To I.G.R.
From J.A.Z.R.
29.1.2001

Twentieth Century Physics
Volume I

Twentieth Century Physics
Volume I

Edited by

Laurie M Brown
Northwestern University

Abraham Pais
Rockefeller University
and
Niels Bohr Institute

Sir Brian Pippard
University of Cambridge

Institute of Physics Publishing
Bristol and Philadelphia

and

American Institute of Physics Press
New York

© IOP Publishing Ltd, AIP Press Inc., 1995

All rights reserved. No part of this publication may be reproduced, stored in a retrieval system or transmitted in any form or by any means, electronic, mechanical, photocopying, recording or otherwise, without the prior permission of the publisher. Multiple copying is permitted in accordance with the terms of licences issued by the Copyright Licensing Agency under the terms of its agreement with the Committee of Vice-Chancellors and Principals. Authorization to photocopy items for internal or personal use, or the internal or personal use of specific clients in the USA, is granted by IOP Publishing and AIP Press to libraries and other users registered with the Copyright Clearance Center (CCC) Transaction Reporting Service, providing that the base fee of $19.50 per copy is paid directly to CCC, 27 Congress Street, Salem, MA 01970, USA

British Library Cataloguing-in-Publication Data
A catalogue record for this book is available from the British Library.

In UK and the Rest of the World, excluding North America:
ISBN   0 7503 0353 0   Vol. I
        0 7503 0354 9   Vol. II
        0 7503 0355 7   Vol. III
        0 7503 0310 7   (3 vol. set)

In North America (United States of America, Canada and Mexico):
ISBN   1-56396-047-8   Vol. I
        1-56396-048-6   Vol. II
        1-56396-049-4   Vol. III
        1-56396-314-0   (3 vol. set)

Library of Congress Cataloging-in-Publication Data are available

Published jointly by Institute of Physics Publishing, wholly owned by The Institute of Physics, London, and American Institute of Physics Press, wholly owned by the American Institute of Physics, New York.

Institute of Physics Publishing, Techno House, Redcliffe Way, Bristol BS1 6NX, UK
Institute of Physics Publishing, Suite 1035, The Public Ledger Building, Independence Square, Philadelphia, PA 19106, USA
American Institute of Physics Press, 500 Sunnyside Boulevard, Woodbury, New York 11797-299, USA

Printed and bound in the UK by Bookcraft Ltd, Bath.

CONTENTS

VOLUME I

Preface	ix
List of contributors	xiii
Biographical captions	xvii

1 **PHYSICS IN 1900**   1
Sir Brian Pippard
 1.1 The community of scientists   3
 1.2 The education of a physicist   8
 1.3 The research physicist   16
 1.4 The support of research   19
 1.5 Black-body radiation   22
 1.6 Experimental facilities   24
 1.7 The world-picture   29
 1.8 The seeds of modern physics   36

2 **INTRODUCING ATOMS AND THEIR NUCLEI**   43
Abraham Pais

 Part 1   ATOMS   43
 2.1 Prelude   43
 2.2 The decade of transition: 1895–1905   52
 2.3 Radioactivity: 1896–1905   56
 2.4 The structure of atoms: 1897–1906   63
 2.5 The birth of quantum physics   65
 2.6 Niels Bohr, father of quantum dynamics   72
 2.7 First the good news. More successes for the old quantum theory   89
 2.8 Now the bad news. The crisis of the old quantum theory   101

 Part 2   NUCLEI   103
 2.9 β-ray spectroscopy, 1906–14   103
 2.10 Models of the nucleus, the beginnings   109
 2.11 1926–32: the years of nuclear paradoxes   115

	2.12 The neutron	119
	2.13 β-spectra: the end of the beginning	126
	2.14 Fission	129
3	**QUANTA AND QUANTUM MECHANICS**	143
	Helmut Rechenberg	
	3.1 Introduction	143
	3.2 Quanta—the empirical foundations (1900–28)	145
	3.3 Origin and completion of quantum mechanics (1913–29)	173
	3.4 The world of microphysics (1925–35)	210
4	**HISTORY OF RELATIVITY**	249
	John Stachel	
	4.1 Introduction	249
	4.2 The special theory of relativity	254
	4.3 The general theory of relativity	286
	4.4 Unified field theories	320
5	**NUCLEAR FORCES, MESONS, AND ISOSPIN SYMMETRY**	357
	Laurie M Brown	
	5.1 The status of physics circa 1930	357
	5.2 *Annus mirabilis*—the new physics of 1932	370
	5.3 Two fundamental theories of nuclear forces	375
	5.4 Cosmic rays in the 1930s: QED, showers, and the mesotron	383
	5.5 Mesotrons, mesons, and the birth of particle physics	394
	5.6 Discoveries during and just after World War II	405
	5.7 Conclusion	409
6	**SOLID-STATE STRUCTURE ANALYSIS**	421
	William Cochran	
	6.1 Crystallography and x-rays before 1912	421
	6.2 The discovery of x-ray diffraction by crystals	424
	6.3 Experimental techniques	432
	6.4 Methods of structure determination	439
	6.5 Accurate structure analysis	450
	6.6 Neutron diffraction	453
	6.7 Electron diffraction	463
	6.8 Surface crystallography	465
	6.9 Imperfect crystals and non-crystalline solids	471
	6.10 The impact of crystal-structure analysis	484
	6.11 Biomolecular structures	496
	6.12 The International Union of Crystallography and related topics	511
7	**THERMODYNAMICS AND STATISTICAL MECHANICS (IN EQUILIBRIUM)**	521
	Cyril Domb	

		Contents
7.1	Introduction—nineteenth century background	521
7.2	Impact of the quantum theory	524
7.3	Formal theoretical developments	531
7.4	The third law of thermodynamics	535
7.5	Phase transitions and critical phenomena	542
7.6	Further topics	574

8 **NON-EQUILIBRIUM STATISTICAL MECHANICS OR THE VAGARIES OF TIME EVOLUTION** 585
Max Dresden
8.1	A phase of transition and consolidation	585
8.2	A history in three parts	592

Illustration acknowledgments	A1
Subject index	S1
Name index	N1

VOLUME II

9	**ELEMENTARY PARTICLE PHYSICS IN THE SECOND HALF OF THE TWENTIETH CENTURY** *Val L Fitch and Jonathan L Rosner*	635
10	**FLUID MECHANICS** *Sir James Lighthill*	795
11	**SUPERFLUIDS AND SUPERCONDUCTORS** *A J Leggett*	913
12	**VIBRATIONS AND SPIN WAVES IN CRYSTALS** *R A Cowley and Sir Brian Pippard*	967
13	**ATOMIC AND MOLECULAR PHYSICS** *Ugo Fano*	1023
14	**MAGNETISM** *K W H Stevens*	1111
15	**NUCLEAR DYNAMICS** *David M Brink*	1183
16	**UNITS, STANDARDS AND CONSTANTS** *Arlie Bailey*	1233

Illustration acknowledgments	A7
Subject index	S1
Name index	N1

Twentieth Century Physics

VOLUME III

17	**ELECTRONS IN SOLIDS** Sir Brian Pippard	1279
18	**A HISTORY OF OPTICAL AND OPTOELECTRONIC PHYSICS IN THE TWENTIETH CENTURY** R G W Brown and E R Pike	1385
19	**PHYSICS OF MATERIALS** Robert W Cahn	1505
20	**ELECTRON-BEAM INSTRUMENTS** T Mulvey	1565
21	**SOFT MATTER: BIRTH AND GROWTH OF CONCEPTS** P G de Gennes	1593
22	**PLASMA PHYSICS IN THE TWENTIETH CENTURY** Richard F Post	1617
23	**ASTROPHYSICS AND COSMOLOGY** Malcolm S Longair	1691
24	**COMPUTER-GENERATED PHYSICS** Mitchell J Feigenbaum	1823
25	**MEDICAL PHYSICS** John R Mallard	1855
26	**GEOPHYSICS** S G Brush and C S Gillmor	1943
27	**REFLECTIONS ON TWENTIETH CENTURY PHYSICS: THREE ESSAYS**	2017
	Historical overview of the twentieth century in physics Philip Anderson	2017
	Nature itself Steven Weinberg	2033
	Some reflections on physics as a social institution John Ziman	2041
Illustration acknowledgments		A11
Journal abbreviations		J1
Subject index		S1
Name index		N1

PREFACE

It is easy enough to find an excuse for celebrating the achievements of physics in the twentieth century. As the year 1900 approached, the splendid edifice of classical physics, founded on the ideas of Newton, Maxwell, Helmholtz, Lorentz, and many others, appeared to be reaching a state of near-perfection; but this very advanced state revealed some structural flaws, which turned out to be more than superficial. The experimental and theoretical discoveries of the years around the turn of the century led to revolutions that transformed the basic outlook of physicists: atomic structure, quantum theory, and relativity. It must be emphasized, however, that the older classical successes were not discarded, but could be seen ultimately as special cases of more general conceptions; the modern physicist must still have a sound appreciation of the old dynamics and electromagnetism. In most technological applications (except for the most advanced 'hi-tech'), it would be futile to drag in relativity or quantum mechanics; with few exceptions, the classical description of everyday events and devices is the best available.

Nevertheless, the switch to what is referred to as Modern Physics, essentially the creation of the present century, enormously enlarged the scope of physical science. Not only was the architecture of the atom and its nucleus, and even that of the nuclear constituents themselves, gradually revealed, but at the other end of the scale of sizes the universe as a whole became accessible to observation, reason, and informed imagination. Quantum mechanics made sense of atomic structure and showed, within a year or two of its invention, that it could explain, at least in principle, the origin of the chemical bond.

In the fifties, the elucidation of some of the simpler proteins and of the DNA double helix by the methods of crystallography transformed the study of biological mechanisms. This is not to imply at all that chemistry and biology are branches of physics—chemists and biologists have their own ways of handling their immensely complex material. Working alone, physicists are no match for them, and can continue with their own adventures, secure in the knowledge that the other sciences are not about to spoil the game by injecting hitherto unsuspected laws of nature, but rather will use the ideas of physics to illuminate their own discoveries.

We recognized from the outset that the best we could hope for in these volumes was a first step toward the writing of this history. At this stage, it cannot be left only to professional historians of science, however much we may hope that they will be inspired later to tackle the job. The first step is for the physicists themselves to point out what they see as important developments in their own fields, and to do their best to strip away the complications which make it hard enough for anyone in the game, let alone outside observers, to see what is going on. We aimed to tell a story that would reveal something of the history to students of physics (including teachers and other science professionals), a story that would lead to enthusiasm, not exasperation. And if we have gone some distance toward achieving this end, serious historians of science will at least have a point of attack for the modern period. Indeed, some of modern physics has already been studied in depth but—without denigrating what has been achieved—these volumes will make clear that the task is only just begun.

At the beginning of the century there were a few leading physicists who kept in touch with almost every active line of research, but now there are none. It is not simply that much more research is being done, but that workers in different areas have little in common beyond what they learnt as students. The theoretical ideas and techniques of fundamental particle physics are only rarely transferable to solid-state physics, and it is several decades since advances in superconductivity stimulated significant contributions to fundamental particles. We had little option but to ask individual specialists to write about their own fields, and hope that they would indicate connections with other fields. The marginal cross-references have picked up a number of these, and should prove helpful in showing where particular ideas have been touched on elsewhere.

Inevitably, however, there are gaps, and we can only apologize to readers who find that their favourite notions have not been mentioned, or their own important contributions overlooked. It is particularly unfortunate that we failed to find an author to tell how electronic circuitry developed from radio communications, through radar and computers, until its technical power came to dominate experimental design, mathematical analysis and computation. To visit a physical laboratory nowadays can make one wonder whether any research would still be possible or worthwhile if the transistor had not been invented. This is only one, though perhaps the most striking, of the technical advances that are inseparable from physics research, and that deserve to be studied in parallel with the history of physical ideas.

We are also aware of having paid scant attention to the social aspects of physics. For example, there is much to be studied about its involvement in war, and about benefits (and possible damage?) that have accrued to the physics profession from military programmes. The politics of funding science, especially 'Big Science', and the communication of scientific results between individuals, laboratories, and countries are other large topics that we have ignored. We have also done very little on the philosophy of science as it relates (as so much of it does) to physics. We do not regard these topics as unimportant—on the contrary they require far more space than we had available in these volumes, and we would like to see them also treated in a comprehensive fashion. Perhaps our work may provide a useful background for such an effort.

Readers accustomed to journalistic hyperbole, and even practising physicists who have noted how many published research letters and articles boast that what has been achieved is for the first time, may be surprised at the low key approach adopted in these accounts. If anyone is entitled to express enthusiasm, it is surely those who have devoted their lives to what they are writing about and have themselves made distinguished contributions. But they are the very people who know what real distinction is. Those who have talked with Einstein or Heisenberg or Feynman (or any other hero) do not lavish praise indiscriminately. Nor do our scientific heroes commonly make grandiose claims; for all the excitement they may feel when a truth is suddenly perceived, and even if they occasionally show off a little in explaining it to others, they know that what they have discovered was there waiting to be discovered. They are (generally speaking) not revolutionaries who have created something where nothing was before—like their forebears they have perceived a flaw in the fabric, found out how to repair it, and take pleasure in the newly revealed beauty.

Regarding the structure of this work, it may be helpful to point out that Volume I mainly covers material up to mid-century, and its chapters are to a large extent written by physicist–historians, that is authors who have previously written on physics and on its history. Volumes II and III have somewhat more specialist flavour and deal with topics more important in the second half-century. Photographs and brief biographies of some of the great physicists of the century will be found scattered among the chapters. They were not chosen by careful deliberation to represent the top fifty or so, but each author was asked to choose a few who had done especially notable work in the writer's field. In this way we present a sample of the architects of modern physics in all its variety. Our history tells how these and myriad others, not all brilliant, but possessed of talent and dedicated to their task, took part in an enterprise that they would not have missed for all the world. It is a story that deserves to be told well, and if it encourages others to tell it better we shall have achieved our aim.

Laurie M Brown
Abraham Pais
Sir Brian Pippard

LIST OF CONTRIBUTORS

EDITORS

Sir Brian Pippard
formerly Cavendish Professor, Cavendish Laboratory, University of Cambridge, Cambridge CB3 0HE, UK

Abraham Pais
Rockefeller University, 1230 York Avenue, New York 10021-6399, USA and Niels Bohr Institute, University of Copenhagen, Blegdamsvej 17, DK-2100 Copenhagen, Denmark

Laurie M Brown
Professor Emeritus of Physics and Astronomy, Northwestern University, 2145 Sheridan Road, Evanston, Illinois 60208-3112, USA

CONTRIBUTORS

Helmut Rechenberg
Max-Planck-Institut für Physik, Werner-Heisenberg-Institut, Föhringer Ring 6, 80805 München, Germany

John Stachel
Department of Physics, Boston University, 590 Commonwealth Avenue, Boston, Massachusetts 02215, USA

William Cochran
Department of Physics and Astronomy, James Clerk Maxwell Building, University of Edinburgh, Mayfield Road, Edinburgh EH9 3JZ, UK

Cyril Domb
Jack and Pearl Resnick Institute of Advanced Technology, Bar-Ilan University, Ramat-Gan, Israel

Max Dresden
Stanford Linear Accelerator Center, PO Box 4349, Stanford, California 94309, USA

Val L Fitch
Joseph Henry Laboratories of Physics, Princeton University, Princeton, New Jersey 08544, USA

Jonathan L Rosner
The Enrico Fermi Institute, University of Chicago, 5640 South Ellis Avenue, Chicago, Illinois 60637-1433, USA

Sir James Lighthill
Department of Mathematics, University College London, Gower Street, London WC1E 6BT, UK

Anthony J Leggett
Department of Physics, University of Illinois at Urbana-Champaign, 1110 W Green Street, Urbana, Illinois 61801, USA

Roger A Cowley
Clarendon Laboratory, University of Oxford, Parks Road, Oxford OX1 3PU, UK

Ugo Fano
James Franck Institute, University of Chicago, 5640 South Ellis Avenue, Chicago, Illinois 60637-1433, USA

Kenneth W H Stevens
Department of Physics, University of Nottingham, Nottingham NG7 2RD, UK

David M Brink
Department of Theoretical Physics, University of Oxford, 1 Keble Road, Oxford OX1 3NP, UK

Arlie Bailey
formerly Division of Electrical Sciences, National Physical Laboratory, Teddington TW11 0LW, UK (*Address for correspondence: Foxgloves, New Valley Road, Milford-on-Sea, Lymington, Hampshire SO41 0SA, UK*)

Robert G W Brown
Department of Electrical and Electronic Engineering, University of Nottingham, Nottingham NG7 2RD, UK (*Address for correspondence: Sharp Laboratories of Europe Ltd, Edmund Halley Road, Oxford Science Park, Oxford OX4 4GA, UK*)

E Roy Pike
Department of Physics, King's College London, The Strand, London WC2R 2LS, and DRA, St Andrews Road, Malvern WR14 3PS, UK

List of Contributors

Robert W Cahn
Department of Materials Science and Metallurgy, University of Cambridge, Pembroke Street, Cambridge CB2 3QZ, UK

Tom Mulvey
Department of Electronic Engineering and Applied Physics, Aston University, Birmingham B4 7ET, UK

Pierre Gilles de Gennes
Ecole Supérieure de Physique et de Chimie Industrielles, 10 rue Vauquelin, 75231 Paris Cedex 05, France

Richard F Post
PO Box 808, Lawrence Livermore National Laboratory, Livermore, California 94550, USA

Malcolm S Longair
Cavendish Laboratory, University of Cambridge, Cambridge CB3 0HE, UK

Mitchell J Feigenbaum
Department of Physics, Rockefeller University, 1230 York Avenue, New York 10021-6399, USA

John R Mallard
formerly Professor of Medical Physics, Department of Bio-Medical Physics and Bio-Engineering, University of Aberdeen, Aberdeen AB9 2ZD, UK (*Address for correspondence: 121 Anderson Drive, Aberdeen AB2 6BG, UK*)

Stephen G Brush
Institute for Physical Science and Technology, University of Maryland, College Park, Maryland 20742-2431, USA

C Stewart Gillmor
Department of History, Wesleyan University, Middletown, Connecticut 06459-0002, USA

Philip Anderson
Joseph Henry Laboratories of Physics, Princeton University, Princeton, New Jersey 08544, USA

Steven Weinberg
Department of Physics, University of Texas at Austin, Austin, Texas 78712-1081, USA

John Ziman
Emeritus Professor of Physics, University of Bristol, Bristol BS8 1TL, UK (*Address for correspondence: 27 Little London Green, Oakley, Aylesbury HP18 9QL, UK*)

BIOGRAPHICAL CAPTIONS

Anderson C D	American	1905–91	Ch 5
Bardeen J	American	1908–91	Ch 11
Blackett P M S	British	1897–1974	Ch 5
Bloembergen N	Dutch	b 1920	Ch 18
Bogoliubov N N	Russian	1900–92	Ch 8
Bohr N	Danish	1885–1962	Ch 2
Boltzmann L E	Austrian	1844–1906	Ch 1
Born M	German	1892–1970	Ch 12
Bragg Sir W L	British	1890–1971	Ch 6
Brockhouse B N	Canadian	b 1918	Ch 12
Chapman S	British	1888–1970	Ch 26
Cottrell A H	British	b 1919	Ch 19
Debye P	Dutch	1884–1966	Ch 12
Eddington A S	British	1882–1944	Ch 23
Einstein A	German/Swiss/American	1879–1955	Ch 4
Fermi E	Italian	1901–54	Ch 15
Feynman R P	American	1918–88	Ch 9
Flory P	American	1910–85	Ch 21
Frank Sir C	British	b 1911	Ch 21
Gabor D	Hungarian	1900–79	Ch 20
Gibbs J W	American	1839–1903	Ch 7
Glazebrook R T	British	1854–1935	Ch 16
Gray L H	British	1905–65	Ch 25
Heisenberg W K	German	1901–76	Ch 5
Herzberg G	German	b 1904	Ch 13
Heycock C T	British	1858–1931	Ch 19
Hubble E P	American	1889–1953	Ch 23
Kamerlingh Onnes H	Dutch	1853–1926	Ch 11
Kubo R	Japanese	b 1920	Ch 8

xvii

Landau L D	Soviet	1908–68	Ch 22
Lawrence E O	American	1901–58	Ch 15
Lorentz H A	Dutch	1853–1928	Ch 1
Maiman T H	American	b 1927	Ch 18
Massey Sir H	Australian	1908–83	Ch 13
Mayneord W V	British	1902–88	Ch 25
Meggers W F	American	1888–1966	Ch 13
Michelson A A	American	1852–1931	Ch 1
Minkowski H	German	1864–1909	Ch 4
Mott N F	British	b 1905	Ch 17
Nernst W H	German	1864–1940	Ch 7
Neville F H	British	1847–1915	Ch 19
Onsager L	Norwegian	1903–76	Ch 7
Planck M	German	1858–1947	Ch 3
Prandtl L	German	1875–1953	Ch 10
Prigogine I	Russian	b 1917	Ch 8
Rosenbluth M N	American	b 1927	Ch 22
Rutherford E	New Zealander	1871–1937	Ch 2
Schawlow A L	American	b 1921	Ch 18
Schrödinger E	Austrian	1887–1961	Ch 3
Seitz F	American	b 1911	Ch 19
Shockley W B	American	1910–90	Ch 17
Stratton S W	American	1861–1931	Ch 16
Strutt J W (Third Baron Rayleigh)	British	1842–1919	Ch 1
Taylor G I	British	1886–1975	Ch 10
Townes C H	American	b 1915	Ch 18
Uhlenbeck G E	Dutch	1900–88	Ch 8
Van Vleck J H	American	1899–1980	Ch 14
von Laue M	German	1879–1960	Ch 6
Wegener A L	German	1880–1930	Ch 26
Weiss P E	French	1865–1940	Ch 14
Wilson K G	American	b 1936	Ch 7
Wilson R R	American	b 1914	Ch 9
Yukawa H	Japanese	1907–81	Ch 5

Chapter 1

PHYSICS IN 1900

Sir Brian Pippard

For a perspective view of the world of physics in 1900 we can hardly do better than chart the growth of research during the present century. *Science Abstracts*, which in its first five years to 1903 covered electrical engineering as well as physics, is a ready source of numerical information since the abstracts, taken from a catholic range of journals from *Annalen der Physik* to *The Horseless Age*, are numbered serially. The total for each year (excluding electrical engineering from the first few), as shown in figure 1.1, is therefore a fairly reliable measure of the level of activity. The influence of the World Wars is obvious, but the point of greatest interest is the explosive rise in the sixties that accompanied the expansion of university education in the Western world and the emergence of Japan as a major power. Most physicists who are still active in research entered the profession after 1960 and have had no first-hand experience of a time when the number of papers relevant to their special interest was not too large to follow closely, and when the leading figures all seemed to know each other personally. If this was so when 10 000 papers appeared each year, how much more so when they numbered less than 2000, as in the early years of the century. The great revolution in physics between 1895 and 1939, the period which saw the discovery of x-rays and x-ray crystallography, radioactivity, quantum theory, the nuclear atom (and its fission), relativity, thermionics and radio communication, was reported in a total of 120 000 papers, rather less than the number published in 1990 alone.

Between 1900 and 1990 the number of published papers per annum increased by a factor of 100; the number of authors who put their names on papers in any one year increased by a factor of 140. It might have been expected, from the modern prevalence of multiple authorship in Big Science, that these two figures would have differed more widely. But if a single paper may represent the achievement of a large team

Twentieth Century Physics

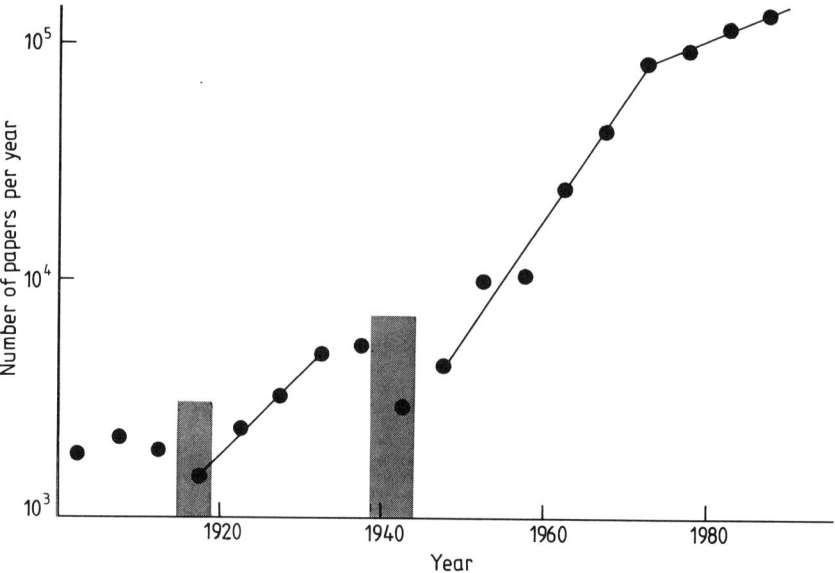

Figure 1.1. *The number of papers in physics published annually during the twentieth century. Each point is the average for a five-year period, 1900–5, 1905–10, etc. The years of the two World Wars are shaded. The straight lines indicate periods during which the publication rate was rising approximately exponentially (the vertical scale is logarithmic) with doubling times of nine years in the interwar period, six years between 1945 and 1980, and 21 years since then.*

(over 430 names on one four-page letter) [1] it is also true that a single name may appear on over 50 publications in a year [2], and the two multiplicities largely cancel each other. Both phenomena are relatively recent; even now the great majority of papers are still written by a single author, or two at most, and few authors publish more frequently than once or twice a year. Lord Kelvin's record of 661 papers and 70 patents in a publishing span of 67 years is unlikely to be bettered [3]. If, in the more spacious air of the past, priority of discovery mattered as much as now, it was usually pursued with decorum and without the benefit of a sensation-hungry press. The press, indeed, was normally indifferent to the obscure activities of physicists or other scientific workers except on such occasions as caught the public fancy—the discovery of argon it might be, or the involvement of a great personage. In 1851, for example, Foucault's pendulum (to which neither he nor his astronomical colleagues attached deep significance) was well covered even by the *Illustrated London News*, a journal dedicated more to political events and fashionable society than to the interests of savants. The wish of the President of France to see a spectacular demonstration must have been responsible. Normally, however, only an eclipse or a natural disaster could deflect this journal's attention to matters where the views of a physicist, sensitively presented, might have made a contribution to mass education†. Unfortunately, where Foucault's pendulum was concerned, conflicting and usually ignorant interpretations probably did more harm than good. Röntgen's discovery of x-rays in 1895 was spared similar confusion since at first, even to experts, the nature of the rays was mysterious, but their astonishing power to reveal bones through living tissue gave splendid scope for a press sensation. Apart from such sporadic excitements, physicists worked in a closed community. They were not chosen as characters in serious fiction nor, apart from the novels of Jules Verne and H G Wells, had the seeds of science fiction yet been sown.

1.1. The community of scientists

In almost every city there were to be found local learned societies, dedicated to self-improvement, where amateurs of arts and science met to discuss their observations and hear about new ideas from visiting lecturers. By 1900 there must have been more than a hundred societies in each of Germany, France and Britain, about 30 each in Italy, Russia, the Netherlands and the United States, and smaller numbers elsewhere [4]. The members of the Manchester Literary and Philosophical Society, founded in 1781, were the first to hear the ideas of John Dalton and James Prescott Joule; and the even older American Philosophical Society (1743) numbered Benjamin Franklin among its founders. But as physical investigations became further removed from direct observation, they featured less frequently at such meetings. This was not true of the Royal Institution of Great Britain, which deserves special mention as being both an instrument of popular education and an outstanding centre of research [5]. Its founder Sir Benjamin Thompson, Count Rumford, intended it to pursue the application of science to public welfare but his successors Sir Humphry Davy and, most famously, Michael Faraday, while not neglecting this task, also carried out a huge and diverse programme of fundamental research in chemistry and physics. At the same time they instituted the occasional lectures and regular Friday Evening Discourses whose technical virtuosity attracted the most influential men and women in London to learn the latest news in science; and at Christmas their children flocked to hear lectures adapted to a 'juvenile auditory'. It was at a Friday Evening Discourse in 1897 that J J Thomson, himself a polished performer, announced his discovery of the electron and left

† *The South African War, which dominated the journal month after month, precluded anything scientific in 1900 until June 16 when Dr Andrew Wilson began a weekly article of 'Science jottings'. His first two articles concerned spiritualism, from which he moved to infant prodigies, the development of flatfish, spiritualism once more, the inheritance of acquired characteristics (for which he quotes decisive evidence) and so on, with no significant scientific content beyond anecdotal natural history.*

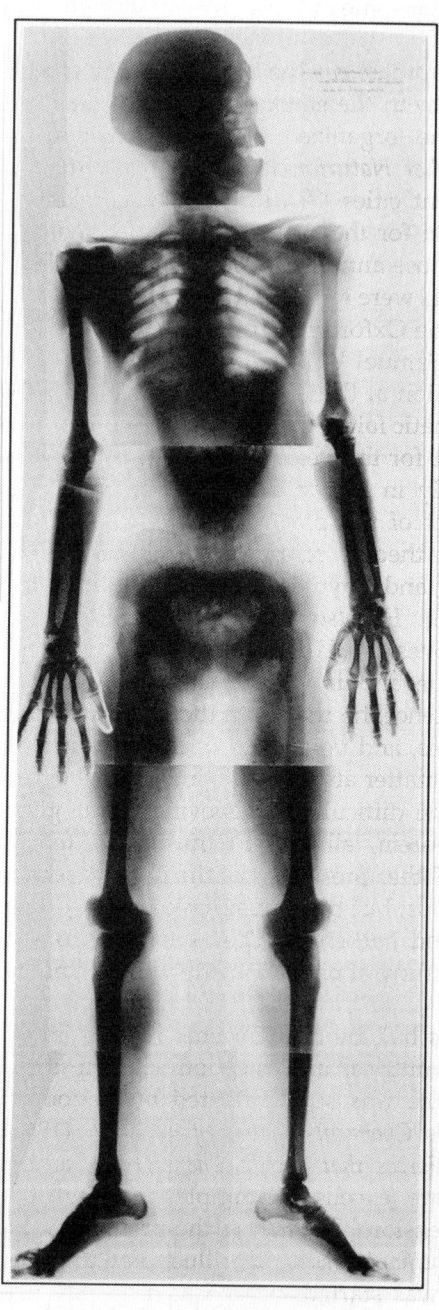

Figure 1.2. *This composite x-ray photograph was made by Ludwig Zehnder in 1897, less than two years after Röntgen's discovery. At least three subjects were involved, and the exposures (one hour for the head) were unacceptably long by modern standards.*

with the feeling that some of those present thought they were the victims of a hoax [6].

A different approach to the dissemination of science was initiated by Lorenz Oken, with the encouragement of Alexander von Humboldt, when in 1822 he organized a German national scientific society, *Gesellschaft deutscher Naturforscher und Ärzte*, which met, usually once a year, in different cities [7]. His idea was taken up in 1831 by The British Association for the Advancement of Science—the British Ass to the facetious—whose annual meetings were attended by thousands and whose proceedings were reported in depth, as they still are, by the quality press. It was at the Oxford meeting in 1860 that Thomas Henry Huxley engaged Bishop Samuel Wilberforce in the sharp exchange that placed Darwinian evolution at the head of the agenda for the conflict between science and dogmatic faith. By 1900 each of several different sections had its own president for the year, whose addresses were followed eagerly and reported fully in *Nature* and other scientific journals. Sir Joseph Larmor, president of the physics section in that year, gave a lengthy review of aether theory, commending the achievements of Maxwell, Hertz, Heaviside and Poynting for the simplicity they had brought to its formulation [8]. He also paid tribute to the work on gas discharges by Schuster, Crookes and J J Thomson, to which Röntgen's discovery had been of great value; and, in what must have been a test of endurance for speaker and audience, he turned to the extension of thermodynamics to chemical processes, and voiced his doubts that it could proceed further to include living matter also. By the time he had expatiated on statistical mechanics and the difficulty of believing in the general validity of the equipartition theorem, all present must have felt they had got their money's worth. Other presidents of the physics section around this time were less ambitious, but they must have assumed they were addressing a fairly specialized audience. They had little to say that would have commended the study of physics to any who had not already been drawn in.

Similar events had by now become regular intellectual carnivals in most Western countries, and the nourishment they provided for the amateur of science was supplemented by periodicals and, of course, books. Faraday's *Chemical History of a Candle* (1861), C V Boys' *Soap Bubbles and the Forces that Mould Them* (1890) and Edwin A Abbott's *Flatland* (1884) are excellent examples in English of popularization without condescension [9]. Among the periodicals, pride of place must surely go to *Scientific American*, 'an illustrated journal of art, science and mechanics' that was started in 1845 as a weekly publication. Though heavily biased towards technology and practical applications of science, it found space for short reports on recent work in pure science, and at the beginning of each year gave a summary of the previous year's discoveries. To be sure, these reports were somewhat sketchy as regards physics, but readers in the first half of 1900 would have learnt, among a host of trifles, that Boys' radiometer could detect a candle at 15 miles and the radiation from Vega and Arcturus (presumably with the aid of a telescope), that Mme Curie had measured the atomic weight (actually the equivalent) of radium, that the radiation from radium, but not polonium, is affected by a magnetic field, that Pupin had greatly improved the performance of telephone cables by inductive loading, that Rowland had found the cause of the Earth's magnetic field (he hadn't) and that the Grand Duke Leopold Ferdinand had seen a purple rainbow at Przemysl. A total eclipse of the Sun, visible in the United States, was prominently discussed beforehand, with obviously expert advice to amateur observers on what to look out for and report, and was described afterwards, with photographs and informed analysis.

Nature (founded 1869) rapidly acquired influence, partly at least through the distinction of its founder-editor, Sir Norman Lockyer, one of the first astrophysicists, of whom it was said that he had a tendency to confuse the Editor of *Nature* with the Author of Nature. It assumed a level of scientific knowledge markedly higher than *Scientific American*, yet contained much to interest the naturalist, the amateur astronomer and the weather-watcher, and it took pains in its book reviews to instruct as well as to criticize. For the specialist it provided authoritative surveys of a wide range of fields, and certainly the physicist could not reasonably have complained of neglect; on the contrary, Lockyer was clearly partial to physics. The correspondence section was given over to discussion of matters arising from earlier articles and reviews, and to odds and ends of ideas and observations. It was not yet, in 1900, a vehicle for preliminary announcements of new results—in this, as in most scientific matters, Germany was the trendsetter. *Physikalische Zeitschrift* (founded 1899), in addition to ephemeral information and abstracts of papers in professional journals, carried from the outset letters describing up-to-the-minute research. Not until Sir Richard Gregory took over the editorship in 1919 did *Nature* begin to follow suit, while *The Physical Review* waited until 1929 before publishing letters (announcing the new practice its editor indicated that an author might expect to see his letter appear in the next number, that is, within two weeks).

The attitude of *Revue Scientifique* (founded 1884) towards the amateur scientist lay somewhere between *Scientific American* and *Nature*. Published weekly, each number contained one or two extended semi-popular articles, together with résumés of papers read at the *Académie des Sciences* and of work published elsewhere; and in the United States *Science* (1883–9; refounded 1895) played a similar role, with rather more emphasis on natural history and popular science, liberally interpreted— a ten-page panegyric of sciosophy [10], for example, by Abner Dean of Angels (commemorated in Bret Harte's *Plain Language from Truthful James* (1870)), and revelations about astral bodies and the successive reincarnations of Alexander the Great. Physics was rather ill-served; during 1900 no significant reference was made to radioactivity. The reports of scientific meetings in America give the impression of a research effort less enterprising than J J Thomson's experiments or European work on the Zeeman effect, both of which were excellently described in long and well-referenced articles. The perceived backwardness of American science forms a persistent undercurrent to the frequent discussions of education, and is explicit in an article [11] describing and commending for emulation the outstanding successes of the Physikalisch-Technische Reichsanstalt, founded in 1887 at Berlin–Charlottenburg, about which more will be said later.

It cannot be doubted that in physics, as in nearly every branch of scholarship, Germany at this time led the world. The universities and their researchers were highly esteemed by their own people, their learned journals were eagerly followed everywhere, and there was an abundance of local societies to meet the needs of amateurs, even if they seem to have been less well served by any periodical of the kind we have noted in other countries. For *Physikalische Zeitschrift* (1899) was from the first aimed at the professional physicist; and he was very well looked after. *Annalen der Physik und Chemie* (1799), which dropped *Chemie* from its title in 1900 when Drude succeeded Wiedemann as editor (and in fact had published virtually no chemistry since Wiedemann took over in 1877) was the premier physics journal of its time, well ahead of *Annales de chimie et de physique* (1789), *Philosophical Magazine* (1798), *Proceedings of the Royal Society* (1856), or its *Philosophical Transactions* (1665), in the number of distinguished papers it contained. Moreover, the systematic diffusion, by means of short abstracts, of work in foreign tongues was far more advanced in Germany than elsewhere. *Fortschritte der Physik* had begun in 1845, and in 1877 Wiedemann instituted a fortnightly supplement of abstracts, *Beiblätter zu den Annalen der Physik*. The English-speaking physicist had to wait until 1897 for *Science Abstracts*.

To conclude this brief sketch of the dissemination of scientific knowledge around 1900 it is worth noting that though the first international conference of research workers was held in Karlsruhe in 1860 [12], such affairs were few in the following fifty years. In 1911 Walther Nernst persuaded Ernest Solvay to provide support for 24 leading physicists, in the hope, by no means unfulfilled, of sorting out the chaos resulting from the new physics of the previous decade [13]. Only after World War II did the conference and the summer school develop into a way of life. Until then it was by reading and discussion of papers at the regular meetings of learned societies, by conversation between colleagues and by voluminous correspondence [14]—such as fell into disuse until the recent rise of facsimile transmission and electronic mail—that ideas were hammered out until there had been achieved the consensus which is the characteristic of progress in science.

1.2. The education of a physicist

Before getting too deeply involved in the professional life of a physicist of earlier times we should remind ourselves again of his rarity. Compared with the present, a much smaller fraction of the population stayed at school until they were of an age to enter college, and of those many gained immediate employment for which further education was unnecessary. In any case public support of students, such as is now taken for granted in many countries (though not the United States), hardly existed and consequently the benefits of college were largely confined to the children of comfortably well-to-do parents. Even in Germany, which until the nineteenth century had taken pride in the equal availability of education to all, it was the prosperous upper and middle classes that supplied the universities with their students. Out of over 12 000 students in Prussian universities around 1890, 170 at most were the sons of working-class fathers [15]. Women in higher education were few indeed except in America, where they were still very much a minority of whom hardly any proceeded to graduate school.

Curricula in the European schools largely followed the traditional humanistic pattern—classics, history, modern languages, mathematics—to which a little science had rather recently been admitted in reluctant response to a general belief that it played an important role in an industrial society. German schools made natural history a significant part of a child's education from an early age, with field expeditions and preserved specimens in the classroom to encourage careful observation [16]. Physics was introduced at about the age of 15, with mechanics as the central theme, but also heat, optics, acoustics, electricity and magnetism. Three hours a week were allocated for this, while chemistry was allowed two and mathematics five†. At first sight this appears reasonably generous, but to a contemporary observer, Professor James Russell of Columbia University, the small proportion of examination questions assigned to science was evidence of its low esteem; and he quotes a German teacher as saying that 'nothing short of a miracle can prevent the promotion of the most deficient member of the class, provided his attainments be satisfactory in other subjects' [17]. At a time when there were few opportunities for pupils to handle equipment themselves, the better physics teachers would set up apparatus in class and encourage discussion on how it might be expected to behave before performing the experiment—an admirable substitute for the practical classes which only a well-endowed school could afford and which, being optional, were attended by very few. In England attempts to introduce practical teaching of science to the labouring classes had fitful success

---

† *These figures are typical of a Prussian* Real-gymnasium, *the class of school that was more inclined towards useful arts than the* Gymnasium *in which the classics were emphasized. Russell [16], writing in 1899, gives much detailed information on the German school system.*

in the first half of the nineteenth century [18], but it was only when the public (independent fee-paying) schools became involved that the process gathered momentum. J M Wilson, first at Rugby School (1860) and later at Clifton College, was the effective pioneer [19]. Clifton, indeed, must surely be unique in having had six members of its teaching staff and one laboratory assistant subsequently elected Fellows of the Royal Society [20]. In one generation a mighty change had overtaken school science, between 1864 when the Public Schools Commission could report that 'Natural Science is practically excluded from the education of the higher classes in England' [21], and 1900 when Arthur Vassall of Harrow School wrote; 'In Public Schools the teaching of science has only recently begun to take reasonable shape, but it is now far enough advanced to make it desirable in Preparatory Schools [for boys under 14].' [22]. Similar developments in the United States may have begun a few years later but moved ahead fast enough for Edwin H Hall to write in 1902: 'On the whole, it appears that the best secondary schools in America, in trying the experiment of teaching physics by means, in part, of laboratory work done by the pupils, have little or nothing to learn from the corresponding schools in France, Germany, or England. For France has apparently never dreamed of such an undertaking, Germany has never seriously considered it, and England is no farther along with it than we are in America, if indeed she is as far.' [23]. Hall, the discoverer of the Hall effect, was a Professor of Physics at Harvard whose thoughts on education deserve to be remembered.

In the colleges and universities the traditional curriculum held sway as strongly as in the schools, but was amplified to include training for the professions—medicine, law and the church. For those who expected to become school teachers, as well as for the general education of a leisured gentleman, mathematics, philosophy and classics could not be bettered. By 1900, however, other branches of learning such as languages and science were enforcing modernization, though at different rates in different countries. This is not an easy subject to penetrate. Statistics [24] of student numbers, hours of formal teaching, financial support, are a good start, but they tell one little about how science was taught or how highly it was valued (as we have just seen in connection with German schools). The custom of basing lecture courses on designated textbooks is rather recent, and to know what books were recommended by a long-dead professor to accompany his course may only indicate what impossible standard of learning he pretended to expect from his class. Even examination papers, when they can be found, may be seriously misleading if one does not know whether they are requests to regurgitate lecture notes or genuine tests of independent thought. In view of these difficulties I shall not pretend to give a comparative account of the growth of higher education in physics, but shall concentrate on the one example, Cambridge University, I know well enough to avoid egregious misrepresentation. Its peculiar merit for this purpose is the ready accessibility of detailed information, a consequence of the tightly structured pattern that led to the award of a degree. In Germany, which otherwise would be a better choice [25], a young man was able to migrate during his often prolonged student days if he was attracted by the teaching of a particular professor. Once matriculated he was free to follow his inclinations until the time he chose to proceed to his doctor's degree by public oral examination [26]; and if Berlin, at least for the physicist, was the leading university there were so many others of distinction, each with its own undocumented customs, as to deter attempts to draw a generalized picture.

In Britain, by contrast, Cambridge was by a long way the leader in mathematics and physics; of 36 Fellows of the Royal Society in 1900 whose academic posts immediately identify them as physicists, 18 had graduated from Cambridge, 4 from London, 3 from Oxford, 5 from the Scottish and 3 from the Irish universities. All the Cambridge Fellows graduated in mathematics [27]. The student with academic ambitions sought entry and financial support by success in the written scholarship examination, resided for three years and gained his degree also by written examination. The examination papers and lists of successful candidates are to be found in the University Library [28], as are the schedules of lectures offered (but it must be remembered that a serious student was also expected to read widely in textbooks and original papers, and to work through many problems for the tutor who gave him personal attention weekly). With all this information it is easy to reconstruct the education of a talented young man†. Moreover, as far as possible, the later career of every graduate has been researched and published [29]. The general picture that emerges will probably serve well enough for other countries besides Britain [30].

A student interested in physics had the choice at the end of his third year of taking his final examination either in natural sciences or mathematics. The Natural Sciences Tripos‡ was a relatively recent creation for which several sciences were taken together, while the Mathematics Tripos was ancient and carried greater prestige than any other university course in the country. In all triposes the successful candidates were graded in three classes, but in mathematics alone they were in addition placed in order of merit. The first-class mathematician was a wrangler; to be Senior Wrangler virtually guaranteed entry

---

† *Women were not admitted to the University. Though they could attend lectures and take the examinations they were awarded only the 'title' of a degree. Very few studied physics.*

‡ *Tripos is the name, peculiar to Cambridge and of great antiquity, given to the final examination for the bachelor's degree. It is believed to have referred originally to the stool on which a disputant sat, but was later transferred to the disputant himself, and only in the last century to the examination.*

into any profession, with rich pickings for those who elected to remain in Cambridge to coach succeeding generations. The pioneering schoolmaster J M Wilson was Senior Wrangler in 1859. A few of the famous names in British mathematical physics, Lord Rayleigh for one, were senior wranglers, but William Thomson (Lord Kelvin), James Clerk Maxwell and Joseph John Thomson were all second wranglers, and many others who later achieved distinction appeared in lower places in the first class. Hope of graduating as a wrangler attracted to Cambridge clever students from all over the land, many of whom were sedulously trained in their last years at school in mastery of the difficult scholarship entrance papers [28]—not a bad introduction to the ceaseless three years' grind that even the brightest had to undergo if they were to acquire the speed and expertise in problem-solving that alone led to success. Those who survived with their imaginations intact were formidably equipped for research in pure or applied mathematics. Something of the same dedication must have inspired corresponding students in other countries, to judge from the fluency with which the special functions of mathematical physics and standard analytical techniques are deployed in so many published papers.

The Natural Sciences Tripos, on the other hand, concentrated on experimental science, with copious practical instruction and minimal demand in the examinations for mathematical skill. One must not imagine these two triposes simply as alternative routes to a career in physics for, as we shall see presently, the route through mathematics was still very strongly the preferred option. Entry to natural sciences by scholarship involved answering questions that nowadays a good student at the age of 17–18 would think fairly elementary—how to measure thermal expansion, form a pure spectrum or construct a telescope, together with problems in mechanics and hydrostatics: by contrast he would be hard put to make a good showing in the mathematics scholarship examination. An example of a mechanics question from each makes the point clearly.

Entrance scholarship examination in physics, 1898
A passenger travelling in a train moving along a straight level railway, wishes to ascertain if the train by which he is travelling is moving with (1) uniform velocity, (2) uniform acceleration. Describe in what way he could obtain this knowledge (a) by observation of the mile-posts and (b) by observation of objects within the carriage only. State in each case how (if at all) he could obtain the arithmetical values of the velocity or acceleration.

Entrance scholarship examination in mathematics, 1899
Determine the position and direction of motion of a projectile at any instant after its projection from a given point in a given manner. Two projectiles are moving in a vertical plane, and h, k

are their distances apart at the beginning and end of an interval during which the line joining them turns through an angle θ. Prove that the shortest distance between them is

$$hk\sin\theta / \sqrt{(h^2 + k^2 - 2hk\sin\theta)}.$$

It will be noted that the second problem suffers, as did many of this type, from an artificiality that suggests that mechanics was often seen by mathematicians as primarily a vehicle for developing manipulative skill. This attitude survived for many decades in the teaching of thermodynamics by mathematicians, for whom it was frequently only an excuse for acquiring proficiency in changing variables by the use of Jacobians [31].

Let us pass rapidly over the student's first three years to the moment when, having graduated, he decides whether to stay on for a further year to specialize in physics. Out of the 40, or thereabouts, obtaining first-class honours each year in the Natural Sciences Tripos, rarely did more than five go further and, of them, one or two at most chose physics. One of these was Owen Richardson, who was awarded the Nobel Prize in 1928 for his work on thermionic emission. Wranglers were fewer in number, 16 in 1900, of whom five (and one of the two women who would have been wranglers but for having been born with the wrong sex) continued for a further year, though none eventually made a career in physics.

What, then, did they do, these intellectual paragons? The answer can be tabulated for the high proportion whose subsequent careers are known [29]. We need not be surprised at the number of medical students, nor at their preponderance in the third class, for many were marking time and enjoying life before their real career started—hospital training followed by their father's practice. It is the number of teachers, and the high standard of their degrees, that marks the difference between those days and ours, and which accounted for the excellent science teaching in many schools, especially those public schools which had embraced science enthusiastically and welcomed a first-class graduate. For a career in higher education and research† a first-class degree was almost imperative, and the same applied to a mathematics graduate, though of the 16 wranglers in 1900 only two became professors, of mathematics not physics (and one of those died young); six became schoolteachers. There still prevailed, and not only in Britain, the conviction that had been unequivocally expressed thirty years before by Isaac Todhunter [32], that research was an occupation for the rare genius, and that young men's heads should not be turned by encouraging them to think themselves thus gifted.

† *The 7 in the table made their careers in botany(2), zoology(2), geology(2) and medicine(1); 3 became Fellows of the Royal Society.*

Physics in 1900

Table 1.1. *Career choices of honours graduates in Natural Sciences Tripos, part 1, 1900.*

	First class	Second class	Third class
Medicine	11	20	25
School teaching	9	6	5
Higher education and research	7	0	0
Colonial administration	4	2	1
Engineering	2	2	3
Other	2	10	4
Not known	5	3	1
Total number	40	43	39

Let us look further at the education received by those few who chose to specialize in physics, either in mathematics or in natural sciences. The examinations at the end of the fourth year show how the differences that were obvious in the scholarship entrance examinations reflect a dichotomy permeating university life, almost everywhere. Physics, alone of the sciences, has been cultivated from two standpoints—from mathematics in a guise we may call *natural philosophy*, and in the laboratory as *experimental physics*—sometimes jealously aloof from each other, sometimes converging in a common pursuit. Finer distinctions may be drawn, but it is enough here to indicate, without embarking on a major digression, that there are historical reasons for the extraordinary difference between the two approaches to professional training, both of which were capable of producing excellent results. Once again, a comparison of examination questions says almost all that is necessary.

Mathematical Tripos, part 2, 1900
Obtain the steady electric and magnetic fields due to a charged particle moving with uniform velocity w in a straight line through the ether. Assuming Poisson's solution of the equation $d^2\phi/dt^2 = V^2\nabla^2\phi$ in the form

$$\phi = \frac{d}{dt}(tw_1) + (tw_2)$$

(where ϕ is the value at a point P at time t, w_1 is the mean value of ϕ when $t = 0$ over the surface of a sphere of radius Vt and centre P, and w_2 is the mean value of $d\phi/dt$ when $t = 0$ over the same surface), show that the effect at a point P of suddenly stopping a small charged sphere when at O is the production of a thin pulse of magnetic force whose intensity is $ew\sin\theta/2ar$. Here e is the charge on the sphere, a its radius, r the distance OP and θ the angle between OP and the direction of the steady velocity

w: also the square of the ratio of w to the velocity of light [V] is neglected. Interpret the result.

Natural Sciences Tripos, part 2 (physics), 1900
Write a general account of the phenomena seen when an electric discharge is taken in a gas between two platinum electrodes, and the pressure of the gas is gradually diminished from atmospheric pressure until the discharge no longer occurs. Describe experiments which support the view that the passage of the discharge is accompanied by electrolytic effects.

It would be easy to infer from these examples—as indeed many mathematicians did—that the experimental physicist was little more than a well-trained technical assistant whose task was to measure and present the facts to the theoretician, and who needed little in the way of exact physical knowledge. In its context this point of view was perhaps forgivable; of the supreme counter-examples, Faraday and Rutherford, the one was long dead and the other only just about to reveal his power. The urgent need at that moment was not seen to be the opening-up of new fields—a task in which an experimenter could excel—but the resolution of the paradoxes arising from successive failures to invent a logically coherent mechanical universe consistent with the known laws of electromagnetism. In any case, many of the older mathematical physicists either had occasion to conduct substantial experimental investigations themselves (Rayleigh, Kelvin) or remembered past heroes (Helmholtz, Hertz, Maxwell) to whom theory and experiment were equally congenial; and they had only to contemplate J J Thomson's transformation from second wrangler to experimentalist to realize that in physics it is intellect and imagination rather than manual dexterity that lead to success.

The experimental physicists themselves, it is hardly necessary to say, did not accept this evaluation, especially in Germany where chairs of physics had been established since the eighteenth century and whatever it was that the professors taught, it was certainly not theoretical physics [33]. With the rise of theoretical physics as a recognizably separate discipline, towards the end of the nineteenth century, the established (*ordinary*) professors of physics jealously guarded their positions against covetous supernumerary (*extra-ordinary*) professors with a theoretical inclination. In Cambridge, on the other hand, the pursuit of theoretical physics sprang directly from the faculty of mathematics which had itself campaigned for the setting up of a department of experimental physics in 1871, but had seen to it that the first three Cavendish professors (Maxwell, Rayleigh, J J Thomson) to whom the teaching of experimental physics was entrusted were wranglers out of their own stable [34]. Probably the mathematics faculty would not have tolerated any greater mathematical content in the physics tripos than the bare minimum revealed in the examination questions such as that quoted above [35].

By various routes, then, a tradition of physics teaching had come about in Germany, Britain and the United States whereby the student spent considerable periods in the laboratory [36]. One German school of physics, Franz Neumann's at Königsberg in East Prussia, has been studied in detail [37]. Between 1826 and 1876 Neumann delivered lecture courses which over the years took in almost every branch of physics that was susceptible to precise mathematical formulation, and a small number of his students undertook experimental investigations. His ethos strikes a modern reader as extraordinarily puritanical—enormous attention was paid to systematic and random errors, apparently because no theory could be deemed secure until it had been shown to agree with observation within stringent and precisely understood tolerances. In this his approach was radically opposed to, say, Faraday's and, it must be said, was markedly unproductive of striking new insights. All the same, one cannot altogether discount a school that produced Kirchhoff and Voigt who, in their turn, handed on the torch to Planck and to Drude. In general the severe German schools, less arid perhaps than Neumann's, turned out considerable numbers of thoroughly competent experimenters and enough physicists of high talent to disarm criticism of their teaching methods†. Not so in France where even in 1916, according to Maurice Caullery (a distinguished and widely learned Professor of zoology at the Sorbonne) the science courses at the prestigious École Polytechnique 'are exclusively theoretical and chiefly mnemonic; one is even tempted to call them psittacizing when one recalls the system of repeated questions on which the students are solely judged' [38]. It must be added, however, that at the Sorbonne itself Jamin, appointed professor in 1869, had established the first university laboratory in France devoted both to teaching and research [39].

The pioneering, and highly influential, textbook of practical physics was that of Friedrich Kohlrausch [40] the Göttingen apostle of precision measurement. It was published in 1870, and thereafter ran through many editions. The first English translation appeared in 1873, and starts 'The numerical value of a physical quantity is affected with error...', as the introduction to a thorough presentation of the theory of errors. It continues with the description of a great variety of experiments concerned with training the student to proficiency in handling equipment and learning all the available techniques. There is no suggestion that a practical class may help the student understand physical ideas. For this we have to thank John Trowbridge and Edwin Hall [41], of Harvard.

---

† *Cahan [30] has suggested that the emphasis on measurement may have been in reaction to the peculiarly German* Naturphilosophie, *a recoil from science in the early years of the last century, not unlike modern 'holistic' aspirations.*

Before them, the astrophysicist Edward C Pickering had produced the first American textbook [42] of practical physics, whose approach follows with less severity that taken by Kohlrausch. A minor point of interest is that he describes how to present results graphically in terms that imply that graph paper was not commercially available at that time. T C Mendenhall [43] writing in 1887, finds it necessary to explain what squared paper is, and even in 1902 Hall, discussing laboratory teaching at secondary schools, makes no mention of graphing results. It was at the Massachusetts Institute of Technology that Pickering worked out the plan of setting up each experiment on its own bench, and having the students tackle each in rotation. This economical use of apparatus and assistants was widely copied, as was the emphasis on Gaussian error theory which, being imperfectly understood by most teachers, acquired the status of an article of faith and exacted the customary worship born of pious incomprehension. The more liberal values recommended by Trowbridge and Hall probably found their way into the secondary schools long before they reached the universities.

1.3. The research physicist

The loose academic structure in Germany which allowed a student to proceed from formal learning to guided research, with no hurdle until the examination for his doctorate, also made it easy for graduates from other countries to embark on research under the most eminent professors. The need to learn German, far from being a pointless burden, gave access to the indispensable source of important results published in that language. Germany was also leading the rest of the world in its conception of a university as the sponsor of new learning, and not solely the guardian and expositor of ancient wisdom. Christa Jungnickel and Russell McCormmach [33] have given a superb account of the development in nineteenth century Germany of the concept of organized research, at a time when most professors in other countries were fully occupied with their teaching duties. It should be stressed that for the most part they did not doubt that teaching was their proper function. It was taken for granted, for example, in many British universities that the courses for the first-year students should be given by the most senior professors; and until well into this century the sentiment prevailed that, where conflict arose between teaching and research, teaching came first. Nevertheless the lure of research was becoming irresistible, and by 1900 other countries, especially the United States, were shifting towards the German ideal. The rise of physics research in the United States has been fully described by Kevles [44].

At this point it is helpful to introduce a little numerical evidence. In compiling the *Dictionary of Scientific Biography* [45] the distinguished historians of science on the editorial board took pains to avoid the unseemly national bias that mars most smaller biographical dictionaries.

Even though they were all American, there is little evidence of partiality in the choice of American scientists with dubious credentials for inclusion among the star performers. If, then, we select from their list those physicists who were alive in 1900 and old enough to have begun making the contributions to research that gave them their fame, we may obtain a rough picture of the influence of different countries. Out of a total of 197, 67 belonged to what we may loosely term the German bloc, made up of 52 Germans, seven Dutch, six Austro-Hungarians and two Swiss†. Next in order came Britain with 35 and France with 34, the United States with 27, Russia with nine, Italy seven, Sweden five, and eight other countries with only 13 in all. The four leaders account for 83% of the top research physicists at that time.

It is small wonder that the United States, which had until very recently developed no tradition of research, should take Germany as its model. Though several universities had instituted PhD programmes in the 1880s or before (J Willard Gibbs obtained his PhD from Yale in 1863), only five of the 27 in our list had taken their doctorates at home, while 10 had gone to Germany (five to Berlin alone); the rest had no PhD, some having worked their way to the top in industry, others having succeeded early to academic posts and therefore having no need of a higher degree. In the whole country only about 15 PhDs a year were awarded in physics [46].

Only two of the 35 British physicists, on the other hand, went to Germany for doctorates, and one of them, Arthur Schuster, was born and schooled there before entering Owens College, Manchester. The most effective single reason for their seeming lack of enterprise was the distinction of Cambridge mathematics and, later, the Cavendish Laboratory, as well as the availability of teaching fellowships at the colleges. A fellowship, with free living quarters and food, an adequate stipend, not too great a teaching load to preclude research, and a good chance in due course of a professorial chair at one of a fast-growing number of universities, offered an unmarried graduate a very acceptable alternative to three or four years in foreign parts; and such fellowships were still frequently the reward of outstanding performance in the tripos, so that they functioned as post-graduate rather than (as now) post-doctoral awards. Out of the 35 in the British group, 14 took this route to preferment. There was no equivalent of the German doctorate, or the modern PhD, but the title Doctor of Science could be earned a few years after graduation by the submission of published work; then, as now, a high standard of originality was expected.

† *Between Germany and Austria there was easy exchange of academic positions; between Germany and Holland less mobility but easy communication. Small in number though the Dutch physicists were, the presence among them of H A Lorentz, H Kamerlingh Onnes, J D van der Waals and P Zeeman gave them great influence. This example warns us against attaching too much weight to numbers alone.*

Hendrik Antoon Lorentz

(Dutch, 1853–1928)

Was the most highly esteemed of all the theoretical physicists of his time. His 'electron theory' was a comprehensive explanation of electromagnetic and material phenomena in terms of charged particles moving through the aether, the medium of their interactions. Light was generated by oscillation of charges, and the effect of a magnetic field (the Lorentz force) was to produce the Zeeman splitting of spectral lines. He, and independently Fitzgerald, proposed to generalize Heaviside's analysis of the foreshortening of the field lines from a particle moving very rapidly, and Lorentz gave mathematical expression to this contraction. His transformation equations survived in relativity theory after the aether was swept away. He chaired the first Solvay Congress in 1911 which highlighted, and began to resolve, the tensions between classical physics and quantum theory, but he himself was never reconciled to the new ideas. Notable for his courtesy, he always presented relativity theory as a remarkable and beautiful source of inspiration, while hoping that it would eventually be reconciled to the classical ideas which he understood better than anyone else.

Similarly, in France, there was no equivalent of the PhD, nor was any demanded as the qualification for an academic post, and a higher doctorate was available as in Britain. After the war of 1870–1, study in Germany held no attractions for a Frenchman, and Paris was the magnet that drew ambitious graduates from other universities. Central government control tended to discourage enterprising research, and old-fashioned teaching syllabuses did not help. In particular, the great tradition of pure mathematics was inimical to theoretical physics. The generally accepted view that French science steadily declined between 1840 and 1900 is unfair except, it seems, in physics where distinction was hardly to be found outside metrology and the design of optical instruments [47].

1.4. The support of research

Before the age of computers the needs of a theoretical physicist were met by study, pencil and paper, good ideas and a room for seminars with his students. An experimenter needed rather more, a laboratory and equipment which by 1900 sometimes stretched his budget, though good work was still possible at a very modest price. More recent topics like chemical physics and electrical conduction in gases might be investigated with standard measuring equipment and a little skill in glass blowing and other workshop crafts. Such string-and-sealing-wax research, evoking sentimental nostalgia in those who found it an ideal way to make fast progress, and did not have to endure its frustrations, is characteristic of a young field before the excitement of discovering new phenomena gives way to systematic measurement. Chemical spectroscopy, whose excitements began around 1860 when Kirchhoff explained the complementarity of absorption (Fraunhofer) and emission lines, and Bunsen and Crookes started to reveal new elements by their spectra, was initially open to all who possessed a simple prism spectrograph and a Bunsen burner. As higher dispersion was found to yield new phenomena, especially following Zeeman's discovery [48] (1896) of the magnetic broadening of spectral lines—soon to be resolved as splitting into many components—the prism gave way to the plane diffraction grating and then (for the fortunate few) one of Rowland's ruled concave gratings [49] (1882). Much valuable work however, was still achieved with simple spectrographs, especially when supplemented by the inexpensive Fabry–Perot interferometer (1897) or the elegant but pricey Lummer–Gehrcke (1902) for fine structure analysis [50]. Provision of these basic necessities and, even more, of the great range of sensitive and/or accurate electrical instruments demanded by both academic and industrial research fuelled the growth of scientific instrument companies and the insatiable demand of scientists for the very latest and best—delights to the mind and often, by virtue of exquisite craftsmanship, to the eye as well.

From this period sounds the cry, echoed and amplified as it comes down to us through the century, for more financial support from the state†. It was not enough to have magnificent laboratories whose donors were satisfied to have left a monument; running costs and the renewal of short-lived equipment has less appeal to the philanthropic pocket. Perhaps only in America, according to Caullery, where 'wealthy classes take an effective interest in the universities, is it easier to build and equip a laboratory than to find a director of the first order for it' [53]. He had good reason to feel envious, for the state-financed universities of France

---

† Rutherford: 'we haven't the money, so we've got to think' [51]; and in Cornford's satirical account (1908) of Cambridge University [52], we meet the Adullamites (i.e. scientists) who 'are dangerous because they know what they want; and that is, all the money there is going'.

lacked the grand laboratories that were springing up elsewhere. The Clarendon Laboratory at Oxford (1872) and the Cavendish Laboratory at Cambridge (1874) owed nothing to state support, and while the latter was mostly equipped at the outset by its donor, the Duke of Devonshire, renewal and replacement costs were hard to come by. In Berlin, Helmholtz's reputation sufficed to persuade the Ministry of Culture to build him a handsome Institute in 1878, but he was soon made aware of the problems of staying abreast of the demands for equipment. In June 1883 and March 1884 he joined with Siemens and others in pointing out [54] 'the advantages likely to accrue to Germany from the maintenance of an imperial institution for research, which should at the same time assume the cognate function of fixing and certifying standards of mechanical and physical measurements'. In drawing attention to the latter function they followed the French example of 1875, when *le Bureau International des Poids et Mesures* [55] was established at Sèvres, a suburb of Paris, 'to develop new metric standards, preserve the international prototypes and make such comparisons as are needed to ensure worldwide uniformity of standards'. From this initiative sprang the dominance of the metric system, first in science and then to an increasing degree in public life.

See also p 1234

Helmholtz and his colleagues, however, had a more ambitious object in mind. 'Other countries, notably England, had enjoyed great renown in science because of the brilliant researches and discoveries of some of the scientific men, who had the good fortune to be possessed of large private means, and the scientific spirit to devote them to investigations demanding both as a *sine qua non*.'†. Those conditions, declared the memorialists, were lacking in the fatherland. Their plea succeeded; there was founded in 1887 the Physikalisch-Technische Reichsanstalt, with Helmholtz as its first director, an institution whose success inspired the consolidation in 1899 of three English laboratories, under the control of the Board of Trade, into one National Physical Laboratory at Teddington [56]‡. At almost the same time (1901) the National Bureau of Standards was set up in Washington, DC [58]. All three began work on the outstanding need to develop and promulgate reliable standards, especially for electrical measurements. The system of absolute electromagnetic units, proposed by Gauss and Weber, and

---

† *For example the Hon. Henry Cavendish (1731–1810), James Prescott Joule (1818–89) and John William Strutt/Lord Rayleigh (1842–1919); to these may be added Michael Faraday (1791–1867) who, though lacking private means, could devote his life to research in the Royal Institution. Among non-physicists Charles Darwin (1809–82) is the outstanding example.*

‡ *Professor H Pellat, of the University of Paris, reporting to the physics congress held there in 1900 [57], extols these initiatives and urges France to follow suit, but with little immediate effect. In his view the United States were already in a strong position since the universities were providing this function.*

John William Strutt, Third Baron Rayleigh

(British, 1842–1919)

Enjoyed private means derived from farming the family estate, and was able to devote himself for much of his life to theoretical and experimental work at his country house, Terling. His *Theory of Sound* is the classic text on vibrations and waves, and is still an invaluable source of mathematical techniques. He published 450 papers in a wide range of topics, especially electromagnetism, hydrodynamics and optics. His practical skill may be judged from his establishment, in collaboration with Mrs Sidgwick, of reliable electrical standards, and from his determination of the relative densities of nitrogen and oxygen, which revealed very small discrepancies and led to the discovery of argon, in collaboration with William Ramsay. By 1900 he had succeeded Kelvin as arbiter of British physics.

based on defining unit current in terms of the force between two current-carrying wires, was of no practical use without reproducible standards of voltage (the standard cell) and resistance (the standard ohm) which could be calibrated in terms of the absolute units. These were essential in industry and law, as well as in pure science, to ensure that components made by different manufacturers were compatible, that disputes about value for money could be resolved in the courts, and that fundamental ideas in physics could be tested by combining results from different laboratories. During the five years that Lord Rayleigh was running the Cavendish Laboratory (1879–84) his principal experimental researches were concerned with establishing electrical standards [59], and work along those lines continued to occupy a large fraction of the Reichsanstalt's time, to judge from Carhart's [11] catalogue of investigations in progress in 1900.

At the same time Helmholtz's authority ensured close connections between the University of Berlin and the Reichsanstalt and enabled extended investigations to be directed by some of the best physicists in the world. Among these was the prolonged study of black-body radiation which led in due course to Planck's formula and his proposal for quantizing the harmonic oscillator. In view of the importance of this work and the opportunity it provides to illustrate the conduct of experimental research at this time, it is worth going into some detail.

1.5. Black-body radiation

Kirchhoff's explanation of the complementary character of absorption and emission spectra was but one corollary of his more basic recognition that the character of the radiation in a uniformly heated cavity must be independent of the nature of the walls; if this were not so, the second law of thermodynamics could be violated by letting radiation pass from one cavity to another at a slightly higher temperature. It is clear, then, that the energy density of the radiation, and the distribution of energy among different wavelengths, must be completely determined by the temperature of the walls. A phenomenon so perfectly independent of the accidents of material bodies must find its explanation solely in terms of fundamental properties of the physical universe, and therefore deserves special attention from experimenters and theorists. A full account of how the theory developed has been given by Klein and Kuhn [60] and we shall concentrate here on experiment.

It is not necessary to study the radiation in a cavity directly if a perfectly black body is available, since Kirchhoff's analysis shows that this emits radiation of the same quality. But although much of the earlier work was done with suitably coated metal strips or wires, doubt about their perfect absorption, particularly in the far infra-red, caused Wien and Lummer to propose [61] (1895) the use of an electrically heated tubular furnace, with a small hole to let radiation out without seriously disturbing the state within the furnace. Designing the furnace so that its surface temperature was uniform was a necessary and relatively simple preliminary; measuring the temperature, however, was a major task, since the only known way of realizing the absolute Kelvin scale was by use of a gas thermometer, which at almost white heat is no small matter. There was, however, a long tradition of extending gas thermometry to ever higher temperatures; Holborn and Wien [62] used a porcelain bulb to calibrate a thermocouple of platinum and platinum–rhodium alloy up to 1700 K, and this was built into the furnace to measure its temperature.

Measurement of radiated power in a narrow spectral range in the infra-red demanded a detector of high sensitivity, for which the pioneering work was that of S P Langley [63], the American astronomer and, later, enthusiast for flying machines. He developed the use of strips of very thin metal foil (first iron and later platinum), blackened to absorb radiation, as bolometers. As one arm of a Wheatstone bridge, incorporating the most sensitive galvanometer available, such a strip changed in resistance by detectable amounts if its temperature rose by 10^{-5} degrees, corresponding to a resistance change of three parts in 10^8. This is a very impressive result, especially in view of the capricious ways of a Thomson moving-magnet galvanometer [64], easily upset by vibrations and by minute fluctuations of the ambient magnetic field. Langley himself provided a dramatic illustration of his bolometer's sensitivity by placing it at the focus of a 13" telescope, and measuring the variation of heat output across the Moon's disc†.

Langley's bolometer and his procedure for measuring refractive index in the far infra-red was adapted by Paschen [67] for the careful investigation that was another essential preliminary to Lummer and Pringsheim's [68] measurement of the spectral distribution of energy in black-body radiation. They arranged for the radiation from a small aperture in the furnace to be dispersed by a fluorite (CaF_2) prism and its intensity scanned with a bolometer. But to interpret the results it was necessary to know very precisely the dispersion of fluorite—how the refractive index varied with wavelength or, in practical terms, how the minimum angle of deviation by the prism varied across the spectrum. This was Paschen's contribution. The prism itself was calibrated by the use of one of Rowland's concave gratings (previously used by Langley) to diffract waves as long as 9.5 μm, the limit of transparency of fluorite. If the source is an electric arc, emitting a wide range of wavelengths, many overlapping spectra are formed. At a chosen angle, where the longest diffracted wavelength is λ_0, wavelengths at integral submultiples λ_0/n also appear. After being selected by a slit they enter the prism, where they are dispersed into a fan of discrete rays. The minimum deviation of each may now be measured to give a set of points on the graph of deviation as a function of wavelength. By working at different diffraction angles these points are supplemented, and a smooth dispersion curve plotted. It is worth remarking that the dispersion is low in the infra-red, so that the slits must be narrow. Very little energy was available and the full sensitivity of the bolometer was required.

These and other measurements proceeding in parallel at the Reichsanstalt well illustrate the service of exact measurement to fundamental science. In this case, by good fortune, Planck was simultaneously working towards his theoretical radiation formula, and a close collaboration soon established remarkable agreement between theory and experiment [68]. It is tempting to see this as the moment when modern physics came into being, but the truth is that for at least a decade after its invention, we find Planck's formula written about as no more than an empirical result of great interest but uncertain significance

---

† Boys' radiometer mentioned earlier [65] had a tiny thermoelectric circuit suspended by a fine fused silica fibre in a magnetic field, and was in effect a d'Arsonval [66] moving-coil galvanometer with its voltage source incorporated in the coil, so that no leads were needed. In fact, as Boys acknowledged later, d'Arsonval had exhibited a similar instrument in 1886.

See also p 6

[69]. Only Einstein, in the years 1905–7, saw at all clearly what Planck's quantum implied for fundamental physics, and his insights also took some years to become known and appreciated.

1.6. Experimental facilities

The spectroscopic equipment used for the study of black-body radiation illustrates the advanced state of technical optics in 1900. The interest of governments had given a strong impetus to the development of telescopes ever since their invention centuries before, and astronomy was an incidental beneficiary. Improvements to the quality, especially uniformity, of optical glass made possible larger refracting telescopes, culminating in the great refractor of 40" aperture erected at Yerkes observatory (Wisconsin) in 1897. Until Foucault's introduction (1858) of chemically silvered glass, reflectors were made of speculum metal whose liability to tarnish was a severe nuisance. The Earl of Rosse's 72" reflector (1848) at Birr Castle in Ireland, in use until 1878, was the largest until the Mount Wilson (California) 100" telescope was mounted in 1912. Most observatories, even those of high distinction, appear to have been able to carry out their work with much smaller instruments, 10" or 12" being typical of the dozens in operation in the early years of the century [70]. The larger telescopes were well beyond the budget of a university, or even a national observatory, and depended on the generosity of rich amateurs.

It is worth mentioning that amateurs have always played a significant part in astronomy. In the nineteenth century many women of private means in Britain devoted themselves to systematic observation with their own telescopes. The Liverpool Astronomical Society (founded 1881) was one of the very few in the physical sciences that admitted women to membership from the start. It was by no means an insignificant provincial society; the Emperor of Brazil was a member [71].

Until the advent of nuclear physics, relativity and quantum mechanics, the principal, perhaps the only, point of contact between astronomy and physics (apart from the mathematical development of Newtonian dynamics and gravitational theory) lay in the study of absorption spectra as indicators of the elements present in the atmospheres of the Sun and stars. The finding (1868) of a hitherto unknown element, helium, in the atmosphere of the Sun, and its identification (1895) with the gas extracted from the uranium mineral, cleveite, is the most famous episode in a long chronicle of important discoveries [72]. The history of spectroscopy, from its earliest days, has been comprehensively written by Kayser [73] in the first of his many volumes tabulating the spectra of the elements. Though a huge amount of information about wavelengths had been amassed by 1900, few regularities had been discerned, of which Balmer's formula (1885) for the lines of atomic hydrogen is the best remembered, and even that

Physics in 1900

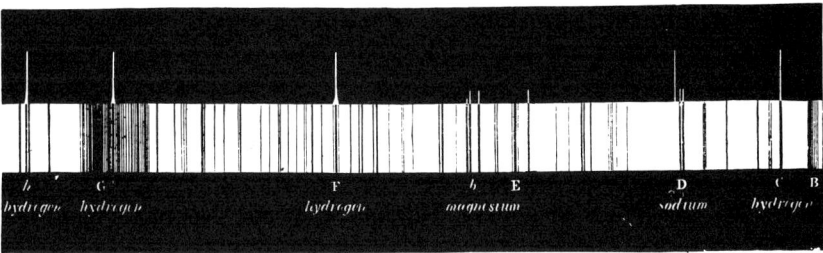

Figure 1.3. *Lockyer (and independently Janssen) focused an image of the edge of the Sun on the input slit of a spectrograph, so that the lower half of the spectrum was formed by the bright photosphere, and the upper half by the chromosphere. This is only visible to the eye during total eclipse, but its light is so concentrated into a few sharp lines that both spectra can be photographed at the same time. These lines appear as emission lines of the chromosphere and as absorption (Fraunhofer) lines in the photosphere. An exception is the bright line just to the left of the sodium doublet (labelled D at the right of the spectrum). Like the hydrogen lines, it is emitted by something that can rise higher in the chromosphere than can the heavier elements Mg and Na, and this led Lockyer and Frankland to propose that there was a hitherto unknown light element in the Sun, which they named helium. (The hydrogen line G, on the left, has wavelength 434 nm and C, on the right, 656 nm.)*

seemed entirely beyond rational explanation at the time. Rydberg had almost arrived at this formula before Balmer, and in about 1890 proposed an extended version which applied rather well to a number of other elements, and is the germ of Ritz's combination principle (1908) [74]. Beyond this, however, spectroscopy remained largely a diagnostic tool in chemistry until the metrologists became interested.

The highest levels of precision in spectroscopy were achieved by Rowland and Michelson in the United States and by the French metrologists Fabry, Perot and Benoît among others. The key to Rowland's success with large concave reflection gratings was his process for grinding screw threads that were immensely superior to anything achieved before [75]. For the first time comparison of wavelengths was possible with errors hardly exceeding one part per million. This invention profoundly impressed the French opticians, Benoît and his colleagues, and stimulated their ingenious advances in interferometry. Among them was the method of exact fractions [76], by which they could avoid having to count the number of wavelengths in a long Fabry–Perot étalon. They then compared the spacings of a series of étalons, each twice as long as the next below, until the largest could be compared directly with the standard metre. In this way they achieved eight-figure accuracy in the measurement of a few selected wavelengths. Michelson had earlier applied his interferometer to the same purpose, showing equal ingenuity in his employment of white-light fringes. This instrument, originally invented to detect the motion of Earth through the aether, became in his hands a remarkably versatile device, of which he gave delightful accounts in his little books, *Lightwaves and their Uses* (1903) and *Studies in Optics* (1927) [77]. It is interesting to note that in the former he dismisses the aether-drift (Michelson–Morley) experiment as a negative result whose prime merit lay in inspiring him to devise his interferometer.

The astronomical telescope and the interferometers of Michelson and of Fabry and Perot were not the only optical instruments which by 1900 had come close to optimal design. Oil-immersion microscope objectives had reached the limit of resolution set by diffraction, and through the work of Rayleigh and Abbe, among others, the general theory had been formulated, though not yet as fully exploited as it was to be in the phase-contrast microscope and other variants. Drude's textbook [78] provides an excellent overview of the advanced state of optical knowledge at the time. It is also one of the first optical texts to present Maxwell's electromagnetic theory of light; the elastic solid theory, or simply an unspecific wave theory, met the needs of most authors.

'My purpose' says Drude 'is attained if these pages strengthen the reader in the view that optics is not an old and worn-out branch of Physics'; from the perspective of 100 years later it appears not indeed worn-out, but certainly mature to a degree that few other branches of experimental physics could claim. To take electrical measurements†, we have already noted the use of the temperamental moving-magnet galvanometer; within a few years this was to be largely replaced by the d'Arsonval type of instrument, where the current flows in a light coil suspended in the strong field of a permanent magnet, so that fluctuations of the ambient magnetic field are of no importance. With the development of improved permanent-magnet materials, compact and relatively cheap instruments became available. Along with the moving-magnet galvanometer, the quadrant and gold-leaf electrometers, of great importance in their time, have become museum pieces whose disappearance from the laboratory was little regretted. The same may be said even more emphatically of a victim of the rise of thermionics, the mechanical oscillograph [80] which, however beautifully designed, was no match either in speed or sensitivity for the cathode-ray oscillographs then beginning to appear. Ferdinand Braun's first instrument [81] (1897) used magnetic deflection of the electron beam in the y-direction, with a spinning mirror to display the time-evolution of the signal along the x-axis. It was Braun also who discovered [82] (1874) the rectifying action

---

† *It is not easy to find instrument makers' catalogues of this period, but the well-illustrated catalogue of the collection in the Whipple Museum, Cambridge* [79] *covers the field fairly thoroughly. A valuable general reference is the* Dictionary of Applied Physics [64], *in which most of the instruments mentioned here, and many others, are treated thoroughly.*

of lead sulphide, but thermionic diodes were introduced [83] (1905) some time before useful solid-state rectifiers. Before then, and indeed for most of the first half of the twentieth century, batteries of electrolytic cells were the standard source of steady current. Trowbridge [84] installed a battery of 20 000 cells in his laboratory at Harvard, giving a steady 40 kV for exciting x-ray tubes, a considerable improvement (he reports) on the induction coil whose intermittent pulses provided the only high-voltage source available to most experimenters.

The study of gas discharges, and particularly the cathode rays emitted during a low-pressure discharge, was an exceptionally fruitful field, with x-rays, the electron and the cathode-ray oscillograph as three important outcomes. Experiments seemed to show [85] that cathode rays could be deflected by a magnetic field, but not by an electric field. This was the cause of widespread confusion, and incidentally explains why Braun used magnetic deflection in his oscillograph. The seat of the trouble was the primitive character of vacuum technology; mechanical pumps (which were usually hand-operated) could not reach a very low pressure, while the mercury-in-glass Sprengel and Toepler pumps [86] were slow and tedious in operation, especially since mercury columns, handled impatiently, are apt at the lowest pressures to slam catastrophically into the sealed-off ends of the glassware. Moreover, it was not fully realized how copiously unbaked glass apparatus can outgas so that, even with the pumps operating continuously, a very low pressure was rarely reached. The residual gas, ionized by the cathode rays, eliminated the electric field between plates which had been provided in the hope of deflecting the rays electrically. It was only by running a discharge for many days in a sealed-off tube that J J Thomson [87] drove enough residual gas off the walls to allow a good vacuum to be maintained. After this the recognition of the electron's existence followed rapidly.

The problem of outgassing is less serious at low temperatures and did not trouble Dewar, whose double-walled vacuum vessels were designed for experiments on liquefied gases, and were first exhibited at the Royal Institution in 1893 [88]. He ingeniously used mercury vapour to expel the air from the interspace before sealing; the residual vapour condensed to a highly reflective film when liquid air was poured in, so that heat gain by radiation was greatly reduced. More complicated vessels, which he needed for his hydrogen liquefier, were made for him in Germany, and thence became more generally available. It was left to a German, R Burger, to appreciate that what can keep liquid cold can also keep it hot, and to acquire, in 1904, patent rights to the Thermos flask [89].

Oxygen and nitrogen had first been indisputably liquefied in 1883. Pictet and Cailletet (1877) independently produced mist in air, but could not collect any liquid. This was achieved some years later by the Cracow physicists, von Wroblewski and Olszewski [90]. By the end of the century the regenerative liquefiers and rectifying columns of Linde and others were providing nitrogen and oxygen for industrial purposes, and liquid air for scientific research [91]. Chemists soon discovered that a liquid-air cooled trap surpassed all other drying agents, and was non-contaminating into the bargain, while Dewar [92] developed the use of coconut charcoal, cooled by externally applied liquid air, to absorb residual gases and produce a really hard vacuum in a sealed vessel. Helium alone resisted absorption, and could thus be prepared, free of other gases, by passage through cooled charcoal. The gases occluded in the monazite sand of Brazil, when treated in this way, provided enough helium for initial studies of its properties. After the liquefaction of hydrogen in 1898 (at 20 K) and its solidification in 1899 (at 12 K) the eventual liquefaction of helium was assured; success came to Kamerlingh Onnes in 1908, at about the same time as it was realized that copious supplies could be extracted from natural gas [72].

A mere recital of significant dates in the production of ever-colder liquid gases would give a false picture of the effort involved, if one failed to appreciate the devoted studies of the $P-V-T$ relations of individual gases which underlay the design of experiments in liquefaction. The demonstration of the critical point in carbon dioxide by Thomas Andrews (1869) was soon followed by van der Waals' proposed equation of state which gave a strong incentive to the study of imperfect gases, and a framework for expressing the data in compact form. According to this equation the inversion temperature of a gas, above which the Joule–Thomson effect (expansion through a nozzle) raises rather than lowers the temperature, is about seven times the critical temperature below which the gas must be cooled if it is to liquefy under pressure. Unreliable though this figure is, it served to inspire measurements at quite modest pressures to locate the inversion temperature, and hence to indicate what preliminary cooling, if any, is needed before regenerative Joule–Thomson cooling will be effective. In addition, of course, it provided an estimate of how far this cooling must proceed before liquefaction is possible. In the last quarter of the nineteenth century much effort went into systematic measurement of the isothermal $P-V$ curves for many different gases, with Regnault and Amagat in France and Kamerlingh Onnes in Holland among the most persistent.

As a final, if minor, example of experimental procedures, the development of fine silica fibres by C V Boys [93] may be cited. His account of how he used a crossbow to draw molten silica into invisible threads, less than $10^{-5}''$ in diameter, makes excellent reading. The fibres, he found, were as strong as steel and, when used in torsion, extraordinarily free from the random drifts which afflicted the silk threads they replaced in moving-magnet galvanometers. Boys' radiometer [65] and his classic determination of the Newtonian constant of gravitation [94] stemmed from this technical innovation. More extensive application of fused silica came later, starting with

See also p 543

Shenstone's [95] introduction (1901) of vitrified silica ware for chemical apparatus.

1.7. The world-picture

Looking back to the beginning of the twentieth century, the physicist is apt to see only the preliminary stirrings of a revolution which swept aside a well-established classical system of thought, based on Newtonian mechanics and Maxwellian electromagnetism. A survey, however, of experimental developments, such as we have just described, shows immediately how much of this seemingly old physics was the product of the preceding quarter-century, often indeed of the decade of the 1890s. Suspicion in continental Europe of Maxwell's theory, fully presented in his great (but difficult) Treatise of 1873, was allayed only by Hertz's demonstrations [96] of electromagnetic waves in 1888, which started a race to transmit and receive signals over ever-greater distances ($\frac{1}{2}$ mile by Rutherford [97] and $1\frac{1}{2}$ miles by Marconi in 1895, across the Atlantic by Marconi [98] in 1901). It should not surprise us, then, that Mendenhall [99], commemorating Rowland in 1901, said 'he lived during a period of almost unparalleled intellectual activity, and his work was done during the last quarter of that century to which we shall long turn with admiration and wonder'. At the same time Oliver Lodge [100], in an elegiac obituary of George Fitzgerald, expressed momentary doubt, only to dismiss it immediately: 'It has been a great epoch this latter half of the nineteenth century ... but it seems closing in. Yet no, there are a few men left, and discoveries are in the air.'. Michelson's famous remark [101], 'Our future discoveries must be looked for in the sixth place of decimals', has sometimes been quoted out of context, as if to imply that he believed all was over for physics. He did, indeed, believe the fundamental laws had been securely established, but not beyond hope of further refinement; and he quoted examples to show that exciting new results (the discoveries of Neptune and argon) had been obtained by observing minute anomalies. In this passage he echoes Maxwell [102] 'I might bring forward instances gathered from every branch of science, showing how the labour of careful measurement has been rewarded by the discovery of new fields of research, and by the development of new scientific ideas'. There is no lack of more recent examples such as the tests of general relativity and of quantum electrodynamics (by the Lamb shift and the g-factor of the electron), but it cannot be denied that the biggest advances came with dramatically new effects like radioactivity which called for talents quite different from Michelson's. It is to the likes of Faraday that we must look to find gifts of the same order as Rutherford's. All the same, the closing-in that Lodge sensed was not quite illusory, as Badash [103] has discussed in some detail. For example, the speculative ferment that gave rise to the variety of hypothetical aether models from Fresnel onwards, and the contrasting German attempts, following Gauss

See also p 1986

and Weber, to adapt action-at-a-distance theories to the latest discoveries in electromagnetism, had largely been abandoned in favour of simpler formal structures of which Lorentz's electron theory [104] was a clear leader.

Before going further, however, into the state of physics theory, a general point is worth making, that the topics encompassed by physics were closely enough related to allow a primarily theoretical physicist to contribute at different times to technical applications and to fundamental speculations, and also to indulge occasionally in serious experimental work; Kelvin, Helmholtz and (later) Einstein are obvious examples. Rayleigh was perhaps the only outstanding physicist to steer clear of fundamentals. Not until the advent of quantum mechanics did a chasm begin to open between those who were excited by the possibility of understanding real materials, and those who were led by the puzzles of the subatomic world to push experiment and theory to increasingly smaller objects and higher energies. Whereas in the 1890s the study of black-body radiation was at the same time an important advance in the measurement, by optical pyrometry, of very high temperatures and an equally important step towards the discovery of the quantum, no one now expects that quarks and chromodynamics will be relevant to the mundane interests of solid-state physicists, chemists and biologists. Such people, being generally satisfied with ideas in quantum mechanics which reached their canonical form sixty years ago, are the spiritual descendants of Rayleigh, for whom the equivalent canons were those of classical mechanics.

The two outstanding preoccupations of physicists during the later decades of the nineteenth century were atoms (or molecules) and the aether. Despite what now seems overwhelming evidence for atoms even then, from the laws of chemical combination and from the kinetic theory of gases, there were doubters, particularly Ernst Mach [105] and Wilhelm Ostwald [106], whose intellectual eminence did not allow their views to be ignored. A few, notably Ludwig Boltzmann, the great pioneer of kinetic theory, may have been disheartened by their antagonism [107], but most (like Planck) [108] responded combatively or (like Poynting) [109] were benignly tolerant of what they saw as an over-reaction to the excessive confidence of some of the atomists. Within a few years the quantitative study of Brownian motion and the discovery of x-ray diffraction had handed victory to the atomists, and we need not trouble to analyse further the influential positivistic views of Mach, though they might be studied with advantage by some modern physicists who dogmatically trumpet their philosophical certainties. We must return later to Ostwald because of his other conflicts with prevailing ideas. On the question of atoms he may be considered as one of the last of the chemists to treat them as no more than useful fictions. In the end, however, he embraced orthodoxy [110].

Ludwig Eduard Boltzmann

(Austrian, 1844–1906)

Showed how Stefan's empirical T^4 law, for the total power radiated by a black body at temperature T, could be derived by the use of thermodynamics. He took the pioneering statistical ideas of Maxwell, whose imaginative insight he greatly admired, and built a systematic theory of gases around the Boltzmann equation, which describes how the distribution of velocities among the molecules develops with time. Various corollaries–the H-theorem, the distribution function, the law of equipartition of energy–are now known by his name, and he was the first to interpret entropy as 'disorder', defined quantitatively. He engaged vehemently with Mach and other influential philosophers of science who sought to abolish unobservable entities, such as atoms, from physics.

The atomists, for all their confidence in basic matters, were far from united in their beliefs about what atoms actually were. W K Clifford's [111] view, to which he was led by the complexity of spectra, that atoms had an elaborate internal structure, was in marked contrast to the conveniently indeterminate model, owing something to the ancient Greeks but more to the eighteenth century mathematician Boscovich, of atoms as indivisible point masses interacting according to some (probably complicated) law of force. This was a good starting point for kinetic theory and was employed powerfully by Maxwell in his analysis of diffusion, viscosity and radiometer forces, even though in his other writings he seems inclined to believe, with Kelvin and many other British physicists, that atoms are not entities in themselves, but manifestations of vortices or other singularities in the aether [112]. By contrast Lorentz and his school assumed the co-existence of matter and aether, which was the medium transmitting the forces of interaction. Individual positively and negatively charged 'electrons' were the elementary material particles which moved freely through the aether without disturbing it. A very full historical account of this complex era of physical thought has been given by Whittaker [104].

Lorentz's theory was the culmination of attempts to resolve the problems arising from the transmission of light through moving transparent media, which in their turn arose from difficulties encountered by the earliest aether models, such as those of Fresnel and Green who imagined it to be an elastic solid. In such models a perpetual headache was the possibility of longitudinal as well as transverse waves, and even as late as 1895, with the discovery of x-rays, there were those, including Röntgen [113] himself, who hailed them as, at long last, the elusive longitudinal vibrations. But such vibrations, it was known, should also be excited when a transverse wave is partially reflected and partially transmitted on striking an interface between media at anything other than normal incidence; there was no evidence for this, nor did the theory give a quantitatively correct account of the reflection coefficient for light at an interface. In any case, before the boundary conditions at the interface can be formulated there must be a theory of dielectric media—for example, does the aether pervade all media uniformly, or has it a different density in dielectrics and in free space? This question arises urgently with moving media, for it determines the extent to which the aether is entrained in the motion. In Fizeau's experiment to test this point, an interferometer was arranged with two beams of light traversing the same path, in one sense moving with a fast water stream, and in the other sense against it. The shift of the interference fringes on reversing the direction of flow was interpreted as showing that when a body of refractive index μ moves with velocity v, the aether drifts through it at a speed v/μ^2 relative to the body. At first sight this seems to dispose of Lorentz's view that the aether is indifferent to the passage of matter, but between Fizeau and Lorentz many developments had occurred. In particular, the Michelson–Morley experiment had failed to detect the expected motion of the Earth relative to the aether. This challenge was picked up independently by Lorentz and by Fitzgerald, who both pointed out [114] that if all bodies suffered a reduction in length by a factor $\sqrt{(1 - v^2/c^2)}$ when moving at a speed v through the aether, the Michelson–Morley experiment would indeed give a null result. This remarkable proposal was not produced out of thin air, but generalized a calculation which showed from Maxwell's theory that the electric and magnetic field patterns produced by a moving charged conducting sphere suffer longitudinal compression at high speeds, and require a corresponding distortion of the sphere into an oblate spheroid with axial ratio $\sqrt{(1 - v^2/c^2)}$. The Lorentz electron model, combined with the Fitzgerald–Lorentz contraction, explained consistently all the experimental results on the optics of moving media.

Looking with modern eyes at the Lorentz theory one wonders why he bothered at all with the aether, since its function is merely to provide

Physics in 1900

Albert Abraham Michelson

(American, 1852–1931)

Was born in Poland but brought as a child to America, where he devoted his life to optical experiments. Over more than fifty years he perfected Foucault's method for determining the velocity of light, and his last measurement, published posthumously, was only superseded as a result of radar developments during World War II. His interferometer, invented to detect the aether drift (the famous Michelson–Morley experiment) proved a versatile instrument which he used to measure wavelengths directly against the standard metre, in the hope of replacing it by a spectroscopic standard, and to study the width of spectral lines and their multiplicity. With a different instrument, the stellar interferometer, he measured for the first time the angular width of a star, Betelgeuse (0.047 seconds of arc).

a medium and a fixed framework in which electromagnetic effects are propagated in accordance with Maxwell's equations. Indeed, it has since been happily abandoned in favour of using the equations and asking no awkward questions. This modern point of view, however, is foreign to the nineteenth century which still nurtured the hope of discovering a completely consistent mechanical model of the Universe, from which inconceivable notions like action-at-a-distance had been banished, and in which all effects were attributed to local causes—electromagnetic forces, for instance, being transmitted from point to neighbouring point by a medium whose mechanical attributes could be defined and made entirely intelligible. It was among the triumphs of Maxwell's theory that the energy possessed by a system of charges and currents could be thought of as distributed throughout space, with (in modern SI notation) an energy density of $\frac{1}{2}ED$ due to the electric field, and $\frac{1}{2}HB$ due to the magnetic field. In addition Poynting's vector $E \times H$ described the flow of energy through space, and tended to confirm the view that energy is something very like a substance in that each element of aether possesses, by virtue of its state of strain, a quantifiable amount of energy which, being conserved, can be followed in its motions and transformations. To be sure, Poynting [115] himself cautioned against dogmatic belief in the actuality of theoretical constructs like aether and its energy, while a little earlier Lord Salisbury [116] (one of the last distinguished politicians to take a keen personal interest in science) had suggested that 'for more than two generations the main, if not the only, function of the word *aether* has been to furnish a nominative case to the verb *to undulate*'. Nevertheless its existence must be reckoned a central article in the belief of most physicists until 1905, when Einstein's relativity theory began to capture the minds of the younger generation and relegate aether to the scrap heap. Many of their elders, including Larmor, were never convinced of the need to change allegiance, and exemplified in a new context Planck's famous remark in his *Scientific Autobiography* [117], referring to his opposition to Mach and Ostwald: 'A new scientific truth does not triumph by convincing its opponents and making them see the light, but rather because its opponents eventually die, and a new generation grows up that is familiar with it.'. To those who yearned for an objective, intelligible world the aether exerted a strong pull, so that as late as 1919 Einstein was stressing that it was not essential to discard it, and in 1955 Dirac (too young to be a survivor of that past age) hoped to revive it in a fully quantized form [118]. But the majority, brought up to abjure metaphysics, were content to have their rules of conduct enshrined in mathematical equations and to seek no deeper meaning such as the aether had once promised to supply. This flight from philosophy, as it usually was, could have claimed support from Karl Pearson [119] who wrote, in *The Grammar of Science* (a book which, written in 1892, casts much light on the climate of scientific thought in its time): 'Hertz's experiments do not seem to me to have logically demonstrated the *perceptual* existence of the aether, but to have immensely increased the validity of the scientific *concept*, aether, by showing that a wider range of perceptual experience may be described in terms of it, than had hitherto been demonstrated by experiment.'. Lorentz himself, lecturing in 1922, was scrupulous to present Einstein's special and general theories as accepted truths, only at the end allowing himself a gentle note of regret: 'As to the aether (to return to it once more), though the conception of it has certain advantages, it must be admitted that if Einstein had maintained it he certainly would not have given us his theory, and so we are very grateful to him for not having gone along the old-fashioned roads.' [120].

The dispute with Ostwald which occasioned Planck's acid comment provides another example of intellectual turmoil at the end of the nineteenth century, arising from the profusion of new ideas. Though the foundations of thermodynamics had been set out in the 1840s and 1850s, and applied by J Willard Gibbs in 1876–8, with superb thoroughness and imaginative power, to chemical processes and other heterogeneous systems, it was still possible for Helm, Ostwald and their followers in the school of energetics [121] to claim that physical processes were almost completely determined by the energy transformations involved. This apotheosis of the first law and disregard for the strict limitations imposed by the second represented a short-lived heresy, but one very offensive to those such as Planck who saw in entropy the clue to the solution of many problems. For him the inspired truth of thermodynamics overrode almost all other concepts, even kinetic theory. For although Boltzmann [122], in his H-theorem†, had provided an exquisite illustration of the law of entropy increase, in Planck's eyes it was no more than a particular case of a law that must transcend particular models, even atomistic theories which were still to be regarded as highly plausible hypotheses rather than established dogma. With hindsight one can appreciate the strength of Planck's position, for Boltzmann's gas theory was, of course, based on Newtonian mechanics, while the law of entropy increase is just as true in the quantum world. As a strict formulation of the basic tendency of all systems left to themselves to become more, never less, disordered, it possesses a universality more profound even than the rules of quantum mechanics.

Of the applications of atomism to explain the properties of matter, the theory of gases was by a long way the farthest advanced. There was, however, a serious difficulty with the specific heats of gases. The ratio $\gamma = C_p/C_v$ should be determined by the number of degrees of freedom possessed by the molecule, being $\frac{5}{3}$ for a monatomic gas such as mercury vapour, argon and neon (the case of argon and neon gave Boltzmann particular satisfaction) [123]. Molecular gases, however, had been an embarrassment from the time of Clausius and Maxwell, since the apparent number of degrees of freedom was almost always less than the expected number, three per atom, and indeed was often fractional; there seemed to be no sensible explanation in terms of internal molecular

---

† *At any instant the distribution of velocities among the molecules of a gas may be specified by the fraction* $f(v_x, v_y, v_z)\, dv_x dv_y dv_z$ *whose velocity components lie within the elementary volume* $dv_x dv_y dv_z$. *Boltzmann defines H as* $\int f \ln f\, dv_x dv_y dv_z$ *and shows that as the molecular velocities are changed by collisions H always decreases. The state of thermal equilibrium is that for which f has the form that minimizes H, and it is in fact the Maxwell distribution of velocities. There is an obvious analogy between* $-H$ *and entropy. A problem arising from this demonstration, one that has been rumbling underground, and occasionally surfacing, ever since, is the fact that strictly according to the laws of mechanics H is as likely to increase as to decrease. Boltzmann's assumption of molecular chaos is enough to eliminate the possibility of increasing H, and is therefore an arbitrary extra imposition. What determines the arrow of time, and whether the law of entropy increase is absolutely or only statistically true, are related questions of which Boltzmann and Planck were very well aware. Insofar as they have been resolved by modern cosmology, information theory or down-to-earth common sense, they belong more to the present century. But they have not completely gone away yet.*

See also p 588

motions. Only opponents of the kinetic theory, among whom was Kelvin [124], took any pleasure in this discrepancy. In due course the necessity of quantizing vibrational modes clarified most of these problems.

When we turn to attempts to understand anything other than thermal properties in atomic terms, the priority assigned to gases is clear. In the *Encyclopaedia Britannica* of 1910 the article *Conduction, electric* devotes nine columns to measurements on metals, with no theory, the same to liquids with serious attention being paid to ionic theories of electrolytic conduction, and 52 substantial columns to gases with a plethora of observations and calculations related to mobility and recombination of ions. On the other hand, 44 columns on *Crystallography* describe in detail the classification of crystal symmetries and the diverse habits of crystallization, without any reference to possible underlying atomic structure. If this edition of the encyclopaedia, with its exceptionally distinguished contributors, is taken to reflect the state of physics in the early years of the century, it must be concluded that beyond the topics already mentioned in this section the compilation of empirical data far outweighed their theoretical interpretation. Newtonian mechanics and electromagnetism had, in general, achieved most of what could be done without excessively complicated analysis. Making sense of the wealth of information on spectra and the properties of matter would have to wait for new ideas beyond the prophetic insights of 1900. While Mendenhall and Lodge did well to extol the advances which we now see as establishing what we call classical physics, the latter's intuition of a changing order was also sound enough, however tentative; neither they, nor any others, can be blamed for having failed to recognize at that early date the immensity of the change that was overtaking the world of physics.

1.8. The seeds of modern physics

In a short space of time four discoveries were made which proved to be of the highest importance to physics and, ultimately, to mankind as a whole. Of these four—x-rays (1895), the Zeeman effect, radioactivity (1896) and the electron (1897)—Röntgen's discovery of x-rays had the most immediate impact both on the popular imagination and on physics. J J Thomson describes in his memoirs [125] how he had a copy of the apparatus set up and looked immediately for any effect on the conductivity of a gas. This was the turning point for him; for the first time he could (as we should put it) ionize the gas without having to pass a hefty high-voltage discharge. From this controllable ionization sprang his systematic study of ions, and in the midst of the frenzy of research thus generated he measured, as already mentioned, the deflection of a beam of cathode rays (for whose production x-rays were in fact unnecessary) by electric and magnetic fields. He was not the first to estimate the ratio, e/m, of charge to mass in the particles of the beam—Wiechert and Kaufman had reached a similar result, as had Lorentz from Zeeman's observation of the magnetic broadening of spectral lines—but he was the first to have the courage to suggest publicly that its very large value implied a charged particle of very small mass; and, of course, recognition of the electron as a universal constituent of matter was a considerable step towards understanding the processes of ionization and recombination which occupied him and his students, not to mention many others elsewhere, for years afterwards; and which still pose problems, and exercise modern students of ionized plasmas.

The discovery of radioactivity in 1896 was a direct consequence of Röntgen's work. Becquerel, who conjectured that x-rays might be a sort of fluorescence, exposed fluorescent salts to sunlight, wrapped them in black paper and placed them on a photographic plate with a cross of copper foil between. After many vain trials he tested a fluorescent uranium salt and at last obtained an image of the cross. Shortly after, however, in a repetition of the experiment the Sun was obscured and he packed away salt and photographic plate in a cupboard, only to find, when he developed the plate days later, an exceptionally strong image. After this happy chance the study of radioactivity proceeded rapidly, so that by 1900 α-, β- and γ-rays had been recognized, and the discovery of 'emanation' (radon) had indicated the possibility of a chain of disintegrations.

These developments, and many events leading to or associated with them, are fully and entertainingly presented by Pais in his book *Inward Bound*, and more briefly in the following chapter. Somewhat earlier, in 1887, the photoelectric effect had been discovered by Hertz. It was explained as the emission of electrons under the influence of light by J J Thomson in 1899, but its importance to fundamental physics was not realized until Einstein's quantum theory of light in 1905. An experimental finding of greater technical than theoretical significance was what was later called thermionic emission [126]. It had long been known that a hot body, placed near an isolated charged object, induced rapid discharge, and a systematic study had been made by Elster and Geitel in 1880. With Thomson's recognition of the electron, and Owen Richardson's exemplary measurements (1903) [127], it became clear that the phenomenon was analogous to the boiling off of electrons from hot metal. Ambrose Fleming [83] showed what thermionic emission held in store when he patented his rectifying valve (diode) in 1904, and at the same time introduced it as a detector of signals from spark transmitters; Lee de Forest's [128] triode in 1906 was followed within a few years by his multistage amplifier and oscillator—the harbinger of the electronic technology which was to transform the world and incidentally, though with leisurely insistence, the techniques of physical measurement.

With these discoveries the golden age of physical research had begun.

References

[1] Aldeva B et al 1990 The L3 collaboration *Phys. Lett.* **236** 109
[2] Hale A 1990 *Physics Abstracts* pp A433, A1770 (Author index)
[3] Thompson S P 1910 *Life of Lord Kelvin* (London: Macmillan) p 1223
[4] Tedder H R 1910 *Encyclopaedia Britannica* vol 25, 11th edn (New York: Encyclopaedia Britannica Inc.) p 309
[5] Caroe G M 1985 *The Royal Institution* (London: Murray)
Thomas J M 1991 *Michael Faraday and The Royal Institution* (Bristol: Hilger)
[6] Thomson J J 1936 *Recollections and Reflections* (London: Bell) p 341
[7] Owen R 1910 *Encyclopaedia Britannica* vol 20, 11th edn (New York: Encyclopaedia Britannica Inc.) p 56
[8] Larmor J 1900 *Report of the 70th meeting of the BAAS* (London: Murray) p 613
[9] Faraday M 1865 *The Chemical History of a Candle* (London: Griffin)
Boys C V 1890 *Soap Bubbles and the Forces that Mould Them* (London: SPCK)
Abbott E A 1884 *Flatland* (London: Seeley)
[10] Dean A 1900 *Science* **11** 763
[11] Carhart H S 1900 *Science* **12** 697
[12] Nye M J 1984 *The Question of the Atom* (Los Angeles, CA: Tomash) p 5
[13] Mehra J 1975 *The Solvay Conferences on Physics* (Dordrecht: Reidel)
[14] See, for example Wilson D B (ed) 1990 *The Correspondence between Sir George Gabriel Stokes and Sir William Thomson* (Cambridge: Cambridge University Press)
[15] Paulsen F (translation Thilly F and Elwang W W) 1906 *The German Universities and University Study* (London: Longmans) p 127
[16] Russell J E 1899 *German High Schools* (New York: Longmans) p 335
[17] Russell J E 1899 *German High Schools* (New York: Longmans) p 349
[18] Layton D 1973 *Science for the People* (London: Allen and Unwin)
[19] Layton D 1973 *Science for the People* (London: Allen and Unwin) p 73
[20] Christie O F 1935 *A History of Clifton College, 1860–1934* (Bristol: Arrowsmith) p 207
[21] Winstanley D A 1947 *Later Victorian Cambridge* (Cambridge: Cambridge University Press) p 190
[22] Smith A and Hall E H 1902 *The Teaching of Chemistry and Physics in the Secondary School* (New York: Longmans) p 361
[23] Smith A and Hall E H 1902 *The Teaching of Chemistry and Physics in the Secondary School* (New York: Longmans) p 370
[24] Forman P, Heilbron J L and Weart S 1975 *Historical Studies in the Physical Sciences* **5**
[25] Thompson S P 1897 *Light, Visible and Invisible* (London: Macmillan) p 262
[26] Mulligen J B 1910 *Encyclopaedia Britannica* vol 27, 11th edn (New York: Encyclopaedia Britannica Inc.) p 748
Paulsen F (translation Thilly F and Elwang W W) 1906 *The German Universities and University Study* (London: Longmans) p 306
[27] Royal Society of London 1900 *Year-Book* (London: Harrison)
[28] The examination papers were published annually by the Cambridge University Press. Examinations results were published in the *Cambridge University Reporter*
[29] Venn J A 1940 *Alumni Cantabrigienses* part 2 (Cambridge: Cambridge University Press)
[30] Cahan D 1985 *Historical Studies in the Physical Sciences* **15**(2) 1
[31] See, for example Margenau H and Murphy G M 1943 *The Mathematics of Physics and Chemistry* (New York: Van Nostrand) ch 1

[32] Todhunter I 1873 *The Conflict of Studies* (London: Macmillan) p 13
[33] Jungnickel C and McCormmach R 1986 *Intellectual Mastery of Nature* vol 1 (Chicago: Chicago University Press) ch 1
[34] Crowther J G 1974 *The Cavendish Laboratory 1874–1974* (London: Macmillan) p 10
[35] Wilson D B 1982 *Historical Studies in the Physical Sciences* **12** 325
[36] Cajori F 1899 *A History of Physics* (New York: Macmillan) p 286
[37] Olesko K M 1991 *Physics as a Calling* (Ithaca, NY: Cornell University Press)
[38] Caullery M (translation Woods J H and Russell E) 1922 *Universities and Scientific Life in the United States* (Cambridge, MA: Harvard University Press) p 247
[39] Paul H W 1985 *From Knowledge to Power* (Cambridge: Cambridge University Press)
[40] Kohlrausch F W G 1870 *Leitfaden der Praktischen Physik* (Leipzig: Teubner) (translation of 2nd edn Wallen T H and Procter H R 1873 *An Introduction to Physical Measurements* (London: Churchill))
[41] Smith A and Hall E H 1902 *The Teaching of Chemistry and Physics in the Secondary School* (New York: Longmans) p 270
[42] Pickering E C 1874–6 *Elements of Physical Manipulation* (London: Macmillan)
[43] Mendenhall T C 1887 *Science* **9** 240
[44] Kevles D J 1978 *The Physicists* (New York: Knopf) ch 1–5
[45] Gillispie C C (ed) 1970 *Dictionary of Scientific Biography* (New York: Scribner)
[46] Caullery M (translation Woods J H and Russell E) 1922 *Universities and Scientific Life in the United States* (Cambridge, MA: Harvard University Press) p 97; see also Kevles D J 1978 *The Physicists* (New York: Knopf) ch 1–5
[47] Fox R and Weisz G (ed) 1988 *The Organization of Science and Technology in France 1808–1914* (Cambridge: Cambridge University Press)
[48] Zeeman P 1913 *Researches in Magneto-optics* (London: Macmillan) p 25
[49] Rowland H A 1883 *Am. J. Sci.* **26** 87
[50] Steel W H 1967 *Interferometry* (Cambridge: Cambridge University Press) pp 116, 201
[51] Mackay A L 1991 *A Dictionary of Scientific Quotations* (Bristol: Hilger) p 214
[52] Cornford F M 1908 *Microcosmographia Academica* (Cambridge: Bowes and Bowes)
[53] Caullery M (translation Woods J H and Russell E) 1922 *Universities and Scientific Life in the United States* (Cambridge, MA: Harvard University Press) p 164
[54] Carhart H S 1900 *Science* **12** 697
Cahan D 1992 *Meister der Messung* (Weinheim: VCH)
[55] Terrien J *et al* 1975 *Le Bureau International des Poids et Mesures 1875–1975* (Sèvres: BIPM)
[56] Pyatt E 1983 *The National Physical Laboratory* (Bristol: Hilger)
[57] Pellat H 1900 *Rapports Présentés au Congrès International de Physique* vol 1 (Paris: Gauthier-Villars) p 101
[58] Kevles D J 1978 *The Physicists* (New York: Knopf) p 66
[59] Strutt R J S (Fourth Baron Rayleigh) 1924 *John William Strutt, third Baron Rayleigh* (London: Arnold) ch 7
[60] Klein M J 1962 *Archive for History of Exact Sciences* **1** 459; see also Kuhn T S 1978 *Black-body Theory and the Quantum Discontinuity 1894–1912* (New York: Oxford University Press)
[61] Wien W and Lummer O R 1895 *Ann. Phys., Lpz* **56** 451

[62] Holborn L and Wien W 1892 *Ann. Phys., Lpz* **47** 107
[63] Langley S P 1881 *Am. J. Sci.* **21** 187; 1886 *Ann. Chim. Physique* **9** 433
[64] Glazebrook R (ed) 1922 *A Dictionary of Applied Physics* vol 2 (London: Macmillan) p 368
[65] Boys C V 1889 *Phil. Trans. R. Soc.* A **180** 159
[66] Gray A 1921 *Absolute Measurements in Electricity and Magnetism* (London: Macmillan) p 412
[67] Paschen F 1894 *Ann. Phys., Lpz* **53** 301
[68] Jungnickel C and McCormmach R 1986 *Intellectual Mastery of Nature* vol 2 (Chicago: Chicago University Press) p 259
[69] See, for example Callendar H L 1910 *Encyclopaedia Britannica* vol 13, 11th edn (New York: Encyclopaedia Britannica Inc.) p 156
[70] Dreyer J L E 1910 *Encyclopaedia Britannica* vol 19, 11th edn (New York: Encyclopaedia Britannica Inc.) p 953
[71] Chapman A 1994 *Yearbook of Astronomy* (London: Sidgwick and Jackson) p 159
[72] Keesom W H 1942 *Helium* (Amsterdam: Elsevier) p 1
[73] Kayser H 1900 *Handbuch der Spectroscopie* vol 1 (Leipzig: Hirzel)
[74] Gillispie C C (ed) 1970 *Dictionary of Scientific Biography* (New York: Scribner) (see entries for Balmer, Rydberg and Ritz)
[75] Rowland H A 1902 *Physical Papers* (Baltimore, MD: Johns Hopkins) p 485
[76] Françon M 1966 *Optical Interferometry* (New York: Academic) p 260
[77] Michelson A A 1903 *Light Waves and their Uses* (Chicago: Chicago University Press); 1927 *Studies in Optics* (Chicago: Chicago University Press)
[78] Drude P 1900 *Lehrbuch der Optik* (Leipzig: Hirzel) (translation Mann C R and Millikan R A 1902 *The Theory of Optics* (London: Longmans))
[79] Lyall K 1991 *Electrical and Magnetic Instruments* (Cambridge: Whipple Museum)
[80] Duddell W D B 1897 *The Electrician* **39** 636
[81] Braun F 1897 *Ann. Phys., Lpz* **60** 552
[82] Braun F 1874 *Ann. Phys., Lpz* **102** 550
[83] Fleming J A 1905 *Proc. R. Soc.* A **74** 476
[84] Trowbridge J 1900 *Am. J. Sci.* **9** 439
[85] Whittaker E 1951 *A History of the Theories of Aether and Electricity* vol 1 (London: Nelson) p 350
[86] Travers M W 1901 *The Experimental Study of Gases* (London: Macmillan) ch 2
[87] Thomson J J 1936 *Recollections and Reflections* (London: Bell) p 334
[88] Dewar J 1893 *Proc. R. Inst.* **14** 1
[89] Burger R 1904 *British Patent* 4421
[90] von Wroblewski S and Olszewski K 1883 *Ann. Phys., Lpz* **20** 243
[91] Ruhemann M and B 1937 *Low Temperature Physics* (Cambridge: Cambridge University Press) ch 2
[92] Dewar J 1905 *Proc. R. Inst.* **18** 177
[93] Boys C V 1887 *Phil. Mag.* **23** 489
[94] Boys C V 1895 *Phil. Trans. R. Soc.* A **186** 1
[95] Shenstone W A 1901 *Proc. R. Inst.* **16** 525
[96] Whittaker E 1951 *A History of the Theories of Aether and Electricity* vol 1 (London: Nelson) p 322
[97] Wilson D 1983 *Rutherford* (London: Hodder and Stoughton) p 89
[98] Gillispie C C (ed) 1970 *Dictionary of Scientific Biography* vol 9 (New York: Scribner) p 98

[99] Mendenhall T C 1902 *Physical Papers of H A Rowland* (Baltimore, MD: Johns Hopkins) p 1
[100] Lodge O J 1901 *The Electrician* **46** 701
[101] Michelson A A 1903 *Light Waves and their Uses* (Chicago: Chicago University Press) p 24; see also Hiebert E N 1990 *Fin de Siècle and its Legacy* (ed) Teich M and Porter R (Cambridge: Cambridge University Press) p 235
[102] Maxwell J C 1890 *Scientific Papers* vol 2 (Cambridge: Cambridge University Press) p 244
[103] Badash L 1972 *Isis* **63** 48
[104] Whittaker E 1951 *A History of the Theories of Aether and Electricity* vol 1 (London: Nelson) ch 13
[105] Brush S G 1976 *The Kind of Motion We Call Heat* vol 1 (Amsterdam: North-Holland) ch 8
[106] Nye M J 1984 *The Question of the Atom* (Los Angeles, CA: Tomash) p 337
[107] Brush S G 1976 *The Kind of Motion We Call Heat* vol 1 (Amsterdam: North-Holland) p 293
[108] Heilbron J L 1986 *The Dilemmas of an Upright Man* (Berkeley, CA: University of California Press) p 44
[109] Poynting J H 1920 *Collected Scientific Papers* (Cambridge: Cambridge University Press) p 607
[110] Brush S G 1976 *The Kind of Motion We Call Heat* vol 1 (Amsterdam: North-Holland) p 299
[111] Clifford W K 1879 *Lectures and Essays* vol 1 (London: Macmillan) p 172
[112] Brush S G 1976 *The Kind of Motion We Call Heat* vol 1 (Amsterdam: North-Holland) p 92
[113] Pais A 1986 *Inward Bound* (Oxford: Clarendon) p 41
[114] Fitzgerald G F 1889 *Science* **13** 390
Whittaker E 1951 *A History of the Theories of Aether and Electricity* vol 1 (London: Nelson) p 404; see also Brush S G 1967 *Isis* **58** 230
[115] Poynting J H 1920 *Collected Scientific Papers* (Cambridge: Cambridge University Press) p 606
[116] Salisbury Marquis of 1894 *Nature* **50** 341
[117] Planck M 1950 *Scientific Autobiography* (London: Williams and Norgate) p 33
[118] Kragh H 1990 *Dirac* (Cambridge: Cambridge University Press) p 200
[119] Pearson K 1892 *The Grammar of Science* (London: Scott) p 214
[120] Lorentz H A 1927 *Problems of Modern Physics* (Boston, MA: Ginn) p 221
[121] Brush S G 1976 *The Kind of Motion We Call Heat* vol 1 (Amsterdam: North-Holland) p 61 and elsewhere
[122] Boltzmann L (translation Brush S G) 1964 *Lectures on Gas Theory* (Berkeley, CA: University of California Press) p 49
[123] Boltzmann L (translation Brush S G) 1964 *Lectures on Gas Theory* (Berkeley, CA: University of California Press) p 216
[124] Smith C and Wise M N 1989 *Energy and Empire* (Cambridge: Cambridge University Press) p 428
[125] Thomson J J 1936 *Recollections and Reflections* (London: Bell) p 325
[126] Richardson O W 1921 *The Emission of Electricity from Hot Bodies* (London: Longmans) ch 1
[127] Richardson O W 1903 *Phil. Trans. R. Soc.* A **201** 497
[128] Gillispie C C (ed) 1970 *Dictionary of Scientific Biography* vol 4 (New York: Scribner) p 6

Chapter 2

INTRODUCING ATOMS AND THEIR NUCLEI

Abraham Pais

PART 1: ATOMS

2.1. Prelude

In 1913 Jean Perrin wrote in his excellent book *Les Atomes* [1] '*La théorie atomique a triomphé. Nombreux encore naguère, ses adversaires enfin conquis renoncent l'un après l'autre aux défiances qui longtemps furent légitimes et sans doute utiles.*'. (The atomic theory has triumphed. Until recently still numerous, its adversaries, at last overcome, now renounce one after another their misgivings, which were, for so long, both legitimate and undeniably useful.)

Perrin's enthusiasm is understandable and justified. Fine experimental work by him and his school, inspired by Einstein's 1905 molecular theory of Brownian motion, had shown that the latter's formula for Avogadro's number, the magnitude of which was already known by then from other arguments, had yielded a sensible experimental value, and thus had verified the molecular/atomic basis of that theory.

Perrin's exaltation on the 'numerous adversaries at last overcome' needs critical scrutiny, however. The fact of the matter is that the acceptance of the reality of atoms and molecules was already widespread as the nineteenth century drew to a close, though there were still some pockets of resistance. It would seem to be a fitting prelude to the present chapter to inquire how and why many scientists had come to some early—limited—understanding about the discrete structure of matter.

The term *atom*, derived from the Greek α, a privative, and $\tau\acute{\epsilon}\mu\epsilon\iota\nu$, to cut, appears first, I am told, in the writings of Greek philosophers of the fifth century BC. Democritus (late fifth century BC) taught that atoms are the smallest parts of matter, though in his view they were not necessarily minute. Empedocles (490–430 BC), physicist, physician, and statesman, held that there are four indestructible and unchangeable elements, fire, air, water and earth, eternally brought into union, and eternally parted from each other by two divine forces, love and discord. Nothing new comes or can come into being. The only changes that can occur are those in the juxtaposition of element with element. Epicurus' (341–270 BC) opinion that atoms cannot be divided into smaller parts by physical means, yet that they have structure, was shared by prominent scientists well into the nineteenth century AD. The Roman poet Lucretius (98–55 BC) was an eloquent exponent of the theory that atoms, infinite in number but limited in their varieties, are, along with empty space, the only eternal and immutable entities of which our physical world is made. Today's scientist will not fail to note that in each of these speculative thinkers' considerations one finds elements that sound curiously modern.

The opposite position, that matter is infinitely divisible and continuous, likewise had its early distinguished proponents, notably Anaxagoras (c 500–c 428 BC) and Aristotle (384–322 BC). The latter's prestige eclipsed the atomist's view until the seventeenth century. Even that late, René Descartes (1596–1650) pronounced that there cannot exist any atoms or parts of matter that are of their own nature indivisible, for though God had rendered a particle so small that it was not in the power of any creature to divide it, He could not, however, deprive himself of the ability to do so [2].

Regarding the understanding of the basic structure of matter, very little changed between the days of speculation by the ancient Greek philosophers and the beginning of the nineteenth century, when, in 1808, the British chemist and physicist John Dalton (1766–1844) commenced publication of his 'New System of Chemical Philosophy'. He had of course illustrious precursors, notably Antoine-Laurent Lavoisier (1743–94). Yet his quantitative theory suddenly could explain or predict such a wealth of facts that he may properly be regarded as the founder of modern chemistry. In a sequel volume (1810) Dalton expressed the fundamental principle of the youngest of the sciences in these words

> I should apprehend there are a considerable number of what may be properly called *elementary* principles, which can never be metamorphosed, one into another, by any power we can control. We ought, however, to avail ourselves of every means to reduce the number of bodies or principles of this appearance as much as possible; and after all we may not know what elements are absolutely indecomposable, and what are refractory, because we do not know the proper means for their reduction. We have already observed that all *atoms of the same kind*, whether simple or compound, must necessarily be conceived to be alike in shape, weight, and every other particular.

The modern nomenclature for elements in terms of the initial one or two letters of their Latin (sometimes Greek) names is due to the Swedish chemist Jöns Jakob Berzelius (1779–1848) who also introduced the notation that exhibits the number of atoms of each element present in a molecule by means of a numerical subscript.

Note that Dalton's compound atom is what we call a molecule. Great confusion reigned through most of the nineteenth century regarding such terminology, one man's molecule being another man's atom. The need for a common language developed, but slowly. Fifty years later, at the first international scientific conference ever held, the steering committee of the 1860 Karlsruhe congress of chemists† still considered it necessary to put at the top of the agenda of points to be discussed the question 'Shall a difference be made between the expressions *molecule* and *atom*, such that a molecule be named the smallest particle of bodies which can enter into chemical reactions and which may be compared to each other in regard to physical properties—atoms being the smallest particles of those bodies which are contained in molecules?' [3]. More interesting than the question itself is the fact that, even in 1860, no consensus was reached.

Especially illuminating for an understanding of science in the nineteenth century are the topics discussed by August Kekulé (1829–96) in the course of his opening address to the Karlsruhe conference. '[He] spoke on the difference between the physical molecule and the chemical molecule, and the distinction between these and the atom. The physical molecule, refers, he said, to the particle of gas, liquid, or solid in question. The chemical molecule is the smallest particle of a body which enters or leaves a chemical reaction. These are not indivisible. Atoms are particles not further divisible' [3]. Both physics and chemistry could have profited if more attention had been paid to the comment by Stanislao Cannizzaro (1826–1910) in the discussion following Kekulé's paper, that the distinction between physical and chemical molecules has no experimental basis and is therefore unnecessary. Indeed, perhaps the most remarkable fact about the nineteenth century debates on atoms and molecules is the large extent to which chemists and physicists spoke at cross purposes when they did not actually ignore each other. This is not to say that there existed one common view among chemists, another among physicists. Rather, in either camp there were many and often strongly diverging opinions which need not be spelled out in detail here. It should suffice to give a few illustrative examples and to note in particular the central themes. The principal point of debate among chemists was whether atoms were real objects or only mnemonic

---

† *The meeting was held 3–5 September 1860. There were 127 chemists in attendance. Participants came from Austria, Belgium, France, Germany, Great Britain, Italy, Mexico, Poland, Russia, Spain, Sweden and Switzerland.*

devices for coding chemical regularities and laws. The main issues for the physicists centred around the kinetic theory of gases; in particular, around the meaning of the second law of thermodynamics.

An early illustration of the dichotomies between the chemists and the physicists is provided by the fact that Dalton did not accept the hypothesis put forward in 1811 by Amedeo Avogadro (1776–1856) that for fixed temperature and pressure, equal volumes of gases contain equal numbers of molecules. Nor was Dalton's position one held only by a single person for a brief time. By all accounts the high point of the Karlsruhe congress was the address by Cannizzaro, in which it was still necessary for the speaker to emphasize the importance of Avogadro's principle for chemical considerations. That conference did not at once succeed in bringing chemists closer together. However, it was recalled by Dmitri Ivanovich Mendeleev (1834–1907) thirty years later that 'the law of Avogadro received by means of the congress a wider development, and soon afterwards conquered all minds' [4].

The tardiness with which Avogadro's law came to be accepted clearly indicates the widespread resistance to the idea of molecular reality. As but one further illustration of this attitude I mention some revealing remarks by Alexander Williamson (1824–1904), himself a convinced atomist. In his presidential address of 1869 to the London Chemical Society, he said 'It sometimes happens that chemists of high authority refer publicly to the atomic theory as something they would be glad to dispense with, and which they are ashamed of using. They seem to look upon it as something distinct from the general facts of chemistry, and something which the science would gain by throwing off entirely. ... On the one hand, all chemists use the atomic theory, and ... on the other hand, a considerable number view it with mistrust, some with positive dislike. If the theory really is as uncertain and unnecessary as they imagine it to be, let its defects be laid bare and examined. Let them be remedied if possible, or let the theory be rejected, and some other theory be used in its stead, if its defects are really as irremediable and as grave as is implied by the sneers of its detractors' [5].

As a final comment on chemistry in the nineteenth century, mention should be made of another regularity bearing on the atomicity of matter and discovered in that period. In an anonymous paper written in 1815, William Prout (1785–1850), a practising physician in London with a great interest in chemistry, claimed to have shown that the specific gravities of atomic species can be expressed as integral multiples of a fundamental unit. In an addendum written the next year, and also published anonymously [6], he noted that this fundamental unit may be identified with the specific gravity of hydrogen 'We may almost consider the $\pi\rho\omega\tau\eta\ \H{\upsilon}\lambda\eta$ of the ancients to be realized in hydrogen'. Yet Prout did not consider his hypothesis as a hint for the reality of atoms 'The light in which I have always been accustomed to consider it [the atomic theory] has been very analogous to that in which I believe most botanists now consider the Linnean system; namely, as a conventional artifice, exceedingly convenient for many purposes but which does not represent nature' [7].

On the whole, molecular reality met with less early resistance in physics than it did in chemistry. That is not surprising. Physicists could already *do* things with molecules and atoms at a time when chemists could, for most purposes, take them to be real or leave them as coding devices. Let us sample a few of the physicists' thoughts.

The insight that gases are composed of discrete particles dates back at least to the eighteenth century. Daniel Bernoulli (1700–82) may have been the first to state that gas pressure is caused by the collisions of particles with the walls within which they are contained. The nineteenth century masters of kinetic theory were atomists—by definition, one might say. In Rudolf Clausius' (1822–88) paper of 1857 entitled *On the kind of motion we call heat* [8], the distinction between solids, liquids, and gases is related to different types of molecular motion. In 1873, James Clerk Maxwell (1831–79) said 'Though in the course of ages catastrophes have occurred and may yet occur in the heavens, though ancient systems may be dissolved and new systems evolved out of their ruins, the molecules [i.e. atoms!] out of which these systems [the Earth and the whole solar system] are built—the foundation stones of the material universe—remain unbroken and unworn. They continue this day as they were created—perfect in number and measure and weight ...' [9].

Ludwig Boltzmann (1844–1906) was less emphatic and in fact reticent at times, but he could hardly have developed his theory of the second law had he not believed in the particulate structure of matter. His assertion that entropy increases almost always, rather than always, was indeed very hard to swallow for those who did not believe in molecular reality. Max Planck (1858–1947), then an outspoken sceptic, saw this clearly when in 1883 he wrote 'The consistent implementation of the second law [i.e. to Planck, increase of entropy as an absolute law] ... is incompatible with the assumption of finite atoms. One may anticipate that in the course of the further development of the theory a battle between these two hypotheses will develop which will cost one of them its life' [10]. This is the battle which Wilhelm Ostwald (1853–1932) joined in 1895 when he addressed a meeting of the Deutsche Gesellschaft für Naturforscher und Ärzte 'The proposition that all natural phenomena can ultimately be reduced to mechanical ones cannot even be taken as a useful working hypothesis: it is simply a mistake. This mistake is most clearly revealed by the following fact. All the equations of mechanics have the property that they admit of sign inversion in the temporal quantities. That is to say, theoretically perfectly mechanical processes can develop equally well forward and backward [in time]. Thus, in a purely mechanical world there could not be a before and an after as we have in our world: the tree could become a shoot and a seed again, the butterfly turn back into a caterpillar, and the old man into a child. No explanation is given by the mechanistic doctrine for the fact that this does not happen, nor can it be given because of the fundamental property of the mechanical equations. The actual irreversibility of natural phenomena thus proves the existence of processes that cannot be described by mechanical equations; and with this the verdict on scientific materialism is settled' [11].

Such were the utterances with which Boltzmann, also present at that meeting, had to cope. We are fortunate to have an eye-witness report of the ensuing discussion from a young physicist who attended the conference, Arnold Sommerfeld (1868–1951). 'The paper on *Energetik* was given by Georg Helm (1851–1923) from Dresden; behind him stood Wilhelm Ostwald, behind both the philosophy of Ernst Mach (1838–1916), who was not present. The opponent was Boltzmann, seconded by Felix Klein. Both externally and internally, the battle between Boltzmann and Ostwald resembled the battle of the bull with the supple fighter. However, this time the bull was victorious over the torero in spite of the latter's artful combat. The arguments of Boltzmann carried the day. We, the young mathematicians of that time, were all on the side of Boltzmann; it was entirely obvious to us that one could not possibly deduce the equations of motion for even a single mass point—let alone for a system with many degrees of freedom—from the single energy equation ...' [12]. As regards the position of Ernst Mach, it was anti-atomistic but of a far more sober variety than Ostwald's: 'It would not become physical science [said Mach] to see in its self-created, changeable, economical tools, molecules and atoms, realities behind phenomena ... the atom must remain a tool ... like the function of mathematics' [13].

Long before these learned *fin de siècle* discourses took place, in fact long before the laws of thermodynamics were formulated, theoretical attempts had begun to estimate the dimensions of molecules. As early as 1816 Thomas Young (1773–1829) noted that 'the diameter or distance of the particles of water is between the two thousand and the ten thousand millionth of an inch' [14]. In 1866 Johann Loschmidt (1821–95) calculated the diameter of an air molecule and concluded that 'in the domain of atoms and molecules the appropriate measure of length is the millionth of the millimetre' [15]. Four years later Kelvin (1824–1907), who regarded it 'as an established fact of science that a gas consists of moving molecules', found that 'the diameter of the gaseous molecule cannot be less than 2×10^{-9} of a centimetre'. In 1873 Maxwell stated that the diameter of a hydrogen molecule is about 6×10^{-8} cm [16]. In that same year Johannes Diderik van der Waals (1837–1923) reported similar results in his doctoral thesis [17]. By 1890 the spread in these values, and those obtained by others, had narrowed considerably. A review of the results up to the late 1880s placed the radii of hydrogen and air molecules between 1 and 2×10^{-8} cm [18], a remarkably sensible range. Some of the physicists just mentioned used methods that enabled them to also determine Avogadro's number N, the number of molecules per mole. For example, Loschmidt's calculations of 1866 imply [15] that $N \approx 0.5 \times 10^{23}$ and Maxwell found [16] $N \approx 4 \times 10^{23}$. The present best value [19] is

$$N \approx 6.02 \times 10^{23}.$$

Towards the end of the nineteenth century, the spread in the various determinations of N was roughly 10^{22} to 10^{24}, an admirable achievement in view of the crudeness—stressed by all who worked on the subject—of the models and methods used. I regret that this is not the place to deal with the sometimes obscure and often wonderful physics contained in these papers, in which the authors strike out into unexplored territory.

So far I have emphasized differences between physics and chemistry in regard to their assessments of atomic reality. I should now mention two points on which these two groups of nineteenth century professionals were in agreement. Interestingly enough, it would become clear in the twentieth century that their shared opinions on both these issues were incorrect.

First point. Until the very last years of the nineteenth century, most if not all scientists who believed in the reality of atoms shared Maxwell's view that these particles remain unbroken and unworn. 'They are ... the only material things which still remain in the precise condition in which they first began to exist' he wrote in his book, *Theory of Heat*, which contains the finest expression of his atomic credo [20]. It is true that many of these same physicists (Maxwell among them) were convinced that something had to rattle inside the atom in order to explain atomic spectra. Therefore, while there was a need for a picture of the atom as a body with structure, this did not mean (so it seemed) that one could take the atom apart. However, in 1899, two years after his discovery of the electron, Joseph John Thomson (1856–1940) announced that the atom had been split 'Electrification [that is, ionization] essentially involves the splitting of the atom, a part of the mass of the atom getting free and becoming detached from the original atom' [21]. By that time it was becoming increasingly clear that radioactive phenomena (first discovered in 1896) also had to be explained in terms of a divisible atom. 'Atoms [of radioactive elements], indivisible from the chemical point of view, are here divisible' Marie Curie wrote in 1900, adding that the explanation of radioactivity in terms of the expulsion of subatomic particles 'seriously undermines the principles of chemistry' [22] (I come back later to Thomson's and Curie's works).

Thus, at the turn of the century, the classical atomists, those who believed both in atoms and in their indivisibility, were under fire from two sides. There was a rapidly dwindling minority of conservatives, led

Twentieth Century Physics

Figure 2.1. *J J Thomson at his work in his laboratory.*

by the influential Ostwald and Mach, who did not believe in atoms at all. At the same time a new breed arose, people such as J J Thomson (figure 2.1), the Curies, and Rutherford, all convinced of the reality of atoms and all—though not always without trepidation, as in the case of Marie Curie—aware of the fact that chemistry was not the last chapter in particle physics. For them, the ancient speculations about atoms had become reality and the old dream of transmutation had become inevitable.

Second point. If there was one issue on which there was agreement between physicists and chemists, atomists or not, it was that atoms, if they existed at all, were too small to be seen. Perhaps no one expressed this view more eloquently than van der Waals in the closing lines of his 1873 doctoral thesis, where he expressed the hope that his work might contribute to bringing closer the time when 'the motion of the planets and the music of the spheres will be forgotten for a while in admiration of the delicate and artful web formed by the orbits of those invisible atoms' [17].

Direct images of atoms were at last produced in the 1950s with the field ion microscope [23].

I conclude this prelude with a comment on another discrete attribute of matter, this one dating back no further than to the nineteenth century: the atomicity of electric charge.

Introducing Atoms and Their Nuclei

The first description of the electric current as a stream of discrete electric charges appeared only in the 1840s. By that time, theoretical speculations of this kind were influenced above all by Michael Faraday's (1791–1867) researches in 1833 on electrolysis. As Maxwell said sometime later in his treatise on electricity and magnetism 'Of all electrical phenomena electrolysis appears the most likely to furnish us with a real insight into the true nature of the electric current, because we find currents of ordinary matter and currents of electricity forming parts of the same phenomenon' [24].

In modern terms Faraday's electrolytic law can be stated as follows. The amount of electricity deposited at the anode by a gram atom of monovalent ions is a universal constant, called the Faraday (F), given by

$$F = Ne$$

where N is again Avogadro's number, and e is the charge of the electron.

'Although we know nothing of what an atom is', Faraday wrote in summarizing his investigations in electrolysis 'yet we cannot resist forming some idea of a small particle, which represents it to the mind ... there is an immensity of facts which justify us in believing that the atoms of matter are in some way endowed or associated with electrical powers, to which they owe their most striking qualities, and amongst them their chemical affinity' [25]. This statement might seem to indicate that he was a believer in the reality of atoms, an atomist; it certainly does indicate an early vision of intra-atomic forces. Yet Faraday's position in the matter of atomic reality was not that unambiguous 'I must confess I am jealous of the term *atom*; for though it is very easy to talk of atoms, it is very difficult to form a clear idea of their nature, especially when compound bodies are under consideration' [26]. That is the true Faraday, exquisite experimentalist, who would only accept what he was forced to believe on experimental grounds.

Maxwell's position on these issues provides another warning against the oversimplified belief that nineteenth century scientists belonged to either of two camps, atomists or anti-atomists. Of course Maxwell believed in the reality of atoms. More than that, he was convinced that an atom was not just a tiny rigid body but had to have some structure. 'The spectroscope tells us that some molecules [our atoms] can execute a great many different kinds of vibrations. They must therefore be systems of a very considerable degree of complexity, having far more than six variables [the number characteristic for a rigid body] ...' he said in a lecture given in 1875 [7]. But he also believed that the structured atom was unbreakable, as already mentioned earlier [28].

In regard to electrolysis, Maxwell's prejudice of atoms unbroken and unworn had to lead him into a quandary. On the one hand, he was willing to concede that electrolysis indicated the existence of 'the most natural unit of electricity' [29], but on the other, he demanded a *dynamical* understanding of the *universality* of this unit. That, however, is possible only if one understands that an ion is a broken atom (or an atom with an atomic fragment tagged on). He therefore *had* to express grave reservations concerning the atomicity of electricity 'If we ... assume that the molecules of the ions within the electrolyte are actually charged with certain definite quantities of electricity, positive and negative, so that the electrolytic current is simply a current of convection, we find that this tempting hypothesis leads us into very difficult ground ... the electrification of a molecule ... though easily spoken of, is not so easily conceived' [29].

Nevertheless, the modern view of the atomicity of electricity gained advocates, the most influential one being Hermann Helmholtz (1824–94), who said in his 1881 Faraday Lecture 'The most startling result of Faraday's law is perhaps this. If we accept the hypothesis that the elementary substances are composed of atoms, we cannot avoid concluding that electricity also, positive as well as negative, is divided into definite elementary portions, which behave like atoms of electricity' [30]—a statement which explains why in subsequent years the quantity e was occasionally referred to in the German literature as '*das Helmholtzsche Elementarquantum*' [31].

Finally, already in 1874 the Irish physicist George Johnstone Stoney (1826–1911) reported to a meeting of the British Association for the Advancement of Science an estimate for e, the first of its kind, based on Faraday's law and the best then available data for F and N. He obtained $e = 3 \times 10^{-11}$ esu, too small by a factor ~ 20, yet not at all bad for a first and very early try [32].

The moral of the story so far is that even before the twentieth century one recognizes results and conjectures that were to survive in later years. At the same time one notes that even some of the greatest scientists were bewildered—as well they should have been.

2.2. The decade of transition: 1895–1905

In March 1905, Ernest Rutherford (1871–1937) delivered the Silliman lectures at Yale. He had chosen radioactive transformations as his topic, and began the first of his talks as follows 'The last decade has been a very fruitful period in physical science, and discoveries of the most striking interest and importance have followed one another in rapid succession ... The march of discovery has been so rapid that it has been difficult even for those directly engaged in the investigations to grasp at once the full significance of the facts that have been brought to light ... The rapidity of this advance has seldom, if ever, been equalled in the history of science'.

The text of Rutherford's lectures [33] makes clear which main facts he had in mind: x-rays, cathode rays, the Zeeman effect, α-, β- and

Introducing Atoms and Their Nuclei

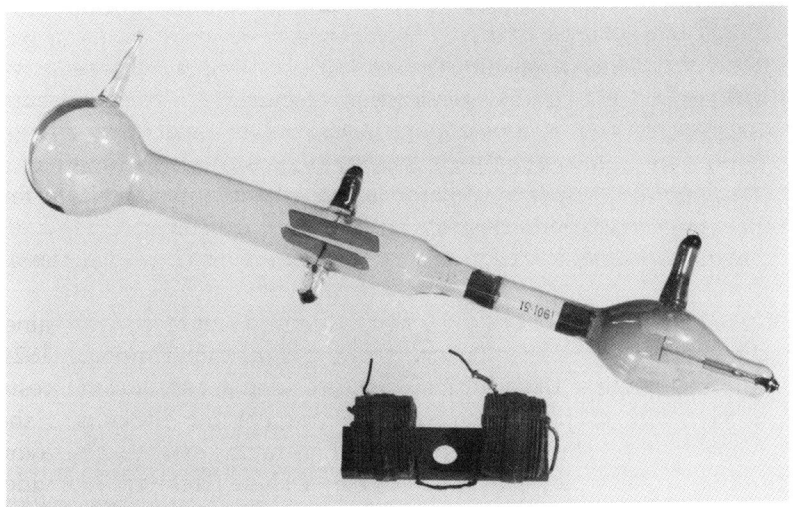

Figure 2.2. *The vacuum tube used by Thomson in his discovery of the electron.*

γ-radioactivity, and the reality as well as the destructibility of atoms. There is no mention, however, of the puzzle posed by Rutherford's own introduction of a characteristic lifetime for each radioactive substance. Nor did he touch upon Planck's discovery of the quantum theory in 1900. He could, of course, not refer to Einstein's article on the light-quantum hypothesis, because that paper was completed on the seventeenth of the very month he was lecturing in New Haven. Nor could he include Einstein's special theory of relativity among the advances of the decade he was reviewing, since that work was completed another three months later. The speed of discovery between 1895–1905 was therefore even greater than the great Rutherford noticed. The following timetable illustrates the rapidity with which one event followed another.

1895. November 8. Wilhelm Röntgen (1845–1923) discovers x-rays. He was so astonished that he said to his wife that when people found out about it they would say *'Der Röntgen ist wohl verrückt geworden'* (Röntgen has probably gone crazy) [34]. For this discovery Röntgen received in 1901 the first Nobel Prize in physics ever awarded.

1896. Antoine Henri Becquerel (1852–1908) observes what he called 'uranic rays', the first phenomenon that opens a new field later called radioactivity. Wilhelm Wien (1864–1928) published his exponential law for black-body radiation, the first quantum law ever written down. *October 31.* Pieter Zeeman's (1865–1934) first paper on the influence of magnetic fields on spectral lines.

1897. Determination of e/m for cathode rays by J J Thomson and others. First mention of a particle lighter than hydrogen (figure 2.2).

Twentieth Century Physics

1898. Rutherford discovers that there are two species of radioactive radiations: α- and β-rays.

1899. J J Thomson measures the electric charge of free electrons, and realizes that in ionization processes atoms are split.

1900. Paul Villard (1860–1934) discovers γ-rays. First determination of a half-life for radioactive decay. Planck discovers the quantum theory.

1905. March. Einstein postulates the light quantum. *June.* His first paper on special relativity theory.

Clearly it was a marvellous period for smart young men to enter physics. Rutherford is a prime example. A native of New Zealand, he first set foot in the Cavendish Laboratory, Cambridge, at age 24, about one month before Röntgen's discovery, worked on x-rays in 1896, and in 1898 began his prolific life's work on radioactivity [35].

It is natural to ask but not easy to answer the question of why so much novelty should be discovered in so short a time span. It is clear, however, that a culmination of advances in instrumentation was crucial. Here are some of them.

(i) *Higher voltages.* Higher voltages were the result of Heinrich Rühmkoff's (1803–74) work, beginning in the 1850s, on an improved version of the induction coil. These were the coils that in 1860 served Gustav Kirchhoff (1824–87) and Robert Bunsen (1811–99) in their analysis of spark spectra; Heinrich Hertz (1857–94) in 1886–88 in his demonstration of electromagnetic waves and his discovery of the photoelectric effect; Röntgen in his discovery of x-rays; Guglielmo Marconi (1874–1937) in his transmission of telegraph signals without wires; Zeeman in his discovery of the Zeeman effect; and J J Thomson in his determination of e/m for electrons. By the turn of the century voltages of the order of 100 000 V were available.

(ii) *Improved vacua.* Improved vacua were achieved in the 1850s, when Johann Geissler (1815–79) began developing the tubes now named after him, and soon was able to reach and maintain pressures of 0.1 mm of mercury. Refined versions of this tube were crucial to the discoveries of Röntgen and J J Thomson.

(iii) *Early versions of the parallel plate ionization chamber.* Parallel plate ionization chambers were first developed in Cambridge in the 1890s. They were used by Rutherford and the Curies in the earliest quantitative measurements of radioactivity.

(iv) *Concave spectral gratings.* Concave spectral gratings were developed starting in the 1880s by Henry Rowland (1848–1901) at the Johns Hopkins University. Their resolving power made Zeeman's discovery possible.

(v) *The cloud chamber.* Work on the development of a cloud chamber was begun in Cambridge in 1895 by Charles T R Wilson (1869–1959). This instrument enabled J J Thomson to measure the electron's charge.

Ernest Rutherford

(New Zealander, 1871–1937)

Ernest Rutherford received his early training in his native New Zealand, then went to Cambridge (1895), where he worked with J J Thomson on gas ionization. In his time as Professor at McGill, Montreal, 1898—1907, where he published *The distinction between α-rays and β-rays* (1899), Rutherford determined the first radioactive half-life (1900), developed the transformation theory (1902), and discovered α-particle scattering (1906). He was awarded the Nobel Prize for chemistry in 1906. Whilst Professor at Manchester, 1907–19, he discovered the nuclear model of the atom. Rutherford was knighted in 1914. He was a member of the Board of Invention and Research which oversaw the British Scientific war effort (1915); several years later (1919) he discovered the first nuclear transmutation. From 1919, until his untimely death in 1937, Rutherford was a Director of the Cavendish Laboratory.

In 1931 he was created Baron Rutherford of Nelson. On 25 October 1937, his ashes were interred in Westminster Abbey, in the presence of family, a representative of the King, several British cabinet members and leaders of science and industry.

Rutherford himself was not only the greatest twentieth century experimentalist but also created a school which produced numerous Nobel laureates. Arthur Eve, his principal biographer, has written of him 'People ask—what was the man like, did he really look like a farmer, what sort of accent did he have, did he believe in immortality? None of these things really matter. Here was a king making his way into the unknown. Who cares what his crown is made of, or what the polish of his boots?'.

It was not only new technology that made possible the wealth of discoveries just listed. Theoretical concepts had to evolve as well. Rutherford summed it all up in terms of an example:

Twentieth Century Physics

> The discovery of the radioactive property of uranium [the element itself was discovered in 1789] might have been made accidentally a century ago [i.e. in the early 1800s] for all that was required was the exposure of a uranium compound on the charged plate of a gold-leaf electroscope ... the discharging property of [uranium] could not fail to have been noted if it had been placed near a charged electroscope. It would not have been difficult to deduce that the uranium gave out a type of radiation capable of passing through metals opaque to ordinary light. The advance would probably have ended there, for the knowledge at that time of the connection between electricity and matter was far too meagre for an isolated property of this kind to have attracted much attention ... If the discovery had been made even a decade earlier [than 1896], the advance must necessarily have been much slower and more cautious [36].

Regarding 'a decade earlier', here Rutherford clearly had in mind the invention of the ionization chamber.

I turn next, one by one, to more details concerning the spectacular discoveries made during the decade of transition.

2.3. Radioactivity: 1896–1905

What were Becquerel's 'uranic rays', his discovery which inaugurated the study of radioactivity? From what we now know it follows that what he originally saw was the effect of β-rays; and that these were not uranic but rather due to the first daughter product, thorium 234. He did not know that his rays were but one of the three distinct types of radioactive radiation; nor that these radiations proceed via a sequence of parent, daughter, granddaughter, etc, elements; nor that β-rays were electrons, as yet undiscovered. In fact he could not even know that radioactive processes emanate from the atomic nucleus, discovered only 15 years later. Indeed, the first observation of a nuclear phenomenon predates the insight that each atom contains a nucleus.

It will be clear that theoretical physicists did not play any role to speak of in the development of radioactivity in its earliest days, both because they were not particularly needed for its descriptive aspects and because the deeper questions were too difficult for that time. Rather, early progress was principally the concern of a fairly small-sized but élite club of experimentalists. Let us see what they were up to.

I will start with Marie Sklodowska (1867–1934) and Pierre Curie (1859–1906) (figure 2.3). Marie, daughter of a Warsaw physics teacher, arrived in Paris in 1891 to study physics at the Sorbonne. She met Pierre in 1894; the next year they married. Pierre's career was already well underway by then. He had done important work on piezo-electricity, on symmetries in crystals, and on magnetism.

Introducing Atoms and Their Nuclei

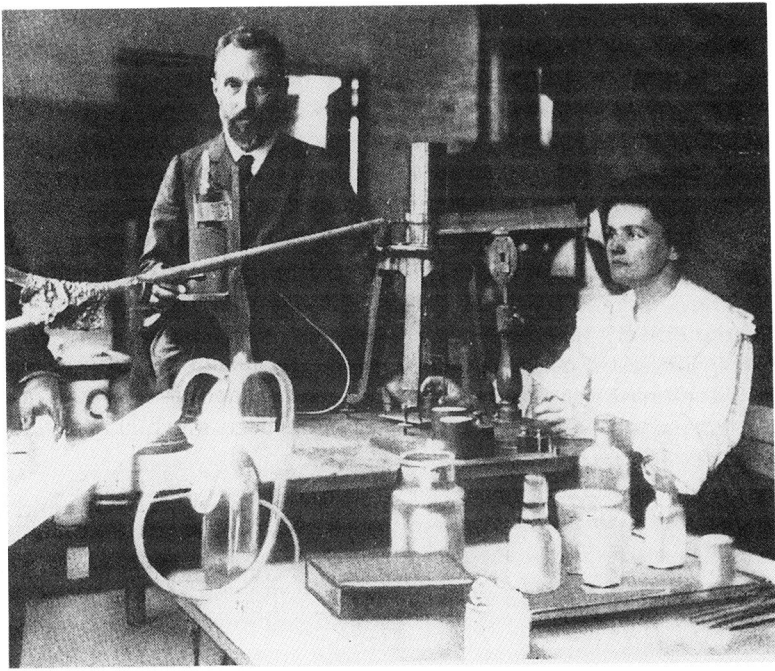

Figure 2.3. *Marie and Pierre Curie in their laboratory.*

When Marie heard of Becquerel's work she discussed it with her husband. 'The study of this phenomenon seemed to us very attractive ... I decided to undertake the study of it ... In order to go beyond the results [which had so far all been qualitative] reached by Becquerel, it was necessary to employ a precise quantitative method. [37]. She is the first of several people we shall meet in this chapter who achieved major results with tools that can only be called primitive by present standards, in her case an ionization chamber of the kind mentioned earlier, with a modest 100 V potential difference between its parallel metal plates, and a sensitive electrometer.

She reported in her first paper (1898) [38] her discovery that thorium was also radioactive, as, unbeknownst to her, Gerhard Schmidt (1865–1949) had also found slightly earlier; and the finding that two minerals which contain uranium, pitchblende and chalcite 'are much more active than uranium itself. This fact is very remarkable and leads one to believe that these minerals contain an element which is much more active than uranium'. With this conjecture young Marie introduced a major novelty: radioactive properties are a diagnostic for the discovery of new substances.

Pierre now joined his wife in her research. Three months later they identified a new element in pitchblende: polonium. Another five months later (December 1898) they, together with Gustave Bémont (1857–1932) isolated radium. In that paper [40] one finds the following reference to her earlier work of that year 'One of us [MC [38]] has shown that radioactivity is an atomic property'. It is the first time in history that radioactivity was explicitly linked to individual atoms.

1898 was the heroic year in the Curies' career. I take leave of them here with an envoi written in 1902 by Rutherford 'I have to keep going, as there are always people on my track. The best sprinters in this road are Becquerel and the Curies in Paris' [41]. Which brings me to Rutherford himself.

I have already mentioned Rutherford's discovery of α- and β-rays. This was the result of absorption experiments which had revealed that one kind (alphas) were readily absorbed while the other (betas) were far more penetrating. But what were these rays?

It took ten years to establish firmly that α-rays consist of doubly ionized helium. Note that the terrestrial presence of helium was only ascertained in 1895, not long before the discovery of α-rays, when Sir William Ramsay (1852–1916), much to his astonishment [42], found helium to be present in a uranium-bearing mineral. 'An examination of the physical and chemical properties [of helium] ... had hardly been completed when Ramsay and Soddy made an examination of the gases liberated from radium and discovered that helium was a product of the transformation of radium' [43]. Because of the presence of helium in several uranium- and thorium-bearing minerals, some connection between α-rays and helium now became plausible but was as yet by no means certain.

For some time α-rays were thought to be neutral, since they apparently did not bend in electric and magnetic fields, but in 1903 Rutherford finally showed that they do deviate in strong fields and carry a positive charge [44]. Next, doubts arose as to whether the e/m for α-particles emitted by polonium on the one hand and by radium on the other were the same. 'Further experimental evidence is required on that important point' Rutherford remarked as late as 1905 [45]. It became clear soon thereafter that the e/m is unique, and, later in 1905, it was stated with certainty that α-particles were charged 'at the moment of their expulsion from the radium atom' [46].

It took still longer to determine the actual value of the α-particle's electric charge. In 1905 the situation was as follows [47] 'Assuming the charge carried by the α-particle to be the same as that carried by the hydrogen atom, the mass of the α-particle is about twice that of the hydrogen atom'. Thus they were closing in on the right e/m value.

Various experiments led to the summary paper of 1908 *Nature of the α-particle* [47], which Rutherford wrote with Hans Geiger (1882–1945), where it was announced: 'On the general view that the charge e carried by a hydrogen atom [i.e. a proton] is the fundamental unit of electricity

... the evidence is strongly in favour of the view that [the charge of the α-particle] $= 2e'$. After nearly a decade of labour, Rutherford was finally prepared to state, italicized, what the α-particle really was 'We may conclude that *an α-particle, after it has lost its positive charge, is a helium atom*'. In a paper together with Thomas Royds (1884–1955) [48], completed in November 1908, he was even more emphatic 'We can conclude with certainty ... that the α-particle is a helium atom [sic]'. They had shown that a discharge sent through a volume in which α-particles from radium had been collected produced the characteristic helium spectrum!

The purpose of relating this story about the nature of the α-particle in such detail is to drive home the fact that much of what we learned in kindergarten was the result of strenuous labours by our ancestors.

The identification of β-rays with electrons was much more rapid, as by 1902 e/m measurements had made the answer obvious. It took 14 years, however, before it was certain that γ-rays are hard photons. In 1902 Rutherford thought that they might be hard β-rays because they were not deflected in magnetic fields [49]. The matter was not definitely settled until 1914 when reflections of γ-rays from crystal surfaces were detected [50].

Rutherford's next major discovery after α- and β-rays came in 1900, after having moved in 1898 from Cambridge to McGill University in Canada. It was that thorium emanation (Rn^{220}) loses half its activity in 60 s (modern value: 56 s) [51]. It is of course no accident that this first observation of a radioactive half-life concerned an element of medium–short life. He was the first to note that if $N(t)$ is the number of active atoms at time t, then the decrease of N with t is well described by

$$dN/dT = -N\lambda \quad \text{or} \quad N(t) = N(0)e^{-\lambda t}.$$

This equation and its generalization to sequential decays was the first of two contributions by Rutherford to theoretical physics, an activity which he did not always hold in the highest esteem. (His second contribution was his theoretical discovery of the central nucleus from the results of scattering experiments.)

The half-life concept enabled Rutherford and Frederick Soddy (1877–1956) in 1902 to formulate their transformation theory [52]. In this 'great theory of radioactivity which these young men sprung on the learned, timid, rather unbelieving, and, as yet, unquantized world of physics of 1902 and 1903' [53] they unabashedly put forward the idea that some atomic species are subject to spontaneous transmutation. Forty years later, a witness to the events characterized the mood of the times as follows 'It must be difficult if not impossible for the young physicist or chemist of the present day to realize how extremely bold it was and how unacceptable to the atomists of the time ... this is a point which must be stressed, for the young generation is more likely to be familiar with the ordered simplicity of the radioactive series as we know them than with the chaotic state which preceded the transformation theory' [54].

The main tenet of the transformation theory is: radioactive bodies contain unstable atoms of which a fixed fraction decay per unit time. The rest of the decayed atom is a new radio-element which decays again, and so forth, till finally a stable element is reached. This theory has remained a principal tool for bringing phenomenological order in the plethora of radioactive phenomena.

It will be clear by now that the period 1896–1905 was rich in experimental advances and phenomenological insights regarding radioactivity. However, on a deeper level of understanding times were not yet ripe, but this did not prevent some physicists from speculating as they always do. Let us see what were some of their thoughts.

In 1910, Marie Curie reminisced as follows about those early days 'The constancy of the uranic radiation caused profound astonishment to those physicists who were the first to be interested in the discovery of H Becquerel. This constancy appears in fact to be surprising; the radiation does not seem to vary spontaneously with time ...' [55]. In order to appreciate this statement fully, three facts should be borne in mind. (1) The radiation emitted by uranium when unseparated from its daughter products does indeed represent, to a very high degree, a steady state of affairs. (2) It took two years from Becquerel's initial discovery until the first parent–daughter separation was effected. (3) It took another two years until it was firmly established that radioactivity does diminish with time.

Speculation on the origin of radioactive energy started with Marie Curie's very first paper on radioactivity (the one in which she announced her discovery of the activity of thorium (1898)). There, cautiously, she suggests the possibility that the energy might be due to an outside source 'One might imagine that all of space is constantly traversed by rays similar to Röntgen rays, only much more penetrating and being able to be absorbed only by certain elements with large atomic weight, such as uranium and thorium' [38]. Shortly afterwards she concluded that an external energy source was a poor excuse 'Any exception to Carnot's principle [first law!] can be evaded by the intervention of an unknown energy which comes to us from space. To adopt such an explanation or to put in doubt the generality of the Carnot principle are in fact two points of view which to us amount to one and the same as long as the nature of the energy here invoked stays entirely *dans le domaine de l'arbitraire*' [56]. She also pointed out that the interior of the atom could be the energy source 'The radiation [may be] an emission of matter accompanied by a loss of weight of the radioactive substances' [56].

Nevertheless experiments were designed to locate a possible outside source of radioactive energy. In an attempt to see whether the Sun could be the cause, the Curies looked for diurnal variations in the activity of uranium. They found no effect [57]. Julius Elster (1854–1920) and Hans Geitel (1855–1923) from Wolfenbüttel reasoned that, if energy were supplied by an x-ray-like radiation which is all pervasive in the atmosphere, then there should be a decrease in activity if the source were placed deep underground. So they requested and obtained permission to do an experiment in the Clausthal mines in the Harz mountains, under 300 m of rock. They found no effect, and concluded (1898) that 'the hypothesis of the excitation of Becquerel rays by radiation pre-existing in space appears improbable to the highest degree (*im höchsten Grade unwahrscheinlich*)' [58]. In a second paper [59] they report on attempts to increase the radioactive emissions by exposing a source to cathode rays, or to sunlight. They found no effect and concluded 'Rather, one will have to derive the source of light [sic] from the atom of the element concerned itself'.

Hence the energy debate might have quieted down were it not that, in March 1903, new fuel was added to it by the discovery that radioactive energy release surpassed in magnitude anything that had been known until then from chemical reactions. In that year Pierre Curie and Albert Laborde (1878–1968) measured the amount of energy released within a Bunsen ice calorimeter by a known quantity of radium [60]. They found that 1 g of radium can heat ~ 1.3 g of water from the melting point to the boiling point in 1 hour!

These results caused a tremendous stir. It was in fact the Curie–Laborde paper which was largely responsible for the worldwide arousal of interest in radium. In a lecture on 'The present crisis of mathematical physics' given in 1904, it led Henri Poincaré (1854–1912) to bring up the energy conservation question '... These principles on which we have built everything, are they about to crumble away in turn? ... When I speak thus, you no doubt think of radium, that grand revolutionist of the present time ... At least, the principle of the conservation of energy still remained with us, and this seemed more solid. Shall I recall to you how it was in its turn thrown into discredit? This [activity of radium] was itself a strain on the principles ... but these quantities of [radioactive] energy were too slight to be measured; at least that was the belief, and we were not much troubled. The scene changed when Curie bethought himself to put radium in a calorimeter; it was then seen that the quantity of heat created incessantly was very notable ...' [61].

The origin of radioactive energy was still being argued in the year in which quantum mechanics was born [62] which brings home the fact that only in the quantum-mechanical era could the definitive proof of the existence of internal mechanisms for energy generation be given, which settled the question for all time.

One fundamental insight regarding the energy content of atoms came much earlier. In the second of his 1905 papers on relativity Einstein stated that 'If a body gives off the energy L in the form of radiation, its mass diminishes by L/c^2 ... The mass of a body is a measure of its energy ... It is not impossible that with bodies whose energy content is variable to a high degree (e.g. with radium salts) the theory may be successfully put to the test'. The level of accuracy of mass measurement was not yet adequate, however, for that test. Even in 1921, in his review of relativity theory, Pauli noted that *'perhaps* the theorem of the equivalence of mass and energy can be checked *at some future date* by observations on the stability of nuclei'* (my italics).

At the turn of the century there already existed a body of knowledge on unstable systems of atomic dimensions. For example, much work had been done at that time on luminescent phenomena. This had made the lifetime concept familiar. However, if these various unstable systems were not amenable to theoretical treatment, they did not appear to pose any manifest paradoxes, principally since one could (so it was thought) ascribe the instability to *external* causes. For example, for a process like luminescence one always had the excuse that some but not all the irradiated matter had been excited. Note also that, in the early days, lifetimes for ordinary light emission by atoms were too short to be observable. Radioactivity, on the other hand, created problems unique for its time. Indeed, the discoveries in 1900 by Rutherford of a half-life and, later that same year, by Planck of the quantum theory signalled the end of the era of classical physics. Neither of them was at the time aware how profoundly his work was to change the course of science. It was only in 1917 that Einstein became the first to grasp that Rutherford's discovery called for no less than a revision of a root concept of classical physics: causality (I will come back to that later). Earlier, however, some wise men had scratched their heads. Lord Kelvin (1824–1907) wrote 'What would be the difference, between radium atoms in a piece of radium bromide, of the atoms which are nearly ripe for explosion, and those which have the prospect of several thousand years of stable diminishing motions before explosion?' [63] and Rutherford wrote 'All atoms formed at the same time should last for a definite interval. This, however, is contrary to the observed law of transformation, in which the atoms have a life embracing all values from zero to infinity' [64]. They were unable to do more than state their puzzlement.

As a fitting conclusion to this account of the early days of radioactivity I note that speculations on the possible good and the possible evil of the atom go back to the founding fathers of radioactivity.

Becquerel, in an early lecture said 'Today the [radioactive] phenomena are of transcendent interest, but in them almost infinitesimal amounts of energy are utilized. Whether ultimately science will have so far advanced as to permit of the practical utilization of the abundant store of energy locked up in every atom of matter is a problem which only the future can answer. Remember, at the dawn of electricity this was looked on as a mere toy, suitable only to amuse children by attracting bits of paper with a stick of rubbed sealing wax' [65].

Pierre Curie, in his 1903 Nobel Prize lecture said 'It can even be thought that radium could become very dangerous in criminal hands, and here the question can be raised whether mankind benefits from the secrets of Nature ...' [66].

Soddy in an early popular book wrote 'If we pause but for a moment to reflect what energy means for the present, we may gain some faint notion as to what the question of transmutation may mean for the future to a fuelless world, once more dependent on a hand-to-mouth method of subsistence. It may still be centuries before this occurs, but neither the application of the discoveries of science nor even their achievement is to be compared with the struggle in winning them' [67].

Also the expression 'atomic energy' entered the language at the very beginning of the twentieth century. The term was first used in 1903 by Rutherford and Soddy [68] *not* just for the energy released by a radioactive element, but much more generally for the energy locked in *any* atom 'All these considerations point to the conclusion that the energy latent in the atom must be enormous compared with that rendered free in ordinary chemical change. Now the radio-elements differ in no way from the other elements in their chemical and physical behaviour. On the one hand they resemble chemically their inactive prototypes in the periodic table very closely, and on the other they possess no common chemical characteristic which could be associated with their radioactivity. Hence there is no reason to assume that this enormous store of energy is possessed by the radio-elements alone. It seems probable that *atomic energy* [my italics] in general is of a similar high order of magnitude, although the absence of change prevents its existence being manifested'. This, truly, is the physics of the twentieth century.

2.4. The structure of atoms: 1897–1906

That basic theory also scored important advances in the 1890s was above all due to the Dutchman Hendrik Anton Lorentz (1853–1928). In 1892 he published his first paper [69] on his atomistic interpretation of electromagnetic theory in terms of charges and currents carried by fundamental particles, which he called charged particles in 1892, ions in 1895, and electrons from 1899 on. In 1895 he introduced the 'Lorentz force' exerted by electromagnetic fields on charged particles [70]. Starting in 1899 he began work on the celebrated transformations that bear his name, achieving their ultimate form in 1904 [71]. More details of his contributions are found elsewhere in this book.

See also p 263

Times were not yet ripe, however, for tackling the theory of atomic structure. 'It is perhaps not unfair to say that for the average physicist of the time, speculations about atomic structure were something like speculations about life on Mars—very interesting for those who like that kind of thing, but without much hope of support from convincing scientific evidence and without much bearing on scientific thought and development' [72].

Nevertheless, J J Thomson began to make serious efforts at building a model for atoms. That he tried is noble, that he failed understandable. Already in his Adams prize essay of 1882 [73] he had included a 'Sketch of a chemical theory', in which he attempted to associate with each atomic species a specific motion of vortices, known for their remarkable indestructibility. As soon as the electron appeared on the scene he betook himself to build atoms out of electrons. In 1897 he referred [74] to 'the hypothesis ... enunciated by Prout [according to which] the atoms of the different elements were hydrogen atoms; in this precise form the hypothesis is not tenable, but if we substitute for hydrogen some unknown primordial substance X, there is nothing known which is inconsistent with this hypothesis'. He left no doubt as to what X had to be '... These primordial atoms, which we shall for brevity call corpuscles' (the way Thomson continued to refer to electrons for some years). In 1899 he wrote 'I regard the atom as containing a large number of ... corpuscles' [75] and in 1903 'The atom of hydrogen contains about a thousand electrons' [76]. In 1904 he wrote '[I] suppose that the mass of an atom is the sum of the masses of the corpuscles it contains' [77]. In 1902 Rutherford referred to 'the view [which] has been put forward that all matter is composed of electrons. On such a view an atom of hydrogen for example is a very complicated structure consisting possibly of a thousand or more electrons' [78].

What agent neutralizes the huge charge within an atom due to all these electrons? And should one not consider that that agent may contribute to the atom's mass?

Thomson assumed that inside the atom the electrons' charge is balanced by a homogeneous positive background which carries no mass! I have found reference to this picture in a Cambridge textbook as late as 1907 'In the present state of our knowledge no certain statement can be attained as to the whole number of electrons in any atom, but the conclusion that it is such that the mass of the atom is the sum of the masses contained in it is so attractive that it seems desirable to accept it provisionally in the absence of any conclusive evidence to the contrary' [79].

Thomson's model raised the central difficulty on which all attempts at an understanding of atoms within the framework of classical physics would ultimately come to grief: the atom's stability. Thomson showed in considerable detail that, given his picture of the positive charge distribution, there do exist electron configurations that are stable—as long as these particles are at rest. He preferred, however, to assume that his electrons move in some sort of closed orbits, his reason being

Introducing Atoms and Their Nuclei

that such motions might explain the magnetic properties of substances [80]. Long before, André Marie Ampère (1775–1836), in his pioneering work on magnetism had proposed that magnetism was caused by electric charges in motion. Could these circuits not simply be electrons moving inside atoms? By a general classical theorem (post-dating Ampère) such orbits are unstable, however, since now the electrons will necessarily lose energy by emission of electromagnetic radiation. I shall not discuss here the ways in which Thomson and others attempted, unsuccessfully, to cope with this extremely troublesome instability [81], suffice it to say for now that quantum mechanics is necessary for the resolution of this fundamental problem.

Nevertheless there was some progress. It was Thomson himself who in 1906 made the discovery, perhaps his greatest as a theoretician, that his early atomic model was fallacious, that in fact 'the number of corpuscles in an atom ... is of the same order as the atomic weight of the substance' [82]. The most famous of his several theoretical reasons for this revision was his theory of the scattering of x-rays by gases. In his primitive, though qualitatively correct, treatment he assumed that this effect is due exclusively to the scattering by x-rays off the interatomic electrons which are treated as free particles. To this day the scattering of low-energy photons by electrons is known as Thomson scattering. From a comparison of his formulae with experimental data Thomson concluded, correctly, that the number of electrons per atom lies somewhere between 0.2 and 2 times the atomic number. What, then, caused the atom to have the mass it has? On that question Thomson remained silent.

2.5. The birth of quantum physics

In the 1890s it came to pass that experimental advances and theoretical stumblings caused developments that, in the year 1900, would lead to the discovery of the quantum theory. In order to set the stage for this evolution it is necessary to probe once again into earlier events.

Once upon a time there lived a man in the town of Heidelberg named Gustav Robert Kirchhoff (1824–87). He was born in Königsberg, the son of a law councillor, a state functionary. He became one of the very few nineteenth century physicists to make fundamental contributions both to experiment and to theory, most of them while being a professor of physics at the University of Heidelberg. He will be best remembered for his activities, experimental as well as theoretical, during the autumn of 1859, which made him the grandfather of the quantum theory.

On October 20 of that year, Kirchhoff submitted his *'unerwartete aufschluss'* (unexpected explanation) [83] of the dark Fraunhofer lines in the solar spectrum: they are due to sodium. He arrived at this conclusion by interposing a flame containing kitchen salt between the solar spectrum and his detector 'If the sunlight is sufficiently damped, then two luminous lines appeared at the position of two dark [solar] D-lines; if the solar intensity surpassed a certain amount, then the two dark D-lines appeared much more pronounced ... the dark D-lines lead one to conclude that there is sodium in the solar atmosphere'. Six weeks later [84] Kirchhoff gave the theoretical interpretation of these observations, showing that they follow from what is now known as Kirchhoff's law, which says the following.

Consider a body in thermal equilibrium with radiation. Let the radiation energy which the body absorbs be converted to thermal energy only, not to any other energy form. Let $E_\nu \, d\nu$ denote the amount of energy emitted by the body per unit time per cm^2 in the frequency interval $d\nu$. Let A_ν be its absorption coefficient for frequency ν. The law states that E_ν/A_ν depends only on ν and the temperature T and is independent of any other characteristic of the body

$$E_\nu/A_\nu = J(\nu, T)$$

Kirchhoff called a body perfectly black if $A_\nu = 1$. Thus $J(\nu, T)$ is the emissive power of a black body. An example of a perfect black body is *Hohlraumstrahlung*, a space enclosed by bodies of equal temperature through which no radiation can penetrate. For this special case the radiation is homogeneous, isotropic and unpolarized so that

$$J(\nu, T) = \frac{c}{8\pi} \rho(\nu, T)$$

where $\rho(\nu, T)$, the *spectral density*, is the energy density per unit volume for frequency ν.

Having discovered that the spectral density depends only on frequency and temperature, the obvious next question is: *how* does it depend on these variables? It was not an easy problem. In Kirchhoff's own words 'It is a highly important task to find this function. Great difficulties stand in the way of its experimental determination. Nevertheless, there appear grounds for the hope that it has a simple form, as do all functions which do not depend on the properties of individual bodies and which one has become acquainted with before now' [84].

'It would be edifying if we could weigh the brain substance which has been sacrificed on the altar of [Kirchhoff's law],' Einstein wrote in 1913 [85]. Forty years would pass between Kirchhoff's discovery that there exists a universal function and the determination of what that function is. Not that, in the intervening years, physicists did not try. Most of those attempts were pure guesswork, with two notable exceptions: the law named after Josef Stefan (1835–93) and Ludwig Boltzmann (1844–1906), according to which the total energy emitted by a black body is proportional to T^4 [86]; and the 'displacement law' of Wilhelm Wien (1864–1928) which states that $\rho(\nu, T)$ must be of the form ν^3 times a

Introducing Atoms and Their Nuclei

function of v/T only [87]. The derivations of both these laws were based on the interplay of two young disciplines: thermodynamics and Maxwell's theory of electromagnetism.

As to the guesswork, all of it is forgettable with one brilliant exception: in 1896 Wien proposed what has since been called his 'exponential law' [88]

$$\rho(v, T) = \frac{8\pi h v^3}{c^3} e^{-hv/kT}$$

where h is Planck's constant, k Boltzmann's constant and c the velocity of light *in vacuo*.

All those fundamental constants did, of course, not appear in Wien's paper†. My reasons for this ahistoric presentation at this point are: firstly, that the exponential law fitted very well with all experimental data known at that time [89], and secondly, that the values of c and k could even then be deduced from other data, so that already in 1896 a good value for Planck's constant could have been obtained!

Although Wien did not know this, his exponential law marked the end of the universal validity of classical physics and the onset of a bizarre turn in science. On the one hand, his law agreed with experiment and therefore had to have something to do with reality. On the other hand, it would have been possible in 1896 to calculate rather than guess at an explicit form for $\rho(v,T)$, namely with the help of Boltzmann's equipartition theorem of statistical mechanics which dates back to the 1870s. Had Wien made that elementary calculation he would have found that ρ is proportional to T, a behaviour violently different from his exponential law and thus, it would then have appeared, in violent disagreement with experiment!

This story has two morals. Firstly, at the turn of the century good theoreticians were not yet familiar with statistical mechanics, a branch of theoretical physics that was indeed less than twenty years old. Secondly, the success of Wien's law indicates, to us, not to Wien, that already in 1896 it should have been clear that there was something seriously amiss with classical physics. That was the situation when Planck appeared on the scene.

Max Planck (1858–1947) was born in Kiel, where his father was professor of jurisprudence at the University. Both his grandfather and great-grandfather had been professors of theology in Göttingen. '[He] would remember, even in his old age, the sight of Prussian and Austrian troops marching into his native town‡ when he was six years old. Throughout his life, war would cause him deep personal sorrow. He lost

---
† Wien wrote $\rho(v, T) = av^3 exp(-bv/T)$.
‡ On their way to do battle with Danish armies.

his eldest son during World War I. In World War II, his house in Berlin burned down during an air raid. In 1945 his other son was executed when declared guilty of complicity in a plot to kill Hitler. In 1867 Planck and his family moved to Munich, where he went to high school. He was a good but not sparkling student, ranking between third and eighth [90]. 'He accepted the authority of the school as later he would the authority of the established corpus of physics' [91].

> See also p 144

Planck himself has described how he came to choose physics as his life's vocation 'The outside world is something independent from man, something absolute, and the quest for the laws which apply to this absolute appeared to me as the most sublime scientific pursuit in life' [92]. He received the first impetus in this direction when a high-school teacher acquainted him with the law of the conservation of energy. 'My mind absorbed avidly, like a revelation, the first law I knew to possess absolute, universal validity, independently of all human agency: the principle of the conservation of energy.' Planck's preoccupation with the absolute and its independence of man's folly is a persistent theme in his writings, never more poignantly expressed than at his moments of discovery. Planck began his university studies in Munich and continued them at the Friedrich Wilhelm University in Berlin, the leading German university of its day. 'I studied experimental physics and mathematics; there were no classes in theoretical physics as yet' [92]. One of his Berlin professors was Kirchhoff—who had moved there from Heidelberg—whom he admired even though he found him dry and monotonous as a teacher. In 1879 he received his PhD in Munich on a thesis dealing with the second law of thermodynamics.

After Kirchhoff's death in 1887, Planck became his successor, first as associate then as full professor in theoretical physics. He had published some 40 papers, mainly on the theory of heat, by the time he discovered the quantum theory.

Planck later wrote how in 1895 he started out on the wobbly road to his monumental achievement: 'At that time I regarded the principle of the increase of entropy as immutably valid ... whereas Boltzmann treated [it] merely as a law of probabilities' [92]. Accordingly he attempted to base his early arguments (1895–98) on the dynamics of Newton/Maxwell. Boltzmann rapidly noted, however, that his reasoning contained serious flaws. There followed a confrontation (in print [93] 1897–98), which ended by Planck frankly admitting that he had been in error.

So Planck started all over again, choosing next the middle ground between dynamics and statistical mechanics: thermodynamics 'my own home territory where I felt myself to be on safe ground'. [92]. He now bestirred himself to prove the universal validity of Wien's exponential law—still, at that time, a viable solution for the Kirchhoff function—from first thermodynamics principles, and in 1899 thought he had done so 'I believe to be forced to conclude that Wien's law [is a] necessary consequence of the application of the entropy increase principle to the theory of electromagnetic radiation and that therefore the limits of validity of this law, if they exist at all, coincide with those of the second law of thermodynamics' [94].

Wrong again.

It was now 7 November 1899. Planck concluded the nineteenth century part of his career with an invocation of the absolute. He had found that the constants a and b in Wien's exponential law together with the velocity of light and the Newtonian constant of gravitation suffice to find a system of natural units 'With the help of a and b it is possible to give units for length [now called the Planck length], mass, time and temperature which, independently of special bodies and substances, retain their meaning for all times and for all cultures, including extraterrestrial and extrahuman ones' [94].

As we now know, Planck's allegations about the limits of validity of Wien's law were once again incorrect. This time it was not a theoretician who found an error. Rather, experimental evidence was presented, in a paper from Berlin, submitted on 2 February 1900, according to which Wien's conjecture did not agree with all the facts 'it has ... been demonstrated that the Wien–Planck spectral equation does not represent the black-body radiation measured by us ... ' [95]. Not that the data on which Wien had based his conjecture were flawed, rather these data had been extended to a hitherto unexplored region, the far infrared. It was there that Wien's law failed.

Planck most probably first wrote down the correct answer, 'Planck's law', on the evening of Sunday 7 October 1900 [96], as the result of new experimental information he had received earlier that day. During the preceding months two groups at the Physikalisch-Technische Reichsanstalt had continued their spectral function measurements which were of excellent quality, as later data confirmed. On the afternoon of that Sunday a colleague had told Planck that in the far infrared this function is proportional to the temperature T rather than to Wien's exponential. So there, in the infrared, appeared the temperature dependence of $\rho(\nu, T)$ predicted by the classical equipartition theorem!

This classical result is now known as the Rayleigh–Einstein–Jeans (REJ) law, after the three men who contributed to its formulation. That law, finally correctly stated in 1905[†], was as little known to Planck in 1900 as it had been to Wien in 1896. It is a curious twist of history that Planck could only discover the first *quantum* law after the *classical* data on infrared radiation had been obtained.

Shortly after 7 October 1900 Planck was informed that his new formula for the spectral density fitted the entire experimental spectrum

[†] *For the history of the* REJ *law see reference* [97].

very well. He publicly stated this result for the first time in a discussion on 19 October [98]. On 14 December he submitted a paper [99] that includes his famous law

$$\rho(\nu, T) = \frac{8\pi h\nu^3}{c^3} \frac{1}{e^{h\nu/kT}-1}.$$

That date marks the birth of quantum physics. Planck's search for the absolute had finally been rewarded. Note that in this paper the constants h and k—mentioned earlier in connection with Wien's work—make their first appearance. Planck's law also makes clear that Wien's law only holds for high frequencies and the REJ law only for low frequencies.

Planck's paper of 14 December 1900 is based partly on solid physical theory, partly on 'a fortunate guess', to use his own words [92]. As a piece of improvization it ranks among the most fundamental works of the twentieth century. It is to Planck's greatest credit and highest glory that, not content with guesswork, he went on to search for a deeper meaning of his law. He later called the resulting labour the most strenuous of his life [100] and 'an act of desparation ... I had to obtain a positive result, under any circumstances and at whatever cost' [101]. I cannot state here all the *ad hoc* assumptions Planck made next [99] except for the most important one, the *quantum postulate*

> The energy of a linear oscillator with frequency ν can only take on *integer* multiples of $h\nu$: the oscillator energy is quantized.

Note: Planck's oscillators were *material* objects, present within the blackbody radiation to ensure radiation mixing.

Einstein wrote later of this work 'Planck's derivation was of unmatched boldness' [102] Planck himself did not at once understand how bold he had been

> I tried immediately to weld the elementary quantum of action somehow in the framework of classical theory. But in the face of all such attempts this constant showed itself to be obdurate ... My futile attempts to put the elementary quantum of action into the classical theory continued for a number of years and they cost me a great deal of effort [92].

Which goes to show that there may exist profound differences between making and understanding a discovery.

Planck's radiation law was rapidly accepted as correct, simply because further experiments in the years immediately after 1900 confirmed his result to ever better accuracy. His derivation caused no stir, however. Peter Debye (1884–1966) has in fact recalled that, soon after its publication, Planck's work was discussed in Aachen, where he was then studying with Arnold Sommerfeld (1868–1951). Planck's work

Introducing Atoms and Their Nuclei

fitted the data well 'but we did not know whether the quanta were something fundamentally new or not' [103].

I count it among Einstein's great achievements that he may well have been the first to realize that the advent of the quantum theory represented a crisis in science 'It was as if the ground was pulled from under one'. He expressed this state of mind shortly after Planck's paper had appeared [104]. It was also Einstein who took the next step in quantum physics—*five* years after the appearance of Planck's paper! His starting point was the experimental validity of Wien's law for high frequencies. His tools were classical statistical mechanics. His finding was the *light-quantum hypothesis*: within the domain of validity of the Wien formula, monochromatic light with frequency v behaves as if it consists of mutually independent quanta with energy hv. So far this statement, referring to free radiation only†, was in the nature of a theorem. Einstein went much further, however, by adding the speculation that the same behaviour of light also holds true for the emission and absorption of light by matter, and gave several experimental tests for this assumption. Later he showed that light-quanta can also be assigned a definite momentum (hv/c). They therefore behave like particles, eventually called photons [105].

Einstein's new proposal that under certain specific circumstances light behaves like particles hit upon very strong resistance. One reason was that, unlike Planck, he could not at once claim experimental support for his predictions; experiment had not yet progressed far enough. Even when later data did confirm his light-quantum predictions, it still took time, well into the 1920s, until the photon was finally accepted as inevitable [106]. What finally settled the issue was the Compton effect (1923): in the process of scattering a light beam off tiny electrically charged particles, light behaves like a stream of particles, photons, in the sense that the collisions between light and particles obey the same laws of energy and momentum conservation as in the collision between billiard balls.

The opposition to Einstein's light-quantum is perhaps most vividly expressed in the recommendation, written in 1913 and signed by four of Germany's most distinguished physicists, accompanying the proposal for Einstein's membership in the Prussian Academy of Sciences. After expressions of highest praise for his achievements, it concludes as follows.

> In sum, one can say that there is hardly one among the great problems in which modern physics is so rich to which Einstein has not made a remarkable contribution. That he may sometimes have missed the target in his speculations, as, for example, in his

---

† *Note a basic difference: Planck had introduced quanta for material resonators, Einstein next did so for light.*

hypothesis of light-quanta, cannot really be held too much against him, for it is not possible to introduce really new ideas even in the most exact sciences without sometimes taking a risk [107].

The primary reason for this nearly two-decades long opposition to the light-quantum is obvious. In the early 1800s the work of Thomas Young (1773–1829) and Augustin Fresnel (1788–1827) had ended the controversy between Christiaan Huyghens' (1629–95) wave description and Isaac Newton's (1642–1727) particle description of light in favour of Huyghens. Along came young Einstein who said that, at least under certain circumstances, light nevertheless behaves as particles. Then what about the interference of light and all other successes of the wave picture? The situation was incomparably more grave then the Newton–Huyghens controversy, where one set of concepts simply had to yield to another. Here, on the other hand, it became clear, as time went by, that *not only could the wave picture lay claim to success, for some phenomena, that excluded the particle picture but also that the particle picture could make similar claims, for other phenomena, that excluded the wave picture.* What was going on?

From the outset Einstein was well aware of these apparent paradoxes, which became ever more strident as time went by. In 1924 he put it like this

> There are ... now two theories of light, both indispensable and—as one must admit despite twenty years of tremendous effort on the part of theoretical physicists—without any logical connection [108].

The resolution came in 1925 when quantum mechanics made clear that Huyghens and Einstein were both right.

During the intervening years, from 1900 to 1925, quantum phenomena were dark and mysterious, and became increasingly important. Planck remarked on this in 1910 '[The theoreticians] now work with an audacity unheard of in earlier times, at present no physical law is considered assured beyond doubt, each and every physical truth is open to dispute. It often looks as if the time of chaos again is drawing near in the theoretical physics' [109].

That very well describes the state of affairs when Niels Bohr's work of 1913 produced further grand successes and added further mysteries to quantum physics.

2.6. Niels Bohr, father of quantum dynamics

2.6.1. Bohr's personal background and early years

See also p 157

Niels Bohr came from an upper-class family. Christian, his father, had been a professor of physiology at the University of Copenhagen, and an excellent teacher with a good mathematical knowledge. During 1905–6

Introducing Atoms and Their Nuclei

he had been Rector of his University and in 1907 and 1908 had been proposed for the Nobel prize in physiology or medicine. Christian's father and grandfather had been high-school teachers, a brother of his great-grandfather had been a member of the Royal Swedish and Norwegian Academies of Science. Thus learning and teaching were major Bohr traditions.

Niels Bohr's mother hailed from a wealthy and influential Danish–Jewish family of bankers. For 14 years her father had been a member of the Danish parliament [110].

Niels started his University studies in 1903. It was an ideal moment for an aspiring young man to enter the field. Half a century of laboratory research had generated an unparalleled backlog of data that demanded understanding. Very recent experiments had brought to light entirely new kinds of physical phenomena. The great twentieth century upheavals that were to rock physics to its foundations had barely begun. The era of classical physics had just come to an end.

With the advantage of almost a century's hindsight, it is not all that difficult to recognize the period at issue as one clearly rich not only in major advances but also in unresolved questions and budding paradoxes. Yet in 1871 James Clerk Maxwell from Cambridge, one of the leading physicists of that time, had found it necessary to sound these words of caution 'The opinion seems to have got abroad that in a few years all the great physical constants will have been approximately estimated, and that the only occupation which will then be left to men of science will be to carry on these measurements to another place of decimals ... but we have no right to think thus of the unsearchable riches of creation, or the untried fertility of those fresh minds into which these riches will continue to be poured' [111]. As had happened before and as will happen again, it was far less obvious to most of those in the midst of events how acute the state of their science was. Perhaps the main cause for this recurrent phenomenon is the physicists' inclination to protect the corpus of knowledge as it exists at any given time, to extend rather than modify the areas where order appears to reign.

Back to Bohr: in his third University year he began his first research, in response to a prize investigation proposed in 1905 by the Royal Danish Academy of Sciences and Letters. The problem posed was to determine experimentally the surface tension of liquids from the surface vibrations of jets emerging from a cylindrical tube. Bohr's studies of that problem, both experimentally and theoretically, gained him a gold metal from the Academy—and his first published scientific paper [112]. This work prolonged his University studies. He received his Master's degree in 1909, and became Doctor of Philosophy in May 1911 on a thesis entitled 'Studies on the electron theory of metals'†.

---

† *Originally in Danish* [113]. *An English translation did not appear until 1972* [114].

In September 1911 Bohr left Denmark for a year's study in England. His original intention was to spend the entire period doing postdoctoral research under J J Thomson, director of the Cavendish Laboratory in Cambridge, hoping to discuss with him the theory of metals, a subject of common interest. Their relations did not develop well, however, mainly because Thomson did not care for Bohr's critique of his own work. Then an event occurred which was to be crucial for Bohr's entire subsequent career: his meeting with Ernest Rutherford (1871–1937) in Cambridge on 8 December 1911.

The preceding May, Rutherford had published his model of the atom [115], consisting of a central body, the nucleus, in which nearly all the atom's mass is concentrated, surrounded by a swarm of electrons. On that December day Rutherford, erstwhile student of Thomson, was in Cambridge to attend the Research Students' Annual Dinner. During that evening Bohr spoke with him, asking whether he might come to Manchester to work in his laboratory, then the world's foremost centre of experimental studies in radioactivity. Rutherford replied that that was all right with him as long as Bohr could settle the matter with Thomson. That arrangement was made and in March 1912 Bohr made his next move.

Bohr now started experimental work but stopped doing so in May to concentrate (with Rutherford's consent) on theoretical issues. In June/July he wrote a draft paper on the structure of atoms and molecules, meant as a sketch for Rutherford's perusal. This document, commonly known as the Rutherford memorandum, was not published until after Bohr's death [116]. I shall come shortly to its content.

On 24 July 1912 Bohr left Manchester for his beloved Denmark. There, on 5 April 1913, he completed his paper on the quantum theory of the hydrogen atom.

That article ushered in a new phase in quantum theory. Planck's as well as Einstein's earlier work on this subject had been based on (classical) statistical mechanics. Bohr's contribution of 1913 should above all be remembered as the first paper ever on quantum *dynamics*, a discipline of which he is the sole father.

In order to appreciate this development it is necessary first to give a sketch of the status of spectroscopy in 1913. This I do next.

2.6.2. Spectroscopy up to 1913

When Newton let sunlight pass through a prism he had observed 'a confused aggregate of rays indued with all sorts of colours'. This spectrum of colours appeared to him to be continuous. The resolving power of his experimental arrangement was not sufficient to show that the solar spectrum actually consists of a huge number of discrete lines interspersed with darkness. The first observation of discrete spectra is apparently due to the Scotsman Thomas Melvill who was born the year before Newton died. He found that the yellow light produced by holding kitchen salt in a flame shows unique refraction, that is, the light is monochromatic. Actually, no single atomic or molecular spectrum is monochromatic, but in the case of kitchen salt the yellow so-called D-line is much more intense than all other lines. Moreover, it was found subsequently that the D-line is in fact a doublet, a pair of close-lying lines. It would also be clear later that the D-line stems from the heated sodium atoms contained in the salt molecules. As to the solar spectrum, I have already noted the importance of the dark, discrete, Fraunhofer lines.

As happens so often in the development of new domains in experimental physics, their origins can be traced to the invention of a new experimental tool. In the case of spectroscopy it was the burner invented in the 1850s by Robert Bunsen (1811–99), now so familiar from elementary laboratory exercises in chemistry. Why was this simple gadget so important to spectroscopy? In order to generate the emission spectrum of a substance, one has in general to heat it. If the heating flame has colours of its own, as with candlelight, then the observation of the spectrum of the substance gets badly disturbed. The virtue of the Bunsen burner is that its flame is non-luminous! And so it came about that analytical spectroscopy started with a collaboration between Kirchhoff and Bunsen, then both professors in Heidelberg. Their tools were simple: a Bunsen burner, a platinum wire with a ringlet at its end for holding the substance to be examined, a prism, and a few small telescopes and scales.

Their results were of the greatest importance. They observed (as others had conjectured earlier) that there is a unique relation between a chemical element and its atomic spectrum. Spectra may therefore serve as visiting cards for new elements 'spectrum analysis might be no less important for the discovery of elements that have not yet been found' [117]. They themselves were the first to apply this insight by discovering the elements caesium and rubidium. Ten more new elements had been identified spectroscopically before the century was over: thallium, indium, gallium, scandium, germanium, and the noble gases helium, neon, argon, krypton, and xenon.

The discovery in 1869 of helium by means of a mysterious yellow line found in the spectrum of the Sun (whence its name) that had no counterpart in terrestrial spectra beautifully illustrated what Kirchhoff and Bunsen had foreseen. '[Spectrum analysis] opens ... the chemical exploration of a domain which up till now has been completely closed ... It is plausible that [this technique] is also applicable to the solar atmosphere and the brighter fixed stars' [117]. The fact that a number of nebulae are luminous gas clouds was another discovery by spectroscopic means. The spectra of these clouds revealed another substance never seen on Earth, accordingly named nebulium, assumed to be a new element. It took 60 years before it was realized that this substance actually is a mixture of metastable oxygen and nitrogen [118]. Another hypothesized stellar element, coronium, turned out to be very highly ionized iron.

Perhaps the most important insight we owe to Kirchhoff and Bunsen is that very minute amounts of material suffice for chemical identification

> The positions occupied [by the lines] in the spectrum determine a chemical property of a similar unchangeable and fundamental nature as the atomic weight ... and they can be determined with almost astronomical accuracy. What gives the spectrum-analytic method a quite special significance is the circumstance that it extends in an almost unlimited way the limits imposed up till now on the chemical characterization of matter [117].

Many more advances in experimental spectroscopy date back to the nineteenth century, such as the discovery that molecules exhibit characteristic 'band spectra', bunches of tightly spaced lines. In fact, by the time Bohr entered the field, the first six volumes of the excellent handbook of spectroscopy by Heinrich Kayser (1853–1940) had appeared, counting in all 5000 pages [119]. I shall, however, conclude this sketch of experimental spectroscopy with just one remark, concerning the spectrum of atomic hydrogen, so essential for what is to follow.

It appears that at least parts of this spectrum were first detected by Anders Ångström (1814–74) 'To my knowledge I was the first who, in 1853, observed the spectrum of hydrogen' [120]. Soon thereafter four (and only four) lines of this spectrum were identified and their frequencies measured, by Julius Plücker (1801–68) [121] and by Ångström (1860) [122]. The latter achieved an accuracy impressive for those days: about one part in ten thousand.

The question of what makes a hot body glow must have come to man's mind long long ago, even long before Newton's conjecture about the mechanism for light emission, found in that utterly remarkable set of open questions, appended to his *Opticks*, which he left for later generations to ponder. Query 8 reads 'Do not all fix'd Bodies, when heated beyond a certain degree, emit Light and shine; and is not this Emission perform'd by the vibrating motions of its parts?'.

The more specific suggestion that these 'parts' are actually parts of atoms or molecules came later, in the nineteenth century. In 1852 George Stokes (1819–1903) suggested 'In all probability ... the molecular vibrations by which ... light is produced are not vibrations in which the molecules move among one another, but vibrations among the constituent parts of the molecules themselves, performed by virtue of the internal forces which hold the parts of the molecules together' [123]. Maxwell's contribution 'Atom' to the 1875 edition of the *Encyclopaedia Britannica* contains the following 'conditions which must be satisfied by an atom ... permanence in magnitude, capability of internal motions or vibration, and a sufficient amount of possible characteristics to account for the difference between atoms of different kinds' [124].

After the discovery of electrons it became extremely plausible that spectra are associated with the motions of these particles inside atoms. One finds several conjectures to this effect in the early years of the twentieth century. Thomson thought (1906), however, that line spectra would be due 'not to the vibrations of corpuscles [i.e. electrons] inside the atom, but of corpuscles vibrating in the field of force outside the atom' [125]. Johannes Stark (1874–1957) believed that band spectra stem from electron excitations of neutral bodies, line spectra from ionized bodies [126]. I mention these incorrect ideas of serious and competent physicists only to illustrate how confused the status of spectra was right up to Bohr's clarifications of 1913.

Towards the end of his life, Bohr was asked what people thought about spectra before 1913. He replied

> One thought [spectra are] marvellous, but it is not possible to make progress there. Just as if you have the wing of a butterfly then certainly it is very regular with the colours and so on, but nobody thought that one could get the basis of biology from the colouring of the wing of a butterfly [127].

In the 1860s, shortly after the first quantitative measurements of spectral frequencies, a new game came to town, less ambitious than trying to find mechanisms for the origin of spectra: spectral numerology, the search for simple mathematical relations between observed frequencies [128]. A textbook published in 1913 [129], at practically the same time that Bohr interpreted the hydrogen spectrum, contains no less than 12 proposed spectral formulae. All these have long been forgotten except for one that will live forever: the Balmer formula for the spectrum of atomic hydrogen.

After receiving a PhD in mathematics, Johann Balmer (1825—1898) became a teacher at a girl's school in Basel, and later also *privat-dozent* at the university there. He was 'neither an inspired mathematician nor a subtle experimentalist, [but rather] an architect ... [To him] the whole world, nature and art, was a grand unified harmony and it was his aim in life to grasp these harmonic relations numerically' [130]. In all he published three physics papers. The first two, completed at age 60, made him immortal; the third, written when he was 72, is uninteresting.

What Balmer did is slightly incredible. Having at his disposal only the four frequencies measured by Ångström, he fitted them with a mathematical expression that predicts an infinity of lines—and his formula is in fact correct! It reads in modernized notation

$$\nu_{ab} = R\left(\frac{1}{b^2} - \frac{1}{a^2}\right)$$

where the symbols have the following meaning: as usual, ν stands for frequency; each distinct pair of values for a and b denotes a distinct frequency, the index b takes on the values 1, 2, 3, ... to infinity and a also runs through the integers but is always larger than b, thus if $b = 2$, then $a = 3, 4, \ldots$; and R is a constant. Balmer found that he could fit the four Ångström lines very well if these are made to correspond to $b = 2$ and $a = 3, 4, 5, 6$, respectively, and if R (itself a frequency, a number of oscillations per second) is given the value $R = 3.291\,63 \times 10^{15}$ per second [131]. Now, a century later, it is known that $R = 3.289\,841\,86 \times 10^{15}$ per second [132], which shows that Balmer's R value was correct to better than one part in a thousand.

Having come that far, Balmer told the professor of physics at Basel University what he had found. This friend told him that actually another 12 lines were known from astronomical observations. Balmer quickly checked that these also fitted his formula, for $b = 2$ and $a = 7$ to 18. Whereupon, he wrote his second physics paper [133], in which he stated that 'this agreement must be considered surprising to the highest degree'.

Balmer's formula has stood the test of time as more hydrogen lines kept being discovered. Soon his results became widely known; they were quoted in the 1912 edition of the *Encyclopaedia Britannica* [134]. For nearly 30 years no one knew, however, what the formula was trying to say.

Then Bohr came along.

2.6.3. Bohr just before March 1913; precursors
Already in the introduction to his doctor's thesis Bohr had opined that mechanics, that is classical physics, does not work inside atoms

> The assumption [of mechanical forces] is not *a priori* self-evident, for one must assume that there are forces in Nature of a kind completely different from the usual mechanical sort; for while on the one hand the kinetic theory of gases has produced extraordinary results by assuming the forces between individual molecules to be mechanical, there are on the other hand many properties of bodies *impossible to explain if one assumes that the forces which act within the individual molecules ... are mechanical also* [my italics]. Several examples of this, for instance calculations of heat capacity and of the radiation law for high frequencies, are well known; we shall encounter another one later, in our discussion of magnetism†.

I have not yet mentioned the 'calculations of heat capacity' made by Einstein in 1906, the first occasion on which the quantum was brought to

---
† See the Danish version [135]. For reasons beyond my comprehension this important passage is not included in the English translation [136].

Niels Bohr

(Danish, 1885–1962)

Niels Henrik David Bohr was a scion of a cultured patrician Danish family. In 1911, after having received the DPhil degree at the University of Copenhagen, he went for post-doctoral research first to J J Thomson, then to Rutherford, the man who became the role model for his personal and scientific style.

Bohr, the founder of quantum dynamics with his theory of the hydrogen atom (1913), was Copenhagen's first professor of theoretical physics (1916). He took the initiative for the creation of the University's first physics institute (now named the Niels Bohr Institute), which opened in 1921. During its first two decades, it was the world's most important theoretical physics centre, counting Heisenberg, Dirac and Pauli among its more than 400 alumni from all over the world. Bohr won the Nobel Prize for physics in 1922.

After the discovery of quantum mechanics (1925), Bohr established its foundations in terms of a new logic which he called complementarity (1927). His other main contributions were the theoretical foundation for the periodic table of elements (1920—22), the theory of the compound nucleus (1936) and the realization that uranium fission by slow neutrons is due to the presence of the isotope 235 (1939). He saw to the development of experimental physics at his institute, and fulfilled numerous obligations regarding social functions of science.

Bohr, one of the most inspiring teachers of his time, died in 1962. It can be said of him what he himself has said of Rutherford 'He was a physicist of the greatest renown. In his life all honours imaginable for a man of science came to him, yet he remained quite simple in all his ways. This, together with the kind interest he took in the welfare of his pupils, caused the spirit of affection he created around him wherever he worked'.

bear on matter rather than radiation. It had been known since the 1870s that at low temperatures the heat capacity of some materials (diamond for instance) is lower than was expected on (classical) theoretical grounds. In showing that this effect was a new manifestation of Planck's quantum in action, Einstein had pioneered a new discipline, the quantum theory of the solid state.

See also p 148

Back to Bohr again: in his Rutherford memorandum he made his first attempt to go beyond classical physics by introducing a new 'hypothesis for which there will be given no attempt at a mechanical foundation (*as it seems hopeless*) [my italics] ...' [116]. Neither the new postulate, essentially a quantum condition on the kinetic energy of an electron inside an atom, nor other details of the memorandum need be discussed here. As Bohr himself said later of this draft paper 'You see, I'm sorry because most of that was wrong' [127]. I shall only note about that work that the simplest system Bohr considered at that time was a hydrogen molecule—the hydrogen atom does not yet appear—and there is no mention at all about spectra.

The year 1913 came and still Bohr's mind was not on spectra. Bohr wrote to Rutherford in January 'I do not at all deal with the question of calculation of frequencies corresponding to the lines in the visible spectrum' [137]. On 7 February he sent to Hevesy a list 'of the ideas I have used as the foundation of my calculations' [138], in which spectra do not appear either.

Shortly after 7 February Bohr heard of the Balmer formula. By 6 March he had completed a paper containing its interpretation. That event marks the beginning of the quantum theory of atomic structure.

On a number of occasions Bohr told others (including me) that he had not been aware of the Balmer formula until shortly before his own work on the hydrogen atom and that everything fell into place for him as soon as he heard about it. The man who told him was Hans M Hansen (1886–1956), one year younger than Bohr, who had done experimental research on spectra in Göttingen. One week before he died Bohr recalled

> I think I discussed with someone ... that was Professor Hansen ... I just told him what I had, and he said 'But how does it do with the spectral formulae?' And I said I would look it up, and so on. That is probably the way it went. I didn't know anything about the spectral formulae. Then I looked it up in the book of Stark (reference [139]) ... other people knew about it but I discovered it for myself. And I found then that there was this very simple thing about the hydrogen spectrum ... and at that moment I felt now we'll just see how the hydrogen spectrum comes [140].

In the course of reminiscing about what he did next, Bohr once remarked 'It was in the air to try to use Planck's ideas in connection with such things' [127]. Let us see what ideas there were about the atom at that time.

First there was the Viennese physicist Arthur Haas (1884–1941) who in 1910 discussed a model for the hydrogen atom consisting of one electron moving on the surface of a positively charged sphere with radius r—not the Rutherford model. He introduced a quantum postulate from which he was able to obtain the correct expression for r, now called the Bohr radius (see further below). There is no mention of spectra in his work [141]. Sommerfeld, commenting on Haas' work during the Solvay Conference of October 1911, thought it plausible 'to consider the existence of molecules as a consequence of the existence of [Planck's constant]' [142]. Bohr said later that he was unaware of Haas' results when he did his own work on hydrogen. However, his paper on that subject does contain reference to Haas [143].

Then there was John Nicholson (1881–1955) from Cambridge, who in 1911 associated spectral lines with various modes of vibration of electrons around their equilibrium orbits in the field of a central charge. He argued that a one-electron atom cannot exist and that the simplest and lightest atoms are, in this order, coronium, with atomic weight about half that of hydrogen, hydrogen, and nebulium, with 2, 3, 4 electrons respectively; helium was considered to be a composite [144]. It was a bizarre collection. Helium is an element, coronium and nebulium are not, and there is no element lighter than hydrogen. For a further insight into the understanding of atoms at that time, it is most significant to note his statement that hydrogen should have three electrons instead of one, in fact there could be no one-electron atom. Hydrogen was therefore still poorly understood.

So far Nicholson's views on atoms are merely passing curiosities, but in a subsequent paper he made a very important comment on angular momentum, a quantity to be called L from here on [145]. L is defined (not in full generality) as follows. Consider a particle with mass m moving with velocity v along a circle with radius r. Then L relative to the circle's centre equals the product mvr. Now, Nicholson noted that this product just equals the ratio of the particle's energy to its frequency. This ratio, Planck had proposed in another context, should equal h. 'If therefore the constant h of Planck has ... an atomic significance, it may mean that the angular momentum of a particle can only rise or fall by discrete amounts when electrons leave or return' [145]. In modern terms, Nicholson had quantized angular momentum. He went on to calculate L for hydrogen, finding an integral multiple of $h/2\pi$ (correct) equal to 18 (weird). Bohr did not think highly of Nicholson's work [127] but did quote him in his paper on hydrogen [143].

Finally, there was the work of Niels Bjerrum dealing with molecules rather than atoms. Bjerrum had been Bohr's chemistry teacher at the University of Copenhagen; later they became good friends. In

1912 Bjerrum published a paper [146], one section of which is entitled *Applications of the quantum hypothesis to molecular spectra*. Those spectra arise as the result not only of electronic but also of nuclear motions. In a diatomic molecule, for example, the vibrations of the two nuclei along the axis joining them generate 'vibrational spectra'. Furthermore, this axis can rotate in space, producing 'rotational spectra'. In 1911 Walther Nernst had pointed out [147] that it is a necessary consequence of the quantum theory that the associated vibrational and rotational energies of molecules must vary discontinuously. Bjerrum was the first to give the explicit expression for the discrete rotational spectral frequencies. That contribution, which does not directly link with what Bohr did next, will be remembered as the first instance of a correct quantum-theoretical formula in spectroscopy.

2.6.4. Bohr's hydrogen atom

On 6 March 1913 Bohr sent a letter [148] to Rutherford in which he enclosed 'the first chapter on the constitution of atoms', asking him to forward that manuscript to *Philosophical Magazine* for publication. Up to that point Bohr had three published papers to his name. These had earned him respect within a small group of physicists. His new paper [143] was to make him a world figure in science and, eventually, beyond.

Two factors decisively influenced Bohr's next advance. First, his insight that 'it seems hopeless' [116] to understand the atom in terms of classical physics, an area of science for which he nevertheless, and with good reason, had the greatest respect, then and later. Secondly, his recent awareness of the Balmer formula. He was convinced (as we have seen) that his answers had to come from the quantum theory (figure 2.4).

To my knowledge, there are no letters nor later interviews describing in detail what Bohr went through in those intense few weeks. Probably he would have been unable to reconstruct later a day by day account of his thoughts during that period. He would, I think, ceaselessly have gone back and forth between trying out, one after the other, new postulates outside classical dynamics and matching those with the guidance provided by the Balmer formula. One finds a residue of these vacillations in the printed paper [116], which contains not one but three derivations of Balmer's expression. The first [149] is based on a hypothesis about which, five pages later, Bohr writes 'This assumption ... may be regarded as improbable', after which it is replaced by another one [150]. Later in 1913 Bohr again treated the Balmer formula differently [151]. Every subsequent version is a distinct improvement over its predecessor. I shall confine myself to a summary of the principal points.

All three treatments have in common the postulate that an electron inside a hydrogen atom can move only on one or another of a discrete set of orbits (of which there are infinitely many), in violation of the tenets of classical physics, which allow a continuum of possible orbits. Bohr

Introducing Atoms and Their Nuclei

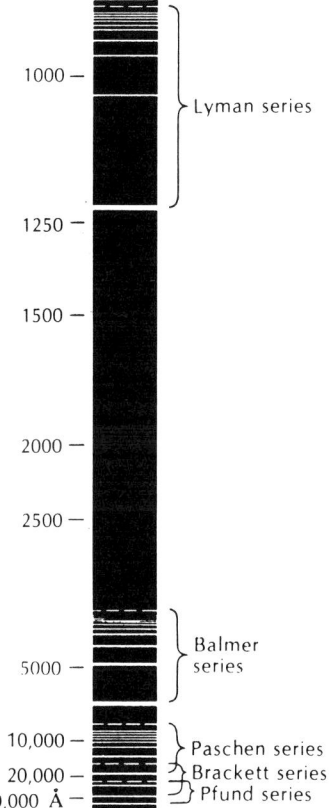

Figure 2.4. *The line spectrum of the hydrogen atom, with wavelengths given in ångstrom units (1 Å= 10^{-8} cm). Bohr's atomic theory gave a fairly accurate account of the line positions, but it could not predict their intensities.*

called his orbits 'stationary states'. Their respective energies, taken in order of increase, will be denoted by E_a, where $a = 1, 2, 3$, and so on. We shall refer to $a = 1$ as the 'ground state', the lowest orbit, closest to the nucleus, which has the lowest energy E_1.

Now, for the first time, I believe, Bohr refers to *radiative instability*. As a result of the energy lost by radiation 'the electron will no longer describe stationary orbits'. In particular, according to classical law, an electron in the ground state will not stay there but will spiral into the nucleus.

Bohr circumvented this disaster by introducing one of the most audacious postulates ever seen in physics. He simply declared that the ground state *is* stable, thereby contravening all knowledge about radiation available up till then!

So much for the ground state. What about the higher stationary states? These are unstable and the electron will drop from a higher to some lower state. Consider two states with energies E_a larger than E_b (thus a is larger than b). Then, Bohr assumed, transitions $a \to b$ are accompanied by the emission of one light-quantum with frequency ν_{ab}, given by

$$E_a - E_b = h\nu_{ab}.$$

(In the first (abandoned) version it was assumed that more light-quanta than one could be emitted.) Thus the discreteness of spectra is a consequence of the discreteness of atomic states. Transitions between these states are the *only* way in which an atom emits (or absorbs) radiation.

Next Bohr returned to his 'new hypothesis' stated in the Rutherford memorandum [116], which now, however, appears in an improved form. Let W_a be the kinetic energy of an electron in orbit a, and ν_a its frequency. (Please keep track of the distinction between ν_a and ν_{ab}. The former is a 'mechanical frequency' of an electron in orbit, the latter an 'optical frequency' of light emitted in a transition.) Bohr now made the explicit proposal

$$W_a = \tfrac{1}{2} a h \nu_a.$$

(If $a = 1$, then $W_1 = \tfrac{1}{2} h \nu_1$, which is what Haas had used.) This equation associates an integer a to each state. It is the first example of a *quantum number* now called the principal quantum number.

It is a straightforward calculation, given in many elementary texts†, to derive the Balmer formula from the last two equations given above and a third one which expresses that the electron is kept in orbit by the balance between the centripetal force which pulls the electron away from and the attractive electric force which pulls it towards the nucleus. One finds that

$$E_a = \frac{hR}{a^2}$$

from which the Balmer formula follows at once.

At this point the frog jumps into the water, as an old saying goes. Bohr was able to predict the value of R!! Let m and $-e$ denote the mass and charge of the electron and Ze the charge of the nucleus ($Z = 1$ for hydrogen of course). Bohr found

$$R_Z = \frac{2\pi^2 m Z^2 e^4}{h^3}$$

where R_1 is our old R. Using the best known experimental values for m, e and h, and putting $Z = 1$, Bohr obtained $R = 3.1 \times 10^{15}$ s^{-1} 'inside the uncertainty due to experimental errors' with the best value of R obtained from spectral measurements.

---

† *These textbook derivations are often imprecise in regard to technical details. I recommend Bohr's own 'third derivation'* [151] *as the superior one.*

Introducing Atoms and Their Nuclei

This expression for R_Z is the most important equation that Bohr derived in his life. It represented a triumph over logic. Never mind that discrete orbits and a stable ground state violated laws of physics which until then were held basic. Nature had told Bohr that he was right anyway, which, of course, was not to say that logic should now be abandoned, but rather that a new logic was called for. That new logic, quantum mechanics, will make its appearance elsewhere in this book.

Bohr was also able to derive an expression for r_a, the radius of the ath orbit:
$$r_a = \frac{a^2 h^2}{4\pi^2 Zme^2}.$$
This yields the 'Bohr radius' $r_1 = 0.55 \times 10^{-8}$ cm for stable hydrogen, in agreement with what was then known about atomic size. Bohr noted further that the larger radii of higher states explain why so many more spectral lines had been seen in starlight than in the laboratory. 'For $a = 33$, the (radius) is equal to 0.6×10^{-5} cm corresponding to the mean distance of the molecules at a pressure of about 0.2 mm mercury ... the necessary condition for the appearance of a great number of lines is therefore a very small density of a gas', more readily available in the atmosphere of stars than on the surface of the Earth.

The capstone of Bohr's work on the Balmer formula is the story of the Pickering lines. In 1896 Charles Pickering (1846–1919) from Harvard had found a series of lines in starlight which he attributed to hydrogen even though this did not fit the Balmer formula. In 1912 these same lines were also found in the laboratory, by Alfred Fowler in London. Bohr pointed out that 'we can account naturally for these lines if we ascribe them to helium' [143], singly ionized helium, that is, a one-electron system with $Z = 2$. According to the formula for R_Z this would give a Balmer formula with R replaced by $R_2 = 4R$. Fowler objected: in order to fit the data the 4 ought to be replaced by 4.0016, a difference which lay well outside experimental error [152]. Whereupon Bohr remarked in October 1913 that the formulation in his paper rested on the approximation in which the nucleus is treated as infinitely heavy compared to the electron; and that an elementary calculation shows that, if the true masses for the hydrogen and helium nuclei are used, then the 4 is replaced by 4.00163, 'in exact agreement with the experimental value' [153]. Up to that time no one had ever produced anything like it in the realm of spectroscopy, agreement between theory and experiment to five significant figures. (This high precision could be attained because the hydrogen/helium ratio R_1/R_2 is independent of the values of e and h.)

Bohr's paper on hydrogen appeared in July 1913. Two sequels followed, one (September) [154] on the structure of atoms heavier than hydrogen and on the periodic table, the other (November) [155] on the structure of molecules. The plethora of experimental data on their spectra available at that time could not then and cannot now be represented as compactly and simply as for hydrogen. One useful formula had been discovered, however, in its most general form by the Swiss physicist Walter Ritz (1878–1909) [156]. It says that the frequencies of a spectrum can be grouped in series, represented by the difference of two functions, each of which depends on a running integer. This, Bohr suggested, indicates an origin of spectra similar to that for hydrogen. 'The lines correspond to a radiation emitted during the passing of the system between two stationary states' [143], a notion that underlies all of spectroscopy to this day.

Following an idea first stated in classical context by J J Thomson, Bohr also correctly identified the origins of x-ray spectra: an electron in an inner ring is knocked out of an atom, after which an electron from an outer ring makes a quantum jump into the unoccupied slot, emitting an energetic light-quantum in the process [154].

For the rest Bohr wisely left spectra alone and, in parts 2 and 3 (references [154, 155]), concentrated on the ground states of atoms and molecules. Electrons are arranged in one or more coaxial and coplanar rings. Here Bohr made another far-seeing remark, insisting that the outermost ring is the seat of most chemical properties of the elements. This observation constitutes the very first step towards quantum chemistry. At this stage Bohr could use another important discovery made in 1913. Henry Moseley (1887–1915) had informed him, in advance of publication, of his experimental conclusion that the number of electrons per atom determines its place in the periodic table [157] (figure 2.5). Clearly then the second and third part of Bohr's trilogy contain nuggets of wisdom—even though much of that work has turned out to rest on incorrect conjectures.

See also p 158

A final remark on Bohr's contributions of 1913. In his third formulation [151] of his theory of the hydrogen atom he argued that for large values of the principal quantum number n the hydrogen levels lie so close together that they form 'almost a continuum', and that therefore the *classical* continuum description for emission of radiation should be very nearly valid for transitions between two very close-lying states both with very large n. This link between the classical and the quantum description was later called by Bohr the *correspondence principle*. Hendrik Kramers (1894–1952), Bohr's first close collaborator, wrote sometime later about this principle 'It is difficult to explain in what [it] consists, because it cannot be expressed in exact quantitative laws and is, on this account, also difficult to apply. [However], in Bohr's hands it has been extraordinarily fruitful in the most varied fields' [158].

2.6.5. Impact of Bohr's ideas

The first response to Bohr's new ideas reached him even before the first paper of his trilogy had appeared in print. After Rutherford had read that article in manuscript form he had written 'There appears to me

Introducing Atoms and Their Nuclei

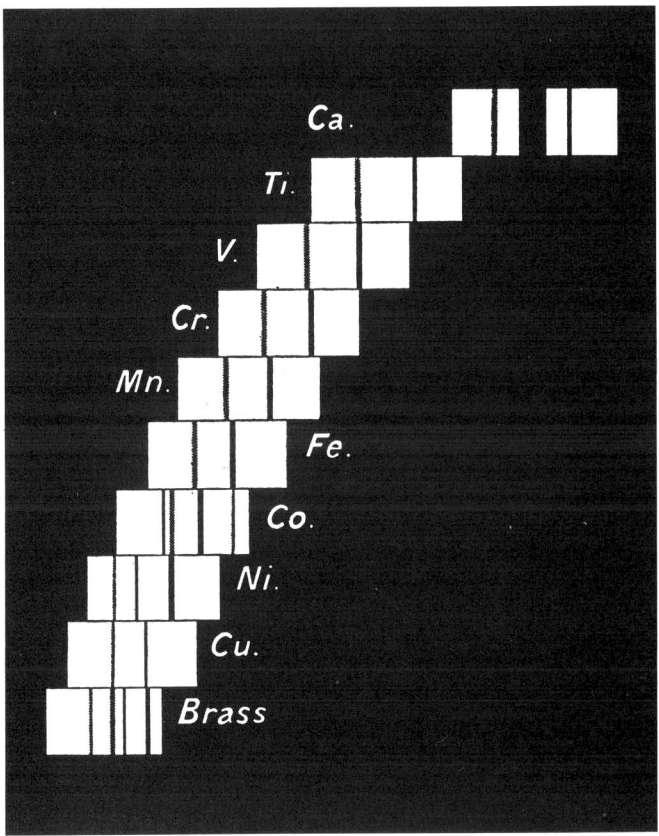

Figure 2.5. *Moseley's plot of his data.*

one grave difficulty in your hypothesis, which I have no doubt you fully realize, namely how does an electron decide what frequency it is going to vibrate at when it passes from one stationary state to the other? It seems to me that you have to assume that the electron knows beforehand where it is going to stop [159]. In typical Rutherford style he had gone right to the heart of the matter by raising the issue of cause and effect, of causality: Bohr's theory left unanswered not only the question of why there are discrete states but also why an individual electron in a higher state chooses one particular lower state to jump into. These questions were to remain unresolved until quantum mechanics gave the surprising answer: they are meaningless.

The older generation did not go along, as was to be expected. When Lord Rayleigh, then past 70, was asked what he thought about Bohr's theory he replied 'I have looked at it, but I saw it was no use to me. I do not say that discoveries may not be made in that sort of way. I think it is very likely they may be. But it does not suit me' [160].

As a final entry from the British side, I note Moseley's brilliant assessment

> Your theory is having a splendid effect on Physics, and I believe when we really know what an atom is, as we must within a few years, your theory even if wrong in detail will deserve much of the credit [161],

and some German responses: Sommerfeld, in a letter to Bohr wrote 'Although I am for the present still a bit sceptical about atomic models, your calculation of the constant [in the Balmer formula] is nevertheless a great achievement' [162]. Friedrich Paschen (1865–1947), the distinguished experimental spectroscopist from Tübingen believed Bohr at once upon hearing of the helium result [163].

Robert Pohl (1884–1976) remembered the reaction in Berlin. 'Shortly after the publication of Bohr's first paper something unusual happened in the *Physikalische Gesellschaft* (Physical Society) in Berlin. Normally, communications at its meetings were original papers, but that time Professor Emil Warburg (1846–1931), a brilliant physicist and teacher, announced a report "on a very important paper", that was Bohr's paper, ... He explained in his dry but clear way ... that this was a real advance, and I believe that the few hundred listeners at once understood: There Bohr has had a stroke of genius (*einen ganz grossen Wurf*), Planck's h proves to be the key for understanding the atom' [164].

Einstein reacted very positively. In September 1913 Hevesy had met him in Vienna and had asked him for his opinion. Einstein replied that Bohr's work was very interesting, and important if right—faint praise, as I know from having heard him make that comment on other occasions. Then, however, Hevesy told him of the helium results, whereupon Einstein said 'This is an enormous achievement. The theory of Bohr must then be right' [165].

In December 1913, Bohr clearly expressed his own reservations 'You understand, of course, that I am by no means trying to give what might ordinarily be described as an explanation ... I hope I have expressed myself sufficiently clearly so that you have appreciated the extent to which these considerations conflict with the admirably coherent group of conceptions which have been rightly termed the classical theory of electrodynamics' [151].

Bohr's trilogy of 1913 appeared at just about midway in the old quantum theory era. During the first half of that period, no more than a handful of papers on quantum theory saw the light of day—but what papers! Very few realized at that time that physics was no longer what it used to be. It was also a rarity in those years for theorists to join the founders in the pursuit of quantum physics. On the experimental side we should note ongoing studies of spectra, and improved measurements of the black-body spectral distributions, of heat capacities at low temperatures and of the photoelectric effect.

All that changed after the appearance of Bohr's papers. Bohr to Rutherford in 1916 'The whole field of work has indeed from a very lonely state suddenly got into a desperately crowded one where almost everybody seems hard at work' [166]. It is not hard, I think, to guess why. Bohr's spectacular successes with the spectra of hydrogen and ionized helium held out promise for an understanding of other spectra as well. Remember that a huge backlog of spectral data amassed in the previous half century were awaiting interpretation. Moreover, at just about the time of Bohr's papers, newly discovered spectral phenomena (to be discussed later) posed fresh challenges. As a result we now observe for the first time that research in quantum physics begins to spread, not only in Europe but also in the United States [167]. In particular we witness the emergence of three schools where the old quantum theory was seriously pursued, in order of appearance in Munich, under Sommerfeld, Copenhagen, under Bohr, and Göttingen, under Born. Each of these centres had its own style. Werner Heisenberg (1901–76), who had spent time at all these places, later said 'I learned optimism from Sommerfeld, mathematics in Göttingen, and physics from Bohr' [168].

2.7. First the good news. More successes for the old quantum theory

2.7.1. The Stark effect
In November 1913 Johannes Stark (1874–1957) announced [169] an important new discovery: when atomic hydrogen is exposed to a static electric field its spectral lines split, the amount of splitting being proportional to the field strength (the linear Stark effect†).

We now encounter for the first time the widening interest in the quantum theory mentioned above. Before the year was out Warburg applied the Bohr theory to the new effect [170]. In 1916 Paul Epstein (1883–1966) from Munich and Karl Schwarzschild (1873–1916) from Göttingen independently showed [171] that the effect in hydrogen is an exactly soluble problem in the old quantum theory. Epstein concluded his papers with these words 'It seems that the efficiency of the quantum theory borders on the miraculous and that it is by no means exhausted'. Their formula, in excellent agreement with experiment, ranks next to the Balmer formula as the greatest quantitative success of the old quantum theory.

2.7.2. The Franck–Hertz experiment
In 1914 James Franck (1881–1964) and Gustav Hertz (1878–1975) published a paper [172] dealing with collisions between electrons and

† *Hydrogen is the only atom that shows a linear Stark effect.*

mercury vapour, a particularly simple substance since its molecules consist of only one atom. This is what they found. If the electron kinetic energy is less than 4.9 eV, the collisions are elastic, that is the electron can change direction but not velocity. When the energy reaches 4.9 eV many collisions become completely inelastic, the electron gives up its entire kinetic energy to the atom. A bit above 4.9 eV many electrons still give 4.9 eV to the atom, then continue with an energy less by that amount.

That was just the kind of behaviour Bohr had predicted. He had suggested that 'an electron of great velocity in passing through an atom and colliding with the bound electrons will loose [sic] energy in distinct finite quanta' [143]. That is precisely what we see in the Franck–Hertz experiment. The energy 4.9 eV corresponds to minus the difference in energy of the most loosely bound electron in the mercury atom (E_1) and the energy of that electron when, due to collision, its energy is raised to the first available discrete excited state (E_2). When the kinetic energy of the impacting electron is less than $E_2 - E_1$, then the bound electron cannot be excited (elastic collisions), when it is a bit more, then all the bound electron can do is to pick up an energy amount $E_2 - E_1$.

According to the Bohr theory the excited electron should eventually drop back to its original orbit under emission of a light-quantum with frequency ν fixed by the Bohr relation

$$h\nu = E_2 - E_1.$$

Now the beauty of Franck and Hertz's work lies not only in the measurement of the energy loss $E_2 - E_1$ of the impinging electron, but they also observed that, when the energy of that electron exceeds 4.9 eV, mercury began to emit ultraviolet light of a definite frequency equal to ν as defined in the above formula. Thereby they gave (unwittingly at first) the first direct experimental proof of the Bohr relation!

2.7.3. Sommerfeld introduces two new quantum numbers. The fine structure of the hydrogen spectrum

In 1887 Albert Michelson (1852–1931) and Edward Morley (1838–1923) had been the first to observe that one of the Balmer lines is actually a 'double line' [173]. In 1916 Sommerfeld gave the quantitative interpretation of this 'fine structure' [174]. It is easiest to divide his reasoning into two stages.

(a) The hydrogen atom, neglecting relativity effects. Bohr, aware of course that the orbits of the electron in that atom are in general ellipses, had tackled the problem only for the subset of circular orbits. It was Sommerfeld who first raised and answered the question: how does one handle ellipses? For circles the only spatial variable (or, as we say, the only degree of freedom) is its radius. Bohr's restriction of the

classical continuum of possible radii to a discrete set had quantized the atom's size. In the elliptic case the atom not only has size but also (loosely speaking) shape, expressed for example by the ratio of the minor axis (length 2b) to the major axis (length 2a) (two degrees of freedom). Sommerfeld showed that the classically possible continuum in regard to both size and shape is restricted to a discrete set of elliptic (including circular) orbits, characterized by two quantum numbers, n and k

$$\frac{b}{a} = \frac{k}{n}.$$

Here n is Bohr's principal quantum number. The quantity k which used to be called the auxiliary or azimuthal quantum number, was a *new* quantum number, also taking on integral values only. Since b at most equals a, it follows that k is restricted to the values $0, 1, 2, \ldots, n$. Circular orbits correspond to $k = n$. Note: $k = l + 1$ where l is the angular momentum quantum number familiar from quantum mechanics.

The inclusion of elliptic orbits—and there is no logical reason to exclude them—shows that Bohr's circular orbits are but a puny subset of all allowed quantum orbits. Then what remains of Bohr's successes regarding hydrogen? The amazing answer is: everything. As Sommerfeld showed, all n states corresponding to fixed n but varying k are 'degenerate', that is they have the same energy. It follows that Bohr, who, of course, knew nothing about this degeneracy, was lucky. His restriction to $k = n$ gave all the representative energy values of the hydrogen levels. Furthermore, again for fixed n and varying k, half the major axes of the corresponding ellipses are equal, hence equal to the radius of the circular orbit, $k = n$. Also Bohr's results for the *sizes* of orbits therefore remained essentially unchanged.

Sommerfeld stressed that the degeneracy he had discovered is unique for the inverse square two-body law in the hydrogen atom and that this degeneracy no longer exists in more complex atoms, multi-body systems in which each of the several electrons is acted on by more complicated forces, due not only to the nucleus but also to the other electrons. This fact is clearly of capital importance for the interpretation of spectra other than that of hydrogen.

(b) The hydrogen atom, including relativity effects. Bohr already knew [175] that relativity leads to precession of the orbits. Sommerfeld duly acknowledged Bohr's point and then went much further by noting that this precession removes the degeneracy of orbits with fixed n, varying k. That fact, he asserted, explains the fine structure. He then proceeded to calculate explicitly the revised formula for the energy levels of hydrogen. The biggest contribution, depending on n only, is the one originally found by Bohr. A much smaller term, depending on both n and k, accounts for the fine structure. Sommerfeld's formula, not given here, is found in any good textbook.

The agreement with experiment, which on the whole was very good, was considered a triumph both for quantum and for relativity theory. It is fitting that from 1916 the quantum rules in atomic physics were called the Bohr–Sommerfeld rules. Yet Sommerfeld's answer has to be considered a fluke. It has been called 'perhaps the most remarkable numerical coincidence in the history of physics' [176]. For more on the fine-structure story see the discussion of the Dirac equation elsewhere in this book.

To conclude this sketch of Sommerfeld's contributions in 1916, I mention his introduction of a third quantum number, which, however, has nothing to do with fine structure. Having quantized the 'shape' of an orbit he was the first to ask next 'The question arises whether the position of the orbit can also be quantized. For that purpose it is necessary, to be sure that at least a preferred direction of reference in space should exist'. That is, the orbital position, given, say, by the direction of the normal to the orbital plane, is only well defined relative to some other fixed direction, such as that of an external electric or magnetic field. For these examples, discussed by Sommerfeld later in 1916, the quantum theory allows only a discrete set of relative directions labelled by a new quantum number generically called m.

Thus the orbits of the electron in hydrogen are now characterized by three quantum numbers, n, k and m. In the presence of an external field, the states with fixed n and k, but varying m, split up. Eventually it became clear that m ranges over the $2l+1$ values $-l, -l+1, \ldots, +l$.

2.7.4. Ehrenfest's adiabatic principle

Paul Ehrenfest studied physics in his native Vienna, where his contact with Boltzmann (under whose guidance he obtained his PhD) was decisive in directing him to his principal scientific devotion, statistical physics. As was noted earlier, that was the branch of physics which had served Planck and Einstein as their prime tool in their earliest work on quantum theory. Ehrenfest studied their papers carefully. As a result he became probably the first after the founders to publish on quantum problems, beginning in 1905 [177]. These early papers already showed what Einstein later called 'his unusually well developed faculty to grasp the essence of a theoretical notion, to strip a theory of its mathematical accoutrements until the simple basic idea emerged with clarity. This capacity made him ... the best teacher in our profession whom I have ever known' [178]. He was respected by all who knew his work except by himself.

Ehrenfest's contribution of interest to us here, his 'adiabatic principle', was inspired by his critical analysis of the contributions by Planck and Einstein, not by those of Bohr, even though, as it turned out, the main applications of his principle were to issues in atomic physics. He published this work [179] in ever more systematic detail in the course of the years 1911–16, most of it from Leiden where, since late 1912, he had been installed as successor to Lorentz.

The gist of the adiabatic principle can be stated as follows. If you give me the quantum rules for a particular system, then I can tell you the rules for a whole class of other systems. The proof is based on the hypothesis that Newtonian mechanics continues to apply as long as systems are in a stationary state, while the quantum theory only comes in to account for jumps from one such state to another. This clearly brings much improved coherence to the old quantum theory: one still did not know why any system is quantized but now one could at least link the quantization of vastly distinct systems.

2.7.5. *Einstein introduces probabilities in quantum physics*

In 1916 Einstein found a new and improved way of understanding Planck's black-body radiation law, which, moreover, linked this law to Bohr's concept of quantum jumps. This work is contained in three overlapping papers [180]. It deals with a system in thermal equilibrium, consisting of a gas of particles 'which will be called molecules' and of electromagnetic radiation with spectral density ρ. Let E_m and E_n, smaller than E_m, denote the energies of two levels of the molecule. Einstein introduced the following new hypothesis. The probability per unit time that the molecule absorbs radiation in making the transition $n \to m$ is proportional to ρ; the probability for emission $m \to n$ consists of the sum of two terms, one proportional to ρ, another, corresponding to 'spontaneous emission', independent of ρ. Combining this hypothesis with some experimental facts about the behaviour of ρ at very high and very low frequencies he found that he could obtain Planck's law if, and only if, the transitions $m \leftrightarrows n$ were accompanied by a *single monochromatic* energy quantum with frequency ν given by $E_m - E_n = h\nu$—Bohr's quantum hypothesis!

The central novelty and lasting feature of this work is the introduction of probabilities in quantum dynamics. I mention two further points Einstein raised in this connection.

Einstein remarked that his mechanism of spontaneous emission of radiation and Rutherford's description, dating back to 1900, of the spontaneous decay of radioactive matter are generically identical 'It speaks in favour of the theory that the statistical law assumed for [spontaneous] emission is nothing but the Rutherford law of radioactive decay'. Ever since 1900 the spontaneous nature of radioactive processes had been a source of bafflement. While Einstein could not explain this phenomenon either, he was the first to note that it could only be understood in a quantum-theoretical context.

Einstein was by no means content with all he had done. He stressed that his theory, which was statistical in nature (it deals only with probabilities), could not predict the direction in which a light-quantum moves after spontaneous emission. Hence his work did not satisfy the causality demand of classical physics: a unique cause leads to a unique effect. That bothered him greatly. In 1920 he wrote to Born 'That business about causality causes me a great deal of trouble ... Can the quantum absorption and emission of light ever be understood in the sense of the complete causality requirement, or would a statistical residue remain? I must admit that there I lack the courage of conviction. However, I would be very unhappy to renounce complete causality' [181].

Quantum mechanics would demand such renunciation. Einstein would never make peace with that.

2.7.6. Selection and polarization rules

In 1918–22, Bohr published a lengthy memoir 'On the quantum theory of line spectra' [182] which contained new applications of the correspondence principle, resulting from an important change in outlook after 1913. At that earlier time his prime concern had been to understand the discreteness of the hydrogen spectrum and, more specifically, the particular frequency values of its spectral lines. Meanwhile, new developments had forced him to reconsider his basic postulate according to which an electron in a state, any state, with energy E_1 can jump to a state with lower energy E_2 accompanied by the emission of a photon with frequency v: $E_1 - E_2 = hv$. In 1913 he had not yet considered the fine structure which splits E_1 and E_2 into various energy levels. Can any of the split E_1 levels go to any of the split E_2 levels? Can any of the split E_1 (or E_2) levels go into a lower split E_1 (or E_2)? The Stark effect also leads to splitting; accordingly the same questions arise there too. It was already known in 1916 that not all possible transitions occur. The number actually seen was considerably smaller. Speculations arose: are some transitions forbidden or allowed but producing lines with an intensity too low to be detectable?

Bohr's memoir centred on the now paramountly important questions of line intensities. His strategy was the use of the correspondence principle outlined earlier (section 2.6.4) but now extended to the case where more than one quantum number appears. Proceeding in this way he obtained a set of rules which, however, hold only in the low-frequency (highly excited states) region where the correspondence principle applies. Next Bohr made the daring and, it has turned out, correct extrapolation that these rules hold for all frequencies. His results were these.

(i) Selection rules for Sommerfeld's quantum number k (or, which is the same, for l): the change Δk for allowed transitions (of the electric dipole type) is restricted to the values ± 1 for one-electron systems (hydrogen, ionized helium) to $0, \pm 1$ for more complicated systems. This result was also obtained independently by Sommerfeld's student Adalbert Rubinowicz (1884–1974) [183].

Introducing Atoms and Their Nuclei

(ii) Bohr correctly guessed the range of admissible m-values. Applying his reasoning to emission in some fixed direction, he obtained a further selection rule: $\Delta m = 0, \pm 1$.

(iii) He also found that $\Delta m = 0(\pm 1)$ corresponds to light polarized parallel (perpendicular) to the direction of an external field present.

2.7.7. The periodic table of the elements
This was a subject dear to Bohr ever since the days of the Rutherford memorandum (section 2.6.3). He returned to this problem in the years 1920–22, adopting a style which he expressed like this 'One of the problems for which one might expect less from a mathematical deductive method than from a physical-inductive approach, the problem of the structure of the atoms and molecules of the elements' [184].

Much had changed since Bohr's discussion of complex atoms in his 1913 trilogy. In particular his pancake picture, according to which electrons move in a set of concentric plane orbits, had run into numerous troubles. From about 1920 ideas moved away from two-dimensional rings to three-dimensional shells of electrons†.

Meanwhile, in 1916, the first successful links, based on physico-chemical reasoning, between the Bohr theory and the periodic table of the elements had been established by Walther Kossel (1888–1956) [186]. His starting point was the striking stability of atoms of the noble gases, manifested by the facts that they are relatively hard to ionize and are unable to form compounds with other atoms. He interpreted these properties to mean that the electron configurations in such atoms consist of 'closed shells', that is they strongly resist turning into positive (negative) ions by giving off (adding on) an electron. These closed shells occur for $Z = 2, 10, 18, 36, 54, 86$ (helium, neon, argon, krypton, xenon, radon; Kossel only considered elements with Z up to 25).

Consider next elements with Z one less than this sequence, 1 (hydrogen), 9, 17, 35, ... (fluorine, chlorine, bromine, ..., the halogens) which are known to turn easily into negative ions by picking up an electron. That is so (Kossel said) because these ions acquire once again noble-gas configurations. For the same reason the alkalis, $Z = 3, 11, 19, \ldots$ (lithium, sodium, potassium, ...) easily turn into positive ions. Arguments like these led Kossel to propose that electrons occupy 'concentric rings or shells on each of which only a certain number of electrons should be arranged', these numbers being (see the Z values of noble gases) 2, 8, 8, 18 ..., counting from inner to outer shells. Further evidence for this behaviour could be read off from other experimental data, among them those concerning optical spectra [187].

These good beginnings of a descriptive picture could not simply be extended to all the elements, however.

---

† *Also from that time date picturesque and forgettable models built out of one or more cubes* [185].

Twentieth Century Physics

By the 1920s areas in the table of elements had been located which did not at all fit the simple periodicities originally postulated by Dimitri Mendeleev (1834–1907) in 1869, when, incidentally, none of the noble gases were yet known. The most striking deviations were a group of scarce elements closely resembling each other in chemical and spectroscopic properties, the rare-earth metals or lanthanides, that name being derived from a Greek verb meaning 'to lie hidden'. Since in 1869 only two of these were known, cerium and erbium, they did not at once draw any particular attention. At the beginning of the twentieth century their number had risen to 13: the 14th and last one, prometheum $Z = 59$ was not found until 1947. The Z values in this group range from 58 to 71.

Efforts to fit these elements into the Mendeleevian scheme only caused trouble, leading some to think that the whole idea of the periodic table might be wrong, others that they should separately be placed in a third dimension relative to the periodic table. In any event it was clear that the rare earths as a group had to occupy an exceptional position in the scheme of things [188].

Thus when in 1920 Bohr turned his full efforts to the periodic table, he had available a mixed bag of promising results and confusion. In 1921 he published two letters to *Nature* on the subject [189]. The most detailed exposé is found in his June 1922 Göttingen lectures, published only after his death [190]. In summary his new picture for the ground states of atoms was the following.

Electrons reside in quantum levels with the lowest available energy. An electron in a given orbit is specified further by a value of k, and is given[†] the overall designation n_k, the range of k being $k = 1, 2, \ldots, n$. An atomic species is fully characterized by a specific set of occupation numbers, the set being called a configuration, which informs us how many electrons are in the n_k orbits $1_1, 2_1, 2_2, 3_1, 3_2, 3_3, \ldots$. The bulk of the gospel according to Bohr consists in the determination of the configuration for each element.

Kramers, the closest witness in Copenhagen, later recalled 'It is interesting to remember that many physicists abroad believed ... that [the theory] was based to a large part on unpublished calculations ... while the truth was that Bohr, with divine vision, had created and deepened a synthesis between spectroscopic and chemical results' [191].

An important tool in Bohr's work is his building-up principle (*Aufbauprinzip*), according to which one may imagine an atom to be formed by the successive capture and binding of the electrons one after another in the field of force surrounding the nucleus. Using this argument Bohr gathered valuable insight on neutral atoms by imagining them to be built up step by step from multiply ionized states. Not until

---

† *Note to the perplexed expert:* $k = 1, 2, 3, \ldots$ *corresponds to s, p, d, ... states.*

1923 did he make fully explicit the crucial postulate that the quantum numbers of the electrons already present are not disturbed by adding a further electron, calling this 'the postulate of invariance and permanence of quantum numbers' [192]. The building-up principle has remained a good though approximate tool used in later improved theories of atomic constitution.

In his Göttingen lectures Bohr took his audience on a guided tour through the periodic system, noting several times on the way how preliminary his reasoning was. Here I shall select only one item, his successful treatment of the rare earths.

Let us first look at a few features of the building-up principle. Up to $Z = 18$, argon, the electrons fill in an orderly way the sequence of states $1_1, 2_1, 2_2, 3_1, 3_2$, reflecting the fact that in this region an electron is bound tighter the smaller n are and, for given n, the smaller k are too. Thereafter 'irregularities' begin because the competition between n and k for lowest energy changes character. Thus the filling of 3_3 is deferred a few steps; in $Z = 19, 20, 4_1$ electrons are added, etc. As we reach lanthanum ($Z = 57$) the outermost electrons are three: two 6_1 and one 5_3. At that stage the 4_4 orbits are still empty. These latter are now added, one by one, Bohr said, as we move through the rare-earth region, from $Z = 58$ to 71, but the 4_4 orbits are smaller than those of the ever-present two 6_1 and one 5_3. Thus the peculiarity of the rare earths is that the building up takes place *inwards not outwards*, all these elements having in common the same three outer (valence) electrons. This, according to Bohr, explains the great chemical similarities of the rare earths. Bohr noted further that another such irregular phenomenon occurs starting right after $Z = 89$, actinium. Beginning with $Z = 90$, thorium, electrons are again added far *inside* the atom (in states 5_4), resulting in a second group of rare earths, the actinides, which is now known to extend into the transuranic region, including neptunium, plutonium, etc—but that is a later story. Bohr's interpretation of the rare earths is the one we use today.

It is not surprising that some of Bohr's results were incomplete, others incorrect. In particular he could not produce a criterion for the famous maximum occupation numbers of closed shells. Nevertheless Bohr's new general strategy of labelling atoms by n_k configurations continues to be fruitful and important. What Sommerfeld put so well in October 1924 still holds true today. 'It will be inevitable that improved spectroscopic data will produce conflicts with Bohr's atomic models. I am nevertheless convinced that in broad outline they are conceptually correct since they so beautifully account for general chemical and spectroscopic traits' [193].

2.7.8. Pauli's exclusion principle
Two months after Sommerfeld had written these lines, Bohr's treatment

Figure 2.6. *Pauli at the blackboard, late 1920s.*

of the periodic table led Wolfgang Pauli (1900–58) (figure 2.6) to make a spectacular discovery which put Bohr's work on its modern footing.

Pauli, son of a medical doctor who later became a university professor, godson of Ernst Mach, began his schooling in his native Vienna. He had already in his high-school years delved deeply into mathematics and physics. When in the autumn of 1918 he enrolled in the University of Munich he brought with him a paper (published shortly afterwards) on general relativity. At the instigation of Sommerfeld, who had, of course, not failed to recognize the brilliance, erudition and ventripotence of his young student, Pauli began in 1920 the preparation of a review article on relativity. After its appearance [194], Einstein wrote 'Whoever studies this mature and grandly conceived work might not believe that its author is a twenty-one year old man' [195]. This article appeared in English translation in 1958 [196], the year of Pauli's death. It is still one of the best presentations of the subject.

On 10 December 1945 Pauli was honoured for his Nobel Prize at a dinner at the Institute for Advanced Study in Princeton (he did not go to Stockholm for the occasion). His late wife told me that Pauli was deeply moved when during an after-dinner toast Einstein said in essence that he considered him as his successor. (I have not seen this in writing.)

Pauli has recalled that during his student days 'I was not spared the shock which every physicist accustomed to the classical way of thinking experienced when he came to know Bohr's basic postulate of quantum theory for the first time' [197]. His first paper on atomic physics dates from 1920. In 1921 he received his PhD *summa cum laude* on a thesis dealing with the quantum theory of ionized molecular hydrogen, whereafter he was for half a year assistant to Born in Göttingen. In 1922 he went to Hamburg first as assistant, then as *privat-dozent*. In 1928 he was appointed professor at the ETH in Zürich, the post he held for the rest of his life.

When in 1922 Pauli went to Göttingen to attend Bohr's lecture series there 'a new phase of my scientific life began when I met Niels Bohr personally for the first time. During these meetings ... Bohr ... asked me whether I could come to Copenhagen for a year' [197]. Pauli accepted and went to Bohr's institute from autumn 1922 to autumn 1923.

I turn now to the exclusion principle. Recall first that the electron states in a hydrogen atom are labelled by three quantum numbers, n, k and m, and that m ranges over the $2k-1$ values $-k-1, \ldots, k-1$. Let us count the total number of states N associated with a given n. For $n=1$: $k=1$, $m=0$, hence $N=1$. For $n=2$: $k=1$, so $m=0$ or $k=2$, so $m=-1,0,1$: hence $N=1+3=4$, etc. For general n: $N=n^2$. Starting from this counting, Pauli introduced three further postulates.

(i) In the spirit of Bohr's central field model, he assigned not only an n and a k but also an m to each electron in a complex atom.

(ii) In his discussion of the anomalous Zeeman effect (to which I shall turn shortly) Pauli had introduced the new hypothesis that the valence electron in alkali atoms exhibits a two-valuedness. Now he assumed the same to be true for all electrons in all atoms, so that each electron is characterized by four quantum numbers, n, k, m and a fourth one capable of two values. It follows that the number of states for given n is not N but $2N$, hence there are two states for $n=1$, eight for $n=2$, 18 for $n=3$: the mystical numbers 2, 8, 18 ruling the periodic table emerge!

(iii) That is all well and good, but why could there not be 17 electrons occupying a state with some fixed value for the four quantum numbers? Pauli decreed 'In the atom there can never be two or more equivalent electrons for which ... the values of all [four] quantum numbers coincide. If there is an electron in the atom for which these quantum numbers have definite values then the state is occupied' full, no more electrons allowed in.

Pauli's decree, called the exclusion principle, is indispensable† not

---

† But of course by no means sufficient. I would in fact mislead the reader if I did not mention just one more point. The first two closed shells, occupied by two and eight electrons, fill all states with $n = 1, 2$, respectively. The third closed shell also has eight electrons, corresponding to $n = 3$, $k = 1, 2$. The fourth period, the first of the 'long periods' has 18 electrons, built out of $n = 3$, $k = 3$, and $n = 4$, $k = 1, 2$. These important details can only be understood from the energy properties of atomic states.

just for the understanding of the periodic table but for ever so much more in modern quantum physics.

2.7.9. The discovery of hafnium [198]
According to the exclusion principle the maximum number of 4_4 states equals $2 \times (2 \times 3 + 1) = 14$, that is the number of rare earths. Had Bohr known this in 1922, he might not have had brief doubts whether that number was 14 or 15. At issue was the nature of the element $Z = 72$. In his Göttingen lecture Bohr had proclaimed 'Contrary to the customary assumption ... the family of rare earths is completed with cassiopeium [$Z = 71$, now called lutetium] ... if our ideas are correct the not yet discovered element with atomic number 72 must have chemical properties similar to those of zirconium and not those of the rare earths' [199].

At this juncture Hevesy persuaded Dirk Coster (1889–1950)—both men were then in Copenhagen—to do an x-ray experiment of their own, there and then, using x-ray equipment acquired by Bohr for his institute and installed in its basement. At first Coster did not feel like it, expecting $Z = 72$ to be rare even for a rare earth. No, said Hevesey, we are not going to look for a rare earth but for an analogue of zirconium. By December they were sure: they had identified $Z = 72$ in all their borrowed zirconium samples. Nor was the new element rare, but in fact as common as tin and a thousand times more plentiful than gold [200]. In their first announcement in print, which appeared on 10 January 1923 they baptized $Z = 72$ 'For the new element we propose the name Hafnium (Hafniae $\equiv$ Copenhagen)' [201].

Bohr himself was absent from Copenhagen on the December day when Coster and Hevesy became convinced of their discovery. He first learned the good news in the following way. 'While Coster telephoned these results through to Bohr, Hevesy took the train to Stockholm to be in time for Bohr's announcement at [his] Nobel lecture' [202].

2.7.10. A fourth quantum number; spin
In 1920 Sommerfeld introduced a fourth quantum number, denoted by j and called the 'inner quantum number' [203]. In doing so he had adopted a reasoning utterly different from the one that had led to the first three quantum numbers, n, k or l, and m. Recall that these had originated from geometrical considerations, the quantization of size, shape and spatial orientation of orbits. In order to make clear how he reasoned this time and to continue separating the good from the bad news, I must anticipate once again the anomalous Zeeman effect which, it had become evident, could not be dealt with in terms of the three quantum numbers known then. Note also that in 1920 it was erroneously thought that this anomaly does not occur for hydrogen but only for the more

complicated motions in many-electron atoms. Sommerfeld assumed that these complex motions are characterized by an additional angular momentum corresponding to 'a hidden rotation'. He associated a new quantum number j with the quantization of this new variable and a new selection rule $\Delta j = 0, \pm 1$. Never mind that these results were obtained by a reasoning now known to be incorrect. The quantum number j and its selection rules have survived all the same. Note the odd stroke of luck: I doubt if Sommerfeld would have introduced j had he known in 1920 that the anomalous Zeeman effect is also anomalous in hydrogen.

There now followed several years during which theoretical physicists experimented with extra quantum numbers and quantum rules. In 1921, Alfred Landé (1888–1975), particularly adept at these games, made another radical proposal: j and m shall take *half*-integer values for certain groups of elements such as the alkalis, the atoms of which consist of one valence electron orbiting around a 'core' of inner electrons [204]. In 1923 Landé proposed, more specifically, that this core performed a 'hidden rotation' with angular momentum $\frac{1}{2}$, in units $h/2\pi$.

Next, in late 1924, followed Pauli's first major discovery. By an ingenious argument he could demonstrate that Landé's model of alkali atoms, a valence electron orbiting around a core with angular momentum $\frac{1}{2}$, was untenable. Yet Landé's results worked well! Pauli found a way out. There *is* a 'hidden rotation' which, however, is not due to the core but to the valence electron itself! The anomalous Zeeman effect 'according to this point of view is due to a peculiar non-classically describable two-valuedness [*Zweideutigkeit*] of the quantum theoretical properties of the valence electron' [205]. (The two-valuedness refers to the values $\pm\frac{1}{2}$ of the m quantum number associated with $j = \frac{1}{2}$.)

About a year later, George Uhlenbeck (1900–88) and Samuel Goudsmit (1902–78) interpreted this two-valuedness as a consequence of an intrinsic angular momentum, a spin, of the electron with value $\frac{1}{2}(h/2\pi)$ [206]. This discovery was made after quantum mechanics had arrived, yet was entirely based on the *classical* analogy with the Earth spinning around an intrinsic axis while orbiting the Sun. I cannot enter here into the complex events surrounding this discovery [207].

2.8. Now the bad news. The crisis of the old quantum theory

2.8.1. Helium

All through the days of the old quantum theory the spectrum of helium remained incomprehensible theoretically; nor was it understood experimentally.

It had been known since the late 1890s that this spectrum consists of two distinct kinds of lines: parahelium, a set of singlets (unsplit lines), and orthohelium, believed to be doublets. Attempts to resolve the doublets remained unsuccessful until January 1927 when it was found that the doublets were actually triplets [208]. 'Within the last few months interest has been revived in these [doublets] by the theoretical work of Heisenberg [209] which predicts a triplet structure ... It is now possible to show that the helium lines really have a structure similar to that predicted by Heisenberg' [210]. That prediction dates from July 1926—and was based on wave mechanics.

In broadest outline, the old quantum theory was a hybrid structure; classical mechanics was supposed to apply to electrons moving *in* stationary orbits, quantum effects were supposed to be confined to jumps *between* orbits. After years of labour on helium, Kramers in Copenhagen concluded that this picture was wrong 'We must draw the conclusion that already in this simple case mechanics is not valid' [211]. Bohr in 1923 on helium stated 'This investigation may ... be particularly suited to provide evidence of the fundamental failure of the laws of mechanics to describe the finer [sic] details of the motion of systems with several electrons' [212], while Sommerfeld in that same year wrote 'All attempts made hitherto to solve the problem of the neutral helium atom have proved to be unsuccessful' [213].

2.8.2. The anomalous Zeeman effect

Zeeman was a young *privat-dozent* in Leiden when in 1897 he discovered that spectral lines split when atoms are placed in a magnetic field [214]. Lorentz at once provided an interpretation in terms of a simple model for an electron moving in an atom [215]. Considering only effects proportional to the first power of the field (the linear Zeeman effect), he showed that a spectral line should split into a doublet or triplet depending on whether the emitted light is parallel or perpendicular to the field direction.

In 1921 Lorentz commented 'Unfortunately, however, theory could not keep pace with experiment and the joy aroused by [this] first success was but short-lived. In 1898 Alfred Cornu (1841–1902) discovered—it was hardly credible at first!—that [a sodium line] is decomposed into a quartet ... Theory was unable to account ... for the regularities observed ... to accompany the anomalous splitting of the lines' [216]. It did not take long to find out that this 'anomalous Zeeman effect' is the rule and the 'normal' effect (Lorentz's prediction) the exception [217].

The year 1916 produced good results when Sommerfeld [218] and independently Debye [219] showed that the normal Zeeman effect fitted nicely in the old quantum theory. The necessary new ingredient was Sommerfeld's 'third quantum number' m. In 1918 Bohr noted [220] that his selection and polarization rules related to m also agreed well for the normal effect. In 1922 he commented as follows on the inexplicable anomalous effect 'The difficulty consists ... in the fact that the ordinary electrodynamic laws can no longer be applied to the motion of the atom in a magnetic field in the same way as seemed to be the case in the theory of hydrogen' [221] still believed to be normal. In those years Bohr repeatedly conjectured that the anomaly might be due to an as yet unclear influence of inner on outer atomic electrons.

All through the years of the old quantum theory the anomalous Zeeman effect remained a mystery†. Sommerfeld in 1919 wrote 'A genuine theory of the Zeeman effect ... cannot be given until the reason for the multiplicities can be clarified' [223]. Pauli in early 1925 reported 'How deep seated the failure of the theoretical principles known till now is can be seen most clearly in the multiplet structure of spectra and their anomalous effect' [224].

2.8.3. The harvest

Having come to the end of my account of what was understood about atoms up till 1925, my readers will, I hope, share my admiration for what had been achieved during the preceding quarter of a century. Those were years in which horse sense, excellent taste for improvisation, and judicious reading of experimental data were at a premium. Logic admittedly was wanting. To recapitulate what has turned out to be of lasting value: the black-body radiation law, the photon, the elements of the quantum theory of the solid state, the spectrum of hydrogen (at least to good approximation), all four quantum numbers and their selection rules in atomic spectroscopy, foundations of the periodic table and hence of quantum chemistry, the adiabatic principle, probabilities introduced in quantum physics, the exclusion principle, and spin. I have not even mentioned that Bose–Einstein and Fermi–Dirac statistics were also initially based on the old quantum theory [225]. As to the helium and anomalous Zeeman effect crises, both were under control by 1926.

Clearly then, we owe high respect to the pioneers who managed to put all this together.

PART 2: NUCLEI

2.9. β-ray spectroscopy, 1906–14

I have already noted that the discovery of the atomic nucleus dates from 1911 (section 2.6.1) and that the first observed nuclear phenomena, radioactive processes, are of earlier vintage (section 2.3). The discussion of these earliest developments, followed up to 1905, will now be continued. The first item of business will be the spectra of radioactive processes.

---

† *To which in 1922 Stern and Gerlach added observations of inexplicable splitting patterns of atomic beams passing through a magnetic field* [222].

Twentieth Century Physics

This subject begins with the observations of William Bragg (1862–1942) and Richard Kleeman (1878–1932) performed in Adelaide, of the range of α-particles emanating from a *thin* layer of radium salt. In 1905 they summarized their findings as follows 'Each α-particle possesses ... a definite range in a given medium the length of which depends on the initial velocity of the particle and the nature of the medium. Moreover, the α-particles of radium which is in radioactive equilibrium can be divided into four groups, each group being produced by one of the first four radioactive changes in which α-particles are emitted. All the particles of any one group have the same range and the same initial velocity' [226]. These conclusions were correct in their essentials and their great interest was described by Soddy in 1905 '... The Daltonian conception that the atoms of the same element are all exactly alike applies even to the velocity with which they expel radiant particles on disintegration. This is probably the most severe experimental test to which this conception has ever been subjected' [227].

When in 1906 Otto Hahn (1879–1968) and Lise Meitner (1878–1968) began their investigations in Berlin (figure 2.7) of the primary β-spectrum, the thought had occurred to them, sensibly enough for its day, that if it is true that α-rays emerge with a unique velocity, then might the same perhaps also be true for β-electrons? They were aware that electrons do not have a clearly marked range in a gaseous medium, so they needed a detection method other than the one used by Bragg–Kleeman. Accordingly they chose to measure the absorption of electrons, adopting as a working hypothesis that in this process, for monoenergetic electrons, absorption follows an exponential fall-off as a function of thickness traversed, a picture widely accepted in 1906–7.

Their first experiments, with thorium, led them to state 'We conjecture from our results that pure β-emitters only emit β-rays of one kind [i.e. one velocity] similar to what happens with α-rays' [228]. In a sequel [229] they made the same claim about actinium. However, beginning with radium, things became more complicated (we are now in September 1909) 'on the grounds of our hypothesis that complex [i.e. non-unique velocity] rays correspond to complex substances ... one must conclude to a complex nature of radium' [230].

Meanwhile, however, their diagnostic, exponential absorption, had begun to fall apart. In 1909 William Wilson (1887–1948) from Manchester showed that an electron beam guaranteed to be monoenergetic does not remotely follow exponential absorption [231].

So what did Hahn do? He used another detection technique. Following Wilson he let β-rays pass through a slit into a space where the action of an homogeneous magnetic field perpendicular to their velocity bent the rays. Unlike Wilson, after traversing a semicircle his electrons were incident on a photographic plate, the blackening of which gave a record of the initial velocity spectrum. This work, done together

Introducing Atoms and Their Nuclei

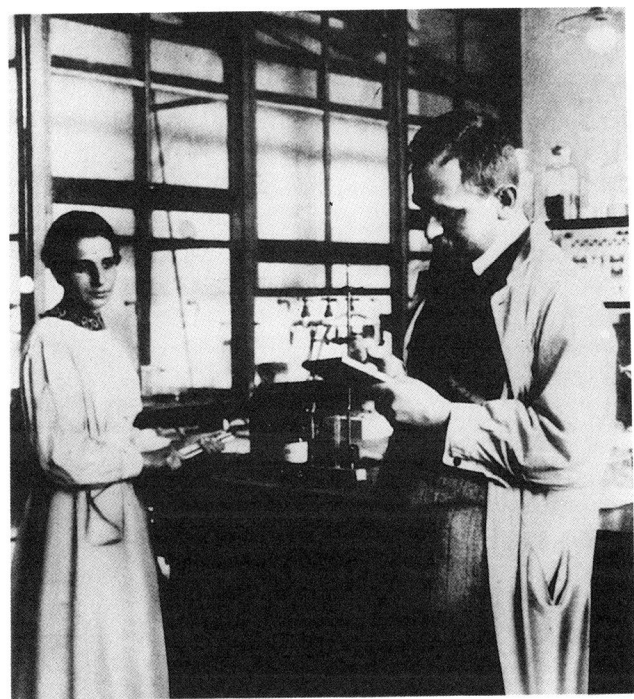

Figure 2.7. *Otto Hahn and Lise Meitner in Hahn's laboratory in Berlin.*

with Otto von Baeyer (1887–1946) led to the conclusion 'The present investigation shows that, in the decay of radioactive substances, not only α-rays but also β-rays leave the radioactive atom with a velocity characteristic for the species in question. This lends new support to the hypothesis of Hahn and Meitner ...' [232].

Thus the erroneous conjecture that a pure β-emitter generates monochromatic electrons was still alive in 1910. It died in April 1911, when Hahn and von Baeyer, together with Lise Meitner, submitted a sequel [233] to the previous paper. Their methods were the same as before. This time, however, they were forced to admit that the *effective* β-spectrum of a pure substance is *inhomogeneous*. Nevertheless, they still held open the possibility that this effective inhomogeneity is a secondary modification of initially monochromatic β-emission 'The inhomogeneity of fast β-rays can have its origin in the fact that the rays were initially emitted by the radioactive substance with unequal velocities ... It is more plausible to look for a secondary cause which renders inhomogeneous the emitted homogeneous rays ... the exponential law [in absorption] cannot be a criterion for the homogeneity of the rays, as was supposed by Hahn and Meitner in contrast to [the view of] other investigators'.

Hahn and Meitner's persistence is impressive, their admission of earlier mistakes forthright. The same frankness strikes me in reading

Hahn's scientific autobiography 'Our earlier opinions were beyond salvage now. It was impossible to assume a separate substance for each beta line. Our original explanation of the exponential absorption had been wrong because we had assumed that we actually had measured the absorption. What we had principally measured was the dispersion and the greater the distance of our preparations from the bottom of the electroscope, the more dispersion we had obtained. In increasing the distance, we had dispersed the weakest beta rays so much that they failed to register, and we had also slowed down the fastest beta rays, so that the average velocity of the non-homogeneous beta rays remained fairly constant over a short length of time ... Though our opinion about the "absorption law" had been wrong, the work had considerably improved our techniques. We had learned how to produce different substances in thin layers and also how to handle them, especially those with short half-lives' [234].

We turn next to the years 1912–13 when the presupposition of monochromatic β-spectra was dropped and it now appeared that β-spectra consist not of one discrete line but of a set of such lines. The first intimations of the line spectra are found in the 1910 paper by von Baeyer and Hahn 'In all cases examined a clearly discontinuous spectrum was obtained' [232]. During the next few years, the study of β-spectra by means of the magnetic separation/photographic detection method continued in several laboratories. All research groups reported complicated discrete spectra. Rutherford also went to work on the problem. Together with Harold Robinson (1889–1955) he studied the spectra of RaB (Pb^{214}) and RaC (Bi^{214}) and found 16 lines for RaB, 48 lines for RaC, the lines falling in seven classes of intensity [235]. However, there were also difficulties in fitting all results by means of line spectra. Already in 1911, Hahn had written to Rutherford 'RaE [Bi^{210}] is the worst of all. We can only obtain a fairly broad band. We formerly thought that it was as narrow as the other bands, but that is not true. It looks as if secondary or such effects had a maximum influence on rays of a medium velocity like RaE' [236]. In a general discussion of the subject, Rutherford did note that continuous bands are sometimes observed. However, he tentatively concluded that 'The continuous β-ray spectrum observed for uranium X [Th^{234}] and radium E [Bi^{210}] may be ultimately resolved in a number of lines' [237]. Thus it came to pass that a well-known physics textbook of the period describes β-radioactivity as a discrete phenomenon and displays elaborate tables of discrete spectral lines [238].

I interrupt the historical account in order to note that in those years it was not yet known that in radioactive decays electrons are produced by two distinct mechanisms: firstly from the primary β-decay process; secondly as a secondary effect accompanying γ-decay. In the latter case, a γ-ray comes out of the nucleus and is next absorbed by a peripheral

Introducing Atoms and Their Nuclei

atomic electron. This process, now called internal conversion, produces discrete electron energies corresponding to the discrete initial γ-energies. It was discovered in 1921 by Charles Drummond Ellis (1895–1980) [239]. I note right away that one year later Ellis introduced another novelty: the first sketch of a nuclear energy level diagram, the first attempt to seek 'support [for] the view that quantum dynamics apply to the nucleus, and that part at least of the structure of the nucleus can be expressed in terms of stationary states' [240].

I return to 1912. In October of that year a new edition of Rutherford's book was ready for press [237]. It came out in 1913. In this volume Rutherford expressed his belief in two distinct causes for radioactive instability '... the instability of the central mass [i.e. the nucleus] and the instability of the electronic distribution. The former type of instability leads to the expulsion of an α-particle, the latter to the appearance of β- and γ-rays ... part of the surplus energy of a ring of [peripheral] electrons is released in the form of a high speed β-particle and part in the form of γ-rays, the division between the two forms of energy depending on factors which are not at present understood' [241].

It was Niels Bohr's first contribution to nuclear physics to observe, in 1913, that 'On the present theory it seems ... necessary that the nucleus is the seat of the expulsion of the high-speed β-particles' [242]. He argued as follows. The chemical properties of elements are dictated by the configurations of their electronic orbits. Isotopes have identical chemical properties, hence identical electron configurations. There are numerous instances in which one isotope emits β-rays with velocities different from another. Thus they are non-identical with regard to β-radioactivity. Hence by exclusion this process must be of nuclear origin.

A propos isotopes: the term was coined in 1913 [243], but already in 1911, even before the discovery of the nucleus, the underlying physical concept had been formulated in its essentials. There is of course nothing startling about that sequence of events. The statement 'there exist elements with identical physical and chemical properties except for atomic weight', which does not contain reference to a nucleus is in need of refinement but will obviously do for the diagnosis of the phenomenon.

I turn to the dénouement of the first phase of β-spectroscopy, the discovery in 1914 of the continuous β-spectrum, made by James (later Sir James) Chadwick (1891–1974), a Rutherford student in Manchester. In 1913 he was awarded an Exhibition of an 1851 Senior Research Studentship. Under its terms he was obliged to carry out research in some place other than Manchester. He chose to go to Berlin, to work with Hans Geiger (1882–1945) at the Physikalisch-Technische Reichsanstalt.

In a letter which Chadwick sent from there to Rutherford we get the first intimation that a major turning point in the history of particle physics was near. The letter, dated 14 January 1914, contains these lines 'We [Geiger and Chadwick] wanted to count the β-particles in the various spectrum lines of RaB+C and then to do the scattering of the strongest swift groups. *I get photographs very quickly easily, but with the counter I can't even find the ghost of a line. There is probably some silly mistake somewhere'.*

The lines which I have italicized contain two key statements. Firstly, Chadwick had been able to obtain line spectra by photographic detection, like everyone else. Secondly, far from having made a silly mistake, he had found that these lines are vastly exaggerated in relative importance. Chadwick gave full details in a paper submitted in April 1914 [244].

On its opening page we find the first expression of concern about the meaning of β-ray line intensities as seen on the photographic plate 'Since ... the photographic action of β-rays with variable velocity is unknown, one does not obtain in this way reliable information about the intensity of the individual groups of rays. It is ... also ... hard to decide whether or not a continuous spectrum is superposed over the line spectrum'. Chadwick then described another detection method. Just as in the Wilson experiment described above, electrons of fixed velocity are bent 180°, then pass through a slit. Thereafter, their intensity is measured by the discharge they cause in an electric potential maintained between a metal plate and a very clean needle with a sharp point, a variant of a Geiger counter. Different velocities are selected by varying the magnetic field strength. Their source was a mixture of radium B (Pb^{214}) and radium C (Bi^{214}). The result: a continuous spectrum on which are superposed four, and only four, lines at the lower energy portion. The position of these lines coincided with some of the strong lines found by others in previous investigations. In addition, he made some tests which convinced him that the continuous energy distribution was not due to secondary scattering effects. Compare these findings with the earlier results of Rutherford and Robinson [235] for the same source: more than 60 discrete lines, no continuous spectrum!

In the same paper Chadwick also gave prescriptions for the production of fake lines 'The difference [with the photographic method] could be explained by the circumstance that the photographic plate is extraordinarily sensitive for small changes in the intensity of the radiation'. For example, he irradiated a photographic plate for 100 minutes, then put a sheet of lead with a narrow slit on top of the plate and re-exposed for an additional 5 minutes. The resulting picture turned out to depend on the details of the development of the plate: normal development generated a sharp line, very slow development, terminated at a convenient time, made it even possible to obtain a nearly black line against a clear background! Thus he drew his final conclusion: the β-rays of RaB and RaC consist of a continuous spectrum. There is an additional line spectrum. With a few exceptions the lines have only a very small intensity.

In the well-known textbook (published in 1930) by Rutherford, Chadwick and Ellis, the only comment on the subject is 'Chadwick showed that the prominence of these groups [of lines] was due chiefly to the ease with which the eye neglects background on a plate' [245].

The discovery of the continuous β-spectrum was Chadwick's first major contribution. What happened to him next is recorded in a letter to Rutherford 'I was in the middle of the experiments on the scattering of β-rays when the war broke out. Radioactivity is naturally not in a flowering condition here ...' The letter was written on 14 September 1915; the sender's address: Barack 10, Engländerlager, Ruhleben. Chadwick had been interned in the stables of a racecourse near Spandau. Yet, under very primitive conditions, he kept working on physics. On 31 March 1917 he wrote to Rutherford from the camp 'I believe I wrote to you before that we had a small lab. A space was granted for scientific work ... you will see that I am not getting rusty for want of work ...' [246]. Chadwick stayed in the camp for the duration of the war. Later we shall catch up with him again, this time at the Cavendish Laboratory.

2.10. Models of the nucleus, the beginnings

2.10.1. The proton–electron (PE) model

Early in 1914 it was known that the nucleus is the seat of all radioactive processes and that a nuclear species is specified by the numbers A and Z. It was believed that the interaction between α-particles and nuclei is purely electromagnetic. Assuming this to be true and then calculating the distance of closest α-particle–nucleus approach, Rutherford had realized [115] by 1911 that the nuclear radius is small: $r \lesssim 3 \times 10^{-12}$ cm. It was obvious that the nucleus of hydrogen is particularly important. 'The hydrogen nucleus is the *positive electron*', Rutherford wrote early in 1914 (his italics) [247]. This nucleus was often called the H-particle in those days, the name 'proton' came later in 1919 [248]. To summarize: early in 1914 one knew some basic facts about radioactivity, about A and Z and r, and about the basic character of the H-particle.

What are nuclei made of?

In February 1914 Rutherford conjectured [249] about the structure of the α-particle itself. 'It is to be anticipated that the helium atom [i.e. the α-particle] contains four positive electrons [H-particles] and two negative', symbolically He $= 4$H $+ 2$e; generally for a given isotopic species X with mass number A and nuclear charge Z (treating A as an integer):

$$X = AH + (A - Z)e.$$

In a Royal Society discussion on 19 March 1914, Rutherford commented further [247] on nuclear structure 'The general evidence indicates that the primary β-particles arise from a disturbance of the nucleus. The latter must consequently be considered as a very complex structure consisting of positive particles and electrons but it is premature (and would serve no useful purpose) to discuss at the present time the possible structure of the nucleus itself'.

Thus Rutherford, though always cautious and averse to speculation, blithely assumed that electrons are nuclear constituents. Actually he would not have conceived of this as an assumption. Was it not self-evident? Did one not see electrons come out of certain nuclei, in β-processes? To Rutherford, as to all physicists of that time, it was equally sensible to speak of electrons as building blocks of nuclei as it was to speak of a house built of bricks, or of a necklace made of pearls.

At about that time we also witness the beginning of a search for substructure within the nucleus. In 1914 Rutherford wrote 'The helium structure nucleus is a very stable configuration which survives the intense disturbances in its expulsion with high velocity from the radioactive atom, and is one of the units of which possibly the great majority of the atoms are composed' [249]. New subunits were proposed: the α'-particle (4H + 4e), the μ-particle (2H + 2e) and others, mainly for reasons of numerology. In 1921 Rutherford wrote 'It is exceedingly easy to write about these matters but exceedingly difficult to get experimental evidence to form a correct decision' [250]. For some time Rutherford himself was led astray by apparent evidence for another subunit, X_3 (3H + e). With some imagination one can see certain of these papers as forerunners of the α-particle model, even the shell model. Nevertheless I do not believe it is unfair to say that none of this work has left any mark on physics.

What forces hold the nucleus together?

Rutherford in 1914 said 'The nucleus, though of minute dimensions, is in itself a very complex system consisting of positively and negatively charged bodies bound closely together by *intense electrical forces*' (my italics) [251]. What else could he say? At that time there simply were no other than electromagnetic and gravitational forces, the latter being manifestly negligible for the problem at hand. Even so, one notes an element of wonder in another of Rutherford's statements that same year '[The nuclear electrons] are packed together with positive nuclei and must be held in equilibrium by forces of a different order of magnitude from those which bind the external electrons' [249].

What about the nucleus and the quantum theory?

In the first edition of *Atombau and Spektrallinien* Sommerfeld ventured the opinion that 'nuclear constitution is governed by the same quantum laws as the [periphery] of atoms [252]. At that time there was not much one could do to verify that idea. If the atomic spectrum of helium would not yield to the quantum theory, who would dare tackle the α-particle?

2.10.2. Binding energy

When in 1905 Einstein derived [253] for the first time his equation

$$E = mc^2$$

he remarked at once 'It is not out of the question that one can devise a test of the theory for bodies the energy content of which is variable to a high degree (as for example for radium salts)'. By 1907 he had reached the conviction [254], however, that it was 'of course out of the question' to reach the experimental precision necessary for his test†; in 1910 he remarked that 'for the moment there is no hope whatsoever' for the experimental verification of $E = mc^2$ [255].

While Einstein was the first to propose a check for that relation in terms of loss of weight in radioactive decays, Planck was the first to draw attention [256] to another test: a bound system should weigh less than the sum of its constituents. As a first example he calculated the mass equivalent of the molecular binding energy for a mole of water. The effect was very small ($\sim 10^{-8}$ g) but the idea was new and excellent.

In 1913 Paul Langevin (1872–1946) applied Planck's remark to the nucleus: 'It seems to me that the inertial mass of internal [i.e. binding] energy is made evident by the existence of certain deviations from the law of Prout' [257]. Unfortunately he did not consider the influence of isotope mixing and therefore overrated binding energy effects.

In order to give a precise meaning to the magnitude of the nuclear binding energy B for a nuclear species X one must of course know what X's constituents are. Until 1932 the answer remained the PE model. As long as one had nothing better one was inevitably stuck with the following incorrect mass formula for nuclei (in obvious notation)

$$m_x = Am_H + (A - Z)m_e - B/c^2.$$

As said, intra-nuclear forces were supposed to be electromagnetic, so therefore was the dynamical origin of nuclear binding B. Rutherford in 1914 wrote 'As Lorentz has pointed out, the electrical mass of a system of charged particles, if close together, will depend not only on the number of these particles, if close together, but on the way their fields interact. For the dimensions of the positive and negative electrons considered, the packing must be very close in order to produce an appreciable alteration of the mass due to this cause. This may, for example, be the explanation of the fact that the helium atom has not quite four times the mass of the hydrogen atom' [249]. Quantitative estimates showed that an assumed electromagnetic origin of nuclear binding would, however, give much too small an effect [258].

So, inconclusively, the best of physicists kept stumbling along. As late as 1921 Pauli remarked 'Perhaps the theorem of the equivalence of mass and energy can be checked *at some future date* [my italics] by observations on the stability of nuclei' [259]. Meanwhile mass spectrographs had begun to pour out good data on isotope masses. In 1927 Frances Aston

† *Radium loses weight at the rate of 1 part per 100 000 per year.*

Twentieth Century Physics

(1877–1945) gave a list of 30 of these [260]. Binding energies are recorded in terms of the packing fraction, the ratio B/A on the scale $A = 16$ for oxygen, so that by definition the packing fraction for oxygen equals zero. It is one of the least transparent ways of representing data I know of, also because oxygen has three stable isotopes.

The book [245] by Rutherford, Chadwick and Ellis, which appeared in 1930, and which for some years was the most influential nuclear physics text, contains the statement that the α-particle has a composition $4H+2e$ and a binding energy of 27 MeV.

The year 1932 brought clarity.

After the discovery of the neutron was published in February, it soon became clear that the nuclear constituents are protons and neutrons (see below). The discovery of deuterium, the hydrogen isotope with mass number two, was announced that same month. Perhaps the last reference to the old H-particle–electron model is found in a paper [261] by Ken Bainbridge (b 1904) entitled *The isotopic weight of H^2* in which he wrote 'On the assumption that the nucleus is composed of two protons and one electron, the energy binding is approximately 2×10^6 electronvolts. If the H^2 nucleus is made up of one proton and one Chadwick neutron of mass 1.0067 then the binding energy of these two particles is 9.7×10^5 electronvolts' (the correct value is $B \simeq 2.15$ MeV). This last value was found from possibly the first application of the correct equation for B

$$m_x = Zm_p + (A - Z)m_n - B/c^2$$

where m_p, m_n are the proton and neutron mass, respectively.

In June 1932, the paper by John Cockcroft (1897–1967) and Ernest Walton (b 1903) on the first nuclear transformation produced by artificially accelerated particles appeared [262]. Their result:

$$\text{Li}^7 + \text{proton} \rightarrow 2\alpha + (14.3 \pm 2.7) \text{ MeV}$$

agreed with $E = mc^2$ within the errors, the masses of all particles appearing in the reaction being well known. This is the oldest example of a new category of tests.

In 1937 a value for the velocity of light, correct to within less than one half of one per cent, was obtained from nuclear reactions in which all relevant masses and kinetic energies were known [263]. Thus the 1930s brought the correct model of the nucleus and advances in experimental techniques necessary for numerous verifications of the energy–mass relation with the help of nuclear phenomena.

Then, of course there was the A-bomb ...

2.10.3. 1919: the first transmutation of an element
In 1917, Rutherford began a series of experiments 'carried out at very irregular intervals, as the pressure of routine and war work permitted'

[264]. He was now Sir Ernest, since 1914, and would be created Baron Rutherford of Nelson in 1931.

Rutherford published his war-time pure research in 1919, in a series of four papers. In the last of these he announced an epoch-making result: bombarding N^{14} nuclei with α-particles yields O^{17}+ a proton—he had observed the first instance of a nuclear transmutation. He concluded this paper as follows 'The results as a whole suggest that if α-particles—or similar projectiles—of still greater energy were available for experiment, we might expect to break down the nucleus structure of many of the lighter atoms' [265]. Note further the remarkable acknowledgment in this paper: he thanks 'Mr William Kay (1879–1961) for his invaluable assistance in counting scintillations'. Thus Rutherford had started a revolution on top of a table, helped by his faithful laboratory steward.

Rutherford was fully aware of the impact of his result. Apologizing to the international anti-submarine warfare committee for his absence at several meetings, he said 'If, as I have reason to believe, I have disintegrated the nucleus of the atom, this is of greater significance than the war' [266]. Yet not even a Rutherford could foresee how far-reaching the discovery of transmutations would become. On 11 September 1933, he remarked in an address given at Leicester: 'The energy produced by the breaking down of the atom is a very poor kind of thing. Anyone who expects a source of power from the transformation of these atoms is talking moonshine ...' [267].

As to further transmutations, note that the Cockcroft–Walton result just mentioned also originated in Rutherford's laboratory, now (since 1919) the Cavendish Laboratory in Cambridge. That experiment had resulted in the first transmutation observed in an accelerator experiment†.

2.10.4. First intimations of a new, nuclear, force

In 1911 Rutherford had deduced the existence of the nucleus from experimental results by Geiger and Ernest Marsden (1889–1970) on the scattering of α-particles by nuclei with high Z (figure 2.8). In 1919 Rutherford did the same experiment but now scattered his 5 MeV α-particles off hydrogen. Since in this case he minimized the Coulomb repulsion (I next briefly anticipate the ultimate interpretation) the intrinsic nuclear forces have the best chance to stand out. As is known from years of subsequent experimentation, 5 MeV is plenty of energy for penetrating the α-hydrogen Coulomb barrier and thus for detecting nuclear force effects.

See also p 366

The result was a great surprise. For α-particles with a range of 7 cm, corresponding to the full 5 MeV energy, he found that 'the number of swift H-atoms produced ... is 30 times greater than the theoretical

† *For an account of the development of accelerators see Chapters 9 and 15.*

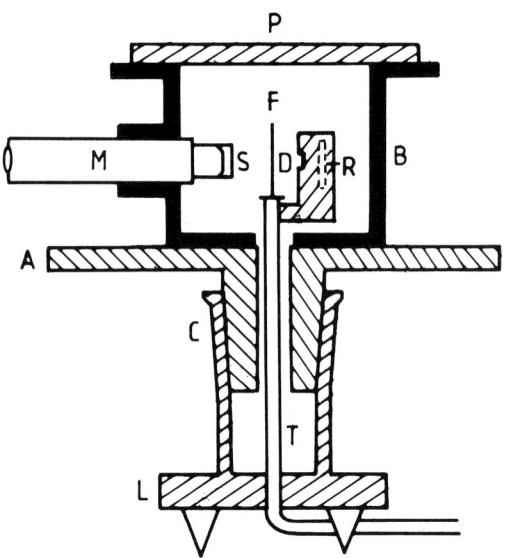

Figure 2.8. *Drawing of Rutherford's α-particle equipment.*

number' [264] not some small correction but a major new effect, which Rutherford could characterize by a range r. He defined r as that distance of closest approach of α- and H-particles at the lowest α-velocity where the Coulomb picture begins to fail. He found the quite sensible value

$$r \simeq 3.5 \times 10^{-13} \text{ cm.}$$

In a subsequent paper Rutherford reported deviations from the 'simple theory' also for α-scattering off nitrogen and oxygen [268].

Rutherford's initial reaction was that the forces were still electromagnetic in the new regime but that the α-particles should not be treated as a point particle, as he had done in 1911. His co-workers Chadwick and Étienne Bieler (1895–1929), who had continued the α-hydrogen scattering study but with improved techniques, were more emphatic in their opinion 'No system of four hydrogen nuclei and two electrons united by inverse square law forces could give a field of force of such intensity over so long an extent. We must conclude either that the α-particle is not made up of four H-nuclei and two electrons, or that the law of force is not the inverse square in the immediate neighbourhood of an electric charge. It is simpler to choose the latter alternative' [269]. They stated their final conclusion in these words

> The present experiments do not seem to throw any light on the nature of the law of variation of the forces at the seat of an electric charge, but merely show that the forces are of very great intensity ... It is our task to find some field of force which will reproduce these effects.

Introducing Atoms and Their Nuclei

I consider this statement, made in 1921, as marking the birth of the strong interactions.

2.11. 1926–32: the years of nuclear paradoxes

We are now entering nuclear physics' third phase. The first, which may be called nuclear physics without nuclei, deals with the discoveries of radioactivity and qualitative properties of its radiations (section 2.3). The second phase covers the discovery of the nucleus, the early exploration of its global properties and the beginning stages of nuclear reactions, just discussed. In all these developments theoretical physicists played no role to speak of. Indeed the two main theoretical contributions during those early days, the transformation theory and the analysis of α-particle scattering, are due to the experimentalist Rutherford. In the third phase, beginning in 1926 and ending with the discovery of the neutron in 1932, theoreticians began to play a role of importance. This marked change was caused by the confrontation between quantum mechanics and nuclear physics.

To set the stage, I turn to the 1931 textbook by George Gamow (1904–68) [270], the first of its kind written by a theoretical physicist. On its first page the author defines his model of the nucleus 'In accordance with the concepts of modern physics we assume that all nuclei are built up of elementary particles—protons and electrons'. Gamow was aware, however, of the difficulties the PE model seemed to be running into. In the course of preparing his manuscript 'Gamow had a rubber stamp made with a skull and crossbones with which he marked the beginning and end of all passages dealing with electrons'. When the Oxford University Press objected to this symbol, Gamow replied 'It has never been my intention to scare the poor readers more than the text itself will undoubtedly do' [271]. The skull and bones were replaced by a boldface sleepy **S**.

These comments by Gamow referred to the sad state of nuclear physics when he was writing his book. In one of its opening pages [272] he stated the situation like this 'The usual ideas of quantum mechanics absolutely fail in describing the behaviour of nuclear electrons; it seems that they may not even be treated as individual particles'. Let us see what caused this opinion.

2.11.1. α-decay explained

Before I start out on tales of woe, it is good to recall first the single success quantum mechanics had scored in the nuclear domain.

In 1928 it was realized by Gamow in Göttingen [273] and independently by Ronald Gurney (1909–53) and Edward Condon (1902–74) [274] in Princeton that α-decay results as a consequence of quantum-mechanical tunnelling through a potential barrier. Moreover, all these authors had a further significant advance to report: the first explanation of the Geiger–Nuttall relation, known phenomenologically [275] since 1912, which establishes a connection between the lifetime of an α-emitter and the range of the produced α-particles.

In the latter to *Nature* by Gurney and Condon, we hear the echoes of a confusing past 'It has hitherto been necessary to postulate some special arbitrary "instability" of the nucleus; but in the following note it is pointed out that disintegration is a natural consequence of the laws of quantum mechanics without any special hypothesis ... Much has been written about the explosive violence with which the α-particle is hurled from its place in the nucleus. But from the process pictured above, one would rather say that the particle slips away almost unnoticed'.

2.11.2. Nuclear size

Imagine that an electron is a little sphere with radius a and that its mass is entirely electromagnetic in origin (both assumptions were widespread until well into the 1920s). Then, classically, $e^2/a \sim mc^2$ or

$$a \sim \frac{e^2}{mc^2} = 2.8 \times 10^{-13} \text{ cm.}$$

As was well known in the 1910s, this length is of the same general order of magnitude as nuclear radii. Question: how is it possible to stuff many electrons this size in a box as small as a nucleus? As Andrade phrased it in 1923 'Coulomb's law cannot account for the stability of a nucleus composed of positive and negative charges which when free would be as large as the nucleus itself' [276].

More interestingly, applications of qualitative quantum mechanics (especially the uncertainty relations) to electrons confined to a nucleus with a typical radius $\sim 5 \times 10^{-13}$ cm showed that it is not the electron's size but rather its de Broglie wavelength λ which causes problems. How can a β-ray with an energy of a few MeV wait inside the nucleus to be released if its λ is large compared with the nuclear radius? (Attempts to treat β-decay by barrier penetration, like α-decay, failed of course.) Moreover, the typical (relativistic) kinetic energy of a confined electron is $\simeq 40$ MeV, implausibly large compared with the average nuclear binding energy per particle [277]. I do not know who first raised these issues.

2.11.3. Nuclear magnetic moments

In 1926 Ralph Kronig (b 1904) pointed out [278] a difficulty related to the magnitude $e\hbar/2mc$ (the Bohr magneton) of the electron's magnetic moment. Since there are supposed to be electrons in the nucleus 'the nucleus too will have a magnetic moment of the order of a Bohr magneton unless the magnetic moments of all the nuclear electrons happen to cancel [the probability for which] seems *a priori* to be very small'. Nuclear magnetic moments had not yet been measured directly

Introducing Atoms and Their Nuclei

at that time. (The first direct measurements of the proton moment date from 1933 [277].) It was clear, however, that nuclear moments of the order of an electron Bohr magneton would give rise to unacceptably large hyperfine splittings of spectral lines, due to the interaction between the magnetic moments of peripheral electrons and of the nucleus.

2.11.4. Nuclear spin

Two years later Kronig spotted another strange consequence of the PE model, this one related to experiments on the intensities in rotational band spectra of N_2^+, singly ionized molecular nitrogen. These spectra consist of two branches [279], one corresponding to transitions $J \to J-1$, the other to $J \to J+1$ where J is the nuclear orbital angular momentum of a rotational level. For the case of two identical nuclei with spin I, the nuclear spin introduces a 'strong to weak' alternating intensity ratio R in each branch, given by

$$R = \frac{I+1}{I}.$$

Measurements had shown [280] that R lies close to 2 for N_2^+, so that the spin of the nitrogen nucleus equals 1. 'One might in the first instance be surprised about this result', Kronig remarked [281]. Indeed. According to the PE model the N^{14} nucleus consists of 14 protons + 7 electrons, that is, of an odd number of spin-$\frac{1}{2}$ particles. The resultant nuclear spin should therefore be half-integer. Then how can it be equal to one? 'Probably one is therefore forced to assume that protons and electrons do not retain their identity to the extent they do outside the nucleus'. [281].

2.11.5. Nuclear statistics

In 1929, Franco Rasetti (b 1901) reported [282] an experimental study of Raman scattering, the inelastic scattering of photons by molecules, for the case of nitrogen. By an argument somewhat similar to the one for rotational band spectra it can be shown [279] that rotational Raman spectra also exhibit alternating intensities which, however, in this case determine the statistics of the nuclei. Rasetti showed that nitrogen nuclei obey Bose statistics.

Immediately afterwards Walter Heitler (1904–81) and Gerhard Herzberg (b 1904) fired off a letter [283] in which they called 'this fact extraordinarily surprising'. According to the PE model the N^{14} nucleus contains an odd number (21) of spin-$\frac{1}{2}$ particles and should therefore obey Fermi–Dirac (FD) statistics. Therefore, they concluded 'This rule is no longer valid in the nucleus ... it seems as if the electron in the nucleus loses, along with its spin [reference to Kronig's earlier conclusion], also its right of participation in the statistics of the nucleus'. Shortly thereafter, hyperfine structure measurements indicated that Li^6 also has the 'wrong' statistics.

Wherever one looked, at nuclear sizes, magnetic moments or statistics, intra-nuclear electrons caused grave problems [284]. The answer was to come soon: there are no electrons in the nucleus, (A, Z) corresponds to Z protons and $(A - Z)$ neutrons. Therefore the number of spin-$\frac{1}{2}$ particles in the nitrogen nucleus is not odd (14 protons + 7 electrons) but even (7 protons + 7 neutrons). It cannot be held against the physicists of the late 1920s that they could not and would not do better than suppose that electrons lose their identity when entering the nucleus.

This concludes the recitation of the nuclear constitution's perplexities which the discovery of the neutron was to resolve. Even after that event constituent electrons were not banned all at once. That postscript can wait until section 2.12. We should next catch up with the developments regarding β-spectra.

2.11.6. β-spectra: 1914–30

From the seven years following Chadwick's discovery of the continuous β-spectrum there is nothing memorable to report on this subject. There was a war going on. Then, in 1921 came Ellis' discovery of internal conversion, as already mentioned.

In 1922, Lise Meitner came forth [285] with a completely different and complicated scheme for β-decays based on the following assumptions: (1) all β-rays start out with a unique energy E inside the nucleus (we are still in the days of the PE model of course); (2) they may escape with this full energy; (3) or they may convert the portion $E - w$ of their energy into a single γ-ray; (4) the γ-ray will give secondary β-rays due to internal conversion in the K, L, M ... shell. I shall not discuss the experimental evidence Meitner claimed for her model—which is of course a variant of the old ideas of Hahn and her of a monochromatic primary β-ray energy. In her paper she refers to Ellis' 'other point of view' but makes no mention of Chadwick's continuous spectrum.

Chadwick and Ellis were not amused. They jointly returned to the continuous spectrum which 'on Meitner's view ... is presumably held to be a fortuitous occurrence due perhaps to the scattering of the homogeneous groups, whereas in our view the continuous spectrum consists of the actual disintegration electrons' [286]. In their paper they report on new experiments which showed that 'The continuous spectrum has a real existence which is not dependent on the spectrum ... Any explanation of it as due to secondary causes is untenable'.

The debate about the primary β-spectrum was nearing its end but was not quite over yet. During 1922–4 several proposals for a secondary origin of the continuous spectrum continued to be published. These were reviewed in 1925 by Ellis and William Wooster (b 1903) who concluded 'There seems to be no doubt that [this spectrum] exists and that the explanation of its occurrence is not to be sought in any ordinary effect' [287]. They also announced the beginnings of a difficult experiment which, they hoped, would further clarify the situation. This was their idea: imagine one could measure calorimetrically the total energy released per individual β-decay. If the continuous spectrum has an 'ordinary' origin, that is if there were a unique primary β-energy which is redistributed by 'ordinary' processes (involving electromagnetic radiation for example) then the energy registered in the calorimeter should be the total primary energy to be identified, they sensibly suggested, with the upper limit of the β-spectrum. Suppose on the other hand that nothing but the electron itself heats up the calorimeter so that the registered energy per decay would be the average over the spectrum of electron energies. That would be 'extraordinary'.

It was.

After two years of hard work Ellis and Wooster announced the result of their experiment [288]. Their source was RaE, chosen because it has practically no line spectrum, so that all complications of internal conversion are avoided. The spectrum has a mean energy $\simeq 0.39$ MeV and an upper limit $\simeq 1$ MeV. The energy detected by the calorimeter was 0.34 MeV with an error of about 10%.

This momentous result ruled out for good any attempt to explain the β-spectrum by 'ordinary' means. The authors stated their main conclusion in these words 'We may safely generalize this result for radium E to all β-ray bodies and the long controversy about the origin of the continuous spectrum of β-rays appears to be settled'.

The news from the Cavendish came as 'a great shock' to Meitner [289]. Good physicist that she was, she and Wilhelm Orthmann (1901–45) repeated the experiment. The resulting paper [290] was completed in December 1929. It handsomely acknowledges Ellis and Wooster, reporting agreement with their results. Both groups remained silent about what it all might mean.

What *does* it all mean?

2.12. The neutron

2.12.1. Chadwick

Hans Bethe (b 1906) has called the 1930s the 'happy thirties in nuclear physics' [291], with good reason: the discovery of the neutron made possible for the first time a rational theory of nuclear phenomena.

In 1930 Walther Bothe (1891–1957) and Herbert Becker reported [292] that the exposure of beryllium to α-particles from polonium generated 'new radiations ... so hard that one can hardly doubt their nuclear origin'. Their assumption that this radiation consists of nuclear γ-rays later turned out to be part of the truth but not the whole truth. There was something odd about this radiation process. Neither its energy balance nor its angular distribution readily fitted the assumption of γ-emission.

Figure 2.9. *James Chadwick.*

Then came the Joliot–Curie communication [293] of 28 January 1932. In this paper entitled *The emission of high energy photons from hydrogenous substances irradiated with very penetrating alpha rays* the authors reported that the alleged γ-rays from the α–beryllium reaction were capable of ejecting protons from paraffin. They further noted that these γ-rays should have an energy ~ 50 MeV if the proton ejection were a Compton effect. They were aware that this was a peculiarly high energy but 'that is not a sufficient reason for rejecting the [Compton effect] hypothesis'. However, on further consideration they changed their minds—not about the γ-rays but about the Compton effect. In a follow-up communication (22 February) [294] they wrote that the mechanism for proton ejection 'corresponds to a new mode of interaction between radiation and matter'.

Meanwhile Chadwick had discovered the neutron (figure 2.9).

When the 28 January paper reached the Cavendish, neither Chadwick nor Rutherford believed its conclusions. Chadwick immediately went to work and on 17 February submitted a paper [295] entitled *Possible existence of a neutron* in which he proposed that the α–Be reaction is (n=neutron)

$$\alpha + \text{Be}^9 \rightarrow \text{C}^{12} + \text{n}$$

where, he noted, the neutron has a mass close to that of the proton.

Introducing Atoms and Their Nuclei

Figure 2.10. *Irene and Frederic Joliot–Curie.*

2.12.2. *Induced radioactivity: the Joliot–Curies*

Fortunately Chadwick's discovery did not lead the Joliot–Curies (figure 2.10) to terminate their experiments. As a result they, too, made a major discovery.

All nuclear reactions known till then had been instantaneous, but when an aluminium foil is irradiated the emission of positrons does not cease immediately when the α-source is removed. The foil keeps emitting positrons; the process has a characteristic half-life ($\sim$ 3 minutes) 'These experiments demonstrate the existence of a new kind of radioactivity with positron† emission' [296]

$$\alpha + Al^{27} \to P^{30} + n$$
$$\searrow$$
$$Si^{30} + e^+.$$

Three weeks later they reported [297] the direct detection by radiochemical means of the respective phosphorous and nitrogen isotopes—which explains why the discovery of β^+ radioactivity was awarded a Nobel Prize in chemistry. (It may also have been a consideration to honour the Joliot–Curies and Chadwick in that same year 1935.)

† *The discovery of the positron in 1932 will be discussed elsewhere in this book.*

Events followed each other very fast now. Another five weeks later Enrico Fermi (1901–54) submitted the first in a series of articles on radioactivity induced by neutron bombardment [298]. With this paper Fermi began his experimental studies in neutron physics which made him perhaps the world's leading expert on that subject during the 1930s.

2.12.3. What is a neutron?
That question took two years to answer: it is a particle as elementary as the proton. I give next some glimpses of the thinking up to that time. All dates to follow refer to 1932.

April 18. Francis Perrin (b 1901) [299] and Pierre Auger (b 1899) [300] suggest that light elements are exclusively built of α-particles, protons, and neutrons, the latter being considered as bound PE-systems, but that, from the naturally radioactive isotope K^{41} on, additional electrons appear as separate nuclear constituents. This proposal was still discussed at the Solvay meeting in October 1933, where, somewhat unexpectedly, it found support from Dirac 'If we consider protons and neutrons as elementary particles we would have three kinds of elementary particles (p, n, e) out of which the nucleus is made up. This number may seem large but, from that point of view, two is already a large number' [301].

April 21. Dmitri Iwanenko (b 1904) must be credited with the first suggestion that the neutron might be something like a proton. 'The chief point of interest is how far neutrons can be considered as elementary particles (something like protons or electrons)' [302]. Not that he banished electrons from the nucleus. Rather, he suggested, these are 'all packed in α-particles or neutrons ... [This] sounds not so improbable if we remember that the nuclei [sic] electrons profoundly change their properties when entering into nuclei'. Thus he believed electrons to be tucked away as constituents of the nucleus—which does, of course, not solve the many problems of the nuclear PE model.

April 18. Chadwick said 'The neutron may be pictured as a small dipole or, perhaps better, as a proton embedded in an electron' [303].

August 17. Iwanenko [304] on the resolution of the spin-statistics paradoxes 'We do not consider the neutron as built up of an electron and a proton but as an *elementary particle*. Given this fact we are obliged to treat neutrons as possessing spin $\frac{1}{2}$ and obeying Fermi–Dirac statistics ... Nitrogen nuclei appear to obey Bose–Einstein statistics'. This now becomes understandable since N^{14} contains just 14 elementary particles [7p+7n], that is an even number, and not 21 [14p+7e]' which is the earliest explicit statement I have found on the resolution of the nitrogen problem.

The year 1932 ended with a stand-off between the old and the new pictures of the neutron.

2.12.4. The first theory of nuclear forces: Heisenberg

Three papers by Heisenberg [305–307] completed in the latter half of 1932 mark the transition to the modern view on nuclear forces. These articles, important though they are, must not be considered as a clean break with the past, however. Heisenberg's nuclear theory is a hybrid of the old and the new. It has the virtue of being based on the proton–neutron model of the nucleus, but the drawback of a proton–electron model for the neutron. In order to understand Heisenberg's position it should be noted that in 1932 he still sided with Bohr in believing that the continuous β-spectrum should be understood as due to a breakdown of the energy conservation law (see section 2.13.1). As he put it in a letter to Bohr 'The basic idea is: shove all the difficulties on to the neutron and practice quantum mechanics inside the nucleus' [308].

See also p 371

Heisenberg's neutron is a particle with spin $\frac{1}{2}$ obeying Fermi–Dirac statistics. 'It will, however, be assumed that under suitable circumstances [the neutron] can break up into a proton and an electron in which case the conservation laws of energy and momentum probably do not apply' [305]. We now grasp Heisenberg's strategy: firstly, let us admit that we do not understand the neutron; secondly, let us see how far we can come with standard non-relativistic quantum mechanics in regard to a nucleus built of protons and neutrons only—the electrons somehow hiding inside the neutrons.

Along with the Coulomb forces between protons Heisenberg postulated new interactions of short range, between protons and neutrons and between neutrons and neutrons, not between protons and protons. A most important novelty of far-reaching consequences is his introduction of 'isotopic spin' in terms of which protons and neutrons are described by a single but two-component wavefunction. (Isotopic spin is discussed in detail elsewhere in this book.)

In spite of shortcomings, Heisenberg's work represents nothing less than a breakthrough, for one single reason: his insistence that the nucleus, considered as a proton–neutron (PN) system, is amenable to non-relativistic quantum mechanics.

2.12.5. The first theory of nuclear reactions: Bohr

The discovery of the neutron was most important not only because it led to the first sensible, though primitive, theoretical description of nuclei but also because this new particle provided an extremely valuable new experimental tool for penetrating the secrets of nuclear matter. Since neutrons carry no electric charge, they can enter and explore the interior of the nucleus without being held back by electrostatic repulsion, as is the case for α-particles. Thus began a new chapter of experimental inquiry, neutron physics, the study of what happens when nuclei are bombarded by neutrons.

See also p 1192

When neutron physics came along, the initial interpretation of data rested on analogies with atomic processes, such as the scattering of electrons by atoms or the ionization of atoms by electrons. A 'one-body model' was postulated, according to which the nucleus as a whole is represented by a rigid body that exerts a fixed overall average force on impinging neutrons. The general framework was of course quantum mechanical.

A few early results looked promising. The model could explain the observed large increase of neutron capture as the neutron moves ever more slowly. Large fluctuations of the capture probability from one element to another could be attributed, it was thought, to corresponding differences in the average force generated by the nucleus.

Very soon severe difficulties arose, however. The theory predicted that the probability for scattering of thermal neutrons should be larger than or comparable with the probability for their capture. Yet already in 1935 experiments showed that for certain elements capture was vastly more probable than scattering, by a factor of about 100 for cadmium for example. Also from that year date the discoveries of 'selective absorption'. These phenomena were discovered by means of experimental arrangements of the following kind. A continuous spectrum of neutrons first impinges on a layer of element X. Some neutrons are absorbed, some pass through. The latter hit a second layer, of element Y, in which further absorption can take place. This second absorption is much smaller if X and Y are the same substance than if they are different.

As a function of neutron energy the absorption exhibits a series of peaks and valleys whatever the absorbing nucleus may be.

All these observations were in conflict with the predictions of the one-body model according to which absorption should exhibit a smooth behaviour (inverse proportionality to the neutron velocity) for any kind of absorbing nucleus.

At this stage Bohr came forth with a radically different approach [309]. His main point was that the idealization of the nucleus as a single rigid body is incorrect. Nuclear reactions should instead be treated, he suggested, as a many-body problem of a special kind: a two-stage reaction.

In the first stage the incoming projectile—which may either be a neutron, a proton or an α-particle, with due regard of course to electrostatic repulsion in the last two instances—merges with the bombarded nucleus into a single unit, the compound nucleus. This compound has many properties in common with a stable nucleus, though it is not a stable body itself. Thus it has a discrete set of energy levels, as had been known to be the case for stable nuclei since the early 1920s (the era that marked the beginnings of 'nuclear spectroscopy'), the analogue for nuclei of the long-familiar situation for atoms and molecules.

In the second stage the compound nucleus disintegrates, either into the particles from which it was formed (elastic scattering), or again into these same particles but with the nucleus turning into one of its excited states (inelastic scattering), or into different particles (nuclear reaction).

A crucial feature of this picture, Bohr continued [309], is that the formation and the break-up of the compound nucleus are well separated in time (always reckoned on the scale of nuclear processes). This separability is a consequence, he thought, of the close packing of nucleons in nuclei. In an encounter between a neutron and a nucleus 'the excess energy of the incident neutron will be rapidly divided among all the nuclear particles with the result that for some time afterwards no single particle will have sufficient kinetic energy to leave the nucleus ... It is therefore clear that the duration of the encounter must be extremely long compared with the time interval, about 10^{-21} s, which the neutron would use in simply passing through a space region of nuclear dimensions' so long that, one might say, the compound nucleus has lost all memory of how it was formed by the time it breaks up. The only chance for break-up is a rare fluctuation in which a single particle gathers most of the available excess energy.

I turn to a few details of Bohr's model. Nuclear photoemission is so pronounced, he reasoned, because it does not depend on the complicated redistribution of energy among nucleons. Any nucleon can at any time get rid of extra energy by emitting a photon. Selective absorption, Bohr argued further, is a consequence of the quantization of the energy levels of the compound nucleus. Quantum mechanics teaches us that the probability of formation of a compound nucleus is quite large if the energy of the incident particle just fits by bringing the whole system into one of the energy levels of the compound nucleus, a phenomenon known as resonance excitation. The probability for compound formation is much smaller if the incident energy does not fit in that way.

Resonance excitation explains the absorption by elements X and Y mentioned before. Absorption by X principally occurs for those incoming neutrons which carry the specific resonant energies corresponding to the formation of the compound nucleus 'X+neutron'. If Y equals X then the neutron spectrum hitting Y is lacking in just those neutrons that can form a compound nucleus in the Y layer. If, on the other hand, Y is different from X, then the incoming continuous neutron spectrum has not suffered depletion of those neutrons that can form the compound 'Y+neutron'. In Bohr's words 'The different selective absorbing elements [show] that the resonance is restricted to narrow energy regions which are differently placed for selective absorbers' [309]. Thus resonance excitation explains both the phenomenon of selective absorption and the general occurrence of peaks and valleys as a function of neutron energy.

By happy coincidence, Bohr's qualitative considerations of these phenomena were greatly helped along by simultaneous and independent quantitative studies of nuclear resonance phenomena undertaken by Gregory Breit (1899–1981) and Eugene Wigner (b 1902) in Princeton. In their paper [310] the authors do not use the compound nucleus model; in fact their main formulae do not depend on any model, and may therefore be applied to Bohr's pictures. This work helped promote 'the profound reordering of the picture of nuclear dynamics implied by Bohr's ideas [which were] rapidly and widely accepted in the nuclear physics community; within months the literature is completely dominated by papers applying, testing, and extending the ideas of the compound nucleus' [311].

It had, of course, been clear to Bohr that analysis with atomic spectra could not be of help to him. Now, together with the brilliant young Danish physicist Fritz Kalckar (1910–38), he suggested [312] that for nuclei a much more proper comparison would be with a drop of liquid†. That analogy should not be taken too literally, the dynamics of a true liquid drop is vastly different from that of nuclei. Yet the comparison, treated cautiously, was tempting and in the event proved fruitful in many respects, particularly in regard to collective motions. A liquid drop in equilibrium can be characterized by forces that would disperse it into smaller parts were it not counteracted by a surface tension. When the drop is excited, it performs vibrations of various kinds: surface waves which do not change the drop volume, and volume waves, a sequence of compressions and dilations. These vibrations, Bohr and Kalckar argued, give a good model for collective motions of nuclei. Vibrational energies, *when quantized*, should describe the spectrum of the compound nucleus. Since the surface waves have frequencies small compared with those of volume waves, the former should be representative of the nuclear spectrum at the relatively low energies at which the processes previously mentioned had so far been observed.

In 1939, Bohr applied his model to nuclear fission. It was to be the compound nucleus' 'finest hour!' [311]. I shall turn to fission shortly.

This concludes my account of the evolution of nuclear physics in the 1930s. The meson theory of nuclear forces, another chapter initiated in the 1930s, will be treated in elsewhere in this book. Two topics remain for me to discuss: the dénouement of the β-spectra puzzle; and fission.

2.13. β-spectra: the end of the beginning

2.13.1. Bohr
Ellis and Wooster's fundamental paper [288] (section 2.11.6) appeared in December 1927. It is my impression that the seriousness of the situation it created was not immediately appreciated. In all the literature of 1928 I found only one (uninteresting) reference to their paper. Pauli's complaint

---

† *In a different context that same thought had occurred earlier to Gamow* [270].

in February 1929 that Bohr was serving him up all kinds of ideas about β-decay 'by appealing to the Cambridge authorities but without reference to the literature' [313] would indicate that not even then had he seen the Ellis–Wooster paper.

The first to take a stab at explaining the continuous nature of β-spectra was Bohr, who proposed that it is due to energy non-conservation in β-decay. The idea had been on his mind since early 1929, as we see from a letter by Pauli to Bohr 'Do you intend to mistreat the poor energy law further?' [314].

Bohr first stated this idea publicly during his Faraday lecture (May 1930) [315], in which he commented on the paradoxes of nuclear constitution as well. Referring to the 4p+2e model of the α-particle he said that 'As soon as we inquire ... into the constitution of even the simplest nuclei the present formulation of quantum mechanics fails completely' and regarding β-decay

> At the present stage of atomic theory we have no argument, either empirical or theoretical, for upholding the energy principle in the case of β-ray disintegrations, and are even led to complications and difficulties in trying to do so ... Still, just as the account of those aspects of atomic constitution essential for the explanation of the ordinary physical and chemical properties of matter implies a renunciation of the classical idea of causality, the features of atomic stability, still deeper-lying, responsible for the existence and the properties of atomic nuclei, may force us to renounce the very idea of energy balance. I shall not enter further into such speculations and their possible bearing on the much debated question of the source of stellar energy. I have touched upon them here mainly to emphasize that in atomic theory, notwithstanding all the recent progress, we must still be prepared for new surprises.

Thus Bohr believed that the β-spectrum *and* the structure paradoxes were indications of limitations of the quantum-mechanical description. Who, in 1929, would have had the temerity to predict that these two ailments demanded two distinct cures: the neutron and the neutrino?

2.13.2. Pauli

In December 1930, Pauli wrote a letter to a physics gathering in Tübingen. Parts of it follow.

> Dear radioactive ladies and gentlemen,
> I have come upon a desperate way out regarding the 'wrong' statistics of the N- and the Li 6-nuclei, as well as of the continuous β-spectrum, in order to save the 'alternation law' of statistics†

---

† *An even (odd) number of spin-$\frac{1}{2}$ particles has integer (half-integer) spin and satisfies Bose–Einstein (Fermi–Dirac) statistics.*

and the energy law. To wit, the possibility that there could exist in the nucleus electrically neutral particles, which I shall call neutrons, which have spin $\frac{1}{2}$ and satisfy the exclusion principle and which are further distinct from light-quanta in that they do not move with light velocity. The mass of the neutrons should be of the same order of magnitude as the electron mass and in any case not larger than 0.01 times the proton mass. The continuous β-spectrum would then become understandable from the assumption that in β-decay a neutron is emitted along with the electron in such a way that the sum of the energies of the neutron and the electron is constant.

There is the further question, which forces act on the neutron? On wave mechanical grounds ... the most probable model for the neutron seems to me to be that the neutron at rest is a magnetic dipole with a certain moment μ. Experiments seem to demand that the ionizing action of such a neutron cannot be bigger than that of a γ-ray, and so μ may not be larger than $e \times 10^{-13}$ cm.

For the time being I dare not publish anything about this idea and address myself confidentially first to you, dear radioactive ones, with the question how it would be with the experimental proof of such a neutron, if it were to have a penetrating power equal to or about ten times larger than a γ-ray.

I admit that my way out may not seem very probable *a priori* since one would probably have seen the neutrons a long time ago if they exist. But only he who dares wins [316].

Let us look at Pauli's letter more closely. His use of 'neutron' (I shall call it the Pauli neutron for a while) is natural, 'our' neutron had not been postulated yet. It is evident that Pauli looked upon his particle as the answer to the problems of nuclear constitution as well as of β-decay. However, he was unable to ban electrons as constituents. The Pauli neutron weighed little; protons still had to supply the nuclear mass, electrons still had to compensate the charge. In its original conception his picture of the nucleus is therefore a 'three species model'. The Pauli neutron had to bind to nuclear matter, whence his need for a magnetic moment.

With regret I must pass by interesting developments during the next few years [317] and move to 1933 when, in April, Pauli noted 'The Italian name [i.e. neutrino] (in contrast to neutron) is made by Fermi' [318], and when, in October, a public confrontation between Pauli and Bohr took place. The setting was the seventh Solvay conference.

Pauli discussed the neutrino '[Bohr's] hypothesis does not seem satisfactory nor even plausible to me'. He insisted on conservation of energy, momentum, angular momentum and statistics 'in all elementary processes'. He regarded it as 'conceivable' that neutrinos had spin $\frac{1}{2}$ and obeyed FD statistics but noted that this has not yet been experimentally verified [319]. There was no longer any mention of the neutrino as a nuclear constituent. Perrin remarked that relativistic kinematics applied to the radium-E spectrum show 'that the neutrino has *zero intrinsic mass*, like the photon' (his italics) [320]. Bohr's comments were rather muted. 'As long as we have no experimental data it is wise not to give up the conservation laws, but on the other hand no one knows what surprises may still await us' [321].

2.13.3. *Fermi*

Fermi was also present during the Solvay discussions on β-decay but kept silent while others were arguing. However, he must have gone to work right after that conference. In December he sent a note on the subject to *Nature*. It was rejected 'because it contained speculations too remote from reality to be of interest to the reader' [322]. An Italian version entitled *'Tentativo di una teoria della emissione di raggi β'* fared better [323]. More detailed accounts [324] appeared early in 1934. Here, at long last, it is stated that the language appropriate to β-decay is the language of quantum field theory 'Electrons (or neutrinos) can be created and can disappear ... The Hamilton function of the system consisting of heavy and light particles must be chosen such that to every transition from neutron to proton there is associated a creation of an electron and a neutrino. To the inverse process, the change of a proton into a neutron, the disappearance of an electron and a neutrino should be associated' [324].

[See also p 374]

In making these assertions Fermi appealed to an analogy with the creation and annihilation of photons. Comparisons like these were in the air. Iwanenko, in August 1932 wrote 'The expulsion of an electron [in β-decay] is similar to the birth of a new particle' [304] and Perrin in December 1933 said 'The neutrino ... does not preexist in atomic nuclei [but] is created when emitted, like the photon' [325]. No one before Fermi had put these ideas in operational form, however. In fact, Fermi was the first to use second quantized spin-$\frac{1}{2}$ fields in particle physics.

Fermi's theory marks one of the high points in the history of particle physics. Its successful predictions constituted indisputable evidence for the proton–neutron picture of the nucleus. From 1934 onwards nuclear electrons were banned for good.

I shall not discuss here the further rich content of Fermi's paper nor the consequences of his theory elaborated by others. I have now in fact fulfilled my task: to show the way through the complexities of β-decay's early years.

2.14. Fission

2.14.1. *The discovery*

On 10 December 1938 Fermi received a Nobel Prize in physics 'for his demonstration of the existence of new radioactive elements produced by neutron irradiation, and for his related discovery of nuclear reactions brought about by slow neutrons'. Among the results Fermi reviewed in his Nobel lecture were those obtained by bombarding uranium (U). Already in June 1934 Fermi and his colleagues had found that neutrons hitting U produce several induced activities, one of which 'suggests the possibility that the atomic number of the [produced] element may be greater than 92' [326], that number being the Z of U.

In his Nobel lecture Fermi summarized these and subsequent findings as follows 'The carriers of some of the activities of U are neither isotopes of U itself nor of the elements lighter than U down to the atomic number 86. We concluded that the carriers were one or more elements of atomic number larger than 92; we, in Rome, use to call the elements 93 and 94 Ausenium and Hesperium respectively' [327]. In early December Fermi's reasoning was eminently logical. If an isotope does not lie in the region $Z = 86$–92, then it must lie above $Z = 92$.

The work of the Rome group was confirmed and extended in other laboratories, notably in Berlin by Hahn, Meitner and Fritz Strassmann (1902–80), who in July 1938 published preliminary evidence for elements with $Z = 93$–96 [328]. In that same paper the authors also announced the discovery of the new unstable isotope $^{239}_{92}U$, seen as a resonance of about 25 electronvolt neutrons. More about that isotope presently.

Thus the realization grew in widening circles that a new chapter in physics and chemistry had begun, that of the transuranic elements.

Wrong.

In the autumn of 1938 Hahn and Strassmann began a most careful radiochemical analysis of the elements produced in neutron–U collisions. Both its motivation and its execution have been summarized in later books by Hahn [329]. I must pass by all these fascinating details and turn directly to their results: among the products of neutron–U interactions they identified three isotopes of barium, which have $Z = 56$—roughly half that of U!

It was a discovery which can be called staggering. Nothing like it had either been seen or dreamed of before; until that time nuclear reactions had never produced changes in Z larger than 2.

Their results appeared in print on 6 January 1939 [330]. The authors were awed by their own findings 'We publish rather hesitantly due to [these] peculiar results ... As chemists we should actually state that the new products are barium ... However, as nuclear chemists, working very close to the field of physics, we cannot bring ourselves yet to take such a drastic step which goes against all previous experience in nuclear physics' [330].

Before 1938 was over it became clear how to interpret the Hahn–Strassmann results. From Otto Frisch's (1904–79) recollections 'Lise Meitner was lonely in Sweden and, as her faithful nephew, I went [from Copenhagen] to visit her at Christmas. There, in a small hotel ... I

Introducing Atoms and Their Nuclei

found her at breakfast brooding over a letter from Hahn' [331]. Frisch was sceptical until Meitner convinced him of Hahn's superior methods.

During the two or three days the two spent together that Christmas 'the idea gradually took shape' that Bohr's liquid-drop model of a nucleus could explain what was going on 'Perhaps a drop could divide itself into two smaller drops by first becoming elongated, then constricted and finally being torn' [331]. They realized that the key was an interplay between two opposing forces. There is an attractive nuclear force, represented in the drop model by a surface tension holding the drop together, similar to what happens to a drop of mercury. There is also a repulsive electrostatic force between protons, tending to pull the nucleus apart. Both forces increase with nuclear size but the repulsive force increases faster. They could estimate that the latter force would win for Z close to 100. Thus 'the uranium nucleus might be a very wobbly unstable drop ready to divide itself at the slightest provocation, such as the impact of a neutron'. They estimated that the kinetic energy of the two fragments was huge: 200 million electronvolts, about ten times larger than in any nuclear reaction previously observed.

Their findings were submitted on 16 January in a letter to *Nature* [332].

From 26–28 January 1939, a small theoretical physics conference was held in Washington, DC, attended by Bohr and by Fermi, who weeks earlier had fled Italy[†], and who, immediately upon hearing the news, had set to work on fission. At that conference both men reported on fission. 'The whole matter, was quite unexpected to all present' [333]. Word about fission immediately appeared in newspapers, for example in the 29 January issue of the *New York Times*.

The January meeting in Washington marked the opening of the widespread hunting season for fission. A research group in Washington reported 'At the conclusion of the conference on 28 January we were privileged to demonstrate [fission] to Professors Bohr and Fermi' [333]. In 1939 over a hundred papers on the subject were published [334]. Among the topics studied were the dependence of fission probability on the incoming neutron energy, the various modes of U fission—by the end of 1939 16 different elements had been identified as U fission fragments, some represented by several isotopes—and the fissionability of other elements. All this was experimental. The only theoretical contributions in that period were made by Bohr and by Bohr and John Wheeler (b 1911).

2.14.2. *Bohr on uranium 235*

When, one day in early February, George Placzek (1905–55) came to see Bohr in Princeton, the latter remarked that now all the confusion about

---

[†] *Due to the recently enacted anti-semitic laws which affected his wife who was of Jewish descent.*

transuranic elements belonged to the past. No, said Placzek, there is still a mystery: as a function of energy, neutron capture by U shows a sizable resonance peak at neutron energies of about 25 eV—the formation of the compound nucleus ^{239}U found earlier by Hahn, Meitner and Strassmann [328]. One must therefore expect, said Placzek, that fission, a decay mode of ^{239}U, should also show a peak at 25 eV—of which, however, there was no sign. That discussion took place over breakfast at the Princeton Club. 'When Bohr heard that he became restless. Let us go to Fine Hall, he said. During the five minutes it took to get there, Bohr was silent, deep in thought. When he came to his office, joined by Placzek, Rosenfeld and Wheeler, he said 'Now hear this, I have understood everything' [335].

Bohr explained: the dominant U isotope with weight 238, has the resonance, but at that energy fission does not take place; the rare isotope with weight 235, making up only 0.7 per cent of natural U, becomes fissile at that energy but does not resonate. A short note to this effect was published on 15 February [336]. It should be stressed that there existed as yet no experimental foundation for Bohr's idea of the role of 235. In fact, at that time 'very few physicists accepted Bohr's explanation. Fermi in particular disagreed strongly' [337]. A direct test would necessitate the enrichment of 235 in a natural U sample which then should exhibit an enhanced rate of fission by slow neutrons. That confirmation came only in March 1940 [338].

During 1939 Bohr collaborated with Wheeler on detailed theoretical studies of fission processes. In this work Bohr met with two old loves: the compound nucleus of 1936 and the study of the relation between surface tension and surface vibrations, the subject of his prize essay written in 1905–6. Their labours resulted in two important papers [339].

Bohr's work with Wheeler was the last major contribution to physics he made. His years of scientific productivity had spanned the period from 1905 to 1939. It had begun with classical physics, vibrations in liquid jets, gone on to the old quantum theory, then to the foundations of quantum mechanics, and finally to nuclear physics. He has put his mark on twentieth century science.

Many years later Bethe wrote 'The compound nucleus dominated the theory of nuclear reactions at least from 1936 to 1954 ... At Los Alamos when we tried to get [probabilities] we used the compound nucleus model and usually our predictions were quite reasonable. The compound nucleus could explain many phenomena ... ' [340]. He should know, since in the war years he was the head of the theory division at the Los Alamos Laboratory.

2.14.3. Postscript: pre-war thoughts on atomic energy from fission
It was already known in 1939 that free neutrons are released in fission processes, and it was at once evident that this might lead to chain reactions which in turn, might lead to atomic bombs.

Introducing Atoms and Their Nuclei

All these early results were published in the open scientific press. So were speculations about their possible implications. To give but one example, in June 1939 there appeared a German article entitled *Can the energy content of atomic nuclei be utilized in technology?* [341]. The press also kept the general public informed from the start, as the following sample of headlines (all from 1939) illustrates: the *New York Times* 5 February 'Don't expect uranium atoms to revolutionize civilisation'; the *New York Herald Tribune* 12 February 'In the realm of science: tempest in cyclotron isn't likely to smash earth'; and 7 May 'In the realm of science: practical development of atomic power is now just a matter of time'; the *Washington Post* 29 April 'Physicists here debate whether experiments will blow up 2 miles of the landscape'.

Headlines such as these may help sell newspapers. Scientists knew well, however, that manna was not about to fall from the heavens. As was written more soberly in the *Scientific American* in October 1939 'Power production by means of nuclear fission would not be beyond the realms of possibility. But under present conditions the process is as inefficient as removing sand from a beach a grain at a time' [342].

For better or for worse, conditions have changed.

But that is another story.

References

[1] Perrin J 1913 *Les Atomes* (Paris: Librairie Alcan) (Engl. Transl. Hammick D L 1916 *Atoms* (New York: Van Nostrand))
[2] Descartes R 1955 *Principles of Philosophy* part 2, principle 20 (see, for example, the translation by Haldane E and Ross G 1955 (New York: Dover))
[3] Compare with de Milt C 1948 *Chymia* **1** 153
[4] Mendeleev D 1891 *The Principles of Chemistry* vol 1 (London: Greenaway) p 315
[5] Williamson A W 1869 *J. Chem. Soc.* **22** 328
[6] Prout W 1815 *Ann. Phil.* **6** 321; 1816 *Ann. Phil.* **7** 111 (reproduced in Alembic Reprints 1932 (London: Gurney & Jackson))
[7] Quoted by Brock W H and Knight M 1965 *Isis* **56** 5
[8] Clausius R 1857 *Ann. Phys., Lpz.* **10** 353
[9] Maxwell J C 1952 *Collected Works* vol 2 (New York: Dover) pp 376–77
[10] Planck M 1883 *Ann. Phys., Lpz.* **19** 358
[11] Ostwald W 1895 *Verh. Ges. Deutsch. Naturf. Ärzte* **1** 155
[12] Sommerfeld A 1944 *Wiener Chem. Z.* **47** 25
[13] Mach E 1910 *Popular Scientific Lectures* (Chicago: Open Court) p 207
[14] Young T 1972 *Miscellaneous Works, Johnson Reprint* vol 9 (New York: Johnson) p 461
[15] Loschmidt J 1866 *Wiener Ber.* **52** 395
[16] Maxwell J C 1952 *Collected Works* vol 2 (New York: Dover) p 361
[17] van der Waals J D 1873 *Over de continuiteit van den gas- en vloeistof toestand* (Leiden: Sÿthoff)

[18] Rücker A W 1888 *J. Chem. Soc.* **53** 222
[19] Deslattes R D 1980 *Ann. Rev. Phys. Chem.* **31** 435
[20] Maxwell J C *Theory of Heat* ch 22 (Westport, CT: Greenwood)
[21] Thomson J J 1899 *Phil. Mag.* **48** 565
[22] Curie M 1900 *Rev. Scientifique* **14** 65
[23] Müller E W 1956 *Phys. Rev.* **102** 624; 1956 *J. Appl. Phys.* **27** 474; 1957 *J. Appl. Phys.* **28** 1; 1957 *Sci. Am.* June 113
[24] Maxwell J C 1873 *A Treatise on Electricity and Magnetism* (Oxford: Clarendon) p 307
[25] Faraday M 1839 *Experimental Researches in Electricity* section 852, (London: Quaritch)
[26] Reference [25] section 869
[27] Reference [9] vol 2, p 418
[28] Reference [9] vol 2, p 361
[29] Reference [24] part 2, ch 4
[30] Compare with Kahl R (ed) 1971 *Selected Writings by Hermann Helmholtz* (Middletown, CT: Wesleyan University Press) p 409
[31] Compare with, for example, Richarz F 1894 *Ann. Phys. Chem.* **288** 385
[32] His paper was not published until 1881: Stoney G J 1881 *Phil. Mag.* **11** 381
[33] Rutherford E 1906 *Radioactive Transformations* (London: Constable) especially pp 1, 16
[34] Nitske W R 1971 *The Life of Roentgen* (Tucson, AZ: University of Arizona Press) p 100
[35] Rutherford's contributions were gathered in 1962–5 *The Collected Papers of Lord Rutherford of Nelson* (London: Allen and Unwin)
[36] Reference [33] p 17
[37] Curie M 1929 *Pierre Curie* (New York: MacMillan)
[38] Sklodowska M 1898 *C. R. Acad. Sci., Paris* **126** 1101
[39] Schmidt G C 1898 *Ann. Phys., Lpz.* **65** 141
[40] Curie P, Curie S and Bémont G 1898 *C. R. Acad. Sci., Paris* **127** 1215
[41] Quoted in Eve A S 1939 *Rutherford* (Cambridge: Cambridge University Press) p 80
[42] Moore R B 1918 *J. Franklin Inst.* **186** 29
[43] Reference [33] p 18
[44] Rutherford E 1903 *Phil. Mag.* **5** 177
[45] Rutherford E 1905 *Radioactivity* 2nd edn (Cambridge: Cambridge University Press)
[46] Rutherford E 1905 *Phil. Mag.* **6** 193; reference [45], p 156
[47] Rutherford E and Geiger H 1908 *Proc. R. Soc.* A **81** 162
[48] Rutherford E and Royds T 1909 *Phil. Mag.* **17** 281
[49] Rutherford E 1902 *Phys. Z.* **3** 517
[50] Rutherford E and da Costa Andrade E N 1914 *Phil. Mag.* **27** 854
[51] Rutherford E 1900 *Phil. Mag.* **49** 1
[52] Rutherford E and Soddy F 1902 *Phil. Mag.* **4** 370, 569
[53] Russell A S 1951 *Proc. Phys. Soc.* **64** 217
[54] Robinson H R 1943 *Proc. Phys. Soc.* **55** 161
[55] Curie M 1910 *Traité de Radioactivité* vol 1 (Paris: Gauthier-Villars) ch 3
[56] Curie M 1899 *Rev. Gen. Sci. Pures Appl.* **10** 41
[57] Reference [55] vol 1, p 129
[58] Elster J and Geitel H 1898 *Ann. Phys., Lpz.* **66** 735
[59] Elster J and Geitel H 1903 *Ann. Phys., Lpz.* **69** 83
[60] Curie P and Laborde A 1903 *C. R. Acad. Sci., Paris* **136** 673
[61] Poincaré H 1913 *The Foundations of Science* (New York: Science) ch 8
[62] Briner E 1925 *C. R. Acad. Sci., Paris* **180** 1586

Introducing Atoms and Their Nuclei

[63] Rayleigh Lord 1969 *The Life of Sir J J Thomson* (London: Dawsons) p 141
[64] Rutherford E 1911 *Radioactive Transformations* (Newhaven, CT: Yale University Press) p 267
[65] Quoted in Crookes W 1910 *Proc. R. Soc.* A **83** xx
[66] Curie P 1967 *Nobel Lectures in Physics 1901–1922* (New York: Elsevier) p 78
[67] Soddy F 1912 *Matter and Energy* (New York: Holt)
[68] Rutherford E and Soddy F 1903 *Phil. Mag.* **5** 576
[69] Lorentz H A 1934 *Collected Papers* vol 2 (The Hague: Nÿhoff) p 164
[70] Lorentz H A 1934 *Collected Papers* vol 5 (The Hague: Nÿhoff) p 1
[71] Lorentz H A 1934 *Collected Papers* vol 5 (The Hague: Nÿhoff) p 172
[72] da Costa Andrade E N 1958 *Proc. R. Soc.* A **244** 437
[73] Thomson J J 1883 *A Treatise on the Motion of Vortex Rings* (London: MacMillan) (reprinted 1968 (London: Dawsons))
[74] Thomson J J 1897 *Phil. Mag.* **44** 293, especially p 311
[75] Thomson J J 1897 *Phil. Mag.* **48** 547, especially p 565
[76] Thomson J J 1904 *Electricity and Matter* (New York: Scribner) p 114
[77] Thomson J J 1904 *Phil. Mag.* **7** 237
[78] Rutherford E 1902 *Trans. R. Soc. Canada* **8** 79
[79] Campbell N R 1907 *Modern Electrical Theory* (Cambridge: Cambridge University Press) p 251
[80] Thomson J J 1903 *Phil. Mag.* **6** 673
[81] For further details see Pais A 1986 *Inward Bound* (Oxford: Oxford University Press) ch 9 section (c) pt 3
[82] Thomson J J 1906 *Phil. Mag.* **11** 769
[83] Kirchhoff G 1859 *Ber. Berliner Akad.* 662 (Engl. Transl. 1860 *Phil. Mag.* **19** 193)
[84] Kirchhoff G 1859 *Ber. Berliner Akad.* 783; 1859 *Ann. Phys. Chem.* **109** 275
[85] Einstein A 1913 *Naturwissenschaften* **1** 1077
[86] Hasenöhrl F (ed) 1968 *Wissenschaftliche Abhandlungen von L Boltzmann* vol 3 (New York: Chelsea) pp 110, 118
[87] Wien W 1893 *Ber. Preuss. Akad. Wiss.* 55; also 1894 *Ann. Phys.* **52** 132
[88] Wien W 1896 *Ann. Phys., Lpz.* **58** 662
[89] See especially Paschen W 1897 *Ann. Phys., Lpz.* **60** 662
[90] Heilbron J L 1986 *The Dilemma of an Upright Man* (Berkeley, CA: University of California Press)
[91] Hermann A 1973 *Planck* (Reinbek: Rowohlt)
[92] Planck M 1950 *Scientific Autobiography and other Papers* (Engl. Transl. F Gaynor 1950 (London: William and Norgate) p 7)
[93] Reference [86] pp 615, 618, 622
[94] Planck M 1900 *Ann. Phys., Lpz.* **1** 69 especially p 118
[95] Lummer O and Pringsheim E 1900 *Verh. Deutsch. Phys. Ges.* **2** 163
[96] Hettner G 1922 *Naturwissenschaften* **10** 1033; this account differs slightly from Planck's own recollections, written in his late eighties Planck M 1958 *Physikalische Abhandlungen and Vorträge* vol 3, ed M von Laue (Braunschweig: Vieweg) p 374
[97] Pais A 1982 *Subtle is the Lord* (Oxford: Oxford University Press) ch 19 section (b)
[98] Planck M 1900 *Verh. Deutsch. Phys. Ges.* **2** 207
[99] Planck M 1900 *Verh. Deutsch. Phys. Ges.* **2** 237
[100] Planck M 1967 *Nobel Lectures in Physics 1901–1921* (New York: Elsevier) p 407
[101] Planck M, letter to R W Wood, 7 October 1931 *Am. Inst. Phys. Archives, New York*
[102] Einstein A 1916 *Verh. Deutsch. Phys. Ges.* **18** 318

[103] Benz U 1975 *Arnold Sommerfeld* (Stuttgart: Wissenschaftliche Verlagsgesellschaft) p 74
[104] Einstein A 1949 autobiographical notes in *Albert Einstein, Philosopher-Scientist* ed P A Schilpp (New York: Tudor)
[105] For the evolution of the photon concept see reference [97], ch 21
[106] Reference [97] ch 19 section(f) and ch 21 section (f)
[107] Kirsten G and Körber H 1975 *Physiker über Physiker* (Berlin: Akademie) p 201
[108] Einstein A 1944 *Berliner Tageblatt* April 20
[109] Planck M 1910 *Phys. Z.* **11** 922
[110] For more details on Bohr's family background see Pais A 1991 *Niels Bohr's Times* (Oxford: Oxford University Press) ch 2
[111] Reference [9] vol 2, p 241
[112] Bohr N 1909 *Trans. R. Soc.* **209** 281; reprinted in *Niels Bohr, Collected Works* vol 1 (Amsterdam: North-Holland) p 29
[113] 1952 *Niels Bohr, Collected Works* vol 1 (Amsterdam: North-Holland) p 167
[114] Reference [113] p 291
[115] Rutherford E 1911 *Phil. Mag.* **21** 669
[116] 1952 *Niels Bohr, Collected Works* vol 2 (Amsterdam: North-Holland) p 136
[117] Kirchhoff G and Bunsen R 1860 *Ann. Phys. Chem.* **110** 160; also 1861 *Ann. Phys. Chem.* **113** 337 (Engl. Transl. 1861 *Phil. Mag.* **20** 89; 1862 *Phil. Mag.* **22** 329–448)
[118] Reference [81] pp 168–170
[119] Kayser H G J 1900–1912 *Handbuch der Spektroskopie* 6 vols (Leipzig: Hirzl) (more volumes appeared later)
[120] Ångström A 1872 *Ann. Phys. Chem.* **144** 300
[121] Plücker J 1859 *Ann. Phys. Chem.* **107** 497, 637
[122] Ångström A 1868 *Recherches sur le Spectre Solaire* (Uppsala: Uppsala University Press)
[123] Stokes G G 1901 *Mathematical and Physical Papers* vol 3 (Cambridge: Cambridge University Press) p 267
[124] Reference [9] vol 2, p 445
[125] Thomson J J 1906 *Phil. Mag.* **11** 769
[126] Stark J 1911 *Prinzipien der Atomdynamik* vol 2 (Leipzig: Hirzl) sections 19, 25
[127] Bohr N, interviewed by T S Kuhn, L Rosenfeld, E Rüdinger and A Peters, 7 November 1962 transcript in Niels Bohr, Archive, Copenhagen
[128] For details, see reference [119] vol 1, pp 123–7; reference [81] pp 171–4
[129] Konen H 1913 *Das Leuchten der Gase und Dämpfe* p 71ff (Braunschweig: Vieweg)
[130] Hagenbach A 1921 *Naturwissenschaften* **9** 451
[131] Balmer J 1885 *Verh. Naturf. Ges. Basel* **7** 548
[132] Zhao P, Lichten W, Layer H and Bergquist J 1987 *Phys. Rev. Lett.* **58** 1293
[133] Balmer J 1885 *Verh. Naturf. Ges. Basel* **7** 548
[134] Schuster A 1911 *Encyclopaedia Britannica* (London: Encyclopaedia Britannica Company Ltd) entry 'Spectroscopy'
[135] Reference [113] p 175
[136] Maxwell J C 1952 *Collected Works* vol 1 (Amsterdam: North-Holland) p 300
[137] Bohr N, letter to E Rutherford 31 January 1913, see reference [116] p 579
[138] Bohr N, letter to G von Hevesy 7 February 1913, see reference [116] p 529
[139] Stark J in reference [126] vol 2, p 44
[140] Reference [127] interviews on 31 October and 7 November 1962

[141] Haas A 1910 *Wiener Ber. IIa* **119** 119; 1910 *Jahrb. Rad. Elektr.* **7** 261; 1910 *Phys. Z.* **11** 537
[142] Sommerfeld A 1912 *Théorie de Rayonnement et les Quanta* (Paris: Gauthier-Villars) p 362
[143] Bohr N 1913 *Phil. Mag.* **26** 1, see reference [116] p 159
[144] Nicholson J W 1911 *Phil. Mag.* **22** 864
[145] Nicholson J W 1912 *Mon. Not. R. Astron. Soc.* **72** 677
[146] Bjerrum N 1912 *Nernst Festschrift* p 90 (Halle: von Knapp) (Engl. Transl. Bjerrum N *Selected Papers* (Copenhagen: Munksgaard) p 34)
[147] Nernst W 1911 *Z. Elektrochem.* **17** 265
[148] Bohr N, letter to E Rutherford 6 March 1913, see reference [116] p 581
[149] Reference [116] section 2
[150] Reference [116] section 3
[151] Bohr N 1914 *Fysisk Tidsskr.* **12** 917 see reference [116] p 303)
[152] Fowler A 1913 *Nature* **92** 95
[153] Bohr N 1913 *Nature* **92** 231, see reference [116] p 273
[154] Bohr N 1913 *Phil. Mag.* **26** 476, see reference [116] p 187
[155] Bohr N 1913 *Phil. Mag.* **26** 857, see reference [116] p 215
[156] Ritz W 1908 *Phys. Z.* **9** 521
[157] Moseley H G J 1913 *Nature* **92** 554; 1913 *Phil. Mag.* **26** 1024; 1914 *Phil. Mag.* **27** 703
[158] Kramers H A and Holst H 1923 *The Atom and the Bohr Theory of its Structure* (New York: Knopf) p 139
[159] Rutherford E, letter to N Bohr 20 March 1913, see reference [116] p 583
[160] Strutt R J 1968 *Life of John William Strutt, Third Baron Rayleigh* (Madison, WI: University of Wisconsin Press) p 357
[161] Moseley H G J, letter to N Bohr 16 November 1913, see reference [116] p 544
[162] Sommerfeld A, letter to N Bohr 4 September 1913, see reference [116] p 603
[163] Gerlach W interview with T S Kuhn 18 February 1963 (transcript in Niels Bohr Archive)
[164] Pohl R W interview with T S Kuhn and F Hund 25 June 1963 (transcript in Niels Bohr Archive)
[165] von Hevesy G, letter to N Bohr 23 September 1913, see reference [116] p 532
[166] Bohr N, letter to E Rutherford 6 September 1916 (transcript in Niels Bohr Archive)
[167] For the United States see especially K R Sopka 1980 *Quantum Physics in America, 1920–1935* (New York: Arno)
[168] Heisenberg W 1984 *Gesammelte Werke* part C, vol 1 (Munich: Piper) p 4
[169] Stark J 1913 *Sitz. Ber. Preuss. Akad. Wiss.* 932; 1913 *Nature* **92** 401
[170] Warburg E 1913 *Verh. Deutsch. Phys. Ges.* **15** 1259
[171] Epstein P S 1916 *Phys. Z.* **17** 148
Schwarzschild K 1916 *Sitz. Ber. Preuss. Akad. Wiss.* 548
[172] Franck J and Hertz G 1914 *Verh. Deutsch. Phys. Ges.* **16** 457
[173] Michelson A A and Morley E W 1887 *Phil. Mag.* **24** 463
[174] Sommerfeld A 1916 *Sitz. Ber. Bayer. Akad. Wiss.* 425, 459; in more detail in 1916 *Ann. Phys.* **51** 1, 125
[175] Bohr N 1915 *Phil. Mag.* **29** 332, see reference [116] p 377
[176] de L Kronig R 1960 *Theoretical Physics in the Twentieth Century* ed M Fierz and V Weisskopf (New York: Interscience) p 50
[177] Compare with Klein M 1970 *Paul Ehrenfest* (Amsterdam: North-Holland) ch 10

[178] Einstein A 1950 *Out of my Later Years* (New York: Philosophical Library) p 236
[179] Complete references are found in Ehrenfest P 1923 *Naturwissenschaften* **11** 543
[180] Einstein A 1916 *Verh. Deutsch. Phys. Ges.* **18** 318; 1916 *Mitt. Phys. Ges. Zürich* **16** 47; 1917 *Phys. Z.* **18** 121
[181] Einstein A, letter to M Born 27 January 1920 reprinted in 1971 *The Born–Einstein Letters* (New York: Walker) p 23
[182] Bohr N 1918 *Kong. Dansk. Vid. Selsk. Skrifter* 1; 1918 *Kong. Dansk. Vid. Selsk. Skrifter* 37; 1922 *Kong. Dansk. Vid. Selsk. Skrifter* 101 reprinted in 1952 *Niels Bohr, Collected Works* vol 3 (Amsterdam: North-Holland) pp 67, 103, 167)
[183] Rubinowicz A 1918 *Naturwissenschaften* **19** 441, 465
[184] 1952 *Niels Bohr, Collected Works* vol 3 (Amsterdam: North-Holland) p 19
[185] See, for example, Heilbron J L 1977 *Lectures on the History of Atomic Physics 1900–1920* (New York: Academic) p 40
[186] Kossel W 1916 *Ann. Phys.* **49** 229
[187] Kossel W and Sommerfeld A 1919 *Verh. Deutsch. Phys. Ges.* **20** 240
[188] For the history of the rare earths see van Spronsen J W 1969 *The Periodic Table* (New York: Elsevier)
[189] Bohr N 1921 *Nature* **107** 104; 1921 *Nature* **108** 208, reprinted in 1952 *Niels Bohr, Collected Works* vol 4 (Amsterdam: North-Holland) pp 71, 175
[190] 1952 *Niels Bohr, Collected Works* vol 4 (Amsterdam: North-Holland) p 341
[191] Kramers H A 1935 *Fys. Tidsskr.* **33** 82
[192] Bohr N 1923 *Ann. Phys.* **71** 228, see reference [190] p 611
[193] Sommerfeld A 1924 *Atombau und Spektrallinien* 4th edn (Braunschweig: Vieweg) p VI
[194] Pauli W 1964 *Collected Scientific Papers* vol 1 (New York: Interscience) p 1
[195] Einstein A 1922 *Naturwissenschaften* **10** 184
[196] Pauli W 1958 *Relativity Theory* (Engl. Transl. G Field) (London: Pergamon)
[197] Pauli W reference [194] p 1073
[198] For more on hafnium see reference [185]
[199] Reference [190] p 405
[200] Coster D 1923 *Physica* **5** 133
von Hevesy G 1951 *Arch. Kemi* **3** 543
[201] Coster D and von Hevesy G 1923 *Nature* **111** 79
[202] See Robertson P 1979 *The Early Years* (Copenhagen: Akademisk Forlag)
[203] Sommerfeld A 1920 *Ann. Phys., Lpz.* **63** 221
[204] Landé A 1921 *Z. Phys.* **5** 231
[205] Pauli W 1925 *Z. Phys.* **31** 373
[206] Uhlenbeck G E and Goudsmit S A 1925 *Naturwissenschaften* **13** 953; 1926 *Nature* **117** 264
[207] For details see Pais A 1991 *Niels Bohr's Times* (Oxford: Oxford University Press) p 241
[208] Houston W 1927 *Proc. Natl Acad. Sci. USA* **13** 91
Hansen G 1927 *Nature* **119** 237
[209] Heisenberg W 1926 *Z. Phys.* **39** 499
[210] Houston W 1927 *Proc. Natl Acad. Sci. USA* **13** 91
[211] Kramers H A 1923 *Z. Phys.* **13** 312
[212] Reference [192]; see especially reference [190] p 643, footnote
[213] Sommerfeld A 1923 *Rev. Sci. Instrum.* **7** 509
[214] Zeeman P 1897 *Phil. Mag.* **43** 226; 1897 *Phil. Mag.* **55** 255
[215] Lorentz H A 1897 *Ann. Phys., Lpz.* **63** 279
[216] Lorentz H A 1921 *Physica* **1** 228

Introducing Atoms and Their Nuclei

[217] For more historical details see Pais A 1986 *Inward Bound* (Oxford: Oxford University Press) ch 4
[218] Sommerfeld A 1916 *Ann. Phys., Lpz.* **51** 125
[219] Debye P 1916 *Phys. Z.* **17** 507
[220] Reference [182] part of 1918
[221] Reference [182] part of 1922, appendix
[222] Stern O and Gerlach W 1922 *Phys. Z.* **23** 476
[223] Reference [193] 1919 1st edn, p 439
[224] Pauli W, reference [194] p 437
[225] For the history of the origins of quantum statistics see Pais A 1982 *Subtle is the Lord* (Oxford: Oxford University Press) ch 23
[226] Bragg W and Kleeman R 1905 *Phil. Mag.* **10** 318
[227] Soddy F 1975 *Radioactivity and Atomic Theory* ed J Trenn (New York: Wiley) p 80
[228] Hahn O and Meitner L 1908 *Phys. Z.* **9** 321
[229] Hahn O and Meitner L 1908 *Phys. Z.* **9** 697
[230] Hahn O and Meitner L 1909 *Phys. Z.* **10** 741
[231] Wilson W 1909 *Proc. R. Soc.* A **82** 612; 1910 *Proc. R. Soc.* **84** 141
[232] von Baeyer O and Hahn O 1910 *Phys. Z.* **11** 488
[233] von Baeyer O, Hahn O and Meitner L 1911 *Phys. Z.* **12** 273
[234] Hahn O 1968 *A Scientific Autobiography* (Engl. Transl. W Ley) (New York: Scribner) p 57
[235] Rutherford E and Robinson H 1913 *Phil. Mag.* **26** 717
[236] Hahn O, letter to E Rutherford 11 January 1911, on microfilm at the Niels Bohr Library, American Institute of Physics, New York
[237] Rutherford E 1913 *Radioactive Substances and their Radiations* (Cambridge: Cambridge University Press) p 256
[238] Müller J and Pouillet C 1914 *Lehrbuch der Physik* 10th edn, vol 4, pp 1272–4 (Braunschweig: Vieweg)
[239] Ellis C D 1921 *Proc. R. Soc.* A **99** 261
[240] Ellis C D 1922 *Proc. R. Soc.* A **101** 1
[241] Reference [237] p 622
[242] Reference [154] section 6
[243] Soddy F 1913 *Nature* **92** 400
[244] Chadwick J 1914 *Verh. Deutsch. Phys. Ges.* **16** 383
[245] Rutherford E, Chadwick J and Ellis C D 1930 *Radiations from Radioactive Substances* (Cambridge: Cambridge University Press)
[246] The Rutherford–Chadwick correspondence is found on microfilm in the Niels Bohr Library, American Institute of Physics, New York
[247] Rutherford E 1914 *Phil. Mag.* **27** 488
[248] See 1920 *Nature* **106** 357
[249] Rutherford E 1914 *Proc. R. Soc.* A **90** insert after p 462
[250] Rutherford E, letter to B Boltwood 28 February 1921 (reprinted in Badash L 1969 *Rutherford and Boltwood* (Newhaven, CT: Yale University Press) p 343)
[251] Rutherford E 1914 *Scientia* **16** 337
[252] Reference [223] p 540
[253] Einstein A 1905 *Ann. Phys., Lpz.* **18** 639
[254] Einstein A 1907 *Jahrb. Rad. Elektr.* **4** 411
[255] Einstein A 1910 *Arch. Sci. Phys. Nat.* **29** 5, 125, see especially p 144
[256] Planck M 1906 *Verh. Deutsch. Phys. Ges.* **4** 136; 1908 *Ann. Phys.* **26** 1
[257] Langevin P 1913 *J. Physique* **3** 553
[258] Lenz W 1920 *Naturwissenschaften* **8** 181
[259] Reference [196] p 123

[260] Aston F W 1927 *Proc. R. Soc.* A **115** 487
[261] Bainbridge K T 1932 *Phys. Rev.* **42** 1
[262] Cockcroft J and Walton E 1932 *Proc. R. Soc.* A **137** 229
[263] Braunbek W 1937 *Z. Phys.* **107** 1
[264] Rutherford E 1919 *Phil. Mag.* **37** 537
[265] Rutherford E 1919 *Phil. Mag.* **37** 581
[266] Wilson D 1983 *Rutherford* (Cambridge, MA: MIT Press) p 405
[267] *New York Herald Tribune* 12 September 1933
[268] Rutherford E 1919 *Phil. Mag.* **37** 571
[269] Chadwick J and Bieler E S 1921 *Phil. Mag.* **42** 923
[270] Gamow G 1931 *Constitution of Atomic Nuclei and Radioactivity* (Oxford: Clarendon)
[271] Casimir H 1983 *Haphazard Reality* (New York: Harper and Row) p 117
[272] Reference [270] p 5
[273] Gamow G 1928 *Z. Phys.* **51** 204
[274] Gurney R W and Condon E 1928 *Nature* **122** 439; 1929 *Phys. Rev.* **33** 127
[275] Geiger H and Nuttall J M 1912 *Phil. Mag.* **23** 439
[276] da Costa Andrade C N 1923 *Proc. Phys. Soc.* **36** 202
[277] Bethe Cf H A and Bacher R 1936 *Rev. Mod. Phys.* **8** 82 sections 3, 38
[278] de L Kronig R 1926 *Nature* **117** 550
[279] For these and further details see Herzberg G 1950 *Molecular Spectra and Molecular Structure* 2nd edn (New York: van Nostrand) p 169
[280] Ornstein L S and van Wyk W R 1928 *Z. Phys.* **49** 315
[281] de L Kronig R 1928 *Naturwissenschaften* **16** 335; 1930 *Naturwissenschaften* **18** 205
[282] Rasetti F 1929 *Proc. Natl Acad. Sci. USA* **15** 515; 1929 *Nature* **123** 757; 1929 *Phys. Rev.* **34** 367
[283] Heitler W and Herzberg G 1929 *Naturwissenschaften* **17** 673
[284] These are discussed in more detail in Pais A 1986 *Inward Bound* (Oxford: Oxford University Press) ch 14
[285] Meitner L 1922 *Z. Phys.* **9** 131
[286] Chadwick J and Ellis C D 1922 *Proc. Cambridge Phil. Soc.* **21** 274
[287] Ellis C D and Wooster W A 1925 *Proc. Cambridge Phil. Soc.* **22** 849
[288] Ellis C D and Wooster W A 1927 *Proc. R. Soc.* A **117** 109
[289] Frisch O 1970 *Biogr. Mem. Fell. Roy. Soc.* **16** 408
[290] Meitner L and Orthmann W 1930 *Z. Phys.* **60** 143
[291] Bethe H A 1979 *Nuclear Physics in Retrospect* ed R H Stuewer (Minneapolis, MN: University of Minnesota Press) p 99
[292] Bothe W and Becker H 1930 *Naturwissenschaften* **18** 705
[293] Curie I and Joliot F 1932 *C. R. Acad. Sci., Paris* **194** 273
[294] Curie I and Joliot F 1932 *C. R. Acad. Sci., Paris* **194** 708
[295] Chadwick J 1932 *Nature* **129** 312
[296] Curie I and Joliot F 1934 *C. R. Acad. Sci., Paris* **198** 254; 1934 *Nature* **133** 201
[297] Curie I and Joliot F 1934 *C. R. Acad. Sci., Paris* **198** 559
[298] Fermi E 1934 *Ric. Scient.* **5** 283 (Engl. Transl. 1962 *E Fermi's Collected Papers* vol 1 (Chicago, IL: University of Chicago Press) p 674)
[299] Perrin F 1932 *C. R. Acad. Sci., Paris* **194** 1343
[300] Auger P 1932 *C. R. Acad. Sci., Paris* **194** 1346
[301] Dirac P A M 1934 *Proc. Seventh Solvay Conf.* (Paris: Gauthier Villars) p 328
[302] Iwanenko D 1932 *Nature* **129** 798
[303] Chadwick J 1932 *Proc. R. Soc.* A **136** 735
[304] Iwanenko D 1932 *C. R. Acad. Sci., Paris* **195** 439

Introducing Atoms and Their Nuclei

[305] Heisenberg W 1932 *Z. Phys.* **77** 1 (Engl. Transl. Brink D M 1965 *Nuclear Forces* (Oxford: Pergamon) p 144)
[306] Heisenberg W 1932 *Z. Phys.* **78** 156
[307] Heisenberg W 1933 *Z. Phys.* **80** 587 (Engl. Transl. Brink D M 1965 *Nuclear Forces* (Oxford: Pergamon) p 155)
[308] Heisenberg W, letter to N Bohr 20 June 1932 (Niels Bohr Archive)
[309] Bohr N 1936 *Nature* **137** 344, reprinted in 1952 *Niels Bohr, Collected Works* vol 9 (Amsterdam: North-Holland) p 152
[310] Breit G and Wigner E P 1936 *Phys. Rev.* **49** 519
[311] Mottelson B 1986 *The Lessons of Quantum Theory* ed J de Boer *et al* (New York: Elsevier) p 79
[312] Bohr N and Kalckar F 1937 *Dansk. Vid. Selsk. Mat.-Fys. Medd* **14** no 10; reprinted in 1952 *Niels Bohr, Collected Works* vol 9 (Amsterdam: North-Holland) p 223
[313] Pauli W, letter to O Klein 18 February 1929 (reprinted 1979 *Pauli Scientific Correspondence* vol 1 (New York: Springer) p 488
[314] Pauli W, letter to N Bohr 5 March 1929 reference [313] p 493
[315] Bohr N 1932 *J. Chem. Soc.* **135** 349, reprinted in 1952 *Niels Bohr, Collected Works* vol 9 (Amsterdam: North-Holland) p 91
[316] Pauli W 1964 *Collected Scientific Papers* vol 2, p 1313
[317] For these see reference [217] pp 316–320
[318] Pauli W, letter to P M S Blackett 19 April 1933 reference [313] vol 2, p 158
[319] 1934 *Structure et Propriétés des Noyaux Atomiques* (Paris: Gauthier-Villars) pp 324–5
[320] Reference [319] p 327
[321] Reference [319] p 328
[322] Rasetti F in reference [298] vol 1, p 450
[323] Reference [298] vol 1, p 538
[324] Reference [298] vol 1, pp 559, 575
[325] Perrin F 1933 *C. R. Acad. Sci., Paris* **397** 1625
[326] Fermi E 1934 *Nature* **133** 898
[327] Fermi E 1965 *Nobel Lectures in Physics, 1922–1941* (New York: Elsevier) p 414
[328] Hahn O, Meitner L and Strassmann F 1938 *Naturwissenschaften* **26** 475
[329] Hahn O 1966 *Mein Leben* (Munich: Bruckmann) p 150; 1966 *A Scientific Autobiography* (New York: Scribner's)
[330] Hahn O and Strassmann F 1939 *Naturwissenschaften* **27** 11 (Engl. Transl. Graetzer H G 1964 *Am. J. Phys.* **32** 9)
[331] Frisch O 1967 *Phys. Today* **20** November 43
[332] Meitner L and Frisch O 1939 *Nature* **143** 239
[333] Roberts R B *et al* 1939 *Phys. Rev.* **55** 416
[334] Reviewed by Turner L A 1940 *Rev. Mod. Phys.* **12** 1
[335] Wheeler J A 1967 *Phys. Today* **20** November 49
[336] Bohr N 1939 *Phys. Rev.* **55** 418, reprinted in 1952 *Niels Bohr, Collected Works* vol 9 (Amsterdam: North-Holland) p 343
[337] 1952 *Niels Bohr, Collected Works* vol 9 (Amsterdam: North-Holland) p 66
[338] Nier A O *et al* 1940 *Phys. Rev.* **57** 546, 748
[339] Bohr N and Wheeler J A 1939 *Phys. Rev.* **56** 426, 1056, see also reference [337] pp 363, 403
[340] Bethe H A reference [291] p 11
[341] Flügge S 1939 *Naturwissenschaften* **27** 492
[342] Harrington J 1939 *Sci. Am.* **161** October 214

Chapter 3

QUANTA AND QUANTUM MECHANICS

Helmut Rechenberg

3.1. Introduction

Before Max Planck entered his studies at the University of Munich in 1874, he asked Philipp von Jolly, Professor of Physics, about the prospects in his field; he obtained the answer that physics was more or less a completed science offering little future prospect. Planck, fortunately, was not put off by this pessimistic view. New facts were soon discovered, e.g. electromagnetic waves (which *had* been anticipated), as well as some phenomena that did not all fit into the scheme of physics known in the late nineteenth century; namely, x-rays, radioactivity and the electron. Even a well-established topic like the kinetic theory of gases, if studied in more detail, revealed serious difficulties, such as those addressed by Lord Rayleigh (1842–1919), who remarked at the turn of the century [1]:

> The difficulties connected with the application of the law of equal partition of energy to actual gases have long been felt ... The value [for the ratio of specific heats] applicable to the principal diatomic gases gives room for three kinds of translation and for two kinds of rotation. Nothing is left for the rotation round the line joining the atoms, nor for the relative motion of the atoms in this line. Even if we regard the atoms as mere points, whose rotation means nothing, there still must exist energy of the last-mentioned kind, and its amount (according to the law) should not be inferior. We are here brought face to face with a fundamental difficulty, relating not to the theory of gases merely, but rather to general dynamics.

Lord Kelvin (1824–1907) called this fundamental difficulty 'a cloud that has obscured the brilliance of the molecular theory of heat and

Max Planck

(German, 1858–1947)

Max Planck, the creator of quantum theory, served for 48 years as academic teacher of theoretical physics at the universities of Munich (1880–85), Kiel (1885–89) and Berlin (1889–1927). He also eagerly participated in the work of the Prussian Academy of Sciences, being elected Fellow in 1894 and acting as permanent secretary from 1912 to 1943. As president of the Kaiser Wilhelm-Gesellschaft (1930–37, 1945–46) he defended German science in the difficult times of the Third Reich and the first post-war years.

Planck acquired a professional reputation with a thorough analysis of the second heat theorem, which he also applied in pioneering papers on the dissociation processes in dilute solutions. Thermodynamic reasoning then opened the path to his law of black-body radiation and quantum theory. Later he extended Einstein's special relativity to mechanics and thermodynamics (1907), but mainly he contributed to quantum theory by using the phase-space approach (zero-point energy 1911, many-degrees of freedom systems 1915, statistics of identical particles 1925). When the new quantum mechanics emerged, he argued vigorously in favour of a causal rather than a probability interpretation.

---

light during the last quarter of the nineteenth century' [2]. This statement was made in April 1900, and a couple of months later Lord Rayleigh treated an interesting problem in that theory, namely the law of what he called 'complete radiation' and others 'black-body' or 'cavity radiation', which at that time was studied experimentally at the Physikalisch-Technische Reichsanstalt in Berlin [3]. Less than half a year later Max Planck, professor of theoretical physics at the University of Berlin, performed *the* crucial step in dissipating Lord Kelvin's cloud: in order to explain the observed black-body radiation intensity, he introduced a concept that gave rise to the quantum theory, which has revolutionized our understanding of Nature. Werner Heisenberg, in 1934, called the quantum theory one of the greatest adventures of

See also p 67

mankind, comparable to Columbus' discovery of America [4], but it took about three decades to evolve, as the details were very tricky. Friedrich Hund (b 1896), at present the oldest surviving witness and contributor to that spiritual adventure, in 1975 asked: 'Could the history of quantum theory have gone by a different path?'. After having analysed various possible starting points and developments—chemical experience, hard (x-)radiation, low-temperature phenomena, spectral laws of atoms and molecules, and matter–wave interferences—he arrived at this conclusion [5]:

> In looking back we state that impressive and quantitatively informative experiments clearly contradicting classical concepts were necessary to give rise to quantum mechanics. The experiences from chemistry were impressive but little quantitative, the experiences from hard radiation were likewise not very quantitative, nor were the phenomena at low temperatures. The spectra ... and also the cavity radiation possessed a high degree of information.
>
> *That quantum theory should have begun with either cavity radiation or spectral laws, seems 'necessary'; Planck's interest in general questions and his understanding of entropy seemed to have 'accidentally' coincided with the measurements at the Physikalisch-Technische Reichsanstalt in Charlottenburg. The lateness of the study of spectral laws, and perhaps of the conception of matter waves, also appears to be accidental.*

In this chapter the complex story of quantum theory, in which many fields of physical sciences become interwoven, has been organized into three parts: (i) the empirical foundations (1900–28); (ii) the origin and completion of quantum mechanics (1913–29); (iii) the world of microphysics (1925–35). They focus, respectively, on three main aspects: the immediate empirical evidence, the mathematical–physical theory, and the interpretation of atomic and molecular phenomena. More details of the historical development of quantum theory may be found in other books [6].

3.2. Quanta—the empirical foundations (1900–28)

In his presidential address opening the first Solvay Congress at Brussels on 30 October 1911, Hendrik Antoon Lorentz (1853–1928) drew attention to the 'serious difficulties' of the kinetic theory, describing the motion of the smallest constituents of matter. He said [7]:

> In this stage of affairs there appeared to us like a wonderful ray of light the beautiful *hypothesis of energy elements* which was first expounded by Mr *Planck* and then extended by Messrs *Einstein*, *Nernst*, and others to many phenomena. It has opened for us unexpected vistas; even those, who consider it with a certain

suspicion, must admit its importance and fruitfulness. Hence it well deserves to be the main subject of our discussions, and certainly also the author of this new hypothesis and those, who have contributed to its development, deserve our sincere gratitude.

Lorentz did not fail to mention that the new theories derived from Planck's hypothesis by no means possessed the exactness and conceptual clarity of the former classical theories; still, a thorough analysis of the various quantum phenomena, separating essential from accidental facts, appeared to be necessary.

The first Solvay Congress of autumn 1911 (see figure 3.1) reviewed the ideas of quanta and their applications since 1900, and the publication of the reports and discussions were vital in propagating these ideas to a larger scientific public, especially outside Germany. Thus the investigation of quantum phenomena became a major occupation of physicists and physical chemists in Europe and America during the following decades. An increasing number of classically non-explainable effects was discovered in atomic and molecular physics, which provided a most impressive and solid basis for the 'new mechanics'—called for in 1911 by Lorentz and named 'quantum mechanics' by Arnold Eucken (1884–1952), the editor of the German proceedings of the Solvay Congress [8].

We can divide the main period of the search for specifically quantum phenomena into three periods. During the years from 1900 to 1913, effects were found in many physical and chemical problems that could be related to or explained by Planck's hypothesis, although no general theoretical scheme connected the various quanta involved in these phenomena (section 3.2.1). The empirical laws of spectra, whose recognition had started far back in the nineteenth century, gave rise, in the years after 1912, to a more or less systematic theoretical organization into a semi-classical quantum scheme, the (empirical) principles of which are described in section 3.2.2. In the last period, i.e. roughly from 1922 to 1928, several crucial discoveries yielded the final proof that no extension of the classical theories, however clever, was able to account for the essential features of atomic and radiation phenomena (section 3.2.3).

3.2.1. Radiation and quanta (1900–13)

It is often said that revolutionary new science is most likely to be produced by young, fresh minds, people just beginning their professional careers. The case of quantum theory does not support this opinion. Max Planck (1858–1947), professor of theoretical physics at the University of Berlin since 1889 (earlier: PhD 1879 and *Habilitation* 1880 in Munich, and *Extraordinarius* 1895 in Kiel), had worked successfully on problems of thermodynamics and electrolytic dissociation, before turning in 1896 to deal with black-body radiation. When he introduced the quantum

Quanta and Quantum Mechanics

Figure 3.1. *The First Solvay Conference (Brussels, 30 October to 3 November 1911). Standing, from left to right: Goldschmidt, Planck, Rubens, Sommerfeld, Lindemann, De Broglie, Knudsen, Hasenöhrl, Hostelet, Herzen, Jeans, Rutherford, Kamerlingh Onnes, Einstein, Langevin. Seated, left to right: Nernst, M Brillouin, Solvay, Lorentz, Warburg, Perrin, Wien, Curie, Poincaré.*

hypothesis, he was 42 years old, and he followed the development of quantum theory actively for the next three decades. After retiring from his chair in 1927, he became President of the main German research organization, the Kaiser Wilhelm-Gesellschaft (1930–37, 1945–46).

The first advocate of the quantum concept, Albert Einstein (1879–1955), however, fitted perfectly the standard opinion. Upon studying at the Swiss Polytechnical Institute (ETH) in Zurich (1896–1900), he took a position at the Swiss Patent Office in 1902. While completing his doctoral dissertation, he published in 1905 three outstanding papers: the best known was on the theory of special relativity, another was on the kinetic explanation of Brownian motion, and the first submitted introduced the light-quantum hypothesis. Einstein continued to work on these three important topics (i.e. on relativity theory, quantum theory and statistical mechanics) during the following 20 years: after that he restricted himself to creating a unified classical field theory of matter and radiation. Though his scientific contributions were quickly acknowledged, notably by Planck, he entered the academic ladder only in 1909 with an extraordinary professorship for theoretical physics at the University of Zurich; from there he moved as full professor to Prague (1911), back to Zurich (1912 ETH) and then Berlin (1914), where he also became director of the newly established Kaiser Wilhelm-Institut für Physik (1917–33). The ascendency of the Nazis in Germany forced him to emigrate to the USA, and he worked from 1933 to his death in 1955 at the Institute for Advanced Study in Princeton, New Jersey.

Planck's and Einstein's ideas were picked up and extended by two further pioneers of quantum theory. The senior of them, Walther Nernst (1864–1941), had studied from 1883 to 1887 in Zurich, Berlin, Graz and Würzburg (with Ludwig Boltzmann, Fritz Kohlrausch and Svante Arrhenius), then collaborated with Wilhelm Ostwald in Leipzig, until he came to Göttingen (professor from 1891) and Berlin (1905) to occupy chairs in physical chemistry. With a high reputation as one of the fathers of theoretical chemistry and electrochemistry and a discoverer of experimental and theoretical laws—e.g. of what he called 'the third law of thermodynamics' (1905)—he began to deal with quantum effects at low temperatures around 1910. He remained a stout adherent and contributor to quantum theory throughout his life; while changing position twice: in 1922 to be President of the Physikalisch-Technische Reichsanstalt (PTR) and in 1924 to succeed Heinrich Rubens in the Berlin University chair of experimental physics (1933 emeritus).

At the end of his scientific career Johannes Stark (1874–1957) also directed the PTR (1933–39). When he first joined the quantum physicists, he had just left a dozentship in Göttingen (since 1900) to accept a more promising one at the Technische Hochschule (TH) Hannover (1906). Three years later he became full professor in Aachen, later in Greifswald (1917) and finally in Würzburg (1920–22). Extremely versatile, the young Stark worked on the most current topics of experimental and theoretical physics. In 1905 he won great esteem by demonstrating the Doppler effect with canal rays, and towards the end of 1913 he found the splitting of spectral lines in electric fields. Between these dates he promoted quantum concepts more boldly than anybody else. His ambitious, at times superficial, theoretical explanations brought him into conflict with his formerly admired colleagues, such as Arnold Sommerfeld and Einstein. Stark turned into a stubborn enemy of the old quantum theory of atomic structures of Niels Bohr and Sommerfeld, and he later opposed modern atomic theory or quantum mechanics, which he denounced (after 1933) as 'Jewish physics'.

3.2.1.1. Heat or black-body (cavity) radiation [9] *(Planck 1900).* In 1860, Gustav Kirchhoff (1824–87) of Heidelberg University proposed the existence of a 'completely black body', which absorbs and emits *all* radiation falling on it [10]. The ratio of emissivity to absorptivity ρ_λ, called 'Kirchhoff's function' and identical to the energy density of black-body radiation, should depend only on the wavelength λ and the absolute temperature T; it was determined with increasing accuracy by using platinum wire blackened by carbon, until Friedrich Paschen (1865–1947) at the TH Hannover obtained the empirical relation [11]

$$\rho_\lambda = c_1 \lambda^{-\alpha} \exp\left(-\frac{c_2}{\lambda T}\right) \qquad (1)$$

with c_1, c_2 and α denoting constants and $\alpha \simeq 5$. Meanwhile, the theory of black-body radiation had been developed: thus Ludwig Boltzmann (1844–1906) combined electrodynamics and thermodynamics in 1884 to derive the temperature dependence of the integrated intensity ($\rho =$ constant T^4, Stefan's law 1879); Willy Wien (1864–1928) used in addition relations of kinetic gas theory (Maxwell–Boltzmann velocity distribution) to obtain (without knowing Paschen's result!) the law (1) [12].

From 1897 to 1899 Planck developed a systematic theory of black-body radiation, based on the absorption and emission of electrodynamic radiation by the 'resonators' of the body and on the laws of thermodynamics [13]. From electrodynamics he derived the relation between the average energy of a resonator U_ν and the density ρ_ν of heat radiation (c is the velocity of light *in vacuo* and ν is the frequency)

$$U_\nu = \frac{c^3}{8\pi \nu^2} \rho_\nu. \qquad (2)$$

Thermodynamics led to a relation for the entropy S of the resonator (of frequency ν, which we suppress from now on in the formulae)

$$\frac{d^2 S}{dT^2} = -\frac{\text{constant}}{U}. \qquad (3)$$

Upon integrating with respect to T and inserting for the resonator energy U, according to equation (2), Planck arrived at the Paschen–Wien law (1), which he considered to be the one singled-out by theory. However, less than a year later improved measurements at the PTR by Otto Lummer (1860–1925) and Ernst Pringsheim (1859–1917), and especially by Heinrich Rubens (1865–1922) and Ferdinand Kurlbaum (1857–1927)—performed with the most perfect black body, a 'cavity' proposed earlier by Wien and Lummer [14]—could be fitted by a different radiation law, namely [15]

$$\rho_\lambda = \frac{c_1}{\lambda^5} \frac{1}{\exp(c_2/\lambda T) - 1}. \qquad (4)$$

Exactly this law resulted if Planck generalized his expression (3) for the entropy to

$$\frac{d^2 S}{dT^2} = -\frac{\alpha}{U(\beta + U)} \qquad (5)$$

with two constants α and β [16]. He proposed this on 19 October 1900 at a Berlin meeting of the German Physical Society, but his argumentation was arbitrary and needed deeper justification.

Planck later, in a letter of 7 October 1931 to Robert W Wood (1868–1955), characterized 'the whole procedure as an act of despair', because 'a theoretical interpretation had to be found at any price, however high it might be'. The price was to introduce considerations of statistical mechanics (which he had avoided so far), notably Boltzmann's relation between entropy S and probability P of 1877, which Planck now wrote for his resonator case as

$$S = k \log P = k \left[\left(\frac{U}{\varepsilon} + 1 \right) \log \left(\frac{U}{\varepsilon} + 1 \right) - \frac{U}{\varepsilon} \log \frac{U}{\varepsilon} \right] \qquad (6)$$

where k is now known as the Boltzmann constant. According to the statistical interpretation, the energy packets ε were distributed among the resonators such that for the equilibrium situation the probability P assumed a maximum. Planck then derived the empirical black-body law (4) (see also figure 3.2), provided he assumed *finite* energy packets of size

$$\varepsilon = h\nu \qquad (7)$$

where ε was identical with β of equation (5) and the constant h assumed the experimental value

$$h = \frac{c_1}{c_2} = 6.55 \times 10^{-27} \text{ erg s}. \qquad (8)$$

Since h had the dimension of an 'action', he later called it the 'quantum of action' (*Wirkungsquantum*).

Planck presented the above consideration on 14 December 1900, again at a meeting of the Berlin Physical Society, which became the birthday of quantum theory [17]. He also calculated there for the first time the constant k of Boltzmann's entropy-probability relation as

$$k = \frac{c_1}{c^2 c_2} = 1.346 \times 10^{-16} \text{ erg deg}^{-1}. \qquad (9)$$

With it he could obtain Loschmidt's or Avogadro's constant and, in particular, the elementary charge (of ions or electrons) from Faraday's constant F and the universal gas constant R,

$$e = \frac{Fk}{R} = 4.69 \times 10^{10} \text{ electrostatic units} \qquad (10)$$

in fair agreement with the available data.

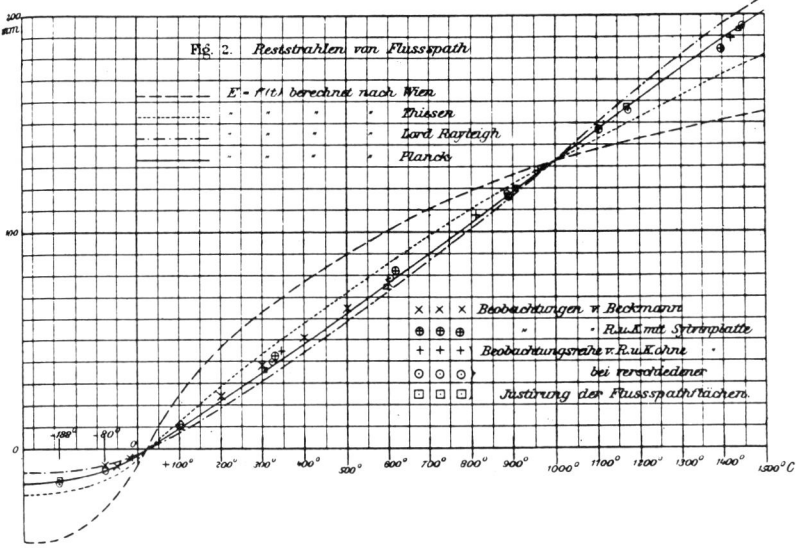

Figure 3.2. *Black-body radiation laws* [18].

3.2.1.2. Light-quanta (Einstein 1905). In 1904 the young Einstein, searching for examples of fluctuations in the kinetic theory of matter, turned to the problem of heat radiation. He found that his fluctuation expression yielded Wien's displacement law relating the wavelength $\lambda_{\max}$ of the maximum in the energy distribution of black-body radiation to the temperature T as $\lambda_{\max} = 0.293/T$ [19]. A year later he dealt with the energy distribution of black-body radiation. While for small frequencies the law

$$\rho_\nu = \frac{8\pi \nu^2}{c^3} kT \qquad (11)$$

followed, the author noted: 'for short wavelength and low radiation densities these foundations completely failed'. Instead, a statistical analysis of Wien's law—into which Planck's law passed over for short wavelength and low density—yielded the result: 'Monochromatic radiation ... behaves, in a thermodynamic sense, as if it consisted of mutually independent radiation quanta of magnitude $R\beta'\nu/N_0$.' [20]. When rewriting Einstein's constants ($R/N_0 = k$ and $\beta' = h/k$), one finds that his 'light-quanta' of frequency ν assumed the energy given by Planck's equation (7).

Einstein immediately applied his new 'heuristic point of view' (as he called his hypothesis) to processes involving the 'transmutation of light'. Thus he showed that the well-known Stokes' rule—i.e. visible fluorescent light (frequency ν_1) was emitted by incoming ultraviolet radiation with frequency $\nu_2 > \nu_1$—followed from energy conservation.

More importantly, he explained the main features of Philipp Lenard's (1862–1947) fundamental observations of the photoelectric effect: (i) only incident light of a certain frequency or higher was able to produce photoelectrons from a given metallic surface; (ii) the number, but not the velocity, of the emitted electrons depended on the intensity of the incident radiation [21]. Einstein's formula, relating the kinetic energy (E_{kin}) of the photoelectrons to the quantum energy of the incident radiation (($R/N_0)\beta\nu$) and the characteristic potential ('*work function*') of the metallic surface (P) as

$$E_{kin} = \frac{R}{N_0}\beta\nu - P = h\nu - P \qquad (12)$$

was checked in greatest detail by Robert A Millikan (1868–1953) about ten years later, providing a 'photoelectric precision value' for Planck's constant $h (= 6.57 \times 10^{-27}$ erg s) [22].

3.2.1.3. Specific-heat quanta (1906–13).

In 1872 Heinrich Friedrich Weber (1843–1912) investigated the specific heat of diamond: at room temperature he found a value much smaller than 6 cal mol^{-1}, the usual result for monatomic substances; but it increased slowly with rising temperature and nearly reached the standard value at 1300 °C [23]. Towards the end of 1906, Einstein (whose physics professor at the ETH Weber had been) knew the reason for this anomalous behaviour. 'If Planck's theory strikes the heart of the matter' he argued 'we must expect contradictions also in other areas of the heat theory [than heat radiation] between the molecular theory of heat and experience, which may be removed by the method proposed.' [24].

Einstein's consideration was straightforward: he extended Planck's *ansatz* of the radiation law (4) to the molecular oscillations caused by heat motion in a monatomic (crystal) solid, associating to each frequency the average energy

$$E_\nu = 3k \frac{h\nu}{\exp(h\nu/kT) - 1} \qquad (13)$$

and assuming for simplicity all oscillations of the N atoms to have the *same* frequency ν. Thus he obtained an expression for the specific heat decreasing exponentially below the standard value at low temperatures T, namely: 'If $T/\beta\nu [= kT/h\nu] < 0.1$, then the [oscillating] object does not contribute appreciably to the specific heat. In between the expression increases, first rapidly and then more slowly.' [25].

This new application of Planck's hypothesis now indeed removed Lord Kelvin's 'dark cloud', but it was hardly noticed, until three years later when the Berlin physico-chemist Walther Nernst systematically studied the specific heats of substances at low temperatures, being motivated by a consequence of his new heat theorem [26]. He summarized the results in February 1910 as [27]:

> If one plots the obtained values graphically, then in most cases one obtains straight lines, which frequently fall off very rapidly at low temperatures; thus one gets the impression that at very low temperatures the specific heats assume a value zero or at least very low values ... This result agrees qualitatively with the theory developed by Einstein.

More refined observations of Nernst and his collaborators produced deviations from Einstein's one-frequency formula. Nernst tried with Frederick A Lindemann (1886–1957) a two-frequency equation (with frequencies v and $v/2$) which seemed to fit the observed data quantitatively [28], and simultaneously received some justification through Planck's 'second quantum hypothesis' of 1911 [29]. These results and ideas were heavily debated, but not resolved, at the first Solvay Congress in Brussels, especially by Einstein, Nernst and Planck. Essential progress, however, came soon afterwards, arising from two separate places.

In March 1912, Peter Debye (1884–1966) of Zurich presented the first results of his calculation of the energy spectrum of a crystal solid, using a continuum approximation, which he cut off at a maximum frequency v_{max} (determined by the condition that a monatomic solid of N atoms should only have $3N$ degrees of freedom). He obtained a closed integral expression for the specific heat c_v, yielding in particular the formula for low temperature T

$$c_v = \text{constant} \frac{T^3}{\theta^3} \quad \text{with } \theta = \frac{h v_{max}}{kT} \qquad (14)$$

(the subscript v denoting constant volume) [30]. Independently, Max Born (1882–1970) and Theodore von Kármán (1881–1963) in Göttingen worked out a detailed dynamical theory of the complex vibrations in crystal lattices. Their formula for the specific heats of monatomic solids contained an integral over all possible frequencies; it also yielded deviations from Einstein's original equation (in the sense shown by the experiments of Nernst and collaborators) and could be extended to polyatomic substances [31]. Because of the nearly simultaneous discovery of the lattice structure of crystals by Max von Laue (1879–1960) and his Munich team, their theory appeared particularly convincing, in spite of its mathematical complexity.

Different aspects of the specific-heat problem arose from the data obtained with diatomic gases at low temperatures; notably, Nernst's student Eucken found a behaviour similar to crystals in the case of hydrogen [32]. Einstein, with Otto Stern (1888–1969), suggested describing this result by giving the rotating molecules a zero-point energy (following from Planck second quantum hypothesis), i.e. [33]

> See also p 973

$$E_\text{rot} = \frac{h\nu}{\exp(h\nu/kT) - 1} + \frac{h\nu}{2}. \quad (15)$$

Although many scientists worked on this problem, a satisfactory solution came about only after the completion of quantum mechanics and the inclusion of the nuclear spins in the hydrogen molecule [34].

3.2.1.4. Chemical and other quanta (1907–13). In late 1907 Stark (see figure 3.3) began to use Planck's hypothesis to account for details of his investigations of the Doppler effect of canal rays (i.e. the positive atomic ions) of hydrogen. He claimed that the presence or absence of emitted lines can be explained by a quantum law of the form

$$E_\text{kin}\alpha = h\nu \quad (16)$$

with α a numerical factor depending on the frequency and the nature of the ion involved. In particular [35]:

> From the above law there follows directly the important postulate that by the impact of a moving ion a characteristic electron oscillation will only then be excited and the corresponding series line be emitted, if the kinetic energy of the atomic ion (canal ray) exceeds a certain threshold value which is characteristic of the ion in question. Hence, if one increases the translational velocity of an atomic ion, starting from the value zero, then a given series line will initially possess no intensity at all; but when the threshold value of the translational velocity has been surpassed, suddenly the emission of a series line takes place.

Stark applied this concept of 'elementary quantum of energy' to the 'chemistry of band spectra' as well as to photochemical reactions in substances, assuming that the latter consist of ions and bound electrons. Although often arriving at keen, later substantiated, conclusions on the structure of matter, he at times claimed more than he could prove; this brought conflict with more cautious colleagues, especially theorists, e.g. a fight with Sommerfeld on the origin of the bremsstrahlung spectrum of x-rays in 1909–10.

Another approach to introduce quanta into chemistry started from Nernst's already mentioned work containing the new heat theorem [26]. The same paper provided an expression for the so-called 'kinetic constant' K_c (entering the fundamental Guldberg–Waage law of mass-action in chemical reactions). Nernst also introduced the 'chemical constant J', for which his student Otto Sackur (1880–1914) and later Stern derived various quantum-theoretical equations [36].

Quanta and Quantum Mechanics

Figure 3.3. *Johannes Stark (1874–1957), a pioneer of quantum physics.*

Early in 1911 Nernst proposed: 'If we generalize the quantum hypothesis ... in such a way that the energy is always absorbed in fixed quanta not only in the case of oscillations around an equilibrium position but also in the case of arbitrary rotation of a mass, then we arrive at further conclusions, which may perhaps be able to explain certain contradictions of the old theory.' [37]. His suggestion was quite favorably received at the Solvay Congress, especially by Einstein and Lorentz. Niels Bjerrum (1879–1958), guest at Nernst's institute, developed a first quantum theory of the infrared absorption spectra of diatomic molecules, which was soon confirmed experimentally by Eva von Bahr [38].

Planck also thought about further quantum phenomena; thus he tried to explain the fact that electrons in metals do not contribute to their specific heats by assuming quantized velocities of electrons in matter [39]. Nernst picked up this idea and claimed that the equation of state for non-atomic gases should deviate, at low temperatures, from the classical equation of state [40]. Lorentz, Planck and others discussed the consequences of this 'gas degeneracy' at a meeting in April 1913 in Göttingen ('*Kinetischer Gaskongreß*'). The problem became settled only a dozen years later within the new types of quantum statistics of Bose–Einstein and Fermi–Dirac.

3.2.2. Atomic structure and spectral lines (1913–21)

Ever since William H Wollaston (1766–1828) discovered the dark lines in the solar spectrum (1802) and Joseph Fraunhofer (1787–1826) noticed the agreement of the position of certain dark lines with some bright lines in the spectrum of flames (1814), the question arose as to their origin. It was essentially settled in 1859 by the cooperation of the Heidelberg physicist Kirchhoff and chemist Robert Bunsen (1811–99): the dark lines correspond to the absorption of discrete spectral lines by the solar atmosphere. Two facts remained to be shown: (i) did atoms or aggregates of atoms (molecules) absorb (and also emit, as Kirchhoff assumed) these lines? (ii) what was the mechanism of absorbing (and emitting) discrete spectral lines? Although considerable efforts were undertaken to answer both questions, they could not be considered as settled when Planck introduced the quantum hypothesis in 1900. A year later, Planck considered 'the problem concerning the nature of light in spectral lines to count among the most difficult and complicated ones, which have ever been posed in optics and electrodynamics' [41]. Might it also be solved with the help of the quantum concept?

'Since the discovery of spectral analysis no knowledgeable person could doubt that the problem of atoms will be solved, if one has learned to understand the language of spectra .... It is the mysterious instrument, on which nature performs the spectral music and organizes the structure of atoms and nuclei according to its rhythm', Arnold Sommerfeld (1868–1951) wrote in the preface of his famous book on *Atomic Structure and Spectral Lines* [42]. But first one had to get some information about the constituents of matter and the nature of laws that should be applied to them.

The recognition of the electron and positive ions (canal rays) as constituents of matter allowed for the first time the construction of realistic models of atoms. In that respect, the British physicist Joseph John Thomson (1856–1940) and the New Zealander Ernest Rutherford (1871–1937) proposed quite different conceptions. Thomson's atom of 1904 consisted of negatively charged electrons, distributed regularly in a sphere of uniformly distributed positive charge; Rutherford's atom of 1911 resembled a planetary system, with electrons moving in circular orbits around a centrally condensed positive charge—the atomic nucleus. By 1910 the Austrian Arthur Erich Haas (1884–1941) had already shown, on the basis of Thomson's model, that the size, i.e. the radius a, of the hydrogen atom could be obtained by the quantum formula

$$h = 2\pi e \sqrt{m_e a} \tag{17}$$

with e and m_e the (absolute) charge and mass of the electron [43]. Another attempt a couple of years later by John William Nicholson (1881–1955) of Cambridge to account for spectra of celestial objects also involved the quantum hypothesis [44]. The decisive breakthrough came in 1913: Niels Bohr (1885–1962), a Danish visitor at Rutherford's Manchester institute, began with the planetary atom that Rutherford had proposed. His approach turned out to the most successful for three reasons: firstly, Bohr selected the most suitable atomic model; secondly, he constructed a dynamics based on mechanical and electrodynamical relations *plus* a quantum hypothesis; thirdly, he could compare the results of the theory with the empirical spectroscopic data.

3.2.2.1. The empirical basis of Bohr's atomic model (1913–14). In April 1913, Bohr communicated the first part of a three-part essay *On the constitution of atoms and molecules* to *The Philosophical Magazine*, adding a new level to the understanding of the structure of matter [45]. In it Bohr proposed a detailed quantum-theoretical model of atoms and molecules, based on the following assumptions:

(i) Rutherford's planetary model of electron orbits around atomic nuclei provides the basic picture.

(ii) Classical dynamics is supplemented by the quantum condition for the angular momentum p_φ (taken over from Nicholson)

$$p_\varphi = nh \qquad (18)$$

which leads to 'stationary states' of energy W_n for the one-electron atoms with nuclear charge Ze, i.e.

$$W_n = -\frac{2\pi^2 m_e e^4 Z^2}{n^2 h^2} \qquad (19)$$

See also p 72

where $n = 1, 2, 3, \ldots$ denotes the quantum number.

(iii) Classical electrodynamics yields the attractive (Coulomb) force between electrons (mass m_e, charge $-e$) and the nucleus (mass M, charge $+e$). However, it is violated in two points: firstly, the electrons do not radiate in Bohr's 'stationary' orbits; secondly, atomic radiation arises from transition between two stationary orbits, its frequency being given by the Planck–Einstein relation

$$h\nu_{nm} = |W_n - W_m|. \qquad (20)$$

Bohr's model immediately explained the size (= diameter) of the lowest ($n = 1$-orbit) of hydrogen ($2a = 1.1 \times 10^{-8}$ cm) and the value of the Rydberg constant R

$$R = \frac{2\pi^2 m_e e^4}{h^3}. \qquad (21)$$

157

Bohr also soon succeeded not only in predicting new hydrogen series (by starting from levels other than $n = 2$, which corresponded to the well-known Balmer series) but also in clarifying the situation with the Pickering series—which he attributed to the ionized helium atom with a slightly different Rydberg constant R_{He^+} ($= 4M_{He}/(M_{He} + m_e)$, with M_{He} the mass of the helium nucleus with charge $2e$!) [46].

The central assumption that 'stationary electron orbits' exist in atoms (and molecules) received justification in the crucial experiments of James Franck (1882–1964) and Gustav Hertz (1878–1975) in Berlin. In 1911 they began investigations on relating the ionization potentials of different gases (helium, neon, hydrogen and oxygen) to quantum theory. Continuing this programme of having electrons of definite kinetic energy collide with atoms of inert gases, they arrived in April 1914 at an unexpected result, which they summarized as follows [47]:

> 1. It will be shown that electrons in mercury vapor undergo elastic collisions with the molecules until they obtain a critical velocity. 2. ... This velocity is equivalent to the one obtained by electrons that have gone through a potential of 4.9 volts. 3. It will be shown that the energy of a 4.9 volt beam equals the energy quantum of the mercury lines 253.6 μm.

Though the authors claimed until 1916 that the 4.9 V represented the ionization potential of mercury, Bohr argued rather: 'It seems that their experiments may possibly be consistent with the assumption that this voltage corresponds only to the transition from the normal state to some stationary state of the neutral atom.' [48]. Hence he considered the Frank–Hertz experiment to give strong support to his atomic theory, and the experimentalists finally agreed with him. A decade later they received the Nobel Prize for physics 'for their discovery of the law governing the impact of an electron upon an atom' and in particular for their verification of Bohr's hypothesis of stationary states and the frequency condition (figure 3.4).

3.2.2.2. X-ray spectra and atomic number (1913–14). X-ray spectroscopy arose from the observation of Charles G Barkla (1877–1944) and his associate Charles A Sadler. They discovered homogeneous x-radiation emerging when x-rays of continuous frequencies are scattered by matter [49]. After 1912 the interference-pattern method of von Laue *et al* and the alternative reflection method of William H Bragg (1862–1943) and his son William L Bragg (1890–1971) gave accurate x-ray wavelengths. In 1913, Henry G J Moseley (1887–1915), a member of Rutherford's Manchester institute, modified Bragg's method and obtained the principal lines of the x-ray spectra of most elements by registering their ionization and photographic images. In November of that year he reported his results to Bohr as confirming the new theory of atomic constitutions and being

Quanta and Quantum Mechanics

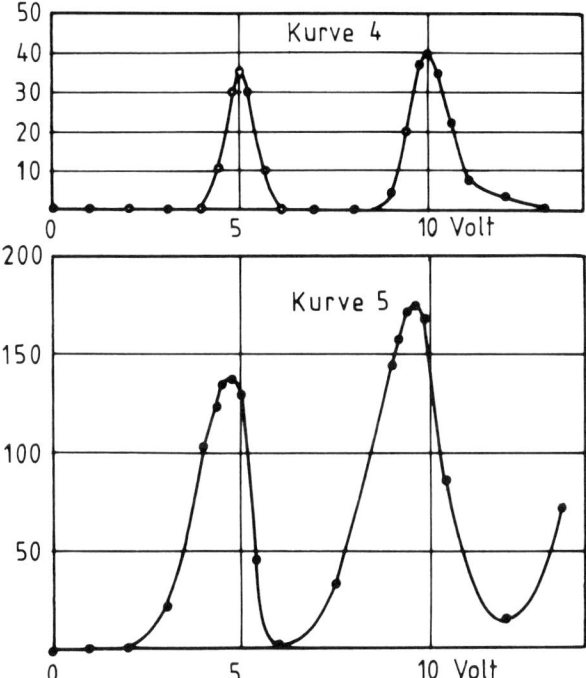

Figure 3.4. *The Frank–Hertz 'ionization voltage curve' confirming Bohr's stationary atomic states* [47].

'extremely simple'. Thus he obtained for the shortest x-ray lines of atoms of molecular charge Ze, the so-called K_α-lines, the formula

$$\nu_\alpha = \nu_0 (Z-1)^2 \left(\frac{1}{1^2} - \frac{1}{2^2} \right) \tag{22}$$

with ν_0 denoting a constant [50]. He also succeeded in correcting the sequence of transition elements to be Fe–Co–Ni according to increasing 'atomic number' Z (rather than to their atomic weight A). That is, the neutral nickel atom possessed a higher nuclear charge and one electron more than the neutral cobalt atom, despite the fact that it had a smaller atomic weight.

In a further paper, presented in spring 1914, Moseley studied another 30 pure chemical substances (aluminium to gold), registering this time the softer L_α lines (because the K_α lines of these higher-Z elements were *too* short for him to observe). He found the frequency dependence described by the formula

$$\nu_\alpha = \nu_0 (Z - 7.4) \left(\frac{1}{2^2} - \frac{1}{3^2} \right) \tag{23}$$

159

where the term 7.4 represented the screening effect of the innermost K-shell of electrons [51]. While Moseley soon had to go to military service, where his death was one of the heaviest losses for science in World War I, the experimental study of x-ray spectra was eagerly continued in Paris by Maurice de Broglie (1875–1960), in Lund by Manne Siegbahn (1886–1977) and in Munich by Röntgen's collaborator Ernst Wagner (1876–1928). Also in Munich, a detailed theory of x-ray spectra was first developed by Walther Kossel (1888–1956) and Sommerfeld. Many important results about the details of atomic structure on the basis of the extended Bohr model were thus achieved in the late 1910s and early 1920s.

3.2.2.3. Sommerfeld's extension of Bohr's model (1915–16). Sommerfeld, professor of theoretical physics at the University of Munich, had been active in the field of quantum theory since 1911; he also had attended the first Solvay Congress and remained alert in looking for quantum phenomena. But his treatment of spectroscopic problems before 1914—e.g. the anomalous Zeeman effects—did not involve quanta. This changed in winter 1914–15, when he delivered a course of lectures on Bohr's theory and its extension. In late 1915 he started to publish the most important results in two principal papers, entitled *On the theory of the Balmer series* [52] and *The fine-structure of the hydrogen and hydrogen-like spectra* [53]. Sommerfeld's work gave Bohr's model of atomic constitution an enormous push forward, as it enabled physicists to deal quantitatively with non-hydrogen-like spectra, i.e. the spectra of atoms having more than one electron, as well as with the action of electric and magnetic field on spectral lines.

Sommerfeld essentially introduced two 'kinematic' generalizations of the mechanical part of Bohr's model:

(i) The (one-dimensional) circular motion of the electrons was replaced by a general elliptical one in a plane, determined by two quantum numbers n_1 and n_2. (A little later Sommerfeld proposed, in dealing with the anomalous Zeeman effect, a third 'spatial quantum number' [54].)

(ii) For particular cases, like some corrections and the action of a magnetic field, Newton's kinematics had to be replaced by Einstein's kinematics of special relativity. (Sommerfeld published these results only in late 1915, after being informed by Einstein that general relativity would not influence them noticeably.)

With generalization (i), Sommerfeld immediately demonstrated why the hydrogen spectrum was simple enough to be represented by Bohr's formulae (19) and (20) of 1913: the different term types and the corresponding series observed with many-electron atoms (and created by terms depending on two quantum numbers, n_1 and n_2) coincided in the case of the hydrogen-like (one-electron) atoms with one whose terms depended only on the sum $n_1 + n_2$. Relativity kinematics, on the other hand, gave rise to a 'fine-structure' of each hydrogen term: thus the ($n_1 + n_2 = 2$) terms became split into doublets, the ($n_1 + n_2 = 3$)-terms into triplets, etc, whose separation was determined by the dimensionless factor $\alpha = (2\pi e^2)/(hc) \simeq 0.7 \times 10^{-3}$, later called (Sommerfeld's) 'fine-structure constant'.

'My measurements [on ionized helium lines] are now finished, and they agree everywhere most beautifully with your fine-structures' Paschen wrote on 25 May 1916 to Sommerfeld [55]. Bohr, the originator of the whole scheme, had already confessed in March 1916 to the Munich professor:

> I thank you so much for your paper, which is so beautiful and interesting. I do not think that I have ever read anything which has given me so much pleasure.

3.2.2.4. Zeeman and Stark effects in the Bohr–Sommerfeld theory (1916–21).

Pieter Zeeman's (1865–1943) discovery of the splitting of spectral lines in magnetic fields (1896) and Hendrik Lorentz' subsequent explanation (1897), for which these physicists shared the Nobel Prize in 1902, provided a triumph for the assumption of electrons being bound in atoms. Each, spectral line was, in the simplest case, originally single then separated into three components, with an unshifted middle line and two others displaced by the amount

$$\Delta \nu = \pm \frac{eH}{4\pi m_e c} \qquad (24)$$

where H denotes the strength of the magnetic field, e and m_e the charge and mass of the electron, respectively. More complex splittings, soon called 'anomalous Zeeman effects', were found in 1898 by Albert A Michelson (1852–1931) and Thomas Preston (1860–1900). The latter stated 'Preston's law', namely that all spectral lines of a given series type (doublet, triplet, etc) yielded the same kind of Zeeman splitting (1899). Later Carl Runge (1856–1927) established numerical rules for the anomalous Zeeman effects (1902, 1907), and finally Paschen and Ernst Back (1881–1959) observed that for high magnetic fields the anomalous separations changed into the normal Zeeman triplets (Paschen–Back effect 1912, 1913). Of these rich and complex phenomena, the Bohr–Sommerfeld theory succeeded only to derive the simplest triplets, in agreement with Lorentz' classical treatment [56]. However, it immediately described another effect: the splitting of spectral lines in an electric field.

On 20 November 1913, Johannes Stark submitted a paper to the Prussian Academy telling of a new observation: if a strong electric field, up to 31 000 V cm^{-1}, acted on radiating hydrogen canal rays, the brightest lines H_α and H_β, when observed in a direction perpendicular to the electric field, were split into five polarized components [57]. A few weeks later Stark and Georg Wendt announced the 'longitudinal effect', i.e. the finding of a triplet, when one looked in the direction of the field vector [58]. The discovery of this electric analogue of the Zeeman effect, for which scientists had searched for nearly two decades, came as a great surprise, because classical electron theory could not account for it. However, Emil Warburg (1846–1931), President of the Reichsanstalt (PTR) and already an eager supporter of quantum theory, provided an estimate of the electric separation in one of the first applications of Bohr's model, obtaining a shift of $hn^2/4\pi m_e e E$, with n the quantum number of the excited hydrogen state and E the strength of the electric field [59]. (Ironically, Stark began just at that time to turn against quantum theory [60].)

In early 1916, Paul Sophus Epstein (1883–1966), a Russian former student of Sommerfeld and at that time (loosely) interned in Munich as an enemy alien, succeeded in calculating the Stark effect completely on the basis of his master's scheme [61]. The electric term splittings followed from the equation

$$W_{el} = -\frac{3h^2 E}{8\pi^2 m_e Z e}(n_1 - n_2)n \tag{25}$$

as expressed with the help of the two quantum numbers (n_1, n_2) of suitably chosen (parabolic) coordinates—with n the principal (Bohr) quantum number of the hydrogen state. By imposing certain selection rules for the quantum numbers n_1, n_2 and n, he reproduced Stark's data perfectly. Independently, and at about the same time, the Berlin astrophysicist Karl Schwarzschild (1873–1916) arrived at an identical treatment [62].

In late 1919 Sommerfeld took the first serious step towards attacking the problem of anomalous Zeeman effects in the Bohr–Sommerfeld theory, which he called in a first note a 'number mystery' [63]. Upon a detailed analysis of the empirical data he formulated five 'rules' for the description and he introduced the 'inner' or 'hidden quantum numbers' of the atom [64]. He did not know its 'geometrical significance', but his former student Alfred Landé (1888–1975), who specialized in anomalous Zeeman effects during the next five years, interpreted it as 'the total quantum number of the atom around its invariant axis' (Landé to Bohr, 16 February 1921)—later called the 'total angular momentum'. The story of deciphering the observation went on until 1925, with Landé, Sommerfeld and his students Werner Heisenberg (1901–76) and Wolfgang Pauli (1900–58) being the main actors on the side of theory, and Ernst Back on the experimental side. It was often a story of errors and arbitrary assumptions (to achieve agreement between theory and experiment), but it paved the way to the final, successful solution.

3.2.3. Quantum-mechanical effects (1922–28)

In January 1909 Einstein submitted a paper to the *Physikalische Zeitschrift*, a contribution to the then ongoing debate on the validity and the significance of Planck's black-body law, in which the experimental pioneers Lummer and Pringsheim and the theoreticians Lorentz, James Hopwood Jeans (1877–1946) and Walther Ritz (1878–1909) participated. Einstein now calculated the energy fluctuations of Planck's law (which he assumed to represent the data correctly) and obtained for the average value of the squared deviation from the mean radiation energy $\bar{E}$ of the frequency interval v and $v + dv$ the result

$$\overline{\varepsilon^2} = \overline{(E - E)^2} = hvE_0 + \frac{c^3}{8\pi v^2 \, dv} \frac{\bar{E}^2}{V} \tag{26}$$

where V is the volume of the cavity containing black-body radiation. The second term on the right-hand side corresponds to the fluctuations provided by the (classical electromagnetic) radiation theory. The first term—which in the case of visible radiation is the larger contribution to the energy fluctuation—showed 'that also the fluctuations of radiation and radiation pressure behave as if radiation consisted of quanta of the size indicated [hv]' [65].

The fluctuation formula did not persuade the community of physicists, even the supporters of quantum theory, to accept the light-quantum hypothesis; it was called 'a speculation that missed the target' (Planck and others in the 1913 letter recommending Einstein's acceptance in the Prussian Academy) and 'an unbearable physical theory' (Millikan in his 1916 paper establishing Einstein's equation for the photoelectric effect). Still, not only did an experiment of 1922 prove the existence of the light-quantum, but the even more important message of the 1909 fluctuation formula, the simultaneous presence of undulatory and corpuscular properties of radiation, also received experimental confirmation in the 1920s. These fundamental results provided the empirical basis of the final atomic theory, quantum mechanics.

A further crucial phenomenon, going beyond the so-called 'older quantum' theory of Bohr and Sommerfeld, was the experimental observation of 'space quantization', the Stern–Gerlach effect, in 1922. The strange statistical behaviour of light-quanta and electrons as obtained, respectively, by the Indian Satyendra Nath Bose (1894–1974) in 1924 and the Italian Enrico Fermi (1901–55) in 1926, disclosed further quantum-mechanical characteristics of atomic objects. Last, but not least, the quantum-theoretical analysis of dispersion of light by atoms, from 1921 to 1925, led not only to formulae substantiated by experiment, but also to the discovery of the Raman effect in 1928.

3.2.3.1. The Stern–Gerlach and Compton effects (1922–23). While many quantum-theory experts considered Sommerfeld's spatial quantization,

Twentieth Century Physics

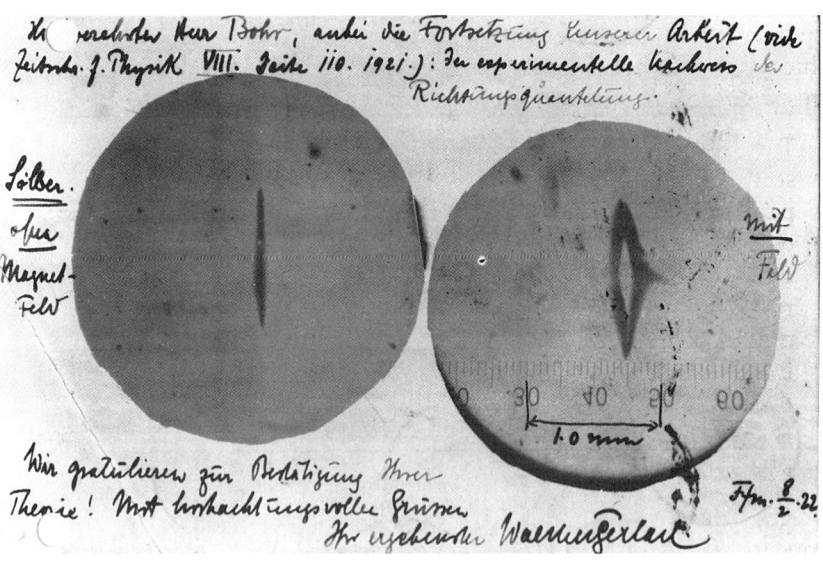

Figure 3.5. *The Stern–Gerlach effect (postcard from Gerlach to Bohr, dated 8 February 1922).*

i.e. the discrete orientation in space of electronic orbits, to be only a mathematical tool, in 1921 Otto Stern in Frankfurt proposed an empirical test: if an electron in an atom carried a magnetic moment of about 1

Bohr magneton ($= eh/4\pi m_e$), an atomic beam of silver atoms should split in passing through a magnetic field of strong inhomogeneity. He performed the experiment with the help of Walther Gerlach (1889–1979), and in February 1922 a splitting of the silver beam into two separate beams indeed showed up (see figure 3.5) [66]. This result then proved space quantization as well as once again confirming the existence of stationary states. However, it also went far beyond the Bohr–Sommerfeld theory, as Einstein and Paul Ehrenfest (1880–1933) demonstrated by their calculation of the time needed by the silver atoms to be oriented in the two directions: they estimated a time for the (classical) magnetic interaction of 10^{11} s, in contrast to the observation—which gave a time of less than 10^{-4} s [67]. Thus one of the bases of the existing atomic theory broke down.

Another basis, namely the classical description of radiation in Bohr's model, was shaken by the experiment of Arthur H Compton (1892–1962) of Washington University, St Louis (see figure 3.6). The American physicist, an expert on x-ray physics, wrote in 1922 a report on the status of *Secondary radiation produced by x-rays*, which he delivered in October of that year to the National Research Council. While he found no definite arguments for 'the quantum idea of [x-ray] scattering' in the available data, he went on to examine the situation further. At the Chicago meeting of the American Physical Society in early December he presented a paper, whose abstract read [68]:

See also p 362

> The hypothesis is suggested that when an x-ray quantum is scattered it spends all of its energy and momentum upon some particular electron. This electron in turn scatters the ray in some definite direction. The change in momentum of the x-ray quantum due to the change in its direction of propagation results in a recoil of the scattering electron. The energy in the scattered quantum [with scattering angle θ] is much less than the energy in the primary quantum by the kinetic energy of the recoil of the scattering electron. The corresponding increase in the wavelength of the scattered beam in $\lambda_\theta = \lambda_0(1 + 2\alpha \sin^2(\theta/2))$ where $\alpha = h/m_e c \lambda_0$.

Compton wrote full details in his *Physical Review* paper, which appeared in May 1923. His 'convincing' result 'that the radiation quanta carries with it directed momentum as well as energy' was attacked heavily in the following years by William Duane (1872–1935) of Harvard University, another x-ray specialist. On the other hand, Debye proposed in Zurich a similar formula for x-ray scattering, although he did not, in contrast to Compton, insist on the existence of the light-quantum [69].

The Compton–Debye description of x-ray scattering stimulated new activities among the quantum physicists. While Einstein, Sommerfeld, Pauli and a few others took the side of the light-quantum, Bohr

Figure 3.6. *Arthur H Compton with his experimental apparatus.*

specifically fought against it, because it did not seem to fit into general quantum-theoretical philosophy based on the correspondence principle (which demanded a smooth transition from classical electromagnetism into a quantum description of radiation). Thus in early 1924, he composed a paper jointly with Hendrik Kramers (1894–1952) and John Slater (1900–76), in which they tried to explain Compton's observations on the assumption that, in any interaction between atoms and radiation, energy is only statistically conserved [70]. To decide between the Copenhagen and the Compton–Debye explanations of the Compton effect, Walther Bothe (1891–1957) and Hans Geiger (1882–1945) of the PTR in Berlin proposed to check the simultaneous occurrence of the scattered x-ray and the recoil electron in a counter experiment. In April 1925 they announced their result [71]:

> The experiments yielded one coincidence per 11 counts approximately, after deducting the counts expected purely accidentally. ...Our experiments, therefore, decide in favor of the older [light-quantum] conception.

A couple of months later Compton and Alfred Simon reached the same conclusion, on the basis of cloud-chamber observation of the recoil electron and the scattered x-ray [72]. The light-quantum was reality and was later baptized the 'photon' [73].

3.2.3.2. *Statistics of light-quanta and electrons (1924–26).* Einstein's idea of light-quanta, though established experimentally by the Compton effect, did not seem to account for the observed radiation phenomena. This is already obvious from the undulatory term in Einstein's fluctuation formula (26)—even if one wants to forget the classical interference patterns, etc. Thus the quantum physicists, notably Einstein himself, Debye, Ladislas Natanson (1864–1937), Ehrenfest, Mieczysław Wolfke (1883–1947) and Bothe, sought to introduce some correlation between independent light-quanta, which would create the interference effects. The solution, however, came from India. On 4 June 1924 Bose wrote a letter to Einstein stating:

> Respected Sir, I have ventured to send you the accompanying article for your perusal and opinion. ... You will see that I have tried to deduce the coefficient $8\pi v^3/c^3$ in Planck's law [equation (2)] independent of classical electrodynamics, only assuming that the ultimate elementary region in the phase space has the content h^3.

He then requested Einstein's help in publishing the paper, which the latter gave immediately, since he recognized that Bose had removed a major objection against light-quanta [74].

Bose started by assuming light-quanta of a given frequency v_s to possess energy hv_s and momentum hv_s/c, as Einstein had done before. He then defined the probability distribution of N_s light-quanta among A_s cells in phase space (where A_s was given, as had been earlier assumed, e.g. by Debye in 1910 [75], by $(8\pi/c^3)Vv_s^2\,dv_s$) through the equation

$$\text{probability} = A_s!/(p_0^s!p_1^s!p_2^s!\ldots) \qquad (27)$$

where p_0^s counts the number of vacant cells (of size h^3), p_1^s those with one light-quantum, etc. In the state of equilibrium the sum of all probabilities (over the index s for different frequency quanta) must assume a maximum for a given total energy ($E = \sum_{r,s} rp_r^s hv_s$). From this the author directly derived Planck's black-body energy distribution law of 1900.

Einstein immediately generalized Bose's statistics to material particles in a series of papers dealing with the quantum theory of the ideal gas [76]. He believed that electrons might also obey the new statistical laws, but new evidence from spectroscopy demanded differently.

In 1924 a combined analysis of several difficulties in explaining the x-ray spectra, the many-line spectra of complex atoms, and their anomalous Zeeman effects by experimental and theoretical experts—especially Landé, Millikan and Ira Sprague Bowen (1898–1973), Louis de Broglie (1892–1987) and Alexandre Dauvillier (1892–1979)—led to a different organization of electronic levels in atoms than so far assumed by Bohr's scheme; this was formulated in some detail by Edmund C Stoner (1889–1973) of Leeds University [77]. Pauli, having criticized Bohr's earlier justification of the closed electron shells (and thus of the periodic system of elements), was able to derive Stoner's organization from a simple assumption. He associated with the electron a 'classically non-describable duplicity of the quantum-theoretical properties', giving each electron bound in atoms four, instead of the usual three, quantum numbers (principal, two azimuthal quantum numbers and a 'magnetic' quantum number m) and claimed his 'exclusion principle' to be valid, namely [78]:

> There can never exist two or more equivalent electrons in the atom for which [...] the values of all [four] quantum numbers coincide.

By letting the fourth quantum number assume the values $+\frac{1}{2}(h/2\pi)$ and $-\frac{1}{2}(h/2\pi)$, Pauli indeed successfully explained the organization of data on complex spectra and their Zeeman effects, as well as the closure of electron shells (with the maximal number of electrons being 2 in the K-shell, 8 in the L-shell, 18 in the M-shell, and 32 in the N-shell, corresponding to the principal quantum numbers $n = 1, 2, 3, 4, 5$).

The exclusion principle not only helped to organize atomic spectra properly, it also gave rise to two important conclusions. First, Fermi employed it to derive the quantum-theoretical statistics obeyed by electrons: he put them in quantum cells containing either zero or one electron, and found distribution resembling the Planck–Bose one, except that the denominator $[\exp(h\nu_s/kT) - 1]$ had to be replaced by $[\exp(h\nu_s/kT) + 1]$ in the Fermi case [79].

Second, George Uhlenbeck (1900–88) and Samuel Goudsmit (1902–83) in Leyden interpreted Pauli's fourth quantum number as a general property of the electron, an 'intrinsic rotation' of the particle later called 'spin' [80]. The half value (in terms of $h/2\pi$) accounted for an old puzzle in atomic theory, the anomaly of the gyromagnetic ratio (of angular momentum over magnetic moment); the discrepancy by a factor 2, as observed in the experiments [81], now came out naturally, and similarly certain factor-of-two discrepancies in explaining data on anomalous Zeeman effects and complex spectra disappeared as well.

3.2.3.3. Wave–particle duality for material objects (1922–27). Bringing together the concepts of particle and wave in atomic physics interested the young Louis de Broglie (see figure 3.7), who worked in the spectroscopic laboratory of his brother Maurice, Duc de Broglie in Paris. In 1922 he suggested associating a small mass m_0 with a photon at rest, in order to treat it statistically like a material particle [82]. Louis de Broglie recalled in an interview for the *Archive of Sources for the History of Quantum Physics* of 7 January 1963 the continuation of his investigations:

> As in my conversations with my brother we always arrived at the conclusion that in the case of x-rays one had [both] waves and corpuscles, thus suddenly—... it was certainly in the course of summer 1923—I got the idea that one had to extend this duality to material particles, especially to electrons. And I realized that, on the one hand, the Hamilton–Jacobi theory pointed somewhat in that direction, for it can be applied to particles and, in addition, it represents a geometrical optics; on the other hand, in the quantum phenomena one obtains quantum numbers, which are rarely found in mechanics but occur very frequently in wave phenomena and in all problems dealing with wave motion.

He formulated the main ideas in the three notes presented in September and October 1923 to the Paris Academy of Sciences [83]:

(i) Due to the relativistic quantum principle ($h\nu_0 = m_0 c^2$), each particle of (rest) mass m_0 obtains an internal frequency ν_0.

(ii) For particles moving with velocity v, two frequencies $\nu(= \nu_0/\sqrt{1 - v^2/c^2}$ and $\nu_1(= \nu_0\sqrt{1 - v^2/c^2})$ exist, which are in phase (since $\nu_1 = \nu(1 - v^2/c^2)$!). They describe a 'fictive wave' having the phase velocity $v_{ph} = c^2/v$, which is associated with each moving particle.

(iii) By applying this 'fictive' or 'phase wave' to Bohr's simplest atomic model, with an electron moving in a circle around the atomic nucleus, the stationary states are represented by standing waves having n nodes, where n is identical to the principal quantum number.

De Broglie concluded that the phase wave guides the motion of the material particle like the propagation of light-quanta, i.e.: 'The rays of the phase waves coincide with the trajectories that are dynamically possible.' [84].

In summer 1924, de Broglie completed the presentation of the concepts in a further note, entitled *On the general definition of the correspondence between wave and [particle] motion* [85]. There he wrote down the fundamental relations expressing energy E, momentum p, and velocity v of a relativistic particle, namely

$$E = h\nu \qquad p = \frac{h\nu}{v_{ph}} \qquad v = \frac{\partial \nu}{\partial(\nu/v_{ph})} = v_{gr} \qquad (28)$$

where v_{ph} and v_{gr} denote phase and group velocities of the 'matter wave'. At the same time, he completed his thesis giving detailed derivations and motivations of all aspects of the theory [86]. There we find also

Figure 3.7. Louis de Broglie (1892–1987), father of matter waves.

the famous equation relating the wavelength λ to the velocity v of a non-relativistic particle of mass m_0, i.e.

$$\lambda = \frac{h}{m_0 v}. \tag{29}$$

Before de Broglie embarked on his theoretical adventure, Clinton J Davisson (1881–1958) and Charles H Kunsman (1890–1970) of the Western Electric Company started a series of experiments, in which they reflected electrons of kinetic energies up to 1500 eV from metallic surfaces; especially at lower speeds they finally observed maxima and minima of the reflected beam intensity depending on the crystal orientation, which they explained by assuming particular atomic models [87]. In early 1925, Friedrich Hund reported on these experiments in Göttingen, and a student, Walter Elsasser (1904–91) became interested in the problem. He thought in particular: 'What if Davisson and Kunsman's maxima and minima were diffraction phenomena similar to those produced when x-rays go through crystals? In this case the electron would not go through the platinum plate but would penetrate ever so slightly and then return; but then the diffraction pattern, if such it was, would be similar.' [88]. Elsasser guessed that these phenomena were related to de Broglie's ideas; he checked the order of magnitude and found it to be right; finally he connected also the Ramsauer effect (i.e. the anomalous behaviour of the electron cross section for small velocities in noble gases) with matter waves and published a short note in *Naturwissenschaften* [89]. He tried to perform an experiment to confirm the guess quantitatively, but he did not succeed and gave up after several months.

A year later, in summer 1926, Davisson stayed in England, where he met Born, Elsasser's thesis advisor, and learned about matter waves and wave mechanics. After his return he started a new series of experiments with Lester H Germer (1896–1971); in March 1927 they reported details of their observations on the diffraction of electrons from monocrystals of nickel, which indeed confirmed the existence of de Broglie waves [90]. Another, independent experimental proof was published shortly afterwards by George Paget Thomson (1892–1975) and Alexander Reid in England: they obtained Debye–Scherrer rings, similar to those observed in x-ray scattering, when electrons of several 10^4 eV passed through thin foils of celluloid [91]. Three of the pioneers of matter waves, de Broglie (1929) and Davisson and Thomson (1937) received the Nobel Prize for physics.

3.2.3.4. Dispersion formulae and Raman effect (1924–28). The quantum-theoretical treatment of dispersion phenomena began in 1921 with Rudolf Ladenburg (1882–1952) replacing amplitude and eigenfrequency in the classical dispersion formula by Einstein's emission coefficient and Bohr's transition frequency [92]. During the following years Ladenburg, Fritz Reiche (1886–1969), and Charles G Darwin (1887–1962) continued on similar lines; but it was Bohr's assistant and collaborator Hendrik Kramers, who opened the path to a quantum-mechanical treatment in spring 1924. Having in the back of his mind the new Bohr–Kramers–Slater radiation theory [70], notably the concept of 'virtual oscillators' emitting and absorbing the quantum-theoretical transition frequencies, he systematically reformulated the classical dispersion formula, which describes the amplitude P of secondary wavelets of frequency v. Thus he replaced the equation

$$P = E \sum_i f_i \frac{e^2}{m_e} \frac{1}{4\pi^2(v_i^2 - v^2)} \tag{30}$$

(with E the electric field strength, and f_i the coupling strengths associated with the eigenfrequencies v_i of the dispersing atoms) in quantum theory by [93]

$$P = E \left(\sum_i A_i^a \tau_i^a \frac{e^2}{m_e} \frac{1}{4\pi^2[(v_i^a)^2 - v^2]} - \sum_j A_j^e \tau_j^e \frac{e^2}{m_e} \frac{1}{4\pi^2[(v_j^e)^2 - v^2]} \right) \tag{31}$$

with A_i^a and τ_i^a the absorption probabilities and corresponding damping times, and A_j^e and τ_j^e the emission probabilities and damping times.

The second term occurs only in quantum theory and yields negative coupling ($-f = -A_j^e \tau_j^e$); it took years before Ladenburg demonstrated experimentally its existence in the special case of anomalous dispersion of the hydrogen atom in the frequency region around the H_α and H_β lines [94].

The development of the dispersion-theoretical approach, in which many quantum physicists (such as Born and Jordan in Göttingen, Reiche and Thomas in Breslau, and Kuhn in Copenhagen) partook in 1924 and 1925, went in two directions. Firstly, Kramers and Heisenberg generalized the dispersion formula (31) to include incoherent scattering [95]. In their treatment—the results of which remained untouched by the subsequent changes brought about by quantum mechanics, as their work was indeed to some extent a forerunner of quantum mechanics—an effect showed up, which had been predicted before by Adolf Smekal (1895–1959) on the basis of the light-quantum hypothesis [96]: the secondary, scattered radiation from atoms emitting spectral lines of frequency v_{nm} also contains radiation of the shifted frequencies

$$v' = v \pm v_{nm} \tag{32}$$

where v is the incident radiation frequency. In early 1928 in Calcutta Chandrasekhara Venkata Raman (1888–1970) discovered this effect in liquids [97]; independently, Georgi Landsberg (1890–1957) and Lew Mandelstam (1879–1944) in Leningrad (now St Petersburg) observed the same effect with quartz [98].

Secondly, the calculation of the intensities of spectral lines was also attacked using the dispersion theoretical approach by Born in Göttingen and John H Van Vleck (1899–1980) in Minnesota, while Leonard S Ornstein (1980–41) and his Utrecht students provided data on intensity ratios of multiplets late in 1924. Even more important consequences emerged from a consideration of sum rules for spectral-line intensities. The Swiss Werner Kuhn (1899–1963) formulated the first and simplest one for the hydrogen case, notably [99]

$$\sum_i f_i = 1. \tag{33}$$

Reiche's Breslau student Willy Thomas independently arrived at the generalization [100]

$$\sum_i f_i^a - \sum f_j^e = s/3 \tag{34}$$

where s denotes the degree of periodicity of the atomic system. Equations (33) and (34) permitted an easily perceptible interpretation: their right-hand sides corresponded to the number of dispersion electrons. Furthermore, equation (33) played an important role in Heisenberg's path to quantum mechanics, as we discuss below.

3.3. Origin and completion of quantum mechanics (1913–29)

When Max Planck in 1900 introduced energy quanta and Albert Einstein in 1905 arrived at the hypothesis of light-quanta, these physicists did not intend to break with the traditional theoretical description of Nature. Both saw, of course, that their procedure contradicted certain features of conventional schemes, i.e. statistical mechanics and electrodynamics. Nevertheless they expected that the quanta would eventually fit into some of the (classical) theories, which were completed in the early years of the twentieth century. The first to reach a final state was statistical mechanics, emerging from the work of James Clerk Maxwell (1831–79) and Ludwig Boltzmann (1844–1906). It received a canonical form in Josiah W Gibbs' (1839–1903) book of 1903, and its consequences were studied between 1900 and 1905 in detail by Lord Rayleigh, James Jeans (1877–1946) and Einstein. Using these methods, Jeans (and Einstein even a little earlier) arrived in 1905 at the 'classical' black-body law (11), which obviously contradicted Planck's law. Three years later, Hendrik Lorentz stated this fact clearly [101]:

> Well, if one compares the theory of Planck with the one of Jeans, one finds that both possess their merits and faults. Planck's theory is the only one which has given us a formula in agreement with the experimental results, but we cannot accept it without profoundly changing our fundamental ideas about electromagnetic phenomena ... Jeans' theory, on the other hand, forces us to ascribe the agreement ... to a coincidence.

When Lorentz expressed this view, he also knew for certain that the disagreement between Jeans' theory and the Berlin experiments could not be removed by the new dynamics of the electron, i.e. the relativity theory presented by Einstein and Henri Poincaré (1854–1912) in 1905. Evidently, Planck had introduced with the *finite energy quantum* a new concept, alien to classical kinetic theory as well as to classical electrodynamics. Planck tried to clarify the nature and meaning of his concept in a lecture course on heat radiation, given in the winter semester 1905–6. In particular, he noted that the phase-space points—i.e. the points representing the position and momentum of an oscillating mass at a given time—of the resonator lie on discrete curves (ellipses). Thus: 'the elementary quantum of action h acquires a new meaning, namely it gives the area of an elementary region in the phase plane of a resonator [i.e. the area between two ellipses determined by the quantum numbers n and $n+1$ of the resonator], no matter what its frequency' [102]. This situation contradicted the classical theory, according to which the phase-space points should fill the phase space continuously. By a closer mathematical analysis of the situation, Poincaré, in his last scientific work of 1911, arrived at the conclusion that Planck's law involved an essential element of discontinuity. Namely, this renowned French scholar (who had also participated in the first Solvay Congress) demonstrated the probability-density function $w(\eta)$ of the resonator in black-body radiation to be a *discontinuous function* of the energy η, i.e. it must be zero for all values of η, except when $\eta = 0, h\nu, 2h\nu$, etc; hence the black-body law and the quantum laws for related phenomena (such as specific heats, etc) could not be described by classical theory or any other intrinsically continuous mathematical formalism [103].

In the years after 1906, Planck tried to bridge the obvious gap between quantum behaviour and classical description. Since he did 'not see any reason to abandon the assumption of the absolute continuity of the free ether and all processes in it' (Planck to Lorentz, 7 October 1908), he wished to dissociate the real absorption and emission processes totally from the quantum hypothesis; in 1914 he declared that the entire quantum action would only 'occur between the oscillators and the free particles (molecules, ions, electrons) which exchange energy on colliding with the oscillators' [104]. Also in the first dynamical quantum theory of atomic structure, as developed by Niels Bohr and Arnold Sommerfeld since 1913, a conservative point of view with respect to the nature of dynamical laws prevailed: the 'revolutionary' aspects were rather packed in additional hypotheses, especially restrictions such as the frequency condition. The theory served for the next decade to explain a considerable part of the spectra of atoms and molecules, until serious discrepancies with data emerged, leading to a complete collapse of what was later called the 'old quantum theory' (section 3.3.1). From a thorough analysis of all the failures, which was mainly carried out at the centres of atomic theory (like Copenhagen and Göttingen), there arose in the summer of 1925 the first consistent 'quantum mechanics'. The concepts of this new atomic theory involved, as had been demanded in 1911 by Poincaré, genuine discontinuities. Only half a year later, however, Erwin Schrödinger (1887–1961) based his own new quantum theory on Louis de Broglie's matter–wave concept which looked completely different; it seemed to work, like all classical dynamical theories, with differential equations and other continuous mathematical tools (section 3.3.2).

Quantum and wave mechanics, although analogues of the corresponding classical theory, exhibited features alien to classical theories. Thus, after the formal equivalence of the two new atomic theories had been proved, Max Born arrived at the probability interpretation of Schrödinger's wavefunction (1926) and Werner Heisenberg discovered the uncertainty relations (1927). David Hilbert's (1862–1943) theory of linear integral equations (1904–12) could now be generalized to give quantum mechanics a sound mathematical foundation. The physicists further succeeded in developing the framework of a relativistic extension, quantum field theory, which allowed the description of atomic particles and electromagnetic radiation in a unified manner (section 3.3.3).

3.3.1. Principles and failures of the 'old quantum theory' (1913–24)

'That this insecure and contradictory foundation was sufficient to enable a man of Bohr's instinct and tact to discover the major laws of the spectral lines and of the electron shells of atoms together with their significance for chemistry appeared to me like a miracle—and appears to me as a miracle even today' wrote Einstein in his reminiscences [105]. He thus characterized the main atomic theory of the years 1913 to 1923 and especially its creator, Niels Bohr, who dominated this decade—in spite of the fact that a number of quantum physicists of the older generation, including Planck, Sommerfeld, Ehrenfest and Einstein, and some members of the younger generation, contributed most aspects of the first quantum-dynamical scheme. The scheme consisted essentially of the laws of classical mechanics, to which quantum conditions had been added, of some electrodynamics—violated by the energy–frequency relation and the existence of radiationless stationary electron orbits—and of a set of additional rules or principles to increase the number of systems treatable by the scheme and to estimate the intensities of atomic radiation. By a most careful—or 'tactful', as Einstein said—application of this contradictory and composite scheme, Bohr finally succeeded in 1921 in establishing a 'theory' of the periodic system of elements, i.e. a description of the physical and chemical properties of atoms having one to 92 electrons and their corresponding nuclear charge Z. At nearly at the same time, several authors began to show that the 'old quantum theory' not only failed to work in the cases of the two-electron helium atom and of more complex atomic systems, but eventually failed even to explain the behaviour of the hydrogen atom under the influence of particular forces.

3.3.1.1. The dynamical theory of the Bohr–Sommerfeld atomic model (1913–16).

In searching for the meaning of his quantum of action, Planck had observed in his winter lectures of 1905–06 that the two-dimensional phase space of the oscillator was divided into regions bounded by ellipses, such that the phase-space areas assumed discrete values, namely,

$$\iint dp\, dq = nh \qquad \text{with } n = 1, 2, 3, \ldots \qquad (35)$$

and q and p denoting position and momentum. Later, Nicholson assumed that the angular momentum of electrons in atoms consists of integer multiples of Planck's constant divided by 2π [44] and Bohr took this over as a condition for the stationary orbits in his atomic model [45]. William Wilson (1875–1965) in England [106], Jun Ishiwara (1881–1947) in Japan [107] and Arnold Sommerfeld in Germany [52] took over the Nicholson–Bohr quantum rule—which agrees with Planck's equation (35) in the case of the resonator—as the phase integral

$$\int p\, dq = nh \qquad (36)$$

and generalized it (as did Planck [108]) to atomic systems having several degrees of freedom. This enabled Sommerfeld to develop a dynamical description of many-electron atoms, and he thus created the tools for a quantum-theoretical treatment of the wealth of empirical data on atomic spectra.

Sommerfeld's theory rested essentially on two assumptions:

(i) for each degree of freedom a quantum condition (36) is valid,
(ii) the classical dynamical laws for multiply periodic systems apply.

The latter laws had been formulated in the nineteenth century by William Rowan Hamilton (1805–65) and Carl Gustav Jacob Jacobi (1804–51). Paul Sophus Epstein [61], Karl Schwarzschild [62] and Peter Debye [56] used them when calculating, following Bohr and Sommerfeld, the Stark and Zeeman effects of the hydrogen atom.

Sommerfeld completed the dynamical theory of the Bohr atom by not only working with the Hamilton–Jacobi partial-differential equation involving the action function S (depending on the action and angle variables, J_i and w_i) but also using the powerful and elegant mathematical method of complex integration [54]. Thus he obtained for the radial-phase integral the result

$$nh = \int \frac{\partial S}{\partial r} \, dr = \int \sqrt{A + 2\frac{B}{r} + \frac{C}{r^2}} \, dr = -2\pi i \left(\sqrt{C} - \frac{B}{\sqrt{A}} \right). \quad (37)$$

This method could be applied to a great variety of atomic problems, relativistic and non-relativistic ones, with and without external electric and magnetic fields. Of course, to achieve the success it was necessary that the Hamilton–Jacobi equation of the system could be split into a set of equations, each depending only on one pair J_i and w_i, of dynamical variables ('separable systems').

3.3.1.2. Principles of atomic theory (1913–18). Late in 1917 Bohr completed the first part of his comprehensive essay dealing with the quantum theory of line spectra, which he subtitled *On the general theory* [109]. He mentioned the recent work on the theory, achieved by Sommerfeld, Einstein, Schwarzschild and Debye, and then continued [110]:

> In spite of the great progress involved in these investigations many difficulties of fundamental nature remain unsolved, not only as regards the limited applicability of the methods used in calculating the frequencies of the spectrum of a given system but especially as regards the question of the polarization and intensity of the emitted spectral lines. These difficulties are intimately connected with the radical departure from the ordinary ideas of mechanics and electrodynamics involved in the main principles of

Quanta and Quantum Mechanics

the quantum theory, and with the fact that it has not been possible hitherto to replace these ideas by others forming an equally consistent and developed structure. Also in this respect, however, great progress has recently been obtained by the work of Einstein and Ehrenfest. On this state of the theory it might therefore be of interest to make an attempt to discuss the different principles from a uniform point of view, and especially to consider the underlying assumptions in their relation to ordinary mechanics and electrodynamics.

The 'great progress' addressed here by Bohr was contained in Ehrenfest's principle of 'adiabatic imvariance' and Einstein's statistical treatment of radiation processes.

In 1913 Boltzmann's Austrian student Ehrenfest, who was a Leyden professor and successor of Lorentz, published an equation valid for two periodic systems A and B, which could be transformed into each other by an infinitesimally slow (i.e. 'adiabatic') change of one or several parameters; he found in particular for the average kinetic energy $\bar{T}_A$ and $\bar{T}_B$ the relation [111]

$$\bar{T}_A/\nu_A = \bar{T}_B/\nu_B \tag{38}$$

which allowed quantum theory to deal with all systems B that could be 'adiabatically' obtained from an already-solved quantum system A.

In his attempts to provide a consistent derivation of Planck's blackbody law, Einstein arrived in July 1916 at a new possibility [112]. He assumed that the equilibrium in a cavity came about through the emission and absorption processes of characteristic frequencies of Bohr atoms, which occurred *statistically* according to the laws of radioactive decay. If he included in the considerations not only absorption and emission of the atomic lines but also 'stimulated emission' by incident radiation of the same frequency, and took into account the correct limiting behaviour, i.e. for infinite equilibrium temperature an infinite radiation density must exist, he obtained the desired Planck law.

> See also p 1398

Evidently, Einstein's procedure of 1916 helped Bohr to approach the problem of calculating line intensities; that is, the intensities obtained in classical electrodynamics, at least for large values of n, 'will on the quantum theory determine the probability of spontaneous transition from a given stationary state ... to a neighboring state' [113]. Here he used, as Einstein had done in 1916 (and Planck even earlier in 1906 [114]), the quite reasonable assumption that the quantum-theoretical relations had to pass, in a suitable limit (notably, for high quantum numbers), into the corresponding classical relations. Such an assumption Bohr had already applied successfully in 1913 to obtain the correct hydrogen spectrum [115]. After 1918 he would call this the 'correspondence principle', and it would play an essential role in atomic theory, starting with Hendrik Kramers' calculations of Stark-effect line intensities for his thesis [116].

In 1918 Sommerfeld's Polish assistant Adalbert (Wojcech) Rubinowicz (1889–1974) derived the polarization of spectral lines by observing that because of a conservation law the angular momentum of an atomic state (electron orbit) may change in the transition by one unit of $h/2\pi$ maximally—this he called the 'selection rule' *(Auswahlprinzip)* [117]. His assumption that both the atom *and* the emitted or absorbed radiation exhibited quantum structure was disliked by Bohr, who arrived at a result similar to that of Rubinowicz by a more involved argument [118]. Indeed, he hoped to solve the polarization problem by invoking the correspondence principle [119]. In spite of the disagreement about possible justification, other selection rules were applied in many atomic problems during the following years; later they were shown to be always connected with symmetry principles.

3.3.1.3. The Aufbauprinzip (1921) and three schools of atomic theory. 'If we admit the soundness of the quantum theory of spectra, the principle of correspondence would seem to afford perhaps the strongest inducement to seek an interpretation of the other physical and chemical properties of the elements on the same line as the interpretation of the series spectra.' [120]. Such strong words were used by Bohr to introduce, in early 1921, his most ambitious programme, namely to provide a theory of the periodic system of chemical elements (from which all physical and chemical properties of the pure substances would follow).

The idea was to construct the many-electron atoms by starting from the hydrogen atom and adding one electron after the other in the available quantum orbits, simultaneously increasing the charge of the nucleus. Bohr explained [121]:

> Thus by means of a closer examination of the progress of the binding process this [correspondence] principle offers a simple argument of concluding that these electrons are arranged in groups in a way which reflects the periods exhibited by the chemical properties of the elements within the sequence of increasing numbers. In fact, if we consider the binding of a large number of electrons by a nucleus of higher positive charge, this argument suggests that after the first two electrons are bound in one-quantum orbits, the next eight electrons will be bound in two-quanta orbits, the next eighteen in three-quanta orbits, the next thirty-two in four-quanta orbits.

Bohr enlarged on the idea in a further short note and presented a very detailed treatment in his Göttingen lectures of June 1922 [122]. One of the attentive members of the audience, Werner Heisenberg, recalled 'Bohr knew the whole periodic system. At the same time one could easily see

... that he had not proved anything mathematically, that he just knew that this was more or less the connection.' [123]. But still the scheme scored a brilliant triumph in late 1922, when its founder obtained the physics Nobel Prize for his atomic theory. It explained the properties of the just discovered element number 72, hafnium, as being the analogue of zirconium, in contrast to other claims [124].

> See also p 99

The quantum physicists (see figure 3.8), though first welcoming enthusiastically the new ideas of Bohr, reacted differently depending on their taste and education. By that time essentially three schools had been formed, directed by Bohr in Copenhagen, Born in Göttingen and Sommerfeld in Munich. Heisenberg having gone through all these schools described the characteristic qualities of their leaders by saying: 'From Sommerfeld I acquired the optimism, from the Göttingen people the mathematics, and from Bohr I learned physics.'. The most senior of these great teachers, Arnold Sommerfeld, was born on 5 December 1868 in Königsberg, where he studied mathematics and received his PhD in 1891. Then he became Felix Klein's (1868–1925) assistant at the University of Göttingen, obtained a mathematics chair at the *Bergakademie* Clausthal (1897) and later at the *Technische Hochschule* Aachen (1900), from where he proceeded in 1906 to Munich to occupy the professorship for theoretical physics. By 1911 Sommerfeld had already turned to quantum theory—dealing first with the problem of β- and γ-emission—and grew into an active supporter even before adopting Bohr's atomic theory in 1915, which he then made the centre of his scientific research for the following decade. At Sommerfeld's institute, Epstein and Rubinowicz worked out their fine contributions, while his former students, Debye in Göttingen and Landé in Berlin and later Frankfurt, picked up his ideas. In Munich Walther Kossel of the Technische Hochschule collaborated with Sommerfeld on x-ray spectra in the later 1910s. After World War I Sommerfeld educated a new generation of talented students, all in atomic theory. Among them were Wilhelm Lenz (1888–1957), Adolf Kratzer (1893–1980), Gregor Wentzel (1898–1978) and Otto Laporte (1902–71). Although they all obtained university chairs for theoretical physics, the best known of Sommerfeld's students were Wolfgang Pauli and Werner Heisenberg.

Bohr, the creator of the first successful quantum theory of atomic constitution, was born on 7 October 1885 in Copenhagen and educated at the university there (PhD 1911). After having spent several years (1911–12 and 1914–16) with Ernest Rutherford in Manchester, he was called back to Copenhagen in 1916 as professor of theoretical physics. There he accepted Hendrik Kramers of Holland as his first student. In 1918 the Swede Oskar Klein (1894–1977) arrived, followed in 1920 by the Norwegian Svein Rosseland (b 1894). After World War I, Copenhagen housed, for a shorter or longer stay, guests from all over the world: until 1925 notably from Germany, for example the seniors James Franck (1920),

Twentieth Century Physics

Figure 3.8. *The great teachers of quantum theory. Top: left—A Sommerfeld (1868–1951), right—Niels Bohr (1885–1962) in 1919; bottom: Max Born (1882–1970) in 1925.*

George de Hevesy (1920–26) and the juniors Pauli (1922–23), Heisenberg (1924–25, 1926–27), and from England Ralph Fowler (1925), from the United States John Slater (1923–24), and from Japan Yoshio Nishina (1923–28). Copenhagen remained the Mecca of postdoctoral studies in atomic theory even in the late 1920s and during the 1930s.

The last of the school founders, Max Born (born on 11 December 1882) originated from Breslau (now Wrocław, Poland); he had studied physics and mathematics in Breslau and later Göttingen, where he received his PhD in 1906. Afterwards he continued, apart from a short stay in Cambridge, England, in academic positions in both places, until he moved in early 1915 to Berlin to take up an extraordinary professorship for theoretical physics at the university. In 1919 he became *Ordinarius* at Frankfurt am Main and in 1921 at Göttingen. Although he had already adopted quantum theory in 1912 (in the work with Theodore von Kármán on the specific heats of solids), he turned to the Bohr–Sommerfeld theory of atomic structure only in 1922 (shortly before Bohr presented his lectures there). He first collaborated with his new assistant Pauli on a quantum-theoretical perturbation scheme, then later also with Heisenberg on the same topic; and he had around him Friedrich Hund and Pascual Jordan (1902–80). They all eagerly picked up atomic problems, applying mostly quite rigorous mathematical methods in the Göttingen tradition. Even before the discovery of quantum mechanics, guests from abroad came to Born, e.g. Enrico Fermi in 1923–24 and Ehrenfest in 1925. Still the main bulk of Born's earlier students worked on problems of crystal lattices, but this changed after 1925, when a considerable number of visitors, especially from the USA (where Born lectured in 1925–26) but also from other countries, poured into Göttingen. The great period of atomic physics ended in 1933, when Born and Franck had to leave Germany. (Born went to Great Britain, returned in 1953 and died on 4 January 1970 in Göttingen.)

Besides these centres, the institutes of Ehrenfest in Leyden, Ralph Fowler (1889–1944) in Cambridge and Lenz in Hamburg served as contributors to atomic theory in the early 1920s (and afterwards), all of them keeping close contacts with Copenhagen, Göttingen and Munich.

3.3.1.4. Failure of atomic theory (1922–24). In the summer of 1921 the Viennese theorist Pauli, who had studied since autumn 1918 at the University of Munich, completed his doctoral dissertation under Sommerfeld; it was devoted to the evaluation of the hydrogen molecule [125]. The author studied all possible orbits and calculated, in particular, the ground state, obtaining a binding energy $W = -0.5175 Rh$ (with R and h denoting Rydberg's and Planck's constant, respectively). In 1921 this seemed to disagree with the experiments, but in 1922—the date of publication—the situation was less clear. Not until 1923 could it be certain that Pauli's calculation demonstrated a clear failure of the celebrated Bohr–Sommerfeld theory. At that time further examples of failure existed.

See also p 101

Perhaps the first atomic system causing difficulties was that of molecular hydrogen. In his pioneering memoir of 1913, Bohr tried a dumb-bell model, but later discussions (of specific heats and the

Zeeman effect) spoke against it, hence Born suggested in 1922 a model in which the electrons possessed more subtle orbits, but without presenting any detailed calculations [126]. Instead he pointed to the case of a related, but simpler two-electron system, the helium atom, whose ground state was calculated at that time (after preliminary work by Landé in 1919) by Kramers in Copenhagen and John H Van Vleck (1899–1980) in Cambridge, Massachusetts. Kramers summarized: 'The calculations have led to the conclusion that in the model for the normal state of the helium atom ... the energy of the atom turns out to be too high by 3.9 volts in comparison to experience.' [127]. The quantum orbit in his calculation, quite similar to Born's suggestion for the hydrogen model, also did not meet the condition of stability.

While the ground-state evaluations of Van Vleck and Kramers did not seem, because of such deficiencies, fully conclusive, the treatment of the states of excited helium by Born and Heisenberg gave a crystal-clear result, as the assumed model (with one very eccentric orbit) appeared straightforward and unproblematic. Still the authors arrived at definitely wrong values for the energy terms, and their conclusion—'either the quantum conditions are wrong, or the motion of the electrons does not satisfy the mechanical equations even in the stationary states' [128]—meant that the semi-classical quantum theory of atomic structure failed. After all, it was the most refined and reliable method of the perturbation theory for multiply-periodic systems (developed by Born, Pauli and Heisenberg) which provided the basis of the helium calculation [129].

The difficulty showed up in multi-electron atoms even more obviously than in the cases of the simple atomic systems. It began with the problem of evaluating the anomalous Zeeman effects, for which Sommerfeld, Landé and Heisenberg in particular tried to establish quantum-theoretical schemes [130] which accounted also for the transformation of the splittings in strong magnetic fields [131]. Their efforts at a complete description of the multiplet and Zeeman phenomena culminated in the 'vector model' of Sommerfeld and Landé [132]. The latter presented his famous formula for the gyromagnetic g factor (determining the pattern and splitting of the spectral lines)

$$g = 1 + \frac{1}{2} \frac{J^2 - \frac{1}{4} + R^2 - K^2}{J^2 - \frac{1}{4}} \tag{39}$$

with quantum numbers K, J and R representing the length of the angular momentum vectors of the series electron, atomic core and the total atom, respectively. When Pauli tried to derive this vector formula, a typical result of 'Sommerfeld's optimism', with the help of the rigorous 'mathematics of the Göttingen people' and the physical argumentation of Bohr, he failed: the phenomenological description of the complex multiplets and their Zeeman effects could not be justified on the basis of the existing atomic theory [133]. 'It's a great misery with the theory of anomalous Zeeman effect and altogether with the atoms containing more than one electron' he complained in a letter to Sommerfeld, written on 19 July 1923.

At about the same time the old quantum theory was about to be further destroyed by the consideration of a rather subtle atomic problem; the simultaneous action of crossed electric and magnetic fields on the hydrogen atom. In 1918 Bohr had suspected that the resulting electron orbits might not be periodic at all, but Epstein five years later gave a different result [134]. The Viennese Otto Halpern (1899–1982) confirmed Epstein's conclusion, and so did detailed investigations of Oskar Klein, then in Minnesota, and Lenz in Hamburg [135]. Pauli, who followed the problem with interest, discussed the result in his handbook article on the old quantum theory and commented [136]:

> Hence the 'allowed' orbits ... pass over into the 'forbidden' orbits ... and *vice versa* ... An escape from this difficulty can be achieved only by a radical change in the foundations of the theory.

3.3.2. Göttingen quantum mechanics and Schrödinger's wave mechanics (1925–26)

> Before 1925, people were working with Bohr orbits ... They provided a pretty satisfactory nonrelativistic picture of atoms in which one electron essentially was the important one ... But there were great difficulties in understanding how two electrons would interact ... and they showed up most clearly when people tried to account for the spectrum of helium. The young people in those days were trying ... by setting up a theory for interaction of Bohr orbits, and there is no doubt that they would have continued along these lines if it had not been for Heisenberg and Schrödinger.

Thus, in April 1970, Paul Dirac (1902–84), one of the prominent actors in the game of finding the new atomic mechanics, recalled the situation he met in the first half of the 1920s [137]. In spite of many deficiencies plaguing the application of the old Bohr–Sommerfeld quantum theory—which were superficially covered by more or less arbitrary hypotheses, e.g. the introduction of half-integral quantum numbers in the description of complex spectra and the associated Zeeman effects, or the assumption that atomic cores or series electrons acted under 'mechanically non-interpretable stresses'—this game of skilful excuses might have been carried on to further refinement for many years, bringing further apparent successes, had not the new mechanics of Heisenberg and Schrödinger radically removed the old picture of electron orbits and their semi-classical treatment.

Indeed, the process of abandoning the concepts of the old quantum theory had already started before 1925, when the idea of a systematic 'discretization' was applied to several characteristic relations used to describe the behaviour of atoms, notably in expressions for the intensities of spectral lines, a procedure which was then called the 'dispersion-theoretical approach' (1923–25). In particular, in Göttingen Born and Heisenberg had been working, since the autumn of 1923, on their programme of a 'discrete quantum mechanics' (executing somehow Poincaré's demand of 1911). When Heisenberg arrived, in summer 1925, at his new 'quantum-theoretical re-interpretation of kinematic and mechanical relations', Born and Jordan in Göttingen, as well as Dirac in Cambridge, England, succeeded quickly in formulating a consistent, discrete quantum mechanics, in which all equations of classical mechanics were replaced by analogous ones, either in the matrix calculus or in the more general mathematical scheme of Dirac's 'q-numbers'. This discrete quantum mechanics was, after several months of existence, confronted with a powerful competitor, namely Schrödinger's apparently 'continuous wave mechanics'—an event which caused some puzzlement in the community of atomic physicists: all of a sudden one possessed, instead of none, *two* obviously *correct atomic theories*.

3.3.2.1. Discretization in atomic dynamics: dispersion theory (1923–25).
On 19 July 1923 Pauli wrote in a letter to Sommerfeld about the problem of the anomalous Zeeman effect and remarked in discussing the formulae for the g factor (from the Bohr–Sommerfeld theory and the data, equation (39), respectively) [138]:

> The structure of the expressions is rather similar but one *cannot* make the difference disappear by changing the normalization of the indefinite additive constants involved in the R, K and J. ... One may also say that the two expressions are related to one another like the differential quotient $(d/dJ)(1/J)$ and the difference quotient $1/J - 1/(J - 1)$, which seems to hint at something that is non-mechanical.

This observation can be considered as the first definite, written proposal to replace a classical differential expression by a difference expression in the available atomic theory. Probably independently of Pauli's proposal, Born gave in the autumn of 1923 a new programme: he called for a systematic 'discretization of physics' or a 'discrete quantum theory'. His assistant Heisenberg, who had finally come to Göttingen, after completing his PhD under Sommerfeld in July 1923, followed it immediately in his next approach to the anomalous Zeeman-effect problem, which yielded the observed half-integral quantum numbers as a natural consequence [139]. (This paper also served its author as his *Habilitation* thesis in summer 1924.) Born, on his part, wrote a paper entitled *On quantum mechanics*, in which he put the new Copenhagen 'dispersion-theory approach' on a sound basis [140].

It had been known for some years that the classical equations describing the dispersion of electromagnetic radiation by matter can be transcribed into quantum formulae, provided one replaced in the classical expressions certain quantities (e.g. the resonance frequency and the number of dispersion electrons), by the corresponding quantum-theoretical expressions, such as the atomic transition frequencies and the Einstein absorption and emission coefficients [141]. In the spring of 1924 Kramers, guided by the Bohr–Kramers–Slater radiation theory, wrote down an improved dispersion formula, equation (31) above [93]. Immediately afterwards, Born derived this formula in the following way [140]: he wrote down, for the ratio P/E (with P denoting the electric polarization induced in the scattering atoms by the action of the electric field strength E of the incident radiation) in the case of coherent scattering, the classical formula as a sum of differential expressions $(\partial \phi / \partial J)$, with respect to the action variables J of the dispersing atom. Then he replaced the differential quotients by difference quotients, i.e.

$$\partial \phi / \partial J \to \frac{\phi(n+\tau) - \phi(n)}{\tau h} \tag{40}$$

where $n+\tau$ and n denote the quantum numbers of two neighbouring stationary states of the atom. (Actually the atom has several degrees of freedom, which are classically denoted by action variables $J_k, k = 1, 2 \ldots$.)

In autumn 1924, Kramers and Heisenberg worked out, according to the prescriptions of Born's previous paper, the general case of dispersion of light, i.e. they considered the coherent scattering of Kramers' formula as well as the incoherent scattering of light [95]. This incoherent scattering consists of radiation having frequencies different from those contained in the incident electromagnetic wave (field strength E), notably the frequencies $\nu + \nu_e$ and $\nu - \nu_e$, as predicted earlier by Adolf Smekal on the light-quantum hypothesis [96]. In the calculations of Kramers and Heisenberg there appeared sums of terms (over all intermediate states R) of the characteristic form

$$M(P, Q; R) = \frac{1}{4h} \left[\frac{A_q(E.A_p)}{\nu_p + \nu} - \frac{A_p(E.A_q)}{\nu_q + \nu} \right] \exp[2\pi(\nu_p + \nu_q + \nu)t] \tag{41}$$

with ν_p and ν_q denoting the transition frequencies from an initial state P to an intermediate state R and from R to the final state Q, respectively, and A_p and A_q the corresponding *transition amplitudes*. Obviously the terms (41) not only possessed the structure of Born's 'quantum-mechanical' expressions, but also exhibited the characteristic feature that they were dependent on three quantum states (hence they sum over intermediate states on two quantum states!) of the atom. This formal feature would help Heisenberg in the decisive step to his quantum-theoretical reformulation scheme.

3.3.2.2. Quantum-theoretical reformulation and non-commuting variables (May–November 1925). Heisenberg had spent the months from September 1924 to April 1925 in Copenhagen. In the discussions with Bohr and Kramers he learned to combine the systematic mathematical formalism of Born with the physical insight of Bohr. He called the resulting procedure used in several papers, the 'sharpened correspondence principle', because it put the original correspondence arguments into a definite quantitive form [142]. Having returned to Göttingen, he used this procedure to achieve a final breakthrough in atomic theory. After an unsuccessful attempt to obtain the line intensities of the hydrogen atom, he first developed a prescription for translating the formulae for classical multiply periodic systems into quantum-theoretical ones (see figure 3.9). He replaced the classical Fourier series—describing the motion of, say of an electron around an atomic nucleus in terms of Fourier amplitudes a_τ and frequencies $\tau \nu$ (with ν the frequency of revolution)—by what he called 'quantum-theoretical Fourier series'—with transition amplitudes $a(n, n - \tau)$, depending on two quantum numbers n and $n - \tau$ of the atomic states involved in the transition, and the corresponding frequencies $\nu(n, n - \tau)$. The classical product term $b_2 \cdot \exp(2\pi i 2 \nu t) = [a_1 \exp(2\pi i \cdot \nu t)]^2$ became in quantum theory

$$b(n, n - 2) \exp[2\pi i \nu(n, n - 2)t]$$
$$= a(n, n - 1)a(n - 1, n - 2) \times \exp[2\pi i \cdot \nu(n, n - 2)t] \quad (42)$$

due to Walther Ritz's combination rule of quantum frequencies

$$\nu(n, n - 1) + \nu(n - 1, n - 2) = \nu(n - 2). \quad (43)$$

When he demanded that all variables in atomic theory should be reformulated as such Fourier sums, $\sum a(n, n - \tau) \cdot \exp[2\pi i \nu(n, n - \tau)t]$, Heisenberg immediately arrived at the surprising consequence that two quantum-theoretical variables, x and y, would not automatically commute (like they always do in classical theory), or in general,

$$x \cdot y \neq y \cdot x. \quad (44)$$

The path-breaking paper submitted in July 1925 contained, as a particular example worked out by the author in full detail, the treatment of anharmonic oscillators along the lines of the new theory [143]. Heisenberg wrote down the equation of motion and integrated it with the help of two conditions (fixing, so-to-speak, the integration constants): (i) the assumption of a state of lowest energy and quantum number n_0, thus that the transition amplitudes $q(n_0, n_0 - 1)$, $q(n_0, n_0 - 2)$, etc disappear; (ii) the reformulated 'quantum condition' ($J = \int p \, dq$), which the author

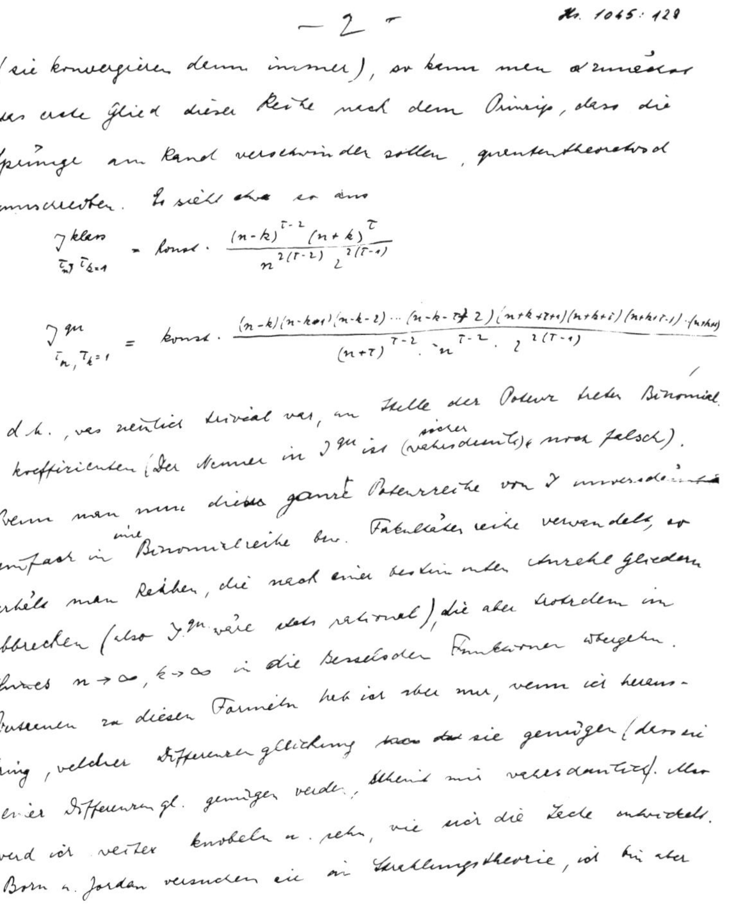

Figure 3.9. Heisenberg's quantum-theoretical reinterpretation (from a letter of Heisenberg to Ralph Kronig, dated 8 May 1925).

took over from the classical equation expressed in the standard form of equation (40) as

$$\frac{\partial}{\partial J}J = 1 = \frac{d}{dn}\frac{1}{h}\int m\left(\frac{dq}{dt}\right)^2 dt = \frac{2\pi m}{h}\sum_{\tau=-\infty}^{+\infty}\frac{d}{dn}[2\pi\nu\tau|a_\tau|^2] \quad (45a)$$

and rewrote in quantum theory as

$$1 = \frac{4\pi m}{h}\sum_{\tau=0}^{\infty}[|a(n,n+\tau)|^2 2\pi\nu(n,n+\tau) - |a(n,n-\tau)|^2 2\pi\nu(n,n-\tau)]. \quad (45b)$$

Thus he computed the solution for the anharmonic oscillator, in which stationary (i.e. time-independent) states existed, whose energy was in the lowest approximation (for the harmonic oscillator)

$$W_n = (n + \tfrac{1}{2})h\nu_0 \qquad n = 0, 1, 2, \ldots . \qquad (46)$$

Evidently, the oscillator energy states showed formally half-integer quantum numbers, $(n + \tfrac{1}{2})$, as the data of molecular spectra had earlier suggested [144]. Furthermore, the author sketched a preliminary theory of the rotator, obtaining results consistent with the intensity rules derived previously for some molecular and complex atomic spectra [145].

After having studied Heisenberg's paper, Born 'began to ponder about the symbolic multiplication' of quantum-theoretical variables [146]:

> I was soon so involved in it that I thought the whole day and could hardly sleep at night .... And one morning .... I suddenly saw light: Heisenberg's symbolic multiplication was nothing but the matrix calculus, well known to me since my student days from the lectures of Rosanes in Breslau.

Then he found, by introducing the canonical-conjugate variables, position q and momentum $p = (m dq/dt)$, that Heisenberg's equation (45b) might be written in terms of infinite matrices **q** and **p**

$$\mathbf{pq} - \mathbf{qp} = \frac{h}{2\pi i}\mathbf{1} \qquad (47)$$

with **1** denoting the unit matrix (having infinitely many diagonal elements 1 and zero elsewhere). This 'commutation relation' was proved by Jordan, who assisted Born in working out what was called 'matrix mechanics'. In their joint paper of September 1925, both outlined the first consistent quantum-mechanical scheme, which replaced the old quantum theory of periodic systems [147]; together with Heisenberg, they generalized the scheme to arbitrarily many degrees of freedom and developed a quantum-mechanical perturbation theory and the new formalism of angular momentum [148].

Independently of Born and Jordan, Dirac (see figure 3.10) created a different formulation of quantum mechanics [149]. He replaced Heisenberg's non-commuting quantum-theoretical Fourier series by what he called 'q-numbers', and introduced as the fundamental relation his translation of the classical Poisson brackets (of the Hamilton–Jacobi action-angle theory, with the canonically conjugate variables J_k and w_k), namely

$$\sum_k \left(\frac{\partial x}{\partial w_k} \frac{\partial y}{\partial J_k} - \frac{\partial y}{\partial J_k} \frac{\partial x}{\partial w_k} \right) \to \frac{2\pi}{ih}[xy - yx] \qquad (48)$$

where x and y denote any two variables of the atomic system under consideration (having f degrees of freedom, $k = 1, 2, \ldots, f$). Thus he replaced, like the Göttingen group, all relations of classical mechanics

Quanta and Quantum Mechanics

Figure 3.10. *Two creators of quantum mechanics: left—Paul Dirac (1902–84) and right—Werner Heisenberg (1901–76) (photograph taken 1929 in Chicago).*

by the corresponding ones in quantum mechanics; indeed, his q-number theory could be applied to a wider range of atomic problems than matrix mechanics [150].

3.3.2.3. Quantum mechanics as reformulated classical mechanics (1925/26). From 14 November 1925 to 22 January 1926, Born delivered at the Massachusetts Institute of Technology two series of lectures, namely: *I. The structure of the atom* and *II. Lattice theory of rigid bodies* [151]. In spite of the already recognized 'superiority of the new [quantum-mechanical] methods to the old', the lecturer presented in the first nine lectures of *Series I* the old Bohr–Sommerfeld quantum theory 'as an application of classical mechanics'. As he said in the preface of the publication, this procedure was to stress not only 'Bohr's great achievement', but at the same time the close analogy of the classical and quantum-mechanical relations: indeed, the latter fulfilled what the Copenhagen pioneer had envisioned by his correspondence principle.

The analogy is evident in the formally identical equations of the Hamiltonian scheme, i.e.

$$\frac{dq_k}{dt} = \frac{\partial H}{\partial p_k} \qquad \frac{dp_k}{dt} = -\frac{\partial H}{\partial q_k} \qquad (49)$$

for a system having arbitrarily many degrees of freedom ($k = 1, 2, 3, \ldots$) and being described by the Hamiltonian H (a function of the canonical variable pairs q_k and p_k, normally the position and momentum coordinates). In matrix mechanics, where the variables (q_k, p_k, H, etc) are expressed by matrices, the right-hand side of equation (49) could be replaced by the commutator expressions, $(2\pi i/h)[\mathbf{H}, \mathbf{q}_k]$ and $(2\pi i/h)[\mathbf{H}, \mathbf{p}_k]$, respectively. (The commutator $[\mathbf{A}, \mathbf{B}]$ of two matrices $\mathbf{A}$ and $\mathbf{B}$ is $\mathbf{AB} - \mathbf{BA}$.) Also the time derivative of all other variables $\mathbf{X}$ was given by the equation

$$\frac{d\mathbf{X}}{dt} = \frac{2\pi i}{h}(\mathbf{HX} - \mathbf{XH}). \tag{50}$$

To these fundamental equations of quantum mechanics one must add the quantum conditions or commutation relations

$$\mathbf{p}_k \mathbf{q}_l - \mathbf{q}_l \mathbf{p}_k = \frac{h}{2\pi i} \delta_{kl} \mathbf{1} \quad (\delta_{kl} = 1 \text{ for } k = l; \ = 0 \text{ otherwise}) \tag{51}$$

(any pair $\mathbf{p}_k, \mathbf{p}_l$ or $\mathbf{q}_k, \mathbf{q}_l$ commutes!).

Furthermore, the Göttingen crew noticed that the matrices representing dynamical variables in quantum mechanics were Hermitian matrices whose (real) *eigenvalues* turn out to be the *observable values* of these variables. They can be obtained by the analogue of a canonical transformation in the classical Hamilton–Jacobi scheme, i.e. a ' similarity transformation' with a unitary matrix $\mathbf{S}$ [148]. In the case of the Hamiltonian matrix $\mathbf{H}$, one arrives at the diagonal matrix $\mathbf{W}$

$$\mathbf{W} = \mathbf{SHS}^{-1} \tag{52}$$

where the diagonal elements denote the energy values.

As in classical mechanics, it is possible to write down the equations of perturbation theory to treat systems disturbed by small perturbing forces (indicated by the small parameter λ). That is, the Hamiltonian matrix $\mathbf{H}$ can be written as

$$\mathbf{H}(\mathbf{p}_0, \mathbf{q}_0) = \mathbf{H}_0(\mathbf{p}_0, \mathbf{q}_0) + \lambda \mathbf{H}_1(\mathbf{p}_0, \mathbf{q}_0) + \lambda^2 \mathbf{H}_2(\mathbf{p}_0, \mathbf{q}_0) + \ldots \tag{53}$$

where the unperturbed $\mathbf{H}(\mathbf{p}_0, \mathbf{q}_0)$ is already a diagonal matrix (i.e. its energy states are known) and the perturbation terms are expressed by a power series in λ. Then the diagonalization of the matrices $\mathbf{H}_1, \mathbf{H}_2$, etc, is obtained in successive steps with the unitary matrix $\mathbf{S}(= 1 + \lambda \mathbf{S}_1 + \lambda^2 \mathbf{S}_2 + \ldots)$ in an expansion according to powers of the parameter λ, namely through the set of equations,

$$\mathbf{H}_0(\mathbf{p}_0, \mathbf{q}_0) = \mathbf{W}^{(0)}$$
$$\mathbf{S}_1 \mathbf{H}_0 - \mathbf{H}_0 \mathbf{S}_1 + \mathbf{H}_1 = \mathbf{W}^{(1)}$$
$$\mathbf{S}_2 \mathbf{H}_0 - \mathbf{H}_0 \mathbf{S}_2 + \mathbf{H}_0 \mathbf{S}_1^2 - \mathbf{S}_1 \mathbf{H}_0 \mathbf{S}_1 + \mathbf{S}_1 \mathbf{H}_1 - \mathbf{H}_1 \mathbf{S}_1 + \mathbf{H}_2 = \mathbf{W}^{(2)} \tag{54}$$
$$\vdots$$

Equations (54) constitute the suitable analogues of the classical differential equations of the Hamilton–Jacobi perturbation theory involving the 'characteristic function' S and the action angle variables J and w (we drop the indices denoting the degrees of freedom), i.e.

$$H_0 = W^{(0)}$$
$$\frac{\partial H}{\partial J_0}\frac{\partial S_1}{\partial w_0} + H_1 = W^{(1)}$$
$$\frac{\partial H_0}{\partial J}\frac{\partial S_2}{\partial w_0} + \frac{1}{2}\frac{\partial^2 H}{\partial J^2}\left(\frac{\partial S_1}{\partial w_0}\right)^2 + \frac{\partial H_1}{\partial J}\frac{\partial S_1}{\partial w_0} + H_2 = W^{(2)} \quad (55)$$
$$\vdots$$

In fact, with the help of Dirac's prescription for replacing the classical Poisson brackets, it is easy to obtain equations (54), considering the Göttingen matrices as a special case of q-numbers.

Evidently, the formal analogy of classical and quantum mechanics becomes even more visible in Dirac's scheme. Dirac had no problems in taking over all Poisson-bracket-like equations of the Hamilton–Jacobi mechanics [149]. Thus he wrote, for example

$$\frac{dJ_k}{dt} = [J_k, H] = 0$$
$$\frac{dw_k}{dt} = [w_k, H] = \frac{\partial H}{\partial J_k} \quad (56)$$
$$[J_k, w_l] = \delta_{kl}, [J_k, J_k] = [w_k, w_l] = 0$$

with $[x, y] = (2\pi/ih)[xy - yx]$ in quantum mechanics (being identical with the corresponding Poisson brackets in classical mechanics). Furthermore, he simply replaced the classical (angular) frequencies of multiply periodic systems by the quantum-mechanical frequencies

$$(\tau\omega) = \omega\left(J, J - \tau\frac{h}{2\pi}\right) = \left[H(J) - H\left(J - \frac{\tau h}{2\pi}\right)\right] / \frac{h}{2\pi}. \quad (57)$$

While in the matrix scheme there were problems in dealing with the solution of complex atomic problems, Dirac had no difficulty, in principle, with any multiply-periodic system and even with an aperiodic collision situation, such as Compton scattering [150].

3.3.2.4. The creation of wave mechanics (November 1925–July 1926). In the third chapter of the long three-man paper on matrix mechanics, Born introduced the most general form of the Göttingen theory, applying his teacher's mathematical theory of linear integral equations. David Hilbert had given a detailed treatment of infinite quadratic forms and matrices, and a method to obtain the eigenvalue spectrum, which included discrete and continuous values [152]. Born took over Hilbert's methods which seemed to offer the hope of dealing with, say, the hydrogen atom, but he did not get anywhere. However, Pauli was able to calculate the discrete Balmer terms, without reference to the 'Göttingen erudition' (*Gelehrsamkeitsschwall*) [153].

In spite of the failure, Born continued to think about a generalization of matrix mechanics that also allowed the treatment of atomic systems having continuous energy states, such as scattering problems. Thus he suggested, in a paper written together with the MIT mathematician Norbert Wiener (1894–1964), a new formulation of quantum mechanics, based on time-dependent operators of the type [154]

$$q(t) = \lim_{\tau \to \infty} \frac{1}{2\tau} \int_{-\tau}^{+\tau} ds\, q(t,s). \tag{58}$$

They could be made to follow the rules of quantum mechanics, and so to satisfy the commutation relations. With the help of harmonic analysis, the matrix elements of the Born–Heisenberg–Jordan scheme were obtained, i.e.

$$q(V,W) = \lim_{\tau \to \infty} \int_{-\tau}^{+\tau} \exp\left(-\frac{2\pi i}{h}Vt\right) q(t) \exp\left(\frac{2\pi i}{h}Wt\right) dt \tag{59}$$

where V and W denote two energy states of the discrete or continuous spectrum (which in the former case yield discrete quantum numbers). The authors applied their 'operator mechanics' to the cases of the harmonic oscillator (reproducing the results of the matrix method) and to the uniform motion of an electron; in the latter case they observed that, in spite of the commutation relation (which established the quantization), the energy spectrum was continuous (as expected from the classical theory and experiment).

What Born and Wiener did not realize was that Hilbert's old integral equation theory offered the possibility of generalizing matrix mechanics even further. This was shown by Cornelius Lanczos (1893–1974) of Frankfurt University in a paper submitted at nearly the same time [155]. Inverting Hilbert's original procedure, he replaced the matrix equations of Born and his co-workers by integral equations; their kernels then represented the dynamical variables of quantum mechanics and acted on a complete set of 'eigenfunctions'. Since he followed different, partly cosmological goals, Lanczos did not turn to practical consequences; furthermore, he refrained from going over to the more practical differential equation method, otherwise he might have arrived at the Schrödinger equation.

So far the inventors of the new quantum mechanics came from the tradition of Bohr and Sommerfeld, whose theory of atomic structure

Erwin Schrödinger

(Austrian, 1887–1961)

Erwin Schrödinger, the creator of wave mechanics, taught physics in Vienna (*Habilitation* 1912), Jena (1920), Stuttgart, Breslau and Zurich (1921–27). His early work dealt with radioactivity (experimental), kinetic theory (dynamics of crystal lattices) and fluctuation phenomena (e.g. proof of the statistical nature of radioactive decay, 1918). He contributed substantially to physiological optics, notably colour theory (1919–25), before he discovered the wave-mechanical equations for atoms and molecules (1925–26).

Succeeding Planck to the theoretical chair at the University of Berlin, he left Germany in 1933 for Oxford (1933–36) and Graz (1936–38). Fleeing a second time from the Nazi government, he settled in Dublin, Ireland, as director at the Institute for Advanced Studies (1939–55). After his retirement he came back to Austria. In later years Schrödinger was interested in general relativity and unified (classical) field theory, and he worked on problems of meson physics. His book *What Is Life* (1944) stimulated progress in biology.

---

they tried to replace by new concepts and mathematics without losing the analogy or correspondence to the old description. Two convictions guided the breakthrough in 1925: first, the distrust of the old pictures of atomic theory, notably electron orbits and the detailed behaviour of atomic objects due to the laws of classical mechanics; second, the necessity to incorporate the discreteness observed in quantum phenomena directly into the new quantum-mechanical laws. Both guidelines could be traced back to Bohr's first ideas of 1913.

Schrödinger, born on 12 August 1887 in Vienna, was only two years younger than Bohr and had gone through a distinguished career as a physicist—having contributed importantly to the kinetic theory of solids and their specific heats, radioactivity, statistical phenomena and to physiological optics, notably colour vision and theory. After leaving Vienna in 1920, he occupied professorships in Jena, Stuttgart and

Breslau, before settling in late 1921 in Zürich in the university's chair for theoretical physics. Since 1921 he had occasionally treated problems of atomic structure, e.g. suggesting the penetration of eccentric orbits of the outer electron into inner shells (1921) or criticizing a Born–Heisenberg result on the theory of hydrogen-like atoms (1925). From 1924 onwards he focused on a programme dealing with quantum statistics, especially with the proper method of counting identical atomic particles. In the course of his work he encountered, in November 1925, the doctoral thesis of Louis de Broglie [86], and it put him quickly on the road to a new atomic theory.

The concept of matter waves helped Schrödinger first to solve a problem in Einstein's new gas theory [156], but, and more importantly, it also set him to work on de Broglie's geometrical phase-wave construction of electron orbits in atoms. This aspect seemed to have a connection with some earlier work of Schrödinger [157], who now tried to extend the construction to the orbits in the Stark and Zeeman effect cases without getting very far. Then he began a different approach by writing down a relativistic wave equation for the electron (mass m, charge $-e$) in a hydrogen atom (i.e. in the Coulomb field of its nucleus), namely (with $\Delta = (\partial^2/\partial x^2) + (\partial^2/\partial y^2) + (\partial^2 \partial z^2)$, and $r = \sqrt{x^2 + y^2 + z^2}$)

$$\Delta \psi + \frac{4\pi^2 m^2 c^4}{h^2}\left[\left(\frac{h\nu}{mc^2} + \frac{e^2}{m^2 c^2 r}\right)^2 - 1\right]\psi = 0. \tag{60}$$

The solution by the standard method of eigenvalues yielded energy states which disagreed with those of Sommerfeld's fine-structure formula of 1915 [53], hence Schrödinger abandoned equation (60); however, he soon (around Christmas 1925) discovered that the non-relativistic approximation yielded the correct Balmer spectrum. He worked out the whole theory in some detail, and on 27 January 1926 the *Annalen der Physik* received his paper containing an exhaustive treatment of the hydrogen problem [158].

In motivating his new non-relativistic hydrogen wave equation, he started from the classical Hamilton–Jacobi equation

$$H\left(q, \frac{\partial S}{\partial q}\right) = E \tag{61}$$

with E and H denoting the energy and the Hamiltonian of the one-electron problem, the latter being a function of its position coordinates q and the canonical momentum p, expressed as a derivative of the action function S. He then replaced S through the transformation

$$S = K \log \psi \tag{62}$$

with K being a constant (of the dimension of an action) and ψ a function of the position variables. This led to the wave equation

$$H\left(q, \frac{K}{\psi}\frac{\partial \psi}{\partial q}\right) = E. \tag{63}$$

Now Schrödinger inserted the classical, non-relativistic action function into the expression for H, and derived the Euler–Lagrange equation for the corresponding variational problem. Thus he finally obtained the hydrogen equation, written in Cartesian position coordinates x, y, z as

$$\frac{\partial^2 \psi}{\partial x^2} + \frac{\partial^2 \psi}{\partial y^2} + \frac{\partial^2 \psi}{\partial z^2} + \frac{2m}{K^2}\left(E + \frac{e^2}{r}\right)\psi = 0 \tag{64}$$

a non-relativistic approximation to equation (60), provided he chose $K = h/2\pi$.

The solutions of this second-order differential equation, which he demanded to be finite and unique everywhere, could be separated into two types:

(a) for negative 'energy eigenvalues' only discrete solutions existed, i.e.

$$E_n = -\frac{2\pi^4 m e^4}{h^2 n^2} \qquad n = 1, 2, 3 \ldots \tag{65}$$

(b) all positive energy values $E > 0$ gave the continuous solutions of equation (64). The whole solution always provided two pieces of information: firstly, the eigenfunctions, ψ_n or ψ_E and, secondly, the eigenvalues.

In the following months, Schrödinger sought to obtain wave equations also for many-electron atoms and molecules. He first developed a 'mechanical-optical analogy', working with wave propagation in a Riemannian space (determined by a line element derived from the kinetic energy expression for the classical system), and found that the spatial part of the wavefunction had to satisfy the equation [159]

$$\text{div grad}\,\psi_q + \frac{8\pi^2}{h^2}(E - V)\psi_q = 0. \tag{66}$$

Soon afterwards, in March 1926, he noticed that equation (66) would also follow by translating the classical Hamilton function $H(q, p)$ into a quantum-mechanical operator, $H(q, (h/2\pi i)(\partial/\partial q))$, i.e. replacing the momentum p by $h/2\pi i$ times the partial derivative with respect to the position coordinate q and having it act on the wavefunction ψ, notably [160]

$$H\left(q, \frac{h}{2\pi i}\frac{\partial}{\partial q}\right)\psi(q) = E\psi(q). \tag{67}$$

Equation (67) could also be generalized to describe time-dependent, even relativistic systems [161].

The author and many other physicists soon started to apply the Schrödinger equation to many problems of atomic theory: band spectra (Erwin Fues, Schrödinger), intensities of x-ray spectra (Gregor Wentzel), Stark effect (Schrödinger, Ivar Waller, Wentzel), the helium problem (Heisenberg), collision phenomena (Born) and the Compton effect (Walter Gordon). The methods of differential equations in wave mechanics proved to be much more powerful and easier to handle than the clumsy matrix method or the tricky q-number formalism, especially since the theoreticians were accustomed to solving differential equations.

3.3.3. Physical interpretation and mathematical foundation (1926—1933)

As mentioned earlier, Lanczos, the Hungarian outsider in quantum theory, recognized towards the end of 1925 the 'closest connection' between the matrix scheme of quantum mechanics and a formulation in terms of integral equations: that is, the equations of the Göttingen people could be rewritten in what he called a 'field-like representation' and 'depending on the nature of the mathematical problem, one might preferably choose one or the other representation' [162]. Schrödinger's wave mechanics, published soon afterwards, surpassed Lanczos' integral method by far with respect to its applicability and efficiency in solving physical problems. Again it exhibited a close connection to matrix mechanics, as shown by Schrödinger himself in March 1926 [160]. In spite of 'fundamentally different starting points, presentation, methods, and in fact the whole apparatus', the author then demonstrated explicitly that [163]:

> To each function of the position- and momentum-coordinates [in wave mechanics] there may be related a matrix in such a way that these matrices, *in every case satisfy* the formal calculation rules of Born and Heisenberg. ... The solution of the natural *boundary-value problem* of this differential equation [in wave mechanics] is completely equivalent to the solution of Heisenberg's algebraic problem.

The crucial point to be established consisted in fixing the proper, detailed relations between the matrix and the wavefunction representation of any given physical quantity.

Independently of Schrödinger, Pauli in Hamburg (in a letter to Jordan) and Carl Eckart (1902–73) in Pasadena derived the desired equivalence relations [164]. Pauli, at the end of his letter, also pointed to deeper conceptional questions that had to be answered, such as:

> The problem of the asymptotic linkage [of the new quantum or wave mechanics] with the usual pictures in space and time for the limiting case of large quantum numbers remains unsolved.

Yet it is definite progress to be able to see the problems [of atomic theory] from two different sides [i.e. from matrix mechanics and wave mechanics]. It seems that one also sees now, from the point of view of quantum mechanics, how the contradistinction between 'point' and 'set of waves' fades away in favor of something more general.

While Wentzel, Léon Brillouin (1889–1979) and Kramers soon worked out the asymptotic linkage [165]—and also, for example, recovered the Bohr–Sommerfeld quantum conditions—it took more than a year, until the end of 1927, before an answer was found for the more general and important question of the physical interpretation; the probability interpretation of the wavefunction by Born and the uncertainty relations of Heisenberg being crucial steps on the way to Bohr's all-embracing 'principle of complementarity'. At the same time, the mathematical scheme for a unified quantum mechanics was perfected by the so-called 'transformation theory' of Dirac and Jordan and the development of more rigorous methods, e.g. for dealing with non-bounded operators, by the mathematicians around Hilbert, especially by Johannes (John) von Neumann (1903–57).

Last, but not least, from the conceptual viewpoint there remained the problem of applying quantum theory to electromagnetic radiation (and, of course, also to relativistic particles). For this reason, by the time of the Born–Jordan paper of 1925, a first attempt had been made at the quantization of the free electromagnetic field. (Indeed, the self-consistency of quantum theory could not be considered as reasonably established until the late 1940s, when the renormalization theory of quantum electrodynamics was formulated.) The early trials of field quantization (including the papers of Jordan and Oskar Klein and of Jordan and Eugene Wigner which will be discussed below) were very much in Bohr's mind when he wrote down his ideas on complementarity.

Before going into these important developments occurring between 1926 and 1928, let us sketch briefly the biographies of some of the actors. Apart from Bohr and Born, only a few of the experienced quantum theorists played a leading role, the most senior of these being Kramers, the others Heisenberg and Pauli, who were then still youngsters (consequently the new physics used to be called '*Knabenphysik*'). Kramers was born in 1894 in Rotterdam and joined Bohr during World War I after studies with Paul Ehrenfest in Leyden. In Copenhagen, over many years, he gradually improved his master's atomic models. Although his original programme eventually failed, many of Kramers' results, especially the dispersion formula of 1924, remained valid. While occupying professorships in Utrecht (1926–33) and Leyden (1933–52, here succeeding Ehrenfest), he contributed further remarkable work in the new atomic theory and quantum electrodynamics. However, in terms of brilliance, Bohr's previous top collaborator was overtaken by the newcomers of the 1920s, not only by Pauli and Heisenberg, but also by Dirac, Jordan and von Neumann, all born between December 1901 and December 1903.

The Englishman Paul Dirac, after studying electrical engineering at Bristol, went to do research in theoretical physics under Fowler in Cambridge. In summer 1925, he seized on Heisenberg's quantum-mechanical ideas and gave them his own original formulation. From then on, every one of his papers revealed important insights, applications and extensions of quantum mechanics: in 1926, for example the relation between statistics and the symmetry of the wavefunction under the exchange of identical particles and the transformation theory; in 1927, crucial steps towards quantum electrodynamics; in 1928, the relativistic electron equation. From the latter he created, in 1931, the concept of the anti-particle. In 1932 he assumed the Lucasian Chair at Cambridge University, once occupied by Isaac Newton.

Born's student Pascual Jordan, who received his PhD in 1924, became famous when, in 1925, he helped his professor to establish matrix mechanics. In 1926, in parallel with Dirac, he proposed the transformation theory of the quantum-mechanical schemes. By 1928, he had become professor at Rostock and in 1944, he succeeded Max von Laue in Berlin. After World War II he taught at the University of Hamburg.

The Hungarian John (originally Janos, in German Johannes) von Neumann, the youngest of the newcomers, studied mathematics in Berlin and Zurich, and received his PhD in Budapest in 1926. Then he occupied junior positions at Berlin and Hamburg, before going as a professor in 1930 to the Princeton Institute of Advanced Study (he stayed in Princeton until his death in 1957). As a guest of David Hilbert in Göttingen from 1926, he completed essential parts of the mathematical foundations of quantum mechanics.

3.3.3.1. Born's statistical interpretation and the Dirac–Jordan transformation theory (June–December 1926). On 25 June 1926, the *Zeitschrift für Physik* received a preliminary note from Göttingen, entitled *On the quantum mechanics of collision processes*; four weeks later there followed a detailed paper on the same subject [166]. Their author Born had been trying to treat these processes involving 'aperiodic motion' in quantum theory for years, without really getting to grips with it [154, 167]. Now Schrödinger's wave mechanics allowed an attack on this problem with greater success, because it included continuous energy eigenvalues in a natural way. Born therefore studied, in first-order perturbation approximation, the scattered wave of, say, an electron at asymptotic distances from the scattering object, either an atom or a potential. If Ψ_m^0 denoted the initial wavefunction of the problem, with m indicating the original quantum state of the scattering atom, then the asymptotic scattered wave in the direction (α, β, γ) could be written as (ω is the solid angle)

$$\psi_{n\tau}^{(1)} = \sum_m \int \int \Phi_{nm}^{(\tau)}(\alpha, \beta, \gamma) \sin\left(\frac{2\pi}{\lambda_{nm}^{(\tau)}}(\alpha x + \beta y + \gamma z + \delta)\right) \psi_m^{(0)} \, d\omega \quad (68)$$

where $\lambda_{nm}^{(\tau)}$ gives the de Broglie wavelength of the scattered particle (derived from the energy $h\nu_{nm}^{(0)} + \tau$), with $\nu_{nm}^{(0)}$ denoting the m transition frequency of the undisturbed scattering atom and τ the energy of the incident particle. Born commented [168]:

> If one wants to interpret the result in a corpuscular manner, then only one interpretation is possible: the probability is proportional to the square of the quantity Φ_{nm}, for the case that the electron coming from the z-direction is scattered into the direction determined by [the cosines] α, β and γ (with a phase shift δ), its energy τ being increased by a quantum $h\nu_{nm}^{(0)}$ at the expense of the atomic energy.

This result now led the author to important conceptual consequences, as he quickly added:

> Schrödinger's quantum mechanics yields a very definite answer on the [scattering] problem; but one does not obtain a causal relation ... One only obtains an answer to the question: 'How probable is a given effect of the collision?' ... Here the whole problem of determinism arises. From the point of view of our quantum mechanics no quantity exists which in a single collision fixes its effect causally. ... I myself tend to the opinion that in the world of atoms, determinism has to be abandoned.

Twenty-eight years later, in December 1954, Born was awarded the Nobel Prize for physics 'for his fundamental work in quantum mechanics and especially for his statistical interpretation of the wavefunction'.

The statistical interpretation received quick acceptance among most quantum physicists (but Schrödinger violently, and Einstein more gently, disagreed). Born himself derived some consequences when he worked on the adiabatic hypothesis in quantum mechanics: in particular, he demonstrated (as Ehrenfest showed in the old quantum theory) how the classical concept of forces can be taken over to a large extent into the new theory [169]. The statistical interpretation played an even more important role in Jordan's quantum-mechanical 'transformation theory'. The idea of applying canonical transformations to solve dynamical problems had entered the Bohr–Sommerfeld quantum theory by 1916, via the Hamilton–Jacobi theory, where it served to calculate the energy states of atoms. Canonical transformations plus the associated perturbation scheme could be taken over into matrix mechanics and wave mechanics.

The formal equivalence of matrix and differential equations, as shown by Schrödinger, Pauli and Eckart, then suggested in spring 1926 the possibility of a different application of canonical transformations: to change matrix relations into wave-mechanical ones, and vice versa. Fritz London worked out the first detailed transformation theory of that type [170].

Later, in December 1926, Dirac submitted his complete demonstration of the equivalence of all existing quantum-mechanical schemes; his transformation scheme included the use of the singular delta-function, which is zero everywhere except at the origin where it becomes infinite [171], for example he transformed the equation for the diagonalization of the energy matrix

$$H(p,q)S(q,W) = S(q,W)W \tag{69}$$

via the intermediate step

$$H\left(\frac{h}{2\pi i}\frac{\partial}{\partial q}, q\right)S(q,W) = WS(q,W) \tag{70}$$

into the Schrödinger equation

$$H\left(\frac{h}{2\pi i}\partial q, q\right)\psi_W(q) = W\psi_W(q). \tag{71}$$

He concluded [172]:

> The eigenfunctions of Schrödinger's wave equation are just the transformation functions (or the elements of the transformation matrix [S]) that enable one to transform from the (q) scheme of matrix representation into a scheme in which the Hamiltonian is a diagonal matrix.

In section 7 of his paper, Dirac showed 'that the present method is in agreement with the assumption formerly used that the square of the amplitude of the wavefunction in certain cases determines a probability'. A much more explicit role was assigned by Jordan to Born's interpretation in his 'new foundation of quantum mechanics', which he submitted for publication in December 1926 [173]. He started directly from 'probability amplitudes' ϕ, which he assumed (on Wolfgang Pauli's suggestion) to depend on two variables x and y, such that $|\phi(x,y)|^2 dx$ is the probability for the variable x to assume values between x_0 and $x_0 + dx$ if y has the fixed value y_0. The author gave the rules (postulates), which were obeyed by these amplitudes, and he found in the case of the canonical pair of variables, p and q, the following result: for q assuming a fixed value, all values for p become equally probable. He also showed that $\phi(x,y)$ satisfied in special cases the Schrödinger equation. After introducing suitable linear operators in his scheme, he finally proved the equivalence between a field-like matrix *a la* Lanczos and the wave-mechanical equivalent.

3.3.3.2. Uncertainty relations and complementarity (1927). At the end of his transformation-theory paper, Dirac wrote [174]:

> If one describes the state of a system at an arbitrary time by giving numerical values to the coordinates and momenta, one cannot actually set up a one–one correspondence between the values of these coordinates and momenta initially and their values at a subsequent time.

This appears to have been a forecast of what Heisenberg would work out in his fundamental paper of March 1927 and what Bohr would discuss generally in terms of 'complementarity' later in autumn of the same year.

The genesis of this development, which provided *the* basis for the physical interpretation of quantum mechanics, can be traced back to events following April 1926. At that time, Heisenberg gave a talk on quantum mechanics in the Berlin colloquium and afterwards discussed the principles of his theory with Einstein. Unexpectedly, the latter criticized the speaker's philosophy of using only observable quantities, claiming: 'The theory decides what we can observe.'. This rejection of Einstein's own previous philosophy (used especially in 1905 to obtain special relativity) surprised Heisenberg. Then, in July 1926 he listened in Munich to Schrödinger's colloquium on wave mechanics and criticized in the discussion the speaker's continuum interpretation. Two months later, Schrödinger came to Copenhagen, where he debated heatedly with Bohr and Heisenberg the question whether quantum mechanics contained an essential discontinuity ('quantum jumps'). In a paper submitted soon afterwards, Heisenberg found 'an argument for the fact that a continuous interpretation of the quantum-mechanical formalism, and also of the de Broglie–Schrödinger waves would not fit into the nature of the known formal relations.' [175].

Although he had used wave-mechanical methods in the helium problem, Heisenberg now returned to pure matrix theory as the most adequate expression of the quantum discontinuities. On the other hand, if that theory really accounted for all atomic processes, it had also to describe the apparently continuous paths of electrons in cloud chambers. In endless discussions with Bohr in Copenhagen, and in letters to Wolfgang Pauli in Hamburg, Heisenberg exposed vague ideas, such as: 'It is meaningless to speak of a position of a particle with a definite velocity. But if one does not take too seriously the accuracy of using the notions of velocity and position, then it may well make sense.' [176].

When Dirac (like Heisenberg, then in Copenhagen) and Jordan developed their transformation schemes, they provided Heisenberg with the mathematical tools to prove his ideas. In particular he started from a Gaussian distribution of width q_1 for the position q of an atomic particle

$$S(\eta, q) \sim \exp\left(-\frac{(q - q')^2}{2q_1^2} - \frac{2\pi i}{h} p'(q - q')\right) \qquad (72)$$

where η represents other properties. Then he evaluated the momentum distributions, $S(\eta, p) = \int_{-\infty}^{+\infty} S(\eta, q) S(p, q)\, dq$, by inserting Jordan's ansatz, $S(p, q) = \exp(2\pi i p q / h)$, obtaining

$$S(\eta, p) \sim \exp\left(-\frac{(p - p')^2}{2p_1^2} + \frac{2\pi i q'(p - p')}{h}\right). \qquad (73)$$

This also described a Gaussian distribution of width p_1 for the momentum and q_1, for the position if the product of the uncertainties for position and momentum assumed the value

$$q_1 p_1 = \frac{h}{2\pi}. \qquad (74)$$

Of course, this is usually written $\Delta q \Delta p = h/2\pi$ (see figure 3.11 [177]).

In his published paper, Heisenberg demonstrated the validity of equation (74) in the cases of the Stern–Gerlach effect, the width of atomic spectral lines, and some *Gedanken* experiments of Bohr and others [178]. He had, in fact, guessed it already from discussing the simultaneous p and q measurement of an electron by short-wavelength radiation (γ-ray microscope). Bohr, who was absent when Heisenberg completed and submitted the manuscript, criticized this discussion; thus Heisenberg supplied an addendum in the proofs stating [179]:

> In first place, the uncertainty of the observation does not rest entirely on the occurrence of discontinuities, but is directly connected with the demand to do justice simultaneously to the different experiences, which are expressed by the corpuscular theory, on the one hand, and by the undulatory theory, on the other.

In particular, the divergence (angle ε) of the illuminating γ-ray (of wave length λ) led to a resolving power

$$\Delta q = \frac{\lambda}{\sin \varepsilon} \qquad (75)$$

following only from the wave theory, while the Compton recoil determined the momentum uncertainty $\Delta p (\approx (h/\lambda) \sin \varepsilon)$.

Even at the time of the greatest successes of his atomic theory, Niels Bohr had kept in mind its difficulties and deficiencies; he was

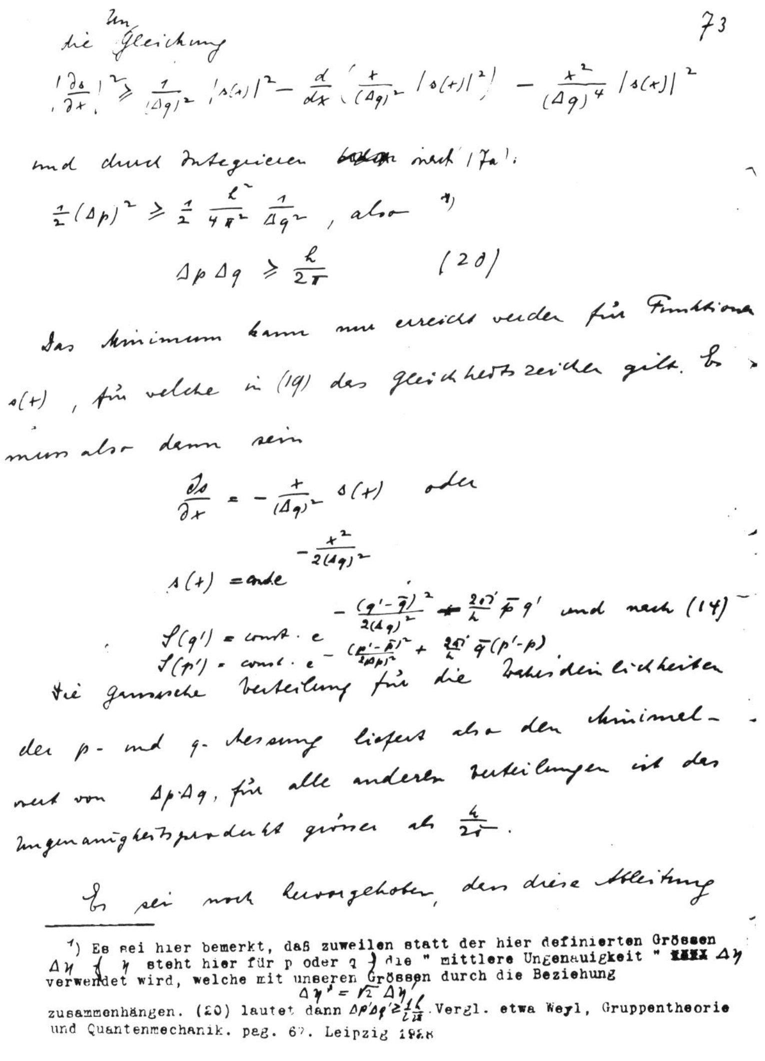

73

die Gleichung

$$\left|\frac{\partial s}{\partial x}\right|^2 \geq \frac{1}{(\Delta q)^2}|s(x)|^2 - \frac{d}{dx}\left(\frac{x}{(\Delta q)^2}|s(x)|^2\right) - \frac{x^2}{(\Delta q)^4}|s(x)|^2$$

und durch Integrieren nach (7a):

$$\frac{1}{2}(\Delta p)^2 \geq \frac{1}{2}\frac{h^2}{4\pi^2}\frac{1}{\Delta q^2}, \text{ also }^{1)}$$

$$\Delta p\,\Delta q \geq \frac{h}{2\pi} \qquad (20)$$

Das Minimum kann nur erreicht werden für Funktionen $s(x)$, für welche in (19) das Gleichheitszeichen gilt. Es muss also dann sein

$$\frac{\partial s}{\partial x} = -\frac{x}{(\Delta q)^2}s(x) \quad \text{oder}$$

$$s(x) = \text{konst}\, e^{-\frac{x^2}{2(\Delta q)^2}}$$

$$S(q') = \text{const}\cdot e^{-\frac{(q'-\bar{q})^2}{2(\Delta q)^2} + \frac{2\pi i}{h}\bar{p}q'} \quad \text{und nach (14)}$$

$$S(p') = \text{const}\cdot e^{-\frac{(p'-\bar{p})^2}{2(\Delta p)^2} + \frac{2\pi i}{h}\bar{q}(p'-p)}$$

Die Gaussche Verteilung für die Wahrscheinlichkeiten der p- und q-Messung liefert also den Minimalwert von $\Delta p\cdot\Delta q$, für alle anderen Verteilungen ist das Ungenauigkeitsprodukt grösser als $\frac{h}{2\pi}$.

Es sei noch hervorgehoben, dass diese Ableitung

¹) Es sei hier bemerkt, daß zuweilen statt der hier definierten Grössen $\Delta\eta$ (η steht hier für p oder q) die " mittlere Ungenauigkeit " $\Delta\eta'$ verwendet wird, welche mit unseren Grössen durch die Beziehung $\Delta\eta' = \sqrt{2}\,\Delta\eta$ zusammenhängen. (20) lautet dann $\Delta p'\Delta q' \geq \frac{h}{2\pi}$. Vergl. etwa Weyl, Gruppentheorie und Quantenmechanik. pag. 67. Leipzig 1928

Figure 3.11. *Derivation of the uncertainty relation.*

especially aware of the 'formal nature' of the old quantum theory, which put into question the possibility of 'forming a consistent picture of phenomena' [180]. He felt the weight of the inconsistencies again in April 1925, after the Bothe–Geiger experiment led to a complete refutation of the interpretation of the Compton effect in the Bohr–Kramers–Slater radiation theory. In a note written a little later, Bohr said [181]:

It must, however, be stressed that the question of a coupling [as the Bothe–Geiger experiment demanded] or of an independence [as the Bohr–Kramers–Slater theory suggested] of the individually observable atomic processes cannot simply be regarded as a choice between two well defined conceptions concerning the propagation of light in empty space, corresponding perhaps to a corpuscular or undulatory theory of light. It involves rather the problem how far spacetime pictures in terms of which physical phenomena have hitherto been described can be applied to atomic processes.

The new quantum-mechanical theories of the years 1925 and 1926 excited constant debates in Copenhagen among Bohr and his associates and guests. The Schrödinger visit in September 1926 confirmed Bohr's opinion that one had to abandon the space-time pictures of classical physics. The wearying discussions, especially with Heisenberg, led Bohr to take off for a winter vacation in Norway in February 1927, from which he returned with the outlines of a 'complementarity view' concerning atomic phenomena. In the autumn of 1927 he formulated his ideas in two lectures [182], presented at the Volta Centenary Congress at Lake Como in Italy in September and at the Solvay Congress in Brussels in October (see figure 3.12).

The basic point, which Bohr displayed in these lectures, was that he did not focus on the detailed formulations of quantum mechanics but rather on an analysis of the underlying concepts and their contradictory features. The two relations for energy and momentum of radiation

$$E = h\nu \qquad \text{and} \qquad p = \frac{h}{\lambda} \qquad (76)$$

intimately tied together the classical concepts of particles and waves. Similarly, he detected in the description of atomic processes both corpuscular and undulatory aspects, which he now declared to be necessary in forming the full physical picture of microscopic phenomena in general. Heisenberg's relation (74) then determined quantitatively the range of the 'complementary' aspects by assigning limits to the simultaneous measurement of physical quantities which do not commute. With these elements Bohr put down the foundation of a new 'theory of measurement' in atomic physics.

3.3.3.3. Completion of the mathematical apparatus (1926–29). The mathematical tools of quantum mechanics, which the physicists had adopted so far, came from Hilbert's theory of linear integral equations, notably the treatment of infinite matrices. Still this theory lacked some elements and extensions needed for the new atomic theory: firstly, the *statistical* nature of the wavefunctions, which could be related to 'Hilbert space vectors', had not yet been included in the mathematical scheme; secondly, the matrices or (integral and differential) operators in quantum

Figure 3.12. *The great reunion of old and new quantum theorists (Fifth Solvay Conference at Brussels in October 1927). Back row, left to right: A Piccard, E Henriot, P Ehrenfest, Ed Herzen, Th De Donder, E Schrödinger, E Verschaffelt, W Pauli, W Heisenberg, R H Fowler, L Brillouin. Centre row, left to right: P Debye, M Knudsen, W L Bragg, H A Kramers, P A M Dirac, A H Compton, L de Broglie, M Born, N Bohr. Front row, left to right: I Langmuir, M Planck, M Curie, H A Lorentz, A Einstein, P Langevin, Ch E Guye, C T R Wilson, O W Richardson. (Absent: W H Bragg, H Deslandres and E Van Aubel.)*

mechanics were not of the 'bounded' type, for which most of the previous results of Hilbert and his followers were valid. Now after the physicists had provided the task and a preliminary treatment of the problem, the senior Göttingen mathematicians again offered their help. In the winter semester 1926–27, Hilbert delivered a lecture course on the mathematical methods of the new quantum theory, the main results of which he published in a joint paper with his assistant Lothar Nordheim (born 1899) and his young guest von Neumann [183]. The paper started by giving the 'physical axioms' for Jordan's probability amplitude; then the authors wrote down the properties and relations of linear operators that should eventually represent the physical quantities, and defined canonical transformations on them; finally, they gave explicit expressions for the physical, Hermitian operators, such as those describing the position and momentum of atomic objects, and formulated the stationary and the time-dependent Schrödinger equations as

$$\left\{ H\left(\frac{h}{2\pi i}\frac{\partial}{\partial x}, x\right) - W \right\} \psi(xW; qH) = 0 \tag{77}$$

and
$$\left\{\left(\frac{h}{2\pi i}\frac{\partial}{\partial x}, x\right) + \frac{h}{2\pi i}\frac{\partial}{\partial t}\right\}\psi(xt;qT) = 0. \tag{78}$$

Here $\psi(xW;qH)$ and $\psi(xt;qT)$ for a given energy value W (of the Hamiltonian H) and a given time t (of the 'time operator' T), respectively, denote the probability amplitudes of the position variable q to assume the value x.

All the quantities introduced and their relations were defined with care, but Hilbert (figure 3.13) and collaborators made free use of Dirac's delta-function in their formalism. In his first contribution, presented on 20 May 1927 to the Göttingen Academy, von Neumann, who continued to deal with the mathematical foundations of quantum mechanics, expressed 'severe mathematical doubts' against this 'improper eigenfunction' (exhibiting so 'absurd properties') [184]. In order to avoid this tool, he proposed to restrict the mathematical function spaces for discrete and continuous spectra, working with 'bounded Hilbert sequences' (i.e. $u_1, u_2, \ldots$ satisfying the condition $\sum |u_n|^2 = 1$) and square integrable functions (satisfying $\int |\psi(q)|^2 \, dq = 1$) and applied the Fischer–Riesz theorem to establish a one-to-one correspondence between the two spaces. Then he presented the axioms for the thus defined 'abstract Hilbert space' (i.e. complete and separable infinite-dimensional vector space having a positive-definite metric), in which the vectors satisfied a Schwarz inequality and Hermitian operators could be defined. He also introduced projection operators P (with the property $P^2 = P$) for writing down the eigenvalue problems. He finally removed the limitation of boundedness in a paper published in 1929 [185].

Besides von Neumann, the Leipzig mathematician Aurel Wintner (1903–58) dealt with a generalization of the older calculus of bounded (infinite) matrices to non-bounded ones [186]. In spite of the fact that many subtle questions, e.g. the problem of normalizing the wavefunctions for continuous eigenstates, remained open (and are still being discussed today), the mathematical foundation of the formalism of non-relativistic quantum mechanics was considered to be firmly established, and in von Neumann's book of 1932 one could find all the necessary details [187].

3.3.3.4. Quantum field theory—non-relativistic and relativistic (1926–33). In his handbook article on general principles of wave mechanics, Pauli wrote [188]:

> Now, we must speak about a peculiar mathematical method which comes from Jordan and Klein [case of symmetrical states] and Jordan and Wigner [case of antisymmetrical states] and may be called repeated quantization of waves in the usual three-dimensional space. This method arose from considering the

Figure 3.13. *Two mathematicians who made essential contributions to quantum mechanics: left—David Hilbert (1862–1943) and his former student Hermann Weyl (1885–1955) in Göttingen, mid-1920s.*

analogy between material particles with symmetrical states, on the one hand, and light-quanta of radiation, on the other hand. One may doubt, whether one is dealing with a really deep-going physical analogy, and it is also certain that all results of wave mechanics can also be obtained without applying this method. But it must be mentioned at least as a method of calculation.

The sceptical judgment on 'repeated' or 'second quantization' (as the partisans would baptize it) might be understood in the case of non-relativistic particle systems; but for relativistic systems this method opened the path to a quite essential and genuinely fundamental scheme, quantum field theory. The first hint of the new method can be traced back to the last section of the Göttingen three-man paper on matrix mechanics, where Jordan suggested a quantization of electromagnetic eigenvibrations in a given volume for the purpose of deriving Einstein's fluctuation formula (26) [189]. In early 1927, Dirac went a step further when he treated the emission and absorption of radiation by (non-relativistic) atoms [190]. He evaluated the perturbation problem of light interacting with atoms with the help of complex operators b_r and b_r^* (whose absolute square, $b_r^* b_r$, defines the probability of the system being in the state r), which follow the commutation relations

$$b_r b_s^* - b_s^* b_r = \delta_{rs} = \begin{cases} 1 \text{ for } r = s \\ 0 \text{ else} \end{cases} \text{ and } b_r b_s - b_s b_r = b_r^* b_s^* - b_s^* b_r^* = 0. \tag{79}$$

He then applied the new formalism to the problem of interaction of an assembly of light-quanta with an atom and showed that Einstein's laws for the emission and absorption of radiation follow.

In a series of papers, written alone or with collaborators in Copenhagen and Göttingen, Jordan generalized Dirac's method [191]. The idea guiding his efforts was, on the one hand, to demonstrate the necessity of a quantization of waves (against Schrödinger's opinion) and, on the other hand, to develop from relativistic one-particle equations the theory of relativistic many-body systems by quantizing the electromagnetic potentials and the wavefunctions in the equations. Thus the non-relativistic approximation, in which the finite velocity of the field propagation is neglected, served only as a preparation for the final, higher goal.

The procedure of Jordan and collaborators consisted of expanding the field functions ψ in terms of the eigenfunctions $u_n(q)$—the latter depending only on the configuration-space variables, i.e. positions and spin. Then the time-dependent probability coefficients, $a_n(t)$, and their complex-conjugates, $a_n^*(t)$, satisfied the commutation (or anti-commutation) relations

$$a_n a_m^* \mp a_m^* a_n = \delta_{nm} \qquad a_n a_m \mp a_m a_n = a_n^* a_m^* \mp a_m^* a_n^* = 0 \tag{80}$$

where the minus sign applies to Bose particles and the plus sign to Fermi particles, respectively. (It should be noted that the variables a are connected with the particle number operator as $N = a_m^* a_m$.) The complex field functions ψ and ψ^* themselves obey similar relations, namely

$$\psi(q)\psi^*(q') \mp \psi^*(q')\psi(q) = \delta(q - q')$$
$$\psi(q)\psi(q') \mp \psi(q')\psi(q) = \psi^*(q)\psi^*(q') \mp \psi^*(q')\psi^*(q) = 0 \tag{81}$$

with $\delta(q - q')$ denoting the product of a delta-function in the space variables and a Kronecker delta-function in the spin variables.

In all cases the authors proved the equivalence of their 'second quantization method' with the method of the n-particle Schrödinger equation (with $n = \sum \int \psi \psi^* dv$, i.e. taking the space integral and the spin-sum). Later Vladimir Fock (1898–1974) in Leningrad formulated second quantization as 'configuration (Fock)-space method'; he constructed systematically from the Bose- or Fermi-type operators a and a^* the time-dependent wavefunction $\Phi(N_1, N_2, \ldots t)$ of $N_1, N_2, \ldots$ particles in the quantum states labelled by 1,2,..., which followed a time-dependent Schrödinger equation [192]

$$-\frac{h}{2\pi i}\frac{\partial \Phi}{\partial t} = H\Phi(N_1, N_2, \ldots, t). \quad (82)$$

The next step involved, as mentioned above, the relativistic one-particle equations, notable the Klein–Gordon equation [193]

$$\Delta \psi - \frac{\partial^2}{\partial t^2}\psi - \frac{4\pi^2 m^2 c^2}{h^2}\psi = 0 \quad (83)$$

for the Bose case, and the Dirac equation for the Fermi case. The latter was proposed by Dirac early in 1928, in an attempt to escape from a difficulty in the probability interpretation of the Klein–Gordon theory: if one introduced new types of field, the four-component spinors ψ_μ, obeying the linear differential equation

$$\left(i\sum_{\nu=1}^{4}\gamma_\nu \frac{\partial}{\partial x_\nu} + \frac{m}{h/2\pi}\right)\psi_\mu = 0 \quad (84)$$

(with the four 4 × 4 gamma matrices γ_ν), the difficulty disappeared and the electron spin $\frac{1}{2}h/2\pi$ followed automatically [194]. The author later noticed that equation (84) actually described not only the electron but also a positively charged Fermi particle, which he identified first with the proton (the hydrogen nucleus); in 1931 however, Dirac claimed it to be the yet unknown 'anti-particle' of the electron having the same mass [195].

The (second) quantization of relativistic fields was again begun by Jordan. In December 1927 he proposed with Pauli a 'quantum electrodynamics of chargeless (i.e. free) fields [196]. Then it took Pauli and Heisenberg more than a year to write down the equations of relativistic quantum theory of interacting fields in a Lagrangian scheme for the field quantities ψ and their canonically conjugate momenta $\pi = \partial L/(\partial \psi/\partial t)$ (with the Lagrangian L), notably the equations of motion plus the relativistic quantization conditions (with a generalized delta-function) [197]. In dealing with interaction terms, they immediately found divergent expressions, e.g. for the self-energy of the electrons. The attempts to remove them would occupy the physicists for the following two decades.

The question of extending the uncertainty relations for the case of relativistic fields turned out to be much less problematic. After preliminary statements of Heisenberg, Lev Landau (1908–68) and Rudolf Peierls (b 1907) claimed the puzzling result that the field-strength uncertainty product for large times becomes arbitrarily small [198], but Bohr and Léon Rosenfeld (1904–74) settled the problem soon in favour of (the standard) finite uncertainties for non-commuting variables of the electromagnetic fields, as one would expect from Heisenberg's principle [199].

3.4. The world of microphysics (1925-35)

> We now know—what the Greek philosophers have already guessed before—that matter which we can see and touch is composed of enormously many, invisibly small corpuscles called atoms. Only in our century what previously had been guess, speculation or phantasy has achieved the state of scientific knowledge substantiated by immediate experimental proof. Hence one can call our century, as far as physics is concerned, the century of atomic research.

Thus Pascual Jordan introduced the first chapter of his book *Physics of the Twentieth Century*, in which he separated the topics of physics into two classes [200]:

> There are large and rich fields in which the atomic constitution of matter does not play a role. ... We now are used to talk about macroscopic physics (or 'macrophysics') if referring to these investigations, in which the existence of atoms cannot be recognized. On the other hand, we call 'microphysics' the research which delves deeply into the atomistic subtleties of matter.

The fields of microphysics have been reviewed in numerous articles and books addressed to a varied public [201]. Max Born's lectures on 'modern physics' of 1933, which appeared two years later in English as the book *Atomic Physics*, provided a fair account of the topics of microphysics. It started with the kinetic theory of gases, introduced the recently recognized 'elementary particles'—i.e. electron, proton, neutron and cosmic-ray particles, then discussed the structure of atoms and their radiation, electron spin and the two quantum statistics and finally the constitution of molecules [202].

Born wrote in the preface of his book that he would not cover all the topics of modern physics, especially not the latest fields of research, nuclear and cosmic-ray physics. Similarly, with one exception, they will not be included in our discussion of the applications of quantum mechanics (section 3.4.1). On the other hand, Born's book and comparable ones on atomic physics, published especially by English speaking authors, avoid the discussion of the conceptual and philosophical implications of quantum mechanics, which Jordan and some other middle-European colleagues often emphasized in their writings [203]. The debates on these important epistemological consequences, conducted between the representatives of the 'Copenhagen interpretation' of microphysics and their opponents (notably Albert Einstein and Erwin Schrödinger), will be reviewed in section 3.4.2. In spite of the enormous success of quantum mechanics, achieved in the late 1920s and early 1930s, new experimental findings stimulated the impression, especially among the great pioneers of the

Quanta and Quantum Mechanics

theory, that a still further revolution in theoretical physics had to be accomplished if one wanted to explain the phenomena observed in nuclear and high-energy physics. Although later developments have shown that a suitably formulated relativistic quantum mechanics is able to describe satisfactorily so far *essentially all* high-energy problems in microphysics, many physicists still hope for a new theory 'beyond the standard model' (section 3.4.3).

3.4.1. Applications of quantum mechanics (1925–32)
The discovery of quantum mechanics started, according to Paul Dirac 'the golden age in theoretical physics, and for a few years after that it was easy for any second rate student to do first rate physics' [204]. A similar feeling was expressed in 1929 by John Slater, another actor in the game, even more optimistically, when he wrote: 'The time has come when physics [i.e. the new atomic theory] is in a position to explain all the properties of matter.' [205]. Three large fields received particular efforts by the quantum physicists (not all 'second rate students', see figure 3.14), who were indeed able to solve most problems connected with the atomic constitution of matter: atomic and molecular physics, solid-state physics and scattering problems in atomic and nuclear physics.

3.4.1.1. Atomic and molecular structure and the group-theoretical model (1925–32). As we have mentioned above, the story of the 'old quantum theory' ended with the latter being unable to explain the behaviour of even the simplest atoms. After the discovery of the new quantum-mechanical schemes, these atomic and molecular problems were again attacked and successfully solved. Thus the matrix-mechanical treatment of the angular momentum served Werner Heisenberg, Born and Jordan to establish the intensity rules for atomic and molecular spectra (previously obtained through semi-empirical guesses), and Pauli solved with it the crossed-field problem of the hydrogen atom [206]. By including the spin property of the electron, Heisenberg and Jordan obtained perhaps the greatest triumph of matrix mechanics: they were able to derive all observed phenomena connected with the anomalous Zeeman effect [207].

Even greater practical successes were scored by those theoreticians who had been using, since February 1926, the wave-mechanical methods to calculate molecular band spectra (Schrödinger, Erwin Fues) and the Stark effect of atoms (Schrödinger, Paul Epstein). While most of these results were familiar from earlier semi-empirical considerations, this was not so in the case of the helium atom. On 5 May 1926, Heisenberg reported from Copenhagen to Pauli [208]:

> We have found a rather decisive argument that your exclusion of equivalent orbits [of electrons in atoms] is connected with the singlet-triplet separation. Consider the energy written as a

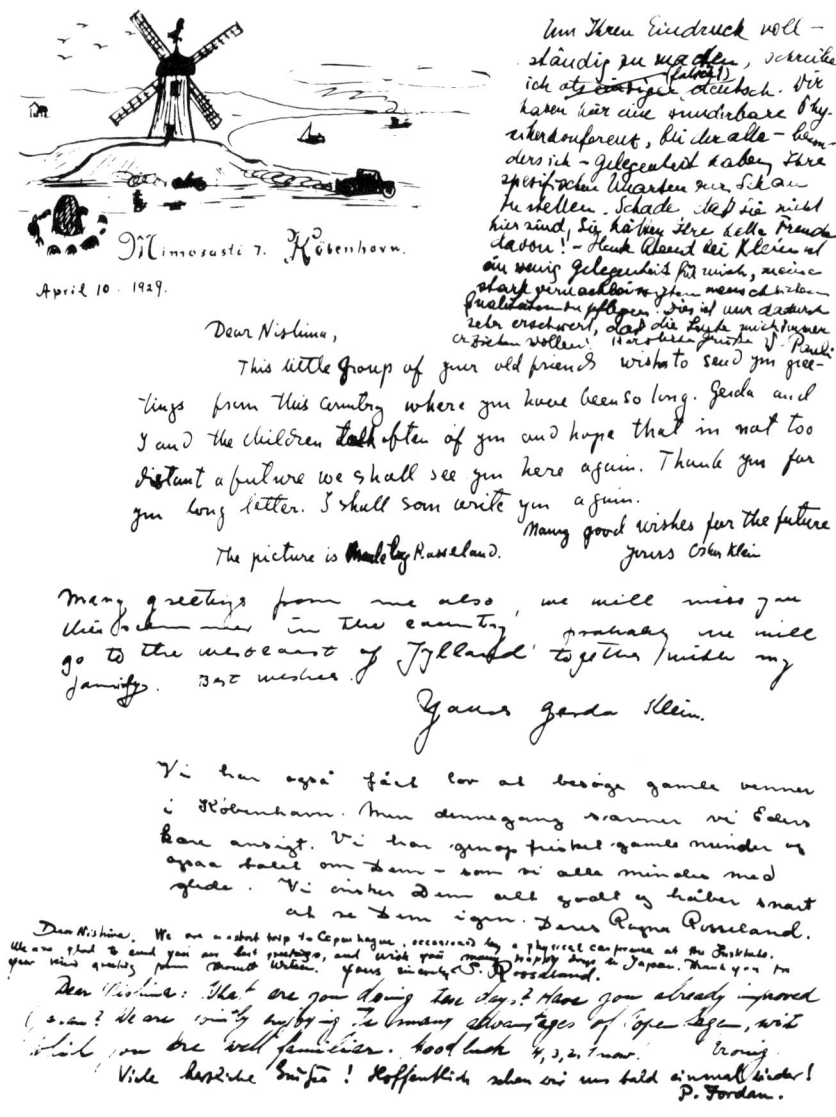

Figure 3.14. *Letter to Yoshio Nishina in Tokyo (dated 10 April 1929, written from Copenhagen and signed by W Pauli, G Klein, O Klein, R Rosseland, S Rosseland and P Jordan).*

function of the transition probabilities: Then a large difference results if one has the transition to 1S, or if, according to your ban, one puts them equal to zero. That is, para and ortho [helium] have different energies, independently of the interaction between magnets [i.e. small interaction between the spin magnetic moments of the electrons].

In his published papers [209] Heisenberg expressed the physical idea underlying his solution in two ways: first, the matrix-mechanical perturbation treatment of the two-electron system, starting from two degenerate states, led to a splitting of the energy terms having the order of magnitude of the Coulomb energy; second, in wave mechanics this splitting was created by the overlap of the wavefunctions of the two electrons (which could be expressed by an 'exchange integral').

Heisenberg's exchange integral paved the breakthrough to understanding the nature of chemical binding forces. Previously, one had been able only to explain polar (or ionic) binding of positively and negatively charged ions, but failed in all cases of molecules that did not consist of ions, such as hydrogen, nitrogen, etc. Walter Heitler (1904–82) and Fritz London (1900–54) now used the exchange integral (or the overlap of the wavefunctions) to derive the 'covalent' binding force [210]. Their application of wave mechanics and slightly different methods—developed especially by Friedrich Hund, Erich Hückel (1896–1980), Robert Mulliken (1896–1986) and Linus Pauling (1901–94)—founded 'quantum chemistry', i.e. the mathematical (or semi-mathematical) treatment of molecules in inorganic and organic chemistry [211] (see figure 3.15 [212]). London and E Eisenschütz then succeeded in 1930 in providing a quantum theory of the weakest forces in chemistry, the Van der Waals forces acting between the molecules of inert gases [213].

Between 1928 and 1932 three books were published (first in German) on the topic 'group theory and quantum mechanics', written by the mathematicians Hermann Weyl (1885–1955) and Bartel Van der Waerden (b 1903) and the physicist Eugene Wigner (1902–95) [214]. As Heisenberg wrote in his review of the first book [215]:

> The quantum mechanics of many-body problems led the physicists into mathematical difficulties which could only be solved with the help of group theory, hence a book, which supplies the mathematical tools in connection with the physical problems, is of utmost importance for the physicist.

He knew quite well what he was talking about, because he had, together with Wigner, pioneered the first explicit applications of group-theoretical methods to the physical problems of many-electron systems [216]. Their example was immediately followed in papers on atomic and molecular problems by Hund, Heitler, London and Weyl; such rich fruits were obtained that the books mentioned above were highly welcomed by the physicists. With respect to the main groups of interest, Weyl stated [217]:

> Two groups, the *group of rotation in 3-dimensional and the permutation groups*, play here the principal role, for the laws governing the possible electronic configurations grouped about the stationary nucleus of an atom or an ion are spherically symmetric with respect to the nucleus, and since the various

Twentieth Century Physics

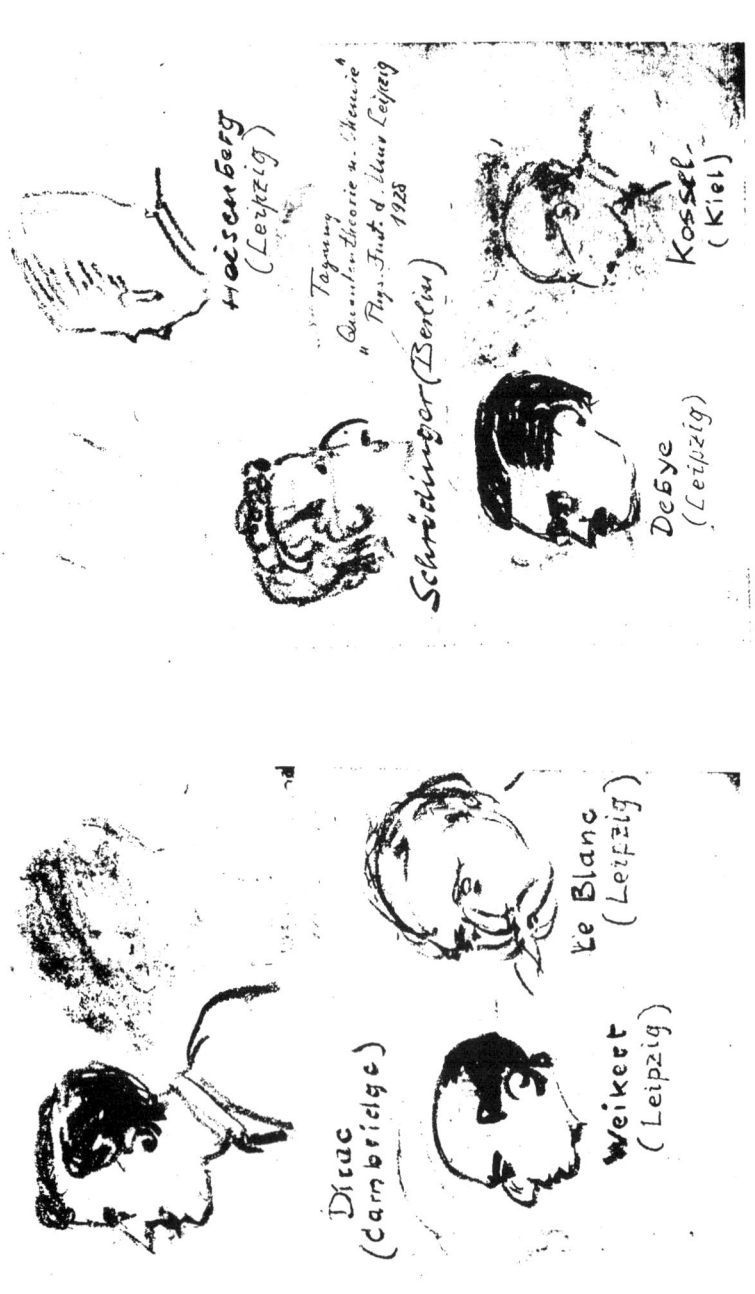

Figure 3.15. *From the early days of quantum chemistry (drawings by W Buchheim at the first Leipzig 'Universitätswoche' 1928).*

electrons of which the atom or ion is composed are identical, these possible configurations are invariant under a permutation of the individual electrons. *All quantum numbers, with the exception of the so-called principal quantum number, are indices characterizing representations of groups.*

Besides the two most obvious ones other groups also entered the treatment of quantum-mechanical problems in due course, since (according to the mathematical theory) all conservation laws in microphysics can be associated with symmetry groups: thus for energy and momentum conservation of non-relativistic atomic systems, one has the groups of time and space translations, and in relativistic systems besides these groups, the group of Lorentz transformations also plays a decisive role [218]. However, the excessive use of group theory also stimulated protests by some quantum physicists. 'It has been rumoured that the "group pest" is gradually being cut out by physics' mentioned Weyl in the preface to the second edition of his book and continued [219]:

> This is certainly not true, in so far as the rotation and Lorentz groups are concerned; as for the permutation group, it does indeed seem possible to avoid it with the help of the Pauli exclusion principle.

On the other hand, John Slater fought against the permutation group by proposing an alternative method, the so-called 'Slater determinant' [220]. Still, as Weyl argued, the group-theoretical method did not disappear from the physicists' papers at all; in the 1930s it gained impetus with the discovery of new symmetry properties in microphysics, notably the isotopic spin symmetry in nuclear structure and high-energy collisions.

3.4.1.2. The electron theory of metals and other solids (1926–33). Quantum mechanics, together with electron spin, gave rise to enormous progress in the theory of the solid state. Ralph Fowler was the first to use Fermi statistics for explaining some properties of the high-density matter in stars [221]. Pauli, on the other hand, used it to solve the first problem in metal physics. He obtained the temperature-independent paramagnetism of alkaline earths (which had resisted all previous attempts on the basis of the classical Langevin theory or the old quantum theory) [222]. He in turn stimulated Arnold Sommerfeld to study systematically the problem of metal electrons, treated as free Fermi particles. Sommerfeld immediately scored an important breakthrough. By considering the electrons as an *ideal degenerate Fermi gas at normal temperatures*, he removed several difficulties of the previous classical Lorentz theory: thus he calculated the correct factor in the electric to thermal conductivity ratio; and he derived reasonable voltaic potentials and equations for the thermionic emission of electrons, the thermoelectric effects, and the behaviour of electric conductivity in magnetic fields (Hall effect) [223].

Sommerfeld's 'free-electron gas hypothesis' needed, of course, justification, since the real situation of electrons in metals is quite different. However, when Felix Bloch (1905–83) discussed the properties of the eigenfunctions of the electrons moving in periodic potentials (as provided by the metal ions in crystal lattices), he found them to account for a certain free mobility, enough to guarantee the results obtained on the free-electron assumption [224]. His considerations were refined over the next three years by Hans Bethe (b 1906), Peierls, Philip Morse (1903–85), Léon Brillouin and Ralph Kronig (b 1904). Kronig, for example, in 1931 presented the popular one-dimensional model calculation showing the following features: the energy of an electron in a crystal lattice cannot take on every value but is restricted to a series of more or less wide 'energy bands', separated from each other by gaps [225]. Peierls and Brillouin, on the other hand, worked out the details of Bloch's ideas in the case of three-dimensional crystal lattices [226]. Brillouin, in particular, gave an easily visualizable construction for the propagation of electrons in crystal lattices by drawing critical 'Brillouin lines' and 'Brillouin zones' in the reciprocal lattice of the 'electron wave vectors'; i.e. if this vector of length $2\pi/\lambda$ (with λ the de Broglie wavelength) and with direction of the propagation of the electron assumes values on the critical line or in the critical zones, the electron is strongly reflected and cannot propagate.

Herbert Fröhlich (1905–92), another main contributor to the topic, summarized the situation of metallic conductivity in his book as follows [227]:

> The creation of an electron current by an electric field is possible only if the crystal possesses energy bands which are not filled up completely. The difference between metals and insulators consists in the fact that the latter contain only filled-up energy bands, while the former possess at least one band which is only partially occupied [by the electrons] ... Closed electron shells [of atoms] always contain an even number of electrons filling a certain number of bands just completely. Atoms with one valence electron ... therefore also represent metals (the alkaline elements, copper, gold, silver). In order that atoms with two valence electrons be metals, it is necessary that the energy band of these electrons overlaps partially with a higher band.

This theory of electrons in solid crystals not only explained the behaviour of metals and insulators, but also the properties of a class of materials in between, the semiconductors: their bands do not overlap at zero temperature but do so increasingly for higher temperatures, hence their electric conductivity has a positive instead of a negative temperature coefficient [228].

The weak binding of electrons in their crystals determines the electric and thermal conductivity of metals. For the explanation of another property, ferromagnetism, a tight binding of the electron spins at the lattice points has to be considered. Heisenberg assumed exchange forces to act between the electrons at neighbouring lattice points and noted that under certain conditions the lowest energy state of a crystal occurs when all electron spins are parallel; he thus provided the first theoretical estimate for the strength of the inner 'Weiss field' and solved the main riddle of ferromagnetism [229]. His assistant Bloch succeeded in deriving fine details of the low-temperature behaviour of ferromagnets by introducing the formalism of spin waves [230]. Although later investigations revealed a much richer complexity than could be described by the original ideas of Heisenberg and Bloch, their work established the firm basis of a consistent theory of ferromagnetism.

3.4.1.3. Scattering problems (1928). Scattering experiments, such as the studies of electron impacts on atoms by James Franck and Gustav Hertz, had played a crucial role in establishing quantum theory. Because the first quantum-mechanical schemes dealt with multiply periodic problems, they were not adapted to describe scattering problems. However, by the spring of 1926 Dirac had already found a clever way of treating the Compton effect when extending his q-number method to a relativistic scheme [231]. Wave mechanics then provided a more suitable description of scattering phenomena, as Born demonstrated in his two fundamental papers of June and July 1926 (which we discussed above in section 3.3.3) [166, 232]. Soon afterwards Walter Gordon (1893–1940) in Berlin and Gregor Wentzel in Leipzig studied the specific problems of the Compton effect and the photoelectric effect with wave-mechanical methods [233].

In dealing with molecular problems, Hund showed that an electron cannot be kept in a potential well if it is not infinitely high [234]. This first observation of what was later called 'tunnel effect' stimulated Bloch's theory of electrons in crystal lattices of 1928 and the nearly simultaneous theory of the nuclear alpha-decay, obtained independently by George Gamow (1904–68) and by Ronald Gurney (1898–1953) and Edward Condon (1902–74) [235]. Upon studying the behaviour of an alpha-particle moving in the potential well of a heavy atomic nucleus, these authors found a 'scattering solution': it corresponded to an exponentially decaying state (with the alpha-particle penetrating through the barrier of the nuclear potential and moving away from the nucleus), and the decay constant λ was obtained to have the right order of magnitude and to satisfy the empirical Geiger–Nuttal rule (relating log λ to the energy E of the alpha-particle).

Quantum-mechanical scattering theory also contributed in another way to understanding processes in nuclear physics. The classical

Rutherford formula of 1911, which fitted the observed alpha-particle scattering from atomic nuclei extremely well, could be reproduced in wave-mechanical calculations [236]. Deviations were expected for larger energies of the alpha-particles, because they are able to penetrate into regions where the Coulomb law is not valid. They were indeed observed experimentally, at the same time as another quantum effect: if two alpha-particles collide, one must take into account the Bose statistics; that is, their wavefunction is, according to Dirac's analysis [237], symmetric in the coordinates of the two particles, i.e.

$$\psi_{mn}(1,2) = \psi_m(1)\psi_n(2) + \psi_m(2)\psi_n(1). \tag{85}$$

(In the case of Fermi particles the wavefunction $\psi_{mn}(1,2)$ would be antisymmetric!) As a consequence, the classical formula for that particular case has to be expanded by additional terms; thus the Rutherford factor was replaced in quantum mechanics by

$$\csc^4\theta \to \csc^4\theta + \sec^4\theta + 2\csc^2\theta \sec^2\theta \cos u \tag{86}$$

where $u = (8/137)(c/v)\log(\cot\theta)$, with θ the scattering angle and v the velocity of the alpha-particle. This result of Mott's scattering theory was substantiated by experiments of his Cambridge colleagues [238].

Finally, we return to the problem of Compton scattering, because the previous treatment of 1926 (using the Klein–Gordon equation) had not taken into account the electron spin. This task was accomplished by Oskar Klein and Yoshio Nishina (1890–1951), who in autumn 1928 used the Dirac equation, and derived a new equation for the intensity I of the radiation scattered by the angle θ

$$I = I_0 \frac{e^4}{m_e^2 c^4 r^2} \frac{\sin^2\delta}{[1+\alpha(1-\cos\theta)]^3} \left\{ 1 + \frac{\alpha^2(1-\cos\theta)^2}{2\sin^2\delta[1+\alpha(1-\cos\theta)]} \right\} \tag{87}$$

with I_0 denoting the intensity of the incident radiation, δ the angle of observation and α the quantity $h\nu/m_e c^2$ (m_e and e mass and charge of the electron, ν frequency of the incident radiation) [239]. This expression deviated by the factor within the curly brackets from the former expression with 'spin-zero electrons', i.e. by terms of the order α^2. (The deviation from the classical Thomson-scattering formula was even larger, i.e. of the order α.)

The Klein–Nishina formula constituted, besides the calculation of the relativistic hydrogen spectrum (by Charles G Darwin and Gordon in 1928), the only acceptable early consequence derived from Dirac's relativistic electron theory. As Klein showed immediately afterwards, the reflection of Dirac electrons from a reasonably high potential barrier yielded a paradoxical result: the electrons, in spite of the electric forces acting against them, penetrate through and arrive at the other side with negative kinetic energy [240]. Klein's paradox originated from the fact that the theory included negative energy states, which Dirac interpreted as 'holes' in a filled up '(Dirac) sea of negative energy states' identifying them first as protons (1930) and finally as new 'anti-particles' to the electron (1931) [241]. While the relativistic theory plagued the experts with further paradoxes and inconsistencies for the next 15 years, the application of the hole concept in non-relativistic atomic and solid-state physics provided many good results [242].

3.4.1.4. The expanding community of quantum physicists (1925–33). At the turn of the year 1927 to 1928 a great revision occurred in the middle-European university chairs for theoretical physics, involving several of the pioneers of quantum mechanics: Werner Heisenberg, from Copenhagen, obtained the professorhsip at the University of Leipzig, Gregor Wentzel, from Leipzig, succeeded Schrödinger (who went to Berlin replacing Planck) at the University of Zurich, and Wolfgang Pauli went to the Zurich ETH (which Peter Debye left in favour of the experimental professorship in Leipzig). Leipzig and Zurich now collaborated very closely with each other and with the previous centres of atomic theory in Munich, Copenhagen and Göttingen. Besides producing joint papers, the new professors exchanged students and assistants: thus Felix Bloch came to Leipzig from Zurich, got his PhD with Heisenberg and later occupied assistant and lecturer's positions, spending some time in between at Utrecht (with Hendrik Kramers 1929–30) and Copenhagen (with Bohr 1931–32); Rudolf Peierls left Munich in 1928 for Leipzig but completed his PhD with Pauli in Zurich, before visiting Rome and Cambridge (1932–33) on a Rockefeller Fellowship. Indeed, quite a few students and guests oscillated between Zurich and Leipzig, stopping at Göttingen, Leyden, Cambridge and Copenhagen *en route*.

Sommerfeld's institute, where Pauli, Wentzel and Heisenberg had obtained their introduction in atomic theory, witnessed a particular revival. Because of the old master's ability to adapt and teach the new wave-mechanical methods [243], numerous visitors from all over the world flooded into Munich. For example, in 1926 Condon and Pauling arrived, and in 1927 William Houston (1900–68) and Carl Eckart. The latter two, in particular, participated actively in the last major original contribution of Sommerfeld to quantum physics, the electron theory of metals; together with visitors in Leipzig and Zurich, like John Slater, these Americans took the new topics home, where their own schools began to flourish in the early 1930s and have continued to do so ever since [244]. Sommerfeld also stimulated another field of 'applied quantum mechanics' astrophysics, as his student Albrecht Unsöld (b 1905) joined older spectroscopists, like the American Henry

Norris Russell (1877–1957), in 1928 and helped establish the physics of stellar atmospheres [245].

Chemistry, metal physics and astrophysics were the immediate beneficiaries of non-relativistic quantum mechanics. In 1932 nuclear physics joined them, but simultaneously the geographical centre of research moved away from central Europe to the West and America, by a singular political event: when the Nazi Party in Germany took over the government in 1933, Jews were declared to be unworthy to occupy public positions, including those at universities. Heisenberg lost Bloch, Sommerfeld lost Bethe, and Göttingen was abandoned by the professors Born and James Franck and a considerable fraction of their most talented collaborators and students (see figure 3.16). It took great efforts of their colleagues and friends abroad, notably in Copenhagen, Cambridge, America and even the Soviet Union, to place the numerous emigrants from Germany. Ultimately, these scholars helped to establish further centres in western Europe and America, for example, Neville Mott's (b 1905) new Bristol laboratory for metal research (where Bethe and Fröhlich went) or Princeton University (where von Neumann and Wigner got positions). As mentioned earlier, former American students had already started new research centres in the United States: Pauling a school of quantum chemistry at Caltech in Pasadena, and Slater, with several of Sommerfeld's visitors, the solid-state group at MIT in Cambridge, Massachusetts; and after 1933, Isidor Rabi (1898–1988) transferred Otto Stern's molecular-ray method to Columbia University. The most decisive role was played by the German emigrants as theoreticians of nuclear physics, both in England and the USA [246]. As a consequence of these migrations, quantum mechanics and its applications became a genuinely international science, in which more scholars collaborated than in any previous topic of physical theory.

3.4.2. Causality, complementarity and reality in quantum mechanics (1926–35)

After Born suggested giving up causal determinism in quantum mechanics, heated discussions took place, especially in Germany and neighbouring countries, about the basic concepts in physical theory. These discussion included renewed analysis of the meaning of probability in atomic physics and reformulation of a generalized causality principle, as first steps. In later stages, consequences were derived from Bohr's general concept of complementarity, and criticism of Einstein was rejected. The debate cooled off after 1935, but was renewed in the 1950s and continues up to the present day.

3.4.2.1. Probability and causality in atomic theory (1926–30). At the end of a talk on 'causal and statistical laws in physics', presented at the Prague

Quanta and Quantum Mechanics

Figure 3.16. *The international community of quantum theorists: top—1931 in Heisenberg's Leipzig seminar (W Heisenberg with R Peierls in the front row, G C Wick, F Bloch and V Weisskopf in the back) and bottom—1937 in Copenhagen (with N Bohr, W Heisenberg, W Pauli, O Stern and L Meitner in front).*

meeting of German physicists and mathematicians in 1929, the expert on probability calculus, Richard von Mises (1883–1953), said [247]:

> The strict determinism, which is usually associated with classical physics as described by differential equations, is only an *apparent* feature; it cannot be kept up if one considers a [physical] theory in principle only together with the experiments testing it, i.e. one restricts oneself to *what is perceptible with human senses* or 'observable in principle'. In the macroscopic world, the indeterministic aspects are partly contained in the *object* of observation; and partly they sneak in by the *measuring process*; every microphysical object carries the statistical element in itself, since this element alone guarantees the *transition* to a frequently repeated phenomenon (*Massenerscheinung*), as is represented already by *each measurement*.

Although the mathematician was leaning to the positivistic views adopted in those days by many scientists and philosophers of science, he addressed an important difference between the classical theories and the new quantum mechanics. The fundamental equations of the classical dynamical theories were differential equations, and the physicists had believed in a 'deterministic hypothesis' of the type: if one knows at a given moment *the initial values of all* parameters describing the system considered, then one can calculate the values of these parameters *for all future times*. Obviously, this hypothesis worked in classical (Newtonian) mechanics and (Maxwell) electrodynamics, and it also could be taken over into the relativistic mechanics and (Einstein's) gravitation theory of 1905 and 1916, respectively, as Hilbert had stated [248]:

> From knowing the values [of the physical state variables], all predictions of the future values follow in a necessary and unique way, in so far as they make sense in physics.

Even classical statistical mechanics did not seem to disturb the 'deterministic hypothesis': the probabilistic description might be considered only a trick to calculate in a comfortable, simple manner the gross properties of a large assembly of particles; an infinitely clever 'Maxwell demon' would, however, be able to disentangle all individual particle tracks, as they followed deterministic classical laws.

The situation changed drastically since 1900, when Planck, in deriving his black-body law, fixed the finite 'size' of molecules and of radiation quanta. Then, in 1905 Egon von Schweidler (1873–1948) expressed the hypothesis that the decay law of radioactive substances was an essentially statistical relation leading to fluctuations, which were substantiated experimentally about 12 years later by the joint effort of Erwin Schrödinger and Elisabeth Bormann [249]. Albert Einstein applied a similar statistical hypothesis to the emission process of radiation from atomic states (which he studied in 1916 for the purpose of deriving Planck's law—see section 3.3.1.2 above). Bohr, Sommerfeld and their followers took over Einstein's procedure and talked about transition probabilities in connection with the emission and absorption of radiation. Bohr even attempted, in the Bohr–Kramers–Slater theory of 1924 (see section 3.2.3.1), to establish a fully probabilistic description of atomic processes by claiming that energy and momentum are only statistically conserved for interactions between atoms and radiation. This claim was rejected the following year by the Bothe–Geiger experiment. While the conservation laws were thus re-established on an atomic level, the statistical hypothesis entered the new quantum mechanics at exactly the place foreseen by Einstein in 1916, namely in the transition amplitudes for atomic radiation or, as Max Born demonstrated in 1926, in the Schrödinger wavefunctions: Born now alos spoke explicitly about a breakdown of the deterministic hypothesis for atomic processes.

Born's statistical interpretation of the wavefunction became an ingredient of the transformation theories of Dirac and Jordan, from which Heisenberg derived his uncertainty relations. The last implied that momentum and position of an atomic particle cannot be measured simultaneously with arbitrary accuracy (see section 3.3.3.2), hence [250]:

> The [classical] causal law loses its content. Since one never knows accurately the initial condition [of a given atomic system], one can never calculate accurately the mechanical behaviour. Each new observation selects from the wealth of possibilities a particular one and restricts the possibilities for later observations.

Heisenberg later formulated the failure of the classical causal law a little more precisely, when he stated [251]:

> The uncertainty relations show first that an exact knowledge of the essential quantities, which are necessary to fix the causal connection in the classical theory, cannot be obtained in quantum theory. The second consequence of the uncertainty consists in the fact that also the future behaviour of such an inaccurately known system can only be predicted in an inaccurate way, namely statistically.

He, therefore, claimed that classical determinism loses sense in microphysics, and thus cannot be applied there. Heisenberg and his friends defended this point of view against all later criticism, such as that expressed in 1932 by Max von Laue and Max Planck (who argued that the uncertainty relation only demonstrated a limitation of the *corpuscular* description in microphysics, but indicated no other limit to the physical knowledge there) [252].

3.4.2.2. Complementarity and measurement process (1927–32). As we have mentioned earlier (in section 3.3.3.2), Bohr gave a different interpretation of the laws of quantum mechanics from that of Heisenberg and Born (who stressed the essential importance of quantum jumps as being explicitly contained in the commutation relations and the uncertainty relations). He rather focused on doubts in the validity of the classical space-time description, which he formulated more definitely in the 1927 discussion of radiation processes [253]:

> On the one hand, in attempting to trace the laws of the time-spatial propagation of light, we are confined to statistical considerations. On the other hand, the fulfilment of the claim of causality for the individual light process, characterized by the quantum of action, entails a renunciation as regards the space-time description.

Bohr considered the mutual exclusion of space-time description and causal description of atomic processes to be an integral part of his 'complementarity theory'. Heisenberg took over this physical interpretation of microphysical phenomena immediately and explained it to a wider public [254]:

> In order to make a statement concerning an [atomic] object, one has to observe it. The observation implies an interaction between observer and object which changes the object ... The causal description of a system is complementary to the space-time description of the same system. Because one has to *observe* in order to establish a space-time description, and this observation disturbs the system. If we disturb the system, we cannot cleanly follow anymore its causal evolution.

In more technical terms, there existed a mathematical description of atomic phenomena which satisfied formally the classical causality, namely with the help of the Schrödinger equation (which had the same differential equation structure as the equations of classical dynamics). However, this description could not be interpreted in space and time without involving the statistical hypothesis (i.e. Born's interpretation of the wavefunction).

As Heisenberg accepted Bohr's complementarity view, Bohr accepted Heisenberg's recognition of the fact that atomic systems are disturbed by certain measurements. He spoke explicitly about 'a new uncontrollable element' introduced by the observation. In later talks, Bohr tried to discuss other fields of science (and even the humanities) in the view of complementarity. In particular, he elaborated in a famous lecture, entitled *Light and life* and given at the 1932 *International Congress on Light Therapy* in Copenhagen, on an application of complementarity to biological systems. Two of his theses impressed some of the experts and were debated in the following years. First, the view 'that we should doubtless kill an animal if we try to carry the investigation of its organs so far that we could tell the part played by the single atom in vital functions'; i.e. a full description of the structure of the living animal excludes that of its biological functions. Second, the view 'that the freedom of the will is to be considered as a feature of conscious life which corresponds to functions of the organism that not only evade a causal mechanical description but resist even a physical analysis, carried to the extent required from an unambiguous application of the statistical laws of atomic mechanics', or in short: free will is complementary to the causal determination of man's action [255].

At another place, in his Faraday lecture of 1930, Bohr addressed the science closest to physics, chemistry, which had now become merged with quantum physics. He found in particular that 'the indeterminacy in the use of classical concepts defining the state of a [microphysical, atomic] system at a given time implies an essential irreversibility in the physical interpretation of this [quantum-mechanical] symbolism.'. Still the nature of this irreversibility was different from that of the irreversibility known from classical thermodynamics: while in the classical description (by statistical mechanics) the mechanical concepts still were thought to describe the microscopic details, in atomic theory the same laws worked only together with the statistical hypothesis. 'In this sense' Bohr argued 'quantum mechanics may be said to represent the next step in the development of our tools of an adequate description of natural phenomena.' [256].

One of the main features of this description consisted of a new concept of the process of measurement. Already within classical theory, the Hungarian Leo Szilard (1898–1964) had concluded that the decrease of entropy caused through the observation of a thermodynamical system (by an intelligent being) must be compensated by an increase of entropy imposed on the observed system through the procedure of measurement [257]. This consideration helped his friend John von Neumann, when he tried to formulate the mathematical theory of measurement in microphysics [187]. While the Schrödinger equation allows for a causal propagation of the wavefunctions describing an evolving system, the intervention of the observer creates instantaneously a discontinuous, acausal, change: it selects from a mixture of states a particular state k ('reduction of the wave packet'). Neumann's procedure of coupling the microphysical system S with the macroscopic apparatus M and separating it again to give as a final answer the state k of S ('von Neumann section') has often been criticized, notably since 1950 in the extended discussion of the quantum-mechanical measurement theory. Many physicists, mathematicians and philosophers of science have attempted to improve on the formulation, but a really convincing alternative procedure has not yet arisen from these efforts.

3.4.2.3. Completeness and reality of the quantum-mechanical description (1935). In struggling with the problem of whether quantum mechanics

Twentieth Century Physics

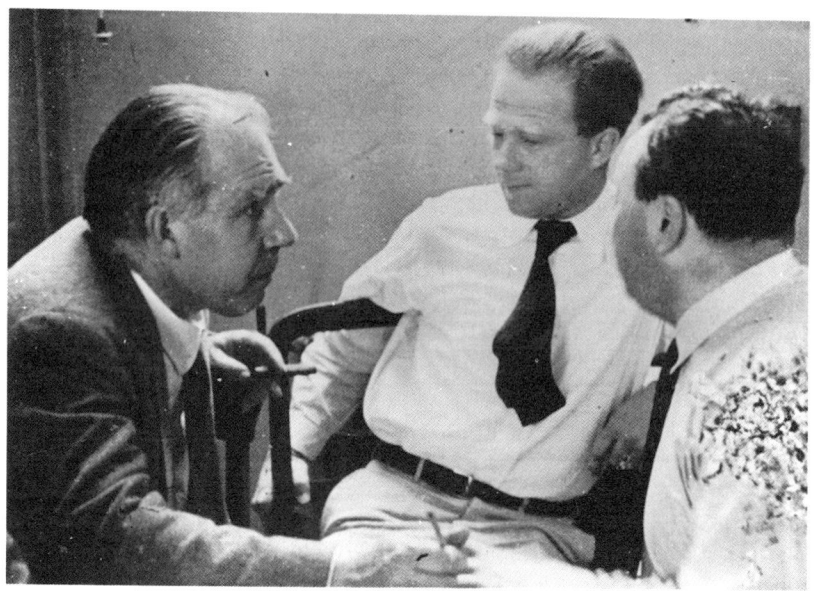

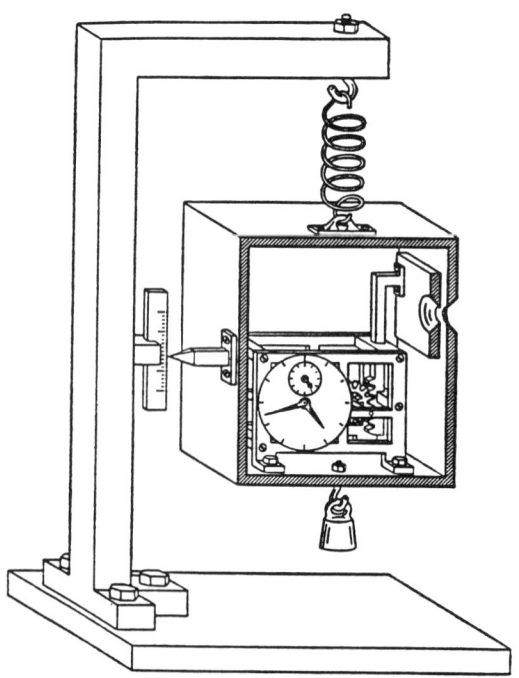

Figure 3.17. *The interpretation of quantum mechanics. Top, the Copenhagen 'mafia', from the left: N Bohr, W Heisenberg and W Pauli; bottom, the discussion of Einstein's thought experiment claiming a violation of the energy–time uncertainty relation.*

yields a correct description of atomic processes, the critics raised two questions: (i) 'Is the new atomic theory a consistent scheme?' and (ii) 'Does the new theory give a complete description of all phenomena observed?' Einstein, who frequently expressed unhappiness about the assumption that 'God plays dice', became the main opponent of the partisans of modern atomic theory in the debates conducted at the Fifth and Sixth Solvay Conferences at Brussels in 1927 and 1930 [258]. After Bohr and his associates (including Heisenberg and Pauli) had beaten back his refined attacks on the validity of the uncertainty relations, e.g. in a system composed of a clock and a gravitating mass, Einstein and other sceptics agreed that quantum or wave mechanics always supplied correct results, wherever this could be applied (see figure 3.17). The second question refers to a deeper conceptual level, and has to do with the ability of quantum mechanics to describe all aspects of microphysics. Since Einstein later shortened this to the question of whether the theory describes 'physical reality', we should discuss briefly the development of this concept after the arrival of the new atomic theory.

Before 1925 there existed only the concept of physical reality underlying the classical theories; it included the existence of phenomena independent of their observation, hence the assumption that it is possible to remove the influence of the observation procedure on the observed object more or less completely. During the first year of wave mechanics Schrödinger had sought to take over essential features of the classical reality concept, e.g. the continuum interpretation of atomic events in space and time. Erwin Madelung (1881–1972) went in the same direction: upon rewriting the time-dependent Schrödinger equation for the one-electron problem in the form of a hydrodynamical continuity equation, he suggested that 'there is a chance to treat the quantum theory of atoms on this basis' [259]. He was not able to develop this idea further (e.g. for many-electron atoms no such simple space-time picture can be given). Also Louis de Broglie did not get very far with his causal extensions of wave mechanics, i.e. the so-called 'double-solution theory' and the 'pilot-wave theory' [260]. Some conservative physicists saw a possibility that hitherto unknown coordinates, so-called 'hidden parameters', might give rise to the apparently acausal behaviour—just as one had previously explained the statistical behaviour of large assemblies of molecules in classical kinetic gas theory, including irreversible effects, as being due to the lack of knowledge about all the dynamical variables of the individual molecules. Von Neumann carried out in the fourth chapter of his *Mathematical Foundations* an analysis of the statistical relations in quantum mechanics (he considered especially the situations of 'homogeneous assemblies' or 'pure cases' and of 'mixtures' in detail) and concluded [261]:

> Nor would it help it if there existed other, as yet undiscovered, physical quantities, in addition to those represented by the

operators in quantum mechanics. The present system of quantum mechanics would have to be objectively false, in order that another description of the elementary processes than the statistical one be possible.

A little later, Jacques Solomon (1908–42) in Paris arrived at the same result by a different argument (which involved, however, the assumption of the uncertainty relations for his quantum-theoretical variables) [262].

Even these demonstrations did not finish the debate on the physical reality question. In 1935 Einstein again formulated what he thought to be a fundamental objection against considering quantum mechanics a complete theory. Together with Boris Podolsky (1896–1966) and Nathan Rosen (b 1909) he assumed as a necessary property for a physical theory to be complete that 'every element of the physical reality must have a counterpart in the physical theory', and defined [263]:

> If, without in any way disturbing a system, we can predict with certainty (i.e. with probability equal to unity) the value of a physical quantity, then there exists an element of physical reality corresponding to this physical quantity.

The authors then discussed the physical variables of two relativistic systems, which had interacted in the past within a finite time interval, and showed that the eigenvalues of non-commuting quantities were not *completely described*, according to their chosen definition of physical reality: 'When the operators corresponding to two physical quantities do not commute, the two quantities cannot have simultaneous reality.'.

Niels Bohr immediately replied: 'A criterion of reality like that proposed by the named authors contains—however cautious its formulation may appear—an essential ambiguity when it is applied to the actual problems with which we are here concerned.'. By analysing the example of Einstein and his colleagues, he found that the quantum-mechanical situation was entirely different, because the conditions (which define the possible types of predictions regarding the future behaviour of the system) contradicted the reality definition. Hence he concluded [264]:

> Since these conditions constitute an inherent element of the description of any phenomenon to which the term 'physical reality' can be properly attached, we see that the argumentation of the above mentioned authors does not justify their conclusion that the quantum-mechanical description is essentially incomplete.

As well as Bohr, Pauli and Heisenberg were also concerned with rejecting the 'Einstein–Podolsky–Rosen (EPR) paradox', and Pauli wrote in a letter to Schrödinger [265]:

> One can*not*, however—as the conservative, old gentlemen wish—declare the statistical results of quantum mechanics as being *correct*

and *nevertheless base* this on a hidden causal mechanism. In this sense the system of quantum-mechanical laws appears to be logically closed (complete in the sense of axiomatics)—in contrast to kinetic gas theory.

Schrödinger, on the other hand, did not content himself with Pauli's verdict; in his opinion the standard atomic theory indeed had to live with rather paradoxical situations, such as he described in the case of the coupling of a radioactively decaying substance with an apparatus killing a cat: as a result, the quantum-mechanical cat had to be considered half-dead and half-alive ('cat paradox') [266].

The EPR paradox inspired many authors afterwards; in particular, discussion emerged on the revival of the hidden-parameter idea by David Bohm (b 1917) and others after the early 1950s [267]. John Bell's (1928–90) analysis of the situation in the 1960s showed that hidden variables resulted in an inequality (for the 'local condition') which could be tested by experiment and found *not* to be satisfied [268].

A different line of causal theories was pursued by Jean Paul Vigier in Paris, who picked up in the 1950s Louis de Broglie's pilot-wave idea of 1927, but his causal formulation of the matter–wave theory did not lead to a real breakthrough. Perhaps Heisenberg was right when he had stated back in 1935 [269]:

> Rather follows the fact that there exists a certain field of experiences, which can be described by Schrödinger's wave mechanics, however, not by classical mechanics, i.e. also the non-perceptible (*unanschaulichen*) features of the quantum-mechanical laws will have to constitute forever an integral part of science.

Causal classical mechanics, on the one hand, and quantum or wave mechanics, on the other, represented two completely different physical schemes—Heisenberg considered them both as consistent 'closed theories'—which rested on different concepts and accounted for different fields of phenomena, those of macrophysics and microphysics, respectively.

3.4.3. Beyond quantum mechanics (1932 to the present)
The decade following the discovery of quantum and wave mechanics brought such rich fruits in successful applications that one would have expected all physicists, especially the pioneers, to be extremely satisfied with the new atomic theory, hoping that it would serve as the physical description of the structure of matter for years to come. However, quite the contrary happened when the quantum theorists hit obstacles in understanding certain phenomena of nuclear and cosmic-ray physics, which they attributed to difficulties of the relativistic quantum field theory. Instead of patiently improving on the formalism, people like Bohr and Heisenberg were soon ready to give up the splendid

accomplishments of quantum mechanics. In a preliminary theoretical analysis of cosmic rays, given in February 1932, Heisenberg said that one would hardly be able to understand the radiation created in high-energy scattering processes 'since the failure in principle of the corresponding radiation theory of Dirac and of the equivalent quantum electrodynamics is already certain because of other reasons' [270]. The other reasons included divergent theoretical results for physical quantities, like the self-mass of electrons and the inability to explain nuclear structure, beta-decay and other nuclear phenomena (such as the so-called Meitner–Hupfeld effect). Hence Heisenberg hoped, and with him Bohr and Pauli, that a new theory, as revolutionary as quantum mechanics in 1925, would soon replace quantum field theory in the domain of nuclear and high-energy physics.

However, at the same time as Heisenberg was writing his cosmic-ray review, a new era of physical theory began, which was characterized not by changing the existing relativistic quantum scheme, but rather by introducing new elementary constituents of matter. This first led to a fuller understanding of nuclear structure and of one part of the high-energy phenomena in cosmic radiation—the (soft) electromagnetic cascade showers. Another part, notably the 'hard' component, could be grasped only after the introduction of further elementary particles. The procedure of increasing the substructure of atoms and nuclei went on throughout the 1950s, 1960s, 1970s until the early 1980s and resulted finally in the present 'standard model'.

The original expectation of the great theoreticians for a completely new fundamental theory beyond quantum mechanics remained unfulfilled. Certainly, the relativistic field theories were developed further by incorporating new features, such as renormalization, symmetry breaking and all that—yet they retained all the characteristics and concepts already known in 1930. However, since then several new ideas have been suggested, and it may well be that they will eventually find a place in a future theory replacing relativistic quantum mechanics.

3.4.3.1. New structures within the 'standard theory' [271]. The standard model of the constitution of matter before 1932 included three fundamental particles: protons and electrons built up the atomic nucleus, the atoms and the molecules, and the light-quantum or photon occurred in processes of atomic and nuclear transformations. Theoretically, one understood fully the structure of atoms and the processes of emission, absorption and scattering of light by atoms. However, serious difficulties showed up in describing the structure of atomic nuclei and their transition processes, except for alpha-decay. Some nuclei, if composed of protons and electrons, showed the wrong (quantum) statistical behaviour, and the continuous spectra of electrons emerging from beta-decay hinted at a violation of the conservation laws for energy, momentum and angular momentum. The discovery of the neutron in February 1932 removed the obstacles to a quantum-mechanical theory of nuclear structure; Pauli's neutrino hypothesis, proposed originally in 1930 and adopted by Enrico Fermi in late 1933 for formulating a quantum field theory of beta-decay, solved the other main problem of nuclear physics.

The positively charged light particle discovered in August 1932 by Carl Anderson (1905–91) of Caltech perfected another essential aspect of the 'standard model', namely Dirac's relativistic electron theory: Dirac himself had first (in 1930) interpreted the 'holes' as protons; but, on theoretical objections (e.g. by Weyl because of the widely different masses, and the prediction of considerable, yet unobserved, annihilation of proton–electron pairs) he predicted in 1931 the existence of an 'anti-electron', having the same mass and opposite charge. The cosmic-ray particle, discovered by Carl Anderson and found copiously in electron–positron pairs by Patrick Blackett (1897–1974) and Guiseppe Occhialini (1917–94) in Cambridge, represented this predicted Dirac particle. Many of the cosmic-ray phenomena (constituting the 'soft component') could be explained with the help of cascades arising from such pairs produced by high-energy photons, which in turn originated as bremsstrahlung quanta. The 'hard component' of the cosmic rays, on the other hand, was satisfactorily interpreted in 1937 as being due to a 'heavy electron' or 'mesotron', with a mass intermediate between those of electrons and protons.

Such an intermediate-mass object could account also for the forces acting between protons and neutrons in nuclei. Heisenberg had introduced exchange forces in 1932 and thus been able to establish a theory of binding energies in nuclei; and Hideki Yukawa (1907–81) in 1934 inferred from the finite range of these forces the existence of a new, intermediate-mass, particle. The discrepancies between the strong forces, predicted by Yukawa for his 'heavy electron', and the weak interaction exhibited by the cosmic-ray mesotron were explained in 1947, when Cecil Powell (1903–69) and his Bristol group identified a second type of intermediate-mass particles in cosmic radiation; the new 'pi-meson' or 'pion' was responsible for the strong nuclear forces, while its decay product, the 'mesotron' of Anderson and Seth Neddermeyer (1907–88)—now called 'mu-meson' or 'muon'—showed only weak beta-decay-like forces. A period lasting 35 years began, in which hundreds of elementary particles were discovered in high-energy reaction, first in cosmic-ray process and later in accelerator experiments, starting with the so-called 'strange particles' and ending with the 'weak' charged and neutral 'W' and 'Z' bosons.

The explosion of the number of elementary particles forced the physicists to organize them eventually with the more elementary scheme of quarks, leptons and exchange bosons, the constituents of the present-day 'standard model'. Again, this can be formulated, apart from some new features (like quark confinement and the symmetry-breaking Higgs mechanism), as a quantum field theory, in fact as two separate theories of that type: the 'electroweak theory' unifying quantum electrodynamics and the extended Fermi theory of weak interactions, with the photon and the weak bosons as intermediate particles exchanging forces between the leptons; and 'quantum chromodynamics', in which 'gluons' establish the strong forces between the 'quarks', the fractionally charged constituents of the proton and the other 'hadrons'. The particle theorists even seek to unite the electroweak theory and quantum chromodynamics in 'grand unified theories', which again are quantum field theories, having to a large extent known features.

Finally, we should mention that the last unexplained solid-state property, superconductivity, was finally embedded into a quantum-theoretical scheme by John Bardeen (1908–91), Leon Cooper (b 1930) and John Schrieffer (b 1931); again it involved a new object, a non-relativistic 'quasi-particle' consisting of a pair of bound electrons.

3.4.3.2. Aspects of a theory beyond quantum mechanics. Perhaps the most significant property of the quantum-mechanical schemes is the linearity of its fundamental equation, which allows the linear combination of solutions and the mathematical representation by linear Hilbert spaces. Many deviations from simple ('ground state') solutions can be obtained as linear, first-order perturbations—a procedure regularly applied in problems of non-relativistic and relativistic quantum theory. On the other hand, the physicists interested in a fundamental description of matter have involved, since Gustav Mie (1868–1957) in 1912, non-linear field theories [272]. Einstein's general relativity and gravitation theory of 1915 became the prototype of an essentially non-linear theory; and the same author sought, in many attempts until his death, to develop from it a unified theory of matter and forces. While Einstein hoped to derive the quantum-theoretical effects from his classical field theory, his successors, such as Born, Heisenberg and Pauli, started right away with a genuine quantum field theory. Heisenberg and Pauli, in particular, debated, for nearly 35 years, the possibility of a unified description of all elementary particles and their interactions [273].

Neither Einstein's nor Heisenberg's unified-field theory proposals achieved the ambitious goal envisaged, but they did contain elements which eventually may enter a more satisfactory future theory. Such an element could be the non-linearity known from Einstein's general relativity, which Heisenberg also recognized as an essential feature of relativistic quantum theories and the description of high-energy phenomena [274]. (Actually, quantum electrodynamics is a non-linear theory, when sources are included, and so are the Yang–Mills schemes, currently used in elementary particle physics.) Also the scattering-matrix theory of elementary particles proposed by Heisenberg in the 1940s, which was revived in 1958 as the S-matrix theory of strong interactions, established intrinsically non-linear relations [275].

From the very beginning, non-linear theories raised great difficulties when one wanted to introduce quantization. Therefore, Heisenberg and Pauli had, since the early 1930s, repeatedly resorted to lattice calculations, which implied formally a quantization of space. Heisenberg even saw in this mathematical possibility a fundamental property of nature: he advocated in the 1930s in many papers the existence of a 'fundamental length' determining the high-energy behaviour of elementary particles; he also hoped that it might play a decisive role—as the third elementary constant besides Planck's h and the vacuum velocity of light c—in a future fundamental theory beyond quantum mechanics [276]. The idea of a granulated or quantized space structure has occurred again and again in discussions up to the present day without, however, finding a more widely accepted formulation (except perhaps as a tool for performing model calculations in a lattice space).

Finally, we should address a different point. The quantum theory of measurement involves, as mentioned above in section 3.4.2.2, an element of irreversibility. We may now ask the question whether this suffices to explain the well-known irreversibility in thermodynamical processes. The debate, carried on over the past 60 years, tends to exclude this attractive possibility. Some experts on the topic, like Ilya Prigogine (b 1917), argue that the dynamical equations of quantum mechanics must be extended, so as to contain an essential element of irreversibility. A new 'supermechanics' of that type would then be able to account for the wealth of phenomena described by equilibrium and non-equilibrium thermodynamics. It is evident that quantum mechanics, in spite of its completeness in mathematical structure and conceptual foundation, *cannot represent the endpoint of all physical theories.* It rather opens new vistas for future generations of scientists interested in unravelling the deep secrets of Nature.

References

[1] Lord Rayleigh 1900 The law of partition of kinetic energy *Phil. Mag.* **49** 98–118, especially p 117

[2] Lord Kelvin 1901 Nineteenth century clouds over the dynamical theory of heat and light *Phil. Mag.* **2** 1–40, especially p 40

[3] Lord Rayleigh 1900 Remarks upon the law of complete radiation *Phil. Mag.* **49** 539–40

[4] Heisenberg W 1934 Wandlungen in den Grundlagen der exakten Naturwissenschaft in jüngster Z. *Angew. Chem.* **47** 697–702, especially p 702

[5] Hund F 1975 Hätte die Geschichte der Quantentheorie auch anders ablaufen können? *Phys. Blätter* **31** 29–35, especially p 35

[6] See especially Hund F 1984 *Geschichte der Quantentheorie* 3rd edn (Mannheim: Bibliographisches Institut)
Jammer M 1966 *The Conceptual Development of Quantum Mechanics* (New York: McGraw-Hill)
Mehra J and Rechenberg H 1982 *The Historical Development of Quantum Theory* (New York: Springer) vol 1–4; 1987 *The Historical Development of Quantum Theory* (New York: Springer) vol 5; *The Historical Development of Quantum Theory* (New York: Springer) vol 6, in preparation
Pauli W 1926 Quantentheorie *Handbuch der Physik* vol 23, ed H Geiger and K Scheel (Berlin: Springer) pt 1 pp 1–278
Reiche F 1921 *Die Quantentheorie. Ihr Ursprung und ihre Entwicklung* (Berlin: Springer) (Engl. Transl. H S Hatfield and H L Brose *The Quantum Theory* (London: Methuen))

[7] Lorentz H A 1914 Ansprache *Die Theorie der Strahlung und Quanten* ed A Eucken (Halle: Knapp) pp 5–7, especially p 5

[8] Eucken A 1914 Anhang *Die Theorie der Strahlung und Quanten* ed A Eucken (Halle: Knapp) p 373

[9] For the history of the black-body law see also Kangro H 1970 *Vorgeschichte des Planckschen Strahlungsgesetzes* (Weisbaden: Steiner) (Engl. Transl. 1976 *Early History of Planck's Radiation Law* (London: Taylor and Francis))
Kuhn T W 1978 *Black-Body Theory and the Quantum Discontinuity 1894–1912* (Oxford: Oxford University Press)
We do not share the view expressed in the last book which postpones the birth of quantum theory to about 1906. Apart from not presenting historical proof for his case, Kuhn does not distinguish between two different steps in the development of the theory: first, the discovery of the existence of the quantum of action; and second, the observation that this concept implies a certain discreteness in the description of atomic phenomena. The second step could not have been taken in 1900, when essential parts of the classical theory, notably the dynamics of the electron, were not completed. Only after 1905 was one able to draw the conclusion that the quantum concept did not follow from an extended classical scheme, but implied a new, probably discrete, mathematical description.

[10] Kirchhoff G 1860 Über das Verhältnis zwischen dem Emissionsvermögen und dem Absorptionsvermögen der Körper für Wärme und Licht *Ann. Phys., Lpz* **109** 275–301

[11] Paschen F 1896 Über die Gesetzmäßigkeiten in den Spektren fester Körper *Ann. Phys., Lpz* **58** 455–92

[12] Wien W 1896 Über die Energieverteilung im Emissionsspektrum des schwarzen Körpers *Ann. Phys., Lpz* **58** 662–9

[13] Planck M 1897 Über irreversible Strahlungsvorgänge. 1.–5. Mitteilung *Sitzungsber. Preuß. Akad. Wiss. (Berlin)* 493–504, 505–7, 508–31; 1898 *Sitzungsber. Preuß. Akad. Wiss. (Berlin)* 449–76; 1899 *Sitzungsber. Preuß. Akad. Wiss. (Berlin)* 560–600

[14] Wien W and Lummer O 1895 Methoden zur Prüfung des Strahlungsgesetzes absolut schwarzer Körper *Ann. Phys., Lpz* **56** 451–6

[15] Rubens H and Kurlbaum F 1900 Über die Emission langwelliger Wärmestrahlen durch den schwarzen Körper bei verschiedenen Temperaturen *Sitzungsber. Preuß. Akad. Wiss. (Berlin)* 929–41

- [16] Planck M 1900 Über eine Verbesserung der Wien'schen Spektralgleichung *Verh. Deutsch. Phys. Ges.* **2** 202–4
- [17] Planck M 1900 Zur Theorie des Gesetzes der Energieverteilung im Normalspektrum *Verh. Deutsch. Phys. Ges.* **2** 237–45
- [18] Rubens H and Kurlbaum R 1901 Anwendung der Methode der Reststrahlen zur Prüfung des Strahlungsgesetz *Ann. Phys., Lpz* **4** 649–66, especially p 659
- [19] Einstein A 1904 Zur allgemeinen molekularen Theorie der Wärme *Ann. Phys., Lpz* **14** 354–62
- [20] Einstein A 1905 Über einen die Erzeugung und Verwandlung des Lichtes betreffenden heuristischen Gesichtspunkt *Ann. Phys., Lpz* **17** 132–48, especially pp 135 and 143
- [21] Lenard P 1902 Über lichtelektrische Wirkung *Ann. Phys., Lpz* **8** 149–98
- [22] Millikan R A 1916 A direct photoelectric determination of Planck's constant '*h*' *Phys. Rev.* **7** 355–88
- [23] Weber H F 1872 Die spezifische Wärme des Kohlenstoffs *Ann. Phys., Lpz* **147** 311–9
- [24] Einstein A 1906 Die Plancksche Theorie der Strahlung und die Theorie der spezifischen Wärme *Ann. Phys., Lpz* **22** 180–90, especially p 184
- [25] Einstein A 1906 Die Plancksche Theorie der Strahlung und die Theorie der spezifischen Wärme *Ann. Phys., Lpz* **22** 180–90, especially pp 186–7
- [26] Nernst W 1906 Über die Berechnung chemischer Gleichgewichte aus thermischen Messungen *Nach. Ges. Wiss. Göttingen* 1–40
- [27] Nernst W 1910 Untersuchungen zur spezifischen Wärme bei tiefen Temperaturen. II *Sitzungsber. Preuß Akad. Wiss. (Berlin)* 262–82, especially p 276
- [28] Nernst W and Lindemann F A 1911 Untersuchungen zur spezifischen Wärme bei tiefen Temperaturen. V *Sitzungsber. Preuß. Akad. Wiss. (Berlin)* 494–501
- [29] Planck M 1911 Eine neue Strahlungshypothese *Verh. Deutsch. Phys. Ges.* **13** 138–48
- [30] Debye P 1912 Les particularités des chaleurs specifiques à basse température *Arch. Sci. Phys. Natur. (Genève)* **33** 256–8; Zur Theorie der spezifischen Wärmen *Ann. Phys., Lpz* **39** 789–839
- [31] Born M and von Kármán T 1912 Über Schwingungen von Raumgittern *Phys. Z.* **13** 297–309; 1913 Zur Theorie der spezifischen Wärme *Phys. Z.* **14** 15–9; 1913 Über die Verteilungen der Eigenschwingungen von Punktgittern *Phys. Z.* **14** 65–71
- [32] Eucken A 1912 Die Molekularwärme von Wasserstoff bei tiefen Temperaturen *Sitzungsber. Preuß. Akad. Wiss. (Berlin)* 141–51
- [33] Einstein A and Stern O 1913 Einige Argumente für die Annahme einer molekularen Agitation beim absoluten Nullpunkt *Ann.Phys.* **40** 551–60
- [34] Dennison D M 1927 A note on the specific heat of the hydrogen molecule *Proc. R. Soc.* A **115** 483–6
- [35] Stark J 1907 Beziehung des Doppler-Effektes bei Kanalstrahlen zur Planckschen Strahlungstheorie *Phys. Z.* **8** 913–919, especially p 914
- [36] Sackur O 1912 Die 'chemischen Konstanten' der zwei- und dreiatomigen Gase *Ann. Phys., Lpz* **40** 87–106
  Stern O 1913 Zur kinetischen Theorie des Dampfdruckes einatomiger freier Stoffe und über die Entropiekonstante einatomiger Gase *Phys. Z.* **14** 629–32
- [37] Nernst W 1911 Zur Theorie der spezifischen Wärme und über die Anwendung der Lehre von den Energiequanten auf physikalisch-

[38] chemische Fragen überhaupt *Z. Elektrochem.* **17** 265–75, especially p 270
Bjerrum N 1912 Über die ultraroten Absorptionsbanden der Gase *Festschrift Walther Nernst* (Halle: Knapp) pp 90–8
von Bahr E 1913 Über die ultrarote Absorption der Gase *Verh. Deutsch. Phys. Ges.* **15** 710–30

[39] Planck M 1911 Eine neue Strahlungshypothese *Verh. Deutsch. Phys. Ges.* **13** 138–48, especially p 147

[40] Nernst W 1912 Der Energieinhalt der Gase *Phys. Z.* **13** 1064–69

[41] Planck M 1902 Über die Natur des weißen Lichtes *Ann. Phys., Lpz* **7** 390–400, especially p 400

[42] Sommerfeld A 1919 *Atombau und Spektrallinien* (Braunschweig: Vieweg), especially p VIII

[43] Haas A 1910 Über die elektrodynamische Bedeutung des Planckschen Strahlungsgesetzes und über eine neue Bestimmung des elektrischen Elementarquantums und der Dimensionen des Wasserstoffatoms *Sitzungsber. Akad. Wiss. (Wien)* 119–44

[44] Nicholson J W 1912 The constitution of the solar corona. II *Mon. Not. R. Astron. Soc.* **72** 677–92

[45] Bohr N 1913 On the constitution of atoms and molecules. Part I, II, III *Phil. Mag.* **26** 1–25, 476–502, 857–75

[46] Bohr N 1913 On the spectra of helium and hydrogen *Nature* **92** 231–2

[47] Franck J and Hertz G 1914 Über Zusammenstöße zwischen Elektronen und den Molekülen des Quecksilberdampfes und die Ionisierungsspannung derselben *Verh. Deutsch. Phys. Ges.* **16** 457–67, especially p 467

[48] Bohr N 1915 On the quantum theory of radiation and the structure of the atom *Phil. Mag.* **30** 394–415, especially p 410

[49] Barkla C G and Sadler C A 1908 Homogeneous secondary Röntgen radiations *Phil. Mag.* **16** 550–84

[50] Moseley H G J 1913 The high-frequency spectra of elements *Phil. Mag.* **26** 1024–34

[51] Moseley H G J 1914 The high-frequency spectra of elements. Part II *Phil. Mag.* **27** 703–13

[52] Sommerfeld A 1915 Zur Theorie der Balmerschen Serie *Sitzungsber. Bayer. Akad. Wiss. (München)* 425–58

[53] Sommerfeld A 1915 Die Feinstruktur der Wasserstoff- und der wasserstoffähnlichen Linien *Sitzungsber. Bayer. Akad. Wiss. (München)* 459–500

[54] Sommerfeld A 1916 Zur Theorie des Zeemaneffektes der Wasserstofflinien mit einem Anhang über den Starkeffekt *Phys. Z.* **17** 491–507

[55] Paschen F 1916 Bohrs Heliumlinien *Ann. Phys., Lpz* **50** 901–40

[56] Debye P 1916 Quantenhypothese und Zeeman-Effekt *Nach. Ges. Wiss. Göttingen* pp 142–153; see also reference [53]

[57] Stark J 1913 Beobachtungen über den Effekt des elektrischen Feldes auf Spektrallinien *Sitzungsber. Preuß. Akad. Wiss. (Berlin)* 932–46

[58] Stark J and Wendt G 1914 Beobachtungen über den Effekt des elektrischen Feldes auf Spektrallinien. II Längseffekt *Ann. Phys., Lpz* **43** 983–90

[59] Warburg E 1913 Bemerkung zur Aufspaltung der Spektrallinien im elektrischen Feld *Verh. Deutsch. Phys. Ges.* **15** 1259–66

[60] Stark J 1914 Schwierigkeiten für die Lichtquantenhypothese im Falle der Emission von Spektrallinien *Verh. Deutsch. Phys. Ges.* **16** 304–6

[61] Epstein P S 1916 Zur Theorie des Starkeffekts *Phys. Z.* **17** 148–50; 1916 *Ann. Phys., Lpz* **50** 489–521

[62] Schwarzschild K 1916 Zur Quantenhypothese *Sitzungsber. Preuß. Akad. Wiss. (Berlin)* 548–68

[63] Sommerfeld A 1920 Ein Zahlenmysterium in der Theorie des Zeeman-Effektes *Naturwissenshaften* **8** 61–4

[64] Sommerfeld A 1920 Allgemeine spektroskopische Gesetze, insbesondere ein magnetooptischer Zerlegungssatz *Ann. Phys., Lpz* **63** 221–63

[65] Einstein A 1909 Zum gegenwärtigen Stand des Strahlungsproblems *Phys. Z.* **10** 185–93, especially p 191

[66] Gerlach W and Stern O 1922 Der experimentell Beweis der Richtungsquantelung *Z. Phys.* **9** 349–52

[67] Einstein A and Ehrenfest P 1922 Quantentheoretische Bemerkungen zum Experiment von Stern und Gerlach *Z. Phys.* **11** 31–4

[68] Compton A H 1923 A quantum theory of the scattering of x-rays by light experiments *Phys. Rev.* **21** 207; full paper in 1923 *Phys. Rev.* **21** 483–502

[69] Debye P 1923 Zerstreuung von Röntgenstrahlen und Quantentheorie *Phys. Z.* **24** 161–6

[70] Bohr N, Kramers H and Slater J 1924 The quantum theory of radiation *Phil.Mag.* **47** 785–822

[71] Bothe W and Geiger H 1925 Experimentelles zur Theorie Bohrs, Kramers und Slater *Naturwissenshaften* **13** 440–1, especially p 441

[72] Compton A H and Simon A W 1925 Directed quanta of scattered x-rays *Phys. Rev.* **26** 289–99

[73] Lewis G N 1926 The conservation of photons *Nature* **118** 874–75

[74] Bose S N 1924 Plancks Gesetz und Lichtquantenhypothese *Z. Phys.* **26** 178–81

[75] Debye P 1920 Der Wahrscheinlichkeitsbegriff in der Theorie der Strahlung *Ann. Phys., Lpz* **33** 1427–37

[76] Einstein A 1924 Quantentheorie des einatomigen idealen Gases. *Sitzungsber. Preuß. Akad. Wiss. (Berlin)* 262–67; 1925 Zweite Abhandlung *Sitzungsber. Preuß. Akad. Wiss. (Berlin)* 3–14; 1925 Quantentheorie des idealen Gases *Sitzungsber. Preuß. Akad. Wiss. (Berlin)* 18–25

[77] Stoner E C 1924 The distribution of electrons among atomic levels *Phil. Mag.* **48** 719–36

[78] Pauli W 1925 Über den Zusammenhang des Abschlusses der Elektronengruppen im Atom mit der Komplexstruktur der Spektren *Z. Phys.* **31** 765–83

[79] Fermi E 1926 Zur Quantelung des idealen einatomigen Gases *Z. Phys.* **36** 902–12

[80] Uhlenbeck G and Goudsmit S 1925 Ersetzung der Hypothese vom unmechanischen Zwang durch eine Forderung bezüglich des inneren Verhaltens jeden einzelnen Elektrons *Naturwissenshaften* **13** 953–4

[81] Beck E 1919 Zum experimentellen Nachweis der Ampèreschen Molekularströme *Ann. Phys., Lpz* **60** 109–48

Arvidsson G 1920 Eine Untersuchung der Ampèreschen Molekularströme nach der Methode von A Einstein und W J de Haas *Phys. Z.* **21** 88–91

[82] de Broglie L 1922 Rayonnement noir et quanta de lumière *J. Phys. Rad.* **3** 422–8

[83] de Broglie L 1923 Ondes et quanta *C. R. Acad. Sci., Paris* **177** 507–10; 1923 Quanta de lumière, diffraction et interférences *C. R. Acad. Sci., Paris* **177** 548–30; 1923 Les quanta, la théorie cinétique de gaz et le principe de Fermat *C. R. Acad. Sci., Paris* **177** 630–32

[84] de Broglie L 1923 Les quanta, la théorie cinétique de gaz et le principe de Fermat *C. R. Acad. Sci., Paris* p 632

[85] de Broglie L 1924 Sur la définition générale de la correspondence entre onde et mouvement *C. R. Acad. Sci., Paris* **179** 39–40

[86] de Broglie L 1925 Recherche sur la théorie des quanta *Ann. Phys., Paris* **3** 22–128
[87] Davisson C J and Kunsman C H 1923 The scattering of low speed electrons by platinum and magnesium *Phys. Rev.* **22** 242–58
[88] Elsasser W 1978 *Memoirs of Physicist in the Atomic Age* (London: Science History Publications and Hilger)
[89] Elsasser W 1925 Bemerkungen zur Quantenmechanik freier Elektronen *Naturwissenshaften* **13** 711
[90] Davisson C J and Germer L H 1927 The scattering of electrons by a single crystal of nickel *Nature* **119** 558–60
[91] Thomson G P and Reid A 1927 Diffraction of a cathode ray by a thin film *Nature* **119** 890
[92] Ladenburg R 1921 Die quantentheoretische Deutung der Zahl der Dispersionselektronen *Z. Phys.* **4** 451–68
[93] Kramers H A 1924 The law of dispersion and Bohr's theory of spectra *Nature* **113** 673–4
[94] Carst A and Ladenburg R 1928 Untersuchungen über anomale Dispersion angeregter Gase. IV. Teil. Anomale Dispersion des Wasserstoffs: wahres Intensitätsverhältnis der Wasserstofflinien H_α and H_β *Z. Phys.* **48** 192–204
[95] Kramers H A and Heisenberg W 1925 Über die Streuung von Strahlung durch Atome *Z. Phys.* **31** 681–708
[96] Smekal A 1923 Zur Quantentheorie der Dispersion *Naturwissenshaften* **11** 873–5
[97] Raman C V 1928 A new radiation *Indian J. Phys.* **2** 387–98
[98] Landsberg G and Mandelstam L 1928 Über die Lichtzerstreuung in Kristallen *Z. Phys.* **50** 769–80
[99] Kuhn W 1925 Über die Gesamtstärke der von einem Zustande ausgehenden Absorptionitinien *Z. Phys.* **33** 408–12
[100] Thomas W 1925 Über die Zahl der Dispersionselektronen, die einem stationären Zustande zugeordnet sind *Naturwissenshaften* **13** 627
[101] Lorentz H A 1908 Le partage de l'énergie entre la matière pondérable et l'ether *Nuovo Cimento* **16** 5–34; also in 1937 *Collected Papers* vol VII (The Hague: Nijhoff) pp 317–41, especially p 341
[102] Planck M 1906 *Vorlesungen über die Theorie der Wärmestrahlung* (Leipzig: Barth) p 156
[103] Poincaré H 1912 Sur la théorie des quanta *J. Physique* **2** 5–34
[104] Planck M 1914 Eine veränderte Formulierung der Quantenhypothese *Sitzungsber. Preuß. Akad. Wiss. (Berlin)* 918–23, especially p 919
[105] Einstein A 1949 Autobiographical notes *Albert Einstein: Philosopher Scientist* ed P A Schilpp (New York: Tudor) pp 45, 47
[106] Wilson W 1915 The quantum theory of radiation and line spectra *Phil. Mag.* **29** 795–802
[107] Ishiwara J 1915 Die universelle Bedeutung des Wirkungsquantums *Tokyo Sugaku Buturigakkakiwi Kizi* **8** 106–16
[108] Planck M 1915 Die Quantenhypothese für Molekeln mit mehreren Freiheitsgraden *Verh. Deutsch. Phys. Ges.* **17** 407–18, 431–51
[109] Bohr N 1918 On the quantum theory of line spectra. I., II. *Kgl. Danske Vid. Selsk. Skrifter, 8. Raekke* **IV.1** 1–36, 37–100
[110] Bohr N 1918 On the quantum theory of line spectra. I., *Kgl. Danske Vid. Selsk. Skrifter, 8. Raekke* **IV.1** p 4
[111] Ehrenfest P 1913 A mechanical theorem of Boltzmann and its relation to the theory of quanta *Proc. Kon. Akad. Wetensch. (Amsterdam)* **16** 591–97;

1916 On adiabatic changes of a system in connection with the quantum theory *Proc. Kon. Akad. Wetensch. (Amsterdam)* **19** 576–87
[112] Einstein A 1916 Strahlungs-Emission und -Absorption nach der Quantentheorie *Verh. Deutsch. Phys. Ges.* **18** 318–23; 1917 Zur Quantentheorie der Strahlung *Phys. Z.* **18** 121–8
[113] Bohr N 1918 On the quantum theory of line spectra. I., *Kgl. Danske Vid. Selsk. Skrifter, 8. Raekke* **IV.1** 1–36, especially p 15
[114] Planck M 1906 *Vorlesungen über die Theorie der Wärmestrahlung* (Leipzig: Barth) p 178
[115] Bohr N 1913 On the constitution of atoms and molecules. Part I *Phil. Mag.* **26** pp 1–25, especially p 14
[116] Kramers H A 1919 Intensities of spectral-lines—on the application of the quantum theory to the problem of the relative intensities of the components of the fine structure and of the Stark effect of these lines of the hydrogen spectra *Kgl. Danske Vid. Selsk. Skrifter, 8. Raekke* **III.3**
[117] Rubinowicz A 1918 Bohrsche Frequenzbedingung und Erhaltung des Impulsmomentes. I, II *Phys. Z.* **19** 441–5, 465–74
[118] Bohr N 1918 On the quantum theory of line spectra. I *Kgl. Danske Vid. Selsk. Skrifter, 8. Raekke* **IV.1** 1–35, especially pp 34–5
[119] Rubinowicz A 1921 Zur Polarisation der Bohrschen Strahlung *Z. Phys.* **4** 343–6

Bohr N 1921 Zur Frage der Polarisation der Strahlung in der Quantentheorie *Z. Phys.* **6** 1–9
[120] Bohr N 1921 Atomic structure *Nature* **107** 104–7, especially p 104
[121] Bohr N 1921 Atomic structure *Nature* **107** 104–7, especially p 105
[122] Bohr N 1921 Atomic structure *Nature* **108** 208–9; 1977 Sieben Vorträge über Atombau (Seven lectures on the theory of atomic structure) *Bohr N: Collected Works, vol 4* (Amsterdam: North-Holland) pp 341–419
[123] Mehra J and Rechenberg H 1982 *The Historical Development of Quantum Theory* vol 1 (New York: Springer) p 357
[124] Coster D and de Hevesy G 1923 On the missing element of atomic number 72 *Nature* **111** 79
[125] Pauli W 1922 Über das Modell des Wasserstoffmolekülions *Ann. Phys., Lpz* **68** 177–240
[126] Born M 1922 Über das Modell der Wasserstoffmolekel *Naturwissenshaften* **10** 677–8
[127] Kramers H A 1923 Über das Modell des Heliumatoms *Z. Phys.* **13** 314–41, especially p 339
[128] Born M and Heisenberg W 1923 Die Elektronenbahnen im angeregten Heliumatom *Z. Phys.* **16** 229–43, especially p 243
[129] Born M and Pauli W 1922 Über die Quantelung gestörter mechanischer Systeme *Z. Phys.* **10** 137–58

Born M and Heisenberg W 1923 Über Phasenbeziehungen bei den Bohrschen Modellen von Atomen und Molekeln *Z. Phys.* **14** 44–55
[130] Landé A 1921 Über den anomalen Zeemaneffekt. I *Z. Phys.* **5** 231–41; 1921 Über den anomalen Zeemaneffekt. II *Z. Phys.* **7** 398–405

Sommerfeld A 1922 Quantentheoretische Umdeutung der Voigtschen Theorie des anomalen Zeeman-Effektes vom D-Linientypus *Z. Phys.* **8** 257–72

Heisenberg W 1922 Zur Quantentheorie der Linienstruktur und der anomalen Zeemaneffekte *Z. Phys.* **8** 273–97
[131] Paschen F and Back E 1912 Normale und anomale Zeemaneffekte *Ann. Phys., Lpz* **39** 897–932; 1913 Nachtrag *Ann. Phys., Lpz* **40** 960–70

[132] Sommerfeld A 1923 Über die Deutung verwickelter Spektren (Mangan, Chrom) nach der Methode der inneren Quantenzahlen *Ann. Phys., Lpz* **70** 32–62
Landé A 1923 Termstruktur und Zeemaneffekt der Multipletts *Z. Phys.* **15** 189–205
[133] Pauli W 1924 Zur Frage der Komplexstrukturterme in starken und schwachen äußeren Feldern *Z. Phys.* **20** 371–87
[134] Bohr N 1918 On the quantum theory of line spectra. II. *Kgl. Danske Vid. Selsk. Skrifter, 8. Raekke* **IV.1**, especially p 93
Epstein P 1923 Simultaneous action of an electric and a magnetic field on a hydrogen-like atom *Phys. Rev.* **22** 202
[135] Halpern O 1923 Über den Einfluß gekreuzter elektrischer und magnetischer Felder auf das Wasserstoffatom *Z. Phys.* **18** 287–303
Klein O 1924 Über die gleichzeitige Wirkung von gekreuzten homogenen elektrischen und magnetischen Feldern auf das Wasserstoffatom *Z. Phys.* **22** 109–18
Lenz W 1924 Über den Bewegungsverlauf und die Quantenzustände der gestörten Keplerbewegung *Z. Phys.* **24** 197–207
[136] Pauli W 1926 Quantentheorie *Handbuch der Physik, vol 23* ed H Geiger and K Scheel (Berlin: Springer) pt 1 pp 1–278, especially pp 163–4
[137] Dirac P A M 1972 Relativity and quantum mechanics *Fields and Quanta* **3** 139–64, especially pp 147–8
[138] von Meyenn K *et al* (ed) 1978 *Wolfgang Pauli: Wissenschaftlicher Briefwechsel/Scientific Correspondence, vol I: 1919–25* (Berlin: Springer) pp 105–7, especially pp 106–7
[139] Heisenberg W 1924 Über eine Abänderung der formalen Regeln der Quantentheorie beim Problem der anomalen Zeemaneffekte *Z. Phys.* **26** 291–307
[140] Born M 1924 Über Quantenmechanik *Z. Phys.* **26** 379–95
[141] Ladenburg R 1921 Die quantentheoretische Deutung der Zahl der Dispersionselektronen *Z. Phys.* **4** 451–68
Darwin C G 1922 A quantum theory of optical dispersion *Nature* **110** 841–2
Ladenburg R and Reiche F 1923 Absorption, Zerstreuung und Dispersion in der Bohrschen Atomtheorie *Naturwissenshaften* **11** 584–98
[142] Heisenberg W 1925 Über eine Anwendung des Korrespondenzprinzips auf die Frage der Polarisation des Fluoreszenzlichtes *Z. Phys.* **31** 617–28, especially p 617
[143] Heisenberg W 1925 Über die quantentheoretische Umdeutung kinematischer und mechanischer Beziehungen *Z. Phys.* **33** 879–93
[144] Kratzer A 1922 Störungen und Kombinationsprinzip im System der violetten Cyanbanden *Sitzungsber. Bayer. Akad. Wiss. (München)* 107–18
Mulliken R S 1925 The isotope effect in band spectra. I, II *Phys. Rev.* **25** 119–38, 259–94
[145] Kronig R de L 1925 Über die Intensität der Mehrfachlinien und ihrer Zeemankomponenten, I *Z. Phys.* **31** 885–97; 1925 II *Z. Phys.* **33** 261–72
[146] Born M 1978 *My Life. Recollections of a Nobel Laureate* (London: Scribner) p 217
[147] Born M and Jordan P 1925 Zur Quantenmechanik *Z. Phys.* **34** 858–88
[148] Born M, Heisenberg W and Jordan P 1926 Zur Quantenmechanik. II *Z. Phys.* **35** 577–615

[149] Dirac P A M 1925 The fundamental equations of quantum mechanics *Proc. R. Soc.* **A109** 642–53
[150] Dirac P A M 1926 Quantum mechanics and a preliminary investigation of the hydrogen atom *Proc. R. Soc.* A **110** 561–79; 1926 The elimination of nodes in quantum mechanics *Proc. R. Soc.* **111** 281–305; 1926 Relativity quantum mechanics with an application to Compton scattering *Proc. R. Soc.* **111** 405–23
[151] Born M 1926 *Problems of Atomic Dynamics* (Cambridge, MA: MIT Press)
[152] Hilbert D 1906 Grundzüge einer allgemeinen Theorie der linearen Integralgleichungen (Vierte Mitteilung) *Nach. Ges. Wiss. Göttingen* 157–227
[153] Pauli W 1926 Über das Wasserstoffspektrum vom Standpunkt der neuen Quantenmechanik *Z. Phys.* **36** 336–63
[154] Born M and Wiener N 1926 A new formulation of the laws of quantization of periodic and aperiodic phenomena *J. Math. Phys.* **5** 84–98
[155] Lanczos C 1926 Über eine feldmäßige Darstellung der neuen Quantenmechanik *Z. Phys.* **35** 812–30
[156] Schrödinger E 1926 Zur Einsteinschen Gastheorie *Phys. Z.* **27** 95–101
[157] Schrödinger E 1922 Über eine bemerkenswerte Eigenschaft der Quantenbahnen eines einzelnen Elektrons *Z. Phys.* **12** 13–23
[158] Schrödinger E 1926 Quantisierung als Eigenwertproblem (Erste Mitteilung) *Ann. Phys., Lpz* **79** 361–76
[159] Schrödinger E 1926 Quantisierung als Eigenwertproblem (Zweite Mitteilung) *Ann. Phys., Lpz* **79** 489–527
[160] Schrödinger E 1926 Über das Verhältnis der Heisenberg-Born-Jordanschen Quantenmechanik zu der meinen *Ann. Phys., Lpz* **79** 734–56
[161] Schrödinger E 1926 Quantisierung als Eigenwertproblem (Vierte Mitteilung) *Ann. Phys., Lpz* **81** 109–39
[162] Lanczos C 1926 Über eine feldmäßige Darstellung der neuen Quantenmechanik *Z. Phys.* **35** 812–30, especially p 819
[163] Schrödinger E 1926 Über das Verhältnis der Heisenberg-Born-Jordanschen Quantenmechanik zu der meinen *Ann. Phys., Lpz* **79** 734–56, especially p 735–6
[164] Pauli W, letter to P Jordan, 12 April 1926, in *Wolfgang Pauli: Scientific Correspondence, vol I: 1919–25* pp 315–320
Eckart C 1926 Operator calculus and the solution of the equations of quantum dynamics *Phys. Rev.* **28** 711–26
[165] Wentzel G 1926 Eine Verallgemeinerung der Quantenbedingungen für die Zwecke der Wellenmechanik *Z. Phys.* **38** 518–29
Brillouin L 1926 La mécanique ondulatoire de Schrödinger; une méthode générale de resolution par approximations successives *C. R. Acad. Sci., Paris* **183** 24–6
Kramers H A 1926 Wellenmechanik und halbzahlige Quantelung *Z. Phys.* **39** 828–40
[166] Born M 1926 Zur Quantenmechanik der Stoßprozesse (Vorläufige Mitteilung) *Z. Phys.* **37** 863–7; 1926 Quantenmechanik der Stoßprozesse *Z. Phys.* **38** 807–27
[167] Born M and Jordan P 1925 Zur Quantentheorie aperiodischer Vorgänge *Z. Phys.* **33** 479–505
[168] Born M 1926 Zur Quantenmechanik der Stoßprozesse (Vorläufige Mitteilung) *Z. Phys.* **37** 863–7, especially p 866
[169] Born M 1926 Das Adiabatenprinzip in der Quantenmechanik *Z. Phys.* **40** 167–92

[170] London F 1926 Über die Jacobischen Transformationen der Quantenmechanik Z. Phys. **37** 915–25; 1926 Winkelvariable und kanonische Transformationen in der Undulationsmechanik Z. Phys. **40** 193–210
[171] Dirac P A M 1926 The physical interpretation of quantum dynamics Proc. R. Soc. A **113** 621–41
[172] Dirac P A M 1926 The physical interpretation of quantum dynamics Proc. R. Soc. A **113** 621–41, especially p 635
[173] Jordan P 1927 Über eine neue Begründung der Quantenmechanik Z. Phys. **40** 809–38
[174] Dirac P A M 1926 The physical interpretation of quantum dynamics Proc. R. Soc. A **113** 621–41, especially p 641
[175] Heisenberg W 1926 Schwankungserscheinungen und Quantenmechanik Z. Phys. **40** 501–6, especially p 506
[176] Heisenberg W to Pauli W, 28 October 1926 *Wolfgang Pauli: Wissenschaftlicher Briefwechsel/Scientific Correspondence, vol I: 1919–25* (Berlin: Springer) pp 349–51, especially p 350
[177] Heisenberg W 1930 *The Physical Principles of the Quantum Theory* (from the German manuscript)
[178] Heisenberg W 1927 Über den anschaulichen Inhalt der quantentheoretischen Kinematik und Mechanik Z. Phys. **43** 172–98
[179] Heisenberg W 1927 Über den anschaulichen Inhalt der quantentheoretischen Kinematik und Mechanik Z. Phys. **43** 172–98, especially pp 197–8
[180] Bohr N 1924 On the application of the quantum theory to atomic structure. Part I. The fundamental postulates Proc. Camb. Phil. Soc. Suppl. 1–42, especially p 34 (Engl. Transl. of a paper first published in Z. Phys. **13** 117–65)
[181] Bohr N 1957 Über die Wirkung von Atomen bei Stößen Z. Phys. **34** 142–57, especially p 154 of the addendum (*Nachschrift*)
[182] Bohr N 1928 The quantum postulate and the recent development of atomic theory *Atti del Congresso Internazionale dei Fisici, vol 2* (Bologna: Zanichelli) pp 565–98; 1928 Le postulate des quanta et le nouveau développement de l'atomistique *Electrons et Photons. Rapports et Discussion de Cinquième Conseil de Physique* Institut International de Physique Solvay (ed) (Paris: Gauthier-Villars) pp 215–47
[183] Hilbert D, von Neumann J and Nordheim L 1928 Über die Grundlagen der Quantenmechanik Math. Ann. **98** 1–30
[184] von Neumann J 1927 Mathematische Begründung der Quantenmechanik Nach. Ges. Wiss. Göttingen 1–57. See also his later 1927 papers: Wahrscheinlichkeitstheoretischer Aufbau der Quantenmechanik Nach. Ges. Wiss. Göttingen 245–72; Thermodynamik quantentheoretischer Gesamtheiten Nach. Ges. Wiss. Göttingen 273–91
[185] von Neumann J 1929 Allgemeine Eigenwerttheorie Hermitischer Funktionaloperatoren Math. Ann. **102** 49–131
[186] Wintner A 1929 *Spektraltheorie unendlicher Matrizen* (Leipzig: Hirzel)
[187] von Neumann J 1932 *Mathematische Grundlagen der Quantenmechanik* (Berlin: Springer) (Engl. Transl. *Mathematical Foundations of Quantum Mechanics* (Princeton, NH: Princeton University Press))
[188] Pauli W 1933 Die allgemeinen Prinzipien der Wellenmechanik *Handbuch der Physik, vol 33 pt 1*, ed H Geiger and K Scheel (Berlin: Springer) pp 83–272, especially p 198 (Engl. Transl. Achuthan P and Venkatesan K 1980 *General Principles of Quantum Mechanics* (Berlin: Springer))
[189] Born M, Heisenberg W and Jordan P 1926 Zur Quantenmechanik. II Z. Phys. **35** 577–615, especially pp 605–15

[190] Dirac P A M 1927 The quantum theory of emission and absorption of radiation *Proc. R. Soc.* A **114** 243–65
[191] Jordan P 1927 Zur Quantentheorie der Gasentartung *Z. Phys.* **44** 473–80; 1927 Über Wellen und Korpuskeln in der Quantenmechanik *Z. Phys.* **45** 766–75
Jordan P and Klein O 1927 Zum Mehrkörperproblem in der Quantenmechanik *Z. Phys.* **45** 751–65
Jordan P and Wigner E 1928 Über das Paulische Äquivalenzverbot *Z. Phys.* **47** 631–51
[192] Fock V 1932 Konfigurationsraum und zweite Quantelung *Z. Phys.* **75** 622–47
[193] Klein O 1926 Quantentheorie und fünfdimensionale Relativitätstheorie *Z. Phys.* **37** 895–906; 1927 Elektrodynamik und Wellenmechanik vom Standpunkt des Korrespondenzprinzips *Z. Phys.* **41** 407–42
Gordon W 1926 Der Comptoneffekt nach der Schrödingerschen Theorie *Z. Phys.* **40** 117–32;
Schrödinger E 1926 Quantisierung als Eigenwertproblem (Vierte Mitteilung) *Ann. Phys., Lpz* **81** 109–39
[194] Dirac P A M 1928 The quantum theory of the electron. I *Proc. R. Soc.* A **117** 610–24; 1928 II *Proc. R. Soc.* A **118** 351–68
[195] Dirac P A M 1931 Quantized singularities in the electromagnetic field *Proc. R. Soc.* A **133** 60–72
[196] Jordan P and Pauli W 1928 Zur Quantenelektrodynamik ladungsfreier Felder *Z. Phys.* **47** 151–73
[197] Heisenberg W and Pauli W 1929 Zur Quantendynamik der Wellenfelder *Z. Phys.* **56** 1–61; 1930 Zur Quantentheorie der Wellenfelder II *Z. Phys.* **59** 168–90
[198] Landau L and Peierls R 1931 Erweiterung des Unbestimmtheitsprinzips für die relativistische Quantentheorie *Z. Phys.* **69** 56–69
[199] Bohr N and Rosenfeld L 1933 Zur Frage der Meßbarkeit der elektromagnetischen Feldgrößen *Kgl. Dansk. Vidensk. Selsk. Math.-Phys. Medd.* **12** No 8
[200] Jordan P 1935 *Die Physik des 20. Jahrhunderts* (Braunschweig: Vieweg) p 1–2
[201] The earliest reviews for physicists one finds in volume 24 of Geiger H and Scheel K *Handbuch der Physik pt I: Quantentheorie* ed A Smekal (Berlin: Springer) (with contributions by A Rubinowicz, W Pauli, H Bethe, F Hund, G Wentzel and N F Mott), and in *pt II: Aufbau der zusammenhängenden Materie* (with contributions by K F Herzfeld, R de L Kronig, A Sommerfeld and H Bethe, M Born and M Goeppert-Mayer, A Smekal, H Grimm and H Wolf). The textbooks for students include Ruark A E and Urey H C 1930 *Atoms, Molecules and Quanta* (New York: McGraw-Hill); Kemble E C 1937 *The Fundamental Principles of Quantum Mechanics with Elementary Applications* (New York: McGraw-Hill) and Pauling L and Wilson E B 1935 *Introduction to Quantum Mechanics* (New York: McGraw-Hill)
[202] Born M 1935 *Atomic Physics* (New York: Hafner)
[203] Bohr N 1934 *Atomic Physics and the Description of Nature* (Cambridge: Cambridge University Press); 1934 *Atomic Physics* (New York: Wiley); 1958 *Human Knowledge* (New York: Wiley);
Heisenberg W 1930 *The Physical Principles of the Quantum Theory* (Chicago: University of Chicago Press); 1935 (and later editions) *Wandlungen in den Grundlagen der Naturwissenschaften* (Leipzig: Hirzel)

[204] Dirac P A M 1968 When a golden age started *From a Life of Physics. Evening Lectures at the ICTP, Trieste* (Vienna: IAEA) p 32
[205] Quoted in Schweber S 1990 The young John Clark Slater and the development of quantum chemistry *Hist. Stud. Phys. Sci.* **20** 339–406, especially p 361
[206] Heisenberg W 1925 Über die quantentheoretische Umdeutung kinematischer und mechanischer Beziehungen *Z. Phys.* **33** 879–93
Born M, Heisenberg W and Jordan P 1926 Zur Quantenmechanik. II *Z. Phys.* **35** 577–615
Pauli W 1926 Über das Wasserstoffspektrum vom Standpunkt der neuen Quantenmechanik *Z. Phys.* **36** 336–63
[207] Heisenberg W and Jordan P 1926 Anwendung der Quantenmechanik auf das Problem der anomalen Zeemaneffekte *Z. Phys.* **37** 263–77
[208] Heisenberg W to Pauli W 1926 In *Wolfgang Pauli: Wissenschaftlicher Briefwechsel/Scientific Correspondence, vol I: 1919–25* ed K von Mayenn *et al* (Berlin: Springer) p 321
[209] Heisenberg W 1926 Mehrkörperproblem und Resonanz in der Quantenmechanik *Z. Phys.* **38** 411–26; Über die Spektra von Atomsystem mit zwei Elektronen *Z. Phys.* **39** 499–518
[210] Heitler W and London F 1927 Wechselwirkung neutraler Atome und homöopolare Bindung nach der Quantentheorie *Z. Phys.* **44** 455–72
[211] See, for example, the lectures at the Hirzel S 1928 *Leipziger Universitätswoche—Quantentheorie und Chemie* (Leipzig); 1929 *Dipolmoment und chemische Struktur* (Leipzig); 1931 *Molekülstruktur* (Leipzig) and the status reviews of Peter Debye, John E Lennard-Jones, Ralph Fowler, Werner Heisenberg and Max Born for the 'Discussion on the structure of simple molecules' (in Hefter W (ed) 1931 *Chemistry at the Centenary (1931) of the British Association for the Advancement of Science* (Cambridge) pp 204–56), as well as Pauling L 1939 *The Nature of the Chemical Bond* (Ithaca, NY: Cornell University Press)
[212] Lea E and Wiemers G 1993 Professor für Theoretische Physik *Werner Heisenberg in Leipzig 1927–1942* ed C Kleint and G Wiemers (Berlin: Akademic Verlag) pp 181–215, especially pp 187, 188
[213] London F and Eisenschütz E 1930 Über das Verhältnis der van der Waalsschen Kräfte zu den homöopolaren Bindungskräften *Z. Phys.* **60** 491–527
[214] Weyl H 1928 *Gruppentheorie und Quantenmechanik* (Leipzig: Hirzel) (Engl. Transl. of the second German edition as: 1931 *Group Theory and Quantum Mechanics* (London: Methuen))
Wigner E 1931 *Gruppentheorie und ihre Anwendung auf die Quantenmechanik der Atomspektren* (Braunschweig: Vieweg)
van der Waerden B L 1932 *Die gruppentheoretische Methode in der Quantenmechanik* (Berlin: Springer)
[215] Heisenberg W 1928 Hermann Weyl, Gruppentheorie und Quantenmechanik *Deutsche Litaraturzeitung* **49** 2473–4
[216] Heisenberg W 1927 Mehrkörperprobleme und Resonanz in der Quantenmechanik. II *Z. Phys.* **41** 239–57
Wigner E 1926 Über nichtkombinierende Terme in der neueren Quantentheorie *Z. Phys.* **40** 883–92; 1927 Einige Folgerungen aus der Schrödingerschen Theorie für die Termstrukturen *Z. Phys.* **43** 624–52
[217] Weyl H 1928 *Group Theory and Quantenmechanics* (Engl. Transl. of the second German edition as: 1931 *Group Theory and Quantum Mechanics* (London: Methuen)) p XXI

[218] As we have already learned from the transformation theory in quantum mechanics, the relation between conservation laws and symmetry transformation (described by unitary matrices or operators) can be described by simple algebraic equations, namely the commutator of the Hamiltonian matrix or operator of the atomic system with the unitary group transformation.

[219] Weyl H 1928 *Group Theory and Quantenmechanics* (Engl. Transl. of the second German edition as: 1931 *Group Theory and Quantum Mechanics* (London: Methuen)) p X

[220] Slater J 1929 The theory of complex spectra. *Phys. Rev.* **34** 1293–322

[221] Fowler R M 1926 On dense matter *Mon. Not. R. Astron. Soc.* **87** 114

[222] Pauli W 1927 Über Gasentartung und Paramagnetismus *Z. Phys.* **41** 81–102

[223] Sommerfeld A 1927 Zur Elektronentheorie der Metalle *Naturwissenshaften* **15** 825–32; 1928 *Naturwissenshaften* **16** 374–81; 1928 Elektronentheorie der Metalle und des Volta-Effekts nach der Fermischen Statistik *Atti Congresso Int. dei Fisici, (Como-Pavia-Roma, September 1927), vol II* (Bologna: Zanichelli) pp 449–73

[224] Bloch F 1928 Über die Quantenmechanik der Elektronen in Kristallgittern *Z. Phys.* **52** 555–600

[225] Kronig R de L and Penney W G 1931 Quantum mechanics of electrons in crystal lattices *Proc. R. Soc.* A **130** 499–513

[226] Peierls R 1930 Zur Theorie der elektrischen und thermischen Leitfähigkeit von Metallen *Ann. Phys., Lpz* **4** 129–48

Brillouin L 1931 *Die Quantenstatistik und ihre Anwendung auf die Elektronentheorie der Metalle* (Berlin: Springer)

[227] Fröhlich H 1936 *Elektronentheorie der Metalle* (Berlin: Springer) especially p 75

For a more technical earlier review, see the handbook article of vol 24 of Geiger H and Scheel K: Sommerfeld A and Bethe H Elektronentheorie der Metalle *Handbuch der Physik Part II Aufbau der zusammenhängenden Materie* pp 333–622

[228] Wilson A H 1931 The theory of electronic semi-conductors *Proc. R. Soc.* A **133** 458–91

[229] Heisenberg W 1928 Zur Theorie des Ferromagnetismus *Z. Phys.* **49** 619–36

The idea that ferromagnetism was an exchange phenomenon was indicated already in a slightly earlier publication by Frenkel J 1928 Elementare Theorie magnetischer und elektrischer Eigenschaften der Metalle beim absoluten Nullpunkt der Temperatur *Z. Phys.* **49** 31–45

[230] Bloch F 1930 Zur Theorie des Ferromagnetismus *Z. Phys.* **61** 206–19

[231] Dirac P A M 1926 Relativity quantum mechanics with an application to Compton scattering *Proc. R. Soc.* **111** 405–23

[232] Also Schrödinger, in his fourth communication (reference [161]), treated at about the same time a scattering problem, the dispersion of light by atoms.

[233] Gordon W 1926 Der Comptoneffekt nach der Schrödingerschen Theorie *Z. Phys.* **40** 117–32;

See also E Schrödinger 1927 Über den Comptoneffekt *Ann. Phys., Lpz* **82** 257–64

Wentzel G 1926 Zur Theorie des photoelektrischen Effektes *Z. Phys.* **40** 574–89; 1927 Über die Richtungsverteilung der Photoelektronen *Z. Phys.* **41** 828–32

[234] Hund F 1927 Zur Deutung der Molekelspektren. I *Z. Phys.* **40** 742–64

[235] Gamow G 1928 Zur Quantentheorie der Atomkerne *Z. Phys.* **51** 204–12;

Gurney R W and Condon E U 1928 Wave mechanics and radioactive decay *Nature* **122** 439

[236] Wentzel G 1926 Zwei Bemerkungen über die Zerstreuung korpuskularer Strahlen als Beugungserscheinung *Z. Phys.* **40** 590–93

Gordon W 1928 Über den Stoß zweier Punktladungen nach der Wellenmechanik *Z. Phys.* **48** 180–91

See also Mott N F 1929 The solution of the wave equation for the scattering of particles by a Coulombian centre of field *Proc. R. Soc.* A **118** 542–9

[237] Dirac P A M 1926 On the theory of quantum mechanics *Proc. R. Soc.* A **112** 661–77

[238] Mott N F 1929 The exclusion principle and aperiodic systems *Proc. R. Soc.* A **125** 222–30; 1930 The collisions between two electrons *Proc. R. Soc.* **126** 259–67

Blackett P M S and Champion F C 1931 The scattering of slow alpha particles by helium *Proc. R. Soc.* A **130** 380–88

[239] Klein O and Nishina Y 1929 Über die Streuung von Strahlung durch freie Elektronen nach der neuen relativistischen Quantendynamik von Dirac *Z. Phys.* **52** 853–68

[240] Klein O 1929 Die Reflexion von Elektronen an einem Potentialsprung nach der relativistischen Dynamik von Dirac *Z. Phys.* **53** 157–65

[241] Dirac P A M 1930 A theory of electrons and protons *Proc. R. Soc.* A **126** 360–5 and reference [195]

[242] See, for example, Heisenberg W 1931 Zum Paulischen Ausschließungsprinzip *Ann. Phys., Lpz* **10** 888–904

[243] See Sommerfeld A 1929 *Atombau und Spektrallinien Wellenmechanischer Ergänzungsband* (Braunschweig: Vieweg)

[244] For an account see, for example, Eckert M 1993 *Die Atomphysiker. Eine Geschichte der theoretischen Physik am Beispiel der Sommerfeldschule* (Braunschweig: Vieweg)

[245] Unsöld A 1938 *Physik der Sternatmosphären* (Berlin: Springer)

[246] See, for example, Brown L M and Rechenberg H 1991 The development of vector meson-theory in Britain and Japan (1937–38) *Br. J. Hist. Sci.* **24** 405–33

[247] von Mises R 1930 Über kausale und statistische Gesetzmäßigkeit in der Physik *Naturwissenshaften* **18** 145–53, especially p 153

[248] Hilbert D 1916 Die Grundlagen der Physik. II *Nach. Ges. Wiss. Göttingen* 53–76, especially p 61

[249] von Schweidler E 1905 Über die Schwankungen der radioaktiven Umwandlung *Comptes Rendus du Premier Congrès International pour l'Etude de la Radiologie et de l'Ionization (Liège, 1905)* ed H Dunot and E Pinat

Schrödinger E 1918 Über ein in der experimentellen Radiumforschung auftretendes Problem der statistischen Dynamik *Sitzungsber. Akad. Wiss. (Wien)* **127** 237–62; 1919 Wahrscheinlichkeitstheoretische Studien, betreffend Schweidler'sche Schwankungen, besonders die Theorie der Meßanordnung *Sitzungsber. Akad. Wiss. (Wien)* **128** 177–237

Bormann E 1918 Zur experimentellen Methodik der Zerfallserscheinungen *Sitzungsber. Akad. Wiss. (Wien)* **127** 2347–407

[250] Heisenberg W 1927 Über die Grundprinzipien der 'Quantenmechanik' *Forschungen und Fortschritte* **3** 83

[251] Heisenberg W 1931 Kausalgesetz und Quantenmechanik. *Erkenntnis* **2** 172–82, especially p 177

[252] von Laue M 1932 Über Heisenbergs Ungenauigkeitsbeziehungen und ihre

erkenntnistheoretische Bedeutung *Naturwissenshaften* **20** 915–6
Planck M 1932 *Der Kausalbegriff in der Physik* (Leipzig: Barth)
[253] Bohr N 1928 The quantum postulate and the recent development of atomic theory *Nature* **121** 580–90, especially p 581
[254] Heisenberg W 1928 Erkenntnistheoretische Probleme der modernen Physik *Werner Heisenberg: Gesammelte Werke/Collected Works* vol CI, ed W Blum *et al* pp 22–8, especially pp 26–7
[255] Bohr N 1958 Light and life *Atomic Physics and Human Knowledge* (New York: Wiley) pp 3–22, especially p 9 and p 11
[256] Bohr N 1932 Chemistry and the quantum theory of atomic constitution *J. Chem. Soc.* 349–84, especially pp 376–7
[257] Szilard L 1929 Über die Entropieverminderung in einem thermodynamischen System bei Eingriffen intelligenter Wesen *Z. Phys.* **53** 840–56
[258] Bohr N 1949 Discussion with Einstein on epistemological problems in atomic physics *Albert Einstein. Philosopher-Scientist* ed P A Schilpp (New York: Tudor) pp 32–66
[259] Madelung E 1926 Quantentheorie in hydrodynamischer Form *Z. Phys.* **40** 322–6
[260] de Broglie L 1927 La mécanique ondulatoire et la structure de la matière et du rayonnement *J. Physique Radium* **8** 225–41; 1928 La nouvelle dynamique des quanta *Electrons et Photons. Rapports et Discussion de Cinquième Conseil de Physique* (Paris: Gauthier-Villars) pp 105–32
[261] von Neumann J 1932 *Mathematische Grundlagen der Quantenmechanik* (Berlin: Springer) (Engl. Transl. *Mathematical Foundations of Quantum Mechanics* (Princeton, NH: Princeton University Press)) p 325
[262] Solomon J 1933 Sur l'indéterminisme de la mécanique quantique *J. Physique* **4** 34–7
[263] Einstein A, Podolsky B and Rosen N 1935 Can quantum-mechanical description of physical reality be considered complete? *Phys. Rev.* **47** 777–80, especially p 777
[264] Bohr N 1965 Can quantum-mechanical description of physical reality be considered complete? *Phys. Rev.* **48** 690–702, especially p 697 and p 700
[265] Pauli W to Schrödinger E, 9 July 1935 1985 *Pauli W: Wissenschaftliche Korrespondenz/Scientific Correspondence, Volume II: 1930–39* ed K van Meyenn (Berlin: Springer) pp 419–22, especially p 421
[266] Schrödinger E 1935 Die gegenwärtige Situation in der Quantenmechanik *Naturwissenshaften* **33** 807–12, 823–8, 844–9, especially p 812
[267] Bohm D 1951 *Quantum Theory* (Englewood Cliffs, NJ: Prentice-Hall), especially chapter 22; 1952 A suggested interpretation of the quantum theory in terms of 'hidden variables' 1952/1953 I, II *Phys. Rev.* **85** 166–79, 180–93; and many other publications
[268] Bell J 1964 On the Einstein–Podolsky–Rosen paradox *Physica* **1** 195–200
Aspect A, Dalibard J and Roger G 1982 Experimental verification of Einstein–Podolsky–Rosen–Bohm *Gedankenexperiment*; 1982 A new violation of Bell's inequalities *Phys. Rev. Lett.* **49** 91–4
[269] Heisenberg W 1936 Prinzipielle Fragen der modernen Physik *Neuere Fortschritte der exakten Wissenschaften* (Leipzig: Deuticke) pp 91–112, especially p 98
[270] Heisenberg W 1932 Theoretische Überlegungen zur Höhenstrahlung *Ann. Phys., Lpz* **13** 430–52, especially p 452
[271] The details of the discovery of new particles in nuclear and cosmic-ray phenomena are discussed in greater detail in Chapter 5 by L M Brown and in Chapter 9 by V L Fitch and J L Rosner.

[272] Mie G Grundlagen einer Theorie der Materie. I, II, III. *Ann. Phys., Lpz* **37** 511–34; **39** 1–40; 1912 *Ann. Phys., Lpz* **40** 1–66

[273] See, for example, Rechenberg H 1993 Heisenberg and Pauli. Their programme of a unified quantum field theory *Werner Heisenberg: Gesammelte Werke/Collected Works* vol AIII, ed W Blum *et al* (Berlin: Springer) pp 1–19

[274] Euler H and Heisenberg W 1936 Folgerungen aus der Diracschen Theorie des Positrons *Z. Phys.* **98** 714–32

Heisenberg W 1936 Zur Theorie der 'Schauer' in der Höhenstrahlung *Z. Phys.* **101** 533–40; 1939 Zur Theorie der explosionsartigen Schauer in der kosmischen Strahlung. II *Z. Phys.* **113** 61–86; 1949 Über die Entstehung von Mesonen in Vielfachprozessen *Z. Phys.* **126** 569–82; 1952 Mesonenerzeugung als Stoßwellenproblem *Z. Phys.* **133** 65–79

See also Hagedorn R Meson showers and multiparticle production (1949–52). In *Werner Heisenberg: Collected Works* vol AIII, ed W Blum *et al* (Berlin: Springer) pp 75–85

[275] Heisenberg W 1943 Die beobachtbaren Größen in der Theorie der Elementarteilchen. I, II *Z. Phys.* **120** 513–38, 673–702; 1944 III *Z. Phys.* **123** 93–112

See also Chew G F 1961 *S-Matrix Theory of Strong Interactions* (New York: Benjamin)

[276] Heisenberg W 1938 Über die in der Theorie der Elementarteilchen auftretende universelle Länge *Ann. Phys., Lpz* **32** 20–33

Chapter 4

HISTORY OF RELATIVITY

John Stachel

4.1. Introduction

'Theory of relativity' is an umbrella term, covering two quite distinct theories, usually called 'the special' and 'the general' theory. The exact nature of the relation between the two is still a subject of controversy, and both might well not bear the common appellation 'relativity' if not so closely associated with the work of one man, Albert Einstein, who developed the second in attempting to extend the first to include gravitation.

Like the other major pillar of twentieth century theoretical physics, 'the' quantum theory, each relativity theory is actually a cluster of somewhat varying physical ideas and associated mathematical formalisms. Each member of the cluster consists of a core of primary concepts and principles and a periphery of derived concepts, theorems, models, etc. The content of the core and periphery varies from interpretation to interpretation and may change over time [1]. However, certain features are common to all members of the cluster, entitling them to a common appellation.

Einstein based his account of the special theory on two core principles, the relativity principle and the light principle (discussed below) [2]. Like those of thermodynamics, with which Einstein often compared them [3], the principles are based on a wealth of empirical evidence, which they generalize at a high level of abstraction. Certain constraints are derived from the principles that all models of physical phenomena must obey. (Of course, this is only true insofar as the phenomena in question fall within the range of validity of the principles. For example, we shall soon see that, according to the general theory, gravitational phenomena do not fall within the range of validity of the special theory.) This is usually guaranteed by imposing these constraints on the dynamical theories purporting to explain the behaviour (mechanical,

Albert Einstein

(German/Swiss/American, 1879–1955)

Albert Einstein is generally regarded as the outstanding theoretical physicist of the twentieth century. The child of German–Jewish parents, he completed his secondary and polytechnical education in Switzerland. Unable to find an academic position, he worked at the Swiss Patent Office from 1902–9. During this time he wrote some of his most important papers, including his work on the special theory of relativity. Growing recognition of his work led to teaching posts at universities in Zurich  and Prague before he moved in 1914 to a research post in Berlin, where he completed work on the general theory of relativity. In 1919, confirmation of his prediction of gravitational light deflection led to worldwide fame, and he subsequently became active in a number of social and political causes, such as pacifism and Zionism. Hitler's access to power in 1933 led him to emigrate to the United States, where he spent the rest of his life at the Institute for Advanced Studies in Princeton, NJ.

In physics, he made the preeminent contribution to the development of the special theory of relativity and single-handedly worked out the foundations of the general theory. He also made major contributions to classical physics, notably his theory of Brownian motion; and to the development of the quantum theory, notably his formulation of the photon hypothesis, the concept of spontaneous transitions between quantum levels, and the development of Bose–Einstein statistics. Never satisfied that quantum mechanics provided a complete explanation of the properties of microsystems, he spent the last three decades of his life in an unsuccessful search for a unified field theory of gravitation and electromagnetism that would also explain quantum phenomena.

optical, electromagnetic, etc) of physical systems [4]. It is the constraints that constitute the common feature of all versions of the special theory.

Various choices of core principles and mathematical formalisms lead to the same set of constraints.

The constraints of the special theory are kinematical, stipulating a certain space-time structure and requiring any dynamical theory to be consonant with it. The so-called Minkowski space-time structure (see below) consists of two elements: an inertial structure governing the motion of force-free bodies (described mathematically by an affine structure), and a chronometry and geometry, or chronogeometry for short, governing the behaviour of (ideal) measuring rods and clocks (described mathematically by a pseudo-metrical structure). Given the chronogeometry, compatibility conditions between the two fix the inertial structure [5]. Minkowski space-time is characterized by its ten-parameter symmetry group [6], called the inhomogeneous Lorentz or Poincaré group. Any set of dynamical variables must form a representation of this group and obey dynamical equations that transform appropriately under it. More precisely, the set of dynamical variables at each point must form a representation space of the homogeneous Lorentz group. For fields, the dynamical equations must guarantee that these representations together form a representation of the inhomogeneous Lorentz (Poincaré) group. Theories obeying these constraints are said (with a slight abuse of language) to be Lorentz- or Poincaré-invariant.

For reasons discussed later, Einstein concluded that the space-time structure of the special theory would have to be generalized to include gravitation. In the resulting so-called *general* theory of relativity [7], gravity is not treated as an external force, but as part of the space-time structure. The fixed inertial structure of the special theory, governing the motion of a body in the absence of gravitation, is replaced by a dynamically determined inertio-gravitational field that includes the effect of gravitation on that motion. For the first time in the history of physics, the space-time structure is not specified *a priori*, but becomes a dynamical physical field. Newtonian gravitation theory can be so reformulated; but in fact this was only done after the development of the general theory and in response to it (see below).

The compatibility conditions still allow the inertial structure to be derived from the chronogeometry, which also becomes a dynamical field, the (pseudo-)metric field. The dual role of this field—defining the chronogeometry of space-time, as well as determining its inertio-gravitational structure [8]—is a central feature of general relativity. Another central feature is the principle of general covariance, a constraint on the dynamical equations permitted by the theory. The symmetry group of the four-dimensional space [9], on which all physical fields are defined, consists of all diffeomorphisms (sufficiently smooth point transformations of the space); so all dynamical equations must be invariant under this group. Applied to the equations obeyed by the metric field, the principle of general covariance singles out Einstein's

Hermann Minkowski

(German, 1864–1909)

Hermann Minkowski is the mathematician who developed the four-dimensional approach to the theory of relativity. Born in Russia of German parents who soon returned to Germany, he was educated in Königsberg, receiving his doctorate in 1885. He taught in Bonn, Königsberg, Zürich (where Einstein was a student in several of his courses) and from 1902 on in Göttingen. Most of his work was in pure mathematics, notably the introduction of geometric methods in various fields such as the theory of quadratic forms and of continued fractions. After coming to Göttingen, his interest in mathematical physics was heightened by contact with Hilbert's group. Participation in a 1905 seminar on electron theory familiarized him with the current problems of the electrodynamics of moving bodies. By 1907 he realized that the work of Lorentz and Einstein could best be understood by uniting space and time into a four-dimensional space-time with a non-Euclidean geometry, now usually called Minkowski space. He developed a mathematical formalism appropriate for this space-time and worked out in detail the four-dimensional treatment of electrodynamics. This four-dimensional approach soon became a standard tool for work on the special theory of relativity. Only after Einstein adopted this approach was he able to complete his work on the general theory.

gravitational field equations.

While there is a broad consensus among physicists on the interpretation of the special theory, this is not true of the general theory. Is the general relativity 'relativistic' in the same (or some analogous) sense as is the special theory [10]? What is the correct formulation of the principle of general covariance [11]? Is it a purely mathematical requirement or does it have a physical content; and if so, what is it [12]? Such questions are still topics of intense discussion among physicists and philosophers of science. Any account, including that given here, must include controversial features.

While well satisfied with the general-relativistic theory of the free gravitational field—indeed, regarding it as his major accomplishment [13]—Einstein was not satisfied with its treatment of matter and non-gravitational fields. Each of these must be described by an appropriate stress–energy tensor, constructed from non-gravitational dynamical variables (as well as the metric tensor) and subject to generally covariant dynamical equations, which must usually be postulated quite independently of the Einstein equations. The principle of 'minimal gravitational coupling', stipulating that—with proper choice of the dynamical variables—the stress–energy tensor should not contain derivatives of the metric, serves to delimit, but not always fix uniquely, possible generalizations of a special-relativistic stress–energy tensor. (In certain cases, the covariant conservation law for the stress–energy tensor, which is an integrability condition for the gravitational field equations, implies these dynamical equations.) In particular, the electromagnetic field, together with the charges and currents that are its sources, must be so introduced. Einstein expected that just as the principle of general covariance singles out the gravitational field equations when applied to the metric field, when applied to some appropriately chosen generalization of that field it would single out equations describing both electromagnetism and gravitation. He hoped that such a unified field theory, as it came to be called, would also provide the key to a deeper understanding of quantum phenomena [14], some of which, such as the mass spectrum of elementary particles, remain unexplained to this day, while quantum mechanics provides explanations of others, such as the energy spectrum of bound systems. While accepting its accomplishments, Einstein regarded quantum mechanics as provisional and incomplete [15], and spent much of his later life in an ultimately unsuccessful search for such a unified theory.

This preliminary account suggests subdividing the history of relativity into the following sections.

(i) The origins and development of the special theory; its application to various branches of physics, such as particle and continuum mechanics, electrodynamics including optics, quantum field theory, etc; and its experimental testing and technological applications.

(ii) The origins and development of the general theory; its application to the gravitational aspects of terrestrial, astronomical and astrophysical phenomena (for other astrophysical and all cosmological applications, see Chapter 23); its observational and experimental testing; the nature of its relation to quantum mechanics, often called the problem of 'quantum gravity'; its considerable philosophical influence, and its impact on popular consciousness.

Figure 4.1. *Galileo Galilei.*

(iii) Attempts to formulate a unified field theory of gravitation and electromagnetism, and—after research on elementary particles led to their introduction—to include the weak and strong force fields as well.

4.2. The special theory of relativity

4.2.1. Origins of the theory: mechanics

The special theory of relativity arose from attempts to reconcile classical mechanics with the optics and electrodynamics of moving bodies [16]. Classical mechanics embodied a mechanical principal of relativity, described by Galileo as follows

> Shut yourself up with some friend in the main cabin below decks on some large ship, and have with you there some flies, butterflies, and other small flying animals. Have a large bowl of water with some fish in it; hang up a bottle that empties drop by drop into a wide vessel beneath it. With the ship standing still, observe carefully how the little animals fly with equal speed to all sides of the cabin. The fish swim indifferently in all directions; the drop falls into the vessel beneath; and, in throwing something to your friend, you need throw it no more strongly in one direction than another, the distances being equal; jumping with your feet together, you pass equal spaces in every direction. When you have observed all these things carefully (though there is no doubt that when the ship is standing still everything must happen in this way), have the ship proceed with any speed you like, so long as the motion is uniform and not fluctuating this way and that.

History of Relativity

Figure 4.2. *Isaac Newton.*

You will discover not the least change in all the effects named, nor could you tell from any of them whether the ship was moving or standing still [17].

Generalizing these observations, we can state the Galilean form of the relativity principle: mechanical phenomena do not allow an absolute distinction between different states of uniform (i.e. unaccelerated) motion. In his *Principia*, Newton states the principle in these words

The motions of bodies included in a given space are the same among themselves, whether that space is at rest, or moves uniformly forwards in a right [i.e. straight] line without any circular motion [18].

Newton's version differs from Galileo's in two respects. Its reference to 'rest' is based on Newton's concept of absolute space, which, like absolute time, is independent of all physical processes [19] and, while Galileo presented it as a generalization from observations, Newton deduced the principle from his three laws of motion. We shall discuss each of these points in turn.

Absolute rest or motion is defined with respect to absolute space; such motion is uniform if equal displacements in absolute space take place in equal absolute times. A 'space [that] ... moves uniformly forwards' without acceleration is now called an inertial frame of reference, a term introduced by Ludwig Lange in 1885 [20]. I shall so denote any frame of reference in which Newton's laws or their special-relativistic generalizations hold. Newton distinguished between absolute quantities, which are inaccessible to direct observation, and relative measures of space, time and motion, which are more or less directly accessible [21].

255

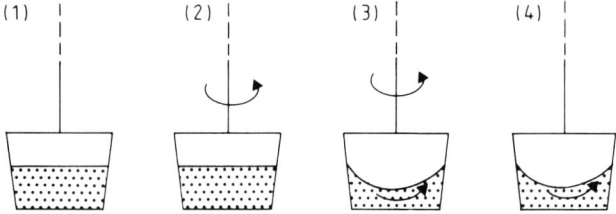

Figure 4.3. *Newton's rotating bracket. (1) A bucket containing water is suspended with its axis of symmetry vertical; the bucket and water are both stationary with respect to the laboratory; the surface of the water is observed to be flat. (2) The bucket is then set in rotation around the axis and relative to the laboratory; at this stage the water remains flat and at rest with respect to the laboratory. (3) Gradually, the water creeps up the sides of the bucket, as it begins to spin with the bucket; maximum concavity of the water is reached when the bucket and water are at rest with respect to each other and spinning with respect to the laboratory. (4) The bucket is then stopped suddenly but the water continues to swirl around, rotating with respect to both bucket and laboratory, the surface of the water remaining maximally concave before receding slowly down the sides of the bucket. The shape of the water's surface indicates whether or not the water is in 'absolute' rotation, and whether or not the water and pail are at 'relative' rest with respect to each other.*

By measuring relative quantities and using his laws of motion, he hoped to attain his real goal: knowledge of the absolute motions of bodies and their causes.

His famous rotating bucket thought experiment [22] gives a simple way of distinguishing between absolute and relative *rotations* (figure 4.3), and the Foucault pendulum provides a more practical method. But his laws of motion do not allow a similar method of distinguishing between absolute and relative *uniform motions (translations)*. Indeed, his relativity principle shows that no observation of the motions of bodies relative to some inertial frame of reference can possibly determine whether that frame is at absolute rest. The Newtonian form of the relativity principle simultaneously *assumes* the existence of a privileged (absolute) frame of reference and *denies* the possibility of observing it: if Newton's laws hold with respect to absolute space, the Newtonian theory of space and time implies that they hold equally well in *any* inertial frame. To get around this difficulty, Newton assumed that the solar system (which he called 'the system of the world') is uniquely privileged: absolute space is simply identified with that inertial frame in which its centre of mass is at rest [23].

By the end of the nineteenth century, some physicists had concluded that the concept of absolute space is not really needed. Since there is no basis for singling out any inertial frame, they used the law of inertia (Newton's first law) to define the entire class of inertial frames [24]. Purged of the concept of absolute space, Newton's laws do single out the class of inertial frames of reference, but assert their complete equality for the description of all mechanical phenomena. This mechanical relativity principle, as we shall call it, constitutes the theoretical explanation, within Newtonian mechanics, of the observations and experiments summed up in the Galilean relativity principle.

As long as the laws of Newtonian mechanics were considered to provide a foundation for the ultimate explanation of all physical phenomena (the mechanical world-view) [25], the mechanical principle of relativity could claim universal validity. Nevertheless, particular mechanical phenomena may single out certain preferred frames of reference for their simplest description. For example, the laws of propagation of sound waves in a medium take a particularly simple form in the rest frame of this medium. In applying the relativity principle to such phenomena, one must include the motion of the medium.

We shall later refer to an important consequence of Galilei–Newtonian kinematics [26]: the vector law of addition of relative velocities. Consider any event E. If we fix a spatial origin O in an inertial system I, the position of E relative to I is fixed by its displacement vector r relative to O; the time t of E is similarly fixed relative to some temporal origin $t = 0$. In Newtonian theory, there is an absolute time, so we need no further specification of t. (In special relativity, as we shall see, the time must be specified relative to the inertial frame.) The same event can be described relative to a second inertial system I' by its displacement vector r' relative to its spatial origin O'; since the time is absolute in Galilei–Newtonian kinematics, the event will occur at the *same* time t. (We assume that the same spatial and temporal units are used in both systems, and the same temporal origin is chosen.) If I' is moving with velocity V with respect to I, then r' and r are related by the so-called Galilean transformation law: $r' = r - Vt + r_0$, where r_0 is the displacement vector between the two origins at the moment $t = 0$. The vector addition rule for relative velocities follows by differentiation with respect to the absolute time: if a body moves with velocity v' with respect to I', then its velocity with respect to I is $v = v' + V$. Note that the velocities r and r' are relative to *different* inertial frames, so this rule is by no means trivial: its proof depends crucially on the existence of an absolute time, common to *both* inertial frames.

Now we turn to an apparent challenge to the mechanical principle of relativity that arose from optics and electrodynamics.

4.2.2. Origins of the special theory: optics and electrodynamics

Two rival theories, corpuscular and wave, were proposed in the seventeenth century to explain the propagation of light [27]. Newton's corpuscular hypothesis, implying that the velocity of light corpuscles, like the speed of bullets, is constant relative to the source that emits them, is entirely consistent with the mechanical relativity principle.

Huygens proposed a wave theory, based on an analogy between light and sound, and assuming that light waves require a medium for their propagation. The speed of light then should be independent of the motion of its source, but constant relative to its medium. Since starlight propagates with no difficulty through interstellar space—an excellent vacuum as far as ordinary matter is concerned—a subtle medium, often called the luminiferous ether, was assumed to permeate all of space [28].

By the mid-nineteenth century, the wave theory had triumphed [29], and the properties of the luminiferous ether became an important topic of study, including the following questions.

(i) Can the propagation of light waves be explained mechanically?; i.e., is it possible to ascribe mechanical properties to the ether such that waves propagating in it possess all the properties of light? Many such models were developed, but ultimately none was fully satisfactory [30].

(ii) What is the relation between ether and ordinary matter? In particular, is the ether completely immobile when matter moves through it, or is the ether totally (or partially) dragged along [31]?

If the ether were completely dragged along by matter, then the relativity principle could be extended from mechanics to optics without explicit reference to the state of motion of the ether. If it were not dragged along, the relativity principle could only be upheld by taking into account motion relative to the ether; in particular it should be possible to detect the motion of the Earth relative to the ether by optical means. The phenomenon of aberration indicated the latter alternative. Stellar aberration is the effect of the Earth's velocity on the apparent direction of light from a star, and (figure 4.4) [32] Fresnel succeeded in explaining it on the basis of the wave theory by assuming the ether to be unaffected by the motion of the Earth [33]. Numerous first-order optical experiments† to detect the motion of the Earth through the ether were carried out [35]; Mascart in 1874 summarized their results

> The translational motion of the Earth has no appreciable influence on optical phenomena produced by a terrestrial source, or light from the Sun, [so] these phenomena do not provide us with a means of determining the *absolute* motion of a body, and *relative* motions are the only ones that we are able to determine [36].

Mascart, in generalizing the results of optical experiments, was extending the Galilean (observational) form of the relativity principle to what we shall call an optical principle of relativity, so far established only for first-order effects.

† *'First-order' experiments are sensitive to effects that depend on the ratio (v/c), where v is a typical velocity of some body (presumably with respect to the ether) in the experiment and c is the velocity of light. Second-order experiments are those that are only sensitive to effects that depend on $(v/c)^2$, etc [34].*

History of Relativity

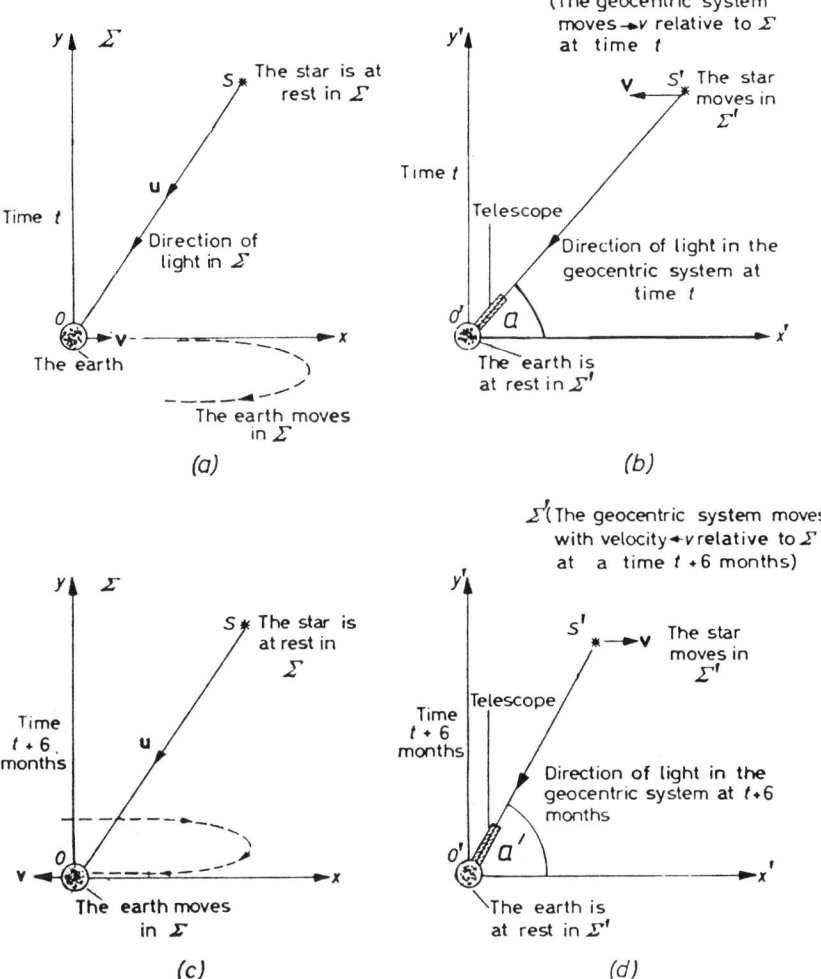

Figure 4.4. *Stellar aberration. Bradley observed a variation in the apparent position of a star at different times of the year. In (a), the Earth is moving with velocity +v relative to the Sun, in the inertial frame Σ. (b) Gives the geocentric system, in which the Earth is at rest. The telescope must be set at an angle a if the light from the star is to enter the telescope normally in the geocentric system. (c) Six months later, the Earth is moving in the opposite direction to the Sun, and in the geocentric system (d), the telescope must now be set at a different angle a' if the light is to enter the telescope normally in the new geocentric system.*

Fresnel had already provided a partial theoretical underpinning for this first-order principle [37]. In order to explain why the Earth's motion had no influence on refraction, he assumed that light waves in an optical medium are partially dragged along by the motion of the medium. In such a medium with index of refraction n at rest in the ether, light

259

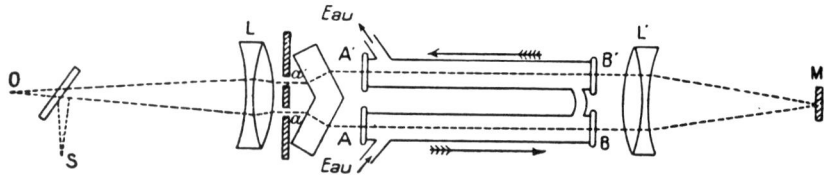

Figure 4.5. *Fizeau's experiment to measure the speed of light in moving water.*

travels with speed $c' = c/n$ (c is the speed of light in vacuum). If the medium moves through the surrounding ether with a velocity v in the same direction as a light ray, then Fresnel asserted that the speed of the light ray with respect to the ether is $c' + (1 - 1/n^2)v$. Fresnel's formula can be reconciled with the Galilean law of addition of relative velocities if it is assumed that, for some reason, a light wave travels with a speed $c' - v/n^2$ with respect to the moving medium, and the formula was so interpreted throughout the nineteenth century. The problem was then seen as finding a theoretical explanation for this behaviour of light waves; but, in retrospect, the formula can be seen as the first indication of a breakdown of the Galilean law of addition for velocities close to c (see section 4.2.6).

Fresnel derived his formula from certain hypotheses about the behaviour of the ether inside a moving body that ultimately proved untenable. Nevertheless, the formula correctly predicted the results of Fizeau's 1851 experiment on the velocity of light in moving water (figure 4.5); and with it Veltmann proved that, to first order, the Earth's motion has no effect on any optical phenomena, including interference effects [38]. Henceforward, any theory of the optics of moving bodies had to derive Fresnel's formula; but by the 1870s, the theories being considered were electrodynamical. Maxwell had reduced optics to the study of electromagnetic waves of a certain frequency range, and the ether had become the carrier of all electric and magnetic fields [39].

Starting with Maxwell, attempts continued for some time to account for the electromagnetic properties of the ether by mechanical models, but their only lasting achievement was to broaden the concept of a dynamical system to include the electromagnetic field (and later other fields). Poincaré emphasized this point and proved that, if one mechanical model of the electromagnetic ether exists, then an infinity of such models must also exist [40]. Physicists gradually accepted the idea that the electromagnetic field represents a dynamical system completely characterized by Maxwell's equations and in need of no mechanical explanation. Some physicists now even attempted to deduce the laws of mechanics from those of electrodynamics (electromagnetic world-view) [41].

Several alternative electrodynamics of moving bodies based on Maxwell's theory were proposed [42]; but by the end of the nineteenth century, H A Lorentz's version was widely accepted because of its success in explaining almost all known electrodynamic and optical phenomena [43]; in particular, it was the only theory from which the Fresnel formula (see above) could be derived. Lorentz assumed the following.

(i) All space, including the interior of ordinary matter, is permeated by the ether, which is completely immobile and the seat of all electric and magnetic fields. Maxwell's equations in their original form hold in the ether rest frame, assumed to be an inertial frame.

(ii) Ordinary matter consists (at least in part) of charged particles, which generate the electric and magnetic fields in the ether.

(iii) These fields in turn exert electric and magnetic forces on the charged particles, which then move through the ether in accord with Newton's laws.

Maxwell's equations thus single out the ether frame as that inertial frame, relative to which the speed of light is everywhere constant and isotropic; and it follows from the vector addition law for relative velocities (see above) that it is the *only* such inertial frame. Lorentz argued that this privileged frame of reference is really not the same as absolute space: as in the case of sound (see above), it is not absolute motion but motion with respect to a medium—here the ether—that is physically relevant [44]. However, there is a major difference between light and sound: it is easy to distinguish between a sound medium and empty space, which does not transmit sound, while even 'empty space' transmits light; so how can one make a distinction between ether and empty space?

In any case, Lorentz's original theory [45] faced a difficulty: in the 1880s experiments began to indicate that the optical relativity principle holds not only to first order but also to second order. The first such experiment was conducted by A A Michelson (1881), based on optical interferometric techniques (figure 4.6), and repeated with greater accuracy in collaboration with E W Morley in 1887 [46]. While able to explain successfully the failure of all first-order experiments to detect the Earth's motion by means of the theorem of corresponding states (see section 4.2.3), Lorentz's electrodynamics of moving bodies predicted a positive outcome for the Michelson–Morley experiment [45].

The problem confronting theorists may be summarized as follows: on the basis of Newtonian mechanics, one can prove the mechanical relativity principle, but Lorentz's theory predicts the relativity principle should fail for second-order optical and electromagnetic phenomena. Experimentally, the acceleration of a body relative to the class of inertial frames is detectable by means of first-order mechanical, optical and

Twentieth Century Physics

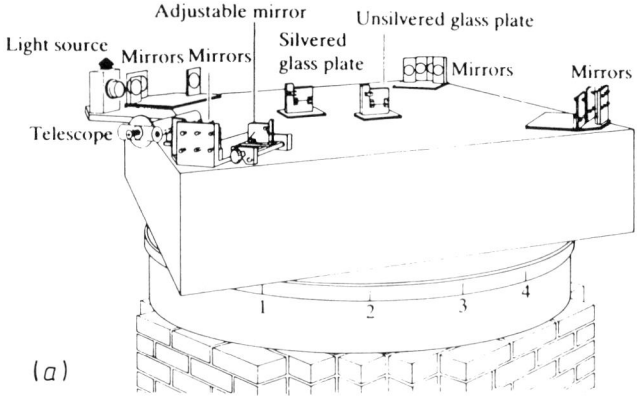

Figure 4.6. (a) The Michelson–Morley experiment. (b) Diagrammatic representation of the Michelson–Morley experiment. A beam of light from the source S is split at the beam-splitter O into two beams l_1 and l_2 travelling along and perpendicular to the direction of the Earth's orbit in its motion. The velocity and direction of the Earth's orbital motion relative to the ether is indicated by the arrow labelled v. Beams l_1 and l_2 reflect from mirrors M_1 and M_2 and return to the beam-splitter, where they are reflected and refracted to the telescope T. A distinctive interference pattern is observed. (c) The movement of the mirrors as the light propagates gives rise to a difference in the time taken for the light to travel along the two perpendicular arms of the interferometer. If the light was propagating through a stationary ether, rotation of the interferometer through 90° should change this time difference and hence should cause a change in the interference pattern. The sensitivity of the instrument would have revealed even a minute shift, but no change was observed.

electromagnetic experiments† but no such experiment, even to second order, singles out any inertial frame that could be identified with the ether. An explanation for such failures to first order could be given within Lorentz's theory, but theorists now had to account for the experimental verification of the relativity principle for optical and electromagnetic phenomena up to second order (at least).

4.2.3. Formulation of the special theory

Lorentz, Poincaré and Einstein made the most significant contributions to the solution to this problem adopted by the physics community. Debate over their respective contributions persists [48], largely because of differing views on what are the central features of special relativity [49]. The following discussion is based on the characterization of the special theory in the Introduction (section 4.1).

Lorentz's method of resolving the discrepancy between theory and experiment was to account for the failure of all attempts to detect the

---

† It was soon verified that rotational motion is detectable by purely optical means. The Sagnac experiment [47], for example, provides an optical analogue to the Foucault pendulum.

History of Relativity

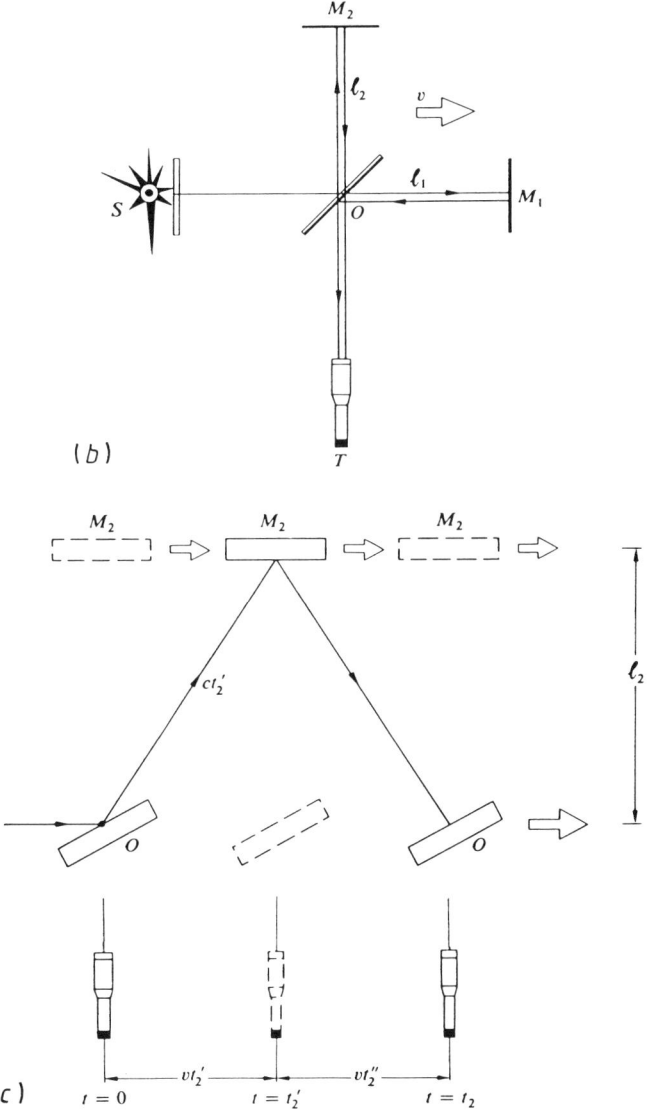

Figure 4.6. *Continued.*

motion of the Earth through the ether by invoking a series of physical effects that just compensate for the expected violations of the relativity principle [50]. He finally achieved compensation to all orders, but only after over a decade of work on the problem [51].

Lorentz used the theorem of corresponding states to show that his theory implied the negative result of attempts to detect motion through the ether: to every possible state (solution of Maxwell's equations) of an electromagnetic system in the ether frame with coordinates r, t, there exists a corresponding state—i.e. a *different* physical solution of Maxwell's equations, but with the *same* mathematical form in an inertial frame moving with velocity v with respect to the ether, when the solution is expressed in terms of suitably transformed variables, r', t' in the moving system. Lorentz used this theorem, in conjunction with certain physical compensations (see below), to deduce optical and electrodynamic versions (first approximate, later exact) of the relativity principle.

The original 1892 approximate version of the theorem explained the failure of all first-order optical experiments. In addition to the Galilei transformation law (see above) from r to r', its proof required introduction of a transformation from the absolute time t to what Lorentz called the local time $t' = t - (v, r)/c^2$. (Throughout this chapter, the scalar product of two vectors A and B will be symbolized by (A, B).) For Lorentz this was just a mathematical device, not a physical compensation. He never related the local time to clock readings in the moving system.

However, in order to explain the negative result of the second-order Michelson–Morley experiment, Lorentz had to introduce such a physical compensation [52]: a contraction by the factor $\sqrt{1 - (v/c)^2}$ in that dimension of a body in the direction of its motion through the ether, for which he gave a dynamical explanation on the assumption that the forces holding matter together are either purely electromagnetic or at least behave in the same way when transformed to a moving frame. This is usually called the Lorentz contraction, and sometimes the Fitzgerald–Lorentz contraction since the same explanation had already been proposed by Fitzgerald [53].

Lorentz further extended the theorem of corresponding states by introducing non-Galilean spatial transformations between r and r', thus defining spatial quantities in the moving system that 'correspond' to those in the ether frame [54]. The new version of the theorem, together with the contraction compensation and a new compensation hypothesis about the variation of the mass of a charged particle with its velocity, enabled him to prove compensation to second-order for all physical systems. By further modifying his transformation laws, he was able to prove exact compensation (i.e. to all orders in v/c) for all the effects of uniform motion through the ether [51]†. The set of transformations of the space and time variables are now called the Lorentz transformations, but he still regarded them as just a mathematical device. Only after

---

† *This statement is not quite accurate. Lorentz's transformation law for the charge density makes it appear that the compensation is not complete for the inhomogeneous Maxwell equations. Poincaré and Einstein gave the correct transformation law for charge density in 1905.*

studying Einstein's work did Lorentz suggest that moving observers would actually measure the 'corresponding' quantities [55].

Like absolute space for Newton, the ether for Lorentz was something real and physically efficacious. The validity of the relativity principle for optics and electrodynamics without the need to take into account the motion of ordinary matter relative to the ether is the remarkable result of 'real' compensating effects—length contraction and mass increase—that arise dynamically just because of such motions through the ether. It was still possible that some non-compensated effect might ultimately confirm the reality of motion through the ether.

Poincaré's approach [56] to the relativity principle was different [57]. The undetectability of motion with respect to the ether extended to all physical phenomena, and he ranked the principle with a small number of other principles—such as the two laws of thermodynamics—based on generalizations from experience (the French word expérience does not make a distinction between experience and experiment) and having a much wider import than any theories that might be used to explain the validity of the principles in particular contexts.

Many other contributions of Poincaré have been incorporated into the generally accepted version of the special theory. He called attention to the conventional element in any definition of the simultaneity of distant events [58], and was the first to give a physical interpretation of Lorentz's local time [59]: clocks at rest in an inertial frame synchronized by means of light signals, without taking into account the motion of the frame relative to the ether, show the local time rather than the true (ether frame) time. As this example shows, he usually took the ether seriously, but on occasion he was agnostic, even suggesting in 1902 that it is just a convenient hypothesis that someday might be discarded. He proved that the Lorentz transformations (which he named) form a group, under which Maxwell's equations with sources are invariant [60].

However, there were limits to his insights. By and large he accepted the ether and its dynamical effects, his formulation of the relativity principle still distinguishing between observers at rest in the ether and moving observers, as does his interpretation of the local time. He never organized his many brilliant insights into a coherent theory that resolutely discarded the ether and the absolute time or transcended its electrodynamic origins to derive a new kinematics of space and time based on a formulation of the relativity principle that makes no reference to the ether.

Einstein was the first to do so. There is very little contemporary documentation of Einstein's work on relativity before his now-classic 1905 paper [61]. Nevertheless, there is an extensive literature attempting to establish how Einstein arrived at his formulation of the theory, based on the scant evidence of the 1905 paper itself and later papers and reminiscences by Einstein and others, which were often supplemented, it must be said, by a good deal of unsubstantiated conjecture [62]. Given the dearth of data and the plethora of possibilities, it is hardly surprising that there is no consensus on answers to such questions as: what had Einstein read by 1905 and how did his readings influence him? Was his work guided primarily by a detailed study of electrodynamics or by more heuristic considerations (such as his well-known *Gedankenexperiment* about chasing a light ray)? What was the place of experimental evidence, such as (but not only) the Michelson–Morley experiment, in his thinking? Did philosophical considerations play an important role, and if so, what philosophy and what role? The following brief account will stick to relatively non-controversial matters [63].

Einstein came to the theory of relativity in the course of his efforts to understand the nature of matter and radiation, in particular his attempt to reconcile the electrodynamics of moving bodies with the optical relativity principle. Letters to his fiancée Mileva Maric [64] show that by 1899 he was seriously studying the electrodynamics of moving bodies [65] and expressing scepticism about the ether

> The introduction of the term 'ether' into theories of electricity leads to the notion of a medium of whose motion one can speak without, I believe, being able to associate any physical meaning with such a statement [66].

He soon supplemented criticism of Hertz's electrodynamics (from a viewpoint remarkably similar to that of Lorentz) with proposals for experiments to detect the motion of the Earth through the ether. By the end of 1901, he was preparing a major work on the electrodynamics of moving bodies, embodying ideas on relative motion that are unfortunately not further detailed in his letters [67]; but he published nothing on the topic until 1905. Practically no contemporary evidence from the crucial years between 1901 and 1905 is known, but the 1905 paper and later reminiscences supply some evidence. Abberation and electromagnetic induction, Fizeau's experiment on the velocity of light in a moving medium and the body of experimental evidence for the optical version of the relativity principle (including, but not confined to the Michelson–Morley experiment) played important roles in convincing him of the universal validity of the relativity principle.

Its evident compatibility with this principle led him to seriously reconsider the long-rejected possibility of an emission theory of light. But the explanation of such things as the reflection of light from a moving mirror and of Fizeau's experiment on the basis of an emission theory forced him to make an ever-expanding series of complicated and non-unique supplementary hypotheses. So he abandoned such attempts [68], and turned to the problem of whether Lorentz's theory (which he knew only in its 1895 version) could be made compatible with the relativity principle.

History of Relativity

His philosophical readings, especially of Hume, Poincaré and Mach helped to liberate his thought from the straight-jacket of classical concepts of space and time [69]. This was a precondition for his crucial 1905 insights: the relativity of simultaneity and the consequent possibility of basing electrodynamics—indeed all of physics—on a new kinematical foundation. Within about six weeks of these insights he completed his now-famous relativity paper [70] (figure 4.7). A great deal of prior work, of which we have no record, must have enabled this rapid *dénouement* of a search that had lasted almost a decade [71].

After resolutely discarding the concept of the ether†, the 1905 paper makes clear the need for a new kinematical foundation for electrodynamics

> The theory to be developed—like every other electrodynamics—is based upon the kinematics of rigid bodies, since the assertions of any such theory concern relations between rigid bodies (coordinate systems), clocks, and electromagnetic processes. Insufficient consideration of this circumstance is at the root of the difficulties with which the electrodynamics of moving bodies currently has to contend [72].

He based the new kinematics on two principles. The first extends the mechanical and optical forms of the relativity principle (see above) to all phenomena

> The laws of nature are independent of the state of motion of the frame of reference, as long as the latter is acceleration free [i.e. inertial] [73].

The second principle, which he calls the principle of the constancy of the velocity of light, asserts that

> ... light in empty space always propagates with a definite velocity, independent of the state of motion of the emitting body [70].

Einstein takes these principles, drawn from previous theory and experiment, as postulates, from which he deduces criteria that must be satisfied by every physical theory. The principles are justified *a posteriori* by the success of theories satisfying these criteria—success in explicating known results, establishing new connections between previously unrelated phenomena and predicting new ones. The first deduction from the principles is the new kinematics. The light principle (taken together with an appropriate definition of simultaneity, see below) requires the speed of light to have the constant and isotropic value c

---

† *Later, he was more precise: it is the concept of an ether with a definite state of motion that is incompatible with the theory of relativity.*

3. *Zur Elektrodynamik bewegter Körper;*
von A. Einstein.

Daß die Elektrodynamik Maxwells — wie dieselbe gegenwärtig aufgefaßt zu werden pflegt — in ihrer Anwendung auf bewegte Körper zu Asymmetrien führt, welche den Phänomenen nicht anzuhaften scheinen, ist bekannt. Man denke z. B. an die elektrodynamische Wechselwirkung zwischen einem Magneten und einem Leiter. Das beobachtbare Phänomen hängt hier nur ab von der Relativbewegung von Leiter und Magnet, während nach der üblichen Auffassung die beiden Fälle, daß der eine oder der andere dieser Körper der bewegte sei, streng voneinander zu trennen sind. Bewegt sich nämlich der Magnet und ruht der Leiter, so entsteht in der Umgebung des Magneten ein elektrisches Feld von gewissem Energiewerte, welches an den Orten, wo sich Teile des Leiters befinden, einen Strom erzeugt. Ruht aber der Magnet und bewegt sich der Leiter, so entsteht in der Umgebung des Magneten kein elektrisches Feld, dagegen im Leiter eine elektromotorische Kraft, welcher an sich keine Energie entspricht, die aber — Gleichheit der Relativbewegung bei den beiden ins Auge gefaßten Fällen vorausgesetzt — zu elektrischen Strömen von derselben Größe und demselben Verlaufe Veranlassung gibt, wie im ersten Falle die elektrischen Kräfte.

Beispiele ähnlicher Art, sowie die mißlungenen Versuche, eine Bewegung der Erde relativ zum „Lichtmedium" zu konstatieren, führen zu der Vermutung, daß dem Begriffe der absoluten Ruhe nicht nur in der Mechanik, sondern auch in der Elektrodynamik keine Eigenschaften der Erscheinungen entsprechen, sondern daß vielmehr für alle Koordinatensysteme, für welche die mechanischen Gleichungen gelten, auch die gleichen elektrodynamischen und optischen Gesetze gelten, wie dies für die Größen erster Ordnung bereits erwiesen ist. Wir wollen diese Vermutung (deren Inhalt im folgenden „Prinzip der Relativität" genannt werden wird) zur Voraussetzung erheben und außerdem die mit ihm nur scheinbar unverträgliche

Figure 4.7. *Opening page of Einstein's first paper on special relativity.* Annalen der Physik Volume 17 p 891 (1905).

relative to some inertial frame. Since there is no preferred (ether) frame, the relativity principle implies that it must have the same value c relative to every inertial frame, clearly incompatible with the vector addition law for relative velocities. Since Galilei–Newtonian kinematics leads to this law (see above), it must be replaced by a new kinematics, which Einstein found by deriving space and time transformation equations from his two principles.

In order to understand the role of the time coordinate, he carefully analysed the concept of time. Noting the element of convention in any definition of the simultaneity of spatially separated events, he defined simultaneity relative to an inertial frame by requiring clocks at rest in the frame be synchronized so that the speed of light is constant and isotropic relative to that frame[†]. It follows from this definition and the light postulate that time is no longer absolute: rather, two events simultaneous with respect to an inertial frame I (as defined by I-synchronized clocks), are non-simultaneous with respect to another inertial frame I' (as defined by I'-synchronized clocks) moving with respect to I.

Einstein then derived the transformation laws between the spatial and temporal coordinates of an event with respect to frames I and I'

$$r' = (r - \gamma V t) + (\gamma - 1)(V, r)V/V^2$$

$$t' = \gamma[t - (V, r)/c^2]$$

$$\gamma = 1/[\sqrt{1 - (V/c)^2}]$$

where V is the velocity of I' relative to I. These transformations agree formally with those Lorentz introduced in 1904 for quite different reasons (see above) [75]. Poincaré soon showed that the set of transformations, which he attributed to Lorentz without mention of Einstein, form a group [76]. He named it the Lorentz group. The inhomogeneous group, which includes spatial and temporal translations as well as Lorentz transformations proper, is now called the Poincaré group.

The new kinematics replaces the Galilean rule for addition of relative velocities with a relativistic formula for the composition of velocities[‡] that resolves the paradox noted above: the speed of light compounded with any lesser speed remains unchanged in magnitude. Its direction and frequency change, of course, and Einstein derived relativistic formulae for aberration and the Doppler effect in this way. He showed that the

---

[†] *Einstein made it clear [74] that any signal can be used to synchronize clocks at rest in an inertial frame, as long as its propagation is isotropic and its speed constant with respect to that frame. The advantage of light is that is has these properties in every inertial frame.*

[‡] *One often speaks of the relativistic law of addition of velocities; but 'addition' is not used here in the sense of arithmetic or vectorial addition, and the two velocities being composed are relative to different inertial frames of reference.*

vacuum Maxwell equations, without and with sources, are invariant under the new kinematics, and derived relativistic equations of motion for a charged particle ('electron').

In 1905, Einstein argued that the inertial mass and energy content of all bodies are proportional, leading to the now famous equation $E = mc^2$ [77] (figure 4.8). Others had suggested a similar relation between mass and electromagnetic energy in special cases [78]; but Einstein was the first to suggest its universal validity for all forms of mass and energy [79].

His later work on the special theory includes: a discussion of the transverse Doppler effect [80], two papers on the electrodynamics of macroscopic media, written in collaboration with Jakob Laub [81], two published review articles [82]; and an extensive review, written about 1912 but only recently published [83]. His two books [84], and many of his survey articles on general relativity include discussions of the special theory, but after 1912 he published no new results in this area.

As noted, Einstein developed the special theory in the course of his search for a theory explaining the nature of matter and radiation, a search that continued throughout his life.

> When we say that we have succeeded in understanding a group of natural processes, we always mean that a constructive theory has been found that embraces the processes in question [85].

His presentation of the special theory bears distinct traces of its origins. Mesmerized by the problem of the nature of light[†], he ignored the discrepancy between the relativity principle, kinematical in nature and universal in scope, and the light principle, taken from 'Lorentz's theory of the immobile light ether' [87] and concerned with a particular type of electrodynamical phenomenon[‡]. Such traces of its electrodynamical origins exposed the new kinematics to avoidable misunderstandings and attacks over the years. Most recently, advocates of 'protophysics', having rightly observed that it should be possible to give a purely kinematical foundation to any theory of kinematics, failed to notice that special-relativistic kinematics has already meet this test (see section 4.2.5.4). Einstein later recognized such defects in his original arguments.

> The special theory of relativity grew out of the Maxwell electromagnetic equations. But it came about that even in the

---

† *In 1951 he wrote 'The whole fifty years of conscious brooding have not brought me nearer to the answer to the question, "What are light quanta?"* ' [86].

‡ *The oddity of this reference to light would be even more apparent if one replaced it by a reference to some other massless field, basing the theory, for example, on the principle of the constancy of the speed of neutrinos. One is reminded of Born's comments on formulations of the second law of thermodynamics that invoke heat engines and refrigerators 'new and strange conceptions, obviously borrowed from engineering'* [88].

13. Ist die Trägheit eines Körpers von seinem Energieinhalt abhängig?
von A. Einstein.

Die Resultate einer jüngst in diesen Annalen von mir publizierten elektrodynamischen Untersuchung[1]) führen zu einer sehr interessanten Folgerung, die hier abgeleitet werden soll.

Ich legte dort die Maxwell-Hertzschen Gleichungen für den leeren Raum nebst dem Maxwellschen Ausdruck für die elektromagnetische Energie des Raumes zugrunde und außerdem das Prinzip:

Die Gesetze, nach denen sich die Zustände der physikalischen Systeme ändern, sind unabhängig davon, auf welches von zwei relativ zueinander in gleichförmiger Parallel-Translationsbewegung befindlichen Koordinatensystemen diese Zustandsänderungen bezogen werden (Relativitätsprinzip).

Gestützt auf diese Grundlagen[2]) leitete ich unter anderem das nachfolgende Resultat ab (l. c. § 8):

Ein System von ebenen Lichtwellen besitze, auf das Koordinatensystem (x, y, z) bezogen, die Energie l; die Strahlrichtung (Wellennormale) bilde den Winkel φ mit der x-Achse des Systems. Führt man ein neues, gegen das System (x, y, z) in gleichförmiger Paralleltranslation begriffenes Koordinatensystem (ξ, η, ζ) ein, dessen Ursprung sich mit der Geschwindigkeit v längs der x-Achse bewegt, so besitzt die genannte Lichtmenge — im System (ξ, η, ζ) gemessen — die Energie:

$$l^* = l \frac{1 - \frac{v}{V} \cos \varphi}{\sqrt{1 - \left(\frac{v}{V}\right)^2}},$$

wobei V die Lichtgeschwindigkeit bedeutet. Von diesem Resultat machen wir im folgenden Gebrauch.

---

1) A. Einstein, Ann. d. Phys. 17. p. 891. 1905.
2) Das dort benutzte Prinzip der Konstanz der Lichtgeschwindigkeit ist natürlich in den Maxwellschen Gleichungen enthalten.

Figure 4.8. *Opening page of Einstein's paper in which he argued that the inertial mass of a body and its energy content are proportional, leading to the famous equation* $E = mc^2$. Annalen der Physik *Volume 18 p 639 (1905).*

derivation of the mechanical concepts and their relations the considerations of those of the electromagnetic field played an essential role. The question as to the independence of these relations is a natural one because the Lorentz transformation, the real basis of the special-relativity theory, in itself has nothing to do with the Maxwell theory [89].

But he never discussed derivations of the Lorentz transformations that do not require the light postulate (see below).

The assimilation of Einstein's work, and that of Lorentz and Poincaré, by the physics community constitutes an integral part of the history of the theory. The creative process does not end with the work of a talented individual, but includes its assessment by a community of experts in the appropriate field, and its incorporation—often in altered form—into the accepted canon [90]. The development of a theory is not over until its assimilation has led to a broad consensus of opinion, which usually includes varying interpretations of 'the' theory. Often, there is even a later reinterpretation of an established theory (see, for example, the earlier discussion of Newton's mechanics and the subsequent discussion of Maxwell's electrodynamics).

To appreciate the choices involved in the formation of the current consensus about the special theory, it is instructive to consider some hypothetical alternative scenarios. The work of Lorentz, Poincaré, and others suggests that, without Einstein's contribution, the consensus version might not have made a clear distinction between kinematic and electrodynamic effects, but interpreted such things as length contraction, time dilation and increase of mass with velocity as dynamical effects, caused by motion relative to the ether frame†. Emphasis would then have been placed on factors leading to the undetectability of absolute velocity, rather than on the complete equivalence of all inertial frames.

On the other hand, if Einstein's emphasis on the light postulate had been recognized as introducing an unnecessary non-kinematical element, and had an approach based entirely on the relativity postulate (see below) been adopted, a more fully kinematical foundation for the special theory than Einstein's might have become canonical.

Turning back to the actual course of events, the theoretical physics community assimilated certain elements of Einstein's work fairly rapidly, albeit with varying emphases on what constituted its central features. In particular, there was often a failure to understand the differences between Lorentz's and Einstein's views (for many years references to the Lorentz–Einstein theory were common), and attempts were made to combine acceptance of relativity with continued adherence to the ether;

---

† In 1908, Laub wrote Einstein 'If your work were not available, then at best we would be taking the same standpoint about Minkowski's time transformation equation (as concerns its physical interpretation) as to Lorentz's "local time"' [91].

History of Relativity

there was also significant resistance to the whole relativity enterprise [92]. Studies of the theory's assimilation by various physics communities have brought out many such differences, and it has been suggested that they are attributable in large part to differing national styles of 'doing physics' [93], but others have argued that this was not the most important factor [94].

4.2.4. Later development of the theory
The further history of the special theory may be roughly classified into the following sections.

(i) The development of varying formulations of the theory, involving alternate fundamental concepts and postulates, and/or the use of different mathematical structures and techniques.

(ii) Application of the theory to various branches of physics, such as: (a) kinematics, (b) particle dynamics, (c) electrodynamics (including optics), (d) thermodynamics, (e) statistical mechanics, (f) continuum mechanics, (g) quantum mechanics and elementary particle theory, and (h) gravitation.

(iii) Experimental tests of the principles and various consequences of the theory, and technological applications.

Systematic research has not yet been done on the history of any these topics, so the following account, while discussing the period immediately after 1905 in somewhat more detail, perforce remains fragmentary and unsystematic [95].

4.2.5. Alternate formulations and formalisms

4.2.5.1. Formalisms for the special theory. Many mathematical formalisms have been used: matrices, vectors, tensors, quaternions, spinors, twistors, etc. The essential requirement is some representation of the Lorentz group—that is, the four-dimensional real pseudo-orthogonal group—or some group isomorphic to it; however, such representations were introduced rather unsystematically.

Einstein's original formulation used Cartesian spatial coordinates relative to various inertial frames; but his results can easily be put into three-dimensional vectorial form, with the time relative to each frame as an additional variable (see above). Such so-called three-plus-one formalisms have the disadvantage that the transformation formulae from one inertial frame to another for many physical quantities are rather complicated. Mathematicians soon developed four-dimensional formalisms that provide major simplifications.

4.2.5.2. Four-dimensional formulation. Poincaré reformulated Lorentz's electrodynamics in a four-dimensional geometrical form [96]†; Minkowski carried out a similar four-dimensional reformulation of Einstein's theory, which had great influence on the subsequent development of relativity [97]. His concept of a world of events in a four-dimensional space-time (often called Minkowski space) brought not only considerable formal simplification, but also conceptual clarification of the interrelations between many physical quantities in the special theory.

The four-dimensional approach suggests a natural generalization of the vector algebra and analysis used in three-dimensional Euclidean space to four-dimensional space-time. Let r be the displacement vector of a point relative to the origin of some inertial frame. The fundamental *spatial* invariant of the frame's Euclidean geometry is the *distance* $\Delta\sigma$ between two points at r and r': $(\Delta\sigma)^2 = (|r' - r|)^2$; it is invariant under the six-parameter (inhomogeneous) Euclidean group, consisting of spatial rotations and translations. If we represent a point in space by the column matrix $\mathbf{r}$, and a rotation by the (real) orthogonal 3×3 matrix $\mathbf{R}$, then $\mathbf{Rr}$ represents the effect of a rotation on $\mathbf{r}$. (A real matrix is orthogonal if it obeys the equation $\mathbf{R}^T\mathbf{R} = \mathbf{I}$, where $\mathbf{R}^T$ is the transpose of $\mathbf{R}$ and $\mathbf{I}$ is the identity matrix.)

Now combine the temporal and spatial coordinates t, r of an event to form a point x in a four-dimensional *space-time*, Minkowski space. The fundamental invariant of this space is the *interval* Δs between two events $x = (t, r)$ and $x' = (t', r')$: $\Delta s^2 = (\Delta\sigma)^2 - (c\Delta t)^2$, where $\Delta t = t' - t$. The action of a Lorentz transformation from one inertial frame to another is represented by a pseudo-rotation of the points of this space. (The time coordinate must be multiplied by c in order to make it dimensionally comparable to the spatial coordinates.) The 'pseudo' in pseudo-rotations refers to the fact that Lorentz transformations are not actually rotations, because the signature of the space-time metric is not positive definite. However, rotations may be used, as Poincaré [96] and Minkowski [97] showed, by taking a pure imaginary time coordinate it, and representing Lorentz transformations by matrices with some pure imaginary components. The interval is invariant under these pseudo-rotations, as well as under the usual spatial rotations and translations in space and time, i.e. under the ten-parameter Poincaré (or inhomogeneous Lorentz) group. Representing an event by a column matrix $\mathbf{x}$, and a Lorentz transformation by a 4×4 pseudo-orthogonal matrix $\mathbf{L}$‡, $\mathbf{Lx}$ represents the effect of a pseudo-rotation—that is, a Lorentz transformation—on the event. It is possible to give two interpretations

---

† *Of course, the introduction of time as a fourth dimension in analytical mechanics long antedates the theory of relativity.*

‡ *The real matrix $\mathbf{L}$ is pseudo-orthogonal if it obeys the equation $\mathbf{L}^T \eta \mathbf{L} = \mathbf{I}$, where $\mathbf{L}^T$ is the transpose of $\mathbf{L}$ and η is the matrix with three $+1$ and one -1 along the diagonal, all non-diagonal elements equalling 0.*

to Lorentz transformations (and indeed all other transformations acting on an abstract space) active and passive. An active transformation takes a point of the space—here an event—into another point, while a passive transformation describes the same point from a different coordinate system—here an inertial frame of reference.

Distances between distinct points in Euclidean space are necessarily positive; but intervals between events in Minkowski space-time fall into three disjoint classes: *space-like* if $(\Delta s)^2 > 0$, *null* (or *light-like*) if $(\Delta s)^2 = 0$ and *time-like* if $(\Delta s)^2 < 0$ (figure 4.9(a)). In the latter case, $\Delta \tau = \sqrt{-(\Delta s)^2}$ is called the proper time between the two events. (Conventions differ as to whether the overall sign of the interval is defined to make the square of space-like intervals positive (as we have) or negative.) This classification defines the light-cone structure of any point O of Minkowski space: the set of events null-separated from O form a cone; all points space-like separated from O lie in the exterior of the cone; all points time-like separated from O lie in the interior (see figure 4.9(b)).

In accord with Minkowski's somewhat grandiose name for space-time—'the world', the path of a point in space-time corresponding to a possible particle motion is called a (time-like) world-line; it is invariantly parametrized by the proper time τ along it, $x = x(\tau)$. Four-dimensional kinematic quantities, defined by derivatives with respect to τ (the four-velocity $V = dx/d\tau$ for example, is the tangent vector to the world-line) are related to combinations of three-plus-one quantities (the components of V in an inertial frame, for example, are $(\gamma, \gamma V)$, where V is the three-velocity). Many physical quantities that appear unrelated in three-plus-one formalisms turn out to be components of the same four-dimensional entity, reflecting the physical relations between these quantities in the special theory (see examples from mechanics and electrodynamics below).

Einstein had already implicitly defined the proper time, and assumed that clocks carried along a time-like world-line read the proper time [77]. Just as the distance between two points in space is not independent of the path between them, the proper time between two events is not independent of the world-line between them—the difference being that a straight line is the shortest path between two points, while the longest proper time elapses along a straight (inertial) world-line between them. This is the basis of the famous 'twin paradox,' implicit in Einstein's presentation but first formulated explicitly by Langevin in 1911 [98], which since then has occasioned considerable controversy [99].

The concept of Minkowski space as the site of all physical processes, with the concomitant use of geometrical concepts and language to describe such processes, was soon widely adopted and expressed in a variety of mathematical formalisms. Minkowski used a four-dimensional matrix formalism, and a quaternion formalism was used by Conway [100] and Silberstein [101]. However, the four-dimensional

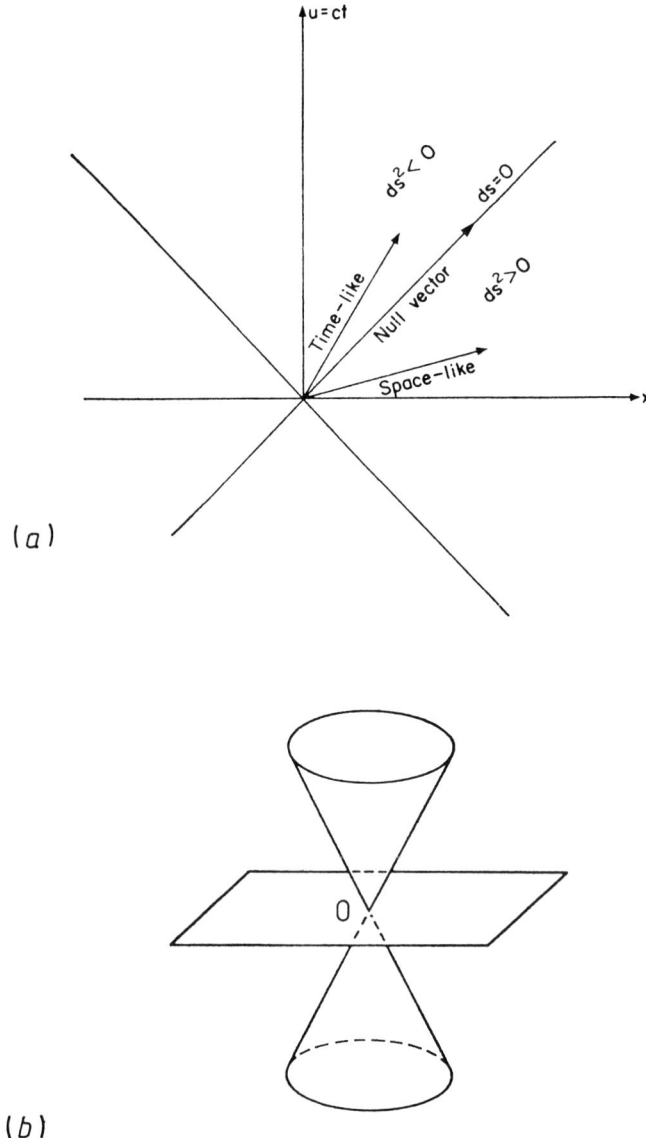

Figure 4.9. (a) Space-like, null and time-like vectors. (b) The light cone.

quasi-Cartesian vector and tensor formalism developed by Sommerfeld enabled physicists already familiar with the corresponding three-dimensional formalism to assimilate the new ideas most readily, and the formalism soon prevailed in applications of the theory [102].

4.2.5.3. Spinor and twistor formalisms. In order to treat electron spin in quantum mechanics, Pauli introduced [103] two-component (non-relativistic) and Dirac introduced [104] four-component (relativistic) wavefunctions. It was soon recognized that these involve non-tensorial representations of the rotation and Lorentz groups, respectively, which are needed for the description of fermion fields (particles with half-integral spin). While classifying linear representations of the orthogonal groups, Cartan in 1913 had introduced these spinorial representations, and in 1929 Van der Waerden developed the required spinor algebra and analysis [105].

Thirty years later, Penrose showed that spinorial techniques could also be applied to advantage in the study of boson fields (particles with integral-spin), such as the Maxwell and linearized gravitational fields [106], and also developed the twistor formalism for relativistic fields, based on representations of the Lorentz group in four-dimensional complex projective spaces [107].

4.2.5.4. Formulations without the light postulate. In 1910, Ignatowski showed that the special theory does not require the light postulate [108]. The relativity principle, together with some additional kinematical assumptions, implies that space-time is invariant under a one-parameter family of space-time transformation groups. For any finite value of the parameter, the group is isomorphic to the Lorentz group (as the parameter becomes infinite, the group contracts to the Galilei group). The parameter has the dimensions of a velocity, and its value is fixed by the light postulate in Einstein's formulation. But purely kinematical assumptions not involving electrodynamics may also be used for this purpose [109]. Refinements and variations of Ignatowski's results have been rediscovered many times since [110]. Einstein never gave up the light postulate (see previous section), and accounts of the theory tended to follow Einstein's lead. Only recently have derivations of the Lorentz transformations similar to Ignatowski's begun to appear in standard textbooks [111].

4.2.5.5. Causal formulations. In 1913–14, Robb proposed another, radically different foundation for the special theory [112]. He was dissatisfied with Einstein's use of inertial-frame-dependent concepts [113], and resolved to develop the theory in absolute—i.e. inertial-frame-independent form. Accordingly, he based his account on an absolute relation between events that he called conical order: of two distinct events, either one precedes the other, or the relation does not hold between them, i.e. it is a partial-order relation. By imposing suitable postulates on this partial-order relation, Robb developed the geometry of Minkowski space in a coordinate-free, purely geometrical manner [114]. Reversing Robb's logic and starting from Minkowski space, his order relation can be defined in terms of the conformal (i.e. light-cone) structure of Minkowski space-time. The relation holds between two events if the interval between them is time-like or null, but fails to hold if it is space-like. Robb's approach did not receive widespread attention until the 1960s [115] with the work of Zeeman and others on the causal structure of Minkowski space-time, and its subsequent generalization by Kronheimer and Penrose to the concept of causal space-times [116].

4.2.6. Relativistic velocity (kinematic) space

In 1907 Laue showed that Fresnel's dragging coefficient, which had previously been explained electrodynamically (see section 4.2.2) can be explained as a purely kinematical effect, without any 'dragging' [117]. Fresnel's formula is just the first-order approximation to the relativistic law of composition of relative velocities as applied to the velocity of light in a moving medium. (Of course, the ether frame plays no role in the relativistic argument. The motion of medium and light may be taken relative to any inertial frame.)

The formulae expressing this law are quite complicated algebraically, especially if the two velocities do not lie in the same direction. Sommerfeld found a simple interpretation of the law [118]: the two velocities that are composed and their resultant form the sides of a spherical triangle on a sphere of imaginary radius ic. Varicak [119] soon gave an interpretation of this result in terms of the geometry of velocity space, sometimes called kinematic space [120]. In special relativity, this space obeys three-dimensional Lobachevskian geometry, i.e. it is a space of constant *negative* curvature, with radius c. (The corresponding Galilei–Newtonian velocity space is flat, leading to the Galilean law of vector addition of relative velocities.) Relative velocities are represented by pairs of points in velocity space, and their composition amounts to solving Lobachevskian (pseudo-spherical) triangles. The geometry of such triangles being formally equivalent to that of spherical triangles of imaginary radius, Sommerfeld's results follow at once [121].

The non-commutativity of the composition of relative velocities in different directions follows immediately from the curvature of relativistic velocity space. Another immediate consequence was noted by Borel [122]. When a vector is parallel-transported around a closed path, it is rotated relative to its original direction. Physically, a torque-free angular momentum vector undergoes parallel transport along its path in velocity space. So, when it moves around a closed orbit at speeds comparable to c, an angular momentum vector undergoes a precession. This effect, for the case of the spin of an electron in an atomic orbit, was rediscovered by Thomas and the resulting spin precession is now called the Thomas effect [123].

4.2.7. Particle dynamics
Planck (1906–7) showed how to derive special-relativistic dynamics from a Lagrangian variational principle (often called a principle of least action, although for dynamical paths the extremum is actually a maximum). The usual techniques applied to the Lagrangian led to simpler definitions of the relativistic force and momentum than those originally proposed by Einstein [124], who soon adopted them [125], and they became standard.

Minkowski showed that, under the Lorentz group, relativistic three-momentum **p** and energy E combine naturally to form a 'four-momentum' vector; and relativistic three-force **f** and rate of work done by it (**f**, **v**) combine to form a 'four-force'.

In 1909, Lewis and Tolman reversed the direction of the previous arguments [126]. Instead of proving the conservation laws for energy and momentum of a system of particles from their equations of motion, they derived the relativistic expressions for the energy and momentum of a particle from the assumption that the conservation laws hold for a system of colliding but otherwise non-interacting particles. Instead of assuming all four conservation laws hold in a single inertial frame, one may assume conservation of energy holds in all inertial frames [127].

In 1903, Schwarzschild derived Maxwell's equations from a variational principle that involved only the charged particles, defining the Maxwell field by the retarded solution for the latter [128]. He thus provided (*before* the advent of the special theory!) the first example of a Lorentz-invariant theory of particles interacting directly but non-instantaneously, and since then numerous other Lorentz-invariant directly interacting particle theories have been proposed [129]. A major attraction of such theories is that, in contrast to field-mediated interactions, such systems involve only a finite number of degrees of freedom [130]. In 1949 Dirac showed that, in addition to the then-known class of such theories, which he called the instant form, there are two other classes, the point and front forms [131]. Considerable work has been done on all three forms, which, however, present serious problems of physical interpretation. One such problem is that a Hamiltonian formulation of the dynamics of a system of relativistic particles, in which the position coordinates of the particles transform appropriately under the Poincaré group, is only possible if the particles are non-interacting [132].

4.2.8. Rigid motions and continuum mechanics
A special-relativistic definition of a rigid body was given by Born [133]. Herglotz and Noether showed that, in contrast to the six degrees of freedom of a rigid body in Newtonian theory, in relativity such a body would only have three [134]; and Laue soon showed that in relativity an extended body cannot have a finite number of degrees of freedom [135]. (If it did, forces applied at a finite number of points of the body would

have to set all of its points into motion instantaneously.) What Born had defined were *rigid motions* of non-rigid bodies, and attention shifted to the dynamics of such non-rigid bodies.

Herglotz gave an elegant special-relativistic formulation of the theory of perfectly elastic media [136]; the special case of relativistic hydrodynamics was also treated by Ignatowski, and by Lamla [137]. Starting with Herglotz, a number of Lagrangian variational principles for adiabatic motions of elastic media, and of Eulerian variational principles for fluids, were formulated [138]. Starting with Eckart [139], attempts have continued to go beyond the adiabatic case and develop a relativistic thermomechanics of continuous media [140].

4.2.9. Electrodynamics

In 1904 Lorentz showed that the vacuum Maxwell equations without sources are invariant under Lorentz transformations, and in 1905 Poincaré and Einstein showed that the equations with sources are also Lorentz-invariant (see above). Minkowski treated relativistic vacuum electrodynamics, utilizing his four-dimensional approach [141] (see above), showing that under the Lorentz group the three-dimensional electric field vector **E** and the magnetic induction vector **B** combine to form a four-dimensional 'six-vector' (as it was then common to call an antisymmetric second-rank tensor), thereby giving a natural mathematical expression to Einstein's demonstration that the division of the electromagnetic field into electric and magnetic fields is relative to an inertial frame. Minkowski also united the three-dimensional electromagnetic energy density scalar, momentum density (Poynting) vector and Maxwell stress tensor into a four-dimensional second-rank symmetric tensor, now known as the stress–momentum–energy tensor (often shortened to stress–energy tensor, energy–momentum tensor or simply stress tensor) of the electromagnetic field. The four-divergence of this tensor is equal to minus a four-vector, whose components are the density of rate of work done by the field and the ponderomotive force density exerted by the field (Lorentz force vector). Minkowski's work suggested to Abraham [142] that a stress–energy tensor be defined for any mechanical system, and Laue [143] showed that the total (electromagnetic plus mechanical) stress–energy tensor of a closed system is conserved. The total stress–energy tensor proved especially important in the general theory, where it serves as the source term in the gravitational field equations (see section 4.3).

Minkowski also developed a phenomenological electrodynamics of macroscopic continuous media, at rest or in motion, introducing a second 'six-vector' that unites the electric displacement **D** and the magnetic field vector **H**, and defining a (non-symmetric) stress–energy tensor and a ponderomotive force density for the electromagnetic field in the medium.

His work started an ongoing discussion on the correct macroscopic treatment of electromagnetic fields in such media. Einstein and Laub proposed different expressions for the stress–energy tensor and ponderomotive force [144]; as did Abraham, whose proposed stress–energy tensor was symmetric [145]. Disputes over the correct forms of the ponderomotive force term and the stress–energy tensor have continued over the years [146]. The reason for such disputes now seems clear: the division of the *total* stress–energy tensor of a *complete* system, consisting of electromagnetic field plus material medium, into matter and field terms is to a large extent arbitrary. Similarly, only the total force acting on a body in a medium is uniquely defined; its division into a ponderomotive term due to the field and a mechanical term due to the medium is non-unique. Furthermore, the ponderomotive term itself cannot be divided uniquely into a force acting on the body's charge and current densities, and another force acting on its polarization and magnetization densities. Used with appropriate care, each of the proposed stress–energy tensors or ponderomotive forces leads to the same physical predictions [147].

Before the advent of the special theory, Lorentz derived a set of macroscopic equations for a material medium from his microscopic electron theory, the medium being modelled as an ensemble of charged particles. His procedure, averaging the fields and particles over small spatial regions at a fixed time, led to macroscopic equations that are not Lorentz-invariant [148]. Dällenbach showed that a similar approach, but based on an invariant averaging procedure over four-dimensional regions in space-time, could be used to derive Minkowski's macroscopic equations [149].

4.2.10. Relativistic thermodynamics

Mosengeil studied the thermodynamics of black-body radiation, with results that are consistent with the special theory [150] but do not require its use: indeed, any system consisting entirely of electromagnetic radiation can be treated adequately on the basis of Maxwell's equations, since the latter are relativistically invariant.

Basing himself on Mosengeil's work, Planck formulated a special-relativistic thermodynamics [151], arguing that the entropy of a system is invariant under Lorentz transformations, while the temperature *decreases* by the factor γ. Einstein derived the same results by a different method [152], and the Planck–Einstein formulation was commonly accepted [153]. But Ott reopened the topic, arguing that the temperature should *increase* by the factor γ [154], setting off a discussion in which the merits of various transformation laws for temperature and other thermodynamic variables were debated with considerable heat [155]. Early in the debate Anderson noted the element of artificiality in all such transformation laws, which depend on little more than a choice of convention [156]. Thermodynamic variables, by their nature, characterize the states of a system in equilibrium, and such a state singles out a unique (inertial) rest frame, in which such variables can be defined and measured.

4.2.11. Relativistic statistical mechanics
A relativistic system of free particles interacting only through collisions has a finite number of degrees of freedom, and can be treated by methods analogous to those used in the kinetic theory of such non-relativistic systems. Jüttner derived the relativistic form of the Maxwell–Boltzmann distribution law [157], and showed that the equation of state of such a gas remains unchanged [158].

The real difficulties in formulating a relativistic kinetic theory for a system of particles begin when other interactions are introduced. Relativistic action-at-a-distance theories (see above) can be used to describe systems that still have only a finite number of degrees of freedom; but when retardation effects are introduced the equations of motion become integro-differential, and one cannot apply the usual methods of kinetic theory. If it is assumed that interactions are mediated by fields, e.g. interactions between charged particles via the electromagnetic field, the number of degrees of freedom of the system becomes infinite. Field degrees of freedom could be eliminated by assuming half-advanced, half-retarded interactions; but retarded interactions are needed to describe the abundant radiation emitted by charged particles interacting at relativistic speeds. The absorber theory of radiation was developed in response to this problem [159]. The application of statistical methods to a system of particles-plus-field with an infinite number of degrees of freedom is extremely difficult, and while certain approximations have proved useful no fully satisfactory treatment has yet been developed [160].

4.2.12. Quantum theory and elementary particles
Probably the most important applications of the special theory have been in high-energy physics, where it is essential for the treatment of particles with speeds that are appreciable fractions of c, and in particular for zero rest-mass particles, whose speed is always c. Indeed, the photon is the first elementary particle for which a fully relativistic description was developed. Einstein introduced the light-quantum as a heuristic concept to describe discrete aspects of energy transfer by the electromagnetic field, but did not find reasons to ascribe momentum to it until 1916 [161]. Not until after the discovery of the Compton effect in 1923 did most theoretical physicists accept the light-quantum (renamed the photon by Lewis).

Reversing Einstein's reasoning, De Broglie argued that, if the electromagnetic field is associated with particles, then other particles, such as the electron, should be associated with a wave field [162]. Attempting to set up a wave equation for this field, Schrödinger first considered a relativistic equation (now called the Klein–Gordon equation after two later rediscoverers); but he could only derive the hydrogen spectrum using the non-relativistic equation now bearing his name [163]. Pauli showed how to incorporate spin into the non-relativistic theory [103] and Dirac formulated a relativistically invariant theory of spin $\frac{1}{2}$ particles (see above) [104]. Problems with the physical interpretation of the Klein–Gordon and Dirac equations as single-particle equations (for particles of spin 0 and $\frac{1}{2}$ respectively) led to their field-theoretical interpretation (so-called second quantization), with the corresponding particles interpreted as quanta of the field.

The extension of quantum mechanics from particles to fields was first discussed by Jordan [164]. Jordan and Pauli applied a Lorentz-invariant method of field quantization to Maxwell's equations [165], which Heisenberg and Pauli generalized to a Lorentz-invariant quantum theory of fields [166]. During the 1930s, quantum electrodynamics, a relativistically invariant theory of the interacting Maxwell and Dirac fields, was developed. The resulting formalism suffered from severe defects: its equations cannot be solved exactly, and series expansion solutions in powers of the fine-structure constant diverge at each order. In the 1940s and 1950s renormalization techniques were successfully developed for covariantly bypassing these divergences at each order and extracting finite numerical results. When applied to problems involving the interaction of matter and electromagnetic radiation, the resulting theory has given numerical results in remarkable agreement with experiment [167].

As more and more elementary particles were discovered, the theory of representations of the Lorentz and Poincaré groups provided the basic theoretical framework for their classification and for setting up the wave equations satisfied by free particles. Dirac (1936), Fierz (1939) and Pauli (1939) were among those who worked on these problems [168], but the systematic classification of all representations of the Poincaré group and corresponding wave equations was first worked out by Wigner in 1939 [169].

A striking application of the special theory is Pauli's spin-statistics theorem: relativistic free fields corresponding to particles of integral spin can only be quantized consistently using commutators (Bose–Einstein statistics), while fields corresponding to particles of half-integral spin require anti-commutator quantization (Fermi–Dirac statistics) [170].

4.2.13. Gravitation theories

Poincaré made the first attempt to set up a special-relativistic theory of gravitation; it would now be called a theory of directly interacting particles (see above) [76]. From the outset, Einstein adopted the field approach [171], but concluded that gravitation could not be incorporated into the special theory without profound modifications of the latter (see section 4.3). Nevertheless, many physicists continued to investigate special-relativistic (or even non-relativistic) theories of gravitation. During the first two decades of this century, such theories were developed by Abraham, Mie and Nordström [172], the last's so-called second (1913) theory being the most interesting such attempt [173]. Einstein and Fokker showed that it could be reinterpreted geometrically as the theory of a conformally flat space-time [174], the conformal factor representing the gravitational potential†. Nordström succeeded in unifying his gravitational theory with electromagnetism, thereby creating the first five-dimensional unified field theory [175] (see section 4.4).

Since then, many special-relativistic theories of gravitation—scalar, vector and tensor—have been and continue to be proposed as competitors of general relativity. In a sense, the problem is too easy: Given a special-relativistic theory with several free parameters, it is rather easy to choose their values so that the theory gives the same predictions as general relativity for the three 'classic tests' of the latter [176].

More interesting are the derivations of the general-relativistic gravitational equations that start from a special-relativistic basis (see section 4.3.4). A major motivation for such efforts has been the difficulty of reconciling the conceptual framework of the general theory of relativity with the framework used in the rest of physics, in particular with that of quantum mechanics. Indeed, many physicists, especially those whose work has been in quantum field theory, have attempted to give a special-relativistic interpretation of the general theory. The fundamental issue remains unresolved: should gravitation theory be modelled after the rest of physics or do the unique features of gravity require fundamental modifications elsewhere in physics? Most physicists still favour the former alternative, but recent successes of gauge theories of the weak and strong forces, combined with the failure of conventional attempts at quantization of general relativity based on its treatment as just another special-relativistic field theory, have made the latter alternative more plausible to many (see section 4.3.15).

4.2.14. Experimental tests and applications
All experiments, e.g. those of Michelson–Morley and Trouton–Noble, originally interpreted as failing to show effects of the motion of ordinary matter with respect to the ether, were reinterpreted by advocates of the special theory as evidence supporting the principle of relativity [177]; later repetitions and variations provided further such support.

† *If the flat Minkowski metric is symbolized by the symmetric second-rank tensor field $\eta_{\mu\nu}$ then the metric of a conformally flat space-time is represented by $\Phi^2 \eta_{\mu\nu}$ with the scalar function Φ playing the role of gravitational potential in Nordström's theory.*

Particularly noteworthy was the Kennedy–Thorndike experiment [178], which ruled out a dependence on the Earth's translational motion of the round-trip time of light around any closed terrestrial path.

As for the light principle, attempts to pursue emission theories, in which the velocity of light is relative to that of its source, continued after 1905. (Einstein had earlier considered the possibility of combining the relativity principle with an emission theory of light; see above.) Ritz proposed a Galilei-invariant emission theory of light, and Ehrenfest noted the absence of conclusive experimental evidence against such a hypothesis [179]. De Sitter finally produced an argument based on observations of the orbits of double-star systems that, taken together with earlier difficulties, more or less disposed of attempts to invalidate the light principle [180]. Nevertheless, emission theories are still proposed occasionally as alternatives to special relativity. In the special theory, there is no problem with an emission theory of light since the velocity of light relative to its source is always c; indeed, the photon hypothesis is the quantum version of such a theory.

The first experimental tests of special-relativistic dynamics concerned the relativistic formula for the variation of (inertial) mass with velocity, then often called the Lorentz–Einstein formula since Einstein's prediction is the same as that of Lorentz's electron theory [51, 54]. (In later years, Einstein felt it better to reserve the term mass for proper mass, and to speak of a variation of kinetic energy with velocity.) Kaufmann had been looking into the problem experimentally since 1901 [181], with results (1905–6) that seemed to argue against the Lorentz–Einstein formula. However, over the course of the next decade, experimental evidence accumulated in support of this formula, contributing to a gradual but ultimately decisive shift of the consensus of opinion in the physics community towards the special theory [182].

Numerous other tests of the kinematical, dynamical, optical and other consequences of the special theory have been carried out over the years. The theory has survived them all [183] and become an everyday working tool of experimental and applied as well as theoretical physicists [184]. Without use of the special theory, for example, it is as impossible to design modern particle accelerators as it is to classify and formulate theories of the many particles discovered with their aid.

The mass–energy equivalence relation was first verified quantitatively by Cockcroft and Walton in a nuclear reaction [185]. With the help of this relation, Bethe showed in detail how nuclear reactions solve the long-standing problem of the source of stellar energy [186]. The uses of what is popularly know as 'atomic' energy—that is, energy from either nuclear fission (1938) or fusion (1949) reactions—for both constructive and destructive purposes provide examples of technological applications, the notoriety of which make $E = mc^2$ the best known equation of our time.

4.3. The general theory of relativity

Einstein started to look for a relativistic theory of gravitation in 1907 (see section 4.2.13). While not alone in this search, he was the only physicist who generalized the (special) relativity principle in order to solve the problem [187]. Thus, in contrast to the many people who contributed to the development of the special theory, the development of the general theory is the story of his quest [188].

Another contrast is the relative abundance of primary source material. Einstein's publications from 1908–15 document successive stages of his work, and surviving research notes and letters provide additional evidence. Detailed studies of this material are becoming available as historians, philosophers of science and relativists have begun to turn their attention to the history of the general theory [189].

The evidence allows a fairly detailed reconstruction, with a minimum of conjecture, of the main steps in the process. Einstein, in retrospect, subdivided his work into three major phases [190].

(i) Formulation (1907) of the equivalence principle, which he interpreted as showing the need to generalize the special theory. (He did not begin to refer to the earlier theory as the 'special' theory until 1915 [191].)

(ii) Transition (1912) from a scalar gravitational potential to a use of a pseudo-Riemannian metric tensor to represent the gravitational field†.

(iii) Formulation (1915) of the generally covariant gravitational field equations for the metric tensor.

4.3.1. Equivalence principle

After trying to set up a special-relativistic generalization of Newton's gravitational theory, Einstein abandoned the attempt after concluding that no such theory could satisfy the equivalence principle [192]. This principle is based on the equality of inertial and gravitational mass, which explains Galileo's observation that test bodies‡ of differing inertial mass and velocity at the same point in a gravitational field all fall with the same acceleration. Newton gave evidence arguing for equality to about 1% accuracy. Eötvös's experiments [193] indicated equality to about 1 part in 10^9 (figure 4.10), but Einstein was apparently not aware

---

† *A Riemannian metric, properly speaking, has a positive-definite signature, so that the distance between any two neighbouring points is always positive. As in the special theory, in general relativity the square of the interval between two events can be either positive (space-like), negative (time-like) or zero (null), and the corresponding tensor field is called a pseudo-Riemannian metric; however, the 'pseudo' is often dropped, especially in the physics literature.*

‡ *An object may be treated as a 'test body' with respect to some field if its passive charge (passive gravitational mass in this case) is big enough to let the field act upon it, while its active charge (active gravitational mass) is so small as to make a negligible contribution to that field.*

History of Relativity

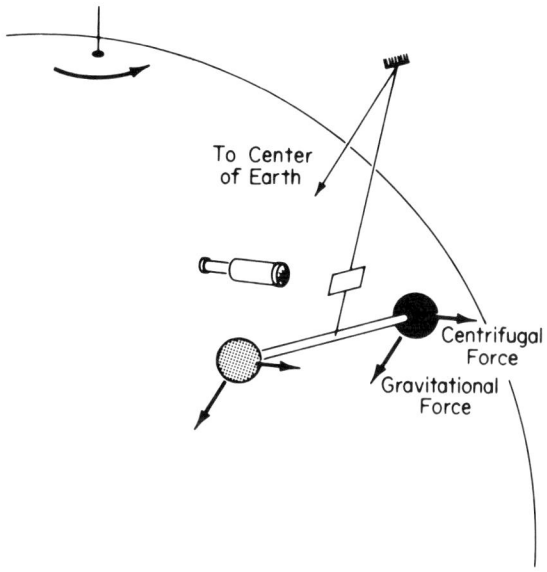

Figure 4.10. *The Eötvos experiment. Fibre supporting the rod does not hang exactly vertically because of the centrifugal force from the Earth's rotation, so the downward gravitational force on the balls is not parallel to the fibre. If gravity pulls one material more strongly than the other, the rod will rotate about the fibre axis. If the entire apparatus is rotated so that the two balls are interchanged, the resulting rotation will be in the opposite sense. The rotation is detected by observing light reflected from a mirror attached to the fibre.*

of this work in 1907. Einstein's conclusion that a special-relativistic theory cannot meet this criterion was incorrect; but (fortunately for the development of the general theory) he did not realize this in 1907. In 1913 he agreed that Nordström's second theory, which he regarded as the only serious rival to his own, incorporates the equivalence principle [194].

What we may call the mechanical principle of equivalence follows from the equality of gravitational and inertial mass. No mechanical experiment can distinguish (at least locally) between:

(i) an *unaccelerated* (inertial) frame of reference in which there is a constant gravitational field, and

(ii) a *uniformly accelerated* frame of reference—with acceleration equal in magnitude but opposite in direction to that produced by the gravitational field of (i)—in which there is no gravitational field present.

The mechanical equivalence principle implies that, once gravitation is taken into account, no mechanical experiment can distinguish between inertial and linearly accelerated frames of reference (figure 4.11). (Since

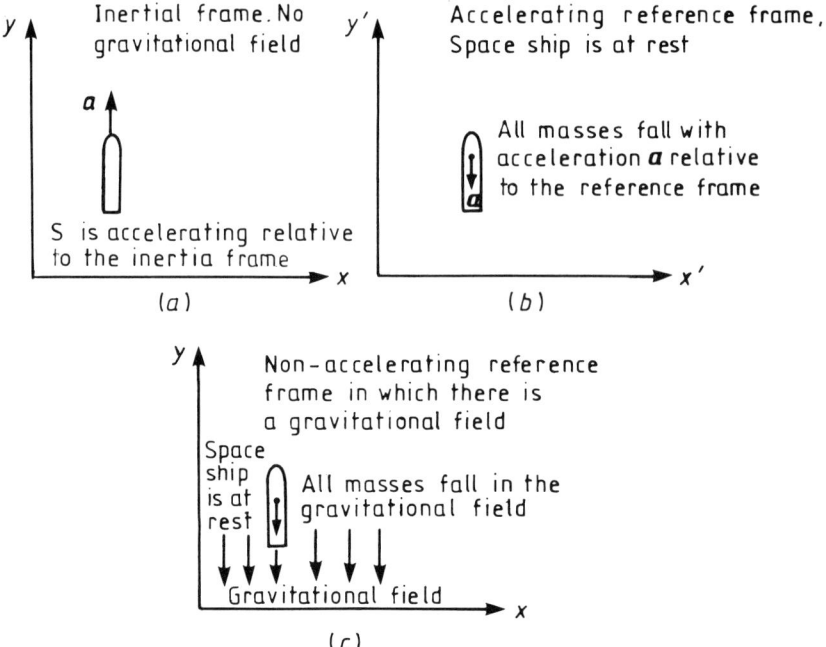

Figure 4.11. *The mechanical principle of equivalence. (a) A body S accelerating with uniform acceleration **a** in an unaccelerated, or inertial, frame of reference, with no gravitational field. (b) A body S at rest within an accelerating reference frame, with uniform acceleration **a**. According the principle of equivalence, measurements carried out in the accelerating reference frame give the same results as experiments carried out in the non-accelerating reference frame in (c), in which there is a gravitational field opposite in direction to **a** and of such a strength that the acceleration due to the force of gravity is numerically equal to **a**.*

acceleration is an absolute quantity in Newtonian theory (i.e. the same for all inertial frames), we may speak of a frame as accelerated without specifying with respect to what inertial frame.) Einstein extended the equivalence principle from mechanics to all physical phenomena, asserting the complete equivalence of the above situations (i) and (ii) for all phenomena—mechanical, electromagnetic, etc. It follows that the absolute distinction between accelerated and unaccelerated frames of reference—as valid in special relativity as is was in Newtonian theory—no longer holds for any physical phenomenon. Taking this principle as the key to a relativistic theory of gravitation, he concluded that such a theory must generalize the original (special) relativity principle for unaccelerated frames of reference, by giving equal status to at least some accelerated frames.

Einstein later put his argument in these terms: the equivalence principle shows that inertia and gravitation are essentially the same, forming a single inertio-gravitational field. The division of this field into inertial and gravitational components is *relative* to the frame of reference chosen†. He suggested that a dynamical theory of gravitation might even show that the inertial mass of a body is entirely due to a gravitational interaction with all the remaining bodies in the Universe [195]. For some time he favoured the programme of explaining the inertio-gravitational field entirely in terms of the behaviour of material bodies, an idea that he later called 'Mach's principle' [196]. It played a considerable motivating role in his work on gravitation, particularly the development of his cosmological model [197]; but he ultimately abandoned it in favour of the unified field theory programme that aimed to explain the nature of matter in terms of the behaviour of fields (see below).

Einstein first generalized the relativity principle to include uniformly accelerated frames [198], often referring to this generalization as the equivalence principle [199]. With it, he was able to predict new gravitational effects, for the effect of a constant gravitational field on any physical phenomenon must be the same as if gravity were absent, but the phenomenon were analysed with respect to a uniformly accelerated frame of reference. On this basis, he predicted that in a gravitational field clocks slow down (gravitational red shift) and light rays bend (gravitational light deflection). (While this way of describing things reproduces Einstein's viewpoint in 1907, and is still often used in accounts of these effects, it does not adequately describe the way these effects are explained by the general theory.)

The Earth's gravitational field is not strong enough to produce effects then measurable; but radiating atoms at a star's surface act like 'clocks,' and Einstein predicted their spectral lines should be shifted to measurably lower frequencies (red shifts) by the star's gravitational field as compared with the corresponding lines at the Earth's surface. Freundlich [200] began the long effort to disentangle the gravitational red shift of solar and other stellar spectral lines from the many astrophysical processes that also produce spectral frequency shifts [201]. Over the next five decades some astronomers claimed that, when suitably corrected, such observations were in quantitative accord with the predicted gravitational effect; while others equally vehemently denied such claims. The same astronomer—notably St John [202]—sometimes switched sides. Only after the then recently discovered Mössbauer effect

---

† *There is an analogy with Einstein's treatment of electromagnetism, where the electromagnetic field divides differently into electric and magnetic fields relative to different inertial (unaccelerated) frames of reference, but the inertio-gravitational field divides differently into inertial and gravitational fields relative to differently accelerated frames. The concept of affine connection, the adequate mathematical expression of Einstein's idea of an inertio-gravitational field, was only developed after the formulation of the general theory (see later).*

Twentieth Century Physics

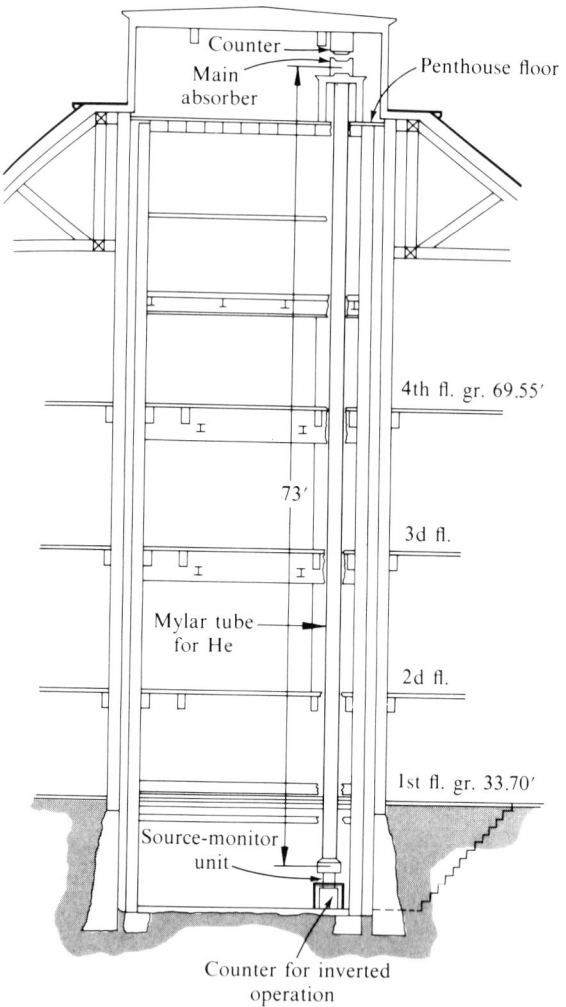

Figure 4.12. *The experiment of Pound and Rebka (1959) and Pound and Snider (1965) on the gravitational red shift of photons rising 22.5 m against gravity through a helium-filled tube in a shaft in the Jefferson Physical Laboratory of Harvard University. The source of Co^{57} had an initial strength greater than a curie. The 14.4 keV gamma-rays had to pass in through an absorber enriched in Fe^{57} to reach the large-window proportional counters. Both source and absorber were placed in temperature-regulated ovens. The velocity of the source consisted of two parts: one steady (v_M), to put the centre of the emission line on the part of the transmission curve that is nearly straight; and the other alternating between $+v_J$ and $-v_J$, to sweep the transmission curve in this straight region; similarly when the steady velocity was $-v_M$. The departure from symmetry between the two cases $+v_M$ and $-v_M$ allows one to determine the offset V_D (the effect of gravitational red shift) from the zero-gravity case of stationary emitter and stationary absorber. The final result for the red shift was (0.9990 ± 0.0076) times the value 4.905×10^{-15} of $2gh/c^2$ predicted from the principle of equivalence (the difference between 'up' experiment and 'down' experiment).*

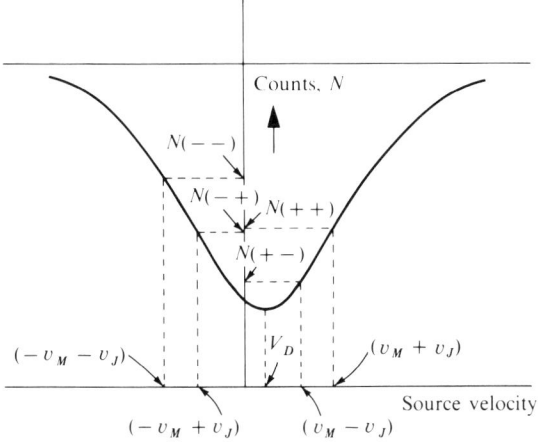

Figure 4.12. *Continued.*

enabled Pound and Rebka (figure 4.12) to observe terrestrial red shifts due to the Earth's gravitational field did serious controversy end [203].

Einstein also predicted (1911) that when its light passed near the Sun's edge, the position of a star's image as observed during a solar eclipse should undergo a measurable gravitational deflection compared with its image on an ordinary night [204]. Measurements were attempted during the solar eclipses of 1912 and 1914, but produced no results because of poor observation conditions and war [205]. By 1916, Einstein had completed the general theory, which predicts a light deflection double that of his 1911 argument, so we shall discuss later observations below.

4.3.2. Metric-tensor field

Einstein developed a scalar field theory of gravitation based on the equivalence principle, first treating the case of a static field [206]. Having previously (1911) derived gravitational light deflection by taking the gravitational potential as an effective index of refraction, which causes the speed of light to vary from point to point, he now (1912) took this varying speed of light $c(r)$ as the scalar field representing gravitation. He argued that since gravitational energy has inertial mass ($m = E/c^2$) and inertial mass equals gravitational mass, gravitational energy must be one source of the gravitational field. Since the gravitational field acts as one of its own sources, $c(r)$ must obey a non-linear equation. This feature of non-linear self-interaction proved to be an abiding feature in the later development of his gravitational theory.

The equations of motion of a particle in a gravitational field provided another clue to the development of the theory. After deriving them by a physical argument, he found they followed from Planck's special-relativistic variational principle for the motion of a free particle, based on the proper time interval $d\tau = \sqrt{c^2 dt^2 - dx^2 - dy^2 - dz^2}$ as Lagrangian, if he simply introduced the variable $c(r)$ into this Lagrangian [207]. The resulting expression for the interval suggested that Minkowski space-time had to be generalized to include gravitation.

An attempt to extend the static theory to stationary gravitational fields provided a third clue. Having studied a uniformly accelerated reference frame in Minkowski space as the equivalent of a uniform gravitational field—the simplest example of a static field—Einstein now studied a uniformly rotating reference frame ('rotating disc') in Minkowski space—the simplest example of a stationary gravitational field. Applied to the relativistic rotating disc, an equivalence principle argument suggests that magnetic-type gravitation effects are associated with spatial curvature [208]. (Of course, neither the uniformly accelerated nor the uniformly rotating frame gives rise to what is sometimes called a permanent gravitational field: since the Riemann tensor vanishes for Minkowski space, its inertio-gravitational fields can be transformed into purely inertial ones by a change of reference frame.)

Towards the end of 1912 Einstein realized that his variational principle could be given a four-dimensional interpretation: the paths of particles in the gravitational field are (time-like) geodesics (i.e. straightest paths) of a four-dimensional space-time with a non-flat metric tensor (a pseudo-Riemannian geometry) [209]. In flat Minkowski space-time, such (time-like) geodesics correspond physically to inertial paths: straight lines in an inertial frame traversed with constant velocity. The geodesics of a non-flat space-time can be interpreted as embodying a generalization of the law of inertia to include gravitation: a particle acted on only by gravitation follows the straightest path in a space-time curved by gravitation. How is such a non-flat space-time characterized mathematically? Einstein recalled his study of Gauss's theory of curved two-dimensional surfaces as a student; his friend, fellow-student and now colleague at the ETH, the mathematician Grossmann, found Riemann had generalized Gauss's theory to an arbitrary number of dimensions, introducing the concepts of the metric-tensor field (metric, for short) that fixes the distance between neighbouring points of space, and the associated curvature tensor (now named after Riemann) built from the metric and its first and second derivatives [210]. Christoffel, Ricci and Levi-Civita had perfected an analytic method†, called the absolute differential calculus but today usually known as tensor analysis, for dealing with Riemannian geometry.

This was the mathematical tool Einstein needed, and his next paper on gravitation, written in collaboration with Grossmann [213], asserts

† *Einstein and Grossmann initially interpreted Riemann's work in the context of the theory of invariants; only after Levi-Civita developed the concept of parallel transport [211] did Einstein and other relativists, notably Weyl [212] begin to emphasize the geometrical interpretation of the general theory.*

that the metric represents both the space-time structure and the inertio-gravitational field. Because of this dual role of the metric, and in contrast to all previous physical theories, the space-time structure of the resulting theory is not fixed *a priori*, but is subject to dynamical equations that link the gravitational field to its sources. This radical break with the previous concept of space-time as the fixed arena of dynamics was the crucial step in the development of general relativity [214].

4.3.3. The field equations

Einstein attributed physical significance to the requirement that a set of equations take the same mathematical form in some set of coordinate systems since every physical frame of reference is associated with some coordinate system. (The converse is not true: not every coordinate system is associated with a physical frame of reference. Hilbert gave conditions [215] (often called the Hilbert conditions) on the components of a metric tensor in a coordinate system that ensure it can be associated with a physical frame of reference.) The equivalence principle (see above) showed that a theory of gravitation had to extend the relativity principle to include at least uniformly accelerated frames of reference, so the question arose: how wide is the class of admissible coordinate systems? If equations take the same mathematical form in every coordinate system, they are called generally covariant. So in a generally covariant theory, all the laws of physics would take the same form in every frame of reference, providing the maximal possible extension of the relativity principle. Could Einstein make his entire gravitational theory generally covariant?

He managed to take into account the effect of the inertio-gravitational field on all non-gravitational processes by means of generally covariant equations. There remained the question of the effect of these non-gravitational processes on the inertio-gravitational field; in other words, the problem of finding the still-unknown gravitational field equations. Analogy with Newtonian theory suggests the gravitational field term (left-hand side) in these equations should involve second-order derivatives of the metric tensor, the relativistic generalization of the Newtonian gravitational potential; and that the source term (right-hand side) should be the total stress–energy tensor, the relativistic generalization of the Newtonian mass. Riemann's curvature tensor is the only tensor that can be formed from the metric tensor and its first and second derivatives, and Grossmann suggested its contraction, the Ricci tensor, as just about the only generally covariant candidate for the gravitational field term.

Einstein rejected this candidate in 1912, and continued to do so for the next three years; his reasons are still a subject of debate among historians of the theory [216]. The once-prevalent view that his difficulties stemmed from ignorance of how to use coordinate conditions to simplify the field equations seems implausible after a recent study of Einstein's notebooks [217].

Initially, he appears to have rejected generally covariant field equations in the belief that they cannot give Newton's theory in the static, weak-field limit. On the basis of his earlier static theory, he adopted a metric for static space-times that is indeed incompatible with the Newtonian limit of generally covariant equations. Einstein only found the correct static metric after returning to general covariance late in 1915 [218].

However, in 1913 he developed what seemed a much stronger argument against generally covariant equations for the gravitational field. According to this 'hole argument', such equations cannot uniquely determine the gravitational field in a matter-free region (hole). The 'hole argument' is not based on an elementary misunderstanding of the significance of coordinate transformations [219]; rather, it turns on the question of whether or not the points of the hole are physically individuated *before* the specification of the metric-tensor field inside the hole [220]. Einstein had tacitly assumed that they are, in which case the 'hole argument' is valid. Only in late 1915, after other difficulties led him to reconsider general covariance, did Einstein realize that the hole argument fails because all physical properties of the points of the hole depend on the metric [221].

But we have jumped ahead in the story. Having abandoned the search for generally covariance, Einstein (1913) proposed a set of non-generally covariant gravitational field equations for the metric. Over the next few years, he attempted to justify these equations by showing that they were uniquely picked out by various physically and/or mathematically desirable criteria. In particular, he claimed that these equations were invariant under the widest possible group of coordinate transformations that avoid the hole argument. But his claims proved fallacious [222], and by mid-1915, mounting problems with the non-generally covariant equations led him to reexamine the question of the covariance group of his field equations. Within a few months he was led, via a rather tortuous series of steps and mis-steps, to adopt generally covariant field equations that differ only slightly from those he had rejected almost three years earlier. To use modern terminology, the Einstein tensor replaces the Ricci tensor on the left-hand side of the field equations [223].

Shortly before Einstein presented the final version of his theory, Hilbert, who was in contact with Einstein throughout this period, derived similar equations from a variational principle, but with rather different motivation. Hilbert was looking for an electromagnetic theory of matter on the basis of Mie's theory of electrodynamics. This theory, an offshoot of the electromagnetic world-view, tried to explain the electron as an intense concentration of electromagnetic field energy that was a solution to a non-linear, non-gauge-invariant generalization of Maxwell's equations. Mie's theory was based on special relativity, but Hilbert took it a step further by including gravitation, adopting Einstein's representation of the latter by the metric-tensor field.

While Hilbert limited himself to a combined theory of the electromagnetic and gravitational fields, Einstein's approach was not tied to a particular model of matter or field. By means of the appropriate stress–energy tensor, any model of matter or (non-gravitational) field could be introduced as a source of his gravitational field equations.

4.3.4. Alternative scenarios

Few followed Einstein's metric-tensor approach to gravitation at the time he was developing it, and it is most unlikely that anyone else would have developed it independently without his lead. But it is interesting to speculate on possible alternative solutions to the problem of reconciling gravitation with the special theory that might have been adopted. While no pressing discrepancies between observation and the predictions of Newtonian gravitational theory impelled the search for a better theory of gravitation†, there was a serious theoretical problem—the discrepancy between the kinematic foundations of Newton's gravitational theory and those of special relativity—which impelled theorists such as Mie and Nordström to propose special-relativistic gravitational theories.

Someone might have discovered (as did Einstein and Fokker in 1913) that Nordström's theory could be reformulated as a dynamical theory of the conformal factor in a conformally flat space-time, thereby suggesting that gravitation and curvature of space-time are connected. But Nordström's theory does not predict any interaction between electromagnetism and gravitation (no influence of electromagnetic energy on the gravitational field, no influence of gravitation on the path of a light ray), nor does it account for the anomalous precession of Mercury's orbit; so it had little to recommend it observationally, and at some point the search might well have begun for a non-scalar theory of gravitation.

An alternate derivation of the Einstein equations starting from flat space-time, first proposed by Weyl for the linearized equations [224] and then by Kraichnan for the exact field equations [225], suggests how they might have emerged from such a search based on special-relativistic field theory [226]. Since masses attract each other gravitationally, such a theory must be based on a field with even spin; since it is an infinite-range force, the field must be massless. So the next candidate after a scalar field is a massless, spin-two field. The equivalence of inertial

---

† The only anomaly seriously discussed at the time, that in the precession of the perihelion of Mercury's orbit, amounted to less than 1% of its total precession due to perturbations by other planets; proposed explanations within the Newtonian framework abounded and the astronomical community had lived with the anomaly for many decades.

mass and energy, together with the equality of inertial and gravitational mass, imply that the field should be self-interacting; that is, the special-relativistically defined stress–energy tensor of the gravitational field should act as one of its own sources. If (by analogy with charge conservation in Maxwell's theory) one further requires that energy–momentum conservation for its sources follow from the field equations themselves, one is led to a special-relativistic theory, the non-linear field equations of which are formally identical to the Einstein equations. However, their physical interpretation is different: space-time 'really' has the Minkowski structure, but this is unobservable because gravity exerts universal forces on clocks and measuring rods, thereby 'distorting' their measurements so that space-time 'appears' to have a non-flat Riemannian metric. (The analogy with Lorentz's interpretation of the Lorentz transformations as dynamical effects of motion through a 'real' ether, discussed earlier, is obvious).

Many theorists who hope to solve the problem of quantum gravity by analogy with the quantization of special-relativistic field theories adopt such an interpretation of general relativity, treating it as a non-linear special-relativistic field theory that happens to have a particularly nasty gauge group. Without here discussing the merits of such a special-relativistic interpretation, it is important to distinguish it from Einstein's geometrical interpretation of the field equations.

Another route that might have led from the special to the general theory is based on the representation of gravitation by an affine connection. This concept, which is the appropriate mathematical realization of Einstein's idea of an inertio-gravitational field, was actually developed after the general theory and largely in response to it (see section 4.3.6). But if the mathematical concept had been formulated independently, an argument based on the equivalence principle might have led to a four-dimensional formulation of Newtonian gravitation theory, based on amalgamation of Galilean inertia and Newtonian gravitational force into a Newtonian inertio-gravitational connection. (In fact, this was done after general relativity by Cartan [227] and Friedrichs [228].) In such a four-dimensional formulation, the equation for the Newtonian gravitational potential becomes an equation for the Ricci tensor of this connection, remarkably similar in form to the Einstein equations. The attempt to bring this four-dimensional geometrical formulation of Newtonian gravitation theory into accord locally with the kinematics of special relativity might then have led to the formulation of general relativity.

4.3.5. Later work on the general theory
The further history of the general theory may be roughly classified into the following sections.

(i) The development of various formulations of the theory involving alternate fundamental concepts and postulates, and/or the use of different mathematical structures and techniques.

(ii) The search for solutions to the field equations, either exact or based on various approximation methods; and the closely connected problem of motion in general relativity.

(iii) Use of the theory to predict novel terrestrial, astronomical and astrophysical phenomena, or significant quantitative corrections to known ones. The study of such phenomena often involves development of general-relativistic versions of existing physical theories, such as particle and continuum mechanics, thermodynamics, electrodynamics, etc.

(iv) Observational and experimental tests of the foundations of the theory and of its detailed predictions.

(v) Attempts to understand the relation between general relativity and quantum mechanics, often referred to as the problem of quantum gravity.

(vi) The influence of general relativity on philosophy of science in general and in particular on the philosophy of space and time, and the impact of the theory on the broader public.

Insufficient systematic historical work has yet been done on any these topics, but the situation is starting to improve [229], and is already somewhat better than in the case of the special theory. Nevertheless, the following account remains fragmentary and unsystematic [230].

Before going into detail, a broad characterization of the later vicissitudes of the general theory is useful. At the time of its birth, the major attention of the still-consolidating theoretical physics community [231] was already shifting towards understanding of the new quantum phenomena, and in particular the structure of the atom. At first, work on the general theory was seen as providing another avenue of approach to the foundations of microphysics, and many prominent mathematicians and physicists worked in both areas. (See the discussion above of Hilbert's work on the general theory. Einstein soon discussed the role of gravitation in the structure of matter [232].) In particular, the search for a unified field theory based on general relativity, and the search for a new quantum mechanics were initially seen as complementary [233], but as quantum mechanics took definitive shape during the later half of the 1920s, a definite rupture between the two fields developed. General relativity and unified field theories more and more became the province of a small band of physicists and mathematicians, whose work was of less and less relevance—or interest—to the mainstream of physics, the low point being reached in the late 1930s and early 1940s [234]. After World War II, a modest revival of interest in the field began, and small groups of younger relativists began to form around a few outstanding figures of the pre-war and immediate post-war generation, such as Bergmann

(Syracuse), Wheeler (Princeton) and DeWitt (Chapel Hill) in the United States, Bondi (London), Lichnerowicz (Paris), Fok (Leningrad), Infeld (Warsaw) and Synge (Dublin). After the first (Berne, 1955) international conference on general relativity [235], contacts between these groups increased and a distinct international community of general relativists emerged and was institutionalized with the foundation of the Society for General Relativity and Gravitation, which publishes a journal and sponsors triennial international meetings on the subject.

By the late 1960s, major improvements in experimental techniques and new discoveries in astrophysics, combined with a general growth in the size of the theoretical physics community fostered by the increased availability of funding for basic research (even on general relativity [236]), started to move the subject back into the mainstream of physics. Today, several journals are devoted to the subject, up to 1000 researchers attend international relativity conferences, relativistic astrophysics is a recognized subdiscipline, observational general relativity has recently become 'big science', and the problem of quantum gravity is considered by many physicists the major conceptual challenge in contemporary physics. Major US government funding of the laser interferometer gravitational-wave observatory or LIGO project to search for gravitational waves [237], and the award of the 1993 Nobel Prize in physics to Taylor and Hulse for their work using data from the binary pulsar PSR 1913+16 [238] to test general relativity, attest to the growing recognition of observational general relativity.

4.3.6. Alternative formulations and foundations
Work on the mathematical and physical foundations of the general theory since 1916 may be classified under the following headings.

(i) Introduction of other mathematical objects to replace or supplement the metric tensor in the formulation of the theory. Notable examples include: the affine connection, formulated by Weyl in 1918 [239]; tetrad fields and differential forms, first used in general relativity by Cartan in 1923 [227]; spinorial fields, first applied to general relativity by Infeld and van der Waerden in 1932 [240], twistor fields, introduced by Penrose [241].

(ii) Extensions of the space-time manifold, especially those that take this as the base manifold of a fibre bundle, in particular jet bundles. Such methods are especially useful for the study of global problems in general relativity, and have brought out many common features shared by general relativity and Yang–Mills theories [242].

(iii) Complexifications of the manifold and/or of various fields on it [243]. This has led to methods for generating new exact solutions to the Einstein equations from old ones; and to the introduction of new sets of gravitational variables that radically simplify the Hamiltonian formulation of the theory [244] (see also section 4.3.15).

4.3.6.1. Alternative foundations. As noted above, the concept of affine connection was developed after the formulation of general relativity. This concept made possible a new approach to the foundations of the theory, based on the recognition that the chronogeometrical structure, described by the metric, and the inertio-gravitational structure, described by the affine connection, are conceptually distinct, even though the physical interpretation imposes certain compatibility conditions between the two: the requirement that the chronogeometry determine the inertio-gravitational structure is the hallmark of general relativity. In large part as a result of Weyl's classic exposition of the theory [245], during the 1920s the physical significance of the affine connection began to be more or less well understood by relativists, including Einstein.

A notable technical advance based on the affine connection was the introduction of the so-called Palatini variational method [246], in which the metric and affine connection are varied independently, yielding not only the field equations but also the compatibility conditions between metric and affinity (vanishing of the covariant derivative of the metric).

Another alternative foundation of the theory does not postulate the metric or affine structure, but derives their existence from two weaker space-time structures. The conformal structure, a weakening of the metric structure, determines the light cone at each point of the manifold (see discussion of Robb above); the projective structure, a weakening of the affine structure, determines the paths of test particles without assigning them preferred (affine) parametrizations. Weyl formulated consistency conditions between the conformal and projective structures that entail the existence of a unique metric, from which they can be derived [247]. More recently Ehlers, Pirani and Schild gave an axiomatic formulation of general relativity based on this idea; not only the metric but the differentiable structure of the manifold are deduced from their axioms [248].

By correlating various space-time structures with (ideal) physical entities—metric with chronogeometry (clocks and measuring rods), conformal structure with light fronts or rays, projective structure with paths of (monopole) test particles, and affine structure with such paths endowed with a preferred time—alternative foundations of the theory can be given an 'operationalist' cast. It has sometimes been suggested that one such foundational scheme is superior to another because the associated entities are more 'realistic' (e.g. light rays as opposed to measuring rods), but all such foundational entities are ideal—i.e. constructs within a certain theoretical framework—and the question of their relation to real measuring devices is by no means trivial. Moreover, classical space-time theories are not basically operationalist: they assert the existence of various space-time structures, which is why the question of their relation to experiment and observation arises in the first place.

4.3.7. The problem of gravitational energy

Einstein had always associated energy with the gravitational field, and in the course of the development of the general theory defined various gravitational analogues of the stress–energy tensor of matter and electromagnetic fields. As long as his field equations were not generally covariant, the fact that these gravitational energy complexes were not tensors was not too troubling. Soon after formulating the generally covariant field equations, Einstein gave a variational derivation of them, based on a first-order variational principle [249], and used this Lagrangian to define an energy–momentum complex for the gravitational field. He showed that the conservation law for the combined gravitational energy complex and non-gravitational stress–energy tensor is generally covariant, even though (as its other name, gravitational energy pseudo-tensor, suggests) the gravitational energy complex is not a tensor (except under linear transformations). Critics soon began to question the interpretation of its components as the energy, momentum and stress densities of the gravitational field [250]. In curvilinear coordinates its components do not vanish for Minkowski space, which one intuitively expects to have no gravitational energy; and in certain coordinate systems its components vanish for some non-flat metrics, which one intuitively expects to have gravitational energy. Nevertheless, Einstein was able to show that, for asymptotically flat space-times in which the field falls off sufficiently fast as one goes to infinity along a space-like hypersurface (intuitively, for non-radiating solutions), the total energy and momentum, as defined by integrals over the relevant components of the pseudo-tensor, behave appropriately (that is, like the components of a free vector) under coordinate transformations between any systems of coordinates, so long as they are asymptotically quasi-Cartesian; and that these integrals are time-independent [251]. Thus began the still-continuing discussion about the concept of gravitational energy (and similarly of momentum and angular momentum) in general relativity [252]. Does it make sense to define a localized gravitational energy density? If not, why not? If so, how should one choose from the infinity of competitors? The major argument against such a complex is that all the first derivatives of the metric tensor, out of which one would expect the components of such a complex to be built, can be made to vanish at any point (indeed along any world-line, as Fermi showed) by an appropriate coordinate transformation. The major argument for defining such a local quantity is that there should be some way of defining how much energy can be extracted from the gravitational field in a given region and converted into other forms of energy that have an unambiguous definition. Over the years various invariant ways of defining a local energy density for the gravitational field have been proposed [253] that seem to have some physical significance in certain cases, such as static or stationary gravitational fields. Freud [254] and Landau–Lifshitz [255] introduced energy complexes that allowed the transformation of the energy volume integrals over all space into surface integrals over the two-sphere at infinity. J Goldberg [256] proved that there are an infinite number of possible energy complexes. There is more agreement about the existence of total, globally defined energy, momentum and angular momentum of the gravitational field in certain asymptotically flat space-times. Just what quantities should be integrated to define the global quantities, and whether and when to take the limit going to infinity along a space-like or null hypersurface, were much debated. Recently a consensus has emerged favouring the ADM (for Arnowitt–Deser–Misner) mass on space-like hypersurfaces, and the Bondi mass on null hypersurfaces. Invariant formulations of each have been developed and the relation between them established [257].

4.3.8. Physical interpretation of general relativity

The difficulties in defining gravitational energy illustrate the type of problem often encountered in the physical interpretation of the general theory, a task that often has proved more subtle and less tractable than was the case for previous physical theories. (This is one feature, at least, that general relativity shares with quantum mechanics.) Basically, such difficulties arise from the fact that, in general relativity, hitherto fixed space-time structures become dynamical entities, so that no 'background' kinematical structures exist for use in the physical interpretation of the theory. For a long time it was customary to evade such problems by using what may be called 'parasitic' interpretations. General relativity has two important but distinct limits: special-relativistic (Minkowski) space-time and Newtonian space-time (including Newtonian gravitation). Parasitic interpretations take advantage of the smallness, in many cases of practical interest, of predicted general-relativistic deviations from special-relativistic results on the one hand, or from those of Newtonian gravity on the other. Concepts developed for the interpretation of the two limit theories are often applied uncritically to the interpretation of small general-relativistic effects. To give just one example: it has become customary to speak of general relativity as predicting a gravitational light deflection. But a deflection only has meaning relative to an undeflected path, that is one uninfluenced by gravitation. This presupposes the existence of a space-time structure that is independent of gravitation, or at least the possibility of a (unique) comparison between two space-time structures, one with and one without the gravitational field. While these ideas make good Newtonian sense, they are meaningless within the context of general relativity, where gravity cannot be invariantly separated from inertia, the path of a light ray is the straightest possible (null geodesic) in the space-time, and there is no unique way of comparing points in two different space-times. But general relativity, as a complete theory of space-time and gravitation, does not require the *ad hoc* introduction of external conceptual elements for its interpretation. Indeed, contrary to the traditional practice, to the extent that special-relativistic or Newtonian concepts are applicable, their use should be justified from within the conceptual framework of general theory. An example of how this can be done is the demonstration of how the usual coordinate-dependent definitions of planetary orbits and their relativistic precession can be given invariant significance [258].

4.3.9. Exact solutions and approximation schemes
The search for exact solutions to the field equations, with or without sources, utilizing techniques adapted to one or another of the mathematical objects mentioned above, started immediately after the formulation of the general theory. The classic technique for finding such solutions is based on the assumption of a symmetry group, generated by one or more Killing vectors, to restrict the class of admissible metrics and thereby simplify the solution of the Einstein equations (for example, by using coordinates adapted to the symmetries). Important solutions found by this method and mentioned below include the spherically symmetric Schwarzschild and collapsing dust solutions, and the plane and cylindrically symmetric gravitational wave solutions.

More recent techniques include: the development of software programs to compute the Riemann, Ricci and Einstein tensors [259]; algebraic classifications of Riemann and Ricci tensors, with the introduction of canonical forms for the metric tensor adapted to various subclasses [260]; the study of space-times with preferred congruences of time-like or null curves having certain simple geometrical properties, and of those with similarly preferred congruences of subspaces of two or three dimensions.

Even so, comparatively few exact solutions have been found [261], and these are not always of great physical interest; so methods of generating approximate solutions have been developed. These may be classified into:

(i) fast approximation methods, based on the special-relativistic limit of general relativity; in particular, the first-order correction to the Minkowski metric is called the linear approximation [262];

(ii) slow approximation methods, based on the Newtonian limit of general relativity; successive corrections to the Newtonian solution are called post-Newtonian, post-post-Newtonian, etc [263];

(iii) asymptotic methods, based on expanding the field of a space-time that is asymptotically flat in the neighbourhood of infinity in inverse powers of a suitably defined radial distance function [264];

(iv) eikonal methods, based on expanding the gravitational radiation field in the high-frequency (eikonal) limit about some suitably defined non-radiating background metric [265].

Applications of all but the last method are discussed below.

4.3.10. Equations of motion
By 1920, several people, including Einstein Eddington, De Donder and Weyl, realized that the equations of motion of the sources of the gravitational field are not independent of the gravitational field equations; and in certain cases are even completely determined by the latter [266]. Einstein and Grommer, in good part motivated by the vain hope that resulting restrictions on the motion might explain quantum phenomena, worked on this problem [267]; but did not really go beyond what had already been done. Returning to the problem with Infeld and Hoffmann [268], Einstein developed a slow approximation method for generating post-Newtonian corrections to the Newtonian equations of motion of a system of bodies of comparable masses interacting gravitationally and moving with speeds small compared with the speed of light (the EIH method) [269]. But what they called 'the new approximation method' had actually been developed in 1917 by Lorentz and Droste, who applied it to the derivation of the equations of motion of extended bodies modelled as continua [263].

More novel was EIH's method of deriving post-Newtonian equations of motion without use of a physical model of the sources of the field. They did this by deriving from the empty-space field equations conditions on certain surface integrals surrounding, but outside, the sources [268]. Shortly thereafter, Fock (1939) used the slow approximation method to derive equations of motion using a fluid model for the sources [270]. His method utilizes the vanishing of the divergence of the stress–energy tensor of the sources and involves much simpler calculations than those of EIH; in the 1950s Infeld introduced a variant of Fock's method, based on the use of so-called 'good delta functions' to model the sources and thereby simplify the EIH-type calculations [271].

4.3.11. The Schwarzschild solution and the classical tests
Solving his new field equations in the linear approximation for the spherically symmetric field around a central mass, Einstein showed that the elliptical orbit of a test particle in this gravitational field precesses in the direction of the planet's motion. Inserting the mass of the Sun and the parameters of Mercury's orbit in his formula for the rate of precession gives very good quantitative agreement with the well-known anomaly in the observed precession of that planet's orbit [272] (figure 4.13). Shortly afterwards, Schwarzschild found the exact spherically symmetric solution to the field equations, now known by his name even though discovered almost simultaneously by Droste [273]; it gives the same numerical result for Mercury.

In 1967, this result was challenged by Dicke, who suggested, on the basis of visual measurements of solar oblateness, that the Sun had a

Twentieth Century Physics

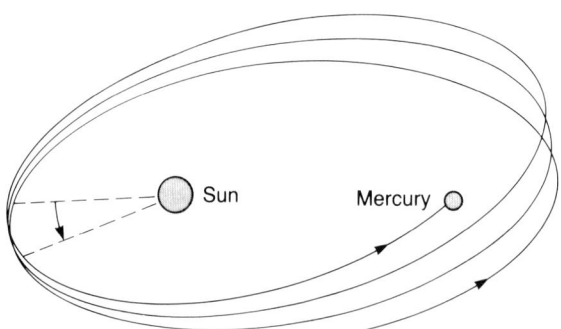

Figure 4.13. *Perihelion advance of Mercury. The point of closest approach of Mercury to the Sun, the perihelion, advances by about 574" per century. Of this, 531" are caused by gravitational perturbations from the other planets, primarily Venus, Earth and Jupiter. The difference, 43" per century, was accounted for by general relativity.*

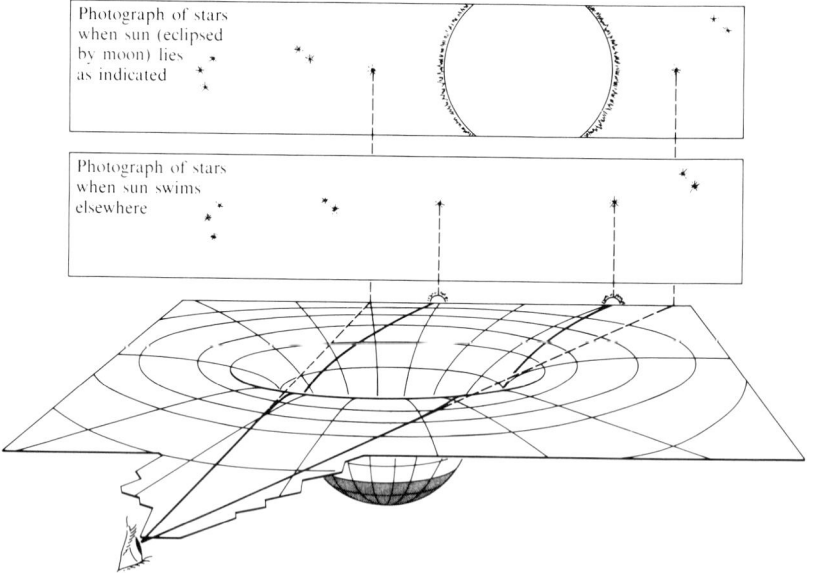

Figure 4.14. *Bending of light by the Sun depicted as a consequence of the curvature of space near the Sun. The ray of light pursues geodesic, but the geometry in which it travels is curved (actual travel takes place in space-time rather than space; the correct deflection is twice that given by above elementary picture). Deflection inversely proportional to angular separation between star and centre of Sun.*

mass quadrupole moment sufficient to significantly affect the Einstein prediction [274]. Since 1980, limits placed on the solar quadrupole moment by more reliable methods (notably helioseismology) seem to rule out such a large effect, leaving the observed perihelion advance of

History of Relativity

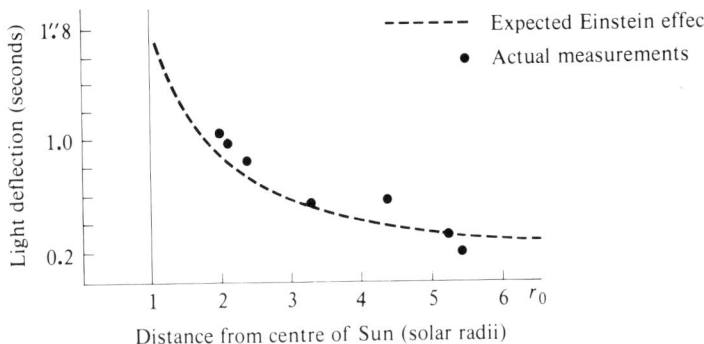

Figure 4.15. *Eddington's results from the 1919 eclipse expedition, showing a close correlation with Einstein's theory.*

Mercury in excellent agreement with the general-relativistic prediction [275].

Einstein also used his approximate solution to calculate the so-called gravitational deflection of a light ray grazing the Sun's limb, getting double the deflection he had predicted in 1911 on the basis of an equivalence principle argument (see figure 4.14). A first attempt to test the new prediction during the partial solar eclipse of 1918 was inconclusive [276], but the 1919 British solar eclipse expeditions led by Eddington and Dyson reported reasonably good agreement between their observations and Einstein's prediction [277] (figure 4.15).

An institute was established in Berlin to test the predictions of the theory. It was headed by Freundlich, the first astronomer to attempt observational tests of general relativity, whose earlier attempts to prove the theory had made him powerful enemies among the German astronomers [278]. He carried out a series of eclipse observations between 1922 and 1929; the results of the last led him to reverse his earlier judgement in favour of general relativity, causing some consternation among relativists [278]. Continuing controversy over the effect, based on optical eclipse observations that at best could claim no more than 10%–20% accuracy, finally dissipated in the late 1960s when radio-telescopic observations of stellar occultations began (figure 4.16), with results that agree with the general-relativistic predictions to an accuracy that has now reached 0.1% [279].

Shapiro proposed a so-called fourth astronomical test of the Schwarzschild metric [280], which really belongs together with the two just discusssed (the gravitational red shift, discussed above, is often grouped with these, but it actually tests only the equivalence principle). It is based on a delay predicted by general relativity in the time it takes for a radio signal, sent from a space probe or a planet, to reach the Earth when the signal passes near the Sun, as compared with the time it would take for the signal to traverse the same path in the absence of the Sun's

Twentieth Century Physics

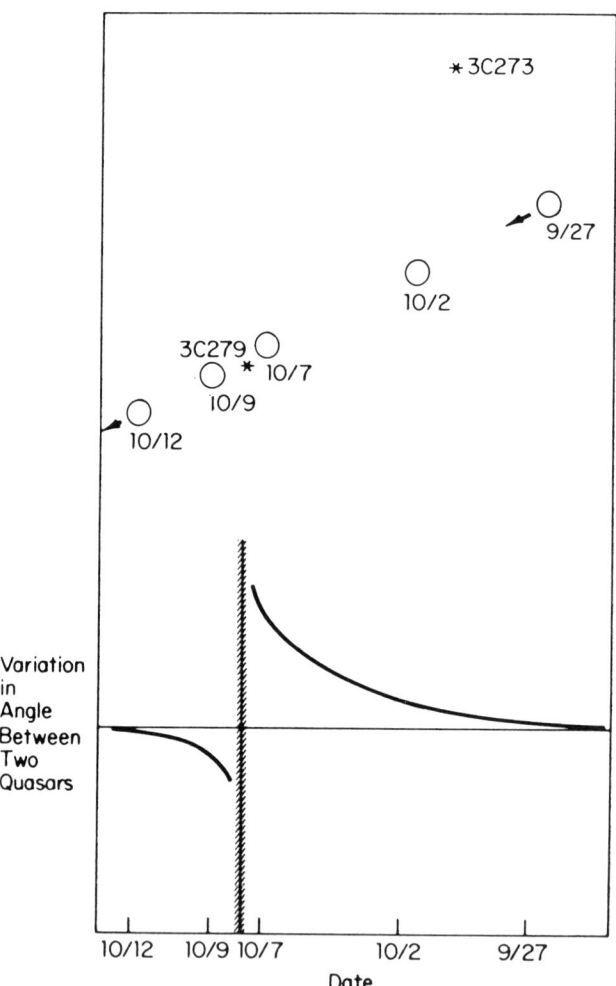

Figure 4.16. *Quasar light deflection measurements. The upper portion of the diagram shows the two quasars 3C273 and 3C279, and the apparent path of the Sun. The lower portion shows the angular variation between the two quasars as the Sun passes in front of 3C279. This variation is due to the deflection of light from 3C279 by the Sun.*

gravitational field (figure 4.17). Results of this test are also now in 0.1% agreement with the general-relativistic predictions [279] (figure 4.18).

4.3.12. Black holes, gravitational collapse and singularities

From the early 1920s, much effort went into attempts to understand the behaviour of the Schwarzschild solution near the Schwarzschild radius, then often also called the Schwarzschild singularity [281]. In the originally used coordinate system, there is a singularity at this radius. Eddington found a coordinate transformation that could remove this

History of Relativity

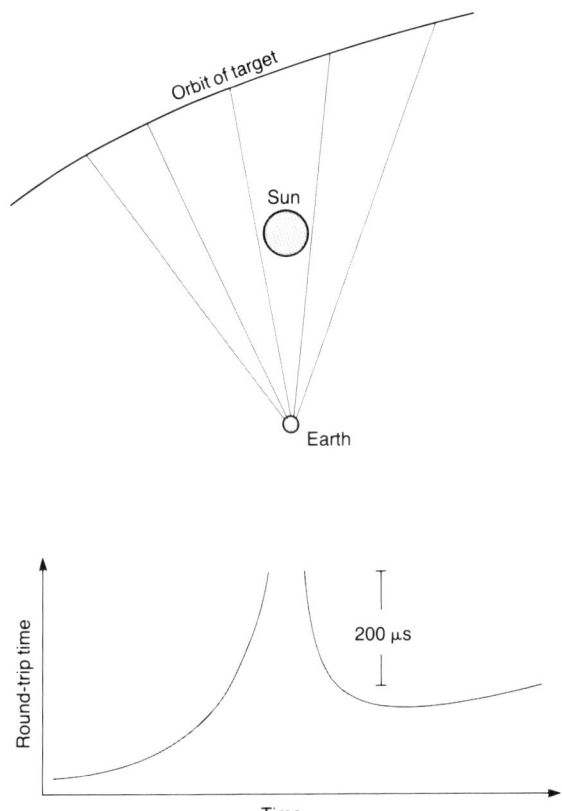

Figure 4.17. *The time delay of light. A target such as planet or a spacecraft moves from left to right on the far side of the Sun while periodic radar tracking signals are sent to it from Earth. As the signals pass the Sun at close range, they suffer an additional delay of up to several hundred microseconds over and above the expected round-trip travel time, which for Mars would be around 42 minutes. Shown in the lower half of the figure is a schematic (and exaggerated) plot of the observed round-trip travel time as a function of time, showing the excess delay for rays that pass near the Sun.*

singularity [282], Lemaître emphasized the non-singular nature of this Schwarzschild horizon [283] and Synge found coordinates that covered the entire Schwarzschild manifold (see below) [284]; but knowledge of their work was not widespread at the time. In 1935, Einstein and Rosen constructed the so called 'bridge' version of the Schwarzschild solution as a classical model of a particle [285], by introducing a topologically non-trivial manifold consisting of two asymptotically flat regions with a 'bridge' connecting them. They had to modify the field equations slightly to get this static interpretation, since the complete Schwarzschild solution is actually non-static (see below). (Topological complications in cosmological models were first discussed by Klein in connection with

Twentieth Century Physics

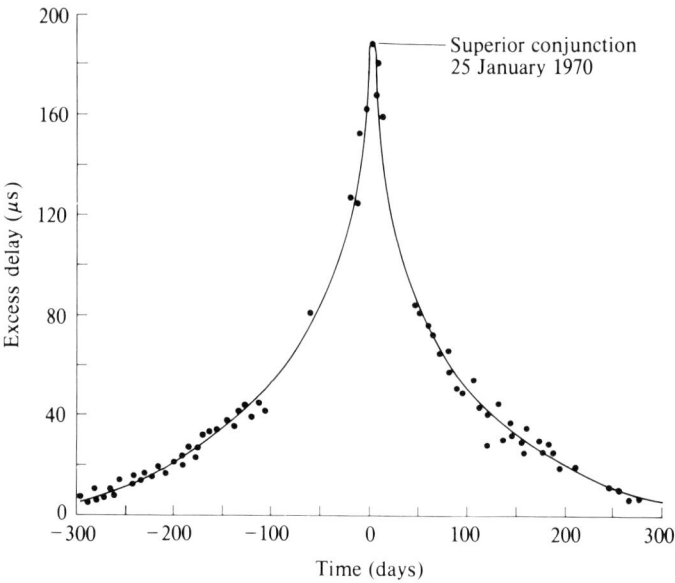

Figure 4.18. *The fourth test of general relativity: time delay of radar echoes from Venus, showing maximum delay when the Sun's edge touches the line between Earth and Venus.*

the elliptic interpretation of the Einstein static universe [286].)

Attempting to show the physical irrelevance of the Schwarzschild singularity, Einstein discussed a static solution representing the interior and exterior (Schwarzschild) gravitational field of a sphere of rotating dust particles (figure 4.19). He showed that, as the radius of the sphere was made smaller and smaller, particles at its surface reached the speed of light well before the sphere reached its Schwarzschild radius [287]. But, as was the case two decades earlier in cosmology, he failed to consider non-static solutions, and in the very same year Oppenheimer and Snyder constructed a model [288] of a collapsing sphere of dust that passes beyond its Schwarzschild radius (figure 4.20), thus providing a powerful argument for taking the solution seriously at and beyond this radius. It was not until two decades later that the fate of gravitationally collapsing stars started to receive widespread attention. Finkelstein called attention to the non-singular nature of the Schwarzschild singularity [289], and Kruskal [290] and Szekeres [291] rediscovered the complete analytic continuation of the Schwarzschild solution. The complete solution is non-static and can be interpreted as having a bridge, wormhole or throat (to use Wheeler's colourful imagery) between two asymptotically flat regions, which constricts until it reaches the true (curvature) singularity (figure 4.21).

It was also shown that sufficiently massive spherically symmetric objects are ultimately unable to resist gravitational collapse beyond their

History of Relativity

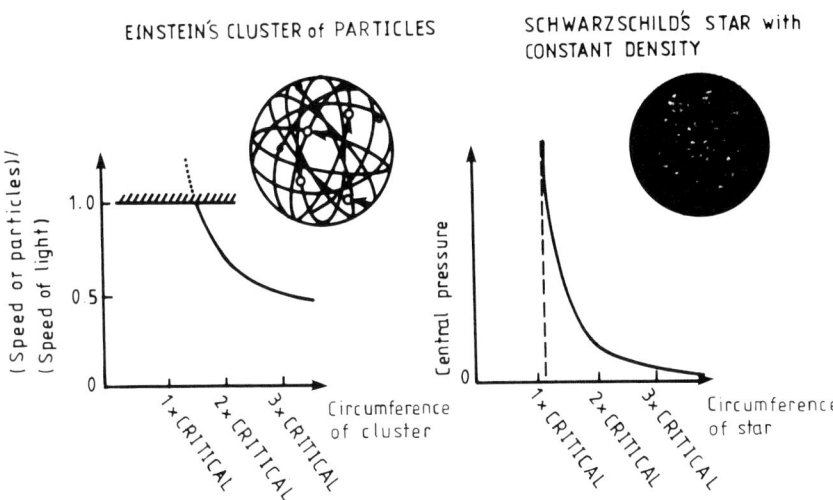

Figure 4.19. *Einstein's evidence that no object can ever be as small as its critical circumference. Left; if Einstein's spherical cluster of particles is smaller than 1.5 critical circumferences, then the particles' speeds must exceed the speed of light, which is impossible. Right; if a star with constant density is smaller than (9/8)–1.125 critical circumferences, then the pressure at the star's centre must be infinite, which is impossible.*

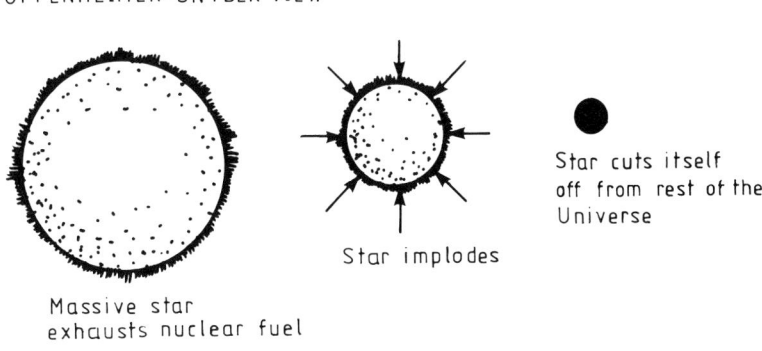

Figure 4.20. *Oppenheimer's view of the fates of large masses.*

Schwarzschild radius and onto the curvature singularity (figure 4.22). Wheeler [292] coined the name 'black holes' for such gravitationally collapsed systems, from which nothing—at least classically (see section 4.3.15)—not even light, can escape [293]. The study of gravitational collapse and black-hole physics have become major theoretical industries, and observations of astrophysical black holes have been claimed [294] (see also Chapter 23). The very names gravitational collapse and black holes, as well as the exotic physical effects associated with them, captured the imagination of physicists and laymen alike, bringing general

309

Twentieth Century Physics

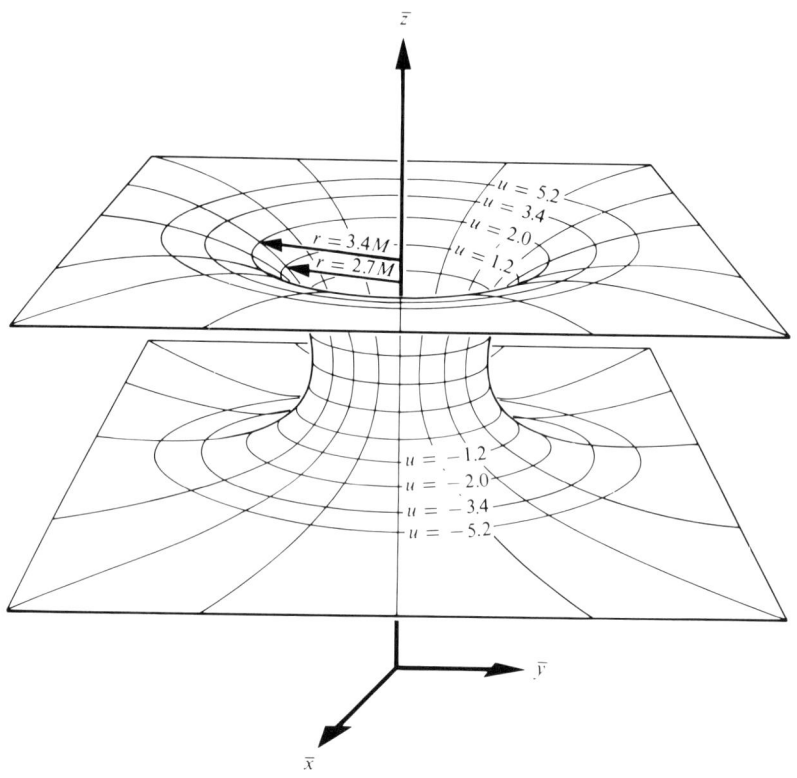

Figure 4.21. The Schwarzschild space geometry at the 'moment of time' $t = v = 0$, with one degree of rotational freedom suppressed ($\theta = \pi/2$). To restore that rotational freedom and obtain the full Schwarzschild 3-geometry, one mentally replaces the circles of constant $\bar{r} = (\bar{x}^2 + \bar{y}^2)^{1/2}$ with spherical surfaces of area $4\pi\bar{r}^{-2}$. Note that the resultant 3-geometry becomes flat (Euclidean) far from the throat of the bridge in both directions (both 'universes').

relativity a much-needed publicity boost in an age when scientific showmanship has become *de rigeur* in the battle for public attention—and funding.

While spherically symmetric collapse beyond the Schwarzschild horizon leads inevitably to curvature singularities, there was considerable discussion about the occurrence of such singularities under more general circumstances. Lifshitz and Khalatnikov [295], on the basis of an analysis of perturbations of the Einstein equations, claimed that the generic solution was non-singular, suggesting that singularities were associated with a high degree of symmetry in the solution. Penrose [296], on the basis of powerful new methods of global differential geometry [297], proved the first of many singularity theorems, which suggested that symmetry had nothing to do with the question. Once an apparent horizon forms (i.e. a

History of Relativity

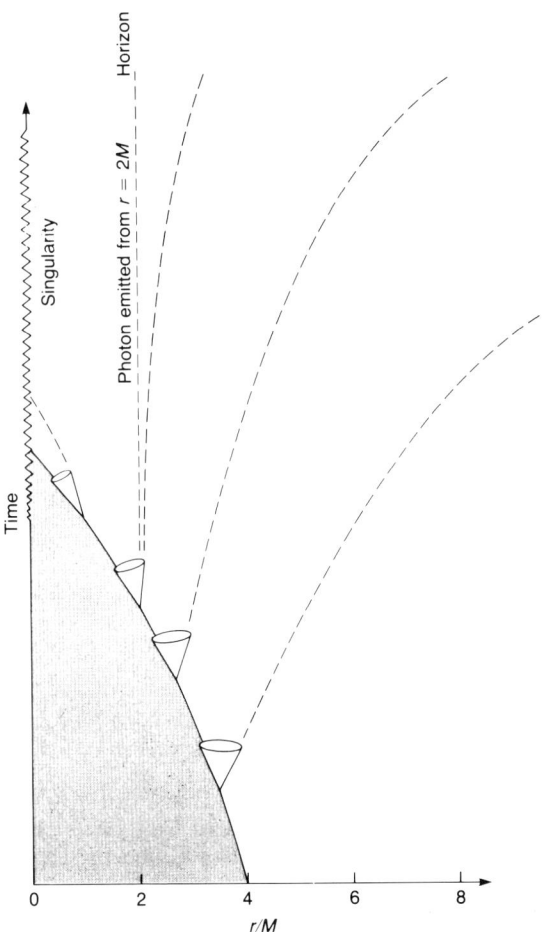

Figure 4.22. *Spherical gravitational collapse to a black hole. Radial distance is plotted horizontally, a variable that measures time is plotted vertically. Here 'M' is shorthand for GM/c^2 which has units of length. The area under the curve is the interior of the collapsing star. Representative 'light cones' are shown. Light cones determine the possible trajectories of light and particles in space-time. Outgoing light rays move along the outward-pointing edge of the cone, while ingoing light rays move along the inward-pointing edge. In the absence of gravity, light rays would move along 45° lines in this diagram. Since particles must move more slowly than light, their trajectories must lie within the opening of the cone. A distant observer sees light rays from an observer falling in on the surface of the star, but as the radius of the star approaches $2M$, the light cones are 'tipped' by space-time curvature so that the rays take longer and longer to reach the distant observer, who sees the process of collapse slow down as $r \to 2M$, the outgoing photon barely escapes, while a photon emitted at $r = 2M$ hovers there, becoming a generator of the event horizon. Inside $r = 2M$, the light cones are tipped over, so that even the 'outgoing' photons actually travel to smaller radii. Everything inside the horizon is forced to reach $r = 0$, where a space-time singularity resides.*

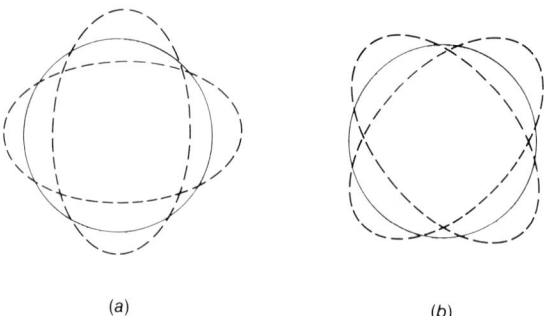

Figure 4.23. *The effect of a gravitational wave on a ring of particles. The wave is propagating out of the page. In (a), the initially circular ring is distorted into an elliptical shape after one-quarter of a cycle of the wave. After half a cycle the ring returns to its circular shape. After three-quarters of a cycle, the ring is distorted into the perpendicular elliptical shape. After one complete cycle of the wave, the ring returns to its initial shape. In (b), the second independent polarization state of the wave causes the same sequence of distortions, with the pattern rotated by 45° relative to that in (a). A general wave causes a superposition of the two distortions.*

surface within which even an 'outgoing' light ray is pulled inwards by the gravitational field), the occurrence of a singularity is inevitable [298]. The controversy was resolved when Khalatnikov and Lifshitz [299] conceded that they had overlooked perturbations leading to singularities.

4.3.13. Gravitational radiation

Einstein [300] found solutions to his field equations, in the linearized approximation, that represented plane gravitational waves; he distinguished between 'real waves,' which carry energy (as defined by his gravitational energy–momentum pseudo-tensor, discussed above) and 'apparent waves', which do not; the latter can be eliminated by a coordinate transformation. Like electromagnetic waves, gravitational waves are transverse with two states of polarization, but instead of being described by a transverse vector they are described by a transverse symmetric, traceless second-rank tensor [301] (figure 4.23).

Eddington first gave an invariant criterion (i.e. valid in any coordinate system) to distinguish real from apparent waves: if the Riemann tensor computed from its metric does not vanish, the wave is physical [302]. The invariant characterization of gravitational fields in terms of their Riemann tensors later proved important, particularly in the identification of radiation fields by such properties [303]. Exact plane wave solutions were given in 1959 by Bondi, Pirani and Robinson [304].

In his papers on linearized gravitation theory, Einstein also investigated the generation of gravitational waves by the sources of the field, deriving the well-known quadrupole moment formula, relating the intensity of the radiation field to the second time derivative of the quadrupole moment of the stress–energy tensor of the sources. In the fast approximation method (see above), Minkowski space-time is the initial solution about which the field is linearized. But the linearized stress–energy tensor, which acts as a source for the linearized field (i.e. first-order correction), obeys special-relativistic conservation laws. So strictly speaking, the original derivation of the quadrupole formula only applies to sources with negligible self-gravitational fields. But in most astrophysical applications, such as gravitationally bound binary star systems, it is just the self-gravitational field that is crucial. Consequently, during the 1940s and 1950s there was considerable controversy over the applicability of the quadrupole formula—and indeed over whether self-gravitating systems emit gravitational radiation at all.

Various ways of attempting to extend the quadrupole formula to self-gravitating systems were proposed, notably those of Landau–Lifshitz [255] and of Fock [305]. Landau and Lifshitz's method is based on a form of the fast motion approximate field equations in which a (non-linear) gravitational energy pseudo-tensor appears alongside the (linearized) non-gravitational energy–momentum tensor as a source of the approximate gravitational field; hence gravitational self-interactions are included in the quadrupole formula calculation. Fock's method is based on the use of a slow approximation (EIH) method for calculating the motions of the sources and the surrounding near field, in which one calculates post-Newtonian corrections to the Newtonian motions of the sources, based on the assumption that v/c is small (where v is a characteristic velocity of the system) and the resulting gravitational fields are weak. If continued, the slow approximation leads to a formal expansion of the field in powers of v/c, in which one solves Poisson-type equations for the field in each order. Gravitational radiation terms do not arise until order $(v/c)^5$, far too high for tractable calculations. So Fock calculated the radiation using the slow motion method in the near (induction) zone, where retardation effects are negligible, to compute the near field; the fast motion method in the far (wave) zone to find the general form of the field there; and matching the two solutions in an intermediate zone to find the gravitational radiation reaction on the sources [306].

Both methods have been developed and improved in response to criticisms; and today the validity of the quadrupole formula, even when applied to self-gravitating systems, is generally accepted.

Bondi, van der Burg and Metzner initiated a new approach to the problem of constructing radiative solutions to the field equations [307]: the study of the asymptotic structure of the gravitational field, i.e. the structure of the field far from its sources along null directions; or, loosely speaking, in the neighbourhood of null infinity. By attaching a conformal boundary to the space-time manifold, Penrose [308] was able to give a precise characterization of the concept of asymptotically flat space-times

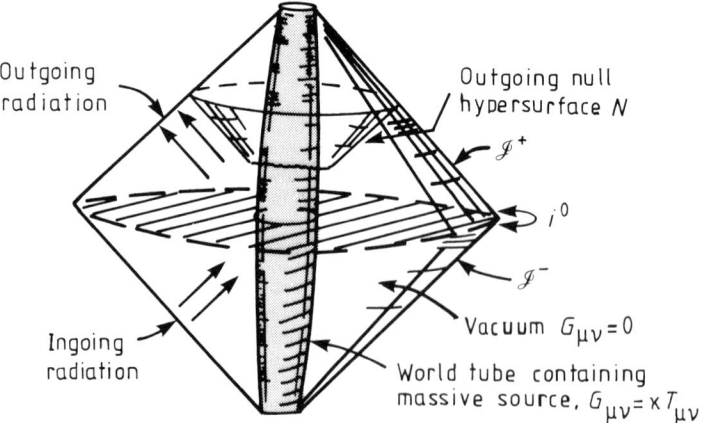

Figure 4.24. *Conjectured global conformal structure of the field of an isolated gravitating system, according to Penrose.*

when the approach to infinity is made on a family of null hypersurfaces (figure 4.24). A different, but analogous, characterization has been given when the approach is on a family of space-like hypersurfaces [309]. Both the null and space-like approaches have been used to give invariant definitions of the global energy, momentum and angular momentum of isolated systems, including the contribution of the gravitational field (see above). A weakness of both approaches so far is the extreme difficulty of connecting the asymptotic solutions with exact near-field solutions, and thus relating the radiated energy as calculated at infinity to the behaviour of some source.

Other important approaches to investigating gravitational radiation include: studies of the Cauchy problem for the Einstein equations, which began with the work of Hilbert in 1917 [310], which led to various canonical $(3 + 1)$ methods for analysing the gravitational field [311]; and studies of the characteristic or null initial value problem for these equations, starting with the work of Darmois in 1927 [312], which led to $(2 + 2)$ approaches to the analysis of the gravitational field [313].

Although originally developed within the framework of the classical theory, these approaches to the study of gravitational radiation—the linearized, the asymptotic, the three-plus-one and the two-plus-two—have been important as starting points of attempts to quantize the gravitational field (see section 4.3.15).

Weber initiated attempts to detect directly gravitational radiation, using the so-called Weber bar [314] (figure 4.25), but his claims have not been verified by other observers [315]. Other methods of detecting gravitational radiation have been proposed, notably the use of laser interferometry [316] (figure 4.26), and the most ambitious effort of this

History of Relativity

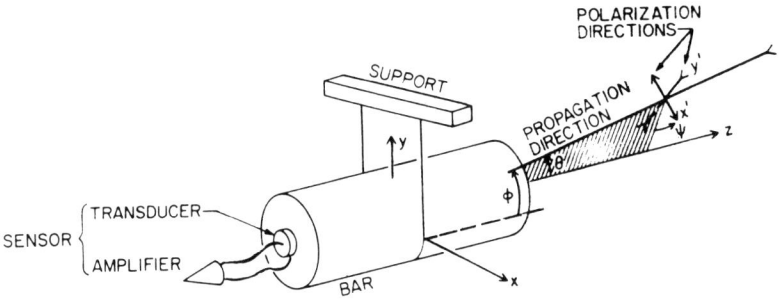

Figure 4.25. *Schematic diagram of a resonant-bar detector for gravitational waves.*

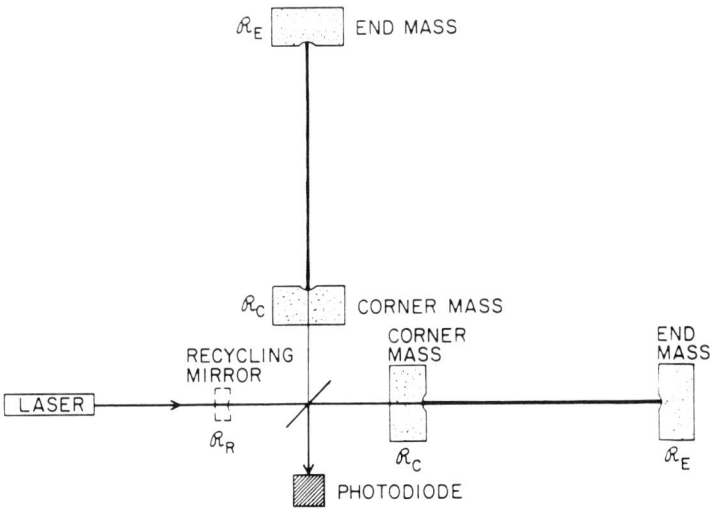

Figure 4.26. *Schematic diagram of a Fabry–Perot-type beam receiver for gravitational waves.*

type, the LIGO project, is currently under construction [237]; but so far there have been no generally accepted direct observations of gravitational waves.

However, good indirect evidence for the existence of gravitational radiation has emerged from analysis of data from the binary pulsar system PSR 1913+16, composed of two neutron stars [317]. Secular changes in the orbit of the pulsar correspond to gravitational radiation damping effects, which are to be expected according to the quadrupole radiation formula [238]. In 1993, Taylor and Hulse were awarded the Nobel Prize in physics for the discovery of and work on this system.

See also p 1757

4.3.14. Recent astronomical and astrophysical applications and tests

For many years attempts at observational verification of the general

Figure 4.27. *The Haystack radar antenna, which Irwin Shapiro and his group have used to collect extensive data on the systematics of the inner part of the solar system. Those data are rapidly becoming the most important source of information about perihelion shifts.*

theory were confined to the three so-called classic tests, all proposed by Einstein in 1916: gravitational red shift (actually a test of the equivalence principle), advance of the perihelion of Mercury and solar light deflection (discussed above). Other predicted general-relativistic effects constituted corrections to the Newtonian results too minute to be observed by available techniques.

Starting in the 1960s, the observational possibilities widened dramatically with the development of new devices such as atomic clocks, radio telescopes (figure 4.27), extraterrestrial satellites tracked by radar and long baseline interferometry detectors, which enable more accurate versions of the older astronomical tests as well as observation of previously inaccessible effects, such as Shapiro's fourth test (see above). So far, all of these tests have given results in very good accord with the predictions of the general-relativity theory [318], which has now become an accepted tool in many branches of astronomy [319].

The discovery of such exotic astrophysical objects as pulsars, quasars, neutron stars and gravitational lenses [320], the theoretical understanding of which involves the development of general-relativistic models, and the search for evidence of the existence of cosmic black holes—general-relativistic objects *par excellence*—have all contributed to the recent renewal of interest in general relativity among astrophysicists, resulting in creation of a new sub-discipline: relativistic astrophysics. The first conference on relativistic astrophysics was held in 1963 [321], and 'Texas Symposiums' have continued regularly ever since [322].

Observational and experimental general relativity is now an active research area [323]. Major US government funding of the LIGO project to search for gravitational waves, and the award of the 1993 Nobel Prize in physics to Taylor and Hulse for their work using a binary pulsar to test general relativity (see above), attest to the growing importance of observational general relativity.

4.3.15. Quantum gravity

Since 1930, there have been numerous attempts to quantize the gravitational field, either in linear approximation or exactly, using either standard methods of (special-relativistic) field quantization, or by developing some alternative to such methods [324]. In the absence of success using any of these methods, in the 1970s considerable attention began to be devoted to the study of quantized non-gravitational fields in background (non-quantized) curved space-times. Work in this area actually started with Schrödinger [325] and Parker [326], but they concentrated on the problem of particle creation in an expanding universe. The newer work was mainly inspired by the hope that such a 'semi-classical' approach would yield new insights into the problematic relation between general relativity and quantum mechanics [327]. Its most dramatic success was Hawking's prediction that black holes emit radiation with a black-body spectrum (Hawking effect) [328].

Returning to quantum gravity proper, in his classic 1916 exposition of general relativity Einstein argued that 'the quantum theory must modify not only Maxwell's electrodynamics but also the new theory of gravitation' [329]. Within a few years, he started work on the unified field theory programme (see section 4.4), intrigued by the hope of thereby finding a classical explanation for quantum phenomena; believing quantum mechanics to be incomplete, he left to others the problem of applying the new quantum-mechanical formalism to the gravitational field.

In the paper which developed special-relativistic quantum field theory (see above), Heisenberg and Pauli remarked optimistically that 'quantization of the gravitational field, which appears to be necessary for physical reasons, may be carried out without any new difficulties by means of a formalism wholly analogous to that applied here' [330]. Rosenfeld first applied this formalism to the linearized gravitational field, introducing the term 'gravitational quanta' for the particles associated with this field [331]. A few years later, Bronsteyn undertook a more critical analysis of linearized quantum gravity [332]. He raised the problem of the measurability of the quantized gravitational field, arguing that, in addition to limits imposed by the uncertainty principle, in general relativity there is an 'absolute limit' to the accuracy with which the components of the linearized affine connection within a given volume can be measured. He suggested that application of the formalism of quantum field theory might not yield the desired fusion of quantum theory and gravitation, calling for 'a radical reconstruction of the theory, and in particular, the rejection of a Riemannian geometry ... and perhaps also the rejection of our ordinary concepts of space and time, modifying them by some much deeper and nonevident concepts' [333].

Thus, by the mid-1930s a line had been drawn between 'conservatives', who feel the problem of quantum gravity is to find the way to apply existing concepts and methods of quantum field theory to the gravitational field equations or some modification of them; and 'radicals', who feel that gravitation has certain unique features (various people have stressed different features) that suggest the need for deep-going conceptual revisions of general relativity and/or quantum mechanics.

However, it was not until the 1950s that interest in the problem of quantum gravity revived, largely through the efforts of the Bergmann group [334], Dirac [335] and Arnowitt, Deser and Misner [336], who worked on canonical methods of quantization, based on singling out a preferred family of space-like hypersurfaces and applying Hamiltonian techniques to the evolution of the three-geometry of these surfaces. After his successes in quantum electrodynamics (see above), Feynman applied the perturbative techniques of special-relativistic quantum field theory to the field equations of general relativity [337]. In contrast to the canonical method, this so-called 'covariant quantization' method introduces a background flat space-time metric, and the difference between the non-flat metric tensor and the background flat metric is treated as a special-relativistic field, obeying non-linear field equations equivalent to the Einstein equations but propagated in accord with the light-cone structure of the flat background [338]. Thus, 'covariant' here does not mean 'generally covariant', but Lorentz covariant, and the intimate union between chronogeometry and inertio-gravitational field that characterized Einstein's theory is lost in this approach. The overwhelming bulk of work on quantum gravity has been done by conservatives following one or other of these approaches, and conservatives still have a majority. But within this majority a shift in attitude has occurred: with the realization that the Einstein field equations are not perturbatively renormalizable, support for the so-called 'covariant quantization' methods has waned; while it has grown for the view that unique features of general relativity require the development of non-perturbative methods based on a canonical formulation of the theory [339].

A growing minority of radicals, led by Penrose [340] and inspired by the failure of attempts at conventional quantization, has gained respectability for efforts to derive general-relativistic space-time itself from some more fundamental set of concepts. Penrose has been engaged in an attempt to construct a non-local theory of space-time based on twistor analysis [107]. But the most popular such attempt in recent years has been string theory, which treats all elementary particles, including the so-far hypothetical gravitons, as quantized excitations of a two-dimensional entity called a string. Most theories of strings consider them to be immersed in some higher-dimensional background (non-dynamical) space-time, which seems self-defeating if the purpose is to construct space-time out of strings; but some theories attempt to meet the challenge of completely eliminating a background metric from their foundations [341].

4.3.16. *Philosophical treatments of relativity and popular reactions*

With the growing acceptance of the special theory by the physics community came growing attention by philosophers and philosophically minded physicists. There were claims that the theory supported one or another philosophical doctrine, such as Petzoldt's plea for positivism [342] and there were attacks on the theory for its supposed incompatibility with one or another philosophical doctrine. However, it was after the announcement in 1919 of the results of the eclipse expeditions testing the general theory (see above) had made relativity a household word (see below) that production of such claims and counterclaims developed into a major philosophical industry [343]. Within a decade, Bergson, Cassirer, Haldane, Myerson, Reichenbach, Russell, Schlick, Whitehead—to name just a few of the more important philosophers—attempted to explain what relativity had done for philosophy and/or what philosophy could do for relativity. During the years between the two World Wars, and more especially after 1945, the philosophy of science was consolidated as a major subdiscipline of philosophy, largely as a result of serious efforts by philosophers to analyse the concepts and structure of the new physics (relativity and quantum theory) rather than try to fit them into a pre-established philosophical mould. Contemporary philosophy of science has itself subdivided, and relativity theory still forms the major area of research in the field called the philosophy of space and time [344].

It is worth noting that the first attacks on Einstein for subverting 'common sense' and 'sound physical intuition' came not from philosophers or political opponents, but from his fellow physicists and started even before he formulated the general theory [345]. In later years, attacks on relativity theory and on Einstein personally came from various quarters. In Germany, these attacks were often politically motivated—Einstein was associated with the democratic and pacifist left during the Weimar Republic—sometimes covered by a scientific or philosophical veneer. The latent anti-semitic animus behind attacks such as those of Lenard and Stark became open during the Nazi era with its attempts to hunt down and ban 'Jewish physics' as well as Jewish physicists. In the Soviet Union, especially during the Stalin era, there were recurrent philosophical attacks, often semi-official, on the theory of relativity as incompatible with dialectical materialism, the officially sanctioned Soviet version of Marxism. Ironically, such attacks alternated with periods of praise for Einstein when his political views happened to coincide with one or another aspect of the current party line. Attacks on Einstein in the western democracies tended to be more restrained, although also not entirely free of anti-semitic animus. A comparative study of the anti-relativity literature in various countries would be a valuable contribution to the historical sociology of science [346].

The extensive popular literature on relativity, which (at its best) attempts to explain the basic ideas of the theory of relativity to a wider, scientifically unsophisticated public, started even before Einstein's own attempt in 1917. It includes books by such outstanding scientists and philosophers as: Bergmann, Born, Borel, Eddington, Freundlich, Geroch, Infeld, Landau, Russell, Schlick, Wald—the list literally goes from A (Ashtekar) to Z (Zee)—and includes a host of books, to say nothing of articles, by more or less eminent (and successful) popularizers. Hardly studied so far, this literature would also repay serious examination by scholars concerned with the diffusion of information about current science to a broader public.

The extensive and intensive preoccupation with Einstein's person as well as his work began in December 1919, when the results of the British eclipse expeditions (see above) were announced with banner headlines in the British, German and United States press, and also featured prominently elsewhere. The resulting sensation soon made Einstein world famous among the general public, a fame that actually increased over the years, making him the world's first scientific superstar. While the public reaction was largely adulatory, it is often forgotten that it was not exclusively positive. Substantial anti-Einstein sentiment also appeared, largely inspired by right-wing political, ultra-nationalistic, and/or anti-semitic motives. The reasons for the origin and persistence of his celebrity, which mystified Einstein, and for the development of the many myths about him [347], have been the subject of much informal speculation over the years, but only recently has the question begun to be studied seriously [348]. A study of those aspects of the Einstein myth relating to his scientific work would also be valuable to scholars concerned with public perceptions and misperceptions of science [349].

4.4. Unified field theories

The search for theories which offer a common explanation for apparently quite distinct physical phenomena has a long history, going back at least to the pre-Socratics. In modern times, the mechanistic world-view, formulated in the wake of the triumphs of Newton's mechanical philosophy and prevalent in physics until the late nineteenth century, assumed that a mechanical explanation could be found for all phenomena; Laplace's 'central force' programme even suggested a concrete way of realizing that goal [350]. The concept of energy provided strong links between all branches of physics, and energeticists suggested that would enable their unification.

But Maxwell's theory provided the first successful example of a field theory that unified previously disparate branches of physics, by explaining all of optics in terms of interacting electric and magnetic fields. The latter were in turn unified by the special theory into a single four-dimensional electromagnetic field. The electromagnetic world-view, espoused by Wien, Abraham and others around the turn of this century (see above), led to attempts to explain the mechanical properties of matter (electrons) on the basis of Maxwell's theory or some suitably modified form of it [351]. Einstein himself engaged in such an attempt around 1909 [352], and for many years Mie searched for a non-linear electrodynamics that would explain the structure of matter [353].

There were also attempts to include gravitation in Mie's programme, starting with his work on a special-relativistic gravitation theory [354]. In 1915, Hilbert attempted to synthesize Mie's approach with Einstein's general-relativity programme, proposing a unified theory with field equations similar to Einstein's (see above) [355]. In 1915 Nordström proposed a five-dimensional unification of Maxwell's theory with his scalar gravitational theory (see above).

After the general theory was completed in 1916, the unified field theory programme took a new turn. Most further attempts were based on the search for a mathematical structure that generalizes that of the general theory in such a way as to include electromagnetism. Weyl proposed a rather natural generalization that introduces electromagnetism in a simple and unforced way by use of a principle which Weyl called gauge invariance [356]. The theory was rejected by Einstein and others for physical reasons; but, in retrospect, it can be seen as the first example of an idea that later proved to be of great importance in field theory: the introduction of a new field into a theory in order to extend its symmetry group from a global to a local gauge group [357].

Weyl's theory introduced the newly developed concept of affine connection (see above) to generalize the metric connection inherent in Einstein's theory. In 1921, Eddington proposed a unified theory based on an even more general (but still symmetric) affine connection [358].

Initially sceptical, Einstein became intrigued with the idea and started to work on it himself [359]. In 1923, Cartan utilized a non-symmetric connection (space with torsion) in another generalization of Einstein's theory of gravitation [227].

In 1919, Kaluza wrote to Einstein suggesting the possibility of unifying general relativity and electromagnetism on the basis of a five-dimensional metric tensor. Einstein encouraged Kaluza to work on this idea, which he published in 1921 [360], and Einstein also worked on this penta-dimensional approach [361]. Both Einstein and Kaluza seemed to have been unaware of Nordström's earlier five-dimensional work, discussed above.

In the late 1920s, Einstein introduced a family of preferred tetrad fields into a Riemannian space-time to characterize spaces with distant parallelism; such spaces served as the basis for a number of attempts at unified field theories [362]. Cartan soon made Einstein aware that he and other mathematicians had already developed the concept of distant parallelism [363].

Einstein's also worked on a theory that employed a complex Hermitian affine connection [364], and with this we have encountered examples of the five basic mathematical structures that have been used in classical (i.e. non-quantum) unified field theories of gravitation and electromagnetism. Broadly speaking, these theories may be classified with the help of these five categories:

(i) theories that utilize spaces of more than four dimensions;

(ii) theories that utilize a metric tensor, symmetric or asymmetric;

(iii) theories that utilize an affine connection, with or without torsion, which is not (completely) deducible from a metric tensor in the theory;

(iv) theories that utilize preferred sets of basis vectors, or the equivalent spinor fields;

(v) theories that utilize complexifications, Hermitian or otherwise, of one or more of categories (ii)–(iv).

These categories are not mutually exclusive. For example, there have been metric-affine theories, five-dimensional metric theories, etc [365].

Einstein, and many others who worked on the unified field theory programme, hoped the requirement that solutions to a properly chosen set of unified field equations be everywhere non-singular would so limit the class of solutions that it would yield a classical explanation of the discreteness associated with so many quantized physical quantities without the need to impose a separate quantization procedure on these equations. Many were sceptical from the beginning [366] and the development of quantum mechanics convinced most physicists that such hopes were Quixotic, but Einstein persisted [367].

After about 1930, the entire search for a unified field theory fell out of fashion among elementary particle physicists. However, this situation started to change in the 1960s when Weinberg and Salam proposed a special-relativistic, locally gauge-invariant field theory unifying the electromagnetic and weak interactions. As a result of this success, the unified field theory programme has experienced a remarkable renaissance in the last two decades. With the development of quantum chromodynamics as a gauge theory of the strong interactions, the goal of the GUT (grand unified theory) programme is now the unification of the strong and electroweak fields, with the ultimate goal of including gravitation as well [368]. However, the modern unification programme differs from Einstein's in at least two major respects: its aim is to unify all four known force fields, not just electromagnetism and gravitation; and it does not attempt to supplant quantum mechanics: formulation of classical unified field equations is just a preliminary to its quantization.

Acknowledgments

Work on this article was partially supported by a grant from the US National Science Foundation, and was completed during a visiting scholarship at the Max Planck Institute for the History of Science, Berlin, Germany.

References

[1] For an attempt to capture this feature in the development of general relativity, see Ray C 1987 Relativity—dead or alive? *The Evolution of Relativity* (Bristol: Adam Hilger) ch 6.
[2] Originally, Einstein only referred to the principle of relativity. 'Theory of relativity' is a term first used by others and the prefix 'special' was added by Einstein after he had generalized his original theory. See Einstein A 1989 *The Collected Papers of Albert Einstein vol 2. The Swiss Years: Writings 1901–1909* (Princeton: Princeton University Press) p 254
[3] Einstein A 1989 *The Collected Papers of Albert Einstein vol 2. The Swiss Years: Writings 1901–1909* (Princeton: Princeton University Press) pp xxi–xxii
[4] For Einstein's distinction between constructive theories, which we call dynamical, and theories of principle, which in the case of the special theory we call kinematical, see Einstein A 1989 *The Collected Papers of Albert Einstein vol 2. The Swiss Years: Writings 1901–1909* (Princeton: Princeton University Press) p xxi, note 28.
[5] See, for example, Stachel J 1994 Changes in the concepts of space and time brought about by relativity *Artefacts, Representations and Social Practice* ed C C Gould and R S Cohen (Dordrecht: Kluwer) pp 141–62
[6] The ten parameters represent the homogeneity and isotropy of space and the homogeneity of time relative to each inertial frame, and the equivalence of all inertial frames. As emphasized by Felix Klein in 1872, a geometry is characterized by its symmetry group. See, for example, Torretti R 1978 *Philosophy of Geometry from Riemann to Poincaré* (Dordrecht: Reidel) pp 137–42; 1983 *Relativity and Geometry* (Oxford: Pergamon) pp 26–7
[7] Actually, the word 'allgemeine' that is translated as 'general' might better be translated as 'universal' in this context, but the English usage is too well-established.
[8] For further discussion of the chronogeometrical, inertial and inertio-gravitational structures, see Stachel J 1994 Changes in the concepts of space and time brought about by relativity *Artefacts, Representations*

and *Social Practice* ed C C Gould and R S Cohen (Dordrecht: Kluwer) pp 141–62.
[9] More accurately, this space is called a differentiable manifold. For a simple discussion, see Kopczynski W and Trautman A 1992 *Spacetime and Gravitation* (New York: Wiley) ch 3.
[10] Fock was the most extreme advocate of the thesis that 'there is no relativity in general relativity'. See Fock V A 1955 *Teoriya Prostranstva Vremeni i Tyagoteniya* (Moscow: Fizmatgiz). For an account of his views, see Gorelik G 1993 Vladimir Fock: philosophy of gravity and gravity of philosophy *Einstein Studies vol 5. The Attraction of Gravitation/New Studies in the History of General Relativity* ed J Earman *et al* (Boston: Birkhäuser) pp 308–31.
[11] For a survey of interpretations of this concept, see Norton J 1994 General covariance and the foundations of general relativity: eight decades of dispute *Rep. Prog. Phys.* **56** 791–858.
[12] For the author's views on this question, see Stachel J 1986 What a physicist can learn from the discovery of general relativity *Proc. Fourth Marcel Grossmann Meeting on General Relativity* ed R Ruffini (Amsterdam: Elsevier) pp 1857–62; 1989 Einstein's search for general covariance, 1912–1915 *Einstein Studies vol 1. Einstein and the History of General Relativity* ed D Howard and J Stachel (Boston: Birkhäuser) pp 63–100; 1993 The meaning of general covariance: the hole story *Philosophical Problems of the Internal and External Worlds/Essays on the Philosophy of Adolf Grünbaum* ed J Earman (Konstanz: Universitätsverlag/Pittsburgh: University of Pittsburgh Press) pp 129–60.
[13] Stachel J 1986 What a physicist can learn from the discovery of general relativity *Proc. Fourth Marcel Grossmann Meeting on General Relativity* ed R Ruffini (Amsterdam: Elsevier) pp 1857–62
[14] Einstein was an early investigator of such phenomena and was preoccupied with their explanation throughout his life. See, for example:
Pais A 1979 Einstein and the quantum theory *Rev. Mod. Phys.* **51** 863–914
Stachel J 1986 Einstein and the quantum: fifty years of struggle *From Quarks to Quasars: Philosophical Problems of Modern Physics* ed R Colodny (Pittsburgh: University of Pittsburgh Press) pp 349–85
[15] For a discussion of Einstein's views on quantum mechanics, see Stachel J 1986 Einstein and the quantum: fifty years of struggle *From Quarks to Quasars: Philosophical Problems of Modern Physics* ed R Colodny (Pittsburgh: University of Pittsburgh Press) pp 349–85.
[16] Tonnelat M-A 1971 *Histoire du principe de relativité* (Paris: Flammarion)
Barbour J 1989 *Absolute or Relative Motion? A study from a Machian point of view of the discovery and the structure of dynamical theories vol 1. The Discovery of Dynamics* (Cambridge: Cambridge University Press)
[17] Galilei G 1632 *Dialogo di Galileo Galilei Linceo Matematico Sopraordinario dello Studio di Pisa. E Filosofo, e Matematico primario del Serenissimo Gr. Duca di Toscano. Doue nei i congressi di quattro giornate si discorre sopra i due Massimi Sistemi del Mondo Tolemaico, e Copernicano; Proponendo indeterminatamente le ragioni Filosofiche, e Naturali tanto per l'una, quanto per l'altra parte* (Florence: Gio Batista Landini) (cited from reimprint (Brussels: Culture et Civilisation, 1966) pp 180–1) (Engl. Transl. Drake S 1967 *Dialogue Concerning the Two Chief World Systems—Ptolemaic and Copernican* (Berkeley: University of California Press) pp 186–7)
[18] Koyré A and Cohen I B (ed) 1972 *Isaac Newton's Philosophiae Naturalis Principia Mathematica: The Third Edition (1726) With Variant Readings*

(in two volumes) (Cambridge, MA: Harvard University Press) fifth corollary to the laws of motion, pp 63–4 (Engl. Transl. Motte A 1729 *Mathematical Principles of Natural Philosophy and His System of the World* revised and edited by F Cajori 1946 (Berkeley: University of California Press) p 20)

[19] For the Leibniz–Clarke controversy over absolute and relational theories of space and time, in which Clarke presented Newton's views, see Alexander H G (ed) 1956 *The Leibniz–Clarke Correspondence. With Extracts from Newton's Principia and Optics* (Manchester: Manchester University Press).

[20] Lange L 1886 *Die geschichtliche Entwickelung des Bewegungsbegriffes und ihr voraussichtliches Endergebnis* (Leipzig: Engelmann)

[21] Koyré A and Cohen I B (ed) 1972 *Isaac Newton's Philosophiae Naturalis Principia Mathematica: The Third Edition (1726) With Variant Readings* (in two volumes) (Cambridge, MA: Harvard University Press) fifth corollary to the laws of motion, pp 46–53 (Engl. Transl. Motte A 1729 *Mathematical Principles of Natural Philosophy and His System of the World* revised and edited by F Cajori 1946 (Berkeley: University of California Press) p 6–12)

[22] Koyré A and Cohen I B (ed) 1972 *Isaac Newton's Philosophiae Naturalis Principia Mathematica: The Third Edition (1726) With Variant Readings* (in two volumes) (Cambridge, MA: Harvard University Press) pp 50–2 (Engl. Transl. Motte A 1729 *Mathematical Principles of Natural Philosophy and His System of the World* revised and edited by F Cajori 1946 (Berkeley: University of California Press) pp 10–11)

[23] Motte A 1729 *Mathematical Principles of Natural Philosophy and His System of the World* revised and edited by F Cajori 1946, Book III (the system of the world), hypothesis I. That the centre of the system of the world is immovable (Berkeley: University of California Press) p 419. Presumably, Newton's attachment to absolute space is connected with his characterization of space as the sensorium of the deity.

[24] See, for example:
Lange L 1886 *Die geschichtliche Entwickelung des Bewegungsbegriffes und ihr voraussichtliches Endergebnis* (Leipzig: Engelmann)
Mach E 1960 *Die Mechanik in ihrer Entwicklung. Historisch-kritisch dargestellt* (Leipzig: Brockhaus) (Engl. Transl. McCormack T J 1960 *The Science of Mechanics* 6th American edn (LaSalle: Open Court))

[25] See, for example, Klein M 1972 Mechanical explanation at the end of the nineteenth century *Centaurus* **17** 58–82

[26] See, for example, Torretti R 1983 Newtonian spacetime *Relativity and Geometry* (Oxford: Pergamon) pp 20–31

[27] See, for example, Sabra A I 1981 *Theories of Light From Descartes to Newton* (Cambridge: Cambridge University Press)

[28] For a brief history of the ether concept, see Swenson L S 1972 *The Ethereal Aether* (Austin, TX: University of Texas Press) ch 1, pp 3–31.

[29] See, for example, Buchwald J Z 1989 *The Rise of the Wave Theory of Light: Optical Experiment and Theory in the Early Nineteenth Century* (Chicago: University of Chicago Press)

[30] For a review of such theories, see Schaffner K F 1972 *Nineteenth-Century Aether Theories* (Oxford: Pergamon).

[31] It was taken for granted that, like ordinary matter, the ether must have a definite state of motion at each point.
For historical accounts of the optics of moving bodies see:
Sesmat A 1937 *Systèmes de référence et mouvements (Physique classique)* vol 6.

L'optique des corps en mouvement (Paris: Hermann)

Hirosige T 1976 The ether problem, the mechanistic world-view, and the origins of the theory of relativity *Hist. Studies Phys. Sci.* **7** sections 2–5

[32] Bradley J 1728 A Letter from the Reverend Mr James Bradley Savilian Professor of Astronomy at Oxford and FRS to Dr Edmund Halley Astronomer Royal, etc, giving an account of a newly discovered motion of the fix'd stars *Phil. Trans. R. Soc.* **35** 637–61

[33] Fresnel A 1818 Lettre de M Fresnel à M Arago, sur l'influence du mouvement terrestre dans quelques phénomènes d'optique *Ann. Chimie Phys.* **9** 57–67

[34] For an account of first-order experiments, see Pietrocola Pinto de Oliveira M 1992 Elie Mascart et l'optique des corps en mouvement *Doctoral Thesis* L'Université Denis Diderot.

[35] For contemporary reviews of such experiments, see Mascart E 1872 Sur les modifications qu'éprouve la lumière par suite du mouvement de la source lumineuse et du mouvement de l'observateur *Ecole Normale Supérieure (Paris) Ann. Sci.* **(2)1** 157–214; 1874 *Ecole Normale Supérieure (Paris) Ann. Sci.* **3** 363–420; 1893 Propagation de la lumière *Traité d'Optique* vol 3 (Paris: Gauthier-Villars) ch 15.

[36] Mascart E 1874 *Ecole Normale Supérieure (Paris) Ann. Sci.* **3** 420

[37] Fresnel A 1818 Lettre de M Fresnel à M Arago, sur l'influence du mouvement terrestre dans quelques phénomènes d'optique *Ann. Chimie Phys.* **9** 57–67

For accounts of Fresnel's work, see:

Hirosige T 1976 The ether problem, the mechanistic world-view, and the origins of the theory of relativity *Hist. Studies Phys. Sci.* **7** 7–8

Torretti R 1983 Newtonian spacetime *Relativity and Geometry* (Oxford: Pergamon) pp 41–2

[38] Fizeau H 1851 Sur les hypothèses relatives à l'ether lumineaux, et sur une expérience qui paraît démontrer que le mouvement des corps change la vitesse avec laquelle la lumière se propage dans leur intérieur *C.R. Acad. Sci., Paris* **33** 349–55

Veltmann W 1873 Ueber die Fortpflanzung des Lichtes in bewegten Medien *Ann. Phys. Chemie* **150** 497–535

[39] See, for example, Buchwald J Z 1985 *From Maxwell to Microphysics/ Aspects of Electromagnetic Theory in the Last Quarter of the Nineteenth Century* (Chicago: University of Chicago Press)

[40] Poincaré H 1890 Introduction *Electricité et Optique vol 1. Les Théories de Maxwell et la Théorie Electromagnetique de la Lumière* (Paris: Georges Carré) pp v–xix

[41] For a discussion of the electromagnetic world-view, see, for example:

McCormmach R 1970 H A Lorentz and the electromagnetic view of Nature *Isis* **58** 459–97

Jungnickel C and McCormmach R 1986 *Intellectual Mastery of Nature vol 2. The Now Mighty Theoretical Physics 1870–1925* (Chicago: University of Chicago Press) pp 227–45

[42] For a review, see Darrigol O 1993 The electrodynamics of moving bodies from Faraday to Hertz *Centaurus* **36** 245–60.

[43] For an influential summary of his views at the turn of the century, see Lorentz H A 1895 *Versuch einer Theorie der elektrischen und optischen Erscheinungen in bewegten Körpern* (Leiden: Brill).

[44] Lorentz H A 1895 *Versuch einer Theorie der elektrischen und optischen Erscheinungen in bewegten Körpern* (Leiden: Brill) p 4

[45] Lorentz H A 1892 La théorie électromagnétique de Maxwell et son

application aux corps mouvants *Arch. Néerlandaises des sciences exactes et naturelles* **25** 363–552

[46] Michelson A A 1881 The relative motion of the Earth and the luminiferous ether *Am. J. Sci.* **22** 120–9

Michelson A A and Morley E W 1887 On the relative motion of the Earth and the luminiferous ether *Am. J. Sci.* **34** 333–45. For a history of the original experiment and later repetitions, see Swenson L S 1972 *The Ethereal Aether* (Austin, TX: University of Texas Press).

[47] Sagnac G 1913 L'éther lumineux démontré par l'effet du vent relatif d'éther dans un interféromètre en rotation uniforme *C. R. Acad. Sci., Paris* **157** 708–10

[48] For early discussions of their respective contributions see:

Pauli W 1921 Relativitätstheorie *Encyklopädie der mathematischen Wissenschaften, mit Einschluss ihrer Anwendungen vol 5. Physik* ed A Sommerfeld (Leipzig: Teubner) part 2, pp 539–775

Kottler F 1924 Considerations de critique historique sur la théorie de la relativité. Première partie: De Fresnel à Lorentz *Scientia* **36** 231–42; Deuxième Partie: Henri Poincaré et Albert Einstein *Scientia* **36** 301–16

Arguments for 'The Relativity Theory of Poincaré and Lorentz' are found in Whittaker E 1953 *A History of the Theories of Aether and Electricity vol II: The Modern Theories* (London: Nelson) ch II.

See also:

Keswami G H 1965 Origin and concept of relativity *Br. J. Phil. Sci.* **15** 286–306; 1966 *Br. J. Phil. Sci.* **16** 19–32

Giedymin J 1982 *Science and Convention: Essays on Henri Poincaré's Philosophy of Science and the Conventionalist Tradition* (Oxford: Oxford University Press)

Zahar E 1983 Poincaré's independent discovery of the relativity principle *Fundamenta Scientiae* **4** 147–75

[49] This is an example of a general principle: in science, many historical questions cannot be formulated properly, let alone answered, without a preliminary critical analysis of the topic. See Grünbaum A 1973 *Philosophical Problems of Space and Time* 2nd edn (Dordrecht: Reidel) ch 12, for other examples from the history of special relativity.

[50] The idea of compensating effects to explain the failure to detect motion through the ether had been invoked much earlier (it is inherent in Fresnel's 1818 paper [33] and see Fizeau H 1851 *C. R. Acad. Sci., Paris* **33** 354).

At about the same time as Lorentz, Larmor developed a theory with many similarities (see Larmor J 1900 *Aether and Matter/ A Development of the Dynamical Relations of the Aether to Material Systems on the Basis of the Atomic Constitution of Matter Including a Discussion of the Influence of the Earth's Motion on Optical Phenomena* (Cambridge: Cambridge University Press)). For accounts of Larmor's theory, see:

Warwick A 1989 The electrodynamics of moving bodies and the principle of relativity in British physics 1894–1919 *PhD Dissertation* Cambridge University; 1991 On the role of the Fitzgerald–Lorentz contraction hypothesis in the development of Joseph Larmor's electronic theory of matter *Arch. Hist. Exact Sci.* **43** 29–91

Darrigol O 1994 The electron theories of Larmor and Lorentz: a comparative study *Hist. Studies Phys. Biol. Sci.* **25** 265–336

[51] Lorentz H A 1904 Electromagnetic phenomena in a system moving with any velocity smaller than that of light *Kon. Neder. Akad. Wet. Amsterdam. Versl. Gewone Vergad. Wisen Natuurkd. Afd.* **6** 809–31

[52] Lorentz H A 1892 The relative motion of the Earth and the ether *Kon. Neder. Akad. Wet. Amsterdam. Versl. Gewone Vergad. Wisen Natuurkd. Afd.* **1** 74–9 (Engl. Transl. in Lorentz H A 1937 *Collected Papers* vol 4 (The Hague: Martinus Nijhoff) pp 219–23); 1895 *Versuch einer Theorie der elektrischen und optischen Erscheinungen in bewegten Körpern* (Leiden: Brill)

[53] Fitzgerald G F 1889 The ether and the Earth's atmosphere *Science* **13** 390

[54] Lorentz H A 1899 Simplified theory of electrical and optical phenomena in moving bodies *Kon. Neder. Akad. Wet. Amsterdam. (Proc. Section of Sciences)* **1** 427–42

[55] Janssen M 1989 *H A Lorentz and the Special Theory of Relativity* unpublished talk at annual meeting of the Nederlandse Natuurkundige Vereniging, April 1989

[56] Poincaré H 1902 *La Science et L'hypothèse* (Paris: Flammarion)

[57] For accounts of Poincaré's work, see:
Miller A I 1973 A Study of Henri Poincaré's 'Sur la Dynamique de l'Electron' *Arch. Hist. Exact Sci.* **10** 207–328
Darrigol O 1994 Henri Poincaré's criticism of fin de siècle electrodynamics *Preprint*

[58] Poincaré H 1898 La mesure du temps *Rev. Métaphys. Morale* **6** 1–13

[59] Poincaré H 1900 La théorie de Lorentz et le principe de réaction *Receuil de travaux offerts par les auteurs à H A Lorentz* ed J Boscha (The Hague: Martinus Nijhoff) pp 252–78

[60] Poincaré H 1905 Sur la dynamique de l'électron *C. R. Acad. Sci., Paris* **140** 1504–8; 1906 Sur la dynamique de l'électron *Circolo Matematico di Palermo. Rendiconti* **21** 129–65

[61] For a survey of the available evidence, see Einstein A 1989 Einstein on the theory of relativity *The Collected Papers of Albert Einstein vol 2. The Swiss Years: Writings 1901–1909* (Princeton: Princeton University Press) pp 252–74.

[62] For much of this literature up to 1981, see Miller A I 1981 Bibliography, secondary sources *Albert Einstein's Special Theory of Relativity: Emergence (1905) and Early Interpretation (1905–1911)* (Reading, MA: Addison-Wesley) pp 434–40; in particular, see the entries under Goldberg, Stanley; Hirosige, Tetu; Holton, Gerald; Klein, Martin; McCormmach, Russell; Miller, Arthur I. Other recent references include:
Earman J, Glymour C and Rynasiewicz R 1983 On writing the history of special relativity *PSA 1982: Proc. 1982 Biennial Meeting of the Philosophy of Science Association* vol 2, ed P D Asquith and T Nickles (East Lansing: Philosophy of Science Association) pp 403–16
Goldberg S 1983 Albert Einstein and the creative act: the case of special relativity *Springs of Scientific Creativity: Essays on the Founders of Modern Science* ed A H Rutherford *et al* (Minneapolis: University of Minnesota Press) pp 232–53; 1984 *Understanding Relativity: Origin and Impact of a Scientific Revolution* (Boston: Birkhäuser)
Gutting G 1972 Einstein's discovery of special relativity *Phil. Sci.* **39** 51–67
Holton G 1980 Einstein's scientific program: the formative years *Some Strangeness in the Proportion* ed H Woolf (Reading, MA: Addison-Wesley) pp 49–65; 1981 Einstein's search for the Weltbild *Proc. Am. Phil. Soc.* **125** 1–15
Miller A I 1982 The special relativity theory: Einstein's response to the physics of 1905 *Albert Einstein: Historical and Cultural Perspectives—The Centennial Symposium in Jerusalem* ed G Holton and Y Elkana (Princeton: Princeton University Press) pp 3–26; 1983 On Einstein's invention of

special relativity *PSA 1982: Proc. 1982 Biennial Meeting of the Philosophy of Science Association* vol 2, ed P D Asquith and T Nickles (East Lansing: Philosophy of Science Association) pp 377–402

Pais A 1982 Relativity, the special theory *Subtle is the Lord, The Science and the Life of Albert Einstein* (Oxford: Oxford University Press) pp 111–74

Renn J 1993 Einstein as a disciple of Galileo: a comparative study of conceptual development in physics *Science in Context* vol 6, ed M Beller *et al* no 1 pp 311–34

Schaffner K F 1983 The historiography of special relativity *PSA 1982: Proc. Biennial Meeting of the Philosophy of Science Association* vol 2, ed P D Asquith and T Nickles (East Lansing: Philosophy of Science Association) pp 417–28

Stachel J 1982 Einstein and Michelson: the context of discovery and the context of justification *Astron. Nach.* **303** 47–53; 1989 *What Song the Syrens Sang: How Did Einstein Discover the Special Theory of Relativity?* (Italian Transl. Lepscky M L 1989 *Quale canzone cantarono le sirene. Come scopri Einstein la teoria speciale della relativita? L'Opera di Einstein, proceedings of the Convegno Internazionale: L'Opera di Einstein, 13–14 Dicembre 1985* ed U Curi (Ferrara: Gabriele Corbo Editore) pp 21–37)

Zahar E 1989 *Einstein's Revolution/A Study in Heuristic* (La Salle: Open Court)

[63] I have speculated more freely on some of these questions in Stachel J 1982 Einstein and Michelson: the context of discovery and the context of justification *Astron. Nach.* **303** 47–53; 1989 *What Song the Syrens Sang: How Did Einstein Discover the Special Theory of Relativity?* (Italian Transl. Lepscky M L 1989 *Quale canzone cantarono le sirene. Come scopri Einstein la teoria speciale della relativita? L'Opera di Einstein, proceedings of the Convegno Internazionale: L'Opera di Einstein, 13–14 Dicembre 1985* ed U Curi (Ferrara: Gabriele Corbo Editore) pp 21–37).

[64] Recently, some currency has been given to claims that Marić, who in 1903 became his wife, played a major creative role in the development of the special theory of relativity. See:

Trbuhović-Gjurić 1969 *U senci Alberta Ajnštajna* (Kruševać: Bagdala) (German Transl. Zimmermann G W (ed) 1983 *Im Schatten Albert Einsteins/ Das tragische Leben der Mileva Einstein-Maric* (Berne: Haupt))

Walker E H and Stachel J 1989 Did Einstein espouse his spouse's ideas *Physics Today* February pp 9–11 (two letters)

Troemel-Ploetz S 1990 Mileva Einstein-Maric: the woman who did Einstein's mathematics *Index on Censorship* **9** 33–6

While the letters show she certainly played a greater role in Einstein's life, including his early scientific work, than acknowledged by previous biographers, the available evidence does not support such claims (see Stachel J 1995 Albert Einstein and Mileva Maric: a collaboration that failed to develop *Creative Couples in Science* ed Pnina Abir-Am and H Pycior (New Brunswick: Rutgers University Press)).

[65] For the text of these letters, see Einstein A 1987 *The Collected Papers of Albert Einstein vol 1. The Early Years, 1879-1902* ed J Stachel *et al* (Princeton: Princeton University Press)

[66] Cited from Stachel J 1987 Einstein and ether drift experiments *Physics Today* May pp 45–7

[67] The editorial note, 'Einstein on the Electrodynamics of Moving Bodies' reference [65], pp 223–5, indicates the letters that discuss each of these topics.

[68] For evidence of his work on an emission theory, see Stachel J 1982

Einstein and Michelson: the context of discovery and the context of justification *Astron. Nach.* **303** 47–53; 1989 *What Song the Syrens Sang: How Did Einstein Discover the Special Theory of Relativity?* (Italian Transl. M L Lepscky 1989 Quale canzone cantarono le sirene. Come scopri Einstein la teoria speciale della relativita? *L'Opera di Einstein, proceedings of the Convegno Internazionale: L'Opera di Einstein, 13–14 Dicembre 1985* ed U Curi (Ferrara: Gabriele Corbo Editore) pp 21–37)

Einstein A 1989 Einstein on the theory of relativity *The Collected Papers of Albert Einstein vol 2. The Swiss Years: Writings 1901–1909* (Princeton: Princeton University Press), pp 253–74, especially pp 263–4

For a review of objections to emission theories, see Pauli W 1958 *Theory of Relativity* ed G Field (Oxford: Oxford University Press) part 2, pp 539–775, with supplementary notes pp 5–9. (Engl. Transl. of Pauli W 1921 Relativitätstheorie *Encyklopädie der mathematischen Wissenschaften, mit Einschluss ihrer Anwendungen vol 5. Physik* ed A Sommerfeld (Leipzig: Teubner))

[69] For a discussion of this period, see Einstein on the Theory of Relativity Einstein A 1989 *The Collected Papers of Albert Einstein vol 2. The Swiss Years: Writings 1901-1909* (Princeton: Princeton University Press) pp 253–74. See pp 258–66, for a number of his later reminiscences; and the 'Introduction to Volume 2' pp xvi–xxix, especially pp xxiii–xxv, for the influence of his readings in philosophy.

[70] Einstein A 1905 Zur Electrodynamik bewegter Körper *Ann. Phys., Lpz* **17** 891–921

[71] For a speculative reconstruction of this work, see Earman J, Glymour C and Rynasiewicz R 1983 On Writing the History of Special Relativity *PSA 1982: Proc. 1982 Biennial Meeting of the Philosophy of Science Association* vol 2, ed P D Asquith and T Nickles (East Lansing: Philosophy of Science Association).

[72] Translation from Stachel J 1994 Changes in the concepts of space and time brought about by relativity *Artefacts, Representations and Social Practice* ed C C Gould and R S Cohen (Dordrecht: Kluwer) p 145.

[73] This formulation is given in Einstein A 1907 Über das Relativitätsprinzip und die aus demselben gezogenen Folgerungen *Jahrb. Radioakt. Elektron.* **4** 411–62

[74] Einstein A 1910 Le principe de relativité et ses conséquences dans la physique moderne *Arch. Sci. Phys. Natur.* **29** 5–28, 125–44

[75] Einstein later stated that he had not read Lorentz' 1904 paper [54] in 1905, but he may have read other papers which utilize the Lorentz transformations, in particular Cohn E 1904 Zur Elektrodynamik bewegter Systeme *König. Preuss. Akad. Wiss. (Berlin). Sitzungsber.* 1294–303, 1404–16. The formal similarity of the transformations led many of their contemporaries—and some of ours—to ignore the differences between Lorentz's and Einstein's interpretations of them (see below). The vector formulation used here was given in Herglotz G 1911 Über die Mechanik des deformierbaren Körpers vom Standpunkt der Relativitätstheorie *Ann. Phys., Lpz* **(4)36** 493–533. Note that as c becomes infinite the spatial transformations approach the Galilean transformations, and t' approaches t.

[76] Poincaré H 1906 Sur la dynamique de l'électron *Circolo Matematico di Palermo. Rendiconti* **21** 129–65

[77] Einstein A 1905 Ist die Tröghirt eines Körpers von seinem Energieinholt abhändgig? *Ann. Phys., Lpz* **18** 639–41

[78] Notably Fritz Hasenöhrl for black-body radiation. See Miller A I 1981 *Albert Einstein's Special Theory of Relativity: Emergence (1905) and Early Interpretation (1905–1911)* (Reading: Addison-Wesley) pp 359–60

[79] For a discussion of Einstein's 'Ist die Tröghirt eines Körpers von seinem Energieinholt abhändgig?' and later criticisms of it, see Stachel J and Torretti R 1982 Einstein's first derivation of mass–energy equivalence *Am. J. Phys.* **50** 760–3. Einstein and others later offered various derivations, the most general of which require only the Lorentz transformations and the laws of conservation of energy and momentum.

[80] Einstein A 1907 Über die Möglichkeit einer neuen Prüfung des Relativitätsprinzips *Ann. Phys., Lpz* **23** 197–98

The effect was first observed by Ives and Stilwell, thereby providing direct evidence for the relativistic time dilation. See Ives H and Stilwell G R 1938 An experimental study of the rate of a moving atomic clock *J. Opt. Soc. Am.* **28** 215–26.

[81] Einstein A and Laub J 1908 Über die elektromagnetischen Grundgleichungen für bewegte Körper *Ann. Phys., Lpz* **26** 532–40; 1908 Über die im elektromagnetischen Felde auf ruhende Körper ausgeübten ponderomotorischen Kräfte *Ann. Phys., Lpz* **26** 541–50

For a discussion, see Einstein A 1989 Einstein and Laub on the Electrodynamics of Moving Media *The Collected Papers of Albert Einstein vol 2. The Swiss Years: Writings 1901–1909* (Princeton: Princeton University Press) pp 503–7.

[82] Einstein A 1907 Über das Relativitätsprinzip und die aus demselben gezogenen Folgerungen *Jahrb. Radioakt. Elektron.* **4** 411–62; corrections in Einstein A 1908 Berichtigungen zu der Arbeit: Über das Relativitätsprinzip und die aus demselben gezogenen Folgerungen *Jahrb. Radioakt. Elektron.* **5** 98–9; 1910 Le principe de relativité et ses conséquences dans la physique moderne *Arch. Sci. Phys. Natur.* **29** 5–28, 125–44

[83] Einstein A 1995 *The Collected Papers of Albert Einstein vol 4. The Swiss Years: Writings 1912–1914* ed M Klein *et al* (Princeton: Princeton University Press) Doc 1

[84] Einstein A 1917 *Über die spezielle und die allgemeine Relativitätstheorie (Gemeinverständlich)* (Braunschweig: Vieweg); 1922 *The Meaning of Relativity: Four Lectures Delivered at Princeton University, May, 1921* (London: Methuen); both went through many editions in Einstein's lifetime and posthumously.

[85] Einstein A 1919 Time, space and gravitation *The Times* (London) 28 November 1919 p 13

[86] Cited from Stachel J 1986 Einstein and the quantum: fifty years of struggle *From Quarks to Quasars: Philosophical Problems of Modern Physics* ed R Colodny (Pittsburgh: University of Pittsburgh Press) p 349

[87] Einstein A 1912 Relativität und Gravitation. Erwiderung auf eine Bemerkung von M Abraham *Ann. Phys., Lpz* **38** 1061

[88] Born M 1949 *Natural Philosophy of Cause and Chance* (Oxford: Oxford University Press) pp 38–9

[89] Einstein A 1935 Elementary derivation of the equivalence of mass and energy *Bull. Am. Math. Soc.* **41** 223–30

[90] Czikszentmihalyi M 1988 Society, culture, and person: a systems view of creativity *The Nature of Creativity: Contemporary Psychological Perspectives* ed R J Steinberg (New York: Cambridge University Press) pp 325–39;

1990 The domain of creativity *Theories of Creativity* ed A Mark *et al* (Newbury Park: Sage)

For further discussion, with examples from the history of relativity, see Stachel J 1994 Scientific discoveries as historical artifacts *Current Trends in the Historiography of Science* ed K Gavroglu (Dordrecht: Reidel) pp 139–48.

[91] Einstein A 1994 *The Collected Papers of Albert Einstein vol 5. The Swiss Years: Correspondence 1902–1914* ed M Klein *et al* (Princeton: Princeton University Press) p 120

[92] Goldberg S 1984 *Understanding Relativity: Origin and Impact of a Scientific Revolution* (Boston: Birkhäuser)

Miller A I 1981 *Albert Einstein's Special Theory of Relativity: Emergence (1905) and Early Interpretation (1905–1911)* (Reading: Addison-Wesley)

[93] For a comparative study of the German, French, British and American receptions, see:

Goldberg S 1984 *Understanding Relativity: Origin and Impact of a Scientific Revolution (Part II, The Early Response to the Special Theory of Relativity, 1905–1911; Part III, From Response to Assimilation, ch 10)* (Boston: Birkhäuser)

See Glick T 1987 *The Comparative Reception of Relativity* (Dordrecht: Reidel) for studies of the reception of the special and general theories in the United States, England, France, Germany, Spain, Italy, Poland, Japan and the Soviet Union.

Glick T 1989 *Einstein and Spain: Relativity and the Recovery of Science* (Princeton: Princeton University Press)

Sánchez Ron J M 1981 *Relatividad Especial, Relatividad General (1905-1923): Orígenes, Desarrollo y Recepción por la Communidad Scientifica* (Barcelona); 1992 The reception of general relativity among British physicists and philosophers *Einstein Studies vol 3. Studies in the History of General Relativity* ed J Eisenstaedt and A J Kox 1992 (Boston: Birkhäuser) pp 57–88

[94] Warwick A 1989 International relativity: the establishment of a theoretical discipline *Studies Hist. Phil. Sci.* **20** 139–49; 1989 The electrodynamics of moving bodies and the principle of relativity in British physics 1894–1919 PhD Dissertation University of Cambridge; 1992 Cambridge mathematics and Cavendish physics: Cunningham, Campbell and Einstein's relativity 1905–1914. Part I: the uses of theory *Studies Hist. Phil. Sci.* **23** 625–56

[95] A good deal of historical information may be found in treatises on the special theory. The books of Henri Arzeliès are particularly useful:

Arzeliès H 1957 *La Dynamique relativiste et ses applications* vol 1 (Paris: Gauthier-Villars); 1959 *Milieux Conducteurs et Polarisables en Mouvement* vol 2 (Paris: Gauthier-Villars); 1963 *Electricité Macroscopique et Relativiste* (Paris: Gauthier-Villars); 1966 *Relativistic Kinematics* (Oxford: Oxford University Press); 1966 *Rayonnement et Dynamique du Corpuscule Chargé Fortement Acceleré* (Paris: Gauthier-Villars); 1967 *Thermodynamique Relativiste et Quantique* (in two volumes) (Paris: Gauthier-Villars).

See also:

Pauli W 1921 Relativitätstheorie *Encyklopädie der mathematischen Wissenschaften, mit Einschluss ihrer Anwendungen vol 5. Physik* ed A Sommerfeld (Leipzig: Teubner) part 2

Lecat M 1924 *Bibliographie de la Relativité* (Bruxelles: Lambertin) for a bibliography up to that date.

[96] Poincaré H 1905 Sur la dynamique de l'électron *C.R. Acad. Sci., Paris*

140 1504–8; 1906 Sur la dynamique de l'électron *Circolo Matematico di Palermo. Rendiconti* **21** 129–65
For studies of Poincaré's work, see:
Miller A I 1973 A study of Henri Poincaré's 'Sur la Dynamique de l'Electron' *Arch. Hist. Exact Sci.* **10** 207–328; Poincaré H 1887 Sur les hypothèses fondamentales de la géometrie *Bull. Soc. Math. France* **15** 203–16, in a discussion of two-dimensional geometries, includes (in addition to Euclid's, Riemann's and Lobachevski's) a 'fourth geometry', which is a two-dimensional version of Minkowski's. Curiously, Poincaré never discussed this 'fourth geometry' in his later work on relativity.
Darrigol O 1994 Henri Poincaré's criticism of fin de siècle electrodynamics *Preprint*

[97] Minkowski H 1908 Die Grundgleichungen für die elektromagnetischen Vorgänge in bewegten Körpern *König. Ges. Wiss. Göttingen. Math.-Phys. Klasse. Nach.* 53–111; 1909 Raum und Zeit *Phys. Z.* **10** 104–11
For studies of Minkowski's work on special relativity, see:
Pyenson L 1977 Hermann Minkowski and Einstein's special theory of relativity *Arch. Hist. Exact Sci.* **17** 71–95 (reprinted in Pyenson L 1985 *The Young Einstein: the Advent of Relativity* (Bristol: Hilger) pp 80–100)
Galison P 1979 Minkowski's space-time: from visual thinking to the absolute world *Hist. Studies Phys. Sci.* **10** 85–121.

[98] Langevin P 1911 L'evolution de l'espace et du temps *Scientia* **10** 31–54
[99] For an account, see Marder L 1971 *Time and the Space Traveller* (London: Allen and Unwin)
[100] Conway A W 1911 On the application of quaternions to some recent developments of electrical theory *Proc. R. Ir. Acad.* A **29** Memoir 1
[101] Silberstein L 1912 Quaternionic form of relativity *Phil. Mag.* **23** 790–809
[102] Sommerfeld A 1910 Zur Relativitätstheorie. I. Vierdimensionale Vektoralgebra *Ann. Phys., Lpz* **32** 749–76; 1910 II. Vierdimensionales Vektoranalysis *Ann. Phys., Lpz* **33** 649–89
Laue M 1911 *Das Relativitätsprinzip* (Braunschweig: Vieweg)
[103] Pauli W 1927 Zur Quantenmechanik des magnetischen Elektrons *Z. Phys.* **43** 601–23
[104] Dirac P A M 1928 The quantum theory of the electron *Proc. R. Soc.* A **117** 610–24
[105] Van der Waerden B L 1929 Spinoranalyse *Ges. Wiss. Göttingen. Math.-Phys. Klasse. Nach* 100–9
For a review of spinor techniques in relativity, see Penrose R and Rindler W 1984 *Spinors and Space-Time vol 1. Two-spinor Calculus and Relativistic Fields* (Cambridge: Cambridge University Press); 1986 *Spinors and Space-Time vol 2. Spinor and Twistor Methods in Space-time Geometry* (Cambridge: Cambridge University Press).
[106] For a review of his work, started in the 1960s, see Penrose R and Rindler W 1984 *Spinors and Space-Time vol 1. Two-spinor Calculus and Relativistic Fields* (Cambridge: Cambridge University Press), which also reviews his work on the application of spinorial techniques in general relativity.
[107] For a review, see Penrose R and Rindler W 1986 *Spinors and Space-Time vol 2. Spinor and Twistor Methods in Space-time Geometry* (Cambridge: Cambridge University Press).
[108] Ignatowski W 1910 Das Relativitätsprinzip *Arch. Math. Phys.* **17** 1–24; 1911 *Arch. Math. Phys.* **18** 17–41
[109] For a discussion of one such approach, with some historical references, see Stachel J 1983 Special relativity from measuring rods *Physics, Philosophy*

and *Psychoanalysis* ed R S Cohen and L Laudan (Boston: Reidel) pp 255–72.

[110] For a review, see Torretti R 1983 *Relativity and Geometry* (Oxford: Pergamon) pp 76–82.

[111] See, for example, Aharoni J 1965 *The Special Theory of Relativity* 2nd edn (Oxford: Clarendon)

[112] Robb A A 1913 *A Theory of Space and Time* (Cambridge: Heffer); 1914 *A Theory of Space and Time* (Cambridge: Cambridge University Press)

[113] Robb objected to the use of relative, frame-dependent concepts as opposed to absolute, frame-independent ones. See the preface to Robb A A 1936 *Geometry of Time and Space* (Cambridge: Cambridge University Press) p vi, 11–13.

[114] Poincaré H 1887 Sur les hypothèses fondamentales de la géometrie *Bull. Soc. Math. France* **15** 203–16 gives a somewhat similar development of a two-dimensional Minkowski space.

[115] But see Weyl H 1923 *Mathematische Analyse des Raumproblems* (Berlin: Springer).

[116] Zeeman E C 1964 Causality implies the Lorentz group *J. Math. Phys.* **5** 490–3

Kronheimer E and Penrose R 1967 On the structure of causal spaces *Proc. Cambridge Phil. Soc.* **63** 481–501

For a discussion of causal space-times, see Torretti R 1983 *Relativity and Geometry* (Oxford: Pergamon) pp 121–9.

[117] Laue M 1907 Die Mitführung des Lichtes durch bewegte Körper nach dem Relativitätsprinzip *Ann. Phys., Lpz* **23** 989–90

[118] Sommerfeld A 1909 Über die Zusammensetzung der Geschwindigkeiten in der Relativtheorie *Phys. Z.* **10** 826–9

[119] Varicak V 1912 Über die nichteukledische Interpretation der Relativtheorie *Deutsche Math. Verein. Jahresber.* **21** 103–27

[120] For a discussion, see Silberstein L 1914 *The Theory of Relativity* (London: MacMillan).

[121] For the relation to Minkowski space see Pauli W 1921 Relativitätstheorie *Encyklopädie der mathematischen Wissenschaften, mit Einschluss ihrer Anwendungen vol 5. Physik* ed A Sommerfeld (Leipzig: Teubner) part 2, pp 539–775, with supplementary notes section 25, note 111 (Engl. Transl. by G Field 1958 *Theory of Relativity* (Oxford: Oxford University Press)).

[122] Borel E 1913 La théorie de la relativité et la cinématique *C. R. Acad. Sci., Paris* **156** 215–8; 1913 La cinématique dans la théorie de la relativité *C. R. Acad. Sci., Paris* **157** 703–5

[123] Thomas L H 1926 The motion of the spinning electron *Nature* **117** 514; 1927 The kinematics of an electron with an axis *Phil. Mag.* **3** 1–22

For an historical account, including the ensuing discussion, see Mehra J and Rechenberg H 1982 *The Historical Development of Quantum Theory vol 3. The Formulation of Matrix Mechanics and its Modification 1925–1926* (Berlin: Springer) pp 270–2.

[124] Planck M 1906 Das Prinzip der Relativität und die Grundgeleichungen der Mechanik *Deutsche Phys. Ges. Verhand.* **8** 136–41; 1907 Zur Dynamik bewegter Systeme *Ann. Phys.* **26** 1–34

For studies of Planck's work on special relativity, see:

Goldberg S 1976 Max Planck's philosophy of Nature and his elaboration of the special theory of relativity *Hist. Studies Phys. Sci.* **7** 125–60

Liu C 1994 Planck and the special theory of relativity, *1993 Meeting of the*

History of Relativity

        History of Science Society, Santa Fe, NM unpublished paper
[125] Einstein A 1907 Über das Relativitätsprinzip und die aus demselben gezogenen Folgerungen *Jahrb. Radioakt. Elektron.* **4** 411–62
[126] Lewis G N and Tolman R C 1909 The principle of relativity and non-Newtonian mechanics *Phil. Mag.* **18** 510–23
[127] Pauli W 1921 Relativitätstheorie *Encyklopädie der mathematischen Wissenschaften, mit Einschluss ihrer Anwendungen vol 5. Physik* ed A Sommerfeld (Leipzig: Teubner) section 30
[128] Schwarzschild K 1903 Zur Elektrodynamik. I. Zwei Formen des Prinzips der Kleinsten Action in der Elektronentheorie *König. Ges. Wiss. Göttingen. Math.-Phys. Klasse Nach.* **128** 126–31; 1903 II. Die elementare elektrodynamische Kraft *König. Ges. Wiss. Göttingen. Math.-Phys. Klasse Nach.* **128** 132–41
        Gauss first suggested that interactions between charged particles are not instantaneous, but involve a finite propagation time. See Larmor J 1900 *Aether and Matter/ A Development of the Dynamical Relations of the Aether to Material Systems on the Basis of the Atomic Constitution of Matter Including a Discussion of the Influence of the Earth's Motion on Optical Phenomena* (Cambridge: Cambridge University Press) p 319
[129] For a review of direct-interaction theories, see Sánchez Ron J M 1978 Action-at-a-distance in XXth century classical physics. Part I: studies of relativistic action-at-a-distance theories *PhD Thesis* University of London ch 2, pp 51–117.
[130] For a review of work on both special and general-relativistic theories of this type, see Havas P 1979 Equations of motion and radiation reaction in the special and general theories of relativity *Isolated Gravitating Systems in General Relativity* ed J Ehlers (Amsterdam: North-Holland) pp 74–155.
[131] Dirac P A M 1949 Forms of relativistic dynamics *Rev. Mod. Phys.* **21** 392–9
[132] For a review of this topic, see Kerner E (ed) 1972 *The Theory of Action-at-a-distance in Relativistic Particle Dynamics. A Reprint Collection* (New York: Gordon and Breach).
[133] Born M 1909 Zur Theorie des starren Elektrons in der Kinematik des Relativitätsprinzips *Ann. Phys., Lpz* **30** 1–56, 840
[134] Herglotz G 1910 Über den vom Standpunkt des Relativitätsprinzips aus als 'starr' zu bezeichnenden Körper *Ann. Phys.* **31** 391–415
        Noether F 1910 Zur Kinematik des starren Körpers in der Relativtheorie *Ann. Phys.* **31** 919–44
[135] Laue M 1911 Zur Diskussion über den starren Körpern in der Relativitätstheorie *Phys. Z.* **12** 85–7; 1911 *Das Relativitätsprinzip* (Braunschweig: Vieweg)
[136] Herglotz G 1911 Über die Mechanik des deformierbaren Körpers vom Standpunkt der Relativitätstheorie *Ann. Phys., Lpz* **36** 493–533
        Nordström G 1911 Zur Relativitätsmechanik deformierbarer Körp *Phys. Z.* **12** 854–7
        For its extension to general relativity, see Nordström G 1917 De gravitatiettheorie van Einstein en de mechanica der continua van Herglotz *Kon. Neder. Akad. Wet. Amsterdam. Versl. Gewone Vergad. Wisen Natuurkd. Afd.* **25** 836–43.
[137] Ignatowski W 1911 Zur Hydrodynamik vom Standpunkte des Relativitätsprinzips *Phys. Z.* **12** 4441–2
        Lamla 1912 Über die Hydrodynamik des Relativitätsprinzips *Ann. Phys., Lpz* **37** 772–96

[138] For a brief review, including the general-relativistic case, see Stachel J 1991 Variational principles in relativistic continuum dynamics *1991 Elba Conference on Advances in Modern Continuum Dynamics* unpublished.
[139] Eckert C 1940 The thermodynamics of irreversible processes III. Relativistic theory of a simple fluid *Phys. Rev.* **58** 919–24
[140] For reviews, which include the extensions to general relativity, see:
Baranov and Kolpascikov 1974 *Relativistic Thermodynamics of Continuous Media* (in Russian) (Minsk: Nauka i Tekhnika)
Sieniutycz S 1994 *Conservation Laws in Variational Thermo-Hydrodynamics* (Dordrecht: Kluwer) ch 12
[141] Minkowski H 1908 Die Grundgleichungen für die elektromagnetischen Vorgänge in bewegten Körpern *König. Ges. Wiss. Göttingen Math.-Phys. Klasse Nach.* 53–111
[142] Abraham M 1909 Zur elektromagnetischen Mechanik *Phys. Z.* **10** 737–41
[143] Laue M 1911 *Das Relativitätsprinzip* (Braunschweig: Vieweg)
[144] Einstein A and Laub J 1908 Über die elektromagnetischen Grundgleichungen für bewegte Körper *Ann. Phys., Lpz* **26** 532–40; 1908 Über die im elektromagnetischen Felde auf ruhende Körper ausgeübten ponderomotorischen Kräfte *Ann. Phys., Lpz* **26** 541–50
[145] Abraham M 1909 Zur Elektrodynamik bewegter Körper *Circolo Matematico di Palermo. Rendiconti* **28** 1–28
For a discussion of this work and the subsequent controversy, see Liu C 1991 Relativistic thermodynamics: its history and foundations *PhD Thesis* University of Pittsburg ch III.
[146] For reviews, see Pauli W 1921 Relativitätstheorie *Encyklopädie der mathematischen Wissenschaften, mit Einschluss ihrer Anwendungen vol 5. Physik* ed A Sommerfeld (Leipzig: Teubner) pp 662–668 (Engl. Transl. Pauli W 1958 *Theory of Relativity* ed G Field (Oxford: Oxford University Press) with supplementary notes pp 106–11, 216)
De Groot S R and Suttorp L G 1972 *Foundations of Electrodynamics* (Amsterdam: North-Holland) ch V, section 7
[147] See, for example:
Pavlov V I 1978 On discussions concerning the problem of ponderomotive forces *Sov. Phys.–Usp.* **21** 171–3
Einstein A 1989 Einstein and Laub on the electrodynamics of moving media *The Collected Papers of Albert Einstein vol 2. The Swiss Years: Writings 1901–1909* (Princeton: Princeton University Press), especially p 507
[148] Lorentz H A 1904 Maxwells elektromagnetische Theorie *Encyklopädie der mathematischen Wissenschaften, mit Einschluss ihrer Anwendungen vol 5. Physik* ed A Sommerfeld (Leipzig: Teubner) part 2, pp 63–144; 1904 Weiterbildung der Maxwellschen Theorie. Elektronentheorie *Encyklopädie der mathematischen Wissenschaften, mit Einschluss ihrer Anwendungen vol 5. Physik* ed A Sommerfeld (Leipzig: Teubner) part 2, pp 145–280
[149] Dällenbach W 1919 Die allgemein kovarianten Grundgleichungen des elektromagnetischen Feldes im Innern ponderabler Materie vom Standpunkte der Elektronentheorie *Ann. Phys., Lpz* **58** 523–548
[150] Mosengeil K 1907 Theorie der stationären Strahlung in einem gleichförmig bewegten Hohlraum *Ann. Phys., Lpz* **22** 867–904, this is a posthumous work edited by Planck.
See also Liu C 1994 Planck and the special theory of relativity *1993 Meeting of the History of Science Society (Santa Fe, NM)* unpublished paper

[151] Planck M 1907 Zur Dynamik bewegter Systeme *Ann. Phys.* **26** 1–34
[152] Einstein A 1907 Über das Relativitätsprinzip und die aus demselben gezogenen Folgerungen *Jahrb. Radioakt. Elektron.* **4** 411–62, section 15. Einstein never returned to the subject in print, but decades later privately criticized this approach.
For further details, see Liu C 1992 Einstein and relativistic thermodynamics in 1952: a historical and critical study of a strange episode in the history of modern physics *Br. J. Phil. Sci.* **25** 185–206.
[153] For a history of relativistic thermodynamics, see Liu C 1991 Relativistic thermodynamics: its history and foundations *PhD Thesis* University of Pittsburg.
[154] Ott H 1963 Lorentz-Transformation der Wärme und der Temperatur *Z. Phys.* **175** 70–104
[155] For critical reviews of the controversy, see:
Arzeliès H 1967 *Thermodynamique relativiste et quantique* (in two volumes) (Paris: Gauthier-Villars)
Liu C 1991 Relativistic thermodynamics: its history and foundations *PhD Thesis* University of Pittsburg; 1994 Is there a relativistic thermodynamics? A case study of the meaning of special relativity *Studies Hist. Phil. Sci.*
[156] Anderson J L 1964 Relativity principles and the role of coordinates in physics *Gravitation and Relativity* ed H-Y Chiu and W F Hoffmann (New York: Benjamin)
[157] Jüttner 1911 Die Dynamik eines bewegten Gases in der Relativtheorie *Ann. Phys., Lpz* **35** 145–61
[158] For discussions of relativistic kinetic theory, see:
Synge J L 1957 *The Relativistic Gas* (Amsterdam: North-Holland)
Ehlers J 1973 Survey of general relativity theory *Relativity, Astrophysics and Cosmology* ed W Israel (Dordrecht: Reidel) pp 1–125
[159] Wheeler J A and Feynman R 1945 Interaction with the absorber as the mechanism of radiation *Rev. Mod. Phys.* **17** 157–81
[160] For reviews, see:
Havas P 1965 Some basic problems in the formulation of a relativistic statistical mechanics of interacting particles *Statistical Mechanics of Equilibrium and Non-Equilibrium* (Amsterdam: North-Holland) pp 1–19
Ehlers J 1975 Progress in Relativistic Statistical Mechanics, Thermodynamics and Continuum Mechanics *General Relativity and Gravitation* ed G Shaviv and J Rose (New York: Wiley) pp 213–32
[161] Einstein A 1905 Über einen die Erzeugung und Verwandlung des Lichtes betreffenden heuristischen Gesichtspunkt *Ann. Phys.* **17** 132–148; 1916 Strahlungs-Emission und -Absorption nach der Quantentheorie *Deutsche Phys. Ges. Verhand.* **18** 318–23
For surveys of Einstein's work on quantum theory, see:
Pais A 1979 Einstein and the quantum theory *Rev. Mod. Phys.* **51** 863–914
Stachel J 1986 Einstein and the quantum: fifty years of struggle *From Quarks to Quasars: Philosophical Problems of Modern Physics* ed R Colodny (Pittsburgh: University of Pittsburgh Press) pp 349–385
[162] de Broglie L 1924 Recherche sur la théorie des quanta *Thèses présentées à la Faculté des Sciences de l'Université de Paris pour obtenir le grade de docteur ès sciences physiques* (Paris: Masson)
For the origins of quantum mechanics, see Mehra J and Rechenberg H 1982 *The Historical Development of Quantum Theory vol 1. The Quantum Theory of Planck, Einstein and Sommerfeld: Its Foundation and the Rise*

of Its Difficulties 1900–1925 (in 2 parts) (Berlin: Springer); 1982 The Historical Development of Quantum Theory vol 3. The Formulation of Matrix Mechanics and its Modification 1925–1926 (Berlin: Springer); 1984 The Historical Development of Quantum Theory vol 4. Part 1: The Fundamental Equations of Quantum Mechanics 1925–1926. Part 2: The Reception of the New Quantum Mechanics 1925–1926 (Berlin: Springer)

For translations of selected papers, see Van der Waerden B L 1967 Sources of Quantum Mechanics (Amsterdam: North-Holland).

[163] Wessels L 1979 Schrödinger's Route to Wave Mechanics Studies Hist. Phil. Sci. **10** 311–340

Mehra J and Rechenberg H 1987 The Historical Development of Quantum Theory vol 5. Part 2: The Creation of Wave Mechanics: Early Response and Applications (Berlin: Springer) pp 426–34

For Schrödinger's earlier work, see Mehra J and Rechenberg H 1987 The Historical Development of Quantum Theory vol 5. Erwin Schrödinger and the Rise of Wave Mechanics. Part 1: Schrödinger in Vienna and Zurich 1887–1925. Part 2: The Creation of Wave Mechanics: Early Response and Applications (Berlin: Springer)

[164] Born M, Heisenberg W and Jordan P 1926 Zur Quantenmechanik Z. Phys. **35** 557–615 (Engl. Transl. in Van der Waerden B L 1967 Sources of Quantum Mechanics (Amsterdam: North-Holland))

For a discussion, see Mehra J and Rechenberg H 1982 The Historical Development of Quantum Theory vol 3. The Formulation of Matrix Mechanics and its Modification 1925–1926 (Berlin: Springer) pp 149–156.

[165] Jordan P and Pauli W 1928 Zur Quantenelektrodynamik ladungsfreier Felder Z. Phys. **47** 151–73

For an account of the early history of quantum field theory, see Schweber S 1994 QED and the Men Who Made It/ Dyson, Feynman, Schwinger and Tomonaga (Princeton: Princeton University Press) ch 1.

[166] Heisenberg W and Pauli W 1929 Zur Quantendynamik der Wellenfelder Z. Phys. **56** 1–61

[167] For selected papers on quantum electrodynamics, see:

Schwinger J (ed) 1958 Selected Papers on Quantum Electrodynamics (New York: Dover)

Miller A I (ed) 1994 Early Quantum Electrodynamics: a Source Book (Cambridge: Cambridge University Press)

For a history through the 1950s, see Schweber S 1994 QED and the Men Who Made It/ Dyson, Feynman, Schwinger and Tomonaga (Princeton: Princeton University Press).

[168] Dirac P A M 1936 Relativistic wave equations Proc. R. Soc. A **155** 447–59

Fierz M 1939 Über die relativistischer Theorie kräftefreier Teilchen mit beliebigem Spin Acta Phys. Helv. **12** 3–37

Fierz M and Pauli W 1939 On relativistic wave equations for particles of arbitrary spin in an electromagnetic field Proc. R. Soc. A **173** 211–232

Pauli W 1941 Relativistic field theories of elementary particles Rev. Mod. Phys. **13** 203–232 (written in 1939)

Others who worked on this problem include Proca and Petiau (see Kragh H S 1990 Dirac: A Scientific Biography (Cambridge: Cambridge University Press)).

[169] Wigner E P 1939 On unitary representations of the inhomogeneous Lorentz group Ann. Math. **40** 149–204

[170] Pauli W 1940 The connection between spin and statistics Phys. Rev. **58** 716–22

[171] Einstein A 1907 Über das Relativitätsprinzip und die aus demselben gezogenen Folgerungen *Jahrb. Radioakt. Elektron.* **4** 411–62
[172] Abraham M 1912 Zur Theorie der Gravitation *Phys. Z.* **13** 1–5, 176
Mie G 1913 Grundlagen einer Theorie der Materie (Dritte Mitteilung) *Ann. Phys., Lpz* **40** 1–66
Nordström G 1912 Relativitätsprinzip und Gravitation *Phys. Z.* **13** 1126–1129
[173] Nordström G 1913 Zur Theorie der Gravitation vom Standpunkt des Relativitätsprinzips *Ann. Phys., Lpz* **42** 533–554
For a discussion of Nordström's theories, see Norton J 1992 Einstein, Nordström and the early demise of scalar, Lorentz-covariant theories of gravitation *Arch. Hist. Exact Sci.* **45** 17–94.
[174] Einstein A and Fokker A D 1914 Die Nordströmsche Gravitationstheorie vom Standpunkt des absoluten Differentialkalküls *Ann. Phys., Lpz* **44** 321–8
[175] Nordström G 1914 Über die Möglichkeit, das elektromagnetische Feld und das Gravitationsfeld zu vereinigen *Phys. Z.* **15** 504–6 (reprinted with Engl. Transl. by P Freund in Appelquist T, Chodos A and Freund P G O 1987 *Modern Kaluza-Klein Theories* (Menlo Park, CA: Addison-Wesley) pp 51–60).
[176] For reviews of special-relativistic theories of gravitation, see:
Whitrow G J and Murdoch G E 1965 Relativistic theories of gravitation *Vistas in Astronomy* vol 6, ed A Beer (Oxford: Pergamon) pp 1–67
Will C M 1993 *Theory and Experiment in Gravitational Physics* 2nd edn (Cambridge: Cambridge University Press)
[177] For early reviews of the literature on experimental and observational results, see:
Laub J J 1910 Über die experimentellen Grundlagen des Relativitätsprinzips *Jahrb. Radioakt. Elektron.* **7** 405–63
Laue M 1911 *Das Relativitätsprinzip* (Braunschweig: Vieweg) (the first monograph on the special theory)
Pauli W 1921 Relativitätstheorie *Encyklopädie der mathematischen Wissenschaften, mit Einschluss ihrer Anwendungen vol 5: Physik* ed A Sommerfeld (Leipzig: Teubner) part 2, pp 539–775
[178] Kennedy R J and Thorndike E M 1932 *Phys. Rev.* **42** 400
Robertson H P 1949 Postulate versus observation in the special theory of relativity *Rev. Mod. Phys.* **21** 378–82
[179] Ritz W 1908 Recherches critiques sur l'électrodynamique générale *Ann. Chimie Phys.* **13** 145–275
Ehrenfest P 1912 Zur Frage der Entbehrlichkeit des Lichtäthers *Phys. Z.* **13** 317–19
[180] DeSitter W 1913 A Proof of the Constancy of the Velocity of Light *Kon. Neder. Akad. Wet. Amsterdam. Versl. Gewone Vergad. Wisen Natuurkd. Afd.* **15** 1297–98; 1913 On the constancy of the velocity of light *Kon. Neder. Akad. Wet. Amsterdam. Versl. Gewone Vergad. Wisen Natuurkd. Afd.* **16** 395–6
For a review of objections to an emission theory, see Pauli W 1958 *Theory of Relativity* ed G Field (Oxford: Oxford University Press) part 2, pp 539–775, with supplementary notes pp 5–9 (Engl. Transl. of Pauli W 1921 Relativitätstheorie *Encyklopädie der mathematischen Wissenschaften, mit Einschluss ihrer Anwendungen vol 5. Physik* ed A Sommerfeld (Leipzig: Teubner))
[181] See Kaufmann W 1905 Über die Konstitution des Elektrons *König. Preuss.*

Akad. Wiss. (Berlin) Sitzungsber. 949–56 and Kaufman W 1906 Über die Konstitution des Elektrons *Ann. Phys., Lpz* **19** 487–53, which refer to the earlier experiments.

[182] For historical reviews of Kaufmann's work and the ensuing discussion, see:
Cushing J T 1981 Electromagnetic mass, relativity, and the Kaufmann experiments *Am. J. Phys.* **49** 1133–49
Miller A I 1981 *Albert Einstein's Special Theory of Relativity: Emergence (1905) and Early Interpretation (1905–1911)* (Reading: Addison-Wesley) pp 334–52

[183] For reviews of experimental evidence, see the relevant sections of:
Pauli W 1958 *Theory of Relativity* ed G Field (Oxford: Oxford University Press) with supplementary notes (Engl. Transl. of Pauli W 1921 Relativitätstheorie *Encyklopädie der Mathematischen Wissenschaften, mit Einschluss ihrer Anwendungen vol 5. Physik* ed A Sommerfeld (Leipzig: Teubner))
Arzeliès H 1957 *La Dynamique Relativiste et ses Applications* vol 1 (Paris: Gauthier-Villars); 1958 *Milieux Conducteurs et Polarisables en Mouvement* vol 2 (Paris: Gauthier-Villars); 1963 *Electricité Macroscopique et Relativiste* (Paris: Gauthier-Villars); 1966 *Relativistic Kinematics* (Oxford: Oxford University Press); 1966 *Rayonnement et Dynamique du Corpuscule Chargé Fortement Acceleré* (Paris: Gauthier-Villars)
Van Bladel B L 1984 *Relativity and Engineering* (Berlin: Springer)

[184] For surveys of technological applications of the special theory see:
Panofsky W 1980 Special relativity theory in engineering *Some Strangeness in the Proportion: A Centennial Symposium to Celebrate the Achievements of Albert Einstein* ed H Woolf (Reading, MA: Addison-Wesley) pp 94–105
Van Bladel B L 1984 *Relativity and Engineering* (Berlin: Springer)

[185] Cockcroft J and Walton E T S 1932 Experiments with high velocity positive ions. II. The disintegration of elements by high velocity protons *Proc. R. Soc.* A **137** 229–42

[186] Bethe H A 1939 On energy generation in stars *Phys. Rev.* **55** 434–56

[187] After coming close to such a generalization (see Abraham M 1912 Zur Theorie der Gravitation *Phys. Z.* **13** 1–5, 176), Abraham abandoned relativity entirely, attacking Einstein for corrupting young physicists (see Abraham M 1912 Relativität und Gravitation. Erwiderung auf eine Bemerkung des Hrn A Einstein *Ann. Phys., Lpz* **38** 1056–1058).
For an account of the Abraham–Einstein controversy, see Cattani C and De Maria M 1989 Max Abraham and the reception of relativity in Italy: his 1912 and 1914 controversies with Einstein 1989 *Einstein Studies vol 1. Einstein and the History of General Relativity* ed D Howard and J Stachel (Boston: Birkhäuser) pp 160–174.

[188] For accounts by Einstein of the development of general relativity, see Einstein A 1921 A brief outline of the development of the theory of relativity *Nature* **106** 782–4; 1933 *Origins of the General Theory of Relativity (Glasgow University Publications)* vol 30 (Glasgow: Jackson, Wylie) (German original Seelig C (ed) 1981 Einiges über die Entstehung der allgemeinen Relativitätstheorie *Mein Weltbild* (Frankfurt: Ullstein Materialien) pp 134–8); 1949 *Autobiographical Notes* (LaSalle: Open Court). First published in Schilpp P A (ed) 1949 *Albert Einstein: Philosopher-Scientist* (LaSalle: Open Court) pp 2–94; 1982 How I created the theory of relativity *physics Today* **35** 45–47 (Translation of notes by Jun Ishiwara of Einstein's 1922 Kyoto lecture, Japanese text in Ishiwara J 1971 *Einstein Kyozyu-Koen-roku* (Tokyo: Kabushika Kaisha) pp 78–88).

For discussions of Einstein's work, see:
Pais A 1982 *Subtle is The Lord, The Science and the Life of Albert Einstein* (Oxford: Oxford University Press) section IV
Stachel J 1982 The genesis of general relativity *Einstein Symposion Berlin aus Anlass der 100. Wiederkehr seines Geburtstages* ed H Nelkowski *et al* (Berlin: Springer)
Torretti R 1983 *Relativity and Geometry* (Oxford: Pergamon)

[189] For an account of the development of general relativity, see Vizgin V P 1981 *Relyativistskaya Teoriya Tyagoteniya (Istori i Formirovanie, 1900–1915)* (Moscow).
See also the proceedings of three conferences on the history of general relativity:
Howard D and Stachel J (ed) 1989 *Einstein Studies vol 1. Einstein and the History of General Relativity* (Boston: Birkhäuser)
Eisenstaedt J and Kox A J (ed) 1992 *Einstein Studies vol 3. Studies in the History of General Relativity* (Boston: Birkhäuser)
Earman J, Jansen M and Norton J D (ed) 1993 *Einstein Studies vol 5. The Attraction of Gravitation/ New Studies in the History of General Relativity* (Boston: Birkhäuser)

[190] Stachel J 1989 Einstein's search for general covariance, 1912–1915 *Einstein Studies vol 1. Einstein and the History of General Relativity* ed D Howard and J Stachel (Boston: Birkhäuser) p 63

[191] Einstein A 1989 *The Collected Papers of Albert Einstein vol 2. The Swiss Years: Writings 1901–1909* (Princeton: Princeton University Press) p 254

[192] For an account of this attempt, see Einstein A 1933 *Origins of the General Theory of Relativity (Glasgow University Publications)* vol 30 (Glasgow: Jackson, Wylie) (German original Seelig C (ed) 1981 Einiges über die Entstehung der allgemeinen Relativitätstheorie *Mein Weltbild* (Frankfurt: Ullstein Materialien)).

[193] Eötvös R V 1889 Über die Anziehung der Erde auf verschiedene Substanzen *Math. Naturwiss. Ber. Ungarn* **8** 65–8

[194] Einstein A 1913 Zum gegenwärtigen Stand des Gravitationsproblems *Phys. Z.* **14** 1249–66
For an account of Einstein's reaction to Nordström's theory, see Norton J 1992 Einstein, Nordström and the early demise of scalar, Lorentz-covariant theories of gravitation *Arch. Hist. Exact Sci.* **45** 17–94

[195] Einstein A 1912 Gibt es ein Gravitationswirkung, die der dynamischen Induktionswirkung analog ist? *Vierteljahrschr. Gerichliche Med. Öffentlich. Sanitätswesen* **44** 37–40

[196] Einstein A 1918 Prinzipielles zur allgemeinen Relativitätstheorie *Ann. Phys., Lpz* **55** 241–4

[197] Einstein A 1917 Kosmologische Betrachtungen zur allgemeinen Relativitätstheorie *König. Preuss. Akad. Wiss. (Berlin) Sitzungsber.* 142–52. See Chapter 23 for a discussion of relativistic cosmology.

[198] Einstein A 1907 Über das Relativitätsprinzip und die aus demselben gezogenen Folgerungen *Jahrb. Radioakt. Elektron.* **4** 411–62

[199] For a discussion of what Einstein meant by the equivalence principle see Norton J 1985 What was Einstein's principle of equivalence? *Studies Hist. Phil. Sci.* **16** pp 203–46.

[200] On Freundlich (later Findlay-Freundlich) and his contributions to the experimental tests of the gravitational red shift and so-called light deflection, see: Hentschel K 1992 *Der Einstein Turm. Erwin F. Freundlich und die Relativitätstheorie—Ansätze zu einer dichten*

Beschreibung von institutionellen, biographischen und theoriengeschichtlichen Aspekten (Heidelberg: Spektrum Akademische); 1994 Erwin Finlay-Freundlich and testing Einstein's theory of relativity *Arch. Hist. Exact Sci.* **47** 143–201.

[201] For accounts of attempts to test the gravitational red shift, see:

Forbes E G 1961 A history of the solar red shift problem *Ann. Sci.* **17** 129–164

Earman J and Glymour C 1980 The gravitational red shift as a test of general relativity: history and analysis *Studies Hist. Phil. Sci.* **11** 175–214

Hentschel K 1992 Grebe/Bachems photometrische Analyse der Linienprofile und die Gravitations-Rotverschiebung: 1919 bis 1922 *Ann. Sci.* **49** 21–46; 1992 *Der Einstein Turm. Erwin F Freundlich und die Relativitätstheorie-Ansïze zu einer dichten Beschreibung von institutionellen, biographischen und theoriengeschichtlichen Aspekten* (Heidelberg: Spektrum Akademische)

[202] Hentschel K 1993 The Conversion of St John: A Case Study in the Interplay of Theory and Experiment *Einstein in Context*, special issue of *Science in Context* vol 6, ed M Beller *et al* no 1 pp 137–94.

[203] Pound R V and Rebka G A 1960 Apparent weight of photons *Phys. Rev. Lett.* **4** 788–803

[204] Einstein A 1911 Über den Einfluss der Schwerkraft auf die Ausbreitung des Lichtes *Ann. Phys., Lpz* **35** 898–908

[205] For accounts of early attempts to use eclipse data to test Einstein's predictions see:

Crelinsten J 1983 William Wallace Campbell and the 'Einstein Problem': An observational astronomer confronts the theory of relativity *Hist. Studies Phys. Sci.* **14** 1–91

Earman J and Glymour C 1980 Relativity and eclipses: The British eclipse expeditions of 1919 and their predecessors *Hist. Studies Phys. Sci.* **11** 49–85

Stachel J 1986 Eddington and Einstein *The Prism of Science, The Israel Colloquium: Studies in History, Philosophy and Sociology of Science* 2 ed E Ullmann-Margalit (Boston: Reidel) pp 225–250

[206] Einstein A 1912 Lichtgeschwindigkeit und Statik des Gravitationsfeldes *Ann. Phys., Lpz* **38** 355–369; 1912 Zur Theorie des statischen Gravitationsfeldes *Ann. Phys., Lpz* **38** 443–58

[207] Einstein A 1912 Zur Theorie des statischen Gravitationsfeldes *Ann. Phys., Lpz* **38** 443–58

[208] For a discussion, see Stachel J 1980 Einstein and the rigidly rotating disc *General Relativity and Gravitation One Hundred Years After the Birth of Albert Einstein* (in two volumes) ed A Held (New York: Plenum) pp 1–15

[209] Einstein seems to have been helped by earlier familiarity with at least some elements of the differential-geometric tradition in nineteenth century mechanics. For this tradition, see Lützen J 1993 *Interactions Between Mechanics and Differential Geometry in the 19th Century* (Københavns Universitet Matematisk Institut, Preprint Series, No. 25).

For Einstein's contacts with this tradition, see Einstein's Research Notes on a Generalized Theory of Relativity in Einstein A 1995 *The Collected Papers of Albert Einstein vol 4. The Swiss Years: Writings 1912–1914* ed M Klein *et al* (Princeton: Princeton University Press)

[210] For reminiscences of his collaboration with Grassmann, on this topic, see Einstein A 1955 Erinnerungen-Souvenirs *Schweizerische*

Hochschulzeitung (Sonderheft) pp 145–153 (reprinted as 1956 Autobiographische Skizze *Helle Zeit–Dunkle Zeit. In Memoriam Albert Einstein* ed C Seelig (Zurich: Europa)).

For a history of nineteenth century geometry, see Torretti R 1978 *Philosophy of Geometry from Riemann to Poincaré* (Dordrecht: Reidel).

For tensor analysis, see Reich K 1994 *Die Entwicklung des Tensorkalküls. Vom absoluten Differentialkalkül zur Relativitätstheorie* (Basel: Birkhäuser).

[211] Levi-Civita T 1917 Nozione di parallelismo in una varieta qualunque *Circolo Matematico di Palermo. Rendiconti* **42** 173–205

[212] Weyl H 1918 *Raum, Zeit, Materie* (Berlin: Springer)

[213] Einstein A and Grossmann M 1913 *Entwurf einer verallgemeinerten Relativitätstheorie und einer Theorie der Gravitation. I. Physikalischer Teil von Albert Einstein. II. Mathematischer Teil von Marcel Grossmann* (Leipzig: Teubner)

[214] Stachel J 1986 What a physicist can learn from the discovery of general relativity *Proc. Fourth Marcel Grossmann Meeting on General Relativity* ed R Ruffini (Amsterdam: Elsevier) pp 1857–1862

[215] Hilbert D 1915 Die Grundlagen der Physik (Erste Mitteilung) *König. Ges. Wiss. Göttingen. Math.-Phys. Klasse Nach.* 395–407

[216] See, for example:
Earman J and Glymour C 1978 Lost in the tensors: Einstein's struggles with covariance principles 1912–1916 *Studies Hist. Phil. Sci.* **9** 251–78
Hoffmann B 1972 Einstein and tensors *Tensor* **6** 157–162
Lanczos C 1972 Einstein's Path From Special to General Relativity *General Relativity: Papers in Honour of J L Synge* ed L O'Raifertaigh (Oxford: Clarendon) pp 5–19
Mehra J 1974 *Einstein, Hilbert and the Theory of Gravitation* (Dordrecht: Reidel)
Stachel J 1989 Einstein's search for general covariance, 1912–1915 *Einstein Studies vol 1. Einstein and the History of General Relativity* ed J Howard and J Stachel (Boston: Birkhäuser)
Vizgin V P and Smorodinskii Ya A 1979 From the Equivalence Principle to the Equations of Gravitation *Sov. Phys.–Usp.* **22** 489–513

[217] Norton J 1984 How Einstein found his field equations: 1912–1915 *Hist. Studies Phys. Sci.* **14** 253–316

[218] Stachel J 1989 Einstein's search for general covariance, 1912–1915 *Einstein Studies vol 1. Einstein and the History of General Relativity* ed J Howard and J Stachel (Boston: Birkhäuser) pp 66–68
Einstein A and Fokker A D 1913 Die Nordströmsche Gravitationstheorie vom Standpunkt des absoluten Differentialkalküls *Ann. Phys., Lpz* **44** 321–328 notes, without further comment, that the original argument against general covariance is wrong.

[219] Stachel J 1989 Einstein's search for general covariance, 1912–1915 *Einstein Studies vol 1. Einstein and the History of General Relativity* ed J Howard and J Stachel (Boston: Birkhäuser)
Norton J 1984 How Einstein found his field equations: 1912–1915 *Hist. Studies Phys. Sci.* **14** 253–316

[220] Stachel J 1986 What a physicist can learn from the discovery of general relativity *Proc. Fourth Marcel Grossmann Meeting on General Relativity* ed R Ruffini (Amsterdam: Elsevier) pp 1857–1862; 1987 How Einstein discovered general relativity: a historical tale with some contemporary morals *General Relativity and Gravitation: Proc. 11th Int. Conf. on General Relativity and Gravitation* ed M A H MacCallum (Cambridge: Cambridge University Press) pp 200–208; 1989 Einstein's

Search for General Covariance, 1912–1915 *Einstein Studies vol 1. Einstein and the History of General Relativity* ed J Howard and J Stachel (Boston: Birkhäuser); 1993 The meaning of general covariance: the hole story *Philosophical Problems of the Internal and External Worlds/Essays on the Philosophy of Adolf Grünbaum* ed J Earman *et al* (Konstanz: Universitätsverlag/Pittsburgh: University of Pittsburgh Press) pp 129–160

[221] In addition to the works on the hole argument cited in [220] see Howard D and Norton J 1993 Out of the Labyrinth? Einstein, Hertz, and the Goettingen answer to the hole argument *Einstein Studies vol 5. The Attraction of Gravitation/ New Studies in the History of General Relativity* ed J Earman *et al* (Boston: Birkhäuser) pp 30–62.

[222] For his claims, see Einstein A 1913 Zum gegenwartigen Stand des Gravitationsproblems *Phys. Z.* **14** 1249–66

For a discussion of Levi-Civita's role in convincing him that his proof was erroneous, see Cattani C and de Maria M 1989 Max Abraham and the reception of relativity in Italy: his 1912 and 1914 controversies with Einstein *Einstein Studies vol 1. Einstein and the History of General Relativity* ed D Howard and J Stachel (Boston: Birkhäuser) pp 160–174.

[223] For details of this story, see:

Norton D 1984 How Einstein found his field equations: 1912–1915 *Hist. Studies Phys. Sci.* **14** 253–316

Stachel J 1989 Einstein's Search for General Covariance, 1912–1915 *Einstein Studies vol 1. Einstein and the History of General Relativity* ed J Howard and J Stachel (Boston: Birkhäuser)

Einstein A 1916 Die Grundlagen der allgemeinen Relativitätstheorie *Ann. Phys., Lpz* **49** 769–822 summarizes his definitive formulation of the theory.

[224] Weyl H 1944 How far can one get with a linear field theory of gravitation in flat space-time? *Am. J. Math.* **66** 591

[225] Kraichnan R H 1955 Special-relativistic derivation of generally covariant gravitation theory *Phys. Rev.* **55** 1118–22

[226] Feynman independently suggested a similar way to discover general relativity if you are not Einstein (see DeWitt C (ed) 1957 *Conf. on the Role of Gravitation in Physics, Proc. W.A.D.C.* Technical Report 57–216; ASTIA Document No. AD 118180 (Wright Air Development Center, Wright-Patterson Air Force Base, OH)).

[227] Cartan E 1923 Sur les variétés à connection affine et la théorie de la relativité généralisée *Ecole Normale Supérieure (Paris) Ann.* **40** 325–412

[228] Freidrichs K 1927 Eine invariante Formulierung des Newtonschen Gravitationsgesetzes und des Grenzüberganges vom Einsteinschen zum Newtonschen Gesetz *Math. Ann.* **98** 566–75

[229] For some recent work, see:

Howard D and Stachel J (ed) 1989 *Einstein Studies vol 2. Einstein and the History of General Relativity* (Boston: Birkhäuser)

Eisenstaedt J and Kox A (ed) 1992 *Einstein Studies vol 3. Studies in the History of General Relativity* (Boston: Birkhäuser)

Earman J, Jansen M and Norton D (ed) 1993 *Einstein Studies vol 5. The Attraction of Gravitation/New Studies in the History of General Relativity* (Boston: Birkhäuser)

[230] A good deal of historical information may be found in some treatises on the general theory. See, for example:

Arzeliès H 1961 *Relativité généralisée. Gravitation vol 1. Principes généraux* (Paris: Gauthier-Villars); 1963 *Relativité généralisée. Gravitation vol 2. Le champ de Schwarzschild* (Paris: Gauthier-Villars)

Pauli W 1958 *Theory of Relativity* ed G Field (Oxford: Oxford University Press) with supplementary notes (Engl. Trans. of Pauli W 1921 Relativitätstheorie *Encyklopädie der Mathematischen Wissenschaften, mit Einschluss ihrer Anwendungen vol 5. Physik* ed A Sommerfeld (Leipzig: Teubner))

Misner C *et al* 1973 *Gravitation* (San Francisco: Freeman)

See also, Lecat M 1924 *Bibliographie de la Relativité* (Bruxelles: Lambertin) for a bibliography up to that date.

[231] For the consolidation of this community, see Jungnickel C and McCormmach R 1986 *Intellectual Mastery of Nature vol 2. The Now Mighty Theoretical Physics 1870–1925* (Chicago: University of Chicago Press).

[232] Einstein A 1919 Spielen Gravitationsfelder im Aufbau der materiellen Elementarteilchen eine wesentliche Rolle? *Preuss. Akad. Wiss. (Berlin) Sitzungsber.* 349–56

[233] Hendry J 1984 *The Creation of Quantum Mechanics and the Bohr–Pauli Dialogue* (Dordrecht: Reidel) ch 2

[234] Eisenstaedt J 1986 La relativité générale a l'étiage: 1925–1955 *Arch. Hist. Exact Sci.* **35** 115–85; 1989 The low water mark of general relativity *Einstein Studies vol 1. Einstein and the History of General Relativity* ed D Howard and J Stachel (Boston: Birkhäuser) pp 277–92. Its influence on mathematics remained much greater.

[235] Mercier A 1956 *Fünfzig Jahre Relativitätstheorie (Helv. Phys. Acta Supplement IV)* (Basel: Birkhäuser)

[236] Goldberg J 1992 US Air Force Support of General Relativity: 1956–1972 *Einstein Studies vol 3. Studies in the History of General Relativity* ed J Eisenstaedt and A J Kox (Boston: Birkhäuser) pp 89–102

[237] Abramovici A *et al* 1992 LIGO: The Laser Interferometer Gravitational-Wave Observatory *Science* **256** 325–33

Thorne K S 1994 LIGO, VIRGO, and the international network of laser-interferometer gravitational wave detectors *Proc. Eighth Nishinomiya Yukawa Symposium on Relativistic Cosmology* ed M Sasaki (Tokyo: Universal Academic)

[238] Taylor J H and Weisberg J M 1989 Further experimental tests of relativistic gravity using the binary pulsar PRS 1913+16 *Astrophys. J.* **345** 434–50

[239] Weyl H 1918 Reine Infinitesiomalgeometrie *Math. Z.* **2** 384–411. Weyl's work drew on the concept of parallel displacement introduced by Levi-Civita (1917) and others. For an account of the development of the concepts of parallel displacement and affine connection, see Reich K 1992 Levi-Civitasche Parallelverschiebung, affiner Zusammenhang, Uebertragungsprinzip: 1916/17–1922/23 *Arch. Hist. Exact Sci.* **44** 77–105.

[240] Infeld L and van der Waerden B L 1933 Die Wellengleichung des Elektrons in der Allgemeinen Relativitätstherorie *Preuss. Akad. Wiss. (Berlin) Phys.-Math. Klasse Sitzungsber.* 380–401

For the applications of spinorial techniques in relativity, see Penrose R and Rindler W 1984 *Spinors and Space-Time vol 1. Two-spinor calculus and relativistic fields* (Cambridge: Cambridge University Press); 1986 *Spinors and Space-Time vol 2. Spinor and Twistor Methods in Space-Time Geometry* (Cambridge: Cambridge University Press).

[241] For a review, see Penrose R and Rindler W 1986 *Spinors and Space-Time vol 2. Spinor and Twistor Methods in Space-Time Geometry* (Cambridge: Cambridge University Press)

[242] For reviews, see:

Hermann R 1975 *Gauge Fields and Cartan-Ehresmann Connections, Part A* (Brookline: Math Sci)

Hermann R 1978 *Yang-Mills, Kaluza-Klein, and the Einstein Program* (Brookline: Math Sci)

Trautman A 1980 Fiber Bundles, Gauge Fields, and Gravitation *General Relativity and Gravitation One Hundred Years After the Birth of Albert Einstein* vol 1, ed A Held (New York: Plenum) pp 287–308

[243] For reviews of complex extensions of Riemannian geometry, see:
Flaherty E J 1976 *Hermitian and Kählerian Geometry in Relativity* (Berlin: Springer)
Held A (ed) 1980 Complex variables in relativity *General Relativity and Gravitation One Hundred Years After the Birth of Albert Einstein* vol 2 (New York: Plenum) pp 207–240

[244] Ashtekar A 1991 *Lectures on Non-Perturbative Canonical Gravity* (Singapore: World Scientific)

[245] Weyl H 1918 *Raum, Zeit, Materie* (Berlin: Springer) which went through five editions with substantial additions up to Weyl H 1923 *Raum-Zeit-Materie. Fünfte, umgearbeitete Auflage* (Berlin: Springer).
For discussions of Weyl's role in the development of general relativity, see: Sigurdsson S 1991 Hermann Weyl, mathematics and physics, 1900–1927 *PhD Dissertation* Harvard University; 1994 Unification, geometry and ambivalence: Hilbert, Weyl and the Goettingen community *Current Trends in the Historiography of Science* ed K Gavroglu (Dordrecht: Reidel)

[246] It was actually Einstein (see Einstein A 1925 Einheitliche Feldtheorie von Gravitation und Elektrizität *Preuss. Akad. Wiss. (Berlin) Phys.-Math. Klasse Sitzungsber.* 414–419) who first introduced independent variations of the connection (see Cattani C 1993 Levi-Civita's influence on Palatini's contribution to general relativity *Einstein Studies vol 5. The Attraction of Gravitation/ New Studies in the History of General Relativity* J Earman *et al* (Boston: Birkhäuser) pp 206–22).

[247] Weyl H 1921 Zur Infinitesimalgeometrie. Einordnung der projektiven und konformen Auffassung *Könij. Ges. Wiss. Göttingen Math.-Phys. Klasse Nach.* 99–112

[248] Ehlers J, Pirani F A E and Schild A 1972 The geometry of free fall and light propagation *General Relativity. Paper in Honour of J L Synge* ed L O'Raifertaigh (Oxford: Clarendon) pp 63–84
Ehlers J 1973 Survey of general relativity theory *Relativity, Astrophysics and Cosmology* ed W Israel (Dordrecht: Reidel) pp 1–125

[249] Einstein A 1916 Hamiltonsches Prinzip und allgemeine Relativitätstheorie *König. Preuss. Akad. Wiss. (Berlin) Sitzungsber.* 1111–16
For discussion of variational principles leading to the Einstein equations, see Kichenassamy S 1993 Variational derivations of Einstein's equations *Einstein Studies vol 5. The Attraction of Gravitation/ New Studies in the History of General Relativity* ed J Earman *et al* (Boston: Birkhäuser) pp 185–205.

[250] For an account of these criticisms, see Cattani C and De Maria M 1993 Conservation laws and gravitational waves in general relativity (1915–1918) *Einstein Studies vol 5. The Attraction of Gravitation/ New Studies in the History of General Relativity* ed J Earman *et al* (Boston: Birkhäuser) pp 63–87.

[251] Einstein A 1918 Die Energiesatz in der allgemeinen Relativitätstheorie *König. Preuss. Akad. Wiss. (Berlin) Sitzungsber.* 448–59

[252] For reviews, see:
Goldberg J 1980 Invariant transformations, conservation laws, and

energy–momentum *General Relativity and Gravitation One Hundred Years After the Birth of Albert Einstein* vol 1, ed A Held (New York: Plenum) pp 469–489

Winicour J 1980 Angular momentum in general relativity *General Relativity and Gravitation One Hundred Years After the Birth of Albert Einstein* vol 2, ed A Held (New York: Plenum) pp 71–96

[253] See, for example, Komar A 1959 Covariant conservation laws in general relativity *Phys. Rev.* **113** 934–6

[254] Freud P von 1939 Über die Ausdrücke der Gesamtenergie und des Gesamtimpulses eines materiellen Systems in der allgemeinen Relativitätstheorie *Ann. Math.* **40** 417

[255] Landau L D and Lifshitz E M 1941 *Teoriya Polya* (Moscow: Nauka)

[256] Goldberg J 1958 Conservation laws in general relativity *Phys. Rev.* **111** 315–25

[257] For the ADM mass, see Arnowitt R, Deser C and Misner C 1962 The Dynamics of General Relativity, *Gravitation: An Introduction to Current Research* ed L Witten (New York: Wiley) pp 227–65.

For the Bondi mass, see:

Bondi H, van der Burg M G J and Metzner A W K 1962 Gravitational waves in general relativity. VII. Waves from axi-symmetric isolated systems *Proc. R. Soc.* A **269** 21–52

Sachs R K 1962 Gravitational waves in general relativity. VIII. Waves in asymptotically flat space-time *Proc. R. Soc.* A **270** 103–26

For invariant definitions of, and the relation between, both, see:

Winicour J 1968 Some total invariants of asymptotically flat space-time *J. Math. Phys.* **9** 861–867

Ashtekar A and Hansen R O 1978 A unified treatment of null and spatial infinity in general relativity. I. Universal structure, asymptotic symmetries, and conserved quantities at spatial infinity *J. Math. Phys.* **19** 1542–66

[258] Infeld L and Plebanski J 1960 *Motion and Relativity* (New York: Pergamon) pp 147–9, 153–5

[259] For a review, see d'Inverno R A 1980 A review of algebraic computing in general relativity *General Relativity and Gravitation One Hundred Years After the Birth of Albert Einstein* vol 1, ed A Held (New York: Plenum) pp 491–537.

[260] For a review, see Petrov A Z 1969 *Einstein Spaces* (Oxford: Pergamon)

[261] For a survey of exact solutions, see Kramer D, Stephani H, MacCallum M and Herlt E 1980 *Exact Solutions of the Einstein Field Equations* (Berlin: Deutscher Verlag der Wissenschaften)

[262] Einstein A 1916 Näherungsweise Integration der Feldgleichungen der Gravitation *König. Preuss. Akad. Wiss. (Berlin) Sitzungsber.* 688–96; 1918 Über Gravitationswellen *König. Preuss. Akad. Wiss. (Berlin) Sitzungsber.* 154–67

[263] Lorentz H A and Droste J 1917 Lorentz, Hendrik Antoon and Droste, Johannes, De beweging van een stelsel lichamen onder den invloed van hunne onderlinge aantrekking, behandeld volgens de theorie van Einstein *Kon. Neder. Akad. Wet. Amsterdam. Versl. Gewone Vergad. Wisen Natuurkd. Afd.* **26** 392–403, 649–60

[264] Bondi H, van der Burg M G J and Metzner A W K 1962 Gravitational waves in general relativity. VII. Waves from axi-symmetric isolated systems *Proc. R. Soc.* A **269** 21–52

[265] Isaacson R 1968 Gravitational radiation in the limit of high frequency, I.

The linear approximation and geometrical optics *Phys. Rev.* **166** 1263–71; 1968 II. Nonlinear terms and the effective stress tensor *Phys. Rev.* **166** 1272–1280

[266] For a discussion of early work on the problem of motion, see Havas P 1989 The early history of the 'problem of motion' in general relativity *Einstein Studies vol 1. Einstein and the History of General Relativity* ed D Howard and J Stachel (Boston: Birkhäuser) pp 234–76.

For discussions of the equations of motion problem that include accounts of later work, see:

Damour T 1987 The problem of motion in Newtonian and Einsteinian gravity *300 Years of Gravitation* ed S Hawking and W Israel (Cambridge: Cambridge University Press) pp 128–198

Havas P 1979 Equations of motion and radiation reaction in the special and general theories of relativity *Isolated Gravitating Systems in General Relativity* ed J Ehlers (Amsterdam: North-Holland) pp 74–155

[267] Einstein A and Grommer J 1927 Allgemeine Relativitätstheorie und Bewegungsgesetz *Preuss. Akad. Wiss. (Berlin) Phys.-Math. Klasse. Sitzungsber.* 2–13

[268] Einstein A, Infeld L and Hoffmann B 1938 The gravitational equations and the problem of motion *Ann. Math.* **39** 65–100

[269] For the development of the EIH method, see Havas P 1989 The early history of the 'problem of motion' in general relativity *Einstein Studies vol 1. Einstein and the History of General Relativity* ed D Howard and J Stachel (Boston: Birkhäuser) pp 234–276.

[270] Fock V 1939 Sur les mouvements des masses finies d'après la théorie de gravitation einsteinienne *J. Physique* **1** 81–116

[271] For a review of the work of Infeld and his school, see Infeld L and Plebanski J 1960 *Motion and Relativity* (New York: Pergamon)

[272] Einstein A 1915 Erklärung der Perihelbewegung des Merkur aus der allgemeinen Relativtätstheorie *König. Preuss. Akad. Wiss. (Berlin) Sitzungsber.* 831–9

For a history of the perihelion problem, see Roseveare N T 1982 *Mercury's Perihelion from LeVerrier to Einstein* (Oxford: Clarendon).

For Einstein's work on it, see Earman J and Janssen M 1993 Einstein's explanation of the motion of Mercury's perihelion *Einstein Studies vol 5. The Attraction of Gravitation/ New Studies in the History of General Relativity* ed J Earman *et al* (Boston: Birkhäuser) pp 129–172.

[273] Schwarzschild K 1916 Über das Gravitationsfeld eines Massenpunktes nach der Einsteinschen Theorie *König. Preuss. Akad. Wiss. (Berlin) Sitzungsber.* 189–96

Droste J 1916 The field of a single centre in Einstein's theory of gravitation, and the motion of a particle in that field *Kon. Neder. Akad. Wet. Amsterdam. Versl. Gewone Vergad. Wisen Natuurkd. Afd.* **19** 197–215

For this and later work on the Schwarzschild solution, see Eisenstaedt J 1982 Histoire et singularités de la solution de Schwarzschild (1915–1923) *Arch. Hist. Exact Sci.* **27** 157–98; 1987 Trajectoires et impasses de la solution de Schwarzschild *Arch. Hist. Exact Sci.* **37** 275–357; 1989 The early interpretation of the Schwarzschild solution *Einstein Studies vol 1. Einstein and the History of General Relativity* ed D Howard and J Stachel (Boston: Birkhäuser) pp 213–33.

[274] Dicke R H and Goldenberg H M 1967 Solar oblateness and general relativity *Phys. Rev. Lett.* **18** 313–6. Dicke's attempt to invalidate general relativity was motivated by his rival scalar-tensor theory of gravitation.

[275] Will C M 1991 *Theory and Experiment in Gravitational Physics* 2nd edn (Cambridge: Cambridge University Press) pp 181–3, 334
[276] Crelinsten J 1983 William Wallace Campbell and the Einstein problem: an observational astronomer confronts the theory of relativity *Hist. Studies Phys. Sci.* **14** 1–91
Earman J and Glymour C 1980 Relativity and eclipses: The British eclipse expeditions of 1919 and their predecessors *Hist. Studies Phys. Sci.* **11** 49–85
[277] Moyer D 1979 Revolution in science: the 1919 eclipse test of general relativity *On the Path of Albert Einstein* ed A Perlmutter and L F Scott (New York:) pp 55–101
Earman J and Glymour C 1980 Relativity and eclipses: the British eclipse expeditions of 1919 and their predecessors *Hist. Studies Phys. Sci.* **11** 49–85
[278] Hentschel K 1992 *Der Einstein Turm. Erwin F. Freundlich und die Relativitätstheorie-Ansätze zu einer dichten Beschreibung von institutionellen, biographischen und theoriengeschichtlichen Aspekten* (Heidelberg: Spektrum Akademische)
[279] Will C M 1993 *Theory and Experiment in Gravitational Physics* 2nd edn (Cambridge: Cambridge University Press)
[280] Shapiro I I 1990 Fourth test of general relativity *Phys. Rev. Lett.* **13** 789–91
[281] Eisenstaedt J 1982 Histoire et singularités de la solution de Schwarzschild (1915–1923) *Arch. Hist. Exact Sci.* **27** 157–98; 1989 The early interpretation of the Schwarzschild solution *Einstein Studies vol 1. Einstein and the History of General Relativity* ed D Howard and J Stachel (Boston: Birkhäuser) pp 213–233
[282] Eddington A S 1924 A comparison of Whitehead's and Einstein's formulas *Nature* **113** 192
[283] Lemaître G 1932 L'univers en expansion *Publication du Laboratoire d'Astronomie et de Geodesié de L'Université de Louvain* **9** 171–205
For a discussion of Lemaître's work, see Eisenstaedt J 1993 Lemaître and the Schwarzschild solution *Einstein Studies vol 5. The Attraction of Gravitation/ New Studies in the History of General Relativity* ed J Earman (Boston: Birkhäuser) pp 353–89
[284] Synge J L 1950 The gravitational field of a particle *R. Ir. Acad. Proc.* A **53** 83–114
[285] Einstein A and Rosen N 1935 The particle problem in general relativity *Phys. Rev.* **48** 73–7
[286] Klein F 1918 Über die Integralform der Erhaltungssätze und die Theorie der räumlich geschlossene Welt *König. Ges. Wiss. Göttingen. Math.-Phys. Klasse Nach.* 394–423
[287] Einstein A 1939 On a stationary system with spherical symmetry consisting of many gravitating masses *Ann. Math.* **40** 922–36
[288] Oppenheimer J R and Snyder H 1939 On continued gravitational contraction *Phys. Rev.* **56** 455–9
[289] Finkelstein D 1958 Past-future asymmetry of the gravitational field of a point particle *Phys. Rev.* **110** 965–7
[290] Kruskal M 1960 Maximal extension of Schwarzschild metric *Phys. Rev.* **119** 1743–5
[291] Szekeres G 1960 On the singularities of a Riemannian manifold *Publ. Math. (Debrecen)* **7** 285–301
[292] Wheeler J A 1968 Our Universe: the known and the unknown *Am. Sci.* **56** 1

[293] For an account of the controversies about gravitational collapse, black holes and the story of their naming, see Thorne K S 1994 *Black Holes and Time Warps: Einstein's Outrageous Legacy* (New York: Norton).
For pre-relativistic ideas about dark stars, see Israel W 1987 Dark Stars: the evolution of an idea *300 Years of Gravitation* ed S Hawking and W Israel (Cambridge: Cambridge University Press) pp 199–276.

[294] For accounts of work on gravitational collapse and black holes, see:
Novikov I D and Frolov V P 1989 *Physics of Black Holes* (Dordrecht: Kluwer Academic)
Thorne K S 1994 *Black Holes and Time Warps: Einstein's Outrageous Legacy* (New York: Norton)

[295] Lifshitz E M and Khalatnikov I M 1960 On the singularities of cosmological solutions of the gravitational equations. I *Zh. Eksp. Teor. Fiz.* **39** 149

[296] Penrose R 1965 Gravitational collapse and space-time singularities *Phys. Rev. Lett.* **14** 57–9

[297] For a review, see Hawking S and Ellis G F R 1973 *The Large Scale Structure of Space-Time* (Cambridge: Cambridge University Press)

[298] For a review, see Tipler *et al* 1980 Singularities and horizons—a review article *General Relativity and Gravitation One Hundred Years After the Birth of Albert Einstein* vol 2, ed A Held (New York: Plenum) pp 97–206

[299] Khalatnikov I M and Lifshitz E M 1970 The general cosmological solution of the gravitational equations with a singularity in time *Phys. Rev. Lett.* **24** 76–9

[300] Einstein A 1916 Näherungsweise Integration der Feldgleichungen der Gravitation *König. Preuss. Akad. Wiss. (Berlin) Sitzungsber.* 688–696; 1918 Über Gravitationswellen *König. Preuss. Akad. Wiss. (Berlin) Sitzungsber.* 154–67

[301] For an extensive review of the properties of gravitational radiation and prospects for its detection, see Thorne K S 1987 Gravitational radiation *300 Years of Gravitation* ed S Hawking and W Israel (Cambridge: Cambridge University Press) pp 330–458.

[302] Eddington A S 1923 The propagation of gravitational waves *Proc. R. Soc.* A **102** 268–82

[303] Pirani F A E 1957 Invariant formulation of gravitational radiation theory *Phys. Rev.* **105** 1089–99

[304] Bondi H, Pirani F A E and Robinson I 1959 Gravitational waves in general relativity. III. Exact plane waves *Proc. R. Soc.* A **251** 519–33

[305] Fock V A 1955 *Teoriya Prostranstva Vremeni i Tyagoteniya* (Moscow: Fizmatgiz)

[306] For a discussion of the problem of motion and radiation reaction in general relativity, see Havas P 1979 Equations of motion and radiation reaction in the special and general theories of relativity *Isolated Gravitating Systems in General Relativity* ed J Ehlers (Amsterdam: North-Holland) pp 74–155.

[307] Bondi H, van der Burg M G J and Metzner A W K 1962 Gravitational waves in general relativity. VII. Waves from axi-symmetric isolated systems *Proc. R. Soc.* A **269** 21–52

[308] Penrose 1963 Asymptotic properties of fields and space-times *Phys. Rev. Lett.* **10** 66–8

[309] For spatial infinity, see Ashtekar A 1980 Asymptotic structure of the gravitational field at spatial infinity *General Relativity and Gravitation One Hundred Years After the Birth of Albert Einstein* (in two volumes)

ed A Held (New York: Plenum).
[310] Hilbert D 1917 Die Grundlagen der Physik (Zweite Mitteilung) *König. Ges. Wiss. Göttingen. Math.-Phys. Klasse Nach.* 55–76
For early work on the Cauchy problem in general relativity, see Stachel J 1991 The Cauchy problem in general relativity: the early years *Einstein Studies vol 3. Studies in the History of General Relativity* ed J Eisenstaedt and A J Kox (Boston: Birkhäuser) pp 405–16
[311] For a survey, see York J W 1979 Kinematics and dynamics of general relativity *Sources of Gravitational Radiation* ed L Smarr (Cambridge: Cambridge University Press)
[312] Darmois G 1927 *Les Équations de la Gravitation* (Paris: Gauthier-Villars)
This problem does not seem to have been discussed again until the independent work of Sachs R K 1962 On the characteristic initial value problem in gravitational theory *J. Math. Phys.* **3** 908–14.
[313] d'Inverno R A and Stachel J 1978 Conformal two-structure as the gravitational degrees of freedom in general relativity *J. Math. Phys.* **19** 2447–60
d'Inverno R A and Smallwood J 1980 Covariant 2 + 2 formulation of the initial value problem in general relativity *Phys. Rev. D* **22** 1233–47
[314] Weber J 1969 Evidence for the discovery of gravitational radiation *Phys. Rev. Lett.* **22** 1320
[315] For accounts of this controversy, see Collins H M 1975 The seven sexes: a study in the sociology of a phenomenon, or the replication of experiments in physics *Sociology* **9** 205–24; 1981 Son of the seven sexes: the social destruction of a physical phenomenon *Social Studies Sci.* **11** 33–62; 1992 Detecting gravitational radiation: the experimenter's regress *Changing Order: Replication and Induction in Scientific Practice* 2nd edn (Chicago: University of Chicago Press) ch 4, pp 79–111.
For the current status of the dispute, see:
Weber J 1992 Supernova 1987A Gravitational wave antenna observations, cross sections, correlations with six elementary particle detectors, and resolution of past controversies *Einstein Studies vol 4. Recent Advances in General Relativity* ed A I Janis and J Porter 1992 (Boston: Birkhäuser) pp 230–240
Thorne K S 1992 On Joseph Weber's new cross section for resonant-bar gravitational wave detectors *Einstein Studies vol 4. Recent Advances in General Relativity* ed A I Janis and J Porter (Boston: Birkhäuser) pp 196–229
[316] For a review, see Thorne K S 1994 *Black Holes and Time Warps: Einstein's Outrageous Legacy* (New York: Norton).
[317] Hulse R A and Taylor J H 1975 Discovery of a pulsar in a binary system *Astrophys. J.* **195** L51–53
[318] For surveys of recent experimental tests, see:
Shapiro I I 1990 Solar system tests of general relativity: recent results and present plans *General Relativity and Gravitation* ed N Ashby *et al* (Cambridge: Cambridge University Press) p 313
Will C M 1993 *Theory and Experiment in Gravitational Physics* 2nd edn (Cambridge: Cambridge University Press); 1993 *Was Einstein Right? Putting General Relativity to the Test* 2nd edn (New York: Basic Books).
[319] For a survey of applications of general relativity in astronomy and geodesy, see Soffel M H 1989 *Relativity in Astrometry, Celestial Mechanics and Geodesy* (Berlin: Springer).
[320] Einstein A 1936 Lens-like action of a star by the deviation of light in the

gravitational field *Phys. Rev.* **49** 404–5
For a survey, see Schneider P, Ehlers J and Falco E 1991 *Gravitational Lenses* (Heidelberg: Springer).

[321] Robinson I, Schild A and Schucking E 1965 *Quasi-Stellar Sources and Gravitational Collapse Including the Proceedings of the First Texas Symposium on Relativistic Astrophysics* (Chicago: University of Chicago Press)

[322] For surveys of relativistic astrophysics, see:
Zeldovich Ya B and Novikov I D 1971 *Relativistic Astrophysics vol 1. Stars and Relativity* (Chicago: University of Chicago Press)
Straumann N 1988 *Allgemeine Relativitätstheorie und relativistische Astrophysik* 2nd edn (Berlin: Springer)
For surveys of gravitational collapse and black-hole physics, see:
Miller J C and Sciama D 1980 Gravitational collapse to the black hole state *General Relativity and Gravitation One Hundred Years After the Birth of Albert Einstein* vol 2, ed A Held (New York: Plenum) pp 359–392
Novikov I D and Frolov V P 1989 *Physics of Black Holes* (Dordrecht: Kluwer Academic)
Thorne K S 1994 *Black Holes and Time Warps: Einstein's Outrageous Legacy* (New York: Norton)

[323] For surveys, see:
Cook A H 1987 Experiments on gravitation *300 Years of Gravitation* ed S Hawking and W Israel (Cambridge: Cambridge University Press) pp 50–79
Will C M 1993 *Theory and Experiment in Gravitational Physics* 2nd edn (Cambridge: Cambridge University Press); 1993 *Was Einstein Right? Putting General Relativity to the Test* 2nd edn (New York: Basic Books)
Everitt C W F 1988 The Stanford Relativity Gyroscope Experiment: a history and overview *Near Zero: New Fontiers of Physics* ed J D Fairbank *et al* (New York: Freeman) pp 587–597; 1992 Background to history: the transition from little physics to big physics in the gravity probe B Relativity Gyroscope Program *Big Science: The Growth of Large-Scale Research* ed P Galison and B Hevly (Stanford: Stanford University Press) ch 8, pp 212–235

[324] For a brief historical account, see Ashtekar's Introduction: the winding road to quantum gravity in Ashtekar A and Stachel J 1991 *Einstein Studies vol 3. Conceptual Problems of Quantum Gravity* (Boston: Birkhäuser) pp 1–9.
For surveys of the various approaches, see:
Ashtekar A and Geroch R 1974 Quantum theory of gravitation *Rep. Prog. Phys.* **37** 1211–56
Ashtekar A and Stachel J 1991 *Einstein Studies vol 3. Conceptual Problems of Quantum Gravity* (Boston: Birkhäuser)
DeWitt B S and Stora R 1984 *Les Houches Session XL: Relativity, Groups and Topology II* (Amsterdam: North-Holland)

[325] Schrödinger E 1939 The proper vibrations of the expanding Universe *Physica* **6** 899–912

[326] Parker L E 1966 The creation of particles in an expanding universe PhD Thesis Harvard University

[327] For surveys of quantum field theory in background space-times, see:
Birrell N D and Davies P C W 1982 *Quantum Fields in Curved Space* (Cambridge: Cambridge University Press)
Fulling S A 1989 *Aspects of Quantum Field Theory in Curved Space-time* (Cambridge: Cambridge University Press)

History of Relativity

[328] Hawking S 1975 Particle creation by black holes *Commun. Math. Phys.* **43** 199–220
For a survey of quantum effects in black-hole physics, see Novikov I D and Frolov V P 1989 *Physics of Black Holes* (Dordrecht: Kluwer Academic).

[329] Einstein A 1916 Die Grundlagen der allgemeinen Relativitätstheorie *Ann. Phys., Lpz* **49** 769–822

[330] Heisenberg W and Pauli W 1929 Zur Quantendynamik der Wellenfelder *Zeit. Physik* **56** 1–61 (Engl. Transl. from Gorelik G 1992 The first steps of quantum gravity and the Planck values *Einstein Studies vol 3. Studies in the History of General Relativity* ed J Eisensaedt and A J Kox (Boston: Birkhäuser) p 370)

[331] Rosenfeld L 1930 Zur Quantelung der Wellenfelder *Ann. Phys., Lpz* **5** 113–152; 1932 La théorie quantique des champs *Institut Henri Poincaré (Paris) Ann.* **2** 25–91

[332] Bronsteyn M P 1936 Quantentheorie schwacher Gravitationsfelder *Phys. Z. Sowjetunion* **9** 140–157; 1936 Kvantovanie gravitatsionnykh voln *Zh. Eksp. Teor. Fiz.* **6** 140–57
For an account of Bronsteyn's work, see:
Gorelik G and Frenkel V J 1985 *M P Bronsteyn i kvantovaya teoriy gravitatsii* in *Eynshteynovskiy Sbornik 1980–1981* (Moscow: Nauka) pp 291–327
Gorelik G 1992 The first steps of quantum gravity and the Planck values *Einstein Studies vol 3. Studies in the History of General Relativity* ed J Eisensaedt and A J Kox (Boston: Birkhäuser)

[333] Bronsteyn M P 1936 Kvantovanie gravitatsionnykh voln *Zh. Eksp. Teor. Fiz.* **6** 140–57 (Engl. Transl. from Gorelik G 1992 The first steps of quantum gravity and the Planck values *Einstein Studies vol 3. Studies in the History of General Relativity* ed J Eisensaedt and A J Kox (Boston: Birkhäuser) p 377)

[334] Bergmann P G, Penfield R, Schiller R and Zatzkis H 1950 The Hamiltonian of the general theory of relativity with electromagnetic field *Phys. Rev.* **80** 81–8

[335] Dirac P A M 1958 The Theory of Gravitation in Hamiltonian Form *Proc. R. Soc. A* **246** 333–43

[336] Arnowitt R, Deser and Misner C W 1962 The dynamics of general relativity *Gravitation: An Introduction to Current Research* ed L Witten (New York: Wiley) pp 227–65

[337] Feynman R P 1963 *Acta Phys. Polon.* **24** 697

[338] For a survey of the canonical and covariant methods and difficulties associated with application of each to general relativity, see Ashtekar A and Geroch R 1974 Quantum theory of gravitation *Rep. Prog. Phys.* **37** 1211–56.

[339] For a discussion of the shift from perturbative to non-perturbative methods in recent years, see Ashtekar's Introduction to Ashtekar A and Stachel J 1991 *Einstein Studies vol 3. Conceptual Problems of Quantum Gravity* (Boston: Birkhäuser).
For a survey of recent progress in the non-perturbative canonical approach, see Ashtekar A 1991 *Lectures on Non-Perturbative Canonical Gravity* (Singapore: World Scientific).

[340] Penrose R 1987 Newton, quantum theory and reality *300 Years of Gravitation* ed S Hawking and W Israel (Cambridge: Cambridge University Press); 1989 *The Emperor's New Mind* (Oxford: Oxford University Press) ch 8

[341] For one such attempt, see Horowitz 1991 String Theory Without Space-Time *Einstein Studies vol 3. Conceptual Problems of Quantum Gravity* ed

A Ashtekar and J Stachel (Boston: Birkhäuser) pp 299–311.
[342] Petzoldt J 1912 Die Relativitätstheorie im erkenntnistheoretischen Zusammenhange des relativistichen Positivismus *Phys. Ges. Berlin Verhand.* **14** 1055–64
[343] For a survey and analysis of the literature on philosophical interpretations of relativity, see Hentschel K 1990 *Interpretationen und Fehlinterpretationen der Speziellen und der Allgemeinen Relativitätstheorie durch Zeitgenossen Albert Einsteins* (Basel: Birkhäuser) ch 4–6.
[344] For a survey of some disputed questions in this area, see Torretti R 1983 *Relativity and Geometry* (Oxford: Pergamon) ch 7.
[345] For early attacks by German physicists, see:
Abraham A M 1912 Relativität und Gravitation. Erwiderung auf eine Bemerkung des Hrn A Einstein *Ann. Phys., Lpz* **38** 1056–8; p 1056 accuses Einstein's theory of exerting a fascinating influence on the most recent mathematical physicists, which threatens to inhibit the healthy further development of theoretical physics.
Gehrke E 1912 *Lehrbuch der Optik von Dr Paul Drude* 3rd edn (Leipzig: Hirzel) pp 446–473
For early critiques from the United States, see:
More L T 1912 The theory of relativity *The Nation* **94** 370–371
Magie W 1912 The primary concepts of physics *Science* **35** 281–93
[346] For surveys of German relativity discussions, pro and con, see Goenner H 1991 The reception of the theory of relativity in Germany as reflected by books published between 1908 and 1945 *Einstein Studies vol 3. Studies in the History of General Relativity* ed J Eisenstaedt and A J Kox (Boston: Birkhäuser) pp 15–38; 1993 The reaction to relativity theory. I. The anti-Einstein campaign in Germany in 1920 *Einstein in Context, Science in Context* vol 6, ed M Beller *et al* no 1 pp 107–133; 1993 The reaction to relativity theory in Germany, III: a hundred authors against Einstein *Einstein Studies vol 5. The Attraction of Gravitation/ New Studies in the History of General Relativity* ed J Earman *et al* (Boston: Birkhäuser) pp 248–273
For discussions of the Russian reactions to relativity, see:
Gorelik G 1993 Vladımir Fock: philosophy of gravity and gravity of philosophy *Einstein Studies vol 5. The Attraction of Gravitation/ New Studies in the History of General Relativity* ed J Earman *et al* (Boston: Birkhäuser)
Graham L 1972 *Science and Philosophy in the Soviet Union* (New York: Knopf)
Joravsky D 1961 *Soviet Marxism and Natural Science 1917–1932* (London: Routledge and Paul)
For discussion of some politically motivated anti-relativity literature, see Hentschel K 1990 *Interpretationen und Fehlinterpretationen der speziellen und der allgemeinen Relativitätstheorie durch Zeitgenossen Albert Einsteins* (Basel: Birkhäuser) ch 3, sections 3.1–3.2.
[347] For some of the myths, see Stachel J 1982 Albert Einstein: the man beyond the myth *Bostonia Mag.* 8–17.
[348] Biezunski M 1987 Einstein's reception in Paris in 1922 *The Comparative Reception of Relativity* ed T Glick (Dordrecht: Reidel) pp 169–188
Missner M 1985 Why Einstein Became Famous in America *Social Studies Sci.* **15** 267–91
[349] Hentschel K 1990 *Interpretationen und Fehlinterpretationen der Speziellen und der Allgemeinen Relativitätstheorie durch Zeitgenossen Albert Einsteins* (Basel: Birkhäuser) ch 2 and 3, has made a good beginning with a

survey and classification of the popular literature.
- [350] For an account of this programme, see Fox R 1974 The rise and fall of Laplacian physics. *Hist. Studies Phys. Sci.* **4** 89–136
- [351] Vizgin V P 1985 *Yedinye Teorii Polia v Pervoi Treti XX Veka* (Moscow: Nauka) (Engl. Transl. Barbour J 1994 *Unified Field Theories in the First Third of the 20th Century* (Basel: Birkhäuser) ch 1)
- [352] Einstein A 1909 Zum gegenwärtigen stand des Strahlungsproblems *Phys. Z.* **10** 185–93
  For a discussion of Einstein's programme, see Stachel J 1986 Einstein and the quantum: fifty years of struggle *From Quarks to Quasars: Philosophical Problems of Modern Physics* ed R Colodny (Pittsburgh: University of Pittsburgh Press) pp 349–385.
- [353] Mie G 1912 Grundlagen einer Theorie der Materie (Erste Mitteilung) *Ann. Phys., Lpz* **37** 511–534; 1913 Grundlagen einer Theorie der Materie (Zweite Mitteilung) *Ann. Phys., Lpz* **39** 1–40; 1913 Grundlagen einer Theorie der Materie (Dritte Mitteilung) *Ann. Phys., Lpz* **40** 1–66
- [354] Mie 1913 Grundlagen einer Theorie der Materie (Dritte Mitteilung) *Ann. Phys., Lpz* **39** 1–40
- [355] Hilbert D 1915 Die Grundlagen der Physik (Erste Mitteilung) *König. Ges. Wiss. Göttingen. Math.-Phys. Klasse Nach.* 395–407
  For discussions, see Mehra J 1974 *Einstein, Hilbert and the Theory of Gravitation* (Dordrecht: Reidel)
  Vizgin V P 1989 Einstein, Hilbert, and Weyl: the genesis of the geometrical unified field theory program *Einstein Studies vol 1. Einstein and the History of General Relativity* ed D Howard and J Stachel (Boston: Birkhäuser) pp 300–14
- [356] Weyl H 1918 Gravitation und Elektrizität *Preuss. Akad. Wiss. (Berlin) Phys.-Math. Klasse. Sitzungsber.* 465–80
- [357] For discussions of Weyl's theory, see Vizgin V P 1985 *Yedinye Teorii Polia v Pervoi Treti XX Veka* (Moscow: Nauka) (Engl. Transl. Barbour J 1994 *Unified Field Theories in the First Third of the 20th Century* (Basel: Birkhäuser) ch 3); 1989 Einstein, Hilbert, and Weyl: the genesis of the geometrical unified field theory program *Einstein Studies vol 1. Einstein and the History of General Relativity* ed D Howard and J Stachel (Boston: Birkhäuser) pp 300–14.
- [358] Eddington A S 1921 A generalization of Weyl's theory of the electromagnetic and gravitational fields *Proc. R. Soc.* A **99** 104–22
- [359] Einstein A 1923 Zur affinen Feldtheorie *Preuss. Akad. Wiss. (Berlin) Phys.-Math. Klasse. Sitzungsber.* 137–40
- [360] Kaluza T 1921 Zum Unitätsproblem der Physik *Preuss. Akad. Wiss. (Berlin) Phys.-Math. Klasse. Sitzungsber.* 966–72
- [361] Einstein A 1927 Zu Kaluzas Theorie des Zusammenhanges von Gravitation und Elektrizität *Preuss. Akad. Wiss. (Berlin) Phys.-Math. Klasse Sitzungsber.* 23–30
- [362] Einstein A 1928 Riemann-Geometrie mit Aufrecherhaltung des Begriffes des Fernparallelismus *Preuss. Akad. Wiss. (Berlin) Phys.-Math. Klasse Sitzungsber.* 217–21
- [363] Einstein A and Cartan E 1979 *Albert Einstein: Letters on Absolute Parallelism 1929–1932* (Princeton: Princeton University Press)
- [364] Einstein A 1945 A generalization of the relativistic theory of gravitation *Ann. Math.* **46** 578–84
- [365] For an account of unified field theories developed during the first third of this century, see Vizgin V P 1985 *Yedinye Teorii Polia v Pervoi Treti XX Veka* (Moscow: Nauka) (Engl. Transl. Barbour J 1994 *Unified Field*

Theories in the First Third of the 20th Century (Basel: Birkhäuser)).

For a survey of later theories, see Tonnelat M-A 1965 *Les Théories Unitaires de l'Electromagnétisme et de la Gravitation* (Paris: Gauthier-Villars).

[366] See, for example, Pauli W 1921 Relativitätstheorie *Encyklopädie der Mathematischen Wissenschaften, mit Einschluss ihrer Anwendungen* vol 5. *Physik* ed A Sommerfeld (Leipzig: Teubner) section 67

[367] For evidence that he also was sceptical about the unified field programme, see Stachel J 1993 The other Einstein: Einstein contra field theory *Science in Context* vol 6, ed M Beller *et al* no 1 pp 275–90.

[368] For a review of these theories, see Gottfried K and Weisskopf V E 1984 *Concepts of Particle Physics* vol 1 (Oxford: Clarendon); 1986 *Concepts of Particle Physics* vol 2 (New York: Oxford University Press).

Chapter 5

NUCLEAR FORCES, MESONS, AND ISOSPIN SYMMETRY

Laurie M Brown

5.1. The status of physics circa 1930

5.1.1. The constituents of matter
According to the American physicist, Robert A Millikan, writing in 1929 [1]

> All atoms are built up out of definite numbers of positive and negative electrons. All chemical forces are due to the attractions of positive for negative electrons. All elastic forces are due to the attractions and repulsions of electrons. In a word, *matter itself is electrical in origin.*

For the modern reader, some interpretation of Millikan's words is required. 'Negative electron' designates the elementary particle whose charge is $-e$, where $e = 1.60 \times 10^{-19}$ C, and whose mass is $m = 9.11 \times 10^{-31}$ kg, in accord with modern usage. However, Millikan's 'positive electron' is the nuclear particle that we call the proton, with positive charge equal to e and mass M, which is 1840 times larger than m. Millikan was using the word 'electron' in an older sense to designate the fundamental unit of charge. Nowadays, we usually reserve the term 'positive electron' to designate the antiparticle of the electron, which has mass m and charge $+e$, and is also known as the positron. In Millikan's picture of the atom, the outer atomic shells consist of negative electrons, in accord with present views, but unlike our current picture, the atomic nucleus was made of 'positive and negative electrons'. In this respect, Millikan was voicing the view commonly held at that time.

At present, physicists speak of a so-called 'standard model', that describes with some success the elementary structures of which matter is composed. It has not yet been fully verified, but there are no experimental results with which it disagrees. It is interesting to note that the 'standard model' that was current in 1930 was based on only two constituent particles, the electron and proton, and it contained only a single type of interaction, namely, electromagnetic. Thus, it was the most fully 'unified theory' since that of the sixth century BC Greek philosopher Anaximenes, who regarded air as the basic substance from which everything else was made [2]. (In this discussion, we have ignored gravitation, because for the small masses at the atomic and subatomic levels, it is too weak to play any significant role.)

In 1925 and 1926, the quantum theory became well-established as the appropriate dynamics of microscopic physics and in 1927 Paul A M Dirac made a relativistic quantum theory of the electromagnetic field in interaction with matter. In this theory, called quantum electrodynamics (QED), the electromagnetic interaction is carried by light quanta, massless particles, for which the chemist G N Lewis in 1926 invented the name photons. Thus, in 1930, physicists viewed the world as constructed out of a fundamental trinity, consisting of two basic constituent particles, the electron and the proton, and one messenger particle, the photon. In contrast to this, the current 'standard model' employs 48 particles (36 quarks and antiquarks and 12 messenger particles). Like the model of 1930, it also fails to include gravitation.

In reviewing the state of knowledge of microscopic physics, a convenient way to broadly classify atomic processes is by the energy scale on which they occur, specified in terms of electronvolts. One electronvolt (eV) is the energy acquired by a particle bearing the electronic charge e when it is accelerated by one volt of potential difference. Thus, for example, the binding energy of the lightest atom, hydrogen, is 13.6 eV, while the inner shells of heavy atoms have typical energies of tens of kiloelectronvolts (keV), as do x-rays. The atomic nucleus has energies measured in millions of electronvolts (MeV), while the primary cosmic rays strike the atmosphere with energies of billions of electronvolts (GeV).

5.1.2. Atomic and molecular (eV) physics in 1930
Werner Heisenberg, one of the great theorists who initiated the quantum-mechanical revolution, wrote in 1969 [3]

> To those of us who participated in the development of atomic theory, the five years following the Solvay Congress in Brussels [in 1927] looked so wonderful that we often spoke of them as the golden age of atomic physics. The great obstacles that had occupied all our efforts in the preceding years had been cleared out of the way; the gate to that entirely new field, the quantum mechanics of the atomic shells stood wide-open, and fresh fruits seemed ready for the plucking.

Nuclear Forces, Mesons, and Isospin Symmetry

Some of those fruits, as explained in more detail in Chapter 3, were: the explanation of the periodic table of the chemical elements, based upon the calculation of the energies of stationary atomic states and the exclusion principle of Wolfgang Pauli; the theory of chemical binding and molecular structure; the theory of elastic and inelastic scattering of charged particles, photons and atomic systems (i.e. atoms and molecules and their ions) on atomic targets; the theory of magnetism; and the theory of radiation from atomic and molecular systems. In addition, a good start had been made upon what would later be called materials science, which deals with the properties of the solid and liquid states of matter.

Among the obstacles that had to be overcome to reach this Golden Age, perhaps the most significant was the psychological reluctance to abandon the seductive semi-classical picture of electrons in orbit around the nucleus, i.e. the atom as a miniature solar system. That picture was first proposed, but in a version that was intrinsically unstable, by Ernest Rutherford in 1911. It was modified by Niels Bohr in 1913, who imposed so-called quantum conditions, which restricted the allowed values of angular momentum to a discrete set and thus stabilized Rutherford's nuclear atom. The quantization procedure was further elaborated by Bohr and by Arnold Sommerfeld over the next decade, and it gave excellent results for the spectrum of hydrogen. However, the Bohr–Sommerfeld model ran into serious problems in the 1920s when applied to other atomic phenomena. In spite of great ingenuity, displayed especially by the young theorists of the generation of Pauli and Heisenberg, born just after the turn of the century, physicists were unable to calculate the spectrum even of helium, the next simplest atom after hydrogen. Neither could they explain the effects of magnetic fields on the wavelengths of most spectral lines (the anomalous Zeeman effect), nor could they predict the intensities of any spectral lines.

The great theoretical breakthrough that swept away these obstacles was quantum mechanics. This theory has forever changed the way in which physics looks at the world—even more, perhaps, than that other great theory of twentieth century physics, relativity. It began with a powerful insight by twenty-three year old Werner Heisenberg in the year 1925. The frequency ν of a photon emitted by an excited atom is given by the Einstein–Bohr frequency condition $h\nu = E_i - E_f$, where h is Planck's constant and E_i and E_f are, respectively, the initial and final energies of the states between which the atomic transition takes place. Heisenberg realized that not only the frequency, but also the probability of transition, which determines the intensity of the spectral line, depends upon both the initial and the final state. He thus focused attention upon a set of quantized probability amplitudes, whose absolute squares are proportional to the line intensities. Not only were these amplitudes complex, but they also obeyed a non-commuting algebra, unlike the amplitudes corresponding to the real classical oscillators, whose existence had been assumed in earlier (and only partially successful) theories of the emission and absorption of light by atoms.

Max Born and Pascual Jordan, in Göttingen, recognized Heisenberg's non-commuting algebra as that obeyed by matrices, and thus was born the form of quantum mechanics known as matrix mechanics, which was further developed by Born, Jordan and Heisenberg in a famous 'three-man paper' [4]. By the end of 1925, Pauli had used this new method to obtain the (non-relativistic) spectrum of the hydrogen atom. A second form of quantum mechanics, known as wave mechanics, was proposed by the Viennese physicist Erwin Schrödinger, working in Zurich, in January 1926. Wave mechanics was soon shown by several physicists, including Schrödinger, to be equivalent in its physical content to matrix mechanics, and it has turned out to be the most commonly used and most convenient way to solve problems in atomic and molecular physics.

Several steps towards the completion of quantum mechanics were taken by the British physicist Dirac, who was working at Cambridge University. The first of these steps was his development of a form of quantum mechanics based upon a generalization of classical Hamiltonian dynamics. Dirac's new method, called transformation theory, was general enough to encompass both matrix mechanics and wave mechanics as special limiting cases. In 1927, Dirac was guided by his method to obtain an equation for the electron that was relativistic, and which at the same time gave correctly the electron's spin of $\frac{1}{2}$ unit (the unit being $\hbar$, or $h/2\pi$, where h is Planck's constant) and its magnetic moment of one Bohr magneton (equal to $e\hbar/2mc$, with c the velocity of light in a vacuum). The spinning electron had been suggested by the Dutch physicists Samuel Goudsmit and George Uhlenbeck in 1925 to help explain the anomalous Zeeman effect. Also in 1927, Dirac proposed the first theory of quantum electrodynamics (QED) and a method of expansion for calculating atomic transition probabilities, the *perturbation theory*, an approximation which relies upon the weakness of the electromagnetic interaction or, in other words, on the small value of the fundamental electronic charge.

By 1929, Dirac could write in the introduction to one of his papers [5]

> The general theory of quantum mechanics is now almost complete, the imperfections that still remain being in connection with the exact fitting of the theory with relativity ideas. These give rise to difficulties only when high speed particles are involved, and are therefore of no importance in the consideration of atomic and molecular structure and ordinary chemical reactions ... The underlying physical laws necessary for the mathematical theory of a large part of physics and the whole of chemistry are thus completely known, and the difficulty is only that the exact

Nuclear Forces, Mesons, and Isospin Symmetry

BOX 5A: QED AND FEYNMAN DIAGRAMS

The Dirac equation for the wavefunction $\psi(x)$ of an electron in interaction with the electromagnetic field is

$$(\gamma_\mu[i\hbar\partial/\partial x_\mu - eA_\mu(x)] - mc)\psi(x) = 0$$

where the γ_μ ($\mu = 0, 1, 2, 3$) are certain 4×4 matrices; A_μ is the electromagnetic potential; ψ is a one-column matrix with entries $\psi_a(x)$, $a = 1, \ldots, 4$; and $\bar{\psi} = \psi^{*T}\gamma_0$ is a one-row matrix of four entries. The $e - \gamma$ interaction is then given by $e\bar{\psi}\gamma_\mu\psi A_\mu$. The various interactions can most easily be visualized in terms of Feynman diagrams, which originated in the mid-1940s. (Applying them in our historic period is anachronistic. However, they provide a compact comparison of various fundamental theories of interest in the 1930s.)

Conventions

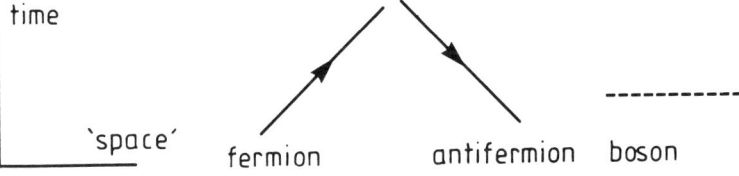

Basic interactions in QED

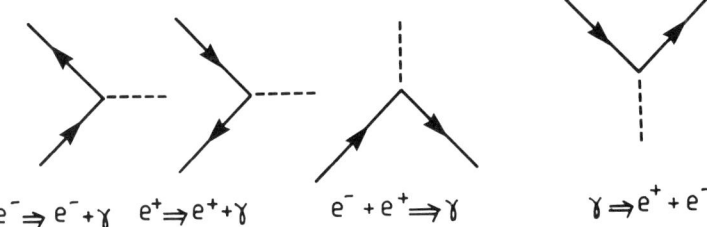

$e^- \Rightarrow e^- + \gamma \quad e^+ \Rightarrow e^+ + \gamma \quad e^- + e^+ \Rightarrow \gamma \quad \gamma \Rightarrow e^+ + e^-$

An example of a derived interaction

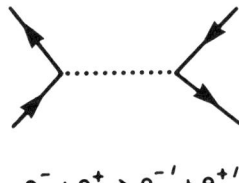

$e^- + e^+ \Rightarrow e^{-\prime} + e^{+\prime}$

361

application of these laws leads to equations much too complicated to be soluble.

5.1.3. X-rays and the Compton effect (keV physics)

During the 1920s three major steps were taken towards combining quantum mechanics with relativity. (We are speaking here of the special theory of relativity, not general relativity, a theory of gravitation which has still, as the century draws to a close, not found a satisfactory quantum formulation.) The first step was the description of x-ray scattering as the collision of an x-ray photon with an almost free electron. This was achieved by Arthur H Compton in 1922, in interpreting the results of his x-ray experiments. The second step was Dirac's relativistic quantum theory of the electron, the Dirac equation of 1927 and his theory of 'holes' in the vacuum, which were interpreted as positive electrons. The third step was QED, the theory of relativistic interaction of the quantized electromagnetic field with the relativistic electron. Again it was Dirac who in 1927 quantized the transverse part of the electromagnetic field. The formulation of QED was completed in 1929–30 by Heisenberg and Pauli. In this section we will consider the first of these three stages, relating to the Compton effect [6].

Wilhelm Röntgen, who discovered x-rays in 1895, was at first unable to find any of the expected light-like properties for the newly discovered x-rays, and so suggested that they might be longitudinal ether vibrations, which were not altogether unexpected in the ether theory of light, although none had previously been observed. In 1897, G G Stokes in England (and independently E Wiechert in Germany) proposed that x-rays were ordinary transverse electromagnetic pulses, produced by the deceleration of individual electrons. This idea was further developed by J J Thomson in his theory of x-ray scattering, and also by his student G C Barkla. The latter demonstrated the transversality of x-rays by a double-scattering experiment and in 1908 discovered the characteristic x-rays, whose homogeneous frequencies were typical of the atoms which produced them. By 1911, Barkla had identified two series of characteristic x-ray lines, which he called K and L. These were studied throughout the periodic system by H J G Moseley, and used to establish the concept of atomic number Z. Moseley's work played a major role in the acceptance of Bohr's atomic theory.

Although the electromagnetic theory was thus preferred, some of Thomson's scattering predictions were contradicted by experiment, especially by x-rays of higher energy and by nuclear γ-rays. The latter were assumed to have the same character as x-rays, i.e. to be electromagnetic wave pulses of still higher frequency. A S Eve had shown in 1904 that, contrary to the Thomson formula, secondary x-rays were much softer (i.e. less penetrating and of lower frequency) than the primary beam. Also contrary to Thomson's theory, there was a forward–backward asymmetry in scattering, as well as discrepancies with the predicted absorption coefficients. For these reasons, W H Bragg in 1907 proposed that x-rays were neutral (possibly composite) particles and this new theory attracted some adherents. The electromagnetic wave theory, however, received a boost in 1912 by the observation by Max von Laue and his collaborators of x-ray diffraction by crystals.

The last result made it more difficult to reconcile the wave theory with the particle properties of x-rays that were expected, according to the light-quantum hypothesis of Einstein. In spite of Einstein's successful use in 1905 of $E = h\nu$ to explain fluorescence and the photoelectric effect, later confirmed by the experiments of R A Millikan, most physicists (including Millikan himself) still remained sceptical. Einstein renewed his interest in light quanta in 1916, when he rederived Planck's law for a 'gas' of light quanta, assuming that [7]

> ... in each energy transfer from radiation to matter the momentum $h\nu/c$ is also transferred to the molecule. Hence we conclude that every such elementary process is a *completely directed* event. With that, the light-quanta must be considered as good as being substantiated.

However, although Einstein's work is compelling in retrospect, the light-quantum hypothesis only began to be accepted by the physics community after it was adopted and utilized by Compton in explaining his x-ray observations. (A similar kinematic analysis was made independently by Peter Debye in Germany and published, as was Compton's, in 1923.)

In late 1922 Compton, who had spent most of his career, beginning with his doctoral thesis at Princeton in 1913, studying the scattering and absorption of x-rays and γ-rays, found that he could explain some very puzzling characteristics of the scattering by an extremely simple model. Namely, he assumed that an x-ray beam of frequency ν is a stream of essentially point-like particles, x-ray photons, each having energy $E = h\nu$ and momentum $p = h\nu/c$. The electron, likewise, he took to be essentially point-like, and assumed that the binding of the electrons played a negligible role, relative to the large quantum energy of the photons. The scattering target (for example, a block of carbon) was thus viewed as merely a box containing free electrons. In his picture an electron, initially at rest (or nearly so), would absorb a single x-ray photon, would recoil, and instantly emit another x-ray photon, generally in a new direction, with the restriction that both energy and momentum were conserved. Since the original photon energy was shared between the emitted photon and the recoiling electron, the secondary photon's energy could be calculated, as well as that of the recoiling electron. Evidently the energy (and hence the frequency) of the secondary photon

See also
p 165

would be less than that of the primary. Thus, unlike the case of Thomson scattering, there was a change (an increase) of the x-ray wavelength.

Except for the use of relativistic kinematics, the calculation was not much more difficult than that of the idealized billiard ball collisions, which are performed by physics students in high school. Yet the results for the dependence of the frequency of the scattered radiation on the direction of scattering were accurate, and together with his experimental work on x-ray scattering and absorption, resolved controversies existing since the discovery of x-rays. This work won Compton the Nobel Prize in Physics for 1927. These struggles over x-rays were related to what became known as the wave–particle paradox, and later (as the 'paradox' became inescapable) simply as 'wave–particle duality'.

Although we have stressed Compton's kinematic analysis of x-ray scattering, which gives the relation between the energy and angular distributions, Compton's theory, like Thomson's, also gave the intensity of the radiation scattered at a given angle (figure 5.1). The relative intensity scattered out of a beam is measured by a quantity called the cross section, which is the effective area subtended by the scatterer, in this case the electron. The total scattering cross section can be related in turn to a quantity called the absorption coefficient μ, and it is this quantity which was most easily measured.

An ideal absorption measurement passes a well-defined beam of homogeneous energy through a relatively thin absorbing layer into a well-shielded and small detector. For such an arrangement, the law of absorption is exponential

$$I(x) = I(0)\exp(-\mu x) \tag{1}$$

where $I(x)$ is the intensity after the beam has traversed a distance x in the absorber. Until about 1932, it was believed that the absorption of x-rays and γ-rays was due entirely to scattering from electrons and to the atomic photoelectric effect. The absorption coefficient per electron (the μ of equation (1), divided by the number of electrons per unit volume) was thus parametrized by a Compton term, independent of Z, and a photoelectric term, proportional to Z^3. For higher-energy photons, the latter term was quite negligible, except for the largest values of Z.

Compton's picture of the scattering was difficult for many physicists to accept, but in 1925 it was confirmed in detail by two experiments showing the simultaneous production of the secondary light quantum and the recoil of the electron. One of them, by Walther Bothe and Hans Geiger in Germany, used for the first time a technique of coincident counting between detectors. The second, by Compton and Alfred Simon, used a Wilson cloud chamber. Both of these techniques became of major importance in the subsequent study of cosmic rays.

In 1926, after the discovery of the new quantum mechanics, Dirac adapted Heisenberg's treatment of radiation from an atom to recalculate

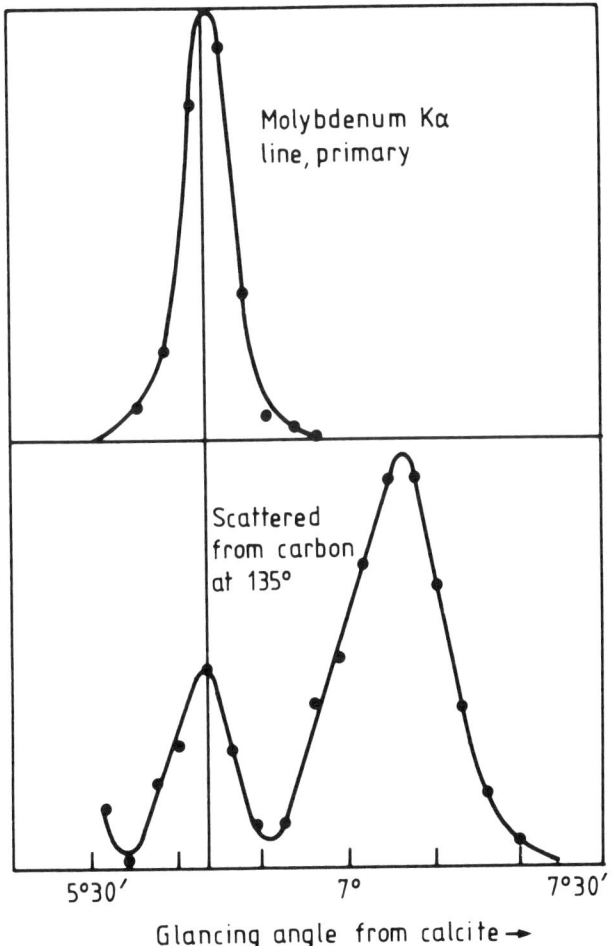

Figure 5.1. *A typical spectrum of scattered x-rays, showing the splitting of the primary ray, originating from a molybdenum target, into an 'unmodified line' having the primary wavelength and a 'modified line' of longer wavelength. The wavelengths are measured by the glancing angle from a calcite crystal.*

the Compton scattering cross section and Walter Gordon obtained the same result using the Schrödinger method. While these were improvements on Compton's 1923 result, neither calculation included the effect of the electron's spin, which was first taken into account in 1929, using Dirac's relativistic electron equation, by O Klein and Y Nishina, working at Bohr's institute in Copenhagen. A series of absorption experiments using γ-rays, between 1924 and 1930, succeeded in showing that the Klein–Nishina formula (KN) was quite exact up to the highest γ-ray energy available (that of Th C", which gave a single narrow γ-ray line at 2.61 MeV). In 1930, some anomalous results were reported by four independent research groups, when high-energy γ-rays and high-Z targets were used. We will discuss those anomalies later.

During the 1920s it was believed that cosmic rays, the primaries striking the atmosphere as well as the high-energy secondary components found at mountain altitude or at sea level, were similar to nuclear γ-rays, but of higher quantum energy, and they were often referred to as *ultra γ-rays*. Since they could not be produced in the laboratory, it was necessary to guess their properties by extrapolation, which meant, for example, that their energies were inferred from their absorption coefficients, theoretically extrapolated by use of scattering formulae (after 1929, this was the KN formula).

5.1.4. α- and β-decay and nuclear systematics (MeV physics)

In Chapter 2 of this volume, Abraham Pais has a section entitled '1926–1932: The years of nuclear paradoxes', in which he has explained the successful treatment of α-decay by the new quantum mechanics, based upon the penetration of the nuclear Coulomb barrier. At the same time, the great puzzle of the continuous β-decay spectrum presented a major paradox to challenge the application of quantum theory to the nucleus. The building-up principle (*Aufbauprinzip*) of the nucleus, unlike that of the atomic shells, also presented insuperable challenges to quantum mechanics, based as it was upon the 1920s standard model that demanded that nuclei be built up out of protons and electrons. As we shall see, many of these problems were solved by enlarging the particle spectrum, but most physicists were not prepared to accept this solution and opted instead for a breakdown of quantum mechanics, supposed to take place at distances smaller than the nuclear radius. Together with the apparent anomaly of γ-ray absorption at high Z, referred to in the previous section, it was clear that MeV physics had a long way to go.

See also p 103

Since most of the present chapter will deal with the nuclear forces, both the strong nuclear binding force and the weak force of β-decay, I would like to consider once again the question of when the first idea of a specific nuclear force surfaced, although this has already been commented upon above by Pais, who attributed the 'first intimations of a new, nuclear, force' to J Chadwick and E S Bieler, associates of Rutherford, who continued Rutherford's experiments on scattering of α-particles on hydrogen and in 1921 confirmed the presence of 'anomalous scattering', that is, deviations from point-like Coulomb scattering (figure 5.2). These Cambridge workers concluded that they could not attribute their results to the usually assumed α-particle structure (consisting of four protons and two electrons, bound by Coulomb forces), but rather that 'the law of force is not the inverse square in the immediate neighborhood of an electric charge' [8].

Nuclear Forces, Mesons, and Isospin Symmetry

Figure 5.2. *A stereoscopic photograph showing α-particle tracks in a cloud chamber. One of the α-particles makes an elastic collision with an oxygen nucleus.*

However, in the relevant scientific literature of the following decade, and specifically in the two great nuclear treatises of 1930 and 1931 [9], there is no evidence that any specifically nuclear force was invoked. In fact, the forces in the nucleus, even those governing β-decay, were always spoken of as electrical, with perhaps a magnetic component due to the high velocities of the nuclear electrons. Even the Chadwick–Bieler paper speaks only of modifying the electric force at small distances, and not of a new nuclear force.

See also p 113

5.1.5. Cosmic rays and Heisenberg's analysis of 1932

Since the early days of the century, there had been evidence for the presence of penetrating radiation in the atmosphere, resembling the nuclear γ-rays discovered in 1900 by Paul Villard in Paris, but of much greater penetrability [10]. At first, these rays were thought to come from radioactive gas or dust. For example, in 1905 A S Eve stated that 'the only ionizing agents' producing the radiation in an ionization chamber were:

(i) radiation due to radioactive matter contained in the air [in the chamber];

(ii) radiations due to active matter on the surface, or in the material of the sides of the vessel;

(iii) penetrating radiation through the sides of the vessel due to radioactive matter in the surrounding bodies [11].

Although Eve was expressing the common view, there had already been suggestions of an extraterrestrial source. For example, C T R Wilson considered the possibility of 'radiation from sources outside our

367

atmosphere, possibly radiation like Röntgen rays or like cathode rays, but of enormously greater penetrating power' [12]. Although Wilson's experiments persuaded him that the Earth's known radioactivity was a more likely source, by 1909 balloon experiments were in progress to determine whether the intensity of the penetrating radiation would increase with altitude, and not decrease with increasing height as would be expected from a radiation originating at the Earth's surface.

The decisive balloon flight was made on 7 August 1912 by Victor F Hess, an Austrian physicist and amateur balloonist, carrying several electrometers to measure the radiation (figure 5.3). At 1500 to 2500 m, the radiation was about as intense as on the ground and it then began a clear rise in ionization until the maximum height of 5350 metres [13]. Other balloon flights by the German physicist Werner Kohlhörster, up to 9000 metres, found an ionization 50 times as intense as at sea level and their absorption coefficient showed that cosmic rays were far more penetrating than any known γ-rays.

During the World War I, research in cosmic rays was discontinued, but it began again in the 1920s. With initial doubts in his mind, Millikan undertook a series of measurements in aeroplanes and on mountain peaks which disagreed in some particulars with those of the European observers. However, he eventually found in experiments comparing the absorption coefficients of cosmic rays in two mountain lakes, that 'the rays do definitely come in from above, and that their origin is entirely outside the layer of atmosphere between the levels of the two lakes' [14].

Up to this point, it had been assumed that the penetrating rays were 'ultra γ-rays', and that their energies could thus be inferred from their penetrability. That is, their absorption coefficient was, in principle, obtainable experimentally by measuring their intensity as a function of depth in the atmosphere (or in another suitable absorber, such as water or lead), and their energy (or frequency) distribution could be inferred from the theory of Compton scattering. All this, of course, assumed that cosmic rays were essentially photons of high energy, possibly with some 'softer' accompaniment. Indeed, Millikan's analysis of the absorption data led him to believe that the primary cosmic rays were the photons released when extraterrestrial protons and electrons combined in space to form elements like helium, oxygen and carbon, and he called cosmic rays, to the delight of the newspapers, 'the birth cries of the elements'.

However, beginning in 1929, evidence accumulated to show that the major part of the cosmic-ray primaries were, in fact, not photons, but charged particles. This was done, using a counter telescope based on the coincidence counting technique, by Bothe and Kolhörster, who showed that the charged particles could penetrate a 4.1 cm gold block. Also, Dmitri Skobeltzyn in Leningrad, directly observed fast charged cosmic-ray particles in a Wilson cloud chamber operated in a magnetic field. This field was strong enough to curve the tracks of electrons from any

Nuclear Forces, Mesons, and Isospin Symmetry

Figure 5.3. *Victor F Hess, who discovered cosmic rays in 1912, pictured as an emeritus professor in his laboratory at Fordham University in New York.*

radioactive source, but it could not bend the charged particles of cosmic rays, thus demonstrating that their energy was much higher [15].

In 1932, only days before Chadwick announced the revolutionary discovery of the neutron, Heisenberg sent to the journal *Annalen der Physik*, a very thorough analysis of the existing cosmic-ray data [16]. Using the quantum-mechanical theories of electron and photon scattering and absorption, he reached the conclusion that 20% of the slowing of fast electrons must be due to electrons in the nucleus. In addition, he concluded from the data on high-energy γ-rays that the nuclear electrons must be scattering coherently, so that the scattering cross section would be proportional to the square of their number, and not linear. These conclusions were based upon the relatively strong scattering and/or absorption of both electrons and photons, and upon the high multiplicity of particle production, reported especially in the multiple coincidence

Twentieth Century Physics

experiments carried out by the Italian physicist Bruno Rossi. It may be that it was partly on account of his familiarity with these high-energy observations that it took some time for Heisenberg to appreciate the true significance of Chadwick's neutron.

5.2. *Annus mirabilis*—the new physics of 1932

5.2.1. The discovery of new particles
Important discoveries made during 1932 led physicists to a new view of the atomic nucleus [17], beginning in January of that year with the report of the concentration and spectroscopic identification of the hydrogen isotope of mass 2 (deuterium), whose nucleus (the deuteron) is to nuclear physics what the hydrogen atom is to atomic physics [18]. This was followed in February by James Chadwick's announcement of the discovery of the neutron, at Rutherford's Cavendish Laboratory, where a particle by that name had been unsuccessfully sought for more than a decade. Rutherford had suggested in 1920 that the e–p model of the nucleus had room for a neutral particle of mass number 1 [19]. Chadwick did not view the neutron as elementary, but rather, as he said at a discussion meeting held at the Royal Society on 28 April 1932 'The neutron may be pictured as a small dipole, or perhaps better as a proton embedded in an electron.'. The first person to suggest that the neutron was a new elementary particle was the Russian physicist Dmitri Iwanenko [20].

> See also p 119

Again at the Cavendish Laboratory, in April of this remarkable year, J D Cockroft and E T S Walton succeeded in accelerating protons up to 600 keV, and used them to produce nuclear reactions in light elements. Electrostatic accelerators of a straight-line type (van de Graaff accelerators) were being built to yield protons of several MeV in Washington DC and also in California. At Berkeley, Ernest Lawrence's circular machine, the cyclotron, began running in late summer and had accelerated protons almost to 5 MeV by the end of the year (figure 5.4).

Equally as important as these developments for the future of nuclear physics was the discovery of two additional elementary particles, the neutrino and the positron. The neutrino was not so much discovered as invented by Pauli, as a way to save the conservation laws apparently violated by the process of β-decay. As explained by Pais in Chapter 2, by the end of 1930 Pauli was convinced, largely by his friend Lise Meitner, that there was a real problem with the continuous β-spectrum requiring, as he said, a 'desperate remedy'. (That was how a new particle was thought of in those days!) During 1931 and 1932 he kept this idea rather private, but at the Seventh Solvay Conference in Brussels in October 1933, he went public with the proposal of a new neutral particle of small mass (possibly massless) and with spin $\frac{1}{2}$. Before Chadwick's discovery,

Nuclear Forces, Mesons, and Isospin Symmetry

Figure 5.4. *Ernest O Lawrence (left) and associates in the machine shop with 85-ton cyclotron.*

Pauli had called his new particle the 'neutron', but at Brussels, following a suggestion of Enrico Fermi, he called it the *neutrino* [21].

As for the positron, although it turned out to be the positive electron, i.e. Dirac's predicted 'hole' in the vacuum, its discovery by Carl Anderson in cosmic rays at Caltech, in August 1932, was purely experimental and without theoretical motivation [22]. Anderson found his 'easily deflectable positives' by using a Wilson cloud chamber equipped with magnetic field and expanded at regular intervals. Quite soon afterwards, P M S Blackett and G P S Occhialini did a similar experiment, but with a Wilson chamber of their design, whose expansion was initiated by the firing of electronic counters—a more efficient detector. In this way, they found that a positron was always produced together with an electron, as predicted by Dirac's theory [23]. Often they found that two or more pairs were produced at one time, foreshadowing the important 'shower' phenomena in cosmic rays.

5.2.2. Heisenberg's n–p model of the nucleus

As we mentioned in section 5.1.5, Heisenberg's analysis of cosmic-ray studies in 1932 strengthened his feeling, based upon β-decay, that electrons must be present in the nucleus, although the idea was in

371

conflict with principles of quantum mechanics, and even contradicted very general conservation laws. However, after the discovery of the neutron, he thought it possible to treat that particle as though it were an elementary constituent of the nucleus. Thus, as he wrote to Bohr 'The idea is to shove all the difficulties of principle into the neutron, and to apply standard quantum mechanics within the nucleus.' [24].

Accordingly, Heisenberg wrote in 1932 a three-part paper on a neutron–proton model, which describes the modern picture of the nucleus, insofar as the structure of the various nuclear species, their binding energies and their stability (aside from questions concerning the neutron itself and β-decay) are treated without reference to nuclear electrons [25]. That is, neither the electrons 'shoved' into neutrons, nor the additional 'free' electrons, which Heisenberg thought were required in order to explain cosmic-ray multiplicities, are considered in the discussion of nuclear systematics [26].

The type of attractive force that Heisenberg introduced to bind the nucleus together was an electron-exchange force, of a type which he had used successfully in his treatment of the helium atom and of ferromagnetism, and which was used to explain homopolar molecular binding. The attractive n–p force in Heisenberg's nuclear model comes from an n–p pair sharing one electron, as in the hydrogen molecular ion H_2^+ (i.e. the neutron is considered to be like a collapsed hydrogen atom). The weaker (in Heisenberg's model) n–n force, also attractive, comes from the sharing, or exchange, of two electrons, as in the hydrogen molecule H_2. The p–p force is simply the repulsive Coulomb interaction of two positive point charges [27].

Clearly, Heisenberg was treating the proton as elementary, and the neutron as composite. Indeed, about half of Heisenberg's paper is concerned with the neutron and the puzzling properties it (and the nuclear electrons) must have when viewed from the standpoint of quantum mechanics. Indeed, in order to take this picture seriously, one must assume that the electrons in the nucleus lose characteristic properties, such as their spin and Bose–Einstein statistics, and that just as in β-decay, they violate local conservation laws within the nucleus.

Nevertheless, Heisenberg's nuclear theory was a great advance, and it opened the door for others to apply quantum mechanics in the nucleus, without necessarily adopting Heisenberg's specific exchange forces [28]. At the Solvay Conference of 1933 on 'The structure and properties of atomic nuclei', referred to above in section 5.2.1, Heisenberg gave the summary talk on nuclear theories. All the theories he discussed were phenomenological; i.e. certain functional forms, such as the exchange potentials, were empirically, not theoretically, derived. Although Pauli proposed his idea of the neutrino in the discussion following Heisenberg's talk, it did not receive general approval. Bohr, for example, recommended a β-decay theory proposed by Guido Beck, which did not

Werner Karl Heisenberg

(German, 1901–76)

The discovery of quantum mechanics in June 1925 by twenty-three year old Heisenberg completely changed the view of the microworld held by physicists and chemists and has had a significant effect on the general intellectual climate of the twentieth century. Born in Würzburg, Germany, into a family of scholars, Heisenberg received all of his early education in Munich and attended the University there, where he studied with the famous researcher and teacher, Arnold Sommerfield. In the Golden Age of quantum theory, between 1925 and 1930, when other versions of the theory were proposed by Erwin Schrödinger and Paul Dirac and many new applications were made, Heisenberg and Niels Bohr developed a new epistemology that was based on the Heisenberg uncertainty principle and Bohr's related principle of complementarity. With Wolfgang Pauli, Heisenberg developed the field theory of quantum electrodynamics. In 1931, he turned his attention to the theory of the nucleus and to the study of cosmic rays. For the rest of his life, these fields, whose high-energy aspect became the new field of elementary particle physics, occupied his main attention. Unlike many prominent German and Austrian physicists of his generation, Heisenberg remained in his position during the Nazi era and in World War II he headed the unsuccessful German Uranium Project. After the war, he was interned in Britain with other leading German scientists, and in 1946 returned to Germany, becoming the director of the Max Planck Institute and professor at the University of Göttingen. Later he and his institute moved to Munich, where he died in 1976.

use the neutrino and conserved neither energy nor angular momentum [29].

5.2.3. Fermi's theory of β-decay

Some physicists were reluctant to abandon the usual conservation laws at very small distance, even when it seemed necessary. Bohr, on the other hand, was quite willing to do so, arguing that energy should be conserved only in a statistical sense, i.e. on the average, and not necessarily in individual elementary processes. This would place the first and second laws of thermodynamics on an equal footing, since the second law (the law of increasing entropy) was certainly valid only statistically. He recalled that quantum theorists had found new laws in passing from the macroscopic to the atomic level, no smaller jump in size than that from atom to nucleus.

However, more conservative physicists (or those more sensitive to the necessary link between conservation laws and symmetries) were unwilling to renounce such basic principles. One of these was Pauli, whose 'desperate remedy', the neutrino, was actually a conservative proposal. Another conservationist was Fermi, who returned to Rome from the Brussels conference with β-decay and the neutrino on his mind.

In Heisenberg's n–p model, β-decay was simply the basic process in which a neutron turned into a proton by emitting an electron, which then left the nucleus. Had the electron been absorbed by another proton in the nucleus, it would have been a strong interaction and not a β-decay. To describe either of these processes, Heisenberg considered that the neutron and the proton were two charge states of a single particle (today called the nucleon), and he introduced operators (now called isospin operators) to change from one state to the other. Fermi also used these isospin operators in his new theory of β-decay.

Fermi wished to modify Heisenberg's theory, so that a neutrino would accompany the emitted electron, as Pauli had suggested; because of the high energy involved in β-decay, the electron e and the neutrino ν had to be treated relativistically. To describe the creation of these particles required a relativistic field theory, similar to QED. Indeed, Fermi's theory bore a striking resemblance to a generalized version of QED, in which the n–p transition played a role similar to an atomic transition producing a photon, and the e–ν pair behaved like the quantum of a four-vector field, not very different in its formal properties from the electromagnetic field.

Fermi's theory obeyed all the standard conservation laws. It also differed from Heisenberg's theory in that the weak interaction strength, determined by a new constant G, was unrelated to the strength of the n–p coupling, the main nuclear binding force in Heisenberg's theory. Working it out during November and December 1933, Fermi sent a brief version of his theory to the British journal *Nature*, whose editor declined to publish it as being 'too remote from physical reality to be of interest to the readers' [30]. However, the main results of the theory were published, still in 1933, in an extended note in the Italian journal *Ricerca*

Scientifica. In his full paper, Fermi also applied the theory to obtain lifetimes for many known examples of β-decay, obtaining remarkable success [31].

5.3. Two fundamental theories of nuclear forces

5.3.1. *The Fermi-field theory*
As we have already implied, Fermi's theory of β-decay did not win immediate acceptance, even though it agreed well with experiment and solved puzzling features of that process. For example, in the theory the e–ν pair is produced at its moment of emission, just as a photon is produced in an atomic de-excitation, and is otherwise not present in the atom. Thus it is possible to exclude electrons from the nucleus and to consider the neutron to be a genuinely elementary particle. Nevertheless, at an international conference held in London and Cambridge in October 1934, which Fermi attended, the only β-decay theory presented was the the energy non-conserving one of Beck and K Sitte. It was even claimed to give a better fit to the currently accepted β-decay electron spectra [32]. (That was not too surprising, since unlike Fermi's theory, the Beck–Sitte theory contained an adjustable function of the electron energy. At the same time, it should be noted that for many years the low-energy end of the electron's β-decay spectrum was distorted by absorption in the source.)

Nevertheless, one by one, the reasons for having electrons in the nucleus were disappearing. They were not needed now for β-decay, nor to explain the electromagnetic cosmic-ray interactions that gave rise to multiple secondary particles. These secondaries were now understood to be produced through pair production and bremsstrahlung (the latter process being the emission of a hard photon by an electron deflected in the field of a nucleus) [33]. In January 1934, Irène Joliot-Curie (the daughter of Marie and Pierre Curie) and her husband, J Frédéric Joliot, announced that they had artificially produced a new form of radioactivity by bombarding aluminium and other light elements with α-particles [34]. The new light radioactive isotopes, having too many protons to be stable, emitted positrons; i.e. they were β^+ emitters, as opposed to the β^- emitters, elements with neutron excess that emitted negative electrons. The discovery of artificial radioactivity thus emphasized the symmetry between neutrons and protons, giving another reason to treat them as equal partners in the theory of nuclear forces. (We note that in the β-processes, e$^-$ is accompanied by an antineutrino $\bar{\nu}$, while e$^+$ is accompanied by a neutrino ν; the two types of neutrino are *not* identical.)

Pauli heard of Fermi's theory, even before it was published, from a former assistant, Felix Bloch, who was on a fellowship with Fermi at Rome. Delighted with this application of his neutrino idea, Pauli wrote an enthusiastic letter about it to Heisenberg, one of his regular correspondents. 'That would be water to drive our mill' he wrote, and Heisenberg replied that he was also very pleased [35]. In the same letter, Heisenberg stated 'If the Fermi matrix elements for the creation of an electron–neutrino pair are correct, they must—just as the possibility of atomic electrons producing light quanta leads to the Coulomb force—give rise in the second approximation to a force between neutron and proton.'.

Heisenberg went on to show how one could obtain either his own nuclear charge-exchange force, or a modification introduced by the Italian theorist Ettore Majorana, which agreed better with the known properties of the deuteron [36]. For the exchange energy itself, Heisenberg obtained

$$J(r) \sim mc^2(10^{-14}/r)^5$$

where mc^2 is the rest energy of the electron (about 0.5 MeV) and r is the separation of neutron and proton in cm. That was terribly (*reichlich*) small, he admitted, but perhaps that was owing to the sloppiness (*Schlampigkeit*) of the calculation! If one sets $r = 2 \times 10^{-13}$ cm, which is about the range of nuclear forces, then $J(r)$ is too small, by a factor of more than 10^9. Results equivalent to Heisenberg's unpublished result for the nuclear exchange potential were obtained in Russia by Igor Tamm and Dmitri Iwanenko, independently, and also in America by Arnold Nordsieck [37].

Heisenberg tried to interest Fermi in his modified nuclear exchange-force model, but Fermi thought it unlikely that the forces could be made strong enough by any acceptable change in his theory of β decay. However, Fermi's assistant, Gian Carlo Wick, wrote to Heisenberg, suggesting that the form of Fermi's interaction might be modified, and Wick also proposed an idea which proved to be influential: namely, he suggested that virtual e–ν currents (i.e. the emission and subsequent reabsorption of such a pair by a nucleon) could explain the 'anomalous' value of the proton's magnetic moment, which had recently been measured as several times the 'Dirac moment', $e\hbar/2Mc$ [38].

In spite of the small force represented by the second-order Fermi interaction, it remained one of the leading fundamental theories of nuclear forces up to the outbreak of World War II [39]. Beginning in 1940, a modified version of the Fermi-field theory (ascribed, however, to Gamow and Teller) was adopted by some physicists, who used it in proposing a two-meson theory to solve certain cosmic-ray puzzles [40]. We may thus understand Heisenberg's desire to push the theory in the mid-1930s, in spite of its shortcomings. In the famous 'Bethe bible' of nuclear physics of 1936, the only field theory of nuclear forces considered was that of the Fermi field, and although the authors estimated the interaction to be too small by a factor of 10^{12}, they gave the following reason for not abandoning the theory [41]:

This highly unsatisfactory result is, of course, due to the extremely small value of the constant g which governs the β-emission. However, the general idea of a connection between β-emission and nuclear forces is so attractive that one would be very reluctant to give it up.

The authors then considered several possible modifications that might improve the agreement.

Heisenberg lectured on his version of the Fermi-field theory in April 1934. In October of the same year, at the London Conference referred to above, Hans Bethe repeated a suggestion for modifying the Fermi theory that he and Rudolph Peierls had made at a conference in Copenhagen a month earlier. The idea was to insert one or more derivatives into the interaction. This would modify the electron spectrum at low energies (which was thought to be necessary), reducing the effective interaction strength at low electron energies (though not at the higher electron energies involved in nuclear binding). Thus, for the same β-decay lifetime, a larger Fermi constant was required, and this increased the predicted nuclear binding strength. Heisenberg was pleased with this idea—but not Pauli, who wrote to him 'The present situation in theoretical physics is this, that a *subtraction physics of electrons and positrons stands face to face with a nuclear physics of arbitrary functions.*' [42].

In February 1935, Heisenberg sent a contribution to a Festschrift in honour of the Dutch physicist Pieter Zeeman, reviewing the Fermi-field theory. He stressed the analogy between the theory of light quanta and that of the Fermi field, pointing out that the relation between the latter and the nuclear potential was similar to the relation between light quanta and the Coulomb field. He also mentioned approvingly the Bethe–Peierls suggestion. (Other physicists in America had independently applied the Bethe–Peierls idea to β decay, but concluded that it did not increase the strong nuclear interaction strength sufficiently [43].)

In 1936, new experiments and new analyses caused a profound change in our view of nuclear forces. This change came about in part from the study of nuclear systematics, but mostly because of new precise measurements of proton–proton collisions. These showed that the p–p nuclear force (after correction for the Coulomb force) was as strong as the n–p force [44]. This, in turn, led to the conjecture known as the charge-independence hypothesis: the n–n, p–p and n–p forces (after correction for Coulomb interaction) are identical [45]. The mathematical formulation of this idea used Heisenberg's isospin operators to write an interaction that was invariant (i.e. unchanged) by the substitution of a proton for a neutron, or vice versa, unless the exchange was forbidden by the Pauli exclusion principle.

In order to extend the idea of exchange forces, additional pairs of light particles were added to the Fermi field, namely, the neutral pairs e^+–e^- and ν–$\bar{\nu}$. Those 'neutral currents', unlike the 'charged currents'

$e^- - \bar{\nu}$ and $e^+ - \nu$, allowed for forces between like nucleon pairs p–p and n–n, as well as n–p. The neutral pairs should also appear in β decay in processes in which the nuclear charge would not change. However, in such a case, electromagnetic processes would be far more probable and would mask the weak interactions [46].

Gregor Wentzel, in Zurich, tried to use the Fermi field, extended by neutral currents, in order to make the nuclear forces charge independent, but concluded that this could be done only if the charged currents made negligible contributions. The key was to couple the two neutral currents in such a way that they interfered destructively for unlike nucleons and constructively for like nucleon pairs. The first successful theory of this type was made by Wentzel's student, Nicholas Kemmer, in London, using the isospin operators to write a quantum field theory interaction which had the invariance property noted by Cassen and Condon [47]. Kemmer's use of isospin invariance to make the first charge-independent meson theory was a development which has played a major role in modern particle physics.

5.3.2. The Yukawa meson theory

In October 1934, Hideki Yukawa proposed a theory of nuclear forces transmitted by a set of conjectured particles, now called mesons, but which he named heavy quanta or *U-quanta*. It became his first published scientific paper [48]. Working at the newly created Science Faculty of the University of Osaka, Yukawa was even more an 'outsider' than Albert Einstein was in 1905, when he created relativity at the Swiss patent office in Bern. After its acceptance in 1937, Yukawa's theory became the dominant fundamental theory of nuclear forces until the present standard model of quarks and gluons became established in the 1970s, and it is still the main way of interpreting intermediate energy physics [49] (figure 5.5).

Yukawa graduated from Kyoto University in 1929, staying on as an unpaid assistant in the Physics Department for several years. During this time he became determined to solve at least one of the two problems of theoretical physics that he considered to be outstanding: first, the construction of a consistent, infinity-free QED; and second, a fundamental theory of nuclear forces. He believed that these two problems were closely related, since he adopted the conventional viewpoint that electrons were nuclear constituents. The problematic electrons of the nucleus were also highly energetic, and it was at very high energy that QED was an especially troubled theory.

After spending two years, with very limited success, on the problems of relativistic QED (which, throughout his lifetime he considered to be the major unsolved problem of physics), Yukawa turned his attention to the 'easier' problem of nuclear forces [50]. In his approach, he was mainly indebted to Heisenberg's three-part paper, discussed above in

Nuclear Forces, Mesons, and Isospin Symmetry

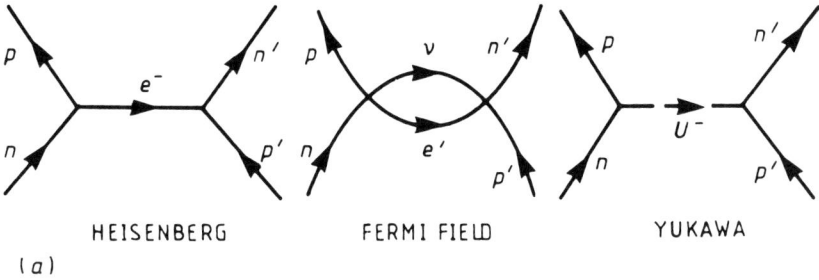

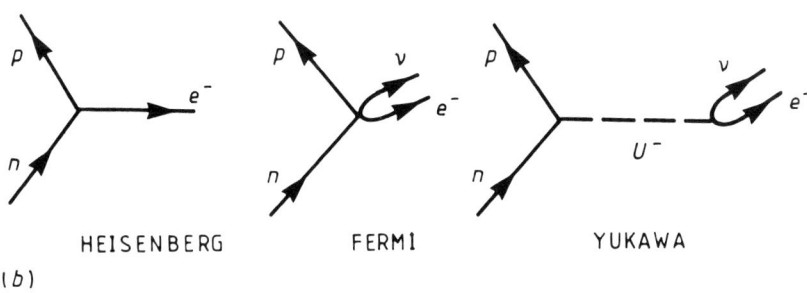

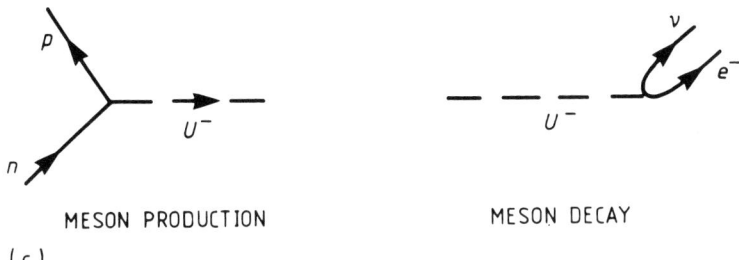

Figure 5.5. *Feynman diagrams (with time as ordinate), comparing Heisenberg, Fermi field, and Yukawa interactions. (a) Models of the strong nuclear charge-exchange force. (b) Models of the (weak) β-decay interactions. (c) Other processes predicted by Yukawa's theory. p, p': protons; n, n': neutrons; e⁻: electron; v: neutrino; U⁻: heavy quantum.*

section 5.2.2, and he prepared a summary and critique of that work, which included this statement [51]:

> Although Heisenberg does not present a definite view on whether neutrons should be seen as separate entities or as combinations of a proton and an electron, this problem, like the β decay problem, cannot be resolved with today's theory. And unless

BOX 5B: QUANTUM FIELD THEORY

Quantum field theories (QFTs) that are used in elementary particle physics are generalizations and adaptations of quantum electrodynamics (QED). Thus, the standard model of the past two decades is built upon elementary spin $\frac{1}{2}$ particles, exchanging spin 1 gauge fields, as is QED.

For simplicity, we consider the quantization of the real scalar field $\phi(x)$, representing the wavefunction of a free neutral spinless particle of mass m and satisfying the equation

$$(\Box + \lambda^2)\phi = 0 \tag{B1}$$

where $\Box = \partial^2/c^2\partial t^2 - \partial^2/\partial x^2 - \partial^2/\partial z^2$ and $\lambda = mc/\hbar$. The four-vector x, with components $x_0 = ct$ and $\boldsymbol{x} = x_1, x_2, x_3$ labels the point in space–time at which the field $\phi(x)$ has a particular value. The position and momentum of the particle are represented by operators q_i, p_i which obey the canonical commutation relations

$$[q_i, p_j] = i\hbar \delta_{ij}. \tag{B2}$$

In quantizing the field $\phi(x)$, its value at x is taken as the coordinate and its canonically conjugate 'momentum' is defined by $\pi(x) = \partial\phi(x)/\partial t$. At a given time, the fields obey

$$[\phi(\boldsymbol{x}), \pi(\boldsymbol{y})] = i\hbar\delta(\boldsymbol{x}-\boldsymbol{y}). \tag{B3}$$

Because $\phi(x)$ was a quantum-mechanical wavefunction upon which we have imposed the quantum conditions (B3), this procedure is sometimes called 'second quantization'. The field $\phi(x)$ now describes any number of particles. (Note that $\boldsymbol{x}$ and $\boldsymbol{y}$ are now c-numbers.)

An important consequence of field quantization is that the number of particles is not constant. The field can be represented by the Fourier transformation

$$\phi(x) = \int \frac{d^3p}{(2\pi)^3} \frac{1}{\sqrt{2E_p}} (a_p e^{-ip\cdot x/\hbar} + \bar{a}_p^{+ip\cdot x/\hbar}) \tag{B4}$$

where E_p is the energy of a particle with momentum p. In this expression a_p and $\bar{a}_p$ are operators that, respectively, destroy and create a particle of momentum p.

CONTINUED

> **BOX 5B:** *CONTINUED*
>
> A quantum field interacts with other quantum fields and it may also have direct self-interaction. During the 1930s, the most important QFTs were: QED, involving the electromagnetic field and the Dirac electron (also applied to the proton); the Fermi theory of β decay, involving the electron, neutrino and the nucleons (also extended as a unified theory to include the strong nuclear force); Yukawa's meson theory, like the Fermi field, a unified theory of the strong and weak nuclear forces.
>
> QED, in the restricted sense, is a theory of electrons and photons in interaction. The free photon field is represented by a real four-vector potential $A_\mu(x)$ satisfying the equation
>
> $$\Box A_\mu(x) = 0. \tag{B5}$$
>
> The supplementary condition $\partial A_\mu/\partial x_\mu = 0$ eliminates the timelike spin 0 component A_0, and because of the masslessness of the photon, it also eliminates the longitudinal component of the field, thus guaranteeing the transversality of free photons.

these problems are resolved, one cannot say whether the view that electrons have no independent existence in the nucleus is correct.

Before he knew of Pauli's suggested neutrino, Yukawa tried to develop a quantum field theory embodying Heisenberg's 1932 electron-exchange force. In Yukawa's theory, the electron was regarded as a field (namely, the Dirac field), with Heisenberg's n–p transition playing the role of 'source current'. The analogy was to an electron transition (e.g. its change of state in an atom) which acts as the source of photon emission. Like Heisenberg's theory, a single basic interaction was intended to account for nuclear binding and scattering, as well as for β decay. Without the equivalent of a neutrino, both theories violated various conservation laws. In 1933, Yukawa made a quantum field theory (QFT) version of Heisenberg's phenomenological theory, and thus he was able to calculate the exchange potential [52]. (He was not trying to predict new physics.) The exchange potential he obtained turned out to be disappointing; he pointed out that it 'has a form like the Coulomb field and does not decrease sufficiently with distance' [53].

After Yukawa became an instructor at Osaka University, a colleague called attention to Fermi's paper on β decay in the Italian journal *Ricerca Scientifica* [54]. In his autobiography, *Tabibito* (The Traveller), dealing with his early years, he described his reaction [55]:

After reading Fermi, I wondered whether the problem of the strong nuclear force could not be solved in the same manner. That is to say, could the neutrons and protons be playing 'catch' with a pair of particles, namely an electron and a neutrino? The 'ball' would be replaced by a pair of particles.

This was the idea of the Fermi-field theory of nuclear forces, as described above in section 5.3.1.

After working on this approach, and after realizing that it gave an inadequate nuclear exchange potential, as published by Tamm and Iwanenko in the British journal *Nature*, Yukawa said (in *Tabibito*), 'it opened my eyes, so that I thought ... if I focus on the characteristics of the nuclear force field, the particle I seek will become apparent'. The key characteristic of the field was the short range of force; Yukawa discovered that this required a new particle, the meson, about 200 times as massive as the electron.

The mechanism of nuclear forces in Yukawa's theory may be understood by a comparison with QED. In the latter theory, the electromagnetic field is represented by its four-vector potential A_μ, which obeys the equation $\Box A_\mu = e j_\mu$, where $\Box$ is the d'Alembertian operator and $e j_\mu$ is the charge-current operator of the electron field†. The basic interaction, written as $e j_\mu A_\mu$, causes the emission or absorption of a photon in first order. In second order (i.e. a photon is emitted by one electron, followed by its absorption by another), it gives rise to the Coulomb potential energy; e^2/r.

Analogously, in Yukawa's theory, the nuclear force field is represented by U, which obeys the equation $(\Box - \lambda^2)U = gJ$, where λ is a new inverse length (related to the range of nuclear force) and gJ is the operator corresponding to the nuclear field (more precisely, to the n–p transition). The basic interaction, written as JU, causes the emission or absorption of a U-quantum in first order. In second order (e.g. a U^+-quantum emitted by a proton and then absorbed by a neutron), it gives rise to the Yukawa potential energy: $g^2 \exp(-\lambda r)/r$. This is Yukawa's expression for Heisenberg's phenomenological charge-exchange energy. (In order to accommodate both U^+ and U^- particles, the U-field must be complex.) The most important result of the theory was the relation between the range parameter λ and the mass m_U of the U-quantum: $\lambda = h/m_U c$. Although this relation is now a commonplace in quantum field theories, it was unknown when Yukawa developed his theory.

In his effort to make a field theoretical version of Heisenberg's n–p model, Yukawa wanted to preserve its character of unifying nuclear force and β decay. To do so, he let the U-field be coupled also to e^-–$\bar{\nu}$ and e^+–ν, with a smaller coupling constant g'. In modern terminology, the U-quanta played the role of 'intermediate bosons'. In his article,

† $\Box = \partial^2/c^2 t^2 \partial t^2 - \partial^2/\partial x^2 - \partial^2/\partial y^2 - \partial^2/\partial z^2$; $j_\mu A_\mu = j_0 A_0 - J_X A_X - J_Y A_Y - J_Z A_Z$.

Yukawa gave the relation between his two constants, g and g' and Fermi's constant G, as follows

$$G = gg'/4\pi.$$

It is worth noting that Yukawa was the first person to clearly distinguish the weak and strong nuclear interactions by assigning them separate coupling strengths.

Yukawa had made an attractive unified theory of nuclear forces and published it, in acceptable English, in a widely available journal. He had exploited the analogy with QED in such a way that it should have been immediately accessible to quantum theorists. Thus it may seem surprising that his theory languished and was not taken seriously, even in his native Japan. It is true that he did not belong to any fashionable school and that he laboured at the geographical periphery; that, doubtless, had some influence on the theory's reception, although science is not supposed to work that way. However, the most important reason seems to have been the reluctance of physicists to consider it valid to postulate new particles without direct observational confirmation. (Pauli's neutrino, his 'desperate solution', was a partial exception, but it too was resisted at first.) As we shall see in the following section, in 1937 it appeared that Yukawa's U-quanta were found in cosmic rays.

5.4. Cosmic rays in the 1930s: QED, showers, and the mesotron

5.4.1. The soft and hard components
As we indicated in section 5.1.5, physicists generally assumed that the primary cosmic rays (those entering the Earth's atmosphere) were 'ultra γ-rays', because of their high penetrating power. However, it was shown in 1928 that the rays at sea level contain charged particles, with enough energy to penetrate an appreciable part of the Earth's atmosphere (equivalent to about 10 m of water). During the 1930s, a worldwide survey of the intensity of cosmic rays as a function of latitude established that the Earth's magnetic field deflects the primaries, which must, therefore, carry electric charge [56] (figure 5.6). These primary particles were then considered by many physicists to be energetic electrons, even though the sign of the primaries appeared to be mainly positive. After World War II, observations with high-flying aeroplanes and balloons showed that the primary cosmic rays were predominantly protons and other atomic nuclei.

However, most of the cosmic rays reaching mountain altitude and sea level are secondaries, consisting of a complex mixture of several kinds of particles with a broad spectrum of energies. Measurements of absorption were made by comparing the intensity of ionization produced by the rays at various heights in the atmosphere, or at different depths in lakes, or by coincidence counting, using solid materials as

Twentieth Century Physics

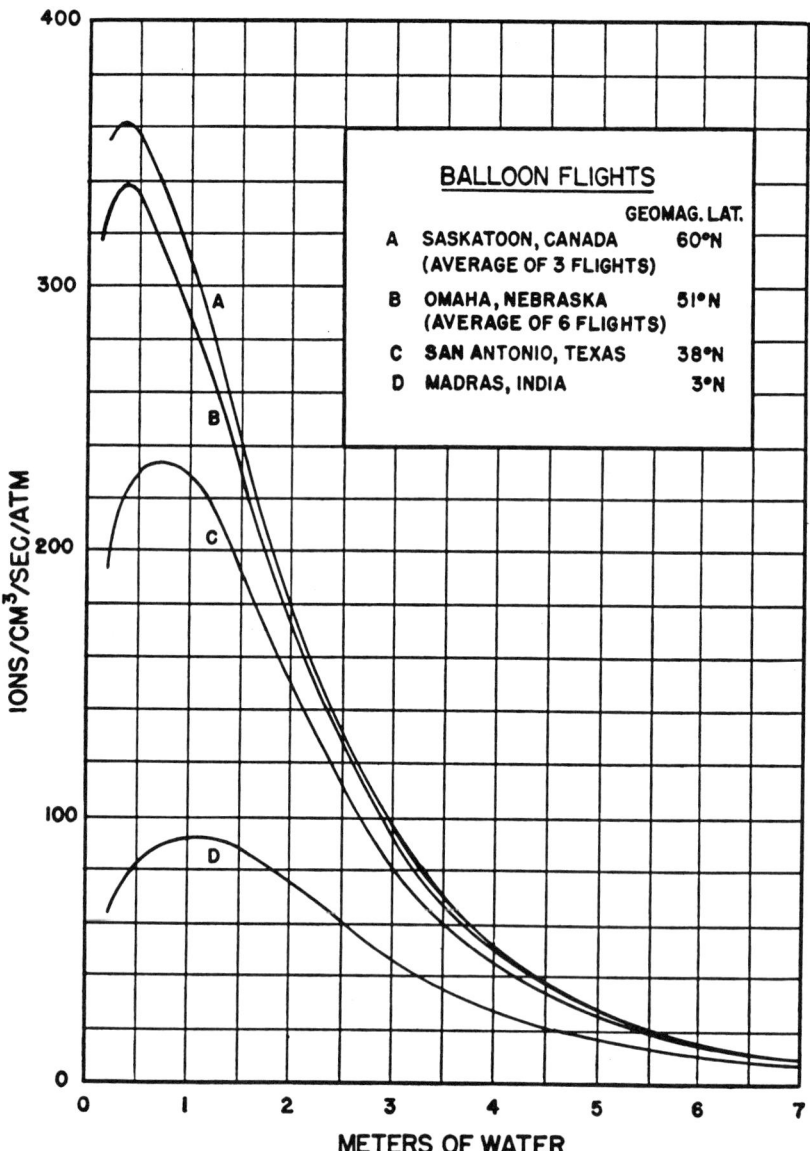

Figure 5.6. *Cosmic-ray intensity, measured by specific ionization, as a function of altitude. The abscissa expresses the altitude as atmospheric pressure in metres of water equivalent (0 m is the top of the atmosphere).*

absorbers. Already in the late 1920s, it became apparent that there is a 'soft' component, easily absorbable, and a 'hard' component, of great penetrating power. The former typically contains many, and sometimes very many, charged particles appearing in groups, forming a cone that can be photographed when produced in or near a Wilson cloud chamber.

The hard component consists mainly of separate singly charged particles, whose small ionizing power and high penetrating power indicate that they have great energy. The intensity of the soft component tends to diminish through absorption, but it is continually regenerated and is present in the atmosphere at all altitudes.

In the 1930s, the only available source of particles with energies greater than a few MeV was cosmic rays; thus they were the natural experimental testing ground for fundamental theories [57]. High particle energy was needed to explore the limits of the theories of QED, relativity and nuclear forces. However, scientific study of cosmic rays was very challenging, and it suffered from severe limitations compared with work with radioactive sources and particle accelerators. One problem was the diversity of the phenomena to be sorted out; a second was the broad spectrum of energy. Both limitations arose from the circumstance that one was observing naturally occurring phenomena, without control of the source, so that it became a matter of observational selection, rather than the more controlled experimentation possible in a laboratory. A related limitation was the relative weakness of the cosmic source, requiring long observation to obtain any sort of precision. Finally, one of the theories to be investigated was itself indispensable for interpreting the observations: namely QED, whose validity had to be assumed, even as it was being tested.

In October 1934, at the international conference on physics in London, cosmic rays were the subject of spirited discussion. One issue, as discussed above, was the question of the particle nature of primaries. Another issue was the validity of QED predictions for high-energy behaviour, which appeared to differ from the observations. In 1934, Bethe and Walter Heitler published their theoretical predictions for bremsstrahlung and e^+–e^- pair production in the field of a nucleus [58]. They found the surprising result that above some energy that was typical of the stopping material (e.g. 20 MeV for lead and 120 MeV for water), almost all the energy loss of an electron would be due to radiative processes, and not to its ionization of the medium through which it passed. Since the radiation intensity is proportional to the squared acceleration of the charge, and hence inversely proportional to the squared mass of the particle, the radiative effects would be about a million times smaller for a proton than for an electron, while its ionization energy loss would be the same at the same velocity. At the London conference, where Carl Anderson and his co-worker Seth Neddermeyer reported on their measurements of absorption in lead plates placed in a cloud chamber, Bethe remarked that 'the radiative energy loss seems far smaller than that predicted by theory' [59]. Bethe's comment was based on the assumption that the particles observed were electrons, and not more massive particles.

That QED might fail at high energy was not in itself surprising.

Carl David Anderson

(American, 1905–91)

Anderson carried out important studies of the particle content of the cosmic rays, using the Wilson cloud chamber operated in a strong magnetic field to allow a measurement of momentum, and discovering several new elementary particles in this way. Anderson did his undergraduate and graduate work at the California Institute of Technology, and continued his research there, originally under the direction of Millikan. Anderson's discovery of the positron in 1932 confirmed the existence of antimatter and the production of electron–positron pairs, predicted by the electron theory of Dirac, although Anderson was not aware of Dirac's theory when he made his discovery. Several years later Anderson and Seth H Neddermeyer found positive and negative particles in the cosmic rays which behaved as heavy electrons, and which they called mesotrons (now called muons). After the discovery of two examples of V-particles by G D Rochester and C C Butler in Manchester in 1947, no further examples were seen for two years until Anderson's group obtained 34 examples in their cloud chamber two years later.

Indeed, it was generally expected to do so, because of QED's 'divergences', i.e. its predictions of infinite mass and charge. The only question in the minds of the theorists was at what energy the theory would 'break down'. Robert Oppenheimer was among those who believed that the breakdown energy could be as low as a few MeV [60]. In fact, most cosmic-ray interactions involve energy transfers not much larger than had already been observed in the laboratory and found to agree with QED. If one views the collision between an electron and a nucleus in the frame of reference in which the electron is at rest, the electric field of the nucleus, moving rapidly in that frame, can be represented, in good approximation, as a superposition of photons,

Nuclear Forces, Mesons, and Isospin Symmetry

the bulk of which have relatively small energy. Their interaction, essentially Compton scattering, is what is seen as bremsstrahlung in the laboratory reference frame [61]. More careful experiments by Anderson and Neddermeyer in 1936, in which electrons were identified and distinguished from the penetrating component did, in fact, verify QED at cosmic-ray energies.

The convincing test of QED was, however, the explanation of the soft component of cosmic rays as a cascade process: an incoming fast electron produces one or more energetic photons (as well as softer ones), and each energetic photon either Compton scatters or produces an e$^+$–e$^-$ pair. These electrons, in turn, interact and produce photons, some of which produce pairs. The result is a 'shower' of many electrons and photons, which continue to multiply in number until the energy is so divided that pair production is no longer possible. The calculation of these shower processes, using the Bethe–Heitler predictions for the fundamental processes, was done in England and in America; the results were in good agreement with each other and with the bulk of the observed shower phenomena [62].

Before the successful treatment of cascade showers, some physicists thought that a shower was produced in a single elementary very high-energy interaction, an interpretation which was consistent with their visual appearance in cloud chambers. Even after it was shown that most showers could be explained as cascades, there were those who believed in the other type, called *explosive* showers, on both experimental and theoretical grounds. On the experimental side, it appeared difficult to account on the cascade theory for certain very large showers, which were called 'bursts', and which continued to puzzle observers as late as 1953, when they were found to be initiated not by photons or electrons, but by energetic nuclear fragments.

On the theoretical side, Heisenberg in June 1936 had proposed a theory of explosive showers arising from strong nuclear interaction. Recall that Heisenberg was an advocate of the Fermi-field theory of nuclear forces (section 5.3.1) [63]. From this he reasoned as follows: the four-fermion Fermi interaction can be written in the form

$$H' = f j_\mu(np) j_\mu(e\nu)$$

where the js represent, respectively, the n–p and the e–ν currents and $f = G_F/\hbar c$, with G_F being the Fermi constant. The strength of the QED interaction is determined by the fine-structure constant $\alpha = e^2/\hbar c \simeq 1/137$. Being *dimensionless*, α does not determine any scale of length. However, the analogous weak coupling coefficient f^2 has the dimensions of the fourth power of a length. Thus the validity of an expansion in powers of f is limited by the requirement that f/λ^2 be small compared to 1, where λ is a the de Broglie wavelength of the electron or the neutrino

[64]. Hence there is a 'characteristic length' $\lambda_0 = \sqrt{f}$, such that $f/\lambda_0^2 = 1$. Since perturbation methods cannot be applied for wavelengths smaller than λ_0, it is possible that high-energy collisions could produce multiple e–ν pairs. Heisenberg conjectured that that was the case, providing a mechanism for explosive showers. Even conceding that most showers might be of the cascade type, Heisenberg and his co-workers insisted that some showers must be of the explosive type [65].

Aside from showers of various types, there remained the puzzle of the penetrating cosmic radiation, the hard component, which could penetrate more than 200 m of water equivalent, or ten atmospheres. These individual high-energy penetrating particles could not be explained by the cascade hypothesis. As Bhabha and Heitler remarked in their shower paper 'We must conclude, either that the extremely hard radiation which penetrates 250 m of water consists of particles of protonic mass, or that the quantum theory of radiation breaks down for radiation of the highest energies if it consists of electrons.' [66]. On the one hand, we recall that it was fashionable to insist on the high-energy breakdown of QED. On the other hand, an (often unstated) objection to the notion that the penetrating particles might be protons is that they were positively and negatively charged in about equal numbers.

Paul Kunze, at the University of Rostock, had studied these very energetic particles with a cloud chamber operated in a strong magnetic field, and had observed charged particles that 'ionize too little for a proton, too much for a positive electron' [67]. Other pioneers in the study of the penetrating radiation were the French group of Louis Leprince-Ringuet and the English group led by P M S Blackett [68]. However, the most extensive data revealing the composition of the penetrating component came from the observations in California of Carl Anderson and Seth Neddermeyer.

Using a counter-controlled cloud chamber with magnetic field, the chamber crossed by a 0.25 cm lead plate, enabling a measurement of the energy lost by shower particles, Anderson and Neddermeyer concluded that '... the experimental energy losses and those calculated from the theory are in agreement up to energies ~ 300 MeV' [69]. However, there were some puzzling tracks that were not associated with showers, and thus unlikely to be electrons, but were also difficult to identify as protons. On 12 December 1936 Anderson, having been awarded the Nobel Prize for his discovery of the positron, which he shared with Victor Hess, the discoverer of cosmic rays, gave an address in Stockholm. Anderson's concluding sentence referred to the puzzling tracks as follows 'These highly penetrating particles, although not free positive and negative electrons, will provide interesting material for future study.'.

Little note was taken of Anderson's comment, nor of the results on penetrating particles that he had published with Neddermeyer. (An important exception was Yukawa, as we shall see presently.)

However, by mid-1937, the existence of new particles, which we shall call mesotrons, had received confirmation from several sources [70]. In a paper published in May 1937, in which they used a 2 cm platinum plate to separate the soft and hard components, the Caltech workers focused their attention on penetrating particles with less than 500 MeV of energy, whose momentum they could analyse magnetically. The particle having the lowest momentum would have ionized 25 times more strongly, had it been a proton. From this and other single tracks, they concluded that they were seeing new particles [71]. Similar results, using a counter telescope and two cloud chambers, were reported in a talk in April by J C Street at a meeting of the American Physical Society [72]. The first mass determinations, yielding a mass of 100 to 200 times the mass of the electron, were made by Street and E C Stevenson and by a Japanese cloud chamber group led by Yoshio Nishina [73].

5.4.2. The new meson theories in Japan and Britain

As mentioned, Yukawa paid careful attention to the experimental studies of the penetrating component of cosmic rays, and this encouraged him to work again on his meson theory in the autumn of 1936. To call attention to a possible connection between his postulated U-quanta and the puzzling cosmic-ray cloud chamber tracks, he wrote a letter to the British journal *Nature*. Yukawa's letter began by pointing out that the Fermi-field theory was unable to account for the very different strengths of nuclear forces and β decay, and it then described his alternative theory of U-quanta. The letter concluded with a reference to the Anderson and Neddermeyer article of 1936 [69], and went on to say [74]

> Now it is not altogether impossible that the anomalous tracks discovered by Anderson and Neddermeyer, which are likely to belong to unknown rays with e/m larger than that of the proton, are really due to such quanta [i.e. U-quanta], as the range-curvature relations of these tracks are not in contradiction to this hypothesis.

Unfortunately, the editor of *Nature* rejected Yukawa's letter; however, he submitted an English language letter of similar content to a Japanese journal, which did publish it [75].

The first response to the mesotron that mentioned the U-quantum hypothesis, other than Yukawa's own letter, was published in the 15 June 1937 issue of the *Physical Review* [76]. The authors, Oppenheimer and Robert Serber, were familiar with the work of the Caltech physicists, and they had discussed the mesotron with Anderson, Neddermeyer and Millikan, all of whom strenuously insisted that the mesotron was a purely experimental discovery. Perhaps for that reason, the theorists' letter, while calling attention to Yukawa's work, nevertheless argued that in his theory of nuclear forces, 'one meets with difficulties hardly

Hideki Yukawa

(Japanese, 1907–81)

Yukawa, the inventor of the meson theory of nuclear forces, was one of the first Japanese scientists to receive his education entirely in Japan and to make an international reputation without leaving the country. Born in Tokyo, he moved with his family to Kyoto as a child (as did his classmate and fellow Nobel Laureate, Sin-itiro Tomonaga). After graduating from Kyoto University in 1929, he stayed on as an unpaid assistant, doing theoretical research on quantum electrodynamics (QED) and nuclear forces, the two most challenging theoretical physics problems of the 1930s. Although he failed to solve the 'divergence' problems of QED (as Tomonaga, influenced by Yukawa, did in the 1940s), at the end of 1934 at Osaka University Yukawa developed a new fundamental theory of nuclear forces, involving the exchange of massive spinless charged particles of both signs, which he referred to as the heavy quanta of a postulated nuclear force field. Observing a connection between the range of the forces and the mass of the field quantum, he predicted that the quanta would appear in cosmic rays, with mass about 200 times that of the electron, as a result of collisions made by fast cosmic-ray particles in the atmosphere. Anderson and Neddermeyer reported the observation of particles fitting this description in 1937, and soon they were widely assumed to be Yukawa's heavy quanta. In fact, those particles (now called muons) are decay products of the Yukawa particles, which are now called pi-mesons or pions, and which were discovered in 1947 by a group from Bristol University led by C F Powell and Giuseppe Occhialini.

less troublesome than in the various forms of electron–neutrino theory which have been proposed [i.e. the Fermi-field theory and its Bethe–Peierls variant]'. Thus, they concluded 'These [Yukawa's] considerations therefore cannot be regarded as the elements of a correct theory, nor serve as any argument whatsoever for the existence of the particles.'.

The second theoretical response, a much more positive one, appeared in the next issue of the *Physical Review*, a letter from the Swiss theorist Ernest Stueckelberg, who stated that he had 'independently arrived at the same conclusion' as Yukawa [77]. After setting forth his own theory (which was a unified theory of strong and *electromagnetic* interactions, involving a five-component field), Stueckelberg concluded

> It seems highly probable that Street and Stevenson, and Neddermeyer and Anderson, have actually discovered a new elementary particle, which has been predicted by theory. This particle is unstable and can only be of secondary origin, its mass being greater than the sum of the masses of electron plus neutrino.

In Yukawa's original paper of 1935, he had stressed the analogy of the heavy quantum with the light quantum of QED, to the extent that he represented the U-field as a relativistic four-vector. Since the sources of the field, i.e. the nucleons, were moving in the nucleus relatively slowly compared with light velocity, the effective nuclear potential was calculated as a generalization of the electric scalar potential, leaving out effects of the magnetic type. The U-quanta thus had zero spin, and were, in that sense 'scalar', but under relativistic transformations behaved as the fourth component of a relativistic four-vector. When Yukawa, at the end of 1936, again turned his attention to the meson theory, with the assistance of his students, Shoichi Sakata and Mituo Taketani (and later Minoru Kobayasi), he formulated two new versions of the theory, their fields transforming, respectively, as relativistic scalar (spin 0) and relativistic four-vector (spin 1) [78].

When he wrote the first meson paper, Yukawa was not aware that Pauli and his assistant Victor Weisskopf had worked out a relativistic quantum theory of an elementary charged scalar field interacting with the electromagnetic field [79]. The scalar theory had many of the characteristic features of Dirac's QED: it described particles and antiparticles of opposite sign and predicted electromagnetic pair production and annihilation, but it did not need the Dirac 'sea' of mostly filled negative energy states. Although the theory did not describe known particles, certainly not electrons, Pauli was very pleased to have made what he later referred to as an 'anti-Dirac' theory. It was the Pauli–Weisskopf theory that Yukawa and Sakata used in formulating the relativistic scalar meson theory in the paper we shall refer to as *Interaction II* [80]. In this paper, they already consider the question of whether neutral, as well as charged, mesons are needed in order to satisfy the charge independence of nuclear forces.

Simultaneously with their work on the relativistic scalar meson theory, Yukawa and Sakata were working with Taketani on a vector meson theory, involving particles of spin 1 [81]. One of their motivations was the discovery that the charge-exchange force in the scalar theory (whose *sign* is fixed by that theory) predicted incorrectly that the ground state of the deuteron should have 0 spin, while the vector theory correctly predicted spin 1. Another reason for preferring spin 1 was their desire to explain the anomalous magnetic moments of the nucleons, and they thought that virtual interactions with scalar mesons could not contribute to the magnetic moments.

Just before sending the third paper, *Interaction III*, for publication in March 1938, Yukawa received a letter (in German) from Herbert Fröhlich and Walter Heitler in Bristol, which initiated a period of cooperation, as well as competition, between the Yukawa school and a group working in Britain. (The other members of the British school were Nicholas Kemmer and Homi Bhabha [82].) The letter began [83]

> Many thanks for the interesting manuscripts that you have been sending us, especially the last one. We quite believe that your theory is correct in principle. We ourselves have considered a great deal about the heavy electron and have formulated a theory of its interaction with the nucleus (together with Kemmer). From the discussion of the spin dependence of the proton–neutron force we have arrived at the conviction that the field [of the heavy electron] must be a vector field, as you have also assumed in your last Japanese note. The wave equation is the same as yours and has already been proposed by Proca (*Journal de Physique* 1936).

Fröhlich and Heitler belonged to the great wave of emigrants fleeing from Germany after the Nazis seized power in 1933. Most of the refugee scientists settled either in Great Britain or in the United States [84]. Both Fröhlich and Heitler had received their doctorates at the Munich institute of Arnold Sommerfeld, a great teacher whose students included Heisenberg, Pauli, Bethe and many other famous physicists. At Bristol, Fröhlich and Heitler studied paramagnetism at low temperature, which had a bearing on the problem of measuring the anomalous magnetic moment of the proton (in connection with the magnetization of solid hydrogen). This, in turn, interested them in a possible theoretical explanation of the proton's 'anomaly'.

Noting the failure of the then dominant Fermi-field theory to account for the proton's magnetic moment, and drawing upon Heitler's knowledge of the new particles, they suggested that the proton spends a fraction α of the time virtually dissociated into a neutron and a positive 'heavy electron', and that a neutron is a fraction α of the time a proton plus a negative 'heavy electron'. Using the measured values of the magnetic moments of the proton and neutron, they found that α was about 0.08 and that the mass of the heavy electron was about 80 times the mass of the electron [85]. Their argument makes a number of assumptions: the nucleon in the virtual state is assumed to have the Dirac moment; the heavy electron is assumed to have no spin or magnetic moment; the basic interaction is assumed to reverse the spin of the nucleon. In their model, the additional magnetic moment arises mainly from the orbital motion of the virtual heavy electron.

The Bristol physicists do not cite Yukawa, but their letter to *Nature* was soon followed by two others which did: from London, a letter from Kemmer, and from Edinburgh, one from Bhabha. Kemmer had been in contact with his former teacher, Wentzel, regarding the formulation of a charge-independent version of the Fermi-field theory (section 5.3.1), and the latter called his attention to Yukawa's and Sakata's *Interaction II*. Citing the difficulty (already noted by the Japanese authors) that the scalar meson theory gave the wrong spin to the deuteron (0, rather than the correct value of 1), Kemmer wrote that 'a more satisfactory relation can be obtained if one admits a vector wave function for the new particle, such as was used by Proca in a different connection' [86].

Immediately following Kemmer's letter in *Nature* was that of Bhabha. As we have noted in section 5.4.1, Bhabha and Heitler formed one of the two teams that had given a theory of the soft component of cosmic rays in terms of cascade showers [62]. Soon afterward, they had separately turned their attention to the hard component [87]. Thanking Heitler for drawing his attention to Yukawa's work during a discussion on cosmic rays, Bhabha advertised the advantages of the vector meson theory over the scalar, both in his letter and in a longer paper which followed [88]. But probably the most important contribution of Bhabha was his emphasis on the consequences of Yukawa's hypothesis for cosmic rays, in particular the meson's instability [89]

> A positive U-particle at rest may disintegrate spontaneously into a positive electron and a neutrino. This disintegration being spontaneous, the U-particle may be described as a 'clock', and hence it follows merely from considerations of relativity that the time of disintegration is longer when the particle is in motion. We believe that this may have to do with the fact observed by Blackett and others that below 2×10^8 eV most cosmic-ray particles are electrons, above this energy heavy electrons. In a previous paper we have shown that the experimental evidence requires that heavy electrons can apparently turn into ordinary electrons. Our U-particles are then to be identified with the heavy electrons, and it follows that most of the heavy electrons have been created either in the Earth's atmosphere or not very far from it.

Both Bhabha and Kemmer pointed out that the charged U-particles should have a neutral partner, Bhabha on the grounds that a strong attractive short-range p–p force was required by experiment, Kemmer on the related grounds of charge independence. Recall that Yukawa and Sakata had claimed as much in *Interaction II*. Bhabha's article appeared in volume A166 of the *Proceedings of the Royal Society*, in which it was followed immediately by an article by Heitler which detailed various 'shower' processes due to the mesotrons that should be observed in cosmic rays [90]. These papers were preceded in the volume by important works on meson theory by Kemmer and by Fröhlich, Heitler and Kemmer, who had met in London at the Royal Society and discovered their common interest [91].

Kemmer's separate paper, which was the more mathematically complete of the two articles, calculated the nuclear potentials resulting from the exchange of U-particles of either spin 0 or 1, each of either parity (i.e. even or odd under a space reflection). Considering then the prototype of nuclear structure physics, the deuteron, Kemmer concluded that the known properties of the ^{3}S ground state of that particle, together with the low-energy scattering data, singled out the vector meson theory. He pointed out that 'The detailed discussion of this fact is the subject of the paper with Fröhlich and others', by which he meant the paper that followed with Heitler and himself. Surprisingly, although Kemmer was the first to write down a charge-independent theory, requiring invariance of the interaction under rotations in isospin space, he overlooked an important consideration. As Pauli pointed out to him in wartime correspondence, a correct fitting of the deuteron's properties, including its quadrupole electric moment, actually requires not a vector but a pseudoscalar meson (i.e. one of spin 0 and odd parity) [92].

Heitler, much later, recalled his association with Kemmer and Fröhlich as follows [93]

> [When Kemmer joined us] he was working in London, but we met often. That gave a very happy balance to our efforts, for Kemmer had mastered the formalism of quantum field theory much better than we. The three of us could then establish a meson theory (as we call it today). It was a vector meson theory, in analogy to the Maxwell theory. Only much later was it recognized that the meson field was a pseudoscalar. We could also predict the existence of a neutral pi-meson (from the proton–proton interaction) which was later discovered. From that Kemmer then developed the charge-symmetric theory, which was the basis of the concept of isospin. The vector meson theory was developed in three different places at the same time and in almost identical ways: by Yukawa, Sakata, and Taketani and by Bhabha [and by ourselves].

5.5. Mesotrons, mesons, and the birth of particle physics

5.5.1. The cosmic-ray mesotron

When new particles of mass intermediate between electrons and protons were found to be present in cosmic rays, there arose the question of

BOX 5C: SPIN AND ISOSPIN

The 'up' and 'down' spin states of a particle with spin $\frac{1}{2}$, such as the electron or the proton, with respect to a direction in space (say, the z axis) can be represented by column matrices

$$\Uparrow = \begin{pmatrix} 1 \\ 0 \end{pmatrix} \quad \text{and} \quad \Downarrow = \begin{pmatrix} 0 \\ 1 \end{pmatrix}.$$

Square arrays (Pauli matrices) are used as quantum operators to represent the spin angular momentum (in units of $\hbar/2$)

$$\sigma_x = \begin{pmatrix} 0 & 1 \\ 1 & 0 \end{pmatrix} \quad \sigma_y = \begin{pmatrix} 0 & -i \\ -i & 0 \end{pmatrix} \quad \sigma_z = \begin{pmatrix} 1 & 0 \\ 0 & -1 \end{pmatrix}.$$

Thus

$$\sigma_z \Uparrow = \Uparrow \qquad \sigma_z \Downarrow = -\Downarrow \qquad \sigma_x \Uparrow = \Downarrow \qquad \text{etc.}$$

We regard the three σs as the components of a vector $\boldsymbol{\sigma}$. The rotationally invariant interaction of the spins of particles a and b is written $\boldsymbol{\sigma}_a \cdot \boldsymbol{\sigma}_b$.

Isospin emerged in the 1930s as an analogue to spin (hence its name). It is a so-called internal quantum number, the first of its kind, that characterizes strongly interacting particles. It was first used by Heisenberg in 1932 in his n–p (neutron–proton) nuclear model, in which p and n are regarded (in a sense) as the charged and neutral states of a single particle (analogous to the two spin states $\Uparrow$ and $\Downarrow$). In the following we shall use the modern term *nucleon* for the nuclear particles p or n, and the notation, differing slightly from Heisenberg's, will also have the modern form.

We define a vector $\boldsymbol{\tau}$ in a three-dimensional abstract space, called charge space. Its components τ_1, τ_2, τ_3 have exactly the same form as the matrices $\sigma_x, \sigma_y, \sigma_z$ and we have

$$p = \begin{pmatrix} 1 \\ 0 \end{pmatrix} \quad \text{and} \quad n = \begin{pmatrix} 0 \\ 1 \end{pmatrix}.$$

The charge-independent nuclear interaction of two nucleons a and b would be written as proportional to $\boldsymbol{\tau}_a \cdot \boldsymbol{\tau}_b$.

In a similar manner, mesons with three charge states (+, − and 0) can be represented by matrices with three rows and three columns, and charge-independent interactions between mesons and nucleons can be constructed analogously.

whether they also played a significant role in the structure of matter. (The search for unity and coherence in the natural world did not end in the ages of myths and religions.) As discussed above, a possible answer was found in the U-particle hypothesis of Yukawa, which initiated a programme to decide whether the 'Yukawa particle' and the 'Anderson particle' were the same. The cosmic-ray discoverers of the new particles had been unaware of Yukawa's proposal, and the first theorists to make the connection differed in their published views: Yukawa and Stueckelberg thought they were the same particle, while Oppenheimer and Serber took a strongly contrary view [76, 77].

Many suggestions were made for naming the new cosmic-ray particle, but by 1939 it came down to roughly an even division for the names 'mesotron' and 'meson', with Millikan an enthusiastic advocate for the former [94]. To those who used other names, especially the detested 'meson', Millikan wrote letters of objection, and most offenders promised to change. An exception was Hans Bethe, who suggested that 'it might be well to keep the name "mesotron" for the experimental thing and "meson" for the theoretical', advice that Millikan considered neither wise nor practical. The experimental thing was, of course, the cosmic-ray particle, and the theoretical one was the 'heavy quantum' of the nuclear force field that Yukawa had proposed. Millikan notwithstanding, we shall adopt Bethe's suggestion and use 'mesotron' and 'meson' in the following, in order to distinguish the two particles.

The task of the cosmic-ray physicists was to determine, by direct observation when possible, and otherwise indirectly from the composition and behaviour of cosmic rays as a function of altitude, latitude, temperature, etc, the mass, mean life, decay scheme and interactions, nuclear and electromagnetic, of the mesotron. From the interactions, it might be possible to infer the spin as well as the coupling strength. The task of the nuclear theorists was to obtain predictions for the range and strength of nuclear forces, based on assumed properties of the mesons, and to compare them with what was known and what was being learned from nuclear systematics and laboratory scattering experiments. Also, by studying radioactive decay, one could investigate the relation between the mesotron mean life, on the one hand, and the mean lives and electron spectra of β-emitters, on the other.

Determining that the mesotron had a mass smaller than the proton and larger than the electron was not very difficult. Since the proton–electron mass ratio is almost 2000, their cloud chamber tracks are very different in appearance; and mesotron tracks did not match either of these profiles. It was quite another matter to determine the mesotron mass accurately with the resources of the 1930s. The momentum of a particle could be deduced from the curvature of its cloud chamber track in a magnetic field, the particle's velocity from its ionization, and the energy from its range, providing that it came to rest before decaying or being captured. Any two different measurements on a track gave a mass determination, but the measurements were difficult to perform except near the end of the range. The first published mass determination on a stopping mesotron was about 130 m_e, made by Street and Stevenson at MIT, and the second was 220 ± 40 m_e, made by Nishina and his co-workers in Tokyo. Because of the uncertainty in the mesotron mass value, and because the complicated nature of the low-energy nuclear forces did not allow a precise definition of 'range of force', there was no sharp confrontation of theory and experiment on this issue [95].

In contrast to the questionable status of the mass comparison, the mesotron lifetime was rich in consequences. Although the first Yukawa meson paper clearly implied the instability (essentially the β decay) of the U-particle, it was not until *Interaction III*, submitted in spring 1938, that the Japanese authors remarked that [96]

> ... a heavy quantum disappears even in the free space by emitting a positive or negative electron and a neutrino or an antineutrino simultaneously ... as already pointed out by Bhabha ... Owing to the small interaction of the heavy quantum with the light particle, the possibility of occurrence of the above process is so small that the mean free path of the high speed heavy quantum in free space is large compared with the dimensions of the measuring apparatus, but is not small enough, in many cases, to make the mean free path larger than the height of the whole atmosphere.

Several implications followed: the probability of a decay within the measuring apparatus was too small for its direct observation in a cloud chamber to be likely. However, the mesotron lifetime was small enough to ensure that the observed mesotrons are produced in the atmosphere, and that they have a high probability of decaying before reaching the ground. The mesotron mean life at rest was estimated by using a weak interaction coupling strength that was obtained from a fit to nuclear β decay. To this a relativistic time-dilation factor was applied, which is essentially the particle energy, measured in units $m_e c^2$. The mean life at rest was given in *Interaction III* as $\tau \simeq 5 \times 10^{-7}$ s, and for a mesotron of energy 10^{10} eV the decay mean free path was 30 km.

Other theoretical estimates of the lifetime, by Heisenberg and Wentzel, agreed with the above value (within a factor of 10) [97]. In a letter to Pauli, Heisenberg noted that such a short decay time 'would be very important for discussion of the cosmic radiation' [98]. This issue was pursued in several papers by Heisenberg's student, Hans Euler, especially in a long review paper with Heisenberg [99]. Assuming that the mesotron was the hard component, and that its decay was responsible for the fast electrons that initiate the soft component, the ratio of the two components at any altitude gave a measure of the mesotron lifetime. From the ratio measured at sea level, Euler inferred that $\tau = (2\pm 1) \times 10^{-6}$ s, and asserted that it agreed with Yukawa's value from β decay, namely 5×10^{-7} s [100].

Another method to determine the mesotron mean life was to compare the absorption of the cosmic-ray intensity, measured by its ionization, under comparable thicknesses of air and water (or other dense material). These thicknesses, conveniently expressed as distance times density, e.g. in units g cm^{-2}, give nearly the same 'absorption' through ionization loss. However, the distance in air corresponding to one such unit is far greater than in water, so that the loss of mesotrons in air because of their decay can be appreciable in air but negligible in water. By comparing the slopes of various absorption curves, Euler and Heisenberg obtained a more exact mean life. As Heisenberg wrote to Yukawa [101]

> Euler has discussed the measurements of [Alfred] Ehmert and similar measurements, on the basis of your hypothesis, and has obtained for the decay time of the heavy electrons 2×10^{-6} s. This value does differ by about a factor 4 from your theoretical value [based upon β decay], but it is certainly questionable how quantitatively certain one can consider experimental numbers in the field of cosmic rays.

Heisenberg asked Yukawa, in the same letter, whether the agreement might be improved if one used the Konopinski–Uhlenbeck form for β decay, which included derivatives in the basic interaction. Yukawa replied that the agreement with Euler's value, using the Fermi form, was rather worse than stated, because of a calculational error. The theoretical value should have been given as 0.25×10^{-6} s. The Konopinski–Uhlenbeck interaction gave a much shorter mean life, by a factor 10^4, 'so that it is impossible to reconcile the cosmic-ray phenomena with the theory of the K-U type' [102].

In an address to the British Association, Section A on 20 August 1938, Patrick Blackett applied the Euler–Heisenberg analysis to observations made in the Holborn station of the London Underground in 1936, as well as to the 1937 work of Ehmert. Blackett also referred to work by Pierre Auger and co-workers in France, bearing on what Blackett called the 'mass absorption anomaly' [103]. After a rough estimate of the mesotron mean life, he continued

> There seems, therefore, to exist definite evidence for the spontaneous decay of the new particle. The accurate determination of this time of decay and of the mass of the particle is now one of the outstanding problems of cosmic ray research.

In addition to absorption experiments in Europe, and also in America, there were attempts to analyse older, pre-mesotron, cosmic-ray measurements which had appeared puzzling at the time. These included experiments carried out in 1934 in Eritrea by Sergio de Benedetti and by Bruno Rossi [104]. The latter reported at a Symposium on Cosmic Rays, held at the University of Chicago in June 1939, that a survey of all available data showed the mean life of the mesotron at rest to be $(2 \pm 1) \times 10^{-6}$ s. At the same time, he pointed to the lack of any direct observation of the decay, and asserted that 'there was no evidence that mesotrons which have been stopped actually disintegrate into an electron and a neutrino' [105]. However, Rossi did not suggest any alternative decay scheme.

At the suggestion of Arthur Compton, Rossi began experiments that summer at various altitudes on Mount Evans, in Colorado. The accurate measurements of his group proved the decay in flight and confirmed the 2 μs mean life. They also verified the relativistic time-dilation effect, its first real test since it was proposed in 1905 by Einstein [106]. The lifetime accuracy was greatly improved by Franco Rasetti, working at Laval University in Quebec. Rasetti, an Italian refugee physicist (as was Rossi), used a 'counter-telescope' of coincidence and anti-coincidence counters with an iron absorber, and was able thus to isolate stopping mesotrons. Using an electronic timing device, he could establish that a mesotron brought to rest emitted an electron a few microseconds later. He found the value $\tau = (1.5 \pm 0.3) \times 10^{-6}$ s [107]. This was further improved to $(2.15 \pm 0.07) \times 10^{-6}$ s by Rossi and Nereson at Cornell, who also reported that 'only half of the absorbed mesotrons undergo disintegration, in agreement with the results of Rasetti' [108].

5.5.2. Mesotron decay and beta decay

By mid-1939 there was great uncertainty about whether the mesotron and meson were identical, and a new idea was beginning to emerge: perhaps there were two different, but nevertheless related particles. Let us review the pros and cons of the situation in 1939.

On the positive side, there clearly existed singly charged particles with mass roughly as expected from meson theory, although the problematic nature of the concept of 'range of force' and the inaccuracy and spread of the mass values made precise comparison out of the question. (Were there, possibly, two or more different masses?) Again on the positive side, the mesotron almost certainly decayed, based initially on indirect cosmic-ray evidence and later on direct observation, and this decay was predicted by the unified Yukawa theory of strong and weak interactions.

On the negative side, one could not be certain that the decay was into one electron (positive or negative) and one neutrino—another neutral particle might also be produced. Also the theoretical lifetime of the meson, based on β decay was always shorter than the observed mean life of about 2 μs, even when the theorist took advantage of the possibility of 'tuning' the theoretical value by modifying the form of the β-decay interaction or choosing a more favourable mass value from the list of

Patrick Maynard Stuart Blackett

(British, 1897–1974)

The son of a London stockbroker, Blackett intended to pursue a naval career and attended naval college. In 1914, at the outbreak of war, he went on active duty as a midshipman and rose to the rank of lieutenant. After the war he resigned his commission and studied physics at Cambridge University. Working under Ernest Rutherford at the Cavendish Laboratory, he used a Wilson cloud chamber in 1924 to visually confirm Rutherford's 1919 observation of the transformation of nitrogen into oxygen through α-particle bombardment (the first observation of a nuclear reaction). In 1932, together with the Italian physicist Giuseppe Occhialini, he constructed a cloud chamber whose expansion was controlled by the coincident signals of a set of Geiger counters, making it possible to choose for observation selected cosmic-ray events in a highly efficient manner. After Anderson's discovery of the positron, the Cambridge investigators used their triggered chamber to show many examples of the production of electron–positron pairs, as predicted by Dirac's relativistic electron theory, and they also demonstrated the important cosmic-ray shower phenomenon. In 1937, Blackett became a professor at the University of Manchester. He was one of the leading scientists to play a role in Britain's World War II effort, especially in the field of anti-submarine warfare. After the war, Blackett returned to Manchester, moving in 1953 to the Imperial College of Science and Technology in London. There he continued his research, and also was active and outspoken on scientific issues related to society.

measured values. One possible solution was to have the β decay proceed by the original Fermi four-fermion interaction, without a meson as intermediary boson. This would decouple β decay from the meson-decay interaction, each being assigned its own coupling strength. However, the meson theory would then lose some of its attractiveness as a unified theory.

Could one use nuclear systematics (i.e. the properties of the stable ground states), low-energy nuclear scattering and nuclear dynamics to determine the mass, spin and coupling strength of the nuclear meson, and from these predict the high-energy behaviour of the cosmic-ray mesotron? On that basis, it was hard to understand why mesotrons did not interact more strongly in flight, scattering or being absorbed in the atmosphere. Why did half of the mesons decay when brought to rest? Finally, the nuclear forces were known to be charge independent. If the charged mesotrons were the quanta of the nuclear charge-exchange force, where were the neutral mesons needed for p–p and n–n interactions? We will consider some attempts to answer these questions, which involve the birth of modern particle physics, in the present section and the next one.

Serber was one of the theorists who took note of the lifetime discrepancy (which had been claimed by Lothar Nordheim to be three orders of magnitude [109]) and pointed out that in the spin 1 meson theory presented in *Interaction III*, the general β-decay interaction had two contributions, one with a meson intermediary and one of the direct four-fermion type. The latter process gave τ proportional to E^7 (E being the energy release) rather than the E^5 dependence of the direct process 'as required by experimental evidence' [110]. Serber continued

> The only hope is to deny the mesotron any role in β decay, and return to direct emission of the light particles by the heavy ones ... One is thus forced to give up any theoretical connection between β decay and mesotron decay: both can take place, but they must be supposed completely independent processes.

Other theorists used the lifetime discrepancy as support for introducing new versions of the meson theory. In Copenhagen, Christian Møller and Leon Rosenfeld introduced a theory of mesons of mixed spin 0 and spin 1. Together with Stefan Rozental, they wrote in *Nature* [111]

> ... it may be expected that it will enable us not only to avoid the discrepancy pointed out by Nordheim, but also to account for such considerable variations of the form of the beta-spectrum and the value of the beta-decay constant from element to element, as are already indicated by the present experimental data.

Bethe was considering a *purely neutral* meson theory, in part to obtain charge-independent forces. Bethe and Nordheim worked on the predictions of the standard charged spin 1 theory, and upon obtaining a meson mean life of about 10^{-8} s wrote [112]

> One has to admit, therefore, that Yukawa's theory in its currently used form does not give a quantitative account of the meson decay. Therefore, in order to obtain a sufficiently rapid nuclear β

decay, it seems necessary to introduce again a direct interaction between heavy and light particles such as the original Fermi interaction ... If this is done, there is, of course, not much point in introducing an additional β decay through the meson field, and we may just as well use the neutral meson theory which does not yield any β decay at all.

In Germany, Sin-itiro Tomonaga, who was on leave from his position in Tokyo to study with Heisenberg in Leipzig, also tried to tackle this problem. His idea was to make the four-fermion interaction the fundamental one, so that the meson decay was a two-step process: first, a virtual decay into a nucleon–antinucleon pair; second, the annihilation of the pair into electron plus neutrino. The difficulty was that the calculation required a 'loop integral', and the result obtained was infinite. Although Tomonaga regretted the time and effort this unsuccessful attempt cost him, after returning to Japan upon the breakout of war in Europe in 1939, he turned his attention to the similar divergences of higher-order calculation in QED, eventually winning a Nobel Prize for his work.

In Japan also, independently of Tomonaga, the idea emerged of making the four-fermion interaction the mechanism for meson decay. Following a suggestion made to him in 1938 by Taketani (at this time in prison for his opposition to the militaristic policy of the Japanese Government), Sakata carried out the calculation and published work based on this idea two years later [113].

5.5.3. The meson and nuclear forces

In a review article in 1939, Rudolph Peierls summarized the situation with regard to the meson theory of nuclear forces [114]. Considering, for simplicity, only the four cases of spin 0 and spin 1, each with even or odd reflection symmetry (parity), he wrote the basic equations of the theories and remarked on the analogy to QED of the spin 1 theory [115]. The latter admitted an interaction analogous to magnetism, so the source particles, the nucleons, could be assumed to have a 'meson moment', coupled to the mesons with a constant f, as well as a charge coupling with a constant g. The scalar theory, however, could have only the charge type of coupling. Thus, Peierls pointed out, 'the details of the meson theory depend upon our assumptions as to (a) the spin of the meson, (b) its symmetry character, (c) the 'meson charge' g and the 'meson moment' f of proton and neutron, (d) the roles of charged and uncharged mesons' [116].

He pointed out some problems of the relativistic scalar theory of *Interaction II*, which included; (i) wrong sign of force in the deuteron ground state, (ii) wrong exchange character of the force for saturation, (iii) no spin–orbit interaction, hence no quadrupole moment of deuteron,

(iv) forces not charge independent [117] and (v) infinite nucleon self-energy. Some of these difficulties can be dealt with by assuming the meson has spin 1 and that the coupling f is non-zero. With suitable choice of g and f one can overcome objections (i), (ii) and (iii). Including neutral mesons, (iv) can be dealt with, and as for the infinite self-energy, this needs a 'cut-off', i.e. a modification of the theory at small distances. A cut-off is required, in any case, in the spin 1 theory, which has a $1/r^3$ singularity in its potential that results in infinite binding energies in attractive states.

One interesting attempt to satisfy the nuclear force requirements with a cut-off at small distances was made by Bethe. Since f must be non-zero, he tried for simplicity $g = 0$, and obtained (with a common cut-off) correct energies for both the spin 1 ground state of the deuteron and for its next higher state, of spin 0. Bethe remarked that 'The meson potential is superior to the old [phenomenological] forces in *predicting a quadrupole moment for the deuteron* because the ground state of this nucleus will now be a mixture of an S and D state rather than a pure S state.' [118]. Such a 'quadrupole moment' of the deuteron had been measured, and the results showed a 'cigar-shaped' nucleus [119]. Bethe found that he could fit that moment, as well as the energy levels, with a 'reasonable' cut-off of about a third of the meson Compton wavelength, providing that he used the purely neutral theory. For Kemmer's symmetric theory, however, the cut-off would be too high ($\sim 1.7\lambda_C$) and worse, the quadrupole moment had the wrong sign (giving a 'pill-box' shape) and it was much too large.

Bethe's stated preference for a purely neutral meson theory was displeasing to many physicists because, as Møller and Rosenfeld (Bohr's associates at Copenhagen) put it [120]

> This ... amounts of course to giving up the remarkable connexions suggested by the symmetrical theory between the problem of nuclear forces and those of cosmic-ray phenomena, beta decay and especially the magnetic moments of proton and neutron.

Referring to their own recent proposal to use a mixture field, they continued [121]

> In view of this unsatisfactory feature of the vector meson theory, we should like to point out that the question of the quadrupole moment of the deuteron, as well as the whole problem of the consistency of the meson theory of nuclear forces in its present provisional form, takes a quite different aspect when ... pseudoscalar meson fields are introduced besides the vector fields in such a way as to cancel all singular terms in the static interaction energy of the heavy particles. In such a theory [using Kemmer's charge-independent interaction], no question of any cutting-off arises.

The reference to pseudoscalar mesons, that is particles of spin 0 and odd parity, turned out to be prophetic, for when Yukawa's mesons were discovered, after World War II, they were found to be of that type.

We close this section by mentioning two other approaches to the meson theory as the 1930s drew to a close. Recall that in section 5.3.1 we discussed how Heisenberg was led, on the basis of the Fermi-field theory of nuclear forces, to the idea that there would be a small 'fundamental length' within which the theory would be greatly modified, resulting in the production of 'explosive showers' or 'bursts'. After the meson theories became accepted, Heisenberg still adhered to the idea of a fundamental length that would limit the application of quantum field theory [122]. Using this idea, he proposed that in treating the multiple production of mesons in high-energy collisions, or their scattering and/or absorption, a useful approach would be to consider the classical meson field (analogous to the use of classical electromagnetic fields when large numbers of photons are involved in a process) [123]. Besides the large, rapidly developing cascade showers that were identified as bursts (as it later turned out), there was another type of shower consisting of strongly interacting particles, initiated by fast protons (or, as it later turned out, by fast nuclei and fast mesons). These were first noted by Marietta Blau and Hertha Wambacher, Viennese physicists, who pioneered in the study of cosmic-rays tracks in photographic emulsion [124].

Finally, a type of theory was proposed that combined features of the Fermi field and Yukawa theories, in order to make use of the mesotrons as the nuclear field quanta and to finesse the absence of neutral mesons [125]. The rationale for this kind of theory was stated this way by Marshak

> The heavy electrons are assumed to be identical with electrons in every respect ('hole' theory, Fermi statistics, etc) except that their rest mass is taken equal to the cosmic-ray meson mass ... The range is directly connected with the rest mass of the heavy electron pair field (in contrast to the Gamow–Teller pair theory). At small r, the potential goes as $1/r^5$ so that one has to cut off in the same way as in the original electron–neutrino theory. The advantage of the heavy electron pair theory is that it deals with particles which can be identified with the cosmic-ray meson.

5.5.4. The penetrating radiation

We have mentioned in section 5.4.1 the very high penetrability of cosmic rays, the hard component being capable of penetrating more than twenty atmospheres. However, it only gradually became apparent that if the bulk of the hard component really consisted of mesotrons with strong nuclear interactions, then they should have scattered strongly (i.e. at large angles) and caused nuclear disruptions that would lead to their slowing, and hence to absorption or decay. On the contrary, charged cosmic rays were observed even in deep railroad tunnels and mines [126].

By 1942, it was apparent that the small nuclear interaction of mesotrons was a serious problem for the theory. As R P Shutt wrote about his observations in a cloud chamber experiment with inserted metal plates [127]

> All theories based upon a mesotron spin of 1 predict a nuclear scattering cross section per neutron or proton between 10^{-25} and 10^{-26} cm^2, values 100–1000 times too large to agree with the experiments ... As Williams and others have shown, because of their intensity and short range, the nuclear forces must scatter into large angles considerably wider than those given by the electrical theory, and this type of scattering would, therefore, be easily recognized.

Although Shutt did observe large-angle scattering in about 2% of the tracks, that was about the same as the estimated proton component in the hard component, so his result was actually compatible with no mesotron nuclear interaction. (It was impossible to distinguish fast protons from fast mesotrons in Shutt's chamber, which in this experiment had no magnetic field.)

5.6. Discoveries during and just after World War II

5.6.1. More doubts about mesotrons: decay versus capture

While impending war loomed, the question 'Does mesotron = meson?' continued to vex the physicists. There were neither logical reasons nor compelling experiments that required a positive answer. Indeed, except for the relationship of mass and nuclear force range (which was not a strong argument), all the other indications pointed to dissimilarity. e.g. the lifetime discrepancy, the lack of a good fit to nuclear β-decay spectra, and the failure to find a neutral meson [128]. Nevertheless, there were powerful teleological and emotional reasons to pursue the possible identity of the two particles. If they were not the mesons promising a unified theory of strong and weak nuclear forces, then why were there mesotrons at all?

The continuing strong interest in pursuing this question in just those terms can be taken as an example of 'social construction'. However, as Steven Weinberg has pointed out, so could, for example, mountain climbing [129]

> Mountain climbers ... may argue over the best path to the peak, and of course these arguments will be influenced by the traditions of mountain climbing and the history and social structure of the expedition. But in the end the expedition will either get to the

peak or they will not, and if they do get there they will know it. No mountaineer would write a book about mountain climbing with a title like 'Constructing Everest'.

As we have indicated in the last section, the lack of evidence for strong nuclear interaction of mesotrons began to emerge as the biggest puzzle in elementary particle physics. A crucial measure of nuclear interaction would be the comparison of the rates of mesotron decay and capture, especially for slow mesotrons. An analysis of this ratio was made by Yukawa and Taisuke Okayama in 1939 [130]. They estimated the capture time in a dense medium, such as lead, to be about 10^{-8} s, much shorter than the stopping time due to ionization, while the decay time was longer (about 10^{-6} s). Thus, they concluded, 'the majority of mesotrons are captured by nuclei after having stopped completely' [131]. In gaseous media, e.g. air, they expected that most mesotrons would decay in flight.

Tomonaga and Gentaro Araki improved the results of Yukawa and Okayama by taking into account the very different effect of the Coulomb field of the nucleus on slow mesotrons of opposite sign. They found that negative mesotrons should almost always be captured, while positive ones should decay [132]

> Since the probability for negative mesons being captured is seen always to be larger than the probability of disintegration ... the negative mesons will be much more likely captured by nuclei than disintegrate spontaneously, not only in dense materials but also in gases. On the other hand, practically all positive mesons will disintegrate spontaneously because of the extremely small capture probability due to the existence of the [electrostatic] potential barrier.

As mentioned above, the later experiment of Rasetti on the mesotron lifetime did show that half of the slow mesons disintegrated [107]. (That half of the slow mesotrons are captured is, however, not an argument for a strong nuclear interaction, since the weakly interacting negative mesotrons may remain near a nucleus for a time comparable to their decay mean life, a long time on the nuclear time scale.)

The predictions regarding capture and decay did not depend greatly upon the still uncertain spin of the mesotron. One of the approaches to try to settle the spin question was based upon the presumed dependence of the electromagnetic properties of the particles on spin. (The reason that this was a presumption was that, aside from spin $\frac{1}{2}$, there were no independent experimental confirmations, and the theories were known to be seriously flawed at high energy.) For example, two of Oppenheimer's students at Berkeley calculated cross sections for the production of a mesotron pair by a photon and also for bremsstrahlung by a mesotron scattering in the electromagnetic field of a nucleus, and

Nuclear Forces, Mesons, and Isospin Symmetry

they compared them with the observed 'burst' frequency in cosmic rays [133]. Oppenheimer argued that their results established that the mesotron must have either spin 0 or spin $\frac{1}{2}$, although he considered the latter value 'improbable'. Since the analysis of nuclear forces tended to favour the pseudoscalar version of the meson theory, he noted that 'The results of CK [Christy and Kusaka] can thus not be regarded as adding a further difficulty to this in itself highly unsatisfactory theory of nuclear forces.' [134].

We can only mention here some of the many attempts to 'explain' the small nuclear interaction of the mesotron. One of the most important of these attempts was at the same time a response to the evident failure of weak coupling perturbation theory because of the large meson–nucleon coupling constant. This problem was attacked by the so-called *strong coupling theory*, in which probability amplitudes were expanded in inverse powers of the coupling constant [135]. However, as the coupling constant was not really 'large', but rather close to unity, Tomonaga improved the theory so that it could accommodate 'intermediate coupling' [136]. However, his paper did not become known in the West until after the war.

Another Japanese wartime innovation of great significance was the two-meson theory, already mentioned above [137]. The idea, first proposed in 1942, was that the Yukawa meson, interacting strongly with nuclei, is produced at a high altitude in the atmosphere and subsequently decays rapidly (with $\tau \approx 10^{-8}$ s) into the weakly interacting cosmic-ray meson, or mesotron, that is observed at lower altitudes and decays more slowly. A full version was read in September 1943 by Sakata and Takeshi Inoue, and an English language version was published later in an international journal started by Yukawa in 1946 [138].

A definitive experiment that established that the mesotrons observed at sea level could not be the Yukawa particle was carried out in Rome, beginning in 1943 in the basement of a high school (Liceo Virgilio), under dangerous wartime conditions [139]. Using an ingenious double magnet arrangement to focus either positive or negative cosmic-ray particles on an iron absorber, the Rome workers confirmed the prediction of Tomonaga and Araki that positives would decay and negatives would be captured [132]. However, when the iron was replaced by carbon, they found that, contrary to the Japanese prediction, both positives and negatives decayed, rather than being captured. This astonishing result, published in 1947, was analysed theoretically to show that it disagreed with the capture rate expected for Yukawa mesons stopping in carbon, the margin of disagreement being a factor of 10^{12} [140]! This conclusion is not in conflict with the capture of the negatives in iron, because the dependence of capture on atomic number Z goes as Z^4.

See also p 652

5.6.2. The discovery of the pion

The same year, 1947, saw the direct observation of the Yukawa particle, which exhibited itself in a way that confirmed the prediction of the two-meson theory of Sakata and Inoue [141]. The technique that revealed the nuclear force meson, and showed it to be different from the mesotron, was a refinement of the photographic emulsion method pioneered by Blau and Wambacher for the study of cosmic rays in 1937. Using microscopic examination and measurement of charged particle tracks in emulsion, it was first put to use as a quantitative detector in cyclotron experiments by Cecil F Powell at the University of Bristol in 1943 [142]. Subsequently, a programme to increase the sensitivity of the emulsions and increase their thickness was set up 'under Joseph Rotblat including chemists from the photographic firms of Ilford Ltd and Kodak Ltd' [143]. Powell was joined from Brazil in 1945 by the Italian expatriate physicist Giuseppe Occhialini, an experienced cosmic-ray physicist who had worked earlier with Blackett. (Occhialini then arranged a year later for the invitation to Bristol of two Brazilian colleagues, Cesare Lattes and Ugo Camerini.) Using the improved emulsions, Donald Perkins, in London, and Occhialini and Powell, in Bristol, observed a small number of events in which a stopping meson produced a 'star', that is, several secondary tracks originating from the point at which the meson came to rest. These were interpreted as the capture of a negative meson, which gave up its rest energy to the capturing nucleus [144].

Much more surprising (as negative mesotrons were already known to be captured) was the observation of 'double meson' tracks, with one intermediate mass particle coming to rest and decaying into another particle of somewhat smaller mass, but not an electron [145]. Soon, using newer electron-sensitive emulsions, the Bristol group was able to observe a two-step decay cascade, consisting of a positive Yukawa meson, christened π (or pion), decaying into a second particle, christened μ (or muon), which also came to rest with a track of fixed length (showing that it was a two-body decay) (figure 5.7). The muon in turn decayed into an electron and one or two other neutral particles. (It has later turned out to be a three-body decay: into a positron and two different neutrinos.)

5.6.3. More particle discoveries

As though the year 1947 were not rich enough in discoveries, astonishing new events were found in cloud chamber observations in Manchester. Using a large counter-controlled Wilson chamber that had been built by Blackett and used for the study of penetrating showers by the Hungarian expatriate physicist Lajos Jánossy, George Rochester and Clifford Butler found two similar events (a half year apart) consisting of 'forked tracks'. The first of these was of inverted-V shape, originating below a thick metal plate. It was interpreted as the decay of a new neutral particle, which analysis showed must be heavier than the pion. The second V

Nuclear Forces, Mesons, and Isospin Symmetry

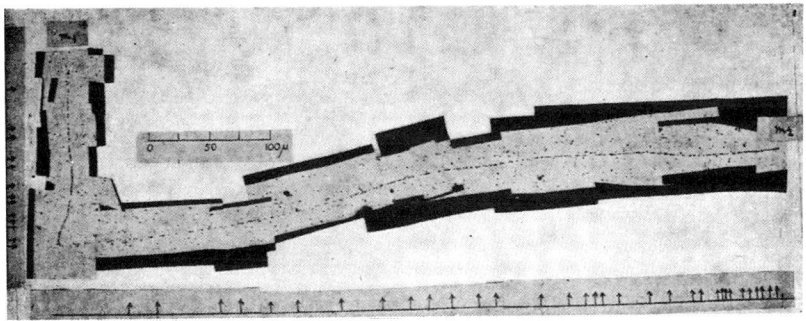

Figure 5.7. *One of the first observed decays of a pion (vertical track) decaying into a muon (horizontal track), after coming to rest in nuclear emulsion.*

particle track, as the new objects came to be known, was interpreted as a charged particle decay [146]. The V particles were later described as 'strange', because they were produced strongly (at high altitudes, not sea level, where as noted, they were scarce), but they decayed with a long mean life, i.e. weakly. The strange particles presaged an 'explosion' in the number of new particles, as discussed in Chapter 9.

See also p 656

Although practically all the particle discoveries that we have discussed were made in cosmic rays, progress in building new accelerators for the production of controlled beams of particles of relativistic energy made rapid progress after the war, and by the mid-1950s largely took over what had come to be called high-energy nuclear physics. Already by 1948, Yukawa mesons were artificially produced at the Berkeley synchrocyclotron (although it took a cosmic-ray physicist from Bristol to demonstrate the nuclear emulsion technique there) [147]; and to emphasize this point: the neutral pion was observed first at another Berkeley accelerator, the electron synchrotron, and only afterwards was detected in cosmic rays [148].

5.7. Conclusion

With the discovery of the neutral pion, it appeared that the way was clear for a complete elementary particle description of matter: atoms made of electrons and nuclei, nuclei made of protons and neutrons interacting with charge-independent forces mediated by the exchange of charged and neutral pions. In various radioactive decays, there appeared neutrino(s) and positive and negative electrons. All this was very satisfactory indeed.

But what was one to make of the muons? (Isadore Rabi is supposed to have asked, 'Who ordered them?'), and what was the role of the 'strange particles'? We know now, of course that they indicated the beginning of the second of a total of three particle 'generations'. Although the questions posed in the 1930s were largely answered in a satisfactory way during the 1940s, many new and urgent questions were posed as mid-century was reached.

References

[1] Millikan R A 1929 *Encyclopaedia Britannica* 14th edn, vol 8, p 340 (original emphasis)
[2] According to Aristotle, Thales of Miletus (seventh century BC) believed that everything was made of water. See Russell B 1945 *A History of Western Philosophy* (New York: Simon and Schuster) p 26
[3] Heisenberg W 1971 *Physics and Beyond* (New York: Harper and Row) p 93
[4] Born M, Heisenberg W and Jordan P 1926 Zur quantenmechanik II *Z. Phys.* **35** 557–615
[5] Dirac P A M 1929 Quantum mechanics of many-electron systems *Proc. R. Soc.* A **123** 714–33
  The prototype hydrogen molecule had been treated by W Heitler and F London in 1928, based on the 1927 approach of M Born and J R Oppenheimer, in which the nuclei are treated as quasi-stationary.
[6] A comprehensive study of Compton's early ideas leading up to his theory of the Compton effect can be found in Stuewer R H 1975 *The Compton Effect* (New York: Science History Publications). A good shorter account is in Mehra J and Rechenberg H 1982 *The Historical Development of Quantum Theory* vol 1 (New York: Springer) pp 512–32
[7] Letter from Einstein to Michele Besso, 6 September 1916. Quoted in Mehra J and Rechenberg H 1982 *The Historical Development of Quantum Theory* vol 1 (New York: Springer) p 515
[8] Chadwick J and Bieler E S 1921 The collisions of α particles with hydrogen nuclei *Phil. Mag.* **42** 923–40
[9] Rutherford E, Chadwick J and Ellis C D 1930 *Radiations from Radioactive Substances* (Cambridge: Cambridge University Press)
  Gamow G 1931 *Constitution of Atomic Nuclei and Radioactivity* (Oxford: Clarendon)
[10] Xu Q and Brown L M 1987 The early history of cosmic ray research *Am. J. Phys.* **55** 23–33
[11] Eve A S 1905 On the radioactive matter present in the atmosphere *Phil. Mag.* **10** 98–112
[12] Wilson C T R 1901 On the ionization of atmospheric air *Proc. R. Soc.* A **68** 151–61
[13] Hess V F 1912 Über Beobachtung der durchdringenden Strahlung bei seiben Freiballonfahrten *Phys. Z.* **13** 1084–91
[14] Millikan R A and Cameron H G 1926 High frequency rays of cosmic origin. III. Measurements in snow-fed lakes at high altitudes *Phys. Rev.* **28** 851–68
[15] For more details, see Skobeltzyn D 1983 The early stage of cosmic ray particle research *The Birth of Particle Physics* ed L M Brown and L Hoddeson (Cambridge: Cambridge University Press) pp 111–9. Also printed in Sekido Y and Elliot H (ed) 1985 *Early History of Cosmic Ray Studies* (Dordrecht: Reidel)
[16] Heisenberg W 1932 Theoretische Überlegungen zur Höhenstrahlung *Ann. Phys.* **13** 430–52
[17] Weiner C 1972 1932—moving into the new physics *Phys. Today* **25** 40–9

See also Brink D M 1965 *Nuclear Forces* (Oxford: Pergamon) for many aspects of this chapter

Brown L M and Rechenberg H 1988 Nuclear structure and beta decay *Am. J. Phys.* **56** 982–8

[18] Urey H C, Brickwedde F G and Murphy G M 1932 A hydrogen isotope of mass 2 *Phys. Rev.* **39** 164–5

[19] Rutherford E 1920 Nuclear constitution of atoms *Proc. R. Soc.* A **97** 374–400

Chadwick J 1932 Possible existence of a neutron *Nature* **129** 319

[20] Iwanenko D 1932 Sur la constitution des noyeaux atomiques *C. R. Acad. Sci., Paris* **195** 236–7

[21] For more on the Pauli's neutrino proposal, see Brown L M 1978 The idea of the neutrino *Phys. Today* **31** 23–8. The neutrino was not detected until the 1950s.

[22] For an historical reminiscence of this discovery, see Anderson C D and Anderson H L 1983 Unraveling the particle content of cosmic rays *The Birth of Particle Physics* ed L M Brown and L Hoddeson (Cambridge: Cambridge University Press) pp 131–54

[23] Blackett P M S and Occhialini G P S 1933 Some photographs of the tracks of penetrating radiation *Proc. R. Soc.* A **139** 699–727

[24] Letter from Heisenberg to Bohr, 20 June 1932 (Bohr Archives, Copenhagen).

[25] Heisenberg W 1932 Über den Bau der Atomkerne *Z. Phys.* **77** 1–11; 1932 *Z. Phys.* **78** 156–64; 1933 *Z. Phys.* **80** 587–96

[26] The connection with cosmic-ray phenomena is that the large production of secondary particles by high-energy electrons and photons implied a strong radiative interaction, which required electric charges of small mass (hence, easily accelerated), i.e. electrons. The light charges are actually the virtual electron–positron pairs present in the strong nuclear electric field, according to Dirac's QED.

[27] In order to account for the approximate equality of the number of neutrons and protons in the lighter nuclei, one must assume that the n–p force is the dominant one.

[28] Majorana E 1933 Über den Kerntheorie *Z. Phys.* **82** 137–45

Wigner E 1933 On the mass defect of helium *Phys. Rev.* **43** 252–7

[29] Beck's position was the following 'It has been suggested that the [lost mechanical] properties be ascribed to an unknown particle which it is proposed to call a "neutrino". There is, however, at present no need to assume the real existence of a neutrino, and the assumption of its existence would even be an unnecessary complication of the description of the β-decay process'—extract from Beck G 1933 Conservation laws and β-emission *Nature* **132** 967

[30] The quote is taken from the account of one of Fermi's associates, F Rasetti in *Enrico Fermi: Collected Papers* vol 1, ed E Segré (Chicago, IL: University of Chicago Press) p 540. For another historical account, see Segré E 1979 *Nuclear Physics in Retrospect* ed R Stuewer (University of Minnesota Press)

[31] Fermi E 1934 Tentativo di una teoria dei raggi β *Nuovo Cimento* **11** 1–19; 1934 Versuch einer Theorie der β *Z. Phys.* **88** 161–77

[32] Beck G and Sitte K 1933 Zur Theorie des β-Zerfalls *Z. Phys.* **86** 105–19; 1934 Bemerkung zur Arbeit von E Fermi *Z. Phys.* **89** 259–60

The London conference proceedings are 1935 *Int. Conf. Physics (London, 1934)* (Cambridge: Cambridge University Press)

[33] Bethe H and Heitler W 1934 On the stopping of fast particles and the

creation of positive electrons *Proc. R. Soc.* A **146** 83–112

[34] Curie I and Joliot F 1934 Une nouveau type de radioactivité *C. R. Acad. Sci., Paris* **198** 254–6

[35] Letter from Pauli to Heisenberg, 7 January 1934; Letter from Heisenberg to Pauli, 12 January 1934; both letters are in von Meyenn K (ed) 1985 *W Pauli, Scientific Correspondence* vol II (Berlin: Springer)

[36] Majorana E 1933 Über die Kerntheorie *Z. Phys.* **82** 137–45; 1933 Sulla teoria dei nuclei *Ric. Scientifica* **4** 559–65

[37] Tamm Ig 1934 Exchange forces between neutrons and protons and Fermi's theory *Nature* **133** 981

Iwanenko D 1934 Interaction of neutrons and protons *Nature* **133** 981–2

Nordsieck A 1934 Neutron collisions and the beta-ray theory of Fermi *Phys. Rev.* **46** 234–5

[38] Frisch R and Stern O 1933 Über die magnetische Ablenkung von Wasserstoffmolekülen und das magnetische Moment des Protons, Part I *Z. Phys.* **85** 4–16

Estermann I and Stern O 1933 Über die magnetische Ablenkung von Wasserstoffmolekülen und das magnetische Moment des Protons, Part II *Z. Phys.* **85** 17–24

Wick G C 1935 Teoria dei raggi β e momento magnetico del protone *Rend. Accad. Lincei* **21** 170–3

[39] Wigner E P, Critchfield C L and Teller E 1939 The electron–positron field theory of nuclear forces *Phys. Rev.* **56** 530–9

Critchfield C L 1939 Spin-dependence in the electron–positron theory of nuclear forces *Phys. Rev.* **56** 540–7

[40] Marshak R E 1940 Heavy electron pair theory of nuclear forces *Phys. Rev.* **57** 1101–6

Gamow G and Teller E 1937 Some generalizations of the β transformation theory *Phys. Rev.* **51** 289

Marshak R E and Bethe H A 1947 On the two-meson hypothesis *Phys. Rev.* **72** 506–9

Another two-meson theory having the correct spin assignments was made some years earlier in Japan: Sakata S and Inoue T 1942 On the relation between the meson and the Yukawa particle *Bull. Phys.-Math. Soc. Japan* **16** 232–4 (in Japanese) (Engl. Transl. 1946 *Prog. Theor. Phys.* **1** 143–50)

[41] Bethe H A and Bacher R F 1936 Nuclear physics, A, stationary states of nuclei *Rev. Mod. Phys.* **8** 82–229, especially p 203

[42] Letter from Pauli to Heisenberg, 1 November 1934 (original emphasis) in reference [35] pp 357–8

[43] Konopinski E J and Uhlenbeck G E 1935 On the Fermi theory of β-radioactivity *Phys. Rev.* **48** 7–12

[44] White M G 1936 Scattering of high energy protons in hydrogen *Phys. Rev.* **49** 309–16

Tuve M A, Heydenburg N P and Hafstad L R 1936 The scattering of protons by protons *Phys. Rev.* **49** 806–25

Theoretical analysis was made by Breit G, Condon E U and Present R D 1936 Theory of scattering of protons by protons *Phys. Rev.* **49** 825–45

[45] Breit G and Feenberg E 1936 The possibility of the same form of specific interaction for all nuclear particles *Phys. Rev.* **49** 850–6

Cassen B and Condon E U 1936 On nuclear forces *Phys. Rev.* **49** 846–9

[46] In referring to 'currents' in the Fermi-field interactions, we are being anachronistic. Such language only became common in the late 1950s.

[47] See references [45], which did not specify a mechanism for the nuclear

forces, but used an arbitrarily specified potential. Kemmer's was a genuine QFT; see Kemmer N 1937 Field theory of nuclear interaction *Phys. Rev.* **52** 906–10

[48] Yukawa H 1935 On the interaction of elementary particles. I *Proc. Phys.-Math. Soc. Japan* **17** 48–57

[49] For Yukawa and the meson theory, see the following: Brown L M 1981 Yukawa's prediction of the meson *Centaurus* **25** pp 71–132
Brown L M 1989 Yukawa in the 1930s: a gentle revolutionary *Historia Scientiarum* **36** 1–21
Darrigol O 1988 The quantum electrodynamic analogy in the early nuclear theory or the roots of Yukawa's theory *Rev. Histoire Sci.* **XLI** 26–297
Mukherji V 1974 A history of the meson theory of nuclear forces from 1935 to 1952 *Arch. History Exact Sci.* **13** 28–100

[50] After World War II, the renormalization theory QED was formulated, and has given extremely accurate results. Sin-itiro Tomonaga, one of the creators of this theory, was closely associated with Yukawa, and has acknowledged the latter's influence upon his work. However, Yukawa never believed that the solution was entirely satisfactory.

[51] Yukawa H 1933 Introduction to W Heisenberg, Über der Bau der Atomkerne *J. Phys.-Math. Soc. Japan* **7** 195–205 (in Japanese)

[52] Almost all of Yukawa's early unpublished work is in the Yukawa Hall Archival Library at the University of Kyoto.

[53] See Brown L M 1985 How Yukawa arrived at the meson theory *Prog. Theor. Phys. Suppl.* **85** 13–19, especially p 16

[54] Brown L M *et al* (ed) 1991 Elementary Particle Theory in Japan, 1930–1960 *Prog. Theor. Phys.* supplement **105** 80

[55] Yukawa H *Tabibito* (Engl. Transl. Brown L M and Yoshida R 1982 (Singapore: World Scientific) p 201)

[56] This issue gave rise to a famous controversy involving Arthur Compton, on the side of charged particles as primaries, and Robert Millikan, who championed γ rays as primaries, well into the 1930s. See Kargon R H 1982 *The Rise of Robert Millikan* (Ithaca, NY: Cornell University Press) pp 154–61

[57] See reference [22], Mukherji [49] and the following:
Brown L M and Rechenberg H 1991 Quantum field theories, nuclear forces, and the cosmic rays (1934–1938) *Am. J. Phys.* **59** 595–605
Rechenberg H and Brown L M 1990 Yukawa's heavy quantum and the mesotron (1935–1937) *Centaurus* **33** pp 214–52
Cassidy D C 1981 Cosmic ray showers, high energy physics, and quantum field theories *Hist. Stud. Phys. Sci.* **12** 1–39
Galison P 1983 The discovery of the muon and the failed revolution against quantum electrodynamics *Centaurus* **26** pp 262–316

[58] Bethe H and Heitler W 1934 On the stopping of fast particles and the creation of positive electrons *Proc. R. Soc.* A **146** 83–112
Heitler W 1936 and 1944 *The Quantum Theory of Radiation* (Oxford: Oxford University Press)

[59] 1935 *Int. Conf. Physics (London, 1934)* vol I (Cambridge: Cambridge University Press) p 250

[60] Serber R 1983 Particle physics in the 1930s: a view from Berkeley *The Birth of Particle Physics* ed L M Brown and L Hoddeson (Cambridge: Cambridge University Press) pp 206–21, especially p 208

[61] von Weizsäcker C F 1934 Ausstrahlung bei Stössen sehr schneller Elektronen *Z. Phys.* **88** 612–45

Williams E J 1935 Nature of high energy particles of penetrating radiation and status of ionization and radiation formulae *Phys. Rev.* **48** 49–54
[62] Bhabha H J and Heitler W 1937 The passage of fast electrons and the theory of cosmic ray showers *Proc. R. Soc.* A **159** 432–58
Carlson J F and Oppenheimer J R 1937 On multiplicative showers *Phys. Rev.* **51** 220–31
[63] Heisenberg W 1936 Zur Theorie der 'Schauer' in der Hohenstrahlung *Z. Phys.* **101** 533–40
[64] The de Broglie wavelength of a particle of momentum p is h/p; thus it is related to the particle's energy.
[65] In addition to the burst phenomena, there was still some doubt about the validity of QED at high energy.
[66] Bhabha H J and Heitler W 1937 The passage of fast electrons and the theory of cosmic ray showers *Proc. R. Soc.* A **159** 455
[67] Kunze P 1933 Untersuching der Ultrastrahlung in der Wilsonkammer *Z. Phys.* **80** 1–18
[68] For the French group, see Leprince-Ringuet L 1983 The scientific activities of Leprince-Ringuet and his group on cosmic rays *The Birth of Particle Physics* ed L M Brown and L Hoddeson (Cambridge: Cambridge University Press) pp 177–82. For the English group, see Wilson J G 1985 The new 'magnet house' and the muon *Early History of Cosmic Ray Studies* ed Y Sekido and H Elliot (Dordrecht: Reidel) pp 145–59
[69] Anderson C D and Neddermeyer S 1936 Cloud chamber observations of cosmic rays at 4300 meters elevation and near sea-level *Phys. Rev.* **50** 263–71, especially p 270
[70] Other names used were heavy electron, barytron, Yukawa particle, Yukon, x-particle, etc.
[71] Neddermeyer S H and Anderson C D 1937 Note on the nature of cosmic ray particles *Phys. Rev.* **51** 884–86
[72] Street J C and Stevenson E C 1937 Penetrating corpuscular component of the cosmic radiation *Phys. Rev.* **51** 1005
For more on the work of Street at MIT, see Galison P 1983 The discovery of the muon and the failed revolution against quantum electrodynamics *Centaurus* **26**
[73] Nishina Y, Takeuchi M and Ichimiya T 1937 On the nature of cosmic ray particles *Phys. Rev.* **52** 1198–9
Street J C and Stevenson E C 1937 New evidence for the existence of a particle of mass intermediate between the proton and the electron *Phys. Rev.* **52** 1003–4
Although published later, the Japanese paper was submitted first.
[74] Brown L M, Kawabe R, Konuma M and Maki Z (ed) 1991 Elementary Particle Theory in Japan, 1930–1960 *Prog. Theor. Phys.* (Supplement 105) 182–185
[75] Yukawa H 1937 On a possible interpretation of the penetrating component of the cosmic ray *Proc. Phys.-Math. Soc. Japan* **20** 712–3
[76] Oppenheimer J R and Serber R 1937 Note on the nature of cosmic ray particles *Phys. Rev.* **51** 113. See also references [60] and Brown and Hoddeson [14] pp 287–8 for discussions of this letter.
[77] Stueckelberg E C G 1937 On the existence of heavy electrons *Phys. Rev.* **52** 41–2
[78] Yukawa H and Sakata S 1937 On the interaction of elementary particles II *Proc. Phys.-Math. Soc. Japan* **19** 1084–1093
Yukawa H, Sakata S and Taketani M 1938 On the interaction of elementary particles III *Proc. Phys.-Math. Soc. Japan* **20** 319–40

Yukawa H, Sakata S, Kobayasi M and Taketani M 1938 On the interaction of elementary particles IV *Proc. Phys.-Math. Soc. Japan* **20** 720–45
[79] Pauli W and Weisskopf V 1934 Über der Quantisierung der skalaren relativistischen Wellengleichung *Helv. Phys. Acta* **7** 709–31
[80] Yukawa H and Sakata S 1937 On the interaction of elementary particles II *Proc. Phys.-Math. Soc. Japan* **19** 1084–1093 (referred to in the text as *Interaction II*)
[81] Yukawa H, Sakata S and Taketani M 1938 On the interaction of elementary particles III *Proc. Phys.-Math. Soc. Japan* **20** 319–40 (referred to in the text as *Interaction III*)
[82] Brown L M and Rechenberg H 1991 The development of the vector meson theory in Britain and Japan (1937–38) *Br. J. Hist. Sci.* **24** 405–33
[83] Letter from W Heitler and H Fröhlich to H Yukawa, 5 March 1938, in Yukawa Hall Archival Library at the University of Kyoto (see reference [52])
The Proca reference is Proca A 1936 Sur la theorie ondulatoire des électrons positifs et négatifs *J. Physique Rad.* **7** 347–53. Although a theory of particles of spin 1, note that Proca proposed it as a theory of the electron.
[84] See, for example,
Hoch P 1990 Flight into self absorption and xenophobia. The plight of refugee theorists among British and American experimentalists in the 1930s, etc *Phys. World* January 23–6
Stuewer R H 1984 Nuclear physicists in a new world. The émigrés of the 1930s in America *Ber. Wiss.* **7** 23–40
[85] Fröhlich H and Heitler W 1938 Magnetic moments of the proton and the neutron *Nature* **141** 37–8. A 'virtual state' is a temporary one, too short, according to the uncertainty principle, to be observable.
[86] Kemmer N 1938 Nature of the nuclear field *Nature* **141** 116-7
[87] Heitler W 1937 On the analysis of cosmic rays *Proc. R. Soc.* A **161** 261–83
Bhabha H J 1938 On the penetrating component of cosmic rays *Proc. R. Soc.* A **164** 257–93
[88] Bhabha H J Nuclear forces, heavy electrons, and the β-decay *Nature* **141** 117–8; On the theory of heavy electrons and nuclear forces *Proc. R. Soc.* A **166** 501–27
The other important vector meson papers at this time were: H Fröhlich, Heitler W and Kemmer M 1938 On the nuclear forces and the magnetic moments of the neutron and the proton *Proc. R. Soc.* A **166** 154–71; Stueckelberg E C G 1938 Die Wechselwirkungkräfte in der Elektrodynamik und in der Feldtheorie der Kernkräfte *Helv. Phys. Acta* **11** 225–44 and 299–328
[89] Bhabha H J Nuclear forces, heavy electrons, and the β-decay *Nature* **141** 118
[90] Heitler W 1938 Showers produced by the penetrating cosmic radiation *Proc. R. Soc.* A **166** 529–43
[91] Kemmer N 1971 Some recollections from the early days of particle physics *Hadronic Interactions of Leptons and Photons* ed J Cumming and H Osborn (London: Academic) pp 1–17
[92] Kemmer N 1971 Some recollections from the early days of particle physics *Hadronic Interactions of Leptons and Photons* ed J Cumming and H Osborn (London: Academic) pp 16, 17
[93] Heitler W 1973 Errinerungen an die gemeinsame Arbeit mit Herbert Fröhlich *Cooperative Phenomena* ed H Haken and M Wagner

(Heidelberg: Springer) pp 421–4, especially pp 422–423

[94] Although Anderson had preferred 'mesoton', Millikan insisted on the r, and he got his way. See reference [70] for some other suggested names.

[95] See references [73]. A review article by Peierls in 1939 gives mass values from five sources, ranging from $39m_e$ to $569m_e$, and he remarks that 'there is as yet no conclusive proof that all mesons have the same mass, though this is probable': Peierls R 1939 The meson *Rep. Prog. Phys.* **6** 78–94

[96] *Interaction III* (reference [81]) p 337. The Bhabha reference is reference [88].

[97] See, for example, Wentzel G 1938 Schwere Elektronen und Theorien der Kernvorgänge *Naturwissenschaften* **26** 273–9, especially p 276

[98] Letter from Heisenberg to Pauli, 14 April 1938, in K von Meyenn (ed) 1985 *Wolfgang Pauli, Scientific Correspondence, Vol. II: 1930–1939* (Berlin: Springer)

[99] Euler H 1938 Über die durchdringende Komponente der Höhenstrahlung und die von ihr erzeugtenaten Hoffmannischen Stösse *Z. Phys.* **110** 692–716

Euler H and Heisenberg W 1938 Theoretische Gesichtspunkte zur Deutung der kosmischen Strahlung *Ergeb. Exakt. Naturwiss.* **17** 1–69

[100] Reference [98] p 692

[101] Letter from Heisenberg to Yukawa, 16 June 1938 (Yukawa Hall Archival Library at the University of Kyoto).

[102] Letter from Yukawa to Heisenberg, 6 August 1938 (copy in Yukawa Hall Archival Library at the University of Kyoto).

[103] Blackett P M S 1938 High altitude cosmic radiation *Nature* **142** 692–3

[104] For a more detailed history, see Brown L M and Rechenberg H Decay of the meson—experiment versus theory (1937–1941) *Preprint* MPI-Ph/92-47 of the Werner Heisenberg Institute for Physics, Munich, Germany

[105] Rossi B 1939 The disintegration of mesons *Rev. Mod. Phys.* **11** 296–303, especially p 296

[106] Rossi B, Hilbury N and Haag J B 1940 The variation of the hard component of cosmic rays with height and the disintegration of mesotrons *Phys. Rev.* **57** 461–9

[107] Rasetti F 1941 Disintegration of slow mesotrons *Phys. Rev.* **60** 198–204

[108] Nereson N and Rossi B 1943 Further measurements of the disintegration curve of mesotrons *Phys. Rev.* **64** 199–201

[109] Nordheim L W 1939 Lifetime of the Yukawa particle *Phys. Rev.* **55** 506

[110] Serber R 1939 Beta-decay and mesotron lifetime *Phys. Rev.* **56** 1065

[111] Møller C, Rosenfeld L and Rozental S 1939 Connexion between the lifetime of the meson and the beta-decay of light elements *Nature* **144** 609

[112] Bethe H A and Nordheim L W 1940 On the theory of meson decay *Phys. Rev.* **57** 998–1006

[113] Sakata S 1940 Connection between the meson decay and the β decay *Phys. Rev.* **58** 576; 1940 On the theory of the meson decay *Proc. Phys.-Math. Soc. Japan* **23** 283–291

[114] Reference [95]

[115] The four types of spin 0 and spin 1 theories were discussed in Kemmer N 1938 Quantum theory of Bose-Einstein particles and nuclear interaction *Proc. R. Soc.* A **166B** 127–53. The spin 1 theories were discussed in [89], in Parts III and IV in [78]; and in H Fröhlich, Heitler W and Kemmer N 1938 On the nuclear forces and the magnetic moments of the neutron and the proton *Proc. R. Soc.* A **166** 154–71. They were discussed by many others later.

[116] Reference [95], p 85
[117] To have a charge-independent theory, neutral mesons are required. Either all nuclear force mesons must be neutral and interact equally with both nucleons, or else the two charged mesons and one neutral meson form an isospin triplet, which can be coupled in a charge-independent way to the nucleon isospin doublet. The latter theory was given in Kemmer N 1938 The charge dependence of nuclear forces *Proc. Camb. Phil. Soc.* **34** 354–64
[118] Bethe H A 1939 The meson theory of nuclear forces *Phys. Rev.* **55** 1261–3, especially p 1261
[119] Kellog J M, Rabi I I, Ramsey N F and Zacharias J R 1939 An electrical quadrupole moment of the deuteron *Phys. Rev.* **55** 318–9
[120] Møller C and Rosenfeld L 1939 The electric quadrupole moment of the deuteron and the field theory of nuclear forces *Nature* **144** 476–7
[121] The work referred to was Møller C and Rosenfeld L 1939 Theory of mesons and nuclear forces *Nature* **143** 241–2. This was followed by a detailed paper: Møller C and Rosenfeld L 1940 On the field theory of nuclear forces *K. Danske Vidensk. Selskab (Math.-fys. Meddelsen)* **17** 1–72
[122] Heisenberg W 1938 Über die in der Theorie der Elementarteilchen auftretende universelle Länge *Ann. Phys.* **32** 20–23; 1938 Die Grenzen der Anwendbarkeit der bisherigen Quantentheorie *Z. Phys.* **110** 251–66
[123] Heisenberg W 1939 Zur Theorie der explosionsartigen Schauer in der kosmischen Strahlung II *Z. Phys.* **113** 61–86. See also Bhabha H J 1939 Classical theory of mesons *Proc. R. Soc.* A **172** 384–409
[124] Blau M and Wambacher H 1937 Disintegration process by cosmic rays with the simultaneous emission of several heavy particles *Nature* **140** 585
The same effect was seen in cloud chambers: Brode R B and Starr M A 1938 Nuclear disintegrations produced by cosmic rays *Phys. Rev.* **53** 3–5
[125] See reference [40] and the following:
Critchfield C L and Lamb W E Jr 1940 Note on a field theory of nuclear forces *Phys. Rev.* **48** 46–49
Marshak R E and Weisskopf V F 1941 On the scattering of mesons of spin $\frac{1}{2}$ by atomic nuclei *Phys. Rev.* **59** 130–5
[126] See, for example, Nishina Y, Sekido Y, Miyazaki Y and Masuda T 1941 Cosmic rays at a depth equivalent to 1400 meters of water *Phys. Rev.* **59** 401
[127] Shutt R P 1942 On the electrical and anomalous scattering of mesotrons *Phys. Rev.* **61** 6–13
[128] A neutral meson in a charge-symmetric theory would be expected to decay into $e^+ + e^-$ and also into $\nu + \bar{\nu}$ both as a free meson and as an intermediary in β decay. However, the former mode could also be electromagnetic (which would dominate the weak decay) and the latter mode would be practically invisible. (The main decay of the neutral pion is into two γ rays.)
[129] Weinberg S 1992 Opening talk at the *Third Int. Symp. History of Particle Physics (Stanford Linear Accelerator Center, June 24, 1992)*, to appear in *Rise of the Standard Model* 1995 (Cambridge: Cambridge University Press)
[130] Yukawa H and Okayama T 1939 Note on the absorption of slow mesotrons in matter *Sci. Papers Inst. Phys. Chem. Res.* **36** 385–9
[131] Yukawa H and Okayama T 1939 *Sci. Papers Inst. Phys. Chem. Res.* **36** 153

[132] Tomonaga S and Araki G 1940 Effect of the nuclear Coulomb field in the capture of slow mesons *Phys. Rev.* **58** 90–1
[133] Christy R F and Kusaka S 1941 The interaction of γ-rays with mesotrons *Phys. Rev.* **59** 405–14; 1941 Burst production by mesotrons *Phys. Rev.* **59** 414–21
[134] Oppenheimer J R 1941 *Phys. Rev.* **59** 462
[135] The first of these papers was: Wentzel G 1940 Zur Problem des statischen Mesonfeldes *Helv. Phys. Acta* **13** 269–308. Others were:
Oppenheimer J R and Schwinger J 1941 On the interactions of mesotrons and nuclei *Phys. Rev.* **60** 1066–7
Pauli W and Dancoff S M 1942 The pseudoscalar meson field with strong coupling *Phys. Rev.* **62** 85–107
Serber R and Dancoff S M 1943 Strong coupling meson theory of nuclear forces *Phys. Rev.* **63** 143–61
Pauli W and Kusaka S 1943 On the theory of a mixed pseudoscalar and a vector meson field *Phys. Rev.* **63** 400–16 and others
[136] Tomonaga S 1941 Zur Theorie des Mesotrons *Sci. Papers Inst. Phys. Chem. Res.* **39** 247–66
[137] See reference [40]. Also see Hayakawa S 1983 The development of meson physics in Japan *The Birth of Particle Physics* ed L M Brown and L Hoddeson (Cambridge: Cambridge University Press) pp 82–107, especially pp 98–102
[138] Sakata S and Inoue T 1946 On the correlations between mesons and Yukawa particles *Prog. Theor. Phys.* **1** 143–9. (Note that 'meson' in the title refers to the particle we have been calling 'mesotron' and is now called muon.)
[139] Conversi M, Pancini E and Piccioni O 1945 On the decay process of positive and negative mesons *Phys. Rev.* **68** 232; 1947 On the disintegration of negative mesons *Phys. Rev.* **71** 209–10. Accounts of the circumstances under which this important experiment was carried out are given in Chapters 13 and 14 of *The Birth of Particle Physics* ed L M Brown and L Hoddeson (Cambridge: Cambridge University Press) by Oreste Piccioni and Marcello Conversi, respectively.
[140] Fermi E, Teller E and Weisskopf V F 1947 The decay of negative mesotrons in matter *Phys. Rev.* **71** 314–5
[141] See reference [138]. Another two-meson theory was proposed shortly before the direct observation; Marshak R E and Bethe H A 1947 On the two-meson hypothesis *Phys. Rev.* **72** 506–9. This theory, formulated independently of Sakata-Inoue, interchanged the spins of the mesotron and meson.
[142] Powell C F and Fertel F 1939 Energy of high velocity neutrons by the photographic method *Nature* **144** 115
[143] Perkins D H 1989 Cosmic ray work with emulsions in the 1940s and 1950s *Pions to Quarks* ed L M Brown *et al* (Cambridge: Cambridge University Press) pp 89–123
See also Powell C F, Fowler H and Perkins D H 1959 *The Study of Elementary Particles by the Photographic Method* (New York: Pergamon)
[144] Perkins D H 1947 Nuclear disintegration by meson capture *Nature* **159** 126–7
Occhialini G P S and Powell C F 1947 Multiple disintegration processes produced by cosmic rays *Nature* **159** 93–4
[145] Lattes C M G, Muirhead H, Occhialini G P S and Powell C F 1947 Processes involving charged mesons *Nature* **160** 694–7
Lattes C M G, Occhialini G P S and Powell C F 1947 Observations on the

tracks of slow mesons in photographic emulsions *Nature* **159** 453–6 and 486–92
[146] Rochester G D and Butler C C 1947 Evidence for the existence of new unstable elementary particles *Nature* **160** 855–7

See also Rochester G D Cosmic-ray cloud-chamber contributions to the discovery of the strange particles in the decade 1947–1957', in L M Brown *et al*, reference [143], pp 57–88
[147] Gardner E and Lattes C M G 1948 Production of mesons by the 184-inch Berkeley cyclotron *Science* **107** 270–1
[148] Steinberger J, Panofsky W K H and Steller J 1950 Evidence for the production of neutral mesons by photons *Phys. Rev.* **78** 802–5

Chapter 6

SOLID-STATE STRUCTURE ANALYSIS

William Cochran

6.1. Crystallography and x-rays before 1912

Crystallography developed primarily as a branch of mineralogy. P von Groth edited the journal *Zeitschrift für Kristallographie und Mineralogie* from 1877 to 1920 in 55 volumes; five volumes of his *Chemische Kristallographia* (1906–19) included no fewer than 3342 drawings and diagrams of crystals. This record of the form, preparation and physical properties of crystals, by no means all of mineral origin, had a considerable influence on the development of the new methods with which we are concerned in this chapter.

The most important instrument for the study of crystal morphology was the optical goniometer, used to measure the angles between the normals of crystal faces, and the polarizing microscope was an important tool for the investigation and identification of mineralogical specimens. The fundamental law of crystal morphology is the law of rational indices [1]. Guided by the external form of the crystal, the normals to three of its faces are chosen to give the directions OA, OB, OC of the crystal axes, as in figure 6.1 (For many crystals it is possible to choose these to be mutually perpendicular). A fourth face is now chosen which cuts all three axes, and the relative lengths OA, OB and OC define the crystallographic axes a, b and c. Since they are relative only, we may take $b = 1$. It is then found that any face of the crystal is such that it is parallel to a plane making intercepts a/h, b/k, c/l on the axes, where h, k, l are small whole numbers called the Miller indices, after W H Miller who introduced them. If the crystal is now represented by the set of normals to its various faces, all passing through the point O, it may be found to possess a number of symmetry elements such as mirror planes and axes of twofold, threefold, fourfold and sixfold rotational

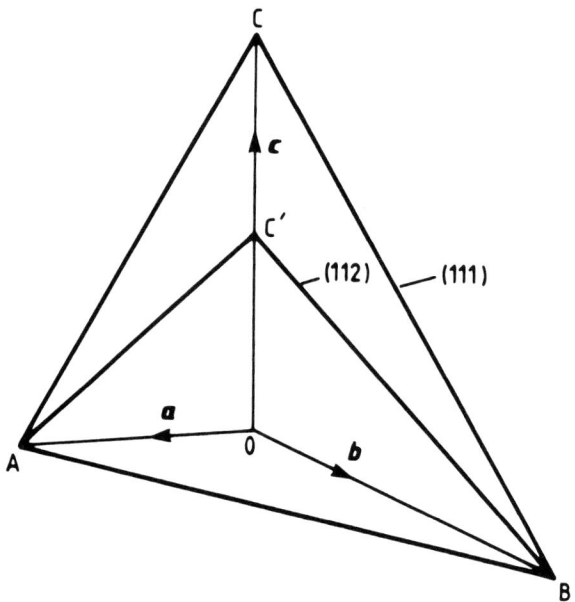

Figure 6.1. *[111] and [112] as examples of (hkl) planes.* $OC' = \frac{1}{2}OC$. *a, b, c are the crystallographic axes.*

symmetry. The latter, for example, is such that rotation of the crystal (idealized by its set of normals) about an angle $2\pi/6$ brings it into self-coincidence. Whether the crystal possesses a centre of symmetry cannot be determined from the symmetry of (the normals to) its faces but this may be decided from other physical properties, for example a crystal which exhibits the piezoelectric effect (e.g. quartz) cannot have a centre of symmetry. The group of symmetry elements is called the point group and a crystal can be assigned to one of 32 crystal classes, based on one of seven systems, depending on its point group. For example, a crystal belonging to the cubic system may have the point-group symmetry of an ideal cube, with three fourfold, four threefold and six twofold axes, a centre of symmetry and mirror planes perpendicular to the fourfold and the twofold axes respectively. Sodium chloride is an example. On the other hand, while belonging to the cubic system it may, like iron pyrites FeS_2, lack some of these symmetry elements, such as the centre of symmetry and even the fourfold axes.

The fact that crystals with fivefold or sevenfold axes never occur naturally found an explanation in Haüy's lattice theory of the structure of crystals [2] (1784). A crystal has associated with it a lattice of points such that if the origin is taken on one such point, an identical point should be found at

$$r_l = l_1 a + l_2 b + l_3 c$$

Solid-State Structure Analysis

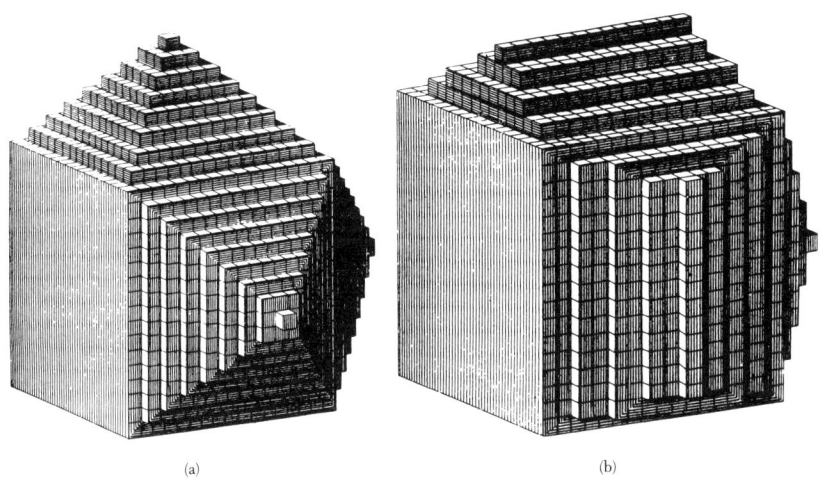

Figure 6.2. *Relation of the external faces of crystals to the building blocks of Haüy.*

where l_1, l_2, l_3 are integers and a, b, c are the fundamental lattice translations or axes of the lattice (and the crystal). The axes a, b, c enclose a parallelepiped called the unit cell which Haüy took to be the building block of the crystal and to be of submicroscopic dimensions. Different crystal faces are developed according to the arrangement of the building blocks, as in figure 6.2.

The problem of finding all possible symmetry types of such a solid was solved in steps associated with the names of A Bravais (1848), L Sohncke (1867), A Schoenflies and E von Federov (both 1891). Bravais' contribution was to show that there are 14 types of lattices of differing geometry and symmetry. Schoenflies and von Federov showed independently that there are 230 space groups (a space group is an arrangement of symmetry elements in space, self-consistent and consistent with the translational symmetry of the lattice). An atom or molecule inserted in this array will be reproduced by the symmetry elements until a system of equivalent particles extends throughout the crystal. The unit of structure, nowadays called the asymmetric unit, is generally smaller than the content of one unit cell, but the symmetry elements multiply it up to fill a unit cell and the crystal is completed by the translational symmetry of the lattice. The nature of the asymmetric unit, or *molécule intégrante* of Haüy, remained obscure following the work on space groups.

A different and more intuitive approach to the problem of crystal structure was initiated by W Barlow [3] who investigated ways of close-packing equal and unequal spheres. His method of close-packing equal spheres gave what later proved to be two of the actual structures of certain metallic elements, and packing two sets of spheres, equal in number but of different radii, was later found to give both the NaCl

423

and the CsCl types of structures. In the early days of structure analysis Barlow's suggested structures provided ideas which could be tested in trying to explain x-ray diffraction patterns. At the time, however, his ideas were not widely accepted, particularly as his structures for simple binary compounds contained no diatomic molecules.

A Cauchy developed a theory of the elasticity of crystals which was based on the lattice theory. However, the number of independent elastic constants for a crystal of the (triclinic) class of lowest symmetry was predicted to be 15 and not the 21 expected from more general considerations. For a cubic crystal, which is characterized by the three elastic constants C_{11}, C_{12} and C_{44}, Cauchy's theory required that $C_{12} = C_{44}$ and this was not confirmed by experiment [4], a discrepancy which helped to prevent the general acceptance of the lattice theory by physicists.

X-rays were discovered by W Röntgen in 1895. While he looked for evidence of reflection, refraction and diffraction, the characteristic features of wave phenomena, he could find none. On the other hand, C G Barkla's experiments on the scattering of x-rays by materials such as paraffin wax showed rather conclusively that the radiation was transversely polarized. He also identified characteristic radiations of two different penetrating powers from the same metal target (anode), which he called the K and the L characteristic radiations, the former being the more penetrating of the two. The greater the atomic weight of the metal target of the x-ray tube, the more penetrating were the K and L radiations.

Despite Barkla's evidence for the wave nature of x-rays, there also seemed to be strong evidence against it in that even a weak source of x-rays can eject electrons from gas molecules. Classical physics could not explain how the energy in a wave spreading out in space could be concentrated to produce this relatively large amount of energy at a point. W H Bragg deduced that x-rays were neutral particles, possibly having an electric dipole moment [5].

6.2. The discovery of x-ray diffraction by crystals

Circumstances in Munich were favourable for the important discovery which was made there in 1912. Groth, the world authority on crystallography, was professor of Mineralogy and Crystallography, Röntgen was professor of Experimental Physics and A Sommerfeld of Theoretical Physics. M von Laue joined Sommerfeld's group in 1909, he had been a pupil of M Planck and was interested in the theory of radiation and in wave-optics. P P Ewald had been given the problem by Sommerfeld of investigating the effect of an incident electromagnetic wave on an array of resonators, that is of isotropically polarizable atoms or molecules arranged on a lattice. Would such a system account for the optical properties of a crystal? Ewald eventually succeeded in showing

Solid-State Structure Analysis

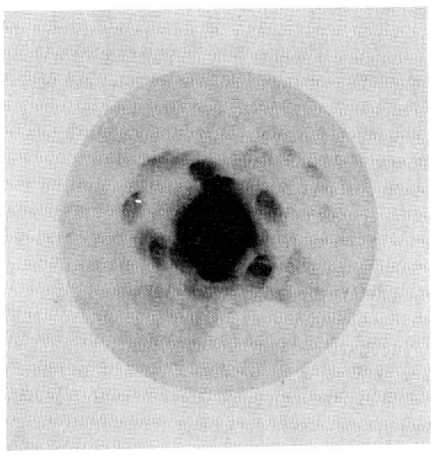

Figure 6.3. *The first successful photograph of x-ray diffraction by a crystal of copper sulphate.*

that it would. In discussing a particular point with Laue, the latter enquired about the likely dimensions of such a lattice if it existed in a crystal, an idea which was apparently novel for him. Then he asked 'what would happen if you assumed very much shorter waves [than ordinary light] to travel in the crystal?'. Ewald replied that this could be worked out exactly by the methods given in his thesis, but Laue was 'listening in a slightly distracted way'. Ewald continued with the completion of his thesis, while Laue discussed with his colleagues the idea that had occurred to him, that x-rays might be diffracted by a crystal. On the whole they were not encouraging, pointing out in particular that thermal vibration of the atoms would destroy the order required. Two of his colleagues, W Friedrich and P Knipping decided nevertheless to try the experiment. A narrow beam of x-rays was passed through a crystal of copper sulphate and impinged on a photographic plate. Eventually diffracted beams were recorded; the first successful photograph is shown as figure 6.3. When a crystal of zinc sulphide, ZnS, which belongs to the cubic system, was used with the incident beam directed along a fourfold axis, the symmetrical photograph shown in figure 6.4 was obtained. A detailed account of the circumstances of this discovery has been given by Ewald [6].

The first paper by von Laue and his two colleagues [7] included conditions (see Box 6A) for an x-ray beam of wavelength λ to be diffracted by a crystal.

Laue found that he could account for the positions of the spots in figure 6.4 with small integral values of h_1, h_2 and h_3, and five separate values for λ/a ranging from 0.038 to 0.143. However, the agreement was not perfect and there was no explanation of the fact that diffracted beams

Max von Laue

(German, 1879–1960)

M von Laue was born near Coblenz, and attended the Universities of Strasbourg, Göttingen and Munich. While acting as assistant to A Sommerfield in Munich he had the idea that x-rays might be diffracted by a crystal and this was confirmed experimentally by his colleagues W Friedrich and P Knipping who obtained the first x-ray diffraction photograph, by a crystal of copper sulphate, in 1912. Laue's interpretation of these and later results was improved on and simplified by W L Bragg. This was Laue's best known contribution to theoretical physics, but he made important contributions to other topics, including relativity, and was awarded the Nobel Prize for physics in 1914. He was a professor of physics successively in the Universities of Zurich (1912–14), Frankfurt (1914–19) and finally Berlin where he was Director of the Institute for Theoretical Physics until his retirement in 1943. After World War II he took an active part in the work of reconciliation and helped to found the International Union of Crystallography (IUCr).

appeared for some values of h_1, h_2, h_3 but not for others.

The scene now shifts to England where W H Bragg was Professor of Physics in Leeds and his son W L Bragg was a research student in the Cavendish Laboratory, Cambridge. News of the discovery in Munich soon reached the Braggs, and the younger Bragg at first tried to interpret the Laue photographs in terms of particles moving down 'avenues' or 'channels' in the crystal. In this he was influenced by his father's ideas, but the attempt was not successful. Two clues were given by a study of the Laue photographs. The first was the way in which the shape of the spots changed as the recording film was moved back from the crystal, the second was that when the crystal was turned by 3° about a vertical axis, spots on a horizontal line through the intercept with the direct beam moved by 6° and changed considerably in intensity. This suggested to

Solid-State Structure Analysis

BOX 6A: LAUE'S EQUATIONS FOR X-RAY DIFFRACTION

Laue's equations for a crystal based on mutually perpendicular axes a, b, c (e.g. ZnS): $(\alpha - \alpha_0)a = h_1\lambda$, $(\beta - \beta_0)b = h_2\lambda$, $(\gamma - \gamma_0)c = h_3\lambda$ where $\alpha_0\beta_0\gamma_0$ and $\alpha\beta\gamma$ are respectively the direction cosines of the incident and diffracted beams and h_1, h_2, h_3 are integers. The first equation is the same as that which applies for the diffraction of a beam of light by a ruled grating of spacing a, while the second applies when there are two 'crossed' gratings with rulings at right angles to one another, having spacings a and b respectively. From this point of view, Laue had only to add the third equation to ensure that all unit cells in the crystal scattered in phase with one another to produce a diffracted beam. When the incident beam is parallel to the c axis, and with $a = b = c$ for a cubic crystal, the equations reduce to $\alpha = h_1\lambda/a$, $\beta = h_2\lambda/a$, $\gamma - 1 = h_3\lambda/a$.

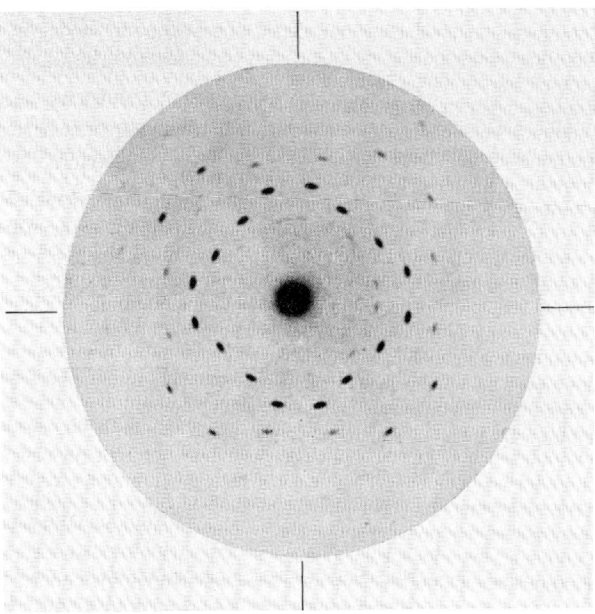

Figure 6.4. *The photograph obtained when the incident x-ray beam is parallel to the fourfold axis of cubic zinc sulphide.*

W L Bragg that the radiation was being reflected from sets of parallel planes in the crystal. This was confirmed by reflecting x-rays over a

427

range of angles from a cleavage plane of mica [8]. He now found that the photographs of ZnS could be explained by the reflection of the x-ray beam, assumed to consist of 'white' radiation covering a continuous range of wavelengths, from various (hkl) planes of the crystal, where the indices (hkl) are the same as Laue's $h_1 h_2 h_3$. The condition for a set of planes to scatter in phase is now known as Bragg's Law,

$$n\lambda = 2d \sin \theta$$

where θ is the glancing angle and d is the spacing of the planes, of indices (hkl). Second-order reinforcement, $n = 2$, for example from planes (100), is equivalent to first-order, $n = 1$, from planes (200). In other words n can be incorporated in d so that Bragg's Law becomes

$$\lambda = 2d \sin \theta. \tag{1}$$

This interpretation is equivalent to Laue's, with the important difference that Bragg found that if ZnS was assumed to be based on a face-centred cubic lattice, and not a simple cubic one, reflections were predicted for (hkl) all even (such as 200, 222) or all odd (such as 111, 135) but not otherwise [9]. The defects in Laue's interpretation were remedied, without at this stage enabling Bragg to discover the arrangement of atoms in the crystal.

This he first achieved by taking Laue photographs of NaCl, KCl and KI [10], on the suggestion of W J Pope, Professor of Chemistry in Cambridge, that alkali halides were likely to have one of the structures predicted by Barlow. The photographs had points of similarity and of difference. The simplest were for KCl, where the K and Cl atoms scatter so nearly equally that the atoms appear to form a simple cubic lattice of lattice constant $a/2$, the distance between nearest neighbours. In KI the iodine atoms dominate and the pattern can be interpreted as that from iodine atoms arranged on a face-centred cubic lattice, with a lattice constant somewhat greater than that of KCl. For NaCl the pattern is more complicated since neither atom dominates, but the structure was deduced to be the same (figure 6.5). These conclusions were confirmed by measurements made with W H Bragg's recently constructed crystal spectrometer (figure 6.6), and it was deduced that the wavelength of an L characteristic radiation from platinum was $\lambda = 1.10 \times 10^{-8}$ cm while the lattice constant of NaCl was $a = 5.60 \times 10^{-8}$ cm, the first of such reliable determinations, [11].

Other investigators, notably G Wulff in Russia, reached similar conclusions and Wulff independently formulated Bragg's Law. An elegant interpretation of the interaction of x-rays with a crystal was given by Ewald [4] in terms of the reciprocal lattice, but did not find application until later.

Sir (William) Lawrence Bragg

(British, 1890–1971)

W L Bragg was born in Adelaide, the son of Sir William Henry Bragg. Partly in collaboration with his father, he founded x-ray crystallography by reinterpreting von Laue's photographs in terms of the reflection of x-rays by planes in the diffracting crystal. He made the first crystal-structure determination, that of rock salt (NaCl). Father and son shared the Nobel Prize for physics in 1915. He was professor of physics at the Victoria University, Manchester, from 1919 to 1937 and succeeded Rutherford as Cavendish Professor in Cambridge (1938–53). Like his father, he became Director of the Royal Institution in London (1954–65), where he did much to popularize science, particularly by his lectures to schoolchildren. As Cavendish Professor he supported and participated in the work of M F Perutz and J C Kendrew on the crystal structures of proteins and also interested himself in the work of F H C Crick and J D Watson on the structure of DNA, thereby assisting in the foundation of molecular biology.

W H Bragg initially used his crystal spectrometer to analyse the radiation from an x-ray source. He found that 'white' radiation was reflected from a particular crystal face over a range of angles, and superimposed on this was a spectrum of sharp peaks of monochromatic x-rays. His observations were greatly extended in a brilliant series of measurements by H J Moseley [12], who founded the subject of x-ray spectroscopy, and introduced the concept of atomic number.

See also p 86

Later the Braggs joined forces and used the crystal spectrometer to determine crystal structures, having their first success with diamond [13]. By January 1915, when their *X-rays and Crystal Structure* was sent to the publisher, they were able to report structures of nine types: sodium chloride, diamond, sphalerite (ZnS), wurtzite (ZnO), caesium chloride,

Twentieth Century Physics

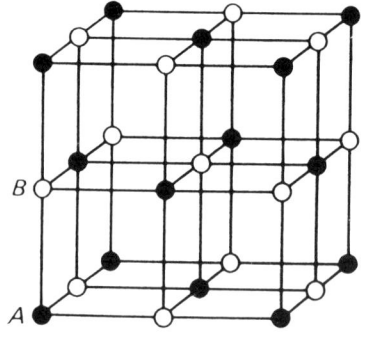

For NaCl
$AB = 2 \cdot 8 \times 10^{-8}$ cm

Figure 6.5. *The structure of sodium chloride.*

Figure 6.6. *W H Bragg's crystal spectrometer. The incident beam is limited by slits at A and B. The crystal at C can be rotated about a vertical axis. The reflected beam is detected in the tubular ionization chamber and the current is measured in the electroscope at E.*

copper, fluorite (CaF_2), pyrites (FeS_2) and calcite ($CaCO_3$). In most of these the positions of the atoms are determined by symmetry, but symmetry alone does not fix the distance between the sulphur atoms in FeS_2, which lie on a threefold axis. This parameter was determined by trial [14], so that there was agreement between measured and calculated x-ray reflection intensities. In a postcard to Ewald, W L Bragg described the structure as 'terribly complicated'. In this early work the scattering power of an atom (now called the atomic scattering factor f) was taken to be proportional to its atomic weight. The concept of a structure factor $F(hkl)$ appeared in an early paper by Laue, but it was apparently A Sommerfeld [15] who first defined it clearly. We take it meantime to be a measure of the amplitude scattered by the electrons in any one unit cell, determined by the coordinates and atomic scattering factors of the atoms in the unit cell, and the indices (hkl) which define the reflecting planes.

W L Bragg and his collaborators [16] found that the intensity of reflection from a particular crystal face depended very much on the degree of perfection of the crystal—on whether the face was a cleaved surface or had been lightly ground, for example. The range of reflection could be anything from a fraction of a degree to two or three degrees. C G Darwin [17] had earlier shown that the intensity of reflection from a perfect crystal is total over an angular range, proportional to the structure factor $|F(hkl)|$, which is no more than seconds of arc in magnitude. However, for a 'mosaic' crystal, i.e. one consisting of crystallites slightly misoriented to one another (as is now known to be caused by dislocations), the integrated intensity is proportional to $|F(hkl)|^2$ and the range of reflection depends on the mosaic structure or relative misorientation. Integrated intensity and experimental aspects of this topic will be considered further in section 6.3.

Fundamental work was done at an early stage by P Debye [18] on another important problem, the effect of thermal vibration of the atoms on the intensity of reflection. He found that the atomic scattering factor for a fixed atom, f, (which is a function of $\sin\theta/\lambda$,) should be replaced by an effective scattering factor

$$f(T) = f \exp(-B \sin^2\theta/\lambda^2) \qquad (2)$$

where B in this 'temperature factor' is equal to $8\pi^2 u^2$, u^2 being the mean square displacement of the atom in any direction due to thermal vibration. At temperatures comparable with the characteristic temperature θ_D, which appears in Debye's theory of the specific heat of a crystal, B is proportional to T and inversely proportional to θ_D^2. Debye's work was extended and corrected by I Waller [19], who treated the matter more rigorously from a quantum-mechanical point of view, showing that the Bragg-reflected radiation is elastically scattered (i.e. without change of energy or wavelength) but that x-rays are also inelastically scattered with slight change of energy in directions not allowed by Bragg's law, the energy being taken up by the thermally-excited waves in the crystal. This connected x-ray scattering with lattice dynamics, as discussed in Chapter 12.

The Nobel Prize for physics was awarded to von Laue in 1914 and was shared by the Braggs in 1915, while Debye was awarded the prize for chemistry in 1936.

6.3. Experimental techniques

We must now define the atomic scattering factor $f(\sin\theta/\lambda)$ more carefully. We take the amplitude scattered by a single electron as the unit. The amplitude scattered by an atom when account is taken of interference between wavelets scattered by different parts of the atom is then given by [20]

$$f(S) = \int 4\pi r^2 \rho(r) \frac{\sin 2\pi r S}{2\pi r S} \, dr.$$

The atom is assumed to be spherically symmetric with electron density $\rho(r)$ such that there are $4\pi r^2 \rho(r)\,dr$ electrons in the thin spherical shell between r and $r + dr$, and $S = 2\sin\theta/\lambda$. The more compact the electron distribution in the atom, the more slowly f falls off with S (mathematically, $f(S)$ is the Fourier transform of $\rho(r)$). Atomic distributions $\rho(r)$ were at first determined by the self-consistent field method of D R Hartree [21]. Later the advent of fast computers enabled greater accuracy to be achieved for the $\rho(r)$ and therefore for the f-curves.

[See also p 1036]

Having dealt with the effect of interference within one atom, we must now consider how the scattered waves from different atoms interfere with one another. This is where the structure factor comes in. It is the (complex) amplitude scattered by all the atoms in one unit cell, and is given by

$$F(hkl) = \sum_{j=1}^{N} f_j \exp 2\pi i (hx_j + ky_j + lz_j). \qquad (3)$$

Here f_j is the atomic scattering factor of the jth atom (they are labelled $j = 1, 2, \ldots, N$) and its position in the unit cell is defined by the vector r_j, expressed by

$$r_j = x_j \mathbf{a} + y_j \mathbf{b} + z_j \mathbf{c}$$

where (xyz) are evidently fractional coordinates. In terms of a vector $\mathbf{H}$ which is perpendicular to the reflecting (hkl) planes and has magnitude $H = 1/d(hkl)$, where $d(hkl)$ is the spacing of the planes (so that also $H = 2\sin\theta/\lambda$, by Bragg's law), $F(hkl)$ can be written more succinctly as

$$F(\mathbf{H}) = \sum_{j=1}^{N} f_j(\mathbf{H}) \exp 2\pi i \mathbf{H} \cdot \mathbf{r}_j. \qquad (4)$$

Other properties of $\mathbf{H}$ are not important at this stage; the reader may think of $\mathbf{H}$ as shorthand for hkl, and $\mathbf{H} \cdot \mathbf{r}$ as shorthand for $hx + ky + lz$.

When the crystal is centrosymmetric, the origin is taken on a centre of symmetry, and for every atom at r_j there is an equivalent one at $-r_j$ so that

$$F(hkl) = \sum_{j=1}^{N} f_j \cos 2\pi(hx_j + ky_j + lz_j) \equiv F(\mathbf{H}) = \sum_{j=1}^{N} f_j(\mathbf{H}) \cos 2\pi \mathbf{H} \cdot \mathbf{r}_j.$$

The structure-factor formula may be made less mysterious by the following considerations. From the definition of an (hkl) plane as one which makes intercepts a/h, b/k, c/l on the axes, (see section 6.1), its equation must be

$$hx + ky + lz = 1.$$

The parallel plane through the origin has $hx + ky + lz = 0$, the one on the other side has $hx + ky + lz = 2$, and so on. An atom j which has its centre precisely on such a plane scatters with phase $\varphi_j = 0$, (or, what is effectively the same thing, $\pm 2\pi$, $\pm 4\pi$ etc). One which falls just midway between planes scatters with phase π, and following this line of argument one concludes that the phase φ_j is proportional to the perpendicular distance $\mathbf{H} \cdot \mathbf{r}_j$ of the atom from a plane. Combining the N contributions by means of an amplitude–phase diagram gives the structure-factor formula.

By using a crystal spectrometer, the intensity of x-ray reflection can be measured on an absolute scale. As was noted in section 6.2, the range of reflection in 2θ depends on the degree of perfection of the crystal, but the integrated intensity is independent of this and proportional to $|F(hkl)|^2$.

Measurements were made on crystals of sodium chloride by R W James et al [22] at temperatures down to that of liquid nitrogen and were corrected for 'secondary extinction' by replacing μ by an effective value (see section 6.5). It was found that the f-curves for Na$^+$ and Cl$^-$ deduced from their measurements agreed well with theoretical ones obtained by Hartree [21], provided account was taken of both the thermal motion and the zero-point motion of the ions as in Debye's theory. It is only for such fundamental results that x-ray intensities on an absolute scale are essential, for crystal-structure analysis relative values of $|F(hkl)|^2$ generally suffice and can be obtained by simpler methods.

Crystal spectrometers were soon replaced by photographic methods. When the crystal is mounted with the c axis, for example, vertical, and is rotated about this axis, $(hk0)$ reflections are produced which are recorded on an equatorial line on a film in a cylindrical holder, the axis of which is also vertical. These are referred to as 'zero-layer reflections'. The next line up contains reflections for which $l = 1$ (first layer) and so on. Since reflections may be recorded on the same layer line, which are distinct but have by chance the same value of $\sin\theta$, there will generally

> ## BOX 6B: DARWIN'S FORMULA FOR THE 'INTEGRATED INTENSITY'
>
> When the beam is intercepted by a face of a mosaic crystal, Darwin [17] deduced that the integrated intensity is given by
>
> $$\frac{E\omega}{I} = \frac{Q}{2\mu}$$
>
> where
>
> $$Q = \left(\frac{e^2}{mc^2}\right)^2 \left(\frac{F(hkl)}{v}\right)^2 \lambda^3 \frac{1 + \cos^2 2\theta}{2 \sin 2\theta}.$$
>
> Here E is the energy reflected into the detector as the crystal is run with angular velocity ω through the reflecting position while I is the energy per second in the incident beam, μ is the absorption coefficient and v the volume of the unit cell. e^2/mc^2 has the dimensions of length and is approximately 10^{-12} cm, often called the classical radius of the electron. The factor $\frac{1}{2}(1 + \cos^2 2\theta)$ is present when the incident beam is unpolarized.

be some overlap of spots on the film, but this can be avoided, or at least minimized, by oscillating the crystal through a limited range of angles. This method was first used by M de Broglie, and was fully developed by J D Bernal [23] who introduced a standard design of camera and constructed charts to facilitate indexing of the reflections. Moving-film methods were introduced by K Weissenberg [24] in 1924. In a Weissenberg camera a metal screen with a slit is arranged so that only reflections belonging to one layer line at a time fall on the film. As the crystal is oscillated, typically through 180°, the film is moved in synchronism parallel to the axis of rotation. Instead of being concentrated on a line, the reflections are well separated, there is no possibility of overlap and indexing is easy. In the precession camera, invented by M J Buerger [25], an even simpler photograph of one layer at a time is obtained.

These methods have in common the fact that a single crystal, measuring some fraction of a millimetre in dimensions, is required. It was found by P Debye and P Scherrer [26], and independently by A W Hull [27], that considerable information can be obtained from the x-ray photograph of a specimen of crystalline powder. Experimental arrangements which differ only in the layout of the recording strip of film are shown diagrammatically in figure 6.7, while figure 6.8 shows powder photographs of KCl and of NaCl. Such photographs contain less

Solid-State Structure Analysis

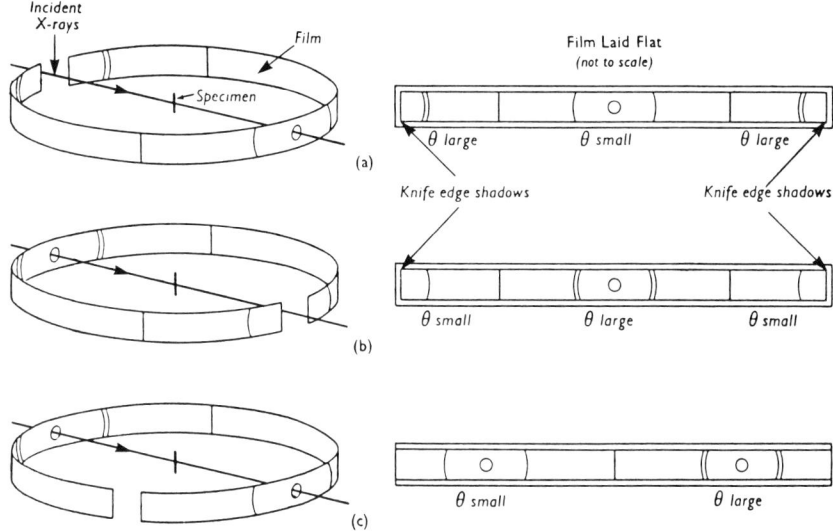

Figure 6.7. *Possible layouts of the recording strip of film in taking a powder photograph.*

information than do those of single crystals, because of the overlap of reflections, and the information is more difficult to extract. The method has, however, been widely used, and Hull in particular was able to determine the crystal structures of a number of metallic elements which had not been studied by single-crystal methods. In the hands of experts such as J Westgren, G Phragmen and A W Bradley, the powder method made it possible in the 1920s and 1930s to determine the structures of alloys almost as complex as any subsequently studied by single-crystal methods.

Photographic techniques [28] have in common that the specimen is immersed in the incident beam, which may have been monochromated by reflection from a crystal, although in many experiments greater intensity is obtained, but with some admixture of white radiation, by passing the beam through a suitable filter. Intensities can be measured with an accuracy of about 10% by eye comparison with a scale of intensity, and more accurately by photometry, which is a tedious process unless the equipment is automated.

A diffractometer [29] operates in somewhat the same way as the Bragg spectrometer, but the incident radiation is filtered, or more often monochromated, the specimen is small and completely immersed in the beam, the detector is a photon counter, and there is a much greater degree of control of the orientation of the specimen. Most importantly, the position of the detector and the orientation of the specimen, as well as the recording of counts representing integrated intensity, are controlled by computer. The commercial development of instruments for single

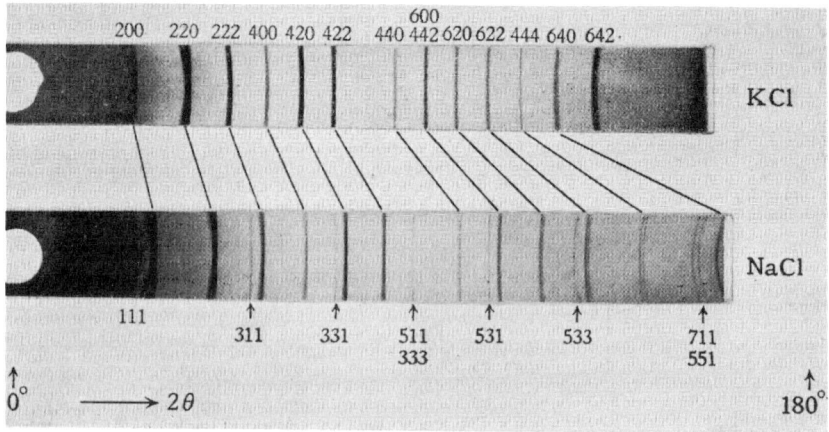

Figure 6.8. *Powder photographs of KCl and NaCl.*

crystals and for powders owes much to T C Furnas and W Parrish, respectively.

Reflections can be measured, with corrections for background etc, at the rate of 100 per hour, typically. Rotating-anode tubes [30] allow 5 to 10 times greater dissipation of energy and/or a smaller area of focal spot. Demountable tubes, with or without a rotating anode, have the advantages that the anode can be replaced to give a selected wavelength, and a burnt-out cathode can be replaced by a new one. They are, however, more troublesome to operate than sealed-off tubes. The development of synchrotron sources has made available greatly increased x-ray intensities [31]. In the synchrotron DESY [32] at Hamburg the range of x-ray wavelengths produced is about the same as from an x-ray tube under typical operating conditions, and for 5 GeV electrons is centred on 1.0 Å. The radiation is polarized in the plane of the orbit and emitted tangentially with a very narrow angular aperture. The spectral luminosity of this particular source was (in 1974) estimated to be some 100 times greater for radiation of wavelength 1.54 Å than from a rotating-anode tube with a Cu anode, and there was the prospect of increasing this ratio by a further factor of perhaps 30. Measurements on a protein crystal [33] were in practice found to be made about 50 times faster than with a rotating anode tube, and by using a monochromator with improved focusing geometry the ratio was increased to about 125.

Many different methods have been reported for recording data from a single crystal or a powder, for both photographic and diffractometer techniques, at high and at low temperatures. The simplest method of cooling to near liquid nitrogen temperatures is to direct a continuous flow of cold gas on to the specimen, with an outer 'curtain' of warm air to prevent condensation. For lower temperatures, helium cryostats

Solid-State Structure Analysis

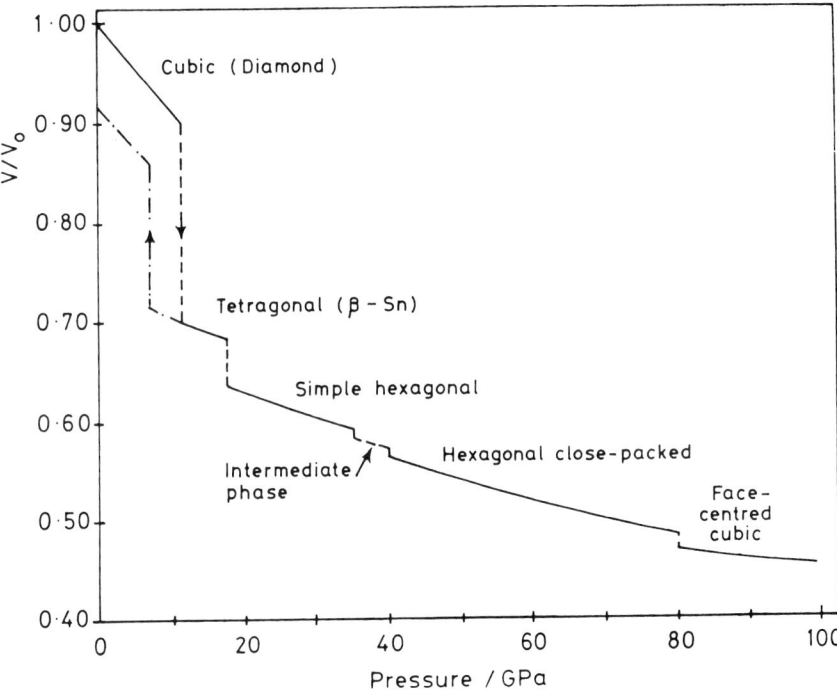

Figure 6.9. *The relative volume V/V_o of silicon over a wide range of pressure.*

have been developed [34]. A design by P Debrenne *et al* [35] allows temperatures up to 2500 °C while the pressure at the specimen is maintained at 10^{-8} Torr.

Equipment for use at high pressures is designed with less difficulty for a neutron source, since neutrons are much less heavily absorbed in the necessary containing vessels. The first use of a diamond anvil high-pressure cell for x-ray powder diffraction was by J C Jamieson *et al* [36]. Since then single-crystal studies at pressures up to 300 kbar with temperatures up to 450 °C have been made [37]. I L Spain [38] reported volume changes in silicon at pressures up to 100 GPa (1000 kbar). Figure 6.9 shows transitions through five main phases, from the cubic (diamond) structure through tetragonal, simple hexagonal and close-packed hexagonal to face-centred cubic. The history of high-pressure diffraction techniques has been reviewed by S Block and G Piermarini [39].

U W Arndt *et al* [40] have described a single-crystal oscillation camera for crystals having large unit cells, in the range 75 to 250 Å. The photograph is taken on a flat film which is subsequently scanned and indexed by a computer-controlled microdensitometer. Figure 6.10 shows an oscillation photograph of a crystalline virus. 1800 reflections could be recorded and measured per hour. Arndt [41] has reviewed

Twentieth Century Physics

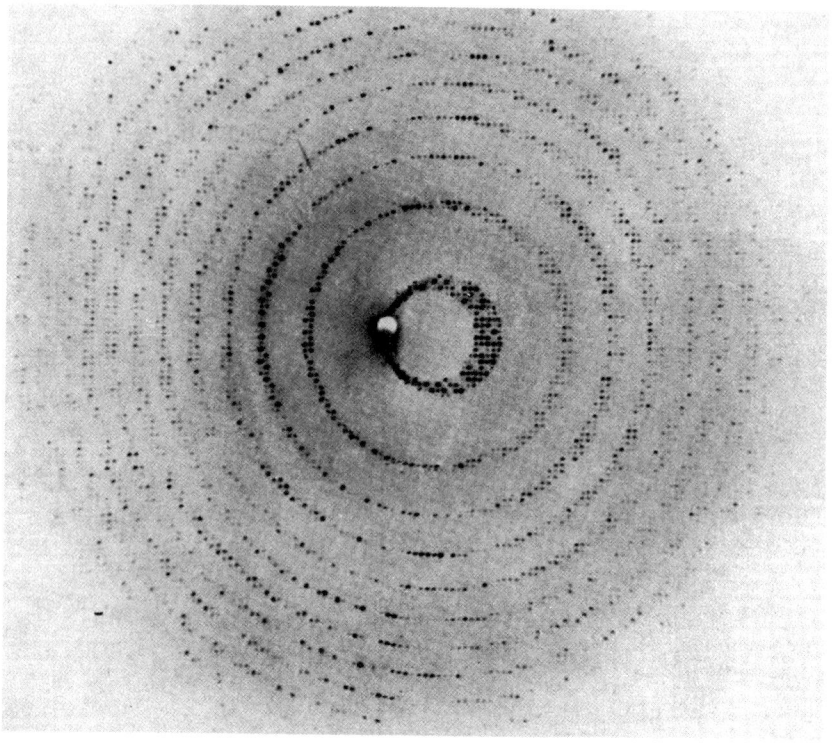

Figure 6.10. *Oscillation photograph of a crystalline virus.*

the processes which can be used to record x-ray data using position-sensitive detectors, while D Bilderback *et al* [42] have reported that an area detector, making use of Kodak storage phosphor technology, gave improved signal-to-noise and increased sensitivity, although less spatial resolution in comparison with film, in recording data from protein crystals. Storage phosphors enabled more data to be acquired per crystal, with less radiation damage.

There has lately been renewed interest in the Laue technique [43]. Many macromolecules retain substantial chemical activity in the crystalline state. The fact that x-ray Laue diffraction patterns can be recorded in a time of 100 ps or less with synchrotron sources and either photographic or area detector recording has opened up the possibility of time-resolved crystallography [44]. A time-resolved experiment has three key components: reaction initiation, monitoring of the reaction through measurement of changes in x-ray intensities, and data analysis. In some instances initiation by a light pulse is possible, and another simple method is by temperature jump. Preliminary studies of protein unfolding in lysozyme have been made by K Moffat *et al* [45].

6.4. Methods of structure determination

In the equation for the structure factor the atomic scattering factor for each atom is known, at least to a good approximation, and $|F(H)|$ can be determined over a range of values of H. 'The phase problem', which is the central problem of crystal-structure analysis, arises from the fact that it is $|F(H)|$ that is measured and not $F(H)$ itself. If $F(H)$ is written in the form

$$F(H) = |F(H)| \exp(i\alpha(H))$$

the phase angles $\alpha(H)$ are not directly accessible to experimental measurement, but if they were all known the atomic coordinates could be found as the maxima in an electron density map

$$\rho(r) = \frac{1}{v} \sum_H |F(H)| \cos(2\pi H \cdot r - \alpha(H)).$$

It was W H Bragg [46] who realized that the structure factors are the Fourier coefficients for the electron density. A brief history of the development of this topic has been given by R W James [47].

The existence of the phase problem was scarcely appreciated in the early crystal-structure determinations mentioned in section 6.2, nor was there much need for the results of space-group theory until structures involving several parameters began to be studied. The connection between space groups and x-ray diffraction was exposed by P Niggli and in more detail by W T Astbury and K Yardley [48]. Their work was extended and incorporated in *International Tables for X-ray Crystallography* which first appeared in 1935 and has since been continually revised and extended. As an illustration of how x-ray data can be used to determine the space group of a crystal, suppose the space-group symmetry is that denoted by the Hermann–Mauguin symbol $P2_1$. This indicates that it belongs to the monoclinic system with a twofold screw axis parallel to b, that is, for every atom at x_j, y_j, z_j there is an equivalent atom in the same unit cell at $-x_j, y_j+1/2, -z_j$. It is then easy to show from the expression for the structure factor that $F(0k0)$ is identically zero when k is odd. If the space group is $P2$ on the other hand, with equivalent points x_j, y_j, z_j and $-x_j, y_j, -z_j$, then there are no 'systematic absences'. The topic is considered in more detail in volume 1 of *The Crystalline State* [1].

Not all space groups can be distinguished by a study of the crystal symmetry and the systematically absent structure factors, but considerable progress was made in this direction by A J C Wilson and his collaborators [49]. First we note from (4) that

$$\overline{|F(H)|^2} = \sum_{j=1}^{N} f_j^2(H)$$

where the average is over a sufficient number of structure factors in the range about H. Since values of $f_j(H)$ are known, this result allows relative values of $|F(H)|$ to be replaced by absolute values, and is referred to as 'Wilson's scaling method'. When the crystal has a centre of symmetry, so that

$$F(H) = \sum_{j=1}^{N} f_j(H) \cos 2\pi H \cdot r_j$$

the value of $\overline{|F(H)|^2}$ is the same, but the *statistical distribution* of values of $|F(H)|$ is changed. This fact forms the basis of practical tests for the presence of a centre of symmetry. Certain other symmetry elements also affect the statistical distribution of x-ray intensities, so that while systematic absences serve to determine unambiguously 49 space groups, statistical methods extend the number to 215.

However, if at least one atom per asymmetric unit scatters 'anomalously', there is no need to resort to statistical methods. When the incident x-ray photon energy is just short of the energy at which there is an absorption edge for a particular element, the scattering factor of the corresponding atom is a complex number

$$f_a = f + i\Delta f''.$$

$\Delta f''$ is appreciable (a few per cent of f) only for comparatively heavy atoms. Friedel's law, which states that $|F(H)| = |F(-H)|$, is still valid when atoms scatter anomalously provided that the crystal has a centre of symmetry, but not otherwise except in special cases. As before, a total of 215 of the 230 space groups can be distinguished [50].

Until the 1930s structure determinations continued to be made by trial and error, postulating a structure which was consistent with the space-group symmetry and which was stereochemically feasible, then testing the agreement between calculated and observed structure factors. A striking example of the success of this method is provided by the analysis of the structure of beryl, $Be_3AlSi_6O_{18}$, by W L Bragg and J West [51] (figure 6.11). The space group was determined from the systematic absences. Bragg's account of the structure determination makes it sound easy, but in fact it represented a triumph of skill and experience which made use of the unit-cell dimensions, space-group symmetry and packing considerations.

The possible use of the Fourier series method for structure determination was demonstrated by W L Bragg [52] in 1929. The structure of diopside, $CaMgSi_2O_6$, is shown in figure 6.12(a). The structure is centrosymmetric in projection on the (010) plane, and the projected electron density is given by

$$\rho(xz) \equiv \int_0^b \rho(xyz)\, dy = \frac{1}{A} \sum_{hl} F(h0l) \cos 2\pi(hx + lz).$$

Solid-State Structure Analysis

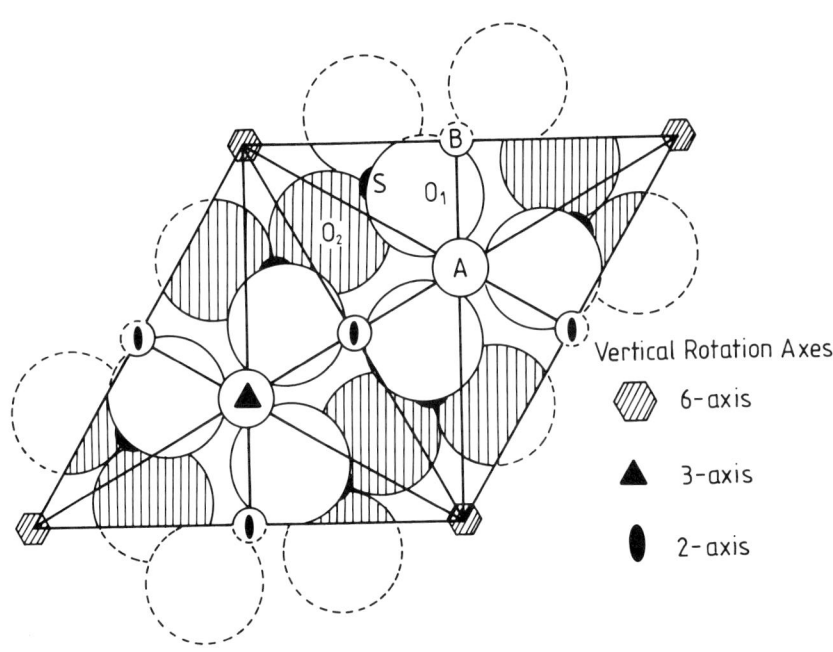

Figure 6.11. The structure of beryl, $Be_3Al_2Si_6O_{18}$. There are six oxygen atoms (shaded) in a compact ring around each sixfold axis. The remaining twelve are shown by the unshaded circles, each of which represents an oxygen atom superposed on its image atom in a horizontal mirror plane. Beryllium atoms are at B positions, aluminium atoms at A positions while silicon atoms appear as small black segments at S positions.

A two-dimensional Fourier series involves a relatively limited number of terms and is less formidable to evaluate. Each $F(h0l)$ is either positive or negative, but in this instance superposed Ca and Mg atoms at the origin make all Fs positive. Figure 6.12(b) shows contours of equal electron density in which silicon and oxygen atoms also show up clearly. The first use of a two-dimensional Fourier series to complete a structure determination was by C A Beevers and H S Lipson [53] in 1934, for $CuSO_4.5H_2O$. The signs of the structure factors were taken to be those of the joint contributions of the copper and the sulphur atoms, whose positions were known, and the oxygen atoms then showed up clearly in the projected electron density.

An important advance was made in 1934 by A L Patterson [54], developing an idea of F Zernike and J Prins [55] who had shown that the radial distribution of the atoms surrounding any given atom in a liquid could be determined from the x-ray diffraction pattern of the liquid. Patterson realized that there must be a corresponding relation connecting the distances between atoms in a crystal with the values of $|F(H)|^2$. He defined what is now known as the Patterson function

441

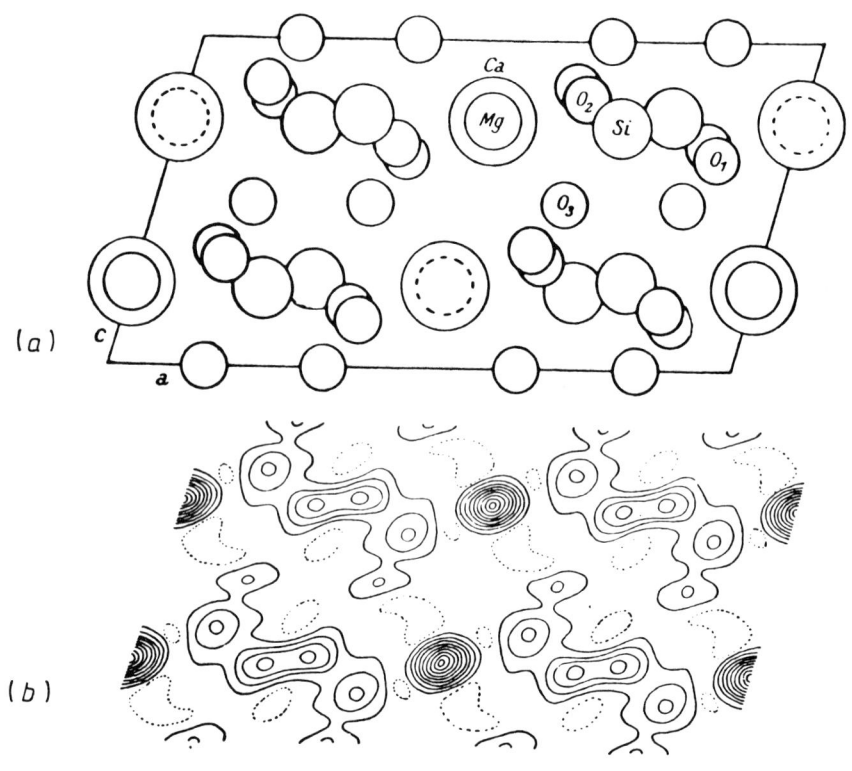

Figure 6.12. (a) The structure of diopside, $CaMgSi_2O_6$, in projection down the b axis, together with (b) a map of the projected electron density obtained by Fourier synthesis.

$$P(r) = \frac{1}{v} \sum_H |F(H)|^2 \cos 2\pi H \cdot r \qquad (5)$$

which does not involve the phase angles $\alpha(H)$, and showed that for every pair of atoms at r_1 and r_2 there is a peak in $P(r)$ at $r_2 - r_1$; $P(r)$ is thus a map of interatomic vectors. If there are N atoms in the unit cell, there are N^2 peaks in $P(r)$, of which N coincide at the origin. The weight of a peak involving atoms of atomic number Z_1 and Z_2 is $Z_1 Z_2$, which is also an approximate measure of the height of this peak. When N is large not many of the $N^2 - N$ peaks will be separately resolved, a fact which diminishes its practical value.

A rather straightforward use of the Patterson function in projection [56], is illustrated by the determination of the crystal structure of salicylic acid. A molecule is shown in figure 6.13(a) together with the interatomic vectors between near-neighbour atoms, which will give peaks near the origin of the Patterson function. This region, in projection down the relatively short c axis, is shown in figure 6.13(b). Molecules occur in projection in two orientations related by a mirror plane perpendicular to

Solid-State Structure Analysis

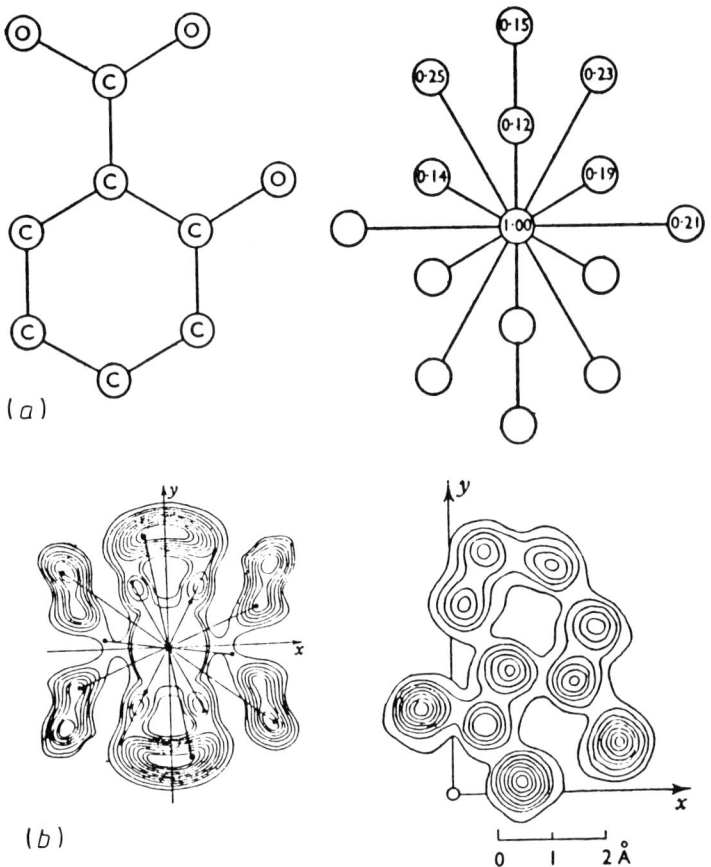

Figure 6.13. (a) A molecule of salicylic acid with the relative positions and weights of peaks near the origin of the Patterson function. (b) Peaks near the origin in the experimental Patterson function of salicylic acid, projected down the c axis, and for comparison the projected electron density for one molecule.

the plane of the diagram. The only way in which the two corresponding sets of interatomic vectors can be fitted to the Patterson function is shown in figure 6.13(b). Having found the orientation of a molecule in projection it was not difficult to find its position in the projected unit cell, and the projected electron density in one molecule is also shown in figure 6.13.

It was shown by D Harker [57] that for certain space groups a plane section through the three-dimensional Patterson function can give direct information about atomic coordinates. In practice, however, the main use of the Patterson–Harker section has been to determine the coordinates of a small number of heavy atoms, that is atoms of comparatively great atomic number, in a crystal structure.

The question 'Can a structure be recovered, at least in principle,

443

Figure 6.14. *The projected electron density in cholesteryl iodide, the sign of each Fourier coefficient having been taken as that of the iodine contribution. This map suggested the positions indicated for the other atoms.*

from its Patterson function?' was answered in the affirmative by D M Wrinch [58]; her paper did not attract much attention at the time but the same conclusion was reached by other authors a decade or so later. For example, consider a set of points at $\pm(r_1, r_2, \ldots, r_N)$ that is, with a centre of symmetry at the origin. The corresponding vector set has points of unit weight at $\pm(2r_1, 2r2, \ldots, 2r_N)$ while all others, apart from that at the origin, have double weight. Thus the points of unit weight give immediately the points of the original set. Two difficulties arise in attempting to apply this argument to the Patterson function. First of all, the latter is made up of peaks which are usually poorly resolved, or overlapping, so that individual peaks are seldom identifiable. More fundamentally, the difficulty arises that atoms at $x_j y_j z_j$ and at $x_j+1/2, y_j, z_j$ etc result in peaks at the *same* point in the Patterson function. The same difficulty prevents the immediate interpretation of a Patterson–Harker section, discussed above. However, methods have been developed [59], which in practice enable a structure to be recovered from its Patterson function once the positions of a small number of constituent atoms have been found. Suppose the positions $r_1, r_2, \ldots, r_n$ of n atoms are known. The structure need not have a centre of symmetry. Now set the Patterson function down with its origin at each of these n points in turn. Where n peaks coincide will be the position of another atom. Somewhat unexpectedly, the best measure of successful n-fold coincidence is to take the minimum value of an individual Patterson function at the point of coincidence. This has been shown by M J Buerger [60] to be the procedure least likely to produce spurious atomic positions by chance coincidences.

One method which utilizes the Patterson function, known as the heavy-atom method, leads in favourable circumstances directly to the structure. Cholesteryl iodide [61] serves as an example. There are two molecules in the unit cell, and the projected Patterson function is dominated by a peak due to the two iodine atoms, whose coordinates are thus fixed. The sign to be attached to each $|F(h0l)|$ was taken to be that of the dominant iodine atom contribution, and the corresponding electron density projection $\rho(xz)$ is shown in figure 6.14 together with the outlines of the two molecules.

There is a related method, known as the method of isomorphous replacement, which makes use of data from *two* crystal structures which are the same, apart from the replacement of one atom (and its symmetry-related equivalents) by another of different atomic number. The two data sets are put on the same absolute scale by Wilson's scaling method, outlined earlier in this section. We consider first the case where the structure is centrosymmetric, or is so in projection. If then the two structure factors are denoted by $F_1(hkl)$ and $F_2(hkl)$, we have that

$$F_2(hkl) - F_1(hkl) = 2(f_2 - f_1)\cos 2\pi(hx_R + ky_R + lz_R).$$

The right-hand side can be evaluated for each (hkl) when the coordinates $x_R y_R z_R$ of the replaceable atom have been determined from the Patterson function, or by a comparison of the two Patterson functions if neither of the replaceable atoms is particularly heavy. Then from the known values of $F_2 - F_1$, $|F_1|$ and $|F_2|$ for each (hkl), the correct signs of F_2 and of F_1 can be found for almost all cases. The isomorphous replacement method has been much used in determining biomolecular structures, such as proteins, which cannot have a centre of symmetry. In this instance it turns out that *three* sets of data are required, from the original crystal, and from the original with an atom replaced by a heavy atom in each of *two different* positions. Fortunately this exacting condition can often be satisfied in biomolecular structures. In practice, because of experimental errors and lack of exact isomorphism, it is helpful to have triple (or more) isomorphous replacement. The method was first suggested by J M Bijvoet and his co-workers [62] and was further elaborated by D Harker [63] with an eye to its application to protein structures. When the replacing atom scatters anomalously with a sufficiently large value of $\Delta f''$, one isomorphous replacement suffices to obtain $\alpha(H)$ from the experimental values of $|F(H)|$, $|F_1(H)|$ and $|F_1(-H)|$ [50].

In biomolecular structures, particularly of proteins and of viruses, a molecule or a virus particle may be built up from several smaller units which although identical or nearly so, are not related by space-group symmetry. Examples will be met with in section 6.11. Also, particularly in protein crystallography, the same or closely related molecules may occur in crystals with different unit cells and space-group symmetry. For example, four molecules of the protein myoglobin make an assembly which is very closely similar to the unit-cell content of haemoglobin. In such circumstances the method of molecular replacement, which developed from a paper by M G Rossman and D M Blow [64] can be used for *ab initio* structure determination as well as to improve and extend phase angles from the stage where such a structure has only been determined at low resolution. The method is too elaborate to describe in detail here; reviews of its different aspects have been published by P Argos and M G Rossman [65] and by M G Rossman [66].

See also p 496

| 76 | C 7 | 76 | 56 | 8 | $\overline{45}$ | $\overline{74}$ | $\overline{66}$ | $\overline{23}$ | 31 | 69 | 72 | 38 | $\overline{16}$ | $\overline{61}$ | $\overline{76}$ | $\overline{51}$ | 0 |
| 19 | S 6 | 0 | 11 | 18 | 18 | 11 | 0 | $\overline{11}$ | $\overline{18}$ | $\overline{18}$ | $\overline{11}$ | 0 | 11 | 18 | 18 | 11 | 0 |

Figure 6.15. *Two examples of Beevers–Lipson strips showing the values of $76\cos 7n6°$ and $19\sin 6n6°$ for $n = 0, 1, \ldots, 15$. The strips facilitated calculation of Fourier syntheses.*

The computational work involved in evaluating a three-dimensional Fourier series is so great that it was not often attempted before the advent of electronic computers, which have made it a simple matter. Even a two-dimensional series took some hours to evaluate and plot out as a figure field on which contour lines were drawn. The work was made more practicable by the availability of Beevers–Lipson strips [67] (figure 6.15) which were commercially available and widely used, not only in crystallography, for many years.

Several special purpose analogue computers were developed before they were made obsolete in the early 1950s by the advent of general purpose electronic computers. The most useful of these was the XRAC, designed by R Pepinsky [68]. Figure 6.16 shows a map of the projected electron density in a molecule of phthalocyanine as produced by XRAC. In the late 1940s and early 1950s XRAC made Pepinsky's laboratory something of a Mecca for crystallographers and a centre for the exchange of ideas on many topics in x-ray crystallography.

The term 'direct methods' is used for those methods which attempt to derive the sign, or the phase angle $\alpha(\mathbf{H})$ directly from the magnitudes of $F(\mathbf{H})$ and other structure factors. Helped by the computer revolution, such methods have undergone spectacular development in the past two decades or so and are now responsible for the majority of structure determinations. (The reader should be warned that the next few pages are quite mathematical, after which the account again becomes more descriptive.)

We begin with some necessary definitions. First of all, structure factors $F(\mathbf{H})$ are rescaled to give 'unitary' structure factors $U(\mathbf{H})$ which correspond to scattering by point atoms, according to the definition

$$U(\mathbf{H}) = F(\mathbf{H}) \bigg/ \sum_{j=1}^{N} f_j(\mathbf{H}) \equiv \sum_{j=1}^{N} f_j(\mathbf{H}) \cos 2\pi \mathbf{H} \cdot \mathbf{r}_j \bigg/ \sum_{j=1}^{N} f_j(\mathbf{H})$$

when the structure is centrosymmetric. The average value of a unitary structure factor does not decrease with increasing H, the maximum possible value of $U(\mathbf{H})$ (when all N atoms scatter in phase) is unity, and an idea of the scale of the Us is got by noting that $\overline{U^2(\mathbf{H})} = 1/N$ when all N atoms are equal. Lastly, we write $s(\mathbf{H})$ for the sign of $U(\mathbf{H})$, which is of course also that of $F(\mathbf{H})$.

Solid-State Structure Analysis

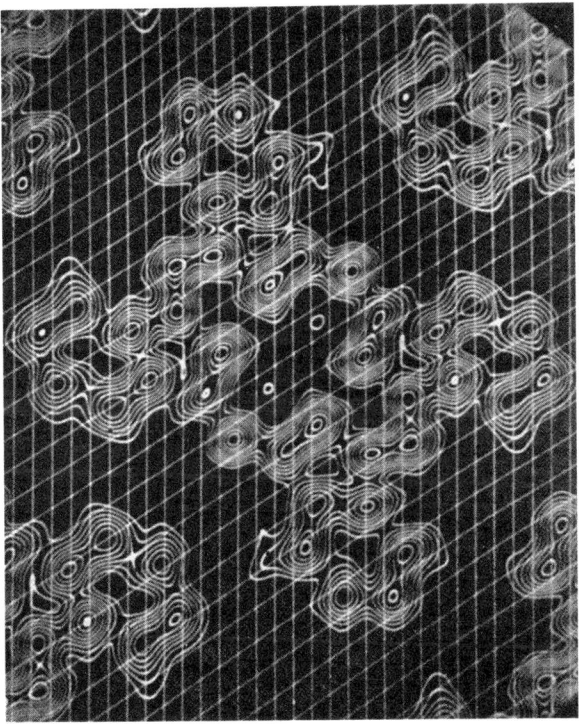

Figure 6.16. *Contours of the projected electron density in phthalocyamine produced on Pepinsky's XRAC.*

The first direct method to be of practical value was discovered by D Harker and J S Kasper [69] and took the form of *inequality relations* between unitary structure factors. Their physical basis is that all atoms scatter x-rays with the same sign and have (approximately) spherical symmetry, but we shall not give any derivations. The simplest Harker–Kasper inequality is

$$U^2(H) \leqslant \tfrac{1}{2}(1 + U(2H)) \tag{6}$$

which, when $|U(H)|$ and $|U(2H)|$ are large enough, requires $U(2H)$ to be positive. Another inequality can be used, again when $|U(H)|$, $|U(H')|$ and $|U(H+H')|$ are large enough, to show that $s(H) = s(H')s(H+H')$, i.e. that $s(H)s(H')s(H+H')$ must be positive.

J S Kasper *et al* [70] determined the crystal structure of dekaborane, $B_{10}H_{14}$, which has four molecules in a unit cell, using inequality relations to find the signs of the larger structure factors. Values of $|U(H)|$ were derived from those of $|F(H)|$, which had not been measured on an absolute scale, using Wilson's scaling method, outlined earlier in this section. The practical limitation of Harker–Kasper inequalities is that unless N is small, no $|U|$s are large enough for the relations to be restrictive.

The condition of positive electron density *by itself* imposes certain conditions, first set out by J Karle and H Hauptman [71]. They take the form of determinantal inequalities, $D \geqslant 0$ where

$$D = \begin{vmatrix} F(0) & F(-\boldsymbol{H}_1) & F(-\boldsymbol{H}_2) & \cdots & F(-\boldsymbol{H}_n) \\ H(\boldsymbol{H}_1 & F(0) & F(\boldsymbol{H}_1 - \boldsymbol{H}_2) & \cdots & F(\boldsymbol{H}_1 - \boldsymbol{H}_n) \\ \vdots & & & & \\ F(\boldsymbol{H}_n) & F(\boldsymbol{H}_n - \boldsymbol{H}_1) & \cdots & \cdots & F(0) \end{vmatrix}.$$

The indices $\boldsymbol{H}$ are different from one another but are otherwise arbitrary. When the additional condition of atomicity is imposed, Fs are replaced by Us.

When the structure is centrosymmetric, $U(\boldsymbol{H}) = U(-\boldsymbol{H})$, and a 3×3 determinant, with non-negative minors, leads to (6), the simplest of the Harker–Kasper inequalities. As M M Woolfson [72] has remarked 'much subsequent development in the theory of direct methods has been found to be an aspect of what determinantal inequalities contain—albeit that the connection is not always blindingly obvious'.

An important advance was made by D Sayre [73], starting from the fact that when the electron density is squared, it still consists of positive peaks in the same positions and the peaks are of equal weight if the atoms are equal. Sayre's equation has its simplest form when written in terms of unitary structure factors, namely

$$U(\boldsymbol{H}) = N\overline{U(\boldsymbol{H}')U(\boldsymbol{H} - \boldsymbol{H}')} \tag{7}$$

where the bar denotes an average over a range of values of $\boldsymbol{H}'$. When the structure is centrosymmetric the above relation can equally well be written as

$$U(\boldsymbol{H}) = N\overline{U(\boldsymbol{H}')U(\boldsymbol{H} + \boldsymbol{H}')}.$$

For the result to be exact, the average over $\boldsymbol{H}'$ must cover a range sufficient to produce equal resolved peaks in an electron density map, or involve a representative selection of such unitary structure factors.

While Sayre was successful in determining the signs of some of the larger Us for a projection of an organic structure, prospects for practical application did not appear bright because of the large number of terms involved on the right-hand side of the equation. It was, however, noticed by W Cochran [74], following a discussion with Sayre, that for a number of known structures $U(\boldsymbol{H})$ and any *one* product $U(\boldsymbol{H}')U(\boldsymbol{H} + \boldsymbol{H}')$ had almost always the same sign when three of the larger $|U|$s were involved. In other words the relation $s(\boldsymbol{H}) = s(\boldsymbol{H}')s(\boldsymbol{H} + \boldsymbol{H}')$ was satisfied although not compelled to be so by a Harker–Kasper inequality, nor did this depend on equality of the atoms present. Cochran was not able to give a quantitative explanation of this observation, but it proved possible to determine the structure of glutamine (in projection) using the triple-product sign relation, starting from ten or so of the largest Us. A broadly similar procedure was used by W H Zachariasen [75] to determine the structure of metaboric acid, H_3BO_3, using three-dimensional data. This was in fact a more difficult structure determination since the asymmetric unit consists of three molecules.

In 1953 H Hauptman and J Karle [76] published a monograph entitled *The Solution of the Phase Problem. I. The Centrosymmetric Crystal.* The title was over-optimistic, and some of the relations which they put forward were not entirely correct, but this publication has been deservedly influential. They established the concept of *probability relations* between structure factors and gave for example an approximate formula for the probability that $s(H)s(H')s(H+H')$ is positive. A more accurate result was obtained by M M Woolfson [77], namely that when there are N equal atoms per unit cell

$$P_+ = \tfrac{1}{2} + \tfrac{1}{2}\tanh(N|U(H)U(H')U(H+H')|)$$

gives the probability that $s(H)s(H')s(H+H')$ is positive. The formula is readily generalized to apply when the atoms are not equal. Results which are in principle more accurate were later obtained by A Klug [78] and others, but the hyperbolic tangent formula is sufficiently accurate for practical purposes. In the 1950s and early 1960s quite a few centrosymmetric structures were solved using sign relations.

When a structure or its projection is not centrosymmetric Sayre's equation still holds in the form (7). The approach initiated by Woolfson was used by Cochran [79] to show that when $U(H')$ and $U(H-H')$ are known for one value of H', the expected value of $\alpha(H)$ is

$$\langle \alpha(H) \rangle = \alpha(H') + \alpha(H - H')$$

and an expression for the probability distribution of $\alpha(H)$ about $\langle \alpha(H) \rangle$ was found. When there are a number of separate indications of the value of $\alpha(H)$, this result becomes

$$\langle \alpha(H) \rangle = \text{phase of} \sum_{H'} U(H')U(H - H').$$

J Karle and H Hauptman [80], following a different route, derived a formula for $\tan\langle\alpha(H)\rangle$ (the tangent formula) which has been particularly valuable. In discussing his results Cochran was decidedly pessimistic about the possibility of phase relations being useful in an *ab initio* structure determination, nevertheless a decade or so later I L Karle and J Karle [81] succeeded in determining the structure of arginine dihydrate using phase relations only. This was a notable and influential advance, in which they did not make much use of a computer. The move towards complete automation of the process of structure determination was initiated by G Germain and M M Woolfson [82] and has been continued, notably by Woolfson and his collaborators, with considerable success. The timing of Woolfson's review [72] felicitously marked the award of a Nobel Prize to J Karle and H Hauptman. The philosophy behind the computer programmes of Woolfson and his co-workers, such as MULTAN (multiple tangent formula method) was that the involvement of the user was to be reduced to a minimum, ideally none at all. From a starting set, several probable sets of phases are derived, each with a figure of merit attached. The corresponding electron density maps are evaluated and printed out with the positions of the highest peaks marked, and these are examined for a plausible molecular configuration. Details can be found in Woolfson's and other reviews [83].

While the phase angles $\alpha(H)$ are not directly accessible to experimental measurement, information about a 'structure invariant' such as $\alpha(H) - \alpha(H') - \alpha(H - H')$ can be obtained by making use of the Renninger effect, as has been shown by B Post and by K Hümmer and H Billy. It is not yet clear how useful this technique will prove to be in practice. Experimental conditions are exacting and will best be met by using a synchrotron source. However, any restrictions that can be placed on the phases in a 'starting set' can be valuable.

6.5. Accurate structure analysis

The structure determination of a biomolecular crystal is often at comparatively low resolution and may not give atomic positions with a standard deviation better than 0.2 Å or so. This, however, is quite sufficient for some purposes, such as the elucidation of the amino-acid sequence of a protein. To solve other problems of stereochemistry, such as the equality or non-equality of certain bond lengths, requires a much smaller standard deviation, while an accurate electron density map is required to locate hydrogen atoms or give information on the bonding electron distribution. The procedure from where a structure has been determined in outline to the point at which the maximum amount of information has been extracted is referred to as the refinement of the structure.

Experience indicates that a structure has probably been correctly determined when the R-factor or agreement index is less than about 0.4, where

$$R = \sum ||F_o| - |F_c|| \bigg/ \sum |F_o|$$

and F_o and F_c denote observed and calculated structure factors. Until the 1960s it was seldom possible to refine a structure beyond $R = 0.1$, but improvements in experimental and computational techniques since then have reduced this figure by a factor of three or more. Since not all measurements are equally reliable, it is preferable to use

$$R_1 = \left(\sum w(|F_o| - |F_c|)^2 \Big/ \sum w|F_o|^2\right)^{1/2}$$

where $w = 1/\sigma^2$ and σ is the variance of a particular $|F_o|$. Many corrections to the measured structure factors may be necessary, of which only a few deserve discussion here, the rest being of interest only to expert practitioners. Extinction is a topic of considerable mathematical difficulty and something of a headache for practical crystallographers [47,84]. The name dates from a 1922 paper by C G Darwin [85]. Notable contributions to both the theory and to methods of correction for extinction have been made by W H Zachariasen [86]. A perfect crystal of volume V and linear dimensions which are not more than a few thousand times greater than a unit-cell dimension gives (see Box 6B) an integrated intensity QV. As the thickness of the crystal is now imagined to increase in the direction of the normal to the reflecting planes, the upper planes reflect away a significant proportion of the incident beam and eventually the lower planes will be completely screened. The intensity will no longer be proportional to V, and since Q is proportional to $|F(H)|^2$ the screening may be complete for strong reflections when it is still unimportant for weak ones. This effect was designated 'primary extinction' by Darwin. Consider next a mosaic crystal in which individual blocks are characterized by a linear dimension too small to give rise to primary extinction. These blocks have a range of misorientation and consequent small random displacements so that phase coherence is lost between the beams from different blocks, and their intensities should be added, not their amplitudes. In these circumstances there will still be shielding of the lower layers when blocks in the upper layers which happen to be in the right orientation reflect away significant energy from the incident beam, attenuating it as if the absorption coefficient had been increased. Darwin designated this effect 'secondary extinction' and his analysis was extended by Zachariasen to derive a usable formula for correcting the measured structure factors.

In a refinement of the structure of hambergite, $Be_2BO_3(OH)$, the R-factor was reduced by this process to 0.041, the largest structure factors having been corrected by a factor of up to two. The weakness of Zachariasen's procedure is that the correction can only be made as part of the refinement, and not of the measurement process, and it is clear that correction for extinction should be kept to a minimum; this can be done by using as small a specimen as possible and a short wavelength of x-rays.

The effect of thermal vibration on the Bragg intensities, as given by (2), is not only the result given by Debye's approximate theory of lattice dynamics, but follows also from the more exact theory introduced by M Born and T von Kármán, provided the harmonic approximation is valid, that is, when the lattice waves obey the principle of superposition. While our discussion so far has assumed that the thermal vibration of each atom is isotropic, in practice it may be anisotropic and allowance for this effect requires the introduction of at most five additional parameters per atom at the refinement stage.

At temperatures comparable with the Debye characteristic temperature θ_D the lattice vibrations cannot be treated as harmonic, the theory must be extended and becomes quite elaborate. Crystallographers have found that it is generally adequate to assume, as in Einstein's model, that each atom vibrates independently, but in a potential that includes terms which are cubic, quartic etc in the atomic displacement. The effect on the temperature factor has been discussed in some detail by B Dawson [87] whose review is concerned with most aspects of structure refinement and the accurate measurement of electron density; unfortunately its completion was prevented by the author's untimely death. M J Cooper et al [88] investigated anharmonicity in BaF_2 using neutron diffraction and found good agreement with the approximate theory.

When the parameters of a refined structure are deduced from an electron density map the termination of the Fourier series introduces ripples in the atomic density, leading to systematic errors that may well be greater than the effect of random errors in the $|F_0|$s [89]. These can be almost completely eliminated by using the data to derive, not the electron density itself, but the difference between the observed and calculated electron densities. Use of this 'difference synthesis' was first advocated by C W Bunn [90] as an aid to structure determination at the stage where some atoms might be incorrectly placed, and by W Cochran [91] as a means of eliminating series termination errors and locating hydrogen atoms. The difference density has also been widely used as a means of measuring the bonding electron density in organic molecules.

Series termination errors can be avoided by using the method of least squares at the refinement stage, as first advocated by E W Hughes [56]. This is now the standard method of refinement and consists in systematically altering the atomic parameters so as to minimize the quantity R defined earlier.

Impressive accuracy has been attained in measuring the electron distribution in organic compounds such as oxalic acid $(COOH)_2 2H_2O$ [92]. Four x-ray and five neutron data sets were collected in different laboratories. The specimen temperatures were about 100 K in each case, $MoK\alpha$ radiation was used in the x-ray studies and neutron wavelengths were in the range 0.525 to 1.07 Å. Atomic positions agreed to better than 0.001 Å, but anisotropic temperature factors did not agree as well as expected from the quality of the data, possibly because no correction was made for scattering by acoustic modes, and different atomic scattering factors were used with different x-ray data sets. In difference density maps the temperature factors were deduced from the neutron data only, since the distribution of bonding electrons could simulate the effect of anisotropic vibration. Figure 6.17(a) shows ($\rho_o - \rho_c$) for three different sets of x-ray data, they are in good general agreement. The positions from which atoms have been subtracted are indicated. Figure 6.17(b) shows theoretical difference density maps; these derive from molecular orbital wavefunctions based on self-consistent field methods. The three maps in figure 6.17(b) correspond to somewhat different parameters of the theory. The agreement between theory and experiment, while not completely quantitative, is quite striking.

6.6. Neutron diffraction [84, 93]

That neutrons would be diffracted by a crystal was realized at least from 1936, when the topic was discussed by W M Elsasser [94]. Figure 6.18 shows diagrammatically the apparatus used by D P Mitchell and P N Powers [95] to demonstrate neutron diffraction experimentally. Fast neutrons from a radium–beryllium source were moderated by paraffin wax to provide thermal neutrons with a maximum in the wavelength distribution estimated as 1.6 Å. Several large crystals of MgO with their (100) planes at the appropriate angle were arranged as shown in the figure. When they were rotated from this position there was a significant reduction in the recorded intensity.

The potential of the method could not be realized until the advent of reactors as sources of thermal neutrons. The first 'pile', constructed in Chicago to the design of E Fermi, began operation in 1942. By 1944 W H Zinn [96], working at the Argonne Laboratory, had investigated reflections from LiF and measured the energy spectrum of neutrons from the reactor. His apparatus is now a museum piece there. By 1946 the work of E O Wollan and C G Shull at Oak Ridge Laboratory was preeminent. Quantitative measurements were made of powder diffraction by NaCl, and diffraction patterns obtained from liquids. Experimental work verifying the predictions of F Bloch and of L Néel (which we discuss later) was begun by C G Shull and J S Smart [97] on magnetic materials, opening up a new field inaccessible to x-rays.

With this brief survey of early work we turn now to a comparison of the theories of x-ray and of neutron scattering, which are quite similar. An important difference results from the fact that while x-rays are scattered by the atomic electrons, occupying a region comparable in size to the wavelength, neutrons are primarily scattered by nuclei of minute dimensions. As a result the atomic scattering factor, f, in (2), has no angular variation; only the temperature factor prevents the neutrons from being scattered isotropically. In place of $f(T)$ we have $-b(T)$, the *bound scattering length* defined for a nucleus that is not free to recoil. This result applies only to a nucleus with zero spin. When

Twentieth Century Physics

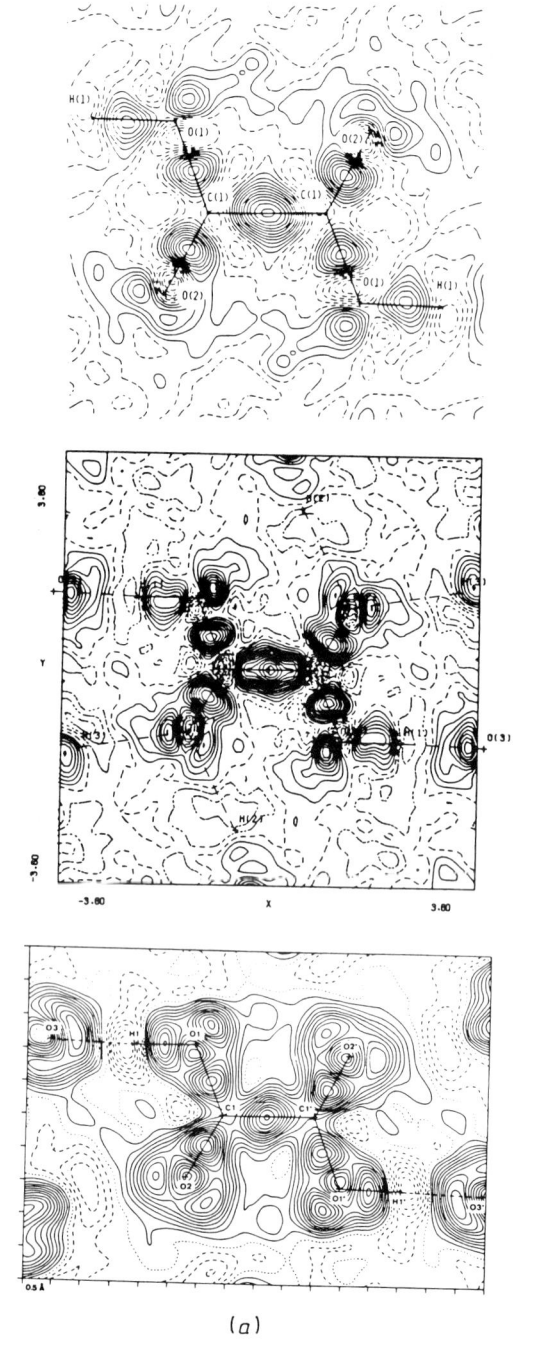

(a)

Figure 6.17. (a) The difference electron density in oxalic acid, from three independent sets of experimental data. (b) Three theoretical maps of the same density corresponding to somewhat different parameters of the theory.

Solid-State Structure Analysis

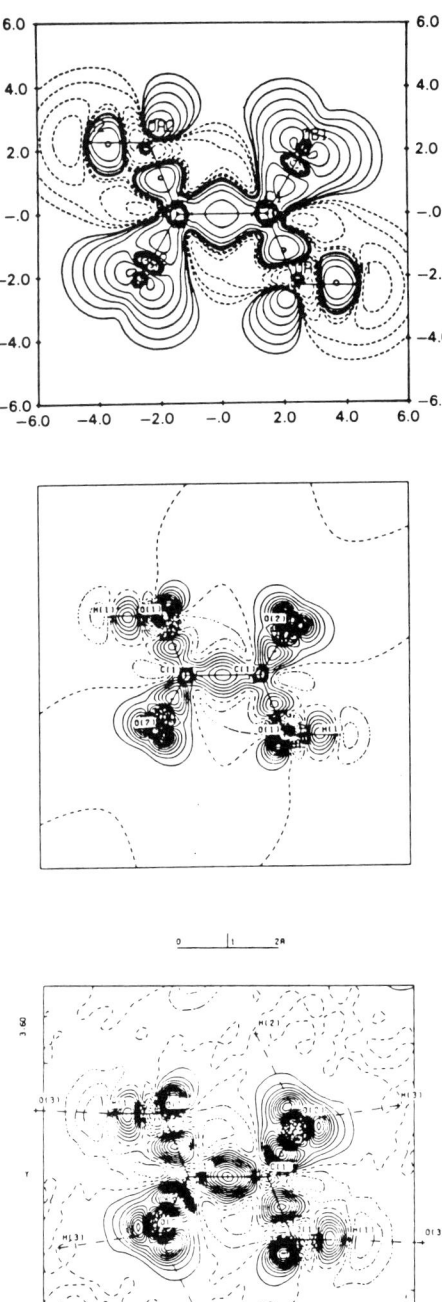

(b)

Figure 6.17. Continued.

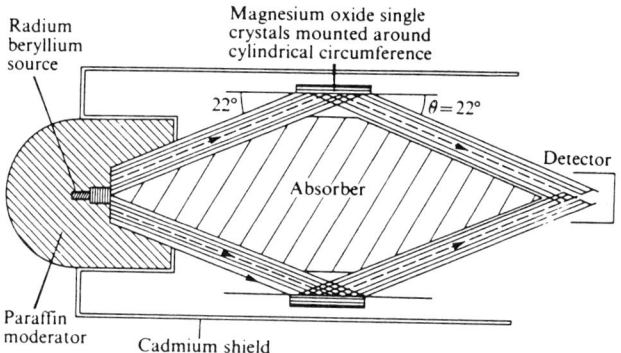

Figure 6.18. *The layout of the apparatus first used to demonstrate neutron diffraction by crystals of MgO (see text).*

the nucleus has a spin it has in effect two scattering lengths b_+ and b_- which may be quite different, and which refer to the two orientations of the neutron spin relative to the nuclear spin. A weighted mean of the two replaces b to give the amplitude of coherently scattered radiation, and there will also be incoherent scattering. There is a further source of incoherent scattering in that isotopes of an element generally have different scattering lengths; the value of b which is effective for coherent scattering is again a weighted mean, $\bar{b}$. In neutron diffraction by a crystal, incoherent scattering produces a slowly varying background on which the Bragg peaks from coherent scattering are superposed. For most elements $\bar{b}$ is positive, increasing slowly with the radius of the nucleus, but is subject to a further contribution which may have the opposite sign when the neutron energy is not far from a resonance level in the nucleus. As a result $\bar{b}$ is negative for H, Mn, Ti, for example, and is almost zero for V. The strong scattering of neutrons by protons makes neutron diffraction a particularly useful tool for showing the position of hydrogen atoms. The scattered amplitude from an atom is not much different for x-rays and neutrons, with Q (Box 6B) about an order of magnitude less for neutron than for x-ray diffraction. Since the low intrinsic absorption of thermal neutrons in most materials allows larger specimens to be used, extinction can be important in neutron diffraction. Indeed early measurements by E Fermi and L Marshall [98] gave the result that the integrated intensity from a number of crystals was nearly proportional to $|F|$ rather than to $|F|^2$, despite their mosaic character. G E Bacon and R Lowde [99] showed that this was to be expected with the large specimens used.

In earlier reactors the isotropic flux of thermal neutrons was at most 10^{12} cm^{-2} s^{-1}, giving a collimated monochromatic beam of some 10^8 cm^{-2} s^{-1}, much less than the photon flux in the beam from an x-ray tube, so that specimens measured in centimetres had to be used to compensate. The neutron intensity from a high-flux reactor designed for experiments on condensed matter, such as that at Brookhaven in USA or at the Institut Laue-Langevin in France, is greater by a factor of about 10^3, while the introduction of guide tubes has further enhanced the intensity of the collimated beam. Consequently specimens not much greater than those suitable for x-ray work can be used, with extinction no more of a problem.

While experimental techniques for neutron diffraction are broadly similar to those used in x-ray diffraction, the scale of the equipment, dictated largely by shielding requirements, is much more massive; capital and running costs are correspondingly great. The limit of thermal neutron flux in the reactors at Brookhaven and Grenoble is not likely to be surpassed, although pulsed reactors may produce a peak thermal flux of about 10^{16} cm^{-2} s^{-1}. Electrons impinging on a target such as tungsten produce γ-rays and make fast neutrons by γ–n reactions, while energetic protons striking a similar target produce neutrons by spallation reactions. The spallation source ISIS has been in operation at the Rutherford Laboratory in England since 1985.

The wavelength λ of a neutron beam is related to the momentum $m_n u$ of a neutron by the de Broglie relation $\lambda = h/m_n u$, and for neutrons in equilibrium with a moderator at temperature T, $\frac{1}{2}m_n \bar{u}^2 = \frac{3}{2}kT$. The wavelengths corresponding to the RMS velocities appropriate to $T = 0\,°\text{C}$ and $100\,°\text{C}$ are 1.55 Å and 1.33 Å, respectively, while neutrons of velocity $u = 2$ km s^{-1} have $\lambda = 2.0$ Å and $E = 0.02$ eV. When neutrons are scattered inelastically by a crystal the energy exchanged with the crystal, which is in units of the phonon energy $h\nu_i(q)$, is generally comparable with the energy of the incident neutrons, so that the energy (and wavelength) of the scattered neutrons is significantly changed. In neutron spectroscopy experiments, measurement of the altered wavelength enables the relation $\nu_i(q)$ between frequency and wavevector of phonons to be determined, as described in Chapter 12. The use of the term *neutron diffraction* usually implies that *neutron spectroscopy* is not included. With this proviso, x-ray diffraction and neutron diffraction are quite similar. However, a neutron beam can be monochromated by using a *chopper* to select neutrons of a particular velocity, and velocity can be measured by a time-of-flight technique. For example, figure 6.19 shows diagrammatically an arrangement by which the diffraction pattern of a powder can be recorded by having as variables in the Bragg equation λ and d for a fixed 2θ, rather than the customary 2θ and d for a fixed λ [100]. This is particularly advantageous when the specimen is enclosed, for example in a pressurized vessel, since provision of only fixed entrance and exit windows is required. A disadvantage is that the energy spectrum of the incident beam has to be known.

The temperature of the moderator is usually between 300 and 400 K

Twentieth Century Physics

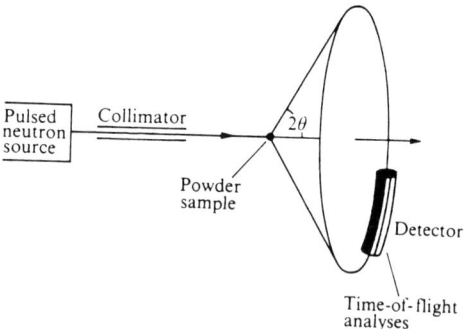

Figure 6.19. *Diagrammatic sketch of apparatus for recording the diffraction pattern from a powder using a time-of-flight method.*

with the peak of the wavelength spectrum in the region 1.0 to 1.5 Å and comparatively few neutrons outside the range 0.7 to 2.8 Å. The number of long wavelength or 'cold' neutrons can be increased by having a cold source in the reactor. For example, a vessel of volume about 300 cm^3 containing liquid hydrogen at $T = 20$ K increases the number of neutrons in the range 5 to 10 Å by an order of magnitude. Conversely a hot source will increase the number of neutrons at the short wavelength end of the spectrum.

The neutron guide [101], which depends on the total reflection of slow neutrons inside a metal tube, can be used to enhance the intensity of a collimated beam, while channelling it into an area of low background. The Grenoble high-flux reactor has ten such guide tubes.

Although photographic techniques have been used [102] to record Laue photographs, the usual detector of thermal neutrons is a cylinder filled with $^{10}BF_3$ gas which acts as a proportional counter. Instead of scanning a single counter over a range of 2θ to record a powder pattern, a linear array of counters can be arranged to record simultaneously. From this, position-sensitive detectors have been developed which may be regarded as multi-cell BF_3 counters. In one such design there are 400 cells, each of width 1 cm arranged around the sample so that each subtends about 0.2°. The cells can also be arranged in a two-dimensional array. The detector of a small-angle scattering instrument at the Grenoble reactor has an array of 64×64 cells each 1×1 cm^2. Other designs utilize a scintillation screen made of lithium-loaded glass coupled to a channel-plate electron-multiplier.

We now review a few experimental studies which exploit the special features of neutron diffraction. An x-ray study of KHF_2 by R M Bozorth [103] had shown that the F–H–F ion was probably linear with a centrally situated hydrogen atom, which was, however, not detected in this or in subsequent x-ray work. S W Peterson and H A Levy [104] used single-crystal neutron diffraction data to calculate a projection of the nuclear density on the (001) plane. In figure 6.20 hydrogens show up as a negative peak midway between positive fluorine peaks which are 2.26 Å apart (two hydrogens coincide in projection). This was the first application of Fourier methods in neutron crystallography. The structure of KH_2PO_4 was studied by G E Bacon and R S Pease [105]. Figure 6.21(a) shows the nuclear density in projection on the (001) plane. Hydrogen atoms are represented by negative peaks with a distribution which is centred on the hydrogen bond between two oxygens and is elongated along it. In figure 6.21(b) the difference Fourier synthesis shows the hydrogens only, when the crystal temperature is 20 °C. At −180 °C the crystal is ferroelectric with the spontaneous polarization along the c axis, perpendicular to the plane of the diagram. From figures 6.21(b) and 6.21(c) it is apparent that each hydrogen is now closer to one oxygen than to the other. When the polarization is reversed each hydrogen moves to the alternative site. This was a direct and elegant demonstration of the atomic movements which take place in a ferroelectric transition, the first of many such studies.

Instances in which neutrons have been used to locate 'light' in the presence of 'heavy' atoms include studies of carbon and nitrogen in metals where x-ray measurements are insensitive to the carbon positions, particularly in rare-earth dicarbides. Neutron diffraction has also played a valuable role in distinguishing metals which are close to one another in atomic number, for example the brasses (Cu–Zn) and the stainless steels (Fe–Mn–Cr etc), and in studying order–disorder transitions in metals [106].

Interaction with the nucleus, though normally the strongest, is not the only mechanism for neutron scattering. In certain materials the magnetic moment of the neutron interacts with the electron spin or orbital moment of an atom to give an additional source of scattering. Much of the theory of neutron scattering by magnetic materials was worked out before the advent of reactors made experimental work possible; W C Koehler [107] has remarked that the classic paper by O Halpern and M H Johnson [108] contained some predictions which were only verified 30 years later. In an elegant experiment reported in 1940, L W Alvarez and F Bloch [109] produced a beam of weakly polarized neutrons by magnetic scattering from a specimen of iron, and by a nuclear resonance experiment obtained a value for the neutron magnetic moment. In what follows, we assume that the magnetic moment of an atom arises solely from the spin of its unpaired electrons, as is the case for atoms of the first transition series in a crystalline environment.

In an ideal paramagnetic material the moments of individual atoms are randomly oriented and uncorrelated, so that only incoherent magnetic scattering occurs. The angular distribution of scattered neutrons then gives information on the distribution of unpaired electrons

Twentieth Century Physics

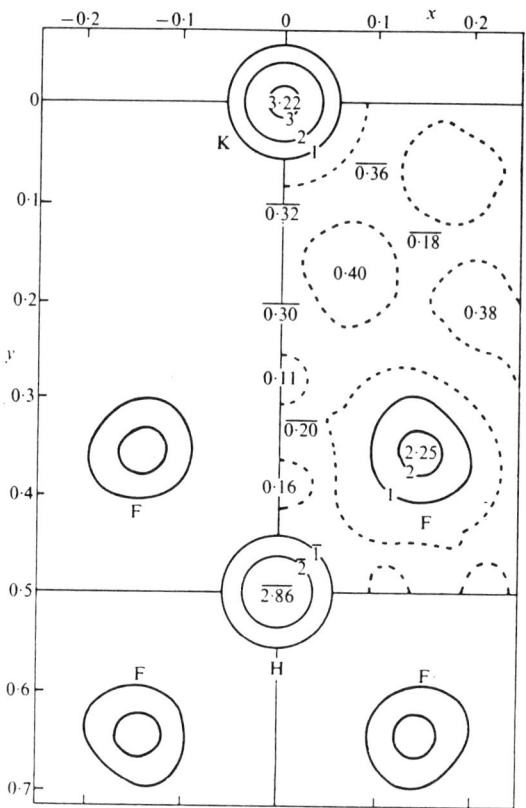

Figure 6.20. *The nuclear density in* KHF$_2$ *projected down the c axis (see text).*

within the atom [110].

A study of the powder pattern of MnO at a temperature of 80 K by C G Shull and J S Smart [97] showed that the dimensions of the 'magnetic' unit cell are twice those of the cubic unit cell determined by x-rays. This results from the antiferromagnetic structure which has anti-parallel alignment of the magnetic moments, as predicted by L Neél (see Chapter 14) The correctness of Neél's model for the ferromagnetic material Fe$_3$O$_4$ was confirmed by C G Shull et al [111] some two years later, and this work laid the foundation for production of polarized beams. An unpolarized beam can be regarded as having two components with spins parallel and antiparallel to the magnetization of a ferromagnetic (or a ferrimagnetic) crystal on which it is incident, with the two components having scattering lengths $b + p$ and $b - p$ respectively, p being the magnetic scattering length. If for a particular reflection it happens that $b = p$, only the one component will be reflected, as a completely polarized beam. This condition is nearly satisfied for the

See also p 1148

See also p 1130

Solid-State Structure Analysis

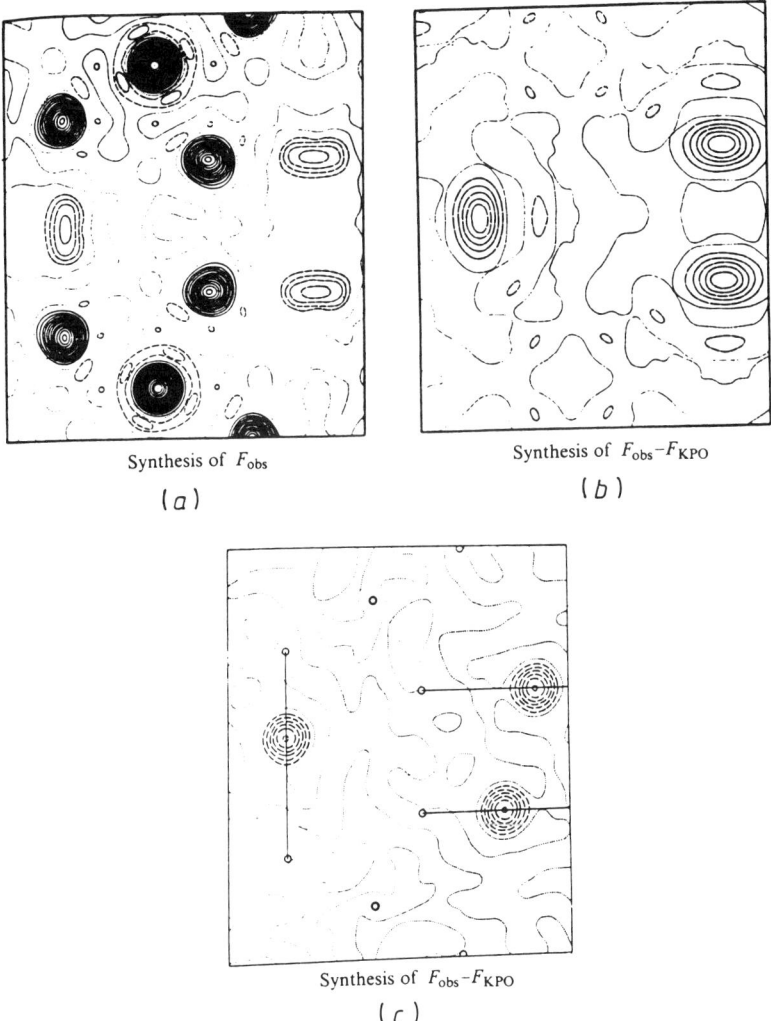

Figure 6.21. *(a) The nuclear density in KH_2PO_4 projected down the c axis. (b) The projected density for hydrogens only, at 20 °C. (c) The projected density for hydrogens only, at −180 °C (see text).*

(220) plane of Fe_3O_4 (magnetite) which can thus be used as a polarizing monochromator.

Figure 6.22 shows schematically a polarized neutron apparatus [112]. The purpose of the magnetic collimator is to retain the polarization direction. The radio frequency in the flipping coil is tuned to the Larmor precession frequency of the neutron in the field and the neutron completes 180° of precession in passing through it. The coil can thus be used to switch the polarization direction. Measurements are usually

461

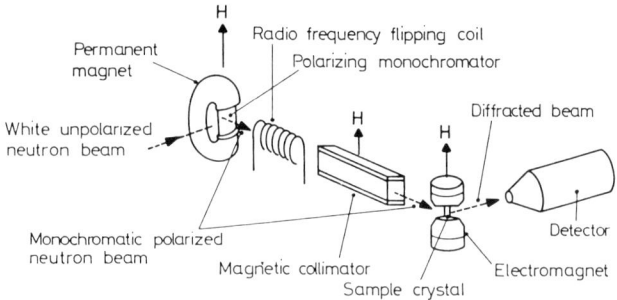

Figure 6.22. *Diagrammatic sketch of apparatus for production of polarized neutron beams.*

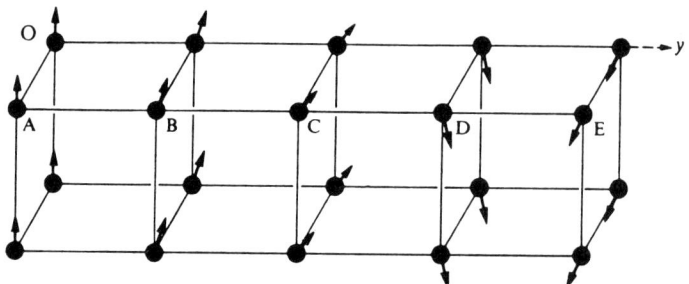

Figure 6.23. *'Helimagnetic' arrangement of magnetic moments in Au_2Mn.*

made by setting the specimen and the detector to the maximum of a Bragg reflection and recording the ratio of the counting rates for the two directions of polarization, from the difference in which the scattering due to the oriented spins is isolated from other scattering [106]. The spin density in the (001) plane of iron was obtained by C G Shull [113] in this way.

The alloy Au_2Mn has an antiferromagnetic structure referred to as 'helimagnetic', illustrated in figure 6.23. There are ferromagnetic sheets of atoms perpendicular to the direction O_y, and the direction of magnetization rotates by about 51° from one sheet to the next. This also provides an example of an incommensurate structure, i.e. the spatial period of the rotation is not commensurate with the corresponding unit-cell dimension. The neutron diffraction pattern then includes pairs of satellite reflections of magnetic origin which accompany the nuclear reflections, in the appropriate range of temperature.

The extension of space-group theory to include magnetic structures has been achieved mainly by Russian crystallographers [114]. New symmetry elements such as anti-translation and anti-reflection increase the number of Bravais lattices from 14 to 36, and of space groups from 230 to 1651, called Shubnikov groups. This treatment of symmetry, incidentally, does not include that of incommensurate structures.

The magnetic structures of rare-earth metals have been found to be both varied and complicated. Some of them absorb neutrons, which has impeded experimental work, much of which has been done by W C Koehler and his colleagues at the Oak Ridge Laboratory. Most of these elements show two or more ordered magnetic structures, in different temperature ranges, usually below 200 K, while the crystal structure remains hexagonal. Koehler [115] has given illustrations of magnetic structures including ferromagnetic with magnetic moments in the plane perpendicular to the c axis, helical antiferromagnetic, conical spiral and antiferromagnetic with a sinusoidal modulation parallel to the c axis.

6.7. Electron diffraction [116]

The discovery of electron diffraction in 1927 ranks in importance for its fundamental significance and its applications with that of x-ray diffraction in 1912. The early history is told in Chapter 3 and we may take it up at the time when G P Thomson heard of Davisson's work at the British Association meeting in the summer of 1926 [117]. On his return to Aberdeen, where he was Professor of Natural Philosophy, he encouraged a research student, A Reid, to modify an existing apparatus so that a beam of electrons with energy in the keV range was transmitted through a film of celluloid and recorded on a photographic plate. They were encouraged by finding diffuse halos. Improved apparatus was constructed and photographs taken with very thin foils of Al and Au as targets. The resulting patterns (figure 6.24) were very similar to x-ray powder photographs, and there was quantitative agreement with an electron wavelength $\lambda = h/\sqrt{2mE}$. Davisson and Thomson shared the Nobel Prize for physics in 1937, which made possible the quip that J J Thomson had been awarded the prize for showing that the electron was a particle and his son received the award for showing that it was not.

See also p 168

A photograph obtained by S Nishikawa and S Kikuchi [118] in 1928 by transmission of 34 keV electrons through mica marked the beginning of the study of crystal structure by electron diffraction. The subject subsequently expanded to include the study of surfaces, electron diffraction by gaseous molecules, electron microscopy etc. In the remainder of this section, however, attention will be confined to structural studies by the transmission technique.

When an electron beam falls on an atom the amplitude of the scattered beam at unit distance is the scattering factor for electrons, f_e, and is proportional to the Fourier transform of the electrostatic potential $\Phi(r)$ in the atom. f_e is related to f, the atomic scattering factor for x-rays by

$$f_e(S) = \frac{2me^2}{h^2} \frac{(Z - f(S))}{S^2}$$

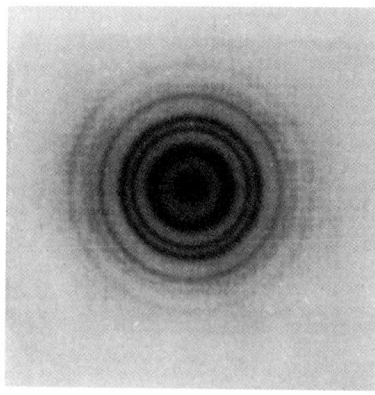

Figure 6.24. *Electron diffraction pattern obtained from a sputtered gold film, which is very similar to an x-ray powder photograph.*

where Z is the atomic number. Use of the Thomas–Fermi method for the electron distribution gives the result that $f_e(S)$ increases approximately as $Z^{1/3}$. The potential near the centre of an atom therefore increases more slowly with Z than does the electron density so that light atoms are relatively more prominent in a map of $\Phi(r)$ than in a map of the electron density $\rho(r)$. The greatest difference in practice between electron diffraction and x-ray or neutron diffraction arises from the much larger scattering factor—f_e is about 10^4 times as great as the other fs. The kinematic theory of electron diffraction, which corresponds to the theory we have met so far for x-rays and neutrons and neglects multiple scattering, is not valid unless the relevant specimen dimension is less than about 1000 Å for materials containing light atoms, or about 100 Å when heavy atoms are involved [119].

Three-dimensional data can be obtained by reorienting thin single-crystal specimens, with considerable overlap from powder specimens, or by using the 'oblique texture' method of Zvyagin. The three-dimensional potential distribution $\Phi(r)$ in diketopiperazine shown in figure 6.25 is an example of Vainshtein's work. The standard deviation in a bond length not involving hydrogen was estimated to be 0.012 Å, but greater deviations were found when comparisons were later made with the results of an x-ray investigation. Examples of other structural studies are given by P Goodman [120]. Multiple scattering can result in non-zero intensities for reflections which correspond to space-group absences, and to a breakdown of Friedel's law [121]. Clearly there are several reasons why the techniques of x-ray or neutron crystallography gave more accurate results at that time.

With the facilities for calculation now available, it is possible to perform computations based on the dynamical theory involving

Solid-State Structure Analysis

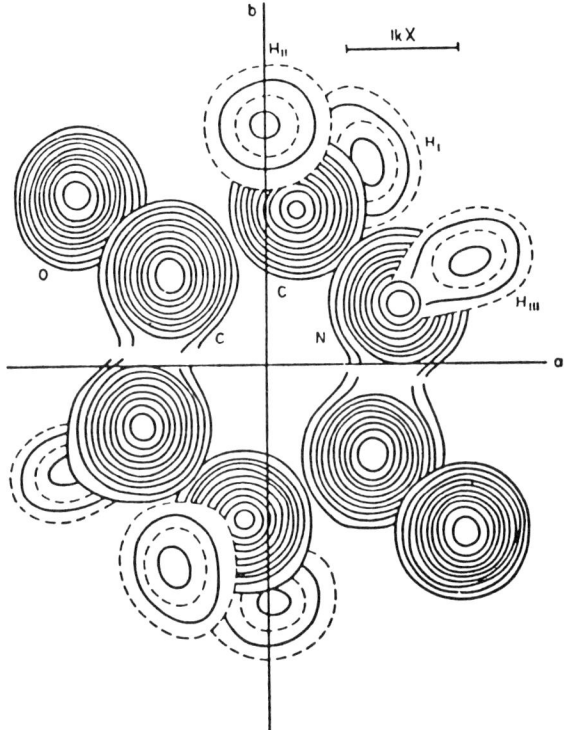

Figure 6.25. *The three-dimensional potential distribution $\Phi(r)$ in a diketopiperazine molecule as determined by electron diffraction.*

hundreds of interacting beams, so that multiple scattering effects are no longer a problem. Indeed, via the interaction analogous to the Renninger effect in x-ray diffraction (see section 6.4), they can be put to advantage, allowing the determination of structure-invariant relations between phase angles $\alpha(H)$ for certain reflections to better than 1°. Experimental techniques for the study of thin crystal specimens have also undergone notable improvement. In using a focused beam for quantitative convergent-beam electron diffraction, a very small volume of specimen is involved, so that inclusions and domains can be studied. From the relation between electron density $\rho(r)$ and electrostatic potential $\Phi(r)$, accurate x-ray structure factors $|F(H)|$ can be derived from the electron diffraction structure factors, and this has become an accurate method of studying the distribution of bonding electrons in inorganic materials with small unit cells such as Ge and GaAs [122].

6.8. Surface crystallography
A knowledge of the structure of the exposed surface of a crystal can

be important in understanding crystal growth, thermionic emission of electrons, catalysis and the formation of metal–semiconductor contacts. This has led to an increased interest in surface crystallography since the early 1960s. Some twenty techniques (and their acronyms) for the investigation of crystal surfaces were listed by D P Woodruff and T A Delchar [123], and others have since been developed. The most important of these for the investigation of structures with which we shall be mainly concerned in this section have been low-energy electron diffraction (LEED), reflection high-energy electron diffraction (RHEED) and in-plane x-ray diffraction. The invention in 1982 of the tunnelling electron microscope by G Binnig et al [124] has given particularly direct and elegant results on the atomic structure of surfaces.

It is not usually worth studying surfaces unless they are uncontaminated. Unless the pressure is very low, contamination by the air is almost instantaneous. Even at a pressure of 10^{-6} Torr (mm of mercury) it takes only about one second to form a monolayer of nitrogen on a metal surface. An uncontaminated surface can be maintained long enough to be studied experimentally only when ultra-high vacuum techniques are available to maintain a pressure of 10^{-10} Torr or better, as has been possible since about 1960 [125]. Detailed comparison of theory with experiment when the LEED technique is used requires calculations which also were scarcely possible before then.

There are five Bravais lattices in two dimensions, while the number of space groups is 17 [123]. The structure of the surface may be unchanged from that of the bulk material, there may be some displacement outwards or inwards of atoms at and near the surface, or there may be a reconstruction involving the outer planes of atoms. Finally atoms or molecules of a different species may be absorbed on to the surface, forming a new structure with a two-dimensional lattice which has a unit cell (or mesh, in 2D) related to that of the substrate.

The electron energy in LEED experiments is in the range 20 to 1000 eV. The beam current is typically 1 μA and the beam is focused to about 1 mm diameter at the specimen. The grids shown in figure 6.26 are spherical and concentric with the recording fluorescent screen and are held at potentials such that electrons which have been scattered inelastically do not reach the screen. Apparatus of this type evolved from a design by W Ehrenberg [126]. Electrons in the LEED range of energy do not penetrate beyond the first three or four layers of atoms since the cross section for elastic scattering is large; furthermore the beam is rapidly attenuated by inelastic scattering. The resulting loss of resolution means that Bragg's law is relaxed, and for a fixed position of the crystal and direction of the incident beam, several diffracted beams can be produced simultaneously. Thus in figure 6.27 the surface structure has a square two-dimensional lattice (or net) of side a, and the incident beam makes an angle φ with the vertical. Three diffracted beams S_5, S_6 and S_7 are

Solid-State Structure Analysis

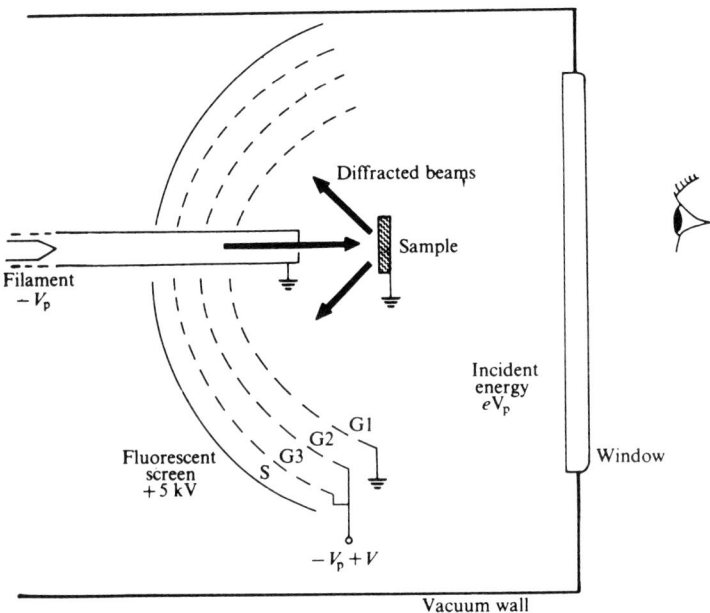

Figure 6.26. *Schematic arrangement of apparatus for* LEED *experiments (see text).*

shown directed into the specimen (and therefore not recorded), while the diffracted beams S_1 to S_4 will reach the recording screen. There is enough information to allow the symmetry and dimensions of the mesh to be worked out.

When a surface is clean and is not reconstructed, this can be deduced from the diffraction pattern. Otherwise, it is necessary to measure the intensity of several beams as a function of electron energy. Figure 6.28 shows the resulting graphs of $I(V)$ measured for angles of incidence φ from $7°$ to $21°$ for a Cu (100) surface. Since the incident beam penetrates some few layers into the crystal we might expect $I(V)$ to be a maximum when the Bragg condition is satisfied for the three-dimensional structure but with broad peaks because of the short effective extension in one direction. The predicted positions of three such peaks (for $\varphi = 0$) are shown by the arrows in figure 6.28. While peaks occur near these positions, there are others which are sometimes larger. The discrepancy comes about because of multiple scattering on the three or four contributing planes of atoms, an effect which can usually be neglected in x-ray or neutron scattering. Figure 6.29 shows how the data may be analysed by comparison with calculated $I(V)$ curves assuming different relaxations of the top layer of atoms. Agreement is best for zero relaxation. Such calculations would be practically impossible without elaborate computer programmes, although it is worth noting that H A Bethe [127] made a beginning in 1928!

Twentieth Century Physics

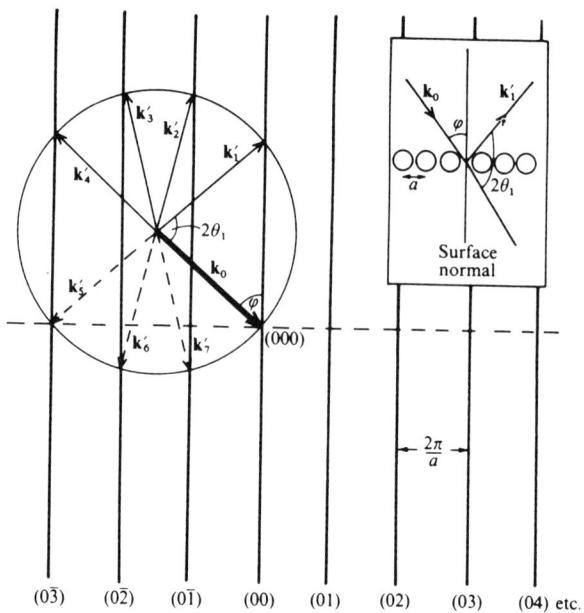

Figure 6.27. *Geometry of diffracted electron beams from a surface having a square net (see text).*

The electron energy in RHEED experiments is typically 100 keV, corresponding to $\lambda = 0.037$ Å. Penetration normal to the surface is small only because the beam is incident at grazing incidence. The spread in λ and in angle of incidence results in a diffraction pattern which is a series of streaks (figure 6.30). The intensity distribution along the streak, and its variation with φ or λ is difficult to calculate and this has not often been attempted.

Neutron beams have rarely been used in surface physics, but J W White [128] succeeded in studying the formation of monolayers of N_2 and Ar on metal surfaces and on graphite. A study of the reflection of monoenergetic molecular beams from the surfaces of crystals was made as early as 1911 by L Dunoyer [129]. Atoms or molecules with kinetic energy not more than about 0.1 eV are very suitable as surface probes since they are sensitive to the structure of the outermost layer only. Other early work, for example that of I Estermann and O Stern [130] was confined to alkali halides but more recent studies of metal and semiconductor surfaces have also been made [123].

From about 1980 a technique was developed by means of which x-rays can be used to investigate surface structure. The first work of this kind was a study of the Ge (100) surface using x-rays at grazing incidence [131]. The geometry of the diffraction process is similar to that used in the RHEED technique, and is referred to as 'in-plane' x-ray diffraction.

Solid-State Structure Analysis

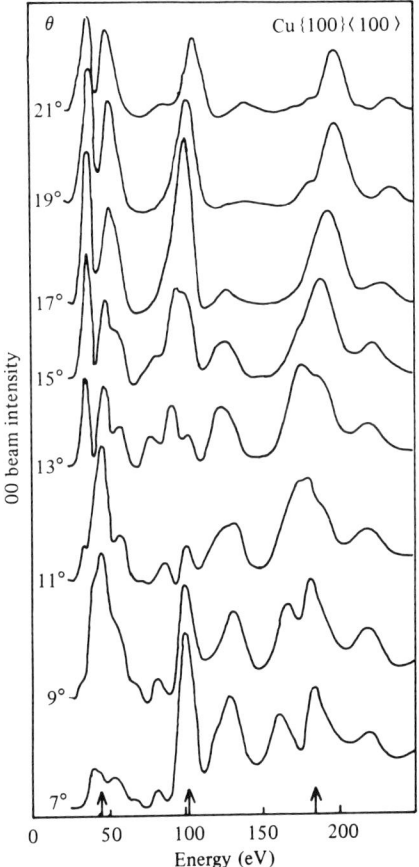

Figure 6.28. *Intensity–energy curve $I(V)$ from a Cu (100) surface. The arrows give the positions of predicted 'Bragg peaks'.*

The incident and diffracted beams, with $\lambda \approx 1.0$ Å, make small angles of order 1° with the crystal surface and the detector is positioned to record the diffracted beam. That a diffracted beam is produced at all in a direction which allows it to be intercepted by the detector depends on the relaxation of Bragg's law which follows from the loss of resolution resulting from the small penetration of the crystal by a beam at grazing incidence. I K Robinson and D J Tweet [132] have given a comprehensive review of variants of this technique, with numerous examples.

An example of two-dimensional x-ray crystallography which made use of Fourier techniques is provided by the work of J Bohr *et al* [133], who investigated the reconstructed (111) surface of InSb. Figure 6.31 shows views of the original structure and the reconstructed 2×2 surface unit cell.

469

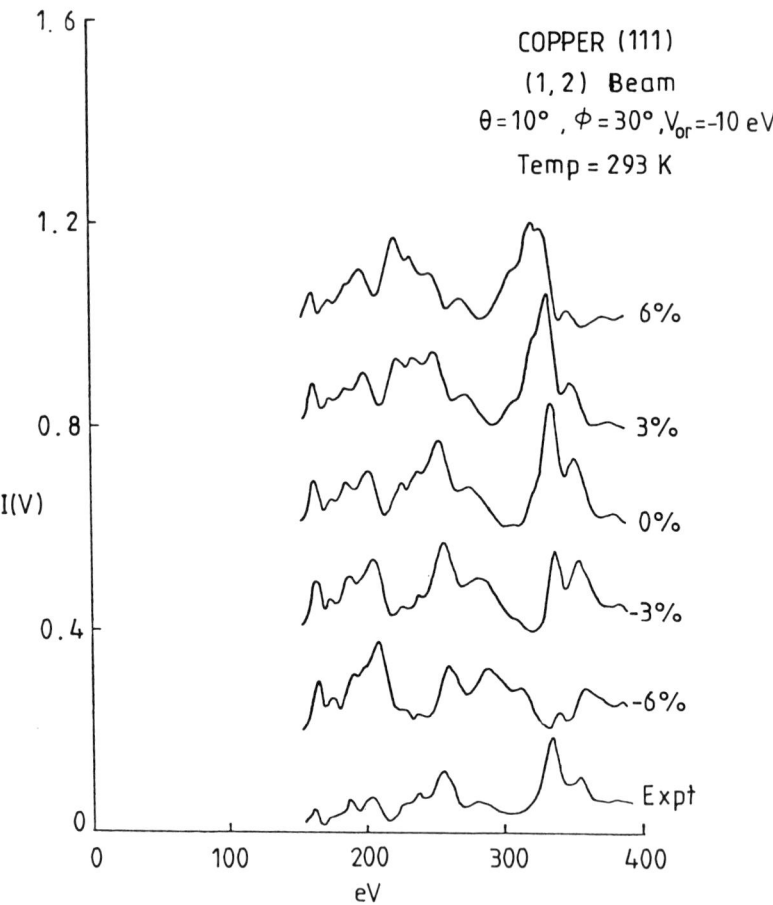

Figure 6.29. *I(V) curve for a [111] surface of Cu compared with calculated curves for different degrees of relaxation of the top layers of atoms.*

The semiconductors Si and Ge have topological properties of their covalent bonds which lead to a great variety of surface reconstructions. The Si (111) 7 × 7 structure in particular has been studied by a variety of techniques, sufficient to warrant a review by D Haneman [134]. It was first reported by R J Schlier and H G Farnsworth [135], one of the pioneers of LEED. At the present time the combined use of tunnelling electron microscopy and in-plane x-ray diffraction probably provides the most powerful means of surface structure determination. There are advantages in using a synchrotron source of x-rays since the high resolution makes the diffracted beams more likely to stand out against the background of thermal diffuse scattering.

Solid-State Structure Analysis

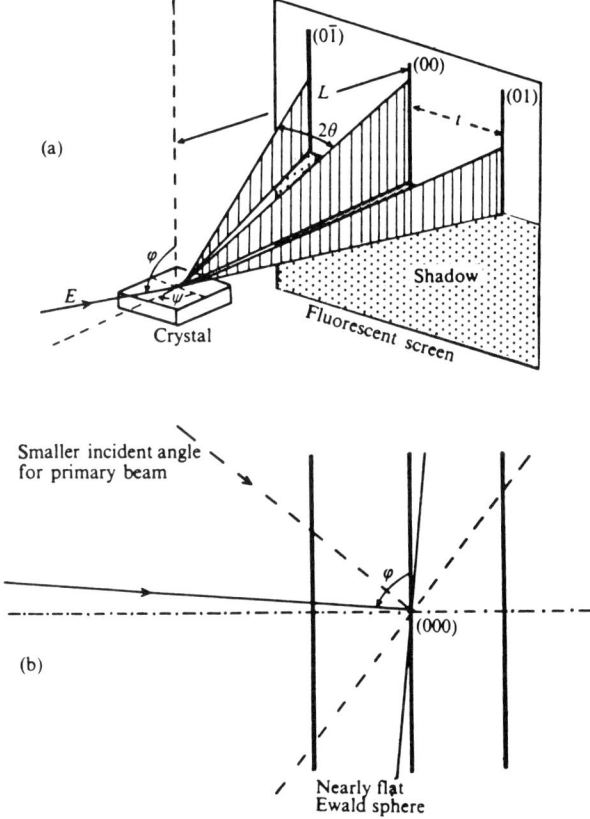

Figure 6.30. *Electron diffraction pattern obtained using the* RHEED *technique. The dimension of the square mesh is given by* $a = (h^2 + k^2)^{1/2} \lambda L/t$ *where L and t are as shown.*

6.9. Imperfect crystals and non-crystalline solids [136]

6.9.1. Line broadening

An otherwise perfect crystal is imperfect in that it has surfaces, and even when these are unreconstructed there is a measurable effect on the diffraction pattern. Consider x-ray reflection from planes parallel to the surface of a thin crystal of thickness t. The loss of resolution resulting from t not being much greater than λ leads to a spread β in the scattering angle 2θ, given by $\beta = \lambda/t \cos\theta$, or $4\lambda/3D \cos\theta$ for a sphere of diameter D. In figure 6.32 powder photographs of gold foil and of colloidal gold are compared; there is a marked broadening of the latter.

In the early 1940s there was controversy over the origin of line broadening shown in photographs of metals which had been severely deformed by hammering or other forms of cold working. The two main schools of thought were (i) that the metal is broken into crystallites,

471

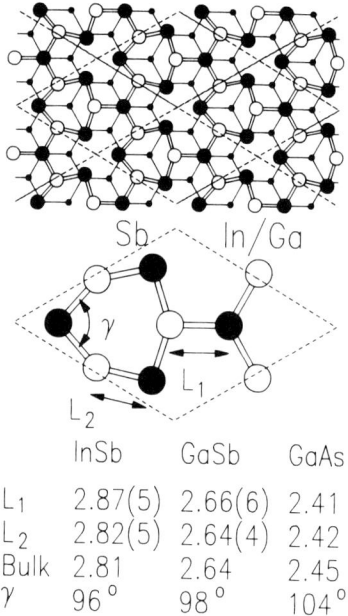

Figure 6.31. *The reconstructed (111) surface of InSb.*

10^{-5} to 10^{-6} cm in linear dimensions, and hence shows diffraction broadening and (ii) that the small crystals are 10^{-4} cm or more in linear dimensions but are elastically distorted. A J C Wilson [136] showed that the experimental evidence favoured the second possibility. This was confirmed when J N Kellar *et al* [137] used an x-ray beam only 10^{-4} cm in diameter so that spotty rather than continuous rings were produced in powder photography of cold-worked aluminium. They deduced a particle size of about 2×10^{-4} cm.

6.9.2. Faults in layer structures
Mistakes that occur during the growth of crystals are particularly prevalent in layer structures such as graphite or cobalt. Cobalt has a hexagonal structure in which layers of close-packed atoms lie in the basal plane (figure 6.33). The first layer is indicated by A, each atom of the second layer B fits into the depression between three atoms of A, while the atoms of the third layer are vertically above those of the first. The sequence ABAB... continued indefinitely gives a perfect crystal. There is, however, another possible position for an added layer, denoted C, which preserves close-packing, and with each added layer there may be a fault, to give ...ABABCBCB... or ...ABABACAC.... Wilson has shown that such faults leave some reflections unaffected, but others exhibit a spread of intensity, the extent of which depends on the probability of a fault.

Solid-State Structure Analysis

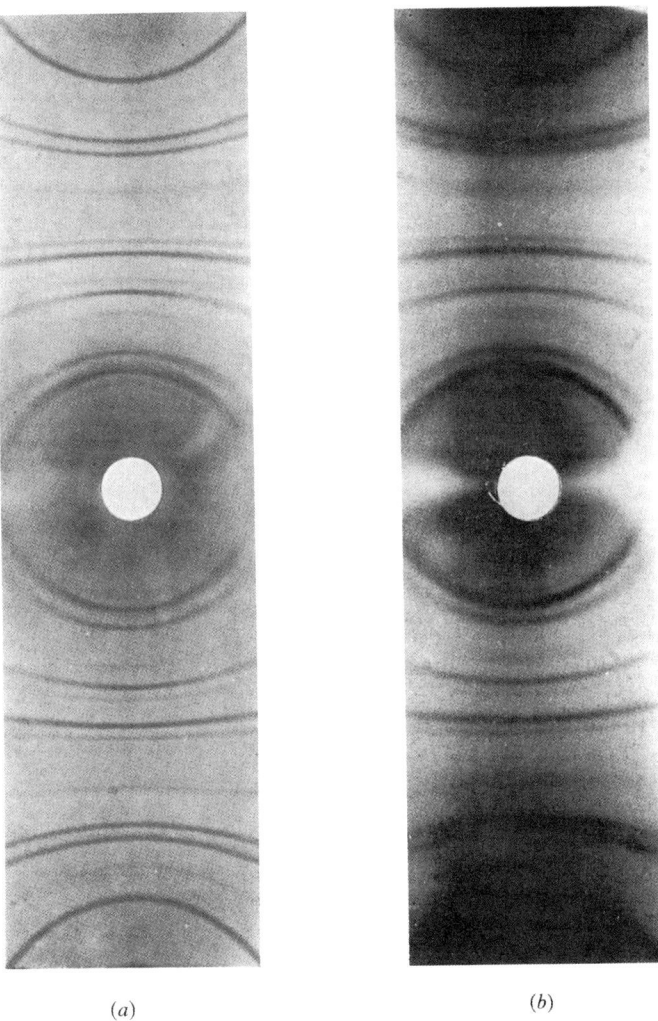

(a) (b)

Figure 6.32. *Powder photographs of (a) gold foil and (b) colloidal gold.*

For a particular specimen of cobalt O S Edwards and H S Lipson [138] deduced that about one layer is ten showed a stacking fault.

6.9.3. Order–disorder transitions

A phenomenon of far wider significance, first revealed by x-ray studies of alloys, is the order–disorder transition. In β-brass, for example, copper and zinc atoms in equal numbers occupy lattice sites indiscriminately at high temperatures, but segregate into a regular alternation of copper and zinc on cooling. The first systematic theory was given by W L Bragg and

Twentieth Century Physics

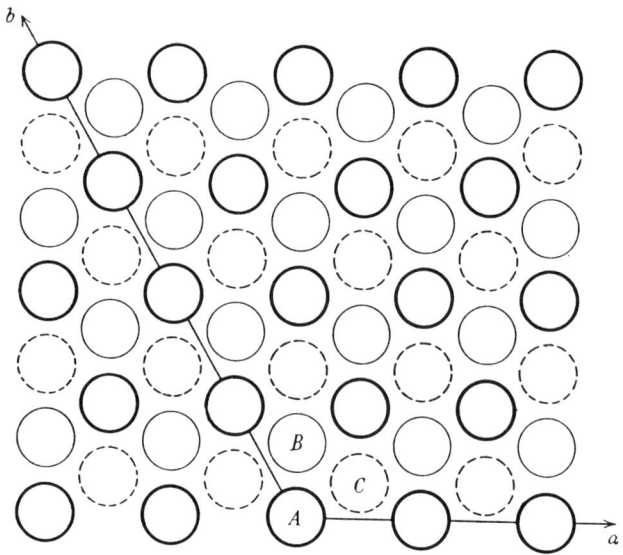

Figure 6.33. *Relative atomic positions in the close-packed layers A, B and C referred to in the text.*

E J Williams [139], whose paper contains references to the experimental origins of the work. The topic is dealt with in greater detail in Chapter 7.

See also p 548

6.9.4. *The structure of ice*

Besides disorder in composition a crystal may have disorder involving the displacement of atoms, or the orientation of molecules. Of many examples, one that particularly deserves mention is the structure of ice. Early investigations have been summarized by W H Barnes [140], whose x-ray studies of single crystals led eventually to the structure shown in figure 6.34, where broken lines outline the hexagonal unit cell. There are puckered sheets of water molecules lying perpendicular to the *c* axis, and each water is tetrahedrally bonded to four near neighbours through hydrogen bonds. This and later x-ray work gave the positions of oxygen atoms only and various proposals were put forward for the hydrogen positions. Majority opinion favoured a structure in which every oxygen has two covalently bound hydrogens, each directed towards neighbouring oxygens, and two more distant ones directed towards it from its other two neighbours, the arrangement of hydrogen on the hydrogen bonds being, however, random. This structure is consistent with the infrared spectrum of ice which shows bands at nearly the same frequencies as for the vapour phase. J D Bernal and R H Fowler [141] summarized the evidence for a structure in which 'ice would be crystalline only in the position of its molecules but glass-like in their orientation'. The first neutron diffraction study was made by

Solid-State Structure Analysis

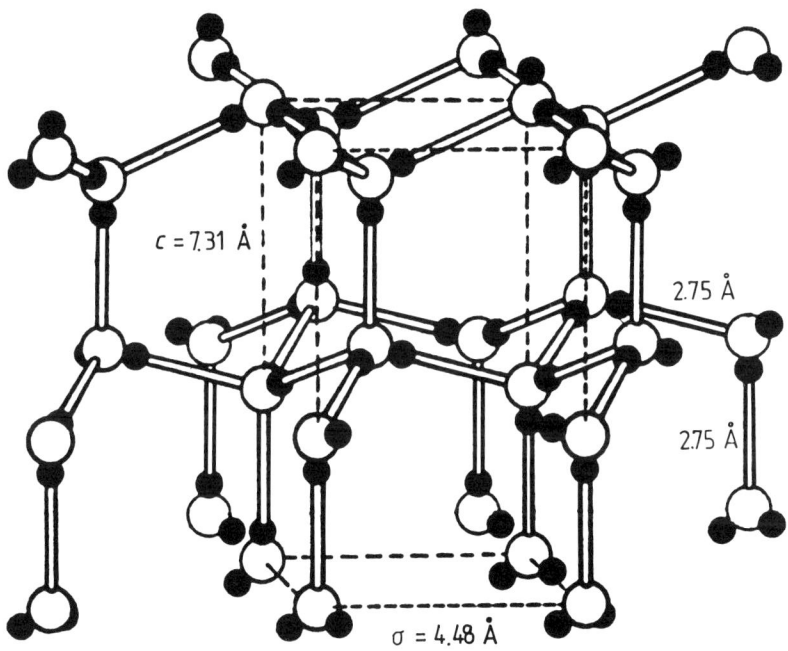

Figure 6.34. *The structure of ice: the unit cell is shown by the broken lines.*

E O Wollan *et al* [142] using powdered D_2O ice at 163 K. They compared the observed intensities with calculated ones for four different structures and found best agreement for the structure with disordered deuterium positions. This was confirmed by later single-crystal studies [143] which showed that hydrogen positions remained disordered at 77 K. Conclusive evidence that disorder persists to zero temperature had in fact been provided by L Pauling [144] who calculated the residual entropy as $S_o = Nk \log(3/2) = 0.4055Nk$ from the number of possible configurations in a crystal of N molecules. This is to be compared with the observed value of $(0.41 \pm 0.03)Nk$.

6.9.5. Dislocations in crystals

We turn now to a different type of imperfection, namely dislocations in crystals, which are particularly important in determining the mechanical properties of materials and are more fully dealt with in Chapter 19. Figure 6.35 shows an edge dislocation in a crystal which has a simple cubic unit cell. It appears as an extra layer inserted into the upper half of the crystal. Away from the lower end of the extra layer the structure is nearly indistinguishable from that of a perfect crystal, but a circuit made around the edge, which would be a closed circuit in a perfect crystal, will fail to close by an amount known as the Burgers vector, in this instance a horizontal lattice translation a. The concept of the edge dislocation was

See also p 1520

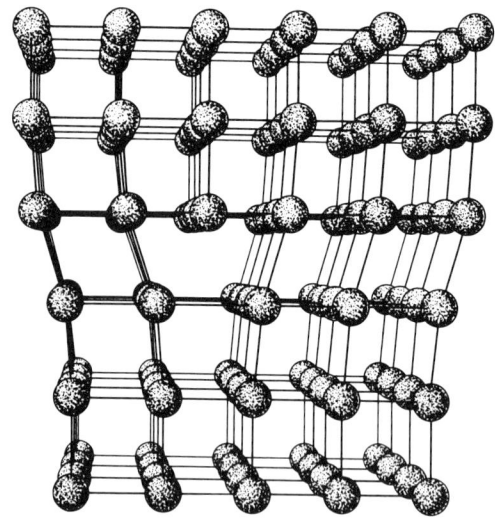

Figure 6.35. *An edge dislocation in a crystal which has a simple cubic structure.*

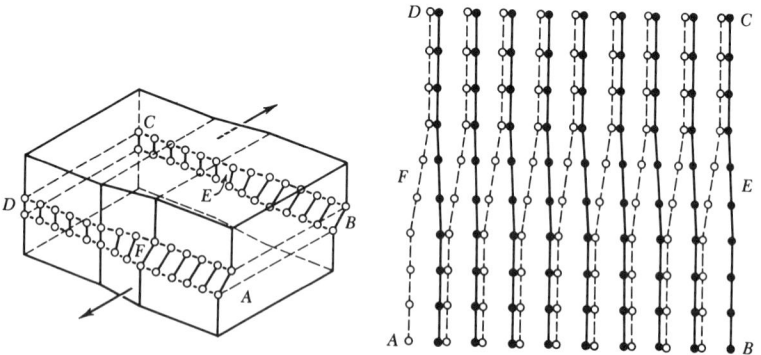

Figure 6.36. *Two views of a screw dislocation.*

introduced independently by E Orowan, M Polanyi and G I Taylor [145]. Figure 6.36 shows two views of a screw dislocation, in which the Burgers vector is parallel to the line of the dislocation. A dislocation line cannot end in a crystal but must continue to a boundary, unless it can form a closed loop. Historical accounts of the development of dislocation theory have been given by W G Burgers and F R N Nabarro, while A K Seeger has drawn attention to papers, some of them from the early 1900s, which foreshadow dislocation theory [146].

The point at which a dislocation ends on the surface of a crystal can show up as a small pit on etching the surface [147]. In certain materials impurity atoms diffuse into the channel at the end of an edge dislocation, thereby 'decorating' it so that its course shows up in visible light under a microscope. More dramatic results, including electron micrographs of moving dislocations in thin metal foil, were first obtained by P B Hirsch [146] and his co-workers in 1956. If an aluminium foil is oriented in an electron beam so as to produce a diffracted beam, an image can be formed in the electron microscope utilizing the intensity in the direct beam (bright field illumination) or in the diffracted beam (dark field illumination). With a perfect crystal, neither image will show any feature of interest. If, however, an edge dislocation is present the local change of orientation may be sufficient to strengthen the diffracted beam and weaken the transmitted one, in which case the dislocation will show up as a dark line in the bright field image [147]. Stress in the crystal combined with an increase in temperature may cause the dislocation to move across the field of view. The theory of contrast produced by dislocations (and other types of defect) is in detail a difficult branch of electron optics in which much of the progress was due to P B Hirsch and his colleagues.

In 1955 J M Menter [148], who had collaborated with this group, realized that the resolving power of his new Siemens electron microscope would be sufficient to resolve the lattice planes in crystals with comparatively large unit cells, and he was able to obtain micrographs which showed edge dislocations directly (figure 6.37) in platinum phthalocyanine, for lattice planes of spacing 12 Å.

6.9.6. Amorphous structures

The term 'amorphous' has no very precise or mutually agreed meaning, but D Weaire [149] suggests 'not crystalline on any significant scale' and quotes G E Morey's [150] definition of a glass 'A glass is an inorganic substance which is continuous with, and analogous to, the liquid state of that substance but which as a result of having been cooled from a fused condition, has attained so high a degree of viscosity as to be for all practical purposes rigid'. F Yonezawa [151] distinguishes a glass as the special group of amorphous solids that have been prepared from the melt by quenching. The process may involve cooling the liquid in a gas (to produce, for example, silicon glass) at a rate of $1-10$ K s^{-1}, cooling in a liquid (to produce alloy wire) at a rate of 10^2-10^5 K s^{-1}, or cooling in contact with a cold solid (to produce alloy ribbon) at a rate of 10^6-10^8 K s^{-1}. Electrodeposition, sputtering and irradiation techniques have also been used.

A successful model, the *continuous random network model*, of amorphous structures for covalently bonded materials was put forward by W H Zachariasen [155]. The network of covalent bonds is random in that there is no long-range order, and there are no broken or 'dangling' bonds. The number and approximate relative orientation of bonds around each atom must be the same as in the corresponding crystal.

BOX 6C: X-RAY DIFFRACTION BY AN AMORPHOUS SOLID

Diffraction measurements using x-rays, neutrons or electrons give information in terms of the radial distribution function, first used by P Debye and H Mencke [152] to investigate the distribution of atoms in liquid mercury. In what follows we discuss the radial distribution function for an elemental material. The generalization necessary to include compound and alloy materials has been given, for example, by B E Warren [153]. Let $I(S)$ be a function which can be obtained from the intensity scattered by an amorphous specimen, measured in much the same way as from a powdered material. The radial distribution function $g(r)$ is defined in terms of the atomic density $n(r)$ at distance r from a given atom, averaged over all actual positions of that atom, so that $g(r) = 4\pi r^2 n(r)$, and the number of atoms within a distance r of the origin atom is $\int_0^r g(r)\,dr$. The relation between $n(r)$ and $I(s)$ is

$$n(r) - n_o = \int 4\pi S^2 I(S) \frac{\sin 2\pi r S}{2\pi r S} dS$$

where n_o is the average atomic density. The integral may be recognized as the Fourier transform of $I(S)$ and takes this form since $I(S)$ is spherically symmetric. The function $G(r) = 4\pi r(n(r) - n_o)$ is sometimes used rather than $g(r)$ and is referred to as the reduced radial distribution function. Finally we note that it is customary to define $I(S) = (I_m(S) - f^2)/f^2$, where $I_m(S)$ is the measured intensity on a scale such that $I_m(S)$ tends to f^2 for large values of S. The effect of subtracting f^2 from $I_m(S)$ is to remove the peak which would otherwise occur in $g(r)$ at the origin, and dividing by f^2 has the effect of 'sharpening' the peaks in $g(r)$. In practice corrections must be made for absorption in the specimen, for Compton scattering etc, and account must be taken of the termination of the Fourier integral at a finite value of S. Details have been given by R J Temkin et al [154] who made a very careful study of amorphous Ge using MoKα x-radiation. They were able to check the accuracy of their work by repeating the measurements and calculations using data from powdered crystalline Ge, where $g(r)$ must satisfy known conditions, such as the areas and positions of the resolved peaks.

D E Polk [156] constructed a ball and spoke model of amorphous Ge containing some 500 atoms. Figure 6.38 shows the model of R J Bell and

Solid-State Structure Analysis

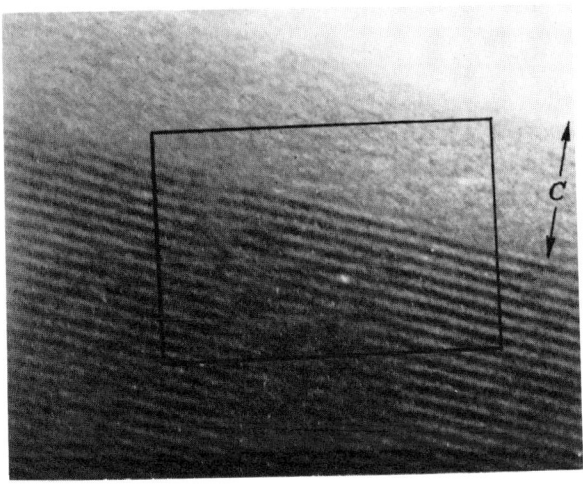

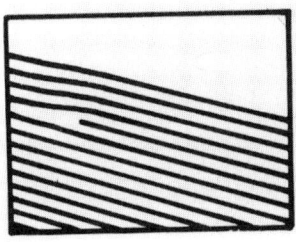

Figure 6.37. *An electron microscope photograph in which resolved planes are seen edge on, and which includes a dislocation.*

P Dean [157] for amorphous SiO$_2$. Figure 6.39 shows $g(r)$ for amorphous Ge determined by Temkin *et al* [154] and compares it with two variants of the random network model [149], with which it is in good agreement.

A dense random packing of hard spheres can be constructed by mixing balls in such a way as to avoid a periodic arrangement. The concept was introduced by J D Bernal [158] and elaborated by his co-worker J L Finney [159] who made measurements of the density and coordination in models consisting of several thousand balls. This model has given better agreement than any other with results obtained on metallic alloy glasses [160]. Figure 6.40 compares experimentally deduced values of $G(r)$, the reduced radial distribution function for Ni$_{26}$P$_{24}$ glass, with the corresponding function derived for Finney's structure by Yonezawa [151]. The agreement is good, and only this model has been able to account for the splitting of the second peak. (The material is not elemental but the atomic radii of Ni and P are not very different.)

Figure 6.38. *A ball and spoke model of amorphous SiO_2.*

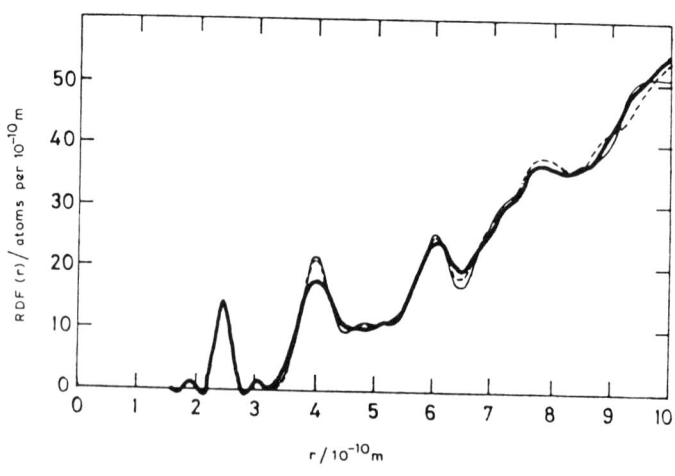

Figure 6.39. *The radial distribution function g(r) for amorphous germanium determined by experiment (heavy line). Calculated curves are also shown for two variants of a random network model.*

Solid-State Structure Analysis

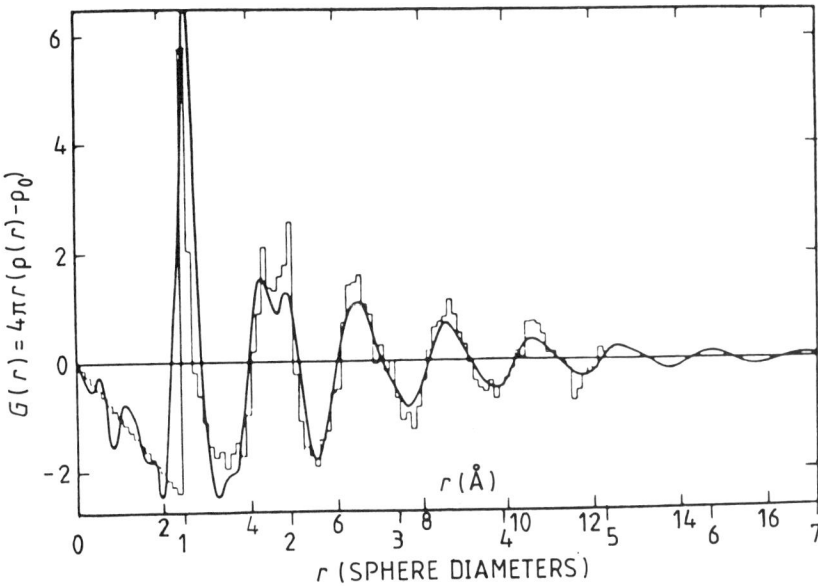

Figure 6.40. *The reduced radial distribution function G(r) determined by experiment for amorphous $Ni_{26}P_{24}$, compared with that calculated for a dense random packing model.*

6.9.7. Quasicrystals

As is remarked in the introduction to a review by D Gratias [161] 'The observation of objects opined to be impossible inevitably requires a revision of traditional theoretical approaches'. Such a situation arose in 1984 when D Schechtman *et al* [162] announced experimental evidence for a metallic alloy with very unusual properties. The alloy of Al and Mn, containing about seven atomic per cent of Mn, was obtained by very rapid cooling from the melt, when an amorphous material might have been expected in the form of a ribbon about 10^{-5} cm thick. However, under the electron microscope pentagonal 'flowers' of the alloy were seen to be surrounded by more Al-rich areas (figure 6.41). The electron diffraction pattern from anywhere within a flower-like area is remarkable in that it has a tenfold axis of rotational symmetry, which is forbidden for a crystal. Nevertheless the sharpness of the pattern shows that the alloy is not amorphous but has long-range order. Further investigation showed that the structure had full icosahedral symmetry, which includes fivefold axes of rotational symmetry. We thus have the paradox of an object with long-range order, yet possessing a symmetry incompatible with the translational periodicity which characterizes the crystalline state. The term quasicrystal was introduced to describe this new state of matter. Not surprisingly it did not find immediate acceptance, and L Pauling [163] suggested that the apparent icosahedral symmetry was

Twentieth Century Physics

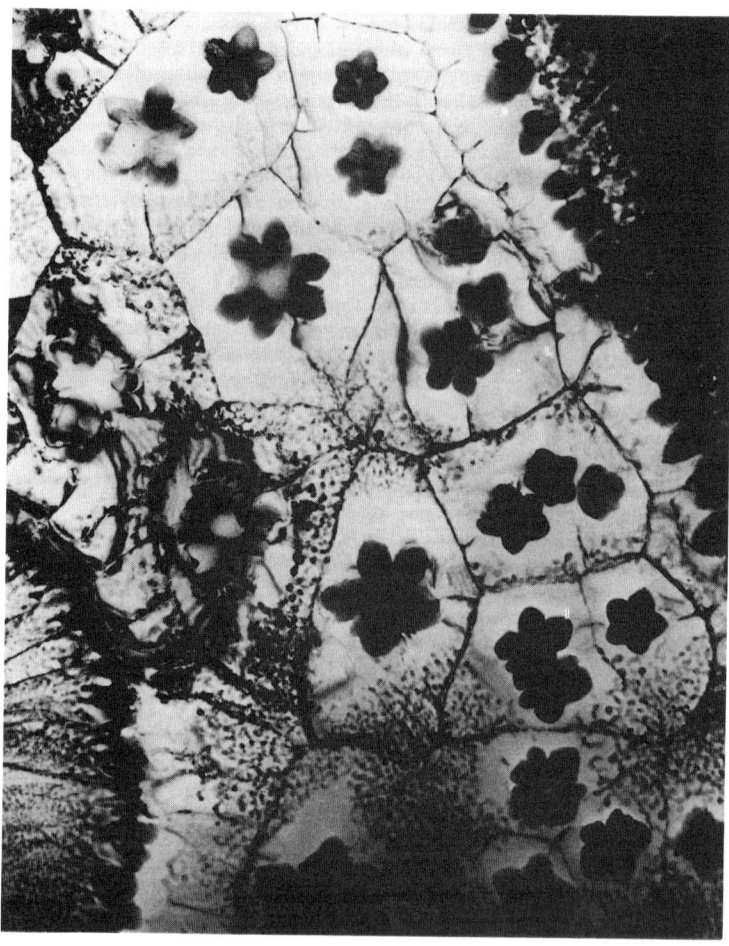

Figure 6.41. *Electron micrograph pattern from an alloy of Mn and Al. The diffraction pattern from the material in the darker areas showed an 'impossible' symmetry (see text).*

an effect produced by multiple twinning of cubic crystallites. A unit cell containing some 2000 atoms was required, not in itself impossible, but Pauling's model cannot explain fully either the electron diffraction pattern or high-resolution electron micrographs [161].

It was not known to crystallographers at the time that the mathematicians H Bohr [164] and A S Besicovich had shown, long before, that strict periodicity is not a necessary condition for obtaining coherent diffraction, with well-defined sharp reflections. While it is impossible to cover a two-dimensional space using regular pentagons or indeed by repetition of any but one of the five two-dimensional unit cells mentioned in section 6.8, it was shown by R Penrose [165] that an irregular assembly of two types of lozenge-shaped figures, one with

Solid-State Structure Analysis

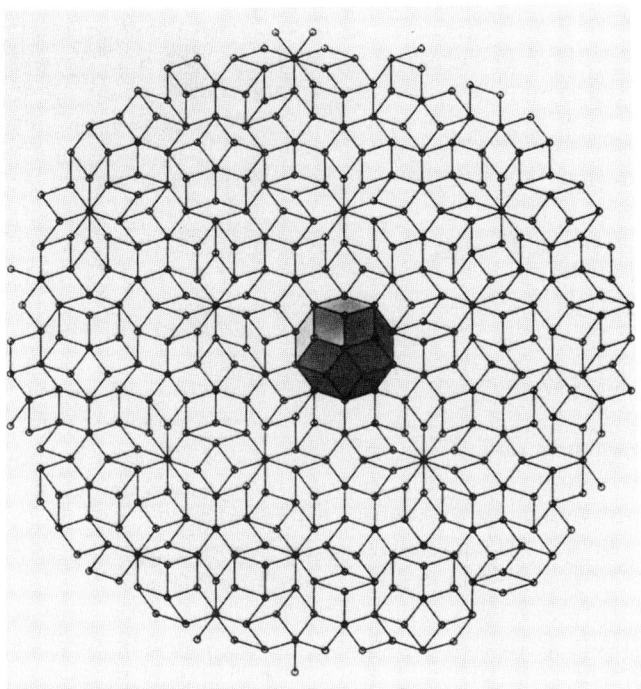

Figure 6.42. *A Penrose tiling in which a plane is covered by an irregular assembly of two lozenge-shaped figures.*

an internal angle of 36°, the other of 72° (e.g. figure 6.42), produces the equivalent in two dimensions of the three-dimensional quasicrystal alloy of Al and Mn. A simpler Penrose tiling [161] which uses three simple parallelograms is shown in figure 6.43. It is quasi-periodic and no displacement will superpose it exactly on itself. It will be recognized, however, as a perspective view of a strictly periodic stacking of cubes in a three-dimensional space. A slice is taken through the stack at an angle whose slope is irrational with respect to the three mutually perpendicular directions in the stack, and all cubes whose tops are included in this slice have their outlines projected in a direction perpendicular to the plane of the cut. In this way we get a quasi-periodic tiling in a plane from a periodic network in three dimensions. The same is true of the original Penrose tilings, which are plane projections of a slice through a simple cubic network, but this time in five-dimensional space! In the same way, Euclidean spaces of dimension greater than three can be used to describe incommensurate phases which can therefore be treated on the same basis as quasicrystals [166].

Three other types of quasicrystals are now known, having rotation axes of order 8, 10 and 12, respectively. In all some forty examples of quasicrystalline alloys were known by the end of 1991, when more

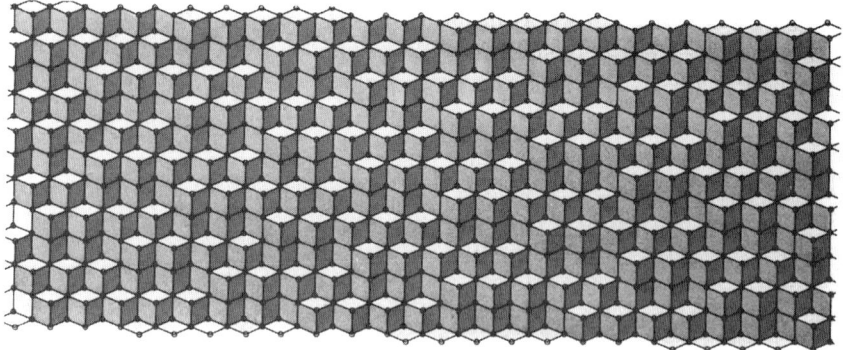

Figure 6.43. *A Penrose tiling involving three parallelograms.*

than a thousand publications had appeared [167]. Less is known about their actual atomic structures, there is of course no unique structure for a quasicrystal and the most that can be asked for are models for the atomic arrangements analogous to the random network model or the dense random packing model for amorphous materials [168].

6.10. The impact of crystal-structure analysis

The discovery of x-ray diffraction was made by physicists and led to the development of new topics within mainstream physics such as x-ray spectroscopy and x-ray optics. The potential for crystal-structure determination, at first exploited mainly by the Braggs and their co-workers, was soon also appreciated by chemists, mineralogists and metallurgists, and to some extent x-ray crystallography became a separate discipline with its own journals, societies and conferences. Of course it continued to influence those sciences from which its participants were drawn, and in physics the fields of ferroelectricity and superconductivity in particular have had important contributions made to them by crystallographic studies.

6.10.1. Cohesive energy and elasticity

The theory of lattice dynamics and of the specific heat of crystals had already been founded by A Einstein's 1907 paper and by those of M Born and T von Kármán and of P Debye, both published in 1912. These papers had at least as much bearing on the emerging quantum theory as they did on the dynamics of atoms in crystals. A knowledge of the crystal structures of the alkali halides enabled Born and his co-workers to develop the theory of the cohesive energy and the elastic constants of these crystals. The discrepancy with Cauchy's theory, mentioned in section 6.1, was cleared up in that it was shown that for cubic crystals such as the alkali halides, the Cauchy condition $C_{44} = C_{12}$ depends on the ions being situated on centres of symmetry and interacting with central

See also p 969

forces. Even when these conditions are satisfied the result cannot be expected to hold exactly since it depends on the validity of the harmonic approximation. These topics are considered in more detail in Chapter 12.

6.10.2. Optical and dielectric properties

The refractive indices of a crystal could be understood in terms of its structure by ascribing an atomic polarizability α to each atom or ion in the unit cell. In an applied field E the electric dipole moment is $p_j = \alpha_j(E + E_{lj})$, where E_{lj} is the local field at site j contributed by the other dipoles. In simple structures such as the alkali halides, E_{lj} is independent of j and equal (in cgs units) to $4\pi P/3$, where P is the polarization, given by the sum of the dipole moments per unit volume. These considerations lead to the result

$$\frac{\varepsilon(\infty) - 1}{\varepsilon(\infty) + 2} = \frac{4\pi}{3} \sum_j \frac{\alpha_j}{v}$$

the well-known Clausius–Mossotti relation. Here $\varepsilon(\infty)$ is the dielectric constant at optical frequencies, equal to the square of the optical index of refraction. Values of α_j have been given for the positive and the negative ions of the alkali halides which account within a few per cent for the values of $\varepsilon(\infty)$ for these crystals [169].

Calcite ($CaCO_3$) is the classic example of a doubly refracting crystal. W L Bragg [170] could account very well for the principal refractive indices of $CaCO_3$ (and the isomorphous $NaNO_3$) on the basis of the crystal structure and assumed values for the atomic polarizabilities.

The static or low-frequency dielectric constant $\varepsilon(0)$ of an ionic crystal is greater than $\varepsilon(\infty)$ because the ions are displaced by a low-frequency field and give an additional contribution to the polarization. The assumption that the ions in an alkali halide carry charges $\pm e$, which Born had found to be satisfactory in evaluating the cohesive energy and elastic constants, does not lead to a value for $\varepsilon(0)$ in agreement with experiment. Even worse, when R H Lyddane and K F Herzfeld [171] used the approach outlined above to calculate the frequencies of the normal modes of vibration of an alkali halide, some of the frequencies came out as imaginary, indicating that the structure is unstable. The source of these discrepancies was traced by B Szigeti [172] to the assumption that the electronic dipole moment of an ion depends only on the field to which it is subjected; it also depends on the configuration of neighbouring ions, particularly nearest neighbours. A shell model put forward by B J Dick and A W Overhauser [173] was found to account satisfactorily for the dielectric properties. In this model the outermost electrons do not move rigidly with the core, and their relative displacement gives an electronic dipole moment which depends both on the field and on the overlap forces with neighbouring ions. This was the situation when

B N Brockhouse and his co-workers [174] made the first measurements of the frequencies of certain modes of vibration in NaI and in KBr, using the technique of neutron spectroscopy. They found that the shell model accounted satisfactorily for the results, which also confirmed the Lyddane–Sachs–Teller formula, $v_L^2/v_T^2 = \varepsilon(0)/\varepsilon(\infty)$, relating the dielectric constants $\varepsilon(0)$ and $\varepsilon(\infty)$ to the frequencies v_L and v_T of long-wavelength optic modes of vibration.

This formula can be generalized to apply when there are more than two types of ion involved. For a crystal such as BaTiO$_3$, which makes a ferroelectric[†] phase transition from a cubic to a tetragonal structure at a temperature, T_c, of 120 °C, and for which in the cubic phase $\varepsilon(0) \propto (T - T_c)^{-1}$, we may expect that $v_T^2 \propto (T - T_c)$ from the Lyddane–Sachs–Teller formula. From this point of view the phase transition at T_c is an instability of the structure when the restoring force associated with one of the transverse optic modes of vibration, often referred to as a *soft mode*, goes to zero. This approach to the theory of ferroelectricity has quite a long history [175] but it was not developed until the connection between dynamical and dielectric properties was clarified by the neutron spectroscopic measurements.

6.10.3. Ferroelectricity

Two main types of ferroelectric phase transition are to be distinguished, displacive and order–disorder. In the former, atoms are displaced from more symmetrical positions only in the ferroelectric phase, and there is a soft mode; in the latter, atoms are already displaced in the high-temperature phase, forming domains of short-range order (see section 6.9) whose extent increases as T_c is approached from above. There was some controversy as to the category to which BaTiO$_3$ and other perovskites belonged. Traditionally BaTiO$_3$ was believed to undergo a displacive transition, but R Comes et al [176] found that it and KNbO3 exhibited the intensity streaks characteristic of crystals with displacement disorder (figure 6.44). More recently it has been concluded [177] that these materials share the characteristics of both types of phase transition; which is the more prominent depends on the temperature and the method used to probe the dynamics of the crystal.

The structural crystallography of ferroelectrics is the subject of several textbooks [178]. The anomalous dielectric properties of BaTiO$_3$ were discovered in ceramic specimens around 1943, and investigated in more detail by A von Hippel et al and by B Wul and I M Goldman [179]. Above 120 °C BaTiO$_3$ has the cubic perovskite structure shown in figure 6.45. At this temperature there is a transition to the ferroelectric tetragonal phase, the unit-cell dimensions changing with temperature as shown in figure 6.46 [180]. At 5 °C there is a transition to an orthorhombic

---

† *Ferroelectricity refers to the analogy with ferromagnetism, not to Fe.*

Solid-State Structure Analysis

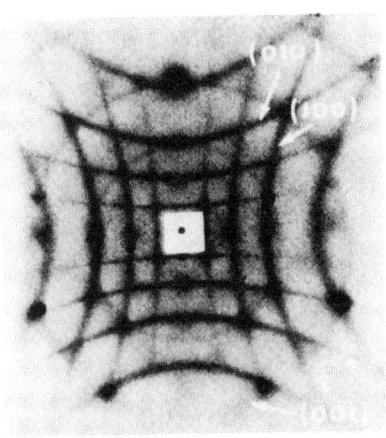

Figure 6.44. *The streaks in this x-ray photograph of KNbO₃ in the cubic plane phase are indicative of displacement disorder in the structure.*

phase (for which $a \neq b \neq c$) and then at -90 °C a final transition to a rhombohedral phase (for which $a = b = c$, $\alpha = \beta = \gamma \neq 90°$). In all three phases the crystal structures are distorted versions of that shown in figure 6.45. The rather complicated sequence is associated with changes in the direction of the spontaneous polarization P_s. In the tetragonal phase it is parallel to the c axis, which is one of the original cubic $\langle 100 \rangle$ directions, in the orthorhombic phase it switches to become parallel to a cubic $\langle 110 \rangle$ direction, while in the rhombohedral phase it is parallel to a cubic $\langle 111 \rangle$ direction. This means that in the ferroelectric phases experimental problems are posed by the existence of domains. In measurements of the dielectric constants W Merz [181] could obtain single-domain crystals only in the tetragonal phase.

Despite several attempts, the structure of the tetragonal phase has not been determined accurately with x-rays, mainly because the barium atoms dominate the scattering. Neutron diffraction studies [182] showed that the octahedral framework of oxygen atoms is only slightly distorted, while titanium and oxygen atoms move relatively in the c direction by about 0.12 Å. The spontaneous polarization is always in the direction of displacement of the Ti ion [183].

Of other perovskites the dielectric and crystallographic properties of NaNbO₃ are particularly complicated, but as always ions are not displaced by more than about 0.1 Å from the symmetrical positions which correspond to the ideal structure of figure 6.43. There are five phases with symmetry ranging from cubic above 640 °C to monoclinic below -200 °C. In the orthorhombic phase at room temperature there are anti-parallel displacements of successive Nb ions and the structure is therefore described as antiferroelectric. Other phases are non-polar

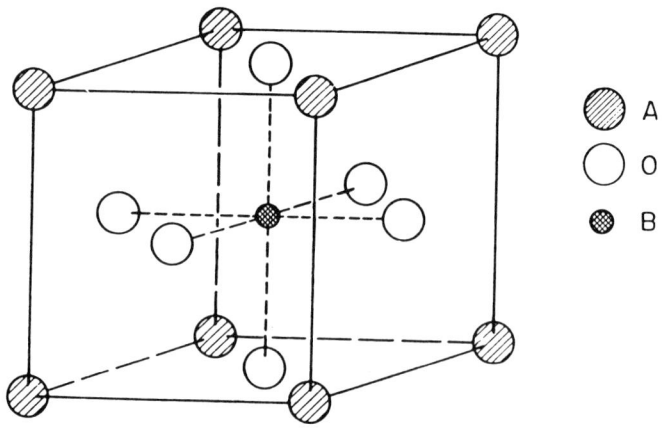

Figure 6.45. *The ideal perovskite structure.*

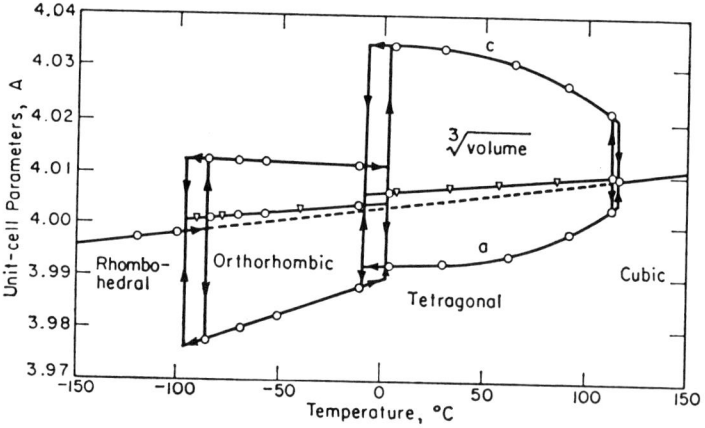

Figure 6.46. Variation of the unit-cell dimensions of $BaTiO_3$ with temperature (see text).

except that which occurs below −200 °C, which is ferroelectric. The results of studies of many other perovskite structures are set out by Jona and Shirane [178].

Potassium dihydrogen phosphate, KH_2PO_4, crystallizes in the tetragonal system at room temperature. G Busch and P Scherrer [184] reported the occurrence of ferroelectricity in this material in 1935, and the physical properties of this and related materials such as $NH_4H_2PO_4$, which has an antiferroelectric phase, have since been extensively investigated. We have already in section 6.6 given an account of a structural study of the phase transition which occurs at 123 K and is accompanied by an ordering of the hydrogen atoms. Neutron diffraction has also been used by R J Nelmes and his co-workers in a series of accurate structure determinations over a wide range of temperature and pressure. The consensus view had been that the large increase in T_c brought about by the replacement of hydrogen by deuterium was attributable to the change in isotopic mass and consequent alteration of tunnelling probability between two sites. The work by Nelmes and his colleagues showed that this cannot be the case.

The material in which ferroelectricity was first discovered, by J Valasek [185] in 1921, was Rochelle salt, $NaKC_4H_4O_6.4H_2O$ and its structure was first investigated by C A Beevers and W Hughes [186] using x-rays. The introduction to this section, on the connection between dynamical and dielectric properties, is immediately relevant to perovskite-type ferroelectrics, less so to the dihydrogen phosphates and perhaps not at all relevant to the theory of ferroelectricity in Rochelle salt and other molecular crystals. That is why they have been discussed here in reverse chronological order.

6.10.4. *Superconductivity*

Crystal-structure investigations did not play an important part in the search for superconductors which followed the discovery of superconductivity in mercury by H Kamerlingh Onnes. Most elemental superconductors have structures in the cubic or hexagonal systems which were already known, and the structures of compound superconductors such as NbC, Nb_3Ge and $CeCu_2Si_2$ are not complicated [187]. Conclusions which could be drawn from the experimental work of B T Matthias and his colleagues have been summarized as 'Matthias' rules' [188]. In particular a structural phase transition frequently occurs at a temperature not much above T_c, the critical temperature for the onset of superconductivity. This is an indication of strong electron–phonon interaction, which also favours superconductivity, but while the two tend to occur together, there is not a definite causal relation.

The BCS theory has been found to account satisfactorily for the properties of these materials, now known as the 'conventional' superconductors, of which Nb_3Ge has the highest T_c of 23.2 K. There are a few perovskite-type materials which should probably not be included in this category, including oxygen-deficient $SrTiO_3$ ($T_c < 1$ K) and $(K_{0.4}Ba_{0.6})BiO_3$ ($T_c = 30$ K) [189]. Despite the existence of the latter material and a few others, such as the crystals of the spherical-shell molecule C_{60} intercalated with K, Rb or Cs ($T_c \lesssim 30$ K), the term *high-temperature superconductor* is always reserved for the class of materials discovered since 1986 to which we now turn our attention, and where x-ray and neutron diffraction have played an essential role.

See also p 937

A new field of research, in which literally thousands of papers have been published and which has attracted as much attention as any discovery in physics since that of nuclear fission, was opened by the

See also p 957

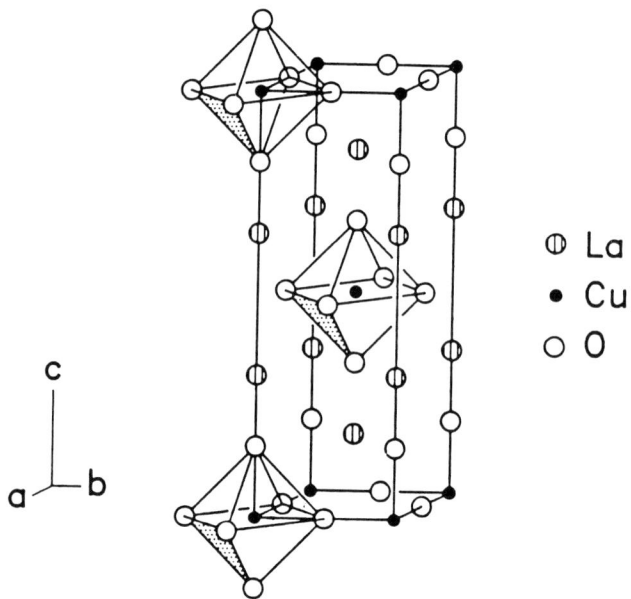

Figure 6.47. *The (ideal) structure of the high-temperature superconductor La_2CuO_4.*

discovery by J G Bednorz and K A Müller [190] of superconductivity in a class of cuprates, La_2CuO_4 in which ions of Ba or Sr can replace La. Their sample contained more than one phase, and the superconducting material was later identified as $(La_{2-x}Ba_x)CuO_{4-\delta}$. T_c depends on x and is a maximum 35 K for $x = 0.15$. A detailed structure was determined for this composition by neutron diffraction [191] (figure 6.47). The ideal structure ($\delta = 0$; all oxygen sites occupied) can be regarded as made by stacking alternate layers of perovskite-type $LaCuO_3$ and LaO. However, the precise structure and resulting properties depend sensitively on processing and particularly on any departure from stoichiometric composition.

Burns [189] has emphasized that in this, as in other high-T_c superconductors, there is another way of describing and categorizing the structure. Figure 6.48(a) shows a plane emphasizing the square-planar Cu–O_2 bonding. In the La_2CuO_4 structure, as represented in figure 6.48(b), planes of this type shown by the solid lines are separated in the c direction by two 'isolation' La–O planes, shown by dashed lines. These separate the nearest-neighbour Cu–O planes by about 6.6 Å. The shorthand notation $n = 1$ which can be used for this structure is extendable as we shall see shortly.

The discovery of $YBa_2Cu_3O_{7-\delta}$ [192], with $T_c = 94$ K showed that the La material was not unique. The structure can also be represented [193] as in figure 6.48(c), where two adjacent Cu–O planes, indicated

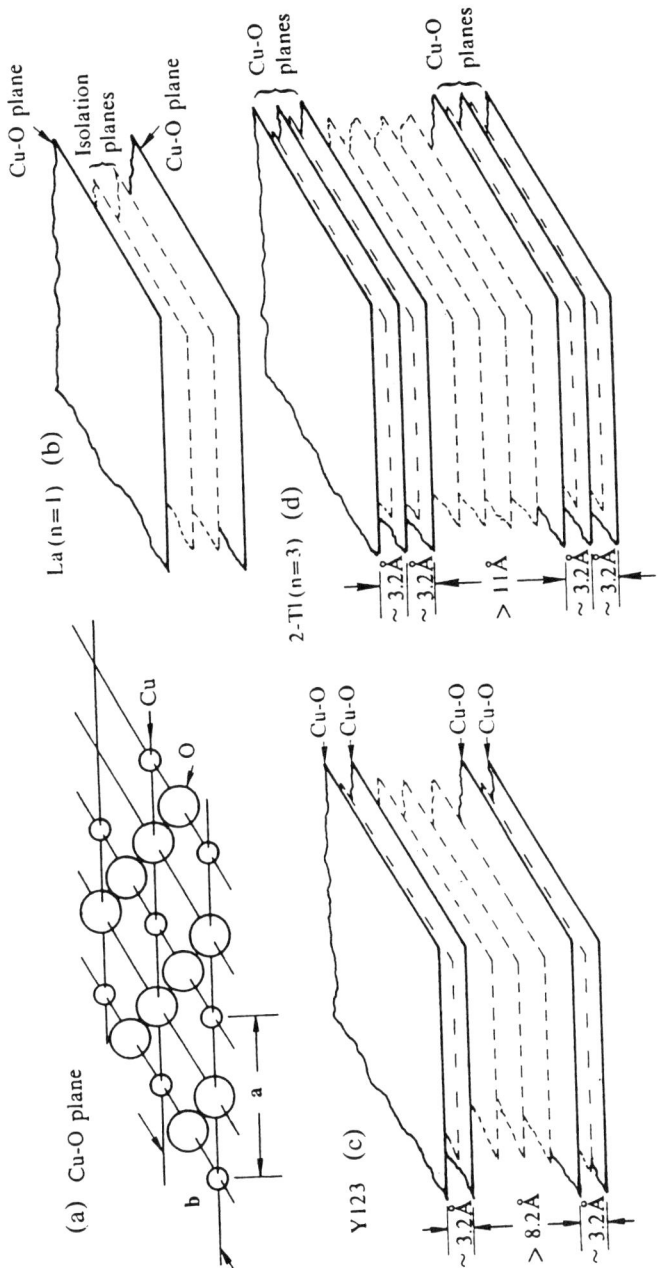

Figure 6.48. Features of the structures of high-temperature superconductors.

by the heavy lines, are separated by a plane sparsely occupied by Y atoms (lightly broken lines), while the three isolation metal–O planes separating pairs of Cu–O planes are indicated by heavier dashed lines. In the notation of the previous paragraph this is therefore an $n = 2$ structure.

At present the material $Tl_2Sr_2Cu_3O_{10}$ has the highest reported value of T_c, about 125 K. It has sets of three immediately adjacent Cu–O planes with four metal–O isolation planes as shown in figure 6.48(*d*), and is therefore an $n = 3$ structure. This topic is further treated in Chapter 11.

6.10.5. Inorganic chemistry

Mineralogy and crystallography were for a long time almost one subject. To quote W L Bragg 'we had to turn to the mineralogists to teach us our crystallography, and they have supplied us with much of the crystalline material which we have studied'. Over the years the debt has probably been more than repaid, particularly by the work of the Bragg school. Minerals comprise an enormous variety of elements and compounds and exhibit every type of chemical bonding. Here we shall concentrate on some general principles, particularly as illustrated by the structures of silicates.

The structure analysis of inorganic compounds has shown that the component atoms or groups of atoms are ionized in the solid state. When two ions approach, the repulsive force between them sets in abruptly, sufficiently so to validate the concept of 'ionic radius', and interatomic distances are usually within 0.1 Å of the sum of the two appropriate ionic radii. The first tables of ionic radii were given by W L Bragg [194] and were subsequently revised and extended by V M Goldschmidt [195] in particular. Rules which govern the structures of inorganic crystals were put forward by L Pauling [196], and minerals afford many examples of Pauling's rules. The first is most simply expressed in terms of lines of force which begin on a positive ion, in number proportional to its valency, and end on neighbouring negative ions, while for each negative ion the sum of the contributions from its neighbouring positive ions is in turn equal to its valency. This simple rule imposes quite rigorous conditions on the geometrical configuration of a structure. For example, W L Bragg [197] considers the structure of silicates 'each silicon atom is surrounded by four oxygen atoms. These atoms have half their valency satisfied by the silicon and so are left with an electrostatic charge which is unity on our valence scale (i.e. it is $-e$). Aluminium within an octahedral group of six oxygens contributes one half to each. Magnesium or ferrous iron contributes one third. Hence we may link a corner of a silicon tetrahedron to another silicon tetrahedron, to two Al octahedra or three Mg octahedra. Proceeding in this way, we find that very few alternatives which obey Pauling's law remain open to a mineral of given composition, and one of these alternatives always turns out to be the actual structure of the mineral.'. A second rule of Pauling's states that the groups around the smaller and more highly charged cations tend to share corners only, whereas edges and even faces are shared by larger groups around weak cations, with the result that highly charged cations tend to be as far apart as possible in the structure.

It is difficult to investigate silicates by chemical methods. The majority exist only as solids and the composition of a particular species is often variable, so that a standard formula cannot be assigned; this created a problem of classification before their structures were known. The types of silicon–oxygen groupings which occur are shown in figure 6.49 and comprise (*a*) closed groups, (*b*) chains and bands, (*c*) sheets and (*d*) three-dimensional nets. Olivine, $(Mg,Fe)_2SiO_4$, is an example of a structure containing SiO_4, while the structure of beryl, $Be_3Al_2Si_6O_{18}$, contains rings formed of six tetrahedra with shared oxygen atoms, although this is not obvious from figure 6.11. With this brief account of the complexity of certain silicates and the simplifications that a knowledge of their structures has made possible, we move to even wider fields.

The first crystal structures to be determined were all inorganic, and no new principles or additions to Pauling's rules emerged from the structure analysis or classification of further inorganic compounds, although in this connection it is worth noting work on an unusual group of compounds, the boron hydrides, whose various crystal and molecular structures were determined almost single-handedly by W N Lipscomb, work which gained him a Nobel prize in 1976. Before moving on to consider organic compounds it is worthwhile pausing to note that the new information provided by x-ray analysis was by no means eagerly welcomed by all inorganic chemists. As late as 1927 H E Armstrong [198] wrote in *Nature* 'Some books are lies frae end to end, says Burns. Scientific (save the mark) speculation would seem to be well on the way to this state! Professor W L Bragg asserts that in sodium chloride there appear to be no molecules represented by NaCl. The equality in number of sodium and chlorine atoms is arrived at by a chess-board pattern of these atoms, it is a result of geometry and not a pairing-off of the atoms. This statement is more than repugnant to common sense. It is absurd to the *n*th degree, not chemical cricket ...'. However, I do not suggest that this view was widely shared!

6.10.6. Organic chemistry
Most organic molecules belong to space groups of low symmetry, with atoms in general positions and a number of parameters required to define their structures. However, the atoms are linked by covalent bonds and the shape of the molecule can often be inferred from the known bond lengths and bond angles involved. It was some decades before x-ray analysis could do more than confirm what stereochemistry could predict, although bond lengths and angles were more precisely

Twentieth Century Physics

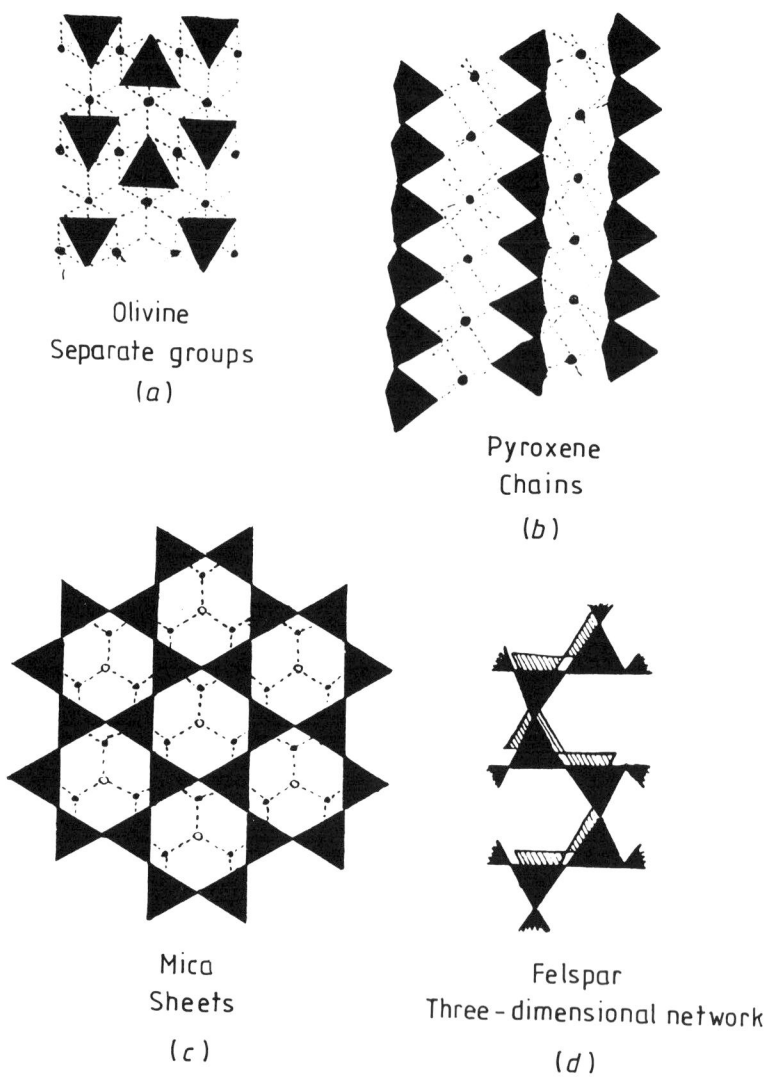

Figure 6.49. *Types of silicon–oxygen groupings which occur in different silicate minerals.*

measured and classified in the process. W H Bragg [199] measured the unit-cell dimensions of two-ringed naphthalene, $C_{10}H_8$, and of three-ringed anthracene, $C_{14}H_{10}$. He concluded that in both crystals the molecules have their longest axis in the c direction, the difference of 2.49 Å in the c dimension of the two unit cells being just enough to accommodate the extra ring of naphthalene. This was confirmed a decade or so later [200]. However '... when attempts were made to go further and postulate detailed atomic structures on the basis of cell dimensions and perhaps a few intensity observations, the results were in general disappointing and even misleading ... nearly all the work on organic crystal structures published before 1932 is quite unreliable in so far as it attempts to describe detailed atomic arrangements.' [201]. A notable exception, and the first organic crystal structure to be correctly and accurately determined, was hexamethylene tetramine, $C_6H_{12}N_4$, by R G Dickinson and A L Raymond [202]. In this instance the unit cell is body-centred cubic, with two molecules per unit cell, and the carbon and nitrogen positions involve only two parameters. An important structure determination, which did not owe anything to space-group symmetry, was that of hexamethyl benzene by K Lonsdale [203]. The unit cell is triclinic with one molecule in the unit cell. The evidence from the x-ray intensities was that each molecule lay in the (001) plane with the centre of the benzene ring at a centre of symmetry in the crystal. These restrictions, taken together with the intensity measurements, made possible quite an accurate structure determination. In 1932 J D Bernal [204], without making a complete structure determination, was able to contribute to the solution of the problem of the correct stereochemical formulae of the sterols.

The development of Fourier methods, (usually confined to two dimensions) and the discovery of the Patterson function opened up a new era in the determination of organic structures in the 1930s. The isomorphous replacement and heavy-atom methods were first applied by J M Robertson [205] to determine the structures of the phthalocyanines. J M Bijvoet and his colleagues [206] were able to determine the structure of strychnine by investigating the isomorphous sulphate and selenate.

It was Bijvoet [207] who first established the absolute configuration of dextro (d) and laevo (l) compounds, of which one is the mirror image of the other, and a solution of one rotates the plane of polarization of light in one sense and that of the other in the opposite sense. Organic compounds are interrelated such that if a conventional way is chosen for any one it will decide the form of all the remainder, but there was no way of telling whether the l-configuration allocated to a particular asymmetric molecule was the actual one or its mirror image, the d-configuration. As we have seen in section 6.4, when the energy of the incident x-rays is just less than an absorption edge of the specimen, there is a breakdown in Friedel's law that $|F(H)| = |F(-H)|$ and the specimen is said to scatter anomalously. For his study of sodium rubidium tartrate, Bijvoet used ZrKα radiation which gave strong anomalous scattering. By noting which of $|F(H)|$ or $|F(-H)|$ was the greater for a number of reflections he was able to prove that the absolute configuration of l-tartaric acid fortunately agreed with Fischer's chemical convention [56].

The structure of vitamin B_{12}, $C_{62}H_{88}N_{14}O_{14}PCo$ was, at the time of its determination in 1957 [208], by a considerable margin the most complicated to have been successfully tackled. Together with the earlier successful determination of the structure of penicillin [209] it earned D C Hodgkin a Nobel Prize for chemistry. The determination of the B_{12} structure was all the more remarkable in that it proceeded from no more than an initial knowledge of the position of the cobalt atom, obtained from the three-dimensional Patterson function, and the knowledge from parallel work in chemistry that this was the centre of 'a large planar group'.

The determination of the structures of molecules containing 100 or so non-hydrogen atoms, with no heavy atoms, is now almost a routine matter, as we noted in a brief review of direct methods in section 6.4. It has become a standard method of determining the structures of 'small' organic molecules of unknown or partly known stereochemistry. Reviews of the impact of x-ray analysis on inorganic chemistry by L Pauling [210], on organic chemistry by J M Robertson [201], and on both by W L Bragg [211] have guided this account.

6.11. Biomolecular structures

The term 'molecular biology' appears to have first been used in print by W Weaver, Director of the Rockefeller Foundation, in his 1938 report to the trustees. J Witkowski [212] marked the 50th anniversary with a paper listing the notable achievements in the subject year by year†. Several papers by x-ray crystallographers feature in this list, as one might expect.

The x-ray investigation of naturally occurring fibres is restricted by their generally poor crystallinity, but useful information has been obtained for cellulose, which is a polymer of anhydro-β-glucose, and for rubber, a polymer of isoprene. These pioneering studies have been reviewed from an historical point of view by R W G Wyckoff and by M Polanyi [213].

Proteins, which form one of the principal groups of molecules found in living matter, are polymers of amino acids linked together to form a polypeptide chain. An amino acid has the formula $NH_2CHR.COOH$, where R represents a group of atoms which distinguishes one amino acid from another and varies from H in glycine to ring systems in tyrosine, tryptophan, etc. There are some 20 commonly occurring amino acids. When a polypeptide chain is formed, amino acids are linked together with elimination of water so that the formula of a residue can be written as in figure 6.50. All amino acids (except glycine) are optically active and only the l-configuration is found in Nature. There are two classes of proteins, fibrous and globular. In the former the polypeptide chains are

---

† *I am grateful to H R Wilson for bringing this paper to my attention.*

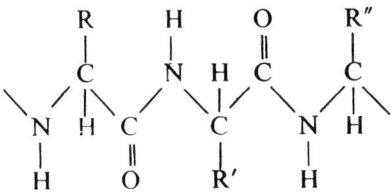

Figure 6.50. *Three amino acid residues forming a short length of polypeptide chain.*

parallel to one another, or nearly so, while in the latter they are folded to give compact molecules.

The first x-ray studies of fibrous proteins were made by R O Herzog and W Jancke [214] who investigated muscle, hair and silk. R Brill [215] made a more detailed study of the latter material and his results and conclusions were extended by K H Meyer and H Mark [216]. It was suggested that extended polypeptide chains with a repeat of 7.0 Å lay parallel to the fibre axis, an essentially correct conclusion. An important contribution was made by W T Astbury [217] and his co-workers who found that fibrous proteins could be classified into two main groups from their x-ray photographs, the keratin group (including myosin, epidermin and fibrinogen) and the collagen group. Some members of the former group had what was called the α-conformation while others existed in the β-conformation. Astbury's suggestion that in the β-conformation extended polypeptide chains were bonded into sheets has been confirmed by later work. It was Astbury, incidentally, who did much to propagate the term 'molecular biology'.

Many globular proteins can be purified and crystallized. The first x-ray photographs from a single crystal of the globular protein, pepsin, were obtained by J D Bernal and D M Crowfoot [218] (later, D C Hodgkin) in 1939, from insulin by D M Crowfoot [219] and from haemoglobin by J D Bernal *et al* [220]. Progress towards the structure determination of globular proteins was very slow until the 1950s when automatic diffractometers and electronic computers became available, and M F Perutz and co-workers [221] discovered that the phase problem could be overcome by the use of the isomorphous replacement method.

Of almost equal importance was the theoretical work of L Pauling and R B Corey [222] who proposed new models for the structures of polypeptide chains. Stereochemical limitations on chain conformations had already been discussed by M L Huggins [223] and by W L Bragg *et al* [224], but Pauling and Corey made use of further restrictions deduced from work by their colleagues on the structures of amino acids and peptides. In particular each amide group, as shown in figure 6.51, should be planar, each nitrogen atom should form a hydrogen bond with an oxygen atom of another peptide group, and the number of groups per turn of a helical structure need not be an integer. The most

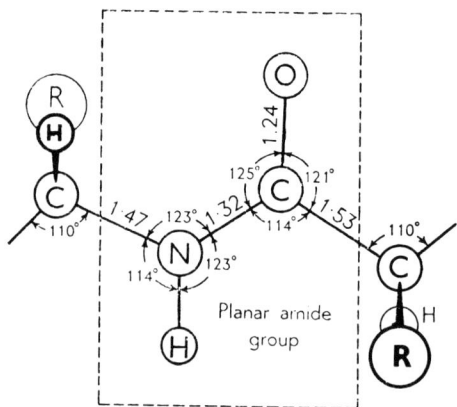

Figure 6.51. *Bond lengths and angles in a polypeptide chain.*

satisfactory model meeting these conditions was the α-helix (figure 6.52). This had a pitch of 5.4 Å and contained 3.6 amide residues per turn, with a vertical distance between residues of 1.5 Å, and hydrogen bonds which are shown as broken lines in the figure. Other structures proposed by Pauling and Corey included the parallel-chain pleated sheet and the antiparallel-chain pleated sheet (figure 6.53) in which hydrogen bonds are formed between chains, which are in the extended or β-conformation, much as Astbury had suggested.

The theory of x-ray diffraction by a continuous helix was worked out by V Vand and was then extended to apply to atoms or groups of atoms regularly repeated in a helical structure, independently by W Cochran and F H C Crick. This work was published jointly [225]. Cochran and Crick [226] applied the theory to interpret x-ray photographs of a synthetic polypeptide, poly-γ-methyl-l-glutamate and found that the polypeptide chain was indeed in the α-helical conformation. Several other synthetic polypeptides have been found with this structure, although the number of amide residues per turn of the helix varies slightly from one to another.

Certain differences between the diffraction patterns from naturally occurring and synthetic α-polypeptides were largely accounted for by F H C Crick [227] on the assumption that the former have a coiled-coil structure, two examples of which are shown schematically in figure 6.54. Such structures apparently make it easier for the side-groups R of one helix to fit between those of its neighbours. The theory of x-ray diffraction by helical structures and their variants has been given succinctly by H R Wilson [228]. As we shall see shortly, the α-helix structure is also found in globular proteins.

Structures in which extended polypeptide chains occur are referred to as β-structures. Astbury's proposal that in β-keratin the chains are

Solid-State Structure Analysis

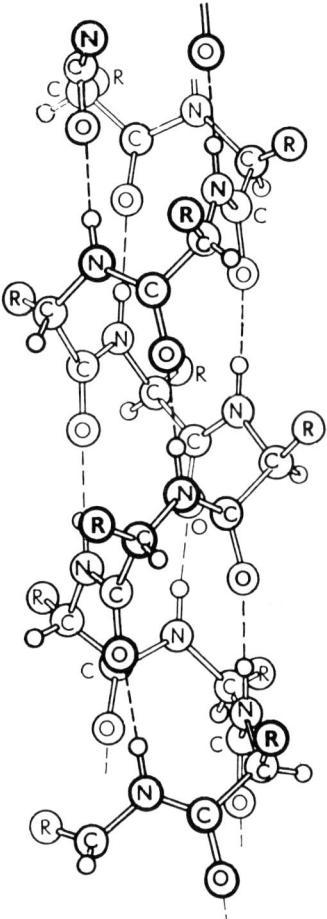

Figure 6.52. *The α-helix of Pauling and Corey.*

bonded to form sheets was modified somewhat by Pauling and Corey when they put forward the structures shown in figure 6.53. Silk fibres are composed of the β-protein, fibroin, and two varieties of this have been shown to have structures built up from antiparallel pleated sheets [229].

There has been much interest in the structure of collagen, a fibrous protein found mainly in connective tissue and tendon. It has an unusual amino acid composition and the hydrogen bonds which occur in the α-helix and the β-sheets cannot be formed. There is now agreement on a three-stranded coiled-coil model, variants of which have been considered and compared by R S Bear [230].

In globular proteins the amino acid sequence along a polypeptide

Twentieth Century Physics

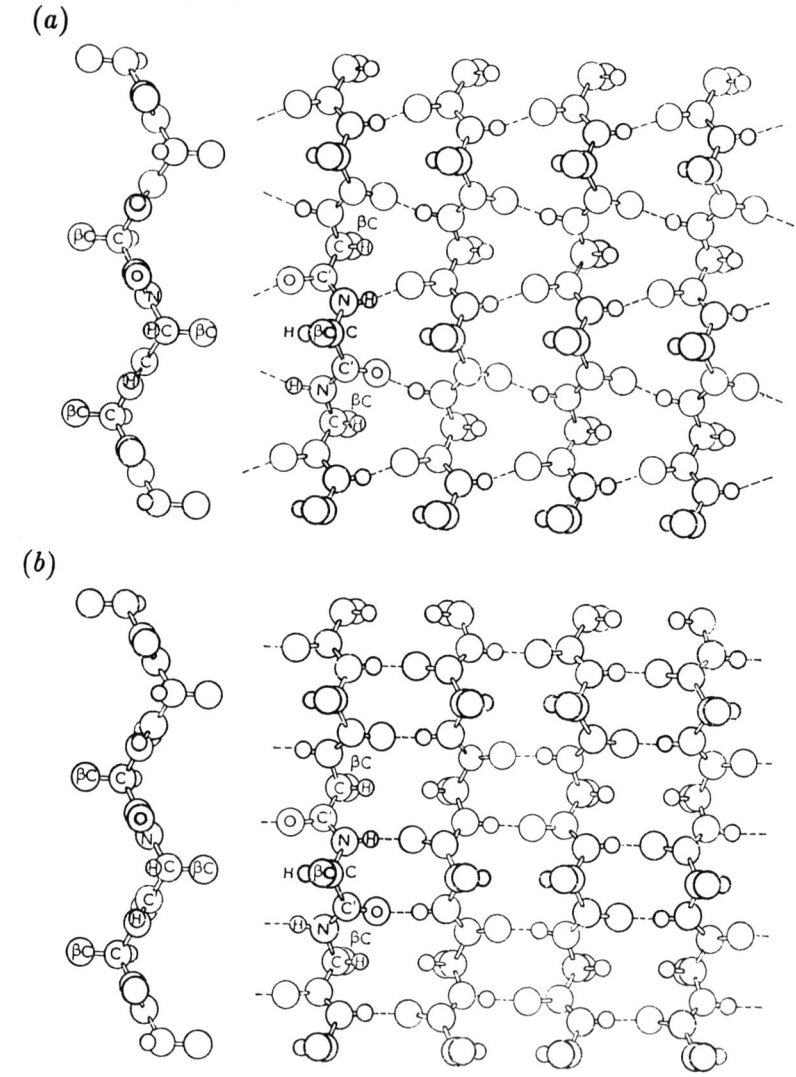

Figure 6.53. *Pleated sheet structures: (a) parallel-chain and (b) antiparallel-chain.*

chain is called the primary structure, the spatial relationship between the amino acid residues is the secondary structure (e.g. the α-helix), the folded course of the chain is the tertiary structure and if a number of folded units together form a larger structure this is the quaternary structure of the molecule. A lengthy study of haemoglobin by M F Perutz and his co-workers [221] led to the successful application of the isomorphous replacement method, which J C Kendrew and his collaborators [231] also used in the first determination of the structure of a globular protein, myoglobin, a related but smaller molecule, using this method. Perutz and Kendrew shared a Nobel Prize in 1962. Here

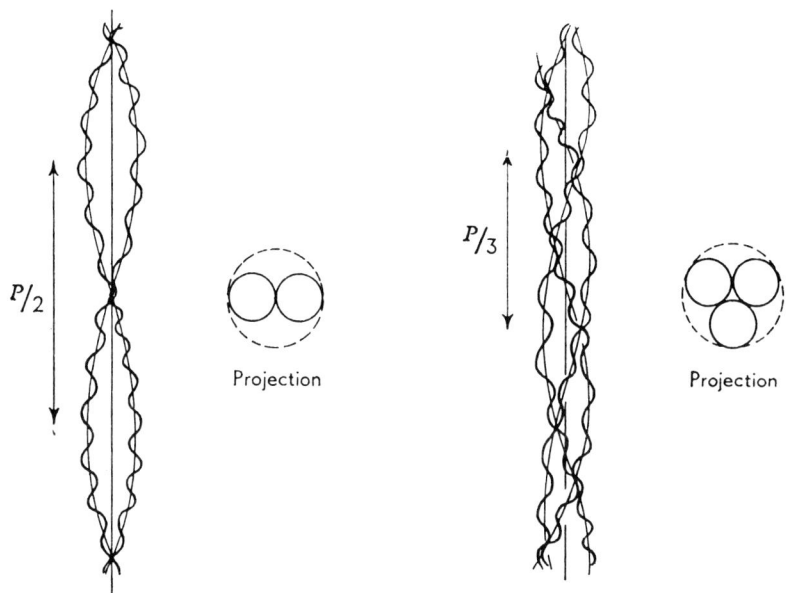

Figure 6.54. *Two-stranded and three-stranded coiled-coils.*

we shall describe only the structure determination of myoglobin in any detail, but since the work of Perutz and Kendrew the structures of some 400 globular proteins have now been determined to atomic resolution, the majority of them by isomorphous replacement or variants of this method.

A molecule of myoglobin consists of one polypeptide chain of 153 amino acid residues and a haem group. About half of the volume of the crystal is occupied by solvent, some of the water molecules being bonded to the protein but the majority of them in a disordered array between molecules. The haem group, the oxygen store in muscle tissue, consists of four five-membered rings in a plane with a central iron atom. Several heavy-atom derivatives were prepared by crystallizing from solutions which contained HgI_4 ions, $AuCl_4$ ions or Ag ions. Heavy-atom coordinates were found from Patterson functions or their projections, and these were used as described in section 6.4 to determine phase angles $\alpha(H)$ for the native protein. The analysis proceeded in several stages. In the first, phase angles were determined for 400 reflections, to a resolution of 6 Å. The corresponding Fourier synthesis showed the course of the polypeptide chain (figure 6.55(a)). The second stage, to a resolution of 2 Å, involved about 10 000 reflections, and showed more clearly the course of the chain and the position of the haem group (figure 6.55(b)). Individual atoms were not resolved but the map showed the secondary structure of most of the chain together with the shape of

Twentieth Century Physics

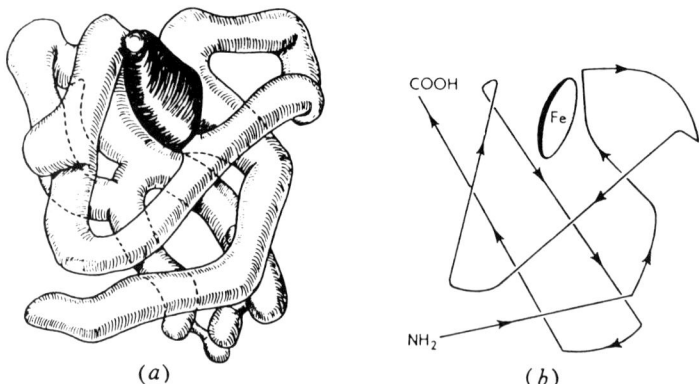

Figure 6.55. *Drawings representing (a) a myoglobin molecule and (b) the course of the polypeptide chain.*

most of the side groups R. Straight rod-like regions of the chain had a high electron density which followed a helical path with the pitch of 5.4 Å to be expected for an α-helix. The positions of 825 non-hydrogen atoms of the 1260 to be found in a molecule were then used to recalculate phase angles. Successive calculations of phase angles and electron density maps pushed the refinement to a resolution of 1.4 Å, as described by Kendrew [231]. More than two-thirds of the chain are in the α-helical conformation, the remainder is non-helical and forms the bends of the chain. Polar side groups are mainly found on the outside of the molecule, non-polar ones inside. Figure 6.56 shows a portion of the three-dimensional electron density map and its interpretation in terms of a helical region bonded to the haem group through a histidine group. When myoglobin is oxygenated the water molecule seen on the other side of the haem group is replaced by an oxygen molecule.

From the long story of protein crystallography [232] I indicate very briefly two topics of research opened up by the possibility of sequence and structure determination in haemoglobins. M F Perutz and H Lehmann [233] investigated the 'molecular pathology' of human haemoglobin. Nearly 100 different mutant haemoglobins are known and structure analysis opened the possibility of studying the part played by amino acid residues that are replaced or deleted, as was first done when V Ingram [234] showed that sickle cell anaemia results from the change of a single residue from valine to glutamine. Most harmful replacements affect residues which are common to all mammalian haemoglobins of known sequence; these have a pattern of sites always occupied by the same residues and include almost all those which make haem contacts or which can disrupt some essential contact between neighbouring segments of polypeptide chains. Mutations at these sites affect respiratory function by decreasing the molecular stability, leading

Solid-State Structure Analysis

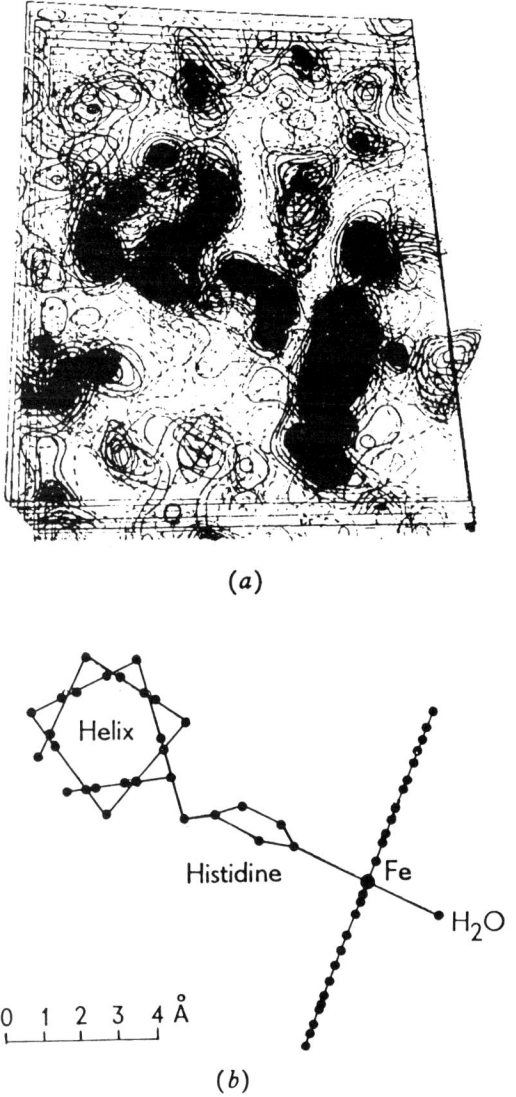

Figure 6.56. *(a) A section of a three-dimensional electron density map of myoglobin at a resolution of 2 Å and (b) its interpretation.*

to the precipitation of the haemoglobin in the red blood cells. M F Perutz has also reviewed work on the regulation of the oxygen affinity of haemoglobin through the steric restraint of the structure of the enclosing globin part of the molecule on the iron atom of the haem group [235].

Nucleic acids are long polymers of nucleotides and play a fundamental part in processes which occur in living cells. Each nucleotide is composed of three components, a base, a sugar and a phosphate group. Nucleic acids fall into two groups depending on the nature of the sugar component; ribonucleic acid (RNA) contains D-ribose while deoxyribonucleic acid (DNA) contains deoxy-D-ribose. There are five commonly occurring bases. In RNA they are the purines adenine (A) and guanine (G) and the pyrimidines cytosine (C) and uracil (U), while in DNA the same two purines occur but the two pyrimidines are cytosine and thymine (T). Their formulae are shown in figure 6.57, while the way in which the sugars are linked through phosphate groups to give polymers is shown schematically in figure 6.58. Much of the chemical work which established the primary structure shown in figure 6.58 was by A R Todd and his collaborators [236]. In 1944 O T Avery et al [237] had shown that hereditary traits could be transmitted from one bacterial cell to another by DNA, strongly suggesting its importance for molecular genetics. In the cell it is combined with protein to form chromosomes, which are replicated when a cell divides. It is now known that DNA controls the functioning of the cell by determining which protein molecules are synthesized there, and the amino acid sequence in the proteins is in turn determined by the sequence of bases in the DNA polymer [228].

DNA has a molecular weight of several million and can be extracted from the cell in the form of a salt. Pioneering studies were made by W T Astbury and F O Bell [238] in 1938 but not enough was known of the chemistry of DNA at that time for much to be deduced from their fibre photographs. A period of rapid advance in x-ray work on DNA began with the experimental work of the team led by M H F Wilkins, and of R E Franklin, and culminated in 1953 in the proposal by Watson and Crick [239] of the famous double-helix structure. Watson's account of this discovery and its background deservedly became a best-seller [240], and in 1962 Crick, Watson and Wilkins shared a Nobel Prize—a prize for which Franklin would have been considered but for her untimely death. Watson's view that they were 'racing Pauling for the Nobel Prize' turned out to be mistaken—Pauling had already qualified and was awarded the Nobel Prize for chemistry in 1954 for his contributions to molecular and biomolecular theory.

The Watson–Crick model, shown in figure 6.59 with its authors, was arrived at from a general knowledge of the experimental x-ray work and by imposing stereochemical limitations, particularly the condition that an adenine base in one chain should form hydrogen bonds with a thymine base in the second chain of the double helix, while guanine and cytosine bases were similarly hydrogen bonded across the axis of the double helix. This A–T and G–C bonding, to give what are now called Watson–Crick base pairs, satisfies Chargaff's rules [241] that the molar ratios of A to T and of G to C are very close to unity in all DNA samples, whatever the sequences of bases. 'This elegant proposal has proved to be one of the most fruitful ideas in the history of biology' [228], but we

Solid-State Structure Analysis

BASES

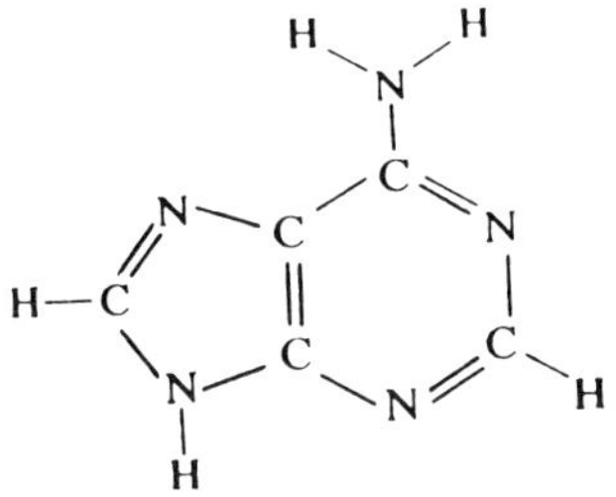

Adenine

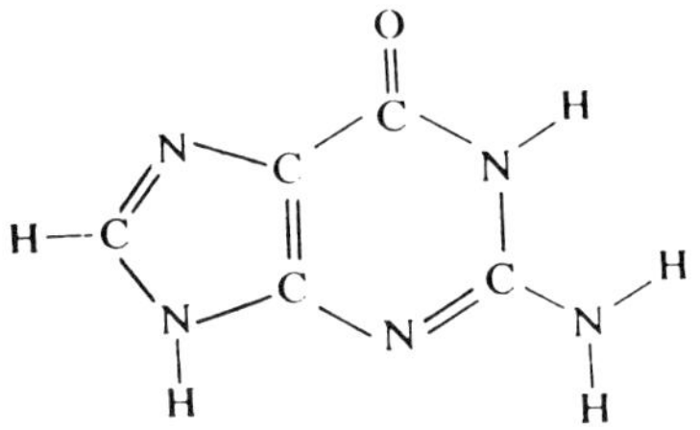

Guanine

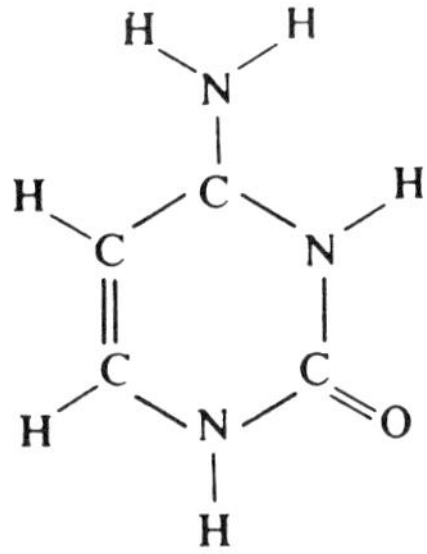

Cytosine

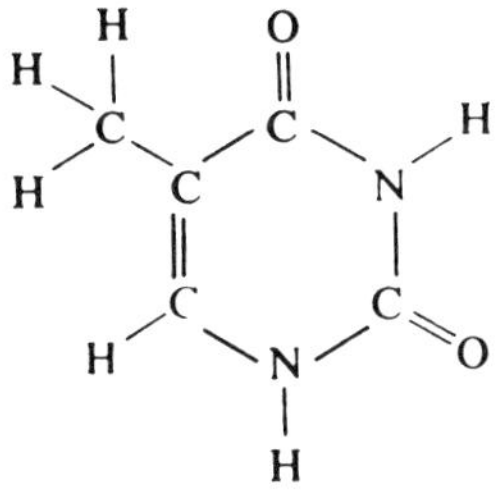

Thymine

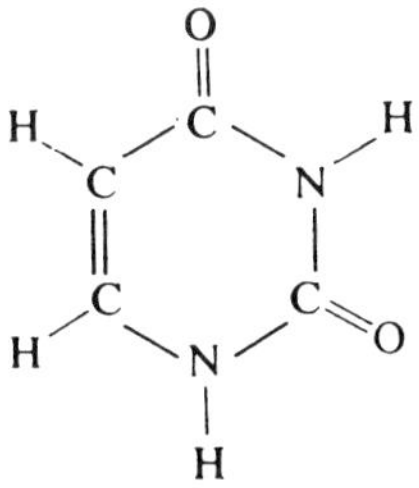

Uracil

SUGARS

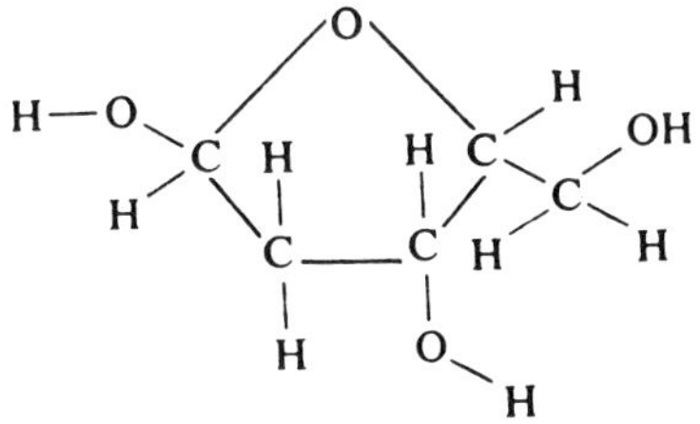

Deoxyribose

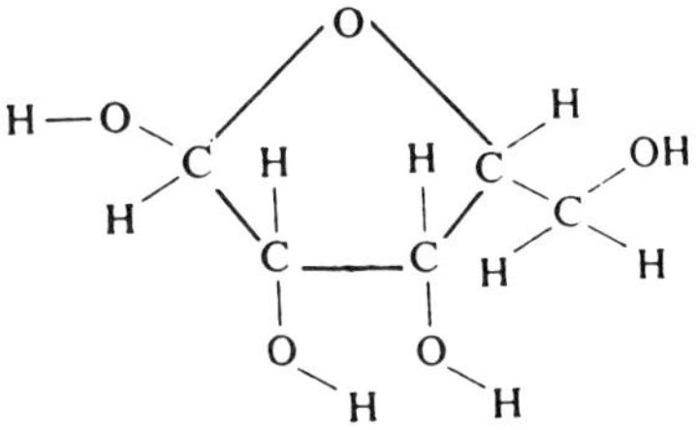

Ribose

PHOSPHORIC ACID

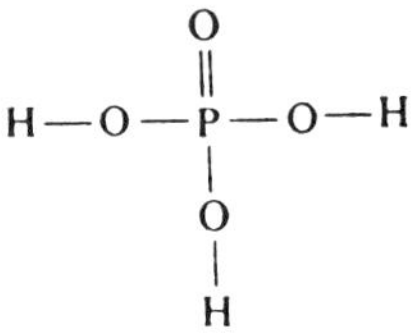

Figure 6.57. *Formulae of the bases adenine (A), guanine (G), cytosine (C), thymine (T) and uracil (U) which are components of nucleic acids. Sugars and phosphoric acid components are also shown.*

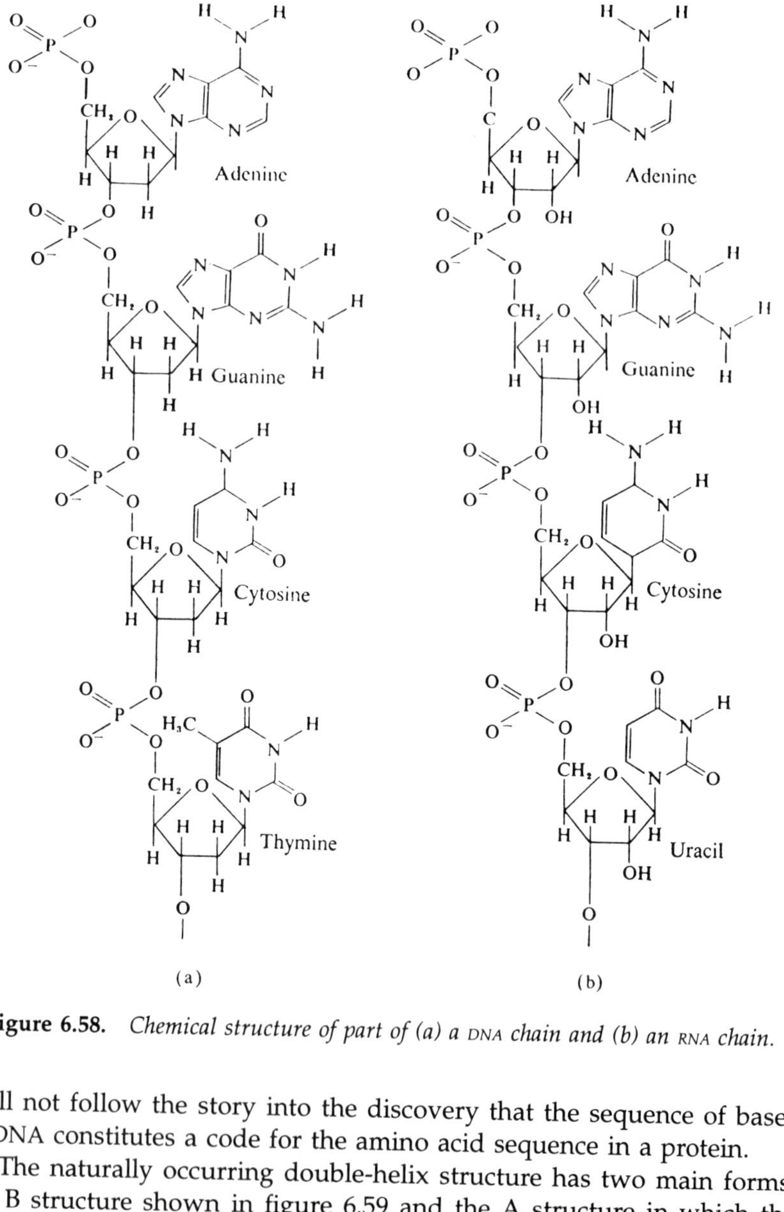

Figure 6.58. Chemical structure of part of (a) a DNA chain and (b) an RNA chain.

shall not follow the story into the discovery that the sequence of bases in DNA constitutes a code for the amino acid sequence in a protein.

The naturally occurring double-helix structure has two main forms, the B structure shown in figure 6.59 and the A structure in which the bases are tilted and the hydrogen bonds are less nearly perpendicular to the axis of the helix. Which occurs depends not on chemical composition but on the degree of hydration.

Nucleotide chains composed of some four to twelve individual nucleotides are referred to as oligonucleotides and the availability since about 1980 of synthetic oligonucleotides of defined base sequence has paved the way for detailed single-crystal structure analysis [242]. The

Figure 6.59. *Watson and Crick with their model of* DNA.

phase problem has been overcome in much the same way as for globular proteins, that is by molecular replacement, isomorphous replacement, and anomalous scattering. The x-ray data do not usually allow a resolution better than 2 Å and crystals usually contain about 50% by volume of solvent, some of it hydrogen-bonded to phosphate groups. Oligonucleotide structures fall into three categories of double helix: the right-handed A and B forms, as in naturally occurring DNA, and the left-handed Z form. Examples are $(GGGGCCCC)_2$, $(CGCGAATTCGCG)_2$ and $(CGCGCG)_2$ which have the A, B and Z forms respectively with base pairs which are all of the Watson–Crick type (figure 6.60).

The genetic code contained in the sequence and structure of DNA is under attack from the environment, and biosynthetic deoxyoligonucleotides containing a variety of errors have provided structural information on the nature of these errors and the implications for error recognition and processing *in vivo* [243]. The inclusion of non-Watson–Crick base pairs in the double helix is the most common error in replication—although the system of checking which has evolved in Nature allows such an error only about once in 10^9 instances!

The simplest viruses consist of nucleic acid and protein, with the protein acting as an outer covering for the nucleic acid, which is the active part of the virus particle. When the nucleic acid enters a living cell a process is initiated which results in the manufacture of more viruses by

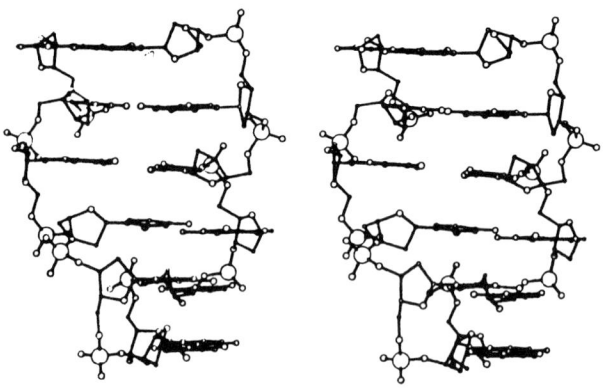

Figure 6.60. *Stereoview of the hexamer oligonucleotide* (CGCGCG)$_2$ *which has a double-helix structure of the Z-form.*

the cell. There are structurally two main types of virus, rod-shaped and spherical. The rod-shaped tobacco mosaic virus (TMV), which is about 95% protein and 5% RNA, has been the subject of several x-ray studies since the first by J D Bernal and I Fankuchen [244] in 1937. All have been made, not on crystalline material, but on oriented gels. The virus particles are about 3000 Å in length and 150 Å in diameter. While they are oriented with their long axes parallel to one another, this is the only element of order present, and the diffraction pattern is the cylindrically averaged diffraction pattern from one virus particle. Following the development of the theory of diffraction by helical structures, Watson [245] was able to show that the TMV pattern could be analysed on the basis of a helical arrangement of subunits, the pitch of the helix being 23 Å and with an integral number $3n + 1$ of subunits in three turns of the helix, i.e. in 69 Å. The value of n could not be determined at that time, but a later study of a mercury-substituted TMV by R E Franklin and K C Holmes [246] gave $n = 16$. Figure 6.61 is a drawing which represents part of the quaternary structure of TMV as derived by A Klug and D L D Caspar [247]. The primary structure of a subunit is known from sequence analysis and there was evidence that part of the polypeptide chain has an α-helical conformation. Part of the RNA chain is shown exposed in the upper part of the figure and it is known that there are three nucleotides per protein subunit.

Despite the fact that only the cylindrical average of the square of the Fourier transform of a TMV particle is obtained experimentally, improved techniques of data collection and the development of new methods of locating the replacing atoms in heavy-atom derivatives has enabled the subunit structure to be determined, first at a resolution of 10 Å and later of 4 Å [248]. Recognizable α-helices can be seen in the protein part, and the sugar and phosphate electron density peaks of the RNA part of

Solid-State Structure Analysis

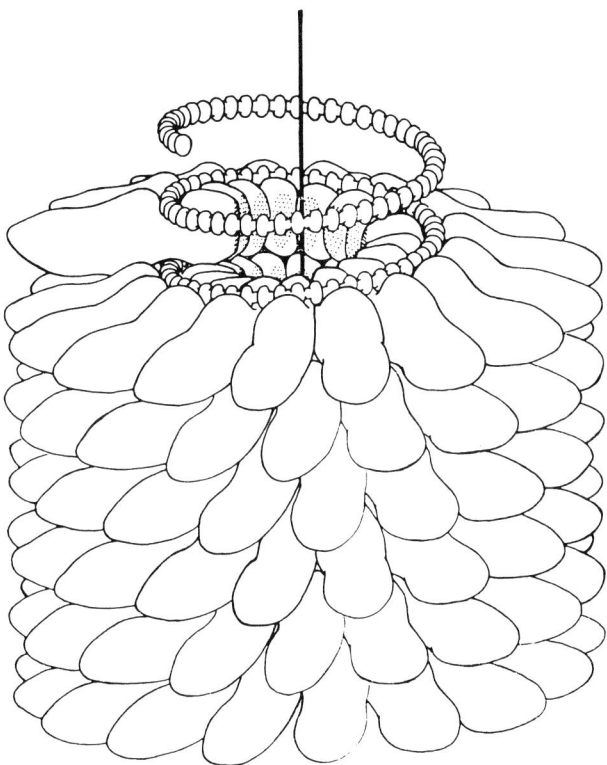

Figure 6.61. *Part of the structure of* TMV. *Part of the* RNA *chain is shown without the surrounding protein subunits.*

a subunit. The RNA follows the virus helix at a radius of 40 Å with considerable puckering, and the protein chain can be traced over about 70% of its length.

We turn now to spherical viruses, which have been even more successfully studied. They form true single crystals, and x-ray photographs of tomato bushy stunt virus (TBSV) were first obtained by D M Crowfoot and G N J Schmidt [249] in 1945. They showed the unit cell to be cubic with $a = 386$ Å; to obtain good x-ray photographs a crystal had to be contained in a capillary tube with some solvent. The large number of reflections and the susceptibility of the crystals to radiation damage posed considerable problems which synchrotron sources have partially overcome more recently.

Crick and Watson [250] considered how protein subunits might be arranged in both the rod-shaped and the spherical viruses. In the former, a helical arrangement makes all subunits equivalent and able to form similar bonds with neighbouring subunits. For the latter an equivalent arrangement around a point is required, i.e. each with the same environment on a spherical surface. The only allowed symmetry operation is that of rotation, and it was concluded that the most probable arrangements of rotation axes through a point would be one of three cubic point groups referred to as having tetrahedral, octahedral and icosahedral symmetry, respectively. Caspar [251] showed that in TBSV the unit cell is indeed cubic and each virus particle has icosahedral symmetry. The occurrence of this non-crystallographic symmetry requires a virus particle to contain at least 60 subunits; if these are identical and equivalently arranged the number must be 60, it can be a multiple of 60 if they are identical but not quite equivalently arranged. Similar conclusions were reached in studies of turnip yellow mosaic virus (TYMV) by A Klug et al [252] and of polio virus by Finch and Klug [253].

A further twist to the story was supplied by electron microscopy studies of spherical viruses. Studies of adenovirus by R W Horne et al [254] showed 252 'morphological units' on the surface of the virus. In subsequent studies of TYMV, 32 morphological units were seen, while for other viruses the number was variously 12, 42, 72, 90, 92, ... In no instance was it 60 or a multiple of 60; clearly the morphological units could not be identified with the subunits. A successful theory of the relation between them was put forward by Caspar and Klug [255]. They took the subunits to be identical in structure but only quasi-equivalent, and proposed that the relation between morphological units and subunits is that the latter cluster to produce the former, and it is the former that are seen in the electron microscope. Both TYMV and TBSV have 180 subunits, but while TYMV has 32 morphological units as a result of hexamer and pentamer clustering, TBSV has 90 morphological units as a result of dimer clustering. Results for other viruses are also consistent with the Caspar–Klug theory of virus architecture, which has been fully confirmed by the determination of several virus structures to near atomic resolution. Klug was awarded the Nobel Prize for chemistry in 1982 in recognition of his contributions to x-ray crystallography and electron microscopy.

The structure of TBSV has been determined, initially to 30 Å resolution, and eventually to 3 Å resolution by S C Harrison et al [256]. Oscillation photographs of 0.5° range were taken of the native virus and of derivatives containing U and Hg as the heavy atoms, recording some 200 000 reflections. Phase determination was improved and extended by averaging electron density maps over the equivalent subunits and by smoothing the density in regions occupied by solvent before recalculating the phase angles from the modified electron density, and so on in a cyclic process. Virus structures determined by the group at Purdue University led by M G Rossman also include that of the common cold virus [257], which has a cubic unit cell with $a = 445$ Å. All four distinct polypeptide chains in a subunit could be traced at 3.5 Å resolution; at 3 Å resolution many water molecules were identified and amino acid sequences could be matched. In a model, 6320 atoms were built into 811 amino acid residues in the four polypeptide chains.

From the structure of sodium chloride to the structures of viruses has taken about three-quarters of the twentieth century.

6.12. The International Union of Crystallography and related topics

X-ray crystallographers have traditionally formed something of an international fellowship. A first informal conference was held in the house of P Ewald's mother, on the Ammersee in 1925. Among those attending, apart from Ewald himself, were M von Laue, W L Bragg, C G Darwin and P J W Debye. In 1929, after a meeting organized by the Faraday Society in London, committees were set up to encourage international cooperation in setting up an abstracting service, in preparing space-group tables and in standardizing notation. This bore fruit when the first edition of *International Tables for the Determination of Crystal Structures* was published in 1935, the result of a collaboration of 19 persons, C Hermann being the editor.

From 1929 the Deutsche Gesellschaft für Technische Röntgenkunde organized annual meetings, in the USA the American Society for X-ray and Electron Diffraction was formed in 1941 and in the UK the X-ray Analysis Group of the Institute of Physics was set up in 1943. Although there were discussions on the formation of an International Society, no progress was made until after World War II, when in 1946 an international meeting was held in London and the decision was taken to form an International Union [258].

At Ewald's suggestion, it assumed responsibility for the publication of an international journal—the *Zeitschrift für Kristallographie* having ceased publication in 1944. *Acta Crystallographica* was launched in 1948 and is now published in four parts, while the *Journal of Applied Crystallography* has appeared since 1968. Another regular publication, *Structure Reports*, took the place of the *Strukturbericht*, which had been published as a supplement to *Zeitschrift für Kristallographie* between 1928 and 1941, and presents abstracts of crystal-structure investigations. The *International Tables*, already mentioned, also became the responsibility of the IUCr; with the appearance of successive editions they have changed from tables of space groups and atomic scattering factors, etc, to a series of authoritative chapters on many aspects of structural crystallography. The *World Directory of Crystallographers* has appeared in successive editions, each larger than the preceding one. The current edition, the eighth, lists 9589 names from 70 countries. The first issue of a Newsletter appeared early in 1993 and has been sent to all those in the *World Directory*.

Apart from regular publications, several books sponsored by the IUCr have appeared, notably *Fifty Years of X-ray Diffraction*, edited by P P Ewald [4], to which we have made frequent reference in earlier sections, and the *Historical Atlas of Crystallography* edited by

J Lima de Faria [259] with six other contributors. The latter traces the history of crystallography from the seventeenth century to the 1970s and includes portraits of prominent crystallographers, not, however, including any women, not even Dame Kathleen Lonsdale or Nobel Prizewinner Dorothy Hodgkin! *Crystallography in North America* [260] was published in 1983. *Out of the Crystal Maze* [261], a contribution to the history of solid-state physics, is not primarily concerned with the topics which we have considered in this chapter.

Since the first General Assembly and Congress at Harvard University in 1948, when there were over 300 participants, similar meetings have been held in pleasant places at three year intervals. The 16th was held in Beijing in 1993. At the 15th, held in Bordeaux, France, an estimated 1750 papers were presented under 16 headings, ranging from instrumentation to crystallographic teaching and history. These were mostly presented in the form of posters with the author(s) in attendance at advertised times. The two largest categories were Biological Macromolecules, and Physical and Chemical Properties in Relation to Structure. In 1991 six smaller conferences, including the 13th European Crystallographic Meeting, and six specialist courses or summer schools, were also organized through the IUCr.

There are 14 commissions of the IUCr at present. Their role is not precisely defined but it is generally to stimulate interest by means of specialist conferences, summer schools and occasional publications and collaborative experiments, such as that on the electron distribution in an organic molecule which was described in section 6.5.

The future of *Structure Reports* is under review because it overlaps with the work of four independent Crystallographic Data Centres. These are the Protein Data Bank at Brookhaven, USA, the Cambridge Structural Database in England, the Inorganic Crystal Structure Database in Bonn and the Metals Crystallographic Data File in Ottawa. These held 427, 73 893, 28 406 and 11 000 entries, respectively, in 1989. The work of the Cambridge Database System has been reviewed by F H Allen *et al* [262]. The International Centre for Diffraction Data, which is also an independent organization, has a file of 170 000 entries for the identification of crystalline powders, in both book and retrieval software form. The Polycrystal Book Service of Dayton, Ohio is an independent book distributor associated with the American Crystallographic Association (the successor to ASXRED mentioned above), specializing in crystallographic books and journals, and some related topics. Their 1992 catalogue listed some 2200 titles.

Acknowledgments

I am indebted to the following persons for helpful advice and for supplying me with references and reprints: F H Allen, U W Arndt,

T Brown, D W J Cruickshank, P D Hatton, J R Helliwell, O Kennard, A Klug, K Moffat, R J Nelmes, G S Pawley, M F Perutz, I K Robinson, M G Rossman, D Sayre, A J C Wilson, H R Wilson, B T Willis, M M Woolfson.

References

[1] Bragg W L 1933 *The Crystalline State* vol 1, ed W H Bragg and W L Bragg (London: Bell) p 6
[2] Haüy R J 1784 *Essai d'une Théorie sur la Structure des Cristaux* (Paris)
[3] Barlow W 1897 *Proc. R. Dublin Soc.*
See also reference [1] p 270
[4] Ewald P P 1962 *Fifty Years of X-ray Diffraction* ed P P Ewald (Utrecht: Oosthoek) ch 3
[5] Bragg W L 1939 (reprint) reference [1] p 272
[6] Ewald P P 1962 reference [4] ch 4
[7] Friedrich W, Knipping P and M von Laue 1912 *Sitz. Math. Phys. Kgl. Bayer. Akad. Wiss. München* 303
[8] Bragg W L 1912 *Nature* **90** 410
[9] Bragg W L 1913 *Proc. Camb. Phil. Soc.* **17** 43
[10] Bragg W L 1913 *Proc. R. Soc.* A **89** 248
[11] Bragg W H and Bragg W L 1913 *Proc. R. Soc.* A **88** 428
[12] Moseley H J 1913 *Phil. Mag.* **26** 1024; 1914 *Phil. Mag.* **27** 703
Siegbahn M 1962 *Fifty Years of X-ray Diffraction* ed P P Ewald (Utrecht: Oosthoek) p 265
[13] Bragg W H and Bragg W L 1913 *Nature* **91** 557
[14] Bragg W L 1939 *The Crystalline State* vol 1, ed W H Bragg and W L Bragg (London: Bell) p 59
[15] Sommerfeld A 1913 *Solvay Conf.*
See also reference [4] p 78
[16] Bragg W L, James R W and Bosanquet C H 1921 *Phil. Mag.* **41** 309
[17] Darwin C G 1914 *Phil. Mag.* **27** 315, 675
[18] Debye P 1913 *Verh. Deutsch. Phys. Ges.* **15** 678
[19] Waller I 1923 *Z. Phys.* **17** 398
[20] Kittel C 1986 *Introduction to Solid State Physics* 6th edn (New York: Wiley)
[21] Hartree D R 1928 *Proc. R. Soc.* A **121** 166
[22] James R W, Waller I and Hartree D R 1928 *Proc. R. Soc.* A **118** 334
[23] Bernal J D 1926 *Proc. R. Soc.* A **113** 118
[24] Weissenberg K 1924 *Z. Phys.* **23** 229
[25] Buerger M J 1944 *The Photography of the Reciprocal Lattice* (New York: ASXRED)
[26] Debye P and Scherrer P 1916 *Nach. Gött. Ges.* **1** 16
[27] Hull A W 1917 *Phys. Rev.* **10** 661
[28] Henry N F, Lipson H S and Wooster W A 1961 *The Interpretation of X-ray Diffraction Photographs* (London: MacMillan)
[29] Arndt U W and Willis B T 1966 *Single Crystal Diffractometry* (Cambridge: Cambridge University Press)
[30] Muller A 1931 *Proc. R. Soc.* A **132** 646
[31] Coppens P 1992 *Synchrotron Radiation Crystallography* (New York: Academic)
[32] Leigh J B and Rosenbaum G 1974 *J. Appl. Crystallogr.* **7** 117

[33] Wilson K S et al 1983 *J. Appl. Crystallogr.* **16** 28
[34] Rudman R 1976 *Low Temperature X-ray Diffraction* (New York: Plenum)
[35] Debrenne P, Laughier J and Chaudet M 1970 *J. Appl. Crystallogr.* **3** 493
[36] Jamieson J C, Lawson A W and Nachtrieb N D 1959 *Rev. Sci. Instrum.* **30** 1016
[37] Jayaraman A 1983 *Rev. Mod. Phys.* **55** 65
[38] Spain I L 1987 *Contemp. Phys.* **28** 523
[39] Block S and Piermarini G 1983 *Crystallography in North America* ed D MacLachan and J P Glusker (New York: American Crystallography Association) p 265
[40] Arndt U W, Champness J N, Phizackerley R P and A J Wonacott 1973 *J. Appl. Crystallogr.* **6** 457
[41] Arndt U W 1986 *J. Appl. Crystallogr.* **19** 145
[42] Bilderback P et al 1988 *Nucl. Instrum. Methods Phys. Res.* A **266** 636
[43] Cruickshank D W J, Helliwell J R and Moffat K 1987 *Acta Crystallogr.* A **43** 656
[44] Moffat K and Helliwell J R 1989 *Topics in Current Chem.* **151** 61
[45] Moffat K, Bilderback D, Schildkamp W and Volz K 1986 *Nucl. Instrum. Methods* A **246** 627
[46] Bragg W H 1915 *Phil. Trans. R. Soc.* A **215** 253
[47] James R W 1948 *The Crystalline State* vol 2, ed W L Bragg (London: Bell) ch 7
[48] Astbury W T and Yardley K 1924 *Phil. Trans. R. Soc.* A **224** 221
[49] Wilson A J C 1962 *Fifty Years of X-ray Diffraction* ed P P Ewald (Utrecht: Oosthoek) p 677
[50] Srinivasan R 1972 *Adv. Struct. Res. Diffraction Methods* **4** 105
[51] Bragg W L and West J, 1926 *Proc. R. Soc.* A **111** 691
[52] Bragg W L 1929 *Z. Krist.* **70** 488
[53] Beevers C A and Lipson H S 1934 *Proc. R. Soc.* A **146** 570
[54] Patterson A L 1934 *Phys. Rev.* **46** 372
[55] Zernike F and Prins J 1927 *Z. Phys.* **41** 184
[56] Lipson H S and Cochran W 1966 *The Crystalline State* vol 3, ed W L Bragg (London: Bell)
[57] Harker D 1936 *J. Chem. Phys.* **4** 381
[58] Wrinch D M 1939 *Phil. Mag.* **27** 98
[59] Reference [56] ch 7
[60] Buerger M J 1950 *Proc. Natl Acad. Sci.* **36** 376 and 738
[61] Carlisle C H and Crowfoot D M 1945 *Proc. R. Soc.* A **184** 64
[62] Bokhoven C, Schoone J C and Bijvoet J M 1951 *Acta Crystallogr.* **4** 275
[63] Harker D 1956 *Acta Crystallogr.* **9** 1
[64] Rossman M G and Blow D M 1962 *Acta Crystallogr.* **15** 24
[65] Argos P and Rossman M G 1980 *Theory and Practice of Direct Methods in Crystallography* ed M Ladd and R Palmer (New York: Plenum)
[66] Rossman M G 1990 *Acta Crystallogr.* A **46** 73
[67] Lipson H S and Beevers C A 1936 *Proc. R. Soc.* A **48** 702
[68] Pepinsky R 1947 *J. Appl. Phys.* **18** 601
[69] Harker D and Kasper J S 1948 *Acta Crystallogr.* **1** 70
[70] Kasper J S, Lucht C M and Harker D 1950 *Acta Crystallogr.* **3** 436
[71] Karle J and Hauptman H 1950 *Acta Crystallogr.* **3** 181
[72] Woolfson M M 1987 *Acta Crystallogr.* A **43** 593
[73] Sayre D 1952 *Acta Crystallogr.* **5** 60
[74] Cochran W 1952 *Acta Crystallogr.* **5** 65
[75] Zachariasen W H 1952 *Acta Crystallogr.* **5** 68

[76] Hauptman H and Karle J 1953 *Solution of the Phase Problem. I. The Centrosymmetric Crystal* (New York: Polycrystal Book Service)
[77] Woolfson M M 1954 *Acta Crystallogr.* **7** 61
[78] Klug A 1958 *Acta Crystallogr.* **11** 515
[79] Cochran W 1955 *Acta Crystallogr.* **8** 473
[80] Karle J and Hauptman H 1956 *Acta Crystallogr.* **9** 635
[81] Karle I L and Karle J 1964 *Acta Crystallogr.* **17** 835
[82] Germain G and Woolfson M M 1968 *Acta Crystallogr.* B **24** 91
[83] Karle J 1989 *Acta Crystallogr.* A **45** 765
   Giacovazzo C 1980 *Direct Methods in Crystallography* (London: Academic)
[84] Bacon G E 1975 *Neutron Diffraction* 3rd edn (Oxford: Clarendon)
[85] Darwin C G 1922 *Phil. Mag.* **43** 800
[86] Zachariasen W H 1963 *Acta Crystallogr.* **16** 1139
   Zachariasen W H 1967 *Acta Crystallogr.* **23** 558
[87] Dawson B 1975 *Adv. Struct. Res. Diffraction Methods* **6** 1
[88] Cooper M J, Rouse K D and Willis B T 1968 *Acta Crystallogr.* A **24** 484
[89] Cruickshank D W J 1949 *Acta Crystallogr.* **2** 65
[90] Bunn C W 1949 *The X-ray Crystallographic Investigation of the Structure of Penicillin* ed D Crowfoot *et al* (Oxford: Oxford University Press)
[91] Cochran W 1951 *Acta Crystallogr.* **4** 81, 408
[92] Coppens P *et al* 1984 *Acta Crystallogr.* A **40** 184
[93] Marshall W and Lovesey S W 1971 *Theory of Thermal Neutron Scattering* (Oxford: Oxford University Press)
   Lovesey S W 1985 *Theory of Neutron Scattering from Condensed Matter* in 2 volumes (Oxford: Oxford University Press)
   Bacon G E (ed) 1986 *Fifty Years of Neutron Diffraction* (Bristol: Hilger)
[94] Elsasser W M 1936 *C. R. Hebd. Séanc. Acad. Sci.* **202** 1029
[95] Mitchell D P and Powers P N 1936 *Phys. Rev.* **50** 486
[96] Zinn W H 1947 *Phys. Rev.* **71** 752
[97] Shull C G and Smart J S 1949 *Phys. Rev.* **76** 1256
[98] Fermi E and Marshall L 1947 *Phys. Rev.* **71** 666
[99] Bacon G E and Lowde R 1948 *Acta Crystallogr.* **1** 303
[100] Buras B and Leciejewicz J 1963 *Nukleonika* **8** 75
[101] Christ J and Springer T 1962 *Nukleonika* **4** 23
[102] Wollan E O and Shull C G 1948 *Phys. Rev.* **73** 830
   Wang S P and Shull C G 1962 *J. Phys. Soc. Japan* **17** (Supplement B3) 340
[103] Bozorth R M 1923 *J. Am. Chem. Soc.* **48** 2128
[104] Peterson S W and Levy H A 1952 *J. Chem. Phys.* **20** 704
[105] Bacon G E and Pease R S 1955 *Proc. R. Soc.* A **230** 359
[106] Brown P J 1979 *Neutron scattering* ed G Kostorz (London: Academic) p 69
[107] Koehler W C 1986 *Fifty Years of Neutron Diffraction* ed G E Bacon (Bristol: Hilger) p 169
[108] Halpern O and Johnson M H 1939 *Phys. Rev.* **55** 898
[109] Alvarez L W and Bloch F 1940 *Phys. Rev.* **57** 111
[110] Shull C G, Strauser W A and Wollan E O 1951 *Phys. Rev.* **83** 333
[111] Shull C G, Wollan E O and Koehler W C 1951 *Phys. Rev.* **84** 912
[112] Nathans R, Shull C G, Shirane G and Andresen A 1959 *J. Phys. Chem. Solids* **10** 138
[113] Shull C G 1963 *Electronic Structure and Alloy Chemistry of the Transition Elements* ed P A Bock (New York: Wiley) p 69
[114] Belov N V, Neronova N N and Smirnova T S 1957 *Sov. Phys. Crystallogr.* **2** 311
[115] Koehler W C 1965 *J. Appl. Phys.* **36** 1078

[116] Thomson G P and Cochrane W 1939 *Theory and Practice of Electron Diffraction* (London: MacMillan)
Pinsker Z G 1949 *Diffraktsia Elektronov* (Engl. Transl. 1953 *Electron Diffraction* (London: Butterworth))
Vainshtein B K 1956 *Strukturnaya Elektronografiya* (Engl. Transl. *Structure Analysis by Electron Diffraction* (Oxford: Pergamon))
Zvyagin B B 1967 *Electron Diffraction Analysis of Clay Minerals* (New York: Plenum)
[117] Thomson G P 1965 *Contemp. Phys.* **9** 1
[118] Nishikawa S and Kikuchi S 1928 *Nature* **121** 1019
[119] Blackman M 1939 *Proc. R. Soc.* A **173** 68
[120] Goodman P 1981 (ed) *Fifty Years of Electron Diffraction* (Dordrecht: Reidel)
[121] Miyaka S and Ueda R 1950 *Acta Crystallogr.* **3** 314
[122] Cowley J M 1992 *Techniques of Transmission Electron Diffraction* (Oxford: Oxford University Press)
Spence J C H and Zuo J M 1992 *Electron Microdiffraction* (New York: Plenum)
[123] Woodruff D P and Delchar T A 1986 *Modern Techniques of Surface Science* (London: Cambridge University Press)
[124] Binnig G, Rohrer H, Gerber C and Weibel E 1982 *Appl. Phys. Lett.* **40** 178
[125] Prutton M 1983 *Surface Physics* (Oxford: Oxford University Press)
[126] Ehrenberg W 1934 *Phil. Mag.* **18** 878
[127] Bethe H A 1928 *Ann. Phys.* **87** 55
[128] White J W 1977 *Dynamics of Liquids by Neutron Scattering* (Heidelberg: Springer)
[129] Dunoyer L 1911 *Radium* **8** 142
[130] Estermann I and Stern O 1930 *Z. Phys.* **61** 95
[131] Eisenberger P and Marra W C 1981 *Phys. Rev. Lett.* **46** 1081
[132] Robinson I K and Tweet D J 1992 *Rep. Prog. Phys.* **55** 599
[133] Bohr J, Reidenhansl R, Nielsen M, Toney M, Johnson R L and Robinson I K 1985 *Phys. Rev. Lett.* **54** 1275
[134] Haneman D 1987 *Rep. Prog. Phys.* **50** 1045
[135] Schlier R J and Farnsworth H E 1957 *Adv. Catalysis* **9** 434
[136] Wilson A J C 1949 *X-ray Optics; the Diffraction of X-rays by Finite and Imperfect Crystals* (London: Methuen)
[137] Kellar J N, Hirsch P B and Thorp J S 1950 *Nature* **165** 554
[138] Edwards O S and Lipson H S 1940 *Proc. R. Soc.* A **180** 268
[139] Bragg W L and Williams E J 1934 *Proc. R. Soc.* A **145** 699
[140] Barnes W H 1929 *Proc. R. Soc.* A **125** 670
[141] Bernal J D and Fowler R H 1933 *J. Chem. Phys.* **1** 515
[142] Wollan E O, Davidson W L and Shull C G 1949 *Phys. Rev.* **75** 1348
[143] Chamberlain J S 1971 *Thesis* University of New England
[144] Pauling L 1935 *J. Am. Chem. Soc.* **57** 2680
[145] Orowan E 1934 *Z. Phys.* **89** 605
Polanyi M 1934 *Z. Phys.* **89** 660
Taylor G I 1934 *Proc. R. Soc.* A **145** 388
[146] Burgers W G 1980 *Proc. R. Soc.* A **371** 125
Nabarro F R N 1980 *Proc. R. Soc.* A **371** 131
Seeger A K 1980 *Proc. R. Soc.* A **371** 173
Hirsch P B 1980 *Proc. R. Soc.* A **371** 160
[147] Amelinckx S 1964 The direct observation of dislocations *Supplement to Solid State Physics* ed F Seitz and D Turnbull (New York: Academic)
[148] Menter J M 1956 *Proc. R. Soc.* A **236** 119
[149] Weaire D 1976 *Contemp. Phys.* **17** 173

[150] Morey G E 1938 *The Properties of Glass* (New York: Reinhold)
[151] Yonezawa F 1991 *Solid State Phys.* **45** 179
[152] Debye P and Mencke H 1931 *Ergebn. Tech. Röntgenk.* **2** 1
[153] Warren B E 1969 *X-ray Diffraction* (Reading, MA: Addison Wesley)
[154] Temkin R J, Paul W and Connell G A N 1973 *Adv. Phys.* **22** 581
[155] Zachariasen W H 1932 *J. Am. Chem. Soc.* **54** 3841
[156] Polk D E 1971 *J. Non-Cryst. Solids* **5** 365
[157] Bell R J and Dean P 1972 *Phil. Mag.* **25** 1381
[158] Bernal J D 1959 *Nature* **183** 141
[159] Finney J L 1970 *Proc. R. Soc.* A **319** 479
[160] Cargill C S 1975 *Solid State Phys.* **30** 227
[161] Gratias D 1987 *Contemp. Phys.* **28** 219
[162] Schechtman D, Black I, Gratias D and Cahn J W 1984 *Phys. Rev. Lett.* **53** 1951
[163] Pauling L 1985 *Nature* **317** 471
[164] Bohr H 1924 *Acta Math.* **45** 29
Besicovich A S 1932 *Almost Periodic Functions* (Cambridge: Cambridge University Press)
[165] Penrose R 1979 *Math. Intell.* **2** 32
[166] de Wolff P M 1977 *Acta Crystallogr.* A **33** 493
[167] Guyot P, Kramer P and de Boissieu M 1991 *Rep. Prog. Phys.* **54** 1373
[168] Guyot P and Audier M 1985 *Phil. Mag.* B **52** 215
[169] Tessman J, Kahn A and Shockley W 1953 *Phys. Rev.* **92** 890
[170] Bragg W L 1933 *The Crystalline State* vol 1, ed W H Bragg and W L Bragg (London: Bell) p 180
[171] Lyddane R H and Herzfeld K F 1938 *Phys. Rev.* **54** 846
[172] Szigeti B 1950 *Proc. R. Soc.* A **204** 51
[173] Dick B J and Overhauser A W 1958 *Phys. Rev.* **112** 80
[174] Woods A D B, Brockhouse B N and Cochran W 1960 *Phys. Rev.* **119** 980
Woods A D B, Brockhouse B N and Cowley R A 1963 *Phys. Rev.* **131** 1025
[175] Cochran W 1981 *Ferroelectrics* **35** 3
[176] Comes R, Lambert M and Guinier A 1970 *Acta Crystallogr.* A **26** 244
[177] Müller K A 1982 *Nonlinear Phenomena at Phase Transitions and Instabilities* ed T Riste (New York: Plenum) p 1
[178] Megaw H D 1957 *Ferroelectricity in Crystals* (London: Methuen)
Jona F and Shirane G 1962 *Ferroelectric Crystals* (Oxford: Pergamon)
Lines M E and Glass A M 1977 *Principles and Applications of Ferroelectrics and Related Materials* (Oxford: Oxford University Press)
[179] von Hippel A, Breckenridge R G, Chesley F C and Tisza L 1946 *Ind. Eng. Chem.* **38** 1097
Wul B and Goldman I M 1945 *C. R. Acad. Sci. USSR* **46** 139
[180] Kay H F and Vousden P 1949 *Phil. Mag.* **40** 1019
[181] Merz W 1949 *Phys. Rev.* **76** 1221
[182] Frazer B C, Danner H R and Pepinsky R 1955 *Phys. Rev.* **100** 745
Harada J, Pedersen T and Barnea Z 1970 *Acta Crystallogr.* A **26** 336
[183] Okaya Y and Pepinsky R 1961 *Computing Methods and the Phase Problem in X-ray Crystal Analysis* ed D W J Cruickshank (Oxford: Pergamon)
[184] Busch G and Scherrer P 1935 *Naturwissenschaften* **23** 737
[185] Valasek J 1921 *Phys. Rev.* **17** 475
[186] Beevers C A and Hughes W 1941 *Proc. R. Soc.* A **177** 251
[187] Matthias B T, Geballe T H and Compton V B 1963 *Rev. Mod. Phys.* **35** 1
[188] Hulm J K and Blaugher R D 1972 *Superconductivity in d- and f-band Metals* ed D H Douglas (New York: AIP)

[189] Burns G 1992 *High Temperature Superconductivity, An Introduction* (New York: Academic)
[190] Bednorz J G and K A Müller 1986 *Z. Phys.* B **64** 189
[191] Jorgensen J G, Schuttler H B, Hinks D G, Capone D W, Zhang K and Brodsky M B 1987 *Phys. Rev. Lett.* **58** 1024
[192] Cava R J, Batlogg B, van Dover R B, Murphy D W, Sunshine S, Siegrist T, Remeika J P, Rieman E A, Zakurak S and Espinosa G P 1987 *Phys. Rev. Lett.* **58** 1676
[193] Beech F, Miraglia S, Santoro A and Roth R S 1987 *Phys. Rev.* B **35** 8778
[194] Bragg W L 1920 *Phil. Mag.* **40** 169
[195] Goldschmidt V M 1929 *Trans. Farad. Soc.* **25** 253
[196] Pauling L 1929 *J. Am. Chem. Soc.* **51** 1010
[197] Bragg W L 1937 *The Atomic Structure of Minerals* (New York: Cornell University Press)
[198] Armstrong H E 1927 *Nature* **120** 478
[199] Bragg W H 1921 *Proc. Phys. Soc.* **34** 33
[200] Robertson J M 1933 *Proc. R. Soc.* A **140** 79
[201] Robertson J M 1962 *Fifty Years of X-ray Diffraction* ed P P Ewald (Utrecht: Oosthoek) p 147
[202] Dickinson R G and Raymond A L 1923 *J. Am. Chem. Soc.* **45** 22
[203] Lonsdale K 1928 *Nature* **128** 810
[204] Bernal J D 1932 *Nature* **129** 2177 and 721
[205] Robertson J M 1935 *J. Chem. Soc.* 615; 1936 *J. Chem. Soc.* 1195
[206] Bokhoven C, Schoone J C and Bijvoet J M 1951 *Acta Crystallogr.* **4** 275
[207] Bijvoet J M 1949 *Koninkl Nederland Akad. Wetenschap.* **52** 313
[208] Hodgkin D C, Kamper J, Lindsey J, MacKay M, Pickworth J, Robertson J H, Shoemaker C B, White J G, Prosen R J and Trueblood K N 1957 *Proc. R. Soc.* A **242** 228
[209] Hodgkin D C, Bunn C W, Rogers-Low B W and Turner-Jones A 1949 *The Chemistry of Penicillin* (Princeton, NJ: Princeton University Press)
[210] Pauling L 1962 *Fifty Years of X-ray Diffraction* ed P P Ewald (Utrecht: Oosthoek) p 136
[211] Bragg W L 1975 *The Development of X-ray Analysis* (London: Bell)
[212] Witkowski J 1988 *Trends Biochem. Tech.* **6** 234
[213] Wyckoff R W G *Fifty Years of X-ray Diffraction* ed P P Ewald (Utrecht: Oosthoek) p 212
Polyani M *Fifty Years of X-ray Diffraction* ed P P Ewald (Utrecht: Oosthoek) p 629
[214] Herzog R O and Jancke W 1920 *Ber. Deutsch. Chem. Ges.* **53** 2162
[215] Brill R 1923 *Liebigs Ann.* **434** 204
[216] Meyer K H and Mark H 1928 *Ber Deutsch. Chem. Ges.* **361** 192
[217] Astbury W T and Street A 1931 *Phil. Trans. R. Soc.* A **230** 75
Astbury W T 1938 *Trans. Farad. Soc.* **34** 378
[218] Bernal J D and Crowfoot D M 1934 *Nature* **133** 794
[219] Crowfoot D M 1935 *Nature* **135** 591
[220] Bernal J D, Fankuchen I and Riley D P 1938 *Nature* **142** 1075
[221] Green D W, Ingram V M and Perutz M F 1954 *Proc. R. Soc.* A **225** 287
[222] Pauling L and Corey R B 1951 *Proc. Natl Acad. Sci. USA* **37** 235–82
[223] Huggins M 1943 *Chem. Rev.* **32** 195
[224] Bragg W L, Kendrew J C and Perutz M F 1950 *Proc. R. Soc.* A **203** 321
[225] Cochran W, Crick F H C and Vand V 1952 *Acta Crystallogr.* **5** 581
[226] Cochran W and Crick F H C 1952 *Nature* **168** 684
[227] Crick F H C 1952 *Nature* **270** 882; 1953 *Acta Crystallogr.* **6** 685

[228] Wilson H R 1966 *Diffraction of X-rays by Proteins, Nucleic Acids and Viruses* (London: Arnold)
[229] Marsh R E, Corey R B and Pauling L 1955 *Acta Crystallogr.* **8** 710
[230] Bear R S 1956 *J. Biophys. Biochem. Cytol.* **2** 363
[231] Kendrew J C, Bodo G, Dintzig H M, Parrish R G, Wyckoff H and Phillips D C 1958 *Nature* **181** 662
Kendrew J C 1963 *Science* **139** 1259
[232] Blundell T and Johnson L N 1976 *Protein Crystallography* (New York: Academic)
[233] Perutz M F and Lehmann H 1968 *Nature* **219** 902
[234] Ingram V 1957 *Nature* **180** 326
[235] Perutz M F 1979 *Ann. Rev. Biochem.* **48** 327
[236] Brown D M and Todd A R 1955 *The Nucleic Acids* vol 1, ed E Chargaff and J N Davidson (New York: Academic)
[237] Avery O T, Macleod C and McCarty M 1944 *J. Exp. Med.* **79** 137
[238] Astbury W T and Bell F O 1938 *Cold Spring Harbor Symp. Quant. Biol.* **6** 109
[239] Watson J D and Crick F H C 1953 *Nature* **171** 737 and 964
[240] Watson J D 1968 *The Double Helix* (London: Wiedenfeld and Nicolson)
[241] Chargaff E 1950 *Experimentia* **6** 201
[242] Kennard O and Hunter W N 1991 *Angew. Chem. Int. Ed. Engl.* **30** 1254
[243] Hunter W N, Leonard G A and Brown T 1993 *Chem. Britain* **29** 484
[244] Bernal J D and Fankuchen I 1937 *Nature* **139** 923
[245] Watson J D 1954 *Biochim. Biophys. Acta* **13** 10
[246] Franklin R E and Holmes K C 1958 *Acta Crystallogr.* **11** 213
[247] Klug A and Caspar D L D 1960 *Adv. Virus Res.* **7** 225
[248] Barrett A N et al 1972 *Cold Spring Harbor Symp. Quant. Biol.* **36** 433
Stubbs G, Warren S and Holmes K 1977 *Nature* **267** 216
[249] Crowfoot D M and Schmidt G M J 1945 *Nature* **155** 504
[250] Crick F H C and Watson J D 1956 *Nature* **177** 473
[251] Caspar D L D 1956 *Nature* **177** 475
[252] Klug A, Finch J T and Franklin R E 1957 *Biochim. Biophys. Acta* **25** 242
[253] Finch J T and Klug A 1959 *Nature* **183** 1709
[254] Horne R W, Brenner S, Waterson A P and Wiedy P 1959 *J. Mol. Biol.* **1** 84
[255] Caspar D L D and Klug A 1962 *Cold Spring Harbor Symp. Quant. Biol.* **27** 1
[256] Harrison S C 1971 *Cold Spring Harbor Symp. Quant. Biol.* **36** 495
Harrison S C, Olson A J, Shutt C E, Winkler F K and Bricogine G 1978 *Nature* **276** 368
[257] Arnold E, Vriend G, Luo M, Griffith J P, Kamer G, Erickson J W, Johnson J E and Rossman M G 1987 *Acta Crystallogr.* A **43** 346
[258] Kamminga H 1989 *Acta Crystallogr.* A **45** 58
[259] Lima de Faria J (ed) 1990 *Historical Atlas of Crystallography* (Dordrecht: Kluwer)
[260] McLachlan D and Glusker J P (ed) 1983 *Crystallography in North America* (New York: American Crystallographic Association)
[261] Hoddesdon L, Braun E, Teichmann J and Weart S 1992 *Out of the Crystal Maze: Chapters from the History of Solid State Physics* (Oxford: Oxford University Press)
[262] Allen F H, Davies J E, Galloy J J, Johnson O, Kennard O, Macrae C F, Mitchell E M, Mitchell G F, Smith J M and Watson D G 1991 *J. Chem. Inf. Comput. Sci.* **31** 187

Chapter 7

THERMODYNAMICS AND STATISTICAL MECHANICS (IN EQUILIBRIUM)

Cyril Domb

7.1. Introduction—nineteenth century background [1]

The first two laws of thermodynamics were triumphs of nineteenth century science. James Prescott Joule [2] found that, however work was dissipated, a given amount of work resulted in the development of the same quantity of heat. He therefore concluded that heat is a form of energy, and the first law of thermodynamics, which embodies this result, is part of the wide general principle of the conservation of energy.

The second law, dealing with the limitations which arise when attempts are made to convert heat into work, can be traced to a remarkable memoir by Sadi Carnot [3] which derived results of permanent significance even though the contemporary theory of the nature of heat was totally incorrect. Independent formulations of the second law were given about 1850 by Rudolf Clausius and William Thomson (for details see Brush [1] section 14.3). Clausius introduced the concept of entropy [4], but its nature remained obscure until it received a statistical-mechanical interpretation.

Thermodynamics soon became established as 'a science with secure foundations, clear definitions, and distinct boundaries' [5], especially in the hands of Josiah Willard Gibbs [6] who applied it to an amazing variety of fundamental problems in what must be one of the most comprehensive scientific memoirs ever published.

In 1859 Gustav Kirchhoff [7] introduced a new question for discussion, the properties of radiation in a cavity whose walls are kept at a constant temperature T. The immense consequences of this suggestion,

Twentieth Century Physics

Josiah Willard Gibbs

(American, 1839–1903)

Willard Gibbs, for most of his life, was more appreciated by Europeans, especially by Clerk Maxwell, than by his American countrymen. As unpaid professor of mathematical physics at Yale he systematized the application of thermodynamics to a wide range of problems in physics and chemistry. His long and difficult papers came in due course, years after his death, to be prescribed as essential reading for American students of thermodynamics. The 'phase rule', a simple and basic deduction from his theory of heterogeneous systems, is known by his name and has played a central role in the study of chemical reactions and of the phase diagrams of alloys. But it is only one of many contributions to chemistry.

Like everything else he did, his last work was slow in achieving recognition. Starting with Boltzmann's ideas he constructed a general theory of statistical mechanics which came to be seen as a foundation for all work in this field during the twentieth century. It is therefore appropriate to recognize him as a great pioneer of modern physics.

He lived quietly with his sisters, having inherited enough money for his and their needs, and never married. In 1901 the Royal Society gave him the Copley medal, their highest award.

---

which ultimately led Max Planck to propose the quantum hypothesis in 1900, are outlined in Chapter 1.

[See also p 22]

Insofar as thermodynamics was concerned to formulate the necessary relationships between measured quantities, without regard to any detailed explanation, the kinetic theory of gases was a separate field of study until its development prompted the enquiry, chiefly by Maxwell and Boltzmann, into the molecular origins of the second law. It is usual to attribute the foundation of the modern kinetic theory of gases to two papers by Clausius [8], especially the second which was concerned with molecular collisions, with the new parameter of *mean free path* characterizing the average distance travelled by molecules between successive collisions.

Surprisingly, Clausius assumed that the molecules all had equal velocities, a defect brilliantly remedied by Maxwell, who later emphasized [9] that the concepts of probability and statistics are those appropriate to a model involving molecular collisions. For this reason it is right to credit him with the transition from the *kinetic theory of gases* to *statistical mechanics*, although the latter term was first coined by Gibbs. Maxwell derived a fundamental result which was known afterwards as the theorem of equipartition of energy. This states that in the equilibrium state of a gas consisting of molecules of different masses and shapes, the mean kinetic energy of each translational and rotational degree of freedom will be the same.

Ludwig Boltzmann entered the field a few years after Maxwell, and from 1868 onwards he dealt with the statistical mechanics of an ideal gas comprising a large number, N, of molecules whose interactions may be neglected, but which may be in a gravitational or other potential field [10]. The total energy E of the system is constant and can be distributed in a large number of different ways among the N molecules. A microstate of gas is defined by specifying the energies of the individual molecules, and Boltzmann made an important new basic assumption that all microstates (which he called *complexions*) have the same *a priori* probability. (These terms, and some of the details of the following argument, will be explained in section 7.2.3.) A state in which a specific molecule has energy ε_r can be called a macrostate. The probability P_r of the macrostate is then proportional to the number of microstates in which the remaining energy $E - \varepsilon_r$ is distributed among the other $(N-1)$ molecules. This is a very large number whose form Boltzmann was able to establish in the relation which carries his name

$$P_r \propto \exp(-\varepsilon_r/kT). \qquad (1)$$

His other contributions to equilibrium statistical mechanics were his derivation of the second law of thermodynamics from the principles of mechanics, his clarification of the statistical nature of the second law (for which Maxwell's demon [11] provided conceptual help, see section 7.6), and the statistical identification of the entropy in the famous relation

See also p 575

$$S = k \ln W. \qquad (2)$$

Here S is the entropy, and W the number of complexions of the macrostate. Although relation (2) is engraved on Boltzmann's tombstone, it was actually first enunciated by Max Planck [12] (all the relevant basic information was indeed provided by Boltzmann). Planck called k Boltzmann's constant; it is written k_B when the letter k is needed for other purposes.

It was this background which prepared Planck for his crucial contribution to the theory of black-body radiation at the dawn of the twentieth century. He was deeply influenced by Clausius, and most of his research during the following decade was devoted to different aspects of thermodynamics. In 1889 he moved to Berlin which was then the centre of theoretical and experimental activity on black-body radiation as described in Chapter 1. The steps by which Planck moved tentatively towards his revolutionary quantum hypothesis are explained in Kuhn's detailed account [13].

We turn finally to the challenge to Boltzmann's derivation of the second law of thermodynamics from the principles of mechanics, which took place in the last decade of the nineteenth century and threatened to block the progress of statistical mechanics. The differential equations of Newtonian mechanics are reversible in time, and Henri Poincaré proved [14] that if a mechanical system is in a given state, it will return an indefinite number of times to a point infinitesimally close to this state. From this result Ernst Zermelo [15] drew the conclusion that a process irreversible in time, like the second law, is impossible in a purely mechanical system. Boltzmann replied [16] that indeed Poincaré's result is correct if *one waits long enough*, but the recurrence time is so long that there is no possibility of ever observing it. The controversy continued for some time, but eventually Boltzmann was completely vindicated. The way was clear for Gibbs to lay the foundation for the modern development of statistical mechanics [17] (section 7.3.1)

See also p 589

7.2. Impact of the quantum theory

Many of the advances in statistical mechanics, particularly the incorporation of quantum theory into the scheme, were made with the intention of understanding specific mysteries, rather than as generalized ideas. They are therefore to be found in appropriate chapters of these volumes and are only epitomized here, except for matters which happen to have been overlooked or cursorily dealt with elsewhere.

7.2.1. Black-body radiation

Planck was inspired to consider, for the first time, the relation between energy U and entropy S in black-body radiation, and proposed [18] the simple differential equation, $\partial^2 S/\partial U^2 = -\alpha/U$, α being a constant. He was led to this equation by its consistency with Wien's law

$$U(\lambda, T) = b\lambda^{-5} \exp(-a/\lambda T)$$

for the dependence on temperature of energy density at wavelength λ. But when improved measurements showed weaknesses in Wien's law, Planck suggested [19] that $\partial^2 S/\partial U^2 = -\alpha/U(\beta + U)$ and its consequence

$$U(\lambda, T) = d\lambda^{-5} / \left[\exp\left(\frac{D}{\lambda T}\right) - 1\right]$$

Thermodynamics and Statistical Mechanics (In Equilibrium)

which fitted the data extremely well, and which in due course he rewrote [20] in its traditional form, in terms of frequency v instead of wavelength

$$U(v, T) = \frac{8\pi v^2}{c^3} \frac{hv}{[\exp(hv/kT) - 1]}.$$

The reasoning that led to this relation is involved, and is usually presented in the form given by Einstein who made clear that it depends on assuming that oscillators can possess energy only in multiples of hv—an idea that Jeans dismissed as a mathematical artefact. The question is discussed more thoroughly in Chapter 3.

See also p 148

7.2.2. Vibrational specific heat of solids

This is a central theme of Chapter 12 which follows the development of the theory from Einstein's introduction of quantum ideas through Debye's wonderful simplification of an immensely complex problem—the enumeration of possible vibrational modes of real crystals—to the more detailed analyses of Born and von Kármán and their successors. The essential piece of quantum statistics is an extension of the original Planck–Einstein idea, that an oscillator has energy only in multiples of hv (more precisely, with an extra $\frac{1}{2}hv$ of zero-point energy which plays no significant part in the specific heat).

7.2.3. Classical and quantum statistics

The dynamics of a classical system, however complicated, can be described in terms of the position (q_i) and momentum (p_i) of each constituent. If they amount to r in number, there are $2r$ separate coordinates and one may define the energy as a function of $2r$ variables. Geometrically considered, and purely as an abstract conception, a single point in a phase space of $2r$ dimensions serves to define the dynamical behaviour of the total system, and as the system evolves the point which represents it moves in phase space. As a basis for his treatment Boltzmann [21] made use of an older theorem of Liouville, which states that any region of phase space occupied by a group of representative points does not change its volume as the system evolves. The motion in phase space thus corresponds to that of an incompressible fluid. If this fluid is at any one time distributed uniformly over the whole of phase space, it will maintain this distribution ever after. If the energy of the system as a whole is fixed at some value E, the only parts of phase space that may be visited are those whose coordinates confer the correct total energy. Boltzmann assumed a uniform distribution over the whole of accessible phase space (*ergodic surface*), and that average properties of the system are to be obtained by giving equal weight to equal volumes.

So far we have considered the $2r$-dimensional phase space, but there is a more restricted phase space which describes only one of all the constituents of the system. If one is dealing with a single particle, its phase space has six dimensions (three of position, three of momentum). Boltzmann assumed he could divide this atomic phase space into small regions of typical volume ω_r, in which each particle would have energy ε_r. Then if one defines the number n_r of particles occupying the rth volume, the system as a whole occupies a volume $\omega_1^{n_1} \omega_2^{n_2} \ldots \omega_r^{n_r}$ in the system phase space, and the product measures the fraction of time the system spends in this precise condition.

Consider now a macrostate with n_1 systems in cell 1, n_2 systems in cell 2, ..., n_r systems in cell r. Boltzmann defined the number of complexions as the number of different arrangements of the atomic systems which give rise to this macrostate. Elementary combinatorial analysis shows this to be

$$\frac{N!}{n_1! n_2! \ldots n_r!}. \tag{3}$$

Following Boltzmann and Gibbs [22], the probability weight assigned this macrostate is

$$W = \frac{N!}{n_1! n_2! \ldots n_r!} \omega_1^{n_1} \omega_2^{n_2} \ldots \omega_r^{n_r} \tag{4}$$

where the n_i satisfy the conditions

$$n_1 + n_2 + \ldots + n_r = N \tag{5}$$

$$n_1 \varepsilon_1 + n_2 \varepsilon_2 + \ldots + n_r \varepsilon_r = E. \tag{6}$$

To derive the values of $n_1, n_2, \ldots, n_r$ in the equilibrium state, identified by Boltzmann as the most probable state, W is maximized subject to the restrictive conditions (5) and (6), to give

$$\langle n_i \rangle \propto \omega_i \exp(-\varepsilon_i / kT) \tag{7}$$

where $\langle \rangle$ denotes the most probable value. In the final step the size of the cells was allowed to tend to zero and any series in the analysis were replaced by integrals.

The quantum theory postulated discrete energy levels ε_r, and the whole of the above analysis could be used without reference to continuous phase space, and without replacing sums by integrals. Moreover the ω_i now had a natural counterpart in the degeneracy, g_i, of the ith energy level. In practice the values of g_i are small, but it is convenient for the moment to group together a substantial number of levels in a small range about ε_i, so that g_i is a large number to which asymptotic formulae can be applied. The need for this device will be removed in section 7.3.1 when the Gibbs *canonical ensemble* is introduced.

Thermodynamics and Statistical Mechanics (In Equilibrium)

In order to derive thermodynamic behaviour from (7) it is particularly convenient to introduce the *partition function*

$$Z = \sum_{j=1}^{r} g_i \exp\left(-\frac{\varepsilon_i}{kT}\right) \tag{8}$$

which is simply related to the Helmholtz free energy F

$$F = -NkT \ln Z. \tag{9}$$

All the properties of the gas of atomic systems can be derived from (9). The sum (8) was introduced by Planck [23] in 1924 following the corresponding integral used by Gibbs in his classical treatment [17]. Planck used the letter Z because of the German term *Zustandsumme*—sum over states—and it has subsequently become standard notation. The above treatment is termed *quantum theory with classical statistics*, and is concerned with distinguishable systems, the interchange of which gives rise to independent states; an alternative description is Maxwell–Boltzmann statistics.

In June 1924 Einstein received a letter from a young Bengali, Satyendra Nath Bose, whose new derivation of Planck's law had been rejected. Einstein was impressed by Bose's paper, translated the paper personally into German, and submitted it with a strong positive recommendation. Publication of the paper [24] transformed Bose's career. His derivation of Planck's law was based on a particle picture, in which the number of particles is not conserved. Bose was led to Planck's formula, even though he does not seem to have been aware that he was using anything different from Boltzmann's statistics [25].

See also p 167

The paper by Bose was followed by two papers by Einstein [26], extending Bose's treatment to material particles whose number is conserved. He clarified the nature of the statistics which had been used for the *identical particles*, showing that the Boltzmann combinatorial factor (4) (with g_i instead of ω_i) is replaced by

$$W = \prod_{i=1}^{r} \frac{(g_i + n_i - 1)!}{n_i!(g_i - 1)!}. \tag{10}$$

Then one finds

$$\langle n_i \rangle = \frac{g_i}{\lambda \exp(\varepsilon_i/kT) - 1} \tag{11}$$

for conserved materials, whilst $\lambda = 1$ for non-conserved particles of light. Bose used a simple and direct argument to evaluate g_i, and deduced that (11) was identical to Planck's law.

The combinatorial formula (10) applies to indistinguishable particles with no limit on the number that can occupy any energy level; such particles are said to satisfy Bose–Einstein statistics. In 1925 Wolfgang Pauli [27] introduced his exclusion principle, which states that no two electrons can occupy the same quantum energy level. Enrico Fermi [28] and Paul Dirac [29] realized independently that this hypothesis would lead to different statistics for electrons, the combinatorial formula (10) being replaced by

$$W = \prod_{i=1}^{r} \frac{g_i!}{n_i!(g_i - n_i)!} \qquad (12)$$

by use of which (11) is replaced by

$$\langle n_i \rangle = \frac{g_i}{\lambda \exp(\varepsilon_i/kT) + 1}. \qquad (13)$$

Equation (12) applies to indistinguishable particles, no two of which can occupy the same level; such particles are said to satisfy Fermi–Dirac statistics.

Quite remarkably, Gibbs had anticipated the need to discriminate between gases consisting of identical particles 'If two phases differ only in that certain entirely similar particles have changed places with one another are they to be regarded as identical or different phases? If the particles are regarded as indistinguishable, it seems in accordance with the spirit of the statistical method to regard the phases as identical' [17]. He went on to point out that if two identical fluid masses are in adjacent chambers and the dividing barrier is removed, there will be no change of entropy, whereas if the fluids are different there will be a change of entropy. The need to distinguish the two cases came to be known as the *Gibbs Paradox*, and was clarified with the introduction of quantum statistics. The theorems of classical mechanics, used by Boltzmann, remain unchanged when the particles are assumed to be identical. In quantum mechanics, on the other hand, there is an absolute distinction between identical and non-identical particles, however similar to one another the latter may be. There is indeed a change of entropy when classical statistics (3) are used; there is no change of entropy when quantum statistics (10) or (12) are used.

7.2.4. Specific heats of gases

The energy levels ε of the molecules of a gas can, to a good approximation, be divided into independent contributions corresponding to translation, vibration, rotation and electronic modes. The partition function Z then reduces to a product, and the free energy to a sum of independent contributions from each degree of freedom.

The equipartition theorem of the classical theory (Maxwell and Boltzmann) assigned a contribution to the energy of $\frac{3}{2}NkT$ for the translational modes, NkT for each vibrational mode, and $\frac{1}{2}NkT$ for each rotational degree of freedom. The specific heat at constant volume, C_V, of a gas of monatomic molecules should be $\frac{3}{2}Nk$, for a gas of diatomic molecules it should be $\frac{7}{2}Nk$, and it should have a much higher value for a gas of polyatomic molecules. Since $C_p - C_v = Nk$, it follows that the corresponding values of the ratio $\gamma = C_p/C_v$ should be $5/3 = 1.67$, $9/7 = 1.29$ and very close to unity.

Only the first result agreed with experiment; experimental results were about 1.40 for diatomic and linear polyatomic molecules and about 1.33 for non-linear polyatomic molecules. Maxwell was very concerned about this discrepancy, which he spoke of as 'the greatest difficulty yet considered by the molecular theory' [30].

Quantum theory removed this discrepancy by allowing for *frozen* modes which make no significant contribution to the specific heat. For example, the energy levels of a vibrational mode of frequency v, are $\frac{1}{2}hv$, $\frac{3}{2}hv$, ... and the classical mean energy kT per vibrator is only attained if all levels can be thermally excited.

For most molecular systems, however, hv is so large that at ordinary temperatures most vibrators remain in the ground state and contribute nothing to the specific heat. Only at higher temperatures do the vibrational modes begin to contribute to the specific heat. By contrast the level separation for rotational modes is usually so small that the classical equipartition theorem remains valid. The energy levels for translation modes are even closer, and classical theory is always valid, but the characteristic temperature for electronic levels is usually high, and their contribution is therefore small. Quantum theory not only removed the general discrepancy, but provided a mechanism for calculating the variation of specific heat with temperature for individual gases if the appropriate numerical physical constants were used.

However, one anomaly arose in relation to the specific heats of gaseous hydrogen, whose rotational energy levels are so far apart that rotation is severely restricted at low temperatures while the gas has still not condensed; on warming up one expects the specific heat to show a progression to the classical form, as in figure 7.1 (curve 1), contrary to experiment (curve 3).

Molecular spectroscopy gave a clear indication that nuclear symmetry must be taken into account [31], and a distinction made between states with symmetrical (S) wavefunctions (para-hydrogen) and anti-symmetrical (A) wavefunctions (ortho-hydrogen). S and A states alternate, the latter being three times as numerous as the former. When the calculation is corrected for this effect, the disagreement with experiment is even worse than before! (curve 2).

The correct interpretation was advanced by Dennison [32]. Spectroscopic evidence shows virtually no transitions between S and A wavefunctions; gaseous hydrogen should therefore be treated as a *mixture* of independent gases, ortho- and para-hydrogen in the ratio 3:1, not as a

See also p 1048

Twentieth Century Physics

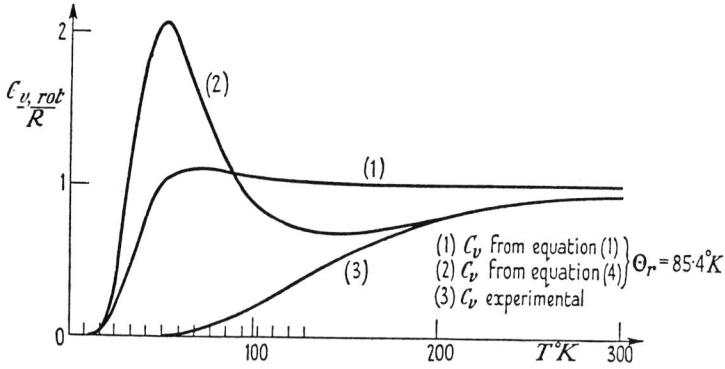

Figure 7.1. *Metastable equilibrium in gaseous hydrogen. (1) c_v, assuming equilibrium, taking no account of nuclear symmetry. (2) c_v, assuming equilibrium taking account of nuclear symmetry. (3) Experiment. The curve is well fitted by taking account of nuclear symmetry but assuming a mixture of independent gases.*

system in thermal equilibrium.

The specific heat then gives excellent agreement with the experimental curve 3, and the system provides an example of *metastable equilibrium*. To achieve thermal equilibrium and curve 2 would require spending very much longer than the experimental time available.

There is no outstanding disagreement between theory and experiment in relation to the specific heats of gases.

7.2.5. Bose–Einstein condensation

For conserved material particles Einstein had derived equation (11); the parameter λ is determined by the number of particles in the system, but cannot be less than 1. Einstein observed that if the total number of particles is found by integrating over a presumed continuous distribution of states, rather than summing over the true states, there is a maximum number N_c of particles that can be accommodated. What happens if N is greater than N_c? The process of integration is impermissible.

What does this mean physically? Einstein's answer was [33]

> I maintain that, in this case, a number of molecules steadily growing with increasing density goes over the first quantum state (which has zero kinetic energy) while the remaining molecules distribute themselves accordingly to the parameter value $\lambda = 1$... A separation is effected; one part condenses, the rest remains a saturated ideal gas.

This exact statistical derivation of a phase transition is now called the Bose–Einstein condensation. Questions were raised about the validity of Einstein's derivation [34], but eventually justifications [35] were forthcoming and, as noted in Chapter 11, F London proposed the

See also
p 927

Thermodynamics and Statistical Mechanics (In Equilibrium)

condensation as an explanation of the transition at 2.2 K of liquid helium into a superfluid.

7.2.6. *Applications of Fermi–Dirac statistics*

At absolute zero, systems obeying Fermi–Dirac statistics take up the lowest available state, while in Bose–Einstein systems all particles occupy the lowest single-particle state; this is not possible for Fermi–Dirac systems since only one particle can occupy each state. They therefore fill up all lowest states, N in total, and the most energetic may possess considerable energy. As a result, there is little thermal excitation until kT approaches the energy of the highest occupied level. The low-temperature distribution is called *degenerate*.

Very shortly after the discovery of Fermi–Dirac statistics, Ralph Fowler [36] published his most original paper in which he suggested that the material in white dwarf stars consists of a degenerate gas. The important astrophysical consequences of this suggestion are discussed in length in Chapter 23.

| See also p 1711 |

Fowler told Milne [37] that he could have kicked himself for not seeing at the same time the application to electrons in metals, which was pointed out by Sommerfeld some two years later, as discussed in Chapter 18. This marks the point at which earlier problems in the physics of metals began to be resolved and the modern quantum theory of conduction began.

| See also p 1288 |

7.3. Formal theoretical developments

7.3.1. *Gibbs ensembles*

Boltzmann's approach described above could be used for assemblies of quasi-independent systems. In 1902 Gibbs introduced the concept of an ensemble [17], a collection of physical bodies following a statistical distribution which is chosen to simulate a particular thermodynamic situation. The properties of the body are calculated by taking a statistical average over the ensemble, and fluctuations above this average can be calculated using standard statistical procedures. It could be anticipated that if the number of constituents, N, of the body becomes large, the fluctuations will decrease, and the averages can then be identified with thermodynamic properties in equilibrium.

Gibbs first took an ensemble of macroscopic assemblies, the number with energy E_n being proportional to $\exp(-\beta E_n)$ and β being related to the temperature T as before. He showed that this ensemble has the important property that if two different ensembles are connected, having the same β, their equilibrium behaviour is unchanged by the connection. Hence, this ensemble simulates a body at constant temperature. Gibbs used classical theory with integral averages, but his ideas could immediately be adapted to the quantum theory, and the average energy could be written

$$\langle E_n \rangle = \frac{\sum_n E_n \exp(\beta E_n)}{\sum_n \exp(-\beta E_n)} = -\frac{\partial}{\partial \beta}(\ln Z_N) \tag{14}$$

where Z_N is the partition function (PF)

$$Z_N = \sum_n \exp(-\beta E_n) \tag{15}$$

$\langle E_n \rangle$ is then identified with the thermodynamic internal energy U.
Similarly the pressure is defined by the average

$$\langle P \rangle = \frac{-\sum_n (\partial E_n / \partial V) \exp(-\beta E_n)}{\sum_n \exp(\beta E_n)} = \frac{1}{\beta} \frac{\partial}{\partial V}(\ln Z). \tag{16}$$

Other forces are defined similarly.

Gibbs showed that by identifying $-kT \ln Z_N$ with the free energy of Helmholtz, $F(T, V, N)$, results were obtained which coincided with thermodynamics. He termed this the canonical ensemble and in principle it provides a mechanism for calculating the equilibrium thermodynamic behaviour of any macroscopic body, solid, liquid, gas, plasma, etc, in terms of the properties of its constituent molecules. The requirement is the knowledge of the different possible energies E_n of the macroscopic body, corresponding to all the different possible arrangements of the constituent molecules.

Boltzmann's treatment corresponds to a constant energy ensemble, and Gibbs noted that his results could be derived from the canonical ensemble by taking a probability distribution which is zero outside the range $(E_n, E_n + dE_n)$ (nowadays this would be described as a δ-function). He called this the *micro-canonical ensemble*; the statistical averages are more difficult to handle, as might have been anticipated in view of the difficulty of applying thermodynamics when the energy is an independent variable rather than the temperature.

In dealing with the thermodynamics of heterogeneous substances, Gibbs had found it useful to introduce the *chemical potential* μ_i of each species i which plays a similar role in controlling chemical equilibrium to that of the temperature in controlling thermal equilibrium. Use of the temperature as an independent thermodynamic variable rather than the energy is paralleled by the use of the chemical potential of a given species as a variable rather than its concentration. Gibbs introduced the *grand canonical ensemble* to simulate a thermodynamic body with constant chemical potentials, the number of assemblies with ν_1 molecules of species 1, ν_2 of species 2, ... ν_r of species r, being proportional to

Thermodynamics and Statistical Mechanics (In Equilibrium)

$\exp(\mu_1 \nu_1 + \mu_2 \nu_2 + \ldots + \mu_r \nu_r - \beta E_n)$. By analogy with (15) Gibbs introduced the grand partition function (GPF)

$$\zeta = \prod_{\nu_1, \nu_2, \ldots, \nu_r, n} \exp(\mu_1 \nu_1 + \mu_2 \nu_2 + \ldots + \mu_r \nu_r - \beta E_n). \tag{17}$$

ζ could readily be identified with an appropriate thermodynamic function. Use of the GPF greatly simplifies the statistical mechanics of heterogeneous assemblies. For example, the formulae (11) and (13) for Bose–Einstein and Fermi–Dirac statistics are derived immediately by using the GPF.

It took several decades for the scope and power of Gibbs' approach to be appreciated. Fowler's treatise on *Statistical Mechanics* [38], which served for many years as the authoritative reference in the field, makes only a scant reference in the introduction to the canonical ensemble and does not use it at all in the rest of the book. A lucid series of lectures by Schrödinger [39] in Dublin in 1944 may well have been responsible for putting Gibbs' ideas at the centre of the map, and once this was achieved the practical consequences of Gibbs' ensemble concept, and the PF and GPF to which it naturally led, formed the basis of statistical-mechanics calculations for the rest of the century.

7.3.2. Einstein's treatment of fluctuations

Einstein was unaware of Boltzmann's formula (2), which related entropy to statistical weight, and he obtained the same result independently, but Einstein's approach to the formula differed significantly from that of Boltzmann and was really complementary to it.

Boltzmann's aim was to establish a theory of matter based on probability and statistics, and formula (2) served as the bridge by which he could get back to thermodynamics. Einstein in his studies on Brownian movement had been concerned with fluctuations and their experimental observation. Rewriting (2) in the form $W = \exp(S/k)$, and making use of the thermodynamic formula expressing the entropy as a function of state, he could calculate the probability of deviations from equilibrium values. He applied this to the calculation of fluctuations in temperature, density, pressure, etc, in a finite system.

See also p 595

Consider a small region of a large reservoir, whose temperature T_0 and pressure P_0 are assumed constant (the subscript 0 always refers to the reservoir); for convenience we fix the number of molecules N in the small region but allow its volume to fluctuate. Fluctuations in the small region of temperature (ΔT), volume (ΔV), entropy (ΔS) and internal energy (ΔU) have a probability w, given by $\exp(\Delta S + \Delta S_0)/k$, if the total system, reservoir + small region, is isolated at constant energy and volume. Thus, $\Delta U + \Delta U_0 = 0$, $\Delta V + \Delta V_0 = 0$. But

$$\Delta S_0 = \left(\frac{\Delta Q}{T_0}\right)_{\text{reservoir}} = \frac{\Delta U_0 + P_0 \Delta V_0}{T_0} = \frac{\Delta U + P_0 \Delta V}{T_0}. \tag{18}$$

Hence

$$w \propto \exp[-\beta(\Delta U - T_0 \Delta S + P_0 \Delta V)]. \tag{19}$$

To determine the fluctuations the exponent in (19) must be expanded as far as second-order terms (The first-order terms vanish since the region is in equilibrium). For the variables ΔT and ΔV, Einstein derived the relation

$$w \propto \exp\left[-\frac{C_v}{2kT^2}(\Delta T)^2 + \frac{1}{2kT}\left(\frac{\partial P}{\partial V}\right)_T (\Delta V)^2\right] \tag{20}$$

and from this he concluded that the fluctuations are Gaussian with $\langle \Delta T^2 \rangle = kT^2/C_v$ and $\langle \Delta V^2 \rangle = kTVK_T$ where K_T is the isothermal compressibility; also the temperature and volume fluctuations are uncorrelated.

When $(\partial P/\partial V)_T$ is zero, as happens at a critical point, the volume fluctuations become large. Details of the behaviour in this case will be discussed in section 7.5.

7.3.3. Mathematical background to the second law: Carathéodory's method

The opening paragraphs of the Introduction (section 7.1) referred briefly to the development of the second law of thermodynamics during the mid-nineteenth century. The derivation followed an exceedingly logical pattern, the key step in the process being the demonstration that it is possible to find a function T with the property that dQ/T is a *perfect differential* i.e. that $\oint dQ/T$ around a reversible cycle is zero.

The arguments of Clausius and Thomson drew on the properties of idealized heat engines, and showed that no engine can be more efficient than that proposed by Carnot. Some mathematical physicists were unhappy about the reference to heat engines, and felt that an alternative development should be possible in which the condition for the existence of an integrating factor for dQ played the central role. This led Carathéodory [40] to develop a treatment based on the alternative hypothesis 'In the neighbourhood of any equilibrium state of a system there are states which are inaccessible by an adiabatic process'. From this hypothesis it was possible to demonstrate the existence of an integrating factor for dQ, and to develop thermodynamics as a mathematical discipline.

Carathéodory also provided a mathematical basis for the concept of temperature as follows. It is an experimental fact that if two bodies are in thermal equilibrium a relationship must exist between their thermodynamic parameters; in addition, if bodies 1 and 2 are in thermal equilibrium, and bodies 2 and 3 are in thermal equilibrium then 1 and 3 are also in thermal equilibrium. These facts are sufficient to establish the existence of an empirical temperature which is the same for all three bodies in equilibrium. This came to be known as the zeroth law of thermodynamics. The same result could be established from first principles using Boltzmann's approach to statistical mechanics.

7.3.4. Method of mean values in statistical mechanics (Darwin–Fowler)
Boltzmann focused attention on the most probable state, and assumed that in equilibrium the contributions of all other states are negligible. This assumption was based on general experience in probability and statistics. In 1922 Darwin and Fowler [41] used an alternative approach in which average values of all quantities of interest are calculated exactly taking all states into account, and these averages are identified with thermodynamic properties in equilibrium. Fluctuations about these averages can also be calculated exactly, which is more satisfying mathematically. The mathematics, however, is hard to explain briefly, and a reference to a specialized text [38] must suffice.

7.4. The third law of thermodynamics [42]

7.4.1. Historical survey
In 1906, soon after Nernst had moved from Göttingen to Berlin he published a paper [43] 'On the calculation of chemical equilibrium from thermal measurements', to which he attached such importance that he ordered 300 reprints, an enormous number at that time. He was concerned with gaseous reactions which attracted much attention at this period because of important industrial and military applications. The concentrations of the resulting gases in equilibrium are governed by the constant K arising in the law of mass action, and van't Hoff [44] in his reaction isochore equation had calculated, from thermodynamics, the variation of K with temperature. But in his formula there remained a constant of integration whose role was important and which defied attempts at calculation.

Nernst felt that the key to the problem lay in assessing the behaviour at the absolute zero. It was difficult to say anything meaningful about the gaseous phases, but he switched to the condensed phases, which had so far not attracted the attention of other chemists. Various arguments led him to suggest that the entropy difference† between the different states of the system (represented by the two sides of the reaction equation) tends to zero as $T \to 0$. It should be noted that there was no reference to the quantum theory, and Nernst assumed that at absolute zero all specific heats have their classical values in accordance with the equipartition theorem.

He next set out to measure the specific heats of solids at low temperatures, and by 1909 he had convinced himself that Einstein's

---

† Nernst did not think in terms of entropy, and couched his statement in terms of the derivatives of energy and free energy.

Walther Hermann Nernst

(German, 1864–1940)

Born into a prominent Prussian family, Nernst studied with leading German physicists before being appointed to W Ostwald in Leipzig. His experimental and theoretical work in electrochemistry made his reputation as a physical chemist and thermodynamicist. On moving to Göttingen he established a large and vigorous research school, and by incorporating Boltzmann's atomic theories into his widely used text *Theoretical Chemistry*, he distanced himself from Ostwald's philosophical preference for treating chemistry without making use of the concept of atoms.

His *New Heat Theorem* (third law of thermodynamics) made the thermodynamics of chemical reactions accessible to experiment, especially through measurements of specific heat at low temperatures. The exploitation of this idea, and the resolution of its problems, occupied many students, including the gifted F E Simon who joined F A Lindemann (later Lord Cherwell) one of Nernst's earlier students, at Oxford and succeeded him as Professor. Thus Nernst's school, sadly depleted by Nazi anti-Jewish measures, flourished in exile. Nernst himself remained in Germany but resisted Nazi policies. Nernst persuaded Ernest Solvay in 1911 to invite leaders of physics, under Lorentz's chairmanship, to a conference on the crisis resulting from Planck's quantum theory. Since then the Solvay conferences have regularly chosen for discussion some of the crucial problems in physics and chemistry, and summoned a limited number of distinguished scientists to share their views.

prediction that they should tend to zero as $T \to 0$ was correct. This led to a new formulation [45] of his theorem as the law of unattainability of absolute zero. It is easy to show that if there are finite differences of entropy between states of a system at $T = 0$, a Carnot cycle can be devised which will enable absolute zero to be reached.

Nernst considered that his theorem was a consequence of the second law. The efficiency of a Carnot cycle between temperatures T_1 and T_2 is equal to $(1 - T_1/T_2)$. He argued that if T_1 can be zero it is possible to convert heat into work with 100% efficiency, contrary to the second law. This proposed cycle of Nernst generated much controversy and criticism, centring around the nature of thermodynamic equilibrium at the absolute zero [46]. Although the question is still open, it is nowadays regarded as semantic; certainly statistical mechanics shows only that dQ/T is a perfect differential, and does not include the third law.

The advent of quantum theory enabled statistical-mechanical calculations to be made of the entropy of gases. The phenomenon of gas degeneracy in quantum statistics, the Bose–Einstein condensation, and the vanishing contribution of electrons to the specific heat provided support for the third law. Already in 1914 Nernst had anticipated such a development [47], but his proposals met with incredulity.

In statistical-mechanical terminology the third law indicated that at absolute zero all systems are in a state of perfect order, but there are substances like glasses and solutions which obviously remain disordered. How did the third law relate to them? It was the work of Franz (later Sir Francis) Simon, both experimental and theoretical, which did much to clarify the situation. He re-formulated the third law as follows 'At absolute zero the entropy differences disappear between all those states of a system between which reversible transitions are possible at least in principle' [48]. In 1930 he expressed the idea in a slightly different form 'the entropy differences disappear between all those states of a system which are in internal thermodynamic equilibrium' [49]. Simon emphasized that a law of thermodynamics could not be expected to apply to a system like glass which was not in equilibrium but in an unstable *frozen-in* state. Despite Simon's elucidation confusion continued for some years, but eventually he succeeded in winning over his critics, and the standard text *Statistical Thermodynamics* by Fowler and Guggenheim [50] contains a much more sympathetic treatment of the third law.

Simon's reformulation subsequently gained wide acceptance and no experimental results were discovered which contravened it. However, it should be pointed out that no theoretical proof of the third law has yet been forthcoming and it must still be regarded as empirical. It is easy to find theoretical [51] models which are not in accordance with the third law; the conclusions must be that such models are idealized and do not represent reality.

In the remainder of this section we discuss briefly the major applications of the third law.

7.4.2. *Phase equilibrium as* $T \to 0$

According to the third law the entropy change ΔS in any phase transition tends to 0 as $T \to 0$. A striking example is provided by the transition

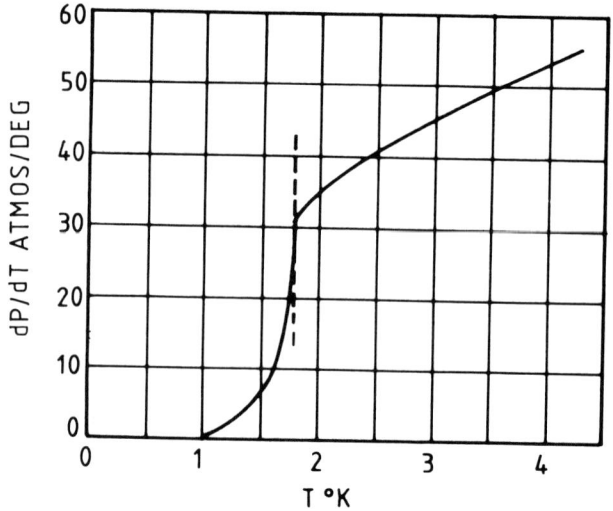

Figure 7.2. *Helium melting curve: dP/dT as a function of T.*

from liquid to solid helium. At atmospheric pressure helium is liquid at absolute zero, but it can solidified by applying a pressure of 25 atmospheres. The liquid, referred to as He II, differs from all other liquids since it is not disordered, and the properties of the solid as well as the mechanism of melting are dominated by zero-point fluctuations [52]. From the Clausius–Clapeyron equation, $dP/dT = \Delta S/\Delta V$, we see that dP/dT should tend to zero, since ΔV remains finite. Figure 7.2 shows measurements of dP/dT by Simon and Swenson [53], which indicate that ΔS tends to zero as T^7.

A second example of the application of the third law is the determination of the equilibrium curve between diamond and graphite. This reaction takes place extremely slowly at normal temperature, and at normal pressure diamond is in the unstable form, owing its existence to the extreme slowness of the transformation. Assuming from the third law that there is no difference of entropy between the two phases at $T = 0$, the free-energy phase diagram can be derived from specific heat measurements down to the lowest temperatures, and the heat of transformation found by normal calorimetric methods. The resulting curve due to Berman and Simon [54] is shown in figure 7.3. It is of particular interest in relation to the production of artificial diamonds from graphite [55] by pressure. In order to check the validity of their derivation Berman and Simon took their measurements to 0.4 K to ensure that there was no specific heat anomaly which could lead to a substantial change in entropy; these measurements, via the third law, have a direct bearing on the reaction at 3000 K.

Thermodynamics and Statistical Mechanics (In Equilibrium)

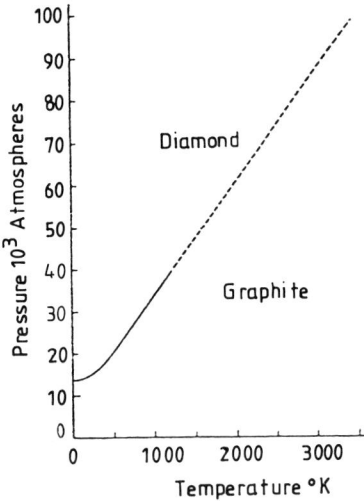

Figure 7.3. *Diamond–graphite equilibrium diagram. Full curve: measured to 1200 K, broken curve: linear extrapolation.*

7.4.3. Calorimetric and statistical estimates of the entropy

Entropy differences between two states of a substance at different temperatures can be derived from calorimetric measurements by use of the second law. Thus, the differences of entropy between the gaseous phase at boiling point and the solid phase at the lowest temperature attained can be calculated from measurements of the specific heats of the liquid and solid phases and of the latent heats of melting and vapourization, together with the latent heats of any transitions which may occur. This is denoted by S_{cal}.

The entropy of the gaseous phase at the boiling point can be calculated directly from statistical mechanics (a small correction can be made for non-ideality of the gas). The molecular data on vibration and rotation levels needed for the calculation can be obtained to high accuracy from spectroscopy; hence this estimate, S_{stat}, is sometimes referred to as a spectroscopic estimate and denoted by S_{spec}. If these two independent estimates of entropy are nearly equal we may legitimately conclude that there is little disorder present in the solid at the lowest temperature of measurement. It is reasonable to extrapolate the solid to $T = 0$ by means of a standard Debye approximation. As typical examples for oxygen [56] O_2, $S_{cal} = 30.85$, $S_{stat} = 30.87$; for methane [57] CH_4, $S_{cal} = 36.53$, $S_{stat} = 36.61$ (the measurements are in entropy units).

However, if we find a significant difference between the two estimates, we can conclude that at the lowest temperatures of measurement there is still significant disorder in the system. Two alternative possibilities exist in relation to further lowering of

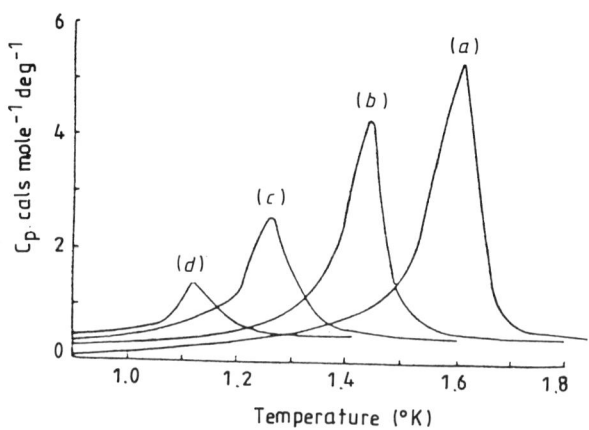

Figure 7.4. λ-anomalies in the specific heat of solid hydrogen: (a) 72.9%, (b) 69.0%, (c) 65.0%, (d) 62.7% ortho-hydrogen.

temperature. The system may become frozen in a non-equilibrium disordered state, or it may stay in equilibrium and manifest a specific-heat anomaly or phase transition which will remove the excess entropy.

As examples of the first type of behaviour we may quote ice H_2O (S_{cal} = 44.29, S_{stat} = 45.10) and carbon monoxide CO (S_{cal} = 37.2, S_{stat} = 38.32) [58]. In the case of ice the residual entropy arises from frozen-in orientations of the water molecule (a detailed model is discussed in section 7.5.12)). In the case of CO the molecules are frozen into an almost random parallel–antiparallel arrangement [59].

A striking example of the second type of behaviour is provided by solid hydrogen in various ortho–para concentrations (section 7.2.4). For pure para-hydrogen [60] there is no residual entropy (S_{cal} = 14.83, S_{stat} = 14.76). This is because para-hydrogen has no net nuclear spin, and no spin entropy; it consists solely of one nuclear species and has no entropy of mixing. However, ortho-hydrogen has a nuclear spin and the lowest energy state is triply degenerate. Hence it is not surprising that for a normal 3:1 mixture of ortho–para hydrogen the statistical entropy exceeds that of pure para-hydrogen by 4.29, and there is a substantial residual entropy at 2 K. The specific-heat anomaly which results when measurements are extended to lower temperatures is shown in figure 7.4 for a variety of ortho–para concentrations. In fact this anomaly removes the threefold degeneracy arising from the rotations but not from the nuclear magnetic moment.

7.4.4. Attaining very low temperatures

Any source of disorder which can be removed adiabatically can be used to cool a system to lower temperatures. The fact that some paramagnetic salts exhibit a susceptibility which follows Curie's law even down to

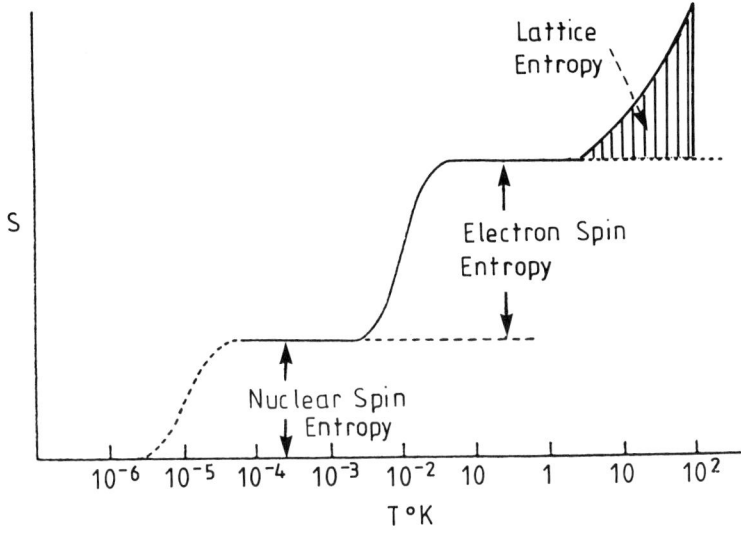

Figure 7.5. *The entropy of a typical paramagnetic salt.*

temperatures as low as 1 K indicates the presence of an electronic spin system which is still fully disordered and has an entropy corresponding to the number of positions that the magnetic ions can take up. By applying a magnetic field sufficiently strong to line up the magnetic ions, and then demagnetizing, temperatures of the order of 10^{-2} K can be attained. In general, if the interaction between spins or between the spin and lattice which ultimately removes the degeneracy has energy ε, the temperature that can be achieved is something like $\sim \varepsilon/k$.

The interaction energies for nuclear spin systems are much smaller than for paramagnetic spin systems, and they order only in the region 10^{-3} to 10^{-6} K. Much higher magnetic fields are needed to align them and great practical difficulties arise because of poor heat transfer at these temperatures. However, these difficulties have been overcome [61] and temperatures as low as 0.3×10^{-9} K have now been attained [62].

See also p 1175

Figure 7.5 shows how the entropy of a paramagnetic salt whose nuclei have a magnetic moment varies with temperature. It also demonstrates that the concept of absolute entropy is not helpful and that it is better to deal separately with the entropy arising from each particular disordering mechanism. At temperatures where significant entropy is distributed by magnetic ions, there is little point in paying attention to the disorder due to nuclear spins. We can choose a zero point of entropy which neglects this latter source of disorder.

7.4.5. Negative temperatures

Experiments on the magnetic resonance of the nuclei of certain solids have shown that it is sometimes possible to induce states of a system which appear to behave, both statistically and thermodynamically, as if they were at negative temperatures. Lithium has a nuclear spin of $\frac{3}{2}$, and if a crystal of lithium fluoride is placed in a magnetic field, its ground state is split into four spin states whose population follows a Boltzmann distribution $n_i \propto \exp(-\beta \varepsilon_i)$, the states of higher energy being less populated. However, the nuclear spins take an unusually long time to come into equilibrium with the lattice. Pound, Ramsey and Purcell [63] exploited this property and reversed the magnetic field in a time sufficiently short for the spins to be unable to follow the field. They remained in their original orientation, but since the energies of the states changed sign, the states of higher energy were now more populated; the system followed a Boltzmann distribution with β corresponding to a negative temperature.

Simon pointed out [64] that such a system does not violate the third law. The negative temperature is achieved by becoming 'hotter than infinity' and not colder than zero, and $T = 0$ is equally unattainable from the negative side. In fact statistical mechanics shows that it is more natural to characterize temperature by means of the variable $\beta = 1/kT$; there is nothing unusual about $\beta = 0$, but $\beta = \pm\infty$ is unattainable.

7.5. Phase transitions and critical phenomena

7.5.1. Introduction

The thermodynamic properties of matter in equilibrium can be considered to fall into two groups—those which vary smoothly and those which have sharp discontinuities. As examples of the first group we may cite the properties of ideal or nearly ideal gases (energy, entropy, specific heat, equation of state), of ideal or nearly ideal solids, of ideal or nearly ideal mixtures of gases or solids; paramagnetism and diamagnetism; and properties of electrons and phonons in normal metals. The second group is usually associated with phase transitions of various types; liquid–vapour equilibrium and the critical point, the melting of solids, phase separation in liquid or solid mixtures or solutions, order–disorder transitions in alloys, ferromagnetism, antiferromagnetism, superconductivity and λ-point anomalies (e.g. liquid helium).

Standard statistical mechanics can handle the first group with relative ease; for example, the treatment of ideal systems when the interactions between gas molecules or between phonons are ignored. Equilibrium in the first-order phase transition, between the crystal and vapour phases, and the form of the vapour pressure curve can then be determined by thermodynamics.

For a slightly non-ideal gas, interaction between molecules can be taken into account by perturbation theory; a chosen thermodynamic property is expressed as a series of ascending powers of a parameter which measures the strength of the interaction. As long as only a finite number of terms are considered the continuity of the thermodynamic properties cannot be destroyed. Discontinuous behaviour can be introduced only by taking the perturbation series to infinity. In fact the problem to be tackled in dealing with phase transitions is the *strong interaction* problem in which the interactions can no longer be treated as a small perturbation, but play a dominant role in the calculations and in the resulting physical properties.

We shall be largely concerned with behaviour near critical points of fluids, magnets and solutions, and with various types of λ-point transition. The second half of the twentieth century saw great progress in the theoretical understanding of how this behaviour depends on the nature of the intermolecular interactions, e.g. their symmetry and range. Some of the mathematics involved is specialized and sophisticated, but I shall try to convey the results and their physical significance without entering into mathematical details.

7.5.2. Liquid–gas critical point

The relation between gases and liquids was the subject of much attention during the first half of the nineteenth century. It was realized quite early that the liquid phase ceased to exist above a certain temperature, but the general assumption was that it disappeared into the gaseous phase. Thus Faraday talked of the 'disliquifying point', while Mendeleev used the term 'absolute boiling temperature' for the point at which the latent heat of evaporation becomes zero. It was Thomas Andrews [65] in his Bakerian Lecture of 1869 who established the true relationship between the liquid and gaseous phases. The term *critical point* was used for the first time in his lecture, and, in addition to a description of careful and accurate experiments on carbon dioxide, he focused attention on the 'close and intimate relations which subsist between the gaseous and liquid states of matter'; the two phases merged at the critical point into one fluid phase, 'but if any one should ask whether it is now in the gaseous or liquid state, the question does not, I believe, admit of a positive reply'. He emphasized this feature in the title of his lecture *On the continuity of the gaseous and liquid states of matter*, and pointed out how it was possible, by suitable choice of path, to pass from the liquid to the gaseous phase without any discontinuity.

Only four years elapsed before van der Waals [66] used the newly developing ideas on kinetic theory of gases to give a plausible theoretical explanation of Andrews' experimental data. He assumed that a gas is made up of molecules with a hard core and a mutual attraction whose range is long compared with the mean free path. The attractive forces give rise to a negative *internal pressure*, which he calculated as $-a/V^2$ by using the virial theorem introduced by Clausius (1870) only a few years before [67]. For the hard core he made the simplest assumption that the

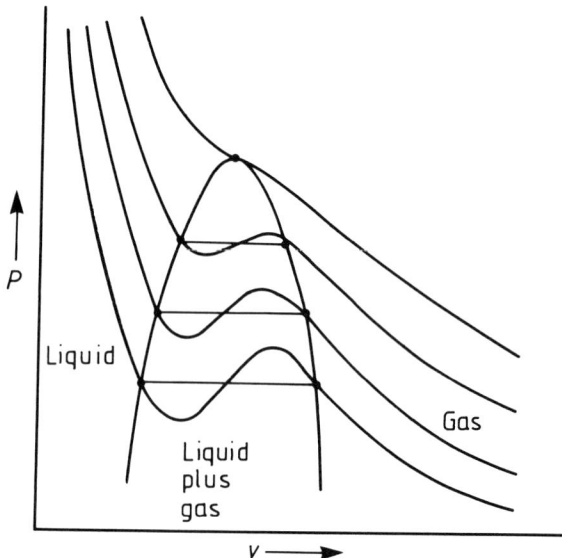

Figure 7.6. *Maxwell's equal area construction which supplemented van der Waals' equation, and enabled the equilibrium state (horizontal lines) to be determined.*

available volume is reduced from V to $V - b$. Hence the equation he put forward was

$$P = P_{\text{internal}} + \frac{RT}{(V-b)} = -\frac{a}{V^2} + \frac{RT}{(V-b)}. \qquad (21)$$

Both Maxwell and Boltzmann were impressed by van der Waals' work. They were reasonably happy with the attractive a/V^2 term, but realized that the repulsive term represented a crude approximation. In a lecture to the Chemical Society in 1875 Maxwell [68] introduced his famous 'equal area' construction, with the help of which it was possible to secure general agreement with Andrews' experimental measurements (figure 7.6).

The algebraic form of the critical isotherm represents the first example of what we shall later call *classical critical exponents*. It was widely assumed that they accurately represented experimental critical behaviour, and van der Waals' theory was not seriously challenged until the 1930s. An analysis of experimental data by Verschaeffelt [69] in 1900, which claimed to find disagreement with experiment, was ignored and took another 45 years to be vindicated.

The new concept of internal pressure which van der Waals introduced was to bear fruit 30 years later in a completely different area of physics.

7.5.3. Curie point of a ferromagnet

The fact that a magnet loses its magnetic power at high temperatures was noted by Gilbert [70] in his famous treatise *De Magnete* in 1600.

Thermodynamics and Statistical Mechanics (In Equilibrium)

More detailed quantitative investigations in the nineteenth century led to Hopkinson's introduction [71] in 1889 of the term *critical temperature* for the temperature at which the magnetism disappears. But the definitive paper for magnets, comparable to that of Andrews for fluids, was written by Pierre Curie [72] in 1895 before he started his more famous investigations of radioactivity.

One of the most interesting new ideas put forward in this paper is the analogy between magnets and fluids. Taking pressure P as the analogue of magnetic field, and density ρ as the analogue of magnetization, Curie points out the close similarity between the P–ρ and M–H isothermals. The paramagnetic state at high temperatures corresponds to the gaseous phase, and the ferromagnetic state at low temperatures to the liquid phase. Curie poses the question whether there exists a precisely defined critical point with associated critical constants for a ferromagnet analogous to a fluid.

It was this analogy which led Pierre Weiss [73] in 1907 to postulate his *molecular field* hypothesis, which simulated the mutual interactions between the molecules by a uniform field nM proportional to the magnetization, M, in the same direction. He states 'One may give to nM the name *internal field* to mark the analogy with the internal pressure of van der Waals'.

See also p 1117

For the magnetic susceptibility of a gas Curie had discovered experimentally the inverse temperature dependence, $\chi \propto 1/T$, and Paul Langevin [74] had used the Boltzmann relation to explain this result theoretically. For an ideal gas of molecules, each having magnetic moment m, Langevin derived the magnetic equation of state (analogous to $PV = RT$ for an ideal gas)

$$M = L\left(\frac{mH}{kT}\right) \tag{22}$$

where $L(x)$ is the Langevin function, $\coth x - 1/x$.

For a ferromagnet Pierre Weiss put forward the simple modification

$$M = L\left(\frac{m(H + H_{\text{int}})}{kT}\right) \quad (H_{\text{int}} = nM) \tag{23}$$

and this led to far-reaching conclusions. There is indeed a sharply defined critical point analogous to that for a fluid; Weiss and Kamerlingh Onnes [75] later termed this the *Curie point* in memory of Curie who had been killed in a street accident in 1906. Below the Curie temperature there is a non-zero spontaneous magnetization; in the paramagnetic state above the Curie temperature, T_c, Curie's inverse temperature relation for the susceptibility is modified to $\chi \propto 1/(T - T_c)$, and this is usually called the Curie–Weiss law.

Twentieth Century Physics

Near the Curie temperature the spontaneous magnetization falls to zero with a parabolic law, and the critical $M-H$ isotherm at T_c follows a cubic law in complete analogy with van der Waals.

The magnet-fluid analogy which proved so fruitful seems to have been forgotten for more than 30 years until it was rediscovered in the lattice-gas model by Cernuschi and Eyring [76]. It was fruitful again in subsequent developments after 1944.

7.5.4. Microscopic critical behaviour of fluids: critical opalescence
In the opening remarks of his 1869 Bakerian lecture, Andrews quoted from an earlier communication 'On partially liquefying carbonic acid by pressure alone, and gradually raising at the same time the temperature to 88 °Fahrenheit, the surface of demarcation between the liquid and gas became fainter, lost its curvature, and at last disappeared. The space was then occupied by a homogenous fluid, which exhibited, when the pressure was suddenly diminished or the temperature slightly lowered, a peculiar appearance of moving or flickering striae throughout its entire mass'.

This phenomenon, subsequently termed critical opalescence, attracted the attention of a number of experimentalists in the late nineteenth and early twentieth centuries. Avenarius [77] described the experimental results in great detail noting striking changes of colour and onset of turbidity in carbon disulphide, ether, carbon dichloride and acetone. Others [78] investigated the phenomenon more precisely without referring to the observations of Andrews or Avenarius. It is indeed striking to observe a colourless transparent fluid suddenly becoming opaque and changing colour in a narrow band of temperatures around T_c. As the temperature is lowered, the fluid splits into colourless liquid and gas with a meniscus separating them. The phenomenon was observed in so many fluids that it could reasonably be regarded as universal. Doubts were expressed as to whether the simple van der Waals theory was capable of accounting for the observations.

Smoluchowski [79] and Einstein [80] were the first to identify the source of the opalescent bands. Fluctuations in the density of the fluid give rise to fluctuations in the refractive index, and hence to light scattering. We have already seen how Einstein used the Boltzmann relation to calculate statistical fluctuations of thermodynamic quantities in equilibrium. For the density he derived the formula $\langle(\Delta\rho)^2\rangle = \rho^2 kT K_T/V$ where K_T is the isothermal compressibility

$$K_T = \frac{1}{\rho}\left(\frac{\partial \rho}{\partial P}\right)_T = -\frac{1}{V}\left(\frac{\partial V}{\partial P}\right)_T. \qquad (24)$$

Assuming that a change in density is accompanied by a change of refractive index according to the Clausius–Mossotti law $(\varepsilon - 1)/(\varepsilon + 2) = A\rho$, and that scattering takes place randomly, Einstein came to the conclusion that the scattering of light of wavelength λ should be proportional to K_T/λ^4. Since the van der Waals equation leads to an infinite value of K_T at T_c, this seemed to provide a satisfactory explanation of critical opalescence.

It was remarkably perceptive of Ornstein and Zernike [81] to note an inconsistency in the treatment arising from the assumption that the fluctuations in all elements of volume are independent of one another, and to point out that there must be a correlation between different elements which increases indefinitely in range as the critical point is approached. To treat this correlation they introduced a basic new function, afterwards called the *pair distribution function*, which has played a central role in the theory of liquids ever since. Let $v(\mathrm{d}r)$ be a random variable representing the number of molecules in a volume $\mathrm{d}r$ centred at r. Since $\mathrm{d}r$ is very small, the probability of occupation by more than one particle, being of order $\mathrm{d}r^2$, can be neglected. Hence we can write $\langle v(\mathrm{d}r) \rangle = n_1(r)\mathrm{d}r$, where $n_1(r)$ is the density, which they took to be a constant ρ. Similarly for the correlation between particles at points r_1, r_2

$$\langle v(\mathrm{d}r_1)v(\mathrm{d}r_2) \rangle = n_2(r_1, r_2)\,\mathrm{d}r_1\,\mathrm{d}r_2 \tag{25}$$

and for an homogeneous isotropic fluid $n_2(r_1, r_2)$ is of the form $n_2(r)$ ($r = |r_1 - r_2|$).

Ornstein and Zernike introduced the function $g(r) = n_2(r)/\rho^2$, which approaches 1 at large distances where the correlation becomes negligible. They used the fluctuation relation quoted above to derive a fundamental identity between K_T and $g(r)$, and also obtained an integral equation for $g(r)$ which they were able to solve explicitly.

The basic physical idea behind Ornstein and Zernike's treatment was to differentiate between the direct influence of molecular interactions, which should be short-ranged, represented by a function $f(r)$, and the correlation between densities, represented by $g(r)$ above, which should become long-ranged as the critical temperature is approached. The integral equation relates $g(r)$ and $f(r)$.

The original paper of Ornstein and Zernike makes difficult reading, and certain aspects of their treatment are obscure (the whole subject has been clarified in a review paper by M E Fisher [82]). Nevertheless, their contribution provided great insight into the nature of critical behaviour.

Their calculations suggested that the correlations fall off asymptotically as $\exp(-\kappa r)/r$, where the value of κ can be determined from the van der Waals equation ($\kappa \sim (T - T_c)^{1/2}$). Their detailed conclusions about light scattering differed from those of Einstein; near the critical temperature the Rayleigh ($\sim \lambda^{-4}$) dependence on wavelength, such as is responsible for the blue of the sky, ceases to be valid, and there is a whitening of the scattered light. By temperature T_c the wavelength dependence has become of the form λ^{-2}.

7.5.5. Critical behaviour of binary alloys

The early years of the twentieth century saw the development of x-ray diffraction as a powerful tool in the investigation of crystal structures, and compounds such as NaCl were found to have a regular ordered structure. However, the ionic bond between Na and Cl is so strong that no significant disordering occurs when the temperature is raised; the crystal melts before it disorders.

In 1919 Tammann suggested that similar ordering might occur in metallic alloys, and this was demonstrated experimentally a few years later [83] by the existence of superlattice lines in the x-ray diffraction pattern of copper–gold alloys. But for such systems there were soon indications that significant disordering takes place with increase in temperature, and that this is accompanied by an anomalous specific heat.

The mathematical description of the disordering process is usually associated with the names of Bragg and Williams. In their first paper [84] in 1934 they introduced a parameter S to characterize the *degree of order*, and used Boltzmann's principle, equation (1), to calculate its behaviour as a function of temperature. They found a pattern closely analogous to the Weiss theory of ferromagnetism, with S falling rapidly to zero at a critical temperature T_c, and remaining zero for $T > T_c$. In fact for a binary alloy with equal concentrations of constituents the equation derived for S was

$$S = \tanh(ST_c/T) \tag{26}$$

which is of the same form as (23) with $H = 0$ and $\tanh x$ replacing the Langevin function $L(x)$.

They elaborated their ideas in a second paper [85]. Because of the analogy with ferromagnetism, they called T_c the Curie temperature of the alloy; S was now termed long-distance order to differentiate it from another parameter which had been introduced by Bethe [86] to characterize the *short-range order* which persists above T_c. They also apologized for having ignored the work of other investigators [87] who had been thinking along similar lines.

Shortly before Sir Lawrence Bragg's death, he told me that in 1933 he had given a seminar in Manchester describing qualitatively how he thought ordering takes place in binary alloys. E J Williams was in the audience, and at the end of the seminar he presented Sir Lawrence with pencilled notes on a sheet of paper which, he claimed, gave numerical substance to the qualitative ideas. Sir Lawrence was naturally impressed, but suggested that if the mathematics was really so simple, someone must surely have done it before, but since no-one in the audience knew of any such calculation in the literature, Bragg and Williams wrote a paper which was presented to the Royal Society.

A few days after the corrected proofs of the paper had been returned, Sir Lawrence on clearing up his desk was shocked to find a preprint by Borelius developing ideas very similar to those which he had just sent off. He told me that he had been very embarrassed in subsequent years to find that all the credit for the development had been given to him and Williams; he would be pleased if a correct perspective could be introduced. He added that they probably obtained the credit because the Bragg–Williams papers described the ideas more clearly than those of any of the other investigators (as is readily confirmed). The important new concept of long-range order is clearly described in the second paper; and a significant distinction is drawn between long-range order and long-range forces. It is stated clearly that short-range forces can give rise to long-range order, a conclusion exactly parallel to that of Ornstein and Zernike for fluids, of which Bragg and Williams were apparently unaware.

7.5.6. Landau theory of second-order transitions: universality

In 1937 L D Landau [88] attempted to provide a unified description for all second-order transitions. In addition to the phenomena described above, experimental evidence was accumulating about specific-heat anomalies in liquid helium, ammonium chloride and a number of other substances which were termed λ-point transitions; the superconducting transition was also in this category. Ehrenfest [89] had introduced a thermodynamic classification of higher-order transitions, but there were difficulties associated with his treatment [90], and we shall avoid describing λ-point transitions as second-order phase transitions although this terminology is common.

Landau generalized the ideas introduced in the theory of alloys to all systems manifesting λ-point transitions, meaning thereby discontinuities in specific heat, without latent heat. He suggested that for every such system one must identify an order parameter analogous to the long-range order in an alloy, which would be zero on the high-temperature side of the transition and non-zero on the low-temperature side. He emphasized the role which symmetry plays in phase transitions, and suggested that the important features of behaviour in the vicinity of a λ-point could be determined by expanding the free energy in a power series as a function of the order parameter η. In the ferromagnetic transition the order parameter is the spontaneous magnetization, in the fluid it is the density difference between liquid and gas; Landau argued that for reasons of symmetry the form of the expansion of the Gibbs function will be

$$\phi(P,T,\eta) = \phi_0(P,T) + A(P,T)\eta^2 + B(P,T)\eta^4 + \ldots \qquad (27)$$

On cooling through the Curie temperature, A passes from positive to negative, while B remains positive. Above the Curie temperature, when $A > 0$, the solution corresponding to a minimum of ϕ is $\eta = 0$; this is the phase of higher symmetry. For example, in an order–disorder transition the disordered phase has the symmetry of the original crystal lattice when an average is taken over all configurations at a given temperature; whereas the ordered phase has the lower symmetry of the superlattice. Below the Curie temperature the minimum of ϕ corresponds to a non-zero value of η given by $\eta^2 = -A/2B$, and this is the phase of lower symmetry.

Landau's theory made it possible to understand why all the systems discussed previously had the same essential pattern of critical behaviour; even though equations of state like (21) and (23) look very different, they both conform to the expansion (27). Critical exponents are the same for all λ-point transitions, and in later terminology the transitions would be described as universal.

The theory predicts the following behaviour of typical thermodynamic quantities for a ferromagnet near $T_c (\tau = T/T_c - 1)$

$$\begin{aligned}
&\text{Spontaneous magnetization} \quad M_0 \sim (\tau)^{1/2} \quad (\tau < 0) \\
&\text{Initial susceptibility} \quad \chi_0 \sim \tau^{-1} \quad (\tau > 0) \\
&\text{Critical isotherm} \quad H \sim M^3 \quad (\tau = 0) \\
&\text{Derivative of susceptibility} \quad \frac{d\chi_0}{dH} \sim \tau^{-4} \quad (\tau > 0).
\end{aligned} \quad (28)$$

The specific heat tends to one value C_- on approaching the transition from a lower temperature, and to another C_+ from above. The transcription to appropriate thermodynamic variables can be readily made for other systems.

We shall see shortly that the expansion (27), on which the above development depends, was invalidated by Onsager's work, but the general principles of Landau's work have retained their validity, and played a major role in subsequent theoretical work. In the opening paragraph of his classic paper [88], he emphasized that the continuity of state between liquid and gas at sufficiently high temperatures is possible only because the liquid and gas phases have the same symmetry. A transition between two phases of different symmetry cannot be continuous; elements of symmetry are either present or absent and no intermediate case is possible. It is surprising that this simple and convincing argument was by-passed for at least 15 years, with discussion continuing on the possibility that the solid–fluid melting transitions might end in a critical point [91]. A summary of Landau's major ideas had appeared in English [92], but was either overlooked or disregarded.

Landau proceeded to a detailed analysis of the role of symmetry in λ-point and Curie-point transitions, which he developed further in his books with Lifshitz [93]. He considered in detail different types of symmetry possible in ordered and disordered phases of the lattice, and allowed for a vector order parameter with different components.

Lars Onsager

(Norwegian, 1903–76)

Born and educated in Norway, Onsager introduced himself to Debye in Zürich by criticizing his theory of strong electrolytes. After a few years with Debye he moved to America where for most of his active life he was at Yale. However much he wished to explain his ideas, his immensely wide learning and mathematical skill got in the way of intelligibility—his lectures on statistical mechanics were known to students as 'Advanced Norwegian'— but in discussions on research problems he would patiently try different approaches in the hope of making intellectual contact with less penetrating minds.

In 1931 he published two papers on the theory of irreversible processes that had little impact at first, but were gradually recognized as fundamental and won him the Nobel Prize in chemistry 37 years later. His second outstanding achievement, the complete solution of the two-dimensional Ising lattice problem, astonished those few who could follow it at the time, and laid the foundation of modern studies of phase transitions.

These were his most remarkable successes, but his contributions to theories of liquid helium, magnetism, metals, ice and turbulence were all influential and characterized by an originality that justifies his being acclaimed as among the greatest of modern scientists.

See also p 601

A subsequent treatment of critical fluctuations by Ginzburg and Landau was adapted by Wilson for use in the renormalization group and will be described later.

See also p 564

7.5.7. Statistical mechanics of a condensing gas: Mayer–Yvon treatment
We have already noted the reservations of Boltzmann and Maxwell in relation to van der Waals' treatment of the hard-core repulsion. If we expand the hard-sphere equation of state in the form

$$\frac{Pv}{RT} = 1 + \frac{B}{v} + \frac{C}{v^2} + \frac{D}{v^3} + \ldots \tag{29}$$

the van der Waals equation requires $B = b$, $C = b^2$, $D = b^3$, etc. Boltzmann had calculated C and found it to be $\frac{5}{8}b^2$, and as a result of the evaluation of a key integral by van Laar [94] in 1899, D was also known to be [95] $0.2869b^3$. Thus it was clear to Boltzmann, and presumably to van der Waals, that the equation could be regarded as rigorously correct only for dilute gases.

There began a period of collection of accurate experimental data initiated by Kamerlingh Onnes [96] who expressed the equation of state in the form

$$\frac{P}{RT} = \frac{1}{v} + \frac{B(T)}{v^2} + \frac{C(T)}{v^3} + \frac{D(T)}{v^4} + \ldots \tag{30}$$

where $B(T)$, $C(T)$, $D(T)$, ... were called the second, third, fourth ... virial coefficients. This then became the standard procedure for recording data on imperfect gases. A formidable challenge was posed to theoreticians to find a theoretical description of virial coefficients.

Then in 1937 it seemed as if one of the great triumphs of statistical mechanics had occurred. J E Mayer [97] developed an elegant formalism which enabled him to express all the virial coefficients as integrals over the intermolecular potentials. The formula he derived was

$$\frac{P}{k_B T} = \rho - \sum_{k=1}^{\infty} \frac{k}{k+1} \beta_k \rho^k \tag{31}$$

where the β_k, which he termed *irreducible cluster integrals*, could be represented simply and diagrammatically as *multiply connected* graphs. For example the first three of the β_k are represented as follows

$$\beta_1 = \bullet\!\!-\!\!\!-\!\!\bullet$$

$$2!\beta_2 = \triangle \tag{32}$$

$$3!\beta_3 = 3\,\square + 6\,\boxtimes + \boxtimes$$

If $\phi(r_{ij})$ is the intermolecular potential between molecules i and j, and we write f_{ij} as $\exp[-\beta\phi(r_{ij})] - 1$, a diagram connecting k points is a $3(k-1)$-dimensional integral, with an appropriate f_{ij} for each pair of points connected by a line in the diagram. For example

$$\boxtimes = \frac{1}{V} \int\!\!\int\!\!\int\!\!\int f_{12} f_{23} f_{34} f_{41} f_{13}\, d\mathbf{r}_1\, d\mathbf{r}_2\, d\mathbf{r}_3\, d\mathbf{r}_4. \tag{33}$$

An equivalent treatment based on correlations between the particles in a fluid had been developed independently by Yvon [98].

Professor M H L Pryce has told me that he was present at the colloquium in Cambridge when Mayer's results were first presented. Fowler, the outstanding world authority on statistical mechanics, was deeply impressed by Mayer's work, and said that he would not have believed that such progress could have been achieved in his lifetime. The way seemed open for a detailed explanation in atomic terms of the liquid and gaseous phases, the critical point and the striking discontinuities that occur in the equation of state. Uhlenbeck has noted [99] that at the great international congress in 1937 to celebrate the centenary of van der Waals' birth, the van der Waals equation was mentioned only once, and all attention was centred on the newly emerging 'rigorous' theories.

Mayer tried to assess the relevant analytical properties of the series (31), combining knowledge of the properties of fluids with conjectures regarding the asymptotic behaviour of β_k. He was strongly supported by Born, who in a joint paper [100] with Fuchs said 'we believe that we have succeeded in showing rigorously, and in a somewhat simpler way than Mayer himself, that his statements are completely correct'.

Subsequent developments dissipated this optimism. From the experimental point of view accurate measurements near the critical point present three major difficulties: (a) the effect of gravity causing a density inhomogeneity in a region of very high compressibility; (b) the long times needed to reach equilibrium and the problem of eliminating temperature gradients; (c) the effect of impurities. The fact that gravity might be important in the critical region was pointed out by Gouy [101] as long ago as 1892. However, in the 1930s it was generally assumed that, in the region then available to experiment, the effect was negligible. The importance of gravity was first clearly demonstrated by Schneider and his co-workers [102], and Schneider showed that Mayer's conclusions on the nature of the liquid–gas co-existence curve disagreed with experiment.

About the same time Yang and Lee [103] derived a number of rigorous results for the *lattice gas model* (which we shall introduce shortly) and concluded that the series (31) related only to the gaseous phase, and could not account for gas–liquid equilibrium. Finally, Uhlenbeck and his co-workers [104] introduced graph theory into statistical mechanics and revealed the true nature of the problem to be tackled in evaluating β_k. Although the number of integrals contributing to β_k is small at first, it grows as $2^{(1/2)k(k-1)}/k!$ and becomes large extremely rapidly. Each integral of order k is in a space of $3k$ dimensions. Even for the simplest hard-sphere potential only six terms have been calculated using the most powerful modern computers; to account for condensation it is essential to take account of the attractive part of the intermolecular forces, and even fewer terms are then available. Despite the elegant formalism of

Twentieth Century Physics

the theory, it provides little practical information on phase transitions and critical points.

7.5.8. Ising model: the Onsager revolution

The Weiss theory described in section 7.5.3 is empirical and does not use a microscopic model involving elementary interactions. In 1925 W Lenz, who had been looking for a simple model which might serve to explain ferromagnetism, put the following suggestion to his graduate student E Ising. Assume that each atom possesses a spin and hence a magnetic moment, μ_0, which can orient either parallel or antiparallel to an external magnetic field H. There is an interaction between nearest-neighbour spins in the lattice, parallel spins having an interaction energy $-J$ and antiparallel spins $+J$ ($J > 0$). Lenz hoped that the model might manifest a non-zero spontaneous magnetization, i.e. that as $H \to 0$ the ratio

$$\frac{(N_2 - N_1)}{(N_2 + N_1)} \tag{34}$$

might tend to a non-zero value. Here N_1, N_2 are the numbers of spins anti-parallel and parallel to the field H.

Ising was able to solve the problem in one dimension only [105] where the statistical problem is quite simple, and the solution did not possess a spontaneous magnetization at any temperature above zero. He concluded wrongly that the model would not give rise to a spontaneous magnetization in higher dimensions. In 1936 Peierls [106] was the first to demonstrate convincingly that for a two-dimensional Ising model, at sufficiently low but non-zero temperatures, the ratio (34) does tend to a non-zero value as $H \to 0$. Ising published no other papers in this field, but by now thousands of papers have discussed the properties of what should be called the *Lenz–Ising model*. (For an historical review of the model see Brush [107]).

In 1928 Heisenberg [108] suggested that the origin of the large interactions giving rise to ferromagnetism lay in the quantum-mechanical exchange forces which can be represented by a vector coupling of spins $-J s_i s_j = -J(s_{xi}s_{xj} + s_{yi}s_{yj} + s_{zi}s_{zj})$. Here s_x, s_y, s_z are quantum-mechanical non-commuting operators representing the components of spin. The internal field of Weiss acting on an atom at a given site is really a fluctuating field arising from the interactions with its neighbours; and it is easily shown that the results of the Weiss theory can be retrieved if this fluctuating field is replaced by its mean value. The Weiss theory is therefore an approximation which one would expect to be valid if the number of interacting neighbours is large. Approximations of this kind were later termed *mean-field approximations*.

In 1944 Onsager [109] achieved a triumph when he published an exact solution of the partition function of the Ising model for the two-dimensional simple quadratic lattice in zero-field. The result was a

Thermodynamics and Statistical Mechanics (In Equilibrium)

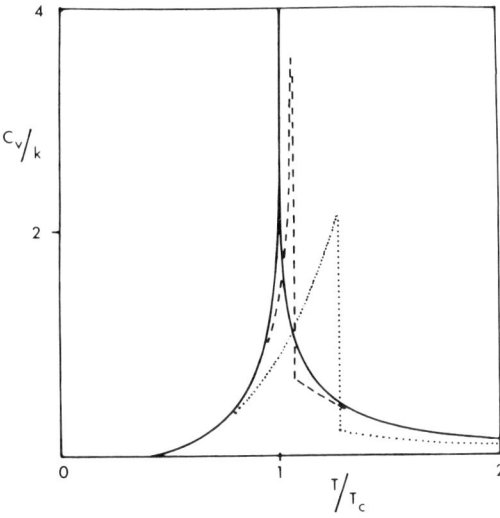

Figure 7.7. Onsager's classic calculation (1944) of the specific heat of the spin $\frac{1}{2}$ Ising model on a simple quadratic lattice. The full curve represents the true specific heat which is logarithmically infinite. The dotted curve represents the Bethe closed form approximation, and the dashed curve an improved closed form approximation due to Kramers and Wannier.

shattering blow to classical theory. The specific heat (figure 7.7) was not discontinuous as required by (28) but logarithmically infinite (this had been conjectured previously by Kramers and Wannier [110]). More importantly, the partition function was non-analytic at T_c so that an expansion of the type used by Landau was completely invalid.

In a subsequent paper with B Kaufman [111], Onsager calculated the correlations and they did not fit in with the results of the Ornstein–Zernike treatment. Finally he was able to determine the spontaneous magnetization [112] (later calculated independently by C N Yang [113]) and this was of the form $(-\tau)^{1/8}$ in the critical region, very different from $(-\tau)^{1/2}$ of the classical Weiss theory (8) (figure 7.8).

Experimental evidence on critical behaviour also began to accumulate in conflict with classical predictions. In 1945 Guggenheim [114] undertook a critical analysis of experimental data on the coexistence curves of a number of gases. According to van der Waals' theory these gases should obey a law of corresponding states i.e. if reduced units T/T_c, ρ/ρ_c are used (where ρ_c, T_c are the density and temperature at the critical point) the coexistence curves for the different gases should fall on a single universal curve. Guggenheim found that the data provided good support for such a law of corresponding states (figure 7.9), but the form of the curve was cubic, $\Delta\rho \sim (-\tau)^{1/3}$, rather than quadratic as required by van der Waals' theory. We have noted a similar conclusion

555

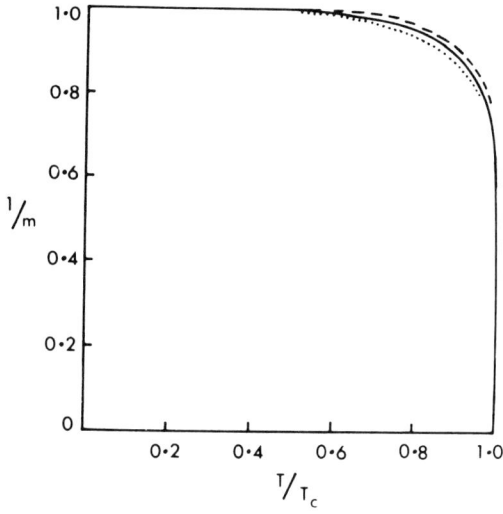

Figure 7.8. *Spontaneous magnetization of the two-dimensional Ising model (full curves simple quadratic lattice, dashed curve honeycomb lattice, dotted curve triangular lattice). The magnetization drops to zero very steeply at $T_c (\sim (T_c - T)^{1/8})$; closed form approximations give $(T_c - T)^{1/2}$.*

by Verschaeffelt [69] in 1900.

In 1954 Habgood and Schneider [115] published precise measurements of the isotherms of xenon near the critical point. They found that the critical isotherm is much flatter than the cubic curve predicted by van der Waals (figure 7.10), and they suggested that the third and fourth derivatives of P with respect to ρ are zero at the critical point.

Onsager's calculations on the Ising model disagreed with classical theory and also with experiment. The latter was not unreasonable since the calculations applied only in two dimensions; it was therefore important to obtain theoretical results for three-dimensional systems, but the exact techniques of Onsager were specific to two dimensions, and to the partition function and its derivative in zero field.

During the two decades that followed the publication of Onsager's solution, the critical properties of model systems had to be established on an individual basis—each model and each property entailed a separate calculation. The most useful tool turned out to be the generation of lengthy perturbation series expansions at high and low temperatures. The idea of generating lengthy series expansions and correlating the coefficients with critical behaviour was first advanced by Domb [116]. Onsager's solution applies only in zero field, and the first target was to explore the properties of the model in a non-zero field.

A few general remarks are necessary to clarify the basis of the method. In normal statistical work it is hazardous to attempt to extrapolate asymptotic behaviour from a finite number of terms. However, in the

Thermodynamics and Statistical Mechanics (In Equilibrium)

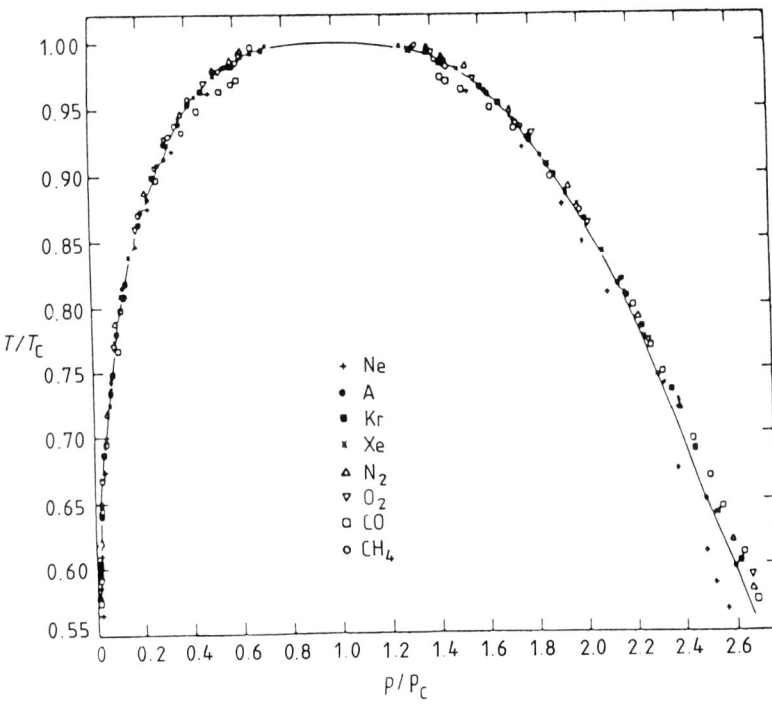

Figure 7.9. *Reduced densities of co-existing liquid and gas phases for a number of simple molecular fluids* [114]. *The experimental points support a law of corresponding states, but the universal curve is cubic rather than quadratic as required by van der Waals' theory.*

case of a ferromagnet one could anticipate on physical grounds a pattern of expected critical behaviour, and use the coefficients of the series to obtain the best fit to parameters such as critical temperature and critical exponents in the assumed asymptotic form for the coefficients. Moreover, the methods of analysis and fitting could be tested on Onsager's exact solution to see how well they worked and to get an idea of the error to be expected. The method proved capable of furnishing important reliable information and, as numerical techniques improved, its use became widespread. M F Sykes played a major role in generating and interpreting such series expansions.

I quote as a typical example the high-temperature expansion of the initial magnetic susceptibility χ_0 for the simple Ising model on the face-centred-cubic lattice

$$\frac{kT\chi_0}{\mu_0^2} = 1 + 12w + 132w^2 + 1404w^3 + 14652w^4 + 151116w^5$$
$$+ 1546322w^6 + 15734460w^7 + 59425580w^8 + \ldots$$
$$(w = \tanh J/kT). \tag{35}$$

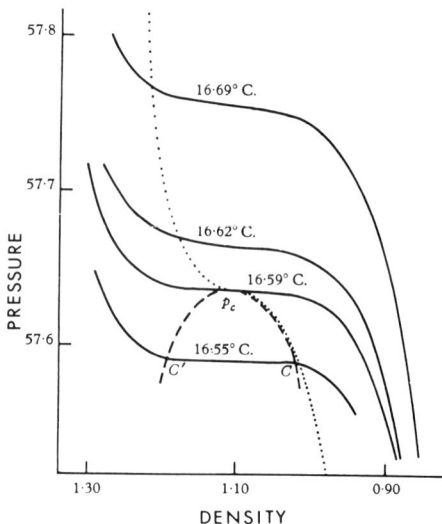

Figure 7.10. *Isotherms of xenon near the critical point* [115]. *The broken curve marks the region of co-existent phases. The dotted curve is the critical isotherm according to van der Waals' equation to be contrasted with the measured 16.59 °C isothermal.*

If we can estimate the asymptotic pattern from these terms we shall be estimating the critical behaviour of χ_0.

There were a number of novel features in the series approach [117]. The actual generation of terms required a close familiarity with graph theory, and the programming of sophisticated graphical enumeration problems on computers. Then there were problems of interpretation, and here the Onsager solution served as a valuable guide. It was usual to assume branch point singularities, i.e. that the critical behaviour of specific heat, magnetic susceptibility, etc, is of the form $(1 - T_c/T)^{-\theta}$, and when the terms were consistent in sign, the asymptotic behaviour of the coefficients could be related directly to critical points, exponents and amplitudes. But in many important cases, particularly for three-dimensional models, the coefficients were not consistent in sign, and spurious unphysical singularities masked the true critical behaviour. G A Baker's application of the Padé Approximant [118], a piece of mathematics which had lain dormant since the end of the nineteenth century, led to remarkable progress, and sparked off similar applications in many other fields concerned with perturbation expansions [119].

Gradually a body of reliable information was assembled on the critical properties of different theoretical models. Their behaviour differed from classical theory, but the differences were much less marked in three dimensions than in two dimensions. Also the theoretical predictions of critical exponents were much closer to experimental results than those of classical theory [120].

Thermodynamics and Statistical Mechanics (In Equilibrium)

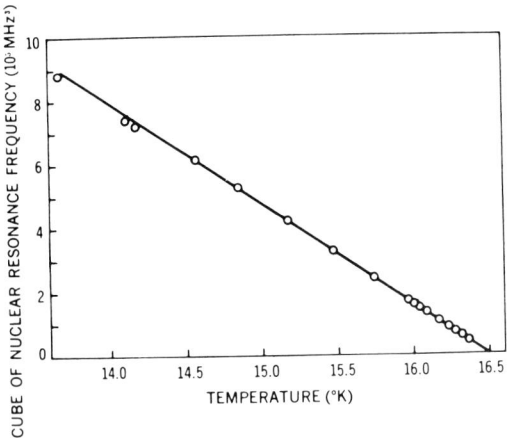

Figure 7.11. *Temperature dependence of the cube of the Eu^{153} in EuS nuclear resonance frequency between 13.60 and 16.33 K. In this range the law $M_0 \sim D(1-t)^{1/3}$ is accurately obeyed* [123].

These theoretical developments now stimulated a new wave of precise experimental measurements, making use of new techniques and new magnetic materials to attain a much greater accuracy than had been available previously [121]. In 1957 Fairbank, Buckingham and Kellers [122] published the result of a new investigation into the nature of the λ-point of liquid He^4. Improved low-temperature techniques enabled them to get very close to the λ-point, and domains and grain boundaries which limit the growth of specific heats of crystals are absent in liquid helium. They concluded that the specific heat is logarithmically infinite on both sides of the λ-point and is in many respects similar to the Onsager specific heat depicted in figure 7.7. Of course they did not suggest that there was any basic reason for the similarity between a three-dimensional quantum fluid and a two-dimensional Ising ferromagnet, but at least they demonstrated that logarithmically infinite specific heats are a reality.

Conventional ferromagnets like iron, nickel and cobalt have high Curie temperatures, and accurate measurements of critical behaviour are difficult. However, new ferromagnets like europium sulphide ($T_c \sim 16.50$ K) are much more amenable to accurate measurements, and nuclear magnetic resonance can be used to measure the fraction of spins oriented in a particular direction at a given temperature, and hence to deduce the spontaneous magnetization. Figure 7.11 is an example from which Heller and Benedek [123] deduced that the cube of the spontaneous magnetization is very nearly linear. They had previously [124] obtained a similar result for the sublattice magnetization of the antiferromagnet manganese fluoride, MnF_2.

It was gratifying to find the new experimental results fitting reasonably well to the theoretical calculations, even though it was well understood that the theoretical models were simple and crude, and did not adequately map the true physical interactions [125]. It seemed as if critical behaviour was insensitive to the details of the interaction mechanism.

As the available theoretical data increased, certain regularities were noted empirically. Domb [126] noted that critical behaviour depended very significantly on dimension, but only slightly on lattice structure in a given dimension. This was borne out by further exact calculations in two dimensions. Extensions of Onsager's work to other lattices gave identical results for critical exponents and small differences in critical amplitudes. Domb and Sykes [127] noted that critical exponents varied from the Ising to the Heisenberg model but seemed to be independent of spin value for a given model (later Jasnow and Wortis [128] suggested more generally that critical exponents might be determined by the symmetry of the ordered state).

An important question raised in the early 1960s was the relationship between the exponents characterizing critical behaviour for different thermodynamic properties. Were they independent or did they satisfy any restrictive conditions? In studying the properties of Fisher's droplet model, Essam and Fisher [129] observed that the critical exponent α', characterizing the low-temperature specific heat ($C_H \sim (-\tau)^{-\alpha'}$), β the spontaneous magnetization ($M_0 \sim (-\tau)^\beta$) and γ the initial high-temperature susceptibility ($x_0 \sim \tau^{-\gamma}$), satisfied the exact relation

$$\alpha' + 2\beta + \gamma' = 2. \tag{36}$$

This relation was also satisfied by the exponents of the two-dimensional Ising model.

This work stimulated Rushbrooke [130] to investigate whether anything could be said purely from thermodynamics (which would then apply to all models). He found that there was indeed a thermodynamic relation

$$\alpha' + 2\beta + \gamma' \geqslant 2. \tag{37}$$

Several other thermodynamic inequalities were later discovered [131].

Two other exact calculations of the partition function for particular model systems became available during this period. The first considered the effect of weak long-range forces, conveniently characterized in d dimensions, at great distances, by $J\lambda^d e^{-\lambda r}$, as $\lambda \to 0$. For a continuum fluid model [132] they reproduce exactly in one dimension† the results of van der Waals with the associated Maxwell construction. Lattice models also reproduce [133] exactly the results of Weiss for ferromagnets and

---

† The $V - b$ excluded volume term (equation (21)) is correct only in one dimension. In two and three dimensions the appropriate hard-sphere partition function must be used.

Thermodynamics and Statistical Mechanics (In Equilibrium)

of Bragg and Williams for alloys. Classical theory was thus partially reinstated as much more than an arbitrary empirical approximation. It is a valid theory for very long-range forces, but does not accord with experiment because intermolecular forces in Nature are usually of shorter range.

The second was a mathematical exercise [134] involving an Ising model in which the spins could have any value subject to the condition that the sum of their squares remained equal to N. An exact solution for this, the 'spherical model', was possible; despite its artificiality, this model, in common with other unrealizable models, was to play a significant part in achieving a proper understanding of critical behaviour.

7.5.9. *Reconciliation: empirical derivation of scaling and universality*

The first international conference devoted specifically to critical phenomena took place in April 1965 in Washington at the National Bureau of Standards. It could reasonably be termed the founding conference of critical phenomena, since for the first time all the different strands of the subject were woven into a coherent fabric, and the major obstacles to progress were clearly delineated. The conference set the pattern for other meetings on critical phenomena which served as important catalysts to progress.

It is particularly fascinating to quote one of the concluding paragraphs of Uhlenbeck's keynote address:

> If there is such a universal, but not-classical behaviour, then there must be a universal explanation which means that it should be largely independent of the nature of the forces. The only corner where this can come from is I think the fact that the forces are not long range. The Onsager solution gives I think a strong hint. It may well be so that away from the critical points the classical theories give a good enough description but that they fail close to the critical point where the substance remembers so to say Onsager. I think that to show something like this is the central theoretical problem. One can call it the reconciliation of Onsager with van der Waals [135].

Uhlenbeck was looking for a new type of universality for short-range forces which would replace the universality of the classical theories which are valid only for very long-range forces.

The first requirement for reconciliation was a coherent description of the 'non-classical universality' analogous to van der Waals' description of classical behaviour. Such a description in the form of a non-classical equation of state, valid in the critical region, emerged within a few months. It was suggested independently by three different groups who used totally different approaches: Widom [136] in the USA searched for a generalization of van der Waals' equation which could accommodate

Twentieth Century Physics

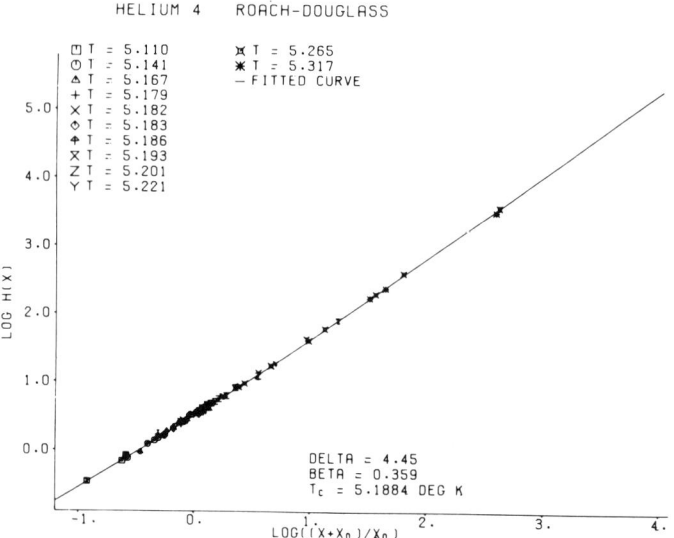

Figure 7.12. *Scaling relation for the fluid He^4.*

non-classical exponents; Domb and Hunter [137] in the UK analysed the behaviour of series expansions of higher derivatives with respect to the magnetic field at the critical point; and Patashinskii and Pokrovskii [138] in the USSR considered the behaviour of multiple-correlations near the critical point. The results were tied together neatly by Griffiths [139] in the form

$$H = M^\delta h(\tau M^{-1/\beta})$$

or for both signs of M

$$H = M|M|^{\delta-1} h(\tau|M|^{-1/\beta}) \tag{38}$$

as the equation of state of a ferromagnet. (For a fluid H was replaced by $P - P_c$ or $\mu - \mu_c$ where μ is the chemical potential and M by $v - v_c$ or $\rho - \rho_c$.) Here β and δ are two parameters which determine all the exponents, and $h(x)$ is an analytic function. Classical theory corresponds to $\delta = 3$, $\beta = \frac{1}{2}$, $h(x)$ linear. Non-classical results could be accommodated with different values of δ, β and a different function $h(x)$.

The two characteristic features of (38) are that critical relations like (36) between exponents are indeed satisfied exactly and that critical data satisfy a 'scaling' relation, i.e. if $HM^{-\delta}$ is plotted against $\tau M^{-1/\beta}$, the two-dimensional data will all fall on a single curve $h(x)$. These two predictions were tested against experimental data for a large number of systems, both fluids and magnets, and were found to be well satisfied (figure 7.12).

However, (38) says nothing about possible values of β, δ and $h(x)$. Empirical information about regularities in the pattern of critical exponents was collected together in the *hypothesis of universality* put forward by Kadanoff [140]; an analogous *smoothness* postulate was advanced independently by Griffiths [141]. This non-classical universality is more sophisticated than had been envisaged by Uhlenbeck; different classes are defined by the space-dimension d and the spin-dimension D (later designated n), and within a given class critical behaviour is universal. If a third parameter σ is introduced to take account of the range of the intermolecular forces, both classical and non-classical behaviour can be taken into account in a wider pattern of universality. Once d, n and σ have been specified the exponents δ, β and the function $h(x)$ are determined, and the behaviour is smooth and universal within the class. Changes in critical exponents with accompanying discontinuities occur in the *crossover* between one universality class and another (e.g. in the transition from a two-dimensional to a three-dimensional system by the introduction of a new interaction).

Two highly theoretical ideas were now pursued which, although remote from physical reality, were subsequently to make an important contribution. Stanley [142] investigated the behaviour of Ising models as a function of spin dimension n. An increase of n gives greater freedom to the spin and corresponds to a decrease of 'co-operative strength'. He found that as $n \to \infty$ the spherical-model solution [134] is retrieved. Hence this model has a place in the general framework.

Joyce [143] had already examined the behaviour of the spherical model with varying dimensions and ranges of force. He found that classical behaviour results with long-range forces in low space-dimensions d, but also with short-range forces if $d > 4$ (with possible logarithmic correction terms). Hence one could reasonably conclude that the same result will also hold for finite n since the co-operation is then stronger. This conclusion was demonstrated convincingly by diagrammatic methods [144] which give precise results when $d > 4$.

We have noted that the behaviour of critical correlations as determined from model calculations [120] differs from the classical results of Ornstein and Zernike [81], but the non-classical correlation exponents and functions fit in with the universality class picture above. Are the correlation exponents determined by δ and β of (38)? In regard to this question the treatment of Patashinskii and Pokrovskii [138] went further than others and suggested the relation

$$d\nu = \beta(\delta + 1) \tag{39}$$

where ν is the non-classical exponent which characterizes the range of correlation ($\kappa \sim \tau^\nu$; see [82]). But the argument was put more cogently by Kadanoff [140] in a key paper in which he tried to find a theoretical basis for the scaling properties that had been discovered in critical behaviour.

Kadanoff noted that near T_c the coherence length ξ becomes very large, and hence it is possible to find a length L, large compared with the lattice spacing a but small compared with ξ. He then considered replacing the interaction of individual spins by the interaction of blocks of L^d spins. One might perhaps expect that in each block the spins would be nearly all up or nearly all down, and the original Ising model of spin σ and interaction J could then be replaced by a new model with block spin $\tilde{\sigma}$ and interaction $\tilde{J}$. If the new block spin model was effectively the same as the original model, the free energies would be related by

$$f(h, \tau) = L^{-d} f(\tilde{h}, \tilde{\tau}) \qquad (h = \beta H). \tag{40}$$

But how would $\tilde{h}$ and $\tilde{\tau}$ be related to h and τ? Kadanoff suggested that

$$\tilde{h} = L^x h \qquad \tilde{\tau} = L^y \tau. \tag{41}$$

It is then easy to show that all the exponents (including ν) can be expressed in terms of x and y, and that both (38) and (39) result.

Relation (39) is satisfied by the Ising model in two dimensions, yet numerical calculations for the three-dimensional model showed a small but persistent discrepancy. Also there are several features of Kadanoff's argument that do not stand up to severe examination, but his ideas stimulated the next major theoretical advance which completed the reconciliation sought by Uhlenbeck.

7.5.10. *Renormalization group (RG); respectability*

At the time of Uhlenbeck's 70th birthday (in 1969), the situation was summarized as follows 'From the point of view of practical calculations of the behaviour near the Curie point we are well on the way to satisfying the needs of the experimentalist for most models of interest. Unifying features have been discovered which suggest that the critical behaviour of a large variety of theoretical models can be described by a simple type of equation of state. But the rigorous mathematical theory needed to make the above developments "respectable" is still lacking' [145].

The application of the renormalization group (RG) which K G Wilson introduced so effectively a year or two later was not rigorous by mathematical standards, and the search for a proper mathematical description of his ideas still continues, but he was able to account convincingly for the striking empirical discoveries of the previous section and the 'respectability' which his work achieved is adequate for most physicists.

A suggestion that the RG could be relevant to critical phenomena was made [146] at a summer school in 1970, but no precise indication was forthcoming as to how it should be used. It had been developed nearly 20 years earlier [147] in connection with field theory, but had no apparent practical consequences, and had not been taken very seriously.

Kenneth Geddes Wilson

(American, b 1936)

K G Wilson had a distinguished scientific family background, his father being an early pioneer of theoretical chemistry. He worked for a doctorate with Murray Gell-Mann at Caltech, and this gave him an opportunity to familiarize himself with all aspects of quantum field theory and elementary particle physics. It also introduced him to the particular formulation in 1954 of the renormalization group idea by Gell-Mann and Low.

In 1963 Wilson accepted an offer of an Assistant Professorship in Cornell. In 1965 he was given a permanent appointment as Associate Professor; the appointment reflects great credit on the University, since even in 1968, 12 years after graduating from Harvard, he had only three publications to his name. The choice of Cornell was very fortunate, since Ben Widom and Michael Fisher, two of the leading international authorities on critical phenomena, ran a joint seminar which attracted Wilson's attention. There could not have been a better place at which to learn first hand of the exciting new developments in critical phenomena.

Wilson's first efforts were directed to clarifying precisely the analogy between quantum field theory and critical phenomena. He worked alone to overcome this extremely difficult challenge, and his successful treatment of the problem represents the crux of his achievement. It provided a theoretical explanation of the pattern of scaling and universality which had been discovered empirically. Kadanoff's 'block scaling' ideas provided a useful pointer, as Wilson acknowledged. Wilson's broad and deep formulation of the RG goes far beyond anything envisaged by the field theorists. His papers proved extraordinarily fertile, and initiated an explosion of significant publications.

He was awarded the Nobel prize for Physics in 1982.

Wilson saw that the theory was capable of providing an understanding of universality, and a framework for the detailed calculation of critical behaviour. He acknowledged his debt to Kadanoff for setting his thoughts in the right direction, but he converted Kadanoff's vague block spin mappings into a precise tool for calculation.

The RG does not produce exact solutions of the Onsager type, and its application involves quite drastic approximations. But unlike the approximations of the classical theory, which do not reflect physical behaviour near T_c except in a small number of specific cases, RG endeavours to retain all the essential physical characteristics of the problem near T_c whilst rejecting insignificant features. This requires continuous thought and attention—in Wilson's graphic language [148] 'One cannot write a renormalization cook book'—and in his Nobel Prize lecture, he describes [149] how hard it was in the initial stages to find approximations which would be computable in practice.

No reliable method has been found of assessing directly the magnitude of the error at any stage of approximation of an RG treatment, and comparison with the results of exact solutions and series expansion estimates played a key role in the successful application of the RG. The methodology described in the previous sections of this chapter did not become redundant as a result of the emergence of the RG—on the contrary its value was enhanced.

Before discussing the RG approach, we shall outline (following Wilson [150]) the nature of the problem presented by critical behaviour, and then discuss qualitatively how the RG deals with it. If we consider the behaviour of a fluid like water, far below the critical point there are microscopic density fluctuations on an atomic scale. If the temperature and pressure are increased towards the critical point, fluctuations become important at longer wavelengths. Sufficiently close to T_c and P_c there are fluctuations on the scale of 1000–10 000 Å which scatter ordinary light and give rise to critical opalescence, making the water look milky. However, the microscopic fluctuations still remain. Close to the critical point there are fluctuations at all wavelengths from 1 Å up to the correlation length ξ which tends to infinity at the critical point. The essence of the problem is how to deal with an assembly with so many scales of length.

The aim of the RG approach is to reduce in a systematic way the large number of degrees of freedom and the large number of length scales of the assembly. A transformation is set up in the parameter space of the Hamiltonian

$$H' = R(H) \qquad (42)$$

which preserves the dimensionality and symmetries of the assembly, and reduces the correlation length by a factor $b > 1$, and the number of degrees of freedom from N to $N' = N/b^d$. The transformation is chosen so that the partition function is preserved

Thermodynamics and Statistical Mechanics (In Equilibrium)

$$Z_{N'}(H') = Z_N(H). \tag{43}$$

A wide choice of transformations is possible, which can be characterized in *real space* or in *momentum space*. In real space the transformation usually follows Kadanoff's idea of introducing block spin variables or eliminating some of the spins by partial summation (decimation transformation). However, the edge of the block is not long compared with the lattice spacing, but contains one or two lattice spacings; also the Kadanoff picture must be amplified, and the relation between the original spin and the block spin must be described precisely. In momentum space the aim is to eliminate high momentum variables which correspond to short wavelength fluctuations.

The transformation is iterated

$$H' = R(H) \qquad H'' = R(H') \qquad \ldots \tag{44}$$

and the universality properties follow from the limiting behaviour of such iterative processes. Elementary iterative processes of the form

$$x_{n+1} = f(x_n) \tag{45}$$

are well known in numerical analysis as a useful method of approximating to a root of the equation

$$x = f(x). \tag{46}$$

They have the advantage that the final solution is within wide limits independent of the starting point. A root of (46) is called a *fixed point*, x^*, of the transformation (45). If (46) has a number of different roots, some of them will correspond to *stable fixed points* (a small perturbation leads back to the fixed point) and the remainder to unstable fixed points. Each of the stable fixed points has a characteristic 'range of attraction' i.e. range of starting points within which convergence will occur to the particular root. Equation (44) involves the generalization of (45) to a multi-dimensional space

$$K_{n+1} = f(K_n) \tag{47}$$

corresponding to a number of parameters in the Hamiltonian, but many of the general characteristics remain the same, including the possibility of convergence to fixed points K^*. The stability pattern is more complex and one must allow for partially stable fixed points which are stable in an appropriate sub-space.

We can now see qualitatively how the above properties provide a basis for the explanation of universality classes. Each fixed point K^*

corresponds to one universality class, and Hamiltonians with a wide variety of parameters all converge to the same fixed point.

Critical behaviour is determined by the behaviour of $f(K)$ near K^*, and if a linear expansion is undertaken near this point, a linear operator can be derived whose eigenvalues are related to critical exponents. A scaling equation of state equivalent to (38) can readily be deduced. The non-analytic critical behaviour arises from analytic functions $f(K)$ as a result of the limiting procedure as the number of iterations tends to infinity.

Although Kadanoff's idea led naturally to real space renormalization, the major progress in actual calculations was achieved by applying perturbation methods which had been developed in quantum field theory to momentum space renormalization. To make this application possible Wilson went back to Landau's ideas, not in the original macroscopic form of equation (27), but in the later microscopic Ginzburg–Landau formulation [151]. This treatment focused attention on the local free energy which varies from point to point because of fluctuations, and can be expanded for a magnetic system, in the form

$$\Delta a(m,t) = B(\nabla m)^2 + C(\delta m)^2 + D(\delta m)^4. \qquad (48)$$

Wilson used a model with a spin s which could vary continuously, and by analogy with (48) introduced the Hamiltonian (for $n = i$)

$$H = (\nabla s)^2 + Rs^2 + Us^4 - Hs \qquad (49)$$

where R and U are analytic functions of T. For general n, s^2 is replaced by $\sum_{i=1}^{n} s_i^2$ and s^4 by $(\sum_{i=1}^{n} s_i^2)^2$ where the s_i are components of the n-vector spin. This was known subsequently as the Landau–Ginzburg–Wilson Hamiltonian. Few who remembered the failure of the macroscopic Landau theory would have been prepared to believe that an expansion of the form (49) was sufficient to lead to non-classical results of Ising-model type. But Wilson was able to indicate that higher-order terms in s^6, s^8, etc, make no significant contribution to critical behaviour (the coefficients are irrelevant parameters).

The parameter chosen [152] for the expansion of critical exponents and scaling functions was $\varepsilon = 4 - d$, where d is the dimensionality of space; the coefficients are functions of n, the spin dimension. $\varepsilon = 0$ corresponds to four dimensions for which classical theory is valid.

Only a few terms of the ε expansion were available; it turns out to be an asymptotic expansion and the value $\varepsilon = 1$ corresponding to three-dimensional systems cannot be assumed small, but by good fortune the convergence is rapid and the results were in satisfactory agreement with those of series expansions (later work derived more terms and devised improved methods of summation [153]). As mentioned at the end of the

Thermodynamics and Statistical Mechanics (In Equilibrium)

previous section, there were some small discrepancies with the results of series expansions, but these have now been fairly well eliminated [154].

Real space renormalization methods worked well in two dimensions, and were able to reproduce the Onsager solution to a high degree of accuracy [155]. Later methods which can be applied in three dimensions contributed useful numerical information [153].

Wilson has described his own work as a second stage of the Landau theory, and has shown precisely where the original Landau theory breaks down [156]. The microscopic Hamiltonian (49) is correct. But the macroscopic form of the free energy (27) obtained by averaging over a region is incorrect since it ignores the variation of R and U with L, the size of the region, which is non-analytic (as had been suggested previously by Kadanoff [140]).

The RG theory did not stop at explaining the empirical results of the previous section. Like all good scientific theories it suggested new avenues of exploration, and was able to deal with problems for which previous methods had not been successful. The most notable are the corrections [157] to the equation of state (38), which incidentally enabled critical behaviour as derived from series expansions to be interpreted more precisely, and systems with long-range dipolar interactions [158] for which the derivation of series expansions is an extremely difficult task.

7.5.11. Self-avoiding walks and configurations of polymers

When the structure of polymer molecules was first clarified [159] in the 1920s and 1930s it posed a number of interesting statistical problems. A polymer chain consists of a molecular unit repeated N times, where N might be thousands or tens of thousands. In a simple model of a polymer in dilute solution each unit could be represented by a bond of fixed length, the relative orientations of successive bonds being random. Features of interest, for example, are the mean square end-to-end length, $\langle R_N^2 \rangle$, the mean square radius of gyration, $\langle S_N^2 \rangle$, and the probability distribution of the end-to-end length, $f(u)du$ ($u = R/\langle R_N^2 \rangle^{1/2}$).

See also p 1596

The earliest research workers used the analogy of a random walk, which leads to the conclusion that asymptotically $\langle R_N^2 \rangle$ and $\langle S_N^2 \rangle$ are proportional to N, and $f(u)$ is Gaussian, $\exp(-u^2)$. This conclusion is unchanged for a modified model in which the angle between successive units is kept constant.

However, it was soon realized that the above models ignore an important physical feature of the polymer, the volume of each molecular unit, and the impossibility of occupying a region of space more than once by any part of the chain. This property, *the excluded volume*, was investigated theoretically by Flory [159] using approximations of mean-field type described in section 7.5.8. He concluded that the effect of the excluded volume is to change the asymptotic behaviour of $\langle R_N^2 \rangle$ and $\langle S_N^2 \rangle$ to $N^{6/5}$.

For theoretical purposes it is often convenient to introduce an artificial lattice structure into continuum problems, and a lattice model which proved fruitful in clarifying the nature of the excluded volume effect was the self-avoiding walk (SAW). A SAW on a lattice is a random walk subject to the condition that no lattice site may be visited more than once in the walk. The prohibition of multiple occupation of the sites provides a convenient representation of the excluded volume effect. The term SAW was first used by Hammersley and Morton [160], although it is not clear who introduced the model [161]. It does not take much effort to convince oneself that the standard mathematical techniques, which are so powerful in elaborating the properties of random walks, are inapplicable to SAWs.

An analogy was early noted [162] between the configurational problems posed by the SAW model and those arising in the development of high-temperature series expansions for the Ising model. We have seen that the critical behaviour of the magnetic susceptibility for this model is determined by the asymptotic behaviour as a function of N of the coefficients in the power series (35). A parallel formalism can be set up for $\langle R_N^2 \rangle$ and $\langle S_N^2 \rangle$.

The precise connection between the SAW model and magnetic models was revealed by de Gennes [163] who pointed out that the SAW configurations correspond to a ferromagnetic model of spin dimension n when n is allowed to take the value zero. The RG calculation of critical exponents can thus be used immediately for the calculation of SAW exponents. The scaling concepts which helped to clarify critical behaviour also have their polymer counterpart [164].

7.5.12. Models with other features of interest

We have seen that exact solutions played a role of special importance in the theory of critical phenomena. For about 20 years after Onsager's work there were no new solutions for realistic interactions which differed in a significant way from that of Onsager. Then in 1967 Lieb [165] produced a variety of new solutions for two-dimensional ferroelectric models which followed a completely different pattern from the Ising model. The models originated in attempts to explain ferroelectric and antiferroelectric phase transitions in hydrogen-bonded crystals like potassium dihydrogen phosphate KH_2PO_4. The structure of the crystal is tetragonal, with every phosphate group surrounded tetrahedrally by four other phosphate groups. The hydrogen atoms are located between each pair of phosphate groups. They are, however, not centrally placed but are to be found near one or other end of the phosphate groups: a 'state' of the crystal is characterized by a specification of the hydrogen positions. If a hydrogen atom near a vertex is denoted by an arrow pointing along

Thermodynamics and Statistical Mechanics (In Equilibrium)

the bond to the vertex, and a hydrogen atom distant from the vertex by an arrow pointing away from the vertex, there will be 16 possible vertex configurations. Onsager [166] suggested that the ferroelectric transition in KH_2PO_4 is connected with an ordering of these vertex configurations, with an energy assigned to each configuration. Configurations must, however, be compatible with their neighbours along the bonds of the lattice, and a partition function must be constructed which takes all possible configurations into account and gives them their correct energy weighting.

It can be shown that the general 16 vertex problem is equivalent to an Ising model with 2, 3 and 4 spin interactions in an external magnetic field. Lieb's exact solutions correspond to particular cases in which ten of the 16 vertex configurations are disallowed (six-vertex models). A simple case allows an exact calculation of the residual entropy of ice at absolute zero, a problem which had previously engaged the attention of Bernal and Fowler, and Pauling [167].

Lieb was followed by Baxter, whose solution of a more complex ferroelectric model [168] (the eight-vertex model) gave rise to exponents which vary continuously with the strength of the interaction. This was a major challenge to the concept of universality, which was resolved when Kadanoff and Wegner [169] demonstrated that a special symmetry of the model gives rise to the continuously varying exponents; this specific feature would not be expected to occur in normal physical systems.

Two-dimensional models with $n = 2$ gave rise to an interesting new type of phase transition. For such models it can be shown rigorously [170] that the spontaneous magnetization (i.e. the order parameter) is zero when $T > 0$. Nevertheless, series expansions gave evidence of a non-zero critical temperature [171]. In 1973 Kosterlitz and Thouless [172] put forward the idea of a new type of ordering, which they called topological long-range order related to the presence of vortices, and this is relevant for the model. They showed that there exist metastable states corresponding to vortices which are closely bound in pairs below some Curie temperature while above it they become free. This idea proved fruitful [173] and the transition is now considered to be well established.

A theoretical model which burst into prominence after being ignored for nearly 20 years is the Potts [174] model. It was first put forward as a generalization of the Ising model having q orientations but only two different energies of interaction; the critical point could be located exactly. A number of physical systems have been identified for which the Potts model is a reasonable representation, and the difference in symmetry between the n-vector ferromagnetic model and the q-component Potts model gives rise to important differences in critical behaviour. Although no exact solution analogous to Onsager's exists for the Potts model, a good deal of exact information is available on critical behaviour and critical exponents [175].

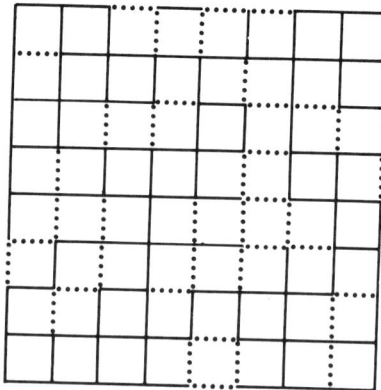

Figure 7.13. *Typical configuration of a bond percolation process. Full lines: free channels, broken lines: blocked channels.*

7.5.13. Percolation processes [176]

The term *percolation process* was first used by Broadbent and Hammersley [177] to describe the flow of a fluid through a random *medium* in contrast to a diffusion process in which the randomness is associated with the particles of the *fluid*. Consider a network, which for convenience we take to be a crystal lattice, with the bonds of the lattice serving as connecting channels; each channel has a finite probability $(1 - p)$ of being blocked. Fluid is introduced at a particular point, and we wish to calculate the probabilities of other points in the network being wet or dry (figure 7.13). In particular we are interested in the probability that the fluid will spread to infinity, and Broadbent and Hammersley were able to establish the existence of a probability of blocking, above which a fluid starting at one point of the lattice would not spread to infinity.

In the above model randomness arises in the bonds of the lattice. An alternative model [178] introduces randomness into the sites of the lattice. Consider a lattice with a random mixture of A and B atoms occupying its sites, the ratio of concentrations being $p : (1 - p)$. The neighbouring sites occupied by like atoms are regarded as connected, and we are interested in the distribution of *connected clusters*. When p is small the A atoms form small islands in a sea of B atoms (figure 7.14). But as the concentration p increases, a critical concentration p_c is reached at which there is non-zero probability that the islands connect to form an 'infinite' cluster spanning the network. If the B atoms are non-magnetic and the A atoms magnetic, we might expect p_c to be associated with the onset of ferromagnetism [179].

Hammersley [180] and his collaborators pioneered the use of Monte Carlo methods for *bond percolation* processes, whilst Domb and Sykes [178] showed that series expansions for *site percolation* processes could be developed in powers of p, and that the techniques used for

Thermodynamics and Statistical Mechanics (In Equilibrium)

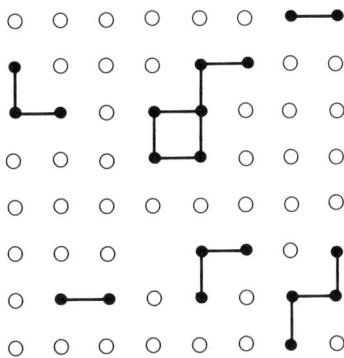

Figure 7.14. *Typical configurations of a site percolation process. Nearest-neighbour black atoms are connected by bonds.*

the exploration of critical behaviour in magnets could be applied to them. The relationship between bond and site percolation processes was clarified by Fisher [181] who showed that every bond process is equivalent to a site process on a different lattice which he called the *covering lattice*.

During the 1960s the world of physics showed little interest in percolation phenomena and only a handful of workers contributed to the literature. There were clear analogies with the magnetic critical phenomena described in the earlier sections of this chapter. For example, the behaviour of $P(p)$, the number of sites belonging to the infinite cluster, closely parallels the behaviour of the spontaneous magnetization [182], and the probability that two sites belong to a single cluster, the *pair connectedness*, parallels the pair correlation function [183]. Sykes and Essam [184] conjectured the exact values of p_c for a number of two-dimensional lattices, recalling the Kramers and Wannier conjecture [110] for the Ising model. Mean-field solutions had their analogue in closed form solutions for tree and cactus structures [185].

The origin of this analogy was revealed when Kasteleyn and Fortuin [186] showed that the bond percolation model is identical with the q-component Potts model with $q = 1$. Hence the general theoretical description provided by the RG applies to percolation but the symmetry is different, and the appropriate universality class in not in the n-vector framework. Toulouse [187] showed that for percolation a mean-field solution is attained only when the space dimension d becomes 6; hence an ε-expansion is not a very useful tool for calculating critical behaviour in three dimensions.

As an interesting aside, Temperley and Lieb [188] connected the percolation model (as well as a number of other models in lattice statistics) to a mathematical problem formulated by Whitney in 1932. They pointed out that the determination by Sykes and Essam of the

573

exact value of p_c for two-dimensional lattices was a rediscovery of a result obtained by Whitney.

The literature on percolation grew extensively in the 1970s and 1980s and a large number of new ideas were injected into the field. They provided a natural application in statistical mechanics of the novel and far-reaching ideas of B Mandelbrot, which we now discuss in the final contribution to this section.

7.5.14. Self-similarity—fractals

In my first year as a mathematics student at Cambridge we were taught to distinguish between continuity and differentiability. It is very easy to see that a function with a sharp change in slope at a given point is continuous but not differentiable. But we were also told of a function discovered by Weierstrass in the 1870s that was everywhere continuous but nowhere differentiable. This was a much more sophisticated idea, and even the very good students who were able to visualize such a function considered it to be a pathological invention of an ingenious mathematician, with no possible practical applications. Mandelbrot [189] demonstrated that things very like these curves do occur in Nature (coastlines are a striking example) and in statistical mechanics. He focused attention on the *self-similar* characteristic of these curves; however much the scale is reduced, the same twists and turns always remain.

A second mathematical idea with no apparent practical application was formulated by Hausdorff in 1929 and developed by Besicovitch in the 1930s. The *dimension* of a space had always been considered as discrete. But if the dimension d of a set of points is defined by means of the relation $N(R) = AR^d$, where $N(R)$ is the number of points of the set within a radius R, the dimension d can take on fractional values. The triangles arranged in figure 7.15 are known as a Sierpinski gasket and the self-similar properties are manifest. Using the above definition of dimension, it is easy to show that the boundary between black and white has dimension $\ln 3 / \ln 2 = 1.5849$. Mandelbrot introduced the term *fractal* to describe such sets, and the non-integral value of d appropriate to them he called their *fractal dimension*.

He identified fractals among the curves and sets which are of interest in statistical mechanics, e.g. random walks, self-avoiding walks, percolation clusters, and their fractal dimensions were determined (or estimated). The literature on fractals has burgeoned in the past few years with applications in many different areas.

7.6. Further topics

The exponential growth which physics has experienced in this century makes it impossible to deal with all topics in thermodynamics and statistical mechanics in which significant progress has been achieved. A choice had to made, but two additional topics deserve brief mention.

Thermodynamics and Statistical Mechanics (In Equilibrium)

Figure 7.15. *Self-similarity illustrated by a Sierpinski gasket. The boundary between black and white has fractal dimension ln 3/ln 2.*

The following passage appears under the heading 'Limitation of the second law of thermodynamics' near the end of Maxwell's *Theory of Heat*

> But if we conceive a being whose faculties are so sharpened that he can follow every molecule in its course, such a being, whose attributes are still as essentially finite as our own, would be able to do what is at present impossible to us. For we have seen that the molecules in a vessel full of air at uniform temperature are moving with velocities by no means uniform, though the mean velocity of any great number of them, arbitrarily selected, is almost exactly uniform. Now let us suppose that such a vessel is divided into two portions, A and B, by a division in which there is a small hole, and that a being, who can see the individual molecules, opens

and closes this hole, so as to allow only the swifter molecules to pass from A to B, and only the slower ones to pass from B to A. He will thus, without expenditure of work, raise the temperature of B and lower that of A, in contradiction to the second law of thermodynamics.

Maxwell's Demon [190] proved to be extraordinarily fruitful in provoking discussion which led to important extensions of the concept of entropy. He wished to highlight the statistical nature of the second law, and really thought that it could be contravened. Kelvin (who introduced the term *demon*) agreed, but emphasized that the demon must be an intelligent being endowed with free will, and could not be purely mechanical. Smoluchowski in 1913 envisaged the arrangements which must be made for the practical use of such a demon. He affirmed that automatic operation is impossible, but thought that even intelligent beings might not be able to violate the second law.

A crucial step was taken by Leo Szilard in 1929 who suggested that the *memory* needed for the demon to operate could be indissolubly connected with the production of entropy; this opened the door to a connection between entropy and information, an idea later developed and amplified by Shannon, Wiener and Brillouin. Szilard illustrated his thesis that memory can entail a continued reduction of entropy by devising a cyclic process which could convert heat into work (the Szilard Engine). If the second law were to retain its validity, the processes of measurement themselves must imply a production of entropy.

Maxwell's Demon cannot operate was the title of a paper published by Brillouin in 1951, in which he argued as follows [191]

> Before an intelligent being can use its intelligence it must perceive its objects, and that requires physical means of perception. Visual perception in particular requires the illumination of the object. Seeing is essentially a non-equilibrium phenomenon; the cylinder in which the demon operates is a closed black body, and to identify particular molecules the observer must use a lamp that emits light of a wavelength not well represented in the black-body radiation. The eventual absorption of this light increases the entropy of the system.

Brillouin went on to show mathematically that this increase of entropy more than balances the decrease of entropy which the demon can effect.

Although this seemed to exorcise the demon, the last word had not yet been spoken. Early in the 1960s Rolf Landauer investigated the thermodynamics of data processing, and concluded that one cannot clear a memory register without generating heat and adding to the entropy of the environment. He pointed out that clearing a memory is a thermodynamically irreversible operation. Charles Bennet then proposed that these ideas provided the basis for Szilard's contention. The last step of the cycle involved resetting the engine's memory to a blank state, and this could not be done without adding at least one bit of entropy to the environment. The correct thermodynamic interpretation is that the engine increases the entropy of its memory in order to decrease the entropy of its environment.

An important phase transition not discussed in section 7.5 is the first-order melting transition. Although there was no dramatic breakthrough comparable to that relating to critical points, important progress was achieved in understanding the nature of the transition.

The simplest statistical-mechanics model which might cast some light on the situation is that of a hard-sphere gas. Does it manifest a phase transition and if so what is the character of the transition? Despite much theoretical effort during the first half of the century no clear answer was forthcoming. Then in the 1950s Alder and Wainwright pioneered a new approach which gave an unequivocal answer—there is a first-order phase transition whose entropy is comparable to the entropy of melting of heavy rare gas fluids. They [192] made use of the tremendous improvement in computing facilities to solve the Newtonian equations of a few hundred spheres confined to a box, trace out their progress after a given initial state, and hence calculate all their thermodynamic properties. This method they termed *molecular dynamics*. A more economical approach (Monte Carlo method) made use of representative statistical samples. Theoretical physicists were a little taken aback at these pragmatic methods, but during the past few decades they have come to play a role of increasing importance in providing insight and information on otherwise intractable problems, and in checking theoretical work which makes use of uncontrolled approximations.

Insight into the nature of the phase transition discovered by Alder and Wainwright and the nature of the liquid state which underlies it was provided by Bernal [193]. Investigating what happens to an assembly of random hard spheres as the volume is steadily decreased, Bernal introduced two important new concepts, random loose packing (RLP) and random close packing (RCP). If pressure is applied to the assembly without any attempt to adjust or shake the spheres, one can reach a maximum density of about 0.60 (RLP); if a shaking procedure is now adopted it is possible to reach a density of 0.637 (RCP, but it is not possible to reach the crystal close packed density of 0.74 without a complete rearrangement).

See also p 619

Experimentally a good deal of interesting information on melting had come from experiments at high pressures [194], and since the density is increased by a substantial factor it is clear that a soft repulsive potential is involved. The following formula was suggested empirically by Simon [195] for the relation between melting temperature and melting pressure,

$$\frac{P}{P_i} = \left(\frac{T}{T_0}\right)^c - 1. \tag{50}$$

Here T_0 is the normal melting temperature, P_i the internal pressure of the solid and c is a constant. This formula was found to fit a wide variety of solids of different types from solidified inert gases to metals. Domb [196] showed from quite elementary considerations that if melting were considered as a generalized order–disorder transformation, a formula of type (50) would be derived, the constant c being related to the repulsive power of the intermolecular potential.

Acknowledgments

Advice and help from L Muldawer on the early history of binary alloys, from N Kurti on nuclear cooling and nuclear refrigeration and from M E Fisher on the early history of the RG is gratefully acknowledged.

References

[1] A general reference on the material of this section is Brush S G 1976 *The Kind of Motion We Call Heat* (Amsterdam: North-Holland)
[2] See, for example, Dampier W C 1966 *A History of Science* (Cambridge: Cambridge University Press) p 226
[3] Sadi Carnot N L 1824 *Reflexions sur la Puissance Motrice du Feu* (Paris: Bachelier) (Engl. Transl. Mendoza E 1960 (New York: Dover))
[4] Clausius R 1865 *Pogg. Ann.* **125** 400
[5] Maxwell J C 1878 *Nature* **17** 257
[6] Gibbs J W 1928 *1875–8 Transactions of the Connecticut Academy; The Collected Works of Willard Gibbs* (New York: Longmans Green) (Dover edition, 1960)
[7] Kirchhoff G R 1859 *Mon.-Ber. Akad. Wiss. Berlin* **783**; 1860 *Ann. Phys.* **109** 275
[8] Clausius R 1857 *Ann. Phys.* **100** 353; 1858 *Ann. Phys.* **105** 259 (Engl. Transl. 1870 *Phil. Mag.* **40** 122)
[9] Maxwell J C 1860 *Report of the 29th Meeting of the British Association* (reproduced in Harman P M (ed) 1990 *The Scientific Letters and Papers of James Clerk Maxwell* vol 1 (Cambridge: Cambridge University Press) p 615); 1860 *Phil. Mag.* **19** 19; 1860 *Phil. Mag.* **20** 21, 33
[10] Boltzmann L 1868 *Wien. Ber.* **58** 517; 1877 *Wien. Ber.* **76** 373
[11] Maxwell J C 1872 *Theory of Heat* 3rd edn (London: Longmans Green) p 308 (reprinted 1970 (Connecticut: Greenwood))
[12] See discussion by Pais A 1982 *Subtle is the Lord* (Oxford: Oxford University Press) p 60–5
[13] Kuhn T S 1978 *Black-Body Theory and the Quantum Discontinuity 1894–1912* (Oxford: Clarendon)
[14] Poincaré H 1890 *Acta Mathematica* **13** 1
[15] Zermelo E 1896 *Ann. Phys., Lpz.* **57** 485
[16] Boltzmann L 1896 *Ann. Phys., Lpz.* **57** 773

Thermodynamics and Statistical Mechanics (In Equilibrium)

[17] Gibbs J W 1902 *Elementary Principles in Statistical Mechanics* (New York: Scribner)
[18] Planck M 1900 *Ann. Phys., Lpz.* **1** 719
[19] Planck M 1900 *Verh. D. Phys. Ges.* **2** 202
[20] Planck M 1900 *Verh. D. Phys. Ges.* **2** 237; 1901 *Ann. Phys.* **4** 564
[21] Boltzmann L 1964 *Lectures on Gas Theory* (Engl. Transl. Brush S G) (California: University of California Press)
[22] For further details see reference [12] and Ehrenfest P and Ehrenfest T 1911 *Enz. Math. Wiss.* vol 4, part 2 (Leipzig: Teubner) (Engl. Transl. Moravcsik M J 1959 *The Conceptual Foundations of the Statistical Approach in Mechanics* (Ithaca, NY: Cornell University Press))
[23] Planck M 1924 *Ann. Phys., Lpz.* **75** 673
[24] Bose S N 1924 *Z. Phys.* **26** 178
[25] Pais A 1982 *Subtle is the Lord* (Oxford: Oxford University Press) p 246
[26] Einstein A 1924 *Sitz. Preuss. Akad. Wiss.* **3** 261; 1925 *Sitz. Preuss. Akad. Wiss.* **3**
[27] Pauli W 1925 *Z. Phys.* **31** 765
[28] Fermi E 1926 *Z. Phys.* **36** 902
[29] Dirac P A M 1926 *Proc. R. Soc.* A **112** 661
[30] Niven W D (ed) 1890 *The Scientific Papers of J C Maxwell* vol 2 (Cambridge: Cambridge University Press) p 433 (reprinted 1965 (New York: Dover))
[31] Hund F 1927 *Z. Phys.* **42** 93
     Hori 1927 *Z. Phys.* **44** 834
[32] Dennison D M 1927 *Proc. R. Soc.* A **115** 483
[33] Pais A 1982 *Subtle is the Lord* (Oxford: Oxford University Press) p 430
[34] Uhlenbeck G E 1927 *PhD Thesis* (The Hague: Nyhoff)
[35] See, for example, Kahn B and Uhlenbeck G E 1938 *Physica* **4** 399
[36] Fowler R H 1926 *Mon. Not. R. Astron. Soc.* **87** 114
[37] Milne E A 1945 *Obituary Notices of Fellows of the Royal Society* **5** 73
[38] Fowler R H 1936 *Statistical Mechanics* 2nd edn (Cambridge: Cambridge University Press)
[39] Schrödinger E 1946 *Statistical Thermodynamics* (Cambridge: Cambridge University Press)
[40] Carathéodory C 1909 *Math. Ann.* **67** 355; 1925 *S. B. Preuss Akad. Wiss.* **39**
     See also Born M 1921 *Phys. Z.* **22** 218, 249, 282
[41] Darwin C G and Fowler R H 1922 *Proc. Camb. Phil. Soc.* **21** 391
[42] General references for this section are:
     Nernst W 1926 *The New Heat Theorem* (Engl. Transl. Barr G) (London: Methuen)
     Simon F E 1956 *Yearbook of the Physical Society (40th Guthrie Lecture)* vol 1
     Wilks J 1961 *The Third Law of Thermodynamics* (Oxford: Oxford University Press)
[43] Nernst W 1906 *Kgl. Ges. Wiss. Gott.* **1**
[44] See, for example, Glasstone S 1948 *Textbook of Physical Chemistry* 2nd edn (London: Macmillan) p 828
[45] Nernst W 1912 *Ber. Kon. Preuss. Acad.* February
[46] See, for example, Einstein A 1913 *Congrès Solvay (Paris 1921)* 293
[47] Nernst W 1914 *Z. Elektrochem.* **20** 397
[48] Simon F E 1927 *Z. Phys.* **41** 806
[49] Simon F E 1930 *Ergeb. Exakt. Naturw.* **9** 222
[50] Fowler R H and Guggenheim E A 1939 *Statistical Thermodynamics* (Cambridge: Cambridge University Press)
[51] See, for example, Fisher M E 1960 *Proc. R. Soc.* A **256** 502
[52] Domb C and Dugdale J S 1957 Solid helium *Progress in Low Temperature*

Physics vol 2, ed C J Gorter (Amsterdam: North-Holland) ch 11
[53] Simon F E and Swenson C A 1950 *Nature* **165** 829
[54] Berman R and Simon F E 1955 *Z. Elektrochem.* **59** 333
[55] Bundy F P, Hall H T, Strong H M and Wentorf R H 1955 *Nature* **176** 51
See also Davies G 1984 *Diamond* (Bristol: Hilger) ch 6
[56] Giauque W F and Johnston H L 1929 *J. Am. Chem. Soc.* **51** 2300
[57] Frank A and Clusius K 1937 *Z. Phys. Chemie* B **36** 291
[58] Giauque W F and Stout J W 1936 *J. Am. Chem. Soc.* **58** 1144
[59] Clayton J O and Giauque W F 1932 *J. Am. Chem. Soc.* **54** 2610
[60] Hill R W and Ricketson B W A 1954 *Phil. Mag.* **45** 277
[61] Kurti N, Robinson F N H, Simon F E and Spohr D A 1957 *Nature* **178** 450
Chapelier M, Goldman M, Chau V H and Abragam A 1969 *C. R. Acad. Sci., Paris* B **268** 1530
[62] Ehnholm G J, Ekstrom J P, Jacquinot J F, Loponen M T, Lounasmaa O V and Soini J K 1979 *Phys. Rev. Lett.* **42** 1702; 1980 *J. Low Temp. Phys.* **39** 417
Hakonen P J, Vuorinen A T and Martikainen J E 1993 *Phys. Rev. Lett.* **70** 2818
It should be noted that in the above case, termed nuclear cooling, only the nuclear spin system reaches the low temperature whilst the surroundings—conduction electrons in the metal etc, may be a thousand times higher. In nuclear refrigeration the spin system cools its surroundings but the lowest temperatures attained are much higher. A recent reference on nuclear refrigeration is Enrico N P, Fisher S N, Guénault A M, Miller I E and Pickett G R 1994 *Phys. Rev.* B **49** 6339
[63] Pound R V 1951 *Phys. Rev.* **81** 156
Purcell E M and Pound R V 1951 *Phys. Rev.* **81** 279
Ramsey N F and Pound R V 1961 *Phys. Rev.* **81** 278
[64] Simon F E 1955 *Temperature—Its Measurement and Control in Science and Industry* vol 2 (New York: Reinhold) p 9
[65] Andrews T 1869 *Phil. Trans. R. Soc.* **159** 575
[66] van der Waals J H 1873 Over de continuiteit van der gas en vloeisoftoestand *Thesis* Leiden
[67] Clausius R 1870 *Ann. Phys.* **141** 124 (Engl. Transl. *Phil. Mag.* **40** 122)
[68] Maxwell J C 1875 *Nature* **11** (Niven W D (ed) 1890 *The Scientific Papers of J C Maxwell* vol 2 (Cambridge: Cambridge University Press) p 418, reprinted 1965 (New York: Dover))
[69] Verschaeffelt J E 1900 *Versl. Kon. Akad. Wetensch. Amsterdam* **5** 94 (*Leiden Comm.* **28**)
[70] Gilbert W 1600 *De Magnete* (Engl. Transl. Mottelay P I 1893 (New York) and 1900 for the Gilbert Club (London)) p 66
[71] Hopkinson J 1889 *Phil. Trans. R. Soc.* A **180** 443
[72] Curie P 1895 *Ann. Chim. Phys.* **5** 289
[73] Weiss P 1907 *J. Physique* **6** 661
[74] Langevin P 1905 *J. Physique* **4** 678; 1905 *Ann. Chim. Phys.* **5** 70
[75] Weiss P 1907 *J. Physique* **6** 662
Weiss P and Kamerlingh Onnes H 1910 *J. Physique* **9** 555
[76] Cernuschi F and Eyring H 1939 *J. Chem. Phys.* **7** 547
[77] Avenarius M 1874 *Ann. Phys. Chem.* **151** 306
[78] Altschul M 1893 *Z. Phys. Chem.* **11** 578
von Wesendonck K 1894 *Z. Phys. Chem.* **15** 262
Travers M W and Usher F L 1906 *Proc. R. Soc.* A **78** 247
Young S 1906 *Proc. R. Soc.* A **78** 262
[79] Smoluchowski M S 1908 *Ann. Phys., Lpz* **25** 205; 1912 *Phil. Mag.* **23** 165

Thermodynamics and Statistical Mechanics (In Equilibrium)

[80] Einstein A 1910 *Ann. Phys., Lpz* **33** 1275
[81] Ornstein L S and Zernike F 1914 *Proc. Akad. Sci. Amsterdam* **17** 793; 1916 *Proc. Akad. Sci. Amsterdam* **18** 1520
[82] Fisher M E 1964 *J. Math. Phys.* **5** 944
[83] Bain E C 1923 *Trans. Am. Inst. Min. (Metall.) Eng.*
      Johansson C H and Linde J O 1925 *Ann. Phys., Lpz* **78** 439
[84] Bragg W L and Williams E J 1934 *Proc. R. Soc.* A **145** 699
[85] Bragg W L and Williams E J 1935 *Proc. R. Soc.* A **151** 540
[86] Bethe H A 1935 *Proc. R. Soc.* A **216** 45
[87] Gorsky W 1929 *Z. Phys.* **50** 84
      Borelius G 1934 *Ann. Phys., Lpz* **20** 57
      Dehlinger U 1930 *Z. Phys.* **64** 359; 1932 *Z. Phys.* **14** 267; 1933 *Z. Phys.* **83** 832
[88] Landau L D 1937 *Phys. Z. Sovietunion* **11** 26, 545
      Landau L D and Lifshitz E M 1938 *Statistical Physics* (Oxford: Oxford University Press)
[89] Ehrenfest P 1933 *Proc. Kon Akad. Wetenschap Amsterdam* **36** 147
[90] Pippard A B 1957 *Classical Thermodynamics* (Cambridge: Cambridge University Press) ch 9
[91] Domb C 1951 *Phil. Mag.* **42** 1316
      Munster A 1952 *Comptes Rendus de la deuxième Reunion Chimie Physique Paris* p 21 (and discussion p 28)
[92] Landau L D 1936 *Nature* **138** 840
[93] Landau L D and Lifshitz E M 1959 *Statistical Physics* (revised edition) (London: Pergamon)
[94] van Laar J J 1899 *Proc. Acad. Sci. Amsterdam* **1** 273
[95] Boltzmann L 1899 *Verlag. Gewone Vergardering Natuur K. Netherlands. Akad. Wetensch.* **7** 484
[96] Kamerlingh Onnes H 1901 *Verslagen Kon. Akad. Wetensch. Amsterdam* **10** 136
[97] Mayer J E 1937 *J. Chem. Phys.* **5** 67
[98] Yvon J 1937 *Actualites Scientifiques et Industrielles* (Paris: Herman) p 542
[99] Uhlenbeck G E 1966 *Critical Phenomena* (NBS Miscellaneous Publications 273) ed M S Green and J V Sengers (Washington, DC: NBS) p 5
[100] Born M and Fuchs K 1938 *Proc. R. Soc.* A **266** 391
[101] Gouy A 1892 *C. R. Acad. Sci., Paris* **115** 720
[102] Schneider W G 1952 *Changements de Phases Comptes Rendus de la deuxième Reunion Annuelle, Société de Chimie Physique* p 69
      Weinberger M A and Schneider W G 1952 *Canad. J. Chem.* **30** 422
[103] Yang C N and Lee T D 1952 *Phys. Rev.* **87** 404 and 410.
[104] Uhlenbeck G E and Ford G W 1962 *Studies in Statistical Mechanics* vol 1, ed J deBoer and G E Uhlenbeck (Amsterdam: North-Holland) p 123
[105] Ising E 1925 *Z. Phys.* **31** 253
[106] Peierls R E 1936 *Proc. Camb. Phil. Soc.* **32** 477
[107] Brush S G 1967 *Rev. Mod. Phys.* **39** 883
[108] Heisenberg W 1928 *Z. Phys.* **49** 619
[109] Onsager L 1944 *Phys. Rev.* **65** 317
[110] Kramers H A and Wannier G H 1941 *Phys. Rev.* **60** 252, 263
[111] Kaufman B and Onsager L 1949 *Phys. Rev.* **76** 1244
[112] Onsager L 1949 *Proc. Florence Conf. on Statistical Mechanics Nuovo Cimento* **6** 261 (result announced with no details of the calculation)
[113] Yang C N 1951 *Phys. Rev.* B **5** 808
[114] Guggenheim E A 1945 *J. Chem. Phys.* **13** 253
[115] Hapgood H W and Schneider W G 1954 *Canad. J. Chem.* **32** 98

[116] Domb C 1949 *Proc. R. Soc.* A **196** 36 and 199; 1974 *Phase Transitions and Critical Phenomena* vol 3, ed C Domb and M S Green (London: Academic) ch 6
[117] Domb C and Martin J L 1974 *Phase Transitions and Critical Phenomena* vol 3, ed C Domb and M S Green (London: Academic) ch 1 and 2
[118] Baker G A 1961 *Phys. Rev.* **124** 768
[119] Baker G A and Gammel J L 1970 *The Padé Approximant in Theoretical Physics* (London: Academic)
[120] Fisher M E 1967 *Rev. Prog. Phys.* **30** 615
[121] Heller P 1967 *Rep. Prog. Phys.* **30** 731
[122] Fairbank W M, Buckingham M J and Kellers C F 1957 *Proc. 5th Int. Conf. on Low Temperature Physics* (Madison, WI: University of Wisconsin Press)
[123] Heller P and Benedek G 1965 *Phys. Rev. Lett.* **14** 71
[124] Heller P and Benedek G 1962 *Phys. Rev. Lett.* **8** 428
[125] Domb C and Miedema A R 1964 *Progress in Low Temperature Physics* vol 4, ed C J Gorter (Amsterdam: North-Holland) ch 4
de Jongh L J and Miedema A R 1974 *Adv. Phys.* **23** 1
[126] Domb C 1960 *Adv. Phys.* **9** 149, 245
[127] Domb C and Sykes M E 1962 *Phys. Rev.* **128** 168
[128] Jasnow D and Wortis M 1968 *Phys. Rev.* **176** 739
[129] Essam J W and Fisher M E 1963 *J. Chem. Phys.* **38** 147
[130] Rushbrooke G S 1963 *J. Chem. Phys.* **39** 842
[131] Stanley H E 1971 *Introduction to Phase Transitions and Critical Phenomena* (Oxford: Oxford University Press) ch 4
[132] Kac M, Uhlenbeck G E and Hemmer P C 1963 *J. Math. Phys.* **4** 216
[133] See, for example, Hemmer P C and Lebowitz J L 1976 *Phase Transitions and Critical Phenomena* vol 5b, ed C Domb and M S Green (London: Academic) ch 2
[134] Berlin T H and Kac M 1952 *Phys. Rev.* **86** 821
[135] Uhlenbeck G E 1965 *Critical Phenomena (NBS Miscellaneous Publications 273)* ed M S Green and J V Sengers (Washington, DC: NBS) p 3
[136] Widom B 1965 *J. Chem. Phys.* **43** 3898
[137] Domb C and Hunter D L 1954 *Proc. Phys. Soc.* **86** 1147
[138] Patashinskii A Z and Pokrovskii V L 1966 *Zh. Eksp. Teor. Fiz.* **50** 439; *Sov. Phys.–JETP* **23** 292
[139] Griffiths R B 1967 *Phys. Rev.* **158** 176
[140] Kadanoff L P 1966 *Physics* **2** 263
[141] Griffiths R B 1970 *Phys. Rev. Lett.* **24** 1479
[142] Stanley H E 1968 *Phys. Rev.* **176** 718
[143] Joyce G S 1966 *Phys. Rev.* **146** 349
[144] Larkin A I and Khmelnitskii D E 1969 *Sov. Phys.–JETP* **29** 1123
[145] Domb C 1971 *Statistical Mechanics at the Turn of the Decade* ed E G D Cohen (New York: Dekker)
[146] de Pasquale F, di Castro C and Jona-Lasinio G 1971 *Proc. Enrico Fermi School on Critical Phenomena* ed M S Green (London: Academic) p 113
[147] Stueckelberg E C G and Peterman A 1953 *Helv. Phys. Acta* **26** 499
Gell-Mann M and Low F E 1954 *Phys. Rev.* **95** 1300
[148] Wilson K G 1975 *Adv. Math.* **16** 170
[149] Wilson K G 1983 *Rev. Mod. Phys.* **55** 583
[150] Wilson K G 1975 *Rev. Mod. Phys.* **47** 773
[151] Ginzburg V L and Landau L D 1950 *Sov. Phys.–JETP* **20** 1064
[152] Wilson K G and Fisher M E 1972 *Phys. Rev. Lett.* **28** 240
[153] Le Guillou J C and Zinn-Justin J 1977 *Phys. Rev. Lett.* **39** 95; 1980 *Phys. Rev.* B **21** 3976

[154] Nickel B 1981 *Proc. 14th Int. Conf. on Statistical Mechanics, Edmonton Canada Physica* A **106** 40; 1991 *Physica* A **177** 189
[155] Neimeijer T and Van Leeuwen J M J 1976 *Phase Transitions and Critical Phenomena* vol 6, ed C Domb and M S Green (London: Academic) ch 7
[156] Wilson K G 1974 *Physica* **73** 119
[157] See, for example, Brezin E, Le Guillou J C and Zinn-Justin J 1976 *Phase Transitions and Critical Phenomena* vol 6, ed C Domb and M S Green (London: Academic) ch 3
[158] Aharony A 1976 *Phase Transitions and Critical Phenomena* vol 6, ed C Domb and M S Green (London: Academic) ch 6
[159] Flory P J 1953 *Principles of Polymer Chemistry* (Ithaca, NY: Cornell University Press)
[160] Hammersley J M and Morton K W 1954 *J. R. Stat. Soc.* B **16** 23
[161] Domb C 1990 *Disorder in Physical Systems* ed G R Grimmett and D J A Welsh (Oxford: Oxford University Press) p 33
[162] Fisher M E and Sykes M F 1959 *Phys. Rev.* **114** 45
       Domb C and Sykes M F 1961 *J. Math. Phys.* **2** 63
[163] de Gennes P G 1972 *Phys. Lett.* A **38** 339
[164] de Gennes P G 1979 *Scaling Concepts in Polymer Physics* (Ithaca, NY: Cornell University Press)
       The pioneer in the application of field theoretical methods to polymer problems was S F Edwards, see Edwards S F 1965 *Proc. Phys. Soc.* **85** 1
[165] Lieb E H 1967 *Phys. Rev. Lett.* **18** 692; 1967 *Phys. Rev. Lett.* **18** 1046; 1967 *Phys. Rev. Lett.* **19** 108; 1967 *Phys. Rev.* **162** 162
[166] Onsager L 1939 *Discussion at Conference on Dielectrics* (New York: New York Academy of Sciences)
[167] Bernal J D and Fowler R H 1933 *J. Chem. Phys.* **1** 515
       Pauling L 1935 *J. Am. Chem. Soc.* **57** 2680
       See also Giauque W F and Stout J W 1963 *J. Am. Chem. Soc.* **58** 1144
[168] Baxter R J 1971 *Phys. Rev. Lett.* **26** 832; *Ann. Phys., NY* **70** 193
[169] Kadanoff L P and Wegner E J 1971 *Phys. Rev.* B **4** 3989
[170] Mermin N D and Wagner H 1966 *Phys. Rev. Lett.* **17** 1133
[171] Stanley H E and Kaplan T A 1966 *Phys. Rev. Lett.* **17** 913
       See also Stanley H E 1974 *Phase Transitions and Critical Phenomena* vol 3, ed C Domb and M S Green (London: Academic) ch 7
[172] Kosterlitz J M and Thouless D J 1973 *J. Phys. C: Solid State Phys.* **6** 1181
[173] Nelson D R 1983 *Phase Transitions and Critical Phenomena* vol 7, ed C Domb and J L Lebowitz (London: Academic) ch 1
[174] Potts R B 1952 *Proc. Camb. Phil. Soc.* **48** 106
       For an historical account see Domb C 1974 *J. Phys. A: Math. Nucl. Gen.* **7** 1335
[175] See, for example, Nienhuis B 1987 *Phase Transitions and Critical Phenomena* vol 11, ed C Domb and J L Lebowitz (London: Academic) ch 1; reference [173] vol 11, ch 1
[176] Stauffer D 1985 *Introduction to Percolation Theory* (London: Taylor and Francis) (2nd edn with A Aharony 1991)
[177] Broadbent S R and Hammersley J M 1957 *Proc. Camb. Phil. Soc.* **53** 629
[178] Domb C and Sykes M F 1961 *Phys. Rev.* **122** 77
[179] Elliot R J, Heap B R, Morgan D J and Rushbrooke G S 1960 *Phys. Rev. Lett.* **5** 366
[180] Frisch H L, Sonnenblick E, Vyssotsky V A and Hammersley J M 1961 *Phys. Rev.* **124** 1020
       Vyssotsky V A, Gordon S B, Frisch H L and Hammersley J M 1962 *Phys. Rev.* **123** 1566

[181] Fisher M E 1961 *J. Math. Phys.* **2** 620
[182] Sykes M F, Glen M and Gaunt D S 1974 *J. Phys. A: Math. Nucl. Gen.* **7** L105
[183] Essam J W 1972 *Phase Transitions and Critical Phenomena* vol 2, ed C Domb and M S Green (London: Academic) ch 6
[184] Sykes M F and Essam J W 1964 *J. Math. Phys.* **5** 1117
[185] Fisher M E and Essam J W 1961 *J. Math. Phys.* **2** 609
[186] Kasteleyn P W and Fortuin C 1969 *J. Phys. Soc. Japan (Suppl.)* **26** 11
[187] Toulouse G 1974 *Nuovo Cimento* B **23** 234
[188] Temperley H N V and Lieb E H 1971 *Proc. R. Soc.* A **322** 251
[189] Mandelbrot B B 1977 *Fractals Form Chance and Dimension*; 1982 *The Fractal Geometry of Nature* (San Francisco: Freeman). These books contain a lucid and fascinating account of the mathematical background and history.
[190] A comprehensive review of this topic with detailed references and reprints of important papers is provided in Leff H S and Rex A F (ed) 1990 *Maxwell's Demon, Entropy, Information, Computing* (Bristol: Hilger)
[191] Ehrenberg W 1967 *Sci. Am.* **217** 103
[192] Alder B J and Wainwright T E 1958 *Transport Processes in Statistical Mechanics* ed I Prigogine (New York: Interscience); 1960 *J. Chem. Phys.* **33** 1439; 1962 *Phys. Rev.* **127** 359
[193] See, for example, Collins R 1972 *Melting and Statistical Geometry of Simple Liquids* vol 2 of the series of reference [117]
[194] See Bridgman P W 1949 *The Physics of High Pressure* (Bell)
[195] Simon F E 1937 *Trans. Farad. Soc.* **33** 65
[196] Domb C 1951 *Phil. Mag.* **42** 1316; 1958 *Nuovo Cimento (Suppl.)* **9** 9

Chapter 8

NON-EQUILIBRIUM STATISTICAL MECHANICS OR THE VAGARIES OF TIME EVOLUTION

Max Dresden

8.1. A phase of transition and consolidation

8.1.1. The incredible initial decade
The first decade of twentieth century physics is associated with such great names as Rutherford, the Curies, Planck, Gibbs and especially Einstein. Although less well known than his monumental contributions in relativity and quantum theory, he made fundamental contributions to statistical mechanics in a series of papers [1]. Einstein succeeded in obtaining the laws of thermodynamics, including irreversibilities, from mechanics and his notion of the probability of a configuration. These same statistical considerations also led him to a study of fluctuations in physical systems; thus the fluctuation from the mean of the energy in a thermal system is

$$(\Delta E)^2 \equiv \langle (E - \langle E \rangle)^2 \rangle = kT^2 \frac{d}{d\tau} \langle E \rangle \tag{1}$$

where $\langle E \rangle$ stands for the mean thermal energy.

This fluctuation relation played a fundamental role in the later development of quantum theory especially as used by Einstein and Lorentz. Einstein showed uncanny insight in looking for systems where fluctuations are significant. This was one of the motivations for studying the motion of a particle suspended in a stationary medium [2]. Thus statistical considerations played a fundamental role in analysing both quantum phenomena and Brownian motion.

In 1902 Gibbs' monumental treatise of statistical mechanics appeared [3]. The idea of an ensemble, which was later used by Einstein and had been contained somewhat implicitly in Boltzmann's earlier work, was made the central notion of statistical mechanics. The thermodynamic behaviour of a system was to be understood in terms of the average behaviour of an appropriately constructed representative ensemble. It was a version of statistical mechanics markedly different from Einstein's or Boltzmann's. Its relation to random processes was rather indirect, but Gibbs obtained the identical fluctuation formula (1) as did Einstein. In 1911, the Ehrenfests [4] gave an incisive, penetrating discussion of Boltzmann's (and Gibbs') statistical researches, their successes and weaknesses, which defined and determined the future directions of research in Boltzmann's approach.

A little earlier (1905) the notion of a random walk was introduced as a query initiating a novel and most active field of statistical physics. Just how the random walks were included as a subclass of random processes, and their close relationship to Brownian motion, was first explicitly stated by M von Smoluchowski and Einstein. Only two years later Langevin devised a novel procedure to analyse these processes, by explicitly introducing a fluctuating force into the equations of motion.

Still in this same decade, Lorentz and Rayleigh exploited and analysed the intimate relation between random processes and statistical mechanics by investigating models, leading to differential equations, especially appropriate for the study of conductivity questions. All these contributions were recognized at the time; they had a substantial impact on the succeeding development. Not recognized, however, was a study by one of Poincaré's students, Bachelier who, stimulated by Poincaré's lectures on probability, defended his thesis (19 March 1900) on 'The mathematical theory of speculation'. He developed a stochastic model of the prices of commodities traded on the stock market, and he independently constructed a theory of random processes based on a probability function $P(x, t)$ (the probability that a commodity at time t has the value x). In so doing he obtained and anticipated many of the results found later by Einstein and Smoluchowski, but his work was ignored; it had no influence on physics, economics or his life.

There is a touch of irony in the realization that in the statistical treatment of the stock market known to Poincaré, but probably not known or appreciated by Einstein, methods, ideas, equations and concepts were developed of direct relevance in physics, astronomy and chemistry. Then, after close to a century of extensive use, it came as a surprise that these very methods were directly applicable to the stock market and economic problems. However, the directions the researches of Einstein, Gibbs, Lorentz, Rayleigh, Boltzmann (via Ehrenfest), Planck, Langevin and Smoluchowski had conferred on statistical mechanics, shaped and formed its future well into the twentieth century. It would be difficult to find a decade of comparable innovations and discoveries; but then the contributors were hardly average scientists. So it is not surprising that their concerns, techniques and ways of viewing basic questions would dominate the succeeding development.

8.1.2. The legacy of the nineteenth century

The investigations alluded to in section 8.1.1 were the continuation of the kinetic theory of gases, which occupied the physicists of the nineteenth century, notably Clausius, Maxwell and Boltzmann. There were two main problems. The first was how to obtain specific physical properties of a gas by considering it as a mechanical system, consisting of N particles of mass m, contained in a volume Ω_0, with a presumably known interaction potential U between the molecules. Quite often auxiliary quantities such as the mean free path were introduced to facilitate the discussion. The second problem was to derive the general laws of thermodynamics from this same mechanical picture. The basic ingredient in both problems was the one-particle distribution function $f(x, v, t)$, which specified the number of molecules at a prescribed point, x, with a prescribed velocity v, at a time t within prescribed ranges as $f(x, v, t) \, d^3x \, d^3v$.

Boltzmann actually preferred a discrete description, where the six-dimensional velocity–position space (called μ-space, after Ehrenfest) is split into cells of finite size ω_i. Because the cell i specifies both a location and a velocity, the state of the gas is described by a set of numbers n_i at time t. Furthermore a cell i also defines the energy ε_i, of a molecule. It is always assumed that $n_i(t) \ll N$; the occupation number of a given cell is always much smaller than the total number of molecules. It is often convenient to mix the discrete and continuous notation (Boltzmann did it all the time). For example, the total number of molecules can be written in two ways

$$N = \sum_i n_i(t) = \int\int d^3v \, d^3x \, f(x, v, t). \tag{2}$$

One of the important legacies of the nineteenth century is the equation Boltzmann derived for the function $f(x, v, t)$ in 1872. It can be written in the form

$$\frac{\partial f}{\partial t} + Sf = C(f, f). \tag{3}$$

The term Sf, the streaming term, describes the changes in f due to the motion of the molecules and the action of the external fields. Sf follows directly from Newton's laws

$$Sf \equiv \sum_\alpha \left(\frac{\partial f}{\partial x_\alpha} v_\alpha + \frac{\partial f}{\partial v_\alpha} \frac{X_\alpha}{m} \right) \qquad (4)$$

α is a Cartesian index, X_α the external force.

The interactions between the molecules are contained in the collision term $C(f, f)$. Although it is in principle determined by the dynamics, Boltzmann needed an extra assumption to obtain an explicit expression for C. In terms of the discrete notation, he assumed that the number of collisions per unit time in which molecules in the cells i, j collide and are scattered to k, l is given by $A_{ij \to kl}$, where

$$A_{ij \to kl} = n_i n_j a_{ij \to kl}. \qquad (5)$$

In (5) $a_{ij \to kl}$ can depend on the dynamics, the collision configuration but not on the time, nor on n_i, n_j, n_k or n_l. The combination of (3), (4) and (5) leads in the continuous notation to

$$\frac{\partial f}{\partial t} + \sum \left(v_\alpha \frac{\partial f}{\partial x_\alpha} + \frac{X_\alpha}{m} \frac{\partial f}{\partial v_\alpha} \right) = \int\!\!\int d^3 v_1 \, d\Omega \, g I(g, \theta)(f_1' f' - f_1 f). \qquad (6)$$

The collision envisaged in (6) is one where molecules with velocities v and v_1, collide and yield molecules with velocities v' and v_1'.

In (6) $g = |g| = |v - v_1|$. $I(g, \theta)$ is the collision cross section, θ is the angle through which the relative velocity g is turned by the collision, and $d\Omega$ is the solid angle.

In principle, the main physical properties of the system should be derivable from a general time-dependent solution of the Boltzmann equation. Obtaining such a solution in full generality is out of the question, but two remarkable conclusions can be drawn. In the case of equilibrium, i.e. when $\partial f / \partial t = 0$, it follows that Sf and $C(f, f)$ from equation (3) are separately zero. This is not obvious but follows in turn from the fact that there is a function H which always decreases. H is defined by

$$H(x, t) \equiv \int d^3 v \, f(x, v, t) \log f(x, v, t) \qquad (7)$$

$$\frac{dH}{dt} \leqslant 0. \qquad (7a)$$

Equation (7a) is not obvious either, but follows directly by manipulating (7) and using (6). It further follows that if $dH/dt = 0$, $\partial f / \partial t = 0$ also, which implies

$$Sf = 0 \qquad (8a)$$

$$C(f, f) = 0 \qquad (8b)$$

which implies equilibrium. Equations (8a) and (8b) can be solved immediately to yield the equilibrium solution, which in discrete notation is

$$\bar{n}_i = A\omega_i \exp\left(-\frac{\varepsilon_i}{kT}\right) \qquad (9a)$$

A is proportional to the density, k is the Boltzmann constant and T the absolute temperature. In the continuous notation the equilibrium solution is

$$f^{(0)} = \frac{n^{(0)}}{(2\pi mkT)^{3/2}} \exp\left(-\frac{1}{kT}\left(\tfrac{1}{2}mv^2 + U\right)\right). \qquad (9b)$$

Not only can all the equilibrium properties of the gas be obtained from the Maxwell–Boltzmann distribution, it also appears, as if by a miracle, that the formulae (7), (8) and (9) describe a universal approach to thermodynamic equilibrium. All that is necessary is to relate H to the entropy, and the equilibrium distribution to a thermodynamic equilibrium state, and this was achieved without solving the forbidding-looking integro-differential equation (6).

It very soon became clear that the unidirectional decrease of H, according to (7), is incompatible with the time reversibility of the underlying dynamics. Since the only assumptions, (3), (4) and (5), are presumably mechanical in nature and thus preserve the reversibility, a strictly irreversible result cannot possibly be a rigorous consequence of that scheme. This conflict and the reconciliation of the intrinsically contradictory features of reversible mechanics and irreversible thermodynamics is another major legacy of the nineteenth century.

In practically all instances this reconciliation is effected by the judicious introduction of probabilities. It is most convenient to illustrate this by using the discrete description of the system. The state of the system (usually called Z), is described by specifying the occupation numbers, energies and volumes of each cell

$$Z \equiv \begin{matrix} \omega_1 & \omega_2 \ldots & \omega_i \ldots \\ n_1 & n_2 \ldots & n_i \ldots \\ \varepsilon_1 & \varepsilon_2 \ldots & \varepsilon_i \ldots \end{matrix} \qquad (10)$$

The most obvious way to assign a probability to a state is to imagine random processes, a lottery where N points are thrown in a random manner over the six-dimensional μ-space, (of volume Ω). Assuming no a priori preference for any part of the space, the probability for an arrangement Z (as given by (10a)) is

$$W(Z) = \frac{N!}{n_i! n_2! \ldots} \left(\frac{\omega_1}{\Omega}\right)^{n_1} \ldots \left(\frac{\omega_i}{\Omega}\right)^{n_i} \ldots \qquad (11)$$

The most probable state is obtained by maximizing $W(Z)$ under the constraints

$$\sum n_i = N$$
$$\sum n_i \varepsilon_i = E = \text{total energy}. \qquad (12)$$

Once again the Maxwell–Boltzmann distribution emerges. The fact that the infinite time limit of a time-dependent distribution is identical with the most probable distribution of a suitably selected random process, singles out the equilibrium distribution as possessing especially simple properties. Of course, employing the ideas of probability has up to this point completely divorced the dynamics of the system from its statistical behaviour. One way to establish some such connection (the last nineteenth century legacy) is to introduce the phase space (called Γ-space) of the dynamical system. In the velocity–position space (the μ-space), the system is represented by a cloud of N points. In the course of time these points move, and because of the collisions also jump in pairs from two cells to two others. The phase space is a $6N$-dimensional space, whose coordinates are $3N$ positions, $x_1, \ldots, x_N$, and $3N$ momenta $p_1, \ldots, p_N$.

One point now represents the detailed state of the complete system at a given moment. In the course of time, this phase point traces out a trajectory. If the system is conservative the trajectory lies on the $(6N-1)$-dimensional surface of constant energy. The motion of the phase point in the constant-energy surface is exact—neither approximations nor probabilities are involved. This is to be contrasted with the earlier Boltzmann description, where it is only known how many molecules are in a cell at a given time, and even if it were known that a particular molecule is in a particular cell, that knowledge would specify its position and momenta in an incomplete manner, since the cells are finite. Thus, corresponding to a Boltzmann state, there would be a finite $(6N - 1)$-dimensional piece of the energy surface. In its motion, the phase point would move from one volume to the next, corresponding to the changes in $n_i(t)$.

This visualization immediately suggests the study of the trajectories on the energy surface. Boltzmann, and Einstein later, assumed that the time spent in a region is proportional to the volume of that region. Boltzmann at one time assumed that a trajectory would actually pass through every point on the energy surface (the ergodic hypothesis), but this was later shown to be untenable. The investigations of the trajectories all had, as one of their main goals, the establishment of relationships between geometrically defined quantities and the temporal behaviour, as, for example, the relations between phase-space averages and time averages.

Thus the nineteenth century contributed the notion of a distribution function, a Boltzmann equation for that function, a sharp distinction between equilibrium and non-equilibrium states (the Boltzmann lottery describing the equilibrium state as the most probable state) and the suggestion that physics could be learned from the study of the set of trajectories on the energy surface. It is against that background that the twentieth century must be seen.

8.1.3. The emerging definition of a discipline
In the early development of statistical mechanics (or kinetic theory) no sharp distinction was made between equilibrium and non-equilibrium phenomena. Nor was their manner of treatment all that different. However, especially through the researches of Boltzmann, the equilibrium and non-equilibrium areas began to separate, in goals as well as in methods. Clearly the time-dependent distribution function required different calculational techniques from the Maxwell–Boltzmann equilibrium distribution. It was in the temporal, irreversible behaviour that the need for a probability interpretation first became evident. The ergodic questions, the trajectories of the phase point on the energy surface, were all stimulated by efforts to gain understanding and control of the time evolution. There is no equilibrium counterpart to many of these questions. Generally speaking, any system not describable by the Maxwell–Boltzmann distribution (temperature and density independent of position and time) or the canonical ensemble or any ensemble density function depending on just the energy (see section 8.2.1.1) belongs to the non-equilibrium category. Systems with temperature or density gradients are the simplest examples of such systems.

The most general problem of non-equilibrium statistical mechanics (in fact the whole subject) is the detailed description of the temporal evolution of a physical (chemical, astronomical) system. This formulation is clearly much too general—as phrased it suggests that there might exist methods and procedures valid for all systems in all states, and that can hardly be expected. From this viewpoint it is even remarkable that there does exist an equilibrium description, the Maxwell–Boltzmann distribution law, which possesses a system-independent universality. It would seem that the only non-equilibrium feature of comparable generality is the approach to equilibrium.

However, the irreversible approach to equilibrium has only been demonstrated from the Boltzmann equation and there the approach has to be in probability terms. Since the Boltzmann equation itself is an approximate equation, valid only for dilute systems and binary collision, this raises a number of questions. Are there arguments to show that *every* physical system approaches a thermal equilibrium, at least in a probability interpretation? If that is indeed the case, can anything be said about the rate at which different systems and different observables approach equilibrium? Is the approach to equilibrium monotonic? Since probability notions are presumably essential in establishing irreversibility, does it make a *physical* difference where and how these notions are introduced? There is clearly a compatibility question—the probabilities introduced must be consistent with the dynamical laws governing the system. In this connection it should be recalled that the thermodynamic irreversibility is not only in conflict with mechanical reversibility, but also with the mechanical recurrence properties, first demonstrated by Poincaré. Consequently it must also be required that the probabilities introduced should resolve the Poincaré–Zermelo recurrence conflict.

Apart from these questions of principle, there are large classes of problems that require a non-equilibrium treatment. The subject of fluctuation phenomena, of which Brownian motion was an early example, demands a control of the temporal behaviour. In these studies the role of the probability postulate is more explicit—and it can indeed be shown that large classes of stochastic systems do approach equilibrium.

A quite different class of phenomena, again demanding a non-equilibrium treatment, are the transport phenomena, diffusion, heat conduction, galvanomagnetic effects, relaxation phenomena and the responses to external fields. It is already clear that non-equilibrium statistical mechanics is not a monolithic field with a single methodology and one universal approach. It ranges from the fluctuating forces on stars to thermal conductivity. All questions of temporal evolution are in principle included. As such, universality is not to be expected, it is remarkable enough that the equilibrium description is universal and system-independent. It is even tempting to conjecture that extreme chaotic situations would again show a similar, universal, system-independent behaviour. Unfortunately this chapter will deal only with the intermediate domain, where little universality can be expected.

8.2. A history in three parts

The discussion so far has shown that non-equilibrium statistical mechanics had developed into an autonomous field, pretty much separate from its equilibrium background. Although autonomous, it is by no means homogeneous, but consists of a number of distinct, almost unrelated areas. To obtain an intelligible overview, it is best to subdivide the century into three distinct epochs. Each epoch will contain a brief discussion of seven or eight distinct topics. The identical topics are re-examined in each epoch; their significance varies from epoch to epoch. This method of presentation of a heterogeneous, somewhat disjointed field has the great advantage that the successes, modifications, redirections, conceptual and methodological shifts can be exhibited sequentially. In this way we achieve an overall understanding of the continually changing diversity with its accompanying shifting emphasis.

The first epoch starts at the beginning of the century with the seminal investigations of Gibbs, Planck and Einstein. The discussion is carried as far as 1940, just before the beginning of World War II. At that time a basic type of 'Master equation' was proposed by Uhlenbeck. The Boltzmann equation is of the Master type. In this first epoch a large number of individually important investigations were carried out, but the absence of a well-defined central theme emphasized the rather disjointed character of the field. Furthermore, non-equilibrium statistical mechanics itself was not in the mainstream of physics. For example, the books written during that period (with the notable exception of reference [5] and a few others), hardly dealt with non-equilibrium processes.

The second epoch covers the developments from 1940–75. Many advances during that period were technical and formal; however, the systematic use of the Liouville equation as the basis for non-equilibrium studies was a dominant theme. The most surprising, even stunning development was the gradual realization that for transport parameters such as viscosity no systematic expansion in the density can exist. By the end of this epoch, the ideas of chaos began to be prevalent in physics, particularly statistical mechanics. Chaos is treated Chapter 24, but its existence did (and does) affect non-equilibrium processes in a direct way. Hence it is appropriate to close the second epoch at the time when the prevalence of chaos began to be appreciated.

The third epoch covers the period from about 1975 to the present. The single most important innovation was undoubtedly the extensive, even ubiquitous, use of large computers. Computers are central in all branches of physics, but especially in statistical mechanics, for the generation of information about model systems. Thus computer data sometimes preempts, sometimes complements, experimental data and analytical discussions. This latest epoch is also characterized by an enormous increase in the variety of topics treated, including stochastic particle acceleration, neural networks and the mixing of galaxies.

8.2.1. Epoch I: from the Boltzmann equation to the Master equation

8.2.1.1. Topic 1: from Boltzmann to Gibbs. In 1902 Gibbs' monumental treatise [3] provided a powerful new method for the derivation of the statistical (macroscopic) properties of physical systems. The basic innovation was the representation of a single system by a collection of identical systems, in different dynamical configurations. An ensemble was described in terms of a density of points in phase space (Γ-space); $\rho(p, q, t) \, d\Gamma$ was the fraction of the ensemble numbers in a volume $d\Gamma$ of phase space. By the Liouville equation of mechanics $\rho(p, q, t)$ satisfies

$$\frac{\partial \rho}{\partial t} + \sum_i \left(\frac{\partial \rho}{\partial q_i} \frac{\partial H}{\partial p_i} - \frac{\partial \rho}{\partial q_i} \frac{\partial H}{\partial p_i} \right) = 0 \qquad (13)$$

See also p 531

G E Uhlenbeck

(Dutch, 1900–88)

George Eugène Uhlenbeck was born in Jakarta (Indonesia) but was educated in the Netherlands. An interest in Lorentz's lectures on physics led to a fascination with Boltzmann's 'Vorlesungen über Gastheory', which stayed with him all his life. He studied theoretical physics in Leiden with Ehrenfest, himself a true disciple of Boltzmann, and obtained his PhD in 1927 on a topic which organized and systematized the statistical notions in the quantum context.

Of his many contributions to physics, the most spectacular was the discovery of the electron spin, which he made together with Goudsmit, in 1925, before either had obtained their PhD degree. But statistical mechanics remained his primary interest. In two classic papers he established the theory of random processes and Brownian motion. He gave the definitive formulation of the stationary Gaussian–Markov processes (now called the Uhlenbeck–Ornstein processes), and with Uehling constructed a quantum version of the Boltzmann equation.

After World War II, he used Bogoliubov's ideas to realize a lifelong scientific ambition: to extend the Boltzmann equation to dense gases. This culmination of Boltzmann's programme, combined with Sinai's result of the ergodicity of the hard sphere, seemed at the time to vindicate Boltzmann and Ehrenfest's approach.

Because of Uhlenbeck's continual insistence on clarity and logic, and his intense dislike of pretension and pomposity, he was highly respected as an arbiter. It was a great shock for him when it was shown that the divergence in the transport coefficients precluded the completion of Boltzmann's programme. He never quite came to terms with that impossibility and probably did not really believe it.

His many other contributions to physics were characterized by clarity and a control of the details, so that he exerted a calming and thoughtful influence on statistical mechanics, contributing greatly to the structure this disjointed field eventually acquired.

where p_i and q_i are the momenta and coordinates and H is the Hamiltonian of the system. The macroscopic observable values of any quantity Q are

$$\langle Q \rangle \equiv \frac{\int \rho Q \, d\Gamma}{\int \rho \, d\Gamma} \qquad d\Gamma \equiv dp_1 \ldots dq_N. \tag{14}$$

The initial ensemble $\rho(p, q, 0)$ must be so selected as to incorporate the knowledge of the system, but must be random otherwise. In equilibrium $\partial \rho / \partial t = 0$. Equation (13) shows that in that case ρ is an arbitrary function of H. Gibbs in a masterly analysis, showed that if the time-independent ensemble is chosen as the canonical ensemble

$$\rho(p, q) = A \exp\left(-\frac{H}{\theta}\right) \tag{15}$$

the resulting ensemble averages formally reproduce all the thermodynamic relations. Gibbs did not really derive the laws of thermodynamics; he constructed thermodynamic analogies. By identifying the parameters in the ensemble density functions the physical thermodynamics was reconstructed, and through (15) and (14) related to the Hamiltonian.

Again the Gibbs method worked much better in equilibrium than in non-equilibrium situations, where both the construction of appropriate initial ensembles and the solution of (13) were formidable obstacles. The ensemble procedure became eventually the method of choice for equilibrium processes. It is clear that in the Gibbs formalism, the approach to equilibrium and ergodic problems play a minor role. The separation between equilibrium and non-equilibrium methods became much more pronounced after Gibbs' work. Planck liked Gibbs' approach, Einstein, Lorentz and later Bohr were very impressed, but to Boltzmann and especially Ehrenfest it appeared that Gibbs had side-stepped all subtle questions of the temporal behaviour by a cheap trick.

8.2.1.2. Topic 2: Brownian motion; random processes; the Fokker–Planck and Langevin equations. In his fundamental paper on Brownian motion Einstein obtained a relation between the positional fluctuations of a colloidal particle and the time as

$$\langle x^2 \rangle = \frac{RT}{3\pi N a \eta} t \tag{16}$$

where R is the gas constant, N Avogadro's number, η the viscosity and a the radius of the particle. This formula connects a fluctuation to a dissipative mechanism (viscosity). It was probably the first fluctuation-dissipation relation. In the derivation of this result Einstein also stressed the relation between diffusion and a random process—Brownian motion.

Random processes are described by a number of probability functions, which successively describe the process in more detail. If y is a fluctuating variable, $W_1(y, t)$ is the probability that at time t the variable has the value y; $W_2(y_1 t_1, y_2 t_2)$ is the joint probability that at time t_1 the variable had the value y_1 and at time t_2 the value y_2. Another important function is the conditional probability $P_2(y_1 t_1 | y_2 t_2)$, the probability that if y had the value y_1 at t_1, it will have the value y_2 at t_2. The classification of random processes was formulated in this epoch. If W_3 and P_3, and all higher Ws and Ps, contain no new information, the process is a Markov process. If all information is already contained in W_1 the process is a purely random process.

For a Markov process P_2 satisfies the Chapman–Kolmogoroff equation

$$P_2(y_1 t_1 | y_3 t_3) = \int P_2(y_1 t_1 | y_2 t_2) \, P_2(y_2 t_2 | y_3 t_3) \, dy_2. \tag{17}$$

(This equation is actually contained in Bachelier's thesis.) For the applications to physics, certain specializations and approximations are useful. For many purposes the information contained in W_1 (which is of course less than the information in W_2 or P_2) is sufficient. These equations have the generic form

$$\frac{\partial W_1(y, t)}{\partial t} = -\frac{\partial}{\partial y}(A(y) W_1(y, t)) + \tfrac{1}{2} \frac{\partial^2}{\partial y^2}(B(y) W_1(y, t)). \tag{18}$$

The coefficients A and B are determined by the system investigated; they in turn determine the nature of the random process. Equations of this class are referred to as Fokker–Planck equations [6,7]. They were discovered between 1914 and 1917, though Smoluchowski had obtained the same type of equation ten years earlier [8].

A quite different method to analyse stochastic processes was developed by Langevin [9]. The basic idea of the Langevin method is to deal with the ordinary equations of motion (say of a Brownian motion particle), but add a time-dependent fluctuating force to the equation of motion. Physically such a fluctuating force represents or simulates a stochastic environment. If a particle is subject to a friction force ηv, and an external force X, the Langevin equation is

$$m \frac{dv}{dt} = -\eta v + X + F(t). \tag{19}$$

$F(t)$ is the fluctuating force; it is not given explicitly, but its average properties are presumed to be known. For example, $F(t)$ may be independent of v and have mean value zero; in addition its variations may be taken to occur on such a short time-scale that there is no significant correlation between F at different times. Since $F(t)$ is only given statistically, one cannot obtain a definite value for v, but one can find the probability distribution function for (say) $W_1(v, t)$. Whether or not the Langevin method and the Fokker–Planck method are equivalent depends on the precise assumptions made about $F(t)$. In this first epoch many examples of the Langevin equation, the Fokker–Planck equation and Markov processes were investigated, but they tended to be isolated examples. They provided useful illustrations, but a more systematic approach had to wait till the second epoch.

8.2.1.3. *Topic 3: the transport coefficients and the Chapman–Enskog method.* Perhaps the most important, certainly the most typical problem in non-equilibrium statistical mechanics was the calculation of the transport coefficients from the Boltzmann equation. The relevant coefficients are always defined from empirical macroscopic laws. Thus the diffusion coefficient is obtained from Fick's law, which relates the particle current i_x in the x direction to the x-component of the density gradient

$$i_x = -D \frac{\partial n}{\partial x}. \tag{20}$$

i_x and $\partial n/\partial x$ can be obtained directly from the distribution function as

$$\int d^3v\, f(\boldsymbol{x}, \boldsymbol{v}, t) = n(\boldsymbol{x}, t) \tag{21a}$$

$$\int d^3v\, v_x f(\boldsymbol{x}, \boldsymbol{v}, t) = i_x(\boldsymbol{x}, t). \tag{21b}$$

For strict equilibrium f is an even function of v so that $i_x = 0$, i.e. there is no particle current. For a homogeneous system, n does not depend on x, so $\partial n/\partial x = 0$. For systems not in equilibrium, neither i_x nor $\partial n/\partial x = 0$ will be zero; hence a knowledge of f, combined with the macroscopic law, allows the calculation of D. All calculations of transport coefficient follow this pattern. The main point therefore is the calculation of f. Practically all schemes devised are approximations since the exact Boltzmann equation is too complicated. The basic physical idea is to investigate configurations 'near' equilibrium. This implies that f can be expanded as

$$f = f^{(0)} + f^{(1)} + f^{(2)} + \ldots. \tag{22}$$

$f^{(0)}$ is the equilibrium solution, $f^{(1)}$ is not; however, by assumption $f^{(1)}$ is small compared with $f^{(0)}$, so that terms such as $(f^{(1)})^2$ can be neglected. Substituting (22) in the Boltzmann equation yields

$$C(f^{(0)}, f^{(0)}) = 0. \tag{23}$$

This is the same equation as (8b) which yields only the velocity dependence of the distribution function. The coefficients in general depend on x and t, and must be selected so that $Sf^{(0)} = 0$. The resulting solution takes the form

$$f^{(0)} = n \left(\frac{m}{2\pi kT}\right)^{3/2} \exp\left(-\frac{m}{2\pi kT}(v-u)^2\right). \tag{24}$$

In (24), n, T and u, are so far undetermined functions of x and t. Equation (24) is a local Maxwell–Boltzmann distribution. Continuing this procedure gives a linear integral equation for $f^{(1)}$. The requirement that it should have solutions provides the necessary equations to determine n, T and u in that order. The physical meaning of u is the local mass velocity (*not* the molecular velocity). This procedure was originated and worked out in great detail by Chapman [10] and Enskog [11]. It is an arcane, involved method requiring the solution of integral equations, and did much to isolate non-equilibrium statistical mechanics from the rest of physics. From these approximate solutions, it could be determined that both the heat conductivity and viscosity coefficients are independent of the pressure in first order (a result discovered much earlier by Maxwell). One striking unanticipated physical feature emerged from this formal analysis. When applied to a mixture of two gases (with an even more elaborate formalism) it was noted that a temperature gradient in the z direction will produce a particle current for either species

$$i_z = kD \frac{1}{T}\frac{dT}{dz} \tag{25}$$

where K is the thermal conductivity and D the ordinary diffusion coefficient. The existence of this thermal diffusion can only be inferred from the full formalism; mean free path arguments do not explain the phenomenon. It is remarkable that when, in 1938, it became desirable (or necessary) to separate different uranium isotopes, thermal diffusion, a rather obscure feature of kinetic theory, became of major industrial importance [12].

Further developments, after the initial studies of Chapman and Enskog, consisted primarily of detailed numerical calculations for specific molecular interactions, such as hard spheres, rough spheres or Maxwell molecules (possessing a repulsive $1/r^5$ interaction). The agreement with experiment was generally fair (5–10%). Burnett [13] calculated the second-order $f^{(2)}$ approximation by expanding in Sonine polynomials. The formulae became very involved, but he did prove that the series development actually converges, something that Chapman and Enskog never did. Enskog made a very nearly desperate attempt to alter the Boltzmann collision term, so that the calculated viscosity would reproduce the observed pressure dependence, but the method was too contrived and the results too marginal.

8.2.1.4. Topic 4: ergodic theorems, mathematical and conceptual questions, the parting of the ways. In 1911 the Ehrenfests [4] gave an incisive analysis of the conceptual and logical status of Boltzmann's efforts to explain thermodynamics and irreversibility on a mechanical basis. In so doing, they made extensive use of the phase-space representation of mechanical systems—the behaviour of systems as expressed by the time a trajectory spends in certain regions of the energy surface. They also clarified how conjectures about this behaviour translated into physical statements: the original ergodic theorem would justify Boltzmann's procedure, while the quasi-ergodic theorem (the statement that the trajectory comes arbitrarily near any point) would not be sufficient to do so.

The demonstrations [14, 15] that such systems were impossible showed, among other things, the rapidly increasing level of mathematical sophistication, as well as the need for a redirection or reinterpretation of Boltzmann's methods. Ehrenfest's analysis made crystal clear that the introduction of suitable probability notions was essential for a consistent implementation of Boltzmann's programme.

In 1931 and 1932 Birkhoff [16] and von Neumann [17] gave a mathematical reorientation of the geometrical problems Boltzmann and Einstein had encountered in describing physical behaviour in terms of trajectories in the energy surface. The basic procedure was to describe the motion of the phase point along a trajectory, from P (the initial point) to P_t some time later, as a mapping. This mapping was then studied rigorously. They obtained mathematically impressive results. For example, the time average of a function $\phi(p, q, t)$ defined by a limit (26) exists

$$\bar{\varphi}_t = \lim_{T \to \infty} \frac{1}{2T} \int_{t-T}^{t+T} \varphi(P_\tau) \, d\tau. \tag{26}$$

The demonstration that a limit exists is not usually a result that thrills physicists. Of greater interest was the demonstration that for a special class of systems, *metrically transitive* systems, the (now existing) time average $\bar{\phi}_t$ was the same as the Gibbs ensemble average $\langle \phi \rangle$ (see (4))

$$\langle \varphi \rangle = \bar{\varphi}_t. \tag{27}$$

A system is metrically transitive if the energy surface cannot be split up into two finite regions, having the property that trajectories starting in one region will stay there for all time. It could be described as 'orbital monogamy'. Unfortunately it is very difficult to decide whether a system is metrically transitive—examples are hard to come by. This is unfortunate, because metrically transitive systems possess exactly the time evolution envisaged by Einstein and Boltzmann. Starting with these studies, ergodic considerations became a part of mathematics. For 50–60 years most physicists considered these investigations and results irrelevant. Now these or similar studies are modern and relevant.

8.2.1.5. Topic 5: the Wiener–Khintchine theorem; the Onsager relations.

In many random processes, the observed information is contained in the temporal record of a fluctuating variable. If observations over an extended period are made on the position of a Brownian particle, (or a fluctuating voltage or an electroencephalogram) a great deal of information is contained in the often irregular graph of the variable y versus t. The basic problem is to extract this information from the observations. The most convenient procedure is to make a Fourier decomposition of the time dependence of the random variable according to

$$y(t) = \int_{-\infty}^{+\infty} d\nu\, A(\nu) e^{2\pi i \nu t}. \tag{28}$$

To avoid mathematical subtleties it is best to assume a long, but finite period of observation from $-T$ to $+T$. A time average is defined in the usual way

$$\langle y^2(t) \rangle_t = \lim_{T \to \infty} \frac{1}{2T} \int_{-T}^{+T} y^2(t)\, dt. \tag{29}$$

Straightforward evaluation yields

$$\langle y^2(t) \rangle_t = \int_0^\infty d\nu\, G(\nu) \tag{30}$$

where $G(\nu)$ is the spectral density of the random process, defined by

$$G(\nu) \equiv \lim_{T \to \infty} \frac{2}{T} |A(\nu)|^2. \tag{31}$$

The spectral density is a measure of the intensity of the νth harmonic in the random variable. The fundamental theorem is the Wiener–Khintchine theorem [18, 19] which relates the autocorrelation function to the spectral density

$$\langle y(t) y(t+\tau) \rangle_t \equiv \lim_{T \to \infty} \frac{1}{T} \int_{-\infty}^{+\infty} y(t) y(t+\tau)\, dt$$
$$= \int_0^\infty d\nu\, G(\nu) \cos 2\pi \nu \tau. \tag{32}$$

It is interesting and important that a correlation function, which describes the process in terms of two times, can be obtained via (31) and (28) from the single time observations. Applying this type of analysis to the fluctuating voltage in an LC circuit leads to an especially simple version of (30) (recall the fluctuations are caused by the thermal fluctuations in the resistor)

$$\langle v_R^2 \rangle = 4RkT \Delta\nu \tag{33}$$

where $\Delta\nu$ is the bandwidth of the circuit and R the resistance. This formula, usually described as the *Johnson noise fluctuations* agrees excellently with experiment. The fluctuations were first observed by Johnson [20]; the derivation of (33) was given by Nyquist [21].

This is yet another example of the intimate relation between the fluctuations of a quantity (the voltage) and a dissipative mechanism (the resistance). There is a general class of such fluctuation dissipation theorems. Macroscopic laws such as Fick's law of diffusion, Fourier's law of thermal conduction, necessarily involves fluctuations. The existence of fluctuations is a necessary consequence of the atomistic substructure. If the macroscopic system is not in equilibrium, so that there are thermal density gradients, or external fields, thermal or electrical or matter currents will be produced. For a class of phenomena, not too far from equilibrium, the currents should be linearly related to the gradients, (often called forces or fluxes) according to

$$J_i = \sum L_{ij} X_j. \tag{34}$$

The L_{ij} are empirical parameters, such as magneto resistance or thermal conductivity; the J_i are the currents, which as argued before will fluctuate.

In a brilliantly original paper Onsager [22] derived relations such (34) by assuming that near equilibrium the entropy (which is a maximum at equilibrium) could be written as

$$S = S_0 - \tfrac{1}{2} \sum S_{ij} \alpha_i \alpha_j \tag{35}$$

where α_i is the deviation of variable i from its equilibrium value. The S_{ij} are determined by the entropy of the system written in terms of the macroscopic variables. Physically the fluctuation of the macroscopic variables drive the system towards equilibrium. Onsager not only obtained an equation of the form (34), but he obtained the incredible bonus that

$$L_{ij} = L_{ji} \tag{36}$$

(which in turn got him the Nobel prize). These Onsager relations established numerical relations between quite different and apparently unrelated processes. There are a large number of physical situations where the Onsager relations provide surprising relations, as in the mechanocaloric effect, where the mass and energy currents of a gas passing through a hole in a partition are connected through L-type coefficients. In chemical reactions, there is a coupling between heat flow and diffusion. In rotating systems and in systems subject to magnetic fields, a generalized version of the Onsager relation yields additional connections. Perhaps the simplest example is that the effect of a thermal gradient on an electric current is the same as the effect of an electric field on a thermal current. Many of the various viscosities in the superfluid description of liquid helium are equal because of the Onsager symmetry. In epoch II (section 8.2.2) the analysis and search for new Onsager-type relations became a major industry.

8.2.1.6. Topic 6: quantum modifications.

The influence of quantum mechanics on equilibrium statistical mechanics was enormous, with the recognition that there are two possible equilibrium distributions, Einstein–Bose and Fermi–Dirac, instead of one single distribution which holds for all objects. However, the formal structure of quantum statistical mechanics parallels the Gibbsian version practically word for word. Instead of a phase-space density function $\rho(p,q,t)$ which satisfies the Liouville equation, there is a density matrix, which satisfies a comparable equation. Macroscopic quantities are obtained by the traces of a matrix product, instead of integrals over phase space [23, 24]. Formally each ensemble member, denoted by α, has a wavefunction $\psi^\alpha(q,t)$. Each such wavefunction can be expanded in a fixed orthonormal set $\phi_n(q)$

$$\psi^\alpha(q,t) = \sum_n a_n^\alpha(t)\,\varphi_n(q) \tag{37}$$

where $|a_n^\alpha(t)|^2$ is the probability that ensemble number α at time t is in state n. If the ensemble has N_e members, the density matrix (relative to the phase ϕ) is defined by

$$\rho_{mn}(t) = \frac{1}{N_e}\sum_\alpha a_m^\alpha (a_n^\alpha)^*. \tag{38}$$

ρ_{mn} is the density matrix written in the base of ϕ_n. From (38) and the Schrödinger equation it follows that the matrix ρ, satisfies the so-called von Neumann equation, which replaces the Liouville equation

$$i\hbar\,\frac{d\rho}{dt} = [H, \rho]. \tag{39}$$

The basic interpretative postulate is that the observed macroscopic value of an observable is the ensemble average of the quantum-mechanical expectation value, leading to

$$Q_{\text{obs}} = \text{Tr}(\rho Q). \tag{40}$$

It should be emphasized that because of the trace in (40) Q_{obs} is independent of the set ϕ_n. This version of quantum statistical mechanics was not extensively discussed in this epoch. Its time would still come. On a more pragmatic level Uhlenbeck and Uchling [25], in 1933, modified the collision term in the Boltzmann equation, so that the Einstein–Bose or Fermi–Dirac statistics were properly taken into account in the collision terms. They further replaced Boltzmann's cell i by quantum state i, and the size of the cell ω_i by the statistical weight g_i of level i. The original Boltzmann collision number assumption

$$A_{ij \to kl} = n_i n_j a_{ij \to kl} \tag{41}$$

was for Fermi particles replaced by

$$A_{ij \to kl} = n_i n_j (1 - n_k)(1 - n_l) a_{ij \to kl}. \tag{42}$$

As written, $n_i n_j$ can only assume the values 0 or 1. Equation (42) clearly incorporates the Pauli principle, if either n_k or $n_l = 1$ the collisions cannot proceed. With this altered collision-number assumption a good deal of the Chapman–Enskog development can be repeated. The results are especially important for conductivity studies.

8.2.1.7. *Topic 7: models.* Models in statistical mechanics are like études in music. They could be finger exercises to improve one's technique, or they could be like Chopin's études and acquire a life and importance (and charm) of their own.

It was Ehrenfest who in his fundamental study employed models to point out inconsistencies, or to clarify subtle points of interpretation. The first of these, the *wind tree model*, was designed to test the validity and consistency of Boltzmann's collision-number assumption, that in each time interval the number of (i, j) (k, l) collisions is precisely given by

$$A_{ij \to kl} \Delta t = n_i n_j a_{ij \to kl} \Delta t. \tag{43}$$

In this model, molecules would move with constant speed c in just four possible directions 1, 2, 3, 4.

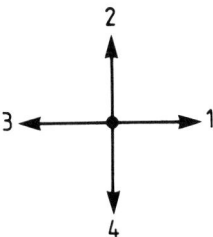

The only dynamics was contained in the scattering of molecules, from fixed obstacles (squares) of side a, so oriented that the only possible effect would be a change in direction.

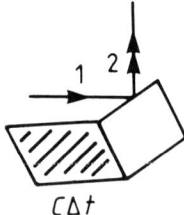

If the averaged exposed surface of a square is B the precise analogue of (43) is

$$A_{i \to j} \Delta t = n_i S_{ij} B. \qquad (44)$$

Here S_{ij} is the area of a parallelogram of sides a and $c\Delta t$. From (44), equations for dn_i/dt can be directly obtained and equilibrium, as expected, is given by

$$n_1 = n_2 = n_3 = n_4 = N/4. \qquad (45)$$

However, from the time-dependent solution it can be directly seen that (44) cannot hold for both a direct (12) collision and a reverse (34) collision where all the velocities are reversed. This example was instrumental in showing that the collision-number assumption had to be interpreted in a probability sense.

The paradoxical aspects of the H theorem were clarified substantially by another Ehrenfest model, 'the dog flea model' [4]. Consider two boxes A and B and a set of $2N$ balls (or two dogs and a set of numbered fleas). Pick a number, and the corresponding ball (or flea) has to jump from wherever it is to the other dog. Put the number back, shuffle and pick again. The question is, what is the temporal behaviour of $\Delta(t) = |n_B(t) - n_A(t)|$ the 'flea excess'? One might hope (expect) that $\lim_{t \to \infty} \Delta(t) = 0$. The transition probability that n_A in one drawing would change from l to m, called $Q(l|m)$, is

$$Q(l|m) = \frac{1}{2N} \delta(l-1, m) + \frac{2N-1}{2N} \delta(l+1, m). \qquad (46)$$

The state after t drawings is described by the Chapman–Kolmogoroff–Smoluchowski–Bachelier equation, for the conditional probability function $P(n|m, t)$ which is (the n is omitted in this notation)

$$P(m|s) = \frac{m+1}{2N} P(m+1, s-1) + \frac{2N-m+1}{2N} P(m-1, s-1). \qquad (47)$$

In spite of its simple appearance, this equation could not be discussed in detail for 50 years. The *average* of $\Delta(t)$ could be computed [26]; it did go to zero as $t \to \infty$.

Non-Equilibrium Statistical Mechanics

In 1926 Kohlrausch and Schrödinger (who had other things on his mind) actually played the Ehrenfest lottery with $N = 40$. They did observe a decrease of Δ in the average, fluctuations around the average and a rare recurrence for an initial state where $\Delta = 10$.

8.2.2. Epoch II: from the Master equation to the onset of chaos 1940–75

In 1940, Nordsieck, Lamb and Uhlenbeck [27] published a paper on the development of cosmic-ray showers, in which they introduced the probabilities for the various possible processes in an explicit manner. This type of equation in which the dynamics is *ab initio* stochastically defined became eventually known as a Master-type equation. It has to be distinguished sharply from equations which are strictly deterministic in character, but where the subsequent interpretation might involve probabilities or averages.

The Liouville equation is a strictly deterministic equation, a Fokker–Planck equation belongs to the Master category. The Boltzmann equation through its explicit dependence on the collision-number assumption, which requires a probability interpretation, also belongs to the Master category. The name Master equation came from the fact that a general probability function as used in the 1940 cosmic-ray paper, allows the calculation of all probabilities, all averages, all correlation functions—in short all physically interesting results can be obtained from such a function (a young aspiring physicist, several years later, looked in vain for references to the presumably famous physicist, Sam Masters).

8.2.2.1. Topic 1: the BBKGY hierarchy and the Master equation–Liouville equation dichotomy.

The Master equation description of a gas was first formulated in an important paper by Siegert [28]. Using the discrete characterization, the state of a gas is described by a probability Master function $W(n_1 n_2, \ldots, n_i, \ldots, t)$ giving the joint probability that they are n_i molecules in cell i (or state i). The temporal evolution of this probability function is determined by the collision-number assumption, itself given in probability terms: the probability for an $(i, j) \to (k, l)$ collision is given by $n_i n_j a_{ij \to kl}$.

The resulting equation for W is rather complicated, but from W all averages can be obtained, such as $\bar{n}_i(t)$ $\overline{n_i(t) n_j(t)}$. Thus

$$\bar{n}_i(t) = \sum_{n_1} \cdots \sum_{n_i} \cdots n_i W(n_1 \ldots n_i \ldots t) \tag{48a}$$

$$\overline{n_i(t) n_j(t)} = \sum_{n_1} \cdots \sum_{n_i} \cdots n_i n_j W(n_1 \ldots n_i \ldots t). \tag{48b}$$

One- and many-particle distribution functions can also be obtained from W

$$F_1(n_1, t) \equiv \sum_{n_2} \cdots \sum_{n_i} \cdots W(n_1 n_2 \ldots n_i \ldots t). \tag{49}$$

From the Master equation it is possible to derive equations for $d\bar{n}_i/dt$ for $\partial F_1/\partial t$, etc. The fundamental significance of the Master-equation category is that the probability functions, such as W or F, all approach equilibrium. The Master-equation category describes irreversible behaviour. In fact Siegert showed from his equation that as $t \to \infty$, W becomes exactly the probability of the Boltzmann lottery (11). In the Master hierarchy W, F_1, F_2, the dynamics is contained implicitly in the expression for $a_{ij\,,kl}$.

In the Gibbs version of statistical mechanics, dynamical behaviour is explicitly expressed through the Liouville equation (13), written here as

$$\frac{\partial f_N}{\partial t} + \sum_i \frac{p_i}{m} \frac{\partial f_N}{\partial q_i} - \sum_i \frac{\partial V}{\partial q_i} \frac{\partial f_N}{\partial p_i} = 0 \qquad (50)$$

where f_N is the ensemble density function, previously called ρ. Just as from the Master function W a hierarchy of distribution functions $F_1, F_2, \ldots$ could be obtained by summation, a hierarchy of functions $f_1, f_2, \ldots$ can be obtained from f_N by integration

$$f_s(\mathbf{p}_1\ldots\mathbf{p}_s, \mathbf{q}_1\ldots\mathbf{q}_s, t) = \Omega_0^s \int \ldots \int d^3 p_{s+1} \ldots d^3 q_N\, f_N \qquad (51)$$

where Ω_0 is the volume. For additive potentials

$$V(\mathbf{q}_1 \ldots \mathbf{q}_N) = \sum_{ij} V(\mathbf{q}_i - \mathbf{q}_j). \qquad (52)$$

Integration of the Liouville equation, in the limit that $\Omega_0 \to \infty$, $N \to n$, $n = N/\Omega$, gives for f_1

$$\frac{\partial f_1}{\partial t} + \frac{p_\alpha}{m} \frac{\partial f_1}{\partial q_\alpha} + X_\alpha \frac{\partial f_1}{\partial p_\alpha} = n \int\int d^3 p_1\, dq_1\, \frac{\partial V(\mathbf{q}-\mathbf{q}_1)}{\partial q_\alpha} \frac{\partial f_2}{\partial p_\alpha}. \qquad (53)$$

This is the first equation of the celebrated BBKGY [29] hierarchy. This hierarchy was named after Bogoliubov, Born, Kirkwood, Guen and Yvon. This particular designation was first used by Uhlenbeck in his (unpublished) Higgins lectures in Princeton (1954). Long as the list of discoveries is, it is still incomplete. The hierarchy was found independently by J de Boer and by M Dresden.

Clearly (53) shows a seductive similarity to the Boltzmann equation (6), but it must be emphasized that neither the Liouville equation nor the Boltzmann-like equation (53) contain any probability whatsoever. The Liouville hierarchy is precisely equivalent to mechanics, and as such reversible, while the Master hierarchy yields an irreversible approach to equilibrium. If the Liouville equation is solved with a δ-function initial condition—the subsequent ensemble density remains a δ-function, moving along the classical trajectory. If the Master equation is solved with an initial δ-function for W, the solution at a later time spreads—it does not remain a δ-function. So the actual question becomes; what, if anything, has to be adjoined to the reversible Liouville description to obtain a relevant irreversible Master description? There are an enormous number of different suggestions (since the final result is known beforehand, they all work). They include, (1) approximate f_2 in (53) by a product of $f_1 f_1$ (the superposition approximations), (2) start from an initially factored distribution and see whether the factorization can be maintained, (3) carry out additional temporal or spatial averaging (4) construct a quantum-mechanical perturbation theory for infinite systems, to arbitrary order. The arbitrariness in deriving the Master equation from the Liouville hierarchy was especially stressed by van Kampen [30]. He preferred to make the necessary additional assumptions totally explicit and early in the formal development. 'One cannot escape from this fact (the reversibility–irreversibility dichotomy) by any amount of mathematical funambulism.' (This rather unusual word seems to be peculiar to van Kampen and its only use in physics is confined to reference [30].) Nevertheless this dichotomy dominated this epoch.

8.2.2.2. Topic 2: Brownian motion and the Kramers equation. Even though during this second epoch there was no single dominating direction of research in these areas, there were a number of highly significant individual contributions. In 1940, Kramers [31] estimated the escape probability of a particle caught in a potential well produced by a fluctuating environment. The precise shape of the potential well was not specified, but it was supposed to have at least one maximum and one minimum. The problem was treated by constructing a Fokker–Planck-type equation for the distribution $f(x, v, t)$ starting from the Langevin equation. The resulting equation for f (assuming a particle of unit mass) is

$$\frac{\partial f}{\partial t} + v\frac{\partial f}{\partial x} + X\frac{\partial f}{\partial v} = \gamma \left(\frac{\partial}{\partial v}(vf) + kT\frac{\partial^2 f}{\partial v^2} \right) \tag{54}$$

γ is the friction coefficient of the medium (often called viscosity) and X the outside force, with the usual meaning of the other symbols. This Kramers equation is similar in structure to the Boltzmann equation with a streaming term (left-hand side) and a collision term (right-hand side). It could be shown in general for times $t \gg \gamma^{-1}$, that f assumes the form

$$f \simeq \sigma(s, t) \exp\left(-\frac{v^2}{2kT}\right) \tag{55a}$$

$$\frac{\partial \sigma}{\partial t} = -\frac{\partial}{\partial x}\left(\frac{X}{\gamma} - \frac{kT}{\gamma}\frac{\partial \sigma}{\partial x}\right). \tag{55b}$$

Thus, for large viscosity, equilibrium is reached in two stages, a rapid approach to a Maxwellian velocity distribution, followed by a slow diffusion (55b). The main problem is to find tractable expressions for the escape probability both for large and small viscosity. This paper was totally ignored for about fifteen years [32]. It is therefore quite remarkable that since about 1990 there has been a veritable explosion of papers, meetings, conferences and summarizing articles (containing more than 200 references) on the Kramers problem. Many phenomena can be described in terms of the escape of a particle (or many particles) from a potential well, or a metastable state under the influence of a fluctuating environment. Examples can be found in chemical kinetics, electron transport in semiconductors, Josephson junctions, noise activated escape in driven systems. There is also a quantum version of Kramers' problem, which can provide information about the energy distribution of particles escaping from a well, or the population relaxation in double wells.

Kramers' derivation of (55) is a marvel of ingenuity—his method (systematized and not as clever) has become the 'elimination of the fast variable' technique—where an initial averaging over the fast variables reduces the problem to a diffusion-type problem.

Other phenomena appear to require non-local versions of the Langevin equation, such as

$$\ddot{x}(t) = X - \gamma \int_0^t F(\tau) x(t-\tau) \, d\tau + F(t). \tag{56}$$

These non-local modifications can explain the time decay of the velocity-correlation function $\langle v(0)v(t) \rangle$ observed in the temporal behaviour of isomeration processes. The processes described by equations such as (56) are no longer Markov processes.

The overall status of stochastic processes during this epoch was summarized in two famous papers. Chandresekhar's [33] paper treats many details of random walks, with and without boundaries; the experiments in colloid statistics are compared with the theoretical predictions of Smoluchowski. It is one of the few places where fluctuation ideas are applied to astronomical problems. Wang and Uhlenbeck [34] give an exhaustive complete discussion of the Brownian motion of coupled harmonic oscillators. This summary also contains the most systematic and complete classifications of random processes—which has become the standard for all succeeding discussions.

8.2.2.3. Topic 3: the new improved Boltzmann equation; the road to kinetic equations. Apart from the formal structure which the BBKGY method had conferred on statistical mechanics, the crucial physical question was whether equations such as (53) could generate physical results not contained in the Boltzmann equation. One specific class of questions was whether this formalism could produce expressions for the transport parameters of dense systems, going beyond the Boltzmann level. In equilibrium dense, non-ideal gases are always treated by an expansion in the density, such as the virial expansion of the pressure. It was widely, in fact generally, believed that a 'beyond Boltzmann' discussion would yield a similar expansion for the viscosities, heat conduction etc.

Bogoliubov [29] suggested a new viewpoint which led to a systematic, genuine extension of the Boltzmann equation. The basic observation was that in the temporal evolution, there are three distinct stages, governed by three characteristic times. The shortest time t_D is the duration of a collision, $t_D = r_0/\langle v \rangle$, ($r_0$ is the range of the intramolecular forces, and $\langle v \rangle$ some average thermal velocity, giving for t_D the value 10^{-12}–10^{-13} s). The second time t_c is the time between collisions. (If l is the mean free path, of order 10^{-4} cm for a not too dense system, $t_c \simeq 10^{-8}$–10^{-9} s). Finally the longest time, or macroscopic time $t_m \simeq L/v_s$, or equivalently $t_m = L/\langle v \rangle$; L is a macroscopic length and v_s the velocity of sound. Clearly

$$t_D \ll t_c \ll t_m \tag{57a}$$

$$r_v \ll l \ll L. \tag{57b}$$

It was Bogoliubov's insight to exploit the existence of these widely separated time-scales and to construct different equations for each stage. For times $t \ll t_D$, the detailed molecular dynamics is clearly essential, while for times $t \gg t_m$, the behaviour will be largely independent of the detailed molecular structure. This suggests a stagewise scenario.

For $t \lesssim t_D$, the system is hardly describable in purely statistical terms, as it will require all the distribution functions, all of which change rapidly. For $t \gtrsim t_D$, the one-particle distribution function f_1 will change rather slowly (f_1 depends on the intramolecular potential through the collision term, which for moderate n is not large compared with f_2, f_3). This is the motivation for the functional assumption that for $t \gtrsim t_D$ the distribution functions f_2, f_3 depend on the time through f_1. In this kinetic stage ($t_d < t < t_c$)

$$f_s(x_1 \ldots x_s, t) \to \Phi(x_1 \ldots x_s, f_1(t)). \tag{58}$$

If this equation is substituted in the first BBKGY hierarchy, a functional equation for f_1 results:

$$\frac{\partial f_1}{\partial t} = A(x_1|f_1) \tag{59}$$

This is the kinetic equation of Bogoliubov, in which x stands for p_1, q_1 in (59).

Of course neither Φ nor A are known *a priori*. To proceed further Bogoliubov assumes a density expansion for A, as well as for the fs distribution functions. For example

$$A_1(x|f_1) = A_1^{(0)}(x|f_1) + nA_1^{(1)}(x|f_1) + n^2 A_1^{(2)} + \ldots . \tag{59a}$$

N N Bogoliubov

(Russian, 1900–92)

N N Bogoliubov was born in 1900 in Nizhnii-Novgorod (Russia) but began his mathematical studies in Kiev under the direction of N M Krylov. His first scientific paper in 1923 was the start of a career that won him many international prizes. He established a school of non-linear mechanics in Kiev and was instrumental in organizing a department of theoretical physics in Moscow.

His research activities were about evenly divided between mathematics, theoretical physics and non-linear studies. He continued the significant investigations, initiated by Liapunov and Krylov, on the stability of non-linear systems, and then turned to purely mathematical questions involving the calculus of variations, quasi-periodic functions and differential equations.

One of his major innovations was an unusual and powerful investigation of non-linear oscillations, which eventually led to the new field of non-linear mechanics, culminating in a general formulation of abstract dynamical systems.

His work in theoretical physics was (at least initially) related to his mathematical studies of the asymptotic properties of differential equations. Of special importance for statistical mechanics was his decisive study of the set of distribution functions describing a general statistical system. (He is the first 'B' in the BBKGY hierarchy). The incisive analysis of the kinetic equations which follow from this system belong to the most important results of statistical mechanics. Bogoliubov's insight that the approach to equilibrium proceeds by different stages exerted an enormous influence on succeeding researches.

Quite different, but probably equally important, were Bogoliubov's contributions to the theory of superconductivity (and superfluidity). The transformations he constructed helped very greatly to develop the theoretical ideas of Fröhlich, Bardeen, Cooper and Schrieffer.

Non-Equilibrium Statistical Mechanics

Inserting these developments in the BBKGY hierarchy leads to an extremely involved but (with patience and skill) manageable system of equations. To second order the equation for f_1 is

$$\frac{\partial f_1}{\partial t} + \frac{p_{1\alpha}}{m}\frac{\partial f_1}{\partial q_{1\alpha}} = n \int\int dy_1\, dy_2\, D_2(q_1|y_1 y_2) f_1(y_1,t) f_1(y_2,t)$$
$$+ n^2 \int\int\int dy_1\, dy_2\, dy_3\, D_3(q_1|y_1 y_2 y_3) f_1(y_1,t) f_1(y_2,t) f_1(y_3,t). \tag{60}$$

The beautiful aspect of this equation is that D_2, D_3, depend exclusively on the two- and three-body dynamics while the statistics is contained in f_1. Of course (60) is based on an expansion in n, and is valid only for $t_D < t < t_c$, but it is an unquestioned advance, if not a simple equation!

8.2.2.4. *Topic 3a: an unwelcome surprise and a welcome new method.* The result of the Bogoliubov analysis for the f_1 equation can be written symbolically as

$$\frac{\partial f_1}{\partial t} + v_\alpha \frac{\partial f_1}{\partial x_\alpha} = J(f_1, f_1) + K(f_1, f_1 f_1) + L(f_1 f_1 f_1 f_1) \tag{61}$$

where J is proportional to n and depends on the dynamics of two isolated particles, K is proportional to n^2 and depends on the dynamics of an isolated group of three particles, etc. It is therefore perhaps not so surprising that the viscosity η can be expanded in a similar way

$$\eta(n, T) = \eta_0(T) + n\eta_1(T) + n^2\eta_2(T) + \cdots \tag{62}$$

where η_0 contains the contributions of binary collisions, η_1 those of triple collisions, etc. The coefficients η_1, η_2, etc, can be formally expressed as the time integral over all collision configurations involving three particles (for η_1) and over all configurations in which just four particles participate (for η_2) etc. These integrals are hard to evaluate in general, but η_1 has been calculated for hard spheres and the results are in reasonable agreement with experiment. Thus the viscosity is density-dependent because of multiple collisions, as expected from earlier studies. It therefore came as a shocking surprise that even for hard spheres η_2, the coefficient of n^2, diverged logarithmically with t. Higher terms in the expansion, η_3 etc, diverged more strongly [35]. It was especially through the investigations of Cohen [36] and Green [37] that the understanding of the origin of the divergences (and indeed their inevitability within the Bogoliubov scheme) was achieved. The divergences arise because in η_2 the time integrals are over all collision configurations—the integrals must include all the collision possibilities of the set of four molecules, considered as isolated autonomous systems. For a set of three hard spheres, there are four possible successive binary collisions between the three molecules. One of these collision sequences may stretch out over a long time, with the last collision occurring a large distance (larger than a mean free path) from the first. Under these circumstances, it is totally unphysical to consider these three as an isolated system; the last collision in the sequence will be influenced or perhaps eliminated because of intervening collision with molecules not belonging to the original set. Thus the expansion in autonomous clusters, one of the ingredients in the Bogoliubov procedure, cannot be valid for all time. These 'pathological' collision sequences cause the divergences in the time integration. Even the determination of the possible collision sequences is highly non-trivial. For example, the number of binary collisions of n identical hard spheres in d dimensions is not known. The few results known are scattered and remarkably inelegant. But it is interesting (and perhaps just lucky) that the Bogoliubov procedure does go beyond the Boltzmann description ($\eta_1(T)$ in (62) is finite).

In 1957 Kubo [38] developed an important technique, which allowed the formal calculation of many transport coefficients without the use of a kinetic equation. Suppose a system is described by a Hamiltonian; its equilibrium thermodynamic properties are determined by the equilibrium density matrix $\rho^{(0)}$. This system is now subject to an external influence described by $AF(t)$, where A is an arbitrary operator and F an arbitrary C-number function of the time. The temporal behaviour of the system is governed by the density matrix $\rho(t)$, which satisfies the von Neumann equation

$$i\hbar \frac{\partial \rho}{\partial t} = [H_t, \rho] \qquad (63)$$

with

$$H_t \equiv H + AF(t) \qquad (63a)$$

So far everything is traditional and exact. The fundamental approximation made in calculating $\Delta\rho \equiv \rho(t) - \rho^{(0)}$, is to solve (63), neglecting quadratic terms in either $\Delta\rho$ or A. This allows the equation for $\Delta\rho$, to be solved in that approximation. The average of a quantity B in this new time-dependent state $\rho_0 + \Delta\rho$ assumes a very elegant form

$$\langle B(t) \rangle = B_0 + \text{Tr}(B\Delta\rho) = B_0 + \int_0^\infty dt'\, \varphi_{BA}(t-t') F(t') \qquad (64)$$

B_0 is the equilibrium value of B and ϕ_{BA} the response function

$$\varphi_{BA}(\tau) = \frac{1}{i\hbar} \text{Tr}\rho^0 [A, B(\tau)] \qquad (65a)$$

R Kubo

(Japanese, b 1920)

Ryogo Kubo was educated primarily at the University of Tokyo, where he acquired the reputation as the best student ever to attend the highly prestigious institution. He was a professor at that University from 1948 until 1981, when he moved to the research institute in Kyoto.

Among his many honours and awards was the Boltzmann medal—arguably the greatest honour that can be conferred on an investigator in statistical mechanics.

One of Kubo's main contributions is in linear response theory. Through his seminal paper of 1957, it became possible to give compact expressions for many of the transport parameters without the need for intervening kinetic equations. Since such equations require separate analysis and justification this was a great advance. The Kubo formulism led naturally to the definition of a number of useful functions, in addition to the response functions, and to a very general formulation of the fluctuation dissipation theorem, exhibiting the general corrections between dissipative parameters and time correlation functions. A further analysis uncovered the direct relationship between the Kubo relations and a generalized version of the Onsager relations.

Kubo's approach to statistical physics, as presented in the second volume of his important book with Toda and Hashitsame, emphasizes the description of many physical phenomena in terms of stochastic processes, which leads to interesting and surprising results. In another study, devoted to quantum statistical mechanics, he introduced the appropriate boundary condition for Green's functions describing the physical system. This is not trivial, for the boundary conditions must be formulated in terms of an imaginary time, which requires the stipulation of the boundary conditions in the complex plane. This same boundary condition was employed by Martin and Schwinger in their field-theoretic formulations of quantum statistical mechanics.

$$B(\tau) \equiv \exp\left(\frac{i}{\hbar}H\tau\right) B \exp\left(-\frac{i}{\hbar}B\tau\right). \tag{65b}$$

This formalism is characteristic of the linear response. If B is chosen to be the electric current caused by the electric field A and $F(t)$ is $e^{i\omega t}$, these formulae yield, after considerable manipulation, the frequency-dependent conductivity tensor $\sigma_{\mu\nu}$

$$\sigma_{\mu\nu} = \beta \int_0^\infty dt\, e^{i\omega t} \langle j_\nu j_\mu(t)\rangle. \tag{66}$$

where $\beta = 1/kT$. The definition of the conductivity, via the linear law relating currents and fields, has been used in (66). This expression for the conductivity tensor typifies the relations obtained by Kubo's procedures. Transport parameters emerge as time integrals over correlation functions. The derivation of these formulae as given is completely formal. More physical arguments, based on the Langevin equation, showed earlier [39], that the friction coefficient can be obtained as a time integral over the average correlation functions of the fluctuating term in the Langevin equation. A more rigorous and more general derivation leads to results possessing the same basic structure: a transport coefficient L can (in a linear response approximation) be written as the thermodynamic limit of a time integral over an averaged correlation function

$$L = \lim_{\varepsilon \to 0} \lim_{N,\Omega \to \infty} \int_0^\infty dt\, e^{-\varepsilon t} \langle J(0)J(t)\rangle. \tag{67}$$

The average in (67) is the average over the canonical ensemble. Calculations of the transport coefficients starting from (67) are more tractable than those obtained from the Bogoliubov analysis, although, as Resibois [40] showed, the results are the same (they show the same divergence and both are approximate).

8.2.2.5. Topic 4: rigorous results; classification of dynamical systems. In 1949 the translation of a book by Khintchine appeared, on *Mathematical Formulation of Statistical Mechanics* [41]. It was extremely critical of Boltzmann's treatment of probability notions which were considered naive and simplistic, and Gibbs was taken to task for not defining his probability notions. Khintchine not only complained about the lack of mathematical rigour (that, he argued, could always be fixed by a mathematician), but he was adamant that the physical formulation was so imprecise that a mathematician could not do much with the subject. His purpose in writing the book was to call attention to rigorous limit theorems and available mathematics (Birkhoff's theorems) to provide a firm foundation for the subject. Most physicists were not impressed by Khintchine's arguments; to them a well-defined

Non-Equilibrium Statistical Mechanics

implementable algorithm was far more important than rigour. Excessive rigour was viewed as an expensive luxury and a persistent irritation. There was nevertheless a considerable amount of excitement when Sinai [42] announced that he had demonstrated that a system of hard spheres enclosed in a box was metrically transitive. Previously metrical transitivity had been shown for contrived physical systems, but Sinai's result was the first legitimate physical system for which this was established. This meant that the original Boltzmann programme for a hard sphere system was completed. As such it created a stir, but there was a surprising by-product as well; Sinai's proof, a true mathematical *tour de force*, applied to any number of spheres n, as long as $n \geqslant 2$. It had always been tacitly assumed that ergodicity (or rather transivity) required a many-body system to provide the necessary irregularity for a valid statistical treatment. Since a system of two or three is ergodic (by Sinai's result), many physicists did not accept ergodic behaviour as a sufficient (or even important) foundation for statistical mechanics [43]. Some considered ergodic studies 'an exercise in sterility'.

Even so Sinai's result forced physicists and mathematicians to reconsider and redefine what are the appropriate notions which describe the different types of time evolution. Each member of an ensemble of points in phase space in the course of time will trace out a complicated trajectory. What are the different patterns of the spreading and redistribution of the ensemble members in the course of time? There exists a hierarchy of behaviours, some extremely regular, others extraordinarily irregular.

(i) In non-ergodic or recurrent systems, a trajectory returns to a given neighbourhood an infinite number of times, without visiting all parts of the energy surface. A phase-space volume moves through the phase space without cumulative distortion.

(ii) In an ergodic (metrically transitive) flow the shape of the phase-space volume changes smoothly while the volume remains constant. In an infinite time a phase-space volume crosses every domain.

(iii) In a mixing flow a phase-space volume is violently distorted, developing fractal-like shapes, spreading out (in the average) uniformly over the phase space. Sinai actually showed that the hard sphere gas is mixing.

(iv) The so called K systems, named after Kolmogoroff [42], possess a very complicated trajectory pattern. Such systems have a positive Kolmogoroff–Sinai entropy. Nearby trajectories separate exponentially in time. The Kolmogoroff–Sinai entropy is a measure of the mean exponential separation coefficient.

These ideas and classifications, far from being esoteric and pathological, are essential for a description and understanding of chaos, as well as for an appreciation of the temporal behaviour in statistical mechanics.

8.2.2.6. Topic 5: the Onsager relations. In contrast to the important discoveries of the Onsager relations and the fluctuation–dissipation theorem in the first epoch, the second epoch was more a period of analysis, consolidation and rethinking. Furthermore many of the Onsager-type relations, as well as the fluctuation dissipation theorems, were subsumed under the more general Kubo relations. A considerable amount of insight was achieved through three new derivations of the Onsager relations [44–46]. Onsager based his early derivation on the assumption that the macroscopic laws of motion hold for the averages in the microscopic domain, even when these exhibit fluctuations. In all derivations it is important to distinguish the usual phase space of the microscopic mechanical variables, and the 'little phase-space gamma', which describes the macroscopic state in terms of macroscopic variables $Q_1, \ldots, Q_n$ ($n \ll N$). It was Einstein who suggested that fluctuation of the macroscopic variables could be described as a kind of Brownian motion. In terms of the γ-space, the macroscopic phase point would perform a Brownian motion or a random walk in that space. The entropy difference between a particular state and the equilibrium state, ΔS is, as before, a quadratic function

$$\Delta S = -k \sum S_{ik} Q_i Q_k. \tag{68}$$

Thus by the relations between entropy and probability, the probability for a particular configuration in the γ-space is

$$W(Q) = \tfrac{1}{2} \exp\left(-\sum S_{ik} Q_i Q_k\right). \tag{69}$$

The temporal change of the fluctuation (in harmony with the Brownian motion picture) can be described in terms of a transition probability per unit time $A_{Q \to Q'}$. If it is further assumed [45] that this is a Gaussian random process, the Onsager reciprocal relations follow. It is also possible to assume a variant of the Onsager regression hypothesis in the form

$$\int dQ' \, Q'_i A_{Q' \to Q} = \sum L_{ij} Q_j \tag{70}$$

which means that the average value of the macroscopic variables is a linear function of the initial values. Onsager's original assumption was the differential form of (70)

$$\frac{d}{dt} \langle Q_i \rangle = \sum_j L_{ij} \langle Q_j \rangle \tag{71}$$

(70) is a weaker form of Onsager's assumption, but it is sufficient to produce the Onsager relations. It should be recalled, nonetheless, that both methods of obtaining the reciprocal relation are approximate. The Onsager relations themselves continue to be experimentally verified with great precision.

8.2.2.7. Topic 6: quantum modifications. Quantum statistical mechanics in this second epoch took a decided turn towards quantum field theory. All aspects of the quantum field-theoretic formalism were applied to statistical problems. The equilibrium problems were primarily treated, starting from the grand canonical ensemble, by exploiting the field-theoretic techniques, Green's functions, diagrammatic developments and time- or temperature-ordered products. This formalism was applied to a number of different physical problems, superconductivity and many-body theory in both nuclear and solid-state physics. Although the majority of applications dealt with equilibrium situations, the field-theoretic methodology was also applied to non-equilibrium studies. In a fundamental study, Martin and Schwinger [47] investigated the temporal behaviour of a general system, acted on by an external perturbation (which need not be small). They constructed a sequence of coupled equations for the time-dependent Green's functions, using field-theoretic techniques. A typical non-equilibrium feature is the selection of an appropriate boundary condition. The formalism is general and imposing; the equations for the Green's function contain, at least in principle, field-theoretic effects such as self-energies and exchange effects. To obtain physically transparent equations numerous approximations have to be made. It is possible to construct equations for the response of this general system; it is even possible to obtain the usual Boltzmann equation, after enough approximations. The formalism is powerful, but its utility is limited by the need for several approximations and the remaining formidable analytical difficulties. There are very few, if any, instances where quantum field-theoretic effects can be made explicit in transport phenomena or other non-equilibrium processes.

It was shown much earlier by Wigner [48] that it is possible to construct a function of momenta and coordinates which mimics many of the properties of the classical Boltzmann distribution function (or the Gibbs phase-space density function). In terms of wavefunctions the Wigner function is defined by

$$W(p,q) = \left(\frac{1}{2\pi\hbar}\right)^n \int \cdots \int d^n q' \, \psi(q-q')\psi^k(q+q')\exp\left(-\frac{2i}{\hbar}pq'\right). \quad (72)$$

It is actually better to define W in terms of the density matrix

$$\rho(q'',q') = \sum_j p_j \psi_j(q'')\psi_j^k(q'). \quad (73)$$

p_j is the probability that an ensemble member is in a state described by a wavefunction ψ_j. The definition of the Wigner function is then

$$W(p,q,t) = \left(\frac{1}{2\pi\hbar}\right)^n \int \cdots \int d^n q' \rho(q-q', q+q') \exp\left(-\frac{2i}{\hbar}pq'\right). \quad (74)$$

The Wigner function was widely ignored for more than 50 years, apart from a few occasional reformulations in a second-quantized form. It was realized that the Wigner function would be useful for the investigation of the precise relation between classical and quantum mechanics. As a consequence the Wigner function has become a major tool in the study of quantum chaos. It is amusing, and perhaps contains a moral, that an old observation, long ignored, reappears as one of the few useful techniques to study a new field.

8.2.2.8. Topic 7: models. As was to be expected, the number and sophistication of models increased sharply during this epoch. One of the notable events was the complete solution of the Ehrenfest dog-flea model by M Kac [49]. His formalism was powerful enough to draw a number of interesting additional conclusions. For example, it could be shown that each state inevitably recurs, which is the direct analogue of the Poincaré recurrence theorem. It was also possible to calculate the mean time of recurrence, Θ_n, of a state where one dog has $2n$ more fleas than the other (the total number of fleas is $2N$).

> See also p 604

$$\Theta_n = \frac{(N+n)!(N-n)!}{(2N)!} 2^{2N}. \quad (75)$$

The mean time of recurrence for $N = 10^4$ and $n = 10^4$ (if there is 1 jump per second) is 10^{6000} years. It is also possible to calculate the physically important fluctuations in the recurrence times; but even for this model, those calculations become quite difficult.

It is remarkable that even a slight change in a model can produce quite different behaviour. In 1972, the Ehrenfest model was modified to make the probability for a jump from dog A to dog B equal to p_n, when there are n fleas on dog A. The Master equation for $P(n,t)$, the probability that there are n fleas on dog A at time t is still quite simple

$$P(n, t+1) = p_{n+1} P(n+1, t) + (1 - p_{n-1}) P(n-1, t) \quad (76)$$

but this model shows a remarkable richness of behaviour. For certain choices of p_n metastable states are formed, where the flea distribution on the two dogs remains constant, except for minor fluctuations. The metastable states are not equilibrium states, and in the limit $t \to \infty$ the equilibrium state is reached, but not monotonically.

Probably the most important study for the succeeding development was the numerical investigation by Alder and Wainwright [50]. Using a computer simulation, they studied a dense gas of hard disks, moving and colliding in a plane. The quantity computed was the autocorrelation function average of a tagged particle: $\langle v(0)v(t)\rangle$, where $v(t)$ is the velocity of the tagged particle at time t. The average is the canonical equilibrium average over the momenta and coordinates of all other particles. This quantity can also be calculated in many different theories, all assuming some version of molecular chaos. All such theories predict an exponential decay with time. For the Enskog theory adapted to hard disks of diameter σ, this correlation function is

$$\langle v(0)v(t)\rangle \simeq \exp(-t/\tau_E) \tag{77}$$

$$\tau_E = \sqrt{\frac{m}{kT}} \frac{1}{n\sigma^2} \tag{77a}$$

where n is the number density of the disks and τ_E is the so-called Enskog time. Alder and Wainwright did find an exponential decay for times $t < \tau_E$. However, for $t > \tau_E$ the data showed a much slower decay. In two dimensions the data could be fitted with a slow decay

$$\langle v(0)v(t)\rangle \sim \left(\frac{t}{\tau_E}\right)^{-1}. \tag{77b}$$

This empirical discovery led to a great deal of new theoretical work [51] which reproduced the data extremely well—and suggested new computer experiments. This was the beginning of an immensely successful interplay between experiments, computer simulations and theoretical studies.

8.2.2.9. Topic 8: Prigogine and the Brussels School. A quite distinct series of developments in statistical mechanics took place in two major centres devoted primarily to non-equilibrium studies, Brussels and Austin (Texas), under the leadership of Prigogine. Even though the Brussels school dealt with many of the same problems in non-equilibrium theory as did others, their approach, terminology, philosophy and diagrammatic techniques were so different that communication has become quite difficult. The Brussels school, like all scientific enterprises went through a number of stages.

Stage 1: 1956–64. This first phase concentrated primarily on an alternative analysis of the Liouville equation and came to a certain measure of conclusion with the publication of Prigogine's book [52]. The first step in the analysis is a Fourier series decomposition of the ensemble density function (the system is enclosed in a finite box).

$$f_N(1\ldots N, t) = \sum_{\bar{k}} a_{\bar{k}}(p_1\ldots p_N, t)\, u_{\bar{k}}(x_1\ldots x_N) e^{-i w_{\bar{k}} t}. \tag{78}$$

Here $\bar{k}$ stands for the collection of wave numbers $k_1 \ldots k_N$, and

$$u_{\bar{k}} = \left(\frac{1}{L}\right)^{3N/n} \exp\left(i \sum_{i=1}^{N} k_i x_i\right) \qquad k_i = \frac{2\pi}{L} n_i \quad (n_i = 0, 1, \ldots). \qquad (78a)$$

Now a remarkable regrouping of f_N is carried out, which allows a separation of the sum over k into separate sums over just one wave number, over pairs of wave numbers, triples, etc. The corresponding coefficients are functions of the momenta; a_0 is a function of $p_1 \ldots p_N$; a_1 a function of k $p_1 \ldots p_N$; a_2 depends on k_1, k_2 and $p_1 \ldots p_N$ etc. This expansion is substituted in the Liouville equation and produces a complicated set of coupled differential equations for the Fourier amplitudes. The solution of this set is an iteration procedure in terms of the molecular interaction λV_{int}. The coupled equations for the amplitudes a allow a convenient graphical representation of the successive terms. The system is studied in the thermodynamic limit and for large (infinite) times. In this process additional approximations are made—omitting the diagrams for processes that vary slowly in time. Even though the formalism, graphs, language and approximations are different, the results obtained do not appear very different. Prigogine, like Uhlenbeck and Bogoliubov, exploits the existence of widely separated time-scales. When applied to a homogeneous plasma the Prigogine–Balescu equations to lowest order in e^2 and all orders in the density give the same results as the more conventional methods.

Stage 2: 1967–75. The period 1967–75 saw a considerable deepening of the Brussels approach (see especially the book by Balescu [53]) with the introduction and exploitation of the notion of sub-dynamics. The general idea is to adhere to Liouville's equation but to split the space of functions f_N into two classes. The functions in one class presumably describe the equilibrium and transport properties, the others the fluctuations, irregular and chaotic behaviour. Each class is governed by a separate equation—one should satisfy a Markovian kinetic equation, the other the full Liouville equation. The macroscopic description should exclusively depend on the variable in the kinetic subspace (which is governed by the Markovian kinetic equation). This splitting is to be accomplished by the introduction of suitable projection operators. Although interesting and suggestive the implementation is not at all simple. The kinetic and chaotic subspaces are still coupled, and not totally autonomous. Furthermore, the distinction between fluctuations and smooth behaviour depends on the scale of precision required. There are serious mathematical differences between projection operators in a space of functions and in a space of operators (Prigogine refers to them as super operators, although they were introduced as early as 1932 by Koopman). The sub-dynamics approach has its stout defenders [53] and severe critics [54].

I Prigogine

(Russian, b 1917)

Ilya Prigogine, born in Moscow, received most of his early training at the 'Free' University of Brussels. He studied both physics and chemistry and in 1941 he obtained a doctor's degree in chemistry from the University at Brussels, where he later became a professor. His many honours include the Nobel Prize in 1977.

His earliest work was in irreversible thermodynamics, following the directions of Gibbs and de Donder, and then turned to an analysis and generalization of the Onsager relations and the molecular theory of solutions (leading to a monograph in 1957).

Prigogine became one of the main contributors to the field on non-equilibrium statistical mechanics, and developed a highly individual approach, carried out with great effectiveness by his students and collaborators at two active and idiosyncratic centres in Brussels and in Austin, Texas. Prigogine later emphasized the need for introducing notions which would characterize states very far from equilibrium, especially the generation of these dissipative structures such as are presumed to play an important role in biological processes. Fluctuations and instabilities play a particularly important role in the temporal evolution as viewed by Prigogine.

In more recent investigations, Prigogine has adopted an even more unconventional view of irreversibility, involving a proper description of entropy. It is bold and suggestive and very different from the more conventional approaches.

Stage 3: 1980–90s. Some of these ideas now became part of a more radical revision of the fundamental ideas of statistical mechanics [55]. There was a marked shift towards more speculative considerations. Ideas from non-linear dynamics—Liapunov functions—began to play a central role. It was strongly hinted that extreme non-equilibrium configurations played a major role in the establishment of structure and order in physics and chemistry as well as biology.

(i) Instead of attempting to derive irreversibility from mechanical and probability considerations, Prigogine insisted that the irreversibility notion was fundamental and irreducible; it should be incorporated in the laws of Nature but not derived.

(ii) Entropy should be considered as a particular kind of Liapunov functional Θ of the physical configuration which satisfies

$$\Theta \geqslant 0 \qquad \frac{d\Theta}{d\tau} \leqslant 0 \qquad (79)$$

(iii) There exists, or should exist, a microscopic entropy operator M which satisfies

$$M \geqslant 0 \qquad (80a)$$

$$\frac{dM}{dt} = iLM \qquad (80b)$$

where L is the Liouville operator, and

$$-i[L, M] = D \leqslant 0 \qquad (80c)$$

where D is the microscopic entropy production operator.

(iv) Prigogine objected to the 'double duty' of the energy operator, which determines both the time evolution and the quantized levels. He suggested that L determines the time evolution and X determines the energy levels where X is determined by

$$\begin{aligned} L\rho &= H\rho - \rho H \\ X\rho &= H\rho + \rho H \end{aligned} \qquad (81)$$

These ideas are certainly unconventional and thought-provoking. Within the physics community they produce a mixture of scepticism, elation and concern: an ideal mixture for further progress.

8.2.3. Epoch III: 1975–90s

8.2.3.1. Consolidation—diversification and computers. One of the benefits of a chronological survey is that the changes in interest and the varying fashions become almost painfully obvious. Since this last epoch overlaps the late twentieth century, it is really impossible to achieve a balanced perspective. For some topics the transition from epoch II to epoch III is so gradual that a mere indication of the subjects and the results achieved is sufficient to acquire a 'short time' insight into the development.

8.2.3.2. Topic 1: the BBKGY hierarchy. The activity in the general area of the BBKGY hierarchy and the associated kinetic equations diminished sharply in this epoch. Most of the investigations were formal extensions, and occasionally the formalism was applied to new systems. For example, the Bogoliubov procedure was applied to binary mixtures; the formalism worked, but there were not many novel physical insights. This is the general character of many of the extensions and elaborations, which require considerable effort and technical ingenuity. Unfortunately the physical results are not commensurate with the effort expended. There are very few new conceptual insights and few (if any) suggestions of new physical phenomena.

An exception is perhaps a paper by Horwitz *et al* [56]. They constructed a manifestly covariant version of the BBKGY hierarchy. In a relativistic context, new insights might be gained from a confrontation between causality, time order and irreversibility.

8.2.3.3. Topic 2: Brownian motion. It is striking how in the period 1975–94 random walks became of central interest. It is not an exaggeration that there was a veritable explosion of random-walk-type models, representing an extraordinary variety of physical, chemical, astronomical and ecological situations. The general features are well known. A walker particle is constrained to move in a prescribed, fixed lattice. Its temporal behaviour is specified by the probability $M(l, l')$ of a jump from l to l' in unit time. The probability of reaching position l after j steps is $P_j(l)$. It satisfies the equation

$$P_{j+1}(l) = \sum_{l'} M(l, l') P_j(l'). \tag{82}$$

If the transition probability depends on $l - l'$, (82) can easily be solved by a Fourier series. For an infinite lattice in s dimensions the solution is

$$P_j(l) = \left(\frac{1}{2\pi}\right)^s \int \cdots \int_{-\pi}^{+\pi} \left(\tilde{M}(k)\right)^s e^{-ilk} \, dk \tag{83}$$

where k stands for s variables and $\tilde{M}$, the structure function of the random walk, is the Fourier transform of M. To calculate quantities of physical interest (first passage times, recurrence times) the explicit form (83) is often not the most convenient; it is better to use a generating function (often called the lattice Green's function) which is defined by

$$G(l, z) \equiv \sum_{j=0}^{\infty} P_l(j) z^j$$

$$= \left(\frac{1}{2\pi}\right)^s \int \cdots \int d^s k \, \frac{\exp(-ilk)}{1 - z\tilde{M}(k)}. \tag{84}$$

Twentieth Century Physics

The Green's function satisfies the equation

$$G(l, z) - z \sum_{l'} M(l - l')G(l', z) = \delta_{l,0}. \tag{85}$$

With these techniques many random walks have been investigated—random walks with traps, where the walkers get stuck, and random walks where particles (positrons) disappear. Random walks using special lattices, e.g. Bethe lattices, fractal lattices, have been analysed in great detail. Random walks, with many walkers, which can be friendly or unfriendly, can be used as models for fluid mixtures. Of special importance (surprisingly enough for field theory) are the self-avoiding walks.

See also p 569

The shape problem, the geometrical characterization of the points visited (usually just once), is important (especially in two and three dimensions) for random walks with boundary conditions. It is surprising that the shape problem for many random walkers (which could be friendly, unfriendly or indifferent) was not treated seriously until 1992 [57]. The resulting geometric patterns are of great importance for the territorial structure of ecology. The method of treatment varies from problem to problem, but is almost always a judicious combination of analytical methods with extensive and exhaustive computer simulations.

In 1986, van Kampen [58] raised an interesting new question by investigating random walks in a manifold. One of the surprising results was that the various equivalences between random walks, Brownian motion, Fokker–Planck equations and the Langevin equations no longer held. This suggestion was not followed up very vigorously, which seems somewhat unfortunate since this is one of the few conceptual (rather than computational) innovations.

8.2.3.4. Topic 3: the Boltzmann equation. It might well appear that with the Bogoliubov generalization of the Boltzmann equation to include triple collisions, interest in the original Boltzmann equation would decline. Surprisingly this was not the case; in this last epoch there was actually more activity in analysing, discussing and approximating the 'old fashioned' Boltzmann equation than in the BBKGY hierarchy. Part of this continued and renewed interest was stimulated by its applicability to plasma physics. Of course approximations to and (in Schwinger's language) mutilations of the Boltzmann equation are an old story.

A widely used simplification [59] was introduced in 1954

$$\frac{\partial f}{\partial t} + v_\alpha \frac{\partial f}{\partial x_\alpha} + \frac{X_\alpha}{m} \frac{\partial f}{\partial v_\alpha} = -\frac{1}{\tau}(f - nf^{(0)}) \tag{86}$$

where $f^{(0)}$ is the equilibrium distribution, n the number density and τ the mean free time for molecules of velocity v; τ depends on v and n, and for simple cases $1/\tau = n/\sigma_0$, σ_0 is the differential scattering cross section, so the collision term is non-linear. For long times and dilute systems (86) is a pretty good approximation, and not unrelated to the relaxation-time approximation for the collision term frequently used in solid-state physics. There the right-hand side of (86) is written as $(f - f^{(0)})/\tau'$; τ' is similar to but not the same as τ, and τ' depends on v but not on n which makes the equation linear and much simpler than (86).

Even earlier [60] Grad employed an expansion of the velocity-dependence of the distribution function in Hermite functions. In this expansion the coefficients of the Hermite polynomials carry the space and time dependence. They are directly related to the moments of the distribution function. By carrying out the expansion to a certain order, Grad could obtain a solution depending on a (any!) finite number of moments. In this manner, he reproduced in a much more transparent manner the earlier results of Enskog, Chapman and Burnett. The period after Grad's article in *Handbuch der Physik* [60] saw a large number of special features for special systems, such as the construction of a Boltzmann equation for polyatomic molecules. Many highly particular matters were studied: the effect of triple collisions in one dimension, the symmetries of the Enskog solutions, the nature of asymptotic solutions. A more systematic study was undertaken by Ernst [61] and his collaborators. These studies all explored the mathematical and physical consequences of non-linearity. For example, they studied the Boltzmann equation for homogeneous systems

$$\frac{\partial f(v,t)}{\partial t} = \int d^3w \int g[I(g,\theta)] d\Omega (f(v',t)f(w',t) - f(v,t)f(w,t)). \quad (87)$$

For Maxwell molecules, they could find exact solutions in closed form.

It is possible in the general case to reduce (87) to an equation for a single scalar function. For an isotropic velocity dependence $f(v,t) = F(|v|,t)$ the resulting equation is

$$\frac{\partial F}{\partial t} = \int_0^\infty du \int_0^u dy \big(K(xy|u) F(y,t) F(u-y,t) - K(yx|u) F(x,t) F(u-x,t) \big) \quad (88)$$

Molecular interactions through the cross section are contained in K. For special molecular models, this non-linear equation can be solved exactly. More interesting are some of the non-intuitive features.

(i) The approach to equilibrium is not uniform.

(ii) Initial configurations exist where the distribution function F oscillates around its equilibrium value

(iii) If an initial distribution is a superposition of equilibrium distributions at different temperatures, this form is maintained for all time with time-dependent parameters.

625

The old-fashioned Boltzmann equation is not dead, it is not even old!

8.2.3.5. Topic 4: rigorous results. The considerable interest in analytical and computational aspects of the old-fashioned Boltzmann equation in this epoch extended to existence and uniqueness proofs. In 1989 a book [62] was devoted entirely to the mathematics of the old Boltzmann equation. There was nothing careless or slipshod about the treatment nor were the mathematical problems routine exercises; on the contrary, they required sophisticated and unusual techniques. This of course separated the practitioners of rigour from the pragmatically oriented physicists. Because of the difficulty of dealing with the thermodynamic limit in a systematic rigorous manner, a number of physicists and mathematicians developed methods (or adapted existing mathematical procedures) which would allow *ab initio* discussion of infinite systems. These so-called C^* methods were first used in connection with field theory [63]. Compared with these, the Birkhoff and von Neumann ergodic theorems are applied physics. Opinions differ widely on the importance and relevance of these results, of which the best available is due to Lanford [64]. He showed that there is a limit (called the Boltzmann–Grad limit) in which the BBKGY hierarchy converges to the Boltzmann equation. In his analysis Lanford considers a system made up of N spheres of diameter d. The Boltzmann–Grad limit is $N \to \infty$ and $d \to 0$, with Nd^2 finite and of course $Nd^3 = 0$. Thus Lanford's result only applies to a system of zero density. Furthermore, the proof guarantees a high (even overwhelming) probability of Boltzmann-like behaviour only for a short time (about $\frac{1}{5}$ of the time between collisions). It is unclear whether this result should be interpreted as a success or a limitation of the C^* method.

8.2.3.6. Topic 5: the Onsager relations and linear response. The activities in this area consisted mainly of re-examination and critical analysis of linear response theory, the Onsager relations and the Langevin equation. There were a number of extensions and generalizations as well, but the main thrusts were in the rethinking of basic ideas. It is in fact quite straightforward to carry out the expansion of the response theory beyond the linear domain. However, consistency in the next approximation requires that the Joule heating be included in the theory of electrical conduction. This changes the physical picture radically, and no completely satisfactory description, taking both the heating and conductivity aspects into account, based on first principles seems to exist. A closely related comment in a paper by Lindhard [65] stresses that the Kubo procedure is an approximation to a reversible theory—hence its results should be reversible as well. Yet the resistance is an explicit manifestation of irreversibility and it is one of the leading results of Kubo theory, while the heating, in the next approximation, is as irreversible as it can be. Van Kampen [66], in an important paper, raised major objections to the linear response assumption. He stresses that the trajectories in phase space are extremely complicated, unstable and sensitive to perturbations. Consequently a perturbation calculation (to first order) of a microscopic dynamical quantity has a very limited range of validity. Yet the macroscopic linear laws deduced by Kubo have a considerable domain of validity. Thus van Kampen claims that some averaging or coarse-graining is implicitly assumed in the derivation of these results. He insists that a sharp distinction be made between microscopic and macroscopic linearity. This is a separate assumption, also made in the derivation of the Onsager relations, and it requires a separate justification.

The canonical answer to van Kampen's objection is that the linear response is not applied to a single unstable trajectory, but to a rather carefully selected ensemble of trajectories, or to a class of smooth phase-space functions. This would imply that the initial ensemble is either in or near equilibrium, and that the perturbations are not too large or long-lived, and finally that fluctuations to unstable configurations are either unlikely or are averaged somewhere in the formalism. However, the validity of the linear response method has not really been precisely displayed.

It is interesting that this same question of assuming linear macroscopic laws occurs not only in linear response theory, in the Onsager relation, but also in the Langevin formalism. There is little argument that a Langevin equation of the form

$$\dot{y} = A(y) + F(t) \tag{89}$$

(where $F(t)$ is the usual fluctuating force) yields a probability distribution $P(y, t)$ which satisfies a Fokker–Planck equation of the type of (18)

$$\frac{\partial P(y,t)}{\partial t} = -\frac{\partial}{\partial y}(A(y)P) = \tfrac{1}{2}\Gamma \frac{\partial^2 P}{\partial y^2} \tag{90}$$

but there is a great deal of argument, indeed a minor controversy, about a similar result for the non-linear Langevin equation

$$\dot{y} = A(y) + C(y)F(t). \tag{91}$$

The difficulty lies in the proper definition of the product of a function $C(y)$ and a fluctuating function F, whose correlation functions are δ-functions.

8.2.3.7. Topic 6: quantum aspects. Most of the quantum-mechanically oriented studies in statistical mechanics continued to employ field-theoretic procedures. Apart from a few investigations of the formal properties of density matrices, most applications remained in condensed matter physics and superconductivity. There was a somewhat surprising resurgence of interest in quantum Brownian motion and in the quantum Langevin equation [67]. The Langevin equation for a single harmonic oscillator in quantum theory has the deceptively simple form (the mass is put equal to 1)

$$\frac{d^2x}{dt^2} + \eta \frac{dx}{dt} + kx = F(t). \tag{92}$$

However, $x(t)$ is the time-dependent position operator and $F(t)$ is an operator-valued random force, so that $F(t_1)$ and $F(t_2)$ do not commute.

$$[F(t_1), F(t_2)] = 2i\hbar\eta\delta'(t_1 t_2) \tag{92a}$$

This leads to an interesting formalism, providing the complete quantum theory of the Brownian motion of a system of linearly coupled oscillators. A similar analysis for more general systems can only be carried out in an approximate manner—but at the time of writing this area is of great concern. In what appears to be a most fundamental and under-appreciated paper, Ford and Lewis made an attempt to construct a quantum version of the classification of random processes classically treated in the paper by Wang and Uhlenbeck [34]. It is remarkable that no previous investigations had attempted such a classification when the underlying dynamics is quantum mechanical rather than classical.

The quantum hierarchy, like the classical hierarchy, operates with an infinite set of joint distribution functions. However, in the quantum theory it is necessary (because of the ubiquitous non-commutativity) to invoke a time-ordering of the functions. It can no longer be assumed, and is in fact not true, that $W_2(y_1 t_1, y_2 t_2) = W_2(y_2 t_2, y_1 t_1)$. Even this single alteration changes the formalism drastically. At the time of writing it appears that the study of quantum stochastic processes is on the verge of becoming a major new fashion in laser physics and femtosecond chemistry. However, important conceptual problems remain—furthermore a number of new precision experiments are now becoming available, which will require novel interpretations [68].

8.2.3.8. Topic 7: models. It was almost inevitable that in this epoch, when computers became common in every office, and computing skills of physicists a more common commodity than quantum mechanics, the study of models would become a booming sub-field of statistical mechanics. The computer simulations, computer graphics and Monte Carlo techniques allowed detailed information and visualization not available in earlier epochs. Even so old models never die; there were a number of variations of the Rayleigh–Lorentz models, treated analytically, describing different aspects of the delocalization of the initial states [68].

The discretization process, the replacement of a continuous system by a discrete model, became very important. The Boltzmann equation, the hydrodynamical equations and quantum field theory can all be formulated on a discrete lattice. Apart from mathematical and sometimes conceptual advantages the computer clearly requires a discrete formulation. Typical equations for such discrete models have the generic form

$$\eta_p(t+1) = f\{\eta_p(t)\} + X_p + F_p(t) \tag{93}$$

where η_p is a vector, describing the state at time t and location p, p describes the discrete structure, f is a functional of a set of neighbours of p, here indicated as $\eta_p|t$, F_p is a fluctuation force and X_p a systematic force. Apart from systems described by (93), many different models have been constructed for specific purposes. The ingenious game 'life' invented by Conway was given a statistical–mechanical treatment by Schulman [69]. It was shown earlier that this game and a more general class of game rules can be subsumed in the category of models described by Dresden [70]. A surprising connection was uncovered by Lebowitz [71], who investigated a weakly asymmetric lattice gas, with a stochastic dynamics (F in (93) was not zero), and was led to the non-linear Burgers equation, well known in hydrodynamics as an equation which exhibits turbulence.

Various features of a discrete Lorentz–Rayleigh model have been combined with elements of the wind-tree model, to produce a very interesting and important class of models, the Lorentz lattice-gas models, which show a surprisingly rich behaviour. Particles move in prescribed directions on a lattice, and scatter from obstacles, so placed and of such geometrical shape that the succeeding motion is again in one of the allowed directions in the lattice. These models were extensively studied, numerically and analytically by the Utrecht School [72]. As always the velocity correlation function and the diffusion parameters provide a sensitive tests for the comparisons between theory and experiment. The analysis is not simple. One average has to be performed over the different trajectories, for a fixed arrangement of scatterers; there follows a second average over the allowed scatterers' arrangements. The results are striking; in general there is no diffusion-type regime; the autocorrelation function inevitably shows long tails.

8.2.3.9. Topic 8: new directions, cellular automata and no conclusions. One new direction, which in point of fact is no longer so new, is the field of cellular automata, in which both space and time coordinates are discrete. The dynamical behaviour is governed by an updating rule, which is a discrete form of an equation of motion. It should be evident that (93) represents such an updating rule. It is also evident that once such a rule is given and the discrete lattice is specified a computer can proceed to trace out the time evolution. In fact the temporal development of the system is precisely reflected in the iterative properties of the computer program. Using appropriate rules, cellular automata can be used to simulate many quite different systems (physical, chemical, ecological etc). To implement such schemes, it is necessary to identify certain quantities within the cellular-automaton description with suitable physical notions such as momentum, energy, etc [73]. By use of a hexagonal lattice [74] many properties of real fluids were modelled, including the Navier–Stokes equations [75]. This is a stunning result and it is therefore not surprising that cellular-automaton techniques are widely applied. It is in particular clear that the Lorentz gas is a cellular automaton, so it has become the custom to refer to this class of systems as LGA (Lorentz gas automata). A complete issue of *Physica* D [76] was devoted to LGA, heralded as 'A new field in statistical mechanics'. Cohen [77] studied diffusion in a Lorentz gas while Ernst [78] investigated the hydrodynamics and correlation function for cellular automata, showing how intertwined these subjects have become.

At present, in the mid-1990s, cellular automata are certainly at the centre of interest in non-equilibrium thinking, but there are a large number of interesting and promising subjects waiting in the wings. The quantum stochastic processes have already been mentioned as likely viable future subjects. In 1989 there was a conference [79] on the unusual topic 'Non-equilibrium theory and quantum kinetic equations in curved spaces'. In 1987 a meeting was held in Santa Barbara discussing transient behaviour in neural nets [80]. Stochastic particle acceleration was the subject of a paper by Khoklov [81], while Khalatnikov [82] investigated 'Stochasticity in relativistic cosmology'.

Astrophysics has its share of statistical notions; the entropy of black holes, the loss of information in black-hole formation, the evaporation of black holes, the speed with which galaxies mix—all these processes involve some aspect of the temporal behaviour of the system and as such belong to non-equilibrium statistical mechanics. But even this medley in no way exhausts the areas where time-dependent statistical processes play a significant role. In 1983 Percus and Yevick [83] suggested a four-dimensional random walk model to study the effectiveness of clinical trials. Wiegel studied the statistical mechanics of red blood cells [84].

Fractals and fractal random walks appear to be relevant to the investigation of flows through porous media, coagulation problems and cloud formation: the subject of this chapter keeps on expanding.

In the background of this frantic activity there is the ubiquitous, but not completely understood role of chaotic phenomena. There is no conclusion, if for no other reason that this summary has shown how difficult it is to conclude exciting and long-time consequences from short-time behaviour. It has been an exciting century, but in the spirit of statistical mechanics it is overwhelmingly likely that the future will be very different from current expectations.

References

[1] Einstein A 1902 *Ann. Phys.* **9** 417–33; 1903 *Ann. Phys.* **11** 170–87; 1904 *Ann. Phys.* **14** 354–70
[2] Einstein A 1905 *Ann. Phys.* **17** 549–67
[3] Willard Gibbs J 1902 *Elementary Principles of Statistical Mechanics* (New Haven: Yale University Press)
[4] Ehrenfest P and Ehrenfest T 1911 Begriffliche Grundlagen der statistishen Anfassing in der Mechanik *Encycl. Math. Wiss.* **4** No. 32
[5] Chapman S and Cowling T G 1939 *The Mathematical Theory of Non-uniform Gases* (Cambridge: Cambridge University Press)
[6] Fokker A D 1914 *Ann. Phys.* **43** 810–23
[7] Planck M 1917 *Sitz. Berich. Preuss. Acad. Wiss.* (collected in Planck M *Physikalische Abhandlungen und Vorträge (PAV)* II p 435)
[8] Smoluchowski M 1906 *Ann. Phys.* **21** 756
[9] Langevin P 1908 *C. R. Acad. Sci., Paris* **146** 530
[10] Chapman S 1916 *Phil. Trans. R. Soc.* A **216** 279; 1917 *Phil. Trans. R. Soc.* A **217** 115
[11] Enskog D 1921 *Svenska Vetensch. Akad. Ark. Matem., Astron., Fys.* **16** 1
[12] Clusius K and Dickel G 1939 *Z. Phys. Chem.* B **44** 397
[13] Burnett D 1935 *Proc. London Math. Soc.* **39** 385; 1935 *Proc. London Math. Soc.* **40** 383
[14] Plancherel M 1913 *Ann. Phys.* **42** 1061
[15] Rosenthal A 1913 *Ann. Phys., Lpz* **42** 796
[16] Birkhoff G D 1932 *Proc. Natl Acad. Sci.* **10** 279
[17] von Neumann J 1932 *Proc. Natl Acad. Sci.* **10** 70, 263
[18] Wiener N 1930 *Acta Math.* **55** 117
[19] Khintchine A I 1934 *Math. Ann.* **109** 604
[20] Johnson J B 1928 *Phys. Rev.* **32** 97
[21] Nyquist H 1928 *Phys. Rev.* **32** 110
[22] Onsager L 1931 *Phys. Rev.* **37** 405; 1931 *Phys. Rev.* **38** 2265
[23] Dirac P A M 1930 *The Principles of Quantum Mechanics* (Oxford: Clarendon)
[24] von Neumann J 1932 *Mathematische Grundlagen der Quantum Mechanik* (Berlin: Springer)
[25] Uhlenbeck G E and Uchling E A 1933 *Phys. Rev.* **43** 552
[26] Uhlenbeck G E and Ornstein L S 1930 *Phys. Rev.* **36** 823
[27] Nordsieck A, Land W E and G E Uhlenbeck 1940 *Physica* **7** 344
[28] Siegert A J F 1940 *Phys. Rev.* **76** 1708
[29] Born M and Guen H S 1949 *A General Kinetic Theory of Liquids* (Cambridge: Cambridge University Press)
Bogoliubov N N 1946 *J. Phys. USSR* **10** 265
Kirkwood J 1940 *J. Chem. Phys.* **14** 180; 1947 *J. Chem. Phys.* **15** 73
Yvon J 1935 *La Theorie Statistiques des Fluides et l'Equation d'Etat* (Paris: Paris Actualités Scientifique et Industrielle)
[30] van Kampen N 1962 *Statistical Mechanics of Irreversible Processes: Fundamental Problems in Statistical Mechanics* (Amsterdam: North-Holland)
[31] Kramers H A 1940 *Physica* **7** 284
[32] Brinkman H C 1956 *Physica* **22** 29, 149

[33] Chandresekhar S 1942 *Rev. Mod. Phys.* **15** 1
[34] Wang Ming Chen and Uhlenbeck G E 1945 *Rev. Mod. Phys.* **17** 323
[35] Dorfman J R and Cohen E G D 1965 *Phys. Lett.* **16** 24
Weinstock J 1965 *Phys. Rev.* A **140** 460
Frieman E and Goldman R 1967 *J. Math. Phys.* **8** 1410
Dorfman J R and Cohen E G D 1967 *J. Math. Phys.* **8** 282
[36] Cohen E G D 1967 *Lectures in Theoretical Physics IX, C* (Boulder, CO) p 279
[37] Green M S 1958 *Physica* **24** 393
[38] Kubo R 1957 *J. Phys. Soc. Japan* **12** 570
[39] Kirkwood J 1946 *J. Chem. Phys.* **14** 180; 1947 *J. Chem. Phys.* **15** 72
[40] Resibois P 1964 *J. Chem. Phys.* **41** 2919
[41] Khintchine A I 1943 *Mathematical Formulations of Statistical Mechanics* (Engl. Transl. Gamov 1949 (New York: Dover))
[42] Sinai Ja 1964 *Am. Math. Soc. Transl.* **39** 83; 1968 *Am. Math. Soc. Transl.* **60** 34; 1963 *Sov. Mat. Dokl.* **4** 1818
Kolmogoroff A N 1959 *Dokl. Akad. Nauk SSSR* **124** 754
[43] Balescu R 1963 *Statistical Mechanics of Charged Particles* (New York: Interscience) p 5
[44] Wigner E 1954 *J. Chem. Phys.* **22** 1912
[45] Onsager L and Mechlup S 1953 *Phys. Rev.* **91** 1505
[46] Casimir H B G 1945 *Rev. Mod. Phys.* **17** 343
[47] Martin P and Schwinger J 1959 *Phys. Rev.* **115** 1342
[48] Wigner E 1932 *Phys. Rev.* **40** 749
[49] Kac M 1947 *Am. Math. Mon.* **54** 369
[50] Alder B J and Wainwright T 1967 *Phys. Rev. Lett.* **10** 988; 1968 *Phys. Soc. Japan* (Supplement 26) 267; 1970 *Phys. Rev.* A **1** 18
[51] Dorfman J R and Cohen E D G 1970 *Phys. Rev. Lett.* **25** 1257
Ernst M H, Hange E H and van Leeuwen J M 1971 *Phys. Rev.* A **2** 2055
[52] Prigogine I 1962 *Non-equilibrium Statistical Mechanics* (New York: Wiley–Interscience)
[53] Balescu R 1975 *Equilibrium and Non-equilibrium Statistical Mechanics* (New York: Wiley–Interscience)
[54] Skarka V 1989 *Physica* A **162** 210
[55] Prigogine I 1980 *From Being to Becoming* (San Francisco: Freeman)
Prigogine I 1981 *Order and Fluctuations in Equilibrium and Non-equilibrium Statistical Mechanics* (New York: Wiley) pp 35–77
[56] Horwitz L 1989 *Physica* **161** 300
[57] Larralde H, Trunfio P, Havlin S, Stanley E and Weiss G H 1992 *Phys. Rev.* A **45** 7128
[58] van Kampen N 1986 *J. Stat. Phys.* **44** 1
[59] Bhatnagar P L, Gross E P and Krook M 1954 *Phys. Rev.* **94** 511
[60] Grad H 1949 *Commun. Pure Appl. Math.* **2** 331; 1958 *Handbuch Phys.* **12** 205
[61] Ernst M 1979 *Phys. Rep.* **78**
[62] Cercignani C 1988 *The Boltzmann Equation and its Applications* (New York: Springer)
[63] Haag R and Kastler D 1964 *J. Math. Phys.* **5** 848
[64] Lanford O 1975 *III ix Proc. 1974 Battelle Rencontu as Dynamical Systems (Lecture Notes in Physics 35)* ed J Misor (Berlin: Springer) p 1
[65] Lindhard J 1993 *J. Stat. Phys.* **72** 539
[66] van Kampen N 1981 *Stochastic Processes in Physics and Chemistry* (Amsterdam: North-Holland) pp 237, 241–4, 250
[67] Ford G W, Kac M and Mazur P 1965 *J. Math. Phys.* **6** 504
[68] Ernst M H and van Velzen G A 1989 *J. Stat. Phys.* **57** 255
[69] Schulman L S 1978 *J. Stat. Phys.* **19** 273

[70] Dresden M 1975 *Proc. Natl Acad. Sci.* **72** 956
[71] Lebowitz 1988 *J. Stat. Phys.* **50** 841
[72] Ernst M 1987 *J. Stat. Phys.* **48** 645; 1988 *J. Stat. Phys.* **51** 312
van Velzen B 1990 *PhD Thesis* University of Utrecht
[73] Hardy J, de Pazzis O and Pomeau Y 1976 *Phys. Rev.* A **13** 1949
[74] Pomeau Y, Frisch U, de Humières D, Hasslacher B, Lallemand P and Rivet J P 1987 *Complex Systems* **1** 648
[75] Hasslacher B 1987 *Los Alamos Science*
[76] 1990 Lattice gas automata *Physica* D **45** November issue
[77] Cohen 1991 *J. Stat. Phys.* **62** 73
[78] Ernst R R 1990 *J. Stat. Phys.* **50** 57
[79] 1989 *Physica* **158** May issue
[80] 1982 Conference on neural nets, Santa Barbara *J. Stat. Phys.* **51** 741
[81] Khoklov 1982 *J. Stat. Phys.* **28** 793
[82] Khalatnikov 1985 *J. Stat. Phys.* **38** 97
[83] Percus and Yevick 1983 *J. Stat. Phys.* **30** 755
[84] Wiegel 1982 *J. Stat. Phys.* **29** 813

ILLUSTRATION ACKNOWLEDGMENTS

Chapter 1

H A Lorentz—AIP Emilio Segrè Visual Archives, Lande Collection
J W Strutt—AIP Emilio Segrè Visual Archives, Physics Today Collection
L E Boltzmann—University of Vienna. Courtesy AIP Emilio Segrè Visual Archives
A A Michelson—AIP Emilio Segrè Visual Archives, W F Meggers Collection
1.2 Courtesy Deutsches Museum, Munich

Chapter 2

E Rutherford—Gawthron Institute, Nelson, New Zealand. Courtesy American Institute of Physics, Emilio Segrè Visual Archives
N Bohr—AIP Niels Bohr Library
2.1 Science Library, London
2.2 Science Library, London
2.3 AIP Niels Bohr Library, E Scott Barr Collection
2.5 Reprinted from *Physics*, Second Edition by Hans C Ohanian, with the permission of W W Norton & Company, Inc. Copyright © 1989, 1985 by W W Norton & Company Inc.
2.7 AIP Niels Bohr Library, Goudsmit Collection
2.8 *A Scientific Autobiography* by Otto Hahn, New York: Ch. Scribners & Sons, 1966
2.9 Nobel Foundation. Courtesy AIP Niels Bohr Library
2.10 French Embassy Information Division. Courtesy AIP Emilio Segrè Visual Archives

Chapter 3

M Planck—AIP Niels Bohr Library, W F Meggers Collection
E Schrödinger—AIP Niels Bohr Library, photograph by Francis Simon
3.1 Instituts Internationaux de Physique et de Chimie, Bruxelles
3.2 Editions de Physique, Les Ulis, France
3.4 From *Verhandlungen der Deutschen Physikalischen Gesellschaft* (2) **16**, 467 (1914), J Frank and G Hertz
3.5 Niels Bohr Archives, Copenhagen
3.6 AIP Niels Bohr Library
3.7 AIP Meggers Gallery of Nobel Laureates
3.8 Top: Niels Bohr Archives, Copenhagen. Bottom: The MIT Museum
3.9 Heisenberg Archive, WHI, Munich
3.10 Heisenberg Archive, WHI, Munich

A1

3.11 Heisenberg Archive, WHI, Munich
3.12 Instituts Internationaux de Physique et de Chimie, Bruxelles
3.13 From *David Hilbert* by C Reid, 1970. Copyright 1970 Springer-Verlag GmbH & Co KG
3.14 Nishina Memorial Foundation, Tokyo
3.15 *Werner Heisenberg in Leipzig* by E Lea and G Wiemers, Sächs. Akad. Wiss., ed Akademie Verlag, Berlin 1993
3.16 (*a*) Heisenberg Archive, Munich. (*b*) Niels Bohr Archives, Copenhagen
3.17 (*a*) Niels Bohr Archives, Copenhagen. (*b*) From *Albert Einstein, Philosopher-Scientist* by P Schlipp, Open Court Publishing, 1970

Chapter 4

A Einstein—Permission granted by the Albert Einstein Archives, The Hebrew University of Jerusalem, Israel
H Minkowski—H A Lorentz, H Minkowski, Das Relatitatsprinzig 1915. Courtesy AIP Emilio Segrè Visual Archives
4.3 IOP Publishing Ltd, Bristol
4.4 From W G V Rosser, *Introductory Relativity*. Reprinted by permission of Plenum Press, New York
4.6 From Eugene Hecht, *Optics*, Second Edition. © 1987 by Addison-Wesley Publishing Company, Inc. Reprinted by permission of the publisher
4.10 *Was Einstein Right? Putting Relativity to the Test*, Second edition by Clifford M Will. Copyright 1986, 1993 by Clifford M Will. Reprinted by permission of BasicBooks, a division of HarperCollins Publishers, Inc.
4.11 From W G V Rosser, *Introductory Relativity*. Reprinted by permission of Plenum Press, New York
4.12 From *Gravitation* by C W Misner, K S Thorne and J A Wheeler. Copyright © 1988 W H Freeman and Company. Used with permission
4.13 From *Gravitation* by C W Misner, K S Thorne and J A Wheeler. Copyright © 1988 W H Freeman and Company. Used with permission
4.14 From *The New Physics*, edited by P C W Davies. Reprinted with the permission of Cambridge University Press
4.15 From *Einstein: A Centenary Volume*, edited by A P French, The International Commission on Physics Education
4.16 *Was Einstein Right? Putting Relativity to the Test*, Second edition by Clifford M Will. Copyright 1986, 1993 by Clifford M Will. Reprinted by permission of BasicBooks, a division of HarperCollins Publishers, Inc.
4.17 From *The New Physics*, edited by P C W Davies. Reprinted with the permission of Cambridge University Press
4.18 From *Einstein: A Centenary Volume*, edited by A P French, The International Commission on Physics Education
4.19 From *Black Holes and Time Warps* by K S Thorne, W W Norton and Co, Inc., New York
4.20 From *Black Holes and Time Warps* by K S Thorne, W W Norton and Co, Inc., New York
4.21 From *Gravitation* by C W Misner, K S Thorne and J A Wheeler. Copyright © 1988 W H Freeman and Company. Used with permission
4.22 From *The New Physics*, edited by P C W Davies. Reprinted with the permission of Cambridge University Press
4.23 From *The New Physics*, edited

Illustration Acknowledgments

	by P C W Davies. Reprinted with the permission of Cambridge University Press
4.25	From *Relativitätstheorie in elementarer Darstellung mit Aufgaben und Lösungen* by Prof Dr H Melcher, VEB Deutscher Verlag der Wissenschaften
4.26	From *Relativitätstheorie in elementarer Darstellung mit Aufgaben und Lösungen* by Prof Dr H Melcher, VEB Deutscher, Verlag der Wissenschaften
4.27	Photograph courtesy of Lincoln Laboratories, MIT

Chapter 5

W K Heisenberg—AIP Emilio Segrè Visual Archives
C D Anderson—AIP Emilio Segrè Visual Archives, E Scott Barr Collection
H Yukawa—AIP Emilio Segrè Visual Archives, W F Meggers Collection
P M S Blackett—AIP Emilio Segrè Visual Archives, W F Meggers Collection

5.1	From *Nobel Lectures in Physics, 1942–62* by permission of the Nobel Foundation, Stockholm
5.2	© Nobel Foundation 1965
5.3	© Nobel Foundation 1964
5.4	Cal-Tech Public Affairs Office
5.5	From *Yukawa's Prediction of the Meson* by L M Brown, Centaurus 1981, Munksgaard International Publishers, Copenhagen, Denmark
5.6	Reprinted with permission from I S Brown, R A Milikan and H V Neher, 1938, *Phys. Rev.* **53**, 856. Copyright 1938. The American Physical Society
5.7	Reprinted with permission from *Nature* **159**, 695, figure 1. Copyright 1947 Macmillan Magazines Ltd
5.8	California Institute of Technology Public Affairs Office

Chapter 6

M von Laue—AIP Emilio Segrè Visual Archives
W L Bragg—AIP Emilio Segrè Visual Archives, W F Meggers Collection

6.2	From *Introduction to Solid State Physics*; 6th Edition by C Kittel, John Wiley, 1986. Reprinted by permission of John Wiley & Sons Inc.
6.3	From *50 Years of X-Ray Diffraction* by P Ewald, International Union of Crystallography 1962
6.4	From *50 Years of X-Ray Diffraction* by P Ewald, International Union of Crystallography 1962
6.5	From *The Development of X-Ray Analysis* by W L Bragg, HarperCollins
6.6	From *The Development of X-Ray Analysis* by W L Bragg, HarperCollins
6.7	From *Interpretation of X-Ray Diffraction Photographs* by N Henry *et al*, Macmillan Publishers Ltd, London
6.8	From *Crystalline State Volume 1* by W L Bragg, HarperCollins
6.9	L Spain *Contemporary Physics* **28** 523, figure 14, Taylor and Francis
6.10	V W Arndt *Journal of Applied Crystallography* **6** 457, figure 2, International Union of Crystallography
6.11	From *The Development of X-Ray Analysis* by W L Bragg, HarperCollins
6.12	From *The Development of X-Ray Analysis* by W L Bragg, HarperCollins
6.13	From *Crystalline State Volume 3: The Determination of Crystal Structures* by Lipson and Cochran, HarperCollins
6.14	From *Crystalline State Volume 3: The Determination of Crystal Structures* by Lipson and Cochran, HarperCollins

A3

6.15 From *Crystalline State Volume 3: The Determination of Crystal Structures* by Lipson and Cochran, HarperCollins
6.16 From *Crystalline State Volume 3: The Determination of Crystal Structures* by Lipson and Cochran, HarperCollins
6.17 International Union of Crystallography
6.18 From *Neutron Diffraction* by G Bacon, Clarendon Press, Oxford
6.19 From *Neutron Diffraction* by G Bacon, Clarendon Press, Oxford
6.20 From *Neutron Diffraction* by G Bacon, Clarendon Press, Oxford
6.21 From *Neutron Diffraction* by G Bacon, Clarendon Press, Oxford
6.22 From G Kostorz, *Neutron Scattering*. Reprinted with permission from Academic Press, Orlando
6.23 From *Neutron Diffraction* by G Bacon, Clarendon Press, Oxford
6.24 From G B Thomson, *Contemporary Physics* **9** 1, figure 1, p11, Taylor and Francis
6.25 From *Crystalline State Volume 3: The Determination of Crystal Structures* by Lipson and Cochran, HarperCollins
6.26 From *Surface Physics* by M Prutton, Oxford University Press, Oxford
6.27 From *Surface Physics* by M Prutton, Oxford University Press, Oxford
6.28 From *Modern Techniques in Surface Science* by D P Woodruff and T A Delchar. Reprinted with the permission of Cambridge University Press, Cambridge, UK
6.29 From *Surface Physics* by M Prutton, Oxford University Press, Oxford
6.30 From *Surface Physics* by M Prutton, Oxford University Press, Oxford
6.31 IOP Publishing Ltd, Bristol
6.32 From *Crystalline State Volume 1* by W L Bragg, HarperCollins
6.33 O S Edwards and H S Lipson, *Proc. Roy. Soc.* **A180**, (1940), 268, figure 4, The Royal Society, London
6.35 From *Introduction to Solid State Physics*; 6th Edition by C Kittel, John Wiley, 1986. Reprinted by permission of John Wiley & Sons Inc.
6.36 From *Introduction to Solid State Physics*; 6th Edition by C Kittel, John Wiley, 1986. Reprinted by permission of John Wiley & Sons Inc
6.37 J M Menter, *Proc. Roy. Soc*, **A236**, (1956), 128, figure 7, The Royal Society, London
6.38 D S Weaire *Contemporary Physics* **17** 173, figure 8, p185, Taylor and Francis
6.39 D S Weaire *Contemporary Physics* **17** 173, figure 2, p177, Taylor and Francis
6.40 F Yonezawa, *Solid State Physics* **45** 179, figure 6b, p191. Reprinted with permission from Academic Press, Orlando
6.41 D Gratias, *Contemporary Physics* **28**, figure 1, p219, Taylor and Francis
6.42 D Gratias, *Contemporary Physics* **28**, figure 6, p219, Taylor and Francis
6.43 D Gratias, *Contemporary Physics* **28**, figure 7, p219, Taylor and Francis
6.44 R A Cowley, *Advances in Physics* **29**, p32, figure 1, Taylor and Francis
6.45 From *Ferroelectric Crystals* by Jona and Shirane, Pergamon Press, Oxford
6.46 From *Ferroelectric Crystals* by Jona and Shirane, Pergamon Press, Oxford
6.47 R Beyers and G Shaw, *Solid State Physics* **42** 135, figure 24, p195, reproduced by permission of Drs Beyers and Shaw
6.48 From G Burns, *High Temperature Superconductors; An Introduction*, figures 1.1a, b, c & d, p4. Reprinted with permission from Academic Press, Orlando
6.49 From *The Development of*

Illustration Acknowledgments

X-Ray Analysis by W L Bragg, HarperCollins

6.50 From *Diffraction of X-Rays by Proteins, Nucleic Acids and Viruses* by H R Wilson. Reproduced by permission of Edward Arnold (Publishers) Ltd, London

6.51 From *Diffraction of X-Rays by Proteins, Nucleic Acids and Viruses* by H R Wilson. Reproduced by permission of Edward Arnold (Publishers) Ltd, London

6.52 From *Diffraction of X-Rays by Proteins, Nucleic Acids and Viruses* by H R Wilson. Reproduced by permission of Edward Arnold (Publishers) Ltd, London

6.53 From *Diffraction of X-Rays by Proteins, Nucleic Acids and Viruses* by H R Wilson. Reproduced by permission of Edward Arnold (Publishers) Ltd, London

6.54 From *Diffraction of X-Rays by Proteins, Nucleic Acids and Viruses* by H R Wilson. Reproduced by permission of Edward Arnold (Publishers) Ltd, London

6.55 From *Diffraction of X-Rays by Proteins, Nucleic Acids and Viruses* by H R Wilson. Reproduced by permission of Edward Arnold (Publishers) Ltd, London

6.56 From *Diffraction of X-Rays by Proteins, Nucleic Acids and Viruses* by H R Wilson. Reproduced by permission of Edward Arnold (Publishers) Ltd, London

6.57 From *Diffraction of X-Rays by Proteins, Nucleic Acids and Viruses* by H R Wilson. Reproduced by permission of Edward Arnold (Publishers) Ltd, London

6.58 From *Diffraction of X-Rays by Proteins, Nucleic Acids and Viruses* by H R Wilson. Reproduced by permission of Edward Arnold (Publishers) Ltd, London

6.59 From *The Double Helix* by J D Watson, Wiedenfeld & Nicholson, London

6.60 Kennard and Hunter *Agn. Chem. Ed. Engl.* **30** 1254 (1991), figure 3, p1256, V C H Verlagsgesellschaft mbH, Weinheim

6.61 From *Diffraction of X-Rays by Proteins, Nucleic Acids and Viruses* by H R Wilson. Reproduced by permission of Edward Arnold (Publishers) Ltd, London

Chapter 7

J W Gibbs—Burndy Library. Courtesy AIP Emilio Segrè Visual Archives

W H Nernst—Rijksmuseum voor de Geschledehis de Natuurwetenschappen Leiden. Courtesy AIP Emilio Segrè Visual Archives

L Onsager—AIP Emilio Segrè Visual Archives

K G Wilson—AIP Meggers Library of Nobel Laureates

7.1 From *Statistical Mechanics* by G S Rushbrooke, Oxford University Press, 1949

7.2 From *Yearbook of the Physical Society* by F E Simon, Physical Society, 1956

7.3 From *Yearbook of the Physical Society* by F E Simon, Physical Society, 1956

7.4 From *Yearbook of the Physical Society* by F E Simon, Physical Society, 1956

7.5 From *Yearbook of the Physical Society* by F E Simon, Physical Society, 1956

7.6 From *Yearbook of the Physical Society* by F E Simon, Physical Society, 1956

7.7 C Domb, *Contemporary Physics* **26** 49, figure 4 (1985), Taylor and Francis

7.8 C Domb, *Contemporary Physics* **26** 49, figure 5 (1985), Taylor and Francis

7.9 C Guggenheim *J. Chem. Phys.* **13** 253, 1945, American Institute of Physics

7.10 C Domb, *Contemporary Physics* **26** 49, figure 7 (1985), Taylor and Francis

7.11 C Domb, *Contemporary Physics* **26** 49, figure 8 (1985), Taylor and Francis

7.12 C Domb, *Contemporary Physics* **26** 49, figure 9 (1985), Taylor and Francis

7.13 *Annals Is. Phys. Soc.* **5** 21, figure 1 (1983)

7.14 *Annals Is. Phys. Soc.* **5** 22, figure 2 (1983)

7.15 From *Fractals Form Chances & Dimension* by B B Mandelbrot, 1977, and *The Fractal Geometry of Nature* by B B Mandelbrot, 1982, W H Freeman, San Francisco

Chapter 8

G E Uhlenbeck—AIP Niels Bohr Library
N N Bogoliubov—AIP Niels Bohr Library
R Kubo—AIP Niels Bohr Library
I Prigogine—AIP Niels Bohr Library

SUBJECT INDEX

Note—volume and page numbers in **bold type** refer to items in boxes

A0620-00 x-ray binary source, III.1757
Abbe optical theory, III.1578–84
Abbe primary image, III.1583
Abbe secondary image, III.1583
Abelian gauge theory, II.693, II.714
Aberdeen MK1 NMR Imager, III.1926
Aberdeen Section Scanner, III.1918
Aberration, I.258, I.266, I.269
 see also Lens
Ability of the 200" telescope to discriminate between selected world models, The, III.1793
Abrikosov vortex lattice, II.935
Absorption, III.1404
Absorption coefficient, I.364, I.368, II.1052
Absorption spectra, I.24
Academia Europaea, III.2055
Académie des Sciences, I.6, II.1233
Accelerators, II.657, II.687–8, II.725–44, II.1201–6, II.1225, III.1883, III.1997, III.2036
 colliding beams, II.738–42
 fixed-field alternating-gradient, II.739
 strong focusing, II.733–4
Accretion disc, III.1756, III.1779
Accretion radius, III.1756
Acoustic energy, II.873

Acoustic modes, II.1004
Acoustic paramagnetic resonance (APR), II.1146–7
Acoustic resonance, III.1357–8
Acta Crystallographica, I.511
Acta Metallurgica, III.1556
Actinides, I.97, II.955
 magnetism, II.1166
Actinium, II.1227
Activation energy, III.1531
Active galactic nuclei, III.1777
 and general relativity, III.1777–81
 non-thermal phenomena in, III.1781–4
Adenine, I.504
Adiabatic demagnetization, II.1125–6
Adiabatic model, III.1802
Adiabatic principle, I.92–3
Adler–Weisberger relation, II.679
Administration of Radioactive Substances Advisory Committee, III.1910
AEG (Allgemeine Elektrizitäts Gesellschaft) Research Institute, III.1577–8
Aerobee (rocket), III.1682, III.2005
Aerodynamics, II.821
 heating phenomenon, II.874
 lift theory, II.817

Aeronautics, II.869–87
Aeronomy, III.1982
Aether-drift, I.26
Age-hardening, III.1546, III.1547
AIDS, III.2029
Aircraft, fixed-wing, II.817
Aircraft wings, II.821, II.824
Airy diffraction limit, III.1467
Alcator, III.1649
ALICE neutral-beam injection experiments, III.1662
Alkali metals, III.1345
Allende meteorite, III.1963
Allgemeine Elektrizitäts Gesellschaft Research Institute, see AEG
Allis-Chalmers Company, III.1871
Along-crest energy propagation, II.863–9
α-decay, I.115–16, I.366–7, II.1189
α-particle energies, II.1201
α-particle model, II.1190
α-particles, I.58, I.59, I.104, I.110, I.112, I.113, I.116, I.119, I.122, I.123, I.367, II.638, II.1189, II.1226, III.1706
α-radioactivity, I.52
α-rays, I.37, I.54, I.58, I.104
β-alumina, III.1539
Aluminium alloys, III.1533
Aluminium–copper alloys, III.1533, III.1546
Aluminium–manganese alloy, I.481–3
Alzheimer's disease, III.1922
American Association of Physicists in Medicine, III.1864
American Crystallographic Association, I.512
American Philosophical Society, I.3
American Physical Society, III.1558, III.2026
American Science and Engineering Group (AS&E), III.1739
American Society for X-ray and Electron Diffraction, I.511
Americium, II.1225
Amino acids, I.496, I.497
Amorphous cement, III.1518
Amorphous magnetism, II.1170–1
Amorphous materials, III.1541
Amorphous metal hypothesis, III.1518
Amorphous solids, x-ray diffraction, **I.478**
Amorphous structures, I.477–9
Ampere (unit), II.1235, II.1254, II.1257
  absolute determination, II.1255
Amplification without inversion (AWI), III.1473
Anaesthetic equipment, III.1897
Anderson–Brinkman–Morel (ABM) state, II.953
Andromeda Nebula, III.1717, III.1718, III.1788
Angiography, III.1930
Anglo-Australian Telescope, III.1744
Angstrom (unit), II.1242
Angular momentum, I.81, I.162, I.359, II.818, II.1124, III.1410, III.1756, III.1779, III.1962, III.1964
Angular momentum algebra, **II.1038**
Anharmonic effects, II.1002–4
Anharmonic oscillators, III.1434
Anisotropic superfluid, II.950–1
Annalen der Physik und Chemie, I.7
Annales de chimie et de physique, I.7
Anomalous skin effect, III.1346, III.1347, III.1351, III.1357
Anthracene, crystal structure, I.494
Anthropic Principle, III.2039–40
Antiferroelectric structure, I.487
Antiferromagnetism, I.559, II.1016, II.1113, II.1129–32, II.1147–8
  exchange interactions, II.1143

resonance, II.1151
structures, II.1011
Antimony, III.1332, III.1335, III.1544
Antiparticles, II.638, II.655
Antiprotons, II.655
Antiscreening. *See* Screening and antiscreening
Anyon mechanism, II.962
Applications of Phase Transitions in Material Science, III.1557
Applied optics, III.1474–81
Applied physics, III.2050
Arginine, structure of, I.449–50
Argon, I.75
Argus experiment, III.1683
Army Operational Research Group, III.1736
Arnowitt–Deser–Misner (ADM) mass, I.301
Arsenic-72 and -74, III.1904
Arteries, artificial, III.1898
Arteriosclerosis, III.1898
Artificial limbs, III.1899
ASDEX tokamak, *see* Tokamak
Associated production, II.657
Association of Universities for Research in Astronomy (AURA), III.1744
Asthenosphere, III.1978
Astigmatism, III.1580
ASTRON, III.1645
Astronomical photography, III.1692
Astronomical spectroscopy, III.1691
Astronomical timescales, II.1227
Astronomical Unit (AU), II.1242
Astronomy, I.24, II.757–62, III.1453–4, III.2046
changing perspective, III.1734–45
general theory of relativity, I.315–17
optical, III.1744–5

Astrophysical Journal, III.1762
Astrophysics, I.630, II.757–62, III.1618, III.1691–821, III.2031
birth of, III.1691
general theory of relativity, I.315–17, III.1776
high-energy, III.1774–84
impact of quantum mechanics, III.1706–8
since 1945, III.1746–807
statistical mechanics in, III.1711
Asymmetric unit, I.423
Asymptotic freedom, II.714
Atmosphere, II.899
Atmospheric physics, III.1983
Atmospheric science, III.1983
Atomic bomb, II.1197, III.1630, III.1891, III.1980
Atomic clock, II.1244, II.1247
Atomic dynamics, discretization in, I.184–5
Atomic energy, I.63, I.113, I.132–3, I.285, III.1628
Atomic Energy Research Establishment, III.1901
Atomic model, I.157–8
Atomic number, I.65, I.158–60, I.407, I.429, II.1057, II.1222, II.1223
Atomic physics, I.31, I.358–62, II.755–6, II.1025–49
Atomic pile, *see* Nuclear reactors
Atomic processes, current overviews, II.1101–6
Atomic scale, II.1026–7
Atomic scattering factor, I.431, I.432, I.439, I.453
Atomic spectra, II.1032–9
Atomic standards of measurements, II.1095
Atomic structure, I.63–5, I.156–62, I.211–15, II.1026–7
Atomic systems, optical actions, II.1097–101
Atomic theory, I.43, I.46, I.227

S3

causality in, I.220–3
   failure of, I.181–3
   principles of, I.176–8
   probability in, I.220–3
   schools of, I.178–81
Atomic weight, I.65
Atoms, I.43–103
   role in metrology and instrumentation, II.1094–7
Atoms for Peace Conference (1955), III.1642
Atoms for Peace Conference (1958), III.1655
AT&T, III.2022
Au$_2$Mn, magnetic moments in, I.462
Aufbauprinzip (building-up principle), I.96–7, I.366
Auger emissions, inner-shell phenomena, II.1090–1
Augmented-plane-wave (APW) method, III.1317
Aurorae, III.2006–7
Auroral phenomena, III.2003
Auswahlprinzip, selection rule, I.178–81
Autocorrelation functions, III.1418
Auto-ionization, II.1061, II.1090, II.1102
Automatic Plate Measuring Machine (APM), III.1745
Avalanche photodiode (APD), III.1419, III.1476
Avogadro's constant, I.150
Avogadro's law, I.46
Avogadro's number, I.43, I.49, I.51, II.1237, II.1269
Azbel'–Kaner cyclotron resonance, III.1357

Balian–Werthamer (BW) state, II.952
Balloon flight, I.368
Balmer formula, I.24, I.78, I.80–5, I.88

Balmer series, I.160
Band gaps, III.1367
Band structure, II.1047, II.1166, III.1314–17, III.1344, III.1530
Band-structure diagram, III.1417
Band theory, II.1165–7, III.1367
Bar-code scanners, III.1480–1
Bardeen–Pines interaction, II.938
Barium, contrast media, III.1863
Barium titanate, III.1538
Baroclinic instability, II.902, II.907
Baroclinic mode, II.896
Baryogenesis, II.761–2
Baryon, II.660, II.661, II.675, II.685, III.1800
Baryon–antibaryon asymmetry, II.760, III.1806
Baryon number, II.658
Basic Orthopaedic BioMechanics, III.1896
BaTiO$_3$, ferroelectric phase transition, I.486
Batteries, I.27
   lithium–iodine, III.1895
   zinc–mercury, III.1895
BBKGY hierarchy, I.605, I.623
BCS theory, I.489, II.935, II.937, II.937–41, II.961–5
Beevers–Lipson strips, I.446
Beiblätter zu den Annalen der Physik, I.7
Bell inequalities, III.1389
Bell Telephone Laboratories, III.1790, III.1944, III.2022, III.2026
Berkelium, II.1225
Bernoulli's equation, II.822, II.885
Beryl, structure of, I.440
Beryllium, I.119, I.120, III.1360
β-decay, I.106, I.366–7, I.370–2, I.375, I.398–402, II.639, II.664, II.665, II.667, III.1706, III.1707
   Fermi's theory, I.374–5
   limits on neutrino masses, II.755

β-effect, II.902
β-electrons, I.104
β-particles, I.107, I.109
β-radioactivity, I.52
β-ray spectroscopy, I.103–9
β-rays, I.37, I.54, I.58, I.59, I.104–9
β-spectra, I.106, I.109, I.118–19, I.126–9
β-structures, I.498
Betatron, II.732, II.1204–5, III.1871
B-factories, II.767–8
Big Bang, III.1784–6, III.1790–3, III.2040
'Big Physics', III.2054–5
'Big Science', III.2023–7, III.2054
Billions of electronvolts (GeV), I.358
Binary alloys, critical behaviour of, I.548–9
Binary collisions, I.611, I.612
Binding energy, I.110–12
Biochemical assay, III.1909
Biomedical technology, III.2031
Biomolecular crystals, I.450–3
Biomolecular structures, I.445, I.496–512
Biomolecules, II.1084
Biophysics, III.2030, III.2046
Biot–Savart law, II.818
Bipolar outflow, III.1769
Birge–Bond diagram, II.1268
Bismuth, III.1282–3, III.1339
Bismuth germanate, III.1921
Bistable optical switching devices (SEEDs), III.1469
Bjorken–Glashow hypothesis, II.700
Black body, I.66
Black-body law, I.173, I.177
Black-body radiation, I.22–4, I.30, I.69, I.144, I.148–50, I.174, I.524- 5, II.758, II.972, II.973, II.982, III.1387, III.1393, III.1402

Black holes, I.306–12, II.760–1, III.1754–7, III.1778–80
  no-hair theorem for, III.1778
BL-Lacertae (or BL-Lac) objects, III.1777, III.1784
Bloch equation, II.1174
Bloch functions, II.1166
Bloch theorem, III.1291–301
Bloch theory, III.1342
Bloch wave, **III.1294–5**, III.1320
Bloch wavelength, III.1301
Bloch wavenumber, III.1294, III.1299
Bloch wave-vector, III.1294–6
Block copolymers, III.1611
Blood flow, III.1897–8, III.1914, III.1920
Blood pressure, monitoring, III.1893–4
B-meson, II.721, II.724–5, II.767–8
Bohm diffusion, III.1630, III.1631
Bohm diffusion formula, III.1632
Bohm diffusion law, III.1655
Bohm–Pines theory, III.1362, III.1363
Bohr–Kramers–Slater radiation theory, I.185, I.203
Bohr magneton, I.116
Bohr orbit, III.1319
Bohr radius, I.85, II.1026
Bohr relation, I.90
Bohr–Sommerfeld atomic model, I.175–6
Bohr–Sommerfeld rules, I.92
Bohr–Sommerfeld theory, I.161
Bohr theory, II.1124
Bolometer, I.23
Boltzmann collision number, I.603
Boltzmann combinatorial factor, I.527
Boltzmann constant, I.67, I.150, I.523, I.589, II.1270, III.1285, III.1387, III.1397
Boltzmann distribution, I.542
Boltzmann equation, I.523, I.533,

I.588, I.591, I.593–5, I.597, I.602, I.605, I.608–11, I.617, I.624–6, III.1303, III.1344, III.1701, III.1992
Boltzmann–Grad limit, I.626
Boltzmann law, III.1287
Bond percolation, I.572
Bone fractures, III.1897
Bootstrap programme, II.681
Born–Oppenheimer approximation, II.1043–6
Bose–Einstein condensation, I.530–1, I.537, II.927, II.930
Bose–Einstein distribution, III.1402
Bose–Einstein statistics, I.155, I.533, II.687, III.1387
Bose statistics, I.117, III.1394
Boson, I.382, II.693, II.694, II.699, II.1218–19
Bound scattering length, I.453
Boundary conditions, II.800–1, II.804
Boundary layer, II.804, II.807, II.812–46
Boundary-layer equations, II.804, II.806, II.843
Boundary-layer solution, II.802
Boundary-layer theory, II.797
Boundary-value problem, I.196
Boys' radiometer, I.28
Brachytherapy, III.1877, III.1890, III.1891
Bragg–Gray theory, III.1882
Bragg Law, I.428, I.431, I.469
Bragg peak, III.1887
Bragg reflection, II.990, III.1367
Brain imaging, III.1906, III.1922, III.1929, III.1932
Brain tumour, III.1871, III.1904, III.1905, III.1919
Bravais lattices, I.462, I.466, III.1513
Bravais model, II.984
Breit–Wigner formula, II.1194
Bremsstrahlung, III.1766

Brillouin frequency shift, III.1398
Brillouin scattering, III.1405
Brillouin zone, II.987, II.1011, III.1293, III.1295, III.1296, III.1300, III.1305, III.1311–13, III.1315, III.1328, III.1331, III.1348, III.1359
British Association for the Advancement of Science, I.5, I.52, II.1252, III.1705
British Broadcasting Company (BBC), III.1988
British Institute of Radiology, III.1861
British Journal of Radiology, III.1867
British Medical Bulletin, III.1901
British Radio Research Board, III.1987–8
Brittle fracture, III.1520
Broadcasting, III.1987, III.1994
Brownian motion, I.30, I.43, I.147, I.533, I.592, I.595–7, I.607–8, I.616, I.623, I.628, III.1441–2
Brussels school, I.619–22
BTH Company, III.1583
Bubble chamber, II.659, II.753
Bubble domains, II.1159–60
Building-up principle (*Aufbauprinzip*), I.96–7, I.366
Bulbous bow, II.885
Bulk-carriers, II.885
Bunsen burner, I.19, I.75
Buoyancy force, II.844
Bureau International de l'Heure (BIH), II.1247, II.1248
Bureau International des Poids et Mesures, I.20
Burgers vector, I.475–6
Busch's lens theory, III.1570
verification of, III.1569–70

Cabibbo–Kobayashi–Maskawa (CKM) matrix, II.674, II.723–5
Caesium, I.75

Subject Index

Caesium-137, III.1892
Caesium chloride, I.429
Caesium clock, II.1245–8
Calcite, I.430, I.485
Calcium tungstate, III.1537, III.1901
Californium, II.1225, II.1227
Calorimetry, II.968, III.1883–4
Calutron isotope separation, III.1630
Cambridge Database System, I.512
Cambridge Electron Accelerator (CEA), II.741
Cambridge Structural Database, I.512
Cambridge University, I.9–15, I.17, I.20, I.21, I.54
Cancer, III.1857, III.1886, III.1917
Candela, II.1251
Canonical ensemble, I.526
Cantor set, III.1846, III.1847, III.1852
Capacitance, II.1255
Capacitors, II.1256
Carbon, formation of, III.1746
Carbon bond, II.1043
Carbon dioxide, atmospheric, II.908
Carbon monoxide in interstellar medium, III.1763–4
Carbon–nitrogen–oxygen cycle (CNO cycle), III.1707–8
Cardiac function, III.1898
Cardiac pacemaker, III.1895
Cardiovascular devices, III.1897–8
Carl Zeiss, III.1410
Carnot cycle, I.536–7, II.1249
Carnot principle, I.60
Cascade showers, I.387
Cassiopeia A, III.1737, III.1775, III.1782
Catalogue of Selected Compact Galaxies and of Post-Eruptive Galaxies, III.1713
Cathode ray, I.27, I.36, I.52, I.53

Cathode-ray oscilloscope, III.1989
Cathode-ray tube, III.1530, III.1568, III.1909
Cauchy condition, I.484
Cauchy integral, III.1404
Cauchy model, II.967
Caustics, III.1466
Cavendish Laboratory, Cambridge, I.17, I.20, I.21, I.54
Cavity radiation, I.144, I.145
CCDM, II.1240
Cellular automata, I.629–31
Cellulose fibres, III.1597
Centaurus X-3 (Cen X-3), III.1754
Centimetre–gram–second (CGS) system, II.1252
Central-axis depth-dose (CADD) curves, III.1872, III.1873
Centrosymmetric structure, I.448, I.449
Cepheids, III.1714, III.1716, III.1719
Ceramics, III.1534
  conductors, III.1537, III.1540
  crystalline glazes on, III.1537
  electrical, III.1536–42
  electro-optical, III.1536–42
  functional (or 'fine'), III.1538
Čerenkov counter, II.749–51, III.1446
Čerenkov detector, II.758, II.765
Čerenkov effect, III.1408
Čerenkov radiation, III.1363
CERN, III.2024, III.2036, III.2054, III.2055, III.2057
Cerro Tololo International Observatory (CTIO), III.1744
Cesium-137, III.1879
C-field, III.1787
CGS units, II.1235
Challenger research ship, III.1973
Challenger space disaster, III.1742
Chandrasekhar limit, III.1759
Chandrasekhar mass, III.1711

Channel classification, prototype of, II.1060–3
Channel coupling, empirical parametrization, II.1073
Chaos, I.593, III.1466–7, III.1823, III.1826, III.1831, III.1848
Chaotic motions, II.815
Chapman–Enskog method, I.597–8
Chapman–Kolmógoroff equation, I.596
Chapman–Kolmógoroff–Smoluchowski–Bachelier equation, I.604
Chapman layer model, III.1994
Characteristic impedance of free space, II.1261
Charge-coupled devices (CCD), II.1248, III.1454, III.1476, III.1586, III.1590, III.1745, III.1770, III.1796
Charged particles
  detection of, II.1200
  measuring devices and detectors, II.1206–8
Charm hypothesis, II.699–700, II.702–4
Charmed particles, II.700–1, II.704–6, **II.705**
Charmonium, II.701–2, II.704
Charnley low-friction arthroplasty, III.1896
CHART fractionation schedule, III.1888
Chemical constant, I.154
Chemical defect, III.1505
Chemical elements, I.45
  formation of, III.1787
  origin of, III.1746–7, III.1785
  synthesis of, III.1747, III.1784–5
Chemical potential, I.532
Chemical quanta, I.154–5
Chemical Rubber Company (now CRC Press), III.1513
Chemical spectroscopy, I.19

Chew–Goldberger–Low (CGL) equations, III.1637
Chew–Low theory, II.680
Chiral invariance, II.676
Chiral smectic C, III.1604, III.1605
Chiral symmetry, II.676
Cholesteryl iodide, I.444
CIE (Commission Internationale de l'Éclairage) chart, II.1251, III.1477
Cinematography, III.1410
Clarendon Laboratory, I.20
Classical statistics, I.525–8
Clausius–Mossotti law, I.485, I.546
Climate, II.899–912
Clocks, II.1242–6
Close collisions, II.1076, II.1077
  by ions and atoms, II.1041–2
  of slow electrons, II.1040–1
Closed trapped surface, III.1778
Cloud chamber, I.54, I.367, I.384, I.388, I.389, I.396, I.408, II.120, II.751–2, II.1200
Cloudy crystal ball model, II.1212
Clusters, II.1084
  masses, III.1772
Cobalt-60, III.1858, III.1873, III.1874, III.1879, III.1892
Cobalt-bomb, III.1872
Cobalt particles, III.1611
Cobalt structure, I.472
COBE satellite, III.1804
Cockcroft–Walton generator, II.639, II.727
Co–Cr–Mo alloys, III.1897
CODATA, II.1261, II.1262, II.1266, II.1268–72
Coherence theory, III.1408, III.1439–41
  cross-spectral purity in, III.1433
Coherent anti-Stokes Raman scattering (CARS), III.1448
Coherent imaging, III.1418
Coherent potential approximation, II.1006

Coherent regeneration, II.674
Cohesive energy, I.484–5
Cold dark matter, III.1804, III.1805
Cold emission, **III.1307**
Collagen, structure of, I.499
Collective Atomic Recoil Laser, III.1456
Collectivization, III.2052–4
Colliding-pulse mode-locking, III.1439
Collision sequences, I.612
Collisions between atoms or ions, II.1039–42, II.1074–81
  collision speed comparable to electron velcities, II.1079–80
  highly stripped ions, II.1080–1
  projectiles faster than atomic electrons, II.1076–7
  projectiles slower than atomic electrons, II.1077–9
Colloids
  definition, III.1610–13
  dilute systems, III.1610
  future lines of research, III.1612–13
  instability, III.1610
  protection, III.1610, III.1611
  recent progress, III.1612–13
Colour centres, III.1529, III.1531, III.1541
Colour flow imaging, III.1914
Coma cluster, III.1732, III.1772, III.1800
Combination scattering, III.1404–6
Combinatorial analysis, I.526
Combinatorial formula, I.527–8
Commission Internationale de l'Éclairage chart, see CIE
Communications
  radio, III.1985–6, III.2000
  theory, III.1417
Compact optical-discs, III.1462, III.1464
Complementarity, I.201–4, I.223–5, III.1395

Complex fluids, III.1596
Complex susceptibility function, III.1404
Complexions, I.523, I.526
Compositeness, II.762–3
Compound nucleus, II.1219
Compound parabolic concentrator (CPC), III.1446
Compton effect, I.120, I.163–72, I.362–6, II.638, III.1302, III.1399, III.1404, III.1862, III.1865
Compton Gamma-Ray Observatory, III.1741, III.1784
Compton scattering, I.218, I.365, I.368, II.991, II.1057, III.1413, III.1773, III.1801
Compton wavelength, II.756, II.1026
Computational fluid dynamics (CFD), II.881–2, II.906–7
Computed tomography (CT), III.1858, III.1873, III.1898, III.1916–23
Computer-aided design (CAD), III.1899
Computer-aided manufacture (CAM), III.1899
Computer analysis, II.906
Computer-generated physics, III.1823–53
Computer graphics, I.628
Computer simulation, I.619, I.628, III.1670–3, III.2044
Computerized treatment-planning systems, III.1875
Computers, I.593, I.622
  early, III.1828–9
  in medical physics, III.1858
  in optical sciences, III.1475–6
  supercomputer, III.2045
Concave spectral gratings, I.54
Concorde, II.877, II.878

Condensation sequence model, III.1965
Condensed matter, II.756, III.2051
Condenser-objective lens, Einfeldlinse, III.1576
Condensing gas, statistical mechanics, I.551–4
Conduction band, III.1319, III.1329, III.1334, III.1340, III.1354
Conduction electrons, II.1168, III.1362
Conductors, magnetism, II.1133–6
Conference on Radioisotope Techniques, III.1901
Configuration space, III.1395
Confocal microscope, III.1450–1
Connected clusters, I.572
Conservation laws, I.374
Conserved vector current (CVC), II.667–9
Constant-energy surfaces, III.1297
Constant-Q technique, II.994
Consultative Committee for Electricity (CCE), II.1254, II.1256–60, II.1262, II.1271, II.1273
Consultative Committee for Photometry (CCP), II.1251
Consultative Committee on the Definition of the Second (CCDS), II.1247
Consultative Committee on Thermometry (CCT), II.1250
Contact-lens technology, III.1414
Contact potential, III.1283, III.1308
Continental drift, III.1967, III.1969, III.1972, III.1973
Continuous phase transitions, II.1009, II.1015
Continuous random network model, I.477
Continuum problems, III.1828
Contraction theory, III.1945

Controlled fusion research, III.1617, III.1633
Controlled fusion research—an application of the physics of high-temperature plasmas, III.1641
Convergent–divergent nozzle, II.879, II.880
Cooling of atoms, II.1099
Cooper pairs, II.938–42, II.946–8, II.950–2, II.954, II.956, II.962–4
Coordinated Universal Time (UTC), II.1247, II.1248
Copernicus satellite, III.1741, III.1765, III.1791
Copper, I.430, III.1515, III.1531, III.1543
Copper–gold alloys, III.1516
Copper oxide superconductors, II.959
Copper–silver solid solution, III.1543
Copper–tin diagram, III.1509
Corona, III.1441
Coronary arteries, III.1898
Coronium, I.76, I.81
Correlation diagram, II.1044
Correlation energy, III.1315
Correspondence principle, I.86
COS-B satellite, III.1740
Cosmic Background Explorer (COBE), II.758, III.1791
Cosmic microwave background, II.758
Cosmic rays, I.364, I.367–70, II.655–7, II.692
 intensity measurement, I.384
 penetrating radiation, I.388, I.404–5
 physics, II.760, III.2004–5
 soft and hard components, I.383–94
 see also Mesotron

Cosmical Electrodynamics, III.1636, III.1685
Cosmological constant, III.1723, III.1724, III.1730, III.1734, III.2039
Cosmological principle, III.1729
Cosmological problem (1939), III.1733–4
Cosmology, II.757–62, III.1691–821
  astrophysical, III.1784–93
  birth of, III.1691
  physical, III.1721–2
  relativistic, III.1721–34
  research since 1945, III.1746–807
  steady-state, III.1786–8
COSMOS High-Speed Measuring Machine, III.1745
COSTAR mission, III.1461
Coster–Kronig transitions, II.1089
Coulomb attraction, III.1331
Coulomb excitation, II.1217
Coulomb field, II.1217, III.1316
Coulomb force, I.123, I.376, III.1315
Coulomb interactions, II.1040, II.1125, III.1290, III.1316, III.1359, III.1362–4, III.1375, III.1551
Coulomb law, I.116, II.641, III.1361
Coulomb repulsion, III.1322
Covering lattice, I.573
Cowling's theorem, III.1958
CP violation, II.671–5
Crab Nebula, III.1737, III.1739, III.1752, III.1774
Critical behaviour, I.568
  binary alloys, I.548–9
Critical current, II.916
Critical density, III.1731
Critical fields, II.920
Critical fluctuations, II.1010, II.1016
Critical magnetic field, II.916
Critical model, III.1731
Critical opalescence, I.546–7

Critical phenomena, I.570, III.1595
Critical point, I.543, III.1947
Critical temperature, I.545, I.547, II.975, II.976
Critical velocity, II.928
Cross-spectral purity in coherence theory, III.1433
Cryostats, helium, I.436–7, III.2025
Crystal boundaries, III.1528
Crystal defects, III.1320, III.1531, III.1535
Crystal diffraction spectroscopy, II.1207
Crystal field, rare-earth elements, II.1164–5
Crystal field splittings, II.1164
Crystal field theory, II.1126–9, II.1136, II.1139, II.1141, II.1144–5
Crystal growth, I.466
Crystal lattice, III.1520
Crystal morphology, I.421
Crystal rectifier, III.1318
Crystal spectrometer, I.430, I.433
Crystal structure, I.423, III.1513–16, III.1529
  analysis, I.484–96
  and magnetic properties, II.1156–8
  determination methods, I.439–50
  faults in layer, I.472–3
  imperfections in, III.1505–6
Crystal surfaces, *see* Surface crystallography
Crystalline cleavage facets, III.1517
Crystalline virus, I.437
Crystallites, III.1280
Crystallographic analysis, III.1546
Crystallographic Data Centres, I.512
Crystallography, I.421–4, III.1515–17, III.1547
  history of, I.512
Crystallography in North America, I.512

Crystals
  dielectric properties, I.485–6
  dislocations, I.475–7
  doubly refracting, I.485
  imperfect, I.471–84
  optical properties, I.485–6
  see also under specific crystal types
CuAu, III.1515
Cu₃Au, III.1515
Cu–Ni–Fe alloy, III.1548
Curie law, I.540–1, II.1116, II.1133
Curie point, I.544–6
Curie temperature, I.545–6, I.549, I.550, I.559, III.1970, III.1976
Curie–Weiss Δs, II.1128–9
Curie–Weiss law, II.1116, II.1120, II.1128
Curium, II.1225
Curve of growth technique, III.1702
CuZn, III.1534
Cyclotrons, II.729–30, II.1199, II.1201, II.1206, II.1221, III.1871
  frequency, III.1334, III.1336
  resonance, III.1336, III.1337, III.1357
Cygnus A, III.1737, III.1775, III.1783, III.1788
Cygnus X-1, III.1754, III.1757
Cytosine, I.504

Daguerreotype process, III.1692
d'Alembert's Paradox, II.796, II.803–8
d'Alembert's Theorem, II.803–8, II.824, II.870, II.886
Danish Atomic Energy Commission, III.1643
Dark matter, II.761, II.768
Darwin–Fisher hypothesis, III.1948
Data collection, II.906
Data processing, thermodynamics of, I.576

de Broglie relation, I.457, III.1296
de Broglie waves, I.199, I.387, II.946, II.975, II.976, II.991, III.1292
Debye characteristic temperature, I.452
Debye equivalent, II.977
Debye formula, II.973
Debye function, II.972
Debye–Hückel theory, III.1521
Debye temperature, II.972, II.975, II.976, III.1305
Deceleration parameter, III.1794–7
Deep inelastic scattering, II.714–15
Defibrillation, III.1895
de Haas–van Alphen effect, II.1166, III.1348–56
δ-function, I.607
δ-rays, I.37
De Magnete, I.544, III.1856
Dementias, III.1932
Dendritic growth, III.1542
Density matrix, I.602, I.617
Density of states, II.986, II.1006
Density parameter, III.1734, III.1799–800
Deoxyribonucleic acid (DNA), I.504, I.507
Depletion layer, III.1325
Detectors, II.744–54
Deterministic hypothesis, I.222
Deuterium, I.112, I.370, III.1627, III.1706, III.1800
Deuterium-tritium, III.1651, III.1674
Deuteron, I.370, I.403
Deutsche Gesellschaft für Metallkunde, III.1553
Deutsche Gesellschaft für Naturforscher und Ärzte, I.47
Deutsche Gesellschaft für Technische Röntgenkunde, I.511

Diabatic potential energy functions, II.1079
Diagnostic machines, III.1897
Diagnostic medicine, III.1900–10
Diagnostic radiology, III.1859–63
Diagnostic ultrasonics, III.1910–11
Diamagnetism, II.1113–15
Diamond, I.429
Diamond–graphite equilibrium diagram, I.539
Diaphanography (light transmission imaging), III.1932–3
Dictionary of Applied Physics, II.977
Dictionary of Scientific Biography, I.16
Dielectric breakdown, III.1336–7
Dielectric constant, III.1595
Dielectric properties of crystals, I.485–6
Dielectric response, II.1055
Dielectric waveguide, III.1397, III.1420
Differential equations, III.1826
Diffractals, III.1466
Diffraction gratings, III.1401–2, III.1466
Diffraction microscopy, III.1582
Diffraction pattern, III.1412
Diffraction theory, III.1420
Diffractometer, I.435
Diffusion, II.818–19, III.1531–5, III.1548
  by continuous movements, II.830
  Fick's law of, I.597, I.601
Diffusion coefficient, I.597
Diffusion–convection balance for scalar quantities, II.841–6
Diffusion creep, III.1534
Diffusion inhibitor, III.1629
Digital subtraction angiography, III.1863
Diketopiperazine, I.464
Dimensional analysis, II.838

Diopside, structure of, I.440
Dipole–dipole interaction, II.987
Dipole vibration, II.1213
Dirac delta-function, I.206
Dirac equation, **I.361**, I.362
Dirac–Jordan transformation theory, I.198–201
Dirac theory, II.651
Directional distribution patterns, **II.1035**
Disabled persons, technology for, III.1899
Disciplinism, III.2042
Discretization process, I.629
Dislocations, I.475–7, III.1507, III.1520, III.1521, III.1525, III.1526, III.1528
Disordered materials, III.1365–9
Dispersion, III.1404
  curves, II.988–90, II.998
  formulae, I.171–2
  relations, II.680
  theory, I.184–5
Dispersive waves, II.856–60
Displacement law, I.66
Distorted wave Born approximation (DWBA), II.1220, II.1221
Distribution of charges and current in an atom consisting of many electrons obeying Dirac's equations, The, III.1622
D-lines, I.66
DNA, *see* Deoxyribonucleic acid (DNA)
Dolen–Horn–Schmid duality, II.683
Doping (and dopants), III.1529, III.1544, III.1593
Doppler effect, I.269, I.270, II.1094, III.1781, III.1784
Doppler shift, III.1398, III.1710
Doppler ultrasound, III.1914
Double-helix structure, I.506

Double strand breaks (DSBs), III.1885
Dreams of a Final Theory, III.2020
Drell–Yan process, II.701–2
Drift chambers, II.749, II.754
Drude–Lorentz model, III.1288
Duality, II.682–4
Dulong and Petit's law, II.970, II.972, II.980
Dwarf stars, III.1696, III.1698
Dye molecules, II.1087
Dynamic hologram, III.1458
Dynamical electroweak symmetry breaking, II.763
Dynamical friction, III.1774
Dynamical symmetry breaking, II.957
Dynamical systems, III.1850
  classification of, I.614–16
Dynamo theory, III.1985

Earth
  core, III.1952–60
  core size, III.1954
  fluid envelope, II.887–909
  gaseous-interior model, III.1947
  Laplace–Herschel–Kelvin scenario, III.1949
  liquid–gas boundary, III.1947
  magnetic field, III.1622–3, III.1952–60, III.1973
  origin and age (post-1935), III.1960–6
  origin and age (pre-1935), III.1944–52
  primaeval, III.1949
  structure, III.1946, III.1954, III.1956, III.1957
Earth–Moon system, III.1948
Earth sciences, III.1967, III.2046
Earthquakes, III.1956, III.1978–80
  zones, III.1976
Echo-sounding data, III.1973
École Polytechnique, I.15

Eddington–Lemaître model, III.1732–3
Eddington luminosities, III.1780
Edge dislocation, I.475–6
Education, I.8–16
  curricula, I.8
  higher, women in, I.8, I.10
EFFI, III.1673
Ehrenfest dog-flea model, I.618
Eigenfunctions, III.1432–7
Eigenvalue problem, III.1848
EIH method, I.303
Einfeldlinse (condenser-objective lens), III.1576
Einstein–Bohr frequency condition, I.359
Einstein–Bose statistics, I.602–3
Einstein–de Sitter model, III.1730–1
Einstein–Podolsky–Rosen(–Bohm) (EPR(B)) paradoxes, I.228–9, III.1457
Einstein X-ray Observatory, III.1798
Einsteinium, II.1224
E layer, III.1994, III.2002
Elasser–Bullard theory, III.1960
Elastic constants, II.983
Elastic scattering, II.687
Elasticity, I.484–5
Electric charges, I.51
Electric fields, I.27, I.261, III.1283, III.1362, III.1434, III.1435
Electric polarization, II.1031
Electrical conduction in metals, III.1284–7
Electrical conductivity, III.1279, III.1530
Electrical counting methods, II.1200
Electrical impedance tomography, III.1932
Electrical stimulation, III.1895
Electrical surges (*Wanderwelle*), III.1566

Electrical units and standards, II.1252–60
Electricity, I.51, I.52
Electricity supply industry, II.1255
Electro-absorption, III.1423
Electrocardiogram/graph, III.1893, III.1896
Electrodynamics, I.88, I.266, I.267, I.270
  in special theory of relativity, I.257, I.280–1
Electroluminescence, III.1417
Electrolysis, I.51
Electromagnetic field, I.358, III.1434, III.2039
Electromagnetic gauge invariance, II.693
Electromagnetic induction, I.266, II.1112
Electromagnetic interactions, I.391
Electromagnetic radiation, I.197, II.1264, III.2018, III.2025
Electromagnetic spectrum, III.1734–45
Electromagnetic theory, I.362
Electromagnetic waves, I.260, III.1391, III.1622, III.1985
Electromagnetism, II.640, II.693, III.1856, III.2000
Electron beams, III.1456, III.1565–91
Electron charge/mass ratio (e/m_e), II.1271
Electron–deuteron scattering, II.698
Electron density, I.440, I.446, I.448, I.452, I.454, I.502, III.1316
  regions, III.2009, III.2010
Electron diffraction, I.463–5, I.481
Electron distribution, I.452
Electron–electron collisions, II.739–40, III.1316
Electron energy
  LEED experiments, I.466
  RHEED experiments, I.468
Electron energy-loss spectrometer (EELS), III.1580
Electron $g - 2$ factor, II.647
Electron gas, III.1361, III.1362, III.1371–5
  quantum theory of, III.1287–91
Electron–hole pairs, II.999
Electron layer, III.1993
Electron magnetic moment, II.644–5
Electron magnetic resonance, III.1859
Electron mass, II.1271
Electron microprobe analyser, III.1548–9
Electron microscopes, I.481, II.931, III.1526, III.1569, III.1570–4
  see also under specific types of electron microscope
Electron mobility, III.1368
Electron multiplication, III.1419
Electron–neutrino pair, II.639
Electron pair, II.1028
Electron paramagnetic resonance (EPR), II.1137–52
  3d ions, II.1139–41
  4f ions, II.1141–3
  gedanken experiment, III.1457, III.1467
  iron group ions, II.1137–9
  rare earths, II.1141–3
Electron–phonon collisions, III.1316
Electron–phonon coupling, II.1000
Electron–phonon interaction, II.941, II.999, III.1303
Electron–phonon scattering, III.1305
Electron physics, III.2000
Electron–positron annihilation, II.690–1, II.715, II.717, II.740
Electron–positron collisions, II.691, II.711, II.740, II.741, II.768
Electron–positron pair production, III.1784, III.1786, III.1806

S15

Electron probe x-ray microanalysis, III.1580
Electron–proton collider (HERA), II.766–7
Electron scattering, II.1033, II.1221–2
  cross section, III.1996
Electron scattering factor, I.463–4
Electron spin, I.215
Electron spin resonance (ESR), II.1137–8, III.1923–33
Electron synchrotrons, II.731–2
Electron theory, I.30
  of metals, I.215–17
Electronic devices, therapeutic, III.1896
Electronics, medical. *See* Medical electronics
Electronics technology, III.2031
Electrons, I.36, I.51, I.53, I.59, I.64, I.77, I.81–90, I.95–9, I.115, I.158, I.167–8, I.369, I.377, I.457, II.637–8, II.641, **III.1298–9**
  free, I.372
  hot, III.1336–41
  in medical physics, III.1857
  in solids, III.1279–383
  see also Valence electrons
Electronvolt (eV), I.358
Electrostatic effects, charge effects, III.1610
Electrostatic electron microscopes, III.1577–8
Electrostatic generators, II.727–9
Electrosurgery, III.1894
Electroweak theory, II.640, II.762
Electroweak unification, II.692–709
Elementary charge (e), II.1271
Elementary excitations, II.928, II.929
Elementary particles, I.115, I.231–2, I.370–1, I.391, I.408–9
  in special theory of relativity, I.282–3
  physics of, II.635–794
  Standard Model of, III.2030, III.2033–9
  see also under specific particles
Elementary phenomena, II.1029–32
Elements, *see* Chemical elements
Elements of Physics, III.1857
El Niño/Southern oscillation phenomenon (ENSO), II.908
Elsasser–Bullard dynamo theory, III.1973
EMI-Scanner, III.1922
Emission electron microscope (EEM), III.1569, III.1577
Emulsions, III.1611
Encyclopaedia Britannica, I.36
Encyclopedia of Materials Science and Engineering, III.1513
Endocrinology, III.1930
Energy bands, III.1300
Energy density, III.1393
Energy fluctuation, I.163
Energy gap, II.932, II.952, III.1329, III.1433
Energy loss, 'hidden', II.846–9
Energy principle for hydromagnetic stability problems, An, III.1672
Enskog time, I.619
Entanglement, III.1394
Entropy, I.47, I.523, I.601
  calorimetric and statistical estimates, I.539–40
Eötvös experiments, I.286–7
Ephemeris Time, II.1247
Epilepsy, III.1922
  diagnosis, III.1916
Epoch of recombination, III.1801, III.1803
Equations of motion, I.48, I.186, I.303
Equilibrium angular momentum, II.932
Equilibrium state, II.920

Subject Index

Equilibrium theory, III.1785
Equipartition theorem, I.5, I.523, I.528
Equivalence principle, I.286–91
Equivalent width concept, III.1702
Erbium-doped optical-fibre amplifier (EDFA), III.1447
Ergodic surface, I.525
Ergodic theorems, I.599, I.626
Essen Ring, II.1244
ETA-BETA I, III.1658
ETA-BETA II, III.1658
Euler–Lagrange equation, I.195
Euratom-CNR facility, Padua, III.1658
European Federation of Medical Physics (EFOMP), III.1934
European Optical Society, III.1474
European research system (or science base), III.2055
European Science Foundation, III.2055
European Southern Observatory (ESO), III.1744
European Space Agency, III.1742, III.2055
Europium sulphide, I.559
Ewald transformation, II.988
Exchange interactions, II.1129, II.1143–4
Excitation channels, II.1059–73
Exciton absorption spectra, III.1423
Exciton theory, III.1423
Exclusion, **II.1028–9**
Exclusion principle, I.97–100, I.168, I.603, II.686, II.1116
Exluded volume, I.569
Experimental facilities, I.24–9
Experimental physics, I.13, I.14
Explorer I, III.2007
Explorer II, III.1740
Explorer III, III.2007
Explorer IV, III.1683, III.2007, III.2008

Explorer satellites, III.1739
Explosive nucleosynthesis, III.1747–8
Explosive showers, I.387
Exponential law, I.67
Extinction coefficient, III.1396
Extremely low-frequency radio physics, III.1995
Eye tumours, III.1876
Eyewall, II.904

F1 layer, III.1994
Fabry–Perot étalons, II.1240
Fabry–Perot interferometer, I.19, III.1412
Fabry–Perot resonators, III.1422, III.1423, III.1429
Fabry–Perot-type beam receiver, I.315
Face-centred-cubic lattice, I.557, III.1295
Facsimile (FAX), III.1476
Far infrared radiation, II.1054
Faraday (unit), I.51, II.1269
Faraday's constant, I.150, II.1237
Faraday's law, I.52
Fast charged particles, II.1039–40
Fast collisions, II.1076
Fast processes, II.1085
F-centre (Farbzentrum), III.1320–1
Fe–Ag–B alloys, III.1543
FECO interferometry, III.1412
Fellgett advantage, III.1416
Fellows of the Royal Society, I.10, I.12
Femtosecond pulses, III.1462
Fermi–Dirac statistics, I.117, I.123, I.155, I.531, I.533, I.603, II.687, III.1706, III.1711
Fermi distribution, II.1168, III.1301, III.1302, III.1308
Fermi energy, III.1295, III.1302, III.1309, III.1364
Fermi-field theory, I.375–8, I.382, I.392, I.393, I.404

S17

Fermi gas, II.1209, **III.1289**, III.1301, III.1308, III.1316
Fermi particles, I.603
Fermi statistics, II.1190, III.1394
Fermi surface, II.999, II.1002, III.1305, III.1308, III.1345–59, III.2025
Fermi theory, II.651
Fermilab, III.2054
Fermions, II.676, II.694, II.955, II.1168–70, III.1355, III.1804
Fermium, II.1224
Ferrimagnetic insulators, II.1156
Ferrimagnetism, II.1113, II.1152–65
  magnetic properties, II.1157
Ferroelastic crystals, III.1538
Ferroelectricity, I.486–9, III.1538
Ferrofluids, III.1611
Ferromagnetic model, I.570
Ferromagnetic resonance, II.1152–3
Ferromagnetism, I.559, I.562, II.1015, II.1113, II.1120–3, II.1134–6, II.1152–65, III.1279
Feynman diagrams, **I.361**
Fibre-optic light, III.1894
Fibre optics, III.1419–21
  see also Optical fibre
Fick's law of diffusion, I.597, I.601
Field cycling behaviour, II.1121
Field emission microscope, III.1569
Field ion microscope, I.50
Fierz interference, II.664
Finalization, III.2048–50
  pre-paradigmatic phase, III.2048
  revolutionary phase, III.2048–9
Fine structure, **II.1036–7**
  constant, II.640, II.1026, II.1262, II.1271
Finite energy quantum, I.173
Finnegans Wake, II.685
First Antarctic Expedition, III.2005
First International Congress of Medical Physics, III.1909, III.1923
First International Congress of Radiology, III.1881
First law of thermodynamics, I.374, I.521
First-order Fermi acceleration, III.1782
First-order melting transition, I.577
First Texas Symposium on Relativistic Astrophysics, III.1776
Fischer–Riesz theorem, I.206
Fitzgerald–Lorentz contraction, I.32, I.264
Fixed point, I.567, III.1832
Fixed-wing aircraft, II.817
Flash-photolysis, II.1085
Flash-spectroscopy, II.1085
Flight efficiency, II.869–73
Flood waters, II.897
Flory temperature, III.1598
Flow
  in pipes, II.814–15, II.825
  resistance, II.815
Fluctuation equation, III.1393, III.1402
Fluid envelope, II.860, II.887–909
Fluid mechanics, II.795–912
Fluid particles, I.572
Fluids, microscopic critical behaviour, I.546–7
Fluorescence, III.1537
Fluorite, I.430
Flying-spot microscope, III.1416
Focal conic defects, III.1605
Fokker–Planck equation, I.595–7, I.624, I.627
Forbidden transitions, III.1762
Forced convection, II.843, II.844
Form factors, II.660, II.687
FORTRAN, III.1843
Fortschritte der Physik, I.7
Foucault's pendulum, I.2–3
Fourier imaging theory, III.1418–19

Fourier series, I.441, I.623
Fourier synthesis, I.501
Fourier techniques, reconstructed (111) surface, I.469
Fourier-transform, III.1418, III.1924
  image processing, III.1413
  spectrometer, III.1416
  spectroscopy, II.1055
Fourier's law of thermal conduction, I.601
Fractals, I.574, I.630
Fractured bones, III.1897
Fragile structures, III.1595
Fragmentation, II.1102, II.1103
  dynamics of, II.1105–6
Franck–Condon rule, II.1046
Franck–Hertz experiment, I.89
Free convection, II.843, II.844
Free-electron gas hypothesis, I.216
Free electrons, I.372
Free-energy phase diagram, I.538
Freedman–Clauser experiment, III.1458, III.1467
Frequency
  dissemination of, II.1248–9
  standards, II.1243–6
Fresnel biprism, III.1585
Fresnel diffraction fringes, III.1580–2
Friedel's law, I.440, I.464, I.495
Friedman world model, III.1725, III.1731, III.1733, III.1784, III.1807
Froome interferometer, II.1264
Frustration, II.1170
FSU, III.2028
Fullerene crystals, II.956
Functional electrical stimulation (FES), III.1900
Functionally graded materials, III.1542
Fundamental observers, III.1729
Fundamental physics, III.2034–5
Fundamental Theory, The, III.1700, III.1787

Fusion bomb, III.2022
Fusion research, III.1617, III.1618, III.1627–9, III.2024
  declassification, III.1640–5
  principal sites for, III.1636
  see also Magnetic fusion research
FWHM, III.1904, III.1922

Gadolinium, magnetic properties, II.1162
Galactic cannibalism, III.1774
Galactic Centre, III.1740, III.1744, III.1763
Galaxy(ies), III.1769–71
  active, III.1797–8
  clusters of, III.1771–4
  crossing times, III.1772
  dark halo, III.1771
  dark matter, III.1771
  distribution of, III.1771–2
  elliptical, III.1771
  formation, III.1800–5
  luminosity function for, III.1769–70
  masses, III.1769
  structure of, III.1714–21
  velocity curves, III.1771
  violent events in nuclei of, III.1777
Galena crystal, III.1318
Galilean transformation law, I.257
Galilei–Newtonian kinematics, I.257
GALLEX experiment, III.1749
Gallium, I.75, III.1749
Gallium arsenide (GaAs), III.1371, III.1375, III.1464
Gamma-10, III.1669
Gamma camera, III.1907–9, III.1918, III.1920, III.1921
Gamma-decay, I.106
Gamma-radioactivity, I.53
Gamma-ray astronomy, III.1738–41
Gamma-ray detectors, III.1740

Gamma-ray imaging, III.1904
Gamma-ray wavelengths, II.1207
Gamma-rays, I.54, I.367, I.457, II.1207, III.1784, III.1879, III.1902
Gamow–Teller type interactions, II.664
Gargamelle Collaboration, II.697
Gas constant (R), II.1269
Gas discharges, I.27
Gas of corpuscles, III.1390–2
Gaseous Electronics Conference, III.1644
Gases, I.47
  ionization phenomena in, III.1644
  kinetic theory of, I.30, I.35, I.46, I.522, I.523, I.587, III.1284
  liquid, I.28, I.543–4
  specific heat, I.528–30
Gaskugeln, III.1695
Gauge theory, III.1414
Gaussian field-distribution profile, III.1429
Gaussian–Hermite (laser) beam propagation, III.1447
Geiger(–Müller) counter, I.108, II.747, II.1200, II.1207, III.1413, III.1900–2, III.2008
Geiger–Nuttall relation, I.116
GEKKO series of lasers, III.1676
Gell-Mann–Okubo mass relation, II.661
General Catalogue of Nebulae, III.1717
General Conference on Weights and Measures (CGPM), II.1234–6, II.1247, II.1248, II.1251, II.1257, II.1262, II.1264
General Electric Company (GEC) Research Laboratories, II.1204
General Electric (GE) Research Laboratory, New York State, III.1528
General theory of relativity, I.249, I.286–320, III.1722–4
  alternative formulations and foundations, I.298–9
  alternative scenarios, I.295–6
  and active galactic nuceli, III.1777–81
  approximation schemes, I.302
  astronomical and astrophysical applications and tests, I.315–17, III.1776
  classical tests, I.303–6
  equivalence principle, I.286–91
  exact solutions, I.302
  field equations, I.293–5, I.313
  gravitational energy, I.300–1
  later work on, I.296–8
  metric-tensor field, I.291–3
  physical interpretation, I.301
  Schwarzschild solution, I.303–6
  software programs, I.302
  tests, III.1757–8
Generalized form factor, II.1040
Genetic code, I.507
Geneva Conference (1958), III.1640–5, III.1642, III.1643
Geniac, III.1828
GENIUS—The Life of Richard Feynman, III.1519
Geology, III.2046
Geomagnetic dynamo theory, III.1958–9
Geomagnetic field, reversals of, III.1973
Geomagnetic field lines, III.2010
Geomagnetic storm, III.2002–3
Geomagnetism, III.2000, III.2001
Geometrical collective model, II.1218
Geophysical observing stations, III.1998
Geophysics, III.1943–2016, III.2046
  history, III.1943–4
  use of term, III.1943

Georan, II.1242
Geospace, III.1981, III.1997–2001
  future study propsects, III.2011
  pictorial models, III.2010
Germanium, I.75, I.470, II.982,
    II.1208, III.1281, III.1301,
    III.1319, III.1322, III.1324,
    III.1332–6, III.1340, III.1544,
    III.1749
Gesellschaft deutscher Naturforscher und Ärzte, I.5
Giant branch, III.1698
Giant dipole resonance (GDR), II.1213–14
Giant molecular clouds, III.1765
Giant stars, III.1696, III.1698, III.1708–9
Gibbs ensembles, I.531–3
Gibbs free energy, III.1543, III.1548
Gibbs function, I.549
Gibbs Paradox, I.528
Ginzburg–Landau theory, I.568, II.934, II.935, II.940
Glashow, Iliopoulos and Maiani (GIM) mechanism, II.700
Glashow model, II.694
Glashow–Weinberg–Salam theory, II.695, II.697, II.698, II.707
Glass industry, III.1517
Glasses, III.1533, III.1600–2
  phonons in, II.1004–7
Glitches, III.1754
Global Positioning Satellite System, II.1249
Glomar Challenger, III.1976, III.1980
Gluons, II.636, II.637, II.684, II.714, II.716
Glutamine, structure of, I.449
God Particle, The, III.2020
God principle, III.2019–20
Gold, III.1515, III.1543
Gold-198, III.1905
Goodyear, III.1593, III.1597
Göttingen quantum mechanics, I.183–96

Gradient-index lenses, III.1391
Grain boundaries, III.1518–20, III.1528
Grain size distribution, III.1549
Grand canonical ensemble, I.532
Grand partition function (GPF), I.533
Grand unified theory, I.323, II.763–4
Grating spectroscopy, III.1391
Gravitation, II.757–62, III.2037
  and quantum theory, I.318
  in special theory of relativity, I.283–4
  Newtonian constant of, I.28
  theory of, I.283–4, II.762
Gravitational collapse, I.306–12
Gravitational constant (G), II.1265–7
Gravitational energy, I.300–1
Gravitational field equations, I.253
Gravitational radiation, I.312–15
Gravitational waves, I.312, I.315
Gravity fields
  Moon, II.888
  Sun, II.888
Gray (unit), III.1867
Grazing collisions by ions and atoms, II.1041–2
Great Debate, III.1716–21
Green statistics, II.687
Greenhouse effect, II.899
Green's function, I.617, I.623–4
Ground-state energy, II.1045
Group-theoretical model, I.211–15
Group velocity, II.856
Guanine, I.504
Guided wave optics, III.1458
Guy's Hospital, III.1864
Gyromagnetic effect, II.1153
Gyromagnetic ratio, II.1237, II.1257
Gyroscope
  femtosecond-pulsed dye-laser ring, III.1433

rotation-sensing ring-laser, III.1433

Habilitation thesis, I.184
Hadron colliders, II.707–8, II.742–4, II.766–7
Hadrons, II.686, II.690–1, II.699–700, II.704, II.713, II.715–17
Haemoglobin, I.500, I.502–3
Hafnium, I.100
Half-life concept, I.59
Hall constant, III.1334, III.1367
Hall effect, I.9, II.1096, III.1282, III.1284–6, III.1296, III.1298, III.1299, **III.1331**
Hall field, III.1373
Hall potential, II.1096
Hall resistance, II.1258
Hambergite, structure of, I.451
Hamilton–Jacobi equation, I.176, I.194
Hamilton–Jacobi theory, I.169, I.191
Hamiltonian, I.566–9, I.612
Hamiltonian matrix, I.190
Hamiltonian scheme, I.189
Hammersmith Hospital, III.1871, III.1873, III.1922
Handbook of Chemistry and Physics, III.1509–13
Handbuch der Physik, III.1513
Harker–Kasper inequality, I.447, I.448
Harmonic oscillator, II.973
Harrison construction, III.1313
Harrison–Zeldovich spectrum, III.1802, III.1807
Harvard Radio Research Laboratory, III.2026
Hayashi limit, III.1767
Heart valves, artificial, III.1898
Heat capacity, I.78–80
Heat conduction theory, III.1945
Heat Theorem, II.968, II.975

Heat-treatment, III.1519
Heaviside layer, III.1987
Heavy-atom derivatives, I.501
Heavy-atom method, I.444
Heavy element catastrophe, III.1785
Heavy fermions, II.955, II.1168–70
Heavy hydrogen, III.1627, III.1628
Heavy-Ion Medical Accelerator, Chiba, Japan, III.1876
Heisenberg–Dirac (H–D) theory, II.1128–9, II.1148–9
Heisenberg exchange integral, I.213
Heisenberg magnet, III.1595
Heisenberg n–p model, I.382
Heisenberg systems, II.1014
Helicity, II.667
Helicons, III.1358–9
Helimagnetic structure, I.462
Helioseismology, III.1749–51
Helium, I.58, I.75, I.81, I.101–2, I.110, I.182, I.211, II.974, II.1004, III.1707
and discovery of cosmic microwave background radiation, III.1790–3
cryostats, I.436–7, III.2025
liquid, II.913–15, II.921–32, III.1341–5
superfluid phases, II.949–55
synthesis, II.1226
α-helix, I.498, I.508
Helmholtz free energy, I.527, I.532
Henry Draper Catalogue of Stellar Spectra, III.1694, III.1695
HERA (electron–proton collider), II.766–7
Hercules X-1 (Her X-1), III.1755
Hermann–Mauguin symbol, I.439
Hermite functions, I.625
Hermite polynomials, I.625
Hermitian affine connection, I.322
Hermitian matrices, I.190
Hermitian operators, I.205

Subject Index

Hertzsprung–Russell diagram, III.1692–8, III.1746, III.1793–4
Her X-1, III.1756
Hess–Fairbank effect, II.944–6, II.955, II.956
Hg–Ba–Ca–Cu–O compound, II.959
Hierarchical clustering, III.1804
Higgs boson, II.695, II.762
Higgs fields, III.1807
Higgs mechanism, II.694
Higgs particle, III.2036
High Energy Astrophysical Observatory A and B (HEAO-A and B), III.1740
High Energy Particle Physics, III.2051
High-energy physics, III.2056, III.2058
High-resolution images, III.1477
High-temperature superconductor, I.489–90
High Voltage Engineering (HVE), II.1203
Highest occupied molecular orbital (HOMO), II.1083
Hilbert conditions, I.293
Hilbert space, II.1025, III.1395
Hip replacement, III.1896
History of Metallography, A, III.1556
History of the University in Europe, III.2041
H-mode operation, III.1649
Hohlraumstrahlung, I.66
Holes, **III.1289–9**
  concept of, III.1297
Holograms, III.1415, III.1438, III.1582–4
Holography, III.1437–9, III.1580–1
  history of, III.1414–17
  off-axis optical beam, III.1585
  principle of, III.1415
Hooker 100" telescope, III.1704, III.1726

Hospital Physicist's Association, III.1864
Hot Big Bang models of Universe, III.1778
Hot Dark Matter, III.1803, III.1804
Hot electrons, III.1336–41
  phonon interaction with, III.1339
Hot shortness, III.1519
Hot-wire techniques, II.831
Hovercraft, II.886
HP65 programmable calculator, III.1829, III.1842, III.1846
H-theorem, I.35
Hubbard U, II.1166
Hubble space telescope, III.1461, III.1474, III.1741–2, III.1745
Hubble's constant, III.1730, III.1731, III.1787, III.1788, III.1793–4, III.1799
Human movement studies, III.1899–900
Hume-Rothery rule, III.1346
Hydraulic jump, II.858–9
Hydrodynamic forces on ships, II.882
Hydrofoil boat, II.886
Hydrogen, I.64, I.77, I.78, I.80–6, I.160, I.359
  molecular, I.181, II.1045, **II.1048**, II.1069–73, III.1766
  ortho, **II.1048**
  ortho–para, I.540
  para, **II.1048**
Hydrogen atoms, I.90–2, I.211, I.360, I.570, II.641
Hydrogen bombs, III.1633, III.1673
Hydrogen bond, II.1043
Hydrogen isotope, I.370
Hydrogen spectrum, I.90
Hydroxyapatite, III.1897
Hyperfine interactions, II.649–50
Hyperfine structure, **II.1036–7**, II.1214
Hyperon, II.657

Hypersonic aerodynamics, II.879–80
Hyperspherical coordinates, II.1103–5
Hypothesis of energy elements, I.145
Hysteresis, II.918, II.1121

Ice, structure of, I.474–5
Illinois University, III.1871
Illustrated London News, I.2
Image processing, Fourier-transform, III.1413
Imaging theory, III.1399–401, III.1467–8
IMB-3 Detector, II.759
Imperfections in Nearly Perfect Crystals, III.1535
Implant surgery, III.1896
Incoherent backscatter, III.1995
Incoherent scatter radar (ISR), III.1995, III.1996
Independent-particle model, III.1359–65
Index Catalogues, III.1717
Indian ink, III.1611
Indirect-exchange-coupling theory, II.1012
Indium, I.75
Indium antimonide (InSb), reconstructed (111) surface, I.469
Induced transitions, III.1398–9
Induction coil, I.54
Ineffectiveness concept, III.1347
Inelastic neutron scattering, II.1162
Inelastic scattering, II.687–90, II.697, II.1221
Inequality relations, I.447
Inertial confinement fusion (ICF), III.1673–7
Inflation, II.761–2
Information retrieval, electronic methods, III.1896
Information technology, III.2044–6

Infrared absorption, II.988
Infrared astronomy, III.1742–4
Infrared Astronomy Satellite (IRAS), III.1744
Infrared radiation, laser sources of, II.1054
Infrared reflectivity, II.1003
Inner quantum number, I.100
Inner-shell phenomena, II.1089–92
  Auger emissions, II.1090–1
  spectra, II.1038–9
  x-ray studies, II.1089–90
Inorganic compounds, structure analysis, I.492–3
Inorganic Crystal Structure Database, I.512
In-plane x-ray diffraction, I.466
Instant camera, III.1417
Institut für Metallphysik und Allgemeine Metallkunde, III.1555
Institute for Defense Analysis, III.2023
Institute for the Study of Metals, University of Chicago, III.1556
Institute Laue-Langevin, III.2054
Institute of General Metallurgy (Allgemeine Metallkunde), III.1554–5
Institute of Metals Research, III.1555
Institute of Physics, III.1558, III.2042
Institution of Physics and Engineering in Medicine and Biology, III.1864
Instrumentation, role of atoms, II.1094–7
Integrated intensity, **I.434**
Integrated optics, III.1449–50
Interaction II, I.391, I.393, I.402
Interaction III, I.392, I.397
Interdisciplinarity, III.2042–4
Interference effects, III.1369–70

Subject Index

Interference pattern, III.1395
Interferometric spectroscopy, III.1410
Interferometry, I.25, I.26, III.1410, III.1738
  see also Optical interferometry techniques
Intermediate state, II.924
Intermediate valence, II.1168–70
Intermetallic compounds, III.1515
Intermetallic Phases Data Bank, III.1513
Internal Constitution of the Stars, The, III.1700, III.1702
Internal conversion, I.107
Internal field, I.545
Internal gravity waves, II.866
Internal pressure, I.543
International Ampere, II.1254, II.1257
International Astronomical Union (IAU), II.1242, II.1247
International Atomic Energy Agency (IAEA), III.1643, III.1648, III.1649, III.1908
  Scanning Conference, III.1916
International Atomic Time (TAI), II.1247, II.1248
International Bureau of Weights and Measures (BIPM), II.1234, II.1236, II.1240, II.1248, II.1251, II.1257, II.1259, II.1272
International Centre for Diffraction Data, I.512
International Commission on Radiation Protection (ICRP), III.1910
International Commission on Radiation Units and Measurements (ICRU), III.1867, III.1881, III.1910, III.1920
International Commission on Radiological Protection, III.1890
International Committee of Weights and Measures (CIPM), II.1234, II.1240, II.1242, II.1247, II.1249–51, II.1257, II.1264, II.1271
International Conference on Medical Electronics, III.1913
International Council for Scientific Unions, III.2006
International Critical Tables, II.1267, III.1509
International Earth Rotation Service (IERS), II.1248
International Federation of Medical and Biological Engineering (IFMBE), III.1934
International Geophysical Year (IGY), III.1973, III.1989, III.2005–6
International Medical Electronics Conference, III.1904
International Ohm, II.1254, II.1257
International Organization of Medical Physicists (IOMP), III.1934
International Polar Year (1882–83), III.2005
International Practical Temperature Scale (IPTS-48), II.1250
International System of Units (SI system), II.1235
International Tables for the Determination of Crystal Structures, I.511
International Tables for X-ray Crystallography, I.439
International Temperature Scale, II.1250
International Toroidal Experimental Reactor) device, *see* ITER

S25

International Ultraviolet Explorer, III.1741
International Ultraviolet Observatory, III.1780
International Union for Geodesy and Geophysics, III.2006
International Union for Scientific Radio, III.2006
International Union of Crystallography, I.511–12
International Union of Physical and Engineering Sciences in Medicine (IUPESM), III.1934
International Union of Pure and Applied Physics (IUPAP), II.1268
International Union of Radio Science (URSI), II.1273
International Units, II.1255, II.1260
International Volt, II.1254, II.1257
International Weights and Measures Organization, II.1234
International X-ray and Radium Protection Commission, III.1890
Internationale Zeitschrift für Metallographie, III.1556
Internationalization, III.2055–6
Inter-Service Ionosphere Bureau (ISIB), III.1998–9
Interservice Radio Propagation Laboratory, III.1998
Interstellar chemistry, III.1765
Interstellar medium, III.1761–9
  soft x-ray background, III.1766
  two-phase model, III.1765
  violent, III.1766
Interstitials, III.1529, III.1531, III.1533
Introduction to the Ionosphere and Magnetosphere, An, III.2001
Introduction to the Study of Physical Metallurgy, An, III.1517

Inverse-Compton catastrophe, III.1781
Inverse-Compton scattering, III.1781
Inverse-square law, II.1266
Inversion layer, III.1373
Investigations of the solar spectrum and the spectra of the chemical elements, III.1692
Iodine, contrast media, III.1863
Iodine-123, III.1920
Iodine-131, III.1900, III.1903–6
Iodine-132, III.1905
Ion–atom collisions, *see* Collisions between atoms or ions
Ion guns, III.1997
Ionic crystals, II.997–8, III.1528–9, III.1531, III.1541
Ionization chamber (thimble chamber), III.1866–7, III.1881
Ionization detectors, II.746–9
Ionization energy, III.1867
Ionization equilibrium, III.1701
Ionization Phenomena in Gases, III.1644
Ionization threshold, II.1069–73
Ionization thresholds, spectral connections at, II.1058–9
Ionizing radiation, III.1857, III.1889
Ionosonde, III.1989, III.1991
Ionosphere, III.1619–23, III.1682–7, III.1981–2001
  'E' layer, III.1621
  electron content, III.1995
  'F' layer, III.1621
Ionospheric physics, III.1944
IRAS satellite, III.1767
Iridium-192, III.1879, III.1892
Iron–carbon alloy, III.1524
Iron-shrouded lens, III.1567
Irreducible cluster integrals, I.552
Irvine–Michigan–Brookhaven (IMB) experiment, III.1760
Ising model, I.554–61, III.1554

Isobars, II.680
Isomeric levels, II.1210
Isomorphous replacement method, I.445
Isospin, **I.395**
Isothermal compressibility, I.546, III.1397
Isotope effect, II.933
Isotopes, I.107
Isotopic charge, II.676
ITER (International Toroidal Experimental Reactor) device, III.1651
ITS-27 temperature scale, II.1250
ITS-90 temperature scale, II.1250, II.1260

J/ψ, II.701–2
Jacobians, I.12
Japanese Atomic Energy Agency, III.1650
JASON, III.2023
Jeans criterion for collapse, III.1768
Jellium, III.1303–5
JET (Joint European Torus), III.1650, III.1651, III.2055
Jets, II.715–16
jj-coupling shell model, II.1209, II.1210
Johnson noise fluctuations, I.601
Joint European Torus, *see* JET
Joint replacement, III.1896
Jones zone, III.1313
Josephson constant, II.1258, II.1271
Josephson effect, II.941–2, II.1096, II.1257, II.1258
Josephson junction, II.1095
Joule–Thomson effect, I.28
Journal of Applied Crystallography, I.511
Journal of Materials Research, III.1556
Journal of Nuclear Materials, III.1534
Julia series, III.1841

Kamioka detector, II.765
Kamiokande neutrino-scattering experiment, II.759, III.1748
Kaons, II.650–4, II.656–7, II.660, II.678
Kapitza's law, III.1283
Kármán vortex street, II.851
K-distributions, III.1466
Kelvin (degree), II.1250
Kelvin Circulation Theorem, II.817, II.822
Kelvin–Helmholtz time-scale, III.1694
Kennedy–Thorndike experiment, I.285
Kennelly–Heaviside layer, III.1621, III.1987
Kerr metric, III.1778
Kerst–Serber theory, II.1206
Kevlar polymeric fibres, III.1552
Keyhole surgery, III.1895
KHF_2, structure of, I.458
KH_2PO_4
 ferroelectricity, I.488
 nuclear density, I.461
 structure of, I.459
Kiloelectronvolt (keV), I.358
Kilogram (unit), II.1233–8
Kinematic shock wave, II.899
Kinematic wave, II.898
Kinematics, relativistic, I.364
Kinetic constant, I.154
Kinetic energy, I.90, II.837
Kinetic equations, I.608–11
Kinetic theory of gases, I.30, I.35, I.46, I.522, I.523, I.587, III.1284
Kirchhoff's function, I.148
Kirchhoff's law, I.66
Kitt Peak National Observatory (KPNO), III.1744
Klein–Gordon equation, I.209, I.283
Klein–Gordon theory, I.209
Klein–Nishina formula, I.218, I.365

Klystrons, II.1206
K-meson, see Kaon
KMS Fusion, III.1676
KNbO$_3$ ferroelectric phase transition, I.486
Knowledge domains, III.2041
Kogelnik–Li equations, III.1430
Köhler illumination, III.1390
Kolmógoroff–Sinai entropy, I.615
Kolmógorov dissipation length, II.838
Kondo effect, II.1167–9
Kondo problem, III.1343
Kondo theory, III.1343
Korteweg–de Vries equation, II.859
Kramers equation, I.607–8
Kramers–Heisenberg effect, III.1405
Kronecker delta-function, I.208
Krypton, I.75
K systems, I.615
Kuhnian revolution, III.2049
Kuiper Airborne Observatory, III.1743
k-values, III.1364

Laboratory of Biological Physics of the College de France, III.1894
LAK theory, III.1359
Lamb-dip, III.1443
Lamb–Retherford shift, III.1421
Lamb shift, II.641–3, II.649
Lamellar phases, III.1608
Laminar flow, II.827
Landau anti-damping, III.1625
Landau critical velocity, II.928
Landau damping, III.1624, III.1625
Landau expansion, II.1009
Landau–Ginzburg–Wilson Hamiltonian, I.568
Landau theory of second-order transitions, I.549–51, II.1009
Landolt–Börnstein series, II.1165
Landolt–Börnstein Tables, III.1513
Lane–Emden equation, III.1695, III.1711
Langevin equations, I.595–7, I.608, I.614, I.624, I.626, I.628
Langevin function, I.548, II.1116
Langley's bolometer, I.23
Laplace–Herschel nebular hypothesis, III.1945, III.1962
Large Electron–Positron Collider (LEP), II.741, III.1792
Large Hadron Collider, III.2036
Large Magellanic Cloud, II.758, III.1760
Large Space Telescope, III.1741
Large-scale-integration, see LSI
Laser anemometry, III.1442
Laser beam
  particle trapping, III.1462
  self-trapping, III.1459
  wavefronts, III.1458
Laser diode, III.1476, III.1480
Laser-fusion, III.1676
Laser-fusion-generated electrical power, III.1479
Laser-fusion pellet, III.1675
Laser interferometer gravitational-wave observatory (LIGO) project, I.298, I.317
Laser Interferometric Gravitational Observatory, III.2024
Laser-light-scattering, III.1441–5
Laser-printing, III.1541
Laser read-out systems, III.1462
Laser sources of infrared radiation, II.1054
Laser speckle, III.1453
Laser standards, II.1242
Laser velocimetry, III.1442
Lasers, III.1417–18, III.1424–52
  applications, III.1478–9
  argon, III.1894
  argon-ion, III.1453
  carbon dioxide, III.1442, III.1677, III.1894

Subject Index

chemical, III.1445
continuous-wave single-mode silica-fibre, III.1447
excimer, III.1453
free-electron, III.1456
GEKKO series, III.1676
glass, III.1677
HCN and H$_2$O action, III.1442–3
helium–cadmium, III.1435, III.1445
helium–neon, III.1427, III.1435, III.1439
history of, III.1411, III.1423
in medicine, III.1480
instabilities, III.1466–7
molecular, III.1442
Nd:glass, III.1479
Nd:YAG, III.1435, III.1471, III.1479, III.1894
noise, III.1444
Nova, III.1464, III.1676
rare-earth ion, III.1438
rare-gas ions, III.1443
ruby, III.1424, III.1426, III.1429, III.1430, III.1432, III.1479, III.1894
soliton, III.1469
spatial mode patterns, III.1431
titanium–sapphire, III.1448
tunable, III.1472
wavelength-tunable dye, III.1448
wire, III.1472
x-ray, III.1464–6, III.1479
see also under specific types and applications
Lasing without inversion (LWI), III.1473
Lattice conduction, II.981
Lattice dynamics, II.967–91, II.997–1007
Lattice gas model, I.553
Lattice gauge theory, II.756
Lattice vibrations, I.452
Lau effect, III.1414
Lau interferometer, III.1414

Laue equations for X-ray diffraction, **I.427**
Laue photographs, I.458
Lead, II.999
 crystals, III.1544
 isotopes, III.1960, III.1961
Lead sulphide detectors, III.1743
Learned societies, I.3
Lehrbuch der Metallographie, III.1554
Length, II.1239–42
Lens aberrations, III.1458
Lens-pupil functions, III.1418
Lenses
 contact, III.1414
 gradient-index, III.1391
 iron-free, III.1568
 magnification ratio, III.1569
 properties of, III.1389
 see also under specific types of lens
Lenz–Ising model. See Ising model
Lepton number, II.664, II.755
Leptons, II.635–6, II.692–5, II.700, II.704, II.719–25
Leukaemia, III.1891
Levitation theory, III.1543
Levshin–Perrin law, III.1400
Liapunov functions, I.621–2
Lidar systems, III.1479
Lift–drag ratio, II.870, II.871, II.881
Light, speed of, I.260, I.261, I.266, II.1240, II.1262–4
Light-beating spectroscopy, III.1421
Light-emitting diode (LED), III.1474, III.1476
Light microscope, resolving power, III.1574
Light-quantum hypothesis, I.71–2, I.151–2, I.167–8, I.185, I.363
Light rays, I.260
Light-scattering spectroscopy, III.1408
Light transformations by atoms, II.1101

S29

Light transmission imaging (diaphanography), III.1932–3
Light waves, I.258
Lightning-induced electron precipitation, III.1686
Line broadening, I.471–2
Line of section, III.1836
Linear accelerators, II.734–8, II.1205, III.1874, III.1875
Linear energy transfer (LET), III.1885
Linear response theory, I.626–7
Linnean system, I.47
Liouville equation, I.593, I.602, I.605–7, I.619, I.620
Liposomes, III.1608
Liquid crystal display (LCD), III.1476, III.1540–1, III.1593, III.2029
Liquid crystal television display, III.1476
Liquid crystals, II.1084, III.1539–42, III.1552–3, III.1603–5, III.2031
Liquid-drop model, II.1186–91, II.1213
Liquid gases, I.28
  critical point, I.543–4
Liquid helium cryostat, III.2025
Liquid metals, III.1366–7
Lithosphere, III.1978
Liverpool Astronomical Society, I.24
Lloyd Triestino, III.1711
LMC X-1, III.1757
LMC X-3, III.1757
Local gauge invariance, II.693
Local gauge transformation, II.693
Lockheed C-141 transport aircraft, III.1743
Lodestone, II.1112, II.1156, III.1856
London Chemical Society, I.46
London equation, II.925, **II.926–7**
Long-chain systems, III.1596–7
Longitudinal waves, III.1956
Long-range collisions, II.1076

Long-range effects, **II.1066–7**
Long-range interactions, II.1063–5
Long-range parameters, **II.1074–5**
Lorentz contraction, I.264
Lorentz–Einstein formula, I.285
Lorentz electron model, I.32
Lorentz factor, III.1783
Lorentz force, III.1622
Lorentz force equation, III.1671
Lorentz–Rayleigh model, I.629
Lorentz transformation, I.264, I.265, I.272, I.274, I.275, I.281
Loschmidt's constant, I.150
Low-energy electron diffraction (LEED), I.466
Low-energy electron microscope (LEEM), III.1577
Lowest unoccupied molecular orbital (LUMO), II.1083
LSI (large-scale-integration), III.1371
Luminescence, III.1537
Luminosity function for galaxies, III.1769–70
Luminosity–spectral class diagram (Russell diagram), III.1696, III.1698, III.1710
Luminous stars, III.1696
Lummer–Gehrcke plate, III.1389
Lummer–Gehrcke spectrograph, I.19

Mach number, II.813, II.814, II.821–4, II.880
Mach's principle, III.1723
Macroscopic wavefunctions, II.943
Macrostate, I.523, I.526
MAFCO, III.1673
Magellanic Clouds, II.758, III.1714, III.1715, III.1719, III.1744, III.1760, III.1788
Magic numbers, II.1225
Magnesium, formation of, III.1747
Magnesium oxide (MgO) crystals, I.453

Magnetic anisotropy, II.1134, II.1154
Magnetic behaviour, types of, II.1112
Magnetic breakdown, III.1359
Magnetic bubbles, II.1161
Magnetic compass, III.1952
Magnetic dipole, II.1111, III.1958, III.2004
  interactions, II.1011
Magnetic domain, II.1154–5
  patterns, II.1135
Magnetic electron lens, III.1566–9
Magnetic equation of state, I.545
Magnetic excitons, II.1162
Magnetic field, I.27, I.102, I.108, I.261, I.541, II.1111, II.1125, II.1126, II.1132, II.1172, III.1282, III.1345, III.1352, III.1358, III.1373
  Earth's, III.1622–3, III.1952–60, III.1973
  in sunspots, III.1958
Magnetic flux, II.919, II.1121
Magnetic fusion research, III.1619, III.1629–31
  classification era, III.1631–40
  post-1960, III.1645–70
Magnetic induction, II.1111
Magnetic insulators, co-operative phenomena in, II.1147–8
Magnetic intensity, II.1111
Magnetic interactions, II.1010
Magnetic lens, III.1568, III.1580
Magnetic materials, I.559
Magnetic mirror, III.1658–70
Magnetic models, I.570
Magnetic moments, I.376, II.1115, II.1117, II.1123, II.1166–7
  in Au$_2$Mn, I.462
Magnetic monopoles, II.766
Magnetic phase transitions, II.1014–18
Magnetic pinch, III.1636
Magnetic properties, II.1112
  and crystal structures, II.1156–8
Magnetic recording, II.1158–9
Magnetic recording, II.1158
Magnetic resonance, I.541–2
Magnetic resonance imaging (MRI), II.1176, III.1873, III.1910, III.1923-33
Magnetic resonance spectroscopy (MRS), III.1930, III.1931
Magnetic saturation, II.1145
Magnetic spectrometers, II.1206–8
Magnetic structures, I.462, I.463, II.991
Magnetic susceptibility, I.545, I.557, I.558, III.1595
Magnetism, I.65, II.1111–81, III.1832
  conductors, II.1133–6
  developments—prior to twentieth century, II.1111–13
  developments—1900–25, II.1113–23
  developments—1925–50, II.1123–5
  developments—1950 onwards, II.1136–7
  molecular field theory, II.1014
  quantum concepts, II.1115–16
Magnetization, I.555, I.559
  intensity of, II.1111
  spontaneous, I.573, II.1118
Magnetoencephalography, III.1932
Magneto-hydrodynamic (MHD) theory, III.1648, III.1650, III.1658, III.1663, III.1670, III.1672, III.1673, III.1684, III.1685
Magneto-ionic theory, III.1990, III.1992
Magnetometer, III.1970
Magneto-optical effect, III.1389
Magnetopause, III.2011
Magnetoresistance, III.1356–7
Magnetosphere, III.1752, III.1981, III.1983, III.1984, III.2001–11

Magnons, II.1162
Manchester Literary and Philosophical Society, I.3
Mandelstam–Brillouin doublet, III.1408
Mandelstam–Brillouin scattering, III.1397
Mandelstam–Brillouin spectral shifts, III.1398
Manganese fluoride, I.559
Manganese oxide (MnO), powder pattern, I.460
Manhattan Project, III.1531, III.1556, III.1630, III.2023, III.2025
Markovian kinetic equation, I.620
Masers, III.1417–18, III.1422
Mass, II.1236–8
Mass absorption anomaly, I.398
Mass–energy absorption coefficients, III.1882
Mass–energy equivalence relation, I.285
Mass spectrographs, I.111
Massachusetts Institute of Technology, I.16
Massey criterion, II.1077
Master equation, I.593, I.605–7, I.618
Master function, I.606
Master-oscillator power amplifiers (MOPAs), III.1474
Materials
  characterization, III.1506
  high-quality, III.1280–1
  physics of, III.1505–64
  science, III.2046–7
  technology, III.2022
  terminology, III.1553
  see also under specific materials
Materials Research Laboratories (MRLs), III.1557
Materials Research Society (MRS), III.1556–7

Materials Science and Technology, III.1513
Mathematical Theory of Relativity, III.1700
Mathematical theory of speculation, I.586
Mathematical tools, III.2030
Mathematics Tripos, I.10, I.13
Matrix mechanics, I.360
Matter, I.44, I.47
  condensed, II.756
  constituents of, I.357–8
  new forms of, II.650–62
  thermodynamic properties, I.542
Max Planck Institute, III.1555
Maxwell–Boltzmann distribution, I.589–91, I.598
Maxwell–Boltzmann law, III.1285
Maxwell–Boltzmann statistics, I.527
Maxwell equations, I.33, I.260–1, I.264, I.265, I.270, I.279, I.283, III.1394, III.1622, III.1625, III.1671, III.2018
Maxwellian distribution of protons, III.1707
Maxwellian electromagnetism, I.29
Maxwell's Demon, I.576
Maxwell's electromagnetic theory of light, I.26
Maxwell's theory, I.29, I.321
Mean field approximations, I.554
Mean free path, I.522, III.1549
Mean square radius of gyration, I.569
Mean thermal energy, I.585
Measurement
  applications, II.1240–2
  capacities, III.2025
Mechanical oscillograph, I.26
Mechanical principle of equivalence, I.287–8
Mechanocaloric effect, II.922
Medical electronics, III.1893–6
Medical imaging, III.1901, III.1912

new techniques, III.1932–3
Medical mechanics, III.1896–900
Medical physics, III.1855–941, III.2022
  branches of, III.1855–9
  chronological evolution, III.1855–9
  computers in, III.1858
  modern applications, III.1857
Medical Research Council (MRC), III.1871, III.1877, III.1891, III.1924
Medical ultrasound, III.1910–15
  clinical perspective, III.1915
Medicine
  lasers in, III.1480
  nuclear, III.1858, III.1892, III.1900–10
Meetings, III.2028
Meissner (or Meissner–Ochsenfeld) effect, II.919–20, II.924, II.941, II.957
Mekometer, II.1240
Men and Decisions, III.1640
Mesomorphic phases, III.1603
Meson Factory, II.1205
Mesons, I.378–83, I.389–94, I.402–4, II.640, II.643, II.661, II.686, II.703, II.721
  B-meson, II.721, II.724–5, II.767–8
Mesophere, III.1982
Mesoscopic systems, III.1370
Mesotrons, I.389, I.404–5
  cosmic-ray, I.394–9
  decay, I.399–402
  decay versus capture, I.405–7
  lifetime, I.397–8, I.406
  nuclear interaction, I.407
Metaboric acid, structure of, I.449
Metal–dielectric interference filters, III.1407
Metallography, III.1468, III.1507, III.1554
Metallurgy, III.1505, III.1554

  pioneers in, **III.1511**
Metals
  electrical conduction in, III.1284–7
  electron theory, I.215–17
  groups I–III, III.1354
  individual properties, III.1345–59
  liquid, III.1366–7
  phonon dispersion curves, II.999
  phonons in, II.998–9
  physics of, III.1518
  post-war research, III.1341–5
  resistivity variation with temperature, III.1281–2
Metals Crystallographic Data File, I.512
Metals Handbook, III.1513
Metals Reference Book, III.1513
Metalworking, III.1505
Metastable effect, II.946
Metastable equilibrium, I.530
Metastable states, II.918
Meteorites, III.1965
Methoden der Mathematischen Physik, III.2043
Methylene di-phosphonate (MDP), III.1920
Metre (unit), II.1235, II.1239
Metre Convention, II.1236
Metric system, II.1233, II.1235
Metrically transitive systems, I.599
Metrology, II.1233, II.1273
  role of atoms, II.1094–7
Metropolitan Vickers Company, III.1871, III.1873
MFTF (Mirror Fusion Test Facility), III.1664, III.1668, III.1669
Michelson interferometer, III.1704
Michelson–Morley experiment, I.32, I.262, I.264, I.266, III.1401
Michelson's interferometer, III.1399

Micro-canonical ensemble, I.532
Microphysics, I.210–33
Microscopes, I.26, III.1475
  see also under specific types of microscope
Microscopic physics, I.358
Microstructology—Behavior and Microstructure of Materials, III.1550
Microstructure, III.1542–3
  quantitative measurement of, III.1549
  study of, III.1506–7
  see also under specific materials
Microtechnology, III.1899
Microwave electron linear accelerators, III.1873
Microwave radiation, II.1053–4
Microwaves
  applications, II.1273
  frequency standards, II.1055
  parametric amplifier for, II.1158
Mid-Atlantic Ridge, III.1973, III.1976
Middlesex Hospital, III.1877
Milky Way, III.1714, III.1716, III.1718
Miller indices, I.421
Millions of electronvolts (MeV), I.358
Mineralogy, III.1514
Minkowski space, I.292
Minor impurity, III.1505
Mirror Fusion Test Facility, see MFTF
Mirror machine, III.1648, III.1662
Mirrors, properties of, III.1389
MIT Radiation Laboratory, III.2026
Mixed state, II.920
MKSA system, II.1235, II.1257
Mobility edge, III.1368
Models, I.603–5, I.618–19, I.628–9
Mohorovičić discontinuity ('Moho'), III.1954
Moiré grating interference, III.1414

Molecular beam analysis, II.1033
Molecular biology, I.496, I.497, III.2020, III.2030, III.2046
Molecular bonds, II.1042–9
Molecular collisions, I.588, II.827
Molecular dissociation, II.1067
Molecular dynamics, I.577
Molecular field hypothesis, I.545
Molecular fluids, I.557
Molecular gases, I.35
Molecular hydrogen, I.181, II.1045, **II.1048**
  in interstellar medium, III.1766
  photoabsorption by, II.1069–73
Molecular line astronomy, III.1762–5
Molecular physics, I.358–62, II.1025–49, II.1081–9
Molecular resonances, II.1084
Molecular sheets, III.1595
Molecular spectra, II.1046–7
Molecular spectroscopy, I.529, II.1085–7
Molecular structure, I.211–15
Molecular systems, perturbations in, III.1593
Molecular theory, I.43
Molecular viscosity, II.826
Molecules, I.45, I.47–9, I.81–2, I.85, I.587, I.588
Molybdenum-99, III.1907
Momentum space, I.567
Momentum transport, II.853
Monomers, III.1596
Monotone curve, III.1837
Monsoon phenomena, II.907–8
Monte Carlo methods, I.572, I.628, III.1560
Moon
  formation of, III.1967
  gravity fields, II.888
  orbit, III.1948
  origin of, III.1952, III.1965, III.1966

Morgan, Keenan and Kellman (MKK) system, III.1698
MOSFET, III.1371, III.1372, III.1375
Moss–Burstein shift, III.1421
Mott transition, III.1365, III.1368
Mount Palomar, III.1745, III.1758
Mount Vernon Hospital, III.1867
Moving-coil experiment, II.1238
MSS (paper), III.1839, III.1843, III.1849, III.1850
MULTAN (multiple tangent formula method), I.450
Multi-channel formulation, II.1067–73
Multi-electron fragmentation, II.1065–7
Multi-electron spectra, II.1035–7
Multi-particle system, III.1394
Multiphase interstellar medium, III.1765–6
Multi-photon processes, II.1099–100
Multiple-beam interferometry, III.1412
Multiple collisions, I.611
Multiple-laser-beam systems, III.1479
Multiple sclerosis (MS), III.1928, III.1932
Multiplex optical spectroscopy, III.1416
Multiply connected graphs, I.552
Multi-TeV hadron colliders, II.767
Muon $g - 2$ factor, II.647–9
Muons, II.640, II.650–4
Myoglobin, I.445
  structure, I.500–1

Nambu–Goldstone boson, II.694, II.695
Nambu–Goldstone mode, II.694
Nambu–Jona-Lasinio model, II.756
$NaNbO_3$, dielectric and crystallographic properties, I.487

Naphthalene, crystal structure, I.494
National Institute of Standards and Technology (NIST), II.1235
National laboratories, role of, II.1254–5
National Measurement Accreditation Service (NAMAS), II.1272
National Physical Laboratory (NPL), I.20, II.1234, II.1243, III.1517, III.1867, III.1881, III.1890, III.1988
National Radium Commission, III.1877
National Science Foundation, III.1980
National Standards Laboratories, III.1867
Natural philosophy, I.13
Natural rubber, III.1593
Natural Sciences Tripos, I.10–14
Nature, I.5, I.6, I.96, I.116, I.129
Naval Research Laboratory (NRL), III.1739, III.1988, III.1992
Navier–Stokes equations, III.1834, III.1835
Near-infrared radiation, II.1053–4
Near-space environment, III.1981
Nebular hypothesis, III.1944–5, III.1950
Nebulium, I.75, I.81
Néel temperature, II.1129, II.1151, II.1152, II.1159
Negative dispersion, III.1399–401
Negative-resistance device, III.1340
Negative temperatures, I.541–2
Neodymium, III.1438
  paramagnetic susceptibility, II.1162
Neodymium–iron borates, II.1165
Neon, I.75, III.1747
Neptunium, II.1223

Nernst–Lindemann formula, II.972
NeuroECAT III, III.1921
Neurological disorders, III.1899
Neutral currents, II.695–7, II.699
Neutral hydrogen, III.1762–5
Neutral kaons, II.658–9
Neutral weak currents, II.695
Neutrino masses, β-decay limits on, II.755
Neutrinoless double-β decay, II.755
Neutrinos, I.129, I.370, I.371, II.639, II.640, II.663–4, II.669–71, II.758–60, III.1748–9
  helicity, **II.670**
  oscillations, II.763
  supernova, II.758
  see also Solar neutrinos
Neutron beams, I.468
Neutron bombardment, I.122
Neutron capture, II.1193, II.1223, II.1227, III.1785
Neutron decay, III.1785
Neutron diffraction, I.452–63, I.487, III.2025
Neutron flux, II.994–6
Neutron guide, I.458
Neutron irradiation, I.130, III.1871, III.1872
Neutron–neutron forces, II.1190
Neutron–proton forces, II.1190
Neutron–proton model, I.371–3, II.1190
Neutron–proton ratio, III.1792
Neutron scattering, I.453, II.991–6, II.1011, II.1166, III.2025
Neutron sources, II.1199, II.1206
Neutron stars, III.1712–13, III.1751–4, III.1756
Neutrons, I.112, I.118–26, I.369–71, I.379, II.639, II.991
  lifetime and decay properties, II.754–5
New General Catalogue of Nebulae and Clusters of Stars, III.1717

Newton Victor Maximar, III.1864
Newtonian constant of gravitation, I.28
Newtonian mechanics, I.29, I.36, I.257, I.261
Newton's equations, III.1622
Newton's rings, III.1386
NGC2264 star cluster, III.1767
$NH_4H_2PO_4$, antiferroelectricity, I.488
Nicholson–Bohr quantum rule, I.175
Nickel–titanium alloys, III.1539
Nobel Prizes, III.2021
Noble metals, III.1343
No-hair theorem for black holes, III.1778
Noise problems in upper-atmospheric physics research, III.1995
Noise radiation, II.872–3
Non-abelian gauge theory, II.693
Non-classical light, III.1388
Non-crystalline solids, I.471–84
Non-dimensional parameters, II.812–16
Non-equilibrium theory, I.630
Non-imaging light concentrators, III.1446
Non-linear differential equations, III.1826
Non-linear effects, II.808, II.846–69
Non-linear equations, III.1826
Non-linear field equations, II.800
Non-linear field theories, II.798
Non-linear optics, II.1100, III.1424–53, III.1458–61
Non-linear oscillations, III.1837–8
Non-linear relations, I.233
Non-linear science, III.2030
Non-linear systems, III.1835, III.2030
Non-linear theory, II.810
Non-linearity, II.856–60, III.1403
Non-metallic materials, III.1528–31

Non-perturbative calculations, III.1825
Non-stationary state, II.1029
Nordström's theory, I.295
Nova laser, III.1464, III.1676
Nuclear atom, discovery of, III.2018
Nuclear binding, I.111, II.1208, II.1210
Nuclear bombs, III.1886
Nuclear demagnetization, II.1175
Nuclear dynamics
  as many-body problem, II.1191–5
  background, II.1183–91
  developments in, II.1183–232
  effects of World War II, II.1196–9
  technical advances, II.1199–208
Nuclear energy, I.107, II.1191, III.1879
Nuclear fission, I.126, I.129–33, II.1195, II.1197
Nuclear forces, I.123, I.378–83, I.402–4, II.1208
  fundamental theories, I.375–83
Nuclear fusion, see Fusion research
Nuclear magnetic moments, I.116–17, II.1171–2
Nuclear magnetic resonance (NMR), II.1094, II.1171–5, III.1859, III.2026
Nuclear magnetic resonance (NMR) imaging, III.1898, III.1923–33
Nuclear magnetism, II.1171–5
Nuclear masses, II.1186–91
Nuclear matter, density of, II.1189, II.1190
Nuclear medicine, III.1858, III.1892, III.1900–10
Nuclear physics, I.115, I.119, II.754–62, III.1871
  defence laboratories, II.1196
  research laboratories, II.1198

Nuclear reactions, I.123–6, II.1219–22, III.1629
  direct reaction mechanism, II.1220–1
  nucleus theory of, II.1192
Nuclear reactors, II.1197, III.1871
Nuclear scattering, II.1012, II.1219–22
Nuclear size, I.116
Nuclear spin, I.117, I.541, II.1171, III.1410–11
Nuclear statistics, I.117–18
Nuclear systematics, I.366–7
Nuclear volume, II.1189, II.1208
Nuclear weapons, III.1980
Nucleation process, III.1547
Nucleic acids, I.503–4, I.507
Nucleon isobars, II.659
Nucleon–nucleon force, II.1212
Nucleons, II.639, II.658, II.660, II.675, II.1218
Nucleosynthesis, II.758–60, III.1746–7, III.1792
Nucleotide chains, I.506
Nucleus, I.103–33, I.370
  collective motion in, II.1213–19
  compound, II.1219
  Heisenberg's n–p model, I.371–3
  low-energy collective modes, II.1214–18
  models, I.109–15
  shell structure in, II.1208–13
Numerical calculations, III.1828
Numerical solution, III.1826

Oak Ridge DCX experiments, III.1661–2
Ocean current, II.862
  patterns, II.892, II.896
Ocean engineering, II.869–87
Ocean ridges, III.1978, III.1981
Ocean tides, II.888
Ocean waves, II.857, II.860
Offshore structures, II.886–7
OGRA mirror machine, III.1662

Ohm
  absolute determination, II.1255
  international, II.1254, II.1257
  monitoring, II.1258
  reference standards, II.1256
Ohm's law, III.1336
Oligonucleotide structures, I.507
On the Principles of Geometry, III.1722
Onsager–Feynman quantization condition, II.944–5
Onsager regression hypothesis, I.616
Onsager relations, I.600–2, I.615–16, I.626–7
Ophthalmology, III.1930
Optical actions, atomic systems, II.1097–101
Optical astronomy, III.1744–5
Optical bistability, III.1449–50
Optical coatings, III.1407
Optical coherence theory, III.1392, III.1407–10, III.1461
Optical comparators, II.1240
Optical computing, III.1461
Optical fibre
  communication, III.1447–9
  sensors, III.1471–2
  telecommunications, III.1471–2, III.1480
  see also Fibre optics
Optical filtering, III.1418
Optical glasses, III.1541
Optical goniometer, I.421
Optical imaging, III.1432–7
Optical information processing, III.1417
Optical instruments and techniques, I.26, III.1406–24
Optical interferometry techniques, III.1411, III.1422
Optical Kerr effect, III.1444–5
Optical maser, III.1443
Optical micrography, III.1508

Optical microscopes, III.1405, III.1422, III.1520, III.1526
Optical parametric oscillator (OPO), III.1435
Optical phonons, III.1338
Optical physics, III.1385–504
Optical properties, crystals, I.485–6
Optical pumping, II.1097–9, III.1415
Optical research, World War II, III.1410–14
Optical sciences, computers in, III.1475–6
Optical signal and data processing, III.1461
Optical Society of America, III.1445, III.1474
Optical spectroscopy, III.1421–2
Optical stethoscopy, III.1406
Optical systems
  polarization in, III.1411–12
  signal-to-noise ratio in, III.1418
Optical theory, Abbe's, III.1578–84
Optical transistor, III.1450
Optical video discs, III.1462
Optics
  classical, III.1385–6
  in special theory of relativity, I.257
  linear, III.1434
  new results and applications (1970–94), III.1452–72
  non-linear, III.1427–32
  towards the twenty-first century, III.1472–81
Optimal filtering, III.1418
Optoelectronic communications, III.2051
Optoelectronic devices, III.1472, III.1476
Optoelectronic industries, III.1476
Optoelectronic integrated circuits (OEICs), III.1449, III.1476
Optoelectronic neural networks, III.1461

Optoelectronic physics, III.1385–504
Optoelectronics
  early research, III.1419
  semiconductors, III.1417
Orbital magnetic moments, II.1012
Orbiting Astrophysical Observatories (OAO), III.1741
Orbiting Solar Observatory (OSO-III), III.1740
Order–disorder transition, I.459, I.473–4, I.578, II.1009, II.1016, III.1515
  lambda transition, II.915, II.922–4, II.927, II.944
Ordinary differential equations (ODEs), III.1826–7, III.1829, III.1835- 9, III.1851
Organic compounds, structure analysis, I.493–6
Orion Nebula, III.1767
Orthogonalized-plane-wave (OPW) method, III.1317
Orthopaedic surgery, III.1896–7
Orthotics, III.1899
Oscillation, II.896, II.907–8, II.973
Oscillator strengths, II.1058
Otolaryngology, III.1930
Out of the Crystal Maze, I.512
Oxalic acid, I.452, I.454
Oxford University, I.20
Oxygen, formation of, III.1747

Pacemakers, III.1895, III.1897
Pacific floor, measurements, III.1974
Padé Approximant, I.558
Pair connectedness, I.573
Pair distribution function, I.547
Palatini variational method, I.299
Palomar Atlas, III.1745
Palomar telescope, III.1721, III.1775, III.1793
Pangaea (giant land-mass), III.1968

Paraelastic crystals, III.1538
Parallel plate ionization chamber, I.54
Parallel shear flow, II.828
Paramagnetic insulators, II.1116–20
Paramagnetic materials, I.459
Paramagnetic salts, I.540–1
Paramagnetism, II.1015, II.1113, II.1119, II.1125–33, III.1354, III.1362
Parametric fluorescence, III.1448–9
Parametrization technique, III.1351
Parastatistics, II.687
Paris Symposium on Radio Astronomy, III.1789
Parity, II.654, **II.1028–9**
  conservation, II.665
  violation, II.639, II.698, II.699
Partial coherence, III.1392
Partial conservation of axial current (PCAC), II.676
Partial differential equations (PDEs), III.1827, III.1834, III.1851
Particle accelerators, *see* Accelerators
Particle content, II.763
Particle energies, II.1201
Particle physics, III.1833, III.2038
  see also Elementary particles; and under specific particle types
Particle size distribution, III.1549
Partition function, I.527, I.532
Parton hypothesis, II.689–90, II.715
Paschen–Wien law, I.149
Patterson functions, I.441–5, I.495, I.501
Pauli exclusion principle, I.97–100, I.168, I.603, II.686, II.1116
Pauli spin-statistics theorem, I.283
Peltier heat, III.1284
Penicillin, structure of, I.496
Penning traps, II.1099

Penrose tiling, I.483
Percolation processes, I.572–4
Perfect cosmological principle, III.1787
Period–luminosity relation, III.1714–15
Periodic table, I.95–7, I.359, II.635
Permanent magnets, II.1165, II.1238
Permittivity of free space, II.1261
Perovskites, I.486–7, I.489, II.1156
Perturbation
  in molecular systems, III.1593
  singular theory, II.800–3, II.829
  strong responses to, III.1593–694
  theory, I.360, III.1830, III.1833
PET (positron emission tomography), III.1904, III.1910, III.1920, III.1922, III.1931
Phase angles, I.439, I.442, I.450, I.502
Phase conjuguation, III.1460
Phase-contrast microscopy, III.1409–10
Phase diagrams, III.1506, III.1508, III.1509, III.1542
  free-energy, I.538
Phase equilibria, III.1506
Phase matching, III.1434
Phase Rule, III.1508, III.1542
Phase transformations, I.542–74, I.577, II.1007–10, III.1542–50
  microstructural and microcompositional analysis in, III.1548–50
Phaseonium, III.1473
Philosophical Magazine, I.7
Philosophical Transactions of the Royal Society, I.7, III.1516–17
PHOBOS space probe, III.1750
Phonon dispersion curves, II.1000, II.1002
Phonon drag, III.1310, III.1311
Phonon exchange, II.953

Phonon interaction with hot electrons, III.1339
Phonons, II.981, II.982, II.990, II.998–9, II.1004, III.1302, III.1338
  in crystals containing defects and in glasses, II.1004–7
Phosphorescent materials (phosphors), III.1403, III.1530, III.1536, III.1537
Phosphors, *see* Phosphorescent materials (phosphors)
Photoabsorption
  by molecular hydrogen, II.1069–73
  spectra, II.1052, II.1053, II.1054, II.1056
Photo-assisted tunnelling, III.1423
Photocathode, III.1439
Photo-cells, III.1321
Photochromic spectacle lenses, III.1542
Photochron streak camera, III.1439
Photoconductivity, III.1321, III.1327–9
Photocopying, III.1541
Photodetection process, III.1441
Photodetectors, III.1433
Photodiode, III.1439
Photoelectric effect, I.37, III.1286
Photoelectronic imaging devices, III.1476
Photo-emission spectroscopy, III.1391
Photographic plates, II.1200
Photography, III.1389, III.1392, III.1410
Photoionization, II.1052, III.1762
Photometry, II.1251–2
Photomultiplier tubes, II.1207, III.1409, III.1411, III.1920
Photon antibunching, III.1459–62
Photon bunching, III.1468
Photon-correlation spectroscopy (PCS), III.1454–8

protein analyser, III.1471
Photon counting statistics, III.1424
Photon energy, II.1052
   conservation process, III.1435
Photonic bandgaps, III.1472
Photonic integrated circuit (PIC), III.1476
Photons, I.117, I.386, II.637, II.638, II.641, II.693, II.1028, III.1302, III.1393, III.1402, III.1403, III.1410, III.1413, III.1451–2, III.1994
Photophone experiments, III.1422
Photoprocesses, II.1102
Photorefractivity, III.1461
Photosensitive devices, III.1321
Photo-voltaic effect, III.1321
Physical constants, II.1233, II.1260–72
   linked sets, II.1267–72
Physical metallurgy
   birth of, III.1516–19
   quantitative theory, III.1519–28
Physical Review, The, I.6
Physical Society, III.2042
Physics
   as domain of knowledge, III.2041
   as social institution, III.2041, III.2056
   definition, III.2041
   methodology of, III.2049
   scope of, III.2041
   sociology of, III.2031
Physics and Chemistry of Solids, III.1556
Physics of Fully Ionized Gases, III.1631
Physikalische Gesellschaft, I.88
Physikalische Metallkunde, III.1555
Physikalische-Technische Reichsanstalt (PTR), I.7, I.20, I.69, I.107, II.916, II.1234, II.1254
Physikalische Zeitschrift, I.6, I.7

Piezoelectric properties, III.1538
Piezophotonic effects, III.1459
π-meson, see Pions
Pioneer III, III.2007–8
Pion–nucleon scattering, II.659
Pion–pion scattering, II.681
Pions, I.408, II.650–5, II.658, II.665, II.675–6, II.711
Pipes, flow in, II.825
Pippard coherence length, II.947
Pitchblende, III.1876
Planck constant, I.67, I.152, I.359, I.360, II.638, II.1026, II.1271, III.1387
Planck energy, II.972
Planck formula, II.970, II.971
Planck law, I.70, I.151, I.363, I.527, III.1399
Planetary dynamics, III.1830–1
Planetary systems, III.1950
Planets
   formation, III.1948–9, III.1963
   infinitesimal, III.1950
Plasma, III.1361
   diagnostic measurements, III.1644
   frequency, III.1623
   instability, III.1630
   oscillations, III.1362, III.1623–5
   radio wave propagation in, III.1987
   temperature, III.1617
   theory, III.1623–5
Plasma physics, III.1617–90, III.1991–2, III.1994
   developments—circa 1950 to the present, III.1627–45
   developments—first half of twentieth century, III.1619–25
   history of, III.1617, III.1618
   landmark experiments, III.1677–81
   research, III.1617, III.1618
   space, III.1682–7

S41

Plasma Physics and the Problem of Controlled Thermonuclear Reactions, III.1642
Plasmapause, III.2009, III.2010
Plasmasphere, III.1982
Plastic deformation, III.1520, III.1526, III.1530
Plate tectonics, III.1956, III.1961, III.1976, III.1978–80, III.2022
Platinum–iridium alloy, II.1237, II.1239
Pleated sheet structures, I.500
Plunging breakers, II.862
Plutonium, II.1223–5
  discovery, II.1197
  weapons-grade, III.1633
p–n junction, III.1325, III.1417
Poincaré–Bendixson theorem, III.1835
Poincaré map, III.1836, III.1851
Poincaré recurrence theorem, I.618
Poincaré–Zermelo recurrence conflict, I.592
Point defects, III.1528–31, III.1533, III.1534
Poisson equation, III.1625
Poisson formula, III.1424
Poissonian statistics, III.1393
Polar Years, III.2005
Polaritons, II.988, II.990
Polarization in optical systems, III.1411–12
Polarization rules, I.94–5
Polarized light microscopy, III.1412
Polarizing microscope, I.421
Polarons, III.1321
Polepiece lens, III.1567
Polonium, I.119, II.1200, II.1206
Polycrystal Book Service, I.512
Polycrystalline metals, III.1518
Polycrystalline samples, III.1280
Polyethylene, III.1602
Polymerization, III.1550, III.1551
Polymerization index, III.1597

Polymers, II.1084, III.1596–603
  blends, III.1553
  chain-folding, III.1552
  colloid protection, III.1611
  configurations, I.569–70
  crystallization, III.1602
  dynamics of entangled systems, III.1600
  fibres, III.1898
  future prospects, III.1603
  gels, III.1602–3
  high-technology, III.1550
  modified, III.1597
  overlapping, III.1599
  physics of, III.1550–3
  reptation process, III.1601
  semicrystalline nature of, III.1551
  solid phases, III.1600
  solutions, III.1599
  statics, III.1599
  structure, I.569–70
Poly-τ-methyl-l-glutamate, I.498
Polymethylmethacrylate (PMMA), III.1602, III.1896
Polyoxyethylene, III.1593
Polypeptide chains, I.496, I.497, I.499–501, I.510
Polysaccharides, III.1611
Polystyrene, III.1602
Polytetrafluoroethylene (Teflon), III.1896
Pomeranchuk trajectory (Pomeron), II.682
Pomeron (Pomeranchuk trajectory), II.682
Positive electron, I.109
Positively charged light particle, I.231
Positron emission tomography, *see* PET
Positronium, II.645–6, II.649, II.650, II.702
Positrons, I.370, I.371, I.377, II.638
Post Office, III.1988

Potassium dihydrogen phosphate (KDP), I.570–1, III.1432
Potential-energy curves, II.1045
Potential flow, II.803, II.804, II.887
Potential vorticity, II.895
Potts model, I.571
Powder diffraction, I.453
Powder photographs, I.435
Poynting's vector, I.33
Precious metals, III.1281
Precipitation-hardening, see Age-hardening
P representation, III.1440
Primaeval atom, III.1785
Primary image, III.1578
Primordial nucleosynthesis, III.1785, III.1786
Princeton Model C stellarator, III.1656
Principal quantum number, I.84, I.91, I.215
Principle of complementarity, I.197
Principle of Relativity, III.2039
Probabilities in quantum physics, I.93–4
Probability amplitude manipulation, II.1064
Probability relations, I.449
Proceedings of the Royal Society, I.7
Programmable calculators, III.1829
Project Sherwood, III.1631–40
Project Sherwood—The US Program in Controlled Fusion, III.1640
Projection electron microscope, III.1584
Proportional counters, II.1207
Protein crystallography, I.445, I.502
Protein Data Bank, I.512
Proteins, I.496
  fibrous, I.497, I.499
  globular, I.497
Proton–antiproton colliders, II.743–4

Proton decay, II.764–5
Proton–electron (PE) model, I.109–10
Proton linear accelerators, II.1205
Proton–neutron (PN) system, I.123
Proton–proton chain, III.1708
Proton–proton colliders, II.742–3
Proton–proton collisions, I.377
Proton synchrotrons, II.732–3
Protons, I.113, I.115, I.118, I.370, II.639, II.701
  in medical physics, III.1876
  Maxwellian distribution of, III.1707
Prototype phenomena, II.1060–3
Pseudopotential, III.1353, III.1367
PSR 1913+16 binary pulsar, III.1757
Publications, III.2028
Pulsars, III.1752
Pulse-echo imaging, III.1911–13
Pulse-echo technique, III.1989
Pump/oxygenator, III.1897
Pump-probe configuration, III.1445
P–V–T relations, I.28
Pyrites, I.430

Q-switching, III.1430, III.1436, III.1438
Quadrature amplitudes, III.1468
Quadrupole moments, I.403, II.1214, II.1215
Qualitative quantum mechanics, I.116
Quanta, I.378, II.638, III.1390
  and radiation, I.146–55
  empirical foundations, I.145–72
  specific heat, I.152–4
Quantitative metallography, III.1549
Quantitative theory, I.44
Quantum chemistry, theory and computations, II.1087–8

Quantum chromodynamics (QCD),
I.322–3, II.636, II.692,
II.709–19, II.756, II.769,
III.2020, III.2031
Quantum communication, III.1472
Quantum conditions, I.359
Quantum cryptography, III.1472
Quantum dot, III.1472
Quantum dynamics, I.72–89
Quantum electrodynamics (QED),
I.358, **I.361**, I.362, I.374, I.378,
I.385, I.386, I.391, II.640–50,
II.646, II.740, II.769, III.1414,
III.1446, III.1472
Quantum energy, III.1361
Quantum field theories (QFTs),
I.206–9, I.318, **I.380–1**, I.617,
II.640, III.2033, III.2037,
III.2038
Quantum formula, I.156
Quantum gravity, I.317–19
Quantum Hall effect, II.1096–7,
II.1257, II.1259, III.1372,
III.1375
Quantum mechanical effects,
I.163–72
Quantum mechanical many-body
problem, II.1212
Quantum mechanical parameters,
II.1063
Quantum mechanical
reformulation, II.1065
Quantum mechanical scattering
theory, I.217–19
Quantum mechanical
transformation theory, I.199
Quantum mechanical tunnelling,
III.1706, III.1707
Quantum mechanical
wavefunction, II.934
Quantum mechanics, I.30, I.61,
I.65, I.72, I.115, I.125,
I.143–248, I.359, I.360,
I.602–3, I.627–8, II.638, II.917,
II.1026, II.1034, II.1058,
III.1280, III.1325, III.1403,
III.1467, III.1830, III.2049
applications, I.211–20, I.229
as reformulated classical
mechanics, I.189–91
aspects of theory beyond,
I.232–3
causality in, I.220–9
complementarity in, I.220–9
completeness and reality of,
I.225–9
in astrophysics, III.1706–8
mathematical foundation,
I.196–208
new structures within 'standard
theory', I.230–2
non-relativistic, I.220
origin and completion of, I.173
physical interpretation, I.196–208
present situation, I.229–30
reality in, I.220–9
Quantum Mechanics, III.2043
Quantum non-demolition (QND)
measurements, III.1469
Quantum numbers, I.84, I.90–100,
I.157, I.162, I.182, I.215,
II.657, II.658, II.1034, II.1036,
II.1124
Quantum of circulation, II.931
Quantum optics, III.1387–406,
III.1468, III.1471, III.1472
Quantum phenomena, I.253
Quantum physics, I.65–72, I.89
probabilities in, I.93–4
Quantum postulate, I.70, I.81
Quantum statistics, I.525–8, I.537
Quantum-theoretical
reformulation and
non-commuting variables,
I.186–9
Quantum theory, I.37, I.54, I.62,
I.65, I.68, I.71, I.80, I.88,
I.89, I.101–3, I.524–31, I.537,
III.1285, III.1387, III.1393,

III.1395, III.1402–4, III.1962, III.2018
and gravitation, I.318
evolution, I.144–5
in special theory of relativity, I.282–3
of electron gas, III.1287–91
of ions, II.1123–5
of measurement, I.223–5, I.233
with classical statistics, I.527
Quarkonium decays, II.716–17
Quarks, II.635–6, II.677, II.684–7, II.699, II.719–25
  b-quarks, II.722–3
  charmed, II.716
  colours, II.687, II.691, II.709, II.711–13
  flavours, II.687
  free, II.691–2
  statistics, II.709–10
Quartz clock, II.1244, II.1247, II.1248
Quasars, III.1776–7, III.1798
  discovery of, III.1789
Quasi-classical distribution function, III.1441
Quasicrystals, I.481–4
Quasiparticles, III.1363, III.1364
Quasi-phase matching, III.1434–5
Quasi-stellar radio sources, III.1776
Quenching, III.1533, III.1542

Radar, III.1944
Radial distribution function, I.480
Radiation, II.1028–32
  and quanta, I.146–55
  see also under specific types of radiation
Radiation accidents, III.1887
Radiation actions, II.1049–59
  spectrum of, II.1055–8
Radiation biology, III.1857
Radiation biophysics, III.1857, III.1884–9

Radiation damage, III.1531, III.1534, III.1884–5
Radiation-dose distributions, III.1872
Radiation dosimetry, III.1880–4, III.1887–9, III.1890
  calorimetry in, III.1883–4
  chemical, III.1884
  ionization methods, III.1880–3
  radionuclides, III.1909
Radiation energy, III.1885
Radiation hazards, III.1886–7, III.1895
Radiation injury, latency of expression, III.1886
Radiation isodose curves, III.1865
Radiation protection, III.1857, III.1871, III.1889–93
Radiation shielding, III.1891
Radiation Trapped in the Earth's Magnetic Field, III.1683
Radiationless transitions, II.1086–7
Radiative instability, I.83
Radio, III.1987
Radio astronomy, III.1736–8, III.1774–6, III.1944
Radio communications, III.1985–6, III.2000
Radio galaxies, III.1795, III.1797
Radio interference, III.2003
Radio propagation, III.1619–23, III.2000, III.2007
Radio pulsars, III.1752, III.1754, III.1757–8
Radio quasars, III.1751, III.1797
Radio research, III.1985
Radio scintillations, III.1751
Radio signals, III.1987, III.1988
Radio sources, III.1776, III.1782, III.1788–90, III.1797–9
Radio telescopes, III.1776
Radio waves, III.1985–8
  attenuation of, III.1995
  propagation, III.1990

Radioactive decay, I.93, I.106, II.639, III.1960
Radioactive isotopes, II.1222
Radioactive Substances Act, III.1910
Radioactive tracers, III.1858
Radioactivity, I.36, I.37, I.49, I.52–3, I.56–63
  artificially produced, III.1900–10
  in medical physics, III.1857
Radiobiology, III.1884–9
Radiochemical Centre, III.1901, III.1907
Radiofrequency energy, II.1057
Radiofrequency spectroscopy, II.1047
Radiography, III.1878
Radiometric dating, III.1969, III.1973
Radionuclides, III.1871, III.1904
  dosimetry, III.1884
  imaging, III.1916
  in medical physics, III.1857
  radiation dosimetry, III.1909
  scanning, III.1898
Radiotherapy, III.1857, III.1858, III.1862–80
  computer-aided dose planning, III.1887
Radium, I.58, I.61, I.104–8, III.1876–9
  bomb, III.1877
  dosimetry, III.1880–1
  implants, III.1878
  needles, III.1877, III.1878
Radon, II.1206, III.1876, III.1879, III.1916
  seeds, III.1877
Raman effect, I.163, I.171–2, II.996, III.1404–6
Raman–Nath diffraction, III.1409
Raman scattering, I.117, II.988, II.990, II.996–7, II.1011, II.1013, III.1397, III.1436
RAND project, III.1741

Random close packing (RCP), I.577
Random loose packing (RLP), I.577
Random processes, I.595–7, I.600
Random walks, I.569–70, I.586, I.624, I.630, III.1597
Rapid solidification processing, III.1543
Rare-earth elements, I.97, II.955
  crystal fields, II.1164–5
  electron paramagnetic resonance (EPR), II.1141–3
Rare-earth ion lasers, III.1438
Rare-earth metals and alloys, I.463, II.1012, II.1168
  magnetic properties, II.1160–5
Rayleigh criterion, III.1467
Rayleigh–Einstein–Jeans (REJ) law, I.69
Rayleigh–Jeans equation, III.1393
Rayleigh–Jeans law, III.1390
Rayleigh–Lorentz models, I.628
Rayleigh number, II.844–5
Rayleigh scattering, II.1057
Rayleigh–Taylor instability, III.1782
Reactive collisions, II.1048–9, II.1088–9
Real space, I.567
  renormalization, I.568, I.569
Realm of the Nebulae, The, III.1720, III.1726
Reciprocal lattice, III.1295, III.1296
Recrystallization, III.1526
Rectifiers, III.1322–7
Red blood cells, I.630
Red giants, III.1704, III.1708–9
Red stars, III.1696
Redshift–distance relation, III.1726
Redshift–magnitude relation, III.1794–7
Redshifts, III.1725, III.1777, III.1798
Reduced radial distribution function, I.481
Reflection high-energy electron diffraction (RHEED), I.466

Refractive index, I.546, III.1449
Regge poles, II.681–2, II.718
Regge trajectories, II.682
Relativistic cosmology, III.1721–34
Relativistic kinematics, I.364
Relativistic statistical mechanics, I.282
Relativistic thermodynamics, I.281–2
Relativistic velocity (kinematic) space, I.278
Relativity, III.2049
  kinematics, I.161
  philosophical treatments and popular reactions, I.319–20
  theory of, I.62, I.249–356, II.1266
  see also General theory of relativity; Special theory of relativity
Relaxation processes, II.1132–3, II.1145–6
Renninger effect, I.465
Renormalizability, II.694, II.695, III.2034, III.2036
Renormalization, I.564–9, II.646–7, II.713, II.756, II.1009, II.1015, III.1829–30, III.1832, III.1834, III.1844, III.1848
Renormalization rule, III.1832
Research
  career prospects, III.2028
  funding, III.2058
  laboratories, III.2050
  linking academy with industry, III.2050–1
  management, III.2049
  needs, I.19–22
  performance, III.2052
  physics, I.16–18
  strategic (or pre-competitive), III.2051
  technologies, III.2044–5
Researching and/or teaching, III.2047–8
Residual resistance, III.1281

Resistivity, **III.1331**, III.1334
  absolute magnitude of, III.1303
  mechanisms of, III.1301–4
  temperature variation of, III.1302, III.1342
Resonance
  effects, II.1059–73
  excitation, I.125
  phenomena, II.659–60, II.1029, II.1060–3, II.1067–73
Resonant absorption of radiation energy, II.1057
Resonant electron excitations, II.1082–4
Reststrahlen bands, II.983
Reversed-field pinch (RFP), III.1656–8
Reviews of Modern Physics, III.1290
Revue Scientifique, I.6
Reynolds number, II.807, II.815, II.816, II.825, II.828, II.834, II.837, II.841, II.844, II.850, III.1852
Reynolds stress, II.826, II.827, II.830, II.837, II.853
R-factor, I.450–1
RF SQUID ring, II.942
Ribonucleic acid (RNA), I.504, I.509
Richardson–Lucy iterative deconvolution algorithm, III.1462
Riemann tensor, I.292, I.312
Riemannian metric, I.286
Riemannian space-time, I.322
River water, II.897–8
RKKT theory, II.1170
RKKY interaction, II.1164
R matrices, II.1068–9, **II.1070**
RNA, *see* Ribonucleic acid (RNA)
Robertson–Walker metric, III.1728–30
Rochelle salt, III.1538
  ferroelectricity, I.489
Rockets, III.1682, III.1738, III.1997, III.1999, III.2005

Rocks
  ages of, III.1960–1
  magnetization, III.1970
  remanent magnetism of,
    III.1970–2
Rod-vision, III.1416
Röntgen (unit), III.1881–2, III.1890
Röntgen Society, III.1861
Rossby waves, II.893, II.894, II.902
Rotating-disc microscopes, III.1451
Rotational motion, II.1216
Rotational spectra, II.1217
Rotational therapy, III.1868
Rotons, II.928
Royal Greenwich Observatory,
  II.1248
Royal Institution of Great Britain,
  I.3, I.27
Royal Society, II.1233
  Fellows of, I.10, I.12
  Philosophical Transactions, I.7,
    III.1516–17
  Proceedings of, I.7
'Rubber Bible', III.1513
Rubbers, III.1551
  vulcanization of, III.1551,
    III.1593
Rubidium, I.75
Running coupling constant, II.713
Russell diagram, III.1696, III.1698,
  III.1710
Russell–Saunders (or L–S)
  coupling, III.1702
Rydberg constant, I.157–8, II.1271
Rydberg energy, II.1026
Rydberg formula, II.1034

Sachs–Wolfe effect, III.1802,
  III.1805
Sacramento Peak Solar
  Observatory, III.1744
Safronov–Wetherill model, III.1963
Safronov–Wetherill theory, III.1966
Sagnac effect, III.1433
Sagnac interferometer, III.1398

Saha equation, III.1699, III.1701
St Bartholomew's Hospital,
  III.1871, III.1901
Salicylic acid, I.443
Salt, *see* Sodium chloride
San Andreas fault, III.1978
Sanduleak-69 202, III.1760
Satellites, III.1683, III.1741, III.1765,
  III.1791, III.1997, III.1999,
  III.2031
  see also under named satellites
Saturation analysis technique,
  III.1909
Scalar quantities,
  diffusion-convection balance
  for, II.841–6
Scale factor, III.1730, III.1731
Scaling, II.688–9
  empirical derivation, I.561–4
  equation of state, I.568
  violation, II.714–15
Scandium, I.75
Scanning electron microscope
  (SEM), III.1451, III.1571,
  III.1586–90
Scanning transmission electron
  microscope (STEM), III.1569,
  III.1587–90
Scattering phenomena, I.217–19
Schizophrenia, III.1922
Schlieren photography, II.863,
  II.880
Schmidt camera, III.1406
Schmidt corrector plate, III.1406
Schmidt telescope, III.1406,
  III.1745, III.1758
Schönberg–Chandrasekhar limit,
  III.1709
Schottky defect, III.1530
Schrödinger equation, I.196, I.205,
  I.209, I.225, I.602, II.798,
  II.862, III.1292, III.1293,
  III.1299, III.1315, III.1317,
  III.1320, III.1366, III.1458
Schrödinger wavefunction, I.174

Subject Index

Schrödinger wave mechanics, I.183–96
Schwarzschild metric, III.1778
Schwarzschild singularity, I.306
Science, I.6
Science Abstracts, I.1, I.7
Scientific American, I.5, I.6
Scintillation counters, II.749–51, III.1901, III.1904, III.1916
Scintillation crystal, III.1907
Scintillation detectors, II.1207–8, III.1901, III.1904
Sco X-1 x-ray source, III.1739, III.1755
Screening and antiscreening, II.1091–2, III.1361, III.1363
Screw dislocation, I.476
Scyllac device, III.1645
SDI (Star Wars enterprise), III.2024
Sea-floor spreading, III.1975, III.1976
Sea transport, II.882–7
Search for Structure, A, III.1556
Second (unit), II.1235, II.1247–8
Second Antarctic Expedition, III.2005
Second-harmonic generation (SHG), III.1429
Second International Congress of Radiology, III.1890
Second International Polar Year (1932–33), III.2005
Second law of thermodynamics, I.46, I.68–9, I.374, I.521, I.524, I.534- 5, I.575
Second-order transitions, I.549–51
Secondary image, III.1578
Seebeck coefficient, III.1310
Seebeck effect, III.1284, III.1309
Seebeck voltage, III.1310
Seismology, III.1954, III.1955, III.1956, III.1980
Selection rules (*Auswahlprinzip*), I.94, I.178–81
Selenium, III.1541

Selenium-75, III.1905, III.1909
Selenium rectifier, III.1318
Self-avoiding walk (SAW), I.569–70, III.1598
Self-consistency, III.1519
Self-interstitials, III.1533
Self-phase modulation, III.1448–9
Self-similarity, I.574
Semi-classical theory, III.1388–9
Semiconductors, II.1002, III.1297, III.1301, III.1323, III.1367–8, III.1419, III.2025
  crystalline, III.1368
  detectors, II.1207
  devices, III.2022
  industry, III.1281
  laser diode, III.1464
  lasers, III.1474
  molten, III.1368
  optoelectronics, III.1417
  particle detectors, II.1208
  physical investigation, III.1318
  physics, III.1330–6
  post-war years, III.1317–22
  stimulated emission in, III.1433
  yellow light emission from, III.1392
Semi-leptonic decays, II.675
Senior Wrangler, I.10–11
Sequential resonant tunnelling, III.1470
Seyfert galaxies, III.1777, III.1780, III.1784
Shallow-water theory, II.890, II.892, II.896
Shannon number, III.1418
Shape-memory alloys, III.1539
Shell model, II.997, II.1214
Shell structure, II.1027, II.1208–13
Ship design, II.884
Ship hulls, II.884, II.885
Ship model testing, II.886
Ship–wave pattern, II.884
Ships, hydrodynamic forces on, II.882

S49

Shock waves, II.798, II.808–12, II.847–9, II.872–82
Shoenberg magnetic interaction, III.1359
Short-range interactions, II.1063–5
Short-range parameters, II.1073, **II.1074–5**
Short Wave Wireless Communication, III.1619
Shortt clock, II.1243
Shubnikov groups, I.462
Sidereal time, II.1247
Siegbahn double-focusing beta spectrometer, II.1206
Siemens-Schuckertwerke, III.1573, III.1574
Signal-to-noise ratio in optical systems, III.1418
Silicon, I.470, III.1281, III.1301, III.1526
Silicon on sapphire (SOS) interface, III.1590
Silver, solid solutions, III.1543
Silver chloride, III.1526
Silver halides, III.1529
Single crystals, I.435–7, I.506, III.1280–1, III.1318, III.1360, III.1520
Single dish antenna, Arecibo, Puerto Rico, III.1996
Single photon emission computed tomography, *see* SPECT
Singular perturbation theory, II.800–3, II.829
Singularities, I.306–12
Site percolation, I.572
Sixteen lectures on controlled thermonuclear reactions, III.1660
Size effect, III.1343
Skin effect, II.1157
Skin-temperature distribution, III.1932
SLAC-LBL Collaboration, II.701
Slender-body aerodynamics, II.877

Sliding charge density-wave conduction, II.936
Sloan-Kettering Memorial Hospital, III.1871
Small Astronomical Satellite (SAS-2), III.1740
Small Magellanic Cloud, III.1715
Small Science, III.2025–31
S-matrix theory, II.680, III.2038
Smectic phases, III.1604
Smith chart, II.680
Snoek pendulum, III.1523
Society of Nuclear Medicine, III.1907
Society of Photo-Optical Instrumentation Engineers, III.1474
Sociological phenomena, III.2029
Sociology of physics, III.2031
Sodium, I.65, II.999, III.1530
Sodium chloride, I.429, I.430, III.1514
Sodium iodide
  crystals, II.1207, III.1903
  scintillation counters, II.1210
Sodium pertechnetate, III.1919
Sodium potassium tartrate, *see* Rochelle salt
Soft matter, III.1593–615
  use of term, III.1593–696
Soft-mode theory, II.1007–8
Solar cell, III.1476
Solar energy, III.1446, III.2005
Solar flares, III.1997
Solar gas, III.2009
Solar neutrinos, II.757, II.1227, III.1748–9
Solar oscillations, III.1749
Solar particles, III.2009
Solar radiation, II.899, II.900
Solar spectrum, spectroscopic measurements of, III.1691–2
Solar stream, III.2002, III.2003
Solar system, III.1824, III.1831, III.1950

dualistic theory, III.1962, III.1963
isotopic anomalies, III.1963-4
monistic theory, III.1962, III.1963
origin of, III.1965
tidal theory, III.1950, III.1951, III.1952
Solar–terrestrial physics (STP), III.1982
Solar–Terrestrial Physics, III.2001
Solar–terrestrial space, III.2001-11
Solar wind, III.1997, III.2009
Solar zenith angle, III.1985
Solenoids, iron-free, III.1568
Solid–solid phase transformations, III.1542, III.1544
Solid solutions, III.1543
Solid-state physics, II.1165-6, II.1169, III.1279, III.1526, III.1529, III.2019, III.2025
  emergence of schools, III.1313-14
Solid-state structure analysis, I.421-519
Solidification, III.1542
Solids
  electrons in, III.1279-383
  properties of, III.1279
Soliton pulses, III.1469
Sorbonne, I.15
Sound from flows, II.850-6
Sound intensity, II.851
Sound waves, II.798, II.846, II.857, II.873, II.950
  interaction of, **II.978-9**
Southern oscillation, II.907-8
Space exploration, III.1479
Space groups, III.1513
Space physics, III.1618, III.1982
Space plasma physics, III.1682-7
Space-probe studies, III.1995
Space research, III.1943, III.1965, III.2007-9, III.2024
Space Shuttle, III.1742
Space technology, III.2022
Space-time, I.251, III.1730

Space Transportation System (STS), III.1742
Spark chambers, II.753-4
Spatial-light modulator technology, III.1461
Special Principle of Relativity, III.2034
Special theory of relativity, I.54, I.249, III.2039
  alternate formulations and formalisms, I.273-8
  causal formulations, I.277-8
  constraints, I.249-51
  continuum mechanics, I.279-80
  core principles, I.249
  electrodynamics in, I.257, I.280-1
  elementary particles in, I.282-3
  experimental tests and applications, I.284-5
  formulation, I.262-73
  formulation without light postulate, I.277
  four-dimensional formulation, I.274-6
  gravitation theories in, I.283-4
  later development, I.273
  origins of, I.254-62
  particle dynamics, I.279
  quantum theory in, I.282-3
  relativistic statistical mechanics, I.282
  relativistic thermodynamics, I.281-2
  relativistic velocity (kinematic) space, I.278
  rigid motions, I.279-80
  space-time structure, I.251
  spinor formalism, I.277
  twistor formalism, I.277
Specific gravity, I.46, II.967
Specific heat, I.550, I.555, I.558-60, II.967-73
  lattice theories, II.982-91
  of gases, I.528-30
  quanta, I.152-4

Speckle interferometry, III.1454
Speckle pattern, III.1441
Speckle phenomena, III.1453
SPECT (single photon emission computed tomography), III.1910, III.1916, III.1917, III.1918, III.1920, III.1931
Spectral analysis, I.156
Spectral connection at ionization thresholds, II.1058–9
Spectral density, I.66, I.600
Spectral fingerprints, II.1092–4
Spectral frequencies, I.77
Spectral function, I.69
Spectral lines, I.156–62
Spectral numerology, I.77
Spectrometer, II.992, II.994
Spectroscopy, I.24, I.74–8, II.1039
  equipment, I.24
  interferometric, III.1410
  parallaxes, III.1698
Speculation, mathematical theory of, I.586
Speed of light, I.260, I.261, I.266, II.1240, II.1262–4
Sphalerite, I.429
Spheromak, III.1670
Spilling breakers, II.862
Spin, I.100–1, **I.395**, I.402, II.654
Spin-blocking, III.1831
Spin-fluctuation-exchange, II.953
Spin-Hamiltonian, II.1139–41, II.1151
Spin-lattice relaxation time, III.1923
Spin model, I.564
Spin–orbit coupling, II.1012, II.1126, II.1127, II.1136, II.1210
Spin–orbit interaction, II.1127, II.1212
Spin-warp technique, III.1924, III.1926
Spin waves, II.1010–14, II.1150–2

Spinels, crystallographic data, II.1156
Spinodal, III.1547–8
Spiral nebulae, III.1716–21
Sponge model, II.921
Sponge phases, III.1608
Spontaneous emission, I.93–4
Spontaneous transitions, III.1398–9
SPRITE detector, III.1477
Sputnik I, III.1738, III.2007
Sputnik II, III.1738, III.2007, III.2008
Sputnik III, III.2008
Squeezed light, III.1468–71
Squeezing parameter, III.1468
SQUIDs, III.1932
SSC debate, III.2020, III.2024, III.2031
Stability/instability, II.828–9
Stable fixed points, I.567
Stacking faults, I.473
Standard Model of elementary particles, III.2030, III.2033–9
Standards, II.1233, II.1234, II.1236–60
Stanford Linear Accelerator Center (SLAC), II.687–8, II.741
Star clusters, III.1696
Stark effect, I.89, I.161, III.1469
Stars, III.1691–713
  early-type, III.1698
  formation, III.1766–9, III.1945
  interstellar medium, III.1761–9
  island universes, III.1714, III.1717, III.1718
  late-type, III.1698
  luminosities, III.1698
  main-sequence, III.1698, III.1705
  mass–luminosity relation, III.1705
  photographic parallax programme, III.1696
  research since 1945, III.1746–60
  see also Stellar
Stationary states, I.83, I.86
Stationary waves, II.875

Statistical mechanics, I.523, I.524, I.542, I.586, III.1830, III.1962
  condensing gas, I.551–4
  development of, I.591
  in astrophysics, III.1711
  method of mean values, I.535
  models in, I.603–5
  non-equilibrium, I.585–633
Steering of atoms, II.1099
Stefan–Boltzmann law, III.1696, III.1756
Stellar aberration, I.258, I.259
Stellar Atmospheres, III.1701
Stellar distances, measurement of, III.1691
Stellar evolution, III.1691–713
  research since 1945, III.1746–60
Stellar interferometer, III.1400
Stellar luminosity function, III.1714
Stellar masses, determination of, III.1698
Stellar optical interferometry, III.1475
Stellar spectra, classification of, III.1692–4
Stellar structure, III.1694–5, III.1699
  theory of, III.1702–5
Stellarator, III.1648, III.1654–6
Stereology, III.1549
Stern–Gerlach effect, I.163–72, I.202
Stern–Gerlach magnets, II.1032, II.1047
Stigmator electrostatic astigmatism corrector, III.1580
Stimulated Brillouin scattering (SBS), III.1444
Stimulated Raman scattering (SRS), III.1432, III.1436
Stimulated Rayleigh-wing scattering, III.1445
Stokes theory, III.1411
Storage-ring synchrotron, II.739
Strain-aging, III.1521, III.1523
'Strange particles', II.655–9, II.677–9
Stratosphere, II.900, III.1982
String theory, II.682–4, II.756, II.766, III.2037–9
Stripping reactions, II.1220
Stroke, III.1922
Strömgren spheres, III.1762
Strong coupling theory, I.407
Strong interactions, II.680–4
Strong responses to perturbations, III.1593–694
Strongly correlated electron systems, II.1168
Strouhal number, II.850
Structure analysis, I.450–3
Structure-factor formula, I.433
Structure factors, I.439, I.441, I.447, I.449
Structure functions, II.690
Structure Reports, I.511, I.512
Strukturbericht, I.511, III.1513
SU(2), II.660, II.694, II.725
SU(3), II.660–2, II.677, II.679, II.684
Subscidy (*s*ubsidiary *sc*ientific domain), III.1507, III.1520
Substance X, I.64
Sudden ionospheric disturbance (SID), III.1995
Sulphur, III.1541
  formation of, III.1747
Sum rules, II.1058–9
Sun
  gravity fields, II.888
  internal structure, III.1694
  luminosity of, III.1708
  magnetic structure, III.1958
  see also Solar
Sunspots, magnetic fields in, III.1958
Sunyaev–Zeldovich effect, III.1773
Supercomputer, III.2045
Superconducting alloys, II.920

Superconducting magnets,
  III.1927, III.1929
Superconducting Super Collider,
  III.2036
Superconductivity, I.232, I.489–92,
  III.1279, III.1347, III.2025
  BCS theory and its aftermath,
    II.937–41
  developments—pre-1933,
    II.915–17
  developments—1945, II.917–21
  developments—1945–56, II.932
  discovery of, II.915
  high-temperature, II.957–65
  modern unified picture, II.943–9
  new developments, II.949
  pre-BCS attempts at microscopic
    theory, II.936–7
  related phenomena, II.955–7
  theoretical developments
    1933–45, II.923–9
  type-I, II.920, II.948
  type-II, II.920–1, II.941, II.948
Supercritical fluid, III.1947
Superfluid density, II.944
Superfluidity, II.913–15, II.922,
  III.2025
  irrotational component, II.929
  modern unified picture, II.943–9
  new developments, II.949
  related phenomena, II.955–7
Superionic conductors, III.1539
Super-Kamiokande detector,
  II.765, II.768, III.2036
Superlattices, III.1470
Supernova neutrinos, II.758
Supernova trigger hypothesis,
  III.1963–4
Supernovae, II.1227–8, III.1712–13,
  III.1718, III.1758–60, III.1766,
  III.1963–4
  light curve, III.1760
  SN1987A, III.1760
Superplasticity, III.1534
Super-resolution, III.1467

Supersonic aerofoil, II.880
Supersonic boom, II.876
Supersonic speeds, II.872, II.874,
  II.877, II.879
Superstring theories, II.684
Supersymmetry, II.762–3
Supra-heat-conductivity, II.922
Surface crystallography, I.465–70
Surface of section, III.1836
Surface waves, II.883
Surfactants, III.1606–8
  colloid protection, III.1611
Surgery
  diathermy, III.1894
  ultrasound in, III.1915
Symmetry, I.430, I.462, **II.1028–9**
Symmetry breaking, II.694, II.762
Symmetry elements, I.423
Symmetry principle, III.2034,
  III.2035, III.2039
Synchrocyclotrons, II.730–1, II.1201
Synchrotron light, II.1050–3
  spectroscopy, II.1052
Synchrotrons, II.639, II.732–3
  DESY, I.436
Synthetic sapphire crystal, II.982
Systems development, III.2030

Tandem accelerator, II.1203
Tandem-mirror facility, III.1669
Tandem Mirror Experiment, see
  TMX
Tandem scanning microscope,
  III.1451
Tankers, II.885
Taylor thickness, II.811
Teaching and/or researching,
  III.2047–8
Technetium-99m, III.1905–7,
  III.1909, III.1919
Technicare Neptune, III.1927
Technische Hoogeschool Delft,
  III.1579
Teflon (polytetrafluoroethylene),
  III.1896

Subject Index

Telecobalt therapy, III.1872
Telecommunication networks, II.906
Telecommunications by optical-fibre technology, III.1480
Telecommunications Research Establishment (TRE), III.1327–8
Telescopes, III.1734, III.1735, III.1738
  see also under specific types and applications
Teletherapy, III.1877, III.1879, III.1890
Television cameras, III.1412
Television industry, III.1476, III.1536
Teller–Northrop theory, III.1683
Temperature coefficients of resistivity, III.1538
Temperature gradients, II.901
Temperature measurement, II.1249–51
Temperature scales, II.1249–51
  ITS-90, II.1260
Temperature variation of resistivity, III.1302, III.1342
Terrella (magnetized sphere), III.2003
Terrestrial Magnetism and Atmospheric Electricity, III.2000
Tetramethyl manganese ammonium chloride (TMMC), II.1014
TFTR, see Tokamak Fusion Test Reactor (TFTR)
Thallium, I.75
Theorem of equipartition of energy, see Equipartition theorem
Theoretical Structural Metallurgy, III.1522

Theory of the Electron, The, III.1992, III.2039
Thermal analysis, III.1508
Thermal capacity, II.967, III.1354
Thermal conductivity, II.977–82, III.1285, III.1290
Thermal energy, I.66
Thermal equilibrium, I.530
Thermal expansion, II.975–7
Thermal vibrations, II.984
Thermal wind equation, II.901
Thermals, II.845
Thermionic emission, I.37
Thermodynamic equilibrium, III.1391
Thermodynamic internal energy, I.532
Thermodynamics, I.5, I.35, I.149
  as science, I.521
  first law of, I.374, I.521
  nineteenth century background, I.521–4
  of data processing, I.576
  of heterogeneous substances, I.532
  relativistic, I.281–2
  second law of, I.46, I.68–9, I.374, I.521, I.524, I.534–5, I.575
  special-relativistic, I.281
  third law of, I.535–7, I.537–42, II.968
Thermoelastic transformation, III.1539
Thermoelectric effects, III.1283–4, III.1306–11
Thermoelectric (Seebeck) voltage, III.1310
Thermography, III.1932
Thermometers, II.1249
Thermonuclear explosion, II.1224
Thermonuclear fusion, III.1618, III.1629
Thermonuclear reactions, II.1226
Thermos flask, I.27
Thermosphere, III.1982, III.2009

S55

Thimble chamber (ionization chamber), III.1866–7, III.1881
Thin-film devices, III.1444
Thin films, II.1170–1
Third International Polar Year (1957–58), III.2006
Third law of thermodynamics, II.968
  applications, I.537–42
  historical survey, I.535–7
Thomas–Fermi method, I.464, II.1190, II.1191
Thomson moving-magnet galvanometer, I.23
Thomson scattering, I.65
Thomson–Lampard theorem, II.1255
Thorium, I.59, I.60
Threshold effects, II.1063–5, **II.1066–7**
Thymine, I.504
Thyroid gland, III.1900–3
Tidal currents, II.888
Tidal kinetic energy, II.891
Tide gauges, II.891–2
Tide-raising forces, II.892, II.896
Tide tables, II.890, II.892
Tight-binding approximation, III.1292
Tight-binding model, III.1295, III.1297
Time
  dissemination of, II.1248–9
  measurement, II.1242–6
Time–position recording, III.1913–14
Time-resolved crystallography, I.438
Time-scales, II.1247–8
Tin oxide, III.1541
Titanium and titanium alloys, III.1281, III.1897
Tl–Ba–Ca–Cu–O compound, II.959
$Tl_2Sr_2Cu_3O_{10}$, I.492

TMX (Tandem Mirror Experiment), III.1667–8
Tobacco mosaic virus (TMV), I.508
Tokamak Fusion Test Reactor (TFTR), III.1646–54, III.1663, III.1677
Tollmien–Schlichting waves, II.832
Tomato bushy stunt virus (TBSV), I.509, I.510
Tomography. *See* Computed tomography
TOPCON EM002B 200 kV column, III.1588
Torsional oscillation, III.1523
Transcurrent faults, III.1980
Transdisciplinary disciplines, III.2045–7
Transform faults, III.1979
Transformation laws, I.269
Transformation theory, I.59, I.60, I.360
Transistors, III.1322–7
  commercial applications, III.1317
  invention of, III.1317
Transition
  abrupt, II.833, II.834
  amplitudes, I.185, **II.1064**
  restrained, II.833, II.834
  state, II.1049
  taxonomy, II.832–41
  types of, II.825–32
Transition metals, III.1343, III.1354
Transmission electron microscope (TEM), III.1526, III.1548, III.1569, III.1570, III.1573
  aberration-free atomic resolution, III.1585–6
  accelerating voltage, III.1590
  and Abbe's optical theory, III.1578–84
  astigmatism corrector, III.1580
  commercial prototype, III.1575–6
  high-resolution commercial, III.1576
  holographic, III.1584

Subject Index

over-focus and under-focus fringes, III.1580
RCA, III.1581
resolution, III.1590
selected area diffraction, III.1579
serially produced, III.1575–7
specimen preparation and radiation damage, III.1574–5
two-stage, III.1574
vertical column, III.1575
Transonic flight, II.882
Transport coefficients, I.597–8, I.612, I.614
Transport parameters, I.593, I.614
Transport phenomena, I.592
Transuranium elements, II.1222–5
Transverse temperature gradient, III.1282
Transverse voltage, III.1282
Transverse waves, III.1956
Trapping of atoms, II.1099
Triplet coupling, II.1086
Triplets (antitriplets), II.684–5
Triton, II.1208
Tropical cyclones, II.904–6
Troposphere, II.900, III.1982, III.1984
T-Tauri stars, III.1766–7
Tubeless syphon, III.1593–4
Tumours. *See* Cancer; Radiotherapy
Tungsten, III.1355
Tunnel effect, I.217
Turbo-fan engine, II.873
Turbulence, II.843, III.1834
  isotropic, II.832
  taxonomy, II.832–41
  types of, II.825–32
Turbulent flow, II.825
Turnip yellow mosaic virus (TYMV), I.510
Twinned crystals, III.1435
Twisted nematic cell, III.1604
Two-fluid hydrodynamics, II.929
Two-level-atom model, III.1459

UHURU satellite, III.1739, III.1754, III.1766
UK Infrared Telescope (UKIRT), III.1743
Ultra high-molecular-weight polyethylene (UHMWPE), III.1896
Ultra low-frequency radio physics, III.1995
Ultrasonic imaging, III.1914–15
Ultrasonic industrial flaw detector, III.1911
Ultrasonic probe, III.1913
Ultrasonic radiation, therapeutic, III.1915
Ultrasound
  diagnostic, III.1910–11
  in medicine, III.1910–15
  in surgery, III.1915
  in therapy, III.1915
  scanning, III.1859, III.1913, III.1915
Ultraviolet astronomy, III.1741–2
Ultraviolet energy, III.1992
Ultraviolet light, II.1050–3, III.1396
Ultraviolet radiation, III.1994, III.2009
Umklapp processes, II.980, III.1304–6
Uncertainty relations, I.174, I.201–4, I.209, I.228, III.1394
Uniaxial ferroelectrics, II.1009
Unified field theories, I.320–3
Unit cell, I.423
Unitary symmetry, II.660–2
Units, II.1236–60
Universal gas constant, I.150
Universal Time (UT), II.1247
Universality, I.549–51, I.563, II.664, II.677, III.1833
  empirical derivation, I.561–4
Universe
  age of, III.1732, III.1793–4
  dark matter, III.1731
  dynamics of, III.1730, III.1787

S57

Einstein's, III.1722–4
  expansion of, III.1786, III.1787, III.1806
  geometry of, III.1723
  Hot Big Bang models of, III.1778
  hot early phases, III.1788, III.1793
  large-scale structure, III.1714–34, III.1805
  mean baryonic density, III.1791–2
  mean density, III.1800
  mean mass density, III.1719
  temperature history, III.1786
  total mass density, III.1800
  very early, III.1805–7
U-particles, I.393, I.394, I.396, I.397
Upper atmosphere, III.1981–4
  geophysics, III.1943–4, III.1982
  ionization of gases, III.1981
Upper Atmosphere, The, III.2000
Upsilon, II.721
Upstream wakes, II.868–9
U-quanta, I.378, I.382, I.389, I.391
Uracil, I.504
Uranium, I.56, I.60, I.61, I.106, II.1199, II.1223, II.1227, III.1630, III.1960
Uranium-235, I.131–2
US Atomic Energy Commission, III.1633
US Bureau of Standards, III.1988
US National Aeronautics and Space Administration (NASA), III.1739, III.1965
US National Bureau of Standards (NBS), I.20, II.1235, II.1236, III.1881, III.1998
U-sequence, III.1839
UTC, *see* Coordinated Universal Time (UTC)
Utrecht School, I.629

Vacancies, III.1505, III.1529, III.1531, III.1533

Vacuum polarization, II.641, II.1031
Vacuum technology, I.27
Vacuum tubes, I.53, III.1987
Valence band, III.1317, III.1319, III.1326, III.1329, III.1334
Valence electrons, III.1314, III.1317, III.1366
Van Allen belts, III.1683–4
van de Graaff generator, II.727–8, II.1203, III.1871, III.1883
van der Waals equation, I.28, I.547, I.552, I.561
van der Waals law, II.973
van der Waals theory, I.555, I.557
Varistors, III.1537
Väsälä–Brunt frequency, II.864
V–A theory, II.667–9
Vavilov–Čerenkov effect, III.1408
Vela satellite, III.1740, III.1741
Velocity–distance relation, III.1726, III.1727, III.1728
Velocity–position space (μ-space), I.587, I.590
Veneziano amplitude, II.684
Vertical-cavity surface-emitting laser (VCSEL), III.1464
Very high-resolution imaging systems, III.1477
Very long baseline interferometry (VLBI), III.1738
Very low-frequency radio physics, III.1995
Very low temperatures, I.540–1
Vesicles, III.1608
Vibration, II.967–1010
Vibrational excitation, II.1073
Vibrational motion, II.1044
Vibrational specific heat of solids, I.525
Vibrational transitions, II.1073
Videophone, III.1477
Viking (rocket), III.2005
Vine–Matthews theory, III.1976

Virial theorem, II.1026, III.1731, III.1772, III.1800
Virtual reality, III.1476
Viruses, I.507–11
  rod-shaped, I.508
  spherical, I.509
Viscosity, II.807, II.813–14
  stabilizing effect of, II.828
Viscous diffusion, II.827
Vitamin B_{12}, structure of, I.495
Vitamin D, isolation of, III.1529
Vlasov–Boltzmann equation, III.1625
Volt
  absolute determination, II.1256
  reference standards, II.1256
Volta effect, III.1283
Voltaic cell, III.1284
von Klitzing constant, II.1259, II.1271
von Neumann equation, I.602, I.612
Vortex lines, II.820, II.822, II.823
Vortex-wake, II.876
Vorticity, II.816–24, II.889, II.890, II.892, II.894, II.901
V-particles, I.408–9
Vulcanization, III.1551, III.1593

Wake fluid, II.821
Wakes, II.812–16
Wanderwelle, electrical surges, III.1566
Water-bath ultrasonic simulator, III.1912
Water waves, II.798
Watson–Crick model, I.504
Wave, *see* under specific wave types
Wave action, II.861
Wave energy, II.862
Wave equation, I.195, I.283, III.1393
Wave frequency, II.863

Wave functions, I.204, **I.361**, I.617, II.941–3, II.949, II.1124
Wave generation, II.846–69, II.882–5, II.900
Wave-interaction effect, III.1994
Wave mechanics, I.174, I.191–6, I.217, I.360
Wave momentum, II.866
Wave numbers, II.838, II.839, II.861, II.864, II.866–8, II.885, II.893, II.984
Wave packets, II.836, III.1366
Wave–particle duality, I.168–71, I.364, III.1394, III.1395
Wave propagation, II.846–69, II.877, III.1407, III.1955
Wave theory, I.258, I.363
Waveforms, II.848, II.858, II.861
Wavelength, II.883, II.902
  standards, II.1239–40, II.1242
Wave-making resistance, II.867
Wavevectors, II.988–90, II.1001
W-bosons, II.637, II.706–9
Weak interactions, II.663–71, II.692–3
Weak universality, II.664
Weather, II.899–912
  forecasting, II.888, II.899, II.906–9
Weber bar, I.314
Weinberg–Salam model, II.695, II.762
Wet-bulb temperature, II.843
Weyl's postulate, III.1729
Wheatstone bridge, I.22
Whiskers, III.1351
Whistler waves, III.1622, III.1623, III.1997, III.2005
  measurements, III.2009
  modelling, III.1999
Whistling atmospherics, III.1995
White dwarfs, III.1709–12
Wide Field Planetary Camera, III.1745

Wiedemann–Franz law, III.1286, III.1287, III.1316
Wiedemann–Franz ratio, III.1290
Wien–Planck spectral equation, I.69
Wiener–Khintchine theorem, I.600–2
Wien's law, I.69, I.151, I.524, III.1393
Wigner function, I.617–18, III.1441
William Herschel Telescope, III.1744
Wilson cloud chamber, see Cloud chamber
Wilson's scaling method, I.440, I.445, I.447
Wind effects, II.901, II.903–4
Wind instruments, II.851–2
Wind speed, II.843
Wind speed/sound speed ratio, II.813
Wind tree model, I.603
Wind-tunnels, II.830, II.879
Winding number, II.955
Wireless. See Radio
Wireless telegraphy, III.1986
Wollaston wire, II.831
Women in higher education, I.8, I.10
Work-function, III.1286, III.1308
Work-hardening, III.1518
World Directory of Crystallographers, I.511
W-particle, III.2036
Wranglers, I.10–12
Wurtzite, I.429

Xenon, I.75, I.556
Xerographic copying, III.1541
X-ray absorption, II.1051
X-ray analysis, III.1514
X-ray Analysis Group, I.511
X-ray angiography, III.1904
X-ray apparatus, II.1051

X-ray astronomy, III.1738–41, III.2005
X-ray binaries, III.1754–7
X-ray contrast media, III.1862–3
X-ray crystallography, I.446, I.469, I.484, III.1515
X-ray diffraction, I.30, I.424–32, I.456, I.457, I.484, I.498, II.983, III.1583
 amorphous solids, **I.478**
 in-plane, I.466
 Laue's equations for, **I.427**
X-ray dose-rate measurements, III.1871
X-ray emission, III.1630
X-ray equipment, III.1861
X-ray imaging, III.1916, III.1922–3
X-ray intensities, I.440
X-ray irradiation, III.1529
X-ray lasers, III.1464–6
X-ray measurements, II.1010
X-ray microscope, III.1583
X-ray optics, I.484
X-ray photons, I.363
X-ray pulsations, III.1756
X-ray pulses, III.1752
X-ray scattering, I.165, I.362, I.365, I.453, II.990, III.1867
X-ray sets, III.1864, III.1877
X-ray sky, III.1798
X-ray sources, III.1755–7
X-ray spectra, I.86, I.158–60
X-ray spectrometer, III.1580
X-ray spectroscopy, I.158–60, I.429, I.484
X-ray studies, inner-shell phenomena, II.1089–90
X-ray telescopes, III.1740
X-ray tomography (X-CT), III.1918
X-ray tubes, III.1859, III.1861, III.1862, III.1865
X-rays, I.3, I.32, I.36, I.52, I.53, I.65, I.362–6, I.421–4, II.1050–3, III.1514
 discovery, I.424

in diagnosis, III.1859–63
in medical physics, III.1857
see also Radiotherapy

Yang–Mills theory, II.692–4, II.713, II.714, II.763
YBa$_2$Cu$_3$O$_7$, II.959
YBa$_2$Cu$_3$O$_{7-\delta}$, I.490–1
Yerkes system, III.1698
Yield-stress, III.1520–1, III.1523
Yin-Yang coil, III.1664
Young's double-slit experiment, III.1402
Yttrium–iron–garnet, II.1159
Yukawa meson theory, I.378–83, II.651–2

Z-bosons, II.637, II.706–9
Zeeman effect, I.7, I.36, I.52, I.54, I.99–103, I.160–2, I.182, I.183, I.184, I.211, II.1114, II.1115, II.1149
Zeeman splittings, II.1120
Zeeman sub-state, II.953
Zeitschrift für Kristallographie, I.511
Zeitschrift für Metallkunde, III.1556
Zenith angle, III.1992
Zernike's circle polynomials, III.1409
Zero-point motion, II.973–5, II.1004
Zero-resistance state, II.961
Zero sound, II.950, III.1364
ZETA (Zero-Energy Thermonuclear Apparatus) toroidal pinch experiment, III.1641–2, III.1645, III.1657
Zhukovski's theorem, II.822
Zinc, III.1360
Zinc oxide, III.1537
Zinc sulphide scintillators, II.1200
Zincblende, III.1514
Zone-levelling, III.1543, III.1544
Zone-refining, III.1543, III.1544, **III.1545**
Zweig's rule, II.685

NAME INDEX

Note—volume and page numbers in **bold type** refer to items in boxes

Aaronson, M, Hubble's constant, III.1793
Abbott, E A, *Flatland*, I.5
Abelson, P H, transuranium element, II.1223
Åberg, T, secondary excitations, II.1090
Abragam, A
  hyperfine structures, II.1139
  nuclear magnetic resonance, II.1175
  resonance research, II.1143
Abraham, electron microscope, III.1573
Abraham, M
  special-relativistic theories, I.284
  stress–energy tensor, I.280, I.281
Abramowicz, M A, self-consistent thick-disc model, III.1780
Abrikosov, A A, superconductivity, II.935
Accardo, C A, p–n junction, III.1417
Adams, M C, dye-laser pulses, III.1439
Adams, W C
  A stars, III.1710
  chemical abundances, III.1702
  luminosity indicators, III.1698
Adams, W S, Sirius B, III.1710, III.1723

Ademollo, M, Dolen–Horn–Schmid duality, II.683
Adlard, photon, III.1451–2
Adler, S, soft photon emission, II.679
Aharonov, Y
  field strength, II.693
  interference patterns, III.1370
Akasofu, S-I, *Solar–Terrestrial Physics*, III.2001
Akhmanov, S I, Nd:YAG laser, III.1435
Albrecht, A, world model, III.1807
Alder, B J
  autocorrelation function of tagged particle, I.619
  molecular dynamics, I.577
Alder, K, Coulomb excitation experiments, II.1217
Alexander, S, fractons, II.1007
Alfvén, H
  baryon symmetric models, III.1806
  Cosmical Electrodynamics, III.1637, III.1685
  Galactic radio emission, III.1737–8
  MHD theory, III.1684, III.1685
  monistic theory, III.1962, III.1963
  ring current around the Earth,

III.2003
Allen, J F, superleak flow, II.946
Allis, W P, electron diffusion rates, III.1644
Alpher, R A
　abundances of elements, III.1785–6
　black-body radiation, II.758
　primordial nucleosynthesis, III.1785, III.1786, III.1790
　proton–neutron ratio, III.1786
Als-Nielsen, J, order–disorder transition in β-brass, II.1009, II.1016
Altar, W, wave propagation, III.1990–1
Altarelli, G, asymptotic freedom, II.715
Al'tshuler, B L, scattered wavelets, III.1370
Alvarez, L W
　neutron diffraction, I.459
　particle interactions, II.659
　proton linear accelerator, II.1205
　proton machines, II.737
　60" cyclotron, II.1202
　tandem accelerator, II.1203
Ambartsumian, V A, T-Tauri stars, III.1766
Amelinckx, S, dislocations, III.1526
Ampère, A M
　electromagnetic interactions, III.1953
　magnetism, I.65
Amrein, W O, point photon, III.1451
Anaxagoras, matter, I.44
Anaximenes, air as basic substance, I.358
Anderson, C D, **I.386**
　absorption in lead plates, I.385
　electron–positron pairs, I.231
　elementary particles, I.391
　mesotron, I.231, I.389
　muon, II.650

positron, III.1706
shower particles, I.388
standard model, I.231
Wilson cloud chamber, I.371
Anderson, J A, angular diameter of Betelgeuse, III.1705
Anderson, J L, relativistic thermodynamics, I.281
Anderson, P W
　bosons, II.695
　disordered materials, III.1365
　electron density, III.1316
　exchange interactions, II.1144
　glasses, II.1006
　phase transitions, II.1008
　phonon exchange, II.953
Anderson, W C
　speed of light, II.1263
　white dwarfs, III.1711
Andrade, E N, crystal diffraction spectroscopy, II.1207
Andrew, E R
　magnetic resonance imaging, III.1924
　nuclear magnetic resonance, II.1175
Andrews, T
　critical point in carbon dioxide, I.28
　critical point phenomenon, III.1947
Anger, H, gamma camera, III.1907
Ångström, A J
　normal solar spectrum, II.1242
　spectrum of hydrogen, I.76
Antonucci, R R, Seyfert galaxies, III.1784
Appleton, E V
　radio receiving apparatus, III.1989
　solar radiation, III.2002
　upper atmosphere, III.1988
　wireless propagation, III.1622
Araki, G
　cosmic-ray particles, I.407

slow mesotrons, I.406
Argos, P, structure determination, I.445
Arima, A, collective motion in nuclei, II.1218
Aristotle, matter, I.44
Armenteros, R, hyperon, II.657
Armstrong, H E, organic compounds, I.493
Arndt, U W, single-crystal oscillation, I.437
Arnett, W D, element abundances, III.1748
Arnott, N
  Elements of Physics, III.1857
  medical physics, III.1856–7
Arp, H, quasars, III.1777
Artsimovich, L
  Atoms for Peace Conference, III.1640
  mirror-type systems, III.1643
  tokamak, III.1648
Asaro, F, alpha-decay spectra, II.1217
Ash, E, optical microscopy, III.1406
Ashkin, A
  argon-ion laser beams, III.1453
  He–Ne laser, III.1434
  optical binding of particles, III.1472
  self-trapping of laser beam, III.1459
Askaryan, G A, self-focusing, III.1436
Aspect, A, Bell inequalities, III.1467
Astbury, W T
  DNA, I.504
  fibrous proteins, I.497
  x-ray diffraction, I.439
Aston, F W
  isotope masses, I.111–12
  stellar energy source, III.1706

Astrom, E O, plasma physics, III.1992
Atkinson, R d'E
  nuclear reactions, II.1225, III.1707
  thermonuclear reactions, III.1627
Auger, P, neutrons, I.122
Austin, L W, radio propagation, III.1987
Autler, S H, 'dressed' atomic states, III.1421
Avenarius, critical opalescence, I.546
Avery, Q T, DNA, I.504
Avogadro, A, hypothesis, I.46
Avrami, M, phase transformations, III.1554
Axford, I, solar wind plasma, III.2010
Ayers, G R, blind deconvolution, III.1461
Azbel, M Ya
  canonical theory, III.1356
  cyclotron resonance, III.1357

Baade, W
  Andromeda Nebula, III.1788
  medium-mass elements, II.1228
  Palomar 200" Telescope, III.1737
  radio sources, III.1775
  supernovae, II.1227–8, III.1712, III.1718, III.1759, III.1775
Bacher, R F
  atomic energy levels, II.1093
  nuclear structure theory, II.1208
  shell model, II.1209
  Thomas–Fermi model, II.1191
Back, E
  total quantum number, I.162
  Zeeman splitting, I.161
Bacon, G E
  neutron diffraction, I.456
  structure of KH_2PO_4, I.459
Badash, L, I.29

Baez, A V, optical microscopy, III.1406
Bahcall, J N
  solar model, III.1749
  solar neutrino flux, III.1748
Bailey, V A, broadcast modulation, III.1994
Bain, E, bainite, III.1515
Bainbridge, K
  H-particle–electron model, I.112
  mass spectrometry, II.1187
Baker, D W, ultrasonic imaging, III.1914
Baker, G A, Padé Approximant, I.558
Baldwin, D, thermal barrier, III.1667
Baldwin, G C, dipole vibration, II.1213
Balescu, R, *Equilibrium and Non-equilibrium Statistical Mechanics*, I.620
Balluffi, R, diffusion, III.1531
Balmer, J, R value, I.77–8
Banerjee, M K, proton inelastic scattering, II.1221
Bangham, vesicles (or liposomes), III.1608
Barber, C, electron storage rings, II.739
Bardakci, K, singularities in complex angular momentum, II.682
Bardeen, J, **II.939**
  BCS theory, II.938
  impurity effects, III.1281
  interatomic forces, II.999
  ion potential, III.1303
  point-contact transistor, III.1326
  solid-state physics, III.1313
  superconductivity, I.232, II.937, III.1364
  theory of phonons, III.1362
  transistor, II.744
  xerography, III.1541

Bardeen, J M, black holes, III.1779
Barham, P M, coherence parameter, III.1412
Barkhausen, H
  magnetic flux, II.1121
  radio waves, III.1622
  whistler waves, III.1622–3
Barkla, C G
  x-ray scattering, I.362, I.424
  x-ray spectroscopy, I.158
Barlow, W, crystal structure, I.423–4
Barnes, W H, structure of ice, I.474
Barrett, C, diffusive transformations, III.1547
Bartels, J, *Geomagnetism*, III.2000
Barthel, P D, radio quasars and radio galaxies, III.1784
Bartlett, J A, shell models, II.1209
Barut, A O, singularities in complex angular momentum, II.682
Basov, N G
  excimer laser, III.1453
  semiconductor maser, III.1423
Bass, M, sum-frequency generation, III.1429
Bate, G, magnetic materials, II.1159
Bates, D R
  ionospheric research, III.1999
  probability of electronic excitations, II.1080
  solar x-rays, III.1994
Bauer, high-reflection coatings, III.1407
Baxter, R J, ferroelectric model, I.571
Bay, Z, electron multiplier, III.1541
Baym, G, neutron stars, III.1754
Bear, R S, three-stranded coiled-coil model, I.499
Beck, C S, defibrillators, III.1895
Beck, G, β-decay theory, I.372, I.375

Beck, K, single crystals of ferromagnets, II.1123
Becker, H, neutron, I.119
Becker, R, nucleation kinetics, III.1547
Becklin, E E
  formation of stars, III.1769
  Orion Nebula, III.1743, III.1767
Becquerel, A H
  radioactivity, I.37, III.1857, III.1876
  uranic rays, I.53, I.56–7
Bednorz, J G
  Meissner effect, II.957–8
  superconductivity, I.490, II.961
Beevers, C A
  Rochelle salt, I.489
  structure determination, I.441
Behn, U, specific heat, II.968
Bell, A G, Photophone, III.1422
Bell, F O, DNA, I.504
Bell, J S
  EPR paradox, I.229
  inequalities, III.1389
Bell, P R, scintillators, II.750
Bell, R J, amorphous SiO_2, I.478–9
Bell, S J
  neutron stars, III.1751
  sky surveys, III.1752
Bémont, G, radium, I.58
Bénard, H
  sound from flows, II.850
  surface tension, II.845
Benedek, G
  'light-beating' spectroscopy, III.1454
  spontaneous magnetization, I.559
Benioff, H, earthquake oscillations, III.1958
Bennet, W R Jr, helium–neon laser, III.1427
Bennett, F, tandem accelerator, II.1203

Bennewitz, K, zero-point motion, II.973–4
Bentley, R, physical cosmology, III.1721
Beran, M J, partial coherence, III.1392
Berkner, L
  IGY, III.2006
  ionospheric research, III.2005–6
  ionospheric sounding, III.1998
  Third International Polar Year, III.2006
Berlinger, W, oxygen octahedra, II.1009
Berman, R
  diamond–graphite equilibrium diagram, I.538
  thermal conductivity, II.981
Bermen, S M, hadrons, II.690
Bernal, J D
  amorphous structure, I.479
  crystal oscillation, I.434
  globular proteins, I.497
  haemoglobin, I.497
  random loose packing (RLP) and random close packing (RCP), I.577
  residual entropy of ice, I.571
  stereochemical formulae, I.495
  structure of ice, I.474
  tobacco mosaic virus (TMV), I.508
Bernard, M G A, stimulated emission in semiconductors, III.1433
Bernoulli, D
  Bernoulli equation, II.806
  gas pressure, I.47
Bernstein, I B
  collisionless shocks, III.1684
  plasma oscillations, III.1625
Berreman, D W, liquid-crystal optics, III.1452
Berry, B, anelastic relations, III.1523

Berry, M V
  diffractals and caustics, III.1466
  topological phase, III.1470
Berson, S, saturation analysis technique, III.1909
Bertero, M, imaging theory, III.1467–8
Bertolotti, M, lasers, III.1423
Berzelius, J J, nomenclature for elements, I.45
Besicovich, A S, quasicrystals, I.482
Bessel, F, stellar distances, III.1691
Beth, R A, angular momentum of photons, III.1410
Bethe, H A
  abundances of elements, III.1785–6
  bremsstrahlung, I.385
  carbon–nitrogen–oxygen cycle (CNO cycle), III.1707–8
  face-centred cubic lattice, III.1293
  fast charged particles, II.1040
  Fermi theory, I.377
  First Shelter Island Conference, II.645
  mesons, I.401, I.403, II.654
  near-field/far field solution, III.1416
  neutron capture, II.1193
  neutron stars, III.1754
  nuclear physics, II.1184
  nuclear reactions, I.132, II.1192, III.1629
  nuclear structure theory, II.1208
  origin of the elements, II.1225
  pp process, II.1226
  quantum fluctuations, II.1013–14
  shell model, II.1209
  spacing of states in highly excited nucleus, II.1213
  theory of metals, III.1290
  Thomas–Fermi model, II.1191

Twelfth Solvay Conference, Brussels, II.646
weak-interaction-treatment, II.1040
Beynon, W, ionosphere, III.1984
Bhabha, H J
  cosmic rays, I.393
  fusion reactions, III.1642
  showers, I.388
  U-particles, I.393
Bhar, J N, ionospheric physics, III.2000
Bhatia, pseudopotential concept, III.1367
Bieler, É S
  α-particles, I.114
  anomalous scattering, I.366
Bier, N, scleral lenses, III.1414
Biermann, L, ionized comet tails, III.2009
Bijvoet, J M
  dextro (d) and laevo (l) compounds, I.495
  strychnine, I.495
Bilby, B A, yield-stress and strain-aging, III.1523
Bilderback, D, area detector, I.438
Billy, H, Renninger effect, I.450
Biloni, H, solidification, III.1542
Binnig, G
  scanning tunnelling microscope, III.1549
  tunnelling electron microscope, I.466
Birch, F, Earth's core, III.1957
Birdsall, C K, stream-type instabilities, III.1673
Birge, R T
  Birge–Bond diagram, II.1268
  constants, II.1271
  least-squares adjustment, II.1270
  speed of light, II.1263
Birkeland, K
  geomagnetic storms, III.2003
  Terella device, III.1682

Birkhoff, G D, energy surface trajectories, I.599
Birkinshaw, M, Sunyaev–Zeldovich effect, III.1773
Bishop, A S, *Project Sherwood—The US Program in Controlled Fusion*, III.1640
Bitter, F, antiferromagnetism, II.1129
Bjerrum, N
 Applications of the quantum hypothesis to molecular spectra, I.82
 molecules, I.81–2
 quantum theory, I.155
Bjorken, B J
 charmed quark, II.700
 quark–lepton analogy, II.700
Bjorken, J D
 hadrons, II.690
 parton hypothesis, II.690
 point-like constituents, II.689
Bjorkholm, J E, self-trapping of laser beam, III.1459
Bjornholm, S, fission isomers, II.1218
Black, J, medical physics, III.1857
Black, J C, ultrasonic properties, II.1006
Blackett, P M S, **I.400**
 cloud chambers, II.651
 Earth's magnetic field, III.1969–70
 electron–positron pairs, I.231
 nuclear reactions, II.1201
 penetrating radiation, I.388
 The craft of experimental physics, II.745
 Wilson cloud chamber, I.371
Blackman, M
 BvK theory, II.984
 perturbation theory, II.1002
 specific heat, II.985–7
 toroidal fusion device, III.1634

Blair, J S
 alpha particles, II.1221
 diffraction theory, II.1221
Blanc-Lapierre, A, polychromatic fields, III.1421
Blau, M, cosmic-rays tracks, I.404
Bleaney, B, resonance research, II.1143
Bleuler, E, two-photon correlation, III.1413
Blewett, J P
 accelerators, II.725–7
 Particle Accelerators, II.1201
Bliss, W R, cross-sectional imaging, III.1912
Bloch, F
 electron gas, III.1291–301
 gyroscopic motion, II.1172
 magnetism, II.1137
 neutron diffraction, I.453, I.459
 nuclear magnetic resonance, II.1094
 quantum theory, I.221, III.1361
 spin waves, II.1010–11
Block, M, parity states, II.665
Block, S, diffraction techniques, I.437
Bloembergen, N, **III.1430**
 anharmonic oscillators, III.1434
 laser spectroscopy, III.1466
 non-linear optics, II.1100, III.1434
 Nonlinear Optics, III.1445
Blow, D M, structure determination, I.445
Bludman, S A, scalar Higgs fields, III.1806
Boas, W, plastic deformation, III.1526
Bobeck, A H, bubble domains, II.1160
Bodin, H, reversed-field pinch (RFP) research, III.1657
Boersch, electron holograms, III.1582–3

Bogoliubov, N N, **I.610**
  BBKGY hierarchy, I.606, I.623, I.624, I.626
  kinetic equation, I.609
  properties of dilute Bose gas, II.930
Bohm, D
  Bohm diffusion, III.1630, III.1631, III.1632
  electrical discharges in gases, III.1361
  EPR paradox, I.229
  field strength, II.693
  First Shelter Island Conference, II.645
  interference patterns, III.1370
  plasma oscillations, III.1625
Bohr, A, II.1216
  compound nucleus theory, II.1219
  Coulomb excitation experiments, II.1217
  geometrical collective model, II.1218
  quadrupole moment, II.1215
Bohr, H, quasicrystals, I.482
Bohr, J, Fourier techniques, I.469
Bohr, N, **I.79**, I.180
  atomic theory, I.157, II.1117, III.1401
  β-decay theory, I.372
  β-spectra, I.126–7
  Balmer formula, I.80, I.82
  Bohr radius, I.85
  building-up principle (*Aufbauprinzip*), I.96–7
  Cockcroft–Walton accelerator, II.1188
  complementarity theory, I.224
  compound nucleus, II.1192, II.1194
  correspondence principle, I.86
  dumb-bell model, I.181–2
  dynamical quantum theory, I.174
  Einstein–Podolsky–Rosen (EPR) paradox, I.228
  electron collisions, II.1042
  fast charged particles, II.1039
  Fifth Solvay Conference (1927), I.205
  fission process, II.1195
  high-speed β-particles, I.107
  hydrogen atom, I.82–6, I.91
  impact of theory, I.86–9
  Light and life, I.224
  liquid drop, I.131, II.1213
  m-values, I.95
  magnetic properties, II.1115
  mechanical forces, I.78
  'new hypothesis', I.84
  nuclear collective motions, II.1213
  nuclear fission, I.126
  nuclear reactions, I.123–6
  On the constitution of atoms and molecules, I.157
  On the general theory, I.176
  On the quantum theory of line spectra, I.94
  periodic system, I.97
  periodic table, I.95
  principal quantum number, I.91
  principle of complementarity, I.197
  quantum conditions, I.359
  quantum dynamics, I.72–82
  quantum hypothesis, I.93
  quantum mechanics, I.225, I.226, I.228
  quantum optics, III.1397
  quantum theory, I.80, I.166, I.175, I.221, II.1115, III.1397
  radiative instability, I.83
  spacing of states in highly excited nucleus, II.1213
  spectra, I.77
  superconductivity, II.917
  theory of atomic structure, III.1699

theory of metals, III.1286
theory of nuclear reactions, II.1192
Twelfth Solvay Conference, Brussels, II.646
uranium-235, I.131–2
x-ray spectra, I.86
Bolsterli, M, computers, III.1843
Bolt, R H, ventriculography, III.1911
Bolton, J G, radio sources, III.1737
Boltzmann, L E, **I.31**
  battle with Ostwald, I.48
  black body, I.66
  black-body radiation, I.149
  energy surface trajectories, I.590
  entropy, I.47
  equipartition theorem, I.67
  H-theorem, I.35
  kinetic theory of gases, I.587
  Liouville theorem, I.525
  process irreversibility, I.524
  second law of thermodynamics, I.524
  specific heat, II.970
  statistical mechanics, I.173, I.523
Bom, N, imaging, III.1914
Bömmel, acoustic resonance, III.1357–8
Bond, W N, constants, II.1267–8
Bondi, H
  chemical elements, III.1788
  field equations, I.313
  plane wave solutions, I.312
  steady-state cosmology, III.1787
Bonhoeffer, K F, molecular hydrogen, II.1048
Bonifacio, R, Collective Atomic Recoil Laser, III.1456
Booker, H G, magneto-ionic theory, III.1992
Bormann, E, decay law of radioactive substances, I.222
Born, M, I.180, **II.985**
  adiabatic approximation, II.999

Atomic Physics, I.210
BBKGY hierarchy, I.606, I.623, I.624, I.626
crystal lattice vibrations, I.153
discrete quantum mechanics, I.184
Dynamical Theory of Crystal Lattices, II.988
Fifth Solvay Conference (1927), I.205
harmonic-oscillator decomposition, III.1402
lattice dynamics, I.452
Lattice theory of rigid bodies, I.189
matrix mechanics, I.360
method of long waves, II.1003–4
On quantum mechanics, I.184
On the quantum mechanics of collision processes, I.198
perturbation theory, II.1002
properties of fluids, I.553
quantum mechanics, I.185, II.638
quantum theory, I.181, I.188
Raman effect, II.996, III.1404
rigid body, I.279
Schrödinger's wavefunction, I.174
self-consistent renormalized harmonic approximation, II.1004
specific heat, II.971–2
spectral line intensity, I.172
The structure of the atom, I.189
theory of ionic crystals, II.997
vibrational specific heat of solids, I.525
Bose, S N
  black-body radiation, III.1402
  light-quanta, I.167
  Planck's law, I.527
  radiation law, III.1288
Bothe, W
  coincident counting, I.364
  cosmic rays, II.748
  neutron, I.119

x-ray scattering, I.166
Bouguer, gravitation, II.1265
Bouwkamp, C J, diffraction theory, III.1420
Bowen, I S, nebulium lines, III.1762
Bowen, R, Gordon Conference (1976), III.1850
Bowles, K L, incoherent scatter radar (ISR), III.1996
Bowley, A R, nuclear medicine, III.1916
Boyd, G D, resonator modes, III.1429
Boyde, A, tandem scanning microscope, III.1451
Boyle, B J
  Newton's rings, III.1386
  quasars, III.1798
Boys, C V
  fine silica fibres, I.28
  gravitation, II.1266
  radiometer, I.6, I.23, I.28
  Soap Bubbles and the Forces that Mould Them, I.5
Bozorth, R M, x-ray study of KHF_2, I.458
Bradley, A W, single-crystal methods, I.435
Bradley, D, dye-laser pulses, III.1439
Bradley, J, stellar aberration, I.259
Bragg, Sir William (Henry), I.429
  α-particles, I.104
  Bragg–Gray theory, III.1882
  crystal spectrometer, I.430
  naphthalene, I.494
  reflection method, I.158
  structure factors, I.439
  x-ray diffraction, I.426, I.429
  x-rays, I.363
Bragg, Sir (William) Lawrence, **I.429**
  binary alloys, I.548
  doubly refracting crystal, I.485

Fifth Solvay Conference (1927), I.205
Fourier series method, I.440
inorganic compounds, I.492
order–disorder transitions, I.473
reflection method, I.158
sodium chloride, I.493
structure of beryl, I.440
Twelfth Solvay Conference, Brussels, II.646
x-ray analysis, I.496
x-ray diffraction, I.426, I.427–8, I.431
x-ray microscope, III.1414
Braginsky, V B, quantum non-demolition (QND) measurements, III.1469
Brahe, T
  supernova, III.1712
  supernovae, II.1228
Brandt, E H, levitation theory, III.1543
Brant, H L, two-photon correlation, III.1413
Brattain, W H
  point-contact transistor, III.1326
  transistor, II.744
Braun, E
  growth spiral, III.1528
  solid-state physics, III.1521
Braun, F, cathode-ray oscillograph, I.26, I.27
Braun, K F, wireless communication, III.1986
Bravais, A, symmetry types, I.423
Breit, G
  First Shelter Island Conference, II.645
  ionospheric conducting layer, III.1988
  ionospheric plasmas, III.1621
  ionospheric sounding, III.1995
  magneto-ionic theory, III.1990
  neutron capture, II.1194
  nuclear resonance, I.126

Brethenton, F P *(cont.)*
  radio receiving equipment, III.1990
  reaction cross sections, II.1219
  Tesla coil, II.1185
Bretherton, F P, momentum transfer, II.867
Breuckner, K, laser fusion programme, III.1676
Brewer, R G
  self-phase modulation, III.1449
  stimulated Brillouin scattering, III.1444
Brickwedde, F G
  deuterium, III.1706
  deuteron, II.650, II.1183
Bridges, W B, argon-ion laser, III.1443
Bridgman, P W
  crystal growth, III.1280
  thermoelectric effects, III.1306
Briggs, B H, ionosphere, III.1984
Brill, R, biomolecular structures, I.497
Brillouin, L
  asymptotic linkage, I.197
  Brillouin lines and Brillouin zones, I.216, III.1311–13
  Fifth Solvay Conference (1927), I.205
  First Solvay Conference (1911), I.147
  Maxwell's Demon cannot operate, I.576
  quantum theory, III.1397
Brindley, G W, zero-point motion, II.974
Brinkman, W F, phonon exchange, II.953
Broadbent, S R, percolation processes, I.572
Brobeck, W, neutral beam injection, III.1661
Brockhouse, B N, **II.993**
  dispersion curves of germanium, II.1002
  interatomic forces, II.999
  neutron spectroscopy, I.486
  phonon dispersion relations, II.993–4
  phonons, II.999
  shell model, II.997
  spin-wave energies, II.1012
  triple-axis crystal spectrometer, II.992
Broer, L J F, nuclear resonance, II.1172
Bromberg, J L, lasers, III.1478
Bronsteyn, M P, linearized quantum gravity, I.317
Brossel, J, optical pumping, III.1415
Brown, B R, computer-generated spatial filter, III.1447
Brown, H, fusion power, III.1673
Brown, R G W, spatial coherence, III.1471
Brown, S, electron diffusion rates, III.1644
Brown, T G, pulse-echo ultrasonic scanner, III.1913
Brüche
  emission electron microscope (EEM), III.1569
  Geometrische Elektronenoptik, III.1578
Brueckner, K A, correlation effects, III.1362
Brunhes, B, Earth's magnetic field, III.1973
Brush, S G
  Lenz–Ising model, I.554
  planetary physics, III.2012
Buchdahl, H A, Langrangean aberration theory, III.1458
Buckingham, M J, λ-point of liquid He^4, I.559
Budden, K G, radio waves, III.1992
Budker, G I
  cooling method, II.743
  magnetic mirror, III.1659

mirror concept, III.1665
mirror machine, III.1662
storage rings, II.741
Buechner, W W, electrostatic generator, II.1203
Buerger, M J, precession camera, I.434
Bullard, E C
  Earth's magnetic field, III.1960
  Pacific floor measurements, III.1974
Bullen, K, Earth's core, III.1957
Bunemann, O, stream-type instabilities, III.1673
Bunn, C W, difference synthesis, I.452
Bunsen, R
  Bunsen burner, I.75–6
  spectral emission lines, III.1692
  spectral lines, I.156
Burbidge, E M
  Evidence for the occurrence of violent events in the nuclei of galaxies, III.1777
  nucleosynthesis in stars, III.1788
  supernovae, II.1227, II.1228
  synthesis of elements, III.1747
Burbidge, G R
  Evidence for the occurrence of violent events in the nuclei of galaxies, III.1777
  inverse-Compton catastrophe, III.1781
  nucleosynthesis in stars, III.1788
  quasars, III.1777
  radio sources, III.1775
  supernovae, II.1227, II.1228
  synthesis of elements, III.1747
  x-ray sources, III.1755
Burger, R, Thermos flask, I.27
Burgers, W G, dislocation theory, I.476
Burke, P G, *R*-matrix calculations, II.1069

Burnham, D C, photon coincidence, III.1435
Burns, G, superconductors, I.490
Burrow, P D, molecular resonances, II.1084
Burrows, angular distributions, II.1220
Burrus, C A, fibre lasers, III.1427
Burstein, E, absorption coefficient of semiconductor, III.1421
Busch, H
  electron motion, III.1568–9
  thin magnetic lens theory, III.1570
Buschow, K H J, permanent magnet materials, II.1165
Butler, C C
  elementary particles, II.654
  forked tracks, I.408
  kaons, II.656
Butler, S T, theory for (d,p) reactions, II.1220
Byer, R L, laser frequency, III.1435
Byrd, R
  First Antarctic Expedition, III.2005
  Second Antarctic Expedition, III.2005

Cabannes, J, combination light-scattering in gases, III.1405
Cabibbo, N, charge-changing weak current, II.677
Cady, properties of quartz, II.1244
Cagniard de la Tour, C, critical point phenomenon, III.1947
Cahan, D, measurement, I.15
Cahn, J, spinodal concept, III.1548
Callan, C, point-like constituents, II.690
Callaway, J, valence/conduction bands, III.1328–9
Cameron, A G W

explosive nucleosynthesis, III.1747
pp chain, II.1227
solar neutrinos, III.1748
supernova trigger hypothesis, III.1964
synthesis of elements, III.1747
Campbell, L L, resistivity variation with temperature, III.1282
Cannizzaro, S
  Karlsruhe congress, I.46
  molecules, I.45
Cannon, A J, stellar spectra, III.1693, III.1694
Capasso, F, photoconductivity, III.1470
Carathéodory, C
  second law of thermodynamics, I.534
  temperature concept, I.534
Carlson, C, xerography, III.1541
Carnot, S, second law of thermodynamics, I.521
Caroe, O, *William Henry Bragg*, III.1514
Carothers, W, polymer industry, III.1550
Carpenter, D, whistler measurements, III.2009–10
Carpenter, H C H
  crystal growth, III.1280
  plastic deformation, III.1526
Carrington, A, vibrational spectrum, II.1067
Carruthers, L, computational expert, III.1844
Cartan, E, non-symmetric connection, I.321
Carter, B, black holes, III.1778
Casagrande, F, laser theory, III.1444
Casimir, H B G, two-fluid model, II.924, II.925
Caspar, D L D
  tobacco mosaic virus (TMV), I.508
  tomato bushy stunt virus (TBSV), I.510
Cassels, J M, critical fluctuations, II.1016
Cassen, B
  automatic rectilinear scanner, III.1901
  isotope invariance, I.378
Castaing, R
  electron microprobe analyser, III.1548
  electron probe x-ray microanalysis, III.1580
Cauchy, A, elasticity of crystals, I.424
Caullery, M
  École Polytechnique, I.15
  laboratory equipment, I.19
Cave, H M, ionization chambers, II.1200
Cavendish, gravitation, II.1266
Caves, C M, quantum non-demolition (QND) measurements, III.1469
Cercignani, C
  BBKGY hierarchy, I.626
  The Boltzmann Equation and its Applications, I.626
Čerenkov, P A
  charged particles, II.750
  radiation from relaxation of non-uniform polarization, III.1408
Chadwick, Sir James, I.120
  α-particles, I.114, II.1190
  anomalous scattering, I.366
  continuous spectrum, I.107–8, I.118
  neutron, I.119–20, I.122, I.369, I.370, II.639, II.1183, III.1706, III.1712
  positron, II.650

Possible existence of a neutron,
 I.120
*Radiations from Radioactive
 Substances,* I.108, I.112,
 II.1199, II.1200
Chalmers, B, new journal, III.1556
Chamberlin, T C
 nebular hypothesis, III.1950
 tidal theory, III.1962
Chambers, K C, radio galaxies,
 III.1796
Chambers, R G
 boundary scattering, III.1343–4
 cyclotron resonance, III.1357
 electron free paths, III.1344
Chandler, P J, waveguide laser,
 III.1471
Chandrasekhar, S
 astrophysics, III.1706
 degenerate stars, III.1711
 dynamical friction, III.1774
 *Eddington: the Most Distinguished
  Astrophysicist of his Time,*
  III.1702–3
 equilibrium theory, III.1785
 liquid crystals, III.1540
 Radiative Transfer, III.1415–16
 Schönberg–Chandrasekhar limit,
  III.1709
Chandresekhar, S, random walks,
 I.608
Chapin, D M, solar cell, III.1421
Chapman, S, **III.1993**
 diamagnetic currents, III.1684
 geomagnetic field, III.2003
 geomagnetic storms, III.2003
 Geomagnetism, III.2000
 ionospheric layers, III.1992–4
 Solar–Terrestrial Physics, III.2001
 statistical mechanics, I.598
 upper atmosphere, III.1982,
  III.1985
Charnley, Sir John, hip
 replacement, III.1896
Charpak, G

drift chambers, II.754
 multi-wire proportional
  chambers, II.749
Cherwell, Lord, *see* Lindemann, F
 A (later Lord Cherwell)
Chester, G V, disordered
 materials, III.1366
Chevalier, R, light curve of
 SN1987A, III.1761
Chew, G F
 pion–nucleon resonance, II.680
 pion–pion scattering, II.681
 Regge trajectories, II.682
 Twelfth Solvay Conference,
  Brussels, II.646
Chiao, R Y
 optical fibres, III.1470
 stimulated Brillouin scattering,
  III.1444
Chodorow, M, high-power pulsed
 klystron, II.736
Christensen-Dalsgaard, J, low
 degree modes, III.1750
Christenson, J, *CP* violation, II.672
Christian, J, phase transformations,
 III.1547
Christoffel, tensor analysis, I.292
Christofilos, N C
 alternating-gradient focusing,
  II.734
 Argus experiment, III.1683
 ASTRON, III.1645
Christy, R F
 meson theory, I.407
 supernovae, II.1228
Cini, M, Twelfth Solvay
 Conference, Brussels, II.646
Clark Jones, R, polarization in
 optical systems, III.1411
Clauser, J F, Einstein–Podolsky
 –Rosen(–Bohm) (EPR(B))
 paradoxes, III.1457
Clausius, R
 heat engines, I.534

kinetic theory of gases, I.522–3, I.587, III.1284
On the kind of motion we call heat, I.47
second law of thermodynamics, I.521
virial theorem, I.543
Clayton, D D
  element abundances, III.1748
  isotopic anomalies, III.1964
Clifford, W K, atoms, I.31
Cline, D, proton–antiproton colliders, II.743
Clothier, W K, voltage measurement, II.1256
Cochran, W
  determinational inequalities, I.448
  helical structure, I.498
  phase transitions, II.1008
  probability relations, I.449
  pseudopotential theory, II.999
  shell model, II.997, II.1002
  structure determination, I.452
Cockcroft, Sir John
  atomic energy project, II.1198
  Atoms for Peace Conference, III.1638
  disintegration experiments, II.1184
  disintegration of lithium, II.729
  electrostatic generators, II.727
  fusion reactions, III.1628
  nuclear disintegration, II.1183
  nuclear reaction, I.285
  nuclear transformation, I.112
  proton acceleration, I.370
Coensgen, F
  deuterium plasma, III.1659
  neutral-beam injectors, III.1663
  plasma gun, III.1661
Cohen, E G D, collision configurations, I.611
Cohen, M

Landau phonon–roton dispersion relation, II.994
Landau spectrum, II.930
Cohen, M H, galactic nuclei, III.1783
Colclough, A R, gas constant (R), II.1269
Cole, E R, blind deconvolution, III.1461
Coleman, S, running coupling constant, II.713
Coles, D
  law of the wake, II.840
  law of the wall, II.840
Colgate, S
  neutral beam injection, III.1661
  neutron star, III.1759
Colle, R, molecular dissociation, II.1067
Collier, R J, holography, III.1438
Collins, M F, spin-wave modes, II.1151
Collins, S C, liquid helium, III.1341
Compton, A H, I.166
  cloud chamber, I.364
  cosmic-ray studies, III.2005
  Fifth Solvay Conference (1927), I.205
  Secondary radiation produced by x-rays, I.165
  x-ray scattering, I.362, I.364, III.1867
  x-rays, I.363
Compton, K T, photoelectric effect, III.1390
Condon, E U
  α-decay, I.115, I.217
  atomic energy levels, II.1093
  electrostatic generators, II.727
  isotope invariance, I.378
  Nature, I.116
  quantum-mechanical theory of alpha decay, II.1184
  The Theory of Atomic Spectra, II.1037

Conrady, A E, chromatic aberration, III.1390
Conversi, M, mesons, II.652
Conway, 'life' (game), I.629
Cook, A H, measurements in physic, II.1273
Coolidge, W D, cathode high-vacuum tube, III.1861
Cooper, L
  BCS theory, II.938
  superconductivity, I.232, II.939, III.1364
Cooper, M J, anharmonicity in BaF_2, I.452
Corey, R B
  biomolecular structures, I.497–8
  pleated sheet structures, I.499
  polypeptide chains, I.497
Cork, C, spark chamber, II.753
Cornford, F M, Cambridge University, I.19
Coster, D, hafnium, I.100
Cotton, Sir Alan, magneto-optical effect, III.1389
Cottrell, Sir Alan, **III.1522**
  discontinuous yield-stress in mild steel, III.1520–1
  dislocation theory, III.1523
  physical metallurgy, III.1555
  yield-stress and strain-aging, III.1523
Coulomb, C A, electrical and magnetic experiments, III.2004
Coulson, C H, computations, II.1087
Courant, *Methoden der Mathematischen Physik*, III.2043
Courant, E D
  accelerators, II.738
  synchrotrons, II.734
Cowan, C, neutrino, II.663
Cowling, T G, axially symmetric field, III.1958

Cox, A, geomagnetic field, III.1973
Cox, D P, interstellar medium, III.1766
Craik, D J, domain techniques, II.1155
Crasemann, B, secondary excitations, II.1090
Crawford, B H, human eye, III.1407
Crewe, A, high brightness field emission sources, III.1589
Crick, F H C
  DNA model, I.507
  double-helix structure, I.504
  helical structure, I.498
  polypeptides, I.498
  protein subunits in viruses, I.509
Critchfield, pp process, II.1226
Croat, J J, neodymium–iron borates, II.1165
Croce, P, coherent imaging, III.1418
Crommelin, eclipse expeditions, III.1722
Cronin, J W
  CP violation, II.672
  spark chamber, II.753
Crookes, W
  cathode rays, II.1271
  phosphorescence of zinc sulphide, II.745
Cross, L E, ceramics, III.1538
Crowfoot, D M
  globular proteins, I.497
  tomato bushy stunt virus (TBSV), I.509
Cummerow, R L, electron spin resonance, II.1138
Cummins, H Z, Doppler shift, III.1442
Curie, I, radioactivity, I.121
Curie, M, I.57
  atoms, I.49
  Fifth Solvay Conference (1927), I.205

Name Index

First Solvay Conference (1911), I.147
  radioactivity, I.56–8, I.60
  radium, I.6, II.1272, III.1863, III.1876
  uranium, I.61
Curie, P
  diamagnetism, II.1116
  ferromagnets, II.1118
  magnetism, I.545, II.1113
  Nobel Prize lecture, I.63
  radioactivity, I.56–8
  radium, III.1876
  uranium, I.61
Curien, H, x-ray scattering, II.990
Curle, N
  jet-edge system, II.852
  sound waves, II.851
Curtis, H D
  Great Debate, III.1718
  novae, III.1717
Cusack, N E, liquid metals, III.1366
Cvejanovic, quantum-mechanical reformulation, II.1065
Cvitanovic, P, universal function, III.1848
Czochralski, single-crystal wires, III.1281

Dabbs, J W T, germanium junctions, II.1208
Dagenais, M, photon antibunching, III.1460
Daguerre, L-J-M, daguerreotype process, III.1692
Dainty, J C, blind deconvolution, III.1461
Dällenbach, W, invariant averaging procedure, I.281
Dallos, J, scleral lenses, III.1414
Dalrymple, G B, geomagnetic field, III.1973
Dalton, B J, laser fluctuations, III.1473
Dalton, J
  learned societies, I.3
  New System of Chemical Philosophy, I.44
Damadian, R, nuclear magnetic resonance, III.1923
Damm, C C
  Alice mirror machine, III.1679
  neutral beam injection, III.1661
Danby, G, two-neutrino experiment, II.672
Daniel, V, copper–nickel–iron alloy, III.1548
Darmois, gravitational field, I.314
Darrow, K K, First Shelter Island Conference, II.645
d'Arsonval, surgical diathermy, III.1894
Darwin, C G
  biological evolution, III.1947
  dispersion formula, I.171
  extinction, I.451
  integrated intensity formula, I.434
  magneto-ionic theory, III.1992
  Raman effect (combination scattering), III.1404
  relativistic hydrogen spectrum, I.218
  statistical mechanics, I.535
  structure factor, I.431
Darwin, G H
  Moon–Earth hypothesis, III.1951–2
  Moon model, III.1966
  Moon's orbit, III.1948
Das Gupta, M K, Cygnus A, III.1775
Dash, W C, diffusion of copper impurity, III.1526
Dashen, R, axial–vector coupling constant and pion–nucleon scattering, II.679
Daunt, energy gap, II.932

Davidovits, confocal scanning laser microscope, III.1450
Davies, D
  detection of impulses, III.1957
  Earth's core, III.1980
da Vinci, L, medical mechanics, III.1856
Davis, D J M Jr, ring-laser gyroscope, III.1433
Davis, D R, Moon model, III.1966
Davis, E A, *Electron processes in Non-crystalline Materials*, III.1369
Davis, R, solar neutrinos, II.757, III.1748
Davisson, C
  electron diffraction, I.463
  electron reflection, I.170
Davy, Sir Humphry, learned societies, I.3
Dawson, B, temperature factor, I.452
Dawson, J, fusion systems, III.1673
Day, A L, report to Philosophical Society of Washington, III.1387
Dean, P, amorphous SiO_2, I.479
de Boer, J H
  critical temperature, II.975
  electrical properties, II.1156
  law of corresponding states, II.974
  semiconductors, III.1330
Debrenne, P, cryostat, I.437
de Broglie, L, I.170
  elementary particles, I.282–3
  Fifth Solvay Conference (1927), I.205
  First Solvay Conference (1911), I.147
  matter-wave theory, I.174, I.229
  On the general definition of the correspondence between wave and (particle) motion, I.169
  wave mechanics, I.227
  wave–particle duality, I.168–9
de Broglie, M
  crystal oscillation, I.434
  x-ray spectra, I.160
Debye, P, **II.971**
  adiabatic demagnetization, II.1125
  dielectric waveguide, III.1397
  dynamical laws, I.176
  electron screening, III.1316
  energy spectrum, I.153
  Fifth Solvay Conference (1927), I.205
  Planck's law, I.70
  screening length, III.1610
  single crystal, I.434
  specific heat, II.971–2, II.977, II.980, II.982, II.1011, III.1398
  thermal conductivity, II.977–8
  thermal vibration, I.431
  vibrational specific heat of solids, I.525
  x-ray scattering, I.165
  x-rays, I.363
  Zeeman effect, I.102
de Forest, L
  triode, I.37
  upper ionized layer, III.1987
de Gennes, P G
  liquid crystals, II.1084, III.1540
  SAW model, I.570
  soft physics, III.2030
de Haas, W J
  Kondo effect, II.1167
  potassium chromic alum, II.1126
de Hevesy, G, quantum theory, I.180
Dehmelt, H G, charged particles, II.1099
Delaunay, C-E, Earth–Moon system, III.1948
Delchar, T A, surface crystallography, I.466
Dellinger, J H, sudden ionospheric disturbance (SID), III.1995

Demirkhanov, R, Atoms for Peace Conference, III.1638
Democritus, atoms, I.43, III.2034–5
Denisyuk, Yu N, holograms, III.1438
Dennison, D M, wavefunctions, I.529
Derjagin, B V, charge effects, III.1611
Derrida, B, delta (δ), III.1849
De Rújula, A, charmed particles, II.704
des Cloizeaux, J
  excitations in $s = 1/2$ Heisenberg antiferromagnet, II.1014
  polymer solutions, III.1599
De Salvo, L, Collective Atomic Recoil Laser, III.1456
Desclaux, J P, shell electrons, II.1091
de Sitter, W
  cosmological constant, III.1724
  double-star systems, I.285
  Einstein–de Sitter model, III.1730–1
  spherical four-dimensional space-time, III.1723
  Universe, III.1724
Destriau, G, electroluminescence, III.1417
Deutsch, M, positronium, II.646
de Vaucouleurs, G, Hubble's constant, III.1793
de Vries, G, weak non-linear and weak dispersive effects, II.859
Devonshire, A F, ferroelectric transition in $BaTiO_3$, II.1007
Dewar, J, liquefied gases, I.28
Dick, B J, shell model, I.485, II.997
Dicke, R H
  Big Bang, III.1790
  Fabry–Perot resonant cavity, III.1423
  optical bomb, III.1419
  solar oblateness, I.303
  world models, III.1806
Dickinson, R G, hexamethylene tetramine, I.495
Dietrich, O W, order–disorder transition in β-brass, II.1009, II.1016
di Francia, T, lens-pupil functions, III.1418, III.1457
Dillon, J F
  bubble domains, II.1159
  domain techniques, II.1155
Dimov, G I, tandem-mirror concept, III.1665–6
Dingle, R, superlattice minibands, III.1459
Dirac, P A M, I.189
  atomic mechanics, I.183
  electromagnetic field, I.358
  electron spin, II.1125
  electron statistics, I.528
  Fifth Solvay Conference (1927), I.205
  four-component wavefunctions, I.277
  free particles, I.283
  particle dynamics, I.279
  properties of the Universe, III.1787
  quantization of free field, III.1402
  quantum electrodynamics (QED), I.360, I.362
  quantum mechanics, I.188, I.198, I.211
  Quantum Mechanics, III.2043
  relativistically invariant theory, I.283
  transformation theory, I.201, I.360
  Twelfth Solvay Conference, Brussels, II.646
  wavefunctions, I.200
Dixon, J M, crystal fields, II.1165

N19

Doell, R, geomagnetic field, III.1973
Doi, M
 entangled systems, III.1600
 monodisperse melt, III.1600
Dolen, R, pion–nucleon charge exchange, II.683
Domb, C
 critical behaviour, I.556, I.560
 critical exponents, I.560
 magnetic field at critical point, I.562
 order–disorder transformation, I.578
 site percolation processes, I.572
Dombrovski, V A, optical continuum of nebula, III.1774–5
Domen, S R, calorimetry, III.1883
Donald, I, diagnostic ultrasound, III.1912–13
Donder, Th De, Fifth Solvay Conference (1927), I.205
Döring, W, nucleation kinetics, III.1547
Dorsey, N E, speed of light, II.1263
Drell, S D, lepton pair, II.690
Dresden, M, game rules, I.629
Dresselhaus, G
 phonons, III.1423
 resonant frequencies, III.1334
Dreyer, J L E
 Index Catalogues, III.1717
 New General Catalogue of Nebulae and Clusters of Stars, III.1717
Droste, J, field equations, I.303
Drude, P K L
 electron optics, III.2000
 electron vibration, II.983
 Hall effect, III.1282
 magneto-ionic theory, III.1990
 metallic conduction, III.1284
 school of physics, I.15
 The Theory of Optics, III.1385
Dryden, H L
 closed-circuit wind-tunnels, II.830–1
 turbulence, II.798
Duane, W
 quantum physics, III.1401–2
 radiation quanta, I.165
 radon, III.1876
 röntgen, III.1867
Duffieux, P M, *The Fourier Transform and its Applications in Optics*, III.1413
Dumke, P W, semiconductor lasers, III.1433
DuMond, J W M, curved crystal focusing spectrometer, II.1207
Dumontet, P, polychromatic fields, III.1421
Dungey, J, magnetosphere, III.2010
Dunlop, J S, radio sources, III.1798
Dunning, J R, fission experiments, II.1196
Dunoyer, L, surface crystallography, I.468
Duraffourg, G, stimulated emission in semiconductors, III.1433
Duwez, P, quenching of hot alloys, III.1542–3
Dyson, F J
 spin-wave theory, II.1011
 Twelfth Solvay Conference, Brussels, II.646
Dyson, J
 Fresnel fringe system, III.1584
 holograms, III.1584

Eberly, J H, unstable and chaotic emission, III.1466
Eccles, W H, radio waves, III.1622, III.1987
Eckart, C
 continuous media, I.280
 equivalence relations, I.196

Eckart, M, theory of metals,
  III.1290
Eckersley, T L
  long-distance communications,
    III.1999
  magneto-ionic theory, III.1992
  solar x-rays, III.1994
  upper ionized layer, III.1987
  whistler waves, III.1622–3,
    III.1995, III.2002
Eddington, A S, **III.1700**
  eclipse expeditions, III.1722
  Eddington–Lemaître model,
    III.1732–3
  energy transport, III.1699
  fine structure constant, II.1262
  Fundamental Theory, III.1787
  gravitational waves, I.312
  internal structure of the stars,
    III.1703
  interstellar medium, III.1761
  mass–luminosity relation,
    III.1705
  Mathematical Theory of Relativity,
    III.1700
  red giants, III.1704
  The Fundamental Theory, III.1700
  *The Internal Constitution of the
    Stars*, III.1700, III.1702
  theory of stellar structure,
    III.1702–6
  virial theorem, III.1731–2
  white dwarfs, III.1710
Edgerton, H E, stroboscopic
  photography, III.1410
Edler, I, time–position recordings,
  III.1913–14
Edwards, O S, stacking faults,
  I.473
Edwards, S F
  entangled systems, III.1600
  monodisperse melt, III.1600
  quantum-mechanical particles,
    III.1599

Egger, confocal scanning laser
  microscope, III.1450
Ehlers, J, general relativity, I.299
Ehrenberg, W, LEED experiments,
  I.466
Ehrenfest, P
  adiabatic principle, I.92–3
  dog-flea model, I.604, I.618
  Fifth Solvay Conference (1927),
    I.205
  higher-order transitions, I.549
  periodic systems, I.177
  phase transitions, II.923–4
  spherical four-dimensional
    space-time, III.1723
  statistical mechanics, I.586
  thermodynamics and
    irreversibility, I.599
  wind tree model, I.603
Ehrenfest, T
  statistical mechanics, I.586
  thermodynamics and
    irreversibility, I.599
Eindhoven, W, medical electronics,
  III.1893
Einstein, A, **I.250**
  anno mirabile, III.2018
  binding energy, I.110–11
  black-body radiation, I.163
  Bohr's theory, I.88
  Brownian motion, I.586
  conservation of energy, III.1390
  cosmological constant, III.1723,
    III.1730
  dielectric constant, III.1396–7
  Einstein–de Sitter model,
    III.1730–1
  energy surface trajectories, I.590
  Fifth Solvay Conference (1927),
    I.205
  First Solvay Conference (1911),
    I.147
  fluctuations, I.533–4
  gas of corpuscles, III.1390–2
  general theory of relativity,

I.251, I.253, I.286–320, III.1722, III.1723
gravitation, I.292
heat capacity, I.78
light-quanta, I.54, I.71, I.151–2, I.167, I.173, I.363, III.1396, III.1452
nucleation of liquid droplets, III.1547
opalescent bands, I.546
optical microscopy, III.1406
quanta, II.969
quantum concept, I.71, I.147, I.228, I.585, III.1398–9
quantum physics, I.93–4
radiation in thermodynamic equilibrium, III.1392–3
radiation law, I.152
relativity theory, I.34, I.61, I.173
space-time structure, I.251
special theory of relativity, I.54, I.249, I.251, I.262–85, III.2039
specific heat, I.153–4, II.970
statistical mechanics, I.173
stress–energy tensor, I.281
unified-field theory, I.232
Ekins, R P, saturation analysis technique, III.1909
Elam, C I
  crystal growth, III.1280
  plastic deformation, III.1526
Elias, G J, solar x-rays, III.1994
Elias, P, communications theory, III.1417
Eliashberg, G M, electron–phonon interaction, II.941
Elliott, R J, phonons, III.1423
Ellis, C D
  continuous spectrum, I.118, I.126
  internal conversion, I.107
  Radiations from Radioactive Substances, I.108, I.112, II.1199, II.1200
Ellis, R E, mirror machine, III.1679
Elsasser, W M

Earth's magnetic field, III.1969–70
electron reflection, I.170
geomagnetic dynamo, III.1958
neutron diffraction, I.453
shell models, II.1209
toroidal (doughnut-shaped) fields, III.1958
Elster, phosphorescence of zinc sulphide, II.745
El-Sum, H M A, holography, III.1415
Emanuel, K A, tropical cyclones, II.904
Emden, R, *Gaskugeln*, III.1695
Emmons, H W
  laminar flow, II.825
  randomly originating turbulent spots, II.835
Empedocles, elements, I.44
Enskog, D, statistical mechanics, I.598
Eötvös, R V
  equivalence principle, I.286
  gravitation, II.1267
Epicurus, atoms, I.44
Epstein, J B, energy in the Sun, III.1748
Epstein, P S
  dynamical laws, I.176
  quantum theory, I.89
  Stark effect, I.162
Ernst, M, non-linearity, I.625
Ertel, H, tropical cyclones, II.905
Esaki, L
  coupling of quantum wells, III.1459
  germanium, III.1340
  quantum-well superlattices, III.1453
Eshelby, J, physical metallurgy, III.1555
Essam, J W
  critical exponents, I.560
  pair connectedness, I.573

two-dimensional lattices, I.573
Essen, L, speed of light, II.1263
Estermann, I, surface
 crystallography, I.468
Eucken, A, II.972
 calorimeter, II.968
 liquid-air temperature, II.977
 quantum mechanics, I.146
 specific-heat problem, I.153–4
Evans, E, liposomes, III.1608
Eve, A S
 ionizing agents, I.367
 radium, III.1881
 secondary x-rays, I.362
Ewald, P P
 crystal structures, III.1514
 Fifty Years of X-ray Diffraction,
  I.511
 reciprocal lattice, I.428
 x-ray diffraction, I.424
Ewan, H I, hyperfine line
 radiation, III.1763
Ewing, Sir Alfred, plastic
 deformation, III.1516
Ewing, D H, compound nucleus,
 II.1219

Faber, T E, liquid metals, III.1366
Fabrikant, V A, light amplification,
 III.1411
Failla, G, medical physics, III.1864
Fairbank, W M
 λ-point of liquid He^4, I.559
 superfluidity, II.932
Fankuchen, I, tobacco mosaic
 virus (TMV), I.508
Fano, U, resonance, III.1473
Faraday, M
 Chemical History of a Candle, I.5
 conductivity, III.1318
 disliquifying point, I.543
 electric current, III.1958
 electrical properties, II.1113
 electrolysis, I.51

electromagnetic interactions,
 III.1953
electromagnetism, III.1856,
 III.2000
experimental physics, I.14
geomagnetic variations, III.1984
learned societies, I.3
magnetic properties, II.1112,
 II.1113
Farnsworth, H G, surface
 crystallography, I.470
Faust, W L, CO_2 laser, III.1442
Feher, E R, crystal field theory,
 II.1144
Fellgett, P, multiplex optical
 spectroscopy, III.1416
Felten, J, luminosity function,
 III.1770
Fermi, E, **II.1187**
 β-decay theory, I.374–5, I.375
 electron statistics, I.528
 neutron bombardment, I.122
 neutron capture, II.1186, II.1193
 neutron diffraction, I.456
 neutron irradiation, I.129–30
 neutron sources, II.1199, II.1206
 quantum mechanics, II.1031
 quantum states, III.1288
 Raman scattering, II.996
 synchrocyclotron, II.733
 *Tentativo di una teoria della
  emissione di raggi β*, I.129
 theory of β-decay, III.1706
Fermi, L, *Atoms in the Family*,
 II.1191
Ferraro, V C A
 diamagnetic currents, III.1684
 geomagnetic field, III.2003
Ferrel, W, Earth–Moon system,
 III.1948
Feshbach, H
 compound nucleus, II.1219
 First Shelter Island Conference,
  II.645

projection-operator technique, II.1221
proton scattering, II.1212
Feynman, R P, **II.648**
  First Shelter Island Conference, II.645
  Landau phonon–roton dispersion relation, II.994
  Landau spectrum, II.930
  neutrinos, II.667
  parity states, II.665
  point-like constituents, II.689
  quantum electrodynamics, II.643, II.740, III.1446
  quantum mechanics, III.1395, III.1830
  renormalization, II.646
  scientific imagination, III.1519
  Twelfth Solvay Conference, Brussels, II.646
  weak vector current, II.668
Field, G B, interstellar medium, III.1765
Fienup, J R, optical data processing, III.1461
Fierz, M, free particles, I.283
Finch, J F, polio virus, I.510
Finkelstein, D, Schwarzschild singularity, I.308
Finney, J L, amorphous structure, I.479
Firestone, F A, ultrasonic flaw detector, III.1911
Fischer, E, polyethylene, III.1552
Fischer, K, AgCl, II.998
Fisher, I Z, neutron frequency, III.1418
Fisher, M E
  covering lattice, I.573
  critical exponents, I.560
  field theory, III.1830
  pair distribution function, I.547
Fisher, O, Earth–Moon system, III.1948
Fitch, V, *CP* violation, II.672

Fitzgerald, G F
  Fitzgerald–Lorentz contraction, I.264
  wireless wave propagation, III.1986
Fizeau, A H L, velocity of light, I.266
Fleming, J A, *Terrestrial Magnetism and Atmospheric Electricity*, III.2000
Flory, P, **III.1598**
  excluded volume, I.569
  polymer chains, III.1551
  polymer solutions, III.1599
  self-avoiding walk, III.1598
Fock, V
  configuration (Fock)-space method, I.208
  equations of motion, I.303
  self-gravitating systems, I.313
Foëx, G, *Le Magnetisme*, II.1119
Foley, electric quadrupole moments, II.1214
Follin, J W, proton–neutron ratio, III.1786
Fomichev, V I, K-shell absorptions, II.1083
Foord, R, photon correlation spectroscopy, III.1454
Ford, F C, mirror machine, III.1679
Försterling, K, wave propagation, III.1991
Fortuin, C, bond percolation model, I.573
Foucault, J B L, pendulum, I.2, I.3
Fowler, R H
  electron gas, III.1711
  Fifth Solvay Conference (1927), I.205
  high-density matter in stars, I.215
  ionization, III.1701
  quantum theory, I.180
  residual entropy of ice, I.571
  statistical mechanics, I.535

Statistical Mechanics, I.533
Statistical Thermodynamics, I.537
stellar spectra, III.1693
structure of ice, I.474
white dwarf stars, I.531
Fowler, T K, tandem-mirror concept, III.1665–6
Fowler, W A
  Big Bang, III.1792
  nuclear interactions between light nuclei, III.1791
  nucleosynthesis in stars, III.1788
  origin of the elements, II.1225
  pp chain, II.1227
  solar neutrinos, III.1748
  supernovae, II.1227, II.1228
  synthesis of elements, III.1747
Fox, A G, Fabry–Perot resonator, III.1429
Franceschini, W, delta (δ), III.1851
Franck, J
  Franck–Hertz experiment, I.89–90
  quantum theory, I.179
  scattering experiments, I.217
  stationary electron orbits, I.158
Frank, A, rare-earth ions, II.1120
Frank, Sir Charles, **III.1606**
  crystal growth, III.1527
  crystallinity, III.1551–2
  line singularities in nematics, III.1605
Frank, I, energy loss formula, II.750
Frank, I M, visible light, III.1409
Frank, N H, theory of metals, III.1290
Franken, P A, ruby laser, III.1427–9
Franklin, B, learned societies, I.3
Franklin, R E
  DNA, I.504
  tobacco mosaic virus (TMV), I.508

Franz, W, photo-assisted tunnelling, III.1423
Fraunhofer, J
  solar spectrum, III.1691–2
  spectral lines, I.156
Frautschi, S C, Regge trajectories, II.682
Frazier, E N, solar oscillations, III.1749–50
Freedman, S J, Einstein–Podolsky–Rosen(–Bohm) (EPR(B)) paradoxes, III.1457
Freeman, A J, rare-earth elements, II.1164
Freidmann, A A, world models, III.1724–5
Frenkel, J
  phonon, III.1302
  semiconductors, III.1319–20
  sound waves, III.1328
  superconductivity, II.917, II.936
  vacancies, III.1529
Fresnel, A
  diffraction, III.1386
  wave description of light, I.72
Freud, P von, energy complexes, I.301
Freundlich, E F
  general relativity, I.305
  red shift, I.289
Fricke, H, chemical dosimetry, III.1884
Friedel, G
  focal conic defects, III.1605
  liquid crystals, III.1540
  mesomorphic phases, III.1603
Friedman, H
  Crab Nebula, III.1739
  rocket experiments, III.1738
  rocket studies, III.2005
  world models, III.1784, III.1807
Friedman, J I, parity violation, II.666
Friedrich, W
  crystal structures, III.1514

x-ray diffraction, I.425
Frisch, O R
  fission hypothesis, II.1195
  neutron–U collisions, I.130
Fritzsch, H, quarks, II.711
Fritzsche, H, semiconductors, III.1330
Fröhlich, H
  electron–phonon interactions, III.1337–8
  heavy electrons, I.392
  lattice deformation, III.1321
  meson theory, I.394
  metallic conductivity, I.216
  superconductivity, II.936
Froome, K D, speed of light, II.1263
Fry, D W, linear accelerators, II.1206
Fry, W J, ultrasonic surgery, III.1915
Fuchs, K, properties of fluids, I.553
Fukui, S, spark chamber, II.753
Fuller, C S, solar cell, III.1421
Fuller, L, upper ionized layer, III.1987
Furnas, T C, powder photographs, I.436
Furry, W, positrons in electron self-energy, II.641

Gabor, D, **III.1567**
  diffraction microscopy, III.1582
  electron microscope, III.1414–15, III.1581
  holography, III.1456, III.1582–4
  magnetic electron lens, III.1566–9
  super-lens, III.1411
Gagarin, Y, orbital flight, III.1738
Gaillard, J-M, two-neutrino experiment, II.672
Galileo Gallilea, I.254
  mechanical principal of relativity, I.254
  principle of inertia, III.1827
  speed of light, II.1262
Galison, P, magnetic moment and angular momentum, II.1123
Gamow, G
  abundances of elements, III.1785–6
  α-decay, I.115, I.217, II.1184
  α-particles, II.1189
  α-scattering, II.1189
  Atomic Nuclei and Nuclear Transformations, II.1225
  Big Bang, III.1784–6
  black-body radiation, II.758
  Constitution of Atomic Nuclei and Radioactivity, I.115
  cosmological constant, III.1730
  electrostatic generators, II.727
  fusion reactions, III.1628
  heavier elements, II.1226, II.1227
  liquid drop, I.126, II.1186
  neutrons, III.1713
  quantum-mechanical tunnelling, III.1706
  The Constitution of Atomic Nuclei and Radioactivity, II.1186
Garmire, E, stimulated Brillouin scattering, III.1444
Garrett, C G B
  anharmonic oscillators, III.1434
  two-photon absorption, III.1429
Garwin, R, parity violation, II.666
Gauss, C F, geomagnetism, III.1984
Geffken, W, metal–dielectric interference filters, III.1407
Gehrcke, E, Lummer–Gehrcke plate, III.1389
Geiger, H
  α-particles, I.113, II.745–6
  α-scattering, II.1200
  β-spectrum, I.107
  coincident counting, I.364
  electrical counting methods, II.1200

electron charge, II.747
Nature of the α-particle, I.58
x-ray scattering, I.166
Geissler, J, improved vacua, I.54
Geitel, phosphorescence of zinc sulphide, II.745
Geller, M, bright galaxies, III.1772
Gell-Mann, M, II.678
  axial coupling constant, II.676
  axial–vector coupling constant and pion–nucleon scattering, II.679
  baryons, II.661
  β-decay, II.677
  correlation effects, III.1362
  isotopic spin multiplets, II.658
  mass relations among baryons and mesons, II.661
  neutrinos, II.667
  quark–lepton analogy, II.700
  quarks, II.711
  renormalization group, III.1830, III.1832
  strangeness, II.662
  SU(3) triplets, II.685
  The Quark and the Jaguar, II.770
  Twelfth Solvay Conference, Brussels, II.646
  unitary symmetry, II.677
  weak hadronic current, II.678
  weak vector current, II.668
Genna, S, calorimetry, III.1883
Georgi, H, charmed particles, II.704
Gerlach, W, beam splitting, I.165
Germain, G, structure determination, I.450
Germer, L H, de Broglie waves, I.171
Gershtein, S S
  neutrinos, III.1803
  weak vector current, II.668
Gerstenkorn, H, Moon model, III.1966

Ghosh, S P, ionospheric physics, III.2000
Giacconi, R, Sco X-1, III.1739
Giaever, I, energy gap, II.933
Giauque, W F
  adiabatic demagnetization, II.1125
  gadolinium sulphate, II.1126
Gibbs, H M
  optical bistability, III.1449
  Optical Bistability, III.1450
  thermal bistable effects, III.1450
Gibbs, J W, **I.522**
  canonical ensemble, I.526
  Elementary Principles of Statistical Mechanics, I.586, I.593
  ensembles, I.531–3
  heterogeneous systems, I.34–5
  identical particles, I.528
  partition function, I.527
  phase equilibria, III.1507–8
  PhD, I.17
  statistical mechanics, I.173, I.523, I.606
  thermodynamics, I.521
Gibson, angular distributions, II.1220
Gibson, G, beta-ray experiment, III.1679–80
Gilbert, W
  De Magnete, I.544, III.1855–6
  Earth's magnetism, III.1952–3
  electromagnetism, II.1252
  magnetic forces, III.1952
  magnetism, II.1111
Gilliland, T R, ionosonde, III.1989, III.1991
Ginsburg, V, Čerenkov effect, III.1409
Gintzon, E L
  high-power pulsed klystron, II.736
  linear accelerators, II.736
Ginzburg, D M
  energy gap, II.932

N27

superconductivity, II.934–5
Ginzburg, V L, Galactic radio emission, III.1737–8
Giorgi, G, MKSA system, II.1235
Gires, F, pump-probe configuration, III.1445
Gittelman, B, electron storage rings, II.739
Gittus, J, radiation damage, III.1534
Glaser, D A, bubble chamber, II.659, II.753
Glaser, W, electron optics, III.1575
Glashow, S L, II.696
  charmed particles, II.704
  charmed quark, II.700, II.705
  neutral currents, II.700
  quark–lepton analogy, II.700
  SU(2) x U(1) group, II.694
Glauber, R J
  neutron frequency, III.1418
  photodetection, III.1441
  quantum theory of coherence, III.1439–41
  radiation field, III.1388
Glazebrook, R T (later Sir Richard), **II.1253**
Gleick, J, *GENIUS—The Life of Richard Feynman*, III.1519
Gold, A V, de Haas–van Alphen effect, III.1352
Gold, T
  magnetosphere, III.1981
  pulsars, III.1752
  Relativistic Astrophysics Symposium, III.1776
  steady-state cosmology, III.1787
Goidberg, H, triplets, II.685
Goldberg, J, conservation laws, I.301
Goldberger, M L
  beta-decay, II.676
  pion decay constant, II.675
  Twelfth Solvay Conference, Brussels, II.646

Goldhaber, A S, photon mass, III.1452
Goldhaber, G, charmed meson, II.704
Goldhaber, M
  dipole vibration, II.1213–14
  neutrino helicity, II.668, II.670
  shell model, II.1215
Goldman, I M, dielectric properties of BaTiO$_3$, I.486
Goldschmidt, V M
  First Solvay Conference (1911), I.147
  inorganic compounds, I.492
Goldsmith, D W, interstellar medium, III.1765
Goldsmith, H H, neutron sources, II.1199
Goldstein, S
  energetic particles, III.1682
  magneto-ionic theory, III.1990
Goldstone, J, general theorem, II.676
Golovin, I N
  beam-fed mirror experiments, III.1662
  magnetic mirror, III.1659
Goodman, C, Coulomb excitation experiments, II.1217
Goodman, P, electron diffraction, I.464
Gordon, J P
  maser, III.1419
  relativistic hydrogen spectrum, I.218
  resonator modes, III.1429
Gordon, W E
  incoherent scatter radar (ISR), III.1996
  Schrödinger method, I.365
Gor'kov, L, BCS theory, II.940
Gorter, C J
  magnetic field, II.1133
  nuclear resonance, II.1172
  relaxation time, II.1133

superconductivity, II.921
Twelfth Solvay Conference, Brussels, II.646
two-fluid model, II.924, II.925
Gorter, E W, ferrimagnetism, II.1157
Görtler, H, boundary layers, II.832–3
Gott, J R, distribution of galaxies, III.1772
Gottleib, M, Atoms for Peace Conference, III.1638
Goudsmit, S
  atomic energy levels, II.1093
  electron spin, I.101
  magnetic moment, II.1123
  quantum number, I.168
  spinning electron, I.360
Gough, D O, low degree modes, III.1750
Gould, G, lasers, III.1423
Gould, L, polar research, III.2006
Goulianos, K, two-neutrino experiment, II.672
Goutier, Y, perturbation theory, III.1473–4
Gouy, A, gravity effects, I.553
Gowing, M, *Britain and Atomic Energy 1939–45*, II.1198
Grad, H
  magnetic cusp, III.1645
  non-linearity, I.625
Gratias, D, quasicrystals, I.481
Gray, L H, **III.1869**
  Bragg–Gray theory, III.1882
  radiotherapy, III.1867
Green, H S, parastatistics, II.687
Green, M B, string theories, II.684
Green, M S, collision configurations, I.611
Green, R F, quasars, III.1798
Greenberg, O W
  quarks, II.687
  SU(3) symmetry, II.709

Greenburger, D M, three-photon interferometry, III.1467
Greene, C H
  critical behaviours, II.1067
  R matrices, II.1069
Greene, J M, plasma oscillations, III.1625
Greening, J R, dosimetry, III.1884
Gregory, Sir Richard, *Nature*, I.6
Greinacher, H, α-particle, II.748
Gribov, V N, singularities in complex angular momentum, II.682
Griffin, J, growth spiral, III.1528
Griffith, A A, crack growth, III.1559
Griffiths, J H E, ferromagnetic resonance, II.1152–3
Griffiths, R B, smoothness postulate, I.563
Gringauz, K I
  plasmapause, III.2009
  solar wind, III.2009
Grodzins, L, neutrino helicity, II.668, II.670
Grondahl, L O
  photo-voltaic effect, III.1321
  semiconductors, III.1318
Gross, D
  point-like constituents, II.690
  running coupling constant, II.713
Gross, E, spectral shift, III.1398
Gross, E P, plasma oscillations, III.1625
Grossmann, M, gravitation, I.292
Groth, crystallography, I.424
Grüneisen, E, thermal expansion, II.975–7
Gubanov, long-range order, III.1368
Guckenheimer, J, MSS, III.1850
Gudden, B
  photo-ionization, III.1320
  single crystals, III.1318–19

temperature variation of resistivity, III.1330
Guen, H S, BBKGY hierarchy, I.606, I.623, I.624, I.626
Guggenheim, E A
  coexistence curves, I.555
  Statistical Thermodynamics, I.537
Guggenheimer, J, shell models, II.1209
Guinier, A, age-hardening in aluminium–copper alloys, III.1546
Gull, S F, supernova remnants, III.1782
Gunn, J E
  electron acceleration, III.1752–3
  mass–luminosity ratio, III.1800
  phase space constraints, III.1804
Gurney, R
  α-decay, I.115, I.217, II.1184
  colour centres, III.1530
  electrostatic generators, II.727
  Nature, I.116
  semiconductors, III.1319
  solid-state physics, III.1314
Gurvich, defibrillators, III.1895
Gusyenov, O H, binary sources, III.1756
Gutenberg, B
  earthquake waves, III.1956
  seismological analysis, III.1954
Guth, A, Universe model, III.1806
Guye, Ch E, Fifth Solvay Conference (1927), I.205
Gyulai, Z, colour centres, III.1529–30

Haake, F, laser theory, III.1444
Haas, A E
  Bohr radius, I.81
  quantum formula, I.156
Haasen, P, *Physikalische Metallkunde*, III.1555
Habgood, H W, isotherms of xenon, I.556

Habing, H J, interstellar medium, III.1765
Hadravský, M, confocal scanning laser microscope, III.1450–1
Hafstad, L, ionospheric plasmas, III.1621
Hahn, E L, resonant absorption, II.1174
Hahn, J, stellar mass, III.1699
Hahn, O, I.105
  β-ray deflection, II.1201
  β-spectrum, I.104, I.105, I.106
  elements with $Z = 93$–96, I.130
  neutron bombardment, II.1194, II.1195
  neutron sources, II.1206
  neutron–U collisions, I.130
  uranium isotope, II.1223
  uranium-239, I.132
Haine, M E, Fresnel fringe system, III.1584
Haldane, F D M, $s = 1$ Heisenberg chains, II.1014
Hale, G E, magnetic fields in sunspots, III.1958
Hale, J E, Palomar 200" telescope, III.1721
Hall, E H
  physics in education, I.9
  practical classes, I.15
  resistivity variation with temperature, III.1282
Halley, E, Earth's magnetic field, III.1953
Halliday, D, electron spin resonance, II.1138
Hallwachs, W L F, photoelectric effect, III.1390–1
Halperin, B I, ultrasonic properties, II.1006
Halpern, O
  neutron diffraction, I.459
  quantum theory, I.183
Hamilton, W R, dynamical laws, I.176

Hammersley, J M
  bond percolation processes, I.572
  percolation processes, I.572
  self-avoiding walk (SAW), I.570
Han, M Y
  quarks, II.687
  SU(3) symmetry, II.709
Hanbury Brown, R
  optical spectroscopy, III.1421
  radio interferometer research, III.1422
Haneman, D, surface crystallography, I.470
Hanna, R C, two-photon correlation, III.1413
Hansen, M, binary systems, III.1509
Hansen, W W
  electron linear accelerators, II.1205
  linac, II.736
  linear accelerators, II.736
Hanson, D
  physics of materials, III.1555
  plastic deformation, III.1526
Hanssen, K-J, atomic resolution, III.1585–6
Hara, Y, quark–lepton analogy, II.700
Hariharan, P, holography, III.1438
Harker, D
  inequality relations, I.447
  Patterson function, I.443
Harper, S, strain-aging law, III.1521
Harrison, E R, structures in the Universe, III.1802
Harrison, S C, tomato bushy stunt virus (TBSV), I.510
Harte, B, *Plain Language from Truthful James*, I.6
Hartley, spherical micelles, III.1606
Hartmann, J, wavefront aberrations, III.1389

Hartmann, W K, Moon model, III.1966
Hartree, D R
  magneto-ionic theory, III.1992
  multi-electron spectra, II.1036
  self-consistent field method, I.432, I.433
  wireless propagation, III.1622
Harvey, E N, luminescence, III.1537
Hasenöhrl, First Solvay Conference (1911), I.147
Hasinger, G, x-ray source counts, III.1798
Hastie, R J, plasma pressures, III.1661
Hauptman, H
  determinantal inequalities, I.448
  tangent formula, I.449
  The Solution of the Phase Problem. I. The Centrosymmetric Crystal, I.449
Hauser, W, compound nucleus, II.1219
Haüy, R J, lattice theory, I.422–3
Hawking, S
  black holes, I.317, II.761, III.1778
  communication aids, III.1899
  hot big bang models, III.1778
Haxel, O
  nuclear binding energies, II.1210
  shell models, II.1209
Hayakawa, S, x-ray sources, III.1755
Hayashi, C, quasi-static models of stars, III.1767
Hayes, W, two-neutrino experiment, II.672
Hayes, W C, *Basic Orthopaedic BioMechanics*, III.1896
Hazard, C, radio sources, III.1776
Heaviside, O
  ionospheric layer, III.1621
  radio waves, III.1987

wireless wave propagation, III.1986
Hebel, P Zum, BCS theory, II.940
Heisenberg, W K, I.189, **I.373**
  atomic theory, I.179
  cosmic-ray review, I.230
  discrete quantum mechanics, I.184
  dispersion formula, I.172
  doublets, I.102
  Einstein–Podolsky–Rosen (EPR) paradox, I.228
  electron spin, II.1125
  equation of motion, I.186
  exchange forces, I.231
  exchange integral, I.213
  Fermi-field theory, I.377
  ferromagnetism, I.554
  Fifth Solvay Conference (1927), I.205
  harmonic-oscillator decomposition, III.1402
  holes, III.2018
  lattice calculations, I.233
  meson theory, I.404
  microphysical phenomena, I.224
  neutron–neutron, neutron–proton forces, II.1190
  neutron–proton model, II.1184
  nuclear charge-exchange force, I.376
  nuclear forces, I.123
  nuclear volume, II.1190
  nucleus, I.371–3, II.1191
  periodic system, I.178
  QED, I.362
  quantum mechanics, I.226, I.358, II.638, II.1058–9
  quantum theory, I.144, I.180, I.187, I.221
  relativity, I.359
  spontaneous emission, III.1401
  superconductivity, II.932, II.934, II.936
  The Physical Principles of the Quantum Theory, III.1394
  theory of explosive showers, I.387–8
  total quantum number, I.162
  Twelfth Solvay Conference, Brussels, II.646
  two-electron system, I.213
  uncertainty relations, I.174
  unified-field theory, I.232
Heitler, W
  bremsstrahlung, I.385
  cosmic rays, I.393
  covalent molecular bond, II.1042
  heavy electrons, I.392
  meson theory, I.394
  nuclear statistics, I.117
  showers, I.388
  The Quantum Theory of Radiation, II.1031, III.1420–1
  Twelfth Solvay Conference, Brussels, II.646
Heller, P, spontaneous magnetization, I.559
Hellmann, H, pseudopotential model, III.1352–3
Hellwarth, R W
  photon, III.1451–2
  Q-switching, III.1430, III.1432, III.1436
Helm, G, *Energetik*, I.48
Helmholtz, H von
  age of the Sun, III.1949
  electricity, I.52
  laboratory equipment, I.20
  medical physics, III.1857
  Physikalisch-Technische Reichsanstalt, I.20, II.1254
  Sun's time-scale, III.1694
Henisch, H K, semiconductors, III.1325
Henkel, superfluidity, II.932
Hénon, M, chaotic behaviour, III.1831

Henrick, L R, equilibrium theory, III.1785
Henriot, E, Fifth Solvay Conference (1927), I.205
Henyey, L G
  radio astronomy, III.1736
  study of the stars, III.1746
Herbert, R, thyroid gland, III.1900
Herbert, R J T, scintillation detector, III.1901
Herbig, G H, T-Tauri stars, III.1766
Herbst, J F, neodymium–iron borates, II.1165
Herglotz, G
  elastic media, I.280
  rigid body, I.279
Herlofson, N, Galactic radio emission, III.1737–8
Herman, F
  photoconductivity, III.1329
  valence/conduction bands, III.1328–9
Herman, R C
  black-body radiation, II.758
  primordial nucleosynthesis, III.1786, III.1790
  proton–neutron ratio, III.1786
Herring, C
  diffusion creep, III.1534
  jellium, III.1303–4
  pseudopotential model, III.1352–3
  solid-state physics, III.1313
  spin waves, II.1012
Herriott, D R, helium–neon laser, III.1427
Herschel, J, *General Catalogue of Nebulae*, III.1717
Herschel, W
  Galaxy, III.1714
  infrared radiation, III.1742
  nebulae, III.1717, III.1945
Hertz, C H, time–position recordings, III.1913–14
Hertz, G
  Franck–Hertz experiment, I.89–90
  scattering experiments, I.217
  stationary electron orbits, I.158
Hertz, H, electromagnetic waves, III.1985
Hertzsprung, E
  diameter of Arcturus, III.1699
  main sequence, III.1698
  mass–luminosity relation, III.1699
  Small Magellanic Cloud, III.1715
  star clusters, III.1696
  stellar sizes, III.1696
Herzberg, G, **II.1072**
  ionization threshold of H_2, II.1069–72
  molecular mechanics, II.1044
  nuclear statistics, I.117
  spectra of diatomic molecules, II.1046
Herzen, E
  Fifth Solvay Conference (1927), I.205
  First Solvay Conference (1911), I.147
Herzfeld, K F, vibration of alkali halide, I.485
Herzog, R O, fibrous proteins, I.497
Hess, G B, superfluidity, II.932
Hess, H
  ocean ridges, III.1974
  sea-floor spreading, III.1976
Hess, V F, I.369
  balloon flight, I.368
  cosmic ray particles, III.1712
  radiation, II.639
Hess, W N, magnetospheric physics, III.2001
Hevesy, G
  Bohr's theory, I.88
  hafnium, I.100
Hewish, A
  neutron stars, III.1751

survey of the sky, III.1788
Hey, J S, radio astronomy, III.1736
Heycock, C T, **III.1511**
　multiphase alloys, III.1548
　phase diagram, III.1508–9
Heydenburg, M P, Coulomb
　excitation experiments,
　II.1217
Heyl, P R, gravitation, II.1266
Higgs, P W
　bosons, II.695
　dynamical symmetry breaking,
　II.957
Higinbotham, W A, diode
　coupling, II.748
Hilbert, D, I.207
　gravitational field, I.314
　Hilbert's space, II.1025
　infinite quadratic forms and
　　matrices, I.191–2
　*Methoden der Mathematischen
　　Physik*, III.2043
　Schrödinger equations, I.205
　theory of linear integral
　　equations, I.174
　variational principle, I.294
Hildebrand, R,
　charge-independence, II.654
Hill, R D, shell model, II.1215
Hillert, M
　atomic segregation/atomic
　　ordering, III.1548
　spinodal concept, III.1548
Hilliard, J, spinodal
　decomposition, III.1548
Hillier, J
　contour phenomena, III.1581–2
　Fresnel diffraction, III.1582
　magnetic lenses, III.1580
Hines, C O
　ionosphere, III.1984
　solar wind plasma, III.2010
Hinks, A, photographic parallax
　programme, III.1696
Hirakawa, K, K_2CoF_4, II.1016

Hisano, K, fine structure, II.1003
Hittorf, J W
　conductivity, III.1318
　rarefied gases, III.1985
Hoag, A A, quasars, III.1798
Hoar, T P, microemulsions,
　III.1608
Hockham, G A, optical
　telecommunication, III.1447
Hodgkin, D C, penicillin, I.496
Hofmann, S, superheavy elements,
　II.1225
Hofstadter, R
　electron scattering, II.1221
　nuclear structure studies, II.1206
　sodium iodide crystals, II.1207
　sodium iodide scintillator,
　　II.750–1
Hohenberg, P
　delta (δ), III.1851
　phonon frequency, II.1002
Holborn, L, black-body radiation,
　I.22
Holmes, A
　Earth's crust, III.1960
　radioactive minerals, III.1969
Holmes, K C, tobacco mosaic
　virus (TMV), I.508
Holstein, T, lattice polarization,
　III.1322
Holt, angular distributions, II.1220
Honda, K, ferromagnetism, II.1134
Hondros, D, dielectric waveguide,
　III.1397
Hooke, R, Newton's rings, III.1386
Hopkins, H H
　coherence parameter, III.1412
　diffraction pattern of a point
　　object, III.1412
　fibre optics, III.1420
　image formation, III.1419
　polarized light, III.1412
　spatial-frequency response of
　　optical systems, III.1422
Hopkins, W

Earth structure, III.1946
earthquake waves, III.1956
Hopkinson, J, *critical temperature*, I.545
Hopwood, medical physics, III.1864
Hora, H, laser systems, III.1464
Horn, D, pion–nucleon charge exchange, II.683
Horn, F H, zone-refined copper, III.1544
Horne, R W, adenovirus, I.510
Horton, self-consistent renormalized harmonic approximation, II.1004
Horwitz, L, BBKGY hierarchy, I.623
Hoselet, First Solvay Conference (1911), I.147
Houston, W V
  de Broglie waves, III.1291–2
  Lamb shift, II.641
  theory of metals, III.1290
Houtermans, F
  nuclear reactions, II.1225
  nuclear relations, III.1707
Houtermans, F G
  age of the Earth, III.1960–1
  thermonuclear reactions, III.1627
Howry, D H
  cross-sectional imaging, III.1912
  prototype scanner, III.1912
  pulse-echo ultrasound, III.1911–12
Hoxton, L G, diffusion cloud chamber, II.752
Hoyle, F
  Big Bang, III.1792
  excited state of ^{12}C, III.1746
  helium synthesis, III.1790
  inverse-Compton catastrophe, III.1781
  nuclear interactions between light nuclei, III.1791
  nucleosynthesis in stars, III.1788
  quasars, III.1777
  red giant stars, II.1226–7
  solar x-rays, III.1994
  steady-state cosmology, III.1787
  supernovae, II.1227, II.1228
  synthesis of elements, III.1747, III.1792
Huang, K
  dipole–dipole forces, II.988
  Dynamical Theory of Crystal Lattices, II.988
  Raman effect, II.996
Hubbard, J
  correlation effects, III.1362
  Hubbard U, II.1166
Hubble, E P, **III.1720**
  counts of faint galaxies, III.1728
  spiral nebulae, III.1719
  The Realm of the Nebulae, III.1720, III.1726
  tuning-fork diagram, III.1721
  velocity–distance relation, III.1726, III.1727
Huchra, J, bright galaxies, III.1772
Hückel, double/triple bonds, II.1043
Hückel, E, wave mechanics, I.213
Hückel, G
  electron screening, III.1316
  screening length, III.1610
Hudson, R P, magnetic cooling, II.1126
Hueter, T F, ventriculography, III.1911
Huggins, M L, chain conformations, I.497
Huggins, W, nebulae, III.1717
Hughes, A L, photoelectric effect, III.1390
Hughes, E W, least squares method, I.452
Hughes, V W, parity violation, II.698
Hughes, W, Rochelle salt, I.489
Hugoniot, A, conservation equations, II.812

Hulburt, E O, solar ultraviolet energy, III.1992
Hull, A W, single crystal, I.434
Hulse, J M
  binary pulsar, I.317
  quadrupole radiation formula, I.315
Hulse, R, binary pulsars, I.298, III.1757
Hulthen, magnetization, II.1131–2
Humason, M L
  galaxy NGC7619, III.1726
  redshift–magnitude relation, III.1788, III.1793
Hümmer, K, Renninger effect, I.450
Hund, F
  analysis of spectra, II.1124
  electron reflection, I.170
  molecular mechanics, II.1044
  quantum theory, I.145
  rare-earth ions, II.1120
  tunnel effect, I.217
  wave mechanics, I.213
Hunter, D L, magnetic field at critical point, I.562
Huntingdon, H B, diffusion, III.1531
Hutson, A R, electron–phonon interaction, III.1339
Hüttel, A, speed of light, II.1263
Huus, T, Coulomb excitation experiments, II.1217
Huxley, T H, Darwinian evolution, I.5
Huygens, C
  light polarization, III.1386
  wave description of light, I.72
  wave theory, I.258
Hyde, E K, *Nuclear Properties of the Heavier Elements*, II.1223

Iachello, F, collective motion in nuclei, II.1218
Iben, I, study of the stars, III.1746

Ibser, H W, neutron sources, II.1199
Ignatowsky, V S, vector theory, III.1412
Ikeda, H, K_2CoF_4, II.1016
Ikeda, K, unstable and chaotic emission, III.1466
Iliopoulos, J
  charmed quark, II.700, II.705
  neutral currents, II.700
Inglis, D R, Cranking Formula, II.1218
Ingram, V, sickle cell anaemia, I.502
Inoue, T, meson theory, I.407, I.408
Inui, T, bulbous bow, II.885
Ioffe, A F
  energy levels, III.1368
  Semiconductor Thermoelements and Thermoelectric Cooling, III.1311
Ioffe, M
  mirror experiment, III.1679
  mirror-type systems, III.1643
Iona, M, KCl, II.988
Irwin, M, redshift quasars, III.1798
Ishiwara, J, quantum rule, I.175
Ising, E
  Ising model, III.1554
  magnetic moment, I.554
Ising, G, linear accelerator, II.736
Israelashvili, J N, measurement of forces between surfaces, III.1612
Iwanenko, D
  neutrons, I.122
  nuclear exchange potential, I.376
Iyengar, P K, dispersion curves of germanium, II.1002

Jackson, K, Materials Research Society (MRS), III.1557
Jacobi, C G J, dynamical laws, I.176

Jacobs, E
  magnetic fusion, III.1629–30
  tokamak, III.1646
Jacobsen, E H, acoustic paramagnetic resonance (APR), II.1146
Jakeman, E
  antibunched light, III.1460
  K-distribution, III.1462
  optical spectrum, III.1454
James, R W
  sodium chloride crystals, I.433
  structure factors, I.439
  zero-point motion, II.974
Jamgochian, E, p–n junction, III.1417
Jamieson, J C, x-ray powder diffraction, I.437
Jancke, W, fibrous proteins, I.497
Jánossy, L, forked tracks, I.408
Jansky, K, radio astronomy, III.1736
Jauch, J M, photon, III.1451
Javan, A, helium–neon laser, III.1427
Jeans, First Solvay Conference (1911), I.147
Jeans, J H
  black-body radiation, I.163
  galaxy formation, III.1800
  gravitational collapse, III.1768
  internal structure of the stars, III.1703
  spiral nebulae, III.1950
  statistical mechanics, I.173
  tidal theory, III.1962
Jefferts, K B, carbon monoxide molecule, III.1763–4
Jeffreys, H
  Earth model, III.1956
  Earth–Moon system, III.1948
  Earth's core, III.1956
  geomagnetism, III.1958
  spiral nebulae, III.1950
  tidal theory, III.1962

Jenkin, F, wire-wound resistors, II.1252
Jennison, R C, Cygnus A, III.1775
Jensen, J H D, II.1211
  Breit–Wigner formula, II.1194
  nuclear binding energies, II.1210
  shell model, II.1209, II.1210, II.1214, II.1220
Johannson, H, emission electron microscope (EEM), III.1569, III.1577
Johansson, C H, crystal structure, III.1515
Johns, H E, telecobalt therapy, III.1872
Johnson, F A, optical-fibre communication, III.1447
Johnson, F M, antiferromagnetism, II.1148
Johnson, J B, Johnson noise fluctuations, I.601
Johnson, L F, rare-earth ion lasers, III.1438
Johnson, M H, neutron diffraction, I.459
Johnson, W H, metre standards, II.1239
Joliot-Curie, F, I.121
  radioactivity, I.121, I.375, II.1186, II.1222
Joliot-Curie, I, I.121
  neutron bombardment, II.1194
  radioactivity, I.375, II.1186, II.1222
Jona-Lasinio, G
  dynamical symmetry breaking, II.957
  pion bound state model, II.756
Jones, H
  Brillouin zone, III.1313
  disordered materials, III.1365
  Fermi surface, III.1295
  Seebeck coefficient, III.1310
  solid-state physics, III.1314
Jordan, P

field quantization, I.283
harmonic-oscillator decomposition, III.1402
matrix mechanics, I.198, I.360
Physics of the Twentieth Century, I.210
quantization of waves, I.208
quantum electrodynamics, I.209
transformation scheme, I.201
Jordan, W C, beta-ray experiment, III.1679–80
Josephson, B, Josephson effect, II.941–2
Jost, vacancies, III.1529
Joule, J P
first law of thermodynamics, I.521
learned societies, I.3
Joy, A H, T-Tauri stars, III.1766
Joyce, G S, long-range and short-range forces, I.563
Julian, B R
detection of impulses, III.1957
Earth's core, III.1980
Jungnickel, C, organized research, I.16
Justi, superconductivity, II.921
Jüttner, Maxwell–Boltzmann distribution law, I.282

Kadanoff, L P
block spin variables, I.567
delta (δ), III.1849
hypothesis of universality, I.563
phase transitions, III.1830
scaling properties, I.563–4
spin-blocking, III.1831, III.1832
universality, I.571
Kadomtsev, B B, fusion-plasma research, III.1642
Kae, M, Ehrenfest dog-flea model, I.618
Kaganov, canonical theory, III.1356

Kaiser, W, two-photon absorption, III.1429
Kalckar, F
liquid drop, I.126, II.1213
spacing of states in highly excited nucleus, II.1213
Källén, G, Twelfth Solvay Conference, Brussels, II.646
Kallman, H
fluorescent light pulses, II.1207
scintillation, II.750
Kaluza, T, general relativity and electromagnetism, I.321–2
Kamerlingh Onnes, H, **II.914**
critical point, I.545
First Solvay Conference (1911), I.147
liquid helium, II.913
superconductivity, I.489, II.915, III.1284
Kaner, E A, cyclotron resonance, III.1357
Kant, I, island universes, III.1714
Kantrowitz, A
magnetic fusion, III.1629–30
tokamak, III.1646
Kao, K C, optical telecommunication, III.1447
Kapany, N S, fibre optics, III.1420
Kapitza, P L
helium liquefier, II.920
law of linear magnetoresistance, III.1356
strong-impulsive-field generator, III.1282–3
superfluidity, II.922
superleak flow, II.946
Kapron, F P, chemical vapour deposition, III.1447
Kapteyn, J C
Galaxy, III.1714, III.1715
luminosity function, III.1714
stellar spectra, III.1696
Kapur, P L, theory of nuclear reactions, II.1219

Karagioz, O V, gravitation, II.1266
Karle, I L, structure of arginine dihydrate, I.449–50
Karle, J
  determinantal inequalities, I.448
  structure of arginine dihydrate, I.449–50
  tangent formula, I.449
  The Solution of the Phase Problem. I. The Centrosymmetric Crystal, I.449
Karpushko, F V, optical bistability, III.1450
Kasai, C, flow imaging, III.1914
Kasper, J S
  crystal structure of dekaborane, I.447
  inequality relations, I.447
Kasteleyn, P W, bond percolation model, I.573
Kastler, A, optical pumping, II.1097, III.1415, III.1447
Kasuya, T, RKKY interaction, II.1164
Kay, W, scintillation counting, I.113
Kaya, S, ferromagnetism, II.1134
Kaye, radiation protection, III.1890
Kayser, H
  handbook of spectroscopy, I.76
  history of spectroscopy, I.24
Keck, D B, chemical vapour deposition, III.1447
Keeler, J, spiral nebulae, III.1950
Keenan, P C
  MKK system, III.1698
  radio astronomy, III.1736
Keesom, A P, liquid helium, II.914
Keesom, W H
  atomic vibrations, III.1287
  liquid helium, II.914, II.922
  Seebeck coefficient, III.1310
Keffer, F, spin-wave modes, II.1151

Kekulé, A, Karlsruhe conference, I.45
Keldysh, L V, photo-assisted tunnelling, III.1423
Kelker, H, liquid crystals, III.1540
Kellar, J N, line broadening, I.472
Keller, A, polyethylene, III.1552
Kellermann, E W, NaCl, II.987–9
Kellers, C F, λ-point of liquid He^4, I.559
Kelley, P L, photodetection, III.1441
Kellman, E, MKK system, III.1698
Kelly, A, Walter Rosenhain, III.1517–19
Kelly, M, cuprous oxide rectifiers, III.1323
Kelvin, Lord
  conducting layer, III.1984
  cooling Earth, III.1945
  Earth structure, III.1946
  Earth's rigidity, III.1953
  gaseous molecule, I.48
  law of equal partition of energy, I.143–4
  Maxwell's Demon, I.576
  papers and patents, I.2
  radium atoms, I.62
  second law of thermodynamics, I.521
  second wrangler, I.11
  Seebeck effect, III.1284
  ship–wave pattern, II.883
  Sun's time-scale, III.1694
  temperature scales, II.1249
Kemmer, N
  meson theory, I.394
  new particles, I.393
  quantum field theory interaction, I.378
  symmetric theory, I.403
  U-particles, I.393
Kendrew, J C
  globular proteins, I.501

isomorphous replacement method, I.500
phase angles, I.502
Kennedy, J W, plutonium isotope, II.1223
Kennelly, A E
　ionosphere, III.1621, III.1985
　radio waves, III.1986, III.1987
Kepler, J, supernovae, II.1228
Kerr, A, electro-optical double refraction, III.1389
Kerr, F J, neutral hydrogen, III.1763
Kerr, R, Kerr metric, III.1778
Kerst, D W
　betatron, II.732, II.1204, III.1871
　fixed-field alternating-gradient accelerators, II.739
　high-intensity beams, II.738
　multipole geometry, III.1645
　orbit calculations, II.732
Keuffel, J W, spark counters, II.753
Kevles, D, physics research, I.16
Khalatnikov, I M
　curvature singularities, I.310
　Stochasticity in relativistic cosmology, I.630
Khandros, L G, metal physics, III.1539
Khintchine, A I, *Mathematical Formulation of Statistical Mechanics*, I.614
Khoklov, stochastic particle acceleration, I.630
Khriplovich, I B, Yang–Mills theory, II.713
Kibble, B P
　fundamental constants, II.1260
　NPL moving-coil experiment, II.1238
Kidder, R
　laser fusion programme, III.1675
　radiation implosion, III.1674
Kihara, T, clusters of galaxies, III.1771

Kikuchi, S, electron diffraction, I.463
Kilby, J St C, integrated circuits, III.1326
Killian, J R Jr, stroboscopic photography, III.1410
Kim, Y K, shell electrons, II.1091
Kimble, H J
　photon antibunching, III.1460
　squeezed light, III.1471
Kinman, T D, helium abundance, III.1790
Kinoshita, S, charged particle tracks, II.1200
Kip, A F, resonant frequencies, III.1334
Kippenhahn, R, study of the stars, III.1746
Kippenhauer, K O, Galactic radio emission, III.1737–8
Kirchheim, R, physics of materials, III.1555
Kirchhoff, G R
　black-body radiation, I.22, I.66, I.148
　collaboration with Bunsen, I.75–6
　induction coil, I.54
　Investigations of the solar spectrum and the spectra of the chemical elements, III.1692
　quantum theory, I.65
　radiation at constant temperature, I.521
　school of physics, I.15
　spectral lines, I.156
Kirkpatrick, H A, curved crystal focusing spectrometer, II.1207
Kirkpatrick, P, holography, III.1415
Kirkwood, J
　BBKGY hierarchy, I.606, I.623, I.624, I.626
　polymer solutions, III.1597

Kirshner, R P, supernovae, III.1759
Kittel, C
  antiferromagnetic resonance, II.1151
  ferromagnetic resonance, II.1153
  ferromagnets, II.1147
  resonance, II.1147
  resonant frequencies, III.1334
  RKKY interaction, II.1164
  spin waves, II.1012
Klaiber, G S, dipole vibration, II.1213
Kleeman, R, α-particles, I.104
Klein, M J, black-body radiation, I.22
Klein, O
  baryon symmetric models, III.1806
  Dirac electrons, I.218
  Dirac equation, I.218
  quantum theory, I.179
  Twelfth Solvay Conference, Brussels, II.646
Kleiner, W H
  crystal field theory, II.1144
  photodetection, III.1441
Kleinman, D A, anharmonic oscillators, III.1434
Klemens, P G, electrons and holes, III.1297
Klug, A
  polio virus, I.510
  tobacco mosaic virus (TMV), I.508
  turnip yellow mosaic virus, I.510
Knight, P L
  laser fluctuations, III.1473
  optical harmonic generation, III.1473
Knipping, P
  crystal structures, III.1514
  x-ray diffraction, I.425
Knoll, M
  Crantz Colloquium, III.1571–2
  electron microscope, III.1571
  high-speed oscillograph, III.1570
  iron-free lenses, III.1568
  scanning electron microscope, III.1587
  transmission electron microscope (TEM), III.1569–70
Knudsen, M
  Fifth Solvay Conference (1927), I.205
  First Solvay Conference (1911), I.147
Kobayashi, M
  CP violation, II.701
  quarks, II.720, II.723
Kocharovskaya, O, x-ray lasers, III.1473
Koehler, T R, self-consistent renormalized harmonic approximation, II.1004
Koehler, W C
  neutron diffraction, I.459
  rare-earth metals, I.463
Koester, C J, optical-fibre laser, III.1427
Kogelnik, H
  Gaussian–Hermite (laser) beam propagation, III.1447
  resonator modes, III.1429
Kogut, J, hadrons, II.690
Kohan, M, *High Performance Polymers*, III.1550
Köhler, A, illumination system for microscopy, III.1390
Kohlhörster, W, balloon flights, I.368
Kohlrausch, F, *Leitfaden der Praktischen Physik*, I.15
Kohlschütter, A, luminosity indicators, III.1698
Kohn, W
  Fermi surface, III.1364
  phonon frequency, II.1002
  wavevectors, II.999

Kolmogoroff, A N, K systems, I.615
Kolmogorov, A N, turbulence, II.798, II.837
Komarov, L I, neutron frequency, III.1418
Kompaneets, A, induced Compton scattering, III.1801
Kondo, J, electron scattering, III.1343
Koo, D C, sky surveys, III.1798
Koppe, superconductivity, II.932, II.934, II.936
Korteweg, D J, weak non-linear and weak dispersive effects, II.859
Koschmieder, E L, free convection, II.846
Koshelev, K, laser amplification, III.1466
Kossel, W
  periodic table, I.95
  x-ray spectra, I.160
Kosterlitz, J M, topological long-range order, I.571
Krakowski, R A, reversed-field pinch (RFP) research, III.1657
Kramers, H A
  asymptotic linkage, I.197
  atomic theory, I.197
  correspondence principle, I.86
  dispersion formula, I.171, I.172, I.185
  dispersion relations, III.1404
  Fifth Solvay Conference (1927), I.205
  Kramers' theorem, II.1125
  particle escape probability, I.607–8
  periodic table, I.96
  potassium chromic alum, II.1126
  quantum theory, I.166, I.179
  specific heat, I.555
  spontaneous emission, III.1401
  two-electron system, I.182

Kraner, H
  crystallization, III.1537
  synthetic zinc silicate, III.1537
Kratzer, A, atomic theory, I.179
Krishnan, K S
  pseudopotential concept, III.1367
  Raman effect (combination scattering), III.1404–6
  Raman scattering, II.996
Kronberg, M L, special orientation relationships, III.1528
Kronig, R L
  dispersion relations, III.1404
  magnetic field, II.1133
  nuclear magnetic moments, I.116
  nuclear spin, I.117
  superconductivity, II.917
Kruger, R, semiconducting crystals, II.1006
Kruskal, M D
  coordinate system, III.1778
  plasma column instabilities, III.1641
  plasma oscillations, III.1625
Kubo, R, **I.613**
  lattice impurities, III.1366
  spin waves, II.1011
  transport coefficients, I.612, I.626–7
Kuhl, D E, tomography, III.1916
Kuhn
  random walk, III.1597
  rubber elasticity, III.1597
Kuhn, T S
  black-body radiation, I.22, II.973
  helium atom, III.1982
  life cycle of field of study, III.2048
  quanta, II.969
Kuhn, W, spectral line intensity, I.172
Kuiper, G P, solar system, III.1963
Kukhtarev, N V, photorefractivity, III.1461

Kukich, superfluidity, II.932
Kulsrud, R, collisionless shocks, III.1684
Kunsman, C H, electron reflection, I.170
Kunze, P, charged particles, I.388
Kurchatov, I V, fusion research, III.1641
Kurdjumov, G, metal physics, III.1539
Kurlbaum, F, black-body radiation, I.149, III.1387
Kurnakov, crystal structure, III.1515
Kurti, N, low-temperature physics, II.920
Kusaka, S, meson theory, I.407
Kyu, T, polymer blends, III.1548

Labeyrie, A E
  holograms, III.1438
  optical astronomy, III.1453
Laborde, radium, I.61
Ladenburg, E, photoelectric effect, III.1390
Ladenburg, R W
  dispersion formula, I.171
  emission and absorption, III.1400
Ladner, A W, *Short Wave Wireless Communication*, III.1619, III.1620
Lallemand, P, stimulated Rayleigh-wing scattering, III.1445
Lallier, E, waveguide laser, III.1471
Lamb, W E
  First Shelter Island Conference, II.645
  hydrogen spectrum, III.1421
  Lamb shift, II.641–3
  laser light, III.1443
  Master-type equation, I.605
  optical maser, III.1443
  photon counting, III.1388
Lambert, J, island universes, III.1714
Lamerton, L F, depth-dose data, III.1867
Lanczos, C
  field equations, III.1724
  matrix mechanics, I.192
  quantum theory, I.196
Land, E H, sheet optical polarizers, III.1417
Landau, L D, **III.1626–7**
  abrupt and restrained transition, II.833–4
  charge effects, III.1611
  damping theory, III.1678
  diamagnetism, II.1133–4
  electron density, III.1316
  electron–electron collisons, III.1316
  energy complexes, I.301
  field-strength uncertainty product, I.209
  gravitational collapse, III.1712
  interacting particles, III.1363
  ionic crystals, III.1321
  light-scattering spectroscopy, III.1408
  magnetization, II.1134
  neutrino, II.667
  neutron stars, III.1713
  non-linear stability theory, II.826
  phenomenological theory, II.1007
  plasma oscillations, III.1623–5
  quantum liquid, II.928
  second-order transitions, I.549–51
  self-gravitating systems, I.313
  spectrum, II.930
  spin-1 particle, II.654
  superconductivity, II.924, II.934–6
  superfluidity, II.929
  surf-riding, III.1362

two-fluid hydrodynamics, II.929, II.944
zero sound, II.950
Landauer, R, thermodynamics of data processing, I.576
Landé, A
quantum numbers, I.101
two-electron system, I.182
Zeeman effect, I.162
Landsberg, G S
Raman effect (combination scattering), III.1404–6
Raman scattering, II.996
spectral shift, III.1397, III.1398
Landweber, L, optical data processing, III.1461
Lane, A M, nuclear reactions, II.1219
Lane, J H, sun's structure, III.1694
Lange, L, inertial frame of reference, I.255
Langenberg, D N, Determination of e/h, II.1268
Langevin, P
Boltzmann relation, I.545
diamagnetism, II.1117
Fifth Solvay Conference (1927), I.205
First Solvay Conference (1911), I.147
internal energy, I.111
stochastic processes, I.596
Zeeman effect/diamagnetism, II.1114
Langley, S P, black-body radiation, I.22
Langmuir, I
Fifth Solvay Conference (1927), I.205
plasma oscillations, III.1623–5
Langsdorf, A Jr, diffusion cloud chamber, II.752
Lankard, J R, dye lasers, III.1446
Laplace, P S
black holes, II.760
central force programme, I.321
fluid-mechanical analysis, II.889
nebular hypothesis, III.1945
sound waves, II.808
Laporte, O, atomic theory, I.179
Larmor, Sir Joseph
aether theory, I.5
radio propagation, III.1622
self-exciting dynamo, III.1958
upper atmosphere, III.1988
Larson, R, formation of stars, III.1769
Laskar, J, chaotic evolutions, III.1823
Lassen, H
upper atmosphere, III.1992
wave propagation, III.1991
Lassettre, E, molecular resonances, II.1084
Lassus St Genies, A H J, lens arrays, III.1411
Latham, R, critical fluctuations, II.1016
Lau, E, Lau effect, III.1414
Laub, J, stress–energy tensor, I.281
Lauer, E J, beta-ray experiment, III.1679–80
Laughlin, J, calorimetry, III.1883
Launay, J M, vibrational excitation, II.1089
Lauterbur, P C, NMR image, III.1924
Laval, J, x-ray scattering, II.990
Lawrence, E O, I.371, **II.1185**
calutron isotope separation, III.1630
cyclotron, II.639, II.729, II.736, II.1201
direct reaction theory, II.1220
Lawson, J D
fusion reactions, III.1628–9
$n\tau$ product, III.1649
Lax, M, laser light, III.1443
Leavitt, H S, Magellanic Clouds, III.1714

Le Berre, M, *Photons and Quantum Fluctuations*, III.1441
Lebowitz, stochastic dynamics, I.629
Lederman, L M
 Drell–Yan process, II.701
 high-mass lepton pairs, II.721
 parity violation, II.666
 two-neutrino experiment, II.672
Lee, B W, renormalizability, II.695
Lee, E W, transverse Kerr effect, II.1155
Lee, T, supernova trigger hypothesis, III.1963
Lee, T D
 lattice gas model, I.553
 neutrinos, II.671
 parity conservation, II.665
 parity violation, II.666
Lee, Y T, cross-beam collision products, II.1089
Leff, H S, *Maxwell's Demon*, I.576
Legvold, S, magnetic configurations, II.1162
Lehmann, H, haemoglobin, I.502
Lehmann, I, Earth structure, III.1956-7
Lehmann, O
 cholesteric esters, III.1540
 liquid crystals, III.1540
Lehovec, K, p–n junction, III.1417
Leibacher, J W, solar oscillations, III.1750
Leighton, R B
 sky survey, III.1743
 solar oscillations, III.1749–50
Leith, E N
 holography, III.1415, III.1437
 off-axis optical beam holography, III.1585
Leksell, L, flaw detector, III.1911
Lemaître, G
 Eddington–Lemaître model, III.1732–3
 primaeval atom, III.1785

Schwarzschild horizon, I.307
spherically symmetric perturbations, III.1801
Universe, III.1725
Lenard, P
 photoelectric effect, I.152, III.1390
 photoelectron emission, III.1396
Lengyels, B A, *Lasers*, III.1425
Lenz, W
 atomic theory, I.179
 ferromagnetism, I.554
Leontovich, M A, *Plasma Physics and the Problem of Controlled Thermonuclear Reactions*, III.1642
Le Pichon, X
 Earth's surface, III.1978
 plate tectonics, III.1978
Le Poole, J B, selected area diffraction, III.1579
Leprince-Ringuet, L, K-meson (or kaon), II.656
Letokhov, V S
 laser amplification, III.1466
 laser beams, III.1462
Le Verrier, U J J, precession of perihelion of Mercury, III.1722
Levi-Civita, T, tensor analysis, I.292
Levine, B F, superlattice device, III.1471
Levinson, C A, proton inelastic scattering, II.1221
Levshin, V L
 absorption of light by uranium glass, III.1403
 Levshin–Perrin law, III.1400
 polarization of fluorescent light, III.1400
 Raman effect, III.1400
Levy, H A, neutron diffraction, I.458–9
Lévy, M

axial coupling constant, II.676
β-decay, II.677
weak hadronic current, II.678
Lewis, G N
  direct reaction theory, II.1220
  particle dynamics, I.279
  photons, I.358, III.1302, III.1403
Lhéritier, M, K-meson (or kaon), II.656
Li, T
  Fabry–Perot resonator, III.1429
  Gaussian–Hermite (laser) beam propagation, III.1447
Libchaber, A, dynamical systems3.1852
Lichte, H, atomic resolution microscopy, III.1585
Lichten, W, potential curves for Ar_2^+ complex, II.1079
Lide, D R Jr, HCN laser line, III.1443
Lieb, E H
  delta (δ), III.1845
  percolation model, I.573
  two-dimensional ferroelectric models, I.570–1
Lifshin, E, electron microprobe analyser, III.1549
Lifshitz, E M
  curvature singularities, I.310
  energy complexes, I.301
  galaxy formation, III.1801
  self-gravitating systems, I.313
  Statistical Physics, I.550
Lifshitz, I M
  canonical theory, III.1356
  crystal defects, II.1004
  crystal surface, II.1005
  Fermi surface, III.1351
  quantum theory, III.1350
Liftshitz, E, magnetization, II.1134
Liley, B, tokamak, III.1647
Lilly, S J, radio galaxies, III.1795–6
Lima de Faria, J, *Historical Atlas of Crystallography*, I.511–12

Linde, A D
  scalar Higgs fields, III.1806
  world model, III.1807
Linde, J O, crystal structure, III.1515
Lindemann, F A (later Lord Cherwell)
  Clarendon Laboratory, II.920
  crystal lattice, III.1315
  First Solvay Conference (1911), I.147
  two-frequency equation, I.153
Lindhard, J
  electron gas, III.1362
  Kubo procedure, I.626
Linfoot, E H, diffraction images, III.1419
Lippmann, G, interference within a thick film, III.1392
Lipscomb, W N, boron hydrides, I.493
Lipson, H, copper–nickel–iron alloy, III.1548
Lipson, H S
  stacking faults, I.473
  structure determination, I.441
Livingston, M S
  accelerators, II.725–7, II.738, II.1201
  cyclotron, II.736
  direct reaction theory, II.1220
  nuclear physics, II.1184
  Particle Accelerators, II.1201
  synchrotrons, II.734
Lobachevsky, N I, *On the Principles of Geometry*, III.1722
Lockyer, Sir Norman
  Nature, I.6
  spectrograph, I.25
  stellar evolution, III.1695
  stellar spectra, III.1693
Lodge, O J, obituary of George Fitzgerald, I.29
Logan, B G, thermal barrier, III.1667

Logan, G, tandem-mirror concept, III.1665–6
Lohmann, A W
  computer-generated spatial filter, III.1447
  holography, III.1415
London, F
  covalent molecular bond, II.1042
  Josephson junction, II.1095
  penetration depth, II.935
  superconductivity, II.924–5
  Superfluids, II.934
London, H
  anomalous skin effect, III.1346–7
  superconductivity, II.921, II.924–5
  superfluidity, II.932
  thermoelectric effects, II.922
Long, R R, upstream wakes, II.868
Longair, M S, radio galaxies, III.1795–6
Longuet-Higgins, M S, wave energy, II.860–1
Lonsdale, K
  hexamethyl benzene, I.495
  x-ray scattering, II.990
Lonzarich, G G, saturation magnetization, II.1167
Lorentz, H A, **I.18**
  anomalous Zeeman effect, I.102
  electric dipoles, II.1117
  electrodynamics of moving bodies, I.261
  electromagnetic theory, I.63
  electron collisions, III.1290
  electron optics, III.2000
  electron theory, II.1114, III.1284
  Fifth Solvay Conference (1927), I.205
  First Solvay Conference (1911), I.147
  general theories of relativity, I.34
  induced precession, II.1115
  magnetism, III.1390

Maxwell equations, I.280
  microscopic electron theory, I.281
  quantum theory, I.145, I.585
  random processes, I.586
  special theory of relativity, I.34, I.262–85
  splitting of spectral lines, I.161
  statistical mechanics, I.173
  The Theory of the Electron, III.1992, III.2039
  thermoelectric effects, III.1306, III.1307–8, III.1308
Loschmidt, J
  Avogadro's number, II.1269
  dimensions of molecules, I.48
Lossew, band-structure diagram, III.1417
Loudon, R, virtual electronic transitions, II.996–7
Louisell, W, laser light, III.1443
Love, A E H
  crystal surface, II.1005
  The Mathematical Theory of Elasticity, II.967
Lovelace, C, pion–pion scattering amplitude, II.684
Loveland, W D, transuranic elements, II.1223
Lovell, B, development of astronomy, III.1735
Low, F E
  bolometers, III.1743
  electric quadrupole moments, II.1214
  pion–nucleon resonance, II.680
  pion–pion scattering, II.681
  renormalization group, III.1830
Lowde, R, neutron diffraction, I.456
Lowman, C E
  magnetic materials, II.1159
  Magnetic Recording, II.1158
Lowry, R A, gravitation, II.1266

Lu, K T, photoabsorption spectrum of xenon, II.1073
Luce, J, energetic particle injection, III.1661
Lücke, K, *Lehrbuch der Metallkunde*, III.1554
Lucretius, atoms, I.44
Ludwig, G, Geiger counter, III.2007
Lugiato, L A, laser theory, III.1444
Lukirskii, A P
  bremsstrahlung, II.1050
  x-ray apparatus, II.1051
Lukosz, W, optical images, III.1439
Lummer, O R, black-body radiation, I.22, I.149, III.1387
Lummer, U, reflection interference fringes, III.1389
Lunbeck, R J, critical temperature, II.975
Lundmark, K, Andromeda Nebulae, III.1718
Luneberg, R K
  imaging theory, III.1418
  Mathematical Theory of Optics, III.1412
Luttinger, J M, Fermi surface, III.1364
Lyddane, R H, vibration of alkali halide, I.485
Lynden-Bell, D
  accretion discs, III.1964
  black holes, III.1779
  violent relaxation, III.1774
Lynn, J E, fission isomers, II.1218
Lyons, H, properties of quartz, II.1244
Lyubimov, V A, electron neutrino, III.1803

McAfee, K B, semiconductors, III.1419
McCall, S L, thermal bistable effects, III.1450
McClelland, C L, Coulomb excitation experiments, II.1217
McClung, F J, Q-switching, III.1430, III.1436
McConnell, A J, photon mass, III.1452
McCormmach, R, organized research, I.16
McCrea, W H, cosmological models, III.1730
Macdonald, resistance of sodium wires, III.1344
McDonald, W, mirror systems, III.1660
MacDougall, D P, gadolinium sulphate, II.1126
McDougall, I, geomagnetic field, III.1973
Macek, W M, ring-laser gyroscope, III.1433
McFarland, B B, laser development, III.1448
McFarlane, C G B, CO_2 laser, III.1442
McFee, J H, electron–phonon interaction, III.1339
McGowan, F K, Coulomb excitation experiments, II.1217
Mach, E
  atoms, I.48
  kinetic theory of gases, I.30
  Mach number, II.813
MacInnes, D, First Shelter Island Conference, II.645
McIntyre, P, proton–antiproton colliders, II.743
McKay, K G, semiconductors, II.1208, III.1419
McKee, C, neutron star, III.1759
McKellar, A, excited states, III.1791
McKenzie, D, plate tectonics, III.1978

Mackintosh, A R, exchange interactions, II.1164
McLennan, liquid helium, II.922
McMahon, R G, redshift quasars, III.1798
McMillan, E M
  electron synchrotron, II.731
  frequency modulation, II.1202
  ion acceleration, II.730
  neptunium, II.1224
  60″ cyclotron, II.1202
  transuranium element, II.1223
Madansky, L, spark counters, II.753
Madelung, E
  electron vibration, II.983–4
  quantum theory, I.227
Madey, J M J, free-electron laser, III.1456
Maeda, K, ionosphere, III.1984
Maiani, L
  charmed quark, II.700, II.705
  neutral currents, II.700
Maiman, T H, **III.1425**
  ruby laser, III.1424–6, III.1894
Majorana, E
  deuteron, I.376
  nuclear stability, II.1190
Maker, P D, pump-probe configuration, III.1445
Maki, A G, HCN laser line, III.1443
Maki, Z, quark–lepton analogy, II.700
Mallard, J R, gamma camera, III.1907
Malmberg, J H, magneto-electrostatic confinement, III.1681
Mandel, L
  coherence theory, III.1433
  optical coherence, III.1392
  photon antibunching, III.1460
  photon beams, III.1424
  quantum theory of coherence, III.1439
  sub-Poissonian statistics, III.1460
Mandelbrot, B
  fractals, I.574
  self-similarity, I.574
Mandelstam, L I
  first-order Raman effect, II.996
  Mandelstam–Brillouin scattering, III.1397
  Raman effect (combination scattering), III.1404–6
  Raman scattering, II.996
  secondary scattered radiation, I.172
Mandelstam, S, Twelfth Solvay Conference, Brussels, II.646
Manley, J M, non-linear optics, III.1424
Mansfield, P
  magnetic resonance imaging, III.1924
  nuclear magnetic resonance, II.1175
Mansur, L K, nuclear materials research, III.1534
Marconi, G
  electromagnetic waves, I.29
  radio waves, III.1995
  telegraph signals, I.54
  wireless transmission, III.1619–20, III.1986
Marechal, A, coherent imaging, III.1418
Margon, B, binary sources, III.1756
Mark, H
  biomolecular structures, I.497
  single crystals, III.1527
Marsden, E
  α-particles, I.113
  α-scattering, II.1200
  medical physics, III.1864
Marshak, R E
  Conceptual Foundations of Modern Particle Physics, II.770

First Shelter Island Conference, II.645
mesons, II.654
neutrinos, II.667
Marshall, L, neutron diffraction, I.456
Marshall, T W, stochastic electrodynamics, III.1388
Martin, J, x-ray diffraction, III.1547
Martin, P, temporal behaviour of a general system, I.617
Marton, transmission microscope, III.1574–5
Martyn, D F
  broadcast modulation, III.1994
  ionosphere, III.1984
Marx, G, neutrinos, III.1803
Mascart, E, motion of the Earth, I.258
Mash, D I, stimulated Rayleigh-wing scattering, III.1445
Mashkevich, V S, covalent crystals, II.1002
Masing, G, *Lehrbuch der Metallkunde*, III.1554
Maskawa, T
  CP violation, II.701
  quarks, II.720, II.723
Massey, Sir Harrie, **II.1024**
  ionospheric research, III.1999
  The Theory of Atomic Collisions, II.1024, II.1039
Mathewson, C, plastic flow and crystal glide, III.1544
Matsuoka, M, x-ray sources, III.1755
Matthews, D, magnetic variations of ocean floors, III.1975
Matthews, T A, stellar spectra, III.1776
Matthias, B T
  ferroelectric crystals, III.1538
  Matthias' rules, I.489

Matuyama, M, Earth's magnetic field, III.1973
Mauguin, C, helical structure, III.1397
Maurer, R D, chemical vapour deposition, III.1447
Maxwell, A, Pacific floor measurements, III.1974
Maxwell, J C
  Atom, I.76
  atoms, I.51
  electrolysis, I.51
  electromagnetic waves, I.260
  electromagnetism, III.1856
  equal area construction, I.544
  equipartition of energy, I.523
  field theory, I.321
  kinetic theory of gases, I.587
  molecules, I.47
  new fields of research, I.29
  second law of thermodynamics, I.523
  second wrangler, I.11
  sodium yellow line, II.1239
  statistical mechanics, I.173
  Theory of Heat, I.575
  wire-wound resistors, II.1252
Mayall, N U, redshift–magnitude relation, III.1788, III.1793
Mayer, J E, statistical mechanics, I.551–4
Mayer, M G, II.1211
  shell model, II.1209, II.1210, II.1214, II.1220
Mayneord, W V, **III.1868**
  depth-dose data, III.1867
  nuclear medicine, III.1901
  thimble ionization chamber, III.1867
Meggers, W F, **II.1030–1**
  atomic energy levels, II.1093
Mehl, R, diffusive transformations, III.1547
Meinesz, F A V, ocean floor, III.1974

Meissner, W, superconductivity, II.916–19, II.946
Meitner, L, I.105
  β-decays, I.118
  β-ray deflection, II.1201
  β-spectrum, I.104, I.105, I.119
  elements with $Z = 93$–96, I.130
  neutron bombardment, II.1194
  neutron sources, II.1206
  quantum theory, I.221
  uranium isotope, II.1223
  uranium-239, I.132
Melvill, T, discrete spectra, I.74–5
Mendeleev, D I
  absolute boiling temperature, I.543
  Avogadro's law, I.46
  periodic table, I.96
Mendelssohn, K
  energy gap, II.932
  low-temperature physics, II.920
Mendenhall, T C, squared paper, I.16
Menter, J M, electron microscope, I.477
Menzel, D H
  abundances of elements, III.1702
  planetary nebulae, III.1761
Merton, R, norms of scientific ethos, III.2057–8
Merz, W, dielectric constants, I.487
Metzner, A W K, field equations, I.313
Meyer, K H, biomolecular structures, I.497
Meyer, R B, smectic phase, III.1604
Michell, J, gravitation, II.1266
Michelson, A A, **I.33**
  Balmer lines, I.90
  cadmium red line, II.1239–40
  future discoveries, I.29
  hydrogen atom, II.641
  interferometer, I.25
  Lightwaves and their Uses, I.26
  optical instruments, III.1391
  optical interferometry, III.1704
  optical relativity principle, I.261
  spectroscopy, I.25
  speed of light, II.1263
  stellar interferometer, III.1400
  Studies in Optics, I.26
  Zeeman splitting, I.161
Mie, G
  non-linear field theories, I.232
  special-relativistic gravitation theory, I.284, I.321
Migdal, A B
  electron–phonon interaction, II.941
  interacting Fermi particles, III.1364
  phonons, II.999
Migneco, E, fission isomers, II.1218
Milburn, G J, quantum non-demolition (QND) measurements, III.1469
Miley, G H, laser systems, III.1464
Miley, G K, radio galaxies, III.1796
Miller, D A B, indium antimonide, III.1450
Miller, D C, Michelson–Morley experiment, III.1401
Miller, H, wedge fields, III.1872
Miller, J S, Seyfert galaxies, III.1784
Miller, S E, integrated optics, III.1444
Miller, W, nebulae, III.1717
Miller, W H, Miller indices, I.421
Millikan, R A
  absorption data, I.368
  constituents of matter, I.357
  electron charge, II.747
  mesotron, I.389
  new cosmic-ray particle, I.396
  oil drop method, II.1271
  photoelectric effect, I.363, III.1390–1
Millington, G, long-distance communications, III.1999

Mills, B, radio surveys, III.1789
Mills, R L
 isotopic spin symmetry, II.693
 local gauge invariance, II.693
 self-interacting fields, II.661
Milne, E A
 cosmological models, III.1730
 electrons in metals, I.531
 ionization, III.1701
 solar atmosphere, III.2003
 stellar spectra, III.1693
 theory of kinematic relativity, III.1787
Minkowski, H, **I.252**
 four-dimensional formulation, I.274
 particle dynamics, I.279
 phenomenological electrodynamics, I.280
 relativistic vacuum electrodynamics, I.280
Minkowski, R
 medium-mass elements, II.1228
 Palomar 200" Telescope, III.1737
 radio galaxy, III.1776
 radio sources, III.1775
 supernovae, II.1227–8
Minnaert, M, equivalent width, III.1702
Minogin, V G, laser beams, III.1462
Minsky, M
 confocal optical microscopy, III.1422
 confocal scanning laser microscope, III.1450
Misener, superleak flow, II.946
Mistry, N, two-neutrino experiment, II.672
Mitchell, D P, neutron diffraction, I.453
Mitchell, J W
 dislocation lines in silver chloride, III.1526
 solid-state physics, III.1530

Mitra, S K
 rocket studies, III.2001
 The Upper Atmosphere, III.2000
Mittelstaedt, O, speed of light, II.1263
Miyamoto, S, spark chamber, II.753
Moffat, K, protein unfolding, I.438
Mögel, H, sudden ionospheric disturbance (SID), III.1995
Mohorovičić, A, Mohorovičić discontinuity (Moho), III.1954
Mollenauer, L F, soliton-pulse scheme, III.1458, III.1471
Möllenstedt, G, Fresnel electron biprism, III.1585
Møller, C
 meson theory, I.401
 neutron capture experiments, II.1192
 S-matrix, II.680
 symmetrical theory, I.403
 Twelfth Solvay Conference, Brussels, II.646
Monck, W H S, stellar spectra, III.1696
Montroll, E W
 atomic models, II.987
 crystal defects, II.1004
 specific heat of small crystals, II.1005
Moore, C, chemical abundances, III.1702
Moore, W S, whole body imager, III.1924
Morey, G E, amorphous structures, I.477
Morgan, J, plate tectonics, III.1978
Morgan, W W, MKK system, III.1698
Moriya, T, lattice distortions, II.1152
Morley, E W
 Balmer lines, I.90

hydrogen atom, II.641
optical relativity principle, I.261
Morris, P, nuclear magnetic
    resonance, II.1175
Mort, J, photocopying, III.1541
Morton, K W, self-avoiding walk
    (SAW), I.570
Moseley, H G J
    atomic number, I.159–60
    Bohr's theory, I.88
    characteristic x-rays, II.1038
    periodic system, I.362
    periodic table, I.86
    x-ray spectra, I.158
Mosengeil, K, relativistic
    thermodynamics, I.281
Moss, T S, absorption coefficient
    of semiconductor, III.1421
Mott, N F, **III.1333**
    Brillouin zone, III.1313
    colour centres, III.1530
    disordered materials, III.1365
    Electron processes in
        Non-crystalline Materials,
        III.1369
    paramagnetism, II.1136
    resistivity of dilute alloy, III.1316
    Seebeck coefficient, III.1310
    semiconductors, III.1319,
        III.1330–4
    solid-state physics, III.1313–14
    The Theory of Atomic Collisions,
        II.1039
Mottelson, B, II.1216
    collective motion, II.1215
    Coulomb excitation, II.1217
    Coulomb excitation experiments,
        II.1217
    geometrical collective model,
        II.1218
    surface oscillations, II.1215
Mould, J, Hubble's constant,
    III.1793
Moulton, F R
    nebular hypothesis, III.1950

    tidal theory, III.1962
Mouton, H, magneto-optical effect,
    III.1389
Mow, V C, *Basic Orthopaedic*
    BioMechanics, III.1896
Mueller, K A, oxygen octahedra,
    II.1009
Mügge, crystals of copper, gold
    etc, III.1517
Mukherjee, A K, plastic
    deformation, III.1534
Mulders, G, equivalent width,
    III.1702
Müller, A
    Meissner effect, II.957–8
    superconductivity, II.961
Muller, C A, hyperfine line
    radiation, III.1763
Müller, E, field emission
    microscope, III.1569
Muller, E W A, solar x-rays,
    III.1994
Müller, K A, superconductivity,
    I.490
Mulliken, R
    molecular mechanics, II.1044
    wave mechanics, I.213
Munro, G, ionosphere, III.1984
Murphy, G M
    deuterium, III.1706
    deuteron, II.650, II.1183
Mushotzky, R F, quasars, III.1780
Mussell, L, isocentric couch,
    III.1873
Mutscheller, A, radiation
    protection, III.1890
Myers, W D, liquid-drop mass
    formula, II.1225

Nabarro, F R N
    diffusion creep, III.1534
    dislocations, I.476, III.1526
    physical metallurgy, III.1555
Nagamiya, T, resonance, II.1147

Nakajima, S, pulse oximeter, III.1894
Nambu, Y
 chiral symmetry, II.676
 decay reaction, II.657
 dynamical symmetry breaking, II.957
 pion bound state model, II.756
 quarks, II.687
 SU(3) symmetry, II.709
 Twelfth Solvay Conference, Brussels, II.646
Nasanow, L H, self-consistent renormalized harmonic approximation, II.1004
Nath, N S N, diffraction of light, III.1409
Neddermeyer, S H
 absorption in lead plates, I.385
 cosmic rays, I.386
 elementary particles, I.391
 mesotron, I.231, I.389
 muon, II.650
 shower particles, I.388
Néel, L
 antiferromagnetism, II.1129, II.1132
 ferrimagnetism, II.1157
 neutron diffraction, I.453
Ne'eman, Y, II.678
 isospin, II.661
 strangeness, II.662
 triplets, II.685
Neidigh, R, plasma column, III.1678
Nelmes, R J, neutron diffraction, I.489
Nereson, N, nuclear forces, II.652
Nernst, W H, **I.536**
 calorimetry, II.968–9
 chemical equilibrium, I.535
 First Solvay Conference (1911), I.147
 new physics, I.7
 quantum hypothesis, I.155
 quantum theory, I.82, I.148, I.154
 specific heat, I.152–3, II.970
 thermodynamic theory, III.1699
 two-frequency equation, I.153
Nethercot, A H
 antiferromagnetism, II.1148
 thermal conductivity, II.982
Neugebauer, G
 formation of stars, III.1769
 Orion Nebula, III.1743, III.1767
 sky survey, III.1742–3
Neumann, F
 school of physics, I.15
 semiconductors, III.1419
Neville, F H, **III.1511**
 multiphase alloys, III.1548
 phase diagram, III.1508–9
Newnham, R E, ceramics, III.1538
Newton, I, I.255
 corpuscular hypothesis, I.257
 first law of motion, I.256
 law of gravitation, II.1265
 Opticks, I.76
 particle description of light, I.72
 physical cosmology, III.1721
 Principia, I.255
 rotating bracket, I.256
 second law of motion, II.806
 third law of motion, II.821
Ng, W K, ruby laser, III.1432
Nichols, G A, optical microscopy, III.1406
Nichols, H W, radiowave propagation, III.1990
Nicholson, J W
 angular momentum, I.175
 electron vibration, I.81
 spectra of celestial objects, I.156–7
Nier, A, lead isotopes, III.1960
Nieto, M M, photon mass, III.1452
Niggli, P, x-ray diffraction, I.439
Nishijima, K
 decay reaction, II.657
 isotopic spin multiplets, II.658

Nishikawa, S, electron diffraction, I.463
Nishina, Y
  Dirac equation, I.218
  letter, I.212
  new particles, I.389
Nishizawa, J, semiconductor maser, III.1423
Niu, K, charmed particles, II.700
Noether, F, rigid body, I.279
Nordheim, L
  disordered materials, III.1365
  Schrödinger equations, I.205
Nordsieck, A
  First Shelter Island Conference, II.645
  Master-type equation, I.605
  nuclear exchange potential, I.376
Nordström, G, special-relativistic theories, I.284
Northrop, G A, thermal conductivity, II.982
Northrop, T G
  adiabatic confinement, III.1679–80
  trapped particles in magnetic-mirror fields, III.1683
Nouchi, photon, III.1451–2
Novikov, I D
  black holes, III.1779, III.1780
  structure in the Universe, III.1801
Nowick, A, anelastic relations, III.1523
Nuckolls, J
  direct-drive approach to inertial fusion, III.1676
  fusion power, III.1673
  radiation implosion, III.1674

Obreimow, crystal growth, III.1280
Occhialini, G P S
  cloud chambers, I.371, II.651
  electron–positron pairs, I.231
O'Dell, C R, helium abundance, III.1790
Oehme, R
  parity violation in β-decay, II.665
  singularities in complex angular momentum, II.682
Oersted, H C
  electric current, III.1958
  electromagnetic interactions, III.1953
Ohkawa, T, multipole geometry, III.1645
Ohnuki, Y, quark–lepton analogy, II.700
Okayama, T, slow mesotrons, I.406
Oke, B, supernovae, III.1759
Oke, I, energy in the Sun, III.1748
O'Keefe, J A, optical microscopy, III.1406
Oken, L, *Gesellschaft deutscher Naturforscher und Ärzt*, I.5
Okubo, S, mass relations among baryons and mesons, II.661
Oldham, R D
  earthquake waves, III.1956
  seismic wave paths, III.1954
Olszewski, K, liquefied gases, I.27
O'Neill, E L, communications theory, III.1417
O'Neill, G K, *Storage-ring synchrotron: device for high-energy physics research*, II.739
Onsager, L, **I.551**
  entropy relations, I.600–2
  Fermi surface, III.1351
  ferroelectric transition in KH_2PO_4, I.571
  hypothetical vortices, II.931
  Ising model, I.554–61, II.1018
  Onsager relations, I.616–17, I.626
  quantum theory, III.1350
Oort, J H
  cosmological density, III.1799

hyperfine line radiation, III.1763
neutral hydrogen, III.1763
Palomar 200" Telescope, III.1775
radio astronomy, III.1762
Öpik, E
　giant stars, III.1708–9
　origin of elements, III.1747
　triple-α reaction, III.1746
Oppenheimer, J R
　adiabatic approximation, II.999
　collapsing sphere of dust, I.308
　direct reaction theory, II.1220
　First Shelter Island Conference, II.645
　general-relativistic analysis, III.1712
　mesotron, I.389
　μ-meson, II.652
　muon, II.651
　neutron stars, III.1713
　QED, I.386
　Twelfth Solvay Conference, Brussels, II.646
Orbach, R, fractons, II.1007
Ornstein, L S
　critical correlations, I.563
　pair distribution function, I.547
　spectral line intensity, I.172
Orowan, E
　dislocations, III.1520
　edge dislocation, I.476
Orthmann, W, β-spectrum, I.119
Ortolani, S, ETA-BETA I devices, III.1658
Oseen, C W, wave propagation in cholesteric liquid crystals, III.1407
Osheroff, D, superfluidity, II.950–1
Osmer, P S, quasars, III.1798
Osmond, block copolymers, III.1611
Osterberg, H, thin-film devices, III.1444
Osterbrock, D E, helium problem, III.1790

Ostriker, J P
　binary sources, III.1756
　electron acceleration, III.1752–3
　galactic cannibalism, III.1774
　spiral galaxies, III.1771
Ostwald, W
　dispute with Planck, I.34
　kinetic theory of gases, I.30
　natural phenomena, I.47–8
　quantum theory, I.148
Overbeek, charge effects, III.1611
Overhauser, A W, shell model, I.485, II.997

Pacini, F, neutron stars, III.1752
Pais, A
　First Shelter Island Conference, II.645
　Inward Bound, I.37, II.1183
　Twelfth Solvay Conference, Brussels, II.646
Pancini, E, mesons, II.652
Pankey, supernovae, III.1759
Panofsky, W K H
　electron storage rings, II.739
　Mark III linac, II.737
　proton linear accelerator, II.1205
　proton machines, II.737
　Stanford Linear Accelerator Center (SLAC), II.687–8
Papanastassiou, D A, supernova trigger hypothesis, III.1963
Parisi, G, asymptotic freedom, II.715
Parker, E N
　collisionless shocks, III.1684
　solar wind, III.1685, III.2009
Parker, H M
　dosimetry, III.1881
　radium needles, III.1877
Parker, W H, determination of e/h, II.1268
Parrent, G B, partial coherence, III.1392

Parrish, W, powder photographs, I.436
Paschen, F
  black-body radiation, I.149
  Bohr's theory, I.88
  bolometer, I.23
  far-infrared spectrum, II.982–3
  Zeeman splitting, I.161
Paschos, E A, parton hypothesis, II.690
Pasternack, S, Lamb shift, II.641
Pastukhov, V P, mirror approach, III.1667
Patashinskii, A Z
  correlation components, I.563
  multiple-correlations near critical point, I.562
Patel, C K N, CO_2 laser, III.1442
Paterson, R
  dosimetry, III.1881
  radium needles, III.1877
Patterson, A L, x-ray diffraction pattern, I.441
Patterson, C C, age of the Earth, III.1961
Pauli, W
  atomic dynamics, I.184
  atomic theory, I.179, I.181
  continuous β-spectrum, I.127–8
  Einstein–Podolsky–Rosen (EPR) paradox, I.228
  electromagnetic field, I.391
  electron gas, III.1288
  equivalence relations, I.196
  exclusion principle, I.98–100, I.359, I.528, III.1711
  field quantization, I.283
  Fifth Solvay Conference (1927), I.205
  free particles, I.283
  lattice calculations, I.233
  Lorentz invariance, III.1451
  neutrino, I.371, I.375, III.1706
  non-relativistic theory, I.283
  paramagnetism of normal metals, II.1133
  quantum electrodynamics, I.209, I.362
  quantum mechanics, I.226
  quantum theory, I.180, I.221
  spin-statistics theorem, I.283
  stability of nuclei, I.111–12
  thermal conductivity, II.979–81
  total quantum number, I.162
  two-component wavefunctions, I.277
  wave mechanics, I.206
  Zeeman effect, I.103
Pauling, L
  biomolecular structures, I.497–8
  First Shelter Island Conference, II.645
  inorganic chemistry, I.496
  inorganic compounds, I.492
  pleated sheet structures, I.499
  polypeptide chains, I.497
  quasicrystals, I.481
  residual entropy of ice, I.571
  structure of ice, I.475
  wave mechanics, I.213
Payne, C H
  abundances of elements, III.1701
  Stellar Atmospheres, III.1701
Payne-Gaposchkin, C, spectrum analysis of stars, III.1962
Peacock, J A, radio sources, III.1798
Pearson, impurity effects, III.1281
Pearson, G L, solar cell, III.1421
Pearson, J J, excitations in $s = 1/2$ Heisenberg antiferromagnet, II.1014
Pearson, K, *The Grammar of Science*, I.34
Pease, F G, angular diameter of Betelgeuse, III.1705
Pease, R S
  structure of KH_2PO_4, I.459
  ZETA experiment, III.1642

Pedersen, P O, radio waves, III.1992
Peebles, P J E
 clusters of galaxies, III.1771
 cold dark matter, III.1804
 spiral galaxies, III.1771
 synthesis of heavy elements, III.1792
 world models, III.1806
Peierls, R E
 exciton absorption spectrum, III.1410
 Fermi theory, I.377
 field-strength uncertainty product, I.209
 holes, III.2018
 Ising model, I.554
 lattice distortion, II.1000
 meson theory of nuclear forces, I.402
 resistivity at low temperatures, III.1306
 Schrödinger equation, III.1299–300
 theory of nuclear reactions, II.1219
 thermal conductivity, II.979–81
 Twelfth Solvay Conference, Brussels, II.646
 Umklapp processes, III.1304–6
Peimbert, M, helium abundance, III.1790
Pekeris, C L, M_2 tide, II.892
Pelzer, H, lattice deformation, III.1321
Penrose, R
 closed trapped surface, III.1778
 global differential geometry, I.310
 hot big bang models, III.1778
 quasicrystals, I.482, I.483
 resonance experiments, II.1139
 space-time, I.313, I.318–19
 spinorial techniques, I.277
Penzias, A A
 carbon monoxide molecule, III.1763–4
 microwave background radiation, III.1790
 microwave radiometer, II.758
 radiation, III.1786
Pepinsky, R
 ferroelectric crystals, III.1538
 XRAC, I.446
Pepper, D M, optical-beam transmission, III.1461
Percus, random walk model, I.630
Perkins, D H
 neutrino scattering, II.697
 photographic emulsions, II.653
Perl, M, *Evidence for the existence of anomalous lepton production*, II.719
Perlman, I
 α-decay spectra, II.1217
 Nuclear Properties of the Heavier Elements, II.1223
Peron, J, controlled fusion, III.1628
Peronneau, P A, ultrasonic imaging, III.1914
Perrin, F
 First Solvay Conference (1911), I.147
 Levshin–Perrin law, III.1400
 neutrons, I.122
 relativistic kinematics, I.129
 Twelfth Solvay Conference, Brussels, II.646
Perrin, J
 Avogadro's number, III.1610
 Les Atomes, I.43
Pershan, P S, non-linear optics, III.1434
Pert, G J, x-ray laser, III.1464
Perutz, M F
 globular proteins, I.501
 haemoglobin, I.500, I.502, I.503
 isomorphous replacement method, I.497
 x-ray analysis, III.1514

Peterson, S W, neutron diffraction, I.458–9
Pethick, C J, neutron stars, III.1754
Petit, R, diffraction theory of gratings, III.1466
Petley, B W
　fundamental constants, II.1261
　gravitation, II.1266, II.1267
Petrán, M, confocal scanning laser microscope, III.1450–1
Petschek, H, collisionless shocks, III.1684
Petzoldt, J, positivism, I.319
Pfann, W G, zone-refining, III.1281, III.1543–5
Pfirsch, D, stellarator-type research, III.1655
Pfund, high-reflection coatings, III.1407
Phillips, M, direct reaction theory, II.1220
Phillips, W A, glasses, II.1006
Phragmen, G, single-crystal methods, I.435
Piccard, A, Fifth Solvay Conference (1927), I.205
Piccioni, O, mesons, II.652
Pickering, C, hydrogen lines, I.85
Pickering, E C
　Elements of Physical Manipulation, I.16
　stellar spectra, III.1693, III.1694
Pidd, R W, spark counters, II.753
Pierce, G W
　radio propagation, III.2000
　semiconductors, III.1318
Pierce, J R, photomultipliers, II.750
Piermarini, G, diffraction techniques, I.437
Pierre, F M, charmed meson, II.704
Pike, E R
　imaging theory, III.1467–8
　optical spectrum, III.1454
Pines, D
　electrical discharges in gases, III.1361
　superconductivity, II.937
Pippard, A B
　optics, III.1467
　penetration depth, II.933–4
Pippin, J E, crystal structures, II.1158
Pirani, F A E
　general relativity, I.299
　plane wave solutions, I.312
Piron, C, photon, III.1451
Placzek, G
　light-scattering spectroscopy, III.1408
　phonon dispersion relations, II.993
　transuranic elements, I.131–2
Planck, M, **I.144**
　black-body radiation, I.144, I.524–5, II.973, II.1269, III.1402
　bound system, I.111
　energy density, III.1387
　energy quanta, I.173
　entropy, I.523
　Fifth Solvay Conference (1927), I.205
　finite energy quantum, I.173
　First Solvay Conference (1911), I.147
　oscillators, I.70
　particle dynamics, I.279
　partition function, I.527
　Planck's law, I.67–70
　quantum optics, III.1387
　quantum theory, I.62, I.146–7, I.156, I.522, III.1385
　school of physics, I.15
　Scientific Autobiography, I.34
　second law, I.47
　special-relativistic thermodynamics, I.281
　thermodynamics, I.68
Platzman, R L, continuous spectra

from synchrotron light sources, II.1050
Plücker, J, spectrum of hydrogen, I.76
Podolsky, B, quantum mechanics, I.228
Pohl, R
  Bohr's theory, I.88
  colour centres, III.1529, III.1530
  disordered systems, II.1006
  ionic crystals, III.1291
Pohlhausen, E, free convection, II.843–5
Pohlhausen, K, conduction of heat, II.842
Poincaré, H
  energy conservation, I.61
  First Solvay Conference (1911), I.147
  perturbation theory, III.1830–31
  radio waves, III.1986–7
  recurrence theorem, I.618
  relativity principle, I.265
  relativity theory, I.173
  special theory of relativity, I.262–85
  statistical mechanics, I.524
Pokrovskii, V L
  correlation components, I.563
  multiple-correlations near critical point, I.562
Polanyi, M
  biomolecular structures, I.496
  dislocations, III.1520
  edge dislocation, I.476
  single crystals, III.1526–7
Polder, D, electromagnetic waves, II.1157
Polikanov, S, fission isomers, II.1218
Politzer, H D, running coupling constant, II.713
Pollack, H O
  eigenfunctions and eigenvalues, III.1468

optical imaging, III.1432
Pomeau, Y, delta (δ), III.1849
Pomeranchuk, I Ya
  electron–electron collisons, III.1316
Pomeranchuk trajectory, II.682
Pontecorvo, B
  neutrinos, II.671
  solar neutrinos, II.757
Popaloizou, J C B, thick discs, III.1781
Porte, G, lamellar phase, III.1608
Porter, A B, imaging theory, III.1391
Porter, C, proton scattering, II.1212
Post, B, Renninger effect, I.450
Post, R F
  Atoms for Peace Conference, III.1639
  Controlled fusion research—an application of the physics of high-temperature plasmas, III.1641
  loss-cone modes of mirror-confined plasmas, III.1662
  mirror machine, III.1679
  Sixteen lectures on controlled thermonuclear reactions, III.1660
Potts, R B, Potts model, I.571
Poulsen, magnetic recording, II.1158
Pound, R V, red shifts, I.291
Powell, C, π-meson, I.231
Powell, C F, photographic emulsions, II.653
Powers, P N, neutron diffraction, I.453
Poynting, J H
  gravitation, II.1266
  theoretical constructs, I.34
Prandtl, L, **II.797**
  boundary-layer equations, II.805, II.806

boundary-layer theory, II.796–8, II.803, II.869
drag due to lift, II.824
drag on a non-lifting body, II.869
elliptic distribution of circulation, II.824
heat transfer, II.842
lift–drag ratios L/D, II.870
perturbation theory, II.829
singular perturbation, II.800
supersonic flow, II.812
triple deck theory, II.829
turbulence, II.815
vortex lines, II.822
vortex wake, II.822
vorticity, II.820
wind-tunnel testing at supersonic speeds, II.879
Prendergast, K H, x-ray sources, III.1755
Prescott, C Y, parity violation, II.698
Press, W H, mass function, III.1804
Preston, T, Zeeman splitting, I.161
Priestley, J, magnesium, III.1359
Prigogine, I, **I.621**
  Brussels school, I.619–22
  Non-equilibrium Statistical Mechanics, I.619
  Twelfth Solvay Conference, Brussels, II.646
Primakoff, H, neutral pion, II.655
Pringle, J E
  accretion discs, III.1964
  thick discs, III.1781
Pringsheim, E, black-body radiation, I.149, III.1387
Prins, J, x-ray diffraction pattern, I.441
Prout, W, atomic species, I.46
Pryce, M H L
  hyperfine structures, II.1139
  quantum theory, III.1413
Purcell, E M
  hyperfine line radiation, III.1763
  nuclear magnetic resonance, II.1094
Pusey, P N, K-distribution, III.1462

Quimby, E
  medical physics, III.1864
  radiotherapy, III.1867
  radium implants, III.1878
Quinn, T J, acoustic interferometer, II.1270

Rabi, I I
  fine/hyperfine structures, II.1047
  First Shelter Island Conference, II.645
  precessional motion, II.1153
Racette, G W, He–Ne laser, III.1435
Radon, J, computed tomography, III.1916
Raff, A D, sea-floor spreading, III.1975
Rainwater, J, II.1216
  quadrupole moment, II.1215
Rajchman, J A, photomultipliers, II.750
Rakshit, H, ionospheric physics, III.2000
Ramachandran, G N, speckle-pattern formation, III.1442
Raman, Sir Chandrasekhara
  Brownian motion, III.1442
  diffraction of light, III.1409
  Lectures in Physical Optics, III.1442
  phase transitions, II.1008
  Raman scattering, II.996, III.1404–6
  secondary scattered radiation, I.172
  speckle, III.1399
Ramberg, E G
  contour phenomena, III.1581–2

magnetic lenses, III.1580
Ramsauer, noble gas atoms, II.1041
Ramsay, Sir William, helium in uranium-bearing mineral, I.58
Rarity, J G, Bell inequalities, III.1471
Rasetti, F
　mesotron lifetime, I.406
　Raman scattering, I.117, II.996
Ratcliffe, J A
　An Introduction to the Ionosphere and Magnetosphere, III.2001
　ionosphere, III.1999
Ray, E, Geiger counter, III.2007
Rayleigh, Lord, **I.21**
　acoustics, III.1910
　aeolian tones, II.850
　black-body radiation, I.30, II.972
　Bohr's theory, I.87
　boundary conditions, II.845
　catastrophe theory, II.798
　Cavendish Laboratory, I.21
　crystal surface, II.1005
　Faraday measurement, II.1269
　grating spectroscopy, III.1391
　law of equal partition of energy, I.143–4
　optics, III.1390
　radio waves, III.1986–7
　random processes, I.586
　senior wrangler, I.11
　shock wave, II.811
　singular perturbation theory, II.811
　sound from flows, II.850
　sound propagation, II.810
　sound waves, II.858
　statistical mechanics, I.173
　The Theory of Sound, I.21, II.808, II.855, III.1910
　wave energy, II.858
　zero-viscosity limit, II.828

Raymond, A L, hexamethylene tetramine, I.495
Raynor, G, physical metallurgy, III.1555
Read, F H, quantum-mechanical reformulation, II.1065
Reber, G
　continuum radio map, III.1762
　radio astronomy, III.1736
Rebka, G A, red shifts, I.291
Rees, M J
　non-thermal phenomena, III.1781
　radio sources, III.1782, III.1783
Regel, A R, energy levels, III.1368
Regener, E, scintillation method, II.1200
Regge, T, Regge poles, II.681–2
Reiche, F, dispersion formula, I.171
Reickhoff, K E, stimulated Brillouin scattering, III.1444
Reid, A
　de Broglie waves, I.171
　electron diffraction, I.463
Reid, J M, pulse-echo ultrasound, III.1911–12
Reimann, B, non-linear analysis, II.857
Reines, F, neutrino, II.663
Reinitzer, F, cholesteric esters, III.1539–40
Reppy, J D, superfluidity, II.932
Retherford, R C, Lamb shift, II.642, II.643
Reusch, rocksalt, III.1517
Revelle, R, Pacific floor measurements, III.1974
Rex, A F, *Maxwell's Demon*, I.576
Reynolds, O
　diffusion of momentum, II.826
　flow in pipes, II.814
　pipe flow, II.825
　Reynolds analogy, II.843
　turbulence, II.798, II.825, III.1834

Rhines, F N
  grain growth, III.1550
  Microstructology—Behavior and Microstructure of Materials, III.1550
Ricci, C G, tensor analysis, I.292
Richards, B, Airy disc, III.1412
Richardson, O W
  electron emission, I.37
  electron optics, III.2000
  Fifth Solvay Conference (1927), I.205
  photoelectric effect, III.1390
  thermionic emission, I.12
Richter, B, electron storage rings, II.739
Richter, R, controlled fusion, III.1628
Riecke
  electrical and thermal conductivities, III.1285
  Hall effect, III.1282
  metallic conduction, III.1284
Riemann, B
  general relativity, III.1722
  propagation speed, II.809
  waveforms, II.810
Risken, H Z
  laser light, III.1443
  laser noise, III.1444
Riste, T, dynamic response of $SrTiO_3$, II.1010
Ritchey, G W, spiral nebulae, III.1717
Ritson, D, b-quark, II.723
Ritter, A, stellar evolution, III.1694–5
Ritz, W
  black-body radiation, I.163
  combination principle, I.25, III.1391
  combination rule, I.186
  spectral frequencies, I.86
Robb, A A, special theory of relativity, I.277–8

Roberts, F, flying-spot microscope, III.1416–17
Roberts, L D, germanium junctions, II.1208
Robertson, H P
  Robertson–Walker metric, III.1729–30
  synchronization of clocks, III.1729
Robertson, J M
  organic chemistry, I.496
  phthalocyanines, I.495
Robinson, D, High Voltage Engineering (HVE), II.1203
Robinson, F N H, anharmonic oscillators, III.1434
Robinson, H
  α-particles, II.1201
  β-ray spectrometer, II.1201
  RaB, RaC and RaE spectra, I.106
Robinson, I K
  plane wave solutions, I.312
  surface crystallography, I.469
Rochester, C D
  elementary particles, II.654
  kaons, II.656
Rochester, G, forked tracks, I.408
Rodionov, S N, tritium-beta experiment, III.1679
Rogers, G L, optical holography, III.1415
Rogerson, J B
  helium problem, III.1790
  interstellar deuterium, III.1791
Rohrer, H, scanning tunnelling microscope, III.1549
Röntgen, W C, x-rays, I.3, I.32, I.53, I.54, I.362, I.424, II.1271, III.1529, III.1857, III.1861
Roosevelt, President, atomic energy research, II.1196
Roozeboom, B, phase diagrams, III.1508
Rosa, E B, speed of light, II.1263

Rosbaud, P, *Progress in Metal Physics*, III.1556
Rose, P H, tandem accelerator, II.1203
Rosen, N, quantum mechanics, I.228
Rosenbluth, M N, **III.1664–5**
  International Summer Course in Plasma Physics 1960, III.1643
  loss-cone modes of mirror-confined plasmas, III.1662
  mirror machine, III.1679
  mirror systems, III.1660
Rosenfeld, L
  gravitational field, I.317
  meson theory, I.401
  symmetrical theory, I.403
  Twelfth Solvay Conference, Brussels, II.646
Rosenhain, W
  amorphous metal hypothesis, III.1518
  An Introduction to the Study of Physical Metallurgy, III.1517
  glass industry, III.1517
  grain boundaries, III.1519, III.1528
  physical metallurgy, III.1516
  slip lines, III.1517
  yield-stresses of single metallic crystals, III.1520
Rosenthal, A H, Sagnac effect, III.1433
Rossby, C G
  potential vorticity concept, II.905
  wave-like current patterns, II.892–5
Rosseland, S, quantum theory, I.179
Rossi, B
  cosmic rays, I.399, II.748
  First Shelter Island Conference, II.645
  nuclear forces, II.652
Rossman, M G
  structure determination, I.445
  virus structures, I.510
Rotblat, angular distributions, II.1220
Roth, W L, superionic conductors, III.1539
Round, H J
  band-structure diagram, III.1417
  yellow light emission, III.1391–2
Roux, D
  lamellar phase, III.1608
  sponge phase, III.1608
Rowan-Robinson, M, radio quasars, III.1797–8
Rowe, H E, non-linear optics, III.1424
Rowland, H, concave spectral gratings, I.54
Royds, T, α-particle, I.59
Rozental, S, meson theory, I.401
Rubbia, C
  field particles W and Z, II.744
  proton–antiproton collider, II.707, II.743
Rubens, H
  black-body radiation, I.149, III.1387
  far-infrared spectrum, II.982–3
  First Solvay Conference (1911), I.147
Rubin, H, magnetic cusp, III.1645
Rubin, V C, rotation curves of galaxies, III.1770
Rubinowicz, A
  polarization of spectral lines, I.178
  selection rules, I.94
Rubinstein, H, Dolen–Horn–Schmid duality, II.683
Rüdenberg, R, electron microscope, III.1573–4
Ruderman, M

indirect-exchange-coupling theory, II.1012
RKKY interaction, II.1164
scalar Higgs fields, III.1806
Rühmkoff, H, higher voltages, I.54
Rumford, Count, *see* Thompson, Sir Benjamin
Runcorn, S K
  continental drift, III.1972
  magnetic North Pole, III.1972
Runge, C, Zeeman splitting, I.161
Rushbrooke, G S, thermodynamic relation, I.560
Ruska, E
  Crantz Colloquium, III.1571–2
  electron microscope, III.1570–1
  electronic lenses, III.1569
  iron-free lenses, III.1568, III.1571
  polepiece lens, III.1567
  transmission electron microscope (TEM), III.1569–70, III.1575–7
Ruska, H, specimen preparation, III.1575–6
Ruskol, E, Moon model, III.1966
Russell, H N
  luminosity–spectral class diagram, III.1696
  photoexcitation, III.1761
  photographic parallax programme, III.1696
  solar atmosphere, III.1701–2
  solar system, III.1962
  tidal theory, III.1951
  white dwarfs, III.1709–10
Russell, J E
  German school system, I.8
  physics in education, I.8
Rutgers, superconductivity, II.924
Rutherford, E (Baron Rutherford of Nelson), **I.55**
  actinium, II.1227
  age of the Earth, III.1694
  α-particles, I.59, I.115, II.638, II.745–6, II.1201
  α-rays, I.54, I.58
  atomic energy, I.63
  atomic nuclei, II.1183
  atomic structure, I.74, I.156
  β-ray spectrometer, II.1201
  β-rays, I.54, I.58
  crystal diffraction spectroscopy, II.1207
  electrical counting methods, II.1200
  electron charge, II.747
  electrons, I.64
  experimental physics, I.14
  First Solvay Conference (1911), I.147
  half-life, I.62
  hydrogen nucleus, I.109
  meeting with Bohr, I.74
  Nature of the α-particle, I.58
  nuclear structure, I.110, I.113, III.2018
  positive ion accelerators, II.1184
  quantum conditions, I.359
  RaB, RaC and RaE spectra, I.106
  Radiations from Radioactive Substances, I.108, I.112, II.1199, II.1200
  radioactive decay, III.1949
  Radioactive Substances and their Radiations, I.107
  Raman effect (combination scattering), III.1404
  Silliman lectures, I.52
  The distinction between α-rays and β-rays, I.55
  thorium emanation, I.59
  transformation theory, I.59
  wireless transmission, III.1986
Ryle, M
  Cassiopeia A, III.1737
  radio astronomy, III.1738
  radio sources, III.1788
  survey of the sky, III.1788
Ryutov, D, mirror physics, III.1666

Sackur, O, quantum-theoretical equations, I.154
Sadler, C A, x-ray spectroscopy, I.158
Safronov, V, solar system, III.1963
Sagdeev, R Z, fusion-plasma research, III.1642
Sagitov, M V, gravitation, II.1266
Sagnac, G, measurement of rotation, III.1398
Saha, M N
  ionospheric physics, III.2000
  stellar atmospheres, III.1699
  stellar spectra, III.1693
Sakata, S
  fundamental triplet, II.685
  meson theory, I.407
  spin-1 particle, II.654
  two-meson theory, I.408
Sakharov, A D
  baryon–antibaryon asymmetry, III.1806
  baryon asymmetry, II.765
  CP violation, II.675
  Plasma Physics and the Problem of Controlled Thermonuclear Reactions, III.1642
  tokamak, III.1630
Salam, A, II.696
  electroweak unification, II.695
  neutrino, II.667
  Twelfth Solvay Conference, Brussels, II.646
Saleh, B E A, sub-Poissonian Franck–Hertz source, III.1460
Salpeter, E E
  α–α scattering, II.1226
  black holes, III.1779
  nuclear reactions, III.1709
  origin of elements, III.1747
  triple-α reaction, III.1746
Sampson, R A, energy transport, III.1699
Samuelson, E J, K_2CoF_4, II.1016

Sandage, A R
  Evidence for the occurence of violent events in the nuclei of galaxies, III.1777
  Herzsprung–Russell diagrams, III.1793
  Hubble's constant, III.1793
  quasars, III.1777
  redshift–magnitude relation, III.1788, III.1793
  stars in globular clusters, III.1746
  stellar spectra, III.1776
  The ability of the 200" telescope to discriminate between selected world models, III.1793
Sargent, W L W, inverse-Compton catastrophe, III.1781
Satomura, S, Doppler ultrasound, III.1914
Saunders, F A, Russell–Saunders (or L–S) coupling, III.1702
Savedoff, M P, interstellar gas, III.1765
Sayre, D, electron density, I.448
Schafroth, superconductivity, II.937
Schawlow, A L, **III.1428**
  Fabry–Perot resonator, III.1429
  lasers, III.1423, III.1424
  laser spectroscopy, III.1466
  maser, III.1422
  Microwave Spectroscopy, III.1428
Schechter, P, mass function, III.1804
Schechtman, D, quasicrystals, I.481
Scheffer, T J, liquid-crystal optics, III.1452
Schelleng, J C, radiowave propagation, III.1990
Schenectady, R T, electrical discharges in gases, III.1361
Scherk, J, quantum gravity, II.684
Scherrer, P, single crystal, I.434
Scherzer

axially symmetric electron
    lenses, III.1579
Geometrische Elektronenoptik,
    III.1578
Scheuer, P A G, radio sources,
    III.1782, III.1789
Schild, A, general relativity, I.299
Schilling, W, point defects, III.1534
Schlichting, H,
    Tollmien–Schlichting waves,
    II.830
Schlier, R J, surface
    crystallography, I.470
Schlüter, A, stellarator-type
    research, III.1655
Schmid, C, pion–nucleon charge
    exchange, II.683
Schmid, E
  plastic deformation, III.1526
  single crystals, III.1527
Schmidt, B V
  Schmidt corrector plate, III.1406
  Schmidt telescopes, III.1745
Schmidt, E, free convection, II.843
Schmidt, G, radioactivity, I.57
Schmidt, G N J, tomato bushy
    stunt virus (TBSV), I.509
Schmidt, M
  optical spectrum of 3C 273,
    III.1776
  quasars, III.1798
  radio quasars, III.1797–8
Schmidt, O, monistic theories,
    III.1963
Schneider, W G
  isotherms of xenon, I.556
  liquid–gas co-existence curve,
    I.553
Schoenflies, A, symmetry types,
    I.423
Schönberg, M,
    Schönberg–Chandrasekhar
    limit, III.1709
Schottky, W
  colour centres, III.1530
  electric potential, III.1289–90
  vacancies, III.1529
Schrieffer, J R
  BCS theory, II.938
  MOSFET, III.1372
  superconductivity, I.232, II.939,
    III.1364
Schrödinger, E, **I.193**
  decay law of radioactive
    substances, I.222
  entangled photon states,
    III.1394–5
  Fifth Solvay Conference (1927),
    I.205
  matter waves, I.194
  non-relativistic action function,
    I.195
  photon mass, III.1452
  quantum mechanics, II.638
  quantum theory, I.174
  relativistic equation, I.283
  wave equations, I.195, I.200
  wave mechanics, I.360
Schubauer, G, boundary layer,
    II.835
Schubnikow
  crystal growth, III.1280
  zone-melting, III.1281
Schulkes, J A, ferrimagnetism,
    II.1157
Schulman, J, microemulsions,
    III.1608
Schulman, L S, 'life' (game), I.629
Schulz, liquid metals, III.1366
Schulz, G J, He atoms, II.1061–2
Schumacher, E E, deformation of
    lead crystals doped with
    antimony, III.1544
Schuster, A
  doctorate, I.17
  electrical discharges in gases,
    III.1985
  energy transport, III.1699
Schwartz, M
  neutrinos, II.671

two-neutrino experiment, II.672
Schwarz, J
  quantum gravity, II.684
  string theories, II.684
Schwarzschild, K
  curvature of space, III.1722
  dynamical laws, I.176
  energy transport, III.1699
  field equations for a point mass, III.1777–8
  Maxwell's equations, I.279
  metric, III.1778
  quantum theory, I.89
  Stark effect, I.162
Schwarzschild, M
  Herzsprung–Russell diagrams, III.1793
  plasma column instabilities, III.1641
  stars in globular clusters, III.1746
Schweber, S
  hierarchical structure, III.2020
  process of emergence, III.2019
Schwers, F, II.972
Schwinger, J
  bosons, II.694–5
  First Shelter Island Conference, II.645
  quantum electrodynamics, II.643, II.740, III.1446
  renormalization, II.646
  temporal behaviour of a general system, I.617
  Twelfth Solvay Conference, Brussels, II.646
Schwitters, R, Mark I Detector, II.703
Scully, M O
  laser light, III.1443
  phaseonium, III.1473
  x-ray lasers, III.1473
Seaborg, G T
  Nuclear Properties of the Heavier Elements, II.1223
  plutonium isotope, II.1223
  transuranic elements, II.1223
Seaton, M J
  chanel coupling formulation, II.1073
  exchanges of angular momentum, II.1060
  R matrices, II.1068
  radial density function, II.1066
Secchi, A, stellar spectra, III.1692–3
Seddon, J A, wave-like patterns, II.897
Seeger, A K, dislocation theory, I.476
Segrè, E
  neutron bombardment, II.1191
  plutonium isotope, II.1223
Seidel, H, optical bistability, III.1449
Seitz, F, **III.1532**
  band structure, III.1314–17
  Brillouin zone, III.1313
  colour centres, III.1531, III.1536
  crystal defects, III.1531
  diffusion, III.1531
  electron band structure of sodium, III.1530
  Imperfections in Nearly Perfect Crystals, III.1535
  materials research, III.1558
  Schrödinger's equation, III.1315
  solid-state physics, III.1313
Selenyi, P, xerography, III.1541
Sen, H K, magneto-ionic theory, III.1992
Sen, N R, ionospheric physics, III.2000
Serber, R
  β-decay, I.401
  direct reaction theory, II.1220
  First Shelter Island Conference, II.645
  mesotron, I.389
  μ-meson, II.652
  muon, II.651

orbit calculations, II.732
Seyfert, C K, spiral galaxies, III.1777
Shaknov, I, scintillation counters, III.1414
Shakura, N, thin accretion discs, III.1779
Sham, L J
  phonon frequency, II.1002
  pseudopotential theory, II.999
Shannon, R R, US optical industry, III.1476
Shapiro, I I, Schwarzschild metric, I.305
Shapiro, J, pion–pion scattering amplitude, II.684
Shapley, H
  Galaxy, III.1714, III.1716, III.1717
  globular clusters, III.1716
  Great Debate, III.1718
  novae, III.1717
  stellar masses, III.1698
Sharp, P F, imaging procedure, III.1920
Sharvin, D Yu, resistance of thin films, III.1370
Sharvin, Yu V, resistance of thin films, III.1370
Shenstone, W A, vitrified silica ware, I.29
Sheppard, R M
  detection of impulses, III.1957
  Earth's core, III.1980
Shimizu, F, self-phase modulation, III.1449
Shklovsky, I S
  Crab Nebula, III.1774
  x-ray sources, III.1755
Shlyaptsev, V, laser amplification, III.1466
Shockley, W B, **III.1324**
  cuprous oxide rectifiers, III.1323
  cyclotron resonance, III.1337–8
  Electrons and Holes in Semiconductors, III.1317
  magnetoresistance, III.1335–6
  photomultipliers, II.750
  transistor, II.744
Shoenberg, D
  de Haas–van Alphen effect, III.1348–52
  superconductivity, II.925
Shore, B W, optical harmonic generation, III.1473
Short, R, sub-Poissonian statistics, III.1460
Shortley, G, *The Theory of Atomic Spectra*, II.1037
Shu, F J, formation of stars, III.1769
Shubnikov, A, superconductivity, II.920
Shuleikin, M V, ionospheric layers, III.1992
Shull, C G
  Fe_3O_4, I.460
  magnetic structures, II.1148
  neutron diffraction, I.453
  powder pattern of MnO, I.460
  spin density, I.462
Shutt, R P
  diffusion chamber, II.752
  heavy particles, II.657
  mesotrons, I.405
Sibbett, W, dye-laser pulses, III.1439
Sidgwick, Faraday measurement, II.1269
Siegbahn, K
  double-focusing beta spectrometer, II.1206
  electron spectrometers, II.1090
Siegbahn, M, x-ray spectra, I.160
Siegert, A J F, Master equation description of gas, I.605
Sievert, R M
  radiation protection, III.1890
  thimble ionization chamber, III.1867

Silin, V P, electrons in metals, III.1363
Silk, J, sound waves, III.1802
Simon, A
  cloud chamber, I.364
  plasma column, III.1678
Simon, Sir Francis
  diamond–graphite equilibrium diagram, I.538
  low-temperature physics, II.920
  third law of thermodynamics, I.537, I.542
  zero-point motion, II.973–4
Simonen, T, TMX experiment, III.1667
Simpson, J A, II.1062
Sinai, Ja, hard spheres system, I.615
Sinitsyn, G V, optical bistability, III.1450
Sitte, K, β-decay theory, I.375
Sitterly, C M, atomic energy levels, II.1093
Skinner, H W B, x-ray emission, III.1361
Sklodowska, M, *see* Curie, M
Slater, J C
  crystal field theory, II.1144
  properties of metals, III.1345
  quantum mechanics, I.211
  quantum theory, I.166, I.180
  Slater determinant, I.215
  Solid State and Molecular Theory, III.1314
  solid-state physics, III.1313
Slee, O B
  radio sources, III.1737
  radio surveys, III.1789
Slepian, D
  eigenfunctions and eigenvalues, III.1468
  optical imaging, III.1432
Slepian, J
  electron emissions, II.749
  secondary emission, III.1399

Slichter, C P, BCS theory, II.940
Slipher, V M, spiral galaxies, III.1725
Sloan, D H, cyclotron, II.736
Slusher, R E, squeezed states of light, III.1469–70
Smale, S, mathematician, III.1840
Smart, J S
  neutron diffraction, I.453
  powder pattern of MnO, I.460
Smekal, A, light-quantum hypothesis, I.172, I.185
Smith, B W, interstellar medium, III.1766
Smith, C S
  A History of Metallography, III.1556
  A Search for Structure, III.1556
  Acta Metallurgica, III.1556
  Institute for the Study of Metals, III.1556
Smith, F G
  Cassiopeia A, III.1737
  radio sources, III.1737
Smith, K, soliton pulse communication, III.1471
Smith, L W, thin-film devices, III.1444
Smith, M G, quasars, III.1798
Smith, N, long-distance communications, III.1998
Smith, P L, *Needs, Analysis and Availability of Data for Space Astronomy*, II.1093
Smith, R A, photoconductivity, III.1327
Smith, T, optical systems, III.1446
Smoluchowski, M S
  dielectric constant, III.1392, III.1396–7
  Maxwell's Demon, I.576
  opalescent bands, I.546
Smoot, G F, cosmic microwave background radiation, III.1804

Name Index

Smythe, *Atomic Energy for Military Purposes*, II.1223
Snitzer, E, optical-fibre laser, III.1427
Snoek, J L
  ferrimagnetism, II.1155
  strain-aging law, III.1521
Snyder, H
  collapsing sphere of dust, I.308
  general-relativistic analysis, III.1712
Snyder, H S
  accelerators, II.738
  synchrotrons, II.734
Sochor, V, HCN/H$_2$O lasing, III.1443
Soddy, F
  atomic energy, I.63
  radioactive decay, III.1949
  transformation theory, I.59
Soffer, B H, laser development, III.1448
Sohncke, L, symmetry types, I.423
Solomon, J, uncertainty relations, I.228
Solvay, E
  First Solvay Conference (1911), I.147
  new physics, I.7
Somer, J C, imaging, III.1914
Sommerfeld, A, I.180
  astrophysics, I.219
  Atombau and Spektrallinien, I.110
  Atomic Structure and Spectral Lines, I.156
  battle between Boltzmann and Ostwald, I.48
  Bohr's theory, I.88
  Drude–Lorentz model, III.1288
  dynamical quantum theory, I.174
  electric potential, III.1289
  extension of Bohr's model, I.160–1
  fine structure, I.90

First Solvay Conference (1911), I.147
free-electron gas hypothesis, I.216
inner quantum number, I.100–1
metal electrons, I.215
Planck's law, I.70
quantization, I.359
quantum model, III.1308
quantum numbers, I.90–2
quantum rule, I.175
quantum theory, I.179, III.1285, III.1361
radio waves, III.1987
structure factor, I.431
theory of metals, III.1290
thermoelectric effects, III.1308
x-ray diffraction, I.424
x-ray spectra, I.160
Zeeman effect, I.102, I.162
Sorby, H
  metallography, III.1507
  microstructure, III.1506
Sorokin, P P
  dye lasers, III.1446
  four-level laser, III.1427
Southworth, G C, waveguide propagation, III.1322
Spain, I L, single-crystal studies, I.437
Spedding, F H, magnetism, II.1143
Spinrad, H, radio galaxies, III.1795
Spitzer, L
  Atoms for Peace Conference, III.1638
  interstellar gas, III.1765
  magnetic fusion, III.1636
  Physics of Fully Ionized Gases, III.1631
  rotational transform, III.1654
  stellarator, III.1630, III.1648, III.1654, III.1655
Sproule, D O, diagnostic ultrasonics, III.1910–11

N71

Sproull, R, materials research, III.1558
Standley, K J
  relaxation time, II.1133
  spinels, II.1156
Stanley, G J, radio sources, III.1737
Stanley, H E, Ising models, I.563
Stark, J
  band spectra, I.77
  chemical and other quanta, I.154
  longitudinal effect, I.162
  quantum physics, I.148, I.155
  Stark effect, I.89
  Zeeman splitting, I.161
Staub, H, element synthesis, II.1226
Staudinger, H, long-chain systems, III.1597
Steenbeck, M, visit to Max Knoll, III.1572–3
Stefan, J, black body, I.66
Stein, P
  computers, III.1843
  mathematician, III.1839
Stein, R F, solar oscillations, III.1750
Steinberger, J
  decay of neutral pion, II.711
  neutral pion, II.655
  two-neutrino experiment, II.672
Steinhardt, P J, world model, III.1807
Steller, J, magnetic mirror, III.1659
Stelson, P H, Coulomb excitation experiments, II.1217
Stephens, W E, element synthesis, II.1226
Stern, J, quantum-theoretical equations, I.154
Stern, O
  electronic orbits, I.164
  ionic crystals, III.1531
  quantum theory, I.221
  specific-heat problem, I.153–4
  surface crystallography, I.468

Sternheimer, R, inner-shell electrons, II.1092
Stevenson, E C
  elementary particles, I.391
  mesotron, I.397
  muon, II.650
  new particles, I.389
Stevenson, M J, four-level laser, III.1427
Stewart, A T, phonons, II.999
Stewart, B
  Earth's magnetic field, III.1621
  geomagnetic variations, III.1984–5, III.1985
  wire-wound resistors, II.1252
Stiles, W G, human eye, III.1407
Stockmayer, W, polymer solutions, III.1597
Stoicheff, B P, stimulated Brillouin scattering, III.1444
Stokes, G G
  molecular vibrations, I.76
  x-rays, I.362
Stoletov, A G, photoelectric effect, III.1390
Stone, J, fibre lasers, III.1427
Stoner, C R, *Short Wave Wireless Communication*, III.1619, III.1620
Stoner, E, white dwarfs, III.1711
Stoner, E C
  electronic levels in atoms, I.168
  spin waves, II.1012
Stoney, G J, estimate for e, I.52
Storey, L R O, whistler studies, III.1995
Stormer, C
  Earth's magnetic field, III.1682
  geomagnetic storms, III.2003
  magnetic-mirror effect, III.1637
  single charged particle, III.2004
Strassmann, F
  elements with $Z = 93$–96, I.130
  neutron bombardment of uranium, II.1195

neutron sources, II.1206
neutron–U collisons, I.130
radiochemical analysis, II.1195
uranium isotope, II.1223
uranium-239, I.132
Stratton, S W, **II.1241**
Straubel, P, imaging theory, III.1418
Strauss, L L
 fusion research, III.1637–40
 Men and Decisions, III.1640
Street, J C
 elementary particles, I.391
 mesotron, I.397
 muon, II.650
 new particles, I.389
Strehl, K, aberration effects, III.1389
Strnat, K J, rare-earth–cobalt alloys, II.1165
Strnat, R M W, rare-earth–cobalt alloys, II.1165
Stroke, G W, holograms, III.1438
Strömgren, B
 interstellar medium, III.1762
 Strömgren spheres, III.1762
Strong, J, anti-reflection coatings, III.1407
Strutinsky, V M, fission isomers, II.1218
Strutt, J W, *see* Rayleigh, Lord
Strutt, R J, thorianite, III.1949
Stueckelberg, E, heavy electrons, I.391
Stuewer, R, nuclear physics, II.1192
Stump, D M, polarized light, III.1412
Sucksmith, W, ferromagnetism, II.1134
Sudarshan, E C G
 neutrinos, II.667
 quasi-classical distribution function, III.1441
 radiation field, III.1388

Suess, E, ocean basins, III.1967
Suess, H E
 chemical elements, III.1747
 nuclear binding energies, II.1210
 shell models, II.1209
Suhl, H, crystal structures, II.1158
Sunyaev, R A
 induced Compton scattering, III.1801
 photon scattering, III.1773
 thin accretion discs, III.1779
Sunyar, A, neutrino helicity, II.668, II.670
Sutherland, P, neutron stars, III.1754
Swedenborg, E, island universes, III.1714
Sweetman, D
 magnetic mirror, III.1659
 neutral beam injection, III.1661
Swiatecki, W T, liquid-drop mass formula, II.1225
Sykes, M E
 critical exponents, I.560
 series expansions, I.557
 site percolation processes, I.572
 two-dimensional lattices, I.573
Synge, E H, optical microscopy, III.1405–6
Szalay, A S, neutrinos, III.1803
Szigeti, B, electronic dipole moment, I.485
Szilard, L
 entropy, I.225
 Maxwell's Demon, I.576
Szöke, A, optical bistability, III.1449
Szymborski, K
 colour centres, III.1529
 solid-state physics, III.1320

Tabor, D, measurement of forces between surfaces, III.1612
Tainter, S, Photophone, III.1422

Talbot, F, daguerreotype process, III.1692
Talleyrand, C M de, metric system, II.1233
Talmi, I, shell model, II.1213
Tamm, I E
  energy loss formula, II.750
  first-order Raman effect, II.996
  nuclear exchange potential, I.376
  Plasma Physics and the Problem of Controlled Thermonuclear Reactions, III.1642
  tokamak, III.1630
  visible light, III.1409
Tammann, G
  binary alloys, I.548
  crystal growth, III.1280
  Hubble's constant, III.1793
  Lehrbuch der Metallographie, III.1554
  liquid crystals, III.1540
  phase diagrams, III.1509, III.1554
Tapster, P R, Bell inequalities, III.1471
Tayler, R J
  helium synthesis, III.1790
  synthesis of heavy elements, III.1792
Taylor, B N
  determination of e/h, II.1268
  magnetic helicity, III.1657
  resonance frequency, II.1095
Taylor, G I, **II.799**
  catastrophe theory, II.798
  diffraction pattern, III.1396
  Diffusion by continuous movements, II.830
  dislocations, III.1520
  edge dislocation, I.476
  equilibrium (wet-bulb) temperature, II.843
  potential-flow forces, II.887
  singular perturbation theory, II.811
  sound waves, II.858
  stabilizing effect of viscosity, II.828
  tidal friction, II.891
  turbulence, II.798, II.832
  vortices, II.833
Taylor, J B
  binary pulsars, III.1757
  plasma pressures, III.1661
Taylor, J H
  binary pulsar, I.298, I.317
  quadrupole radiation formula, I.315
Tebble, R S, domain techniques, II.1155
Teich, M C, sub-Poissonian Franck–Hertz source, III.1460
Teichmann, J
  colour centres, III.1529
  solid-state physics, III.1320
Telegdi, V L, parity violation, II.666
Teller, E
  adiabatic confinement, III.1679–80
  Atoms for Peace Conference, III.1639
  dipole vibration, II.1213–14
  First Shelter Island Conference, II.645
  trapped particles in magnetic-mirror fields, III.1683
Temkin, R J
  amorphous germanium, I.479
  amorphous structure, I.478
Temmer, G M, Coulomb excitation experiments, II.1217
Temperley, H N V, percolation model, I.573
Terhune, R W, Raman Stokes and anti-Stokes rings, III.1436
Ter-Martirosyan, Coulomb excitation, II.1217

Name Index

Ter-Pogossian, M M, positron emission tomography (PET), III.1920
Tesla, N, radio and high-frequency technology, III.1620
Tharp, M, Mid-Atlantic Ridge, III.1973
Thellung, A, disordered materials, III.1366
Theobald, J P, fission isomers, II.1218
Thiessen, A, direct-drive approach to inertial fusion, III.1676
Thom, R, dynamical systems, III.1850
Thomas, L H, Thomas effect, I.278
Thomas, R G, nuclear reactions, II.1219
Thomas, W, spectral line intensity, I.172
Thompson, Sir Benjamin (Count Rumford), learned societies, I.3
Thompson, G B H, semiconductor lasers, III.1437
Thompson, R, hyperon, II.657
Thompson, S, radioactivity, III.1876
Thompson, W, *see* Lord Kelvin, Lord
Thomson, Sir George
  controlled fusion, III.1636
  de Broglie waves, I.171
  electron diffraction, I.463
  finite electron temperatures, III.1624–5
  toroidal fusion device, III.1634
Thomson, J J, I.50
  atomic model, I.64, I.65
  atomic structure, I.49, I.156
  deflection measurements, II.1271
  determination of e/m, I.53, I.54
  discovery of electron, I.3, III.1565
  electrical conductivity, III.1986
  electrical discharges in gases, III.1985
  electron, III.1857
  electron charge, II.747
  electron diffraction, I.463
  electron emission, I.37
  electronic theory of magnetism, II.1114
  experimental physics, I.14
  finite electron temperatures, III.1624–5
  free electrons, I.54
  gravitation, II.1266
  heat engines, I.534
  line spectra, I.77
  metallic conduction, III.1284
  photoelectron emission, III.1396
  Recollections and Reflections, I.27, I.36
  second wrangler, I.11
  x-ray scattering, I.362
Thomson, R, materials physics, III.1558
Thomson, W, *see* Kelvin, Lord
Thonemann, P C
  fusion reactions, III.1642
  ZETA experiment, III.1641
Thorne, K S, binary sources, III.1756
Thouless, D J, topological long-range order, I.571
Tiktopoulos, singularities in complex angular momentum, II.682
Ting, S C C
  dilepton mass spectrum, II.701
  J resonance, II.702
Tisza, L, two-fluid model, II.927
Tizard, Sir Henry, atomic energy project, II.1198
Todd, A R, nucleic acids, I.504
Todd, Lord, polymerization, III.1550
Todhunter, I, research and genius, I.12

Toennies, J P, helium atom scattering, II.1006
Tolansky, S
  FECO interferometry, III.1412
  microtopography, III.1412
  optical interferometry, III.1411
  rings lens-plate experiment, III.1412
Tollmien, W
  parallel flows, II.828
  stability calculation, II.829
Tolman, R C
  particle dynamics, I.279
  quantum physics, III.1401–2
  spherically symmetric perturbations, III.1801
Tolpȳgo, K B, covalent crystals, II.1002
Tomonaga, S
  β-decay, I.402
  cosmic-ray particles, I.407
  coupling theory, I.407
  quantum electrodynamics, II.643, II.740, III.1446
  renormalization, II.646
  slow mesotrons, I.406
  Twelfth Solvay Conference, Brussels, II.646
Tonks, L, plasma oscillations, III.1624
Totsuji, H, clusters of galaxies, III.1771
Toulouse, percolation processes, I.573
Touschek, B, storage rings, II.740
Townes, C H, **III.1420**
  'dressed' atomic states, III.1421
  electric quadrupole moments, II.1214
  Fabry–Perot resonator, III.1429
  lasers, III.1423
  maser, III.1419, III.1422
  stimulated Brillouin scattering, III.1444
Townsend, A, 'big eddies', II.839

Townsend, J S, ionization detectors, II.746
Toya, T, dispersion curves in sodium, II.999
Trahin, M, perturbation theory, III.1473–4
Treiman, S B
  A Century of Particle Theory, II.770
  β-decay, II.676
  pion decay constant, II.675
Treloar, L R G, rubber elasticity, III.1551
Tremaine, S D
  galactic cannibalism, III.1774
  phase space constraints, III.1804
Trimble, V L, binary sources, III.1756
Trowbridge, J
  electrolytic cells, I.27
  practical classes, I.15
Trump, J G
  electrostatic generator, II.1203
  High Voltage Engineering (HVE), II.1203
Tsoi, V S, bismuth crystal, III.1344–5
Tsu, R
  coupling of quantum wells, III.1459
  quantum-well superlattices, III.1453
Tuck, J L
  magnetic cusp, III.1645
  magnetic pinch, III.1636
Turlay, R, *CP* violation, II.672
Turnbull, D
  age-hardening, III.1533
  nucleation kinetics, III.1547
Turner, R C, flaw detector, III.1911
Tuve, M A
  ionospheric conducting layer, III.1988
  ionospheric plasmas, III.1621
  ionospheric sounding, III.1995

radio receiving equipment, III.1990
van de Graaff generator, II.1203, II.1204
Tweet, D J, surface crystallography, I.469
Twiss, R Q
  optical spectroscopy, III.1421–2
  radio interferometer research, III.1422
Tyson, A, galaxies, III.1797

Uchling, E A, collision term in Boltzmann equation, I.602
Uhlenbeck, G E, **I.594**
  Brownian motion coupled harmonic oscillators, I.608
  collision term in Boltzmann equation, I.602
  Critical Phenomena, I.553
  electron spin, I.101
  First Shelter Island Conference, II.645
  graph theory in statistical mechanics, I.553
  magnetic moment, II.1123
  Master equation, I.593–605
  quantum number, I.168
  random processes, I.628
  spinning electron, I.360
  universality, I.561
Ullmaier, H, point defects, III.1534
Ulrich, R K
  solar model, III.1749
  solar oscillations, III.1750
Underwood, E, *Quantitative Stereology*, III.1549–50
Unsöld, A, hydrogen abundance, III.1701
Upatnieks, J, holography, III.1415, III.1437
Upatnieks, Y J, off-axis optical beam holography, III.1585
Urey, H C
  chemical elements, III.1747
  deuterium, III.1706
  deuteron, II.650, II.1183
  solar system, III.1963

Valasek, J, Rochelle salt, I.489
Vallarta, M S, magnetic-mirror effect, III.1637
Van Allen, J A
  Geiger counter, III.2007
  magnetospheric physics, III.2001
  Radiation Trapped in the Earth's Magnetic Field, III.1683
  rocket sounding experiments, III.1682–3
  Third International Polar Year, III.2006
  Van Allen belts, III.1636, III.1683, III.1684
van Cittert, P H, optical coherence, III.1407–8
Van, V, x-ray diffraction, I.498
van de Graaff, R J
  electrostatic generator, II.727–8, II.1186, II.1203
  High Voltage Engineering (HVE), II.1203
van de Hulst, H, radio astronomy, III.1762–3
van der Burg, M G J, field equations, I.313
van der Meer, S
  antiprotons, II.707
  field particles W and Z, II.744
  stochastic cooling, II.743
van der Waals, J D
  doctoral thesis, I.48
  kinetic theory of gases, I.543
  quantum mechanics, II.1042
  spinodal, III.1547–8
Van der Waerden, B, group theory, I.213
van Heel, A C S, fibre optics, III.1419–20
van Hove, L

magnetic correlation theory, II.1016
neutron frequency, III.1418
phonon dispersion relations, II.993
Twelfth Solvay Conference, Brussels, II.646
Van Kampen, N, linear response assumption, I.627
van Laar, J J, statistical mechanics, I.551–4
van Leeuwen, J H
  magnetic properties, II.1115
  theory of metals, III.1286–7
van Maanen, A
  bright spirals M31 and M33, III.1717–18
  nebulae, III.1718
van Rhijn, P J, luminosity function, III.1714
Van Vleck, J H, **II.1131**
  Bohr theory, II.1123–4
  crystal field theory, II.1132
  ferromagnetic resonance, II.1153
  First Shelter Island Conference, II.645
  H–D exchange, II.1140
  magnetic ions, II.1128–9
  molecular orbital theory, II.1143
  Quantum Principles and Line Spectra, II.1131
  rare-earth ions, II.1120
  relaxation theory, II.1146
  spectral line intensity, I.172
  spin–orbit interactions, II.1136
  Theory of Electric and Magnetic Susceptibilities, II.1131
  two-electron system, I.182
van't Hoff, J H, reaction isochore equation, I.535
Varicak, V, kinematic space, I.278
Vashakidze, M A, optical continuum of nebula, III.1774–5

Vassall, A, physics in education, I.9
Vaughan, R A, relaxation time, II.1133
Vavilov, S I
  absorption of light by uranium glass, III.1403
  Milrostructura Sveta, III.1403
  non-linear optics, III.1403
  polarization of fluorescent light, III.1400
  Raman effect, III.1400
  visible light, III.1409
Veksler, V, ion acceleration, II.730
Velikovsky, I, *Worlds in Collision*, III.2013
Veltman, M, electroweak parameters, II.725
Veneziano, G
  Dolen–Horn–Schmid duality, II.683
  scattering amplitude, II.683–4
Venkatesen, T N C, optical bistability, III.1449
Verne, J, science fiction, I.3
Verschaeffelt, J E, classical critical exponents, I.544
Verschaffelt, E, Fifth Solvay Conference (1927), I.205
Verwey, E J W
  charge effects, III.1611
  electrical properties, II.1156
  semiconductors, III.1330
Vigier, J P, matter-wave theory, I.229
Villard, P
  γ-rays, I.54
  ionization, III.1880
  nuclear γ-rays, I.367
Vine, F, magnetic variations of ocean floors, III.1975
Vinen, W F, vibrating-wire experiment, II.931
Vinogradov, A V, laser amplification, III.1466

Virasoro, M, Dolen–Horn–Schmid duality, II.683
Vlasov, A, Landau damping, III.1624, III.1625
Voigt, W
  electron optics, III.2000
  electronic theory of magnetism, II.1114
  magneto-ionic theory, III.1990
  school of physics, I.15
Volkoff, G, neutron stars, III.1713
Volmer, M, nucleation kinetics, III.1547
von Ardenne, scanning transmission microscope (STEM), III.1569, III.1587–9
von Baeyer, O
  β-ray deflection, II.1201
  β-spectrum, I.106
  radioactive decay, I.105
von Borries, B
  polepiece lens, III.1567
  transmission electron microscope (TEM), III.1575–7
von Braun, W, rocket technology, III.1738
von Federov, E, symmetry types, I.423
von Groth, P, crystallography, I.421
von Gutfeld, R J, thermal conductivity, II.982
von Hippel, A, dielectric properties of BaTiO$_3$, I.486
von Humboldt, A, *Gesellschaft deutscher Naturforscher und Ärzt*, I.5
von Kampen, N
  BBKGY hierarchy, I.624
  random walks, I.624
von Kármán, T
  BvK theory, II.985
  crystal lattice vibrations, I.153
  lattice dynamics, I.452
  sound from flows, II.850
  specific heat, II.971–2
  turbulence, II.832
  vibrational specific heat of solids, I.525
von Klitzing, K
  plateaux studies, III.1375
  semiconductors, II.1258
von Laue, M, **I.426**
  crystal structures, III.1514
  electromagnetic wave theory, I.363
  Fresnel's dragging coefficient, I.278
  imaging theory, III.1418
  lattice structure, I.153
  partial coherence, III.1392
  superconductivity, II.918–19
  x-ray diffraction, I.424, I.425
von Neumann, J
  energy surface trajectories, I.599
  First Shelter Island Conference, II.645
  materials research, III.1558
  Mathematical Foundations, I.227
  microphysics, I.225
  non-bounded operators, I.197
  quantum mechanics, I.198, I.206
  Schrödinger equations, I.205
von Rebeur-Paschwitz, E, earthquake vibrations, III.1953–4
von Schweidler, E, decay law of radioactive substances, I.222
von Smoluchowski, M, Brownian motion, I.586
von Weizsäcker, C F
  carbon–nitrogen–oxygen cycle (CNO cycle), III.1707–8
  monistic theories, III.1963
  nuclear energies, II.1191
von Wroblewski, S, liquefied gases, I.27
Voss, H D, *Lightning-induced electron precipitation*, III.1686–7

Wagner, vacancies, III.1529
Wagner, E, x-ray spectra, I.160
Wagoner, R V
  Big Bang, III.1792
  nuclear interactions between light nuclei, III.1791
Wahl, A C, plutonium isotope, II.1223
Wainwright, T E
  autocorrelation function of tagged particle, I.619
  molecular dynamics, I.577
Wakefield, J, holograms, III.1584
Wald, G, rod-vision, III.1416
Wali, K C, *Chandra: a Biography of S Chandrasekhar*, III.1711
Walker, A G
  Robertson–Walker metric, III.1729–30
  synchronization of clocks, III.1729
Walker, Sir Gilbert, systematic shifts of climate, II.907
Walker, J G
  antibunched light, III.1460
  optical data processing, III.1461
Walker, M F, star cluster NGC2264, III.1766–7
Walkinshaw, W, linear accelerators, II.1206
Walls, D F, quantum non-demolition (QND) measurements, III.1469
Walraven, T, Palomar 200" Telescope, III.1775
Walter, W F, germanium junctions, II.1208
Walton, E T S
  disintegration experiments, II.1183, II.1184
  disintegration of lithium, II.729
  electrostatic generators, II.727
  fusion reactions, III.1628
  nuclear reaction, I.285
  nuclear transformation, I.112

proton acceleration, I.370
Wambacher, H, cosmic-rays tracks, I.404
Wandel, A, quasars, III.1780
Wang, C C, He–Ne laser, III.1435
Wang, Ming Chen
  Brownian motion coupled harmonic oscillators, I.608
  random processes, I.628
Wannier, G
  multi-electron fragmentation, II.1065
  optoelectronic properties, III.1410
  semiconductors, III.1320
Wannier, G H, specific heat, I.555
Warburg, E
  First Solvay Conference (1911), I.147
  quantum number, I.162
Ward, F A B, ionization chambers, II.1200
Ward, J C, quantum theory, III.1413
Ward, W R, Moon model, III.1966
Wardlaw, R S, crystal fields, II.1165
Ware, A A, ZETA experiment, III.1641
Warren, B E, amorphous structure, I.478
Wasserburg, G J, supernova trigger hypothesis, III.1963
Watanabe, Y, semiconductor maser, III.1423
Waterman, P C, electromagnetic scattering, III.1446
Watson, G N, radio waves, III.1987
Watson, J D
  DNA model, I.507
  double-helix structure, I.504
  protein subunits in viruses, I.509
Watson-Watt, R, upper atmosphere, III.1988
Watt, R W, ionosphere, III.1981

Weaire, D, amorphous structures, I.477
Weaver, W, molecular biology, I.496
Weber, A, nucleation kinetics, III.1547
Weber, H F, specific-heat quanta, I.152
Weber, R L, maser principle, III.1417
Weber, W E
  magnetic moments, II.1114
  Zeeman effect, II.1114
Wegener, A L, **III.1968**
  continental drift, III.1967–9
Wegner, E J, universality, I.571
Weidlich, W, laser theory, III.1444
Weigert, F, polarization of fluorescent light, III.1400
Wein, First Solvay Conference (1911), I.147
Weinberg, D L, photon coincidence, III.1435
Weinberg, S, II.696
  Dreams of a Final Theory, II.770
  electroweak unification, II.695
  social construction, I.405
Weinrich, M, parity violation, II.666
Weis, P E, **II.1119**
  Le Magnetisme, II.1119
Weiss, M T, crystal structures, II.1158
Weiss, P
  critical point, I.545
  magnetic fields, II.1117–18
  molecular field hypothesis, I.545
Weissenberg, K, moving-film methods, I.434
Weisskopf, V F
  compound nucleus, II.1219
  electromagnetic field, I.391
  First Shelter Island Conference, II.645

positrons in electron self-energy, II.641
proton scattering, II.1212
quantum theory, I.221
The Joy of Insight, II.770
Welford, W T, compound parabolic concentrator (CPC), III.1446
Wells, H G, science fiction, I.3
Wells, P N T, ultrasonic imaging, III.1914
Wendt, G, longitudinal effect, I.162
Wentzel, G
  asymptotic linkage, I.197
  atomic theory, I.179, I.219
  Fermi field, I.378
  Twelfth Solvay Conference, Brussels, II.646
Wenzel, W A, spark chamber, II.753
West, J, structure of beryl, I.440
Westerhout, G, neutral hydrogen, III.1763
Westgren, J, single-crystal methods, I.435
Weyl, H, I.207
  eigenfunctions with continuous spectra, II.1025
  gauge invariance, I.321
  group theory, I.213
  Weyl's postulate, III.1729
Weymann, R J, electron scattering, III.1801
Whaling, W, resonance state of nucleus, III.1746–7
Wharton, C B, microwave interferometer, III.1659
Wheeler, J A
  First Shelter Island Conference, II.645
  fission process, I.132, II.1195
  hydrogen bomb, III.1636
  S-matrix theory, II.680
  two-photon correlation, III.1413
Whitcomb, R T, area rule, II.881

White, D L, electron–phonon interaction, III.1339
White, J W, neutron beams, I.468
Whitham, G B
  action-conservation principle, II.862
  'equal area' law, II.847
  'exploding wire' phenomenon, II.848
  water motions, II.860
Wick, G C
  nuclear exchange-force model, I.376
  quantum theory, I.221
  Twelfth Solvay Conference, Brussels, II.646
Wideröe, R
  resonant cyclic accelerators, II.729
  The Infancy of Particle Accelerators, II.770
Widom, B
  phase transitions, III.1830
  van der Waals equation, I.561
Wiechert, E
  cathode ray tube, III.1566
  Earth model, III.1953
Wiedlich, W, laser theory, III.1444
Wiegel, red blood cells, I.630
Wieman, C, parity violation, II.699
Wien, W
  atomic vibrations, III.1287
  black-body radiation, I.22, I.53, I.149
  exponential law, I.66–7
Wiener, N
  Brownian process, III.1830
  quantum mechanics, I.192
Wiersma, E C, potassium chromic alum, II.1126
Wiese, W L, *Needs, Analysis and Availability of Data for Space Astronomy*, II.1093
Wightman, A S, Twelfth Solvay Conference, Brussels, II.646

Wigner, E
  band structure, III.1314–17
  electron band structure of sodium, III.1530
  electron gas, III.1361–2
  neutron capture, II.1194
  nuclear resonance, I.126
  reaction cross sections, II.1219
  Schrödinger's equation, III.1315
  Twelfth Solvay Conference, Brussels, II.646
  Wigner function, I.617–18
Wilberforce, S, Darwinian evolution, I.5
Wilczek, F, running coupling constant, II.713
Wild, J J, pulse-echo ultrasound, III.1911–12
Wilkins, M H F, DNA, I.504
Wilkinson, D H
  Ionisation Chambers and Counters, II.1207
  pulse height conversion to pulse time, II.751
  shell-model theory, II.1214
Williams, E J, order–disorder transitions, I.474
Williams, R C, Lamb shift, II.641
Williamson, A, atomic theory, I.46
Wilson, A, science jottings, I.3
Wilson, A J C
  line broadening, I.472
  structure factors, I.439
Wilson, C T R
  alpha/beta particle tracks, II.1200–1
  cloud chamber, I.54
  droplet formation, II.751–2
  Fifth Solvay Conference (1927), I.205
  radiation, I.367
Wilson, H A, electron charge, II.747
Wilson, H F, special orientation relationships, III.1528

Wilson, H R, helical structures, I.498
Wilson, J M
  physics in education, I.9
  senior wrangler, I.11
Wilson, J T, plate tectonics, III.1978
Wilson, K
  computers, III.1834
  field theory, III.1830, III.1833–34
  fixed point, III.1832, III.1833–34
  magnetism, III.1834
  non-perturbative calculations, III.1825
Wilson, K G, **I.565**
  renormalization group, I.564–9
Wilson, R, microwave background radiation, III.1790
Wilson, R R, **II.735**
  Fermilab accelerator, II.734
Wilson, R W
  carbon monoxide molecule, III.1763–4
  microwave radiometer, II.758
  radiation, III.1786
Wilson, W
  electron beams, I.104
  quantum rule, I.175
Winston, R, compound parabolic concentrator (CPC), III.1446
Winther, A, Coulomb excitation experiments, II.1217
Wintner, A, bounded matrices, I.206
Wirtz, C W, spiral galaxies, III.1725
Wisdom, J
  chaotic evolutions, III.1823
  computing, III.1831
Witkowski, J, biomolecular structures, I.496
Wolf, D, melting, III.1533–4
Wolf, E
  Airy disc, III.1412
  coherence properties of partially polarized light, III.1411
  diffraction images, III.1419
  energy transport, III.1459
  optical coherence theory, III.1392, III.1421
  quantum theory of coherence, III.1439
Wolfe, J P, thermal conductivity, II.982
Wolfke, M
  two-stage imaging, III.1583
  x-ray microscope, III.1414
Wollan, E O
  neutron diffraction, I.453
  structure of ice, I.475
Wollaston, W H
  medical physics, III.1857
  solar spectrum, III.1691
  spectral lines, I.156
Wollfson, M M, determinantal inequalities, I.448
Wood, R W
  black-body radiation, I.150, III.1387
  diffraction gratings, III.1397
  direct-drive approach to inertial fusion, III.1676
  fluorescence and photochemistry of materials, III.1401
  Physical Optics, III.1391
  resonance radiation, III.1390
Woodbury, E J, ruby laser, III.1432
Woodruff, D P, surface crystallography, I.466
Woods, A D B, shell model, II.997
Woodward, J J, linear accelerators, II.736
Woolfson, M M
  probability relations, I.449
  structure determination, I.450
Wooster, W, continuous spectrum, I.118, I.126
Wright brothers, powered flight, II.873
Wright, T, island universes, III.1714

Wrinch, D M, Patterson function, I.444
Wu, C S
　parity violation, II.666
　scintillation counters, III.1414
Wu, L-A, squeezed light, III.1471
Wu, S L, gluon jets, II.716
Wu, T T, *CP* violation, II.673
Wül, B, dielectric properties of $BaTiO_3$, I.486
Wyckoff, R W G
　biomolecular structures, I.496
　crystal structures, III.1513
Wyler, A A, magneto-ionic theory, III.1992
Wynn-Williams, C E
　binary counting circuits, II.748
　ionization chambers, II.1200

Xiao, M, squeezed light, III.1471

Yalow, R, saturation analysis technique, III.1909
Yamaguchi, Y, decay reaction, II.657
Yan, T-M, lepton pair, II.690
Yang, C N
　CP violation, II.673
　isotopic spin symmetry, II.693
　lattice gas model, I.553
　local gauge invariance, II.693
　neutrinos, II.671
　parity conservation, II.665
　parity violation, II.666
　self-interacting fields, II.661
　spin-1 particle, II.654
Yardley, K, x-ray diffraction, I.439
Yariv, A
　guided wave optics, III.1458
　optical waveguides, III.1459
　Quantum Electronics, III.1436
Yeh, Y, Doppler shift, III.1442
Yevick, random walk model, I.630

Yonezawa, F, amorphous structures, I.477, I.479
York, D G, interstellar deuterium, III.1791
York, H
　fusion research, III.1637
　magnetic mirror, III.1659
Yosida, K, RKKY interaction, II.1164
Young, angular distributions, II.1220
Young, F Jr, materials physics, III.1558
Young, J Z, flying-spot microscope, III.1416–17
Young, T
　dimensions of molecules, I.48
　medical physics, III.1857
　wave description of light, I.72
Yuen, H P, minimum-uncertainty packets, III.1469
Yukalov, V I, melting, III.1533–4
Yukawa, H, **I.390**
　exchange forces, I.231
　meson theory, I.378–83, I.389, II.640, II.651–2, II.669
　neutral pions, II.654
　slow mesotrons, I.406
　spin-1 boson, II.692
　Twelfth Solvay Conference, Brussels, II.646
　U-particle hypothesis, I.396
Yuniev, defibrillators, III.1895
Yvon, J, BBKGY hierarchy, I.606, I.623, I.624, I.626

Zachariasen, W H
　amorphous structures, I.477
　extinction, I.451
　structure of metaboric acid, I.449
Zahradnicek, J, gravitation, II.1266
Zanstra, H, planetary nebulae, III.1762

Zavoisky, E, paramagnetic resonance, II.1137–8
Zeeman, P
  magnetism, III.1390
  splitting of spectral lines, I.161
  Zeeman effect, I.54
Zehnder, L, x-ray photograph, I.4
Zeiger, H J, maser, III.1419
Zel'dovich, B Ya, phase conjugation, III.1458
Zeldovich, Ya B
  Big Bang, III.1790
  binary sources, III.1756
  black holes, III.1779, III.1780
  collapsing cloud, III.1803
  cosmological constant, III.1723, III.1806
  induced Compton scattering, III.1801
  neutrinos, III.1803
  photon scattering, III.1773
  photon to baryon–antibaryon ratio, III.1806
  structures in the Universe, III.1801, III.1802
  synthesis of heavy elements, III.1792
  weak vector current, II.668
Zeller, R C, disordered systems, II.1006
Zener, C
  Fermi surface, III.1295
  valence conduction bands, III.1337
Zenneck, J
  radio propagation, III.2000
  radio waves, III.1987
Zermelo, E, process irreversibility, I.524
Zernike, F
  circle polynomials, III.1409
  critical correlations, I.563
  optical coherence, III.1407–9
  pair distribution function, I.547
  phase-contrast microscopy, III.1409–10, III.1419, III.1475
  x-ray diffraction pattern, I.441
Zheludev, N I, polarization symmetry breaking, III.1467
Zherikhin, A, laser amplification, III.1466
Zienau, S, lattice deformation, III.1321
Ziman, J M
  electrons and holes, III.1297
  Electrons and Phonons, III.1305
  lattice conduction of heat in metals, II.981
  pseudopotential concept, III.1367
Zimkina, T M, K-shell absorptions, II.1083
Zimm, B, polymer solutions, III.1597
Zimmerman, G, direct-drive approach to inertial fusion, III.1676
Zinn, W H, neutron diffraction, I.453
Zinn-Justin, J, renormalizability, II.695
Zobernig, G, gluon jets, II.716
Zoll, P
  cardiac pacemakers, III.1895
  defibrillators, III.1895
  electrical stimulation, III.1895
Zupancic, C, Coulomb excitation experiments, II.1217
Zvyagin, B B, oblique texture method, I.464
Zwanziger, singularities in complex angular momentum, II.682
Zweig, G, fractionally charged constituents of matter, II.685
Zwicky, F
  Catalogue of Selected Compact Galaxies and of Post-Eruptive Galaxies, III.1713

clusters of galaxies, III.1731
Coma cluster, III.1732
Schmidt telescope, III.1745
supernovae, III.1712, III.1718, III.1758, III.1759, III.1775

Zworykin, V K
  phosphors, III.1536–7
  photomultipliers, II.750
  synthetic zinc silicate, III.1537
  xerography, III.1541

To I.G.R.
From J.A.Z.R.
29.1.2001

Twentieth Century Physics
Volume II

Twentieth Century Physics
Volume II

Edited by

Laurie M Brown
Northwestern University

Abraham Pais
Rockefeller University
and
Niels Bohr Institute

Sir Brian Pippard
University of Cambridge

Institute of Physics Publishing
Bristol and Philadelphia

and

American Institute of Physics Press
New York

© IOP Publishing Ltd, AIP Press Inc., 1995

All rights reserved. No part of this publication may be reproduced, stored in a retrieval system or transmitted in any form or by any means, electronic, mechanical, photocopying, recording or otherwise, without the prior permission of the publisher. Multiple copying is permitted in accordance with the terms of licences issued by the Copyright Licensing Agency under the terms of its agreement with the Committee of Vice-Chancellors and Principals. Authorization to photocopy items for internal or personal use, or the internal or personal use of specific clients in the USA, is granted by IOP Publishing and AIP Press to libraries and other users registered with the Copyright Clearance Center (CCC) Transaction Reporting Service, providing that the base fee of $19.50 per copy is paid directly to CCC, 27 Congress Street, Salem, MA 01970, USA

British Library Cataloguing-in-Publication Data
A catalogue record for this book is available from the British Library.

In UK and the Rest of the World, excluding North America:
ISBN   0 7503 0353 0   Vol. I
       0 7503 0354 9   Vol. II
       0 7503 0355 7   Vol. III
       0 7503 0310 7   (3 vol. set)

In North America (United States of America, Canada and Mexico):
ISBN   1-56396-047-8   Vol. I
       1-56396-048-6   Vol. II
       1-56396-049-4   Vol. III
       1-56396-314-0   (3 vol. set)

Library of Congress Cataloging-in-Publication Data are available

Published jointly by Institute of Physics Publishing, wholly owned by The Institute of Physics, London, and American Institute of Physics Press, wholly owned by the American Institute of Physics, New York.

Institute of Physics Publishing, Techno House, Redcliffe Way, Bristol BS1 6NX, UK

Institute of Physics Publishing, Suite 1035, The Public Ledger Building, Independence Square, Philadelphia, PA 19106, USA

American Institute of Physics Press, 500 Sunnyside Boulevard, Woodbury, New York 11797-299, USA

Printed and bound in the UK by Bookcraft Ltd, Bath.

CONTENTS

VOLUME II

9 **ELEMENTARY PARTICLE PHYSICS IN THE SECOND HALF OF THE TWENTIETH CENTURY**   635
  Val L Fitch and Jonathan L Rosner
  - 9.1 Introduction   635
  - 9.2 Preludes (before 1940)   637
  - 9.3 Quantum electrodynamics   640
  - 9.4 New forms of matter up to the mid-1960s   650
  - 9.5 Interactions up to the mid-1960s   663
  - 9.6 The quark revolution   684
  - 9.7 Electroweak unification   692
  - 9.8 Quantum chromodynamics   709
  - 9.9 Three families of quarks and leptons   719
  - 9.10 Accelerators   725
  - 9.11 Detectors: from Rutherford to Charpak   744
  - 9.12 Overlaps with other subjects   754
  - 9.13 Unsolved problems and hopes for the future   762
  - 9.14 Conclusions   769

10 **FLUID MECHANICS**   795
  Sir James Lighthill
  - 10.1 Yet another great success for twentieth century physics   795
  - 10.2 Boundary layers and wakes; instability and turbulence; heat and mass transfer   812
  - 10.3 Non-linear effects on the generation and propagation of waves   846
  - 10.4 Transforming the human condition through aeronautics and ocean engineering   869
  - 10.5 Dynamics of the Earth's fluid envelope, and its forecasting applications   887

11 SUPERFLUIDS AND SUPERCONDUCTORS — 913
A J Leggett
- 11.1 Introduction — 913
- 11.2 The period 1945–70 — 930
- 11.3 New developments — 949

12 VIBRATIONS AND SPIN WAVES IN CRYSTALS — 967
R A Cowley and Sir Brian Pippard
- 12.1 The beginnings of lattice dynamics — 967
- 12.2 New experimental techniques — 991
- 12.3 The development of lattice dynamics — 997
- 12.4 Structural phase transitions — 1007
- 12.5 Spin waves — 1010
- 12.6 Magnetic phase transitions — 1014

13 ATOMIC AND MOLECULAR PHYSICS — 1023
Ugo Fano
- 13.1 Introduction — 1023
- 13.2 Atomic–molecular physics at mid-century — 1025
- 13.3 Completing the spectrum of radiation actions — 1049
- 13.4 Excitation channels and resonance effects — 1059
- 13.5 Collisions between atoms or ions — 1074
- 13.6 Molecular physics — 1081
- 13.7 Inner-shell phenomena — 1089
- 13.8 Spectral fingerprints of atoms and molecules — 1092
- 13.9 The role of atoms in metrology and instrumentation — 1094
- 13.10 Optical handling of atomic systems and transformation of light by atoms — 1097
- 13.11 A current overview — 1101

14 MAGNETISM — 1111
K W H Stevens
- 14.1 Introduction — 1111
- 14.2 The period 1900–25 — 1113
- 14.3 The period 1925–50 — 1123
- 14.4 Paramagnetism — 1125
- 14.5 Conductors — 1133
- 14.6 1950 onwards — 1136
- 14.7 Electron paramagnetic resonance — 1137
- 14.8 Ferromagnetism and ferrimagnetism — 1152
- 14.9 The changing pattern — 1165
- 14.10 Nuclear magnetism — 1171
- 14.11 Concluding remarks — 1175

15	**NUCLEAR DYNAMICS**	1183
	David M Brink	
	15.1 Background	1183
	15.2 Nuclear dynamics as a many-body problem	1191
	15.3 The effects of World War II	1196
	15.4 Technical advances	1199
	15.5 Shell structure in nuclei	1208
	15.6 Collective motion in nuclei	1213
	15.7 Nuclear scattering and reactions	1219
	15.8 New isotopes and new elements	1222
	15.9 Creation of the elements	1225
16	**UNITS, STANDARDS AND CONSTANTS**	1233
	Arlie Bailey	
	16.1 Introduction	1233
	16.2 Units and standards	1236
	16.3 Physical constants	1260
	16.4 Applications	1272

Illustration acknowledgments	A7
Subject index	S1
Name index	N1

VOLUME I

Preface	ix
List of contributors	xiii
Biographical captions	xvii

1	**PHYSICS IN 1900**	1
	Sir Brian Pippard	
2	**INTRODUCING ATOMS AND THEIR NUCLEI**	43
	Abraham Pais	
3	**QUANTA AND QUANTUM MECHANICS**	143
	Helmut Rechenberg	
4	**HISTORY OF RELATIVITY**	249
	John Stachel	
5	**NUCLEAR FORCES, MESONS, AND ISOSPIN SYMMETRY**	357
	Laurie M Brown	

6	**SOLID-STATE STRUCTURE ANALYSIS** *William Cochran*	421
7	**THERMODYNAMICS AND STATISTICAL MECHANICS (IN EQUILIBRIUM)** *Cyril Domb*	521
8	**NON-EQUILIBRIUM STATISTICAL MECHANICS OR THE VAGARIES OF TIME EVOLUTION** *Max Dresden*	585
	Illustration acknowledgments	A1
	Subject index	S1
	Name index	N1

VOLUME III

17	**ELECTRONS IN SOLIDS** *Sir Brian Pippard*	1279
18	**A HISTORY OF OPTICAL AND OPTOELECTRONIC PHYSICS IN THE TWENTIETH CENTURY** *R G W Brown and E R Pike*	1385
19	**PHYSICS OF MATERIALS** *Robert W Cahn*	1505
20	**ELECTRON-BEAM INSTRUMENTS** *T Mulvey*	1565
21	**SOFT MATTER: BIRTH AND GROWTH OF CONCEPTS** *P G de Gennes*	1593
22	**PLASMA PHYSICS IN THE TWENTIETH CENTURY** *Richard F Post*	1617
23	**ASTROPHYSICS AND COSMOLOGY** *Malcolm S Longair*	1691
24	**COMPUTER-GENERATED PHYSICS** *Mitchell J Feigenbaum*	1823
25	**MEDICAL PHYSICS** *John R Mallard*	1855
26	**GEOPHYSICS** *S G Brush and C S Gillmor*	1943

27 **REFLECTIONS ON TWENTIETH CENTURY PHYSICS: THREE ESSAYS**   2017

    Historical overview of the twentieth century in physics   2017
    Philip Anderson

    Nature itself   2033
    Steven Weinberg

    Some reflections on physics as a social institution   2041
    John Ziman

Illustration acknowledgments   A11

Journal abbreviations   J1

Subject index   S1

Name index   N1

Chapter 9

ELEMENTARY PARTICLE PHYSICS IN THE SECOND HALF OF THE TWENTIETH CENTURY

Val L Fitch and Jonathan L Rosner

9.1. Introduction

The past 50 years of elementary particle physics have witnessed an explosion of data, followed by simplifications based on classification and solid theory. Attempts to describe the fundamental interactions from a more unified point of view have borne fruit in a combined theory of weak and electromagnetic interactions based on self-interacting quantum fields and a similarly based theory of the strong interactions.

The understanding of the periodic table of the elements bears some similarity to the story of particle physics. An initial systematization of data was followed by firmer theoretical efforts, culminating in the advent of quantum mechanics. The vast variety of atoms and isotopes could be understood in terms of fundamental protons, neutrons and electrons interacting via electromagnetic (well understood) and strong (poorly understood) forces.

In the 1960s, a scheme for classifying the strongly interacting particles based on the group SU(3) began to make sense of the rapidly proliferating spectrum. Eventually, the success of SU(3) and related symmetries was traced to the existence of a few constituents—the *quarks*. Now we are confronted with a proliferation of quarks and *leptons* (the electron, muon, tau and their respective neutrinos) for which a deeper explanation is still lacking. These are summarized in table 9.1.

As more and more fundamental building blocks of matter were being uncovered, the way in which fundamental forces were described also

Table 9.1. *The quarks and leptons as of 1994.*

	Leptons			Quarks	
Symbol	Name	Charge	Symbol	Name	Charge
ν_e	Electron neutrino	0	u	Up	2/3
e^-	Electron	−1	d	Down	−1/3
ν_μ	Muon neutrino	0	c	Charmed	2/3
μ^-	Muon	−1	s	Strange	−1/3
ν_τ	Tau neutrino[a]	0	t	Top	2/3
τ^-	Tau	−1	b	Bottom	−1/3

[a] Not yet directly observed.

evolved. The unification of forces has a long tradition, dating from Newton's synthesis of terrestrial and celestial gravity and Maxwell's synthesis of electricity and magnetism. In this century it included a detailed understanding of the weak interactions and their violation of mirror symmetry, and culminated in the unified theory of weak and electromagnetic interactions of Glashow, Weinberg and Salam, and the discovery of the predicted carriers of the weak force, the W and Z. Still to be understood at the deepest level is the violation of the combined symmetry of charge reversal and mirror reflection, discovered in 1964.

The success of the electroweak theory was particularly heartening because it took place in the context of quantum field theory, previously thought to be useful only for describing electromagnetic processes. A parallel development, also relying on quantum field theory, was the emergence of a theory of strong interactions, now known as *quantum chromodynamics* (QCD). This theory describes why quarks are different from leptons (quarks have a new kind of charge dubbed *colour*, while leptons are colourless) and gives quantitative predictions for their interactions with one another via the exchange of quanta known as *gluons*. It explains, through the dependence of interaction strength on distance, why it makes sense to speak of quarks at all, even though they appear to be permanently bound to one another.

A chart of the carriers of strong and electroweak forces is given in table 9.2. The picture of particle physics as consisting of quarks and leptons interacting via exchanges of photons, gluons, Ws and Zs has come to be known as the 'standard model'.

A symposium has been devoted to the emergence of the standard model [1], and an extensive book [2] treats the whole period with which we are concerned. Specific chapters in particle physics, in the 1930–50s [3] and the period 1947–64 [4], are also the subject of excellent historical reviews. In this chapter, we touch on some high points of the progress made in this fruitful field in the past 50 years. Our hope is to give some

Elementary Particle Physics

Table 9.2. *Carriers of the strong and electroweak forces.*

Symbol	Name	Force carried	Mass (GeV/c^2)
γ	Photon	Electromagnetic	0
g	Gluon	Strong	0
$W^{\pm}$	W-boson	Weak (charged)	80.3 ± 0.2
Z^0	Z-boson	Weak (neutral)	90.189 ± 0.004

flavour of how far it has come, and where we might expect it to lead in the future.

We do not wish to give the impression that progress in elementary particle physics, any more than in any other field, is an orderly process. In the interest of space, our story omits many blind alleys and wrong experiments. We have chosen to speak of discoveries and ideas that have had some lasting value. At the same time, we cannot claim to be comprehensive in our treatment. There is necessarily some choice of subjects involved, for which we take full responsibility.

We begin in section 9.2 with a few key points of early twentieth century particle physics, to set the stage for our later discussion. This period is dealt with more extensively in reference [5], which may be consulted for citations. We then (section 9.3) describe progress in quantum electrodynamics (QED), which until the advent of electroweak unification and QCD was our only example of a successful, relevant quantum field theory. Except for the treatment of QED, we break our discussion at the mid-1960s, treating properties of matter (section 9.4) and forces (section 9.5) before proceeding further.

The description of strongly interacting particles, or *hadrons*, in terms of quarks (section 9.6) marks a turning point in particle physics in the latter half of this century. With quarks taken seriously, the way was paved for extension of electroweak unification (section 9.7) from its original province of leptons to the whole range of elementary particles. Moreover, the route was now established for the development of QCD (section 9.8). New forms of matter, in the form of the third family of quarks and leptons (section 9.9), could be accommodated without much difficulty in the new framework.

Almost all of the results in particle physics in the past 50 years have been crucially dependent on continued progress in the development of accelerators (section 9.10) and detectors (section 9.11). Elementary particle physics has profited immensely from its overlap with other fields (section 9.12). We mention some puzzles and hopes in section 9.13 and we give our conclusions in section 9.14.

9.2. Preludes (before 1940)

The first 'elementary particle' identified as such was the electron, whose charge-to-mass ratio was first measured by J J Thomson in 1897. The discreteness of the charge itself was demonstrated somewhat later by Millikan.

Experiments by Rutherford in the early part of the twentieth century showed that alpha particles underwent scattering from matter at much greater angles than one might have anticipated. A popular model of the atom at that time envisioned material spread uniformly through it, whereas Rutherford's scattering experiments pointed towards an intense concentration of most of the matter over less than 10^{-4} of the atom's linear size. Niels Bohr, a young visitor at Rutherford's laboratory in Manchester in 1911–12, was inspired by Rutherford's experiments to attempt to construct a model of the atom based on negatively charged electrons orbiting a positively charged nucleus. He was forced to introduce new physics, foreshadowing quantum mechanics, in order to keep the orbits from decaying by emission of radiation. He was not initially motivated by data on the spectra of light emitted by hydrogen, but when he learned of the Balmer spectrum the whole problem became clear to him and his solution was presented within a month.

The Bohr atom made use of an analogy with previously known ideas, such as orbits. A break with the past occurred in the mid-1920s with the fully fledged development of quantum mechanics by Heisenberg, Schrödinger, Born and others. A crucial aspect of quantum mechanics was the scale set by Planck's constant h, with dimensions of (energy) × (time) or (momentum) × (length); another was the identification by de Broglie of the connection between waves and particles: (wavelength) = h/(momentum), confirmed experimentally by Davisson and Germer using electrons. Similarly, though people had been accustomed since Maxwell's time to view electromagnetic radiation in terms of waves, Einstein's explanation of the photoelectric effect in 1905 indicated that light could also be regarded as composed of *quanta*, or discrete units, with (energy) = h (frequency). This idea was confirmed by the discovery of the *Compton effect*, the scattering with change of wavelength of electromagnetic quanta (*photons*) on electrons.

The version of quantum mechanics developed in the mid-1920s applied to particles with velocities small compared with that of light. In seeking an equation of motion for particles not subject to this limitation, Dirac introduced new degrees of freedom. His equation applied to a quantity with a total of four components. A twofold multiplicity allowed one to describe particles such as the electron which have two possible directions of spin. However, an additional twofold multiplicity was a necessary consequence of invariance of Dirac's equation under the transformations of special relativity. Dirac interpreted this additional doubling to imply the existence of *antiparticles*, with opposite charge and the same mass as particles. Thus, there should exist a positively charged version of the electron. This particle, the *positron*, was identified by

Anderson in cosmic radiation in 1932.

The comparison between the charges and masses of atomic nuclei, and the detailed study of their spins, made it clear that one could not build the nucleus merely out of protons, nor of protons and electrons. A new building block was needed, similar in mass to the proton but electrically neutral. This particle, the *neutron*, was discovered by Chadwick in 1932. Its existence made the picture of atomic nuclei fall into place. The charge (Z) counted the number of protons, while the mass number (A) counted the total number of protons and neutrons. The mass of the nucleus was slightly less than the sum of the masses of individual neutrons and protons (*nucleons*), because of the effects of binding energy.

Radiation from outer space had been identified by V Hess and others in the early part of the twentieth century. By the mid-1930s, cosmic rays were a subject of some experimental interest, and were recognized as providing a useful source of highly accelerated particles, as were products of radioactive decay. However, these sources were soon joined by a number of devices invented for artificially accelerating particles, which were then focused with the help of electric and/or magnetic fields. These devices included the Cockcroft–Walton generator, the van de Graaff generator, and the cyclotron. The last was pioneered by Ernest O Lawrence, and was extensively used for particle physics studies until the mid-1950s. At that time, various versions of the *synchrotron* became available.

The prediction of infinite quantities, such as the field energy (self-interaction energy) of the electron, meant that a consistent quantum-mechanical description of the interaction of radiation with matter still had not been found in the 1930s. While it was possible during this period to calculate many processes by an approximation valid to lowest order in the interaction strength, a self-consistent description to all orders was lacking.

The continuous nature of the energy spectrum of electrons or positrons emitted in beta decay, and the balance between initial and final particles, implied that an unseen agent was carrying off momentum and angular momentum in the decay. This particle, dubbed the *neutrino*, always accompanied the electron in beta decay. The fundamental process was then n $\rightarrow$ pe$^-\bar{\nu}_e$, where n is the neutron, p is the proton, e$^-$ is the electron and $\bar{\nu}_e$ is an antineutrino. In a heavy nucleus with a large proton excess, the process p $\rightarrow$ ne$^+\nu_e$, could occur instead, even though forbidden by energy conservation for a free proton and neutron. Both processes were described by an interaction, postulated by Fermi, which was almost, but not quite, correct. It made no provision for the violation, discovered in the 1950s, of mirror symmetry (*parity violation*) in the beta-decay interaction. In its description of the production of the electron–neutrino pair, the Fermi theory was one of the first applications of *quantum field theory*, which makes provision for the production and annihilation of particles.

The extremely short range of the nuclear force led Yukawa to postulate the existence of a new particle, the *meson*, whose exchange would give rise to a short-range interaction. In 1937, a new type of particle was seen in cosmic radiation. Charged, and with a mass very close to that predicted by Yukawa, this new particle (the *muon*) was initially identified as Yukawa's meson. However, if the muon were really the carrier of the strong nuclear force, it should interact strongly with matter. Its persistent failure to do so led to the gradual realization that the muon was *not* Yukawa's meson. The particle predicted by Yukawa was yet to be discovered, as we shall see in section 9.4.

The picture of elementary particle physics as of 1940 was fairly simple and self-consistent. The atom consisted of a nucleus built of neutrons and protons, bound to electrons via electromagnetism. Neutrinos were hypothetical particles emitted in beta decay. The 'four forces of nature' were already in place: strong (holding nuclei together), electromagnetic, weak (associated with beta decay) and gravitational. The elements in the periodic table of the elements up to uranium had almost all been seen and elements heavier than uranium were starting to be discovered. There were few intimations of the rich variety of particles or the progress in understanding of forces that would characterize the next 50 years.

9.3. Quantum electrodynamics

One of the early triumphs of theory in describing elementary particle physics lay in the realm of the purely electromagnetic interactions. This area, *quantum electrodynamics*, or QED, evolved through an interplay of experiment and theory. For many years, its success was regarded as an exception, to be contrasted with much more phenomenological descriptions of the weak and strong interactions. With hindsight, we now know that those theories have followed a route related to that pioneered by QED. Indeed, the weak interactions have now been unified with QED into an *electroweak* theory, and strong forces are described by a theory which could well be unified in the future with the electroweak interactions. For historical purposes, however, it is appropriate to trace the development of QED in its own right. Even today, progress is continuing to be made in calculations, and some interesting puzzles remain for the hardy experimentalist and theorist.

We shall discuss primarily purely electromagnetic processes, not affected by uncertainties of weak- or strong-interaction physics, taking examples mainly from the interaction of photons with electrons or muons. The calculations we shall describe are organized as a series in increasing powers in $\alpha = e^2/\hbar c \approx 1/137$, the *fine-structure constant*.

9.3.1. Infinite quantities in the theory

Although the interactions of photons and electrons were described successfully to lowest order in α, as in photon–electron scattering or electron–positron pair production by a photon in an intense external field, it was realized quite early that higher orders in α led to difficulties [6], manifested in a series of calculations which gave infinite answers.

One can see that classical electromagnetic theory is plagued by infinite quantities just by calculating the field energy surrounding a point electron. The energy diverges as $1/r_0$, where r_0 is the minimum distance to the electron taken in the integral over energy density. Could a proper relativistic quantum-mechanical treatment cure this problem?

The description of electrons in a way compatible with special relativity entails also the existence of positrons [7]. With the help of W Furry, Weisskopf [8] showed that the inclusion of contributions from positrons in the electron self-energy calculation reduced the degree of divergence to $\ln(1/r_0)$, where r_0 again represents a minimum cut-off distance. Thus positrons were a partial, but not sufficient, help.

Another infinite quantity occurring in quantum electrodynamics arises as a result of the production of virtual electron–positron pairs by a photon. Like the electron self-energy, this *vacuum polarization* divergence depends on the logarithm of a cut-off parameter. Despite the infinite nature of vacuum polarization, it was possible to calculate its effect on the Coulomb interaction, for instance in a hydrogen atom, by comparing the interaction at large and shorter distances. The result [9] was a prediction that the $2P_{1/2}$ level of hydrogen should lie 27 MHz above the $2S_{1/2}$ level, whereas the Dirac theory predicts them to be degenerate.

9.3.2. Early experimental developments

9.3.2.1. Lamb shift.

The fine structure in the Balmer Series spectrum of the hydrogen atom was first observed in the H_α line ($n = 3 \to 2$) by Michelson and Morley in 1887. By the time the Dirac equation was available in the late 1920s more than 15 spectroscopic measurements were available for comparison with the theory. A difficulty was immediately encountered. The intensity ratios of the observed lines were not those expected; even more importantly, the splitting of the lines was different from that predicted. Already in 1933 a letter had been published in *Physical Review* addressing the issue of this discrepancy [10]. It was entitled *On the breakdown of the Coulomb law for the hydrogen atom*. Measurements later in the decade on the D_α line by Houston and Williams [11] sharpened the disagreement with the observation of a third line (deuterium shows less Doppler broadening). This work stimulated Pasternack [12] to observe that the observations could be accounted for if the 2S level were shifted upwards by about 0.03 cm^{-1} (900 MHz in units used subsequently). The vacuum polarization correction alone [9] was much too small and in the wrong direction to account for the discrepancy. This was the situation until after World War II.

In the US during wartime many physicists worked in one of the Radiation Laboratories devoted to radar development, or on the Manhattan Project concerned with nuclear weapons. Willis Lamb, at Columbia University, was originally denied necessary security clearance to work at the Columbia Radiation Laboratory (CRL) because his wife was not a US citizen. Instead, he taught physics, including atomic physics, to Navy students, thus becoming familiar with the problems associated with the H_α and D_α spectra discussed above. Eventually he was allowed to work at the CRL on high-frequency magnetrons in which capacity, though nominally a theoretical physicist, he built one of the first continuous wave magnetrons with his own hands. It operated at 2.7 cm which, not accidentally, was just the frequency of the fine-structure splitting of hydrogen. Immediately after the war, with a graduate student, R C Retherford (who had himself developed an expertise in high-vacuum techniques as well as in the measurement of tiny currents during the war), Lamb proceeded to mount an experiment designed to answer definitively the questions posed by the hydrogen fine structure. He exploited many of the new techniques and instrumentation developed during the war years. It was a brilliant effort and the experiment succeeded beyond all of Lamb's dreams.

The fine-structure splitting between the 2P and 2S levels in hydrogen was in the range of Lamb's 3 cm magnetron. Nominally, the 2S level is metastable with a lifetime sufficiently long to survive a reasonably long path through the apparatus. However, any small stray electric field will mix the 2S and 2P levels and shorten the lifetime of the 2S state, perhaps to the point where no atoms initially in the 2S state could survive the trip through the apparatus. To address these and other problems the apparatus was composed of five distinct elements. First, the source was an oven enclosing a hot tungsten surface which dissociated the molecular hydrogen to atoms. On exiting from the oven, a beam of atomic hydrogen was formed by collimation. Second, an electron beam was positioned to cross the beam of hydrogen atoms to excite at least some of the atoms to the 2S state. This process was very inefficient; only about 1 in 100 million atoms were so excited, but it was enough. Third, the atomic beam was passed through a radio-frequency field to induce transitions from the 2S to various 2P levels. Atoms in the P states decay so rapidly to the ground state that they travel less than 10^{-3} cm before they are lost to the beam of metastable atoms. Fourth, a uniform magnetic field enveloped the whole apparatus to remove the (possible) near degeneracy between the 2P and 2S levels through the Zeeman effect, and thereby minimize the danger of stray electric field which would shorten the lifetime of the beam. Finally, the beam ended with a detector designed to selectively sense the hydrogen atoms in the 2S state and reject all others. Following on the work of Massey and Oliphant, Lamb and Cobas had previously calculated that metastable hydrogen atoms impinging on tungsten would de-excite with the emission of electrons from the tungsten and thereby produce a current. Utilizing this fact, the detector consisted of a plate of tungsten connected to the most sensitive current meter available at the time, an FP 54 electrometer.

The experimental measurement consisted of setting the radio frequency and varying the magnetic field until a dip occurred in the detector current. This corresponded to RF-induced transitions from the 2S to one of the 2P states. By extrapolating to zero magnetic field, Lamb and Retherford found that the transitions occurred at a frequency 1000 MHz less than that deduced from theory, just as expected if the $2S_{1/2}$ level were shifted by this amount. They also reported seeing directly the transitions between the $2S_{1/2}$ and the $2P_{1/2}$ levels at this frequency. In the Dirac theory, these levels have exactly the same energy. The results [13] from this very first experiment, obtained 16 April 1947, are shown in figure 9.1. In this one elegant measurement the speculations of Pasternack had been shown to be correct, but the effect was removed far beyond the realm of speculation and now deserved the most serious attention.

Before World War II there had been considerable theoretical effort directed towards the question of the self-energy of the electron. However, because of the war, interest had remained dormant. Now, with the stimulus of the results of Lamb and Retherford the latent interest developed into a major attack by theoretical physicists, and within a few years the problem was solved to the satisfaction of nearly everyone. (To the end of his life, however, Dirac maintained that any theory involving the subtraction of infinities was ugly, unsatisfactory and surely incomplete.)

Lamb first announced the results, which he had obtained only five weeks earlier, at a conference held on Shelter Island in Peconic Bay, Long Island, New York, 2–4 June 1947. Sponsored by the National Academy of Sciences and organized by Robert Oppenheimer, it was attended by most of the leading theoretical moguls at the time in the US (see figure 9.2, from the second of references [14] p 380). At that conference, not only were the results of Lamb and Retherford made known, but also R Marshak first suggested that there were two kinds of mesons, and H Kramers laid the groundwork for reinterpreting infinite quantities in quantum field theory by means of 'renormalization' [14].

Within a few days after the conference, Bethe had calculated the 'Lamb shift' to be 1040 MHz using old-fashioned non-relativistic methods but with ingenious subtractions of infinite terms [15]. It was a calculation that led to many refinements [16], ending within three years with the fully developed theory of quantum electrodynamics by Feynman, Schwinger and Tomonaga [17].

Twentieth Century Physics

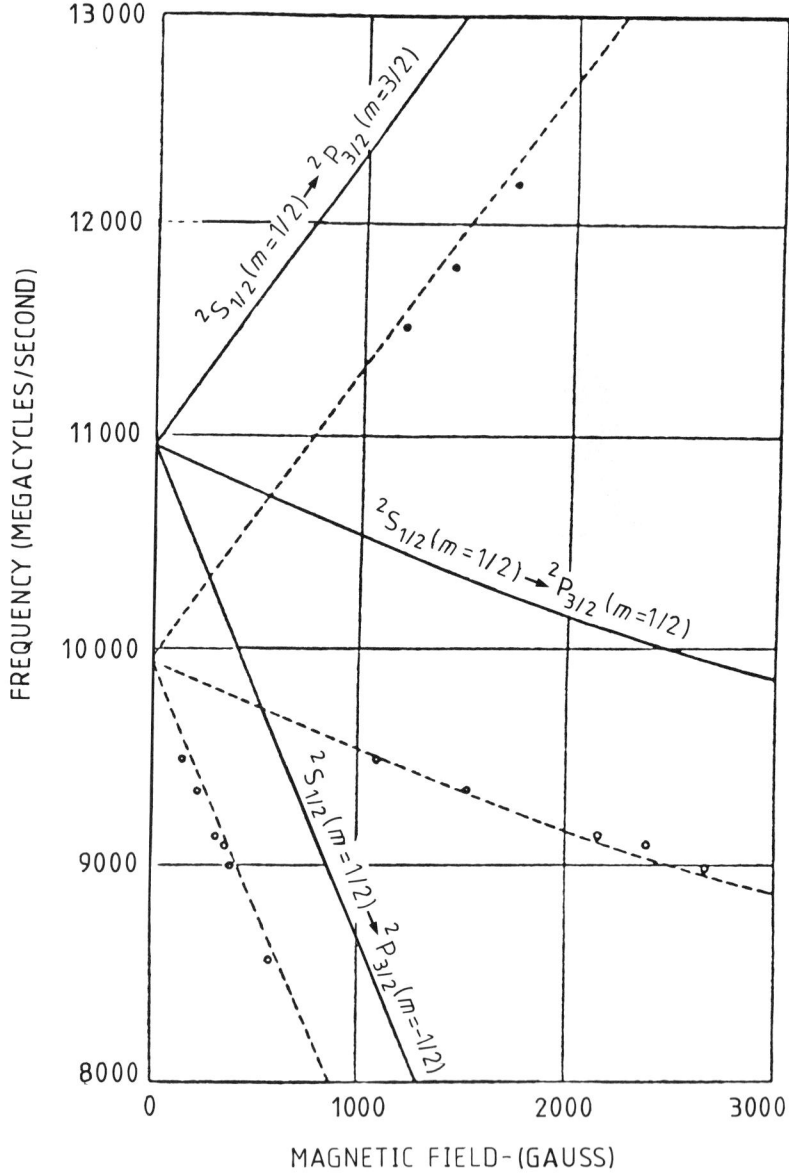

Figure 9.1. *Evidence for the Lamb shift* [13]. *The solid lines are three of the expected values of level splittings as a function of magnetic field in the absence of a level shift; the broken lines are those which would be expected in the presence of a 1000 MHz shift.*

9.3.2.2. *Electron magnetic moment.* The Dirac theory [7] of the electron predicts the factor g in the relation $\mu = geS/(2m)$ between the spin S and the magnetic moment μ to be exactly 2. Comparison of the fine-structure and hyperfine interactions in sodium and gallium suggested

Elementary Particle Physics

Figure 9.2. *Participants at the first Shelter Island conference, 2–4 June 1947: (1) I I Rabi; (2) L Pauling; (3) J Van Vleck; (4) W E Lamb Jr; (5) G Breit; (6) D Mac Innes (National Academy of Sciences); (7) K K Darrow; (8) G E Uhlenbeck; (9) J Schwinger; (10) E Teller; (11) B Rossi; (12) A Nordsieck; (13) J von Neumann; (14) J A Wheeler; (15) H A Bethe; (16) R Serber; (17) R E Marshak; (18) A Pais; (19) J R Oppenheimer; (20) D Bohm; (21) R P Feynman; (22) V F Weisskopf; (23) H Feshbach. Photograph provided by D Mac Innes.*

a departure from this value [18]: $g - 2 = 0.00229 \pm 0.00008$. This result was announced at the 1947 Shelter Island conference. Schwinger [19] calculated the effect of the electron self-interaction on this quantity, finding $g - 2 = \alpha/\pi = 0.00232$ in agreement with experiment. The stage was set for a more thorough understanding of such effects. How could one get results in accord with experiment from a theory beset by infinite quantities?

9.3.2.3. Positronium. We have previously discussed ideas lying dormant during the war and blossoming quickly afterwards. Thus as early as December 1945, a few months after the end of the war, Purcell, Torrey and Pound at Harvard, and Bloch and Packard at Stanford had independently discovered nuclear magnetic resonance. In 1946 Wheeler published a paper [20] on an idea he had developed during the war in which he worked out the details of bound states involving one or more electrons (e⁻) and positrons (e⁺). The simplest system, the e⁺e⁻

645

Figure 9.3. *Participants in the Twelfth Solvay Conference, Brussels, 1961. From left to right, front row: S Tomonaga, W Heitler, Y Nambu, N Bohr, F Perrin, J R Oppenheimer, Sir W Lawrence Bragg, C Møller, C J Gorter, H Yukawa, R F Peierls, H A Bethe. Second row: I Prigogine, A Pais, A Salam, W Heisenberg, F J Dyson, R P Feynman, L Rosenfeld, P A M Dirac, L Van Hove, O Klein. Back row: A S Wightman (slightly to front), S Mandelstam, G Chew, M L Goldberger, G C Wick, M Gell-Mann, G Källén, E P Wigner, G Wentzel, J Schwinger, M Cini.*

bound state, we now call positronium. He included the rates of annihilation from the singlet and triplet states and the relative polarization of the two photons from the singlet-state annihilation. This paper was published in an unlikely journal, the *Annals of the New York Academy of Sciences*, because they were offering a prize for the best paper, which Wheeler won. Experimental confirmation required the development of new tools and instrumentation. The predicted bound state was finally discovered in 1951 by Martin Deutsch [21].

9.3.3. Advent of renormalization

The banishment of infinite quantities from quantum electrodynamics has been traced in references [14] and [17]. Many of the participants in this effort are shown in figure 9.3. Early leaders included Heisenberg, Dirac, Oppenheimer and Stückelberg. The procedure of reinterpreting infinite quantities in terms of physical ones was proposed by Kramers at the 1947 Shelter Island Conference [14] and developed systematically by Sin-Itiro Tomonaga, Julian Schwinger and Richard P Feynman (see box). The proof that this method was consistent to every order in perturbation theory was given by Dyson and Salam and Ward.

Three major sorts of infinities occur in quantum electrodynamics. The first, associated with the electron's infinite energy of interaction with its own electromagnetic field, is removed by redefining its mass to be the physical value, order by order, in perturbation theory. The second can be removed by demanding that a free electron produced at a given point in space be detectable with unit probability at some distant point at a later time. The third, related to the polarization of the vacuum pairs by a test charge, can be removed by redefining the electron's charge as its value as seen by a distant observer.

9.3.4. Higher-order corrections and experimental confirmation

The successive improvements in theory and experiment in quantum electrodynamics have been a long and mainly happy story [22].

9.3.4.1. Electron $g-2$ factor. The measurement of the electron's anomalous magnetic moment benefited greatly from the ability to investigate magnetically confined free electrons [23]. The most recent experiments make use of traps employing both electric and magnetic fields [24]. A single electron has been kept in solitary confinement for nine months in such a trap!

The results for electrons and positrons for $a \equiv (g-2)/2$ are [25] $a(e^-) = (1159\,652\,188.4 \pm 4.3) \times 10^{-12}$, $a(e^+) = (1159\,652\,187.9 \pm 4.3) \times 10^{-12}$, where the error is a combination of statistical and systematic uncertainties. The theoretical prediction [26] is

$$a(e) = \alpha/(2\pi) - 0.328\,478\,965\,(\alpha/\pi)^2 + C_3(\alpha/\pi)^3 + C_4(\alpha/\pi)^4 + \delta a(e). \quad (1)$$

The first two terms have been calculated analytically, while only numerical evaluations of $C_3 = 1.176\,11 \pm 0.000\,42$ and $C_4 = -1.434 \pm 0.138$ have been performed. The term $\delta a(e) = 4.46 \times 10^{-12}$ arises from electroweak interactions and from Feynman graphs involving internal muons, τ-leptons and quarks.

Several determinations of α exist, the most precise of which, $\alpha^{-1}(\text{QHE}) = 137.035\,9979 \pm 0.000\,0032$ makes use of the quantum Hall effect [27]. Using this value, one predicts [26]

$$a(e) = (1159\,652\,140 \pm 5.3\,[C_3] \pm 4.1\,[C_4] \pm 27.1\,[\alpha]) \times 10^{-12}. \quad (2)$$

This result agrees with experiment within errors. Until the uncertainty in α is reduced, one is not really testing the $(\alpha/\pi)^4$ term incisively.

9.3.4.2. Muon $g-2$ factor. The lowest-order Feynman graphs leading to a difference between $a(e)$ and $a(\mu)$ contribute at second-order in α. Instead of $-0.328\ldots$ for the coefficient of $(\alpha/\pi)^2$, one obtains $+0.754\ldots$ [28].

The earliest measurement of $a(\mu)$ confirmed that it was within 5% of zero [29]. A pioneering series of experiments was begun at CERN

Richard P Feynman

(American, 1918–88)

The contributions of Richard P Feynman to theoretical physics range from the low temperatures of liquid helium to the highest-energy collisions of elementary particles. Feynman developed a relativistic theory of quantum electrodynamics using a path integral approach, aided by diagrams that could be identified with actual physical processes. For this work he shared the Nobel Prize in 1965 with Julian Schwinger and Sin-Itiro Tomonaga. Not only have Feynman diagrams become the standard language of elementary particle theory, but they have been widely applied in other areas, such as nuclear physics and condensed matter theory.

Another contribution of Feynman which gave him great pleasure was a description of the weak interactions (the V–A theory, formulated in collaboration with Murray Gell-Mann and independently by George Sudarshan and Robert E Marshak) which incorporated the newly discovered violation of mirror symmetry. Still another was his recognition, along with J D Bjorken, that experiments at Stanford were probing point-like structures (called 'partons' by Feynman, but soon to be identified as quarks) in the proton.

Feynman was born in Brooklyn, New York, and went to public high school there. He received his undergraduate training at the Massachusetts Institute of Technology and his PhD from Princeton in 1942. During World War II he headed the Scientific Computing division at Los Alamos. After several years on the faculty at Cornell University, he joined the California Institute of Technology in 1950, where he remained intensely active in physics. The *Feynman Lectures in Physics,* based on an introductory course at Caltech, remain one of the foremost guides to novel and simple ways to view our Universe.

in the late 1950s, involving the containment of muons first in a long dipole magnet and then in storage rings [30]. The most recent results of these latter experiments are: $a(\mu^+) = (1165\,910 \pm 11) \times 10^{-9}$, $a(\mu^-) = (1165\,936 \pm 12) \times 10^{-9}$ or, combining the two results, $a(\mu) = (1165\,923 \pm 8.5) \times 10^{-9}$, where the total (statistical and systematic) errors are shown.

The theoretical value contains contributions from QED, intermediate states involving strongly interacting particles ('hadrons') and weak interactions [31]

$$a(\mu)|_{\text{QED}} = (1165\,846.955 \pm 0.046 \pm 0.028) \times 10^{-9}$$
$$a(\mu)|_{\text{hadron}} = (70.27 \pm 1.75) \times 10^{-9} \quad (3)$$
$$a(\mu)|_{\text{weak}} = (1.95 \pm 0.10) \times 10^{-9}$$

leading to a total of

$$a(\mu)|_{\text{theor}} = (1165\,919.18 \pm 1.76) \times 10^{-9}. \quad (4)$$

The errors in $a(\mu)|_{\text{QED}}$ reflect the theoretical uncertainty and that due to the measurement of α. The dominant error in $a(\mu)|_{\text{hadron}}$ and hence in (4) comes from uncertainty in hadronic vacuum polarization effects in the $\mathcal{O}(\alpha^2)$ contribution.

An experiment is being mounted at Brookhaven National Laboratory to measure $a(\mu)$ about 20 times more sensitively [30]. Its interpretation will require reduction of the error in the hadronic vacuum polarization contribution by means of more precise e^+e^- annihilation experiments. With that error reduced, one will then be able to test the weak contribution $a(\mu)|_{\text{weak}}$ to about 20% of its expected value.

9.3.4.3. *Lamb shift.* The most recent measurement of the $2S_{1/2} - 2P_{1/2}$ splitting in atomic hydrogen is $(1057\,851.4 \pm 1.9)$ kHz [32], to be compared with the theoretical prediction [33] of $(1057\,853 \pm 14)$ kHz or $(1057\,871 \pm 14)$ kHz, depending on how the proton's structure is described. As in the case of $a(\mu)$, there is satisfactory agreement with experiment, but a hadronic measurement is needed to reduce theoretical uncertainty.

9.3.4.4. *Hyperfine interactions.* A history of atomic hyperfine structure experiments is given by Ramsey [34]. The most precise value for the splitting between the $^3S_1$ and $^1S_0$ levels in hydrogen is $\Delta\nu_H = (1420\,405\,751.7667 \pm 0.0009)$ Hz. The theoretical prediction is within about 1 kHz of this value, but is affected by unknown proton structure effects. While experimental accuracy far outstrips theory, the agreement is still impressive.

Purely leptonic systems also exhibit hyperfine structure. The experimental value for muonium agrees satisfactorily with theory within the errors associated with uncertainty in the muon mass. In positronium, the observed hyperfine splitting is $\Delta\nu^{exp} = (203\,389.10\pm0.74)$ MHz, to be compared with the prediction $\Delta\nu^{theor} = (203\,404.5\pm9.3)$ MHz, where the dominant source of error comes from uncalculated corrections of order α^2 with respect to the leading result. The calculation of these corrections is a challenge to the most enthusiastic theorist.

9.3.4.5. Positronium decay: a current puzzle. The annihilation of positronium ground states into photons is governed entirely by QED. The $^1S_0$ (singlet) state decays into two photons, while the $^3S_1$ (triplet) state decays into three. The singlet rate [35] is found to be $\lambda_s = (7.994 \pm 0.011)$ ns^{-1}, to be compared with the prediction [20, 35, 36] $\lambda_s = (7.986\,654 \pm 0.000\,001)$ ns^{-1}, whose error is governed by that on α. The agreement is satisfactory. On the other hand, the value measured in vacuum [37] of the triplet rate, $\lambda_t = (7.0482 \pm 0.0016)$ μs^{-1}, is more than six standard deviations from the calculated value [38] $\lambda_t = (7.03831 \pm 0.00007)$ μs^{-1}. One would need a large $\mathcal{O}(\alpha^2)$ correction to this value to bring theory into accord with experiment. The evaluation of this term represents a frontier of QED.

9.4. New forms of matter up to the mid-1960s

The number of known 'elementary' particles grew by an enormous factor in the two decades after World War II. In this section we describe how that growth took place. The penetrating component of cosmic radiation, the muon, was eventually understood as being distinct from Yukawa's particle, the pion. The antiproton, predicted by Dirac's theory, was eventually found. Many 'strange' particles were discovered, and evidence was gathered for resonant particles, some living less than 10^{-23} s. Classification schemes during this period made widespread use of symmetry principles. The understanding of several hundred particles in terms of a few simple constituents came later.

9.4.1. The 1930s and 1940s: muons, pions and kaons
Modern particle physics starts with the discovery of the deuteron in 1931 by Urey, Brickwedde and Murphy [39]; the neutron in 1932 by Chadwick [40], the positron in 1932 by Anderson [41] (to be confirmed soon after by Blackett and Occhialini in an experiment discussed below) and what is now called the muon (then, the mesotron or meson) by Neddermeyer and Anderson, Street and Stevenson, and the Nishina group in 1937 [42].

Though Dirac's paper [7] predated the discovery of the positron by three years and Yukawa's meson [43] came before the muon by two years, they played no role in stimulating experimental searches. For example, Anderson observed that a person accepting the Dirac theory at face value could have discovered the positron in an afternoon. 'However, history did not proceed in such a direct and efficient manner, probably because the Dirac theory, in spite of its successes, carried with it so many novel and seemingly unphysical ideas, such as negative mass, negative energy, infinite charge density, etc. Its highly esoteric character was apparently not in tune with most of the scientific thinking of the day.' In addition, while today the Dirac theory is held as a monument to deductive reasoning, a splendid example of the 'unreasonable effectiveness of mathematics in physics', it was not accepted for a long time. Leading detractors included the likes of Pauli and Heisenberg. While the discovery of the positron had much to do with the acceptance of the Dirac theory, the theory in no way stimulated the activity leading to the discovery.

In the case of Yukawa's meson theory, Japanese scientific literature was not widely disseminated in the West, and the average physicist simply was unaware of the Yukawa particle. Only after the muon's discovery did Oppenheimer and Serber [44] discuss (with reservations) the possibility of identifying it with the Yukawa particle, making the first reference to Yukawa's paper in Western literature [45].

During the 1930s the only theoretical construct that took hold was the neutrino [46], forced on the community of physicists by energy conservation. This apparent independence of experimental and theoretical work contrasts strikingly with the close communication and interplay that developed between theory and experiment by the 1960s.

Blackett and Occhialini [47] not only beautifully confirmed the observations of Anderson, but gave photographic cloud-chamber evidence of cosmic-ray induced showers of particles. In addition, their experiment initiated one of the most important new techniques of the decade, the use of counter-controlled cloud chambers. The previous practice was to expand the cloud chamber and, after an optimum time, take a photograph. Consequently, most of the photographs were of an empty chamber, or showed old tracks fuzzed out by diffusion. With counter-controlled expansions, more than 75% of the pictures showed ionizing tracks. Never again, relative to the passage of particles, would cloud chambers be expanded randomly in time. The neutral pion excepted, for the next 20 years all of the new particles were discovered through the study of cosmic rays via visual techniques: by the use of counter-controlled cloud chambers and, after World War II, photographic emulsions.

The Yukawa meson theory [43] (1935) and the Fermi theory [48] (1934) of beta decay were patterned on electrodynamics, where the force between charges is viewed as arising from photon exchange. In the case of radiation from atoms, the photons did not previously exist in the atom but were created at the moment of radiation. Likewise, in Fermi's beta-decay theory, the electron and neutrino did not pre-exist in the nucleus but were spontaneously created. This enormous conceptual advance immediately resolved many questions, including the nature of the neutron. From its discovery until the advent of the Fermi theory,

Twentieth Century Physics

> See also p 119

a hotly debated topic had been whether the neutron was a particle with the same status as the proton or whether it was an electron–proton composite. With the Fermi theory, beta decay was simply the transmutation of the neutron to proton, electron and neutrino. The lifetime of the neutron was not measured until 1948 but in 1934 the Fermi theory estimated it to be $\sim 10^3$ s. (It is now known to be very close to 15 min: 887 ± 2 s [49].)

In the case of the Yukawa theory the particles mediating the interaction were given mass to produce a short-range 'Yukawa potential', $V(r) \sim \exp[-r(mc/\hbar)]/r$. Choosing a range then known to be characteristic of nuclear forces, $\sim 10^{-13}$ cm, one gets a mass around 200 MeV. Oppenheimer and Serber asked whether this was the same particle as the μ-meson (which they prophetically referred to as a heavy electron). However, this association almost immediately ran into conflict with experiment.

If the meson were to account for nuclear forces it should interact strongly with nuclear matter, so that the cross section for interaction should approach the geometric cross section with a mean free path of about 100 g cm^{-2}. It quickly became apparent that the penetrating component of the cosmic radiation interacted with a much smaller cross section. Furthermore, Rossi and Nereson [50] showed in 1940 by studying the flux of this component as a function of altitude that the attenuation through the atmosphere could only be accounted for by including a decay component as well as nuclear interactions. They were able to measure the lifetime to be $(2.15 \pm 0.07) \times 10^{-6}$ s, roughly 100 times longer than the lifetime of the Yukawa particle estimated on theoretical grounds. Shortly afterwards, a meson was seen to decay to an electron plus neutral particles in a cloud chamber [51].

Early on it became apparent that positively and negatively charged mesons could be expected to behave quite differently on coming to rest in matter. The positive particle would very quickly come to thermal energies and decay at its natural rate. The negative particle would come to rest, find itself attracted to a positive nucleus, cascade down through the various atomic levels by emitting radiation or Auger electrons, and then, depending on the nature of the particle, either decay or interact with the nucleus. The whole process, stopping and atomic transitions to the 1S state, was estimated to take less than 10^{-11} s, a time much shorter than the expected decay times.

> See also p 407

The experiment which definitively showed that the meson discovered in 1937 was not Yukawa's particle was performed by Conversi, Pancini and Piccioni [52] (1945). Using magnetized iron as a charge selector, they showed that both positive and negative mesons appeared to decay when coming to rest in carbon, while if the particles came to rest in iron only the positive ones decayed. Since negative mesons with Yukawa-like properties would be expected to interact even in elements as light

Elementary Particle Physics

Figure 9.4. *The Pic du Midi, site of early cosmic-ray experiments.*

as carbon, the evidence was unmistakable that the mesons could not be the strongly interacting Yukawa type.

In the meantime, significant advances were taking place in the production of photographic emulsions sufficiently sensitive to detect lightly ionizing particles. Following initial applications in nuclear physics, in 1946 Perkins at the Imperial College in London and Powell and his group at Bristol first exposed the new emulsions to cosmic rays [53]. (The history of the photographic emulsion technique and its contributions to particle physics has been beautifully recorded in reference [54].)

Though the first emulsions were not sensitive to minimum ionizing particles, the technique almost immediately showed its value. Perkins found an example of a meson coming to rest and depositing its rest energy into a 'star' of heavily ionizing fragments in an emulsion exposed in an aeroplane flown at 30 000 ft. The Bristol group exposed their emulsions at the Pic du Midi laboratory in the French Alps (figure 9.4), quickly finding examples of π mesons decaying to μ-mesons with characteristic tracks 600 microns long. (This group first labelled the two kinds of mesons as π and μ, the only two Greek letters on Powell's typewriter.) The π-meson decay was clearly two-body. (The emulsions were not yet sufficiently sensitive to see the electron track from the subsequent μ-decay.)

See also p 408

The paper of the Powell Group at Bristol, generally marked as announcing the discovery of the π-meson or pion [55], appeared in

Nature of 24 May 1947. The existence of two types of mesons had been anticipated by Sakata's group in Japan, whose work only became known after several years, and was discussed in the West by R E Marshak and H Bethe [56].

Seven months later in the same journal appeared a paper *Evidence for the existence of new unstable elementary particles* by C D Rochester and C C Butler [57], reporting on the first two examples of 'strange' particle decay in a cloud chamber. We would now call them beautiful examples of K-mesons, the neutral K_S and a charged K^+. Elementary particle physics would be occupied for the next quarter century with the properties of pions, muons and the 'strange' particles.

9.4.2. Pion properties

Nuclear forces were expected to be *charge-independent*, obeying a symmetry which amounted to invariance under rotation in an abstract *isotopic spin* or *isospin* space [5]. As a consequence, Yukawa's theory entailed not only charged but also neutral pions [58], first seen around 1950 in various accelerator and cosmic-ray experiments [59].

9.4.2.1. Spin and parity. The spin of charged pions was established by comparing the rates for $p + p \to \pi^+ + d$ and $\pi^+ + d \to p + p$. The ratio of the rate for the first process to that for the second is proportional to $2S_\pi + 1$, where S_π is the spin of the pion. The result of measurements of both reactions [60] led to the conclusion that $S_\pi = 0$.

Particles can be characterized by an intrinsic *parity*, describing the behaviour of their fields under space inversion. The negative parity of charged and neutral pions was deduced by comparing [61] rates for the process $\pi^- + p \to n + \gamma$ with $\pi^- + p \to n + \pi^0$, and $\pi^- + d \to 2n$ with $\pi^- + d \to 2n + \gamma$ and $\pi^- + d \to 2n + \pi^0$, and by comparing [62] the yields of neutral and charged pions in proton collisions on light nuclei. An early test of charge-independence by comparing the cross sections for $n + p \to d + \pi^0$ and $p + p \to d + \pi^+$ was performed by R Hildebrand [63].

9.4.2.2. Charged and neutral pion decays. The lifetime of the charged pion was first measured using pions produced artificially at an accelerator [64]. Within a couple of years, precise electronic timing techniques had yielded a result [65] very close to that known at present (26 ns), and considerably shorter than that of the muon. The dominant channel is $\pi^\pm \to \mu^\pm \nu$.

The nature of the neutral pion was still open to question. It was seen to decay to two photons, but direct measurements of its spin, parity and lifetime were needed. S Sakata, L D Landau and C N Yang [66] showed that a spin-1 particle could not decay to two photons, and showed how to determine the parity by comparing the linear polarizations of the two photons. Parallel polarizations implied even parity, while perpendicular polarizations implied odd parity. Later measurements [67] showed the parity to be odd, in agreement with the result of reference [61].

The lifetime of the neutral pion was determined by direct measurements [68] to be less than 10^{-13} s. Estimates [69] by J Steinberger suggested the possibility of an even shorter lifetime than this upper bound, outside the reach of direct detection methods at the time. It was proposed by H Primakoff [70] to use the rate of neutral pion photoproduction in the Coulomb field of a nucleus to measure the decay rate indirectly. This method was first applied a number of years later, yielding values slightly below 10^{-16} s [71]. Direct methods, now the most accurate, also evolved for measuring such short lifetimes using sandwiches of foils [72].

9.4.3. Antiprotons

The Dirac theory of antiparticles, so stunningly confirmed by the positron's discovery, also predicted a negative version of the proton, the *antiproton* or $\bar{p}$. The antiproton and proton were expected, though not unanimously, to have equal masses. The energy of a new accelerator under consideration at Berkeley was chosen to lie above the threshold for the reaction $p + p \rightarrow p + p + p + \bar{p}$. This accelerator, the Bevatron, not only discovered the antiproton, but a wealth of other particles, ushering in a new era of high-energy physics.

Several groups were searching for antiprotons at the Bevatron. Key players included O Chamberlain, O Piccioni, E Segré, W Wenzel, C Wiegand and T Ypsilantis. The construction of a beam of antiprotons was a crucial feature of the discovery. Protons in the machine struck an internal target, and the resulting negatively charged particles were focused into a well-collimated external beam with a narrow momentum range. Antiprotons were distinguished from the much more abundant negative pions by precise time-of-flight measurements and with the help of focusing Čerenkov counters, a tool that contributed greatly to the experiment's success. Knowing both the momentum and the velocity of particles in the beam, one could measure their masses. A clear signal was seen of negatively charged particles with the mass of a proton [73]. Their identity as examples of antimatter was confirmed by the observation of their annihilations in nuclear emulsions [74].

9.4.4. Strange particles

The story of the discovery and classification of 'strange particles', as they were called, affords a lovely example of order emerging from chaos [75].

9.4.4.1. Discoveries in cosmic rays. Two techniques for studying cosmic rays were in use during the 1940s: nuclear emulsions, in which the passage of particles exposed a three-dimensional photographic image,

Twentieth Century Physics

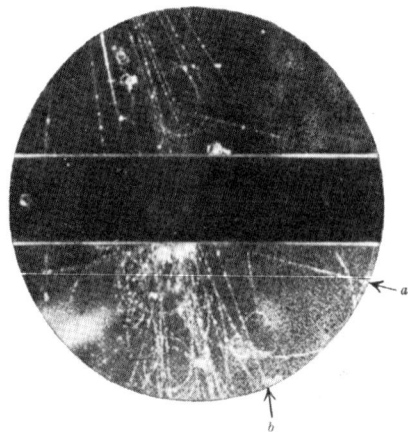

Figure 9.5. *First evidence for a neutral strange particle, from the cloud-chamber experiment of Rochester and Butler [57]. The particle was produced in a lead plate (shown at centre) and decayed to two charged particles (a and b).*

and cloud chambers, in which charged particles left a trail of droplets in a supercooled vapour. Both yielded evidence for new particles besides the positron, muon and pion.

By studying elastic scattering of charged particle tracks on electrons in emulsions, Leprince-Ringuet and Lhéritier [76] deduced the existence of a new particle of mass about 990 times the electron's mass. In retrospect, this appears to have been the first measurement of the mass of the charged *K-meson* or *kaon*.

Using a cloud chamber at Manchester, Rochester and Butler [57] saw two events which we would now characterize as decays of kaons, as mentioned in section 9.4.1. The neutral particle led to the emergence of a 'forked' pair of tracks (figure 9.5), while the charged one was seen decaying to an electron and one or more missing neutral particles. A long 'dry spell' for the Manchester group followed, during which an emulsion experiment [77] saw the decay of a charged kaon to $\pi^+ + \pi^+ + \pi^-$, permitting a decisive measurement of its mass. Meanwhile confirmation of the existence of Rochester and Butler's events came from a cloud chamber operating primarily at high altitude in California [78]. The Manchester group then set up operation on the Pic du Midi in the Pyrenees, where many more events were found.

The neutral kaon was seen to decay to $\pi^+ + \pi^-$. Because of the equal masses of the two pions, the momentum distributions of their tracks were equal, and so were the typical angles they made with the direction of the incident kaon. However, in 1950, an event was seen in which the positive track appeared to be a proton [79]. In the decay of a neutral particle to $p + \pi^-$, the proton, because of its larger mass, tends to carry most of the momentum, leading to an asymmetric configuration. The Pic du Midi group [80], led by R Armenteros, and a cloud chamber group at Indiana [81] led by R Thompson resolved the decay $K^0 \to \pi^+ + \pi^-$ from what we now understand to be $\Lambda \to p+\pi^-$. The Λ was the first example of a *hyperon*, a particle heavier than the proton. Cosmic-ray studies also yielded the first evidence for charged hyperons, the Ξ^- and the Σ^+. The Ξ^- was called the 'cascade' particle because it was seen [82] to decay to $\Lambda + \pi^-$, with the subsequent decay of the Λ. The Σ^+ was seen [83] decaying to $n + \pi^+$ and $p + \pi^0$.

In contrast to the neutral kaon, the charged kaon was found to have many different decay modes. Sorting out all these results took some time and effort, in which emulsion studies were of tremendous help.

9.4.4.2. Discoveries in accelerators. The Brookhaven Cosmotron, which began operation in 1952, belonged to a new generation of accelerators constructed to study fundamental interactions at unprecedented energies [84]. Its attention was quickly turned to strange particles. In a series of experiments performed with a hydrogen diffusion cloud chamber, Ralph Shutt and his group [85] were able to demonstrate what cosmic-ray experiments had been unable so far to reveal: the new heavy particles were produced in pairs, as in the reaction $\pi^- + p \to \Lambda^0 + K^0$, confirming earlier hypotheses by several groups in Japan [86] and by A Pais [87]. In addition to the associated production of Λs just mentioned, Shutt's group was the first to produce and observe neutral and negative Σ hyperons.

9.4.4.3. Associated production and strangeness. The new heavy particles were produced fairly copiously (in about 1% of all cosmic-ray events), but decayed very slowly (with lifetimes about 10^{12} times longer than one might expect if their production and decay were governed by the same interaction). A clear statement of this paradox was given very early by Nambu, Nishijima and Yamaguchi in their second paper of reference [86]

> ... production and decay are not inverse processes and/or some kind of selection rules (in a very general sense) are at work in the decay reaction).

One option for such a rule was a quantum number ('isotopic parity') taken to be $+1$ for the nucleon, pion and muon, and -1 for the new ones (corresponding to what we now call the K and Λ). This quantum number was taken to be conserved *multiplicatively* in production processes, but could be violated in the much weaker decay processes. Thus, a kaon always had to be produced in association with a Λ, a rule that came to be known as 'associated production'.

Associated production as formulated required one to assign an odd isotopic parity to the Σ hyperons, but did not prohibit reactions like $p+p \to \Sigma^+ + \Sigma^+$ or $n+n \to \Lambda + \Lambda$ that were not seen. Moreover, associated production ran into difficulty with the negative cascade particle, Ξ^-, which decays to $\Lambda\pi^-$ with a typical 'slow' lifetime of the order of 10^{-10} s. By the above scheme we would then assign even isotopic parity to the Ξ^-. But then what would prohibit the decay $\Xi^- \to n + \pi^-$, a process that has not been seen to this day?

The strong interactions of pions and nucleons conserve isospin symmetry, mentioned at the beginning of section 9.4.2. Pions have isospin $I = 1$ and nucleons have $I = 1/2$. The charge of a pion is the third component of its isospin, while that of the nucleon is displaced by half a unit. For both particles, one can write $Q = I_3 + B/2$, where the *baryon number B* is 0 for pions and 1 for nucleons.

M Gell-Mann [88] and K Nishijima [89] generalized this relation to the new particles by postulating 'displaced' isotopic spin multiplets, with $I = 0$ for the Λ, $I = 1$ for the Σ^+ and Σ^- (thereby predicting the Σ^0), and $I = 1/2$ for charged and neutral kaons. The conservation of charge and I_3 in strong interactions then required every multiplet to be characterized by an additional quantum number, called *strangeness*, also conserved in the strong interactions.

The relation between the charge and the third component of isospin could now be written $Q = I_3 + Y/2$, where the *hypercharge* $Y = B + S$, with a new additive quantum number S assigned to every strongly interacting particle, described the displacement of the isospin multiplet. The K^0 and K^+ were taken to have $S = 1$ and the Λ and Σ were assigned $S = -1$. The reactions $p + p \to \Sigma^+ + \Sigma^+$ and $n + n \to \Lambda + \Lambda$ were forbidden by this scheme. Decays such as $K^0 \to \pi^+ + \pi^-$ and $\Lambda \to p + \pi^-$, violating S by one unit, were allowed to proceed, but only weakly.

In order that $\Xi^- \to \Lambda + \pi^-$ be a weak decay violating strangeness by one unit, but to prevent the occurrence of $\Xi^- \to n + \pi^-$, one had to take $S(\Xi^-) = -2$. Then $K^- + p \to K^+ + \Xi^-$ should proceed strongly, as was later confirmed. Moreover, $I_3(\Xi^-)$ should equal $-\frac{1}{2}$, entailing the existence of a neutral partner Ξ^0 with $I_3 = +\frac{1}{2}$, a particle eventually seen in a bubble chamber [90].

9.4.4.4. Neutral kaons and their lifetimes. The existence of kaons with both signs of charges had been known for some time before Gell-Mann and Nakano and Nishijima proposed that these mesons had isospin $\frac{1}{2}$. But that proposal implied the existence also of two kinds of *neutral* kaons, the K^0 (with strangeness $S = 1$ like the K^+) and the $\bar{K}^0$ (with strangeness $S = -1$ like the K^-). How could one tell them apart? This question was raised by Fermi in June of 1954 at a seminar by Gell-Mann describing the strangeness scheme.

The resolution of this problem [91] provided a beautiful application of quantum mechanics. The K^0 and $\bar{K}^0$ are two degenerate states; nothing specifies what combination of them corresponds to states of definite mass and lifetime. However, if invariance under charge reflection is assumed in the K decay process (as it was in the mid-1950s), one linear combination of K^0 and $\bar{K}^0$ could decay to $\pi^+ + \pi^-$, while the other could not. These two states were denoted K_1^0 and K_2^0, respectively. The K_1^0 had already been seen, and the K_2^0 was a newly predicted particle that should live much longer than the K_1^0, since its decay to $\pi^+ + \pi^-$ was forbidden.

An experiment at the Brookhaven Cosmotron [92] soon confirmed the existence of the predicted neutral kaon with its characteristic long lifetime and dominant three-body decay mode. Supporting evidence came from four events which, most plausibly, originated from interactions of K_2^0 particles in photographic emulsions [93].

9.4.5. Resonances

9.4.5.1. Pion–nucleon scattering. The isospin of a combined pion–nucleon system can only be $\frac{1}{2}$ or $\frac{3}{2}$, entailing simple relations among pion–nucleon scattering amplitudes [94]. (Other early applications of isotopic spin are noted in references [95].) As artificially produced pions began to be available in the late 1940s because of increased cyclotron energies, attention was turned to the properties of pion–nucleon scattering. The development of electron synchrotrons capable of reaching energies of several hundred MeV led to production of high-energy photons, also useful in producing pions [96]. A crucial result of these experiments was the discovery of the first *isobar*, an excited nucleon state.

9.4.5.2. Nucleon isobars. With increasing pion and photon energies, the cross sections for pion–nucleon elastic scattering and single pion photoproduction grew in a very specific pattern [97]. Theorists [98] proposed that this pattern might signal the existence of a short-lived excited state of the nucleon with $I = J = \frac{3}{2}$. With still higher energies [99], the cross sections traced out a well-defined pion–nucleon resonance, called the (3, 3) resonance because of its spin and isospin. Its current appellation is $\Delta(1232)$, where the number (here and later) denotes the mass in MeV/c^2. The phase shift in the corresponding channel was seen to pass through 90°, as appropriate for a resonance.

Still higher energies led to the discovery of additional peaks in cross sections, corresponding to states with well-defined spins and isospins. By the late 1950s, several of these pion–nucleon resonances had been identified [97].

9.4.5.3. The resonance explosion. Invention of the bubble chamber [100] permitted the study of particle interactions in much greater detail during the mid-1950s. Several groups, including Luis Alvarez and collaborators [101], introduced automated methods to scan photographs of interactions. Pions, nucleons and strange particles all were found to participate in the formation of resonant states with well-defined masses and spins. Widths of these states were typically tens of MeV, as expected for states formed and decaying via the strong interactions [102].

Some of the earliest resonances had been anticipated from spin and isospin properties of nuclear forces. A 'mixture of mesons', not just Yukawa's proposed particle, appeared to be needed [103]. Experiments on the *form factors* of nucleons [104], to be discussed in section 9.6.2.1, could be interpreted as though the photon coupled to nucleons through a spin-1, isoscalar particle [105], decaying to three pions. An $I = J = 1$ resonance decaying to $\pi^+ + \pi^-$ also was suggested [106] on the basis of form-factor experiments. This resonance, the $\rho(776)$, was the first meson resonance to be discovered [107]; the three-pion isoscalar resonance $\omega(783)$ followed shortly thereafter [108].

Resonance searches utilized two major techniques: 'formation', or varying the beam energy and observing a peak in the cross section as in the discovery of the (3, 3) resonance, and 'production', or grouping particles in the final state into combinations whose 'effective mass' spectrum could be studied for enhancements. We already noted that formation experiments in pion–nucleon and photon–nucleon collisions led to a fertile yield of resonances. The advent of beams of negatively charged kaons revealed many resonances of negative strangeness formed in K$^-$p and K$^-$n reactions. Excited states of the Λ and Σ hyperons thus began to be uncovered. The production experiments had unique ability to see meson resonances, but also were the first to detect several excited baryon states. By the early 1960s, the $\rho(776)$, $\omega(783)$, another three-pion resonance, the $\eta(547)$ (reference [109]), a $J = 1$ 'excited kaon,' K*(892), and a $J = \frac{3}{2}$ version of the Σ at 1385 MeV had appeared. The 'elementary particles' had thus progressed from a handful to a veritable zoo in the course of 20 years [102].

9.4.6. Unitary symmetry

9.4.6.1. Preludes. By 1960, a wide variety of baryons with spin $\frac{1}{2}$ had been identified. Isospin had proved a successful guide to the classification of all the strongly interacting particles, but the existence of particular multiplets and the masses of the baryonic states remained a mystery. Similarly, it appeared that kaons were strange versions of pions, with the same spin and intrinsic parity.

9.4.6.2. SU(3). Initial efforts to unify non-strange and strange particles [110] included a 'global symmetry' of the strong interactions, composite models of mesons, and searches for various Lie groups containing isospin [SU(2)] as a subgroup. The group SU(3) was proposed [111] as a symmetry of strongly interacting particles in 1959, with the proton, neutron and Λ assigned to a triplet representation. The mesons π, K and an as yet undiscovered eighth meson (now known as the η), all composites of p, n, Λ and their antiparticles, were to form an octet.

Murray Gell-Mann had been searching for several years for higher symmetries containing isospin which could unify the baryons. In 1960 he realized that the octet representation of SU(3) was a perfect home for the spin-$\frac{1}{2}$ states N, Λ, Σ and Ξ. His conclusions on this 'eightfold way' were originally published only in preprint form [112]. Independently, Yuval Ne'eman, an Israeli military attaché in London working for his PhD in physics, was asked by his supervisor, Abdus Salam, to find an appropriate group containing isospin to classify the observed hadrons. He generalized the work of Yang and Mills [113] on self-interacting fields to higher symmetries, and realized the importance of SU(3). He, too, suggested that the spin-$\frac{1}{2}$ baryons belonged to an octet [114]. Gell-Mann's approach was part of a larger scheme in which *currents* (such as the electromagnetic current) played a crucial role [115]. We shall return to that aspect of SU(3) in section 9.5.

9.4.6.3. Consequences for masses; verification. The SU(3) scheme, when combined with an assumption about the nature of symmetry breaking, led Gell-Mann and Okubo [116] to mass relations among the baryons and among the mesons:

$$M(N) + M(\Xi) = [3M(\Lambda) + M(\Sigma)]/2 \qquad M(K) = [3M(\eta) + M(\pi)]/4.$$

The relation for the baryons is well satisfied: with the observed masses, we have 2257 MeV ≈ 2270 MeV. For the mesons, the symmetry breaking is greater, and we have 495 MeV ≈ 446 MeV. The success of the mass formula for baryons and the discovery [109] of the predicted η were notes of encouragement for the SU(3) scheme. Another higher symmetry, based on the group G_2, had room for only seven mesons and seven baryons.

When SU(3) was proposed, an isospin-$\frac{3}{2}$ non-strange multiplet (the Δ(1232)) and an isospin-1 strange multiplet (the Σ(1385)) were known. At the 1962 International Conference on High Energy Physics in Geneva an excited state of the Ξ at 1530 MeV/c^2, decaying to Ξ + π, was announced. The SU(3) scheme and another contender, the G_2 group mentioned above, had very different predictions for the remaining particles if the Δ(1232), Σ(1385) and Ξ(1530) all belonged to the same multiplet.

A ten-dimensional SU(3) representation was the smallest which could accommodate the Δ and its partners. The Gell-Mann–Okubo mass relation predicted equal spacings between the masses of its members. The Ξ(1530) obeyed this prediction. It was expected to have isospin $\frac{1}{2}$. Nine states of the ten would then have been seen: four Δs, three Σs

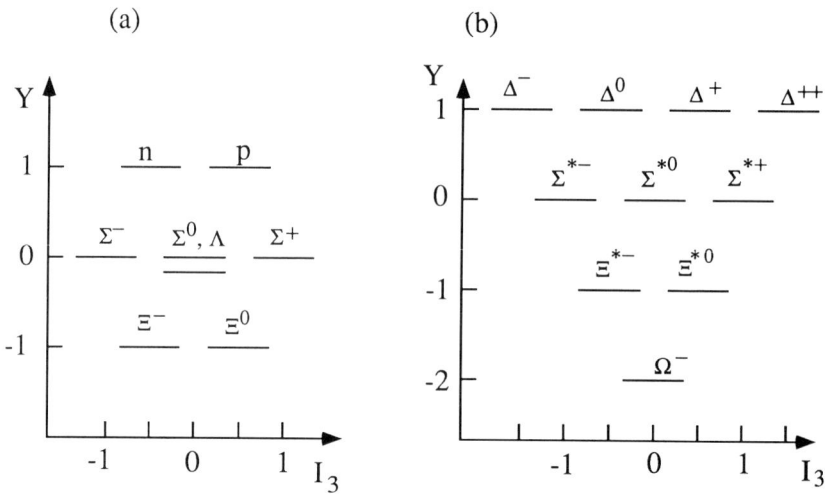

Figure 9.6. *Low-lying baryonic states as classified in SU(3): (a) octet, (b) decimet.*

and two Ξs. The missing tenth state was expected to have strangeness −3 and a mass of about 1675 MeV. With this mass, it would be *stable* with respect to the strong interactions, decaying instead weakly to such channels as $\Lambda + K^-$. The members of the lowest octet and decimet of baryons are shown in figure 9.6.

The next largest representation of SU(3) which could decay to a pair of octets and could hold a Δ is 27-dimensional. It contains resonances of positive strangeness, decaying to such states as $K^+ + p$ and $K^+ + n$. However, no such resonances had been seen [117].

The smallest G_2 multiplet accommodating the Δ was 14-dimensional. Although it had room for a state of strangeness −3 (like the 10-plet of SU(3)), it had an isoscalar positive-strangeness resonance. The absence of such a resonance thus weighed against the G_2 scheme, and favoured the 10-plet (rather than the 27-plet) assignment of the multiplet within SU(3).

The SU(3) prediction of a strangeness −3 state was stressed by Gell-Mann and Ne'eman at the 1962 Geneva Conference [118]. Shortly thereafter, a new 80″ bubble chamber at Brookhaven was exposed to a beam of K^--mesons. The story of how the predicted particle, the Ω^-, was discovered [119] in 1964 is one of the exciting chapters in the history of particle physics. Its mass was almost exactly the predicted value, and the first event (figure 9.7) was a particularly clean signature in which all the decay products revealed themselves.

Elementary Particle Physics

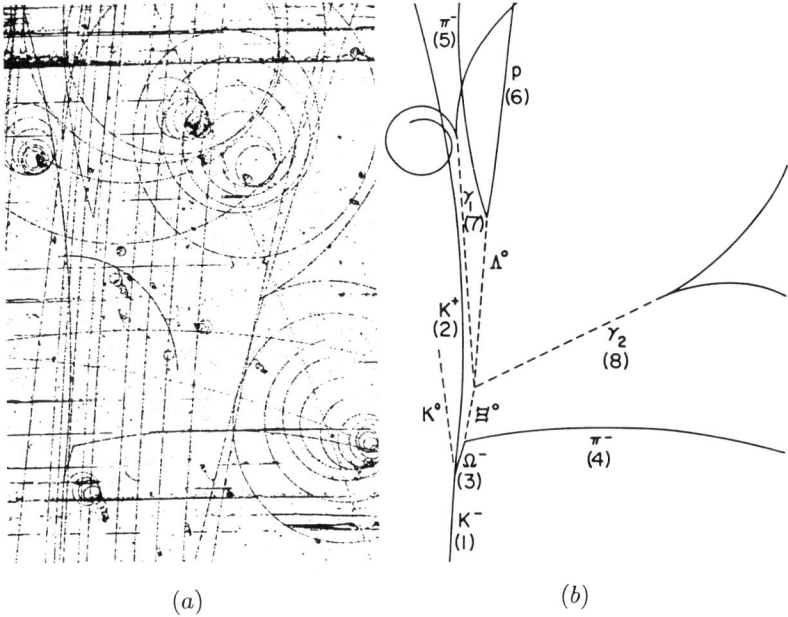

Figure 9.7. *First evidence for the Ω^- baryon* [119]. *(a) Bubble-chamber photograph; (b) interpretation.*

9.5. Interactions up to the mid-1960s

9.5.1. Weak interactions and V–A theory

9.5.1.1. Observation of the neutrino. The neutrino was postulated in the early 1930s to carry off the unseen energy and momentum in the beta-decay process $n \to p + e^- + \nu$ (see section 9.2). However, no direct evidence for it had been found. After World War II, F Reines and some of his colleagues considered using a nuclear explosion as a copious source of neutrinos [120]. Eventually he and Clyde Cowan decided to use neutrinos produced at a nuclear reactor, eliciting an approving letter from Fermi that the new method should be much simpler to carry out and had 'the advantage that the measurement can be repeated any number of times'.

The inverse of the beta-decay reaction, the process $\bar{\nu} + p \to n + e^+$, was studied. A very large target, well shielded against accidental interactions, was required to observe this process, which is about 10^{18} times less frequent than a proton–proton interaction.

The initial experiment was performed at the Hanford reactor in 1953. Signals with the reactor on and off were compared. The large reactor-off signal, eventually traced to cosmic-ray interactions, prevented

663

a definitive observation [121]. The experiment was then moved to Savannah River, yielding a signal sufficiently convincing [122] for Reines to send a telegram to Pauli in June of 1956 informing him that the particle he predicted more than 25 years earlier had been found.

The observation of the reaction $\bar{\nu}+p \to e^+ + n$, without any examples of $\bar{\nu} + n \to e^- + p$, was the first proof of *lepton number conservation* in the weak interactions. The electron and the neutrino, by convention, are assigned lepton number $L = 1$. The neutral lepton accompanying an electron in beta decay (the source of reactor neutrinos) must then be an antineutrino with $L = -1$, giving rise only to positrons (with $L = -1$).

9.5.1.2. Early classification of beta decay. Fermi's 1934 theory of beta decay [5], mentioned in section 9.2, was based on an interaction of the *vector* (V) type between pairs of fermions, one of five possibilities consistent with Lorentz invariance and mirror symmetry. The others are scalar (S), tensor (T), axial vector (A) and pseudoscalar (P), whose names refer to each fermion pair's properties under Lorentz transformation.

The non-relativistic limit of the S and V interactions corresponds to a unit operator evaluated between spinors. Beta-decay transitions of this type were called *Fermi* transitions; many were identified experimentally. The non-relativistic limits of the T and A interactions involve instead the Pauli matrix σ. Transitions of this *Gamow–Teller* type [123] also were identified quite early, through their characteristic change of one unit of nuclear spin, suggesting complements to Fermi's hypothesis.

The energy dependence of the beta-decay spectrum depends on the form of couplings. When S and V or T and A interactions coexist, terms arise of the form m_e/E_e, where m_e and E_e are the electron's mass and energy. Such *Fierz interference* terms [124] were not seen, restricting the allowed interactions.

Electron–neutrino correlations in beta-decay experiments allow one to sort out different interaction types, with distributions proportional to $1 + b\hat{p}_e \cdot \hat{p}_\nu$, with [125] ($b = -1, 1, \frac{1}{3}, -\frac{1}{3}$) for pure (S, V, T, A) interactions, respectively. In the mid-1950s, experiments on $^6\text{He} \to {}^6\text{Li} + e^- + \bar{\nu}$, $^{19}\text{Ne} \to {}^{19}\text{F} + e^+ + \nu$ and $^{35}\text{Ar} \to {}^{35}\text{Cl} + e^+ + \nu$ gave conflicting results, which were soon to be sorted out.

9.5.1.3. Universality hypothesis. The weak interactions, though initially involved only in nuclear beta decay, were soon seen to participate with equal strength in $\mu \to e\nu\bar{\nu}$ and muon capture by nuclei. This observation led to the hypothesis [126] of *weak universality*, according to which any of the pairs (p, n), (e, ν), and (μ, ν) would interact equally with itself or any other pair. The charged pion's decay proceeds at a rate consistent with weak universality if dominated by a nucleon–antinucleon intermediate state. The decays of strange particles appeared to be due to an interaction similar to but somewhat weaker than the beta-decay interaction.

9.5.1.4. Overthrow of parity conservation.
As decays of charged strange mesons were identified in the early 1950s, the masses of particles decaying in various modes all clustered more and more closely about a single value, now known to be $m(K) = 493.6$ MeV/c^2, with lifetimes showing a similar approach to a common value [127]. However, two distinct decays, called $\theta^+ \to \pi^+ + \pi^0$ and $\tau^+ \to \pi^+ + \pi^+ + \pi^-$, could not correspond to the same particle if the weak interactions were invariant under mirror symmetry.

An examination [128] of the kinematic distribution of the three-pion ('τ^+') decays favoured zero relative angular momentum between the two like pions and between them and the unlike pion. In that case, the τ^+ had to have spin J equal to zero and negative intrinsic parity P (since the parity of each pion was identified as negative; see section 9.4.2.1). A mixture of internal relative angular momenta leading to $J^P(\tau^+) = 2^-$ was also possible.

The two-pion ('θ^+') decay can only occur for a particle with $P = (-1)^J$. Thus, since the above analysis favoured $J^P(\tau^+) = 0^-$ or possibly 2^-, the τ^+ and θ^+ could not be the same *if parity was conserved in weak decays*.

The possibility of parity violation in the weak interactions began to be debated informally in seminars early in 1956. At the Sixth International Conference on High Energy Physics in Rochester, New York, in the early spring of that year, Feynman quoted a question by Martin Block asking whether the τ and θ could be different parity states of the same particle with no definite parity, implying that parity is not conserved [129]. Returning from the conference and reviewing all the evidence, T D Lee and C N Yang (figure 9.8) realized that no conclusive test had yet been performed of parity conservation in the weak interactions; such a test should measure the expectation value of an operator odd under spatial reflection, such as the scalar product of a momentum and a spin vector. They suggested several such tests [130]. (For two complementary historical accounts, see references [131].)

Before Lee and Yang's paper was published, R Oehme pointed out to them [132] that one could not see the effects of parity (P) violation in a beta-decay experiment if charge-conjugation invariance (C) and time-reversal symmetry (T) were preserved. Moreover, in the absence of significant final-state interactions, even C invariance alone prevented the observation of a P-violating asymmetry. Thus, Lee and Yang's suggestion of P violation in beta decay implied as well the violation of C. The published work incorporated this suggestion, and was followed by a joint paper [133] which also pointed out that the equality of masses and lifetimes for particles and antiparticles followed just from CPT invariance, a consequence of Lorentz invariance and local field theory [134]. Thus, for example, the observed equality of positive and negative muon lifetimes could not be used as evidence of invariance under charge

Figure 9.8. T D Lee (left) and C N Yang in 1957. Photograph by Alan W Richards.

conjugation since it followed from much weaker assumptions.

The Lee–Yang suggestion was taken up by C S Wu at Columbia and her collaborators at the National Bureau of Standards in Washington, and by V L Telegdi at Chicago and his student J I Friedman. Wu and her colleagues found that parity was indeed violated [135] in the beta decay of polarized ^{60}Co. Friedman and Telegdi began a search, which ultimately bore fruit [136], for parity violation in the decay $\pi \to \mu \to e$ in emulsions.

Wu returned to Columbia early in 1957 with the preliminary news of her results. Within a weekend Garwin, Lederman and Weinrich [29] were able to devise and perform an experiment searching for parity violation in $\pi \to \mu \to e$. The story is told in reference [137] by one of the participants.

Lee and Yang suggested that one parity-violating observable could be an up–down asymmetry in the weak decays of polarized Λ hyperons. The observation of this asymmetry [138] was a further confirmation that, indeed, parity was violated in the weak interactions.

The weak interactions as now written down could be described by a theory which violated spatial reflection (P) and charge conjugation (C) symmetries, but was invariant under time reversal (T), the product CP thereby satisfying the requirement of a CPT-invariant theory. The

Elementary Particle Physics

conservation of T by the weak interactions, as newly formulated, explained why the neutron did not appear to have an electric dipole moment. Experiments placed quite stringent limits on such a moment [139]. Its absence had been thought to be evidence against parity violation, but it was now realized that T invariance alone would forbid an electric dipole moment for any elementary particle.

9.5.1.5. V–A theory and conserved vector current (CVC). In the initial experiment demonstrating parity violation in beta decay, the electrons were found to be emitted in a direction correlated with the spin of the initial nucleus. Soon thereafter, a number of experiments found that the spins of beta-decay electrons and positrons were correlated with their own velocities. Electrons were emitted spinning in a predominantly left-handed manner, or with left-handed *helicity*, while positrons emerged primarily with right-handed helicity.

Before the observation of parity violation in the weak interactions, Salam and Landau had proposed that a neutrino could exist in a single helicity state [140]. Indeed, Weyl had considered spinors with two components much earlier, but this possibility was rejected by Pauli in the 1930s for the neutrino because it did not conserve parity [141]. The ordinary spin-$\frac{1}{2}$ particle satisfying the Dirac equation has four degrees of freedom—two for each spin of both particle and antiparticle. When the rest mass is zero, however, the four-component solution breaks apart into two two-component ones, the first describing a left-handed particle and a right-handed antiparticle and the second describing a left-handed antiparticle and a right-handed particle.

A two-component neutrino was tailor-made for the large degree of parity violation observed in beta-decay experiments [142]. If a neutrino had one helicity and its antiparticle the opposite helicity, an electron released with velocity v in beta decay had to have polarization equal to $\pm v/c$, while a positron with the same velocity had to have the opposite polarization. This was indeed verified. Electrons always had polarization $P(e^-) = -v/c$, while positrons had polarization $P(e^+) = v/c$.

In view of the apparent two-component nature of the neutrinos participating in beta decay and the beauty and simplicity of the two-component formalism, Feynman and Gell-Mann and Sudarshan and Marshak proposed [143–145] that *all* particles participate in the weak interactions in the same two-component manner. As a consequence, each particle pair (such as (e, ν), (μ, ν), and (n, p)) had to participate in the weak interaction with a definite combination of vector (V) and axial (A) strengths, by convention called V–A. The scalar (S), tensor (T) and pseudoscalar (P) interactions vanish under the two-component hypothesis.

The universal V–A interaction agreed with many results at the time it was proposed. In addition to predicting the polarizations of electrons and positrons in beta decay, it reproduced the observed muon lifetime and the electron–neutrino correlations in most beta-decay processes. However, it seemed incompatible with some observations, including the apparent dominance of S and T interactions in ^{6}He beta decay [146] and the apparent absence [147] of the decay $\pi \to e\nu$. In due course, a new ^{6}He experiment [148] confirmed the V–A prediction, and the $\pi \to e\nu$ decay was eventually found at the predicted level [149]. The theory also predicted parity violation in purely *non-leptonic* processes, thus resolving the $\tau - \theta$ puzzle mentioned earlier. It implied that the neutrino emitted in beta decay was left-handed, as was verified in an elegant experiment (see Box 9A) by M Goldhaber, L Grodzins and A Sunyar [150] using the apparatus described in figure 9.9. A more detailed discussion of these and other tests of the V–A theory appears in reference [151].

A general test for Lorentz structure of couplings in muon decay had been proposed in 1950 [152]. The electron energy spectrum in muon decay could be described in terms of a parameter ρ and the ratio x of the electron energy in the muon rest frame to its maximum value. Normalized to unit integral over the range $0 \leqslant x \leqslant 1$, the spectrum has the form $N(x) = 6x^2[2(1 - x) + (4/9)\rho(4x - 3)]$. The V–A theory implied that labelling electrons as particles but muons as antiparticles yields $\rho = 0$, while labelling both muons and electrons as particles yields $\rho = \frac{3}{4}$. Experiments eventually converged on the latter value.

The universal V–A interaction was also generalized to strange particles. Notable among the results [143] was the linkage of strangeness S and charge Q in beta-decay interactions, represented by the empirical rule $\Delta S = \Delta Q$. If $\Delta S = -\Delta Q$ transitions also had been allowed, they would have led to unobserved processes such as $\Xi^- \to \pi^- + n$.

The universal weak (current) × (current) interaction, where each V–A current carries unit charge, implies that all Fermi-type interactions in beta decay stem from the nucleon's V current, while all Gamow–Teller transitions arise from A. A remarkable feature of these currents, particularly of the vector current, is their universal strength. For example, in Fermi transitions between nuclei of zero spin, as in ^{14}O beta decay, the coupling strength appears to be almost identical with that in muon decay, despite the presence of strong interactions which in principle could modify the nucleon's current.

To explain the universality of the weak vector current, Feynman and Gell-Mann [143] and Gershtein and Zel'dovich [153] suggested that the charged weak vector current belonged to an isotopic spin *triplet* of conserved vector currents. This *conserved vector current* hypothesis identified the vector weak current with the generator of the isotopic spin symmetry, whose matrix elements are set by simple isospin considerations, independently of strong interaction details. For example,

Elementary Particle Physics

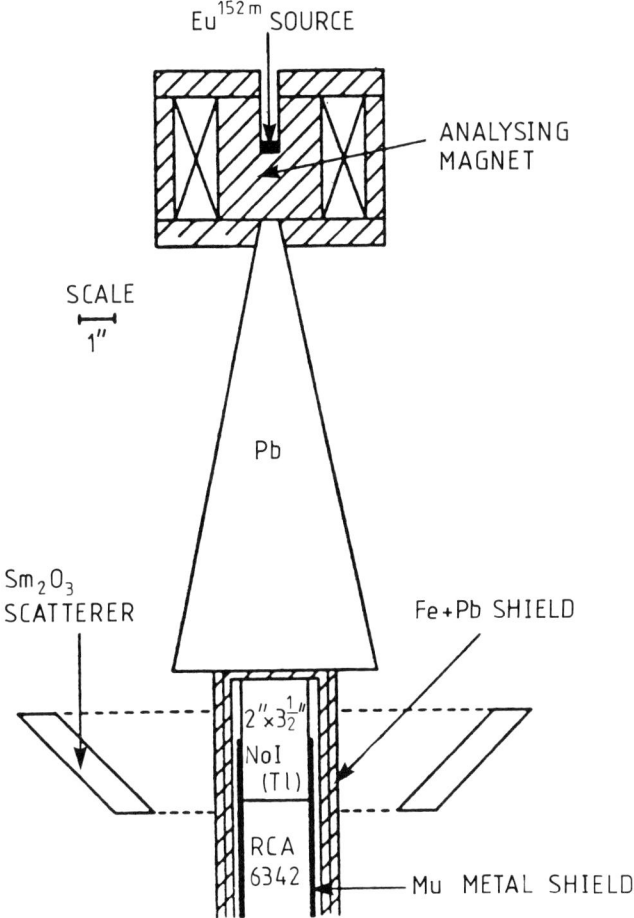

Figure 9.9. *Apparatus in the experiment of Goldhaber, Grodzins and Sunyar* [150] *to determine the neutrino helicity by measuring circular polarization of resonant scattered γ-rays. The apparatus has cylindrical symmetry about the vertical axis.*

the rate for the decay $\pi^+ \to \pi^0 + e^+ + \nu_e$ was predicted precisely [143]; the process was observed, with the predicted rate, several years later [154].

9.5.1.6. Two-neutrino hypothesis and its confirmation. The current–current form of the weak interactions, with a Fermi coupling strength $G_F \approx 10^{-5} m_p^{-2}$, leads to cross sections which violate the unitarity of the S-matrix at high energies. One cure for this problem was the proposal, first appearing in Yukawa's theory of the meson and in various forms thereafter [143, 155], that the weak interactions arose from the exchange of a meson, in analogy with the pion for the strong force and the photon for electromagnetism. The V–A nature of the weak interaction required

BOX 9A: THE HELICITY OF THE NEUTRINO

Everyone knows that the neutrino is a most elusive particle requiring tons of material just to detect. Who then would have the daring to even contemplate measuring its spin direction relative to its direction of motion, its helicity? M Goldhaber, L Grodzins and A Sunyar (GGS) (reference [150]) succeeded in doing this in an ingeniously conceived and executed experiment. It was an experiment that required encyclopaedic knowledge of nuclear isotopes as well as cunning experimental technique.

The nucleus ^{152}Eu decays to ^{152}Sm by electron capture from the K-shell with the emission of a neutrino with energy 0.840 MeV. The ^{152}Eu has spin zero but the ^{152}Sm which accompanies the neutrino is in an excited state, Sm* with spin 1. It decays in 3×10^{-14} s to the spin-zero ground state with the emission of a gamma ray. How did GGS put all this together to measure the helicity of the neutrino? By conservation of angular momentum, the spin of the neutrino plus the Sm* must equal the spin of the captured electron, forcing the spin of the Sm* to be opposite to that of the neutrino. Since the Sm* nucleus is recoiling away from the neutrino, the *helicity of the Sm* must be the same as the neutrino*. Measuring the helicity of the neutrino has been reduced to measuring the helicity of the recoiling Sm*.

The angular momentum of the Sm* will be carried away by the emitted gamma ray. It is possible to measure the circular polarization of gamma rays by passing them through magnetized iron since the absorption cross section depends on the spin direction of the scattering electrons. To get the helicity of the Sm* it is still necessary to determine its direction of travel relative to the direction of the gamma. GGS note that the gamma-ray energy is 0.960 MeV, not far from the neutrino energy. If the gamma ray is emitted in the direction the Sm* is travelling, the energy will be boosted just enough so that the gamma will be scattered resonantly (hence, with a large cross section) from Sm in its ground state if the scattering angle is about 90°. This then determines that the gamma ray has been emitted in the direction of the Sm*.

This brilliant, complex scheme was implemented in the apparatus shown schematically in figure 9.9. By measuring the gamma-ray intensity in the NaI crystal as a function of the direction of the magnetization of the iron, the neutrino was found to be *left-handed*!

this particle to have spin 1, i.e. to be a *vector* boson. The modern name for this boson is W.

The vector boson proposal predicted the process $\mu \to e\gamma$ to occur with a branching ratio far above the upper limits set in the early 1960s. The process involved a virtual transition of the muon to some charged intermediate state and a neutrino. The charged intermediate state would then radiate a photon and reabsorb the neutrino to become an electron. However, the predicted decay would not occur if the muon and electron coupled to *separate* neutrinos.

An experiment was performed at Brookhaven National Laboratory by the participants shown in figure 9.10 to study the interactions of neutrinos produced in the decay of pions in the process $\pi^\pm \to \mu^\pm \nu$. Neutrinos distinct from those emitted in beta decay would produce only muons, not electrons, in reactions such as $\nu+n \to \mu^-+p$ and $\bar{\nu}+p \to \mu^++n$. The result was conclusive: the muon and electron neutrinos were different [156].

The history of the two-neutrino experiment is told in reference [157]. The development of neutrino beams had been first proposed by B Pontecorvo and M Schwartz [158], with resulting physics anticipated in a prescient paper by Lee and Yang [159].

One goal of early neutrino experiments was the search for the W-boson mentioned above. The results in experiments on both sides of the Atlantic [160] were negative, placing a lower limit of 2 GeV/c^2 on the mass of the W, which is now known to be much heavier. Neutrino interactions were so rare in those early days (1963) that a bottle of champagne was opened each time an interaction was seen at the European experiment, performed at CERN.

9.5.2. CP violation

9.5.2.1. Implications of P and C violation for neutral kaon systems. The V–A theory of weak interactions violated both parity (P) and charge-conjugation invariance (C), while preserving time-reversal invariance and the product CP. Now, the original prediction of two types of neutral kaons, the short-lived kaon decaying to $\pi\pi$ and the long-lived one forbidden from doing so [91], was based on the assumption of C invariance of the weak interactions. Similar results were shown to follow [161] from combined invariance under the product CP. Hence, the K_2^0, mentioned in section 9.4.4, was still expected not to be able to decay to $\pi\pi$ if CP invariance was valid in the weak interactions.

9.5.2.2. Initial experiments. Searches for the decay $K_2^0 \to \pi^+\pi^-$ during the late 1950s and early 1960s failed to find any signal at a level of a part in 300. With the advent of spark chambers permitting selective triggering on events of a given type, a more sensitive search was mounted by

Figure 9.10. *Participants in the two-neutrino experiment at Brookhaven* [156]. *From left to right: J Steinberger, K Goulianos, J-M Gaillard, N Mistry, G Danby, W Hayes, L Lederman, M Schwartz. Top: then; bottom: recently.*

J Christenson, J Cronin, V Fitch and R Turlay at Brookhaven National Laboratory. The early history and motivation for this experiment are described in reference [162]. A signal was indeed seen (figure 9.11), corresponding to one event for every 500 decays [163].

Theoretical analyses [133, 164] had set the stage for a description of CP violation in decays of neutral kaons. After its discovery,

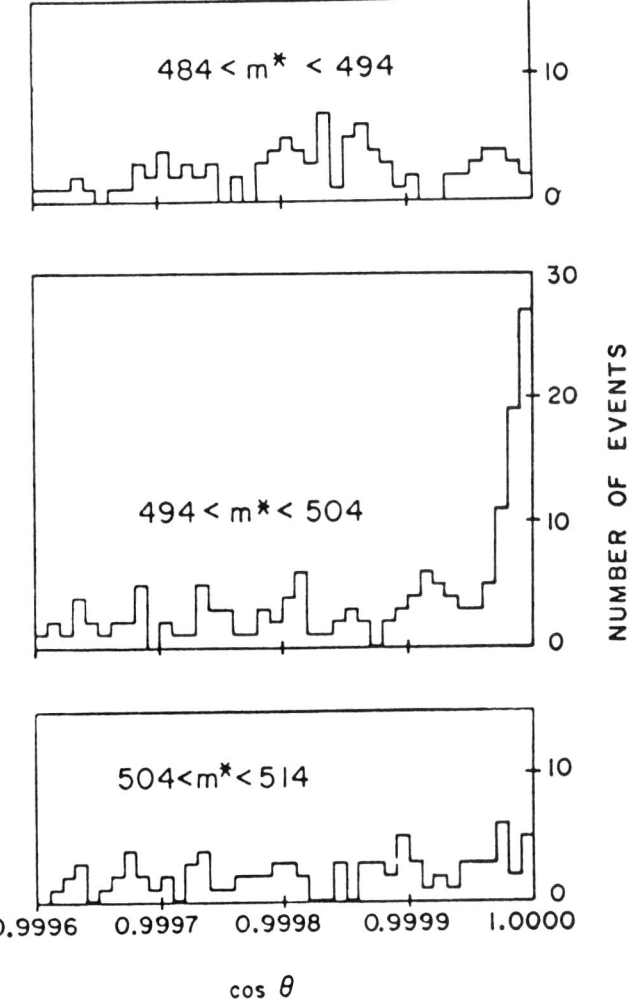

Figure 9.11. *Evidence [163] for the decay $K_L \to \pi^+\pi^-$. The angular distribution of the two-pion system with respect to the beam is shown for three mass ranges, of which the middle range corresponds to the signal.*

Wu and Yang suggested further experiments [165], establishing many conventions still in use today. Since both K_1^0 and K_2^0 were now seen to decay to pairs of pions, their names were changed to denote states of definite lifetime: K_S for the short-lived neutral kaon, with lifetime 0.09 ns, and K_L for the long-lived neutral kaon, with lifetime 52 ns.

The presence of the decay $K_L \to \pi^+\pi^-$ indicated CP violation in the weak interactions. Assuming the validity of the CPT theorem, this implied also T violation, a result not to be taken lightly. The

673

experimenters spent more than six months analysing their signal before concluding that their data were indeed evidence for CP violation. Detailed analyses (see, for example, reference [166]) without the assumption of CPT invariance later showed that indeed the data were also signalling T violation.

A crucial test confirming that the $\pi^+\pi^-$ final state indeed originated from the long-lived kaon was the observation [167] of interference between the decays $K_L \to \pi^+\pi^-$ and $K_S \to \pi^+\pi^-$. The K_S was produced by *coherent regeneration* [164, 168]. As a K_L beam passes through matter, the phase and magnitude of the beam's K^0 and $\bar{K}^0$ components change in different ways, inducing a mixture of K_S. As a result, one can measure not only the magnitude but also the phase of the ratio of $K_L \to 2\pi$ and $K_S \to 2\pi$ amplitudes.

The evidence for CP violation has so far been confined exclusively to the system of neutral kaons. Reviews of major results since the discovery of CP violation can be found in references [166] and [169].

9.5.2.3. Decays to charged and neutral particles. The origin of CP violation in neutral kaon decay still is not understood in terms of a fundamental theory. A question arose immediately whether the observed effect results entirely from mixing, or stems in part from direct CP violation in the decay process.

The states $K^0_{S,L}$ of definite mass and lifetime can be regarded as mixtures of states $K^0_{1,2}$ with even and odd CP. In a CPT-conserving theory [133], one can write

$$|K_S\rangle = \frac{1}{\sqrt{(1+|\epsilon|^2)}}(|K_1\rangle + \epsilon|K_2\rangle) \qquad |K_L\rangle = \frac{1}{\sqrt{(1+|\epsilon|^2)}}(|K_2\rangle + \epsilon|K_1\rangle)$$

where the complex parameter ϵ denotes the effect of the mixing. The amplitudes for the decays of K_L to $\pi^+\pi^-$ or $\pi^0\pi^0$ may be characterized by their ratios to the corresponding K_S decay amplitudes

$$\frac{A(K_L \to \pi^+\pi^-)}{A(K_S \to \pi^+\pi^-)} \equiv \eta_{+-} \qquad \frac{A(K_L \to \pi^0\pi^0)}{A(K_S \to \pi^0\pi^0)} \equiv \eta_{00}$$

where $\eta_{+-} = \eta_{00} = \epsilon$ if the only CP violation takes place through mixing. In general one can write $\eta_{+-} = \epsilon + \epsilon'$ and $\eta_{00} = \epsilon - 2\epsilon'$. Here ϵ' describes the effect of direct CP violation (i.e. that not due to mixing) in the decay amplitude, leading to a final state with isospin 2.

The leading candidate for a theory of CP violation in the neutral kaon system, the presence of phases in the Cabibbo–Kobayashi–Maskawa matrix (section 9.9), predicts that ϵ' should have a phase very close to that of ϵ, with $\mathrm{Re}(\epsilon'/\epsilon)$ a few parts in 10^4. However, a model with $\epsilon' = 0$, in which CP violation in neutral kaons is due entirely to a $\Delta S = 2$

'superweak' interaction [170] mixing K^0 and $\bar{K}^0$, still is not conclusively excluded by present data.

One experiment [171] finds $\text{Re}(\epsilon'/\epsilon) = (7.4 \pm 5.9) \times 10^{-4}$, consistent with zero, while another [172] finds $\text{Re}(\epsilon'/\epsilon) = (23 \pm 6.5) \times 10^{-4}$, non-zero at the level of more than three standard deviations. There is not a serious contradiction between the two results if $\text{Re}(\epsilon'/\epsilon)$ is around 14×10^{-4}. Improved versions of both experiments are planned. The theory can also be tested by studies of B-mesons (section 9.9).

9.5.2.4. Asymmetries in semi-leptonic decays. By comparing rates for neutral kaons to decay to states such as $\pi^- e^+ \nu_e$ and $\pi^+ e^- \bar{\nu}_e$, one can learn the relative admixture of K^0 and $\bar{K}^0$ in each neutral kaon's wavefunction [173]. The $\Delta S = \Delta Q$ rule of strangeness-changing weak interactions mentioned above implies that only negative pions emerge from K^0 and only positive from $\bar{K}^0$. CP violation implies an asymmetry proportional to $2\text{Re}\epsilon$ in rates of K_L decay to these final states. This asymmetry was searched for and found; the results [174] are in perfect accord with expectations.

9.5.2.5. CP violation and baryon asymmetry in the Universe. Shortly after CP violation was discovered, Sakharov [175] pointed out that it was a key ingredient in understanding why the observed Universe contains more baryons than antibaryons. Other crucial features were interactions which violate baryon-number conservation, and a period in the early Universe when reactions governed by these interactions were out of thermal equilibrium. Those conditions have been realized in subsequent models [176], to which we shall return in sections 9.12 and 9.13.

9.5.3. Current algebras

9.5.3.1. The role of the pion. In the newly proposed V–A theory, the weak interactions of strongly interacting particles remained mysterious. While the vector weak current of nucleons seemed to be nearly unaffected by the strong interactions, the axial vector coupling strength, parametrized by $G_A \approx 1.25$, deviated significantly from the value $G_A = 1$ expected if nucleons behaved like leptons. Moreover, decays of strongly interacting particles to purely leptonic final states were characterized by arbitrary constants, such as the pion decay constant f_π for $\pi \to \mu + \nu$.

Armed with experience in both strong and weak interactions, M L Goldberger and S B Treiman [177] embarked on a calculation of f_π. Their result involved a crucial role for the nucleon–antinucleon intermediate state and hence for the pion–nucleon coupling constant $g_{\pi NN}$. They obtained the relation $f_\pi = m_N G_A / g_{\pi NN}$, which is accurate to a few per cent. Here m_N is the mass of the nucleon.

The vector current's time component, when integrated over all space, is the *isotopic charge*, or generator of isotopic spin transformations. The conservation of vector current is thus associated with isotopic spin invariance. The space integral of the *axial* current's time component also performs isotopic spin transformations, rotating left-handed and right-handed fermions oppositely in isospin space. Invariance under separate isospin rotations of left-handed and right-handed fermions is known as *chiral invariance*.

But is the axial current conserved? Apparently not; the divergence of the matrix element of the axial current between one-nucleon states is proportional to m_N, and so cannot vanish unless the nucleon is massless. Under such circumstances, one indeed would have a chiral invariance. Chiral symmetry would be realized in the *Wigner–Weyl* sense, by manifest invariance of both the equations of motion and their solution. One implication of manifest chiral symmetry is the prediction that for every particle with a given parity, there should be a degenerate particle with opposite parity. This certainly did not seem to be so.

In a sequel to their first paper on the pion decay constant, Goldberger and Treiman [178] showed that the requirement of a conserved axial current entails a zero-mass pole in beta-decay amplitudes. Inspired by this result and by an analogy with superconductivity, Y Nambu [179] discovered another mode in which chiral symmetry could be realized. If the *vacuum itself* were not chirally symmetric, so that the part of the symmetry generated by the axial current was *spontaneously broken*, there was no need for parity doubling in the spectrum of states like the nucleon. Instead, a massless pseudoscalar particle, the pion, miraculously appeared; the parity partner of the nucleon was then a state with a nucleon and a pion. This behaviour illustrated a general theorem of J Goldstone [180]: spontaneous breakdown of a global, i.e. space-independent symmetry (such as chiral symmetry), necessarily leads to massless particles in the spectrum. For systems of pions and nucleons, chiral symmetry is said to be realized in the *Nambu–Goldstone* sense: via spontaneous symmetry breaking. In Nambu's approach, the pion satisfied the Goldberger–Treiman relation as a natural consequence of its role in the breakdown of chiral invariance. The zero-mass pole which Goldberger and Treiman had discovered could be, in fact, identified with the pion in the limit of exact chiral invariance.

M Gell-Mann and M Lévy were motivated to understand the deviation of the axial coupling constant G_A from unity by identifying the divergence of the axial current with the pion field itself [181]. They constructed several field-theoretic models in which this was so. The Goldberger–Treiman relation was a welcome consequence. The fact that the divergence of the axial current was non-zero, but identified with the pion field, was called *partial conservation of axial current* (PCAC).

9.5.3.2. Gell-Mann–Lévy discussion of universality. Although the vector current was not expected to be renormalized by the strong interactions, the Fermi constant G as measured in ^{14}O beta decay appeared to be about 3% lower than G_μ as measured from muon decay. Gell-Mann and Lévy proposed [181] that the missing 'strength' was made up by the beta-decay coupling of the proton to the Λ: in modern notation, the hadronic beta-decay current then involved the charge-changing transitions p ↔ n$\cos\theta + \Lambda\sin\theta$. With $\sin^2\theta \approx 0.06$, one was then able to understand both $G/G_V \approx 0.97$ and the relative suppression of strangeness-changing weak decays (particularly notable for the process $\Lambda \to $ pe$^-\bar{\nu}$ [182]).

9.5.3.3. Gell-Mann's algebra of currents. In search of a dynamical principle underlying the success of unitary symmetry, Gell-Mann [183] (figure 9.12) realized that both isotopic spin (the group SU(2)) and its generalization to SU(3) were generated by *charges*—space integrals of time components of vector currents. These charges F_i ($i = 1,\ldots,8$) obeyed an algebra of the form $[F_i, F_j] = \mathrm{i} f_{ijk} F_k$, where the f_{ijk} are totally antisymmetric *structure constants*.

The spatial integrals of time components of axial vector currents give rise to charges F_i^A, which necessarily transform as vectors under the symmetry: $[F_i, F_j^A] = \mathrm{i} f_{ijk} F_k^A$. The existence of an octet of light pseudoscalar mesons is the reason the axial charges are not conserved; when acting on the vacuum, they produce a pion, kaon or η.

When one has a set of generators of a symmetry, it is natural to ask for the behaviour of *all* commutators. Gell-Mann postulated that the algebra was completed by the simplest possibility for the commutator of two *axial* charges: $[F_i^A, F_j^A] = \mathrm{i} f_{ijk} F_k$. This is indeed the case for the leptons. It could be true for strongly interacting particles as well if they were made up of more fundamental entities—at least, one could conceive of interactions not affecting the basic relations. These entities could be identified with *quarks* [184]—members of the triplet representation of SU(3). We shall discuss them in section 9.6.

The combinations $(F_i + F_i^A)/2 \equiv F_i^R$ and $(F_i - F_i^A)/2 \equiv F_i^L$ then would obey two independent SU(3) algebras. This SU(3) × SU(3) structure has led to successful predictions for many quantities previously regarded as the province of intractable strong-interaction physics.

9.5.3.4. Cabibbo theory of strange-particle decays. If vector and axial vector currents transformed as generators of SU(3), their matrix elements between various states of SU(3) multiplets could be related to one another. By 1963, data on beta decays of various hyperons in the baryon octet had been accumulating, and the time was ripe for such an analysis.

N Cabibbo [185] postulated that the charge-changing weak current behaved as a member of an SU(3) octet, consisting of a linear combination of a piece, transforming like a charged pion (with coefficient $\cos\theta$) and a

Figure 9.12. *Murray Gell-Mann (top); Yuval Ne'eman (bottom).*

charged kaon (with coefficient $\sin\theta$). While the angle θ is the one Gell-Mann and Lévy proposed for rescuing universality of the weak hadronic current, we shall henceforth refer to it as θ_C.

The conservation of vector current specifies uniquely the vector current's matrix element between baryon states. However, the non-

Elementary Particle Physics

conservation of axial current allows for two types of matrix elements for the axial current. In SU(3) there are two ways to couple an octet current to initial and final octet baryons, called F (totally antisymmetric) and D (totally symmetric).

One combination of F and D coupling for the axial current is the axial charge of the nucleon, $F + D = G_A \approx 1.25$. The Cabibbo theory described decays such as $K \to \pi e \nu$, $\Lambda \to p e \nu$, $\Sigma^- \to n e^- \nu$, and $\Sigma^- \to \Lambda^0 e^- \nu$ with a single value of θ_C and a self-consistent value of F/D. In modern fits, which also include data on several other hyperon beta decays (and on the ratios of axial to vector couplings in some of them) one finds [186] $\sin \theta_C \approx 0.22$ and $F/D \approx 2/3$. This last value turns out to be close to what one expects in a quark picture of baryons.

9.5.3.5. *Adler–Weisberger relation.* The PCAC hypothesis permitted calculations of soft pion emission [187] in a manner analogous to calculations of soft photon emission using general principles of electromagnetism [188]. Stephen Adler, a skilled practitioner of this technique, used it to study a series of processes, including low-energy pion–nucleon scattering [189]. Hearing of the proposal by Murray Gell-Mann and his student Roger Dashen to apply the commutator of two axial isospin generators to obtain a relation between the axial–vector coupling constant and pion–nucleon scattering, Adler realized that his experience was ideal for the problem. In very short order he related G_A to the difference in total cross sections of positive and negative pions on nucleons, integrated over energy [190]. At the same time, W I Weisberger produced a similar calculation, and subsequently generalized the result to relate $|\Delta S| = 1$ transitions to kaon–nucleon scattering [191].

The Adler–Weisberger relation in the zero-pion-mass limit may be written

$$G_A^2 = 1 - 2\frac{f_\pi^2}{\pi} \int_0^\infty \frac{d\nu}{\nu} (\sigma^{\pi^- p}(\nu) - \sigma^{\pi^+ p}(\nu))$$

where ν stands for the laboratory energy of the pion. Its prediction for G_A depends to some extent on the treatment of corrections for massive pions, but the resulting value $G_A \approx 1.2$ was sufficiently close to experiment that the power of current algebra was immediately recognized.

9.5.3.6. *Other current algebra relations.* The Adler–Weisberger relation exploits the nonlinearity of the current commutation relations to normalize axial charges. The commutator between axial and vector charges also contains useful information, providing a relation [192] between the semi-leptonic process $K \to \pi e \nu$ and the purely leptonic decay $K \to \mu \nu$. Even purely hadronic processes such as pion–pion scattering could be attacked by such methods, as shown by Weinberg [193]. Many other successes of current algebra were chronicled in contemporary texts and reviews [194].

9.5.4. Strong-interaction schemes
While the theory of weak interactions enjoyed tremendous progress during the 1950s, the strong interactions remained a mystery. Symmetry arguments sufficed for many results, but the underlying forces were not understood until the advent of quantum chromodynamics in the 1970s. Still, the efforts to understand strong forces bore fruit in a number of areas even in the absence of a fundamental theory.

9.5.4.1. S-matrix theory and dispersion relations. A unitary matrix relating outgoing to incoming scattering states was introduced by Wheeler in the 1930s [195] in the context of non-relativistic nuclear reactions, with a fully relativistic treatment developed independently by Heisenberg [196]. The scattering matrix, or S-matrix, as it came to be called, had an interesting history during World War II [197]. News of Heisenberg's work was brought to Japan by German submarine in the form of a letter from Heisenberg to Nishina. A unitary S-matrix appeared in Japanese literature of the 1940s in analyses of microwave junctions by Tomonaga and his group [198]. The US wartime microwave work also employed the S-matrix in descriptions of junctions [199]. Around this time Stückelberg independently introduced an analogue of the S-matrix [200]. In the physics of antenna impedance matching, a transformation very similar to Stückelberg's had been proposed even earlier, forming the basis of the frequently used *Smith chart* [201]. The development of the relativistic S-matrix in the post-war years owed much to the work of Møller [202].

As the strong coupling of pions to nucleons became clear in the 1950s, physicists despaired of describing the strong interactions by quantum field theory, so successful for QED. It was hoped [203] that by characterizing the singularities of the S-matrix, a theory could avoid expanding amplitudes as a perturbation series, which fails because of the large pion–nucleon coupling constant. Thus began a period of intense study of analytic properties of scattering amplitudes and of 'bootstrap' theories in which the known particles were all viewed as composites of one another. An early success of this programme was the development of dispersion relations for scattering amplitudes [204], relating the real parts of amplitudes to integrals over total cross sections. Precise measurements of pion–nucleon total cross sections and of real parts of amplitudes [205] provided one impressive verification of these relations. Dispersion relations could also be written simultaneously in variables corresponding to relativistic generalizations of energy and momentum transfer [206], providing insight into parameters governing the range of strong forces.

9.5.4.2. Chew–Low theory and isobars. The first pion–nucleon resonance, the $\Delta(1232)$ mentioned above, was described by Chew and Low [207] in a theory that viewed the force between a pion and a baryon (the nucleon

Elementary Particle Physics

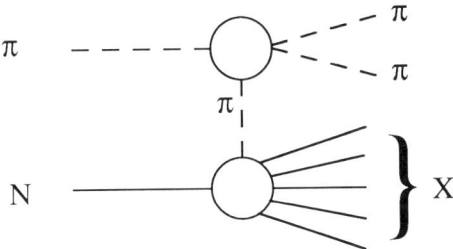

Figure 9.13. *Diagram describing the Chew–Low proposal [208] for studying pion–pion scattering.*

or Δ) as due primarily to baryon exchange. Indeed, the singularities in the scattering amplitude most important for its low-energy behaviour are just those associated with baryons in the crossed channel.

9.5.4.3. Pion–pion scattering and the bootstrap programme. The idea that strongly interacting particles could all be viewed as composites of one another (encouraged by Chew and Low's result for the Δ, nucleon and pion) grew in the 1950s into a fully fledged 'bootstrap' programme, dealing only with the self-consistent singularities of the S-matrix actually found in Nature. The key ingredients in this programme were the unitarity of the S-matrix and the analyticity and *crossing symmetry* of the scattering amplitude. This last feature relates such processes as $A + B \to C + D$ to *crossed reactions* such as $\bar{C} + B \to \bar{A} + D$.

A simple test of the bootstrap programme was provided by pion–pion scattering. Since pions are not available in the laboratory, how would one employ them as targets? The answer was provided by Chew and Low [208]: use virtual pions (see figure 9.13). In the reaction $\pi + p \to X + n$, for example, the scattering amplitude has a pole in the invariant momentum transfer, corresponding to exchange of a pion, which dominates the scattering for small enough values of momentum transfer.

Many bootstrap calculations of pion–pion scattering were performed [203], with varying degrees of success in reproducing the dominant low-energy features. These included a prominent P-wave resonance in the $I = 1$ channel, the $\rho(770)$ (the number refers to the mass in MeV/c^2), and a slow increase of the S-wave phase shift in the $I = 0$ channel. Once the low-energy behaviour predicted by current algebra [193] was specified, successful bootstrap calculations were indeed displayed [209]. However, analyticity, unitarity and crossing symmetry proved inadequate to specify scattering amplitudes uniquely.

9.5.4.4. Regge poles. A paper by Tullio Regge [210] dealing with scattering in a non-relativistic potential provided an important step in the description of scattering in particle physics. The S-matrix has poles in complex angular momentum which *change their position as a function of energy*, solving a knotty problem that had plagued field-theoretic descriptions of scattering amplitudes.

Unitarity and the short-range nature of the strong force allowed Froissart to show that total cross sections can grow no more rapidly at high energy than $\sigma_T \sim (\log s)^2$, where s is the square of the centre-of-mass energy [211]. For exchange of any particle of spin J, the total and elastic cross sections should grow as $\sigma_T \sim s^{J-1}$ and $\sigma_{el} \sim s^{2J-2}$, respectively, exceeding the Froissart bound (and violating $\sigma_{el} \leqslant \sigma_T$) for $J > 1$. Particles of high spin had been observed, but cross sections showed no sign of violating the Froissart bound.

Adapting Regge's result to relativistic scattering, Chew and Frautschi proposed [212] that particles lie on *Regge trajectories*, with specific states corresponding to integer or half-integer values of angular momentum $J = \alpha(s)$. Since all the known trajectories had values of $\alpha(t)$ less than or equal to 1 for the scattering regime $t \leqslant 0$, where t represents the invariant momentum transfer variable, there was no conflict with the Froissart bound. Understanding the approach of total cross sections at high energy to an approximately constant value required the introduction of another trajectory with $\alpha(0) = 1$, called the *Pomeranchuk trajectory* in honour of Pomeranchuk's description [213] of the high-energy behaviour of total cross sections.

The derivation of singularities in the complex angular momentum plane from field theory using dispersion relations was performed by Gribov, Bardakci, Barut and Zwanziger, and Oehme and Tiktopoulos in 1962 [214].

The hypothesis of Regge pole exchange also implies a definite phase of the scattering amplitude. Evidence for this phase confirmed the hypothesis. The result relies on the asymptotic energy dependence of the scattering amplitude and its properties with regard to analyticity and crossing. Regge trajectories, an example of which is shown in figure 9.14, display a high degree of linearity in $s = m^2$, a feature to be discussed in section 9.8.3.

9.5.4.5. Duality and forerunners of string theory. Progress in one area of theory frequently leads to results in another. Thus, attempts to understand the strong interactions eventually led to a candidate theory for quantum gravity.

During the 1960s, it was popular to describe scattering processes at high energies by sums of Regge pole exchanges. For processes with no exchange of quantum numbers, the Pomeranchuk trajectory, or *Pomeron*, for short, plays the dominant role. Differences between various such processes are governed by contributions of the non-leading trajectories. If quantum numbers are exchanged, the Pomeron cannot contribute. For example, in pion–nucleon charge exchange, $\pi^- + p \to \pi^0 + n$,

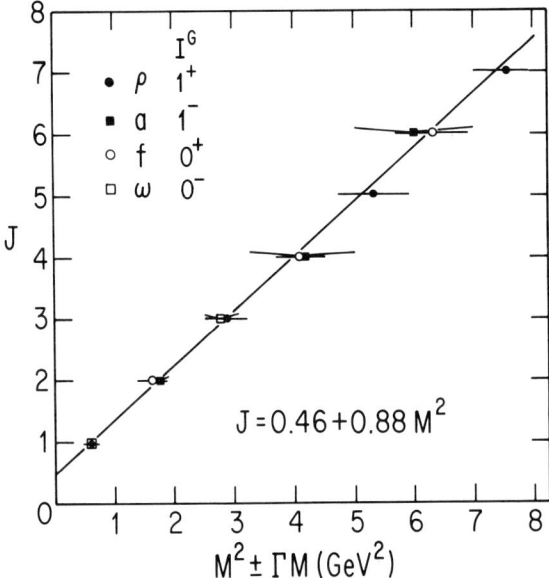

Figure 9.14. *Example of Regge trajectories for non-strange mesons. Points denote values of (squared) mass; error bars denote effects of natural widths. The trajectories for isospins I = 0 and 1 and for positive and negative G-parities (charge-conjugation eigenvalue of the neutral member of the isospin multiplet) are seen to coincide in this case.*

the dominant trajectory is the one on which the ρ-meson lies (see figure 9.14), with energy dependence at high energy well described by ρ trajectory exchange. But what about the behaviour at low energies, when resonances lead to rapid oscillations of the amplitude?

R Dolen, D Horn and C Schmid [215] found that Regge and resonance descriptions of pion–nucleon charge exchange are not to be added to one another, but instead are complementary descriptions of the same physics, the Regge trajectory providing an *average description of the resonant behaviour*. Moreover, the average description of the resonances entails a phase of the scattering amplitude similar to that expected at higher energies from Regge pole exchange. This *duality* between resonances and Regge behaviour applies only to non-Pomeranchuk trajectories [216], and can be visualized in terms of quark graphs [217].

M Ademollo, H Rubinstein, G Veneziano and M Virasoro studied the Dolen–Horn–Schmid duality result for the process $\pi + \pi \to \pi + \omega$, which is particularly simple under crossing symmetry. Over the course of a year they hammered together a model scattering amplitude with increasing features of duality [218]. Finally Veneziano, *en route* by boat from Israel to Italy and his first postdoctoral job at CERN, found a remarkable formula for a scattering amplitude with both poles and Regge behaviour in two kinematic variables [219]

$$A(s,t) \sim B(1-\alpha(s), 1-\alpha(t)) \equiv \frac{\Gamma(1-\alpha(s))\Gamma(1-\alpha(t))}{\Gamma(2-\alpha(s)-\alpha(t))}.$$

The Veneziano amplitude created quite a stir among physicists accustomed to the intractability of the strong interactions. C Lovelace and J Shapiro [220] proposed a model pion–pion scattering amplitude very similar to the one just quoted, with many attractive features, including a low-energy limit in agreement with current algebra results [193]. Attempts to construct a theory yielding the Veneziano amplitude led to the forerunners of string theory [221] and of supersymmetry [222].

String theories of the hadrons were originally viewed as models of the strong interactions, perhaps valid in some limit. One of their seemingly unattractive features was their need to exist in 26 space-time dimensions, rather than the usual four. In 1974 J Scherk and J Schwarz [223] realized that these theories possess a massless spin-2 particle—a good candidate for the graviton—and, together with a few other devotees, began to study them as possible versions of quantum gravity. Then, in 1984, M Green and J Schwarz [224] discovered that supersymmetric versions of certain string theories could exist in 10, rather than 26, dimensions, and the extra six dimensions began to be interpreted in terms of internal symmetries of particles. These *superstring* theories became the subject of intense theoretical activity, which we shall describe further in section 9.13.

All in all, despite the absence of a fundamental theory, much was learned in the 1950s and 1960s from the study of the strong interactions. We shall return to them in section 9.8 in the context of quantum chromodynamics.

9.6. The quark revolution

During the 1960s and early 1970s, particle physics underwent a transformation. Hundreds of strongly interacting resonances and a bewildering hodgepodge of interactions were gradually understood in terms of a few basic building blocks interacting via Yang–Mills fields (to be described in section 9.7). Thus was the physics of quarks and gluons born. In this section we describe the building blocks; we return to the interactions in sections 9.7 and 9.8.

9.6.1. The quark model

9.6.1.1. Triplets as a basis for SU(3). A good deal of the behaviour of a symmetry group such as SU(2) or SU(3) may be learned from its action on its *fundamental* representation, corresponding to a triplet of particles for SU(3). One can build any SU(3) representation out of sufficiently many triplets or antitriplets.

A fundamental triplet, consisting of the proton, neutron and Λ as building blocks, was employed by Sakata [225] for constructing models of elementary particles, and by his colleagues [111] whom we have mentioned earlier in the context of SU(3). However, with the advent of the eightfold way (section 9.4.6.2), the proton, neutron and Λ no longer could be regarded as fundamental; they belonged instead to an *octet* of particles, along with three Σs and two Ξs. Moreover, there appeared to exist a 10-plet of baryons of spin $\frac{3}{2}$, including the Δ(1238), the Σ(1385), the Ξ(1530) and the predicted Ω(1675). Triplets were employed quite early to construct such states by Goldberg and Ne'eman [226].

Could one regard all the baryons (and the emerging multiplets of spin-0 and spin-1 mesons) as composites of more fundamental entities? During a visit to Columbia University in 1963, Murray Gell-Mann proposed taking SU(3) triplets seriously as fundamental subunits of the hadrons [184, 227]. Calling the subunits *quarks* (as in the passage 'Three quarks for Muster Mark' from *Finnegans Wake* [228]), he found that all baryons could be identified with states of three spin-$\frac{1}{2}$ quarks, while mesons could be represented as quark–antiquark pairs. The members of the SU(3) triplet were an isospin doublet (u, d) and a (strange) singlet s. They had to have fractional charges: $Q(u) = \frac{2}{3}$, $Q(d) = Q(s) = -\frac{1}{3}$, but this was a heavy price, since fractionally charged entities had never been observed in Nature. Working independently at CERN, George Zweig developed a picture of the same fractionally charged constituents of matter [229], which he called 'aces'. In the choice of a name, quarks won out over aces, poetry over poker.

Zweig wished to understand why some decays were allowed and others were forbidden. A spin-1 meson known as the ϕ, conjectured to exist [230] as a mixture of a singlet and octet in SU(3), had been observed [231] decaying to K^+K^- and $K^0\bar{K}^0$. Its decay to $\rho\pi$, though allowed by the combined symmetry of charge conjugation and isospin known as *G-parity* [232], seemed suppressed. This fact was hard to appreciate group-theoretically but could be comprehended immediately through quark diagrams. (See figure 9.15.) The suppression of decays in which the initial quarks must annihilate one another rather than appearing in the final particles is now known as *Zweig's rule*.

The quark picture immediately explained why baryons appeared only in singlets, octets and 10-plets of SU(3), since these are the states that can be formed of three triplets. Mesons occurred only in singlets and octets, states that can be formed of a triplet and antitriplet. For states without orbital angular momentum, one could make positive-parity baryons of spin $\frac{1}{2}$ and spin $\frac{3}{2}$, and negative-parity mesons of spin 0 and 1. These coincided exactly with the lowest-mass observed families of strongly interacting particles. Additional orbital angular momentum would yield states of higher spin and even or odd parity.

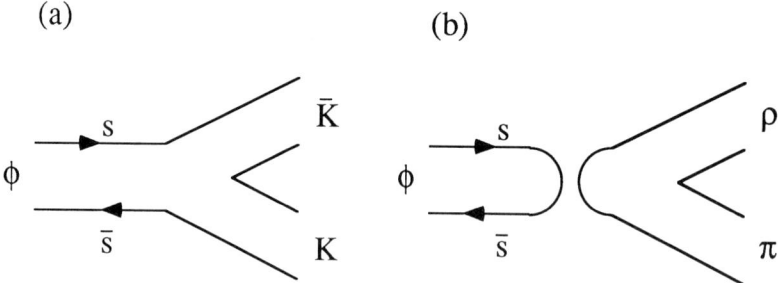

Figure 9.15. Diagrams illustrating Zweig's rule. (a) Allowed decay; (b) forbidden decay.

9.6.1.2. *The spectrum of resonant particles.* The rules for combining quarks into hadrons are simple. For mesons, a quark and antiquark can form states of spin $S = 0$ or 1. For baryons, three quarks can form states of spin $S = \frac{1}{2}$ or $\frac{3}{2}$. These quark spins combine vectorially with an orbital angular momentum L to form states with various total angular momenta J. The intrinsic parities P of these states are also well defined in terms of orbital configurations. Since the relative parities of a quark and antiquark are opposite, the mesons should have parities $P = (-1)^{L+1}$. Indeed, the lowest-lying mesons have $J^P = 0^-$ and 1^-, as predicted for $L = 0$ states. The next-lowest mesons have $J^P = 0^+$, 1^+ and 2^+, as predicted. The lowest-lying baryons (the octet and decimet shown in figure 9.6) indeed have $J^P = (\frac{1}{2})^+$ and $(\frac{3}{2})^+$, as expected for $L = 0$ states, while their first excitations appear to have negative parity and $J = \frac{1}{2}, \frac{3}{2}$ and $\frac{5}{2}$ as expected for $L = 1$ states.

The quark model allowed one to describe masses and magnetic moments of elementary particles with remarkable success [233]. A more algebraic formulation embodying many successes of the quark model was based on the group SU(6) (formed from the product of SU(3) and an SU(2) for quark spin) [234].

There is evidence for a substantial number of orbital excitations and some radial excitations as well. However, the nature of the baryon ground states posed a problem best visualized by referring to the Δ.

9.6.1.3. *The quark statistics problem.* The $\Delta(1232)$ multiplet has isospin $I = \frac{3}{2}$ and spin $J = \frac{3}{2}$. It is a natural candidate for an S-wave state of three quarks. It is symmetric in isospin (containing, for example, the Δ^{++} with three u-quarks). In the absence of orbital angular momentum among the quarks, it is symmetric in spin (containing, for example, the $J_z = \frac{3}{2}$ state with all three quarks aligned along the $+z$ axis). It is symmetric in space if it is the spatial ground state, as is likely, but this behaviour is unacceptable for a state composed of three identical fermions, which must obey the Pauli exclusion principle.

It was shown under certain assumptions [235] that particles of integral spin should obey Bose–Einstein statistics, while particles of half-integral spin should obey Fermi–Dirac statistics. These properties are expressed in terms of commutation relations for boson field operators and anticommutation relations for fermion fields. In 1953, H S Green [236] discovered a generalization of these rules, known as *parastatistics*, based on the assumption of new internal degrees of freedom of the fields. In 1964, Greenberg [237] suggested that quarks obeyed Green statistics, behaving as parafermions of order 3.

The parafermion hypothesis for quarks was equivalent to imagining quarks to occur in three versions, baryons consisting of one of each. A wavefunction antisymmetric in this new degree of freedom was symmetric in all the remaining ones. Greenberg was then able to classify the symmetry of all three-quark wavefunctions, finding a close correspondence with many known baryon states.

A more concrete proposal of an extra degree of freedom for quarks was made by Han and Nambu [238]. The quarks' apparent fractional charges arose from averaging over three species with *integral* charges. Thus, two u-quarks would have charge 1 and one would have charge 0, making an average of $\frac{2}{3}$, while two d-quarks would have charge 0 and one would have charge -1, making an average of $-\frac{1}{3}$. Again, baryons would have one quark of each type.

We shall see the eventual outcome of the quark statistics problem when discussing quantum chromodynamics in section 9.8. The internal 'triplicity' proposed by Greenberg and by Han and Nambu has come to be known as 'colour'. Whereas more 'flavours' of quarks are now known than the u, d and s of the 1960s, the number of colours has remained three. Thus, the quote 'Three quarks for Muster Mark' remains as appropriate as ever.

9.6.2. Deep inelastic scattering

9.6.2.1. Elastic scattering and form factors. The advent of new electron accelerators and sensitive detection techniques after World War II enabled the study of electron elastic scattering on protons and neutrons at momentum transfers large enough to probe nucleon structure [104]. The proton and neutron were found to be structures with radii of the order of 1 fm (10^{-13} cm). Their *form factors* (the Fourier transforms of their charge and magnetic moment distributions [239]) were found to fall off at large momentum transfers q roughly as q^{-4}.

9.6.2.2. Design and operation of SLAC. In the 1950s, planning began for a large linear accelerator to be located west of the Stanford University campus. The persistent efforts of W K H Panofsky and others eventually

Twentieth Century Physics

Figure 9.16. *Aerial view of the Stanford Linear Accelerator Center in 1969.*

led to the construction of the Stanford Linear Accelerator Center (SLAC) (see figure 9.16), which began operation in 1967.

Opinion was divided as to whether the SLAC accelerator was worth the cost and effort [240]. When the machine was conceived, many people thought that the proton form factor would continue to drop with increasing momentum transfer. A similar behaviour was anticipated for each form factor for excitation of specific resonances like the Δ. If true, this behaviour would mean that the high-energy scattering of electrons on protons would prove a barren desert.

9.6.2.3. Discovery and interpretation of scaling. An experiment to study the *inelastic* scattering of electrons on protons was mounted by an MIT-SLAC collaboration. One simply detected the scattered electron, inferring some properties of the final hadronic state from the direction and energy of the electron alone.

Elementary Particle Physics

The results [241] were as surprising as those obtained by Rutherford more than half a century earlier. The number of high-angle scatterings was far in excess of what one expected on the basis of rapidly decreasing form factors, as if the proton itself contained point-like constituents. A 1966 paper by Bjorken [242], predicting such behaviour on the basis of current algebra, had anticipated this viewpoint.

Two relativistically invariant quantities characterize the deep inelastic scattering of a lepton on a target of mass M: the square of the invariant momentum transfer, Q^2, and the product $2M\nu$, where $\nu = E - E'$ is the difference between the initial and final lepton energies E and E' in the laboratory system. Bjorken's result predicted that, aside from well-defined kinematic factors, the scattering could be described in terms of a *scaling* variable $x \equiv Q^2/2M\nu$, i.e. in terms of the *ratio* of the two quantities just mentioned. Observations confirmed this behaviour. The decrease of form factors for elastic scattering and for excitation of specific resonances with increasing Q^2 was compensated by the ability to excite more and more states of high mass.

9.6.2.4. Initial neutrino experiments. In the early plans for Fermilab, it was recognized that deep inelastic scattering of neutrinos on protons and neutrons could play a role complementary to that of the SLAC experiments. Whereas deep inelastic electron scattering is governed by the electromagnetic interactions, and hence is sensitive to charges of the nucleon's constituents, the neutrino reactions known to occur at the time were sensitive to the presence of constituents able to change their charges by ±1 unit.

Two experiments were mounted to study deep inelastic scattering of neutrinos on nucleons in the earliest days of operation of Fermilab: a Harvard–University of Pennsylvania–Wisconsin collaboration [243] (E-1) and a Caltech–Columbia–Fermilab–Rockefeller collaboration [244] (E-21). Experiments at CERN with bubble chambers such as Gargamelle [245], using neutrinos produced at the lower energies of the CERN Proton Synchrotron (PS), also made early studies of deep inelastic neutrino scattering. As in the SLAC experiments, scaling was dramatically confirmed. Moreover, a comparison of the electromagnetic and weak scattering experiments made it possible to determine the charges of constituents [246], confirming the hypothesis advanced several years earlier that the constituents of protons and neutrons had charges of $\pm\frac{1}{3}$ and $\pm\frac{2}{3}$.

9.6.2.5. Parton hypothesis and its successes. Bjorken noted that his current algebra results, which had predicted the scaling behaviour observed at SLAC and in inelastic neutrino interactions, could be interpreted in terms of point-like objects in the proton. Feynman [247] took the idea of point-like constituents seriously, interpreting the scaling

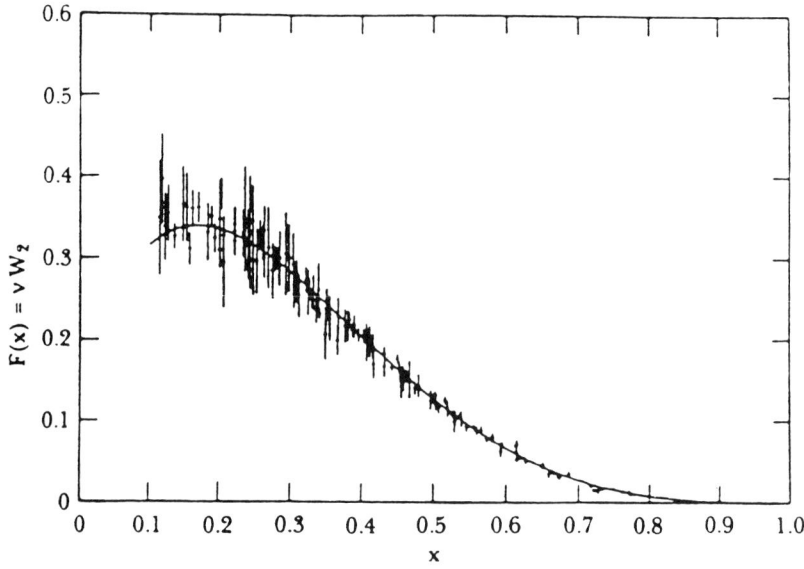

Figure 9.17. *Proton structure function versus the scaling variable x.*

variable $x = Q^2/2M\nu$ as the fraction of a proton's momentum carried by the constituent, or *parton*, that absorbs the momentum and energy from the scattered lepton. A test for the spin of the constituents proposed by C Callan and D Gross [248] soon supported the idea that the point-like objects in the proton indeed were quarks.

A systematic exploration of the parton hypothesis, undertaken by Bjorken and Paschos [249], developed a consistent description of the momentum distributions of partons in protons and neutrons for all deep inelastic scattering data known at the time. Those distributions are parametrized by *structure functions*, an example of which is shown in figure 9.17 (from reference [240] p 152).

Structure functions are also probed when a parton in one hadron annihilates an antiparton in another to produce a lepton pair. That process, first proposed by Drell and Yan in 1971 [250], still plays a major role today in the study of parton distributions and in tests of QCD, and is central to the production of W- and Z-bosons to be discussed in section 9.7. Another effect of point-like constituents in hadrons, predicted by Berman, Bjorken and Kogut [251] in 1971 and observed in a series of experiments at CERN and Fermilab [252], was the production of hadrons at high transverse momenta as a result of parton–parton collisions.

9.6.3. Electron–positron annihilation

The cross section for $e^+ + e^- \to$ hadrons can be normalized by defining a ratio $R \equiv \sigma(e^+ + e^- \to \text{hadrons})/\sigma(e^+ + e^- \to \mu^+ + \mu^-)$. In the quark

Elementary Particle Physics

model this ratio measures just the sums of the squares of the produced charges, and would be expected to increase when the energy crosses the threshold for new kinds of quarks. For example, just above the threshold (about 300 MeV) for the production of pairs of pions which are composed of up and down quarks, this ratio should be $R = 3(q_u^2 + q_d^2)/q_\mu^2$, where q_u, q_d and q_μ (the charge of the up quark, the down quark, and the muon) are $+\frac{2}{3}$, $-\frac{1}{3}$ and -1, respectively. The factor '3' represents the number of quark 'colours', while the muon occurs in only one variety. R should equal $\frac{5}{3}$ if these daring conjectures are correct, i.e. if the assigned quark charges are correct and if quarks have three colours. (The presence of sharp resonances in the cross section prevents this average behaviour from being realized in practice.) As the energy of the colliding particles crosses the threshold for strange quark production as manifested by pairs of kaons, at about 1 GeV, the ratio R should rise to 2. As resonances become broader and start to overlap at higher energies, this average behaviour should be observable.

The study of electron–positron collisions was pioneered by several groups starting in the late 1950s (see section 9.10.9), leading to the construction of machines at Stanford, Orsay (France), Frascati (Italy) and Novosibirsk (USSR). In the late 1960s, a group at Frascati brought into operation the ADONE collider, with a centre-of-mass energy of up to 3 GeV, the first capable of studying the process $e^+ + e^- \rightarrow$ hadrons above the region where prominent resonances led to fluctuations in the cross section. This machine, with an energy sufficient to produce pairs of u, d and s-quarks, should have yielded $R = 2$. Results were consistent with this, though with wide errors [253], as shown in figure 9.18.

In an attempt to reach higher collision energies, the Cambridge Electron Accelerator, or CEA, was adapted for electron–positron collisions in the early 1970s (see section 9.10.9). Above the threshold (slightly below 4 GeV) for 'charmed' quark production ($q_c = +2/3$), the quark model predicted that R should rise to $3\frac{1}{3}$. However, when higher energies of 4 and 5 GeV were reached at CEA, the value of R was found to rise to around 5 [254]! These results were eventually confirmed by the Stanford electron–positron collider SPEAR (sections 9.9.1, 9.10.9). New physics in the form of a τ-lepton (section 9) had unexpectedly appeared, the decay products of the τ being counted as hadrons. (The values of R in figure 9.18 have this contribution subtracted out.) It took some years to straighten out this initially confusing situation [255], but eventually quark counting by means of the ratio R proved a very useful technique.

9.6.4. *Searches for free quarks*

The absence of free quarks was a continuing source of concern to many. Searches for fractionally charged particles (for a review, see reference [256]) were mounted as soon as proposed; even the early observations of Millikan were examined with quarks in mind. Claims

Twentieth Century Physics

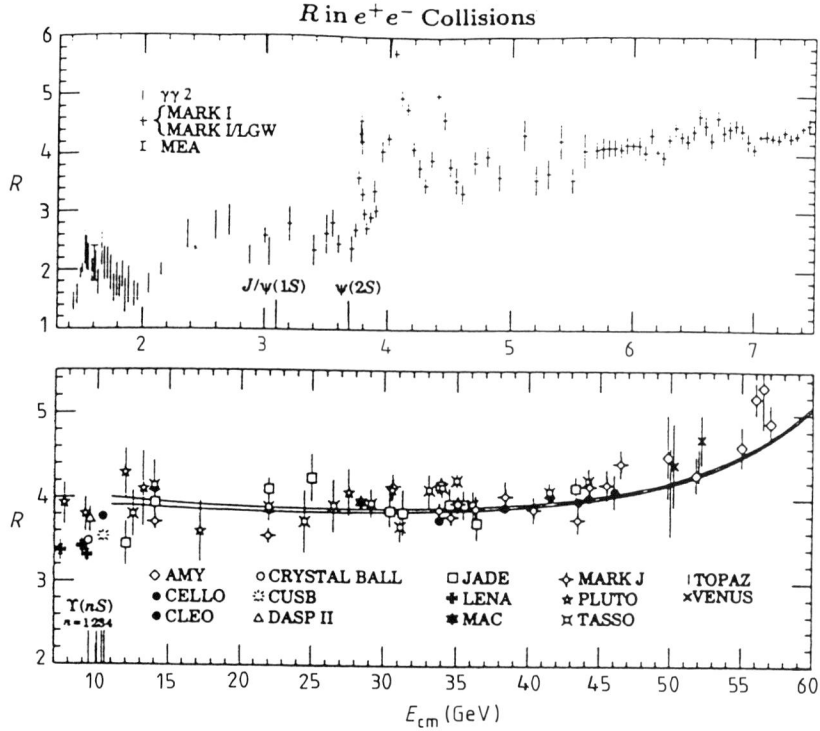

Figure 9.18. Behaviour of the ratio $R \equiv \sigma(e^+e^- \to hadrons)/\sigma(e^+e^- \to \mu^+\mu^-)$ as a function of centre-of-mass energy E_{CM}.

for detection of free quarks in cosmic rays [257] were not supported by further evidence [256]. Physicists looked in such diverse places as oysters (where substances with odd chemical properties might be concentrated [258]) and samples from the Moon [259]. Excitement over free quarks peaked in the late 1970s with experiments of Fairbank and collaborators [260] which seemed to show fractional charges on niobium balls heated on a tungsten substrate. However, other experiments [256, 261] failed to see the effect, which was eventually understood to result from inhomogeneities on the surfaces of the balls.

To this day, free quarks have not been observed. Quantum chromodynamics provides a plausible reason why they should be permanently confined in hadrons, as we shall see in section 9.8.

9.7. Electroweak unification

9.7.1. A theory of leptons

9.7.1.1. Yang–Mills fields. As mentioned in section 9.5.1.6, Yukawa [43] and subsequent authors [143 155] suggested that the weak interactions

Elementary Particle Physics

stem from exchange of a spin-1 boson, in analogy with the exchange of photons in electromagnetism. A crucial step laid the ground for a self-consistent theory based on this idea. In 1954 C N Yang and R L Mills proposed [113] that isotopic spin symmetry arises in the same way as electromagnetic gauge invariance. Their theory of fields which interact with *one another* as well as with external matter turned out to serve not only to unify electromagnetic and weak interactions, but also to describe strong forces. (In this context see also reference [262].)

The following discussion owes much to an exposition by Yang [263]. In electromagnetism, consider a slowly moving test charge undergoing a virtual displacement by a four-vector dx_μ. The change in the phase of its wavefunction (aside from any $p \cdot x$ contribution, where p is its four-momentum) is $eA^\mu(x)dx_\mu$ (in units with $\hbar = c = 1$), where $A^\mu(x)$ is the vector potential and e is the electric charge of the particle in question.

According to *local gauge invariance*, the phase convention for the particle can be set independently at each space-time point x, corresponding to changing the vector potential by the divergence of a scalar quantity which depends on x. This *local gauge transformation* changes only the difference in phases at the endpoints when a particle is taken along a path from one point to another. When the endpoints coincide, so that the particle is taken on a voyage along a *closed path* in space-time, this phase difference vanishes.

Every experiment corresponds to either taking a particle on a closed path in space-time, or comparing the results of two *different* paths from a point x_1 to another x_2. Thus, one can take a particle through either of two slits or around either side of a solenoid (as in a classic experiment proposed by Aharonov and Bohm [246]). One measures not $A^\mu(x)$, but the spatial integral of the *field strength* $F^{\mu\nu} \equiv \partial^\nu A^\mu - \partial^\mu A^\nu$ over an area enclosed by the particle's closed path in space-time (or by the area between the two paths from x_1 to x_2). The field strength is invariant under a local gauge transformation. Since changes of phase commute with one another, such a gauge theory is said to be *abelian*. The group of phase changes is U(1), the unitary group in one dimension.

Yang and Mills' generalization of local gauge invariance involves not only transformations of phase, but local rotations in isotopic spin space. Thus, the quantity $A^\mu(x)$ induces isotopic spin transformations, and acts as a matrix in isospin space. Its dimension depends on the representation chosen for isotopic spin: two-dimensional for proton and neutron, three-dimensional for π^+, π^0 and π^- and so on. Since rotations about different axes do not commute with one another, the corresponding field strength has an additional term

$$F^{\mu\nu} \equiv \partial^\nu A^\mu - \partial^\mu A^\nu + g[A^\mu, A^\nu].$$

Here g is the gauge coupling strength, analogous to the electric charge e. Such a theory, said to be *non-abelian*, has fields which interact with

693

one another as a result of the additional term in $F^{\mu\nu}$. The isotopic spin group is denoted by SU(2); the 2 stands for the dimension of its smallest non-trivial representation.

9.7.1.2. Glashow's model. The first attempts [143, 155] to unify the charge-changing weak interactions with electromagnetism assumed that charged intermediate bosons (the Ws, mentioned in section 9.5.1.6) together with the (neutral) photon formed an SU(2) triplet. However, the Ws couple only to left-handed fermions, while the photon couples to both left- and right-handed fermions. In order to allow for this difference, Glashow [265] extended the SU(2) to an SU(2) × U(1) group. Gauge bosons corresponding to SU(2) would consist of W^+, W^- and a neutral W^0, while an additional neutral boson (B^0 in today's notation) would correspond to U(1) gauge transformations. The W^0 and B^0 would be mixed by an interaction also giving rise to masses of charged Ws. Two linear combinations of fields then emerged: the massless photon γ, with field $A^\mu(x)$, and a *massive* neutral vector boson Z with field $Z^\mu(x)$

$$A = B\cos\theta + W^0 \sin\theta \qquad Z = -B\sin\theta + W^0 \cos\theta.$$

The mass of the Z was predicted in terms of θ and the mass of the W by $M_Z = M_W / \cos\theta$. Weak processes involving exchange of a Z ('neutral-current interactions') were thus predicted by this theory.

9.7.1.3. Symmetry breaking and the Higgs mechanism. The origin of W and Z masses remained unexplained in Glashow's theory. In an electroweak theory based on a *renormalizable* Yang–Mills theory (one in which divergent quantities could be defined away by redefinitions of fundamental parameters like masses and coupling constants), it appeared that gauge bosons had to remain massless. The *ad hoc* introduction of gauge-boson masses destroys renormalizability, gauge invariance or both. How could one break the SU(2) × U(1) symmetry without spoiling the attractive properties of the theory?

It became apparent in the late 1950s and early 1960s that symmetries could be manifested in two ways (see section 9.5.3.1). In the *Wigner–Weyl* realization, symmetry is manifest through degeneracies in the spectrum. In the *Nambu–Goldstone* mode [179, 180], the vacuum breaks the symmetry but the equations of motion retain it. In this mode, each broken symmetry operation corresponds to a massless, spinless particle, known as a *Nambu–Goldstone boson*.

In a theory with massless vector bosons, such as electromagnetism, only two polarization states are allowed: the two directions transverse to the boson's direction of propagation, or, equivalently, left and right circular polarization. A massive spin-1 particle has, in addition, a longitudinal polarization state. It was noted by J Schwinger,

P W Anderson, Peter Higgs and others [266] that in a gauge theory containing Nambu–Goldstone bosons, such bosons could act as the needed third component of a massive vector boson. One can describe the massless scalar particles as being 'eaten' by the gauge bosons, which then become heavy. This process is now known as the *Higgs mechanism*.

9.7.1.4. The Weinberg–Salam model. The Higgs mechanism was soon employed in the service of electroweak unification by Weinberg and Salam [267, 268] (see figure 9.19). The simplest version of the theory introduced only an SU(2) doublet (ϕ^+, ϕ^0) and its complex conjugate $(\phi^-, \bar{\phi}^0)$. With an appropriate self-interaction of the fields, the combination $\eta \equiv (\phi^0 + \bar{\phi}^0)/\sqrt{2}$ would acquire a non-zero vacuum expectation value v, thereby breaking SU(2) × U(1) down to U(1) of electromagnetism as desired. The fields $\phi^\pm$ and $(\phi^0 - \bar{\phi}^0)/\sqrt{2}$, corresponding to the Nambu–Goldstone bosons of the broken symmetry, would be 'eaten' by the $W^\pm$ and Z, while the difference $H \equiv \eta - v$, often referred to as 'the' Higgs boson, would remain in the spectrum as a massive particle.

9.7.1.5. Renormalizability. The Weinberg–Salam theory remained a curiosity for several years, referred to very little even by its inventors. It predicted unobserved *neutral weak currents*, such as neutrino interactions in which a neutrino rather than a charged lepton emerged, and strangeness-changing decays such as $K^+ \to \pi^+ \nu \bar{\nu}$. Limits on such decays were particularly stringent, lying far below the level of ordinary weak interactions. That is why Weinberg entitled his model 'A theory of leptons'.

It was, furthermore, not clear at the outset that the Higgs mechanism solved the problem of the renormalizability of the electroweak theory, until the proof was supplied in 1971 by Gerard 't Hooft [269], a student of Martinus Veltman in Utrecht (see reference [270] for some historical perspectives on this work) and extended by Benjamin W Lee and Jean Zinn-Justin [271].

The proofs of the self-consistency of the Glashow–Weinberg–Salam electroweak theory were a major factor in its acceptance, and sent the particle physics community into a state of excitement. Weinberg immediately realized the significance of the result [272]. Early calculations of rates for neutral-current processes began to appear [273]. Discussions with experimentalists began in earnest: had neutral currents been overlooked? Variants of the model without neutral currents but with new particles were proposed [274]. Attempts were renewed to extend the theory to hadrons; these will be discussed in section 9.7.3.

9.7.2. Experimental confirmation of neutral currents
With the growth of interest in the Glashow–Weinberg–Salam theory in the early 1970s, the search for weak neutral currents entered a more

Twentieth Century Physics

(a)

(b)

(c)

Figure 9.19. *(a) S Glashow; (b) S Weinberg; (c) A Salam.*

Elementary Particle Physics

serious phase [275]. Previous upper bounds on such effects were re-examined in many processes. Particularly stringent bounds appeared to exist for reactions initiated by neutrinos.

9.7.2.1. Neutrino scattering. The Gargamelle collaboration at CERN had been investigating deep inelastic neutrino scattering during the late 1960s and early 1970s. Its limits on neutral current events were reported by Perkins at the 1972 International Conference on High Energy Physics [246]. Deep inelastic neutrino scattering was also beginning to be studied by experiments at Fermilab, as mentioned in section 9.6.2.4.

Neutrinos were produced at accelerators mainly from the decays $\pi \to \mu\nu$ and $K \to \mu\nu$, and hence were of the muonic type. When interacting in matter via the charged current, they gave rise to hadronic showers and to a clearly identified muon in the final state. The presence of a muon rather than an electron was the basis of the claim that the muon and electron neutrinos were distinct (see section 9.5.1.6). Occasional hadronic showers without an accompanying muon, observed even in the earliest neutrino experiments [276], were usually ascribed to contamination of the beam with neutral particles (particularly neutrons). One could check this possibility by studying in detail the distribution of events with respect to distance along the detector or inward from the lateral boundaries. Neutron-induced events would become rarer as either distance increased, while the rate for neutrino-induced events would be independent of distance.

The likelihood that the Gargamelle Collaboration was seeing neutral-current events was discussed informally at CERN in the early months of 1973. First, however, a detailed check of backgrounds had to be made [277]. Once the experimenters were satisfied that neutrons and other backgrounds could not be responsible for their signal, they announced the discovery of neutral currents in weak interactions of neutrinos. The effects were seen both in deep inelastic scattering [278] and in a 'golden' neutrino–electron elastic scattering event [279]. The E-1A Collaboration at Fermilab also saw deep inelastic neutral-current interactions of neutrinos [280]. Their signal was present at an early stage, but an attempt to understand it better by reconfiguring the detector led to a temporary loss of the effect [281].

How could neutral-current effects have been overlooked for so long? One source of the difficulty in identifying them can be seen in figure 9.20. Define R_ν and $R_{\bar{\nu}}$ as the ratios of neutral-current to charged-current events for deep inelastic neutrino scattering. The predictions of the Glashow–Weinberg–Salam theory as a function of the angle θ (section 9.7.1.2) yield a nose-like curve relating R_ν and $R_{\bar{\nu}}$. The present data, shown as the plotted point [282], sit at the bottom of the 'nose'. The ratio R_ν is expected to be quite small for a large range of values of $\sin^2\theta$, while the ratio $R_{\bar{\nu}}$, though larger than R_ν, is about as small as it

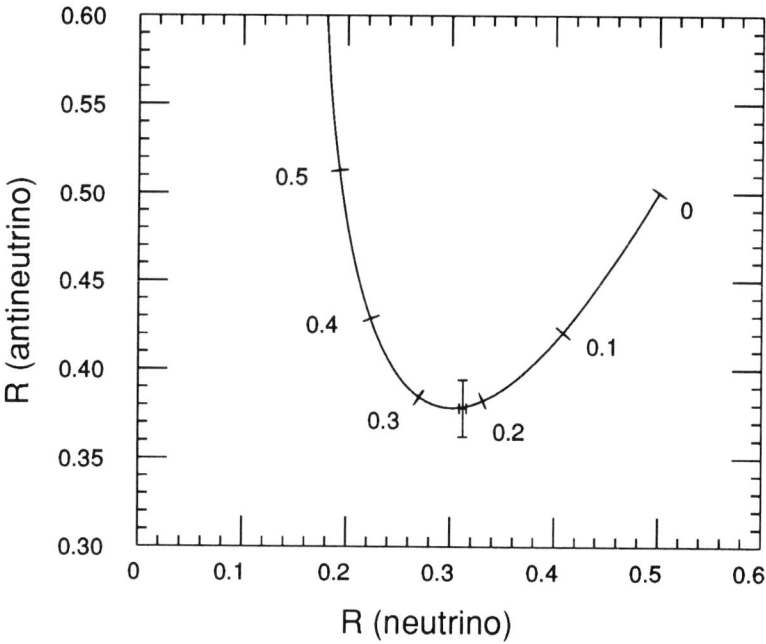

Figure 9.20. *The neutral-current to charged-current ratios R for deep inelastic scattering of neutrinos and antineutrions, plotted parametrically as functions of $\sin^2\theta$ (labels on curve). The plotted point shows an average of recent data.*

can be. Neutral currents were also expected (and eventually found [283]) in *elastic* neutrino–proton scattering.

9.7.2.2. Parity violation in electron–deuteron scattering. The Glashow–Weinberg–Salam theory also predicted neutral-current weak processes involving charged leptons. These processes are normally masked by the much stronger electromagnetic interactions. However, in contrast to electromagnetism, the neutral weak force does not conserve parity. Despite its weakness, its effects can be observed through *interference* of the electromagnetic and weak interactions.

See also p 665

A collaboration led by V W Hughes and C Y Prescott searched at SLAC for the predicted parity violation in deep inelastic scattering of polarized electrons [284]. A difference in interactions of left-circularly polarized and right-circularly polarized electrons would confirm the predicted effect. It was necessary to prepare a source of polarized electrons, and to arrange their arrival at the target with desired (and known) polarization. The choice of target was also important: the predicted effect turns out to be much smaller in hydrogen than in deuterium, so both were investigated.

The results were announced in 1978. The electron had neutral weak interactions, just as the theory had predicted. One could observe the interference of photon and Z^0 exchange. This confirmation was a major source of excitement at the 1978 International Conference on High Energy Physics in Tokyo [285].

9.7.2.3. Parity violation in atoms. A fundamental violation of parity in the electron–nucleus interaction can lead to such effects as optical rotation in atoms [286]. Initial experiments (chronicled in the first of references [284] and the reviews of reference [287]) did not see the expected effects, leading to some bizarre variations on the simplest version of the theory. Eventually laser stability was brought under control, and calculations of atomic physics effects to high orders were performed. The first experiments to claim the observation of parity violation were performed in atomic bismuth, with other groups who had previously not seen a signal then finding one.

More recent experiments [288] have attained ever-improving accuracy in measuring atomic parity violation. The most recent results in caesium by C Wieman's group, when taken in conjunction with atomic physics calculations [289], not only confirm the electroweak theory at better than the 3% level, but provide a useful constraint on physics beyond the standard model [290].

9.7.2.4. Other neutral-current processes. Many other neutral-current effects have been observed over the years since they were first predicted. They include forward–backward asymmetries in electron–positron reactions, parity violation in polarized-electron scattering on nuclei other than hydrogen and deuterium, detailed studies of $\nu_\mu e$ and $\bar{\nu}_\mu e$ elastic scattering and interference between charged and neutral currents in $\nu_e e$ and $\bar{\nu}_e e$ elastic scattering. An excellent source for tracking how these data have steadily improved is the series of reviews of reference [291].

9.7.3. Extension to hadrons and the charm hypothesis
The Cabibbo theory of strange particle decays, discussed in section 9.5.3.4, can be re-expressed in terms of quarks and W-bosons. A charged W can be emitted or absorbed in transitions between a u-quark and the linear combination $d' \equiv d \cos\theta_C + s \sin\theta_C$, with $\sin\theta_C \approx 0.22$. The charge-conjugate couplings also exist.

The charged currents carrying a left-handed u-quark into d' can be taken as part of a triplet. The properties of the neutral member of this triplet can be deduced by recalling the commutation relations for SU(2). The raising and lowering operators $J_\pm$ of angular momentum, for example, are related to the third component J_3 by $[J_+, J_-] = 2J_3$. The normalization of the currents then is specified by the property $[J_3, J_\pm] = \pm J_\pm$. But a neutral current defined in this way contains strangeness-changing pieces. How could these be avoided?

Each charged lepton (electron and muon) couples to its own variety of neutrino by the charge-changing current (the analogue of $J_\pm$). The corresponding neutral current is then *diagonal* in neutrino species. Moreover, even if one were to define linear combinations of neutrinos coupled to electrons and muons, neutral currents are avoided as long as these linear combinations are orthogonal to one another.

A quark–lepton analogy, drawn in 1964 by Gell-Mann [184], Bjorken and Glashow [292], Y Hara [293], and Z Maki and Y Ohnuki [294], proposed that as long as each charged lepton had its own neutrino, the d-quark and s-quark should not have to share the u-quark in weak transitions. Instead, one combination $d' = d\cos\theta_C + s\sin\theta_C$ would couple to u-quarks, while the orthogonal combination $s' = -d\sin\theta_C + s\cos\theta_C$ would couple to a new quark, which Bjorken and Glashow called *charmed*. This charmed quark c was to be heavier than the u-quark, but, like it, would have charge $\frac{2}{3}$. The neutral currents as defined by the commutator of two charged currents then are automatically diagonal in flavour.

Several years later, Glashow, J Iliopoulos and L Maiani realized that not only did the Bjorken–Glashow hypothesis guarantee that neutral currents preserve strangeness in lowest order, but strangeness-changing neutral-current effects in higher-order weak processes were also drastically suppressed [295]. The mixing between a neutral kaon and its antiparticle is one such process. Since this mixing appears to be no stronger than the product of two first-order weak transitions (as one would have, for example, in the two-step process $K^0 \leftrightarrow \pi\pi \leftrightarrow \bar{K}^0$), it was possible to deduce an upper limit to the mass of the charmed quark. Glashow, Iliopoulos and Maiani (GIM) predicted that the charmed quark lay below about 2 GeV/c^2 and proposed ways to search for it.

The GIM mechanism was applied by Gaillard and Lee [296] to a systematic study of rare decays of kaons, setting the stage for experimental investigations that continue to this day. Many suggestions for finding charmed particles began to be made [297], particularly after evidence for the predicted neutral currents began to appear. The role of a charmed quark in the cancellation of *triangle anomalies* (appearing in higher-order electroweak calculations) was stressed [298].

9.7.4. Experimental confirmation of charm

9.7.4.1. Early hints. Charmed particles were taken seriously very early in Japan. Some of the initial proposals of quartet models of hadrons had been made by Japanese authors [293, 294]. An event seen in emulsion by K Niu and his colleagues at Nagoya University was interpreted in 1971 as a possible candidate for a charmed particle [299]. It appeared to decay to $\pi^+\pi^0$ with a mass of about 2 GeV/c^2 and a lifetime of about 10^{-14} s. Niu's event (and others like it) suggested that quarks always occurred in pairs: u with d and c with s. Kobayashi and Maskawa [300] recognized that with a third pair one could describe CP violation. Both of the quarks they proposed have now been seen (section 9.9).

9.7.4.2. Discovery of the J/ψ and the charmonium spectrum. An early study of the Drell–Yan process (section 9.6.2.4) by Leon Lederman and collaborators observed a shoulder in the effective mass spectrum of muon pairs [301], which could have been interpreted as a new particle viewed with poor mass resolution. Stimulated by experience in the study of lower-mass dilepton pairs and by this result, Samuel C C Ting and collaborators mounted an experiment to measure the dilepton mass spectrum with mass resolution far better than ever achieved before. The experiment consisted in colliding an intense beam of protons from the Brookhaven Alternating-Gradient Synchrotron (AGS) with a beryllium target, and observing electron–positron pairs with a two-arm spectrometer with approximately 30° between the two arms. Čerenkov counters filled with hydrogen served to identify electrons, with bending magnets and eleven planes of proportional chambers providing precise momentum measurements.

By September of 1974, it was clear to Ting's group that they had indeed come upon something either very wrong or very exciting. A peak in the e^+e^- effective mass was showing up at 3.1 GeV/c^2 with almost no background on either side. While a number of cross-checks were performed, Ting spoke of his result discreetly to colleagues, who generally urged him to publish it [240]. Meanwhile, on the West Coast, studies of electron–positron annihilations with the SPEAR detector (section 9.6) had been continuing. Cross sections had been measured in steps of 0.2 GeV in centre-of-mass energy. The values obtained at 3.2 and 4.2 GeV seemed a bit high in comparison with those at other energies. The cross section was remeasured in those regions in June 1974, but no structure was noticed.

The SLAC–LBL Collaboration resumed its scrutiny of the energy-dependence data in October 1974. The cross section at 3.1 GeV, upon closer examination, seemed not to be reproducible. Two out of ten runs at that energy gave much larger values than the others. Measurements were resumed over the weekend of 9 and 10 November in order to check the anomaly.

Ting arrived at SLAC on 11 November for a programme committee meeting with the news of his discovery [302]. The peak he and his colleagues were seeing, dubbed the 'J' (figure 9.21), was almost free of background. He was greeted with news of an electron–positron annihilation cross-section peak seen in the Mark I detector (figure 9.22) by the SLAC–LBL Group, who called their effect [303] the ψ. The dual name J/ψ has persisted for the particle with mass 3.1 GeV/c^2. Within ten days, the SLAC-LBL group had raised the beam energy of SPEAR,

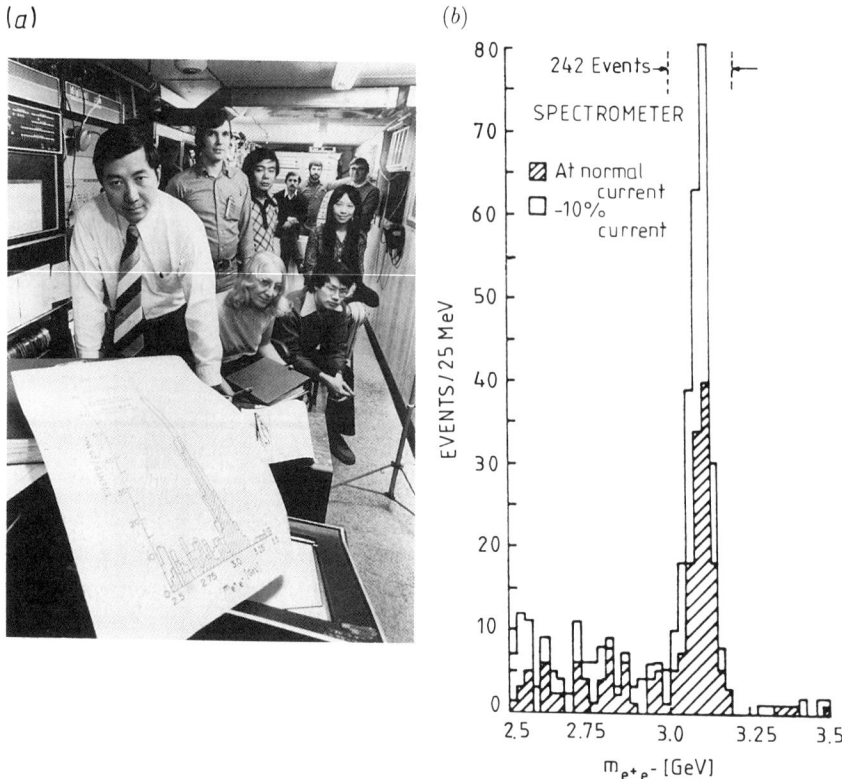

Figure 9.21. (a) S C C Ting and his group, with a plot of the J resonance. (b) The mass spectrum of e^+e^- pairs, showing the J peak [302].

finding another bump—the ψ'—at 3.7 GeV [304]. Electron–positron rings at Frascati in Italy and Hamburg in Germany were able to confirm the existence of the ψ in a matter of days [305].

The J/ψ and ψ' could be viewed as the ground state and first radially excited state of a charmed quark and a charmed antiquark with zero relative orbital angular momentum and parallel spins. In analogy with the name *positronium*, for the bound state of an electron and a positron, this system was called *charmonium*. Several sets of authors published interpretations of the new particles based on the charm hypothesis [306]. However, some questions had to be settled before that interpretation could be regarded as established and alternative schemes (see, for example, references [307]) laid to rest.

9.7.4.3. Early confusion in the search for charm. The charm hypothesis predicted P-wave $c\bar{c}$ states lying between the J/ψ and the ψ'. The ψ' should be able to decay to them electromagnetically. A detector at

Elementary Particle Physics

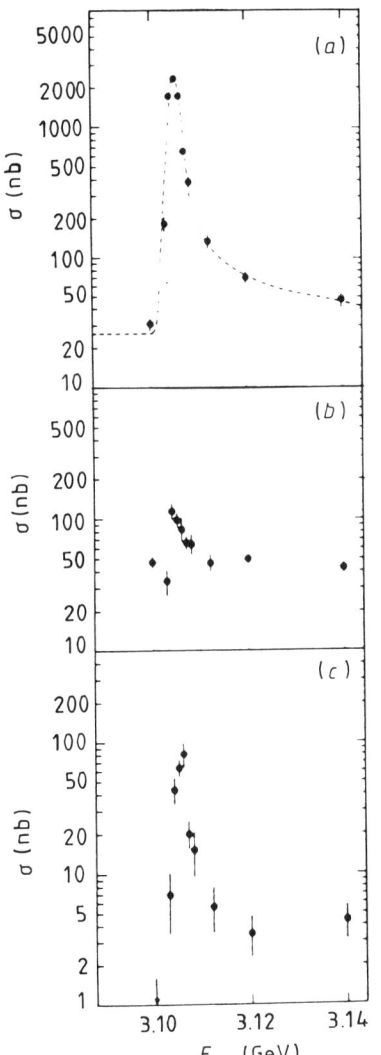

Figure 9.22. Left: the cross section for e^+e^- annihilations to various final states as measured by the Mark I detector [303]: (a) multihadrons, (b) e^+e^-, (c) $\mu^+\mu^-$. Right: the Mark I Detector with Roy Schwitters.

SPEAR (located at the other side of the ring from the SLAC–LBL set-up) initially did not see these transitions. They were first found at the DORIS storage ring, and confirmed at SLAC [308].

Mesons were expected containing one charmed quark and one light (u or d) antiquark, decaying to such final states as $\bar{K}\pi$. Early searches were not turning up such decays [309].

703

The cross section for $e^+e^- \to$ hadrons was indeed increasing above a centre-of-mass energy of 4 GeV in a manner indicating new particle production. However, the increase was actually *too large* for what was expected on the basis of charm. Moreover, an increase in the number of kaons, expected since the charmed quark's dominant decay should be to strange quarks, was not materializing.

The resolution of this confusion is traced in references [255]. Not only were charmed particles being produced in pairs, but a new lepton, the τ, was also being formed in pairs at roughly the same energy. Many of the signals for charm were counterbalanced by signals of the new lepton! It was not until the identification of the new lepton [310] that the 'inclusive' signals for charm could be sorted out.

9.7.4.4. Observation of charmed particles. By the spring of 1976, the absence of a signal for charmed mesons in the SPEAR experiment was becoming a source of concern. Calculations by De Rújula, Georgi and Glashow [311] indicated that the lightest charmed mesons should lie between 1.8 and 1.86 GeV/c^2, and the lightest charmed baryons should have a mass of about 2.2 GeV/c^2. The predictions for baryons were borne out by the discovery of two candidates for charmed baryons, a Σ_c^{++} (= uuc) decaying to a Λ_c^+ (= udc) and a pion, in a single neutrino interaction [312]. Observation of additional leptons in neutrino experiments [313] and direct leptons in hadronic reactions [314] also hinted strongly at the existence of charmed particles. Gerson Goldhaber and F M Pierre combed through the Mark I data to see if a charmed-meson signal was lurking there. Indeed it was; the lightest charmed meson, called the D (for doublet [292]) was found [315] in both predicted charge states, $D^0 = c\bar{u}$ and $D^+ = c\bar{d}$, with branching ratios close to those originally estimated in reference [295].

Glashow had exhorted a conference of meson spectroscopists in 1974 to go out and find charm [316]. If they did not do so by the next conference, he would eat his hat. If 'outlanders' (non-meson-spectroscopists) found charm, the spectroscopists were to eat their hats. The organizers of the next (1977) conference graciously distributed candy hats to all participants [317].

9.7.4.5. Spectroscopy of charmonium and charmed particles. The study of charmonium and charmed particles has come a long way since the mid-1970s. The charmonium spectrum (see figure 9.23) is already richer than that of positronium, while numerous non-strange and strange charmed mesons and baryons have been discovered. These systems, the lightest for which ideas of perturbative quantum chromodynamics (section 9.8) begin to hold, thus form one of the testing grounds for the new understanding of the strong interactions that arose in the 1970s.

BOX 9B:
THE DISCOVERY OF CHARMED PARTICLES

Glashow, Iliopoulos and Maiani [295] described how the puzzling absence of strangeness-changing neutral currents could be neatly explained if there existed a new fourth quark, the charmed quark. Immediately experimentalists started to ask: 'How can we find it? What would be the experimental signature of such an object?'. Gradually, the details were fleshed out [297,316]. By the summer of 1974, the masses of charmed mesons and baryons were predicted, as well as their decay modes and their branching ratios. When the J/ψ was discovered in November 1974 it fitted neatly into the scheme as the $c\bar{c}$ analogue of the ϕ_s, the vector meson. The search for charmed mesons and baryons began at laboratories around the world.

At hadron accelerators, searches were started for associated production of charmed mesons and baryons in analogy with strange-particle production more than 20 years earlier, mainly using the predicted decay of the pseudoscalar charmed meson, $D^0 \rightarrow K^-\pi^+$, as an identifying signature. The SLAC group, working with the e^+e^- collider detector, also started a search for charmed particles. However, the first clear-cut example of a charmed baryon came from a bubble-chamber event, produced by a high-energy neutrino [312]. The search for charm at SPEAR was complicated by the appearance of the τ lepton. Not until this confusion was unravelled was the D-meson clearly identified [315] in 1976, a year and a half after the discovery of the J/ψ.

In the meantime, searches for charmed particles at the hadron machines had been singularly unsuccessful, with only upper limits for the production cross sections. The charmed particles, even though produced in great numbers, were swamped by the high attendant backgrounds in hadronic interactions. It was not until the development of silicon strip detectors with their excellent spatial resolution and tolerance for high counting rates that charmed particles were detected at hadron machines [318]. This approach was highly successful. Almost immediately samples of data at the electron–positron colliders, previously counted in tens, were counted in thousands.

At present, experiments at hadron machines are recording millions of events carrying charm. The study of charmonium itself has been extended via proton–antiproton annihilation to states unavailable in e^+e^- collisions.

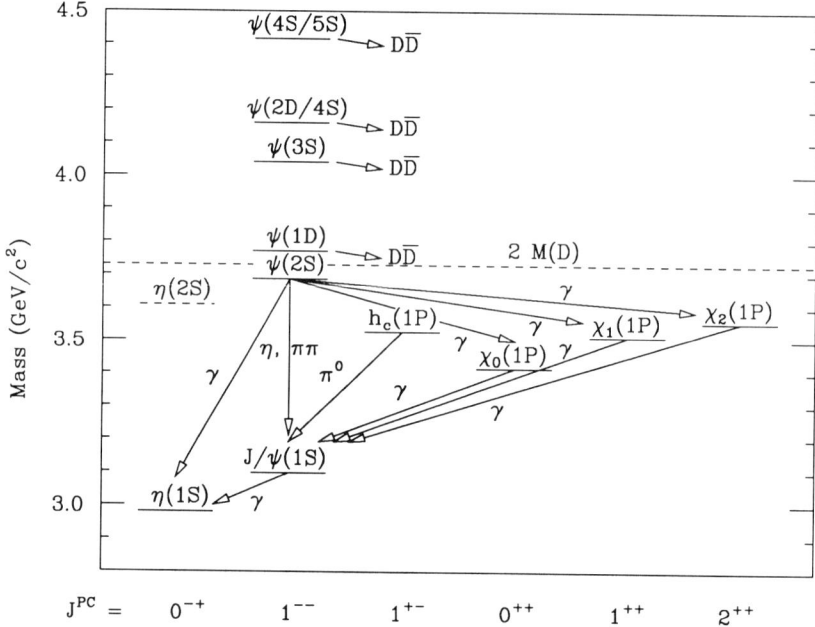

Figure 9.23. The charmonium spectrum. Observed and predicted levels, as quoted in reference [319], are denoted by solid and broken horizontal lines, respectively. Arrows are labelled by particles emitted in transitions.

9.7.4.6. Charmed-particle decays. Since charmed quarks couple to strange quarks with a strength proportional to the cosine of the Cabibbo angle (section 9.7.2), the lifetimes of charmed particles are very short, ranging between about 10^{-14} and 10^{-13} s. The *differences* between these lifetimes arising from the strong interactions are a topic of much current interest.

9.7.5. The W and Z

9.7.5.1. Discovery. In a theory with an intermediate vector boson W, the Fermi coupling constant $G_F = (1.16639 \pm 0.00002) \times 10^{-5}$ GeV^{-2} can be re-expressed as a constant of order one times the square of a coupling constant divided by the square of the W mass. The mass thus follows from the value of the coupling constant.

Early searches for W-bosons, whose coupling constant was unknown, explored mass ranges of a few GeV [160, 320], as mentioned in section 9.5.1.6. Part of the motivation for studies of deep inelastic neutrino scattering at Fermilab was to search for indirect effects of W exchange (see, for example, references [243] and [244]). Sensitivity could thus be achieved to masses up to several tens of GeV.

The Glashow–Weinberg–Salam electroweak theory relates the Fermi coupling constant, the SU(2) coupling strength g, and the W mass M_W by $G_F/\sqrt{2} = g^2/8M_W^2$. Moreover, the relation between g and the electric charge e depends on the angle θ: $e = g \sin\theta$. The combined results predict $M_W = 37.2$ GeV$/\sin\theta$ (with $\alpha = e^2/(4\pi\hbar c) = 1/137$ for the electromagnetic fine-structure constant).

The study of deep inelastic neutrino scattering and parity-violating interactions of polarized electrons in the late 1970s gradually yielded a value of $\sin\theta$ slightly lower than $\frac{1}{2}$, implying a W mass above 75 GeV$/c^2$. The electroweak theory also predicted the Z mass to be slightly higher than that of the W: $M_Z = M_W/\cos\theta$. With these estimates of W and Z masses, one could anticipate the cross sections for their production in quark–antiquark annihilations. Proton–proton or proton–antiproton collisions with centre-of-mass energies of several hundred GeV would provide quarks with energies sufficient to produce the desired particles. Schemes to collide protons with antiprotons at Fermilab and in the CERN Super Proton Synchrotron (SPS), to be described in greater detail in section 9.10.10.2, began taking shape in the late 1970s. A major step in each project was the collection of sufficient antiprotons in a narrow range of space and momentum. Important steps in this 'cooling' procedure were taken by Simon van der Meer, working at CERN [321]. The effort required to turn the CERN SPS into a proton–antiproton collider was coordinated by Carlo Rubbia [322]. First collisions were seen in 1982. The Fermilab project, requiring the construction of a ring of superconducting magnets, came into operation several years later.

Two detectors were constructed at CERN to study the debris of collisions. The UA1 Collaboration (UA stands for Underground Area) was led by Rubbia, while the spokesman for UA2 was Pierre Darriulat. In January of 1983 the first signals of W production [323] were seen by both groups: an electron of high transverse momentum, with apparent imbalance of transverse momentum in the opposite direction, as expected for a W decaying to an electron and a neutrino. Within six months, both groups also reported the observation of the Z (figure 9.24), decaying to pairs of charged leptons [324].

The W and Z masses are slightly higher than anticipated on the basis of the lowest-order theory (with $\alpha = 1/137$). At the short distances characteristic of the W and Z Compton wavelengths, the vacuum polarization effects mentioned in section 9.3 predict $\alpha \approx 1/128$. With this simple correction, the W and Z masses are in remarkable accord with theory.

9.7.5.2. W and Z properties in hadron colliders. The CERN discoveries of the W and Z were followed by studies of their production and decays with ever increasing accuracy, by the UA1 and UA2 Collaborations and by the Collider Detector Facility (CDF) at Fermilab (figure 9.25).

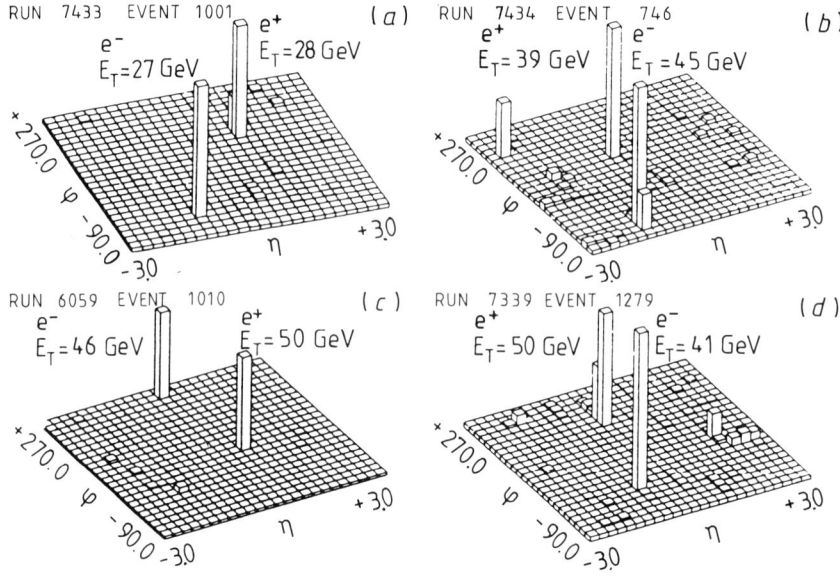

Figure 9.24. Early evidence for Z production from the first of references [324]. Heights of bar graphs denote energy transverse to the beam deposited in various regions of the UA1 detector.

A precise determination of the Z mass was made [325], shortly before electron–positron colliders began to refine that quantity further. Production cross sections were measured; comparison of lepton yields in W and Z decays yielded indirect measurements of the ratio of W and Z widths [326], helping to specify the open W decay channels and thereby place an indirect lower bound on the top quark's mass. We shall return to the top quark in section 9.9.

Recently the D0 Detector (figure 9.25) has begun operation at Fermilab. (The name refers to its location around the accelerator ring.) Its fine-grained calorimetry and large angular coverage aim at a precise measurement of the W mass.

9.7.5.3. Results from SLAC and LEP on the Z. The Z mass was predicted by the electroweak theory to be about 90 GeV/c^2. At SLAC and CERN, plans began to develop in the late 1970s and early 1980s to build electron–positron colliders ('Z factories') with at least that energy in the centre of mass.

The Stanford Linear Collider (SLC) involved upgrading the SLAC linear accelerator to a beam energy of 50 GeV, adding a positron source, installing positron and electron cooling, and building two arcs to bring the electrons and positrons into collision (see figure 9.26 (top)). The first results [327], obtained in 1989, included precise measurements of the Z mass and width. More recently, collisions of longitudinally polarized electrons with positrons have provided a precise measurement of $\sin^2\theta$ through the difference between cross sections for left-handed and right-handed electrons [328].

In the LEP (Large Electron–Positron) Collider, lying on the French–Swiss border at the base of the Jura Mountains (see figure 9.26 (bottom) and reference [330] for an early progress report), positrons and electrons injected from the CERN site are detected at four symmetrically placed locations around the ring. The acronyms for the detectors are ALEPH (Apparatus for LEP Physics), DELPHI (Detector with Lepton, Photon and Hadron Identification), L3 (internal CERN numbering of experiment) and OPAL (Omni-Purpose Apparatus for LEP).

LEP measured the Z width to an accuracy establishing its invisible decay modes to be the three known pairs of neutrinos $\nu_e\bar{\nu}_e$, $\nu_\mu\bar{\nu}_\mu$, and $\nu_\tau\bar{\nu}_\tau$. Precise measurements of the Z mass, of its total width, of branching ratios to various final states and of asymmetries of various sorts, test not only the electroweak theory, but higher-order corrections stemming from the top quark and Higgs boson [331]. The top quark was thereby anticipated to have a mass below 200 GeV/c^2, as indeed transpired [332] (section 9.9).

9.8. Quantum chromodynamics

9.8.1. Early suggestions of colour-triplicity

In section 9.6 we mentioned briefly that quarks appear classified by three 'colours', as evidenced by several circumstances.

9.8.1.1. Quark statistics. In baryons, the product of the quarks' spin, space and 'flavour' (u, d, s, etc) wavefunctions appears to be symmetric under interchange of any two quarks. Since quarks have spin $\frac{1}{2}$ and thus should obey Fermi statistics, one would expect their total wavefunction to be antisymmetric under interchange of any two quarks. This goal is achieved by adding a new degree of freedom in which all the quarks in a baryon are antisymmetric.

According to Greenberg [237] and Han and Nambu [238], the three-quark structure of baryons suggested a new SU(3) symmetry, under which quarks would transform as triplets (3), while all known hadrons would be singlets. Since

$$3 \times \bar{3} = 1 + 8 \qquad 3 \times 3 \times 3 = 1 + 8 + 8 + 10$$

the known mesons would be the singlets of $3 \times \bar{3}$ (i.e. of quark–antiquark pairs), while the baryons would be the singlets of $3 \times 3 \times 3$ (three-quark states). Nambu [333] showed it plausible that the only states manifest

Twentieth Century Physics

(a)

(b)

Figure 9.25. *(a) The CDF detector; (b) the central calorimeter modules of the D0 detector; (c) a recent aerial view of the Fermilab site.*

Elementary Particle Physics

(c)

Figure 9.25. *Continued*

in Nature are singlets, and discussed the possibility that this new SU(3) corresponded to a gauge theory.

9.8.1.2. Decay of the neutral pion. The decay $\pi^0 \to \gamma\gamma$ was attributed by Steinberger in 1949 [69] to a diagram involving a virtual proton (see figure 9.27). Nearly 20 years later, a similar calculation (involving quark loops instead) was set on a firmer footing by Adler, Bell and Jackiw. The amplitude for this process was found to be proportional to the number of quarks travelling around the loop, indicated to be three by the experimental $\pi^0 \to \gamma\gamma$ rate [334].

9.8.1.3. Cross section for electron–positron annihilations. The study of electron–positron collisions (see section 9.6.3) provided a further confirmation of three quark 'colours'. The cross section $\sigma(e^+ + e^- \to \text{hadrons})$ is proportional to the number N_c of quark colours. Once the charges of quarks produced at any given energy were understood, the data indicated that $N_c = 3$.

9.8.2. Requirements of a gauge theory of strong interactions
Murray Gell-Mann and Harald Fritzsch [335] stressed the colour-triplet nature of quarks in 1972, summarizing the evidence for a threefold degree

Twentieth Century Physics

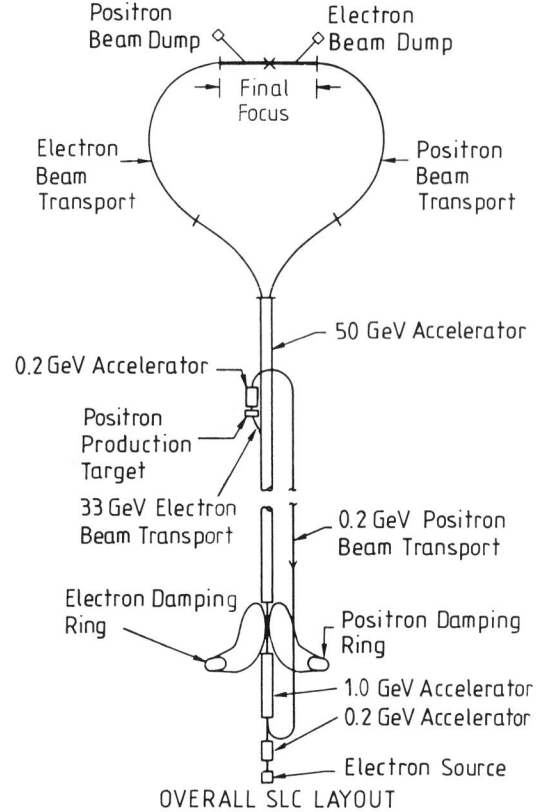

Figure 9.26. Top: sketch of the Stanford Linear Collider, from reference [329]. Bottom: photograph of the LEP site. The small and large solid circles indicate the position of the SPS and LEP rings, while the dotted line denotes the French–Swiss border.

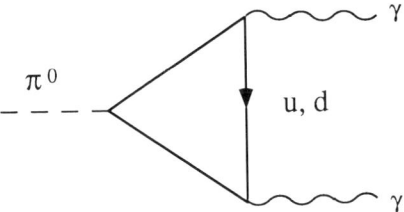

Figure 9.27. *Graph responsible for the decay $\pi^0 \to \gamma\gamma$. The original calculation by Steinberger [69] involved protons instead of u- and d-quarks travelling around the loop.*

of freedom and emphasizing the need for a gauge theory of the strong interactions which would not destroy the successes of current algebra.

A vector-like interaction, similar to that of electromagnetism, was favoured, strong enough at large distances to bind quarks into hadrons yet weak enough at short distances to let the quarks appear quasi-free in deep inelastic scattering. Theories like electromagnetism behave differently. Their vacuum polarization effects (see section 9.3) augment the strength of electromagnetic interactions at short distances. The stage was thus set for a systematic examination of quantum field theories suitable to describe the strong interactions.

9.8.3. Asymptotic freedom and infrared slavery

Vacuum polarization effects in Yang–Mills theories were first calculated correctly, using non-covariant methods, by Khriplovich in 1969 [336]. The self-interaction of the Yang–Mills field led to an important contrast with quantum electrodynamics, the force in a Yang–Mills theory becoming *weaker* at short distances as required for a theory of the strong interactions.

Khriplovich's result seems to have escaped general attention until after its rediscovery in the early 1970s. At that time, several people began examining Yang–Mills theories as candidates for the strong interactions [337]. Two manifestly covariant calculations of the coupling strength's dependence on distance appeared in 1973 [338]. A student of Sidney Coleman at Harvard, H David Politzer, and David Gross and his student Frank Wilczek at Princeton both found a result for SU(N) gauge theories which relates a *running coupling constant* α_N measured at a momentum scale μ_1 to its value at another scale μ_2

$$[\alpha_N(\mu_1)]^{-1} = [\alpha_N(\mu_2)]^{-1} + \frac{1}{4\pi}\left(\frac{11}{3}N - \frac{2}{3}n_f\right)\log\frac{\mu_1^2}{\mu_2^2}.$$

Here n_f is the number of fermion species contributing to the vacuum polarization for a gauge boson. This result made use of the *renormalization group* concept, which specifies the dependence of quantum field theory parameters on changes of scale [339].

The coefficient of the logarithm depending on N stems from gauge-boson loops. Its positive sign means that the interaction in a Yang–Mills theory weakens at large momentum scales (equivalently, at short distances), a property dubbed *asymptotic freedom*. In order for it to hold, the number of fermions n_f in the loop must be small enough for the term proportional to N to dominate.

For an abelian gauge theory such as electromagnetism, the term proportional to N is absent; fermion loops contribute a negative coefficient to the logarithm, and the interaction becomes *stronger* at large momentum scales or short distances.

The *long-distance* behaviour of the coupling constant also differs markedly in abelian and non-abelian theories. In an abelian theory, since the coupling strength ceases to 'run' at scales μ much lower than the mass of the lightest fermion contributing to vacuum polarization, one can measure a well-defined charge at long distances. In non-abelian theories, the logarithmic dependence on μ associated with gauge-boson loops eventually leads the coupling constant to diverge at some small value of μ. The interaction then becomes strong, and perturbation theory ceases to be valid.

It was proposed quite early that the Yang–Mills theory of strong interactions would lead to interaction energies proportional to the interquark separation; such a force between quarks and antiquarks at long distances would keep them from ever being torn apart from one another. This suggestion leads [340] to families with angular momentum J linear in M^2, and is supported by the spectrum of particles lying along the highest Regge trajectories (cf section 9.5.4.4 and figure 9.14). As a counterpart to asymptotic freedom, the confinement of quarks by an ever-increasing potential has sometimes been called *infrared slavery*.

The theory of the strong interactions based on a Yang–Mills quantum field acting on the colour degree of freedom has come to be known as *quantum chromodynamics* or QCD. The quanta of QCD are called *gluons*, since they provide the 'glue' holding hadrons together.

9.8.4. Scaling violation in deep inelastic scattering

Although deep inelastic lepton scattering experiments (section 9.6) reveal point-like constitutents in the proton, the details of nucleon structure depend slightly on the momentum transferred to the target by the leptons. Figure 9.28 shows modern data [49] illustrating this behaviour, predicted by QCD. Quarks can emit or absorb gluons, thereby shifting their momenta. A proton probed with very high momentum transfers thus appears to have 'softer' and more numerous constituents. The scaling predicted by Bjorken is violated slightly, to a degree that sheds light on the strong coupling constant $\alpha_s(\mu)$ (the subscript denotes 'strong') at any chosen momentum scale μ. Several quantitative treatments of this feature appeared in the mid-1970s, starting with the

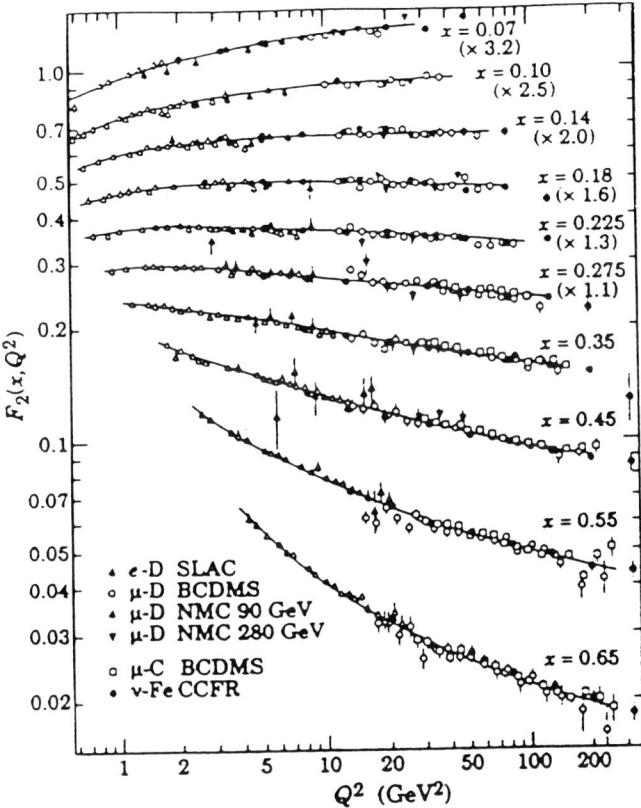

Figure 9.28. *Non-scaling behaviour of structure functions.*

initial discoveries of asymptotic freedom and expressed transparently in the work of Altarelli and Parisi [341]. Since many measurements of α_s are performed at the mass of the Z-boson, a convenient reference point is $\mu = M_Z$, even though the deep inelastic scattering experiments probe lower momentum scales. A recent analysis of deep inelastic scattering [342] finds $\alpha_s(M_Z) \approx 0.12$, consistent with other determinations.

9.8.5. Jets and other high-$p^\perp$ phenomena

An early application of the parton picture, mentioned in section 9.6.2.5, predicted [251] hadron production at high transverse momenta, confirmed by experiments at the CERN ISR (Intersecting Storage Rings) and Fermilab [252]. The behaviour of cross sections at high transverse momenta indicated that the constituents of colliding protons were interacting with one another at a fundamental level.

Direct evidence for gluons emerged from the study of three-jet events in electron–positron annihilations [343], as detailed in reference [240]. In quantum chromodynamics, the process $e^+ + e^- \to$ hadrons evolves in two stages. First (essentially instantaneously), a quark–antiquark (q$\bar{\text{q}}$) pair is produced by the electromagnetic current. Then, over a longer time scale, these quarks materialize into hadrons through production of additional quark–antiquark pairs, with transverse momenta small with respect to the original quarks. The original quark and antiquark thereby define a direction for two *jets* of particles, which become more and more clearly defined with increasing energy. Because QCD interactions grow stronger at momentum scales below 1 GeV/c, it took several GeV in the centre of mass to begin seeing two-jet events. Such jets were identified in e^+e^- annihilations [344] and hadronic collisions [345] in the mid-1970s.

At the 1978 International Conference on High Energy Physics in Tokyo, a much-discussed particle was the Υ (section 9.9), a composite of the fifth (b) quark and its antiquark in the same way that the J/ψ is a charm–anticharm bound state. Had the three-gluon decay of the Υ been seen? It was not clear whether the data were really showing three distinct gluon jets, or just effects of data selection. The higher energies of two new electron–positron colliders, PETRA in Hamburg and PEP at Stanford, allowed a much crisper search for gluon jets. QCD predicted a *third* jet, corresponding to the emission of a gluon by one of the final-state quarks, in a fraction of e^+e^- annihilations into hadrons. This gluon jet should be identifiable if emitted at a sufficient angle with respect to one of the quarks.

A technique for identifying gluon jets [346] was proposed by Wu and Zobernig, members of the TASSO Collaboration at PETRA. A pretty picture of a three-jet event (see figure 9.29) was presented at international conferences during the early summer of 1979 [347]. At the Lepton–Photon Symposium at Fermilab in August 1979, all four groups working at PETRA presented evidence for three-jet events.

As available centre-of-mass energies increased, spectacular particle jets at large angles with respect to the initial beams were observed at the CERN SPS Collider [348] and the Fermilab Tevatron [349]. The decays of Z-bosons to hadrons seen at LEP have yielded vast samples of events with three, four and even more jets. The rate for $n+1$ jets is related to that for n jets by one power of α_s, permitting an estimate $\alpha_s(M_Z) \approx 0.12 \pm 0.01$ from multi-jet production rates compatible with that from deep inelastic scattering.

9.8.6. Other applications

9.8.6.1. Quarkonium decays. Bound states of charmed quarks such as the J/ψ, mentioned in section 9.7, provided the first laboratory for application of perturbative QCD to decays [350]. These studies were helped greatly by the discovery of the fifth quark b and of the corresponding b$\bar{\text{b}}$ bound states such as the Υ (section 9.9). One can

Elementary Particle Physics

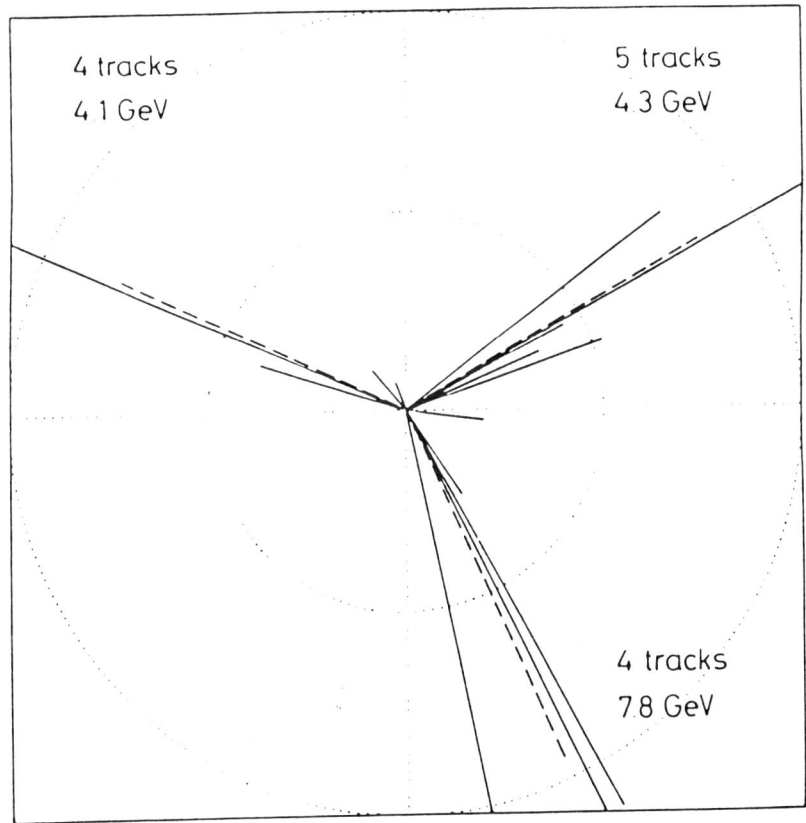

Figure 9.29. *Sketch of particle tracks in an early three-jet event in electron–positron annihilation.*

measure $\alpha_s(m_b)$, for instance, by comparing decays such as $\Upsilon \to 3g$ (g = gluon) and $\Upsilon \to 2g + \gamma$. With higher-order QCD corrections, the result [342] is $\alpha_s(m_b) \approx 0.19$, corresponding to $\alpha_s(M_Z) \approx 0.11$.

9.8.6.2. Hadron production in electron–positron annihilations. The lowest-order QCD correction to the ratio R characterizing hadron production in electron–positron annihilations (section 9.6.3) amounts to a factor $1 + (\alpha_s/\pi)$. Data taken over a wide range of centre-of-mass energies, ranging from several GeV to M_Z, are indeed consistent with this correction.

9.8.6.3. Counting rules. Differential cross sections at high energies and fixed angles [351] behave in ways which may be derived from QCD by tracing the flow of high momentum transfers through diagrams for scattering at the quark level.

717

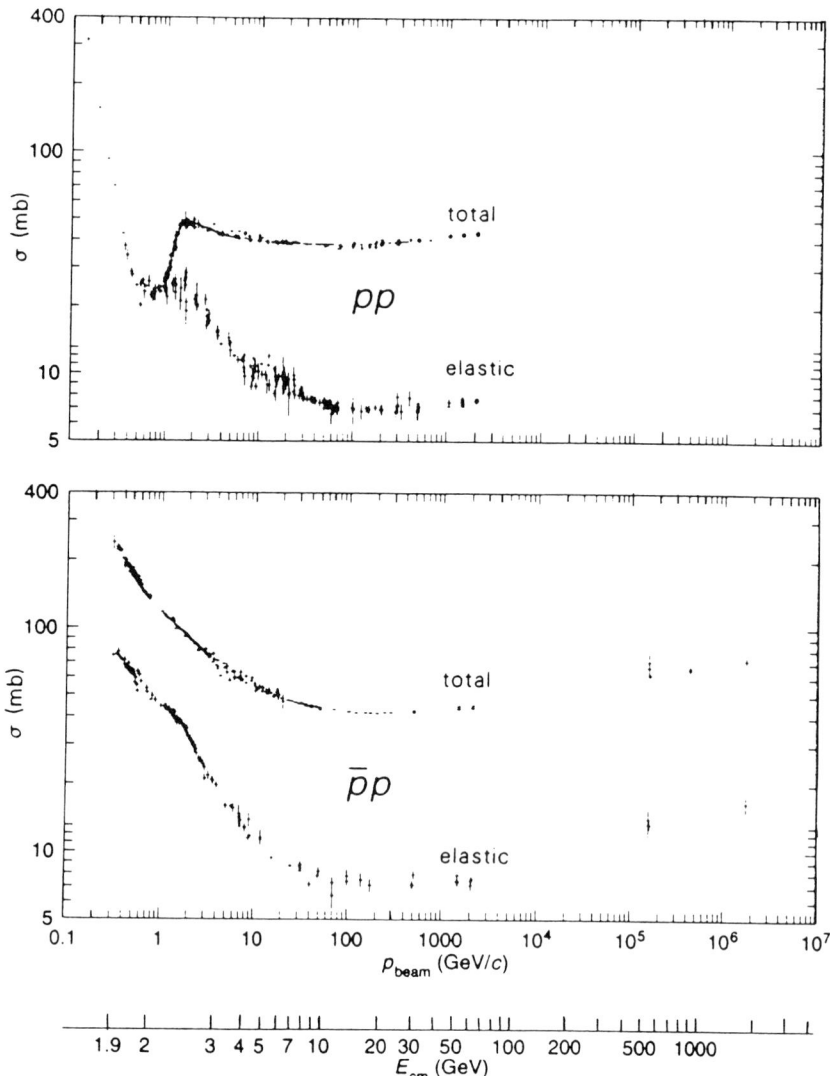

Figure 9.30. *Total and elastic pp and $\bar{p}p$ cross sections as functions of energy.*

9.8.6.4. Non-perturbative effects. Not all phenomena in high-energy collisions are understood quantitatively within the framework of QCD. Scattering at small momentum transfers is still most economically described within the framework of Regge poles (section 9.5.4). The Pomeranchuk trajectory may reflect exchanges of a pair (or more) of gluons [352]. The rise of total cross sections at high energies [353] is a topic of interest, with recent data [354] summarized in figure 9.30.

Multiple particle production [355] in hadronic interactions displays simple regularities understood mainly on a phenomenological basis.

The *rapidity* of a particle may be defined as $y \equiv (\frac{1}{2})\log[(E + p_z)/(E - p_z)]$, where E is its energy and p_z its momentum along the beam axis. A hadronic collision leads to a number of particles per unit rapidity growing only slightly with increasing energy. A qualitative understanding of this behaviour exists on the basis of quark fragmentation [356].

QCD has thus proved able to describe a wide variety of strong-interaction phenomena. Although QCD only permits perturbative calculations at energies and momenta above one or several GeV, this aspect should not be held against its validity. Methods for dealing with the non-perturbative regime, including lattice gauge theories [357] and QCD sum rules [358], are being actively pursued.

9.9. Three families of quarks and leptons

9.9.1. The τ-lepton

9.9.1.1. Discovery. The repetitive structure of the leptons (electron and its neutrino, muon and its neutrino) stimulated searches for additional 'sequential' doublets, consisting of a heavy charged lepton and its neutrino, with negative results up to the early 1970s [359]. Around that time, several theorists [360] worked out the consequences of heavy charged leptons. The process $e^+ + e^- \to l^+ + l^-$, under scrutiny at new electron–positron colliders, was one way to produce these leptons.

The Cambridge Electron Accelerator (see section 9.6.3) reported in the summer of 1973 that the ratio $R = \sigma(e^+ + e^- \to \text{hadrons})/\sigma(e^+ + e^- \to \mu^+\mu^-)$ at centre-of-mass energies of 4 and 5 GeV exceeded the value of 2 expected for u-, d- and s-quarks [254]. These results were confirmed by the SPEAR Collider later that year [361]. The charmed quark, discovered in 1974 (section 9.7), accounted for part of the rise in R. It contributed $\frac{4}{3}$ to R, leading to a total of $3\frac{1}{3}$. However, the SPEAR results above 4 GeV indicated values of R well in excess of 4.

Other curious features of the SPEAR data at and above 4 GeV appeared not to be attributable to charmed-particle production. Events with an electron, a muon of opposite charge, no other charged particles, and missing energy suggested production of a pair of new leptons: $e^+ + e^- \to \tau^+ + \tau^-$, with one of them decaying to $e\nu\bar{\nu}$ and the other to $\mu\nu\bar{\nu}$. This signature had been sought earlier at the ADONE electron–positron collider at Frascati [359]. However, in retrospect, the energy was too low for τ pair production.

In 1975, Martin Perl and his collaborators published a paper entitled *Evidence for the existence of anomalous lepton production*, which they concluded with the statement 'A possible explanation for these events is the production and decay of new particles each having a mass in the range of 1.6 to 2 GeV/c^2.'. Two years later they were able to quote a mass of 1.80 ± 0.045 GeV/c^2 for the τ-lepton [310]. Initial failure to detect a crucial decay mode, $\tau \to \pi\nu$, at the DORIS storage rings at DESY in Hamburg led to some scepticism as late as 1977, but confirmation of this and other modes by the DESY experiments followed soon thereafter [362].

The τ was a truly new discovery in not having been anticipated by any direct theoretical prediction. It was the first member of a third family of quarks and leptons.

9.9.1.2. *Properties.* The τ decays to a ν_τ (whose existence remains inferred, having not been detected directly) and a virtual W-boson, materializing into $u\bar{d}$, $u\bar{s}$ (with reduced rate), $e\bar{\nu}_e$ or $\mu\bar{\nu}_\mu$. Decay rates for these final states are consistent with the standard weak-interaction theory, leading to an overall lifetime of about 0.3 ps. The ratio of the $u\bar{d}$ rate to the $e\bar{\nu}_e$ or $\mu\bar{\nu}_\mu$ rate, a factor of three times a small correction, provides further evidence for three colours of quarks. Reference [363] provides a sample of the physics that may be learned from the increasing precision in the measurement of τ properties.

9.9.2. *The fifth quark*

9.9.2.1. *Kobayashi and Maskawa's suggestion and its implications.* Spurred by hints of charmed particles in nuclear emulsions [299] (section 9.7.4.1), M Kobayashi and T Maskawa [300] asked 'As long as there seemed to be two families of quarks and leptons, why not three?'. A third family of quarks would permit the parametrization of CP violation by introduction of a non-trivial phase in the charge-changing weak couplings of quarks.

The search for the charmed quark had been partly motivated by the existence of two families of leptons, a family structure being particularly appropriate for cancelling triangle anomalies [298] (section 9.7.3). Thus, a third doublet of quarks would imply a third doublet of leptons and vice versa. Following the τ's discovery, it was then natural to expect a third quark of charge $-\frac{1}{3}$ (in addition to d and s) named b for 'bottom,' and a third quark of charge $\frac{2}{3}$ (in addition to u and c) named t for 'top' (in analogy to 'down' and 'up'). The quantum number carried by the b-quark has been referred to as 'beauty'.

Some false alarms and harbingers of the true signal marked the search for the third family of quarks. An apparent anomaly in deep inelastic antineutrino scattering [364] suggested that a threshold for production of the b-quark had been crossed. (This effect, not confirmed in other experiments, was later understood in terms of instrumental bias.) A search at Fermilab for particles heavier than the J/ψ decaying to lepton pairs yielded an apparent peak ('Υ') at 6 GeV in the e^+e^- channel, not confirmed in the $\mu^+\mu^-$ channel. However, a peak at 9.5 GeV/c^2 in that channel tempted John Yoh, then a postdoctoral fellow working on the experiment, to put a bottle of Mumm's champagne labelled '9.5' (to be opened presently) in the group's refrigerator [365]. Another group, also studying muon pairs at Fermilab, had a single event at 9.5 GeV/c^2, labelled affectionately 'Big Mac'. The $\mu^+\mu^-$ spectrum in the 1976 experiment of reference [366] showed a small anomaly around 10 GeV/c^2.

9.9.2.2. *Discovery of the upsilon family.* The study of hadronically produced leptons at Fermilab evolved through several stages, culminating in Experiment E-288, dedicated to the study of high-mass lepton pairs, under the leadership of Leon Lederman. After the false alarm of the Υ at 6 GeV in 1976 and a fire in the spring of 1977 whose damage was quickly repaired, the group began running with high intensity and improved resolution in the late spring of 1977. The data soon proved that John Yoh's bottle of champagne had been correctly labelled. Not only was there a peak at 9.5 GeV/c^2, but another smaller one seemed to be riding on its tail, about 0.6 GeV/c^2 higher [367] (see figure 9.31(a)), in a manner reminiscent of charmonium, for which the ψ' lies about 0.6 GeV/c^2 above the J/ψ. In contrast to charmonium, however, there appeared to be *three* narrow peaks [368]. Called $\Upsilon(1S)$, $\Upsilon(2S)$, and $\Upsilon(3S)$, they correspond to the first, second and third S-wave systems of a heavy quark bound to the corresponding antiquark. (Figure 9.32(b) shows a recent spectrum.)

The partial width of the Υ for decay into pairs of leptons, measured at the DORIS storage rings [369], suggested that this particle consisted of quarks with charge $-\frac{1}{3}$, a conclusion strengthened upon observation of the second peak [370], the Υ', whose hadronic parameters were predicted with greater confidence. Once the threshold for production of mesons containing a single heavy quark was passed, the ratio R in e^+e^- rose by $\frac{1}{3}$, confirming the new quark's charge of $-\frac{1}{3}$ and identifying it with the b. The Υ and its excited levels were $b\bar{b}$ states.

9.9.2.3. *Beauty mesons.* The 12 GeV Cornell Electron Synchrotron, first operated in 1967, was converted in the late 1970s to the Cornell Electron Storage Ring (CESR), with beam energies of up to 8 GeV for electrons and positrons. This machine could not have come at a more opportune time, detecting the $\Upsilon(1S)$, $\Upsilon(2S)$ and $\Upsilon(3S)$ in short order. At higher energy a fourth Υ level was seen, with a natural width [371] indicating its decay into pairs of b-flavoured mesons, now called B-mesons. These new mesons were first identified by the presence of a lepton and additional kaons (corresponding to the weak decay of a b-quark) in their decay products [372]. Reconstruction of specific decay channels [373] yielded a B-meson mass around 5.28 GeV/c^2. Since the threshold for production of a pair of B-mesons is twice this value, the $\Upsilon(4S)$, at a mass of 10.575 GeV/c^2, is thus ideal for producing pairs of B-mesons nearly at rest.

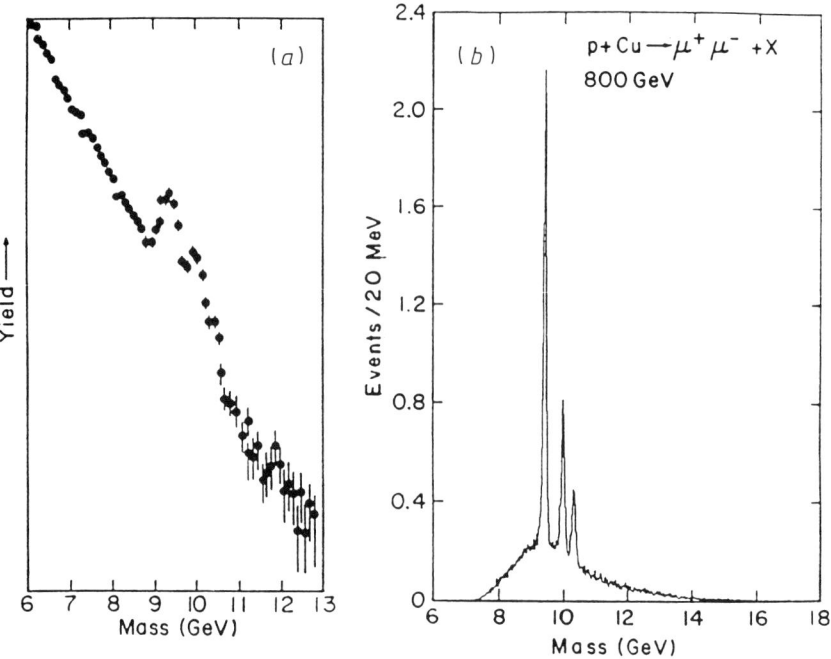

Figure 9.31. *Evidence for the* Υ *resonances in hadronic interactions.* (a) *From reference* [367]; (b) *a recent spectrum from the second of references* [368].

9.9.2.4. Spectroscopy of the b-quark. The present state of our knowledge about b$\bar{\text{b}}$ levels is summarized in figure 9.32. At least six sets of S-wave levels and two groups of P-wave levels have been identified, all spin-triplets thus far. Their electromagnetic transitions to spin-singlet levels require a b-quark spin to flip, and are expected to occur with rates below present levels of sensitivity.

The b$\bar{\text{b}}$ system exhibits the interplay of short-range and long-range effects in QCD [374, 375]. The interquark potential combines the short-range Coulomb-like behaviour expected from single gluon exchange with the long-range linear behaviour proposed by Nambu [340].

9.9.2.5. Decays of particles with b-quarks. Once the b-quark was identified, its charge-changing weak decays were seen to favour the c-quark and not the lighter u-quark [376], a conclusion reached on the basis of the emitted leptons' momentum spectrum. The decay b → ul$\bar{\nu}_l$, initially not detected at all, was eventually identified at the level of about 1–2% of the b → cl$\bar{\nu}_l$ process [377]. Considering the different phase space available for the two decays b → cl$\bar{\nu}_l$ and b → ul$\bar{\nu}_l$, the result implies a ratio of b → u and b → c couplings of between 0.05 and 0.1.

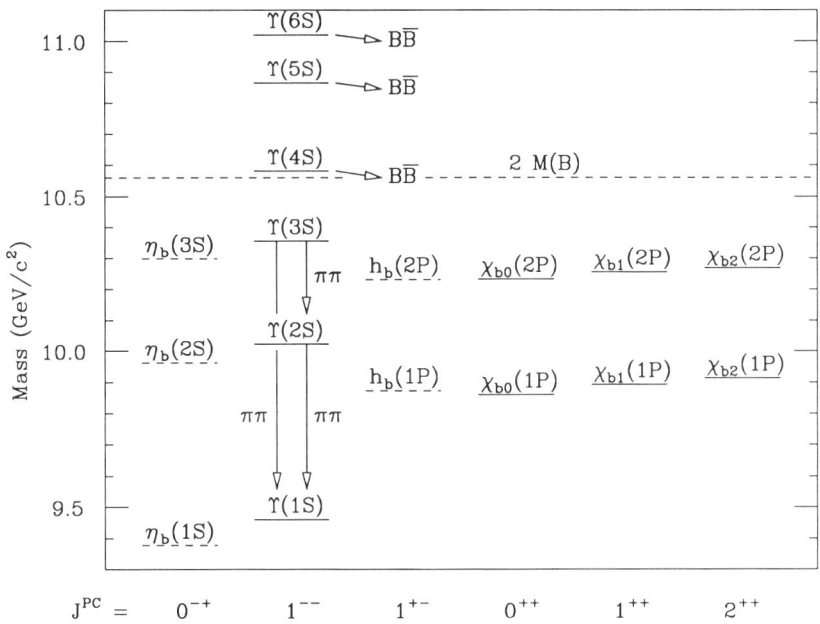

Figure 9.32. *The spectrum of $b\bar{b}$ states. Observed and predicted levels, as quoted in reference [319], are denoted by solid and broken horizontal lines, respectively. In addition to the transitions denoted by arrows, numerous electric dipole transitions have been seen between the Υ and χ_b states, e.g. $3S \to 2P \to 2S \to 1P \to 1S$, $3S \to 1P$ (very weak) and $2P \to 1S$.*

Another peculiar feature of b-quark decays lies in the relative weakness of the b → c weak coupling. Charmed mesons, with masses below 2 GeV/c^2, live for about 0.4 to 1 ps, depending on the light quark they contain. By contrast, mesons with a b-quark live for about 1.5 ps on average [378], in spite of their much larger masses. This curiously long life of the b-quark was first observed by the MAC Collaboration at SLAC under the leadership of David Ritson [379].

9.9.3. The Cabibbo–Kobayashi–Maskawa (CKM) matrix

9.9.3.1. Role in CP violation. The charge-changing couplings of quarks in the three-family model of Kobayashi and Maskawa [300] can be expressed in terms of a 3×3 matrix with rows pertaining to the charge $-\frac{1}{3}$ quarks d, s, b and columns to the charge $\frac{2}{3}$ quarks u, c, t. The unitarity of this matrix, required by its role in the diagonalization of quark mass matrices, implies the absence of flavour-changing neutral currents, just as for the two-family model.

For n quark families, a unitary $n \times n$ matrix has n^2 real parameters. For $n = 2$, a single angle (the Cabibbo angle) suffices to specify the matrix, the

Twentieth Century Physics

remaining three parameters setting relative quark phases. Three families require four real parameters in addition to five arbitrary relative quark phases.

A convenient parametrization of the CKM matrix [380] is

$$\mathbf{V}_{CKM} = \begin{bmatrix} V_{ud} & V_{us} & V_{ub} \\ V_{cd} & V_{cs} & V_{cb} \\ V_{td} & V_{ts} & V_{tb} \end{bmatrix}$$

$$\approx \begin{bmatrix} 1 - \lambda^2/2 & \lambda & A\lambda^3(\rho - i\eta) \\ -\lambda & 1 - \lambda^2/2 & A\lambda^2 \\ A\lambda^3(1 - \rho - i\eta) & -A\lambda^2 & 1 \end{bmatrix}$$

with A, ρ and η of order unity. Only the leading orders of the matrix elements are displayed here. The matrix is unitary as written up to order λ^4. The violation of CP invariance is represented by the non-zero value of η, at least three families of quarks being needed for this purpose.

9.9.3.2. Experimental information on matrix elements. The parameter $\lambda \approx 0.22$ is the sine of the Cabibbo angle, well known from the analysis of strange-particle decays. Information on the b-quark lifetime places A around 0.8, while the measurement $|V_{ub}/V_{cb}| = 0.05 - 0.1$ implies $(\rho^2 + \eta^2)^{1/2}$ between $\frac{1}{4}$ and $\frac{1}{2}$.

Individual values of η and ρ are less well known. Values of ρ between -0.4 and 0.4 and of η between 0.2 and 0.6 are compatible with present data. Indirect information is provided by CP violation in the kaon system and by mixing of the neutral B-meson $\bar{B}^0 \equiv \bar{b}d$ with its antiparticle $B^0 \equiv b\bar{d}$, first discovered in 1987 [381]. Uncertainties plaguing the estimates of η and ρ from these diagrams include errors in the top quark mass ($m_t = 180 \pm 12$ GeV/c^2; see below), and in various hadronic matrix elements, for which estimates from lattice gauge theories [357] and QCD sum rules [358] are beginning to be useful.

A 'superweak' theory [170] (independent of η in the CKM matrix) can still account for all the CP-violating phenomena observed in the kaon system. Two main checks of the CKM theory of CP violation hinge on rare decays of kaons and on the search for CP-violating B decays [382].

Detection of a difference between the ratios of the amplitudes for decay of K_L and K_S into pairs of charged and neutral pions would provide evidence against the superweak model, as mentioned in section 9.5.2.3. Searches for this difference have achieved a level of better than 0.2%. Two experiments [171,172] give somewhat different answers on this detection; improved versions of both experiments are planned.

The study of B-meson decays has been pioneered by detectors in e^+e^- colliders (see figure 9.33) such as ARGUS at DESY and CLEO at CESR. More luminous e^+e^- sources of B-mesons, with an energy asymmetry useful in studying time-dependent decay effects, are now being planned at several laboratories. It is also becoming possible to isolate B hadron signals from intense backgrounds in hadron colliders.

9.9.4. On the trail of the top quark; observation

Although the family structure of quarks and leptons is simple and repetitive, the b-quark could have broken away from this pattern, just as the transition metals are associated with variation in the periodic table of the elements. However, several pieces of data indicated that the b-quark had to be a member of a doublet of weak SU(2).

If the b-quark were a singlet, flavour-changing neutral decays such as $b \to s\mu^+\mu^-$ would occur with a rate much higher than anticipated in standard electroweak theory [383]. No such enhanced rate has been observed. Furthermore, the cross section and forward–backward asymmetry in the reaction $e^+ + e^- \to b + \bar{b}$, over a range of energies, imply that the b is a doublet member, just like the d and s. The self-consistency of calculations based on box diagrams for mixing of neutral B-mesons with their antiparticles also supports the existence of a heavy top quark.

Ever since the discovery of the b, searches for the top continued at electron–positron colliders such as PEP (Stanford), PETRA (Hamburg), TRISTAN (Japan) and LEP (CERN), and hadron colliders at CERN. With the higher hadronic collision energies available at Fermilab, in a *tour de force* of analysis involving several different decay modes, the CDF Collaboration presented evidence in 1994 for a top quark [332], confirming it in 1995 and measuring its mass to be $m_t = 176 \pm 8 \pm 10$ GeV/c^2. The D0 Collaboration has also observed a statistically significant signal [384], quoting $m_t = 199^{+19}_{-21} \pm 22$ GeV/c^2.

The effect of a large top quark mass on electroweak parameters was first pointed out by Veltman [385]. Virtual top quark contributions to W and Z self-energies alter the lowest-order predictions of the ratio $M_W^2/M_Z^2 = \cos^2\theta$ and the Z mass itself. Since independent (very accurate) determinations of the electroweak angle θ and the Z mass exist, for instance from LEP experiments, the top quark had been anticipated to lie below about 200 GeV/c^2. The measurement of its mass within these bounds was a striking confirmation of the electroweak theory, and indirectly places bounds on other effects such as Higgs bosons.

9.10. Accelerators

Our knowledge of the nucleus and, subsequently, of elementary particles has closely parallelled the development of particle accelerators at higher and higher energies. An excellent and detailed review of accelerator development up to the time of colliding beams (ca 1960) has been given by two of the principal participants in this effort, Livingston and Blewett

See also p 1201

Twentieth Century Physics

Figure 9.33. *The ARGUS (top) and CLEO (bottom) detectors.*

[386]. For each type of machine, we give here only a few examples of those actually constructed around the world.

9.10.1. Electrostatic generators
The first nuclear reactions produced by protons accelerated to high voltages bombarding a target at rest in the laboratory were observed by Cockcroft and Walton [387]. Their work was directly stimulated by the theoretical work of Gurney and Condon and of Gamow on barrier penetration. To produce the required voltage, they developed a scheme to multiply a transformer's output voltage through an ingenious arrangement of rectifying diodes and capacitors. Whereas nearly any multiplying factor can be achieved, the technical advantage lay in the fact that if the transformer's secondary produced a voltage V, no capacitor or rectifier was required to withstand a voltage greater than $2V$.

The more difficult technical problem at the time was posed by the need to provide an ion source at the high voltage and a tube evacuated of air through which the ions could be accelerated. The tube had to be insulated to withstand the high potential between the ion source and the target, with suitable arrangements for a uniform potential gradient along the tube to minimize spurious electrical discharges.

Cockcroft and Walton's incredible technical feat was recognized with a Nobel Prize in 1951. Until the advent of radio-frequency quadrupoles, their accelerator served as the low-voltage stage of every proton accelerator. While theirs was not the first scheme for voltage multiplication, it was superior, their device becoming universally known as the Cockcroft–Walton generator.

The other static voltage source widely used for the acceleration of nuclear particles was the electrostatic generator of van de Graaff [388]. While a Rhodes scholar at Oxford in 1927–28 van de Graaff had become interested in the possibility of producing high voltages by belt-charging. On his return to the United States as a National Research Council Fellow at Princeton University, he constructed his first models (figure 9.34), one for positive and one for negative potentials. Each consisted of a motor at the base turning a pulley which drove an insulating belt. The belt passed vertically about 7 feet into a copper sphere about 24" in diameter which contained the return pulley. The sphere and return pulley arrangement was supported on a glass tube about 2" in diameter. Charge was 'sprayed' on the belt at the bottom and removed in the sphere at the top with brushes.

Van de Graaff's first two generators were used for many years at Princeton in freshman physics demonstration lectures. Eventually one of them was given to the Smithsonian Institution for exhibit on the condition that they supply a duplicate to allow the demonstrations to continue. Subsequently, the duplicate model (made with modern

Figure 9.34. *Robert van de Graaff and one of the earliest versions of his generator.*

materials) has not performed as well as the original, a wonderful testimony to the inventor's skill.

The van de Graaff generators were immediately successful at producing high voltages with relative ease. However, as with the Cockcroft–Walton generator, combining the source of high voltage with an ion source and accelerating column presented a daunting technical problem. With van de Graaff's success, a number of groups became interested in adapting the generator to particle acceleration. Among these were Ray Herb and his group at Wisconsin, Merle Tuve and his group at the Carnegie Institution in Washington, and Compton and Van Atta at MIT, who, joined by van de Graaff [389], produced a machine which developed a potential of 5.1 MV. At more modest voltages the groups at Wisconsin [390] and the Carnegie Institution [391] produced the first practical van de Graaff generators for accelerating particles. With their machine the Carnegie group made the first excitation measurements showing sharp nuclear resonances.

Electrostatic generators have been adapted to accelerate electrons and to produce high-energy x-rays. In this capacity they have extensive

Elementary Particle Physics

industrial and medical usage. In nuclear physics, they have long had the advantage of excellent voltage stability, permitting good control of beam energy in precision measurements of nucleon–nucleon scattering. More recently, the development of negative hydrogen ion sources has permitted the construction of so-called tandem generators. These machines accelerate negative hydrogen ions from ground to a high positive potential where a thin foil strips them of their two electrons; the positive ions are then accelerated back to ground, thereby doubling their energy. Despite these developments, machines relying on high static voltages have been limited to around 50 MeV, restricting them to nuclear physics studies.

9.10.2. The cyclotron

Concurrently with the development of the static high-voltage machines, work began on resonant cyclic accelerators. The first of these, a linear machine designed and constructed by Wideröe [392], consisted of three cylindrical tubes positioned end to end and electrically insulated from each other. A radio-frequency voltage was applied to the central tube, properly phased to accelerate an ion travelling from the first to the second tube. By adjusting the radio frequency to reverse the voltage during the ion's transit down the middle tube, the ion was again accelerated when passing from the second to the third tube. Reading about this device in the library at Berkeley, E O Lawrence realized that the ions might be recycled through the radio-frequency voltages again and again by bending the particles back on their original path with a magnetic field. He quickly saw that the angular frequency of revolution of an ion in a circular path in a magnetic field, $\omega = eB/mc$, remains constant, independent of radius and momentum! In such a way the cyclotron and a whole class of magnetic resonant machines were born [393].

| See also p 1201 |

The first cyclotron applied to physics research had magnetic pole pieces 10″ in diameter, producing protons over 1 MeV in energy. It came into operation in 1932 a few months after the disintegration of lithium by protons was announced by Cockcroft and Walton, confirming their results immediately [394].

A number of features important to successful cyclotron operation were not recognized at the beginning. Happily, the device worked anyway. The cyclotron angular frequency, $\omega = eB/mc$, remains constant only as long as B and m do not change. Countering the requirement of constant B is the need for some vertical focusing to prevent ions with some initial vertical velocity from striking the pole tips of the magnet before undergoing a significant number of rotations. Vertical focusing requires the field to fall off with increasing radius. Fortunately for the cyclotron inventors, any magnet with circular pole tips will, barring heroic attempts at field shaping, naturally show a small decrease of field with radius, thus providing some vertical focusing. At the same time, the change of field with radius should not be so great as to violate the cyclotron equation grossly. This too was the case in the first cyclotron. In addition, relativistic effects cause the mass to increase with energy. Again providence smiled on the inventor, since this effect does not become limiting until a proton's energy reaches about 25 MeV.

After the cyclotron's success at Berkeley, many similar machines were constructed throughout the world. In the 1930s every research university in the country arranged to have a cyclotron for nuclear physics studies. Berkeley, of course, showed the way, with larger and larger cyclotrons. This activity culminated with the construction of a 184" giant, completed only after World War II.

9.10.3. Phase stability and the synchrocyclotron

Shortly after the war, McMillan at Berkeley and Veksler in the Soviet Union independently showed that the conflict between requiring the field to decrease with radius for vertical focusing and increase with radius to compensate for the relativistic mass increase could be resolved by changing the frequency during ion acceleration [395]. To quote from McMillan's paper

> The device proposed here makes use of a 'phase stability' possessed by certain orbits in a cyclotron. Consider, for example, a particle whose energy is such that its angular velocity is just right to match the frequency of the electric field. This will be called the equilibrium energy. Suppose further that the particle crosses the accelerating gaps just as the electric field passes through zero, changing in such a sense that an earlier arrival of the particle would result in an acceleration. This orbit is obviously stationary. To show that it is stable, suppose that a displacement in phase is made such that the particle arrives at the gap too early. It is then accelerated; the increase in energy causes a decrease in angular velocity, which makes the time of arrival tend to become later. A similar argument shows that a change of energy from the equilibrium value tends to correct itself. These displaced orbits will continue to oscillate, with both phase and energy varying about their equilibrium values.
>
> In order to accelerate the particles it is now necessary to change the value of the equilibrium energy, which can be done by varying either the magnetic field or the frequency. While the equilibrium energy is changing, the phase of the motion will shift ahead just enough to provide the necessary accelerating force; the similarity of this behaviour to that of a synchronous motor suggested the name of the device ...

The advantages of phase stability far exceeded its single disadvantage of requiring the frequency to change during the time the particles are acquiring energy. The ions are injected at the high-frequency end of the cycle, and the frequency is reduced during the acceleration. The ions are finally brought to an internal target or ejected from the machine at the highest energy (lowest frequency). The frequency cycle is then repeated. Since ions can only be accepted for acceleration during a small part of the whole frequency cycle, the beam intensity will accordingly be much lower, about 1% of that in conventional cyclotrons.

It did not take long for the fundamental soundness of the idea to be lastingly recognized. The ideas expressed by McMillan have been invoked in every subsequent accelerator. Immediately after World War II the 184" cyclotron, after some modelling studies, was converted to a 'synchrocyclotron'. It first came into operation in November 1946, accelerating deuterons to 190 MeV and, later, protons to 350 MeV. It was thus an enormous success from the beginning, followed between 1948 and 1952 by five additional synchrocyclotrons with energies ranging from 150 to 450 MeV constructed at various universities in the US and a further six elsewhere in the world. The physics pouring from these accelerators was immense, primarily because their energy exceeded in most cases the 290 MeV threshold for creating pions. With these machines the properties of the pion and the muon were elucidated in detail. Pionic and muonic atoms were discovered and quantitative measurements on these new kinds of atoms were made. Parity violation was observed and thoroughly explored in the pion–muon–electron decay chain. It was a fantastically productive period in the history of particle physics.

9.10.4. Electron synchrotrons

The concept of phase stability was also exploited in electron accelerators, leading to the first electron synchrotrons. Synchrotrons accelerate particles in orbits of constant radius. After an electron's energy has reached 4–5 MeV its speed is essentially that of light so that the transit time around an orbit of fixed radius is nearly constant, requiring minimal change in the frequency of the accelerating voltage. Therefore, use of a separate system such as a van de Graaff generator for pre-acceleration to a few MeV simplifies the subsequent synchrotron design enormously.

The acceleration of electrons in circular guide fields is accompanied by significant loss of energy to radiation. This loss implies a smaller orbit and a smaller transit time around the ring. If the particles are accelerated on the falling side of the radio-frequency voltage, those particles arriving earlier will be accelerated by higher voltages thereby compensating for the radiation loss. Under the guidance of McMillan the first electron synchrotron was started in Berkeley in 1946 and completed in 1949, producing electrons of 335 MeV. Experiments using the photons from the electrons impinging on Pb were immediately started and the first photoproduced charged pions were quickly detected [396]. A major result from this machine was the discovery of the neutral pion [68] in 1950 (cf section 9.4.2). As with the synchrocyclotrons, electron synchrotrons were constructed at many universities, most of them in the range of 300 MeV. They were responsible for tracing out the full peak of the famous (3, 3) resonance, mentioned in section 9.4.5.2.

9.10.5. *The betatron*

Independently of all the activity devoted to electron synchrotrons, in fact back in 1940, the first successful betatron was developed by D W Kerst [397] with an energy of 2.3 MeV. The electric field associated with a changing magnetic field served to accelerate electrons, while the same magnetic field guided the particles, a combination which would seem fairly obvious. However, the attendant technical problems turned out to be severe, thwarting many attempts before Kerst's. His machine followed the detailed magnet design based on elaborate orbit calculations by Kerst and Serber [398]. An interesting technical point about the betatron has become a favourite examination question in courses in electricity and magnetism: it is easily shown that, for a constant orbit radius, the magnetic field at the orbit must be just one-half the average magnetic field linked by the orbit. One problem associated with extrapolating the betatron principle to high energies lies in maintaining this factor of 2 in the face of iron saturation, etc. The development of the betatron culminated with the construction at the University of Illinois of a 300 MeV device, completed in 1950 [399]. We have already noted the desirability of pre-acceleration of electrons in a synchrotron by, for example, van de Graaff generators. Another even more popular method was pre-acceleration of the electrons in the synchrotron itself using the betatron principle.

9.10.6. *Proton synchrotrons*

By 1948 it had become clear that achieving still higher energy protons than accessible by synchrocyclotrons would hinge on the synchrotron principle. Cyclotrons require the quantity of magnetic iron to increase roughly as the cube of the maximum proton momentum, whereas synchrotron requirements scale only linearly with the momentum. The crossover had already been reached with cyclotrons under construction. The US Atomic Energy Commission planned to construct two machines; one on Long Island at the newly created Brookhaven National Laboratory in the energy range of 2–3 GeV and the other with 5–6 GeV at Berkeley.

The previous considerations of electron synchrotrons make it clear why physicists were more timid in proceeding with proton machines, whose required frequency change was larger. Furthermore, it was necessary to synchronize the frequency closely with the instantaneous magnetic field. At the Brookhaven machine (ultimately named, somewhat grandiosely, the Cosmotron) the protons were injected from a van de Graaff generator with 4 MeV energy. The initial machine design involved a magnetic field of 311 gauss and a frequency of rotation around the magnet ring of 0.4 MHz at injection. At the full energy, 3 GeV, the average field reached 13 kilogauss with a frequency of 4.2 MHz. In contrast to the early electron machines, whose acceleration time was typically 5 ms, the period of acceleration was 1 s, allowing the final energy of 3 GeV to be reached with only about 2 kV of accelerating voltage per turn.

The Cosmotron started operating in 1952 at 2.3 GeV. Almost immediately it began producing noteworthy physics, such as the first observation of associated strange-particle production (cf section 9.4.4). Early in 1954 it started operating at its full design energy of 3 GeV.

The Berkeley synchrotron, called the Bevatron (a name based on the earlier abbreviation for GeV), reached its design energy of 5.4 GeV in October 1954. It could be pushed to 6.4 GeV at a lower cycling rate. Ostensibly the design energy of the Bevatron was chosen to exceed the antiproton production threshold, which is 5.6 GeV for protons striking protons at rest. Appropriately, antiprotons produced in the Bevatron were found by the Chamberlain–Segré group [73] in 1955 (cf section 9.4.3).

9.10.7. Strong focusing

At the 1952 January meeting of the American Physical Society, held on the campus of Columbia University in New York City, a highlight was an invited talk by Enrico Fermi [400], describing research in progress at the University of Chicago's 450 MeV synchrocyclotron which had started operation the preceding year. It was the highest energy accelerator then in operation. Near the end of his talk Fermi described, in jest, the ultimate accelerator of the future, a machine circling the Earth.

Another talk at this same meeting contained the seeds of an idea that made ultra-high-energy accelerators possible. Quoting from the abstract [401] of a paper describing the original pair of quadrupoles devised to focus the external beam of the Princeton cyclotron

> Under these conditions, because of the curvature of the fringing field, their focal lengths in the plane of the field are very nearly equal, but opposite in sign, to their focal lengths in the plane of the pole faces. However, if two of these astigmats are spaced a distance comparable with their focal lengths with their field directions orthogonal, a point image of a point source may be obtained. If the source distance, the separation of the astigmats, and the desired image distance are specified, the focal lengths required for double focusing can be reliably calculated from the thin lens equation.

This is the first published reference to a principle eventually called strong focusing. This idea was quickly exploited by the early summer of 1952 to focus external pion beams at Columbia's Nevis cyclotron [402].

Strong magnetic focusing was in the air. In a highly influential paper, Courant, Livingston and Snyder showed how that principle could be used with great advantage in the design of synchrotrons [403], demonstrating in particular that strong focusing, implemented by alternating the gradient of the magnetic field guiding the particles, could be maintained in the closed orbits required in accelerators, that phase stability would persist, and that the large 'momentum compaction' would result in relatively small beam displacements in spite of rather large momentum spreads. This paper also created a magnetic quadrupole design which would not produce a net bending of the beam, a disadvantage of the original design by the Princeton group.

It developed that Nicholas Christofilos, an electrical engineer working in Greece, had applied in 1950 for a US patent on a conceptual design for an accelerator using alternating-gradient focusing. His seminal contribution went unrecognized because his work was never published, but his patent (awarded 28 February 1956) is contained in reference [404].

Following the prescriptions supplied in the Courant, Livingston and Snyder paper, designs of two alternating-gradient synchrotrons to operate at about 30 GeV, one at Brookhaven and one at CERN, were immediately started. However, the first accelerator to employ the alternating-gradient strong-focusing principle, producing 1.1 GeV electrons in 1954, was constructed under the direction of R R Wilson at Cornell University. The 28 GeV machine at CERN (called the Proton Synchrotron or PS) started operations in 1959, while at Brookhaven experiments were begun in 1960 with the machine (the AGS) operating at 33 GeV and still in service at the time of writing in the mid-1990s.

For many years, proton synchrotrons became the machines of choice in the push towards ever-increasing energies. An accelerator at Serpukov in Russia began operation at 76 GeV in 1971. The Fermilab accelerator, planned for 200 GeV, was constructed by Robert R Wilson (see box) ahead of schedule and under budget to operate at 400 GeV. It revealed the Υ (section 9.9.2) and produced a wide variety of other physics results. It has now been outfitted with a ring of superconducting magnets which permit beam energies of 0.9–1 TeV and can collide protons with antiprotons, yielding centre-of-mass energies of up to 2 TeV (see section 9.10.10.2 below). A machine similar in scope to the original Fermilab accelerator, the SPS, was constructed at CERN, with a beam energy of 400 GeV. It achieved proton–antiproton collisions a few years ahead of Fermilab, making possible the discovery of the W and Z (section 9.7.5).

9.10.8. Linear accelerators

As we have seen, special problems arise with injection and extraction in accelerators such as cyclotrons employing magnetic confinement. These problems are not nearly as challenging for linear machines such as van de Graaff generators. However, large fixed potentials are impossible to deal

Robert R Wilson

(American, b 1914)

When Robert Rathbun Wilson was chosen to lead the construction of a 200 GeV accelerator in 1967 he was given 10 square miles of farmland west of Chicago and a budget of 240 million dollars. By taking high technical risks and eschewing conventional engineering practices, Wilson finished, under budget, a machine yielding twice the energy originally specified. Along with the machine was a great laboratory which attracted particle physicists from around the world. Wilson, an amateur sculptor, had a passionate interest in architecture as well as accelerator design. Every construction on the site bore his stamp of originality.

Wilson was born in 1914 in Frontier, Wyoming, where his family had a cattle ranch. Educated through high school in local schools, he went to the University of California at Berkeley where he obtained an undergraduate degree followed by a PhD working under E O Lawrence. In 1940 he went to Princeton as an instructor in the physics department (where Richard Feynman was still a graduate student). Almost immediately, because of World War II, he headed a project intended to separate the isotopes of uranium, based on a device of his own invention, the isotron. Moving to Los Alamos in 1943 at the insistence of Robert Oppenheimer who had known him in Berkeley, Wilson became head of the experimental nuclear physics division.

After the war, following a brief period at Harvard, Wilson moved to Cornell University where he was director of the Laboratory of Nuclear Studies. In this capacity he supervised the design and construction of a series of electron accelerators of ever-increasing energy, one of which was the very first machine to use alternating-gradient strong focusing. With his considerable experience in experimental physics and accelerator construction it was natural for him to be selected in 1967 to create, out of the farmlands of Illinois, the Fermi National Accelerator Laboratory.

with beyond about 25 MV. Linear accelerators or linacs, were developed to maintain the ease of beam injection and extraction of fixed potential machines without the limiting high potentials, avoided by using cyclic resonant voltages.

The first proposal for a linear accelerator powered by radio-frequency voltages was by G Ising in Sweden in 1925. However, no attempt was made to implement its general idea until Wideröe succeeded in accelerating potassium and sodium ions in 1928. As noted above, Wideröe's device provided inspiration for the cyclotron of E O Lawrence.

Lawrence was clearly driven to make accelerators to do physics. Concurrently with his development of the cyclotron with his student, M S Livingston, he had another student, D H Sloan, working on a linear device more in keeping with Wideröe's scheme. This early machine consisted of a number of cylindrical tubes arranged in a line with a spacing between tubes sufficient to prevent breakdown of the radio-frequency voltage applied between adjacent sections. Acceleration of the particles occurred only between the tubes. The ions drifted down the axis of each tube for a time sufficient for the alternating voltage to change sign. Progressing down the accelerator, the tubes were made progressively longer to compensate for the ions' acceleration. By 1931 Lawrence and Sloan had, with ten tubes, accelerated mercury ions to 1.25 MeV. Unfortunately, RF generators then available did not permit operation at high frequencies. These early machines were restricted to accelerating slow-moving heavy ions, too slow and highly charged to overcome the Coulomb barrier and initiate nuclear reactions in target materials.

The modern linac stems from the work of William W Hansen at Stanford University starting in the mid-1930s. The initial work centred on developing a single cavity for accelerating electrons, requiring large amounts of RF power at high frequency. Correspondingly, Hansen, with the Varian brothers, set about inventing an appropriate RF generator. This activity yielded the klystron, which played an important role for radar in World War II. In the meantime, Kerst's successful development of the betatron raised doubts whether a linear device could ever compete.

The radar developments during the war, especially with respect to powerful RF sources, raised hopes. After the war, Hansen, with E L Gintzon and J J Woodward, concluded that the magnetrons then available would be suitable for linear accelerators of a few MeV, but that higher energies would require the further development of RF power sources. Hansen pursued the idea of accelerating the electrons with an electromagnetic field travelling in a cylindrical waveguide held in phase with the electrons by 'loading' the waveguide, leading to the first (Mark I) Stanford linac with an energy of 6 MeV.

Concurrently, work initiated by Chodorow and Gintzon was proceeding on a high-power pulsed klystron. After a successful test

Elementary Particle Physics

Figure 9.35. *Wolfgang K H Panofsky, the builder and first director of the Stanford Linear Accelerator Center.*

in 1949, plans went forward for a GeV electron linac, completed in 1952 (the Stanford Mark III) [405]. Unfortunately, W W Hansen, who had been the inspirational leader of the whole project, died prematurely in 1949 after living to see the successful test of the high-powered pulsed klystron.

The Mark III linac was the progenitor of the two-mile Stanford Linear Accelerator, constructed on the Stanford campus in the early 1960s under the direction of W K H Panofsky (figure 9.35). Initially operated at 20 GeV with an average current of 50 μA, this machine now regularly operates at 50 GeV, corresponding to an average accelerating field of about 150 000 V cm^{-1}.

The post-war development of electron linacs was parallelled by work on proton machines, principally at Berkeley under the direction of L Alvarez and W K H Panofsky [406]. Since protons of equal energy travel much more slowly than electrons, far lower RF frequencies (of the order of 200 MHz) are required, making the accelerators considerably more cumbersome. In fact, the technique of accelerating electrons with a travelling wave in the waveguide cannot be successfully applied to protons, necessitating methods more akin to the old drift tube idea. In particular, what emerged was a resonant cavity 40 ft long and 39″ in diameter, operated in the TM$_{010}$ mode at 202.5 MHz and filled with 42 drift tubes. This pioneering work at Berkeley yielded a machine which accepted protons from a 4 MeV van de Graaff generator and accelerated them to 31.5 MeV.

The advantage of a high beam current with low angular divergence has retained for proton linacs a unique place in the accelerator hierarchy. They are used as injectors to the large proton synchrotrons, typically following a Cockcroft–Walton machine or radio-frequency quadrupole. An elaboration of the original standing-wave design of the Berkeley group has been highly successful at the Los Alamos Meson Factory, LAMPF, where protons are accelerated to 800 MeV with an average beam current of 1 mA.

9.10.9. Colliding beams

As noted above, the trail-blazing paper of Courant, Livingston and Snyder showed the feasibility of accelerators using the strong-focusing principle. It also introduced a new era in the sophistication of machine design. In particular, designs could be contemplated providing a new level of precision in beam handling and transport, whether around the closed orbits of an accelerator, or in beam lines external to the machine.

In 1956 it was first recognized that high-intensity beams, positioned and focused with high precision and directed against beams of equal and opposite momenta, could achieve reaction rates sufficient for interesting physics results. Just to contemplate this possibility is somewhat astonishing since the density of target particles in a beam is vastly smaller than in more usual targets such as liquid hydrogen or solid materials. It was Kerst [407] who first recognized that beams stored in fixed orbits could be made to collide with each other repetitively, thereby compensating for their diaphanous nature. All previous accelerators had directed their beams against targets at rest in the laboratory, either within the machine itself, or in an external beam extracted after the completion of the acceleration cycle.

Two beams of equal and opposite energy deposit all of their energy in the centre-of-mass (CM) system, in contrast to the collision of a particle with another at rest, for which the requirement of momentum conservation decreases rather spectacularly the energy available in the CM system at high energies. For example, two protons directed against one another, each with energy E, have an energy $2E$ in the CM system, whereas the CM energy of a proton of energy E striking another at rest is $\sqrt{2m(m+E)}$, where m is the proton mass. Energy is not wasted, of course, in fixed target machines; it yields high-energy beams of secondary particles, such as mesons and neutrinos. However, achieving the maximum energy for the production of new particles hinges on the colliding-beam technique. In 1956, when the notion of colliding beams was being advanced, the 30 GeV proton accelerators at Brookhaven and CERN were under construction. With 30 GeV on a fixed target, the available CM energy is 7.6 GeV; arranging these machines to collide beams with one another would yield almost 8 times as high a CM energy.

The proposal by Kerst *et al* *Attainment of very high energies by means of intersecting beams of particles* [408], recognized that two fixed-field alternating-gradient accelerators [409] could be arranged to circulate their high-energy beams in opposite directions over a common path in a straight section common to the two accelerators. G K O'Neill, at about the same time, published a paper entitled *Storage-ring synchrotron: device for high-energy physics research* [410]. (A non-technical review of the origin of the ideas and the development of the first practical storage rings has been provided in reference [411].) Since beams had already been extracted, highly successfully, from the Cosmotron, he proposed that such an extracted beam be stored in two magnet rings with the stored beams rotating in opposite directions. If the two rings were tangent at one point, the beams could be made to collide. Some details of the O'Neill proposal relating to injection were unworkable, but the general scheme has been followed in every one of the many colliding beam machines constructed subsequently.

Making proposals and suggesting ideas is easy and fun. A rare person is sufficiently convinced of their ultimate value to make a very major commitment of time and effort over a number of years. O'Neill, then an instructor in the Princeton University Physics department, was such a person. He quickly realized that storage rings might be easier to implement with electrons than with protons. With electrons, radiation quickly damps out transverse oscillations, reducing the cross-sectional area of the beams, a major advantage since the beam particles' interaction probability depends inversely on the area. (In those days no mechanism was known for damping the transverse oscillations in proton machines.) Furthermore, radiation automatically provides the energy loss mechanism necessary at injection for the particles to end up in stable orbits.

With a tentative design for an electron–electron collider in hand, O'Neill went to Stanford where the linear electron accelerator already provided an ideal injector. He convinced the director of that facility, W K H Panofsky, as well as Burton Richter and Carl Barber, of the proposal's worth. Somewhat later, O'Neill recruited B Gittelman, also a Princeton faculty member, to work on the project. After a year of detailed design and after funding had been secured from the office of Naval Research, construction was begun in 1959 on a pair of 0.5 GeV electron storage rings at Stanford. The team of Barber, Gittelman, O'Neill and Richter did the pioneering work.

And pioneering work it was. As soon as the first electrons were stored, totally new and unanticipated difficulties arose. To quote O'Neill 'We were crudely reminded that Nature is an ingenious troublemaker.'. For example, the large peak currents induced in the vacuum chamber electromagnetic fields that reacted back on the beam destructively. One by one these problems were solved, with many people contributing to their solution. Finally, in 1965, the first electron–electron scattering results were obtained. The quantum electrodynamics of Feynman, Schwinger and Tomonaga survived an especially clean test. As Feynman has said, 'The test of all knowledge is experiment. Experiment is the sole judge of scientific truth.'.

Electron–electron colliders were limited to studies and tests of electrodynamics. It was clear from the beginning that electron–positron collisions offered a much richer opportunity for interesting physics. Electron–positron annihilations proceed through a virtual photon and thus are ideal for creating vector particles as well as particle–antiparticle pairs. This option had been considered by the Princeton–Stanford group but rejected because of the worry that the positron intensity could never reach the level needed to get physics results.

A most important contributor to the development of storage rings was a group working under the direction of B Touschek at the Frascati laboratory in Italy near Rome. Early in 1960 they decided to make an electron–positron storage ring of 0.25 GeV. An immediate simplification is associated with electrons and positrons which, travelling in opposite directions, can be guided by the same magnetic field and vacuum chamber. However, it was the spirit of the Italian group that guaranteed this electron–positron collider (called ADA, for *aniello d'accumulazione*, or accumulation ring) would be a useful test facility.

Originally constructed and tested at Frascati only for electrons, ADA was loaded on a truck and transported to Orsay near Paris in 1962 with its ultra-high-vacuum chamber intact. The development of storage rings was characterized by the appearance of unexpected instabilities whenever the operation was pushed into a new realm of intensity or energy. Here, true to the pattern, more intense beams could be injected by a 1 GeV linear electron accelerator, and new instabilities (the Touschek effect) associated with the new intensities were found. The source of the new difficulties was eventually identified and the problems brought under control. The first electron–positron annihilations were recorded in ADA in 1963. While their intensity was too low for significant physics to emerge, it was apparent that the problem of low positron intensity could be overcome. Indeed, this was the case and every collider constructed since has been of this type.

ADA had a glorious career as a testing ground for e^+-e^- colliders and was suitably retired in 1965. Early in 1965 the Italian group started the construction of a much larger machine, ADONE, which was to have an energy of 1.5 GeV per beam. The first beam–beam interactions from this machine were observed early in 1967. The maximum energy of ADONE turned out to be a most unfortunate choice. The J/ψ, produced in 1974 at an energy of 3.1 GeV at SLAC, could have been uncovered much earlier at ADONE had that machine been designed to reach, say, 1.6 GeV per beam.

Elementary Particle Physics

In the early 1960s a French group at Orsay started constructing an electron–positron machine with an energy of 0.385 GeV per beam. Activity in storage-ring design and construction was also initiated in Russia at Novosibirsk under the direction of G Budker. At the Electron–Photon Conference at Stanford University in 1967, experimental results were reported from three storage rings [412]: Orsay (385 MeV), Novosibirsk (510 MeV) and Stanford (550 MeV). The Orsay and Novosibirsk machines were both electron–positron colliders.

Electron–positron colliders were also on the drawing boards in the US. Competitive proposals were received by government agencies from groups at the Cambridge Electron Accelerator (CEA) and the Stanford Linear Accelerator Center (SLAC). The SLAC proposal won out but no funding was forthcoming. Finally, the Stanford Positron–Electron Collider, SPEAR, was constructed as an experiment, albeit a somewhat expensive one, with a maximum energy of 3.5 GeV per beam, starting operation in 1972. In the meantime, the group at CEA, still wanting to build a collider, started by inventing a 'low-β interaction region', a scheme which increases the interaction rates by factors of 10 to 100. This idea, combined with an ingenious and difficult beam handling arrangement, converted the CEA's electron synchrotron into a colliding beam machine, yielding collisions at energies much higher than could be achieved elsewhere and indicating the high value of R already mentioned in sections 9.6.3 and 9.9.1.

The CEA results were largely discounted at the time. A popular impression viewed the collider operation of the CEA as exceedingly difficult and thus unlikely to yield good physics. In the end the CEA group was vindicated; their results were shown to be correct by the more extensive measurements at the SPEAR collider.

We have already shown the behaviour of R as a function of energy in figure 9.18. Below the mass of the top quark the value of R from simple quark and colour counting should approach $3\frac{2}{3}$. This it does, rising at the highest energies as the low-energy tail of the Z^0 is approached.

After SPEAR many electron–positron colliders have been constructed throughout the world, each with a name derived from some appropriate acronym. A listing of those existing near the end of the twentieth century is given in table 9.3.

This activity has culminated in the construction of LEP, the Large Electron Positron collider at CERN, with a circumference of 27 km, operating, in its first phase, at single beam energies up to about 50 GeV. The energy of this machine, completed in 1989, was designed to produce the Z^0, with a mass of 91 GeV. In the second phase of operation, the beam energy will be raised to about 90 GeV, to produce $W^+ - W^-$ pairs. This energy represents the practical limit for electron–positron storage rings, because energy loss to synchrotron radiation is proportional to the fourth power of the beam energy divided by the radius of curvature of

Table 9.3. *Electron–positron colliders with CM energies above 10 GeV.*

Machine	Beam Energy (GeV)	Location
DORIS	4–5.3	DESY (Germany)
VEPP-4	5	Novosibirsk (Russia)
CESR	8	Cornell
PEP	17	Stanford
PETRA	23	DESY (Germany)
TRISTAN	35	KEK (Japan)
LEP	50–100	CERN

the bending magnets. When operating at 50 GeV per beam it amounts to 1.6 MW at the design luminosity, rising to 14 MW at the higher energy. This energy loss must be made up by the considerable radio-frequency power delivered to the beam.

To avoid the limitation of synchrotron radiation a daring proposal was made at SLAC in 1979 [413] to accelerate alternate pulses of electrons and positrons down the length of the two-mile long machine to 46 GeV. Each type of particle is then separately steered to arrange for their head-on collision. Clearly, for these collisions to take place with an appreciable chance of interaction, the beams must be incredibly small (of the order of two wavelengths of visible light) and must be steered with even better accuracy. To appreciate this technical *tour de force,* imagine a needle-shaped bunch of electrons with the diameter of the needle, one-fifth that of a human hair and perhaps a centimetre long, colliding with a similar bunch of positrons moving in the opposite direction, each bunch of particles having travelled through a semi-circle half a mile in radius. This project, the SLC (for Stanford Linear Collider), was successfully completed; while its flux of Z^0s is significantly lower than obtained in LEP, its electrons and positrons can be polarized, adding important information to the results. It is generally agreed that still higher energy electron–positron collisions will require the further development of the idea of the linear collider.

9.10.10. Hadron colliders

9.10.10.1. Proton–proton colliders. As noted above, machines intended to collide protons are intrinsically much more difficult to design. The absence of radiation damping complicates considerably both injection and storage. As has been noted, every proton remembers its history forever in the absence of damping. The first proton storage-ring-collider complex (the ISR or Intersecting Storage Rings) was constructed at CERN using the 28 GeV protons provided by the PS synchrotron, first operating in 1971. It was, technically, a very complex device

Elementary Particle Physics

with elegant solutions to many difficult problems concerning beam handling and proton stacking and storage. Its most significant new technical development was the invention by Simon van der Meer [321] of stochastic cooling, which replaces the radiation cooling in electron machines and has made possible all of the subsequent hadron colliders. This invention recognizes that protons' radial positions oscillate about a central orbit at the so-called betatron frequency. The technique samples the departure of the mean beam position from its nominal radius at a particular point on the ring of magnets, feeding back the amplified signal across a chord to make up for delays in amplification and signal transmission. The amplified signal is applied to correcting electrodes appropriately positioned relative to the pickup electrodes in terms of betatron wavelengths so as to deflect the protons towards their central orbit. Since the pickup electrodes only sense the average position of the beam, some of the protons will be 'heated', but, on the average, the beam is cooled; hence the name 'stochastic cooling'.

Concurrently, a different cooling method was developed by G Budker at Novosibirsk. His method enveloped a beam of protons with an intense beam of electrons moving with the same speed and in the same direction. Any proton with a velocity different from the electrons will scatter from them, losing energy. This technique is most useful for low-energy beams and has been applied successfully to cyclotrons.

9.10.10.2. Proton–antiproton colliders. The cooling method of van der Meer became the crucial feature in the next round of accelerators, the proton–antiproton colliders mentioned in section 9.7.5. Early in 1976 a proposal was prepared by C Rubbia, P McIntyre and D Cline [414] to convert the existing highest-energy accelerators to proton–antiproton colliders, with a physics motivation explicit from their proposal's title 'Producing massive neutral intermediate bosons with existing accelerators'. This proposal was complete with estimates of the production cross sections of $W^{\pm}$ and W^0s (called Z^0s by the time of their discovery). It required beam handling techniques considerably beyond current methods

> The main elements are (1) an extracted proton beam to produce an intense source of antiprotons at 3.5 GeV/c, and (2) a small ring of magnets and quadrupoles that guides and accumulates the $\bar{p}$ beam, (3) a suitable mechanism for damping the transverse and longitudinal phase spaces of the $\bar{p}$ beam (either electron cooling or stochastic cooling), (4) an RF system that bunches the protons in the main ring and in the cooling ring, and (5) transport of the 'cooled' RF bunched $\bar{p}$ beam back to the main ring for injection and acceleration.

The authors also provided a sketch of the proposal's implementation at the Fermi National Accelerator Laboratory (Fermilab) in Batavia,

Illinois. This proposal implied that the projected production cross sections would permit the construction of a proton–antiproton collider with sufficient luminosity to produce a reasonable rate of detected Ws and Zs. When budgetary constraints at Fermilab did not allow the laboratory to proceed with these plans, despite the importance of the possible results, the authors took their proposal to CERN where a machine of similar energy (the SPS) was also in operation. Here their plans were accepted and construction was started on this grand experiment. As mentioned in section 9.7.5, it resulted in the discovery of the $W^{\pm}$ early in 1983 and of the Z^0 several months later.

In 1984 Simon van der Meer and Carlo Rubbia were awarded the Nobel Prize 'for their decisive contributions to the large project, which led to the discovery of the field particles W and Z, communicators of weak interaction' [321, 322].

In the meantime, plans at Fermilab were evolving in the late 1970s to make a $p\bar{p}$ collider of much higher energy using superconducting rings piggy-backed on the original iron magnet ring. This machine is another technical *tour de force* owing to the additional complications associated with the use of superconducting magnets in a pulsed mode around a 6.3 km ring. It was brought into operation in 1985. Each beam has an energy close to 1 TeV, hence its name, the Tevatron. From its initial operation it has been the lead accelerator in the search for the top quark, culminating in its observation (section 9.9.4).

9.11. Detectors: from Rutherford to Charpak

Newtonian mechanics, Maxwell's equations and Einstein's relativity are all recognized as monuments to deductive reasoning. Their place in history both in time and in the development of our understanding of Nature are well known and are part of the textual lore of physics. Conclusions drawn from experimental data which have changed the way we view things—Rutherford scattering, the Bohr atom, the Hubble radius—are likewise part of every textbook. Only rarely do major conceptual developments in the history of science go unnoticed. One example is the recognition that the Sun itself is a star [415]. This was surely a most profound contribution to understanding the Universe, but who first proposed it and how it came to be accepted is totally unknown. Similarly, some developments in instrumentation, in the tools physicists use which have had profound consequences, have had such diffuse origins that it is difficult to identify the individuals who might be credited. Bardeen, Brattain and Shockley are credited for the transistor. Dating from the late 1940s, this is recognized quite properly as a momentous invention. It had a greater impact on the nature of physics research in the last half of the twentieth century than any other invention. But a further development, which truly brought about a revolution not only in physics research but in nearly every aspect of life, was the large-scale integration of transistor circuits on single pieces of silicon. This profound development, occurring in the 1960s, came from incremental advances in many different laboratories, and has a more diffuse history. Near the end of the century commercially available devices are available which have four million transistors on a single piece of silicon!

In a charming essay [416] entitled *The craft of experimental physics*, P M S Blackett begins 'More has been written of what the experimental physicist has discovered than of how he has discovered it. Because he has changed the technique of living by his intense curiosity to find out about obscure things, many of his discoveries have become common knowledge. But his method of experimental discovery, how he works and thinks, is much less known.'.

Also, many of the tools which the experimental physicist has invented and developed for the purpose of doing experiments have not been accorded the attention they deserve. Since these investigations are not an end in themselves, there is nothing which is forgotten faster than an outdated technique no matter how clever or original its conception. From 1920 to 1960 an important device for the physicist was the vacuum tube. Today's students are hard pressed to recognize one, and the inventors are forgotten. In the present section we review the development of some of the tools and techniques which have most contributed to our present knowledge of particle physics.

At the last 'turn of the century', in 1900, the electron had just been discovered. So too with radioactivity. The immediate problem was to unravel the nature of the alpha, beta and gamma rays that had just been discovered. The principal tools consisted of electroscopes and electrometers for measuring ionization.

In 1903 Crookes and also Elster and Geitel discovered the phosphorescence of zinc sulphide on exposure to alpha particles. It had also been discovered rather early that pure zinc sulphide did not fluoresce; some impurity, e.g. copper, was necessary to make it work. A thin layer of finely powdered zinc sulphide spread on a glass plate and exposed to an alpha source such as radium or polonium could be seen to scintillate, with dark-adapted eyes, through a low-powered microscope. The scintillation technique served not only to count particles but also to locate their positions in space. This method became famous later when used to study the scattering of α-particles, leading to Rutherford's discovery of the atomic nucleus. Earlier Rutherford and Geiger had become involved in studying the nature of the α-particle itself. The question of its charge was crucial. By measuring the total charge deposited on an electrometer from a particular source within a set period of time and knowing the flux of particles from the number of scintillations with the same source in the same period of time, the charge on each particle could be obtained. Of course, this method assumes an efficiency of 100% for α-particles producing scintillations, an assumption that had never been tested. In a classic experiment published in 1908 Rutherford and Geiger made an invention to resolve the question [417].

9.11.1. Ionization detectors

The ionization produced by individual α-particles did not provide a sufficient signal in an electrometer. To quote Rutherford and Geiger

> Some preliminary experiments to detect a single α-particle by its direct ionisation were made by us, using specially constructed sensitive electroscopes. As far as our experience has gone, the development of a certain and satisfactory method of counting the α-particles by their small direct electrical effect is beset with numerous difficulties.
>
> We then had recourse to a method of automatically magnifying the electrical effect due to a single α-particle. For this purpose we employed the principle of production of fresh ions by collision. In a series of papers, Townsend [418] has worked out the conditions under which ions can be produced by collisions with the neutral gas molecules in a strong electric field.

As implemented by Rutherford and Geiger, the high electric field was produced by applying voltage on a fine wire running down the centre of a conducting cylindrical tube, 25 cm long and 1.7 cm in diameter. They operated the device with a gain of only a few hundred but it was enough to give distinguishable kicks ('throws') to their electrometer.

The paper ends with six conclusions.

(i) By employing the principle of magnification of ionization by collision, the electrical effect due to a single α-particle may be increased sufficiently to be readily observed by an ordinary electrometer.

(ii) The magnitude of the electrical effect due to an α-particle depends upon the voltage employed, and can be varied within wide limits.

(iii) This electric method can be employed to count the α-particles expelled from all types of active matter which emit α-rays.

(iv) Using radium C as a source of α-rays, the total number of α-particles expelled per second from 1 gram of radium have been accurately counted. For radium in equilibrium, this number is 3.4×10^{10} for radium itself and for each of its three α-ray products.

(v) The number of scintillations observed on a properly prepared screen of zinc sulphide is, within the limit of experimental error, equal to the number of α-particles falling upon it, as counted by the electric method. It follows from this that each α-particle produces a scintillation.

(vi) The distribution of the α-particles in time is governed by the laws of probability.

After these conclusions the authors go on to observe 'Calculation shows that under good conditions it should be possible by this method to detect a single β-particle, and consequently to count directly the number of β-particles expelled from radio-active substances.'.

In connection with this paper two comments are of interest. The first is that Rutherford and Geiger took the charge of the electron to be $e = 3.6 \times 10^{-10}$ esu, vastly different from the currently accepted value of 4.8×10^{-10} esu. This was obtained by averaging the results of J J Thomson (3.4), H A Wilson (3.1) and R A Millikan (4.06) (times 10^{-10} esu), all of which were obtained by observing the effect of an electric field on charged water droplets. In a subsequent paper [419] Rutherford and Geiger report the charge of the α-particle to be 9.3×10^{-10} esu. Independently, they conclude that the charge of the α-particle must be $2e$ and hence $e = 4.65 \times 10^{-10}$ esu, close to the currently accepted value. They then go on to identify the source of the error in the previous determinations of e, i.e. the evaporation of the water droplets during the measurements. Millikan eventually avoided this problem by using oil drops.

The second comment relates to conclusion (vi), which is correct but not supported by their data. Discussing the interval distribution between successive α-particles the authors show, very qualitatively, a curve which starts from zero, rises, and then comes down. They appear to have neglected to allow for the dead-time of the apparatus. We now know that the interval distribution must be, from the Poisson distribution, purely exponential.

And so was born the proportional counter, in general form not dissimilar to that used today. This device continued to be one of the most important detectors of charged particles through the rest of the century. The counter was invented to answer a particular physics question. Necessity is indeed the mother of invention.

It is perhaps surprising that it took 20 more years before the so-called Geiger–Müller (G-M) counter emerged [420]. Of course, World War I removed four years from this period since, unlike in World War II, counting nuclear particles was hardly at the top of the wartime priority list. Nonetheless, it is still surprising that it took so long. In retrospect it is difficult to imagine the circumstances under which experimental physicists laboured and the relatively primitive materials available. It was literally the era of string and sealing wax. It is a sobering experience to read, for example the chapter on Geiger counters in reference [421]. Making G-M tubes was apparently still something of a black art ten years after their invention.

The Geiger–Müller (or 'Zählrohr') counters were an immediate success, with many applications. They were sensitive over the area projected by the outer cylinder, a large area at the time, producing signals sufficiently large to require a minimum amount of vacuum tube amplification to drive mechanical registers for tabulating counts. The G-M counters were indiscriminate, responding to any and all types of ionizing radiation, a serious disadvantage in many applications. However, when arranged in time coincidence the counters could be used to define beams of particles. In this mode, initially in the hands of Bothe [422] and Rossi, these devices became a principal tool in the study of cosmic rays. In 1954, W Bothe was awarded the Nobel Prize in Physics 'for the coincidence method and his discoveries made therewith'. (Rossi's circuit was a very significant improvement: a triple coincidence device, totally symmetric in all the inputs, whereas the original Bothe circuit was twofold and asymmetric. Coincidence circuits, now called logical 'AND' circuits, have continued to be incrementally improved in timing resolution ever since, first in the vacuum tube form and now in transistor versions.)

Somewhat earlier, physicists had started using vacuum tube amplifiers in conjunction with ionization chambers. In 1926 Greinacher [423] showed it possible to amplify the current due to an α-particle sufficiently to register as a click in a headphone. Subsequently, he also detected single protons by the same method. High-voltage power supplies utilizing transformers and vacuum tube diodes became the norm, largely replacing the banks of batteries used previously.

The mechanical counting registers used in conjunction with G-M tubes were limited to a maximum rate of about ten per second. Correspondingly, a counting loss of the order of 10% occurred for rates as low as one per second, stimulating the invention of electronic 'prescaling' of the counts by divide-by-two circuits before registering on mechanical counters. The first of these binary counting circuits, devised by Wynn-Williams in 1932 [424], consisted of two vacuum tubes coupled so that only one could be conducting at a time. The driving pulse would turn off the conducting tube, but at the end of the driving pulse, in the absence of special provisions, either tube could become conducting whereas it was desired that the tubes conduct alternately on successive driving pulses. Memory of the previous conducting state, a necessary intrinsic feature, was provided by an RC circuit with a time constant long enough to more than cover the duration of the driving pulse. The divide-by-two circuits could be cascaded to any power of two. Binary division with six stages (by 64) before recording on mechanical registers was common.

A very significant improvement in the reliability of binary dividers was made in the early 1940s by W A Higinbotham using diode coupling between stages. While Higinbotham neither published nor patented his contribution, it is included in reference [425]. The early vacuum tube circuits could respond in the microsecond range, thus avoiding the counting-loss problem of mechanical counters. The essential ingredients of these circuits, bistable elements with intrinsic memory, persist in the highly refined binary transistor circuitry of today.

In the 1960s large-scale integration of transistors was accompanied by a striking decrease in the cost of complex circuits. Their reliability, as measured by the mean time between failures, made it possible to contemplate experiments with massive quantities of electronics. It was in this technological environment that G Charpak, working at CERN, started a revolution in particle detectors by constructing large-area multi-wire proportional chambers. (See also Thompson in the first of references [3], p 274.)

Ever since the advent of scintillation counters, proportional counters had almost become extinct, but now they were resurrected on a grand scale. Multi-wire chambers had been used earlier, but never approaching the magnitude pursued by Charpak. Each wire had its own solid-state amplifier and discriminator to produce the signals necessary to log on magnetic tapes for later computing processing. Charpak further refined the idea of the drift chamber: the time between the passage of a particle through a chamber and the appearance of the signal on the 'sense' wire indicated the distance of the particle's trajectory from the wire. In certain gases the electron drift velocity was remarkably independent of the electric field, making the device very linear. Layers of chambers with the wires at fixed angles to each other provided x–y position information, critically dependent on the position of the sense wires themselves. While the scheme required unprecedented electronic circuitry, such circuitry was inexpensive, and perhaps more importantly, reliable. These devices were fast, acquiring great quantities of data in a short time and providing in ideal form the sort of data required for analysis in a modern digital computer. The drift chamber is now a standard part of the particle physicist's tool kit. For this work Charpak was awarded the 1992 Nobel Prize in Physics.

9.11.2. Scintillation and Čerenkov counters

By 1903 zinc sulphide was known to emit visible flashes of light when bombarded by α-particles, but was relatively insensitive to the gamma and beta radiations most often accompanying the α-particles. The arduous business of counting visible scintillations in dark rooms with dark-adapted eyes was the technique of choice for many years, yielding much information about α-radiating sources, the nature of the α-particles and Coulomb scattering. It was only the invention of photomultipliers capable of detecting the scintillations and converting them to electrical signals that afforded full exploitation of the scintillation phenomenon.

See also p 1200

In the early 1900s certain materials when bombarded with electrons of rather low energy (about 100 volts) showed a propensity to emit more electrons than struck the surface. These surfaces were thus acting as electron multipliers! The first patent on using these materials to amplify small currents, granted in 1919 to Joseph Slepian [426], did not result in practical applications. Only later [427], in 1936, were multiple-stage devices, using both magnetic and electrostatic focusing, invented and applied to the amplification of photoelectric currents. Even though the early photomultipliers were scarcely practical, they were recognized as presenting a unique noise problem, which Shockley and Pierce addressed [428].

The first device resembling those in current use was described by Zworykin and Rajchman in 1939 [429]. With careful attention to electrostatic focusing, these inventors managed to avoid the severe space-charge limitations of the early models. In the nomenclature of the RCA company, the new device became the 931A photomultiplier. This tube had commercial applications in sensing the sound track on movie film, and was also used in World War II, far from its original purpose, as a generator of white noise for masking radio and radar signals. Tubes selected for exceptional sensitivity were labelled as 1P21.

In the meantime a new effect associated with the energy loss of charged particles passing through material had been discovered. In 1934 Čerenkov [430] had observed 'feeble visible radiation' from β-rays passing through a clear liquid. In 1937 Frank and Tamm [431] developed an energy loss formula which included radiation at a particular angle relative to the track of the particle. That this radiation was emitted at a characteristic angle was very shortly afterwards confirmed by Čerenkov [432]. Photomultipliers had still another application: detecting Čerenkov radiation.

While zinc sulphide was almost ideal for detecting slow-moving α-particles, it appeared difficult, if not impossible, to produce the large crystals needed for electron and γ-ray detection. For this reason, the discovery in 1947 by Kallman [433] of scintillation by various organic crystals proved enormously important, immediately stimulating investigations of various materials. P R Bell [434] found that anthracene and later stilbene were good scintillators and could be grown in large sizes. These materials, with very short recovery times, were quickly exploited by experimentalists. Viewed by the RCA 1P21 phototubes, they made sensitive counters with exceedingly fast response that could cover relatively large areas. These devices played a pivotal role in the discovery of the neutral pion, the early lifetime measurements on positronium, and the measurement of the relative polarization of the two γ-rays from positronium annihilation.

Certain fluorescing chemicals were also found which, when added to organic liquids, would result in a solution that would scintillate with good efficiency [435]. Such liquid scintillators proved extremely useful for large volume detectors. Later, in the mid-1950s, it was found that a plastic such as polystyrene could serve as the solvent. Now, towards the end of the century, plastic is the scintillation material of choice in every application except γ-ray spectroscopy, being available commercially in all shapes and sizes.

Parallelling the developments in organic scintillators in the late 1940s, Hofstadter [436] found that sodium iodide activated by thallium impurities proved a very efficient scintillator. Unlike zinc sulphide, it could be grown into very large transparent crystals, and despite its highly deliquescent nature could be packaged into quite stable forms. The high atomic number of the iodine gave the crystal a high sensitivity to γ-rays. Indeed, it became possible, depending on the γ energy and the size of the crystal, to capture the full γ-ray energy. When coupled to a photomultiplier and a device for measuring the resulting spectrum of pulse heights, this crystal led to an enormously valuable spectrometer for γ-rays and mesic x-rays.

Again necessity stimulated invention. The RCA 1P21 phototube had a high sensitivity to photons, but it was difficult to channel the light from a crystal efficiently to photocathodes in the tube's interior. To correct this disadvantage, the RCA company, as well as EMI, developed the first end-window phototubes, the 5819 (RCA) and the 5060 (EMI) (see reference [437]). These new phototubes proved critical for the new spectrometers, essentially making them feasible.

The electronic developments associated with pulse-height analysers (kick-sorters in England) had started at Los Alamos during World War II for use with ionization chambers and proportional counters. In general, they were rather cumbersome and complicated devices, notoriously difficult to maintain. The threshold for each channel would drift relative to its neighbour, which in turn would distort pulse amplitude spectra. It was Wilkinson [438] who first hit on the solution to this problem, by 'laying the pulse on its side', i.e. by converting a pulse height to a pulse time. He charged a capacitor with the input pulse, then discharged it linearly in time, measuring the time between input and the end of the discharge by simply using an oscillator. The resulting time interval was then proportional to the original pulse amplitude. Electronically, it was much easier to keep time intervals constant than to keep voltage amplitude thresholds constant. Today, these circuits, called ADCs ('analogue-to-digital converters'), are ubiquitous throughout the electronics industry. The old technique of setting channel intervals with voltage comparators is still alive because of its speed advantage. When applied to transistor electronics, the device is called a 'flash ADC'.

9.11.3. The visual techniques

We have already mentioned (section 9.4.1) the key role played by photographic emulsions in detecting tracks of elementary particles. Cloud chambers, bubble chambers and spark chambers were also of importance.

9.11.3.1. Cloud chambers. Even before the new century began, C T R Wilson had been studying the process of droplet formation in supersaturated water vapour. It is not a simple process. Even in a supersaturated vapour, tiny droplets will inevitably evaporate in the absence of a formation nucleus, such as a dust particle, which in effect provides a flat surface for condensation. Alternatively, Wilson learned that charged molecular ions also provide centres for droplet formation since the repulsion of the charges causes the droplet to become larger, overcoming the tendency to evaporate away.

Experimentalists were quick to seize on the effects uncovered by Wilson. As we have already noted, in the early 1900s the study of the behaviour of charged water droplets in an electric field provided the first measurements of the electron charge. It was not until 1911, however, that Wilson first observed and photographed [439] the formation of tracks of droplets along the paths of single charged particles in a gas of water vapour supersaturated by a sudden expansion of its volume. Therein was born one of the most important tools in particle physics, the expansion cloud chamber. In 1927 Wilson was awarded the Nobel Prize 'for his method of making the paths of electrically charged particles visible by condensation of vapor'.

The application of this technique to the study of cosmic rays was considerably enhanced when the tracks of ions were found to persist for a time sufficiently long to permit the chamber's expansion and the growth of droplets along the tracks. The expansion could be triggered by signals from counters [47] (cf section 9.4.1). The discoveries of the positron, the muon and strange particles were all made using the expansion cloud chamber.

It occurred to a number of people in the 1930s to arrange for a cloud chamber to be continuously sensitive. A first crude device was constructed by Hoxton [440], but its general principle was elaborated by Langsdorf into a device called a 'diffusion cloud chamber' that became the model for all subsequent chambers [441] (see also reference [442]). A vertical temperature gradient was arranged in a gas such that the vapour was unsaturated at the top but highly saturated at the bottom. In between, droplets would form in a narrow horizontal layer. The sensitive region tended to be rather thin, making the device unsuitable for vertically moving particles such as cosmic rays but presenting clear advantages for the horizontally moving particles from accelerators.

Ralph Shutt's group at Brookhaven exploited the diffusion chamber, initially at the Nevis cyclotron [443] and later at Brookhaven. The chamber could be made of a size ideal for studying the production and decay of the recently discovered strange particles. Traditionally the practice had been to place sheets of material, copper or lead, in the chamber to serve as sources of the new particles when struck by other particles incident from outside the chamber. The ideal target material was hydrogen. To this end, the diffusion chamber of Shutt et al contained 21 atmospheres of hydrogen with methanol vapour. When the chamber was exposed to 1.5 GeV negative pions at the Cosmotron at BNL, the associated production of strange particles was discovered.

Elementary Particle Physics

9.11.3.2. Bubble chambers. Using the diffusion chamber both as a target and as a detector was a clear advantage. However, even when operating at a pressure of 21 atmospheres, the probability of a pion interacting in the chamber was only about one in a thousand. Even with as many as ten incident particles in each photograph, it was still necessary to take 100 pictures per observed interaction. Therefore, when Glaser's invention of the bubble chamber [100] (see section 9.4.5.3) proved that liquid hydrogen (with a density 50 times that of gaseous hydrogen at 21 atmospheres) would work [444], the diffusion chamber was doomed as a tool for doing physics.

The explosive development of the Glaser bubble chamber in its liquid hydrogen version culminated in the 1950s with the Berkeley 72″ chamber. A development of the Alvarez group, it was to be the source of discovery of many unstable particles, as discussed in section 9.4.

Appropriately, the Nobel Prize was awarded in 1960 to Donald Glaser 'for the invention of the bubble chamber', and in 1968 to Luis Alvarez [101] 'for his decisive contributions to elementary particle physics, in particular the discovery of a large number of resonant states, made possible through his development of the technique of using the hydrogen bubble chamber and data analysis'.

9.11.3.3. Spark chambers. An effort was made in the late 1940s to find a counting device that would improve on the rather poor timing resolution characteristic of Geiger counters. Keuffel [445] and Pidd and Madansky [446] constructed 'spark counters' consisting of parallel conducting plates between which high voltages were applied. An electrical discharge would occur whenever a charged particle traversed the gap between the plates. While Keuffel observed that the discharge occurred along the path of the particle, the emphasis at the time was on electronic timing, so the feature of track delineation was ignored. Shortly after, scintillation counters with superb timing characteristics were invented, removing the principal motivation for further spark-counter development. In addition, spark counters with steady voltages were plagued with spurious discharges not associated with the passage of particles. When pulsed voltages were used [447] it was found that spurious discharges could be avoided, but only single tracks could be recorded.

The spark chamber was first proved viable for recording particle tracks by Fukui and Miyamoto [448]. Using pulsed voltages, initiated by scintillation counters, with a neon–argon mixture as the gas between the plates, they showed that tracks of more than one particle could be recorded. The receipt of the preprint of their paper in the US initiated a flurry of activity devoted to further development. Foremost among the leaders in this development were Cronin at Princeton [449] and Cork and Wenzel at Berkeley [450]. By 1960 these groups were doing highly interesting physics using spark chambers. Guided by their work, a Columbia–Brookhaven group built a large, massive spark chamber to detect neutrino interactions at the AGS at Brookhaven (section 9.5.1.6).

Spark chambers had a clear advantage over bubble chambers in that the interaction and tracking of particles could be separated, and the multiple Coulomb scattering, which limited the momentum resolution in bubble chambers, could be reduced significantly. Furthermore, spark chambers could be triggered on interesting events, as determined by accessory information from counters. These advantages were quickly exploited and at least two major discoveries (the existence of two neutrinos [156] and CP violation [163]) were shortly made with this technique.

Initially, information from spark chambers was obtained photographically. The spark positions were digitized from film records and processed on large mainframe computers. Subsequently, in some experiments, the thunder, not the lightning, was recorded using acoustical detectors. Later, the chamber electrodes were made of wire crossing magnetostrictive lines. The discharge current would generate an acoustical signal in the line, to be sensed at its end. The arrival time provided a measure of the spark's position. Such acoustical schemes lent themselves to full automation in a computer environment. Variations developed in the 1960s included the wide gap chamber and the streamer chamber. The spark chamber, operated in a non-visual mode, became the detector of choice in many experiments until replaced by the multi-wire proportional chambers and by Charpak's drift chambers [451], able to acquire data at far higher rates.

9.12. Overlaps with other subjects

As emphasized earlier, particle physics was very different in the periods before about 1960 and after about 1973. Before the 1960s, properties of individual particles were studied without the unifying themes of quarks and leptons. Similarly, the understanding of strong and weak interactions underwent a significant change with the employment of non-abelian gauge theories after the early 1970s.

During these developments, particle physics drew on a number of other areas for insight and experimental results. We mention a few of these.

9.12.1. Nuclear physics

9.12.1.1. Neutron lifetime and decay properties. Nuclear reactors provided copious sources of neutrons, whose mean lifetime and ratio of axial–vector to vector couplings were important ingredients in the evolution of weak-interaction theory. In view of the difficulty of confining a neutron for its mean lifetime (about 15 minutes), an accurate measurement of this

Elementary Particle Physics

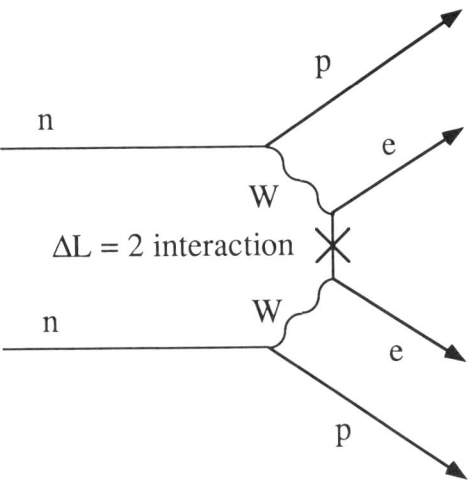

Figure 9.36. *Diagram contributing to neutrinoless double-beta decay. The line labelled by a cross is a Majorana neutrino, containing components with lepton number $L = 1$ and $L = -1$.*

quantity remained elusive for many years. 'Magnetic bottle' techniques to contain neutrons led to significant progress. The present value [49] of 887 ± 2 s provides not only useful information on the axial–vector coupling constant [452] (see sections 9.5.1 and 9.5.3), but also constrains the rate at which hydrogen was incorporated into helium in the early Universe (section 9.12.4.4).

9.12.1.2. Neutrinoless double-beta decay. All processes observed so far conserve *lepton number*, an additive quantum number equal to 1 for the negatively charged leptons and their neutrinos and −1 for the corresponding antiparticles. However, since neutrinos can, in principle, mix with their antiparticles [453], an interaction changing lepton number by two units is conceivable. Such a *Majorana mass* can lead to neutrinoless double-beta decay through the graph illustrated in figure 9.36.

Present searches for neutrinoless double-beta decay have observed no signal for this process, leading to an effective upper limit of a few eV on the relevant neutrino Majorana masses [454].

9.12.1.3. Beta-decay limits on neutrino masses. Details of the electron spectrum in nuclear beta decay can reveal distortions stemming from neutrino masses. No such effects have been observed, leading to upper limits of several electronvolts on the electron neutrino mass [455].

9.12.2. Atomic physics

We referred in section 9.7.3.3 to experiments searching for parity violation in atomic physics due to the neutral-current interaction. Many of these

755

results rely on detailed atom-trapping methods developed since the earliest post-World War II days. The trapping methods have also been of great help in measuring anomalous magnetic moments of leptons (section 9.3), in studying neutron decays (see section 9.12.1.1), and in searching for the neutron's electric dipole moment (so far, without success).

9.12.3. Condensed matter

9.12.3.1. Nambu–Jona-Lasinio model. The bound state model of the pion proposed by Nambu and Jona-Lasinio [179] was inspired by analogy with superconductivity [456]. In both cases, a pair of fermions can form a bound state, leading to an energy gap and a zero-mass excitation.

9.12.3.2. Renormalization group. The behaviour of physical quantities under changes in scale was studied not only in elementary particle physics [339], but also in condensed-matter problems, where 'real-space' aspects of scale changes emerged particularly explicitly [457].

9.12.3.3. Lattice gauge theory. Non-relativistic field theories have been formulated on discrete lattices, such as occur in actual solids, yielding many properties of elementary and collective excitations. For example, one can determine the presence or absence of phase transitions, the masses and spatial extent of bound states, and bulk properties such as critical temperatures and specific heats. The methods for describing these phenomena are readily adapted to continuum field theories by approximating space-time with a discrete lattice. This programme [357] has been particularly vigorously applied in constructing a lattice version of quantum chromodynamics (QCD), aimed at understanding all of low-energy hadron physics. The lattice spacing must then be much less than a hadron's size (10^{-13} cm). One current obstacle to this programme is its inability to reproduce the pion well. In the limit of exact (spontaneously broken) chiral symmetry, the pion would be massless, with an infinite Compton wavelength, requiring for its representation a lattice of infinite extent. The actual pion has a Compton wavelength exceeding 10^{-13} cm, demanding many lattice points (and hence larger and faster computers) to reproduce correctly both pions and the rest of hadron physics.

9.12.3.4. String theory and soluble two-dimensional models. We mentioned dual models in section 9.5.4.5, and will come to their present incarnation (string theories) in section 9.13.5. Such theories also have been found relevant to two-dimensional models of critical behaviour in condensed-matter systems [458].

9.12.4. *Astronomy, astrophysics, gravitation and cosmology*

The interface between elementary particle physics and astrophysics is a vast subject (see, for example, the seminal text of reference [459]), for which we can only touch on a few selected topics. Particle physics is at the heart of many astrophysical processes (such as supernova evolution), while astrophysical constraints have proved useful in anticipating elementary particle properties (such as the number of types of light neutrino). For many problems in astrophysics, solutions based on fundamental particle physics are among several options.

9.12.4.1. Solar neutrinos. The detection of neutrinos from the Sun was proposed long ago by Pontecorvo [460]. The first search for solar neutrinos was mounted by R Davis and collaborators [461], employing the reaction $\nu + {}^{37}Cl \rightarrow e^- + {}^{37}Ar$ with chlorine in the form of cleaning fluid contained in a large tank located in the Homestake Gold Mine in South Dakota to shield against cosmic-ray backgrounds. The detection of a signal relies on the extraction from the tank of less than one atom of argon per day, whose radioactive decay is the signature of its production.

Expectations for Davis' experiment were worked out in detail by Bahcall and collaborators [462]. The rate observed in the chlorine experiment was lower than expected and has remained so over more than 20 years. The present ratio of experiment to theory is about 0.3.

More recently, several other experiments, sensitive to different neutrino energies, have also indicated a solar-neutrino rate lower than expected. An experiment in the Kamioka mine in Japan, sensitive only to the highest-energy neutrinos (above 5 MeV), sees a rate about half that expected. Its detector consists of a large tank of very pure water, viewing neutrino–electron interactions via Čerenkov light [463]. Two experiments sensitive to a lower neutrino energy threshold [464] rely on the reaction $\nu + {}^{76}Ga \rightarrow e^- + {}^{76}Ge$, with extraction of the radioactive germanium and subsequent detection of its decay. These also show a signal somewhat lower than predicted by the standard solar model.

Do these results, if correct, indicate a shortcoming of solar models (with all their attendant details of nuclear physics), or do they point to new elementary particle physics [465]? For example, do electron neutrinos (the type emitted in the Sun) undergo oscillations [466] to other species undetected in the above experiments? Such oscillations could either take place in vacuum or be induced by interaction with the Sun's matter.

Further experiments are in the construction or planning stage [467]. An ideal detector would be sensitive to (i) a wide range of neutrino energies, ranging from less than a few hundred keV to many MeV, (ii) the direction of the neutrinos' source, as in the Kamioka experiment, and (iii) interactions of neutrinos other than ν_e through neutral-current effects.

9.12.4.2. Supernova neutrinos. The Kamioka detector was one of a number set up in the early 1980s for the entirely different goal of searching for proton decay (see section 9.13.3). Another large water Čerenkov detector had been set up in the Morton Salt Mine northeast of Cleveland, Ohio, with the same purpose (see figure 9.37 [468]). After several years of operation, neither detector had seen any signal for proton decay. However, on 23 February 1987, both saw bursts of counts induced by neutrinos emerging from the explosion of a supernova in the Large Magellanic Cloud [469]. The observation by Kamioka was particularly fortunate since the detector was just a minute away from a scheduled shutdown. The supernova was the first seen in 1987, and hence was denoted SN1987A.

This observation confirmed basic predictions of the life cycle of a supernova [469, 470], which in turn relied crucially upon the presence of neutral as well as charged currents in the weak interactions. Furthermore, the neutrinos arrived at the Earth within a few seconds of one another despite travelling about 160 000 light years, allowing estimates of upper limits on their masses [471].

If and when a supernova explodes in our own galaxy (whose diameter is about 60 000 light years), the neutrinos emitted will produce a stronger signal than SN1987A. Several other detectors, to be mentioned in section 9.13.5.5, may be sensitive to this signal.

9.12.4.3. Cosmic microwave background. In 1948, George Gamow, R A Alpher and R C Herman predicted [472] that the black-body radiation remaining from the birth of the Universe should have a temperature of a few degrees K. This expectation was verified by Penzias and Wilson in 1964, whose microwave radiometer revealed a persistent noise signal [473]. A Princeton group that had been looking for the same effect supplied an explanation [474].

The microwave background has proved remarkably homogeneous and isotropic, with a black-body temperature of 2.74 K, but displays a distortion caused by the velocity of our galaxy of some 600 km s^{-1} towards a point in Virgo, as well as fluctuations of order 10^{-5} recently identified by the Cosmic Background Explorer (COBE) Satellite and other experiments [475]. These fluctuations are a relic, looking backwards in time, of the physics generating the much larger fluctuations now seen as galaxies and clusters.

9.12.4.4. Nucleosynthesis and number of neutrinos. The neutron lifetime, as mentioned earlier [452], affects the rate at which hydrogen was incorporated into helium in the early Universe. Another crucial variable in that calculation lies in the number of light neutrino species. The microwave background radiation provides a calibration of the mass density of the Universe contributed by each neutrino species at the time

Elementary Particle Physics

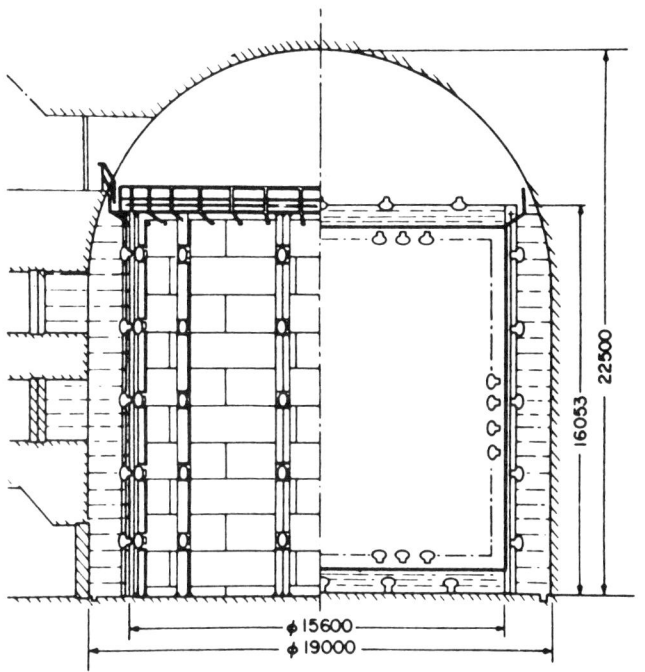

Figure 9.37. *Top: sketch of the Kamiokande II Detector. Bottom: wide-angle photograph of the IMB-3 detector. The hemispheres partially visible in the square plastic sheets are large photomultiplier tubes. The plastic squares, called waveshifters, catch a portion of the light which does not directly strike the photomultipliers. Also visible is a diver. The photograph is taken through about 20 metres of water, showing the remarkable clarity of the purified water.*

of helium nucleosynthesis. This density affects the rate of the Universe's expansion and hence the competition between the decay of neutrons and their incorporation into helium. Too many neutrino species would imply too much helium. An upper limit was thus set [476] of no more than four species; even this number started to look unlikely when limits on the neutron lifetime became tighter. Theorists thus breathed a sigh of relief (and at least one theorist won a case of wine) when the study of Z decays at LEP (section 9.7.5.3) indicated that indeed three was the correct number.

9.12.4.5. Baryon asymmetry of the Universe. The observed part of the Universe contains far more baryons than antibaryons. As mentioned in section 9.5.2.5, the groundwork for understanding this asymmetry was laid by A Sakharov [175] in 1967. The necessary ingredients, still important in any current theory [176], included:

(i) CP violation;
(ii) baryon-number violation (to be discussed in section 9.13.3.2);
(iii) a period in which the Universe was out of thermal equilibrium.

The possibilities for realizing these conditions are quite varied, but the form of CP violation incorporated into the phases of the CKM matrix seems insufficient. Rather, the CKM phases may be only one manifestation of a broader role for CP violation.

9.12.4.6. Cosmic-ray physics. Cosmic rays played an important role in elementary particle physics until the advent of accelerators in the mid-1950s, as mentioned in section 9.4. Even today, however, information on fundamental interactions continues to emerge from cosmic-ray studies. As one example, total cross sections for particle interactions appear to rise with increasing energies beyond the limits of terrestrial accelerators [477].

A new field of cosmic-ray physics has been spawned by the possibility of observing point sources of gamma rays at TeV energies and beyond. Čerenkov detection of TeV air showers has pinpointed emissions from the Crab Nebula (the remnants of a supernova seen by Chinese astronomers in 1054), and possibly even from extragalactic sources [478]. The search for point sources of gamma rays at higher energies and for the highest-energy cosmic rays is being undertaken using extensive arrays, some as large as 10^8 m^2 in area [479].

9.12.4.7. Black holes. A star may form with a mass so great that its own gravitational field prevents everything, even light, from escaping from its surface. To quote Laplace [480] (1798)

> A luminous star, of the same density as the Earth, and whose diameter should be two hundred and 50 times larger than that

of the Sun, would not, in consequence of its attraction, allow any of its rays to arrive at us; it is therefore possible that the largest luminous bodies in the Universe may, through this cause, be invisible.

Astronomical evidence has now been accumulating that such objects, now called [480] 'black holes', do, indeed, exist.

The relation between black holes and elementary particle physics is one of the fascinating unsolved problems of the present century. Stephen Hawking showed a black hole to be capable of producing pairs of particles in its vicinity, one of which is ejected and one of which falls into the black hole [481]. The ejected particle carries off energy which the black hole must supply by losing mass. Thus, a black hole eventually evaporates. Is this behaviour the source of a paradox? A pure quantum-mechanical state is thereby converted to a mixed one, leading to numerous suggestions for modifying the laws of classical gravitation, quantum mechanics, or both. As candidate quantum theories of gravitation, string theories (section 9.13.4) are being used to model such processes.

9.12.4.8. Dark matter. There is ample evidence that not all matter in the Universe takes the form of visible stars, gas or dust. One can measure the mass in our galaxy (or others), for example, by plotting the velocity of rotation of objects around the centre as a function of distance from the centre [482]. Much more mass is inferred than is seen.

Is the 'dark matter' of the Universe in the form of failed stars (Jupiter-size objects) or is it more exotic? Many proposals have been made, including a variety of as yet unseen elementary particles such as very light spinless particles (for example, particles known as 'axions' [483]), particles predicted by supersymmetry (see section 9.13.1.2) and massive neutrinos (see section 9.13.2). Each species seems to be able to account for some but not all of the desired properties of dark matter. For instance, a neutrino with a mass of about 10 eV, stable on the time-scale of the Universe's lifetime, can provide just the right amount of mass for the Universe to neither expand indefinitely nor contract back to a point after a finite time. But a 10 eV neutrino does not form the seeds for small-scale structure seen in galaxies, requiring another mechanism for galaxy formation.

9.12.4.9. Inflation and baryogenesis. The study of the early phases of the Universe led in the early 1980s to a remarkable observation that persisted in essence while undergoing various revisions in detail [484]. It is likely that at one or more stages, the Universe underwent an exponential increase in scale, wiping out all fluctuations. We infer this behaviour from the remarkable isotropy and homogeneity of the Universe today. A consequence of this 'inflationary scenario' is that the Universe should be exactly on the boundary between open (expanding forever) and closed (collapsing back to a point). This observational question remains to be settled by the study of the Universe's dark matter mentioned above.

9.12.4.10. Searches for deviations from Einstein's gravitation theory. The general theory of relativity does not distinguish between the gravitational force felt by any object; all that matters is its mass. Early tests [485] confirming this 'equivalence principle' had been refined to considerable precision by the mid-1960s [486]. A reanalysis [487] of the original experiments seemed to show a departure from the equivalence principle in the mid-1980s, leading to a flurry of improved tests and searches for a 'fifth force'. No evidence for it has survived these improved experiments [488].

9.13. Unsolved problems and hopes for the future

9.13.1. Electroweak theory: symmetry-breaking sector

9.13.1.1. Hunting for the Higgs boson. The electroweak theory described in section 9.7, though immensely successful in reproducing current data, is incomplete. We do not know how the masses of the W, Z, quarks or leptons (all of which break the original symmetry) actually arise.

The Weinberg–Salam mechanism for breaking electroweak SU(2) × U(1) involves the existence of a *Higgs boson H*, corresponding to the fluctuations of a neutral field with respect to its vacuum expectation value v. The non-zero value of v gives rise to the W and Z masses, with $v = 2^{-1/4} G_F^{-1/2} = 246$ GeV fixed by the value of the Fermi coupling constant. Masses of quarks and leptons arise through their Yukawa couplings to the Higgs field.

The *mass* of the Higgs boson, not specified by the electroweak theory, is arbitrary, lying anywhere between the experimentally determined lower limit [489] (about 60 GeV/c^2 at present) and about 1 TeV. The upper limit reflects a requirement that the scattering of two longitudinally polarized gauge bosons preserve S-matrix unitarity [490]. A similar requirement for pion–pion scattering is met by the observed spectrum of $\pi\pi$ resonances.

The search for the Higgs boson is one reason often quoted in support of multi-TeV hadron colliders (see section 9.13.5). A straightforward Higgs signature would be a resonance in the W^+W^- and ZZ channels; this and many others are discussed, for example, in references [491] and [492].

9.13.1.2. Alternatives for the Higgs sector: supersymmetry, compositeness. Two main streams of theoretical thought concern the underlying nature of the Higgs boson. One of them views the Higgs boson as an elementary scalar particle whose mass must be 'protected' by some mechanism from acquiring large (and uncontrollable) radiative corrections. *Supersymmetry* [493], the most popular of such schemes, postulates for every particle of spin S the existence of a partner of spin $S\pm\frac{1}{2}$, whose presence in radiative corrections would exactly cancel the contribution of its 'superpartner' if exactly degenerate with it in mass. Supersymmetry, of course, would be verified by observing superpartners, not too different in mass from their ordinary counterparts to facilitate the desired cancellation. Most current versions of supersymmetric theories predict that at least some superpartners should exist below a mass of 1 TeV.

The other class of theories views the Higgs boson as a composite of more fundamental objects [494]. Such theories of *dynamical electroweak symmetry breaking* require the Higgs boson to reveal its structure by an energy of one or two TeV, implying a rich spectrum of resonances in W–W, Z–Z, and W–Z scattering starting at such energies.

9.13.2. Neutrino mass

9.13.2.1. Direct searches. Searches for electron-neutrino masses in beta-decay experiments have substantially improved over the past few years, as mentioned in section 9.12.1.3, with hopes for further advances [495]. Accelerator-based experiments are expected to make modest but not spectacular gains over present bounds on muon and tau neutrino masses.

9.13.2.2. Oscillations. A major hope for future observation of neutrino masses rests upon the phenomenon of neutrino oscillations (section 9.12.4.1), detectable in several ways.

(i) Further experiments could confirm the proposed role of neutrino oscillations in solar neutrino physics by measuring the *spectrum* of neutrinos from the Sun and comparing it with predictions of solar models.

(ii) Studies of the fluxes of different neutrino types generated by cosmic rays in the atmosphere can test for oscillations. At present the observed ratio of muon neutrinos to electron neutrinos seems somewhat smaller than anticipated theoretically [496], but further data are needed.

(iii) Direct searches for oscillations at accelerators [497] are probing new ranges of masses and mixing parameters. Experiments also have been proposed in which an accelerator neutrino beam is directed at an underground target several hundred or more km away [498].

9.13.3. Grand unified theories

9.13.3.1. Particle content. The description of electroweak and strong interactions through two separate Yang–Mills theories in the early 1970s

Twentieth Century Physics

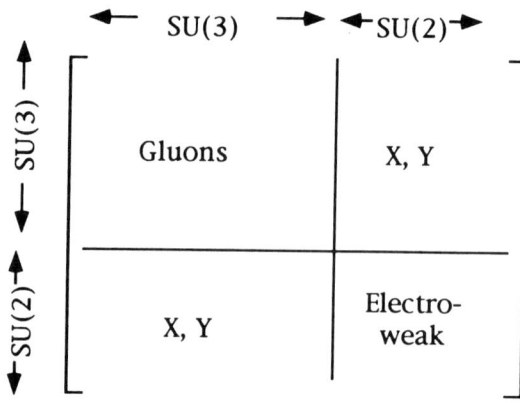

Figure 9.38. *Gauge bosons contained within the SU(5) grand unified group.*

suggested that a *single* such theory might describe particle physics. The 'grand unified' group would have to include $SU(3)_{QCD} \times SU(2)_{weak} \times U(1)_{weak}$ as a subgroup. Whereas the coupling constants of these groups differ at low energies, their different dependences on changes of momentum scale suggest that they approach one another at high energies. At such a 'unification energy', quarks and leptons behave very similarly to one another [499].

Grand unified groups proposed quite early included SU(5) (reference [500]) and SO(10) (reference [501]). The structure of SU(5), in particular, is very easy to visualize, involving the 3 × 3 matrices of SU(3) and 2 × 2 matrices of SU(2) arranged in block diagonal form, as shown in figure 9.38. The U(1) fits in naturally as a diagonal SU(5) matrix commuting with both SU(3) and SU(2). A minor blemish on the theory lies in its slightly arbitrary choice of quarks and leptons as members of 5- and 10-dimensional representations of SU(5). This aspect is handled more elegantly in the SO(10) scheme, where a single 16-dimensional representation suffices for each family of quarks and leptons. The presence of small but non-zero masses for neutrinos in SO(10) can be arranged, since the theory includes both left-handed and right-handed neutrinos [502].

Coupling constants in SU(5) do not approach a common value at high energy, a feature which can be bypassed by making the theory supersymmetric [503] or by permitting unification at more than one mass scale [504]. These possibilities are compared with the standard SU(5) scheme in figure 9.39.

9.13.3.2. Proton decay; experiments. Grand unification predicts that the proton can decay, with a pair of quarks turning into an antiquark and a lepton, by the exchange of a gauge boson of the extended group (see figure 9.40). This generic feature [499] was anticipated earlier by

Elementary Particle Physics

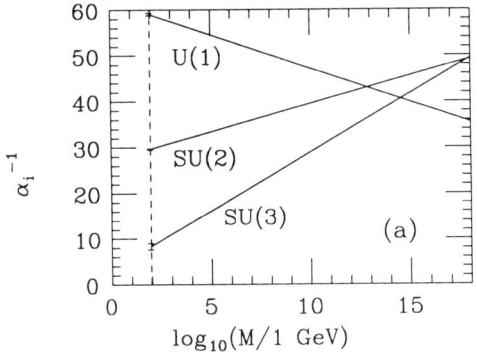

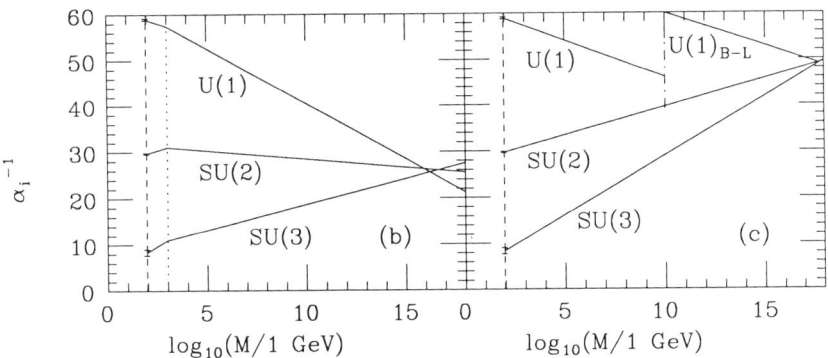

Figure 9.39. *Behaviour of coupling constants predicted by the renormalization group in various grand unified theories [504]. Error bars in plotted points denote uncertainties in coupling constants measured at $M = M_Z$ (broken vertical line). (a) SU(5), showing failure of coupling constants to meet at a point; (b) supersymmetric SU(5) scheme [503] with superpartners at 1 TeV (dotted line); (c) example of a two-scale SO(10) model with an intermediate mass scale (dot–dashed vertical line).*

Sakharov as a way to generate the baryon asymmetry of the Universe [175] in the presence of CP-violating interactions, as mentioned in sections 9.5.2.5 and 9.12.4.5. Many experiments were mounted beginning in the late 1970s to search for proton decay. The simplest SU(5) grand unified theory predicted a proton lifetime lower than about 10^{30} years [176], accessible to experiments with detectors of several tens of tons.

As of now, several multi-kiloton detectors have set limits [505] on the proton lifetime near $\tau_p > 10^{32}$ years. The two largest, the Kamioka and IMB water Čerenkov detectors mentioned in section 9.12.4.2, did record neutrinos from SN1987A. A 50-kiloton version of the Kamioka detector ('Super-Kamiokande') is now under construction [506], aimed at extending the lifetime limit to about 10^{34} years.

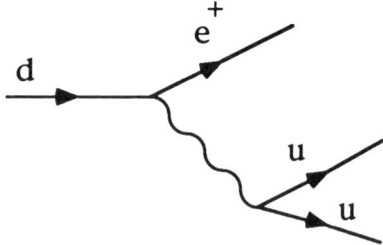

Figure 9.40. *Example of a process contributing to proton decay.*

9.13.3.3. Magnetic monopoles. The transition in the early Universe from a symmetric phase to one with broken symmetry can generate large numbers of magnetic monopoles [507]. Limits on their existence are very stringent [508], thus requiring means to greatly dilute their abundance or to avoid their generation. The apparent dearth of monopoles contributed to the inflationary universe scenario mentioned in section 9.12.4.9.

9.13.4. String theories

Some aspects of dual resonance models and their connection to the physics of strings have already been mentioned in section 9.5.4.5. The connection of string theories to condensed-matter problems was mentioned in section 9.12.3.4. A 'heterotic' version, with both 26- and 10-dimensional properties, gave the promise of realistic grand unification schemes [509]. It was hoped that the observed quarks and leptons could be explained in terms of singularities on surfaces generated when the extra dimensions curled up into unseen structures [510].

The predictive power of string theories for physics at the 100 GeV scale was soon recognized to be limited, as expected for theories whose fundamental scale is the Planck mass $m_P \equiv (\hbar c/G_N)^{1/2} \approx 10^{19}$ GeV/c^2, where G_N is Newton's gravitational constant. Nonetheless, construction of string theories with implications for grand unification schemes continues. Attention has also recently been devoted in string theories to problems of their internal self-consistency and to the construction of models of quantum gravitation. As mentioned in section 9.12.4.7, string theories are being used at present to investigate the behaviour of quantum-mechanical information loss accompanying black-hole evaporation [511].

9.13.5. Future facilities

As questions in elementary particle physics shift to higher energy domains and different particles, new facilities are under construction or under consideration to investigate the new domains. These include the first electron–proton collider (HERA), multi-TeV hadron colliders,

Elementary Particle Physics

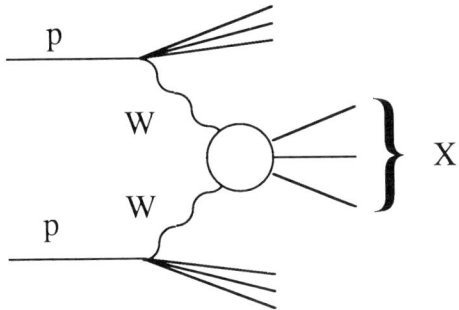

Figure 9.41. *W–W scattering induced by a hadronic collision.*

facilities for producing B-mesons copiously ('B-factories'), and large electron–positron colliders.

9.13.5.1. HERA. The first electron–proton collider began operating in 1991 at the DESY laboratory in Hamburg, Germany [512], colliding 27 GeV electrons with 820 GeV protons at a total centre-of-mass energy of about 300 GeV. So far, HERA has measured the photon–proton total cross section, has produced its first data on proton structure functions, and has observed the expected reaction $e^- + p \rightarrow \nu_e +$ (anything). It will be a welcome instrument for measuring the behaviour of structure functions (section 9.6.2) in new kinematic regions and in searching for new particles.

9.13.5.2. Multi-TeV hadron colliders. In the early 1980s, with the discovery of the W and Z anticipated, plans were laid to explore the mechanism of electroweak symmetry breaking through multi-TeV hadronic collisions. These plans crystallized into a proposal for a Large Hadron Collider (LHC) in the LEP tunnel at CERN and a project to construct the Superconducting Supercollider (SSC) in Texas. A good sampling of the physics accessible to these machines may be found in reference [491]. The LHC is designed for a centre-of-mass energy of 14 TeV and a luminosity of 10^{34} cm^{-2} s^{-1}, affording the study of such processes as W–W scattering (figure 9.41). The SSC was to have 40 TeV in the centre of mass and about ten times less luminosity, but support for it was withdrawn after several years of construction [513].

9.13.5.3. B factories. The standard electroweak theory (described for three families of quarks and leptons in section 9.9) predicts B-meson decays to exhibit CP-violating asymmetries which can be quite large, depending on values of parameters still being determined. However, since B-mesons decay to any given final state with a typical probability of less than 10^{-3}, many B-mesons are needed in order to permit such studies: probably at least 10^8 B$\bar{\text{B}}$ pairs, out of the reach of present electron–positron colliders.

The purest source of B-mesons, lying just above the B$\bar{\text{B}}$ threshold, is the $\Upsilon(4S)$ resonance, produced in e^+e^- collisions with a cross section of about 1 nb. An upgraded version of the CESR e^+e^- collider at Cornell, operating at the $\Upsilon(4S)$ resonance for several years with a luminosity of several times 10^{33} cm^{-2} s^{-1}, could produce 10^8 B$\bar{\text{B}}$ pairs. The most clear-cut CP-violating signatures at the 4S resonance require one to observe the time-dependence of both B-mesons' decays, most easily accomplished using electrons and positrons of different energies. Asymmetric B-factories are being constructed at KEK in Japan and at SLAC in the United States.

9.13.5.4. Large linear electron–positron colliders. The likely existence of a top quark below 200 GeV, the hopes that Higgs bosons and superpartners may exist with masses below several hundred GeV/c^2 and the sheer technical challenge of taking the next step in energy and luminosity have led to planning for a large linear electron–positron collider with a centre-of-mass energy of 500 GeV and a luminosity of 10^{33} cm^{-2} s^{-1}. Workshops on this facility have been held now for several years. In contrast to the Stanford Linear Collider, whose beams are bent in arcs to achieve collisions, this machine would be truly linear, with dimensions set by the attainable accelerating gradients; a total length of about 20 km is typical of plans.

9.13.5.5. Non-accelerator facilities. Several large facilities under construction or under consideration will shed light on particle physics without any need for an accelerator. The 'Super-Kamiokande' detector mentioned earlier will take a further step towards observing a proton's decay or setting improved limits on its lifetime. It will also be sensitive to neutrino interactions, as will a number of other facilities located underground (Soudan [514] in Minnesota, MACRO [515] in Italy, the Sudbury Neutrino Observatory [SNO] [516] in Canada), underwater (DUMAND [517] off the coast of Hawaii), or even under ice (AMANDA [518] at the South Pole). The MACRO detector will also search for monopoles at unprecedented sensitivities.

Searches for 'dark matter' will take on a new dimension with the development of cryogenic techniques, able, in somewhat simplified terms, to detect 'things that go bump in the night', otherwise invisible [519].

Large surface arrays are envisioned to detect cosmic-ray air showers. It may be possible to synchronize electronically the operation of an array as large as 5000 km^2 in area [520]. More modest surface arrays at higher elevations than currently in use are also planned [521].

9.14. Conclusions

Elementary particle physics in the second half of this century has oscillated between complexity and simplicity, between confusion and order.

The number of known 'elementary' particles grew enormously in the years after World War II thanks to new production and detection techniques. While some physicists despaired in the 1950s and 1960s of ever comprehending this burgeoning zoo of particles and their interactions, symmetries such as SU(3) began organizing these particles into families; the advent of the quark model introduced further regularity. However, not until a genuine theory of the strong interactions based on quantum chromodynamics (QCD) appeared in the early 1970s could one understand these successes in a fundamental way.

The weak interactions at the midpoint of the century were in a similar muddle. Failure to identify the most fundamental beta-decay processes made it hard to deduce the structure of the basic interaction. The discovery of parity non-conservation in the weak interactions lifted blinders from physicists' eyes, helped identify some experiments as right and some as wrong, and led almost immediately to a satisfactory description based on the so-called V–A theory. Still, the weak interactions remained incompletely understood. The V–A theory could not be used for calculations of higher precision or at higher energies. It took the synthesis of weak interactions with electromagnetism to construct a theory with these features.

Our present understanding of both strong and electroweak interactions relies on quantum field theory. At the midpoint of the century, field theory appeared to apply to a small area of particle physics, namely, quantum electrodynamics (QED), a limitation that vanished entirely in the course of 20 years.

Now that we understand the strongly interacting particles as consisting of quarks, held together with gluons, and understand the basis for both strong and electroweak forces, new questions have arisen. One of the earliest, arising just seven years after the discovery of parity violation, concerns the violation of combined charge-reflection (C) and parity (P) symmetries. A proposal that this violation requires at least three families of quarks has been supported by the discovery of the third family, but we still do not know whether the observed CP violation arises from this source.

Still mystifying, and possibly related to the origin of CP violation, is the pattern of masses and electroweak couplings of quarks and leptons. Quark masses necessarily break the electroweak symmetry. So do the masses of the weak force carriers, the charged Ws and the neutral Z. The source of this symmetry breaking has been generically termed the 'Higgs mechanism' and vague predictions exist for a 'Higgs boson', whose discovery would undoubtedly shed light on the electroweak symmetry-breaking mechanism. However, lurking in the background of any scenario involving the Higgs boson are many other new particles or effects, to be probed in the next generation of experiments.

The study of elementary particle physics in the past 50 years has been a noble enterprise, complete with its share of successes and setbacks. Let us hope that the progress in the next century is as rich as that in the present one.

9.14.1. Additional sources

In addition to the conference proceedings and historical works mentioned at the beginning of this article [1–5], and in the course of our discussion (see, for example, references [14, 129, 137, 227, 240, 275, 359]), several other useful references are worth mentioning. They include an eminently readable account by Jeremy Bernstein [522], a brief report by the United States Department of Energy [523], a thoughtful book by Steven Weinberg [524], a memoir spanning seven decades by Victor Weisskopf [525], a concise account of results in *Physical Review* by Sam Treiman [526], a new work by Murray Gell-Mann [527], Robert Marshak's last opus [528], and an early history of particle accelerators based on the life and work of Rolf Wideröe [529]. A biennial compilation of relevant data is produced by the Particle Data Group [49], carrying on a long tradition begun in the 1950s by Arthur H Rosenfeld and collaborators.

Acknowledgments

We thank J Bernstein, J D Bjorken, M M Block, L M Brown, U Fano, P Freund, H Frisch, R Hildebrand, Y Nambu, R Oehme, A Pais, M Perl, P Ramond, R Sachs, M M Shapiro, F C Shoemaker, V Telegdi, and A Tollestrup for helpful advice, and Gordon Fraser and Adrienne Kolb for help in obtaining illustrations. Ugo Fano provided indispensable editorial advice in the course of a careful reading of the manuscript. This work was performed in part at the Aspen Center for Physics, and was supported in part by the United States Department of Energy under Grant No. DE FG02 90ER 40560. JLR wishes to thank the Theory Group of Fermilab for their hospitality during the completion of this chapter.

References

[1] Brown L M, Dresden M, Hoddeson L and Riordan M (ed) 1995 *Third Int. Symp. on the History of Particle Physics: The Rise of the Standard Model (Stanford, CA, June 24–27, 1992)* (Cambridge: Cambridge University Press)

[2] Pais A 1986 *Inward Bound* (Oxford: Clarendon)

[3] Brown L M and Hoddeson L (ed) 1983 *Int. Symp. on the History of Particle Physics: The Birth of Particle Physics* (Cambridge: Cambridge University Press)

Colloque International sur l'Histoire de la Physique des Particules (International Colloquium on the History of Particle Physics) (Paris, 21–23 July, 1982) (Les Ulis, France: Les éditions de Physique); 1982 *J. Physique Coll.* **43** C8, supplement 12

[4] Brown L M, Dresden M and Hoddeson L (ed) 1989 *Pions to Quarks: Particle Physics in the 1950s (Proc. Second Int. Symp. on the History of Particle Physics (Fermilab, 1985))* (Cambridge: Cambridge University Press)

[5] Brown L M 1995 Nuclear forces, mesons and isospin symmetry *Twentieth Century Physics* (Bristol: Institute of Physics) Chapter 5

[6] Heisenberg W and Pauli W 1929 *Z. Phys.* **56** 1–61; 1930 *Z. Phys.* **59** 168–90
Oppenheimer J R 1930 *Phys. Rev.* **35** 461–77
Heisenberg W 1934 *Z. Phys.* **90** 209–231
Dirac P A M 1934 *Proc. Camb. Phil. Soc.* **30** 150–63

[7] Dirac P A M 1928 *Proc. R. Soc.* A **117** 610–24; 1928 *Proc. R. Soc.* A **118** 351–61; 1930 *Proc. R. Soc.* A **126** 360–5; 1931 *Proc. R. Soc.* A **133** 60–72

[8] Weisskopf V F 1939 *Phys. Rev.* **56** 72–85

[9] Uehling E A 1935 *Phys. Rev.* **48** 55–63
Serber R 1935 *Phys. Rev.* **48** 49–54

[10] Kemble E C and Present R D 1933 *Phys. Rev.* **44** 1031–2

[11] Houston W V 1937 *Phys. Rev.* **51** 446–9
Williams R C 1938 *Phys. Rev.* **54** 558–67

[12] Pasternack S 1938 *Phys. Rev.* **54** 1113

[13] Lamb W E Jr and Retherford R C 1947 *Phys. Rev.* **72** 241–3
Lamb W E Jr in the first of references [3] ch 20

[14] Schweber S S 1984 *Relativity, Groups and Topology II (1983 Les Houches Lectures)* ed B S de Witt and R Stora (Amsterdam: North-Holland) pp 37–220
See also Marshak R E in the first of references [3] pp 376–401

[15] Bethe H A 1947 *Phys. Rev.* **72** 339–41

[16] Kroll N M and Lamb W E 1949 *Phys. Rev.* **75** 388–98
Feynman R P 1949 *Phys. Rev.* **76** 769–89 (see in particular footnote 13)
Schwinger J 1949 *Phys. Rev.* **75** 898–9
French J B and Weisskopf V F 1949 *Phys. Rev.* **75** 388, 1240–8
Nambu Y 1949 *Prog. Theor. Phys.* **4** 82–94

[17] Schwinger J (ed) 1958 *Selected Papers on Quantum Electrodynamics* (New York: Dover); see also the second of references [3] pp C8-409–23

[18] Foley H M and Kusch P 1947 *Phys. Rev.* **72** 1256–7; 1948 *Phys. Rev.* **73** 412

[19] Schwinger J 1948 *Phys. Rev.* **73** 416–7; Erratum 1949 *Phys. Rev.* **76** 790–817

[20] Wheeler J A 1946 *Ann. NY Acad. Sci* **48** 219–38
See also Pirenne J 1944 Thesis University of Paris; 1946 *Arch. Sci. Phys. Nat.* **28** 273; 1947 *Arch. Sci. Phys. Nat.* **29** 121

[21] Deutsch M 1951 *Phys. Rev.* **82** 455–6

[22] Kinoshita T (ed) 1990 *Quantum Electrodynamics* (Singapore: World Scientific)

[23] Schupp A R, Pidd R W and Crane H R 1961 *Phys. Rev.* **121** 1–17

[24] Dehmelt H 1990 *Rev. Mod. Phys.* **62** 525–30
Van Dyck R S Jr 1990 in reference [22] ch 8

[25] Van Dyck R S Jr, Schwinberg P B and Dehmelt H G 1987 *Phys. Rev. Lett.* **59** 26–29

[26] Kinoshita T and Yennie D 1990 in reference [22] ch 1

[27] Cage M E *et al* 1989 *IEEE Trans. Instrum. Meas.* **38** 284–9

[28] Suura H and Wichmann E 1957 *Phys. Rev.* **105** 1930–1
Petermann A 1957 *Phys. Rev.* **105** 1931

[29] Garwin R, Lederman L M and Weinrich M 1957 *Phys. Rev.* **105** 1415–7

[30] Farley F J M and Picasso E in reference [22] ch 11
[31] Kinoshita T and Marciano W J in reference [22] ch 10
[32] Pal'chikov V G, Sokolov Yu L and Yakovlev V P 1983 *Pis. Zh. Eksp. Teor. Fiz.* **38** 347–9 (Engl. Transl. *JETP Lett.* **38** 418–20)
[33] Sapirstein J R and Yennie D R in reference [22] ch 12
[34] Ramsey N F in reference [22] ch 13
[35] Mills A P and Chu S in reference [22] ch 15
[36] Harris I and Brown L M 1957 *Phys. Rev.* **105** 1656–61
[37] Nico J S, Gidley D W, Rich A and Zitzewitz P W 1990 *Phys. Rev. Lett.* **65** 1344–7
[38] Ore A and Powell J L 1949 *Phys. Rev.* **75** 1696–9
Caswell W G and Lepage G P 1979 *Phys. Rev.* A **20** 36–43
Adkins G S 1983 *Ann. Phys., NY* **146** 78–128
[39] Urey H, Brickwedde F G and Murphy G M 1932 *Phys. Rev.* **39** 164–5, 864
[40] Chadwick J 1932 *Nature* **129** 312; 1932 *Proc. R. Soc.* A **136** 692–708, 744–8
[41] Anderson C 1932 The positive electron *Science* **76** 238–9; the first of references [3] ch 7
[42] Neddermeyer S H and Anderson C D 1937 *Phys. Rev.* **51** 884–6
Street J C and Stevenson E C 1937 *Phys. Rev.* **51** 1005
Nishina Y, Takeuchi M and Ichimaya T 1937 *Phys. Rev.* **52** 1198–9
[43] Yukawa H 1935 *Proc. Phys.-Math. Soc. Japan* **17** 48–57
[44] Oppenheimer J R and Serber R 1937 *Phys. Rev.* **51** 1113
[45] Pais A in reference [2] p 433
[46] Pauli W as described in reference [2] p 315
[47] Blackett P M S and Occhialini G P S 1933 *Proc. R. Soc.* A **139** 699–726
[48] Fermi E 1934 *Nuovo Cimento* **11** 1–19; 1934 *Z. Phys.* **88** 161–71
[49] Montanet L *et al* (Particle Data Group) 1994 *Phys. Rev.* D **50** 1173–1825
[50] Rossi B and Nereson N 1942 *Phys. Rev.* **62** 417–22
Nereson N and Rossi B 1943 *Phys. Rev.* **64** 199–201
Rossi B in the first of references [3] ch 11
[51] Williams E J and Evans G R 1940 *Nature* **145** 818–9
[52] Conversi M, Pancini E and Piccioni O 1947 *Phys. Rev.* **71** 209–10
Piccioni O in the first of references [3] ch 13
Conversi M in the first of references [3] ch 14
[53] Perkins D H in reference [4] ch 5
Perkins D H 1947 *Nature* **159** 126–7
Occhialini G P S and Powell C F 1947 *Nature* **159** 93–4
[54] Powell C F, Fowler P H and Perkins D H 1959 *The Study of Elementary Particles by the Photographic Method* (New York: Pergamon)
[55] Lattes C M G, Muirhead H, Occhialini G P S and Powell C F 1947 *Nature* **159** 694–7
Lattes C M G, Occhialini G P S and Powell C F 1947 *Nature* **160** 453–6, 486–92
[56] Sakata S and Inoue T 1946 *Prog. Theor. Phys.* **1** 143–50
Tanikawa Y 1947 *Prog. Theor. Phys.* **2** 220–1
Marshak R E and Bethe H A 1947 *Phys. Rev.* **72** 506–9
[57] Rochester G D and Butler C C 1947 *Nature* **160** 855–7
[58] Kemmer N 1938 *Proc. R. Soc.* A **166** 127–53
Sakata S 1941–2 Unpublished correspondence with S Tomonaga (Y Nambu, private communication)
[59] Bjorklund R, Crandall W E, Moyer B J and York H F 1950 *Phys. Rev.* **77** 213–18
Steinberger J, Panofsky W K H and Steller J 1950 *Phys. Rev.* **78** 802–5
Steinberger J in reference [4] ch 20

Elementary Particle Physics

- Carlson A G, Hooper J E and King D T 1950 *Phil. Mag.* **41** 701–24
- [60] Clark D C, Roberts A and Wilson R 1951 *Phys. Rev.* **83** 649
- Durbin R, Loar H and Steinberger J 1951 *Phys. Rev.* **83** 646–8
- [61] Panofsky W K H, Aamodt R L and Hadley J 1951 *Phys. Rev.* **81** 565–74
- [62] Hales R W, Hildebrand R H, Knable N and Moyer B 1952 *Phys. Rev.* **85** 373–4
- [63] Hildebrand R H 1953 *Phys. Rev.* **89** 1090–2
- [64] Richardson J R 1948 *Phys. Rev.* **74** 1720–1
- [65] Chamberlain O, Mozley R F, Steinberger J and Wiegand C 1950 *Phys. Rev.* **79** 394–5
- [66] Sakata S 1942 (unpublished) (Y Nambu, private communication)
- Landau L D 1948 *Dokl. Akad. Nauk SSSR* **60** 207–9
- Yang C N 1950 *Phys. Rev.* **77** 242–5
- [67] Samios N P, Plano R, Prodell A, Schwartz M and Steinberger J 1962 *Phys. Rev.* **126** 1844–9
- [68] Kaplon M F, Peters B and Bradt H L 1950 *Phys. Rev.* **76** 1735–6
- [69] Steinberger J 1949 *Phys. Rev.* **76** 1180–6
- [70] Primakoff H 1951 *Phys. Rev.* **81** 899
- [71] Tollestrup A V, Berman S, Gomez R and Ruderman H 1960 *Proc. 1960 Ann. Int. Conf. on High Energy Physics at Rochester (Rochester, August 25–September 1, 1960)* ed E C G Sudarshan *et al* (New York: Interscience [University of Rochester]) pp 27–30
- Ruderman H A 1962 *PhD Thesis* Caltech (unpublished)
- Bellettini G, Bemporad C, Braccini P L and Foà L 1965 *Nuovo Cimento* A **40** 1139–70
- [72] von Dardel G, Dekkers D, Mermod R, Van Putten J D, Vivargent M, Weber G and Winter K 1963 *Phys. Lett.* **4** 51–4
- Atherton H W *et al* 1985 *Phys. Lett.* **158B** 81–4
- [73] Chamberlain O, Segrè E, Wiegand C and Ypsilantis T 1955 *Phys. Rev.* **100** 947–50
- Goldhaber G in reference [4] ch 16
- Chamberlain O in reference [4] ch 17
- Piccioni O in reference [4] ch 18
- [74] Chamberlain O *et al* 1956 *Phys. Rev.* **102** 921–3
- Barkas W H *et al* 1957 *Phys. Rev.* **105** 1037–58
- [75] Gell-Mann 1982 in the second of references [3] pp C8-395–408; reference [2] ch 20
- Perkins D H in reference [4] ch 5
- Rochester G D in reference [4] ch 4
- [76] Leprince-Ringuet L and Lhéritier M 1944 *C. R. Acad. Sci., Paris* **219** 618–20; 1946 *J. Phys. Radium* **7** 65–9
- See, however, the criticism by Bethe H A 1946 *Phys. Rev.* **70** 821–31
- [77] Brown R, Camerini U, Fowler P H, Muirhead H, Powell C F and Ritson D M 1949 *Nature* **163** 82–7
- See also Fowler P H, Menon M G K, Powell C F and Rochat O 1951 *Phil. Mag.* **42** 1040–9
- [78] Seriff A J, Leighton R B, Hsiao C, Cowan E W and Anderson C D 1950 *Phys. Rev.* **78** 290–1
- [79] Hopper V D and Biswas S 1950 *Phys. Rev.* **80** 1099–1100
- [80] Armenteros R, Barker K H, Butler C C, Cachon A and Chapman A H 1951 *Nature* **167** 501–3
- Armenteros R, Barker K H, Butler C C and Cachon A 1951 *Phil. Mag.* **42** 1113–35
- [81] Thompson R W, Cohn H O and Flum R S 1951 *Phys. Rev.* **83** 175

[82] Thompson R W, Buskirk A V, Etter L R, Karzmark C J and Rediker R H 1953 *Phys. Rev.* **90** 329–30. An account of subsequent work by this group is given by Thompson R W in the first of references [3] ch 15.
Armenteros R, Barker K H, Butler C C, Cachon A and York C M 1952 *Phil. Mag.* **43** 597–612
Leighton R B, Cowan E W and van Lint V A J 1953 *Proc. Conf. Int. Ray. Cosmique (Bagnères de Bigorre)* (Toulouse: University of Toulouse) pp 97–101
Anderson C D, Cowan E W, Leighton R B and van Lint V A J 1953 *Phys. Rev.* **92** 1089
[83] York C M, Leighton R B and Bjornerud E K 1953 *Phys. Rev.* **90** 167–8
Bonetti A, Levi Setti R, Panetti M and Tomasini G 1953 *Nuovo Cimento* **10** 345–6, 1736–43
[84] Blewett J P in reference [4] ch 10
Livingston M S and Blewett J P 1962 *Particle Accelerators* (New York: McGraw-Hill)
Blewett M H (ed) 1953 *Rev. Sci. Instrum.* **24** 723–870
[85] Fowler W B, Shutt R P, Thorndike A M and Whittemore W L 1953 *Phys. Rev.* **90** 1126–7; 1953 *Phys. Rev.* **91** 1287; 1954 *Phys. Rev.* **93** 861–7; 1955 *Phys. Rev.* **98** 121–30
See also Walker W D 1955 *Phys. Rev.* **98** 1407–10
For an historical account see Fowler W B in reference [4] ch 22.
[86] Nambu Y, Nishijima K and Yamaguchi Y 1951 *Prog. Theor. Phys.* **6** 615–19, 619–22
Miyazawa H 1951 *Prog. Theor. Phys.* **6** 631–3
Oneda S 1951 *Prog. Theor. Phys.* **6** 633–5
[87] Pais A 1952 *Phys. Rev.* **86** 663–72; 1953 *Physica* **19** 869–87
[88] Gell-Mann M 1953 *Phys. Rev.* **92** 833–4; 1953 On the classification of particles (unpublished)
Gell-Mann M and Pais A 1955 *Proc. 1954 Glasgow Conf. on Nuclear and Meson Physics* ed E H Bellamy and R G Moorhouse (London: Pergamon)
Gell-Mann M 1956 *Nuovo Cimento* **4** (Supplement) 848–66
[89] Nakano T and Nishijima K 1953 *Prog. Theor. Phys.* **10** 581–2
Nishijima K 1954 *Prog. Theor. Phys.* **12** 107–8; 1955 *Prog. Theor. Phys.* **13** 285–304
[90] Alvarez L W, Eberhard P, Good M L, Graziano W, Ticho H K and Wojcicki S G 1959 *Phys. Rev. Lett.* **2** 215–9
The history of this and related discoveries is recounted by Alvarez L 1969 *Science* **165** 1071–91.
[91] Gell-Mann M and Pais A 1955 *Phys. Rev.* **97** 1387–9
[92] Lande K, Booth E T, Impeduglia J and Lederman L M 1956 *Phys. Rev.* **103** 1901–4 (an account of this experiment is given by Chinowsky W in reference [4] ch 21)
[93] Fry W F, Schneps J and Swami M S 1956 *Phys. Rev.* **103** 1904–5
[94] Heitler W 1946 *Proc. Ir. Acad.* **51A** 33–9
Nambu Y and Yamaguchi Y 1951 *Prog. Theor. Phys.* **6** 1000–6
[95] Brueckner K A and Watson K M 1951 *Phys. Rev.* **83** 1–9
Watson K M 1952 *Phys. Rev.* **85** 852–7
Adair R K 1952 *Phys. Rev.* **87** 1041–3
[96] Walker R L in reference [4] ch 6
[97] Bishop A S, Steinberger J and Cook L J 1950 *Phys. Rev.* **80** 291
Steinberger J, Panofsky W K H and Steller J 1950 *Phys. Rev.* **78** 802–5
Chedester C D, Isaacs P, Sachs A and Steinberger J 1951 *Phys. Rev.* **82** 958–9

Anderson H L, Fermi E, Long E A, Martin R and Nagle D E 1952 *Phys. Rev.* **85** 934–5

Fermi E, Anderson H L, Lundby A, Nagle D E and Yodh G B 1952 *Phys. Rev.* **85** 935–6

Anderson H L, Fermi E, Long E A and Nagle D E 1952 *Phys. Rev.* **85** 936

[98] Marshak R E 1950 *Phys. Rev.* **78** 346

Fujimoto Y and Miyazawa H 1950 *Prog. Theor. Phys.* **5** 1052–4

Brueckner K A and Case K M 1951 *Phys. Rev.* **83** 1141–7

Brueckner K A 1952 *Phys. Rev.* **86** 106–9

Brueckner K A and Watson K M 1952 *Phys. Rev.* **86** 923–8

Many of these works made reference to earlier schemes by Wentzel G 1940 *Helv. Phys. Acta* **13** 269–308; 1941 *Helv. Phys. Acta* **14** 633–5; 1947 *Rev. Mod. Phys.* **19** 1–18.

Pauli W and Dancoff S M 1942 *Phys. Rev.* **62** 85–108

[99] Walker R L, Oakley D C and Tollestrup A V 1953 *Phys. Rev.* **89** 1301–2

Anderson H L, Fermi E, Martin R and Nagle D E 1953 *Phys. Rev.* **91** 155–68

Yuan L C L and Lindenbaum S J 1953 *Phys. Rev.* **92** 1578–9

Ashkin J, Blaser J P, Feiner F, Gorman J and Stern M O 1954 *Phys. Rev.* **93** 1129–30

[100] Glaser D A 1952 *Phys. Rev.* **87** 665; 1953 *Phys. Rev.* **91** 496, 762–3
[101] Alvarez L A 1969 *Science* **165** 1071–1091; reference [4] ch 19
[102] A brief history of these discoveries is given by Samios N P in reference [1].
[103] Moller C and Rosenfeld L 1940 *Kong. Danske Vid. Selsk., Matt-fys. Medd.* **17** no 8 pp 1–72

Schwinger J 1942 *Phys. Rev.* **61** 387

Rosenfeld L 1948 *Nuclear Forces* (New York: Interscience) p 322

[104] Hofstadter R in reference [4] ch 7 pp 126–43
[105] Nambu Y 1957 *Phys. Rev.* **106** 1366–7
[106] Frazer W R and Fulco J R 1959 *Phys. Rev. Lett.* **2** 365–8
[107] Stonehill D C, Baltay C, Courant H, Fickinger W, Fowler E C, Kraybill H, Sandweiss J, Sanford J and Taft H 1961 *Phys. Rev. Lett.* **6** 624–5

Erwin A R, March R, Walker W D and West E 1961 *Phys. Rev. Lett.* **6** 628–30

[108] Maglic B C, Alvarez L W, Rosenfeld A H and Stevenson M L 1961 *Phys. Rev. Lett.* **7** 178–82
[109] Pevsner A *et al* 1961 *Phys. Rev. Lett.* **7** 421–3
[110] These are documented in the introduction to Gell-Mann M and Ne'eman Y 1964 *The Eightfold Way* (New York: Benjamin).
[111] Ikeda M, Ogawa S and Ohnuki Y 1959 *Prog. Theor. Phys.* **22** 715–24; 1960 *Prog. Theor. Phys.* **23** 1073–99
[112] Gell-Mann M 1961 The eightfold way *Caltech Report* CTSL-20, reprinted in Gell-Mann M and Ne'eman Y 1964 *The Eightfold Way* (New York: Benjamin) p 11
[113] Yang C N and Mills R L 1954 *Phys. Rev.* **96** 191–5
[114] Ne'eman Y 1961 *Nucl. Phys.* **26** 222–9
[115] Gell-Mann M 1962 *Phys. Rev.* **125** 1067–84
[116] Okubo S 1962 *Prog. Theor. Phys.* **27** 949–66

Gell-Mann M 1962 *Proc. 1962 Int. Conf. on High Energy Physics at CERN (Geneva, 4–11 July, 1962)* ed J Prentki (Geneva: CERN) p 805

[117] Goldhaber S, Chinowsky W, Goldhaber G, Lee W, O'Halloran T, Stubbs T F, Pjerrou G M, Stork D H and Ticho H 1962 Presented at the conference by Goldhaber G *Proc. 1962 Int. Conf. on High Energy Physics at CERN (Geneva, 4–11 July, 1962)* ed J Prentki (Geneva: CERN) pp 356–8

[118] Gell-Mann M 1962 *Proc. 1962 Int. Conf. on High Energy Physics at CERN (Geneva, 4–11 July, 1962)* ed J Prentki (Geneva: CERN)
Ne'eman Y 1962 Personal conversation with G Goldhaber at 1962 CERN Conference (Geneva, 4–11 July, 1962)
Goldhaber G in reference [1]
[119] Barnes V E *et al* 1964 *Phys. Rev. Lett.* **12** 204–6
See also Barnes V E *et al* 1964 *Phys. Lett.* **12** 134–6
Abrams G S *et al* 1964 *Phys. Rev. Lett.* **13** 670–2
[120] Reines F 1979 *Science* **203** 11–16; reference [4] ch 24
[121] Reines F and Cowan C L Jr 1953 *Phys. Rev.* **92** 830–1
[122] Cowan C L Jr, Reines F, Harrison F B, Kruse H W and McGuire A D 1956 *Science* **124** 103–4
Reines F, Cowan C L Jr, Harrison F B, McGuire A D and Kruse H W 1960 *Phys. Rev.* **117** 159–70
[123] Gamow G and Teller E 1936 *Phys. Rev.* **49** 895–9
[124] Fierz M 1937 *Z. Phys.* **104** 553–65
[125] See, for example, Gasiorowicz S 1966 *Elementary Particle Physics* (New York: Wiley) p 502
[126] Pontecorvo B 1947 *Phys. Rev.* **72** 246–7
Klein O 1948 *Nature* **161** 897–9
Puppi G 1948 *Nuovo Cimento* **5** 587–8
Lee T D, Rosenbluth M and Yang C N 1949 *Phys. Rev.* **75** 905
Tiomno J and Wheeler J A 1949 *Rev. Mod. Phys.* **21** 144–52, 153–65
[127] Fitch V and Motley R 1956 *Phys. Rev.* **101** 496–8
Motley R and Fitch V 1957 *Phys. Rev.* **105** 265–6
Alvarez L W, Crawford F S, Good M L and Stevenson L 1956 *Phys. Rev.* **101** 503–5. Other references are cited in reference [2] sec 20(c).
[128] Dalitz R H 1953 *Proc. Conf. Int. Ray. Cosmique (Bagnères de Bigorre)* (Toulouse: University of Toulouse) p 236; 1953 *Phil. Mag.* **44** 1068–80; 1954 *Phys. Rev.* **94** 1046–51
Fabri E 1954 *Nuovo Cimento* **11** 479–91
A much fuller account is given by Dalitz R H in reference [4] ch 30.
[129] Polkinghorne J *Rochester Roundabout: The Story of High Energy Physics* (New York: Freeman) p 57
[130] Lee T D and Yang C N 1956 *Phys. Rev.* **104** 254–8
[131] Lee T D 1971 *Elementary Processes at High Energy (International School of Subnuclear Physics, Erice, Italy, 1970)* ed A Zichichi (New York: Academic Press) pp 827–40
Yang C N *C N Yang: Selected Papers 1945–80, With Commentary* (San Francisco: Freeman) pp 26–31
[132] Oehme R *C N Yang: Selected Papers 1945–80, With Commentary* (San Francisco: Freeman) pp 32–33
Lee T D and Yang C N *C N Yang: Selected Papers 1945–80, With Commentary* (San Francisco: Freeman) pp 33–34
[133] Lee T D, Oehme R and Yang C N 1957 *Phys. Rev.* **106** 340–45
See also Ioffe B L, Okun' L B and Rudik A P 1957 *Zh. Eksp. Teor. Fiz.* **32** 396–7 (Engl. Transl. 1957 *Sov. Phys.–JETP* **5** 328–30)
[134] Schwinger J 1953 *Phys. Rev.* **91** 713–28 (see especially p 720ff); 1954 *Phys. Rev.* **94** 1362–84 (see especially equation (54) on p 1366 and p 1376ff)
Lüders G 1954 *Kong. Danske Vid. Selsk., Matt-fys. Medd.* **28** 5; 1957 *Ann. Phys., NY* **2** 1–15
Pauli W (ed) 1955 *Niels Bohr and the Development of Physics* (New York: Pergamon) pp 30–51

Bell J S 1955 *Proc. R. Soc.* A **231** 479–95
[135] Wu C S, Ambler E, Hayward R W, Hoppes D D and Hudson R P 1957 *Phys. Rev.* **105** 1413–15
[136] Friedman J I and Telegdi V L 1957 *Phys. Rev.* **105** 1681–2; 1957 *Phys. Rev.* **106** 1290–3
[137] Lederman L M 1993 *The God Particle: If the Universe is the Answer, What is the Question?* (Boston: Houghton Mifflin) pp 256–73
[138] Crawford F S Jr, Cresti M, Good M L, Gottstein K, Lyman E M, Solmitz F T, Stevenson M L and Ticho H 1957 *Phys. Rev.* **108** 1102–3
Eisler F *et al* 1957 *Phys. Rev.* **108** 1353–5
Leipuner L B and Adair R K 1958 *Phys. Rev.* **109** 1358–63
[139] Purcell E M and Ramsey N F 1950 *Phys. Rev.* **78** 807
Smith J, Purcell E M and Ramsey N F 1951 (unpublished)
Smith J 1951 PhD Thesis Harvard University (unpublished) as quoted by Ramsey N F 1953 *Nuclear Moments* (New York: Wiley) p 8
[140] Salam A 1957 *Nuovo Cimento* **5** 299–301
Landau L 1957 *Zh. Eksp. Teor. Fiz.* **32** 407–8 (Engl. Transl. 1957 *Sov. Phys.–JETP* **5** 337–8); 1957 *Nucl. Phys.* **3** 127–31
[141] Pauli W 1933 *Quantentheorie (Handbuch der Physik 24/1)* (Berlin: Springer) pp 83–272
[142] Lee T D and Yang C N 1957 *Phys. Rev.* **105** 1671–5
Jackson J D, Treiman S B and Wyld H W 1957 *Phys. Rev.* **106** 517–21
[143] Feynman R P and Gell-Mann M 1958 *Phys. Rev.* **109** 193–8
[144] Sudarshan E C G and Marshak R E 1958 *Proc. Int. Conf. on Mesons and Newly Discovered Particles (Padua–Venice, 22–28 September, 1957)* ed N Zanichelli (Bologna: Società Italiana di Fisica), reprinted in Kabir P K (ed) 1963 *The Development of Weak Interaction Theory* (New York: Gordon and Breach) pp 118–28; 1958 *Phys. Rev.* **109** 1860–2
[145] Sakurai J J 1958 *Nuovo Cimento* **7** 649–60
[146] Rustad B M and Ruby S L 1955 *Phys. Rev.* **97** 991–1002
[147] Anderson H L and Lattes C M G 1957 *Nuovo Cimento* **6** 1356–81
[148] Allen J S, Burman R L, Herrmannsfeldt W B, Stähelin P and Braid T H 1959 *Phys. Rev.* **116** 134–43
[149] Fazzini T, Fidecaro G, Merrison A W, Paul H and Tollestrup A V 1958 *Phys. Rev. Lett.* **1** 247–9
Impeduglia G, Plano R, Prodell A, Samios N, Schwartz M and Steinberger J 1958 *Phys. Rev. Lett.* **1** 249–51
[150] Goldhaber M, Grodzins L and Sunyar A W 1958 *Phys. Rev.* **109** 1015–17
[151] Telegdi V L in reference [4] ch 32
[152] Michel L 1950 *Proc. Phys. Soc.* **63** 514–31
[153] Gershtein S S and Zel'dovich Ya B 1955 *Zh. Eksp. Teor. Fiz.* **29** 698–9 (Engl. Transl. 1956 *Sov. Phys.–JETP* **2** 576–8)
[154] Depommier P, Heintze J, Mukhin A, Rubbia C, Soergel V and Winter K 1962 *Phys. Lett.* **2** 23–6
Depommier P, Heintze J, Rubbia C and Soergel V 1963 *Phys. Lett.* **5** 61–3
[155] Klein O 1939 *Les Nouvelles Théories de la Physique* (Paris: Inst. Int. de Coöperation Intellectuelle) pp 81–98
Schwinger J 1957 *Ann. Phys., NY* **2** 407–34
Bludman S A 1958 *Nuovo Cimento* **9** 433–45
Feinberg G 1958 *Phys. Rev.* **110** 1482–3
Glashow S L 1959 *Nucl. Phys.* **10** 107–17
Gell-Mann M 1959 *Rev. Mod. Phys.* **31** 834–8
Lee T D and Yang C N 1960 *Phys. Rev.* **119** 1410–19
[156] Danby G, Gaillard J M, Goulianos K, Lederman L M, Mistry N,

Schwartz M and Steinberger J 1962 *Phys. Rev. Lett.* **9** 36–44
[157] Schwartz M 1989 *Rev. Mod. Phys.* **61** 527–32
[158] Pontecorvo B 1959 *Zh. Eksp. Teor. Fiz.* **37** 1751–7 (Engl. Transl. 1960 *Sov. Phys.–JETP* **10** 1236–40)
Schwartz M 1960 *Phys. Rev. Lett.* **4** 306–7
[159] Lee T D and Yang C N 1960 *Phys. Rev. Lett.* **4** 307–8
[160] Block M M *et al* 1964 *Phys. Lett.* **12** 281–5
Bernardini G *et al* 1964 *Phys. Lett.* **13** 86–91
Burns R, Goulianos K, Hyman E, Lederman L, Lee W, Mistry N, Rehberg J, Schwartz M, Sunderland J and Danby G, 1965 *Phys. Rev. Lett.* **15** 42–5
[161] Landau L 1957 *Zh. Eksp. Teor. Fiz.* **32** 405–6 (Engl. Transl. 1957 *Sov. Phys.–JETP* **5** 336–7); 1957 *Nucl. Phys.* **3** 127–31
[162] Fitch V L 1981 *Rev. Mod. Phys.* **53** 367–71
[163] Christenson J H, Cronin J W, Fitch V L and Turlay R 1964 *Phys. Rev. Lett.* **13** 138–40
[164] Sachs R G 1963 *Ann. Phys., NY* **22** 239–62
[165] Wu T T and Yang C N 1964 *Phys. Rev. Lett.* **13** 380–5
[166] Bell J S and Steinberger J 1966 *Proc. Oxford Int. Conf. on Elementary Particles (19–25 September, 1965)* ed T R Walsh (Chilton: Rutherford High Energy Laboratory) pp 193–222
Cronin J W 1981 *Rev. Mod. Phys.* **53** 373–83
[167] Fitch V L, Roth R F, Russ J S and Vernon W 1965 *Phys. Rev. Lett.* **15** 73–6
[168] Pais A and Piccioni O 1955 *Phys. Rev.* **100** 1487–9
Case K M 1956 *Phys. Rev.* **103** 1449–53
Good M L 1957 *Phys. Rev.* **106** 591–5
[169] Lee T D and Wu C S 1966 *Ann. Rev. Nucl. Sci.* **16** 511–90
Kleinknecht K 1976 *Ann. Rev. Nucl. Sci.* **26** 1–50
Kabir P K 1968 *The CP Puzzle: Strange Decays of the Neutral Kaon* (New York: Academic)
Sachs R G 1987 *The Physics of Time Reversal* (Chicago: University of Chicago Press)
Jarlskog C (ed) 1989 *CP Violation* (Singapore: World Scientific)
[170] Wolfenstein L 1964 *Phys. Rev. Lett.* **13** 562–4
[171] Gibbons L K *et al* 1993 *Phys. Rev. Lett.* **70** 1203–1206
[172] Barr G D *et al* 1993 *Phys. Lett.* **317B** 233–42
[173] Treiman S B and Sachs R G 1956 *Phys. Rev.* **103** 1545–49
Wyld H W Jr and Treiman S B 1957 *Phys. Rev.* **106** 169–70
[174] Lüth V 1974 *Thesis* University of Heidelberg
Geweniger C *et al* 1974 *Phys. Lett.* **48B** 483–6
[175] Sakharov A D 1967 *Pis. Zh. Eksp. Teor. Fiz.* **5** 32–5 (Engl. Transl. *JETP Lett.* **5** 24–7)
[176] Langacker P 1981 *Phys. Rep.* **72** 185–385 and references therein
[177] Goldberger M L and Treiman S B 1958 *Phys. Rev.* **110** 1178–84; 1958 *Phys. Rev.* **111** 354–61
Treiman S B in reference [4] ch 27
[178] Goldberger M L and Treiman S B 1958 *Phys. Rev.* **110** 1478–9
[179] Nambu Y 1960 *Phys. Rev. Lett.* **4** 380–2
Nambu Y and Jona-Lasinio G 1961 *Phys. Rev.* **122** 345–58; 1961 *Phys. Rev.* **124** 246–54
Nambu Y in reference [4] ch 44
[180] Goldstone J 1961 *Nuovo Cimento* **19** 154–64
Goldstone J, Salam A and Weinberg S 1962 *Phys. Rev.* **127** 965–70
[181] Gell-Mann M and Lévy M 1960 *Nuovo Cimento* **16** 705–25

[182] Crawford F S Jr, Cresti M, Good M L, Kalbfleisch G R, Stevenson M L and Ticho H K 1958 *Phys. Rev. Lett.* **1** 377–80
Nordin P, Orear J, Reed L, Rosenfeld A H, Solmitz F T, Taft H T and Tripp R D 1958 *Phys. Rev. Lett.* **1** 380–2
[183] Gell-Mann M 1962 *Phys. Rev.* **125** 1067–84; 1964 *Physics* **1** 63–75
[184] Gell-Mann M 1964 *Phys. Lett.* **8** 214–5
[185] Cabibbo N 1963 *Phys. Rev. Lett.* **10** 531–3
[186] Bourquin M *et al* 1983 *Z. Phys.* C **21** 27–36
Leutwyler H and Roos M 1984 *Z. Phys.* C **25** 91–101
Donoghue J F, Holstein B R and Klimt S W 1987 *Phys. Rev.* D **35** 934–8
[187] Nambu Y and Lurié D 1962 *Phys. Rev.* **125** 1429–36
[188] Bloch F and Nordsieck A 1937 *Phys. Rev.* **52** 54–9
Low F E 1958 *Phys. Rev.* **110** 974–7
[189] Adler S L 1965 *Phys. Rev.* B **137** 1022–33; 1965 *Phys. Rev.* B **139** 1638–43
[190] Adler S L 1965 *Phys. Rev. Lett.* **14** 1051–5; 1965 *Phys. Rev.* B **140** 736–47; 1966 *Phys. Rev.* **149** 1294(E)
[191] Weisberger W I 1965 *Phys. Rev. Lett.* **14** 1047–51; 1966 *Phys. Rev.* **143** 1302–9
[192] Callan C G and Treiman S B 1966 *Phys. Rev. Lett.* **16** 153–7
[193] Weinberg S 1966 *Phys. Rev. Lett.* **16** 879–83; 1966 *Phys. Rev. Lett.* **17** 616–21
[194] Adler S L and Dashen R F 1968 *Current Algebras* (New York: Benjamin)
Gasiorowicz S and Geffen D A 1969 *Rev. Mod. Phys.* **41** 531–73
Lee B W 1972 *Chiral Dynamics* (New York: Gordon and Breach)
[195] Wheeler J A 1937 *Phys. Rev.* **52** 1107–27
[196] Heisenberg W 1943 *Z. Phys.* **120** 513–38, 673–702; 1944 *Z. Phys.* **123** 93–112
[197] Rechenberg H in reference [4] ch 39
[198] Tomonaga S-I 1947 *J. Phys. Soc. Japan* **2** 151–71; 1948 *J. Phys. Soc. Japan* **3** 93–105 (reprinted in Miyazima T (ed) 1971–6 *Scientific Papers of Tomonaga* vol 2 (Tokyo: Misuzu Shobo) pp 1–48)
[199] Dicke R H 1948 *Principles of Microwave Circuits* vol 8, ed C G Montgomery (New York: McGraw-Hill) ch 5 pp 130–161
[200] Stückelberg E C G 1943 *Helv. Phys. Acta* **16** 427–8; 1944 *Helv. Phys. Acta* **17** 3–26
[201] Smith P H 1939 *Electronics* **12** 29–31; 1944 *Electronics* **17** 130–133, 318–325
[202] Møller C 1945 *Kong. Danske Vid. Selsk., Matt-fys. Medd.* **23** no 1 pp 1–48; 1946 *Kong. Danske Vid. Selsk., Matt-fys. Medd.* **22** 19
[203] See, for example, Zachariasen F 1964 *Strong Interactions and High Energy Physics: Scottish Universities' Summer School 1963* ed R G Moorhouse (Edinburgh: Oliver and Boyd) pp 371–409
Chew G F 1968 *Science* **161** 762–5
[204] Toll J S 1952 *PhD Thesis* Princeton University (unpublished)
Gell-Mann M, Goldberger M L and Thirring W E 1954 *Phys. Rev.* **95** 1612–27
Goldberger M L, Miyazawa H and Oehme R 1955 *Phys. Rev.* **99** 986–8
Goldberger M L 1955 *Phys. Rev.* **97** 508–10; *Phys. Rev.* **99** 979–85
Goldberger M L, Nambu Y and Oehme 1956 *Ann. Phys., NY* **2** 226–82
Bremermann H J, Oehme R and Taylor J G 1958 *Phys. Rev.* **109** 2178–90
Bogoliubov N N, Medvedev B V and Polivanov M V 1958 *Voprosy Teorii Dispersionnykh Sootnoshenii* (Moscow: Fizmatgiz)
[205] Anderson H L, Davidon W C and Kruse U E 1955 *Phys. Rev.* **100** 339–43
Foley K J, Jones R S, Lindenbaum S J, Love W A, Ozaki S, Platner E D, Quarles C A and Willen E H 1967 *Phys. Rev. Lett.* **19** 193–8, 622(E); 1969 *Phys. Rev.* **181** 1775–93
[206] Mandelstam S 1958 *Phys. Rev.* **112** 1344–60; 1959 *Phys. Rev.* **115** 1741–51, 1752–62

[207] Chew G F and Low F E 1956 *Phys. Rev.* **101** 1571–9, 1579–87
[208] Chew G F and Low F E 1959 *Phys. Rev.* **113** 1640–8
[209] See, for example, Brown L S and Goble R L 1968 *Phys. Rev. Lett.* **20** 346–9; 1971 *Phys. Rev.* D **4** 723–5
Basdevant J L and Lee B W 1970 *Phys. Rev.* D **2** 1680–1701
[210] Regge T 1959 *Nuovo Cimento* **14** 951–76; 1960 *Nuovo Cimento* **18** 947–56
[211] Froissart M 1961 *Phys. Rev.* **123** 1053–7
[212] Chew G F and Frautschi S C 1961 *Phys. Rev. Lett.* **7** 394–7
Blankenbecler R and Goldberger M L 1962 *Phys. Rev.* **126** 766–86
[213] Pomeranchuk I Ya 1958 *Zh. Eksp. Teor. Fiz.* **34** 725–8 (Engl. Transl. 1958 *Sov. Phys.–JETP* **7** 499–501)
[214] Oehme R 1964 *Strong Interactions and High Energy Physics: Scottish Universities' Summer School 1963* ed R G Moorhouse (Edinburgh: Oliver and Boyd) pp 129–222 and references [20–23] therein
[215] Dolen R, Horn D and Schmid C 1967 *Phys. Rev. Lett.* **19** 402–7; 1968 *Phys. Rev.* **166** 1768–81
[216] Freund P G O 1969 *Phys. Rev. Lett.* **20** 235–7
Harari H 1968 *Phys. Rev. Lett.* **20** 1395–8
[217] Harari H 1969 *Phys. Rev. Lett.* **22** 562–5
Rosner J L 1969 *Phys. Rev. Lett.* **23** 689–92
[218] Ademollo M, Rubinstein H R, Veneziano G and Virasoro M A 1967 *Phys. Rev. Lett.* **19** 1402–5; 1968 *Phys. Lett.* **27B** 99–102; 1968 *Phys. Rev.* **176** 1904–25
See also Mandelstam S 1968 *Phys. Rev.* **166** 1539–52
[219] Veneziano G 1968 *Nuovo Cimento* **57A** 190–7
[220] Lovelace C 1968 *Phys. Lett.* **28B** 264–8
Shapiro J A 1969 *Phys. Rev.* **179** 1345–53
[221] Nambu Y 1970 *Symmetries and Quark Models, Int. Conf. on Symmetries and Quark Models (Wayne State University, 18–20 June, 1969)* ed R Chand (New York: Gordon and Breach) pp 269–78
Susskind L 1969 *Phys. Rev. Lett.* **23** 545–7; 1970 *Phys. Rev.* D **1** 1182–6
Fubini S, Gordon D and Veneziano G 1969 *Phys. Lett.* **29B** 679–82
[222] Ramond P 1971 *Phys. Rev.* D **3** 2415–8
[223] Scherk J and Schwarz J H 1974 *Phys. Lett.* **52B** 347–50
[224] Green M B and Schwarz J H 1984 *Nucl. Phys.* B **243** 475–536; 1984 *Phys. Lett.* **149B** 117–122
[225] Sakata S 1956 *Prog. Theor. Phys.* **16** 686–8
[226] Goldberg H and Ne'eman Y 1963 *Nuovo Cimento* **27** 1–5
[227] Gell-Mann M 1987 *Symmetries in Physics (1600–1980), Proc. First Int. Meeting on the History of Scientific Ideas (Sant Feliu de Guíxols, Catalonia, Spain, 20–26 September, 1983)* ed M G Doncel *et al* (Barcelona: Servei de Publicacions, Universitat Autònoma de Barcelona) pp 473–97
[228] Joyce J 1939 *Finnegans Wake* (New York: Viking) p 383
[229] Zweig G 1964 CERN reports 8182/TH 401 and 8419/TH 412 (unpublished); second paper reprinted in Lichtenberg D B and Rosen S P (ed) 1980 *Devlopments in the Quark Theory of Hadrons* vol 1 (Nonantum, MA: Hadronic) p 22
[230] Gell-Mann M in the first of references [183]
Sakurai J J 1962 *Phys. Rev. Lett.* **9** 472–5
[231] Bertanza L *et al* 1962 *Phys. Rev. Lett.* **9** 180–3
Schlein P, Slater W E, Smith L T, Stork D H and Ticho H K 1963 *Phys. Rev. Lett.* **10** 368–71
Connolly P L *et al* 1963 *Phys. Rev. Lett.* **10** 371–6
[232] Lee T D and Yang C N 1956 *Nuovo Cimento* **3** 749–53

Elementary Particle Physics

Goldhaber M, Lee T D and Yang C N 1958 *Phys. Rev.* **112** 1796–8
[233] Dalitz R H 1967 *Proc. XIII Int. Conf. on High Energy Physics* (Berkeley, CA: University of California Press) pp 215–36
Zel'dovich Ya B and Sakharov A D 1966 *Yad. Fiz.* **4** 395–406 (Engl. Transl. 1967 *Sov. J. Nucl. Phys.* **4** 283–90)
Lipkin H J 1973 *Phys. Rep.* **8** 173–268 and references therein
Sakharov A D 1975 *Pis. Zh. Eksp. Teor. Fiz.* **21** 554–7 (Engl. Transl. *JETP Lett.* **21** 258–9)
De Rújula A, Georgi H and Glashow S L 1976 *Phys. Rev.* D **12** 147–62
Lipkin H J in reference [1]
[234] Gürsey F and Radicati L 1964 *Phys. Rev. Lett.* **13** 173–5
Pais A 1964 *Phys. Rev. Lett.* **13** 175–7
Gürsey F, Pais A and Radicati L 1964 *Phys. Rev. Lett.* **13** 299–301
Bég M A B, Lee B W and Pais A 1964 *Phys. Rev. Lett.* **13** 514–7
Sakita B 1964 *Phys. Rev.* **136** B1756–60; 1964 *Phys. Rev. Lett.* **13** 643–6
[235] Reference [2] p 528
[236] Green H S 1953 *Phys. Rev.* **90** 270–3
Greenberg O W and Messiah A M L 1965 *Phys. Rev.* B **138** B1155–67
See also Gentile G 1940 *Nuovo Cimento* **17** 493–7
[237] Greenberg O W 1964 *Phys. Rev. Lett.* **13** 598–602
[238] Han M Y and Nambu Y 1965 *Phys. Rev.* **139** B1006–10
[239] Sachs R G and Wali K C in reference [4] ch 8 pp 144–6
[240] Riordan M 1987 *The Hunting of the Quark* (New York: Simon and Schuster)
[241] Bloom E D *et al* 1969 *Phys. Rev. Lett.* **23** 931–4
Breidenbach M, Friedman J I, Kendall H W, Bloom E D, Coward D H, DeStaebler H, Drees J, Mo L W and Taylor R E 1969 *Phys. Rev. Lett.* **23** 935–8
[242] Bjorken J D, Friedman J I, Kendall H W, Bloom E D, Coward D H, DeStaebler H, Drees J, Mo L W and Taylor R E 1966 *Phys. Rev.* **148** 1467–78; 1967 *Phys. Rev.* **160** 1582(E); see also 1966 *Phys. Rev. Lett.* **16** 408; 1967 *Phys. Rev.* **163** 1767–9; 1969 *Phys. Rev.* **179** 1547–53
[243] Benvenuti A *et al* 1973 *Phys. Rev. Lett.* **30** 1084–7
[244] Barish B C *et al* 1973 *Phys. Rev. Lett.* **31** 565–8
[245] Budagov I *et al* 1969 *Phys. Lett.* **30B** 364–8
[246] Perkins D H 1972 *Proc. XVI Int. Conf. on High Energy Physics (Chicago and Batavia, IL, September 6–13, 1972)* vol 4, ed J D Jackson *et al* (Batavia, IL: Fermilab) pp 189–247
[247] Feynman R P 1969 *Phys. Rev. Lett.* **23** 1415–7; 1972 *Photon–Hadron Interactions* (Reading, MA: Benjamin)
[248] Callan C G and Gross D J 1969 *Phys. Rev. Lett.* **22** 156–9
[249] Bjorken J D and Paschos E A 1969 *Phys. Rev.* **185** 1975–82
[250] Drell S D and Yan T-M 1970 *Phys. Rev. Lett.* **25** 316–20, 902(E)
[251] Berman S M, Bjorken J D and Kogut J 1971 *Phys. Rev.* D **4** 3388–3418
[252] Büsser F W *et al* 1973 *Phys. Lett.* **46B** 471–6
Cronin J W, Frisch H J, Shochet M J, Boymond J P, Piroué P A and Sumner R L 1973 *Phys. Rev. Lett.* **31** 1426–9; 1975 *Phys. Rev.* D **11** 3105–23
Appel J A *et al* 1974 *Phys. Rev. Lett.* **33** 719–22
Albrow M G *et al* 1978 *Nucl. Phys.* B **135** 461–85; *Nucl. Phys.* B **145** 305–48
[253] See, for example, Richter B in reference [1]
[254] Litke A *et al* 1973 *Phys. Rev. Lett.* **30** 1189–92, 1349(E)
Tarnopolsky G, Eshelman J, Law M E, Leong J, Newman H, Little R, Strauch K and Wilson R 1974 *Phys. Rev. Lett.* **32** 432–5
[255] Gilman F J 1976 *Proc. 1975 Int. Symp. on Lepton and Photon Interactions (Stanford University, August 21–27, 1975)* ed W T Kirk (Stanford, CA:

Stanford Linear Accelerator Center) pp 131–54
Harari H 1976 *Proc. 1975 Int. Symp. on Lepton and Photon Interactions (Stanford University, August 21–27, 1975)* ed W T Kirk (Stanford, CA: Stanford Linear Accelerator Center) pp 317–53
Harari H in reference [1]

[256] Jones L W 1977 *Rev. Mod. Phys.* **49** 717–52
[257] McCusker C B A and Cairns I 1969 *Phys. Rev. Lett.* **23** 658–9
Cairns I, McCusker C B A, Peak L S and Woolcott R L S 1969 *Phys. Rev.* **186** 1394–1400
[258] Rank D M 1968 *Phys. Rev.* **176** 1635–43
[259] Stevens C M, Schiffer J P and Chupka W 1976 *Phys. Rev.* D **14** 716–27
[260] LaRue G S, Fairbank W M and Hebard A F 1977 *Phys. Rev. Lett.* **38** 1011–14
[261] Marinelli M, Gallinaro G and Morpurgo G 1981 *Nucl. Instrum. Methods* **185** 129–40
Morpurgo G 1987 *Fundamental Symmetries: Proc. Int. School of Physics with Low-Energy Antiprotons (Erice, Italy, 27 September–3 October, 1986)* ed P Bloch *et al* (New York: Plenum) pp 131–159
[262] Sachs R G 1948 *Phys. Rev.* **74** 433–41 (see especially eqution (24))
[263] Yang C N 1974 *Phys. Rev. Lett.* **33** 445–7
[264] Aharonov Y and Bohm D 1959 *Phys. Rev.* **115** 485–91
[265] Glashow S L 1961 *Nucl. Phys.* **22** 579–88
[266] Schwinger J 1962 *Phys. Rev.* **125** 397–8
Anderson P W 1963 *Phys. Rev.* **130** 439–42
Higgs P W 1964 *Phys. Lett.* **12** 132–3; 1964 *Phys. Rev. Lett.* **13** 508–9; 1966 *Phys. Rev.* **145** 1156–63
Englert F and Brout R 1964 *Phys. Rev. Lett.* **13** 321–3
Guralnik G S, Hagen C R and Kibble T W B 1965 *Phys. Rev. Lett.* **13** 585–7
Kibble T W B 1967 *Phys. Rev.* **155** 1554–61
[267] Weinberg S 1967 *Phys. Rev. Lett.* **19** 1264–6
[268] Salam A 1968 *Proc. of the Eighth Nobel Symp.* ed N Svartholm (Stockholm: Almqvist and Wiksell/New York: Wiley) pp 367–77
See also Salam A and Ward J C 1964 *Phys. Lett.* **13** 168–71
[269] 't Hooft G 1971 *Nucl. Phys.* B **33** 173–99; 1971 *Nucl. Phys.* B **35** 167–88
't Hooft G and Veltman M 1972 *Nucl. Phys.* B **44** 189–213
[270] Veltman M J G in reference [1]
't Hooft G in reference [1]
Veltman M J G 1994 *Neutral Currents: Twenty Years Later (Paris, 6–9 July, 1993)* ed U Nguyen-Khac and A M Lutz (River Edge, NJ: World Scientific)
[271] Lee B W 1972 *Phys. Rev.* D **5** 823–35
Lee B W and Zinn-Justin J 1972 *Phys. Rev.* D **5** 3121–37, 3137–55, 3155–60
[272] Weinberg S 1971 *Phys. Rev. Lett.* **27** 1688–91
[273] 't Hooft G 1971 *Phys. Lett.* **37B** 195–6
Weinberg S 1971 *Phys. Rev.* D **5** 1412–7
[274] Georgi H and Glashow S L 1972 *Phys. Rev. Lett.* **28** 1494–7
[275] Galison P 1983 *Rev. Mod. Phys.* **55** 477–507; 1987 *How Experiments End* (Chicago: University of Chicago Press); 1994 *Discovery of Weak Neutral Currents: The Weak Interaction Before and After (AIP Conf. Proc. 300)* ed A K Mann and D B Cline (New York: AIP) pp 244–86
[276] Galison P in the second of references [275] p 208
[277] Fry W F and Haidt D 1973 CERN-TCL, Technical memorandum, 22 May, as quoted in reference [275]
Haidt D 1994 *Discovery of Weak Neutral Currents: The Weak Interaction Before and After (AIP Conf. Proc. 300)* ed A K Mann and D B Cline

[278] Hasert F J et al 1973 *Phys. Lett.* **46B** 138–40; 1974 *Nucl. Phys.* B **73** 1–22
[279] Hasert F J et al 1973 *Phys. Lett.* **46B** 121–4
[280] Benvenuti A et al 1974 *Phys. Rev. Lett.* **32** 800–3
Aubert B et al 1974 *Phys. Rev. Lett.* **32** 1454–60
[281] Cline D B 1994 *Discovery of Weak Neutral Currents: The Weak Interaction Before and After (AIP Conf. Proc. 300)* ed A K Mann and D B Cline (New York: AIP) pp 175–86
Mann A K 1994 *Discovery of Weak Neutral Currents: The Weak Interaction Before and After (AIP Conf. Proc. 300)* ed A K Mann and D B Cline (New York: AIP) pp 207–243
[282] Rosner J L 1992 *IV Mexican School of Particles and Fields (Oaxtepec, Mexico, 3–14 December, 1990)* ed M J L Lucio and A Zepeda (Singapore: World Scientific) pp 355–405
The data are from Bogert D et al (FMM Collaboration) 1985 *Phys. Rev. Lett.* **55** 1969–72.
Allaby J V et al (CHARM II Collaboration) 1987 *Z. Phys.* C **36** 611–28
Blondel A et al (CDHSW Collaboration) 1990 *Z. Phys.* C **45** 361–79
Reutens P et al (CCFRR Collaboration) 1990 *Z. Phys.* C **45** 539–50
[283] Lee W et al 1976 *Phys. Rev. Lett.* **37** 186–9
[284] Prescott C Y in reference [1]
Prescott C Y et al 1978 *Phys. Lett.* **77B** 347–52; 1979 *Phys. Lett.* **84B** 524–8
[285] Baltay C 1979 *Proc. 19th Int. Conf. on High Energy Physics (Tokyo, 1978)* ed S Homma et al (Tokyo: Physical Society of Japan) pp 882–903
Weinberg S 1979 *Proc. 19th Int. Conf. on High Energy Physics (Tokyo, 1978)* ed S Homma et al (Tokyo: Physical Society of Japan) pp 907–18
[286] Zel'dovich Ya B 1959 *Zh. Eksp. Teor. Fiz.* **36** 964–6 (Engl. Transl. 1959 *Sov. Phys.–JETP* **9** 682–3)
Bouchiat M A and Bouchiat C 1974 *Phys. Lett.* **48B** 111–4; 1974 *J. Physique* **35** 899–927; 1975 *J. Physique* **36** 493–509
[287] Commins E D and Bucksbaum P H 1980 *Ann. Rev. Nucl. Part. Sci.* **30** 1–52
Fortson E N and Lewis L L 1984 *Phys. Rep.* **113** 289–344
Khriplovich I B 1991 *Parity Nonconservation in Atomic Phenomena* (Philadelphia: Gordon and Breach)
Stacey D N 1992 *Phys. Scr.* **T40** 15–22
Sandars P G H 1993 *Phys. Scr.* **T46** 16–21
[288] Drell P S and Commins E D 1984 *Phys. Rev. Lett.* **53** 968–71; 1985 *Phys. Rev.* A **32** 2196–2210
Tanner C E and Commins E D 1986 *Phys. Rev. Lett.* **56** 332–5
Bouchiat M A, Guéna J, Pottier L and Hunter L 1986 *J. Physique* **47** 1709–30
Noecker M C, Masterson B P and Wieman C E 1988 *Phys. Rev. Lett.* **61** 310–3
Macpherson M J D, Zetie K P, Warrington R B, Stacey D N and Hoare J P 1991 *Phys. Rev. Lett.* **67** 2784–7
Wolfenden T M, Baird P E G and Sandars P G H 1991 *Europhys. Lett.* **15** 731–6
Meekhof D M, Vetter P, Majumder P K, Lamoreaux S K and Fortson E N 1993 *Phys. Rev. Lett.* **71** 3442–5
[289] Blundell S A, Johnson W R and Sapirstein J 1990 *Phys. Rev. Lett.* **65** 1411–4
Blundell S A, Sapirstein J and Johnson W R 1992 *Phys. Rev.* D **45** 1602–23
Dzuba V A, Flambaum V V and Sushkov O P 1989 *Phys. Lett.* **141A** 147–53
[290] Sandars P G H 1990 *J. Phys. B: At. Mol. Phys.* **23** L655–8
Marciano W J and Rosner J L 1990 *Phys. Rev. Lett.* **65** 2963–6; 1992 *Phys. Rev. Lett.* **68** 898(E)

Peskin M E and Takeuchi T 1992 *Phys. Rev.* D **46** 381–409
[291] Kim J E, Langacker P, Levine M and Williams H H 1981 *Rev. Mod. Phys.* **53** 211–52
Amaldi U, Böhm A, Durkin L S, Langacker P, Mann A K, Marciano W J, Sirlin A and Williams H H 1987 *Phys. Rev.* D **36** 1385–1407
Langacker P, Luo M and Mann A K 1992 *Rev. Mod. Phys.* **64** 87–192
[292] Bjorken B J and Glashow S L 1964 *Phys. Lett.* **11** 255–7
[293] Hara Y 1964 *Phys. Rev.* **134** B701–4
[294] Maki Z and Ohnuki Y 1964 *Prog. Theor. Phys.* **32** 144–58
[295] Glashow S L, Iliopoulos J and Maiani L 1970 *Phys. Rev.* D **2** 1285–92
[296] Gaillard M K and Lee B W 1974 *Phys. Rev.* D **10** 897–916
[297] Carlson C E and Freund P G O 1972 *Phys. Lett.* **39B** 349–52
Snow G 1973 *Nucl. Phys.* B **55** 445–54
Gaillard M K, Lee B W and Rosner J L 1975 *Rev. Mod. Phys.* **47** 277–310
[298] Bouchiat C, Iliopoulos J and Meyer P 1972 *Phys. Lett.* **38B** 519–23
Georgi H and Glashow S L 1972 *Phys. Rev.* D **6** 429–31
Gross D J and Jackiw R 7 *Phys. Rev.* D **4** 7–93
[299] Niu K, Mikumo E and Maeda Y 1971 *Prog. Theor. Phys.* **46** 1644–6
[300] Kobayashi M and Maskawa T 1973 *Prog. Theor. Phys.* **49** 652–7
Kobayashi T in reference [1]
[301] Christenson J H, Hicks G S, Lederman L M, Limon P J, Pope B G and Zavattini E 1970 *Phys. Rev. Lett.* **25** 1523–6; 1973 *Phys. Rev.* D **8** 2016–34
Lederman L M in reference [1]
[302] Aubert J J et al 1974 *Phys. Rev. Lett.* **33** 1404–6
[303] Augustin J-E et al 1974 *Phys. Rev. Lett.* **33** 1406–8
[304] Abrams G S et al 1974 *Phys. Rev. Lett.* **33** 1453–5
[305] Bacci C et al 1974 *Phys. Rev. Lett.* **33** 1408–10
Braunschweig W et al 1974 *Phys. Lett.* **53B** 393–6
[306] Appelquist T and Politzer H D 1975 *Phys. Rev. Lett.* **34** 43–5
De Rújula A and Glashow S L 1975 *Phys. Rev. Lett.* **34** 46–9
Borchardt S, Mathur V S and Okubo S, 1975 *Phys. Rev. Lett.* **34** 38–40
Callan C G, Kingsley R L, Treiman S B, Wilczek F and Zee A 1975 *Phys. Rev. Lett.* **34** 52–6
Appelquist T, De Rújula A, Politzer H D and Glashow S L 1975 *Phys. Rev. Lett.* **34** 365–9
Eichten E, Gottfried K, Kinoshita T, Lane K D and Yan T-M 1975 *Phys. Rev. Lett.* **34** 369–72
Gaillard M K, Lee B W and Rosner J L 1975 *Rev. Mod. Phys.* **47** 277–310
[307] Goldhaber A S and Goldhaber M 1975 *Phys. Rev. Lett.* **34** 36–7
Schwinger J 1975 *Phys. Rev. Lett.* **34** 37–8
Barnett R M 1975 *Phys. Rev. Lett.* **34** 41–3
Nieh H T, Wu T T and Yang C N 1975 *Phys. Rev. Lett.* **34** 49–52
Sakurai J J 1975 *Phys. Rev. Lett.* **34** 56–8
[308] Braunschweig W et al 1975 *Phys. Lett.* **57B** 407–12
Feldman G J et al 1975 *Phys. Rev. Lett.* **35** 821–4
Tanenbaum W et al 1975 *Phys. Rev. Lett.* **35** 1323–6; 1978 *Phys. Rev.* D **17** 1731–49 and references therein
[309] Boyarski A M et al 1975 *Phys. Rev. Lett.* **35** 196–9
[310] Perl M L et al 1975 *Phys. Rev. Lett.* **35** 1489–92; 1976 *Phys. Lett.* **63B** 466–70; 1977 *Phys. Lett.* **70B** 487–90
[311] De Rújula A, Georgi H and Glashow S L 1976 *Phys. Rev.* D **12** 147–62
[312] Cazzoli E G, Cnops A M, Connolly P L, Louttit R I, Murtagh M J, Palmer R B, Samios N P, Tso T T and Williams H H 1975 *Phys. Rev. Lett.* **34** 1125–8

[313] Benvenuti A et al 1975 Phys. Rev. Lett. **34** 419–22
Blietschau J et al 1976 Phys. Lett. **60B** 207–10
von Krogh J et al 1976 Phys. Rev. Lett. **36** 710–3
Barish B C et al 1976 Phys. Rev. Lett. **36** 939–41
[314] See, for example, Boymond J P, Mermod R, Piroué P A, Sumner R L, Cronin J W, Frisch H J and Shochet M J 1974 Phys. Rev. Lett. **33** 112–5
Appel J A et al 1974 Phys. Rev. Lett. **33** 722–5
[315] Goldhaber G et al 1976 Phys. Rev. Lett. **37** 255–9
Peruzzi I et al 1976 Phys. Rev. Lett. **37** 569–71
[316] Glashow S L 1974 *Experimental Meson Spectroscopy—1974* ed D A Garelick (New York: AIP) pp 387–392
[317] Riordan M in reference [240] p 321
[318] Anjos J C et al 1989 Phys. Rev. Lett. **62** 513–6
[319] Eichten E and Quigg C 1994 Phys. Rev. D **49** 5845–56
[320] Lederman L M and Pope B G 1971 Phys. Rev. Lett. **27** 765–8
[321] van der Meer S 1985 Rev. Mod. Phys. **57** 689–97
[322] Rubbia C 1985 Rev. Mod. Phys. **57** 699–722
[323] Arnison G et al 1983 Phys. Lett. **122B** 103–116; 1983 Phys. Lett. **129B** 273–82
Banner M et al 1983 Phys. Lett. **122B** 476–85
[324] Arnison G et al 1983 Phys. Lett. **126B** 398–410
Bagnaia P et al 1983 Phys. Lett. **129B** 130–40
[325] Abe F et al (CDF Collaboration) 1989 Phys. Rev. Lett. **63** 720–3
[326] Abe F et al (CDF Collaboration) 1993; D0 Collaboration 1993, as reported by Swartz M 1994 *Lepton and Photon Interactions: XVI Int. Symp. (Ithaca, NY, August, 1993) (AIP Conf. Proc. 302)* ed P Drell and D Rubin (New York: AIP) pp 381–424
[327] Abrams G S et al (Mark II Collaboration) 1989 Phys. Rev. Lett. **63** 724–7
[328] Abe K et al 1993 Phys. Rev. Lett. **70** 2515–20
[329] Seeman J T 1991 Ann. Rev. Nucl. Part. Sci. **41** 393
[330] Schopper H 1985 *Proc. Int. Symp. on Lepton and Photon Interactions at High Energy (Kyoto, August 19–24, 1985)* ed M Konuma and K Takahashi (Kyoto: Kyoto University) p 769
[331] Swartz M 1994 *Lepton and Photon Interactions: XVI Int. Symp. (Ithaca, NY, August, 1993) (AIP Conf. Proc. 302)* ed P Drell and D Rubin (New York: AIP) pp 381–424
[332] Abe F et al (CDF Collaboration) 1994 Phys. Rev. Lett. **73** 225–31; 1994 Phys. Rev. D **50** 2966–3026; 1995 Phys. Rev. Lett. **74** 2626-31
Abachi S et al (D0 collaboration) 1995 Phys. Rev. Lett. **74** 2632–7
[333] Nambu Y 1966 *Preludes in Theoretical Physics in Honor of V F Weisskopf* ed A De-Shalit A et al (Amsterdam: North-Holland/New York: Wiley) pp 133–42
[334] Adler S L 1969 Phys. Rev. **177** 2426–38
Bell J S and Jackiw R 1969 Nuovo Cimento **60A** 47–61
Okubo S 1970 *Symmetries and Quark Models, Int. Conf. on Symmetries and Quark Models (Wayne State University, 18–20 June, 1969)* ed R Chand (New York: Gordon and Breach) pp 59–79
[335] Fritzsch H and Gell-Mann M 1972 *Proc. XVI Int. Conf. on High Energy Physics (Chicago and Batavia, IL, September 6–13, 1972)* vol 2, ed J D Jackson et al (Batavia, IL: Fermilab) pp 135–65
See also Gell-Mann M 1972 Acta Phys. Austriaca Suppl. **IX** 733–61
Bardeen W A, Fritzsch H and Gell-Mann M 1973 *Scale and Conformal Symmetry in Hadron Physics* ed R Gatto (New York: Wiley) p 139
[336] Khriplovich I B 1969 Yad. Fiz. **10** 409–24 (Engl. Transl. 1970 Sov. J. Nucl. Phys. **10** 235–42)

[337] 't Hooft G in reference [1]
Gross D in reference [1]
[338] Gross D J and Wilczek F 1973 *Phys. Rev. Lett.* **30** 1343–6; 1973 *Phys. Rev.* D **8** 3633–52; 1974 *Phys. Rev.* D **9** 980–93
Politzer H D 1973 *Phys. Rev. Lett.* **30** 1346–9; 1974 *Phys. Rep.* **14C** 129–80
[339] Stückelberg E C G and Petermann A 1953 *Helv. Phys. Acta* **26** 499–520
Gell-Mann M and Low F E 1954 *Phys. Rev.* **95** 1300–12
Bogoliubov N N and Shirkov D V 1955 *Dokl. Akad. Nauk SSSR* **103** 203–6; 1956 *Nuovo Cimento* **3** 845–63
Wilson K 1969 *Phys. Rev.* **179** 1499–1512; 1970 *Phys. Rev.* D **2** 1438–72; 1971 *Phys. Rev.* D **3** 1818–46
Callan C G Jr 1970 *Phys. Rev.* D **2** 1541–7
Symanzik K 1970 *Commun. Math. Phys.* **18** 227–46
Wilson K G and Kogut J 1974 *Phys. Rep.* **12C** 75–199
[340] Nambu Y 1974 *Phys. Rev.* D **10** 4262–8
[341] Altarelli G and Parisi G 1977 *Nucl. Phys.* B **126** 298–318
[342] Bethke S 1993 *Proc. XXVI Int. Conf. on High Energy Physics (Dallas, TX, August 6–12, 1992)* ed J R Sanford (New York: AIP) pp 81–113
Voss R 1994 *Lepton and Photon Interactions: XVI Int. Symp. (Ithaca, NY, August, 1993) (AIP Conf. Proc. 302)* ed P Drell and D Rubin (New York: AIP) pp 144–171
[343] Newman H et al (Mark J Collaboration) 1979 *Proc. Int. Symp. on Lepton and Photon Interactions at High Energies (Fermilab, August 23–29, 1979)* ed T B W Kirk and H D I Abarbanel (Batavia, IL: Fermilab) pp 3–18
Berger Ch et al (PLUTO Collaboration) 1979 *Proc. Int. Symp. on Lepton and Photon Interactions at High Energies (Fermilab, August 23–29, 1979)* ed T B W Kirk and H D I Abarbanel (Batavia, IL: Fermilab) pp 19–33
Wolf G et al (TASSO Collaboration) 1979 *Proc. Int. Symp. on Lepton and Photon Interactions at High Energies (Fermilab, August 23–29, 1979)* ed T B W Kirk and H D I Abarbanel (Batavia, IL: Fermilab) pp 34–51
Orito S et al (JADE Collaboration) 1979 *Proc. Int. Symp. on Lepton and Photon Interactions at High Energies (Fermilab, August 23–29, 1979)* ed T B W Kirk and H D I Abarbanel (Batavia, IL: Fermilab) pp 52–69
[344] Hanson G J et al 1975 *Phys. Rev. Lett.* **35** 1609–12
[345] Selove W 1979 *Proc. 19th Int. Conf. on High Energy Physics (Tokyo, 1978)* ed S Homma et al (Tokyo: Physical Society of Japan) pp 165–70
McCarthy R L 1979 *Proc. 19th Int. Conf. on High Energy Physics (Tokyo, 1978)* ed S Homma et al (Tokyo: Physical Society of Japan) pp 170–1
Cool R L 1979 *Proc. 19th Int. Conf. on High Energy Physics (Tokyo, 1978)* ed S Homma et al (Tokyo: Physical Society of Japan) pp 172–3
Clark A G 1979 *Proc. 19th Int. Conf. on High Energy Physics (Tokyo, 1978)* ed S Homma et al (Tokyo: Physical Society of Japan) pp 174–6
Hansen K H 1979 *Proc. 19th Int. Conf. on High Energy Physics (Tokyo, 1978)* ed S Homma et al (Tokyo: Physical Society of Japan) pp 177–81
Nakamura K 1979 *Proc. 19th Int. Conf. on High Energy Physics (Tokyo, 1978)* ed S Homma et al (Tokyo: Physical Society of Japan) pp 181–3; summarized by Sosnowski R 1979 *Proc. 19th Int. Conf. on High Energy Physics (Tokyo, 1978)* ed S Homma et al (Tokyo: Physical Society of Japan) pp 693–705.
[346] Wu S L and Zobernig G 1979 *Z. Phys.* C **2** 107–10
[347] Wiik B 1979 *Proc. Neutrino 79, Int. Conf. on Neutrinos, Weak Interactions, and Cosmology (Bergen, June 18–22, 1979)* ed A Haatuft and C Jarlskog (Bergen: University of Bergen) pp 113–54
Söding P 1980 *Proc. European Physical Society Int. Conf. on High Energy*

Physics (Geneva, 27 June–4 July, 1979) vol 1, ed W S Newman (Geneva: CERN) pp 271–81
[348] Banner M et al (UA2 Collaboration) 1982 Phys. Lett. **118B** 203–10
Bagnaia P et al (UA2 Collaboration) 1983 Z. Phys. C **20** 117–34
[349] Abe F et al (CDF Collaboration) 1993 Phys. Rev. Lett. **70** 1376–80
[350] Novikov V A, Okun L B, Shifman M A, Vainshtein A I, Voloshin M B and Zakharov V I 1978 Phys. Rep. **41C** 1–133
[351] Gunion J F, Brodsky S J and Blankenbecler R 1972 Phys. Rev. D **6** 2652–8; 1973 Phys. Rev. D **8** 287–312
Brodsky S J and Farrar G R 1973 Phys. Rev. Lett. **31** 1153–6; 1975 Phys. Rev. D **11** 1309–30
Blankenbecler R, Brodsky S J and Gunion J F 1975 Phys. Rev. D **12** 3469–87
[352] Nussinov S 1975 Phys. Rev. Lett. **34** 1286–9
Low F E 1976 Phys. Rev. D **12** 163–73
[353] Block M M and Cahn R N 1985 Rev. Mod. Phys. **57** 563–98; 1990 Czech J. Phys. **40** 164–75
[354] Hikasa K et al (Particle Data Group) 1992 Phys. Rev. D **45** S1-S584
[355] Dremin I M and Quigg C 1978 Science **199** 937–41
[356] Field R D and Feynman R P 1978 Nucl. Phys. B **136** 1–76
[357] Wilson K 1974 Phys. Rev. D **10** 2445–59
Mackenzie P B and Kronfeld A S 1993 Ann. Rev. Nucl. Part. Sci. **43** 793–828
[358] Shifman M A, Vainshtein A I and Zakharov V I 1979 Nucl. Phys. B **147** 385–447, 448–518, 519–534
Shifman M A 1983 Ann. Rev. Nucl. Part. Sci. **33** 199–233
Reinders L J, Rubinstein H and Yakazi S 1985 Phys. Rep. **127** 1–97
[359] Alles-Borelli V, Bernardini M, Bollini D, Brunini P L, Massam T, Monari L, Palmonari F and Zichichi A 1970 Lett. Nuovo Cimento **4** 1156–9
Bernardini M, Bollini D, Brunini P L, Fiorentino E, Massam T, Monari L, Palmonari F, Rimondi F and Zichichi A 1973 Nuovo Cimento **17A** 383–9
Perl M in reference [1]; 1994 Stanford Linear Accelerator Center Report SLAC-PUB-6584 (to be published in Proc. Int. Conf. on the History of Original Ideas and Basic Discoveries in Particle Physics (Erice, Sicily, 29 July–4 August, 1994))
[360] Tsai Y-S 1971 Phys. Rev. D **4** 2821–37
Thacker H B and Sakurai J J 1971 Phys. Lett. **36B** 103–5
[361] Augustin J-E et al 1975 Phys. Rev. Lett. **34** 764–7
[362] Alexander G et al (PLUTO Collaboration) 1978 Phys. Lett. **78B** 162–6
[363] Drell P and Patterson J R 1993 Proc. XXVI Int. Conf. on High Energy Physics (Dallas, TX, August 6–12, 1992) ed J R Sanford (New York: AIP) pp 3–32
Schwarz A S 1994 Lepton and Photon Interactions: XVI Int. Symp. (Ithaca, NY, August, 1993) (AIP Conf. Proc. 302) ed P Drell and D Rubin (New York: AIP) pp 671–94
[364] Benvenuti A et al 1976 Phys. Rev. Lett. **36** 1478–82
[365] Lederman L in reference [1]
[366] Kluberg L, Piroué P A, Sumner R L, Antreasyan D, Cronin J W, Frisch H J and Shochet M J 1976 Phys. Rev. Lett. **37** 1451–4
[367] Herb S W et al 1977 Phys. Rev. Lett. **39** 252–5
Innes W R et al 1977 Phys. Rev. Lett. **39** 1240–2, 1640(E)
[368] Ueno K et al 1979 Phys. Rev. Lett. **42** 486–9
Lederman L 1989 Rev. Mod. Phys. **61** 547–60
[369] Berger Ch et al (PLUTO Collaboration) 1978 Phys. Lett. **76B** 243–5
Darden C W et al (DASP Collaboration) 1978 Phys. Lett. **76B** 246–8
[370] Bienlein J K et al 1978 Phys. Lett. **78B** 360–3
C W Darden et al (DASP Collaboration) 1978 Phys. Lett. **78B** 364–5

[371] Andrews D A et al (CLEO Collaboration) 1980 Phys. Rev. Lett. **45** 219–20
Finocchiaro G et al (CUSB Collaboration) 1980 Phys. Rev. Lett. **45** 222–5
[372] Bebek C et al (CLEO Collaboration) 1981 Phys. Rev. Lett. **46** 84–7
Chadwick K et al (CLEO Collaboration) 1981 Phys. Rev. Lett. **46** 88–91
Spencer L J et al (CUSB Collaboration) 1981 Phys. Rev. Lett. **47** 771–4
Brody A et al (CLEO Collaboration) 1982 Phys. Rev. Lett. **48** 1070–4
Giannini G et al (CUSB Collaboration) 1982 Nucl. Phys. B **206** 1–11
[373] Behrends S et al (CLEO Collaboration) 1983 Phys. Rev. Lett. **50** 881–4
Giles R et al (CLEO Collaboration) 1984 Phys. Rev. D **30** 2279–94
[374] Buchmüller W and Cooper S 1988 *High Energy Electron–Positron Physics* ed A Ali and P Söding (Singapore: World Scientific) pp 410–87
[375] Rosner J L 1991 *Testing the Standard Model (Proc. 1990 Theoretical Advanced Study Institute in Elementary Particle Physics (Boulder, CO, 3–27 June, 1990))* ed M Cvetič and P Langacker (Singapore: World Scientific) pp 91–224
[376] Spencer L J et al (CUSB Collaboration) 1981 Phys. Rev. Lett. **47** 771–4
[377] Fulton R et al (CLEO Collaboration) 1990 Phys. Rev. Lett. **64** 16–20
Albrecht H et al (ARGUS Collaboration) 1990 Phys. Lett. **234B** 409–16; 1991 Phys. Lett. **255B** 297–304
[378] Venus W 1994 *Lepton and Photon Interactions: XVI Int. Symp. (Ithaca, NY, August, 1993) (AIP Conf. Proc. 302)* ed P Drell and D Rubin (New York: AIP) pp 274–91
[379] Fernandez E et al (MAC Collaboration) 1983 Phys. Rev. Lett. **51** 1022–5
[380] Wolfenstein L 1983 Phys. Rev. Lett. **51** 1945–7
[381] Albrecht H et al (ARGUS Collaboration) 1987 Phys. Lett. **192B** 245–52
[382] Ellis J, Gaillard M K, Nanopoulos D V and Rudaz S 1977 Nucl. Phys. B **131** 285–307
Carter A B and Sanda A I 1980 Phys. Rev. Lett. **45** 952–4; 1981 Phys. Rev. D **23** 1567–79
Bigi I I and Sanda A I 1981 Nucl. Phys. B **193** 85–108
Winstein B and Wolfenstein L 1993 Rev. Mod. Phys. **65** 1113–47
[383] Kane G L and Peskin M E 1982 Nucl. Phys. B **195** 29–38
[384] Abachi S et al (D0 Collaboration) 1995 Phys. Rev. Lett. **74** 2632–7
[385] Veltman M 1977 Nucl. Phys. B **123** 89–99
[386] Livingston M S and Blewett J P 1962 *Particle Accelerators* (New York: McGraw-Hill)
[387] Cockcroft J D and Walton E T S 1932 Proc. R. Soc. A **136** 619–30; 1932 Proc. R. Soc. A **137** 229–42
[388] van de Graaff R J 1931 Phys. Rev. **38** 1919–20
For earlier work using pulsed electrostatic generators see Breit G and Tuve M A 1928 Nature **121** 535–6 and other authors noted in reference [2] pp 405–6.
[389] van de Graaff R J, Compton K T and Van Atta L C 1933 Phys. Rev. **43** 149–57
Van Atta L C, Northrop D L, Van Atta C M and van de Graaff R J 1936 Phys. Rev. **49** 761–76
[390] Herb R G, Parkinson D B and Kerst D W 1935 Phys. Rev. **48** 118–24
[391] Tuve M S, Hafstad L R and Dahl O 1935 Phys. Rev. **48** 315–37
[392] Wideröe R 1928 Arch. Elektrotech. **21** 387, 486 (Engl. Transl. Livingston M S (ed) 1966 *The Development of High-Energy Accelerators* (New York: Dover) pp 92–114)
[393] Lawrence E O and Livingston M S 1931 Phys. Rev. **37** 1707; 1931 Phys. Rev. **38** 834; 1931 Phys. Rev. **40** 19–35
[394] Lawrence E O, Livingston M S and White M G 1932 Phys. Rev. **42** 150–1

[395] McMillan E M 1945 *Phys. Rev.* **68** 143–4
Veksler V 1945 *J. Phys. (USSR)* **9** 153–8, reprinted in Livingston M S (ed) 1966 *The Development of High Energy Accelerators* (New York: Dover) pp 202–10
[396] McMillan E M, Peterson J M and White R S 1949 *Science* **110** 579–83
[397] Kerst D W 1940 *Phys. Rev.* **58** 841; 1941 *Phys. Rev.* **60** 47–53
[398] Kerst D W and Serber R 1941 *Phys. Rev.* **60** 53–8
[399] Kerst D W, Adams G D, Koch H W and Robinson C S 1950 *Phys. Rev.* **78** 297
[400] Fermi E 1952 *Phys. Rev.* **86** 611
[401] Shoemaker F C, Britton R J and Carlson B C 1952 *Phys. Rev.* **86** 582
[402] Fitch V L and Rainwater J 1953 *Phys. Rev.* **92** 789–800
[403] Courant E D, Livingston M S and Snyder H S 1952 *Phys. Rev.* **88** 1190–6
[404] Christofilos N 1956 *US Patent* No 2,736,799, reprinted in Livingston M S (ed) 1966 *The Development of High-Energy Accelerators* (New York: Dover) pp 270–80
[405] Chodorow M, Ginzton E L, Hansen W W, Kyhl R L, Neal R B and Panofsky W K H, 1955 *Rev. Sci. Instrum.* **26** 134–204
[406] Alvarez L W, Bradner H, Franck J V, Gordon H, Gow J D, Marchall L C, Oppenheimer F, Panofsky W K H, Richman C and Woodyard J R 1955 *Rev. Sci. Instrum.* **26** 111–33
[407] Kerst D W 1955 (unpublished). This was first discussed at a conference sponsored by the Midwest Universities Research Association, MURA, in late 1955 (Shoemaker F C, private communication).
[408] Kerst D W, Cole F T, Crane H R, Jones L W, Laslett L J, Ohkawa T, Sessler A M, Symon K R, Terwilliger K M and Nilsen N V 1956 *Phys. Rev.* **102** 590–1
[409] Symon K R, Kerst D W, Jones L W, Laslett L J and Terwilliger K M 1956 *Phys. Rev.* **103** 1837–59
[410] O'Neill G K 1956 *Phys. Rev.* **102** 1418–9
[411] O'Neill G K 1966 *Sci. Am.* **215** 107–116
[412] Marin P 1967 *Proc. Third Int. Symp. on Electron and Photon Interactions (SLAC, 1967)* ed S M Berman (Stanford, CA: SLAC) pp 376–86
[413] Schwarzschild B M 1980 *Phys. Today* **33** January pp 19–21
[414] Rubbia C, McIntyre P and Cline D 1977 *Proc. Int. Neutrino Conf. (Aachen, 1976)* ed H Faissner (Braunschweig: Vieweg) pp 683–7
[415] See, for example, Manchester W R 1992 *A World Lit Only by Fire* (Boston: Little Brown) p 294
[416] Blackett P M S 1933 *Cambridge University Studies* ed H Wilson (London: Nicholson and Watson) pp 67–96
[417] Rutherford E and Geiger H 1908 *Proc. R. Soc.* A **81** 141–61
[418] Townsend J S 1901 *Phil. Mag.* **1** (Ser. 6) 198–227; 1902 *Phil. Mag.* **3** 557–76; 1903 *Phil. Mag.* **5** 389–98; 1903 *Phil. Mag.* **6** 358–61, 598–618
[419] Rutherford E and Geiger H 1908 *Proc. R. Soc.* A **81** 162–73
[420] Geiger H and Müller W 1928 *Phys. Z.* **29** 839–41
[421] Neher H V 1938 (seventeenth printing in 1952) *Procedures in Experimental Physics* ed J Strong *et al* (New York: Prentice-Hall) pp 259–304
[422] Bothe W 1929 *Z. Phys.* **59** 1–5
Rossi B 1930 *Nature* **125** 636
[423] Greinacher H 1926 *Z. Phys.* **36** 364–73
[424] Wynn-Williams C E 1932 *Proc. R. Soc.* A **136** 312–24
[425] Elmore W C and Sands M 1949 *Electronics* (New York: McGraw-Hill) p 210
[426] Slepian J 1919 *US Patent* No 1,450,265 (April 3, 1919)
[427] Zworykin V K, Morton G A and Mather L 1936 *Proc. IRE* **24** 351–75

[428] Shockley W and Pierce J R 1938 *Proc. IRE* **26** 321–32
[429] Zworykin V K and Rajchman J A 1939 *Proc. IRE* **27** 558–66
[430] Čerenkov P A 1934 *C. R. Acad. Sci. URSS* **2** 451–4
[431] Frank I and Tamm I 1937 *C. R. Acad. Sci. URSS* **14** 109–14
[432] Čerenkov P A 1937 *Phys. Rev.* **52** 378–9
[433] Kallman H 1947 *Nat. Tech.* July
[434] Bell P R 1948 *Phys. Rev.* **73** 1405–6
[435] Reynolds G T, Harrison F B and Salvini G 1950 *Phys. Rev.* **78** 488
[436] Hofstadter R 1948 *Phys. Rev.* **74** 100–1; reference [4] ch 7 pp 126–143
[437] Morton G A 1949 *RCA Rev.* **10** 525–53
[438] Wilkinson D H 1950 *Proc. Camb. Phil. Soc.* **46** 508–18
[439] Wilson C T R 1911 *Proc. R. Soc.* A **85** 285–8; 1912 *Proc. R. Soc.* A **87** 277–92
[440] Hoxton L G 1933–4 A continuously operating cloud chamber *Proc. Virginia Acad. Sci.* **9** 23
[441] Langsdorf A Jr 1936 *Phys. Rev.* **49** 422; 1939 *Rev. Sci. Instrum.* **10** 91–103
[442] Vollrath R E 1936 *Rev. Sci. Instrum.* **7** 409–10
[443] Shutt R P, Fowler E C, Miller D H, Thorndike A M and Fowler W B 1951 *Phys. Rev.* **84** 1247–8
[444] Hildebrand R H and Nagle D E 1953 *Phys. Rev.* **92** 517–8
[445] Keuffel J W 1948 *Phys. Rev.* **73** 531
[446] Pidd R W and Madansky L 1949 *Phys. Rev.* **75** 1175–80
[447] Cranshaw T E and DeBeer J F 1957 *Nuovo Cimento* **5** 1107–17
[448] Fukui S and Miyamoto S 1959 *Nuovo Cimento* **11** 113–5
[449] Cronin J W 1967 *Bubble and Spark Chambers* vol 1, ed R P Shutt (New York: Academic) pp 315–405
[450] Wenzel W A 1964 *Ann. Rev. Nucl. Sci.* **14** 205–38
[451] Charpak G 1993 *Rev. Mod. Phys.* **65** 591–8
[452] Freedman S J 1990 *Comments Nucl. Part. Phys.* **19** 209–20
[453] Majorana E 1937 *Nuovo Cimento* **14** 171–84
[454] Kayser B, Gibrat-Debu F and Perrier F 1989 *The Physics of Massive Neutrinos* (Teaneck, NJ: World Scientific)
Boehm F and Vogel P 1987 *Physics of Massive Neutrinos* (Cambridge: Cambridge University Press)
Bilenky S M and Petcov S T 1987 *Rev. Mod. Phys.* **59** 671–754; 1989 *Rev. Mod. Phys.* **61** 169(E)
[455] Kawakami H *et al* 1991 *Phys. Lett.* **256B** 105–11
Robertson R G H, Bowles T J, Stephenson G J Jr, Wark D J, Wilkerson J F and Knapp D A 1991 *Phys. Rev. Lett.* **67** 957–60
Decman D and Stoeffl W 1992 *Bull. Am. Phys. Soc.* **37** 1286
Holzschuh E, Fritschi M and Kündig W 1992 *Phys. Lett.* **287B** 381–8
Weinheimer Ch *et al* 1993 *Phys. Lett.* **300B** 210–6
[456] Bardeen J, Cooper L N, and Schrieffer J R 1957 *Phys. Rev.* **108** 1175–1204
Anderson P W 1958 *Phys. Rev.* **112** 1900–16
[457] See, for example, Kadanoff L P 1966 *Physics* **2** 263–72; 1975 *Phys. Rev. Lett.* **34** 1005–8
Kadanoff L P, Götze W, Hamblen D, Hecht R, Lewis E A S, Palciauskas V V, Rayl M, Swift J, Aspnes D and Kane J 1967 *Rev. Mod. Phys.* **39** 395–431
Kadanoff L P and Houghton A 1975 *Phys. Rev.* B **11** 377–86
Wilson K G 1971 *Phys. Rev.* B **4** 3174–83, 3184–3205; 1975 *Rev. Mod. Phys.* **47** 773–840
Wilson K G and Kogut J 1974 *Phys. Rep.* **12C** 75–199
Fisher M E 1974 *Rev. Mod. Phys.* **46** 597–616

Elementary Particle Physics

[458] Friedan D, Qiu Z and Shenker S 1984 *Phys. Rev. Lett.* **52** 1575–8; 1985 *Phys. Lett.* **151B** 37–43
[459] Weinberg S 1972 *Gravitation and Cosmology: Principles and Applications of the General Theory of Relativity* (New York: Wiley)
[460] Pontecorvo B 1946 *Chalk River Laboratory Report* PD-205 (unpublished)
[461] Davis R Jr, Mann A K and Wolfenstein L 1990 *Ann. Rev. Nucl. Part. Sci.* **39** 467–506
[462] Bahcall J N 1989 *Neutrino Astrophysics* (Cambridge: Cambridge University Press)
[463] Hirata K S *et al* 1991 *Phys. Rev. D* **44** 2241–60; 1992 *Phys. Rev. D* **45** 2170(E)
Totsuka Y 1992 *Rep. Prog. Phys.* **55** 377–430
[464] Abazov A I *et al* 1991 *Phys. Rev. Lett.* **67** 3332–5
Gavrin V *et al* 1993 *Proc. XXVI Int. Conf. on High Energy Physics (Dallas, TX, August 6–12, 1992)* ed J R Sanford (New York: AIP) pp 1101–10
Anselmann P *et al* 1992 *Phys. Lett.* **285B** 376–89, 390–7; 1993 *Phys. Lett.* **314B** 445–8
[465] Bludman S, Hata N, Kennedy D C and Langacker P 1993 *Phys. Rev. D* **47** 2220–33
[466] Pontecorvo B 1967 *Zh. Eksp. Teor. Fiz.* **53** 1717–25 (Engl. Transl. 1968 *Sov. Phys.–JETP* **26** 984–8)
Wolfenstein L 1978 *Phys. Rev. D* **17** 2369–74
Mihkeev S P and Smirnov A Yu 1985 *Yad. Fiz.* **42** 1441–8 (Engl. Transl. *Sov. J. Nucl. Phys.* **42** 913–7); 1986 *Nuovo Cimento* **9C** 17–26; 1987 *Usp. Fiz. Nauk* **153** 3–58 (Engl. Transl. 1987 *Sov. Phys.–Usp.* **30** 759–90)
[467] Norman E B *et al* 1992 *The Fermilab Meeting DPF 92 (Proc. 1992 Division of Particles and Fields Meeting (Fermilab, 10–14 November, 1992))* ed C H Albright *et al* (Singapore: World Scientific) pp 1450–2
Raghavan R S and Pakvasa S 1988 *Phys. Rev. D* **37** 849–57
[468] Hirata K S *et al* 1988 *Phys. Rev. D* **38** 449
[469] Hirata K *et al* 1987 *Phys. Rev. Lett.* **58** 1490–3
Bionta R M *et al* 1987 *Phys. Rev. Lett.* **58** 1494–6
[470] Colgate S A and White R H 1966 *Astrophys. J.* **143** 626–81
Arnett W D 1982 *Astrophys. J. Lett.* **263** L55–7
[471] Bahcall J and Glashow S L 1987 *Nature* **326** 476–7
Arnett W D and Rosner J L 1987 *Phys. Rev. Lett.* **58** 1906–9
Abbott L F, De Rújula A and Walker T P 1988 *Nucl. Phys.* B **299** 734–56
[472] Gamow G 1948 *Phys. Rev.* **74** 505–6; 1948 *Nature* **162** 680–2
Alpher R A and Herman R C 1948 *Nature* **162** 774–5; 1949 *Phys. Rev.* **75** 1089–95 and references therein
[473] Penzias A A and Wilson R W 1965 *Astrophys. J.* **142** 419–21
[474] Dicke R H, Peebles P J E, Roll P G and Wilkinson D T 1965 *Astrophys. J.* **142** 414–9
Dicke R H and Peebles P J E 1966 *Nature* **211** 574–5
[475] Smoot G F *et al* 1992 *Astrophys. J. Lett.* **396** L1–5
[476] Steigman G, Schramm D N and Gunn J E 1977 *Phys. Lett.* **66B** 202–4
Walker T P, Steigman G, Schramm D N, Olive K A and Kang H S 1991 *Astrophys. J.* **376** 51–69
[477] Yodh G B *First Aspen Winter Physics Conf.* ed M M Block (*Ann. NY Acad. Sci.* **461** 239–59)
[478] Weekes T C 1988 *Phys. Rep.* **160** 1–121
Weekes T C *et al* 1989 *Astrophys. J.* **342** 379–95
Punch M *et al* 1992 *Nature* **358** 477–8
[479] Alexandreas D E *et al* 1991 *Phys. Rev. D* **43** 1735–8; 1992 *Nucl. Instrum. Methods* A **311** 350–67

Cronin J W et al 1992 *Phys. Rev.* D **45** 4385–91
Nagano M et al 1992 *J. Phys. G: Nucl. Phys.* **18** 423–42
Chiba N et al 1992 *Nucl. Instrum. Methods* **A311** 338–49
Bird D J et al 1993 *Phys. Rev. Lett.* **71** 3401–4

[480] Misner C W, Thorne K S and Wheeler J A 1973 *Gravitation* (San Francisco: Freeman) p 872
[481] Hawking S 1975 *Commun. Math. Phys.* **43** 199–220; *Quantum Gravity: An Oxford Symposium, 1975* ed C J Isham et al (Oxford: Clarendon) pp 219–67; 1981 *Encyclopedia of Physics* ed R G Lerner and G L Trigg (Reading, MA: Addison-Wesley) pp 81–83
[482] Trimble V 1987 *Ann. Rev. Astron. Astrophys.* **25** 425–72
Primack J R, Seckel D and Sadoulet B 1988 *Ann. Rev. Nucl. Part. Sci.* **38** 751–807
[483] Peccei R D and Quinn H R 1977 *Phys. Rev. Lett.* **38** 1440–3; 1977 *Phys. Rev.* D **16** 1791–7
Weinberg S 1978 *Phys. Rev. Lett.* **40** 223–6
Wilczek F 1978 *Phys. Rev. Lett.* **40** 279–82
[484] Guth A 1980 *Phys. Rev.* D **23** 347–56
Linde A 1983 *Phys. Lett.* **129B** 177–81 and references therein
[485] Eötvös R v, Pekár D and Fekete E 1922 *Ann. Phys., Lpz* **68** 11–66
[486] Dicke R H, Roll P G and Krotkov R 1964 *Ann. Phys., NY* **26** 442–517
[487] Fischbach E, Sudarsky D, Szafer A, Talmadge C and Aronson S H 1986 *Phys. Rev. Lett.* **56** 3–6, 1427(E)
[488] Adelberger E G, Heckel B R, Stubbs C W and Rogers W F 1991 *Ann. Rev. Nucl. Part. Sci.* **41** 269–320
[489] For a recent reference see, for example, Buskulic D et al (ALEPH Collaboration) 1993 *Phys. Lett.* **313B** 299–311.
[490] Veltman M 1977 *Acta Phys. Pol.* **B8** 475–92; 1977 *Phys. Lett.* **70B** 253–4
Lee B W, Quigg C and Thacker H B 1977 *Phys. Rev. Lett.* **38** 883–5; 1977 *Phys. Rev.* D **16** 1519–31
[491] Eichten E, Hinchliffe I, Lane K and Quigg C 1984 *Rev. Mod. Phys.* **56** 579–707; 1986 *Rev. Mod. Phys.* **58** 1065(E)
[492] Dawson S, Gunion J F, Haber H E and Kane G L 1990 *The Higgs Hunter's Guide* (Redwood City, CA: Addison-Wesley)
[493] Wess J and Bagger J 1983 *Supersymmetry and Supergravity* (Princeton, NJ: Princeton University Press)
Freund P 1986 *Introduction to Supersymmetry* (Cambridge: Cambridge University Press)
[494] Weinberg S 1976 *Phys. Rev.* D **13** 974–96; 1979 *Phys. Rev.* D **19** 1277–80
Susskind L 1979 *Phys. Rev.* D **20** 2619–25
[495] Decman D and Stoeffl W in reference [455]
[496] Hirata K S et al 1992 *Phys. Lett.* **280B** 146–52
Beier E W et al 1992 *Phys. Lett.* **283B** 446–53
Fukuda Y et al 1994 *Phys. Lett.* **335B** 237–45
[497] Arik E et al (CHORUS Collaboration) 1991 CERN Experiment WA-95, approved September 1991, K Winter, spokesperson
Kadi-Hanifi M et al (NOMAD Collaboration) 1991 CERN Experiment WA-96, approved September 1991, F Vannucci, spokesperson
[498] Bernstein R H and Parke S J 1991 *Phys. Rev.* D **44** 2069–78 and references therein
[499] Pati J C and Salam A 1974 *Phys. Rev.* D **10** 275–89
[500] Georgi H and Glashow S L 1974 *Phys. Rev. Lett.* **32** 438–41
[501] Georgi H 1975 *Proc. 1974 Williamsburg DPF Meeting* ed C E Carlson (New York: AIP) pp 575–82

Fritzsch H and Minkowski P 1975 *Ann. Phys., NY* **93** 193–266
[502] Gell-Mann M, Ramond P and Slansky R 1979 *Supergravity* ed P van Nieuwenhuizen and D Z Freedman (Amsterdam: North-Holland) pp 315–21
Yanagida T 1979 *Proc. Workshop on Unified Theory and Baryon Number in the Universe* ed O Sawada and A Sugamoto (Tsukuba, Japan: National Laboratory for High Energy Physics)
[503] Amaldi U, Bohm A, Durkin L S, Langacker P, Mann A K, Marciano W J, Sirlin A and Williams H H 1987 *Phys. Rev.* D **36** 1385
Amaldi U, de Boer W and Fürstenau H 1991 *Phys. Lett.* **260B** 447–55
Langacker P and Polonsky N 1993 *Phys. Rev.* D **47** 4028–45
[504] Rosner J L 1994 *DPF 94 Proc. DPF 94 Meeting (Albuquerque, NM, August, 1994)* (Singapore: World Scientific)
[505] Seidel S *et al* 1988 *Phys. Rev. Lett.* **61** 2522–25
Hirata K S *et al* 1989 *Phys. Lett.* **220B** 308–16
[506] Totsuka Y 1992 *Rep. Prog. Phys.* **55** 377–430
[507] Preskill J P 1979 *Phys. Rev. Lett.* **43** 1365–8
[508] See, for example, Adams F C, Fatuzzo M, Freese K, Tarlé G, Watkins R and Turner M S 1993 *Phys. Rev. Lett.* **70** 2511–14
[509] Gross D J, Harvey J A, Martinec E and Rohm R 1985 *Phys. Rev. Lett.* **54** 502–5; 1985 *Nucl. Phys.* B **256** 253–84; 1986 *Nucl. Phys.* B **267** 75–124
[510] Candelas P, Horowitz G T, Strominger A and Witten E 1985 *Nucl. Phys.* B **258** 46–74
Witten E 1985 *Nucl. Phys.* B **258** 75–100
[511] Callan C G Jr, Giddings S, Harvey J A and Strominger A 1992 *Phys. Rev.* D **45** 1005–9
[512] See, for example, Derrick M *et al* (ZEUS Collaboration) 1992 *Phys. Lett.* **293B** 465
Ahmed T *et al* (H1 Collaboration) 1993 *Phys. Lett.* **299B** 374–84, 385–93
[513] Ritson D 1993 *Nature* **366** 607–10
[514] Ayres D S *et al* 1991 *Proc. 25th Int. Conf. on High Energy Physics (Singapore, August 2–8, 1990)* ed K K Phua and Y Yamaguchi (Singapore: World Scientific [South East Asia Theoretical Physics Association and the Physical Society of Japan]) pp 480–1
Thron J L 1993 *Proc. XXVI Int. Conf. on High Energy Physics (Dallas, TX, August 6–12, 1992)* ed J R Sanford (New York: AIP) pp 1232–7
[515] Calicchio M *et al* 1988 *Nucl. Instrum. Methods* A **264** 18–23
Ahlen S P *et al* 1993 *Nucl. Instrum. Methods* A **324** 337–62
[516] Norman E B *et al* 1992 *The Fermilab Meeting DPF 92 (Proc. 1992 Division of Particles and Fields Meeting (Fermilab, 10–14 November, 1992))* ed C H Albright *et al* (Singapore: World Scientific) pp 1450–2
[517] Roberts A 1992 *Rev. Mod. Phys.* **64** 259–312
[518] Barwick S *et al* 1993 *Proc. XXVI Int. Conf. on High Energy Physics (Dallas, TX, August 6–12, 1992)* ed J R Sanford (New York: AIP) pp 1250–3
[519] Sadoulet B 1990 *Proc. First Int. Symp. on Particles, Strings, and Cosmology (Boston, MA, 27–31 March, 1990)* ed P Nath and S Reucroft (Teaneck, NJ: World Scientific) pp 147–84
[520] Cronin J W 1992 *Proc. Symp. on the Interface of Astrophysics with Nuclear and Particle Physics (Zuoz, Switzerland, 11–18 April, 1992)* ed M P Locher (Villigen: Paul Scherrer Institute) pp 341–3
[521] Klein S *et al* 1992 *The Fermilab Meeting DPF 92 (Proc. 1992 Division of Particles and Fields Meeting (Fermilab, 10–14 November, 1992))* ed C H Albright *et al* (Singapore: World Scientific) pp 1364–6
[522] Bernstein J 1989 *The Tenth Dimension: An Informal History of High Energy*

[523] United States Department of Energy 1990 *The Ultimate Structure of Matter: The High Energy Physics Program from the 1950s through the 1980s* (Washington, DC: USDOE Office of Energy Research)
[524] Weinberg S 1992 *Dreams of a Final Theory* (New York: Pantheon)
[525] Weisskopf V 1991 *The Joy of Insight* (New York: Basic Books)
[526] Treiman S B 1993 *A Century of Particle Theory*
[527] Gell-Mann M 1994 *The Quark and the Jaguar: Adventures in the Simple and the Complex* (New York: Freeman)
[528] Marshak R E 1993 *Conceptual Foundations of Modern Particle Physics* (River Edge, NJ: World Scientific)
[529] Waloschek P (ed) 1994 *The Infancy of Particle Accelerators: Life and Work of Rolf Wideröe* (Braunschweig: Vieweg)

Chapter 10

FLUID MECHANICS

Sir James Lighthill

10.1. Yet another great success for twentieth century physics

10.1.1. A parallel revolution in fluid mechanics
Much of this book has been concerned with those profound changes in the pure and applied physical sciences that were initiated at the outset of the twentieth century through revolutionary discoveries which included radioactivity, quantum theory, the nuclear atom and relativity, together with thermionics and x-rays and their many applications. These were discoveries that transmuted the nineteenth century worldview of physics while addressing with scintillating success many of its recognized failures.

This chapter recalls yet another success story of twentieth century physics which, like the others, had profound implications for the human condition. It was based on brilliant experiments but depended above all on revolutionary approaches to interpretation, which both accounted for some spectacular failures of earlier approaches and, for the first time, placed upon sound foundations a certain major branch of physics: fluid mechanics.

Our environment on Earth confers a special human importance on two fluids: air and water; and for this principal reason we limit the present history to the mechanics of these two fluids. Indeed, the failures of nineteenth century fluid mechanics were particularly notorious for these familiar fluids of low viscosity—whereas, for example, the lubricating action of liquids of much higher viscosity in bearings under load had by 1886 been reliably accounted for with a sound fluid-mechanical theory [1].

The low viscosity of air, on the other hand, had tempted nineteenth century physicists to try to relate its dynamics to that of an 'ideal' fluid without any viscosity at all. For such a fluid there existed a most

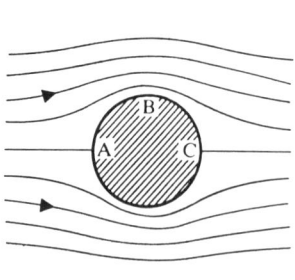

Figure 10.1. *Contrast between the potential field of flow around a circular cylinder (left) and an experimentally observed field [3] (right). Note: in both fields the oncoming flow is steady (not varying with time) but in the observed field the 'wake' (behind the cylinder) is highly unsteady.*

elegant and extensive theory [2], constructed—sometimes, in response to suppositions about the ether!—by many of the century's finest physicists. Yet the true motions of air, when disturbed by the movement of a typical body through it, bore in reality no relation whatsoever to the predictions of this ideal-fluid theory.

Admittedly, in acknowledging the theory's failure, the scientists concerned paid excessive attention to just one erroneous prediction—the famous d'Alembert's Paradox, that a body moving at uniform velocity through fluid would experience no resistive force—instead of admitting that the predicted flow field as a whole was quite different from that actually occurring. This predicted flow field was just such a regular 'potential' field, derived from a scalar potential, as is familiar from electrostatics or magnetostatics, and was in total contrast (see figure 10.1) to the real field, with its swirling wake behind the body [3]. And yet another false prediction of the theory was that fixed-wing aircraft would be unable to fly, because the air could not exert any force at all—resistive or lifting force—on a steadily moving body.

Just one scientist, Ludwig Prandtl, must be given the main credit for the brilliant discovery [4] which resolved these anomalies. Indeed his revolutionary discovery of the boundary layer in 1904 had the same transforming effect on fluid mechanics as Einstein's 1905 discoveries had on other parts of physics.

Such a very special level of recognition is due to Prandtl for two principal reasons. In theoretical terms his solution, besides representing an extremely early example of a singular perturbation (see section 10.1.2 below), was successfully applied to field equations of a fully non-linear type. In practical terms, Prandtl's new insights brought about the

Ludwig Prandtl

(German, 1875–1953)

Ludwig Prandtl was born on 4 February 1875 in Freising, Bavaria and studied Mechanical Engineering in Munich under August Foeppl. His doctor's thesis appeared in 1900 with some immediately acclaimed discoveries on the torsion of beams. However, by the following year, when appointed Professor of Mechanics in Hannover, he had already begun his path-breaking career in fluid mechanics. It was to the Third International Congress of Mathematicians (Heidelberg, 1904) that his epoch-making discovery of the boundary layer was first presented. That year, the famous Göttingen mathematician Felix Klein secured for Prandtl a chair in applied physics (later renamed applied mechanics) at Göttingen; which was to become the base from which he progressively revolutionized the understanding of aerodynamic drag and lift and of fluid flows at high Mach number. Many of his later contributions were made as leader of a fine team whose other members (including A Betz, W Tollmien, M Munk, J A Ackeret, H Schlichting, A Busemann) would themselves all win international recognition. In 1909 he married Gertrude Foeppl, by whom he had two daughters. His superb guide to fluid mechanics ultimately appeared in an extended 1952 English-language edition as *Essentials of Fluid Dynamics*. He died in Göttingen on 15 August 1953. His complete works [4] were published by Springer in 1963.

introduction of 'streamlined' shapes which, by permitting just modest deviations from potential flow fields, would experience very low resistive forces; while Prandtl himself went on to generate the first quantitative understanding of lift (and drag) forces on fixed-wing aircraft based on sound physical principles [5].

The twentieth century history of many crucially important developments that were to evolve from boundary-layer theory is sketched in section 10.2 below. These included developments in which Prandtl and scientists of his Göttingen school (A Betz, H Schlichting and above all Theodore von Kármán, who would soon set up new schools first at Aachen and later at Caltech) combined the new ideas with ideas from elsewhere, including the work of Zhukovski in Russia, to develop knowledge on boundary layers and wakes, and on the aerodynamic forces associated with them. Also, they included important studies on the instability of such shear layers (where early advances were made by physicists distinguished in other contexts, including Rayleigh and Sommerfeld) and on their tendency to develop the chaotic form of fluid motion known as turbulence—on which key papers by Osborne Reynolds, concentrating on turbulence in pipe flow, had first appeared in the 1880s and where important later advances would be made not only by Prandtl but also by H L Dryden in the United States, G I Taylor in England and A N Kolmogorov in Russia.

Here all of that extended historical account is preceded by a succinct introduction which begins with a quite simple explanation (section 10.1.2) of the idea of a singular perturbation. This leads on to a brief indication (section 10.1.3) of what the very earliest discoveries on boundary layers and wakes meant for those new aeronautical developments that were, in due course, to transform the human condition in ways which are outlined in section 10.4.

Then a first sketch is given (section 10.1.4) of that parallel revolution in knowledge of non-linear effects in the generation and propagation of waves in fluids—including sound waves and water waves—which began from elucidation of the physics of shock waves. These, indeed, offered another outstanding example of a singular perturbation successfully applied to field equations of a non-linear type, where Rayleigh [6] and Taylor [7] made independently the key discovery in two papers published in 1910. The whole scientific study constituted an early example of what would later be given the generic name of 'catastrophe theory'—just as the fluid mechanics of turbulence became an important forerunner to later generic concepts of chaos. Moreover, the same consistent involvement of twentieth century exponents of fluid mechanics in non-linear field theories led them to pioneer many new concepts—from solitons to uses of non-linear Schrödinger equations—that, after first appearing in the advanced study of water waves, would contribute later to other areas of the physical sciences.

The two great revolutions in twentieth century fluid mechanics were combined in many later developments that are outlined in section 10.4. For example, aircraft flying at speeds near, or greater than, the speed of sound may exhibit complex interactions of boundary layers with shock waves. Again, the mechanics of bodies near the ocean surface—whether ships or offshore structures—may often involve a similar mixture of considerations from the modern non-linear studies of both water waves and shear layers.

Geoffrey Ingram Taylor

(British, 1886–1975)

Geoffrey Ingram Taylor—a grandson of the mathematical originator George Boole —was born in London on 7 March 1886. He studied Mathematics and Physics at Cambridge; where, after successful early experimental researches within quantum optics, he produced in 1910 his revolutionary paper on the physics of shock waves. After his 1911 appointment to a newly established Readership in Dynamical Meteorology at Cambridge, he began to publish pioneering experimental investigations of turbulence in the atmosphere. Moreover, he demonstrated the importance of turbulent dissipation of energy in tidal streams (1919) and pursued fundamental investigations on the motion of rotating fluids (1922) and on hydrodynamic stability (1923); while his 1921 paper *Diffusion by continuous movements* crucially contrasted turbulence with apparently analogous kinetic-theory phenomena and led in 1935 to massive developments in statistical theory of turbulence. Appointment in 1923 to a Royal Society Research Professorship allowed him thereafter to concentrate his energies on personal researches. Alongside crucial discoveries in solid mechanics (including the beginnings of dislocation theory), he continued to expand knowledge of shock waves and of atmospheric processes (e.g. buoyant plumes and thermals) while making also key initial studies (1951–52) on the fluid mechanics of the biosphere. He married Stephanie Ravenhill in 1925. His complete works in 4 volumes were published by Cambridge University Press (1958–71). He died in Cambridge on 27 June 1975.

Beyond our history of responses of fluid mechanics to such engineering challenges, we offer in section 10.5 an enormously larger-scale view of the mechanics of the Earth's whole fluid envelope including oceans, rivers and the atmosphere. The twentieth century has seen great advances in the successful application of fluid mechanics to the understanding, and to the prediction, both of disasters such as storms and floods, and moreover of weather and climate in general. Yet, many of these advances, too, can trace their lineage back to the revolution in fluid mechanics achieved in 1904 by Ludwig Prandtl.

10.1.2. An extremely simple example of a singular perturbation

Prandtl's introduction of what we now call a singular perturbation [8] was applied to field equations of a fully non-linear type. However, the essential idea can be illustrated with an exceedingly simple application to an ordinary differential equation which is linear with constant coefficients, and where the correctness of the method is easily verified by comparison with an exact solution.

Suppose that, in seeking to solve the differential equation

$$\varepsilon y'' + y' + ky = 0 \quad \text{for } 0 < x < 1 \tag{1}$$

subject to the boundary conditions

$$y = 0 \text{ at } x = 0, \qquad y = 1 \text{ at } x = 1 \tag{2}$$

we know that the second differential coefficient y'' is multiplied by an extremely small factor ε (which might be the viscosity in a fluid-mechanical application). The ordinary perturbation theory which is used throughout physics would look first for the solution with $\varepsilon = 0$ and then improve it by a process of successive approximation; for example, one involving an expansion in powers of ε.

This approach seems at first to make good progress because when $\varepsilon = 0$ equation (1) takes the still simpler form

$$y' + ky = 0 \tag{3}$$

which has the well known general solution

$$y = ae^{-kx} \tag{4}$$

with a an arbitrary constant. However, it is impossible to find a value of a which satisfies both the boundary conditions given in (2); for example, the choice $a = e^k$ satisfies the second condition, giving

$$y = e^{k(1-x)} \quad \text{with } y = 1 \text{ at } x = 1 \tag{5}$$

but this solution fails to satisfy the first condition since

$$\text{the value of } y \text{ at } x = 0 \text{ is } e^k \tag{6}$$

instead of being equal to zero. Moreover it is easily verified that no procedure of moving to higher approximations through the usual expansion in powers of ε yields any improvement at all.

The difficulty arises of course because, although two boundary conditions as in (2) are appropriate to a second-order differential equation (1), nevertheless the solution of a first-order equation (3) is uniquely determined by just a single boundary condition as in (5). The 'extra boundary condition' $y = 0$ at $x = 0$ (which might be the no-slip condition at a solid boundary in the fluid-mechanical application) makes completely impossible the usual perturbation approach.

If, however, we probe just what has gone wrong, we must recognize that the solution (5) which meets the second boundary condition should in general satisfy the true differential equation (1) to quite a close approximation because the term $\varepsilon y''$ is very small compared with the other terms in the equation. By contrast, things become very different as the solution (5) approaches $x = 0$ because it must encounter a region of very rapid change between the value (6) and the true value of $y = 0$ at $x = 0$.

In such a region the rate of change y' can become very large and its own rate of change y'' can become enormously larger so that even the term $\varepsilon y''$ in equation (1) takes a large value. If $\varepsilon y''$ and y' are both far bigger than the term ky, then equation (1) can be approximated as

$$\varepsilon y'' + y' = 0 \tag{7}$$

so that $\varepsilon y' + y$ has zero rate of change and must take a constant value c. But the solution of

$$\varepsilon y' + (y - c) = 0 \tag{8}$$

is

$$y - c = A e^{-x/\varepsilon} \tag{9}$$

where A is another arbitrary constant; here A replaces a while $1/\varepsilon$ replaces k in the former general solution (4) of equation (3).

Now, given that $y = 0$ at $x = 0$ (the first boundary condition), the value of A is determined as $-c$, so that

$$y = c(1 - e^{-x/\varepsilon}) \tag{10}$$

in the region of rapid change. The broken line in figure 10.2 shows, however, that such rapid change soon disappears as y tends to an asymptotic value c.

In its simplest form, the idea of singular perturbation theory is that equation (7) with its solution (10) is just appropriate to the 'boundary layer' region of very rapid change with x extremely small, while equation

801

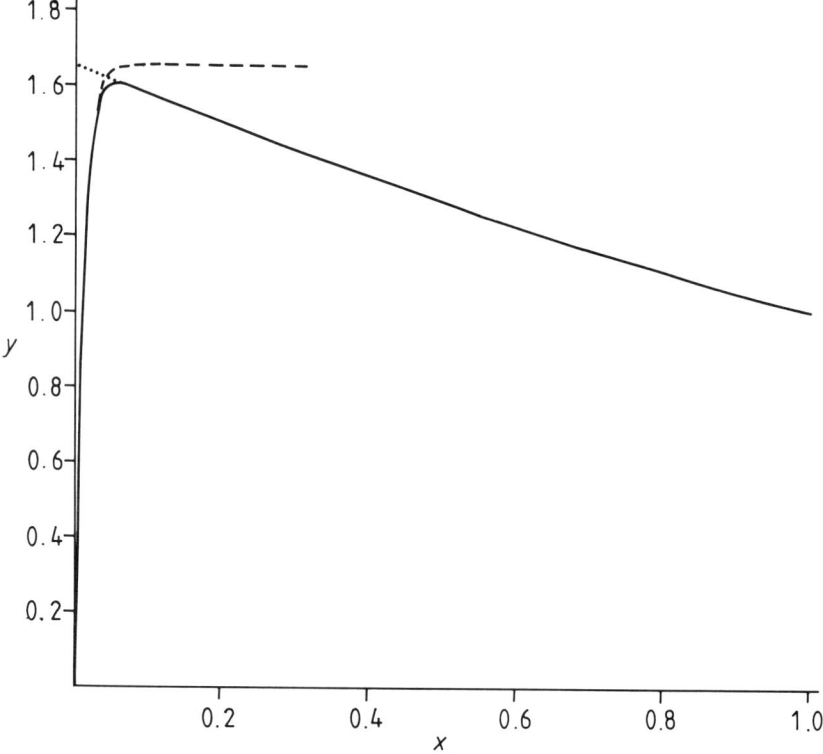

Figure 10.2. *Solid line, exact solution of equation (1) subject to conditions (2). Dotted line, 'external' solution (5). Broken line, 'boundary-layer' solution (10).*

(3) with its solution (4) is appropriate to all other regions. Moreover the two solutions are readily matched together provided that

$$c \text{ is given the value } e^k \qquad (11)$$

so that the value c of y just outside the region of rapid change coincides with the limit (5) of y as it approaches the region around $x = 0$.

Figure 10.2 assesses the accuracy of this approach. The solid line shows the exact solution of equation (1) subject to the conditions in (2) while the broken line shows the 'boundary-layer' solution (1), with c as in (11), and the dotted line shows the solution (5) for the region outside the boundary layer. The idea that the true solution makes, quite simply, a smooth transition between its two approximate forms is well substantiated in this case†.

† *More advanced forms of singular perturbation theory [8] are able to achieve any desired degree of accuracy by matching a series expansion for y in powers of ε with coefficients functions of x/ε in the region of very rapid change to another expansion in powers of ε with coefficients functions of x outside that region.*

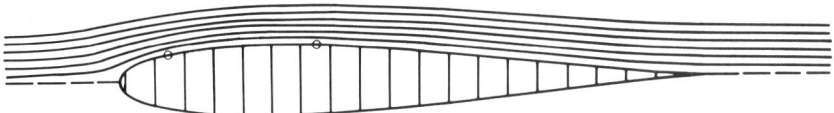

Figure 10.3. *For specially designed shapes, potential flow can be achieved (as illustrated here for flow, symmetrical about the broken line, around a blade whose cross-section is a Zhukovski [10] aerofoil) except in a very thin boundary layer. For magnified illustrations of the fluid motion in the two circular regions near the flow boundary, see figure 10.4 below.*

The above introductory account of a singular perturbation has concentrated on an extremely simple example, in order that readers may easily appreciate the idea and why it can work so well. It is a case where an exact solution (figure 10.2) was readily obtained; on the other hand, we shall see in the next two sections that the revolution in fluid mechanics came from applications of the same idea to field equations (that is, to partial differential equations) of a fully non-linear type where progress would have been impossible without the use of such a radically new approach.

10.1.3. How d'Alembert's Paradox became d'Alembert's Theorem

Of the two remaining introductory sections this first is devoted to outlining Prandtl's early discoveries about the boundary layer, and to sketching the resulting transformation of ideas about how a body moving with uniform velocity disturbs the air around it. In section 10.1.1 we recalled d'Alembert's Paradox, that the potential flow field suggested by theories of an 'ideal' fluid without any viscosity would generate no resistive force on such a body [9]. Prandtl's work explained why, for typical body shapes, the true flow field is so vastly different (see figure 10.1) in spite of the extremely low viscosity of air; but, above all, it gave the first indications that for specially designed body shapes the flow field could be much closer to a potential field—in such a way that the resistive force while not zero would become very small (see figure 10.3).

This was a development in which, effectively, d'Alembert's Paradox became d'Alembert's Theorem: an encouraging affirmation that, if ever the flow around a steadily moving body could be made quite close to a potential flow, then the resistive force should likewise become quite close to that zero force which an exactly potential flow would exert. These considerations prompt, of course, the question 'just what may be needed for the flow of an only slightly viscous fluid to become close to a potential flow?'.

Although (as has been indicated more than once) the field equations for the flow of such a fluid are 'of a fully non-linear type'—with non-linear terms as large and important as the linear terms in those equations—nevertheless it turns out that a potential flow field does exactly satisfy these full non-linear equations for a viscous fluid. On the other hand a potential flow field can satisfy only one boundary condition at a solid surface instead of the two which a real fluid satisfies. The genius of Prandtl showed itself in his bold introduction of a revolutionary approach—to which the reader has already been introduced, under its modern name singular perturbation theory—in recognition of this deficiency in the number of boundary conditions that a potential flow field can satisfy.

Actually, the flow field around a body moving with uniform velocity through still air is easiest to describe in a frame of reference with the body at rest, in which the problem becomes the exactly equivalent problem of how a stationary body disturbs a uniform wind. The classical potential-flow solution to this problem satisfies just one boundary condition; namely that, because there can obviously be no flow across the solid surface, the flow velocity at that surface must be directed along it (that is, tangentially).

A real fluid, however, must satisfy a second boundary condition, stating that the magnitude of that flow velocity tends to zero as the surface is approached. At the solid surface, indeed, departures from local thermodynamic equilibrium must be just as small as in the rest of the fluid (owing—in both locations—to the huge frequency of molecular collisions); so that fluid in contact with the surface satisfies the equilibrium conditions of zero velocity relative to the surface and temperature equal to that of the surface.

It is impossible for this second boundary condition to be satisfied by any potential flow field; for which, rather, the velocity magnitude is always substantial on the solid surface (and, actually, attains its maximum at a certain point of the surface). As in the simple example of section 10.1.2, however, it remains possible that the external potential flow is separated from the body's surface by a thin boundary layer [4, 11] in which the velocity falls steeply from its potential-flow value to a zero value at the solid surface (see figure 10.4).

Simplified forms of the field equations can be expected to apply in such a boundary layer (if it exists) just as equation (1) within its boundary layer takes the simplified form (7). Here, we merely describe in physical terms the nature of these simplified field equations for the boundary layer; which are interesting primarily because

(i) their solutions for certain shapes of body represent boundary layers which remain thin and attached to the surface—these are the 'streamlined' shapes which experience low resistive force because the flow around them differs only moderately from a potential flow—whereas

(ii) for a much wider range of shapes, solutions to these boundary-layer equations simply cease to exist downstream of a certain point on

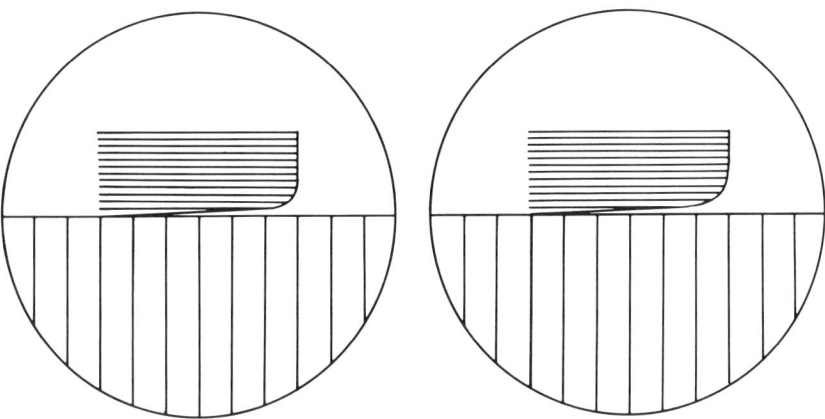

Figure 10.4. *Magnified illustration of flow (from left to right) in the two circular regions of figure 10.3. Each horizontal line near the flow boundary has length equal to the local flow speed. The boundary layer, within which the speed tends to zero, is very thin in the left-hand region. Moreover, in the right-hand region it remains attached to the boundary even though diffusion of momentum has caused it to grow a little thicker.*

the body surface, near which the flow along that surface can be described as undergoing 'separation' from the solid surface.

It is of course this flow separation that in case (ii) may produce a prominent swirling wake behind a bluff body; whereas the attached boundary layer in case (i) may emerge as a thin, far less prominent wake behind a streamlined body.

These distinctions make plain the crucial importance of those boundary-layer equations [4, 11] which Prandtl derived already in 1904. The physical effects represented in those equations are as follows.

First of all, the very steep gradient of velocity across a boundary layer (from a potential-flow value, say V, at its outer edge down to zero at the solid surface) allows viscous effects to become important even for fluids of very low viscosity; indeed, the thickness δ of the layer adjusts itself to the viscosity μ of the fluid. Velocity gradients of the order of V/δ produce viscous stresses of the order $\mu V/\delta$; here, a stress is of course a force per unit area, so that forces per unit volume in a layer of thickness δ are of the order $\mu V/\delta^2$. Simply from a balance of these viscous forces per unit volume against any other effects (see below), it is already clear that the small thickness δ of a boundary layer for fluids of different viscosity μ must vary as the square root $\mu^{1/2}$ of the viscosity.

For extremely small values of μ, then, boundary layers may be very thin indeed. At the same time, the actual viscous stresses in these layers (of order $\mu V/\delta$) may themselves be small (of order $\mu^{1/2}$); so that, in cases where the boundary layer remains attached, these stresses may contribute only a modest augmentation to the zero resistive force on the

805

body suggested by d'Alembert's Theorem.

Those other effects, independent of viscosity, against which the viscous forces per unit volume have to be balanced are two in number. Newton's second law of motion, of course, equates the total force on any particle to its mass times its acceleration, so that force per unit volume on a fluid is equal to its density ρ times its acceleration. This is the non-linear term in the field equations of fluid mechanics; thus, in steady flow around a body, the acceleration of any particle of fluid along its (curved) path has to be written as $v \, dv/ds$ in terms of the distance s along that path—because its velocity v is the rate ds/dt of increase of s—and this gives a mass-acceleration $\rho v \, dv/ds$ per unit volume.

At the same time, the total force per unit volume which this has to balance includes not only the viscous force but also any gradient $(-dp/ds)$ of the fluid pressure† along the path (with negative sign because pressures increasing with s oppose the motion). Outside the boundary layer, where v assumes its potential-flow value V, this gradient $(-dp/ds)$ is the sole force balancing the mass-acceleration term $\rho V \, dV/ds$, giving the famous Bernoulli equation

$$p + \tfrac{1}{2}\rho V^2 = \text{constant} \qquad (12)$$

explained in 1738 (for such steady motion without viscous effects) by Daniel Bernoulli [12].

Prandtl brilliantly recognized that the pressure would be given throughout the boundary layer by this same equation (12) depending on the velocity V outside it (briefly, any pressure change across the very thin boundary layer must be negligible since pressure gradients at right angles to the flow are limited to their centrifugal values). It followed that the pressure gradient in the flow direction, $(-dp/ds)$, takes a value $\rho V \, dV/ds$ which, in combination with viscous forces, has to balance the mass-acceleration $\rho v \, dv/ds$ within the boundary layer. It is this balance

$$\rho v \frac{dv}{ds} - \rho V \frac{dV}{ds} = \text{viscous force per unit volume} \qquad (13)$$

which—in physical terms—constitutes Prandtl's boundary-layer equations, governing how the velocity v falls from its value V just outside the thin boundary layer to a zero value at the solid surface. (Here, because the viscous stress takes a value $\mu \, dv/dn$, proportional to a gradient of v normal to the flow, each particle of fluid is subjected to a viscous force per unit volume $\mu \, d^2v/dn^2$, associated with the difference of stresses

---

† Here, more strictly, p should be taken as the 'excess pressure' (any excess of fluid pressure over its purely hydrostatic value), so that the gradient $(-dp/ds)$ accounts also for any component of gravity along the path. In the case of air, however, that distinction is generally unimportant.

acting on both sides of it—and, as foreshadowed earlier, with order of magnitude $\mu V/\delta^2$.)

These equations were of revolutionary significance because they focused attention on that thin layer whose dynamics was really determining the character of the whole flow. In mathematical terms they are quite complicated non-linear field equations of 'parabolic' type which would be subjected to over half a century's intensive study because of their complexity and because of their importance—see (i) and (ii) above—for a wide range of body shapes [11]. Although section 10.2 includes some account of these analyses that confirmed how the equations can be solved by a systematic progression in the downstream direction, which may either proceed regularly as in (i) or else abruptly cease as in (ii), we reproduce here just the essential physical insight into those alternatives that was included in Prandtl's 1904 paper.

Viscosity, of course, is the physical process by which fluid momentum is diffused down its gradient. This has different effects in three different types of boundary layers: (a) when the external flow speed V is increasing in the direction of flow (dV/ds positive); (b) when it is slightly decreasing (dV/ds negative but small); and (c) when it is falling more steeply (dV/ds negative but not small). In case (a) the pressure gradient $\rho V dV/ds$ is accelerating the fluid in the boundary layer (which therefore remains thin) while any excess momentum is transferred by diffusion to the solid surface. In case (b) the pressure gradient is gradually reducing the momentum of fluid near the surface, which would be brought to rest but for the fact that diffusion from the external flow replenishes that momentum. In case (c) such diffusion is insufficient to keep this fluid moving and a stagnant region forms near the wall from which the main flow separates.

Briefly, then, the shapes that can benefit from d'Alembert's Theorem by retaining a thin, attached boundary layer are those shapes for which the external potential flow field combines regions of types (a) and (b) only and avoids regions of type (c). The flow speed V increases over the front part of the body to a maximum where flow lines are most packed together, and is then reduced only gradually over the smoothly tapering rear portion (see figure 10.3). For example, when a herring after a burst of swimming glides forward rigidly, its rate of loss of speed is low because its shape satisfies this condition.

Widespread engineering application of these ideas is described in section 10.2, where their further development takes, in essence, three forms. First, imprecise phrases like 'very low viscosity' are abandoned in favour of that quantitative comparison of a product of flow quantities with μ which 'Reynolds number' gives; so that boundary layers become thinner (relative to body size) with increasing Reynolds number.

Next, the tendency as Reynolds number becomes still larger for the interior of a boundary layer to become turbulent is explained along with several implications. Above all, increased diffusion in the boundary layer assists in countering the tendency to separation—as Prandtl himself was to find—so that a rather wider range of boundary layers with retarded external flow becomes comprised within case (b) above.

Finally, boundary-layer theory was also developed by Prandtl in another, even more dramatic, form [5]. This explained how the nature of thin wakes from well designed wing shapes could allow wings to experience large forces ('lift') at right angles to the flow, even while resistive forces would continue to be low in the general spirit of d'Alembert's Theorem.

10.1.4. The physics of shock waves

This last introductory section is devoted to one further topic, completely different from that of boundary layers, where a long-standing enigma in fluid mechanics was resolved early in the twentieth century by an application of, yet again, the essential idea (section 10.1.2) that is now called singular perturbation theory. It was the resolution of that enigma which first elucidated the physics of shock waves.

The present brief account of this major achievement may serve to introduce that wide subject—non-linear effects in the generation and propagation of waves in fluids—to which section 10.3 is devoted. There are many different kinds of waves in fluids; amongst which sound waves in air may be most familiar to our ears, and gravity waves on a water surface to our eyes. Non-linear effects include interaction of such waves with fluid flows, and indeed this interaction is the principal feature of waves in fluids which makes their physics so different from that of other types of waves.

Linear theories of waves in general became well developed during the nineteenth century, and Rayleigh's superb treatise *The Theory of Sound* [13] expounded comprehensively an enormous body of knowledge obtained from linear wave theory as applied to sound generation and propagation. Yet in section 253 of this great work Rayleigh brilliantly showed how consideration of just the simplest possible problem in the non-linear theory of sound posed an enigma which it was quite impossible to resolve with the knowledge then available. Briefly, in any large-amplitude sound wave, higher values of the pressure may propagate faster than lower values and 'catch up' with them so that, apparently, continuous motion ceases to be possible.

Here, the enigma is explained for air, treated as a perfect gas with constant specific heats in a ratio γ which is close to 1.4. Ever since a famous 1816 paper of Laplace [14] it had been appreciated that, in sound waves, those changes of pressure p and density ρ whose ratio $dp/d\rho$ is the square of the sound speed c are 'adiabatically' related; this means that a particle of fluid experiences no heat input (or, indeed, output) so that when it expands it cools—by doing work—and the pressure falls more than it would in an isothermal process. For air with undisturbed pressure p_0 and density ρ_0 this adiabatic relationship takes the form

$$\frac{p}{p_0} = \left(\frac{\rho}{\rho_0}\right)^\gamma \text{ giving } c^2 = \frac{dp}{d\rho} = c_0^2 \left(\frac{\rho}{\rho_0}\right)^{\gamma-1} \text{ with } c_0^2 = \frac{\gamma p_0}{\rho_0}. \tag{14}$$

On linear theory sound is propagated at the undisturbed speed of sound c_0, whereas on non-linear theory higher pressures travel at an increased sound speed, amounting on a first approximation to

$$c = c_0 + \frac{\gamma - 1}{2} \frac{p - p_0}{\rho_0 c_0}. \tag{15}$$

But, this is the propagation speed relative to motions of the air, whose velocity u in the direction of propagation is given on linear theory by a well known relation

$$p - p_0 = \rho_0 c_0 u. \tag{16}$$

(Indeed, in a wave travelling at the linear-theory speed c_0, the acceleration of any particle of fluid is $-c_0 du/dx$ so that equation (16) allows the mass-acceleration per unit volume ($-\rho_0 c_0 du/dx$) to balance the pressure force $-dp/dx$.)

These arguments suggest that the speed $c + u$ at which pressure changes may be propagated on non-linear theory can, by equations (15) and (16), be written

$$c_0 + \frac{\gamma + 1}{2} u \quad (= c_0 + 1.2u \quad \text{for } \gamma = 1.4). \tag{17}$$

Note that the excess propagation speed for air is about $1.2u$, out of which just one-sixth is due to the increase (15) in c while five-sixths is due to convection of sound at the air velocity u; here, already, there is an important interaction of a fluid flow with waves.

Although readers have just been introduced by crude approximate arguments to expression (17) for the propagation speed, a brilliant 1859 analysis by the great mathematician Bernhard Riemann [15] had proved it to be absolutely accurate for plane sound waves of any amplitude propagated one-dimensionally into still air under adiabatic conditions. Briefly, the relationship $c = c_0 + \frac{1}{2}(\gamma - 1)u$ is exact; the expressions for pressure and density may be derived from c by equations (14); and, most important of all, each value of u is propagated at precisely the speed (17).

These conclusions were well known to Lord Rayleigh, who recognized also their sensational implications. Figure 10.5 shows these in the case of a single pulse of positive excess pressure, represented as an initial plot (solid line) of fluid velocity u against distance. On a linear theory of one-dimensional sound waves, each value of u would

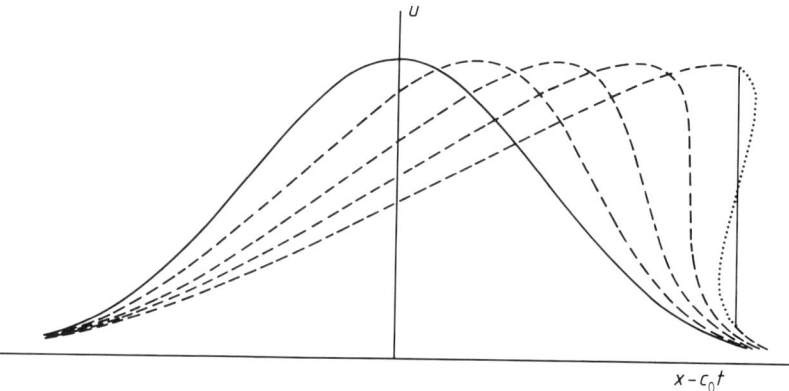

Figure 10.5. Solid line, initial waveform (graph of fluid velocity u against $x - c_0 t$). Broken lines, changed waveforms at later times, expected since values of u are propagated not at speed c_0 but at speed $c_0 + 1.2u$. Dotted line, a still later waveform similarly derived, yet manifestly impossible. Vertical line, tentative resolution of the enigma proposed by Riemann [15].

be propagated at speed c_0, so that the shape of the pulse would remain unchanged when plotted (as here) against $x - c_0 t$.

On the exact non-linear theory, however, each value of u is propagated at speed $c_0 + 1.2u$, so that after a time t it has travelled a distance $(c_0 + 1.2u)t$. In the representation of figure 10.5, then, where the values of u are plotted against $x - c_0 t$, each value of u has been shifted a distance $1.2ut$ to the right. Small values of u have hardly moved at all, while large values have moved much more—permitting them, as suggested earlier, to 'catch up' with smaller values.

These distorted pulse shapes are shown in figure 10.5 (broken lines) for a sequence of increasing values of t until a time has been reached where the pulse shape has a vertical tangent. Pulse shapes at still later times continue to be predicted by the theory; however, parts of these shapes are shown in figure 10.5 as dotted lines in recognition of the clear impossibility of the fluid velocity u taking three different values at one and the same point!

A tempting idea for resolving this enigma is to suppose that the solution develops a discontinuity. Riemann himself noticed that a discontinuity—indicated in figure 10.5 by the vertical solid line—could be inserted in place of the dotted line in such a way that total mass and momentum for the system are conserved, while all continuous parts of the curve (broken lines) still satisfy exact equations for sound propagation under adiabatic conditions. Rayleigh, however, objected that this idea not only (i) left unexplained how a discontinuity would arise but also, still more seriously, (ii) failed to satisfy overall conservation of energy.

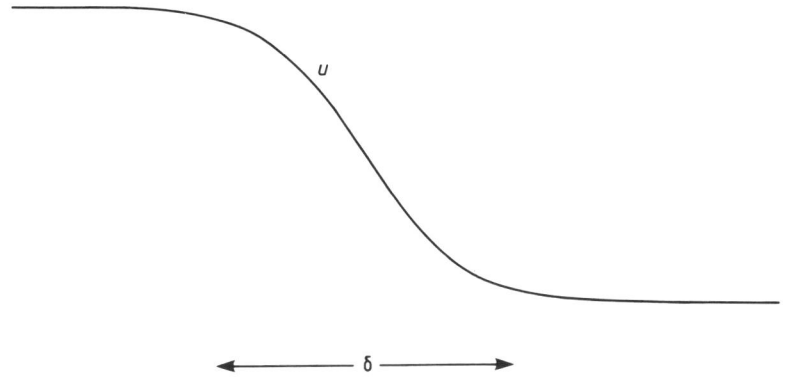

Figure 10.6. *The abrupt change of velocity u within the extremely small 'Taylor thickness' δ of a (not too strong) shock wave takes up this distribution, such that diffusion balances non-linear effects.*

Much later, in 1910, the enigma was resolved both by Rayleigh himself [6] and by the young G I Taylor [7] in two independent studies published in a single number of the Proceedings of the Royal Society. Essentially, the idea that would later be called 'singular perturbation theory' was again used so as to overcome both the objections (i) and (ii) above.

In our first simple illustration (section 10.1.2) and in the aerodynamic application (section 10.1.3) of this idea, it was a discontinuity between external trends and a boundary condition that necessitated the introduction of 'a region of very rapid change' at the boundary, with gradients large enough for 'diffusion' effects to become important. In the present application no such 'boundary layer' can appear; instead, a discontinuity is emerging in the midst of the fluid; yet it once more gives rise to a region with gradients large enough for diffusion to become important (and the analyses of Rayleigh and Taylor showed that diffusion of momentum as in section 10.1.3 is of quite comparable importance with thermal diffusion).

It is because the effects of diffusion (being proportional to gradients) can, if the region of very rapid change is thin enough, attain any required level that they are able to cancel out those effects of excess propagation speed which tend to produce the unrealistic 'overturning' of the waveform represented by the dotted curve in figure 10.5. Instead, there is formed something close to the discontinuous solution shown as a solid vertical line; where, however, the discontinuity possesses a definite thickness δ ('the Taylor thickness') and an associated internal structure (figure 10.6) which allow those 'overturning' and 'diffusion' effects to be in precise balance [16].

While countering Rayleigh's objection (i) by explaining how this nearly discontinuous wave—called a shock wave—arises, the above indication of important diffusion effects within it also demolishes any idea that the process is adiabatic. On the contrary, the fluid traversed by a shock wave experiences a departure from adiabatic conditions, which of course—in accordance with the second law of thermodynamics—involves an increase in entropy (so that p/p_0 rises above the value given by (14) for given ρ/ρ_0). This counters objection (ii) because the shock wave achieves energy conservation by abandoning entropy conservation—instead of the other way round.

Shock waves are much thinner than boundary layers; essentially, because the main non-linear term $\rho v dv/ds$ in the field equations of fluid mechanics includes a rate of change d/ds in the flow direction—that is, across the shock wave's small thickness δ. Therefore, while the balance (13) in a boundary layer equates a diffusion term of order $\mu V/\delta^2$ to other terms independent of δ (the flow being directed along the boundary), a similar balance in a shock wave equates a diffusion term to a non-linear term of order $\rho V^2/\delta$. Accordingly, the shock wave thickness δ comes out to be directly proportional to the relevant diffusivity rather than to its square root, and typical values of δ in atmospheric air vary from a millimetre for quite weak shock waves to a few micrometres for much stronger ones.

For practical engineering purposes, then, it often suffices to treat shock waves as sharp discontinuities satisfying conservation equations for mass, momentum and energy. These equations had actually been written down in 1889—for fluids in general—by A Hugoniot [17] who, on the other hand, had not been in a position to appreciate why, in most fluids including air, the physics of shock-wave formation (see above) permits only the appearance of compressive discontinuities.

Important applications of such approaches to supersonic aerodynamics are outlined in section 10.4. They are above all founded on a deep study of supersonic flow in shaped nozzles [18] which Ludwig Prandtl published in the same year as his boundary-layer theory: 1904, the *annus mirabilis* that transformed fluid mechanics.

10.2. Boundary layers and wakes; instability and turbulence; heat and mass transfer

10.2.1. The liveliest of non-dimensional parameters
This section 10.2 traces the further development of many ideas—and especially those related to boundary layers and wakes—that have been briefly introduced in section 10.1.3. The historical developments here described were centred around problems of how stationary solid objects disturb a steady wind. (As stated on p 804, these problems are identical to questions of how still air is disturbed by the steady movement of a solid body through it; when, indeed, air motions are most conveniently considered as relative to the movement of the body.)

Non-dimensional parameters proved useful in the twentieth century development of several branches of physics. Their particular importance in fluid mechanics may be illustrated by two contrasting examples, related to assessments of the significance for airflows of different physical properties: compressibility and viscosity.

The air density ρ, according to equation (14), should change by just a very small fraction of its undisturbed value ρ_0—so that compressibility might be neglected—provided that the pressure p changed by a very small fraction of p_0. But, if the wind speed is U, Bernoulli's equation (12) suggests that changes in pressure may be of the order of changes in $\frac{1}{2}\rho V^2$ where flow velocities V are of the same order of magnitude as U; so that changes in pressure are very small compared with the undisturbed pressure p_0 if the ratio

$$\tfrac{1}{2}\frac{\rho_0 U^2}{p_0} = \tfrac{1}{2}\gamma \left(\frac{U}{c_0}\right)^2 \tag{18}$$

is very small.

Extensive experimental studies confirmed the importance of the ratio of wind speed to sound speed

$$\frac{U}{c_0} = M \tag{19}$$

as a non-dimensional parameter small values of which (say, values less than 0.2) could be used to diagnose the absence of any influence of compressibility. The flow field is then 'solenoidal'—like a magnetic field—with flow speeds increasing wherever flow lines come together because the volume flux through a tube of flow lines remains constant.

The name 'Mach number' given to M honours Ernst Mach who in 1887 interpreted photographs of bullets in flight, showing (see section 10.4.2 below) how the pattern of airflow undergoes an exciting change [19], with the prominent appearance of shock waves, when M exceeds 1. Here, however, we are entirely concerned with the 'unexciting' tendency of compressibility to lose all importance for those flows, with small values of M, to which section 10.2 is devoted.

Actually, various other non-dimensional parameters of fluid mechanics [11] share this relatively unexciting property that their smallness implies the negligibility of a certain physical property. By contrast there are, as explained in section 10.1.3, absolutely no airflows around solid bodies for which the effects of viscosity can be neglected, since these are always important in that boundary layer whose development critically influences the character of the entire flow.

Nevertheless a non-dimensional parameter involving viscosity does prove important for a wide variety of reasons other than any misguided

See also p 873

aim of diagnosing when viscosity might be negligible. This parameter depends not only on the wind speed U but also on the body's linear dimension ℓ.

Thus, for all bodies of a well-defined shape but different sizes (commonly described as 'geometrically similar' bodies) the dimension ℓ in the direction of the wind is usually taken as a measure of size. Then the parameter

$$\frac{\rho U \ell}{\mu} = R \tag{20}$$

is non-dimensional since the viscosity μ is defined as the ratio of a stress (which, like a pressure, has the dimension ρU^2) to a velocity gradient (with dimension U/ℓ).

Whenever the Mach number (19) is small, so that compressibility is negligible, the pattern of disturbances to the wind caused by the presence of the stationary body can depend only on the (effectively uniform) air density ρ, on the viscosity μ, on the wind speed U and on the body size ℓ. But the general principles of dimensional analysis tell us that this apparent dependence on four variables is really only a dependence on the single non-dimensional parameter (20). (Briefly, any changes in ℓ, U, ρ and μ which preserve the value of R are readily seen to be equivalent to simple changes in the fundamental units of length, time and mass, which leave unaffected both the laws of mechanics and the relation of viscous stress to velocity gradient.)

Accordingly, the airflow around a large body is just a scaled-up version of the flow around a smaller, but geometrically similar, model of the body at the same value of R. This principle has many applications to wind-tunnel experiments on airflow; for example, a scale-model with a reduced value of ℓ may be tested in a wind-tunnel that uses air at high pressure or low temperature (respectively, increasing ρ or decreasing μ) so that the ratio R is unchanged.

An even more important use of R, the non-dimensional parameter (20), is to characterize which out of several possible types of flow will occur in a particular case. Thus the flows loosely described in section 10.1 as motions of 'a fluid of very low viscosity' are more correctly described as flows with high values of R; indeed, it is for $R \geqslant 10^3$ (about) that a thin boundary layer can occur, and any dependence (section 10.1.3) of its thickness δ on the square root $\mu^{1/2}$ of viscosity is better expressed as the non-dimensional statement [11] that a thickness-to-length ratio

$$(\delta/\ell) \text{ tends to vary as } R^{-1/2}. \tag{21}$$

However, such thinning of a boundary layer as R grows larger and larger is by no means indefinitely continued.

For his fine 1880s studies of flow in pipes, the Manchester engineer Osborne Reynolds [20] had used a ratio just like (20) but with the length

ℓ replaced by the pipe's dimension normal to the flow (the internal diameter), and had discovered that the flow tended to become turbulent (that is chaotic) when that ratio exceeded about 2000. In his honour, the ratio (20) is named Reynolds number by aerodynamicists, who moreover use R_δ to mean the corresponding ratio

$$R_\delta = \frac{\rho U \delta}{\mu} = \left(\frac{\delta}{\ell}\right) R \tag{22}$$

based on the boundary layer's dimension δ normal to the flow. This flow within a boundary layer tends (see section 10.2.3) to become turbulent when such a ratio (22) reaches a critical value whose rough order of magnitude is again 10^3; however, since R_δ tends—by equation (21)—to vary as $R^{1/2}$, it may be for values of R in the broad neighbourhood of 10^6 rather than of 10^3 that a boundary layer becomes turbulent.

Prandtl recognized very early [21] that such a transition to turbulence in boundary layers played three important roles, all associated with the increased diffusion of momentum which results from chaotic motions. Thus,

(i) the boundary layer becomes considerably thicker, while

(ii) resistance receives a substantial increase in consequence of greater diffusion of momentum towards the solid surface, and yet

(iii) this same diffusion makes the layer relatively less prone to separation;

that is, able (see p 807) to remain attached to the surface at relatively higher values of the rate of decrease $(-dV/ds)$ of external flow velocity with distance s along the surface.

Prandtl saw, too, that (ii) and (iii) are in an interesting conflict. For example, an 'aerofoil' shape may need to be as thin as that in figure 10.3 if separation is to be avoided at $R = 10^4$, when the boundary layer takes a steady, 'laminar' form (figure 10.4). Yet, at Reynolds numbers around 10^6 or more, aerofoil sections almost twice as thick (with $(-dV/ds)$ almost twice as large) will avoid separation as a consequence of transition to turbulence in the boundary layer; and then the associated increase (ii) in resistance, though substantial, is far less than would have been the excess resistance associated with separated flow.

Moreover [21], the resistance to flow over an aerodynamically 'bad' shape like a sphere is reduced by a large factor (around 3) when the boundary layer makes such a transition—from laminar to turbulent—before it separates (figures 10.7 and 10.8). This transition may be generated either (a) by an increase in R to rather large values (in excess of 10^5); or, at lower values of R, (b) by turbulence artificially generated, as shown in some dramatic photographs [22] of flow around a sphere with and without a 'trip wire' which promotes transition. It is, of course,

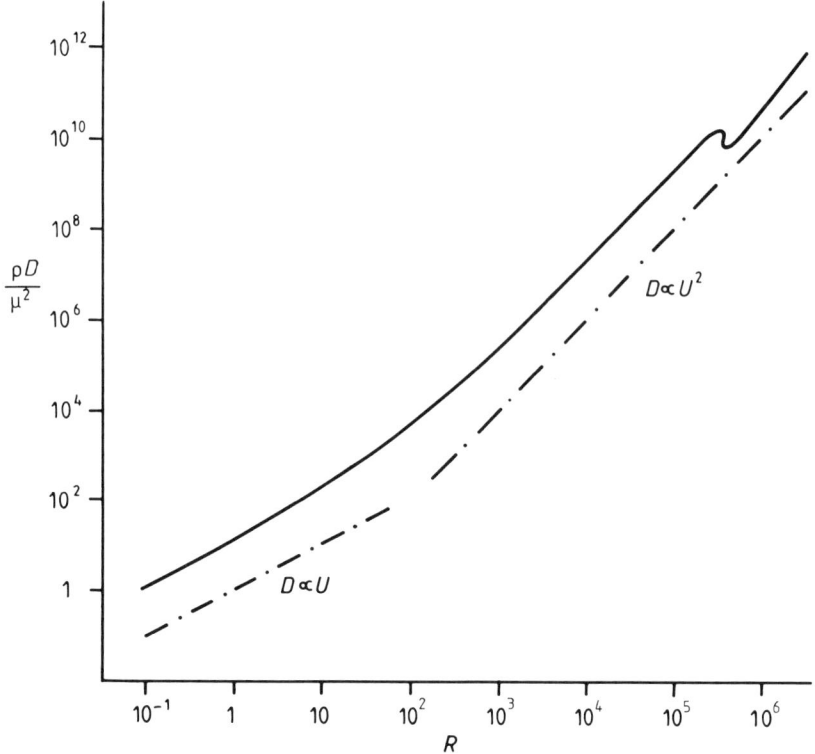

Figure 10.7. *A first illustration of the 'liveliness' of dependence on Reynolds number $R = \rho U \ell / \mu$. The drag force D on a smooth sphere of diameter ℓ in a steady wind of speed U, while broadly varying as U^2 (due to pressures of order ρU^2) for $R > 10^3$ and yet as U (due to viscous stresses of order $\mu U/\ell$) for $R < 10$, experiences [29] a threefold drop, due to transition to turbulence in the boundary layer, at around $R = 3 \times 10^5$.*

well known that many ball games are made interestingly complicated by such sensitive dependence of forces on a moving sphere to both speed (or, more correctly, Reynolds number) and various surface disturbances.

Reynolds number has been described [11] as 'the liveliest of non-dimensional parameters' because régimes of flow vary so strikingly over an extended range of orders of magnitude of R. Such variations are found both for increases in R (as briefly illustrated above) from 10^3 to a sequence of higher orders of magnitude and also for decreases to low values, even less than 1, as revealed under the microscope in the marvellous world of the swimming of small organisms [23].

10.2.2. New roles for vorticity

It is a paradox that the most efficacious twentieth century tool for understanding the forces (including the aeronautically crucial lift force) that may act on a stationary body in a wind was a concept, VORTICITY,

Fluid Mechanics

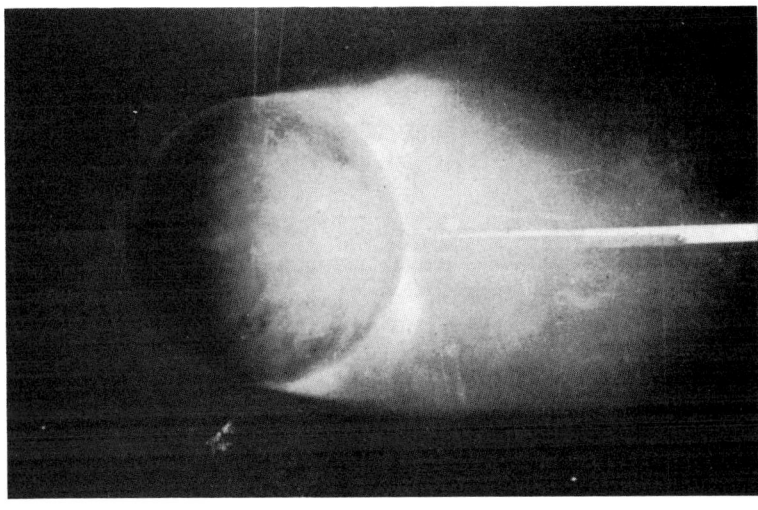

Figure 10.8. *Flow (from left to right) around a smooth sphere (a) without, and (b) with, a trip wire* [22].

which—far from having been previously neglected—was a specially favoured tool of nineteenth century fluid-mechanics theorists! Obsessed with vorticity, they nevertheless had used it to pursue many fruitless quests (for example, the representation of atoms as vortices in the ether) which need not be recalled here, whilst one of their most penetrating general results—Kelvin's Circulation Theorem, destined to play a key role in aerodynamic lift theory—was mistakenly viewed by its author as incompatible with any possibility of fixed-wing aircraft sustaining a lift force.

Some quite new roles for vorticity emerged, then, in the twentieth century. These were above all derived from reinterpretation of discoveries on boundary layers and wakes in terms of vorticity distributions that possessed both well defined momentum and associated consequences for aerodynamic forces.

The importance of vorticity derives from the idea that, wherever viscous stresses can be neglected, a small sphere of fluid is subjected only to pressure forces [11]; which, acting through its centre, are not altering its angular momentum (figure 10.9). The sphere's motion can be divided into three parts: (i) uniform translation with the velocity v of the centre, (ii) rigid rotation with angular velocity $\frac{1}{2}\omega$ where ω is the vorticity—parts (i) and (ii) carry, respectively, all the sphere's momentum and angular momentum—and (iii) a symmetrical squeezing or 'straining' motion in which the sphere is instantaneously changing its shape into an ellipsoid of the same volume. Part (iii) involves elongations in some directions and foreshortenings in others, which are, respectively, decreasing or increasing its moment of inertia about these directions. Conservation of angular momentum implies, then, that components of vorticity along axes which are being elongated or foreshortened are, respectively, increased or decreased; and, indeed, the vorticity vector itself is subject to precisely the same changes in length and direction as a line of particles of fluid through the centre of the sphere undergoes in the course of the sphere's straining motions.

This leads to the idea of a 'vortex line' as a line, or 'necklace', of moving particles of fluid that—wherever viscous stresses can be neglected—continues to point always in the direction of the vorticity vector; a vector whose magnitude, moreover, varies in proportion as the necklace is locally stretched. The familiar smoke ring is a visible bundle of vortex lines.

A typical shearing motion, such as a boundary layer (figure 10.4), is a region of strong vorticity, with magnitude ω equal to the gradient dv/dn of velocity. Indeed, pure rotation with angular velocity ω is obtained by vectorially adding such a shearing motion to another at right angles, and this indicates $\frac{1}{2}\omega$ as the rotary component (ii) for each (figure 10.10).

The above considerations lead to the classical vector relationship $\omega = \text{curl } v$ expressing vorticity components as a combination of gradients of velocity components. This has two major implications, of which the first is that the vorticity field itself is solenoidal (div $\omega = 0$) so that vortex lines can never end in the fluid. Furthermore, just as the corresponding magnetostatic relationship curl $H = J$ allows us to regard a current distribution J as precisely determining its associated magnetic field H in accordance with the Biot–Savart law, so too the vorticity distribution ω completely determines the fluid flow field v.

Where viscous stresses can by no means be neglected, as in a boundary layer, their effect is to produce diffusion, not only of velocity

Fluid Mechanics

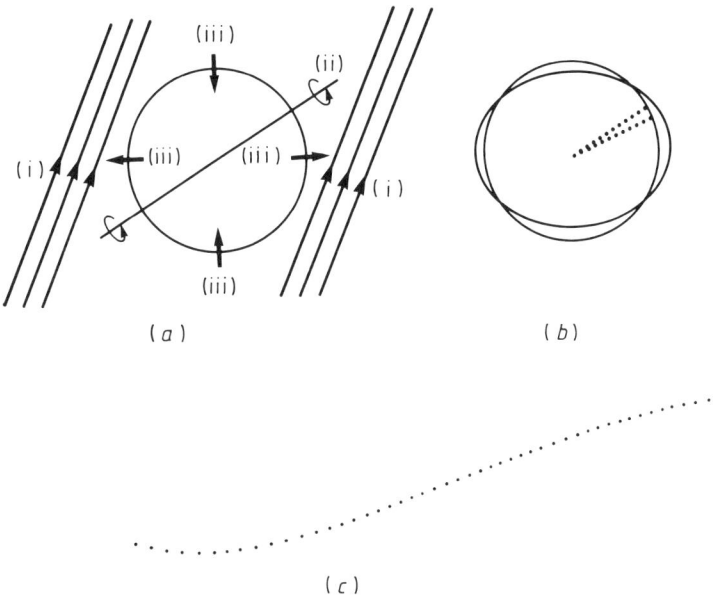

Figure 10.9. *(a) The instantaneous motion of a small sphere of fluid may be divided into three parts, with part (ii)—rotation at angular velocity $\tfrac{1}{2}\omega$—carrying all the angular momentum. Now, pressure forces on the sphere act through its centre, and so are not changing this angular momentum. (b) Yet ω may be changing because the 'straining' motion (iii) is altering moments of inertia about different axes, in such a way that the vorticity vector is subject to the same straining motions as a line of fluid particles. (c) Thus a 'necklace' of fluid particles which coincides with a vortex line at one instant will continue to do so.*

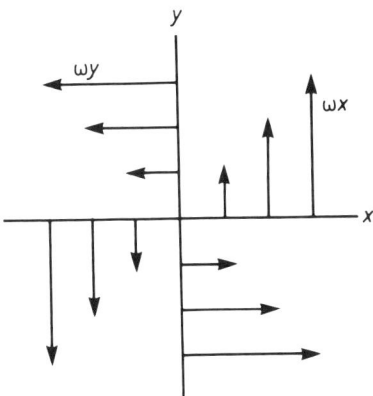

Figure 10.10. *Since these two shearing motions combine vectorially to give pure rotation with angular velocity ω, the rotary component (ii) in each has angular velocity $\tfrac{1}{2}\omega$.*

819

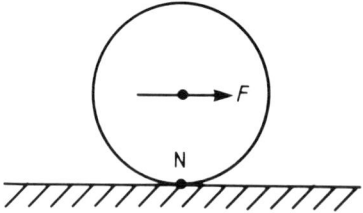

Figure 10.11. *Rotation is generated in a small sphere of fluid, touching a solid surface at N—where it satisfies a no-slip condition—by a force per unit mass F, where (see section 10.1.3) $F = -(1/\rho)(dp/ds) = V(dV/dS)$.*

components (section 10.1.3) but also of velocity gradients, so that the vorticity too undergoes diffusion with just the same value (μ/ρ) of the diffusivity. Vortex lines, then, are subject not only to convection by the fluid motions but also to such diffusion.

When a stationary body disturbs a uniform wind, the solid surface itself is the only available source of vorticity. From this viewpoint, a boundary layer arises when diffusion of the vorticity generated at the surface, together with convection downstream, allows that vorticity to remain confined in a thin boundary layer and wake—outside of which a potential flow (that is, a flow with zero vorticity) is to be found.

The rate of vorticity generation at a solid surface (per unit area) is VdV/ds in the notation of section 10.1.3. A schematic diagram (figure 10.11) readily suggests why this pressure-gradient force per unit mass tends to generate rotation in a small sphere of fluid which—like a rubber ball!—satisfies a no-slip condition at the surface.

From this point of view, boundary-layer regions defined on p 807 as (a) with dV/ds positive, (b) with dV/ds negative but small and (c) with dV/ds negative but not small are characterized by (a) positive rates of vorticity injection, (b) small negative rates (vorticity abstracted gradually enough to be counterbalanced by diffusion), and (c) large negative rates, leading to backflow ($dv/dn < 0$) near the surface. Out of these, only (c) makes the boundary layer separate.

Prandtl, about a decade after his initial explanation of flows that avoid separation (because no region (c) is present), made another extraordinary discovery about such flows [5]. This was that the vorticity shed behind the body exercises a determining influence on both airflow and aerodynamic forces. Here, a nineteenth century theorem proved invaluable: a vorticity distribution ω not only determines the flow field v but also specifies the momentum of that flow field very directly (in mathematical terms, as $\frac{1}{2}\rho$ times the moment of the vorticity distribution).

Admittedly, these effects may be relatively insignificant for a purely symmetrical flow like that of figure 10.3, simply because boundary layers

Fluid Mechanics

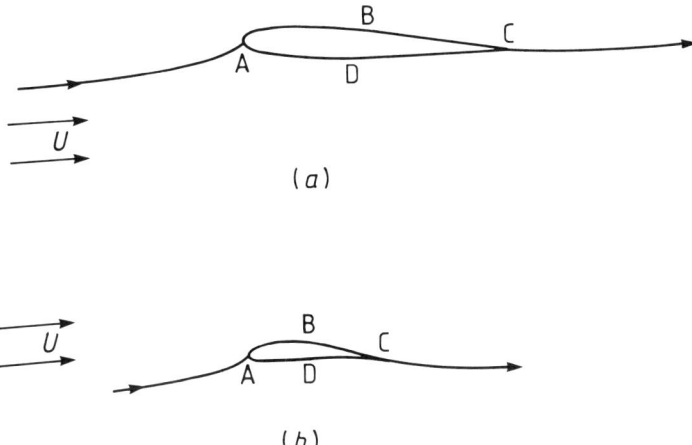

Figure 10.12. *Lift-generating airflows around wing cross-sections. (a) Flow about a symmetrical shape (the same as in figure 10.3) when the 'angle of attack' between the wind vector U and the 'chord' (line joining leading and trailing edges) is 6°. The circulation (amount by which the integral ∫ V ds taken along ABC exceeds its value taken along ADC) is positive because values of the velocity V just outside the boundary layer are greater on the upper surface than on the lower surface. (b) Flow at the same angle of attack about a cambered cross-section. The circulation now includes a contribution due to camber in addition to that due to angle of attack.*

on the body's upper and lower surfaces shed equal and opposite vorticity (clockwise and anticlockwise, respectively). In a unit length of wake, then, there is no resultant vorticity, although it does have a (very small) moment corresponding to that rate of creation of forward momentum in the wake fluid which Newton's third law tells us must accompany the (very small) resistive force acting on the body.

But things are quite different wherever some departure from flow symmetry—arising either when a symmetrical shape like that of figure 10.3 is placed in the wind at a positive 'angle of attack', or when the centreline of its cross-section is not straight but 'cambered' (arched)—removes any such tendency for vorticity on the upper and lower surfaces to be equal (figure 10.12). Then, the wake may carry a large and growing downward momentum, whose rate of increase represents the aerodynamic lift on the body (such an upward force of the air on the body being equal and opposite to the body's rate of communication of downward momentum to the air).

Twentieth century developments in the aerodynamics of aircraft wings—at those relatively low Mach numbers to which section 10.2 is devoted—have concentrated on winning ever greater quantitative precision for each element within Prandtl's fundamental picture [5] of the associated vorticity pattern, which critically involves 'bound' vorticity (any resultant vorticity attached to the wing) as part of a continuous field of vortex lines with the 'trailing' vorticity in the wake. Here this pattern can be sketched in just the simplest terms.

Evidently, the expression $\omega = dv/dn$ for vorticity in a boundary layer, where the velocity v rises from zero at the surface to V at the edge of the boundary layer, implies a resultant (total) vorticity $+V$ per unit area of a wing's upper surface, alongside a similar resultant $-V$ on the lower surface. Bound vorticity, representing a positive resultant (per unit span) for the wing section as a whole, takes therefore the form of any difference between values of the integral $\int V ds$ along the upper and the lower surfaces. Such a difference, called the 'circulation' Γ around the wing section, appears whenever the extent and magnitude of clockwise values of V on its upper surface exceed those for its anticlockwise values on its lower surface (figure 10.12).

Circulation is important because it generates lift; Bernoulli's equation (12) already suggests this—since it makes the pressure greatest where V is least—but a rather general mathematical analysis known as Zhukovski's theorem [10] established the lift as $\rho U \Gamma$ per unit span for potential flow of 'two-dimensional' character [9] over a wing of arbitrary (though uniform) section and very large span. Prandtl's more physical analysis [5], on the other hand, employed fundamental properties of vortex lines

(i) to infer the lift per unit span on each cross-section as $\rho U \Gamma$ for rather general 'three-dimensional' wing shapes; and, moreover,

(ii) to quantify the associated 'penalty', in the form of an extra resistance now commonly described as 'drag due to lift'.

The insight essential to these achievements was a recognition that, since vortex lines cannot end in the fluid, any bound vorticity must be linked in a continuous system with trailing vorticity in the wake.

Figure 10.13(a) offers a crude schematic view of how the bound vorticity Γ, varying with position z along the wing span (and of course falling to zero in the region beyond the wing tips $z = \pm b$ where no source of vorticity exists), is incorporated within such a pattern. (Note: dotted lines in this diagram are deliberately oversimplified indications that the vortex lines must be closed; also, they constitute a recognition—which can be linked to the above-mentioned Kelvin's Circulation Theorem—that, around any wing with positive camber and/or angle of attack, it is possible to establish a smooth flow with attached boundary layers incorporating bound vorticity Γ only if equal and opposite vorticity—sometimes referred to as a 'starting vortex'—has been shed earlier when the motion was initiated. Actually, patterns of trailing vorticity far behind an aircraft tend to 'roll up' into a pair of concentrated vortices such as are often observed on humid days as 'condensation trails'.)

The vortex wake close behind the wing (solid lines) has a flow pattern as indicated in figure 10.13(b)—due to Prandtl [5]—with a downward momentum, given by $\frac{1}{2}\rho$ times the moment of the vorticity distribution,

Fluid Mechanics

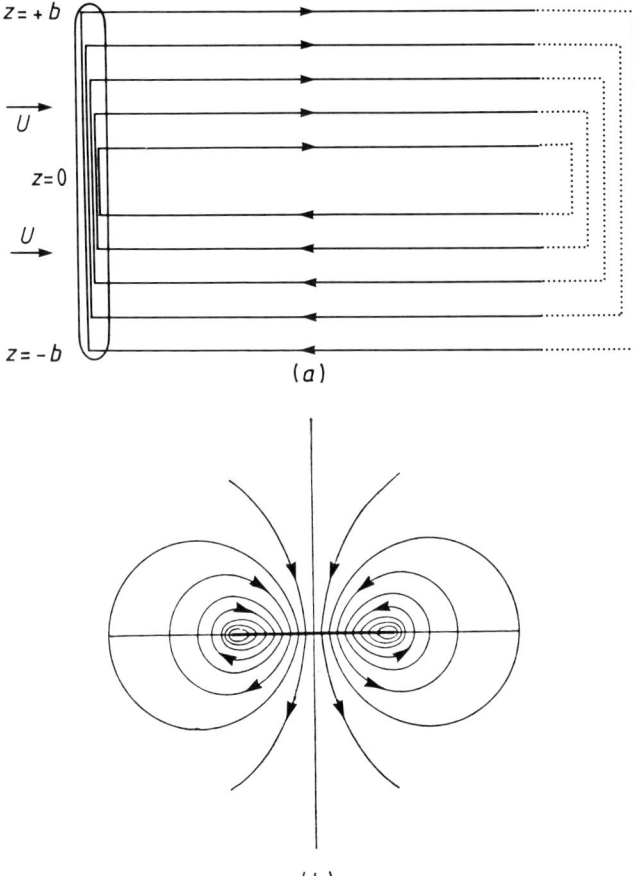

Figure 10.13. (a) Here, solid lines show schematically the overall pattern of vortex lines (incorporating both bound vorticity $\Gamma(z)$—attached to the wing, and greatest where the vortex lines are most thickly packed—and also trailing vorticity) for a lifting wing-pair in a steady airflow. (For dotted lines, see text.) (b) Prandtl's diagram [5] of the pattern of wake flow in a vertical plane just downstream of the wing associated with this vorticity distribution. Forces on the wing arise from the continued shedding of this pattern of airflow; the lift L being its additional downward momentum per unit time while its kinetic energy per unit length of wake is the drag due to lift, D_L.

which increases at a rate

$$L = \rho U \int_{-b}^{b} \Gamma \, dz \qquad (23)$$

representing the lift on the wing—each section of which contributes as in (i) above. It is important, too, that each unit length of the growing wake has a calculable kinetic energy, D_L; which, being necessarily supplied from work done in maintaining the wing's steady relative

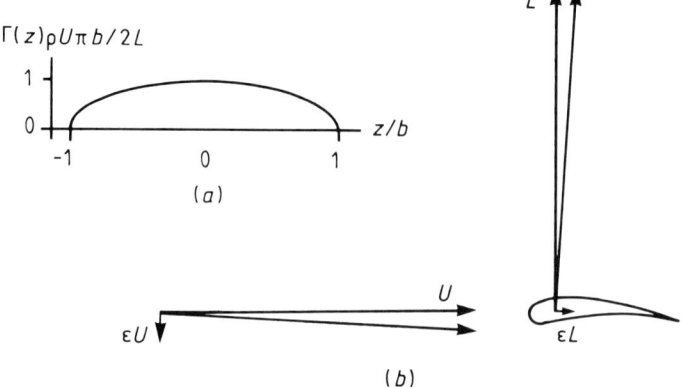

Figure 10.14. (a) This 'elliptic' distribution of circulation was shown by Prandtl to reduce drag due to lift to its minimum value εL, while the associated trailing vorticity would generate at the wing a uniform downward motion εU. (b) Indeed, where a horizontal wind of speed U produces an upward force L, its downward deflexion by the vertical motion εU must give an extra horizontal force εL.

motion through the air per unit distance travelled, is—exactly as in (ii) above—the drag due to lift.

For a given span $2b$, Prandtl successfully demonstrated that the least possible drag due to lift is

$$D_L = \varepsilon L \quad \text{where } \varepsilon = \frac{L}{2\pi \rho U^2 b^2} \tag{24}$$

this minimum being achieved—amongst all distributions of Γ across the span—by the 'elliptic' shape of distribution actually shown in figure 10.14. Prandtl also uncovered yet another instructive way of looking at this drag due to lift: the wake vorticity ω produces a velocity field v which, at the wing itself, takes the form of a downward motion εU. Locally, this is equivalent to the wind being inclined downwards through an angle ε, and then the expected force L normal to this effective wind is inclined backwards at an angle ε and so includes a 'drag' component εL (figure 10.14).

D'Alembert's Theorem does not, of course, apply to these flows which—far from being nearly potential flows—are strongly influenced by a complicated distribution of vorticity. Yet ε is a small number, of the order of magnitude 10^{-2}, for aircraft wing pairs whose span $2b$ greatly exceeds their other dimensions. Then, as remarked at the end of section 10.1.3, the general spirit of d'Alembert's Theorem is retained in these airflows where huge lift forces, capable (see section 10.4) of sustaining hundreds of tonnes of metal in the air, are achievable with only a modest drag penalty.

10.2.3. Types of transition, types of turbulence: (i) struggles before 1940

However justifiably a historian of twentieth century fluid mechanics may adopt a heroic narrative style for outlining key discoveries on boundary layers and wakes made in the century's first quarter, he must acknowledge that such a style would be out of place for describing the course of twentieth century developments in turbulence research. This has been a field where progress in the twentieth century was much more gradual—notwithstanding the benefits of starting from the strong position of Reynolds's 1880s researches [20]. Only during the 1940s, indeed, did understanding begin substantially to progress beyond the point which Reynolds had reached in those researches and in a powerful theoretical analysis [24] dated 1895.

Although (section 10.2.1) his 1880s studies had concentrated on turbulence in pipe flow, they had recognized too the existence of some different, sharply contrasted types of turbulent flow and of transition to turbulence. Yet investigators during the next half century showed an excessive tendency to regard 'turbulence' as a single, integrated phenomenon, research on any aspect of which would illuminate the field as a whole. This led to prolonged struggles up blind alleys; which may have yielded some refinement of techniques, but need not be chronicled in a brief survey.

Famous papers are rarely recalled in their entirety! For pipe flow itself, Reynolds had not just determined an easily remembered minimum Reynolds number for transition to turbulence. A meticulous observer, he had also uncovered two extraordinary features—intermittency, and abruptness—of that transition; which involved the intermittent, sudden appearance of abrupt 'flashes' of intensely chaotic motion in the midst of an otherwise laminar flow. Only in 1951 were these two features rediscovered, by Howard Emmons, in brilliant experiments at Harvard on transition in boundary layers on flat plates [25]; comprising the intermittent, abrupt initiation of growing 'spots' of highly chaotic flow, each separated by a sharp boundary from a laminar-flow environment, but all merging downstream into a fully turbulent boundary layer (figure 10.15).

For a long time, again, little attention was paid to Reynolds's lists of other types of flow which

(i) show a much greater proneness to transition (by comparison with pipe flow), and yet

(ii) make that transition in more restrained ways which he described as 'sinuous'.

These lists included, among parallel flows, those where the vorticity has a maximum in the flow itself (instead of at a solid boundary); and, among curved flows, those with velocity greatest 'on the inside'. Broadly, they are flows where fluid mechanics theory has proved rather

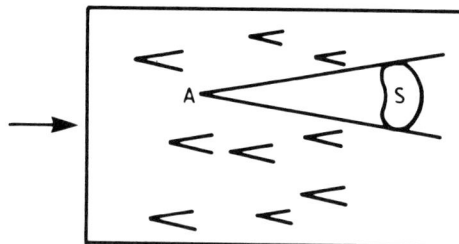

Figure 10.15. *Emmons's observations [25] of turbulent 'spots' in the boundary layer on a flat plate. Different 'spots' originate randomly, each at some point A which is the apex of a wedge within which the spot S subsequently grows as shown. Other wedges (all with about the same included angle) within which spots grow as they are swept downstream are also shown. Turbulent motion is fully developed within each spot. Spots thus randomly initiated merge together, forming ultimately a fully turbulent boundary layer.*

readily able to forecast flow instability (as Rayleigh did first in each case [26, 27]) and where, initially, disturbances predicted to develop unstably are rather similar to those observed. Yet it was only in the second half of the twentieth century that distinctions between 'abrupt' and 'restrained' transition would be properly recognized, and interpreted by ideas from non-linear stability theory—ideas of which the great Russian physicist Landau [28] had sketched the first in 1944.

In the meantime, differences in kind between the diffusion of momentum by molecular viscosity and by turbulence had been well brought out in Reynolds's 1895 paper [24], and may here be illustrated by the case of a parallel airflow with velocity U in the x-direction which varies as a function $U(y)$ of the coordinate y in a perpendicular direction (figure 10.16). If the airflow is turbulent, this $U(y)$ represents the average of a randomly fluctuating fluid velocity, with components

$$U(y) + u, \ v, \ w \quad \text{in the } x, \ y, \ z \text{ directions.} \tag{25}$$

Now, although the velocity fluctuations u, v, w have zero means, a product like uv can have a non-zero mean $\langle uv \rangle$, in which case excess x-momentum (ρu per unit volume) is transported in the y-direction at an average rate

$$\rho \langle uv \rangle \tag{26}$$

now known as a Reynolds stress. Energy is then being extracted from the mean flow, and fed into the turbulence, at a rate

$$-\rho \langle uv \rangle \frac{dU}{dy} \tag{27}$$

per unit volume.

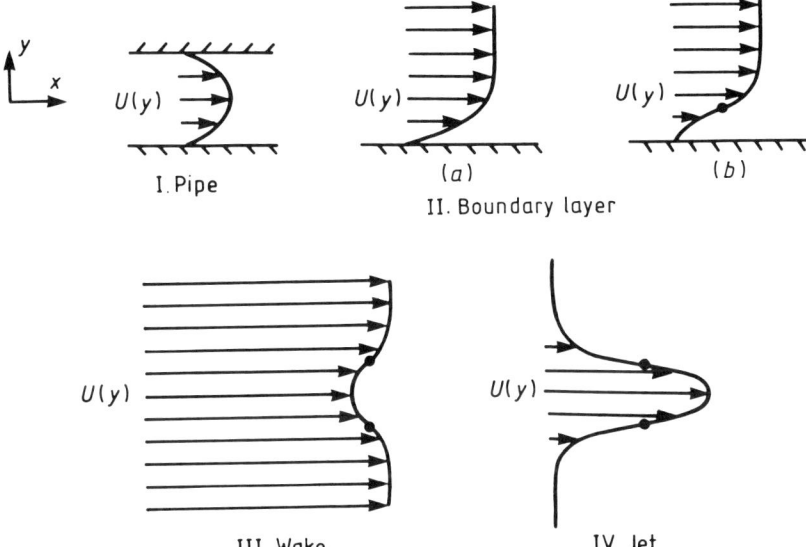

Figure 10.16. *Some (closely) parallel flows: (I) pipe; (II) boundary layer on a flat plate in cases (a) dV/ds positive and (b) dV/ds small and negative; (III) wake; (IV) jet. Vorticity maxima are found at the solid boundary in cases I and II(a) but appear in the flow itself (black circles) in cases II(b), (III) and IV.*

In a laminar flow, by contrast, the only source of momentum transport is viscous diffusion, interpreted of course in an apparently similar way on the kinetic theory of gases—where equations like (25), (26) and (27) may arise (with (25) representing velocities of molecules while angle brackets in (26) and (27) signify weighted means based on molecular weights). These superficial similarities are, however, gravely misleading; molecular velocities u, v, w are enormous (of order the sound speed) but a mean like (26) takes a very small value (the viscous stress, $-\mu dU/dy$) because molecular collisions act constantly to restore thermodynamic equilibrium and it is only over the mean free path between collisions (10^{-4} mm in atmospheric conditions) that a molecule's momentum shows, in a statistical sense, any effective 'persistence'.

Such contrasts not only demonstrate an absolute scale separation between the statistics of molecular velocities in kinetic theory and the statistics of fluid velocities in turbulence, but suggest also that attempts at finding anything analogous to a mean free path in turbulence may be fruitless. Ultimately, those attempts—known as 'mixing length theories' and pursued throughout the century's second quarter [29]—were abandoned in recognition that velocity fluctuations on length scales comparable with the thickness of the entire parallel flow contribute to the Reynolds stress (26).

Interspersed, however, with many less productive attacks on

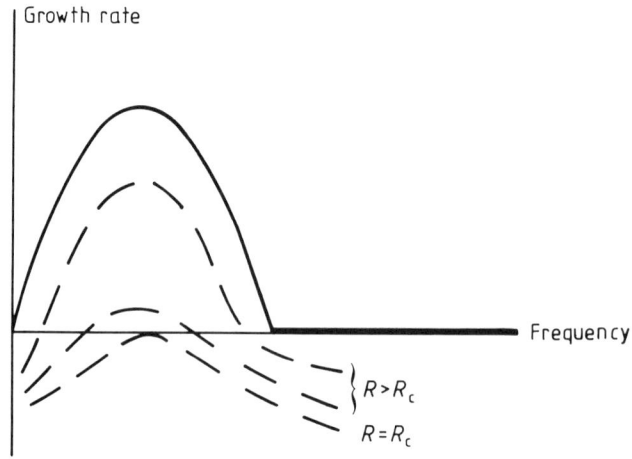

Figure 10.17. *Growth rate of small sinusoidal disturbances to a parallel shear flow: typical results when vorticity maximum is in the midst of the fluid* [11]. *Solid line, vorticity neglected. Broken lines, growth rates for finite Reynolds number R (there are growing disturbances when R exceeds a 'critical' value R_c).*

problems of turbulence and transition, a moderate number of relatively valuable additions to knowledge were made during the period before the 1940s. Tollmien in 1935 completed Rayleigh's work on parallel flows in the limit of zero viscosity by establishing [30] that the existence of a vorticity maximum in the midst of the fluid is not only a necessary, but also a sufficient, condition for their instability to small disturbances. Figure 10.17 shows in a typical case how the rate of growth of disturbances varies with frequency (solid line); while results for finite Reynolds number (broken lines) indicate how, as might be expected, viscous effects diminish that rate of growth [11]. This is a flow where transition to turbulence starts from an initial development of regular, 'sinuous' disturbances at or around the frequency for maximum growth rate.

In (slightly) curved flows, where the condition 'greater velocity on the inside of the bend' for a similar instability to small disturbances in the zero-viscosity limit had in 1916 been demonstrated [27] by Rayleigh, G I Taylor showed seven years later [31] that the stabilizing effect of viscosity could be calculated in an important special case—where experiment and theory were in agreement about the stability boundary. Near the boundary, moreover, the disturbances predicted (rings of vorticity, aligned with the flow direction, of axially alternating sign) were 'robustly' detected in the experiments (figure 10.18).

Instability, of course, was interpreted by practically all scientists before 1940 as just such an instability to small disturbances. It was a paradox for them, therefore, that common flows without any vorticity

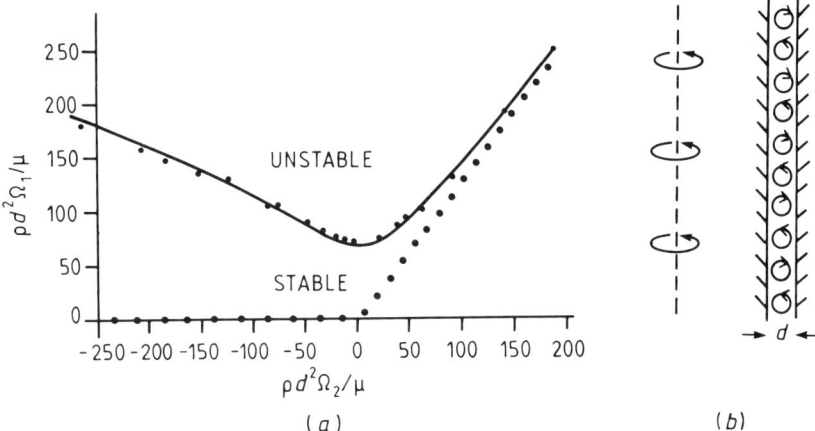

Figure 10.18. *Instability of fluid motion in the annular gap between an inner solid cylinder and an outer hollow cylinder rotating about the broken-line axis at angular velocities Ω_1 and Ω_2, respectively. (a) Solid line, stability boundary calculated by G I Taylor [31], whose experimental points are also shown. Dotted line, stability boundary with viscosity neglected. (b) Form (both predicted and observed) of disturbances.*

maximum in the midst of the fluid—including flow in a pipe or in a boundary layer on a flat plate—are (see above) stable to small disturbances in the zero-viscosity limit; especially, since such stability was presumed to be reinforced by any viscous effects.

The resolution of this paradox emerged in two halves. Prandtl used yet another singular perturbation theory to initiate the first half in 1921; showing [32] that, when the undisturbed vorticity attains its maximum at a solid surface, the effect of viscosity on small disturbances is felt above all within an 'inner surface layer'—considerably thinner than the main boundary layer (or pipe as the case may be)—where viscous action gives the fluctuations in surface frictional force a 45° phase lead over local velocity fluctuations which can tend to destabilize the disturbances. (This phase lead arises from the fact that, for sinusoidal disturbances of frequency ω, an extra term $\rho i \omega v$ on the left-hand side of equation (13) is capable of becoming dominant in such an inner surface layer.) Prandtl's 1921 paper is important in a wider context because it foreshadowed the general 'triple deck' theory (see section 10.4.2) of inner surface layers within main boundary layers.

But once again the powerful analytical skills of W Tollmien were needed to achieve (in 1929) a complete stability calculation [33], taking viscous effects fully into account, for the boundary layer on a flat plate. The unstable disturbances take the form of progressive waves that are exponentially amplified, and the analysis had to allow for viscous effects not only in the inner surface layer of Prandtl but also in a thin 'critical layer' around where the wave speed coincided with the local flow speed. Tollmien's brilliant analysis was given some important detailed refinement four years after by H Schlichting [34], and the waves—whose full significance would emerge only afterwards, with the second half of the resolution of the paradox—are now referred to as Tollmien–Schlichting waves.

Beyond all this work on how transition to turbulence may be initiated, a stride forward in analysing fully developed turbulence was taken by G I Taylor [35] in his 1922 study 'Diffusion by continuous movements'. Here a potent tool, for characterizing how diffusion by chaotic yet continuous fluid flows differs from that achieved by free motions of gas molecules between collisions, was created by extending that idea of a mean product of two velocity fluctuations which appears in the Reynolds stress (26).

The full tensor of Reynolds stresses includes the mean values, not only of uv as in (26) but also of u^2, v^2, w^2, wu and vw; with each mean product (on multiplication by ρ) representing a component of momentum transport. Taylor's innovation was to consider the mean product of a component (u, v or w) of the velocity fluctuation at a point P with its value (or the value of one of the others) at a nearby point Q. For example, if Q lies at a distance r downstream of P, the mean product

$$\langle u_P u_Q \rangle \tag{28}$$

(or 'covariance') coincides with the mean square $\langle u_P^2 \rangle$ (or 'variance') when r is very small; yet the chaotic nature of turbulence implies that, when r becomes large, u_P and u_Q become essentially uncorrelated so that the covariance (28) tends to zero.

The curve of (28) as a function of r gives, then, a first indication of the spatial scales of coordinated motions (or 'eddies') within the turbulence (figure 10.19). Another such indication is given by its Fourier transform $\Phi(k)$ which, by a general statistical theorem, indicates the spectrum of turbulence as a function of the downstream wavenumber k—in the sense that $\Phi(k)\,dk$ represents that part of the variance $\langle u^2 \rangle$ which is contributed to it by sinusoidal components $a\sin(kx+\alpha)$ of u with wavenumbers k in the interval dk.

One type of turbulence to which Taylor's innovations were applied was that studied (in 1929–36) by Hugh Dryden and his colleagues [36] in the course of their development of closed-circuit wind-tunnels. These would become the main test facilities for tackling the central problem (see section 10.2.1) of how stationary objects disturb a steady wind; however, great care was needed to ensure that the wind would indeed be practically steady in the working section of the wind-tunnel (figure 10.20). After the flow downstream of that section is accelerated by a fan, a gradual expansion of cross-section retards it slowly enough (case (b), dV/ds negative but small, of p 820) for separation from the tunnel walls to be avoided. Then its slow passage through a so-called honeycomb

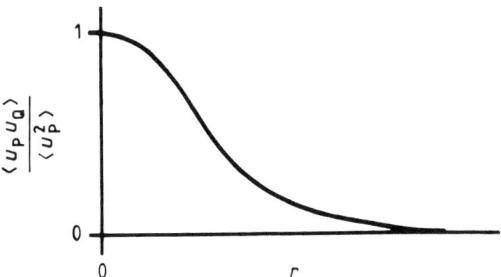

Figure 10.19. *Typical dependence* [39] *of the covariance (28) on the point Q's distance r downstream of P.*

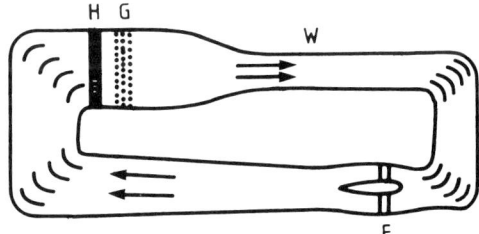

Figure 10.20. *A typical closed-circuit wind-tunnel (schematic). W, working section; F, fan; H, honeycomb; G, gauze screens. Note: the airflow is assisted in turning corners by guide vanes.*

(at Reynolds numbers low enough to retain laminar flow) brings it to a constriction carefully shaped to achieve condition (a), dV/ds positive, so that the wall boundary layers in the working section remain thin while the flow at restored speed between them becomes very nearly uniform. But the ability to demonstrate this demanded the development of refined 'hot-wire' techniques for measurement of the extremely low levels of residual turbulence left in the flow.

Wollaston wire, with a platinum core of radius 5 μm embedded in a silver wire of ten times the radius, can have a very small length of platinum core uncovered (by etching) so that, in an electric circuit, it provides the main resistance—with a value sensitively dependent on the temperature of the platinum. In a wind, this temperature is determined by a balance between ohmic production of heat and its removal by the airflow, whose instantaneous velocity may thus be measured with a suitable circuit. Dryden showed how, provided that this circuit includes compensation for the wire's thermal inertia, such a measurement can have excellent accuracy.

In the meantime the frequency spectrum obtained from the measured velocity fluctuations allows a direct derivation of Taylor's spectrum $\Phi(k)$

831

with respect to downstream wavenumber (since a sinusoidal component $a\sin(kx + \alpha)$ on being swept past the hot wire at velocity U becomes $a\sin(kx - kUt + \alpha)$, with frequency kU). Also, the use of two hot wires at points P and Q permits the covariance (28) to be measured as a function of the distance r between them.

Turbulence of the type which may remain in the working section of a wind-tunnel was found to be 'isotropic'; that is, with statistical properties which were invariant on any rotation (or reflection) of axes. This permitted valuable determinations—to which both Taylor [37] and von Kármán [38] contributed—of the entire tensor of covariances between different components of velocity at nearby points P and Q in terms of the vector separation between them [39], provided that just one component (28) was known as a function of r when Q is a distance r downstream of P. Similarly, the complete energy spectrum $E(k)$ of velocity fluctuations as a function of the magnitude k of the three-dimensional wavenumber vector could be expressed in terms of Taylor's one-dimensional spectrum $\Phi(k)$ by a differential relationship

$$E(k) = k^3 \left[\frac{\Phi'(k)}{k}\right]'. \tag{29}$$

Yet isotropic turbulence seemed merely to be just one more 'type of turbulence', by no means generic; particularly as the component (26) of Reynolds stress is zero in the isotropic case so that no acquisition (27) of energy from a mean flow is possible and the turbulent energy must always decay as a function of time.

10.2.4. Types of transition, types of turbulence: (ii) the new taxonomy

Many of the difficulties outlined above began to be resolved in papers which appeared in the early 1940s yet—because of restricted communication between scientists at that time—would enjoy wide appreciation only later. Here, just brief indications are given of how work initiated in those papers led thereafter to rather systematic developments of the 'taxonomy' both of transition and of fully developed turbulence.

Regarding transition in boundary layers, two very different 1940 papers carried knowledge of their stability to small disturbances far beyond the analysis of Tollmien–Schlichting waves for a boundary layer on a flat plate. On the one hand, for boundary layers treated as parallel flows, Schlichting himself showed [40] how, for slightly retarded flows (type (b), dV/ds small but negative, of p 820, with vorticity abstracted at the wall), the presence of a vorticity maximum in the midst of the layer implied instability to disturbances with a wide range of frequencies as in figure 10.17; moreover, a relatively low Reynolds number sufficed for the onset of this instability. On the other hand, Görtler showed [41] how boundary layers on curved walls, with the curvature concave to the

Fluid Mechanics

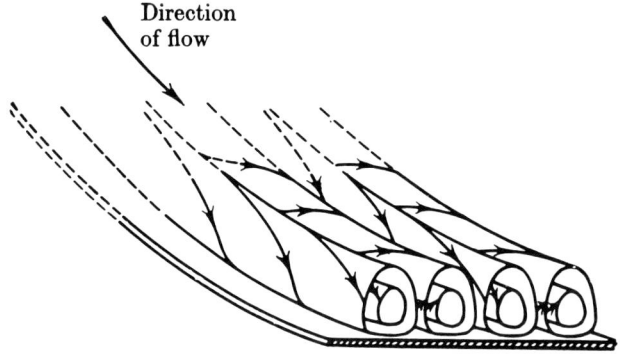

Figure 10.21. *The thin boundary layer on a concave surface is unstable (Görtler [41]) to the onset of Taylor–Görtler vortices illustrated in this (magnified) diagram, and regarded as analogous to the vortices found by Taylor [31] when an inner cylinder rotates in a stationary outer cylinder (figure 10.18 in case $\Omega_2 = 0$).*

flow (so that velocity is greatest on the inside of the bend), are subject to unstably developing disturbances with vorticity aligned to the flow, as had been studied (figure 10.18) by Taylor [31]; such boundary-layer vortices are now called Taylor–Görtler vortices (figure 10.21).

Then, just four years after these significant further contributions to 'the first half' (p 829) of the resolution of the boundary-layer transition paradox, a paper from Russia by L D Landau [28] was to sow the seeds of a great mass of radically new theory, that would branch out far beyond studies of stability to small disturbances in contributing to 'the second half'. Although this mass of (non-linear) theory would, in its later ramifications, become formidably complex, something of its essence may be perceived already in a brief summary of Landau's 1944 paper.

Essentially, he attempted in that paper to model the distinction (p 826) between 'abrupt' and 'restrained' transition. Here, for example, 'abrupt' transition may be found in systems with a vorticity maximum at the wall, which tend to become unstable to small disturbances when R—the Reynolds number (20)—reaches a 'critical' value R_c which is relatively high; R_c being defined so that such small disturbances have an amplitude A which changes *exponentially* like

$$e^{\gamma t} \text{ with } \gamma < 0 \text{ when } R < R_c \text{ but } \gamma > 0 \text{ when } R > R_c. \qquad (30)$$

In these systems, moreover, even positive values of the growth rate γ may be only moderate. Yet any rather large disturbance can cause vorticity to be abstracted at the wall so that the vorticity maximum moves into the fluid, generating momentarily an unstable local motion with a much lower R_c and a far greater amplification of disturbances. By contrast, systems tending to exhibit—at least in early stages of transition—a regular or 'sinuous' form of disturbance, may often become unstable more readily (having much lower values of R_c) but may be subject to an effect by which, at least initially, the appearance of larger amplitudes 'restrains' the exponential growth (30) of disturbances.

Landau's highly simplified mathematical model of these differences takes the form of an equation

$$\frac{d(A^2)}{dt} = 2\gamma A^2 - \alpha A^4 \qquad (31)$$

for a variable 'amplitude squared' proportional to the energy of disturbances to the mean flow. For very small A the right-hand side of (31) is dominated by its first term (evidently, one wholly consistent with the behaviour (30) for small disturbances), but Landau's second term gives a preliminary idea, on an assumption (acknowledged by him as oversimplified) that only the amplitude A of a single mode is important, of how any growth of amplitude may be either restrained ($\alpha > 0$) or enhanced ($\alpha < 0$).

Now it turns out that the substitution $y = (1/A^2) - (\alpha/2\gamma)$ throws equation (31) into the very simple form (3)—with 2γ for k and t for x—of which (4) is the general solution. This means that Landau's equation (31) has the general solution

$$A^2 = \frac{1}{(\alpha/2\gamma) + ae^{-2\gamma t}} \qquad (32)$$

whose form is plotted (figure 10.22) in two cases:

(i) restrained transition ($\alpha > 0$) where the solid lines give the behaviour of disturbances when $\gamma > 0$ (implying, by (30), that $R > R_c$); and

(ii) abrupt transition ($\alpha < 0$) where disturbances with $\gamma < 0$ (with subcritical Reynolds number $R < R_c$) behave as shown by the broken lines.

(Note: all solutions assume one of the shapes shown, though with a possible horizontal shift.)

In case (i), very small disturbances begin by growing exponentially. However, this does not continue indefinitely; after a certain time, a definite level of disturbance is maintained—as in the example (figure 10.18) of Taylor vortices.

In case (ii), by contrast, an 'abrupt' transition becomes possible already at subcritical Reynolds number (with $\gamma < 0$). There is a certain threshold level of energy $A^2 = 2\gamma/\alpha$ such that, for initial disturbances below it, $a > 0$ and so the disturbance energy damps out to zero. On the other hand, for initial energies above that threshold, $a < 0$; and then there is a runaway instability—the 'abrupt' form of transition—such that,

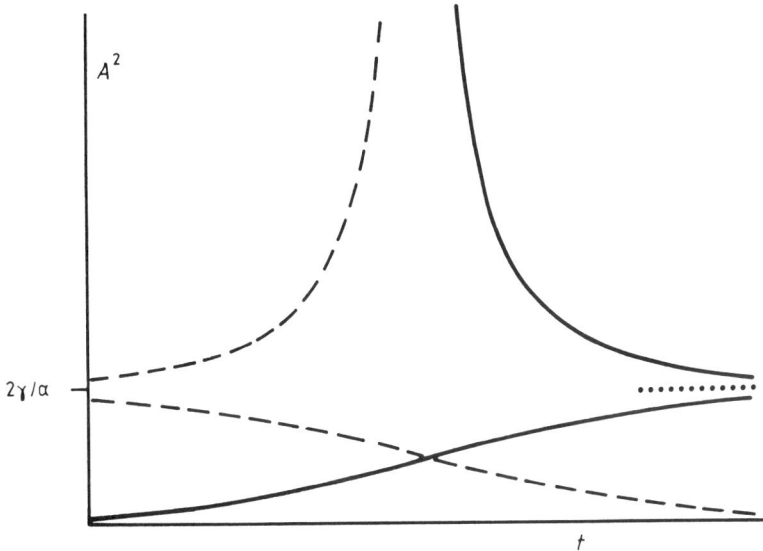

Figure 10.22. *Solutions of equation (31). Solid lines, case where $\alpha > 0$, $\gamma > 0$. Broken lines, case where $\alpha < 0$, $\gamma < 0$.*

in a finite time, the energy has already grown without limit (reaching levels, of course, where other terms besides those retained in equation (31) become important).

For the paradox concerning transition to turbulence in the boundary layer on a flat plate, such considerations opened the way to completing 'the second half' of its resolution, based on the layer's enhanced instability to disturbances above a certain level. Both halves were triumphantly confirmed soon after by Galen Schubauer and his colleagues at the National Bureau of Standards, Washington, DC (the inheritors, and further developers, of Dryden's precision techniques for aerodynamic instrumentation).

They showed on the one hand [42] that, when an enormous effort was made to suppress flow disturbances in the wind-tunnel's working section, and also any roughness elements (common sources of disturbances) on the solid surface, it became possible to excite growing Tollmien–Schlichting waves, at the predicted frequencies, by very small vibrations of a 'ribbon' flush with the surface (figure 10.23). This discovery has since been developed much further in researches detailing all those stages through which the growing waves develop three-dimensional characteristics and, ultimately, the chaotic nature of turbulence.

But the same group had next followed up Emmons's discovery [25] of randomly originating turbulent 'spots' (p 825) with a comprehensive demonstration [43] that these indeed were the typical form of transition

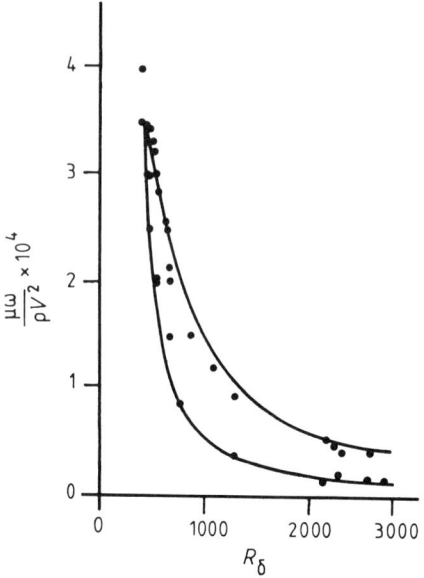

Figure 10.23. *Instability of boundary layer on a flat plate to small wave-like disturbances. These are able to grow, at given Reynolds number R_δ, for frequencies ω between the upper and lower branches of the 'neutral curve' shown. (———, predicted; ●, observed) neutral curve. (Note: here, the value of δ used to define R_δ is the 'displacement thickness' of the layer, defined so that its defect in mass flux is $V\delta$.)*

in boundary layers on flat plates (and also in pipes and channels), with each spot growing in dimensions amid the surrounding laminar flow at quite a regular rate as it is swept downstream. Locally, such a transition takes a literally 'abrupt' form in the sense that within each spot an almost fully developed turbulence is already present; thus, fluid on passing inside a spot adopts almost instantaneously a chaotic form of motion. On the other hand, some averaged quantity, like the frictional stress in the wall, may exhibit just a gradual change through a 'transition region' that extends from where the spots are first making their appearance to where the statistical assemblage of spots has grown to fill the whole layer (figure 10.15).

Subsequent work to uncover those mechanisms by which the spots of turbulence maintain sharp boundaries as they grow within the surrounding laminar flow has made rather slow progress; mainly, by using ideas (cognate to those of section 10.1.4) from the theory of non-linear wave propagation—although spots need of course to be viewed [44] as three-dimensional 'wave packets' (see section 10.3) rather than as anything similar to long-crested waves. Only in the century's concluding decade do answers to this challenging question seem at last to be emerging [45].

Fluid Mechanics

In the meantime, many other theoretical studies of transition at subcritical Reynolds numbers have followed somewhat more closely Landau's idea of improving the equation for development of a small-disturbance mode with a higher-order addition—but, now, one calculated from the equations of fluid mechanics [46, 11]. The most important higher-order effect is that associated with distortion of the mean-flow distribution $U(y)$ by action of a Reynolds stress (26) calculated from values of the small-disturbance velocities u and v. At a given subcritical Reynolds number, this once again yields an amplitude level above which the mode grows without limit.

For restrained transition (case (i) above) theoretical and experimental studies have combined to elucidate that 'spectral evolution' of disturbances which, at sufficiently supercritical Reynolds number, follows any initial amplitude growth at the frequency for which the system is most unstable. By the 1960s already, many papers were demonstrating those sequences of 'period-doubling bifurcations' which would later be acknowledged as standard routes to chaos. These may occur either as a process in time [47, 48] or, alternatively [49], as a gradual development with increase of R beyond R_c in a system like that of figure 10.18. In either case, the end result is fully developed turbulence.

For turbulence itself, just as for transition, a key paper offering the subject increased coherence emerged from Russia in the early 1940s. Kolmogorov's 1941 paper [50] explained how, even though ideas of turbulence 'as a single integrated phenomenon' are erroneous (section 10.2.3), and although turbulence (like transition) exists in a variegated taxonomy of types, nonetheless all those types have their small-scale features in common at very high Reynolds number.

Moreover, because one of the types is isotropic turbulence (section 10.2.3), which obeys certain special rules, including equation (29) for the spectrum $E(k)$ of turbulent energy, and because any small-scale features are associated with properties of spectra at relatively large magnitudes k of the wavenumber vector, this law of Kolmogorov implies that every type of turbulence becomes isotropic at relatively large wavenumbers and obeys those special rules. Isotropic turbulence has, then, this one 'generic' property (see the end of section 10.2.3), that all turbulence exhibits locally isotropic small-scale behaviour at very high Reynolds number.

In brief summary, Kolmogorov's analysis allows for the fact that, while $E(k)$ is the spectrum of the kinetic energy, $\frac{1}{2}(u^2 + v^2 + w^2)$ per unit mass, a corresponding spectrum of the rate

$$\varepsilon \text{ per unit mass} \tag{33}$$

of viscous dissipation of turbulent energy into heat is a spectrum of squares (and products) of gradients of u, v and w which should be proportional to $k^2 E(k)$; it turns out, for example, to be $(2\mu/\rho)k^2 E(k)$ for isotropic turbulence. Now, in a long spectrum, $k^2 E(k)$ reaches its maximum at much higher wavenumbers than $E(k)$, and this leads to the concept of energy being dissipated in 'small eddies' (that is, spectral components of motion with large k) even though the main energy-containing eddies have a much larger scale (smaller k). Also, expression (27) for the rate at which energy is fed into the turbulence includes a product of velocities u and v (not of their gradients) and its spectral peak is close to that of the turbulent energy itself.

Turbulence is a process, then, where the main energy-containing eddies, while receiving energy directly, are passing it down a 'cascade' of eddy sizes to those small eddies which dissipate it into heat. At every stage of the cascade, non-linear effects stemming from the basic non-linearity of momentum transport tend to generate 'overtones' or 'summation tones' of somewhat higher wavenumber; yet many successive stages, all involving random elements, are needed to reach large wavenumbers. After just a few stages, indeed, the wavenumber spectrum $E(k)$ becomes 'statistically decoupled' from the main energy-containing eddies, and assumes a universal 'equilibrium' form, independent of the type of turbulence in which the small eddies are embedded.

The principles of dimensional analysis (section 10.2.1) tell us that this large-wavenumber behaviour of the energy spectrum $E(k)$, dependent only on the four variables k, ρ, μ and ε (see (33) above), must really involve the non-dimensional form

$$\varepsilon^{-2/3} k^{5/3} E(k) \tag{34}$$

of the spectrum in a dependence on just a single non-dimensional variable ηk, where

$$\eta = (\mu/\rho)^{3/4} \varepsilon^{-1/4} \tag{35}$$

is known as the Kolmogorov dissipation length and characterizes the size of the energy-dissipating eddies. Extensive experimental data support this conclusion, with a dependence of expression (34) on ηk close to that shown on a log–log plot (solid line, derived from Pao's 1965 studies [51]) in figure 10.24. It is noteworthy that, within the 'equilibrium' range of wavenumbers where this dependence applies, there is a subrange $\eta k < 0.1$ (commonly called 'inertial' subrange) for which expression (34) is approximately constant, with $E(k)$ decreasing like $k^{-5/3}$; yet the spectrum of energy dissipation, proportional to $k^2 E(k)$, is still increasing (broken line) towards its peak around $\eta k = 0.3$.

In parallel with the researches on these large-wavenumber features common to different types of turbulence, studies of specific aspects of each type at the level of the energy-containing eddies made steady progress. Here, although space does not allow detailed descriptions, it

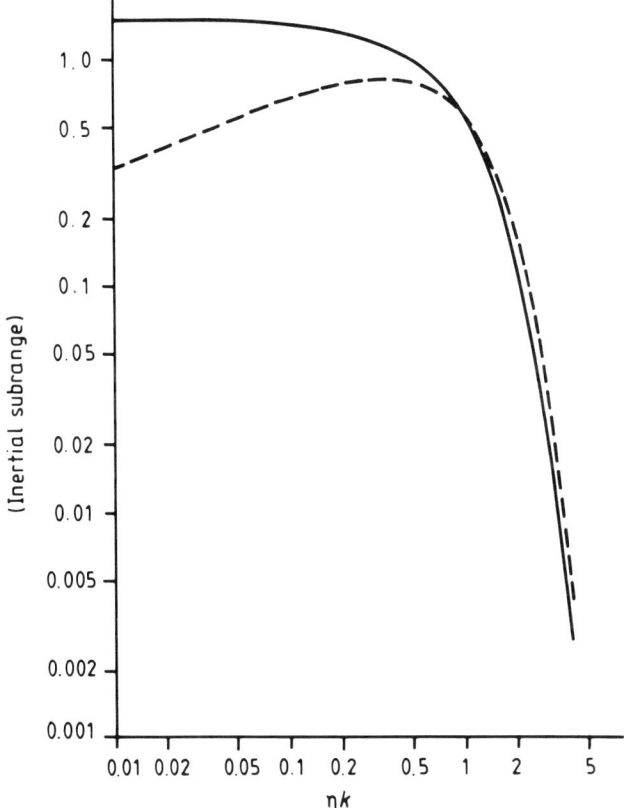

Figure 10.24. *Energy spectrum of turbulence in the equilibrium range. The solid line shows the dependence of expression (34) on ηk, while the broken line plots a quantity $(\varepsilon^{-2/3}\eta^{1/3})k^2 E(k)$, proportional to the spectrum of energy dissipation.*

may again be interesting to note an individual property which all the types have in common, as suggested first by Alan Townsend in his 1956 book [52].

Townsend amassed evidence that a role of some importance in guiding the development of turbulence was played by what he called 'big eddies'. These were substantially larger in scale than the main energy-containing eddies, and tended to fill the whole of any shear layer (boundary layer, wake, jet); in recent years they have often been described as 'coherent structures' within some largely chaotic shear layer.

The full wavenumber spectrum of turbulence is analysed, then, (figure 10.25) into 'big', 'energy-containing' and 'small' eddies—with the equilibrium range of 'small' eddies subdivided further (as the log–log plot of figure 10.24 shows best) into an inertial subrange and the main energy-dissipating eddies. Here, Townsend's graph reproduced as figure

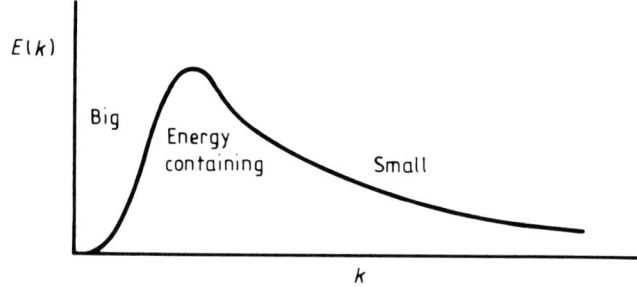

Figure 10.25. *Analysis of the energy spectrum $E(k)$, derived [53] from measurements of isotropic turbulence behind a grid of bars of mesh M in a stream of velocity U, where $\rho U M/\mu = 5300$, into 'big', 'energy containing' and 'small' eddies. (Note: linear scales for k and $E(k)$ are used here; for a continuation of the small-eddy range on a log–log scale, see figure 10.24.)*

10.25 is a spectrum derived (1951) by equation (29) from measurements of $\Phi(k)$ for a particular case of isotropic turbulence [53], but such a 'fourway split' is now seen as characteristic of turbulence in general.

For example, in any type of turbulence adjacent to a solid surface, those big eddies which exercise a guiding role in its development have the form of elongated vortices, their vorticity being aligned to the wind but with crosswind alternations of sign. These, as indicated in figure 10.26 (where the mean flow is perpendicular to the paper), have two important effects on the mean vorticity (here, directed from left to right):

(a) they produce a correlation of vortex-line stretching (p 818) with inflow, which greatly increases vorticity levels near the solid surface [11]; yet

(b) elsewhere, they push vorticity maxima out into the midst of the fluid so that, from this highly unstable configuration, there emerge 'bursts' of intense turbulence which feed the system as a whole [54].

Both effects (a) and (b) influence also the distribution of mean velocity $U(y)$ found near a solid surface (figure 10.27): associated with a given frictional stress at the surface, there is an equilibrium (and approximately logarithmic) distribution of $U(y)$, aptly named 'law of the wall' in the 1955 studies of D Coles [55]—who interpreted it a lot more convincingly than had the investigators (obsessed with 'mixing length' ideas) of the 1930s.

But turbulent flows away from any solid surface—with mean-flow distributions referred to as 'law of the wake' by Coles [56]—are influenced very differently by their big eddies; whose mixing action tends not to intensify differences in the shapes of mean-flow distributions from those characteristic of laminar flow, but to smooth them away. An extreme example of this is the turbulent jet, with its aeronautical importance (section 10.4) for exerting a thrust F equal to the total rate of

Figure 10.26. Tendency of 'big eddies' in a turbulent boundary layer to generate both (a) correlation of vortex-line stretching with inflow and (b) convection of vorticity maxima into the midst of the fluid.

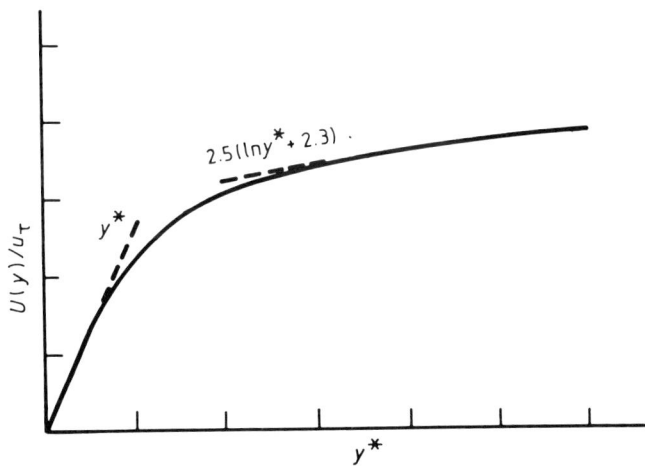

Figure 10.27. Illustrating how, in turbulent flow near a smooth solid surface on which a frictional stress $\tau = \mu U'(0)$ is exerted, the mean velocity $U(y)$ may be related to the 'friction velocity', $u_\tau = (\tau/\rho)^{1/2}$, and to the non-dimensional distance from the wall, $y^* = \rho u_\tau y/\mu$.

momentum transport in the jet. The quantity $(\rho F)^{1/2}$ has the dimensions of viscosity, and the fully developed jet flow has an effective Reynolds number $(\rho F)^{1/2}/\mu$ independent of distance from the orifice (indeed, its diameter increases linearly with distance, in inverse proportion to the diminution in average velocity). Actually, a laminar flow with a viscosity of about

$$0.055(\rho F)^{1/2} \tag{36}$$

would have a velocity distribution (figure 10.28) almost identical to that of the mean flow [58] in a fully developed turbulent jet!—which leads to expression (36) being often referred to as the effective 'eddy viscosity'.

10.2.5. The diffusion–convection balance for scalar quantities

Stationary solid objects in a wind (the main topic of section 10.2), besides being subject to forces, are often cooled and/or dried. In other words, it is not only vector quantities like momentum—or vorticity—that may be subject in boundary layers to the combined action of convection and

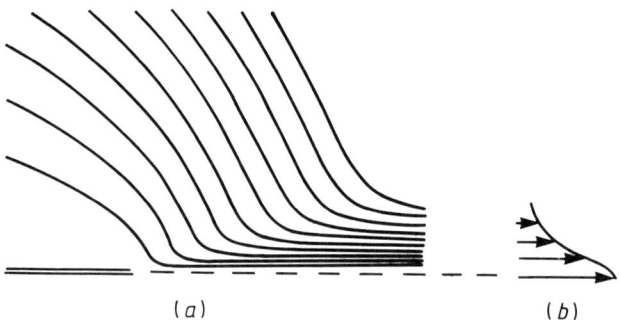

Figure 10.28. (a) Flow lines for an air jet with momentum flux F, as calculated [57] for laminar flow with the viscosity (36). (Flow is symmetrical about the broken lines.) Note the entrainment of external air into the jet, which causes its mass flux (although not its momentum flux) to increase with distance from the orifice. (b) Distribution of velocity in the jet.

diffusion, but also scalar quantities like heat or water vapour. Prandtl himself recognized early [59] how, in particular, this must overturn views on the modes (radiation, conduction, convection) of heat transfer.

It used to be thought that, apart from any radiative loss, a body in a wind would lose heat exclusively by convection. This is impossible, however, since at the solid surface itself the air is stationary (section 10.1.3). Conduction must, therefore, operate to move heat out into the boundary layer before convection can sweep it away.

Accordingly, the rate of heat transfer from unit area of solid surface at temperature T_s into a wind at temperature T_w may be written as

$$k\frac{T_s - T_w}{h} \tag{37}$$

where k is thermal conductivity and h is a certain fraction of the boundary-layer thickness δ (numerically derived for an important case in 1921 by Prandtl's colleague K Pohlhausen [60], and calculated later in many other cases). Just like viscosity, then, conduction of heat can never be disregarded.

Similar remarks apply to water vapour, which is transferred from a wet surface at a rate

$$D\frac{q_s - q_w}{h} \tag{38}$$

here, D is the diffusivity of water vapour, defined as rate of transfer of water mass across unit area divided by the gradient in its volume concentration q—while q is q_w in the wind and takes its saturated value q_s at the surface. Actually, this diffusivity D for water vapour is almost identical numerically with the diffusivity of heat (which is $k/\rho c_p$ because the volume concentration of heat has its gradient in a ratio ρc_p to that of temperature), and it follows that h is the same fraction of δ in both (37) and (38), being determined by an identical diffusion–convection balance in each case [11].

It was G I Taylor (1933) who first saw [61] how this explains why a wet body in a wind reaches the equilibrium ('wet-bulb') temperature

$$T_s = T_w - \frac{L_v D}{k}(q_s - q_w) \tag{39}$$

where L_v is the latent heat of evaporation. Equation (39) is called for by the condition that the rates of transfer of sensible heat (37) and of latent heat (expression (38) multiplied by L_v) add up to zero.

In the case of a turbulent boundary layer, the turbulence itself generates practically all the diffusion; whether of momentum, heat or water vapour (in laminar flow, by contrast, the diffusivity μ/ρ of momentum is about 30% less than that of heat or water vapour). At the extraordinarily early date of 1874, Osborne Reynolds had noticed [62] how, in any turbulent flow, this implied almost identical diffusion–convection balances for momentum and for heat (now known as the Reynolds analogy).

Thus, for a body in a wind of speed U, if τ is the frictional stress at the surface (rate of transfer of momentum to unit area) and Q is the heat transfer rate, then to a close approximation

$$\frac{Q}{\rho c_p (T_s - T_w)} = \frac{\tau}{\rho U} \tag{40}$$

in turbulent flow, because the same diffusion–convection balance governs the right-hand side (momentum transfer divided by momentum per unit volume in the wind) and the left-hand side (heat transfer divided by difference in heat content per unit volume between surface and wind). Equation (40), to which a similar result for water–vapour transfer in turbulent flow may be added, has continued to be widely applied in twentieth century fluid mechanics [29].

Although all further discussion of 'forced convection' (heat transfer from, or to, a body in a wind) is postponed to section 10.4.2, some brief comments are needed here on 'free convection', which, in the absence of any wind, is generated simply by buoyancy forces acting on air warmed by heat transfer from the body. In free convection from a vertical surface (to be described first), it was again a penetrating analysis by Pohlhausen, working in collaboration with the brilliant experimenter E Schmidt, that established a proper boundary-layer description of the phenomenon, and won excellent agreement (see figure 10.29) between theory and experiment [63].

In the boundary-layer equation (13), one term $\rho V dV/ds$ could be suppressed because there was no flow outside the layer, but an additional

Twentieth Century Physics

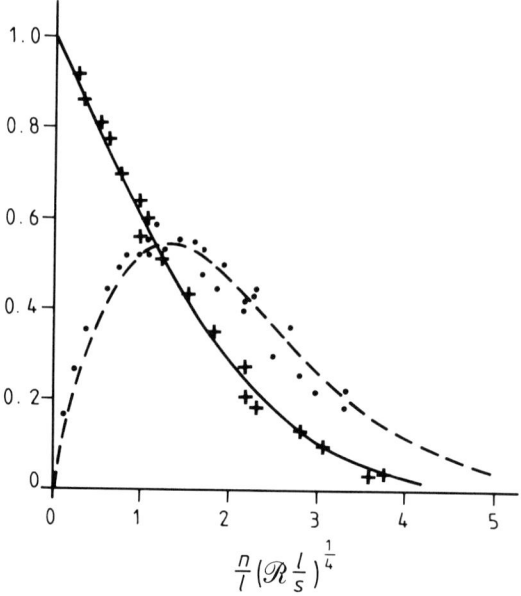

Figure 10.29. *The free-convection boundary layer on a vertical flat plate of height l maintained at a temperature T_1 exceeding the ambient air temperature T_0. Calculated and measured dependence [63] of the temperature T and the velocity v in the boundary layer at a normal distance n from the plate and at a height s above its base.*

force per unit volume—the buoyancy force $\rho g(T - T_0)/T_0$, where T_0 is the ambient air temperature—needed to be incorporated on the right-hand side

$$\rho v \frac{dv}{ds} = \rho g \frac{(T - T_0)}{T_0} + \mu \frac{d^2v}{dn^2} \qquad (41)$$

while the diffusion–convection balance for the temperature T could be similarly expressed as

$$\rho c_p v \frac{dT}{ds} = k \frac{d^2T}{dn^2}. \qquad (42)$$

Pohlhausen had used equation (42) on its own to solve the forced-convection problem [60], with v obtained from an independent solution of (13); but now it was necessary [64] to solve (41) and (42) as two coupled equations to obtain the results of figure 10.29.

In such free-convection problems the absence of any external wind U means that the Reynolds number (20) has to be replaced by a different non-dimensional parameter, known (for reasons indicated below) as the Rayleigh number $\mathcal{R}$. The form

$$\mathcal{R} = g \frac{T_1 - T_0}{T_0} \ell^3 \left(\frac{\rho}{\mu}\right) \left(\frac{\rho c_p}{k}\right) \qquad (43)$$

of $\mathcal{R}$, for a vertical surface of height ℓ maintained at temperature T_1, is suggested by the closely parallel importance of the diffusivities μ/ρ in (41) and $k/\rho c_p$ in (42). For the total rate of heat transfer from unit width of such a vertical surface, Pohlhausen obtained the expression

$$0.52\mathcal{R}^{1/4}k(T_1 - T_0) \tag{44}$$

proportional to the fourth root of the Rayleigh number.

Similar calculations can be made (but with g replaced by its component parallel to the surface) for a flat surface at an angle to the vertical—provided that the angle is acute! By contrast, the nature of free convection above a horizontal surface is altogether different, depending entirely on the instability of fluid heated from below.

Aspects of this problem on the huge length scales ℓ (more precisely, huge values of $\mathcal{R}$) that are relevant in meteorology involve, as described in section 10.5, a highly chaotic motion; with randomly spaced—and timed—ascents through cool air, from a hot ground, of those masses of warm air known as 'thermals'. This main section 10.2, however, is now concluded with a brief account of free-convection instability on a laboratory scale, that follows on rather naturally from earlier discussions of instability and transition.

See also p 899

Laboratory experiments on fluid heated from below are carried out either

(a) on a gas or liquid between two solid horizontal surfaces, or
(b) on a liquid between a solid surface and a free surface.

Some famous 1901 experiments of Bénard [65] had used system (b); much later (1958), however, it was recognized [66] that the instability which they demonstrated was primarily driven not by buoyancy forces, but by the temperature dependence of surface tension! (Warm liquid rising to the surface is pulled onwards by adjacent cold liquid with its greater surface tension.)

For problem (a) on the other hand, coupled equations similar in general character to (41) and (42), but without any simplifying boundary-layer approximation, may be written down and analysed in the small-disturbance limit. This was first done (1916) by Rayleigh [67], who, however, was not able to solve them with realistic boundary conditions. Later workers [68] showed in case (a) that the critical value $\mathcal{R}_c$ (minimum value of $\mathcal{R}$ for instability to small disturbances) is 1700. (In a liquid, the T_0^{-1} factor in (41) and (43) needs to be replaced by the liquid's coefficient of cubic expansion.)

The mode of disturbance to which the system is most unstable takes the form of long 'rolls'; moreover, transition is of the restrained type—corresponding to a positive α in Landau's equation (31)—so that the rolls assume a definite amplitude when $\mathcal{R}$ only moderately exceeds $\mathcal{R}_c$. The

Twentieth Century Physics

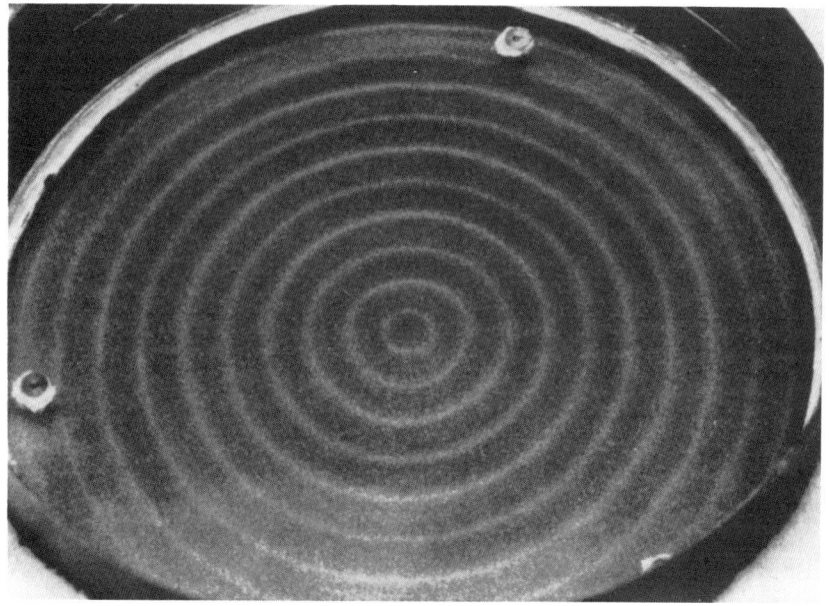

Figure 10.30. *Free convection: motions of fluid heated from below, made visible by Koschmieder [69] in a cylindrical container between two horizontal circular plates of glass.*

1966 experiments of Koschmieder [69] showed that the detailed form of the rolls within a container reflects above all its side-wall geometry. Figure 10.30, indicating the true form of free-convection instability in liquid between two horizontal glass plates, is here offered as a sort of counterpoise to the historical overemphasis on 'Bénard cells'.

10.3. Non-linear effects on the generation and propagation of waves

10.3.1. Waves with hidden energy loss

Waves and flows interact [70]; in a sound wave, for example, values of the air velocity u are propagated (p 809) at a signal velocity $u + c$ (the local sound speed relative to the local air motion) which exceeds the undisturbed sound speed c_0 by an amount $1.2u$ of which five-sixths† results from convection of sound by the air motion u. This section 10.3 outlines twentieth century developments in the understanding of wave/flow interactions, beginning (section 10.3.1) with further consequences of larger values of u 'catching up' with smaller ones

---

† *Sound waves in water (with its different compressibility properties) behave similarly except that the coefficient 1.2 is replaced by about 4, so that only one quarter (instead of five-sixths) of the excess signal velocity results from convection in this case.*

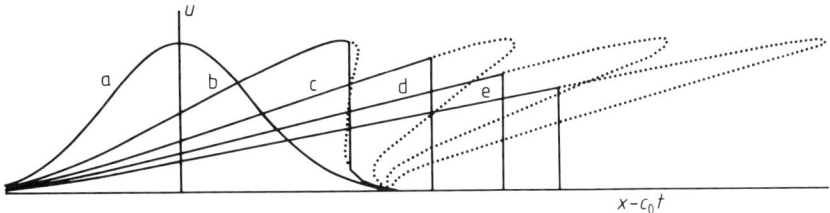

Figure 10.31. *Propagation of plane sound pulse. Curves a and b reproduce from figure 10.5 the initial wave form and an early wave form incorporating a shock wave, while curves c, d and e show—at equal time intervals—the later development of the wave form (derived from the dotted curves by Whitham's equal-area rule).*

and going on (section 10.3.2) to the effects of more complex (e.g. turbulent) flows on sound, before illustrating some very different types of interaction between flows and dispersive waves (those where the small-disturbance wave speed is not a constant c_0 as with sound but may depend e.g. on frequency).

For plane sound waves, consequences of the fundamental 1910 discovery (section 10.1.4) that 'catching up' proceeds until an extremely sharp discontinuity is formed—the shock wave, with thickness proportional to a diffusivity so that diffusion can balance convection within it whilst also dissipating wave energy into heat—were gradually developed during the next half-century. Essentially, the conclusion [70] was that sound propagation outside the shock wave continues to be governed by constant-entropy laws (section 10.1.4) because the very slight entropy inhomogeneities present in air traversed by a shock wave of changing strength have negligible effects on propagation. By contrast, the overall wave energy is progressively diminished by the accumulated effects of energy losses 'hidden' inside such an apparent discontinuity.

Not only energy, but also information, disappears in this process!—as may be illustrated by continued tracing (figure 10.31) of the non-linear propagation, with shock-wave formation, depicted in figure 10.5. At each stage the necessary discontinuity is incorporated at that location for which overall mass conservation is retained; this condition, the 'equal area' law of Whitham [71], requires that the discontinuity leaves unaltered the area under the curve (strictly, area under the corresponding curve of density ρ against distance [70]—although a closely linear relationship of density variations to u allows the law to be applied, with good approximation, to the graph of u itself).

Absolutely all information about the original shape of the compression pulse—except about the area Q under the curve—has disappeared in the later stages of this process, when the waveform has become a right-angled triangle of area Q with a hypotenuse of slope $(1.2t)^{-1}$ (the reciprocal slope $\delta x/\delta u$ increases at a rate 1.2 because a value $u + \delta u$ propagates faster than a value u by a signal-speed excess of

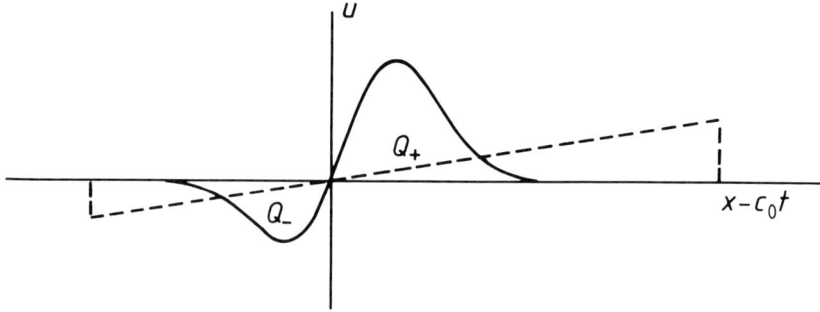

Figure 10.32. *Case of an initial waveform (solid line) with positive lobe of area Q_+ and negative lobe of area Q_-. Broken line, subsequently developing N-wave with unchanged areas for positive and negative lobes.*

$1.2\delta u$). Such a triangular waveform has a height $(5Q/3t)^{1/2}$—the strength of the shock wave—and a length $(12Qt/5)^{1/2}$. Moreover the energy of the wave, proportional to length and to amplitude squared, diminishes as $t^{-1/2}$ (what an enormous contrast to energy attenuation following an exponential law, as found for small-amplitude waves!) and the rate of attenuation, proportional to $t^{-3/2}$, is found to agree exactly with the rate of energy loss 'hidden' inside the discontinuity.

The corresponding conclusion for a sound pulse incorporating negative as well as positive values of u is that it develops asymptotically into an 'N-wave' (figure 10.32) which consists of two right-angled triangles, with areas Q_+ above the $u = 0$ axis and Q_- below it and with aligned hypotenuses of slope $(1.2t)^{-1}$. The shock strengths are $(5Q_+/3t)^{1/2}$ and $(5Q_-/3t)^{-1/2}$ and the overall length is $(12Q_+t/5)^{1/2} + (12Q_-t/5)^{-1/2}$; while, yet again, wave energy diminishes as $t^{-1/2}$ with all energy loss 'hidden' inside the two shock waves.

It is not just in plane waves that such behaviour is found; as may be illustrated (again, following Whitham [71]) by analysing the 'exploding wire' phenomenon that results from instantaneous discharge of a condenser through a very thin wire. A cylindrically expanding sound wave is generated in the surrounding air by the wire's sudden vaporization; however, an interesting difference from the plane-wave case emerges already from a simple linear analysis of the type familiar to nineteenth century physicists: cylindrical wave propagation, on such an analysis, converts the sound source (the outward-pushing vapour) into a wave with both positive and negative values of the outward velocity u. Thus, at radial distances r beyond a certain value r_1

$$u = \left(\frac{r_1}{r}\right)^{1/2} u_1(t - c_0^{-1}r) \qquad (45)$$

where the waveform $u_1(t)$ comprises a positive phase followed immediately by a negative phase of identical area Q (making $\int u_1(t)\,dt = 0$, and thus reconciling the necessary r^{-1} dependence of energy flux to the impossibility of any indefinite growth with r in the total outward displacement $2\pi r \int u \, dt$).

This means that, on non-linear theory, an N-wave (and, indeed, a 'balanced' N-wave with equal areas $Q_+ = Q_- = Q$) is necessarily formed. Whitham's analysis showed how a value $u_1 = (r/r_1)^{1/2} u$ is once more propagated at a signal speed $dr/dt = c_0 + 1.2u$ (replacing the simple c_0 value of (45)), giving to a close approximation

$$\frac{dt}{dr} = c_0^{-1} - 1.2 u c_0^{-2} = c_0^{-1} - 1.2 \left(\frac{r_1}{r}\right)^{1/2} u_1 c_0^{-2} \qquad (46)$$

which can be integrated to give

$$t = c_0^{-1} r - 0.6 (r_1 r)^{1/2} u_1 c_0^{-2} + \text{constant} \qquad (47)$$

as the altered time when a given value of u_1 is found. Thus the temporal waveform, of u_1 as a function of the time t, is now sheared (backwards) in such a way that the reciprocal slope $\delta t / \delta u_1$ takes an asymptotic value

$$-0.6 (r_1 r)^{1/2} c_0^{-2} \qquad (48)$$

whose magnitude increases indefinitely with r.

The ultimate N-wave therefore takes the form of two right-angled triangles of area Q whose aligned hypotenuses have the reciprocal slope (48). Each triangle has a height (discontinuity in u_1) equal to

$$\left[Q c_0^2 / 0.3 (r_1 r)^{1/2} \right]^{1/2} \qquad (49)$$

which varies as $r^{-1/4}$; but, since $u = (r_1/r)^{1/2} u_1$, this yields Whitham's remarkable inverse-three-quarters-power prediction for the strengths of both shock waves—which has been accurately verified in experiments on exploding wires (figure 10.33). The time interval separating the shock waves takes a value

$$\left[1.2 Q (r_1 r)^{1/2} c_0^{-2} \right]^{1/2} \qquad (50)$$

which increases as the fourth root $r^{1/4}$ of the distance travelled, while the wave energy decays as $r^{-1/4}$; with all the energy dissipation hidden, once more, inside the two shock waves.

Although the small part of twentieth century research on shock-wave dynamics that is mentioned in Chapter 10 has been selected with the aim of omitting studies on explosions for warlike purposes, the brief introduction given here (and in section 10.1.4) is further amplified in section 10.4.2. There, in descriptions of aeronautically generated shock waves [72], the 'double bang' described above as an N-wave is again characteristically found.

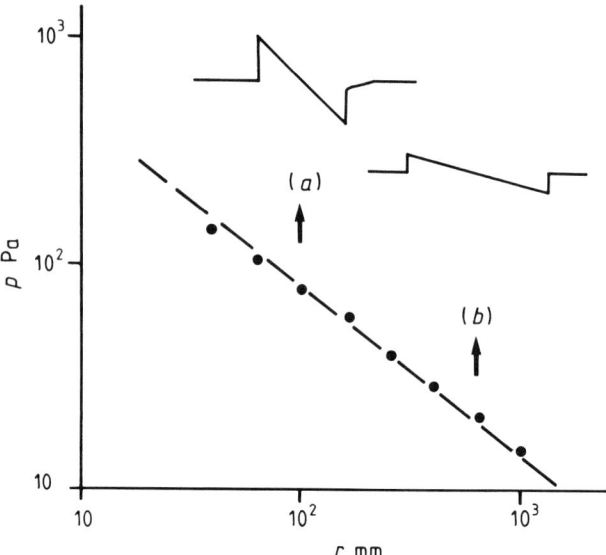

Figure 10.33. *Illustrating how an exploding wire generates pressure–time curves of N-wave shape, (a) and (b), at different radial distances r, and comparing the pressure rise at the initial shock wave with an $r^{-3/4}$ law (broken line).*

10.3.2. Sound from flows

Rayleigh's genius for recognizing when a phenomenon was simply unexplainable on the basis of existing knowledge (see section 10.1.4) was exhibited too in his 1879 paper on the aeolian harp [73], an instrument which responded with 'eerie' or 'ethereal' sounds when placed in a wind. Strouhal had accumulated data on the tones emitted by a wire of diameter ℓ in a wind of speed U, expressing their frequency f in a non-dimensional form

$$S = \frac{f\ell}{U} \tag{51}$$

which under the name Strouhal number is still used in many contexts; but he called them Reibungstöne in a false analogy with the effect of a bow on a violin string. Rayleigh preferred the expression 'aeolian tones' and showed that they could be present with or without vibrations of the wire but that such vibration when it occurred was at right angles to the wind direction. He correctly concluded [13] that production of the aeolian tone must be a consequence of some fluid-mechanical phenomenon still to be discovered.

The challenge was answered in various accounts of experiments (beginning with a—this time, flawless—paper [74] published in 1908 by H Bénard) which Theodore von Kármán would interpret in a famous 1911 study [75]. At Reynolds numbers between about 50 and about 2000, the entire flow past the wire exhibits an instability of the 'restrained' type

Figure 10.34. *The Kármán vortex street in the wake of a thin wire (observations* [76] *at R = 71).*

(section 10.2.4), involving the periodic shedding of vorticity of opposite signs from the upper and lower surfaces of the wire to form a well defined 'Kármán Vortex Street' (figure 10.34) behind it. The wake has a steady rate of growth in its momentum ($\frac{1}{2}\rho$ times the moment of the vorticity distribution) in a direction opposed to the wind, which Kármán could identify with the drag (equal and opposite to the force with which the wire acts on the air); while periodic vortex shedding leaves a periodically varying circulation of alternating sign around the cylinder, with which oscillatory lift is associated. The Strouhal number (51) for the oscillations of lift takes the same value (about 0.2, though falling away towards 0.1 at the lower end of the Reynolds number range) as for the aeolian tones.

Reciprocally, the wire exerts on the air an equal and opposite oscillatory force, such as would be supposed in nineteenth century acoustics to act as a dipole source of sound waves. Gradually, it became recognized that the waves generated are indeed the dipole field associated with this force; a 1955 paper by N Curle first demonstrated [77] that the presence of flow insignificantly affects the radiated sound field. He pointed out too that the radiated power takes a form

$$\langle \dot{L}^2 \rangle / (12\pi \rho c_0^3) \tag{52}$$

in terms of the mean square rate of change of total lift L on the whole length of wire; which at frequency f is $(2\pi f)^2 \langle L^2 \rangle$ so that, as the wind speed U increases, the proportionality of lift to U^2 and the dependence (51) of f on U imply a sixth-power law (like U^6) of dependence of sound radiation on wind speed.

Expression (52) for the sound radiated is perfectly valid whether or not the oscillatory lift causes the wire to vibrate. In the fluctuations of total lift L, on the other hand, considerable cancelling can result from fluctuations in circulation not being perfectly in phase all along the wire. Then a resonant vibration of the wire—as in one of the variously tuned strings of the aeolian harp—may produce an increase in sound intensity simply by bringing all these fluctuations into phase, 'locked on' to the phase of that vibration.

Much more musically important are those many wind instruments which use the interaction of a narrow air jet with a sharp edge to generate

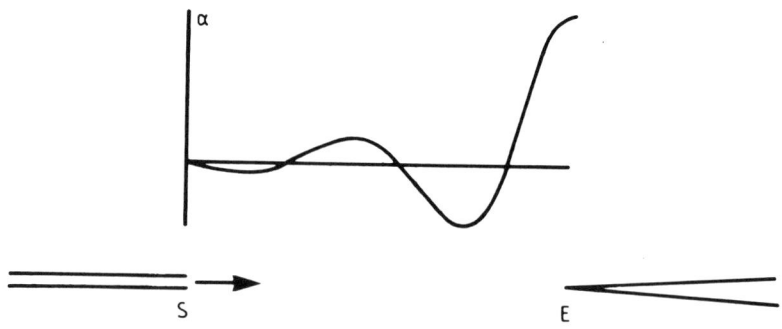

Figure 10.35. When an air jet issues horizontally (at moderate Reynolds number) from a long slit S, variations in the extremely small angle α of upward deflexion of the jet take the form of a travelling sinusoidal wave of exponentially increasing amplitude. At the instant shown, the edge E experiences a positive angle of attack α and exerts a downward force on the fluid, causing positive rate of increase of the deflexion α at S and therefore tending to reinforce the travelling wave.

sound—which, again, is often intensified by such 'locking on' to some resonant frequency (in an adjacent pipe). Curle identified in the jet-edge system a fundamental feedback loop which allows the air to exert on the edge an oscillatory lift force with its related dipole sound field [78].

Specifically, the dipole field produces a minute oscillation in flow direction at the jet orifice, and this disturbance as it travels downstream becomes amplified in consequence of the instability (again, of 'restrained' type) associated with the parallel flow in the jet. Feedback is effective, therefore (figure 10.35), at those frequencies for which the amplified direction change on reaching the edge has the right phase to reinforce the lift fluctuations.

Sound from flows, then, may take a dipole form, with power output (52), if a flow exercises a fluctuating force L on a solid body immersed in it; and at moderate Reynolds number, when 'restrained' instability can produce periodic fluctuations in L, the sound may be a musical note. At higher R, on the other hand, the fluctuations become chaotic, so that they generate acoustic noise; yet continued 'Strouhal scaling' (51) of typical frequencies f in the turbulent flow ensures a continued U^6 dependence of acoustic power output on the wind speed U. The 'roar' of a high wind is generated by fluctuating forces arising in its interactions with solid objects.

Various faster airflows, however, can generate significant noise even when not interacting with solid objects. The problems of estimating the noise that turbulent jets may radiate—without edges (or other objects) placed in them—began to be studied around 1950, when a fruitful method since named 'the acoustic analogy' was introduced [79].

Outside the jet, where any sound generated is just a minor by-product

Fluid Mechanics

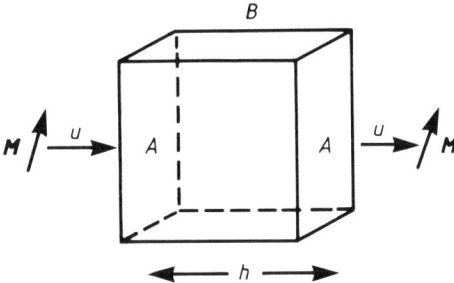

Figure 10.36. *Showing the part of the momentum transport into a box-shaped element B which occurs across one pair of opposite faces, of area A, a distance h apart.*

of the turbulence, the ordinary linear equations of sound may be applied. Essentially these equate (i) the rate of change of mass in a box-shaped volume element B to the net rate of mass flow into B, and (ii) the rate of change of momentum in B to the pressure force acting. (They are a closed system of equations because momentum per unit volume coincides with mass flux, while mass per unit volume is adiabatically related to pressure.)

But in accurate fluid mechanics the rate of change of momentum in B should include also the net rate of momentum flow into the element (figure 10.36). Momentum transport, of course, involves a product of two velocities, of the kind whose mean value appears in the Reynolds stress (26)—although in the acoustic analogy we must concern ourselves with its fluctuations. This product of two velocities is legitimately neglected outside the jet, where the linear equations of sound are applied, but cannot be ignored in the turbulence itself. There, it effectively appears in the second acoustic equation (ii) as an additional force, equal to the net rate of momentum transport into the element; thus, the ordinary equations of sound become correct if—simply, within the turbulent flow—such an additional force is incorporated within them.

These arguments demonstrate the validity of the acoustic analogy between the sound field of the turbulence and the sound field that would be generated in a linear acoustic medium by such a distribution of additional forces. This analogy is fruitful in many ways; not least, by showing how the absence of any solid body in the flow makes an essential difference to the radiated sound.

Indeed, without any resultant action on the fluid of a fluctuating external force L with its associated dipole radiation (52), the generation of sound must depend entirely on the net effect of a distribution (with zero resultant) of internal forces between elements. Now, the net radiation emitted (at an angle θ) by two equal and opposite dipoles separated by a distance h arises entirely (figure 10.37) from the difference $c_0^{-1} h \cos\theta$ in times of emission of signals received simultaneously by a far-off observer.

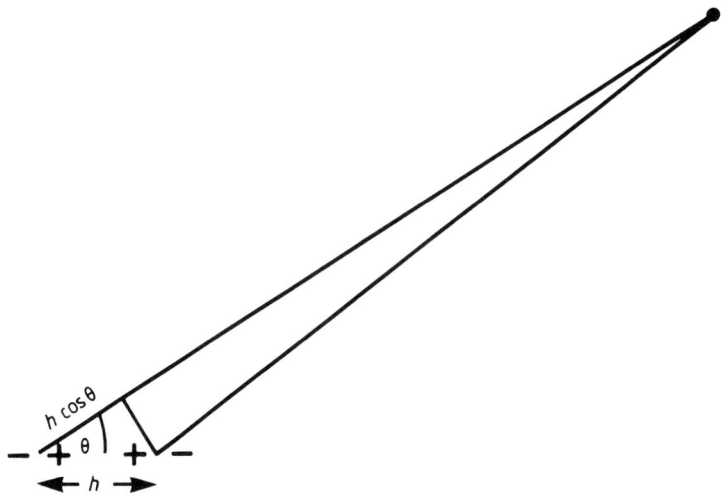

Figure 10.37. For two equal and opposite dipoles, a distance h apart, the far field in a direction making an angle θ with the line joining the dipoles results from the difference $h\cos\theta$ in distances travelled, leading to a difference $c_0^{-1} h\cos\theta$ in times of emission, of signals received simultaneously.

So this radiation—described as that of a quadrupole, of strength equal to h times the dipole strength—takes the form

$$(c_0^{-1} h \cos\theta) \left(\frac{d}{dt}\right) \text{ (dipole radiation).} \qquad (53)$$

Here, the new differentiation with respect to time (additional to that which is already present in the radiation field of a dipole of strength L) produces an extra frequency factor in the sound field; which, with Strouhal scaling (51), and also the c_0^{-1} factor present in (53), yields an additional Mach number factor (U/c_0) in amplitude and the same factor squared in the power radiated. Essentially, this is why flows at low Mach number where fluctuating body force is present generate sound of purely dipole type, compared with which any quadrupole radiation is negligible.

In a jet, however, fluctuations in body force are absent. Then the U^6 dependence of the acoustic power which they would generate is converted by the extra $(U/c_0)^2$ factor into an eighth power dependence (like U^8) of jet noise radiation [80]. The enormous difference between this U^8 law for jet noise and the U^2 dependence of jet thrust (section 10.2.4) was to prove crucial for making possible (section 10.4.1) an environmentally acceptable development of jet engines for air-transport purposes.

Reconsideration of the box-shaped element B gives a simple idea of how quadrupoles arise (figure 10.36). Net transport of momentum into B across any one pair of opposing faces, each of area A, involves an input MuA at one face and $-MuA$ at the other, where u is the velocity normal to those faces and the vector M represents momentum per unit volume. Each rate of change of momentum is equivalent to a force and so to a dipole and the two dipoles together form a quadrupole of strength $MuAh$ in a volume element Ah; thus a quadrupole of strength Mu per unit volume is associated with this pair of faces (with similar associations for the other two pairs).

It is indeed rather closely correct (as suggested by such crude arguments) that the strength per unit volume of the quadrupole sound source is equal to the momentum transport. Of course, while either a force or the equivalent dipole is a vector, with a direction as well as a magnitude, momentum transport involves two directions (the momentum's direction and that in which it is transported) and therefore is a tensor—exactly like a quadrupole strength, which involves both the direction of the component dipoles and that of the displacement between them.

Identification of the effective quadrupole source strength with the momentum transport tensor proved a solid foundation for developments in the second half of the twentieth century in the science of sound from flows—which, in its aeronautical application, has become known as aeroacoustics [81]. The many other considerations that have needed to be taken into account in these developments include, for example, these:

(i) that there is phase coherence between such sources at points P and Q only when they are close enough to permit the covariance plotted in figure 10.19 to be substantial; and

(ii) that, in a jet, all fluctuations are convected by the mean velocity distribution [80].

This brief account of sound from flows may be concluded by an even briefer postscript on flows from sound. Although some aspects of such 'acoustic streaming' were well covered in Rayleigh's *Theory of Sound*, a completely new aspect emerged in the second quarter of the twentieth century with the widespread availability of piezoelectric quartz crystals as powerful ultrasonic sources. Early experimenters were puzzled by 'the quartz wind'!—that was often observed to accompany such sources.

The acoustic energy density in an ultrasonic beam takes the value $\rho \langle u^2 \rangle$ (twice the kinetic energy density because of equipartition between kinetic and potential energies). But this expression is also the mean momentum flux in the beam (transport, per unit area, of u-momentum in the u-direction). Integrating across the area of the beam, we conclude that the total momentum transport in the beam is equal to the acoustic energy per unit length; or to c_0^{-1} times the total energy transport (as indeed might be expected by analogy with light).

Now, the energy of an ultrasonic beam in the megahertz range falls away very steeply from a variety of dissipative effects; yet momentum has to be conserved. Necessarily, then, a mean flow of jet type is formed to carry that part of the beam's original momentum transport which can no longer be present in the beam after its energy losses [82]. Ultimately, the jet takes the form indicated in figure 10.28, corresponding to a total rate of transport F of momentum in the jet equal to c_0^{-1} times the power of the ultrasonic source. Between such a 'turbulent jet generated by sound' and the problem described earlier of 'sound produced by a turbulent jet' there is an intriguing reciprocity—with momentum transport as the key to both phenomena.

10.3.3. Non-linearity competing with dispersion

Waves where, even for very small amplitude, the speed of propagation may vary (e.g. with wavelength) are called dispersive, because different sinusoidal components of an initial local disturbance will travel at different speeds and so become 'dispersed' from one another. In the ocean, for example, a local storm can create an extremely complicated disturbance of the water surface; yet, after a large time t, ocean waves of approximately sinusoidal form are observed as 'swell' (see section 10.5) at distances $c_g t$ from the storm region, where the velocity c_g takes different values for waves of different length λ.

Nineteenth century physicists, who had named c_g the 'group velocity'—or speed of travel of the energy in a group of approximately sinusoidal waves—had exploded the fallacy that this speed is equal to the 'wave velocity' c with which the crests of a sinusoidal wave train move; demonstrating rather [13] that if c varies with λ all the energy in a group of waves travels, not at any such illusory speed c suggested by the movement of crests, but at the group velocity

$$c_g = c - \lambda \frac{dc}{d\lambda}. \tag{54}$$

Twentieth century oceanographers systematically verified this statement by measurements of swell at great distances (many thousands of km) from where it had been generated [83].

For these waves on 'the surface of the deep' (section 10.3.4), satisfying the classical relationship

$$c = \left(\frac{g\lambda}{2\pi}\right)^{1/2} \quad \text{so that } c_g = \tfrac{1}{2}c \tag{55}$$

both velocities vary so steeply with λ that any modest non-linear effects on signal speed, such as were discussed for sound waves in section 10.3.1,

Fluid Mechanics

become unimportant by comparison; indeed, the substantial non-linear effects on ocean waves which have been investigated in the twentieth century are of a markedly different character. Before outlining these studies, however, we devote section 10.3.3 to the interesting topic of wave propagation where there is competition on a roughly equal footing between non-linearity and dispersion in their effects on signal speed [70].

Such propagation is typified by waves in 'shallow' water; that is, water whose mean depth h_0 (say, above a flat horizontal bottom) is very much smaller than the wavelength λ. A superficial, yet illuminating, analogy with sound waves suggests that any local value of the depth h should be propagated horizontally with wave speed $c = (gh)^{1/2}$. Very briefly, what governs the propagation in just two (horizontal) dimensions of these essentially longitudinal waves is the relationship between the vertically integrated values, ρ_v and p_v, of the density and pressure, given by equations

$$\rho_v = \rho h, \quad p_v = \tfrac{1}{2}\rho g h^2 \quad \text{so that} \quad c^2 = \frac{dp_v}{d\rho_v} = \frac{\rho g h \, dh}{\rho \, dh} = gh \quad (56)$$

is the square of the wave speed.

Pursuing the analogy, we see that a first approximation to the increased wave speed with which values of h in excess of h_0 are propagated is

$$c = c_0 \left(1 + 0.5 \frac{h - h_0}{h_0}\right) \quad \text{where} \quad c_0 = (gh_0)^{1/2} \quad (57)$$

is the linear-theory wave speed. Here, of course, c is the velocity of propagation relative to any water motion; whose velocity u is given on linear theory as

$$u = c_0 \frac{h - h_0}{h_0} \quad (58)$$

because any rate of increase of depth in a wave travelling at velocity c_0 must take the value $-c_0 dh/dx$ and be balanced by the downward flux gradient $-d(h_0 u)/dx$. These crude approximate arguments suggest augmented values of the wave speed and the signal speed

$$c = c_0 + 0.5u \quad \text{and} \quad c + u = c_0 + 1.5u \quad (59)$$

respectively; where the factor 1.5, replacing the 1.2 in equation (17), implies that 2/3 of the signal-speed excess is due to convection of waves at the water velocity u and 1/3 to the increase in c.

Once again, a full non-linear analysis by Riemann's methods [15] of such shallow-water propagation—assumed unidirectional as well as longitudinal—proves the equations (59) to be completely accurate for waves of any amplitude; with depth changes related to changes in c by equation (56), and with values of u propagated at precisely the speed $c_0 + 1.5u$. Just as with sound, then, a waveform develops as shown in figure 10.5 (except that 1.2 is replaced by 1.5) and poses the same enigma of how the tendency to form a discontinuity is compatible with the unavoidable loss of wave energy—a loss, per unit mass of water, shown by Rayleigh in a 1914 paper [84] to be

$$\frac{g(h_2 - h_1)^3}{4h_1 h_2} \qquad (60)$$

wherever the depth increases discontinuously from h_1 to h_2.

Much later, however, the resolution of this enigma was found to differ fundamentally from that which Rayleigh [6] and Taylor [7] had revealed for sound waves. Briefly, shallow-water waves have a very slightly dispersive character; which becomes more important as the gradient of the waveform in figure 10.5 is increased and, 'in competition with' non-linear effects, resolves the enigma.

Dispersion is found even in linear theory, where a sinusoidal waveform

$$h - h_0 = a \sin[k(x - c_0 t)] \quad \text{of length } \lambda = 2\pi/k \qquad (61)$$

with the water velocity (58), gives potential and kinetic energies per unit mass of water as

$$\frac{ga^2}{4h_0} \quad \text{and} \quad \frac{c_0^2 a^2}{4h_0^2} \qquad (62)$$

respectively; equipartition being assured by equation (57) for c_0. Yet not all the kinetic energy corresponds to the horizontal component u of water motion; a changing depth involves a vertical velocity varying linearly from dh/dt at the surface to zero at the bottom, which makes an additional contribution $(k^2 c_0^2 a^2/12)$ from vertical motions to the kinetic energy per unit mass. Equipartition now modifies the wave velocity from $c_0 = (gh_0)^{1/2}$ to a reduced value

$$[gh_0/(1 + \tfrac{1}{3}k^2 h_0^2)]^{1/2} \simeq c_0 \left(1 - \tfrac{1}{6}k^2 h_0^2\right). \qquad (63)$$

With this value for c in equation (54), the group velocity c_g becomes

$$c_g = c - \lambda \frac{dc}{d\lambda} = c + k\frac{dc}{dk} = c_0 \left(1 - \tfrac{1}{2}k^2 h_0^2\right) \qquad (64)$$

so that energy travels just a little slower than the wave crests in these slightly dispersive waves.

It follows [70] that a travelling discontinuous increase in depth—or 'hydraulic jump'—can drag along behind it a regular wave train with

Fluid Mechanics

Figure 10.38. *An undular bore on the river Severn photographed by Professor D H Peregrine.*

crests that travel at the same speed; yet whose energy, moving more slowly, is being gradually carried backwards relative to the jump, thus achieving practically all of the energy loss (60). The UK's famous Severn bore, generated as the flood tide propagates up the narrowing Severn estuary and river, often takes this form (figure 10.38).

A properly balanced description of the competition between weak non-linear and weak dispersive effects is obtainable from the equation introduced in 1895 by D J Korteweg and G de Vries [85], yet applied by others only much later to shallow-water propagation as well as to other physical phenomena influenced by these competitive effects. That equation makes the rate of change of u a sum of the term $-c_0 du/dx$ appropriate to constant signal velocity and terms

$$-\tfrac{1}{6}c_0 h^2 \frac{d^3 u}{dx^3} - 1.5u \frac{du}{dx} \qquad (65)$$

which, respectively, give the corrected linear dispersion relation (63) and the non-linear effect (59) on signal speed [70]. The regular waves behind a hydraulic jump are shown by this equation to be not sinusoidal but 'cnoidal'!—with shapes described by the Jacobian elliptic function cn.

The Korteweg–de Vries equation has also the famous 'soliton' solution, observable in shallow water as a propagating 'solitary wave' of locally increased depth whose effective wavelength is small enough for dispersive effects to cancel completely the steepening tendency of figure

859

10.5 so that the waveform remains of permanent type. Solitons, although largely a curiosity in fluid mechanics, have been shown to exhibit intriguing types of mutual interaction which make them of widespread interest to theoretical physicists in general [86].

10.3.4. The surface of the deep

Two great interacting components of the Earth's fluid envelope (section 10.5) are the ocean and the air above it, where winds in a turbulent boundary layer generate transfer of water vapour (section 10.2.5) from ocean to atmosphere, and transfer of heat between them (in either direction); while also transferring momentum from air to ocean, where much of it assumes the form of surface waves. As in other fields of physics, ocean waves of energy E (say, per unit horizontal area) carry momentum E/c in the direction of propagation, with c as the wave velocity [70]. However, when agencies such as wave breaking which diminish E reduce also the momentum E/c carried by the waves—without, of course, any change in total momentum—the remainder is necessarily converted into currents (compare section 10.3.2 on acoustic streaming).

In the absence of such energy dissipation, and of any renewed momentum transfer from winds, ocean waves have in the twentieth century been shown, even at amplitudes where non-linear effects are important, to satisfy conservation laws which are interestingly reminiscent of wave–particle duality principles. These laws are outlined here, though further reference to geophysical implications of knowledge on surface waves is deferred to section 10.5.

Duality relates waves to particles carrying action in specific amounts; similarly, a key concept for ocean waves is the wave action, A per unit horizontal area. For small-amplitude waves on still water, A is defined so that the energy E is $A\omega$ in terms of the frequency ω while the wave momentum is $A\mathbf{k}$ in terms of the vector wavenumber $\mathbf{k}$ (its magnitude, therefore, being E/c as stated above).

These quantities, moreover, can still be defined for regular long-crested waves even at large amplitudes [86], when surface waves (unlike the minimally dispersive waves of section 10.3.3) acquire from non-linear effects no continual steepening of a periodic waveform of length λ but just a limited modification from the sinusoidal shape into one [87] with peakier crests and flatter troughs (figure 10.39). Then $\mathbf{k}$ is simply defined as a vector in the direction of propagation with magnitude $2\pi/\lambda$, and ω as 2π divided by the wave period; while Whitham's fine 1965 studies [88] proved that the water motions possess momentum $A\mathbf{k}$ and kinetic energy $\frac{1}{2}A\omega$—although departures from equipartition due to non-linear effects make the associated potential energy slightly less.

Figure 10.40 shows the results of accurate energy computations made in 1975 by M S Longuet-Higgins [89] for regular long-crested waves of

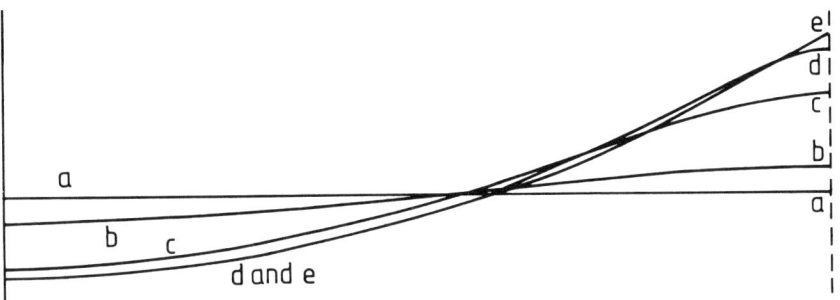

Figure 10.39. *Waveforms [87] for periodic deep-water waves of length λ and of different heights H between crest and trough. The broken line marks the plane of symmetry at each crest. (a) Undisturbed water surface. (b) Waveform for $H = 0.030\lambda$. (c) Waveform for $H = 0.100\lambda$. (d) Waveform for $H = 0.130\lambda$. (e) Waveform for the maximum wave height $H = 0.141\lambda$.*

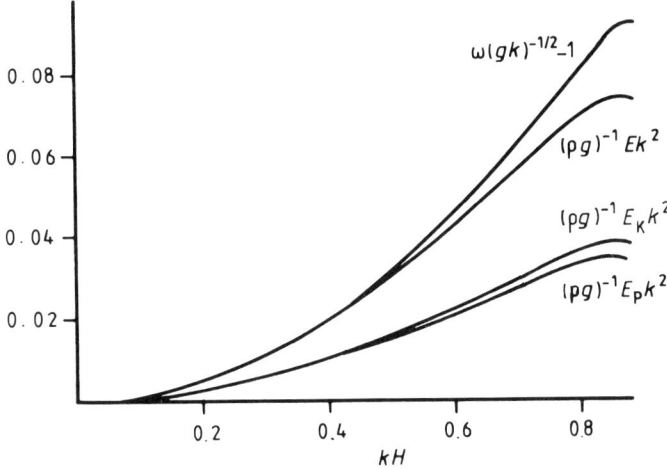

Figure 10.40. *Frequency ω, and wave energy E (divided into kinetic E_K and potential E_P) calculated [89] for deep-water waves of wavenumber k.*

wavenumber k (the slight rise in frequency ω above its small-amplitude value $(gk)^{1/2}$ is also shown). Both the kinetic and potential energies, as well as their sum the total wave energy E, increase with amplitude to a maximum value, beyond which the wave assumes a sharp-crested form (figure 10.39) such that, with further addition of energy, dissipation arises [90] through foaming at the crests (the celebrated 'white horses' phenomenon).

Conservation of wave action holds, on the other hand, when such dissipation is absent. Indeed, even where waves propagate into regions

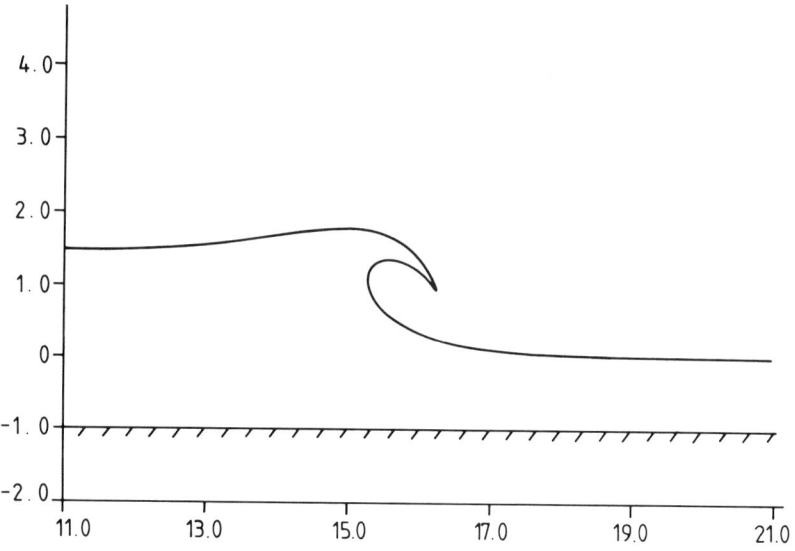

Figure 10.41. *Form of a plunging breaker computed by Professor D H Peregrine.*

of ocean current (so that wave energy need not be conserved because of exchange with the energy of currents), Whitham's action-conservation principle still applies. The kinetic energy of the waves becomes $\frac{1}{2}A(\omega - V \cdot k)$ in a current of velocity V, where $\omega - V \cdot k$ describes a modification to the effective frequency resulting from the familiar Doppler effect. Thus waves which propagate from still water into a region of opposing current (negative $V \cdot k$) may experience substantial gains in wave energy.

Space allows reference to but two other developments in non-linear ocean-wave theory. First, the need to make non-linear propagation studies still more precise (those based on action-conservation laws retain good accuracy only where values of the amplitude change per wavelength are very small) led in the able hands of K Stewartson and his colleagues to the fruitful description of a wavetrain's changing envelope by means of a non-linear Schrödinger equation, excellently adapted for convenient numerical solution [91].

Secondly, the study of 'spilling breakers' (waves in a slowly varying wavetrain which, as described above, attain the amplitude for maximum energy only to lose some of it by foam production at crests) has been valuably complemented by investigations of those 'plunging breakers' that arise when waves move into shallow water. Superb modern developments in computational fluid dynamics for unsteady motions of water with a free surface of complicated shape have permitted great insight into this alternative, and more spectacular, type of breaking to be achieved (figure 10.41).

Fluid Mechanics

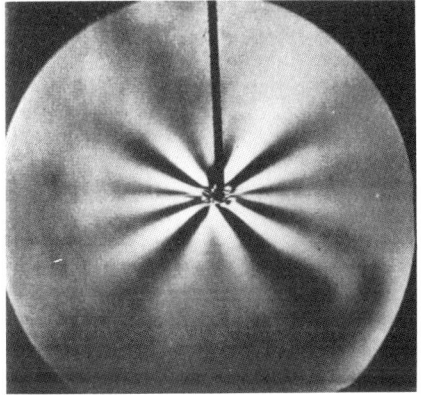 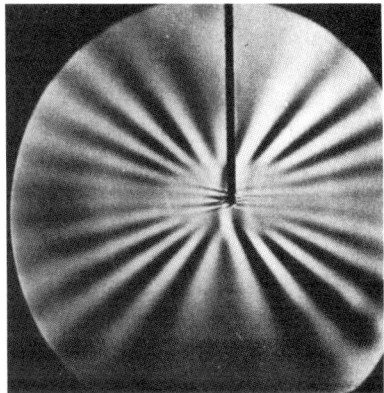

Figure 10.42. *Schlieren photographs [92] of waves with along-crest energy propagation (left, at time t = 10 s; right, at time t = 25 s).*

10.3.5. Along-crest energy propagation

The vector group velocity c_g, or velocity of propagation of the energy of waves in fluids, may differ not only in magnitude but also in direction from the wave velocity c of propagation of crests normal to themselves. This section is devoted to dispersive waves where the two velocities are orthogonal!—so that the wave energy is propagated along crests instead of at right angles to them [70].

Figure 10.42 shows the results [92] of experiments in a uniformly stratified salt solution (where the density reduction with height follows a constant gradient). Here, the vertical rod's function is simply to support at its lowest end an immersed horizontal cylinder which was given a brief horizontal displacement at time $t = 0$. Waves from this source have been made visible by schlieren photography at $t = 10$ s and at $t = 25$ s.

These waves are seen to have crests which—far from being concentric circles, as when energy generated at the source travels at right angles to crests—are arranged radially, because of along-crest energy propagation. A motion picture shows all crests moving normal to themselves, downwards for those above the source but upwards for those below it; yet wave energy, travelling radially outward, has reached considerably farther at $t = 25$ s than at $t = 10$ s.

Fundamental properties of disturbances to stratified fluids are geophysically important [93] because of density stratification in both the atmosphere and the ocean (section 10.5). This section, however, concentrates on the basic physics of along-crest energy propagation.

It arises [70] in any system where the wave frequency ω depends only on the direction and not on the magnitude of the wavenumber k. The simplest formula $c_g = d\omega/dk$ for the group velocity in one-dimensional propagation—which, since $\omega = ck$, agrees with an expression like

863

$c + k \, dc/dk$—can be generalized in three dimensions to a statement that the vector group velocity c_g is the gradient of the frequency in wavenumber space (on duality principles, indeed, waves move like particles, with the Hamiltonian expression for particle velocity as the gradient of energy in momentum space adapted through energy and momentum being associated with frequency and wavenumber). Where ω depends only on the direction of k it must stay constant along the vector k itself—which, therefore, is orthogonal to c_g (as the gradient of ω).

Figure 10.43 shows the nature of plane waves in a stratified liquid like that of figure 10.42, with a straight-line dependence

$$\rho = \rho_0(1 - \varepsilon z) \tag{66}$$

of density on the height z. The effective incompressibility of the liquid prevents the particle motions from having any 'longitudinal' component (a component perpendicular to crests, as in sound waves); they are forced, rather, to lie in surfaces of constant phase—where, being subject to a pressure-gradient force BC perpendicular to those surfaces and a vertical gravity force BD, they must move always up or down a direction BE of steepest descent. Now a particle displaced upwards by a small distance s at an angle θ to the vertical finds itself denser than its surroundings by an amount $\rho_0 \varepsilon (s \cos \theta)$ and subject to a downward gravity force $g \rho_0 \varepsilon s \cos \theta$ per unit volume, so that

$$g \rho_0 \varepsilon s \cos^2 \theta \tag{67}$$

is the resultant restoring force in the direction BE. Therefore such a particle, of mass ρ_0 per unit volume, performs simple harmonic motion with frequency

$$\omega = N \cos \theta \quad \text{where} \quad N = (g\varepsilon)^{1/2} \tag{68}$$

is known as the Väisälä–Brunt [94, 95] frequency. (Note: when such arguments are applied to atmospheric air in section 10.5.2, the only essential modification is a reduction in N that results from a change to the restoring force linked with the particle's adiabatic density drop in response to pressure loss with height.)

As foreshadowed, then, it is only on the direction of the wavenumber vector that the wave frequency ω depends. If moreover θ is slightly reduced to $\theta - \delta\theta$, the frequency increases by

$$(N \sin \theta) \delta \theta = (N k^{-1} \sin \theta) k \delta \theta \tag{69}$$

while the wavenumber k experiences a displacement $k \delta \theta$, which points downwards or upwards (figure 10.43) according as k has an upward or downward component, respectively [70]; so the group velocity vector c_g has

Fluid Mechanics

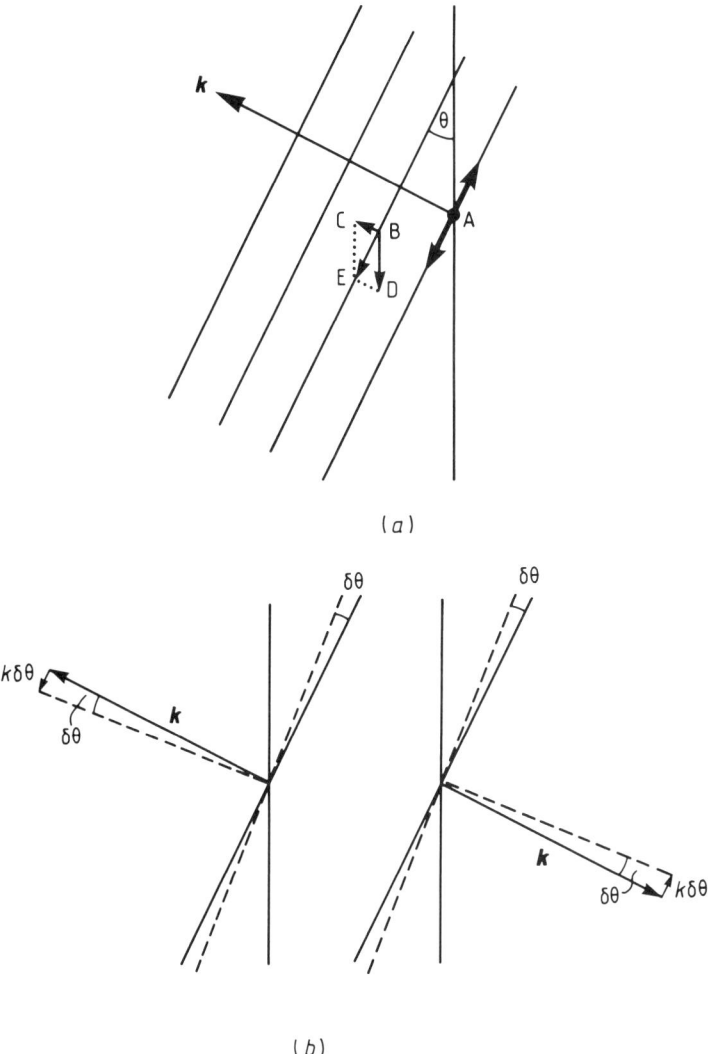

Figure 10.43. *(a) Illustrating plane waves in uniformly stratified fluid, including (at A) particle movement parallel to crests at angle θ with vertical, and (at B) resultant of pressure forces BC and gravity forces BD. (b) When θ is reduced to θ − δθ, the change (of magnitude kδθ) in the vector wavenumber **k** has vertical component of opposite sign to that of **k**.*

(i) magnitude $Nk^{-1}\sin\theta$ (see (69)) and
(ii) direction at an angle θ to the vertical, with
(iii) a vertical component opposite in sign to that of **k**.

Property (iii) explains why waves with the vertical component of

865

c_g positive or negative (found above or below the source in figure 10.42) have downward or upward directions of propagation, respectively. Property (i) indicates that, in terms of the wavelength λ, the radial distance travelled by wave energy in time t is $c_g t = (N\lambda t/2\pi)\sin\theta$ so that any small angle subtended at the source by crests a distance λ apart takes the value

$$2\pi/(Nt\sin\theta) \tag{70}$$

which again agrees with figure 10.42 (the angle varies with t like t^{-1} and with θ like $(\sin\theta)^{-1}$).

These 'internal' gravity waves within a stratified fluid, besides displaying the unusual features just discussed, exhibit interesting interactions between waves and flows: the central theme of section 10.3. Such interactions may be exemplified first in a uniform flow and then in a sheared one.

It is possible [96] for a two-dimensional pattern of internal gravity waves to remain stationary in a uniform stream of velocity U—so enabling them to form part of the steady flow past a long obstacle stretched perpendicularly to it—provided that their wavenumber has magnitude $k = N/U$. Then crests remain stationary because by equation (68) the wave velocity $c = \omega/k$, with which they move normal to themselves, cancels the velocity $U\cos\theta$ which is the wind's opposing component normal to crests. The energy in such waves, if generated at the obstacle, has travelled a distance $Ut\sin\theta$ parallel to crests (by (i) above) as well as being swept downstream by the greater distance Ut (figure 10.44(a)), so simple trigonometry makes the crests normal to a radius vector from the obstacle. Thus they are circular arcs; yet, because the energy's net distance downstream of the obstacle is necessarily positive, they are not full circles but semi-circles (figure 10.44(b)).

Early theoretical studies, having demonstrated that disturbances made by an obstacle in a uniform stream of stratified fluid satisfy equations governing stationary cylindrical waves, wrongly concluded that crests would be circles. This error is underlined here as one representative example out of many hundreds of studies of waves in fluids where determination of the uniquely correct wave pattern has needed application not only of equations of motion but also of the above condition (analogous to Sommerfeld's radiation condition in electromagnetic theory) that energy flow must be directed away from the source.

Accompanying this energy flow–say, E per unit length of obstacle—is a flow

$$(E/c)\cos\theta = E/U \tag{71}$$

of the component of wave momentum in the direction opposed to the wind. This momentum is also generated at the obstacle; which must experience an equal and opposite force, E/U per unit length, known

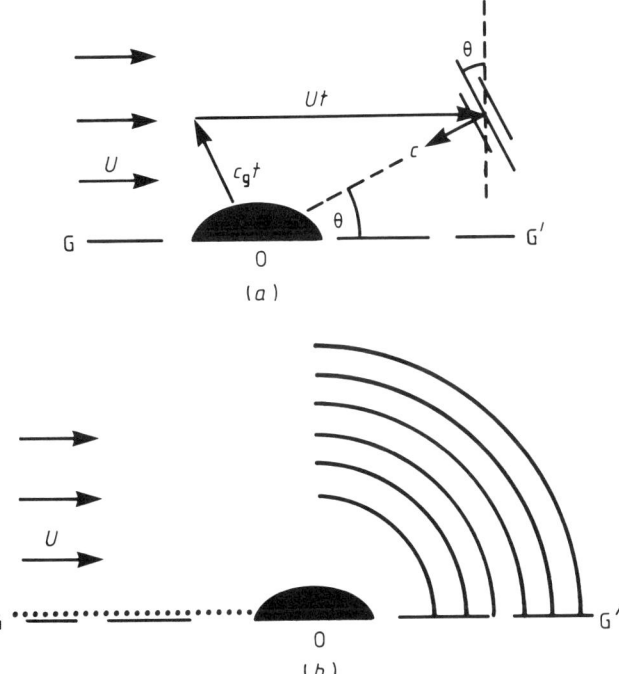

Figure 10.44. (a) In a uniform stream of speed U, internal waves of wavenumber $k = N/U$ have two remarkable properties: (i) crests remain stationary, because their propagation at wave velocity $c = \omega/k = (N\cos\theta)/k$ cancels the stream's component $U\cos\theta$ normal to the crests, while (ii) energy travels from a source (the obstacle O) at group velocity $c_g = Nk^{-1}\sin\theta = U\sin\theta$ at an angle θ to the upward vertical (along-crest propagation), besides being convected at velocity U by the stream, and so is found downstream in a direction making an angle θ with the horizontal; that is, in a direction normal to crests. (b) Consequential circular-arc form of crests. Also (see next page) the 'upstream wake' is indicated. (Note: the plane GG' may represent either a flat ground beneath the wind flow (wherein wave crests are quarter-circles) or else a plane of symmetry for a large flow (in which the wave crests with their mirror images in that plane are semi-circles.)

as wave-making resistance. The associated momentum transfer is not to wake fluid behind (and about on a level with) the obstacle as with frictional resistance; but, as first emphasized by F P Bretherton [97], to fluid vertically remote from it.

Although these conclusions have been explained in one very special case—uniform stratification, uniform stream, long obstacle of uniform cross-section—where simple results are readily derived, conclusions that are essentially similar can be worked out in an enormously wider range of cases, where any or all of the above elements may be non-uniform. In every case, stationary waves are found downstream of (yet vertically remote from) the obstacle, which experiences a wave-making resistance associated with the upstream component of their momentum. Such momentum transfer to vertically remote fluid has to be taken into account in modelling atmospheric flows over mountain ranges (section 10.5.2).

Paradoxically, 'upstream wakes' are possible in all these cases!—as can be seen in the simple case treated above by allowing θ to take its limiting value $\frac{1}{2}\pi$ (horizontal crests; a possibility tacitly ignored in the earlier analysis). Then the frequency (68) vanishes so that waves of any (vertical) wavenumber k can be stationary. Yet their along-crest energy propagation (at velocity Nk^{-1}, by (i) above) may exceed the stream velocity U if $k < N/U$, so that such waves can exist upstream of any obstacle tall enough to generate significant disturbances with vertical wavenumbers $k < N/U$. (Another approach to this surprising phenomenon is to remark that the stream's kinetic energy may be inadequate to supply the gravitational potential energy needed for lower parts of it to surmount a tall obstacle.) Comprehensive non-linear accounts of upstream wakes—with which, again, a resistance is associated—were first given by R R Long [98].

There is one very special capability that along-crest energy propagation gives to internal gravity waves in a sheared wind (one of velocity V varying with altitude z): they can intensify the wind by giving up all their energy to it. Indeed, conservation of wave action (section 10.3.4) tells us that, at a 'critical' altitude $z = z_{cr}$ where the relative frequency $\omega - V \cdot k$ falls to zero, wave energy is reduced to zero through exchange with the flow.

Here, relative frequency means the frequency in a frame of reference with the air at rest; which equation (68) identifies with $N\cos\theta$, so that the angle θ tends to $\frac{1}{2}\pi$ at the critical altitude. Waves, as they propagate through the sheared wind, must maintain constant values of the absolute frequency ω and the horizontal component k_H of wavenumber, but its varying vertical component $k_H \tan\theta$ can become larger and larger, in such a way that the vertical component of group velocity, $Nk_H^{-1}\cos^2\theta\sin\theta$, tends to zero as the square of the distance from the critical level $z = z_{cr}$ where $\omega - Vk_H = N\cos\theta$ vanishes. This allows the energy unlimited time to reach the critical level—so that mere reflection at that level, which occurs in many other physical problems, is impossible (as Booker and Bretherton [99] first showed), being replaced by total absorption. Some high-level jet-like winds are maintained by such absorption of energy from internal gravity waves.

Along-crest energy propagation occurs also in other fluid-mechanical systems; for example, in a homogeneous liquid rotating with constant angular velocity Ω. Here, plane waves with crests making an angle θ to the axis of rotation have frequency $\omega = 2\Omega\sin\theta$, which again depends only on the direction of the wavenumber vector; and various properties of internal gravity waves are repeated with N and θ replaced by 2Ω and

Fluid Mechanics

$\frac{1}{2}\pi - \theta$. Most spectacularly, the analogue of the upstream wake becomes a wake at $\theta = \frac{1}{2}\pi$, the 'Taylor column'; thus, when slow flow at right angles to the axis of rotation encounters a three-dimensional body, it was shown by G I Taylor [100] to become a two-dimensional flow about a column of fluid which amounts to a sort of extension of the body shape in the direction of that axis!

10.4. Transforming the human condition through aeronautics and ocean engineering

10.4.1. Fluid mechanics of flight efficiency

Human life on Earth was transformed in the twentieth century by various technologies, including those that made the Earth 'more of a neighbourhood' through fast and efficient movement of people and valued goods across it. This section 10.4 outlines the critical contribution of fluid mechanics to such developments, especially in the air but also in the ocean; which were achieved in close cooperation with developments (founded on other physical advances) in the worldwide transmission of information.

Aviation contributed decisively to these changes as a direct result of two great discoveries of Prandtl: that avoidance of boundary-layer separation can maintain very low drag on a non-lifting body (p 807) while the extra drag due to wing lift (p 824) can be kept very low for high wing spans. It was above all the 'Earth-shrinking' effects of ever greater range (distance travelled by an aircraft on a single flight) that responded so sensitively to these, as well as to some later, discoveries.

The elementary 'range equation' which give insight into such effects (see Küchemann [101], for example) describes the distance X travelled by a transport aircraft under those 'cruise' conditions—constituting the main part of its flight—where the lift L essentially balances the aircraft weight W while the drag D is balanced by the engine thrust T. The specific fuel consumption of the engine is defined as the ratio

$$s = \frac{\text{weight of fuel consumed per unit time } (-dW/dt)}{\text{useful work done per unit time } (UT)} \quad (72)$$

and it follows that

$$\frac{dW}{dt} = -sUT = -sUD = -sU\frac{D}{L}W \quad (73)$$

so that the cruise range X (the distance $\int U \, dt$ covered, at speed U, as the aircraft's weight W is reduced from W_0 at the start to W_1 at the end of cruise) is

$$X = \int U \, dt = \frac{L}{D} \frac{1}{s} \int_{W_0}^{W_1} \left(-\frac{dW}{W}\right) = \left(\frac{L}{D}\right)\left(\frac{1}{s}\right)\left(\ln \frac{W_0}{W_1}\right) \quad (74)$$

869

on the reasonable assumption that both L/D and $1/s$ are to be kept essentially constant (close, that is, to their maximum values) under cruise conditions.

Although the total range exceeds X (by the much smaller distances covered during the climb to, and descent from, cruise altitude), equation (74) is a valued first indication that achievable range may be dominated by a product of three separate factors, related to respective achievements in the three professional disciplines (aerodynamics, propulsion, structures) that make up aeronautical engineering. Evidently, the aerodynamicist and the propulsion engineer need to secure high maximum values of L/D and $1/s$, respectively. Moreover, successes in the design of a reliably safe structure which eschews unnecessary weight determine how much fuel for cruise, $W_0 - W_1$, can be carried in addition to other components of the all-up weight W_0; which include above all structure weight—along with fuel for climb and descent and for potentially necessary diversions, and of course 'payload' (used as a shorthand for all weights stemming from the aircraft's transport function).

Prandtl's discoveries [4, 5] led to a recognition of how high maximum lift-drag ratios L/D could be achieved. Briefly, the drag includes a 'frictional' element D_f, associated with the (mainly, turbulent) boundary layers on an aircraft's wings and fuselage, together with a drag due to lift, D_L, related to the work done in generating that vortex wake which produces the lift L. Equation (24) shows that the least possible value of D/L is

$$\frac{D_f}{L} + \varepsilon = \frac{D_f}{L} + \frac{L}{2\pi \rho U^2 b^2} \tag{75}$$

which is itself minimized when both terms are equal. This specifies

$$L = \left(2\pi \rho U^2 b^2 D_f\right)^{1/2} \quad \text{and Max}\ \frac{L}{D} = \left(\frac{\pi \rho U^2 b^2}{2 D_f}\right)^{1/2}. \tag{76}$$

Here, given meticulous design to ensure that separation is avoided all over the airframe, the 'frictional' drag D_f takes a form $k_f \rho U^2 S$ linked mainly to turbulent diffusion of momentum over a total area S of wings and fuselage, with k_f as a small constant—its smallness stemming, of course, from d'Alembert's Theorem—of order 10^{-2}. Then equations (75) and (76) yield results

$$\varepsilon = \left(\frac{k_f}{2\pi}\frac{S}{b^2}\right)^{1/2} \quad \text{and Max}\ \frac{L}{D} = \frac{1}{2\varepsilon} = \left(\frac{\pi}{2 k_f}\frac{b^2}{S}\right)^{1/2} \tag{77}$$

which demonstrate the benefits to be derived from sufficiently high wing spans b—or, more precisely, from sufficiently large values of the ratio

b^2/S—when lift–drag ratios around 20 or more (corresponding to values of $\varepsilon < 0.025$) have proved readily achievable [101].

Fluid mechanics, then, contributed greatly towards increases in the cruise range (74) through this achievement of large lift–drag ratios; while structural branches of aeronautical engineering made another big contribution through an increase in W_0/W_1, resulting from designs of reliably safe structures with reduced ratios of structural to all-up weight. Of course, different demands interact: the semi-span b of the wings may, essentially, be determined as a compromise between lift–drag benefits of increasing b and those structure-weight penalties which may result e.g. from the associated increase of bending moment at the wing 'root'.

The third great contribution to increasing the cruise range (74) arose from the evolution of power plants with massively reduced values of the specific fuel consumption s. Alongside refined thermodynamic and engine-design developments, these achievements depended also on advances in fluid mechanics which, yet again, interacted with considerations of lift–drag-ratio maximization.

Paradoxically [101], important reductions in the specific fuel consumption (72) accrue from increasing the aircraft speed U. For (say) a propeller aircraft, if an air mass Q enters the propeller disks per unit time at velocity U and is accelerated to velocity V, then that air's rates of increase of momentum and of kinetic energy are, respectively,

$$T = Q(V - U) \quad \text{and} \quad dE_K/dt = Q\left(\tfrac{1}{2}V^2 - \tfrac{1}{2}U^2\right) \tag{78}$$

so that equation (72) becomes

$$s = \frac{(-dW/dt)}{dE_K/dt}\frac{\tfrac{1}{2}V^2 - \tfrac{1}{2}U^2}{(V-U)U} = \frac{(-dW/dt)}{dE_K/dt}\frac{V+U}{2U}. \tag{79}$$

Here, the first factor is (inversely) related to the thermodynamic efficiency of conversion of the fuel's chemical energy into mechanical energy. But it is the second, more strictly fluid-mechanical, factor which so strikingly depends on the airspeed U, and which is reduced substantially as U increases (whether for given V, or for given $V - U$, or for given $\tfrac{1}{2}V^2 - \tfrac{1}{2}U^2$).

Similar equations for jet aircraft, though a little more complex because the exhaust gas is not air but an air-fuel combustion product, give a fluid-mechanical factor in s which is very similarly reduced (actually, a bit more) as U increases. With typical energy increases $\tfrac{1}{2}V^2 - \tfrac{1}{2}U^2$ derived from combustion, such considerations suggest that valuable increases in s may come from increasing the airspeed U to very large values.

Interactions of this conclusion with lift–drag-ratio maximization emerged in two principal stages. First, the cruise condition (76), with $W = L$, fixes the value of the ratio

$$W/\rho U^2 = (2\pi b^2 k_f S)^{1/2} \tag{80}$$

Figure 10.45. *A Boeing 747/400, with Rolls-Royce RB211-524G turbofan engines, and with all-up weight 395 tonnes, cruise Mach number 0.85 and maximum range 15 400 km.*

so that large airspeeds U demand low values of the air density ρ in cruise, and a further reduction in ρ as W decreases from W_0 to W_1. In other words, cruise must take place at rather high altitudes, increasing slightly during flight. (This cruise phase of flight is of course bracketed by other, briefer phases—takeoff and climb, descent and landing—when values of $W/\rho U^2$ may need to be far higher than the value (80) which maximizes lift–drag ratio.)

Secondly, the potential for increases in U becomes severely limited as U approaches the sound speed c_0. As indicated in section 10.4.2 below, energy losses in aeronautical shock waves represent a potent additional source of drag for aircraft at supersonic speeds; and, even when the Mach number (19) is somewhat less than 1, passage of the airflow over the wing can accelerate it to supersonic speeds and allow shock-wave formation. On the other hand, sweptback wings (figure 10.45) have allowed cruise Mach numbers to be increased to quite high subsonic values [102], around 0.85; essentially, because just the component of airflow perpendicular to the wing is accelerated by passage over it, while the parallel component remains unaltered.

Another fluid-mechanical limit on the early evolution of jet-propelled aircraft proved to be the intolerably high level of noise radiation from the jets themselves. This noise problem threatened to make the development of increasingly powerful jet engines environmentally incompatible with acceptable existences for communities around airports.

It is, perhaps, historically unusual that two big problems are found to have a single common solution! Yet two great obstacles to development of efficient jet aircraft—limitations on the reduction of s placed by an upper limit on cruise Mach number, and environmentally imposed restrictions on noise radiation from jets—were overcome by essentially the same development [80].

The theory of jet noise sketched on p 854 indicates that the acoustic energy radiated varies as the eighth power of the jet exit velocity, written there as U but here rewritten as V. High engine thrusts are needed at takeoff and in the early stages of climb, when the airspeed U is not large so that the thrust for a jet of given cross-section varies as V^2, as opposed to the V^8 dependence of the noise radiated to nearby communities. This contrast points to the possibility of increasing thrust, yet reducing noise, by large increases in jet cross-section accompanied by reductions in V.

The turbo-fan engine has been highly successful in achieving this objective. Although the energy acquired from air-fuel combustion would give its products a very high velocity if they emerged as a simple jet, most of that mechanical energy can first be extracted in order to operate a turbine which, by rotating a very large fan, accelerates huge masses of air that have by-passed the combustion chamber. Values of V for both this air and the combustion products can then be kept down to moderate values, so that the ratio (proportional to V^6) of noise emitted to thrust exerted is enormously reduced: the engine is both quieter and more powerful.

Simultaneously, these developments have led to yet another augmentation of the cruise range (74). Essentially, they allow the fluid-mechanical factor $(V + U)/(2U)$ in the specific fuel consumption (79) to be diminished significantly, even under strict conditions of an upper limit on the cruise Mach number U/c_0, simply because V itself takes a reduced value.

Powered flight—over distances less than 1 km—had first been achieved by the Wright brothers in 1904. Then progress in the science of fluid mechanics, that started in the very same year, made a massive contribution to those twentieth century developments which, by allowing increases in cruise range for transport aircraft to over 15 000 km, with closely associated cost-reduction benefits, would significantly transform the character of human life on Earth.

10.4.2. Aeronautical shock waves

This section 10.4.2 outlines that 'exciting change, with the prominent appearance of shock waves' (p 813) which airflow patterns undergo when the Mach number (19), defined as the ratio of wind speed U to sound speed c_0, exceeds 1. (In aeronautics, of course, this wind speed U means the speed of the air relative to a moving aircraft.) After that change, not only do 'waves and flows interact' (section 10.3.1), but the flow consists almost entirely of waves whose propagation has been annulled (that is, brought to rest) by the flow.

For example, propagation at an angle θ to the wind of very weak sound waves at the undisturbed sound speed c_0 can be annulled by the opposing component $U \cos \theta$ of airflow if

$$U \cos \theta = c_0. \tag{81}$$

Provided that $U > c_0$, an angle θ satisfying (81) exists, and such waves can be a permanent feature of the flow [19].

For waves that are not so weak, the true signal velocity (17) replaces c_0 in equation (81), so that θ is reduced for positive u (or positive excess pressure) and increased for negative u. Moreover, if a shock wave appears, its own speed of propagation replaces c_0 in (81).

An extra source of drag is represented by the 'hidden energy loss' arising (section 10.3.1) in any shock waves [103], such lost energy needing to be restored from additional work done by thrust to overcome this component of drag. So 'aeronautical shock waves' need to be kept as weak as possible.

There is a huge contrast between the low-Mach-number flow around a blade with an 'aerofoil' section appropriate to such flow (figure 10.3) and the pattern of flow around a supersonic aerofoil (figure 10.46). The latter shape must be even thinner, and must also have a sharp leading edge, so that all disturbances to the oncoming supersonic stream remain small and the shock waves generated are weak. Moreover, the whole visible flow pattern takes the form of stationary waves propagating at angles θ satisfying either equation (81) or similar equations with modified right-hand sides [103].

Each wave arises from a disturbance to the incident airflow by the aerofoil surface. The point A where such disturbance is zero ($u = 0$) emits a wave in accordance with equation (81). But the direction of the surface at B is associated, as figure 10.46(b) shows, with propagation of a positive value of u at an increased speed $c_0 + 1.2u$ giving a reduction in θ; while the direction at C is shown in figure 10.46(c) to be compatible with $u < 0$ and an increased value of θ. The excess pressures, positive at B and negative at C, produce a resultant drag, related rather precisely [103] to the hidden energy loss in the N-waves (compare figures 10.46(d) and 10.32) generated by such non-linear propagation. But practically no waves arise in the region behind the rear shock wave, where the incident airflow is undisturbed.

At the same time, just as thin boundary layers are shown in figure 10.3 as attached to the surface, so also (to complete the comparison) boundary layers with broadly similar distributions of the velocity v are present on a supersonic aerofoil. Yet there is an astonishing difference in the temperature distribution from that discussed in section 10.2.5. At supersonic speeds, a cold wind no longer cools a body; it heats it!

This 'aerodynamic heating' phenomenon arises [103] because the energy per unit mass which is diffused in a boundary layer is no longer just the heat energy $c_p T$ but the total energy $c_p T + \frac{1}{2} v^2$ obtaining by adding on the kinetic energy. Because that total energy has the value $c_p T_0 (1 + 0.2 M^2)$ in an incident airflow of temperature T_0, the temperature T at the solid surface (where $v = 0$) tends to rise close to the value

$$T = T_0(1 + 0.2M^2). \tag{82}$$

Fluid Mechanics

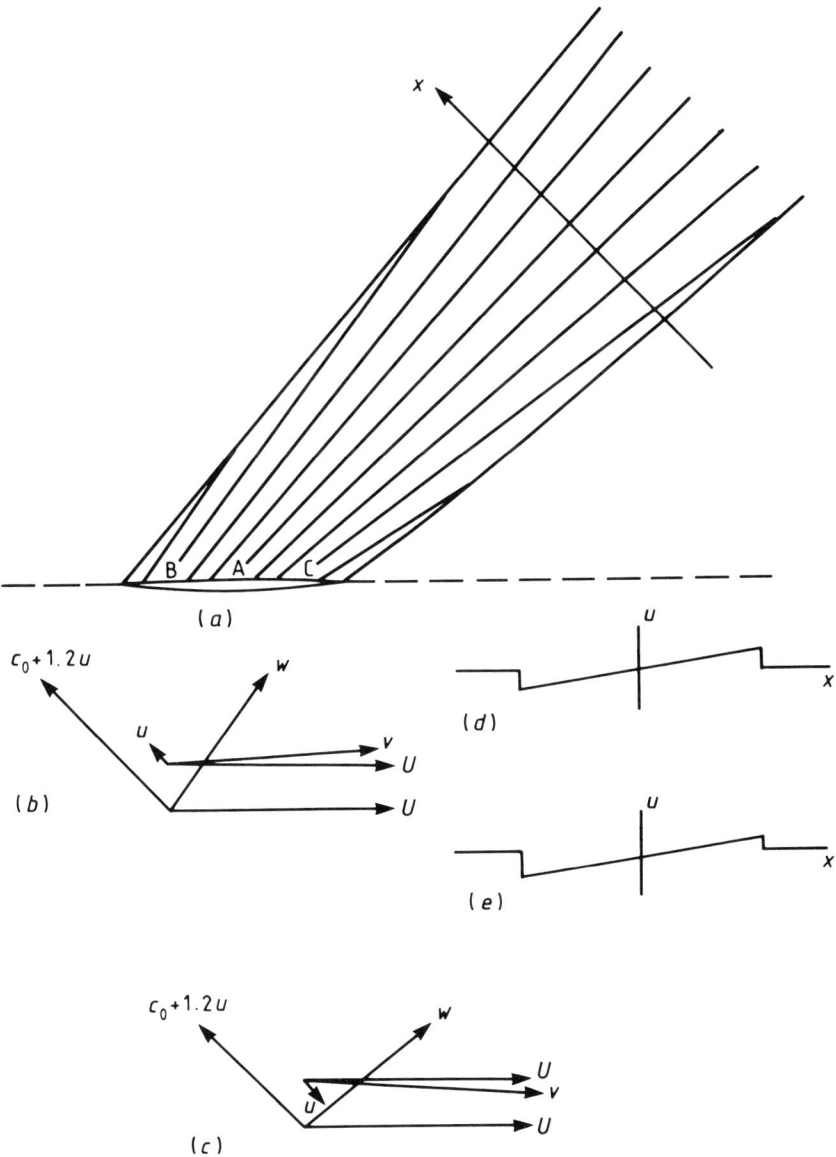

Figure 10.46. (a) Stationary waves generated in supersonic flow (symmetrical about the broken line) around a thin aerofoil. (b) At B, the wind velocity U and a velocity u carried by the waves have a resultant v that slopes upwards (along the tangent to the aerofoil surface). The stationary wave is in the direction w (resultant of U and the signal velocity). (c) At C, the negative value of u causes the resultant v to slope downwards (again along the tangent) and the stationary wave now has a reduced angle to the wind. (d) Form of the balanced N-wave at the section shown. (e) At positive angle of attack, this becomes an unbalanced N-wave, with downward momentum.

875

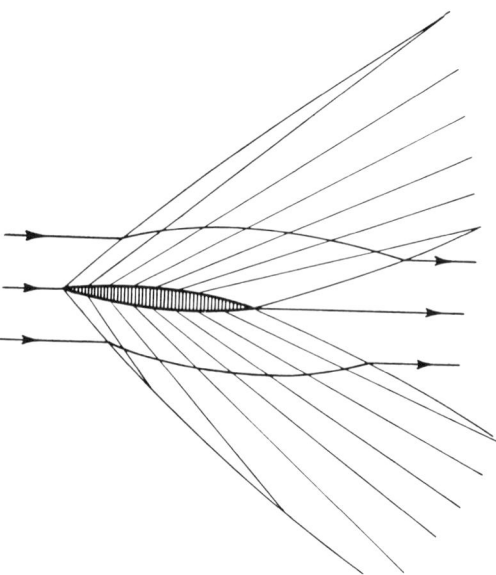

Figure 10.47. *The complete flow field* [103] *around a lifting aerofoil (actually, one twice as thick as that in figure 10.46).*

When, for example, $M = 2$ while T_0 takes a stratospheric value 210 K, this temperature (82) is 378 K, around the boiling point of water!

And one further comparison may be made: just as the symmetrical flow in figure 10.3 changes, at a positive angle of attack, into the lift-generating flow of figure 10.12(a), so also such an angle of attack converts the symmetrical flow of figure 10.46(a) into a flow that gives lift; briefly, because each value of u is decreased on the upper surface while values on the lower surface are increased, so that all pressures on the lower surface exceed the corresponding upper-surface pressures. And the aerofoil exerts once more an equal and opposite force, which confers downward momentum on the fluid. Yet this momentum appears, not in any vortex-wake region behind the aerofoil, but almost exclusively in the flow between the shock waves!—where the balanced N-waves of figure 10.46(d) become unbalanced, as figure 10.46(e) shows, and include [103] a downward component of momentum (see also figure 10.47).

Nevertheless, designers of supersonic aircraft in the twentieth century were increasingly drawn away from wings with such aerofoil sections by hard facts about aeronautical shock waves. The aim of keeping these weak so as to reduce shock-wave drag (and, often, shock-wave noise: the 'supersonic boom' heard at ground level) is achieved best, not by shapes that are thin in just one dimension (figure 10.46), but by shapes that are slender in both dimensions perpendicular to the flow [103]. Accordingly, the rather simple concepts of supersonic aerofoil theory (figure 10.46) required careful extension to involve three-dimensional wave propagation—using ideas outlined in section 10.3.1 above—if essential features of the 'slender-body aerodynamics' needed for $M > 1$ were to be usefully exploited [104, 72].

A good first suggestion of what may be called for is given by the analysis (p 848) of shock waves produced in air by an 'exploding wire'. Here, the source of sound is the appearance in the fluid of foreign matter, the outward-pushing vapour, whose cross-sectional area S increases very suddenly with time; generating, on a linear analysis, air motions (45), which are converted by non-linear effects into an N-wave.

Flight at supersonic speeds through a mass of air similarly introduces foreign matter (the aircraft), whose overall cross-section S in contact with that air mass increases very suddenly with time—though it then falls abruptly to zero. For a slender shape at supersonic speeds, this suggests the possible importance of a function $S(x)$, describing variation in the wind-direction x of the body's cross-sectional area S normal to that direction.

Slender-body aerodynamics showed how shapes where the function $S(x)$ varies rather smoothly and gradually along the aircraft's length have two advantages: (i) shock-wave strengths are kept relatively low; and (ii) they are quite well estimated [71], as for an exploding wire, in a two-stage analysis—linear, and then non-linear—with source strengths depending on the total area $S(x)$ of each cross-section rather than on its detailed shape. Of course the waves spread, not cylindrically, but conically at the angle θ defined by equation (81); as the linear part of the analysis makes clear already because equation (81) represents a 'stationary phase' condition.

This initial, linear analysis can be further improved [105] if the non-directional simple-source radiation is modified by adding a distribution of downward-pointing dipoles of strength $L(x)$, representing (section 10.3.2) forces on the fluid equal and opposite to the lift forces on aircraft cross-sections. This improvement, after the necessary non-linear adaptation, produces modest increases of N-wave strength below, and decreases above, the aircraft.

During the twentieth century, a significant first step in extending air transportation to supersonic speeds was made with Concorde [106], which offers busy passengers the benefit of journey times of only 3 hours over ranges of around 6000 km. Its cruise Mach number $M = 2$, attained at stratospheric altitudes between 15 and 16 km, limits aerodynamic heating to temperatures (82) at which aircraft materials retain safe properties; while its external shape (figure 10.48), with a sharp-edged slender delta wing, is a remarkable compromise between aerodynamic requirements in the supersonic and the low-Mach-number parts of its flight.

Thus, in spite of substantial shock-wave drag, it achieves a supersonic

Figure 10.48. *An Aérospatiale/British Aerospace Concorde with Rolls-Royce/Snecma Olympus 593 turbojet engines, and with all-up weight 185 tonnes, cruise Mach number 2.0 and maximum range 6500 km.*

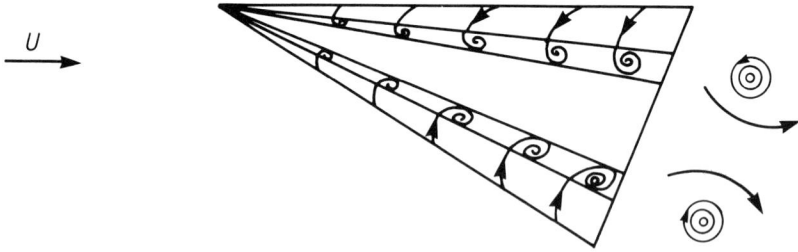

Figure 10.49. *Schematic illustration of lift generation by a sharp-edged delta wing at low Mach number. At a moderate angle of attack, a delta wing like that of Concorde sheds vorticity from the sharp leading edges—which rolls up into a vortex pair similar to that shed from the trailing edge in figure 10.13(b). Creation of downward momentum in the flow associated with this vortex pair balances the weight of the aircraft.*

lift–drag ratio around 10, and the shock waves produced below it are found, after propagation into the much denser air at ground level, to involve pressure jumps of only about a millibar. Yet the most innovative feature of the Concorde wing design is its fluid-mechanical method for achieving lift stably at low Mach number: the vortex pair essential (as figure 10.13(b) suggests) to this purpose is shed, not from the trailing edge, but from the sharp leading edge (figure 10.49), generating downward momentum which supports the take-off weight of Concorde stably at angles of attack of over 20°.

Fluid Mechanics

Development of such a shape needed extensive testing of models, both in ordinary low-speed wind-tunnels (figure 10.20) and also in supersonic wind-tunnels. Actually, it had been studies by Prandtl, initiated [18] in yet another fine paper of 1904, which produced insight into how wind-tunnel testing at supersonic speeds could be achieved.

Although an airflow's *convergence*—that is, reduction in cross-sectional area, giving an increase in ρV, the mass flow per unit area—produces the expected acceleration at subsonic speeds (as in figure 10.20), it has the opposite effect at supersonic speeds. Indeed, the balance (section 10.1.4) of mass-acceleration with pressure gradient

$$\rho V \frac{dV}{ds} = -\frac{dp}{ds} \left(= -c^2 \frac{d\rho}{ds} \text{ in adiabatic motion} \right) \tag{83}$$

makes ρV change with distance s at a rate

$$\frac{d(\rho V)}{ds} = \rho \frac{dV}{ds} + V \frac{d\rho}{ds} = \rho \left(1 - \frac{V^2}{c^2}\right) \frac{dV}{ds} \tag{84}$$

implying, as first recognized in the 1880s by the great Swedish engineer Laval, that acceleration to supersonic speeds needs a convergence followed by a divergence. What else does it need?

Prandtl suggested the answer with a diagram (figure 10.50) where the solid curves show all possible adiabatic distributions of pressure in a certain 'convergent–divergent nozzle', given the upstream pressure p_0. The fainter curves, describing wholly subsonic flow, apply for downstream pressures exceeding a certain value p_1; while the bold curve, describing continuous acceleration to supersonic flow, is achieved for downstream pressure p_2. But for downstream pressures between p_2 and p_1, where no continuous solution exists, Prandtl proposed theoretically and confirmed experimentally that the pressure follows an essentially discontinuous progression from the bold curve to one of those dotted curves which describe solutions at increased entropy. This penetrating discovery indicated how, whereas supersonic motion at a point is often determined just by upstream conditions, there may nonetheless be exceptions where the value of a downstream pressure is highly influential.

Extensions of such 'one-dimensional' analyses, including extensions to unsteady flow, allowed refined design of, and effective 'starting' procedures for, supersonic wind-tunnels; as achieved by those, beginning with A Busemann [107] in Göttingen, who would build up in the twentieth century such a deep experimental knowledge of aerodynamics at $M > 1$. (See figure 10.51 for a fine schlieren photograph [108] taken in an early supersonic tunnel at Göttingen.) But here, rather than pursuing this theme further, or extending it to that so-called 'hypersonic'

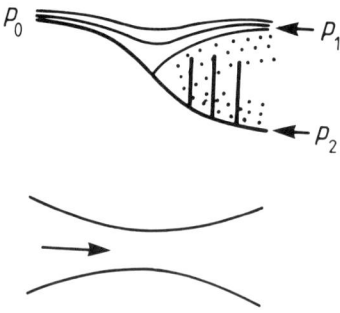

Figure 10.50. *Possible pressure distributions in a convergent–divergent nozzle. Solid lines, adiabatic distributions, of which the bold curve shows acceleration to supersonic speeds and the fainter curves describe wholly subsonic flow. Dotted lines, distributions with higher entropy, towards which the pressure rises after the discontinuities indicated.*

Figure 10.51. *Schlieren photograph of flow around a supersonic aerofoil published in 1927 by J Ackeret, another pioneer (then at Göttingen, but later in Zürich) of supersonic aerodynamics.*

aerodynamics [109] of spacecraft re-entry (with M as high as 20) in which shock-wave drag may actually be a friend—but the enemy is aerodynamic heating—this section 10.4.2 is concluded with comments on transonic flow: the transition régime between subsonic and supersonic aerodynamics [110].

Emphasis was placed in section 10.4.1 on the importance for flight efficiency of achieving high subsonic values of cruise Mach number without incurring shock-wave drag penalties. To help appreciate how shock waves may arise, the contrast between figure 10.46 at $M > 1$ and figure 10.3 at low Mach number may here be 'interpolated' with a diagram (figure 10.52) of symmetrical flow over an aerofoil at high subsonic Mach number.

Here the acceleration of air over the aerofoil surface generates a local region of supersonic flow. On the other hand, as in Prandtl's nozzle

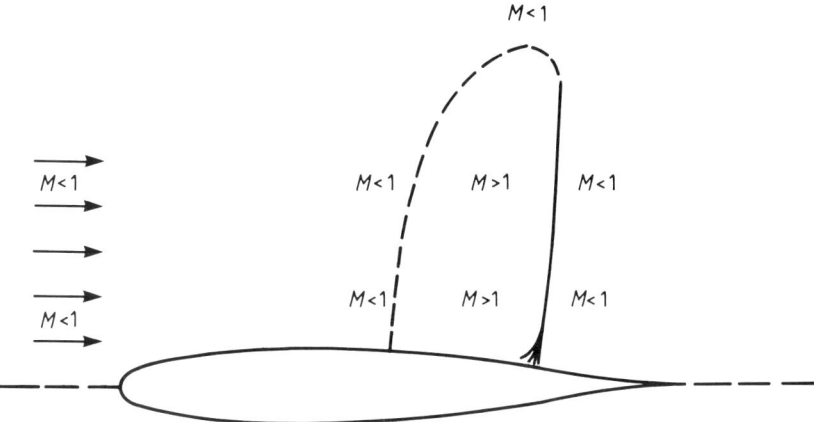

Figure 10.52. *Illustrating a flow (symmetrical about the broken line) around an aerofoil at high subsonic Mach number, when there appears in the flow a limited supersonic region terminated by a 'discontinuity' that incorporates a shock wave/boundary-layer interaction.*

example (figure 10.50), this flow can adjust to the level of downstream pressures only through a 'discontinuity'. Moreover, it is recognized now—in both cases—that this discontinuity involves not just a simple shock wave but a complex 'shock wave/boundary-layer interaction', including upstream influence [111] of the shock wave on the boundary layer (that is, an influence extending much farther upstream that might have been expected). And it is noteworthy that, here, the standard analysis (section 10.1.3) of an air flow into two parts (an exterior flow and a boundary layer) requires modification; often, by means of a 'triple deck' analysis in which the boundary layer is further subdivided, with a very thin inner part of the boundary layer near the wall responding to the external flow with particular sensitivity [112, 113, 114].

In the design of transport aircraft it is above all such shock-wave effects—being harmful both to lift–drag ratio and in other ways—which need to be avoided by ideas such as sweepback (figure 10.45). Other concepts valuable in design for transonic speeds include the celebrated 'area rule' of Whitcomb [115]. Briefly, an aircraft for which $S(x)$, the distribution of cross-sectional area defined above, varies smoothly and gradually with x will experience aerodynamic forces which, to an even greater extent at transonic than at supersonic speeds, are determined almost entirely by this distribution. (For such a determination, however, the earlier two-stage analysis—first linear and then non-linear—needs at transonic speeds to be replaced by a single-stage, fully non-linear analysis.)

But, beyond any such conceptual aids to aerodynamic design, it is above all on advanced computational fluid dynamics (CFD) that aircraft companies rely for detailed analysis of the expected performance of an aircraft configuration in complicated conditions like those characteristic of transonic flight. Modern CFD programs [116] give good representations of the steady flow outside the boundary layer around intricate aircraft shapes in three dimensions; also, where continuous flow is impossible, these programs identify the location and strength of any resulting shock wave—although without describing its interaction with the boundary layer. The data obtained, taken together with experimental knowledge of real interactions of external flows with boundary layers, are found invaluable for all aspects of design, including those directed (section 10.4.1) towards the improvement of flight efficiency.

10.4.3. Fast ships and secure offshore structures
Alongside the revolution in long-distance transport of people and of high-value goods that emerged from aeronautical engineering, some significant transformations of bulk carriage by sea, and of other fields within ocean engineering, were taking place. In section 10.4.3, selected contributions from fluid mechanics to these developments are outlined.

The surface of the deep (section 10.3.4) offers evident advantages for carriage of heavy goods, solid or liquid, in bulk-carriers or tankers whose overall weights can be in hydrostatic balance with the weight of water displaced. However, that surface's ability to propagate waves creates some counterbalancing disadvantages, including both speed and weather limitations.

At sea, relatively fast travel can once again yield benefits—above all, ships and crew complete more journeys each year—but, just as a limit on the speed of economical air transport is placed by the onset of extra drag due to shock-wave generation, so also some limitations on speeds of economical marine transport are placed by the onset of significant drag due to surface-wave generation. Here, methods for pushing back this onset to higher speeds (analogous to the use of sweepback in aeronautics) are sketched. On the other hand modern 'ship routing' methods for minimizing delays from stormy weather depend on advanced forecasting techniques and are postponed to section 10.5.2.

Away from any storms, hydrodynamic forces on ships can be estimated by linearly combining any forces [117] due to ambient ocean waves (especially, swell from far-off storms) with forces which the ship's steady motion at speed U would produce in a calm sea. But such calm-sea forces, on which we concentrate, may be influenced by the ship's own power to generate waves.

Waves so generated remain stationary relative to the ship, so that if they propagate at an angle θ to the ship's motion with wave velocity c they satisfy the equation

$$U \cos \theta = c \tag{85}$$

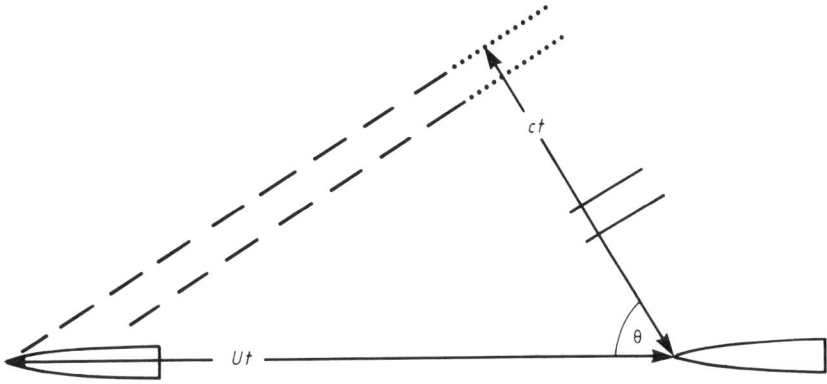

Figure 10.53. *Waves generated by a ship moving at speed U are propagated at various angles θ, but each has velocity c = U cos θ. If they travelled a distance ct during time t, while the ship moves a distance Ut, then by simple trigonometry the crests (dotted) would be on (broken) lines through the new position of the ship. Actually, the energy in the (plain-line) crests has travelled half-way towards these lines at the group velocity $c_g = \frac{1}{2}c$. These arguments, including consideration of all angles θ, lead to the Kelvin ship-wave pattern of figure 10.54 below.*

analogous to (81) for sound waves. With the linear-theory value (55) for c, equation (85) gives the wavelength as

$$\lambda = \frac{2\pi U^2}{g} \cos^2 \theta. \qquad (86)$$

An enormous difference from sound waves arises, however, from the fact that the energy in surface waves travels, not at the wave velocity c with which crests move, but at the group velocity $c_g = \frac{1}{2}c$. Figure 10.53 points this contrast: during time t, while a ship moves a distance Ut, the waves generated travel a distance $c_g t$. If c_g took the value $c = U \cos \theta$ then any wave generated would lie on a line through the ship's present position (as with sound waves in figure 10.46). But waves with $c_g = \frac{1}{2}c$ have only travelled half as far. Trigonometry then shows [70] that—on these linear-theory arguments, first offered in the 1880s by Lord Kelvin [118]—the waves all lie within a 'wedge' of semi-angle

$$\sin^{-1}(1/3) = 19.5° \qquad (87)$$

and take forms sketched in figure 10.54. But we shall see that (i) for a given ship at a given speed, only part of the Kelvin ship-wave pattern is observed, while (ii) refinements of the analysis (including non-linear considerations) produce some broadening of the wedge.

Because the greatest possible wavelength (86) for ship-generated waves is $2\pi U^2/g$ (attained in figure 10.54 by those longer waves which

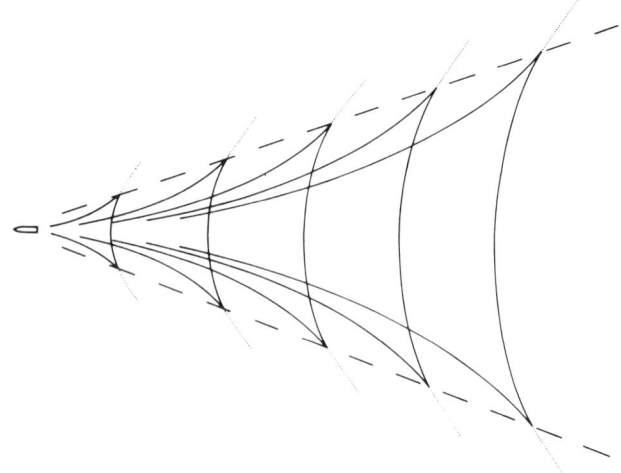

Figure 10.54. *Plain lines, ship-wave pattern suggested by Kelvin [118] from simple group-velocity arguments. Dotted lines, extensions beyond the broken-line 'caustic' derived from exact linear-theory calculations [70]. (Note: non-linear effects produce a further slight broadening of the Kelvin wedge.)*

are propagating at small angles θ), the classical approach to minimizing the energy in such waves (so that wavemaking drag is kept very small) is to ensure that the ship's length, and smoothness of shape along it, are both so great that relatively short waves, with $\lambda \leqslant 2\pi U^2/g$, simply cannot be excited. But the fine Japanese naval architect, T Inui, demonstrated in 1962 an important improvement [119] on this classical approach, which may here be briefly explained [70].

At low enough speeds U, a ship experiences only frictional drag because it generates no waves. Then the flat free surface behaves like a reflection plane for the physicist's 'method of images'; and, indeed, a common experimental approach to estimating frictional drag for a ship design of given shape is to build a 'double model'—a model of the immersed part of the hull together with its reflection in the free surface—and to test this model fully immersed (with the boundary layer made turbulent to simulate full-scale Reynolds number) in a water tunnel. Frictional drags so estimated can later be subtracted from drags measured in flows with a free surface to indicate values of wavemaking drag.

In such a double-model experiment, the half of the flow below the plane of symmetry (which, by symmetry, remains flat) represents the flow below the real ocean surface. However, it fails in just one respect to represent it with full accuracy: the distribution of pressure varies in that plane, instead of taking a constant value (the atmospheric pressure) as it should on a flat surface. Approximately, any waves generated are those

Fluid Mechanics

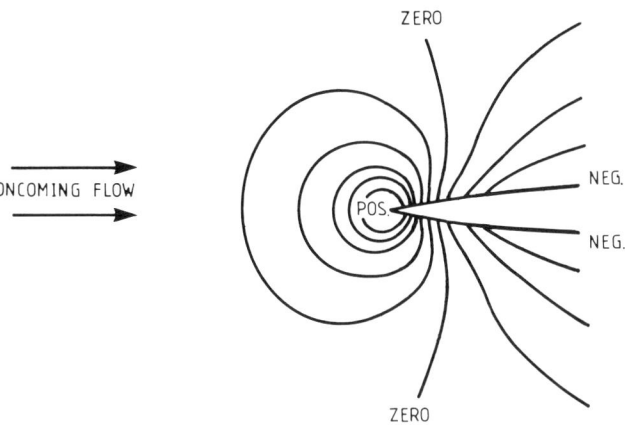

Figure 10.55. *For ship hulls with conventional bows, a typical distribution of excess pressure measured (or computed) in the plane of symmetry of a 'double model' (immersed part of hull plus its reflection in that plane) is as shown, with spectral elements of highest wavenumber arising from the steep positive pressure peak just in front of the bow.*

which this distribution of pressure, moving at speed U, would produce if applied to the free surface; and the value of U at which such waves become important is that for which the highest significant wavenumbers in the spectrum of this distribution with respect to x are around g/U^2 (corresponding to $\lambda = 2\pi U^2/g$).

Both experimental and computational studies of the pressures in the plane of symmetry confirm [70] that these highest wavenumbers are associated with a steep positive pressure peak just ahead of the ship's bow (figure 10.55), related by Bernoulli's equation (9) to a sharp retardation of flow in this region. They might be eliminated, then, by an added hull element which tends to produce a cancelling peak of negative pressure in the same region.

Such an element is the subsurface 'bulbous bow' of Inui [119], projecting ahead of the ship's visible bow. In the corresponding double model, with two bulbous projections, convergence of the flow between them produces in the plane of symmetry a local velocity peak related, by equation (9), to a local negative-pressure peak which can suppress the former positive peak and postpone the former drag rise to higher values of U.

The economics of transport by modern tankers and bulk-carriers benefited, then, from the introduction of (i) longer ships with (ii) subsurface bulbous bows. And thick boundary layers near the stern of such longer ships confer an intriguing fluid-mechanical advantage, because much of the water entering the propeller disks is water from the boundary layers. Briefly, since its velocity relative to the ship takes a reduced value $v < U$, its momentum gain (contribution to thrust) per unit mass when kinetic energy E is added becomes

$$(2E + v^2)^{1/2} - v \tag{88}$$

which is substantially augmented as v is reduced.

Like aeronautical engineers, naval architects combine the use of conceptual aids to design with model testing and progressively refined CFD techniques for analysing flows around ship hulls [120]. These are now able to apply at the free surface, not a linearized, but an accurate boundary condition (with some associated widening of the Kelvin wedge), and have led to further valuable improvements in hull design.

The second half of the twentieth century saw also two useful innovations in the fast marine transport of middle-sized loads over fairly short ranges [121]. Both of these achieve major reductions in frictional drag by lifting the main part of the hull out of the water.

The first is the hovercraft, with an annular jet used to allow creation of a cushion of air at positive pressure which supports the weight of the hull. (Any wave drag is then generated literally, rather than just conceptually as above, by such a distribution of pressure on the water surface, moving at speed U.)

The second is the hydrofoil boat, where a wing-shaped element under the water generates, after the design speed is reached, the main lifting force that holds the hull out of the water. (Then the most important resistance is the wing drag, including the drag due to lift.)

And yet another area of ocean engineering that gained special importance in the twentieth century was the engineering of secure offshore structures, mainly for oil and gas exploration and exploitation. Essentially, this has been a field where applications of fluid mechanics could no longer benefit from d'Alembert's Theorem. Thus, whereas shapes of aircraft or ships can be 'streamlined' (section 10.1.4) so that drag forces are low when the vehicle moves ahead in its designed direction of travel, an offshore structure has no such predetermined direction for the relative motions of water and structure. Essentially, it must withstand the onset of waves from all directions.

These considerations led to the widespread use of structural members with no preferred direction, including (especially) circular cylinders [122]. Some complexities in the interaction of a circular cylinder with just a steady oncoming flow were illustrated already in figure 10.1, but offshore engineering has demanded from fluid mechanics a detailed understanding of the cylinder's far more complicated interaction with very unsteady flow fields in ocean waves. Above all, hydrodynamic loads on the cylinder are associated, of course, with changes of momentum in the fluid flow as affected by such interaction.

Here, two aspects of these changes are emphasized. First, any vorticity that is shed as a result of flow separation in one phase of an oscillatory flow may be convected back over the cylinder in later phases. Deep study of the resulting vorticity distributions led to an impressive level of knowledge of how the associated momentum ($\frac{1}{2}\rho$ times the moment of the vorticity distribution) would vary, with some vital consequences for hydrodynamic forces on the cylinder [123].

But the part of the flow that is without vorticity has its own additional momentum. This is, of course, a potential flow—with a momentum that shows no change with time (corresponding to d'Alembert's Theorem that the associated force is zero) when the oncoming flow is steady. Yet when the cylinder, of radius, a, is placed in a flow whose velocity U varies with time, the potential-flow part of the interaction has a varying momentum associated with a force whose value

$$2\pi a^2 \rho \frac{dU}{dt} \tag{89}$$

was already well established in the nineteenth century.

In 1928, however, G I Taylor foresaw [124] the need to calculate such potential-flow forces when a body is placed in an ambient fluid motion that varies in space as well as in time. He analysed them by means of a profound application of general Hamiltonian dynamics, and also verified the results directly with a more direct, yet far more laborious, calculation of pressure distributions. For example, in the particular case of a circular cylinder, with axis in the z-direction, placed in such a variable flow field with U and V as its x- and y-components, the x-component of the potential-flow force turns out to be

$$2\pi a^2 \rho \left[\frac{dU}{dt} + \frac{1}{2} \frac{d}{dx} (U^2 + V^2) \right]. \tag{90}$$

For many decades, G I Taylor's result (90) was used only rarely. In offshore-engineering applications, moreover, there was almost universal loyalty to the classical expression (89) for the potential-flow force. By the mid-1980s, however it became recognized that the quadratic correction in equation (90) could be important [125]; e.g. for the excitation of resonant modes of motion of structures (such as tension-leg platforms) with a very low natural frequency—which quadratic forces allow to be excited as a 'difference tone' by a spectrum of waves of considerably higher frequencies.

Interactions between offshore engineering and fluid mechanics cover an enormously wide field. Here, we have touched on just a very small sample of that field, chosen above all to illustrate differences in its character from other areas of application of fluid mechanics.

10.5. Dynamics of the Earth's fluid envelope, and its forecasting applications

10.5.1. Wave-like current patterns

The Earth's fluid envelope is a mixture of air and water. In the atmosphere air predominates, yet water (transferred from the ocean) plays a spectacular role. In oceans, rivers and lakes water predominates, yet they teem with life because of dissolved oxygen and CO_2. Not only mass exchange between the different components but also transfers of heat and of momentum are major influences on their dynamics.

Study of that dynamics, besides its scientific interest, has crucially important applications to meeting the requirements of countless industries—agriculture, shipping, aviation, energy-supply, coastal engineering, river-management, etc—for improved weather forecasting; see section 10.5.2, where some account of climate-change forecasting is also given. Practically none of the dynamical or forecasting problems involve only one component of the Earth's fluid envelope in isolation from others; nevertheless, this section 10.5.1 is devoted to certain features of ocean and river dynamics which, while influenced by atmospheric input, assume something of a life of their own because patterns of water flow propagate in a wave-like manner.

Forecasting was first achieved successfully for ocean tides [126]. Superficially, these seem to be periodic vertical motions that alter sea level; yet practically all their kinetic energy is in the horizontal motions: those powerful tidal currents which make an exciting spectacle off many coastlines. The pattern of these currents propagates in a wave-like manner, similar in all ways but one (see below) to the shallow-water propagation described in section 10.3.3.

Deep oceans masquerade, then, as shallow water for propagation of tidal currents. Indeed the ratio $\frac{1}{3}k^2h_0^2$ between kinetic-energy contributions from vertical and horizontal motions is so small that it annuls the dispersion effect (63): at frequency $\omega = kc_0$, the value (57) for c_0 gives

$$kh_0 = \frac{\omega h_0}{c_0} = \omega \left(\frac{h_0}{g}\right)^{1/2} \tag{91}$$

which for the tidal component of lowest period (half a day) is less than 0.005 even in the ocean's deepest parts, making $\frac{1}{3}k^2h_0^2$ less than 10^{-5}.

Newton had recognized that gravitational theory gave the Moon a special influence on tides: relative to the motions of the solid Earth in the gravity fields of Sun and Moon, water nearest to each body would be subject to an excess attraction while water farthest from it would experience a defect; in other words, a relative repulsion. For each body, then, forces tending to raise sea level would reach maxima (figure 10.56) at the rotating Earth's nearest and farthest points—whose positions change with a period close to half a day. But, although the Moon was far less massive than the Sun, its proximity would yield bigger values of those gradients of gravitational force with distance which produce tide-raising forces. In addition, once a fortnight at full or new Moon, the Moon's tide-raising force would be augmented by an almost collinear solar force, giving 'spring' tides (with perfect collinearity at an equinox); while neap tides, with the forces orthogonal, would occur in between.

Fluid Mechanics

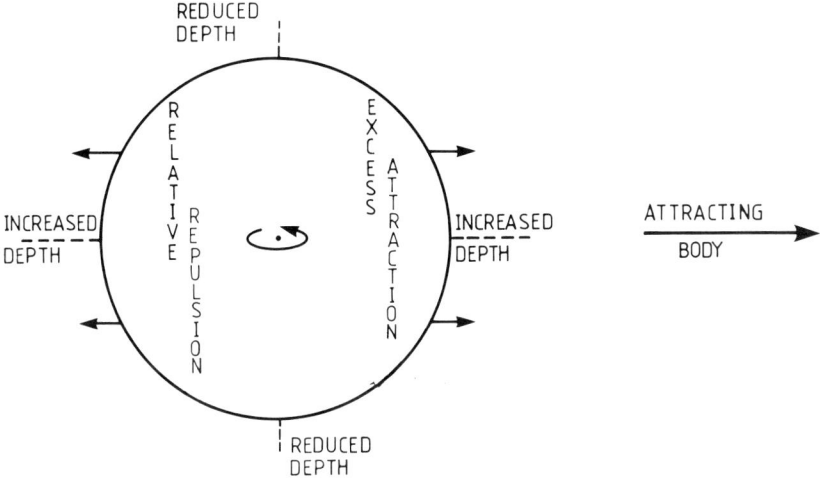

Figure 10.56. *Gravitational forces tending to raise sea level (short arrows) reach maxima at points nearest to or farthest from an attracting body. Newton argued that the resulting ocean depth distribution might be as suggested by the broken lines. (Also, see note (ii): the horizontal components of the forces play a leading role.)*

Fluid-mechanical analysis, however, in the hands of Laplace [127] and his successors, gave a picture differing from that of Newton in three important respects:

(i) it allowed for the wave-like character of the pattern of ocean-current response to tide-raising forces; of which, furthermore,

(ii) only the horizontal components could energize tidal currents (whose convergence, then, would raise sea level); while moreover

(iii) the Earth's rotation, besides determining periods for different tide-raising forces, had also an important dynamic effect.

This effect, first identified for the tides, is important too for most oceanic or atmospheric motions [93]. Vorticity was defined in section 10.2.2 so that a fluid simply rotating with angular velocity Ω has vorticity 2Ω. Accordingly, just the Earth's rotation with axial angular velocity 2π per day gives all terrestrial fluids an axial vorticity of 4π per day. But analysis of essentially horizontal tidal currents requires only a knowledge of the vorticity's vertical component, equal at latitude θ to

$$f = \frac{4\pi \sin\theta}{\text{day}} = \frac{4\pi \sin\theta}{24(3600\,\text{s})} = (1.45 \times 10^{-4} \sin\theta)\,\text{s}^{-1}. \tag{92}$$

The sense of this 'planetary' vorticity is cyclonic: anticlockwise in the northern hemisphere ($\theta > 0$) and clockwise in the southern ($\theta < 0$).

Its importance for tides is that when these raise ocean depth from an undisturbed value h_0 to a new value h, such stretching (see figure 10.9) of a vertical column of water stretches the vertical vorticity component (92) to a new value fh/h_0; of which the part

$$f\left(\frac{h}{h_0} - 1\right) \tag{93}$$

must correspond to effects other than Earth's rotation. Expression (93) represents, then, the vorticity ς of the tidal currents themselves (motions relative to the Earth).

Simple shallow-water theory is shown in equation (56) to be a direct analogue of the theory of sound. Two-dimensional propagation over shallow water whose undisturbed depth h_0 varies with position would, similarly, equate with two-dimensional acoustic propagation in a medium with non-uniform undisturbed sound speed c_0 if, as in sound theory, potential flow (with zero vorticity) could be assumed. However, tidal currents—although propagating effectively over shallow water—are not potential flows since they have a vorticity (93), where h/h_0 represents the depth increase due to convergence of the same currents. This difference is found, very briefly, to make the waves dispersive after all!—with a local dispersion relationship

$$\omega^2 = f^2 + (gh_0)k^2 \tag{94}$$

of frequency ω to wavenumber k for small disturbances (corresponding to wave velocity $\omega/k = (gh_0)^{1/2}$ if f is absent).

All analyses of tides, and all interpretations of tidal observations, start [126] from Newton's elucidation of tide-raising forces and spectrally analyse these. The largest component is known as M_2, the Moon's 'semi-diurnal' influence; actually, with period 12 h 25 min, greater than half a day by 1/28 because the Moon's orbital motion with period 28 days is in the same sense as Earth's rotation. The next largest is S_2, that part of the solar tide-raising force which is exactly semi-diurnal (with period 12 h). There are very many other components, all with considerably longer periods, which need to be allowed for in constructing tide tables; however, it is noteworthy that the basic cycle of spring and neap tides derives already from the beats between M_2 and S_2.

These, moreover, are the two components which allow equation (94) to determine a real wavenumber k for, essentially, all latitudes θ. Thus S_2 has frequency 4π per day, equal to a polar value of f, while M_2 has an only slightly lower frequency, equal to the value (92) of f for $\theta = 75°$. In both cases, the equation simply indicates that k becomes small (that is, tidal wavelengths become large) at relatively high latitudes (figure 10.57). By contrast, all spectral components with lower frequency are faced with a considerably lower maximum latitude for real k; which acts

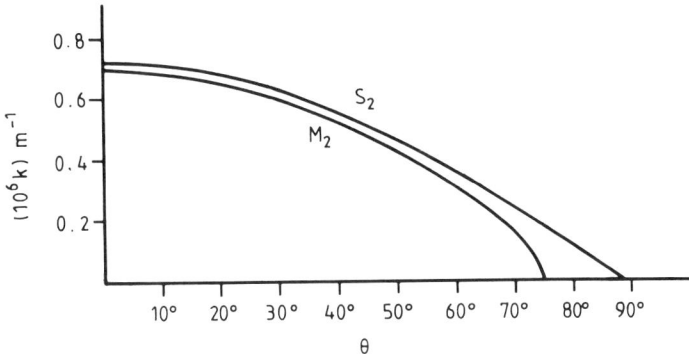

Figure 10.57. *Wavenumber k at latitude θ as indicated for the M_2 and S_2 tides by equation (94), with h_0 taken as 4 km (the oceans' mean depth).*

as a caustic bounding any waves (there is, of course, the usual tailing-off of amplitude beyond the caustic).

On any continental shelf, concentration of the tidal kinetic energy within water of much diminished depth may augment local currents, with two results: (i) dissipation of energy by bottom friction and (ii) various non-linear distortions of waveforms, some of which (section 10.3.3) may involve a 'hidden' energy loss. What is the energy source that compensates for any dissipation?

The energy source for tides is the Earth's rotation itself, which stores energy rather as a flywheel does in some machines. If the Earth rotated so slowly as to turn always the same face to the Moon, there would be no tidal movements at all but just a static elevation of water (as in figure 10.56) at points nearest to or farthest from the Moon. Tides are due to excess rotation (the day being less than the Moon's orbital period) and derive their energy from it. Thus, dissipation of that energy reduces the excess; in short, 'tidal friction lengthens the day'. By 1919 already [128], G I Taylor had successfully verified this statement through a comprehensive comparison of the rate at which Earth's rotation is becoming slower with an estimated rate of dissipation of tidal energy in shallow seas. (For the Moon's own rotation, of course, any analogous excess was long ago annulled by dissipation of tidal motions in the solid Moon itself.) Ocean tides, then, being rather lightly damped forced oscillations, take broadly the form of standing waves, with local wavenumbers given by equation (94); yet, because dissipation is localized in shallow shelf areas, nearby tidal-current patterns assume rather more of a progressive-wave character, with energy flowing towards where it is being dissipated.

Historically, tide gauges have long been abundant in those areas of relatively shallower water where tidal forecasting was needed, with a spectral analysis of their records having allowed the construction of reliable tide tables. But it was only in the second half of the twentieth century [129] that pan-oceanic pictures of tidal current patterns became achievable by numerical analysis. This uses 'shallow-water theory' (see above) with the correct distribution (93) of vorticity, with known ocean-depth distributions, and with continental-shelf boundary conditions that allow for realistic values of energy dissipation.

Figure 10.58 illustrates a pioneering computation of this type for the M_2 tide in the Atlantic Ocean [130], obtained by C L Pekeris in 1969. (It represents part of a computation for the Earth's oceans taken as a whole.) Here, wavelengths follow approximately the trend shown in figure 10.57, while the importance of vorticity is shown especially at 'amphidromic' points around which the phase of the standing wave rotates. This diagram, which is in rather good local agreement with coastal and island measurements, offers a remarkable contrast to the simple Newtonian picture of figure 10.56.

An excess rise of sea level, above that associated with the Moon's and the Sun's tide-raising forces, may occur in stormy weather; which produces an additional horizontal force on the ocean associated with momentum transfer from winds. Such an excess rise, called 'storm surge', is numerically predicted from wind forecasts (section 10.5.2) using shallow-water approximations as above [131]; with some severe coastal inundations being experienced when high tide and peak storm surge coincide.

Nevertheless, much more persistent ocean-current patterns are generated, not by sudden storms, but by the prolonged action of winds at the surface. In fact, most of the circulation of ocean waters is due to this cause [93] (the main exception being the 'thermohaline' part, driven by the sinking of water with high salt concentration generated [126] by ice formation in polar oceans).

For such forcing by slowly varying winds, another mechanism for vorticity change becomes important besides that effect of stretching of a vertical column of fluid which underlies equation (93) for the vertical vorticity ς of the current pattern itself. Even if there were no stretching, nonetheless movement of vortex lines with the fluid (figure 10.9) would imply conservation of the total vertical vorticity $f + \varsigma$ in any northward motion at velocity v; yet such motion makes the value (92) for the planetary vorticity f increase at a rate

$$\beta v \quad \text{where } \beta = (2.3 \times 10^{-11} \cos\theta)\,\text{s}^{-1}\text{m}^{-1} \qquad (95)$$

implying an equal and opposite rate of change, $(-\beta v)$, for ς. In 1940 the gifted oceanographer C G Rossby [132] saw the importance of wave-like current patterns governed by this relationship $d\varsigma/dt = -\beta v$ (making the rate of change of the curl of the current pattern proportional to one component thereof).

Fluid Mechanics

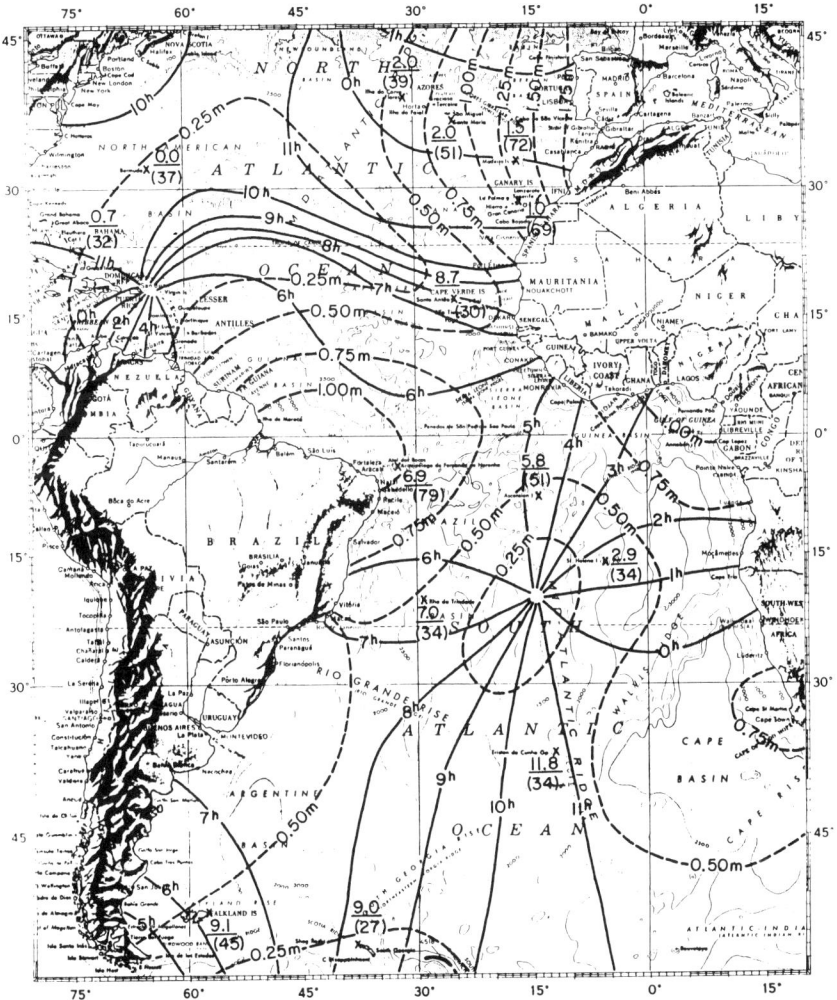

Figure 10.58. *An early computation* [130] *of the Atlantic's* M_2 *tide. The computed cotidal lines (solid) give phase lag of high tide in hours after lunar transit at Greenwich, while corange lines (broken) give tidal range in metres. Underlined figures give observed phase lags, and bracketed figures observed ranges in centimetres.*

The dispersion relationship for such 'Rossby waves' is

$$\omega k^2 = -\beta k_1 \tag{96}$$

where k_1 is the eastward component of the wavenumber vector k. If its northward component is k_2, then the curves of constant ω in a wavenumber plane with k_1 and k_2 as coordinates are circles (figure 10.59). But the group velocity (p 864) is the gradient of ω in this plane, which is directed radially inwards towards the centre of each circle [93].

893

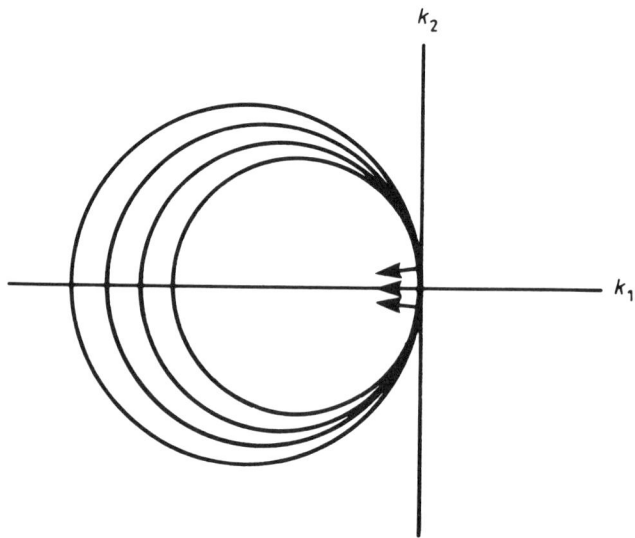

Figure 10.59. *The dispersion relationship (96) for Rossby waves describes circles in the wavenumber plane (k_1, k_2) whose radius $\beta/2\omega$ becomes less as ω increases, so that the gradient of ω points inwards towards the centre of each circle. Arrows in the direction of the group velocity refer to the case of waves with $k_2 \gg k_1$ produced by zonal forcing.*

To give just one illustrative example, the contrast between westerly mean surface winds in middle latitudes and easterly trade winds at relatively lower latitudes tends on average to generate a broad zonal pattern of oceanic vorticity that is anticyclonic (ς of opposite sign to f). Such zonal forcing has k_2 much greater than k_1, giving a westward group velocity (figure 10.59). So all the energy of the resulting current patterns travels westward (along-crest energy propagation, as in section 10.3.5), to become concentrated in a poleward current on the western boundary (the Gulf Stream in the North Atlantic, the Kuroshio current in the North Pacific, the Agulhas current in the South Indian Ocean, the East Australian current in the South Pacific, etc). Figure 10.60 shows schematically (i) the mechanism of westward travel of such a zonally distributed signal and (ii) its tendency to generate a thin boundary current [133]. This has been successfully analysed by boundary-layer methods (allowing [93] for non-linear inertial effects and effects of momentum transport across the layer) similar to those of section 10.1.3; furthermore it may separate like other boundary layers—as, for example, the Gulf Stream separates from the North American coast at Cape Hatteras, becoming a somewhat meandering jet thereafter as it makes it way across the Atlantic.

Actually, Rossby's analysis [132] took account of vorticity changes of both types. If the stretching effect (93) as well as the effect (95) of

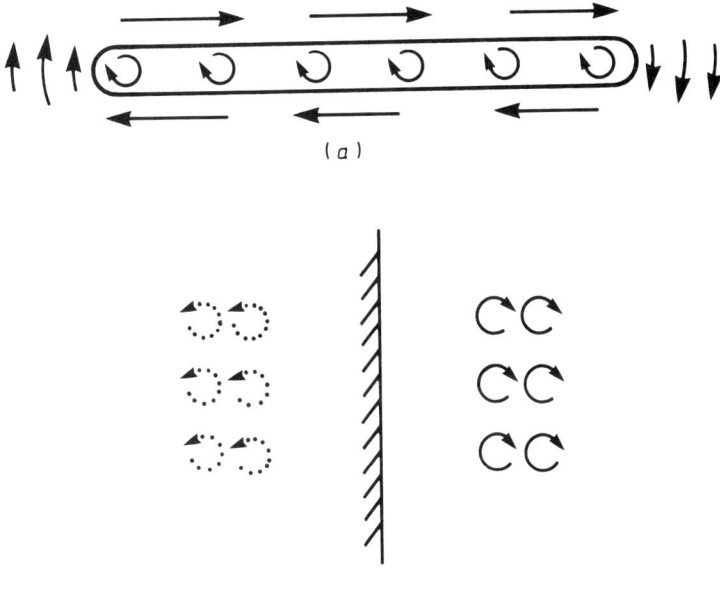

Figure 10.60. (a) Schematic view of the mechanism for westward travel of a disturbance with nearly zonal forcing (northern-hemisphere case). Negative ς induces northward flow to the west, and so generates negative ς there by equation (95). (b) At any western boundary a westward travelling pattern of vorticity, along with its image system (broken lines), generates a northward boundary current.

'advection' by any northward motion v is allowed for, then the total vorticity $f + \varsigma$, instead of being unchanged, must vary in proportion to the depth h. Rossby gave the quantity

$$\frac{f + \varsigma}{h} \tag{97}$$

which remains constant in any advective motion, the imaginative name 'potential vorticity', which was to prove highly influential on both oceanographers and meteorologists (section 10.5.2). The associated dispersion relationship includes the whole of (94)—except that the left-hand side (negligible at very low frequencies) is suppressed—together with (96) to give

$$\omega \left(\frac{f^2}{gh_0} + k^2 \right) = -\beta k_1. \tag{98}$$

With use of (98), the geometry of figure 10.59 is a little modified, yet conclusions on western boundary currents in mid-latitude oceans remain essentially unchanged.

But an important difference between current patterns generated by tide-raising forces distributed uniformly with depth and by wind stresses concentrated at the ocean surface still needed to be addressed. Below a well-mixed layer near the ocean surface, water density shows an increase with depth that is associated with gradients of temperature and salinity (figure 10.61). The gravitational response of water with such a distribution of density can be analysed into modes [133], of which primarily the first two are important:

(i) the 'shallow-water' mode, frequently called 'barotropic', where currents distributed uniformly with depth generate sea-level changes by convergence and

(ii) the 'baroclinic' mode without sea-level changes, where a vertical column of fluid has no net momentum because currents near the surface are opposed by deeper currents, and where convergence simply tilts density contours.

Gravity acting on stratified fluid opposes such tilting; very briefly, the term gh_0 which represents the restoring force for the barotropic mode is retained for the baroclinic mode except that h_0 ceases to be the ocean depth (of many km) but is replaced by a quantity of order 1 m. Then the bracketed expression in (98) is dominated by its first term, so that the equation describes non-dispersive waves propagated to the west at the wave speed

$$\left(-\frac{\omega}{k_1}\right) = gh_0 \frac{\beta}{f^2} \tag{99}$$

which is plotted as a function of latitude in figure 10.62.

Mid-latitude oceans respond very sluggishly in the baroclinic mode. Thus figure 10.62, although merely describing small-amplitude waves, suggests that, with speeds of the order of 1 cm s^{-1}, they might need a decade to cross an ocean; and numerical models confirm that, in these latitudes, distributions of ocean current with depth take at least a decade to respond to forcing by wind—even though such surface forcing has a basic tendency to excite more baroclinic than barotropic response.

But equatorial oceans respond readily in the baroclinic mode. Figure 10.62, besides indicating this, suggests that any simple ray-theory analysis is impossible near the equator because of steep wave-speed gradients; however a detailed wave theory shows how baroclinic waves, trapped near the equator, propagate along it [134, 135]. The different trapped-wave modes, with waveforms as in the Schrödinger equation for the harmonic oscillator, include one that propagates eastwards at velocity $(gh_0)^{1/2}$, around 3 m s^{-1}, while the rest propagate westwards at somewhat lower velocities. The former contributes to that irregular pattern of tropical-meteorology changes known as the Southern Oscillation (section 10.5.2), while the latter modes play a role in the lively

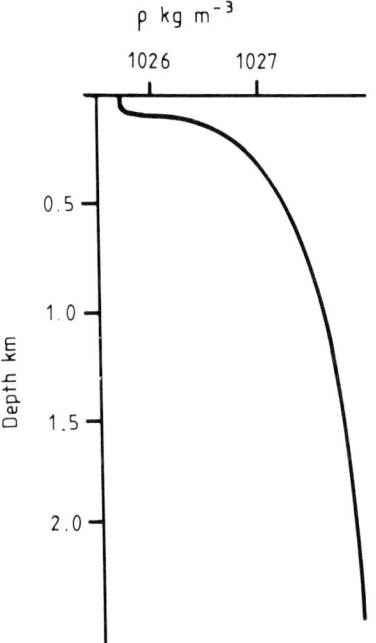

Figure 10.61. *A typical average variation of density ρ in the ocean as a function of depth.*

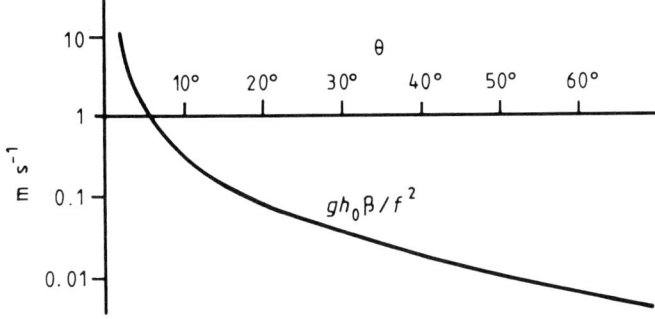

Figure 10.62. *The wave speed (99) for westward propagation of small baroclinic disturbances is plotted (on a logarithmic scale) against latitude θ for $h_0 = 1$ m.*

response of the Indian Ocean currents to onset of the Southwest Monsoon [133].

In another part of the Earth's fluid envelope—the rivers—downstream movements of flood waters generated by intense rainfall may exhibit wave-like patterns that were first elucidated [136] by a fine hydraulic engineer, J A Seddon, for the great rivers of the USA.

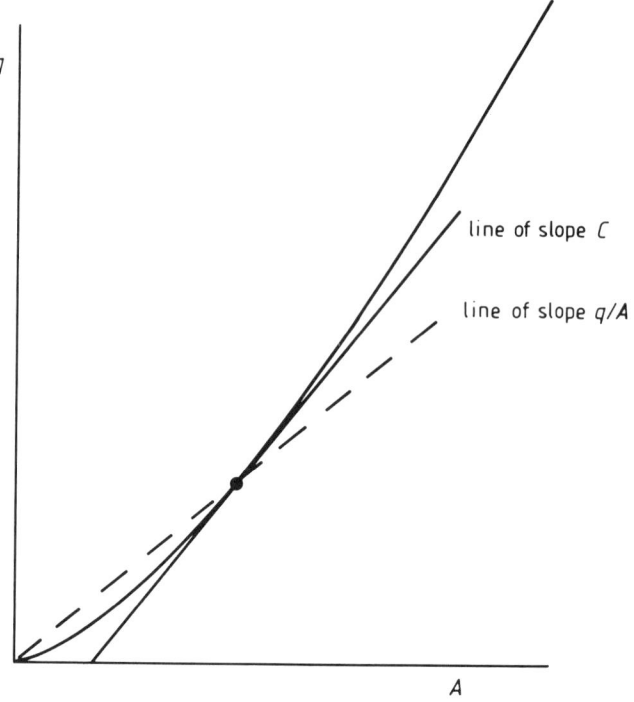

Figure 10.63. *The relationship between volume flow q and cross-sectional area A of river water is concave upwards at any location, with slope $C = dq/dA$ exceeding the mean flow velocity q/A. Multiplying the local rate of change of A by C, we obtain the local rate of change of q in the form $-C dq/dx$, so that values of q travel at velocity C.*

Anywhere on a river, there must be a relationship between the water's cross-sectional area, A, and the volume flow of river water, q. This relationship is concave upwards (figure 10.63) because, where α is the bed slope, the downstream gravity force $\rho g \alpha A$ on a section is balanced by frictional resistance (associated with turbulent flow in the river), which varies as about the square of the mean flow velocity q/A. But, besides this mean velocity, a somewhat higher velocity C defined as the slope $C = dq/dA$ of figure 10.63 plays an important part in determining how flood waters move. At any location, simple kinematics shows that the rate of change of water cross-section A (depending on inflow minus outflow) is minus the gradient of the flow q; so that, on multiplying this relationship by C, we deduce that the rate of change of q is $(-C)$ times the gradient of q itself.

This means that changes in q travel downstream at velocity C, in the form of what has been called a 'kinematic' wave. (Its wave speed C, although greater than the mean flow velocity q/A, is much less than the propagation velocity c_0 for 'dynamic' waves such as gravity waves on shallow water; which, indeed, become powerfully damped in long turbulent rivers.) Moreover, different values of q travel at different speeds C, in such a way that larger values may tend (yet again!) to catch up with smaller values [137]; until an equilibrium waveform— the 'kinematic shock wave', with thickness of order 1 km—is formed from a balance of kinematic and dynamic effects. This then is yet another flow response to atmospheric input which assumes something of 'a life of its own' in the form of a current pattern that propagates in a wave-like manner.

10.5.2. Weather and climate
An interacting system comprising atmosphere, oceans, continents and the month-by-month distribution of incident solar radiation determines the Earth's climate; that is, the average geographical distribution of winds, temperatures, cloudiness and precipitation in each month of the year. It is strongly affected by contrasts between the heat-absorbing properties of the ocean's well-mixed layer (about 50–100 m thick) and of weakly conducting continental surfaces; while evaporation of ocean water has two opposing effects on climate: the 'greenhouse' effect (interception of outgoing long-wave radiation by water vapour), and reductions of incoming solar radiation through reflection from its condensed forms (clouds, snow cover, etc).

However, the system's many unstable and indeed chaotic features bring about an enormous variability about any monthly average; and such variations on small time-scales, of course, constitute the weather. Also some of the instabilities, by generating large-scale horizontal (as well as vertical) fluxes, are influences on climate itself.

The atmosphere, no less than the ocean, is strongly affected [93] by the planetary vorticity (92), by its change (95) in north–south movements, and by a form of Rossby's 'potential vorticity' idea (97). Fundamental applications of these concepts to weather and climate are outlined in section 10.5.2 before (i) describing twentieth century successes in meeting urgent needs of agriculture, transport, civil protection, etc for improved weather forecasting, and (ii) reviewing possibilities for discerning changes of climate over periods exceeding the basic annual cycle. First of all, however, some key influences on atmospheric stratification must be described [138].

Air temperatures normally decrease with altitude, but never at a rate exceeding 1 °C per 100 m; determined as g/c_p, with the air's specific heat c_p at constant pressure around 1000 J kg^{-1} per °C. The explanation is that heat input when pressure as well as temperature changes is

$$c_p \, dT - \rho^{-1} \, dp \quad \text{which is} \quad c_p \, dT + g \, dz \qquad (100)$$

wherever pressure drop is due to an increase in altitude dz. So air rising with zero heat input (that is, adiabatically) loses temperature by 1 °C per 100 m, and would continue to rise—while falling air would continue to fall—if ever the ambient lapse rate (gradient of temperature decrease) exceeded this value. Such excessive lapse rates, in short, are rapidly eliminated by vertical mixing.

On the other hand, any rising air with a fully saturated partial pressure of water vapour (a quantity which becomes less, of course, after a temperature decrease) must as it cools experience condensation, with associated latent heat release, and a consequently more gradual temperature drop—along the 'moist-air adiabatic'—of around 0.5 °C per 100 m. It follows that, wherever the ambient air's lapse rate lies between 0.5 °C and 1 °C per 100 m, any saturated air may experience that vigorous convective mixing, involving vertical motions with associated condensation, which characterizes (say) cumulus clouds.

In the more stably stratified air just above the tops of cumulus clouds, a continued pounding from below by the more vigorously rising parcels of moist air generates internal waves (section 10.3.5). Flight through these feels much like flight through turbulence, and pilots often call them clear air turbulence. The theory of internal waves given in section 10.3.5 for a liquid with the density stratification (66) needs just one modification in the atmosphere [139]: air in an adiabatic upward displacement $s\cos\theta$ suffers a pressure drop $\rho g s \cos\theta$ and a consequent density drop $\rho g c^{-2} s \cos\theta$ where c is the sound speed (14); so it finds itself denser than its surroundings by an amount $\rho(\varepsilon - gc^{-2})s\cos\theta$. Thus it is only when $\varepsilon > gc^{-2}$ that the stratification is stable (this, of course, is just another form of the condition on lapse rate), and then the Väisälä–Brunt frequency N takes a modified form [94, 95]

$$N = \left[g(\varepsilon - gc^{-2})\right]^{1/2}. \tag{101}$$

However, after this change, all aspects of the theory of internal waves given in section 10.3.5 remain unaltered.

Where solar radiation strongly heats land surfaces, the powerful air movements that remove much of the heat by upward convection are described (p 845) as thermals. In each thermal, the fluid flow field is broadly of the nature of a 'vortex ring', with a large upward momentum. A thermal grows as it rises by entraining ambient air and, very gradually, slows down—although, especially in tropical regions, the upward movement may continue to altitudes of very many km.

These and many other mixing processes generate the troposphere, that fairly well mixed region of the atmosphere which extends to a height varying from 16 km over the equator to mid-latitude values around 11 km; and, in polar winters, a level as low as 8 km. Above it, those substantial temperature lapse rates that are associated with fast mixing processes disappear or are reversed, and this region with very stable stratification is known as the stratosphere.

Fluid Mechanics

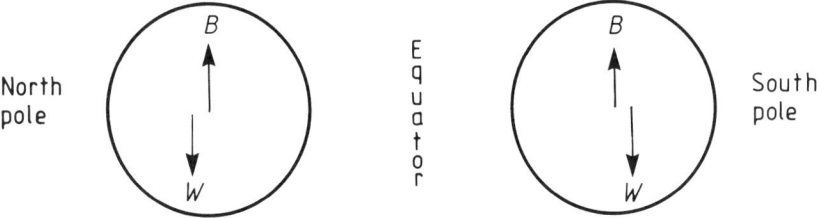

Figure 10.64. *Schematic explanation of the vorticity-generating effect of horizontal temperature gradients. The line of action of a fluid sphere's weight W is displaced to the cold side of its geometric centre through which the buoyancy force B acts.*

The general climatology of global winds is often summarized in terms of surface-wind features (e.g. westerlies in temperate zones, easterly 'trade winds' at lower latitudes), but from the standpoint of basic physics the surface winds are a second-order aspect of climate [93]. The outstanding feature of the distribution of mean winds is the enormous general increase of westerly wind with height at practically all latitudes; for example, the easterly trade winds extend only to altitude 2–3 km, above which they are overlain with increasingly strong westerlies. It is above all the large temperature difference between equator and poles (derived from the distribution of solar radiation) which makes for this general increase of westerly mean wind with height, in a way which can be interpreted [140] in terms of fundamental properties of vorticity.

Horizontal temperature gradients change the essential conclusion (figure 10.9) that, because pressure forces act through a sphere's geometrical centre, they are not altering its angular momentum. Figure 10.64 shows how a sphere, subject to an equatorward temperature gradient T' at a given pressure level, has its centre of gravity displaced polewards so that the couple formed from its weight and the Archimedes buoyancy force on the sphere generates angular momentum. The associated rate of change of vorticity in a direction perpendicular to the paper is gT'/T; and, in equilibrium, it is only the increase with height z of the westerly wind U (the component 'into the paper') which, by tilting the planetary vorticity (93) at a rate dU/dz, can generate horizontal vorticity at a balancing (equal and opposite) rate $f dU/dz$. This balance

$$f\frac{dU}{dz} = g\frac{T'}{T} \tag{102}$$

known as 'the thermal wind equation', accounts in the troposphere for typical increases of mean westerly wind by 3–4 m s^{-1} per km of altitude.

Some idea of how the planetary vorticity's change (95) in north–south motions allows flow patterns to be wave-like was outlined for the oceans in section 10.5.1. Corresponding studies for the atmosphere are a bit more complicated—because of the large vertical gradient (102) of mean velocity—so that there is space here for only a brief summary of the conclusions.

Particularly important atmospheric roles are played by two types of wave, both capable of growing to substantial amplitudes [93]. The two types are on markedly different length-scales, and they also differ in another important respect, reminiscent of the distinction (section 10.2.4) between flow instabilities of the 'abrupt' and the 'restrained' kinds. Yet they interact with each other.

One has wavelengths so large [141] that just two or three lengths embrace the globe; a small-amplitude theory of this wave shows general similarity to that of a Rossby wave, but with capabilities of amplitude growth. Yet the growth, being of a 'restrained' type, retains [142] the general wave-like form (figure 10.65); moreover, within the overall westerly wind pattern, the wave maintains a nearly stationary position which (roughly speaking) can appear 'anchored' to the topography of continents. In the wave, large maximum wind speeds are found in the famous upper-troposphere jet stream. Thus, the climatological mean wind field indicated by the thermal wind equation (102) and the equator–pole distribution of mean temperature represent smoothed-out versions of the fluctuations of an often more intense wind pattern, associated with more concentrated temperature gradients.

Another (smaller-scale) type of growing wave [143], again influenced both by the mean wind shear (102) and by the 'beta effect' (95), has something in common with the growing Tollmien–Schlichting waves found in boundary layers (p 830) and, like them, tends to generate chaotic flow patterns. These are the cyclonic disturbances which randomly appear, especially over mid-latitude oceanic regions, and are amplified—while becoming distorted in various ways—as they travel eastwards. Weather in such regions is powerfully influenced by this chaotic tendency, known to meteorologists as 'baroclinic instability', which moreover has been shown to put some definite limits (see below) on the times over which forecasting can be attempted. Also, the disturbances interact with the large-scale wave system [93]; indeed, part of their energy is capable, in a manner analogous to the absorption of internal waves at a critical level (p 868), of being fed into the jet stream.

Furthermore, climate is itself affected by all of those disturbances to pure east–west motion which have been mentioned. Very briefly, the north–south components of those disturbances produce a net northward transport of the angular momentum of westerly winds about the axis of the Earth [93]. Thus, over and above the tendency (102) for the mean westerly wind to increase with height, mid-latitude regions are gaining some additional 'westerly' angular momentum at the expense of lower latitudes. And so the observed surface wind pattern emerges as what was described earlier as a 'second-order' effect: on the average, the poleward angular momentum transport is balanced by the moments

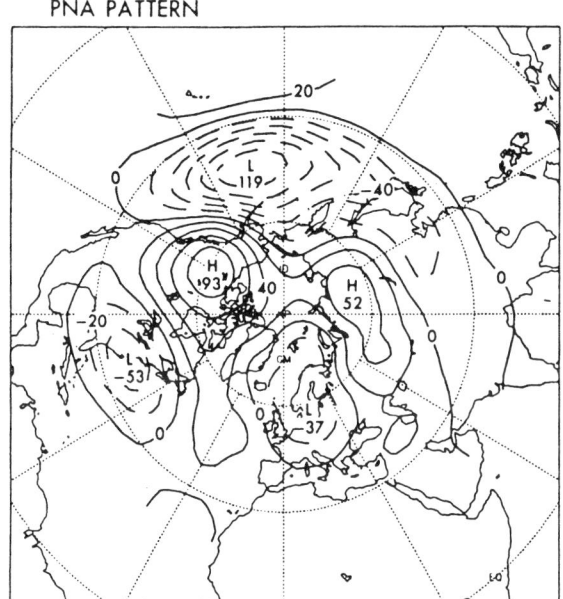

Figure 10.65. *Example of an upper-troposphere stationary-wave pattern in the northern hemisphere* [142]. *Altitude departures (in m) of the 300-millibar pressure surface from zonally averaged levels are shown for the rather common winter condition known as PNA (Pacific–North American anomaly, with a 'high' H over North America and a 'low' L over the Pacific). Tightly packed contours (in other words, unusually steep slopes of the constant-pressure surface) are associated with high-speed upper-troposphere winds.*

of the resistive forces exerted by the Earth's surface on westerly surface winds in middle latitudes and easterly 'trade winds' at lower latitudes.

Climatologically, the trade-wind zone is a relatively stable one, where a slow process of 'subsidence' (sinking of air) necessarily balances the continued rapid ascent of air (especially in 'thermals') over equatorial regions. Such slow subsidence is by no means adiabatic; instead, much of the heat gained in compression is lost by radiation, so the lapse rate is much reduced (also, the descending air is rather dry) and a highly stable stratification results [144]—except in a well-mixed layer close to the surface. Another climatologically significant feature of the trade-wind zone is that poleward transport of heat is preferentially achieved [145], not by atmospheric motions, but by such oceanic flows as western boundary currents (section 10.5.1); a good example of climate's dependence on the Earth's fluid envelope as a whole.

In the field of weather, two examples of discontinuity formation may be mentioned, one from middle latitudes and one from tropical regions. As mid-latitude cyclonic disturbances develop, horizontal gradients of temperature often become progressively sharper until there is formed, as J Bjerknes explained already in 1919, an effective discontinuity. At such a 'front', inclined at a small angle α to the horizontal, both the temperature and the wind velocity parallel to the front make practically discontinuous changes, ΔT and ΔU, such as are permitted by equation (102) to be in balance. (Essentially, the horizontal gradient T' inside the inclined front is $\alpha \mathrm{d}T/\mathrm{d}z$, so that an integrated form of (102) across the front makes $f \Delta U$ equal to $g\alpha \Delta T/T$.) Fronts are striking weather features in temperate zones, with major influences on cloud formation and precipitation.

But it is a tropical weather feature which to the greatest degree can threaten disastrous effects on human populations [146]. The so-called tropical cyclones (known also as typhoons or as hurricanes) are vastly bigger and more intense than mid-latitude cyclones, exhibiting horizontal length scales of very many hundreds of km (figure 10.66) and surface wind speeds of the order of 50 m s^{-1}. They are formed when a wind pattern with substantial cyclonic vorticity comes in contact with a tropical ocean (having surface temperature 26 °C or more) provided that the atmosphere's vertical structure exhibits certain features; including, especially, a lapse rate in that range well above 0.5 °C per 100 m (see above) where fully saturated air is freely able to make vertical motions.

The discontinuity in this case is known as the 'eyewall': a vertical wall of dense convective cloud surrounding an 'eye' which, often, is nearly free of cloud (figure 10.66). The eyewall forms where air, having been continuously accelerated as it followed a long spiral path inwards over the warm ocean, has finally reached its saturation vapour pressure; then it can rise fast all the way to the top of the troposphere (at, say, 15 km altitude) where it tends to spiral outwards in a broadly anticyclonic motion. The eye, by contrast, consists of rather calm air at the same (low) pressure as the fast moving air in the eyewall.

In 1986, K A Emanuel pointed out [147] that tropical cyclones can be regarded as heat engines operating on something close to a Carnot cycle, in which the working fluid is that intimate mix of dry air with water in all its forms (vapour, droplets, ice crystals) which appears in the atmosphere. It follows the 'moist-air adiabatic' (see above) as it rises in the eyewall, after taking in energy at, essentially, the ocean surface temperature—partly as sensible heat but mainly as latent heat associated with an increase in partial pressure of water vapour. (From a physics standpoint, the energy transfer arises because the wind extracts the more energetic water molecules from the ocean surface.) Then, at a much lower upper-troposphere temperature, the working fluid loses energy by radiation and by mixing with its environment. The large difference between the heat-input and heat-output temperatures makes for an efficient heat engine.

Mechanical energy so generated is dissipated mainly by frictional resistance at the ocean surface—which, in a sense, transfers momentum

Fluid Mechanics

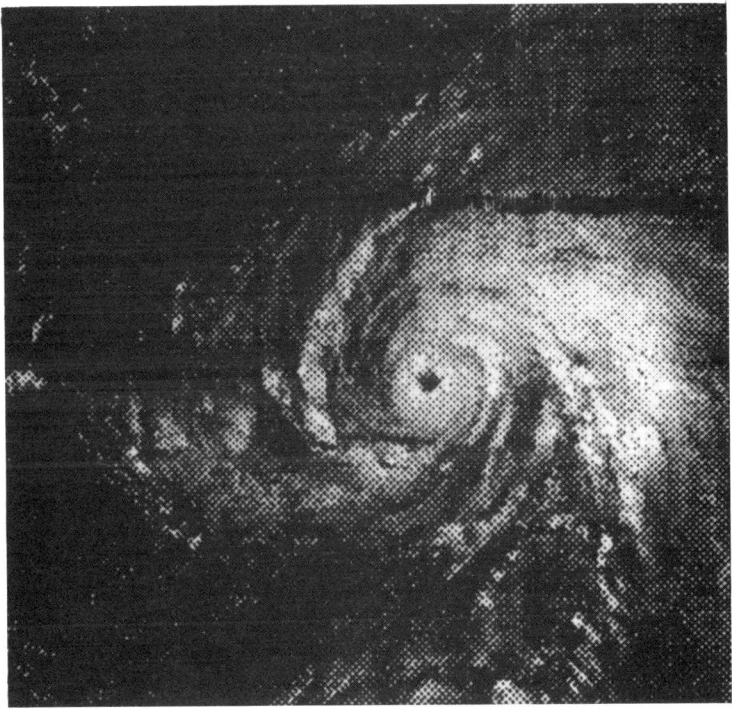

Figure 10.66. *Typical satellite photograph of a northern-hemisphere tropical cyclone with a horizontal extent of about 1000 km. Note the 'eye of the storm'.*

'back' to the ocean. Such transfer, wherever a tropical cyclone approaches a coastline, can threaten a devastating 'storm surge' (p 892) in addition to any more direct threats of disaster from extreme wind speeds—or from extreme rainfall if the storm deposits its huge moisture content over land.

Insight into the initial formation of tropical cyclones as well as of many other important atmospheric phenomena is given [146] by an extension, originally made in 1942 by H Ertel [148], of Rossby's 'potential vorticity' concept (97). Very briefly, Ertel's extension takes h as the vertical distance between neighbouring surfaces of constant entropy; so that, in adiabatic processes where such surfaces move with the fluid, the total vorticity $f + \varsigma$ changes in proportion as the distance h is stretched. Regions with anomalously high values of Ertel's potential vorticity 'have the potential' as it were to develop into large cyclonic structures. Moreover, potential vorticity was recently shown to have properties analogous to those of vorticity proper in aeronautics where (section 10.2.2) the vorticity field entirely determines the flow; similarly, meteorologists found an inversion algorithm which uniquely relates a wind field to its distribution of potential vorticity, so underlining this distribution's diagnostic value.

Weather forecasting [149] needs to be global: ships require advance warning of storms to plan their routes, winds must be forecast at all the levels where aircraft fly (section 10.4), warnings of a dangerous tropical cyclone are needed well in advance of its approach to land. Agriculture too needs forecasts several days ahead, and such forecasts must be global because over a period of a few days the influence of remote weather patterns becomes important. Predictions of both water supply and energy demand need to be based on weather forecasts, which indeed are valued in relation to most human activities.

The last quarter of the twentieth century saw the achievement [150] of truly global forecasting of increasing accuracy by means of computational fluid dynamics (CFD). The computer programs, using horizontal mesh sizes of the order of 100 km, are based on comprehensive descriptions of the dynamics of air motions at many altitude levels (around 20), with separate accounts taken of interactions with land topography and with the ocean surface, of water vapour and its condensation, of moist-air convective movements, of more general cloud physics and precipitation (and subsequent evaporation of falling rain), and of the influence of all these variables on radiative heat transfer. A global data-gathering and telecommunication network provides initial data for the programs, primarily at 00.00 and 12.00 GMT, including information at the different altitudes derived from the release of radiosonde balloons that can telemeter data on wind, temperature, humidity etc. from all levels. Some more two-dimensional, yet still very valuable data are provided from meteorological satellites and from other instrumentation, while various users of forecasts (ships, aircraft) are also communicators of data to the network. Computer analysis assimilates all the data (discarding anomalous values) and puts them into a form compatible with use as initial conditions for the CFD program. This is then used to forecast developments up to 10 days ahead, with outputs generated at very many intermediate forecast times; the output data being appropriately interpreted and communicated in various different forms to the different industries that need it as well as to the general public.

Later, all forecasts are compared with actual outcomes to determine how various measures of error increase with forecast time. Towards the end of the twentieth century, the general utility of 3-day forecasts had become very substantial, while 6-day forecasts were beginning to show a definite value.

In addition, some sort of measure of best possible 'predictability horizon' (allowing for unavoidable errors in initial data) was being estimated [150]. To this end, programs are run with just very small differences in initial data; and, as with other chaotic systems, large differences in the forecasts emerge beyond a certain predictability horizon, which seems to be around 15–20 days. An ingenious method for studying which errors produce most rapid divergences in forecasts has identified them to be errors in data at the western edge of any large mid-latitude ocean; these, of course, are the errors which are most magnified by 'baroclinic instability' effects.

In relation to such regions, efficient methods exist for reconciling the use of a mesh size of the order of 100 km with the physical tendency to form fronts—whose locations and strengths are routinely inferred from output data. In areas prone to tropical cyclones, however, problems arise from the far more drastic discontinuities at eyewalls [146].

Indeed, even the original formation of such a discontinuity remains hard to predict (although improvements may emerge now that plots of potential vorticity are included in some computer outputs); accordingly, forecasting may need to start from an 'eye' already identified by satellite and to impose a vortical structure (centred thereon) upon nearby initial data. Some CFD programs using this technique along with 'movable nested meshes' (where the finest mesh moves along with the eye region) are beginning to offer improved accuracies in forecasting the tracks of tropical cyclones, such as are urgently needed by the inhabitants of vulnerable coastlines [146].

Last of all, possibilities of systematic shifts of climate on time-scales exceeding the basic annual cycle are briefly reviewed. Work on the most important of these was pioneered in the 1920s by the Director General of the India Meteorological Department, Sir Gilbert Walker [151].

Indian agriculture depends critically on summer rainfall generated by the Southwest Monsoon. In the days before long-term storage of grain from high-yield crops, any 'bad' (low-rainfall) monsoon was a grave disaster for India. Sir Gilbert discovered long-distance correlations between monsoon behaviour over India and climatic variations stretching from East Africa to South America, and viewed such 'teleconnections' as evidence for a 'Southern Oscillation', that is, a (somewhat irregular) shift back and forth between extremes of climate over the Earth's more southerly oceans.

Later investigations of monsoon phenomena [152] suggested some of the physical basis for such an 'oscillation'. The familiar South Asian monsoons are part of a system of strong low-level winds—broadly opposed by even stronger high-level winds—on an enormous scale; for example, the Southwest Monsoon involves low-level winds beginning in the southern part of the Indian Ocean and blowing over East Africa and South Asia before moving onwards across most of the Pacific. Yet monsoons continue to be regarded as wind patterns related to large-scale ocean-land contrasts. Indeed, an expansion of either (i) global land cover or (ii) global wind patterns in spherical harmonics includes—alongside the zonally averaged components with $n = 0$—some specially important components with $n = 1$, corresponding (i) to the area of Eurasia/Africa hugely exceeding that of the Americas and (ii) to the related monsoon wind fields.

Further progress on the Southern oscillation required a prolonged collaboration of meteorologists with oceanographers. Very briefly, this irregular cycle emerges [153] from large-scale effects of sea-surface temperature distributions on winds and of wind variability on ocean current patterns. At one phase of the oscillation, for example, a marked slackening of trade winds changes the normal pattern of wind stress on the tropical Pacific and allows a 'trapped' baroclinic wave (p 896) to travel slowly eastwards across it; producing on arrival that event known as 'El Niño', which by warming Peruvian surface waters seriously harms the big local fishing industry. The complex nature of 'ENSO' (the popular name for the El Niño/Southern oscillation phenomenon) illustrates yet again the leading climatological role played by ocean/atmosphere interaction.

It is this intimate coupling which makes CFD modelling of climate change so difficult. It might be thought that, even though a CFD model of weather exhibits beyond the predictability horizon a massive divergence of solutions with only slightly different initial conditions, nonetheless the statistics of those solutions could represent climate. Some initial success with this approach (using relatively coarse-mesh models) was achieved [154] in the 1980s, but it became recognized in the 1990s that models developed primarily for weather forecasting take insufficiently detailed account of oceanic processes to be used for climate-change forecasting. Accordingly, some intricate coupled ocean/atmosphere models began to be adopted—in spite of severe difficulties in quantifying the necessary coupling [155].

A principal objective of such models was to forecast climate changes related to likely increases in atmospheric CO_2. Even though the 'greenhouse effect' (see above) is mainly associated with water vapour, carbon dioxide makes a modest additional contribution to it. Moreover a forecast doubling of atmospheric CO_2 by the middle of the twenty-first century could influence climate, not only through its direct effect on interception of outgoing radiation, but also indirectly through two positive-feedback effects: (i) increased evaporation leading to enhanced interception by water vapour and (ii) reduced snow cover raising the absorption of solar radiation. Current models, by comparing solutions with constant and with increasing CO_2 levels, have indeed estimated so-called 'global warming' as rather non-uniform—appearing preferentially over northern-hemisphere land masses [156].

The fluid mechanics of just air and water were seen in this chapter to have yielded 'a fair share' out of the many exciting developments in twentieth century physics; those mentioned having been chosen for their general importance and interest to non-specialist readers. Any specialist reader, of course, may argue that his or her specialism merited more detailed description; yet the chapter's author can legitimately counter suggestions of personal bias with a comment that his own major area of specialist expertise—fluid mechanics of the biosphere—has been omitted altogether.... Above all, he hopes that the chapter will extend the interests of many generally 'physics-minded' readers into fluid mechanics.

References

[1] Reynolds O 1886 *Phil. Trans. R. Soc.* **177** 157
[2] Basset A B 1888 *Hydrodynamics* (Cambridge: Cambridge University Press)
[3] Tietjens O 1931 *Handbuch der Experimentalphysik* vol 4, part 1, ed W Wien and F Harms (Leipzig: Akademische Verlagsgesellschaft) pp 669–703
[4] Prandtl L 1905 *Verhandlungen des III. Internationalen Mathematischen Kongresses (Heidelberg, 1904)* (Leipzig: Teubner) pp 484–91. Note: this and the other cited papers of Prandtl can be found also in his collected papers: Prandtl L 1961 *Gesammelte Abhandlungen* ed W Tollmien, H Schlichting and H Görtler (Berlin: Springer).
[5] Prandtl L 1918, 1919 Tragflügeltheorie, parts I, II *Nachr. Ges. Wiss. Göttingen, Math.-Phys. Klasse* 151, 107
[6] Rayleigh Lord 1910 *Proc. R. Soc.* A **84** 247
[7] Taylor G I 1910 *Proc. R. Soc.* A **84** 371
[8] Van Dyke M D 1975 *Perturbation Methods in Fluid Mechanics* (Palo Alto, CA: Parabolic)
[9] Lamb H 1932 *Hydrodynamics* 6th edn (Cambridge: Cambridge University Press)
[10] Zhukovski N E 1912 *Teoreticheskie osnovy aerodinamiki* (Moscow: Tekhnicheskaya uchilishcha) (French Transl. Joukowsky N 1916 *Aérodynamique* (Paris: Gauthier-Villars))
[11] Rosenhead L (ed) 1963 *Laminar Boundary Layers* (Oxford: Oxford University Press)
[12] Bernoulli D 1738 *Hydrodynamica* (Strasbourg: Argentorati)
[13] Rayleigh Lord 1878/1896 *The Theory of Sound* 1st/2nd edns (London: Macmillan)
[14] Laplace P S 1816 *Ann. Chim. Phys.* **3** 238
[15] Riemann B 1859 *Abh. Göttingen Ges. Wiss.* **8** 43
[16] Lighthill M J 1956 *Surveys in Mechanics* ed G K Batchelor and R M Davies (Cambridge: Cambridge University Press) pp 250–351
[17] Hugoniot A 1889 *J. de l'Ecole Polytech.* **58** 1
[18] Prandtl L and Pröll A 1961 *Z. Vereines Deutsch. Ing.* **48** 348
[19] Mach E 1887 *Sitzungsber. Wiss. Akad.* **95** 164
[20] Reynolds O 1883 *Phil. Trans. R. Soc.* **174** 935
[21] Prandtl L 1914 *Nachr. Ges. Wiss. Göttingen, Math.-Phys. Klasse* 177
[22] Wieselsberger C 1914 *Z. Mech.* **5** 140
[23] Lighthill J 1976 Flagellar hydrodynamics *SIAM Rev.* **18** 161–230
[24] Reynolds O 1895 *Phil. Trans. R. Soc.* **186** 123
[25] Emmons H W 1951 *J. Aeronaut. Sci.* **18** 490
[26] Rayleigh Lord 1880 *Proc. Lond. Math. Soc.* **19** 67
[27] Rayleigh Lord 1916 *Proc. R. Soc.* A **93** 148
[28] Landau L D 1944 *Dokl. Akad. Nauk. SSSR* **30** 299
[29] Goldstein S (ed) 1938 *Modern Developments in Fluid Dynamics* (2 volumes) (Oxford: Oxford University Press)

[30] Tollmien W 1935 *Nachr. Ges. Wiss. Göttingen, Math.-Phys. Klasse* 79–114
[31] Taylor G I 1923 *Phil. Trans. R. Soc.* A **223** 289
[32] Prandtl L 1921 *Z. Angew. Math. Mech.* **1** 431
[33] Tollmien W 1929 *Nachr. Ges. Wiss. Göttingen, Math.-Phys. Klasse* 21–44
[34] Schlichting H 1933 *Nachr. Ges. Wiss. Göttingen, Math.-Phys. Klasse* 181–208
[35] Taylor G I 1922 *Proc. Lond. Math. Soc. (2)* **21** 196
[36] Dryden H L 1929–36 *Nat. Adv. Comm. Aeronaut. Report* **320** (with A M Kauth) and **581** (with G B Schubauer, W C Mock and H K Skramstad)
[37] Taylor G I 1935 *Proc. R. Soc.* A **151** 429
[38] von Kármán T and Howarth L *Proc. R. Soc.* A **164** 192
[39] Batchelor G K 1953 *The Theory of Homogeneous Turbulence* (Cambridge: Cambridge University Press)
[40] Schlichting H 1940 *Jahrb. Deutsch. Luftfahrtf.* **1** 97
[41] Görtler H 1940 *Nachr. Ges. Wiss. Göttingen, Math.-Phys. Klasse* 1–26
[42] Schubauer G B and Skramstad H K 1947 *Nat. Adv. Comm. Aeronaut. Report* **909**
[43] Schubauer G B and Klebanoff P S 1955 *Nat. Adv. Comm. Aeronaut. Report* **1289**
[44] Gaster M 1968 *J. Fluid Mech.* **32** 173
[45] Smith F T 1992 *Phil. Trans. R. Soc.* A **340** 171
[46] Stuart J T 1958 *J. Fluid Mech.* **4** 1
[47] Sato H 1960 *J. Fluid Mech.* **7** 53
[48] Michalke A 1965 *J. Fluid Mech.* **23** 521
[49] Coles D 1965 *J. Fluid Mech.* **21** 385
[50] Kolmogorov A N 1941 *Dokl. Akad. Nauk. SSSR* **30** 301
[51] Pao Y H 1965 *Phys. Fluids* **8** 1063
[52] Townsend A A 1956 *The Structure of Turbulent Shear Flows* (Cambridge: Cambridge University Press)
[53] Stewart R W and Townsend A A 1951 *Phil. Trans. R. Soc.* A **243** 359
[54] Kim H T, Kline S J and Reynolds W C 1971 *J. Fluid Mech.* **50** 133
[55] Coles D 1955 *Fünfzig Jahre Grenzschichtforschung* ed H Görtler and W Tollmien (Braunschweig: Vieweg) pp 153–63
[56] Coles D 1956 *J. Fluid Mech.* **1** 191
[57] Squire H B 1951 *Q. J. Mech. Appl. Math.* **4** 321
[58] Reichardt H 1942 *Gesetzmässigkeiten der freien Turbulenz* vol 414 (Berlin: Verein Deutscher Ingenieure)
[59] Prandtl L 1910 *Phys. Z.* **11** 1072
[60] Pohlhausen E 1921 *Z. Angew. Math. Mech.* **1** 115
[61] Whipple F J W 1933 *Proc. Phys. Soc.* **45** 307 gives a full account of Taylor's (not separately published) analysis.
[62] Reynolds O 1874 *Proc. Manchester Lit. Phil. Soc.* **14** 7
[63] Schmidt E and Beckmann W 1930 *Tech. Mech. Thermod.* **1** 391
[64] Schmidt E and Beckmann W 1930 *Tech. Mech. Thermod.* **1** 391 gives a full account of Pohlhausen's (not separately published) analysis.
[65] Bénard H 1901 *Ann. Chim. (Phys.)* **23** 62
[66] Pearson J R A 1958 *J. Fluid Mech.* **4** 489
[67] Rayleigh Lord 1916 *Phil. Mag.* **32** 529
[68] Pellew A and Southwell R V 1940 *Proc. R. Soc.* A **176** 312
[69] Koschmieder E L 1966 *Beitr. Phys. Atmos.* **39** 209
[70] Lighthill J 1978 *Waves in Fluids* (Cambridge: Cambridge University Press)
[71] Whitham G B 1956 *J. Fluid Mech.* **1** 290
[72] Whitham G B 1952 *Commun. Pure Appl. Math.* **5** 301
[73] Rayleigh Lord 1879 *Phil. Mag.* **7** 161

[74] Bénard H 1908 *Comptes Rendus* **147** 839
[75] von Kármán T 1911 *Phys. Z.* **13** 49
[76] Homann F 1936 *Forsch. Geb. Ing.-Wes.* **7** 1
[77] Curle N 1955 *Proc. R. Soc.* A **231** 505
[78] Curle N 1953 *Proc. R. Soc.* A **216** 412
[79] Lighthill M J 1952 *Proc. R. Soc.* A **211** 564
[80] Lighthill M J 1963 *Am. Inst. Aeronaut. Astron. J.* **1** 1507
[81] Goldstein M E 1976 *Aeroacoustics* (New York: McGraw-Hill)
[82] Lighthill J 1978 *J. Sound. Vib.* **61** 391
[83] Munk W H, Miller G R, Snodgrass F E and Barber N F 1963 *Phil. Trans. R. Soc.* A **255** 505
[84] Rayleigh Lord 1914 *Proc. R. Soc.* A **90** 324
[85] Korteweg D J and de Vries G 1895 *Phil. Mag.* **39** 422
[86] Whitham G B 1974 *Linear and Nonlinear Waves* (New York: Wiley)
[87] Schwartz L W 1974 *J. Fluid Mech.* **62** 553
[88] Whitham G B 1965 *J. Fluid Mech.* **22** 273
[89] Longuet-Higgins M S 1975 *Proc. R. Soc.* A **342** 157
[90] Banner M L and Peregrine D H 1993 *Ann. Rev. Fluid Mech.* **25** 377
[91] Davey A and Stewartson K 1974 *Proc. R. Soc.* A **338** 101
[92] Stevenson T N 1973 *J. Fluid Mech.* **60** 759
[93] Gill A E 1982 *Atmosphere-Ocean Dynamics* (New York: Academic)
[94] Väisälä V 1925 *Soc. Sci. Fenn. Commentat. Phys.-Math.* **2** 19
[95] Brunt D 1927 *Q. J. R. Meteorol. Soc.* **53** 30.
[96] Yih C S 1980 *Stratified Flows* (New York: Academic)
[97] Bretherton F P 1969 *Q. J. R. Meteorol. Soc.* **95** 213.
[98] Long R R 1970 *Tellus* **22** 471
[99] Booker J R and Bretherton F P 1967 *J. Fluid Mech.* **27** 513
[100] Taylor G I 1923 *Proc. R. Soc.* A **104** 213.
[101] Küchemann D 1978 *The Aerodynamic Design of Aircraft* (Oxford: Pergamon)
[102] Lucas J 1988 *Boeing 747: the First Twenty Years* (London: Taylor & Francis)
[103] Sears W R (ed) 1954 *General Theory of High Speed Aerodynamics* (Princeton, NJ: Princeton University Press)
[104] Ward G N 1955 *Linearized Theory of Steady High-Speed Flow* (Cambridge: Cambridge University Press)
[105] Hayes W D 1971 *Ann. Rev. Fluid Mech.* **3** 269
[106] Morgan M B 1972 *J. R. Aeronaut. Soc.* **76** 1
[107] Busemann A 1931 Gasdynamik *Handbuch der Experimentalphysik* vol 4, ed W Wien and F Harms ch 1 pp 343–460
[108] Ackeret J 1927 Gasdynamik *Handbuch der Physik* vol 7, ed H Geiger and K Scheel (Berlin: Springer) pp 289–342
[109] Hayes W D and Probstein R F 1966 *Hypersonic Flow Theory* (New York: Academic)
[110] Zierep J and Oertel H (ed) 1988 *Symposium Transsonicum III* (Berlin: Springer)
[111] Lighthill M J 1953 *Proc. R. Soc.* A **217** 478
[112] Stewartson K 1969 *Mathematika* **16** 106
[113] Neiland V Ya 1969 *Izv. Akad. Nauk. SSSR Mekh. Zhidk. Gaz.* **4** 33
[114] Messiter A F 1970 *SIAM J. Appl. Math.* **18** 241
[115] Whitcomb R T 1952 *Nat. Adv. Comm. Aeronaut. Memorandum* RN L52 H08
[116] Jameson A, Baker T J and Weatherill N P 1986 *Am. Inst. Aeronaut. Astron. Paper* 86-0103
[117] Newman J N 1991 *Phil. Trans. R. Soc.* A **334** 213
[118] Kelvin Lord 1891 *Popular Lectures* vol 3 (London: Macmillan) pp 450-500
[119] Inui T 1962 *Trans. Soc. Nav. Arch. Mar. Eng.* **70** 282

[120] Wehausen J V and Salvesen N (ed) 1977 *Numerical Ship Hydrodynamics* (Berkeley, CA: University of California)
[121] Trillo R L (ed) 1993–94 *Jane's High-Speed Marine Craft* (London: Jane's Information Group)
[122] Chapman J C 1979 *BOSS '79 (Proc. 2nd Int. Conf. on Behaviour of Off-Shore Structures)* ed H S Stephens and S M Knight (Cranfield: BHRA Fluid Engineering) pp 59–74
[123] Bearman P W and Graham J M R 1979 *BOSS '79 (Proc. 2nd Int. Conf. on Behaviour of Off-Shore Structures)* ed H S Stephens and S M Knight, (Cranfield: BHRA Fluid Engineering) pp 309–22
[124] Taylor G I 1928 *Proc. R. Soc.* A **120** 260
[125] Rainey R C T 1989 *J. Fluid Mech.* **204** 295
[126] Defant A 1961 *Physical Oceanography* (2 volumes) (Oxford: Pergamon)
[127] Laplace P S 1775 *Mém. Acad. R. Sci. (Paris)* 75–182
[128] Taylor G I 1919 *Phil. Trans. R. Soc.* A. **220** 1
[129] Hendershott M and Munk W 1970 *Ann. Rev. Fluid Mech.* **2** 205
[130] Pekeris C L and Accad Y 1969 *Phil. Trans. R. Soc.* A. **265** 413
[131] Jelesnianski C P 1967 *Mon. Weath. Rev.* **98** 740
[132] Rossby C G 1940 *Q. J. R. Meteorol. Soc.* **66** (Supplement) 68
[133] Lighthill J 1971 *Phil. Trans. R. Soc.* A **270** 371
[134] Blandford R R 1966 *Deep-Sea Res.* **13** 941
[135] Matsuno T 1966 *J. Meteorol. Japan* **44** 25
[136] Seddon J A 1900 *Trans. Am. Soc. Civ. Eng.* **43** 179
[137] Lighthill M J and Whitham G B 1955 *Proc. R. Soc.* A **229** 281
[138] Brunt D 1939 *Physical and Dynamical Meteorology* 2nd edn (Cambridge: Cambridge University Press)
[139] Turner J S 1973 *Buoyancy Effects in Fluids* (Cambridge: Cambridge University Press)
[140] Lighthill M J 1966 *J. Fluid Mech.* **26** 411
[141] Charney J G and Eliassen A 1949 *Tellus* **1** 38
[142] Karoly D J, Plumb R A and Ting M 1989 *J. Atmos. Sci.* **46** 2802
[143] Charney J G 1947 *J. Meterol.* **4** 135
[144] Betts A K and Ridgway W 1988 *J. Atmos. Sci.* **45** 522
[145] Vonder Haar T H and Oort A H 1973 *J. Phys. Oceanogr.* **3** 169
[146] Lighthill J, Zheng Z, Holland G and Emanuel K (ed) 1993 *Tropical Cyclone Disasters* (Beijing: Peking University Press)
[147] Emanuel K A 1986 *J. Atmos. Sci.* **43** 585
[148] Ertel H 1942 *Meteorol. Z.* **59** 271
[149] Houghton D D (ed) 1985 *Handbook of Applied Meteorology* (New York: Wiley)
[150] Manabe S 1985 *Issues in Atmospheric and Ocean Modeling. Part B. Weather Dynamics* (New York: Academic)
[151] Walker G 1928 *Q. J. R. Meteorol. Soc.* **54** 79
[152] Lighthill J and Pearce R P (ed) 1981 *Monsoon Dynamics* (Cambridge: Cambridge University Press)
[153] Philander S G 1990 *El Niño, La Niña, and the Southern Oscillation* (New York: Academic)
[154] Washington W M and Parkinson C L 1986 *An Introduction to Three-Dimensional Climate Modelling* (Mill Valley, CA: University Science Books)
[155] Houghton J T, Jenkins G J and Ephraums J J (ed) 1990 *Climate Change: the IPCC Scientific Assessment* (Geneva: World Meteorological Organisation)
[156] Carson D A 1992 *The Hadley Centre Transient Climate Change Experiment* (Bracknell: UK Meteorological Office)

Chapter 11

SUPERFLUIDS AND SUPERCONDUCTORS

A J Leggett

11.1. Introduction

11.1.1. Liquid helium: the early days

If the subject which we now know as low-temperature physics can be said to have a birthday, that day would be 10 July 1908—the date on which Heike Kamerlingh Onnes and his team at the University of Leiden first successfully cooled the element helium (^{4}He) below 4.2 K and thereby liquefied it. For the next 15 years, the only place in the world where liquid helium existed was the Leiden laboratory (now named after Onnes).

If helium, like other elements, became liquid, should it not also, like them, become solid under its own vapour pressure when cooled to sufficiently low temperatures? Onnes certainly expected this, and over the next 15 years reached lower and lower temperatures in an unsuccessful search for the freezing point. By 1922 he was speculating that helium might remain liquid even if cooled down to absolute zero. In fact, after his death, the Leiden team did succeed in inducing freezing, but only at a pressure of 30 atmospheres. Onnes' speculation was correct, the phase diagram of helium is quite different from that of any ordinary element.

In the mid- and late-1920s experimental research on liquid helium accelerated, with its production in a number of laboratories in Europe and North America, and a number of curious but apparently minor anomalies were observed, in particular at a temperature close to 2.2 K, which was eventually interpreted as a transition point between a higher- and a lower-temperature phase of the liquid. These phases were christened He I and He II, respectively. Until 1936, however, it seems that

Heike Kamerlingh Onnes

(Dutch, 1853–1926)

Onnes studied in Groningen and Heidelberg, and in 1882 was appointed to the first chair in experimental physics established in the Netherlands in Leiden. Partially with a view to testing van der Waals' hypothesis of the law of corresponding states, he embarked on a programme to cool various gases to low temperatures, and in particular to liquefy helium, a goal which he achieved in 1908. From then until his death in 1926 the Leiden laboratory was the acknowledged world leader in low-temperature physics; it was there, in 1911, that the phenomenon of superconductivity was discovered, and Onnes and his co-workers subsequently established many of its principal characteristics. He received the Nobel Prize in 1913 for his work on helium. Onnes had a reputation as a far-sighted and careful worker who, despite his considerable theoretical talents, always emphasized the primacy of experiment: his motto was *'door meten tot weten'* ('through measurement to knowledge').

---

liquid helium was regarded as a curiosity of Nature, but mainly because it did not freeze under its own vapour pressure (a property eventually explained to most people's satisfaction in terms of the anomalously large quantum-mechanical zero-point energy). It was not suspected that its behaviour might be qualitatively different from that of any other known liquid. It is astonishing that the characteristic feature of He II which we now know as superfluidity, must have been almost an everyday occurrence in low-temperature laboratories for a quarter of a century, without ever being consciously observed!

An important development came in 1932, when Willem Keesom and his daughter A P Keesom conducted careful experiments on the thermal properties of liquid helium near the apparent transition between He I and He II at 2.2 K. They were surprised to find no latent heat at the

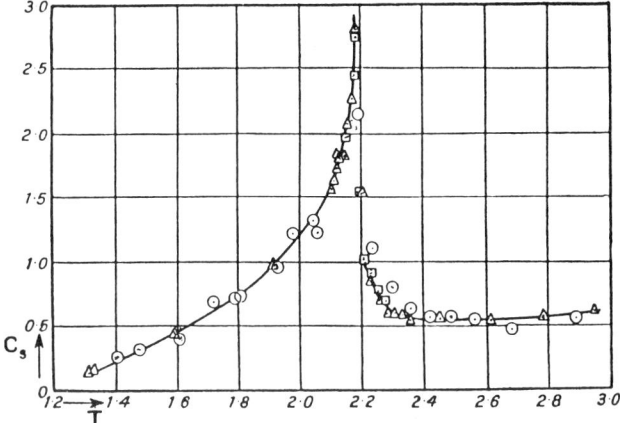

Figure 11.1. *The specific heat of liquid ^{4}He around 2 K. The shape of the curve resembles the Greek letter lambda (λ).*

transition, but a remarkable variation of specific heat (figure 11.1) with an apparent discontinuity at the transition itself. This characteristic shape led to the name lambda-transition, but the nature of the transition and the difference between He I and He II remained unclear until the late 1930s.

11.1.2. Superconductivity up to 1933

Having successfully liquefied helium, and thereby obtained the means to cool other substances to temperatures of a few degrees absolute, Onnes pressed ahead with his assistant Holst to investigate, among other things, the electrical properties of metals in this temperature range. He believed that the resistance of a pure metal should disappear on cooling to absolute zero. After some inconclusive experiments on platinum he decided to try mercury. On the basis of previous work it had been concluded that 'at very low temperatures ... the resistance would, within the limits of experimental accuracy, become zero. Experiment has confirmed the forecast. While the resistance at 13.9 K ... is still 0.034 times the resistance of solid mercury extrapolated to 0 °C, at 4.3 K it is only 0.0013; and at 3 K it falls to less than 0.0001'. Within a few weeks the true significance of what was going on had been appreciated: a second paper, entitled 'Disappearance of the electrical resistance of mercury at helium temperatures' sets an upper limit on the resistance at 3 K of 3×10^{-6} Ω, one ten-millionth of that at 0 °C. Onnes concluded that by this means 'conductors of zero resistance are obtainable': he had discovered the phenomenon of *superconductivity*.

Like liquid helium, superconductivity remained until 1923 exclusive to the Leiden laboratory. Many important features of the phenomenon were soon established. For example, it was found that the onset of the zero-resistance state was abrupt, taking place over an unmeasurably small range of temperature, that this state was destroyed by a sufficiently large current (*critical current*) and that it could also be destroyed by application of a *critical magnetic field*, of a few hundredths of a tesla. These last two observations were unified by Silsbee's conjecture that, as experiment later demonstrated, the critical current was simply the current necessary to produce the critical magnetic field at the surface of the wire. Also, superconductivity was observed in tin and lead as well as mercury (though not in gold or platinum). Finally, it was confirmed that the phenomenon of zero resistance manifested itself not only as a zero potential drop along a current-carrying superconducting wire, but far more sensitively in the failure to decay, over an observable timescale, of a current set up in a ring made entirely of superconducting metal. On observing this stability of a ring current, Onnes commented on the similarity to the molecular currents envisaged by Ampère in his explanation of ferromagnetism.

By the mid-1920s research on superconductivity was started in Toronto and in the Physikalische-Technische Reichsanstalt (PTR) in Berlin as well as Leiden, and many more materials were found to be superconducting; the element with the highest transition temperature, T_c, was niobium (8.4 K). Many alloys and chemical compounds were also found to be superconducting, including copper sulphide and gold–bismuth alloys where neither element was by itself a superconductor; this established that the phenomenon could not be associated with individual atoms.

In 1932 the head of PTR, Walter Meissner, reviewed the current state of experiment and theory in superconductivity, and formulated the most important question regarding superconductivity as follows 'Is the superconducting current just a variation ("*Abart*") of the usual current, or are we dealing with a totally different phenomenon?'. In particular, are the electrons which carry the supercurrent the same ones that carry the current in a normal metal, or are they newly released (e.g. from atomic traps) when the metal becomes superconducting? Meissner gives arguments for each hypothesis, and recounts various experiments intended to shed light on this question, including experiments, at that time inconclusive, on the important question of whether the supercurrent is a bulk or a surface effect. As a result of this work Meissner drew the conclusion that 'the electrons which carry the supercurrent cannot move from the superconductor into a [normal] metal closely attached to it'—an unjustified conclusion from a modern point of view. Towards the end of his discussion he remarks that one of the most important experimental questions is 'whether other physical properties [besides the electrical ones] also undergo a jump on the transition to the superconducting state. In all experiments to date this happens for no other physical property,

Superfluids and Superconductors

which makes superconductivity appear particularly mysterious'. The absence of a jump in the thermal conductivity at T_c seems especially to tell against the first hypothesis above (since if the electrons suddenly cease to be scattered at T_c, why does the thermal conductivity not jump up?) but perhaps 'only a small fraction of the ordinary electrons are superconducting at T_c'. Meissner's final speculation is that a proper explanation of superconductivity would require not only quantum mechanics but also quantum electrodynamics 'which is not yet fully developed'.

Actually, there were few leading theorists of the period who had not tried their hands at the problem, particularly during the years 1928–32 when the recently developed concepts of quantum mechanics held out high promise of a key. Some went so far as to publish papers on it, while others, such as Felix Bloch and Niels Bohr, developed theories which eventually failed to satisfy them and thus were never published. Of the ideas floated in these years, two in particular stand out. The first, originated by Bloch and developed by Landau and Frenkel, was that the ground state of a superconductor is characterized by the existence of spontaneous currents, which, however, flow in random directions and therefore on average cancel until ordered by an externally imposed current. This model, no doubt largely motivated by analogy with a ferromagnet whose different domains are oriented in mutually cancelling ways until aligned by an external field, has not stood the test of time (though it had useful by-products). A second idea, which was advocated by Kronig, Bohr and Frenkel and which strikes more of a resonance today, is that superconductivity must result from electron–electron interactions and consist in a *correlated* motion of all the electrons such that they cannot individually be scattered. However, this correlated state was apparently thought of as *crystalline* in nature, which is far from the modern picture. Over the whole discussion there hung a famous theorem of Bloch, which stated that *no* current-carrying state can be the ground state, so that the supercurrent (as then understood) can only be *metastable*. It is not surprising that little progress was made in this period, since a crucial piece of the puzzle was still missing.

11.1.3. The Meissner effect and other experimental developments in superconductivity up to 1945

The critical year in the history of experimental work in superconductivity is 1933. For the first 20 years it had been almost universally taken for granted that the only significant difference between a superconductor and a normal metal lay in the property of perfect electrical conductivity; consequently theoretical efforts had focused on attempts to show how the collisions which limit the conductivity in the normal state could be switched off. Now, it is easy to show that if a superconductor is no more than a metal with infinite conductivity, then the magnetic flux in the interior of a completely superconducting body cannot change. If one attempts to change it by imposing (or changing) an external magnetic field, the time variation of the magnetic field will give rise to an electric field which, although transient, will set up a circulating current that is just sufficient to cancel the externally imposed field in the interior of the body. By virtue of the infinite conductivity, this current will thereafter never decay (so long, of course, as the external magnetic field is not further changed), and the superconductor will behave as a permanent magnet. A particular and apparently trivial special case is that when the metal is cooled through T_c in a steady magnetic field, the superconductor retains the original flux.

This leads us to ask what is the equilibrium state of a superconducting sphere (for example) at a temperature below T_c and in a weak external magnetic field H ('weak' means less than the critical field at that temperature)? Now, we can reach this state by at least two different paths: in path (a) we first cool the sphere below T_c in zero field and then switch on the field, while in path (b) we first switch on the magnetic field while the temperature is above T_c and only thereafter cool through T_c.

In case (a) it is clear from the above considerations (and had been established by 1933 in numerous experiments) that the magnetic flux cannot penetrate the body, and the field lines behave as in figure 11.2(a). On the other hand, in case (b), when the field is switched on in the normal state, the magnetic field lines penetrate the sample as usual (figure 11.2(b)). When the sample is cooled below T_c, in constant external magnetic field, the above argument indicates that there should be no currents set up and no change in the flux in the interior of the body, so that the final state also should be that represented in figure 11.2(b). If this is correct, the final state of the sphere for the given temperature T and field H is not unique, but is a function of its history. There is nothing intrinsically absurd about this proposition—many other cases are known in physics in which, depending on the history, various 'metastable' states can be generated—but it has the surprising feature that the non-uniqueness persists for arbitrarily small values of the magnetic field, that is, the two different final states can be brought arbitrarily close together without one being able to relax into the other.

As a matter of fact, by 1932 indirect evidence had begun to accumulate, from experiments on the hysteretic behaviour of superconducting wires when the external field on them was cycled, that the above picture might be too simple. Ironically, however, the crucial experiment was actually designed to look for something else. Meissner had at the forefront of his mind the question whether the current in a superconducting wire flowed uniformly over the cross section or only in a thin layer at the surface. Max von Laue, who was a consultant to the PTR, suggested that this question could be resolved by measuring

Superfluids and Superconductors

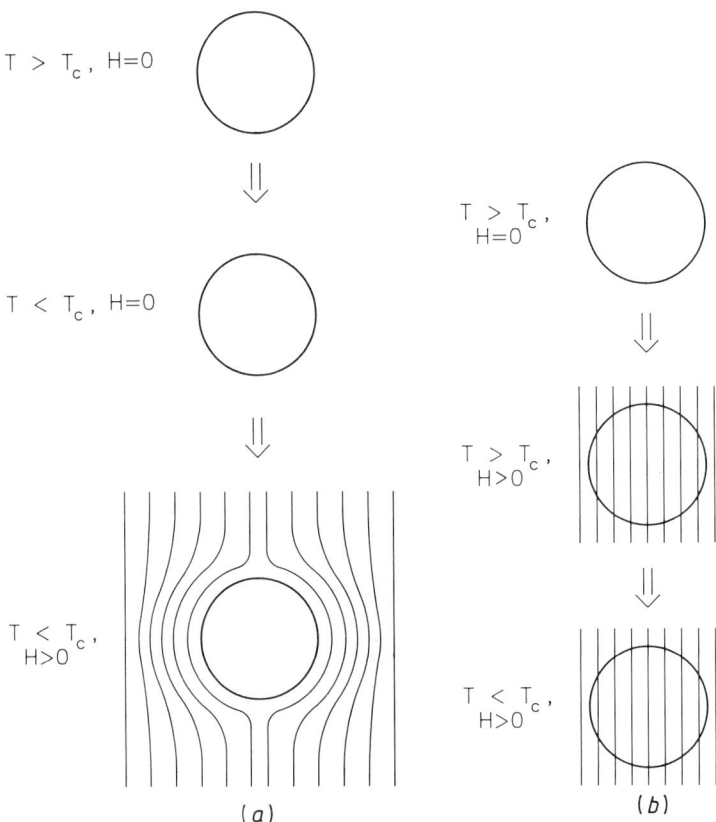

Figure 11.2. *Pre-Meissner expectations about the expulsion of magnetic flux from a superconductor.*

the magnetic field at a small distance from a pair of current-carrying superconducting cylinders. Normally, in such an experiment one would cancel the earth's magnetic field before cooling the sample through T_c (in zero current), but in some of the PTR experiments this was not done (it is still debatable whether on the first occasion this was an accident!) and, as the sample cooled through T_c, a clear jump in the measured field was seen, indicating that all or at least most of the magnetic flux was expelled from the interior of the superconductor: the final state, in this case also, corresponds to figure 11.2(a) and not figure 11.2(b)! (On the other hand, no change was detected in the flux threading a hollow cylinder.)

The fundamental significance of this effect, the Meissner (or Meissner–Ochsenfeld) effect, which was rapidly reproduced at other laboratories, was immediately recognized: superconductivity is more than simply a vanishing of the electrical resistance, it is a state with totally different thermodynamic properties from the normal state of metals. The fact that the magnetic field was actually expelled from the interior of

919

the superconductor shows beyond reasonable doubt that the zero-flux state—that of figure 11.2(*a*)—is the *equilibrium* state of the body, not just a consequence of a failure to relax to true equilibrium. Theorists at last had something to get their teeth into, and progress was not long in coming.

By this time political events in Europe had begun to affect the course of low-temperature physics. Following the accession to power of the Nazi government in Germany, the years 1933–34 saw the exodus of many Jewish physicists, including Kurt Mendelssohn, Franz Simon, Nicholas Kurti and the London brothers, all of whom were attracted to Oxford by F A Lindemann (later Lord Cherwell) at the Clarendon Laboratory, which thereby rapidly became a major centre in experimental low-temperature physics. At the same time superconductivity at the PTR suffered a severe blow when Meissner left for Munich and von Laue was fired by the new Nazi-appointed head. In the next few years politics in the Soviet Union also had an impact. The Soviet experimentalist Pyotr Kapitza, who had almost completed his helium liquefier in Cambridge, was detained on a return trip to Moscow and spent the rest of his long working life there; and the work of the Kharkov laboratory, which despite its recent origin had produced one of the most convincing confirmations of the Meissner effect and pioneered work on superconducting alloys, came to a halt when its leader, A Shubnikov, was arrested in the Stalinist purges of 1937. He died in prison in 1945.

Apart from the Meissner effect, the most significant experimental developments in superconductivity in the years 1933–45 were probably those on superconducting alloys. It had long been appreciated that in some alloys the critical field (the magnetic field beyond which superconductivity disappears) was much larger than in pure elements. It was primarily Shubnikov's group in Kharkov who showed that in many alloys there are *two* 'critical fields' (called by Shubnikov H_{c1} and H_{c2}, a designation that has stuck). Below the 'lower critical field' H_{c1}, the field cannot penetrate (just as in pure elements). However, between H_{c1} and the much larger 'upper critical field' H_{c2} the field can penetrate partially, without destroying superconductivity, which vanishes only when H reaches H_{c2} (figure 11.3). This is the property which, much later, allowed coils made of alloy to become the routine method of producing very strong magnetic fields. The characteristic behaviour of alloys, including marked magnetic hysteresis, is known as type-II superconductivity, in contrast to the type-I behaviour of pure elements, and the region between H_{c1} and H_{c2} is known as the *mixed state*.

The historical development of our understanding of superconductivity would no doubt have been smoother and more logical—though it is not clear that it would have been any faster—had type-II superconductivity never been discovered until a microscopic theory of the type-I version was in place. As it was, these new experimental developments gave rise to a considerable amount of debate and, in retrospect, confusion. It be-

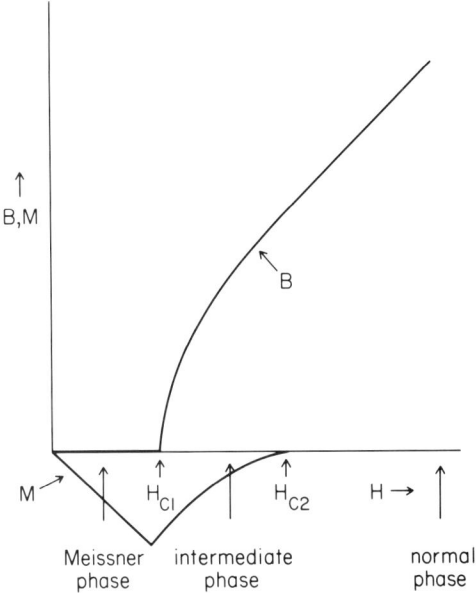

Figure 11.3. *The behaviour of the magnetization (M) and magnetic induction (B) of a 'type-II' superconductor as a function of external magnetic field H.*

came clear only slowly that the mixed state of alloy superconductors between H_{c1} and H_{c2} had essentially nothing in common with the intermediate state of elemental superconductors, in which one also gets partial flux penetration; the latter, unlike genuine type-II superconductivity, can be accounted for in terms of macroscopic energetics. It was also debated whether the characteristic features of alloy superconductivity were really a thermodynamic equilibrium effect, as claimed by Gorter and H London, or rather a consequence of hysteresis, as postulated in the *sponge model* of Mendelssohn. But by the mid-1960s the Gorter–London view was generally accepted.

Other experimental advances in this period will be mentioned later for their theoretical implications, but one discovery must be mentioned for its technical consequences. Justi and his group at the PTR discovered, in 1941, superconductivity in niobium nitride at 15 K, a record transition temperature at the time, and significant because it meant that liquid hydrogen could in principle replace liquid helium as a coolant. This was the forerunner of a whole class of superconducting materials with transition temperatures ~16–24 K and which, though not disordered, show characteristic alloy behaviour.

11.1.4. Liquid helium: the experimental revolution

At the beginning of 1936 liquid helium was still just one more inert-gas liquid, which was peculiar only in a few respects, mainly that it did not solidify under its own vapour pressure and that it appeared to have two different liquid phases, He I and He II, with the curious 'lambda-transition' between them at 2.17 K. Within three years the picture had changed totally, and it was clear that liquid helium had properties qualitatively different from those of any other known substance. While the Leiden and Toronto laboratories were involved in the earlier phase, the main breakthroughs came in the single year 1938, almost entirely from the more recently established low-temperature groups at Oxford and Cambridge in the UK and Moscow and Kharkov in the Soviet Union.

It had been noted by McLennan in Toronto in 1932 that when helium which had been boiling furiously just above the λ-temperature was cooled into the He II phase, the boiling abruptly stopped and the liquid became quiescent. The significance of this phenomenon became clear in 1936, when the Keesoms found that the thermal conductivity of He II was greater than that of He I by a factor of at least 3×10^6, a behaviour they named 'supra-heat-conductivity'. A year later a team in Cambridge found that the heat current in He II was not even proportional to the temperature gradient as in ordinary materials, so that a thermal conductivity in the normal sense could not even be defined.

This might have been explained as convective transport of heat, if the viscosity of He II had been low enough. But although the viscosity was indeed found to decrease rapidly below the λ-point, the reduction was not nearly enough. This, together with some observations of the strange propensity of He II to leak from otherwise sound containers, which had apparently been common knowledge among experimenters but had not previously been thought worthy of systematic investigation, suggested to Kapitza in Moscow and to Allen and Misener in Cambridge that a direct measurement of the viscosity by flow through thin capillaries might be of interest. They were indeed: the two papers, published simultaneously in 1938, showed unambiguously that the viscosity of He II when measured in this way must, if it existed at all, be less than one part in 1500 of that of He I! For this totally unexpected and unique behaviour Kapitza coined, by analogy with superconductivity, the term *superfluidity*.

This was not the only remarkable property of He II that now came to light. If the liquid was placed in an open-ended U-tube with the bottom packed with fine powder (so that it, unlike an ordinary liquid, could by virtue of its superfluidity leak through it) and one side was heated by shining a flashlight on it, it rushed from the cold to the hot side—so violently that a 'fountain' could be produced! (figure 11.4). Later, the inverse of this effect was discovered—an increase of pressure could cause a rise in temperature (*mechanocaloric effect*)†. Another curious effect which

---

† H London applied to these phenomena a thermodynamical argument devised by Kelvin to relate thermoelectric effects, and in this way demonstrated the quantitative connection between them.

Superfluids and Superconductors

Figure 11.4. *The fountain effect in superfluid liquid ^{4}He.*

had been briefly noted by Onnes, but apparently not taken seriously, was rediscovered: if a small beaker is placed inside a larger one, with different levels of fluid in the two, they rapidly equalize, even though the rim of the smaller beaker is well above either surface! It seemed as if a thin film of liquid helium covered the whole surface of the inner beaker, and that below the λ-temperature this film, like the helium in narrow capillaries, became superfluid. This was soon verified, and the thickness of the film estimated to be 20–30 nm—a conclusion which was confirmed directly and elegantly by optical methods a dozen years later.

By the end of 1938 it was clear that superfluid helium possessed at least as many bizarre properties as superconducting metals. The stage was set for a theoretical synthesis.

11.1.5. Theoretical developments, 1933–45
The first hint that the superconductors and liquid He II might have something in common came from a study of the thermodynamics of the two phase transitions. Prompted in part by the observation that while the specific heat appears to be discontinuous across the λ-transition there is no latent heat, i.e. no discontinuity in the entropy, Ehrenfest formulated in 1933 the general concept of a higher-order phase transition, in which the first derivatives of the free energy (entropy, volume) are

continuous but higher-order derivatives could undergo a discontinuity. In Ehrenfest's classification, second-order phase transitions (as identified by the behaviour of the specific heat) included not only the λ-transition of helium but also the transitions of metals such as iron to the ferromagnetic state and of some binary alloys to the ordered state in which the two types of atoms were arranged on regular sublattices. Shortly thereafter, the superconducting transition was added to the list when measurements at Leiden showed a specific heat jump there also. As discussed in Chapter 7, Landau developed a theory of such transitions in terms of an *order parameter* which grew from zero as the temperature was lowered through T_c. But although the nature of the order parameter was clear enough for magnetic transitions and binary alloys, it remained obscure for liquid helium and superconductivity. In the latter case, it was tentatively identified by Landau, in one of the most interesting of the pre-Meissner theoretical papers on superconductivity, as the critical current. In Gorter and Casimir's two-fluid model, the fraction of electrons which acquired superconducting properties rose from zero at T_c to 1 as the temperature decreased to zero, and served as an order parameter. Like Bohr and Kronig, they visualized the superconducting electrons as forming some kind of crystalline lattice.

Ehrenfest's work and the discovery of the Meissner effect gave a fillip to attempts to apply standard thermodynamic arguments to the superconducting transition. Ehrenfest's student Rutgers formulated a relation between the critical field and the specific heat of the normal and superconducting states, and this was rapidly verified experimentally. Shortly thereafter Gorter and Casimir gave a general discussion of the transition in samples of different shapes, showing in particular that under certain conditions it would be energetically advantageous for normal and superconducting regions to be interleaved—the configuration now known as the *intermediate state*.

The first decisive theoretical breakthrough in superconductivity was made in 1935 by the brothers Fritz and Heinz London. They observed that since superconductors do not obey Ohm's law (steady-state current proportional to electric field) it is most natural to suppose that it is the *acceleration* of the current which is proportional to field (Box 11A). This assumption, combined with Maxwell's electromagnetic equations, leads to the conclusion that in a superconductor in equilibrium the magnetic field consists of two parts: a part which dies off exponentially with distance from the surface over a layer (now called the London penetration depth) of thickness $\lambda_L \sim 10^{-5}$ cm† and a *frozen-in* field which is simply the field the sample possessed when cooled into the superconducting

---

† *This conclusion had actually been drawn ten years earlier by de Haas Lorentz, but the paper was in Dutch and not widely read. It was also derived in a 1933 paper by Becker, Heller and Sauter, and slightly later by Braunbek.*

state. But given Meissner's disproof of the frozen-in fields, they proposed to start instead from an equation—now known as the London equation (equation (A3) in Box 11A), which rather than relating the *acceleration* of the current to the *electric* field, relates the *equilibrium* value of the current to the *magnetic* field. This leads (in a simply connected superconductor) to a field that contains only the first of the above two components, and therefore accords with Meissner's observations. They further supposed that the n which enters the definition of λ_L (equation (A2)) is the number of superconducting electrons and is temperature-dependent; in particular the penetration depth should become infinite as T_c is approached. In a second paper the Londons also discuss the case of a *multiply connected* superconducting body (such as the rings used by Onnes), and conclude that in this case a frozen-in flux and persistent current can exist, in accordance with experiment†.

The phenomenological London equation is now generally believed to give a correct description of most of the electromagnetic properties of superconductors in thermodynamic equilibrium. The most interesting and prescient part of the Londons' work, however, is their speculations about the microscopic origins of this equation. They recognized clearly that it would follow, if one made the assumption (see Box 11A) that the wavefunctions of the electrons are unaltered by a magnetic field, somewhat as happens with the electron in a hydrogen atom. The system would, in fact, be behaving exactly like a huge atom!

But why should the wavefunctions of the electrons be unaltered by the magnetic field, quite unlike what happens in normal metals? 'Suppose the electrons to be coupled by some form of interaction, in such a way that the lowest state may be separated by a finite interval from the excited ones. Then the disturbing influence of the field on the eigenfunctions can only be considerable if it is of the same order of magnitude as the coupling forces.' This was to prove a key to the development of a microscopic theory.

The work of the Londons made an immediate impact, and a number of experiments were undertaken to verify its predictions, in particular that the penetration depth should tend to infinity as the transition temperature T_c is approached. In a classic series of experiments on colloidal mercury, Shoenberg determined the temperature variation of $1/\lambda_L^2$ (i.e. the 'number of superconducting electrons'). Some years later his result was shown to fit rather well the formula $1 - (T/T_c)^4$, precisely the temperature-dependence predicted by Gorter and Casimir's two-fluid model. In addition, experiments on (the lack of) thermoelectric effects in the superconducting state conducted on the eve of World War II (but

---

† In their 1935 paper, the Londons come within a hair's breadth of predicting the phenomenon we now know as 'flux quantization' (F London in fact predicted it explicitly in 1948).

BOX 11A: THE LONDON EQUATION

An electron, of mass m and charge e, not subject to collisions, acquires drift velocity v from an electric field E in accordance with the equation
$$\dot{v} = eE/m.$$
In a gas of n electrons per unit volume, the current density J is nev, whence it follows that

$$\dot{J} = ne^2 E/m. \quad (A1)$$

This is the acceleration equation for current. Now in a changing magnetic field, Faraday's law of induction can be written $\mathrm{curl} E = -\mu_0 \dot{H}$. Combined with (A1) this gives

$$\lambda_L^2 \mathrm{curl} \dot{J} + \dot{H} = 0 \quad \text{where } \lambda_L^2 = m/\mu_0 n e^2. \quad (A2)$$

The Londons proposed to drop the time-derivatives and take as the basic equation for the supercurrent

$$\lambda_L^2 \mathrm{curl} J + H = 0 \quad (A3)$$

where λ_L is the London penetration depth.

In a simply connected superconductor (A3) is equivalent to the relation
$$J = -\lambda_L^{-2} \mu_0^{-1} A = -(ne^2/m) A \quad (A4)$$
where A is the electromagnetic vector potential, defined by $\mathrm{curl} A = H$. In their paper the Londons remark that this relation may be intuitively understood as follows: for a single electron the expression for the electric current may be written

$$J = -(i\hbar e/2m)(\psi \nabla \psi^* - \psi^* \nabla \psi) - (e^2/m)|\psi|^2 A \quad (A5)$$

CONTINUED

only reported in full after the war) were interpreted as suggesting the idea of an energy gap, though no direct measurement of its value was made until much later.

Meanwhile, the question of the nature of the He II phase was also of great interest to theorists, particularly after the rash of discoveries in 1938.

BOX 11A: CONTINUED

In a normal metal the wavefunctions ψ deform in response to the application of a potential A in just such a way as to cancel the principal effects of the last term, so that to a first approximation the current is zero. However, for $A \to 0$ this deformation requires, according to the principles of perturbation theory, mixing in to the original ($A = 0$) wavefunction ψ_0 states of arbitrarily low energy. If for some reason no such states exist in the superconductor, then even for finite A we may have $\psi = \psi_0$, $\nabla \psi = \nabla \psi_0 = 0$, and only the second term is left. Since $|\psi|^2$ is the single-electron probability density, summation of (A5) over all the electrons leads directly to (A4).

Once the idea that the λ-transition was an 'order–disorder' transition was established, the most obvious hypothesis was that it corresponded to some kind of short-range positional ordering of the atoms of the liquid, and such a scenario was considered by Fröhlich, Fritz London and others. However, the fact that x-ray diffraction studies seemed to show no difference between the ordering of the atoms in He I and He II was discouraging. In 1938 London came up with a radically new hypothesis: it had been pointed by Einstein in 1924 that a set of non-interacting atoms[†], if cooled to a sufficiently low temperature, would undergo the phenomenon now known as *Bose* (or *Bose–Einstein*) *condensation*, in which a finite fraction of the atoms, which tends to one as the temperature falls to zero, occupies the single state with lowest energy. Possibly in part because of objections raised by Uhlenbeck, this proposal had apparently not been taken very seriously. London now resurrected it and, noting that the temperature T_c estimated for a gas of free atoms at a density equal to that of liquid helium, (3.13 K) was not so different from the observed λ-temperature (2.17 K), proposed to identify the λ-transition with the onset of Bose condensation. He supported this identification by pointing out that the specific heat of the He II phase, while not identical to that expected for a Bose-condensed gas, was qualitatively similar. Tisza quickly followed with his two-fluid model, the condensed and non-condensed atoms acting as two mutually interpenetrating fluids; the condensed atoms are responsible for the characteristic superfluid properties while the rest (*normal*) behave much like atoms in an ordinary gas. Thus 'the viscosity of the system is due entirely to the atoms in excited states'. Tisza recognized clearly that

---

[†] Later it became clear that only atoms of integral total spin, such as ^{4}He but not ^{3}He, obey Bose–Einstein statistics.

in measurements, for example, of the damping of an oscillating cylinder the normal atoms would confer viscosity, whereas only condensed atoms would flow through a superleak, without viscous drag. He also predicted the existence of a 'temperature wave' in which the two sorts of atoms oscillated relative to one another, with no net oscillation in the total density, but the idea was criticized, among other things, because there was no obvious way of inhibiting collisions between the normal and the condensed atoms.

Although the war seriously disrupted low-temperature research in Western Europe and America, rather surprisingly its impact was less in Moscow, where Kapitza continued his experiments and Landau his related theoretical work. In 1941 there appeared a classic paper in which Landau did for superfluid helium approximately what the Londons had done for superconductors. Starting from a general consideration of the quantum mechanics of a liquid of Bose particles at zero temperature, he deduced that the low-lying excited states of such a system could be described in terms of *elementary excitations*—entities which carry a definite momentum p and energy ε (the energy depends on the momentum). Provided there are not too many of them, the total momentum and energy of the system is just the sum of that carried by the elementary excitations. The ground state of the system is free of elementary excitations, but except at zero temperature a certain number will be thermally excited with an energy distribution appropriate to a Bose gas of particles which can be created or destroyed. The concept of an elementary excitation (or *quasiparticle*) is a very general and important one which has been applied to many other condensed-matter systems besides superfluids and superconductors. In one special case, that of quantized lattice vibrations or phonons in crystalline solids, the concept was already implicit in the standard description, but Landau's 1941 paper probably marks its first appearance in a more general context.

Landau deduced that in a quantum liquid the excitations were of two kinds. The first are phonons, quantized sound waves for which ε and p are related by the equation $\varepsilon = cp$, c being the velocity of sound; these dominate the thermodynamics at temperatures well below the λ-point, giving a specific heat proportional to T^3. The second branch of excitations, *rotons*, corresponded to quantized rotational motion (vorticity) and have a finite energy gap Δ: in the original paper Landau assumed that the minimum energy occurred for $p = 0$, so that the spectrum is $\varepsilon(p) = \Delta + p^2/2\mu$ (where μ is a sort of 'effective mass' for the roton), but he was later forced by experiment to modify this to $\varepsilon(p) = \Delta + (p-p_0)^2/2\mu$, i.e. the lowest-lying roton has a finite momentum, p_0. In this paper he gave a famous argument to the effect that if the liquid flows through a capillary with velocity v, it will be unstable against the creation of an avalanche of elementary excitations as soon as v exceeds a *critical velocity* v_L (Landau critical velocity) given by the minimum

See also p 1363

value of $\varepsilon(p)/p$. For the assumed spectrum v_L is finite (it is the smaller of the quantities c and $(2\Delta/\mu)^{1/2}$) and therefore it is possible that the flow is dissipation-free (superfluid) for $v < v_\text{L}$. Landau himself was careful to state that the criterion $v < v_\text{L}$ was a necessary condition for superfluidity but not a sufficient one; this caveat seems not always to have been remembered by subsequent workers.

He now proceeded to derive a conceptually much more satisfactory version of the two-fluid model originally suggested by Tisza. He assumed that one could think of an unexcited part of the liquid and a 'gas' of elementary excitations, like two interpenetrating fluids obeying different laws of motion. However, he stressed that this model should not be taken too literally and was only a manner of speaking. The hydrodynamic equations obeyed by the gas of thermally excited elementary excitations—the *normal component*—were identical to those of an ordinary liquid; in particular the drift velocity v_n vanishes at the walls of the containing vessel. By contrast, the velocity v_s of the unexcited superfluid component was postulated to be *irrotational*—that is, in a simply connected bulk sample the integral of v_s around any closed curve should be zero (cf section 11.2.6). This condition is similar to that satisfied, in the London theory, by the electrical current in a superconductor in the absence of electric and magnetic forces. In a remarkable *tour de force*, Landau succeeded in deriving from these simple considerations a complete 'two-fluid hydrodynamics' for the description of a superfluid liquid. In this description the entropy S is associated entirely with the normal component and can be calculated if one knows the excitation spectrum $\varepsilon(p)$ of the elementary excitations: the apparent infinite thermal conductivity is simply due to convective counterflow of the two components. The fact that the superfluid carries zero entropy offers an immediate explanation of the fountain and mechanocaloric effects, since in those experiments it is only the superfluid which flows through the superleak.

From his two-fluid hydrodynamics, Landau (apparently unaware, because of war-time conditions, of Tisza's detailed work) also predicted a second type of sound wave. However, he did not identify it as a pure temperature oscillation, so that the first attempts to detect it, which used mechanical excitation, were unsuccessful. It was only after Lifshitz pointed out that temperature changes would be a far more effective excitation mechanism that Peshkov, using a heater and thermometer, detected this wave in 1944. Landau also pointed out that the moment of inertia of the helium in a rotating bucket would be proportional to the 'density of normal excitations' ρ_n and hence could be used to measure the latter, as was done in a famous experiment by Andronikashvili at the end of the war period.

11.2. The period 1945–70

11.2.1. Liquid helium

While World War II interrupted low-temperature research in the USA and Western Europe, it had some beneficial side effects. Technological spin-offs from war-time research included sophisticated electronic techniques and the development of the Collins helium liquefier, which made research at helium temperatures accessible to a large number of laboratories. On the theoretical side, Bohm's war-time work on plasma problems related to the isotopic separation of uranium led him, together with Pines and others, to investigate the problem of the electron gas in metals—thus making an essential contribution to the eventually successful theory of superconductivity.

> See also p 1361

The year 1946 not only marked the resumption of large-scale low-temperature research in the West, it is also a landmark year for the first genuinely microscopic theory of either helium or superconductivity that has stood the test of time, namely N N Bogoliubov's theory of the properties of a dilute Bose gas with weakly repulsive interactions. Starting from London's assumption that such a gas would undergo Bose–Einstein condensation he developed a systematic field-theoretic perturbation theory to show that the long-wavelength energy spectrum would have the phonon form predicted by Landau. At shorter wavelength the atoms behave essentially as if free. This work is important not only in its own right, but as the beginning of a spate of theoretical papers which treated various many-body problems by the methods of quantum field theory.

Meanwhile, continued measurements by Peshkov of the velocity of second sound had failed to give particularly good agreement with either Tisza's or Landau's theory, and in 1947 led Landau to modify his spectrum so that the rotons were no longer a different branch, but joined continuously on to the phonon spectrum (figure 11.5). Tisza continued to maintain his own theory, and the issue was not settled in favour of Landau's prediction until 1950.

The physical meaning of the Landau spectrum was much clarified by Feynman and Cohen. Using a variational wavefunction, Feynman was able to show that only the low-momentum excitations have the phonon form postulated by Landau, but that at higher momenta the motion of a particular atom would induce a backflow of the rest of the liquid. Using this idea, he obtained precisely Landau's predicted dip in the energy spectrum at high momenta (figure 11.5). A few years later $\varepsilon(p)$ was measured directly by neutron scattering, and the theoretical model confirmed. It has to be said that despite this work the detailed microscopic nature of the 'roton' excitations has remained somewhat obscure to this day. The general opinion, however, is that they have

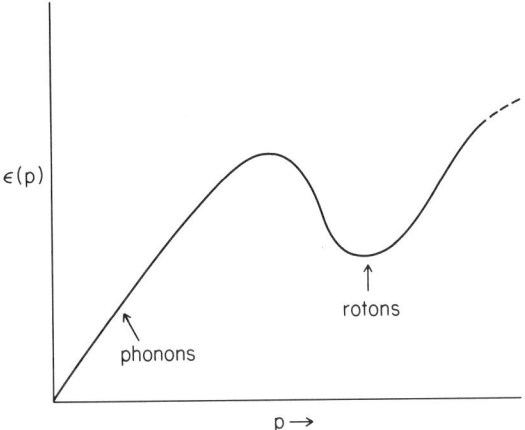

Figure 11.5. *The excitation spectrum of He II, as postulated in Landau's 1947 paper and subsequently confirmed by experiment.*

at most a distant connection with the original ideas about quantized vorticity in which their name originates.

In fact, ideas about vorticity in superfluid helium developed in a quite different direction. According to Landau, the velocity field of the superfluid should be irrotational, which means that at sufficiently low temperature, when there is essentially no 'normal component' left, liquid helium should not be able to rotate at all, and therefore in a rotating bucket would not take up a parabolic surface. It was therefore a considerable surprise when in 1950 the first experiment of this type showed that He II, even at low temperatures, behaved like any other liquid. The explanation was given by Onsager in a famous conference discussion remark, and later in a published paper by Feynman: the liquid acquires vortex lines on which the density of superfluid is zero, and around which the superfluid flows with quantized circulation, as will be described in more detail later. When the container is rotated, the vortices arrange themselves so that the macroscopically averaged velocity is the same as in an ordinary liquid, and the macroscopically averaged surface (all that could be measured in the experiment) is correspondingly identical.

See also p 944

Evidence for Onsager's hypothetical vortices was soon forthcoming: Hall and Vinen showed that the more severe damping of second sound in a rotating container could be explained quantitatively by the presence of vortices, and in 1961 Vinen's vibrating-wire experiment showed the *quantum of circulation* to be close to the expected h/m. In recent years even more direct observations of the pattern of vortices in the rotating liquid have been made by electron microscopy: since free electrons preferentially travel along vortex lines, the pattern of their arrival at

931

a detector placed above the liquid images the pattern of the vortices. These vortices play an important role in many experimental properties of He II; in particular, it is believed that under most conditions its inability to sustain persistent currents above a 'critical velocity' v_c is due to the fact that pre-existing vortices become free to move across the current, thereby reducing it.

Among many other experiments which probed the conceptual foundations of the theory of superfluidity, two are particularly noteworthy: in one, suggested by London and carried out by Hess and Fairbank, the liquid was cooled through the λ-temperature while rotating very slowly (so slowly that not even one vortex line could be formed) and it was observed that the *equilibrium* angular momentum decreased in proportion to the normal density. While the apparently similar experiment of Andronikashvili is the analogue of the failure of magnetic flux to penetrate an already superconducting sample, this experiment of Hess and Fairbank is the analogue, for a neutral system, of the Meissner effect. The second experiment by Kukich, Henkel and Reppy, probed the critical velocity very close to T_λ, and confirmed a theoretical prediction that the circulating-current state should not in fact be perfectly metastable: rather, it should be possible to make transitions, by a process of thermal activation, to lower-energy (and lower-current) states, leading to an observable decay of the current with time.

Despite the very considerable body of circumstantial evidence that accumulated in these years in support of the theoretical picture of superfluidity, it was always a source of unease that the phenomenon held to be the key to it, Bose condensation, had never been directly observed. Neutron scattering experiments have been interpreted as showing the existence of a finite condensate fraction (of the order of 10%), but cannot be said to have demonstrated it beyond reasonable doubt. Indeed, to this day a direct demonstration of the existence and magnitude of the condensate in ^{4}He is lacking.

11.2.2. *Superconductivity—experiment and phenomenology, 1945–56*

The idea that, for the wavefunction of the superconducting state to possess the rigidity required by the London equation, a minimum energy was needed to excite single electrons—the *energy gap*—was emphasized by Ginzburg in his 1946 book. It was also a central result of a microscopic theory by Heisenberg and Koppe, which, although eventually abandoned, had a big influence in the late 1940s. In their 1946 report of their experiments conducted on the eve of the outbreak of war, Daunt and Mendelssohn inferred the existence of such a gap from the absence of a thermoelectric effect (an inference which in the light of modern ideas is not compelling). More convincing evidence came in 1953–54, with experiments on the thermal conductivity and specific heat which indicated rather firmly that the 'density of superconducting electrons' decreased *exponentially* as $T \to 0$, as would be expected if there were an energy gap. Yet more direct evidence came from measurements of electromagnetic absorption in the superconducting state, which occurs above a certain critical frequency, with a fairly abrupt threshold. (To conclude the story, in 1960, when the BCS theory was already established, a classic experiment by Giaever gave what is still the most direct evidence for the gap: when a normal and a superconducting metal are joined by a thin insulating barrier, no current flows until the voltage bias reaches a certain value—that necessary to overcome the energy gap in the superconductor.)

A very critical experiment was on the *isotope effect*—the influence of the isotopic mass of an element upon its superconducting transition temperature. Ironically, this question had been of interest (under the influence of now discarded ideas about ionic 'vibrators') in the very early days of the subject, and Onnes sought an effect in lead, but without success. The development of isotopic separation during and after the war helped, and in 1950 two groups simultaneously announced that the transition temperature of mercury decreased with increasing isotopic mass. It was immediately clear that the mechanism of superconductivity must have something to do not so much with the interaction of electrons with static lattice ions (which would be independent of isotopic mass) as with the dynamics of the lattice (i.e. the phonons). Coincidentally, Fröhlich had developed a theory in which the electron–phonon interaction played a vital role, and which predicted that T_c should be inversely proportional to the square root of the isotopic mass, as was immediately verified for isotopes of tin.

But an even more crucial role in the development of the theory in this period was played by the experiments of Pippard in Cambridge on the penetration depth. Starting from an idea of H London, that a magnetic field should affect the 'number of superconducting electrons' and therefore the London penetration depth λ_L (hereafter λ), Pippard decided to measure the dependence of λ on field (from the change of resonant frequency of a microwave cavity containing the specimen). The result was a surprise: λ varied much less with field than predicted. It appeared that whatever was responsible for the London equation was not itself confined to the London penetration depth! Thus Pippard was led to postulate that there was some kind of 'long-range order' in the superconductor which extended over perhaps 20 times the zero-temperature penetration depth λ_0, or about 10^{-4} cm. Noting that specimens much smaller than this (indeed smaller than λ_0) had been shown to be fully superconducting, Pippard commented that 'the range of order must therefore not be regarded as a minimum range necessary for the setting up of an ordered state, but rather as the range to which order will extend in the bulk material'.

An even more exciting result was found when Pippard went on to show that the penetration depth increased markedly with alloying, while the thermodynamic properties were essentially unchanged. Since in the simple London theory it is the same quantity, n_s, which determines both the thermodynamics and the penetration depth, something had to give. Part of Pippard's PhD thesis had been concerned with the anomalous skin effect in non-superconducting metals which arises when the current is confined to a surface layer thinner than the electronic mean free path. The relation between current and field then becomes *non-local* (see Chapter 17). In his thesis Pippard had noted that a parallel consideration should apply in Heisenberg and Koppe's model of superconductivity. While the London equation relates the current $J(r)$ directly to the vector potential $A(r)$ at the same point, a non-local theory would relate it to some average of A over a region around that point. He now realized the possible significance of this observation: the range over which the non-locality extends is essentially the 'range of order' (*coherence length*) introduced previously. When the mean free path is reduced to much less than this length, e.g. by alloying, only a fraction of the region which 'ought' to be contributing to the response does so, so the 'effective' number of superconducting electrons goes down and the penetration depth correspondingly increases. The microscopic meaning of Pippard's coherence length became beautifully clear with the BCS theory.

Meanwhile, the phenomenology of superconductivity was being developed in other directions. In 1950 Fritz London (now long settled at Duke University in North Carolina) published the first volume on superconductivity of his two-volume series *Superfluids*. As the name implies, he strongly emphasized the parallels between the superconductivity of metals and the superfluidity of He II, and, among other things, reiterated his recent prediction that the magnetic flux through a thick superconducting ring should be *quantized* in units of h/e (h = Planck's constant, e = electron charge). At the time, the anticipated difficulty of the experiment apparently deterred people from trying to verify this remarkable prediction; had they done so, they would have had a shock and the history of the subject might have gone a little differently!

Unknown to most people in Western Europe and America because of Cold War conditions, a momentous development was meanwhile taking place in Moscow. Ginzburg and Landau, building in part on London's ideas, attempted a specific realization for superconductors of Landau's earlier general theory of second-order phase transitions. With another of the inspired guesses which have been characteristic of the history of the theory, they identified the order parameter of the Landau theory with a *quantum-mechanical wavefunction* such as the one invoked intuitively by London—they did not say of what—and proceeded to treat its interaction with a magnetic field just as one would for a single electron. In the Ginzburg–Landau theory there is a sort of kinetic energy associated with the spatial variation of the order parameter (just as there would be for a single electron), and one can therefore define a (temperature-dependent) characteristic length $\xi(T)$, namely the length ξ over which the order parameter has to be 'bent' before the required energy exceeds the condensation energy of the superconducting state. Since the theory also incorporates London's formula for the penetration depth $\lambda(T)$, one can form a dimensionless ratio κ of these two quantities, which turns out to be temperature- independent and is therefore an intrinsic characteristic of the material. Ginzburg and Landau noted that if κ were greater than $1/\sqrt{2}$, the effective surface energy between a superconducting and a normal region of the metal would become negative, which would allow superconducting filaments to persist in the normal metal beyond the thermodynamic critical field. A very similar idea, though phrased in terms of his coherence length, had been put forward by Pippard (and indeed earlier by Gorter) and used to explain the flux penetration in superconducting alloys. However, Ginzburg and Landau surprisingly remark that 'from the experimental data, it follows that κ is always $\ll 1$' and that it was therefore unnecessary to examine what would happen in the case $\kappa > 1/\sqrt{2}$!

This state of affairs did not last long. Studying results obtained by his experimental colleague Zavaritskii on the magnetic behaviour of amorphous thin films, Landau's student, A A Abrikosov, began to wonder whether the restriction $\kappa > 1/\sqrt{2}$ was really sacrosanct, and to think about what would happen were it to be violated. He concluded that the magnetic behaviour would indeed be that characteristic of superconducting alloys (type II), and, eventually, that the way that the magnetic field would penetrate in the mixed state would be in the form of *vortices*, that is, regions with a core which is similar to a region of normal metal, allowing penetration of the field; currents flow around these regions in such a way as to screen the field out from the rest of the metal. These vortices are in fact the exact analogue, for a superconducting (electrically charged) system, of those which occur in superfluid (uncharged) ^{4}He. This theory was published in 1957, more or less simultaneously with the BCS microscopic theory. Ten years later the predicted *Abrikosov vortex lattice* was observed directly by means of magnetic decoration (a technique in which fine magnetic particles deposited on the surface of a superconductor are attracted to the positions of the vortices), finally dispelling lingering doubts about its reality.

The Ginzburg–Landau theory initially attracted little attention in the West, and not all of that favourable. It was only later, after it had been shown to be derivable under appropriate conditions from the more microscopic BCS theory, that it began to be taken seriously and its formidable potential for solving complicated problems of magnetic behaviour was recognized. Nowadays it is a standard item in textbooks on superconductivity.

11.2.3. Pre-BCS attempts at a microscopic theory

It is somewhat remarkable that while the early period of superconductivity saw a number of attempts at microscopic, first-principles theory, the dozen years following Meissner's momentous discovery brought very few—perhaps because it was now obvious that no simple consideration of scattering mechanisms could solve the problem. The late 1930s saw some qualitative work by Slater and a more quantitative attack by Welker, but neither stood the test of time. When, after the war, renewed attempts at a microscopic model were made, they rather surprisingly echoed the pre-Meissner work of Frenkel and Landau in assuming that a superconductor is characterized, even in the absence of external electric and magnetic fields, by the presence of currents flowing in random directions: the role of external fields is mainly to align these currents appropriately (the force of Bloch's theorem seems to have been less than fully realized in its application to these models). Born and Cheng hoped to avoid Bloch's theorem by taking into account the (static) periodic potential of the ions: in their model, the Coulomb interaction transfers electrons from the valence to odd corners of the conduction band, where they are responsible for spontaneous currents. The model of Heisenberg and Koppe started with electrons moving freely (no lattice potential) but, again under the influence of the Coulomb force, rearranged to produce spatially localized wave packets which then moved in a correlated way under the influence of fields. While this theory did not in the end turn out to have much relation to the true explanation of superconductivity, it had two interesting aspects which foreshadowed some features of the BCS theory. Firstly, it predicted an energy gap for one-electron excitations, and, secondly, it caused the authors to observe that the only reasonable explanation, both of the very small condensation energy of the superconducting state ($\sim 10^{-7}$ eV/atom, compared to typical Coulomb energies of a few eV/atom) and of the enormous variation between systems of the value of T_c (a factor of order 100), must lie in the transition temperature being exponentially dependent on material properties.

Fröhlich's attempt lay closer to what turned out to be the successful line of development. He started from the idea that the polarizability of the ions gives rise to an indirect interaction: as one electron passes by, it polarizes the ionic lattice (figure 11.6) and since the ions are heavy their relaxation is slow, so that a second electron passing by some time after will be attracted to the lingering positive charge density. However, while the idea that this attraction would lead to an instability of the Fermi distribution was to play a crucial role in the BCS theory, Fröhlich's specific conjecture for the solution was eventually shown not to give a Meissner effect; so too with his second theory published four years later, which turned out to be a model not of superconductivity but of the *sliding charge density-wave conduction* now known to be characteristic of some quasi-one-dimensional metals such as $NbSe_3$.

Superfluids and Superconductors

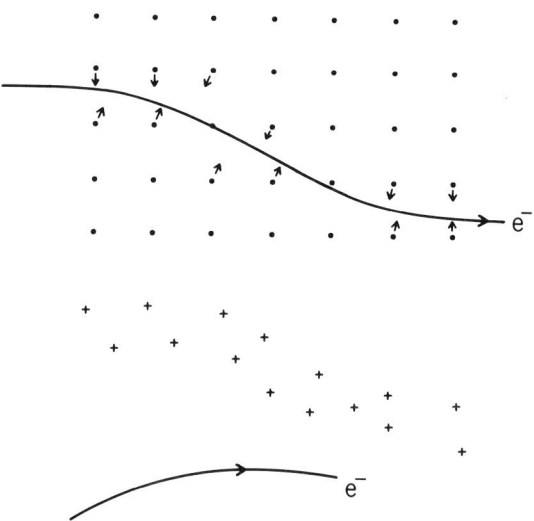

Figure 11.6. *The mechanism of electron–electron attraction in a metal, as envisaged by Fröhlich and subsequently by Bardeen, Cooper and Schrieffer.*

In 1955 Bardeen and Pines went further, taking into account both the ion–electron interaction and the direct Coulomb repulsion between the electrons to produce a more realistic interaction which had a crucial feature: provided the energy exchanged between two electrons near the Fermi surface was not too great, it could be *attractive*. (Fröhlich had already shown that the electron–phonon interaction *alone* would lead to an attraction, but this was not convincing so long as it seemed that it might be outweighed by the Coulomb repulsion.) In retrospect it can be seen that all the pieces were now in place for a definitive microscopic theory of superconductivity.

One piece of work, indeed, which already foreshadowed some of the qualitative features of the BCS theory was that of Schafroth, who in 1954 suggested, in the spirit of London, that superconductivity results from the Bose condensation of quasi-molecules formed by pairs of electrons (an idea which had actually been suggested as long ago as 1946 by R A Ogg in the context of a very specific problem in chemical physics). However, in Schafroth's formulation it was very difficult to calculate specific properties of the system, and in a subsequent paper with Blatt and Butler on the problem of formation of 'molecules' in Fermi systems he confined himself to 'speculations' about superconductivity, without claiming to have a concrete theory.

11.2.4. BCS and its aftermath

The critical breakthrough in our microscopic understanding of superconductivity (at least of the old-fashioned sort!) came in the twelve months July 1956–July 1957, at the University of Illinois at Urbana-Champaign, at the hands of John Bardeen (professor), Leon Cooper (a postdoctoral research associate whom Bardeen had brought from Princeton) and Robert Schrieffer (a graduate student of Bardeen). The first hint came when Cooper, trying out various forms of interaction which he hoped might yield an energy gap, took up the somewhat artificial problem of a single pair of electrons, with total momentum zero, which attracted one another, not in free space but in the presence of the Fermi sea, so that states below the Fermi surface were blocked. He quickly realized that such a pair would form a bound state, that is, a state with energy less than the minimum allowed for freely moving electrons subject to the same restriction. A significant feature was that the magnitude of the binding energy of the molecule-like object so formed depended exponentially on the strength of the interaction—the very feature which Heisenberg had identified as an essential ingredient in any viable theory.

It therefore looked promising to attribute the onset of superconductivity to the formation of many such *Cooper pairs*, and to identify the energy gap, which was becoming increasingly indicated by experiment, with the binding energy of the pair. Clearly, however, the pairs could not be regarded as independent objects, even if one allowed them to undergo Bose condensation *à la* Schafroth: a simple calculation showed that the separation of the pair was so large (typically $\sim 10^{-4}$ cm) that between any two putatively paired electrons there would be millions of others, and the whole notion of identifying 'partners' was meaningless. The critical technical problem, then, was to implement the idea of Cooper pairing quantitatively, taking into account the Fermi statistics obeyed by the electrons. The first step was to guess the nature of the many-body wavefunction; this was taken, in effect, to be simply that of a collection of Cooper pairs all occupying the same state (thus somewhat resembling a Bose condensate) but with the critical proviso that the wavefunction was correctly antisymmetrized to take account of the Fermi statistics. The second, and highly non-trivial, step was to find a way of actually carrying out calculations on a system described by such a wavefunction. This problem was eventually solved with the aid of a trick similar to one used ten years earlier by Bogoliubov for liquid helium and a quantitative theory was born.

The BCS theory of superconductivity, as it is now universally known, starts from the Bardeen–Pines interaction which is attractive at low energies, but it replaces the rather complicated Bardeen–Pines form by a simpler interaction, which is attractive and constant when the electrons are close to the Fermi surface, but otherwise zero (this turned out to be a serendipitous simplification). A guess is then made: the most important interactions are likely to be those between electrons of opposite momentum and spin (as in the Cooper problem), so all others

John Bardeen

(American, 1908–91)

Bardeen studied electrical engineering as an undergraduate at the University of Wisconsin and, after a few years in industry, went on to do his doctorate at Princeton with E P Wigner. He then went back to industrial research, and it was as a member of the technical staff at Bell Telephone Laboratories that he invented, with Brattain and Shockley, the transistor, a discovery which won him the Nobel Prize in 1956. After joining the University of Illinois in 1951 he devoted himself to the problem of finding a microscopic basis for the theory of superconductivity, and in 1956–57 produced, with Leon Cooper and J Robert Schrieffer, the theory which is nowadays universally known as the BCS theory and which won him and them his second Nobel Prize in 1972.

He remained at the University of Illinois until his death in 1991. Bardeen always regarded himself as an electrical engineer as much as a physicist, and at Illinois held chairs in both departments; his theoretical work is characterized by an intimate contact with real experimental systems and a very individual brand of intuition which he sometimes found difficult to convey to his colleagues, but which paid off in many different areas of solid-state physics.

are neglected. The problem is then in essence exactly soluble, and what is found is that indeed at zero temperature the ground state does in some sense represent a Bose condensate of 'diatomic-molecule-like' objects—Cooper pairs—which all have the same wavefunction, corresponding to zero centre-of-mass momentum and, as regards the internal state, zero total spin and orbital angular momentum. To break up a pair and create an excited state takes a minimum energy equal to twice the energy gap, a quantity which turns out to be exponentially dependent on the strength of the interaction and hence on the material parameters. Because of this feature, the electronic wavefunction has indeed the kind of rigidity needed in the London approach to produce the Meissner effect and moreover to guarantee the stability of persistent currents. The Pippard coherence length, in pure metals, turns out to be nothing other than the effective radius of the Cooper pairs: when an electric field is applied to one 'side' of the pair, the effect is automatically propagated over a distance of the order of this radius. At finite temperatures the superconducting state is stable below a temperature T_c, where kT_c is of the same order as the zero-temperature energy gap and depends on the isotopic mass as $M^{-1/2}$. The superfluid density (density of superconducting electrons), and hence the London penetration depth, can be calculated as a function of temperature and turns out to have just the behaviour postulated in the phenomenological London theory.

A very exciting aspect of the BCS theory was that it could make quantitative predictions of various experimental quantities, some of which had not at the time been measured. Particularly striking were the predictions for the ultrasonic attenuation rate and nuclear spin relaxation rate. Both these quantities are in some sense a measure of the number of electrons available to exchange small amounts of energy with other excitations (phonons and nuclear spins, respectively), and because of the energy gap one would not expect the superconducting electrons to contribute; so the naive prediction is that both rates would fall off fast below T_c. However, because of some very specific features of the BCS wavefunction, it turns out that while the naive prediction is correct for ultrasonic attenuation, the nuclear spin relaxation rate is predicted to *rise* just below T_c, before falling at lower temperatures. The confirmation of this counter-intuitive prediction in the spring of 1957 by Bardeen's experimental colleagues Hebel and Slichter was a dramatic event which probably did more than anything to promote the general acceptance of the theory, despite some initial scepticism by the authors of competing theories.

The BCS theory was indeed enthusiastically received and over the next few years spawned literally thousands of theoretical and experimental papers. A particularly noteworthy development occurred in 1959, when Landau's student, Lev Gor'kov, extended the BCS formalism to give a microscopic derivation of the phenomenological Ginzburg–Landau theory (still unappreciated in the West), and identified their order parameter as nothing other than the local energy gap. One striking 'correction', however, had to be made: it was clear that the order parameter has to do with Cooper pairs, and therefore the charge to be associated with the elementary entity is $2e$. An immediate consequence (pointed out explicitly by Byers and Yang on rather more general grounds a year later) was that in a multiply connected superconductor flux should be quantized not in units of h/e—as in London's experimentally untested prediction— but rather in units of half the size, $h/2e$. This prediction was soon confirmed experimentally, and was generally regarded as another striking piece of evidence in favour of the BCS theory. It is amusing to note that Ginzburg had flirted with the idea of an effective charge e^*, but had been warned off by Landau.

The next fifteen years saw increasing refinement and generalization of the theory in different directions. With the help of a general technique for handling the electron–phonon interaction, due to Migdal and Eliashberg in the Soviet Union, detailed quantitative methods were developed to explain, and in one or two cases even predict, the transition temperatures of a large number of materials, even in circumstances where the original BCS model was too crude. This work also helped in understanding why the original model, despite its simplifications, works so well for so many systems. Detailed theories were developed to deal with the effect on superconductivity of both non-magnetic impurities (which have a very small effect, at least as regards the thermodynamics, although they can significantly change the electromagnetic properties) and magnetic impurities (which, because the two electrons of the pair, having different spins, see them differently, are very deleterious). Most importantly, the magnetic behaviour of type-II superconductors was quantitatively understood in the light of the ideas of Ginzburg and Landau and of Abrikosov, which were now recognized as firmly based in microscopic theory. A particularly important point is that type-II superconductors in the mixed state, while they show partial Meissner effect, may not always display zero resistance, because the Abrikosov vortices can move across the current flow and in so doing generate a finite voltage drop. In the development of high-field superconducting magnets—a major technological spin-off—it became essential to find ways of 'pinning' the vortices to the right sort of inhomogeneity. Nowadays superconducting magnets producing fields of up to 20 T, (2×10^5 gauss) with no generation of heat, are almost standard equipment.

11.2.5. The Josephson effect

One development which is of such fundamental conceptual and practical importance that it deserves a section to itself is the work of a young student working in Cambridge in 1962. Brian Josephson was struck by the idea that the 'wavefunction' which describes the centre-of-mass behaviour of the Cooper pairs has not just an amplitude but also a phase, and asked how this would affect the transmission of electrons between two bulk superconductors joined by a thin insulating barrier (such as those used by Giaever for his tunnelling experiments). Performing a calculation along the lines of the BCS theory, he came to the surprising conclusion that one would not only, like Giaever, see tunnelling of the excited single electrons (making allowance now for the existence of an energy gap on both sides) but also tunnelling of the Cooper pairs (now called Josephson tunnelling). This process is a highly coherent one in which the approximately 10^{20} Cooper pairs participate, as it were, as one, and has the remarkable feature that for no bias voltage the current

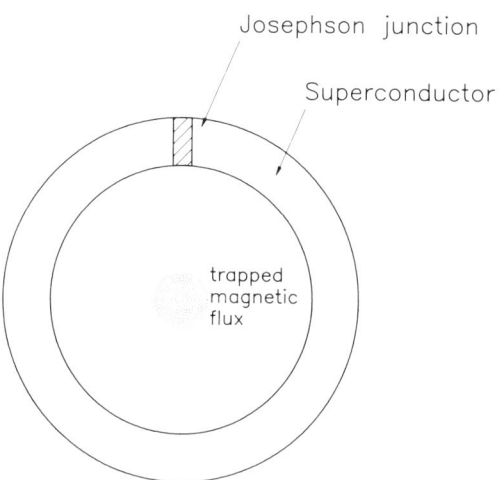

Figure 11.7. *A radio-frequency* SQUID *(superconducting quantum interference device).*

is proportional to the sine of the difference in phase of the Cooper pair wavefunctions in the two superconductors, while for a finite bias voltage V the current oscillates with the Josephson frequency $2eV/h$.

Although Josephson's own attempt to verify his prediction was unsuccessful (it turned out later that, as so often happens, others had probably 'seen' the effect but were unable to interpret it and had therefore discarded the relevant data), confirmation was not long in coming, and the Josephson effect is now the basis of an important class of electronic devices. One reason for its importance is the following: it can be shown that a Josephson supercurrent of maximum magnitude I_c is equivalent to the existence of an energy of magnitude $\hbar I_c/2e$ which depends on the relative phase of the Cooper-pair wavefunctions in the two superconductors. But this relative phase may be regarded as a macroscopic variable, and is associated with various unarguably macroscopic quantities: for example, in a so-called RF SQUID ring (a thick superconducting loop interrupted by a single Josephson junction, as in figure 11.7), there is a close association between the phase difference across the junction and the total magnetic flux trapped in the ring. Since the phase-dependent energy can be very small—of the order of the thermal energy of a single atom at room temperature—we have the extraordinary situation of a *macroscopic* variable which is controlled by a *microscopic* energy. This not only makes possible devices of unprecedented sensitivity but has in recent years allowed various tests to be made of the applicability of quantum mechanics at the macroscopic level—tests which so far the theory has passed with flying colours.

11.2.6. The modern unified picture of superconductivity and superfluidity

The understanding of superconductivity and superfluidity as macroscopic phenomena, which had evolved by the end of the 1960s (and which is still largely accepted) is firmly rooted in the ideas of the Londons and of Landau a generation earlier. In the case of superconductivity the microscopic underpinning was spectacularly provided by the work of BCS, but a microscopic understanding of superfluidity came more gradually and imperceptibly, and even today is not quantitatively as satisfactory as for superconductivity. In this section, I give a brief overview of the current picture, emphasizing the ideas which are common to the two branches of the subject. This section is necessarily somewhat more technical in nature than the rest of the chapter.

Let us start with the superfluidity of He II. The key notion is London's idea that the λ-transition corresponds to the onset of Bose condensation in this system. Generalizing this beyond the original case of thermodynamic equilibrium, we assume that below T_λ, for any state of the liquid, we can find a set of single-particle wavefunctions (in general these will depend on position r, and the choice of them may depend on time) such that exactly one of them—labelled $\chi_0(rt)$—is occupied on average by a number $N_0(t)$ of atoms which is a significant fraction of the total number N, while all the rest will have an occupation of order 1. (It is *not* assumed that N_0 need ever be equal to N, even in thermal equilibrium at zero temperature, because collisions are always scattering atoms between this 'special' state and the rest.) Given this hypothesis, we can define an order parameter (or, as it is sometimes rather misleadingly called, a 'macroscopic wavefunction' of the superfluid) $\Psi(rt)$ by the prescription

$$\Psi(rt) \equiv \sqrt{N_0(t)} \chi_0(rt). \tag{1}$$

Thus $\Psi(rt)$, like the one-particle wavefunction χ_0, is a complex quantity. We analyse it into amplitude and phase

$$\Psi(rt) \equiv A(rt) \exp i\phi(rt) \qquad (A, \phi \text{ real}). \tag{2}$$

By applying the usual expressions for the probability density and particle current of a single particle, we derive the density $\rho_c(rt)$ and mass current $j_c(rt)$ of the condensed particles

$$\rho_c(rt) = (A(rt))^2 \qquad j_c(rt) = (\hbar/m)(A(rt))^2 \nabla \phi(rt). \tag{3}$$

We take the ratio of these quantities to define the superfluid velocity $v_s(rt)$

$$v_s(rt) \equiv j_c(rt)/\rho_c(rt) = (\hbar/m) \nabla \phi(rt). \tag{4}$$

Thus, the superfluid velocity is, apart from a factor, nothing but the gradient of the single-particle wavefunction $\chi_0(rt)$ into which condensation has taken place. It should be emphasized that although the quantity $\chi_0(rt)$ is indeed a perfectly legitimate single-particle wavefunction, its evolution is governed by many-body effects and therefore not described by any simple equation of the Schrödinger type.

From (4) it follows immediately that, for all points in the liquid at which $\chi_0(rt)$ is finite, $\text{curl} v_s = 0$, i.e. the superfluid flow is irrotational. This provides the basis for the two-fluid hydrodynamics of Landau, and from this point on we can essentially just follow his arguments. One thing to note is that the so-called *superfluid density* ρ_s of the liquid, that is, the ratio of the total current associated with v_s to v_s itself, is not in general identical to the condensed fraction N_0. In particular, there are rather general arguments to show that at zero temperature ρ_s is equal to the total density ρ of the liquid, even though N_0 is believed to be less than 10% of N. (In effect, the condensate particles may drag all or, at non-zero temperatures, part of the uncondensed particles along with them.)

Our microscopic identification of v_s, however, permits us to go beyond Landau by considering the case of a multiply connected geometry, e.g. a ring. We cannot now conclude that the integral of v_s around the ring must vanish (since in the space in the middle v_s is, of course, not defined). But the order parameter, like χ_0 in (1), must be single-valued, i.e. its phase $\phi(rt)$ must be well-defined within addition factors of $2n\pi$. This immediately leads to the Onsager–Feynman quantization condition

$$\oint v_s \cdot dl = nh/m \qquad (5)$$

which is nothing but the analogue of the Bohr quantization condition $l = nh$ for an electron in an atom. If it should happen that $\chi_0(rt)$ vanishes along a line within the liquid, then for any contour which encloses this line, condition (5) must be satisfied. In particular, if we consider a straight line and a circular contour of radius R, and take $n = 1$, we have

$$v_s(R) = \hbar/mR. \qquad (6)$$

This is just the classical flow pattern of a vortex line. Note that although the condensate wavefunction χ_0 must vanish at the vortex core ($R = 0$), there is nothing to say that the total density of the liquid must do so. A vortex ring (vortex line joined in a loop) is also a possible pattern.

Let us consider a particular experimental geometry, a fine tube bent into a ring of radius R, which may be rotated at will. We consider two experiments. In the first (the idealized version of the Hess–Fairbank experiment), the liquid is cooled through T_λ while the annulus is rotating with a small angular velocity ω, of the order of the quantity $\hbar/mR^2$. Above the λ-transition the helium behaves much like an ordinary liquid, and rotates with the container. When the liquid is cooled below T_λ,

Superfluids and Superconductors

however, a finite fraction of the atoms have to condense into a single one-particle state, with wavefunction χ_0. Because this is governed by the Onsager–Feynman quantization condition, only discrete choices are possible and the system will choose that single-particle state χ_0 which most nearly corresponds to rotation with the container; in particular, for $\omega < \hbar/2MR^2$ it will come to rest in the laboratory frame. Thus a fraction $\rho_s(T)/\rho$ of the liquid will appear to have ceased to rotate, while the rest will behave much like a normal liquid. While the ring geometry is particularly simple, similar results can be obtained for a simple bucket provided we remember the possibility of vortices.

A second experiment, which at first sight looks similar but is actually conceptually very different, involves starting with the liquid above T_λ in a container rotating rather fast ($\omega \gg \hbar/MR^2$). Once again, the helium comes into rotation with the container. We now cool through T_λ with the container still rotating; when we do so, the state of the liquid does not change perceptibly. This is easy to understand in the light of the above analysis—the liquid had velocity v such that $\oint v \cdot dl$ was already large compared with h/m, so that the Onsager–Feynman condition could be satisfied by a v_s which was imperceptibly different from v. Finally, we stop the rotation of the container, still holding the liquid below T_λ. What we observe experimentally, except when the temperature is extremely close to T_λ, is that the circulating current is persistent and will never die away while the liquid remains below T_λ (though changing the temperature can change its magnitude as more or less of the liquid is found in the condensed state).

It should be emphasized that whereas the behaviour of the liquid in a Hess–Fairbank type of experiment is exactly analogous to that which would be expected (if we could do an analogous experiment!) for a single electron in a hydrogen atom, its behaviour in this second ('persistent-current') experiment is very different. It is clear that the persistent-current state cannot be the state of thermodynamic equilibrium, and in the atomic analogue (the electron excited to a state of high angular momentum) the excited state would rapidly decay to the ground state (s-state). Why does superfluid helium behave so differently? The reason is as follows: if we consider either the wavefunction of the atomic electron, or, in the analogous quantity, the order parameter, for helium, then in both cases for the ground state the 'winding number' n which occurs in the Onsager–Feynman quantization condition, i.e. the number of 'turns' of 2π made by the phase in going around the ring, is non-zero in the initial state but zero in the final state (s-state). It is impossible to go from one state to the other without the wavefunction (order parameter) passing through zero at some point on the ring at some stage. For the atomic electron this can and does happen without difficulty. However, in the case of superfluid helium the strong interactions between the atoms resist the creation of such a zero in the order parameter, and the non-zero n state may remain metastable. One way of changing n without a prohibitive expenditure of energy is to move a pre-existing vortex line—a region where $|\Psi|$ indeed vanishes—across the ring; it can be shown that this can reduce n by 1. Also, extremely close to the λ-transition where the superfluid density and the associated energy are very small, it is possible to create a vortex ring from scratch, by a thermal fluctuation, and then expand it until it fills the whole tube, with the same effect; this is thought to be the mechanism operating in the experiments of Kukich, Henkel and Reppy.

Thus, consideration of Bose condensation plus the effects of interatomic interactions can explain both the Hess–Fairbank result (a thermodynamic equilibrium phenomenon) and the existence of persistent currents (a metastable effect). It is amusing (and rarely noted in the literature of the subject) that the classic experiment on superleak flow as carried out by Kapitza and by Allen and Misener (section 11.1) may correspond, depending on the 'de Broglie wavelength' $\lambda \equiv h/mv_s$ realized in the experiment, to either or both of these effects: if λ is small compared with the length of the superleak, 'superfluidity' must essentially be understood as a metastable phenomenon, while if the opposite condition is realized it is a thermodynamically stable one!

See also p 922

We turn now to superconductivity. The principal complications are (i) that the electrons in metals obey Fermi rather than Bose statistics, so that there is no question of 'Bose condensation' in the usual sense, and (ii) that unlike the atoms of ^{4}He, they are charged. As to the first, BCS showed that the many-body wavefunction of a superconducting system corresponds to a state in which a finite fraction of the electrons (which in their model is unity at zero temperature) pair up to form 'di-electronic molecules' (Cooper pairs) which all have the same two-particle wavefunction ($\chi(r_1\sigma_1 : r_2\sigma_2)$ (where σ_i is the spin projection of the ith electron). Although this is not exactly Bose condensation, it is clear that there is a qualitative similarity. Indeed, for our purposes, as in (1), we may take the 'order parameter' of the superconducting state to be just the 'molecular' wavefunction $\chi_0(r_1\sigma_1 : r_2\sigma_2 : t)$ times the square root of the number of electrons condensed into it

$$\Psi_2(r_1\sigma_1 : r_2\sigma_2 : t) = \sqrt{N_0(t)}\chi_0(r_1\sigma_1 : r_2\sigma_2 : t). \tag{7}$$

Unlike the order parameter of a Bose system, this is a two-particle quantity and, as we shall see in the case of ^{3}He, can have quite a complicated structure. However, it is believed that for superconductors (at least of the pre-1970 sort), the spin wavefunction is just a singlet and moreover the dependence on the relative coordinate ρ is isotropic, with its dependence on $|\rho|$ fixed by the energetics. Thus, if we denote the centre-of-mass coordinate of the pair, i.e. the quantity $(r_1 + r_2)/2$, by r, an adequate approximation to (7) for our purposes is

$$\Psi_2(r_1\sigma_1 : r_2\sigma_2 : t) = \Psi(r, t) \cdot f(|\rho|) \cdot \frac{1}{\sqrt{2}}(\uparrow\downarrow - \downarrow\uparrow) \tag{8}$$

where the last factor is a symbolic representation of the spin-singlet state. It is the centre-of-mass wavefunction $\Psi(r, t)$ which in thermodynamic equilibrium becomes the order parameter of Ginzburg and Landau (and more generally the analogue of the superfluid-helium order parameter defined by (1)). The other terms are constant and may be ignored for present purposes (although, of course, a knowledge of $f(|\rho|)$ is important when we want to discuss the internal properties of the pairs such as the Pippard coherence length.)

We can now proceed exactly as in the helium case, with one important proviso: since electrons, unlike He atoms, are charged, the relation between the superfluid velocity v_s and the phase of the wavefunction χ_0 must be modified to include a possible electromagnetic vector potential $A(rt)$. For a single particle the correct replacement would be $\nabla\phi \to \nabla\phi - eA/\hbar c$; however, since the 'phase' in question is that of the centre-of-mass wavefunction of a *pair*, e should be replaced by $2e$ (and m by $2m$) and we therefore get as the generalization of (4)

$$v_s(r, t) = \frac{\hbar}{2m}\left(\nabla\phi(rt) - \frac{2e}{\hbar c}A(rt)\right). \tag{9}$$

The electric current j associated with the pairs is taken to be proportional to v_s, with a constant of proportionality $\rho_s(T)$ which is roughly speaking the fraction of electrons bound into Cooper pairs (though, as with helium, it is not identical to the N_0 introduced above). Hence, on taking the curl of (9) we find, for any region in which the order parameter is non-vanishing, the London equation

$$\operatorname{curl} j \propto \operatorname{curl} A \equiv H \tag{10}$$

from which the whole of the London phenomenology follows, in particular that in a bulk sample the magnetic field is screened out in a penetration depth λ_L which is proportional to $\rho_s^{-1/2}$. Moreover, if we apply (9) to a thick ring, and consider a path which lies well inside the material, where no field can penetrate, we may take $v_s = 0$ and hence $\nabla\phi = 2eA/\hbar c$ (figure 11.8). Integrating this relation around the loop and using the fact that ϕ must be single-valued up to additive terms of $2n\pi$, we obtain the famous quantization condition for the total flux Φ threading the ring

$$\Phi \equiv \oint A \cdot dl = n\phi_0 \qquad \phi_0 \equiv hc/2e \tag{11}$$

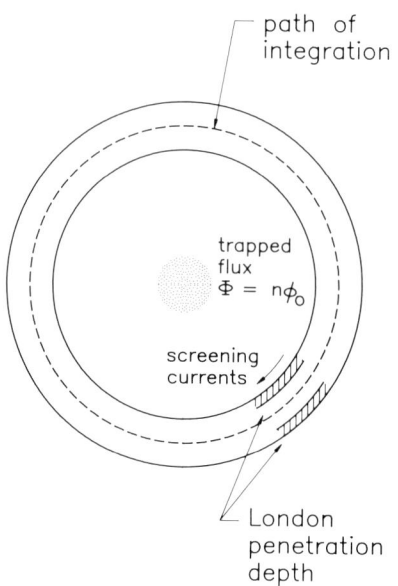

Figure 11.8. *Flux quantization in a superconducting ring.*

as originally predicted by London (with, however, ϕ_0 given by an expression twice as large) and subsequently by Byers and Yang.

Thus, if we compare a charged system (superconductor) with a neutral one such as superfluid helium, we see that the Hess–Fairbank 'no-rotation' effect and the Meissner effect, are analogues, as are the Onsager–Feynman quantization of circulation (5) and the flux quantization (11). We recall that the Ginzburg–Landau parameter κ, which determines the magnetic behaviour of a superconductor, is defined as λ_L/ξ: since, as we let the electric charge tend to zero, λ_L tends to infinity we may say that liquid helium (for which λ_L is of course not strictly defined) corresponds to the limit $\kappa \to \infty$, i.e. an 'extreme type-II' superconductor. Thus, at not too high angular velocities it is always energetically advantageous for rotating helium to admit the penetration of vortices (type-II behaviour): by contrast, as we have seen, superconductors in a magnetic field may show either type-I or type-II behaviour depending on the value of κ. Alloying tends to reduce the superfluid density ρ_s, and hence to increase λ_L and κ and promote type-II behaviour.

The explanation of the metastability of persistent currents in a ring proceeds along the same lines as outlined above for He. In the present case the finiteness everywhere of the order parameter Ψrt, which guarantees the topological stability of the current-carrying states, is due to the attractive interaction which binds the Cooper pairs. To break up a pair so as to reduce Ψ requires an energy of at least twice the energy gap. As in the helium case, it is a delicate question whether 'superconductivity' in the original sense of finite current flow under zero voltage drop is better thought of as essentially a manifestation of an equilibrium phenomenon (Meissner effect) or of a non-equilibrium one (metastability of supercurrent flow). In type-II superconductors, just as in helium, the motion of existing vortex lines across the current can lead to its decay. This a serious problem in the design of type-II superconductors for high-field magnets, but one which has been overcome by, for example, cold-working, to provide defects which will pin the vortices.

Finally, it should be noted that the Josephson and related effects arise because the order parameter is essentially a single-particle wavefunction (in helium) or the centre-of-mass wavefunction of a pair of particles (in a superconductor). These effects, discovered in superconductors in 1962, had to wait until the 1980s for their confirmation in superfluid helium.

11.3. New developments

11.3.1. The superfluid phases of ^{3}He

Superfluid ^{3}He is unusual among the topics of this chapter in that its existence, if not most of its detailed properties, was predicted theoretically before its experimental discovery. The light (mass-3) isotope of helium is extremely rare in nature (about 1 part in 10^7 of naturally occurring helium), and experiments on it only became feasible with the development of nuclear reactors, which produce ^{3}He via the decay of tritium (^{3}H). By the early 1950s it was known that this isotope liquefied at a temperature only a little below that of the common mass-4 isotope, and a few properties of the liquid had been measured. Like ^{4}He, it appeared that it would not solidify under its own vapour pressure.

Since the ^{3}He atom contains an odd number of spin $-1/2$ elementary particles (2 protons, 1 neutron, 2 electrons) it should differ from ^{4}He in obeying Fermi–Dirac statistics; liquid ^{3}He was the first system of this kind which could be cooled without freezing to the point of *degeneracy* where the statistics might play an important role. It was recognized that there was a close parallel with electrons in (normal) metals, which could be understood qualitatively as a system of non-interacting Fermi particles, and in 1956 Landau provided a firmer theoretical base for this notion with his theory of a Fermi liquid, i.e. a strongly interacting Fermi system which is liquid in the degeneracy regime. As in his earlier work on the Bose liquid ^{4}He, Landau introduced the idea of an elementary excitation: only in this case the elementary excitations of the strongly interacting system were in one-to-one correspondence with those of the free gas, and in particular, like them, obeyed Fermi statistics. As a result, the properties of the liquid are mostly qualitatively similar to those of the free gas, the only major exception being the modification, in the strongly degenerate regime, of the ordinary sound waves into a peculiar collective excitation called by Landau *zero sound*. Over the next decade or so measurements down to about 3 mK showed that below 100 mK ^{3}He is indeed very well described by Fermi-liquid theory.

The atoms of liquid ^{3}He in this regime, then, behave qualitatively much like the electrons in a normal metal. But a metal which is normal at temperatures of more than a few degrees may in the very low-temperature regime become superconducting, and once the microscopic (Cooper-pairing) mechanism for this was understood in the work of BCS it became an obvious question whether something similar might not happen in liquid ^{3}He at sufficiently low temperatures. The very earliest speculations assumed that if this happened the Cooper pairs would form in an s-wave, spin-singlet state as in superconductors, i.e. their wavefunction would be of the general form of equation (8), but it was very soon realized that the very strong repulsion between the 'hard cores' of the helium atoms would make this form of pairing energetically unfavourable, and attention shifted to the possibility of pairing with non-zero orbital angular momentum l (which would reduce the effect of the core repulsion because the centrifugal force automatically keeps the atoms from getting too close). Because of the Pauli principle, even-l pairs must be in a spin-singlet state (with the spins of the two atoms anti-parallel) while odd-l pairs must be spin triplets (spins parallel). The majority (though not universal) opinion throughout the 1960s was that the state with $l = 2$ (hence spin-singlet) was likely to be the most stable, and the properties of this state were explored in some detail, in particular in an influential 1962 paper by Anderson and Morel. Since the internal wavefunction of the Cooper pairs in this case does not possess spherical symmetry, neither do the physical properties: the liquid should be anisotropic. While a system of electrically neutral atoms could not be superconducting, it was predicted to display the analogous property of superfluidity. Thus this as yet hypothetical phase of matter became known as an *anisotropic superfluid*. Throughout the 1960s and early 1970s repeated attempts were made by theorists to predict the transition temperature of liquid ^{3}He into the anisotropic superfluid phase, with results ranging from temperatures just below the lowest currently achieved in the liquid all the way down to 10^{-17} K.

In the autumn of 1971, a graduate student at Cornell University, Doug Osheroff, was measuring pressure changes during the cooling of a mixture of liquid and solid ^{3}He. Between 3 and 1.5 mK, the record of pressure against time showed two small but reproducible anomalies: a change in the slope of the curve at 2.6 mK (the *A feature*), and at a rather lower temperature a tiny discontinuity in trace (*B feature*). The first paper in which Osheroff, with his supervisor D M Lee and R C Richardson, reported these results was entitled 'Evidence for a new phase of solid ^{3}He'; however, within weeks magnetic resonance experiments showed

Superfluids and Superconductors

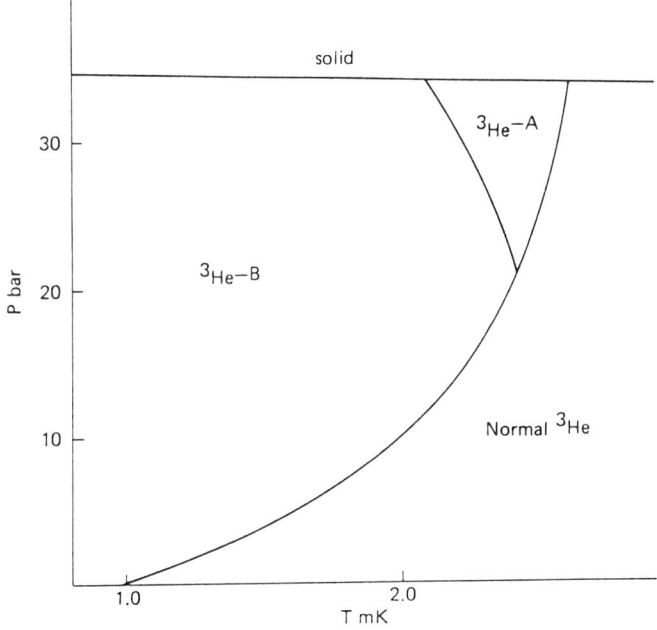

Figure 11.9. *Phase diagram of liquid ^{3}He below 3 mK.*

unambiguously that whatever was going on, at both the A and B features, was going on in the liquid rather than the solid. Was this the long-anticipated anisotropic superfluid phase?

The outcome of the feverish period of experimentation and theoretical analysis which followed could hardly have been anticipated: below 3 mK liquid ^{3}He has not one but three new phases, now called A, B and A_1, each of which is in some sense an 'anisotropic superfluid'. In zero magnetic field the phase diagram is approximately as shown in figure 11.9 (but the A phase can be supercooled considerably below the equilibrium 'AB line'). As the field increases, the A phase fills a larger and larger region of the diagram, suppressing the B phase entirely by about 6 kG, while the second-order A-to-normal transition splits into two, with the A_1 phase a thin sliver between them.

The origin of this surprising result is the fact that the primary mechanism of attraction which binds the Cooper pairs does not arise, as in superconductors, from the polarization of an ionic background (there is none!) but rather through a conceptually similar mechanism involving polarization of the spins of the 'background' liquid. This feature had indeed been anticipated in the 1960s, and it had been appreciated that such a mechanism, if dominant, would favour pairing in a spin triplet (figure 11.10) and hence odd-l state rather than the generally favoured $l = 2$ state. What had not been realized, however, is the richness

Twentieth Century Physics

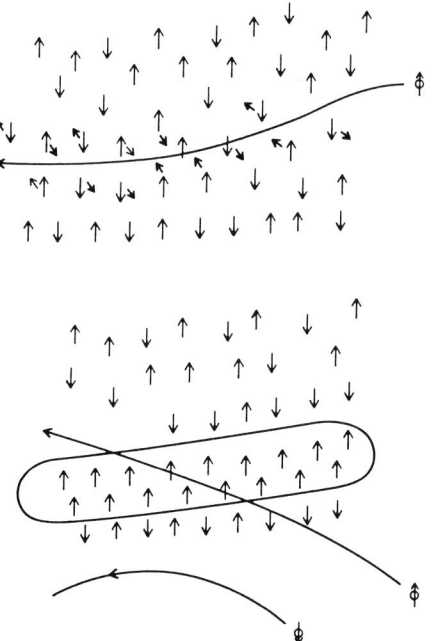

Figure 11.10. *Mechanism of indirect attraction in liquid ³He. An 'up' spin atom creates a region of predominantly 'up' polarization near its path; this region then attracts other up-spin atoms but repels down-spin ones.*

of behaviour which this kind of state permits. Suppose that, as we now believe, the favoured pairing configuration is a p-wave ($l = 1$) triplet, then there are three Zeeman spin sub-states and three orbital ones, and this gives nine possible 'components' of the wavefunction, each of which can enter multiplied by a complex coefficient. Now, if we just follow through the BCS argument, there is precisely one combination which is the most stable over the whole of the region of existence of the superfluid phase: this corresponds to the so-called Balian–Werthamer (BW) state which is now almost universally identified with the experimentally observed B phase of ³He. This state had been described, and its properties discussed in some detail, in an important paper by Balian and Werthamer in 1963. (It had actually been introduced earlier by Vdovin in the Soviet Union, but the paper was published in an obscure conference proceedings and was virtually unknown in the West.) In this phase, which in the terminology of atomic physics would be called ³P₀, the Cooper pairs all have orbital angular momentum $l = 1$ and total spin $S = 1$ (spins 'parallel'), but the vectors l and S are oppositely directed so as to give a total angular momentum J equal to zero (but see below). The energy gap and hence the thermodynamics of such a phase are indistinguishable from those of an ordinary (s-wave) superconductor.

Within this scheme, however, one cannot understand the occurrence of the A phase. In a seminal paper written within a few months of the experimental discovery, Anderson and Brinkman pointed out a crucial difference between the lattice-polarization *phonon exchange* mechanism of attraction between electrons in a superconductor and the spin-polarization *spin-fluctuation-exchange* mechanism believed to operate in liquid ^{3}He. In the former the polarized medium is the ionic lattice whose behaviour is hardly affected by the onset of Cooper pairing of the electrons. By contrast, in ^{3}He the polarized medium is the liquid itself, i.e. the same system as forms the Cooper pairs: pair formation can therefore affect the polarizability, and hence the strength of the interaction which binds the pairs. This puts a premium on those states which retain the full normal-state polarizability, which is possible only if two of the three Zeeman sub-states (corresponding to $S_z = \pm 1$ along some axis, not also $S_z = 0$) are formed. Less obviously, it favours a particular one of this class of states, which had been discussed in the early 1960s by Anderson and Morel and is nowadays known as the Anderson–Brinkman–Morel state (ABM) state. The principal characteristic of this state is that, irrespective of their spin states ($S_z = \pm 1$), all Cooper pairs have an apparent angular momentum $\hbar$ directed along the same axis. Moreover, unlike the BW state or the s-wave pairing state of superconductors, this state has an energy gap which vanishes for certain directions of motion of the quasiparticles. Further implementation of these ideas has given a fairly quantitative account of the relative stability of the A and B phases.

For completeness, it should be added that in the 'sliver' of A_1 phase it is believed that only a single Zeeman sub-state (e.g. $S_z = +1$) is occupied, the two spin directions being no longer equivalent in a magnetic field, so that one may have a slightly higher transition temperature to the Cooper-paired state than the other.

What is especially exciting about the new phases of liquid ^{3}He is that the order parameter (Cooper-pair wavefunction) in each has, as well as the usual centre-of-mass degree of freedom which is possessed also by the electron pairs in superconductors, one or more internal degrees of freedom which are not fixed by the energetics of the pair-forming process (sometimes called 'broken symmetries'). For example, in the A phase the pairs are characterized by two axes: one (conventionally denoted *d*) is perpendicular to the axis (which can actually be chosen to be any axis in a plane) along which pairs with $S_z = \pm 1$ are formed, while the other (denoted *l*) is the axis along which they have relative orbital angular momentum $\hbar$. As far as the 'gross' energies which are responsible for the pair-formation process are concerned, these axes are quite arbitrary. For the B phase the 'broken symmetry' is a bit more subtle; it corresponds to the possibility of starting from the '$^3P_0$' state and performing an arbitrary rotation of the spin coordinates relative to the orbital ones. The resulting state still has an isotropic energy gap, and indeed those of its properties which depend only on the spin coordinates or only on the orbital axes are unaffected. However, the effect on these properties which involve *correlations* between the spins and orbital coordinates, is profound.

The crucial point is that the formation of the superfluid state (the 'Bose condensation' of the Cooper pairs, if one likes to think of it that way) requires that all pairs have the same wavefunction not only with respect to their centre-of-mass coordinate but also with respect to their *internal* degrees of freedom. This feature was, of course, already present for the electrons in superconductors, but without striking consequences, because the relative wavefunction was uniquely fixed by the energy considerations which formed the pairs in the first place. By contrast, for superfluid ^{3}He the effects are spectacular—effects which would be totally negligible in a single molecule can be amplified by the 'Bose condensation' to the point where they dominate the behaviour. The best-known example concerns the interaction of the nuclear magnetic moments of the ^{3}He nuclei. If we think of these nuclei as tiny classical magnets, the usual theory of magnetism tells us that, if constrained to be parallel, they would prefer to lie end to end rather than side by side. When rotating around one another, while pointing in a constant direction, they achieve a favourable configuration for about half the time if the magnets lie in the plane of rotation, rather than normal to it. However, the energy E_d which favours the parallel orientation over the perpendicular is so tiny that even if two ^{3}He atoms in, say, the gas phase were to form a diatomic molecule (which in fact they do not), thermal disorder would completely swamp it, and the molecules of the gas would rotate with random orientations. In the superfluid phase, however, all N of the Cooper pairs must rotate in the *same* way. The energy difference between the two configurations (in-plane and perpendicular) is now not E_d, but NE_d, which is always much larger than the thermal energy kT. Thus the in-plane configuration is always overwhelmingly favoured and, except in special circumstances, is always realized. Furthermore, when this configuration is disturbed, for example by applying an RF magnetic field to the spins, the existence of this coherent dipole energy has a spectacular effect on the resonance behaviour. Historically it was in fact these effects which permitted the earliest convincing identification of the microscopic nature of the three superfluid phases. Various other ultra-weak effects may be similarly amplified (it has even been proposed that it might be possible in this way to manifest the effects on a macroscopic scale of the parity-violating 'weak neutral currents' introduced in the 'Standard Model' of the electroweak interaction in particle physics, on a macroscopic scale).

See also p 695

The above discussion assumes implicitly that the internal degrees of freedom are constant in space. However, spectacular consequences follow also when we consider the possibility of their spatial variation.

Perhaps the most intriguing of all relates to the question of the (meta)stability of supercurrents in an annulus. We saw that in a BCS superconductor the reason for this stability was essentially topological in origin: the number of twists of 2π the phase ϕ undergoes as we go once around the annulus (*winding number*) cannot be changed without at some stage depressing the magnitude $|\Psi|$ of the order parameter to zero. The situation is essentially the same in the B phase of superfluid ^{3}He, but in the A (and A$_1$) phase there is a crucial difference. Suppose we start with the 'orbital vector' l constant in direction throughout (which allows us to define a phase $\phi(r)$ unambiguously), it is then possible to go from winding number n to $n-2$ (though not to $n-1$) via a process in which the magnitude of the order parameter never changes from its original value, provided that at intermediate stages we allow the direction of l to vary in different places. Thus, in an ideal situation where there is no energy to 'pin' l, 'superfluid' ^{3}He-A would actually not be superfluid, at least in the sense of showing metastability of supercurrents. (It would still be expected to show a Hess–Fairbank effect.) That in real life the A phase appears to behave much like ^{4}He, with the usual metastability, is attributed to the combined effects of the nuclear dipole force and the walls in 'pinning' the vector l. This circumstance shows spectacularly the necessity, emphasized earlier, of distinguishing between the 'stable' and 'metastable' manifestations of superfluidity.

11.3.2. Miscellaneous recent developments

In this section, I briefly review a few systems, other than traditional superconductors and the two pure isotopes of helium, in which superconductivity, superfluidity or related phenomena either exist or it is believed that they might occur under conditions not yet experimentally realized. The most spectacular of these, the high-temperature superconductors, are treated in the next section.

The major surprise of the 1970s in the area of superconductivity was its discovery in some members of a class of materials known as the *heavy-fermion* systems. As already mentioned, a good description of many systems of strongly interacting fermions, such as the electrons in ordinary metals in their normal state is given by Landau's Fermi-liquid theory. In this theory the elementary excitations, which behave qualitatively much like free electrons, are characterized by an *effective mass* which for most ordinary metals is of the order of the original electron mass: the electronic specific heat at low temperatures is proportional to this mass. However, some metal compounds containing rare-earth or actinide elements, such as CeAl$_3$, CeCu$_2$Si$_2$ or UPt$_3$, have very anomalous properties at low temperatures, in particular specific heats so huge that if the Landau theory applies the effective mass of the electrons must be hundreds or even thousands of times the real mass (hence the name heavy-fermion). Indeed, at first it was suspected that this enormous specific heat might arise from localized electrons to which the concept of effective mass does not apply, but the picture changed dramatically in 1979, when a superconducting transition was discovered in UBe$_{13}$ and subsequently in various other members of the class. The fact that the specific heat drops rapidly below T_c, as in a typical BCS superconductor, is generally taken to indicate that the electrons contributing to it must be those that form Cooper pairs and thus must be mobile. Both the normal and the superconducting properties of the heavy-fermion systems are the subject of considerable ongoing research, and it is already clear that the normal state cannot be described by any simple variant of the so-called Bloch–Sommerfeld picture which has been so successful for simpler metals. While the behaviour in the superconducting state resembles that of a typical BCS superconductor, there are indications that in UPt$_3$, at least, and possibly in others, the Cooper pairs form in an 'anisotropic' state similar to that which occurs in superfluid ^{3}He. However, the rather complicated effects of the crystal structure, and the absence of any simple diagnostic probe analogous to nuclear magnetic resonance in ^{3}He, has prevented this from being firmly established so far.

A second class of novel superconducting materials is the alkali-doped fullerene crystals. The fullerenes are crystals of hollow polyhedra of carbon—C$_{60}$ is the most studied—and were first synthesized in bulk in 1990. Pure C$_{60}$ is an insulator, but when it is doped with alkali atoms these contribute conduction electrons and the crystal becomes metallic. Moreover, for certain levels of doping it also becomes superconducting, with a transition temperature as high as ~33 K—which, had it been discovered seven years earlier, would at that time have been the world record! While all the evidence is that superconductivity in these materials is due to formation of s-wave Cooper pairs as in the 'old-fashioned' superconductors, it is unclear whether the mechanism which binds the pairs is the usual phonon exchange or (as it is thought to be in the heavy-fermion superconductors) a sophisticated effect of the inter-electron interactions.

It is worth noting briefly that there are several areas outside condensed matter physics where the ideas developed in the theory of superconductivity and superfluidity have made an impact. The idea of Cooper pairing has been applied with considerable success to explain some of the properties of heavier nuclei. Since in this case we are talking about a system of at most a few hundred particles, macroscopic concepts such as superfluidity have no real meaning, but the occurrence of pairing is reflected in the reduced moment of inertia of the nucleus (compare the Hess–Fairbank effect). A more direct, if more speculative, application of BCS theory is to neutron stars in which, depending on the pressure, etc, various superfluid phases, analogous to those of ^{3}He, are predicted to occur. Finally, ideas analogous to those of the BCS theory have found major application in particle physics, in the *dynamical symmetry breaking* scenario of Nambu and Jona-Lasinio and the mechanism, originally recognized by Anderson and elaborated by Higgs, by which the intermediate vector bosons of the unified electroweak theory acquire mass; a concept inspired by the Meissner effect.

Finally, there are systems in which superfluidity might occur but has not yet been demonstrated unambiguously, for example, the dilute ($\lesssim$ 8%) fermion system of ^{3}He atoms which is stable in (itself superfluid) ^{4}He at low temperatures, spin-polarized atomic hydrogen (a Bose gas), and the metastable Bose system formed by excitons in some insulators such as Cu_2O (where indirect manifestations of 'superfluid' behaviour may have been seen in recent experiments). The observation of superfluidity in spin-polarized atomic hydrogen ($H_\uparrow$) would be particularly exciting because, if the mechanism is indeed Bose condensation as is almost universally assumed both here and in He II (see section 11.2.6), then this latter phenomenon, which has to date resisted direct observation in He II, should be detectable almost with the naked eye; a sharp change is predicted to occur, in $H_\uparrow$ under the relevant experimental conditions, in the spatial density profile. In this way we should be able to plug, in an experimental sense, the last loophole in the general argument of section 11.2.6.

11.3.3. *High-temperature superconductivity*

Over the three-quarters of a century following Onnes' 1911 discovery, the maximum temperature achieved for the transition to the superconducting state crept up gradually from the 4.2 K of mercury to an apparent limit in the region ~20–25 K (figure 11.11). True, claims had been made of superconductivity at higher temperatures, for example 60 K for CuCl, but these had all been later discredited. Moreover, many theorists had convinced themselves that the phonon mechanism could never produce a T_c much higher than those observed; and while some, such as Little in the US and Ginzburg in the Soviet Union, speculated that alternative mechanisms might produce a superconducting state at much higher temperatures, perhaps even room temperature, specific attempts to build such a 'high-temperature superconductor' had always ended in failure. Thus, in the summer of 1986 most physicists familiar with the subject would probably have bet 100 to 1 that T_c would never go above 30 K. In the event, within a year virtually every laboratory in the world was almost routinely observing superconductivity above not 30 but 90 K, well above the boiling point of nitrogen; the record today stands at over 150 K, half way to room temperature. Many physicists would regard this as the most significant development in solid-state physics for half a century.

The story begins with a paper from IBM Zurich in November 1986 by Alex Müller and Georg Bednorz, reporting the onset of a partial Meissner effect around 30 K in a compound first made a few years

See also
p 490

Twentieth Century Physics

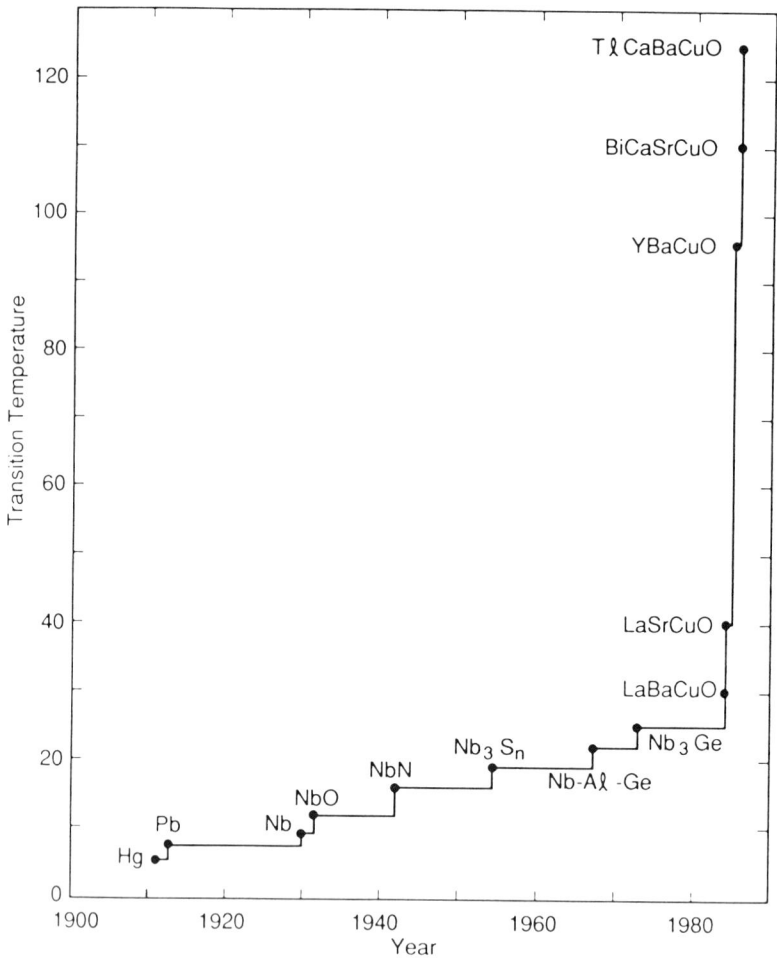

Figure 11.11. *The history of the maximum known temperature for superconductivity to occur (up to 1991).*

earlier, La$_2$CuO$_4$ doped with a small amount of barium. They tentatively (and in view of earlier false alarms, bravely) identified this as indicating a superconducting transition. (In a rare example of the award of the Nobel Prize satisfying the original condition that the work recognized should have been done in the preceding year, Müller and Bednorz received the prize in 1987.) The significance of this work was rapidly appreciated, and over the next few months many laboratories frantically explored the chemical 'neighbours' of this material to try to increase the transition temperature. A notable step, in early 1987, was the discovery by a Houston–Alabama collaboration led by Chu and Wu of superconductivity in YBa$_2$Cu$_3$O$_7$ (usually nowadays known as YBCO or '123') at 90 K, and a year or so later the maximum transition temperature was raised to 125 K in a Tl–Ba–Ca–Cu–O compound ('2223'). At the time of writing, about 100 different compounds are known which have temperatures above the 1986 dream of 30 K (not counting the fullerenes); the current record, about 160 K, is held by a Hg–Ba–Ca–Cu–O compound under pressure.

Although the high-temperature superconductors (HTS) belong to several different classes, they share a number of features[†]. First, they are all chemically complicated (few HTS are known which contain less than four different elements) and have fairly complicated crystal structures with large unit cells, but invariably containing well-separated planes (or pairs or triples thereof) of copper and oxygen atoms (CuO$_2$ planes) (figure 11.12). For this reason they are often referred to as *copper oxide superconductors*. It seems overwhelmingly probable that the mechanism of superconductivity is to be sought in those CuO$_2$ planes, and that the role of the off-plane atoms is mainly to act both as donors of electrons (or, more usually, of holes) to the planes and as spacers between them; this partially explains the enormous number of chemical compounds which show high-temperature superconducting behaviour. Secondly, in most cases superconductivity occurs in a region of the phase diagram close to an antiferromagnetic phase (figure 11.13). Thirdly, the properties of these materials in the normal state are highly anomalous. In particular, the electrical resistivity is proportional to temperature from 1000 K down to the transition temperature. A resistivity proportional to temperature is of course a feature of ordinary metals at temperatures above the Debye temperature θ_D, and one might think that the copper oxide superconductors simply have rather low values of θ_D. However, this idea seems implausible when one finds that Bi$_2$Sr$_2$CuO$_6$ behaves like the HTS in all respects except that its superconducting transitions are much lower (9 K); but resistivity is still linear all the way down to T_c, far below any remotely plausible value of θ_D (which in any case can be measured directly and is not specially low for the HTS). The Hall coefficient and NMR (nuclear magnetic resonance) behaviour are also highly unusual in the normal state of the HTS.

While the HTS are indeed superconductors in that they show, under appropriate conditions, both a Meissner effect and resistance-free flow of current, the difference of their electromagnetic behaviour from that of the old-fashioned superconductors is so marked as to be almost qualitative. In the first place, they show type-II magnetic behaviour in an extreme form: while the lower critical field, H_{c1}, is probably typically of the order

---

[†] *A slight complication is caused by the discovery, since 1986, of superconductivity above 30 K in BaKBiO$_3$. This has none of the characteristics described and is usually thought to be a simple* BCS *superconductor with an anomalously high T_c.*

Twentieth Century Physics

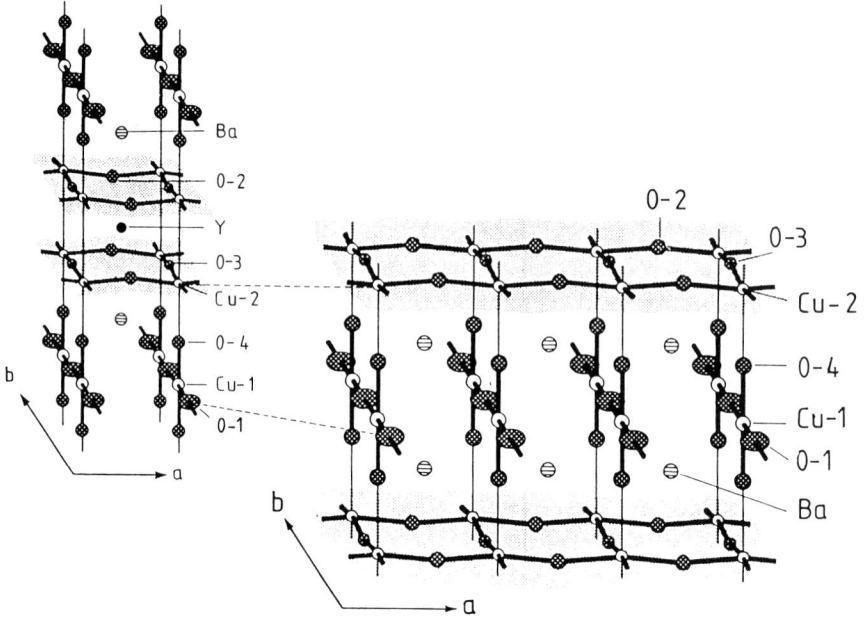

Figure 11.12. *The crystal structure of $YBa_2Cu_3O_7$.*

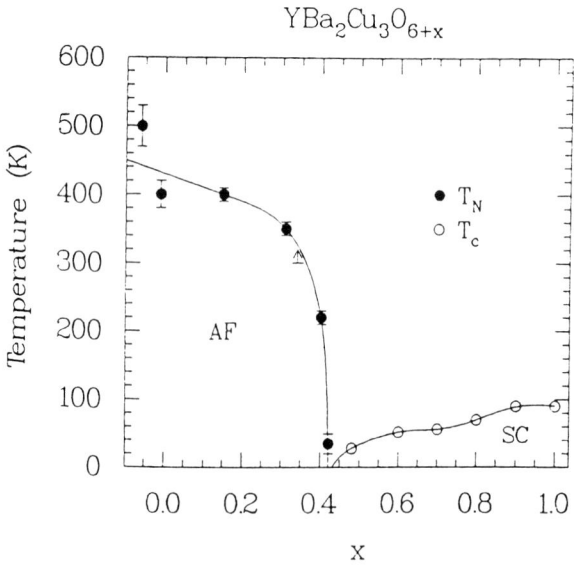

Figure 11.13. *The phase diagram of $YBa_2Cu_3O_{6+x}$, as a function of temperature T and oxygen concentration x. T_N is the temperature of transition to the AF (antiferromagnetic) state and T_c that to the SC (superconducting) state.*

of 100 gauss, the upper one, H_{c2}, is so enormous that for temperatures only just below T_c it cannot actually be reached with existing magnets. Secondly, the *transition to the zero-resistance state* is much less sharp than for conventional superconductors, particularly in a magnetic field (figure 11.14). It is also directionally dependent: recent experiments show that there is sometimes a region of temperature below T_c in which the resistance parallel to the planes drops as expected, but perpendicular to the planes actually increases! Indeed, it is currently a matter of debate whether, in certain regions of the phase diagram, the HTS should really be described as superconducting, since the vortices move around so much more readily than in a traditional superconductor. No doubt this is partly because the order parameter is large only on the CuO_2 planes, so that rather than a continuous line one should think of a vortex as like a loosely connected string of beads, which can 'wobble' or even be broken without too large an energy cost.

Turning from the macroscopic electromagnetic behaviour to the more 'microscopic' properties of the HTS we find a similarity to BCS superconductors which is the more surprising given the qualitatively anomalous behaviour in the normal state. In fact, nearly every feature of the BCS model shows up at least qualitatively, for example, both the spin susceptibility and the electronic specific heat drop off sharply below T_c, and there appears to be a relatively well-defined energy gap Δ, in the sense that most of the electronic excitations lie above a threshold energy which it is natural to call the 'gap'. However, there are two significant differences from the traditional superconductors. First, nearly every method of measuring Δ gives a value at zero temperature whose ratio with T_c is considerably greater than the BCS ratio of 1.75, and moreover in some experiments it is not clear that Δ tends to zero as T approaches T_c. Secondly and perhaps more significantly, a considerable number of experiments (tunnelling, specific heat, NMR, penetration depth) indicate that there are some electronic excitations which have energies small compared to Δ. Apart from these two features, the similarity to the BCS predictions is remarkable.

The question of the mechanism by which superconductivity is induced in the HTS is hotly debated, and any present attempt to review it is likely to be out of date within a few years, if not months. Among the few principles which would probably be generally agreed are: (i) that the mechanism of high-temperature superconductivity is 'universal'—there is not one type of mechanism for the original lanthanum compound of Bednorz and Müller, a different one for YBCO, and so on; (ii) that the dominant role is played by the CuO_2 planes, with the off-plane atoms essentially acting as 'spectators', and the differences in T_c are due in large part to the different numbers of electrons or holes which these atoms donate to the planes; (iii) that the mechanism can reasonably be investigated theoretically by considering only a single

Twentieth Century Physics

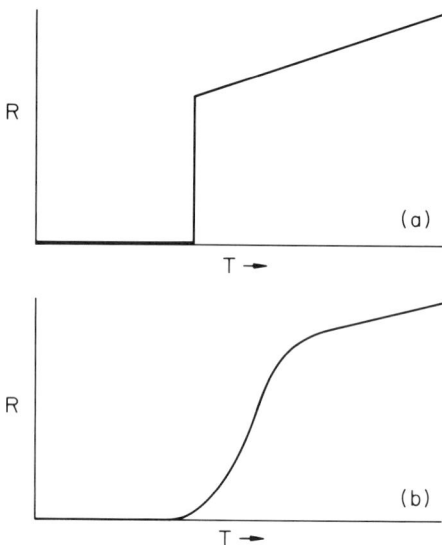

Figure 11.14. *Resistance as a function of temperature for (a) Al in zero magnetic field and (b) $Bi_2Sr_2CaCu_2O_8$ in a field of a few tesla (schematic).*

plane (although one very recent theoretical proposal strongly disputes this), the role of inter-plane interactions being primarily to provide a Josephson-type coupling between the order parameter on neighbouring planes, which prevents the fluctuations becoming so large as to make the situation qualitatively different from that in a typical three-dimensional superconductor, and (iv) (commanding widespread though not universal acceptance) that the proximity to an antiferromagnetic instability cannot be an accident. Beyond this there is little or no agreement.

We may take the question in the following three stages.

(1) Is superconductivity due to the formation of Cooper pairs, as in the old-fashioned superconductors, or is some completely different mechanism at work?

(2) If it is Cooper pairs, what is the attraction responsible for binding them?

(3) What is the symmetry of the pair state?

Regarding (1), since the discovery of HTS a number of alternatives to the Cooper-pair mechanism have been proposed. One, which for a time looked particularly attractive, was the so-called *anyon* mechanism. Anyons are a peculiar type of excitation, with statistics intermediate between those of fermions and bosons, which can exist only in a strictly two-dimensional system, a feature which in view of the strongly 'two-dimensional' character of the CuO_2 planes in the HTS makes their attraction obvious. Moreover, it is often claimed (though the validity of this claim is not obvious to me) that at zero temperature a system of anyons is superconducting even without the need for interactions. Unfortunately, it has proved very difficult to do concrete calculations of experimental properties on the basis of the anyon model, and the few distinguishing features which have been predicted have not yet been reliably observed, so that interest in the idea, at least in the context of HTS, seems to be ebbing. Various other proposals, for example involving the Bose condensation of a fictitious 'gauge boson' associated with the strong hard-core repulsions of the electrons in the CuO_2 plane, have been made. A very strong constraint on any such 'exotic' mechanism (including the anyon one) is that it must be consistent with the observation that the unit of flux quantization in the HTS, as in the BCS sort, is $h/2e$ (and not, for example, h/e): it is not clear whether or not any theory possessing such a property must in the last resort just be another language for describing the process of Cooper-pair formation.

A variant which is similar, but not quite identical, to the Cooper-pair idea is that bound pairs of electrons are already formed in the normal state. Such pairs would be bosons, and the observed transition temperature would then simply correspond to the point of Bose condensation, as in ^{4}He. Some of the experimental observations, for example that in some (but not all) HTS the spin susceptibility begins decreasing at temperatures much higher than T_c, would find a natural explanation in this model, which as yet has not been developed far.

Probably the majority opinion is that the mechanism of superconductivity in the HTS is essentially Cooper pairing. If this is correct, we can tentatively apply the prescriptions of standard BCS theory to find out something of their characteristics. One feature immediately stands out if we do this: whereas in traditional superconductors the ratio of the Pippard coherence length (Cooper-pair radius) to the mean distance between electrons is thousands or even tens of thousands, in the HTS it is less than ten. This would explain why the magnetic behaviour of these materials is so extremely type-II, and would also mean that many of the approximations made in the standard BCS theory need to be re-examined, so that it would not be surprising if the quantitative aspects of its predictions are modified.

Turning to question (2), it is fairly generally agreed (by those who believe in the Cooper-pair mechanism) that the attraction which binds the pairs must be mediated by the exchange of some kind of low-energy excitation obeying Bose statistics; the question is whether this is a phonon as in traditional superconductors or some kind of collective excitation of the electron system, and if the latter, what kind. Arguments against the phonon mechanism include the absence of any appreciable isotope effect in the higher-temperature HTS, and the fact that if, as seems natural, it is collisions with this same boson which are responsible for the normal-state resistance then, as already mentioned, the linearity of the latter in temperatures down to 9 K is incompatible with any reasonable Debye temperature. However, these arguments are not incontrovertible. If the boson is a collective excitation of the electron system, what kind of excitation? Here one finds a variety of answers. One of the most popular scenarios identifies the boson as an antiferromagnetic spin fluctuation, the 'relic' of the nearby antiferromagnetic state, and this makes a striking prediction concerning question (3), the symmetry of the Cooper-pair wavefunction, to which we now turn.

Are the Cooper pairs formed in a simple s-wave state, as in traditional BCS superconductors, or with 'exotic' symmetry, as in superfluid ^{3}He and possibly some of the heavy-fermion superconductors? This question is particularly interesting because whereas, for example, a mechanism of attraction due to phonons usually favours an s-wave state, the scenario based on exchange of antiferromagnetic spin fluctuations indicates rather unambiguously that the pair state should be of the so-called $d_{x^2-y^2}$ type; this state has nodes (points on the Fermi surface where the energy gap is zero) and therefore even at low energies there should exist an appreciable number of electronic excitations, contrary to the simple BCS case where there are exponentially few. Thus, the apparent evidence for such excitations in a variety of experiments (see above) is often taken as supporting the $d_{x^2-y^2}$ hypothesis, as is the direct observation in recent photoemission experiments, of a gap which appears to be large along the x and y axis in the CuO_2 plane but small and possibly zero along the 45° axis—precisely the behaviour of the gap in the $d_{x^2-y^2}$ state. However, this argument is not totally foolproof. To see this, in figure 11.15 we compare schematic representations of the $d_{x^2-y^2}$ state and a strongly deformed s-state—a quite plausible candidate in some alternative scenarios. We see that, except very close to the 45° axis, the magnitude of the gap is almost identical in the two cases (and hence, except at very low temperatures, the number of excitations will be very similar); the principal difference between the two states is in the relative *sign* of the wavefunction in the various lobes. It would be highly desirable to measure this sign directly, and the first measurements of this type (using the Josephson effect) have very recently been made; at the time of writing the situation is confused, and it is not clear that there is *any* symmetry assignment which is compatible, given current theoretical understanding, with the totality of the data.

At present, the technically important business of finding HTS with ever higher transition temperatures is very much a matter of trial and error. Needless to say, a major motivation for trying to understand the mechanism is the hope of identifying those chemical, structural and other features which tend to raise T_c, so as to orient the search among the billions of currently unexplored compounds. While it seems unlikely that the current record will stay unchallenged for long, it is anyone's guess whether genuine 'room-temperature' superconductivity will be achieved in our lifetime (or ever); and whether, if so, the technical difficulties in

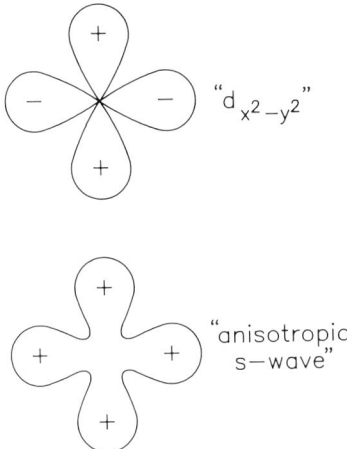

Figure 11.15. *Comparison of the angular dependence of the energy gap for (a) the $d_{x^2-y^2}$ and (b) a very anisotropic s-wave state. The + signs indicate the sign of the pair wavefunction in each state.*

the practical application of the HTS (connected, for example, with the annoyingly high mobility of the vortices) will be overcome to the extent that they may have a revolutionary effect on the technology of everyday life.

11.3.4. Further reading
In writing this chapter I have drawn heavily on references [1,2]. Other books [3–9] also provide well-referenced accounts of related material.

Acknowledgments

This work was supported by the National Science Foundation under grant No NSF-DMR-92-14236. I thank Gordon Baym and Brian Pippard for many useful comments and suggestions.

References

[1] Hoddeson L, Braun E, Teichmann J and Weart S (ed) 1992 *Out of the Crystal Maze* (New York: Oxford University Press)
[2] Dahl P F 1992 *Superconductivity* (New York: American Institute of Physics)
[3] Keesom W H 1942 *Helium* (Amsterdam: Elsevier)
[4] Shoenberg D 1952 *Superconductivity* (Cambridge: Cambridge University Press)
[5] London F 1950 *Superfluids I Macroscopic Theory of Superconductivity* (New York: Wiley); 1954 *Superfluids II Macroscopic Theory of Superfluid Helium* (New York: Wiley)

[6] Gorter C J and Brewer D F (ed) 1955 *Progress in Low Temperature Physics* (in 13 volumes) (Amsterdam: North-Holland)
Note particularly:
Feynman R P 1955 *Applications of Quantum Mechanics to Liquid Helium (Progress in Low Temperature Physics 1)* (Amsterdam: North-Holland) p 17
Bardeen J and Schrieffer J R 1961 *Recent Developments in Superconductivity (Progress in Low Temperature Physics 3)* (Amsterdam: North-Holland) p 170
[7] Wilks J 1967 *The Properties of Liquid and Solid Helium* (Oxford: Clarendon)
[8] Vollhardt D and P Wölfle 1990 *The Superfluid Phases of Helium 3* (London: Taylor and Francis)
[9] Ginsberg D M (ed) 1994 *Physical Properties of High-temperature Superconductors* (in 4 volumes) (Singapore: World Scientific)

Chapter 12

VIBRATIONS AND SPIN WAVES IN CRYSTALS

R A Cowley and Sir Brian Pippard

12.1. The beginnings of lattice dynamics

The elastic solid, considered as a structureless continuum, claimed the attention of many applied mathematicians in the nineteenth century. It was a matter of controversy whether the most general formulation (stress being assumed proportional to strain) demanded 21 independent elastic constants as Green rightly believed, or only 15 as followed from Cauchy's model of material points interacting through central forces. The majority, however, were more concerned with applications of the theory, generally in simplified form, to real problems of wave propagation, or arising from engineering structures. The scale of their achievements may be judged from Love's [1] formidable treatise, which opens with a valuable historical survey before settling into its dense mathematical exposition.

The obvious weaknesses of Cauchy's model—it required Poisson's ratio to be $\frac{1}{4}$ while something around $\frac{1}{3}$ was much more usual—probably discouraged widespread speculation on the atomic structure of solids until x-ray crystallography gave the necessary impetus. Before that, however, Einstein had initiated a different line of enquiry with his theory of the specific heat of solids. The specific heat is so central a topic of crystal lattice dynamics that we must begin with the events that led up to Einstein's seminal paper.

12.1.1. Specific heats

The very term, specific heat, implies something different from the modern approved term, thermal capacity. Just as specific gravity refers the weight of a body to the weight of an equal volume of water, so specific heat refers its thermal capacity to that of an equal mass of water. This was a natural way of interpreting the standard procedure

Twentieth Century Physics

for measuring specific heat by the method of mixtures—dunking the hot body in a water-filled calorimeter—which was good enough to allow Dulong and Petit to formulate their law in 1819. We need not rehearse the early history of this law; by 1900 it was accepted that the atomic heat of most solid elements is about 6 calories per degree. If the chemists, who found it a valuable starting-point for estimating the atomic weight of a new element, wondered what was responsible for this remarkable regularity they seem to have been wise enough not to record their thoughts. They were aware of exceptions, notably diamond, whose specific heat is much smaller than it should be. Only in 1898 did Behn's [2] measurements begin to reveal that other substances showed at lower temperatures the same discrepancy as diamond at room temperature.

On first reading, Behn's technique seems very primitive, in that it still employed the method of mixtures. A block of metal would be cooled in liquid air (-186 °C) or in a mixture of solid CO_2 and alcohol (-79 °C), and transferred to a water calorimeter. In this way he deduced the mean specific heat between $-79°$ and $18°$ and between $-186°$ and $-79°$; for aluminium he found 5.3 and 4.2 cal deg^{-1}, the room-temperature value being 6.0. The demonstration was convincing enough, and he allowed himself to speculate that the specific heat might vanish at absolute zero. Ten years later, thanks to Nernst and Einstein, this idea was on the way to being generally accepted, and it was these two who were principally responsible for initiating a radically new understanding of the whole field.

Although Einstein's contribution came first, in 1907, it made little impact until the experiments of Nernst's school provided massive empirical support. It was his assistant, Eucken [3], who built the first electrically heated calorimeter (figure 12.1), which was elaborated and improved by Nernst and successive members of his team, including Lindemann and Simon [4]. At first the lowest temperature (Nernst mentions -210 °C) was attained by pumping on liquid air, but later liquid hydrogen and liquid helium brought temperatures as low as 1 K into the available range. It should be remembered that in a specific heat measurement a temperature difference, often quite small, must be measured, and this places considerable demands on the calibration of the thermometer against the absolute scale represented by a perfect gas thermometer. If there is a calibration error t which varies with temperature T, it is not the relative error t/T that matters, but the differential dt/dT. Any departure from smoothness in the calibration curve can be serious, and the establishing of a reliable scale, especially at extremes of temperature, has been a major occupation of standards laboratories throughout the century. Compared with this, other technical problems in calorimetry are quite minor.

See also p 1249

Nernst's principal interest in calorimetry stemmed from his Heat Theorem of 1906, now known as the Third Law of Thermodynamics,

See also p 535

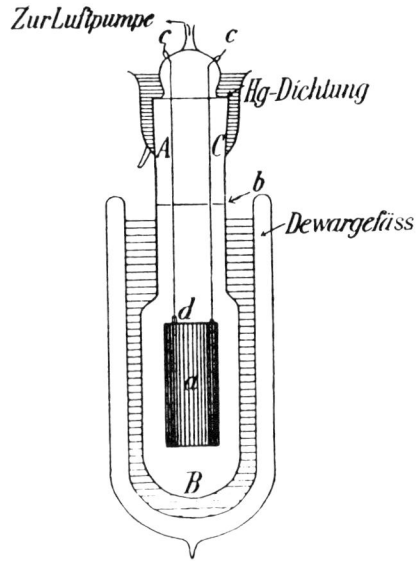

Figure 12.1. *Eucken's vacuum calorimeter. The hollow copper block a, into which samples can be sealed, hangs in an evacuated glass vessel, suspended from the leads to the platinum winding which serves both as heater and as thermometer. The whole is immersed in liquid air. The ground glass stopper is surrounded by mercury to make it airtight. Radiation from outside is minimized by silvering the inside of the vessel, and by the tin-foil-covered mica plate b.*

an essential ingredient in the technically very important process of determining the thermodynamical parameters of chemical reactions from specific heat measurements. Thus much of the emphasis in his school centred on acquiring data of immediate interest to the chemical industry, and the study of specific heats of elements for their own sake was hardly more than a by-product. Nevertheless, a picture soon emerged of the temperature variation of specific heats, remarkably similar for most elementary solids, and largely confirming the ideas put forward by Einstein.

The development of Einstein's thought on quanta has been described in detail by Kuhn [5]. Being convinced that the radiation field was quantized, he found himself forced to accept that the oscillations of charged particles, in dynamic equilibrium with the radiation, were also quantized, and hence that all oscillations, whether of charged or uncharged particles, were quantized. In his 1907 paper [6] he supposed that every atom of a solid vibrates at the same frequency, which he identified with the *reststrahlen* frequencies (in the far infrared) at which Rubens and Nichols had found almost perfect reflection, as will be discussed later. Owing to Einstein's omission of references it has sometimes been supposed that he introduced the concept of $3N$ atomic

See also p 152

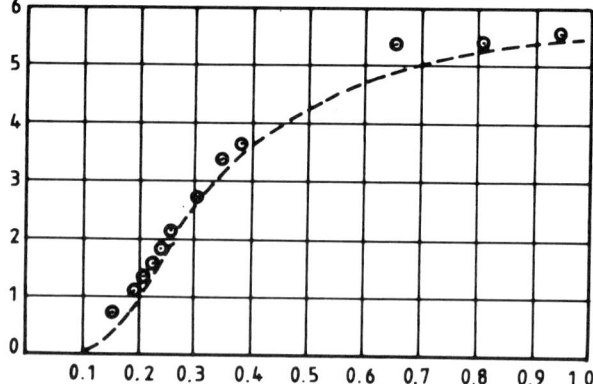

Figure 12.2. *Einstein's comparison of his specific heat formula with the available data for diamond. The abscissa is (in modern terms) $\hbar\omega_E/kT$, where ω_E is a characteristic frequency chosen to give a satisfactory fit to theory. The ordinate is specific heat in calories/gram-atom.*

oscillators—N atoms vibrating in three dimensions—but in fact it was mentioned by Boltzmann [7] in a footnote with which Einstein most certainly would have been familiar.

Boltzmann notes that the specific heat of a solid should be twice that of a perfect gas of the same atoms (Dulong and Petit's law) but Einstein, applying Planck's formula, shows how it should fall as the temperature is lowered, and cites the room-temperature values for three light elements (B, C, Si) in support of this view. Better still, he takes Weber's measurements of the specific heat of diamond between 222 and 1258 K and shows they lie reasonably close to his theoretical curve (figure 12.2) if he assumes the vibration frequency to be that of infrared radiation of 11 μm wavelength. 'Thus' he says, 'according to the theory it is to be expected that diamond shows an absorption maximum at $\lambda = 11\mu$'; but here he had over-reached himself, as he soon realized, for the atomic oscillators need not be electrically charged. So a correction [8] soon appeared: 'diamond either shows an absorption maximum at $\lambda = 11\mu$ or it has no optically demonstrable proper frequency whatsoever'. The latter is now known to be a good approximation to the truth.

By about 1911 Nernst had shown that Einstein's formula was roughly correct, but gave far too small a specific heat at the lowest temperatures. His first attempt, with Lindemann [9], to patch it up involved supposing there were two natural frequencies, one twice the other, and each accounting for half the total number of degrees of freedom. This had no theoretical justification, nor had Nernst's [10] next attempt which was to add a term proportional to $T^{5/2}$. Meanwhile Einstein [11] himself had realized the basic weakness of his own model—a single vibrating atom rapidly radiates an acoustic wave into the surrounding medium, and is

Peter Debye

(Dutch, 1884–1966)

Born and schooled in Maastricht, Peter Debye found it convenient to cross the border into Germany and enter college at Aachen, where Wien and Sommerfeld were professors. Until 1940 he spent most of his time in Germany, but when he found himself under Nazi pressure to adopt German citizenship he moved to Cornell where he spent the last 26 years of his life. He was an astonishingly versatile physicist and physical chemist who laid the foundations of many areas of study—the specific heat of solids, the theory of dipole moments, strong electrolytes, x-ray diffraction from random polycrystals, the Compton effect, adiabatic demagnetization and many others that are described in Mansel Davies' biographical memoir (Davies M 1970 *Biographical Memoirs of Fellows of the Royal Society* (London: Royal Society) p 175). Over a period of 60 years he published nearly 250 papers, despite Heisenberg's view that he 'had a certain tendency to take things easy ... I could frequently see him walking around in his garden and watering the roses even during duty hours of the institute. But the centre of his interest was undoubtedly his science'; and according to Henri Seck 'he will live in our memory as a brilliant scientist, a great teacher, a fatherly and helpful adviser and, above all, as a happy man.'.

in fact a heavily damped oscillator to which the application of Planck's formula must be considered very dubious. In a last-minute addendum to his paper of 1911 he remarked that if the atoms are capable of vibrating with a range of frequencies, each separate frequency contributes a Planck term to the thermal energy. He did not, however, suggest any way to calculate the frequency spectrum.

This was the problem that Debye on the one hand, Born and von Kármán (BvK) on the other, tackled independently. First to publish was

Debye [12] with a characteristically brilliant approximation that could be worked out completely, followed by BvK [13] with an attempt at a complete solution whose realization was quite beyond their powers. They agreed, however, that the vibrations to be quantized were not vibrations of individual atoms, but of the solid as a whole. The enumeration of the number of independent modes of vibration in a chosen range of wavelengths had been given by Rayleigh [14] in his classical treatment of black-body radiation, but Debye was apparently unaware of this and, having acquired from his earlier work on diffraction theory a taste for Bessel functions, he worked it out with considerable labour for a spherical body (Rayleigh had chosen a cube and written down the answer in two lines). BvK introduced the idea of periodic boundary conditions and also found the answer easily. In 1911 Weyl [15] proved, what every physicist takes for granted (though it is in fact a subtle matter), that the shape of the body is immaterial. The number of independent modes with wavenumber between q, defined as $2\pi/\lambda$, and $q + dq$ is proportional to $q^2\, dq$, and if the waves are non-dispersive, so that frequency ω is proportional to q, the density of states $g(\omega)$, being the number in unit frequency range, is proportional to ω^2. This was Debye's assumption, not because he believed it true but because he argued that it would be true at very low temperatures, when only long waves were excited, and that he could satisfy the Dulong–Petit law at high temperatures by cutting off the spectrum at a certain maximum frequency ω_0, so that the total number of modes was three times the number of atoms. If the curve fitted the facts at low and high temperatures, it could not be too bad in between, and in any case he did not see his way to calculating anything more sophisticated. BvK attempted a theory of a real atomic lattice, and laid down the mathematical framework for a solution but could make little progress without severe approximations. We shall return to them later; meanwhile Debye holds the stage. Like everyone else, he assigned the Planck energy, $\hbar\omega/(e^{\hbar\omega/kT} - 1)$, to each mode. There is no need to enter into details which can be found in many textbooks. The principal features of his theory are that the specific heat C is a function with one parameter only, the Debye temperature $\Theta_D = \hbar\omega_0/k$, and that at low temperatures $C \propto T^3$. It is striking that the Nernst–Lindemann formula fits Debye's curve within 2% all the way down to $\Theta_D/12$; only below this does the T^3 law show the superiority of Debye's theory.

[See also p 982]

Nernst and his school espoused Debye's theory enthusiastically even before Eucken and Schwers [16] confirmed the T^3 law in fluorspar. As more substances were investigated the observed agreement seems to have conferred on the theory a status approaching that of Holy Writ. It soon became customary to express the measurements of specific heat C in terms of the variation of Θ_D with temperature; that is, the Debye function $C = D(T/\Theta_D)$ is inverted to $T/\Theta_D = D^{-1}(C)$, or $\Theta_D = T/D^{-1}(C)$. Since for many substances the Debye formula is well obeyed, $\Theta_D(T)$ does not vary over a wide range and a compact representation of the results is achieved, without any implied interpretation. If every substance had shown the same pattern of $\Theta_D(T)$ it would have been clear that this represented the effects of Debye's approximations. When, however, different patterns were found there was a tendency (very reasonably discussed in the Ruhemanns' book [4]) to dissect C into a Debye-like part and an anomalous extra for which some cause ought to be discovered. On the whole this rather arid exercise was kept within bounds, real and fictitious anomalies being clearly discerned, and perhaps the chief consequence of Debye's great success was that generations of students grew up in the belief that it was the last word on the subject. As we shall see later, it would be truer to regard it as almost the first and in itself something of a dead end, since it was the atomic model adumbrated by BvK that developed into the elaborate modern theories.

12.1.2. Zero-point motion

Between 1910 and 1912, as Kuhn [5] has described in detail, Planck turned again to the quantum theory of black-body radiation, with the intention of resolving a deep enigma—how could a material oscillator absorb energy only in discrete quanta? If the radiation was exceedingly weak, it would take a long time to excite an extra quantum of oscillation, and what would happen if the radiation were turned off half way through? We need not follow the steps by which he concluded, entirely contrary to Einstein's view, that the process of absorption was continuous while emission was discontinuously quantified. The end-product of his thoughts was that his original formula for the mean energy of an oscillator, $\hbar\omega/(e^{\hbar\omega/kT} - 1)$, must be replaced by $\frac{1}{2}\hbar\omega/(e^{\hbar\omega/kT} + 1)/(e^{\hbar\omega/kT} - 1)$. All he had done was to add $\frac{1}{2}\hbar\omega$ to the mean energy, but it implies that even at absolute zero an oscillator retains some energy. In due course Schrödinger's theory of the harmonic oscillator produced the same result automatically, and by that time there was enough experimental evidence to make the concept of zero-point energy generally acceptable. It appeared at first to be an elusive effect, having no influence on the specific heat unless, as Einstein and Stern remarked, the oscillator frequency changes with temperature; but convincing examples were lacking. Later Bennewitz and Simon [17] developed an argument that was interesting in itself and had even more interesting consequences. They noted that Trouton's empirical rule broke down in a systematic way with light elements. Trouton's rule states that for all liquids the latent heat of evaporation per gram-molecule bears a constant relation to the boiling point; i.e. the entropy change on evaporation is the same for all. This is an extension of van der Waals' law of corresponding states, and would be exactly true if the law of force between molecules took the same form for all, apart from scale factors defining the strength of the potential, ε, and its characteristic range, σ†, but with argon and hydrogen, and especially helium, the latent heat is markedly smaller than Trouton's rule predicts. If, however, the zero-point energy (which vanishes in the gas) is more or less the same, $\frac{9}{8}k\Theta_D$, in the liquid as in the solid, being particularly large for light atoms, it very satisfactorily accounts for the deficit in latent heat. A striking example of the failure of the law of corresponding states, because of zero-point energy, is provided by the heat of sublimation of solid hydrogen and deuterium. If ever two substances should confirm the law it is these, since the mass of the molecule should be irrelevant; yet sublimation of deuterium takes 50% more energy than hydrogen.

Further support for the existence of zero-point motion came from x-ray diffraction, but this had to wait until wave mechanics had been applied by Hartree to determine the electron density distribution in an atom. James and Brindley [18] were then able to compute absolutely the scattering amplitudes of the potassium and chlorine atoms, and hence the strength of various reflections from a crystal of KCl at liquid-air temperature. They found the measured amplitudes less than expected on the assumption that the atoms were at rest, but in agreement with the assumption of zero-point motion.

To return to Bennewitz and Simon: noting how large was the effect of zero-point motion in liquid helium, on account of the lightness of the atom and the weakness of the cohesive force, they suggested that it acted as an internal pressure, expanding the liquid to a much lower density than it would achieve otherwise—so much lower that the atoms could not hold a rigid structure. This would explain why, even at absolute zero, a pressure of 25 atmospheres was needed to reduce its volume to the point of solidification. Helium has a liquid–vapour critical temperature (5.2 K) but no triple point. The argument was generally accepted and when, in about 1948, the possibility arose of extracting enough of the light isotope, ^{3}He, from the atmosphere, its properties were the subject of lively, mainly informal, discussion. There were those who believed that the even greater zero-point motion would prevent condensation taking place at all, except under pressure, so that the liquid–vapour critical point would also be absent. In the middle of these speculations, de Boer [19] resolved the issue with a dramatic coup. He first restated the rationale of the law of corresponding states—if only two scale factors, ε and σ, are needed to specify the forces between molecules, the equation of state must be the same for all materials when written in terms of dimensionless *reduced* variables; $P^* = P\sigma^3/\varepsilon$, $V^* = V/N\sigma^3$ and $T^* = kT/\varepsilon$, N being

---

† Another example is Lindemann's theory of fusion, according to which a solid melts when the amplitude of atomic oscillations reaches a certain fraction of the spacing between atoms. As Einstein and Kamerlingh Onnes both pointed out, this is a dimensional consequence, like the law of corresponding states, of the assumption of a universal law of force between molecules. It tells one nothing of the mechanism of melting.

the number of molecules in the sample. As was well known, the molecular mass m cannot be incorporated in these expressions and is therefore irrelevant to the equation of state. This is no longer true when quantum effects enter, as de Boer pointed out, for a further dimensionless parameter may be constructed, $\Lambda^* = \hbar/\sigma\sqrt{m\varepsilon}$; apart from a numerical factor, Λ^* is the de Broglie wavelength of a molecule having energy ε, expressed in terms of the range σ of the force field. The reduced Debye temperature, $\Theta^* = \Theta_D/T_c$, and Λ^* involve the same combination of parameters, if zero-point motion plays no part. We therefore expect Θ^* to be a unique function of Λ^*, the two being proportional for heavy atoms. For each elementary gas, ε and σ can be found from its high-pressure behaviour, and de Boer was thus able to prove his point by the curve in figure 12.3(a).

In a second paper he and Lunbeck [20] plotted the reduced critical temperature T_c^* against Λ^* for a number of light elements, and again found a smooth curve (figure 12.3(b)). If ^{3}He differs from ^{4}He only in its mass, ε and σ being the same, Λ^* is greater in ^{3}He by a factor $\sqrt{(4/3)}$. On extrapolating their curve they deduced T_c^* for ^{3}He, and hence the actual critical temperature $T_c = 3.3 \pm 0.2$ K. In the event the limits of error were overcautious, since the measured T_c is 3.32 K.

This example is perhaps not strictly relevant to lattice dynamics, but it is so elegant an illustration of the power of dimensional analysis that it deserves to be remembered. We must now return to an earlier period.

12.1.3. Thermal expansion

Grüneisen [21], working at the Reichsanstalt, pointed out in 1908 that α, the linear thermal expansion coefficient of a solid, varies with temperature in the same way as its specific heat—$\alpha/C = $ constant for a given material. He illustrated the relationship with data from a number of metals over a wide range of temperature, and noted that a similar idea had occurred to Slotte some years earlier†. Soon afterwards he derived this result for solids governed by classical dynamics, making use of Clausius' virial theorem which had served van der Waals well in his theory of gases. Three years later [22] he provided a more straightforward and more general thermodynamical explanation, starting from Nernst's heat theorem. Because of it he could assume the entropy S to vanish at zero temperature, and therefore derive $S(T)$ as $\int C \, dT/T$. The volume expansion coefficient, 3α, follows by use of Maxwell's thermodynamics relation, $3\alpha V = (\partial V/\partial T)_P = -(\partial S/\partial P)_T$. It is necessary to know how the specific heat reacts to the volume changes due to P, and

---

† *Slotte had worked in Helsingfors since before 1893 on a kinetic theory of solids, using much-simplified models and elementary analysis to derive a picture of thermal expansion. Einstein was certainly aware of his work, since he wrote an abstract of one of Slotte's papers in 1904, three years before his own specific heat paper; but it is unlikely that Slotte was an effective stimulus to Einstein's thought.*

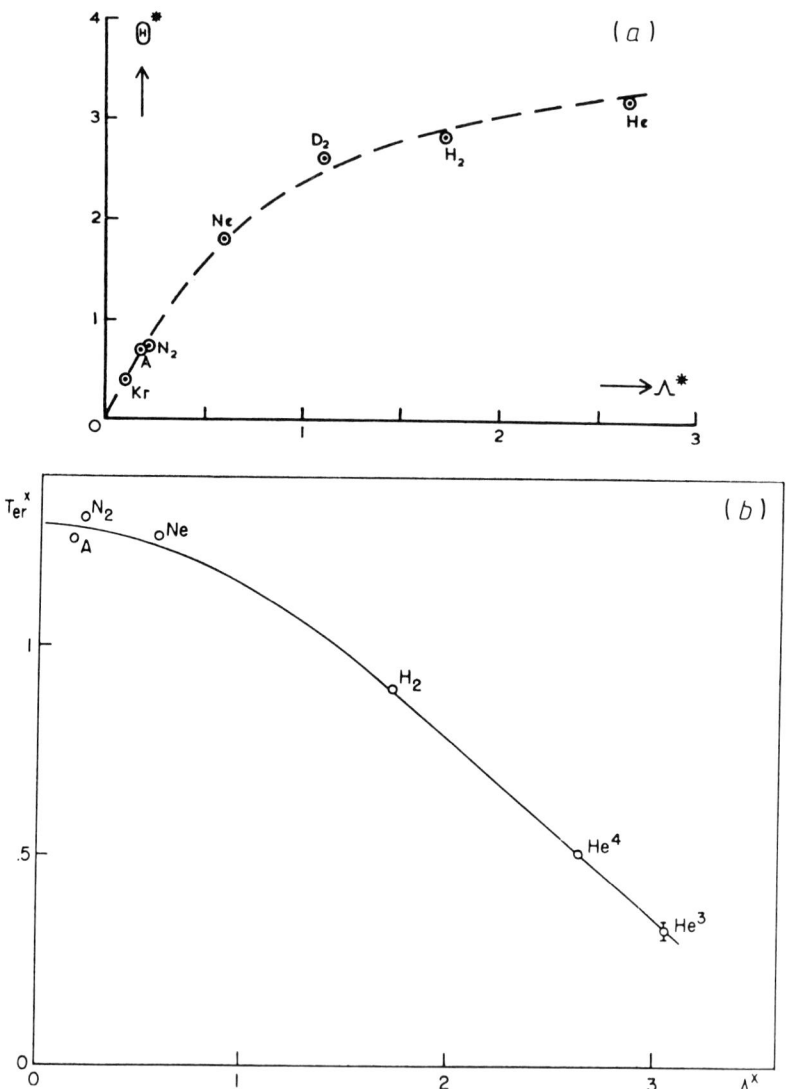

Figure 12.3. (a) The Debye temperature of elementary solids, in reduced terms, is a smooth function of the reduced de Broglie wavelength. (b) The critical temperature is also a smooth function which can be extrapolated to estimate the critical temperature of ^{3}He.

Grüneisen made use of the Einstein and Nernst–Lindemann expressions, in both of which C is a function only of v/T, so that the sole variable involved is the response of the atomic frequency v to volume changes. A few lines of calculation sufficed to show that $\alpha/C = \frac{1}{3}\gamma K$, in which C is the specific heat per unit volume, K is the compressibility and γ

(Grüneisen's constant) is the relative change of atomic frequency with volume, $\gamma = -(V/\nu)\,\mathrm{d}\nu/\mathrm{d}V$.

Shortly after, Debye [23] tackled the same problem from the standpoint of his theory of specific heats. His approach was from first principles, and looks altogether more elaborate in starting from the statistical mechanical partition function, but the end result differs only from Grüneisen's in replacing the single atomic frequency ν by its Debye equivalent, Θ_D.

Grüneisen noted that γ was approximately 2 for several materials (the law of corresponding states would have it the same for all). If the force law between molecules was strictly harmonic (force $\propto$ displacement) there would be no frequency change ($\gamma = 0$) on compression or expansion, and Grüneisen commented that the actual value of γ showed the repulsive force to rise very rapidly at short distances, as indeed van der Waals, and later Mie, had supposed. The vanishing of γ for a strictly harmonic lattice illustrates how anharmonicity and thermal expansion are intimately connected, but the importance of anharmonicity only became apparent when the problem of thermal conductivity in solids came under serious scrutiny.

12.1.4. Thermal conductivity [24]

It is obvious to the touch that metals conduct heat much better than non-metals at room temperature. The explanation of the Wiedemann–Franz law connecting electrical and thermal conductivity in metals was the one great success of the earliest electron theories of metals, as is discussed in Chapter 17. Here we concentrate on non-metals, especially crystals, many of which conduct so poorly that great care is needed, when measuring the conductivity, to avoid serious error from extraneous heat leaks. The once-famous Lees' disc apparatus was more successful than most, and not as crude as perhaps it appears to modern observers who, especially if they are skilled in vacuum and low-temperature techniques, may forget how difficult it was in 1900 to obtain a really hard vacuum. A good overview of the experimental position around 1920 is to be found in the *Dictionary of Applied Physics* [25], even if it has much to say about technical matters such as the heat-insulating properties of powders and fibres, and rather little about measurements down to liquid-air temperature and beyond, such as were pioneered by Eucken [26]. He found the thermal conductivity of a number of salts to vary inversely with temperature over a factor of 4.5 between 373 K and 83 K, and this regularity—about the only one known at the time—was in Debye's mind when he added a paragraph on thermal conductivity to his paper [23] on the expansion coefficient.

Up to this time little interest had been shown in the theory of thermal conduction in non-metals, perhaps through scarcity of information but also because Einstein's model (like earlier atomic models) provided no

See also p 1285

BOX 12A: INTERACTION OF SOUND WAVES

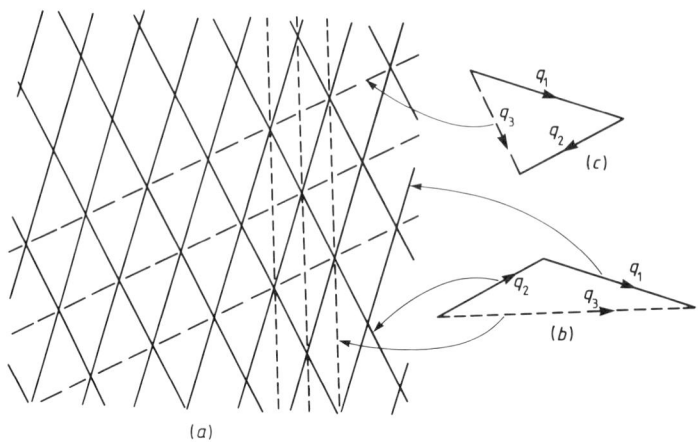

The diagram (a) represents two compressional waves moving from left to right and crossing in a solid. The full lines are wavefronts on which the bonds between atoms are stretched to a maximum. Halfway between these lines the atoms are pushed together. At the intersections the forces causing elongation are added; at the centre of each lozenge the compressive forces are added. Because it is harder to push two atoms together than to pull them apart by the same amount, the elongations exceed the compressions. As a result, along the two sets of broken lines the mean elongation is not zero, as it would be in a harmonic lattice, but positive; and greater than the mean elongation between the broken lines. These broken lines therefore represent a wave-like disturbance in the medium which might serve as a source for a new wave if an essential condition is satisfied: as the original waves progress, the new wavefronts progress with them, and if it happens that they move with the speed of sound waves they can excite a real progressive wave.

CONTINUED

mechanism for energy transport so long as each atom was conceived to vibrate independently. Einstein did indeed, as already noted, recognize that the energy of an oscillator would be rapidly dissipated, but he took the argument no further. Debye found himself in a quandary precisely opposite to Einstein's—so long as the acoustic waves were independent of each other, there was nothing in a perfect crystal to limit the conductivity. He seems (his full account [27] of the work is hard to

BOX 12A: *CONTINUED*

The vector diagrams on the right describe the situation in (*a*). The full lines, drawn normal to the wavefronts, represent the wavevectors of the original waves, while vector addition (*b*) and subtraction (*c*) give the wavevectors of the waves that might be produced by the interaction. The frequencies of the new wave motions are $\omega_1 + \omega_2$ in (*b*) and $\omega_1 - \omega_2$ in (*c*), where ω_1 and ω_2 are the frequencies of the original waves. As Pauli remarked, this is the analogue for waves of the production of combination tones when two musical notes interact in a non-linear medium (e.g. the ear). The process is possible if the new wavevector q_3 and frequency ω_3 are such that ω_3/q_3 is the velocity of a sound wave of frequency ω_3.

The condition for wave interaction can be expressed in a familiar form in terms of phonons, the particle-like quanta of wave motion which have momentum $\hbar q$ and energy $\hbar \omega$. When two phonons combine to give a single third phonon, their energy and momentum must be conserved—$\hbar \omega_3 = \hbar \omega_1 + \hbar \omega_2$ and $\hbar q_3 = \hbar q_1 + \hbar q_2$ in (*b*); in (*c*) a process of stimulated emission occurs, and plus signs are replaced by minus signs. There is, or course, no need to include $\hbar$ in these equations, which then express conservation of frequency and wavevector.

If the waves are non-dispersive, ω/q is constant, and two waves can only interact successfully if they are moving in the same direction. This does not alter the flow of energy and therefore does nothing to limit the transport of heat.

find) to have hoped at first that spontaneous density fluctuations would serve to scatter the waves; but since these fluctuations are themselves no other than thermally generated sound waves they cannot interact with other waves in a truly harmonic medium. He was forced to fall back on anharmonicity as the scattering mechanism (see Box 12A), unaware that this also is ineffective, as Pauli pointed out in 1925. We can jump straight to this moment in the story, since the intervening years are barren of interesting ideas on the subject.

Pauli's account [28] is only an extended abstract of a talk to the German Physical Society and leaves much to the imagination. According to his student Peierls [29] it is also unique among Pauli's works in that the analysis is wrong. What is not wrong is the leading idea that on an atomic lattice, as distinct from Debye's continuum, two waves may interact to produce a third. On Pauli's suggestion, Peierls studied this

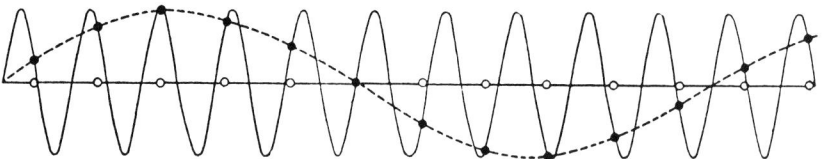

Figure 12.4. *The dark circles represent the displacement of the evenly spaced atoms (open circles) when a transverse wave passes. The full and broken lines are equally valid interpretations of the waveform responsible.*

See also p 1304

problem and arrived at the concept of Umklapp processes. There is no need to repeat the discussion given in Chapter 17, but it should be remarked that it was in connection with the present topic, heat transport in non-metals, that the concept emerged. The essential point is that in a discrete atomic lattice there is no unique specification of wavevectors (figure 12.4); if $e^{iq \cdot r}$ describes the displacement of every atom, r being a typical position vector for an atom, then $e^{i(q+g) \cdot r}$ does equally well if g is any vector of the reciprocal lattice, having the property that $e^{ig \cdot r}$ is the same for every atom. A typical process that conserves wavevectors is expressed by $q_3 = q_1 + q_2 + g$, and this loosening of the conservation law allows waves, when they interact, to conserve energy but not flux of energy, so that a heat current can be dissipated. Momentum in the wave system is not conserved, and the balance is taken up by the lattice as a whole, in units of $\hbar g$.

Peierls' analysis is long and detailed, as befits the first substantial piece of work of a research student, but the principal results can be stated briefly, if anachronistically, in terms of phonons. At high temperatures ($T > \Theta_D$), where Dulong and Petit's law holds, the mean free path of a given phonon, before it is destroyed in collision with another, is inversely proportional to the numbers of other phonons present, and therefore varies as $1/T$. Since the thermal conductivity κ is proportional to the lattice specific heat and to the phonon free path, and since C is constant, $\kappa \propto 1/T$ as found by Eucken and explained by Debye, though the latter's demonstration was wrong in principle. The Umklapp process is the responsible agent, but at lower temperatures it becomes less effective; when all phonons present have values of q significantly less than the smallest g, no Umklapp can satisfy the conservation conditions, and the mean free path rises steeply. In a large perfect crystal there would be nothing to limit the rise as T falls to zero, but Peierls noted that imperfections would ultimately take control by scattering the phonons; and if they did not, the free path would exceed the lateral dimensions of the sample, so that the phonons would ricochet from wall to wall in random directions. The free path would then, in a cylindrical rod, become equal to the diameter of the rod and κ would fall to zero as T^3, according to Debye's specific heat law.

Vibrations and Spin Waves in Crystals

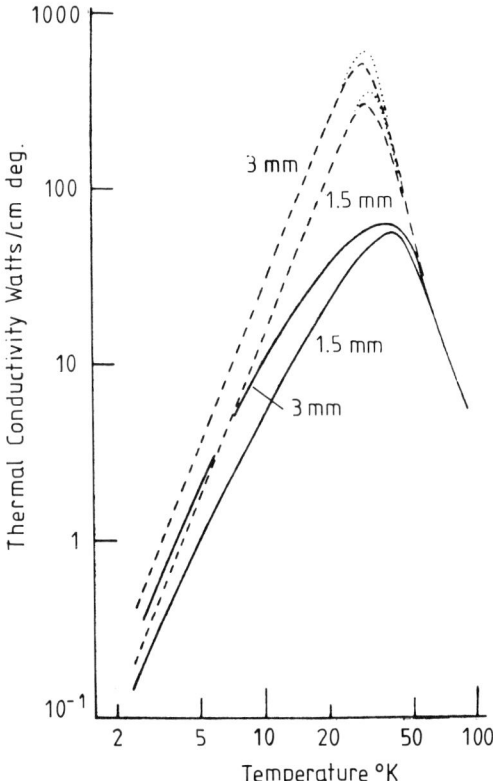

Figure 12.5. *Thermal conductivity of single-crystal sapphire rods (logarithmic scales). The abscissa is temperature and the ordinate conductivity in W cm^{-1} deg^{-1}. As the surface of the rod is roughened the conductivity at the lower temperatures is reduced, since the phonons are scattered by the surface instead of being reflected specularly. Above 50 K scattering within the rod limits the phonon free path, and surface conditions do not matter.*

The first unequivocal verification of Peierls' predictions had to wait until 1951, when Berman began obtaining results such as that shown in figure 12.5. The conductivity at the peak is 40 times that of copper at room temperature. Further work has extended and refined both experiment and theory, without significantly affecting the foundations laid by Peierls. A review by Berman [24], and Ziman's monograph [30], bring the story up to 1960, by which time interest in the topic was declining. Ziman also covers the case of lattice conduction of heat in metals, where usually the electrons dominate, as well as very effectively limiting the phonon free path. This topic is of more interest for what it can reveal of electronic properties than of lattice dynamics, and will not be considered here.

The long free paths of phonons at low temperatures was

Figure 12.6. *Northrop and Wolfe's demonstration of the preferred directions (light image) for ballistic motion of phonons through a germanium crystal.*

demonstrated directly by von Gutfeld and Nethercot [31] in 1964. They evaporated thin films on either face of a synthetic sapphire (Al_2O_3) crystal, one to serve as heater and the other as thermometer. A very short injected heat pulse arrived at the thermometer as two well-resolved pulses, with such time delays as identified them as longitudinal and transverse waves which had travelled ballistically without intervening collisions. On account of the elastic anisotropy a wave-group does not in general travel along the direction of the normal to its wavefront, and in certain directions there is a strong concentration of radiated power. Northrop and Wolfe [32] exhibited this in a striking way by attaching a small heat sensor to one face of a germanium crystal, and scanning a focused laser beam across the opposite face in a television raster. The response of the sensor (figure 12.6) shows immediately the directions of channelling of power, the details of which can be confirmed as consistent with the known elastic constants of germanium. In itself, the experiment tells one nothing about lattice dynamics, since it can be fully explained by continuum elastic theory, but the use of ballistic phonons as a research tool has proved valuable in a variety of other contexts.

12.1.5. Lattice theories of specific heat

The huge triumph of Debye's specific heat theory, and the difficulty of handling three-dimensional lattices, caused the BvK approach to the problem almost to disappear from view. Not entirely, however, and it can now be seen providing the basis of all later developments. As such, it deserves more than the passing mention which is all we have so far accorded it. To appreciate its origins we may start with the systematic studies of the far-infrared spectrum that were pursued by Paschen, Rubens and others [33] at the Reichsanstalt as a consequence of their interest in black-body radiation.

Vibrations and Spin Waves in Crystals

It was found that the dispersion (variation of refractive index with wavelength) of transparent crystals could be explained by the presence of strong absorption lines in both the ultraviolet and the infrared. If such an absorption follows the standard resonance curve, the crystal will reflect very strongly—almost like a metal—wavelengths near the absorption maximum. Rubens and Nichols mounted an experiment in which heat radiation was reflected many (~6) times from crystal faces and then analysed with a reflection-grating spectrograph. They found that only narrow lines (reststrahlen or residual rays) survived the process and determined, for example, that quartz had three absorption lines at 8.50, 9.02 and 20.74 μm. In the course of time many crystals were studied, some showing reststrahlen as far out as 100 μm. As already remarked, diamond is transparent right into the far infrared.

The explanation given by Drude [34] was that, unlike the case in metals whose optical properties he rightly ascribed to itinerant electrons, the electrons in insulators are attached to atoms and give rise to two characteristic types of vibration: at high ultraviolet frequencies the electrons vibrate with respect to their own atom, while vibrations of whole atoms relative to one another take place at lower, infrared, frequencies. Reststrahlen bands therefore reveal the frequencies of the latter type of vibration—an idea taken over by Einstein in his specific heat theory. Two years later Madelung [35] greatly refined this idea, and his work lies at the root of the BvK approach. He recognized that a typical vibration involved not one atom alone (as in Einstein's theory) but the whole crystal, and that there is a natural highest frequency characterized by neighbouring atoms vibrating in antiphase. Madelung was not concerned with specific heats, but with finding a connection between the elastic constants of a crystal and its reststrahlen frequency. He assumed a model, which he thought might suit NaCl, in which the atoms are arranged in a simple cubic lattice, alternately Na and Cl, charged oppositely like the ions in salt solution, according to the ionic hypothesis that by this time was generally accepted. The ions are imagined as connected to all their surrounding neighbours by springs whose strength can be determined from the elastic constants. From the spring constants and the atomic masses the frequency of any chosen vibrational mode can in principle be found. The mode chosen by Madelung as responsible for the infrared absorption was that in which all Na ions on planes normal to a cube diagonal vibrate in phase with one another, but in antiphase with the Cl ions on planes sandwiched between them; he found good agreement between calculated and measured frequencies. In view of the fact that his model is exactly that which the Braggs determined six years later by x-ray diffraction, one might have expected them to mention Madelung, but it is likely enough that they were unaware of his work. They say there is strong evidence for the model they (and Madelung) choose, but are probably referring to the extensive studies of the packing

See also p 428

of spheres by Barlow and Pope. What seems to be new in Madelung's approach is the assumption that the units are ions, not atoms. These conceptions, dating from 1906 to 1909, deserve to be remembered by text-book writers who imply that crystal chemistry started with von Laue and the Braggs.

To progress beyond Madelung's chosen vibrational mode to the general case of a standing wave of arbitrary wavevector q was the essential core of the BvK approach. The difficulties were that too little was known about the interatomic forces, and that the numerical work needed for detailed calculations was too demanding until the brute force of computers could be brought to bear on the problem. A one-dimensional chain of atoms coupled by harmonic elastic springs can be analytically solved, and BvK give the solution, presumably not knowing, as Debye did, that Rayleigh had already discussed the problem in detail. Instead of the angular frequency, ω, of a normal mode being related to its wavenumber q by a constant velocity u, $\omega = uq$, they found $\omega = \omega_0 \sin(aq/2) = \omega_0 \sin(\pi a/\lambda)$, where a is the spacing of the atoms and ω_0 depends on the mass and spring constant; ω_0 is the highest vibrational frequency, attained when $\lambda = 2a$, and alternate atoms vibrate in antiphase. Over the band of wavenumbers q around π/a, the frequency varies only slightly from ω_0, and since the wavenumbers of the modes are evenly spaced in q, there are a large number of normal modes with frequencies close to ω_0. This feature differs from the Debye approximation, and gives rise to the temperature dependence of the Debye temperature deduced from the specific heat, as discussed above.

BvK simplified the three-dimensional problem by assuming that the lattice is isotropic with the same dispersion curve (dependence of ω upon q) for the normal modes in every direction of q. It was then straightforward to obtain the energy of the thermal vibrations, and hence the specific heat. As Blackman comments in a useful review of the early work [36], BvK's expression is more complicated than Debye's, but fits the experimental results no better. At nearly the same time, 1914, Born [37] applied the BvK theory to the specific case of diamond, whose crystal structure had been only recently determined by the Braggs. The theory required a knowledge of the interatomic force constants in diamond, and to obtain these Born developed the idea that the long-wavelength modes had to be consistent with the macroscopic elastic waves which could propagate in the material. As a result of this Born showed that the number of elastic constants obtained from BvK theory of non-Bravais lattices, with forces which depended on the angles between the atomic bonds, agreed with macroscopic theory and not with the predictions of Cauchy's oversimplified Bravais model. These results were then in agreement with the known results for the elastic constants of diamond.

After this the development of the BvK approach rested, apart from some minor improvements, until Born and his group began work again

Max Born
(German, 1892–1970)

Max Born was born in Breslau in December 1892. His father was an embryologist who held a Chair at the University. He went to school in Breslau and then to university where he attended lectures on physics, chemistry, zoology, philosophy, logic and mathematics. He moved to Göttingen and completed his doctorate in 1907 where he studied and worked with Carathéodory, Courant, Hilbert, Minkowski, Schwarzschild and Voigt.

After his marriage in 1913 to Hedwig Ehrenberg, he moved to a Chair in Berlin in 1914 and then, in 1919, exchanged Chairs with Max von Laue at Frankfurt. In 1921 he became Director of the Physical Institute in Göttingen in succession to Debye where his assistants included Pauli, Heisenberg and Jordan, and the Institute was in the forefront of the development of quantum theory. In 1933 Hitler came to power and Max Born was deprived of his Chair and he accepted an invitation to move to Cambridge.

In 1936 he was appointed to the Tait Chair of Natural Philosophy in Edinburgh where he established a research school and was elected a Fellow of the Royal Society. He retired in 1953 and returned to Bad Pyrmont in Germany and became active in the cause of the social responsibility of scientists. He was awarded the Nobel Prize in 1954 for his work on quantum theory.

Throughout his life he was a keen and good musician and he died in Göttingen in 1970 leaving his widow, one son and two daughters.

---

in the thirties. For von Kármán it had been a brief intermission from his life's work in aerodynamics, while Born, reminiscing in 1965, admitted that he lost interest in the field after his work with von Kármán, but that it was a valuable source of problems for research students.

Blackman was one of these students and he tackled the problem

See also p 851

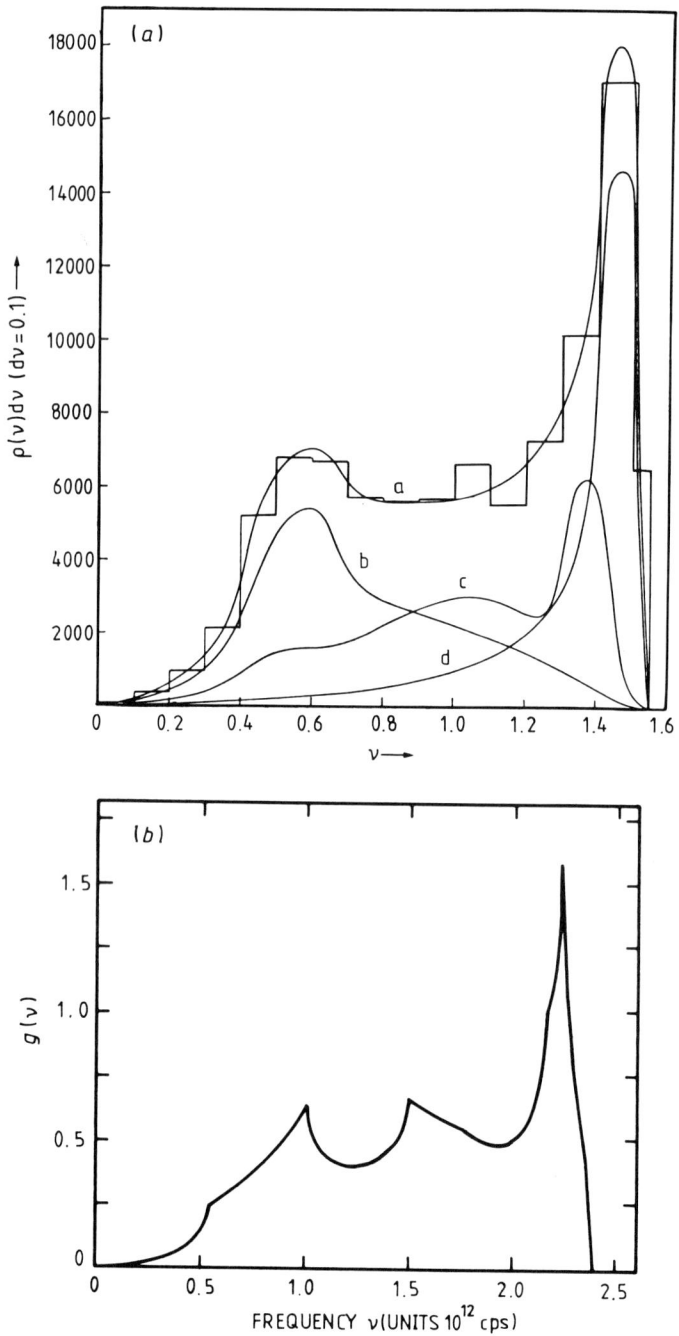

Figure 12.7. (a) The density of states for a simple cubic lattice with nearest-neighbour interactions. Curve 'a' gives the total density of states [36]. (b) The frequency distribution function for potassium at 9 K calculated from 60 466 176 normal modes in the Brillouin zone [38].

of a simple cubic solid with identical atoms connected by elastic springs between the neighbouring atoms along the cube edges and face diagonals. The model was not very realistic, but computing the specific heat [36] with even this simple model was a formidable problem when only hand-operated calculators were available. The formalism of BvK immediately reveals where the hard work arises: the dynamical equations for the displacements of the atoms are, for each wavevector q in three dimensions, a set of three coupled equations. The normal mode frequencies are the eigenvalues of the 3×3 matrix, or equivalently the equation for the frequencies of these modes is cubic and gives three independent normal modes, which in symmetry directions are a longitudinal and two transverse modes. The calculation of the specific heat requires repeating the procedure (which involves numerical solution of the cubic equations) for a large number of different wavevectors throughout the Brillouin zone. In figure 12.7(a) we show the frequency distribution calculated by Blackman with 8000 points in the Brillouin zone, and it is compared with a more recent calculation, figure 12.7(b), which included interatomic forces up to fifth-nearest neighbours and many more points in the Brillouin zone. The calculation of Θ_D from Blackman's density of states then gave qualitatively the features observed in many solids—a minimum of Θ_D at intermediate temperatures which reflects the peak in the density of states from the transverse acoustic modes at the Brillouin zone boundary.

Because of the difficulty of making these calculations, considerable effort was then directed by Montroll [39] and others towards developing analytic methods for a variety of simple two- and three-dimensional atomic models. The results have now largely been superseded by detailed computer calculations, apart from the work of van Hove [40] on the nature of the singularities in the density of states, arising from the maxima, minima and saddle points of the dispersion curves. The singularities are mostly ones in which the derivative of the density of states with respect to frequency is discontinuous at the frequency of the singularity as illustrated in figure 12.7(b).

After his move to Edinburgh in 1936, Born's group continued its work on the theory of lattice dynamics. Of particular note is the work of Kellermann [41] on the alkali halide, NaCl. Kellermann used Madelung's model for alkali halides, of charged ions interacting with electrostatic forces and a short-range repulsive interaction, and computed the vibrational frequencies for 48 different wavevectors in the Brillouin zone. The new difficulty which was overcome in performing this calculation was the evaluation of the electrostatic forces. The displacement of the ions gives rise to electrostatic dipoles which interact with other dipoles, through a dipole–dipole interaction which decreases with distance as $1/r^3$. The summation over all the dipole–dipole interactions is of very long range, and indeed if wavevector q is zero is only conditionally convergent and dependent on the macroscopic shape of the crystal. These problems were solved by Kellermann by the use of the Ewald transformation which converts the long-range dipole–dipole sum into two rapidly convergent sums. The same procedure is still used today with computers, but is much more easily implemented than it was for Kellermann working with a hand calculator.

The dispersion curves for various high-symmetry directions are shown in figure 12.8. The results show longitudinal and transverse modes of both an acoustic and optical nature. The lower-frequency acoustic modes have the Na and Cl ions moving largely in phase, and in the higher-frequency optic modes they are largely out of phase. Of particular interest as the wavevector, q, tends to zero is that the limiting frequency of the transverse and longitudinal optic modes is different. This is a direct consequence of the long-range forces being only conditionally convergent at long wavelengths, and is characteristic of all infrared-active optic modes in ionic crystals.

Kellermann's calculations and similar calculations by Iona [42] for KCl gave good agreement with specific heat measurements. Later Huang [43] extended the work of Kellermann, using Maxwell's electromagnetic equations instead of Coulomb's electrostatic law to calculate the forces between the dipoles. The differences in the results are only important where the wavevector is so small (10^{-4} of the Brillouin zone dimension) that the optic modes and light waves have similar frequencies. For these wavevectors the light waves, photons, interact with the lattice dynamical waves, phonons, to produce coupled waves as illustrated in figure 12.9. Later these results were less elegantly rediscovered, and named *polaritons*, and later again Raman scattering measurements were made which showed the correctness of this theory.

The development of lattice dynamics up until 1954 is very completely summarized in the monograph by Born and Huang [45], which still for many remains the definitive account of the formal theory. In addition to the topics discussed above, the work of Born and his students had laid down the framework of the theory for the optical properties— infrared absorption and Raman scattering—and for the elastic and thermal properties in terms of the interatomic forces. The difficulties at this time were that although the formal theory had been developed, very little was known about the interatomic forces, and the calculations needed to give quantitative results to compare with measurements of, say, the specific heat were excessively lengthy and tedious. Consequently the subject as a whole was somewhat arid and did not attract a great deal of attention. It needed more precise experimental data against which to compare the theory, and more information about the interatomic forces. One approach, to refine the measurements of the specific heat and to deduce from them more information about the frequency distribution, proved unrewarding—thermal expansion and anharmonic effects cause

Vibrations and Spin Waves in Crystals

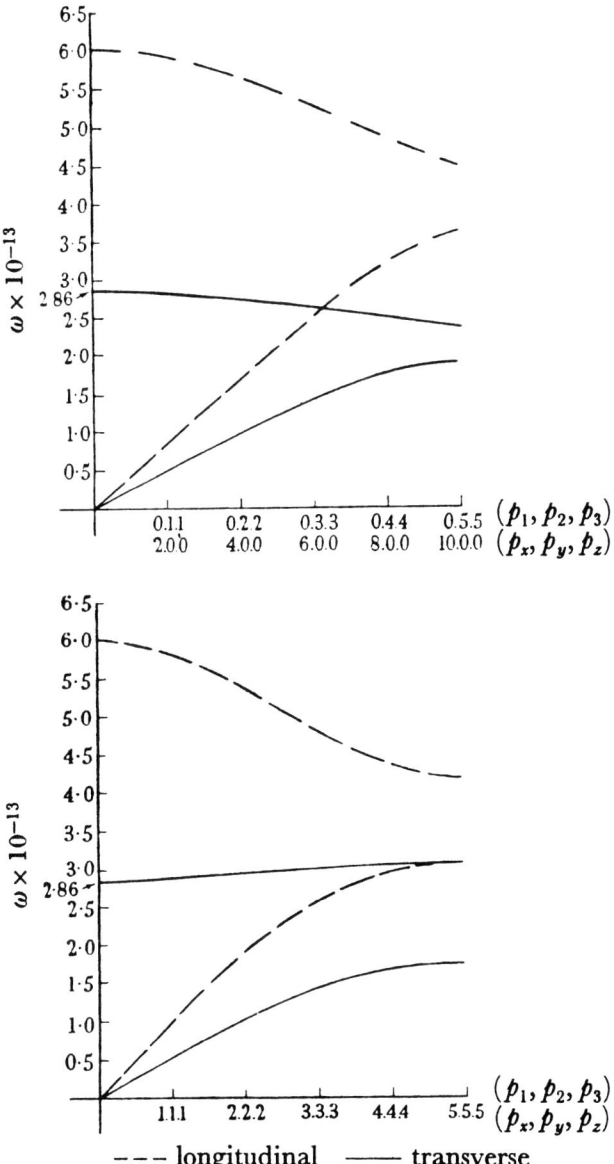

--- longitudinal ——— transverse

Figure 12.8. *The dispersion curves calculated by Kellermann [41] for NaCl. The wavevector $p = 10\,aq/2\pi$, and the figure shows the dispersion curves in the (100) and (111) directions.*

small changes in the specific heat, which are difficult to estimate and can cause substantial errors in the resulting frequency distributions. Clearly the information required was a direct measurement of the frequencies ω for particular wavevectors, q, and preferably the dispersion curves $\omega(q)$

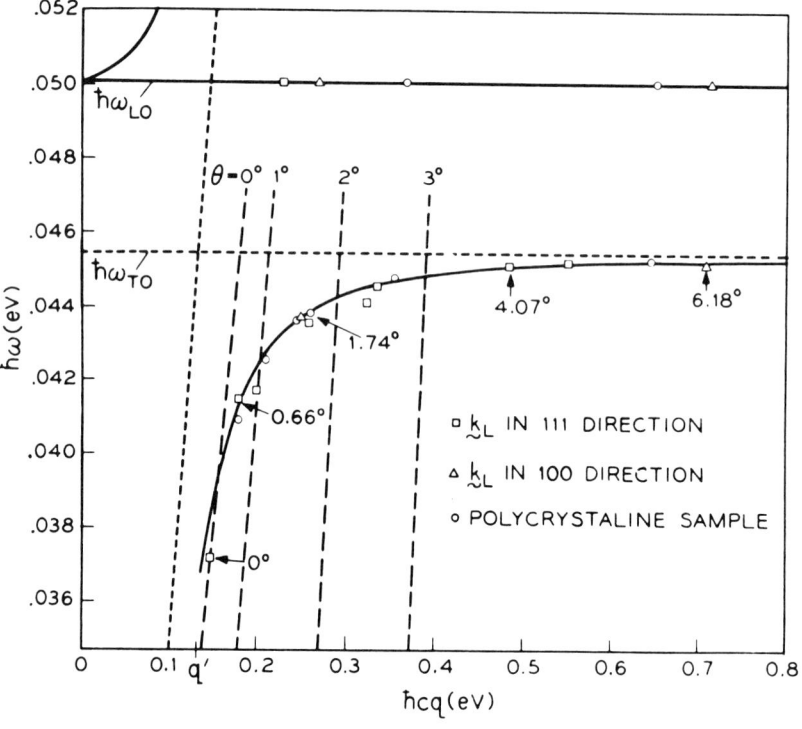

Figure 12.9. *The measured dispersion curves for polaritons, the coupled longitudinal optic and photon excitations. The measurements were made by Raman scattering at small angles and the material was GaP* [44].

as illustrated in figure 12.8.

The first direct measurements of the dispersion curves were made using x-ray scattering techniques. The long and involved history of this topic is recounted at least in part in Chapter 6, and in more detail in a review by Lonsdale [46]. X-ray scattering from a crystal can arise not only from the Bragg reflections, but also when phonons are produced or absorbed. The wavevector transfer in the scattering, Q, is then $q + g$ where g is a reciprocal lattice vector (Bragg reflection) and q is the phonon wavevector. This gives rise to diffuse haloes around each Bragg reflection, due to scattering from the acoustic modes, which were investigated by Laval [47] and more quantitatively by Curien [48]. The detailed theory shows that the intensity of the scattering by the phonon is inversely proportional to the square of the frequency, and dependent on the pattern of the displacements in the normal mode. By making careful quantitative measurements, Walker [49] managed to deduce the dispersion curves for the phonons in aluminium, as shown in figure 12.11. The method is, however, very difficult to apply generally

Vibrations and Spin Waves in Crystals

because corrections are needed for Compton scattering of the x-rays and for multiple phonon scattering. As a result the methods cannot give such accurate measurements as one in which the frequencies are directly measured rather than inferred from intensity measurements.

It was at the end of the 1950s that the technique of neutron scattering became available which enabled detailed measurements of dispersion curves to be made, and these have had a great influence on subsequent developments. At about the same time light scattering methods became far more efficient because of the development of intense monochromatic laser sources, and these two techniques have greatly increased the amount and usefulness of the experimental information available about phonons and interatomic forces.

12.2. New experimental techniques

12.2.1. Neutron scattering

The neutron was discovered in 1932, and soon afterwards it was shown that neutrons were diffracted by crystalline materials. Before progress could be made, however, a more intense source was needed and this was provided by the nuclear reactors built during and after World War II. For these reactors to operate, they needed neutrons that were thermalized in a moderator, and these thermal or moderated neutrons then had a Maxwell–Boltzmann distribution of energies centred around the temperature of the moderator multiplied by Boltzmann's constant. The wavelength of the de Broglie waves for neutrons from a room-temperature moderator is 0.18 nm, and very similar to the wavelength of x-rays and the interatomic spacing in crystals. Diffraction of neutrons by single crystals is then similar to that of x-rays but, as discussed in Chapter 6, neutrons have some advantages over x-rays in that the scattering is related to the nuclear properties rather than the atomic number and so they can readily be used for light atoms and to distinguish atoms with similar atomic numbers. The neutron, having a magnetic moment, can also be used to measure magnetic structures, and is now the accepted tool for the determination of the details of these structures.

> See also p 453

After the development of the first nuclear reactors, experimenters at the Oak Ridge Laboratory began using neutrons to study structures. Nearly monochromatic (i.e. mono-energetic) beams of neutrons were obtained by Bragg reflecting the beams from the reactor with a large single crystal, and these monochromatic beams were then used together with conventional crystallographic techniques to determine a wide variety of crystallographic and magnetic structures. The neutron had, however, more to offer solid-state physics than the determination of structures. The energy of these thermal neutrons is very similar to the quantized energy associated with a normal mode of vibration, $\hbar\omega(q)$. When a neutron is scattered by a single phonon, with energy transfer

Twentieth Century Physics

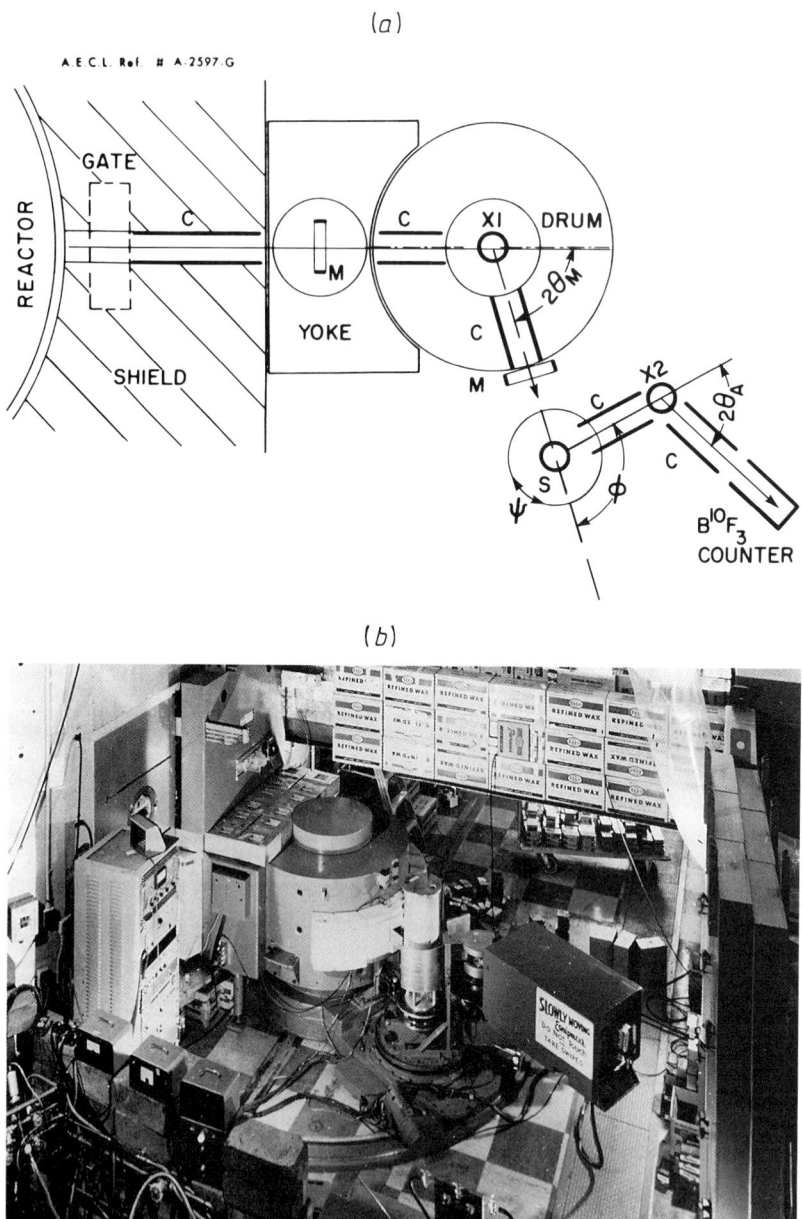

Figure 12.10. (a) A schematic diagram of the original triple-axis crystal spectrometer built by Brockhouse [50]. (b) A photograph of the spectrometer showing the monochromator drum, the specimen cryostat and the analysing spectrometer.

$\hbar\omega(q)$, and wavevector transfer $Q = g + q$, both $\omega(q)$ and Q are in principle measurable because the changes are large in comparison with

Bertram N Brockhouse

(Canadian, b 1918)

Bertram Brockhouse was born in 1918 and served in the Canadian Navy from 1939 to 1945. He obtained his BA from the University of British Columbia in 1947 and then moved to the University of Toronto to study for his PhD.

He married Doris Miller in 1948 and in 1949 became a lecturer at the University of Toronto until he completed his PhD in 1950. After his PhD he joined Atomic Energy of Canada Ltd as a research officer and worked initially with Donald Hurst on neutron diffraction.

Apart from an extended visit to Brookhaven National Laboratory when the NRX reactor at Chalk River was out of use, he remained at Chalk River until 1962. Most of his important experiments were performed between 1956 and 1962 and he was the Branch Head of the Neutron Physics Branch from 1960 to 1962. In 1962 he became Professor of Physics at the McMaster University from which post he retired in 1984. He has been awarded many honours for his work including the Nobel Prize in 1994, Fellowship of the Royal Society in 1965 and the Buckley Prize of the American Institute of Physics in 1962.

He is a Canadian patriot and has four sons and one daughter.

---

the original energy and wavevector of the neutron. This is different from the situation with x-rays where Q is easily measured, but the change in energy, one part in 10^5, is very small.

In the 1940s the theory of thermal neutron scattering by materials was of importance in the design of the moderators for reactors, and little of the early work was directed to basic solid-state physics. By 1950 it was realized in several laboratories that neutron scattering, in principle, provided a tool for measuring phonon dispersion relations, as was explicitly spelt out by Placzek and van Hove [51] in 1954. This possibility attracted several groups, and the most successful experiments were performed at Atomic Energy of Canada by Brockhouse. He realized that the experiments would be difficult and the intensities very low, but had the advantage of the reactor at Chalk River which had the highest flux of any in the world for this type of experiment. He therefore began developing the necessary techniques to measure the energy of the scattered neutrons. He improved the neutron detectors and the shielding around the instruments and, possibly most importantly, the size and reflectivity of the monochromating crystals. These developments enabled him to build the first triple-axis spectrometer (figure 12.10), in which the scattered energy is measured by a second monochromator crystal. The first experiments [52] used a fixed incident neutron energy, and enabled measurements to be made of the phonon dispersion curves in aluminium (figure 12.11). These were the first direct observations of phonons in metals, but the triple-axis crystal spectrometer became a much more useful tool when it was realized that it could be controlled to measure the frequencies at certain predetermined wavevectors, the so-called constant-Q technique. This was implemented in 1958, and in 1960 the spectrometer was redesigned so that it was controlled by computer-generated punched paper tape. A large number of very important measurements were made with this spectrometer, and it is still used in much the same form as when it was developed, though with further improvements in the efficiency of the monochromators and in the direct computer control which makes it far more adaptable and convenient.

In the 1950s most other laboratories chose a different approach to measure the energy transferred by the neutrons. A chopper was used to produce a pulsed monochromatic beam and the time-of-flight between chopper and detector measured to determine the scattered energy. This type of instrument has proved to be not nearly so powerful for single crystals as the triple-axis crystal spectrometer, but for isotropic systems (liquids and amorphous materials) it has proved invaluable, especially with modern computers. The first successful experiment performed with the time-of-flight technique was the measurement of the Landau phonon–roton dispersion relation in superfluid liquid helium by groups at Atomic Energy of Canada, Stockholm and Los Alamos in 1957, following the suggestion by Cohen and Feynman [53].

See also p 931

The first neutron scattering experiments were all performed on reactors built largely for the production of isotopes and material testing. The importance and success of the early experiments led to a rapid increase in the number of groups performing these experiments and to the construction of reactors at Brookhaven and Grenoble solely for the purpose of providing intense beams for neutron scattering. The reactor at Grenoble has hot and cold sources to adjust the temperature of the moderated neutrons, and makes extensive use of guide tubes to transport the neutrons away from the reactor. The instruments have become more complex and refined, but the basic principles have remained similar to those of the early experiments.

The neutron flux of these reactors is only 5–10 times that of the reactor

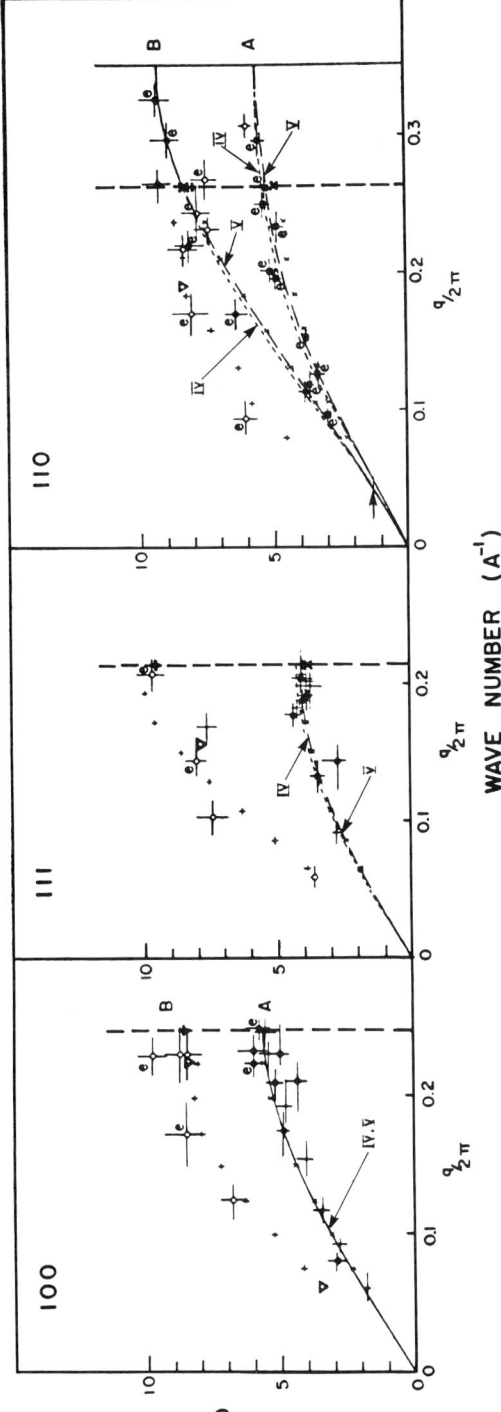

Figure 12.11. *The phonon dispersion curves of aluminium along the three principal symmetry directions; ν is in THz, the lines are various models, the + and × are from x-ray measurements [49] and the ● and ○ from neutron measurements [52].*

used by Brockhouse. The increase in flux is small because it is limited by heat transfer problems in the reactor. As a result new sources of neutrons using proton accelerators and the spallation process are being developed at the Rutherford Appleton Laboratory and other laboratories around the world. Meanwhile the range of problems in condensed-matter science to which neutrons are making a crucial contribution has spread from physics into chemistry, biology and materials science.

12.2.2. Raman scattering

Raman scattering, the inelastic scattering of light by phonons, was first observed in solids by Landsberg and Mandelstam [54] in 1928, at about the same time as Raman and Krishnan [55] observed it in liquids and gases. Landsberg and Mandelstam observed that sharp spectral lines developed sharp satellites, resulting from what was later interpreted as scattering from single optical phonons of long wavelength; in contrast to this Fermi and Rasetti [56] observed in 1931 a continuous spectrum which arises from the simultaneous scattering by two phonons. The Raman effect is very weak, and because the frequency changes in the scattering are typically about 10^{-3} of the incident frequency, the incident light must be highly monochromatic and the scattered light detected by a high-resolution spectrometer. These were difficult conditions when a white-light mercury lamp was all that was available for a source. As a result few experimental groups continued with the study of Raman scattering. The field was completely changed in the early 1960s by the development of laser sources to provide intense monochromatic and highly collimated beams of light, and the technique was rapidly taken up in many laboratories. The high resolution, and relative ease, of performing experiments has made Raman scattering widely available, and a powerful tool for studying the frequency and lifetime of phonons.

The theory of the first-order Raman effect was first discussed classically by Mandelstam in 1930 [57], and in the same year quantum mechanically by Tamm. After Placzek's 1934 review [58], there was little progress until Born and Bradburn [59] discussed the two-phonon spectrum of alkali halides in 1947, and Smith [60] did comparable calculations for diamond. The theory is summarized very clearly in the text by Born and Huang [45]. Full understanding of the Raman effect involves on the one hand a knowledge of the frequencies and lifetimes of the phonon modes which couple to the light, and on the other hand the nature of the coupling between the light and the phonons. The former problem is the basic problem of lattice dynamics discussed in other parts of this chapter. There have been several valiant attempts to develop a theory of the coupling mechanism by Placzek and Born, expanding the polarizability in terms of the displacements of the atoms, and by Loudon [61] using third-order perturbation theory to describe the virtual electronic transitions. Nevertheless these are difficult to apply

See also p 1397

Vibrations and Spin Waves in Crystals

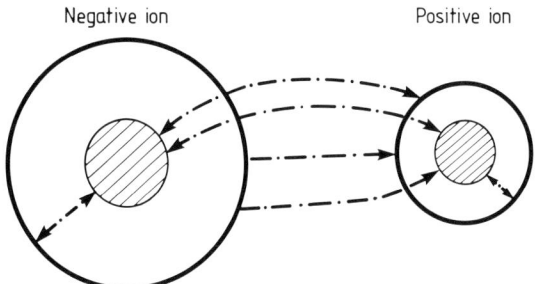

Figure 12.12. *The shell model for ions in crystals (of an alkali halide), showing the different short-range forces and shell–core coupling.*

quantitatively, and this limits the application of an otherwise valuable technique.

12.3. The development of lattice dynamics

12.3.1. Ionic crystals

Born's theory of ionic crystals is unsatisfactory because it neglects the polarizability of the ions, which is why the optical refractive index of crystals is different from that of free space—ions become polarized in the electric field of the light. The early phases of this story have been described in Chapter 6, and we take it up again with the proposal by Dick and Overhauser [62] of a shell model of the ion (figure 12.12). The outer electrons are described by a massless shell which can be displaced from the core to produce a polarization dipole. The shells interact with other cores and shells both by short-range overlap forces and by electrostatic forces, and are attached by a spring to their own cores. Woods, Cochran and Brockhouse [63] extended this model to explain the recently measured dispersion curves for NaI and KBr, and obtained very reasonable agreement (figure 12.13). There was no doubt that the shell model was successful in capturing much of the essential physics, and since then it, and various equivalent models, have been widely used to describe the interatomic forces of ionic crystals.

Despite this success, a careful comparison of the results of the shell model with the experimental results showed that there were still difficulties. In particular the frequencies of some modes could only be correctly obtained when some of the parameters had unphysical values, such as giving the shell of electrons a positive charge! This difficulty was traced to the neglect of another type of electronic distortion. Schröder [64] pointed out that it was necessary to include distortions of the ions in the form of an isotropic radial breathing motion of the electronic distributions. This does not produce a charged dipole and is

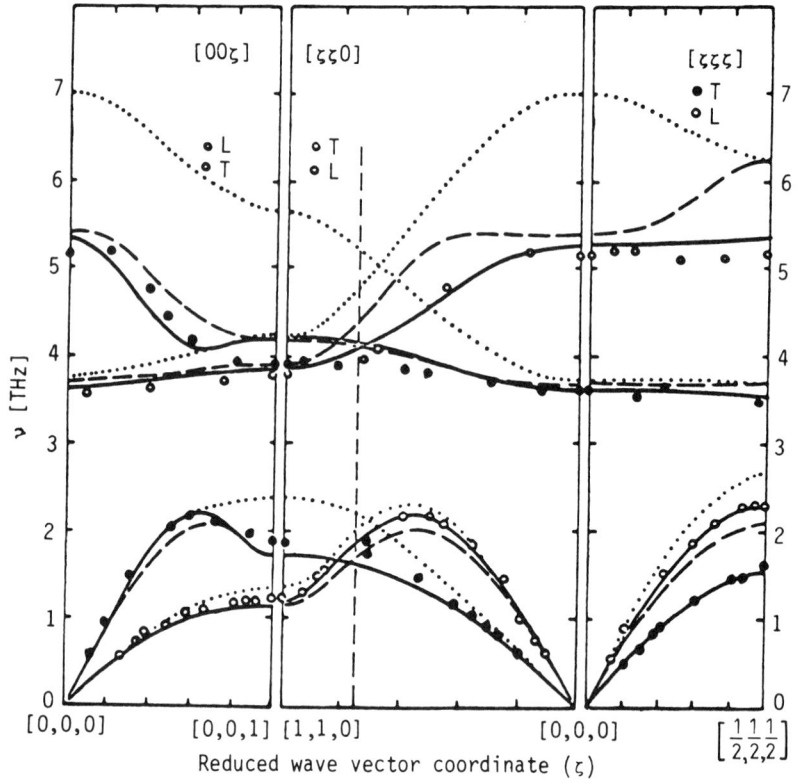

Figure 12.13. *The phonon dispersion curves of NaI [63]. The points are measured frequencies, the dotted lines are a rigid-ion model, the dashed lines a shell model and the solid lines a breathing-shell model.*

not produced by long-range dipole forces, but solely by the short-range forces.

Clearly one must ask whether it is necessary to introduce yet further distortions of the ions. Fischer *et al* [65] have suggested that quadrupolar distortions of the Ag ion are needed to describe the forces in AgCl. In most ionic materials, however, the simple mechanical shell model does provide a very satisfactory account of the interatomic forces with short-range overlap forces only between a few near neighbours. It may seem surprising that a complex electronic system can be so well described by such a model, but the electronic distortions of the spherical ions are presumably related to the nature of the excited electronic states, and for closed shells these have usually a character no more complicated than s or p waves [66].

12.3.2. Metals

Initially the existence of well-defined phonons was less certain in metals than in ionic crystals, because of the possibility that the free electrons might scatter the phonons; or, to express the problem differently, it was unclear that the energy of the crystal could be expressed in terms of a potential function depending only on the atomic coordinates. This is the adiabatic approximation introduced by Born and Oppenheimer [67], but they were unable to show that it was a good approximation for metals. This did not prevent solid-state physicists from taking the existence of phonons for granted, and eventually the situation was clarified by the experiments of Brockhouse and Stewart [52], who showed that they really did exist in aluminium, and by the theoretical work of Migdal [68].

There followed more detailed measurements of the phonon dispersion curves of metals, particularly sodium and lead. The dispersion curves of sodium [69] could be explained by a BvK model including forces (classical springs) between near neighbours. In detail, forces up to fifth neighbours were needed to give agreement with the experimental data. This model showed that the forces must arise through the conduction electrons, as a direct overlap between the atomic cores could not extend as far as the fifth-nearest neighbours; the model was also unsatisfactory in that it involved 13 parameters. At about the same time as these measurements were made, new light was thrown on the coupling between the conduction electrons and the atomic cores through the development of the pseudopotential method [70], which enabled the interatomic forces to be calculated in terms of the electron–phonon interaction. Movement of one atom changes the conduction electron distribution, and thus leads to a force on another atom. This theory of the origin of the interatomic forces goes back at least as far as Bardeen's important paper [71] in 1937, but was first applied in detail by Toya [72] in 1957 to calculate the dispersion curves in sodium, although he did not explicitly use pseudopotential theory. Successful application by Cochran, to sodium, of the formulation developed particularly by Sham [73] demonstrated the power of the pseudopotential approach.

See also p 1353

Sodium is a particularly simple metal, with a near-spherical Fermi surface, and the potential of the ion cores has little effect on the conduction electrons. Lead is a more complex metal with four conduction electrons per atom and with a stronger electron–phonon interaction, for which Brockhouse and his collaborators [74] found that interatomic forces of much longer range were required to explain the measured dispersion curves. This is a consequence of an effect pointed out by Kohn [75] in 1959. For wavevectors q of magnitude greater than the diameter of the Fermi surface, $2k_F$, the screening differs from that with q less than $2k_F$ in that only in the latter case is it possible for phonons to excite electron–hole pairs with both the electron and the hole at the Fermi surface. At the critical value, $q = 2k_F$, the dispersion curve is predicted to show a kink, such as is shown in figure 12.14. In free-electron metals like sodium the electron–phonon coupling is too weak to allow a Kohn

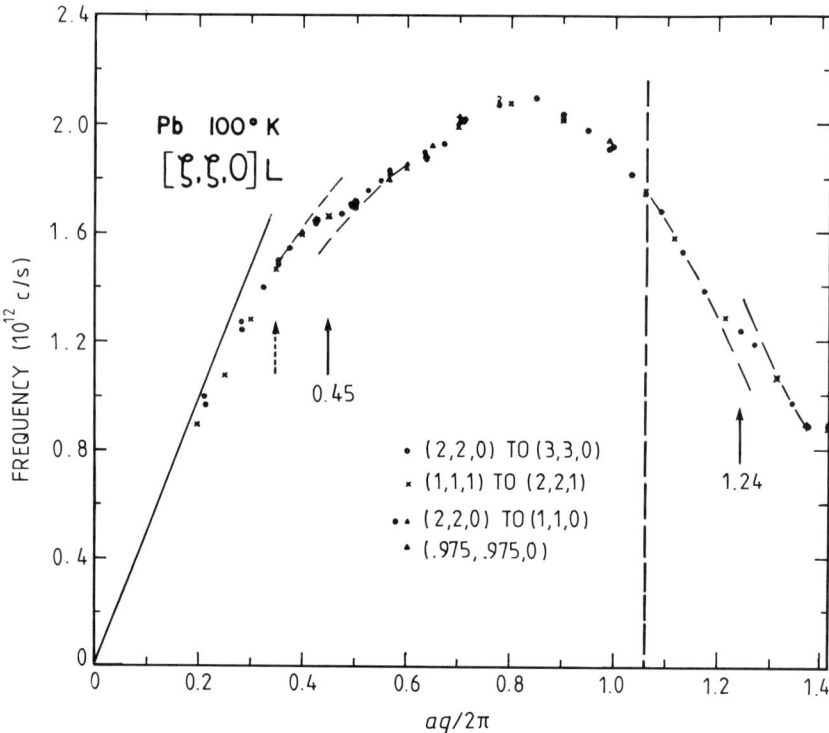

Figure 12.14. *The phonon dispersion curve of lead [74] along the (110) direction. Note the abrupt anomalies for $aq/2\pi = 0.45$ and 1.24.*

anomaly to be observed. As well as the electron–phonon coupling, the strength of the anomaly depends on the density of states at the Fermi surface, which becomes particularly large if the material is quasi-one-dimensional, owing its metallic behaviour to linear chains of conducting material. As first predicted by Peierls [76] in 1928, the lattice is then unstable against a distortion which opens up a gap in the electronic energies at the Fermi surface; in terms of the Kohn anomaly there is an infinite singularity in the dispersion curves at $q = 2k_F$. Although the precise behaviour of this strongly interacting system is still unclear, the phonon dispersion curve [77] of the one-dimensional metal KCP, whose chemical formula is given in figure 12.15, provides good evidence for a very strong Kohn anomaly, such as might be expected in a quasi-one-dimensional material, and which gives rise to a distorted crystal structure.

Since 1960, a large body of experimental information has been built up about the phonon dispersion curves in most simple metals and ordered alloy structures. Theoretically, however, progress has been less satisfactory in that pseudopotential theory has not been extended with

Vibrations and Spin Waves in Crystals

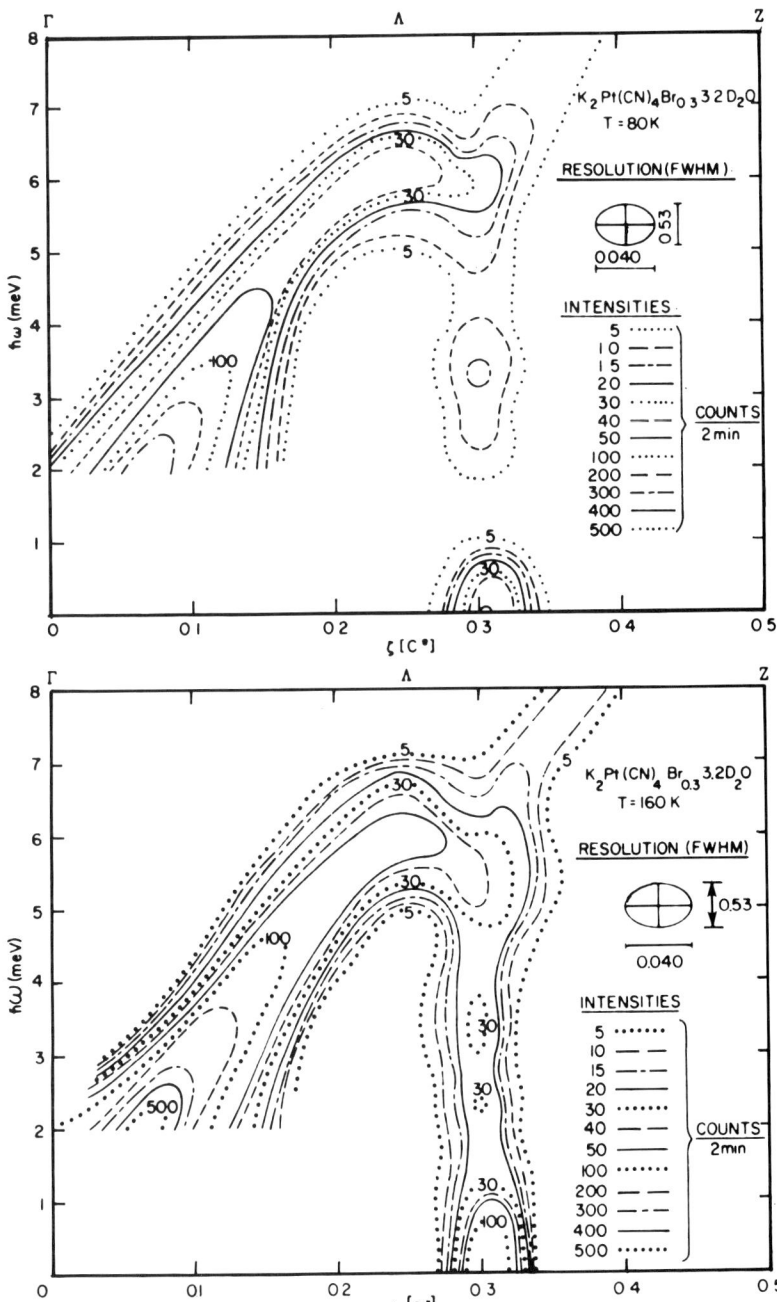

Figure 12.15. *The scattering observed [77] for wavevectors along the chain direction c, $q = \zeta c^*$ in KCP at 80 K and 160 K. The column of scattering for $\zeta = 0.3$ arises from the Kohn anomaly associated with the nearly one-dimensional Fermi surface and at 80 K there is a Peierls instability and elastic Bragg peak.*

any great success to the more complex metals. For these the problem of calculating the screening by the conduction electrons is far from being solved, and there are further calculational difficulties in including the real band structure and Fermi surface.

12.3.3. Semiconductors

The nature of the interatomic forces and phonon dispersion curves in semiconductors is an important question, because they are intermediate between ionic crystals and metals. The initial calculations of H M J Smith used nearest-neighbour forces and the BvK theory to calculate the dispersion curves but, as with ionic crystals and metals, better experiments showed the limitations of this approach. After Brockhouse and Iyengar [78] measured the dispersion curves of germanium in 1959, Herman [79] showed that if one was going to use BvK theory long-range forces, at least up to fifth-nearest neighbours, would be needed.

Mashkevich and Tolpȳgo [80] suggested that induced dipolar forces were important in covalent crystals, and Cochran [81] used the shell model to describe germanium. A nearest- and next-nearest-neighbour shell model gave a surprisingly satisfactory account of the experimental results, but this model is not as satisfactory for semiconductors as it is for ionic crystals. Firstly, the detailed fits to the dispersion relation are less good unless a large number of parameters are introduced, and secondly the covalent binding in semiconductors is directional and it is unclear how this can be modelled by spherical shells. As a result a large number of different models have been introduced, with charges placed at the centre of the bonds between the atoms to describe the bonding electrons. With sufficient adjustable parameters all these models provide a description of some of the experimental results, but none are completely convincing [82].

An alternative approach is to calculate the interatomic forces or phonon dispersion curves from a full microscopic model. Hohenberg, Kohn and Sham [83] used their local density functional formulation to calculate the frequencies of a few phonons by comparing the energy of distorted and undistorted crystals. This procedure has given reasonable agreement for a few phonons in simple semiconductors but is not easily generalized to give the full phonon dispersion curves.

12.3.4. Anharmonic effects

The interatomic potentials in crystals are not harmonic but also include non-linear (anharmonic) terms which, as well as being needed to explain the thermal conductivity and thermal expansion, are also responsible for relatively small corrections ($\sim 10\%$) to the specific heat, the frequencies of the normal modes and the elastic constants. Since the effects are relatively small, many can be treated by perturbation theory, as was used by Born and Blackman [84] to account for the temperature dependence

Vibrations and Spin Waves in Crystals

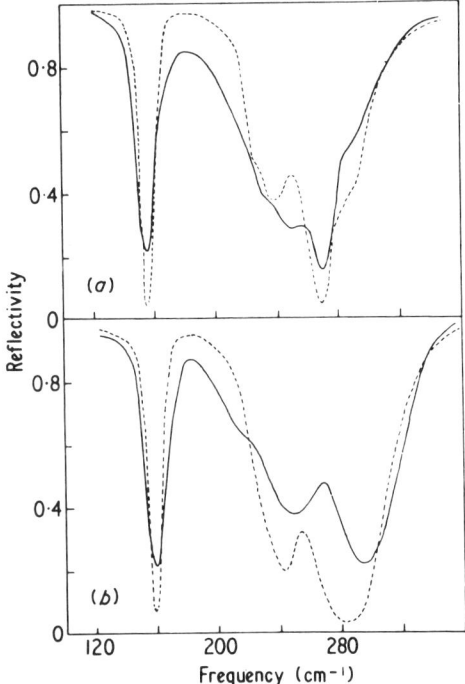

Figure 12.16. *The infrared reflectivity [86] of 2.5 μm NaCl films observed at (a) 45° and (b) 65°. Note the complex shape of the observed peaks (full line) and theory (dotted) due to the frequency dependence of the optic mode self-energy.*

of the linewidths of the normal modes observed in the far-infrared spectrum of ionic crystals. The development of many-body techniques in the late 1950s led to a more elegant recasting of the theory, and its extension to systems where anharmonic effects are large. This has proved particularly fruitful in the theory of the infrared spectrum. The model developed originally by Madelung for alkali halide crystals predicts an uncomplicated resonant absorption curve for each normal mode which can be excited by the infrared, but detailed measurements show considerable fine structure in the wings of the main resonance. The many-body theory describes this structure as, in some sense, an artefact of the method of observation. Since the lifetime [85] of the infrared mode is determined by its decay into pairs of other phonons, it is dependent on the density of two-phonon states, and this varies with the frequency which is used to study the normal mode. In an anharmonic crystal it is not then possible to specify the frequency and lifetime of a mode, without also specifying the frequency used to study the mode. This fine structure was first calculated in 1963 [85], and the results obtained by Hisano *et al* [86] in 1973 are shown in figure 12.16.

The method of long waves developed by Born and his collaborators

1003

was based on the realization that the long-wavelength acoustic modes must be equivalent to the waves obtained from microscopic elasticity theory. In an anharmonic crystal the situation is more complex. At very long wavelengths the period of the wave is much greater than the lifetime of most of the phonons in the material; there are many phonon collisions during each period of the wave and the material is at every point in a state of thermodynamic equilibrium as defined by the local temperature and strain. In contrast, at higher frequencies there is insufficient time for all the phonons to reach thermal equilibrium in each period of the wave, and the acoustic modes propagate in a collisionless mode. In the former case the waves are known as *first sound*, and can be described by thermodynamics in a similar way to the propagation of sound in a gas. In the collisionless regime the waves are known as *zero sound*, and though heavily overdamped in most liquids and gases, they are well defined in a crystal and can be studied by neutron scattering. The frequencies thus measured yield different elastic constants from those determined at low frequencies by, for example, the velocity of ultrasonic waves [87].

Anharmonic effects are particularly large in quantum crystals such as solid helium. The zero-point motion is large, and expands the interatomic distances well beyond the position of the minimum in the interatomic potential; so far, indeed, that the second derivative of the interatomic potential is negative, and according to conventional theory the crystal is unstable. The difficulty was overcome by Koehler and Nosanow [88] who used the self-consistent renormalized harmonic approximation, first suggested by Born [89] and by Horton, in which the second derivatives of the potential are not calculated at the interatomic distance, but are averaged over the distances sampled as the atoms move. The average is then positive, so the crystal is after all stable, and calculated normal mode frequencies are in reasonable accord with the measurements. After further improvement, by including the residual anharmonic effects, the results give very good agreement with experiment for the frequencies, lifetimes and the details of the measured neutron scattering [90].

12.3.5. Phonons in crystals containing defects and in glasses
The changes in the normal modes of vibration of systems caused by the introduction of defects was studied by Rayleigh, who proved a number of important theorems. The subject then lay almost dormant until the work of I M Lifshitz, which began in 1943, and of Montroll and his collaborators in 1955 [91]. Much of this early work was theoretical, and concerned with the mathematical properties of very idealized, usually one-dimensional, models. The basic changes in the phonon spectrum, when a few defects are introduced, are the development of either localized or resonant modes associated with the defect. Localized modes occur when the characteristic vibrational frequency of the defect

is different from that at which any normal mode can propagate through the host crystal; this tends to occur when the defect has a smaller mass than the host. The motions in these modes are then confined to the immediate neighbourhood of the defect. If, however, the frequency of the defect lies within a band of normal mode frequencies of the host crystal, as when a massive atom is introduced, the dispersion curves are perturbed around the frequency of the defect in a characteristic resonant way. The defects also break the translational symmetry of the crystal so both effects may be observed with optical techniques. Localized modes were first observed with infrared techniques by Schaefer [92] who studied colour centres in alkali halides, and resonant perturbations caused by Ag in NaCl and KCl. Subsequently there have been a large number of very detailed optical experiments, and detailed calculations using the shell models described above [93]. The theory has also been tested for metals by using neutron scattering techniques; localized modes were observed when defects (e.g. chromium) were introduced into a tungsten crystal, and resonant perturbation was seen with gold in copper [94].

A different type of defect, but one for which the theory is not too different from that of isolated defects, is the crystal surface. The first treatment of waves on the surface of an isotropic continuous medium was by Rayleigh in 1885, and his work and that of Love in 1911 form the basis of our understanding of seismic and other surface waves. Solid-state physics is, however, concerned with processes and effects at the atomic level, and the extension of surface wave theory to atomic lattices was initiated by Lifshitz and co-workers from 1948 onwards [91]. They used similar techniques to those they had developed for isolated defects, and the results are qualitatively similar. If the surface is perpendicular to z, and periodic in x and y, a surface mode is characterized by a wavevector of only two components, q_x and q_y. Perpendicular to the surface, the amplitude of a surface mode falls off rapidly with distance into the bulk of the crystal. The surface modes are therefore described by dispersion curves in q_x and q_y, and there are a series of branches, depending on whether the modes are polarized in the plane or perpendicular to it, and whether they have optical or acoustic character. In the same way as for isolated defects there are localized or resonant modes depending on whether the mode frequency is outside, or within, the band of frequencies of the host, so now the behaviour of the surface mode with z depends on whether the frequency is within the band of host frequencies with the same q_x and q_y.

The surface modes can influence many of the properties of crystals, particularly if the crystal size is very small. Montroll [95] in 1950 discussed the specific heat of small crystals, and predicted that the surface modes would give rise to a term proportional to the surface area and to T^2. Detailed measurements of the dispersion curves for surface waves had, however, to await the development of suitable experimental techniques. A probe is needed which can transfer the appropriate energy and wavevector, yet interacts only with the surface. A suitable probe is helium atom scattering and the experiment is very similar to that of neutron scattering, except that the source is a beam of helium atoms. The first successful experiments performed with this type of equipment were by Toennies et al [96] in 1983.

The developments just described led to an understanding of the properties of crystals containing a few defects, although there are still problems associated with the determination of the force constants coupling defects to the lattice and the atoms at surfaces. The formal theory of isolated defects is not easily applied when the number of defects becomes large in mixed crystals. One can start by treating the mixed crystals with the *coherent potential approximation* [97] in which the phonons are treated as being scattered by defects arising from the deviations away from the average lattice. The theory has had some success but fails to explain how a sufficient concentration of defects can lead the localized modes to cooperate in forming a band of propagating phonons. Experimentally, optical measurements on mixed ionic and semiconducting crystals have shown two types of behaviour. In 1928 Kruger et al [98] showed that the frequency of the transverse optic mode in a typical mixed alkali halide, KCl_xBr_{1-x} varied continuously from that of KCl to that of KBr. In contrast, in semiconductors such as InP_xAs_{1-x}, there are two frequencies of optical absorption, one close to that of InP and the other to that of InAs [99]. There is no general theory that encompasses both types of behaviour.

One of the unexpected features of very disordered systems came to light in 1971 when Zeller and Pohl [100] measured the specific heat and thermal conductivity of glasses. At low temperatures (<1 K) they found that the specific heats of vitreous silica, vitreous GeO_2 and selenium were proportional to temperature and much larger than for crystalline materials, while the low-temperature thermal conductivities were proportional to T^2. Other properties of glasses such as the velocity of sound, ultrasonic attenuation and dielectric response have also been found to differ in several respects from those observed in crystalline materials, and the anomalies were later found in almost all non-crystalline materials. An explanation for these results was put forward independently in 1972 by Anderson et al [101], and by Phillips [102], who suggested that in glasses there are two-level systems corresponding to parts of the glass structure flipping between two different configurations. The density of states of these levels was further assumed to be independent of energy to explain the specific heat measurements. The theory was further developed by Black and Halperin [103] in 1977, to account for the ultrasonic properties, but there is still no satisfactory understanding of the microscopic nature of these two-level systems.

Vibrations and Spin Waves in Crystals

On a large scale, disordered solids behave like a homogeneous three-dimensional material, in the sense that the number of particles is proportional to the cube of the linear dimension of the sample. More generally, $N \propto l^d$, where l is the size of the system and d is the dimensionality. However, on length scales less than some characteristic L, some disordered systems have $N \propto l^D$ where D is the fractal dimension, less than d. Examples of fractal systems are percolating clusters, polymers, rubbers and gels. The phonon density of states, $g(\omega)$, for modes with a wavelength greater than L has the form $g(\omega) \propto \omega^{d-1}$, but for modes with wavelengths less than L, $g(\omega) \propto \omega^{\bar{d}-1}$, where $\bar{d}$ is known as the fracton dimensionality; the modes in this frequency range were called *fractons* by Alexander and Orbach [104] in 1982, who also suggested that $\bar{d} = 4/3$. Boukenter et al [105] measured the Raman scattering from silica gel and found that over a limited but considerable frequency range, the scattering varied with a power of the frequency transfer very different from that observed in non-fractal glasses, such as As_2S_3. They deduced that $\bar{d} = 1.27 \pm 0.16$ consistent with Alexander and Orbach's conjecture.

See also p 574

12.4. Structural phase transitions

It is very common for the crystal structure of a solid to change as the temperature or pressure is altered, and the methods of x-ray or neutron diffraction enable us to determine what has changed in the arrangement of the atoms, as described in Chapter 6. In a first-order transition there is a latent heat, and the two phases are distinct and may coexist in the same way as water and ice can coexist. More commonly, however, the transition is continuous or nearly continuous, and there is no question of distinct coexisting phases. Indeed, in the neighbourhood of the transition there are small-scale fluctuations of one phase in the other. A full theory of the phase transition must take account of these fluctuations, though they were not included in the very important phenomenological theory put forward by Landau [106] in 1937 in a general form, and rediscovered and applied to the ferroelectric transition in $BaTiO_3$ by Devonshire [107] in 1949. The theory makes use of symmetry to write down the free energy of the crystal in terms of the displacement of the atoms from their positions in the high-temperature phase, and assumes the simplest possible temperature dependence. Close to the transition temperature the theory is equivalent to the molecular field theory of magnetism which also neglects fluctuations, and despite its shortcomings has been very successful in describing a wide range of different phenomena. The Bragg–Williams theory of the order–disorder transitions in alloys was a similar theory, and much thought was put into ways of including the short-range order or fluctuations, as described in Chapter 7.

See also p 549

The understanding of structural phase transitions developed differently, and the next step was the soft-mode theory. Although earlier

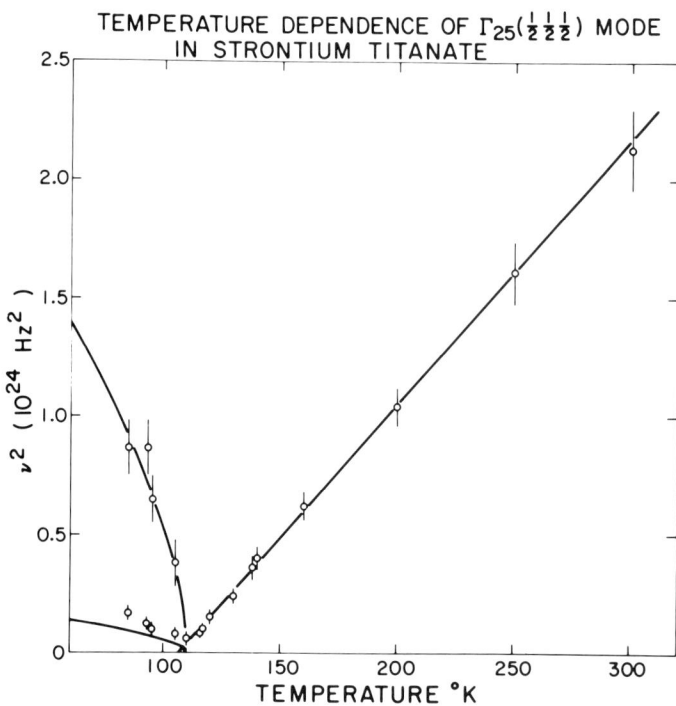

Figure 12.17. *The frequencies of the soft phonon modes with* $q = (2\pi/a)\,(\frac{1}{2}, \frac{1}{2}, \frac{1}{2})$ *in SrTiO$_3$* [109]. *Below the phase transition at 110 K, the distorted structure splits the degeneracy of the mode into two.*

proposed by Raman and by Anderson, it was the work of Cochran [108] in 1960 which led to direct tests of the suggestion that the phase transition resulted from a particular type of instability of the crystal. As the transition is approached the frequency of one of the normal vibrational modes decreases as $(T - T_c)^{1/2}$; at T_c the restoring force associated with this mode of distortion vanishes, and the crystal spontaneously distorts to a new structure. Infrared measurements and neutron scattering measurements showed evidence of a strongly temperature-dependent mode in SrTiO$_3$; although it does not in fact undergo a ferroelectric transition, the measurements lent considerable support to the soft-mode hypothesis. This was further strengthened by the later discovery of the antiferrodistortive transition in SrTiO$_3$ and detailed studies of the associated soft modes by Raman and neutron scattering techniques [109], as shown in figure 12.17.

These measurements showed that at least some structural phase transitions were associated with a soft mode, but there was still no explanation for why this particular normal mode was so strongly temperature dependent. It was shown that the theory of anharmonic crystals could qualitatively account for the temperature dependence assumed for the mode and for the Landau expansion of the free energy, provided that the self-consistent phonon theory was used to remove inconsistencies associated with unstable modes. From detailed calculations of the anharmonic effects it appeared that they could account in a consistent manner for a large number of different properties of SrTiO$_3$, but subsequently it has been questioned whether the detailed model used for the anharmonic interactions is correct. The difficulty lies in our lack of knowledge about the anharmonic part of the interatomic potentials [110].

While this work was progressing, soft modes were being found in many different crystals, but studies of the statistical mechanics of other phase transitions were progressing differently. The distortions occurring below the transition vary, according to Landau theory, as $(T_c - T)^{1/2}$, whereas exact solutions of the two-dimensional Ising model and other phase transitions give $(T_c - T)^\beta$, with an exponent β much less than 0.5.

The discrepancy springs from Landau's neglect of fluctuations. The first experimental evidence of their importance for structural phase transitions came from a detailed study of the order–disorder transition in β-brass by Als-Nielsen and Dietrich [111] who showed that the exponents were clearly different from those given by Landau theory. Thus fluctuations must play a crucial role at least for order–disorder transitions, in which slow diffusion is the mechanism by which phase transformation is achieved. The situation is very different in structural phase transitions like that in SrTiO$_3$, but here also the Landau theory fails in matters of detail, as was found by Mueller and Berlinger [112] who showed that the rotation of the oxygen octahedra below T_c was proportional to $(T_c - T)^{1/3}$. This made clear that critical fluctuations were important at structural phase transitions involving soft modes, as well as those involving the much slower diffusive motions in alloys.

The development of renormalization group techniques in the early 1970s clarified the role of critical fluctuations at phase transitions, and the concepts were applied to the diversity of structural phase transitions. In particular, the results showed that transitions involving elastic distortions as a primary order parameter were well described by Landau theory; the critical fluctuations did not cause changes in the behaviour. Uniaxial ferroelectrics were also described by Landau theory, albeit with logarithmic corrections to the temperature dependencies. We have reached the point where further development would involve the detailed description of particular phase transitions and technical arguments. Instead of following this road we refer the reader to some recent reviews [110].

The modern theory of continuous phase transitions depends on the concept of scaling, which implies that the fluctuations are controlled by a single length scale which diverges as the transition is approached.

Similarly the dynamics of the fluctuations is determined by a single time-scale which also diverges as the transition is approached. Structural phase transitions have provided some evidence that these concepts may need modification. The first evidence came from the experiments performed by Riste *et al* [113] in 1971 which showed that the dynamic response of $SrTiO_3$ above T_c consisted of two components; a quasi-elastic component and the oscillating soft mode. Similar results have now been found for many other transitions. These unexpected results do show that there are two time-scales involved in the fluctuations and there have been many attempts to explain these effects. Some of the explanations [110] concentrate on developing more complex treatments of the critical fluctuations, but these theories have not, as yet, reproduced the large difference between the time-scales. Halperin and Varma [114] suggested that the short time-scale arose from the phonons, while the long time-scale arose from interactions of the phonons with defects. The difficulty is that the appropriate defects have not been identified.

More recently, high-resolution x-ray measurements have identified two length scales [115] for the critical fluctuations. The experimental results are also inconsistent with existing theories of phase transitions, and furthermore the role, if any, of defects has not been clarified. These two results, showing that there are more than one time-scale and length scale for the critical fluctuations, pose severe problems for the understanding of structural phase transitions in terms of the conventional scaling theory of phase transitions. Further work is needed to understand these effects, and to decide whether the results show that defects nearly always play a crucial role at structural phase transitions, or whether the scaling theory of phase transitions is inadequate.

12.5. Spin waves

The spin waves play a similar role for magnetic materials to the one that the normal modes of vibration play for crystal structures: they describe the deviations of the atomic magnetic moments away from the ideal magnetic ordering. They were first introduced for ferromagnetic materials by Bloch [116] who calculated the quantized energy of the excitations as a function of the wavevector, q. He showed that if the magnetic interactions were of Heisenberg form

$$H = -J \sum S_i \cdot S_j$$

then the energies were given for small wavevectors by

$$\hbar\omega(q) = Dq^2$$

where the constant D depends on the exchange constant, J, and the crystal structure.

Since the spin waves describe spin deviations, the thermal excitation of spin waves decreases the ordered magnetic moment and Bloch showed that, for bulk ferromagnetic materials at low temperatures, the magnetization varied with temperature as

$$M(T) = M(0) - CT^{3/2}$$

where C is a constant which depends on J. This result has now been accurately tested experimentally.

A similar analysis to that of Debye for the specific heat shows that the magnetic contribution to the specific heat is proportional to $T^{3/2}$. The difference between the T^3 law of Debye and the $T^{3/2}$ arises from the linear dispersion curve of the normal modes and the quadratic dispersion curve of the spin waves. The difficulty of separating the magnetic $T^{3/2}$ term from the electronic T and normal mode T^3 term seems to have inhibited clear tests of this prediction.

Further theoretical work then concentrated on obtaining higher-order corrections to the magnetization by considering corrections to the low-k behaviour of the dispersion relation, and spin-wave interactions. Dyson [117] in 1956 concluded that linear spin-wave theory with non-interacting spin waves is good enough for all practical purposes.

At long wavelengths the long-range magnetic dipole interactions become important and modify the dispersion curves in a way that depends on the geometry of the sample. The results are essentially similar to those described above for the dielectric properties of crystals but, in practice, were worked out again without awareness of the dielectric results [118]. The theories gave good results in comparison with ferromagnetic resonance measurements.

The spin waves in antiferromagnetic structures were not studied theoretically until the work of Kubo *et al* [119] was done in the early 1950s. The difficulty is that the simple antiferromagnetic structure is not an eigenstate of the Hamiltonian and so spin waves might destroy the long-range order. They showed that, at least in three dimensions, this was not the case and that the antiferromagnetic spin waves had the dispersion curve $\omega = cq$, as for the normal modes of vibration. The spin waves therefore make a similar contribution to the specific heat.

As with the development of the understanding of the normal modes of vibration, further progress was inhibited by the lack of any experimental technique to study the spin waves in detail throughout the Brillouin zone. The solution to this was provided by the development of neutron scattering and Raman scattering techniques. The neutron has a magnetic moment and this interacts with the magnetic fields in a magnetic material giving rise to scattering of the neutron. For what appear to be completely fortuitous reasons, the strength of the scattering from a typical magnetic system is very similar to that of the nuclear scattering. This fortunately enables both types of scattering to be determined. Because the magnetic dipolar interaction is more complex than the Fermi pseudopotential describing the neutron–nucleus interaction, the form of the scattering cross-section is more complex, but the basic concepts of conservation of energy and crystal momentum are the same for spin waves as for normal modes of vibration.

Brockhouse [120] was the first to use neutron scattering to determine the spin-wave energies in the ferrimagnet magnetite and the ferromagnet cobalt [121]. He identified the scattering as arising from spin waves by applying a magnetic field in different directions and ensuring that the cross-section varied as predicted for the magnetic scattering. These measurements showed that short-wavelength spin waves do exist and that exchange constants could be determined experimentally. Shortly after Brockhouse's work, measurements were made on a number of different materials and one of the most complete was on the simple antiferromagnet, MnF_2, for which the dispersion curves are shown in figure 12.18 [122]. These results showed that spin waves did occur in antiferromagnets, and had the linear dispersion curve at small wavevectors as predicted by the theory, apart from the effects of anisotropy.

Since these first measurements, the spin waves have now been measured in a large number of different materials enabling the exchange constants to be determined in many materials. Measurements on simple insulators have shown the essential correctness of the theory of super-exchange, while the longer-range interactions found in rare-earth metals have confirmed the correctness of the indirect-exchange-coupling theory developed by Ruderman et al [123].

A different type of development has been the theory and experimental study of systems in which the electronic/magnetic energy levels are strongly influenced by the local crystalline environment, spin–orbit coupling and orbital magnetic moments. Much of our understanding of these systems comes from spin resonance experiments on the dilute salts. This work has now been extended to concentrated systems, the detailed study of the spin-wave/exciton dispersion relations and the interaction between the excitons on different atoms.

The nature of the spin waves in itinerant metals is also clarified. The theory developed by Stoner [124] suggested that in a magnetic metal the different spin bands had different energies resulting in a different filling of these bands and hence a spontaneous magnetization. Herring and Kittel [125] then showed that there were two types of excitation in these metals: collective spin waves with $\omega \propto Dq^2$, and single electron–hole excitations corresponding to transitions of the electrons from one band to the other. The details depend on the band structure but both types of excitation have been observed in magnetic metals such as the weak ferromagnetic MnSi.

Vibrations and Spin Waves in Crystals

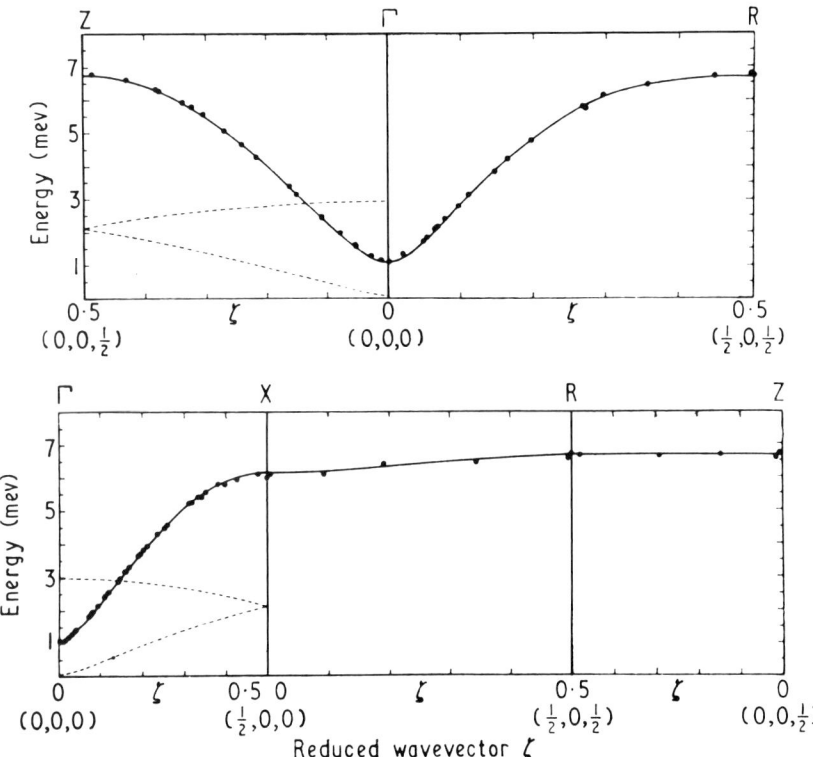

Figure 12.18. *The spin-wave dispersion curves in the antiferromagnet MnF$_2$ [122]. The solid line is a fit with dipolar forces and nearest- and next-nearest-neighbour exchange interactions.*

Raman scattering [126] has also proved to be a useful probe of spin waves, particularly in insulators. The one-magnon scattering gives information about the zone centre magnon frequency. The two-magnon scattering gives information about the two-spin-wave density of states. Since, however, the two spin waves are created on neighbouring sites, the effects of spin-wave—spin-wave interactions give rise to a marked change in the peak and shape of the spectrum.

The effect of the interactions between spin waves is particularly strong in materials that have magnetic chains, or other one-dimensional arrangements of the spins. These interactions are strong for small values of the spin for which the quantum fluctuations are large. In one-dimensional systems long-range order can be destroyed by the presence of a single spin-wave excitation. Since, for temperature $T > 0$, there is always an excitation somewhere on an infinite chain, no one-dimensional systems can have long-range magnetic order except when $T = 0$. In addition, Bethe [127] showed in 1931 that the quantum fluctuations destroy the long-range order of the $s = 1/2$ Heisenberg antiferromagnetic chain even at $T = 0$. The effect of these large fluctuations on the excitations has been studied now for several systems. For large spin, $s = 5/2$, the experiments [128] on tetramethyl manganese ammonium chloride (TMMC) showed fairly well-defined spin waves at low temperatures. The behaviour of the low-spin systems is, however, more complicated because of the quantum effects.

When a neutron, for example, creates a spin excitation at a particular site in a chain, the magnetic bonds must be broken on both sides of the site. These broken bonds are similar to domain walls and, in the case of Heisenberg systems, are known as spinons, and are necessarily created in pairs in a neutron scattering experiment. The excitations in the $s = 1/2$ Heisenberg antiferromagnet were studied by des Cloizeaux and Pearson [129] in 1962; they found the lowest-energy excitation for each wavevector, q. For many years this was interpreted as the energy of a spin-wave dispersion curve but, after computer simulations, further theoretical work and finally experiments [130], it is now appreciated that well-defined spin waves do not occur, despite their calculation in numerous tutorial and examination questions. The excitations are a continuum spectrum resulting from the excitation of pairs of spinons with wavevectors q_1 and $q - q_1$ which, for $s = 1/2$, can propagate largely independently.

The results for $s = 1$ Heisenberg chains are different. Haldane [131] showed that in this case the spinons are bound and so a finite energy is needed to break up this binding. The excitation spectrum then consists of a gap rather than the linear spin-wave excitation predicted by spin-wave theory for Heisenberg systems. This gap has now been observed [132]. Clearly the ability to produce materials containing magnetic chains has enabled the detailed study of quantum fluctuations in low-dimensional systems to be undertaken.

12.6. Magnetic phase transitions

The neutron scattering experiments described above have also enabled detailed measurements to be made of the fluctuations occurring close to magnetic phase transitions. This work has been recently reviewed [133], and so will not be described in full detail. The important feature is that the wavevector transfers and frequency transfers available with neutron scattering techniques enable the spatial extent and time dependence of the critical fluctuations to be studied. The simplest theories of phase transitions, the Landau theory described in section 12.4 or the well-known molecular field theory of magnetism, fail because they neglect the effect of the fluctuations. Most of the recent work on phase transitions has been in developing ways of including the effects of the fluctuations and, since neutron scattering enables measurements to be made of the fluctuations, they have played a crucial role in the development of our understanding of phase transitions.

Vibrations and Spin Waves in Crystals

Table 12.1. *Exponents of $d = 3$ Ising systems.*

System	β	ν	γ	Reference
β-brass	0.305(5)	0.65(2)	1.25(2)	[111]
MnF_2		0.634(40)	1.27(4)	[136]
CoF_2	0.305(30)	0.61(2)	1.21(6)	[137]
Theory	0.312	0.64	1.25	
Mean field	0.5	0.5	1.0	

Table 12.2. *Exponents of $d = 3$ Heisenberg systems.*

System	β	ν	γ	Reference
$RbMnF_3$	0.32(2)	0.701(11)	1.366(24)	[138]
Theory	0.38	0.702	1.375	
Mean field	0.5	0.5	1.0	

Continuous phase transitions, such as the transition from paramagnet to ferromagnet, are characterized by a change of symmetry and the development of an order parameter (ferromagnetic moment) in the low-symmetry phase. If the order parameter is the magnetization, M, then theory predicts that

$$M = M_0(T_c - T)^\beta$$

where M_0 is a constant and β is known as a critical exponent. The critical fluctuations are described by a length scale, ξ, and ξ is described by another exponent ν such that

$$\xi = \xi_0(T - T_c)^{-\nu}$$

and the susceptibility by χ where

$$\chi = \chi_0(T - T_c)^{-\gamma}.$$

Mean field theory or Landau theory give the same values for the exponents for every type of transition: $\beta = 1/2$, $\nu = 1/2$ and $\gamma = 1$. Exact solutions of some simple models, series approximations and experiments showed by the early 1970s that these values were often incorrect, and that better theories were needed. The improved theories made use of the scaling hypothesis and the renormalization group as described in Chapter 7. The scaling hypothesis states that the critical fluctuations are similar at all temperatures close to T_c provided that all lengths are scaled by the correlation length, ξ. Renormalization group theory provided an understanding of scaling theory and why phase transitions show universal behaviour in the sense that the exponents and critical properties are the same for systems having the same symmetry, dimensionality and range of forces.

See also p 562

Scattering experiments directly measure the temperature dependence of the order parameter, while the critical scattering, particularly above T_c, gives the susceptibility from the intensity and the correlation range from the wavevector spread of the scattering. The experiments then yield information about the exponents as well as giving detailed information about the form of the critical scattering.

By now many measurements have been performed and so only a brief review can be given of the results. The first measurement of the critical fluctuations was by Latham and Cassels [134] in 1952, who found that the total cross-section of iron increased at T_c and correctly interpreted this as arising from scattering by the magnetic fluctuations, as is well known from the critical opalescence of light at the critical point of materials. A major step forward [135] was the development of the magnetic correlation theory by van Hove with the use of mean field theory. In our view the first critical experiment that clearly showed the failure of mean field theory and demonstrated the power of neutron scattering techniques was the study of the order–disorder transition in β-brass [111] by Als-Nielsen and Dietrich in 1967. They measured the three critical exponents β, γ and ν and showed that they all differed from mean field theory and were in agreement with the exponents calculated from series expansions.

Soon after this result was obtained, similar experiments were performed on three-dimensional antiferromagnets and particularly careful measurements were made on MnF_2 [136] and CoF_2 [137]. These are anisotropic antiferromagnets and so belong to the $d = 3$ Ising universality class like β-brass: in table 12.1 we compare the measured exponents. The satisfactory agreement between the experiments gave good experimental support for universality. Another very careful experiment was performed [138] on a cubic antiferromagnet, $RbMnF_3$, which is an example from the isotropic Heisenberg $d = 3$ universality class. The exponents for this material are listed in table 12.2. There is reasonable agreement between theory and experiment and both give results that are different from those for the Ising system shown in table 12.1.

One of the most important developments was the growth of materials in which the magnetic ions are in sheets or chains, because they enabled two- and one-dimensional systems to be studied. Some aspects of the one-dimensional systems have been described in section 12.5 and so, in this section, we discuss only two-dimensional systems. Very important experiments were performed in 1973 by Samuelson [139] and by Ikeda and Hirakawa [140] on K_2CoF_4. In this material the square $Co-F_2$ planes are separated from one another by two K–F planes. The magnetic interactions are then strong in the planes but much weaker, by a factor of about 10^5, between the planes. Because of the crystal field effects on the Co ions, the effective interactions between the Co ions are anisotropic and

Vibrations and Spin Waves in Crystals

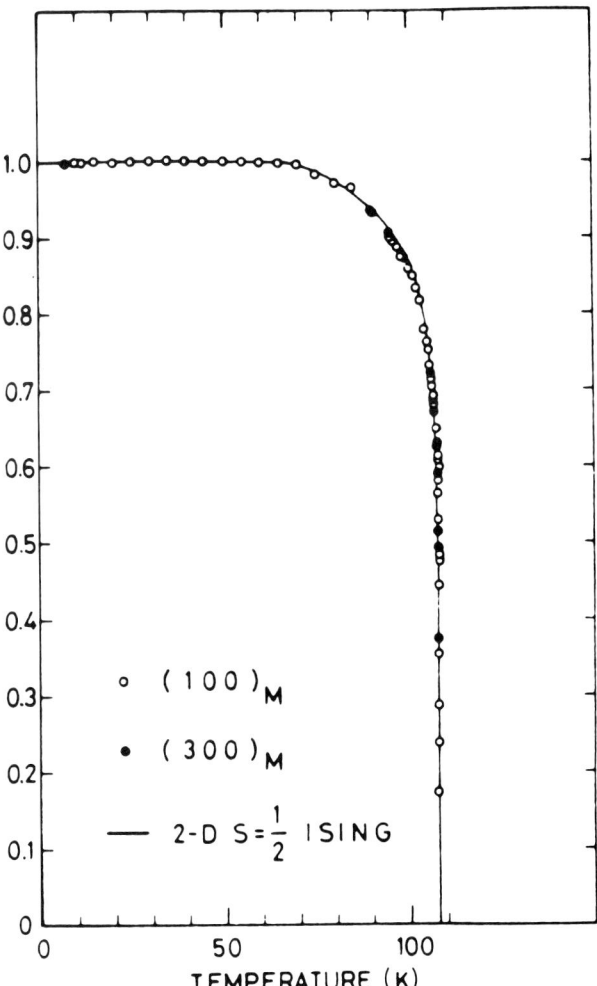

Figure 12.19. *The intensity (square of sublattice magnetization) in K_2CoF_4 which is a model two-dimensional system. The solid line shows Onsager's exact solution* [140].

See also p 554

Table 12.3. *Exponents of $d = 2$ Ising systems.*

System	β	ν	γ	Reference
K_2CoF_4	0.128(4)	0.99(3)	1.77(6)	[140]
Theory	0.125	1.0	1.75	[141]

the spins align antiferromagnetically, perpendicular to the c axis. They are therefore close to model systems for studying the two-dimensional Ising model.

This model is nearly unique in statistical physics because it was solved exactly, by Onsager [141] in 1944. In figure 12.19 we show a comparison of the measured order parameter with Onsager's solution and the agreement is excellent. In table 12.3, we list the measured values for the exponents as well as Onsager's results; the agreement is clearly very good and the values are very different from those of mean field theory.

Since these experiments were performed, the subject has widened to study phase transitions in many different areas. The reviews [133] give details of the way in which the experiments have shown that defects can change the universality class, and that percolation is dominated by the one-dimensional weak links in the backbone of the percolating cluster. Phase diagrams have been measured, and the way disorder and competing interactions in spin glasses and random field problems lead to non-ergodic behaviour and other unexpected features has been studied.

References

[1] Love A E H 1927 *The Mathematical Theory of Elasticity* 4th edn (Cambridge: Cambridge University Press)
[2] Behn U 1898 *Ann. Phys., Lpz.* **66** 237
[3] Eucken A 1909 *Phys. Z.* **10** 586
[4] Ruhemann M and Ruhemann B 1937 *Low Temperature Physics, part 2* (Cambridge: Cambridge University Press) ch 2
[5] Kuhn T S 1978 *Black-body Theory and the Quantum Discontinuity 1894–1912* (Oxford : Clarendon)
[6] Einstein A 1907 *Ann. Phys., Lpz.* **22** 180
[7] Boltzmann L 1964 *Lectures on Gas Theory* (Engl. Transl. S G Brush) (Berkeley, CA: University of California) p 330
[8] Einstein A 1907 *Ann. Phys., Lpz.* **22** 800
[9] Nernst W and Lindemann F 1911 *Z. Elektrochim.* **17** 817
[10] Nernst W 1911 *Ann. Phys., Lpz.* **36** 395
[11] Einstein A 1911 *Ann. Phys., Lpz.* **35** 679
[12] Debye P 1912 *Ann. Phys., Lpz.* **39** 789
[13] Born M and von Kármán T 1912 *Phys. Z.* **13** 297
[14] Lord Rayleigh 1900 *Phil. Mag.* **49** 539
[15] Weyl H 1911 *Math. Ann.* **71** 441
[16] Eucken A and Schwers F 1913 *Verh. Deutsch. Phys. Ges.* **15** 578
[17] Bennewitz K and Simon F 1923 *Z. Phys.* **16** 183
[18] James R W and Brindley G W 1928 *Proc. R. Soc.* A **121** 155
[19] de Boer J 1948 *Physica* **14** 139
[20] de Boer J and Lunbeck R J 1948 *Physica* **14** 510
[21] Grüneisen E 1908 *Ann. Phys., Lpz.* **26** 211
[22] Grüneisen E 1911 *Ber. Deutsch. Phys. Ges.* **13** 426, 491
[23] Debye P 1913 *Phys. Z.* **14** 259
[24] Berman R 1953 *Adv. Phys.* **2** 103
[25] Glazebrook R (ed) 1922 *A Dictionary of Applied Physics* 1 (London: Macmillan) p 428
[26] Eucken A 1911 *Ann. Phys., Lpz.* **34** 185

[27] Debye P 1914 *Vorträge über die Kinetische Theorie der Materie und die Elektrizität* (Leipzig: Teubner)
[28] Pauli W 1925 *Verh. Deutsch. Phys. Ges.* **6** 10
[29] Peierls R E 1980 *Proc. R. Soc.* A **371** 28
[30] Ziman J M 1960 *Electrons and Phonons* (Oxford: Clarendon)
[31] von Gutfeld R J and Nethercot A H 1964 *Phys. Rev. Lett.* **12** 641
[32] Northrop G A and Wolfe J P 1979 *Phys. Rev. Lett.* **43** 1424
[33] Paschen F 1895 *Ann. Phys., Lpz.* **54** 668
Rubens H and Nichols E F 1897 *Ann. Phys., Lpz.* **60** 418
[34] Drude P 1904 *Ann. Phys., Lpz.* **14** 677
[35] Madelung E 1909 *Ges. Wiss. Göttingen Nach. Math.-Phys. Klasse* **1** 100
[36] Blackman M 1935 *Proc. R. Soc.* A **148** 365
[37] Born M 1914 *Ann. Phys., Lpz.* **44** 605
[38] Cowley R A, Woods A D B and Dolling G 1966 *Phys. Rev.* **150** 487
[39] Montroll E W 1947 *J. Chem. Phys.* **15** 575
[40] van Hove L 1953 *Phys. Rev.* **89** 1189
[41] Kellermann E W 1940 *Phil. Trans. R. Soc.* **238** 513; 1941 *Proc. R. Soc.* A **178** 17
[42] Iona M 1941 *Phys. Rev.* **60** 822
[43] Huang K 1951 *Proc. R. Soc.* A **208** 352
[44] Henry C H and Hopfield J J 1965 *Phys. Rev. Lett.* **15** 964
[45] Born M and Huang K 1954 *Dynamical Theory of Crystal Lattices* (Oxford: Oxford University Press)
[46] Lonsdale K 1943 *Rep. Prog. Phys.* **9** 256
[47] Laval J 1954 *J. Phys. Radium* **15** 545
[48] Curien H 1952 *Acta Crystallogr.* **5** 392
Jacobsen E H 1955 *Phys. Rev.* **97** 654
[49] Walker C B 1956 *Phys. Rev.* **103** 547
[50] Brockhouse B N 1961 *Inelastic Scattering of Neutrons in Solids and Liquids* (Vienna: International Atomic Energy Agency) p 113
[51] Placzek G and van Hove L 1954 *Phys. Rev.* **93** 1207
[52] Brockhouse B N and Stewart A T 1955 *Phys. Rev.* **100** 756
[53] Cohen M and Feynman R P 1957 *Phys. Rev.* **107** 13
[54] Landsberg G and Mandelstam L 1928 *Naturwissenshaft* **16** 557
[55] Raman C V and Krishnan K S 1928 *Nature* **121** 501
[56] Fermi E and Rasetti F 1931 *Z. Phys.* **71** 689
[57] Mandelstam L, Landsberg G and Leontowitsch M 1930 *Z. Phys.* **60** 334
[58] Placzek G *Handbuch der Radiologie VI* ed E Marx (Leipzig: Akademische Verlagsgellschaft) p 205
[59] Born M and Bradburn M 1947 *Proc. R. Soc.* A **188** 161
[60] Smith H 1948 *Phil. Trans. R. Soc.* A **241** 105
[61] Loudon R 1964 *Adv. Phys.* **13** 423
[62] Dick B J and Overhauser A W 1958 *Phys. Rev.* **112** 90
[63] Woods A D B, Cochran W and Brockhouse B N 1960 *Phys. Rev.* **119** 980
[64] Schröder U 1966 *Solid State Commun.* **4** 347
[65] Fischer K, Bilz H, Haberhorn R and Weber W 1972 *Phys. Status Solidi* b **54** 285
[66] Further references and details of many materials are given in Cochran W 1973 *The Dynamics of Atoms in Crystals* (London: Arnold) and Bilz H and Kress W 1979 *Phonon Dispersion Relations in Insulators (Springer Series in Solid State Sciences 10)* (Berlin: Springer).
[67] Born M and Oppenheimer R 1927 *Ann. Phys., Lpz.* **84** 457
[68] Migdal A B 1957 *Sov. Phys.–JETP* **5** 333

[69] Woods A D B, Brockhouse B N, March R H, Stewart A T and Bowers R 1962 *Phys. Rev.* **128** 1112
[70] Harrison W 1965 *Pseudopotential in the Theory of Metals* (New York: Benjamin)
[71] Bardeen J 1937 *Phys. Rev.* **52** 688
[72] Toya T 1958 *J. Res. Inst. Catalysis, Hokhaido Univ.* **6** 183; 1959 *J. Res. Inst. Catalysis, Hokhaido Univ.* **7** 60
[73] Sham L 1965 *Proc. R. Soc.* A **283** 33
[74] Brockhouse B N, Arase T, Caglioti G, Rao K R and Woods A D B 1962 *Phys. Rev.* **128** 1099
[75] Kohn W 1959 *Phys. Rev. Lett.* **2** 393
[76] See Peierls R E 1964 *Quantum Theory of Solids* (Oxford: Oxford University Press) p 108
[77] Carneiro K, Shirane G, Werner S A and Kaiser S 1976 *Phys. Rev.* B **13** 4258
[78] Brockhouse B N and Iyengar P K 1958 *Phys. Rev.* **111** 747
[79] Herman F 1959 *J. Phys. Chem. Solids* **8** 405
[80] Mashkevich V S and Tolpȳgo K B 1957 *Sov. Phys.–JETP* **32** 520
[81] Cochran W 1959 *Proc. R. Soc.* A **253** 260
[82] For a review see Bilz H, Strauch D and Wehner R K 1984 *Handbuch der Physik XXV/2d* (Berlin: Springer)
[83] Hohenberg P and Kohn W 1964 *Phys. Rev.* B **136** 864
Sham L J and Kohn W 1966 *Phys. Rev.* **145** 561
[84] Born M and Blackman M 1933 *Z. Phys.* **82** 551
[85] Cowley R A 1963 *Adv. Phys.* **12** 421
[86] Hisano K, Placido F, Bruce A D and Holah G D 1972 *J. Phys. C: Solid State Phys.* **5** 2511
[87] Cowley R A 1967 *Proc. Phys. Soc.* **90** 1127
[88] Koehler T R 1966 *Phys. Rev. Lett.* **17** 89
Nosanow L H 1966 *Phys. Rev.* **146** 120
[89] Born M 1951 *Festschrift der Akademie der Wissenschaften Göttingen*
Choquard P 1967 *The Anharmonic Crystal* (New York: Benjamin)
[90] Horner H 1972 *Phys. Rev. Lett.* **29** 556
Minkiewicz V J, Kitchens T A, Shirane G and Osgood E B 1973 *Phys. Rev.* A **8** 1513
[91] A review is given by Maradudin A A, Montroll E W and Weiss G H 1963 *Theory of Lattice Dynamics in the Harmonic Approximation* (New York: Academic)
[92] Schaefer G 1960 *J. Phys. Chem. Solids* **12** 233
[93] Moller H B and Mackintosh A R 1965 *Phys. Rev. Lett.* **15** 623
[94] Svensson E C and Brockhouse B N 1967 *Phys. Rev. Lett.* **18** 858
[95] Montroll E W 1950 *J. Chem. Phys.* **78** 183
[96] Brusdeylias G, Doak R B and Toennies J P 1983 *Phys. Rev.* B **27** 3662; 1983 *Phys. Rev.* B **28** 2104
[97] Elliott R J and Taylor D W 1967 *Proc. R. Soc.* A **296** 201
[98] Kruger R, Reinkober O and Koch-Holm E 1928 *Ann. Phys., Lpz.* **85** 110
[99] Oswald F 1959 *Z. Naturf.* a **14** 374
[100] Zeller R C and Pohl R O 1971 *Phys. Rev.* B **4** 2029
[101] Anderson P W, Halperin B I and Varma C M 1972 *Phil. Mag.* **25** 1
[102] Phillips W A 1972 *J. Low Temp. Phys.* **7** 351
[103] Black J C and Halperin B I 1977 *Phys. Rev.* B **16** 2879
[104] Alexander S and Orbach R 1982 *J. Physique* **43** L625
[105] Boukenter A, Champagnon B, Duval E, Durnas J, Quinson J F and Serughetti J 1986 *Phys. Rev. Lett.* **57** 2391
[106] Landau L D 1937 *Phys. Z. Sov.* **11** 26

[107] Devonshire A F 1949 *Phil. Mag.* **40** 1040
[108] Cochran W 1960 *Adv. Phys.* **9** 389
[109] Fleury P A, Scott J F and Worlock J M 1968 *Phys. Rev. Lett.* **21** 16
Cowley R A, Buyers W J L and Dolling G 1969 *Solid State Commun.* **7** 181
Shirane G and Yamada Y 1969 *Phys. Rev.* **177** 858
[110] A review is given by Bruce A D and Cowley R A 1980 *Structural Phase Transitions* (London: Taylor and Francis).
[111] Als-Nielsen J and Dietrich O W 1967 *Phys. Rev.* **153** 706, 711, 717
[112] Mueller K A and Berlinger W 1971 *Phys. Rev. Lett.* **25** 734
[113] Riste T, Samuelson E J and Otnes K 1971 *Structural Phase Transitions and Soft Modes* (Oslo: Oslo Universitetsfurlaget) pp 395–408
[114] Halperin B I and Varma C M 1976 *Phys. Rev.* B **14** 4030
[115] Andrews S R 1986 *J. Phys. C: Solid State Phys.* **19** 3721
McMorrow D F, Hamaya N, Shimonura S, Fujii Y, Kishimoto S and Iwasaki H 1990 *Solid State Commun.* **76** 443
[116] Bloch F 1932 *Z. Phys.* **74** 295
[117] Dyson F J 1956 *Phys. Rev.* **102** 1217, 1230
[118] For a review see van Kranendonk J and Van Vleck J H 1958 *Rev. Mod. Phys.* **30** 1
[119] Kittel C 1951 *Phys. Rev.* **82** 565
Anderson P W 1952 *Phys. Rev.* **86** 694
Kubo R 1952 *Phys. Rev.* **87** 568
[120] Brockhouse B N 1957 *Phys. Rev.* **106** 859
[121] Sinclair R N and Brockhouse B N 1960 *Phys. Rev.* **120** 1638
[122] Nikotin O, Lindgard P A and Dietrich O W 1969 *J. Phys. C: Solid State Phys.* **2** 1168
[123] Reviews are given in Stirling W G and McEwan K A 1987 *Methods of Experimental Physics 23 C* (New York: Academic) and Jensen J and Mackintosh A R 1991 *Rare Earth Magnetism* (Oxford: Oxford University Press).
[124] Stoner E C 1938 *Proc. R. Soc.* A **165** 372
[125] Herring C and Kittel C 1951 *Phys. Rev.* **81** 869
Mattis D C 1965 *The Theory of Magnetism* (New York: Harper and Row)
[126] Cottam M G and Lockwood D J 1986 *Light Scattering in Magnetic Solids* (New York: Wiley)
[127] Bethe H A 1931 *Z. Phys.* **71** 205
[128] Hutchings M T, Shirane G, Birgeneau R J and Holt S L 1972 *Phys. Rev.* B **5** 1999
[129] des Cloizeaux J and Pearson J J 1962 *Phys. Rev.* **128** 2131
[130] Tennant D A, Perring T G, Cowley R A and Nagler S E 1993 *Phys. Rev. Lett.* **70** 4003
[131] Haldane F D M 1983 *Phys. Rev. Lett.* **50** 1153
[132] Buyers W J L, Morra R M, Armstrong R L, Hogan M J, Gerlach P and Hirakawa K 1986 *Phys. Rev. Lett.* **56** 371
[133] Collins M F 1989 *Magnetic Critical Scattering* (Oxford: Oxford University Press)
Cowley R A 1987 *Neutron Scattering Part C* ed K Sköld and D L Price (New York: Academic)
[134] Latham R and Cassels J M 1952 *Proc. Phys. Soc.* A **65** 241
[135] van Hove L 1954 *Phys. Rev.* **95** 249, 1374
[136] Schulhof M P, Nathans R, Heller P and Linz N 1970 *Phys. Rev.* B **1** 2034
[137] Cowley R A and Carneiro K 1980 *J. Phys. C: Solid State Phys.* **13** 3281
[138] Tucciarone A, Lau H Y, Corliss L M, Depalme A and Hastings J M 1971 *Phys. Rev.* B **4** 3206

[139] Samuelson E J 1973 *Phys. Rev. Lett.* **31** 936
[140] Ikeda H and Hirakawa K 1974 *Solid State Commun.* **14** 529
[141] Onsager L 1944 *Phys. Rev.* **65** 117

Chapter 13

ATOMIC AND MOLECULAR PHYSICS

Ugo Fano

13.1. Introduction

As recounted in Chapter 3, studies of atomic and molecular phenomena played a central role in the development of quantum mechanics, as well as in its verification, throughout the early 1930s. Mainstream physics then directed its focus to nuclear and other pursuits; yet early instances of novel atomic–molecular phenomena kept emerging. These will be reported in the next section on 'atomic and molecular physics at mid-century', which should serve as a primer for this chapter's subject and is accordingly extensive.

This chapter deals mainly with subsequent developments, powered by the steady injection of new energies (personnel and funding) and stimulated in part by the role of atomic phenomena in astrophysical and laboratory plasmas. Exploration of new areas and collection of data played comparable roles in this process, but the mechanisms of phenomena will be emphasized here, since their understanding proves essential when extrapolating knowledge of atomic processes to new substances and to new environments. It is, of course, trivial that the behaviour of atoms and molecules underlies not only chemistry but the properties of all matter, gaseous or condensed.

Sections 13.3–13.6 will accordingly deal with rather general subjects, namely: the gross response of atomic systems to the whole spectrum of electromagnetic radiation, the more specific excitation channels and resonances identified by this response and by the effects of electron impact, the main features of collisions between atoms, molecules and their ions, and key developments on aggregates (molecules and clusters). Section 13.7 will deal with inner-shell phenomena.

Sir Harrie Massey

(Australian, 1908–83)

Born near Melbourne, Harrie Massey was a student of Rutherford, Lecturer at The Queen's University of Belfast (1933–38) and Professor at University College London for the rest of his life. He was a major figure of atomic physics, providing leadership throughout the broad context of its relevance to astrophysical and upper atmospherical plasmas. His early co-authorship in 1933 of the fundamental *Theory of Atomic Collisions* (Oxford: Clarendon) was followed by the development of increasingly ample editions over several decades. His research, his personal and organizational leadership and his recruitment and training of new leaders, among them the late Sir David Bates, Michael J Seaton, Phillip G Burke and Alex Dalgarno, established the eminence and style of a British School, in the mid-century period, that spread a lasting influence throughout the world. Under his inspiration, the wealth and diversity of atomic and molecular phenomena in the cosmos provided guidance, models and targets to the development of atomic theory and of its chemical applications. (Photograph by D H Rooks.)

The last three sections will report on activities that are more focused but prove nevertheless broadly relevant: spectroscopy's role in collecting and analysing masses of data on energy levels, which serve primarily as diagnostic tools; the invention and development of novel methods and instruments of measurement, including very accurate control of atomic samples, which underlie not only advances to be reported here but also the whole area of modern metrology involving previously undreamed of precision; the optical pumping and precision handling of atoms. A comprehensive view will be finally outlined.

A few remarks may complement this introduction. The development of our subject has been understandably influenced by historical circumstances. On the experimental side, the progress of techniques and instrumentation naturally had a controlling influence, to be outlined in each relevant context, while on the theoretical side quirks and fads have had curious roles, as sketched in the following paragraphs.

The flush of quantum mechanics' success led influential figures to view the remaining tasks of atomic theory as confined to applications of analytical or numerical procedures for solving Schrödinger's equation, much as Newtonian mechanics had seemed to reduce astrophysics to constructing celestial trajectories. This view found particularly fertile ground in Britain where theory is cast as applied mathematics. Indeed Massey's school pioneered the study of collision physics for decades, combining quantum and classical procedures.

The theory of discrete stationary states centred instead on utilizing their symmetries through group theory concepts leading to algebraic procedures, complemented by numerical integrations over radial coordinates. Atomic physics seemed thus to split into two sub-fields, dealing with discrete and continuous spectra, respectively. That the transition between these spectra at ionization and dissociation thresholds is actually smooth, making their separation irrelevant, was soon realized but sank into the general consciousness imperfectly and only after a delay.

A rather poorly appreciated aspect of quantum theory was that its seemingly novel procedures, in terms of operators, eigenvalues, eigenfunctions, etc, are in fact straightforward applications of the unified view of mathematical physics developed in the 1900s by Hilbert's school; only the code name 'Hilbert's space' became popular. In fact, Hilbert had cast his procedures as extensions of Fourier's familiar analysis. Unfamiliarity with this aspect still underlies many theoreticians' reluctance to deal with eigenfunctions with continuous spectra, whose systematics had been developed in the 1911 thesis by Hermann Weyl, a student of Hilbert. This reluctance has contributed to the artificial separation of 'spectra' and 'collisions'.

A diverging trend has emerged from recent studies of atomic states with large quantum numbers, which appear *prima facie* amenable to semiclassical treatments. This interest has been fanned by the lingering (resurging?) preference of many theorists for the study of trajectories, and more generally for exploring the boundary between quantum and classical phenomena. Thereby it is possible to overlook the fact that quantum phenomena, in contrast to trajectories, inherently spreads over all three dimensions of physical space. Studies of chaotic aspects of semiclassical mechanics thus forfeit perceiving manifestations of quantum effects localized in limited portions of the accessible space.

13.2. Atomic–molecular physics at mid-century

The physics of atoms and molecules developed early in this century through many novel experiments, whose interpretation proved altogether puzzling as described in Chapter 2. Rutherford's discovery of the concentration of atomic masses and positive charges in nuclei 10 000 times smaller than atoms raised the most profound questions: which circumstances set the size of atoms? and why don't the negatively charged electrons drop on nuclei, generating currents that radiate energy away?

Analysis of such questions and of their underlying phenomena, by a new brand of physicists guided largely by N Bohr, led eventually to a scientific milestone, namely, the development of quantum mechanics in the mid-1920s. This development lies outside our scope; we shall instead introduce here the new framework of knowledge and concepts that set atomic–molecular physics on its way through the rest of the century.

13.2.1. Atomic scale and structure

Quantum mechanics has interpreted the size of atoms by discovering that confinement of any particle within a volume of radius a boosts its kinetic energy by $\sim \hbar^2/2ma^2$, where $\hbar$ indicates Planck's constant and m the particle's mass. This boost opposes the electrons' attraction towards atomic nuclei, which is represented by the potential energy $-Ze^2/a$ for an electron at distance a from a nuclear charge Ze. For a single electron about this charge, boost and attraction balance out at the Bohr radius $a = \hbar^2/mZe^2 = 0.053/Z$ nm, where the potential energy's magnitude amounts to twice the electron's kinetic energy (Virial theorem). This radius optimizes the electron's bond to the nucleus, represented by the 'Rydberg energy'

$$E = Ze^2/a - \hbar^2/2ma^2 = Z^2 e^4 m/2\hbar^2 = 13.6 Z^2 \text{ eV}. \tag{1}$$

Expression (1) is conveniently resolved into factors

$$E = \tfrac{1}{2} Z^2 (e^2/\hbar c)^2 mc^2 = \tfrac{1}{2}(Z\alpha)^2 mc^2 \tag{1'}$$

whose dimensional element mc^2 represents the electron's (relativistic) 'rest energy'. The numerical coefficient $\alpha = 1/137.036$ represents instead the ratio of the electron's average velocity in the ground state of atomic H to the velocity of light c. Since the magnetic effects of the electron's motion are proportional to its speed, α (or $Z\alpha$) represents the relative roles of magnetic and electric forces in atomic phenomena. The small value of α implies small magnetic effects on energy levels; for this reason α is called the *fine structure* (or *relativistic*) constant. It is widely believed that the value of α is set by fundamental physics and there is a consensus that this piece of theory is beyond our present understanding. Solution of this puzzle might be viewed as determining the magnitude of the unit charge e.

Combinations of the ratio α and of the radius a are also important: the Compton wavelength αa serves as a unit for the wavelength increase

Atomic and Molecular Physics

experienced by x-rays recoiling from an electron. The product $\alpha^2 a = 2.8$ fm represents the electron's classical radius, i.e. the radius of a spherical charge e whose electrostatic field has energy mc^2.

Regarding the 'shell structure' of atoms, probably familiar to readers, we note that the ratio a/Z represents the confinement radius of the innermost ('K') electron pair for an element with atomic number Z. This radius shrinks a little further for heavy elements whose values of $Z\alpha$, the ratio of electron and light velocities, are no longer $\ll 1$, because approach to the velocity of light boosts the electron's inertia and thereby the effective value of its mass m in the definition $a = \hbar^2/mZe^2$.

The confinement radii of other shells result from variants of equation (1), modified by the addition of centrifugal potentials. The latter include the centrifugal repulsion between electrons with parallel 'spin' angular momenta, a phenomenon usually called 'Pauli exclusion' or 'exchange effect' and outlined in Box 13A [1].

The radius of the outermost (valence) electron shell increases slowly with the atomic number, roughly proportional to $Z^{1/3}$. For all neutral atoms, save those of noble gases, the electron motions in this shell have a net non-zero angular momentum. These electrons, free from the constraint of cancelling one another's momentum, are thus prone to combine with electrons of other atoms to form chemical bonds and are accordingly said to be 'chemically unsaturated'—in other words, reactive—as we shall see in section 13.2.5.

'Atoms', without further qualification are implied to be electrically neutral, that is, to include a number of (negative) electrons equal to the atomic number Z which represents the multiple of unit positive charges on their nuclei. Atoms stripped of some of their electrons are called positive ions. Atoms that include supernumerary electrons—generally a single one—are called negative ions.

The structure of the compact form (ground state) of generic atomic species, considered thus far, is complemented by the occurrence of 'excited stationary states' which occupy a larger volume of space with one (or more) electrons less strongly bound than in the ground state. The term 'stationary' means that these states remain unchanged in time, in the absence of external actions. Their stability is ensured by optimization of parameters analogous to the radius a in equation (1). Their diverse properties and classifications will be introduced in section 13.2.3.

A still broader multitude of 'non-stationary' states occurs, whose time-dependence is usually represented by a Fourier-like series of sinusoidal oscillations. Each oscillation frequency equals the energy difference of a pair of stationary states, divided by $\hbar$, much as the oscillation frequency of a light beam's polarization, along its path through an anisotropic medium, reflects differences between alternative refractive indices of that material.

BOX 13A:
SYMMETRY, PARITY AND EXCLUSION

The structure of atoms and molecules frequently displays symmetry elements, typically invariance under reflection through a point, an axis or a plane. This symmetry emerges in the statistical distribution of atomic constituents for stationary states, which exhibit a maximum, minimum (often a zero) or a saddle point at the symmetry point, axis or plane. Quantum mechanics represents such *non-negative* distributions as the squared moduli of 'probability amplitudes' or 'wavefunctions', essentially square roots of probabilities, which may simply reverse their sign at a symmetry element. The absence or presence of such a sign reversal ('even' or 'odd' character) is a parameter of stationary states called *parity*, with reference to a specific symmetry.

The parity of pairs of *identical particles* under reflection through their centre of mass, an operation that leaves the pair *unchanged*, has a major relevance for atomic structure because it leaves its wavefunction unchanged, i.e. with even parity. (The wavefunction's parity for a *single* electron under reflection through a symmetry element may instead be odd.) The wavefunction of the *electron pair* consists of two factors, one pertaining to its spin and one to its position distribution; its even parity requires both factors to have *equal* parity, whether even or odd. The spin factor is even for an antiparallel spin-pair whose zero angular momentum is independent of orientation, but is odd for the parallel spin function which reverses its sign, as a vector does, under a rotation by 180° about the centre of mass and which leaves the pair unchanged. Parallel spins thus imply an odd distribution function in space, which vanishes at the centre of mass, representing an effect of centrifugal repulsion. A striking effect of this symmetry on molecular rotations will be outlined in Box 13E.

CONTINUED

13.2.2. Radiation

'Radiation' will indicate here electromagnetic oscillations of all frequencies; their exchanges of energy with atoms and molecules have provided the main evidence for atomic structure and mechanics. These exchanges proceed by discrete units (photons) proportional to the radiation frequency ω (rad s^{-1}), $\Delta E = \hbar\omega$ [2].

The principal interaction between radiation and an isolated atom or

Atomic and Molecular Physics

BOX 13A: CONTINUED

Symmetry thus prevents electrons with parallel spins from approaching closely. The usual description of this 'exclusion' disregards electron identity initially by attributing labels 1 and 2 to individual electrons, it then requires their combined wavefunction to be odd under permutation of these labels. (The spin-pair's wavefunction, disjoined from position, remains even under this permutation.) Reliance on a pair's symmetry about its centre of mass, stressed by Feynman, is viewed here as preferable conceptually, but electron labelling followed by enforcing odd symmetry under permutation of each pair proves more convenient generally.

---

molecule is represented by a term of their combined energy, namely, the product of a radiation's electric field and of the electric dipole moment of an atomic *non-stationary* state. (Single atoms display no oscillating dipole while remaining in a stationary state, but are forced into non-stationary states by their coupling to radiation.) Energy transfers between radiation and atoms are particularly intense when their oscillation frequencies coincide, i.e. at *resonance*. Stationary normal modes of coupled oscillations of radiation and atoms, in which energy is shared between these partners, occur when both partners are confined in a restricted volume of space.

The most instructive processes occur, however, when the radiation spreads over volumes V far larger than atoms. In this case the unit energy, $\hbar\omega$, of each independent radiation mode is highly diluted with density, $\hbar\omega/V$, and so is its field strength acting on any single atom. This circumstance allows the evaluation of the radiation action as a 'perturbation', *linear* in the field strength A and in the electron charge, i.e. in the parameter $\alpha^{1/2}$ of equation (1'). The density of radiation modes with frequencies capable of resonating with an atomic dipole's becomes, by the same token, very high—indeed proportional to V—thus yielding a significant total contribution.

13.2.2.1. Elementary phenomena. Three alternative 'elementary' processes result from the atom–radiation interaction: (a) the resonant *absorption* of one photon by an atom, whose energy is thereby raised to a higher level, which may be bound or ionized; (b) the reciprocal 'spontaneous' emission of one photon by an atom (resonance being readily afforded by the dense spectrum of radiation modes in a large volume V); and (c) the *scattering* of a photon from one radiation mode to another,

William F Meggers

(American, 1888–1966)

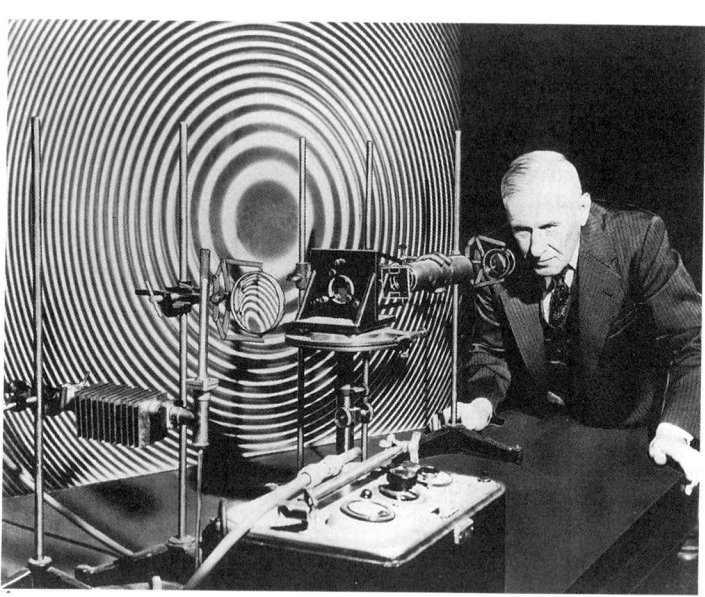

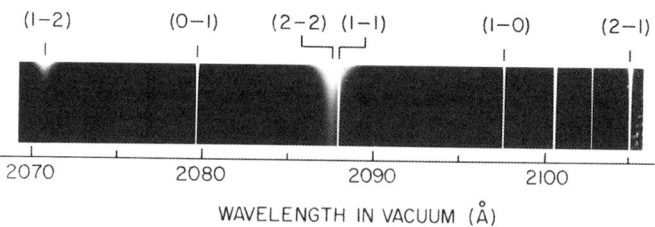

WAVELENGTH IN VACUUM (Å)

William Meggers, of the US National Bureau of Standards (NBS), was a pillar of spectral analysis through the mid-century period focusing on instrumental precision and data analysis. Under his guidance and initiative a team of spectroscopists observed, measured and otherwise collected thousands upon thousands of precise spectral line wavelengths, serving as standards throughout the world. Their lists were then patiently and persistently inspected over decades in the search of patterns that would reveal regularities of atomic mechanics.

CONTINUED

> **CONTINUED**
>
> Cumulative experience developed a flair that enlivened the search. Mechanization of the search started about mid-century, initially by IBM card-sorting devices but eventually through electronic computers scanning lists of spectral frequencies for pairs with equal differences; the results of these scans are embodied in the basic NBS *Tables of Atomic Energy Levels*. The clarity of interference rings beside his picture underlies the accuracy of a Hg 198 wavelength standard. The spectral lines displayed here belong to transitions between levels of two Zn multiplets with *j*-values indicated above each line. Note the differences among line widths reflecting the varying rates of decay of excited states through alternative channels. (Increasing widths near the upper edge reflect an instrumental artefact.)

mediated by an atom through absorption + emission (not necessarily at resonance). Photon scattering may be elastic (Rayleigh) or inelastic (Raman), the latter leaving a fraction of the absorbed photon in the atomic system.

High photon energies afford the 'Compton' variant of inelastic scattering, in which an electron recoils clear out of its atomic system. Another process, manifest mostly at high photon energies, is the emission of a continuous spectrum of radiation (bremsstrahlung) by fast electrons deflected, mainly by nuclear attraction, when traversing atomic fields.

Quantitative treatments of these processes were key elements in the development of quantum mechanics, reviewed in 1932 in Fermi's lectures [3] and more comprehensively in Heitler's book of 1936 [4].

The interaction of electrons with radiations whose photon energy approaches the electron rest energy mc^2 (= 511 keV) requires relativistic treatments that intermix radiation and electrostatic fields. A major result in this range lay in the 1932 discovery of the combination of MeV-range photons with the Coulomb field of high-Z nuclei yielding a 'pair-production', namely, the generation of electron pairs with charges of opposite sign. This process leads, in turn, to the surprising secondary process of 'vacuum polarization' in the space surrounding nuclei or other charges, a high-energy analogue of the usual electric polarization of atoms and atomic aggregates by electrostatic fields.

Electric polarization itself amounts to the non-resonant generation in matter—and also in vacuum—of the same 'dipole' displacements of electric charges as absorb or emit radiation at resonant frequencies.

Twentieth Century Physics

Further atomic polarization effects, corresponding to very high-frequency resonances, were discovered in the late 1940s: 'Lamb shifts' of atomic energy levels and ~0.1% corrections to the magnetic moments of electron spins. A major problem arose in these developments from an inherent divergence of the high-frequency contribution to polarizations in the limit $\omega \to \infty$. Procedures were developed by 1950 to remove these singularities by appropriate circumscribing of the problem, thus affording very successful calculations by expansion into powers of the electron charge parameter α. However, in the view of the author of this chapter, the basic problem persists to this day.

13.2.3. Atomic spectra

Spectroscopes are instruments that resolve a beam of radiation into components of different frequencies, displaying the intensity of the various components on a screen or on a detector's surface. Early displays of the analysis for the radiation emitted or absorbed by monatomic vapours revealed their frequencies to lie at or about discrete values (spectral lines) in, or near, the visible range ($10^{15} < \omega < 10^{16}$ rad s^{-1}). The multitude of lines observed for all elements was soon reduced somewhat by representing them as differences between a smaller set of spectral terms, later identified as equivalent to energies of stationary states of each atom. Extracting the terms from rich spectra tested the skills and devotion of spectroscopists, until the process was computerized.

Observed lines correspond only to pairs of states generating transitions with an electric dipole moment, i.e. pairs with opposite parity (under reflection through the atom's centre) and angular momenta j differing by no more than one unit of $\hbar$. These *selection rules* afford a partial classification of spectral terms. (The 'quantum number' j represents the magnitude of the vector angular momentum $\boldsymbol{j}$, namely, the value of $|\boldsymbol{j}|^2 = j(j+1)\hbar^2$.)

Terms of free atoms are independent of their orientation in space. This 'degeneracy' is broken in the presence of external fields, electric (Stark) or magnetic (Zeeman), which split each term with $j \neq 0$ into a multiplet of levels with different quantum numbers m. Beams of atoms, or molecules, with $j \neq 0$ are resolved into components with different m values by transmission through 'Stern–Gerlach magnets'. The uniform separation of the components measures their magnetic moment parallel to the field (figure 13.1). Elastic scattering of radiation or electrons, the latter to be outlined in section 13.2.4, can form an image of the shape of an atom in a stationary state (figure 13.2).

13.2.3.1. *The simplest spectra.*
The simplest spectra occur for atoms of hydrogen and of monovalent metals, whose single valence electron moves in a field with central symmetry allowing separation of its rotational and radial motions. This electron's vector angular momentum

Atomic and Molecular Physics

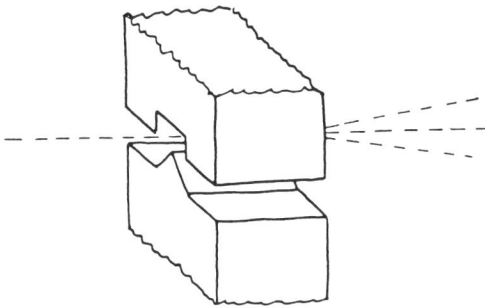

Figure 13.1. *Diagram of molecular beam analysis by a Stern–Gerlach inhomogeneous magnet. The incident beam is resolved here into three components.*

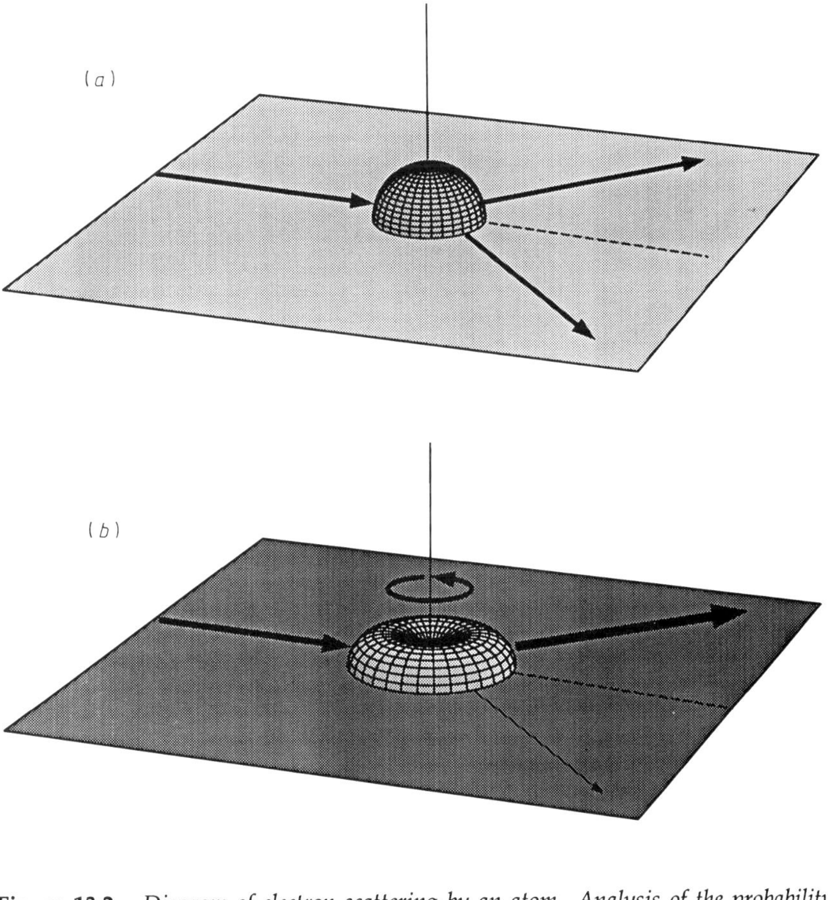

Figure 13.2. *Diagram of electron scattering by an atom. Analysis of the probability of electron deflections by different angles determines the shape of an atom. Asymmetries between deflections to the right and left reflect intra-atomic currents.*

j is accordingly a constant of the motion and might have been expected to vanish in some states, yielding a statistically uniform distribution of electron positions about the atom. The latter feature is indeed verified by certain observations on atomic ground states, but magnetic analysis of monovalent atomic beams showed them to be resolved into two components labelled by $m = \pm\frac{1}{2}$, implying that $j \neq 0$ and that the corresponding quantum number $j = \frac{1}{2}$, thus demonstrating the existence of the electron's spin [5].

The separation of rotational and radial motions is also a factor affecting the density of electron distributions. For the H ground state this density is simply proportional to $\exp(-2r/a)$ with $a = 0.053$ nm as in equation (1). For the sequence of H excited states with spherical symmetry the corresponding density is represented by the product of a similar factor $\exp(-2r/na)$ and of a squared polynomial of degree $n-1$ (i.e. with $n-1$ roots) in the ratio r/na, with $n = 2, 3, \ldots$. The electron binding energy for this sequence of states, easily obtained from the observed H spectrum (see Chapter 2, section 2.6), is also simply related to (1) by the Rydberg formula $E_n = 13.6$ eV/n^2. The basic relation displayed here between the parameter na of an exponential factor of the radial density and the electron's binding energy, with an appropriate value of a, is *common to all atoms*.

Rotational motions of an electron generate directional patterns of its density distribution. Quantum mechanics represents these patterns through a set of mathematical expressions, $|Y_{lm}(\theta, \varphi)|^2$, described in Box 13B. This distribution is complemented by a radial factor $|f_{nl}(r)|^2$, analogous to that described above for the spherical distribution, but with l fewer roots in its polynomials. The energies of excited stationary states with $l \neq 0$ coincide approximately with those of equal n and $l = 0$, but depart from them through a 'fine structure' splitting, described in Box 13C.

The spectra of monovalent metals (alkaline and noble), resulting from excitation of their valence electron, resemble the H spectrum closely because this electron moves in a Coulomb field when outside the atom's (spherical) ionic core. However, this electron's penetration into the core modifies the radial density functions $f_{nl}(r)$ mainly *within* the core. Outside the core, $f_{nl}(r)$ keeps its hydrogen character to within parameter shifts in its exponential and in the roots of its polynomial factor; these are represented effectively by subtracting from the principal quantum number n a new semi-empirical quantum defect parameter μ_l, $n \to n - \mu_l$. The expression 13.6 eV/n^2 for the binding energy of atomic hydrogen levels is thus extended into the generic formula for levels of Rydberg series $E_{nl} = 13.6$ eV$/(n - \mu_l)^2$. The l-dependence of μ_l removes the approximate coincidence of the energies of stationary states with equal n and different l (except for l values at which μ_l is negligible).

BOX 13B: PATTERNS OF DIRECTIONAL DISTRIBUTION

Directions of physical space are generally indicated (by analogy with coordinates on the Earth's surface) with reference to a polar axis and a 0° meridian, by a pair of angles (θ, φ). The 'longitude' φ runs from 0° to 360° (2π radians) around the axis and the 'co-latitude' θ runs from one pole to the other (0° to 180°).

Distributions in longitude are represented through a Fourier series of trigonometric functions $\{\cos m\varphi, \sin m\varphi\}$ (m = any integer) or of their equivalent complex combination $\exp(im\varphi) = \cos m\varphi + i \sin m\varphi$. The rotational motion of an electron, or other system, about the polar axis, with a constant angular momentum component $m\hbar$ along this axis, is represented by a distribution function with the single trigonometric component $\exp(im\varphi)$, whose real and imaginary parts are subdivided into m *meridian* lobes.

Distributions in co-latitude are represented through a series of polynomials of degree $l - m$ in the variable $\cos\theta$, with an additional factor $\sin m\theta$, (associated Legendre polynomials, $P_{lm}(\cos\theta)$). The distribution $|P_{lm}(\theta)|^2$ has $l - m$ lobes (e.g. $P_{00} = 1$, $P_{10} \propto \cos\theta$, $P_{11} \propto \sin\theta$, $P_{20} \propto 3\cos^2\theta - 1$). A combined distribution function $P_{lm}(\theta)\exp(im\varphi)$ is indicated in the text by the spherical harmonic function $Y_{lm}(\theta, \varphi)$. A single function Y_{lm} represents a rotational motion with squared angular momentum $|l|^2 = l(l+1)\hbar^2$ and with a component $m\hbar$ of l along the polar axis. The index l thus represents the degree of directional confinement of rotational motions.

13.2.3.2. Multi-electron spectra. Working out the energy levels of multi-electron atoms looks, at first sight, to be a daunting task. Monovalent atoms have been viewed in section 13.2.3.1 as consisting of a single electron and of a spherical ionic core, without describing how the core-electron mechanics determines the quantum defects μ_l. Other ionic cores are asymmetric (except for monovalent cores like Mg$^+$), so their own structure needs study.

The task is reduced considerably by the algebraic treatment of directional interactions among electrons as outlined in Box 13D, but interactions among radial motions also require attention. These interactions are themselves restricting the electrons of filled shells within ionic cores, because shell filling exhausts the angular degrees of freedom accessible to any single electron. The still current procedure,

BOX 13C: FINE AND HYPERFINE STRUCTURES

The fine structure of atomic spectra stems primarily from the interactions between the magnetic moments of electron spins and those resulting from intra-atomic rotational motions of electrons. The magnetic field experienced by the spin of each electron, owing to that electron's rotational motion, stems actually from the apparent rotation of *other* charges, mainly of the nucleus, about the spin itself.

The nucleus' contribution is proportional to its charge Ze and to the inverse cube of the electron's distance from it. The resulting energy of spin-orbit coupling is accordingly proportional to the integral over $|f_{nl}(r)|^2/r^3$ and to the product of the orbital and spin angular momenta $l \cdot s$. Its net effect on the spectra of monovalent metals is represented by a shift of the quantum defect μ_l proportional to $s \cdot l$ and increasing rapidly with Z, as $Z\langle|f_{nl}(r)|^2/r^3\rangle$. (An early apparent discrepancy between experiment and theory of this phenomenon was quickly removed in 1926 by L H Thomas's accurate evaluation of the field experienced by an orbiting electron.) Contributions of relativistic effects proportional to $l \cdot s$, are disregarded here.

The product $l \cdot s$ depends on the squared magnitude of the vector sum

$$|l + s|^2 = |l|^2 + 2l \cdot s + |s|^2 = |j|^2 = j(j+1)\hbar^2 \qquad (2)$$

that is, on the quantum number $j = l \pm \frac{1}{2}$, in addition to the values of l and $s = \frac{1}{2}$. The resulting energy levels, -13.6 eV$/(n - \mu_{lj})^2$, turned out to be lower for $j = l - \frac{1}{2}$ (antiparallel mutual orientation) than for $j = l + \frac{1}{2}$. This result accounted for the earlier discovery of 'doublets', i.e. of closely spaced pairs of spectral lines. An adequate quantitative understanding of the 'simple spectra' involving a single valence electron was thus achieved by the mid-1920s.

CONTINUED

introduced by Hartree in 1928, exploits this angular confinement by requiring a single radial distribution function $f_{nl}(r)$ to represent the radial distribution of all electrons with the same quantum numbers. This approach, motivated initially by disregarding all electron correlations, determines the $f_{nl}(r)$ values of closed shells 'self-consistently' so as

> **BOX 13C:** *CONTINUED*
>
> Concurrently a still smaller, 'hyperfine', structure of spectral lines was resolved, analysed and evaluated, stemming from the interaction of electronic and nuclear distributions of charges and currents. We do not describe this interaction beyond stating that its energy depends on the product of electronic and nuclear angular momenta, $j \cdot I$, as reflected in the squared combined momentum $|j + I|^2 = |F|^2 = F(F+1)\hbar^2$.
>
> The ground state of the H atom ($j = \frac{1}{2}$, $I = \frac{1}{2}$) resolves into two components with $F = 0$ and 1. Short-wave radio emission at 1420 MHz by transitions between these levels throughout the cosmos has been of great value in assaying the distribution of atomic hydrogen. (See also section 13.2.5.3.)
>
> The combination of orbital and spin momenta that yields energy levels with alternative values of j alters the directional asymmetry of the electron's density distribution illustrated in figure 13.2. This subject belongs to a broader class of studies whose outline and scope are introduced in Box 13D.

to minimize the energy of an entire closed shell core. Its Hartree–Fock variant also accounts for the exchange energy, a side-effect of the centrifugal repulsion of electron pairs with parallel spins mentioned in section 13.2.1 and Box 13A.

Each electron of an incomplete shell, whether in the ground state or excited, is also assigned quantum numbers nl (or nlj) and a radial distribution $f_{nl}(r)$, whose determination is even less straightforward. The electronic structure of ground or excited states is then represented by a 'configuration' formula, $\Pi_i (n_i l_i)^{N_i}$, where N_i indicates the number of electrons assigned to the ith subshell. The distributions $f_{n_i l_i}(r)$ are then determined so as to stabilize the total energy of the stationary state, depending on both radial and angular distributions. Bringing the energy into adequate agreement with experiment generally requires an additional step of 'configuration mixing'.

The state of this art in the mid-1930s was summarized in a treatise by E U Condon and G Shortley [10]. This state was fairly adequate to deal with excited 's' or 'p' electrons (i.e. with $l = 0$ or 1). Full unravelling of the very rich spectra of partially filled 'd' shells ($l = 2$), which accommodate up to $2(2l + 1) = 10$ electrons, began with Racah's work in the 1940s [8]. The f-shell ($l = 3$) spectra of the rare-earths defied analysis until later decades.

BOX 13D: ANGULAR MOMENTUM ALGEBRA

The evaluation of the spin-orbit coefficient $l \cdot s$ through the identity $|l + s|^2 = |j|^2$, equation (2), is a prototype of very extensive procedures central to spectral theory and to other subjects. Problems analogous to this prototype occur in evaluating the repulsion energy between two (or more) electrons with density distributions represented through spherical harmonics $Y_{lm}, Y_{l'm'}$. Electron interactions represented by the Coulomb potential $e^2/|r - r'|$, to be averaged over their combined density distributions, can be expanded into spherical harmonics. We deal here with generalizations of the trigonometry formula $\cos ax \cos bx = [\cos(a+b)x + \cos(a-b)x]/2$ which resolves a product into a sum of terms involving sums and differences of the indices (a, b). The manifold generalizations of this formula rest on the transformations of harmonic functions induced by changes of their coordinates, transformations that form a group and whose expressions in terms of indices $\{l, l' \ldots\}$ are provided by group theory.

The relevant formulations for quantum systems by algebraic functions of angular momentum quantum numbers $\{l, s, j; m, m'\}$ were provided around 1930 in parallel books by E Wigner and L Van der Waerden [6,7]. Thereby *all averagings* over directional distributions are reduced to standard formulae. Further progress was achieved by G Racah [8], and independently by Wigner [9], relating alternative combinations of three or more functions through algebraic functions of the indices $\{l, s, j, \ldots\}$ only, which are independent of coordinate systems.

13.2.3.3. Inner-shell spectra. Inner-shell electrons are readily ejected from atoms by incident projectiles with sufficient energy. The resulting vacancy is then quickly filled by an electron with higher energy, $B_o > B_i$; the transition between these shells releases the energy difference $B_o - B_i$, a process involving a dipole oscillating current with frequency $\omega = (B_o - B_i)/\hbar$ that may, but need not, result in the emission of a photon $\hbar\omega$. Incidentally, this photon emission was anticipated by Bohr in 1913, upon his first development of the 'planetary model'; equation (1) predicted that ejection of a K-shell electron would generate x-rays with $\sqrt{\omega} \propto Z$. This prediction was soon verified by Moseley's discovery of characteristic x-rays, so called because of the contrast between their line spectra and Röntgen's earlier continuous x-ray emission. This verification provided, in turn, early evidence of the assignment of successive values of Z to the

See also
p 158

Atomic and Molecular Physics

nuclear charge Ze for successive elements of Mendeleev's table.

The whole spectrum of characteristic x-rays consists of several series, in different frequency ranges, labelled K, L, M, ... series in correspondence to initial vacancies in inner shells with quantum number $n = 1, 2, 3, \ldots$. The lines of each series are labelled $\alpha, \beta, \gamma \ldots$ depending on the n value of the B_o shell.

The, generally more frequent, alternative to x-ray emission arises by transfer of the energy $B_o - B_i$ to an electron of the shell o, or external to it, evidenced by the detection of monoenergetic (Auger) electrons with characteristic spectra.

13.2.4. Collisions

Collisions of atomic particles have been a major tool of modern physics since Rutherford's α-particle experiments. Spectroscopy, which has provided more details about atoms, also depends largely on collisions by electrons or ions as sources of excited atoms.

'Long-range' collisions transfer energy to atoms through the Coulomb field of fly-by charged particles that remain—often far—outside a target atom or molecule. (A rapid pulse of this field acts much like a continuous spectrum of radiation.) 'Close', short-range collisions lead to formations of a *complex*, usually short-lived, in which projectile and target coalesce to varying degrees. Intermediate 'grazing' collisions, characteristic of most atomic projectiles involve only a small interpenetration of projectile and target.

The angular momentum of the complex (projectile + target combination), about its centre of mass, remains constant throughout a collision. One can take advantage of this conservation by 'partial wave analysis' of the incident beam into components with different orbital momenta. The probability of collision for each component is proportional to $(2l + 1)/v^2$ (v = velocity), thus increasing with l but with a coefficient that generally drops with increasing v. Application of this approach has been mostly confined thus far to a subset of problems, despite its basic significance.

The alternative classification of collisions by 'impact parameter', the least distance of the incident projectile's trajectory from the target, is generally appropriate to all long-range collisions and to most grazing and close collisions by projectiles heavier than electrons.

The first treatise of *The Theory of Atomic Collisions* by N F Mott and H S W Massey appeared as early as 1933 [11]. It centres on a wave-mechanical formulation illustrated by experimental results.

13.2.4.1. The action of fast charged particles.

The action of fast charged particles has been a relevant ingredient of many experiments since Rutherford's, its main elements being assessed by Bohr in 1913. He stressed the contrast between the *impulsive* character, prevailing when the collision lasts less than the longest period $1/\omega$ of atomic oscillations $b\omega/v \ll 1$ (b = impact parameter, v = particle velocity) and the reversible, i.e. *elastic*, character when $b\omega/v \gg 1$. The energy dissipated by the projectile in impulsive collisions amounts to the recoil energy of all atomic electrons within the impulsive range, *as though* these electrons were recoiling freely [12].

A wave-mechanical treatment of this phenomenon, by H A Bethe (1930), added a new feature and a correction to Bohr's ideas. It viewed the Coulomb interaction between a fast particle with charge ze at r and an atomic electron at r_i as 'weak', representing it by a Fourier integral over exponentials

$$\frac{ze^2}{|r - r_i|} = \frac{ze^2}{2\pi^2} \int \frac{dk}{k^2} e^{ik \cdot (r_i - r)}. \tag{3}$$

Each element of this integral represents the probability amplitude of transfer of momentum $\hbar k$ by the projectile to the ith electron. The squared amplitudes $|\exp(-ik \cdot r)|^2$ for momentum loss by the projectile, i.e. the probabilities of its deflection, add incoherently because the deflection is observable. The corresponding amplitudes $\exp(ik \cdot r_i)$ for momentum gain by individual electrons add instead coherently because only the probability of net excitation of an atomic target to its nth level is observable (unless an electron recoils clear out of the target). This probability is proportional to a function $|F_{n0}(\Sigma_i e^{ik \cdot r_i})|^2$ called the 'generalized form factor' of the target. Bethe verified that the resulting average energy absorption, $\Sigma_n E_n |F_{n0}(\Sigma_i e^{ik \cdot r_i})|^2$, equals the sum of the recoil energies of independent electrons, as anticipated by Bohr.

Bethe also noted that the minimum k-value contributing to $\Sigma_n E_n |F_{n0}|^2$ is set by the energy–momentum balance of the projectile, a limit somewhat tighter for fast electrons than Bohr's $b\omega/v \sim 1$. Mott noted that Bethe's procedure applies for massive (ion) projectiles also when $b\omega/v \gg 1$. In 1924 Fermi exploited the correspondence between the pulsed field of fast projectiles and radiation, replacing $\Sigma_n E_n |F_{n0}|^2$ by an equivalent macroscopic dielectric property of the target material.

13.2.4.2. Close collisions of slow electrons. Bethe's weak-interaction (actually impulsive) treatment [13] applies also to close collisions by fast electrons but gives unrealistically high results at incident velocities comparable to those of atomic electrons. Variants intended to remove this discrepancy did not appear desirable.

Incorporation of the incident electron into the target to form a complex with constant angular momentum might have appeared most appropriate to the present problem. However, only its application to noble gas targets with spherical symmetry proved successful in the early days (because of the complications of generic multi-electron complexes, whose treatment started only after mid-century). Within this framework, the density distribution of the colliding electron within the complex, with squared angular momentum $l(l+1)\hbar^2$, is represented by $|f_{kl}(r)Y_{lm}(\theta,\phi)|^2$ as for the valence electron of a monovalent atom in section 13.2.3.1. Here the index k, representing the momentum $\hbar k$ of the incident electron, replaces the quantum number n of the metal's valence electron. The quantum defect μ_l of the incident electron turns here into an equivalent phase shift $\delta_l(k) = \pi\mu_l$.

Experiments on elastic scattering of electrons by noble gas atoms, with unit incident flux and unit solid angle of deflection $d\Omega$, are represented in this case by the cross section

$$d\sigma(\theta) = \frac{4\pi}{k^2}|\Sigma_l(2l+1)^{1/2}\exp(i\delta_l)\sin\delta_l(k)Y_{l0}(\theta,\varphi)|^2\,d\Omega. \quad (4)$$

(The function $f_{kl}(r)$ is in this case, outside the atomic core, a spherical Bessel function $j_l(kr + \delta_l(k))$.) Great surprise was caused by Ramsauer's early discovery that this cross section vanishes, as though atoms of Ar, Kr, Xe become *fully transparent*, at an energy of impact (< 1 eV) characteristic for each atom. It was understood later that attraction by the polarized target prevails at near zero velocity, yielding $\delta_l(k) > 0$, but is overcome at larger energies by the net repulsion of the incident electron by the 'closed-shell' atom. The phase shifts $\delta_l(k)$ vanish at the transition between these energy regimes.

13.2.4.3. Close and grazing collisions by ions and atoms. Atoms and molecules, whether in ionic or neutral states, can be slower than all target electrons and still pack kinetic energies in the keV range. The momenta of their nuclei are similarly large allowing the projectile to bore through targets on a near-straight line (except when passing close to a target nucleus). Their influence on target electrons is mostly elastic, but the aggregate number of excitations and ionizations along their paths is non-negligible. (The speed of thermal atoms is orders of magnitude lower than those of atomic electrons, hence thermal collisions are elastic, barring chemical reactions.) Electron transfers between heavy projectiles and targets have been the object of much study.

A curious aspect of electron transfers, specifically of 'electron capture' by an ion colliding with a neutral target, was pointed out by L H Thomas in the mid-1920s. Head-on collision of the ion with a target electron would push the electron along the ion's path, thus facilitating its capture, but the *resulting* speed of the electron, double the ion's, would prevent its capture. Velocity matching of the electron and ion requires instead electron ejection at 60° from the ion's track. A second, elastic, collision by the electron with a nucleus at rest (i.e. a 'two-shot' process) is required to turn the electron parallel to the projectile's track, thus facilitating its capture. This phenomenon and its variants were observed half-a-century later.

Close and grazing collisions between slow atoms and/or molecules, whether ionic or neutrals, belong properly to the physics of the resulting molecular complex. 'Slow' means, in this context, generally 'slower than valence electrons'. These collisions depart from ordinary molecular mechanics primarily by their high energy of internuclear motion. Their experimental and theoretical study was delayed beyond mid-century by the resulting generally multiple fragmentation, whose study requires elaborate detection and analysis techniques.

The case of grazing collisions presents an easier opportunity, especially in the elementary case when the complex has a single valence electron, whose independent motion can be studied in the field of a pair of slow moving atomic or molecular cores. A theoretical method of 'perturbed stationary states', introduced early to treat this and related phenomena [11] and widely applied later, rests on a basic molecular procedure to be described in section 13.2.5.1.

For collisions at speeds faster than those of valence electrons, a key qualitative rule was formulated early by N Bohr [14], namely, that all electrons slower than the collisions's speed are likely to be swept out of their initial states.

13.2.5. *Molecular bonds and behaviour*

Pairs of atoms and/or molecules experience a generic weak attraction, a function proportional to $1/r^6$ of their distance r, a force whose existence had been inferred from gas properties in the 1800s and which bears the name of its discoverer, van der Waals. Quantum mechanics has traced this force to mutual, synchronously oscillating, electric polarizations of the two particles. This force draws particles together until stronger forces, attractive or repulsive, become dominant upon sufficient overlap of the partners' electron distributions. It even suffices to bind pairs of rare gas atoms at temperatures of the order of 1 K, against their short-range repulsion.

The origin of the basic 'covalent' molecular bond was first established by quantum mechanics, through analysis of the prototype approach of two H atoms by W Heitler and F London (1928). The short-range force between these atoms depends critically on the mutual orientation of their respective electron spins. Parallel orientation leads to repulsion, as noted in Box 13A; opposite orientation favours instead the pair's distribution function $f(r_1, r_2)$ to centre between the two nuclei, thus optimizing the attraction of the electron pair by both nuclei and offsetting the repulsions within each pair of equal charges (four attractions, two repulsions). (The original study, based on separate atomic distributions $f(r_1, r_2) = f_{10}(r_1 - R_1) f_{10}(r_2 - R_2)$ led to an optimum bond energy $2E(H) - E(H_2) \sim 3.7$ eV, but subsequent release of $f(r_1, r_2)$ allowed the calculated energy to approach the experimental value of the H$_2$ bond, 4.65 eV; the corresponding average internuclear distance also approached its value of 0.074 nm.)

Mechanisms substantially equivalent to the H$_2$ bond were soon found to underlie the formation of most chemical bonds. Bonds between atoms with different affinity for additional electrons are asymmetric, shifting the joint distribution $f(r_1, r_2)$ closer to one than to the other atom; typically the O–H bond in H$_2$O draws the bonding pair closer to the O atom. In extreme cases, such as in the NaCl molecule, the bond is called ionic and is indicated by Na$^+$Cl$^-$.

Divalent atoms form bonds with two atoms, on a straight line for group II (e.g. Mg) and at an angle for group VI (e.g. O). Trivalent atoms form three bonds, at 120° on a plane for group III (e.g. B) and in trihedral form for group V (e.g. N). Tetravalent atoms similarly form four bonds in tetrahedral directions, typically in CH$_4$ but also in the crystalline network of tetrahedrally bonded carbon, namely, diamond.

Carbon, and other atoms to a lesser extent, develop an alternative mixed type of bonding which underlies their far richer chemistry. The graphite form of carbon utilizes three electrons per atom to form a plane 120° network of bonds and the remaining one electron per atom to run through this network in quasi-metallic form. 'Aromatic' molecules share these features as do innumerable compounds with intermingling double, or even triple, bonds. A basic treatment of these phenomena was developed by Hückel in the early 1930s.

The ground states of simple, stable molecules often consist of closed shells, with electron spins coupled to cancel each other; a notable exception occurs in the oxygen molecule with net spin $S = 1$ in its ground state and low-energy metastable excitations with $S = 0$. Low excitations with $S = 0$ within the ground-state shell occur in many molecules. Excitations of one electron to higher shells are analogous to those of single atoms, with angular momenta of that electron variously coupled to those of the molecular core.

Important conceptual progress in about 1930 traced the origin of rubber elasticity, and of analogous phenomena, to the thermal behaviour of long-chain molecules (polymers). These molecules consist of sequences of N atomic groups, e.g. CH$_2$, linked by bonds that are free to rotate with respect to one another. The distance of the chain's end points is proportional to N when the chain is stretched, but only to $\sqrt{N}$ for random orientation of successive bonds. Mechanical action, such as a pull by an external force, can thus stretch a length of rubber but thermal agitation tends to restore random orientations with individual molecules wrapped onto themselves.

13.2.5.1. The Born–Oppenheimer approximation. The Born–Oppenheimer approximation, basic to molecular physics, rests on the electrons moving much faster than nuclei, owing to their smaller mass. Accordingly electron distribution functions $f(r_1, r_2, \ldots)$ are evaluated assuming fixed positions of the nuclei $\{R_1, R_2, \ldots\}$ and indicated by $F(\{R_1, R_2, \ldots\}; r_1, r_2, \ldots)$. The energy $E_n(\{R_1, R_2, \ldots\})$ of a stationary electron state with distribution F_n is then viewed as the potential energy governing the slower nuclear motion. 'Non-adiabatic' corrections to the initial results thus obtained are functions of derivatives $\partial F_n/\partial R_\alpha$.

Figure 13.3 illustrates the dependence of a succession of energy levels E_n of the H_2 molecule on the internuclear distance $R = |R_1 - R_2|$. Most of the $E_n(R)$ curves display a deep (near parabolic) minimum which holds the internuclear distance within a range $\sim 10^{-2}$ nm. The *vibrational* motion of the nuclear distance, thus confined, has a sequence of its own stationary states with energies E_{nv}, whose spacings $E_{n,v+1} - E_{nv}$ are roughly uniform and of the order of 0.5 eV for H_2 (but much smaller for heavier molecules). One of the curves $E_n(R)$ has, however, no minimum; which implies that its electron state pushes the nuclei apart, indeed towards 'dissociation', corresponding to an electron pair with parallel spins, whose centrifugal effect keeps the electrons apart thus preventing the bond's formation.

'Potential' surfaces of appropriate dimensions are the analogues of figure 13.3 for polyatomic molecules. Their calculations require elaborate numerical procedures, but have been expanding into a sort of industry with the development of increasingly powerful computers.

An additional element of molecular mechanics consists of the *rotational* motion of each whole molecule about its centre of mass. Its role in the spectra, and the role of vibrations, dominate the molecular spectra to be described in section 13.2.5.2. The energy scales of rotational, vibrational and electronic energy levels differ by 1–2 orders of magnitude, depending on circumstances, thus affording the corresponding motions to proceed quite independently.

The basic items thus outlined were soon developed, by about 1930, into an extensive theoretical treatment of molecular mechanics which interpreted the extensive spectral observations then available. Chief leaders in this process were F Hund, R Mulliken [15] and G Herzberg [16].

Two important concepts emerged in this early period:

(i) The set of potential curves in figure 13.3 constitutes a *correlation diagram* that connects the 'united atom' states of the H_2 molecule at $R = 0$ to its 'separate atom' states at $R = \infty$. The former arise when $R_2 \to R_1$, forming for H_2 a doubly-charged H_2^{++} core whereby the electron stationary states coincide with those of He; the latter represent pairs of separate H atoms in various states.

(ii) Excitation of an electron, for example due to photoabsorption by H_2's electronic ground state $n = 1$ with vibrational state v, changes the initial density distribution function $F_{1v}(R; r_1, r_2)$ into a different

Atomic and Molecular Physics

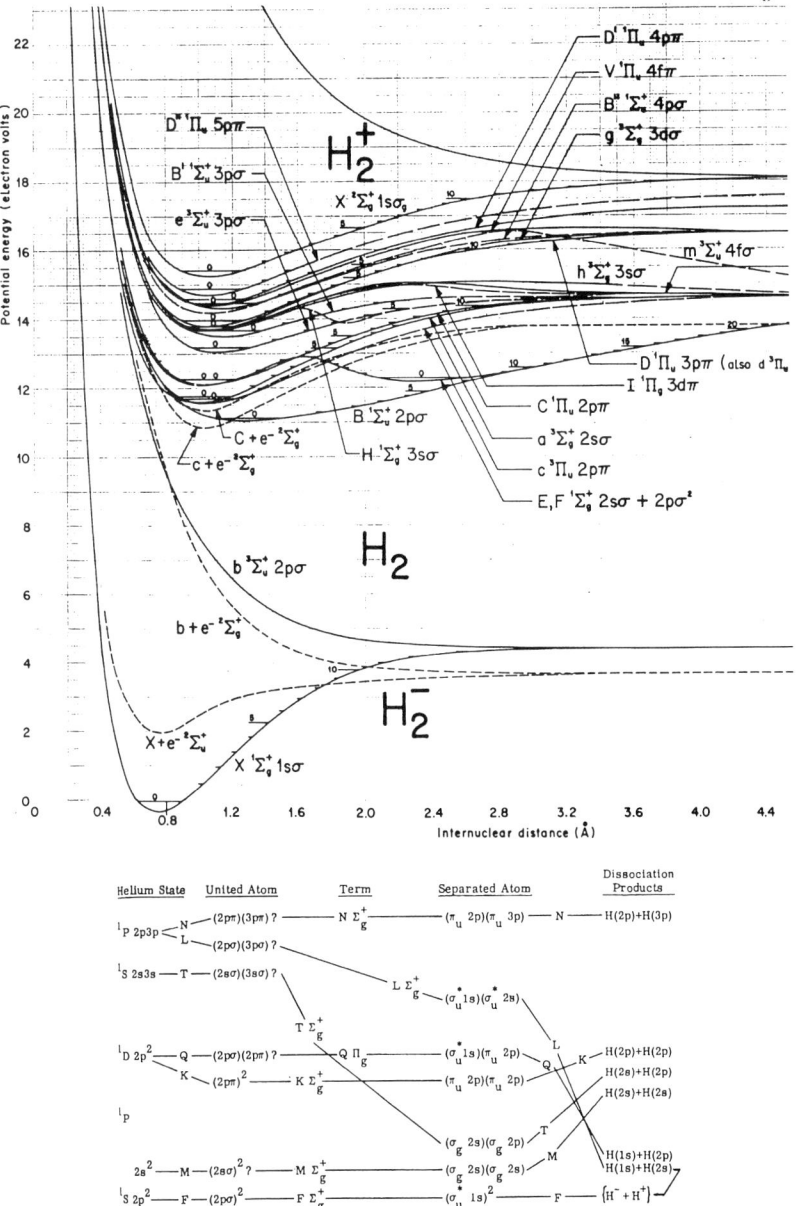

Figure 13.3. *Potential-energy curves of molecular hydrogen and of its ions. The ordinate represents the electron energy level of various stationary states at each value of the internuclear distance R plotted as the abscissa. The minimum of each curve represents the equilibrium value of that distance, the curvature at that point determines the vibration frequency; the energy values at large abscissas represent dissociation thresholds. The zero value of the ordinates corresponds to the ground-state energy of the H_2 molecule, inclusive of its vibrational energy; the curve labelled b^2 represents the repulsion of a pair of H atoms with parallel spins. Dashed curves represent energies of the H_2^- ion. Below: partial diagram of correlations between 'united' and 'separate' atom limits.*

1045

one $F'(R; r_1, r_2)$ that is *not* stationary because the radiation of interest resonates with the electron oscillations only, owing to the larger nuclear mass. The F' distribution resolves, in time, into a number of stationary components F_{nv}. This circumstance, called the *Franck–Condon* rule and extensively verified by experiments, is being utilized currently to manoeuvre the internuclear distance R by repeated appropriate photoabsorptions so as to recast molecular structures into novel, if unstable, shapes of interest.

13.2.5.2. Molecular spectra. The aspect of molecular spectra contrasts with that of atomic spectra through the contributions of vibrational and rotational motions. Each electronic transition, which would give rise to a single spectral line in an atom, is generally accompanied by manifold transitions between alternative pairs of vibrational and rotational energy levels. Only a small subset of these transitions generates oscillating electric dipoles and is thus readily apparent in the spectra, but this subset is nevertheless large.

Typically an oscillating current arises in transitions by 'optically active' molecules in which the centre of mass of the electrons does not coincide with the centre of charge of the nuclei. An example is the water molecule, H_2O, whose electrons lean onto the O atom leaving each H atom with a net positive charge. Further restrictions (selection rules) restrict the effective intensity of dipole oscillation to transitions between successive levels of rotational or vibrational energy in each spectral sequence.

The energy separation of transitions between different pairs of vibrational levels is of the order of 10^{-2} eV and between pairs of rotational levels is still smaller by 1–2 orders of magnitude. Accordingly each vibrational transition generates a band of rotational lines, whose own spacings are linear or quadratic functions of the relevant angular momentum numbers J. The vanishing spacing as $J \to 0$, or at other 'band heads', is prominent (figure 13.4). The art of interpreting these complex manifolds of lines was well developed in the mid-1920s, at least for the diatomic molecules whose stationary states are classified by the constant angular momentum component $\lambda \hbar$ along the internuclear axis. Symmetries under inversion at the centre of mass and under interchange of identical nuclei have also been noted and interpreted.

By 1950 the spectra of numerous molecules, in the visible and near visible range of frequencies, had been studied as reported in G Herzberg's classic texts [16]. His 1950 table reported basic parameters of $\sim$500 individual diatomic molecules: bond energies, internuclear distances, vibrational frequencies, moments of inertia, electron excitation energies, etc. Direct observations of rotational spectra by microwave spectroscopy were just beginning at that time; they will be discussed

Atomic and Molecular Physics

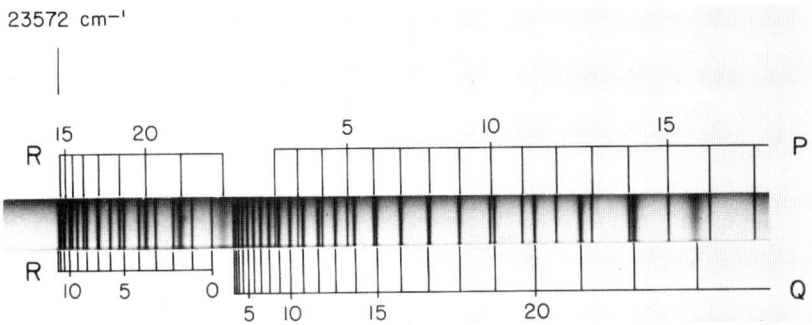

Figure 13.4. *Band structure in the spectrum of electronic and vibrational excitation to states of AlH about 2.5 eV above its ground state. (Abscissa scale covers an energy range of nearly 0.05 eV, increasing towards the left.) The index N of each absorption line labels the initial (ground state) rotational energy level. The Q branch represents transitions in which the N value remained constant but the rotational energy was reduced by the increased moment of inertia. The P branch pertains to transitions N → N − 1 which reduce the rotational energy even further. The R branch pertains to transitions N → N + 1; here the increase of N prevails initially over the increased inertia but is then overwhelmed by it.*

in later sections. A remarkable rotational property of H_2 and other molecules is described in Box 13E.

Excitation of inner-shell electrons gives rise to emission of x-rays or of Auger electrons, much as outlined for atoms in section 13.2.3.3. Note, however, that temporary vacancies generated in inner shells are generally localized in one individual atom within a molecule thus disturbing the equilibrium of valence bonds and enriching the spectra.

Fine and hyperfine structures, analogous to those described for atoms in Box 13C, are also present in molecular spectra. Their initial observation, in the mid-1930s, by I I Rabi's group on beams of molecules—or atoms—in their ground states, subjected simultaneously to a constant magnetic field and to an oscillating field orthogonal to it, ushered in the new field of *radiofrequency spectroscopy*.

Molecules with a specified angular momentum component $j_z = m\hbar$, parallel to a uniform magnetic field $B\hat{z}$, were first sorted out by a Stern–Gerlach magnet, as outlined in section 13.2.3. Their magnetic energy levels in this field are represented in terms of a gyromagnetic (or magnetomechanical) ratio γ by $E_m = -\gamma m B\hbar$. Exposure of the beam to an oscillating field tuned to the precise frequency γB may cause transitions to $m \pm 1$, in which case molecules are rejected by a further analyser. The sensitivity of this selection allowed measurements of parameters γ of electrons and nuclei to unprecedented accuracy; Rabi hailed these results as the 'utilization of single molecules as nuclear laboratories'. This technique even led to the discovery of a charge asymmetry (quadrupole moment) in the nucleus of 'heavy' hydrogen ^2H.

> ## BOX 13E: PARA- AND ORTHO-HYDROGEN
>
> Molecular hydrogen was discovered by K F Bonhoeffer in 1929 to consist of two components that convert into one another only very slowly (in the absence of catalysts), a property shared by many homonuclear diatomics as predicted by D M Dennison. Hydrogen at very low temperature turns entirely into its *para* form.
>
> This observation reflects the exclusion phenomenon (Box 13A) pertaining to its pair of nuclei, identical particles with spin $\frac{1}{2}$ and equivalent to a pair of electrons in this respect. The distribution function in space of hydrogen nuclei with parallel spins must change its sign under rotation of the molecule by 180°, as it does for the analogous electron pair. For the electron pair this change of sign implies a centrifugal repulsion, as it does for the nuclei of H_2, but here the centrifugal effect implies a more obvious manifestation, namely, non-zero rotational energy (more precisely, molecular rotation with an odd value of its angular momentum index J). Accordingly, the *ortho* species of H_2 disappears at low kelvin temperatures, whose minimal thermal agitation does not support even the lowest non-zero rotational level. Rising temperatures allow the resumption of H_2 molecular rotation but the spin arrangement of the nuclei is well insulated from external thermal actions, whereby the onset of odd-J rotation hinges largely on external magnetic actions.
>
> Incidentally the nomenclature 'ortho' and 'para' for parallel and antiparallel spins originates from the empirical systematics of the atomic spectrum of helium, which also appeared to consist of two substances with different spectra, until Heisenberg traced their difference to the mutual electron spin orientation in 1927. Analogous classifications arise for electron and rotational spectra of all systems whose identical particles have alternative spin states.

13.2.5.3. Reactive collisions. Reactive collisions involve a transfer of particles between projectile and target, including the restructuring of a pair of molecules into new entities, the essential feature of chemical reactions. The large mass of molecules, as compared with electrons, makes them candidates for at least partially semi-classical treatments. We outline here a sketch of the widely studied prototype approach to such a treatment, dating from the 1930s and to be expanded in section 13.6.4.

Consider the simple reaction

$$AB + C \rightarrow A + BC \tag{5}$$

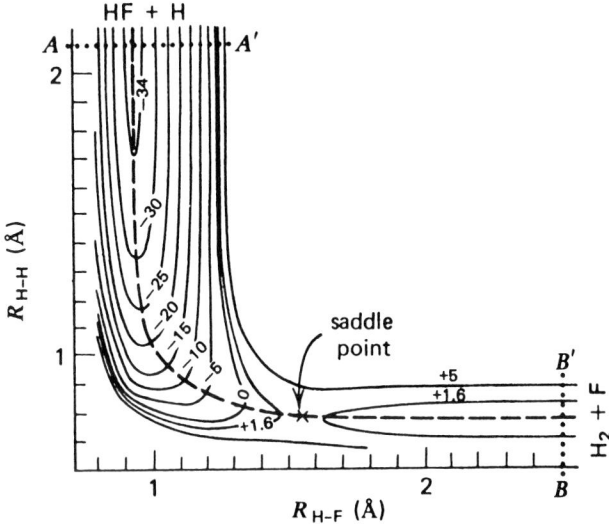

Figure 13.5. 'Contour map' of the potential energy surface for the linear molecular complex F-H-H, corresponding to equation (5) in the text ($F \equiv A$; $H \equiv B, C$). Energy contours are labelled in kilocalories per mole (1 kcal mol^{-1} = 0.043 eV). The zero of energy corresponds to the limit of separate H_2+F. The dashed line indicates the reaction path.

where {A, B, C} indicate atoms or molecular groups, assuming its elements to be restricted to a straight line. This model involves two parameters only, namely the non-negative distances r_{AB} and r_{BC} shown in figure 13.5. Initially r_{AB} oscillates in a potential well analogous to those of figure 13.3, whereas r_{BC} is very large; at the end r_{AB} is large and r_{BC} confined. The rearrangement takes place at $r_{AB} \sim r_{BC}$, that is, astride the potential barrier that separates the entrance and exit valleys. The ABC complex evolving in this region is said to be in its *transition state*. The relevance of this schematic model to the actual three-dimensional phenomenon does not appear to have been discussed at that time.

13.3. Completing the spectrum of radiation actions

Two major gaps remained at mid-century in the study of the radiation action, an action introduced in section 13.2.2 and elaborated in sections 13.2.3 and 13.2.5.2 within a limited context. The information provided in the latter sections dealt primarily with the action of visible light extending into the near infrared on the lower frequency side and somewhat further into the ultraviolet at higher frequencies.

Additional studies had dealt with radiofrequencies generated by electric processes and with the x-rays generated mostly by bremsstrahlung (section 13.2.2.1) and also by nuclei or cosmic rays. Technical difficulties had prevented detailed exploration of the action of radiation with frequencies in the gaps between these three ranges, each gap covering 2–3 decades of the spectrum. These difficulties concerned the generation of radiation with any desired frequency in the gaps, as well as the handling and measurements of these radiations. Typically, radiations between the far ultraviolet and near x-ray ranges are absorbed strongly by all kinds of matter.

Newer technical developments, mainly in the 1960s, have since afforded practically unrestricted exploration throughout both gaps. Their nature and results are reported separately for each gap in subsections 13.3.1 and 13.3.2. Main features of radiation action throughout its spectrum and on elements throughout the periodic system are summarized in sections 13.3.3 and 13.3.4

13.3.1. From ultraviolet to x-rays—synchrotron light
Transmission of ultraviolet light through optical systems is practically limited to photon energies below 10 eV, which is the threshold of absorption by lithium fluoride. Beyond this limit continuous spectra are produced by discharges in noble gases; their extraction into vacuum through orifices hinges, however, on the use of powerful pumps which became practical in the 1960s [17].

On the x-ray side, efficient production of bremsstrahlung was limited to photon energies in the keV range. This limit was overcome in St Petersburg by Lukirskii's clever instrumentation, reaching down to about 50 eV photons in the early 1960s [18] (figure 13.6) which, although a notable achievement, was soon overshadowed by the onset of alternative developments.

Within the gap lay the extensive emission spectra of atomic innershell transitions (section 13.2.3.5), whose discrete photon energies afford fragmentary glimpses of evidence on radiation action. Weissler's 1956 review of photoionization up to ~50 eV reflects the modest progress achieved through such sources [19]. Evidence about photoabsorption spectra from inner shells of atoms seemed at the time compatible with continuation at lower photon energies of the saw-tooth spectra characteristic of the x-ray range (figure 13.7).

R L Platzman stressed in that period that adequate exploration of the 10–1000 eV range would hinge on extensive use of the continuous spectra from 'synchrotron light' sources. This radiation is emitted by electrons travelling on closed paths at speeds approaching the velocity of light c. Slower electrons travelling on closed paths act as radio antennas emitting a single frequency v reciprocal to their period of circulation, in directions above and below their path. At speeds approaching c, however, this emission consists mainly of extremely high multiples of that frequency, focused tangentially to the electron path and forming a practically continuous spectrum of uniform intensity. This spectrum extends to the frequency $(E/mc^2)^3 v$, where E indicates the electron's

Atomic and Molecular Physics

(a)

(b)

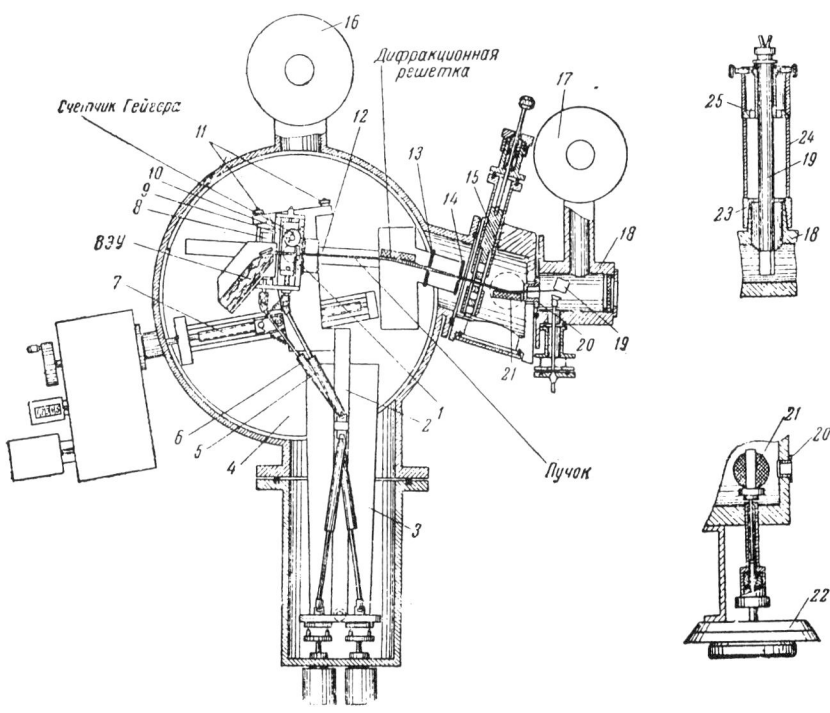

Figure 13.6. *Diagram and photograph (with T M Zimkina) of the x-ray apparatus designed by A P Lukirskii in about 1960, which extended the range of x-ray absorption measurements down to ~50 eV.*

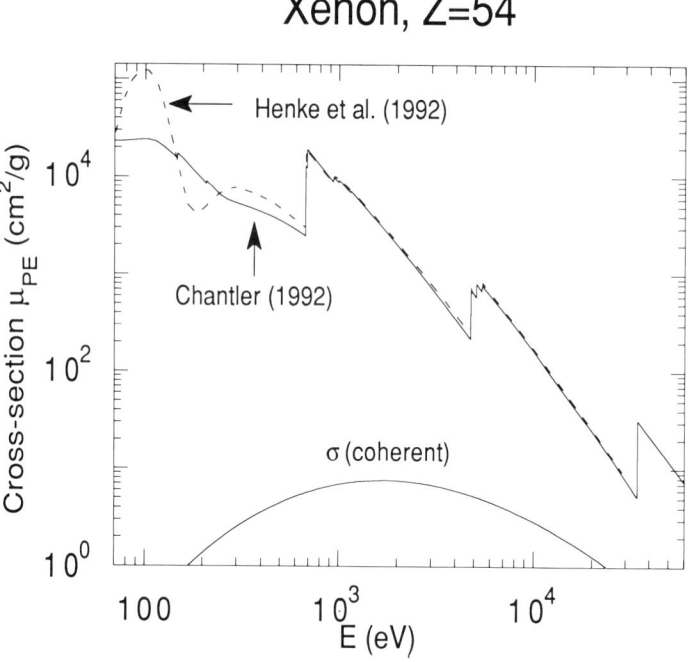

Figure 13.7. *Absorption coefficient of Xe gas as a function of photon energy, showing the sharp onsets of the K, L, M shell contributions (in order from the right-hand side). The units of ordinates are reciprocal to the gas layer thickness expressed in g cm^{-2}. The partial contribution of Rayleigh scattering is indicated by σ.*

kinetic energy and m its mass. (The ratio $(E/mc^2)^3$ amounts to $\sim 10^{12}$ for $E = 5$ GeV.) This remarkable emission, long anticipated by theorists, was first detected in the late 1940s, and applied in 1953 to detect the photoabsorption by K electrons of beryllium at and beyond 110 eV.

The process of perceiving the full impact of synchrotron light spectroscopy and providing for its conversion into a basic tool occupied a whole decade. The striking detection by this method of a novel Rydberg series of He absorption lines near 60 eV in 1963 (figure 13.8) was, however, quickly followed by very extensive experimentation, mapping out the main features of photoabsorption spectra throughout the former gap, as reported comprehensively in a 1968 review [20]. These accomplishments required considerable technical developments for handling and measuring radiation in the new range of frequencies, such as improvements in the reflectance of mirrors and gratings and in the adaptation of photographic emulsions.

The photoabsorption and photoionization spectra of atoms and molecules thus obtained in the photon range of 10–1000 eV depart from the simple hydrogen-like pattern displayed in figure 13.7 in several

Atomic and Molecular Physics

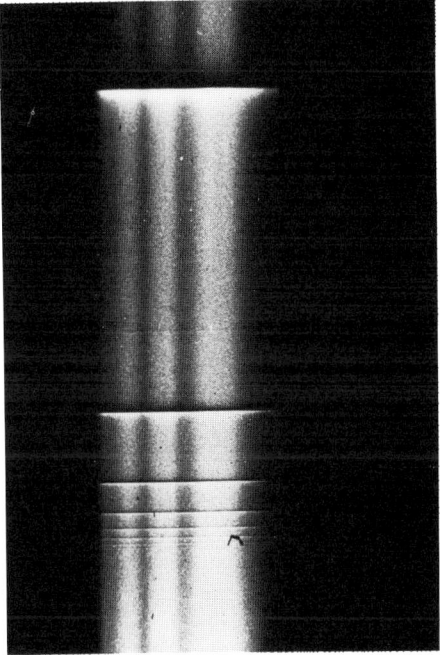

Figure 13.8. *Photoabsorption spectrum of He gas in the range ~60–65 eV (190–210 Å) showing a Rydberg series of resonances, generated by excitation to doubly excited $sp^1 P$ states.*

respects, as illustrated in figure 13.9:

(i) The rise of photoionization at the threshold for ejection from any one shell is often delayed and smoothed out by centrifugal barriers hindering the escape of slow photoelectrons.

(ii) The peaks of photoabsorption above each threshold are often broadened by the mutual repulsions between the numerous electrons of p, d or f subshells.

(iii) Characteristic deep intensity minima occur generally in the photoabsorption by electrons of incomplete shells. (This phenomenon, observed in alkali spectra since 1928, had been viewed as accidental; an indication of it in the Ar spectrum of reference [18] accordingly surprised its discoverers.)

(iv) Ubiquitous peaks often rise above the background of photoionization spectra as in figure 13.9.

13.3.2. *From the near infrared to the microwave range*

Extension to the infrared of the usual spectroscopy techniques meets difficulties. The lower-energy photons fail to activate usual types of

1053

Twentieth Century Physics

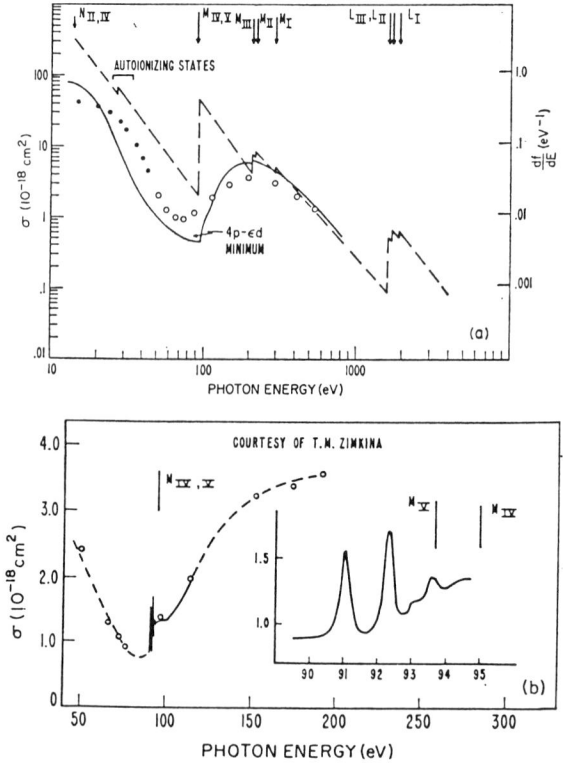

Figure 13.9. *Photoabsorption spectrum of Kr gas near the onset of the $M_{IV,V}$ (i.e. 3d) showing subshell contributions. (a) ○ and ●, experimental values; – – –, hydrogenic approximation; ——, one electron model. (b) Details near the $M_{IV,V}$ threshold.*

photographic or photoelectric detectors; detection relied instead on observing the small temperature rise of 'bolometers', small black objects that absorb the incident radiation energy. Energy emission by intra-atomic currents acting as antennas is reduced in proportion to the cube of their oscillation frequency; its detection thus requires instruments with unprecedented sensitivity. The longer wavelength of infrared radiation boosts its penetration into mirrors or diffraction gratings, thus increasing energy dissipation.

The generation, manipulation and measurement of far infrared radiations of specified frequency and observation of their action hinged therefore on developing novel techniques. Three principal innovations have served this purpose progressively since the early 1960s, namely, laser sources of radiation (Chapter 19), analysis by Fourier-transform spectroscopy, and photoconductor detectors with thresholds in the far infrared (up to ~30 μm wavelengths).

Laser sources of infrared radiation were initially available at specific frequencies. The desired frequency selection was then attained indirectly

Atomic and Molecular Physics

through pairs of tunable dye lasers, whose beat (i.e. difference) frequency can scan the whole infrared range. An alternative consists of 'beating' the precise frequency of CO_2 or CO infrared lasers with the precise frequencies of electrically generated microwaves ($\simeq 10^{12}$ Hz). More recently, tunable lasers have been introduced, covering most of the infrared range.

Fourier-transform spectroscopy served initially to convert into frequencies, using computers without loss of accuracy, measurements of infrared wavelengths that were then more readily accessible by high-precision interferometry [21]. Direct and precise measurements of frequencies have become available more recently, including the generation of very high harmonics of microwave frequency standards.

Application of the principles outlined above have increased both the sensitivity and the resolution of measurements by several orders of magnitude. Thereby they afforded discoveries of molecular spectroscopy as will be reported in sections 13.6 and 13.8. Note that infrared radiation is hardly absorbed by electrons of single atoms in their ground state or of most substances other than metals. Its basic action consists of exciting vibrational and rotational motions of free molecules (and of their analogues in condensed matter). These excitations are restricted by the selection rules indicated in section 13.2.5.2 for single molecules. The restrictions are, however, bypassed in dense gases, through the effect of very frequent collisions that perturb the internal mechanics of each molecule. For example, single H_2 molecules have no independent electric dipole moment on which radiation may act, but absorb and emit radiation very appreciably when impacted by other molecules at high density or when polarized by external fields.

An additional related aspect of the infrared radiation's action in dense gases results from the enhanced ability of each molecule to transfer its excitation energy to surrounding molecules very rapidly. Retention of excitation by any molecule for only a brief interval Δt implies a reciprocal broadening $\Delta \omega \sim 1/\Delta t$ of its characteristic (resonant) absorption frequency ω. This broadening progresses, with increasing gas density, to the point of changing the aspect of absorption spectra radically, as illustrated in figure 13.10. It also lowers progressively the mean $\langle \omega \rangle$ of the frequencies thus absorbed.

13.3.3. The spectrum of radiation actions
The macroscopic action of radiation, with frequency ω and wavenumber k, on bulk matter is represented by the dielectric response function $\epsilon(\omega, k)$, often simply $\epsilon(\omega)$. A gaseous assembly of atoms or molecules is, of course, an example of bulk matter. The dielectric response $\epsilon(\omega, k)$ is a function of complex variables: its real part represents the electric susceptibility χ, i.e. the ratio of the matter's electric polarization to the inducing field strength E; its imaginary part represents the conductivity

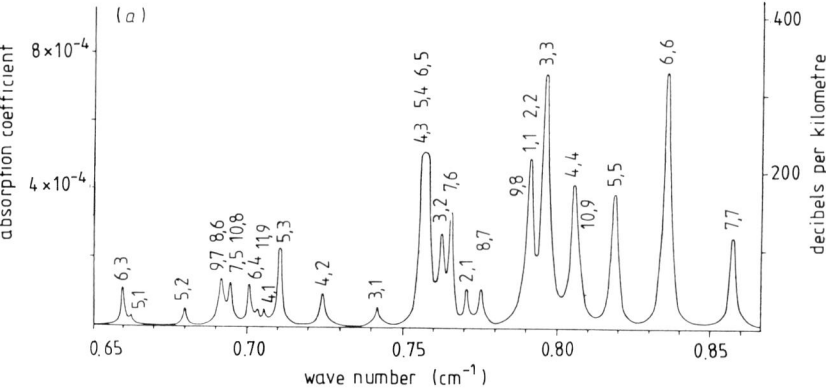

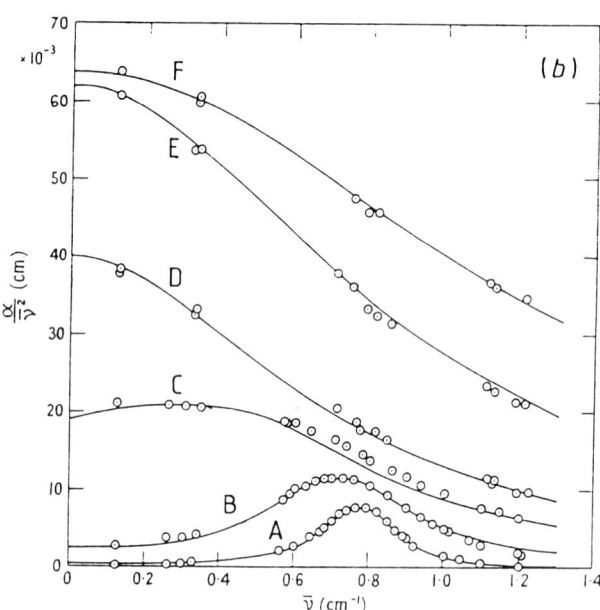

Figure 13.10. *Photoabsorption spectrum generated by excitation of the 'inversion' vibration of the ammonia (NH$_3$) molecule, in which the nitrogen atom oscillates back and forth through the triangular structure of the hydrogen atoms, with a frequency of 24 GHz (photon energy 10^{-4} eV). (a) In the gas at low density the oscillation frequency depends on the rotational state of the molecule, giving rise to manifold absorption peaks. (b) at increasing pressure the spectral lines are broadened by intermolecular collisions that interrupt the orderly vibrations. The entire spectrum shown in (a) has coalesced into the single broad absorption of curve A; at pressures increasing to 6 atmospheres the spectrum evolves into the curve F. The entire evolution of this spectrum has since been represented by a three-parameter formula.*

σ, i.e. the ratio of the current density to the inducing E. The dependencies of χ and of σ on (ω,k) are interrelated. Either of them has a peak slope where the other has a peak value.

The energy dissipated by radiation is proportional to $\sigma(\omega,k)$. The study of this parameter, especially in the range of high photon energies $\hbar\omega$, received great support, about and after mid-century, motivated by the need to shield workers, or even whole populations, from nuclear radiations. Here we survey briefly the spectral dependence of $\epsilon(\omega)$.

Radiofrequency energy is dissipated in conducting materials, i.e. metals, electrolytes and ionized gases (plasmas), mostly through collisions of free particles oscillating in the field. Similarly the rotation of polar molecules (i.e. molecules with electric dipole moments) is excited by radiation and then dissipated through collisions [22]. This process extends to non-polar molecules at high densities where polarity is induced by collisions, as noted in section 13.3.2.

Oscillating currents induced by radiation in non-resonant atoms or molecules act as antennas emitting in all directions (Rayleigh scattering), but these emissions combine to yield a net effect only when their density is non-uniform within a radiation wavelength. Non-uniform density fluctuations in the upper atmosphere thus combine to scatter blue light from the Sun, a process that redistributes radiation energy without absorbing it.

Resonant absorption of radiation energy by atoms and molecules of various species sets in progressively in the infrared, visible or ultraviolet frequency ranges, as described earlier and illustrated in figure 13.11 for H. Departures of this familiar process from the simple pattern of H astride its absorption threshold have been described in section 13.3.1 and are illustrated in figure 13.9. These departures subside, however, at higher frequencies ω; on the other hand the intensity of photoabsorption drops rapidly as ω increases because this process conserves both energy and momentum only within smaller and smaller volumes about the nucleus.

Radiation actions other than photoabsorption become appreciable at photon energies in the x-ray (multi-keV) range whose wavelengths are comparable to single-atom diameters. Here the electron density varies rapidly within a wavelength and non-resonant electron oscillations redistribute radiation energy in a process of elastic (coherent) scattering. This process gradually overshadows photoabsorption as $\hbar\omega$ and the atomic number increase, being increasingly accompanied by the 'incoherent' Compton scattering, also introduced in section 13.2.2.1. Finally, the pair-production process—also introduced in section 13.2.2.1—emerges in the MeV photon range, becoming eventually the main agent of radiation-energy dissipation. The comparative contributions to energy dissipation in different materials and at different frequencies by these alternative processes have been the object of much study, as

13.3.4. Spectral connection at ionization thresholds—sum rules

The onset of resonant photoabsorption by atoms, starting from lower frequencies, occurs when the photon energy $\hbar\omega$ matches the difference between the atom's ground-state binding energy E_g and that of the lowest excited level, that is, at

$$\hbar\omega = E_g - E_n \qquad (n > 1). \tag{6}$$

The value of E_n has been indicated in section 13.2.3.1 as $13.6\ \text{eV}/n^2$ for the H atom, and $13.6\ \text{eV}/(n - \mu_l)^2$ for alkali atoms. In either example the values of E_n converge rapidly to zero as n increases, indeed the photoabsorption spectra of atoms consist generally of a Rydberg series of lines converging to the limit, as $n \to \infty$, $\hbar\omega_n \to E_g$, beyond which the excited electron escapes from the ionic residue with kinetic energy $\hbar\omega - E_g$. Photoabsorption extends beyond this ionization threshold declining rapidly with increasing frequency as described above (section 13.3.3).

The qualitative difference between the discrete line spectrum below the ionization threshold and the absorption continuum above it contrasts with the basic continuity of the photoabsorption through the threshold illustrated in figure 13.11. This continuity, demonstrated mathematically by theorists in about 1960, is implied by the representation of photoabsorption through the imaginary part of the single function $\epsilon(\omega, k)$.

The magnitudes of photoabsorption in the discrete and continuum ranges, related to parameters of the $\epsilon(\omega)$ singularities, have been represented by numerical indices called 'oscillator strengths' since the earliest days of atomic physics. The oscillator strength of the nth line of the discrete spectrum, f_n, equals the ratio of its photoabsorption to that of an electron held near to equilibrium by a spring with force constant $k = m\omega_n^2$. The corresponding ratio applies to the oscillator strength of absorption in the continuum frequency range $\delta\omega$ about ω, $(\mathrm{d}f/\mathrm{d}\omega)\delta\omega$.

The photoabsorption (at $k \sim 0$) integrated over the whole spectrum of any atom or molecule, expressed in the scale of oscillator strengths, equals its number of electrons

$$\Sigma_n f_n + \int_{E_g/\hbar}^{\infty} \mathrm{d}\omega\, \mathrm{d}f/\mathrm{d}\omega = Z. \tag{7}$$

This famous 'Thomas, Reiche, Kuhn' sum rule guided Heisenberg's development of quantum mechanics. Its significance emerges by considering that the integral over the spectrum represents the action of an

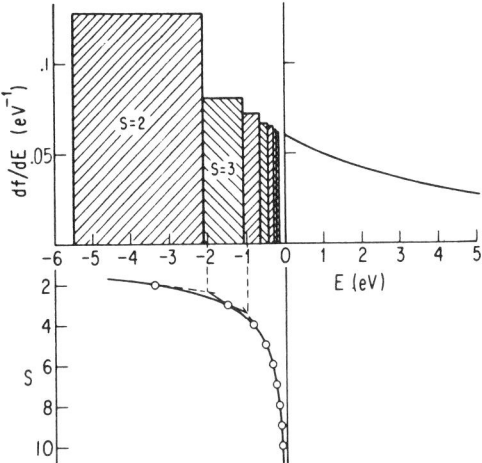

Figure 13.11. *Contributions to the photoabsorption by atomic H by its excitation to the series of discrete energy levels represented by the histogram and to the continuous spectrum of ionized states. The lower part of the diagram relates to the construction of the histogram.*

intense pulse of negligible duration, in which case each electron absorbs energy independently of all others. Useful sum rules analogous to (7) represent the spectral averages of various powers of the frequency [20].

Verification of equation (7) by experimental values of f_n for each material places a stiff requirement on the accuracy and completeness of photoabsorption measurements. An initial attempt to verify (7) for Al metal, in about 1967, fell short of the mark; critical analysis of the experimental data obtained with synchrotron light identified an instrumental defect. Successive data achieved agreement. Verification for other materials is now acknowledged as an important check of photoabsorption measurements, applied with success to elements with Z values up to ~30. Application to higher Z substances is complicated by difficulties in separating photoabsorption adequately from other processes, and by breakdown of the decay law $\omega^{-1-(7/2)}$ for photoabsorption anticipated qualitatively in section 13.3.3.

13.4. Excitation channels and resonance effects

The atomic excitation spectra described in section 13.2.3.1 pertain to unusually simple circumstances, in which a single electron moves about a spherically symmetric ion. Excitations of most atoms involve electron motions about ions with lower (or no) symmetry, often also joint excitations of two or more electrons in which the ion itself is also excited. In the case of molecules, the ionic core is apt to rotate and vibrate thus exchanging energy and angular momentum with external electrons.

Twentieth Century Physics

The analysis of excitation spectra thus involves additional concepts and parameters that constitute the subject of the present section.

The common element of atomic or molecular excitations lies in energy that allows electrons, or other constituents, of a system to separate into fragments. The excitation energy may or may not suffice for the fragments to separate completely; full separation amounts to an ionization or a dissociation of the initial system. Alternative modes of fragmentation are called 'channels', following a custom of nuclear physics. Different stationary states (energy levels) of each channel differ in their total energy but are otherwise characterized by single values of the total angular momentum and total parity and by specified allotments of energy, angular momentum and parity among the fragments.

This allotment is, however, no constant of the motion, since interactions between fragments afford exchanges of energy, angular momentum and parity, especially when fragment pairs approach one another in the course of their motion. Thus, for example, a molecular electron may exchange energy with the ion's rotation, one of many possible effects of 'channel coupling'. The identity of each channel is thus manifest in its stages of large fragment separation, but each channel is *not* independent of other channels. The resulting shuffle of single-channel spectra obscures the simple pattern of Rydberg series described in section 13.2.3.1. However, much of this shuffle has been unravelled since the 1960s, through a major development of atomic theory.

This development started with the qualitative analysis of simple phenomena, to be outlined in section 13.4.1, observed in detail by novel instrumentation. Soon thereafter, in the mid-1960s, M J Seaton developed an analytic treatment of exchanges of angular momentum and of modest amounts of energy between a single electron and an atomic ion core. This treatment was intended for numerical *ab initio* evaluation of its parameters under simple circumstances. Its structure, on the other hand, soon proved applicable to an expanding range of spectra, through empirical fitting of parameters. Its main features will be described in the following subsections. The current need for its further extension will be discussed in section 13.10.

13.4.1. Prototype phenomena

A prototype of channel classification, shown in figure 13.12, rests on the energy levels of an ionic residue, He^+, omitting specification of angular momentum and parity allotment between this ion and an electron. Analysis of channel couplings centres thus on further identification and evaluation of relevant parameters. The spectrum of photoabsorption leading to excitation of He from its ground state into the cross-hatched block of the second row in figure 13.12 has been displayed in figure 13.8. The succession of intensity peaks in that spectrum corresponds to a

Atomic and Molecular Physics

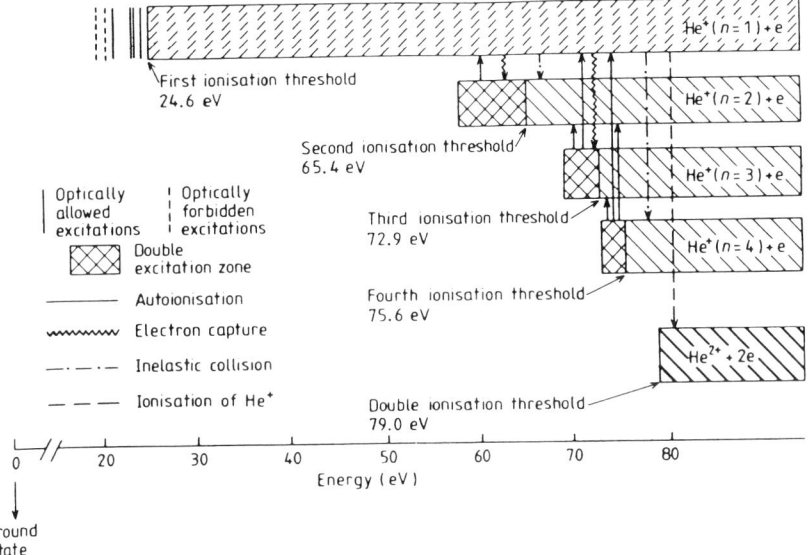

Figure 13.12. *Diagram showing an increasing number of channels in the spectrum of helium and their interconnections by auto-ionization and other processes.*

succession of doubly excited levels classed as $^1P^o$ (singlet pairing of electron spins, total orbital momentum $L = 1$, odd parity). The width and parity of each peak reflect the strength and mode of its coupling to the top channel of figure 13.12, namely, to separate fragments $He^+(1s)$ + free electron with orbital momentum $l = 1$. The relevant parameters are described in the following subsections. Roughly, the spectral width $\Delta\omega$ of each peak, in frequency units, is reciprocal to the time Δt spent by the atom in a doubly excited state of the second row before decaying into an ionized state with equal energy of the first row in figure 13.12.

The inverse process of this decay starts with the collision between an electron with kinetic energy of 35–40 eV and $l = 1$, and a helium ion He^+ in its ground state. The energy range specified here corresponds, in the first row of figure 13.12, to the cross-hatched portion of the second row. If the electron energy happens to match one of the energies of the peaks of figure 13.8, within its peak's width, the first-row ionized state is said to 'resonate' with the doubly excited state of the second row and can transform into it.

Such resonance phenomena were known previously at lower energies but their abundance, spread and sharpness had been widely underestimated. The first glimmer of the actual occurrence of resonances emerged two months prior to the observation of the spectrum shown in figure 13.8, in a measurement by G J Schulz of the elastic electron collisions with He atoms in their ground state, $e(\sim 20 \text{ eV}) + \text{He}$ [24]. The

1061

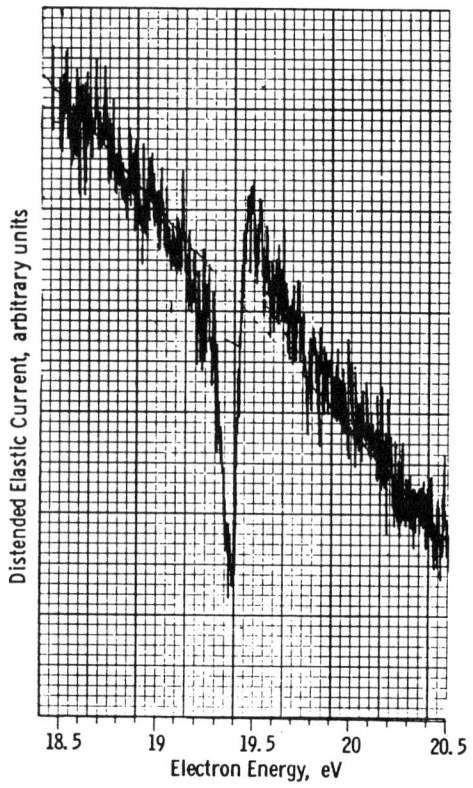

Figure 13.13. *Resonance at 19.37 eV in the spectrum of electrons deflected by 72° in elastic collisions with He.*

initial state of this collision belongs to the first row of the analogue of figure 13.12 for negative ions He⁻. The sharp oscillation in the number of 19.37 eV electrons deflected by 72°, displayed in figure 13.13, implies the temporary ($\Delta t \sim 3 \times 10^{-12}$ s) formation of a discrete resonant level of He⁻, a level classified as $1s2s^2$ ^2S whose existence had been anticipated theoretically.

Schulz's observation, with its newly achieved resolution of the energy scale, was soon enlarged by J A Simpson with a more versatile instrument of equivalent resolution that detected *doublet* resonances in negative ions of all other noble gases. A significant aspect of this discovery lay in the coincidence of the doublet spacing of each *negative* ion resonance with the spacing of the ground-state doublet of the *positive* ion of the same element, implying that the resonant negative ion consists of a positive ion combined with a spinless pair of excited electrons. The versatility of Simpson's instrument yielded a steady stream of new resonance detections, in molecules as well as atoms, throughout 1963 and 1964.

A large amount of knowledge on atomic processes emerged from

Atomic and Molecular Physics

extensions of these pioneering observations of resonances. In particular, the, then novel, subject of negative ion resonances has recently been reviewed [25]. We shall return to this broader scope after introducing relevant theory.

13.4.2. Quantum-mechanical parameters
Our initial description of atomic electron states, in section 13.2.3.1, was cast in terms of radial and angular density distributions, represented as squared moduli of angular $[Y_{lm}(\theta, \varphi)]$ and radial $[f_{nl}(r)]$ 'density functions'. Such functions, generally complex valued, serve to represent electron densities and other observable quantities, much as complex amplitudes of light waves serve to represent their alternative (linear, circular or elliptical) polarizations. Light wave amplitudes are measurable, at least in principle, in macroscopic experiments; the quantum density functions are instead mathematical symbols, called 'probability amplitudes'—or more often wavefunctions—because their squared moduli serve only to represent probabilities of alternative observations of atomic phenomena, as described in Box 13F.

The number of excitation channels of any multi-particle atom or molecule relates to the number of its degrees of freedom and may thus be very large, even though finite. In practice, however, this number has been modest in most of the situations studied thus far, where many parameters remain 'frozen', barring very large energy inputs such as required, for example to destroy a whole atomic structure. Nevertheless the following developments should be viewed as being circumscribed by practical, if often implied, schematizations.

13.4.2.1. Long- and short-range interactions—threshold effects. The interactions that underlie the transitions of fragments between channels extend over a range of inter-fragment distances, even though they vanish at large separations. This distinction is particularly obvious for the simple case of single-electron fragments, where the spherically symmetric Coulomb field of an ionic core predominates at large radial distances r over the dipole, quadrupole, and higher order 2^n-pole field components generated by core asymmetries, whose potentials decay as $r^{-(n+1)}$. Channel coupling stems entirely from multi-pole components. This section will deal primarily with a single-electron fragment and more briefly, in separate subsections, with multi-electron fragments and molecular dissociations.

Even though channel coupling receives contributions from interactions over a range of radial distances, it was found conceptually convenient to parametrize its effect by 'short-range transition matrices', acting at or near core surfaces, in contrast to matrices analogous to equation (8) that represent relationships at the limit of large radial distances. We have

BOX 13F: TRANSITION AMPLITUDES

As a prototype demonstration of probability amplitude manipulation, we describe here the amplitude that forms the core of the cross section (4) for scattering of an electron incident along a z axis into a direction (θ, φ), namely, $\Sigma_l(2l+1)^{1/2} \exp(i\delta_l) \sin \delta_l(k) Y_{l0}(\theta, \varphi)$. Its first factor $(2l+1)^{1/2}$ represents (to within irrelevant factors) the probability amplitude for the incident electron to enter a state with orbital index l, an amplitude indicated in standard notation by $(i|l)$. The second and third factors, $\exp(i\delta_l) \sin \delta_l(k)$, represent a 'transition element', i.e. the probability amplitude $T_l(k)$ for the electron to rebound from a collision with parameter l into a direction other than its initial one. The last factor $Y_{l0}(\theta, \varphi)$, represents the probability amplitude of the electron rebounding in a specified final direction (f) forming an angle θ with the incidence. Thus one writes the probability amplitude formula for the specified collision as

$$(i, \theta_i = 0 | T | f, \theta_f) = \Sigma_l (i|l) T_l(k) (l|f, \theta_f)$$
$$\propto \Sigma_l (2l+1)^{1/2} \exp(i\delta_l) \sin \delta_l(k) Y_{l0}(\theta, \varphi). \qquad (8)$$

The set of amplitudes (8) for all pairs of initial and final directions (i, f) forms a square array, of infinite dimension, familiar in mathematics as a matrix, specifically in this case a *transition matrix*. It will serve us as a prototype for representing the connections among all pairs (i, f) of excitation channels.

The number of excitation channels to be considered in the next sections is generally finite, in contrast to the number of channels i and f of the amplitude (8). In practice, however, this amplitude also involves only a finite number of independent channels because the phase shifts $\delta_l(k)$ vanish generally for larger (e.g. two-digit) values of the orbital index l. Expansion of matrices $(i|T|f)$, connecting alternative excitation channels, into sums of terms with indices analogous to the orbital l of equation (8) is generally possible, but the physical significance of such indices is not apparent at this point of our story. Incidentally, however, we shall deal here with sets of channels having a common value of the total angular momentum index J, analogous to the index l summed over in equation (8).

previously used this approach implicitly in section 13.2.3.1 where the dependence of outer electron levels upon the structure of a multi-electron spherical core was incorporated into a single quantum defect parameter μ_l. A key feature of short-range parameters lies in their dependence on energies at a $\sim$10 eV scale, contrasting with long-range interactions whose effects are sensitive to energies near ionization thresholds at a sub-1 eV scale. Short-range parameters depend instead on energy only on the scale of potential and kinetic energies prevailing at short ranges. This concept, as well as most of its following elaboration and extensions, stems from the work of M J Seaton, often known as 'multi-channel quantum defect theory' [26, 27], partly outlined in Box 13G.

13.4.2.2. Multi-electron fragmentation. This subject was pioneered by G Wannier in a 1953 extension [28] of Wigner's threshold laws, dealing with the threshold for ionization of atoms by electron collision. The final state of this process has a pair of electrons emerging from an ion's field with near-zero energy; Wannier recognized that their joint escape requires the electrons to share their combined energy E above threshold evenly by escaping with nearly equal velocities on opposite sides of the atom. This requirement restricts the probability of joint escape severely; its analysis by classical mechanics identified a tiny bundle of suitable trajectories, accessible with probability $E^{1.127}$, the exponent being the root of a quadratic equation.

This famous result met with widespread resistance by theorists, even after its quantum-mechanical reformulation in 1970, until verified in 1973 by the brilliant experiment of Read and Cvejanovic [29]. A single electron emerging from the collision was detected, with energy ϵ <0.02 eV, as a function of the incident energy $E+24.58$ eV, 24.58 eV being the ionization threshold of the He target. Thereby only a fraction $0.02/E$ of double ionizations could be detected, with a net probability $E^{0.127}$ far easier to identify than the departure of double ionizations from a linear law. Figure 13.14(*a*) displays the results, extending to incident energies $E < 0$. The dip of the scored electrons astride $E = 0$ reflects the low probability of the two emerging electrons sharing their energy nearly evenly. The best fit shows the detection to be proportional to $E^{0.131}$.

The metastability of the approach to double escape identified by Wannier seems paradoxical: whereas the configuration of the two electrons on opposite sides of the ion is stabilized by their mutual repulsion, their maintenance of nearly equal velocities (and equal distances from the nucleus) during the escape seems wholly unstable. Indeed the pair's potential *peaks* at equal radial distances, $r_1 = r_2$, for a constant value of $R^2 = r_1^2 + r_2^2$. Stability is actually ensured by steady motion along R, as implied by Wannier's treatment and confirmed by later developments, but is nevertheless less than obvious.

This stability was then verified by another experiment in Read's laboratory [30]. This experiment measured the production of metastable excited states of He, at $\sim$20 eV above the ground state, by incident

BOX 13G: LONG-RANGE AND THRESHOLD EFFECTS

Seaton's analysis deals with the radial density function (wavefunction) of a single electron in an ion's Coulomb field, initially in the energy range where the fragmentation proceeds to infinite distance. We indicate this function here by $f_{kl}(r)$ as in section 13.2.4.2. The electron's energy in excess of the ionization threshold amounts to $13.6k^2$ eV if the wavenumber k, corresponding to the wavelength $2\pi/k$ of $f_{kl}(r)$'s oscillations at $r \to \infty$, is expressed in atomic units reciprocal to the radius a of equation (1). Specifically the leading term of $f_{kl}(r)$ at large r includes a main factor

$$\sin[kr + (\ln r)/k + \eta_l(k) + \delta_l(k)] \tag{9}$$

whose phase element η_l reflects a long-range influence whereas $\delta_l(k) = \pi \mu_l(k)$ reflects short-range influences, as it did in section 13.2.4.2.

The centrepiece of our parametrization connects the sine factor of $f_{kl}(r)$ with the corresponding factor of the functions $f_{kl}(r)$ pertaining to energy levels *below* the ionization threshold, as indicated in section 13.3.4. These levels, negative when measured from the threshold, were called 'binding energies' with positive values $13.6/(n - \mu_l)^2$ eV in section 13.2.3.1. A negative value of the energy, $13.6k^2$ eV implies an imaginary value of the wavenumber k. This shift to imaginary k affects, in equation (9), the short-range parameter $\delta_l(k)$ hardly at all, but changes the trigonometric dependence $\sin kr$ into a hyperbolic dependence $\sinh |k|r$. Since $\sin kr$ can be represented as the difference of two complex exponential functions, $\exp(\pm ikr)$, each of them becomes real, $\exp(\pm r/\nu)$, if we set $k \to i/\nu$. Expression (9) thus becomes

$$b e^{r/\nu} r^{-\nu} \sin \pi(\nu - l + \mu_l) - b^{-1} e^{-r/\nu} r^{\nu} \cos \pi(\nu - l + \mu_l) \tag{9'}$$

with coefficients b smoothly dependent on ν. The obvious condition, that the probability amplitude $f_{nl}(r)$ remains finite as $r \to \infty$, implies that the value of $\nu - l + \mu_l$ equals *an integer*, as anticipated by the binding energy formula 13.6 eV$/(n - \mu_l)^2$.

Physically, the radial probability distribution $|f_{nl}(r)|^2$ of an electron bound to an ion consists of an *integer* number of lumps separated by 'nodes' of the wavefunction $f_{nl}(r)$.

CONTINUED

Atomic and Molecular Physics

> **BOX 13G:** *CONTINUED*
>
> Smoothness of phenomena across fragmentation thresholds generally applies only to the photoionization of a single electron, and to the analogous particle–ion collisions. Extension to fragmentations with alternative interactions by C H Greene *et al* (cited in references [26, 27]) displays a variety of critical behaviours.

electrons of 20–25 eV. The yield of these excitations, plotted against the incident energy in figure 13.14(*b*), shows a series of peaks diagnosed as doubly excited states of He$^-$ with the pair of excited electrons at comparable distances from the nucleus ($r_1 \sim r_2$). This region of high potential energy has long been called the 'potential ridge' of two-electron systems.

The study of two, or more, electron excitations, of threshold phenomena, and of stability on the potential ridge has drawn extensive effort ever since the mid-1960s. Section 13.11 will return to this subject in a broader context.

13.4.2.3. Molecular dissociation. The dissociation of molecules, into neutral or charged fragments, is analogous to ionization in that it occurs at the limit of a succession of excited states, vibrational excitations in this case. This process has been treated in the framework of quantum defect theory by R Colle.

The succession of vibrational levels terminating at the dissociation threshold is finite, at variance with the infinite Rydberg series of electronic spectra, reflecting the difference of the relevant potentials. The vibrational spectrum has been studied in detail up to threshold in the 1980s mainly for H_2 and HD by A Carrington; photoexcitations to the continuum are beginning to be studied by multi-photon absorption.

The Born–Oppenheimer approximation that governs the vibrational motion of molecular nuclei becomes insecure at the very threshold of dissociation, because the electrons no longer move faster than nuclei in the tails of their distribution functions which are separating out. We will return to this subject in sections 13.6 and 13.11.

13.4.3. Multi-channel formulation—resonances

The theoretical elements introduced in section 13.4.2 can be used, in principle, to treat all multi-channel phenomena. In practice *ab initio* applications have been confined to phenomena with no more than two electrons outside an ionic core. Empirical determination of short-range

Twentieth Century Physics

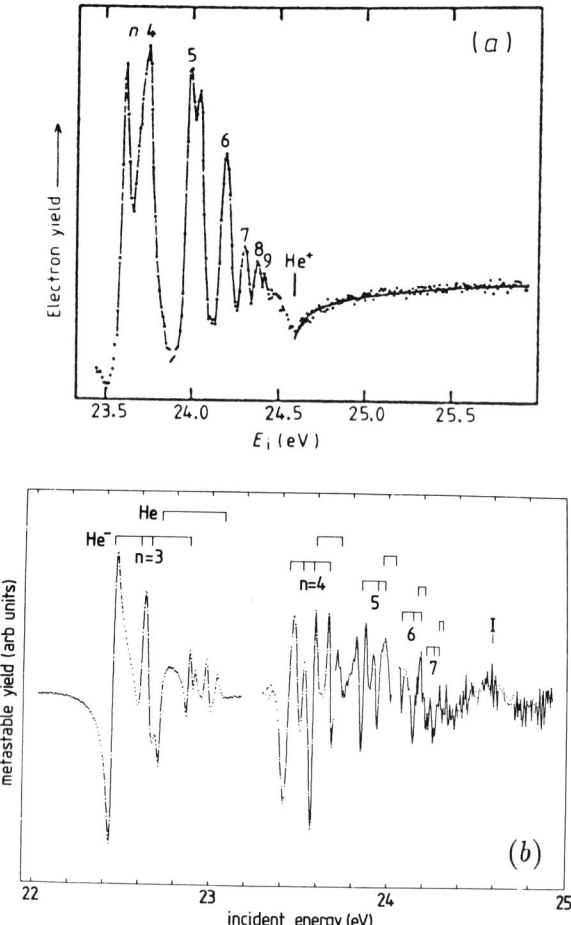

Figure 13.14. *Novel experimental features in the inelastic collision spectrum of 20–25 eV electrons with helium: (a) Yield of very slow electrons emerging from the collision. (b) Yield of metastable excitations of He.*

parameters, to be introduced in section 13.4.3.2, has afforded broader applications.

The main limitation to *ab initio* studies lies in the current want of proficiency in dealing with more than two electrons over radial distances far exceeding those of ionic cores. The problem lies in calculating the relevant transition matrices, $(i|T|f)$, or their equivalent 'R matrices' [31], for larger sets of open channels; these parameters must encompass relevant effects of electron interactions outside the core. Seaton's compendium [27] calculates R matrices within the framework of collision theory, i.e. by generalizing section 13.2.4.2 to include excitations of a single target electron to alternative discrete

1068

energy levels. C H Greene and collaborators have refined more recently effective variational calculations of R matrices in diagonal form [32], that is, with eigenvalues $\tan \delta_{\alpha J}$ (or $\tan \pi \mu_{\alpha J}$) analogous to the phase shifts δ_l of equation (8), and eigenvectors $(i|\alpha J)$ that connect the channels αJ to alternative specified channels i. The calculation utilizes sets of independent-electron orbitals in the core range, i.e. within a sphere of radius r_0, treating electrons of closed shells as forming a spherical potential $v(r)$, as was also done by Seaton. A limitation lies in formulating appropriate boundary conditions for electrons reaching the radius r_0. Analogous, but far more extensive, R-matrix calculations have been performed by P G Burke and a large number of collaborators over the last 20 years [33].

An essential aspect of channel coupling effects lies in the occurrence of 'closed' ionization or dissociation channels $|f\rangle$, whose threshold energy for fragmentation exceeds the initial $E + E_g$. The diverging term of the factor (9') of $|f\rangle$'s density functions, namely, the term with the factor $\exp(r/\nu)$ should, of course, be removed as was done in that single channel example by requiring $\nu - l + \mu_l$ to take an integer value. At variance with removal of the analogous term $\exp(-ik_f r)$ for $E_{ft} < E + E_g$, mentioned in Box 13H, the elimination of each $\exp(r/\nu_f)$ requires a specific concurrence of all terms of the $\Sigma_{\alpha J}$ in equation (10). The appropriate procedure, to be introduced in section 13.4.3.1, constitutes the source of all spectral *resonances* and thereby a central element of the multi-channel quantum defect theory.

13.4.3.1. *Photoabsorption by molecular hydrogen near its ionization threshold. A '2-3 channel' example.*

The workings of multi-channel theory came into sharp focus in 1969 through the successful analysis of a single experiment, even though the channel and resonance concepts had been previously familiar and Seaton's formalism was complete. G Herzberg aimed at observing a line spectrum converging to the ionization threshold of H_2, in order to establish the threshold's position very accurately. To this end he minimized the influence of molecular rotations and vibrations by studying samples of H_2 in its state of lowest energy, boiling off a low-temperature liquid [34]. His goal was, however, frustrated by the irregularity of the spectrum actually observed (figure 13.15(b)).

The origin of the irregularity, soon apparent to Herzberg, is indicated in the level diagram next to the spectrum: photoabsorption actually excites two alternative short-range (αJ) channels of H_2 with $J = 1$, each of them coupled to two different long-range channels, with ionization thresholds (E_{t0}, E_{t2}). The short-range channels correspond to excitations polarized along (σ) and across (π) the molecular axis; this distinction persists with increasing radial distance as long as the electron's motion about the core, with orbital index $l = 1$ (p state), remains anchored to

BOX 13H: R-MATRIX APPLICATIONS

One obvious application of R matrices consists of extending the elastic cross section (4) to include the inelastic channels of interest. For any given energy E and direction of an incident electron i, one determines its relevant admixture of channels $|\alpha\rangle$ with angular momenta J (including spins) of the complex formed by the electron + target combination $(i|\alpha J)$. These are then combined with the R matrices to form the transition probability amplitudes

$$(i|T|f) = \Sigma_{\alpha J}(i|\alpha J) \exp(i\delta_{\alpha J}) \sin \delta_{\alpha J}(\alpha J|f) \qquad (10)$$

to alternative final channels $|f\rangle$. The set $\{|f\rangle\}$ should include all channels that have at least one energy level lower than the sum of the target E_g and the incident E.

Here, as in equation (4), it is understood that the density functions of incident channels $(i|$ include only the 'ingoing' complex portion $\exp -i(k_i r + \ldots)$ of the sine factor (9), and those of $|f\rangle$ only the 'outgoing' portion $\exp i(k_f r + \ldots)$. These conditions are fulfilled automatically by constructing the coefficients $(i|\alpha J)$ and by using the transition element in equation (10).

A second basic application of R matrices deals with the photoabsorption of radiation with frequency ω by a target with ground-state energy E_g and with multi-channel excitations. The photoabsorption event proper takes place within the volume of radius r_o included in the R-matrix calculation. The probability amplitude for excitation into an eigenchannel $(\alpha J|$ of the R matrix with energy $E_g + \hbar\omega$ if often indicated by $D_{\alpha J}$. (The value of J is the vector sum of the ground-state momentum J_g and of the angular momentum, generally unity, of the absorbed photon.) The probability amplitude for excitation into the channel f with energy $E_g + \hbar\omega$ is then represented by a formula akin to (10), namely

$$(i|D|f) = \Sigma_{\alpha J} D_{i\alpha} \exp(i\delta_{\alpha J}) \sin \delta_{\alpha J}(\alpha J|f) \qquad (10')$$

if $E_g + \hbar\omega$ exceeds the fragmentation threshold E_{ft}. This expression needs modifying if $E_g + \hbar\omega$ falls short of E_{ft}, as it does in the case of collisions, by a procedure outlined in the example of section 13.4.3.1.

---

the molecular axis by a quadrupole potential that decreases as r^{-3}. At larger distances the electron motion is instead controlled by its sharing of

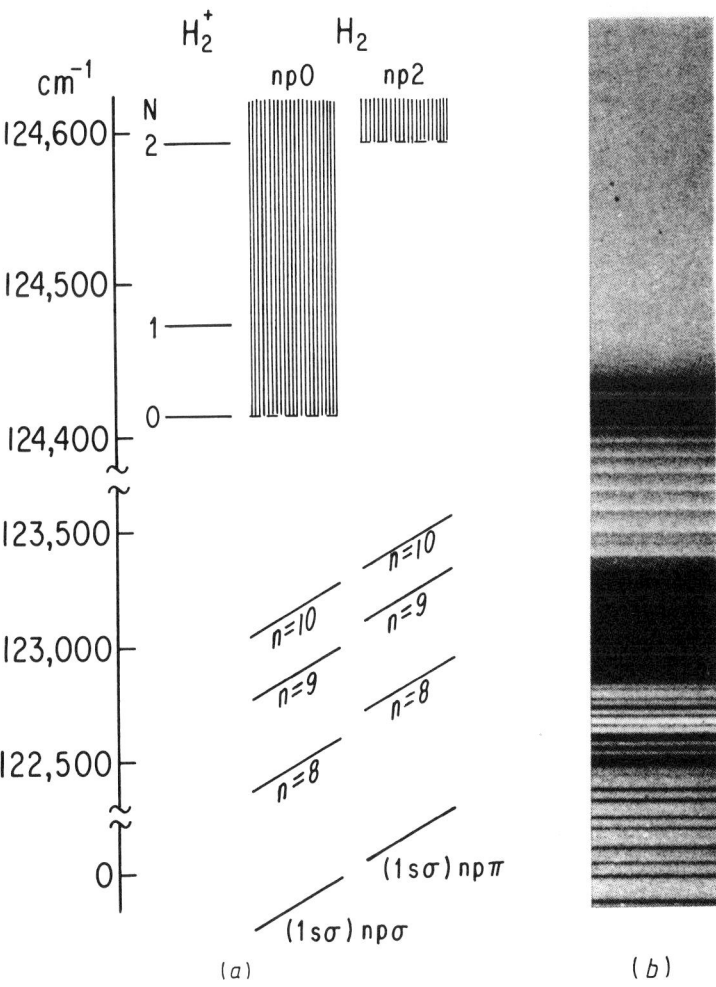

Figure 13.15. *Photoabsorption spectrum of molecular hydrogen in its state of lowest energy* $((1s\sigma)^2, v = J = 0)$. *(a) Schematic diagram of alternative pairs of excitation channels identifiable in the ranges of lower excitation and above ionization thresholds, respectively. The lower excitations are labelled by* $|\alpha J\rangle$, *where* $\alpha \equiv (1s\sigma np\sigma$, *or* $1s\sigma np\pi)$ *and* $J = 1$. *The ionization channels are instead labelled by* $|fJ\rangle$ *with* $f \equiv (N = 0 \text{ or } N = 2)$ *and* $J = 1$. *The lower excitations are represented schematically by three pairs of levels with* $n = 8, 9, 10$ *and quantum defects* $\mu_\sigma = 0.22$ *and* $\mu_\pi = -0.06$. *The ionization levels are represented by continuous bands with different thresholds labelled by* $N = 0$ *or* 2. *(b) Observed spectrum extending over a range of about* 100 cm^{-1} *straddling the* $N = 0$ *threshold. Individual features are represented accurately by solutions* $\{A_\sigma, A_\pi\}$ *of the pair of equations (12) below the* $N = 0$ *threshold and of the single equation with* $N = 2$ *above that threshold.*

Gerhard Herzberg

(German, b 1904)

Gerhard Herzberg was born and educated in Hamburg, and received his PhD in Darmstadt. He has since pioneered in molecular physics for nearly 70 years, producing an extraordinary volume of research, both experimental and theoretical, basic textbooks and the development of a first-class laboratory. Forced out of Germany, he worked and taught at the University of Saskatchewan, at the Yerkes Observatory, and finally at the NRC Institute in Ottawa which now bears his name. He received the Nobel Prize for Chemistry in 1970.

The whole body of molecular physics, primarily through the study of spectra, relies largely on his action, through his invention of experimental approaches, his detection of novel regularities, and his development of texts with extensive tables of data. The monumental series *Molecular Spectra and Molecular Structure* consists of three volumes on *Spectra of Diatomic Molecules*, *Infrared and Raman Spectra* and *Structure of Polyatomic Molecules*.

energy with rotation of the ionic core H_2^+ with angular momenta $N = 0$ or 2; the small rotational energy difference, $E_{t2} - E_{t0} = 0.022$ eV, looms large on the scale in figure 13.15(a) and of the electron motion close to its ionization threshold [35]. The combined action of these circumstances is treated in Box 13I.

The treatment of Herzberg's two-channel example extends readily to atoms and molecules with any number of channels, namely, to the majority of spectra, as noted in Box 13I. It is indeed normal that the state of an electron, or other fragment, emerging from a core still bears the characteristic of the combined system 'core + fragment'. This state evolves, however, as the fragment's radial distance increases, to assume the character of a loose association of fragment and core. It is thus generally appropriate to distinguish long-range channels (f or i) with loose association from short-range eigenchannels αJ.

The characteristic feature of the algebra in Box 13I, introduced by Seaton [27], lies in the occurrence of M-fold products of periodic functions of parameters v_f (or equivalent) which depend in turn on the total energy E. The seeming capriciousness of spectral lines and resonances actually reflects the interplay of the periodic functions in equations (9′), (12) and (13). Knowledge of this feature helps to discriminate intrinsic from accidental features of observations.

A third excitation channel, namely, vibrational excitation, manifests itself in figure 13.15(b) mainly through the interloping dark bands, generated by combinations of lower, shorter-lived electronic excitations and of vibrational transitions. The latter are very intense because the removal of the excited electron from the molecule's ground state relaxes the bond between its atoms, thus inducing a major readjustment of the nuclear motion. The vibrational excitations have been fitted quantitatively within this framework, as detailed in the references quoted above. Additional applications of this approach to molecules, prior to 1985, have been reviewed in reference [37].

13.4.3.2. Empirical parametrization of channel coupling. The use of experimental short-range parameters $\delta_{\alpha J} = \pi \mu_{\alpha J}$ and of the separation of ionization thresholds $E_{t2} - E_{t0}$ in section 13.4.3.1, together with graphical illustrations of the solution of equation (12), opened up an extensive series of spectral applications. The first of these consisted of a detailed analysis of the photoabsorption spectrum of xenon by K T Lu [38]. The Xe atom, and those of all noble gases except He, have a doublet ionization threshold $(t_{3/2}, t_{1/2})$ with easily measurable separation. Accordingly, the set of v_f parameters of the relevant form of equation (13) reduces to two functionally related elements $\{v_{3/2}, v_{1/2}\}$ at all energies much lower than the thresholds for electron excitations of Xe^+. Fitting the wealth of lines observed in this spectrum afforded a quantitative determination of an extended set of parameters, analogues of those in equations (11–13).

It proved particularly effective here as it had in reference [35], to plot the energy of each spectral line, represented by its parameter pair $(v_{3/2}, v_{1/2})$ as a point on a graph with these coordinates. The dependence of equation (13) on trigonometric functions of v_f made it possible to restrict the coordinate scales to their non-integral values, as illustrated by figure 13.16. The features of the curves drawn through the experimental points provided rich physical interpretations.

Spectra with multiple ionization potentials have been similarly analysed using alternative partial two-dimensional plots, combined with parameter fitting by computer operations. Such procedures have now become a major tool of spectral analysis. Indeed it has become practical to distribute spectroscopic information in terms of multi-channel theory.

BOX 13I: INTERPLAY OF SHORT-RANGE AND LONG-RANGE PARAMETERS

The channel coupling formulation of Box 13H, with coefficients $(\alpha J|f)$, turned out to be ideally suited to connect the short- $\{(\alpha J| \equiv (\sigma, 1), (\pi, 1)\}$ and long-range $\{|f) \equiv (N = 0, N = 2)\}$ parameters. The relevant $(\alpha J|f)$ coefficients, determined by simple angular momentum algebra, equal $\{\sqrt{(1/3)}, \pm\sqrt{(2/3)}\}$; the quantum defects $\{\mu_\sigma, \mu_\pi\}$, observed in low excitation spectra, yield the two eigenphases $\delta_{\alpha J} = \pi \mu_{\alpha J}$, whereas the two ν_f parameters $\{\nu_0, \nu_2\}$ are related to the threshold energies E_{t0}, E_{t2}. and to the electron's energy E by

$$\nu_f = \left(\frac{13.6 \text{ eV}}{E_{tf} - E}\right)^{1/2} \qquad f \equiv N = (0, 2). \qquad (11)$$

The occurrence of different pairs of channels, at short and long ranges from the H_2^+ ion core (see section 13.4.1), now plays a decisive role: whereas, in the single channel equation (9'), the parameters ν and μ_l belong to the same channel, here the short-range quantum defect μ_l is replaced by the pair of alternative short-range parameters (μ_σ, μ_π). By the same token, both short-range channels contribute to each long-range channel $|f)$, with respective amplitudes (A_σ, A_π). The condition $\sin \pi (\nu - l + \mu_l) = 0$ that removes the rising exponential in equation (9') is now replaced by the *pair* of conditions

$$\Sigma_{\alpha=\sigma,\pi} A_\alpha (\alpha J|f) \sin \pi (\nu_f - l + \mu_\alpha) = 0 \qquad \text{for } f \equiv N = (0,2) \quad (12)$$

where each sine function depends on parameters (ν_f, μ_α) of *different* channel sets.

This pair of linear equations, in the two amplitudes (A_σ, A_π), is solved by elementary algebra *if, and only if,* its coefficients satisfy the trigonometric condition.

$$\tfrac{1}{3}\sin\pi(\nu_0+\mu_\sigma)\sin\pi(\nu_2+\mu_\pi) + \tfrac{2}{3}\sin\pi(\nu_0+\mu_\pi)\sin\pi(\nu_2+\mu_\sigma) = 0. \qquad (13)$$

CONTINUED

13.5. Collisions between atoms or ions [39]

The study of collisions, introduced in section 13.2.4, has expanded greatly since mid-century. The wide diversity of circumstances presented by

Atomic and Molecular Physics

BOX 13I: CONTINUED

Solutions of equations (12) and (13) determine then the positions and intensities of the spectral lines as well as the intensity spectrum and angular distribution of photoelectrons ([26, 35] and references therein).

The basic equation (13) remains relevant for electron energies E between the two ionization thresholds E_{t0} and E_{t2}, with a minor modification. Here the condition (12) for $f \equiv N = 0$ is no longer relevant, whereby the spectrum of energies E is continuous. The intensity of this spectrum is, however, modulated by equation (12) with $f = 2$, a modulation consisting of *seemingly capricious resonances*, corresponding loosely to the Rydberg series that would converge to the ionization threshold E_{t2} in the absence of channel coupling [36].

Thus equations (11) and (12) apply to larger channel sets $\{|f\rangle\}$, to sums over many eigenchannels $(\alpha J|$ and to larger matrices of coefficients $(\alpha J | f)$.

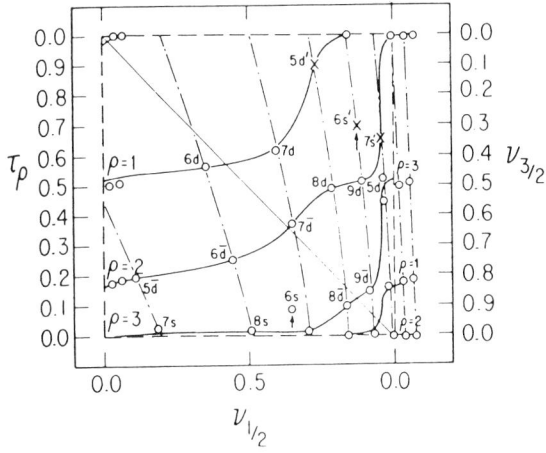

Figure 13.16. *Plot of observed level energies E in the Xe spectrum as open circles, each with a pair of coordinates $(\nu_{1/2}, \nu_{3/2})$ determined from the analogue of equation (11) with reference to the observed pair of series limits $(E_{t3/2}, E_{t1/2})$ as indicated in the text. The large number of points available proved sufficient to identify three lines $(\rho = 1, 2, 3)$ on which all points lie, and thereby to evaluate all parameters of algebraic equations represented by those lines.*

this study nevertheless still leaves it in a rather fragmentary state, a view applying specifically to the subject of the present section. Electron

1075

collisions have been largely included in section 13.4, albeit mostly by implication. Collisions between molecular aggregates of atoms will be touched upon in section 13.6 on molecules.

Collisions between atoms or ions are governed by different circumstances depending on whether the atoms' speed is higher or lower than that of the relevant atomic electrons. (The *energy* of colliding atoms generally exceeds by far that of relevant electrons owing to the far higher mass of nuclei; recall in this context that a heavy particle can impart to a single electron at rest no more than twice its own speed.)

Collisions by very fast atoms are thus unlikely to transfer much energy to electrons. Collisions by very slow atoms or ions are also unlikely to transfer much of their own energy in a single process. Collisions with speed comparable to that of valence electrons have the highest probability of exciting their target; their correct treatment has become reasonably accurate only in the last decade. These three collision classes will be discussed in separate subsections. Collisions involving highly charged (stripped) ions will be dealt with separately in section 13.5.4.

13.5.1. *Projectiles faster than atomic electrons*

Long-range collisions of this class, mediated by the Coulomb field of charged projectiles, were introduced in section 13.2.4.1. Their average energy transfer to the target decreases approximately in inverse ratio to the squared velocity of the projectile owing to the shorter duration of the collision. Specific effects of interest may decrease much faster, of course, as the projectile's speed increases; an extreme example is seen in the electron capture by the projectile which would decrease as v^{-12}, according to elementary theory, but actually does so only as v^{-11} through the Thomas process described in section 13.2.4.3. Neutral projectiles interact only in grazing and close collisions. At high speeds, projectile and target constituents act basically as independent nuclei or electrons colliding with one another in binary processes as follows.

The impulsive character of fast collisions, stressed by Bohr (section 13.2.4.1), is shared by close- and long-range collisions; one or more constituent of the target, recoiling upon absorbing a quick shock, may react independently if the shock is sufficiently violent.

Attention has focused, in recent decades, on the following special processes.

(i) A variant of electron capture, in which a target electron fails to become attached to the projectile but travels with it at nearly equal speed and direction (Rudd–Macek effect).

(ii) Ejection of inner-shell electrons by the impact of fast ions.

(iii) Electron capture by highly charged ions; in this case the electron settles preferentially in excited states where its speed remains more nearly comparable to the ion's velocity.

(iv) Collisions leading to excitation of two (or more) electrons. This process, purely electronic for long-range collisions, may or may not involve repeated actions by the projectile.

(v) In contrast to long-range collisions, whose action is independent of the sign of the projectile's charge, close collisions differ, e.g. for protons or antiprotons which attract or repel the electrons, respectively.

13.5.2. Projectiles slower than atomic electrons
The effectiveness of slower collisions is restricted by the 'Massey criterion', introduced in section 13.2.4.1, which excludes energy transfers whenever the product of the collision's duration and of the transfer's oscillation frequency greatly exceeds unity. This limiting frequency drops, however, sharply for collisions at certain interatomic distances that bring the energies of two electron levels within the *colliding pair* into near coincidence (quasi-degeneracy). The collision is accordingly viewed in this case as a *molecular* process.

The occurrence of quasi-degeneracies, apt to generate collisional energy transfers, becomes obvious upon inspection of the 'Born–Oppenheimer potential curves'—introduced in section 13.2.5.1 and illustrated in figure 13.3—for the 'complex' formed by the colliding atoms. Degeneracies correspond to crossings and quasi-degeneracies to 'avoided crossings' of such curves. Sizeable studies of such excitations resulting from grazing collisions have been developed in recent decades, combining experiments and theory. One specific aim of these studies has been to observe and interpret the orientation and alignment of excited atoms induced by collision [40].

Close collisions tend to be more complex as atoms interpenetrate deeply, displacing whole shells of electrons. A dramatic demonstration of the role of level crossings emerged, however, in the 1960s, in experiments on $Ar^+ + Ar$ collisions at energies of the order of 50 keV, that is, at a speed far lower than the electrons' [41]. In these collisions the nuclei follow a rather definite trajectory, thus correlating the observed projectile deflection with the closest approach, r_0, of the two nuclei. The balance of momentum and energy transfers also determines the total energy Q transferred to electrons. Morgan's plot of the average $\bar{Q}$ as a function of r_0 showed a sharp jump from ~100 eV to 600–700 eV as r_0 traverses the very narrow range from ~0.025 to ~0.023 nm (figure 13.17(a)). This sharp discontinuity was interpreted as the *onset of overlap* between the inner L-shells of the colliding atoms, in apparent contrast to the notion of each shell spreading over a broad range of radial distances.

Further details emerged soon from observations at St Petersburg detecting projectile and recoil ions in coincidence, and thereby resolving the jump into two steps of ~250 eV each, corresponding to the sequential ejection of two L-shell electrons. Connection with theory was then

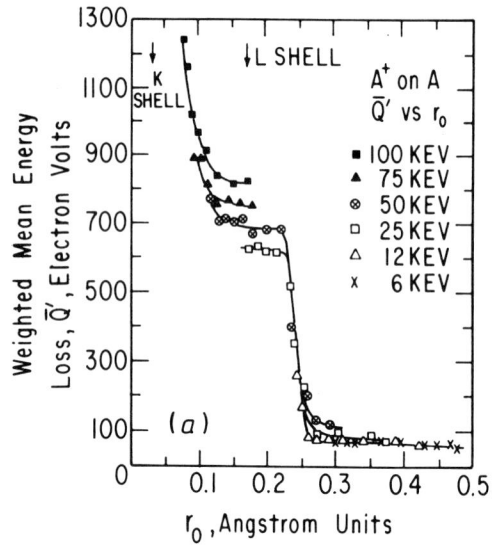
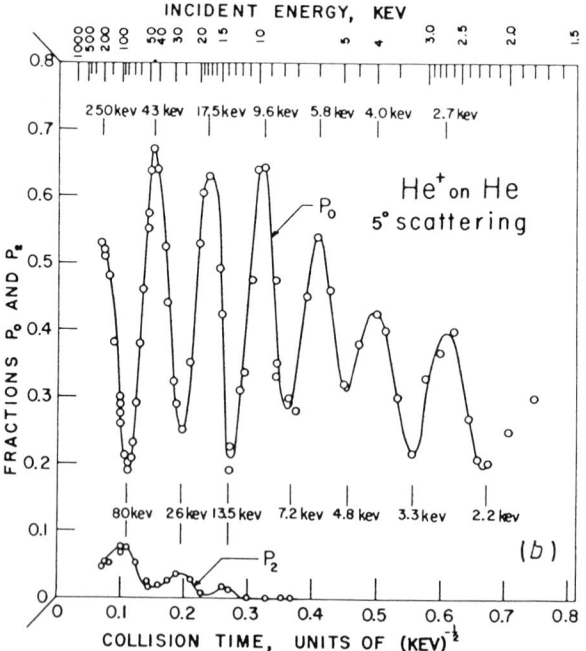

Figure 13.17. Early demonstrations of novel phenomena in ion–atom collisions. (a) Energy transfer to electrons in $Ar^+ + Ar$ collisions as a function of internuclear distance, for various collision energies. (b) Incident ions He^+, colliding with neutral He gas atoms and emerging with 5° deflection, are selected according to their charge after collision: neutral, single or double. The fractions (P_0, P_1, P_2) thus selected are oscillating functions of the collision duration with nearly constant frequency.

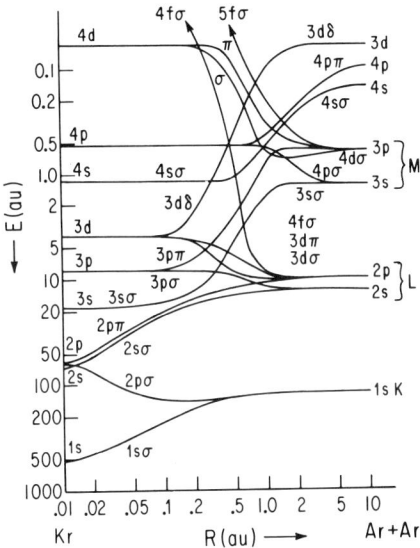

Figure 13.18. *Diabatic potential energy functions for the Ar$_2$ complex.*

established by W Lichten's construction of potential curves for the Ar$_2^+$ complex (figure 13.18), whose 'united atom' limit reduces the pair to a Kr$^+$ ion. Lichten identified a single level, occupied by just two electrons, which belongs to the L shell ($n = 2$) manifold in the 'separate atom' limit Ar+Ar$^+$, but gets 'promoted' [15], as r_0 decreases, to the $n = 4$ manifold of Kr$^+$. At the critical range of r_0, where $\bar{Q}$ jumps, this rising level *crosses* a bundle of descending levels that represent bonding valence states of the Ar$_2^+$ molecular ion; the degeneracy of the crossing thus shifts the electron pair into the valence shell and eventually to full ejection.

The striking phenomenon in figure 13.17(a) appears now as an early example of a multitude of level-crossing effects, ubiquitous in atomic interpenetrations, which promote electrons from inner to outer shells. The notion has been introduced recently that these effects may be usefully traced to a single (or multiple?) singularity of the mathematical representation of collisions as functions of the internuclear distance r_0 viewed as a complex variable [42].

13.5.3. Collision speed comparable to electron velocities
Early expectations that these collisions would favour electron transfers between projectiles and targets led theorists to anticipate that even repeated transfers could occur in a single grazing process. Actual observation of this phenomenon in He$^+$–He scattering with 5° deflection, at Connecticut in 1958, nevertheless attracted considerable attention [43] (figure 13.17(b)). Formation of the molecular complex He$_2^+$, as He$^+$ and

He merge at modest speed (but multi-keV energy), leads one electron to occupy with near equal probability either of two molecular energy levels, the lower one 'bonding' the two atoms, the higher one being instead repulsive (anti-bonding) as displayed in figure 13.3 for H_2^+. The energy gap between these levels rises to a few eV as projectile and target approach, decaying again as they separate. This circumstance leads the electron to oscillate between the two atoms with the beat frequency corresponding to the energy gap, as outlined at the end of section 13.2.1. The number of oscillations displayed in figure 13.17(b) represents the product of the oscillation frequency and of the collision's lifespan.

This notable observation coincided with the introduction, by D R Bates and collaborators, of systematic procedures for calculating the probability of electronic excitations resulting from similar processes, a development motivated at the time by interest in astrophysical applications. Projectiles are treated in these procedures as moving classically on straight, or nearly straight, trajectories. Atomic electrons are treated instead quantum mechanically, by perturbation methods under favourable conditions. More elaborate procedures are required to evaluate sizeable probabilities of single or sequential electron transitions. Specifically one calculates probability amplitudes $a_n(t)$ of observing, at various times t during the collision, the target and/or the projectile in their nth alternative 'closely coupled' state. The collision dynamics is then embodied in a system of differential equations governing the amplitudes $a_n(t)$ [44].

Applications of this approach have expanded progressively, utilizing the improvements in computer performance and responding to the rising requirements of plasma research, which involves highly ionized and colliding gases at thermal energies of the order of keV. Critical to the study of relevant collisions is the selection of distribution functions analogous to the $f_{kl}(r)$ (introduced in section 13.2.4.2 and again in section 13.4) pertaining to the relevant atomic or molecular state of projectiles and targets, and of their collision complexes. A recent review of these studies includes over 350 references [45].

These collision studies have dealt largely with grazing collisions and outer-shell excitations but have also reached deeper into atoms, treating for example the transfer of inner-shell electrons onto fast incident protons or onto highly charged stripped ions, viewed at times as analogues of the simple process of figure 13.17(b). We shall return to this subject in sections 13.5.4 and 13.7.

13.5.4. *Highly stripped ions*

Experimentation with multiply charged atomic ions started in about 1970, utilizing accelerators originally built (but grown obsolescent) for nuclear studies. Low-charge ions thus accelerated lost additional electrons upon traversing a foil; the higher charge thus attained could then amplify the energy gain by further acceleration followed by further stripping, at least in principle. Currently, alternative procedures can strip all electrons even from the atoms with highest nuclear charge.

The accelerated ions with intermediate charge ze (e.g. with $z \sim 10$) available in the 1970s and 1980s served initially as projectiles, more efficient than protons or α-particles, to study inner-electron ejection in ion–atom collisions. Accordingly, the circumstances of these collisions are generally analogous to those described in section 13.5.1. Electrons still clothing a stripped ion are also liable to excitation or ejection by interacting with a target. The study of ion–atom collisions was thereby enriched and pursued extensively. Collisions were classified according to the location and number of electrons actually involved. This multiplicity of alternative channels has hindered, of course, the development of any comprehensive theory.

The more recent availability of ions with high atomic number Z and clothed with only very few electrons has shifted recent attention progressively to a more straightforward task, namely, the collision and spectral study of few-electron systems in very strong Coulomb fields where relativistic conditions prevail. This study presents a specific challenge in that its relativistic multi-particle aspects are only amenable to perturbative, if very accurate, treatment. No unanticipated observation has emerged thus far in this field, but its study warrants attention [46].

Highly stripped ions are also normal constituents of plasmas at stellar temperatures, in the range 1–100 keV. Such plasmas are now generated in laboratories rather easily, if briefly, by the focused action of a pulsed laser on condensed matter surfaces; they are also maintained for longer intervals in devices developed for nuclear fusion. Their radiative emissions are of interest on one hand as useful indicators of the instantaneous state of a plasma and on the other hand as undesirable vehicles of energy dissipation. The spectroscopy of these emissions has expanded greatly in recent decades under the stimulus of plasma research. Collisions among stripped ions and with plasma electrons are, of course, important elements of plasma dynamics and evolution, as reported in Chapter 22.

13.6. Molecular physics [47]

The study of molecules has branched out since mid-century into a wealth of new directions, opened by technical developments. Novel observations of phenomena, specific advances of molecular spectroscopy, giant strides of quantum-chemistry calculations afforded by computer development, and progress on reactive collisions will be dealt with in separate subsections.

13.6.1. New avenues of experimentation
Much of the experimental progress resulted from improvements of seemingly straightforward procedures such as: (a) achieving, maintaining and measuring lower and lower pressures in experimental vessels by means of more powerful pumps, wall materials and seals, by about 1960; (b) producing, maintaining and monitoring more perfect and cleaner surfaces, whose study became possible in the 1970s; (c) improved preparation of material samples; (d) fast electronic controls of valves and of measuring devices (e.g. of mass spectrometers); (e) availability of intense, monochromatic and sharply tunable radiation sources (lasers) for spectroscopy; (f) development of shorter and shorter radiation pulses reaching now into the femtosecond (10^{-15} s) scale.

13.6.1.1. Cross-beam experiments. Chemical reactions had traditionally been produced and studied in bulk matter, generally gaseous or in solution. The production and observation of single collisions at the crossing point of unidirectional, monovelocity [48] beams of reactant molecules opened a new era of chemistry, beginning in the 1950s and fully established after 1960 [49]. Each collision product could then be collected and analysed in appropriate directions, thus determining dynamical parameters of each collision event.

13.6.1.2. Low-temperature beams from nozzle expansions. Sudden (adiabatic) expansion of a compressed gas into vacuum lets it squirt into a jet, converting most of its thermal energy into kinetic energy of unidirectional motion. Development of this technique in the late 1960s made available beams of desired composition at controlled kinetic temperatures of the order of 1 K [50]. At these temperatures, new molecular aggregates (van der Waals molecules and clusters of desired composition) form under the influence of their van der Waals attraction. It proved practical to expand buffer gases of low molecular weight, typically He or Ar, 'seeded' by traces of molecules of interest; fair-sized molecules could thus be observed with a few rare gas atoms stuck to them. These atoms could then be shaken off following spectral excitation of the carrier molecule [51]. A large variety of experiments has since been performed by this method.

13.6.1.3. Resonances. Resonant electron excitations, similar to those of atoms discussed in section 13.4, occur in molecules as well. A separate mechanism is afforded by suitable distributions of atoms and charges occurring in molecular structures, distributions that can hold an additional electron temporarily within the volume of a valence shell. A simple such mechanism rests on the non-negligible length/width ratio of diatomic molecules. This ratio allows a slow electron with orbital momentum $l = 2$ or higher to penetrate the centrifugal barrier

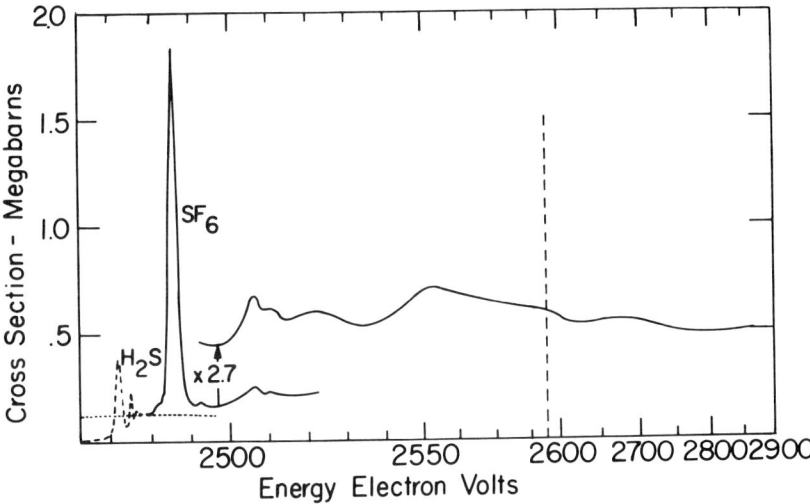

Figure 13.19. Cross sections for photoabsorption by the K shell of the sulphur atom, near its threshold at 2.5 keV, in two alternative molecular combinations. The photoabsorption by SF_6 is boosted greatly by resonance resulting from the obstacle to S-electron escape through the surrounding, negatively charged, fluorine atoms. The integral of S-photoabsorption over the whole spectrum is equal for the two molecules, but its concentration near threshold in the case of SF_6 is outstanding. The vertical dotted line represents each of two different, but accidentally equal, contributions to photoabsorption: (a) the constant value of the K-shell photoabsorption in H_2S above its ionization threshold, (b) the contribution of photoabsorption by the six F atoms of SF_6 in this energy range.

surrounding the molecule along its interatomic axis, thus reaching its internal Coulomb field more readily. The successive electron escape from this internal trap proceeds again more readily along the axis. The formation of such a short-lived molecular state is called a 'shape resonance'; the lowest one is known as the 'lowest unoccupied molecular orbital' (LUMO), a name contrasting with the 'highest occupied molecular orbital' (HOMO). Numerous such resonances have been observed at energies up to 30–40 eV above the valence levels [52, 53].

This phenomenon is magnified in molecules with a central atom surrounded by 'electronegative' atoms that bear more than their share of electrons; the net charge on these atoms hinders the escape of any extra electron injected in their middle. Injection is provided here conveniently by photoexcitation of an inner shell, as illustrated in figure 13.19 by the contrasting K-shell absorptions of sulphur in the molecules SF_6 and SH_2 [54]. This striking observation was soon extended to other molecules, judiciously selected particularly by V I Fomichev and T M Zimkina in St Petersburg, and has recently become a frequent tool for studying the mechanics of molecular valence shells.

1083

Extensive molecular resonances have also been observed in collisions of electrons with molecules, processes discussed in section 13.4 and illustrated in figure 13.12. E Lassettre in the 1950s and 1960s and P D Burrow more recently have been major contributors to these studies, which actually have a bearing on negative ion states of each target molecule.

13.6.1.4. Larger aggregates. The structure and the dynamics of the flimsy molecules held together by van der Waals forces, rather than by chemical bonds, have been studied by microwave spectroscopy, typically along the low-temperature beams within which they originate. The strength of the 'adsorption' bond by which a small molecule hangs onto a large one can thus be measured. New insight is also gained from the geometrical structures thus determined: For example, the dimer of the polar molecule HF might have been expected to form with each positively charged H atom next to the negatively charged F atom of the other molecule. One F atom appears instead to be bonded to both H atoms, tightly to one and loosely to the other, with the two bonds at an angle comparable to that of the water molecule H_2O, whereas the second F atom hangs loose [55].

Clusters, typically of monatomic substances, have been studied extensively since the 1970s, primarily as transition structures between molecules and bulk matter. Their cohesion energy per particle appears to oscillate with increasing size, being largest for atom numbers that afford a high-symmetry structure. A typical question concerns cluster reactions to external disturbances: is it elastic as for solids or yielding as for liquids? It is actually solid-like at lower temperatures and liquid-like at higher ones [56].

Polymers, i.e. aggregates of long and flexible molecular chains, were introduced in section 13.2.5, with reference to the rubber elasticity. They have been studied extensively since mid-century, with diverse compositions and properties, typically with regard to their relaxation after removal of an externally imposed strain.

Liquid crystals, i.e. aggregates of moderate-sized, often rod-like, molecules with important technical applications, have also been synthesized and studied extensively since the 1960–70s. P G de Gennes, a principal theoretician of both liquid crystals and polymers, reports on this subject in Chapter 21.

Biomolecules, primarily proteins and polynucleotides, also consist of chemically bonded chains of molecular groups more highly differentiated than for ordinary polymers. Knowledge of their structure was achieved in the 1950s through dramatic progress of x-ray diffraction of their crystalline form. Understanding of their delicate, critical and still rather mysterious properties—responses to external ionic charges, folding and unfolding in specific manners, efficient and controlled transport of electrons over nanometre ranges—remains fragmentary.

13.6.1.5. *Observation of fast processes.* Molecular processes develop on various, quite different, time-scales. Their observation and study have thus expanded in step with the progress of time-interval resolution of physical measurements. This resolution stood roughly at the nanosecond (10^{-9} s) scale in about 1960, at picosecond (10^{-12} s) in about 1970, and at femtosecond (10^{-15} s) in the late 1980s. The nanosecond scale introduced 'flash-photolysis' and 'flash-spectroscopy' in physico-chemical research. The picosecond scale afforded observations of dissociation processes, of the approach to relaxation of the orientation and energy of molecules in solution; the femtosecond scale afforded fuller scope of relaxation phenomena.

13.6.2. *Molecular spectroscopy*

Molecular spectroscopy continued to develop after mid-century on its earlier path, yet expanded remarkably through the progress reported in sections 13.3, 13.4 and 13.6.1.

The treatment of multi-channel electronic excitations and ionizations, reported in section 13.4, more specifically through its illustration by the example of H_2 photoionization (section 13.4.3.1), proved very feasible for the molecular applications reviewed in reference [37]. Specifically relevant to molecules was the introduction of the quantum defect's dependence on internuclear distances R as the agent of specific transitions between vibrational states with distribution functions $|\chi_v(R)|^2$ and $|\chi_{v'}(R)|^2$

$$(v'|\mu_{\alpha j}|v) = \int_0^\infty dR \chi_{v'}(R) \tan[\pi \mu_{\alpha j}(R)] \chi_v(R). \tag{14}$$

The availability of radiation sources with photons in the range $20 < \hbar\omega < 1000$ eV (section 13.3.1) has facilitated the detection of valence-shell responses to the ejection of inner-shell electrons. A surprising observation thus emerged by comparing the K-shell ionization thresholds of atomic and molecular oxygen, which differ by $\sim$10 eV ($\sim$540 eV for O, $\sim$530 eV for O_2), owing to relaxation of the valence electrons about the K-shell vacancy [57]. (This relaxation delays the vacancy's 'tunnelling' between the O_2 atoms beyond its lifetime, thus 'breaking the symmetry' of the homonuclear diatomic O_2.)

Molecular physics was greatly advanced by removal of the gap between the near infrared and microwave spectral ranges (section 13.3.2), affording full exploration of the vibrational and rotational spectra that serve to identify molecules, e.g. in outer space. Indeed the astrophysical role of molecules was hardly suspected until rotational frequencies appeared in radiofrequency spectra from astrophysical sources, in about 1960. Laboratory studies of molecular microwave spectra hinged initially mainly on measuring the power loss generated by introducing vapours in tunable cavities. The extension of vibrational spectra to approach the dissociation thresholds of H_2 and HD has been noted in section 13.4.2.2. Currently molecular spectra are often studied by passing a beam through waveguides that generate standing waves of accurately determined wavelength in the microwave or infrared range as indicated in section 13.3.2.

In contrast to the easy identification of optical emission or absorption in visible spectra, the detection of resonances in the lower frequency range hinges on preliminary, but quite accurate, knowledge of their probable location, owing to the limited practical range of tunability. A striking example of the difficulty in identifying specific spectra is afforded by the saga of the ion H_3^+, the smallest polyatomic structure, which was identified early in this century by mass spectroscopic analysis of discharges in hydrogen gas. Its spectral 'fingerprint' eluded detection until 1980, when successive estimates of its vibration frequencies and improved techniques led T Oka to its spectral detection at $\lambda \sim 4$ μm. Knowledge of its spectrum led then to abundant detection of H_3^+ in outer space, especially in the outer atmosphere of Jupiter [58].

13.6.2.1. Radiationless transitions. The motions of electrons and nuclei within a molecule have been viewed in section 13.2.5 as basically independent, owing mainly to the different scales of their respective energy spectra. (Electronic excitations are nevertheless generally accompanied by satellite vibrational excitations.) However, the higher-level density of polyatomic molecules, and of all molecules near their fragmentation thresholds, restricts the independence of electronic and nuclear motions, often drastically.

A prototype of radiationless transition between electronic levels, which transfers their energy difference to vibrational motions rather than to radiation, occurs commonly in organic molecules. The spins of ground-state electrons normally cancel each other forming a 'singlet' state, as indicated in section 13.2.5; their excitation by photoabsorption does not affect their spin state directly. On the other hand, the stability accruing to ground states by forming a spin singlet no longer holds for the excited electron. Indeed, a 'spin flip' by this electron—stimulated by the spin-orbit coupling (section 13.2.3.2)—reduces its Coulomb repulsion by the residual unpaired electron, owing to the centrifugal repulsion of 'triplet coupling' (Box 13A). (Excited electron states with very similar distributions in space generally occur with alternative mutual orientations of their spins, higher net spins then correspond to energy levels lower by amounts of the order of 1 eV, according to Hund's rule.) The energy thus released by electron spin flips is readily dissipated to vibrational motion in molecules of sufficient size or excitation.

This phenomenon is typical of organic *dye molecules* which absorb light of specific colours readily and fail to release it promptly by fluorescence coupled with a second spin flip. The energy is trapped instead in triplet or higher spin states, unable to dissipate quickly a larger amount of energy to vibration; it may be re-emitted eventually by delayed photoemission (phosphorescence) or dissipated slowly to other molecules in condensed media.

Analogous processes of radiationless transitions are increasingly common for higher and higher excitations of molecules. These phenomena escaped general attention prior to mid-century, when attention focused mainly on diatomics. Their emergence in the 1950s and 1960s represented a departure from the conventional wisdom viewing electronic and vibrational motions as distinct. General appreciation of the rates of electron energy conversion to vibration, rates that are modest yet frequently larger than for light emission, took hold eventually in the community, closing a chapter of doubts and debates.

13.6.3. Quantum chemistry. Theory and computations

The theory of molecular excitations and transformations (quantum chemistry), laid out in its essentials prior to mid-century (section 13.2.5), has been pursued vigorously. Its applications have expanded greatly, of course, stimulated by the pace of computer technology and by the amount of available effort. The basic ingredient of calculations is built from large sets of independent-electron distribution functions (orbitals) suitably combined with coefficients adjusted to optimize energy levels or analogous parameters. The optimization utilizes the Hartree–Fock method, configuration mixing and other variational procedures, as in the past. A bench-mark was achieved in the 1960s, by the Kolos–Wolniewicz calculation of the ionization threshold of H_2 molecules within 1 part in 10^5 of its experimental value. C Rothaan's introduction of analytic orbitals in the 1950s, affording easier interpretation and handling of electron distributions, leavened the field but wholly numerical descriptions of orbitals remain in use.

Still greater evolution occurred in the diversity of researchers' perceptions and attitudes regarding computations. Numerical results, as provided by current procedures, long reflected mainly the input data rather than the underlying physics. This subject was discussed in the closing lecture by C H Coulson to a 1959 conference, a lecture viewed now as historical [59]. Coulson noted the danger that the quantum-chemistry community would split between practitioners of massive computations and workers looking for greater transparency of results and approaches as well as for greater adherence to physical mechanisms.

This contrast is still apparent but has not led to dire consequences. It has been attenuated by the reduced cost of computing, both in terms of money and of personal effort. This reduction has encouraged heuristic computations as tools to explore, for example, the dependence of results on alternative inputs, an activity that has been expanding in recent years.

The construction of potential surfaces, requiring calculation of bond energies over multi-dimensional lattices of atomic configurations, has been particularly enhanced by computer progress. This activity was confined largely to tri-atomic systems for some years, but has since expanded rapidly to larger systems, for example Truhlar's group has dealt quantitatively with the seven-atom set in the reaction $CH_4 + OH \rightarrow CH_3 + H_2O$ [60]. Only three of these atoms (C, H, O) participate directly in the transfer of a single H atom, but the relaxation of the CH_3 residue is also relevant.

13.6.4. Reactive collisions

Reactive collisions have also been advanced greatly by computer developments, mainly through the construction of the potential surfaces outlined above. Basically, the calculation of a reactive collision consists of studying the evolution of a system of interest along a path on a potential surface, from a starting point where atoms are grouped, e.g. as AB+C (as introduced in section 13.2.5.3) to an end point with the structure A+BC.

The drastic schematization of such a process, described in section 13.2.3.5 as collinear, is now generally dispensed with, but section 13.6.3 noted that calculations extending beyond tri-atomic examples still remain infrequent. The central difficulty of this study lies in satisfying its implication that the potential surface should reflect realistically the deformation of atomic and molecular structure as reactants approach and intermingle into a complex.

A more subtle difficulty lies in verifying compliance with a main feature of the early concept of traversing a transition state in mid-collision, namely, that the traversal would proceed steadily forward without intermediate loops. Deviations from purely classical motion along a reaction path have been allowed for in past decades, for example by allowing for quantum tunnelling through potential barriers. Allowances for effects of quantum fluctuations of a system's motion astride and along a classical path appear to have been scarce.

Some progress towards treating the motion of several atoms quantum mechanically and more visually has been achieved by representing their positions on potential surfaces in polar coordinates [61]. (These coordinates are called 'hyperspherical' for 'surfaces' of more than three dimensions.) Thereby the coming together and eventual separation of reactants is represented by the variation of a single radial variable.

A valuable example treated successfully by this approach is provided by the reaction

$$H_2 + F = HF + H \tag{15}$$

studied by Y T Lee through detailed observation of its cross-beam collision products. The energy released by this exothermic process is found to remain largely concentrated in the vibrational excitation of HF to its $v = 3$ level, whereas H_2 was initially in its ground state. (This finding parallels the electronic excitation of an ion reported as item (iii) of section 13.5.1.) Quantum hyperspherical treatment of this reaction by J M Launay has reproduced the vibrational excitation successfully [62].

13.7. Inner-shell phenomena [63]

The production of vacancies in the inner shells of atoms or molecules, generally by photoabsorption or collision, serves two main purposes: (a) the generation of monochromatic x-rays by the oscillating current of an electron transition that fills the vacancy; (b) the study of secondary processes resulting from a similar, but *radiationless* (section 13.6.2.1) transition.

Vacancies, with the same consequences, can also arise as secondary effects of nuclear processes: (c) electron capture into a nucleus made unstable by excessive positive charge, accompanied by ejection of a neutrino; (d) internal conversion of nuclear energy through electromagnetic interaction analogous to 'photoemission by the nucleus + photoionization' of the inner shell. The occurrence of either process is manifested by the subsequent x-ray emission or radiationless transitions.

The rate of x-ray emission, proportional to the squared ratio of its electron displacement to the x-ray wavelength at lower photon energies, lags behind that of competing radiationless (Auger) transitions, until its proportionality to the cube of frequency prevails at $\hbar\omega > 10$ keV. Auger transitions occur instead generally within ~ 10 fs, regardless of their energy.

Specific aspects of the x-ray emission and Auger processes are outlined in sections 13.7.1 and 13.7.2, respectively. Section 13.7.3 deals instead with the role of inner electrons in properties of whole atoms and molecules encompassed by the generic names 'screening' and 'antiscreening'. The former refers to the reduction of nuclear attraction upon outer electrons by the negative charge of inner electrons. The latter refers to a less obvious influence of inner electrons that amplifies the action of outer electrons on nuclear orientations.

13.7.1. X-ray studies

The broad classification of atomic x-ray spectra, characteristic of each element, into K, L, M ..., series was established long before mid-century (section 13.2.3.5). Coster–Kronig transitions were also identified, connecting subshell levels with the same n and different l and/or j values (as defined in section 13.2.3.2); the energy separation of these levels reflects differences in their closest approach to the nucleus as represented by different values of their quantum defects μ_{lj}. The comparatively high intensity of Coster–Kronig emissions demonstrates the high intensity of the electron currents involved in transitions between levels of the same shell.

Attention has been directed more recently to the 'satellites' of most x-ray lines, that is to sets of lines with frequencies slightly lower than that of a regular line, whose emission is accompanied by one of many alternative secondary excitations. The measurement, classification and theory of these emissions have been advanced mainly by two schools guided by B Crasemann at Oregon and T Åberg in Finland.

The precision determination of characteristic x-ray wavelengths and photon energies has been an important element not only for providing specific standards but for interconnecting measurements performed on diverse scales. It involves, specifically, the x-ray diffraction by crystal lattices, the lattice spacing of the crystals, the Planck constant, etc. A recent paper in this field indicates a current accuracy of the order of 1 part in 10^6 [64]. The intensity of each x-ray emission reflects instead the distribution in space of the radiating current.

Precision experiments and theory on x-ray transitions between levels of highly stripped ions have attracted much attention recently, as reported in section 13.5.4.

'Resonant' production of an inner-shell vacancy, as a preliminary to x-ray emission, consists of exciting an inner-shell electron to a discrete level within, or slightly above, the valence shell of an atom or molecule. This process, facilitated by high-resolution synchrotron radiation, has attracted much attention recently as a tool to observe the influence of the excited electron on the valence shells, particularly in molecules, as noted in section 13.6. This influence is reflected, in part, by the subsequent x-ray emission.

13.7.2. Auger emissions

The Auger phenomenon consists of the escape from atoms of monoenergetic electrons, such as would result from photoionization by characteristic x-rays. Indeed their energy coincides with that of photoelectrons ejected by x-rays from the same atom, but is transmitted by direct interaction with an electron that fills an inner-shell vacancy. This transfer of excitation energy may be viewed as an example of the coupling between 'closed' and 'open' channels described in section 13.4, i.e. as an example of auto-ionization.

Beginning in the 1950s the study of Auger electrons was enhanced by K Siegbahn's design of ESCA electron spectrometers with high resolving power and luminosity and by its observation in gas rather than condensed phase, largely at W Mehlhorn's initiative. The analysis of fine structures and of satellite spectra, analogous to those of x-rays, thus became accessible. The Auger process is initiated, of course, by the production of an inner-shell vacancy, i.e. by photoionization or by collision as for the production of characteristic x-rays.

The combination of vacancy generation and of Auger emission in rapid succession (~10 fs) affords opportunities for further studies of atomic dynamics. Thus, vacancy production with low excess energy generates a rather slow electron that may be overtaken by the following, faster, Auger electron, thus affording exchanges of energy and/or angular momentum between these electrons (post-collision interactions). The production of a specific inner-shell vacancy by an incident beam generally impresses on the vacancy an alignment and/or orientation with reference to the beam direction; these geometric characteristics are then transmitted to the Auger electrons whose study proved more fertile than the corresponding study of characteristic x-rays. The greater flexibility of Auger-electron analysis extends to observing effects of resonant generations of vacancies mentioned at the end of section 13.7.1.

A comprehensive review of recent studies on Auger electrons from free atoms and molecules, including their detection in coincidence with recoil ions, is provided in reference [65]. Since the 1970s Auger emission has also emerged as an analytical tool of the presence and state of atoms or molecules adsorbed on surfaces.

13.7.3. Screening and antiscreening

The electric potential within an atom, at a radial distance r from its nucleus with charge Ze, can be represented as $(Z - s)e/r$ where the number s indicates the screening action of all the atomic electrons at radial distances $r' < r$. The value of s equals the sum over all bound electrons with indices (n, l) of the values of their distribution functions $|f_{nl}(r')|^2$ (introduced in section 13.2.3.1) integrated over the range $0 < r' < r$.

The determination of the functions $f_{nl}(r)$ for atomic ground states has been indicated in section 13.2.3. The resulting values of the screening function $s(Z; r)$ depend smoothly on the atomic number Z, in general. We note here, however, a distinct effect of the relativistic mass boost of K-shell electrons of high-Z elements, mentioned in section 13.2.1, that was identified by J P Desclaux and Y K Kim in the 1970s. This boost draws K electrons closer to the nucleus, thereby increasing their contribution to the screening of other electrons, especially of those in states with $l \neq 0$ that are kept away from the nucleus by the centrifugal force. The resulting modification of screening propagates domino-like from shell to shell, as l increases, as demonstrated by appreciable reductions of the quantum defects μ_l.

The 'antiscreening' effect stems from asymmetries, of electron distribution or of spin orientation, inherent to the structure of incomplete (open) valence shells (or subshells). The Coulomb interaction between each electron of such a shell and each electron of the underlying, spherically symmetric, 'closed' shells perturbs this symmetry. This perturbation, inversely proportional to the third, or higher, power of the distance of each electron pair, propagates—again domino-like—from shell to shell, eventually influencing the orientation of a nuclear spin. Its inverse dependence on electron distances magnifies the initially insignificant magnitude of the perturbation as it propagates to increasingly dense distribution of inner-shell electrons.

This phenomenon, and its name, was identified by R Sternheimer around 1950 in the context of the quadrupole coupling of valence electrons with nuclei, that is, between the elongations (or oblateness) of their respective electric charges, whose weak strength is nevertheless detected by nuclear magnetic resonance. The corresponding magnification of the interaction between the magnetic moments of the valence electrons and nuclei proved even more striking. Roughly speaking, electrons with parallel spins keep more nearly apart than those with antiparallel spins, as noted in section 13.2.1. Thereby electrons of each shell are shifted slightly towards smaller or larger radial distances depending on whether their spins are parallel or antiparallel to the spin of valence electrons. (A similar effect occurs for the magnetism of orbital currents.) In about 1960 it was noted that the magnetic field prevailing in the outer shell of an iron magnet ($\sim 30\,000$ G) is amplified by antishielding to tenfold strength in its action on Fe nuclear spins.

13.8. Spectral fingerprints of atoms and molecules [66]

The yellow colour of a flame has long been recognized as indicating the presence of a sodium salt. Nineteenth century spectroscopy traced this colour to emission of light with wavelength of $\sim$580 nm, which has since been resolved into a doublet with the wavenumbers 16 956.183 and 16 973.379 cm^{-1}. Again in the early nineteenth century, Fraunhofer identified characteristic absorption lines in the solar spectrum; by the turn of the century helium was discovered in the solar spectrum, but the absorption by the H$^-$ ion and the ion's very existence were first noticed there in 1938. The unravelling of rare-earth spectra, to the extent of identifying the ground states of these chemically similar elements, was completed after mid-century.

The present section outlines the general character and the current availability of spectral information for diagnostic purposes.

A notable systematics governs the visible and near-visible spectra of atomic elements, which reflect the structure of their valence shells, particularly the number of their s and p ($l = 0, 1$) electrons. These spectra are accordingly similar for all elements of each column of Mendeleev's periodic system. A different systematics emerges instead for the far more complex spectra of the transition and rare-earth groups, elements that differ from one another by the number of their d or f ($l = 2$ or 3) electrons. The distribution functions $f_{nl}(r)$ of d and f electrons peak at shorter radial distances; these electrons thus bear little influence on the chemical properties of each element but their interaction with s and p electrons enrich and complicate the spectra greatly. These features of atomic spectra bear also on the spectra of molecules containing each element, but systematic features are less apparent in molecules owing to their more complex structure.

Astrophysics has been, of course, both a principal consumer and a provider of spectral information, if limited in the past to the visible and near-visible range. This limitation afforded little evidence of the presence of molecules in space. The actual wealth of astrophysical molecular processes has emerged only since the 1960s through the evidence on rotational and vibrational spectra tapped by joining the infrared and radiofrequency ranges (section 13.3.2).

The progress of spectroscopy, mainly on either side of the optical range, rested until mid-century upon the dedicated effort of a modest number of individuals, who improved their measurements, extended their coverage and gladly relayed their results to users. A first comprehensive publication listing atomic energy levels was published by R F Bacher and S Goudsmit in 1932. The growing number of users led W F Meggers, in the late 1940s, to organize—with the support of E U Condon, Director of the US National Bureau of Standards—a standing programme to collect, analyse and produce extensive reports on that subject [67]. (Its last volume, on rare-earth spectra, appeared in 1978.) Dr Charlotte Moore Sitterly, an astrophysical spectroscopist, led that work to the end of her career.

Since that time, the growth of physics personnel, the extension of spectroscopy to far broader frequency ranges (section 13.3), the onset of plasma physics for nuclear fusion and other purposes (Chapter 22) and the spectacular expansion of astrophysics (Chapter 23) have occurred. These combined factors have transformed the task initiated by Meggers into a sizeable industry, growing on its own and in response to needs and opportunities. The NBS (now the National Institutes of Standards and Technology) still constitutes a pillar of this activity; neither it nor other institutions appear to take a comprehensive responsibility for the whole industry.

A useful survey of the present situation, with regard to astrophysics, has emerged from the International Astronomical Union's convening in 1991 a meeting on the subject, which led in turn to a published study of the *Needs, Analysis and Availability of Data for Space Astronomy* [68]. This study includes a report on the recent international 'Opacity Project', organized by M J Seaton to calculate atomic data bearing on the opacity of stellar envelopes

On the plasma side, support has been provided to extend atomic spectroscopy to the study of variously stripped ions. Here, however, the main diagnostic tool appears to consist of the easily identified and simpler spectra of ions isoelectronic to the (monovalent) alkali atoms introduced in section 13.2.3.1 (see also section 13.5.4).

13.9. The role of atoms in metrology and instrumentation [69]

The role of atoms in metrology and instrumentation has become prominent since mid-century. This role rests on two main considerations: (a) molecules of any chemical substance are *indistinguishable* from one another and so are all of their atomic or subatomic constituents; (b) each of their properties remains constant in time. Moreover, most species of atoms or molecules are readily available to experimenters, thus serving as convenient standards.

> See also p 1233

These features emerged strikingly from the 1946 independent discoveries of nuclear magnetic resonance by F Bloch and by E Purcell. The spins of hydrogen nuclei in a water sample subjected to a magnetic field B precess about the field direction at a uniform rate proportional to B; this rate is readily detected and measured by 'tuning in' an induction circuit or a cavity. A water sample and common instruments can thus serve as standards to determine the absolute strength of unknown magnetic fields, less easily measurable otherwise.

A key circumstance complements here the role of atoms: frequency standards of high accuracy and precision are readily provided not only by electric technology, reaching now into the THz range, but also by atomic radiation sources. The international standard of time intervals has been provided since 1971 by a hyperfine frequency transition of ^{133}Cs nuclei in an atomic beam, namely, 9 192 631 770 Hz. Frequency measurements lend themselves to high precision owing to the linearity of typical electrical phenomena that afford superposing currents or voltages and beating (heterodyning) of their frequencies for easy observation and intercomparison. Access to high frequencies—up to 10^9 Hz at mid-century, far higher by the mid-1990s—provides long trains of oscillations with high-quality periodicity.

The accuracy of a periodic motion depends on the length of its oscillation train, which is extremely high for many spectral lines and is reflected in their observed line width as noted earlier (section 13.3.3). Spectral lines emitted by a gas are, however, broadened by the Doppler effect, i.e. by the apparent frequency shift of emissions by atoms moving at different speeds v in the observer's direction, a shift proportional to v/c. This effect is reduced for emissions from beams at very low temperature directed orthogonally to the observer.

Major efforts have been devoted for decades to the task of sharpening line profiles by controlling the temperature of radiation emitters. A further stage has been reached more recently by slowing down individual atoms through light pressures applied simultaneously from different directions, thus trapping, storing and displacing them as desired [70].

Such increasingly refined controls are currently leading to major strides in metrology.

Atomic standards of measurements have been replacing earlier ones gradually. The ^{133}Cs frequency–time standard was adopted officially when it proved more stable than the Earth's rotation rate [71]. Wavelength standards, anchored to interference fringes of atomic emissions from ^{86}Kr, replaced the original 'metre bar' as length units for a time. They have been replaced, in turn, by the distance travelled by light *in vacuo* in 1 s, thus being anchored to the frequency standard by a numerically defined value of the velocity of light c.

This evolving process reflects the continuing progress of technology in comparing measurements of physical quantities greatly differing in magnitude, often by 6–10 orders of magnitude, i.e. basically in the ratio of laboratory and atomic dimensions. These intercomparisons are done by stages of scaling, by factors of 100 to 1000 at a time. Each stage has two components: (a) the 'scoring' of discrete 'steps' such as the passage of wave crests, the turns of precessing motions, or discontinuities in an intensity–voltage plot; (b) the precise measurement of fractions of a single step. The scoring process reflects the nature of atomic elementary processes (section 13.2.2.1). Fractions of a turn are familiarly measured on graduated circles. Basically similar procedures are reported for precision measurements of widely different phenomena.

This universality of the atomistic underpinning of diverse phenomena has been documented further since the 1960s by the unanticipated contribution of the Josephson junction and of the quantum Hall effect to basic metrology. Both of these phenomena involve frictionless conduction in bulk materials. The Josephson junction connects materials in their superconducting state, which implicitly extends properties of single atoms and molecules to macroscopic systems, as pointed out by F London in 1935; the quantum Hall effect deals with electrons confined in one direction but free to drift across it. These surprising contributions to metrology are outlined below.

A Josephson junction consists of a nanometre thick layer of insulating material separating two superconductors held at a small potential difference of V volts. Complementing this constant potential by an AC component of frequency $2eV/\hbar$ causes electron *pairs*, in a singlet spin state, to tunnel through the junction as predicted by B D Josephson in 1962. However, surprise greeted the later publication of a thorough study by B N Taylor *et al* [72], which showed that the resonance frequency $2\,eV/\hbar$ is *fully independent* of the nature and detailed properties of the two superconducting materials. The study thus documented that the pair of materials behaves just like a *single* atom whose electrons are transferred from one to another stationary state by resonant radiation, fully verifying the atomistic stationary character of superconducting states. (Notice, incidentally, that injection of an electron pair into a

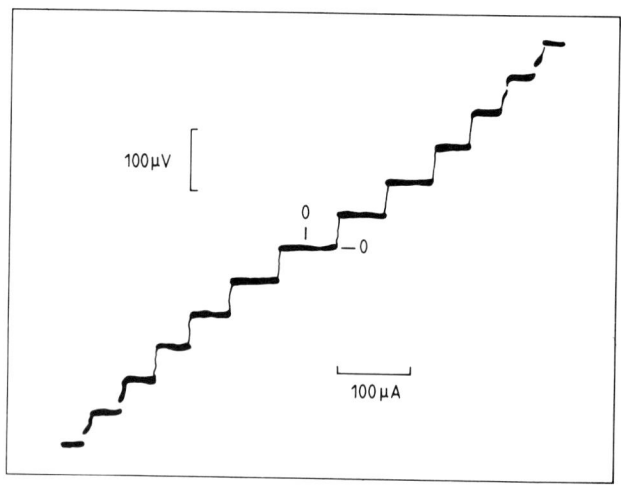

Figure 13.20. *An oscilloscope display of the voltage–current relation for a Josephson junction exposed to 36 GHz microwaves. The sequence of identical supercurrent steps affords precision measurements of the ratio 2e/h.*

system with macroscopic capacitance is hardly perceptible, in contrast to its addition to an atom or to the recently developed 'mesoscopic' systems.) Frequency measurements of the Josephson effect provide the present standard of electric potential.

The Josephson effect can be displayed in alternative fashions, of which the above outline appears conceptually simplest while others appear more striking. If the junction is exposed to microwaves of constant frequency ω and to an increasing potential V that drives through it a supercurrent of intensity i, the $i(V)$ plot rises *stepwise* whenever 2 eV traverses a *multiple* of the photon energy $\hbar\omega$, as shown in figure 13.20. This plot illustrates how the number of successive steps can be scored so as to interconnect quite different potential scales precisely, e.g. the μV scale of this figure and the voltage scale of electrical engineering. Anchoring engineering voltages to the voltage scale of the Josephson phenomenon in this figure, a task performed by Taylor at NBS shortly thereafter, rested on this process.

The normal Hall effect is displayed by an electric current flowing along a conductor strip subjected to a magnetic field orthogonal to the strip. This field pushes the current towards one side of the strip, generating a Hall potential U_H, i.e. a voltage difference across the strip corresponding to a transverse electric field E_H whose strength balances that of the magnetic action. The magnitude and sign of U_H are reciprocal to the concentration and sign of the charge carriers in the conductor; indeed measurements of U_H serve generally to assess that concentration and sign. The quantum Hall effect, which occurs under very special circumstances, measures similarly a fundamental atomic parameter, namely, the fine structure constant α introduced in section 13.2.1.

The carriers in the strip labelled 'surface channel' in the inset of figure 13.21 are electrons squeezed into a narrow layer at the interface of two semiconductors at kelvin temperatures. (The figure is drawn from the original quantum Hall report [73].) The squeezing stems from a 'gate voltage' V_g that controls the electron distribution among energy levels whose spectrum consists of a few energy bands, each band corresponding to one 'Landau level' of motion in a strong magnetic field. The electron density, per unit area, in each filled band is fixed by the magnetic field's strength. Figure 13.21 shows how the Hall voltage U_H remains constant as V_g increases through each gap between a filled and an empty level; the potential U_{pp} *along* the strip remains at or near zero in the gap. Additional electrons move into the layer as V_g enters a new band, causing U_H to decline; U_{pp} rises instead sharply as a new level gets filled. Each *step* value of U_H equals a fixed factor divided by the parameter α and by the number of filled levels, a number that determines the electron density throughout the layer. Here again, as in the Josephson junction studies, we observe the occurrence of level steps of macroscopic quantities, whose value reflects the value of atomic parameters.

Both of these phenomena afford tools for the precise measure of macroscopic variables. The same function is performed by the observation of fixed or moveable sequences of interference or diffraction patterns, whether of microwave, optical or x-ray radiation, whose shift reflects the corresponding fine displacements of mirrors, gratings or crystals. Such displacements are controlled at this time to picometre accuracy. An additional important tool for such fine adjustments has been provided by piezoelectric crystals whose length varies at rates of the order of 0.1 nm V^{-1}.

13.10. Optical handling of atomic systems and transformation of light by atoms

Technologies have arisen since mid-century that channel atomic motions into desired paths through optical actions. Conversely atoms serve currently to transform radiations. These activities have grown into a discipline and an industry loosely identified as 'quantum optics'.

This development was sparked in 1950 by A Kastler's conception of 'optical pumping', an action that transfers light polarization (circular or linear) to atoms in gaseous or condensed phases [74]. More recently one has learned how to steer, cool and trap atomic particles, as anticipated in section 13.9, and to achieve and study multi-photon processes. These several aspects are outlined in the following subsections.

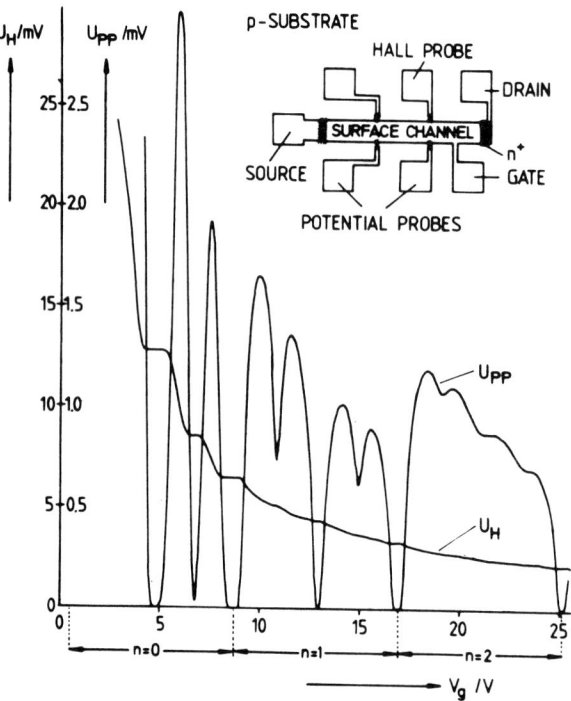

Figure 13.21. Recordings of the Hall voltage U_H and of the voltage drop U_{pp} between the potential probes as functions of the gate voltage V_g at the temperature 1.5 K, constant magnetic field 18 T and source drain current 1 μA. The inset shows a top view of the device, with length 400 μm, width 50 μm and 130 μm distance between the potential probes. The flat portions of the U_H plot play a metrology role analogous to that of the steps in figure 13.20.

13.10.1. Optical pumping

As a prototype process, consider monovalent sodium atoms in their ground state, subjected to a magnetic field pointing in the $\hat{z}$ direction. Atoms with downward spin direction ($j_z = -\hbar/2$) have lower energy than those with opposite spin; thermal collisions nevertheless keep both spin orientations statistically in near balance. Consider now the action of yellow light—in resonance with atomic excitation—propagating along $\hat{z}$ with positive circular polarization, thus imparting to atoms the angular momentum $j_z = \hbar$ per absorbed photon. Each photoexcitation event is followed by re-emission of a photon (fluorescence) with unspecified polarization. The angular momentum imparted by the light thereby shifts the balance of spin orientation towards the $j_z = \hbar/2$ ground-state component, thus 'pumping up' the spin population.

An analogous effect occurs in the absence of the external field, owing to the alternative mutual orientation of the electron's and nuclear spins. If the electron spin was initially pumped in the presence of a magnetic field, removal of this field would allow the hyperfine interaction (Box 13C) to share part of the electron's acquired orientation with the nucleus. Spin polarization of electrons is also transferred to nuclei of other elements, e.g. Xe, in the course of collisions [75].

The ensuing rich phenomenology has been studied extensively [76], achieving widespread application to orient the spins of nuclei and other particles in accelerator beams and in their targets.

13.10.2. Cooling, steering and trapping of atoms

The generic process through which electric fields steer atoms or molecules consists of polarizing them—i.e. of inducing a dipole moment—and thereby attracting them towards positions of lower potential energy [77]. A more specific process imparts to each atom or molecule the momentum $\hbar\omega/c$ of each absorbed photon $\hbar\omega$ in the direction of a light beam; the momenta lost in the successive fluorescence point in average directions thus leaving the absorbers with a net momentum in the beam's direction. 'Cooling' is achieved through this process by tuning the light frequency a little below the resonance of absorbers at rest; the Doppler effect of an absorber's motion causes the light to be absorbed only by atoms moving with more than average thermal velocity opposite to the beam's direction.

A convenient arrangement for steering, and cooling, atoms consists of setting up a standing wave of light resonant or quasi-resonant with an atomic sample. Atoms traversing this wave are strongly polarized and/or excited and thereby driven towards the wave troughs where the light intensity and the atoms' energy are lowest, an arrangement largely developed and utilized in D E Pritchard's laboratory.

'Trapping' is achieved generally by steering atoms into a small volume by converging light beams or static fields. Particularly effective for metrology have proven 'Penning traps' that hold single charged particles (electrons or ions) for long times during which their motion is controlled precisely by oscillating electromagnetic fields. Charge/mass ratios and magnetic moments of such particles have thus been measured with extreme precision, especially by H G Dehmelt and his school.

13.10.3. Multi-photon processes

The phenomena of photo-absorption and -emission by atoms have been introduced in section 13.2.2 as weak processes, whose treatment is elementary, on the grounds that each relevant mode of a radiation field spreads over a volume of space far larger than atoms. The electric field strength corresponding to the 'photon' unit of excitation of each mode is thus extremely small compared with intra-atomic field strengths.

This feature has been altered by the introduction of lasers, whose very principle lies in multiple excitation of specific radiation modes. The action of early lasers, with modest intensity, could still be treated as weak, in the sense that it did allow for the resonant absorption of two or more photons of the same mode in a single elementary process, if only with a small probability.

The introduction of this class of processes nevertheless opened a new field of physics, dubbed *non-linear optics* (Chapter 18), under primary leadership by N Bloembergen in the early 1960s. Simultaneous absorption of two or more photons affords bypassing the selection rules of previous spectroscopy and attaining high excitations by lower-frequency sources. The small probability of multi-photon processes is, however, enhanced whenever the energy of a fraction of the absorbed photons approaches that of an intermediate atomic level, affording thus a 'quasi-resonant' stepping stone.

The 'perturbative' treatment of multi-photon processes grew progressively inadequate as laser intensities increased, particularly with the introduction of short-pulse lasers that release energy stored in a condenser within a short time, currently even in the femtosecond range. An early indication of high-probability multi-photon processes emerged from the discovery of photoionized electrons ejected with alternative kinetic energies differing by one or more photons (above threshold ionization).

The ratio of radiation field strength to the strength of atomic fields (the latter of the order of 100 V nm^{-1} in atomic valence shells) was stated to be exceedingly small in ordinary processes. It has now been boosted even past unity by the combination of several elements: strong laser sources, pulse compression into short intervals, focusing the source image into a spot a few μm across. This ratio attains unity for power fluxes of the order of 10^{17} W cm^{-2}. Still higher fluxes have been seen to pulverize atomic structures.

A distinct process competes with photoabsorption proper as the ratio of field strengths approaches unity, namely, the 'quiver motion' characteristic of the free electrons' response to an AC field of strength E and frequency ω. Its amplitude, $eE/m\omega^2$, is particularly large at the low frequencies of powerful infrared lasers and may even exceed the size of excited atoms. The free-electron aspect of this phenomenon contrasts sharply with the photoabsorption's dependence on intra-atomic forces, which is required to preserve the balance of both energy and momentum (section 13.3.3).

These contrasting features of diverse yet comparably strong actions of intense radiation pulses complicate their joint treatment. Successful, but fragmentary, quantitative evaluations of specific processes have been reported, but no coherent theory has been developed thus far. Whether the more comprehensive and flexible approach of section 13.11.2 will prove adequate remains to be seen.

13.10.4. Transformations of light by atoms

The achievement of multi-photon absorption processes has opened the door to multiplying, and subdividing, radiation frequencies. A transparent example is afforded by the process called 'four-wave mixing'. Basically, an atomic or molecular level excited by absorption of several photons may return to the ground state by emitting a single photon with energy and frequency equal to the sums of those absorbed. This process is, however, restricted by the requirement of parity conservation (Box 13A). Each photoabsorption step generally involves transitions between atomic states of opposite parity that display an oscillating, odd parity, dipole moment. A closed cycle of photo-absorption and photo-emission must thus involve an even number of steps. The earliest opportunity for frequency multiplication is thus afforded by triple absorption followed by a single emission.

Parity conservation may be bypassed by introducing an asymmetry. Frequency doubling is thus typically achieved by light propagation through 'optically active' media, with non-centrosymmetric structure, or—selectively so—at interfaces. An atom, excited to a level $\hbar\omega$ and then *traversing* a cavity (Chapter 18) tuned to support oscillations with frequency $\omega/2$, can yield its excitation to the second excited level of the cavity in a process dubbed 'parametric amplification'.

Combinations of manifold analogous processes have afforded a wealth of applications, many of them bearing on precision metrology or affording novel measurement procedures. These endeavours now occupy a major portion of atomic–molecular science.

13.11. A current overview

Studies of the diverse topics outlined in the previous sections are currently developing over expanding areas and scales which in turn are increasingly being subdivided into smaller branches. This progress, fuelled by the rising input of personnel and of technological innovation, remains constrained conceptually and computationally by reliance on the original independent-particle model of atomic processes. This model implies that interparticle correlations are to be either evaluated and described laboriously for each particle pair, triplet, etc, or else taken into account 'behind the scenes' by computer evaluation of interactions among large basis sets of single-particle 'orbitals'. Such procedures, well suited to their earlier goal of treating a few particles at a time, have grown increasingly laborious and opaque as applied to larger and integrated systems.

The desirability of more flexible models, utilizing appropriate 'collective coordinates', has been apparent for some time. Progress towards this goal has been slow, because innovation implies identifying and overcoming novel hurdles, with benefits that remain elusive and distant for years. Section 13.11.1 will outline phenomenological elements whose theoretical formulation would encompass a large fraction of all atomic–molecular processes. Section 13.11.2 will introduce the hyperspherical approach to multi-particle phenomena, which displays the required flexibility.

13.11.1. A comprehensive phenomenology

An illusory contrast between collision and spectral phenomena has been indicated in section 13.1. To remove its source, consider that the formation and the fragmentation of a complex, in the course of a 'close' collision (section 13.2.4), are actually *reciprocal* processes governed by the *same dynamics*, regardless of differences between the initial and final reactants [78]. The common element of formation and fragmentation lies in the evolution of the complex between any of its compact states and one or another of its alternative fragmented states. This evolution and its inverse are naturally represented by reciprocal parameters.

The excitation of an atom, molecule, or even of a cluster, into a spectrum of alternative levels, following the absorption of a photon, may be viewed as the evolution of a complex towards fragmentation, an evolution powered, for example, by the injection of radiation energy into an initial ground state. Whether this evolution can proceed to complete fragmentation in one channel, A, or be aborted by meeting a potential barrier, depends on whether the available energy does or does not exceed the threshold for full fragmentation in that channel. If it does not, the evolution towards fragmentation in A is inverted, but may again proceed to completion in another channel B with lower threshold.

The formation of standing waves by reflection on a barrier in channel A then generates resonances in the spectrum of fragments in channel B, as described in section 13.4; this phenomenon is called 'auto-ionization' or 'autodissociation' of discrete levels of the 'closed' channel A. The continuity of parameters as functions of energy through thresholds has been stressed in sections 13.3.4 and 13.4.

Photoprocesses may also be viewed as forming a complex consisting of atomic target + radiation energy, much as a negative ion is formed by the combination of an incident electron with a neutral atom. On the other hand, no complex is formed in long-range collisions (section 13.2.4) with large impact parameters, in which the Coulomb field of a charged projectile transfers energy to a distant target much as radiation does (section 13.2.4.1). We thus see that complex formation may or may not involve an extensive structural rearrangement.

The chief common element in the evolution of complexes between their compact and fragmented states lies in the contrasting symmetry features of these states. A complex in its compact state displays central symmetry by rotating freely about its centre of mass. A fragmented state displays instead symmetry about the axis through the centres of mass of separate fragments. This feature was stressed by A R P Rau in the

1970s, jointly with its analogue in the desorption of a particle from a surface. The contrasting symmetry will also lead us to identify the main dynamical element of the evolution of a complex.

13.11.2. Dynamics of the evolution of complexes in the hyperspherical approach

The elements of this dynamics are viewed here as analogues of those introduced in section 13.2 for the dynamics of a single electron. The confinement of an electron within a volume of radius a was stated to boost its kinetic energy by an amount proportional to $1/a^2$. Similarly the confinement of a whole complex into a compact state of radius R about its centre of mass boosts the total kinetic energy of its constituents by a corresponding factor, to be defined in the following. A complex formed by collision or photoabsorption has generally a radius smaller than its equilibrium value. The corresponding excess kinetic energy will drive its subsequent evolution towards fragmentation.

The motion of a single excited electron was resolved, in section 13.2.3, into radial and rotational components. The radial component, represented by a density factor $|f_{nl}(r)|^2$, is governed by the nuclear attraction and by the radial component of the kinetic energy, i.e. by a factor proportional to $1/a^2$ or its equivalent. The rotational component is represented by a directional distribution $|Y_{lm}(\theta,\varphi)|^2$ (Box 13B), whose index l indicates its number of lobes, i.e. in essence the degree of its directional confinement; the kinetic energy associated to this confinement also tends to push the electron outwards, being accordingly often represented as a centrifugal potential proportional to $l(l+1)/r^2$.

The rotational confinement of a multi-particle complex exhibits analogous features, even though more complex than for a single electron. One element of this complication was faced before mid-century in evaluating the rotational energy of multi-electron levels (section 13.2.3.2), a goal attained by the algebraic procedures mentioned in Box 13D. The introduction of hyperspherical coordinates has extended those procedures since the 1960s to evaluate the energy of both rotationally and radially correlated motions of multi-particle states.

13.11.2.1. Hyperspherical coordinates.
Hyperspherical coordinates were introduced in the 1930s (and described in the physico-mathematical text by Morse and Feshbach) mainly to provide a more explicit description of the correlated motion of two electrons in the helium atom. V Fock utilized them in the 1950s to discover a previously overlooked aspect of He dynamics. Figure 13.22 shows how they serve to represent the joint position of the He's electron pair by a hyper-radius $R = (r_1^2 + r_2^2)^{1/2}$ and five angles; four angles identify the directions of the two electrons, while the fifth one represents the ratio $r_2/r_1 = \tan\alpha$ of their respective distances from the nucleus. Wannier's analysis of the double-electron

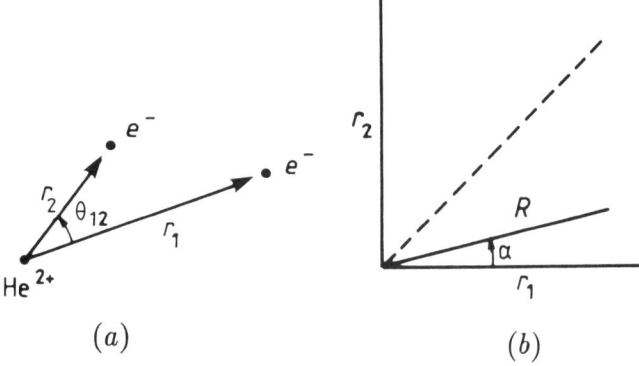

Figure 13.22. *Representation of the joint position of the electron pair in a He atom by five hyperspherical coordinates. The plane containing the nucleus and the electrons is identified by three standard Euler angles (not shown). (a) The two-electron configuration with respect to the nucleus is identified by their radial distances (r_1, r_2) and the angle θ_{12}. (b) The pair of radial distances is identified in polar coordinates (R, α).*

fragmentation of He (section 13.4.2.1) centred on the value $\alpha = \pi/4$ in figure 13.22 as identifying the critical path of the pair towards double ionization at near-threshold energies.

These coordinates allow resolving of the evolution of an He state into a hyper-radial motion, represented by a distribution $|f_{n\lambda}(R)|^2$, and a hyper-rotational motion, represented by harmonic distribution functions in the five angles $|Y_{\lambda\mu'\mu''\mu'''\mu''''}(\theta_1, \varphi_1, \theta_2, \varphi_2, \alpha)|^2$. (The set of indices $\{\mu', \ldots, \mu''''\}$ will be indicated by μ in the following, the set $\{\theta_1, \ldots \alpha\}$ by ω.) Thereby one views radial correlations, represented by the dependence of $|Y_{\lambda\mu}(\omega)|^2$ on α, as fully analogous to their dependence on the angular correlations represented through the angle θ_{12} between the directions (θ_1, φ_1) and (θ_2, φ_2). The harmonics $Y_{\lambda\mu}(\omega)$ are constructed explicitly by standard analytical procedures.

The radial kinetic energy of the electron pair depends on R^2, as it did on the parameter a^2 of equation (1); the rotational kinetic energy is proportional, as for a single electron, to a quadratic polynomial in λ divided by R^2. The Coulomb potential energy between the three helium constituents, averaged over their hyperangular distribution, is represented by an algebraic function of the indices $\{\lambda, \mu\}$ divided by R.

At variance with the case of a single electron, the energies of the hyper-radial and hyper-rotational motions and of the Coulomb interactions do not remain independent as functions of R. Their interdependence has been studied extensively over the last 25 years with notable, if partial, success.

The key effect of the hyperspherical formulation of He mechanics has been to represent implicitly all correlation dynamics among the three constituents of He—one nucleus and two electrons—by algebraic functions of the parameters $\{\lambda, \mu\}$ and of the single radius R.

This representation extends straightforwardly to all atomic–molecular systems.

It suffices to view the positions of N particles as represented by a single point in a $3(N-1)$-dimensional space with the origin of its coordinates at the centre of mass. (He yields $N=3$ and $3(N-1)=6$.) The hyper-radius R needs redefining for particles of different mass; since it represents the radius of inertia for He (regarding the nucleus as infinitely heavy) it may be replaced by the square root $\sqrt{I}$ of the moment of inertia about the systems' centre of mass or, with reference to the total mass M, by $R = (I/M)^{1/2}$. Angular and radial correlations are then represented through harmonics $Y_{\lambda\mu}(\omega)$ of a set $\{\omega\}$ with $3N-4$ elements.

The analytical geometry of these multi-particle spaces was developed largely by Russian nuclear physicists in the 1960s and 1970s, including the all-important transformations among alternative sets $\{\omega\}$, properly viewed as Jacobi coordinates of analytical mechanics [79]. An historical incident appears relevant in this context: molecular theorists, meeting shortcomings of the Born–Oppenheimer approximation (section 13.2.5.1) near a dissociation threshold, bypassed them through an analytical transformation, later identified as leading to hyperspherical coordinates. Similarly, extensive efforts to develop a 'correct' form of the translation factors mentioned in section 13.5.3, are fully bypassed in the hyperspherical frame.

13.11.2.2. The dynamics of the fragmentation process. The dynamics of the fragmentation process hinges primarily on the evolving balance between the complex' kinetic and potential energies, as it does for all mechanical systems. For quantum aggregates of nuclei and electrons, the kinetic energy is proportional to $1/R^2$ and the potential energy to $1/R$, with the 'radius' R defined above as $(I/M)^{1/2}$, because all potential energies are Coulombic. The coefficients of $1/R^2$ and $1/R$ are algebraic functions of the indices λ and $\{\mu', \mu'', \ldots\}$ introduced above. Fragmentation will introduce separate parameter sets $\{R_1, \lambda_1, \mu_1\}$ and $\{R_2, \lambda_2, \mu_2\}$ for the fragments. The total angular momentum J is conserved in the process.

The earmarks of compact states are small values of R and the lowest values of the λ index compatible with a given J value and the maximal correlations—angular and radial—which are characteristic of compactness. (Since the functions $Y_{\lambda\mu}(\omega)$ are, in essence, polynomials of degree λ in the variables ω, their dependence on ω sharpens as λ increases.) Small values of R boost, of course, the kinetic energy factor $1/R^2$ in contrast to the $1/R$ dependence of the potential energy.

The earmark of fragmentation channels—i.e. of each among the alternative channels—consists of a set of $\{\lambda, \mu\}$ indices that makes the potential energy stable—i.e. maximum, minimum or saddle—under small variations of the indices. (This requirement of stability is illustrated by obvious physical examples; for example, dissociation of a molecule leads generally to stable or metastable fragments.) The process of fragmentation is thus described by the evolution of the set of $\{\lambda, \mu\}$ values from the lowest values of λ, earmarks of compactness, to the values of $\{\lambda, \mu\}$ that characterize any specific fragmentation channel. This evolution is paced by the increasing access to larger values of λ afforded, at constant total energy, by the decline of the rotational energy $[\lambda + (3N-5)/2]^2/R^2$ with increasing R.

These features of the evolution, common to most atomic and molecular processes, were pointed out in 1981 [80], in the context of a theoretical procedure that had proved valuable in calculating the lower doubly excited levels of He. However, that procedure failed to yield the observed approach to double ionization along the fragmentation axis $\alpha = \pi/4$ (figure 13.22 and section 13.4.2.1), by disregarding the limited access to high λ values mentioned above [81]. The key role of this limitation for stabilizing states at high potential energies was indeed overlooked until the late 1980s. The progressive access to distribution functions $Y_{\lambda\mu}(\omega)$ with higher λ indices, as R increases, also illustrates how the state of a complex grows progressively less compact, acquiring sharper features in its correlated variables ω. Sharpness of these features is, of course, essential to the contrasting localization of alternative channels on the space of ω variables.

Appreciation of these aspects of the evolution of complexes has encouraged the development of procedures, leading to its description through analytical functions of the $\{\lambda, \mu\}$ indices at increasing values of R. This development, carried out thus far for simple systems, has verified the expectation that the distribution of the indices $\{\lambda, \mu\}$ concentrates along fragmentation channels with characteristic features [82].

References

[1] The evidence for the spin itself, i.e. for an intrinsic angular momentum of each electron, is reported in section 13.2.3.
[2] The occurrence of these units emerges from considerations analogous to those that determine the size of atoms. For monochromatic radiation of frequency ω, the energy within a small volume oscillates periodically between its electric and magnetic forms; it thus mimics the energy of a mechanical oscillator, with mass m, force constant $k = m\omega^2$ and amplitude A, which oscillates between its potential and kinetic peak values $kA^2/2$ and $m\omega^2 A^2/2$. Quantum mechanics sets a lower limit to the energy of *any* variable thus confined, as indicated at the outset of section 13.2.1.
[3] Fermi E 1932 *Rev. Mod. Phys.* **4** 87
[4] Heitler W 1936 *Quantum Theory of Radiation* 5th edn (Oxford: Clarendon)
[5] The splitting of Ag beams into two components was discovered in the early Stern–Gerlach experiments prior to quantum mechanics, its implications remaining unclear.

[6] Wigner E P 1931 *Gruppen Theorie und ihre Anwendungen* (Braunschweig: Vieweg) (Engl. Transl. 1959 *Group Theory* (New York: Academic))
[7] Van der Waerden B L 1932 *Die Gruppentheoretische Methode in der Quantenmechanik* (Berlin: Springer)
[8] Racah G 1949 *Phys. Rev.* **76** 1352; 1965 *Group Theory and Spectroscopy (Springer Tracts in Modern Physics 37)* (Berlin: Springer) and references therein
[9] Wigner E P 1959 *Group Theory* (New York: Academic)
[10] Condon E U and Shortley G H 1935 *The Theory of Atomic Spectra* (Cambridge: Cambridge University Press)
[11] Mott N F and Massey H S W 1933 *The Theory of Atomic Collisions* (Oxford: Oxford University Press) (5th edn 1965)
[12] The probability of impulsive collisions decreases as the projectile's velocity increases because the projectile acts for a shorter time. On the other hand the collision becomes elastic at low velocities. The probability of inelastic collisions thus appears to peak for $b\omega/v \sim 1$. Analogous considerations have often been presented under the name of '*Massey criteria*'. Their simplistic application may, however, be vitiated when the amount of energy $\hbar\omega$ required by a transition within the target depends on the impact parameter b. An instance of such strong dependence is described in section 13.5.2.
[13] Bethe presents his treatment as a weak (Born) approximation that yields a probability amplitude of momentum transfer $\hbar k$ *linear* in the Fourier coefficient $1/k^2$ of equation (3). In this respect his calculation of the projectile's Coulomb field action on each electron is actually exact, in both quantum and classical mechanics. Indeed Rutherford's classical formula yields a probability of momentum transfer between colliding charges proportional to $1/k^4$, with the same coefficient as Bethe's.
[14] Bohr N 1948 The penetration of atomic particles through matter *Kgl. Via. Selsk. Mater.-Fys. Medd.* **18** 8
[15] Mulliken R 1930 *Rev. Mod. Phys.* **2** 60 and 506; 1931 *Rev. Mod. Phys.* **3** 89; 1932 *Rev. Mod. Phys.* **4** 1
[16] Herzberg G 1950 *Spectra of Diatomic Molecules* (Princeton, NJ: Van Nostrand); 1954 *Infrared and Raman Spectra* (Princeton, NJ: Van Nostrand); *Electronic Structure and Spectra of Polyatomic Molecules* (Princeton, NJ: Van Nostrand)
[17] Dibeler V H and Reese R M 1964 *J. Res. NBS* A **68** 409
[18] Lukirskii A P, Rumsk M A and Smirnov L A 1960 *Opt. Spectrosc.* **9** 262
[19] Weissler G L 1956 *Encyclopedia of Physics* vol 21, ed S Fluegge (Berlin: Springer) p 306
[20] Fano U and Cooper J W 1968 *Rev. Mod. Phys.* **40** 441
[21] Comer P 1970 *Ann. Rev. Astron. Astrophys.* **8** 269
[22] Polar molecules in interstellar space receive instead energy from collisions and dissipate it by radiation.
[23] Available from the US National Institute of Standards and Technology, Report NBSIR 87-3597 by M J Berger and J Hubbell.
[24] Schulz G J 1963 *Phys. Rev. Lett.* **10** 104
[25] Buckman S J and Clark C W 1994 *Rev. Mod. Phys.* **6** 539
Compton R N 1993 *Negative Ions* ed V A Esaulov (Cambridge: Cambridge University Press)
[26] Fano U and Rau A R P 1986 *Atomic Collisions and Spectra* (Orlando, FL: Academic) ch 7 and 8
[27] Seaton M J 1983 *Rep. Prog. Phys.* **46** 167 and references therein
[28] Wannier G 1953 *Phys. Rev.* **90** 817

[29] Fano U 1983 *Rep. Prog. Phys.* **46** 258
[30] Buckman S J, Hammond P, King G C and Read F H 1983 *J. Phys. B: At. Mol. Phys.* **16** 4219
[31] The term 'R matrix' was introduced by Wigner, in nuclear physics, in about 1948 to denote 'short-range' adaptations of 'reaction matrices', called K in scattering theory, which contain the same information as the transition matrices $(i|T|f)$ in symmetrized form. The K and T matrices are related by $K = T/(1 + iT)$ and $T = K/(1 - iK)$; the transition element $T_l = \exp(i\delta_l)\sin\delta_l$ of Box 13F corresponds in this notation to $K = \tan\delta_l$, as easily verified. The matrix $(i|K|f)$ is similarly obtained from $(i|T|f)$ by matrix algebra. Short-range K matrices may be denoted by $K^{(s)}$ instead of R.
[32] Greene C H, Fano U and Strinati G 1983 *Phys. Rev.* A **28** 2209; 1988 *Phys. Rev.* **38** 5953; *Rev. Mod. Phys.*
[33] Lutz H O, Briggs J S and Kleinpoppen H (ed) 1985 *Proc. NATO Advanced Study Institute S Flavia, Italy* (New York: Plenum) p 51
[34] A parallel independent experiment was done by S Takezawa.
[35] Herzberg G and Jungen Ch 1972 *J. Mol. Spectrosc.* **41** 425
[36] The behaviour is represented analytically by replacing the product πv_0 in equation (13) by an index $-\Delta$, where Δ represents the phase shift of the photoelectron exiting from the atom in the open channel $f \equiv N=0$. The equation thus modified is readily solved to yield the value of Δ as a function of the energy E or of its equivalent value of v_2, from equation (11). The modulation of the photoelectron intensity is represented by the derivative $d\Delta/d\pi v_2$. The modulations of the density of spectral lines and of their intensities, displayed in figure 13.15 below the lowest threshold E_{t0}, is conveniently viewed as an appendage of the pattern prevailing above the threshold and reflecting the presence of the closed channel $f \equiv N = 2$.
[37] Greene C H and Jungen Ch 1985 *Adv. At. Mol. Phys.* **21** 51
[38] Lu K T 1971 *Phys. Rev.* A **4** 579
[39] The content of this section 13.5 derives largely from conversations with C D Lin (Kansas State University) and from references provided by him.
[40] Andersen N O, Gallagher J W and Hertel I V 1988 *Phys. Rep.* **165** 1
[41] Morgan G H and Everhart E 1962 *Phys. Rev.* **128** 662
Afrosimov V V, Gordeev Yu S, Panov M N and Fedorenko N V 1965 *Zh. Tekhn. Fiz.* **34** 1613ff
Kessel Q C, Russek A and Everhart E 1965 *Phys. Rev. Lett.* **14** 484
[42] Ovchinnikov S U and Solov'ev E A 1986 *Sov. Phys.–JETP* **63** 538
[43] Ziemba F P and Everhart E 1959 *Phys. Rev. Lett.* **2** 299
[44] In the process of electron capture from the target into a projectile's field, the analytical description of the electron's motion must display explicitly its absorption of the necessary momentum. This requirement was met initially by Bates' group by inserting an *ad hoc* translation factor. The correct form of this factor has been discussed for decades in dozens of contributions. Only in the 1980s was the *ad hoc* approach itself found to be inadequate, an observation not yet widely known to which we shall return in section 13.10.
[45] Fritsch W and Lin C D 1991 *Phys. Rep.* **202** 1
[46] Richard P, Stockli M, Cocke C L and Lin C D (ed) *VIth Int. Conf. on Physics of Highly Charged Ions (AIP Conf. Proc. 274)* (New York: AIP)
[47] Discussions, data and critical readings of this section have been provided by R S Berry, D H Levy and other colleagues.

[48] Velocity selection by filtration through rotating wheel slots is costly in terms of beam intensities. Experimental art lies in finding suitable compromises.
[49] Herschbach D R 1966 *Adv. Chem. Phys.* **10** 319
[50] Anderson J B, Andres R P and Fenn J B 1966 *Adv. Chem. Phys.* **10** 275
[51] Smalley R E, Levy D H and Wharton L 1976 *J. Chem. Phys.* **81** 5417
[52] Dehmer J L, Parr A C and Southworth S H 1976 *J. Chem. Phys.* **64** 3266
See also, Nenner I *Handbook on Synchrotron Radiation* ed G V Marr (Amsterdam: North-Holland) p 355
[53] The retention of electrons by centrifugal potentials is also apparent in atomic photoionization of high-l inner shells, typically in $4d \rightarrow 4f$ transitions of rare-earth atoms (section 13.7).
[54] La Villa R and Deslattes R D 1966 *J. Chem. Phys.* **44** 4399
[55] Howard B J, Dyke T R and Klemperon W 1984 *J. Chem. Phys.* **81** 5417
[56] Haberland H (ed) 1994 *Clusters of Atoms and Molecules* (Berlin: Springer)
[57] Bagus P S and Schaeffer H F III *J. Chem. Phys.* **56** 224
[58] Oka T 1992 *Rev. Mod. Phys.* **64** 1141
[59] Coulson C A 1960 *Rev. Mod. Phys.* **32** 170
[60] Truong T N and Truhlar D G 1990 *J. Chem. Phys.* **93** 1761
[61] Manz J 1986 *Commun. At. Mol. Phys.* **17** 91
[62] Launay J M and Lepetit B 1988 *Chem. Phys. Lett.* **144** 346
[63] This section owes much to advice and material provided by Professor W Mehlhorn (Freiburg) and by R D Deslattes (NIST) on x-ray measurements.
[64] Mooney T, Lindroth E, Indelicato P, Kessler E G and Deslattes R D 1992 *Phys. Rev. A* **45** 1531
[65] Mehlhorn W 1990 *X-Ray and Inner Shell Processes (AIP Conf. Proc. 215)* ed T A Carlson *et al* (New York: AIP) p 465
[66] Major contributions to the material of this section and of the preceding section 13.2.3 have been provided by Dr William C Martin (NIST).
[67] Moore C E 1971 *Atomic Energy Levels (NSRDS-NBS 35, vol I-III)* (Gaithersburg, MD: NBS)
Martin W C, Zalubas R and Hagan L *Atomic Energy Levels (NSRDS-NBS 60, vol IV)* (Gaithersburg, MD: NBS)
[68] Smith P L and Wiese W L (ed) 1992 *Atomic And Molecular Data for Space Astronomy (Lecture Notes in Physics 407)* (Berlin: Springer)
[69] The content of this section derives largely from extended conversations with Dr R D Deslattes (NIST) and from references and illustrations provided by him.
[70] Cohen Tannoudji C N and Phillips W D 1990 *Phys. Today* **43** 33
[71] Petley B W 1985 *The Fundamental Physical Constants* (Bristol: Hilger) p 15ff
[72] Taylor B N, Parker W H and Langenberg D N 1969 *Rev. Mod. Phys.* **41** 375
[73] von Klitzing K, Dorda G and Pepper M 1980 *Phys. Rev. Lett.* **45** 494
[74] Kastler A 1950 *J. Phys. Radium* **11** 255
[75] See, for example, Schaefer S R, Cates D G and Happer W 1990 *Phys. Rev. A* **41** 6063
[76] Fano U and Macek J H 1973 *Rev. Mod. Phys.* **45** 553
[77] This process parallels the magnetic steering displayed in figure 13.3, except that electrons have an intrinsic spin dipole magnetic moment, whereas atoms are normally unpolarized. Many molecules, e.g. H_2O, have an intrinsic electrical moment, that is usually averaged out by thermal rotation.
[78] The probabilities of a collision starting from complex formation in channel A followed by fragmentation in channel B and of its reciprocal collision,

         $B \to A$, may nevertheless differ owing to differing densities of states for alternative reactants.
[79] Smirnov Yu F and Shitikova K V 1977 *Fiz. Elem. Chast. At. Yadra* **8** 847 (Engl. Transl. 1977 *Sov. J. Part. Nucl.* **8** 344)
[80] Fano U 1981 *Phys. Rev.* A **24** 2402
[81] This failure has since been remedied by S Watanabe through an extensive calculation of the necessary corrections.
[82] Fano U and Sidky E 1992 *Phys. Rev.* A **45** 4776; 1993 *Phys. Rev.* **47** 2812
Bohn J 1994 *Phys. Rev.* A **49** 3761; *Phys. Rev.* A **50** 2893; *Phys. Rev.* A **51** 1110

Chapter 14

MAGNETISM

K W H Stevens

14.1. Introduction

14.1.1. Magnetism prior to the twentieth century

At the beginning of the nineteenth century the phenomenon of magnetism, which had been known for many centuries, was assumed to be due to the existence of magnetic dipoles, pairs of closely spaced equal and opposite magnetic poles, a concept first introduced by Gilbert in the sixteenth century. By the end of the century this picture had almost entirely disappeared and magnetism was then regarded as due to tiny circulating electric currents, each of which would set up a magnetic field which would be indistinguishable from the field of a magnetic dipole. Thus magnetism and electricity had become different aspects of a unified description, which made free use of the field concept, that at each point in space there would be electric and magnetic fields which would have magnitudes and directions. In fact for both there were two fields, denoted in the case of magnetism by B, the *magnetic induction*, and H, the *magnetic intensity*, two vector quantities related by $B = H + 4\pi M$, where M denoted the *intensity of magnetization*, which also had magnitude and direction. (This is the relation in the cgs system of units then in use.) At any point in a vacuum M was taken to be zero; B and H then became identical, which raised the question of which to use and what to call it. The convention chosen was to use H and call it the magnetic field.

The introduction of B, H and M was the outcome of attempts to formulate the theory without assumptions about microscopic mechanisms, and yet incorporate the idea that the field inside a magnetized specimen is not necessarily the same as the external field creating the magnetization. It has led to a good deal of confusion for teachers and students, not least in reconciling this continuum theory with atomistic models. One of the problems is Lorentz's demonstration that when magnetic effects are ascribed entirely to currents the average magnetic field, h, (taken over all space, inside and outside atoms) is the same as B. Since these various problems are man-made, and evaporate in a consistent application of Lorentz's methods and their successors in atomic physics, I shall avoid using B and H as far as possible, except in free space, where they are identical, and in experimental curves where the originators have used them. The magnetization, M, however, is a fundamental concept and its microscopic interpretation plays a central role in this story.

For most of the nineteenth century magnetic phenomena were regarded as static, and the advent of Maxwell's electromagnetic equations, late in the period, made little difference. Even now most investigations of magnetic phenomena which involve electromagnetic waves use frequencies for which the wavelengths are long compared with some relevant dimension of the sample of interest.

It was mainly due to Faraday that in the second half of the nineteenth century it had at last been realized that there were several different types of magnetic behaviour. Lodestone had long been known for its magnetism and had been turned to practical use in compasses. Iron too had also been found to have magnetic properties, for it was attracted to lodestone, but after these two discoveries centuries were to go by before any further additions were made. The next significant step occurred in the eighteenth century when cobalt and nickel were also found to have magnetic properties similar to those of iron, though it was not until the nineteenth century that there was reasonable certainty that the properties of cobalt could be attributed to the pure element. So at the beginning of the twentieth century there were just three known elements, iron, cobalt and nickel, some of their alloys and lodestone which had strong magnetic properties.

Iron and its alloys had, of course, become extremely important in the development of structural engineering; the fortunate combination of its strength and magnetic properties, together with Faraday's discovery of electromagnetic induction, had resulted in the establishment of the electrical power industry. Thus by 1900 some quite advanced electrical machinery existed along with a range of sensitive instruments for measuring electrical currents and voltages. The combination of high-field electromagnets and improved instrumentation meant that it had become possible to make measurements of magnetic properties which would have been quite impossible in the early part of the previous century.

Faraday, about the middle of the 1840s, also initiated a general study of the magnetic properties of gases, liquids and solids. This work produced no further examples of strong magnetic materials but it did show that weak magnetism appeared to be a universal property. As a result he could begin dividing the weakly magnetic materials into two classes, according to whether they were repelled from a region of high

Magnetism

magnetic field or attracted into it. The former were said to be *diamagnets* and the latter *paramagnets*; a diamagnetic substance, when placed in a magnetic field, develops a magnetic moment in the opposite direction to the field whereas a paramagnetic substance develops a moment in the direction of the field. To these classes we must add the *ferromagnets* (Fe, Co, Ni), but not lodestone, which is nowadays classed as a *ferrimagnetic*, and the *antiferromagnets*, both of which are much later discoveries.

It is also possible to subdivide the various magnetic solids according to whether or not they are electrical conductors. This separates lodestone from the ferromagnetic metals and alloys as well as separating the diamagnets from most of the paramagnets. For the normal metals, those which do not show ferromagnetism, the paramagnetism shows little or no temperature dependence, whereas for most of the insulators it is diamagnetism which shows little temperature dependence. Thus the electrical behaviour separates the paramagnets from the diamagnets except for a small class, mainly the insulating salts of the 3d transition elements and the rare-earths, which show pronounced temperature-dependent paramagnetism.

In describing the historical development a classification based on electrical properties has seemed to be slightly more convenient than that introduced by Faraday. Table 14.1 gives some idea of the range of topics to be covered, and the main theoretical concepts. The century itself will be treated in three periods—before the introduction of quantum mechanics (1900–25), up to the end of World War II and the subsequent years of reconstruction (1925–50), and the rest of the century (1950 onwards). In the third period, as well as the extension of the concept of antiferromagnetism to ferrimagnetism, two major new fields were opened, the study of the rare-earths and the actinides, and nuclear magnetic resonance. The first can be incorporated in the general scheme, but the second is so different that it has seemed best to regard it as a separate topic and describe it separately. Then towards the end of this period developments in technology opened further fields of study, two based on random systems, spin glasses and amorphous metals, and another on the properties of very thin films. These also are best treated as separate topics.

14.2. The period 1900–25

14.2.1. Diamagnetism

The investigations of magnetic properties, begun by Faraday, were gradually extended, particularly by Curie [1], with the result that by the end of the first quarter of this century it had been found that the majority of gases (except O_2, NO, NO_2 and ClO_2), liquids and solids were diamagnetic, the exceptions being the ferromagnets, most compounds containing ions of the 3d elements (Ti, V, Cr, Mn, Fe, Co, Ni and Cu) and

Table 14.1. *The top part of the table lists the topics of special interest in two periods, 1900–50 and 1950 onwards. The bottom part, under 'Theoretical concepts', numbers the main theoretical ideas. Their numbers are attached to the topics in the top half, indicating where they have been used.*

	1900–50	1950 onwards
Insulators	Diamagnetism 7 Curie–Weiss paramagnetism 1, 2	Antiferromagnetism 2, 4 Ferrimagnetism 2, 4
Conductors	Temperature-independent paramagnetism 3 Ferromagnetism 2, 3	Rare-earth metals 2, 3, 4
		Nuclear magnetism 5 Spin glasses 4 Amorphous alloys 6 Thin films 2, 3, 6
Theoretical concepts	1. Crystal fields 2. Exchange interactions 3. Band theory 4. Spin-Hamiltonians 5. Bloch equations 6. Correlated electrons 7. Closed shell ions	

the normal metals, the non-ferromagnetic conductors. It had also been established that their susceptibilities, usually denoted by χ, and defined as the ratio of the magnetization to the applied magnetic field in the weak-field limit, were independent of temperature and of the strengths of the fields then available. Weber, in 1854, had already suggested that molecules could be divided into two classes according to whether or not they had resultant magnetic moments due to circulating currents. On the supposition that an applied field would induce circulating currents in molecules which otherwise had no such currents, the characteristics of diamagnetism could be explained. The molecular model was by no means generally accepted, and Weber's views were fulfilled only with Lorentz's electron theory of matter, initiated in 1892 and given a firm basis by Thomson's discovery, in 1897, of what we now know as the electron.

Among the many phenomena Weber analysed was the Zeeman effect in spectroscopy. Voigt [2] and Thomson [3] had already tried to develop an electronic theory of magnetism, with limited success, and it was Langevin [4] who pointed out that the Zeeman effect and diamagnetism were different aspects of the same phenomenon. The application of a magnetic field would modify all orbital motions by inducing precessions about the field direction, so creating changes in each molecular magnetic moment. Furthermore, the theory could be used to estimate the average sizes of the electron orbits from the experimental results, and Langevin noted that the orbits were small enough to be contained within molecules.

Lorentz went further, suggesting that the induced precession was due to the force on each electron from the electric field set up as the magnetic field was increased from zero, and that there would need to be a massive and immovable positively charged sphere present to maintain electrical neutrality. (It is remarkable that although planetary motion was well understood there seems to have been a marked reluctance to envisage molecules as mostly empty structures.) He also expressed concern that the precession, once it had been set up, seemed to persist indefinitely, which appeared contrary to physical experience: explaining why was to dog magnetic theory until the advent of quantum concepts.

Lorentz's misgivings were only too well founded since, as van Leeuwen [5] showed in 1919, a system of charged particles governed by classical mechanics can have no magnetic properties in a state of thermal equilibrium. This was remarked on by Bohr [6] in his doctoral thesis but became widely known only through van Leeuwen's work.

14.2.2. Quantum concepts

The justification of Langevin's model came with the Bohr [7] theory which supplemented the Rutherford model of the atom, a central, massive and positively charged nucleus about which the electrons were circulating as in a tiny planetary system, with a quantum postulate which restricted the possible atomic energies to a number of discrete values. In particular an atom or ion would have a lowest energy level from which there could be no decay. Thus the problem of motion with no dissipation in energy disappeared.

The initial impact of the Bohr theory was, however, much greater in spectroscopy than in magnetism, where an extensive theoretical structure was constructed and used to interpret a wide range of experimental observations. Three quantum numbers were introduced, one associated with each degree of spatial freedom, with relations between them. (The position of any particle can be described in a variety of ways using three 'coordinates'. Those chosen for an orbiting electron were its distance from the nucleus, and two angles, one of which was in the plane of its motion and the other gave the angle of the normal to the plane relative to some fixed direction.) In due course these were found to be insufficient and a fourth quantum number was added, associated with the assumption that the electron was spinning about its axis.

Magnetic moments had already been associated with two of the original quantum numbers, those related to angular variables, and it was natural to associate magnetic moment with the spin of the electron: the only strange feature, apart from the spin taking a half integer quantum number, was that from the study of the Zeeman effect it appeared that the magnetic moment associated with the spin angular momentum was twice that associated with orbital momentum. To complete the picture it was also necessary to assume [8] that no two electrons could have the same four quantum numbers (Pauli's exclusion principle).

The restriction imposed by this last assumption had direct consequences for magnetism because it led to the picture of the shell structure of atoms and ions and the recognition that in the diamagnetic insulating compounds the electronic structures were those of ions with all the occupied shells completely filled. Thus all the electrons could be regarded as paired off; with neither a resultant orbital nor a resultant spin angular momentum there would be no net magnetic moment. Furthermore, it was realized that the paramagnetic insulators all contained ions with electrons in incompletely filled shells and that non-zero magnetic moments could well be associated with these.

The diamagnetism of closed shells was accounted for by the precession of the orbits in the presence of a magnetic field, an extension of Lorentz's proposal [9]. Then the temperature independence followed because the electronic arrangement was so stable. To unpair any of the electrons by, say, moving one of them to an unoccupied orbital would need far more energy than was available from thermal fluctuations. So such transitions would not occur, a conclusion which also explained why these diamagnets were insulators.

With just the Bohr theory no further progress was possible.

14.2.3. Paramagnetic insulators

In 1895 Curie reported that a number of substances showed susceptibilities which as well as differing from diamagnetism in the direction of the induced moment also differed in having a strong temperature dependence; the results could be fitted to the relation $\chi = C/T$, where T is the absolute temperature and C is a constant which depended on the nature of the sample. The relation is now known as Curie's law. As the temperature range was extended it was found that a better fit was usually obtained with the Curie–Weiss law, $\chi = C/(T - \Delta)$, where Δ was positive for some substances and negative for others. There were no measurements down to temperatures of the order of the magnitudes of the Δs, so the corrections to Curie's law were small. In 1905 Langevin put forward an explanation of the Curie form using the assumption that not only could an external magnetic field induce precessions but it would also tend to line up any permanent magnetic moments. The alignment would not be complete because of the randomizing effects of temperature, which could be taken into account using Boltzmann's statistics. At low field strengths the magnetization M is proportional to the applied field, but in very strong fields it saturates at M_0, when all individual moments are aligned with H. The detailed expression for the variation of M with H is still known as Langevin's function

$$M/M_0 = \coth a - 1/a$$

Magnetism

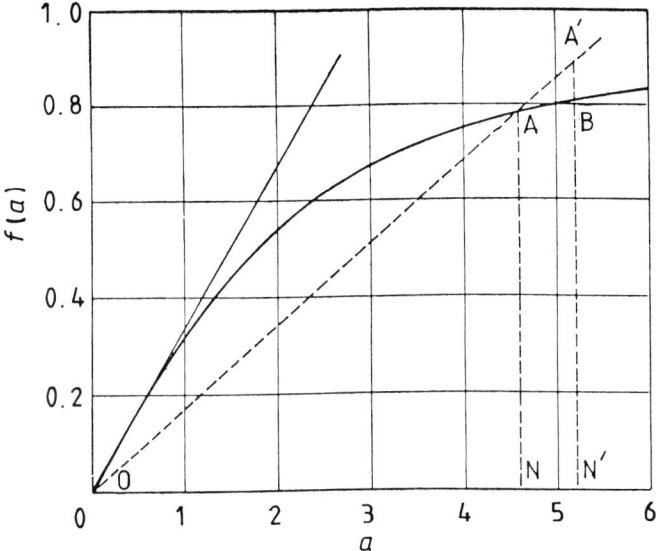

Figure 14.1. *Langevin's graph of the function* $\coth a - 1/a$ *(the curved solid line).*

where $a = \mu H/kT$, μ being the magnetic moment of a molecule (figure 14.1). As with Langevin's theory of diamagnetism this analysis also falls foul of van Leeuwen's theorem, but is restored by Bohr's assumption of stable orbits giving rise to permanent molecular moments.

Debye [10], who had based his theory of polar molecules on Langevin's work, modified the details of the argument to take account of the concept of spatial quantization that had developed from Bohr's atomic theory. The permanent magnetic moment could not be assumed capable of pointing at any angle to H, but only at a few discrete orientations. The theory was expected to be only strictly applicable when the magnetic interactions between different molecules could be neglected, as in gases, but from the observation that the law held more widely it could be inferred that the restriction was probably unnecessary.

This conclusion received strong support from Lorentz's treatment of the interaction of electric dipoles. He showed by direct calculation that if the molecules in a spherical sample are arranged in a cubic array they produce no internal field; each dipole is acted upon by the external field as if all the others were absent. The same was expected to apply to the magnetic case, for small fields and high temperatures, particular as the strength of the magnetization is then usually so small that there is a negligible difference between the spherical sample and a sample of any other shape (the restriction to a cubic array was not expected to be of any significance).

In spite of this result Weiss [11] took the remarkably bold step of extending the Langevin theory by assuming that the field acting on a molecular magnet would be the externally applied field plus a substantial extra field proportional to the magnetization. This led him to substitute $H + \lambda M$ in place of H in the Langevin expression for the paramagnetic moment, taking λ to be positive. The expression for the magnetization for a sample obeying Curie's law, $M = CH/T$, was changed to $M = C(H + \lambda M)/kT$, which could be re-arranged to give $M = CH/(T - \lambda C)$, the same form for the susceptibility, M/H, as that of the Curie–Weiss law, when Δ was identified as λC. Weiss must have realized that if this was to be the way of accounting for the Curie–Weiss law it was quite certain, from the known values of the Δs, that they could not be due to the known interactions between the magnetic dipoles. He was therefore led to postulate that some other interaction, of unknown origin, must be present, which he presciently suggested might arise from the electrical forces between the molecules.

See also p 545

Even more remarkably, when he examined the modified formula of Langevin, without making approximations about the temperature and the strength of the magnetic field, he found that quite a different behaviour emerged at low enough temperatures and with H put to zero. There were two solutions for M. It would either be zero or it would have a finite value which depended on T, and the solution with the finite M would have the lower energy. This suggested that a system obeying the Curie–Weiss law with a positive Δ could be expected to develop a spontaneous moment at low enough temperatures. The theory also showed that this moment would grow from zero as the temperature was further lowered from a critical value (now usually referred to as the Curie temperature).

That a spontaneous magnetization might arise can be inferred by supposing that a uniform magnetization has developed in the absence of an applied field. Weiss's assumption would then imply that there would be an internal field in the direction of this magnetization which would create a magnetization in the direction of the internal field. A self-consistency argument would then suggest that the initial spontaneous magnetization should be identified with the magnetization it actually produced.

It may seem surprising that the basis of Weiss's result can be understood by such a simple argument, but like most of the advances in physics something has to initiate events. Similar reasoning is now used in a variety of contexts, often under a description which includes the term 'self-consistency'. It seems that this was the first example of its use in a solid-state context, where it provided an explanation of a solid-state phase transition from paramagnetism to ferromagnetism.

The implications for experimental physics were considerable. It had already been suggested, by Curie, that ferromagnets would become paramagnetic at high enough temperatures. Now it seemed

Pierre Ernst Weiss

(French, 1865–1940)

Physics has always exploited its discoveries and has also had room for the experimentalist whose interest is in facts rather than the current notions of theoreticians. Pierre Ernst Weiss, who was born in Mullhouse, France, spent almost the whole of his professional life working in magnetism. Faraday had suggested that all materials would have magnetic properties and it was Weiss, using the most sensitive equipment then available, who confirmed this suggestion, exploring, with his students, a large number of substances over extended ranges of temperature. Beginning in Zurich he eventually moved to Strasbourg, in 1919, where he created a centre for magnetic studies which still exists. His discoveries showed the limitations of current theory, which he proceeded to extend, by assumptions which, although they must have been anathema to the theoreticians of the day, explained his observations and produced the first theory to explain how phase transitions could occur in solids. In his ideas he was far ahead of his time. His ideas can be found, along with many experimental results, in *Le Magnetisme* (1931) [12], a book with G Foëx as co-author, which, incidentally, makes no use of quantum mechanics!

that examples were to hand of paramagnets which would become ferromagnetic at low enough temperatures. However, the necessary low-temperature facilities were not available to test this, so advantage was taken of another prediction, that the paramagnetic magnetization would not remain proportional to the field strength as the field increased and that the departures from linearity would be easier to observe as the temperature approached that of the phase transition, which was predicted to occur when the temperature equalled Δ.

All this was a considerable stimulus to the development of low-temperature facilities, and indeed the interest in producing ever lower temperatures and research in magnetism have all along been closely linked. Nevertheless the experimental demonstration of a paramagnetic salt becoming ferromagnetic at low temperatures failed, in spite of there being a number of positive Δs. Weiss and Foëx [12] in 1931 gave a table of values about half of which were positive. In fact it is now known that ferromagnetism is uncommon in the salts. It seems probable that the observations were made on samples containing unidentified impurities. A tiny amount of unrecognized ferromagnetic iron could have invalidated the observations.

Having accounted for the $1/(T - \Delta)$ part of the Curie–Weiss law there remained the question of C, which reflected the magnitudes of the ionic magnetic moments. They should have been obtainable from the Zeeman splittings of the ground states of the ions. However, it was found that in most cases the values obtained for C did not agree with the experimental values. (Nor did the Bohr theory explain the Zeeman observations.)

This was the position prior to the introduction of electron spin and its anomalous magnetic moment. Once it was included the problems over the Zeeman splittings largely disappeared, but this did not clear up the problem of the C values. Then it was realized that much better agreement could be obtained if the magnetic moments in the ionic compounds were assumed to be entirely due to electron spin, so creating the problem of accounting for the disappearance of orbital contributions in the crystals. The Bohr theory gave no answer, though the position was somewhat improved when Hund [13] showed that the Zeeman observations on a range of rare-earth ions could be used to explain most of their Cs. Van Vleck and Frank [14] subsequently showed that his failure on two of the ions had a simple explanation.

A failure to obtain a fit to experimental results because some feature has been accidentally omitted is unfortunate, but when the error is subsequently corrected there is an almost overwhelming impression that one's understanding of what is happening must be correct. Since the rare-earth ions were known to have magnetic properties due to electrons in 4f shells, which were well inside the ions whereas the magnetic properties of the 3d transition ions were known to be due to electrons on the outsides of the ions, it was an obvious step to assume that in some way the difference between the two families arose because the 3d ions were more sensitive to their crystallographic environment. The Bohr theory, however, gave no help.

14.2.4. Ferromagnetism

Although the Weiss theory explained how ferromagnetism might occur when a paramagnet is cooled sufficiently there was a problem with applying this argument, because freshly prepared iron at room temperature has no magnetic moment. To cope with this situation Weiss first assumed that a typical iron sample would consist of a random assembly of single crystals. He then assumed that in each single crystal the direction of the spontaneous moment would be in a direction determined by its crystal structure. A random distribution of crystallites would therefore have no net moment. In a uniform external magnetic field only a few of the crystallites would have their moments lying parallel to this field, so he next considered those crystallites with the opposite alignment assuming that on increasing the field from zero nothing much would happen until the external field reached a critical value, when the magnetization would suddenly completely reverse. In fact, of course, the bulk of the crystallites are not aligned in either of these directions and for these he assumed that those lying near the field directions would not be affected but those lying in the 'wrong' direction would eventually reverse their moments, when the component of the applied field in the direction of their axes reached the critical value. Thus the overall picture, on gradually increasing the external field, would show evidence of moment reversals taking place successively as the field reached the critical value for the different crystallites. Eventually all the wrongly aligned moments would be reversed, after which there would be no further increase in overall moment. On reducing the field nothing much would happen and the system would retain its magnetization until the field reached zero. On increasing it again, in the opposite direction, moment flips would begin as the axial component of the field reached the critical value. In this way Weiss was able to explain how an initially unmagnetized sample would acquire a moment in an applied field and how the moment would change under field cycling [15].

Field cycling behaviour (hysteresis, the term introduced by Ewing in 1881) had long been observed experimentally and although this model, when worked out in detail, did not fully reproduce the observations there was enough similarity between its predictions and what was being observed to give confidence in the basic idea (figure 14.2). For many years, however, there was no direct way of viewing the distribution of the magnetization, though there was evidence of sudden changes in it as an applied field was altered, through the Barkhausen [16] effect—a sudden change in the magnetic flux threading a coil in series with an earphone produced a click.

A complete theory of hysteresis was of much less importance than the need to have information about the hysteresis properties of the ferromagnetic elements and alloys. This was complicated because evidence was accumulating that the magnetic behaviour was probably being affected by the presence of unidentified impurities in both the elements and the alloys, and by inhomogeneities in their structures. (The mechanical properties of iron, for example, are strongly dependent on the amount of carbon present and the heat treatments that have been carried out.) The reason for the interest in hysteresis was a technical one, the need for magnets with optimum properties. There were basically two different requirements, permanent magnets and core materials for transformers. For permanent magnets the need was to have as large a magnetization as possible in high field and a large critical field. With

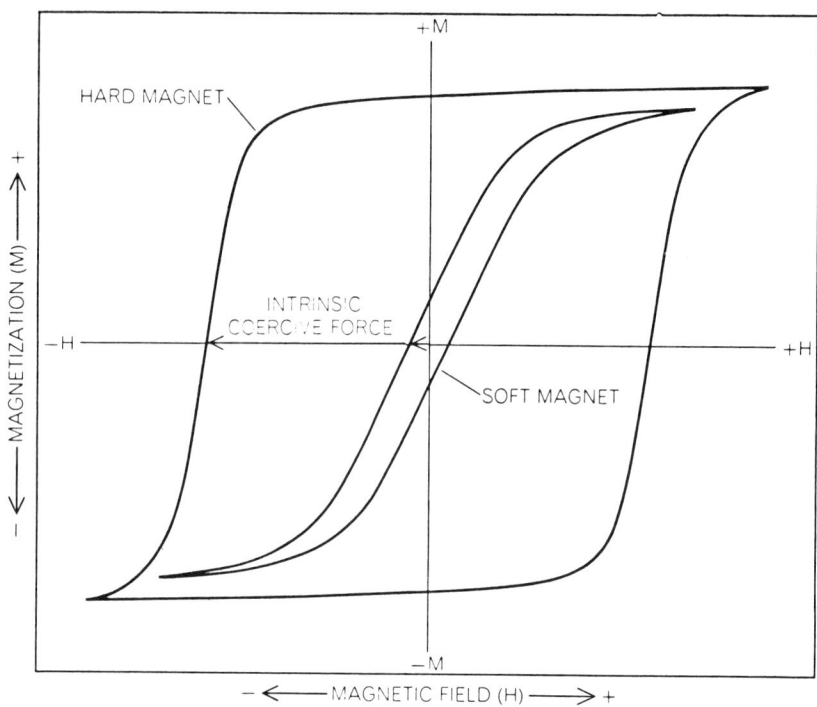

Figure 14.2. *A schematic illustration of the hysteresis behaviour of a hard magnetic material, the large loop, and a soft magnetic material, the small loop. The initial magnetization curve contains a part which joins the origin to the loop.*

the former a large magnetization would then be retained on reducing the high field to zero and the latter would ensure its stability against the random fields to which the magnet would inevitably be exposed. Stability against temperature fluctuations was also a requirement. In the transformer applications the magnetization would be repeatedly taken round a hysteresis cycle and a simple calculation showed that in each cycle there would be an energy loss determined by the area of the hysteresis loop. Apart from heating the core, which meant that some form of cooling might be necessary, it was undesirable from the point of view of efficiency, so the need was to make the area of the hysteresis loop as small, and the magnetization as large, as possible. As a result ferromagnets were broadly classified by their hysteresis loops; 'hard' ferromagnets having large-area hysteresis loops were potentially useful for permanent magnets, and 'soft' ferromagnets having small areas were potentially useful in transformers. The only way to find suitable materials for the different uses was by trial and error, with the result that while many alloys were classified as above, relatively few were used. In the transformer applications the fact that the ferromagnets were also conductors was a disadvantage, because the magnetic flux changes induced current flows in them, resulting in ohmic energy losses. As the number of hysteresis cycles per second and the magnitudes of the induced voltages increase with frequency, ferromagnetically cored transformers have been confined to low-frequency applications.

Although there was considerable interest in the technical applications there were also academic aspects of ferromagnetism, particularly in the relationship between magnetic moment and angular momentum. Indeed it is interesting to read, in an account of this investigation by Galison [17], of Einstein's interest from 1905 onwards and of the controversy which arose amongst the experimentalists, Richardson, Barnett, Einstein and de Haas, Beck and co-workers. It arose over whether or not the *gyromagnetic ratio*, the ratio of the magnetic moment to the angular momentum, in appropriate units, was 1, as expected from Ampère's model of current loops, or some other value. Galison actually gives a diagram in which the values from the different investigations are plotted against their dates, showing the wild fluctuations which were obtained from about 1915 until the middle 1920s, when the ratio more or less settled down near 2. No explanation could be offered until 1925 when Goudsmit and Uhlenbeck [18] suggested that the electron had a spin and that to account for spectroscopic observations the associated magnetic moment had to be twice that expected for ordinary angular momentum. The ratio, in appropriate units, was denoted by g and the value, which is very close to 2, has since been known as the anomalous g-value of the electron spin.

These experiments showed that the magnetic moments of ferromagnets arise almost entirely from the spins of unpaired electrons. Furthermore, from the saturation moments, measured in high magnetic fields when it could be assumed that all the local moments were aligned, the total number of aligned electrons could be determined. Surprisingly the numbers were not simply related to the number of atoms, being for example 0.6 for the unpaired electrons per atom in nickel; and this in a lattice in which all the nickel atoms seem to be identically situated. It was a long time before an explanation was forthcoming.

It was during this period that single crystals of ferromagnets were first made, with Beck [19] the pioneer in the study of their magnetic properties. Little of value was discovered until later, and the important work will be described in due course.

See also p 1134

14.3. The period 1925–50

14.3.1. Quantum theory of ions

By the early 1920s doubts were beginning to creep in about the merits of the Bohr theory, particularly in the context of spectroscopy, where it was being severely tested. A comprehensive 300-page review by Van Vleck [20] in 1926 was probably the last to appear before the theory was entirely replaced by quantum mechanics. This development had enormous consequences for most of physics, including, of course, magnetism. There is no need for a comprehensive account of it, but as some of the results of its application to the spectroscopy of ions had immediate consequences for magnetism an account of these seems desirable.

The description of an ion which it gave was, in some ways, similar to that in the Bohr theory. In a first approximation each electron was described by a wavefunction which involved four quantum numbers, identical to those in the Bohr theory, with the same restriction that no two electrons could have the same set. The energies of the individual electrons were also similar, so the same shell structures emerged; but in the concept of angular momentum differences particularly relevant to magnetism became apparent.

Angular momentum is a difficult concept in quantum mechanics because of the role played by the uncertainty principle. In classical mechanics it is represented by a vector quantity which can be regarded as having components along three orthogonal axes, $0x$, $0y$ and $0z$. In quantum theory the measurement of any one component upsets the others, so an angular momentum cannot be specified as in classical physics. The position is not quite as bad, however, as the above statement might suggest, for although the three components cannot be known simultaneously it is possible to know one component and the total magnitude. The component, in magnetism, is usually taken as that in a direction determined by an external field. There is then no possibility that the moment is completely aligned in this direction, for this would imply that the components at right angles are also known to be zero. So there is always an angle between the direction of the total moment and the direction of the field. In describing a moment as aligned the meaning is that the projection of the total moment onto the field direction has its maximum allowed value. Quantum mechanics confirms the conclusion of the Bohr theory, that an ion which has the right number of electrons to fill the shells up to definite values of n and l, leaving the shells outside this empty, possesses neither orbital nor spin angular momentum and therefore no magnetic moment. Quantum theory was thus able to account, in much the same way as the Bohr theory, for the temperature-independent diamagnetism of a wide range of insulators.

The next application was to the salts showing the Curie–Weiss type of paramagnetic susceptibility, ionic insulators which contain ions of either the iron or the rare-earth groups, with electrons in partially filled shells. Hund's [21] analysis of spectra, even before the advent of quantum mechanics, showed that in the lowest energy arrangements the electrons in the partially filled shells have, as far as possible, aligned spins. (When a shell is more than half-filled the exclusion principle does not allow all the electrons to have parallel spins.) Quantum theory provided an explanation in terms of the Coulomb interactions between the electrons and the effects of the exclusion principle. It also showed how the spectral properties could be used to estimate the energy needed to reverse one spin against the others. This result was far larger than any previous estimates and provided evidence for an interaction of ferromagnetic character between spins, of electrical origin as had been postulated much earlier by Weiss.

About the same time Heisenberg [22] and Dirac [23] independently studied a problem in which N electrons were placed in N different orbital wavefunctions, leaving the spin of each electron free to take either of two spin orientations ($m_s = \pm\frac{1}{2}$, in units of $\hbar$). They were able to show that the energies of such a model would be exactly the same as would be found for a different model, one in which there were N spins of $\frac{1}{2}$ coupled by what is now known as a ferromagnetic Heisenberg exchange interaction, and that the energy would be lowest when all the spins were aligned in the same direction. An external magnetic field would cause all the magnetic moments to align themselves in its direction. Here was another example of a ferromagnetic interaction and again it had arisen from the Coulomb interaction between electrons and the need to satisfy the exclusion principle.

Before ending this account of some of the results of quantum theory mention must be made of Kramers' [24] theorem, because it is frequently used in magnetism. He proved that, if the number of electrons is odd and no external field is present, every energy level will contain an even number of states of the same energy. In many cases this is the only degeneracy which remains and its importance is that if a magnetic field is applied it can be expected that the two levels will diverge linearly. When the splitting is measured it gives the magnetic moment to be associated with each state. (One state has its moment parallel to the field and the other antiparallel, so the level separation gives the energy needed to reverse the moment against the field.)

14.4. Paramagnetism

14.4.1. *Adiabatic demagnetization*

In 1926 Debye [25] and Giauque [26] independently pointed out that the rise in temperature found on applying a magnetic field to paramagnetic salts indicated that the reverse process could be used to obtain cooling in the helium temperature range beyond the limit which had then been reached. The proposal was rapidly followed up, though a good deal of preparatory work was first required. The argument involved no quantum theory, but only thermodynamic reasoning and experimental results. It was more a matter of choosing an optimum material. Kurti and Simon [27] showed that recent results on the specific heat of the rare-earth salt gadolinium sulphate indicated that it would be an even more promising material than they had first expected. This was confirmed by Giauque and MacDougall [28] who used it to reach a temperature of 0.53 K, beginning at 3.4 K with a field of 8000 gauss (the Earth's field is about 1 gauss). Almost immediately afterwards de Haas, Wiersma and Kramers [29] reached 0.05 K using potassium chromic alum.

In zero field and at low temperatures each ion in an assembly of identical Kramers ions will be in one or other of two states, with equal probabilities. On applying a field the states will be symmetrically split; having equal populations is then not the thermal equilibrium arrangement so some ions will move from the upper to the lower energy level and the energy released will appear as heat. It is removed by conduction and the overall energy is reduced. The assembly is now thermally isolated and the field is reduced slowly enough to maintain thermal equilibrium at all times. As the splitting of the states decreases, ions return from the lower to the upper state, for which they need energy. This comes from the thermal energy of the lattice, and so the crystal is cooled.

The experimental demonstration of the cooling was not easily accomplished. A major problem arises from heat leaks, particularly if the time to equalize the populations, at the final stage, is long. A full account has been given by Hudson [30].

14.4.2. Crystal field theory

Adiabatic demagnetization focused attention on the need to understand more about the low-lying energy levels of magnetic ions in crystals and on the way in which thermal equilibrium is established following a disturbance of the level populations. This was met by crystal field theory.

The basic idea [31] was that the electrons of importance in a magnetic ion, those in a partially filled shell, are exposed to an electric field from adjacent negatively charged diamagnetic ions, normally identical, six in number, and arranged octahedrally (figure 14.3). The first theoretical analyses, based on a crystal field which was not that of a regular octahedron, led to problems [32] which were only resolved when Penney decided to use what is now known as a cubic potential, that due to a regular octahedron. This had the required property that at a distance r from the magnetic ion the potential is largest along lines to nearest neighbours and weakest in intermediate directions. Such a potential hinders the free orbital motion of the electrons of an isolated ion and results in their energy levels being split and their states being modified.

The theory was extended to take account of small departures from octahedral symmetry, the presence of spin–orbit coupling and the inclusion of external magnetic fields. The first produced a further lowering of the symmetry, which generally meant that all orbital

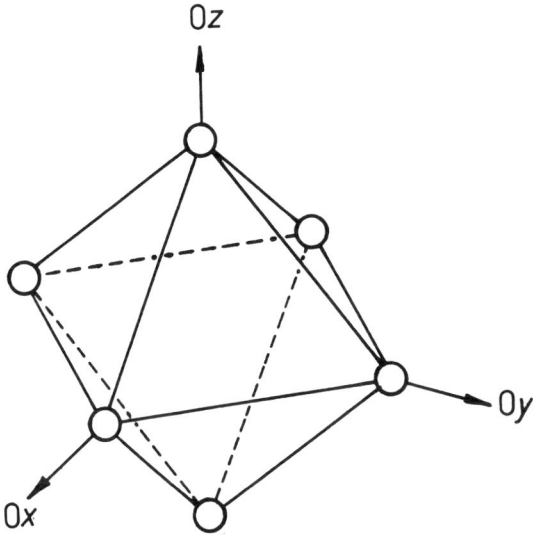

Figure 14.3. *The six circles represent the positions of the nuclei of neighbours in an octahedral arrangement. The axes, which are mutually at right-angles, join the centre of the octahedron to the vertices and pass through the magnetic ion.*

degeneracies of the free ion were resolved and all orbital motion was *quenched*, leaving no orbital magnetic moment. This explained why spin-only moments were observed in the paramagnetism (figure 14.4).

The wavefunctions which go with these energy levels are not easy to represent diagrammatically, for generally they describe many-electron functions. However, when there is only a single 3d electron in the ion they are easier to picture, and an example of how quenching arises is shown in figure 14.5. For a more complicated ion it is simplest to accept that the charge distribution has lobes which in the lowest state are set so as to minimize the energy. Detailed theory is needed to calculate the orientation of least energy.

When the spin–orbit interaction was included it was found that the orbital magnetic moment was then not quite quenched and that the effect was as if the g-value of the electron had been changed by about 10% from the normal value (approximately 2). The theory was able to predict the sign of the departure from 2, which varied with the ion, and also that it would be directionally dependent. That is, the observed Zeeman splittings would vary with the direction of the magnetic field relative to the crystallographic directions.

The same basic theory was also applied to the rare-earth ions, with the assumption that the crystal fields would be much weaker and that the effect of the spin–orbit coupling would be stronger. The result was that crystal field effects were expected to be negligible except at very low temperatures, a conclusion which fitted in well with the knowledge

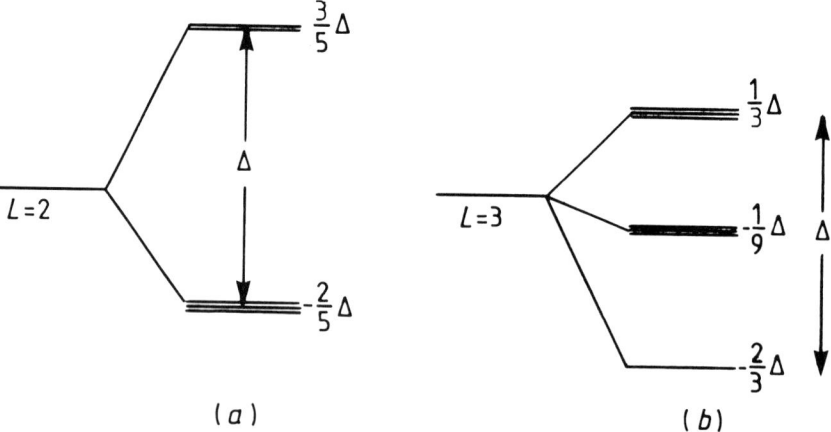

Figure 14.4. *In an octahedral environment the five-fold degeneracy of a free ion with $L = 2$ is split into two coincident states at $(3/5)\Delta$ and three coincident states at $(-2/5)\Delta$, where Δ is the overall splitting. Its sign depends on the number of d electrons and is positive for $3d^1$ and $3d^6$ and negative for $3d^4$ and $3d^9$. (For clarity the levels which are actually coincident are shown with small splittings.) The pattern for $L = 3$ has three coincident states at $(1/3)\Delta$, three at $(-1/9)\Delta$ and a single level at $(-2/3)\Delta$. Δ is positive for $3d^3$ and $3d^8$ and negative for $3d^2$ and $3d^7$. $3d^5$ has $L = 0$ and so shows no splitting. Its pattern has therefore been omitted. The Δ used in this figure should not be confused with the Δ in the Curie–Weiss law.*

that their susceptibilities could be obtained from observations of the free ion moments. When, eventually, low-temperature susceptibility results became available, on the sulphates of praseodymium and neodymium, it was found that χ tended to a constant for the former and to infinity for the latter as the temperature tended to zero. These conclusions were in agreement with the predictions of crystal field theory. So it seemed that it was a viable theory for both the iron and the rare-earth salts. (Later work has shown that six-fold coordination of neighbours is uncommon for rare-earth salts and that an arrangement with nine neighbours is much more common. Thus the satisfactory explanation of the low-temperature susceptibilities of these two salts probably owes less to the choice of crystal field than to the fact that praseodymium is a non-Kramers ion, and so has a singlet lowest state, whereas neodymium is a Kramers ion and so has a doublet lowest.)

The only remaining need, as far as susceptibilities were concerned, was to account for the Curie–Weiss Δs, for the Cs were now explained. Weiss had already attributed the Δs to interactions between ions, and the Heisenberg–Dirac (H–D) theory had produced exchange couplings which could be re-interpreted as arising from the spins of magnetic ions in quenched orbital states. Van Vleck [33] had therefore no hesitation

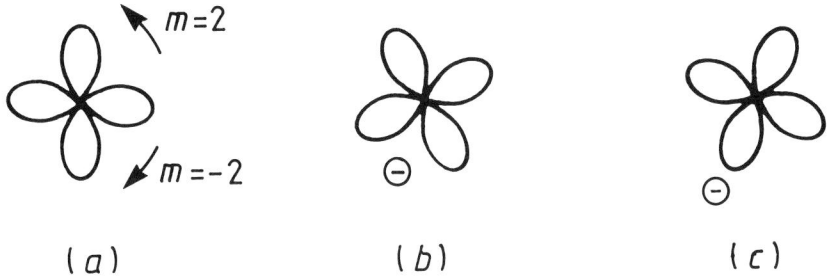

Figure 14.5. (a) The instantaneous projection of a typical d-wave function onto a horizontal plane. The state with 2 units of angular momentum has this shape rotating anticlockwise about the normal to the plane. The state with −2 units is its clockwise rotating companion. (b) and (c) show what happens if a single negative charge is placed somewhere in the horizontal plane. The rotational motion is arrested and in the lowest energy state the electron is in a wavefunction which places it as far away as possible from the hindering charge. That is, in the state shown in (b). As there were initially two states and the number is unaltered by the presence of the charge there is a second, higher energy state, in which the electron is much closer to the negative charge, (c). The states with 2 and −2 units of angular momentum can be regarded as wave motions about the vertical axis. (b) and (c) can be regarded as standing wave patterns constructed from them. Having no angular motion they have no magnetic moments.

in attributing the Δs to exchange interactions between the spins of the magnetic ions. In 1937 he gave a theory for a set of such spins interacting through nearest-neighbour exchange interactions and confirmed that the susceptibility would then take the Curie–Weiss form and that Δ would be proportional to the exchange constant. It would therefore be positive for the ferromagnetic exchange of the H–D theory. So the origins of the Δs were established, except that their signs, on the whole, were turning out to be opposite to those of the H–D theory.

The expression *exchange interaction* seems, by this time, to have become fairly firmly embedded in the literature of magnetism, both in the context of the paramagnetic salts and in the ferromagnetism of iron, cobalt and nickel. This was undoubtedly due to the initial concept of Weiss, though it is curious that where the theory had been most developed the experimental evidence was that the sign was not that of ferromagnetism, and where it was least developed it was.

14.4.3. Antiferromagnetism

The study of salts with negative Δ was dominated, from 1932 onwards, by the work of Néel, who pioneered the concept of antiferromagnetism. The term is now applied to such salts when they are below a phase transition which occurs at the *Néel temperature*, T_c ($\sim -\Delta$). Bitter [34] described the transition by an extension of the Weiss theory of ferromagnetism, using the negative Δs to justify the choice of a model in which the sign of the exchange tended to make neighbouring spins

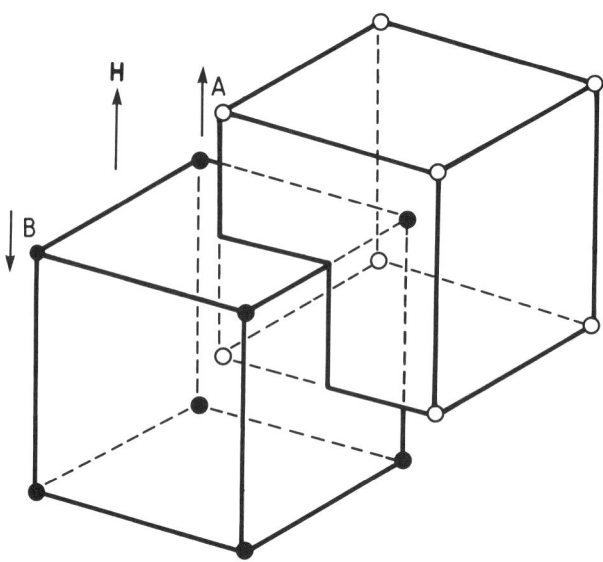

Figure 14.6. *The body-centred cubic lattice can be thought of as two interpenetrating simple cubic lattices, one unit cell of each being shown here. If there is a spin pointing up at each open circle, e.g. A, and a spin pointing down at each solid circle, e.g. B, the result is an antiferromagnetic system. Each spin is surrounded by a shell of 8 oppositely pointing spins.*

align antiparallel rather than parallel to one another. To cope with this difference he assumed that the magnetic lattice could be regarded as composed of two interpenetrating sublattices, A and B, which were magnetized in opposite directions (figure 14.6). In this example each spin is surrounded by 8 others of opposite orientation—A by 8 Bs and B by 8 As.

If the aligning forces are short-ranged the internal field at each spin is due entirely to its neighbours and in the same direction relative to its own moment as in the Weiss theory. In an external field H, weaker than any internal exchange fields, and in the direction of the moment at an A site, the internal field at A is in this same direction and is of the form $H + \gamma M_B$ (M_B and γ are both positive) while at a B site the field in the direction of the B spin is $\gamma M_A - H$, with M_A and γ again positive. The total moment in the direction of H is $M_A - M_B$.

The various moments and fields are then substituted in the Langevin expressions. At high temperatures and small H the theory predicts a Curie–Weiss type of susceptibility with a negative Δ and, at low enough temperatures, a phase change such that M_A and M_B would take finite values even when H is put to zero. Unlike the ferromagnetic case there is no net magnetic moment, and in fact a susceptibility can be defined even below the phase transition. Like the ferromagnetic case the theory does

John Hasbrouck Van Vleck

(American, 1899–1980)

The first book to apply quantum mechanics to solid-state physics was published in 1932 by J H Van Vleck, a Professor of Theoretical Physics at Wisconsin and a tenth-generation American. It was not his first major publication, for in 1926 he had published *Quantum Principles and Line Spectra*, which discussed the application and limitations of Bohr's correspondence principle. Nevertheless it is quite remarkable how quickly he realized the significance of the new theory and set about using it. Indeed, so thorough was his understanding that over 60 years later, his *Theory of Electric and Magnetic Susceptibilities* is still regarded as the 'bible' of magnetic theory.

In 1934 he moved to Harvard, where he was to remain for the rest of his life, and from 1951 to his retirement in 1969 he held the Hollis Chair of Mathematics and Natural Philosophy, the oldest endowed scientific chair in North America. Throughout this period, and afterwards in retirement, the steady flow of important scientific papers never ceased, and it gave great pleasure to his many friends and admirers when his career was crowned in 1977 by the award of the Nobel Prize for Physics.

---

not predict the direction of alignment in zero external field. If, however, there is some interaction, neglected in this analysis, which determines the alignment, the susceptibility in an external field will differ from that predicted, and will depend on the angle between the external field and the direction of alignment. For parallel alignment of moment and field, the susceptibility was predicted to fall to zero linearly with temperature from its value at the phase transition, and for a perpendicular alignment to remain constant. A consequence is that, as an external field is increased from zero, the magnetic energy depends both on the strength of the field and its orientation relative to the magnetization. The lowest energy, in a high enough field, is obtained with the alignment perpendicular to the field direction, as had previously been found by Hulthen [35], who had used a different theory to investigate the way in which the low-temperature alignments would alter with temperature. He had suggested that on gradually increasing a field aligned in the direction of the magnetization a value would be reached at which the moment would change its direction to a perpendicular alignment. Such an effect was not observed until 1961, in MnF_2 [36].

At the time of this analysis no antiferromagnet had been positively identified, though in 1936 Néel [37] had suggested, because of a specific heat anomaly near 350 °C, that manganese, a metal, might be antiferromagnetic. The first positive identification [38], MnO in 1938, was closely followed [39] by Cr_2O_3. The basic experimental difficulty was that nothing very drastic happens at the Néel temperature. No large magnetization develops and there is only a slight change in the susceptibility. There had been indications of phase changes in thermal measurements, but these could hardly be regarded as indicating magnetic phase changes. Agreement with theoretical predictions about susceptibilities were in the end much more convincing.

14.4.4. Relaxation

As a result of the above developments it must have seemed that the magnetism of the transition group salts was well on the way to being understood. Attention was also being given to the question of what determined the rates at which level populations, and therefore magnetic moments, could be changed. The process was well understood in the spectroscopy of ions, where transitions between energy levels are induced by radiation of specific frequencies. But while there was no direct experimental evidence about the very closely spaced energy levels of magnetic ions it could be estimated, by extending the theory to very low frequencies, that thermodynamic equilibrium could not be restored in a reasonable time after demagnetization had the process needed to rely on transitions associated with electromagnetic radiation.

Attention therefore turned to a new mechanism, energy exchanges between ions and lattice vibrations. The result was a sequence of suggestions about detailed mechanisms and estimates of the times which they would give. Each was weeded out in turn, for not giving fast enough relaxation, only to be replaced by an even more complicated theory. The climax was reached in 1940, when Van Vleck [40] extended crystal field theory, with all its ramifications of spin–orbit interactions and external magnetic fields, by including the modulation of the crystal field by vibrational motion of the surrounding ions. This predicted that there should be a wide variation in relaxation properties depending on how the energy level patterns were split by the static crystal field and whether the ions had an odd or an even number of electrons. From his estimates of the rates it seemed that the theory would account for the relaxation process, though its complication and the time of its publication did not encourage detailed tests. It was, in fact, many years before the theory was generally accepted as correct. Apart from the initial delay due to World War II it took some time to realize that the measured relaxation times were often not those of the ion being examined. A given crystal can often contain undetected impurities which relax rapidly, and mechanisms exist whereby the energy of a slow relaxing ion can be rapidly transferred to a fast relaxing ion and thence to the lattice, so short-circuiting its own relaxation processes.

The problem of understanding the phenomenon of relaxation did not inhibit its experimental investigation which was started in 1936 by Gorter and Kronig [41]. The basic idea was to superimpose a time-dependent magnetic field on a steady field applied to a paramagnet. The simplest conceptual arrangement is one in which the additional field is switched periodically from h to $-h$. Then, if the switching rate is slow, the magnetization in the field $H+h$ will settle at $\chi(H+h)$ before there is a change to $H-h$, when it will settle to another steady value, $\chi(H-h)$. As the rate of switching is increased a point will be reached at which the magnetization will have so little time to adjust to its equilibrium value before the field is again switched that it will tend to stay at the average value, χH. Thus by gradually increasing the rate of switching it should be possible to observe the change in behaviour and so obtain an estimate of the time to establish equilibrium as a function of the steady field and of the temperature. In fact at the time there was relatively little experience of step-wise field switching, and sinusoidally oscillating fields were used as a matter of course. At low frequencies the magnetization should then vary in phase with the oscillating field, but as the frequency increases a phase difference should develop along with a decrease in magnitude of the response. From such observations it should be possible to estimate the relaxation time. A limit was set, however, on the range of relaxation times which could be measured by the range of available frequencies. In the first experiments, Gorter used frequencies in the range 10–30 MHz, later extended to 78 MHz, with small amplitude fields (8 gauss). Full accounts of the method were published in 1947 by Gorter [42] and more recently by Standley and Vaughan [43].

14.5. Conductors

14.5.1. Normal metals

Little need be said here about the paramagnetism of normal metals for in 1927 Pauli had used the standard quantum theory treatment of a particle in a box to account for it. At absolute zero all the electrons are spin-paired and at a finite temperature the number of spins which can be freely oriented is proportional to T. So all that is necessary, at least in an elementary version of the theory, is to replace the C in the Curie law by a multiple of T and so obtain a temperature-independent paramagnetism. In 1930 Landau [44] showed that there was a diamagnetic contribution

Twentieth Century Physics

> See also p 1348

because an applied field would also modify the orbits, and there the position rested until interest was revived, after World War II. As this is described in detail in Chapter 17 nothing more need be said here.

14.5.2. Ferromagnets

The advent of quantum mechanics had little initial impact on the study of ferromagnetism, which was a field of considerable technical importance. Many elegant experimental studies were carried out, with a view to throwing light on such phenomena as hysteresis and what were known as easy and hard directions of magnetization (i.e. why the moments lined up in specific crystallographic directions and how far they could be moved into other directions with externally applied magnetic fields). Single crystals were much in demand for such studies. Thus Honda and Kaya [45] reported measurements, in 1928, on carefully prepared oblate spheroids made from single crystals of iron, and similar measurements on disc-shaped crystals of nickel were reported by Sucksmith *et al* [46], and on cobalt by Kaya [47]. While it is difficult to assess the intrinsic value of the results obtained from the careful shaping, which seems to have been dictated by the classical concepts of demagnetizing factors, such experiments gave important technical information about magnetic anisotropy, which had not been included in the original theory. Thus iron was found to have the [100] directions as easy directions of magnetization, nickel the [111] and cobalt its hexagonal axis. It seems that a single crystal was automatically assumed to be a single domain, an assumption that was later questioned by Landau and Liftshitz [48] who pointed out that, in the absence of an external field, a rod-shaped specimen with its easy direction of magnetization along the axis could lower its energy by breaking into two oppositely directed domains. Thus in crossing the rod there would first be a region where the magnetization would be along the axis in one direction, followed by another region in which the magnetization was in the opposite direction. It was not known how abruptly the change occurred, nor what would happen if a field were to be applied along the axis, though it could be expected that the favourably oriented domain would grow at the expense of the unfavourably oriented one. The details of this were clearly important in the context of hysteresis, but how the Bloch wall, the region between the two oppositely oriented domains, would change when the external field altered was difficult to determine.

A new method of investigation emerged from the work of Bitter [49] who had shown that if finely divided magnetic powder was distributed over the surface of a ferromagnet it would tend to be attracted towards any domain boundaries which were present. However, the use of this method called for polished surfaces and suitably chosen suspensions of powders, and before it could come into general use a good deal of effort had to be devoted to optimizing these. Even then the techniques

Figure 14.7. *Many different domain patterns have been observed, of which these are examples.*

offered little scope for observing movements of domain boundaries. Nevertheless the technique eventually became widely used [50], many beautiful domain patterns being seen and analysed (figure 14.7).

Another area of study was the behaviour of ferromagnets based on iron, cobalt and nickel above their Curie temperatures, to determine whether they then showed a Curie–Weiss type of susceptibility and whether the C corresponded to a whole number of spins at each site. As the Curie temperatures were all well above room temperature, so limiting the accessible ranges for paramagnetism, it was difficult to reach firm conclusions. The position was different with gadolinium, which had been shown to be a ferromagnet [51] in 1935. Its Curie temperature is at room temperature and the measurements indicated that it had seven aligned spins at each site in both phases. Its properties could be understood in terms of the Weiss model, a spin of $\frac{7}{2}$ at each site and an H–D form of exchange between the spins at adjacent sites. For iron, cobalt and nickel the Weiss theory continued to find favour, with the ferromagnetic alignment attributed to 'exchange interactions'—a useful phrase to describe the origin of Weiss's macroscopic internal field.

1135

Various attempts were made to provide a microscopic model which would also account for the 0.6 unpaired spins per site in nickel. An early example from 1933 is the work of Stoner [52], who used the concept of energy bands which was proving so useful for normal conductors. The basic idea was that since an external field produced a temperature-independent paramagnetism in a normal metal a strong exchange interaction would, as in the Weiss theory, produce ferromagnetism. Shortly after, Mott [53] suggested that there might be two overlapping bands, one based on the atomic 4s states and the other on the 3d states, and that the electrons were distributed between the bands. The conduction could then be associated with the 4s electrons, which are on the outsides of the atoms, and the magnetism with the electrons in the 3d band, which would be similar to the electrons responsible for the moments in the 3d ions. This idea allowed the possibility of a fractional number of magnetic ions per site, but it left unexplained why they were aligned. Similar ideas were applied to the magnetic alloys as well as the pure metals in attempts to reveal systematic behaviour, with limited success. The stumbling block was that although it seemed fairly certain that the ferromagnetism was due to the Coulomb interactions between the electrons combined with the exclusion principle there was no satisfactory way of showing this. (Even if it had been possible it would not have explained why there are easy directions of magnetization, a comment which also applies to the Weiss theory and its extension to antiferromagnetism.)

Such a deficiency had been rectified in crystal field theory, where the introduction of spin–orbit couplings had produced anisotropic g-values, so showing that magnetic properties can be anisotropic. In an attempt to explain ferromagnetism by an extension of crystal field theory, Van Vleck [54] suggested that the spin–orbit interactions would also modify the form of the H–D exchange interaction and as a result a ferromagnet would have easy and hard directions of magnetization. While this was almost certainly the germ of the explanation of an important property of ferromagnets there was already enough difficulty in dealing with the Coulomb interactions without having to take on the spin–orbit interactions as well. So while both the origin of ferromagnetism and the reason for its anisotropy were thought, in principle, to be understood no fully convincing theoretical explanations were forthcoming.

14.6. 1950 onwards

World War II interrupted most magnetic research, and recovery took much longer in some countries than others, but once recovery had begun there were few further breaks, except those due to the discovery of new techniques and new lines of development in existing topics. In describing the second half of the century it has seemed best to follow particularly interesting lines as long as they were progressing rapidly, rather than describe the general field of magnetism in any particular period, except in the last decade.

A significant change in the attitude to research in magnetism came about after the war. Having known some of the pre-war experimentalists and read about the work of many of the others, I have concluded that on the whole they were much more comfortable with the ideas of classical magnetism than with those of quantum mechanics. Enough satisfaction could apparently be derived from developing experimental techniques, observing properties and applying the result of the classical theory of magnetism. Such interpretations as needed quantum mechanics could be safely left to a small number of theoreticians.

With the ending of the war the number of students in physics departments increased greatly, and this in due course led to considerably increased research in magnetism. The newcomers had usually taken comprehensive courses in quantum mechanics and seemed generally to be unattracted by simply making experimental observations, so much so that it became almost *de rigueur* to attempt a quantum theoretical explanation of all observations. Indeed some of the papers on what might be supposed to have been about experimental work contained so much theory that one is led to wonder whether the distinction between theoretician and experimentalist had almost disappeared. Later the pendulum was to swing even further, for now it is almost impossible to have a paper solely on experimental work accepted, while papers which are entirely theoretical have little difficulty.

The beginning of this period saw the development of two new techniques, magnetic resonance and neutron scattering. Both were destined to have a major impact on research in magnetism. The former could be used to study both electronic and nuclear magnetism and shared a similar theoretical basis, the Bloch [55] equations, which had originally been proposed for use in nuclear resonance. The experimental techniques were, however, somewhat different, with the result that not only did the two fields develop differently but the field of electronic resonance also divided. For convenience, therefore, several fields of study will be described as if they were in separate compartments, as indeed they were from the experimental point of view, though theoretically they had much in common. Since neutron scattering developed into a major tool in wide areas of science it is treated in some detail elsewhere. Here it will be assumed that the reader is familiar with its general features and its impact on magnetism will be mentioned as appropriate.

14.7. Electron paramagnetic resonance

14.7.1. Iron group ions
In 1945, in Kazan, Zavoisky [56] observed what was known initially as *paramagnetic resonance*, though later the description *electron spin resonance*

(ESR) came into use as an alternative. He applied a uniform magnetic field, the main field, to salts containing divalent copper or manganese ions, both of which have an odd number of electrons, so that there were Kramers degeneracies to be lifted. He then applied a small oscillating magnetic field at right angles to the main field, knowing from the quantum-mechanical selection rules, that if the quantum, $h\nu$, of the oscillating field equalled the separation between the split Kramers levels, energy should be absorbed from the oscillator and the loss might be observable. He chose a frequency of 120 MHz for the oscillating field and looked for an absorption as the main magnetic field was varied. Rather surprisingly, for the experimental arrangement was by no means an optimum, he found absorption curves, almost as broad as the range of available main fields. These were the first resonance results to be reported for magnetic ions in crystals and the only experiments which Zavoisky is known to have carried out.

In 1946 Cummerow and Halliday [57] published the results of a similar experiment, also on a manganese salt, using a much higher frequency from a source which had been developed during the war. At much the same time similar work began in Oxford [58], at room temperatures and an even higher frequency ($\sim 10^{10}$ Hz). A number of absorption lines were obtained by slowly changing the main field, which had a magnitude of a few thousand gauss and was produced by an electromagnet. The widths of the lines were of the order of several hundreds of gauss, much narrower than those observed by Zavoisky.

It was also significant that in quite a number of crystals which contained Kramers ions no resonances were found. An explanation came from Van Vleck's theory of spin-lattice relaxation which predicted that if an ion had a set of closely spaced orbital levels the spin-lattice relaxation time could be exceptionally short. It could also be expected, on general grounds, that the linewidth would depend inversely on the relaxation time, so the obvious next step was to lengthen the relaxation time by going to lower temperatures.

This became the province of Bleaney and Penrose, and in a very short time all the missing resonances, except from titanium, had been found. Advantage was taken of the variable valencies shown by these atoms. Thus if resonance could not be observed with one valency it was usually possible to find another compound in which the ion had a different valency. So one way or another each ion could be found with an odd number of electrons and then the Kramers degeneracy came in useful. (Resonances from Ti^{3+}, which has an exceptionally fast relaxation time, were later found.) A survey of the results published [58] in 1948 demonstrated that *electron paramagnetic resonance* (EPR) was an important new technique. For the next few years there was intense activity in its study, particularly at Oxford, where the investigations had got off to a flying start, due to the availability of the necessary microwave equipment. Gradually more groups joined in and now EPR has become a standard technique in chemistry and biology as well as in solid-state laboratories.

From crystal field theory it was found, as expected, that most of the resonances occur at frequencies close to the free electron spin resonance in the same main magnetic field, and that they depended on the orientation of the field relative to the crystallographic axes. (For experimental reasons the frequency was held constant and the main field was varied in magnitude and orientation.) What was surprising was the strange behaviour of the linewidths as the direction of the main field was altered. Then, in 1948, Penrose went to Leiden, where Gorter was anxious to set up resonance work because of the information it could provide about the potential of magnetic crystals for use in adiabatic demagnetization. On returning briefly to England for Christmas, Penrose reported that he had carried out resonance experiments on a non-magnetic magnesium Tutton salt in which about 5% of the zinc ions had been replaced by copper ions. It had been expected that such a dilution would sharpen and weaken any copper resonance lines; sharpen because the interactions between the ions were reduced and weaken because of the smaller number of magnetic ions. Unexpectedly, instead of there being one sharper line there was a pattern of four lines (later recognized to be eight). Penrose had revealed for the first time in a solid a process known to occur in the spectroscopy of free ions, a splitting of energy levels due to an interaction between the magnetic electrons of a Cu^{2+} ion and its nuclear spin. Further investigations showed that the hyperfine splittings, as they are known, varied considerably with the field direction and that it was the underlying unresolved hyperfine structure in the fully concentrated salts which led to the strange behaviour of the linewidths. Unfortunately Penrose died, at the early age of 28, before he could pursue his results and it fell to Gorter to write a brief account of them [59]. The discovery of nuclear hyperfine structures resulted in a major re-direction of the resonance work at Oxford; henceforth almost all the resonance studies were on magnetically dilute crystals.

The first attempt, by Abragam and Pryce [60], to produce a theory of these nuclear hyperfine structures was unsuccessful, for they had not appreciated that the magnetically active spins in the 3d shell induce magnetic moments in the s electrons and so produce extra indirect interactions with the nuclear spin. Once this mechanism was introduced a satisfactory explanation emerged [61].

14.7.2. 3d ions: spin-Hamiltonians and paramagnetic resonance
There are classes of compounds which have similar structure and, particularly with transition-metal ions, such a class is produced when one transition ion is replaced by another, leaving the rest of the constituents unaltered. The class is often named after the first person to describe the

structure (e.g. Tutton) or a particular gem stone that is a member of the same class (e.g. garnet). Resonance work frequently relies on finding a class in which magnetic ions replace non-magnetic ions, for this is a convenient way of obtaining magnetically dilute crystals. Such crystals are usually then described by the magnetic impurity rather than the non-magnetic ion it is replacing. Thus Penrose's sample would be described as a diluted copper rather than a magnesium salt.

The class called Tutton salts have the property that the surrounding six non-magnetic ions do not form a regular octahedron because an opposite pair are further away from the central ion than its other four neighbours. They therefore provided an opportunity to study a range of different magnetic ions in a tetragonal environment and to compare these with another family, such as the fluosilicates, in which two opposite faces of the octahedron are separated to produce an environment having three-fold symmetry. The initial interpretations, though complicated and lengthy, used a direct application of Van Vleck's crystal field theory. However, Van Vleck [62] had shown, along the lines of the H–D exchange calculation, that for one specific magnetic ion it was possible to derive an operator which involved only the angular momentum operators of the total spin and a few parameters to describe the lowest-lying crystal field levels. Since these were just the levels being examined in the resonance experiments it was quickly realized [63] that the same concept could be applied to most of the ions and that the forms of these so-called spin-Hamiltonians could be tentatively written down with just a knowledge of the total spin of the ion and the symmetry of its surroundings. To interpret his measurements the experimentalist needed to know the properties of spin angular momentum operators, but beyond that it was only a matter of choosing a few parameters. This procedure was remarkably successful. For a given ion, resonances could be observed for magnetic fields at virtually all angles to the crystallographic axes and a wealth of results could be obtained and fitted to a consistent model. Thus spin-Hamiltonians, though initially introduced via crystal field theory, came to be established in their own right as an economical, and physically meaningful, description of experimental results.

The spin-Hamiltonians for ions which have $S = \frac{1}{2}$ are generally very simple, for unless there is a nucleus with a spin, only three parameters, usually denoted g_x, g_y and g_z, are needed to account for the resolution of the two-fold Kramers degeneracy by a magnetic field. (With three-fold and four-fold symmetries the form is even simpler, for two of the three are equal.) With nuclear spins present the forms become slightly more complicated, though they take a common form for all nuclear spins. For ions which have electron spins greater than $\frac{1}{2}$ and no nuclear spins, the spin-Hamiltonians usually need extra terms to describe what are known as zero-field splittings, that is small splittings of energy levels which are present even before a magnetic field is applied.

A main role for the theoretician was to explain the origins of these parameters, using crystal field theory, which involved a fair amount of quite complicated mathematics. Thus for a given class it seemed of interest to show that keeping the crystal field parameters much the same would nevertheless give the observed spin-Hamiltonian parameters, which, on the whole, showed considerable variations from ion to ion.

In 1952 I showed [64] that the theory could be greatly simplified by introducing the concept of equivalent operators, a step which eliminated much of the complication of crystal field theory. Indeed it then became very similar to spin-Hamiltonian theory, for crystal field theory can be regarded as a sequence of steps showing how the degenerate states of a low-lying energy level of an isolated ion are first partially resolved to give a low-lying orbital level, and then how the spin degeneracy of this is lifted. Spin-Hamiltonian theory cuts out the first step and assumes that the orbital resolution has already occurred. After that an operator in spin variables suffices to describe the resolution of the remaining degeneracies. The equivalent operator technique deals with the first step by replacing the crystal fields by expressions in orbital momentum operators and so makes it similar to the second step, with orbital replacing spin for the angular momentum variables. This formalism is now widely adopted, so much so that crystal fields as such are seldom given explicitly. Instead an expression in equivalent operators, each multiplied by a parameter, is written down and the parameters are chosen to fit the observations. These parameters are often called crystal field parameters, though it is now established that there is actually no need to have a crystal field model to justify their introduction. This has led to some confusion in the literature over how to describe them, for *crystal field* has been attached both to the parameters in the spin-Hamiltonian and to those in the equivalent operators.

14.7.3. 4f ions: the rare-earths

With the growth in understanding of the 3d ions attention turned to the rare-earths which had become more readily available, as a by-product of intense metallurgical work for the atom-bomb project. The main problem with applying the resonance technique was that these ions generally had extremely short relaxation times so that it was necessary to work at liquid helium temperatures. The programme began with a study of the rare-earth ethyl sulphates, compounds which had already been studied for possible use in adiabatic demagnetization experiments. In a very short time a host of new results were obtained and reviewed [65].

The obvious assumption was that crystal field theory would also account for these, though initially there was little supporting evidence, for the only published work on rare-earths had assumed that, like the iron group ions, the dominant crystal field was due to an octahedron of neighbours, but much weaker in strength. This had seemed satisfactory at the time, but a later attempt to explain the susceptibility (actually the Faraday rotation, to which it is proportional) of some rare-earth ethyl sulphates had created misgivings and indeed Van Vleck told me in a letter dated 7 February 1951, that he doubted whether crystal field theory could be used for the rare-earths. However, new resonance results on a number of Kramers ions were making it clear that the crystal fields were quite different from those of iron group ions and that crystal field theory could be used. In particular the g-values were much more anisotropic and there was no way in which this could occur with an approximately octahedral crystal field. Nor was it consistent with the crystal structure, which showed that the magnetic ions had twelve nearest neighbours. Treating the crystal field of such an arrangement was a major problem, and indeed it was to circumvent this that the equivalent operator version was introduced. After that the algebraic difficulties disappeared. However, there was one particular feature, a sixth-order crystal field parameter B_6^6, which was new and for which there was no direct evidence from most of the resonance results. In due course Gd^{3+} was examined. This is an ion with a high spin, $\frac{7}{2}$, which presents no problem for crystal field theory, for there is no orbital degeneracy in the free state. All that should be found is a straightforward Zeeman splitting of the spin degeneracy into eight equally spaced levels. In fact what was actually found was quite different. In the absence of a magnetic field the eight spin states were split into four Kramers doublets and the resonance experiment showed that there was a six-fold symmetry axis. It did not, at least in any simple way, justify the introduction of a B_6^6 crystal field parameter into the spin-Hamiltonian, for all such parameters should been zero. It did, however, show that indeed the six-fold symmetry of the neighbours existed and it also focused attention, once again, on the problem of understanding the zero field splittings of S-state ions, a problem which had already arisen with the iron group examples and is still not resolved.

Even though the resonance results could be explained using equivalent operators there still remained too much ambiguity in the values of the parameters. This arose because for most of the ions the theory predicted that there would be an extensive pattern of low-lying levels, doublets for the Kramers ions. To determine the parameters precisely it was necessary to known the properties of all the levels, but at liquid helium temperatures it was expected that only the lowest of these would be populated. The best that could then be done was to supplement the resonance measurements with susceptibility results. Later, inelastic neutron scattering proved to be a valuable technique for determining these quite closely spaced energy levels and now the uncertainty in the parameters is much reduced.

The investigation of the rare-earths by resonance methods provided a wealth of information about a group of elements which were previously quite unfamiliar to most physicists, and it is appropriate to acknowledge the contribution made to this field and to magnetism in general by F H Spedding, of Iowa State University. Thanks to him these elements became available in substantial quantities and in separated forms, leading to a growth in interest in all their properties, going far beyond what has so far been described. They are now used extensively in technological applications.

EPR was also used to study ions of the higher transition groups, though on the whole there was not the same diversity of results because many of the ions are found in covalent complexes and show no resonances. An interesting exception was the $IrCl_6$ complex [66] in which the magnetic electrons were found to be interacting with the spins of all seven nuclei, showing that the electronic orbits spread over the whole complex. Each magnetic electron of the nominally iridium ion was estimated to spend approximately 30% of its time near the chlorine sites. Crystal field theory provided no framework for explaining this and it was necessary to turn to a molecular orbital type of theory, such as was first described in 1935 by Van Vleck [67]. This was an indication that all might not be well with crystal field theory. Nevertheless in its spin-Hamiltonian form it continued to meet all requirements, as is readily apparent from Abragam and Bleaney's [68] very full account of the resonance work and its associated theory up to 1970.

14.7.4. Exchange interactions

We now return to salts of the iron group. It had been realized that in a magnetically dilute crystal a statistical accident was likely to produce a few pairs of magnetic ions on adjacent lattice sites. However, before this could be explored by spin resonance Guha [69] suggested that an anomalous susceptibility in copper acetate might be due to the presence of pairs of Cu^{2+} ions. This was confirmed by resonance methods [70]. Resonance showed three closely spaced levels, an excited set with an effective spin of 1, above a non-magnetic singlet ground state. This is just what was to be expected from isolated pairs of Cu^{2+} ions, each with spin of $\frac{1}{2}$, coupled by an antiferromagnetic exchange interactions. Further analysis showed the mechanism to be an exchange interaction, dominantly that of the H–D model though with the opposite sign, with a smaller part due to coupling of the spin directions to the lattice structure.

The understanding of exchange interactions was in its infancy, though ever since the H–D analysis it had been assumed that there would be such interactions between nearest-neighbour magnetic ions, and various observations had been interpreted on this basis. Copper acetate gave direct evidence for it and confirmed that it could be of antiferromagnetic character. Also, at about the same time neutron diffraction showed [71] that exchange interactions could be strongest between magnetic ions which were not nearest neighbours, particularly if they are separated by non-magnetic ions.

Explanations were put forward under a number of descriptions (e.g. direct exchange, superexchange) but a satisfactory and systematic theory had to wait until about 1959 when Anderson wrote several papers, including a review of previous work [72].

14.7.5. Problems with crystal field theory

Although crystal field theory was being assumed necessary for the derivation of spin-Hamiltonians there were one or two clouds on the horizon. The iron group cyanides, covalently bonded complexes, had been investigated, and for these it seemed necessary to have a theory like that used for the $IrCl_6$ complexes. Then Feher [73], using a complicated technique known as *Endor*, produced evidence that the electrons of the so-called ionic magnetic ions were straying far from the nucleus, which again could not be explained by crystal field theory. Another cloud was that Kleiner [74] had calculated the coefficient of the cubic part of the crystal field using quantum mechanics rather than the classical theory used by Van Vleck, and had come out with the opposite sign which, had it been correct, would have ruined all agreement between theory and experiment. Perhaps not unnaturally this result was largely ignored. (A later 'improved version' [75] restored the original sign but with a value which seemed far too small.)

A potentially much more serious criticism was that in crystal field theory electrons are being distinguished, for those on the neighbouring ions are regarded as simply producing contributions to the crystal field whereas those on the magnetic ion are given a full quantum-mechanical treatment. A fundamental property of electrons was therefore being violated. The criticism came to the fore because of the good progress that was being made in understanding metals. While crystal field theory took account only of local variations of potential around the ion to which an electron was attached, in a metal the potential is periodic in space, and Bloch's theorem shows that electrons are no longer confined to the vicinity of any particular nucleus. There is no problem in incorporating their indistinguishability and the exclusion principle into the basic theory. The dilemma was that the fully concentrated magnetic crystals are also periodic, so it seemed only reasonable that their electrons should also move through the whole lattice, a feature that is not contemplated in crystal field theory. Slater [76] expressed the view that it was only a matter of time before band theory replaced crystal field theory. It had already been used with modest success by Stoner and Mott, in an attempt to explain the ferromagnetism of iron, cobalt and nickel, and it seemed reasonable that the paramagnetism of the transition group salts could be understood in similar terms. There would be no need to distinguish between electrons nor to use concepts like the crystal field. Slater had undoubtedly put his finger on a weak spot, but the implication that all would come right through band theory was far too optimistic; it has still not happened [77]. It has, however, proved possible, comparatively recently, to reformulate the theory of magnetic insulators, retaining the periodicity of the lattice and keeping the electrons indistinguishable, and to show that this leads to each ion being described by a spin-Hamiltonian of the usual form together with terms that describe exchange-like interactions between the effective spins [78]. This work did not use the concept of electrons in bands. If electrons are regarded as moving independently in a common periodic potential then bands are inevitably introduced. However, the concept of a periodic potential is an approximation, and another one, presumably better for the insulating magnets, gives results much more in line with observations.

The result of not realizing this was that for some 25 years from the early 1950s magnetism was divided into two camps, one wedded to crystal fields and the other to band theory, with the result that the literature of magnetism is sprinkled with comments about the need to decide, in accounting for some observation, whether to use a localized description (a crystal field model) or a delocalized description (a band theory model). The crunch came when the rare-earth metals were studied, for it was then found that the best way to explain their conductivity was to invoke bands for the outer electrons and, for the magnetism of the 4f electrons, to use crystal field concepts. While there is still no satisfactory theory which allows spin-Hamiltonians for localized moments and band states for conduction, at least each camp is now using the other's concepts.

14.7.6. Relaxation processes

The study of magnetic ions as low-concentration impurities in non-magnetic hosts had really only been possible by working at low temperatures, because the intensity of the absorption increased as the temperature was lowered, so helping to offset the loss due to dilution, but it also raised other difficulties, one being that of saturation. The basic idea of the resonance experiment was to induce absorption of electromagnetic energy by raising an ion in a low energy state to one of higher energy. But if this was all that happened it would not have been long before all the ions would have been lifted to the higher state, in which case no further absorption would occur. That something of this kind could happen had been observed and it had become part of the resonance technique to vary the incident power to check that *saturation*, as it was known, was not occurring. As the resonance technique became established interest grew in the relaxation processes which tended to prevent saturation occurring.

In the spectroscopy of isolated ions there are two processes which can prevent saturation, both associated with electromagnetic radiation and famously described by the Einstein A and B coefficients—A for spontaneous emission and B for the stimulated exchange of energy between the ion and incident radiation of the correct frequency. In optical spectroscopy spontaneous emission usually dominates. However, in EPR the observations gave relaxation times which were far too short to be due to spontaneous emission, nor could stimulated emission be invoked because this would simply return any absorbed energy to the exciting wave and so reduce the overall absorption. Almost from the beginning, therefore, it was assumed that the answer was to be found in Van Vleck's [40] theory of relaxation, which replaced the modes of the electromagnetic spectrum by the spectrum of the lattice vibrations. However, the theory had not been tested in detail, though the prediction that the relaxation times would be temperature dependent was obviously correct.

See also p 177

14.7.7. *Acoustic paramagnetic resonance*

If magnetic ions do relax by stimulated emission of lattice vibrations then it should also be possible to excite transitions by means of monochromatic vibrational waves. There were, however, several problems to be overcome. At a frequency of 10^{10} Hz lattice waves have a wavelength approximating to that of visible light and could be expected to be highly attenuated. Furthermore, no way was known of generating such monochromatic waves, until in 1959 Jacobsen *et al* [79] showed that it was possible to use the piezoelectric properties of quartz to generate monochromatic lattice waves of the same frequency from electromagnetic waves; with these he and his colleagues demonstrated what is now known as *acoustic paramagnetic resonance* (APR). Even in the first experiments resonances were seen which had not been observed by EPR in the same sample. The inference was that some unidentified magnetic centres were present which EPR had missed. The experimental technique was quite different from that used in EPR because of the very short wavelength, but nevertheless it was successfully developed and many resonances that EPR had missed were later detected. (The detection of a resonance, whether by EPR or by APR, does not in itself identify its origin; hyperfine structure, if present, is a great help for it will usually identify the nucleus.)

Looking back at the EPR results it could be seen that most of the information came from Kramers ions, which were predicted by the Van Vleck theory to be much more weakly coupled to the lattice than the non-Kramers ions. So there was an obvious reason why APR should differ, for it would be most effective for non-Kramers ions. The specific detection of such ions confirmed an impression that had already come from work on relaxation processes in EPR, that Kramers ions were relaxing by a process which was not included in the Van Vleck theory—cross-relaxation—which was a well-established process in the field of nuclear resonance [80]. Basically it involves an ion in an excited state dropping to a lower one while a similar ion in the lower state moves to a higher one, a process in which both energy and total spin are conserved. A fast relaxing and different ion can be substituted for one of the ions and, provided their resonance lines overlap, energy can be taken from an excited A ion, a slow relaxing Kramers ion, and used to excite a fast relaxing B ion which then rapidly de-excites by transferring the energy to the lattice. A small number of undetected B ions can completely alter the relaxation characteristics of the A ions. It was the demonstration that such fast relaxing ions were indeed present which explained why the relaxation observations on what should have been slow relaxing ions had played havoc with the testing of the Van Vleck theory. (It is unusual for a theory to exist for something like a quarter of a century before it is proved to be correct.)

14.7.8. Co-operative phenomena in magnetic insulators—antiferromagnetism

Until the early 1950s there were two experimental methods available to investigate antiferromagnetism. The most direct was the measurement of magnetic susceptibility on, if possible, single crystals, as a function of temperature and field direction relative to the crystalline axes. The second used the Faraday effect, the rotation of the plane of polarization of light when it propagates in the direction of the magnetic field, the magnitude of effect having been shown to be proportional to the susceptibility [81]. When the new technique of EPR was developed it was naturally of interest to examine what would happen, in a fully concentrated crystal, to a resonance line as the temperature was lowered through an antiferromagnetic phase transition.

The first experiments [82] were on powdered Cr_2O_3, which has a Néel temperatures near 40 °C. While a resonance was observed above the transition, below the transition it was difficult to decide whether it had disappeared completely or moved outside the range of available fields. The same phenomenon was soon observed in other salts, and explained independently by Kittel [83] and Nagamiya [84], who showed that the resonance had probably moved to a much higher frequency.

Their theory was based on the Bitter model of an antiferromagnet, with the added concept that the sublattice magnetizations were not simply lined up in antiparallel directions but were actually precessing, as for electron spins. Thus the moment of one sublattice would be precessing in the effective internal magnetic field due to its exchange interactions with the other sublattice and vice versa. A similar model had already been introduced for ferromagnets, also by Kittel [85], the difference in the antiferromagnets being that the two-sublattices precess in opposite directions. As a consequence two resonances could be expected, both with the same frequency in zero applied field. This was verified in 1959 by Johnson and Nethercot [86] who took a good deal of care over their choice, MnF$_2$, as a suitable antiferromagnetic. They found resonant absorption, rather as in EPR, except that no external magnetic field was needed. The resonant frequency was temperature dependent because of the temperature dependences of the two sublattice magnetizations. This may be regarded as the first definitive observation of what is now known as antiferromagnetic resonance.

> See also p 460

A large change in the experimental techniques came in 1951, when Shull *et al* [71] used the elastic scattering of neutrons in studies on a number of Mn^{2+} salts, obtaining their magnetic structures, the magnitudes of the magnetic moments and their orientations relative to the crystal axes. In the simplest examples their results confirmed the Néel picture of two oppositely directed interpenetrating ferromagnetically ordered sublattices, as modelled in Bitter's theory. The experiment relied on the neutrons, which have magnetic moments due to their spins, being scattered from the static magnetic array in much the same way as x-rays are Bragg scattered from the electronic charges in a crystal. The neutrons could therefore be used to determine the magnetic structure and, in particular, show whether this was the same or different from the charge structure. It came as no surprise to find the two were not the same in an antiferromagnet, though it was a considerable surprise to find, in MnO for example, that the strongest antiferromagnetic couplings were not between nearest-neighbour magnetic ions but between next nearest neighbours, magnetic ions separated by non-magnetic O^{2-} ions (figure 14.8). The nearest Mn^{2+} ion which has a spin which is antiparallel to that of a chosen Mn^2 ion is reached through an intervening O^{2-} ion.

Since then many antiferromagnets have been discovered, with a variety of sublattice structures, and neutron scattering has become the favoured technique for its examination.

14.7.9. *The Heisenberg–Dirac model and spin waves*

> See also p 1010

Ever since the introduction of the H–D model of exchange coupled spins there has been a continuing interest in using it to show theoretically that such a system would show a phase transition at low enough temperatures. It is deceptively simple-looking and so has been particularly interesting to theoreticians. The discovery of antiferromagnetism, which in the theory simply meant that the ferromagnetic pattern of energy levels was inverted, provided a stimulus to further studies. However, it would be out of place in this account to attempt a review of such progress as there has been, which is quite limited, so I shall content myself with a reference [87] and some observations. There is extremely little evidence that the model applies to any known system. Ferromagnetic insulating salts are quite uncommon and in the antiferromagnets there are known to be alternating magnetic

Magnetism

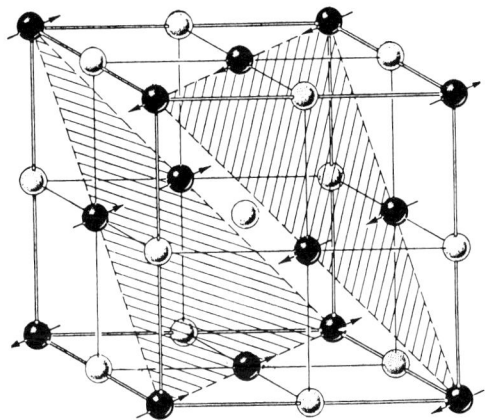

Figure 14.8. *In MnO the moments which are antiparallel have an intervening oxygen ion. The structure shows that the strongest exchange interactions must therefore be between next nearest neighbours and that they are of antiferromagnetic character.*

moments at the sites of the order of magnitude of those of unpaired electrons. The H–D model predicts that at absolute zero there would be no moment at any site, which suggests that a realistic low-temperature model needs to include further (small) interactions. On the other hand the model does seem to have the potential for predicting the existence of a cooperative phase transition and the temperature at which it occurs.

Alongside the rigorous work on the H–D model there has been a large effort devoted to approximate models, which also can be traced back to the H–D model of a ferromagnet, the ground state of which has all the spins aligned parallel to one another (identical m_s values) to create a macroscopic moment. If each s is taken to be $\frac{1}{2}$ and its magnetic moment is taken to be the Bohr magneton μ_B, then for N spins the total moment will have a projected magnitude of $N\mu_B$ which, in accordance with the uncertainty principle, means that the total moment has a magnitude of $[N(N+1)]^{1/2}\mu_B$. For large N, therefore, the total magnitude and its projection are virtually aligned and so are like classical moments. In the absence of any external field they can point in any direction. However, in the presence of a field, H, they can align themselves at any one of a large number of angles to the field direction to give rise to a set of equally spaced energy levels, the separation being $\mu_B H$—a macroscopic Zeeman effect. The total moment precesses like a gyroscope around the field direction (figure 14.9). There is also another set of energy levels with the same spacings but with a slightly smaller total moment. This arises from a spin arrangement in which one of the spins has been reversed, which makes the largest projected value $(N-1)\mu_B$, a value which also occurs for one of the θ values in the case when all the spins are all parallel.

1149

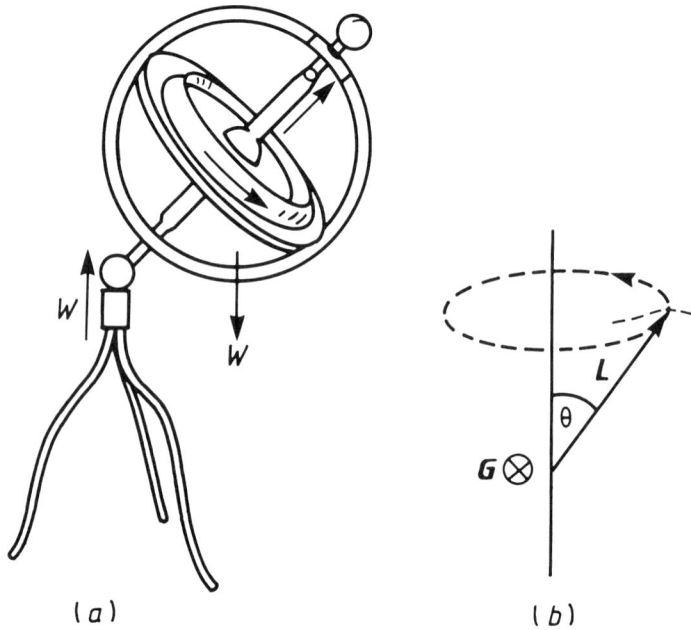

Figure 14.9. In the classical gyroscope (a) the couple, G, is provided by the weight W and an equal and opposite force acting on its point of support. G creates an angular acceleration in its own direction, which is at right angles but in the same plane as the angular momentum L of the fly wheel. The result is that there is a precession about the vertical at an angle θ. Due to frictional forces the angular motion is damped and θ increases until the whole precessional motion collapses. For a magnetic moment (b) the couple G is created by an external magnetic field and the angular momentum L is usually due to aligned electron spins. There is no damping and the precessional angle θ remains constant. However, θ is restricted to a finite number of values, two in the case of a single electron, and the classical damping is replaced by discontinuous changes in θ being allowed. The energy changes associated with these are quantized, for the energy levels are equally spaced.

When the arrangements are examined in detail it is found, in typical quantum-mechanical fashion, that although the moment in the direction of the field is less than the fully aligned value, the spin reversal is not at a specific site. Rather it must be regarded as propagating, with a range of possible velocities, so that it is equally likely to be at any site. A phrase has been invented to describe these motions—*spin deviation waves* or simply *spin waves*—and various attempts have been made to visualize them (figure 14.10).

The mode which is in phase everywhere corresponds to the case when no spin has been reversed, so nothing is being propagated. In an extended system it is, of course, possible to envisage that two or more spin wave-packets have been excited and that they are in different parts of the sample. They should therefore be able to travel a long way before

Magnetism

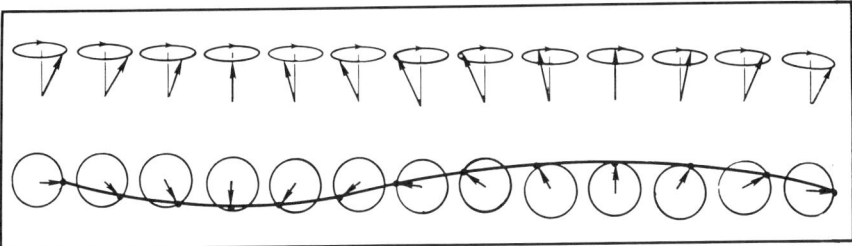

Figure 14.10. *A ferromagnetic spin wave, a normal mode of the spin system, is described semiclassically as a precession of each spin about the magnetization direction with the same frequency. The top diagram shows the spins viewed from the side and the bottom diagram shows the view from above. Due to the rotation the wave appears to move with a wavelength determined by the difference in angle between adjacent spins. Energy is needed to change this angle so the energy associated with a spin wave depends on its wavelength. Most wave motion has a finite number of wavelengths and forms a 'wave-packet'.*

encountering one another. It seems reasonable to suppose that quite a lot of spin waves can exist at the same time without much mutual interference.

With such diagrams it is easy (and sensible) to forget the reasoning which has led to them, because they are physically convincing and therefore likely to be independent of any mathematical, or semi-mathematical derivation. Such waves will probably exist for any ferromagnet and possibly for any cooperative magnetic system, including antiferromagnets. The really relevant thing is to demonstrate their existence.

The Kittel theory of antiferromagnetic resonance can be regarded as an illustration of a theory of two motions which have the same phase throughout each sublattice, so they correspond to the non-propagating modes of the ferromagnet. The first propagating mode was observed [88] by inelastic neutron scattering in 1961. A little later Collins [89] observed 42 spin-wave modes in MnO and used his results and spin-wave theory to estimate the nearest and next nearest exchange interactions, commenting that this was the only technique for observing spin-wave modes. The theory has been fully reviewed by Keffer [90]. A later investigation [91] on CoF_2, which from EPR was expected to have a complicated spin-Hamiltonian, produced evidence for a temperature dependence of the spin-wave modes as the Néel temperature (38 K) was approached, as well as evidence for their existence above the Néel temperature. The latter was unexpected, though the former could have been anticipated from the Kittel theory (but not from the version of the H–D model which takes the fully aligned sublattices as a starting point).

By the end of the 1960s the interest in crystalline antiferromagnetism seems to have been waning, probably because a wide range of other magnetic phenomena was being generated, particularly those connected with the technical uses of new insulating materials which, unlike the antiferromagnets, possessed permanent magnetic moments. So this section will be closed with a brief mention of some special cases of antiferromagnetism.

The rare-earth chromites show two Néel temperatures [92], one corresponding to an antiparallel alignment of chromium moments and a second due to an alignment of rare-earth moments. Then a number of crystals which were expected to be antiferromagnetic, e.g. Fe_2O_3, Ni_2O_3 were found to be weakly ferromagnetic. Two independent mechanisms were introduced [93] to account for this. The two-sublattice concept was retained with the assumption that they did not have antiparallel alignments, either because there were zero-field splittings due to crystal fields with different distortion axes for the different sublattices, or because of the presence of antisymmetric exchange interactions, which might arise from the combination of spin–orbit interactions and lattice distortions. The former were familiar from EPR studies, whereas the significance of the latter was not appreciated until it was invoked by Moriya [94] to account for slight departures from antiparallel alignments.

14.8. Ferromagnetism and ferrimagnetism

It is convenient to begin this section by grouping ferro- and ferrimagnetism together, for the experimentalists had long found difficulty in distinguishing between them. Theoretically there was no difficulty, nor would there have been experimentally if the significance of electrical properties had been appreciated. Rather it seems that the experimentalists, particularly those brought up using pre-war techniques, regarded the advent of ferrimagnetism as an addition to the range of ferromagnets, one which provided a new set of materials to which to apply their existing experimental methods.

14.8.1. *Ferromagnetic resonance*

This position changed when Griffiths returned to Oxford at the end of World War II and began using the microwave equipment there for an investigation of the high-frequency permeabilities of the ferromagnetic metals. Using a resonant cavity, one end wall of which was made of the ferromagnetic metal, he attempted to measure one of the sources of energy loss in a ferromagnet at a much higher frequency than had previously been available. In trying to isolate the loss he decided to apply a steady magnetic field in the plane of the wall and as a result discovered [95] a strong resonance as the magnitude of the field was varied (figure 14.11). This occurred near the value expected from the relation $h\nu = 2\mu_B H$, appropriate for an electron in the internal field of classical magnetic theory. The experiment was basically the same as that

Magnetism

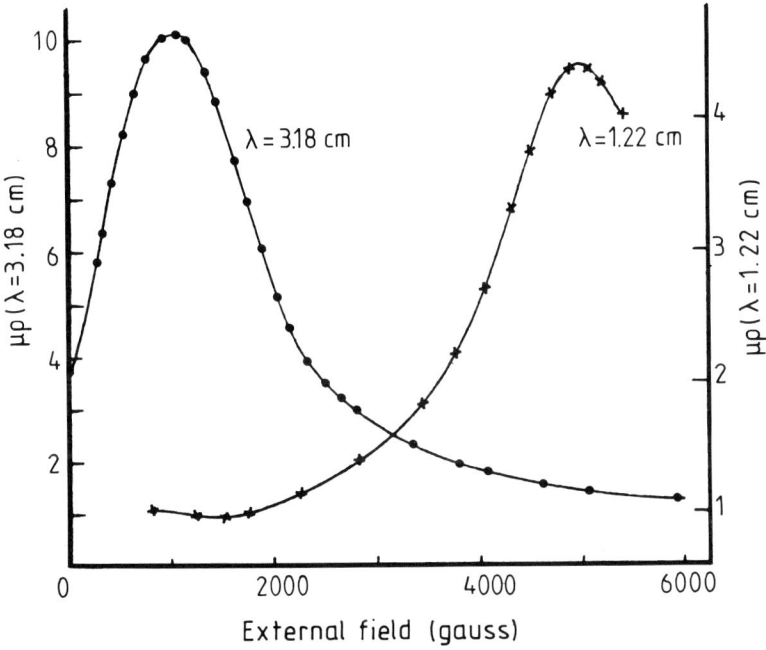

Figure 14.11. *The ferromagnetic resonances first observed by Griffiths.*

being used in the early days of EPR. It was, though, quite clear that the resonance was not due to a set of independent electrons, both from its intensity and because they would not be showing ferromagnetism. Instead, all the spins were imagined to be precessing at the same rate about the direction of the internal field, as in figure 14.10, but remaining parallel to one another as they did so.

Apart from the indirect observation in the gyromagnetic effect, the first direct observation of precessional motion seems to have been in atomic and molecular beams by Rabi [96] in 1937, then (1946) in solids in the macroscopic resonance of Griffiths' ferromagnetic end walls and about the same time in water, in the nuclear moments of protons, by Bloch *et al* [97] and independently, by Purcell *et al* [98]. Resonance could thus be induced in atomic beams, in ferromagnets, in paramagnets and in nuclei, a remarkably wide range.

The original theoretical explanation of Griffiths was corrected by Kittel [85], and then re-derived using microscopic equations by Van Vleck [99]. As a result it was realized that ferromagnetic resonance provided a way of using single crystals to determine the magneto-crystalline anisotropy energies, macroscopic expressions which were introduced to account for the preferred directions of magnetization found in ferromagnetic crystals.

14.8.2. Magnetic domains

The concept of a macroscopic anisotropy energy can probably best be understood by considering another technique for exploring ferromagnets, which had, by the late 1940s, been brought to a mature stage, the study of domain structures by means of Bitter patterns. In a very short time a large number of domain patterns were photographed and showed that a uniform magnetization is rarely to be found in a ferromagnet with no external field applied. Rather, the magnetization consists of domains of uniform magnetization. So two questions came to the fore. What determined the direction of the moment within a given domain and what determined the domain patterns?

See also p 1134

For the most part the theory needed to answer these questions was couched in macroscopic terms, for the quantum theory of ferromagnetism in conductors was still in an insufficiently developed state. This was not regarded as a handicap, but just the opposite, for the ideas of classical magnetism, with a few *ad hoc* assumptions, had been quite successful so far and there was no reason to suppose it would now fail. So the technique was to write down, for a uniformly magnetized region, an energy expression which depended on assumed directions and magnitudes for the magnetization, the magnetic anisotropy energy. The values actually taken up would then be those which minimized the total energy, subject to any forces across the boundaries between the uniformly magnetized regions. There also needed to be terms associated with exchange and since H–D theory gave an energy which depended solely on the angle between two spins it was given a classical interpretation, as an energy which remained constant as the direction of the magnetization rotated and which was proportional to the square of its magnitude. The anisotropic energy already had a classical explanation in terms of demagnetizing factors, the only problem being that the associated energies seemed to be far too small, though the angular dependence seemed to be correct. So the simplest thing was to accept the measured values for their coefficients and attribute them to microscopic anisotropic exchange, while describing them macroscopically. In more sophisticated treatments further energies were included, such as those which couple the dimensions of the domain to its magnetization, for it had been found that magnetization was accompanied by volume changes. Armed with these energy expressions it was possible to use various measurements to sort out the magnitudes of the unknown parameters.

The Bitter patterns showed that in many cases the sample was not a single domain. The model therefore had to be enlarged to include energy changes as one domain grew at the expense of another. Also the domain walls made their contribution to the energy; within these the magnetization changes direction so the isotropic part of the exchange plays a significant role, for adjacent spins are no longer parallel. So complicated did the domain patterns and the theory become that their study, for many of the experimentalists, became more like a study of art forms, while for the theoretician they became more like a nightmare.

The further question of how to understand their time dependence, the changes in shapes and sizes of domains and the motions of domain walls during a hysteresis cycle was even more of a problem. The Bitter technique, being static, was little help. A new avenue was opened by Lee et al [100], who introduced the use of the transverse Kerr effect, which relied on the observation that when a plane-polarized beam of light is reflected from the surface of a magnetized sample the plane of polarization is rotated. The experimental method used changes in the angle of rotation of the polarization of a narrow beam of light to monitor changes in the magnetization at the spot from which the beam was being reflected. It was not entirely what was wanted, because it did not follow the motion, for example, of a domain wall. This was to come later, particularly for ferrimagnets, as will be discussed in due course. Domain techniques have been reviewed by Craik and Tebble [101] and Dillon [102].

14.8.3. Ferrimagnetism and ferromagnetism

It is now convenient to turn to ferrimagnetism, which for many years had hardly been distinguished experimentally from ferromagnetism, a situation which was to continue for long after it had been shown that ferrimagnetism was fundamentally closer to antiferromagnetism. The reason was, of course, that the two shared the important property of having macroscopic magnetic moments, and this was of more importance than the way in which it came about. Thus any property of one which depended on the macroscopic moment could be expected to occur with the other.

In particular ferrimagnets showed [103] the equivalent of ferromagnetic resonance as well as the more classical phenomena of hysteresis and domain structures.

With such a similarity to ferromagnets it might have been expected that some of the ferrimagnets would simply have replaced some of the existing ferromagnetic metals and alloys in specific applications. However, this is not what happened, because the ferrimagnets had one property which made them significantly different from the ferromagnets: they were electrically insulating. Thus instead of rivalling the ferromagnets they complemented them, and so greatly extended the frequency range in which magnetized materials could be used.

The first inkling that ferrimagnets could have valuable uses came in 1946 when Snoek [104], working at the Philips Laboratories in Holland, announced that *ferrites* had been produced with strong magnetic properties, high electrical resistivity and low hysteresis loss. (Ferrite is used to describe a whole class of oxides.) Hilpert had made some of them in 1909 but had concluded that his were of no commercial value, so until

Snoek's announcement there had been little interest in their magnetic properties, though de Boer and Verwey [105] had been interested in their electrical properties (see Chapter 17).

> See also p 1330

14.8.4. Crystal structures

The magnetic properties of the ferrimagnets are strongly linked with their crystal structures. For example, Fe_3O_4 is basically a face-centred cubic lattice of O^{2-} ions with Fe ions in interstitial positions. There are 96 of these in the unit cell, 64 having four and 32 having six adjacent O^{2-} ions, with only 24 Fe ions to be accommodated. Eight of these are present as Fe^{2+}, which has four unpaired electrons, and 16 are present as Fe^{3+}, which has five unpaired electrons. So not only does iron occur with two different magnetic moments, it also has a choice of sites with quite different crystal fields. The actual distribution is determined in the formation of the crystal, and this and the states of ionization of the ions at the various sites can only be determined afterwards, which is not easily done. There is also some evidence that the choice is not the same for all unit cells.

Standley [106] lists the crystallographic data for a range of similar compounds, the spinels (with the general formula $M^{2+}Fe_2^{3+}O_4^{2-}$ or $M^{2+}O^{2-}$, $Fe_2^{3+}O_3^{2-}$, where M is a divalent metal ion) the garnets and the perovskites. Among these there are many examples, including lodestone, that are both ferrimagnetic and insulating.

By this stage the reader will probably have concluded that the oxides form a large family of complicated substances, a view which is not uncommonly held, particularly when it is realized that in addition to their number they are all highly refractory and so difficult to prepare with a given composition. It is therefore reasonable to ask why they are given any attention? The best answer seems to be that in the late 1940s they were indeed of very limited interest. Now they are all around us, though we may not realize it. To give a few examples, we rely on particular members of the ferrites and some closely related families for the internal aerials of our radio sets, the recording media for our tape and video recorders, and the permanent information stores in our computers and various 'cards'. And these are just a few of the uses of specially selected ferrimagnetic insulators.

A perhaps more pertinent question should therefore be, how is a ferrimagnet chosen for a particular application? In the early days such a question could hardly have been asked, let alone answered, for it was only after the properties of many of them had been explored that the possible applications were identified. Now it can be answered: from a pool of painstakingly acquired information. Thus the story of the development of ferrites is a good counter-argument to a claim that research should be consumer directed. Would the need for a radio small enough to be portable ever have produced even one of the necessary components, a ferrite aerial?

For all that the macroscopic magnetic properties of ferrimagnets are similar to those of ferromagnets, microscopically they are closer to antiferromagnets. An interesting consequence is that there is no reason to suppose, even for one which can be regarded as composed of two antiparallel magnetic sublattices, that the moments on the two sublattices should be similar either in magnitude or temperature dependence. Thus the thermal behaviour of the net macroscopic moment of a ferrimagnet may be quite different from that of a ferromagnet. This was described by Néel [107] in an account of his original work, which includes a number of illustrations of what might occur. Probably the most interesting prediction was that as the temperature is raised the net magnetization might decrease to zero and then rise in the opposite direction. While this does not happen in the majority of ferrimagnets it was demonstrated by Gorter and Schulkes [108] in some mixed ferrites. The temperature at which the net moment falls to zero is now known as the compensation point.

For the most part attention will, from now on, be focused on the uses which arose for ferrites, garnets etc, which are electrical insulators.

An immediate consequence of the high resistivity was apparent in the resonance experiments. In ferromagnetic resonance in a conductor the microwave radiation can only penetrate a very small distance (*skin effect*) which is why Griffiths used thin films, supported by the non-magnetic wall of a standard microwave resonator. The number of unpaired spins which could partake in the resonance was therefore limited. With an insulator the position was quite different, for it could be expected that it would be virtually transparent to microwave radiation unless there was unexpectedly large absorption due to its magnetic properties, which might include absorption from domain wall movements and rotations of magnetization, as well as the resonance. The non-resonant effects were generally found to be small which meant that large samples could be studied and, if necessary, placed anywhere in the resonator.

The transparency was exploited in an investigation of the propagation of electromagnetic waves along the direction of an external magnetic field. The theory was developed by Polder [109] who confirmed that the Faraday rotation of the plane of polarization would have the interesting property that if for one direction of propagation through a finite sample it was rotated through a particular angle then on the propagation in the opposite direction the rotation would be through the same angle. That is, if it corresponded to a right-handed screw rotation for one direction of propagation it corresponded to a left-handed screw for the other. This property is exploited in a number of devices [110], such as the *isolator* (figure 14.12). It was also found [111] that when certain ferrites were exposed to high microwave powers they would radiate microwaves at a

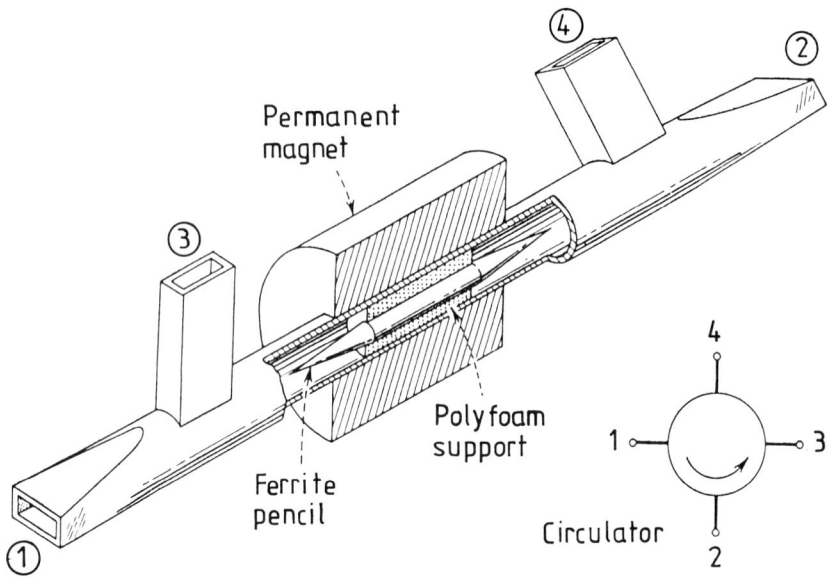

Figure 14.12. *The polarization of a wave incident at 1 is rotated through 45° and emerges at 2. A similar wave incident at 2 emerges at 3. So waves travelling in opposite directions are separated. The device can be used in radar systems, to separate outgoing waves from their reflections. (A rotation of order 90° can be achieved in distances of order 1 cm.)*

frequency double that of the incident radiation, a result which led to a good deal of interest in non-linear effects. For example Pippin [112] soon showed that two incident waves of differing frequencies could produce both sum and difference frequencies, and Suhl [113] showed that it should be possible to use the non-linear properties to produce a parametric amplifier for microwaves, the first demonstration being given by Weiss [114]. Since any amplifier can be made to oscillate by feeding back part of the output to the input this was later turned into an oscillator.

14.8.5. Magnetic recording

As long ago as 1898 Poulsen filed a patent which described an instrument which would record and reproduce sound through the orientation of magnetic domains, and a picture of his device is to be found on p 13 of *Magnetic Recording* by Lowman [115]. When it was demonstrated at the Paris Exposition of 1900 the interest was so great that he was awarded the grand prize. However, the practical obstacles to be faced before magnetic recording reached the present stage were immense, though the basic ideas have hardly changed. Before the ferrites came along thin strips of metallic ferromagnets were investigated; it was hoped that a strip, on emerging from a magnetizing coil, would retain a record of the time-dependent current in the exciting coil in the form of a variation of the local permanent magnetization along its length. Unfortunately, because of the conductivity of the strip, on entering the coil the induced magnetization took time to build up and as it emerged took time to settle to its remanent value, so it did not accurately represent the instantaneous current. Another problem arose because the easiest experimental arrangement was to magnetize the strip perpendicularly to its plane whereas the marked preference of a strip is to have the magnetization in its plane.

The advent of ferrites (which in this context includes all ferri- and ferromagnetic insulators) changed the position, since among them were materials which would respond much more rapidly than metallic films to changes in magnetizing fields. There were plenty of other difficulties, but at least the goal seemed sufficiently near in the late 1940s to justify a good deal of industrial research activity.

A modern tape consists of particles which have anisotropic magnetic properties such that the magnetization is uniaxial. The particles are then dispersed in some suitable non-magnetic medium and supported on a non-magnetic film. In some manufacturing processes, the tapes are exposed to strong magnetic films in the plane of the tape to orient the particles, after which they are cemented in position and demagnetized, for example, by heating to above the Néel temperature. On its passage through the recording head the tape passes through a field which first rises from, and then returns to, zero. It is left with a residual magnetization determined by the field cycle. For good reproduction this needs to be proportional to the current exciting the coil, which is unlikely to occur of its own accord, since the magnetization cycle is highly non-linear. It has been found that the imposition of a much stronger field, which oscillates at a frequency higher than that which is being recorded, gives better reproduction.

There is a good deal of technical knowledge and experience about the best magnetic materials to use, the production of suitable films and the best recording techniques. The range of materials in use is reviewed by Bate [116], and the techniques of recording by Lowman [115].

14.8.6. Bubble domains

In 1957 Dillon [117] reported that it was possible to transmit visible light through thin sections of the ferrimagnet, yttrium–iron–garnet. This opened the possibility of using Faraday rotation to see domain patterns with a microscope. If for one direction of magnetization the rotation of plane-polarized light is right-handed then for the other it is left-handed, and the two should be distinguishable using an analyser. The first photographs showing domain structures obtained by this new technique, in a film which was not in a magnetic field, soon followed [118]. Each domain appeared as a snake-like figure, winding its way across the sample, adjacent 'snakes' being oppositely magnetized in the direction normal to the film. The patterns could readily be changed by the application of a magnetic field and it is noticeable that each domain preferred to end at a boundary of the film rather than to close on itself. From then on the technique received considerable development and there is an extensive literature on it (figure 14.13).

It was suggested that instead of storing information in a static form on a tape and reading it by moving the tape, one might store it on a stationary tape and move the information. Bobeck [119] has given a preliminary account of progress towards this end. A large number of possible ferrimagnets were studied, among which were some where the serpentine domains of Dillon closed on themselves. Then, when a uniform magnetic field was applied, in a direction perpendicular to the tape, the regions with a magnetization in the direction of the field would grow at the expense of those magnetized in the opposite direction. For certain ranges of fields there would appear to be small circular domains with one sign of magnetization immersed in a sea with the opposite magnetization. From their appearance when viewed through the optical system they came to be known as magnetic bubbles. In their physical properties they were similar to small cylindrical magnets, which meant that they could be moved fairly easily by non-uniformities in the applied magnetic field. The study of bubble domains and their dynamics became a field of considerable interest [120]. By applying the correct fields near the edge of the tape, and by other techniques, it was possible to nucleate closed domains and so create bubbles, which could afterwards be moved and arranged as needed. So the problem of moving information along a stationary tape was well on the way to being solved, if the local concentration of bubbles could be varied as required and the whole pattern moved as one.

The direction of the development then changed when it was found that bubbles and their absence offered a way of providing high-density permanent stores for 'noughts' and 'ones', and that it was also possible to add and subtract them and so use them as computing elements [15]. Not much, though, seems to have come of this because of the competition from semiconductors.

14.8.7. Rare-earth metals and their alloys

There had been an interest in the magnetic properties of the rare-earth metals even before World War II, but as they became more readily available, and in much better separated forms, it became clear that the doubtful purity of the materials made a good deal of the pre-war work unreliable. Indeed many of the early EPR studies in insulators showed resonances from rare-earth ions which were not supposed to be there, indicating in particular that gadolinium impurities would have significantly distorted earlier macroscopic magnetic measurements.

Magnetism

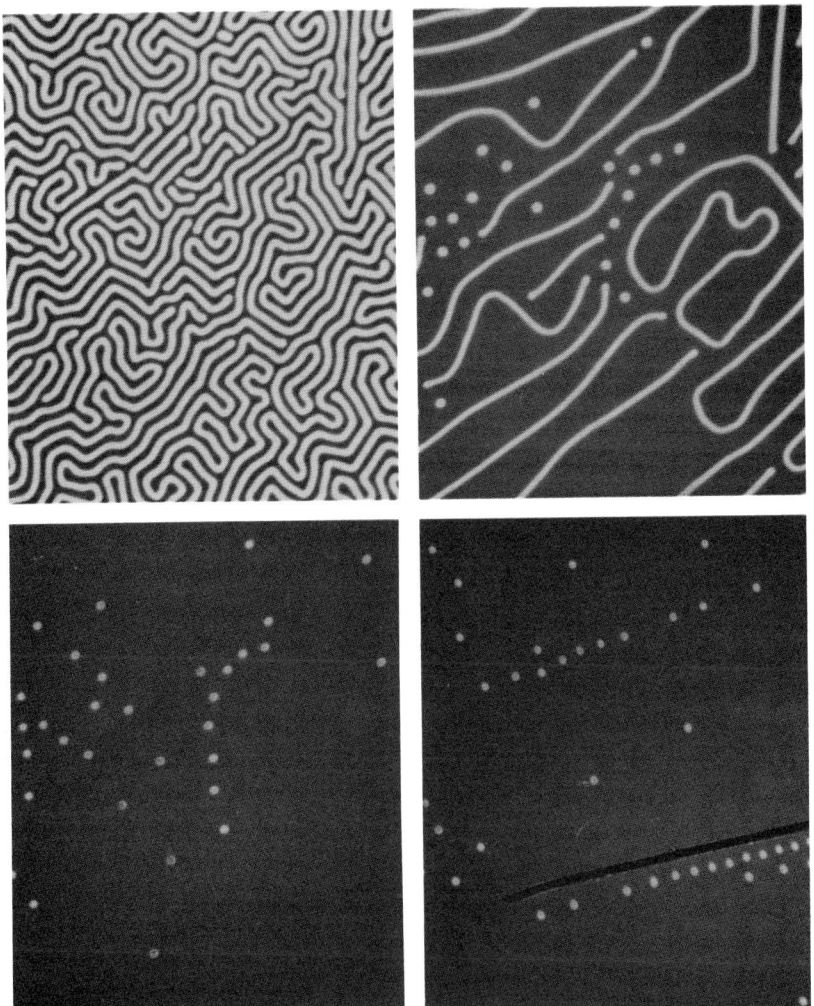

Figure 14.13. *Emergence of magnetic bubbles in a thin wafer of magnetic garnet is demonstrated in this sequence of photomicrographs taken at the Bell Telephone Laboratories. The magnetic domains in the specimen rotate polarized light in different directions depending on whether the internal magnets in the crystal point up or down. By adjustment of a polarizing filter, domains with the same orientation can be made to look either bright or dark. When no external magnetic field is present (top left), the domains form serpentine patterns, with domains of opposite magnetization occupying equal areas. (The apparent departure from equality here is an artifact of the exposure.) When an external magnetic field is applied perpendicularly to the specimen (top right), the domains that are magnetized in the opposite direction shrink and in a few cases contract into bubbles. Although these circular domains are called bubbles, they are actually stubby cylinders viewed from the end. A further increase in the external bias field (bottom left) converts all the remaining 'island' domains into bubbles. With a 'soft' magnetic wire whose magnetization is polarized by the external field one can move the bubbles around freely (bottom right). Because the bubbles repel one another they tend to maintain a certain minimum separation. Nevertheless, they can be packed with a density of more than a million per square inch.*

Nevertheless the pre-war work was indicative of interesting magnetic properties, for gadolinium had been found to be ferromagnetic and it seemed probable that all the members of the family would be of interest. So when they became more readily available, in increasingly pure forms, their study was taken up in earnest.

Unlike most metals, the paramagnetic susceptibility of neodymium varied strongly over a range of temperatures in the vicinity of room temperature, and the variation was very much like that found in salts containing trivalent neodymium ions [121]. Since the explanation of the susceptibilities of these salts was falling nicely into place as a result of the free-ion Zeeman studies and the interpretations of EPR, the results on the metal were similarly interpreted on the assumption that all the magnetism was due to localized inner-shell 4f electrons.

There was no inkling of the complexity yet to come. The first indication came from neutron diffraction studies of polycrystalline samples of erbium and holmium [122]. As the study developed a panoply of exotic magnetic configurations was revealed in the elements and, in due course, in their alloys. Legvold [123] has given a detailed account of this and various other physical properties together with diagrams illustrating the various types of magnetic ordering. Below some critical temperature, which is usually below room temperature, some of the elements (e.g. gadolinium) show moments aligned as in typical ferromagnets, in others (e.g. terbium) there are aligned moments in parallel planes but with the directions rotating through a fixed angle from plane to plane, and in still others (e.g. holmium) there are combinations of the above so that some components of magnetization rotate from plane to plane while other components point perpendicularly to the plane (figure 14.14). As if this was not enough the arrangement in any one element could also change with temperature.

> See also
> p 991

A good deal of information was also found by inelastic neutron scattering, as described in Chapter 12. An incident neutron would lose some of its energy and change its direction of motion, showing that an excitation of known energy and momentum had been created in the magnetic system. While this was, experimentally, similar to the excitation of a spin wave in a more conventional ferromagnet there was some hesitation over using such a description in such complicated magnetic structures. The excitations could hardly be regarded as due to simple reversals of electron spins. They came to be referred to as *magnetic excitons* or *magnons*.

It had long been accepted that the rare-earth ions are unique in having partially filled 4f shells which are well inside the electronic charge distribution. This concept seems to have been readily accepted for the elementary metals. So they too had immobile electrons in partially filled 4f shells, which meant that the conductivity had to be associated with the outer electrons, which were, presumably, in conduction bands. Little

Magnetism

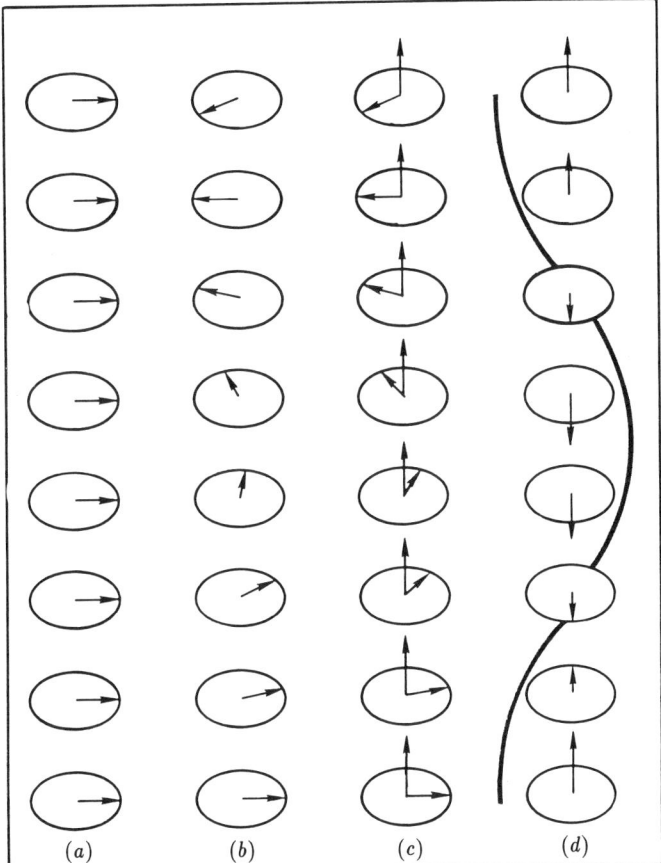

Figure 14.14. *Magnetic structures of the heavy rare-earths. The moments in a particular hexagonal layer are all parallel. In the basal-plane ferromagnet, (a), all moments are aligned along the direction of magnetization. The moments in the helix, (b), rotate by a specific angle between neighbouring planes. The cone, (c), is a combination of a helix and an axial ferromagnetic component, so that the total moment rotates on the surface of a cone. In the longitudinal wave structure, (d), the c-axis component varies sinusoidally, while that in the basal plane is disordered. These structures can all be understood as a combination of the exchange interaction, which produces long-range periodicity along the c axis, and the crystal fields, which tend to orient the moments.*

notice sees to have been taken of Slater's view, at least as far as the 4f electrons were concerned, that in a periodic structure all electrons should be regarded as in band states. Rather the accepted view was that the 4f electrons could be distinguished and treated by crystal field theory, with the 'ions' bathed in a sea of conduction electrons which itself provided a major contribution to the crystal fields.

The description of the metals then became quite similar to those of the salts, for their Curie–Weiss-like susceptibilities at high temperatures

1163

could readily be understood without a need for a detailed description of the exchange interactions. However, explaining the temperatures at which the cooperative phases set in was more of a problem because they are much higher than those of the salts. Some years before there had been an interest in exchange interactions between manganese ions in a non-magnetic host and out of this had come the idea of an indirect exchange interaction due to conduction electrons. This idea was adopted for the rare-earth metals. When a conduction electron is in the vicinity of a localized moment it is assumed that there will be an exchange interaction between its spin and the local moment. Then, as the conduction electron moves away, some memory of the correlation will be retained, and this will be passed on to a second localized moment, so producing an indirect correlation between the two localized moments. The resultant coupling is now usually described as the RKKY interaction (using the initials of four of the main contributors, Ruderman and Kittel [124], Kasuya [125] and Yosida [126]). Unlike exchange in insulators it was predicted that its magnitude would fall off more slowly with the local moment separations and show oscillations in sign. The magnons could then be pictured as excitations of the coupled system of localized moments and induced moments in the cloud of conduction electrons [127].

To find out more about the crystal fields the rare-earth elements were dissolved in non-magnetic host metals of the same structure. This reduced the exchange interactions and allowed the low-lying patterns of crystal field splittings to be observed directly by the inelastic scattering of neutrons [128]. Not only did this work yield the crystal field parameters but it showed there was little variation from one rare-earth to the next.

This work showed clear evidence of unquenched orbital moment and created problems in describing the exchange interactions, which could hardly be expected to have the same form as when the orbital moment is quenched. There was also the complication of taking account of the crystal field splittings. The only hope seemed to be to use the neutron dispersion results to provide detailed descriptions of the exchange interactions and the success achieved has been described by Mackintosh [129].

It seems probable that had the work on the rare-earth elements not been so successful the division of opinion about which description to use, localized or non-localized, would have continued, but with these examples, where there were undoubtedly periodic lattices, it became increasingly difficult, though not impossible (according to Freeman [130] on whom Slater's mantle appears to have fallen) to maintain that they could only be understood on the basis of band theory. The corresponding difficulty with the insulators was resolved without the introduction of bands, and it seems not improbable that the same will occur with the conductors. Undoubtedly the simple concept of crystal fields will disappear, though this has not stopped various workers from trying to estimate them using various electrostatic-like models of host metals, such as gold and aluminium. In most of this work the 4f electrons are distinguished from the rest, though not by Schmitt [131] who found that there was an exchange-like contribution to the crystal field.

I took up the problem of how to obtain equivalent operator forms in place of the crystal fields, while retaining indistinguishability, without success [132]; and though Dixon and Wardlaw [133] later managed to clarify some of the issues, they did not entirely resolve them.

While theoretical progress has been limited the same has not been true experimentally, for it has been found that alloys containing rare-earths have a diversity of uses. On this topic a few remarks and references must suffice. The remarks are based on three articles in *Magnetism in the Nineties*, which look back as well as forward in time. The first, by Strnat and Strnat [134], gives an account of the work on rare-earth–cobalt alloys and lists their uses. Some are of considerable industrial importance, particularly where very powerful permanent magnets are needed and expense is not a restriction. The magnets in use are not simple things—the authors remark that 'Real magnets are more complex than the intermetallics identified as potential hard-magnetic materials. They are not stoichiometric compounds; there are always several phases present; the metallurgical microstructure is typically complex and not in thermodynamic equilibrium. Some magnets are even composites containing nonmagnetic binders.'. The second article, by Herbst and Croat [135], describes another important family, the neodymium–iron borates, and the third, by Buschow [136], describes the permanent magnet materials of type $RT_{12-x}M_x$, where T can be any of Co, Ni and Mn and M any of Al, Si, Ti, V, Cr, Mo, W and Re. All three papers give a large number of references and a further source is a volume in the series known as Landolt-Börnstein [137].

14.9. The changing pattern

By about the middle of the 1970s a good deal of the steam resulting from the immediate post-war developments in electron resonance, ferrites and rare-earths seems to have been dissipated, so although developments in all of them continued, interest was beginning to turn to new directions. For twenty or more years Slater's view, that with periodic systems their understanding would necessarily come from band theory, had been ignored and as a result that part of magnetism which used localized concepts with great success had become divorced from the magnetism which used band theory, which had hardly progressed at all. This was in remarkable contrast with practically all the rest of solid-state physics where band theory had come to be the dominant approach. So the traditional ferromagnets iron, cobalt and nickel had come to be 'the piggy in the middle', explicable neither by band theory nor by localized moments. Furthermore, with the subsequent demonstration that the localized models needed neither to distinguish electrons nor to give up periodicity the common sense approach of staying with concepts which worked seemed to have paid off.

The above remarks have slightly exaggerated the position for although band theory was having enormous success in wide areas of solid-state physics there was a growing sense that something more was sometimes needed. It would be out of place to go into a lot of detail here so a few remarks must suffice. There were two basic problems. The first was that given a band structure, say that of sodium, it was possible to work out the probabilities that, at a given lattice site, the number of 3s electrons would be 0, 1 or 2. The isolated sodium atom has only one such electron and the expectation was that this would be by far the most probable occupation number in the solid, and yet the values derived from band theory were sufficiently different to give cause for concern. The idea was creeping in that it might be necessary to add something to band theory to produce restrictions in the possible local expectation numbers. One technique [138] was to introduce localized functions derived from Bloch functions and characterized by a parameter commonly called the Hubbard U. (A tribute to John Hubbard who wrote a number of important papers on this topic in the 1960s and died at an early age.)

The second problem was that in band theory almost all the electrons are spin-paired, so there was limited scope for magnetism to arise from unpaired electron spins, and yet the number of examples where local moments were being found was increasing. Neutron scattering had indicated that even in chromium and manganese there were small localized spin magnetic moments, with similar results on some of the actinide elements. (The actinides were, at one time, expected to be magnetically similar to the rare-earths, having unpaired electrons in 5f orbitals, but since these electrons are on the outsides of the ions there was the alternative possibility that they would be much more sensitive to their environment. Sorting out what actually occurs has been far from simple, particularly when there is also the possibility that some or all of the electrons might be in the nearby 6d orbitals. The magnetism of the actinides has received a lot of attention, but this will not be reviewed here, except to say that there seems to be little that is systematic about the magnetism of these elements and their compounds.) The de Haas–van Alphen effect, which has proved so powerful in the analysis of electron behaviour in non-magnetic metals, has also given strong support to Stoner's conception of itinerant magnetism. As explained in Chapter 17, the effect provides a detailed picture of the band structure of conduction electrons. Gold was the first to show that the conduction electrons in iron and nickel form two different assemblies, each containing spins of one orientation. The difference in the number of electrons in the two assemblies accounts fairly well for the saturation magnetization. This

See also p 1348

Magnetism

is a complex subject with many unresolved problems concerning the distinction between localized and itinerant electrons, such as to keep alive the issues which troubled Slater. A review has been given by Lonzarich [139].

The problem of describing localized charge distributions, where the changes in magnitude on moving from one site to another are smaller than the electronic charge, led to the introduction of the concept of charge fluctuations, as well as spin fluctuations for describing local magnetic moments when the moments are less than that of a single spin. These are useful expressions but they can also be misleading to the casual reader, for they suggest that charge or moment at a site is changing with time, whereas often the change is with position and is static in time, but as technical terms they are here to stay.

An upshot of the various extensions to band theory has been a growth in interest in the magnetism of conductors, so much so that the study of itinerant magnetism is becoming one of the most active topics in the field; perhaps the most remarkable point is that quite a lot of this is actually about the disappearance of magnetism as the temperature is lowered.

14.9.1. *The Kondo effect*

To give an example, many years ago de Haas *et al* [140] observed a minimum in the electrical resistance of gold at about 5 K, a similar effect being subsequently found in magnesium [141] and other elements. There was no explanation of this quite unexpected effect, though it was suggested that it might be due to impurities. In due course this was confirmed, as it became increasingly clear from work on non-magnetic metals containing deliberately incorporated 3d elements, such as manganese, that localized magnetic moments were involved. (By this time temperature-independent magnetism from band electrons and/or filled inner shells of electrons was regarded as universal in normal metals. So, any conductor which had magnetic properties solely of this kind was regarded as non-magnetic.) The paramagnetism with the impurities present was Curie-like and so was obviously due to localized moments. Then it was found that at about the temperature of the resistance minimum the paramagnetism disappeared.

See also p 1342

This phenomenon led to a lot of theoretical work, under the description *Kondo effect*, which acknowledged the contribution to the explanation which Kondo made in a series of papers in the 1960s [142]. The physical explanation was that as the temperature decreased the arrangement of the conduction electrons would change and that a degree of charge and spin localization would develop near a given impurity, such that the localized magnetic moment in the conduction cloud would actually cancel the localized moment on the impurity. It was as if the spin angular momentum of the impurity had become coupled to an equal spin angular momentum induced in the cloud of conduction electrons to give a total spin of zero, so forming a non-magnetic cluster. Thus an interest which was originally in electrical conduction came to be associated with the disappearance of magnetism and the extension of itinerant magnetism to take account of what became known as *strongly correlated electron systems*.

14.9.2. *Intermediate valence and heavy fermions*

In the early 1970s a rather similar magnetic phenomenon was found, the disappearance of magnetic moments in certain fully concentrated conducting compounds of rare-earths at low temperatures. This too might have been regarded as indicating that the conduction electrons had developed magnetic moments, which precisely cancelled the localized moments from the 4f electrons, had it not been that it was difficult to accept that it could happen at all the rare-earth sites in a crystal. (There is a related question as to what happens to the Kondo effect in a normal metal when the number of magnetic impurities is increased.)

To illustrate the direction taken by the explanation it is useful to go back to the rare-earth metals and ask why there are itinerant conduction electrons as well as electrons localized in partially filled 4f states. The simplest explanation is that, for the postulated electronic arrangement to be in equilibrium, it must require a positive amount of energy to take an electron out of the 4f shell at a given site and place it at the top of the Fermi distribution of the conduction electrons. Similarly it must also take a positive amount of energy to move an electron from the top of the Fermi distribution to a 4f orbital, so increasing the number of 4f electrons at a typical site. In other words it must be energetically unfavourable to alter the number of 4f electrons at any site.

Suppose now that in some rare-earth conductor a small amount of energy could actually be gained by moving a 4f electron from a particular site and placing it at the Fermi level. Then this should happen, for it will lower the energy. But having granted this, why not allow a second electron to move from another site? This too could happen. The transferred electrons are delocalized and as the process is continued and electrons from successive sites are moved to the conduction band the later ones will have to go into increasingly higher-lying band states, for the lower ones, which were previously empty, will now be occupied by electrons which have already been transferred. So less energy will be gained and the process will eventually stop, when the energy needed for the transfer becomes positive. But if the number of electrons so transferred is less than the total number of 4f sites it would seem that there is now a mixture of sites, some of which have lost an electron from the 4f shell and others which have not. There is then the problem of reconciling this picture with the quantum requirement that crystallographically identical sites must have identical charges.

Experiments on some rare-earth sulphides seemed to indicate that such transfers had occurred and that all the rare-earth sites had, somehow, remained crystallographically identical. To add to the difficulty the conduction electrons were then found, at low temperatures, to develop localized magnetic moments which cancelled those in the 4f shells. So there seemed to be something like a Kondo effect in a fully concentrated system. There would have been localized moments had no transfers occurred and probably there still are, at high enough temperatures, even though transfers to conduction states had occurred.

The phenomenon as a whole was described as *intermediate valency* and, whether or not the above physical description is valid, it was quickly recognized that its theoretical treatment presented a major problem in quantum mechanics. The basic phenomenon, the disappearance at low temperatures of a magnetic moment, can hardly be regarded as an exciting event in magnetism, any more than other recent topics involving strongly correlated electrons—Kondo lattices [143] and heavy fermions [144], a rather similar phenomenon which is particularly associated with actinide conductors. Yet magnetism is the soil from which these new developments have sprung, and it is experts in magnetism who are most familiar with the available theoretical techniques.

A problem with most of these new developments is a shortage of experimental techniques which can pin-point what is happening, the realization of which serves to emphasize how fortunate it was, particularly for the insulators, that resonance techniques were available. The position is much less favourable for conductors, for with bulk specimens something which will penetrate a metal is needed. Neutrons can be used and so can elastic waves, though only over a limited frequency range. Electromagnetic radiation is restricted to very low and very high frequencies. X-rays can excite electrons from bound states to high-lying conduction states, so high in fact that the electrons emerge from the conductor with energies that can be measured. This technique, the photoemission of electrons, has been used, but so far these and similar methods have not been as easy to interpret as resonance experiments.

Probably the best way of regarding these efforts to extend itinerant magnetism is to expect that they will result in much of magnetism disappearing as a separate topic in solid-state physics. Indeed it is probably only because magnetism was a well-developed subject before the relevance of band theory was fully appreciated and that, when it was, the main developments were dominated by crystal field theory, that it remained separate. Whether the whole of magnetism will disappear into solid-state physics is another matter, for just as the traditional topics, which focused on crystalline materials, seem to be close to being unified, new fields, involving non-periodic lattices, are being opened. There is much interest in amorphous magnetic materials and very thin films which, though periodic in two dimensions, are far from periodic in the third.

14.9.3. *Amorphous magnetism and thin films*

There is a long history, in magnetism, of the experimental study and use of magnetic alloys, which can be classed under the general heading of random systems. Similarly in EPR the magnetic ions are often widely dispersed in magnetic hosts, and so provide another example of randomness, though one in which the consequences of randomness are not obvious. The position changes as the concentration of magnetic ions is increased, particularly if the exchange interactions are antiferromagnetic, as they often are. For low concentrations there will probably be no phase transition corresponding to the onset of antiferromagnetism, but as the concentration is increased it becomes increasingly probable that some kind of phase transition will occur. The study of some of these systems comes under the heading of spin glasses, which are usually taken, at least in theoretical studies, to imply either a random spatial arrangement of spins interacting with short-range exchange interactions or a random occupation of lattice sites with longer-range exchange interactions which vary in sign, as in the RKKY theory of impurity spins in a conductor. Both models present great theoretical difficulties and have excited a good deal of interest in the last decade, particularly over the question of how to recognize a phase transition if it should occur; or indeed whether there will be a transition at all. A simple example of just three spins mutually coupled antiferromagnetically serves to illustrate the sort of problem that can arise. It is already difficult to decide how these would align at low temperatures because they cannot all be antiparallel, a situation described as *frustration* [145].

The experimentalist can fairly readily be provided with amorphous magnetic samples, for there is a variety of techniques for making them, and they can be either insulating or conducting. So there is no shortage of examples to study. On the other hand the theoretician, while he can invent a variety of models, is not able to get very far with most of them. So both experimentalist and theoretician can be kept occupied, but whether, for any specific case, there is a convincing match between experiment and theory is a matter I do not feel competent to decide. It is probably not important at the present stage, because magnetism has usually progressed quite well simply on the basis of experimental investigations, with theory sometimes providing a stimulus and sometimes an *a posteriori* explanation. Here there is an extra stimulus, a technological interest in producing isotropic magnetic materials.

The other big development is the study of thin films, which may be of the order of 100 atomic layers in thickness, crystalline or amorphous, magnetic or non-magnetic. Such films need to be supported, and a variety of materials can be used as substrates, either for single films or for sandwiches of films having different compositions. The possibilities are almost endless, and enough has already been done to show that a rich variety of new properties will be found. Thus it has been possible to grow a layer of an insulating ferrimagnet on top of a normal non-magnetic conducting layer, and demonstrate that changing the magnetization in the ferrimagnet changes the resistivity of the conductor. Such an arrangement may well find use in the reading heads of tape recorders.

I am aware that no references have been given in this section and for this there are two reasons. One is that so many workers have contributed that it is difficult to decide on priorities and the other is that I find it difficult to make an assessment of which of the many possibilities have the greatest potential for development. I hope, therefore, that my deficiency can be compensated by a reference to the articles in *Magnetism in the Nineties* [144], which is primarily aimed at predicting what is going to happen in magnetism in the rest of this century. For the immediate past there are the papers given at the International Conference on Magnetism, I.C.M. 91 [146], one of a regular sequence held at intervals of approximately three years.

14.10. Nuclear magnetism

14.10.1. Nuclear magnetic resonance (NMR)
At about the time that ferromagnetic and electron paramagnetic resonance were being introduced a related interest was being developed in resonances associated with nuclei. In 1946 two papers appeared [97,98] which were to set the scene for a whole new development in magnetism. It was known that a substantial number of nuclei possess nuclear spins and associated nuclear magnetic moments which, as with the electrons, are tied to the spin directions. So in an external magnetic field it could be expected that such a nucleus would precess about the field direction with a frequency determined by the strength of the field and the nuclear g-value, the ratio of the magnetic moment to the angular momentum in appropriate units. Also that an oscillating transverse field, of the right frequency, would induce transitions between the nuclear spin orientations (the moment projections in the direction of the field). Indeed the basic concepts of the various magnetic resonance phenomena are so similar that they can reasonably be regarded as variants of the same concept. That more unification has not occurred can be attributed to differences in experimental technique. Nevertheless some contact has been maintained, particularly on the theoretical side, so in describing NMR the differences rather than the similarities will be stressed.

The magnitudes of nuclear spins are of the same order as those of the electron, but because the nuclear masses are something like 1000 times greater the nuclear magnetic moments are less than the electronic moment by a similar factor. Thus a magnetic field which gives an EPR resonance in the microwave region, at say 10^{10} Hz, is likely to give a nuclear resonance at a frequency near 10^7 Hz, where the experimental techniques are substantially different. In both, the strength of the absorption depends on the magnitude of the magnetic moment, the total number of spins in the sample and the difference between the number of these which are oriented parallel and antiparallel to the magnetic field, which introduces a factor of $\mu H/kT$, where μ is the magnetic moment and T is the temperature. In introducing any new technique it always seems difficult to observe any responses to begin with, and it is therefore remarkable that nuclear resonances were first observed at about the same time as electron resonances, for all the above numerical factors, except the total number of spins, are far less favourable. The apparent imbalance was largely offset by the higher detection sensitivity at the lower frequency, though the choice of sample was also important.

The first nuclear resonance experiment in a solid, by Gorter and Broer [147] in 1942, was unsuccessful. However, by 1946 resonances due to protons, hydrogen nuclei, had been reported in paraffin and water and once these had been found the way was open to making improvements in the experimental techniques and moving to other nuclei. From the beginning there was an interest in relaxation phenomena, energy exchanges between the spins and energy losses to the environment, which cannot readily be described in terms of the concept of Zeeman split levels. Instead they were described phenomenologically, in the Bloch [55] equation, which is an extension of the classical and quantum-mechanical equations of gyroscopic motion. The experiment is regarded as inducing a change in the precessing magnetization, which can be divided into two parts, a constant part in the direction of the field and another part which is rotating in the perpendicular plane (see figure 14.10). The latter was regarded as the resultant of all the spin precessions, and so it would have a persistent non-zero value as long as the exciting field continued to be present. However, if the excitation is removed both components of magnetization can be expected to change. The steady part should alter because the resonance phenomenon will have changed the initial thermal distribution of up and down moments (the constant part in the field direction). So there will be a change back to the original thermal equilibrium distribution, characterized by a time, T_1, known as the spin-lattice relaxation time. The transverse moment, which would have been zero before the experiment began, will decay back to zero by spin–spin relaxation processes, mutual spin-flips of oppositely aligned spins which destroy the coherences in the precessions of the spins. This decay was characterized by another time, T_2. For a liquid there is, of course, no lattice and the T_1 and the T_2 process are related. Two magnetic dipoles close together interact, and if they are at fixed positions, as in a solid, this interaction accounts for T_2. However, in a liquid the nuclei are moving and the motion of one dipole creates a time-dependent field acting on the others, which is much like that of the oscillating field in the experiment. It can therefore induce spin reversals as well as mutual spin-flips, and so contribute to T_1 as well. Its similarity in this respect to spin-lattice relaxation in EPR may be taken to indicate the great potential of NMR for studying motion in liquids.

The lower frequencies used in NMR give it an advantage over EPR for the study of metals, since the fields penetrate further; NMR has proved valuable here and has found a wide use mainly because the steady field which the nucleus experiences can be different from that which is applied, due to moments induced in the electron clouds. These investigations have helped in the study of local moments in conductors. A similar effect in molecules has been widely taken up in chemistry, for it has provided a means of identifying different chemical complexes. Thus the chemical shift, as it is known, can be used to distinguish between the protons in the CH_3, CH_2 and OH groups, which all occur in ethyl alcohol. The electronic arrangements in the groups are different, so the fields at the protons in the different groups are also different, even though they are all in the same external field. The resonances are very close, so the individual lines must be so narrow that they do not overlap. This presents more of a problem in solids than in liquids, where the motional narrowing, as it is called, helps to keep the lines narrow, but even so the resolution may be lost if the external field is not sufficiently uniform over the sample. Resonances from the same group in different parts of the sample will then occur at different field strengths and artificially broaden the lines. This high-resolution NMR, of which an example is shown in figure 14.15, therefore requires a highly uniform external field. The intensity of each line is proportional to the number of protons in each group, which can also be useful.

The reasoning can be inverted, for if the applied field is non-uniform in a known way, and all the nuclei of interest have identical surroundings, the intensity of the resonance line at a particular field value gives a measure of the number of nuclei in that particular field. If it could be arranged that every tiny cube in the sample experienced a different applied field, the line shape would give the spatial density of the nuclei. Furthermore, if the same nucleus occurs in two different environments and the non-uniform field can be given a time variation there is the possibility of determining the spatial distribution of each. Demonstrating this was a major challenge because the more the nuclei are divided up into packets which resonate at different field values the less they are resonating at any given field strength, and the weaker the resonance lines. Nevertheless the challenge has now been met, but only after a sequence of quite remarkable concepts was proposed and the relevant experiments performed.

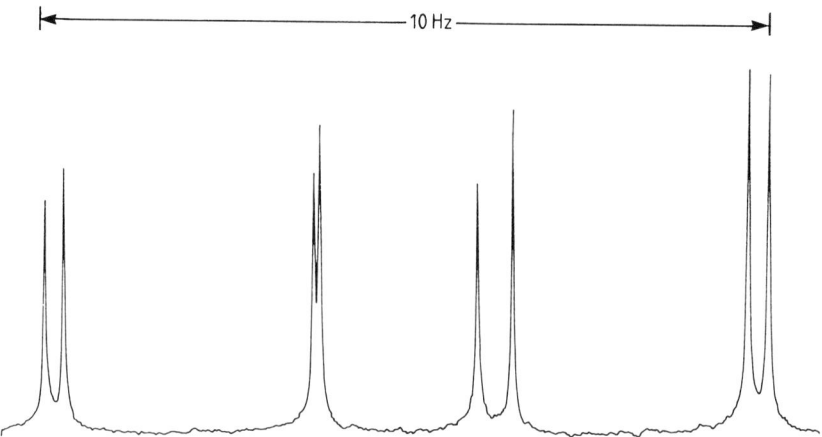

Figure 14.15. *Example of extremely sharp proton resonances at 200 MHz in a magnetic field of 4.7 T. The whole spectrum shown occupies a band of 10 Hz and the lines are typically 0.04 Hz wide at half-peak. The spectrum was obtained with a WP 200 spectrometer (Bruker, Spectrospin Ltd), the sample being a dilute solution of ortho-dichlorobenzene, $C_6H_4Cl_2$ in deuterated acetone.*

The success is largely due to Hahn's [148] idea of substituting pulses for the steady intensity of radiation used up till then for observing the resonant absorption. While this may not seem to be much of a change, or even perhaps a step backwards, because it reduces the energy to which the spins are exposed, the ramifications have been enormous. In the first place the Bloch equation could be solved, to sufficient accuracy, by what amounted to a geometrical type of reasoning, by considering how a rotating observer would describe what was occurring. It then became possible, particularly for Hahn and his associates, to see how to design sequences of pulses to increase the experimental sensitivity. One development was the production of 'echoes'. A spin system, suitably excited would produce an 'echo' at a later time, a response in a receiver coil, which, depending on the sequence, would give values either for T_1 or T_2. Other sequences would transfer spin polarizations between low- and high-abundant spin systems and so allow the resonances of low-abundance spins to be detected indirectly. These techniques, when combined with time variations of the so-called steady main field, produced magnetic resonance imaging, which is now finding wide use in the medical field by revealing the density (and/or the relaxation time) distribution of protons and other nuclei in human organs, by what is thought to be a non-invasive technique. It is confidently expected that, in due course, it will be possible to follow some of the important chemical reactions which take place in the body, by combining magnetic resonance imaging (MRI) and chemical shifts. (A reaction results in changes in the concentrations of different chemical groups.)

The literature on nuclear magnetic resonance is very extensive, and only a brief survey has been possible here. For further reading there are books by Andrew [149], Abragam [150] and Mansfield and Morris [151], which between them cover most of its aspects.

14.10.2. Nuclear demagnetization

After a considerable amount of effort the demagnetization of suitable magnetic crystals produced cooling to temperatures well below those which can be obtained with liquid helium, so extending the available range down to about 0.01 K, with the hope of an extension to 0.001 K [30]. To someone unfamiliar with low-temperature work such a small temperature decrease probably seems a small reward for so much effort, at least until it is realized that in low-temperature physics the important parameter is the reciprocal of the temperature. With the boiling point of liquid helium being near 4 K, a drop to 0.01 K changes $1/T$ from about 0.25 to about 100, which puts the extension in quite a different light. Reaching 0.001 K would give 1000 for $1/T$ and so would represent another major extension. A two-stage demagnetization process was therefore considered by Kurti, with the first using electronic moments and the second nuclear moments [152]. Eventually, however, the first stage became unnecessary when the development of dilution refrigerators, based on the properties of mixtures of helium isotopes allowed starting temperatures of the order of 0.01 K to be maintained indefinitely.

From about 0.01 K the direct demagnetization of nuclei looked promising, for the factors which had favoured electronic demagnetization from an initial temperature of 2 K could then be seen to be similar to those for nuclei starting from 0.01 K, but this was by no means enough, for in working at such low temperatures there were many obstacles to be overcome. It would be out of place to go into all the details, which can be found in two papers, describing work at Lancaster [153] and Bayreuth [154]. Both groups cooled copper down to about 10^{-5} K, and more recently the nuclei in Cu have been cooled to about 10^{-7} K, at which extremely low temperature there was a transition of the nuclear spin array to an antiferromagnetic structure.

14.11. Concluding remarks

A century of magnetism is a long period to describe in an article of limited length, and from direct familiarity with only part of the period. I do not expect that all readers will agree with my selection or with my presentation. I hope though that my deficiencies will not have upset any of the many friends I have made through some forty years of interest in magnetism, and that I have managed to show less expert readers why we have found magnetism so interesting. Those of us who have been privileged to work in the field have found it fascinating, as

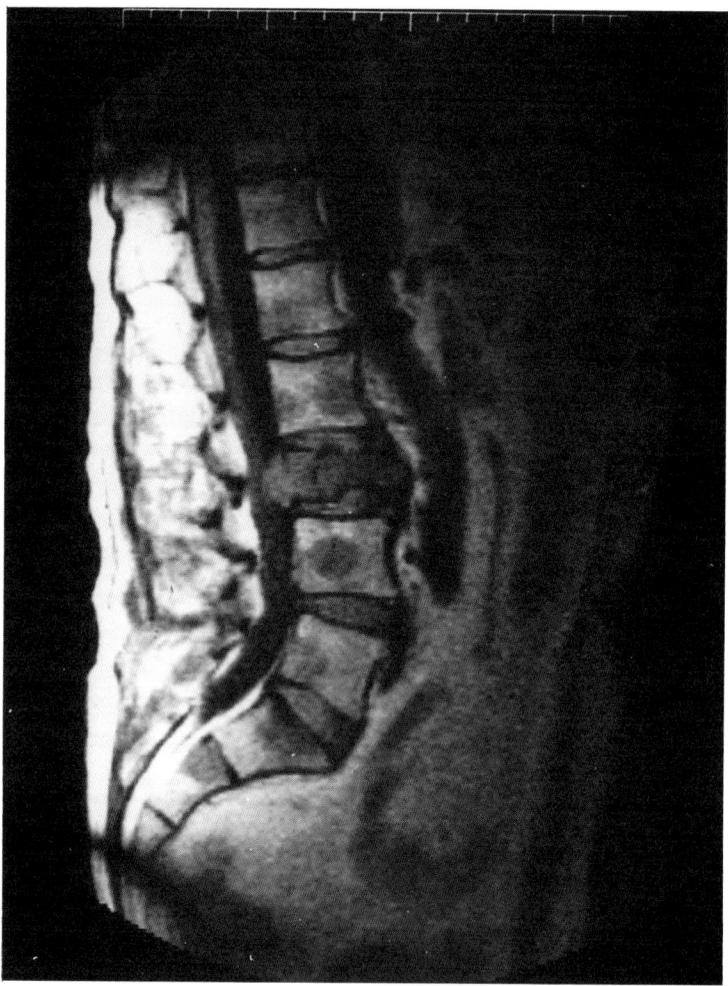

Figure 14.16. *A magnetic resonance image of a human spine, showing (i) a collapsed vertebrae in the centre of the spinal column, and (ii) tumours within vertebrae, where the tumours replacing normal marrow are shown by the dark areas. The spinal cord and nerve roots are also clearly visible to the left of the spinal column.*

well as being a specialism which we can confidently feel has contributed enormously to the well-being of society. From compasses and magnets to pick up pins, it has progressed through the development of the electrical power industry to information storage and retrieval in many forms. Now magnetic resonance imaging seems set to change whole fields of medical practice, without, as far as is known, introducing any undesirable side-effects (figure 14.16).

See also
p 1923

Not a bad record, surely, for something which throughout has been driven by curiosity!

Acknowledgments

I would like to acknowledge the help and interest shown to me by Brebis Bleaney in the early stages of the project, and my colleagues in Nottingham throughout. In particular I would like to thank John Fletcher, who made his substantial collection of early books on magnetism available to me and allowed me to draw on his wide knowledge of physics, John Owers-Bradley who tutored me on adiabatic demagnetization and several other topics, Peter Morris for his guidance about nuclear magnetic resonance and Brian Worthington for figure 14.16.

References

[1] Curie P 1895 *Ann. Chim. Phys.* **5** 289
[2] Voigt W 1902 *Ann. Phys., Lpz* **9** 115
[3] Thomson J J 1903 *Phil. Mag.* **6** 673
[4] Langevin P 1905 *Ann. Chim. Phys.* **5** 70
[5] van Leeuwen J H 1919 *Dissertation* Leiden
[6] Bohr N 1972 *Collected Works* (with English translations) vol 1, ed J R Nielson (North-Holland: Amsterdam)
[7] Bohr N 1913 *Phil. Mag.* **26** 1
[8] Pauli W 1925 *Z. Phys.* **31** 765
[9] Pauli W 1920 *Z. Phys.* **23** 201
[10] Debye P W 1926 *Phys. Z.* **27** 67
[11] Weiss P 1907 *J. Physique* **5** 70
[12] Weiss P and Foëx G 1931 *Le Magnétisme* (Paris: Libraire Armand Colin)
[13] Hund F 1925 *Z. Phys.* **33** 85
[14] Van Vleck J H and Frank A 1929 *Phys. Rev.* **34** 1494
[15] Bobeck A H and Scovil H E D 1971 *Sci. Am.* June p 78
[16] Barkhausen H 1919 *Phys. Z.* **20** 401
[17] Galison P 1987 *How Experiments End* (Chicago: University of Chicago Press) ch 2
[18] Goudsmit S and Uhlenbeck G E 1925 *Naturwissenschaften* **13** 953
[19] Beck K 1918 *Vjschr. Naturf. Ges. Zurich* **63** 116
[20] Van Vleck J H 1926 *Bull. Natl Res. Counc.* **10** 9
[21] Hund F 1927 *Z. Phys.* **43** 788
[22] Heisenberg W 1926 *Z. Phys.* **38** 411
[23] Dirac P A M 1926 *Proc. R. Soc.* A **112** 661
[24] Kramers H A 1930 *Proc. Amsterdam Acad.* **33** 959
[25] Debye P 1926 *Ann. Phys., Lpz* **81** 1154
[26] Giauque W F 1927 *J. Am. Chem. Soc.* **49** 1864
[27] Kurti N and Simon F 1933 *Naturwissenschaften* **21** 178
[28] Giauque W F and MacDougall D P 1933 *Phys. Rev.* **43**
[29] de Haas W J, Wiersma E C and Kramers H A 1934 *Physica* **1** 1

[30] Hudson R P 1972 *Principles and Application of Magnetic Cooling* (Amsterdam: North-Holland)
[31] Van Vleck J H 1932 *Electric and Magnetic Susceptibilities* (Oxford: Oxford University Press)
[32] Van Vleck J H 1950 *Am. J. Phys.* **18** 495
[33] Van Vleck J H 1937 *J. Chem. Phys.* **5** 320
[34] Bitter F 1938 *Phys. Rev.* **54** 79
[35] Hulthen 1936 *Proc. Am. Acad. Sci.* **39** 190
[36] Jacobs I S 1961 *J. Appl. Phys.* **32** 61S
[37] Néel L 1936 *Ann. Phys.* **5** 232
[38] Bizette H, Squire C F and Tsai B 1938 *C. R. Acad. Sci., Paris* **207** 449
[39] Foëx G and Graff M 1939 *C. R. Acad. Sci., Paris* **209** 160
[40] Van Vleck J H 1940 *Phys. Rev.* **57** 426
[41] Gorter C J and Kronig R 1936 *Physica* **3** 1009
[42] Gorter C J 1947 *Paramagnetic Relaxation* (Amsterdam: Elsevier)
[43] Standley K J and Vaughan R A 1969 *Electron Spin Relaxation Phenomena in Solids* (London: Hilger)
[44] Landau 1930 *Z. Phys.* **64** 629
[45] Honda K and Kaya S 1928 *Sci. Rep. Tohoku Univ.* **17** 1157
[46] Sucksmith W, Potter H H and Broadway L 1928 *Proc. R. Soc.* A **117** 471
[47] Kaya S 1928 *Sci. Rep. Tohoku Univ.* **17** 1157
[48] Landau L and Liftshitz E 1935 *Phys. Z.* **8** 153
[49] Bitter F 1932 *Phys. Rev.* **41** 507
[50] Bozorth R M 1951 *Ferromagnetism* (New York: Van Nostrand)
[51] Urbain G *et al* 1935 *C. R. Acad. Sci., Paris* **200** 2132
[52] Stoner E C 1933 *Phil. Mag.* **15** 1018
[53] Mott N F 1935 *Proc. Phys. Soc.* **47** 571
[54] Van Vleck J H 1937 *Phys. Rev.* **52** 1178
[55] Bloch F 1946 *Phys. Rev.* **70** 460
[56] Zavoisky E 1945 *Fiz. Zh.* **9** 211 245
[57] Cummerow R L and Halliday D 1946 *Phys. Rev.* **70** 433
[58] Bagguley D M S, Bleaney B, Griffiths J H E, Penrose R P and Plumpton B I 1948 *Proc. Phys. Soc.* **61** 542, 551
[59] Penrose R P 1949 *Nature* **163** 992
[60] Abragam A and Pryce M H L 1949 *Nature* **163** 992
[61] Abragam A and Pryce M H L 1950 *Proc. Phys. Soc.* A **63** 409; 1951 *Proc. R. Soc.* A **205** 135
[62] Van Vleck J H 1939 *J. Chem. Phys.* **7** 61
[63] Abragam A and Pryce M H L 1951 *Proc. R. Soc.* A **205** 135; 1952 *Proc. R. Soc.* A **206** 164, 173
 Stevens K W H 1963 *Magnetism* vol 1 (New York: Academic) p 1
[64] Stevens K W H 1952 *Proc. Phys. Soc.* **65** 209
[65] Bleaney B and Stevens K W H 1953 *Rep. Prog. Phys.* **16** 108
[66] Owen J and Stevens K W H 1953 *Nature* **171** 836
[67] Van Vleck J H 1935 *J. Chem. Phys.* **3** 807
[68] Abragam A and Bleaney B 1970 *Electron Paramagnetic Resonance of Transition Ions* (Oxford: Clarendon)
[69] Guha B C 1951 *Proc. R. Soc.* A **206** 353
[70] Bleaney B and Bowers K D 1952 *Phil. Mag.* **43** 372; 1952 *Proc. R. Soc.* A **214** 451
[71] Shull C G, Strausen W A and Wollan E O 1951 *Phys. Rev.* **83** 333
[72] Anderson P W 1963 *Magnetism* vol 1 (New York: Academic) ch 2
[73] Feher E R 1956 *Phys. Rev.* **103** 500
[74] Kleiner W H 1952 *J. Chem. Phys.* **20** 1784

[75] Freeman A J and Watson R E 1960 *Phys. Rev.* **120** 1254
[76] Slater J C 1953 *Rev. Mod. Phys.* **25** 199
[77] Cyrot M 1982 *Magnetism of Metals and Alloys* (Amsterdam: North-Holland)
de Haas W J et al 1933 *Physica* **1** 1115
[78] Gondairi K-I and Tanabe Y 1966 *J. Phys. Soc. Japan* **21** 1527
Stevens K W H 1976 *Phys. Rep.* C **24** 1
Brandow B 1977 *Adv. Phys.* **26** 651
[79] Jacobsen E H, Shiren N S and Tucker E B 1959 *Phys. Rev. Lett.* **3** 81
[80] Bloembergen N, Shapiro S, Persham P S and Attman J O 1959 *Phys. Rev.* **114** 445
[81] Van Vleck J H and Penney W G 1934 *Phil. Mag.* **17** 9
[82] Trounson E P et al 1950 *Phys. Rev.* **79** 542
[83] Kittel C 1951 *Phys. Rev.* **82** 565
[84] Nagamiya T 1951 *Prog. Theor. Phys. Japan* **6** 342
[85] Kittel C 1948 *Phys. Rev.* **73** 155
[86] Johnson F M and Nethercot A H Jr 1959 *Phys. Rev.* **114** 705
[87] Caspers W J 1989 *Spin Systems* (Singapore: World Scientific)
[88] Riste T and Wanic A 1961 *J. Phys. Chem. Solids* **17** 318
[89] Collins M F 1964 *Proc. Int. Conf. on Magnetism (Nottingham, 1964)* (London: Institute of Physics/Physical Society) p 319
[90] Keffer F 1966 *Handbuch der Physik Band XVIII/2* p 1
[91] Martel P, Cowley R A and Stevenson R W H 1967 *J. Appl. Phys.* **39** 1116
[92] Aleonard R, Panthenet R, Rebouillat J P and Vevrey C 1968 *J. Appl. Phys.* **39** 379
[93] Moriya T 1963 *Magnetism* (New York: Academic) ch 3
[94] Moriya T 1960 *Phys. Rev.* **117** 91
[95] Griffiths J H E 1946 *Nature* **158** 670
[96] Rabi I I 1937 *Phys. Rev.* **51** 652
[97] Bloch E, Hansen W W and Packard M 1946 *Phys. Rev.* **70** 474
[98] Purcell E M, Torrey H C and Pound R V 1946 *Phys. Rev.* **69** 37
[99] Van Vleck J H 1950 *Phys. Rev.* **78** 266
[100] Lee E W, Callaby D R and Lynch A C 1958 *Proc. Phys. Soc.* **72** 233
[101] Craik D J and Tebble R S 1961 *Rep. Prog. Phys.* **14** 116
[102] Dillon J F 1963 *Magnetism* vol III (New York: Academic)
[103] Bagguley D M S and Owen J 1957 *Rep. Prog. Phys.* **20** 304
[104] Snoek J L 1947 *New Developments in Ferromagnetic Materials* (Amsterdam: Elsevier)
[105] de Boer J H and Verwey E J W 1937 *Proc. Phys. Soc.* A **49** 59
[106] Standley K J 1962 *Oxide Magnetic Materials* (Oxford: Clarendon) ch 3
[107] Néel L 1948 *Ann. Phys., Paris* **206** 49
[108] Gorter E W and Schulkes J A 1953 *Phys. Rev.* **90** 487
[109] Polder D 1949 *Phil. Mag.* **40** 99
[110] Nicolas J 1980 *Ferromagnetic Materials* vol 2 (Amsterdam: North-Holland) 243
[111] Melchor J L et al 1957 *Proc. Inst. Radio Eng.* **45** 643
[112] Pippin J E 1956 *Proc. Inst. Radio Eng.* **44** 1054
[113] Suhl H 1957 *Phys. Rev.* **106** 384
[114] Weiss M T 1957 *Phys. Rev.* **107** 317
[115] Lowman C E 1972 *Magnetic Recording* (McGraw-Hill: New York)
[116] Bate G 1980 *Ferromagnetic Materials* vol 2 (Amsterdam: North-Holland) p 381
[117] Dillon J F 1957 *Phys. Rev.* **105** 759
[118] Dillon J F 1958 *J. Appl. Phys.* **29** 1286
[119] Bobeck A H 1967 *Bell Syst. Tech. J.* **46** 1901

[120] Bobeck A H and Della Torre E 1975 *Magnetic Bubbles* (Amsterdam: North-Holland)
Malozemoff A P and Slonczewski J C 1979 *Applied Solid State Science: Advances in Materials and Device Research (Supplement 1: Magnetic Domain Walks in Bubble Materials)* ed R Wolfe (New York: Academic)
[121] Elliot J F, Legrold S and Spedding F H 1954 *Phys. Rev.* **94** 50
Bates L F, Leach S J, Loasby R G and Stevens K W H 1955 *Proc. R. Soc.* B **68** 181
[122] Koehler W C and Wollan E O 1955 *Phys. Rev.* **97** 1177
[123] Legvold S 1980 *Ferromagnetic Materials* vol 1, ed E P Wohlfarth (Amsterdam: North-Holland) ch 3
[124] Ruderman M A and Kittel C 1954 *Phys. Rev.* **96** 99
[125] Kasuya T 1956 *Prog. Theor. Phys.* **16** 45
[126] Yosida K 1957 *Phys. Rev.* **106** 893
[127] Kasuya T 1966 *Magnetism* vol IIB, ed G T Rado and H Suhl (New York: Academic) p 215
[128] Rathmann O and Toubourg P 1977 *Phys. Rev.* B **16** 1212
Toubourg P 1977 *Phys. Rev.* B **16** 1201
[129] Mackintosh A R 1977 *Phys. Today* June p 23
[130] Freeman A J 1972 *Magnetic Properties of Rare Earth Metals* ed R J Elliott (London: Plenum)
[131] Schmitt D 1979 *J. Phys. F: Met. Phys.* **9** 1759
[132] Stevens K W H 1976 *Magnetism in Metals and Metallic Compounds* ed J T Lopuszanski, A Pekalski and J Przystawa (New York: Plenum) p 1
Stevens K W H 1977 *Crystal Field Effects in Metals and Alloys* ed Furrer (New York: Plenum)
[133] Dixon J M and Wardlaw R S 1986 *Physica* **135** 105
[134] Strnat K J and Strnat R M W 1991 *Magnetism in the Nineties* ed A J Freeman and K A Gescheidner Jr (Amsterdam: Elsevier) p 38
[135] Herbst J F and Croat J J 1991 *Magnetism in the Nineties* ed A J Freeman and K A Gescheidner Jr (Amsterdam: Elsevier) p 57
[136] Buschow K H J 1991 *Magnetism in the Nineties* ed A J Freeman and K A Gescheidner Jr (Amsterdam: Elsevier) p 79
[137] Landolt-Börnstein 1962 *Eigenschaften der Materie in Ihren Aggregatzuständen* (Berlin: Springer) p 9
[138] Hubbard J 1966 *Proc. R. Soc.* A **296** 82
[139] Lonzarich G G 1980 *Electrons at the Fermi Surface* ed M Springford (Cambridge: Cambridge University Press) p 225
[140] de Haas W J, Wiersma E C and Kramers H A 1934 *Physica* **1** 1
[141] MacDonald D K C and Mendelssohn K 1950 *Proc. R. Soc.* A **202** 523
[142] Kondo J 1969 *Solid State Phys.* **23** 183
[143] Lacroix C 1991 *Magnetism in the Nineties* ed A J Freeman and K A Gescheidner Jr (Amsterdam: North-Holland) p 90
[144] Adroja D T and Malik S K 1991 *Magnetism in the Nineties* ed A J Freeman and K A Gescheidner Jr (Amsterdam: North-Holland) p 126
Steglich F 1991 *Magnetism in the Nineties* ed A J Freeman and K A Gescheidner (Amsterdam: North-Holland) p 186
[145] Binder K and Young A P 1986 *Rev. Mod. Phys.* **58** 801
[146] ICM '91 *J. Magn. Magn. Mater.* **104–107, 108**
[147] Gorter C J and Broer L J F 1942 *Physica* **9** 591
[148] Hahn E L 1950 *Phys. Rev.* **80** 580
[149] Andrew E R 1955 *Nuclear Magnetic Resonance* (Cambridge: Cambridge University Press); 1970 *Magnetic Resonance* ed C K Coogan, N S Ham, S N Stuart, J R Pilbrow and G V H Wilson (New York: Plenum) p 163

[150] Abragam A 1961 *The Principles of Nuclear Magnetism* (Oxford: Oxford University Press)
[151] Mansfield P and Morris P 1982 *Advances in Magnetic Resonance* Supplement 2, ed Waugh (New York: Academic)
[152] Kurti N *et al* 1956 *Nature* **178** 450
[153] Bradley D I, Guenault A M, Keith V, Kennedy C J, Miller I E, Mussett S G, Pickett G R and Pratt W P 1984 *J. Low Temp. Phys.* **57** 359
[154] Gloos K, Smeibide P, Kennedy C, Singaas A, Sekowski P, Mueller R M and Pobell F 1988 *J. Low Temp. Phys.* **73** 101

Chapter 15

NUCLEAR DYNAMICS

David M Brink

15.1. Background

From the time of the discovery of the neutron, physicists became increasingly interested in studying the properties of nuclei. This chapter describes the development of nuclear structure physics from about 1936 until 1975. The main focus is on the decade following the end of World War II. Many people contributed to experimental and theoretical advances, but here we concentrate on the most significant developments which opened up new fields of research.

The year 1932 was a turning point in the development of nuclear physics. On 28 April 1932, the Royal Society held an important meeting on the structure of atomic nuclei. Lord Rutherford delivered the opening address [1] reviewing the progress made in the previous three years, and spoke about the discovery of the neutron by James Chadwick [2] and the deuteron by Urey, Brickwedde and Murphy. He also reported the results of the first nuclear disintegration experiments with artificially accelerated ions carried out by J Cockcroft and E Walton [3]. These were the first of a series of dramatic developments which transformed our understanding of atomic nuclei. Some of these advances are described in detail in Chapter 2 by Abraham Pais and Chapter 5 by Laurie M Brown. We mention them again here to provide a background for the subsequent developments in nuclear dynamics. The book *Inward Bound* by Abraham Pais [4] is a fascinating source of information about the history of events of this period.

In 1932, most nuclear physicists supposed that the nucleus of a heavy element consisted mainly of alpha particles with an admixture of a few free protons and electrons. Lord Rutherford spent a part of his lecture to the Royal Society discussing that model and the difficulties associated with it. He suggested that 'an electron cannot exist in a free state in a stable nucleus but must always be associated with a proton or other possible massive units. The indication of the existence of the neutron in certain nuclei is significant in this connection.'.

In June 1932, just four months after Chadwick had published his discovery of the neutron, Werner Heisenberg [5] submitted the first of a series of papers in which he developed the basic ideas of the neutron–proton model of nuclear structure. He assumed that neutrons obey the laws of Fermi statistics and have spin $\frac{1}{2}$, arguing that these assumptions were necessary in order to explain the Bose statistics of the nitrogen nucleus and empirical results on nuclear moments. These developments are described by Abraham Pais in section 2.12 of Chapter 2.

In his address to the Royal Society in 1932, Lord Rutherford described the development of positive ion accelerators by research workers at the Cavendish Laboratory in Cambridge, UK, the Department of Terrestrial Magnetism in Washington DC and the University of California in Berkeley, and announced the results of the first experiments using accelerators for nuclear studies carried out at Cambridge. The other two groups performed similar experiments during the next few months. Until that time most information on the structure of nuclei had come from experiments with alpha particles from natural radioactive sources. Accelerators formed an additional line of attack which had many advantages. The intensities were much greater and the energies of the particles could be varied at will. A contemporary picture of the impact of accelerators on the development of nuclear physics can be obtained from the review article of M Livingston and H Bethe [6] published in 1937. Detailed information about the events described in the next few paragraphs can be found in their review. Pais gives interesting historical insights in section 17b of his book [4].

It had been expected that energies higher than the potential barrier associated with the Coulomb repulsion between the target and projectile would be required to produce a nuclear reaction. In 1928 G Gamow [7] and E Condon and R Gurney [8] introduced the quantum-mechanical theory of alpha decay which predicted that there was a certain probability of penetration of a potential barrier by particles with considerably less energy than required to go over the top of the barrier (cf Chapter 2, section 2.11). Cockcroft and Walton [9] realized that this might make a nuclear reaction possible with relatively low-energy protons.

J Cockcroft and E Walton had begun their work in 1929 at the Cavendish Laboratory. In 1930, using a half-wave rectifier to obtain a constant potential and a single-section accelerator tube, they produced protons with an energy of 300 keV. Then they developed a more powerful machine with a voltage multiplier and a two-section tube and obtained protons with energies up to 700 keV. The first disintegration experiments were performed with this machine in 1932 on the reaction

$$^7\text{Li} + p = {}^4\text{He} + {}^4\text{He}.$$

Ernest O Lawrence

(American, 1901–58)

Ernest O Lawrence was born in Canton, South Dakota, on 8 August 1901 of parents who were Norwegian immigrants. In 1919 he joined the university of South Dakota and graduated in 1922 with a BA degree in Chemistry. Then he spent periods at the Universities of Minnesota, Chicago and Yale. He was awarded his PhD at Yale in 1925 and obtained a National Research Fellowship which allowed him to stay on there for two years. He married Mary Kimerly Blumer in 1923.

In 1927 Lawrence accepted a faculty position as Associate Professor at Berkeley. He was strongly attracted to nuclear physics and was stimulated by the experiments of Rutherford. He saw that new advances could be made if ways could be found to accelerate charged particles to high velocities. As a result of his efforts and those of his colleagues the first cyclotron was operating at Berkeley in 1932. He became the Director of the Radiation Laboratory in 1935 and was awarded the Nobel Prize in 1939.

Lawrence had an intuitive feeling for physics and was a natural inventor. In the end, his unusual powers of leadership, his enthusiasm and personality were more important than his physics. He died on 27 August 1958.

E O Lawrence of the University of California began to work on the design of the cyclotron in 1929. By 1930 he had a small model working and obtained protons with an energy of 80 keV. In 1932 Lawrence and S Livingston had constructed a larger machine which produced protons with energies up to 1 MeV.

Several years earlier, in 1926, G Breit had investigated the Tesla coil as a means of producing high voltages. He was joined by M Tuve at the Department of Terrestrial Magnetism in Washington DC, and in the years up to 1930 made great progress in the design of accelerator tubes. However, the Tesla coil was not satisfactory as a high-voltage source because of the pulsed nature of the potential and the fluctuating value of the peak voltage. At Princeton in 1929, R van de Graaff built the first electrostatic generator of the type named after him. Tuve and his collaborators realized the virtues of the van de Graaff generator as a high-voltage source, and in 1932 they built a small model with a sectional metal and glass accelerating tube which had been used previously in the Tesla coil work. They published the first results of nuclear disintegration experiments with ions accelerated up to 600 keV by this machine in 1932. The results of experiments performed with these accelerators over the next few years are collected together in the 1937 review article of Livingston and Bethe [6].

An event of considerable importance was the discovery of artificial radioactivity by Irène and Frédéric Joliot-Curie [10] in 1934. Not only was this the first example of the creation of a new isotope, it was also a discovery of considerable practical importance. It provided a convenient label for identifying unstable isotopes, and was a precursor of Enrico Fermi's important experiments on neutron capture reactions. It also opened up the possibility for the use of radioisotopes for chemical and medical research (cf Chapter 2, section 2.12).

15.1.1. Nuclear masses and the liquid-drop model
The liquid-drop model was invented by George Gamow to describe a number of basic properties of nuclei. We begin the story with the first chapter of his book *The Constitution of Atomic Nuclei and Radioactivity* [11], which was published in May 1931 just before the discovery of the neutron. The book is a slim volume of 114 pages and the author aimed to give 'as complete an account as possible of our present experimental and theoretical knowledge of the nature of atomic nuclei'. Gamow was well placed to do this job. He had developed the theory of alpha decay in 1928 and had shown that the phenomenon could be understood as a quantum-mechanical tunnelling process. Niels Bohr was impressed with Gamow's work and invited him to spend the academic year 1928–9 in Copenhagen (figure 15.1). Bohr also arranged for him to visit Cambridge in January and February 1929, and he presented his theory at a meeting of the Royal Society in London in February [12]. He spent the academic year 1929–30 in Cambridge and then went back to Copenhagen. He interacted with Bohr, H Casimir, F Houtermans, L Landau and N Mott while he was in Copenhagen.

When Gamow wrote his book, detailed information about nuclear masses was beginning to emerge. In 1919 F Aston published an account of his mass spectrograph. His first results confirmed F Soddy's prediction of the widespread occurrence of isotopes of the elements. His instrument could measure masses with an accuracy of 1 part in 1000. The measurements showed that the atomic mass of every isotope was approximately a whole number (the mass number A) on a scale where the mass of the common isotope of oxygen was 16. Other workers

Enrico Fermi

(Italian, 1901–54)

Enrico Fermi was born in Rome on 29 September 1901. His parents came from the Po valley and his father had a Civil Service position with the Italian Railways. Fermi became committed to physics at the age of 15, and his unusual grasp of the subject amazed his examiners when he was admitted to the Scuola Normale in Pisa. He graduated in 1922 and spent a short time in Göttingen and Leiden before being appointed to the new chair of Theoretical Physics at the University of Rome in 1926. From 1922 onwards he had strong support from Senator Corbino, politician, scholar and scientist who was the Director of the Physics Institute at the University of Rome. He published his work on Fermi statistics in 1926 and his theory of beta decay in 1934.

Fermi started to form a research group in 1927 to study nuclear physics and in 1934 started experiments on neutron-induced radioactivity. In July 1928 he married Laura Capon. He was awarded the Nobel Prize in 1938. He left Italy for Columbia University immediately afterwards and arrived in New York on 2 January 1939. At Columbia he started to study chain reactions. He went to Chicago in 1942 to direct the experiments which lead to the first chain-reacting pile. After the war he was a professor at the Institute of Nuclear Studies at the University of Chicago where he remained until he died on 29 November 1954. Fermi had unusual gifts of concentration and a passion for clarity. In all his work he emphasized the physical content of theories rather than their formal aspects.

entered the field and there were advances in techniques for producing ion sources for heavy metals. Aston had developed his second mass spectrograph by 1927. With the new instrument an accuracy of 1 part in 10 000 could be achieved for light isotopes, and 1 part in 5000 for heavy isotopes. Interesting historical information is contained in an article on mass spectrometry prepared by K Bainbridge in 1951 [13].

Twentieth Century Physics

Figure 15.1. *Niels Bohr, pictured beside a Cockcroft–Walton accelerator.*

By 1928 the atomic weights of about 250 isotopes were known. They were precise enough to measure mass defects (the differences between the exact atomic weights and the mass numbers) to an accuracy of about 20%. The mass defects were interpreted by a model which assumed that a nucleus was composed of protons and electrons. The idea, as stated by Rutherford [14] at a Royal Society meeting in 1929, was as follows 'The difference in mass between the free and nuclear proton is ascribed to a packing effect, i.e. to the interaction of the electromagnetic fields of the protons and electrons in the highly condensed nucleus. On modern views we know that there is a close relation between mass and energy (Einstein's relation $E = mc^2$). The free proton has a mass 1.0073 while the proton in the nucleus has a mass very nearly 1. This apparently small loss of mass means that a large amount of energy has been emitted in the entrance of the free proton into the structure of the nucleus, an amount equal to about 7 MeV.'. Aston's measurements showed that the binding energy per proton was almost constant throughout the entire periodic table. The significance of this important fact was first recognized by Gamow who used it in his formulation of the liquid-drop model.

We know nowadays that the density distribution of matter in a nucleus is approximately constant in its interior and falls rapidly to zero at the nuclear surface. The radii of all nuclei may be represented approximately by the formula

$$R = r_0 A^{-1/3} \qquad (1)$$

where A is the mass number and r_0 is approximately 1.2 fm (1 fm = 10^{-15} m). The nuclear volume is approximately proportional to A or, equivalently, the density of 'nuclear matter' in the interior of a nucleus is nearly constant throughout the whole periodic table.

The position was not nearly so clear in 1931. Something was known about the size of light nuclei from the scattering of alpha particles and heavy nuclei from alpha-decay lifetimes. As early as November 1928 Gamow had pointed out that the radius of the aluminium nucleus obtained from alpha scattering and the radii of heavy nuclei deduced from alpha-decay lifetimes were consistent with the formula in equation (1). There was hardly sufficient data to prove Gamow's hypothesis, but the idea was consistent with the model of the nucleus which he presented at the Royal Society discussion in February 1929.

Gamow proposed that alpha particles played an important role in nuclear structure. He argued that nuclei contained a few 'loose protons' and a number of 'loose electrons' but that most of the protons and electrons in a nucleus were 'packed up' into alpha particles. At the time there was no way to understand the bewildering behaviour of electrons in nuclei. He writes in his book [11] 'For some unknown reason, although electrons in the nucleus behave in a peculiar and obscure way, this does not affect very much the laws governing the motion of the nuclear alpha particles and protons.'. There is more about the 'electron problem', especially in the context of beta decay in the chapters by Abraham Pais and Laurie M Brown of the present book (Chapters 2 and 5). Gamow was able to make progress by ignoring the nuclear electrons. Of course the 'electron problem' disappeared after the discovery of the neutron by Chadwick in 1932.

Gamow assumed that the alpha particles in a nucleus behaved like hard spheres interacting by a strong attractive force with a strength decreasing rapidly with distance. Such a collection of alpha particles would have properties similar to a small drop of liquid. On the basis of this model the binding energy and volume of a nucleus should be proportional to the number of alpha particles, a result which was consistent with Aston's mass measurements.

The alpha-particle model was soon to be superseded. In February 1932, Chadwick's letter giving evidence for the existence of the neutron appeared in *Nature*. Just four months later in June 1932, Heisenberg published the first of a series of three papers in which he put forward the model which formed the basis of later investigations of nuclear structure. He suggested that nuclei were composed of protons and neutrons and that nuclear structure could be described using the laws of quantum mechanics in terms of the interaction between the nuclear particles. But Gamow's ideas about nuclear matter carried over into the new theory. His assumption that the nuclear volume was proportional to the mass number was used by W Heisenberg [5] in 1932 for his discussion of nuclear stability. E Majorana [15] wrote in his 1933 paper on the neutron–proton model of the nucleus 'Thus one finds in the centre of an atom a sort of matter which has the same property of uniform density as ordinary matter. Light and heavy nuclei are built up from this matter and the difference between them depends mainly on their different content of nuclear matter.'.

In his third paper, Heisenberg assumed forms for the neutron–neutron, neutron–proton forces and applied the Thomas–Fermi method to study the properties of nuclear matter. He imagined the nucleus to be a gas of freely moving particles obeying Fermi statistics and held together by the nuclear forces, and tried to find a self-consistent effective potential. Heisenberg's theory was the first application of the independent particle model to nuclear structure and was a precursor of the shell model. Heisenberg used the theory to obtain an approximate expression for the binding energy of nuclear matter as a function of the mass number A. There was a difficulty with the theory: Heisenberg found that the nuclear binding energy as a function of the atomic weight A increased more rapidly than A^2 if the interaction was attractive for all values of the inter-nucleon separation. This result was in contradiction to the mass measurements of Aston. The constant density of nuclear matter and the proportionality of the binding energy to A are important facts which impose strong conditions on nuclear forces. Forces which reproduce these properties satisfy the 'saturation condition'. Heisenberg's force did not produce saturation.

See also p 372

In order to satisfy the saturation condition, Heisenberg had to assume that the nuclear force became repulsive at small distances. Another way out of the difficulty was proposed by Majorana in March 1933. Heisenberg was guided by an analogy when he was trying to find a suitable interaction between nuclear particles. He treated the neutron as a composite particle and assumed an exchange interaction similar to the one responsible for the binding of the hydrogen molecular ion. Majorana doubted the validity of this analogy and preferred an alternative approach. He wanted to find the simplest law of interaction which would produce saturation. He came up with the exchange interaction named after him and showed that it satisfied the above criterion. During the next few years, a number of physicists tried to calculate the properties of nuclear matter using forces derived from the theory of light nuclei. All of these attempts were unsuccessful and in their 1936 review article Bethe and Bacher concluded that 'the statistical [i.e. Thomas–Fermi] model was quite inadequate for the treatment of nuclear binding energies'. In fact it is very difficult to calculate the properties of nuclear matter from a realistic nucleon–nucleon interaction and Bethe and others spent the next half-century grappling with this problem.

In 1935 C von Weizsäcker [16] proposed a semi-empirical method for calculating nuclear energies. He assumed a form for the nuclear energy which was indicated by the statistical model, and which could be derived from simple qualitative considerations based on a liquid-drop description. However, in the formula certain constants were left arbitrary and were determined from experimental data. von Weizsäcker's approach was very successful and gave strong support to the liquid-drop model of nuclear structure.

15.2. Nuclear dynamics as a many-body problem

Heisenberg [5] imagined the nucleus to be a gas of protons and neutrons obeying Fermi statistics, moving independently but held together by strong nuclear forces. This Fermi-gas model was the standard picture of nuclear structure for two or three years, but it went out of favour in the mid-1930s. The change of viewpoint was a consequence of Fermi's neutron capture experiments and their interpretation by Bohr.

Soon after Chadwick had found the neutron and the Joliot-Curies had discovered artificial radioactivity, Fermi decided that neutron-induced reactions would be an ideal tool for studying nuclei. He had collected together a powerful group of young physicists in Rome and they decided to investigate artificial radioactivity produced by neutron bombardment [17]. The story is told in Laura Fermi's book [18] and in an article by E Segrè [19] written after Fermi's death in 1955. Experimental nuclear physics was a new field to the Rome group and several of its members had visited laboratories around the world [19] to learn techniques. The laboratories visited included those of R Millikan in Pasadena, P Debye in Leipzig, O Stern in Hamburg and Lise Meitner in Berlin. The first positive result of their neutron experiments was obtained with a fluorine target in March 1934. By July the Rome group had investigated about 60 elements. Of these, more than 40 could be activated by neutrons. In October the group found that the rate of neutron capture was greatly increased by the presence of materials containing hydrogen. They explained this in terms of the slowing of the neutrons to thermal energies by collisions with protons in the hydrogen and by the rapid increase in reaction cross sections at low velocities.

The first results on reaction rates seemed very erratic until it was realized that there were many sharp resonances in the cross sections. These results were a mystery until Bohr [20] formulated his compound nucleus theory of nuclear reactions. The advent of the compound nucleus model is described by Pais in Volume 1 of this history and by Stuewer in his article on Niels Bohr and nuclear physics [12]. Stuewer describes the movement away from the independent particle model of nuclear structure. In September 1934, Bethe gave a seminar in Copenhagen in which he presented a single-particle theory of nuclear reactions. From the time when the results of the first artificial disintegration experiments came out Bohr had wondered about the reaction mechanism. He was not happy with Bethe's approach. Six months later C Møller, who had just returned to Copenhagen from Rome, reported on the latest neutron capture experiments. Apparently, after about half an hour, Bohr suddenly took to the floor and offered a new interpretation of Fermi's results. By the end of 1935 Bohr had developed his compound nucleus theory of neutron-induced reactions [20].

| See also p 123 |

The change of viewpoint is described in the introduction to Bethe's 1937 review paper [21] on nuclear reactions. Bethe writes 'Bohr was the first to point out that every nuclear process must be treated as a many-body problem. It is not at all permissible to use a one-body approximation, particularly in the case of heavy nuclei.'. He goes on to say that when a neutron or proton falls on a target nucleus, it interacts strongly with the individual nuclear particles in the target. As a result the energy, which is initially concentrated in the incident particle, will soon be distributed among all the particles of the compound nucleus which consists of the particles in the original nucleus plus the incident particle. Bethe expressed the idea as follows 'Each of the particles of the "compound nucleus" will have some energy, but none will have sufficient energy to escape from the rest. Only after a comparatively long time the energy may "by accident" be again concentrated on one particle so that this particle can escape.'.

Although nuclear processes are governed by quantum laws, Bohr recognized that a classical picture could describe the essential physics of compound nucleus formation. He had small models made in the workshop of the Copenhagen Institute to illustrate his theory. They consisted of circular wooden dishes (representing the nucleus) filled with small ball-bearings (representing the nucleons). In 1937 Bohr lectured on the compound nucleus model in a number of universities in the USA, always with the help of his wooden model (figure 15.2). He described how it worked in an article in *Science* [22] in 1937 'If the bowl were empty, then a ball which was sent in would go down one slope and pass out on the opposite side with its original energy. When however there are other balls in the bowl, then the incident one will not be able to pass through freely but will divide its energy first with one of the balls, these two

Nuclear Dynamics

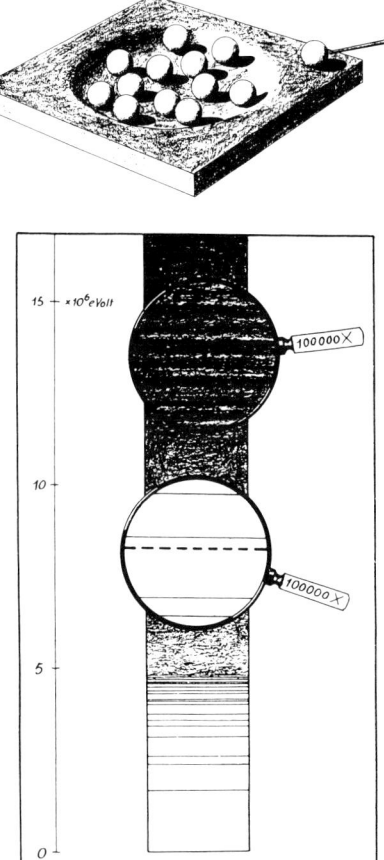

Figure 15.2. *Bohr compound nucleus diagrams.*

will share their energy with others, and so on until the kinetic energy is divided among all the balls.'. According to quantum mechanics a bound nucleus has stationary energy states. Bohr recognized that a nucleus with an excitation energy of several MeV should have a series of closely spaced energy levels. These would have essentially the same character below and above the dissociation energy for break-up into a neutron (or proton) and a residual nucleus.

In 1935 Bethe and Fermi had developed theories of neutron capture based on an independent particle picture. They wanted to understand why slow neutrons could be easily captured by many nuclei. They represented the interaction between a neutron and a nucleus by a potential well and calculated the capture cross section by quantum mechanics. In the limit of low neutron velocities, they found that the capture cross section was inversely proportional to the neutron velocity. This result was in agreement with some aspects of the experimental data but not with others.

Breit and Wigner, working at Princeton in 1936, proposed a mechanism [23] in which the neutron was captured into a quasi-stationary (virtual) energy level, which could then decay either by re-emitting a neutron or by gamma emission. Their result was the famous Breit–Wigner formula for the reaction cross section of a resonance reaction. Bethe described the idea as follows 'If the energy of the incident particle is equal, or nearly equal, to one of the energy levels of the compound nucleus, the probability of the formation of the compound nucleus will obviously be much greater than if the energy of the particle falls in the region between two resonance levels. Therefore we shall find characteristic fluctuations of the yield of every nuclear process with energy, from high values at the resonance energies to low values between resonance levels.'. In spite of the complexity of the compound nucleus, the simple Breit–Wigner formula for the reaction cross section gave a quantitative account of its variation with energy. The resonance energies, total widths and partial widths were parameters which could be fitted to experimental data.

Physicists soon understood that the Breit–Wigner theory brought Bohr's semi-classical picture of the compound nucleus into agreement with the postulates of quantum mechanics. Jensen was in Copenhagen at the time and described the impact of the Breit–Wigner formula in his Nobel Prize Lecture in 1963 'It originated outside Copenhagen, but it soon could be seen on every blackboard of Niels Bohr's institute.'.

The compound nucleus picture fitted in well with the liquid-drop model. It implied a strong coupling between the nucleons in a nucleus and seemed to rule out any treatment in terms of independently moving particles. In his 1936 paper Bohr wrote 'In contrast to the usual view, where the excitation is attributed to an elevated quantum state of an individual particle in the nucleus, we must in fact assume that the excitation will correspond to some quantized collective motion of all the nuclear particles.'. This idea was elaborated by Bohr and Kalkar in 1937. They proposed [24] that there could be excited states of a nucleus corresponding to the surface oscillations of the nuclear 'liquid drop'.

Bohr's compound nucleus model was invented to describe the experimental results of Fermi and his collaborators on neutron capture reactions. The Rome group had set out to study neutron capture in all the elements they could lay their hands on. In particular, by bombarding uranium with neutrons they hoped to produce elements with atomic numbers larger than 92. Irène Joliot–Curie in Paris and Lise Meitner and Otto Hahn in Berlin also took up this line of research. From the beginning the results for uranium were confusing. Many radioactive periods were found and radiochemical methods were used in an attempt to assign atomic numbers to the reaction products without any clear conclusion.

Hahn and Meitner had been collaborators off and on for almost 30 years and had done important work on the beta and gamma spectra of radioactive nuclei. Fritz Strassmann, a young analytical chemist, joined their uranium research project and contributed to the radiochemical analysis. The chemistry was difficult because of the minute size of the samples and also because elements like radium and barium with quite different mass numbers have very similar chemical properties. In June 1938, about four years after the project had started, Meitner, who was Jewish, had to leave Berlin. Until then she had been protected from Nazi persecution because of her Austrian nationality, but when the German army marched into Austria on 12 March 1938 her position changed overnight. She moved to Copenhagen (where her nephew Otto Frisch was working in the Niels Bohr Institute) and then, in July, to Stockholm to join K Siegbahn's laboratory. Hahn and Strassmann continued their studies in Berlin and soon found that a group of at least three radioactive products, formed from uranium under neutron bombardment, were chemically very similar to barium. At first they thought that they had produced isotopes of radium, but further experiments forced them to conclude that isotopes of barium ($Z = 56$) were formed as a consequence of bombardment of uranium ($Z = 92$) with neutrons.

Hahn and Strassmann wrote up their results for publication [25] and sent the news to Meitner on 19 December 1938. She and Frisch spent Christmas together on the west coast of Sweden. They discussed the outcome of the Berlin experiments and arrived at an interpretation which was published [26] in a note to *Nature* on 11 January. They wrote 'On account of their close packing and strong energy exchange, the particles in a heavy nucleus would be expected to move in a collective way which has some resemblance to the movement of a liquid drop. If the movement is made sufficiently violent by adding energy, such a drop may divide itself into two smaller drops.'. After division the drops would repel each other and gain a kinetic energy of about 200 MeV. The whole 'fission' process could be described in an essentially classical way.

See also p 129

On returning to Copenhagen, Frisch conceived an experiment to confirm the 'fission' interpretation. He measured the ionization produced by the highly charged fission fragments. The experiment, which he finished within a week, was consistent with the fission hypothesis, and the results were published in *Nature* [27] on 18 February 1939. Frisch discussed the fission hypothesis with Bohr early in January. Bohr accepted it immediately. A few days later he set off on a visit to the United States taking the news with him. It made an instantaneous impact and on 29 January there was even a note in the New York Times about the discovery. While in America, Bohr collaborated with John Wheeler to produce a detailed theory of the fission process. Their paper [28] was submitted to *Physical Review* at the end of 1939.

15.3. The effects of World War II

World War II had a profound impact on the post-war development of nuclear physics. This section focuses on two points. The first is to outline how the wartime activities led to the creation of national laboratories in the post-war period. The second is to mention some specific areas where there were major advances in experimental techniques and in the understanding of nuclear dynamics resulting from the wartime programme.

Before the war there was a massive migration of scientists. Many physicists, chemists and biologists left Germany, and other continental European countries for the United States, for Britain and for different countries throughout the world. Many of these scientists were later involved in defence work. During the war others moved from the UK, Denmark and France to the USA or Canada to participate in defence programmes. These scientists were based in various laboratories in the USA and Canada and established contacts which persisted and developed into collaborations when the scientists returned to their home institutions at the end of the war. This had a profound influence on the way in which nuclear physics progressed in the post-war period.

Scientists in the United States began to think about a nuclear defence programme early in 1939 [29, 30]. Fermi had moved to the United States at the end of 1938 and joined the Physics Department of Columbia University. After Niels Bohr had brought the news of the discovery of fission to New York in January 1939, Fermi and Herbert Anderson, with the collaboration of the head of the physics department, Professor Pegram and the director of the cyclotron laboratory, J R Dunning, began to make fission experiments with the Columbia University cyclotron [18]. They were joined by the Hungarian physicist Leo Szilard and the Canadian Walter Zinn. Quite soon they became concerned that fission could be the basis of a nuclear weapon and in March 1939 made the first attempts to establish contacts with the government. In October President Roosevelt appointed an Advisory Committee on Uranium. In December 1941, just before Pearl Harbour, there was a decision by the United States to make an all-out effort in atomic energy research. This was soon to become the huge Manhattan Project. This was after the United States Government had been shown, in July 1941, the British Maud Report which demonstrated how and why an atomic bomb was possible [31, 32].

Almost immediately a number of nuclear physics defence laboratories were set up. Already in June 1940 President Roosevelt had established the National Defense Research Council, with Vannevar Bush as its head. In October 1941 MIT was chosen as the site for the new radiation laboratory for radar research. Nuclear programmes started at the Metallurgical Laboratory of the University of Chicago in the first half of 1942, at Los Alamos, New Mexico, early in 1943, at the Clinton

Nuclear Dynamics

Figure 15.3. *The pile in Chicago, in which Fermi achieved, for the first time on 2 December 1942, a man-made chain reaction. (To the left, Dr H W Newson, a member of the original Fermi team.)*

Laboratories in Oak Ridge in late 1943 and at the Hanford Engineering works in Washington State in 1944. From April 1942 until summer 1944 Fermi was based in Chicago.

In 1941 plutonium was discovered at Berkeley. The story of the discovery is described in section 15.8 of this chapter. It was recognized that neutrons produced in the fission of ^{235}U might allow the conversion of natural uranium into plutonium. Work on plutonium was continued at the Berkeley cyclotron until the early part of 1942 and then transferred to the Metallurgical Laboratory at Chicago. At the same time research on reactor physics began. The first atomic pile (nuclear reactor) began operating in a squash court in Chicago in December 1942 (figure 15.3). The trial atomic bomb was exploded at Alamogordo, in New Mexico, on 16 July 1945 and the two nuclear bombs were dropped on Japan in August 1945, just before the end of the Pacific war.

After the end of the war, the United States Government decided to continue nuclear research both for military and civil purposes. Some of the wartime laboratories were converted to National Research Laboratories for the development of nuclear power and other nuclear applications. The Metallurgical Laboratory in Chicago was the precursor of the Argonne National Laboratory. Los Alamos and Oak Ridge also developed into National Laboratories. They had strong physics divisions where important research in nuclear physics was carried out.

Research laboratories were also founded in other countries. The origin of the Chalk River Laboratory in Canada and Harwell in England is particularly interesting. Some British scientists had ideas about nuclear weapons soon after the discovery of fission. Margaret Gowing tells the story in her book *Britain and Atomic Energy 1939–45* [31]. The Maud Committee was set up in 1940 to report on the feasibility of the uranium bomb. There was a period of intense and productive work by British scientists over 15 months which resulted in the presentation of the Maud Report in July 1941. In 1939–40 a French group under J F Joliot were working on ideas for a nuclear reactor project in France and had succeeded in obtaining a supply of heavy water from Norway for use as a moderator. Just before the fall of France two French nuclear physicists, H von Halban and L Kowarksi, succeeded in escaping to England with 185 kg of heavy water.

Contacts between British and American groups began with a visit of Sir Henry Tizard and John Cockcroft to Washington DC in the autumn of 1940. Canada also became involved with the British project as a supplier of uranium and also because Canadian physicists were interested. Canada was considered as a possible site for a British atomic energy project. After the Maud report had been submitted there was a question of cooperation between the British and American projects. There were many difficulties and misunderstandings and the cooperation problem took two years to resolve. The final result was the Quebec Agreement of August 1943. The immediate result was that groups of British scientists joined various American projects at Los Alamos, Oak Ridge and Berkeley. Even before the Quebec Agreement the British and Canadians and the French group started a laboratory in Montreal to continue work on a heavy-water moderated reactor. In the end, after complex negotiations between the United States, Canada and Britain, there was an agreement in April 1944 that a project to build a heavy-water reactor in Canada should go ahead. This led to the establishment of the Chalk River Laboratory which had a staff of Canadian, French and British physicists. It started late, the small pilot reactor ZEEP had not even gone critical by the end of the Pacific war, but the project became the cornerstone of the British and Canadian post-war atomic energy programmes. British scientists gained experience at building and operating reactors. Cockcroft went straight from being director at Chalk River to being director at Harwell. British and Canadian physicists worked closely together over a number of years even after the establishment of the Harwell Laboratory. Harwell was the leading nuclear physics laboratory in Britain for a number of years in the post-war period.

During the war many technical advances were made and new

Nuclear Dynamics

knowledge was acquired. The advances were particularly striking in several fields, notably in neutron physics, and in isotope production and separation. The defence work focused on military applications but after the war the methods evolved could be applied in other areas. Neutron sources developed out of all recognition during the war period. In the early years of the war Fermi and other physicists at the Columbia Cyclotron developed neutron sources 100 000 times more intense than the ones they had used in Rome, by bombarding beryllium with deuterons. Later, in 1943-4, at the Metallurgical Laboratory in Chicago they had neutron sources from the nuclear reactor ('pile') which were 100 000 times stronger than the ones in Columbia. With these sources they were able to produce macroscopic amounts of radioactive isotopes. The emphasis was on the production of plutonium, but the knowledge gained there could be applied later to the production of other radioisotopes for research purposes and for applications in medicine and industry.

High-intensity cyclotrons could be used directly for the production of significant amounts of radioisotopes. This was done first be Seaborg and his collaborators in Berkeley where they produced about 0.5 micrograms of plutonium by bombarding a uranium target with 3500 microampere hours of deuteron current. Later on it became a standard method for producing other radioisotopes.

In the course of the defence programme neutron cross sections were measured for many elements with high-intensity neutron sources. A collection of cross sections was compiled in 1945 for the Manhattan Laboratories by Goldsmith and Ibser [33] at the Metallurgical Laboratory of the University of Chicago. An updated version of this compilation was eventually published in the *Review of Modern Physics* in 1947 [33]. In the fission of uranium many unstable isotopes are produced. About 60 fission chains were identified and studied during the war. Many of the isotopes produced were isolated and their masses and lifetimes were measured. The results were eventually published in the Plutonium Project Report in 1946 [34].

15.4. Technical advances

15.4.1. *Techniques available in 1930*

Theoretical advances in physics have often been associated with the development of new experimental techniques. Since nuclear physics began there has been a continuous effort to find new sources of projectiles for nuclear reactions and new methods for detecting reaction products. This section describes the history of some of the breakthroughs made in the period 1930–65. The book *Radiations from Radioactive Substances* by Rutherford, Chadwick and Ellis [35] gives a detailed account of the techniques available in 1930 just before the discovery of the neutron and the invention of accelerators in 1932.

Until 1932 projectiles used in nuclear reaction studies were from natural radioactive sources. One popular material was polonium (RaF) which could be electrochemically separated from parent materials and supplied a pure source of monoenergetic alphas with an energy of 5.30 MeV. Higher energy alphas (7.68 MeV) were obtained from RaC'. One of the most useful gamma-ray sources was ThC" which has a single strong line of 2.62 MeV.

In 1930 four kinds of devices were in common use for the detection of charged particles. They were photographic plates, zinc sulphide scintillators, Geiger counters and the Wilson cloud chamber. Becquerel made his original discovery of radioactivity by observing its action on photographic plates. In the period up to 1930, and even beyond, photographic plates were used as detectors in various kinds of particle spectrometers. They could also be used in another way. Alpha particles and other charged particles leave tracks in photographic plates, and studies initiated by Kinoshita [36] in 1910 showed that charged particle tracks were made clearly visible by the succession of grains along the track and that reactions occurring in the plate could be studied by measuring the tracks.

The scintillation method for counting alpha particles with zinc sulphide as a scintillator was devised by Regener [37] in 1908. It was used in the original experiments on alpha scattering by Geiger and Marsden [38] in 1908–9 and zinc sulphide was still an important detector in 1930. The scintillations were counted visually using a microscope with a large numerical aperture to increase their apparent brightness. The book of Rutherford, Chadwick and Ellis has a section on methods for determining the efficiency of an observer in counting scintillations.

The first steps towards electrical counting methods were made by Rutherford and Geiger in 1908 [39]. They made use of a property discovered by Townsend, that an ion moving in a strong electric field in a gas at low pressure, produces an avalanche of fresh ions by collision with gas molecules. By suitable adjustment of the conditions the ionization due to a single alpha particle yielded a measurable electric current. Geiger subsequently developed variants of the method which gave much greater sensitivity and the Geiger counter became one of the most important detection devices for charged particles up until about 1950. The importance of the Geiger counter increased when electronic counting methods were introduced at the end of the 1920s. At the same time other kinds of ionization chambers were developed detecting individual charged particles and measuring their energies (Ward, Wynn-Williams and Cave 1929 [40]). The Geiger counter could also detect gamma rays.

The expansion method resulted from the discovery by Wilson [41] that ions produced in gases by the passage of charged particles act as nuclei for the condensation of water vapour when a certain supersaturation condition is reached. Wilson found by an application of this method that tracks of individual alpha and beta particles could be clearly seen and photographed. The Wilson cloud chamber was especially important for the early studies of nuclear reactions by Blackett in 1925 [42] and others. Particle energies could be measured in a cloud chamber by using range–energy relations.

The energies of charged particles were measured by deflection in magnetic fields. The first measurements on alpha particles were made by Rutherford in 1903 and there was a steady improvement in techniques over the next few decades. The discovery of monoenergetic groups of electrons emitted by beta radioactive substances was due to the work of von Baeyer, Hahn and Meitner [43] in 1911. They studied the deflection of beta rays in a weak magnetic field. In fact these homogeneous groups of electrons were not the disintegration electrons at all, but were internal conversion lines associated with gamma transitions. These groups proved to be of the greatest importance for measuring gamma-ray transition energies. The first focusing beta-ray spectrometer was invented by Danysz in 1913 and developed by Rutherford and Robinson [44, 45]. By 1950 many sophisticated beta spectrometers were available. Gamma-ray wavelengths were first measured directly using crystal diffraction by Rutherford and Andrade in 1914 [46].

Initially, alpha-particle energies were more difficult to measure because stronger magnetic fields were needed to produce a significant deflection. Techniques improved and in 1914 Rutherford and Robinson [45] were able to measure the energies of alpha particles from RaC with an accuracy of better than 1%.

15.4.2. Accelerators

Over the years the cyclotron and the synchrocyclotron have been amongst the most important accelerators for nuclear physics research. Their history was reviewed by Livingston in 1952 [47] and more information about the early period can be found in his article. The book by Livingston and Blewett [48] is an interesting source of information about accelerators in general up to 1962.

The cyclotron was developed in Berkeley, California, by Ernest O Lawrence. The first machine built by Lawrence and Livingston in 1931 had pole faces 11" in diameter and produced a 1.2 MeV proton beam. Soon it was superseded by one with poles of 37" diameter which could produce 16 MeV alpha particles. Lawrence and his co-workers continued with the design of larger instruments and in 1939 completed the 60" machine which could produce 8 MeV protons, 16 MeV deuterons and 38 MeV alpha particles (figure 15.4). The Berkeley 60" machine was the first of the big cyclotrons and set the pattern for others of this size. In 1952 there were 14 cyclotrons in England and Europe and 21 in the United States. Most of them were in scientific laboratories and supported

Twentieth Century Physics

Figure 15.4. *The 60" cyclotron as installed at the Radiation Laboratory. Left to right top; Alvarez and MacMillan: left to right bottom; Cooksey, Lawrence, Thornton, Backus, Salisbury.*

research in nuclear physics or produced radioactive tracer isotopes for research in chemistry, biology or medicine.

There is a relativistic limitation on the energy of a fixed frequency cyclotron which was reached in the 60" machines. In 1945 McMillan [49] suggested the use of frequency modulation to overcome the relativistic limitation. Veksler in the USSR discovered important stability properties of the orbits in such a machine. It could be shown that when the applied frequency is decreased slowly the particles follow this change adiabatically so as to gain energy slowly and remain in resonance. The first of the frequency modulated machines was the 184" synchrocyclotron which began operating in Berkeley at the end of 1946. It produced 350 MeV protons, 200 MeV deuterons and 400 MeV alphas. By 1952 there were seven synchrocyclotrons operating throughout the world. At the end of World War II there were four cyclotrons in Japan, but they were destroyed by the United States occupation forces.

There was a further advance in 1959 with the invention of the spiral ridge cyclotron. This is a fixed-frequency machine and relativistic variation of mass with energy is compensated by a magnetic field which varies with angle and radius in a particular way. One of the first of its kind was an 88" machine built in Berkeley in 1962. The average beam intensity in a spiral ridge machine is 100 times larger than in a synchrocyclotron.

Another type of accelerator which has been widely used for nuclear physics research is the van de Graaff electrostatic generator. There is an interesting review by van de Graaff, Trump and Buechner [50] which gives a picture of the state of the art in 1948 just fifteen years after the first experiments were made with a van de Graaff machine. The first large machines were completed at the Department of Terrestrial Magnetism of the Carnegie Institution by Tuve *et al* in 1935 and at MIT by van de Graaff and his colleagues in 1936 (figure 15.5). They had terminal voltages of about 3 MeV. The potential developed in an electrostatic machine is limited by corona breakdown and it was found that the maximum potential developed could be increased by operating the machine under pressure in a gaseous insulating medium. Herb and his associates at Wisconsin obtained remarkable results with a generator operating in an atmosphere containing freon. By 1948 sulphur hexafluoride was the favoured insulating gas.

P H Rose wrote an interesting article on the origin of the tandem accelerator in a book dedicated to L W Alvarez [51]. The information in this paragraph comes from there. Rose writes that the private company High Voltage Engineering (HVE) was founded in 1946 to build and market electrostatic generators. Denis Robinson was the president and Robert van de Graaff and John Trump were associated with it. HVE built a very effective 6.5 MeV machine for the Oak Ridge National Laboratory in 1951. With this success HVE stole the initiative from university and government laboratories and for 20 years was the world leader in direct current accelerators. In the mid-1950s they developed the tandem accelerator which became one of the most widely used machines for low-energy nuclear physics research. In a tandem, negative ions from an ion source are accelerated towards the positive high-voltage terminal. Inside the terminal the ions pass through a region of higher gas pressure where collisions with the gas molecules strip off electrons converting the negative ions into positive ions. The acceleration continues by repulsion of these positive ions away from the high-voltage terminal. The idea of the tandem had been patented by Bennett in 1950. The concept was reinvented by Alvarez who promoted it enthusiastically and encouraged Robinson to develop it as a new HVE product. The first machine was sold to the Chalk River Laboratory and delivered in 1958. It had a terminal voltage of 5 MeV and could accelerate protons to 10 MeV and heavy ions to a higher energy. The machine was simple to install and use. It proved to be very popular and a total of 62 were built and sold to laboratories throughout the world.

Two other kinds of accelerators have had a significant impact on nuclear structure physics. They are the betatron and the linear accelerator. A third, the synchrotron, has been very important for particle

Figure 15.5. *The 2 m van de Graaff generator at the Carnegie Institution. Merle Tuve is in the suit.*

physics, but not as much for nuclear physics.

The betatron was invented by D Kerst (cf reference [48]) at the University of Illinois and in 1940 he built a 2.3 MeV machine in his laboratory there. It was a circular machine with a special magnetic field configuration and particles were accelerated by magnetic induction. The design was based on detailed theoretical calculations by Kerst and Serber. Once the basic principle had been understood these machines were very easy to construct and could accelerate electrons up to much higher energies. Kerst transferred his base of operations to the General Electric Company (GEC) Research Laboratories where a 20 MeV machine was completed in 1942. Westendorp and Charlton of GEC continued the development with a 100 MeV machine in 1945. Commercial production was started at GEC and in other companies. The commercial machines were mainly in the 20 MeV range and were built to meet the rising demand from research laboratories, hospitals and industrial plants. Kerst returned to Illinois to build a 300 MeV model, the largest and possibly the ultimate betatron. Nowadays betatrons are hardly used for nuclear physics research but in the 1940s they provided the first source of high-energy gamma rays. These were x-rays (bremsstrahlung) produced when the high-energy electrons impinged on a tungsten target. The energy could be varied by changing the betatron energy. In 1947 experiments with high-energy gamma rays from the 100 MeV betatron at GEC lead to the discovery of the collective giant dipole resonance in nuclei.

The first proton linear accelerator was built by L Alvarez and his group in Berkeley in 1947. There is a description of the construction of this accelerator by W Panofsky in reference [52]. A low-energy linear accelerator had been built by Sloan and Lawrence in 1931. Interest was renewed towards the end of World War II due to the many advances which had been made in the development of very high frequency power sources. In fact both Lawrence and Alvarez had worked on power sources and transmitters for microwave radar at the MIT Radiation Laboratory from 1940 to 1943 [53]. One of the things which motivated Alvarez to push ahead with linear accelerators were cost estimates for high-energy accelerators. He argued that the cost of a cyclotron would increase roughly with the third power of the energy, while for a linear accelerator the energy–cost relation would be approximately linear. Thus eventually, as one went to higher energies, the linear accelerator would replace the cyclotron on economic grounds. This argument was correct, but irrelevant after the invention of the strong focusing synchrotron. Alvarez's second motivation was the availability of surplus radar equipment after the war. The machine was built and operated successfully to produce a 32 MeV proton beam. It was used for fundamental proton–proton scattering experiments and for studying various nuclear reactions. Several other linear proton machines were built, the most notable being the 'Meson Factory', LAMPF, at Los Alamos which was the leading tool in medium-energy nuclear physics for many years.

Linear accelerators proved to be much more important for electrons than for protons. The linear accelerator is particularly suited to producing high-energy electrons. In a circular machine the continual acceleration of the electrons towards a centre by the magnetic field produces enormous radiation losses. These are avoided in the linear machine. Another advantage of the linear accelerator is the excellent collimation of the emergent beam. Stanford has always been a centre for high-energy electron linear accelerators. Initially the driving force was W W Hansen who in the mid-1930s was thinking about this kind of machine. It became a practical possibility with the advent of magnetrons developed for radar applications. Work on these accelerators also began at the end of World War II and a 2 m machine was constructed in 1948 which produced a 4 MeV electron beam. Progress in those early days was described in a review article by Fry and Walkinshaw in 1948 [54]. The Stanford Mark I machine [48] was built in 1947. It was 12 ft long, was powered by magnetrons and produced a 6 MeV beam of electrons. Then the Stanford team were able to develop high-powered klystrons which had much better phase stability properties. The Stanford Mark II was powered by these klystrons and yielded 35 MeV electrons. This was the prototype for the Mark III, designed for electron energies up to 1 GeV, and consisted essentially of 30 Mark II accelerators arranged in a line. It was finished in 1960 but was built in stages and was used at lower energy for important nuclear structure studies by Hofstadter and collaborators from 1953 onwards.

15.4.3. Neutron sources

Between 1934 and 1942 the standard neutron source was a sealed tube containing Be powder and Ra gas. The radon is alpha active and the alphas produce neutrons from the reaction

$$^{4}\text{He} + {}^{9}\text{Be} \rightarrow {}^{12}\text{C} + \text{n}.$$

Most of the experiments of Fermi and collaborators, Hahn, Meitner and Strassmann and others were all made with this kind of source. Radon and its daughters have a high gamma emission which contaminates the neutron source. This contamination can be removed by replacing the radon by polonium as the source of alpha particles.

The development of the fission reactor in 1942 provided a new, high-intensity neutron source. Cutting a small hole in the shielding of a reactor allows a beam of neutrons to be extracted for experimental purposes. The high neutron flux from a reactor is particularly useful for the production of radioisotopes by neutron capture.

15.4.4. Measuring devices and detectors

Magnetic spectrometers measure the momenta of charged particles by deflection in a suitable arrangement of magnetic fields. Already by the beginning of World War II the designs had been developed to quite a high level of sophistication and used in mass spectrometers and in beta-ray spectrometers. After the war spectrometers began to be used in association with accelerators for the analysis of the products of nuclear reactions produced by the accelerator beams; also solutions to accelerator problems contributed to the design of spectrometers and vice versa. In particular the Kerst–Serber [48] theory of orbit stability in a cyclotron or betatron was used in the design of spectrometers and the principle of the Siegbahn double-focusing beta spectrometer [55] was adapted to the design of spectrometers for accelerators.

Another technique which is now of historical interest is crystal diffraction spectroscopy for measuring gamma-ray wavelengths. The method was first used by Rutherford and Andrade [46] in 1914, and was later developed by other workers. The most successful device was the curved crystal focusing spectrometer proposed by DuMond and Kirkpatrick in 1930 [56]. A spectrometer based on this principle was designed in 1947 and built by DuMond and his collaborators. It could measure wavelengths of gamma rays in the 0–500 keV range with an accuracy of better than 0.01%. Now semiconductor counters can measure gamma-ray energies with comparable accuracy.

Several major advances in detector technology were made after World War II. One of these was the invention of proportional counters. These arose from a refinement of the Geiger counter. If an avalanche counter was carefully designed and operated under controlled conditions, then it could act as a detector for beta or gamma rays and, at the same time measure the energy of the radiation. The status of proportional counters in 1950 is reported in the book by D H Wilkinson [57]. By 1960 this kind of counter had been largely replaced by semiconductor counters.

A method of counting ionizing radiations by photoelectric measurement of the fluorescent light pulses emitted by certain organic substances was first pointed out by Kallman in 1947. It was a development of the old scintillation method, but the scintillations were counted electronically by a photomultiplier rather than the eye. Researchers immediately began to search for scintillating materials which would work well in conjunction with photomultipliers. In 1948 Hofstadter [58] found that sodium iodide crystals activated by thallium could be used as an efficient detector of gamma rays. Hofstadter was not searching at random. He knew the properties which a suitable phosphor should possess. The importance of impurities was known to Rutherford and his colleagues and in section 126 of reference [35] there is a statement 'It is well known that pure zinc sulphide does not scintillate when bombarded with alpha particles nor does it fluoresce under the action of light. In order to sensitize the zinc sulphide, the insertion of a minute quantity of impurity, such as copper, is necessary.'. The phosphorescent properties of alkali halides had already been studied in the 1930s. Sodium iodide which had a good stopping power for gamma rays and could be grown into large crystals seemed to be a promising choice. Until 1949 the Geiger counter had been the most commonly used detector for gamma rays. Within the space of a couple of years it had been replaced by the sodium iodide detectors, and counting efficiencies had been increased by a factor of 1000. Many important experiments made in the early 1950s would have been impossible without the new detectors. Sodium iodide detectors are still in common use as gamma detectors, often in combination with semiconductor detectors.

The scintillation detector combined high detection efficiency with moderate energy resolution. If high energy resolution was required then a beta-spectrometer or a crystal diffraction technique still had to be used. This situation changed with the introduction of semiconductor counters based on silicon or germanium. References [59] and [60] review the development of these detectors. Already in 1949 McKay [61] had made the first steps towards designing a semiconductor counter based on germanium and in 1953 he explored the use of silicon. A group at Purdue [62] was also actively interested. The turning point was a successful application of germanium junctions in a low-temperature nuclear physics experiment by Walter, Dabbs and Roberts in 1958 [63]. Then a number of laboratories in Europe and North America began intensive development of semiconductor particle detectors. These were in common use by 1962. Next came lithium drifted germanium, Ge(Li), gamma-ray detectors which were used widely by 1967. For the past 25 years most high-resolution detectors for charged particles and gamma rays have been based on semiconductor technology. The Ge(Li) detectors for gamma rays are less suitable than sodium iodide scintillators when the highest efficiency is required but the energy resolution is better by more than an order of magnitude and line widths of 1 keV can be obtained. The energy resolution of silicon particle detectors is not as good as for magnetic spectrometers, but for most purposes such high resolution is not required.

15.5. Shell structure in nuclei

The 1936 review article of Bethe and Bacher [64] gives an account of nuclear structure theory just four years after the discovery of the neutron. It was recognized that nuclear forces saturated so that nuclear binding energy and the nuclear volume were approximately proportional to the mass number A. The variation of nuclear binding energies with neutron and proton numbers N and Z was understood at the level of the Weizsäcker mass formula. Physicists knew that at least two different kinds of force could produce saturation. One possibility was a van der Waals type force which was attractive at large distances and repulsive at small distances. Another was an exchange interaction. There was also a pairing effect which made nuclei with N and Z both even somewhat more stable than odd nuclei. The 1936 review discussed the theory of the deuteron and included calculations of neutron–proton scattering, the capture of neutrons by protons and the photoelectric dissociation of the deuteron. Experimental data on these reactions was becoming available from the new accelerators built around 1932. The review also contained a section on the theory of the triton, He^3 and the alpha particle. The authors concluded that the binding energies of these nuclei were consistent with a short-range force between nucleons. The neutron–neutron force had to be approximately equal to the proton–proton force (charge symmetry). Experimental binding energies were available from mass measurements and nuclear reaction data from the new accelerators.

After an analysis of the Fermi-gas model for nuclear matter, Bethe and Bacher discussed a Hartree or 'shell' model for the structure of heavier nuclei. The model assumed independent motion of individual protons and neutrons. The authors wrote 'This assumption can certainly not claim more than moderate success as regards the calculation of nuclear binding energies. However, it is the basis for a prediction of certain periodicities in nuclear structure for which there is considerable experimental evidence.'. This evidence related to the strong binding of O^{16} and to breaks in the abundance of elements when plotted against $I = N - Z$. There was some evidence for anomalies at $N = 82$ and $N = 126$. The regularities were noticed by Bartlett [65], Elsasser [66] and Guggenheimer [67]. In 1933–4 Elsasser considered the energy levels of neutrons and protons in a potential well. The levels were grouped together in 'shells'. Neutrons and protons filled up the levels in order of increasing energy, taking into account the restrictions of the Pauli principle. Whenever a shell was completed a particularly stable nucleus could be expected. When a new shell started the binding energy of the newly added particles should be less than for the particles completing the previous shell.

The Elsasser [66] shell model predicted closed shells for N or $Z = 2$, 8 and 20. These fitted in with the binding energy and abundance data. However, the model failed to give closed shells at $N = 82$ and $N = 126$. Not long after this Niels Bohr formulated his compound nucleus theory. This was a strong interaction model and seemed to be incompatible with independent particle motion. As a result the shell model was abandoned for more than a decade.

The jj-coupling shell model was discovered independently by Maria Goeppert Mayer [68] and Haxel, Jensen, and Suess [69] in 1949. There are interesting historical perspectives in the Nobel Prize Lectures of Mayer [70] and Jensen [71] and in lectures delivered by D Kurath [72] and H A Weidenmüller [73] at a symposium held at the Argonne National Laboratory in 1989 to celebrate the 40th anniversary of the shell model. Maria Mayer studied with Max Born in Göttingen. After taking her doctorate she married an American, Joseph Mayer, and moved to the United States in 1930. There she could not get a job because of antinepotism rules in universities, but managed to continue studying physics with the help of various associates. In 1946 Joseph Mayer moved to the University of Chicago and Maria was able to get a half-time position at the newly established Argonne Laboratory. Hans Jensen studied in Hamburg and during the 1930s spent some time at the Niels Bohr Institute in Copenhagen. After the war he had a position in Hannover and began collaborating with Haxel and Suess.

Around 1947 Maria Mayer worked with Edward Teller on the

origin of the elements and, as a first step, began analysing data on isotopic abundances. Haxel, Jensen and Suess also became intrigued by regularities in nuclear binding energies as evident, for instance, from the abundance distribution of the elements. In 1948 there was much more information available than there had been in 1934 and the picture was clearer. In 1948 Maria Mayer [74] published evidence for the particular stability for the numbers 20, 50, 82 and 126. Part of the evidence was the small neutron absorption cross sections for targets with 82 or 126 neutrons as measured by D J Hughes at Argonne. The ideas of Elsasser and others in the 1930s were revived and the search was on for a suitable shell model. Several ideas were tried out and then Mayer and Jensen came up, independently, with the jj-coupling shell model which fitted the observed 'magic numbers' (figure 15.6). Maria Mayer wrote in her Nobel Prize Lecture [70] 'At that time Enrico Fermi had become interested in the magic numbers ... One day as Fermi was leaving my office he asked: "Is there any evidence of spin–orbit coupling?" Only if one had lived with the data as long as I could one immediately answer "Yes, of course and that will explain everything.". Fermi was sceptical, and left me with my numerology.'. Later in the same lecture she said 'After a week when I had written up the other consequences carefully Fermi was no longer sceptical. He even taught it in his class in nuclear physics.'.

Maria Mayer's 1949 paper already noted some of the predictions of the shell model. One related to the spins and parities of odd-A nuclei. Already in 1949 some were known. In the next few years many spins and parities of both ground and excited states were measured. Gamma-ray angular correlation was one technique for measuring spins of nuclear states. This method was being developed around 1948, but after the advent of sodium iodide scintillation counters the experiments became much more practicable. Measured spins and parities could be compared with shell-model predictions.

One class of excited states was isomeric levels. They had very long lifetimes: hours, days or even years. The explanation was that the spins of the isomeric states and the ground state were very different and the gamma transition rates were hindered by angular momentum selection rules. The pattern of levels in the jj shell model was such that nuclei in particular regions of the periodic table were expected to have isomeric states. These regions were called islands of isomerism. The multipolarities and hence the lifetimes of the de-excitation gamma transitions could be predicted. These predictions of the jj shell model were in striking accord with experimental data.

A necessary condition for the validity of the shell model was that the mean free path of a nucleon inside the nucleus should be large compared with the nuclear radius so that the single-particle orbits could be well defined. This appeared to be in contradiction to Bohr's theory of nuclear

Nuclear Dynamics

Figure 15.6. *Maria Goeppert-Mayer and Hans Jensen.*

reactions. In fact the success of Bohr's theory had encouraged theorists to abandon the shell-model description of nuclear structure. In his 1936 paper [20] Bohr explicitly rejected theories of nuclear structure based on independent particle motion. Referring to the shell model he wrote '... it is, at any rate, clear that the nuclear models hitherto treated in detail are unsuited to account for the typical properties of nuclei for which, as we have seen, energy exchanges between individual nuclear particles is a decisive factor'. He continued later in the same paragraph 'In the atom and in the nucleus we have indeed two extreme cases of mechanical many-body problems for which a procedure of approximation resting on a combination of one-body problems, so effective in the former case, loses any validity in the latter where we, from the very beginning, have to do with essentially collective aspects of the interplay between the constituent particles.'.

In 1949 the shell model was accepted immediately in spite of the apparent conflict with Bohr's views stated in the last paragraph. There were probably two reasons for this: (i) the model could explain the magic numbers and (ii) it gave wavefunctions for nuclear states which could be used to make predictions which could be tested by observations using new experimental techniques. Bohr was also convinced by the new results and in an unpublished note [75] which is reproduced in volume 9 of his collected works he argued that a deeper understanding of the quantum-mechanical many-body problem was needed to reconcile the two points of view.

In the meantime another piece of evidence for independent particle motion emerged. This was the optical model of low-energy neutron or proton scattering proposed by Feshbach, Porter and Weisskopf [76] in 1954. They argued that a neutron incident on a target nucleus might be absorbed to form a compound nucleus, or it might pass through with its motion modified due to its average interaction with the nucleons in the target. The optical model was sometimes called the cloudy crystal ball model. 'Cloudy' referred to absorption, while 'crystal ball' alluded to the refraction of the path of the neutron by the average field. The average field in the optical model had the same origin as the average potential in the shell model and the real part of the optical potential should be similar to the shell-model potential. The absorption was represented by an imaginary part in the optical potential. Bohr's compound nucleus model corresponds to a limiting case where the absorption is strong and the nucleus is 'black'. New experimental results on elastic scattering confirmed the predictions of the optical model and showed that the absorption corresponded to a relatively long mean free path for a nucleon inside a nucleus. It was long enough to permit the definition of shell-model orbits.

Another mystery in the shell model was the origin of the spin–orbit interaction. It had to be there to get the right magic numbers but for several years there was no clear idea about where it came from. The question was partially resolved by advances in the experimental and theoretical understanding of the nucleon–nucleon force. According to Yukawa the force between nucleons was due to meson exchange. The only strongly interacting meson known in 1949 was the pion and the nucleon–nucleon force produced by pion exchange had a tensor component but no spin–orbit part. At the time there were attempts to derive a spin–orbit force as a second-order effect of the tensor force. The resulting spin–orbit effect was too weak. The situation changed in the early 1960s when the rho- and omega-mesons were discovered. These were vector mesons with spin $S = 1$. A generalization of the Yukawa theory to vector mesons produced a nuclear force with a spin–orbit component. Throughout the 1950s there was a systematic effort to measure the nucleon–nucleon force by doing proton–proton and neutron–proton scattering experiments. An analysis by Stapp et al [77] in 1957 found clear evidence for the existence of a spin–orbit component in the nuclear force. This would explain at least a part of the spin–orbit force in the shell model. Even now there is no clear consensus as to whether it is the final answer.

Since 1949 the shell model has been generalized to enable it to describe properties of complex nuclear states and has given rise to tremendous progress in the understanding of nuclear structure and reactions. Some of this has been reviewed in a lecture by Talmi [78] at the 40th anniversary symposium for the shell model in 1989. He wrote 'It is no longer necessary to justify the shell model on "theoretical grounds". The proof of the shell model is in the agreement of its predictions with experiment and that is established without reservation.'. Perhaps not all nuclear physicists would agree with this statement, but it does represent a widely held point of view.

15.6. Collective motion in nuclei

Niels Bohr wrote about nuclear collective motions in several paragraphs of his 1936 paper on the compound nucleus theory. He saw it as a natural feature of a situation where nucleons in a nucleus interacted strongly with each other. He imagined the nucleus as a small drop of liquid, or even as a solid. In either case it could have various vibrational modes. In his 1937 review paper on nuclear dynamics Bethe reported on a calculation by Bohr and Kalckar on the spacing of states in a highly excited nucleus. The excitations were built up from surface and volume vibrational modes. Bohr and Kalckar also calculated the thermal properties of the liquid drop with this model. The liquid-drop model of nuclear fission due to Frisch and Meitner and later to Bohr and Wheeler involved large amplitude collective motion.

15.6.1. Giant resonances

A new kind of collective motion was discovered in 1948. The experiments were made by Baldwin and Klaiber [79] and the interpretation was due to Goldhaber and Teller, and Migdal [80]. They used the term 'dipole vibration' for the collective mode. Later it was called a 'giant dipole resonance' or GDR. The experiments were made at the General Electric Company laboratory at Schenectady, New York with gamma rays from one of the new betatrons built by Kerst and collaborators. Baldwin and Klaiber measured the photo-fission cross sections for uranium and thorium, and the photo-neutron cross sections for carbon and copper. The cross sections had the appearance of a high-frequency resonance. The resonance energies were 30 MeV in carbon, 22 MeV in copper, and 16 MeV in uranium and thorium. The results were remarkable in that nuclei from completely different regions of the periodic table had similar resonances. The resonance energy decreased regularly with increasing A; also the resonances were very similar for reactions with quite different final states.

Goldhaber and Teller [80] suggested that the gamma rays excited a collective motion in the nucleus in which the protons moved in one direction while the neutrons moved in the opposite direction. This was the origin of the name 'dipole vibration'. They also argued that the width of the resonance was due to the transfer of energy from the orderly dipole vibration into other modes. That is the width was due to a process analogous to damping by friction. The reaction had two stages. The first was absorption of a photon to form a compound nucleus, while the second was the decay of the compound nucleus into the final channel. The preferred decay of the compound nucleus was fission for heavy nuclei and neutron emission for lighter ones. Goldhaber and Teller were able to estimate the energy integrated photo-absorption cross section, with results consistent with the experimental data. Experiments on gamma-ray induced reactions made over the next few years confirmed the universal character of the giant dipole resonance.

A few years later Wilkinson [81] proposed a shell-model theory of the giant dipole resonance. He argued that the nuclear states which contribute to the dipole resonance are those formed by promotion of a single nucleon into the next higher shell. In all cases the angular momenta of the promoted particle and that of the 'hole' left behind were coupled to a total $J = 1$. The overall parity was negative. The energies of all these states clustered together around the observed resonance energy. The photon absorption excited a particular linear combination of these shell-model particle–hole states. The coherent superposition of the shell-model states can be shown [82] to correspond closely to the collective Goldhaber–Teller mode. This is an example of a case where the collective and shell-model descriptions are equivalent. In 1961 Thouless [83] showed that the random-phase approximation could give a systematic theory of vibrational states like the giant dipole resonance.

The giant dipole resonance is an isovector mode because the neutrons and protons move out of phase. In 1972 the isoscalar giant quadrupole mode was discovered in proton inelastic scattering experiments [84]. Later, in 1977, the isoscalar giant monopole, or breathing mode, was found using inelastic alpha-particle scattering. The neutrons and protons move in phase in these isoscalar modes. The quadrupole and monopole modes have the same universal properties as the giant dipole resonance. They exist in nuclei throughout the periodic table and their properties vary in a systematic way with N and Z. Spin giant resonances have also been found in many nuclei, but their properties are more dependent on shell structure. They can be excited by proton or electron inelastic scattering. Bertrand and Bertsch reviewed the state of the field in 1981 [85].

15.6.2. Low-energy collective modes

The shell model of Mayer and Jensen was based on the idea that protons and neutrons in a nucleus moved in a spherical shell-model potential representing their average interaction with the other nucleons. In 1949 Townes, Foley and Low (cf reference [87]) collected all the existing information about the electric quadrupole moments of odd nuclei. The quadrupole moments had been measured in the hyperfine structure of atomic spectra. Townes and his co-workers found evidence for shell structure in a plot of Q/R^2 against the atomic number Z (Q is the quadrupole moment and R is the nuclear radius). The plot showed a regular shape related to the filling of the individual particle levels and passed through zero at the closed-shell values. The quadrupole moment was negative just below the closed shell and positive above as predicted by the shell model. The plot had a broad peak between the closed shells $Z = 50$ and $Z = 82$. There was an unexpected result. At its maximum Townes et al estimated that the experimental Q/R^2 was 35 times larger than the value expected for an odd proton.

In a paper published in 1950 James Rainwater [86] argued that the quadrupole moment data gave a strong indication that the basic nuclear shape for these nuclei was not spherical, but corresponded to a distortion of the whole nucleus into a spheroidal shape. He also showed that, while the individual particle motion of the nucleons favours a spherical shape at closed shells, away from closed shells the centrifugal pressure of the odd nucleons tends to produce a deformed equilibrium shape. Rainwater was a professor in the Physics Department of Columbia University and Aage Bohr (Niels Bohr's son) held a research fellowship in the same department during the 1949–50 academic year (figure 15.7). They shared an office during that year [87]. Bohr was also interested in the large quadrupole moments of nuclei between closed shells and studied the possibility that deformed nuclei might have rotational spectra. He worked out some of the consequences of a model for odd nuclei in which the odd nucleon was coupled to the rotations of a deformed core containing all the other nucleons. In particular he studied the influence of deformations on nuclear magnetic moments [88].

Aage Bohr returned to Copenhagen in the autumn of 1950 and continued to work on collective motion. He was joined by Ben Mottelson who came to the Bohr Institute with a fellowship from Harvard. As far as one can judge from the published record, it was in the second half of 1952 that things began to fall into place. In June of that year Mottelson gave a lecture in which he outlined the consequences of a model with surface oscillations coupled to easily excitable single-particle modes. There is a reprint of that lecture in the book of Alder and Winther on Coulomb excitation [89]. Progress at this point was influenced by the appearance in July 1952 of an extensive review on 'Nuclear isomerism and the shell model' by Goldhaber and Hill [90] in which they collected together a mass of information on measured level schemes and electromagnetic transition rates in many nuclei. In November 1952 Bohr and Mottelson submitted a letter to *Physical Review* [91] on electric quadrupole isomeric transitions in a number of even nuclei. They remarked that the measured electromagnetic transition rates were more than a factor of 100 faster than shell-model predictions, but could be explained if the nuclei were deformed. They gave an interpretation in terms of rotational states of

Figure 15.7. *(a) Aage Bohr, (b) Ben Mottelson and (c) James Rainwater.*

deformed nuclei.

Bohr and Mottelson submitted another letter to *Physical Review* in March 1953 [92] which was focused specifically on rotational motion. A deformed rotating nucleus with N and Z even was expected to have a band of states with excitation energies $E = AI(I + 1)$ where the spin $I = 0, 2, 4, 6, \ldots$ and A is a constant proportional to the inverse of the moment of inertia. The spectrum of an odd nucleus could be obtained by coupling single-particle states to the rotating core. In the

March letter they showed the first examples of rotational spectra in two isotopes of hafnium with $A = 178$ and $A = 180$. These and other results were presented in a more extensive paper in 1953 [93]. Within the next year many examples of rotational spectra were found. For example in July 1953 Asaro and Perlman [94] submitted a letter to *Physical Review* reporting patterns in the alpha-decay spectra of even nuclei which could be interpreted by the rotational model.

At the same time the Copenhagen group was thinking of other ways in which the collective model could be studied experimentally. In May 1953 Mottelson [89] suggested that Coulomb excitation would be an ideal experimental tool for studying low-energy collective modes in nuclei. He proposed to 'bombard atomic nuclei with charged particles whose energy is well below the barrier. The probability of penetrating the barrier would not be appreciable, and the nucleus would experience only the Coulomb field. Thus the excitation resulting from the pulse of electromagnetic energy would reflect the properties of the target nucleus free from complication by the nuclear force between target and projectile.'. Coulomb excitation had been considered before, first by Weisskopf in 1938 and later by Ter-Martirosyan (cf reference [89]) in 1951. There were also two experiments by Barnes and Ardine, and by Lark-Horowitz and collaborators in 1939 in which Coulomb excitation had been observed. The different situation in 1953 was that theoreticians and experimentalists were looking for a powerful method for investigating collective motion. Coulomb excitation offered a promising approach.

The first of the new generation of Coulomb excitation experiments were made in 1953 by Huus and Zupancic [89,95] in Copenhagen and by McClelland and Goodman [96] at MIT. Both used protons as projectiles and NaI counters as gamma-ray detectors. Both groups observed strong Coulomb excitation. After this there was an explosion of activity. Temmer and Heydenburg [89,97] at the Department of Terrestrial Magnetism, Carnegie Institution of Washington DC used alphas as projectiles. The Coulomb fields were stronger, the barriers higher and the cross sections larger. McGowan and Stelson at Oak Ridge [89,98] combined Coulomb excitation with gamma-ray angular distributions to measure spins of excited nuclear states. The theory and the first experimental results were set out in a big review article by Alder, Bohr, Huus, Mottelson and Winther in 1956 [89,99]. By then Coulomb excitation experiments had been made on almost 150 different target isotopes. The Nobel Prize Lectures of Bohr [100] and Mottelson [101] contain interesting information about all these developments.

In 1955 [102] Nilsson made an important step towards unifying the rotational model and the shell model. There was already a precursor of the idea in Rainwater's original paper where he calculated the individual nucleon states in a spheroidal potential well. Following Rainwater, Nilsson contended that the shell-model potential representing the average interaction of a nucleon with a nucleus, should be non-spherical in a deformed nucleus. The single-particle wavefunctions calculated in an appropriate deformed potential could be combined with the rotational model to make predictions about the properties of deformed nuclei with odd N or Z. This model, and more refined versions which followed it, were very successful. The Nilsson model gave a static picture of deformed nuclei. Dynamics was introduced by Inglis in 1954 with his remarkable 'Cranking Formula' [103].

The study of low-energy collective motion has been one of the most fruitful fields in nuclear structure physics. It has been characterized by a productive interaction between theory and experiment and many new and interesting phenomena have been found and continue to be found. The theory has been particularly effective in accounting for properties of nuclei with large deformations. A recent development has been the discovery of superdeformed nuclei at the Daresbury Laboratory by Twin and his collaborators in 1985. These are nuclear states where the major-to-minor axis ratio is 2:1. They have some unusual properties which are thought to reflect closed-shell effects which can occur in strongly deformed Fermi systems. Such effects had been predicted by Strutinsky [104] in 1966 and were observed in fission isomers. The first fission isomer was found by Polikanov and his colleagues in Dubna. Strutinsky's predictions about the properties of fission isomers were confirmed by Migneco and Theobald in 1968 and by other groups. The fission isomer story was reviewed by Bjornholm and Lynn [105] in 1980 and references can be found there.

15.6.3. *The interacting boson model*

Bohr and Mottelson's theory of nuclear structure has sometimes been called the geometrical collective model. It had its origins in Niels Bohr's image of nuclei as small liquid drops which had excited states related to the motion of the nuclear surface. In the case of deformed nuclei the motions were rotations and associated vibrations about the non-spherical equilibrium shape. Some nuclei have a spherical equilibrium shape. In the geometrical collective model the excited states would have a vibrational character. According to the ideas of the geometrical model the excited states of even nuclei with a few nucleons outside closed shells have vibrational spectra. Those with many nucleons outside closed shells are rotational. In between there are transitional nuclei with more complicated spectra related to anharmonic vibrations.

In 1977 Arima and Iachello [106] invented a new theoretical approach to the study of collective motion in nuclei. It is called the algebraic collective model as opposed to the geometrical collective model of Bohr and Mottelson. The algebraic model has its origin in the shell model. Nucleons near the Fermi surface in a nucleus interact and form pairs. The pairs act like bosons. Collective states are built out of these bosons.

Group theory is the mathematical apparatus of the interacting boson model, and the theory has several limiting cases where its predictions can be written in a closed form by using algebraic methods. The theory has been very successful in describing the spectra of even nuclei, especially those of the transitional nuclei mentioned in the last paragraph.

15.7. Nuclear scattering and reactions

15.7.1. *The compound nucleus*

A central idea of Bohr's compound nucleus theory was to divide a nuclear reaction into two stages. The first step was the formation of the compound nucleus and the second was its disintegration into the products of the reaction. At moderate excitation energies, reaction cross sections showed a resonance structure. Individual resonances corresponded to quantum states of the compound nucleus. The theory evolved in two ways. One was a statistical approach which averaged over the resonant structure, the other aimed at a complete characterization of the detailed energy dependence.

Weisskopf and Ewing [107] were able to develop a quantitative theory of energy-averaged cross sections based on the compound nucleus idea by assuming that the two stages of the reaction could be treated as independent processes. The mode of disintegration was assumed to depend on the energy, angular momentum and parity of the compound system but not on the specific way in which the compound nucleus was formed. The theory of Weisskopf and Ewing made some simplifying assumptions about energy and angular momentum dependence but was very effective and is still used today. A more complete theory was produced by Hauser and Feshbach in 1952 [108] and formed the basis of sophisticated statistical model theories which have been used to analyse experimental data.

The resonant structure of reaction cross sections was first worked out by Breit and Wigner [23] in 1936. A more detailed approach which formed the basis of a rigorous theory of nuclear reactions was given by Kapur and Peierls in 1938 [109]. This theory separated an internal region where the constituents of the compound nucleus were interacting strongly, and an external region where the reaction products were separating freely. The two regions were connected by boundary conditions. The structure of the compound nucleus could be analysed by focusing on the internal region. An extensive development of the formalism was made by Wigner and his collaborators in 1946 [110]. Wigner's presentation differed from that of Kapur and Peierls in the form of the boundary conditions. There is an extensive review of the subsequent developments by Lane and Thomas [111].

15.7.2. Direct reaction theories

There were new ideas in the air about both nuclear structure and nuclear reactions in the immediate post-war period. The shell model of Mayer and Jensen appeared in 1949, and the first direct reaction theories almost immediately afterwards. A direct reaction is a fast process and does not involve the formation of a compound nucleus intermediate state. The reaction products are produced as soon as the projectile interacts with the target.

A direct reaction mechanism was first suspected in the case of (d,n) reactions by Lawrence, Livingston and Lewis in 1933 [112] and the first direct reaction theory was proposed by Oppenheimer and Phillips in 1935 [113]. There were many other measurements of (d,p) and (d,n) reactions in the pre-war period [21] and they were known to have large cross sections. Experiments made after the war with 200 MeV deuterons from the Berkeley synchrocyclotron gave yields which were much too large to be accounted for by Bohr's compound nucleus model. Other experimenters reported forward peaking in the angular distribution. This did not fit with Bohr's compound nucleus theory which predicted that angular distributions should be symmetric about 90°. Bohr's compound nucleus model was clearly not working and theorists began to think of new approaches to reaction theory. One of the first was suggested by Serber in 1947 [114].

In 1950 and 1951 angular distributions for several (d,p) and (d,n) reactions with resolved final states were measured. Experiments were performed by Burrows, Gibson and Rotblat and by Holt and Young in 1950. They made measurements with ^{16}O and ^{27}Al as targets and found that the angular distributions were characterized by a large forward peak and one or more lesser peaks, very reminiscent of a diffraction pattern. In 1950 Butler [115], working with Peierls in Birmingham, proposed a theory for (d,p) reactions and used it to analyse the data of Burrows and his collaborators. By fitting the angular distribution to the experiment he was able to determine the angular momentum of the transferred neutron in the residual nucleus. In Butler's theory the target nucleus 'stripped' the neutron from the deuteron and the proton continued on its way. For this reason (d,p) and (d,n) reactions came to be called stripping reactions. Butler's theory went through several refinements. In the end the most successful approach was the distorted wave Born approximation (DWBA). Its systematical mathematical formulation was given by Horowitz and Messiah in 1953.

Stripping reactions interpreted by Butler's theory or its successor, the DWBA, were immediately recognized as an excellent spectroscopic tool. The sensitivity to the angular momentum transfer was used to give information about the spins of nuclear states reached by stripping reactions. Moreover the probability for stripping was proportional to the probability of the target nucleus accepting a single neutron or proton, so the magnitude of the cross section gave information about the single-particle structure of nuclear states. Stripping reactions with deuterons and other projectiles have continued to be important for spectroscopic studies until the present day.

Inelastic scattering of nucleons, alpha particles or other projectiles is another class of direct interactions. The Butler theory could be modified to calculate this kind of reaction but was not so successful. A version of the distorted wave Born approximation could be formulated for inelastic scattering and was applied to a case of proton inelastic scattering by Levinson and Banerjee [116] in 1957. It was the first application of the full DWBA theory to a nuclear reaction and was quite successful, though the reason for the success was not fully understood. Physical insight resulted from a series of scattering experiments with 40 MeV alpha particles in 1959. They were made by Blair and his colleagues at the University of Washington cyclotron [117]. The elastic angular distributions showed a series of regularly spaced diffraction maxima as would be expected from a black nucleus model. In 1959 Blair [118] developed a diffraction theory for inelastic scattering with extremely good results. He realized that inelastic scattering angular distributions were sensitive to the transparency of the target nucleus to the projectile. The black nucleus model was a good approximation for alpha particles but not for nucleons. After this insight DWBA could be used with confidence for various kinds of projectiles, provided care was taken in choosing the optical potential which generated the distorted waves.

One of the big problems in compound nucleus theory was to separate the internal compound nucleus region, which was interesting from the nuclear structure point of view, from the external region which was dominated by simple kinematics. Another problem was to provide a unified description of compound nucleus and direct reactions. In 1958 Feshbach [119] invented a new approach based on a projection-operator technique which was convenient from the mathematical point of view and allowed a clear physical separation of the prompt (direct reaction) and time-delayed (compound nucleus) component of a reaction amplitude.

15.7.3. Electron scattering
The first high-energy elastic electron scattering experiments were made with the Stanford linear accelerator by Hofstadter in 1953 [120]. They were continued over the following few years and the results were reviewed by Hofstadter in 1956. The electron scattering is sensitive to the distribution of electric charge in a nucleus. In order to investigate the details of the charge distribution the electrons must have a de Broglie wavelength which is small compared with nuclear dimensions. This means they must have energies in the range 200–500 MeV.

The first Stanford experiments in 1953 were made with 116 MeV electrons. This was soon increased to 180 MeV, and 550 MeV electrons were available by 1956. By that time the techniques were good enough to give accurate values of the radius of the charge distribution and of the surface diffuseness [121].

15.8. New isotopes and new elements

Most nuclei of atoms existing on the Earth are stable. An element may have several stable isotopes and about 270 stable isotopes are known. They are clustered along the valley of stability on a plot of Z against N. A number of radioactive isotopes also exist in nature. Most of them either have a lifetime longer than the age of the earth, or occur in a radioactive decay chain deriving from long-lived isotopes. A few radioactive isotopes are produced in small quantities by cosmic rays. An example is ^{14}C which is used for dating organic materials. Nuclei of isotopes beyond mass number $Z = 82$ are all radioactive and are unstable against alpha decay or spontaneous fission. This instability is due to the electrostatic repulsion between protons, and nuclei become more unstable as Z increases. Lighter nuclei off the stability line may be beta active. On the proton-rich side they decay by positron emission or electron capture. On the neutron-rich side they decay by electron emission. Beta-active isotopes usually have half-lifetimes in the range microseconds to years. They are long enough for their properties to be studied. When nuclei become very proton rich or very neutron rich they become unstable against proton or neutron emission, respectively. Such nuclei are said to lie beyond the proton or neutron drip lines and have very short lifetimes. All the isotopes considered in this section are radioactive, but lie within the proton and neutron drip lines.

The first isotopes, ^{13}C and ^{30}Si, were the artificially radioactive nuclei produced by the Joliot-Curies in 1933. Fermi and the Rome group found many more in their experiments on neutron reactions. By 1937 about 200 artificially radioactive isotopes had been produced using various nuclear reactions. In 1960 this number had increased to about 800. Many of the new isotopes were fission fragments. Their masses and lifetimes had been measured and in many cases there had been studies of the spectrum of excited states.

15.8.1. Transuranic elements

One of the most exciting developments in the history of nuclear physics has been the artificial preparation of transuranium elements, that is elements with atomic numbers larger than 92. They are all unstable and decay by alpha or beta emission or by fission. The lifetimes of some transuranium elements are very long, but not long enough for them to have survived in nature even if high concentrations had existed at the time of the formation of the Earth's crust. The one exception is the isotope ^{239}Pu of plutonium which has been found in minute quantities in pitchblende from Canada and the Congo. It is thought to have been produced by neutron capture in uranium with neutrons produced by spontaneous fission of uranium. There are many historical references in the book of Hyde, Perlman and Seaborg [122]. References to earlier reviews are quoted there. Other interesting and more recent references are Hyde et al [123] and Seaborg and Loveland [124].

The story began with the neutron capture experiments of Fermi and his co-workers in Rome in 1934. They had isolated a 13 min activity from a uranium sample after neutron irradiation and separated it chemically from all elements of atomic number 82 to 92. They concluded that the 13 min activity was an isotope of element 93. Further experiments by the Rome group and by groups in other countries showed that the situation was very complex and the matter was not resolved until after the discovery of fission by Hahn and Strassmann at the end of 1938. Their work cleared the way to further progress on the transuranic elements.

McMillan at the University of California focused on an activity with a 23 min half-life which had been identified with an isotope of uranium by Meitner, Hahn and Strassmann in 1937. Eventually in 1940 McMillan and Abelson [125] were able to announce the discovery of the first transuranium element (figure 15.8). It had atomic number $Z = 93$ and was named neptunium from the planet Neptune which is the first planet beyond Uranus. Klaproth who discovered uranium in 1789 had given the element this name in honour of the discovery of the planet Uranus in 1781 by Herschel. The identification of neptunium was delayed because its chemical properties were different from those which had been expected. The 23 min activity was due to the decay of ^{239}U into ^{239}Np. The ^{239}Np was itself unstable and decayed by beta-decay into the isotope ^{239}Pu of plutonium with $Z = 94$. The positive identification of this isotope of the new element was hampered by its long half-life, but was eventually accomplished by Kennedy, Seaborg, Segrè and Wahl [126] in 1941 in a major experiment with a 16 MeV deuteron beam from the Berkeley 60" cyclotron. The experimenters succeeded in isolating a 0.5 microgram sample of ^{239}Pu and found that it fissioned when bombarded with slow neutrons.

It was recognized that the slow neutron fission of ^{235}U might allow the conversion of appreciable quantities of natural uranium into plutonium. It was then that the secret plutonium project was set up by the Uranium Committee. The history was recorded in the book *Atomic Energy for Military Purposes* by Smythe in 1945. After 1940 all work on transuranium elements was classified and results were not published until after 1945.

Transuranium elements have many isotopes. Some were found in accelerator experiments and others with reactor experiments. Once one isotope had been produced its chemistry could be studied and the information so obtained helped with the identification of other

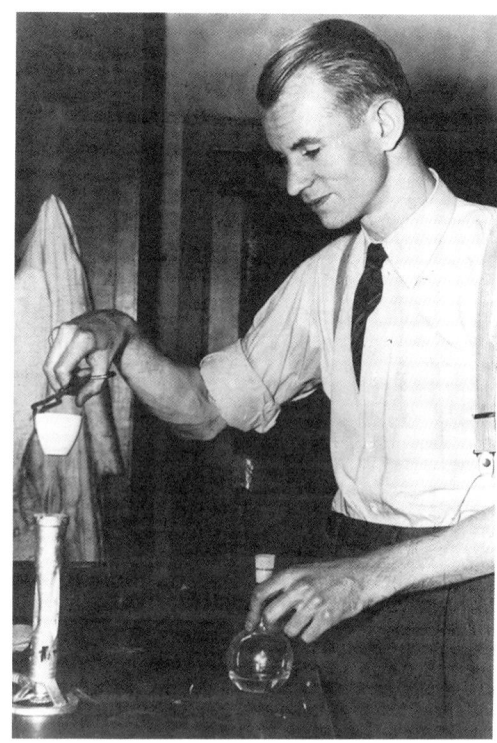

Figure 15.8. *Edwin McMillan in his laboratory at the time of the discovery of neptunium.*

isotopes. The next elements were found in 1944–5 after the advent of nuclear reactors. By that time ^{239}Pu existed in large enough quantities to be used as a target in accelerator or reactor experiments. The first isotope of curium (^{242}Cu, $Z = 96$), named after Marie and Pierre Curie, was prepared by Seaborg, James and Ghiorso in mid-1944 by bombarding a plutonium target with alpha particles at the Berkeley cyclotron. An isotope of americium (^{241}Am, $Z = 95$) was produced in the Chicago Metallurgical Laboratory as a product of neutron irradiation of plutonium.

The general strategy for producing new elements was to select the heaviest target nuclei possible, then add a few nucleons to them. Thus the first isotopes of the next two elements, berkelium ($Z = 97$) and californium ($Z = 98$), were made in 1949–50 by using americium and curium as targets and alpha particles as projectiles. Microgram quantities of the target material were sufficient for the experiments. One of the interesting consequences of the November 1952 'Mike' thermonuclear explosion was the discovery of elements 99 and 100, einsteinium and fermium, in the resulting debris. In addition to those new elements and many new isotopes of plutonium, americium, curium, berkelium and californium [127] were identified.

By 1992 the formation of all elements up to $Z = 109$ had been established and there were some indications that elements with Z up to 112 might be produced [128]. Most of these transuranium elements were first produced at Berkeley, some at Dubna and some at Darmstadt. As Z increases the elements become more and more unstable against fission and alpha emission. For example, the most stable californium isotope ($Z = 98$) has a half-life of 900 yr, while the most unstable isotope of hahnium ($Z = 104$) is about 40 s.

The availability of heavy-ion accelerators has changed the tools and strategies available for the formation of new elements. Until 1955 new elements were made by selecting the heaviest available target and adding a few nucleons. With heavy-ion projectiles, lighter more convenient targets can be used [124]. Element 107 was produced at the GSI Laboratory in Darmstadt by the reaction

$$^{209}\text{Bi} + ^{54}\text{Cr} \rightarrow ^{262}107 + \text{n}.$$

In 1966 Myers and Swiatecki used an elaborate liquid-drop mass formula with empirical shell corrections to study the fission barriers of transuranium elements. Their calculations suggested that nuclei near the magic numbers $Z = 114$ and $N = 184$ should be stabilized by closed-shell effects and have higher fission barriers than lighter nuclei. The general trend for nuclei with larger A to have shorter lifetimes for fission and alpha decay might be reversed and some of these 'superheavy elements' could even be stable. This suggestion started a programme to try to produce superheavy elements in heavy-ion reactions and to search for them in nature. So far none have been found, but the heaviest transuranium elements that have so far been produced are not so very far away from the predicted superheavy region.

At the time of going to press, the production of the element with $Z = 110$ at GSI was reported by Hofmann *et al* [129].

15.9. Creation of the elements

Progress in understanding the origin of the elements is reviewed in the Nobel Prize Lectures of Bethe [130] in 1967 and Fowler [131] in 1983. There are also some recent textbooks [132, 133]. The material for this section comes mainly from references [130, 131]. The question of the origin of the elements is closely connected with the production of energy in stars. In 1929 Rutherford wrote [134] 'It is natural to suppose that the uranium on our Earth has its origin in the Sun.'. In 1929 Atkinson and Houtermans concluded that in the interior of stars the temperature was high enough to allow nuclear reactions to occur. Gamow's book *Atomic Nuclei and Nuclear Transformations* [135] written in 1936 had a chapter on the relative abundance and the origin of the elements. He wrote about thermonuclear reactions in stars and noted that neutron capture would be an effective way of building up heavy elements.

It has been known for a long time that most stars are composed of hydrogen and helium with less than 1% of heavier elements. Thus if nuclear reactions are responsible for the production of energy, then hydrogen and helium must be involved. A scheme (the pp process) for producing helium from protons was suggested by von Weizsäcker in 1937 and calculated by Bethe and Critchfield in 1938. In the first step two protons interacted to produce a deuteron, a positron and a neutrino. This was a weak interaction process with a very small cross section; but enough to be effective on an astronomical timescale. The deuterons quickly interacted further and in the end produced ^{4}He. Stimulated by a conference organized in Washington DC by Gamow in 1938, Bethe produced an alternative scheme, the carbon–nitrogen cycle, for building up helium from hydrogen. In this process the carbon acted as a catalyst and was regenerated at the end of the cycle. It turned out that both processes were important. The pp process is now known to be the mechanism which operates in the sun while the carbon–nitrogen cycle dominates the energy production in a massive star during the hydrogen-burning stage.

In an attempt to find a way of synthesizing heavier elements Gamow and his collaborators [136] suggested that the early universe could have played the role of a gigantic fusion reactor. All atomic nuclei could be produced from an initial mix of protons and neutrons by neutron capture, beta decay and other nuclear reactions. The idea ran into a difficulty because there were no stable nuclei with mass 5 or 8. In 1939 Staub and Stephens [137] working in the Kellogg Radiation Laboratory at Caltech found that ^{5}He was not stable. The same was later shown to be true of ^{5}Li. In 1949 Hemmendinger [138] found that ^{8}Be was also unstable. The mass gaps at $A = 5$ and $A = 8$ were a problem for element synthesis not only in the big bang, but also in the interior of stars.

Helium could be synthesized from hydrogen by the pp process or the carbon–nitrogen cycle. The next step was to find a mechanism for building heavier elements out of helium. There was a problem in bridging the gap between ^{4}He and ^{12}C because of the unstable nuclei at mass 5 and 8. In the end the problem was solved by Fred Hoyle [139]. He was convinced that ^{12}C nuclei could be formed from the fusion of three alpha particles in red giant stars. This reaction required the simultaneous collision of three alpha particles, a very unlikely event except for the fact that it is favoured by a double resonance. The effect of the resonance in alpha–alpha scattering was calculated by Salpeter [140] during a visit to the Kellogg Radiation Laboratory in 1951, but his estimate for the fusion rate for three alpha particles into ^{12}C was still too small. Hoyle realized that the process would be speeded up if ^{12}C had an excited state just above the threshold for break up into three alpha particles. Hoyle visited the Kellogg Laboratory early in 1953 and questioned the staff about the possible existence of his state. At the time it was not known but Dunbar *et al* [141] looked for the state in a nuclear reaction and found it almost exactly where Hoyle had predicted it to be.

Once the mass gaps at $A = 5$ and 8 had been passed, heavier elements could be built up by a variety of nuclear reactions. In Hoyle's classic papers stellar nucleosynthesis of elements up to the iron group was achieved by charged particle reactions; heavier elements could be produced by neutron capture as suggested by Gamow [135, 136]. There are two versions to the neutron capture process. The s-process was suggested by Gamow in 1935. It is slow. The time interval between successive captures is generally longer than the beta-decay lifetimes and the pathway for building up heavier nuclei is along the valley of stability. The r-process is a rapid capture process. Two or more neutrons may be captured before beta decay and it tends to produce neutron-rich nuclei. It was used as a mechanism for heavy-element production in supernovae in the paper of Burbidge, Burbidge, Fowler and Hoyle [142] in 1957.

The solar neutrino problem has been a puzzle for more than 20 years. There are various side branches to the pp chain for hydrogen burning. In 1958 Fowler and Cameron [143] showed that ^{7}B and ^{8}B should be produced in small amounts as byproducts of the pp chain. The beta decay of these nuclei produces neutrinos which should be energetic enough to be detected through their interaction with ^{37}Cl to form radioactive ^{37}Ar. The idea was suggested by B Pontecorvo and L Alvarez in the 1940s. In 1970 an experimental set-up was constructed by R Davis and collaborators. It is located deep underground in the Homestake Gold Mine in Lead, South Dakota and started taking data in 1970. Only about one quarter of the number of neutrinos expected have been found. This is a problem for the standard stellar models.

The idea of using nuclear methods to measure astronomical timescales was proposed by Rutherford [134] in 1929 in a note to *Nature* on the origin of actinium and the age of the Earth. He assumed that actinium was a member of a decay chain originating from actino-uranium, a then unknown isotope of uranium with mass number 235. By making use of the known abundances of isotopes in the different radioactive decay chains he was able to estimate that the half-life of actino-uranium (^{235}U) should be about 0.42 billion years and argued that the Earth could not be older than 3.4 billion years. This was the first attempt to use nuclear processes to measure cosmological timescales.

We conclude this chapter with another example relating nuclear physics and the cosmos. The isotope of californium with $A = 254$ was first identified in the debris of the 1952 thermonuclear explosion. The half-life was measured to be 55 days. Baade and Minkowski studied the decay in the light intensity of supernovae and found that two kinds of behaviour could be distinguished. All type I supernovae had remarkably similar light curves and about 100 days after the birth of the supernova the brightness decreased linearly on a logarithmic scale with a half-life of about 50 days. In 1956 Baade, Burbidge, Hoyle, Burbidge, Christy and Fowler [144] suggested that the energy source for the light could be the decay of ^{254}Cf which is produced by the r-process in the explosion of the supernova. The isotope with $A = 254$ decays by fission and releases a lot of energy. Baade [145] has collected information about observations of historical supernovae and suggested that the supernovae observed by Tycho Brahe in 1572 and Kepler in 1604 had the same decay period. It was soon recognized that the story was more complicated. The isotope ^{254}Cf is indeed produced in supernova explosions, but not in large enough quantities to explain the light curves. Isotopes of some medium-mass elements with half-lives in the region of 50 days give the major contribution to the effect noticed by Baade and Minkowski. The nucleus ^{59}Fe is particularly important. It has a half-life of 45 days and decays by beta emission. Each decay releases an energy of 1.57 MeV. This is small compared with the energy released in the fission of ^{254}Cf, but the ^{59}Fe is produced in much greater quantities [146].

References

[1] Rutherford E 1932 *Proc. R. Soc.* A **136** 735
[2] Chadwick J 1932 *Nature* **129** 312
[3] Cockcroft J D and Walton E T S 1932 *Proc. R. Soc.* **137** 229
[4] Pais A 1986 Possible existence of a neutron *Inward Bound* (Oxford: Oxford University Press)
[5] Heisenberg W 1932 Über den Bau der Atomkerne *Z. Phys.* **77** 1; 1932 *Z. Phys.* **78** 156; 1933 *Z. Phys.* **80** 587
[6] Livingston M S and Bethe H A 1937 Nuclear physics C. Nuclear dynamics, experimental *Rev. Mod. Phys.* **3** 245
[7] Gamow G 1928 *Z. Phys.* **51** 204; 1929 *Z. Phys.* **52** 510
[8] Condon E U and Gurney R W 1928 *Nature* **122** 439; 1929 *Phys. Rev.* **33** 127
[9] Cockcroft J D and Walton E T S 1930 *Proc. R. Soc.* A **129** 477
[10] Curie I and Joliot F 1934 *C. R. Acad. Sci., Paris* **198** 254, 561
[11] Gamow G 1931 Un nouveau type de radioactivité *The Constitution of Atomic Nuclei and Radioactivity* (Oxford: Oxford University Press)
[12] Stewer R 1985 Niels Bohr and nuclear physics *Niels Bohr a Centennial Volume* ed A P French and P J Kennedy (Cambridge, MA: Harvard University Press)
[13] Bainbridge K T 1960 Charged particle dynamics and optics, relative isotopic abundances of the elements, atomic masses *Experimental Nuclear Physics* vol 1, ed E Segrè (New York: Interscience)
[14] Rutherford E 1929 *Proc. R. Soc.* A **123** 373
[15] Majorana E 1933 Über die Kerntheorie *Z. Phys.* **82** 137
[16] von Weizsäcker C F 1935 *Z. Phys.* **96** 431
[17] Fermi E, Amaldi E, D'Agostino O, Rasetti F and Segrè E 1934 *Proc. R. Soc.* **146** 483

[18] Fermi L 1955 *Atoms in the Family* (London: Allen and Unwin)
[19] Segrè E 1955 Fermi and neutron physics *Rev. Mod. Phys.* **27** 257
[20] Bohr N 1936 Neutron capture and nuclear constitution *Nature* **137** 344
[21] Bethe H A 1937 Nuclear physics B. Nuclear dynamics, theoretical *Rev. Mod. Phys.* **9** 69
[22] Bohr N 1937 Transmutations of atomic nuclei *Science* **86** 161
[23] Breit G and Wigner E 1936 Capture of slow neutrons *Phys. Rev.* **49** 519
[24] Bohr N and Kalkar F 1937 On the transmutations of atomic nuclei by impact of material particles, I: general theoretical remarks *Mater.-Fys. Medd. Dan. Vidensk. Selsk.* **14** No. 10
[25] Hahn O and Strassmann F 1939 *Naturwissenshaften* **27** 11
[26] Meitner L and Frisch O R 1939 Disintegration of uranium by neutrons: a new type of nuclear reaction *Nature* **143** 239
[27] Frisch O R 1939 Physical evidence for the division of heavy nuclei under neutron bombardment *Nature* **143** 276
Bohr N 1939 Disintegration of heavy nuclei *Nature* **143** 330
[28] Bohr N and Wheeler J A 1939 The mechanism of nuclear fission *Phys. Rev.* **56** 426
[29] Smyth H D 1946 *Atomic Energy for Military Purposes* (Washington, DC: US Government Printing Office)
[30] Hewlett R G and Anderson O F Jr 1962 *The New World: History of the United States Atomic Energy Commission* (University Park, PA: Penn State University Press)
[31] Gowing M 1964 *Britain and Atomic Energy 1939–45* (New York: Macmillan)
[32] Gowing M 1985 *Niels Bohr, a Centenary Volume* ed A P French and P J Kennedy (Cambridge, MA: Harvard University Press) p 266
[33] Goldsmith H H, Ibser H W and Feld B T 1947 Neutron cross sections of the elements *Rev. Mod. Phys.* **19** 259
[34] Siegel J M (ed) 1946 Plutonium project report *Rev. Mod. Phys.* **18** 513; 1946 *J. Am. Chem. Soc.* **68** 2411
[35] Rutherford E, Chadwick J and Ellis C D 1930 *Radiations from Radioactive Substances* (Cambridge: Cambridge University Press)
[36] Kinoshita S 1910 *Proc. R. Soc.* A **83** 432
[37] Regener E 1908 *Verh. Deutsch. Phys. Ges.* **19** 78, 351
[38] Geiger H 1908 *Proc. R. Soc.* A **81** 174
Geiger H and Marsden E 1909 *Proc. R. Soc.* A **82** 495
[39] Rutherford E and Geiger H 1908 *Proc. R. Soc.* A **81** 141–61
[40] Ward F A B, Wynn-Williams C E and Cave H M 1929 *Proc. R. Soc.* **125** 715
[41] Wilson C T R 1913 *Phil. Trans. R. Soc.* **193** 289
[42] Blackett P M S 1925 *Proc. R. Soc.* A **107** 349
[43] von Baeyer O and Hahn O 1910 *Phys. Z.* **11** 488
von Baeyer O, Hahn O and Meitner L 1911 *Phys. Z.* **12** 273, 378
[44] Rutherford E and Robinson H 1913 *Phil. Mag.* **26** 717
[45] Rutherford E and Robinson H 1914 *Phil. Mag.* **28** 557
[46] Rutherford E and Andrade E N 1914 *Phil. Mag.* **27** 854; 1914 *Phil. Mag.* **28** 263
[47] Livingston M S 1952 *Ann. Rev. Nucl. Sci.* **1** 157, 169
[48] Livingston M S and Blewett J P 1962 *Particle Accelerators* (New York: McGraw-Hill)
[49] Livingston M S 1959 *Phys. Today* **12** 18
McMillan E M 1959 *Phys. Today* **12** 24
[50] van de Graaff R J, Trump J G and Buechner W W 1946 *Rep. Prog. Phys.* **11** 1
[51] Rose P H 1987 The tandem accelerator: workhorse of nuclear physics

Discovering Alvarez; Selected Works of L W Alvarez ed W P Trower (Chicago, IL: University of Chicago Press)
[52] Panofsky W K H 1987 Building the proton linear accelerator *Discovering Alvarez; Selected Works of L W Alvarez* ed W P Trower (Chicago, IL: University of Chicago Press)
[53] Johnston L 1987 The war years *Discovering Alvarez; Selected Works of L W Alvarez* ed W P Trower (Chicago, IL: University of Chicago Press)
[54] Fry D W and Walkinshaw W 1948 Linear accelerators *Rep. Prog. Phys.* **xii** 102
[55] Siegbahn K (ed) 1955 *Beta- and Gamma-ray Spectroscopy* (Amsterdam: North-Holland)
[56] Du Mond J W M and Kirkpatrick H A 1930 *Rev. Sci. Instrum.* **1** 88
[57] Wilkinson D H 1950 *Ionisation Chambers and Counters* (Cambridge: Cambridge University Press)
[58] Hofstadter R 1948 *Phys. Rev.* **74** 100; 1949 *Phys. Rev.* **75** 796
[59] Miller G L, Gibson W M and Donovan P F 1962 *Ann. Rev. Nucl. Sci.* **12** 189
[60] Tavendale A J 1967 *Ann. Rev. Nucl. Sci.* **17** 73
[61] McKay K G 1949 *Phys. Rev.* **76** 1537
[62] Orman C, Fan H Y, Goldsmith G J and Lark-Horowitz K 1950 *Phys. Rev.* **78** 646
[63] Walter F J, Dabbs J W T and Roberts L D 1958 *Bull. Am. Phys. Soc.* **3** 181
[64] Bethe H A and Bacher R F 1936 Nuclear physics A. Stationary states of nuclei *Rev. Mod. Phys.* **8** 82
[65] Bartlett J A 1932 *Phys. Rev.* **41** 370; 1932 *Phys. Rev.* **42** 145
[66] Elsasser W M 1933 *J. Phys. Radium* **4** 549; 1934 *J. Phys. Radium* **5** 389
[67] Guggenheimer J 1934 *J. Phys. Radium* **5** 253, 475
[68] Goeppert Mayer M 1949 *Phys. Rev.* **75** 1969
[69] Haxel O, Jensen J H D and Suess H E 1949 *Phys. Rev.* **75** 1766
[70] Goeppert Mayer M 1973 *Nobel Lectures: Physics, 1963–72* (Amsterdam: Elsevier) p 20
[71] Jensen J H D 1973 *Nobel Lectures: Physics, 1963–72* (Amsterdam: Elsevier) p 40
[72] Kurath D 1990 *Nucl. Phys.* A **507** 1c
[73] Weidenmüller H A 1990 *Nucl. Phys.* A **507** 5c
[74] Goeppert Mayer M 1948 *Phys. Rev.* **74** 235
[75] Bohr N 1977 Comments on atomic and nuclear constitution *Niels Bohr: Collected Works* vol 9 (Amsterdam: North-Holland) p 523
[76] Feshbach H, Porter C and Weisskopf V F 1954 *Phys. Rev.* **96** 448
[77] Stapp H P, Ypsilantis T J and Metropolis N 1957 *Phys. Rev.* **105** 302
[78] Talmi I 1990 *Nucl. Phys.* A **507** 295c
[79] Baldwin G C and Klaiber G S 1947 *Phys. Rev.* **71** 3; 1948 *Phys. Rev.* **73** 1156
[80] Goldhaber M and Teller E 1948 *Phys. Rev.* **74** 1046
Migdal A 1944 *J. Phys. (Moscow)* **8** 331
[81] Wilkinson D H 1959 *Ann. Rev. Nucl. Sci.* **9** 1
[82] Brown G E and Bolsterli M 1959 *Phys. Rev. Lett.* **3** 472
[83] Thouless D J 1960 *Nucl. Phys.* **21** 225; 1961 *Nucl. Phys.* **22** 78
[84] Lewis M B and Bertrand F E 1972 *Nucl. Phys.* A **196** 337
Pitthan R and Walcher Th 1971 *Phys. Lett.* B **36** 563
Fukuda S and Torizuka Y 1972 *Phys. Rev. Lett.* **29** 1109
[85] Bertrand F 1981 *Nucl. Phys.* A **354** 129c
Bertsch G F 1981 *Nucl. Phys.* A **354** 157c
[86] Rainwater J 1950 *Phys. Rev.* **79** 432
[87] Rainwater J 1976 Background for the spheroidal nuclear model proposal

(Nobel Lecture) *Rev. Mod. Phys.* **48** 385
[88] Bohr A 1951 *Phys. Rev.* **81** 134
[89] Alder K and Winther A 1966 *Coulomb Excitation* (New York: Academic) (a collection of reprints with an introductory review)
[90] Goldhaber M and Hill R D 1952 *Rev. Mod. Phys.* **24** 179
[91] Bohr A and Mottelson B 1953 *Phys. Rev.* **89** 316
[92] Bohr A and Mottelson B 1953 *Phys. Rev.* **90** 717
[93] Bohr A and Mottelson B R 1953 *Dan. Mater. Fys. Medd.* **27** No. 16
[94] Asaro F and Perlman I 1953 *Phys. Rev.* **91** 763; 1954 *Phys. Rev.* **93** 1423. See also, 1953 *Phys. Rev.* **92** 694, 1495
[95] Huus T and Zupancic C 1953 *Dan. Mater. Fys. Medd.* **28** No. 1
[96] McLelland C L and Goodman C 1953 *Phys. Rev.* **91** 51
[97] Temmer G M and Heydenburg N P 1954 *Phys. Rev.* **94** 426
[98] McGowan F K and Stelson P H 1955 *Phys. Rev.* **99** 127
[99] Alder K, Bohr A, Huus T, Mottelson B and Winther A 1976 *Rev. Mod. Phys.* **28** 432
[100] Bohr A 1976 Rotational motion in nuclei (Nobel Lecture) *Rev. Mod. Phys.* **48** 365
[101] Mottelson B 1976 Elementary modes of excitation in the nucleus (Nobel Lecture) *Rev. Mod. Phys.* **48** 375
[102] Nilsson S G 1955 *Mater.-Fys. Medd. Dan. Vidensk. Selsk.* **29** No. 16
[103] Inglis D R 1954 *Phys. Rev.* **96** 1059
Thouless D J and Valatin J G 1962 *Nucl. Phys.* **31** 211
[104] Strutinsky V M 1966 *Yad. Fiz.* **3** 614; 1967 *Nucl. Phys.* A **95** 420
[105] Bjornholm S and Lynn J E 1980 *Rev. Mod. Phys.* **52** 725
[106] Arima A and Iachello F 1975 *Phys. Rev. Lett.* **35** 1069
Arima A, Otsuka T, Iachello F and Talmi I 1977 *Phys. Lett.* **66** B 205
[107] Weisskopf V F 1937 *Phys. Rev.* **52** 295
Weisskopf V F and Ewing D H 1940 *Phys. Rev.* **57** 472, 935
[108] Hauser W and Feshbach H 1952 *Phys. Rev.* **87** 366
[109] Kapur P L and Peierls R E 1938 *Proc. R. Soc.* A **166** 277
[110] Wigner E P 1946 *Phys. Rev.* **70** 606
Wigner E P and Eisenbud L 1947 *Phys. Rev.* **72** 29
[111] Lane A M and Thomas R G 1958 *Rev. Mod. Phys.* **30** 257
[112] Lawrence E O, Livingston M S and Lewis G N 1933 *Phys. Rev.* **44** 56
[113] Oppenheimer J R and Phillips M 1935 *Phys. Rev.* **47** 845; 1935 *Phys. Rev.* **48** 500
[114] Serber R 1947 *Phys. Rev.* **72** 1008
[115] Butler S T 1950 *Phys. Rev.* **80** 1095; 1951 *Proc. R. Soc.* A **208** 559
[116] Levinson C A and Banerjee M K 1957 *Ann. Phys., NY* **2** 471, 499; 1958 *Ann. Phys., NY* **3** 67
[117] McDaniels D K, Blair J S, Chen S Y and Farwell G W 1960 *Nucl. Phys.* **17** 614
[118] Blair J S 1959 *Phys. Rev.* **115** 928
[119] Feshbach H 1958 *Ann. Phys., NY* **5** 357; 1962 *Ann. Phys., NY* **19** 287; 1967 *Ann. Phys., NY* **43** 410
[120] Hofstadter R, Fechter H R and McIntyre J A 1953 *Phys. Rev.* **91** 422
[121] Hofstadter R 1956 *Rev. Mod. Phys.* **28** 214
[122] Hyde E K, Perlman I and Seaborg G T 1971 *Nuclear Properties of the Heavy Elements* (New York: Dover) ch 9 p 745
[123] Hyde E K, Perlman I and Seaborg G T 1971 *The Nuclear Properties of the Heavy Elements* vol II (New York: Dover) p 745
[124] Seaborg G T and Loveland W D 1984 *Treatise on Heavy Ion Science* vol 4, ed D A Bromley (New York: Plenum) p 254

[125] McMillan E and Abelson P H 1940 *Phys. Rev.* **57** 1186
[126] Kennedy J W, Seaborg G T, Segrè E and Wahl A C 1946 *Phys. Rev.* **69** 555 (The original report was written in May 1941 but only published after the World War II.)
[127] Fields P R *et al* 1956 Transplutonium elements in thermonuclear test debris *Phys. Rev.* **102** 180
[128] Barber R C *et al* 1992 Discovery of the transfermium elements *Prog. Part. Nucl. Phys.* **29** 453 (This is the report of a working group set up by the IUPAP and IUPAC to consider questions of priority in the discovery of elements with nuclear charge greater than 100.)
[129] Hofmann S, Nivor V, Hessberger E P, Armbruster P, Folger H, Munzenberg G, Schölt H J, Popeko A C, Yeremin A V, Andreyev A N, Saro S, Janik R and Leino M 1995 *Z. Phys.* A (January)
[130] Bethe H A 1973 *Nobel Lectures: Physics, 1963–72* (Amsterdam: Elsevier)
[131] Fowler W A 1983 Nobel Prize Lecture *Nobel Lectures in Physics 1981–90* ed G Ekspong (Singapore: World Scientific) p 172
[132] Rolfs C E and Rodney W S 1988 *Cauldrons of the Universe* (Chicago, IL: Chicago University Press)
[133] Clayton D D 1968 *Principles of Stellar Evolution and Nucleosynthesis* (New York: McGraw-Hill)
[134] Rutherford E 1929 *Nature* **123** 313
[135] Gamow G 1936 *Atomic Nuclei and Nuclear Transformations* (Oxford: Oxford University Press)
[136] Gamow G 1948 *Nature* **162** 680
  Alpher R and Herman R C 1950 *Rev. Mod. Phys.* **22** 153
[137] Staub H and Stephens W E 1939 *Phys. Rev.* **55** 131
[138] Hemmendinger A 1948 *Phys. Rev.* **73** 806; 1949 *Phys. Rev.* **74** 1267
[139] Hoyle F 1946 *Mon. Not. R. Astron. Soc.* **106** 343; *Astrophys. J. Suppl.* **1** 121
[140] Salpeter E E 1952 *Astrophys. J.* **115** 326
[141] Dunbar D N F, Pixley R E, Wenzel W A and Whaling W 1953 *Phys. Rev.* **92** 649
[142] Burbidge E M, Burbidge G R, Fowler W A and Hoyle F 1957 *Rev. Mod. Phys.* **29** 547
[143] Fowler W A 1958 *Astrophys. J.* **127** 551
  Cameron A G W 1958 *Ann. Rev. Nucl. Sci.* **8** 249
[144] Baade W, Burbidge G R, Hoyle F, Burbidge E M, Christy R F and Fowler W A 1956 Supernovae and californium 254 *Proc. Astron. Soc. Pacific* **68** 296
[145] Baade W 1943 *Astron. J.* **97** 119; 1945 *Astron. J.* **102** 309
[146] Hoyle F 1975 *Astronomy and Cosmology* (San Francisco, CA: Freeman) p 383

Chapter 16

UNITS, STANDARDS AND CONSTANTS

Arlie Bailey

16.1. Introduction

The central theme of this chapter is the application of new discoveries in physics to replace arbitrary material standards of measurement by physical constants. This has not been an accidental unplanned process. Metrology is a branch of science which has been consistently sponsored by government since standards of weight, length and capacity were issued by royal decree in Babylon about 2500 BC. It is generally accepted that maintenance and dissemination of standard weights and measures for use in trade ('legal metrology'), often in association with the coinage, is a responsibility which only government can carry†.

However, advanced technology is involved and in the eighteenth century scientific interest in accurate measurement began to develop. The Royal Society in Britain and the Academie des Sciences in France took part in experiments aimed at improving standards, including international comparisons. One conclusion drawn from this was that there would be advantage in having standards related to physical constants which could be reproduced wherever they were required. The length of the seconds pendulum was considered as a possibility but rejected because of its dependence on the value of gravity. When, in 1790, Talleyrand proposed to the French National Assembly the setting up of the metric system, the standards chosen were one ten-millionth of the quadrant of the Earth's meridian for the metre and the mass of one cubic decimetre of water at 4 °C for the kilogram. Much effort

---

† *In this chapter the word 'standard' is used exclusively in the sense of 'measurement standard' (French 'étalon'), not as a 'specification standard' (as in 'British standard'— French 'norme').*

went into determining these, and practical standards were made which became known as the 'Metre and Kilogram of the Archives'. Improved measurement techniques revealed that they were not exact realizations of the defined values so the definitions were abandoned: but the material standards remained in use.

In the nineteenth century, the growth of technology-based industry led to requirements for a wider range of standards and for national and international organizations to support them. In Britain this was largely undertaken by the British Association for the Advancement of Science, founded in 1841, which set up facilities for the testing of thermometers, clocks, lenses and other instruments, and took the initiative in defining electrical units and developing standards to meet the needs of the electricity supply industry.

In 1867 a number of scientists attending the Paris Exhibition discussed the needs of industry for internationally accepted standards. As a result, the French Government called a conference in 1875 at which twenty countries were represented and which led to the signing of the Convention of the Metre. This established the International Weights and Measures Organization including:

(i) The International Bureau of Weights and Measures (BIPM), the laboratory at Sèvres, near Paris, which establishes basic standards, carries out international comparisons, and makes determinations of fundamental physical constants, in cooperation with national laboratories [1].

(ii) The General Conference on Weights and Measures (CGPM), which oversees the administration and financing of BIPM and ratifies important decisions of CIPM†.

(iii) The International Committee of Weights and Measures (CIPM), the scientific committee which executes the decisions of CGPM, recommends changes in the international system of units and standards when needed, provides a forum for the coordination of programmes in national laboratories, and directly supervises the operation of BIPM.

(iv) A number of Consultative Committees and Working Groups which advise CIPM on specific topics and arrange international comparisons.

National laboratories with responsibility for the development and maintenance of standards were set up following the signing of the Convention of the Metre. The first of these, in 1887, was the Physikalisch-Technische Reichsanstalt (now Bundesanstalt, PTB) in Germany. This was followed by the National Physical Laboratory (NPL) in Britain [2],

---

† *Records of the meetings of the General Conference of Weights and Measures (CGPM), the International Committee of Weights and Measures (CIPM) and of its Consultative Committees and Working Groups are published by Le Bureau International des Poids et Mesures, Pavillon de Breteuil F-92313, Sèvres Cedex, France.*

the National Bureau of Standards (NBS—now the National Institute of Standards and Technology, NIST) in the USA [3], and a number of similar laboratories in other countries [4,5]. They have played a crucial part in exploring new physical phenomena which can be used in the realization of standards and the establishment of systems of units.

The metric system slowly found acceptance not only in France but in other countries in Europe and Latin America where it was in use by the mid-nineteenth century: for day-to-day purposes it is of course still by no means universally welcomed in English-speaking countries, but a major factor in its use for scientific purposes was the need for consistent measurement of electrical quantities brought about by the growth of the electric telegraph and supply industries. In 1851 W E Weber proposed a coherent system of units, based on the centimetre, gramme and second as base units, and taking two forms for electrostatic and electromagnetic quantities, based respectively on the inverse-square laws of force between electric charges or magnetic poles. Maxwell showed that the ratio of these units was determined by the speed of light; the first example of the linking of units to a fundamental physical constant. The British Association committee concerned with electrical standards recommended the adoption of the CGS units in 1863.

However, with the expansion of the supply industry it became clear that the magnitudes of the units were not convenient for practical use and in 1881 there was international agreement to use practical units, the volt, the ohm and the ampere, which were respectively 10^8, 10^9 and 10^{-1} times their CGS magnetic equivalents. They were still not easy to realize in terms of their definitions and in 1908 the International Congress in London defined the International Units in terms of material standards— the International Ohm as the resistance of a specified column of mercury, the International Ampere as the current depositing silver from solution at a specified rate and the International Volt as their product or as a given fraction of the voltage of a Weston Cell. These remained in use until 1948 when improved methods of absolute determination allowed the CGPM to replace them with the present definition of the ampere.

Meanwhile, in 1902 G Giorgi made the radical proposal which led to the adoption of the MKSA system in which there are four base units—the metre, the kilogram, the second and the ampere. This gave units of a magnitude more convenient for practical purposes, and the adoption of the ampere removed the distinction between electromagnetic and electrostatic units. The system was coherent, in the sense that other units could all be derived from the base units without multipliers other than unity. The CGPM became increasingly involved in the definition of units as other base units, thermodynamic temperature, luminous intensity and amount of substance, were identified. After much discussion the Eleventh CGPM in 1960 adopted the International System of Units, SI, which is now the recognized basis for physical measurement [6]. In this chapter we first look briefly at how the base units (other than the mole, which is not a physical quantity in the same sense as the others) and certain derived units have evolved. We then consider the determination of fundamental physical constants and how accurate knowledge of their values has reacted back on the process of measurement itself.

16.2. Units and standards

16.2.1. Mass

16.2.1.1. The kilogram. The kilogram is unique. It is the only base unit whose definition has not changed in the past hundred years and the only unit still defined in terms of a material standard. In 1878, following the signing of the Metre Convention, three prototype kilograms were made in the form of cylinders of an alloy of 90% platinum, 10% iridium [1]. They were compared with the kilogram of the archives at the Paris Observatory: the one closest in mass was chosen as the International Kilogram and has been held as such by BIPM ever since (figure 16.1). It was formally adopted by the first CGPM in 1889. A further forty cylinders were ordered in 1882: their masses were adjusted to bring them within 1 mg of the mass of the International Kilogram and 34 of them were allocated for use as national standards by the member states of the Metre Convention. Others have since been made. They are kept in specially protected conditions and used only rarely. There have been two formal international comparisons of the standards, reported to the CGPM in 1913 and 1954.

During the past century a number of special balances for the comparison of masses have been developed, mainly in national laboratories. Key points have been finding ways of avoiding changing loads on the knife-edges, keeping the system thermally stable, and eliminating ground tremors. The best system is a single-arm balance developed by the US National Bureau of Standards [1] in which reference standards are compared in turn with a counterweight. One of these is now in use by BIPM. By this means, standard kilograms can be compared with a precision of about 1 μg, or 1 part in 10^9.

Weighing is, of course, an important operation in trade as well as in physics and secondary standards are maintained nationally for this purpose. In Britain, for example, the National Weights and Measures Laboratory maintains sets of standards for metric, avoirdupois and troy weights, all traceable to the NPL kilogram.

16.2.1.2. Possible alternatives. Much effort has been put into exploring ways in which the kilogram might be expressed in terms of fundamental physical constants rather than as the mass of a material standard. This might be done through the electrical units: the ampere is defined in terms

Units, Standards and Constants

Figure 16.1. *The kilogram. The kilogram is now the only physical unit defined in terms of a material standard—this platinum–iridium cylinder was constructed in 1878 and is held by BIPM.*

of the force between a pair of current-carrying conductors and is thus related to the kilogram. If the process were reversed and the ampere were defined in terms of fundamental constants, the kilogram could be derived.

Determination of the values of the fundamental constants is discussed in section 16.3. Possibilities which have been explored include the use of the gyromagnetic ratio of the proton [7,8], the Faraday constant and Avogadro's number [9]. At present these all have accuracies in the region of 1 in 10^6, which is about three orders of magnitude short of what is needed to replace the material standard.

See also p 1260

The most promising method now appears to be that devised by

Figure 16.2. *The NPL moving-coil experiment—an ongoing attempt to replace the definition of the kilogram in terms of a material standard by relating its value to electrical phenomena.*

Kibble at NPL in which a coil suspended from a balance is placed partly in the field of a permanent magnet (figure 16.2). Two measurements are made: first, a known current flowing through the stationary coil is balanced against a known mass and, second, the coil is moved through the field with known velocity and the induced voltage measured. From these the mass can be related to the electrical quantities, without direct involvement of the fundamental constants and, since the same equipment is used, a number of uncertain factors cancel out [10, 11], but there is still much work to be done before this method is fully evaluated.

16.2.2. Length

16.2.2.1. The International Metre. The metre started in the same way as the kilogram, defined in terms of a material standard, but its subsequent history has been different: after seventy years it was redefined in terms of a physical quantity, wavelength, and subsequently in terms of a fundamental constant, the speed of light.

Using the same platinum–iridium alloy as for the kilogram, a number of prototypes of the metre were constructed and one of these was selected as the international standard metre and ratified by the CGPM at its first meeting in 1889. However, the metre was not formally defined as the length of the international prototype until the Seventh CGPM in 1927. A number of copies were made and distributed to national laboratories and BIPM as reference standards. An international comparison to verify the values of the national metres started in 1921 and continued for some fifteen years, the results being published in 1940.

The metre prototypes consisted of a Pt–Ir bar of x-shaped cross-section of overall length about 120 cm. On the flat central strip three transverse lines were engraved at each end, the distance between the two middle lines defining the metre [1]. The conditions of use had to be closely specified and the 1927 definition in full stated [6]

> The unit of length is the metre, defined by the distance, at 0°, between the axes of the two central lines marked on the bar of platinum–iridium kept at the BIPM and declared Prototype of the metre by the 1st CGPM, this bar being subject to standard atmospheric pressure and supported on two cylinders of at least one centimetre diameter, symmetrically placed in the same horizontal plane at a distance of 571 mm from each other.

BIPM was involved, from its early days, in the calibration of scales for the measurement of lengths other than one metre and of surveying tapes. A 'geodetic baseline' was installed for the latter purpose, using a number of microscopes and travelling scales and capable of calibrating a 24 m wire with an uncertainty of about 10 μm. A number of techniques for accurately comparing the lengths of metre standards were developed in the early part of this century: they are described in some detail by Johnson [12]. Using the Brunner comparator, BIPM could compare two standards with a precision of 0.1 μm [1].

16.2.2.2. Wavelength standards. The idea that the wavelength of light could be used as a measure of length is perhaps not quite as old as the wave theory itself but certainly dates back to the early years of the nineteenth century. In 1859 Maxwell suggested that the sodium yellow line could be used as a standard. BIPM used interferometers to monitor length standards and in 1893 Michelson provided one which he used at

BIPM to measure the wavelength of the cadmium red line, which later (1927) became the length standard for spectroscopy and the basis for the angstrom. In the period up to 1940 some nine determinations of the metre in terms of wavelengths were made in various laboratories. In the process much was learned about the detailed characteristics of various spectral lines and what features had to be taken into account for the most accurate measurements. Doppler effect, hyperfine structure and separation of isotopes were just some of the problems which had to be tackled. After World War II the idea of a wavelength standard was given much consideration by CIPM and its Consultative Committee for the Definition of the Metre. About 1950 the choice of a suitable isotope was narrowed down to cadmium-114, mercury-198 and krypton-84 or -86. BIPM examined in detail the profiles of lines from these and ways in which they could be generated. In the end, CCDM recommended use of the krypton-86 orange line, produced by a cold-cathode discharge lamp cooled to liquid nitrogen temperature [13].

16.2.2.3. The speed of light. Section 16.3.2 describes how the measurement of the value of the speed of light developed until it became one of the most accurately known physical constants. Then in 1983 the Seventeenth CGPM redefined the metre in the following terms

> The metre is the length of the path travelled by light in vacuum during a time interval of 1/299 792 458 of a second.

Since laser interferometry had played a large part in leading up to this decision, most of the facilities needed to relate reference standards to the definition were already in existence and the change enabled improved precision to be achieved with relatively minor alterations in procedures.

16.2.2.4. Applications. There are, of course, innumerable applications where accurate measurement of length is needed. For trade purposes line standards are widely used. In engineering, gauge blocks and similar instruments have to be calibrated. For surveying there are tapes. All these and other requirements have to be met by national services in legal metrology with measurements traceable through the national standards laboratory.

In the early part of the century, optical comparators were in general use [12]. With the change to wavelength standards, interferometric methods arrived and have been adapted to use lasers. In some cases they have influenced the requirements. At BIPM a technique for standardizing 24 m baselines for tape measurement was developed using Fabry–Perot étalons [14], and distance-measuring instruments for surveying have been developed: the first of these was the mekometer [15] which uses light polarization—modulated at typically 500 MHz. By comparing the phases of the transmitted and reflected waves, accuracies of the

Samuel Wesley Stratton

(American, 1861–1931)

Dr Stratton was born in Litchfield, Illinois, the son of a farmer. He became interested in farm machinery and other mechanical devices and went to the University of Illinois where he read mechanical engineering, followed by research. In 1892 he moved to the University of Chicago at the invitation of Albert Michelson, with whom he did research on instrumentation. He also became interested in defence matters and served as a US Navy Lieutenant during the Spanish–American war of 1898.

He was appointed first Director of the National Bureau of Standards when it was founded in 1901. He was responsible for finding a site for the Bureau, for planning the laboratories and for deciding its programme of work. This required constant interaction with Congress and he acquired a reputation as 'a scientific politician'. World War I made great demands on NBS and Stratton's defence background was invaluable in deciding the projects to be undertaken, including aeronautical research and work for the Navy.

After the war, Stratton used his contacts with industry to redirect the Bureau's programme along lines of more commercial interest: this brought him into contact with the Secretary of Commerce, Herbert Hoover, who, in 1923, arranged for Stratton to become President of the Massachusetts Institute of Technology. Again he took a very active interest in steering the programme of work, while maintaining his interest in NBS as a member of the Visiting Committee. In October 1931 he died from a heart attack while dictating an article on Thomas Edison. His broad background in science and engineering, his sound judgement of national needs for research and his political sense made him the ideal first Director of NBS.

order of a millimetre per kilometre are possible. This, and its further developments such as the Georan, which uses two wavelengths, have found applications in surveying, civil engineering and geodesy [16].

For scientific purposes there has been a continuing demand for spectral lines of accurately known wavelength. In 1965 the International Astronomical Union recognized the wavelength of the Kr-86 line used to define the metre and four other wavelengths each from Kr-86, Hg-198 and Cd-114, covering the wavelength range from 435 to 645 nm with uncertainties from 2 to 7 parts in 10^8 [17]. With the development of lasers other more accurate sources have become available. Following the redefinition of the metre in 1983, the CIPM recommended five stabilized laser standards [18], and recognized standards are now available throughout the infrared region.

For specialized purposes, other units have been used. The Astronomical Unit (AU—approximately the radius of the Earth's orbit round the Sun) is the distance at which the Gaussian gravitational constant, k, has the value 0.017 202 098 95. The parsec (from a *parallax* of one *sec*ond of arc) is the distance at which 1 AU subtends an angle of 1 second: it is 3.086×10^{16} m. The light-year (9.46×10^{15} m) is perhaps more easily understood and is used in popular publications on astronomy.

At the other end of the scale we have the angstrom used in spectroscopy and named after the Swedish physicist, A J Ångström who, in 1868, published a diagram of the 'normal solar spectrum' in which the wavelengths were given in ten-millionths of a millimetre. It was defined in terms of the cadmium red line whose wavelength in normal air was 6438.4696 Å. After the change of the definition of the metre to that using the speed of light, the angstrom was redefined as exactly 10^{-10} m. The former 'X unit' or 'Siegbahn unit', about 10^{-13} m, which was formerly used to express the wavelength of x-rays, was abandoned in 1948 in favour of the angstrom.

CIPM does not, of course, encourage the use of units outside the SI system. The X-unit is included in the list of 'units generally deprecated' and the angstrom as 'in use temporarily', though the use of the nanometre (10 Å) is preferred.

16.2.3. Time and frequency

16.2.3.1. Background. The measurement of time dates from prehistory [19]. The earliest clocks were forms of sundials and water clocks. Oscillating systems, linking time to frequency, were not introduced until the middle ages. The earliest of these was the foliot, from about 1300, in which a pivoted bar moved to and fro to allow a toothed wheel to rotate: this was not a sinusoidal oscillator, but perhaps more in the nature of a multivibrator. The pendulum clock, a true oscillator, was invented about

1660 and the spring-controlled balance wheel twenty years later. These have remained the principles of mechanical clocks.

Until the 1960s time scales were considered to be almost exclusively the province of astronomers, although some pragmatic steps were taken to meet everyday needs. Local time was in general use until the mid-nineteenth century, so the time in Bristol, for example, lagged that of London by ten minutes. The coming of the railways led to a common 'railway time' and in 1880 'Greenwich Time' became the legal standard for Britain. At an international conference in Washington in 1884 it was agreed that the world should be split into a number of time zones based on Greenwich time: these have survived with little change to the present day.

The growth of industry in the nineteenth century led to a demand for checking the accuracy of clocks and watches which astronomers were not equipped to meet. In Britain, the British Association took over the disused Kew Observatory and set up facilities for testing these and other instruments. Responsibility for this passed to NPL on its founding in 1900. It is perhaps interesting to note that NPL tested 200 watches for the London Olympic Games in 1948.

16.2.3.2. Time keepers and frequency standards. The first pendulum clocks had a variation of about ten seconds a day in their time-keeping. More accurate operation was needed, particularly for the purposes of navigation on long sea voyages. Sources of error included temperature changes with their effect on the length of the pendulum, the effect of barometric pressure, and the need to couple mechanically or electrically to the pendulum to keep it swinging and to drive the rest of the mechanism. By 1900 much work had been done on these and stabilities of about 0.01 second per day had been achieved. The ultimate development of the pendulum clock was the Shortt clock which reached its final form in 1924 (figure 16.3). It used two pendulums: one was free-swinging except for a minimum period every half-minute when a driving impulse was applied and a synchronizing signal provided for the second, 'slave' pendulum which drove the escapement. This made possible a variation of only about 0.001 second per day. One of these clocks was used at the National Physical Laboratory to provide time signals interpolating between signals from astronomical observatories: this was used for accurate time interval measurement purposes, as in the checking of chronometers.

Another form of clock based on an accurate frequency was developed at NPL using a tuning fork [20], starting in 1922 with an improved version in 1931. It used an elinvar fork between the coils of an electromagnet driven by a thermionic valve. In an airtight enclosure and at constant temperature, it had a stability of about 5 in 10^8 over one

Figure 16.3. *The Shortt clock—the ultimate pendulum clock with master and slave pendulums used by NPL in the 1920s for time interval measurements.*

hour and 3 in 10^7 over a week. It remained in use until superseded by the quartz clock.

The use of the piezoelectric properties of quartz to control accurately the frequency of an oscillatory circuit is well known. Early work was done by Cady in 1921 and the first application to time measurement was made by Marrison, of the Bell Telephone Laboratories, in 1927. Work at NPL led to the development of the 'Essen Ring' which became widely used as a very stable standard of time and frequency. Stabilities of 4 in 10^{10} over one hour and 1 in 10^8 over a month were achieved [21].

Even though they have been superseded by atomic clocks for the most accurate purposes, quartz oscillators are still in widespread use for a great variety of applications, scientific, industrial, and even for cheap wrist watches which are stable to perhaps 10 seconds in a year. The properties of quartz change slightly with age. It has been proposed that this might be overcome by using an external reference. In 1949 H Lyons at NBS developed a system of this kind using an absorption line in ammonia as the reference; a long-term frequency stability of about 1 in 10^8 was demonstrated [22]. This was, in effect, the first atomic clock (figure 16.4). Since then many other possible systems have been investigated [23–26].

In the 1920s and 1930s a number of experiments showed that state transitions could occur in beams of atoms passing through magnetic and

Units, Standards and Constants

Figure 16.4. *The first atomic clock—constructed at NBS in 1949, this used a microwave absorption line in the ammonia molecule to control the frequency of an oscillator to about 1 part in 20 million.*

radio-frequency fields [23]. In 1938 Rabi and his colleagues demonstrated a resonance curve for the transition between two states in the lithium nucleus [27], and in 1950 Ramsey showed that a much sharper resonance could be obtained by using two separate regions of RF field with a drift space between them [28]. This principle was used in the first operational caesium clock developed at NPL by Essen and his colleagues and brought into use in 1955 [29]. In the caesium clock, the ground state is split into two components with an energy difference corresponding to a microwave

Figure 16.5. *The NPL caesium clock of the 1970s.*

frequency of about 9192 MHz (figure 16.5). In the Essen version, the beam of atoms from an oven passed through a magnet which separated the two states. The atoms in one state then passed through a microwave cavity about 50 cm long with a collimating slit in the middle and a longitudinal magnetic field, and finally through a second cavity, coupled to the first and a further magnet designed to deflect any atoms whose state had changed on to a detector. By varying the frequency of the applied microwave field the resonance pattern could be explored: the central peak was found to have a width in the region of 500 Hz. By designing a circuit to lock on to this, a stable source was obtained with a standard deviation of frequency of 1 in 10^{10}.

There have been many further developments in caesium clocks and, as well as being used in national laboratories, they are widely available commercially, some in portable form. The best standards now have accuracies of about 1 in 10^{14} [30–32]. Particularly interesting work has been carried out at NRC in Canada and PTB in Germany [33, 34].

Many other atomic systems have also been studied [26]. These include rubidium standards, which are commercially available, thallium and hydrogen beam systems and a variety of masers and lasers. The hydrogen maser has very good short-term stability and is in use in many laboratories, often supplementing caesium standards.

Units, Standards and Constants

16.2.3.3. Time scales and the second. At the beginning of this century time scales were based entirely on astronomical observations and the second was defined as 1/86 400 of the mean solar day. Because of the movement of the Earth in its orbit round the Sun, the length of the true solar day, the interval between successive crossings of the meridian by the Sun, varies during the year by some seconds, and the time scale based on it differs from a scale based on the mean solar day, averaged over the year, by a maximum of sixteen minutes in November. The difference is defined by 'the equation of time' [35].

For astronomical purposes the position of the Sun is less important and sidereal time is used. This is based on the time at which stars cross the meridian. There are corrections for minor changes in the Earth's axis of rotation: as measurement accuracy improved it was observed that there can be unpredictable changes in the Earth's rate of rotation caused by structural changes which affect its moment of inertia.

A variety of other time scales have been used for specific purposes. In 1952 the International Astronomical Union (IAU) introduced Ephemeris Time for dynamical calculations. Effectively it is based on the length of the tropical year 1900 and its disadvantage for general use is that it is accurately known only years after the event because of the need to process the observations [36–38]. Other scales, following the introduction of the atomic clock, are mentioned below.

As the accuracy of time measurement improved, first with quartz clocks and later with caesium clocks, it became clear that the definition of the second in terms of the mean solar year was no longer satisfactory. In 1956 the CIPM set up its Consultative Committee on the Definition of the Second (CCDS) to study the problem, and after much discussion the CGPM, in 1967-8, adopted the following definition

> The second is the duration of 9 192 631 770 periods of the radiation corresponding to the transition between the two hyperfine levels of the ground state of the caesium 133 atom.

This remains the definition of the second.

From 1955 the Bureau International de l'Heure (BIH) maintained a mean atomic scale known as AM by comparing clocks in different countries by means of low-frequency radio signals. In 1970 the name was changed to International Atomic Time (TAI). By then mean sidereal time had been improved by correcting certain irregularities due to Earth motion and was known as Universal Time (UT). To maintain continuity it was arranged that TAI and UT would coincide on 1 January 1958. For everyday use it is necessary to take account of the Earth's rotation and from 1 January 1972 Coordinated Universal Time (UTC) was established with the second markers coinciding with those of TAI but 'leap seconds' added or subtracted from time to time to keep the hours and minutes aligned with the solar day. This is the scale now generally used for such purposes as broadcast time signals.

BIPM had of course long been concerned with the realization of the second as a unit but after the introduction of TAI it became increasingly concerned also with time scales. CCDS was deeply involved in advising BIH and in 1971 the Fourteenth CGPM endorsed TAI and gave a formal definition of it. The Fifteenth CGPM in 1975 endorsed the use of UTC. In 1987 responsibility for time scales was transferred to BIPM and BIH ceased to exist. The other part of its programme, concerned with Earth rotation, was retained at the Paris Observatory under the new title of the International Earth Rotation Service (IERS). BIPM is continuing to maintain and disseminate TAI [39] and has now set up its own timekeeping unit using caesium clocks.

16.2.3.4. The dissemination of time and frequency. A very important role of observatories has been to make time scales available to those who need to use them. New communication techniques have always had the transmission of time signals as an early task. Just before 1800 the British Admiralty set up a chain of semaphore stations to provide a link from London to the dockyard at Portsmouth and a daily time signal was transmitted (one wonders how they managed on foggy days). When the electric telegraph was developed it was widely used to transmit time signals. Radio was also used at an early stage. In 1908 the Bureau des Longitudes in France proposed that the radiotelegraph station which had been set up on the Eiffel Tower for military purposes should be used to transmit time hourly. At the 1912 conference which led to the setting up of the BIH a programme of time and weather signals from radio stations in various parts of the world was started. This enabled ships at sea to check their chronometers. The stations chosen all operated on the long wavelength of 2500 metres, thus ensuring reception over long distances [40].

These services were primarily for military or technical purposes. From the very early days of broadcasting, time signals were transmitted for reception by the general public. In Britain, the Royal Greenwich Observatory provided the timing for the BBC's 'six pips', referred to its quartz clocks after 1942; and a number of broadcast transmitters in different parts of the world have accurately controlled carrier frequencies. These include the BBC 198 kHz transmitter at Droitwich whose frequency is controlled to within 2 in 10^{12}.

Since the 1920s a number of countries have provided special radio transmissions for accurate time and frequency signals, on a range of frequencies from VLF to HF. But as alternative methods have evolved many of these have been terminated. One development which depends on accurate timing and has provided useful signals has been the Loran-C navigation system. This uses a number of stations in different locations transmitting accurately synchronized pulses on 100 kHz. By receiving these pulses and comparing their times of arrival, a receiver can determine its position with an error of 50–500 m, to a distance of 2000 km or more. At a known position, the signals provide time comparisons to about 1 μs, equivalent to an uncertainty in frequency of about 1 in 10^{12}. This method has been used to combine national scales to form TAI.

To a large extent radio methods have been replaced by satellites as a means of time comparison. Travelling clocks have been used by the US Naval Observatory [41] and it is much involved in GPS, the US Navy's Global Positioning Satellite System, in which each satellite carries atomic clocks and from which time can be derived to about 10 ns. There has also been recent work at NBS [42].

16.2.4. Temperature

16.2.4.1. Background. The first thermometers were invented about 1600 and many different types have followed [43]. These were of practical use but the scientific significance of temperature did not become clear until the nineteenth century. In the words of the Ruhemanns 'Not until Lord Kelvin showed that the second law of thermodynamics supplied us with an absolute scale of temperature was a thermometer anything more than an arbitrary definition of a zero point and a degree' [44]. Kelvin showed that the Carnot cycle could be used to define a thermodynamic temperature scale independent of the material used, and it was later shown that a thermometer using a 'perfect' gas at zero pressure would measure thermodynamic temperature. By the end of the nineteenth century the use of the thermodynamic scale and of the gas thermometer for its realization had been accepted. The present century has seen a wide variety of developments in techniques of temperature measurement and their applications [43]. We have space only to consider a few of the fundamental aspects here.

16.2.4.2. Practical temperature scales. From its foundation in 1875, BIPM was actively involved in thermometry, mainly for the practical reason that its standards of the kilogram and the metre needed to be kept at accurately known temperatures. The hydrogen thermometer was selected as giving the best approximation to the thermodynamic scale and in 1887 the CIPM adopted this for the standard, using as fixed points the temperatures of melting ice and boiling water in specified conditions.

With the development of national standards laboratories early in this century other requirements emerged, in particular the need for a greater temperature range. In 1911 it was proposed that the thermodynamic scale should be officially adopted with a number of additional fixed points to extend the range. The outbreak of World War I delayed matters but, after long discussion, the Seventh CGPM authorized the

International Temperature Scale for temperatures above −190 °C. While accepting the thermodynamic scale as formally defining temperature, it was agreed that a practical way of realizing it should be stated. Six fixed points were specified, from the boiling point of oxygen to the freezing point of gold, together with the instruments to be used (platinum resistance thermometer and Pt/PtRh thermocouple) with their interpolation formulae.

This scale, which became known as ITS-27, was widely welcomed and came into general use. In time the introduction of new techniques showed that some revision was desirable and in 1939 the CIPM set up a Consultative Committee on Thermometry (CCT) to take advice and consider what changes should be made [46]. Again, war intervened, but in 1948 the Ninth CGPM adopted the revised International Practical Temperature Scale, IPTS-48. It also decided to change the name of the unit from the 'degree Centigrade' to the 'degree Celsius'. The scale was very similar to the earlier one, with a few refinements. The differences between them were insignificant below about 600 °C but rose considerably at higher temperatures.

It was realized that for the thermodynamic scale it was anomalous to have a unit based on the interval between two fixed points: all that is needed is an absolute zero and one fixed point. After much discussion the Tenth CGPM decided in 1954

> to define the thermodynamic temperature scale by choosing the triple point of water as the fundamental fixed point, and assigning to it the temperature 273.16 degrees Kelvin, exactly.

This made the degree Kelvin (now simply the kelvin) as nearly equal to the degree Celsius as possible.

In 1968 the CIPM recommended a further revision of the scale and IPTS-68 was formally adopted by the Fifteenth CGPM in 1975 [47,48]. New fixed points were added to extend the range and the interpolation formulae were revised to align the practical scale with the thermodynamic scale as closely as new techniques had made possible.

In 1990 yet another scale was introduced [49], known as ITS-90. The lowest temperature on the scale was reduced to 0.65 K and the temperature assigned to the sixteen fixed points were adjusted again to match thermodynamic temperatures as closely as possible. The structure of the scale illustrates how recent developments in physics have influenced the measurement process:

(i) between 0.65 and 5.0 K, the temperature, T_{90}, is defined in terms of the law relating vapour pressure and temperature for helium-3 and helium-4;

(ii) between 3.0 K and the triple point of neon (25 K), it is defined in terms of a helium gas thermometer;

(iii) between the triple point of hydrogen (14 K) and the freezing point of silver (1235 K) it uses platinum resistance thermometers;

(iv) above the freezing point of silver it uses the Planck radiation law.

It will be observed that some of the defining ranges overlap: this allows consistency of measures over limited ranges without sharp discontinuities. The differences between T_{90} and T_{68} are within 0.02 K up to 200 °C and within 0.3 K to 1000 °C.

16.2.5. Photometry

Even more than thermometry, photometric measurements have application mainly for industrial and domestic purposes and have not made an important contribution to the science of physics. However, the major national standards laboratories have been involved in these measurements from their early days and BIPM more recently: the candela is the SI base unit of luminous intensity. So we will give a very brief account of the way photometry units and standards have developed.

The earliest official standard of illumination was probably the British Parliamentary Candle introduced by the Metropolitan Gas Referees in 1860 for testing gas lighting. Other countries developed different standards and there was little coordination between them, perhaps because what was being sought was not a standard simply of a physical quantity but of its subjective effect. In 1909 agreement was reached between NBS, NPL and LCIE (the French Laboratoire Centrale d'Electricité) to use a common unit based on carbon filament lamps. At a meeting of the Commission Internationale de l'Eclairage (CIE) in 1921, Belgium, Italy, Spain and Switzerland joined in the agreement and the unit was named the 'International Candle'.

Further developments led to CIPM establishing a Consultative Committee for Photometry (CCP) in 1937. They recommended the adoption of a unit, the 'new candle' based on the brightness of a blackbody radiator. In 1948 this was formally approved by the Ninth CGPM, with the name changed to 'candela'. The CGPM also approved the definition of the lumen and when the SI units were adopted in 1960, they included the candela as the base unit and the lumen and the lux as derived units [50–52].

As photometric techniques improved it became possible to relate measures of illumination to objective radiometric measurements and the disadvantages of the black-body source became apparent. In 1979 the Sixteenth CGPM decided that a complete change was needed and adopted an entirely new definition of the candela

> The candela is the luminous intensity, in a given direction, of a source that emits monochromatic radiation of frequency

540×10^{12} Hz and that has a radiant intensity in that direction of 1/683 watt per steradian.

The frequency of 540×10^{12} Hz, corresponding to a wavelength of 555 nm, was chosen as the peak of the sensitivity curve of the human eye [53].

16.2.6. Electrical units and standards

16.2.6.1. Background. The definitions of the electrical units and the availability of standards for their realization have been closely linked with the development of the science and its industrial applications. The phenomena of electromagnetism were first recognized in the sixteenth century by Gilbert, but it was not until Faraday's experiments were followed by Maxwell's equations in the nineteenth century that it really became a major branch of physics. The electric telegraph was introduced in the 1830s and the first supplies of power for electric lighting in the 1880s. In 1861 the British Association for the Advancement of Science set up an Electrical Standards Committee which played the leading part in defining a rational system of units and developing practical standards. It was received with some scepticism by the telegraph industry who felt that 'The gentlemen who constitute the Committee are but little connected with practical telegraphy ... and may be induced simply to recommend the adoption of Weber's absolute units, or some other units of a magnitude ill adapted to the peculiar and various requirements of the electric telegraph.' [54]. In fact the Committee showed great perception in distinguishing between the roles of units and standards and in building a useful practical system on a sound scientific base.

As there was at that time no coherent system for electrical quantities it gave this matter high priority. About ten years earlier Weber had proposed the centimetre–gram–second system and the Committee accepted this as the base while recognizing that multiples of the units to match practical needs would be desirable. It reported back to the Association in 1863 and the practical units were defined as the ohm, the ampere and the volt, being respectively 10^9, 10^{-1} and 10^8 times the CGS electromagnetic units.

The British Association also sponsored a large programme of work to construct reference standards and to make absolute determinations of their value. The first wire-wound resistors were constructed and their values determined in 1863 by Maxwell, Fleeming Jenkin and Balfour Stewart. The method used had been devised by Weber and used a coil rotating in the Earth's field which both caused a current to pass through the resistor and deflected a magnet. The ampere was determined in a similar manner using Weber's electrodynamometer. After some years, the main activity was transferred to the Cavendish Laboratory

Richard Tetley Glazebrook

(British, 1854–1935)

Sir Richard Glazebrook was born in Liverpool. After taking his degree in mathematics at Trinity College, Cambridge, he undertook research in optics at the Cavendish Laboratory, under Maxwell. In 1883 the British Association appointed him secretary of its Committee on Electrical Standards. He was responsible for maintaining the BA standard resistors and capacitors and for absolute determinations of the units.

In 1900, after a year as the Principal of University College, Liverpool, he became the first Director of the National Physical Laboratory. His experience, his clearsightedness and his strategic abilities enabled him to lead NPL successfully through its early years and to impress the Government with the vital part it had to play in developing new technologies, including aerodynamics, ship testing, roads and radio. About 1917 he negotiated with the Treasury new arrangements for the control of NPL as a government establishment in DSIR, the Department of Scientific and Industrial Research.

In 1919 he retired from NPL but continued to advise on its programme. He became Zaharoff Professor of Aviation at Imperial College. He was knighted in 1917 and received many other honours and awards. He was a major figure in setting the pattern of twentieth century research in applied physics in Britain and internationally.

---

in Cambridge and Dr Glazebrook, later Sir Richard and the first Director of NPL, took charge of it [55, 56].

In 1881, at a congress in Paris, the BA system of units was accepted internationally. It was agreed that a column of mercury should be adopted as a reference standard for the ohm, as had been proposed by Siemens, with specified dimensions. This was known as the 'legal ohm'.

Although BIPM was not officially involved in electrical standards at that time, Dr Benoit of the Bureau made a number of copies of the legal ohm for the French government. But, as the value differed slightly from the BA value, it was never recognized in Britain.

This was the first step towards setting up a practical system of standards which could be used in industry and would not be affected by the occasional changes in absolute values which continuing research was revealing. In 1893 an international conference in Chicago took this process further and it culminated in the London conference of 1908. It was here agreed that International Electrical Units would be defined in terms of material standards:

(i) the International Ohm as the resistance of a column of mercury at the temperature of melting ice, 14.4521 grams in mass, of constant cross-sectional area, and of a length of 106.300 centimetres;

(ii) the International Ampere as the current which, when passed through a solution of silver nitrate in water deposits silver at a rate of 0.001 118 00 of a gram per second;

(iii) the International Volt as the pressure to produce a current of 1 ampere through 1 ohm; subsequently an alternative definition was accepted on the basis that a Weston cell at a temperature of 20 °C has an EMF of 1.0183 International Volts.

The London Conference very effectively cleared the air. The International Units provided a satisfactory practical system for industry and other users for many years, while allowing research on absolute units to carry on unhindered.

16.2.6.2. The role of the national laboratories. Although BIPM maintained some facilities for electrical measurements, which in 1907 again provided standards for the French government, they were not formally involved in electrical standards until later. In 1921 the Sixth CGPM amended the Convention of the Metre to include electrical units. In 1927 the Seventh CGPM set up the Consultative Committee for Electricity (CCE): this met the following year and recommended that BIPM should play a full part in international electrical standards.

Up to that time, the national laboratories had been responsible for establishing and coordinating standards. Indeed, the demand for standards to meet the need of the rapidly growing electrical industry played a major part in the decisions to set up national laboratories. The first of these, the Physikalisch-Technische Reichsanstalt (PTR) in Germany, was founded in 1883 with strong support from Siemens and with Helmholtz as its first director.

In Britain the Electric Lighting Acts of 1882 and 1888 made the Board of Trade (originally set up in 1696 to look after the trade of the American colonies and later charged with the oversight of weights and measures) responsible for the regulation of the electrical supply industry. This included the provision of standards and the type approval of electricity meters. It gave legal endorsement to the International Units, provided national standards based on the BA standards and set up a laboratory for the verification of instruments. But it was recognized that it would not be in a position to develop new standards or take part in international activities, so when NPL had established its programme, the Board of Trade standards were handed over to it in 1920: for legal reasons they were maintained as separate standards until about 1950.

Similar developments have taken place in other countries. In the United States, Congress was reluctant to sanction the setting up of a national laboratory, though after Edison established his research laboratory at Menlo Park in 1876 and others followed (but electrical instruments for export had to be sent to PTR in Germany for calibration) the case became overwhelming and NBS was established in 1901. In some countries the work is done jointly by industry and government: in France, for example, LCIE the Laboratoire Centrale des Industries Electriques, is jointly supported.

16.2.6.3. Absolute determinations. While the International Units met the needs of industry they were less satisfactory from the scientific point of view and national laboratories continued to look for better ways of realizing them from their CGS definitions. In the early 1900s, PTR, NBS, NPL and LCIE undertook a coordinated programme using a variety of current balances to determine the ampere in terms of the force between current-carrying coils (figure 16.6). The ohm was measured by the Lorenz machine (in which a conducting disc rotates in the field from a coil of known dimensions) and by Campbell's inductor whose mutual inductance is calculable from its dimensions. The results of these measurements were used to calibrate mercury resistance standards and Weston cells which had travelled between the participating laboratories [55]. Continuing work in this area after World War I led to BIPM's involvement in electrical standards and eventually to the SI system (section 16.2.6.4.).

After World War II further developments took place; of these two of the most important originated in NML in Australia. In 1956 the Thomson–Lampard theorem was published [57, 58]. They showed that if the wall of a conducting cylinder of arbitrary cross-section is divided parallel to the axis into four parts with negligible gaps, the cross-capacitance between opposite pairs of faces is calculable and if they are made symmetrical the capacitance per unit length is given by

$$C = \varepsilon_0 (\ln 2)/\pi \text{ F m}^{-1}.$$
$$= 1.953\,549\,04 \text{ pF m}^{-1}.$$

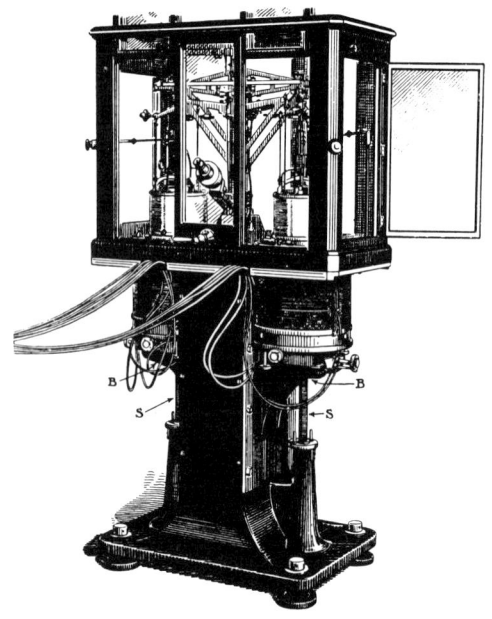

Figure 16.6. *The Ayrton–Jones current balance for the absolute determination of the ampere–initiated by the British Association and completed at NPL in 1907.*

Capacitors using this principle were constructed in NML, NBS, ETL and NPL and used by means of a quad-bridge relating capacitance and resistance to determine the absolute value of the ohm with an uncertainty in the region of 1 part in 10^7, rather better than an order of magnitude improvement on previous determinations [59].

An absolute determination of the volt can be made by measuring the force of attraction between two plane electrodes at different voltages. The force per unit area between two infinite planes with a voltage V between them and d metres apart in vacuum is given by

$$\varepsilon_0 V^2/(4\pi d) \text{ N m}^{-1}.$$

A number of measurements based on this were made using metal plates and a balance at NBS and ETL. In 1965 Clothier, of NML, proposed the use of a plate over a pool of mercury: the rise in the surface level of the mercury when a voltage was applied could be measured with great accuracy. This has achieved an uncertainty in the region of 3 in 10^7 [60, 61].

16.2.6.4. Later developments of electrical units. Following the recommendation of the CCE, BIPM in 1929 set up reference standards of the volt and the ohm and in 1932 the first international comparison took place when four national laboratories sent standards to BIPM. Further comparisons, with a growing number of participants, took place at intervals of two years until they were interrupted by World War II. They resumed later, but from 1957 only took place every three years.

The 1928 CCE concluded that the International Volt, Ohm and Ampere should not be perpetuated but that they should be replaced by an absolute system as soon as the relations between the two sets of units could be established with adequate accuracy. After further work, CIPM decided that the absolute system should come into use on 1 January 1948. The basis was the modified CGS system proposed by Giorgi in 1901 [62, 63]. In this, the MKSA system, the units are of a magnitude more suitable for practical measurements and the awkward 4π was removed from the definition of the ampere. The definitions ratified by the Ninth CGPM are of the ampere in terms of the force between two parallel conductors, the volt as the potential difference between two points of a conductor carrying one ampere and dissipating one watt, and the ohm as the resistance in which one volt produces one ampere [6]. Other units were also defined and they were all taken over into the SI system when that was formally approved by the Eleventh CGPM in 1960, with the ampere designated as the base unit for electrical quantities.

However, the definition of the ampere does not lend itself to easy practical realization, so intercomparisons with BIPM continued and it was accepted by convention that the values of the BIPM standards would be taken as the volt and the ohm. By 1970 some ten national laboratories were sending their standards to BIPM for intercomparison every three years: the spreads in value were about 1 in 10^6 for the ohm and two or three times this for the volt [64]. In the light of the results, national laboratories have sometimes adjusted the values of their standards and BIPM also made an adjustment in 1968.

With the growing workload for BIPM and the limitation of accuracy caused by having to transport reference standards, particularly Weston cells, it became clear that an alternative procedure was desirable. As the work described in section 16.3 of this chapter proceeded, it became clear that physical constants could potentially replace material standards. There have been many suggestions for ways in which electrical standards might be derived. At one time the most promising appeared to be through the gyromagnetic ratio of the proton [7]. In practice, however, this had not achieved any significant improvement and the two effects which are now used are the Josephson effect for the volt and the quantum Hall effect for the ohm.

In 1962 Josephson predicted the effect which now bears his name [65]. If two superconductors are linked by a thin insulating gap irradiated by a frequency f, an applied DC voltage will produce a current and the current–voltage characteristic has voltage steps given by

$$\Delta V = (h/2e)f$$

See also p 941

where h is Planck's constant and e the charge on the electron. This gave rise to a great deal of interest in metrological laboratories because it clearly offered a way of determining the volt in terms of frequency, accurately measurable, and fundamental constants. The difficulty was that our knowledge of the constants was not as precise as the precision with which voltage standards can be compared. But if we write the relation as

$$\Delta V = K_J f$$

where K_J, the Josephson constant, has a value to be agreed, this still, though not now an absolute method, provides a useful way of monitoring voltage standards and comparing them without the need to transport Weston cells. A number of national laboratories made measurements of the Josephson constant in terms of their volts and used it to monitor their standards. At its meeting in 1972, CCE took note of these developments and concluded that the BIPM volt on 1 January 1969, V_{69-BI}, was equal within half a part in a million to the voltage which would be produced by the Josephson effect in a junction irradiated at the frequency of 483 594.0 GHz (figure 16.7). Not all countries were prepared to adopt this value and it became clear that it was not in fact consistent with the SI value. In 1986 CCE set up a Working Group to review the evidence and after it reported back CCE recommended that the value

$$K_{J-90} = 483\,597.9 \text{ GHz V}^{-1}$$

should be adopted from 1 January 1990 and used where possible for monitoring national standards, with an uncertainty estimated to be 4 in 10^7. CCE made it clear that this represented a reference standard and did not imply any change in the SI definition of the units or in the values of the fundamental constants. It expressed the view that no change in the recommended value would be necessary in the foreseeable future.

One practical problem was the small voltage produced by a Josephson junction—about 10 mV for a frequency of 5 GHz, for example. This made it less than easy to monitor a 1 V standard, but work at NIST has since led to the development of arrays of up to 20 000 junctions in series, which gets round this problem [66].

Monitoring the ohm is basically similar. If a semiconductor carries a current I in a direction at right angles to a magnetic field, a voltage V_H is generated at right angles to both. This is known as the Hall effect and the ratio

$$R_H = V_H/I$$

is known as the Hall resistance. It normally depends on the geometry and the strength of the field, but in 1980 von Klitzing demonstrated that in certain semiconductors where the current flow is effectively limited to two dimensions the Hall resistance is quantized and has the value

$$R_H = R_K/i$$

Units, Standards and Constants

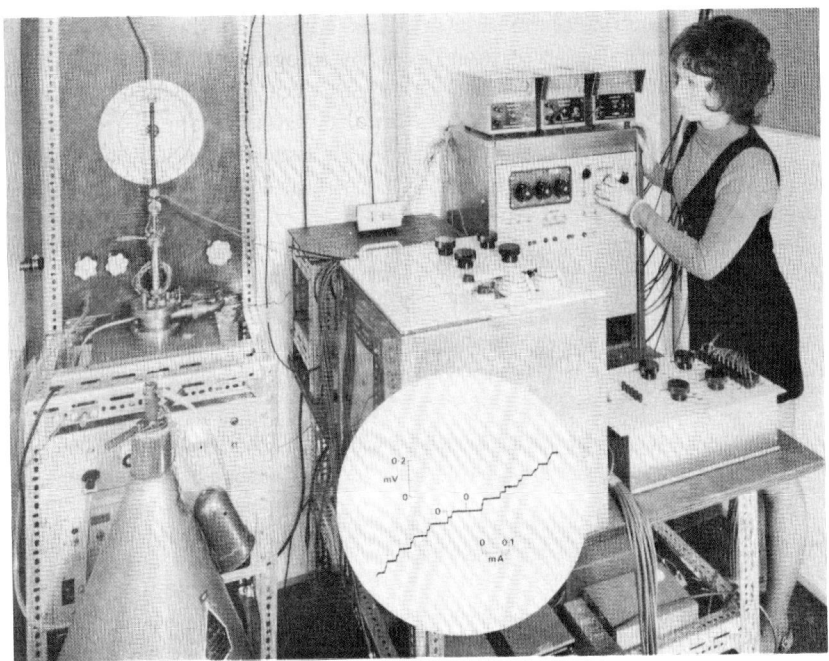

Figure 16.7. *The Josephson junction voltage standard at NPL—inset shows the current–voltage characteristic.*

where $i = 1, 2, 3, \ldots$ (depending on the field strength) and R_K is a constant known as the von Klitzing constant. Its theoretical value is h/e^2 [67–70]. Using this effect it is possible to set up a standard of resistance which can be used to monitor national ohms. As with Josephson, the equipment is not simple—it has to operate at liquid helium temperatures, for example—so that it can hardly be used as an everyday reference, but for comparing standards at the highest level without the need to transport them it has been very useful.

In 1986 the CCE set up another Working Group to consider the quantum Hall effect. After it reported in 1988, CCE recommended that the von Klitzing constant should be used from 1 January 1990 to monitor the values of national ohms and that its value should be

$$R_{K-90} = 25\,812.807 \text{ ohms}.$$

CCE again emphasized that this implied no change in the definition of the units or in the values of h and e but that it was simply a reference value to which no change was likely to be needed in the foreseeable future. The estimated uncertainty was 2 in 10^7.

The changes in connection with the volt and the ohm were announced by BIPM a year in advance of the date when they were to take place [71].

At the same time the new temperature scale, ITS-90, was announced. Advice was given in some detail on the new procedures to be used and on how results should be expressed. This involved the national laboratories in a considerable amount of work not only in changing their own procedures but also in making sure that all laboratories dependent on them for calibrations were aware of the situation. Almost all national volts and ohms were effectively changed in value by some parts in a million and this had to be taken into account in maintaining calibration histories.

In fact, the process seems to have gone reasonably smoothly and we look forward to a more tranquil period for electrical measurements. It is somewhat intriguing that a major shake-up of electrical measurements seems to take place every twenty years: in 1908 we adopted the International Units, in 1928 the first CCE met, in 1948 the 'absolute' units were adopted, in 1968 a major readjustment of the values of the national volts and ohms was decided and in 1988 we were committed to Josephson and von Klitzing. One wonders what 2008 will bring.

These changes have all been accompanied by reductions in uncertainty. At the time of writing we seem to be ahead of the user demand, but this is unlikely to remain so. The periodic review of the values of the fundamental constants will continue and there are still possible new techniques to be explored. A useful survey of the present situation and future trends has been given by Kibble [10].

The other intriguing point is that although the thrust has repeatedly been towards standards based on absolute values and physical constants, we are in fact ending the century as we began it with electrical quantities referred to standards chosen by an international committee.

16.3. Physical constants

16.3.1. Types of constant

We come now to an application of measurement techniques which is fundamental to physics and which reacts back on the process of measurement itself, as we have just seen in relation to the electrical units.

Physical theory depends on a large number of quantities which have values which we believe to be constant over the whole of time and the whole of the universe (although this belief has occasionally been questioned). It is convenient to divide them into several categories:

(a) Mathematical constants, such as e and π whose values are fixed and are calculable from formulae to as many places of decimals as may be needed.

(b) Defined constants, such as the speed of light in vacuum, $c = 299\,792\,458$ m s^{-1} exactly, and the permeability of free space

$$\mu_0 = 4\pi \times 10^{-7} = 1.256\,637\ldots \times 10^{-6} \text{ H m}^{-1}.$$

Units, Standards and Constants

Table 16.1. *Fundamental constants of physics (reference [59] p 2).*

Velocity of electromagnetic radiation in free space	c
Elementary charge	e
Mass of electron at rest	m_e
Mass of proton at rest	m_p
Avogadro constant	N_A
Planck constant	h
Universal gravitational constant	G
Boltzmann constant	k

From these may be derived the permittivity of free space

$$\varepsilon_0 = 1/\mu_0 c^2 = 8.854\,19\ldots \times 10^{-12}\ \text{F m}^{-1}$$

and the characteristic impedance of free space

$$Z_0 = (\mu_0/\varepsilon_0)^{1/2} = 376.730\ldots\,\Omega.$$

(c) Measured constants, which comprise the great majority of physical constants, including, for example, the gravitational constant, G, the Planck constant, h, the charge of the electron, e (not to be confused with $e = 2.718\ldots$ in (a) above), and many others. Their values are determined by measurement and many of them are closely interrelated. For this reason the most accurate values are derived not from individual measurements but by a least-squares treatment of all the available data. This is undertaken from time to time by CODATA, the Committee on Data for Science and Technology of the International Council of Scientific Unions (ICSU). The latest set of data on some hundred quantities was published in 1986 [72]: some changes have already been proposed [73].

These constants are often referred to, by CODATA for instance, as the fundamental physical constants. This can be somewhat confusing, as the links between them mean that the number of independent constants is very much less than the total number of those whose values are published. They are all perhaps fundamental in a sense in some branch of physics, but one may be forgiven for thinking that some are more fundamental than others. There is merit in the list of the fundamental constants of physics given by Petley (reference [59] p 2) and shown in table 16.1.

It is important to remember that constants may not be forever assigned to a particular class. Today π is for us a mathematical constant calculable from a formula to any accuracy we wish. This was not so in the third century BC when Archimedes 'measured the circle by inscribing and circumscribing polygons, increasing the number of sides till the polygons nearly met on the circle' [74]. In this way he arrived at a value between $3\frac{1}{7}$ and $3\frac{10}{71}$ (within three decimal places of the exact value). More recently the speed of light has changed from being a measured constant to a defined constant: after many determinations of its value, the CGPM in 1983 redefined the metre, thus fixing the speed of light (see section 16.2.2.3). Further measurements of its value would thus be meaningless. It is important to distinguish between this case and the apparently similar cases of the Josephson and von Klitzing constants for which CCE gave recommended values in 1988 for monitoring the volt and the ohm (section 16.2.6.4), but no change of definition of the units was involved, the constants remain measurable, and improved values may be recommended at some time in the future.

It is also worth noting that the values for the constants derived from coordinated experiments have not always been universally accepted. For example, Eddington derived a theoretical value of exactly 137 for the fine structure constant, α^{-1}, on the basis of the number of independent components of 'a complete energy tensor' [75]. Experimental values have been in the region of 137 but not exactly equal to it: the latest CODATA figure is 137.035 989 61 [72]. A more profound challenge is offered by cosmological theories which assume physical 'constants' in fact change over periods of time.

In this section we consider first the two 'classical' constants, c and G, whose values have been measured over several hundred years and are not closely linked to those of other constants. Then we discuss the ways in which the determination of linked groups of constants has become organized and finally we give examples of the coordinated measurements which have contributed to this. Many hundreds of such determinations have taken place in this century and clearly we do not have space to discuss all the experimental techniques in detail. We do, however, give some examples of the most significant developments and list references where further details can be found.

16.3.2. The speed of light
We have already mentioned how measurements of the speed of light eventually led to its becoming a defined quantity. These measurements have a long history.

In classical times it was believed that light travelled with an infinite velocity. Galileo doubted this and attempted to measure the velocity by covering and uncovering lanterns a couple of miles apart but of course this did not give a result. The first effective measurement was made by Roemer in 1676 by observing the time of the eclipses of the satellites of Jupiter from different points in the Earth's orbit. In 1849 Fizeau made the first successful measurement on Earth, using a rotating wheel as a timing shutter on a beam of light reflected from a mirror some miles distant. Foucault made similar measurements. But one of the most significant developments about this time was Maxwell's theory of electromagnetic waves, with his conclusion that the speed of light is given by

$$c = (\mu_0 \varepsilon_0)^{-1/2} = \nu\lambda$$

where ν is the frequency, λ the wavelength, and μ_0 and ε_0 are the permeability and permittivity of free space, as before. This opened the way to a variety of methods of measuring c which we will discuss later. Towards the end of the century there was much debate on the existence of the aether. It was expected that if the Earth was moving through the aether the speed of light would depend on the direction of transmission. The Michelson–Morley experiment of 1887 showed that this was not the case. This led to the idea of the Fitzgerald–Lorentz contraction and ultimately to Einstein's theory of relativity in 1905. So, even before the start of the present century, measurement had a profound effect on physical theory.

Throughout the twentieth century there has been a continuing interest in the value of the speed of light and many careful determinations of its value have been made, far too many for us to describe them all in detail here. A variety of methods have been used. In the first half of the century, Michelson made further measurements using the rotating mirror method [76], Mittelstaedt [77,78], Hüttel [79] and Anderson [80,81] used a basically similar method with a Kerr cell as the timing shutter, and Mercier [82] found the velocity of electromagnetic waves at about 50 MHz by observing standing waves on parallel transmission lines. But the most original method was that used by Rosa and Dorsey [83] in 1907 from the ratio of the electrostatic and electromagnetic electric units, which gave much better precision than any other method at that time. The results up to 1940 were reviewed by Birge [84]. He was somewhat critical of the error estimates in some of the earlier measurements and after detailed analysis of the results arrived at

$$c = 299\,776 \pm 4 \text{ km s}^{-1}$$

as the best available value.

The new technologies developed for defence purposes during World War II had profound influence on many aspects of metrology. Radio navigation results showed the need for better values of c and the opening up of new areas of the spectrum in the microwave and infrared regions made new measurements possible. Essen and his colleagues at NPL used a microwave cavity resonator to measure the wavelength of radiation of known frequency [85,86]. His colleague, Froome, made similar measurements at higher frequencies (24 and 72 GHz) using a microwave interferometer [87,88] (figure 16.8). Further Kerr cell measurements were made by a number of people, radar was used, and other geodetic

Twentieth Century Physics

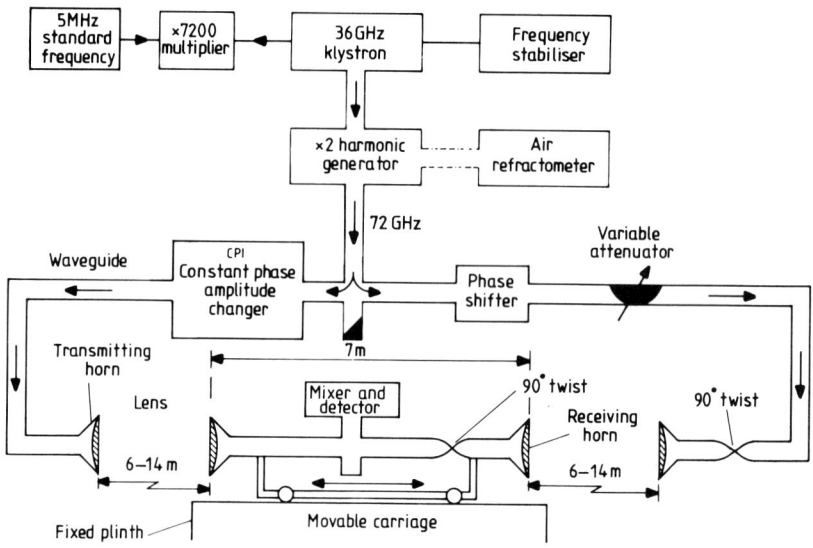

Figure 16.8. *The Froome interferometer for measurement of the speed of electromagnetic radiation at microwave frequencies.*

methods. Froome's results were the most accurate, with a quoted uncertainty of 0.1 km s^{-1} for his 1958 result of 299 792.5 km s^{-1}.

After 1970, the development of lasers with very high spectral purity made even better accuracies possible (figure 16.9). Frequencies were compared with the caesium standard using harmonic mixing and wavelengths were measured in terms of the krypton-86 length standard by means of intermediate calibrated lasers. A coordinated programme of work at NBS [89, 90], NPL [91, 92] and NRC [93] led to results with standard deviations in the region of 1 m s^{-1}. These measurements, and earlier results, are discussed in some detail by Petley [59]. They provide a clear example of how new technology can make possible a significant improvement in measurement accuracy—about two orders of magnitude in this case.

These measurements had a profound effect on the SI system of units, as we have already seen in section 16.2.2. In 1975, after a review of the results available, the CGPM recommended that the value $c = 299\,792\,458$ m s^{-1} should be used for the speed of light, with an uncertainty of $\pm 4 \times 10^{-9}$. Then in 1983 the Seventeenth CGPM used this value to redefine the metre as the length of the path travelled in vacuum in 1/299 792 458 of a second. It also invited the CIPM to draw up rules for the practical realization of the metre, and this was done in terms of lasers stabilized by certain specified molecular transitions. Thus the speed of light ceased to be a measurable constant. No further determinations of its value will be made.

Units, Standards and Constants

Figure 16.9. *The array of lasers used to measure the speed of light at NPL in the mid-1970s.*

16.3.3. The constant of gravitation

The other constant of which independent determinations have been made over several centuries is the gravitational constant, G. In the latter part of the seventeenth century, Newton developed his law of gravitation, starting reputedly from the relation between the fall of an apple and the motion of the planets and published in *Principia*. We can express the law as

$$F = Gm_1m_2r^{-2}$$

where F is the force between two masses m_1 and m_2 a distance r apart. Over the years much work has gone into determining the value of G and whether it really is a constant. Its value is not only of theoretical interest, but it provides a way of determining the mass, and hence the density, of the Earth.

Early determinations of G were made by measuring the deflection from the vertical of a plumbline near a large mountain. The first attempt to do this in 1740 at Mount Chimborazo in Peru by Bouguer was ruined by atrocious weather conditions, and although it was clear that the effect occurred, the results were not numerically significant. Better results were obtained by Maskelyne, then Astronomer Royal, in 1774 on Mount Schiehallion in Perthshire. Others measured G by comparing the values of the acceleration due to gravity at the surface of the Earth and the bottom of a deep mine, but it was clear that none of these methods could produce very accurate results and they were superseded by laboratory measurements of the attractive force between two masses. The first of these used a torsion balance; it was started by Michell, who died before the construction of the apparatus was complete. Cavendish took over the equipment and carried out the experiment, publishing his results in 1798. He obtained the value of 5.448 for the density of the Earth, corresponding to

$$G = 6.754 \times 10^{-11} \text{ m}^3 \text{ kg}^{-1} \text{ s}^{-2}.$$

Many other determinations followed, particularly in the latter half of last century, including the work of Boys, who used a torsion balance with an arm rotating about a vertical axis, similar to that of Cavendish, and of Poynting, who used a balance with an arm rotating in the vertical plane. Two large masses were attached to the ends of the arm, with a central knife-edge so arranged that gravity provided the restoring force. A very large mass was mounted on a turntable so that it could be positioned under either of the two masses. By moving it from one position to the other and observing the deflections, the forces, and hence G, could be calculated. Useful surveys of these measurements of G were given by Boys [94] and by Poynting and Thomson [95]. The former gives a table of the values achieved at the end of last century.

In the present century there have been many further determinations of G, all using some form of the torsion balance. Heyl [96] used an oscillation method and Zahradnicek [97] a resonance method. More recently there have been further results published by Karagioz [98] and Sagitov [99] in the USSR and Luther and Towler [100] in the USA. The results have been reviewed by Lowry et al [101] and by Petley (reference [59] pp 226–36). The CODATA recommended value is now

$$G = 6.67259(85) \times 10^{-11} \text{ m}^3 \text{ kg}^{-1} \text{ s}^{-2}$$

with an uncertainty of about 1 part in 10^4 [73].

Interest in the gravitational constant is not only in its absolute value. Gravitational theory is central to studies of the development of the universe. Newton's simple inverse-square law has in this century been called into question in a number of ways. It is a fascinating story but unfortunately we do not have the space to go into it here in detail. We will try to indicate some of the main directions which have been explored and the measurements which have affected them, and indicate where further details can be found. A major change was brought about by Einstein's theory of relativity which led to the law of gravitation based on tensors [102]. There have also been attempts to develop a quantum theory of gravitation [103].

There has been much interest in trying to determine whether G really is constant over vast distances and long periods of time, an

See also p 286

interest stimulated by the publication in 1937 by Dirac of his theory of multiplicative creation cosmology, in which G and other 'constants' are changing. On this theory the change in G might be in the region of 10^{-11} per year—quite out of range of direct measurements, of course. This led to many attempts to estimate limits to any variation in G, mainly using palaeontological evidence of changes in the Earth's temperature, changes in the orbits of the Earth and other planets, continental drift, clustering of galaxies, and also more recently using radar and lidar measurements of the distances of the Moon and planets, and other phenomena. A survey of the researches and a summary of the results are given by Petley (reference [59] pp 47–51). The recent results have ranged from an uncertainty of less than 10^{-12} per year to about 10^{-10}. Clearly, in relation to Dirac's theory, this does not prove that G is definitely constant, but it is fair to say that it provides no evidence to the contrary.

Another question has been whether gravitational force is a function solely of mass or whether it depends on the nature of the material involved. The classical experiments to investigate this were those of Eötvös [104]. Using a wide variety of substances, he found no evidence to suggest that the composition had any effect and subsequent work has confirmed this (reference [59], pp 241–6). There have also been checks of the inverse-square law and on the possible variation of G with other physical parameters, such as quantum state. So far no clear cut dependence has been found, but further experiments are continuing and seem likely to do so indefinitely. Some surveys of the present situation and recent results are given in reference [105].

16.3.4. Linked sets of constants

The two constants we have just discussed, c and G, are of significance in classical physics and their values have been determined mainly by direct measurement. The other constants that we have to consider are almost all involved in atomic or quantum theory and their values are generally related. It has therefore become the practice to determine the values not just of single constants but of linked groups from series of measurements.

The need for this began to be appreciated early in this century, but the first formal recognition was provided by the publication in the early 1920s of the International Critical Tables. These gave the values of nine 'Accepted Basic Constants' and of a number of other constants derived from them. Raymond Birge, of the University of California, Berkeley, was not entirely happy with the derivation of some of the published values and undertook an extensive critical evaluation of all the available evidence. The results were published in his classic paper of 1929 [106] 'Probable values of the general physical constants (as of January 1, 1929)'. He stressed the importance of using the least-squares method to calculate probable errors. In 1930 and 1931, W N Bond published papers [107, 108] which discussed Birge's results and built on them and other data to give results which he claimed to be '... definitely the most accurate so far obtained by any method'. These were criticized by Birge in his next paper in 1932 'Probable values of e, h, e/m and α' [109]. But Bond had made a very important contribution by using a graphical presentation, consistent with the least-squares method, to display the results of a series of related measurements and to assist in evaluating the optimum values. This was adopted by Birge in his 1932 paper and has become known as the 'Birge–Bond Diagram': an example, with a detailed explanation, is given by Petley (reference [59] p 296). In 1941, Birge published a further review of 'The general physical constants' [84], giving the latest values of some dozen principal constants and a number of derived constants. His principal constants included the speed of light, the gravitational constant, the litre (in cm^3), the volume of ideal gas, the International Ohm and Ampere, some atomic weights, the standard atmosphere, the ice point, the Joule equivalent, the Faraday constant, the Avogadro number, the electronic charge and the Planck constant. It is interesting to note that some of these, the litre and the International Ohm and Ampere, are no longer measurable constants.

In 1969 Taylor, Parker and Langenberg published their paper 'Determination of e/h, using macroscopic quantum phase coherence in superconductors: implications for quantum electrodynamics and the fundamental physical constants' [110]. In this they showed that the improved accuracy in the measurement of e/h as a result of using the Josephson effect made it necessary to recalculate the values of all the related constants. It became clear that revaluations were necessary from time to time. The International Union of Pure and Applied Physics (IUPAP) had set up a Commission on Atomic Masses and Related Constants following a conference in 1956. Further conferences followed at intervals of a few years and, in 1972, the emphasis was changed to Atomic Masses and Fundamental Constants [111]. The parent body of IUPAP, ICSU, has CODATA as a committee now publishing least-squares revisions of the fundamental constants from time to time, as we have seen [72].

16.3.5. Some examples of linked determinations

16.3.5.1. F, R, N_A and k. During the greater part of the nineteenth century, ideas were developing of the atomic and molecular structure of matter and the kinetic theory of gases, together with Maxwell's electromagnetic theory. In particular, it was realized that some properties could usefully be expressed per mole of substance, i.e. in terms of the gram-molecular weight. The quantities in this group all have their roots in this period and precede the discovery of the electron. They are:

F, the Faraday, the electric charge carried by one gram-equivalent of any element;
R, the gas constant, given by $PV = RT$ for one mole of an ideal gas;
N_A, the Avogadro constant, the number of molecules per mole; and
k, the Boltzmann constant, given by $k = R/N_A$.

Direct measurements of the Faraday have been carried out by electrolysis, measuring the quantity of material deposited from an electrolyte when a known current is passed for a known time. The early measurements, including one by Rayleigh and Sidgwick [112], were made using silver. This has the disadvantage that silver does not normally consist of a single isotope. Iodine does and for this reason the measurements made by Washburn and Bates in 1912 using iodine were of particular importance [113]. Subsequent measurements have also used organic compounds, but the reported uncertainties have not greatly changed (reference [59], section 4.7). The currently accepted value is [73]

$$F = 96\,485.309 \text{ C mol}^{-1}.$$

The difficulty in measuring the gas constant, R, is of course that no gas is an ideal gas and Boyle's law is not strictly obeyed. The traditional method has been to measure the density of a real gas, usually oxygen, at various pressures and to extrapolate to zero pressure. More recently, Quinn and his colleagues have used an acoustic interferometer maintained at an accurately known temperature to determine R [114] (figure 16.10). The results have been reviewed by Colclough [115] and the latest value from CODATA is [73]

$$R = 8.314\,510 \text{ J mol}^{-1} \text{ K}^{-1}.$$

In the mid-nineteenth century, Loschmidt attempted to measure Avogadro's number, but the result was wildly in error. Early in the present century, Rutherford measured the charge of an alpha particle by counting the number of particles given off by a sample of radium and the total charge. A little later Millikan measured the charge on the electron directly (see following section). From the known values of ionic charges in electrolytes, this made it possible to calculate N_A. In 1900 Planck had stated his law of black-body radiation which involved k and h. From experimental results, he calculated the values of k and h and derived $N_A = R/k$. More recently, N_A has been determined by accurate measurement of the density of silicon together with its lattice spacing. When allowance is made for the isotopic composition of the samples, this method has provided probably the best values of N_A so far [116]. The latest CODATA value is [73]

$$N_A = 6.022\,1367 \times 10^{23} \text{ mol}^{-1}.$$

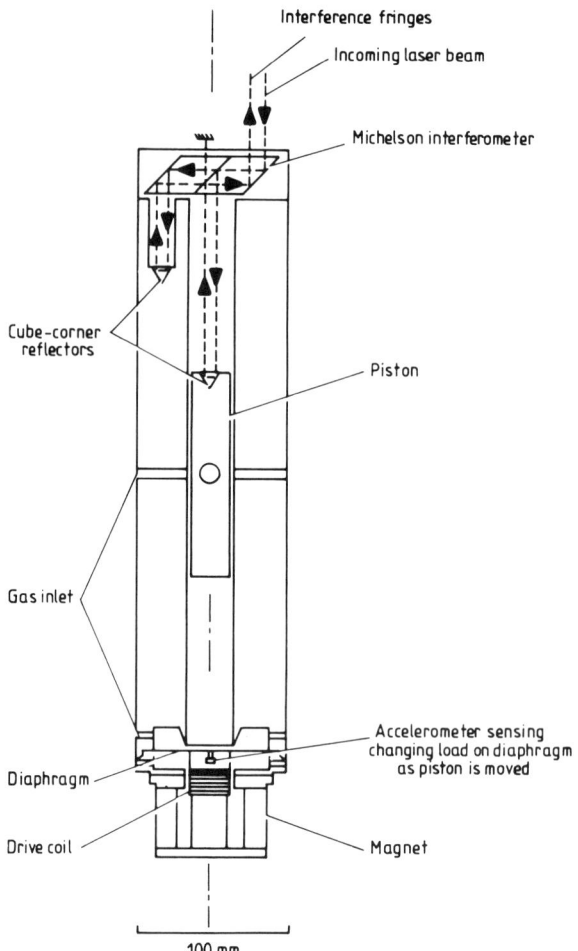

Figure 16.10. *The acoustic interferometer for measurement of the universal gas constant used by Quinn et al [114] in the early 1970s.*

Although Planck derived a value of the Boltzmann constant, k, from radiation measurements, it is now calculated from the values of R and N_A, which are more accurately known by other means, as we have just seen. The CODATA value is [73]

$$k = 1.380\,658 \times 10^{-23} \text{ J K}^{-1}.$$

16.3.5.2. e, h, e/m and α. This group of constants forms the core of Birge's least-squares adjustment of 1932 [109]. They are related to the properties of the electron and indeed measurement of some of them

Units, Standards and Constants

Table 16.2. *Latest CODATA values for physical constants.*

Elementary charge	e	$1.60217733 \times 10^{-19}$ C
Planck constant	h	$6.6260755 \times 10^{-34}$ J s
Electron charge/mass ratio	e/m_e	$1.75881962 \times 10^{11}$ C kg^{-1}
Fine structure constant	α	$7.29735308 \times 10^{-3}$
Rydberg constant	R_∞	10973731.534 m^{-1}
Electron mass	m_e	$9.1093897 \times 10^{-31}$ kg

led to the identification of the electron in the first place. By the mid-nineteenth century, ionic conduction in electrolytes was well understood and when, after their discovery by Röngten in 1895, x-rays were found to induce conduction in gases, it was easy to identify this as caused by ionic conduction in the gaseous state, but cathode rays, which had first been produced around 1870, presented more of a problem. They were first thought to be waves similar to light, but then Crookes showed that they were deflected by a magnetic field and in 1890 Schuster used this effect to show that they must be particles with a charge-to-mass ratio, e/m, many times that of the hydrogen ion. Many further measurements followed: among the most notable were Thomson's measurement of the deflection by electric as well as magnetic fields [117] and Millikan's determinations of the charge by the oil drop method [118] which continued into the 1920s. These measurements established the identity of the electron as a particle with a mass about one two-thousandth of that of the hydrogen ion and an equal but opposite charge [119].

Later determinations of the properties of the electron have made use of spectroscopic and indirect methods, usually involving the values of some other constants including Planck's constant, h, the fine structure constant, α ($= \mu_0 c e^2/2h$), the Rydberg constant R_∞ ($= m_e c \alpha^2/2h$), the Faraday constant and the Avogadro constant. The relations between these were discussed in some detail by Birge [84, 109]. More recently, the discovery of the Josephson and von Klitzing effects has provided alternative routes. The Josephson constant is given by

$$K_J = 2e/h$$

and the von Klitzing constant by

$$R_K = h/e^2 = \mu_0 c/2$$

where μ_0 is the permeability of free space.

The need to provide reference standards for the volt and the ohm (see section 16.2.6.4) led to a concentrated and coordinated programme of measurement of these two constants to the best possible accuracy. The results were reviewed by two working parties of the CCE [120]. Recommended values were agreed by the CIPM. The latest CODATA values for the constants mentioned above are shown in table 16.2 [73].

16.3.5.3. Other constants. The CODATA list gives values of over 100 constants. We have been able to discuss only a minority of these. Most of the others are of a specialized nature: about 60, for example, give values for detailed properties of fundamental particles—the electron, the muon, the proton, the neutron and the deuteron. A few are of more general interest—the Bohr magneton, the Stefan–Boltzmann constant, the first and second radiation constants, and the Wien displacement law constant, for example. One constant which has been the subject of some interest recently has been the gyromagnetic ratio of the proton; it was considered that this might provide an alternative and accurate means for the absolute determination of the ampere and a number of measurements were undertaken with this end in view [8, 9].

Determination of the values of the physical constants is an ongoing process, likely to continue indefinitely as measurement techniques improve. A new least-squares adjustment is scheduled for completion in 1995 [73]. A new set of recommended values will then presumably be issued by CODATA.

16.4. Applications

In this chapter we have looked in some detail at the derivations of the base units of the SI system and of the standards for maintaining them, and also at the determinations of the fundamental physical constants. These are the areas of metrology most relevant to the science of physics and where international cooperation is most deeply involved.

In addition, national standards laboratories have the responsibility of maintaining secondary standards of the base units and of derived units for the support of trade, industry and domestic purposes and of disseminating these through calibration services. In Britain, for example, the National Measurement Accreditation Service, NAMAS, which is a service of the National Physical Laboratory, accredits over a thousand calibration and testing laboratories in industry and elsewhere so that measurements made are traceable to the national and international standards. For the most part there is little basic physics involved in this and details of the services provided in most countries are readily available [4, 5].

There are two areas of derived quantities where the science of metrology has developed mainly in this century, where the establishment of standards has been vital and where BIPM has played a major part in establishing them. These are radioactivity and microwaves. Radioactivity was discovered by Becquerel in 1896 when he observed that uranium salts could affect a photographic plate. Thorium and radium were soon also discovered and in 1911 Marie Curie made the first radium standard which was deposited with BIPM and became the first international standard. The need for further standards eventually led to BIPM setting up new laboratories, opened in 1964 and responsible for the preparation of standard sources, establishment of x-ray and γ-ray dosimetry standards and neutron measurements [1]. They have also led the way to the definition of the SI units of radioactivity including the gray and the sievert [121].

See also p 1863

Microwave applications to communications, radar and navigation grew enormously during World War II. After the war it soon became clear that there was a need for international harmonization for standards of measurement. In 1952 URSI, the International Union of Radio Science, sponsored international comparisons of microwave power, followed by other quantities including noise, attenuation and permittivity. A more formal arrangement than URSI could provide was clearly desirable and in 1965 CCE appointed a Working Group on RF Measurements, GT-RF, which organizes international comparisons. More than a dozen laboratories take part. Frequency bands up to 18 GHz are covered for most quantities, with a few measurements up to about 100 GHz and laser power at 10.6 and 1.06 μm [122, 123]. Lists of quantities offered by national laboratories are given in the URSI Register [5].

Metrology is an interesting example of a subject which serves both pure science and everyday needs. National systems of weights and measures date from at least 3000 BC and have been internationally coordinated at least since Roman times. In the early eighteenth century the Royal Society in Britain and the Academy of Science in France arranged intercomparisons and began to look at the possibility of relating practical standards to physical constants. Since then the development of metrology as a science and its applications in trade and everyday life have gone hand in hand. There can be few branches of science where international cooperation has been so active and harmonious over so many years and where the results have influenced so wide a range of human activity. In this chapter we have emphasized its importance to new developments in physics and we end with references to papers by Cook which give a comprehensive and profound view of this topic [124–126].

References

[1] Page C H and Vigoureux P 1975 *The International Bureau of Weights and Measures 1875–1975* (Washington, DC: US Department of Commerce)
[2] Pyatt E C 1983 *The National Physical Laboratory—A History* (Bristol: Hilger)
[3] Cochrane R C 1966 *Measures for Progress—A History of the National Bureau of Standards* (Washington, DC: US Department of Commerce)
[4] Dobbie B, Darrell J, Poulter K and Hobbs R 1987 *Review of DTI Work on Measurement Standards* (London: Department of Trade and Industry) ch 20
[5] Bailey A E (ed) 1990 *URSI Register of National Standards Laboratories for Electromagnetic Metrology* (Bristol: Hilger)

[6] Bell R J (ed) 1993 *SI—The International System of Units* 6th edn (London: HMSO)
[7] Vigoureux P 1971 The gyromagnetic ratio of the proton—a survey *Precision Measurement and Fundamental Constants* (Gaithersburg, MD: NBS)
[8] Kibble B P and Hunt G J 1979 A measurement of the gyromagnetic ratio of the proton in a strong magnetic field *Metrologia* **15** 5–30
[9] Taylor B N 1973 Determining the Avogadro constant to high accuracy via improved measurements of the absolute ampere *Metrologia* **9** 21–3
[10] Kibble B P 1991 Present state of the electrical units *Proc. IEE* A **138** 187–97
[11] Kibble B P, Robinson I A and Belliss J H 1988 A realisation of the SI watt by the NPL moving-coil balance *NPL Report* DES 88
[12] Johnson W H 1923 Comparators *and* Line standards of length *Dictionary of Applied Physics, III* ed R T Glazebrook (London: Macmillan) pp 232–57, 465–77
[13] 1960 Comptes Rendues *11th CGPM Resolution* 6
[14] Carré P and Hamon J 1966 Mesure interférentielle de la base géodésique du BIPM *Metrologia* **2** 143–50
[15] Froome K D and Bradsell R H Distance measurements by means of a modulated light beam yet independent of the speed of light *Symposium on Electronic Distance Measurement (Oxford, September, 1965)* (London: Hilger-Watts)
[16] Laurila S H 1976 *Electronic Surveying and Navigation* (New York: Wiley)
[17] 1965 *Transactions of the International Astronomical Union* vol XII A
[18] 1983 Procès-Verbaux *CIPM Recommendation* 1
[19] Ward F A B 1970 *Time Measurement—Historical Review* (London: Science Museum)
[20] Dye D W and Essen L 1934 *Proc. R. Soc.* A **143** 285
[21] Essen L 1938 *Proc. Phys. Soc.* **50** 413
[22] Lyons H 1949 Microwave spectroscopic frequency and time standards *Electron. Eng.* **68** 251
[23] Beehler R E 1967 A historical review of atomic frequency standards *Proc. IEEE* **55** 792–805
[24] Audoin C and Vanier J 1976 Atomic frequency standards and clocks *J. Phys. E: Sci. Instrum.* **9** 697–720
[25] Terrien J 1976 Standards of length and time *Rep. Prog. Phys.* **39** 1067–108
[26] Vanier J and Audoin C 1989 *The Quantum Physics of Atomic Frequency Standards* (Bristol: Hilger)
[27] Rabi I I, Zacharias J R, Millman S and Kusch P 1938 A new method of measuring nuclear magnetic moment *Phys. Rev.* **53** 318
[28] Ramsey N F 1950 A molecular beam resonance method with separated oscillatory fields *Phys. Rev.* **78** 695–9
[29] Essen L and Parry J V L 1957 The caesium resonator as a standard of frequency and time *Phil. Trans.* A **250** 45–69
[30] Bowhill S A (ed) 1984 Brief reviews of developments in frequency standards and in time scale formation and comparisons *Review of Radio Science 1981–83* (Brussels: URSI (International Union of Radio Science))
[31] Mungall A G 1986 Frequency and time—national standards *Proc. IEEE* **74** 132–6
[32] See several other papers in the 1986 special issue on radio measurement methods and standards *Proc. IEEE* **74** 137–82.
[33] Mungall A G, Daams H and Boulanger J-S 1980 Design and performance of the new 1-m NRC primary caesium clocks *IEEE Trans. Instrum. Meas.* **IM-29** 291–7
[34] Becker G 1977 Performance of the primary Cs-standard of the

Physikalisch-Technische Bundesanstalt *Metrologia* **13** 99–104
[35] Ward F A B 1970 *Time Measurement—Historical Review* (London: Science Museum) ch I
[36] Guinot G and Seidelmann P K 1988 Time scales: their history, definition and interpretation *Astron. Astrophys.* **194** 304–8
[37] Winkler G M R and Van Flandern T C 1977 Ephemeris time, relativity, etc *Astron. J.* **82** 84–92
[38] See also Kaye and Laby 1992 (reprint) Astronomical and atomic time systems *Tables of Physical and Chemical Constants* 15th edn (London: Longmans) section 1.9.1
[39] Quinn T J 1991 The BIPM and the accurate measurement of time *Proc. IEEE* **79** 894–905
[40] Fleming J A 1913 *The Wonders of Wireless Telegraphy* (London: SPCK)
[41] Winkler G M R 1986 Changes at USNO in global timekeeping *Proc. IEEE* **74** 151–5
[42] Beehler R E and Allan D W 1986 Recent trends in NBS time and frequency distribution services *Proc. IEEE* **74** 155–7
[43] Quinn T J and Compton J P 1975 The foundations of thermometry *Rep. Prog. Phys.* **38** 151–239
[44] Ruhemann M and Ruhemann B 1937 *Low Temperature Physics* (Cambridge: Cambridge University Press) p 52
[45] Hall J A 1967 The early history of the International Practical Scale of Temperature *Metrologia* **3** 25
[46] Hall J A and Barber C R 1967 The evolution of the International Practical Temperature Scale *Metrologia* **3** 78
[47] Barber C R 1969 International Practical Temperature Scale of 1968 *Nature* **222** 929
[48] *The International Practical Temperature Scale of 1968* (London: National Physical Laboratory/HMSO) (amended edn 1975)
[49] 1990 The International Temperature Scale of 1990 (ITS-90) *NPL Leaflet*. See also Kaye and Laby 1992 (reprint) The International Temperature Scale of 1990 (ITS-90) *Tables of Physical and Chemical Constants* 15th edn (London: Longmans) section 1.5.1
[50] Preston J S 1961 Photometric standards and the unit of light *NPL Notes on Applied Science No 24* (London: HMSO)
[51] Crawford B H 1962 Physical photometry *NPL Notes on Applied Science No 29* (London: HMSO)
[52] Jones O C 1975 Optical radiation scales *Contemp. Phys.* **16** 287–310
[53] Jones O C 1978 Proposed changes to the SI system of photometric units *Lighting Res. Technol.* **10** 37–40
[54] Clark Latimer 1862 Letter to the editor *The Electrician* 17 January p 1862
[55] Glazebrook R T 1913 The Ohm, the Ampere and the Volt—a memory of fifty years, 1862–1912: The Fourth Kelvin Lecture, 27 February 1913 *Proc. IEE* **50** 560–92
[56] Paul R W 1936 Electrical measurements before 1886 *J. Sci. Instrum.* **13** 1–8
[57] Thomson A M and Lampard D G 1956 A new theorem in electrostatics with applications to calculable standards of capacitance *Nature* **177** 888
[58] Lampard D G 1957 A new theorem in electrostatics with applications to calculable standards of capacitance *Proc. IEE* **104** C 271–80
[59] Petley B W 1985 *The Fundamental Physical Constants and the Frontier of Measurement* (Bristol: Hilger) pp 142–5
[60] Clothier W K 1965 A proposal for an absolute liquid electrometer *Metrologia* **1** 181–4
[61] Clothier W K, Sloggett G J, Bairnsfather H, Currey M F and Benjamin D J

1989 A determination of the volt with improved accuracy *Metrologia* **26** 9–46

[62] Giorgi G 1901 Unita razionali di elettromagnetismo *Atti Assoc. Elettrotecn.* **5** 402–18
[63] Vigoureux P 1989 Eighty-eight years of Giorgi's MKS units *J. Phys. E: Sci. Instrum.* **22** 671–3
[64] Dix C H and Bailey A E 1975 Electrical standards of measurement, Part 1 DC and low-frequency standards *Proc. IEE* **122** 1018–36
[65] Josephson B D 1962 Supercurrents through barriers *Phys. Lett.* **1** 251. See also 1964 *Rev. Mod. Phys.* **26** 216; 1965 *Adv. Phys.* **14** 419
[66] Burroughs C J and Hamilton C A 1990 Voltage calibration systems using Josephson junction arrays *IEEE Instrumentation and Measurement Conference (13–15 February, 1990)* (New York: IEEE) pp 291–4
[67] Von Klitzing K, Dorda G and Pepper M 1980 New method for high accuracy α determination based on quantised Hall resistance *Phys. Rev. Lett.* **45** 494–7
[68] Von Klitzing K and Ebert G 1985 Application of the quantum Hall effect in metrology *Metrologia* **21** 12–18
[69] Taylor B N 1987 History of the present value of $2e/h$ commonly used for defining national units of voltage and possible changes in national units of voltage and resistance *IEEE Trans. Instrum. Meas.* **IM-36** 659–64
[70] Taylor B N and Witt T J 1989 New international electrical reference standards based on the Josephson and quantum Hall effects *Metrologia* **26** 47–62
[71] Quinn T J 1989 News from the BIPM *Metrologia* **26** 69–74
[72] The 1986 adjustment of the fundamental physical constants: report of the CODATA Task Group on Fundamental Constants *CODATA Bulletin 63* (Oxford: Pergamon). See also 1973 Recommended consistent values of the fundamental physical constants *CODATA Bulletin 11* (Paris: CODATA Secretariat)
[73] Cohen E R and Taylor B N 1992 The fundamental physical constants *Phys. Today* August BG 9-13
[74] Dampier-Whetham W C D 1930 *A History of Science* (Cambridge: Cambridge University Press) p 47
[75] Eddington A S 1946 *Fundamental Theory* (Cambridge: Cambridge University Press)
[76] Michelson A A, Pease F G and Pearson F 1935 *Astrophys. J.* **82** 26
[77] Mittelstaedt O 1929 *Ann. Phys., Lpz* **2** 285
[78] Mittelstaedt O 1929 *Phys. Z.* **30** 165
[79] Hüttel A 1940 *Ann. Phys., Lpz* **37** 365
[80] Anderson W C 1937 *Rev. Sci. Instrum.* **8** 239
[81] Anderson W C 1941 *J. Opt. Soc. Am.* **31** 187
[82] Mercier J 1924 *J. Phys. Radium* **5** 168
[83] Rosa E B and Dorsey N E 1907 *Bur. Stand. Bull.* **3** 433
[84] Birge R T 1942 *Rep. Prog. Phys.* **8** 90
[85] Essen L and Gordon-Smith A C 1948 *Proc. R. Soc.* A **194** 348
[86] Essen L and Froome K D 1951 *Proc. Phys. Soc.* B **64** 862
[87] Froome K D 1958 *Proc. R. Soc.* A **247** 109
[88] Froome K D and Essen L 1969 *The Velocity of Light and Radio Waves* (New York: Academic)
[89] Bay Z and White J A 1972 *Phys. Rev.* D **5** 796
[90] Evenson K M, Wells J S, Peterson F R, Danielson B L, Day G W, Barger R L and Hall J L 1972 *Phys. Rev. Lett.* **29** 1349
[91] Blaney T G, Bradley C C, Edwards G J, Jolliffe B W, Knight D J E,

Rowley W R C, Shotton K C and Woods P T 1977 *Proc. R. Soc.* A **355** 61
[92] Woods P T and Jolliffe B W 1976 *J. Phys. E: Sci. Instrum.* **9** 395
[93] Baird K M, Smith D S and Witford B G 1980 *Opt. Commun.* **12** 367
[94] Boys C V 1923 Earth, the mean density of the, *Dictionary of Applied Physics* ed R T Glazebrook (London: Macmillan)
[95] Poynting J H and Thomson J J 1934 *Textbook of Physics; Properties of Matter* (London: Griffin)
[96] Heyl P R 1930 *Bur. Stand. J. Res.* **5** 1243
Heyl P R and Chrzanowski P 1942 *J. Res. NBS* **29** 1
[97] Zahradnicek J 1933 *Phys. Z.* **34** 126
[98] Karagioz O V, Ismaylov V P, Agafonov N L, Kocheryan E G and Tarakanov Yu A 1976 *Izv. Akad. Sci. USSR* **12** 351
[99] Sagitov M V, Milyukov V R, Monakhov E A, Nazhidinov V S and Tadzhidinov Kh G 1977 *Dokl. Akad. Nauk.* **245** 567
[100] Luther G G and Towler W R 1982 *Phys. Rev. Lett.* **48** 121
[101] Lowry R A, Towler W R, Parker H M, Kuhlthau A R and Beams J W 1972 The gravitational constant G *Atomic Masses and Fundamental Constants 4* ed J H Sanders and A H Wapstra (New York: Plenum) p 521
[102] See, for example, Nunn T P 1923 *Relativity and Gravitation* (London: University of London Press); Hawking S W and Israel W 1979 *General Relativity—an Einstein Centenary Survey* (Cambridge: Cambridge University Press)
[103] Ashtekar A and Geroch R 1974 Quantum theory of gravitation *Rep. Prog. Phys.* **37** 1211-56
[104] Eötvös R V, Pekar D and Feteke E 1922 *Ann. Phys., Lpz* **68** 11
[105] See, for example, Petley B W 1985 *The Fundamental Physical Constants and the Frontier of Measurement* (Bristol: Hilger) ch 7
Gillies G T 1982 *BIPM Rapport* BPM-82/9
Will C M 1984 The confrontation between general relativity and experiment, an update *Phys. Rep.* **113** 345-422
[106] Birge R T 1929 *Rev. Mod. Phys.* **1** 1
[107] Bond W N 1930 *Phil. Mag.* **10** 994
[108] Bond W N 1931 *Phil. Mag.* **12** 632
[109] Birge R T 1932 *Phys. Rev.* **40** 228
[110] Taylor B N, Parker W H and Langenberg D N 1969 *Rev. Mod. Phys.* **41** 375
[111] Sanders J H and Wapstra A H (ed) 1972 *Atomic Masses and Fundamental Constants 4* (New York: Plenum)
[112] Lord Rayleigh and Mrs Sidgwick 1884 *Phil. Trans.* **175** 411
[113] Washburn E W and Bates S J 1912 *J. Am. Chem. Soc.* **34** 1341, 1515
[114] Quinn T J, Colclough A R and Chandler T R D 1976 *Phil. Trans.* **283** 367
[115] Colclough A R 1981 *Precision Measurement and Fundamental Constants 2, NBS Special Publication 635* (Gaithersburg, MD: NBS) p 263
[116] Deslattes R D 1980 *Ann. Rev. Phys. Chem.* **31** 435
[117] Thomson J J 1897 *Phil. Mag.* **44** 293
[118] Millikan R A 1912 *Trans. Am. Electrochem. Soc.* **21** 185
[119] Millikan R A 1917 *The Electron—Its Isolation and Measurement and the Determination of some of its Properties* (Chicago: University of Chicago Press) (2nd edn 1924)
[120] Comité Consultatif d'Electricité 1988 *Report of the 18th meeting*, Appendix E2, *Report from the Working Group on the Josephson Effect*; and Appendix E3, *Report from the Working Group on the Quantum Hall Effect* BIPM. See also Hartland A 1988 Quantum standards for electrical units *Contemp. Phys.* **29** 477

[121] International Commission on Radiological Protection 1980 *Radiation Quantities and Units ICRP Report 33* (Oxford: ICRP)
[122] Bailey A E 1980 International harmonisation of microwave standards *Proc. IEE H* **127** 70–3
[123] Bailey A E, Hellwig H W, Nemoto T and Okamura S 1986 International organisation in electromagnetic metrology and international comparison of RF and microwave standards *Proc. IEEE* **74** 9–14
[124] Cook A H 1972 Quantum metrology–standards of measurement based on atomic and quantum phenomena *Rep. Prog. Phys.* **35** 463
[125] Cook A H 1975 The importance of precise measurement in physics *Contemp. Phys.* **16** 395
[126] Cook A H 1977 Standards of measurement and the structure of physical knowledge *Contemp. Phys.* **18** 393

ILLUSTRATION ACKNOWLEDGMENTS

Chapter 9

R P Feynman—AIP Meggers Gallery of Nobel Laureates
R R Wilson—Courtesy of Fermilab
9.1 Reprinted with permission from *Phys. Rev.* **72**, 241–3. Copyright 1995 The American Physical Society
9.2 AIP Emilio Segrè Visual Archives
9.3 Instituts Internationaux de Physique et de Chimie, Bruxelles
9.5 Reprinted with permission from *Nature* **160** p855. Copyright 1995 Macmillan Magazines Ltd
9.7 Reprinted with permission from *Phys. Rev. Lett.* **12**, 204–6. Copyright 1964 The American Physical Society
9.8 Photograph by Alan W Richards. Courtesy AIP Emilio Segrè Visual Archives
9.9 M Goldhaber, L Grodzins and A W Sunyar *Phys. Rev.* **109**, 1015–7. Copyright 1958 The American Physical Society
9.10 Courtesy of Brookhaven National Laboratory
9.11 J W Cronin, V L Fitch and R Turlay *Phys. Rev. Lett.* **13**, 138–40. Copyright 1964 The American Physical Society
9.12 Courtesy of Fermilab
9.16 Courtesy of Stanford Linear Accelerator Center and US Department of Energy
9.18 From M Riordan, *The Hunting of the Quark*, Simon & Schuster, 1987
9.19 S Glashow and A Salam: photographs courtesy of AIP. Photograph of S Weinberg courtesy of Fermilab
9.21 (*a*) Brookhaven National Laboratory. Courtesy AIP Emilio Segrè Visual Archives (*b*) Reprinted with permission from *Phys. Rev. Lett.* **33** 1404–6. Copyright 1974 The American Physical Society
9.22 Left: (*a*) Reprinted with permission from *Phys. Rev. Lett.* **33** 1406–8. Copyright 1974 The American Physical Society. (*b*) Reprinted with permission from *Phys. Rev. Lett.* **33** 1406–8. Copyright 1974 The American Physical Society. (*c*) Reprinted with permission from *Phys. Rev. Lett.* **33** 1406–8. Copyright 1974 The American Physical Society
9.22 Right: Stanford Linear Accelerator Center and US Department of Energy
9.24 Reprinted with permission from *Phys. Lett.* **122B** 398–410. Copyright 1983 Elsevier Science BV, Amsterdam
9.25 Courtesy of Fermilab
9.26 Top: reproduced with

A7

permission from *Annual Review of Nuclear and Particle Science* Vol 41 © 1991, by Annual Reviews, Inc.

9.28 Reprinted with permission from *Phys. Rev.* **D50** 1173-825. Copyright 1994 The American Physical Society

9.30 Reprinted with permission from *Phys. Rev.* **D45** S1–S584. Copyright 1992 The American Physical Society

9.31 (*a*) Reprinted with permission from *Phys. Rev. Lett.* **39** 1240–2. Copyright The American Physical Society. (*b*) Reprinted with permission from *Rev. Mod. Phys.* **61** 547–60. Copyright 1989 The American Physical Society

9.33 Top: DESY, Hamburg. Bottom: CESR, Cornell University

9.34 AIP Niels Bohr Library

9.35 Courtesy of Fermilab

9.37 Bottom: courtesy of Joe Stancampiano and Karl Luttrell

9.39 Reproduced with permission from *Proceedings DPF 94 Meeting (Albuquerque, NM, August 1994)*, World Scientific, Singapore

Chapter 10

L Prandtl—AIP Emilio Segrè Visual Archives, Lande Collection

G I Taylor—Cambridge University Library. Courtesy AIP Emilio Segrè Visual Archives

Chapter 11

H Kamerlingh Onnes—Burndy Library. Courtesy AIP Emilio Segrè Visual Archives

J Bardeen—AIP Emilio Segrè Visual Archives

11.1 From *Low Temperature Physics* by L C Jackson, John Wiley (US) and Methuen (UK)

11.4 Courtesy of Prof J F Allen FRS

11.9 From *Helium-3* by W P Halperin and L P Pitaevskii, North-Holland, Amsterdam

11.11 From *Superconductivity* by P F Dahl, AIP Press, New York

11.12 Reprinted with permission from *Appl. Phys. Lett.* **51** 57. Copyright 1987 The American Physical Society

11.13 Reprinted with permission from *Phys. Rev.* B **38**, 2477. Copyright 1988 The American Physical Society

Chapter 12

P Debye—AIP Emilio Segrè Visual Archives. Photograph by Francis Simon

M Born—AIP Emilio Segrè Visual Archives, Lande Collection

12.3 (*a*) Reprinted with permission from *Physica* **14** 139. Copyright 1948 Elsevier Science. (*b*) Reprinted with permission from *Physica* **14** 510. Copyright 1948 Elsevier Science

12.5 R Berman, 1953, *Advances in Physics* **2** 103, Taylor and Francis

12.7 (*a*) Reprinted with permission from M Blackman, *Proc. R. Soc.* A**148** 365. Copyright 1935 The Royal Society

12.8 Figure 1: reprinted with permission from E W Kellerman, *Phil. Trans. R. Soc.* **238** 513. Copyright 1940 The Royal Society. Figure 2: reprinted with permission from E W Kellerman, *Proc. R. Soc.* A**178** 17. Copyright 1941 The Royal Society

12.11 Reprinted with permission from *Phys. Rev.* **100** 756. Copyright 1955 The American Physical Society

12.12 Reprinted with permission from *Phys. Rev.* **112** 90. Copyright 1958 The American Physical Society

12.14 Reprinted with permission from *Phys. Rev.* **128** 1099.

Illustration Acknowledgments

Copyright 1962 The American Physical Society
12.15 Reprinted with permission from *Phys. Rev.* B **13** 4258. Copyright 1976 The American Physical Society
12.19 Reprinted with permission from *Phys. Rev.* **65** 117. Copyright 1944 The American Physical Society

Chapter 13

H Massey—Photograph by D H Rooks
W F Meggers—AIP Meggers Collection of Nobel Laureates
G Herzberg—AIP Meggers Collection of Nobel Laureates
13.1 Courtesy of J J McLelland, NIST
13.2 From *Physics of Atoms and Molecules* by U Fano, University of Chicago Press
13.3 T E Sharp, *Atomic Data* **2** 119–69 (1972), Academic Press, San Diego
13.4 A E Douglas and W E Jones, *Can. J. Phys.* **44** 2251 (1966), National Research Council of Canada, Ottawa
13.5 From *Physical Chemistry* by R S Berry, S A Rice and J Ross. Copyright 1980 John Wiley and Sons. Reprinted by permission of John Wiley and Sons, Inc.
13.6 A P Lukirskii *et al*, *Optics and Spectroscopy* **9** 262. Copyright 1960 The Optical Society of America
13.7 Courtesy of R D Deslattes, NIST
13.8 Courtesy of R P Madden, NIST
13.9 Reprinted with permission from *Rev. Mod. Phys.* **40** 456, figure 68. Copyright 1968 The American Physical Society
13.10 Reprinted with permission from A B Bleaney *et al, Proc. Roy. Soc.* A**189**. Copyright 1947 The Royal Society, London
13.11 From U Fano and A R P Rau, *Atomic Collisions and Spectra*, Academic Press, Orlando
13.12 IOP Publishing Ltd, Bristol
13.13 Reprinted with permission from *Phys. Rev. Lett.* **10** 104, figure 2. Copyright 1963 The American Physical Society
13.14 IOP Publishing Ltd, Bristol
13.15 (*a*) G Herzberg and C Jungen, 1972, *J. Mol. Spect.* **41** 425, Academic Press, Orlando
13.16 Reprinted with permission from K T Lu, *Phys. Rev.* A**4**, 579. Copyright 1971 The American Physical Society
13.17 (*a*) Reprinted with permission from *Phys. Rev.* **128** 662, figure 7. Copyright 1962 The American Physical Society. (*b*) Reprinted with permission from P F Ziemba and E Everhart, *Phys. Rev. Lett.* **2** 299, figure 1. Copyright 1959 The American Physical Society
13.18 Reprinted with permission from S U Ovchinnokov and E A Solov'ev, *Sov. Phys. JETP* **63** 538, figure 2. Copyright 1986 American Institute of Physics
13.19 Reprinted with permission from R Villa and R D Deslattes, *J. Chem. Phys.* **44** 4399. Copyright 1966 American Institute of Physics
13.20 B W Petley, *The Fundamental Physical Constants*, Adam Hilger
13.21 Reprinted with permission from K von Klitzing *et al*, *Phys. Rev. Lett.* **45** 494, figure 1. Copyright 1980 The American Physical Society
13.22 IOP Publishing Ltd, Bristol

Chapter 14

P E Weiss—AIP Emilio Segrè Visual Archives, Goudsmit Collection
J H Van Vleck—AIP Meggers Gallery of Nobel Laureates
14.2 From *Magnetic Bubbles*, p84, by A H Bobeck and H E D Scovil. Copyright © June 1971 Scientific American, Inc. All Rights reserved

Twentieth Century Physics

14.7 From R M Bozorth *Ferromagnetism* D Van Nostrand Co., Inc. Copyright Bell Telephones Laboratories, Inc.
14.10 Reprinted with permission from A R Mackintosh, *Physics Today*, June 1977, figure 6, p28. Copyright 1977 American Institute of Physics
14.11 Reprinted with permission from Griffiths, *Nature* **158** 670. Copyright 1946 Macmillan Magazines Ltd
14.12 From J K Standley, *Oxide Magnetic Materials*, figure 9.7, p183, Clarendon Press, Oxford
14.14 Reprinted with permission from A R Mackintosh, *Physics Today*, June 1977, figure 4, p26. Copyright 1977 American Institute of Physics
14.15 From A B Pippard, *Physics of Vibration*. Reprinted with the permission of Cambridge University Press, Cambridge
14.16 Courtesy of Professor B S Worthington

Chapter 15

E O Lawrence—Lawrence Radiation Laboratory. Courtesy AIP Emilio Segrè Visual Archives
E Fermi—AIP Emilio Segrè Visual Archives
15.1 Niels Bohr Institute. Courtesy AIP Emilio Segrè Visual Archives
15.2 Courtesy of Max-Plank-Institut für Kernphysik
15.3 Argonne National Laboratory. Courtesy AIP Emilio Segrè Visual Archives
15.4 Lawrence Berkeley Laboratory. Courtesy of AIP Emilio Segrè Visual Archives
15.5 Department of Terrestrial Magnetism, Carnegie Institute of Washington. Courtesy AIP Emilio Segrè Visual Archives
15.7 Oakland Tribune. Courtesy AIP Emilio Segrè Visual Archives

Chapter 16

R T Glazebrook—Courtesy of Mrs Marguerite Pyatt, Hungerford, Berks
16.2 *Measurement at the Frontiers of Science*, 1991, National Physical Laboratory
16.3 E Pyatt, *National Physical Laboratory—A History*, 1983, Adam Hilger
16.4 Cochrane, *Measures for Progress*, NIST
16.5 E Pyatt, *National Physical Laboratory—A History*, 1983, Adam Hilger
16.6 E Pyatt, *National Physical Laboratory—A History*, 1983, Adam Hilger
16.7 E Pyatt, *National Physical Laboratory—A History*, 1983, Adam Hilger
16.8 B W Petley, *The Fundamental Constants and the Frontier of Measurement*, 1988, Adam Hilger
16.9 E Pyatt, *National Physical Laboratory—A History*, 1983, Adam Hilger
16.10 B W Petley, *The Fundamental Constants and the Frontier of Measurement*, 1988, Adam Hilger

SUBJECT INDEX

Note—volume and page numbers in **bold type** refer to items in boxes

A0620-00 x-ray binary source, III.1757
Abbe optical theory, III.1578–84
Abbe primary image, III.1583
Abbe secondary image, III.1583
Abelian gauge theory, II.693, II.714
Aberdeen MK1 NMR Imager, III.1926
Aberdeen Section Scanner, III.1918
Aberration, I.258, I.266, I.269
  see also Lens
Ability of the 200" telescope to discriminate between selected world models, The, III.1793
Abrikosov vortex lattice, II.935
Absorption, III.1404
Absorption coefficient, I.364, I.368, II.1052
Absorption spectra, I.24
Academia Europaea, III.2055
Académie des Sciences, I.6, II.1233
Accelerators, II.657, II.687–8, II.725–44, II.1201–6, II.1225, III.1883, III.1997, III.2036
  colliding beams, II.738–42
  fixed-field alternating-gradient, II.739
  strong focusing, II.733–4
Accretion disc, III.1756, III.1779
Accretion radius, III.1756
Acoustic energy, II.873

Acoustic modes, II.1004
Acoustic paramagnetic resonance (APR), II.1146–7
Acoustic resonance, III.1357–8
Acta Crystallographica, I.511
Acta Metallurgica, III.1556
Actinides, I.97, II.955
  magnetism, II.1166
Actinium, II.1227
Activation energy, III.1531
Active galactic nuclei, III.1777
  and general relativity, III.1777–81
  non-thermal phenomena in, III.1781–4
Adenine, I.504
Adiabatic demagnetization, II.1125–6
Adiabatic model, III.1802
Adiabatic principle, I.92–3
Adler–Weisberger relation, II.679
Administration of Radioactive Substances Advisory Committee, III.1910
AEG (Allgemeine Elektrizitäts Gesellschaft) Research Institute, III.1577–8
Aerobee (rocket), III.1682, III.2005
Aerodynamics, II.821
  heating phenomenon, II.874
  lift theory, II.817

Aeronautics, II.869–87
Aeronomy, III.1982
Aether-drift, I.26
Age-hardening, III.1546, III.1547
AIDS, III.2029
Aircraft, fixed-wing, II.817
Aircraft wings, II.821, II.824
Airy diffraction limit, III.1467
Alcator, III.1649
ALICE neutral-beam injection experiments, III.1662
Alkali metals, III.1345
Allende meteorite, III.1963
Allgemeine Elektrizitäts Gesellschaft Research Institute, *see* AEG
Allis-Chalmers Company, III.1871
Along-crest energy propagation, II.863–9
α-decay, I.115–16, I.366–7, II.1189
α-particle energies, II.1201
α-particle model, II.1190
α-particles, I.58, I.59, I.104, I.110, I.112, I.113, I.116, I.119, I.122, I.123, I.367, II.638, II.1189, II.1226, III.1706
α-radioactivity, I.52
α-rays, I.37, I.54, I.58, I.104
β-alumina, III.1539
Aluminium alloys, III.1533
Aluminium–copper alloys, III.1533, III.1546
Aluminium–manganese alloy, I.481–3
Alzheimer's disease, III.1922
American Association of Physicists in Medicine, III.1864
American Crystallographic Association, I.512
American Philosophical Society, I.3
American Physical Society, III.1558, III.2026
American Science and Engineering Group (AS&E), III.1739

American Society for X-ray and Electron Diffraction, I.511
Americium, II.1225
Amino acids, I.496, I.497
Amorphous cement, III.1518
Amorphous magnetism, II.1170–1
Amorphous materials, III.1541
Amorphous metal hypothesis, III.1518
Amorphous solids, x-ray diffraction, **I.478**
Amorphous structures, I.477–9
Ampere (unit), II.1235, II.1254, II.1257
absolute determination, II.1255
Amplification without inversion (AWI), III.1473
Anaesthetic equipment, III.1897
Anderson–Brinkman–Morel (ABM) state, II.953
Andromeda Nebula, III.1717, III.1718, III.1788
Angiography, III.1930
Anglo-Australian Telescope, III.1744
Angstrom (unit), II.1242
Angular momentum, I.81, I.162, I.359, II.818, II.1124, III.1410, III.1756, III.1779, III.1962, III.1964
Angular momentum algebra, **II.1038**
Anharmonic effects, II.1002–4
Anharmonic oscillators, III.1434
Anisotropic superfluid, II.950–1
Annalen der Physik und Chemie, I.7
Annales de chimie et de physique, I.7
Anomalous skin effect, III.1346, III.1347, III.1351, III.1357
Anthracene, crystal structure, I.494
Anthropic Principle, III.2039–40
Antiferroelectric structure, I.487
Antiferromagnetism, I.559, II.1016, II.1113, II.1129–32, II.1147–8
exchange interactions, II.1143

Subject Index

resonance, II.1151
structures, II.1011
Antimony, III.1332, III.1335, III.1544
Antiparticles, II.638, II.655
Antiprotons, II.655
Antiscreening. *See* Screening and antiscreening
Anyon mechanism, II.962
Applications of Phase Transitions in Material Science, III.1557
Applied optics, III.1474–81
Applied physics, III.2050
Arginine, structure of, I.449–50
Argon, I.75
Argus experiment, III.1683
Army Operational Research Group, III.1736
Arnowitt–Deser–Misner (ADM) mass, I.301
Arsenic-72 and -74, III.1904
Arteries, artificial, III.1898
Arteriosclerosis, III.1898
Artificial limbs, III.1899
ASDEX tokamak, *see* Tokamak
Associated production, II.657
Association of Universities for Research in Astronomy (AURA), III.1744
Asthenosphere, III.1978
Astigmatism, III.1580
ASTRON, III.1645
Astronomical photography, III.1692
Astronomical spectroscopy, III.1691
Astronomical timescales, II.1227
Astronomical Unit (AU), II.1242
Astronomy, I.24, II.757–62, III.1453–4, III.2046
 changing perspective, III.1734–45
 general theory of relativity, I.315–17
 optical, III.1744–5

Astrophysical Journal, III.1762
Astrophysics, I.630, II.757–62, III.1618, III.1691–821, III.2031
 birth of, III.1691
 general theory of relativity, I.315–17, III.1776
 high-energy, III.1774–84
 impact of quantum mechanics, III.1706–8
 since 1945, III.1746–807
 statistical mechanics in, III.1711
Asymmetric unit, I.423
Asymptotic freedom, II.714
Atmosphere, II.899
Atmospheric physics, III.1983
Atmospheric science, III.1983
Atomic bomb, II.1197, III.1630, III.1891, III.1980
Atomic clock, II.1244, II.1247
Atomic dynamics, discretization in, I.184–5
Atomic energy, I.63, I.113, I.132–3, I.285, III.1628
Atomic Energy Research Establishment, III.1901
Atomic model, I.157–8
Atomic number, I.65, I.158–60, I.407, I.429, II.1057, II.1222, II.1223
Atomic physics, I.31, I.358–62, II.755–6, II.1025–49
Atomic pile, *see* Nuclear reactors
Atomic processes, current overviews, II.1101–6
Atomic scale, II.1026–7
Atomic scattering factor, I.431, I.432, I.439, I.453
Atomic spectra, II.1032–9
Atomic standards of measurements, II.1095
Atomic structure, I.63–5, I.156–62, I.211–15, II.1026–7
Atomic systems, optical actions, II.1097–101
Atomic theory, I.43, I.46, I.227

causality in, I.220–3
failure of, I.181–3
principles of, I.176–8
probability in, I.220–3
schools of, I.178–81
Atomic weight, I.65
Atoms, I.43–103
role in metrology and instrumentation, II.1094–7
Atoms for Peace Conference (1955), III.1642
Atoms for Peace Conference (1958), III.1655
AT&T, III.2022
Au$_2$Mn, magnetic moments in, I.462
Aufbauprinzip (building-up principle), I.96–7, I.366
Auger emissions, inner-shell phenomena, II.1090–1
Augmented-plane-wave (APW) method, III.1317
Aurorae, III.2006–7
Auroral phenomena, III.2003
Auswahlprinzip, selection rule, I.178–81
Autocorrelation functions, III.1418
Auto-ionization, II.1061, II.1090, II.1102
Automatic Plate Measuring Machine (APM), III.1745
Avalanche photodiode (APD), III.1419, III.1476
Avogadro's constant, I.150
Avogadro's law, I.46
Avogadro's number, I.43, I.49, I.51, II.1237, II.1269
Azbel'–Kaner cyclotron resonance, III.1357

Balian–Werthamer (BW) state, II.952
Balloon flight, I.368
Balmer formula, I.24, I.78, I.80–5, I.88

Balmer series, I.160
Band gaps, III.1367
Band structure, II.1047, II.1166, III.1314–17, III.1344, III.1530
Band-structure diagram, III.1417
Band theory, II.1165–7, III.1367
Bar-code scanners, III.1480–1
Bardeen–Pines interaction, II.938
Barium, contrast media, III.1863
Barium titanate, III.1538
Baroclinic instability, II.902, II.907
Baroclinic mode, II.896
Baryogenesis, II.761–2
Baryon, II.660, II.661, II.675, II.685, III.1800
Baryon–antibaryon asymmetry, II.760, III.1806
Baryon number, II.658
Basic Orthopaedic BioMechanics, III.1896
BaTiO$_3$, ferroelectric phase transition, I.486
Batteries, I.27
lithium–iodine, III.1895
zinc–mercury, III.1895
BBKGY hierarchy, I.605, I.623
BCS theory, I.489, II.935, II.937, II.937–41, II.961–5
Beevers–Lipson strips, I.446
Beiblatter zu den Annalen der Physik, I.7
Bell inequalities, III.1389
Bell Telephone Laboratories, III.1790, III.1944, III.2022, III.2026
Berkelium, II.1225
Bernoulli's equation, II.822, II.885
Beryl, structure of, I.440
Beryllium, I.119, I.120, III.1360
β-decay, I.106, I.366–7, I.370–2, I.375, I.398–402, II.639, II.664, II.665, II.667, III.1706, III.1707
Fermi's theory, I.374–5
limits on neutrino masses, II.755

Subject Index

β-effect, II.902
β-electrons, I.104
β-particles, I.107, I.109
β-radioactivity, I.52
β-ray spectroscopy, I.103–9
β-rays, I.37, I.54, I.58, I.59, I.104–9
β-spectra, I.106, I.109, I.118–19, I.126–9
β-structures, I.498
Betatron, II.732, II.1204–5, III.1871
B-factories, II.767–8
Big Bang, III.1784–6, III.1790–3, III.2040
'Big Physics', III.2054–5
'Big Science', III.2023–7, III.2054
Billions of electronvolts (GeV), I.358
Binary alloys, critical behaviour of, I.548–9
Binary collisions, I.611, I.612
Binding energy, I.110–12
Biochemical assay, III.1909
Biomedical technology, III.2031
Biomolecular crystals, I.450–3
Biomolecular structures, I.445, I.496–512
Biomolecules, II.1084
Biophysics, III.2030, III.2046
Biot–Savart law, II.818
Bipolar outflow, III.1769
Birge–Bond diagram, II.1268
Bismuth, III.1282–3, III.1339
Bismuth germanate, III.1921
Bistable optical switching devices (SEEDs), III.1469
Bjorken–Glashow hypothesis, II.700
Black body, I.66
Black-body law, I.173, I.177
Black-body radiation, I.22–4, I.30, I.69, I.144, I.148–50, I.174, I.524- 5, II.758, II.972, II.973, II.982, III.1387, III.1393, III.1402

Black holes, I.306–12, II.760–1, III.1754–7, III.1778–80
 no-hair theorem for, III.1778
BL-Lacertae (or BL-Lac) objects, III.1777, III.1784
Bloch equation, II.1174
Bloch functions, II.1166
Bloch theorem, III.1291–301
Bloch theory, III.1342
Bloch wave, **III.1294–5**, III.1320
Bloch wavelength, III.1301
Bloch wavenumber, III.1294, III.1299
Bloch wave-vector, III.1294–6
Block copolymers, III.1611
Blood flow, III.1897–8, III.1914, III.1920
Blood pressure, monitoring, III.1893–4
B-meson, II.721, II.724–5, II.767–8
Bohm diffusion, III.1630, III.1631
Bohm diffusion formula, III.1632
Bohm diffusion law, III.1655
Bohm–Pines theory, III.1362, III.1363
Bohr–Kramers–Slater radiation theory, I.185, I.203
Bohr magneton, I.116
Bohr orbit, III.1319
Bohr radius, I.85, II.1026
Bohr relation, I.90
Bohr–Sommerfeld atomic model, I.175–6
Bohr–Sommerfeld rules, I.92
Bohr–Sommerfeld theory, I.161
Bohr theory, II.1124
Bolometer, I.23
Boltzmann collision number, I.603
Boltzmann combinatorial factor, I.527
Boltzmann constant, I.67, I.150, I.523, I.589, II.1270, III.1285, III.1387, III.1397
Boltzmann distribution, I.542
Boltzmann equation, I.523, I.533,

I.588, I.591, I.593–5, I.597, I.602, I.605, I.608–11, I.617, I.624–6, III.1303, III.1344, III.1701, III.1992
Boltzmann–Grad limit, I.626
Boltzmann law, III.1287
Bond percolation, I.572
Bone fractures, III.1897
Bootstrap programme, II.681
Born–Oppenheimer approximation, II.1043–6
Bose–Einstein condensation, I.530–1, I.537, II.927, II.930
Bose–Einstein distribution, III.1402
Bose–Einstein statistics, I.155, I.533, II.687, III.1387
Bose statistics, I.117, III.1394
Boson, I.382, II.693, II.694, II.699, II.1218–19
Bound scattering length, I.453
Boundary conditions, II.800–1, II.804
Boundary layer, II.804, II.807, II.812–46
Boundary-layer equations, II.804, II.806, II.843
Boundary-layer solution, II.802
Boundary-layer theory, II.797
Boundary-value problem, I.196
Boys' radiometer, I.28
Brachytherapy, III.1877, III.1890, III.1891
Bragg–Gray theory, III.1882
Bragg Law, I.428, I.431, I.469
Bragg peak, III.1887
Bragg reflection, II.990, III.1367
Brain imaging, III.1906, III.1922, III.1929, III.1932
Brain tumour, III.1871, III.1904, III.1905, III.1919
Bravais lattices, I.462, I.466, III.1513
Bravais model, II.984
Breit–Wigner formula, II.1194
Bremsstrahlung, III.1766

Brillouin frequency shift, III.1398
Brillouin scattering, III.1405
Brillouin zone, II.987, II.1011, III.1293, III.1295, III.1296, III.1300, III.1305, III.1311–13, III.1315, III.1328, III.1331, III.1348, III.1359
British Association for the Advancement of Science, I.5, I.52, II.1252, III.1705
British Broadcasting Company (BBC), III.1988
British Institute of Radiology, III.1861
British Journal of Radiology, III.1867
British Medical Bulletin, III.1901
British Radio Research Board, III.1987–8
Brittle fracture, III.1520
Broadcasting, III.1987, III.1994
Brownian motion, I.30, I.43, I.147, I.533, I.592, I.595–7, I.607–8, I.616, I.623, I.628, III.1441–2
Brussels school, I.619–22
BTH Company, III.1583
Bubble chamber, II.659, II.753
Bubble domains, II.1159–60
Building-up principle (*Aufbauprinzip*), I.96–7, I.366
Bulbous bow, II.885
Bulk-carriers, II.885
Bunsen burner, I.19, I.75
Buoyancy force, II.844
Bureau International de l'Heure (BIH), II.1247, II.1248
Bureau International des Poids et Mesures, I.20
Burgers vector, I.475–6
Busch's lens theory, III.1570
verification of, III.1569–70

Cabibbo–Kobayashi–Maskawa (CKM) matrix, II.674, II.723–5
Caesium, I.75

Caesium-137, III.1892
Caesium chloride, I.429
Caesium clock, II.1245–8
Calcite, I.430, I.485
Calcium tungstate, III.1537, III.1901
Californium, II.1225, II.1227
Calorimetry, II.968, III.1883–4
Calutron isotope separation, III.1630
Cambridge Database System, I.512
Cambridge Electron Accelerator (CEA), II.741
Cambridge Structural Database, I.512
Cambridge University, I.9–15, I.17, I.20, I.21, I.54
Cancer, III.1857, III.1886, III.1917
Candela, II.1251
Canonical ensemble, I.526
Cantor set, III.1846, III.1847, III.1852
Capacitance, II.1255
Capacitors, II.1256
Carbon, formation of, III.1746
Carbon bond, II.1043
Carbon dioxide, atmospheric, II.908
Carbon monoxide in interstellar medium, III.1763–4
Carbon–nitrogen–oxygen cycle (CNO cycle), III.1707–8
Cardiac function, III.1898
Cardiac pacemaker, III.1895
Cardiovascular devices, III.1897–8
Carl Zeiss, III.1410
Carnot cycle, I.536–7, II.1249
Carnot principle, I.60
Cascade showers, I.387
Cassiopeia A, III.1737, III.1775, III.1782
Catalogue of Selected Compact Galaxies and of Post-Eruptive Galaxies, III.1713
Cathode ray, I.27, I.36, I.52, I.53

Cathode-ray oscilloscope, III.1989
Cathode-ray tube, III.1530, III.1568, III.1909
Cauchy condition, I.484
Cauchy integral, III.1404
Cauchy model, II.967
Caustics, III.1466
Cavendish Laboratory, Cambridge, I.17, I.20, I.21, I.54
Cavity radiation, I.144, I.145
CCDM, II.1240
Cellular automata, I.629–31
Cellulose fibres, III.1597
Centaurus X-3 (Cen X-3), III.1754
Centimetre–gram–second (CGS) system, II.1252
Central-axis depth-dose (CADD) curves, III.1872, III.1873
Centrosymmetric structure, I.448, I.449
Cepheids, III.1714, III.1716, III.1719
Ceramics, III.1534
  conductors, III.1537, III.1540
  crystalline glazes on, III.1537
  electrical, III.1536–42
  electro-optical, III.1536–42
  functional (or 'fine'), III.1538
Čerenkov counter, II.749–51, III.1446
Čerenkov detector, II.758, II.765
Čerenkov effect, III.1408
Čerenkov radiation, III.1363
CERN, III.2024, III.2036, III.2054, III.2055, III.2057
Cerro Tololo International Observatory (CTIO), III.1744
Cesium-137, III.1879
C-field, III.1787
CGS units, II.1235
Challenger research ship, III.1973
Challenger space disaster, III.1742
Chandrasekhar limit, III.1759
Chandrasekhar mass, III.1711

Channel classification, prototype of, II.1060–3
Channel coupling, empirical parametrization, II.1073
Chaos, I.593, III.1466–7, III.1823, III.1826, III.1831, III.1848
Chaotic motions, II.815
Chapman–Enskog method, I.597–8
Chapman–Kolmógoroff equation, I.596
Chapman–Kolmógoroff–Smoluchowski–Bachelier equation, I.604
Chapman layer model, III.1994
Characteristic impedance of free space, II.1261
Charge-coupled devices (CCD), II.1248, III.1454, III.1476, III.1586, III.1590, III.1745, III.1770, III.1796
Charged particles
  detection of, II.1200
  measuring devices and detectors, II.1206–8
Charm hypothesis, II.699–700, II.702–4
Charmed particles, II.700–1, II.704–6, **II.705**
Charmonium, II.701–2, II.704
Charnley low-friction arthroplasty, III.1896
CHART fractionation schedule, III.1888
Chemical constant, I.154
Chemical defect, III.1505
Chemical elements, I.45
  formation of, III.1787
  origin of, III.1746–7, III.1785
  synthesis of, III.1747, III.1784–5
Chemical potential, I.532
Chemical quanta, I.154–5
Chemical Rubber Company (now CRC Press), III.1513
Chemical spectroscopy, I.19

Chew–Goldberger–Low (CGL) equations, III.1637
Chew–Low theory, II.680
Chiral invariance, II.676
Chiral smectic C, III.1604, III.1605
Chiral symmetry, II.676
Cholesteryl iodide, I.444
CIE (Commission Internationale de l'Éclairage) chart, II.1251, III.1477
Cinematography, III.1410
Clarendon Laboratory, I.20
Classical statistics, I.525–8
Clausius–Mossotti law, I.485, I.546
Climate, II.899–912
Clocks, II.1242–6
Close collisions, II.1076, II.1077
  by ions and atoms, II.1041–2
  of slow electrons, II.1040–1
Closed trapped surface, III.1778
Cloud chamber, I.54, I.367, I.384, I.388, I.389, I.396, I.408, II.120, II.751–2, II.1200
Cloudy crystal ball model, II.1212
Clusters, II.1084
  masses, III.1772
Cobalt-60, III.1858, III.1873, III.1874, III.1879, III.1892
Cobalt-bomb, III.1872
Cobalt particles, III.1611
Cobalt structure, I.472
COBE satellite, III.1804
Cockcroft–Walton generator, II.639, II.727
Co–Cr–Mo alloys, III.1897
CODATA, II.1261, II.1262, II.1266, II.1268–72
Coherence theory, III.1408, III.1439–41
  cross-spectral purity in, III.1433
Coherent anti-Stokes Raman scattering (CARS), III.1448
Coherent imaging, III.1418
Coherent potential approximation, II.1006

Subject Index

Coherent regeneration, II.674
Cohesive energy, I.484–5
Cold dark matter, III.1804, III.1805
Cold emission, **III.1307**
Collagen, structure of, I.499
Collective Atomic Recoil Laser, III.1456
Collectivization, III.2052–4
Colliding-pulse mode-locking, III.1439
Collision sequences, I.612
Collisions between atoms or ions, II.1039–42, II.1074–81
  collision speed comparable to electron velcities, II.1079–80
  highly stripped ions, II.1080–1
  projectiles faster than atomic electrons, II.1076–7
  projectiles slower than atomic electrons, II.1077–9
Colloids
  definition, III.1610–13
  dilute systems, III.1610
  future lines of research, III.1612–13
  instability, III.1610
  protection, III.1610, III.1611
  recent progress, III.1612–13
Colour centres, III.1529, III.1531, III.1541
Colour flow imaging, III.1914
Coma cluster, III.1732, III.1772, III.1800
Combination scattering, III.1404–6
Combinatorial analysis, I.526
Combinatorial formula, I.527–8
Commission Internationale de l'Éclairage chart, *see* CIE
Communications
  radio, III.1985–6, III.2000
  theory, III.1417
Compact optical-discs, III.1462, III.1464
Complementarity, I.201–4, I.223–5, III.1395

Complex fluids, III.1596
Complex susceptibility function, III.1404
Complexions, I.523, I.526
Compositeness, II.762–3
Compound nucleus, II.1219
Compound parabolic concentrator (CPC), III.1446
Compton effect, I.120, I.163–72, I.362–6, II.638, III.1302, III.1399, III.1404, III.1862, III.1865
Compton Gamma-Ray Observatory, III.1741, III.1784
Compton scattering, I.218, I.365, I.368, II.991, II.1057, III.1413, III.1773, III.1801
Compton wavelength, II.756, II.1026
Computational fluid dynamics (CFD), II.881–2, II.906–7
Computed tomography (CT), III.1858, III.1873, III.1898, III.1916–23
Computer-aided design (CAD), III.1899
Computer-aided manufacture (CAM), III.1899
Computer analysis, II.906
Computer-generated physics, III.1823–53
Computer graphics, I.628
Computer simulation, I.619, I.628, III.1670–3, III.2044
Computerized treatment-planning systems, III.1875
Computers, I.593, I.622
  early, III.1828–9
  in medical physics, III.1858
  in optical sciences, III.1475–6
  supercomputer, III.2045
Concave spectral gratings, I.54
Concorde, II.877, II.878

S9

Condensation sequence model, III.1965
Condensed matter, II.756, III.2051
Condenser-objective lens, *Einfeldlinse*, III.1576
Condensing gas, statistical mechanics, I.551–4
Conduction band, III.1319, III.1329, III.1334, III.1340, III.1354
Conduction electrons, II.1168, III.1362
Conductors, magnetism, II.1133–6
Conference on Radioisotope Techniques, III.1901
Configuration space, III.1395
Confocal microscope, III.1450–1
Connected clusters, I.572
Conservation laws, I.374
Conserved vector current (CVC), II.667–9
Constant-energy surfaces, III.1297
Constant-Q technique, II.994
Consultative Committee for Electricity (CCE), II.1254, II.1256–60, II.1262, II.1271, II.1273
Consultative Committee for Photometry (CCP), II.1251
Consultative Committee on the Definition of the Second (CCDS), II.1247
Consultative Committee on Thermometry (CCT), II.1250
Contact-lens technology, III.1414
Contact potential, III.1283, III.1308
Continental drift, III.1967, III.1969, III.1972, III.1973
Continuous phase transitions, II.1009, II.1015
Continuous random network model, I.477
Continuum problems, III.1828
Contraction theory, III.1945

Controlled fusion research, III.1617, III.1633
Controlled fusion research—an application of the physics of high-temperature plasmas, III.1641
Convergent–divergent nozzle, II.879, II.880
Cooling of atoms, II.1099
Cooper pairs, II.938–42, II.946–8, II.950–2, II.954, II.956, II.962–4
Coordinated Universal Time (UTC), II.1247, II.1248
Copernicus satellite, III.1741, III.1765, III.1791
Copper, I.430, III.1515, III.1531, III.1543
Copper–gold alloys, III.1516
Copper oxide superconductors, II.959
Copper–silver solid solution, III.1543
Copper–tin diagram, III.1509
Corona, III.1441
Coronary arteries, III.1898
Coronium, I.76, I.81
Correlation diagram, II.1044
Correlation energy, III.1315
Correspondence principle, I.86
COS-B satellite, III.1740
Cosmic Background Explorer (COBE), II.758, III.1791
Cosmic microwave background, II.758
Cosmic rays, I.364, I.367–70, II.655–7, II.692
  intensity measurement, I.384
  penetrating radiation, I.388, I.404–5
  physics, II.760, III.2004–5
  soft and hard components, I.383–94
  see also Mesotron

Cosmical Electrodynamics, III.1636, III.1685
Cosmological constant, III.1723, III.1724, III.1730, III.1734, III.2039
Cosmological principle, III.1729
Cosmological problem (1939), III.1733-4
Cosmology, II.757-62, III.1691-821
  astrophysical, III.1784-93
  birth of, III.1691
  physical, III.1721-2
  relativistic, III.1721-34
  research since 1945, III.1746-807
  steady-state, III.1786-8
COSMOS High-Speed Measuring Machine, III.1745
COSTAR mission, III.1461
Coster–Kronig transitions, II.1089
Coulomb attraction, III.1331
Coulomb excitation, II.1217
Coulomb field, II.1217, III.1316
Coulomb force, I.123, I.376, III.1315
Coulomb interactions, II.1040, II.1125, III.1290, III.1316, III.1359, III.1362-4, III.1375, III.1551
Coulomb law, I.116, II.641, III.1361
Coulomb repulsion, III.1322
Covering lattice, I.573
Cowling's theorem, III.1958
CP violation, II.671-5
Crab Nebula, III.1737, III.1739, III.1752, III.1774
Critical behaviour, I.568
  binary alloys, I.548-9
Critical current, II.916
Critical density, III.1731
Critical fields, II.920
Critical fluctuations, II.1010, II.1016
Critical magnetic field, II.916
Critical model, III.1731
Critical opalescence, I.546-7

Critical phenomena, I.570, III.1595
Critical point, I.543, III.1947
Critical temperature, I.545, I.547, II.975, II.976
Critical velocity, II.928
Cross-spectral purity in coherence theory, III.1433
Cryostats, helium, I.436-7, III.2025
Crystal boundaries, III.1528
Crystal defects, III.1320, III.1531, III.1535
Crystal diffraction spectroscopy, II.1207
Crystal field, rare-earth elements, II.1164-5
Crystal field splittings, II.1164
Crystal field theory, II.1126-9, II.1136, II.1139, II.1141, II.1144-5
Crystal growth, I.466
Crystal lattice, III.1520
Crystal morphology, I.421
Crystal rectifier, III.1318
Crystal spectrometer, I.430, I.433
Crystal structure, I.423, III.1513-16, III.1529
  analysis, I.484-96
  and magnetic properties, II.1156-8
  determination methods, I.439-50
  faults in layer, I.472-3
  imperfections in, III.1505-6
Crystal surfaces, *see* Surface crystallography
Crystalline cleavage facets, III.1517
Crystalline virus, I.437
Crystallites, III.1280
Crystallographic analysis, III.1546
Crystallographic Data Centres, I.512
Crystallography, I.421-4, III.1515-17, III.1547
  history of, I.512
Crystallography in North America, I.512

Crystals
  dielectric properties, I.485–6
  dislocations, I.475–7
  doubly refracting, I.485
  imperfect, I.471–84
  optical properties, I.485–6
  see also under specific crystal types
CuAu, III.1515
Cu$_3$Au, III.1515
Cu–Ni–Fe alloy, III.1548
Curie law, I.540–1, II.1116, II.1133
Curie point, I.544–6
Curie temperature, I.545–6, I.549, I.550, I.559, III.1970, III.1976
Curie–Weiss Δs, II.1128–9
Curie–Weiss law, II.1116, II.1120, II.1128
Curium, II.1225
Curve of growth technique, III.1702
CuZn, III.1534
Cyclotrons, II.729–30, II.1199, II.1201, II.1206, II.1221, III.1871
  frequency, III.1334, III.1336
  resonance, III.1336, III.1337, III.1357
Cygnus A, III.1737, III.1775, III.1783, III.1788
Cygnus X-1, III.1754, III.1757
Cytosine, I.504

Daguerreotype process, III.1692
d'Alembert's Paradox, II.796, II.803–8
d'Alembert's Theorem, II.803–8, II.824, II.870, II.886
Danish Atomic Energy Commission, III.1643
Dark matter, II.761, II.768
Darwin–Fisher hypothesis, III.1948
Data collection, II.906
Data processing, thermodynamics of, I.576

de Broglie relation, I.457, III.1296
de Broglie waves, I.199, I.387, II.946, II.975, II.976, II.991, III.1292
Debye characteristic temperature, I.452
Debye equivalent, II.977
Debye formula, II.973
Debye function, II.972
Debye–Hückel theory, III.1521
Debye temperature, II.972, II.975, II.976, III.1305
Deceleration parameter, III.1794–7
Deep inelastic scattering, II.714–15
Defibrillation, III.1895
de Haas–van Alphen effect, II.1166, III.1348–56
δ-function, I.607
δ-rays, I.37
De Magnete, I.544, III.1856
Dementias, III.1932
Dendritic growth, III.1542
Density matrix, I.602, I.617
Density of states, II.986, II.1006
Density parameter, III.1734, III.1799–800
Deoxyribonucleic acid (DNA), I.504, I.507
Depletion layer, III.1325
Detectors, II.744–54
Deterministic hypothesis, I.222
Deuterium, I.112, I.370, III.1627, III.1706, III.1800
Deuterium-tritium, III.1651, III.1674
Deuteron, I.370, I.403
Deutsche Gesellschaft für Metallkunde, III.1553
Deutsche Gesellschaft für Naturforscher und Ärzte, I.47
Deutsche Gesellschaft für Technische Röntgenkunde, I.511

Diabatic potential energy functions, II.1079
Diagnostic machines, III.1897
Diagnostic medicine, III.1900–10
Diagnostic radiology, III.1859–63
Diagnostic ultrasonics, III.1910–11
Diamagnetism, II.1113–15
Diamond, I.429
Diamond–graphite equilibrium diagram, I.539
Diaphanography (light transmission imaging), III.1932–3
Dictionary of Applied Physics, II.977
Dictionary of Scientific Biography, I.16
Dielectric breakdown, III.1336–7
Dielectric constant, III.1595
Dielectric properties of crystals, I.485–6
Dielectric response, II.1055
Dielectric waveguide, III.1397, III.1420
Differential equations, III.1826
Diffractals, III.1466
Diffraction gratings, III.1401–2, III.1466
Diffraction microscopy, III.1582
Diffraction pattern, III.1412
Diffraction theory, III.1420
Diffractometer, I.435
Diffusion, II.818–19, III.1531–5, III.1548
  by continuous movements, II.830
  Fick's law of, I.597, I.601
Diffusion coefficient, I.597
Diffusion–convection balance for scalar quantities, II.841–6
Diffusion creep, III.1534
Diffusion inhibitor, III.1629
Digital subtraction angiography, III.1863
Diketopiperazine, I.464
Dimensional analysis, II.838

Diopside, structure of, I.440
Dipole–dipole interaction, II.987
Dipole vibration, II.1213
Dirac delta-function, I.206
Dirac equation, **I.361**, I.362
Dirac–Jordan transformation theory, I.198–201
Dirac theory, II.651
Directional distribution patterns, **II.1035**
Disabled persons, technology for, III.1899
Disciplinism, III.2042
Discretization process, I.629
Dislocations, I.475–7, III.1507, III.1520, III.1521, III.1525, III.1526, III.1528
Disordered materials, III.1365–9
Dispersion, III.1404
  curves, II.988–90, II.998
  formulae, I.171–2
  relations, II.680
  theory, I.184–5
Dispersive waves, II.856–60
Displacement law, I.66
Distorted wave Born approximation (DWBA), II.1220, II.1221
Distribution of charges and current in an atom consisting of many electrons obeying Dirac's equations, The, III.1622
D-lines, I.66
DNA, *see* Deoxyribonucleic acid (DNA)
Dolen–Horn–Schmid duality, II.683
Doping (and dopants), III.1529, III.1544, III.1593
Doppler effect, I.269, I.270, II.1094, III.1781, III.1784
Doppler shift, III.1398, III.1710
Doppler ultrasound, III.1914
Double-helix structure, I.506

Double strand breaks (DSBs), III.1885
Dreams of a Final Theory, III.2020
Drell–Yan process, II.701–2
Drift chambers, II.749, II.754
Drude–Lorentz model, III.1288
Duality, II.682–4
Dulong and Petit's law, II.970, II.972, II.980
Dwarf stars, III.1696, III.1698
Dye molecules, II.1087
Dynamic hologram, III.1458
Dynamical electroweak symmetry breaking, II.763
Dynamical friction, III.1774
Dynamical symmetry breaking, II.957
Dynamical systems, III.1850
　classification of, I.614–16
Dynamo theory, III.1985

Earth
　core, III.1952–60
　core size, III.1954
　fluid envelope, II.887–909
　gaseous-interior model, III.1947
　Laplace–Herschel–Kelvin scenario, III.1949
　liquid–gas boundary, III.1947
　magnetic field, III.1622–3, III.1952–60, III.1973
　origin and age (post-1935), III.1960–6
　origin and age (pre-1935), III.1944–52
　primaeval, III.1949
　structure, III.1946, III.1954, III.1956, III.1957
Earth–Moon system, III.1948
Earth sciences, III.1967, III.2046
Earthquakes, III.1956, III.1978–80
　zones, III.1976
Echo-sounding data, III.1973
École Polytechnique, I.15

Eddington–Lemaître model, III.1732–3
Eddington luminosities, III.1780
Edge dislocation, I.475–6
Education, I.8–16
　curricula, I.8
　higher, women in, I.8, I.10
EFFI, III.1673
Ehrenfest dog-flea model, I.618
Eigenfunctions, III.1432–7
Eigenvalue problem, III.1848
EIH method, I.303
Einfeldlinse (condenser-objective lens), III.1576
Einstein–Bohr frequency condition, I.359
Einstein–Bose statistics, I.602–3
Einstein–de Sitter model, III.1730–1
Einstein–Podolsky–Rosen(–Bohm) (EPR(B)) paradoxes, I.228–9, III.1457
Einstein X-ray Observatory, III.1798
Einsteinium, II.1224
E layer, III.1994, III.2002
Elasser–Bullard theory, III.1960
Elastic constants, II.983
Elastic scattering, II.687
Elasticity, I.484–5
Electric charges, I.51
Electric fields, I.27, I.261, III.1283, III.1362, III.1434, III.1435
Electric polarization, II.1031
Electrical conduction in metals, III.1284–7
Electrical conductivity, III.1279, III.1530
Electrical counting methods, II.1200
Electrical impedance tomography, III.1932
Electrical stimulation, III.1895
Electrical surges (*Wanderwelle*), III.1566

Subject Index

Electrical units and standards, II.1252–60
Electricity, I.51, I.52
Electricity supply industry, II.1255
Electro-absorption, III.1423
Electrocardiogram/graph, III.1893, III.1896
Electrodynamics, I.88, I.266, I.267, I.270
  in special theory of relativity, I.257, I.280–1
Electroluminescence, III.1417
Electrolysis, I.51
Electromagnetic field, I.358, III.1434, III.2039
Electromagnetic gauge invariance, II.693
Electromagnetic induction, I.266, II.1112
Electromagnetic interactions, I.391
Electromagnetic radiation, I.197, II.1264, III.2018, III.2025
Electromagnetic spectrum, III.1734–45
Electromagnetic theory, I.362
Electromagnetic waves, I.260, III.1391, III.1622, III.1985
Electromagnetism, II.640, II.693, III.1856, III.2000
Electron beams, III.1456, III.1565–91
Electron charge/mass ratio (e/m_e), II.1271
Electron–deuteron scattering, II.698
Electron density, I.440, I.446, I.448, I.452, I.454, I.502, III.1316
  regions, III.2009, III.2010
Electron diffraction, I.463–5, I.481
Electron distribution, I.452
Electron–electron collisions, II.739–40, III.1316
Electron energy
  LEED experiments, I.466
  RHEED experiments, I.468

Electron energy-loss spectrometer (EELS), III.1580
Electron $g - 2$ factor, II.647
Electron gas, III.1361, III.1362, III.1371–5
  quantum theory of, III.1287–91
Electron–hole pairs, II.999
Electron layer, III.1993
Electron magnetic moment, II.644–5
Electron magnetic resonance, III.1859
Electron mass, II.1271
Electron microprobe analyser, III.1548–9
Electron microscopes, I.481, II.931, III.1526, III.1569, III.1570–4
  see also under specific types of electron microscope
Electron mobility, III.1368
Electron multiplication, III.1419
Electron–neutrino pair, II.639
Electron pair, II.1028
Electron paramagnetic resonance (EPR), II.1137–52
  3d ions, II.1139–41
  4f ions, II.1141–3
  gedanken experiment, III.1457, III.1467
  iron group ions, II.1137–9
  rare earths, II.1141–3
Electron–phonon collisions, III.1316
Electron–phonon coupling, II.1000
Electron–phonon interaction, II.941, II.999, III.1303
Electron–phonon scattering, III.1305
Electron physics, III.2000
Electron–positron annihilation, II.690–1, II.715, II.717, II.740
Electron–positron collisions, II.691, II.711, II.740, II.741, II.768
Electron–positron pair production, III.1784, III.1786, III.1806

S15

Electron probe x-ray microanalysis, III.1580
Electron–proton collider (HERA), II.766–7
Electron scattering, II.1033, II.1221–2
　cross section, III.1996
Electron scattering factor, I.463–4
Electron spin, I.215
Electron spin resonance (ESR), II.1137–8, III.1923–33
Electron synchrotrons, II.731–2
Electron theory, I.30
　of metals, I.215–17
Electronic devices, therapeutic, III.1896
Electronics, medical. *See* Medical electronics
Electronics technology, III.2031
Electrons, I.36, I.51, I.53, I.59, I.64, I.77, I.81–90, I.95–9, I.115, I.158, I.167–8, I.369, I.377, I.457, II.637–8, II.641, **III.1298–9**
　free, I.372
　hot, III.1336–41
　in medical physics, III.1857
　in solids, III.1279–383
　see also Valence electrons
Electronvolt (eV), I.358
Electrostatic effects, charge effects, III.1610
Electrostatic electron microscopes, III.1577–8
Electrostatic generators, II.727–9
Electrosurgery, III.1894
Electroweak theory, II.640, II.762
Electroweak unification, II.692–709
Elementary charge (e), II.1271
Elementary excitations, II.928, II.929
Elementary particles, I.115, I.231–2, I.370–1, I.391, I.408–9
　in special theory of relativity, I.282–3
　physics of, II.635–794
　Standard Model of, III.2030, III.2033–9
　see also under specific particles
Elementary phenomena, II.1029–32
Elements, *see* Chemical elements
Elements of Physics, III.1857
El Niño/Southern oscillation phenomenon (ENSO), II.908
Elsasser–Bullard dynamo theory, III.1973
EMI-Scanner, III.1922
Emission electron microscope (EEM), III.1569, III.1577
Emulsions, III.1611
Encyclopaedia Britannica, I.36
Encyclopedia of Materials Science and Engineering, III.1513
Endocrinology, III.1930
Energy bands, III.1300
Energy density, III.1393
Energy fluctuation, I.163
Energy gap, II.932, II.952, III.1329, III.1433
Energy loss, 'hidden', II.846–9
Energy principle for hydromagnetic stability problems, An, III.1672
Enskog time, I.619
Entanglement, III.1394
Entropy, I.47, I.523, I.601
　calorimetric and statistical estimates, I.539–40
Eötvös experiments, I.286–7
Ephemeris Time, II.1247
Epilepsy, III.1922
　diagnosis, III.1916
Epoch of recombination, III.1801, III.1803
Equations of motion, I.48, I.186, I.303
Equilibrium angular momentum, II.932
Equilibrium state, II.920

Equilibrium theory, III.1785
Equipartition theorem, I.5, I.523, I.528
Equivalence principle, I.286–91
Equivalent width concept, III.1702
Erbium-doped optical-fibre amplifier (EDFA), III.1447
Ergodic surface, I.525
Ergodic theorems, I.599, I.626
Essen Ring, II.1244
ETA-BETA I, III.1658
ETA-BETA II, III.1658
Euler–Lagrange equation, I.195
Euratom-CNR facility, Padua, III.1658
European Federation of Medical Physics (EFOMP), III.1934
European Optical Society, III.1474
European research system (or science base), III.2055
European Science Foundation, III.2055
European Southern Observatory (ESO), III.1744
European Space Agency, III.1742, III.2055
Europium sulphide, I.559
Ewald transformation, II.988
Exchange interactions, II.1129, II.1143–4
Excitation channels, II.1059–73
Exciton absorption spectra, III.1423
Exciton theory, III.1423
Exclusion, **II.1028–9**
Exclusion principle, I.97–100, I.168, I.603, II.686, II.1116
Exluded volume, I.569
Experimental facilities, I.24–9
Experimental physics, I.13, I.14
Explorer I, III.2007
Explorer II, III.1740
Explorer III, III.2007
Explorer IV, III.1683, III.2007, III.2008

Explorer satellites, III.1739
Explosive nucleosynthesis, III.1747–8
Explosive showers, I.387
Exponential law, I.67
Extinction coefficient, III.1396
Extremely low-frequency radio physics, III.1995
Eye tumours, III.1876
Eyewall, II.904

F1 layer, III.1994
Fabry–Perot étalons, II.1240
Fabry–Perot interferometer, I.19, III.1412
Fabry–Perot resonators, III.1422, III.1423, III.1429
Fabry–Perot-type beam receiver, I.315
Face-centred-cubic lattice, I.557, III.1295
Facsimile (FAX), III.1476
Far infrared radiation, II.1054
Faraday (unit), I.51, II.1269
Faraday's constant, I.150, II.1237
Faraday's law, I.52
Fast charged particles, II.1039–40
Fast collisions, II.1076
Fast processes, II.1085
F-centre (Farbzentrum), III.1320–1
Fe–Ag–B alloys, III.1543
FECO interferometry, III.1412
Fellgett advantage, III.1416
Fellows of the Royal Society, I.10, I.12
Femtosecond pulses, III.1462
Fermi–Dirac statistics, I.117, I.123, I.155, I.531, I.533, I.603, II.687, III.1706, III.1711
Fermi distribution, II.1168, III.1301, III.1302, III.1308
Fermi energy, III.1295, III.1302, III.1309, III.1364
Fermi-field theory, I.375–8, I.382, I.392, I.393, I.404

Twentieth Century Physics

Fermi gas, II.1209, **III.1289**, III.1301, III.1308, III.1316
Fermi particles, I.603
Fermi statistics, II.1190, III.1394
Fermi surface, II.999, II.1002, III.1305, III.1308, III.1345–59, III.2025
Fermi theory, II.651
Fermilab, III.2054
Fermions, II.676, II.694, II.955, II.1168–70, III.1355, III.1804
Fermium, II.1224
Ferrimagnetic insulators, II.1156
Ferrimagnetism, II.1113, II.1152–65
  magnetic properties, II.1157
Ferroelastic crystals, III.1538
Ferroelectricity, I.486–9, III.1538
Ferrofluids, III.1611
Ferromagnetic model, I.570
Ferromagnetic resonance, II.1152–3
Ferromagnetism, I.559, I.562, II.1015, II.1113, II.1120–3, II.1134–6, II.1152–65, III.1279
Feynman diagrams, **I.361**
Fibre-optic light, III.1894
Fibre optics, III.1419–21
  see also Optical fibre
Fick's law of diffusion, I.597, I.601
Field cycling behaviour, II.1121
Field emission microscope, III.1569
Field ion microscope, I.50
Fierz interference, II.664
Finalization, III.2048–50
  pre-paradigmatic phase, III.2048
  revolutionary phase, III.2048–9
Fine structure, **II.1036–7**
  constant, II.640, II.1026, II.1262, II.1271
Finite energy quantum, I.173
Finnegans Wake, II.685
First Antarctic Expedition, III.2005
First International Congress of Medical Physics, III.1909, III.1923

First International Congress of Radiology, III.1881
First law of thermodynamics, I.374, I.521
First-order Fermi acceleration, III.1782
First-order melting transition, I.577
First Texas Symposium on Relativistic Astrophysics, III.1776
Fischer–Riesz theorem, I.206
Fitzgerald–Lorentz contraction, I.32, I.264
Fixed point, I.567, III.1832
Fixed-wing aircraft, II.817
Flash-photolysis, II.1085
Flash-spectroscopy, II.1085
Flight efficiency, II.869–73
Flood waters, II.897
Flory temperature, III.1598
Flow
  in pipes, II.814–15, II.825
  resistance, II.815
Fluctuation equation, III.1393, III.1402
Fluid envelope, II.860, II.887–909
Fluid mechanics, II.795–912
Fluid particles, I.572
Fluids, microscopic critical behaviour, I.546–7
Fluorescence, III.1537
Fluorite, I.430
Flying-spot microscope, III.1416
Focal conic defects, III.1605
Fokker–Planck equation, I.595–7, I.624, I.627
Forbidden transitions, III.1762
Forced convection, II.843, II.844
Form factors, II.660, II.687
FORTRAN, III.1843
Fortschritte der Physik, I.7
Foucault's pendulum, I.2–3
Fourier imaging theory, III.1418–19

Subject Index

Fourier series, I.441, I.623
Fourier synthesis, I.501
Fourier techniques, reconstructed (111) surface, I.469
Fourier-transform, III.1418, III.1924
  image processing, III.1413
  spectrometer, III.1416
  spectroscopy, II.1055
Fourier's law of thermal conduction, I.601
Fractals, I.574, I.630
Fractured bones, III.1897
Fragile structures, III.1595
Fragmentation, II.1102, II.1103
  dynamics of, II.1105–6
Franck–Condon rule, II.1046
Franck–Hertz experiment, I.89
Free convection, II.843, II.844
Free-electron gas hypothesis, I.216
Free electrons, I.372
Free-energy phase diagram, I.538
Freedman–Clauser experiment, III.1458, III.1467
Frequency
  dissemination of, II.1248–9
  standards, II.1243–6
Fresnel biprism, III.1585
Fresnel diffraction fringes, III.1580–2
Friedel's law, I.440, I.464, I.495
Friedman world model, III.1725, III.1731, III.1733, III.1784, III.1807
Froome interferometer, II.1264
Frustration, II.1170
FSU, III.2028
Fullerene crystals, II.956
Functional electrical stimulation (FES), III.1900
Functionally graded materials, III.1542
Fundamental observers, III.1729
Fundamental physics, III.2034–5
Fundamental Theory, The, III.1700, III.1787

Fusion bomb, III.2022
Fusion research, III.1617, III.1618, III.1627–9, III.2024
  declassification, III.1640–5
  principal sites for, III.1636
  see also Magnetic fusion research
FWHM, III.1904, III.1922

Gadolinium, magnetic properties, II.1162
Galactic cannibalism, III.1774
Galactic Centre, III.1740, III.1744, III.1763
Galaxy(ies), III.1769–71
  active, III.1797–8
  clusters of, III.1771–4
  crossing times, III.1772
  dark halo, III.1771
  dark matter, III.1771
  distribution of, III.1771–2
  elliptical, III.1771
  formation, III.1800–5
  luminosity function for, III.1769–70
  masses, III.1769
  structure of, III.1714–21
  velocity curves, III.1771
  violent events in nuclei of, III.1777
Galena crystal, III.1318
Galilean transformation law, I.257
Galilei–Newtonian kinematics, I.257
GALLEX experiment, III.1749
Gallium, I.75, III.1749
Gallium arsenide (GaAs), III.1371, III.1375, III.1464
Gamma-10, III.1669
Gamma camera, III.1907–9, III.1918, III.1920, III.1921
Gamma-decay, I.106
Gamma-radioactivity, I.53
Gamma-ray astronomy, III.1738–41
Gamma-ray detectors, III.1740

S19

Gamma-ray imaging, III.1904
Gamma-ray wavelengths, II.1207
Gamma-rays, I.54, I.367, I.457, II.1207, III.1784, III.1879, III.1902
Gamow–Teller type interactions, II.664
Gargamelle Collaboration, II.697
Gas constant (R), II.1269
Gas discharges, I.27
Gas of corpuscles, III.1390–2
Gaseous Electronics Conference, III.1644
Gases, I.47
  ionization phenomena in, III.1644
  kinetic theory of, I.30, I.35, I.46, I.522, I.523, I.587, III.1284
  liquid, I.28, I.543–4
  specific heat, I.528–30
Gaskugeln, III.1695
Gauge theory, III.1414
Gaussian field-distribution profile, III.1429
Gaussian–Hermite (laser) beam propagation, III.1447
Geiger(–Müller) counter, I.108, II.747, II.1200, II.1207, III.1413, III.1900–2, III.2008
Geiger–Nuttall relation, I.116
GEKKO series of lasers, III.1676
Gell-Mann–Okubo mass relation, II.661
General Catalogue of Nebulae, III.1717
General Conference on Weights and Measures (CGPM), II.1234–6, II.1247, II.1248, II.1251, II.1257, II.1262, II.1264
General Electric Company (GEC) Research Laboratories, II.1204
General Electric (GE) Research Laboratory, New York State, III.1528
General theory of relativity, I.249, I.286–320, III.1722–4
  alternative formulations and foundations, I.298–9
  alternative scenarios, I.295–6
  and active galactic nuclei, III.1777–81
  approximation schemes, I.302
  astronomical and astrophysical applications and tests, I.315–17, III.1776
  classical tests, I.303–6
  equivalence principle, I.286–91
  exact solutions, I.302
  field equations, I.293–5, I.313
  gravitational energy, I.300–1
  later work on, I.296–8
  metric-tensor field, I.291–3
  physical interpretation, I.301
  Schwarzschild solution, I.303–6
  software programs, I.302
  tests, III.1757–8
Generalized form factor, II.1040
Genetic code, I.507
Geneva Conference (1958), III.1640–5, III.1642, III.1643
Geniac, III.1828
GENIUS—The Life of Richard Feynman, III.1519
Geology, III.2046
Geomagnetic dynamo theory, III.1958–9
Geomagnetic field, reversals of, III.1973
Geomagnetic field lines, III.2010
Geomagnetic storm, III.2002–3
Geomagnetism, III.2000, III.2001
Geometrical collective model, II.1218
Geophysical observing stations, III.1998
Geophysics, III.1943–2016, III.2046
  history, III.1943–4
  use of term, III.1943

Subject Index

Georan, II.1242
Geospace, III.1981, III.1997–2001
  future study propsects, III.2011
  pictorial models, III.2010
Germanium, I.75, I.470, II.982,
  II.1208, III.1281, III.1301,
  III.1319, III.1322, III.1324,
  III.1332–6, III.1340, III.1544,
  III.1749
Gesellschaft deutscher Naturforscher und Ärzte, I.5
Giant branch, III.1698
Giant dipole resonance (GDR), II.1213–14
Giant molecular clouds, III.1765
Giant stars, III.1696, III.1698, III.1708–9
Gibbs ensembles, I.531–3
Gibbs free energy, III.1543, III.1548
Gibbs function, I.549
Gibbs Paradox, I.528
Ginzburg–Landau theory, I.568, II.934, II.935, II.940
Glashow, Iliopoulos and Maiani (GIM) mechanism, II.700
Glashow model, II.694
Glashow–Weinberg–Salam theory, II.695, II.697, II.698, II.707
Glass industry, III.1517
Glasses, III.1533, III.1600–2
  phonons in, II.1004–7
Glitches, III.1754
Global Positioning Satellite System, II.1249
Glomar Challenger, III.1976, III.1980
Gluons, II.636, II.637, II.684, II.714, II.716
Glutamine, structure of, I.449
God Particle, The, III.2020
God principle, III.2019–20
Gold, III.1515, III.1543
Gold-198, III.1905
Goodyear, III.1593, III.1597
Göttingen quantum mechanics, I.183–96

Gradient-index lenses, III.1391
Grain boundaries, III.1518–20, III.1528
Grain size distribution, III.1549
Grand canonical ensemble, I.532
Grand partition function (GPF), I.533
Grand unified theory, I.323, II.763–4
Grating spectroscopy, III.1391
Gravitation, II.757–62, III.2037
  and quantum theory, I.318
  in special theory of relativity, I.283–4
  Newtonian constant of, I.28
  theory of, I.283–4, II.762
Gravitational collapse, I.306–12
Gravitational constant (G), II.1265–7
Gravitational energy, I.300–1
Gravitational field equations, I.253
Gravitational radiation, I.312–15
Gravitational waves, I.312, I.315
Gravity fields
  Moon, II.888
  Sun, II.888
Gray (unit), III.1867
Grazing collisions by ions and atoms, II.1041–2
Great Debate, III.1716–21
Green statistics, II.687
Greenhouse effect, II.899
Green's function, I.617, I.623–4
Ground-state energy, II.1045
Group-theoretical model, I.211–15
Group velocity, II.856
Guanine, I.504
Guided wave optics, III.1458
Guy's Hospital, III.1864
Gyromagnetic effect, II.1153
Gyromagnetic ratio, II.1237, II.1257
Gyroscope
  femtosecond-pulsed dye-laser ring, III.1433

S21

rotation-sensing ring-laser, III.1433

Habilitation thesis, I.184
Hadron colliders, II.707–8, II.742–4, II.766–7
Hadrons, II.686, II.690–1, II.699–700, II.704, II.713, II.715–17
Haemoglobin, I.500, I.502–3
Hafnium, I.100
Half-life concept, I.59
Hall constant, III.1334, III.1367
Hall effect, I.9, II.1096, III.1282, III.1284–6, III.1296, III.1298, III.1299, **III.1331**
Hall field, III.1373
Hall potential, II.1096
Hall resistance, II.1258
Hambergite, structure of, I.451
Hamilton–Jacobi equation, I.176, I.194
Hamilton–Jacobi theory, I.169, I.191
Hamiltonian, I.566–9, I.612
Hamiltonian matrix, I.190
Hamiltonian scheme, I.189
Hammersmith Hospital, III.1871, III.1873, III.1922
Handbook of Chemistry and Physics, III.1509–13
Handbuch der Physik, III.1513
Harker–Kasper inequality, I.447, I.448
Harmonic oscillator, II.973
Harrison construction, III.1313
Harrison–Zeldovich spectrum, III.1802, III.1807
Harvard Radio Research Laboratory, III.2026
Hayashi limit, III.1767
Heart valves, artificial, III.1898
Heat capacity, I.78–80
Heat conduction theory, III.1945
Heat Theorem, II.968, II.975

Heat-treatment, III.1519
Heaviside layer, III.1987
Heavy-atom derivatives, I.501
Heavy-atom method, I.444
Heavy element catastrophe, III.1785
Heavy fermions, II.955, II.1168–70
Heavy hydrogen, III.1627, III.1628
Heavy-Ion Medical Accelerator, Chiba, Japan, III.1876
Heisenberg–Dirac (H–D) theory, II.1128–9, II.1148–9
Heisenberg exchange integral, I.213
Heisenberg magnet, III.1595
Heisenberg n–p model, I.382
Heisenberg systems, II.1014
Helicity, II.667
Helicons, III.1358–9
Helimagnetic structure, I.462
Helioseismology, III.1749–51
Helium, I.58, I.75, I.81, I.101–2, I.110, I.182, I.211, II.974, II.1004, III.1707
  and discovery of cosmic microwave background radiation, III.1790–3
  cryostats, I.436–7, III.2025
  liquid, II.913–15, II.921–32, III.1341–5
  superfluid phases, II.949–55
  synthesis, II.1226
α-helix, I.498, I.508
Helmholtz free energy, I.527, I.532
Henry Draper Catalogue of Stellar Spectra, III.1694, III.1695
HERA (electron–proton collider), II.766–7
Hercules X-1 (Her X-1), III.1755
Hermann–Mauguin symbol, I.439
Hermite functions, I.625
Hermite polynomials, I.625
Hermitian affine connection, I.322
Hermitian matrices, I.190
Hermitian operators, I.205

Subject Index

Hertzsprung–Russell diagram, III.1692–8, III.1746, III.1793–4
Her X-1, III.1756
Hess–Fairbank effect, II.944–6, II.955, II.956
Hg–Ba–Ca–Cu–O compound, II.959
Hierarchical clustering, III.1804
Higgs boson, II.695, II.762
Higgs fields, III.1807
Higgs mechanism, II.694
Higgs particle, III.2036
High Energy Astrophysical Observatory A and B (HEAO-A and B), III.1740
High Energy Particle Physics, III.2051
High-energy physics, III.2056, III.2058
High-resolution images, III.1477
High-temperature superconductor, I.489–90
High Voltage Engineering (HVE), II.1203
Highest occupied molecular orbital (HOMO), II.1083
Hilbert conditions, I.293
Hilbert space, II.1025, III.1395
Hip replacement, III.1896
History of Metallography, A, III.1556
History of the University in Europe, III.2041
H-mode operation, III.1649
Hohlraumstrahlung, I.66
Holes, **III.1289–9**
  concept of, III.1297
Holograms, III.1415, III.1438, III.1582–4
Holography, III.1437–9, III.1580–1
  history of, III.1414–17
  off-axis optical beam, III.1585
  principle of, III.1415
Hooker 100" telescope, III.1704, III.1726

Hospital Physicist's Association, III.1864
Hot Big Bang models of Universe, III.1778
Hot Dark Matter, III.1803, III.1804
Hot electrons, III.1336–41
  phonon interaction with, III.1339
Hot shortness, III.1519
Hot-wire techniques, II.831
Hovercraft, II.886
HP65 programmable calculator, III.1829, III.1842, III.1846
H-theorem, I.35
Hubbard U, II.1166
Hubble space telescope, III.1461, III.1474, III.1741–2, III.1745
Hubble's constant, III.1730, III.1731, III.1787, III.1788, III.1793–4, III.1799
Human movement studies, III.1899–900
Hume-Rothery rule, III.1346
Hydraulic jump, II.858–9
Hydrodynamic forces on ships, II.882
Hydrofoil boat, II.886
Hydrogen, I.64, I.77, I.78, I.80–6, I.160, I.359
  molecular, I.181, II.1045, **II.1048**, II.1069–73, III.1766
  ortho, **II.1048**
  ortho–para, I.540
  para, **II.1048**
Hydrogen atoms, I.90–2, I.211, I.360, I.570, II.641
Hydrogen bombs, III.1633, III.1673
Hydrogen bond, II.1043
Hydrogen isotope, I.370
Hydrogen spectrum, I.90
Hydroxyapatite, III.1897
Hyperfine interactions, II.649–50
Hyperfine structure, **II.1036–7**, II.1214
Hyperon, II.657

S23

Hypersonic aerodynamics, II.879–80
Hyperspherical coordinates, II.1103–5
Hypothesis of energy elements, I.145
Hysteresis, II.918, II.1121

Ice, structure of, I.474–5
Illinois University, III.1871
Illustrated London News, I.2
Image processing, Fourier-transform, III.1413
Imaging theory, III.1399–401, III.1467–8
IMB-3 Detector, II.759
Imperfections in Nearly Perfect Crystals, III.1535
Implant surgery, III.1896
Incoherent backscatter, III.1995
Incoherent scatter radar (ISR), III.1995, III.1996
Independent-particle model, III.1359–65
Index Catalogues, III.1717
Indian ink, III.1611
Indirect-exchange-coupling theory, II.1012
Indium, I.75
Indium antimonide (InSb), reconstructed (111) surface, I.469
Induced transitions, III.1398–9
Induction coil, I.54
Ineffectiveness concept, III.1347
Inelastic neutron scattering, II.1162
Inelastic scattering, II.687–90, II.697, II.1221
Inequality relations, I.447
Inertial confinement fusion (ICF), III.1673–7
Inflation, II.761–2
Information retrieval, electronic methods, III.1896
Information technology, III.2044–6

Infrared absorption, II.988
Infrared astronomy, III.1742–4
Infrared Astronomy Satellite (IRAS), III.1744
Infrared radiation, laser sources of, II.1054
Infrared reflectivity, II.1003
Inner quantum number, I.100
Inner-shell phenomena, II.1089–92
  Auger emissions, II.1090–1
  spectra, II.1038–9
  x-ray studies, II.1089–90
Inorganic compounds, structure analysis, I.492–3
Inorganic Crystal Structure Database, I.512
In-plane x-ray diffraction, I.466
Instant camera, III.1417
Institut für Metallphysik und Allgemeine Metallkunde, III.1555
Institute for Defense Analysis, III.2023
Institute for the Study of Metals, University of Chicago, III.1556
Institute Laue-Langevin, III.2054
Institute of General Metallurgy (Allgemeine Metallkunde), III.1554–5
Institute of Metals Research, III.1555
Institute of Physics, III.1558, III.2042
Institution of Physics and Engineering in Medicine and Biology, III.1864
Instrumentation, role of atoms, II.1094–7
Integrated intensity, **I.434**
Integrated optics, III.1449–50
Interaction II, I.391, I.393, I.402
Interaction III, I.392, I.397
Interdisciplinarity, III.2042–4
Interference effects, III.1369–70

Subject Index

Interference pattern, III.1395
Interferometric spectroscopy, III.1410
Interferometry, I.25, I.26, III.1410, III.1738
  see also Optical interferometry techniques
Intermediate state, II.924
Intermediate valence, II.1168–70
Intermetallic compounds, III.1515
Intermetallic Phases Data Bank, III.1513
Internal Constitution of the Stars, The, III.1700, III.1702
Internal conversion, I.107
Internal field, I.545
Internal gravity waves, II.866
Internal pressure, I.543
International Ampere, II.1254, II.1257
International Astronomical Union (IAU), II.1242, II.1247
International Atomic Energy Agency (IAEA), III.1643, III.1648, III.1649, III.1908
  Scanning Conference, III.1916
International Atomic Time (TAI), II.1247, II.1248
International Bureau of Weights and Measures (BIPM), II.1234, II.1236, II.1240, II.1248, II.1251, II.1257, II.1259, II.1272
International Centre for Diffraction Data, I.512
International Commission on Radiation Protection (ICRP), III.1910
International Commission on Radiation Units and Measurements (ICRU), III.1867, III.1881, III.1910, III.1920
International Commission on Radiological Protection, III.1890
International Committee of Weights and Measures (CIPM), II.1234, II.1240, II.1242, II.1247, II.1249–51, II.1257, II.1264, II.1271
International Conference on Medical Electronics, III.1913
International Council for Scientific Unions, III.2006
International Critical Tables, II.1267, III.1509
International Earth Rotation Service (IERS), II.1248
International Federation of Medical and Biological Engineering (IFMBE), III.1934
International Geophysical Year (IGY), III.1973, III.1989, III.2005–6
International Medical Electronics Conference, III.1904
International Ohm, II.1254, II.1257
International Organization of Medical Physicists (IOMP), III.1934
International Polar Year (1882–83), III.2005
International Practical Temperature Scale (IPTS-48), II.1250
International System of Units (SI system), II.1235
International Tables for the Determination of Crystal Structures, I.511
International Tables for X-ray Crystallography, I.439
International Temperature Scale, II.1250
International Toroidal Experimental Reactor) device, *see* ITER

S25

Twentieth Century Physics

International Ultraviolet Explorer, III.1741
International Ultraviolet Observatory, III.1780
International Union for Geodesy and Geophysics, III.2006
International Union for Scientific Radio, III.2006
International Union of Crystallography, I.511–12
International Union of Physical and Engineering Sciences in Medicine (IUPESM), III.1934
International Union of Pure and Applied Physics (IUPAP), II.1268
International Union of Radio Science (URSI), II.1273
International Units, II.1255, II.1260
International Volt, II.1254, II.1257
International Weights and Measures Organization, II.1234
International X-ray and Radium Protection Commission, III.1890
Internationale Zeitschrift für Metallographie, III.1556
Internationalization, III.2055–6
Inter-Service Ionosphere Bureau (ISIB), III.1998–9
Interservice Radio Propagation Laboratory, III.1998
Interstellar chemistry, III.1765
Interstellar medium, III.1761–9
  soft x-ray background, III.1766
  two-phase model, III.1765
  violent, III.1766
Interstitials, III.1529, III.1531, III.1533
Introduction to the Ionosphere and Magnetosphere, An, III.2001
Introduction to the Study of Physical Metallurgy, An, III.1517

Inverse-Compton catastrophe, III.1781
Inverse-Compton scattering, III.1781
Inverse-square law, II.1266
Inversion layer, III.1373
Investigations of the solar spectrum and the spectra of the chemical elements, III.1692
Iodine, contrast media, III.1863
Iodine-123, III.1920
Iodine-131, III.1900, III.1903–6
Iodine-132, III.1905
Ion–atom collisions, *see* Collisions between atoms or ions
Ion guns, III.1997
Ionic crystals, II.997–8, III.1528–9, III.1531, III.1541
Ionization chamber (thimble chamber), III.1866–7, III.1881
Ionization detectors, II.746–9
Ionization energy, III.1867
Ionization equilibrium, III.1701
Ionization Phenomena in Gases, III.1644
Ionization threshold, II.1069–73
Ionization thresholds, spectral connections at, II.1058–9
Ionizing radiation, III.1857, III.1889
Ionosonde, III.1989, III.1991
Ionosphere, III.1619–23, III.1682–7, III.1981–2001
  'E' layer, III.1621
  electron content, III.1995
  'F' layer, III.1621
Ionospheric physics, III.1944
IRAS satellite, III.1767
Iridium-192, III.1879, III.1892
Iron–carbon alloy, III.1524
Iron-shrouded lens, III.1567
Irreducible cluster integrals, I.552
Irvine–Michigan–Brookhaven (IMB) experiment, III.1760
Ising model, I.554–61, III.1554

Isobars, II.680
Isomeric levels, II.1210
Isomorphous replacement method, I.445
Isospin, **I.395**
Isothermal compressibility, I.546, III.1397
Isotope effect, II.933
Isotopes, I.107
Isotopic charge, II.676
ITER (International Toroidal Experimental Reactor) device, III.1651
ITS-27 temperature scale, II.1250
ITS-90 temperature scale, II.1250, II.1260

J/ψ, II.701–2
Jacobians, I.12
Japanese Atomic Energy Agency, III.1650
JASON, III.2023
Jeans criterion for collapse, III.1768
Jellium, III.1303–5
JET (Joint European Torus), III.1650, III.1651, III.2055
Jets, II.715–16
jj-coupling shell model, II.1209, II.1210
Johnson noise fluctuations, I.601
Joint European Torus, see JET
Joint replacement, III.1896
Jones zone, III.1313
Josephson constant, II.1258, II.1271
Josephson effect, II.941–2, II.1096, II.1257, II.1258
Josephson junction, II.1095
Joule–Thomson effect, I.28
Journal of Applied Crystallography, I.511
Journal of Materials Research, III.1556
Journal of Nuclear Materials, III.1534
Julia series, III.1841

Kamioka detector, II.765
Kamiokande neutrino-scattering experiment, II.759, III.1748
Kaons, II.650–4, II.656–7, II.660, II.678
Kapitza's law, III.1283
Kármán vortex street, II.851
K-distributions, III.1466
Kelvin (degree), II.1250
Kelvin Circulation Theorem, II.817, II.822
Kelvin–Helmholtz time-scale, III.1694
Kennedy–Thorndike experiment, I.285
Kennelly–Heaviside layer, III.1621, III.1987
Kerr metric, III.1778
Kerst–Serber theory, II.1206
Kevlar polymeric fibres, III.1552
Keyhole surgery, III.1895
KHF_2, structure of, I.458
KH_2PO_4
  ferroelectricity, I.488
  nuclear density, I.461
  structure of, I.459
Kiloelectronvolt (keV), I.358
Kilogram (unit), II.1233–8
Kinematic shock wave, II.899
Kinematic wave, II.898
Kinematics, relativistic, I.364
Kinetic constant, I.154
Kinetic energy, I.90, II.837
Kinetic equations, I.608–11
Kinetic theory of gases, I.30, I.35, I.46, I.522, I.523, I.587, III.1284
Kirchhoff's function, I.148
Kirchhoff's law, I.66
Kitt Peak National Observatory (KPNO), III.1744
Klein–Gordon equation, I.209, I.283
Klein–Gordon theory, I.209
Klein–Nishina formula, I.218, I.365

Klystrons, II.1206
K-meson, see Kaon
KMS Fusion, III.1676
KNbO$_3$ ferroelectric phase transition, I.486
Knowledge domains, III.2041
Kogelnik–Li equations, III.1430
Köhler illumination, III.1390
Kolmógoroff–Sinai entropy, I.615
Kolmógorov dissipation length, II.838
Kondo effect, II.1167–9
Kondo problem, III.1343
Kondo theory, III.1343
Korteweg–de Vries equation, II.859
Kramers equation, I.607–8
Kramers–Heisenberg effect, III.1405
Kronecker delta-function, I.208
Krypton, I.75
K systems, I.615
Kuhnian revolution, III.2049
Kuiper Airborne Observatory, III.1743
k-values, III.1364

Laboratory of Biological Physics of the College de France, III.1894
LAK theory, III.1359
Lamb-dip, III.1443
Lamb–Retherford shift, III.1421
Lamb shift, II.641–3, II.649
Lamellar phases, III.1608
Laminar flow, II.827
Landau anti-damping, III.1625
Landau critical velocity, II.928
Landau damping, III.1624, III.1625
Landau expansion, II.1009
Landau–Ginzburg–Wilson Hamiltonian, I.568
Landau theory of second-order transitions, I.549–51, II.1009
Landolt–Börnstein series, II.1165
Landolt–Börnstein Tables, III.1513

Lane–Emden equation, III.1695, III.1711
Langevin equations, I.595–7, I.608, I.614, I.624, I.626, I.628
Langevin function, I.548, II.1116
Langley's bolometer, I.23
Laplace–Herschel nebular hypothesis, III.1945, III.1962
Large Electron–Positron Collider (LEP), II.741, III.1792
Large Hadron Collider, III.2036
Large Magellanic Cloud, II.758, III.1760
Large Space Telescope, III.1741
Large-scale-integration, see LSI
Laser anemometry, III.1442
Laser beam
 particle trapping, III.1462
 self-trapping, III.1459
 wavefronts, III.1458
Laser diode, III.1476, III.1480
Laser-fusion, III.1676
Laser-fusion-generated electrical power, III.1479
Laser-fusion pellet, III.1675
Laser interferometer gravitational-wave observatory (LIGO) project, I.298, I.317
Laser Interferometric Gravitational Observatory, III.2024
Laser-light-scattering, III.1441–5
Laser-printing, III.1541
Laser read-out systems, III.1462
Laser sources of infrared radiation, II.1054
Laser speckle, III.1453
Laser standards, II.1242
Laser velocimetry, III.1442
Lasers, III.1417–18, III.1424–52
 applications, III.1478–9
 argon, III.1894
 argon-ion, III.1453
 carbon dioxide, III.1442, III.1677, III.1894

chemical, III.1445
continuous-wave single-mode silica-fibre, III.1447
excimer, III.1453
free-electron, III.1456
GEKKO series, III.1676
glass, III.1677
HCN and H_2O action, III.1442–3
helium–cadmium, III.1435, III.1445
helium–neon, III.1427, III.1435, III.1439
history of, III.1411, III.1423
in medicine, III.1480
instabilities, III.1466–7
molecular, III.1442
Nd:glass, III.1479
Nd:YAG, III.1435, III.1471, III.1479, III.1894
noise, III.1444
Nova, III.1464, III.1676
rare-earth ion, III.1438
rare-gas ions, III.1443
ruby, III.1424, III.1426, III.1429, III.1430, III.1432, III.1479, III.1894
soliton, III.1469
spatial mode patterns, III.1431
titanium–sapphire, III.1448
tunable, III.1472
wavelength-tunable dye, III.1448
wire, III.1472
x-ray, III.1464–6, III.1479
see also under specific types and applications
Lasing without inversion (LWI), III.1473
Lattice conduction, II.981
Lattice dynamics, II.967–91, II.997–1007
Lattice gas model, I.553
Lattice gauge theory, II.756
Lattice vibrations, I.452
Lau effect, III.1414
Lau interferometer, III.1414

Laue equations for X-ray diffraction, **I.427**
Laue photographs, I.458
Lead, II.999
  crystals, III.1544
  isotopes, III.1960, III.1961
Lead sulphide detectors, III.1743
Learned societies, I.3
Lehrbuch der Metallographie, III.1554
Length, II.1239–42
Lens aberrations, III.1458
Lens-pupil functions, III.1418
Lenses
  contact, III.1414
  gradient-index, III.1391
  iron-free, III.1568
  magnification ratio, III.1569
  properties of, III.1389
  see also under specific types of lens
Lenz–Ising model. See Ising model
Lepton number, II.664, II.755
Leptons, II.635–6, II.692–5, II.700, II.704, II.719–25
Leukaemia, III.1891
Levitation theory, III.1543
Levshin–Perrin law, III.1400
Liapunov functions, I.621–2
Lidar systems, III.1479
Lift–drag ratio, II.870, II.871, II.881
Light, speed of, I.260, I.261, I.266, II.1240, II.1262–4
Light-beating spectroscopy, III.1421
Light-emitting diode (LED), III.1474, III.1476
Light microscope, resolving power, III.1574
Light-quantum hypothesis, I.71–2, I.151–2, I.167–8, I.185, I.363
Light rays, I.260
Light-scattering spectroscopy, III.1408
Light transformations by atoms, II.1101

Light transmission imaging (diaphanography), III.1932–3
Light waves, I.258
Lightning-induced electron precipitation, III.1686
Line broadening, I.471–2
Line of section, III.1836
Linear accelerators, II.734–8, II.1205, III.1874, III.1875
Linear energy transfer (LET), III.1885
Linear response theory, I.626–7
Linnean system, I.47
Liouville equation, I.593, I.602, I.605–7, I.619, I.620
Liposomes, III.1608
Liquid crystal display (LCD), III.1476, III.1540–1, III.1593, III.2029
Liquid crystal television display, III.1476
Liquid crystals, II.1084, III.1539–42, III.1552–3, III.1603–5, III.2031
Liquid-drop model, II.1186–91, II.1213
Liquid gases, I.28
critical point, I.543–4
Liquid helium cryostat, III.2025
Liquid metals, III.1366–7
Lithosphere, III.1978
Liverpool Astronomical Society, I.24
Lloyd Triestino, III.1711
LMC X-1, III.1757
LMC X-3, III.1757
Local gauge invariance, II.693
Local gauge transformation, II.693
Lockheed C-141 transport aircraft, III.1743
Lodestone, II.1112, II.1156, III.1856
London Chemical Society, I.46
London equation, II.925, **II.926–7**
Long-chain systems, III.1596–7
Longitudinal waves, III.1956
Long-range collisions, II.1076

Long-range effects, **II.1066–7**
Long-range interactions, II.1063–5
Long-range parameters, **II.1074–5**
Lorentz contraction, I.264
Lorentz–Einstein formula, I.285
Lorentz electron model, I.32
Lorentz factor, III.1783
Lorentz force, III.1622
Lorentz force equation, III.1671
Lorentz–Rayleigh model, I.629
Lorentz transformation, I.264, I.265, I.272, I.274, I.275, I.281
Loschmidt's constant, I.150
Low-energy electron diffraction (LEED), I.466
Low-energy electron microscope (LEEM), III.1577
Lowest unoccupied molecular orbital (LUMO), II.1083
LSI (large-scale-integration), III.1371
Luminescence, III.1537
Luminosity function for galaxies, III.1769–70
Luminosity–spectral class diagram (Russell diagram), III.1696, III.1698, III.1710
Luminous stars, III.1696
Lummer–Gehrcke plate, III.1389
Lummer–Gehrcke spectrograph, I.19

Mach number, II.813, II.814, II.821–4, II.880
Mach's principle, III.1723
Macroscopic wavefunctions, II.943
Macrostate, I.523, I.526
MAFCO, III.1673
Magellanic Clouds, II.758, III.1714, III.1715, III.1719, III.1744, III.1760, III.1788
Magic numbers, II.1225
Magnesium, formation of, III.1747
Magnesium oxide (MgO) crystals, I.453

Magnetic anisotropy, II.1134, II.1154
Magnetic behaviour, types of, II.1112
Magnetic breakdown, III.1359
Magnetic bubbles, II.1161
Magnetic compass, III.1952
Magnetic dipole, II.1111, III.1958, III.2004
  interactions, II.1011
Magnetic domain, II.1154–5
  patterns, II.1135
Magnetic electron lens, III.1566–9
Magnetic equation of state, I.545
Magnetic excitons, II.1162
Magnetic field, I.27, I.102, I.108, I.261, I.541, II.1111, II.1125, II.1126, II.1132, II.1172, III.1282, III.1345, III.1352, III.1358, III.1373
  Earth's, III.1622–3, III.1952–60, III.1973
  in sunspots, III.1958
Magnetic flux, II.919, II.1121
Magnetic fusion research, III.1619, III.1629–31
  classification era, III.1631–40
  post-1960, III.1645–70
Magnetic induction, II.1111
Magnetic insulators, co-operative phenomena in, II.1147–8
Magnetic intensity, II.1111
Magnetic interactions, II.1010
Magnetic lens, III.1568, III.1580
Magnetic materials, I.559
Magnetic mirror, III.1658–70
Magnetic models, I.570
Magnetic moments, I.376, II.1115, II.1117, II.1123, II.1166–7
  in Au$_2$Mn, I.462
Magnetic monopoles, II.766
Magnetic phase transitions, II.1014–18
Magnetic pinch, III.1636
Magnetic properties, II.1112
  and crystal structures, II.1156–8
Magnetic recording, II.1158–9
Magnetic recording, II.1158
Magnetic resonance, I.541–2
Magnetic resonance imaging (MRI), II.1176, III.1873, III.1910, III.1923-33
Magnetic resonance spectroscopy (MRS), III.1930, III.1931
Magnetic saturation, II.1145
Magnetic spectrometers, II.1206–8
Magnetic structures, I.462, I.463, II.991
Magnetic susceptibility, I.545, I.557, I.558, III.1595
Magnetism, I.65, II.1111–81, III.1832
  conductors, II.1133–6
  developments—prior to twentieth century, II.1111–13
  developments—1900–25, II.1113–23
  developments—1925–50, II.1123–5
  developments—1950 onwards, II.1136–7
  molecular field theory, II.1014
  quantum concepts, II.1115–16
Magnetization, I.555, I.559
  intensity of, II.1111
  spontaneous, I.573, II.1118
Magnetoencephalography, III.1932
Magneto-hydrodynamic (MHD) theory, III.1648, III.1650, III.1658, III.1663, III.1670, III.1672, III.1673, III.1684, III.1685
Magneto-ionic theory, III.1990, III.1992
Magnetometer, III.1970
Magneto-optical effect, III.1389
Magnetopause, III.2011
Magnetoresistance, III.1356–7
Magnetosphere, III.1752, III.1981, III.1983, III.1984, III.2001–11

Magnons, II.1162
Manchester Literary and Philosophical Society, I.3
Mandelstam–Brillouin doublet, III.1408
Mandelstam–Brillouin scattering, III.1397
Mandelstam–Brillouin spectral shifts, III.1398
Manganese fluoride, I.559
Manganese oxide (MnO), powder pattern, I.460
Manhattan Project, III.1531, III.1556, III.1630, III.2023, III.2025
Markovian kinetic equation, I.620
Masers, III.1417–18, III.1422
Mass, II.1236–8
Mass absorption anomaly, I.398
Mass–energy absorption coefficients, III.1882
Mass–energy equivalence relation, I.285
Mass spectrographs, I.111
Massachusetts Institute of Technology, I.16
Massey criterion, II.1077
Master equation, I.593, I.605–7, I.618
Master function, I.606
Master-oscillator power amplifiers (MOPAs), III.1474
Materials
 characterization, III.1506
 high-quality, III.1280–1
 physics of, III.1505–64
 science, III.2046–7
 technology, III.2022
 terminology, III.1553
 see also under specific materials
Materials Research Laboratories (MRLs), III.1557
Materials Research Society (MRS), III.1556–7

Materials Science and Technology, III.1513
Mathematical Theory of Relativity, III.1700
Mathematical theory of speculation, I.586
Mathematical tools, III.2030
Mathematics Tripos, I.10, I.13
Matrix mechanics, I.360
Matter, I.44, I.47
 condensed, II.756
 constituents of, I.357–8
 new forms of, II.650–62
 thermodynamic properties, I.542
Max Planck Institute, III.1555
Maxwell–Boltzmann distribution, I.589–91, I.598
Maxwell–Boltzmann law, III.1285
Maxwell–Boltzmann statistics, I.527
Maxwell equations, I.33, I.260–1, I.264, I.265, I.270, I.279, I.283, III.1394, III.1622, III.1625, III.1671, III.2018
Maxwellian distribution of protons, III.1707
Maxwellian electromagnetism, I.29
Maxwell's Demon, I.576
Maxwell's electromagnetic theory of light, I.26
Maxwell's theory, I.29, I.321
Mean field approximations, I.554
Mean free path, I.522, III.1549
Mean square radius of gyration, I.569
Mean thermal energy, I.585
Measurement
 applications, II.1240–2
 capacities, III.2025
Mechanical oscillograph, I.26
Mechanical principle of equivalence, I.287–8
Mechanocaloric effect, II.922
Medical electronics, III.1893–6
Medical imaging, III.1901, III.1912

S32

Subject Index

new techniques, III.1932–3
Medical mechanics, III.1896–900
Medical physics, III.1855–941, III.2022
  branches of, III.1855–9
  chronological evolution, III.1855–9
  computers in, III.1858
  modern applications, III.1857
Medical Research Council (MRC), III.1871, III.1877, III.1891, III.1924
Medical ultrasound, III.1910–15
  clinical perspective, III.1915
Medicine
  lasers in, III.1480
  nuclear, III.1858, III.1892, III.1900–10
Meetings, III.2028
Meissner (or Meissner–Ochsenfeld) effect, II.919–20, II.924, II.941, II.957
Mekometer, II.1240
Men and Decisions, III.1640
Mesomorphic phases, III.1603
Meson Factory, II.1205
Mesons, I.378–83, I.389–94, I.402–4, II.640, II.643, II.661, II.686, II.703, II.721
  B-meson, II.721, II.724–5, II.767–8
Mesophere, III.1982
Mesoscopic systems, III.1370
Mesotrons, I.389, I.404–5
  cosmic-ray, I.394–9
  decay, I.399–402
  decay versus capture, I.405–7
  lifetime, I.397–8, I.406
  nuclear interaction, I.407
Metaboric acid, structure of, I.449
Metal–dielectric interference filters, III.1407
Metallography, III.1468, III.1507, III.1554
Metallurgy, III.1505, III.1554
  pioneers in, **III.1511**
Metals
  electrical conduction in, III.1284–7
  electron theory, I.215–17
  groups I–III, III.1354
  individual properties, III.1345–59
  liquid, III.1366–7
  phonon dispersion curves, II.999
  phonons in, II.998–9
  physics of, III.1518
  post-war research, III.1341–5
  resistivity variation with temperature, III.1281–2
Metals Crystallographic Data File, I.512
Metals Handbook, III.1513
Metals Reference Book, III.1513
Metalworking, III.1505
Metastable effect, II.946
Metastable equilibrium, I.530
Metastable states, II.918
Meteorites, III.1965
Methoden der Mathematischen Physik, III.2043
Methylene di-phosphonate (MDP), III.1920
Metre (unit), II.1235, II.1239
Metre Convention, II.1236
Metric system, II.1233, II.1235
Metrically transitive systems, I.599
Metrology, II.1233, II.1273
  role of atoms, II.1094–7
Metropolitan Vickers Company, III.1871, III.1873
MFTF (Mirror Fusion Test Facility), III.1664, III.1668, III.1669
Michelson interferometer, III.1704
Michelson–Morley experiment, I.32, I.262, I.264, I.266, III.1401
Michelson's interferometer, III.1399

Micro-canonical ensemble, I.532
Microphysics, I.210–33
Microscopes, I.26, III.1475
  see also under specific types of microscope
Microscopic physics, I.358
Microstructology—Behavior and Microstructure of Materials, III.1550
Microstructure, III.1542–3
  quantitative measurement of, III.1549
  study of, III.1506–7
  see also under specific materials
Microtechnology, III.1899
Microwave electron linear accelerators, III.1873
Microwave radiation, II.1053–4
Microwaves
  applications, II.1273
  frequency standards, II.1055
  parametric amplifier for, II.1158
Mid-Atlantic Ridge, III.1973, III.1976
Middlesex Hospital, III.1877
Milky Way, III.1714, III.1716, III.1718
Miller indices, I.421
Millions of electronvolts (MeV), I.358
Mineralogy, III.1514
Minkowski space, I.292
Minor impurity, III.1505
Mirror Fusion Test Facility, *see* MFTF
Mirror machine, III.1648, III.1662
Mirrors, properties of, III.1389
MIT Radiation Laboratory, III.2026
Mixed state, II.920
MKSA system, II.1235, II.1257
Mobility edge, III.1368
Models, I.603–5, I.618–19, I.628–9
Mohorovičić discontinuity ('Moho'), III.1954
Moiré grating interference, III.1414

Molecular beam analysis, II.1033
Molecular biology, I.496, I.497, III.2020, III.2030, III.2046
Molecular bonds, II.1042–9
Molecular collisions, I.588, II.827
Molecular dissociation, II.1067
Molecular dynamics, I.577
Molecular field hypothesis, I.545
Molecular fluids, I.557
Molecular gases, I.35
Molecular hydrogen, I.181, II.1045, **II.1048**
  in interstellar medium, III.1766
  photoabsorption by, II.1069–73
Molecular line astronomy, III.1762–5
Molecular physics, I.358–62, II.1025–49, II.1081–9
Molecular resonances, II.1084
Molecular sheets, III.1595
Molecular spectra, II.1046–7
Molecular spectroscopy, I.529, II.1085–7
Molecular structure, I.211–15
Molecular systems, perturbations in, III.1593
Molecular theory, I.43
Molecular viscosity, II.826
Molecules, I.45, I.47–9, I.81–2, I.85, I.587, I.588
Molybdenum-99, III.1907
Momentum space, I.567
Momentum transport, II.853
Monomers, III.1596
Monotone curve, III.1837
Monsoon phenomena, II.907–8
Monte Carlo methods, I.572, I.628, III.1560
Moon
  formation of, III.1967
  gravity fields, II.888
  orbit, III.1948
  origin of, III.1952, III.1965, III.1966

Morgan, Keenan and Kellman (MKK) system, III.1698
MOSFET, III.1371, III.1372, III.1375
Moss–Burstein shift, III.1421
Mott transition, III.1365, III.1368
Mount Palomar, III.1745, III.1758
Mount Vernon Hospital, III.1867
Moving-coil experiment, II.1238
MSS (paper), III.1839, III.1843, III.1849, III.1850
MULTAN (multiple tangent formula method), I.450
Multi-channel formulation, II.1067–73
Multi-electron fragmentation, II.1065–7
Multi-electron spectra, II.1035–7
Multi-particle system, III.1394
Multiphase interstellar medium, III.1765–6
Multi-photon processes, II.1099–100
Multiple-beam interferometry, III.1412
Multiple collisions, I.611
Multiple-laser-beam systems, III.1479
Multiple sclerosis (MS), III.1928, III.1932
Multiplex optical spectroscopy, III.1416
Multiply connected graphs, I.552
Multi-TeV hadron colliders, II.767
Muon $g - 2$ factor, II.647–9
Muons, II.640, II.650–4
Myoglobin, I.445
  structure, I.500–1

Nambu–Goldstone boson, II.694, II.695
Nambu–Goldstone mode, II.694
Nambu–Jona-Lasinio model, II.756
$NaNbO_3$, dielectric and crystallographic properties, I.487

Naphthalene, crystal structure, I.494
National Institute of Standards and Technology (NIST), II.1235
National laboratories, role of, II.1254–5
National Measurement Accreditation Service (NAMAS), II.1272
National Physical Laboratory (NPL), I.20, II.1234, II.1243, III.1517, III.1867, III.1881, III.1890, III.1988
National Radium Commission, III.1877
National Science Foundation, III.1980
National Standards Laboratories, III.1867
Natural philosophy, I.13
Natural rubber, III.1593
Natural Sciences Tripos, I.10–14
Nature, I.5, I.6, I.96, I.116, I.129
Naval Research Laboratory (NRL), III.1739, III.1988, III.1992
Navier–Stokes equations, III.1834, III.1835
Near-infrared radiation, II.1053–4
Near-space environment, III.1981
Nebular hypothesis, III.1944–5, III.1950
Nebulium, I.75, I.81
Néel temperature, II.1129, II.1151, II.1152, II.1159
Negative dispersion, III.1399–401
Negative-resistance device, III.1340
Negative temperatures, I.541–2
Neodymium, III.1438
  paramagnetic susceptibility, II.1162
Neodymium–iron borates, II.1165
Neon, I.75, III.1747
Neptunium, II.1223

Nernst–Lindemann formula, II.972
NeuroECAT III, III.1921
Neurological disorders, III.1899
Neutral currents, II.695–7, II.699
Neutral hydrogen, III.1762–5
Neutral kaons, II.658–9
Neutral weak currents, II.695
Neutrino masses, β-decay limits on, II.755
Neutrinoless double-β decay, II.755
Neutrinos, I.129, I.370, I.371, II.639, II.640, II.663–4, II.669–71, II.758–60, III.1748–9
  helicity, **II.670**
  oscillations, II.763
  supernova, II.758
  see also Solar neutrinos
Neutron beams, I.468
Neutron bombardment, I.122
Neutron capture, II.1193, II.1223, II.1227, III.1785
Neutron decay, III.1785
Neutron diffraction, I.452–63, I.487, III.2025
Neutron flux, II.994–6
Neutron guide, I.458
Neutron irradiation, I.130, III.1871, III.1872
Neutron–neutron forces, II.1190
Neutron–proton forces, II.1190
Neutron–proton model, I.371–3, II.1190
Neutron–proton ratio, III.1792
Neutron scattering, I.453, II.991–6, II.1011, II.1166, III.2025
Neutron sources, II.1199, II.1206
Neutron stars, III.1712–13, III.1751–4, III.1756
Neutrons, I.112, I.118–26, I.369–71, I.379, II.639, II.991
  lifetime and decay properties, II.754–5
New General Catalogue of Nebulae and Clusters of Stars, III.1717

Newton Victor Maximar, III.1864
Newtonian constant of gravitation, I.28
Newtonian mechanics, I.29, I.36, I.257, I.261
Newton's equations, III.1622
Newton's rings, III.1386
NGC2264 star cluster, III.1767
$NH_4H_2PO_4$, antiferroelectricity, I.488
Nicholson–Bohr quantum rule, I.175
Nickel–titanium alloys, III.1539
Nobel Prizes, III.2021
Noble metals, III.1343
No-hair theorem for black holes, III.1778
Noise problems in upper-atmospheric physics research, III.1995
Noise radiation, II.872–3
Non-abelian gauge theory, II.693
Non-classical light, III.1388
Non-crystalline solids, I.471–84
Non-dimensional parameters, II.812–16
Non-equilibrium theory, I.630
Non-imaging light concentrators, III.1446
Non-linear differential equations, III.1826
Non-linear effects, II.808, II.846–69
Non-linear equations, III.1826
Non-linear field equations, II.800
Non-linear field theories, II.798
Non-linear optics, II.1100, III.1424–53, III.1458–61
Non-linear oscillations, III.1837–8
Non-linear relations, I.233
Non-linear science, III.2030
Non-linear systems, III.1835, III.2030
Non-linear theory, II.810
Non-linearity, II.856–60, III.1403
Non-metallic materials, III.1528–31

Non-perturbative calculations, III.1825
Non-stationary state, II.1029
Nordström's theory, I.295
Nova laser, III.1464, III.1676
Nuclear atom, discovery of, III.2018
Nuclear binding, I.111, II.1208, II.1210
Nuclear bombs, III.1886
Nuclear demagnetization, II.1175
Nuclear dynamics
  as many-body problem, II.1191–5
  background, II.1183–91
  developments in, II.1183–232
  effects of World War II, II.1196–9
  technical advances, II.1199–208
Nuclear energy, I.107, II.1191, III.1879
Nuclear fission, I.126, I.129–33, II.1195, II.1197
Nuclear forces, I.123, I.378–83, I.402–4, II.1208
  fundamental theories, I.375–83
Nuclear fusion, *see* Fusion research
Nuclear magnetic moments, I.116–17, II.1171–2
Nuclear magnetic resonance (NMR), II.1094, II.1171–5, III.1859, III.2026
Nuclear magnetic resonance (NMR) imaging, III.1898, III.1923–33
Nuclear magnetism, II.1171–5
Nuclear masses, II.1186–91
Nuclear matter, density of, II.1189, II.1190
Nuclear medicine, III.1858, III.1892, III.1900–10
Nuclear physics, I.115, I.119, II.754–62, III.1871
  defence laboratories, II.1196
  research laboratories, II.1198
Nuclear reactions, I.123–6, II.1219–22, III.1629
  direct reaction mechanism, II.1220–1
  nucleus theory of, II.1192
Nuclear reactors, II.1197, III.1871
Nuclear scattering, II.1012, II.1219–22
Nuclear size, I.116
Nuclear spin, I.117, I.541, II.1171, III.1410–11
Nuclear statistics, I.117–18
Nuclear systematics, I.366–7
Nuclear volume, II.1189, II.1208
Nuclear weapons, III.1980
Nucleation process, III.1547
Nucleic acids, I.503–4, I.507
Nucleon isobars, II.659
Nucleon–nucleon force, II.1212
Nucleons, II.639, II.658, II.660, II.675, II.1218
Nucleosynthesis, II.758–60, III.1746–7, III.1792
Nucleotide chains, I.506
Nucleus, I.103–33, I.370
  collective motion in, II.1213–19
  compound, II.1219
  Heisenberg's n–p model, I.371–3
  low-energy collective modes, II.1214–18
  models, I.109–15
  shell structure in, II.1208–13
Numerical calculations, III.1828
Numerical solution, III.1826

Oak Ridge DCX experiments, III.1661–2
Ocean current, II.862
  patterns, II.892, II.896
Ocean engineering, II.869–87
Ocean ridges, III.1978, III.1981
Ocean tides, II.888
Ocean waves, II.857, II.860
Offshore structures, II.886–7
OGRA mirror machine, III.1662

Ohm
　absolute determination, II.1255
　international, II.1254, II.1257
　monitoring, II.1258
　reference standards, II.1256
Ohm's law, III.1336
Oligonucleotide structures, I.507
On the Principles of Geometry, III.1722
Onsager–Feynman quantization condition, II.944–5
Onsager regression hypothesis, I.616
Onsager relations, I.600–2, I.615–16, I.626–7
Ophthalmology, III.1930
Optical actions, atomic systems, II.1097–101
Optical astronomy, III.1744–5
Optical bistability, III.1449–50
Optical coatings, III.1407
Optical coherence theory, III.1392, III.1407–10, III.1461
Optical comparators, II.1240
Optical computing, III.1461
Optical fibre
　communication, III.1447–9
　sensors, III.1471–2
　telecommunications, III.1471–2, III.1480
　see also Fibre optics
Optical filtering, III.1418
Optical glasses, III.1541
Optical goniometer, I.421
Optical imaging, III.1432–7
Optical information processing, III.1417
Optical instruments and techniques, I.26, III.1406–24
Optical interferometry techniques, III.1411, III.1422
Optical Kerr effect, III.1444–5
Optical maser, III.1443
Optical micrography, III.1508

Optical microscopes, III.1405, III.1422, III.1520, III.1526
Optical parametric oscillator (OPO), III.1435
Optical phonons, III.1338
Optical physics, III.1385–504
Optical properties, crystals, I.485–6
Optical pumping, II.1097–9, III.1415
Optical research, World War II, III.1410–14
Optical sciences, computers in, III.1475–6
Optical signal and data processing, III.1461
Optical Society of America, III.1445, III.1474
Optical spectroscopy, III.1421–2
Optical stethoscopy, III.1406
Optical systems
　polarization in, III.1411–12
　signal-to-noise ratio in, III.1418
Optical theory, Abbe's, III.1578–84
Optical transistor, III.1450
Optical video discs, III.1462
Optics
　classical, III.1385–6
　in special theory of relativity, I.257
　linear, III.1434
　new results and applications (1970–94), III.1452–72
　non-linear, III.1427–32
　towards the twenty-first century, III.1472–81
Optimal filtering, III.1418
Optoelectronic communications, III.2051
Optoelectronic devices, III.1472, III.1476
Optoelectronic industries, III.1476
Optoelectronic integrated circuits (OEICs), III.1449, III.1476
Optoelectronic neural networks, III.1461

Optoelectronic physics, III.1385–504
Optoelectronics
  early research, III.1419
  semiconductors, III.1417
Orbital magnetic moments, II.1012
Orbiting Astrophysical Observatories (OAO), III.1741
Orbiting Solar Observatory (OSO-III), III.1740
Order–disorder transition, I.459, I.473–4, I.578, II.1009, II.1016, III.1515
  lambda transition, II.915, II.922–4, II.927, II.944
Ordinary differential equations (ODEs), III.1826–7, III.1829, III.1835- 9, III.1851
Organic compounds, structure analysis, I.493–6
Orion Nebula, III.1767
Orthogonalized-plane-wave (OPW) method, III.1317
Orthopaedic surgery, III.1896–7
Orthotics, III.1899
Oscillation, II.896, II.907–8, II.973
Oscillator strengths, II.1058
Otolaryngology, III.1930
Out of the Crystal Maze, I.512
Oxalic acid, I.452, I.454
Oxford University, I.20
Oxygen, formation of, III.1747

Pacemakers, III.1895, III.1897
Pacific floor, measurements, III.1974
Padé Approximant, I.558
Pair connectedness, I.573
Pair distribution function, I.547
Palatini variational method, I.299
Palomar Atlas, III.1745
Palomar telescope, III.1721, III.1775, III.1793
Pangaea (giant land-mass), III.1968

Paraelastic crystals, III.1538
Parallel plate ionization chamber, I.54
Parallel shear flow, II.828
Paramagnetic insulators, II.1116–20
Paramagnetic materials, I.459
Paramagnetic salts, I.540–1
Paramagnetism, II.1015, II.1113, II.1119, II.1125–33, III.1354, III.1362
Parametric fluorescence, III.1448–9
Parametrization technique, III.1351
Parastatistics, II.687
Paris Symposium on Radio Astronomy, III.1789
Parity, II.654, **II.1028–9**
  conservation, II.665
  violation, II.639, II.698, II.699
Partial coherence, III.1392
Partial conservation of axial current (PCAC), II.676
Partial differential equations (PDEs), III.1827, III.1834, III.1851
Particle accelerators, *see* Accelerators
Particle content, II.763
Particle energies, II.1201
Particle physics, III.1833, III.2038
  see also Elementary particles; and under specific particle types
Particle size distribution, III.1549
Partition function, I.527, I.532
Parton hypothesis, II.689–90, II.715
Paschen–Wien law, I.149
Patterson functions, I.441–5, I.495, I.501
Pauli exclusion principle, I.97–100, I.168, I.603, II.686, II.1116
Pauli spin-statistics theorem, I.283
Peltier heat, III.1284
Penicillin, structure of, I.496
Penning traps, II.1099

Penrose tiling, I.483
Percolation processes, I.572–4
Perfect cosmological principle, III.1787
Period–luminosity relation, III.1714–15
Periodic table, I.95–7, I.359, II.635
Permanent magnets, II.1165, II.1238
Permittivity of free space, II.1261
Perovskites, I.486–7, I.489, II.1156
Perturbation
  in molecular systems, III.1593
  singular theory, II.800–3, II.829
  strong responses to, III.1593–694
  theory, I.360, III.1830, III.1833
PET (positron emission tomography), III.1904, III.1910, III.1920, III.1922, III.1931
Phase angles, I.439, I.442, I.450, I.502
Phase conjuguation, III.1460
Phase-contrast microscopy, III.1409–10
Phase diagrams, III.1506, III.1508, III.1509, III.1542
  free-energy, I.538
Phase equilibria, III.1506
Phase matching, III.1434
Phase Rule, III.1508, III.1542
Phase transformations, I.542–74, I.577, II.1007–10, III.1542–50
  microstructural and microcompositional analysis in, III.1548–50
Phaseonium, III.1473
Philosophical Magazine, I.7
Philosophical Transactions of the Royal Society, I.7, III.1516–17
PHOBOS space probe, III.1750
Phonon dispersion curves, II.1000, II.1002
Phonon drag, III.1310, III.1311
Phonon exchange, II.953

Phonon interaction with hot electrons, III.1339
Phonons, II.981, II.982, II.990, II.998–9, II.1004, III.1302, III.1338
  in crystals containing defects and in glasses, II.1004–7
Phosphorescent materials (phosphors), III.1403, III.1530, III.1536, III.1537
Phosphors, *see* Phosphorescent materials (phosphors)
Photoabsorption
  by molecular hydrogen, II.1069–73
  spectra, II.1052, II.1053, II.1054, II.1056
Photo-assisted tunnelling, III.1423
Photocathode, III.1439
Photo-cells, III.1321
Photochromic spectacle lenses, III.1542
Photochron streak camera, III.1439
Photoconductivity, III.1321, III.1327–9
Photocopying, III.1541
Photodetection process, III.1441
Photodetectors, III.1433
Photodiode, III.1439
Photoelectric effect, I.37, III.1286
Photoelectronic imaging devices, III.1476
Photo-emission spectroscopy, III.1391
Photographic plates, II.1200
Photography, III.1389, III.1392, III.1410
Photoionization, II.1052, III.1762
Photometry, II.1251–2
Photomultiplier tubes, II.1207, III.1409, III.1411, III.1920
Photon antibunching, III.1459–62
Photon bunching, III.1468
Photon-correlation spectroscopy (PCS), III.1454–8

protein analyser, III.1471
Photon counting statistics, III.1424
Photon energy, II.1052
  conservation process, III.1435
Photonic bandgaps, III.1472
Photonic integrated circuit (PIC), III.1476
Photons, I.117, I.386, II.637, II.638, II.641, II.693, II.1028, III.1302, III.1393, III.1402, III.1403, III.1410, III.1413, III.1451–2, III.1994
Photophone experiments, III.1422
Photoprocesses, II.1102
Photorefractivity, III.1461
Photosensitive devices, III.1321
Photo-voltaic effect, III.1321
Physical constants, II.1233, II.1260–72
  linked sets, II.1267–72
Physical metallurgy
  birth of, III.1516–19
  quantitative theory, III.1519–28
Physical Review, The, I.6
Physical Society, III.2042
Physics
  as domain of knowledge, III.2041
  as social institution, III.2041, III.2056
  definition, III.2041
  methodology of, III.2049
  scope of, III.2041
  sociology of, III.2031
Physics and Chemistry of Solids, III.1556
Physics of Fully Ionized Gases, III.1631
Physikalische Gesellschaft, I.88
Physikalische Metallkunde, III.1555
Physikalische-Technische Reichsanstalt (PTR), I.7, I.20, I.69, I.107, II.916, II.1234, II.1254
Physikalische Zeitschrift, I.6, I.7

Piezoelectric properties, III.1538
Piezophotonic effects, III.1459
π-meson, *see* Pions
Pioneer III, III.2007–8
Pion–nucleon scattering, II.659
Pion–pion scattering, II.681
Pions, I.408, II.650–5, II.658, II.665, II.675–6, II.711
Pipes, flow in, II.825
Pippard coherence length, II.947
Pitchblende, III.1876
Planck constant, I.67, I.152, I.359, I.360, II.638, II.1026, II.1271, III.1387
Planck energy, II.972
Planck formula, II.970, II.971
Planck law, I.70, I.151, I.363, I.527, III.1399
Planetary dynamics, III.1830–1
Planetary systems, III.1950
Planets
  formation, III.1948–9, III.1963
  infinitesimal, III.1950
Plasma, III.1361
  diagnostic measurements, III.1644
  frequency, III.1623
  instability, III.1630
  oscillations, III.1362, III.1623–5
  radio wave propagation in, III.1987
  temperature, III.1617
  theory, III.1623–5
Plasma physics, III.1617–90, III.1991–2, III.1994
  developments—circa 1950 to the present, III.1627–45
  developments—first half of twentieth century, III.1619–25
  history of, III.1617, III.1618
  landmark experiments, III.1677–81
  research, III.1617, III.1618
  space, III.1682–7

Plasma Physics and the Problem of Controlled Thermonuclear Reactions, III.1642
Plasmapause, III.2009, III.2010
Plasmasphere, III.1982
Plastic deformation, III.1520, III.1526, III.1530
Plate tectonics, III.1956, III.1961, III.1976, III.1978–80, III.2022
Platinum–iridium alloy, II.1237, II.1239
Pleated sheet structures, I.500
Plunging breakers, II.862
Plutonium, II.1223–5
  discovery, II.1197
  weapons-grade, III.1633
p–n junction, III.1325, III.1417
Poincaré–Bendixson theorem, III.1835
Poincaré map, III.1836, III.1851
Poincaré recurrence theorem, I.618
Poincaré–Zermelo recurrence conflict, I.592
Point defects, III.1528–31, III.1533, III.1534
Poisson equation, III.1625
Poisson formula, III.1424
Poissonian statistics, III.1393
Polar Years, III.2005
Polaritons, II.988, II.990
Polarization in optical systems, III.1411–12
Polarization rules, I.94–5
Polarized light microscopy, III.1412
Polarizing microscope, I.421
Polarons, III.1321
Polepiece lens, III.1567
Polonium, I.119, II.1200, II.1206
Polycrystal Book Service, I.512
Polycrystalline metals, III.1518
Polycrystalline samples, III.1280
Polyethylene, III.1602
Polymerization, III.1550, III.1551
Polymerization index, III.1597

Polymers, II.1084, III.1596–603
  blends, III.1553
  chain-folding, III.1552
  colloid protection, III.1611
  configurations, I.569–70
  crystallization, III.1602
  dynamics of entangled systems, III.1600
  fibres, III.1898
  future prospects, III.1603
  gels, III.1602–3
  high-technology, III.1550
  modified, III.1597
  overlapping, III.1599
  physics of, III.1550–3
  reptation process, III.1601
  semicrystalline nature of, III.1551
  solid phases, III.1600
  solutions, III.1599
  statics, III.1599
  structure, I.569–70
Poly-τ-methyl-l-glutamate, I.498
Polymethylmethacrylate (PMMA), III.1602, III.1896
Polyoxyethylene, III.1593
Polypeptide chains, I.496, I.497, I.499–501, I.510
Polysaccharides, III.1611
Polystyrene, III.1602
Polytetrafluoroethylene (Teflon), III.1896
Pomeranchuk trajectory (Pomeron), II.682
Pomeron (Pomeranchuk trajectory), II.682
Positive electron, I.109
Positively charged light particle, I.231
Positron emission tomography, *see* PET
Positronium, II.645–6, II.649, II.650, II.702
Positrons, I.370, I.371, I.377, II.638
Post Office, III.1988

Potassium dihydrogen phosphate (KDP), I.570–1, III.1432
Potential-energy curves, II.1045
Potential flow, II.803, II.804, II.887
Potential vorticity, II.895
Potts model, I.571
Powder diffraction, I.453
Powder photographs, I.435
Poynting's vector, I.33
Precious metals, III.1281
Precipitation-hardening, *see* Age-hardening
P representation, III.1440
Primaeval atom, III.1785
Primary image, III.1578
Primordial nucleosynthesis, III.1785, III.1786
Princeton Model C stellarator, III.1656
Principal quantum number, I.84, I.91, I.215
Principle of complementarity, I.197
Principle of Relativity, III.2039
Probabilities in quantum physics, I.93–4
Probability amplitude manipulation, II.1064
Probability relations, I.449
Proceedings of the Royal Society, I.7
Programmable calculators, III.1829
Project Sherwood, III.1631–40
Project Sherwood—The US Program in Controlled Fusion, III.1640
Projection electron microscope, III.1584
Proportional counters, II.1207
Protein crystallography, I.445, I.502
Protein Data Bank, I.512
Proteins, I.496
  fibrous, I.497, I.499
  globular, I.497
Proton–antiproton colliders, II.743–4

Proton decay, II.764–5
Proton–electron (PE) model, I.109–10
Proton linear accelerators, II.1205
Proton–neutron (PN) system, I.123
Proton–proton chain, III.1708
Proton–proton colliders, II.742–3
Proton–proton collisions, I.377
Proton synchrotrons, II.732–3
Protons, I.113, I.115, I.118, I.370, II.639, II.701
  in medical physics, III.1876
  Maxwellian distribution of, III.1707
Prototype phenomena, II.1060–3
Pseudopotential, III.1353, III.1367
PSR 1913+16 binary pulsar, III.1757
Publications, III.2028
Pulsars, III.1752
Pulse-echo imaging, III.1911–13
Pulse-echo technique, III.1989
Pump/oxygenator, III.1897
Pump-probe configuration, III.1445
$P–V–T$ relations, I.28
Pyrites, I.430

Q-switching, III.1430, III.1436, III.1438
Quadrature amplitudes, III.1468
Quadrupole moments, I.403, II.1214, II.1215
Qualitative quantum mechanics, I.116
Quanta, I.378, II.638, III.1390
  and radiation, I.146–55
  empirical foundations, I.145–72
  specific heat, I.152–4
Quantitative metallography, III.1549
Quantitative theory, I.44
Quantum chemistry, theory and computations, II.1087–8

S43

Quantum chromodynamics (QCD),
  I.322–3, II.636, II.692,
  II.709–19, II.756, II.769,
  III.2020, III.2031
Quantum communication, III.1472
Quantum conditions, I.359
Quantum cryptography, III.1472
Quantum dot, III.1472
Quantum dynamics, I.72–89
Quantum electrodynamics (QED),
  I.358, **I.361**, I.362, I.374, I.378,
  I.385, I.386, I.391, II.640–50,
  II.646, II.740, II.769, III.1414,
  III.1446, III.1472
Quantum energy, III.1361
Quantum field theories (QFTs),
  I.206–9, I.318, **I.380–1**, I.617,
  II.640, III.2033, III.2037,
  III.2038
Quantum formula, I.156
Quantum gravity, I.317–19
Quantum Hall effect, II.1096–7,
  II.1257, II.1259, III.1372,
  III.1375
Quantum mechanical effects,
  I.163–72
Quantum mechanical many-body
  problem, II.1212
Quantum mechanical parameters,
  II.1063
Quantum mechanical
  reformulation, II.1065
Quantum mechanical scattering
  theory, I.217–19
Quantum mechanical
  transformation theory, I.199
Quantum mechanical tunnelling,
  III.1706, III.1707
Quantum mechanical
  wavefunction, II.934
Quantum mechanics, I.30, I.61,
  I.65, I.72, I.115, I.125,
  I.143–248, I.359, I.360,
  I.602–3, I.627–8, II.638, II.917,
  II.1026, II.1034, II.1058,
  III.1280, III.1325, III.1403,
  III.1467, III.1830, III.2049
  applications, I.211–20, I.229
  as reformulated classical
    mechanics, I.189–91
  aspects of theory beyond,
    I.232–3
  causality in, I.220–9
  complementarity in, I.220–9
  completeness and reality of,
    I.225–9
  in astrophysics, III.1706–8
  mathematical foundation,
    I.196–208
  new structures within 'standard
    theory', I.230–2
  non-relativistic, I.220
  origin and completion of, I.173
  physical interpretation, I.196–208
  present situation, I.229–30
  reality in, I.220–9
Quantum Mechanics, III.2043
Quantum non-demolition (QND)
  measurements, III.1469
Quantum numbers, I.84, I.90–100,
  I.157, I.162, I.182, I.215,
  II.657, II.658, II.1034, II.1036,
  II.1124
Quantum of circulation, II.931
Quantum optics, III.1387–406,
  III.1468, III.1471, III.1472
Quantum phenomena, I.253
Quantum physics, I.65–72, I.89
  probabilities in, I.93–4
Quantum postulate, I.70, I.81
Quantum statistics, I.525–8, I.537
Quantum-theoretical
  reformulation and
  non-commuting variables,
  I.186–9
Quantum theory, I.37, I.54, I.62,
  I.65, I.68, I.71, I.80, I.88,
  I.89, I.101–3, I.524–31, I.537,
  III.1285, III.1387, III.1393,

Subject Index

III.1395, III.1402–4, III.1962, III.2018
and gravitation, I.318
evolution, I.144–5
in special theory of relativity, I.282–3
of electron gas, III.1287–91
of ions, II.1123–5
of measurement, I.223–5, I.233
with classical statistics, I.527
Quarkonium decays, II.716–17
Quarks, II.635–6, II.677, II.684–7, II.699, II.719–25
  b-quarks, II.722–3
  charmed, II.716
  colours, II.687, II.691, II.709, II.711–13
  flavours, II.687
  free, II.691–2
  statistics, II.709–10
Quartz clock, II.1244, II.1247, II.1248
Quasars, III.1776–7, III.1798
  discovery of, III.1789
Quasi-classical distribution function, III.1441
Quasicrystals, I.481–4
Quasiparticles, III.1363, III.1364
Quasi-phase matching, III.1434–5
Quasi-stellar radio sources, III.1776
Quenching, III.1533, III.1542

Radar, III.1944
Radial distribution function, I.480
Radiation, II.1028–32
  and quanta, I.146–55
  see also under specific types of radiation
Radiation accidents, III.1887
Radiation actions, II.1049–59
  spectrum of, II.1055–8
Radiation biology, III.1857
Radiation biophysics, III.1857, III.1884–9

Radiation damage, III.1531, III.1534, III.1884–5
Radiation-dose distributions, III.1872
Radiation dosimetry, III.1880–4, III.1887–9, III.1890
  calorimetry in, III.1883–4
  chemical, III.1884
  ionization methods, III.1880–3
  radionuclides, III.1909
Radiation energy, III.1885
Radiation hazards, III.1886–7, III.1895
Radiation injury, latency of expression, III.1886
Radiation isodose curves, III.1865
Radiation protection, III.1857, III.1871, III.1889–93
Radiation shielding, III.1891
Radiation Trapped in the Earth's Magnetic Field, III.1683
Radiationless transitions, II.1086–7
Radiative instability, I.83
Radio, III.1987
Radio astronomy, III.1736–8, III.1774–6, III.1944
Radio communications, III.1985–6, III.2000
Radio galaxies, III.1795, III.1797
Radio interference, III.2003
Radio propagation, III.1619–23, III.2000, III.2007
Radio pulsars, III.1752, III.1754, III.1757–8
Radio quasars, III.1751, III.1797
Radio research, III.1985
Radio scintillations, III.1751
Radio signals, III.1987, III.1988
Radio sources, III.1776, III.1782, III.1788–90, III.1797–9
Radio telescopes, III.1776
Radio waves, III.1985–8
  attenuation of, III.1995
  propagation, III.1990

S45

Radioactive decay, I.93, I.106, II.639, III.1960
Radioactive isotopes, II.1222
Radioactive Substances Act, III.1910
Radioactive tracers, III.1858
Radioactivity, I.36, I.37, I.49, I.52–3, I.56–63
  artificially produced, III.1900–10
  in medical physics, III.1857
Radiobiology, III.1884–9
Radiochemical Centre, III.1901, III.1907
Radiofrequency energy, II.1057
Radiofrequency spectroscopy, II.1047
Radiography, III.1878
Radiometric dating, III.1969, III.1973
Radionuclides, III.1871, III.1904
  dosimetry, III.1884
  imaging, III.1916
  in medical physics, III.1857
  radiation dosimetry, III.1909
  scanning, III.1898
Radiotherapy, III.1857, III.1858, III.1862–80
  computer-aided dose planning, III.1887
Radium, I.58, I.61, I.104–8, III.1876–9
  bomb, III.1877
  dosimetry, III.1880–1
  implants, III.1878
  needles, III.1877, III.1878
Radon, II.1206, III.1876, III.1879, III.1916
  seeds, III.1877
Raman effect, I.163, I.171–2, II.996, III.1404–6
Raman–Nath diffraction, III.1409
Raman scattering, I.117, II.988, II.990, II.996–7, II.1011, II.1013, III.1397, III.1436
RAND project, III.1741

Random close packing (RCP), I.577
Random loose packing (RLP), I.577
Random processes, I.595–7, I.600
Random walks, I.569–70, I.586, I.624, I.630, III.1597
Rapid solidification processing, III.1543
Rare-earth elements, I.97, II.955
  crystal fields, II.1164–5
  electron paramagnetic resonance (EPR), II.1141–3
Rare-earth ion lasers, III.1438
Rare-earth metals and alloys, I.463, II.1012, II.1168
  magnetic properties, II.1160–5
Rayleigh criterion, III.1467
Rayleigh–Einstein–Jeans (REJ) law, I.69
Rayleigh–Jeans equation, III.1393
Rayleigh–Jeans law, III.1390
Rayleigh–Lorentz models, I.628
Rayleigh number, II.844–5
Rayleigh scattering, II.1057
Rayleigh–Taylor instability, III.1782
Reactive collisions, II.1048–9, II.1088–9
Real space, I.567
  renormalization, I.568, I.569
Realm of the Nebulae, The, III.1720, III.1726
Reciprocal lattice, III.1295, III.1296
Recrystallization, III.1526
Rectifiers, III.1322–7
Red blood cells, I.630
Red giants, III.1704, III.1708–9
Red stars, III.1696
Redshift–distance relation, III.1726
Redshift–magnitude relation, III.1794–7
Redshifts, III.1725, III.1777, III.1798
Reduced radial distribution function, I.481
Reflection high-energy electron diffraction (RHEED), I.466

Refractive index, I.546, III.1449
Regge poles, II.681–2, II.718
Regge trajectories, II.682
Relativistic cosmology, III.1721–34
Relativistic kinematics, I.364
Relativistic statistical mechanics, I.282
Relativistic thermodynamics, I.281–2
Relativistic velocity (kinematic) space, I.278
Relativity, III.2049
  kinematics, I.161
  philosophical treatments and popular reactions, I.319–20
  theory of, I.62, I.249–356, II.1266
  see also General theory of relativity; Special theory of relativity
Relaxation processes, II.1132–3, II.1145–6
Renninger effect, I.465
Renormalizability, II.694, II.695, III.2034, III.2036
Renormalization, I.564–9, II.646–7, II.713, II.756, II.1009, II.1015, III.1829–30, III.1832, III.1834, III.1844, III.1848
Renormalization rule, III.1832
Research
  career prospects, III.2028
  funding, III.2058
  laboratories, III.2050
  linking academy with industry, III.2050–1
  management, III.2049
  needs, I.19–22
  performance, III.2052
  physics, I.16–18
  strategic (or pre-competitive), III.2051
  technologies, III.2044–5
Researching and/or teaching, III.2047–8
Residual resistance, III.1281

Resistivity, **III.1331**, III.1334
  absolute magnitude of, III.1303
  mechanisms of, III.1301–4
  temperature variation of, III.1302, III.1342
Resonance
  effects, II.1059–73
  excitation, I.125
  phenomena, II.659–60, II.1029, II.1060–3, II.1067–73
Resonant absorption of radiation energy, II.1057
Resonant electron excitations, II.1082–4
Reststrahlen bands, II.983
Reversed-field pinch (RFP), III.1656–8
Reviews of Modern Physics, III.1290
Revue Scientifique, I.6
Reynolds number, II.807, II.815, II.816, II.825, II.828, II.834, II.837, II.841, II.844, II.850, III.1852
Reynolds stress, II.826, II.827, II.830, II.837, II.853
R-factor, I.450–1
RF SQUID ring, II.942
Ribonucleic acid (RNA), I.504, I.509
Richardson–Lucy iterative deconvolution algorithm, III.1462
Riemann tensor, I.292, I.312
Riemannian metric, I.286
Riemannian space-time, I.322
River water, II.897–8
RKKT theory, II.1170
RKKY interaction, II.1164
R matrices, II.1068–9, **II.1070**
RNA, see Ribonucleic acid (RNA)
Robertson–Walker metric, III.1728–30
Rochelle salt, III.1538
  ferroelectricity, I.489
Rockets, III.1682, III.1738, III.1997, III.1999, III.2005

Rocks
　ages of, III.1960–1
　magnetization, III.1970
　remanent magnetism of, III.1970–2
Rod-vision, III.1416
Röntgen (unit), III.1881–2, III.1890
Röntgen Society, III.1861
Rossby waves, II.893, II.894, II.902
Rotating-disc microscopes, III.1451
Rotational motion, II.1216
Rotational spectra, II.1217
Rotational therapy, III.1868
Rotons, II.928
Royal Greenwich Observatory, II.1248
Royal Institution of Great Britain, I.3, I.27
Royal Society, II.1233
　Fellows of, I.10, I.12
　Philosophical Transactions, I.7, III.1516–17
　Proceedings of, I.7
'Rubber Bible', III.1513
Rubbers, III.1551
　vulcanization of, III.1551, III.1593
Rubidium, I.75
Running coupling constant, II.713
Russell diagram, III.1696, III.1698, III.1710
Russell–Saunders (or L–S) coupling, III.1702
Rydberg constant, I.157–8, II.1271
Rydberg energy, II.1026
Rydberg formula, II.1034

Sachs–Wolfe effect, III.1802, III.1805
Sacramento Peak Solar Observatory, III.1744
Safronov–Wetherill model, III.1963
Safronov–Wetherill theory, III.1966
Sagnac effect, III.1433
Sagnac interferometer, III.1398

Saha equation, III.1699, III.1701
St Bartholomew's Hospital, III.1871, III.1901
Salicylic acid, I.443
Salt, see Sodium chloride
San Andreas fault, III.1978
Sanduleak-69 202, III.1760
Satellites, III.1683, III.1741, III.1765, III.1791, III.1997, III.1999, III.2031
　see also under named satellites
Saturation analysis technique, III.1909
Scalar quantities,
　diffusion-convection balance for, II.841–6
Scale factor, III.1730, III.1731
Scaling, II.688–9
　empirical derivation, I.561–4
　equation of state, I.568
　violation, II.714–15
Scandium, I.75
Scanning electron microscope (SEM), III.1451, III.1571, III.1586–90
Scanning transmission electron microscope (STEM), III.1569, III.1587–90
Scattering phenomena, I.217–19
Schizophrenia, III.1922
Schlieren photography, II.863, II.880
Schmidt camera, III.1406
Schmidt corrector plate, III.1406
Schmidt telescope, III.1406, III.1745, III.1758
Schönberg–Chandrasekhar limit, III.1709
Schottky defect, III.1530
Schrödinger equation, I.196, I.205, I.209, I.225, I.602, II.798, II.862, III.1292, III.1293, III.1299, III.1315, III.1317, III.1320, III.1366, III.1458
Schrödinger wavefunction, I.174

Schrödinger wave mechanics, I.183–96
Schwarzschild metric, III.1778
Schwarzschild singularity, I.306
Science, I.6
Science Abstracts, I.1, I.7
Scientific American, I.5, I.6
Scintillation counters, II.749–51, III.1901, III.1904, III.1916
Scintillation crystal, III.1907
Scintillation detectors, II.1207–8, III.1901, III.1904
Sco X-1 x-ray source, III.1739, III.1755
Screening and antiscreening, II.1091–2, III.1361, III.1363
Screw dislocation, I.476
Scyllac device, III.1645
SDI (Star Wars enterprise), III.2024
Sea-floor spreading, III.1975, III.1976
Sea transport, II.882–7
Search for Structure, A, III.1556
Second (unit), II.1235, II.1247–8
Second Antarctic Expedition, III.2005
Second-harmonic generation (SHG), III.1429
Second International Congress of Radiology, III.1890
Second International Polar Year (1932–33), III.2005
Second law of thermodynamics, I.46, I.68–9, I.374, I.521, I.524, I.534- 5, I.575
Second-order transitions, I.549–51
Secondary image, III.1578
Seebeck coefficient, III.1310
Seebeck effect, III.1284, III.1309
Seebeck voltage, III.1310
Seismology, III.1954, III.1955, III.1956, III.1980
Selection rules (*Auswahlprinzip*), I.94, I.178–81
Selenium, III.1541

Selenium-75, III.1905, III.1909
Selenium rectifier, III.1318
Self-avoiding walk (SAW), I.569–70, III.1598
Self-consistency, III.1519
Self-interstitials, III.1533
Self-phase modulation, III.1448–9
Self-similarity, I.574
Semi-classical theory, III.1388–9
Semiconductors, II.1002, III.1297, III.1301, III.1323, III.1367–8, III.1419, III.2025
  crystalline, III.1368
  detectors, II.1207
  devices, III.2022
  industry, III.1281
  laser diode, III.1464
  lasers, III.1474
  molten, III.1368
  optoelectronics, III.1417
  particle detectors, II.1208
  physical investigation, III.1318
  physics, III.1330–6
  post-war years, III.1317–22
  stimulated emission in, III.1433
  yellow light emission from, III.1392
Semi-leptonic decays, II.675
Senior Wrangler, I.10–11
Sequential resonant tunnelling, III.1470
Seyfert galaxies, III.1777, III.1780, III.1784
Shallow-water theory, II.890, II.892, II.896
Shannon number, III.1418
Shape-memory alloys, III.1539
Shell model, II.997, II.1214
Shell structure, II.1027, II.1208–13
Ship design, II.884
Ship hulls, II.884, II.885
Ship model testing, II.886
Ship–wave pattern, II.884
Ships, hydrodynamic forces on, II.882

Shock waves, II.798, II.808–12, II.847–9, II.872–82
Shoenberg magnetic interaction, III.1359
Short-range interactions, II.1063–5
Short-range parameters, II.1073, **II.1074–5**
Short Wave Wireless Communication, III.1619
Shortt clock, II.1243
Shubnikov groups, I.462
Sidereal time, II.1247
Siegbahn double-focusing beta spectrometer, II.1206
Siemens-Schuckertwerke, III.1573, III.1574
Signal-to-noise ratio in optical systems, III.1418
Silicon, I.470, III.1281, III.1301, III.1526
Silicon on sapphire (SOS) interface, III.1590
Silver, solid solutions, III.1543
Silver chloride, III.1526
Silver halides, III.1529
Single crystals, I.435–7, I.506, III.1280–1, III.1318, III.1360, III.1520
Single dish antenna, Arecibo, Puerto Rico, III.1996
Single photon emission computed tomography, *see* SPECT
Singular perturbation theory, II.800–3, II.829
Singularities, I.306–12
Site percolation, I.572
Sixteen lectures on controlled thermonuclear reactions, III.1660
Size effect, III.1343
Skin effect, II.1157
Skin-temperature distribution, III.1932
SLAC-LBL Collaboration, II.701
Slender-body aerodynamics, II.877

Sliding charge density-wave conduction, II.936
Sloan-Kettering Memorial Hospital, III.1871
Small Astronomical Satellite (SAS-2), III.1740
Small Magellanic Cloud, III.1715
Small Science, III.2025–31
S-matrix theory, II.680, III.2038
Smectic phases, III.1604
Smith chart, II.680
Snoek pendulum, III.1523
Society of Nuclear Medicine, III.1907
Society of Photo-Optical Instrumentation Engineers, III.1474
Sociological phenomena, III.2029
Sociology of physics, III.2031
Sodium, I.65, II.999, III.1530
Sodium chloride, I.429, I.430, III.1514
Sodium iodide
 crystals, II.1207, III.1903
 scintillation counters, II.1210
Sodium pertechnetate, III.1919
Sodium potassium tartrate, *see* Rochelle salt
Soft matter, III.1593–615
 use of term, III.1593–696
Soft-mode theory, II.1007–8
Solar cell, III.1476
Solar energy, III.1446, III.2005
Solar flares, III.1997
Solar gas, III.2009
Solar neutrinos, II.757, II.1227, III.1748–9
Solar oscillations, III.1749
Solar particles, III.2009
Solar radiation, II.899, II.900
Solar spectrum, spectroscopic measurements of, III.1691–2
Solar stream, III.2002, III.2003
Solar system, III.1824, III.1831, III.1950

dualistic theory, III.1962, III.1963
isotopic anomalies, III.1963–4
monistic theory, III.1962, III.1963
origin of, III.1965
tidal theory, III.1950, III.1951, III.1952
Solar–terrestrial physics (STP), III.1982
Solar–Terrestrial Physics, III.2001
Solar–terrestrial space, III.2001–11
Solar wind, III.1997, III.2009
Solar zenith angle, III.1985
Solenoids, iron-free, III.1568
Solid–solid phase transformations, III.1542, III.1544
Solid solutions, III.1543
Solid-state physics, II.1165–6, II.1169, III.1279, III.1526, III.1529, III.2019, III.2025
  emergence of schools, III.1313–14
Solid-state structure analysis, I.421–519
Solidification, III.1542
Solids
  electrons in, III.1279–383
  properties of, III.1279
Soliton pulses, III.1469
Sorbonne, I.15
Sound from flows, II.850–6
Sound intensity, II.851
Sound waves, II.798, II.846, II.857, II.873, II.950
  interaction of, **II.978–9**
Southern oscillation, II.907–8
Space exploration, III.1479
Space groups, III.1513
Space physics, III.1618, III.1982
Space plasma physics, III.1682–7
Space-probe studies, III.1995
Space research, III.1943, III.1965, III.2007–9, III.2024
Space Shuttle, III.1742
Space technology, III.2022
Space-time, I.251, III.1730

Space Transportation System (STS), III.1742
Spark chambers, II.753–4
Spatial-light modulator technology, III.1461
Special Principle of Relativity, III.2034
Special theory of relativity, I.54, I.249, III.2039
  alternate formulations and formalisms, I.273–8
  causal formulations, I.277–8
  constraints, I.249–51
  continuum mechanics, I.279–80
  core principles, I.249
  electrodynamics in, I.257, I.280–1
  elementary particles in, I.282–3
  experimental tests and applications, I.284–5
  formulation, I.262–73
  formulation without light postulate, I.277
  four-dimensional formulation, I.274–6
  gravitation theories in, I.283–4
  later development, I.273
  origins of, I.254–62
  particle dynamics, I.279
  quantum theory in, I.282–3
  relativistic statistical mechanics, I.282
  relativistic thermodynamics, I.281–2
  relativistic velocity (kinematic) space, I.278
  rigid motions, I.279–80
  space-time structure, I.251
  spinor formalism, I.277
  twistor formalism, I.277
Specific gravity, I.46, II.967
Specific heat, I.550, I.555, I.558–60, II.967–73
  lattice theories, II.982–91
  of gases, I.528–30
  quanta, I.152–4

S51

Speckle interferometry, III.1454
Speckle pattern, III.1441
Speckle phenomena, III.1453
SPECT (single photon emission computed tomography), III.1910, III.1916, III.1917, III.1918, III.1920, III.1931
Spectral analysis, I.156
Spectral connection at ionization thresholds, II.1058–9
Spectral density, I.66, I.600
Spectral fingerprints, II.1092–4
Spectral frequencies, I.77
Spectral function, I.69
Spectral lines, I.156–62
Spectral numerology, I.77
Spectrometer, II.992, II.994
Spectroscopy, I.24, I.74–8, II.1039
 equipment, I.24
 interferometric, III.1410
 parallaxes, III.1698
Speculation, mathematical theory of, I.586
Speed of light, I.260, I.261, I.266, II.1240, II.1262–4
Sphalerite, I.429
Spheromak, III.1670
Spilling breakers, II.862
Spin, I.100–1, **I.395**, I.402, II.654
Spin-blocking, III.1831
Spin-fluctuation-exchange, II.953
Spin-Hamiltonian, II.1139–41, II.1151
Spin-lattice relaxation time, III.1923
Spin model, I.564
Spin–orbit coupling, II.1012, II.1126, II.1127, II.1136, II.1210
Spin–orbit interaction, II.1127, II.1212
Spin-warp technique, III.1924, III.1926
Spin waves, II.1010–14, II.1150–2

Spinels, crystallographic data, II.1156
Spinodal, III.1547–8
Spiral nebulae, III.1716–21
Sponge model, II.921
Sponge phases, III.1608
Spontaneous emission, I.93–4
Spontaneous transitions, III.1398–9
SPRITE detector, III.1477
Sputnik I, III.1738, III.2007
Sputnik II, III.1738, III.2007, III.2008
Sputnik III, III.2008
Squeezed light, III.1468–71
Squeezing parameter, III.1468
SQUIDs, III.1932
SSC debate, III.2020, III.2024, III.2031
Stability/instability, II.828–9
Stable fixed points, I.567
Stacking faults, I.473
Standard Model of elementary particles, III.2030, III.2033–9
Standards, II.1233, II.1234, II.1236–60
Stanford Linear Accelerator Center (SLAC), II.687–8, II.741
Star clusters, III.1696
Stark effect, I.89, I.161, III.1469
Stars, III.1691–713
 early-type, III.1698
 formation, III.1766–9, III.1945
 interstellar medium, III.1761–9
 island universes, III.1714, III.1717, III.1718
 late-type, III.1698
 luminosities, III.1698
 main-sequence, III.1698, III.1705
 mass–luminosity relation, III.1705
 photographic parallax programme, III.1696
 research since 1945, III.1746–60
 see also Stellar
Stationary states, I.83, I.86
Stationary waves, II.875

Statistical mechanics, I.523, I.524, I.542, I.586, III.1830, III.1962
  condensing gas, I.551–4
  development of, I.591
  in astrophysics, III.1711
  method of mean values, I.535
  models in, I.603–5
  non-equilibrium, I.585–633
Steering of atoms, II.1099
Stefan–Boltzmann law, III.1696, III.1756
Stellar aberration, I.258, I.259
Stellar Atmospheres, III.1701
Stellar distances, measurement of, III.1691
Stellar evolution, III.1691–713
  research since 1945, III.1746–60
Stellar interferometer, III.1400
Stellar luminosity function, III.1714
Stellar masses, determination of, III.1698
Stellar optical interferometry, III.1475
Stellar spectra, classification of, III.1692–4
Stellar structure, III.1694–5, III.1699
  theory of, III.1702–5
Stellarator, III.1648, III.1654–6
Stereology, III.1549
Stern–Gerlach effect, I.163–72, I.202
Stern–Gerlach magnets, II.1032, II.1047
Stigmator electrostatic astigmatism corrector, III.1580
Stimulated Brillouin scattering (SBS), III.1444
Stimulated Raman scattering (SRS), III.1432, III.1436
Stimulated Rayleigh-wing scattering, III.1445
Stokes theory, III.1411
Storage-ring synchrotron, II.739

Strain-aging, III.1521, III.1523
'Strange particles', II.655–9, II.677–9
Stratosphere, II.900, III.1982
String theory, II.682–4, II.756, II.766, III.2037–9
Stripping reactions, II.1220
Stroke, III.1922
Strömgren spheres, III.1762
Strong coupling theory, I.407
Strong interactions, II.680–4
Strong responses to perturbations, III.1593–694
Strongly correlated electron systems, II.1168
Strouhal number, II.850
Structure analysis, I.450–3
Structure-factor formula, I.433
Structure factors, I.439, I.441, I.447, I.449
Structure functions, II.690
Structure Reports, I.511, I.512
Strukturbericht, I.511, III.1513
SU(2), II.660, II.694, II.725
SU(3), II.660–2, II.677, II.679, II.684
Subscidy (*subs*idiary *sci*entific *dom*ain), III.1507, III.1520
Substance X, I.64
Sudden ionospheric disturbance (SID), III.1995
Sulphur, III.1541
  formation of, III.1747
Sum rules, II.1058–9
Sun
  gravity fields, II.888
  internal structure, III.1694
  luminosity of, III.1708
  magnetic structure, III.1958
  see also Solar
Sunspots, magnetic fields in, III.1958
Sunyaev–Zeldovich effect, III.1773
Supercomputer, III.2045
Superconducting alloys, II.920

Superconducting magnets, III.1927, III.1929
Superconducting Super Collider, III.2036
Superconductivity, I.232, I.489–92, III.1279, III.1347, III.2025
　BCS theory and its aftermath, II.937–41
　developments—pre-1933, II.915–17
　developments—1945, II.917–21
　developments—1945–56, II.932
　discovery of, II.915
　high-temperature, II.957–65
　modern unified picture, II.943–9
　new developments, II.949
　pre-BCS attempts at microscopic theory, II.936–7
　related phenomena, II.955–7
　theoretical developments 1933–45, II.923–9
　type-I, II.920, II.948
　type-II, II.920–1, II.941, II.948
Supercritical fluid, III.1947
Superfluid density, II.944
Superfluidity, II.913–15, II.922, III.2025
　irrotational component, II.929
　modern unified picture, II.943–9
　new developments, II.949
　related phenomena, II.955–7
Superionic conductors, III.1539
Super-Kamiokande detector, II.765, II.768, III.2036
Superlattices, III.1470
Supernova neutrinos, II.758
Supernova trigger hypothesis, III.1963–4
Supernovae, II.1227–8, III.1712–13, III.1718, III.1758–60, III.1766, III.1963–4
　light curve, III.1760
　SN1987A, III.1760
Superplasticity, III.1534
Super-resolution, III.1467

Supersonic aerofoil, II.880
Supersonic boom, II.876
Supersonic speeds, II.872, II.874, II.877, II.879
Superstring theories, II.684
Supersymmetry, II.762–3
Supra-heat-conductivity, II.922
Surface crystallography, I.465–70
Surface of section, III.1836
Surface waves, II.883
Surfactants, III.1606–8
　colloid protection, III.1611
Surgery
　diathermy, III.1894
　ultrasound in, III.1915
Symmetry, I.430, I.462, **II.1028–9**
Symmetry breaking, II.694, II.762
Symmetry elements, I.423
Symmetry principle, III.2034, III.2035, III.2039
Synchrocyclotrons, II.730–1, II.1201
Synchrotron light, II.1050–3
　spectroscopy, II.1052
Synchrotrons, II.639, II.732–3
　DESY, I.436
Synthetic sapphire crystal, II.982
Systems development, III.2030

Tandem accelerator, II.1203
Tandem-mirror facility, III.1669
Tandem Mirror Experiment, *see* TMX
Tandem scanning microscope, III.1451
Tankers, II.885
Taylor thickness, II.811
Teaching and/or researching, III.2047–8
Technetium-99m, III.1905–7, III.1909, III.1919
Technicare Neptune, III.1927
Technische Hoogeschool Delft, III.1579
Teflon (polytetrafluoroethylene), III.1896

Subject Index

Telecobalt therapy, III.1872
Telecommunication networks, II.906
Telecommunications by optical-fibre technology, III.1480
Telecommunications Research Establishment (TRE), III.1327–8
Telescopes, III.1734, III.1735, III.1738
 see also under specific types and applications
Teletherapy, III.1877, III.1879, III.1890
Television cameras, III.1412
Television industry, III.1476, III.1536
Teller–Northrop theory, III.1683
Temperature coefficients of resistivity, III.1538
Temperature gradients, II.901
Temperature measurement, II.1249–51
Temperature scales, II.1249–51
 ITS-90, II.1260
Temperature variation of resistivity, III.1302, III.1342
Terrella (magnetized sphere), III.2003
Terrestrial Magnetism and Atmospheric Electricity, III.2000
Tetramethyl manganese ammonium chloride (TMMC), II.1014
TFTR, *see* Tokamak Fusion Test Reactor (TFTR)
Thallium, I.75
Theorem of equipartition of energy, *see* Equipartition theorem
Theoretical Structural Metallurgy, III.1522

Theory of the Electron, The, III.1992, III.2039
Thermal analysis, III.1508
Thermal capacity, II.967, III.1354
Thermal conductivity, II.977–82, III.1285, III.1290
Thermal energy, I.66
Thermal equilibrium, I.530
Thermal expansion, II.975–7
Thermal vibrations, II.984
Thermal wind equation, II.901
Thermals, II.845
Thermionic emission, I.37
Thermodynamic equilibrium, III.1391
Thermodynamic internal energy, I.532
Thermodynamics, I.5, I.35, I.149
 as science, I.521
 first law of, I.374, I.521
 nineteenth century background, I.521–4
 of data processing, I.576
 of heterogeneous substances, I.532
 relativistic, I.281–2
 second law of, I.46, I.68–9, I.374, I.521, I.524, I.534–5, I.575
 special-relativistic, I.281
 third law of, I.535–7, I.537–42, II.968
Thermoelastic transformation, III.1539
Thermoelectric effects, III.1283–4, III.1306–11
Thermoelectric (Seebeck) voltage, III.1310
Thermography, III.1932
Thermometers, II.1249
Thermonuclear explosion, II.1224
Thermonuclear fusion, III.1618, III.1629
Thermonuclear reactions, II.1226
Thermos flask, I.27
Thermosphere, III.1982, III.2009

Thimble chamber (ionization chamber), III.1866–7, III.1881
Thin-film devices, III.1444
Thin films, II.1170–1
Third International Polar Year (1957–58), III.2006
Third law of thermodynamics, II.968
  applications, I.537–42
  historical survey, I.535–7
Thomas–Fermi method, I.464, II.1190, II.1191
Thomson moving-magnet galvanometer, I.23
Thomson scattering, I.65
Thomson–Lampard theorem, II.1255
Thorium, I.59, I.60
Threshold effects, II.1063–5, **II.1066–7**
Thymine, I.504
Thyroid gland, III.1900–3
Tidal currents, II.888
Tidal kinetic energy, II.891
Tide gauges, II.891–2
Tide-raising forces, II.892, II.896
Tide tables, II.890, II.892
Tight-binding approximation, III.1292
Tight-binding model, III.1295, III.1297
Time
  dissemination of, II.1248–9
  measurement, II.1242–6
Time–position recording, III.1913–14
Time-resolved crystallography, I.438
Time-scales, II.1247–8
Tin oxide, III.1541
Titanium and titanium alloys, III.1281, III.1897
Tl–Ba–Ca–Cu–O compound, II.959
$Tl_2Sr_2Cu_3O_{10}$, I.492

TMX (Tandem Mirror Experiment), III.1667–8
Tobacco mosaic virus (TMV), I.508
Tokamak Fusion Test Reactor (TFTR), III.1646–54, III.1663, III.1677
Tollmien–Schlichting waves, II.832
Tomato bushy stunt virus (TBSV), I.509, I.510
Tomography. *See* Computed tomography
TOPCON EM002B 200 kV column, III.1588
Torsional oscillation, III.1523
Transcurrent faults, III.1980
Transdisciplinary disciplines, III.2045–7
Transform faults, III.1979
Transformation laws, I.269
Transformation theory, I.59, I.60, I.360
Transistors, III.1322–7
  commercial applications, III.1317
  invention of, III.1317
Transition
  abrupt, II.833, II.834
  amplitudes, I.185, **II.1064**
  restrained, II.833, II.834
  state, II.1049
  taxonomy, II.832–41
  types of, II.825–32
Transition metals, III.1343, III.1354
Transmission electron microscope (TEM), III.1526, III.1548, III.1569, III.1570, III.1573
  aberration-free atomic resolution, III.1585–6
  accelerating voltage, III.1590
  and Abbe's optical theory, III.1578–84
  astigmatism corrector, III.1580
  commercial prototype, III.1575–6
  high-resolution commercial, III.1576
  holographic, III.1584

over-focus and under-focus fringes, III.1580
RCA, III.1581
resolution, III.1590
selected area diffraction, III.1579
serially produced, III.1575–7
specimen preparation and radiation damage, III.1574–5
two-stage, III.1574
vertical column, III.1575
Transonic flight, II.882
Transport coefficients, I.597–8, I.612, I.614
Transport parameters, I.593, I.614
Transport phenomena, I.592
Transuranium elements, II.1222–5
Transverse temperature gradient, III.1282
Transverse voltage, III.1282
Transverse waves, III.1956
Trapping of atoms, II.1099
Triplet coupling, II.1086
Triplets (antitriplets), II.684–5
Triton, II.1208
Tropical cyclones, II.904–6
Troposphere, II.900, III.1982, III.1984
T-Tauri stars, III.1766–7
Tubeless syphon, III.1593–4
Tumours. *See* Cancer; Radiotherapy
Tungsten, III.1355
Tunnel effect, I.217
Turbo-fan engine, II.873
Turbulence, II.843, III.1834
  isotropic, II.832
  taxonomy, II.832–41
  types of, II.825–32
Turbulent flow, II.825
Turnip yellow mosaic virus (TYMV), I.510
Twinned crystals, III.1435
Twisted nematic cell, III.1604
Two-fluid hydrodynamics, II.929
Two-level-atom model, III.1459

UHURU satellite, III.1739, III.1754, III.1766
UK Infrared Telescope (UKIRT), III.1743
Ultra high-molecular-weight polyethylene (UHMWPE), III.1896
Ultra low-frequency radio physics, III.1995
Ultrasonic imaging, III.1914–15
Ultrasonic industrial flaw detector, III.1911
Ultrasonic probe, III.1913
Ultrasonic radiation, therapeutic, III.1915
Ultrasound
  diagnostic, III.1910–11
  in medicine, III.1910–15
  in surgery, III.1915
  in therapy, III.1915
  scanning, III.1859, III.1913, III.1915
Ultraviolet astronomy, III.1741–2
Ultraviolet energy, III.1992
Ultraviolet light, II.1050–3, III.1396
Ultraviolet radiation, III.1994, III.2009
Umklapp processes, II.980, III.1304–6
Uncertainty relations, I.174, I.201–4, I.209, I.228, III.1394
Uniaxial ferroelectrics, II.1009
Unified field theories, I.320–3
Unit cell, I.423
Unitary symmetry, II.660–2
Units, II.1236–60
Universal gas constant, I.150
Universal Time (UT), II.1247
Universality, I.549–51, I.563, II.664, II.677, III.1833
  empirical derivation, I.561–4
Universe
  age of, III.1732, III.1793–4
  dark matter, III.1731
  dynamics of, III.1730, III.1787

Einstein's, III.1722–4
expansion of, III.1786, III.1787, III.1806
geometry of, III.1723
Hot Big Bang models of, III.1778
hot early phases, III.1788, III.1793
large-scale structure, III.1714–34, III.1805
mean baryonic density, III.1791–2
mean density, III.1800
mean mass density, III.1719
temperature history, III.1786
total mass density, III.1800
very early, III.1805–7
U-particles, I.393, I.394, I.396, I.397
Upper atmosphere, III.1981–4
geophysics, III.1943–4, III.1982
ionization of gases, III.1981
Upper Atmosphere, The, III.2000
Upsilon, II.721
Upstream wakes, II.868–9
U-quanta, I.378, I.382, I.389, I.391
Uracil, I.504
Uranium, I.56, I.60, I.61, I.106, II.1199, II.1223, II.1227, III.1630, III.1960
Uranium-235, I.131–2
US Atomic Energy Commission, III.1633
US Bureau of Standards, III.1988
US National Aeronautics and Space Administration (NASA), III.1739, III.1965
US National Bureau of Standards (NBS), I.20, II.1235, II.1236, III.1881, III.1998
U-sequence, III.1839
UTC, *see* Coordinated Universal Time (UTC)
Utrecht School, I.629

Vacancies, III.1505, III.1529, III.1531, III.1533
Vacuum polarization, II.641, II.1031
Vacuum technology, I.27
Vacuum tubes, I.53, III.1987
Valence band, III.1317, III.1319, III.1326, III.1329, III.1334
Valence electrons, III.1314, III.1317, III.1366
Van Allen belts, III.1683–4
van de Graaff generator, II.727–8, II.1203, III.1871, III.1883
van der Waals equation, I.28, I.547, I.552, I.561
van der Waals law, II.973
van der Waals theory, I.555, I.557
Varistors, III.1537
Väsälä–Brunt frequency, II.864
V–A theory, II.667–9
Vavilov–Čerenkov effect, III.1408
Vela satellite, III.1740, III.1741
Velocity–distance relation, III.1726, III.1727, III.1728
Velocity–position space (μ-space), I.587, I.590
Veneziano amplitude, II.684
Vertical-cavity surface-emitting laser (VCSEL), III.1464
Very high-resolution imaging systems, III.1477
Very long baseline interferometry (VLBI), III.1738
Very low-frequency radio physics, III.1995
Very low temperatures, I.540–1
Vesicles, III.1608
Vibration, II.967–1010
Vibrational excitation, II.1073
Vibrational motion, II.1044
Vibrational specific heat of solids, I.525
Vibrational transitions, II.1073
Videophone, III.1477
Viking (rocket), III.2005
Vine–Matthews theory, III.1976

Virial theorem, II.1026, III.1731, III.1772, III.1800
Virtual reality, III.1476
Viruses, I.507–11
 rod-shaped, I.508
 spherical, I.509
Viscosity, II.807, II.813–14
 stabilizing effect of, II.828
Viscous diffusion, II.827
Vitamin B_{12}, structure of, I.495
Vitamin D, isolation of, III.1529
Vlasov–Boltzmann equation, III.1625
Volt
 absolute determination, II.1256
 reference standards, II.1256
Volta effect, III.1283
Voltaic cell, III.1284
von Klitzing constant, II.1259, II.1271
von Neumann equation, I.602, I.612
Vortex lines, II.820, II.822, II.823
Vortex-wake, II.876
Vorticity, II.816–24, II.889, II.890, II.892, II.894, II.901
V-particles, I.408–9
Vulcanization, III.1551, III.1593

Wake fluid, II.821
Wakes, II.812–16
Wanderwelle, electrical surges, III.1566
Water-bath ultrasonic simulator, III.1912
Water waves, II.798
Watson–Crick model, I.504
Wave, *see* under specific wave types
Wave action, II.861
Wave energy, II.862
Wave equation, I.195, I.283, III.1393
Wave frequency, II.863

Wave functions, I.204, **I.361**, I.617, II.941–3, II.949, II.1124
Wave generation, II.846–69, II.882–5, II.900
Wave-interaction effect, III.1994
Wave mechanics, I.174, I.191–6, I.217, I.360
Wave momentum, II.866
Wave numbers, II.838, II.839, II.861, II.864, II.866–8, II.885, II.893, II.984
Wave packets, II.836, III.1366
Wave–particle duality, I.168–71, I.364, III.1394, III.1395
Wave propagation, II.846–69, II.877, III.1407, III.1955
Wave theory, I.258, I.363
Waveforms, II.848, II.858, II.861
Wavelength, II.883, II.902
 standards, II.1239–40, II.1242
Wave-making resistance, II.867
Wavevectors, II.988–90, II.1001
W-bosons, II.637, II.706–9
Weak interactions, II.663–71, II.692–3
Weak universality, II.664
Weather, II.899–912
 forecasting, II.888, II.899, II.906–9
Weber bar, I.314
Weinberg–Salam model, II.695, II.762
Wet-bulb temperature, II.843
Weyl's postulate, III.1729
Wheatstone bridge, I.22
Whiskers, III.1351
Whistler waves, III.1622, III.1623, III.1997, III.2005
 measurements, III.2009
 modelling, III.1999
Whistling atmospherics, III.1995
White dwarfs, III.1709–12
Wide Field Planetary Camera, III.1745

Wiedemann–Franz law, III.1286, III.1287, III.1316
Wiedemann–Franz ratio, III.1290
Wien–Planck spectral equation, I.69
Wiener–Khintchine theorem, I.600–2
Wien's law, I.69, I.151, I.524, III.1393
Wigner function, I.617–18, III.1441
William Herschel Telescope, III.1744
Wilson cloud chamber, *see* Cloud chamber
Wilson's scaling method, I.440, I.445, I.447
Wind effects, II.901, II.903–4
Wind instruments, II.851–2
Wind speed, II.843
Wind speed/sound speed ratio, II.813
Wind tree model, I.603
Wind-tunnels, II.830, II.879
Winding number, II.955
Wireless. *See* Radio
Wireless telegraphy, III.1986
Wollaston wire, II.831
Women in higher education, I.8, I.10
Work-function, III.1286, III.1308
Work-hardening, III.1518
World Directory of Crystallographers, I.511
W-particle, III.2036
Wranglers, I.10–12
Wurtzite, I.429

Xenon, I.75, I.556
Xerographic copying, III.1541
X-ray absorption, II.1051
X-ray analysis, III.1514
X-ray Analysis Group, I.511
X-ray angiography, III.1904
X-ray apparatus, II.1051

X-ray astronomy, III.1738–41, III.2005
X-ray binaries, III.1754–7
X-ray contrast media, III.1862–3
X-ray crystallography, I.446, I.469, I.484, III.1515
X-ray diffraction, I.30, I.424–32, I.456, I.457, I.484, I.498, II.983, III.1583
  amorphous solids, **I.478**
  in-plane, I.466
  Laue's equations for, **I.427**
X-ray dose-rate measurements, III.1871
X-ray emission, III.1630
X-ray equipment, III.1861
X-ray imaging, III.1916, III.1922–3
X-ray intensities, I.440
X-ray irradiation, III.1529
X-ray lasers, III.1464–6
X-ray measurements, II.1010
X-ray microscope, III.1583
X-ray optics, I.484
X-ray photons, I.363
X-ray pulsations, III.1756
X-ray pulses, III.1752
X-ray scattering, I.165, I.362, I.365, I.453, II.990, III.1867
X-ray sets, III.1864, III.1877
X-ray sky, III.1798
X-ray sources, III.1755–7
X-ray spectra, I.86, I.158–60
X-ray spectrometer, III.1580
X-ray spectroscopy, I.158–60, I.429, I.484
X-ray studies, inner-shell phenomena, II.1089–90
X-ray telescopes, III.1740
X-ray tomography (X-CT), III.1918
X-ray tubes, III.1859, III.1861, III.1862, III.1865
X-rays, I.3, I.32, I.36, I.52, I.53, I.65, I.362–6, I.421–4, II.1050–3, III.1514
  discovery, I.424

in diagnosis, III.1859–63
in medical physics, III.1857
see also Radiotherapy

Yang–Mills theory, II.692–4, II.713, II.714, II.763
YBa$_2$Cu$_3$O$_7$, II.959
YBa$_2$Cu$_3$O$_{7-\delta}$, I.490–1
Yerkes system, III.1698
Yield-stress, III.1520–1, III.1523
Yin-Yang coil, III.1664
Young's double-slit experiment, III.1402
Yttrium–iron–garnet, II.1159
Yukawa meson theory, I.378–83, II.651–2

Z-bosons, II.637, II.706–9
Zeeman effect, I.7, I.36, I.52, I.54, I.99–103, I.160–2, I.182, I.183, I.184, I.211, II.1114, II.1115, II.1149
Zeeman splittings, II.1120

Zeeman sub-state, II.953
Zeitschrift für Kristallographie, I.511
Zeitschrift für Metallkunde, III.1556
Zenith angle, III.1992
Zernike's circle polynomials, III.1409
Zero-point motion, II.973–5, II.1004
Zero-resistance state, II.961
Zero sound, II.950, III.1364
ZETA (Zero-Energy Thermonuclear Apparatus) toroidal pinch experiment, III.1641–2, III.1645, III.1657
Zhukovski's theorem, II.822
Zinc, III.1360
Zinc oxide, III.1537
Zinc sulphide scintillators, II.1200
Zincblende, III.1514
Zone-levelling, III.1543, III.1544
Zone-refining, III.1543, III.1544, **III.1545**
Zweig's rule, II.685

NAME INDEX

Note—volume and page numbers in **bold type** refer to items in boxes

Aaronson, M, Hubble's constant, III.1793
Abbott, E A, *Flatland*, I.5
Abelson, P H, transuranium element, II.1223
Åberg, T, secondary excitations, II.1090
Abragam, A
  hyperfine structures, II.1139
  nuclear magnetic resonance, II.1175
  resonance research, II.1143
Abraham, electron microscope, III.1573
Abraham, M
  special-relativistic theories, I.284
  stress–energy tensor, I.280, I.281
Abramowicz, M A, self-consistent thick-disc model, III.1780
Abrikosov, A A, superconductivity, II.935
Accardo, C A, p–n junction, III.1417
Adams, M C, dye-laser pulses, III.1439
Adams, W C
  A stars, III.1710
  chemical abundances, III.1702
  luminosity indicators, III.1698
Adams, W S, Sirius B, III.1710, III.1723

Ademollo, M, Dolen–Horn–Schmid duality, II.683
Adlard, photon, III.1451–2
Adler, S, soft photon emission, II.679
Aharonov, Y
  field strength, II.693
  interference patterns, III.1370
Akasofu, S-I, *Solar–Terrestrial Physics*, III.2001
Akhmanov, S I, Nd:YAG laser, III.1435
Albrecht, A, world model, III.1807
Alder, B J
  autocorrelation function of tagged particle, I.619
  molecular dynamics, I.577
Alder, K, Coulomb excitation experiments, II.1217
Alexander, S, fractons, II.1007
Alfvén, H
  baryon symmetric models, III.1806
  Cosmical Electrodynamics, III.1637, III.1685
  Galactic radio emission, III.1737–8
  MHD theory, III.1684, III.1685
  monistic theory, III.1962, III.1963
  ring current around the Earth,

N1

III.2003
Allen, J F, superleak flow, II.946
Allis, W P, electron diffusion rates, III.1644
Alpher, R A
  abundances of elements, III.1785–6
  black-body radiation, II.758
  primordial nucleosynthesis, III.1785, III.1786, III.1790
  proton–neutron ratio, III.1786
Als-Nielsen, J, order–disorder transition in β-brass, II.1009, II.1016
Altar, W, wave propagation, III.1990–1
Altarelli, G, asymptotic freedom, II.715
Al'tshuler, B L, scattered wavelets, III.1370
Alvarez, L W
  neutron diffraction, I.459
  particle interactions, II.659
  proton linear accelerator, II.1205
  proton machines, II.737
  60" cyclotron, II.1202
  tandem accelerator, II.1203
Ambartsumian, V A, T-Tauri stars, III.1766
Amelinckx, S, dislocations, III.1526
Ampère, A M
  electromagnetic interactions, III.1953
  magnetism, I.65
Amrein, W O, point photon, III.1451
Anaxagoras, matter, I.44
Anaximenes, air as basic substance, I.358
Anderson, C D, **I.386**
  absorption in lead plates, I.385
  electron–positron pairs, I.231
  elementary particles, I.391
  mesotron, I.231, I.389
  muon, II.650

positron, III.1706
shower particles, I.388
standard model, I.231
Wilson cloud chamber, I.371
Anderson, J A, angular diameter of Betelgeuse, III.1705
Anderson, J L, relativistic thermodynamics, I.281
Anderson, P W
  bosons, II.695
  disordered materials, III.1365
  electron density, III.1316
  exchange interactions, II.1144
  glasses, II.1006
  phase transitions, II.1008
  phonon exchange, II.953
Anderson, W C
  speed of light, II.1263
  white dwarfs, III.1711
Andrade, E N, crystal diffraction spectroscopy, II.1207
Andrew, E R
  magnetic resonance imaging, III.1924
  nuclear magnetic resonance, II.1175
Andrews, T
  critical point in carbon dioxide, I.28
  critical point phenomenon, III.1947
Anger, H, gamma camera, III.1907
Ångström, A J
  normal solar spectrum, II.1242
  spectrum of hydrogen, I.76
Antonucci, R R, Seyfert galaxies, III.1784
Appleton, E V
  radio receiving apparatus, III.1989
  solar radiation, III.2002
  upper atmosphere, III.1988
  wireless propagation, III.1622
Araki, G
  cosmic-ray particles, I.407

slow mesotrons, I.406
Argos, P, structure determination, I.445
Arima, A, collective motion in nuclei, II.1218
Aristotle, matter, I.44
Armenteros, R, hyperon, II.657
Armstrong, H E, organic compounds, I.493
Arndt, U W, single-crystal oscillation, I.437
Arnett, W D, element abundances, III.1748
Arnott, N
  Elements of Physics, III.1857
  medical physics, III.1856–7
Arp, H, quasars, III.1777
Artsimovich, L
  Atoms for Peace Conference, III.1640
  mirror-type systems, III.1643
  tokamak, III.1648
Asaro, F, alpha-decay spectra, II.1217
Ash, E, optical microscopy, III.1406
Ashkin, A
  argon-ion laser beams, III.1453
  He–Ne laser, III.1434
  optical binding of particles, III.1472
  self-trapping of laser beam, III.1459
Askaryan, G A, self-focusing, III.1436
Aspect, A, Bell inequalities, III.1467
Astbury, W T
  DNA, I.504
  fibrous proteins, I.497
  x-ray diffraction, I.439
Aston, F W
  isotope masses, I.111–12
  stellar energy source, III.1706

Astrom, E O, plasma physics, III.1992
Atkinson, R d'E
  nuclear reactions, II.1225, III.1707
  thermonuclear reactions, III.1627
Auger, P, neutrons, I.122
Austin, L W, radio propagation, III.1987
Autler, S H, 'dressed' atomic states, III.1421
Avenarius, critical opalescence, I.546
Avery, Q T, DNA, I.504
Avogadro, A, hypothesis, I.46
Avrami, M, phase transformations, III.1554
Axford, I, solar wind plasma, III.2010
Ayers, G R, blind deconvolution, III.1461
Azbel, M Ya
  canonical theory, III.1356
  cyclotron resonance, III.1357

Baade, W
  Andromeda Nebula, III.1788
  medium-mass elements, II.1228
  Palomar 200" Telescope, III.1737
  radio sources, III.1775
  supernovae, II.1227–8, III.1712, III.1718, III.1759, III.1775
Bacher, R F
  atomic energy levels, II.1093
  nuclear structure theory, II.1208
  shell model, II.1209
  Thomas–Fermi model, II.1191
Back, E
  total quantum number, I.162
  Zeeman splitting, I.161
Bacon, G E
  neutron diffraction, I.456
  structure of KH_2PO_4, I.459
Badash, L, I.29

Baez, A V, optical microscopy, III.1406
Bahcall, J N
  solar model, III.1749
  solar neutrino flux, III.1748
Bailey, V A, broadcast modulation, III.1994
Bain, E, bainite, III.1515
Bainbridge, K
  H-particle–electron model, I.112
  mass spectrometry, II.1187
Baker, D W, ultrasonic imaging, III.1914
Baker, G A, Padé Approximant, I.558
Baldwin, D, thermal barrier, III.1667
Baldwin, G C, dipole vibration, II.1213
Balescu, R, *Equilibrium and Non-equilibrium Statistical Mechanics*, I.620
Balluffi, R, diffusion, III.1531
Balmer, J, *R* value, I.77–8
Banerjee, M K, proton inelastic scattering, II.1221
Bangham, vesicles (or liposomes), III.1608
Barber, C, electron storage rings, II.739
Bardakci, K, singularities in complex angular momentum, II.682
Bardeen, J, **II.939**
  BCS theory, II.938
  impurity effects, III.1281
  interatomic forces, II.999
  ion potential, III.1303
  point-contact transistor, III.1326
  solid-state physics, III.1313
  superconductivity, I.232, II.937, III.1364
  theory of phonons, III.1362
  transistor, II.744
  xerography, III.1541

Bardeen, J M, black holes, III.1779
Barham, P M, coherence parameter, III.1412
Barkhausen, H
  magnetic flux, II.1121
  radio waves, III.1622
  whistler waves, III.1622–3
Barkla, C G
  x-ray scattering, I.362, I.424
  x-ray spectroscopy, I.158
Barlow, W, crystal structure, I.423–4
Barnes, W H, structure of ice, I.474
Barrett, C, diffusive transformations, III.1547
Bartels, J, *Geomagnetism*, III.2000
Barthel, P D, radio quasars and radio galaxies, III.1784
Bartlett, J A, shell models, II.1209
Barut, A O, singularities in complex angular momentum, II.682
Basov, N G
  excimer laser, III.1453
  semiconductor maser, III.1423
Bass, M, sum-frequency generation, III.1429
Bate, G, magnetic materials, II.1159
Bates, D R
  ionospheric research, III.1999
  probability of electronic excitations, II.1080
  solar x-rays, III.1994
Bauer, high-reflection coatings, III.1407
Baxter, R J, ferroelectric model, I.571
Bay, Z, electron multiplier, III.1541
Baym, G, neutron stars, III.1754
Bear, R S, three-stranded coiled-coil model, I.499
Beck, C S, defibrillators, III.1895
Beck, G, β-decay theory, I.372, I.375

Beck, K, single crystals of ferromagnets, II.1123
Becker, H, neutron, I.119
Becker, R, nucleation kinetics, III.1547
Becklin, E E
  formation of stars, III.1769
  Orion Nebula, III.1743, III.1767
Becquerel, A H
  radioactivity, I.37, III.1857, III.1876
  uranic rays, I.53, I.56–7
Bednorz, J G
  Meissner effect, II.957–8
  superconductivity, I.490, II.961
Beevers, C A
  Rochelle salt, I.489
  structure determination, I.441
Behn, U, specific heat, II.968
Bell, A G, Photophone, III.1422
Bell, F O, DNA, I.504
Bell, J S
  EPR paradox, I.229
  inequalities, III.1389
Bell, P R, scintillators, II.750
Bell, R J, amorphous SiO_2, I.478–9
Bell, S J
  neutron stars, III.1751
  sky surveys, III.1752
Bémont, G, radium, I.58
Bénard, H
  sound from flows, II.850
  surface tension, II.845
Benedek, G
  'light-beating' spectroscopy, III.1454
  spontaneous magnetization, I.559
Benioff, H, earthquake oscillations, III.1958
Bennet, W R Jr, helium–neon laser, III.1427
Bennett, F, tandem accelerator, II.1203

Bennewitz, K, zero-point motion, II.973–4
Bentley, R, physical cosmology, III.1721
Beran, M J, partial coherence, III.1392
Berkner, L
  IGY, III.2006
  ionospheric research, III.2005–6
  ionospheric sounding, III.1998
  Third International Polar Year, III.2006
Berlinger, W, oxygen octahedra, II.1009
Berman, R
  diamond–graphite equilibrium diagram, I.538
  thermal conductivity, II.981
Bermen, S M, hadrons, II.690
Bernal, J D
  amorphous structure, I.479
  crystal oscillation, I.434
  globular proteins, I.497
  haemoglobin, I.497
  random loose packing (RLP) and random close packing (RCP), I.577
  residual entropy of ice, I.571
  stereochemical formulae, I.495
  structure of ice, I.474
  tobacco mosaic virus (TMV), I.508
Bernard, M G A, stimulated emission in semiconductors, III.1433
Bernoulli, D
  Bernoulli equation, II.806
  gas pressure, I.47
Bernstein, I B
  collisionless shocks, III.1684
  plasma oscillations, III.1625
Berreman, D W, liquid-crystal optics, III.1452
Berry, B, anelastic relations, III.1523

Berry, M V
  diffractals and caustics, III.1466
  topological phase, III.1470
Berson, S, saturation analysis technique, III.1909
Bertero, M, imaging theory, III.1467–8
Bertolotti, M, lasers, III.1423
Berzelius, J J, nomenclature for elements, I.45
Besicovich, A S, quasicrystals, I.482
Bessel, F, stellar distances, III.1691
Beth, R A, angular momentum of photons, III.1410
Bethe, H A
  abundances of elements, III.1785–6
  bremsstrahlung, I.385
  carbon–nitrogen–oxygen cycle (CNO cycle), III.1707–8
  face-centred cubic lattice, III.1293
  fast charged particles, II.1040
  Fermi theory, I.377
  First Shelter Island Conference, II.645
  mesons, I.401, I.403, II.654
  near-field/far field solution, III.1416
  neutron capture, II.1193
  neutron stars, III.1754
  nuclear physics, II.1184
  nuclear reactions, I.132, II.1192, III.1629
  nuclear structure theory, II.1208
  origin of the elements, II.1225
  pp process, II.1226
  quantum fluctuations, II.1013–14
  shell model, II.1209
  spacing of states in highly excited nucleus, II.1213
  theory of metals, III.1290
  Thomas–Fermi model, II.1191
  Twelfth Solvay Conference, Brussels, II.646
  weak-interaction-treatment, II.1040
Beynon, W, ionosphere, III.1984
Bhabha, H J
  cosmic rays, I.393
  fusion reactions, III.1642
  showers, I.388
  U-particles, I.393
Bhar, J N, ionospheric physics, III.2000
Bhatia, pseudopotential concept, III.1367
Bieler, É S
  α-particles, I.114
  anomalous scattering, I.366
Bier, N, scleral lenses, III.1414
Biermann, L, ionized comet tails, III.2009
Bijvoet, J M
  dextro (d) and laevo (l) compounds, I.495
  strychnine, I.495
Bilby, B A, yield-stress and strain-aging, III.1523
Bilderback, D, area detector, I.438
Billy, H, Renninger effect, I.450
Biloni, H, solidification, III.1542
Binnig, G
  scanning tunnelling microscope, III.1549
  tunnelling electron microscope, I.466
Birch, F, Earth's core, III.1957
Birdsall, C K, stream-type instabilities, III.1673
Birge, R T
  Birge–Bond diagram, II.1268
  constants, II.1271
  least-squares adjustment, II.1270
  speed of light, II.1263
Birkeland, K
  geomagnetic storms, III.2003
  Terella device, III.1682

Birkhoff, G D, energy surface trajectories, I.599
Birkinshaw, M, Sunyaev–Zeldovich effect, III.1773
Bishop, A S, *Project Sherwood—The US Program in Controlled Fusion*, III.1640
Bitter, F, antiferromagnetism, II.1129
Bjerrum, N
 Applications of the quantum hypothesis to molecular spectra, I.82
 molecules, I.81–2
 quantum theory, I.155
Bjorken, B J
 charmed quark, II.700
 quark–lepton analogy, II.700
Bjorken, J D
 hadrons, II.690
 parton hypothesis, II.690
 point-like constituents, II.689
Bjorkholm, J E, self-trapping of laser beam, III.1459
Bjornholm, S, fission isomers, II.1218
Black, J, medical physics, III.1857
Black, J C, ultrasonic properties, II.1006
Blackett, P M S, **I.400**
 cloud chambers, II.651
 Earth's magnetic field, III.1969–70
 electron–positron pairs, I.231
 nuclear reactions, II.1201
 penetrating radiation, I.388
 The craft of experimental physics, II.745
 Wilson cloud chamber, I.371
Blackman, M
 BvK theory, II.984
 perturbation theory, II.1002
 specific heat, II.985–7
 toroidal fusion device, III.1634
Blair, J S
 alpha particles, II.1221
 diffraction theory, II.1221
Blanc-Lapierre, A, polychromatic fields, III.1421
Blau, M, cosmic-rays tracks, I.404
Bleaney, B, resonance research, II.1143
Bleuler, E, two-photon correlation, III.1413
Blewett, J P
 accelerators, II.725–7
 Particle Accelerators, II.1201
Bliss, W R, cross-sectional imaging, III.1912
Bloch, F
 electron gas, III.1291–301
 gyroscopic motion, II.1172
 magnetism, II.1137
 neutron diffraction, I.453, I.459
 nuclear magnetic resonance, II.1094
 quantum theory, I.221, III.1361
 spin waves, II.1010–11
Block, M, parity states, II.665
Block, S, diffraction techniques, I.437
Bloembergen, N, **III.1430**
 anharmonic oscillators, III.1434
 laser spectroscopy, III.1466
 non-linear optics, II.1100, III.1434
 Nonlinear Optics, III.1445
Blow, D M, structure determination, I.445
Bludman, S A, scalar Higgs fields, III.1806
Boas, W, plastic deformation, III.1526
Bobeck, A H, bubble domains, II.1160
Bodin, H, reversed-field pinch (RFP) research, III.1657
Boersch, electron holograms, III.1582–3

Bogoliubov, N N, **I.610**
  BBKGY hierarchy, I.606, I.623, I.624, I.626
  kinetic equation, I.609
  properties of dilute Bose gas, II.930
Bohm, D
  Bohm diffusion, III.1630, III.1631, III.1632
  electrical discharges in gases, III.1361
  EPR paradox, I.229
  field strength, II.693
  First Shelter Island Conference, II.645
  interference patterns, III.1370
  plasma oscillations, III.1625
Bohr, A, II.1216
  compound nucleus theory, II.1219
  Coulomb excitation experiments, II.1217
  geometrical collective model, II.1218
  quadrupole moment, II.1215
Bohr, H, quasicrystals, I.482
Bohr, J, Fourier techniques, I.469
Bohr, N, **I.79**, I.180
  atomic theory, I.157, II.1117, III.1401
  β-decay theory, I.372
  β-spectra, I.126–7
  Balmer formula, I.80, I.82
  Bohr radius, I.85
  building-up principle (*Aufbauprinzip*), I.96–7
  Cockcroft–Walton accelerator, II.1188
  complementarity theory, I.224
  compound nucleus, II.1192, II.1194
  correspondence principle, I.86
  dumb-bell model, I.181–2
  dynamical quantum theory, I.174

Einstein–Podolsky–Rosen (EPR) paradox, I.228
electron collisions, II.1042
fast charged particles, II.1039
Fifth Solvay Conference (1927), I.205
fission process, II.1195
high-speed β-particles, I.107
hydrogen atom, I.82–6, I.91
impact of theory, I.86–9
Light and life, I.224
liquid drop, I.131, II.1213
m-values, I.95
magnetic properties, II.1115
mechanical forces, I.78
'new hypothesis', I.84
nuclear collective motions, II.1213
nuclear fission, I.126
nuclear reactions, I.123–6
On the constitution of atoms and molecules, I.157
On the general theory, I.176
On the quantum theory of line spectra, I.94
periodic system, I.97
periodic table, I.95
principal quantum number, I.91
principle of complementarity, I.197
quantum conditions, I.359
quantum dynamics, I.72–82
quantum hypothesis, I.93
quantum mechanics, I.225, I.226, I.228
quantum optics, III.1397
quantum theory, I.80, I.166, I.175, I.221, II.1115, III.1397
radiative instability, I.83
spacing of states in highly excited nucleus, II.1213
spectra, I.77
superconductivity, II.917
theory of atomic structure, III.1699

theory of metals, III.1286
theory of nuclear reactions, II.1192
Twelfth Solvay Conference, Brussels, II.646
uranium-235, I.131–2
x-ray spectra, I.86
Bolsterli, M, computers, III.1843
Bolt, R H, ventriculography, III.1911
Bolton, J G, radio sources, III.1737
Boltzmann, L E, **I.31**
    battle with Ostwald, I.48
    black body, I.66
    black-body radiation, I.149
    energy surface trajectories, I.590
    entropy, I.47
    equipartition theorem, I.67
    H-theorem, I.35
    kinetic theory of gases, I.587
    Liouville theorem, I.525
    process irreversibility, I.524
    second law of thermodynamics, I.524
    specific heat, II.970
    statistical mechanics, I.173, I.523
Bom, N, imaging, III.1914
Bömmel, acoustic resonance, III.1357–8
Bond, W N, constants, II.1267–8
Bondi, H
    chemical elements, III.1788
    field equations, I.313
    plane wave solutions, I.312
    steady-state cosmology, III.1787
Bonhoeffer, K F, molecular hydrogen, II.1048
Bonifacio, R, Collective Atomic Recoil Laser, III.1456
Booker, H G, magneto-ionic theory, III.1992
Bormann, E, decay law of radioactive substances, I.222
Born, M, I.180, **II.985**
    adiabatic approximation, II.999

Atomic Physics, I.210
BBKGY hierarchy, I.606, I.623, I.624, I.626
crystal lattice vibrations, I.153
discrete quantum mechanics, I.184
Dynamical Theory of Crystal Lattices, II.988
Fifth Solvay Conference (1927), I.205
harmonic-oscillator decomposition, III.1402
lattice dynamics, I.452
Lattice theory of rigid bodies, I.189
matrix mechanics, I.360
method of long waves, II.1003–4
On quantum mechanics, I.184
On the quantum mechanics of collision processes, I.198
perturbation theory, II.1002
properties of fluids, I.553
quantum mechanics, I.185, II.638
quantum theory, I.181, I.188
Raman effect, II.996, III.1404
rigid body, I.279
Schrödinger's wavefunction, I.174
self-consistent renormalized harmonic approximation, II.1004
specific heat, II.971–2
spectral line intensity, I.172
The structure of the atom, I.189
theory of ionic crystals, II.997
vibrational specific heat of solids, I.525
Bose, S N
    black-body radiation, III.1402
    light-quanta, I.167
    Planck's law, I.527
    radiation law, III.1288
Bothe, W
    coincident counting, I.364
    cosmic rays, II.748
    neutron, I.119

x-ray scattering, I.166
Bouguer, gravitation, II.1265
Bouwkamp, C J, diffraction theory, III.1420
Bowen, I S, nebulium lines, III.1762
Bowen, R, Gordon Conference (1976), III.1850
Bowles, K L, incoherent scatter radar (ISR), III.1996
Bowley, A R, nuclear medicine, III.1916
Boyd, G D, resonator modes, III.1429
Boyde, A, tandem scanning microscope, III.1451
Boyle, B J
  Newton's rings, III.1386
  quasars, III.1798
Boys, C V
  fine silica fibres, I.28
  gravitation, II.1266
  radiometer, I.6, I.23, I.28
  Soap Bubbles and the Forces that Mould Them, I.5
Bozorth, R M, x-ray study of KHF_2, I.458
Bradley, A W, single-crystal methods, I.435
Bradley, D, dye-laser pulses, III.1439
Bradley, J, stellar aberration, I.259
Bragg, Sir William (Henry), I.429
  α-particles, I.104
  Bragg–Gray theory, III.1882
  crystal spectrometer, I.430
  naphthalene, I.494
  reflection method, I.158
  structure factors, I.439
  x-ray diffraction, I.426, I.429
  x-rays, I.363
Bragg, Sir (William) Lawrence, **I.429**
  binary alloys, I.548
  doubly refracting crystal, I.485
  Fifth Solvay Conference (1927), I.205
  Fourier series method, I.440
  inorganic compounds, I.492
  order–disorder transitions, I.473
  reflection method, I.158
  sodium chloride, I.493
  structure of beryl, I.440
  Twelfth Solvay Conference, Brussels, II.646
  x-ray analysis, I.496
  x-ray diffraction, I.426, I.427–8, I.431
  x-ray microscope, III.1414
Braginsky, V B, quantum non-demolition (QND) measurements, III.1469
Brahe, T
  supernova, III.1712
  supernovae, II.1228
Brandt, E H, levitation theory, III.1543
Brant, H L, two-photon correlation, III.1413
Brattain, W H
  point-contact transistor, III.1326
  transistor, II.744
Braun, E
  growth spiral, III.1528
  solid-state physics, III.1521
Braun, F, cathode-ray oscillograph, I.26, I.27
Braun, K F, wireless communication, III.1986
Bravais, A, symmetry types, I.423
Breit, G
  First Shelter Island Conference, II.645
  ionospheric conducting layer, III.1988
  ionospheric plasmas, III.1621
  ionospheric sounding, III.1995
  magneto-ionic theory, III.1990
  neutron capture, II.1194
  nuclear resonance, I.126

Name Index

radio receiving equipment, III.1990
reaction cross sections, II.1219
Tesla coil, II.1185
Bretherton, F P, momentum transfer, II.867
Breuckner, K, laser fusion programme, III.1676
Brewer, R G
  self-phase modulation, III.1449
  stimulated Brillouin scattering, III.1444
Brickwedde, F G
  deuterium, III.1706
  deuteron, II.650, II.1183
Bridges, W B, argon-ion laser, III.1443
Bridgman, P W
  crystal growth, III.1280
  thermoelectric effects, III.1306
Briggs, B H, ionosphere, III.1984
Brill, R, biomolecular structures, I.497
Brillouin, L
  asymptotic linkage, I.197
  Brillouin lines and Brillouin zones, I.216, III.1311–13
  Fifth Solvay Conference (1927), I.205
  First Solvay Conference (1911), I.147
  Maxwell's Demon cannot operate, I.576
  quantum theory, III.1397
Brindley, G W, zero-point motion, II.974
Brinkman, W F, phonon exchange, II.953
Broadbent, S R, percolation processes, I.572
Brobeck, W, neutral beam injection, III.1661
Brockhouse, B N, **II.993**
  dispersion curves of germanium, II.1002

interatomic forces, II.999
neutron spectroscopy, I.486
phonon dispersion relations, II.993–4
phonons, II.999
shell model, II.997
spin-wave energies, II.1012
triple-axis crystal spectrometer, II.992
Broer, L J F, nuclear resonance, II.1172
Bromberg, J L, lasers, III.1478
Bronsteyn, M P, linearized quantum gravity, I.317
Brossel, J, optical pumping, III.1415
Brown, B R, computer-generated spatial filter, III.1447
Brown, H, fusion power, III.1673
Brown, R G W, spatial coherence, III.1471
Brown, S, electron diffusion rates, III.1644
Brown, T G, pulse-echo ultrasonic scanner, III.1913
Brüche
  emission electron microscope (EEM), III.1569
  Geometrische Elektronenoptik, III.1578
Brueckner, K A, correlation effects, III.1362
Brunhes, B, Earth's magnetic field, III.1973
Brush, S G
  Lenz–Ising model, I.554
  planetary physics, III.2012
Buchdahl, H A, Langrangean aberration theory, III.1458
Buckingham, M J, λ-point of liquid He^4, I.559
Budden, K G, radio waves, III.1992
Budker, G I
  cooling method, II.743
  magnetic mirror, III.1659

mirror concept, III.1665
mirror machine, III.1662
storage rings, II.741
Buechner, W W, electrostatic generator, II.1203
Buerger, M J, precession camera, I.434
Bullard, E C
  Earth's magnetic field, III.1960
  Pacific floor measurements, III.1974
Bullen, K, Earth's core, III.1957
Bunemann, O, stream-type instabilities, III.1673
Bunn, C W, difference synthesis, I.452
Bunsen, R
  Bunsen burner, I.75-6
  spectral emission lines, III.1692
  spectral lines, I.156
Burbidge, E M
  Evidence for the occurrence of violent events in the nuclei of galaxies, III.1777
  nucleosynthesis in stars, III.1788
  supernovae, II.1227, II.1228
  synthesis of elements, III.1747
Burbidge, G R
  Evidence for the occurrence of violent events in the nuclei of galaxies, III.1777
  inverse-Compton catastrophe, III.1781
  nucleosynthesis in stars, III.1788
  quasars, III.1777
  radio sources, III.1775
  supernovae, II.1227, II.1228
  synthesis of elements, III.1747
  x-ray sources, III.1755
Burger, R, Thermos flask, I.27
Burgers, W G, dislocation theory, I.476
Burke, P G, R-matrix calculations, II.1069

Burnham, D C, photon coincidence, III.1435
Burns, G, superconductors, I.490
Burrow, P D, molecular resonances, II.1084
Burrows, angular distributions, II.1220
Burrus, C A, fibre lasers, III.1427
Burstein, E, absorption coefficient of semiconductor, III.1421
Busch, H
  electron motion, III.1568-9
  thin magnetic lens theory, III.1570
Buschow, K H J, permanent magnet materials, II.1165
Butler, C C
  elementary particles, II.654
  forked tracks, I.408
  kaons, II.656
Butler, S T, theory for (d,p) reactions, II.1220
Byer, R L, laser frequency, III.1435
Byrd, R
  First Antarctic Expedition, III.2005
  Second Antarctic Expedition, III.2005

Cabannes, J, combination light-scattering in gases, III.1405
Cabibbo, N, charge-changing weak current, II.677
Cady, properties of quartz, II.1244
Cagniard de la Tour, C, critical point phenomenon, III.1947
Cahan, D, measurement, I.15
Cahn, J, spinodal concept, III.1548
Callan, C, point-like constituents, II.690
Callaway, J, valence/conduction bands, III.1328-9
Cameron, A G W

explosive nucleosynthesis,
III.1747
pp chain, II.1227
solar neutrinos, III.1748
supernova trigger hypothesis,
III.1964
synthesis of elements, III.1747
Campbell, L L, resistivity variation
with temperature, III.1282
Cannizzaro, S
Karlsruhe congress, I.46
molecules, I.45
Cannon, A J, stellar spectra,
III.1693, III.1694
Capasso, F, photoconductivity,
III.1470
Carathéodory, C
second law of thermodynamics,
I.534
temperature concept, I.534
Carlson, C, xerography, III.1541
Carnot, S, second law of
thermodynamics, I.521
Caroe, O, *William Henry Bragg*,
III.1514
Carothers, W, polymer industry,
III.1550
Carpenter, D, whistler
measurements, III.2009–10
Carpenter, H C H
crystal growth, III.1280
plastic deformation, III.1526
Carrington, A, vibrational
spectrum, II.1067
Carruthers, L, computational
expert, III.1844
Cartan, E, non-symmetric
connection, I.321
Carter, B, black holes, III.1778
Casagrande, F, laser theory,
III.1444
Casimir, H B G, two-fluid model,
II.924, II.925
Caspar, D L D

tobacco mosaic virus (TMV),
I.508
tomato bushy stunt virus (TBSV),
I.510
Cassels, J M, critical fluctuations,
II.1016
Cassen, B
automatic rectilinear scanner,
III.1901
isotope invariance, I.378
Castaing, R
electron microprobe analyser,
III.1548
electron probe x-ray
microanalysis, III.1580
Cauchy, A, elasticity of crystals,
I.424
Caullery, M
École Polytechnique, I.15
laboratory equipment, I.19
Cave, H M, ionization chambers,
II.1200
Cavendish, gravitation, II.1266
Caves, C M, quantum
non-demolition (QND)
measurements, III.1469
Cercignani, C
BBKGY hierarchy, I.626
*The Boltzmann Equation and its
Applications*, I.626
Čerenkov, P A
charged particles, II.750
radiation from relaxation of
non-uniform polarization,
III.1408
Chadwick, Sir James, I.120
α-particles, I.114, II.1190
anomalous scattering, I.366
continuous spectrum, I.107–8,
I.118
neutron, I.119–20, I.122, I.369,
I.370, II.639, II.1183, III.1706,
III.1712
positron, II.650

N13

Possible existence of a neutron, I.120
Radiations from Radioactive Substances, I.108, I.112, II.1199, II.1200
Chalmers, B, new journal, III.1556
Chamberlin, T C
  nebular hypothesis, III.1950
  tidal theory, III.1962
Chambers, K C, radio galaxies, III.1796
Chambers, R G
  boundary scattering, III.1343–4
  cyclotron resonance, III.1357
  electron free paths, III.1344
Chandler, P J, waveguide laser, III.1471
Chandrasekhar, S
  astrophysics, III.1706
  degenerate stars, III.1711
  dynamical friction, III.1774
  Eddington: the Most Distinguished Astrophysicist of his Time, III.1702–3
  equilibrium theory, III.1785
  liquid crystals, III.1540
  Radiative Transfer, III.1415–16
  Schönberg–Chandrasekhar limit, III.1709
Chandresekhar, S, random walks, I.608
Chapin, D M, solar cell, III.1421
Chapman, S, **III.1993**
  diamagnetic currents, III.1684
  geomagnetic field, III.2003
  geomagnetic storms, III.2003
  Geomagnetism, III.2000
  ionospheric layers, III.1992–4
  Solar–Terrestrial Physics, III.2001
  statistical mechanics, I.598
  upper atmosphere, III.1982, III.1985
Charnley, Sir John, hip replacement, III.1896
Charpak, G
  drift chambers, II.754
  multi-wire proportional chambers, II.749
Cherwell, Lord, *see* Lindemann, F A (later Lord Cherwell)
Chester, G V, disordered materials, III.1366
Chevalier, R, light curve of SN1987A, III.1761
Chew, G F
  pion–nucleon resonance, II.680
  pion–pion scattering, II.681
  Regge trajectories, II.682
  Twelfth Solvay Conference, Brussels, II.646
Chiao, R Y
  optical fibres, III.1470
  stimulated Brillouin scattering, III.1444
Chodorow, M, high-power pulsed klystron, II.736
Christensen-Dalsgaard, J, low degree modes, III.1750
Christenson, J, *CP* violation, II.672
Christian, J, phase transformations, III.1547
Christoffel, tensor analysis, I.292
Christofilos, N C
  alternating-gradient focusing, II.734
  Argus experiment, III.1683
  ASTRON, III.1645
Christy, R F
  meson theory, I.407
  supernovae, II.1228
Cini, M, Twelfth Solvay Conference, Brussels, II.646
Clark Jones, R, polarization in optical systems, III.1411
Clauser, J F, Einstein–Podolsky–Rosen(–Bohm) (EPR(B)) paradoxes, III.1457
Clausius, R
  heat engines, I.534

kinetic theory of gases, I.522–3, I.587, III.1284
On the kind of motion we call heat, I.47
second law of thermodynamics, I.521
virial theorem, I.543
Clayton, D D
  element abundances, III.1748
  isotopic anomalies, III.1964
Clifford, W K, atoms, I.31
Cline, D, proton–antiproton colliders, II.743
Clothier, W K, voltage measurement, II.1256
Cochran, W
  determinational inequalities, I.448
  helical structure, I.498
  phase transitions, II.1008
  probability relations, I.449
  pseudopotential theory, II.999
  shell model, II.997, II.1002
  structure determination, I.452
Cockcroft, Sir John
  atomic energy project, II.1198
  Atoms for Peace Conference, III.1638
  disintegration experiments, II.1184
  disintegration of lithium, II.729
  electrostatic generators, II.727
  fusion reactions, III.1628
  nuclear disintegration, II.1183
  nuclear reaction, I.285
  nuclear transformation, I.112
  proton acceleration, I.370
Coensgen, F
  deuterium plasma, III.1659
  neutral-beam injectors, III.1663
  plasma gun, III.1661
Cohen, E G D, collision configurations, I.611
Cohen, M

Landau phonon–roton dispersion relation, II.994
Landau spectrum, II.930
Cohen, M H, galactic nuclei, III.1783
Colclough, A R, gas constant (R), II.1269
Cole, E R, blind deconvolution, III.1461
Coleman, S, running coupling constant, II.713
Coles, D
  law of the wake, II.840
  law of the wall, II.840
Colgate, S
  neutral beam injection, III.1661
  neutron star, III.1759
Colle, R, molecular dissociation, II.1067
Collier, R J, holography, III.1438
Collins, M F, spin-wave modes, II.1151
Collins, S C, liquid helium, III.1341
Compton, A H, I.166
  cloud chamber, I.364
  cosmic-ray studies, III.2005
  Fifth Solvay Conference (1927), I.205
  Secondary radiation produced by x-rays, I.165
  x-ray scattering, I.362, I.364, III.1867
  x-rays, I.363
Compton, K T, photoelectric effect, III.1390
Condon, E U
  α-decay, I.115, I.217
  atomic energy levels, II.1093
  electrostatic generators, II.727
  isotope invariance, I.378
  Nature, I.116
  quantum-mechanical theory of alpha decay, II.1184
  The Theory of Atomic Spectra, II.1037

Conrady, A E, chromatic aberration, III.1390
Conversi, M, mesons, II.652
Conway, 'life' (game), I.629
Cook, A H, measurements in physic, II.1273
Coolidge, W D, cathode high-vacuum tube, III.1861
Cooper, L
  BCS theory, II.938
  superconductivity, I.232, II.939, III.1364
Cooper, M J, anharmonicity in BaF_2, I.452
Corey, R B
  biomolecular structures, I.497–8
  pleated sheet structures, I.499
  polypeptide chains, I.497
Cork, C, spark chamber, II.753
Cornford, F M, Cambridge University, I.19
Coster, D, hafnium, I.100
Cotton, Sir Alan, magneto-optical effect, III.1389
Cottrell, Sir Alan, **III.1522**
  discontinuous yield-stress in mild steel, III.1520–1
  dislocation theory, III.1523
  physical metallurgy, III.1555
  yield-stress and strain-aging, III.1523
Coulomb, C A, electrical and magnetic experiments, III.2004
Coulson, C H, computations, II.1087
Courant, *Methoden der Mathematischen Physik*, III.2043
Courant, E D
  accelerators, II.738
  synchrotrons, II.734
Cowan, C, neutrino, II.663
Cowling, T G, axially symmetric field, III.1958

Cox, A, geomagnetic field, III.1973
Cox, D P, interstellar medium, III.1766
Craik, D J, domain techniques, II.1155
Crasemann, B, secondary excitations, II.1090
Crawford, B H, human eye, III.1407
Crewe, A, high brightness field emission sources, III.1589
Crick, F H C
  DNA model, I.507
  double-helix structure, I.504
  helical structure, I.498
  polypeptides, I.498
  protein subunits in viruses, I.509
Critchfield, pp process, II.1226
Croat, J J, neodymium–iron borates, II.1165
Croce, P, coherent imaging, III.1418
Crommelin, eclipse expeditions, III.1722
Cronin, J W
  CP violation, II.672
  spark chamber, II.753
Crookes, W
  cathode rays, II.1271
  phosphorescence of zinc sulphide, II.745
Cross, L E, ceramics, III.1538
Crowfoot, D M
  globular proteins, I.497
  tomato bushy stunt virus (TBSV), I.509
Cummerow, R L, electron spin resonance, II.1138
Cummins, H Z, Doppler shift, III.1442
Curie, I, radioactivity, I.121
Curie, M, I.57
  atoms, I.49
  Fifth Solvay Conference (1927), I.205

First Solvay Conference (1911), I.147
radioactivity, I.56–8, I.60
radium, I.6, II.1272, III.1863, III.1876
uranium, I.61
Curie, P
diamagnetism, II.1116
ferromagnets, II.1118
magnetism, I.545, II.1113
Nobel Prize lecture, I.63
radioactivity, I.56–8
radium, III.1876
uranium, I.61
Curien, H, x-ray scattering, II.990
Curle, N
jet-edge system, II.852
sound waves, II.851
Curtis, H D
Great Debate, III.1718
novae, III.1717
Cusack, N E, liquid metals, III.1366
Cvejanovic, quantum-mechanical reformulation, II.1065
Cvitanovic, P, universal function, III.1848
Czochralski, single-crystal wires, III.1281

Dabbs, J W T, germanium junctions, II.1208
Dagenais, M, photon antibunching, III.1460
Daguerre, L-J-M, daguerreotype process, III.1692
Dainty, J C, blind deconvolution, III.1461
Dällenbach, W, invariant averaging procedure, I.281
Dallos, J, scleral lenses, III.1414
Dalrymple, G B, geomagnetic field, III.1973
Dalton, B J, laser fluctuations, III.1473

Dalton, J
learned societies, I.3
New System of Chemical Philosophy, I.44
Damadian, R, nuclear magnetic resonance, III.1923
Damm, C C
Alice mirror machine, III.1679
neutral beam injection, III.1661
Danby, G, two-neutrino experiment, II.672
Daniel, V, copper–nickel–iron alloy, III.1548
Darmois, gravitational field, I.314
Darrow, K K, First Shelter Island Conference, II.645
d'Arsonval, surgical diathermy, III.1894
Darwin, C G
biological evolution, III.1947
dispersion formula, I.171
extinction, I.451
integrated intensity formula, I.434
magneto-ionic theory, III.1992
Raman effect (combination scattering), III.1404
relativistic hydrogen spectrum, I.218
statistical mechanics, I.535
structure factor, I.431
Darwin, G H
Moon–Earth hypothesis, III.1951–2
Moon model, III.1966
Moon's orbit, III.1948
Das Gupta, M K, Cygnus A, III.1775
Dash, W C, diffusion of copper impurity, III.1526
Dashen, R, axial–vector coupling constant and pion–nucleon scattering, II.679
Daunt, energy gap, II.932

Davidovits, confocal scanning laser microscope, III.1450
Davies, D
  detection of impulses, III.1957
  Earth's core, III.1980
da Vinci, L, medical mechanics, III.1856
Davis, D J M Jr, ring-laser gyroscope, III.1433
Davis, D R, Moon model, III.1966
Davis, E A, *Electron processes in Non-crystalline Materials*, III.1369
Davis, R, solar neutrinos, II.757, III.1748
Davisson, C
  electron diffraction, I.463
  electron reflection, I.170
Davy, Sir Humphry, learned societies, I.3
Dawson, B, temperature factor, I.452
Dawson, J, fusion systems, III.1673
Day, A L, report to Philosophical Society of Washington, III.1387
Dean, P, amorphous SiO_2, I.479
de Boer, J H
  critical temperature, II.975
  electrical properties, II.1156
  law of corresponding states, II.974
  semiconductors, III.1330
Debrenne, P, cryostat, I.437
de Broglie, L, I.170
  elementary particles, I.282–3
  Fifth Solvay Conference (1927), I.205
  First Solvay Conference (1911), I.147
  matter-wave theory, I.174, I.229
  On the general definition of the correspondence between wave and (particle) motion, I.169
  wave mechanics, I.227
  wave–particle duality, I.168–9
de Broglie, M
  crystal oscillation, I.434
  x-ray spectra, I.160
Debye, P, **II.971**
  adiabatic demagnetization, II.1125
  dielectric waveguide, III.1397
  dynamical laws, I.176
  electron screening, III.1316
  energy spectrum, I.153
  Fifth Solvay Conference (1927), I.205
  Planck's law, I.70
  screening length, III.1610
  single crystal, I.434
  specific heat, II.971–2, II.977, II.980, II.982, II.1011, III.1398
  thermal conductivity, II.977–8
  thermal vibration, I.431
  vibrational specific heat of solids, I.525
  x-ray scattering, I.165
  x-rays, I.363
  Zeeman effect, I.102
de Forest, L
  triode, I.37
  upper ionized layer, III.1987
de Gennes, P G
  liquid crystals, II.1084, III.1540
  SAW model, I.570
  soft physics, III.2030
de Haas, W J
  Kondo effect, II.1167
  potassium chromic alum, II.1126
de Hevesy, G, quantum theory, I.180
Dehmelt, H G, charged particles, II.1099
Delaunay, C-E, Earth–Moon system, III.1948
Delchar, T A, surface crystallography, I.466
Dellinger, J H, sudden ionospheric disturbance (SID), III.1995

Demirkhanov, R, Atoms for Peace Conference, III.1638
Democritus, atoms, I.43, III.2034–5
Denisyuk, Yu N, holograms, III.1438
Dennison, D M, wavefunctions, I.529
Derjagin, B V, charge effects, III.1611
Derrida, B, delta (δ), III.1849
De Rújula, A, charmed particles, II.704
des Cloizeaux, J
  excitations in $s = 1/2$ Heisenberg antiferromagnet, II.1014
  polymer solutions, III.1599
De Salvo, L, Collective Atomic Recoil Laser, III.1456
Desclaux, J P, shell electrons, II.1091
de Sitter, W
  cosmological constant, III.1724
  double-star systems, I.285
  Einstein–de Sitter model, III.1730–1
  spherical four-dimensional space-time, III.1723
  Universe, III.1724
Destriau, G, electroluminescence, III.1417
Deutsch, M, positronium, II.646
de Vaucouleurs, G, Hubble's constant, III.1793
de Vries, G, weak non-linear and weak dispersive effects, II.859
Devonshire, A F, ferroelectric transition in $BaTiO_3$, II.1007
Dewar, J, liquefied gases, I.28
Dick, B J, shell model, I.485, II.997
Dicke, R H
  Big Bang, III.1790
  Fabry–Perot resonant cavity, III.1423
  optical bomb, III.1419
  solar oblateness, I.303
  world models, III.1806
Dickinson, R G, hexamethylene tetramine, I.495
Dietrich, O W, order–disorder transition in β-brass, II.1009, II.1016
di Francia, T, lens-pupil functions, III.1418, III.1457
Dillon, J F
  bubble domains, II.1159
  domain techniques, II.1155
Dimov, G I, tandem-mirror concept, III.1665–6
Dingle, R, superlattice minibands, III.1459
Dirac, P A M, I.189
  atomic mechanics, I.183
  electromagnetic field, I.358
  electron spin, II.1125
  electron statistics, I.528
  Fifth Solvay Conference (1927), I.205
  four-component wavefunctions, I.277
  free particles, I.283
  particle dynamics, I.279
  properties of the Universe, III.1787
  quantization of free field, III.1402
  quantum electrodynamics (QED), I.360, I.362
  quantum mechanics, I.188, I.198, I.211
  Quantum Mechanics, III.2043
  relativistically invariant theory, I.283
  transformation theory, I.201, I.360
  Twelfth Solvay Conference, Brussels, II.646
  wavefunctions, I.200
Dixon, J M, crystal fields, II.1165

Doell, R, geomagnetic field, III.1973
Doi, M
 entangled systems, III.1600
 monodisperse melt, III.1600
Dolen, R, pion–nucleon charge exchange, II.683
Domb, C
 critical behaviour, I.556, I.560
 critical exponents, I.560
 magnetic field at critical point, I.562
 order–disorder transformation, I.578
 site percolation processes, I.572
Dombrovski, V A, optical continuum of nebula, III.1774–5
Domen, S R, calorimetry, III.1883
Donald, I, diagnostic ultrasound, III.1912–13
Donder, Th De, Fifth Solvay Conference (1927), I.205
Döring, W, nucleation kinetics, III.1547
Dorsey, N E, speed of light, II.1263
Drell, S D, lepton pair, II.690
Dresden, M, game rules, I.629
Dresselhaus, G
 phonons, III.1423
 resonant frequencies, III.1334
Dreyer, J L E
 Index Catalogues, III.1717
 New General Catalogue of Nebulae and Clusters of Stars, III.1717
Droste, J, field equations, I.303
Drude, P K L
 electron optics, III.2000
 electron vibration, II.983
 Hall effect, III.1282
 magneto-ionic theory, III.1990
 metallic conduction, III.1284
 school of physics, I.15
 The Theory of Optics, III.1385
Dryden, H L
 closed-circuit wind-tunnels, II.830–1
 turbulence, II.798
Duane, W
 quantum physics, III.1401–2
 radiation quanta, I.165
 radon, III.1876
 röntgen, III.1867
Duffieux, P M, *The Fourier Transform and its Applications in Optics*, III.1413
Dumke, P W, semiconductor lasers, III.1433
DuMond, J W M, curved crystal focusing spectrometer, II.1207
Dumontet, P, polychromatic fields, III.1421
Dungey, J, magnetosphere, III.2010
Dunlop, J S, radio sources, III.1798
Dunning, J R, fission experiments, II.1196
Dunoyer, L, surface crystallography, I.468
Duraffourg, G, stimulated emission in semiconductors, III.1433
Duwez, P, quenching of hot alloys, III.1542–3
Dyson, F J
 spin-wave theory, II.1011
 Twelfth Solvay Conference, Brussels, II.646
Dyson, J
 Fresnel fringe system, III.1584
 holograms, III.1584

Eberly, J H, unstable and chaotic emission, III.1466
Eccles, W H, radio waves, III.1622, III.1987
Eckart, C
 continuous media, I.280
 equivalence relations, I.196

Name Index

Eckart, M, theory of metals, III.1290
Eckersley, T L
  long-distance communications, III.1999
  magneto-ionic theory, III.1992
  solar x-rays, III.1994
  upper ionized layer, III.1987
  whistler waves, III.1622–3, III.1995, III.2002
Eddington, A S, **III.1700**
  eclipse expeditions, III.1722
  Eddington–Lemaître model, III.1732–3
  energy transport, III.1699
  fine structure constant, II.1262
  Fundamental Theory, III.1787
  gravitational waves, I.312
  internal structure of the stars, III.1703
  interstellar medium, III.1761
  mass–luminosity relation, III.1705
  Mathematical Theory of Relativity, III.1700
  red giants, III.1704
  The Fundamental Theory, III.1700
  The Internal Constitution of the Stars, III.1700, III.1702
  theory of stellar structure, III.1702–6
  virial theorem, III.1731–2
  white dwarfs, III.1710
Edgerton, H E, stroboscopic photography, III.1410
Edler, I, time–position recordings, III.1913–14
Edwards, O S, stacking faults, I.473
Edwards, S F
  entangled systems, III.1600
  monodisperse melt, III.1600
  quantum-mechanical particles, III.1599

Egger, confocal scanning laser microscope, III.1450
Ehlers, J, general relativity, I.299
Ehrenberg, W, LEED experiments, I.466
Ehrenfest, P
  adiabatic principle, I.92–3
  dog-flea model, I.604, I.618
  Fifth Solvay Conference (1927), I.205
  higher-order transitions, I.549
  periodic systems, I.177
  phase transitions, II.923–4
  spherical four-dimensional space-time, III.1723
  statistical mechanics, I.586
  thermodynamics and irreversibility, I.599
  wind tree model, I.603
Ehrenfest, T
  statistical mechanics, I.586
  thermodynamics and irreversibility, I.599
Eindhoven, W, medical electronics, III.1893
Einstein, A, **I.250**
  anno mirabile, III.2018
  binding energy, I.110–11
  black-body radiation, I.163
  Bohr's theory, I.88
  Brownian motion, I.586
  conservation of energy, III.1390
  cosmological constant, III.1723, III.1730
  dielectric constant, III.1396–7
  Einstein–de Sitter model, III.1730–1
  energy surface trajectories, I.590
  Fifth Solvay Conference (1927), I.205
  First Solvay Conference (1911), I.147
  fluctuations, I.533–4
  gas of corpuscles, III.1390–2
  general theory of relativity,

I.251, I.253, I.286–320,
III.1722, III.1723
gravitation, I.292
heat capacity, I.78
light-quanta, I.54, I.71, I.151–2,
I.167, I.173, I.363, III.1396,
III.1452
nucleation of liquid droplets,
III.1547
opalescent bands, I.546
optical microscopy, III.1406
quanta, II.969
quantum concept, I.71, I.147,
I.228, I.585, III.1398–9
quantum physics, I.93–4
radiation in thermodynamic
equilibrium, III.1392–3
radiation law, I.152
relativity theory, I.34, I.61, I.173
space-time structure, I.251
special theory of relativity, I.54,
I.249, I.251, I.262–85, III.2039
specific heat, I.153–4, II.970
statistical mechanics, I.173
stress–energy tensor, I.281
unified-field theory, I.232
Ekins, R P, saturation analysis
technique, III.1909
Elam, C I
crystal growth, III.1280
plastic deformation, III.1526
Elias, G J, solar x-rays, III.1994
Elias, P, communications theory,
III.1417
Eliashberg, G M, electron–phonon
interaction, II.941
Elliott, R J, phonons, III.1423
Ellis, C D
continuous spectrum, I.118, I.126
internal conversion, I.107
*Radiations from Radioactive
Substances*, I.108, I.112,
II.1199, II.1200
Ellis, R E, mirror machine, III.1679
Elsasser, W M

Earth's magnetic field,
III.1969–70
electron reflection, I.170
geomagnetic dynamo, III.1958
neutron diffraction, I.453
shell models, II.1209
toroidal (doughnut-shaped)
fields, III.1958
Elster, phosphorescence of zinc
sulphide, II.745
El-Sum, H M A, holography,
III.1415
Emanuel, K A, tropical cyclones,
II.904
Emden, R, *Gaskugeln*, III.1695
Emmons, H W
laminar flow, II.825
randomly originating turbulent
spots, II.835
Empedocles, elements, I.44
Enskog, D, statistical mechanics,
I.598
Eötvös, R V
equivalence principle, I.286
gravitation, II.1267
Epicurus, atoms, I.44
Epstein, J B, energy in the Sun,
III.1748
Epstein, P S
dynamical laws, I.176
quantum theory, I.89
Stark effect, I.162
Ernst, M, non-linearity, I.625
Ertel, H, tropical cyclones, II.905
Esaki, L
coupling of quantum wells,
III.1459
germanium, III.1340
quantum-well superlattices,
III.1453
Eshelby, J, physical metallurgy,
III.1555
Essam, J W
critical exponents, I.560
pair connectedness, I.573

two-dimensional lattices, I.573
Essen, L, speed of light, II.1263
Estermann, I, surface crystallography, I.468
Eucken, A, II.972
 calorimeter, II.968
 liquid-air temperature, II.977
 quantum mechanics, I.146
 specific-heat problem, I.153–4
Evans, E, liposomes, III.1608
Eve, A S
 ionizing agents, I.367
 radium, III.1881
 secondary x-rays, I.362
Ewald, P P
 crystal structures, III.1514
 Fifty Years of X-ray Diffraction, I.511
 reciprocal lattice, I.428
 x-ray diffraction, I.424
Ewan, H I, hyperfine line radiation, III.1763
Ewing, Sir Alfred, plastic deformation, III.1516
Ewing, D H, compound nucleus, II.1219

Faber, T E, liquid metals, III.1366
Fabrikant, V A, light amplification, III.1411
Failla, G, medical physics, III.1864
Fairbank, W M
 λ-point of liquid He^4, I.559
 superfluidity, II.932
Fankuchen, I, tobacco mosaic virus (TMV), I.508
Fano, U, resonance, III.1473
Faraday, M
 Chemical History of a Candle, I.5
 conductivity, III.1318
 disliquifying point, I.543
 electric current, III.1958
 electrical properties, II.1113
 electrolysis, I.51
 electromagnetic interactions, III.1953
 electromagnetism, III.1856, III.2000
 experimental physics, I.14
 geomagnetic variations, III.1984
 learned societies, I.3
 magnetic properties, II.1112, II.1113
Farnsworth, H G, surface crystallography, I.470
Faust, W L, CO_2 laser, III.1442
Feher, E R, crystal field theory, II.1144
Fellgett, P, multiplex optical spectroscopy, III.1416
Felten, J, luminosity function, III.1770
Fermi, E, **II.1187**
 β-decay theory, I.374–5, I.375
 electron statistics, I.528
 neutron bombardment, I.122
 neutron capture, II.1186, II.1193
 neutron diffraction, I.456
 neutron irradiation, I.129–30
 neutron sources, II.1199, II.1206
 quantum mechanics, II.1031
 quantum states, III.1288
 Raman scattering, II.996
 synchrocyclotron, II.733
 Tentativo di una teoria della emissione di raggi β, I.129
 theory of β-decay, III.1706
Fermi, L, *Atoms in the Family*, II.1191
Ferraro, V C A
 diamagnetic currents, III.1684
 geomagnetic field, III.2003
Ferrel, W, Earth–Moon system, III.1948
Feshbach, H
 compound nucleus, II.1219
 First Shelter Island Conference, II.645

projection-operator technique, II.1221
proton scattering, II.1212
Feynman, R P, **II.648**
  First Shelter Island Conference, II.645
  Landau phonon–roton dispersion relation, II.994
  Landau spectrum, II.930
  neutrinos, II.667
  parity states, II.665
  point-like constituents, II.689
  quantum electrodynamics, II.643, II.740, III.1446
  quantum mechanics, III.1395, III.1830
  renormalization, II.646
  scientific imagination, III.1519
  Twelfth Solvay Conference, Brussels, II.646
  weak vector current, II.668
Field, G B, interstellar medium, III.1765
Fienup, J R, optical data processing, III.1461
Fierz, M, free particles, I.283
Finch, J F, polio virus, I.510
Finkelstein, D, Schwarzschild singularity, I.308
Finney, J L, amorphous structure, I.479
Firestone, F A, ultrasonic flaw detector, III.1911
Fischer, E, polyethylene, III.1552
Fischer, K, AgCl, II.998
Fisher, I Z, neutron frequency, III.1418
Fisher, M E
  covering lattice, I.573
  critical exponents, I.560
  field theory, III.1830
  pair distribution function, I.547
Fisher, O, Earth–Moon system, III.1948
Fitch, V, *CP* violation, II.672

Fitzgerald, G F
  Fitzgerald–Lorentz contraction, I.264
  wireless wave propagation, III.1986
Fizeau, A H L, velocity of light, I.266
Fleming, J A, *Terrestrial Magnetism and Atmospheric Electricity*, III.2000
Flory, P, **III.1598**
  excluded volume, I.569
  polymer chains, III.1551
  polymer solutions, III.1599
  self-avoiding walk, III.1598
Fock, V
  configuration (Fock)-space method, I.208
  equations of motion, I.303
  self-gravitating systems, I.313
Foëx, G, *Le Magnetisme*, II.1119
Foley, electric quadrupole moments, II.1214
Follin, J W, proton–neutron ratio, III.1786
Fomichev, V I, K-shell absorptions, II.1083
Foord, R, photon correlation spectroscopy, III.1454
Ford, F C, mirror machine, III.1679
Försterling, K, wave propagation, III.1991
Fortuin, C, bond percolation model, I.573
Foucault, J B L, pendulum, I.2, I.3
Fowler, R H
  electron gas, III.1711
  Fifth Solvay Conference (1927), I.205
  high-density matter in stars, I.215
  ionization, III.1701
  quantum theory, I.180
  residual entropy of ice, I.571
  statistical mechanics, I.535

Statistical Mechanics, I.533
Statistical Thermodynamics, I.537
stellar spectra, III.1693
structure of ice, I.474
white dwarf stars, I.531
Fowler, T K, tandem-mirror concept, III.1665–6
Fowler, W A
  Big Bang, III.1792
  nuclear interactions between light nuclei, III.1791
  nucleosynthesis in stars, III.1788
  origin of the elements, II.1225
  pp chain, II.1227
  solar neutrinos, III.1748
  supernovae, II.1227, II.1228
  synthesis of elements, III.1747
Fox, A G, Fabry–Perot resonator, III.1429
Franceschini, W, delta (δ), III.1851
Franck, J
  Franck–Hertz experiment, I.89–90
  quantum theory, I.179
  scattering experiments, I.217
  stationary electron orbits, I.158
Frank, A, rare-earth ions, II.1120
Frank, Sir Charles, **III.1606**
  crystal growth, III.1527
  crystallinity, III.1551–2
  line singularities in nematics, III.1605
Frank, I, energy loss formula, II.750
Frank, I M, visible light, III.1409
Frank, N H, theory of metals, III.1290
Franken, P A, ruby laser, III.1427–9
Franklin, B, learned societies, I.3
Franklin, R E
  DNA, I.504
  tobacco mosaic virus (TMV), I.508

Franz, W, photo-assisted tunnelling, III.1423
Fraunhofer, J
  solar spectrum, III.1691–2
  spectral lines, I.156
Frautschi, S C, Regge trajectories, II.682
Frazier, E N, solar oscillations, III.1749–50
Freedman, S J, Einstein–Podolsky–Rosen(–Bohm) (EPR(B)) paradoxes, III.1457
Freeman, A J, rare-earth elements, II.1164
Freidmann, A A, world models, III.1724–5
Frenkel, J
  phonon, III.1302
  semiconductors, III.1319–20
  sound waves, III.1328
  superconductivity, II.917, II.936
  vacancies, III.1529
Fresnel, A
  diffraction, III.1386
  wave description of light, I.72
Freud, P von, energy complexes, I.301
Freundlich, E F
  general relativity, I.305
  red shift, I.289
Fricke, H, chemical dosimetry, III.1884
Friedel, G
  focal conic defects, III.1605
  liquid crystals, III.1540
  mesomorphic phases, III.1603
Friedman, H
  Crab Nebula, III.1739
  rocket experiments, III.1738
  rocket studies, III.2005
  world models, III.1784, III.1807
Friedman, J I, parity violation, II.666
Friedrich, W
  crystal structures, III.1514

x-ray diffraction, I.425
Frisch, O R
  fission hypothesis, II.1195
  neutron–U collisions, I.130
Fritzsch, H, quarks, II.711
Fritzsche, H, semiconductors, III.1330
Fröhlich, H
  electron–phonon interactions, III.1337–8
  heavy electrons, I.392
  lattice deformation, III.1321
  meson theory, I.394
  metallic conductivity, I.216
  superconductivity, II.936
Froome, K D, speed of light, II.1263
Fry, D W, linear accelerators, II.1206
Fry, W J, ultrasonic surgery, III.1915
Fuchs, K, properties of fluids, I.553
Fukui, S, spark chamber, II.753
Fuller, C S, solar cell, III.1421
Fuller, L, upper ionized layer, III.1987
Furnas, T C, powder photographs, I.436
Furry, W, positrons in electron self-energy, II.641

Gabor, D, **III.1567**
  diffraction microscopy, III.1582
  electron microscope, III.1414–15, III.1581
  holography, III.1456, III.1582–4
  magnetic electron lens, III.1566–9
  super-lens, III.1411
Gagarin, Y, orbital flight, III.1738
Gaillard, J-M, two-neutrino experiment, II.672
Galileo Gallilea, I.254
  mechanical principal of relativity, I.254
  principle of inertia, III.1827
  speed of light, II.1262
Galison, P, magnetic moment and angular momentum, II.1123
Gamow, G
  abundances of elements, III.1785–6
  α-decay, I.115, I.217, II.1184
  α-particles, II.1189
  α-scattering, II.1189
  Atomic Nuclei and Nuclear Transformations, II.1225
  Big Bang, III.1784–6
  black-body radiation, II.758
  Constitution of Atomic Nuclei and Radioactivity, I.115
  cosmological constant, III.1730
  electrostatic generators, II.727
  fusion reactions, III.1628
  heavier elements, II.1226, II.1227
  liquid drop, I.126, II.1186
  neutrons, III.1713
  quantum-mechanical tunnelling, III.1706
  The Constitution of Atomic Nuclei and Radioactivity, II.1186
Garmire, E, stimulated Brillouin scattering, III.1444
Garrett, C G B
  anharmonic oscillators, III.1434
  two-photon absorption, III.1429
Garwin, R, parity violation, II.666
Gauss, C F, geomagnetism, III.1984
Geffken, W, metal–dielectric interference filters, III.1407
Gehrcke, E, Lummer–Gehrcke plate, III.1389
Geiger, H
  α-particles, I.113, II.745–6
  α-scattering, II.1200
  β-spectrum, I.107
  coincident counting, I.364
  electrical counting methods, II.1200

electron charge, II.747
Nature of the α-particle, I.58
x-ray scattering, I.166
Geissler, J, improved vacua, I.54
Geitel, phosphorescence of zinc sulphide, II.745
Geller, M, bright galaxies, III.1772
Gell-Mann, M, II.678
  axial coupling constant, II.676
  axial–vector coupling constant and pion–nucleon scattering, II.679
  baryons, II.661
  β-decay, II.677
  correlation effects, III.1362
  isotopic spin multiplets, II.658
  mass relations among baryons and mesons, II.661
  neutrinos, II.667
  quark–lepton analogy, II.700
  quarks, II.711
  renormalization group, III.1830, III.1832
  strangeness, II.662
  SU(3) triplets, II.685
  The Quark and the Jaguar, II.770
  Twelfth Solvay Conference, Brussels, II.646
  unitary symmetry, II.677
  weak hadronic current, II.678
  weak vector current, II.668
Genna, S, calorimetry, III.1883
Georgi, H, charmed particles, II.704
Gerlach, W, beam splitting, I.165
Germain, G, structure determination, I.450
Germer, L H, de Broglie waves, I.171
Gershtein, S S
  neutrinos, III.1803
  weak vector current, II.668
Gerstenkorn, H, Moon model, III.1966

Ghosh, S P, ionospheric physics, III.2000
Giacconi, R, Sco X-1, III.1739
Giaever, I, energy gap, II.933
Giauque, W F
  adiabatic demagnetization, II.1125
  gadolinium sulphate, II.1126
Gibbs, H M
  optical bistability, III.1449
  Optical Bistability, III.1450
  thermal bistable effects, III.1450
Gibbs, J W, **I.522**
  canonical ensemble, I.526
  Elementary Principles of Statistical Mechanics, I.586, I.593
  ensembles, I.531–3
  heterogeneous systems, I.34–5
  identical particles, I.528
  partition function, I.527
  phase equilibria, III.1507–8
  PhD, I.17
  statistical mechanics, I.173, I.523, I.606
  thermodynamics, I.521
Gibson, angular distributions, II.1220
Gibson, G, beta-ray experiment, III.1679–80
Gilbert, W
  De Magnete, I.544, III.1855–6
  Earth's magnetism, III.1952–3
  electromagnetism, II.1252
  magnetic forces, III.1952
  magnetism, II.1111
Gilliland, T R, ionosonde, III.1989, III.1991
Ginsburg, V, Čerenkov effect, III.1409
Gintzon, E L
  high-power pulsed klystron, II.736
  linear accelerators, II.736
Ginzburg, D M
  energy gap, II.932

superconductivity, II.934–5
Ginzburg, V L, Galactic radio emission, III.1737–8
Giorgi, G, MKSA system, II.1235
Gires, F, pump-probe configuration, III.1445
Gittelman, B, electron storage rings, II.739
Gittus, J, radiation damage, III.1534
Glaser, D A, bubble chamber, II.659, II.753
Glaser, W, electron optics, III.1575
Glashow, S L, II.696
  charmed particles, II.704
  charmed quark, II.700, II.705
  neutral currents, II.700
  quark–lepton analogy, II.700
  SU(2) x U(1) group, II.694
Glauber, R J
  neutron frequency, III.1418
  photodetection, III.1441
  quantum theory of coherence, III.1439–41
  radiation field, III.1388
Glazebrook, R T (later Sir Richard), **II.1253**
Gleick, J, *GENIUS—The Life of Richard Feynman*, III.1519
Gold, A V, de Haas–van Alphen effect, III.1352
Gold, T
  magnetosphere, III.1981
  pulsars, III.1752
  Relativistic Astrophysics Symposium, III.1776
  steady-state cosmology, III.1787
Goldberg, H, triplets, II.685
Goldberg, J, conservation laws, I.301
Goldberger, M L
  beta-decay, II.676
  pion decay constant, II.675
  Twelfth Solvay Conference, Brussels, II.646

Goldhaber, A S, photon mass, III.1452
Goldhaber, G, charmed meson, II.704
Goldhaber, M
  dipole vibration, II.1213–14
  neutrino helicity, II.668, II.670
  shell model, II.1215
Goldman, I M, dielectric properties of BaTiO$_3$, I.486
Goldschmidt, V M
  First Solvay Conference (1911), I.147
  inorganic compounds, I.492
Goldsmith, D W, interstellar medium, III.1765
Goldsmith, H H, neutron sources, II.1199
Goldstein, S
  energetic particles, III.1682
  magneto-ionic theory, III.1990
Goldstone, J, general theorem, II.676
Golovin, I N
  beam-fed mirror experiments, III.1662
  magnetic mirror, III.1659
Goodman, C, Coulomb excitation experiments, II.1217
Goodman, P, electron diffraction, I.464
Gordon, J P
  maser, III.1419
  relativistic hydrogen spectrum, I.218
  resonator modes, III.1429
Gordon, W E
  incoherent scatter radar (ISR), III.1996
  Schrödinger method, I.365
Gor'kov, L, BCS theory, II.940
Gorter, C J
  magnetic field, II.1133
  nuclear resonance, II.1172
  relaxation time, II.1133

superconductivity, II.921
Twelfth Solvay Conference, Brussels, II.646
two-fluid model, II.924, II.925
Gorter, E W, ferrimagnetism, II.1157
Görtler, H, boundary layers, II.832–3
Gott, J R, distribution of galaxies, III.1772
Gottlieb, M, Atoms for Peace Conference, III.1638
Goudsmit, S
  atomic energy levels, II.1093
  electron spin, I.101
  magnetic moment, II.1123
  quantum number, I.168
  spinning electron, I.360
Gough, D O, low degree modes, III.1750
Gould, G, lasers, III.1423
Gould, L, polar research, III.2006
Goulianos, K, two-neutrino experiment, II.672
Goutier, Y, perturbation theory, III.1473–4
Gouy, A, gravity effects, I.553
Gowing, M, *Britain and Atomic Energy 1939–45*, II.1198
Grad, H
  magnetic cusp, III.1645
  non-linearity, I.625
Gratias, D, quasicrystals, I.481
Gray, L H, **III.1869**
  Bragg–Gray theory, III.1882
  radiotherapy, III.1867
Green, H S, parastatistics, II.687
Green, M B, string theories, II.684
Green, M S, collision configurations, I.611
Green, R F, quasars, III.1798
Greenberg, O W
  quarks, II.687
  SU(3) symmetry, II.709

Greenburger, D M, three-photon interferometry, III.1467
Greene, C H
  critical behaviours, II.1067
  R matrices, II.1069
Greene, J M, plasma oscillations, III.1625
Greening, J R, dosimetry, III.1884
Gregory, Sir Richard, *Nature*, I.6
Greinacher, H, α-particle, II.748
Gribov, V N, singularities in complex angular momentum, II.682
Griffin, J, growth spiral, III.1528
Griffith, A A, crack growth, III.1559
Griffiths, J H E, ferromagnetic resonance, II.1152–3
Griffiths, R B, smoothness postulate, I.563
Gringauz, K I
  plasmapause, III.2009
  solar wind, III.2009
Grodzins, L, neutrino helicity, II.668, II.670
Grondahl, L O
  photo-voltaic effect, III.1321
  semiconductors, III.1318
Gross, D
  point-like constituents, II.690
  running coupling constant, II.713
Gross, E, spectral shift, III.1398
Gross, E P, plasma oscillations, III.1625
Grossmann, M, gravitation, I.292
Groth, crystallography, I.424
Grüneisen, E, thermal expansion, II.975–7
Gubanov, long-range order, III.1368
Guckenheimer, J, MSS, III.1850
Gudden, B
  photo-ionization, III.1320
  single crystals, III.1318–19

temperature variation of resistivity, III.1330
Guen, H S, BBKGY hierarchy, I.606, I.623, I.624, I.626
Guggenheim, E A
　coexistence curves, I.555
　Statistical Thermodynamics, I.537
Guggenheimer, J, shell models, II.1209
Guinier, A, age-hardening in aluminium–copper alloys, III.1546
Gull, S F, supernova remnants, III.1782
Gunn, J E
　electron acceleration, III.1752–3
　mass–luminosity ratio, III.1800
　phase space constraints, III.1804
Gurney, R
　α-decay, I.115, I.217, II.1184
　colour centres, III.1530
　electrostatic generators, II.727
　Nature, I.116
　semiconductors, III.1319
　solid-state physics, III.1314
Gurvich, defibrillators, III.1895
Gusyenov, O H, binary sources, III.1756
Gutenberg, B
　earthquake waves, III.1956
　seismological analysis, III.1954
Guth, A, Universe model, III.1806
Guye, Ch E, Fifth Solvay Conference (1927), I.205
Gyulai, Z, colour centres, III.1529–30

Haake, F, laser theory, III.1444
Haas, A E
　Bohr radius, I.81
　quantum formula, I.156
Haasen, P, *Physikalische Metallkunde*, III.1555
Habgood, H W, isotherms of xenon, I.556

Habing, H J, interstellar medium, III.1765
Hadravský, M, confocal scanning laser microscope, III.1450–1
Hafstad, L, ionospheric plasmas, III.1621
Hahn, E L, resonant absorption, II.1174
Hahn, J, stellar mass, III.1699
Hahn, O, I.105
　β-ray deflection, II.1201
　β-spectrum, I.104, I.105, I.106
　elements with $Z = 93$–96, I.130
　neutron bombardment, II.1194, II.1195
　neutron sources, II.1206
　neutron–U collisions, I.130
　uranium isotope, II.1223
　uranium-239, I.132
Haine, M E, Fresnel fringe system, III.1584
Haldane, F D M, $s = 1$ Heisenberg chains, II.1014
Hale, G E, magnetic fields in sunspots, III.1958
Hale, J E, Palomar 200″ telescope, III.1721
Hall, E H
　physics in education, I.9
　practical classes, I.15
　resistivity variation with temperature, III.1282
Halley, E, Earth's magnetic field, III.1953
Halliday, D, electron spin resonance, II.1138
Hallwachs, W L F, photoelectric effect, III.1390–1
Halperin, B I, ultrasonic properties, II.1006
Halpern, O
　neutron diffraction, I.459
　quantum theory, I.183
Hamilton, W R, dynamical laws, I.176

Hammersley, J M
  bond percolation processes, I.572
  percolation processes, I.572
  self-avoiding walk (SAW), I.570
Han, M Y
  quarks, II.687
  SU(3) symmetry, II.709
Hanbury Brown, R
  optical spectroscopy, III.1421
  radio interferometer research, III.1422
Haneman, D, surface crystallography, I.470
Hanna, R C, two-photon correlation, III.1413
Hansen, M, binary systems, III.1509
Hansen, W W
  electron linear accelerators, II.1205
  linac, II.736
  linear accelerators, II.736
Hanson, D
  physics of materials, III.1555
  plastic deformation, III.1526
Hanssen, K-J, atomic resolution, III.1585–6
Hara, Y, quark–lepton analogy, II.700
Hariharan, P, holography, III.1438
Harker, D
  inequality relations, I.447
  Patterson function, I.443
Harper, S, strain-aging law, III.1521
Harrison, E R, structures in the Universe, III.1802
Harrison, S C, tomato bushy stunt virus (TBSV), I.510
Harte, B, *Plain Language from Truthful James*, I.6
Hartley, spherical micelles, III.1606
Hartmann, J, wavefront aberrations, III.1389

Hartmann, W K, Moon model, III.1966
Hartree, D R
  magneto-ionic theory, III.1992
  multi-electron spectra, II.1036
  self-consistent field method, I.432, I.433
  wireless propagation, III.1622
Harvey, E N, luminescence, III.1537
Hasenöhrl, First Solvay Conference (1911), I.147
Hasinger, G, x-ray source counts, III.1798
Hastie, R J, plasma pressures, III.1661
Hauptman, H
  determinantal inequalities, I.448
  tangent formula, I.449
  The Solution of the Phase Problem. I. The Centrosymmetric Crystal, I.449
Hauser, W, compound nucleus, II.1219
Haüy, R J, lattice theory, I.422–3
Hawking, S
  black holes, I.317, II.761, III.1778
  communication aids, III.1899
  hot big bang models, III.1778
Haxel, O
  nuclear binding energies, II.1210
  shell models, II.1209
Hayakawa, S, x-ray sources, III.1755
Hayashi, C, quasi-static models of stars, III.1767
Hayes, W, two-neutrino experiment, II.672
Hayes, W C, *Basic Orthopaedic BioMechanics*, III.1896
Hazard, C, radio sources, III.1776
Heaviside, O
  ionospheric layer, III.1621
  radio waves, III.1987

wireless wave propagation, III.1986
Hebel, P Zum, BCS theory, II.940
Heisenberg, W K, I.189, **I.373**
  atomic theory, I.179
  cosmic-ray review, I.230
  discrete quantum mechanics, I.184
  dispersion formula, I.172
  doublets, I.102
  Einstein–Podolsky–Rosen (EPR) paradox, I.228
  electron spin, II.1125
  equation of motion, I.186
  exchange forces, I.231
  exchange integral, I.213
  Fermi-field theory, I.377
  ferromagnetism, I.554
  Fifth Solvay Conference (1927), I.205
  harmonic-oscillator decomposition, III.1402
  holes, III.2018
  lattice calculations, I.233
  meson theory, I.404
  microphysical phenomena, I.224
  neutron–neutron, neutron–proton forces, II.1190
  neutron–proton model, II.1184
  nuclear charge-exchange force, I.376
  nuclear forces, I.123
  nuclear volume, II.1190
  nucleus, I.371–3, II.1191
  periodic system, I.178
  QED, I.362
  quantum mechanics, I.226, I.358, II.638, II.1058–9
  quantum theory, I.144, I.180, I.187, I.221
  relativity, I.359
  spontaneous emission, III.1401
  superconductivity, II.932, II.934, II.936

The Physical Principles of the Quantum Theory, III.1394
  theory of explosive showers, I.387–8
  total quantum number, I.162
  Twelfth Solvay Conference, Brussels, II.646
  two-electron system, I.213
  uncertainty relations, I.174
  unified-field theory, I.232
Heitler, W
  bremsstrahlung, I.385
  cosmic rays, I.393
  covalent molecular bond, II.1042
  heavy electrons, I.392
  meson theory, I.394
  nuclear statistics, I.117
  showers, I.388
  The Quantum Theory of Radiation, II.1031, III.1420–1
  Twelfth Solvay Conference, Brussels, II.646
Heller, P, spontaneous magnetization, I.559
Hellmann, H, pseudopotential model, III.1352–3
Hellwarth, R W
  photon, III.1451–2
  Q-switching, III.1430, III.1432, III.1436
Helm, G, *Energetik*, I.48
Helmholtz, H von
  age of the Sun, III.1949
  electricity, I.52
  laboratory equipment, I.20
  medical physics, III.1857
  Physikalisch-Technische Reichsanstalt, I.20, II.1254
  Sun's time-scale, III.1694
Henisch, H K, semiconductors, III.1325
Henkel, superfluidity, II.932
Hénon, M, chaotic behaviour, III.1831

Henrick, L R, equilibrium theory, III.1785
Henriot, E, Fifth Solvay Conference (1927), I.205
Henyey, L G
  radio astronomy, III.1736
  study of the stars, III.1746
Herbert, R, thyroid gland, III.1900
Herbert, R J T, scintillation detector, III.1901
Herbig, G H, T-Tauri stars, III.1766
Herbst, J F, neodymium–iron borates, II.1165
Herglotz, G
  elastic media, I.280
  rigid body, I.279
Herlofson, N, Galactic radio emission, III.1737–8
Herman, F
  photoconductivity, III.1329
  valence/conduction bands, III.1328–9
Herman, R C
  black-body radiation, II.758
  primordial nucleosynthesis, III.1786, III.1790
  proton–neutron ratio, III.1786
Herring, C
  diffusion creep, III.1534
  jellium, III.1303–4
  pseudopotential model, III.1352–3
  solid-state physics, III.1313
  spin waves, II.1012
Herriott, D R, helium–neon laser, III.1427
Herschel, J, *General Catalogue of Nebulae*, III.1717
Herschel, W
  Galaxy, III.1714
  infrared radiation, III.1742
  nebulae, III.1717, III.1945
Hertz, C H, time–position recordings, III.1913–14
Hertz, G

Franck–Hertz experiment, I.89–90
scattering experiments, I.217
stationary electron orbits, I.158
Hertz, H, electromagnetic waves, III.1985
Hertzsprung, E
  diameter of Arcturus, III.1699
  main sequence, III.1698
  mass–luminosity relation, III.1699
  Small Magellanic Cloud, III.1715
  star clusters, III.1696
  stellar sizes, III.1696
Herzberg, G, **II.1072**
  ionization threshold of H_2, II.1069–72
  molecular mechanics, II.1044
  nuclear statistics, I.117
  spectra of diatomic molecules, II.1046
Herzen, E
  Fifth Solvay Conference (1927), I.205
  First Solvay Conference (1911), I.147
Herzfeld, K F, vibration of alkali halide, I.485
Herzog, R O, fibrous proteins, I.497
Hess, G B, superfluidity, II.932
Hess, H
  ocean ridges, III.1974
  sea-floor spreading, III.1976
Hess, V F, I.369
  balloon flight, I.368
  cosmic ray particles, III.1712
  radiation, II.639
Hess, W N, magnetospheric physics, III.2001
Hevesy, G
  Bohr's theory, I.88
  hafnium, I.100
Hewish, A
  neutron stars, III.1751

N33

survey of the sky, III.1788
Hey, J S, radio astronomy, III.1736
Heycock, C T, **III.1511**
  multiphase alloys, III.1548
  phase diagram, III.1508-9
Heydenburg, M P, Coulomb excitation experiments, II.1217
Heyl, P R, gravitation, II.1266
Higgs, P W
  bosons, II.695
  dynamical symmetry breaking, II.957
Higinbotham, W A, diode coupling, II.748
Hilbert, D, I.207
  gravitational field, I.314
  Hilbert's space, II.1025
  infinite quadratic forms and matrices, I.191-2
  Methoden der Mathematischen Physik, III.2043
  Schrödinger equations, I.205
  theory of linear integral equations, I.174
  variational principle, I.294
Hildebrand, R, charge-independence, II.654
Hill, R D, shell model, II.1215
Hillert, M
  atomic segregation/atomic ordering, III.1548
  spinodal concept, III.1548
Hilliard, J, spinodal decomposition, III.1548
Hillier, J
  contour phenomena, III.1581-2
  Fresnel diffraction, III.1582
  magnetic lenses, III.1580
Hines, C O
  ionosphere, III.1984
  solar wind plasma, III.2010
Hinks, A, photographic parallax programme, III.1696
Hirakawa, K, K_2CoF_4, II.1016

Hisano, K, fine structure, II.1003
Hittorf, J W
  conductivity, III.1318
  rarefied gases, III.1985
Hoag, A A, quasars, III.1798
Hoar, T P, microemulsions, III.1608
Hockham, G A, optical telecommunication, III.1447
Hodgkin, D C, penicillin, I.496
Hofmann, S, superheavy elements, II.1225
Hofstadter, R
  electron scattering, II.1221
  nuclear structure studies, II.1206
  sodium iodide crystals, II.1207
  sodium iodide scintillator, II.750-1
Hohenberg, P
  delta (δ), III.1851
  phonon frequency, II.1002
Holborn, L, black-body radiation, I.22
Holmes, A
  Earth's crust, III.1960
  radioactive minerals, III.1969
Holmes, K C, tobacco mosaic virus (TMV), I.508
Holstein, T, lattice polarization, III.1322
Holt, angular distributions, II.1220
Honda, K, ferromagnetism, II.1134
Hondros, D, dielectric waveguide, III.1397
Hooke, R, Newton's rings, III.1386
Hopkins, H H
  coherence parameter, III.1412
  diffraction pattern of a point object, III.1412
  fibre optics, III.1420
  image formation, III.1419
  polarized light, III.1412
  spatial-frequency response of optical systems, III.1422
Hopkins, W

Name Index

Earth structure, III.1946
earthquake waves, III.1956
Hopkinson, J, *critical temperature*, I.545
Hopwood, medical physics, III.1864
Hora, H, laser systems, III.1464
Horn, D, pion–nucleon charge exchange, II.683
Horn, F H, zone-refined copper, III.1544
Horne, R W, adenovirus, I.510
Horton, self-consistent renormalized harmonic approximation, II.1004
Horwitz, L, BBKGY hierarchy, I.623
Hoselet, First Solvay Conference (1911), I.147
Houston, W V
  de Broglie waves, III.1291–2
  Lamb shift, II.641
  theory of metals, III.1290
Houtermans, F
  nuclear reactions, II.1225
  nuclear relations, III.1707
Houtermans, F G
  age of the Earth, III.1960–1
  thermonuclear reactions, III.1627
Howry, D H
  cross-sectional imaging, III.1912
  prototype scanner, III.1912
  pulse-echo ultrasound, III.1911–12
Hoxton, L G, diffusion cloud chamber, II.752
Hoyle, F
  Big Bang, III.1792
  excited state of ^{12}C, III.1746
  helium synthesis, III.1790
  inverse-Compton catastrophe, III.1781
  nuclear interactions between light nuclei, III.1791
  nucleosynthesis in stars, III.1788
  quasars, III.1777

  red giant stars, II.1226–7
  solar x-rays, III.1994
  steady-state cosmology, III.1787
  supernovae, II.1227, II.1228
  synthesis of elements, III.1747, III.1792
Huang, K
  dipole–dipole forces, II.988
  Dynamical Theory of Crystal Lattices, II.988
  Raman effect, II.996
Hubbard, J
  correlation effects, III.1362
  Hubbard U, II.1166
Hubble, E P, **III.1720**
  counts of faint galaxies, III.1728
  spiral nebulae, III.1719
  The Realm of the Nebulae, III.1720, III.1726
  tuning-fork diagram, III.1721
  velocity–distance relation, III.1726, III.1727
Huchra, J, bright galaxies, III.1772
Hückel, double/triple bonds, II.1043
Hückel, E, wave mechanics, I.213
Hückel, G
  electron screening, III.1316
  screening length, III.1610
Hudson, R P, magnetic cooling, II.1126
Hueter, T F, ventriculography, III.1911
Huggins, M L, chain conformations, I.497
Huggins, W, nebulae, III.1717
Hughes, A L, photoelectric effect, III.1390
Hughes, E W, least squares method, I.452
Hughes, V W, parity violation, II.698
Hughes, W, Rochelle salt, I.489
Hugoniot, A, conservation equations, II.812

N35

Hulburt, E O, solar ultraviolet energy, III.1992
Hull, A W, single crystal, I.434
Hulse, J M
  binary pulsar, I.317
  quadrupole radiation formula, I.315
Hulse, R, binary pulsars, I.298, III.1757
Hulthen, magnetization, II.1131–2
Humason, M L
  galaxy NGC7619, III.1726
  redshift–magnitude relation, III.1788, III.1793
Hümmer, K, Renninger effect, I.450
Hund, F
  analysis of spectra, II.1124
  electron reflection, I.170
  molecular mechanics, II.1044
  quantum theory, I.145
  rare-earth ions, II.1120
  tunnel effect, I.217
  wave mechanics, I.213
Hunter, D L, magnetic field at critical point, I.562
Huntingdon, H B, diffusion, III.1531
Hutson, A R, electron–phonon interaction, III.1339
Hüttel, A, speed of light, II.1263
Huus, T, Coulomb excitation experiments, II.1217
Huxley, T H, Darwinian evolution, I.5
Huygens, C
  light polarization, III.1386
  wave description of light, I.72
  wave theory, I.258
Hyde, E K, *Nuclear Properties of the Heavier Elements*, II.1223

Iachello, F, collective motion in nuclei, II.1218
Iben, I, study of the stars, III.1746
Ibser, H W, neutron sources, II.1199
Ignatowsky, V S, vector theory, III.1412
Ikeda, H, K_2CoF_4, II.1016
Ikeda, K, unstable and chaotic emission, III.1466
Iliopoulos, J
  charmed quark, II.700, II.705
  neutral currents, II.700
Inglis, D R, Cranking Formula, II.1218
Ingram, V, sickle cell anaemia, I.502
Inoue, T, meson theory, I.407, I.408
Inui, T, bulbous bow, II.885
Ioffe, A F
  energy levels, III.1368
  Semiconductor Thermoelements and Thermoelectric Cooling, III.1311
Ioffe, M
  mirror experiment, III.1679
  mirror-type systems, III.1643
Iona, M, KCl, II.988
Irwin, M, redshift quasars, III.1798
Ishiwara, J, quantum rule, I.175
Ising, E
  Ising model, III.1554
  magnetic moment, I.554
Ising, G, linear accelerator, II.736
Israelashvili, J N, measurement of forces between surfaces, III.1612
Iwanenko, D
  neutrons, I.122
  nuclear exchange potential, I.376
Iyengar, P K, dispersion curves of germanium, II.1002

Jackson, K, Materials Research Society (MRS), III.1557
Jacobi, C G J, dynamical laws, I.176

Jacobs, E
  magnetic fusion, III.1629–30
  tokamak, III.1646
Jacobsen, E H, acoustic paramagnetic resonance (APR), II.1146
Jakeman, E
  antibunched light, III.1460
  K-distribution, III.1462
  optical spectrum, III.1454
James, R W
  sodium chloride crystals, I.433
  structure factors, I.439
  zero-point motion, II.974
Jamgochian, E, p–n junction, III.1417
Jamieson, J C, x-ray powder diffraction, I.437
Jancke, W, fibrous proteins, I.497
Jánossy, L, forked tracks, I.408
Jansky, K, radio astronomy, III.1736
Jauch, J M, photon, III.1451
Javan, A, helium–neon laser, III.1427
Jeans, First Solvay Conference (1911), I.147
Jeans, J H
  black-body radiation, I.163
  galaxy formation, III.1800
  gravitational collapse, III.1768
  internal structure of the stars, III.1703
  spiral nebulae, III.1950
  statistical mechanics, I.173
  tidal theory, III.1962
Jefferts, K B, carbon monoxide molecule, III.1763–4
Jeffreys, H
  Earth model, III.1956
  Earth–Moon system, III.1948
  Earth's core, III.1956
  geomagnetism, III.1958
  spiral nebulae, III.1950
  tidal theory, III.1962

Jenkin, F, wire-wound resistors, II.1252
Jennison, R C, Cygnus A, III.1775
Jensen, J H D, II.1211
  Breit–Wigner formula, II.1194
  nuclear binding energies, II.1210
  shell model, II.1209, II.1210, II.1214, II.1220
Johannson, H, emission electron microscope (EEM), III.1569, III.1577
Johansson, C H, crystal structure, III.1515
Johns, H E, telecobalt therapy, III.1872
Johnson, F A, optical-fibre communication, III.1447
Johnson, F M, antiferromagnetism, II.1148
Johnson, J B, Johnson noise fluctuations, I.601
Johnson, L F, rare-earth ion lasers, III.1438
Johnson, M H, neutron diffraction, I.459
Johnson, W H, metre standards, II.1239
Joliot-Curie, F, I.121
  radioactivity, I.121, I.375, II.1186, II.1222
Joliot-Curie, I, I.121
  neutron bombardment, II.1194
  radioactivity, I.375, II.1186, II.1222
Jona-Lasinio, G
  dynamical symmetry breaking, II.957
  pion bound state model, II.756
Jones, H
  Brillouin zone, III.1313
  disordered materials, III.1365
  Fermi surface, III.1295
  Seebeck coefficient, III.1310
  solid-state physics, III.1314
Jordan, P

field quantization, I.283
harmonic-oscillator
  decomposition, III.1402
matrix mechanics, I.198, I.360
Physics of the Twentieth Century,
  I.210
quantization of waves, I.208
quantum electrodynamics, I.209
transformation scheme, I.201
Jordan, W C, beta-ray experiment,
  III.1679–80
Josephson, B, Josephson effect,
  II.941–2
Jost, vacancies, III.1529
Joule, J P
  first law of thermodynamics,
    I.521
  learned societies, I.3
Joy, A H, T-Tauri stars, III.1766
Joyce, G S, long-range and
  short-range forces, I.563
Julian, B R
  detection of impulses, III.1957
  Earth's core, III.1980
Jungnickel, C, organized research,
  I.16
Justi, superconductivity, II.921
Jüttner, Maxwell–Boltzmann
  distribution law, I.282

Kadanoff, L P
  block spin variables, I.567
  delta (δ), III.1849
  hypothesis of universality, I.563
  phase transitions, III.1830
  scaling properties, I.563–4
  spin-blocking, III.1831, III.1832
  universality, I.571
Kadomtsev, B B, fusion-plasma
  research, III.1642
Kae, M, Ehrenfest dog-flea model,
  I.618
Kaganov, canonical theory,
  III.1356

Kaiser, W, two-photon absorption,
  III.1429
Kalckar, F
  liquid drop, I.126, II.1213
  spacing of states in highly
    excited nucleus, II.1213
Källén, G, Twelfth Solvay
  Conference, Brussels, II.646
Kallman, H
  fluorescent light pulses, II.1207
  scintillation, II.750
Kaluza, T, general relativity and
  electromagnetism, I.321–2
Kamerlingh Onnes, H, **II.914**
  critical point, I.545
  First Solvay Conference (1911),
    I.147
  liquid helium, II.913
  superconductivity, I.489, II.915,
    III.1284
Kaner, E A, cyclotron resonance,
  III.1357
Kant, I, island universes, III.1714
Kantrowitz, A
  magnetic fusion, III.1629–30
  tokamak, III.1646
Kao, K C, optical
  telecommunication, III.1447
Kapany, N S, fibre optics, III.1420
Kapitza, P L
  helium liquefier, II.920
  law of linear magnetoresistance,
    III.1356
  strong-impulsive-field generator,
    III.1282–3
  superfluidity, II.922
  superleak flow, II.946
Kapron, F P, chemical vapour
  deposition, III.1447
Kapteyn, J C
  Galaxy, III.1714, III.1715
  luminosity function, III.1714
  stellar spectra, III.1696
Kapur, P L, theory of nuclear
  reactions, II.1219

Name Index

Karagioz, O V, gravitation, II.1266
Karle, I L, structure of arginine dihydrate, I.449–50
Karle, J
    determinantal inequalities, I.448
    structure of arginine dihydrate, I.449–50
    tangent formula, I.449
    The Solution of the Phase Problem. I. The Centrosymmetric Crystal, I.449
Karpushko, F V, optical bistability, III.1450
Kasai, C, flow imaging, III.1914
Kasper, J S
    crystal structure of dekaborane, I.447
    inequality relations, I.447
Kasteleyn, P W, bond percolation model, I.573
Kastler, A, optical pumping, II.1097, III.1415, III.1447
Kasuya, T, RKKY interaction, II.1164
Kay, W, scintillation counting, I.113
Kaya, S, ferromagnetism, II.1134
Kaye, radiation protection, III.1890
Kayser, H
    handbook of spectroscopy, I.76
    history of spectroscopy, I.24
Keck, D B, chemical vapour deposition, III.1447
Keeler, J, spiral nebulae, III.1950
Keenan, P C
    MKK system, III.1698
    radio astronomy, III.1736
Keesom, A P, liquid helium, II.914
Keesom, W H
    atomic vibrations, III.1287
    liquid helium, II.914, II.922
    Seebeck coefficient, III.1310
Keffer, F, spin-wave modes, II.1151

Kekulé, A, Karlsruhe conference, I.45
Keldysh, L V, photo-assisted tunnelling, III.1423
Kelker, H, liquid crystals, III.1540
Kellar, J N, line broadening, I.472
Keller, A, polyethylene, III.1552
Kellermann, E W, NaCl, II.987–9
Kellers, C F, λ-point of liquid He^4, I.559
Kelley, P L, photodetection, III.1441
Kellman, E, MKK system, III.1698
Kelly, A, Walter Rosenhain, III.1517–19
Kelly, M, cuprous oxide rectifiers, III.1323
Kelvin, Lord
    conducting layer, III.1984
    cooling Earth, III.1945
    Earth structure, III.1946
    Earth's rigidity, III.1953
    gaseous molecule, I.48
    law of equal partition of energy, I.143–4
    Maxwell's Demon, I.576
    papers and patents, I.2
    radium atoms, I.62
    second law of thermodynamics, I.521
    second wrangler, I.11
    Seebeck effect, III.1284
    ship–wave pattern, II.883
    Sun's time-scale, III.1694
    temperature scales, II.1249
Kemmer, N
    meson theory, I.394
    new particles, I.393
    quantum field theory interaction, I.378
    symmetric theory, I.403
    U-particles, I.393
Kendrew, J C
    globular proteins, I.501

isomorphous replacement method, I.500
  phase angles, I.502
Kennedy, J W, plutonium isotope, II.1223
Kennelly, A E
  ionosphere, III.1621, III.1985
  radio waves, III.1986, III.1987
Kepler, J, supernovae, II.1228
Kerr, A, electro-optical double refraction, III.1389
Kerr, F J, neutral hydrogen, III.1763
Kerr, R, Kerr metric, III.1778
Kerst, D W
  betatron, II.732, II.1204, III.1871
  fixed-field alternating-gradient accelerators, II.739
  high-intensity beams, II.738
  multipole geometry, III.1645
  orbit calculations, II.732
Keuffel, J W, spark counters, II.753
Kevles, D, physics research, I.16
Khalatnikov, I M
  curvature singularities, I.310
  Stochasticity in relativistic cosmology, I.630
Khandros, L G, metal physics, III.1539
Khintchine, A I, *Mathematical Formulation of Statistical Mechanics*, I.614
Khoklov, stochastic particle acceleration, I.630
Khriplovich, I B, Yang–Mills theory, II.713
Kibble, B P
  fundamental constants, II.1260
  NPL moving-coil experiment, II.1238
Kidder, R
  laser fusion programme, III.1675
  radiation implosion, III.1674
Kihara, T, clusters of galaxies, III.1771

Kikuchi, S, electron diffraction, I.463
Kilby, J St C, integrated circuits, III.1326
Killian, J R Jr, stroboscopic photography, III.1410
Kim, Y K, shell electrons, II.1091
Kimble, H J
  photon antibunching, III.1460
  squeezed light, III.1471
Kinman, T D, helium abundance, III.1790
Kinoshita, S, charged particle tracks, II.1200
Kip, A F, resonant frequencies, III.1334
Kippenhahn, R, study of the stars, III.1746
Kippenhauer, K O, Galactic radio emission, III.1737–8
Kirchheim, R, physics of materials, III.1555
Kirchhoff, G R
  black-body radiation, I.22, I.66, I.148
  collaboration with Bunsen, I.75–6
  induction coil, I.54
  Investigations of the solar spectrum and the spectra of the chemical elements, III.1692
  quantum theory, I.65
  radiation at constant temperature, I.521
  school of physics, I.15
  spectral lines, I.156
Kirkpatrick, H A, curved crystal focusing spectrometer, II.1207
Kirkpatrick, P, holography, III.1415
Kirkwood, J
  BBKGY hierarchy, I.606, I.623, I.624, I.626
  polymer solutions, III.1597

N40

Name Index

Kirshner, R P, supernovae, III.1759
Kittel, C
  antiferromagnetic resonance, II.1151
  ferromagnetic resonance, II.1153
  ferromagnets, II.1147
  resonance, II.1147
  resonant frequencies, III.1334
  RKKY interaction, II.1164
  spin waves, II.1012
Klaiber, G S, dipole vibration, II.1213
Kleeman, R, α-particles, I.104
Klein, M J, black-body radiation, I.22
Klein, O
  baryon symmetric models, III.1806
  Dirac electrons, I.218
  Dirac equation, I.218
  quantum theory, I.179
  Twelfth Solvay Conference, Brussels, II.646
Kleiner, W H
  crystal field theory, II.1144
  photodetection, III.1441
Kleinman, D A, anharmonic oscillators, III.1434
Klemens, P G, electrons and holes, III.1297
Klug, A
  polio virus, I.510
  tobacco mosaic virus (TMV), I.508
  turnip yellow mosaic virus, I.510
Knight, P L
  laser fluctuations, III.1473
  optical harmonic generation, III.1473
Knipping, P
  crystal structures, III.1514
  x-ray diffraction, I.425
Knoll, M
  Crantz Colloquium, III.1571–2
  electron microscope, III.1571
  high-speed oscillograph, III.1570
  iron-free lenses, III.1568
  scanning electron microscope, III.1587
  transmission electron microscope (TEM), III.1569–70
Knudsen, M
  Fifth Solvay Conference (1927), I.205
  First Solvay Conference (1911), I.147
Kobayashi, M
  CP violation, II.701
  quarks, II.720, II.723
Kocharovskaya, O, x-ray lasers, III.1473
Koehler, T R, self-consistent renormalized harmonic approximation, II.1004
Koehler, W C
  neutron diffraction, I.459
  rare-earth metals, I.463
Koester, C J, optical-fibre laser, III.1427
Kogelnik, H
  Gaussian–Hermite (laser) beam propagation, III.1447
  resonator modes, III.1429
Kogut, J, hadrons, II.690
Kohan, M, *High Performance Polymers*, III.1550
Köhler, A, illumination system for microscopy, III.1390
Kohlhörster, W, balloon flights, I.368
Kohlrausch, F, *Leitfaden der Praktischen Physik*, I.15
Kohlschütter, A, luminosity indicators, III.1698
Kohn, W
  Fermi surface, III.1364
  phonon frequency, II.1002
  wavevectors, II.999

N41

Kolmogoroff, A N, K systems, I.615
Kolmogorov, A N, turbulence, II.798, II.837
Komarov, L I, neutron frequency, III.1418
Kompaneets, A, induced Compton scattering, III.1801
Kondo, J, electron scattering, III.1343
Koo, D C, sky surveys, III.1798
Koppe, superconductivity, II.932, II.934, II.936
Korteweg, D J, weak non-linear and weak dispersive effects, II.859
Koschmieder, E L, free convection, II.846
Koshelev, K, laser amplification, III.1466
Kossel, W
  periodic table, I.95
  x-ray spectra, I.160
Kosterlitz, J M, topological long-range order, I.571
Krakowski, R A, reversed-field pinch (RFP) research, III.1657
Kramers, H A
  asymptotic linkage, I.197
  atomic theory, I.197
  correspondence principle, I.86
  dispersion formula, I.171, I.172, I.185
  dispersion relations, III.1404
  Fifth Solvay Conference (1927), I.205
  Kramers' theorem, II.1125
  particle escape probability, I.607–8
  periodic table, I.96
  potassium chromic alum, II.1126
  quantum theory, I.166, I.179
  specific heat, I.555
  spontaneous emission, III.1401
  two-electron system, I.182

Kraner, H
  crystallization, III.1537
  synthetic zinc silicate, III.1537
Kratzer, A, atomic theory, I.179
Krishnan, K S
  pseudopotential concept, III.1367
  Raman effect (combination scattering), III.1404–6
  Raman scattering, II.996
Kronberg, M L, special orientation relationships, III.1528
Kronig, R L
  dispersion relations, III.1404
  magnetic field, II.1133
  nuclear magnetic moments, I.116
  nuclear spin, I.117
  superconductivity, II.917
Kruger, R, semiconducting crystals, II.1006
Kruskal, M D
  coordinate system, III.1778
  plasma column instabilities, III.1641
  plasma oscillations, III.1625
Kubo, R, **I.613**
  lattice impurities, III.1366
  spin waves, II.1011
  transport coefficients, I.612, I.626–7
Kuhl, D E, tomography, III.1916
Kuhn
  random walk, III.1597
  rubber elasticity, III.1597
Kuhn, T S
  black-body radiation, I.22, II.973
  helium atom, III.1982
  life cycle of field of study, III.2048
  quanta, II.969
Kuhn, W, spectral line intensity, I.172
Kuiper, G P, solar system, III.1963
Kukhtarev, N V, photorefractivity, III.1461

Name Index

Kukich, superfluidity, II.932
Kulsrud, R, collisionless shocks, III.1684
Kunsman, C H, electron reflection, I.170
Kunze, P, charged particles, I.388
Kurchatov, I V, fusion research, III.1641
Kurdjumov, G, metal physics, III.1539
Kurlbaum, F, black-body radiation, I.149, III.1387
Kurnakov, crystal structure, III.1515
Kurti, N, low-temperature physics, II.920
Kusaka, S, meson theory, I.407
Kyu, T, polymer blends, III.1548

Labeyrie, A E
  holograms, III.1438
  optical astronomy, III.1453
Laborde, radium, I.61
Ladenburg, E, photoelectric effect, III.1390
Ladenburg, R W
  dispersion formula, I.171
  emission and absorption, III.1400
Ladner, A W, *Short Wave Wireless Communication*, III.1619, III.1620
Lallemand, P, stimulated Rayleigh-wing scattering, III.1445
Lallier, E, waveguide laser, III.1471
Lamb, W E
  First Shelter Island Conference, II.645
  hydrogen spectrum, III.1421
  Lamb shift, II.641–3
  laser light, III.1443
  Master-type equation, I.605
  optical maser, III.1443
  photon counting, III.1388
Lambert, J, island universes, III.1714
Lamerton, L F, depth-dose data, III.1867
Lanczos, C
  field equations, III.1724
  matrix mechanics, I.192
  quantum theory, I.196
Land, E H, sheet optical polarizers, III.1417
Landau, L D, **III.1626–7**
  abrupt and restrained transition, II.833–4
  charge effects, III.1611
  damping theory, III.1678
  diamagnetism, II.1133–4
  electron density, III.1316
  electron–electron collisons, III.1316
  energy complexes, I.301
  field-strength uncertainty product, I.209
  gravitational collapse, III.1712
  interacting particles, III.1363
  ionic crystals, III.1321
  light-scattering spectroscopy, III.1408
  magnetization, II.1134
  neutrino, II.667
  neutron stars, III.1713
  non-linear stability theory, II.826
  phenomenological theory, II.1007
  plasma oscillations, III.1623–5
  quantum liquid, II.928
  second-order transitions, I.549–51
  self-gravitating systems, I.313
  spectrum, II.930
  spin-1 particle, II.654
  superconductivity, II.924, II.934–6
  superfluidity, II.929
  surf-riding, III.1362

two-fluid hydrodynamics, II.929, II.944
zero sound, II.950
Landauer, R, thermodynamics of data processing, I.576
Landé, A
  quantum numbers, I.101
  two-electron system, I.182
  Zeeman effect, I.162
Landsberg, G S
  Raman effect (combination scattering), III.1404–6
  Raman scattering, II.996
  spectral shift, III.1397, III.1398
Landweber, L, optical data processing, III.1461
Lane, A M, nuclear reactions, II.1219
Lane, J H, sun's structure, III.1694
Lange, L, inertial frame of reference, I.255
Langenberg, D N, Determination of e/h, II.1268
Langevin, P
  Boltzmann relation, I.545
  diamagnetism, II.1117
  Fifth Solvay Conference (1927), I.205
  First Solvay Conference (1911), I.147
  internal energy, I.111
  stochastic processes, I.596
  Zeeman effect/diamagnetism, II.1114
Langley, S P, black-body radiation, I.22
Langmuir, I
  Fifth Solvay Conference (1927), I.205
  plasma oscillations, III.1623–5
Langsdorf, A Jr, diffusion cloud chamber, II.752
Lankard, J R, dye lasers, III.1446
Laplace, P S
  black holes, II.760

central force programme, I.321
fluid-mechanical analysis, II.889
nebular hypothesis, III.1945
sound waves, II.808
Laporte, O, atomic theory, I.179
Larmor, Sir Joseph
  aether theory, I.5
  radio propagation, III.1622
  self-exciting dynamo, III.1958
  upper atmosphere, III.1988
Larson, R, formation of stars, III.1769
Laskar, J, chaotic evolutions, III.1823
Lassen, H
  upper atmosphere, III.1992
  wave propagation, III.1991
Lassettre, E, molecular resonances, II.1084
Lassus St Genies, A H J, lens arrays, III.1411
Latham, R, critical fluctuations, II.1016
Lau, E, Lau effect, III.1414
Laub, J, stress–energy tensor, I.281
Lauer, E J, beta-ray experiment, III.1679–80
Laughlin, J, calorimetry, III.1883
Launay, J M, vibrational excitation, II.1089
Lauterbur, P C, NMR image, III.1924
Laval, J, x-ray scattering, II.990
Lawrence, E O, I.371, **II.1185**
  calutron isotope separation, III.1630
  cyclotron, II.639, II.729, II.736, II.1201
  direct reaction theory, II.1220
Lawson, J D
  fusion reactions, III.1628–9
  $n\tau$ product, III.1649
Lax, M, laser light, III.1443
Leavitt, H S, Magellanic Clouds, III.1714

Le Berre, M, *Photons and Quantum Fluctuations*, III.1441
Lebowitz, stochastic dynamics, I.629
Lederman, L M
  Drell–Yan process, II.701
  high-mass lepton pairs, II.721
  parity violation, II.666
  two-neutrino experiment, II.672
Lee, B W, renormalizability, II.695
Lee, E W, transverse Kerr effect, II.1155
Lee, T, supernova trigger hypothesis, III.1963
Lee, T D
  lattice gas model, I.553
  neutrinos, II.671
  parity conservation, II.665
  parity violation, II.666
Lee, Y T, cross-beam collision products, II.1089
Leff, H S, *Maxwell's Demon*, I.576
Legvold, S, magnetic configurations, II.1162
Lehmann, H, haemoglobin, I.502
Lehmann, I, Earth structure, III.1956–7
Lehmann, O
  cholesteric esters, III.1540
  liquid crystals, III.1540
Lehovec, K, p–n junction, III.1417
Leibacher, J W, solar oscillations, III.1750
Leighton, R B
  sky survey, III.1743
  solar oscillations, III.1749–50
Leith, E N
  holography, III.1415, III.1437
  off-axis optical beam holography, III.1585
Leksell, L, flaw detector, III.1911
Lemaître, G
  Eddington–Lemaître model, III.1732–3
  primaeval atom, III.1785

Schwarzschild horizon, I.307
spherically symmetric perturbations, III.1801
Universe, III.1725
Lenard, P
  photoelectric effect, I.152, III.1390
  photoelectron emission, III.1396
Lengyels, B A, *Lasers*, III.1425
Lenz, W
  atomic theory, I.179
  ferromagnetism, I.554
Leontovich, M A, *Plasma Physics and the Problem of Controlled Thermonuclear Reactions*, III.1642
Le Pichon, X
  Earth's surface, III.1978
  plate tectonics, III.1978
Le Poole, J B, selected area diffraction, III.1579
Leprince-Ringuet, L, K-meson (or kaon), II.656
Letokhov, V S
  laser amplification, III.1466
  laser beams, III.1462
Le Verrier, U J J, precession of perihelion of Mercury, III.1722
Levi-Civita, T, tensor analysis, I.292
Levine, B F, superlattice device, III.1471
Levinson, C A, proton inelastic scattering, II.1221
Levshin, V L
  absorption of light by uranium glass, III.1403
  Levshin–Perrin law, III.1400
  polarization of fluorescent light, III.1400
  Raman effect, III.1400
Levy, H A, neutron diffraction, I.458–9
Lévy, M

axial coupling constant, II.676
β-decay, II.677
weak hadronic current, II.678
Lewis, G N
  direct reaction theory, II.1220
  particle dynamics, I.279
  photons, I.358, III.1302, III.1403
Lhéritier, M, K-meson (or kaon), II.656
Li, T
  Fabry–Perot resonator, III.1429
  Gaussian–Hermite (laser) beam propagation, III.1447
Libchaber, A, dynamical systems3.1852
Lichte, H, atomic resolution microscopy, III.1585
Lichten, W, potential curves for Ar_2^+ complex, II.1079
Lide, D R Jr, HCN laser line, III.1443
Lieb, E H
  delta (δ), III.1845
  percolation model, I.573
  two-dimensional ferroelectric models, I.570–1
Lifshin, E, electron microprobe analyser, III.1549
Lifshitz, E M
  curvature singularities, I.310
  energy complexes, I.301
  galaxy formation, III.1801
  self-gravitating systems, I.313
  Statistical Physics, I.550
Lifshitz, I M
  canonical theory, III.1356
  crystal defects, II.1004
  crystal surface, II.1005
  Fermi surface, III.1351
  quantum theory, III.1350
Liftshitz, E, magnetization, II.1134
Liley, B, tokamak, III.1647
Lilly, S J, radio galaxies, III.1795–6
Lima de Faria, J, *Historical Atlas of Crystallography*, I.511–12

Linde, A D
  scalar Higgs fields, III.1806
  world model, III.1807
Linde, J O, crystal structure, III.1515
Lindemann, F A (later Lord Cherwell)
  Clarendon Laboratory, II.920
  crystal lattice, III.1315
  First Solvay Conference (1911), I.147
  two-frequency equation, I.153
Lindhard, J
  electron gas, III.1362
  Kubo procedure, I.626
Linfoot, E H, diffraction images, III.1419
Lippmann, G, interference within a thick film, III.1392
Lipscomb, W N, boron hydrides, I.493
Lipson, H, copper–nickel–iron alloy, III.1548
Lipson, H S
  stacking faults, I.473
  structure determination, I.441
Livingston, M S
  accelerators, II.725–7, II.738, II.1201
  cyclotron, II.736
  direct reaction theory, II.1220
  nuclear physics, II.1184
  Particle Accelerators, II.1201
  synchrotrons, II.734
Lobachevsky, N I, *On the Principles of Geometry*, III.1722
Lockyer, Sir Norman
  Nature, I.6
  spectrograph, I.25
  stellar evolution, III.1695
  stellar spectra, III.1693
Lodge, O J, obituary of George Fitzgerald, I.29
Logan, B G, thermal barrier, III.1667

Logan, G, tandem-mirror concept, III.1665–6
Lohmann, A W
  computer-generated spatial filter, III.1447
  holography, III.1415
London, F
  covalent molecular bond, II.1042
  Josephson junction, II.1095
  penetration depth, II.935
  superconductivity, II.924–5
  Superfluids, II.934
London, H
  anomalous skin effect, III.1346–7
  superconductivity, II.921, II.924–5
  superfluidity, II.932
  thermoelectric effects, II.922
Long, R R, upstream wakes, II.868
Longair, M S, radio galaxies, III.1795–6
Longuet-Higgins, M S, wave energy, II.860–1
Lonsdale, K
  hexamethyl benzene, I.495
  x-ray scattering, II.990
Lonzarich, G G, saturation magnetization, II.1167
Lorentz, H A, **I.18**
  anomalous Zeeman effect, I.102
  electric dipoles, II.1117
  electrodynamics of moving bodies, I.261
  electromagnetic theory, I.63
  electron collisions, III.1290
  electron optics, III.2000
  electron theory, II.1114, III.1284
  Fifth Solvay Conference (1927), I.205
  First Solvay Conference (1911), I.147
  general theories of relativity, I.34
  induced precession, II.1115
  magnetism, III.1390
  Maxwell equations, I.280
  microscopic electron theory, I.281
  quantum theory, I.145, I.585
  random processes, I.586
  special theory of relativity, I.34, I.262–85
  splitting of spectral lines, I.161
  statistical mechanics, I.173
  The Theory of the Electron, III.1992, III.2039
  thermoelectric effects, III.1306, III.1307–8, III.1308
Loschmidt, J
  Avogadro's number, II.1269
  dimensions of molecules, I.48
Lossew, band-structure diagram, III.1417
Loudon, R, virtual electronic transitions, II.996–7
Louisell, W, laser light, III.1443
Love, A E H
  crystal surface, II.1005
  The Mathematical Theory of Elasticity, II.967
Lovelace, C, pion–pion scattering amplitude, II.684
Loveland, W D, transuranic elements, II.1223
Lovell, B, development of astronomy, III.1735
Low, F E
  bolometers, III.1743
  electric quadrupole moments, II.1214
  pion–nucleon resonance, II.680
  pion–pion scattering, II.681
  renormalization group, III.1830
Lowde, R, neutron diffraction, I.456
Lowman, C E
  magnetic materials, II.1159
  Magnetic Recording, II.1158
Lowry, R A, gravitation, II.1266

Lu, K T, photoabsorption spectrum of xenon, II.1073
Luce, J, energetic particle injection, III.1661
Lücke, K, *Lehrbuch der Metallkunde*, III.1554
Lucretius, atoms, I.44
Ludwig, G, Geiger counter, III.2007
Lugiato, L A, laser theory, III.1444
Lukirskii, A P
  bremsstrahlung, II.1050
  x-ray apparatus, II.1051
Lukosz, W, optical images, III.1439
Lummer, O R, black-body radiation, I.22, I.149, III.1387
Lummer, U, reflection interference fringes, III.1389
Lunbeck, R J, critical temperature, II.975
Lundmark, K, Andromeda Nebulae, III.1718
Luneberg, R K
  imaging theory, III.1418
  Mathematical Theory of Optics, III.1412
Luttinger, J M, Fermi surface, III.1364
Lyddane, R H, vibration of alkali halide, I.485
Lynden-Bell, D
  accretion discs, III.1964
  black holes, III.1779
  violent relaxation, III.1774
Lynn, J E, fission isomers, II.1218
Lyons, H, properties of quartz, II.1244
Lyubimov, V A, electron neutrino, III.1803

McAfee, K B, semiconductors, III.1419
McCall, S L, thermal bistable effects, III.1450
McClelland, C L, Coulomb excitation experiments, II.1217
McClung, F J, Q-switching, III.1430, III.1436
McConnell, A J, photon mass, III.1452
McCormmach, R, organized research, I.16
McCrea, W H, cosmological models, III.1730
Macdonald, resistance of sodium wires, III.1344
McDonald, W, mirror systems, III.1660
MacDougall, D P, gadolinium sulphate, II.1126
McDougall, I, geomagnetic field, III.1973
Macek, W M, ring-laser gyroscope, III.1433
McFarland, B B, laser development, III.1448
McFarlane, C G B, CO_2 laser, III.1442
McFee, J H, electron–phonon interaction, III.1339
McGowan, F K, Coulomb excitation experiments, II.1217
Mach, E
  atoms, I.48
  kinetic theory of gases, I.30
  Mach number, II.813
MacInnes, D, First Shelter Island Conference, II.645
McIntyre, P, proton–antiproton colliders, II.743
McKay, K G, semiconductors, II.1208, III.1419
McKee, C, neutron star, III.1759
McKellar, A, excited states, III.1791
McKenzie, D, plate tectonics, III.1978

Mackintosh, A R, exchange interactions, II.1164
McLennan, liquid helium, II.922
McMahon, R G, redshift quasars, III.1798
McMillan, E M
  electron synchrotron, II.731
  frequency modulation, II.1202
  ion acceleration, II.730
  neptunium, II.1224
  60" cyclotron, II.1202
  transuranium element, II.1223
Madansky, L, spark counters, II.753
Madelung, E
  electron vibration, II.983–4
  quantum theory, I.227
Madey, J M J, free-electron laser, III.1456
Maeda, K, ionosphere, III.1984
Maiani, L
  charmed quark, II.700, II.705
  neutral currents, II.700
Maiman, T H, **III.1425**
  ruby laser, III.1424–6, III.1894
Majorana, E
  deuteron, I.376
  nuclear stability, II.1190
Maker, P D, pump-probe configuration, III.1445
Maki, A G, HCN laser line, III.1443
Maki, Z, quark–lepton analogy, II.700
Mallard, J R, gamma camera, III.1907
Malmberg, J H, magneto-electrostatic confinement, III.1681
Mandel, L
  coherence theory, III.1433
  optical coherence, III.1392
  photon antibunching, III.1460
  photon beams, III.1424
  quantum theory of coherence, III.1439
  sub-Poissonian statistics, III.1460
Mandelbrot, B
  fractals, I.574
  self-similarity, I.574
Mandelstam, L I
  first-order Raman effect, II.996
  Mandelstam–Brillouin scattering, III.1397
  Raman effect (combination scattering), III.1404–6
  Raman scattering, II.996
  secondary scattered radiation, I.172
Mandelstam, S, Twelfth Solvay Conference, Brussels, II.646
Manley, J M, non-linear optics, III.1424
Mansfield, P
  magnetic resonance imaging, III.1924
  nuclear magnetic resonance, II.1175
Mansur, L K, nuclear materials research, III.1534
Marconi, G
  electromagnetic waves, I.29
  radio waves, III.1995
  telegraph signals, I.54
  wireless transmission, III.1619–20, III.1986
Marechal, A, coherent imaging, III.1418
Margon, B, binary sources, III.1756
Mark, H
  biomolecular structures, I.497
  single crystals, III.1527
Marsden, E
  α-particles, I.113
  α-scattering, II.1200
  medical physics, III.1864
Marshak, R E
  Conceptual Foundations of Modern Particle Physics, II.770

First Shelter Island Conference, II.645
mesons, II.654
neutrinos, II.667
Marshall, L, neutron diffraction, I.456
Marshall, T W, stochastic electrodynamics, III.1388
Martin, J, x-ray diffraction, III.1547
Martin, P, temporal behaviour of a general system, I.617
Marton, transmission microscope, III.1574–5
Martyn, D F
  broadcast modulation, III.1994
  ionosphere, III.1984
Marx, G, neutrinos, III.1803
Mascart, E, motion of the Earth, I.258
Mash, D I, stimulated Rayleigh-wing scattering, III.1445
Mashkevich, V S, covalent crystals, II.1002
Masing, G, *Lehrbuch der Metallkunde*, III.1554
Maskawa, T
  CP violation, II.701
  quarks, II.720, II.723
Massey, Sir Harrie, **II.1024**
  ionospheric research, III.1999
  The Theory of Atomic Collisions, II.1024, II.1039
Mathewson, C, plastic flow and crystal glide, III.1544
Matsuoka, M, x-ray sources, III.1755
Matthews, D, magnetic variations of ocean floors, III.1975
Matthews, T A, stellar spectra, III.1776
Matthias, B T
  ferroelectric crystals, III.1538
  Matthias' rules, I.489

Matuyama, M, Earth's magnetic field, III.1973
Mauguin, C, helical structure, III.1397
Maurer, R D, chemical vapour deposition, III.1447
Maxwell, A, Pacific floor measurements, III.1974
Maxwell, J C
  Atom, I.76
  atoms, I.51
  electrolysis, I.51
  electromagnetic waves, I.260
  electromagnetism, III.1856
  equal area construction, I.544
  equipartition of energy, I.523
  field theory, I.321
  kinetic theory of gases, I.587
  molecules, I.47
  new fields of research, I.29
  second law of thermodynamics, I.523
  second wrangler, I.11
  sodium yellow line, II.1239
  statistical mechanics, I.173
  Theory of Heat, I.575
  wire-wound resistors, II.1252
Mayall, N U, redshift–magnitude relation, III.1788, III.1793
Mayer, J E, statistical mechanics, I.551–4
Mayer, M G, II.1211
  shell model, II.1209, II.1210, II.1214, II.1220
Mayneord, W V, **III.1868**
  depth-dose data, III.1867
  nuclear medicine, III.1901
  thimble ionization chamber, III.1867
Meggers, W F, **II.1030–1**
  atomic energy levels, II.1093
Mehl, R, diffusive transformations, III.1547
Meinesz, F A V, ocean floor, III.1974

Meissner, W, superconductivity, II.916–19, II.946
Meitner, L, I.105
  β-decays, I.118
  β-ray deflection, II.1201
  β-spectrum, I.104, I.105, I.119
  elements with $Z = 93$–96, I.130
  neutron bombardment, II.1194
  neutron sources, II.1206
  quantum theory, I.221
  uranium isotope, II.1223
  uranium-239, I.132
Melvill, T, discrete spectra, I.74–5
Mendeleev, D I
  absolute boiling temperature, I.543
  Avogadro's law, I.46
  periodic table, I.96
Mendelssohn, K
  energy gap, II.932
  low-temperature physics, II.920
Mendenhall, T C, squared paper, I.16
Menter, J M, electron microscope, I.477
Menzel, D H
  abundances of elements, III.1702
  planetary nebulae, III.1761
Merton, R, norms of scientific ethos, III.2057–8
Merz, W, dielectric constants, I.487
Metzner, A W K, field equations, I.313
Meyer, K H, biomolecular structures, I.497
Meyer, R B, smectic phase, III.1604
Michell, J, gravitation, II.1266
Michelson, A A, **I.33**
  Balmer lines, I.90
  cadmium red line, II.1239–40
  future discoveries, I.29
  hydrogen atom, II.641
  interferometer, I.25
  Lightwaves and their Uses, I.26
  optical instruments, III.1391

optical interferometry, III.1704
optical relativity principle, I.261
spectroscopy, I.25
speed of light, II.1263
stellar interferometer, III.1400
Studies in Optics, I.26
Zeeman splitting, I.161
Mie, G
  non-linear field theories, I.232
  special-relativistic gravitation theory, I.284, I.321
Migdal, A B
  electron–phonon interaction, II.941
  interacting Fermi particles, III.1364
  phonons, II.999
Migneco, E, fission isomers, II.1218
Milburn, G J, quantum non-demolition (QND) measurements, III.1469
Miley, G H, laser systems, III.1464
Miley, G K, radio galaxies, III.1796
Miller, D A B, indium antimonide, III.1450
Miller, D C, Michelson–Morley experiment, III.1401
Miller, H, wedge fields, III.1872
Miller, J S, Seyfert galaxies, III.1784
Miller, S E, integrated optics, III.1444
Miller, W, nebulae, III.1717
Miller, W H, Miller indices, I.421
Millikan, R A
  absorption data, I.368
  constituents of matter, I.357
  electron charge, II.747
  mesotron, I.389
  new cosmic-ray particle, I.396
  oil drop method, II.1271
  photoelectric effect, I.363, III.1390–1
Millington, G, long-distance communications, III.1999

Mills, B, radio surveys, III.1789
Mills, R L
  isotopic spin symmetry, II.693
  local gauge invariance, II.693
  self-interacting fields, II.661
Milne, E A
  cosmological models, III.1730
  electrons in metals, I.531
  ionization, III.1701
  solar atmosphere, III.2003
  stellar spectra, III.1693
  theory of kinematic relativity, III.1787
Minkowski, H, **I.252**
  four-dimensional formulation, I.274
  particle dynamics, I.279
  phenomenological electrodynamics, I.280
  relativistic vacuum electrodynamics, I.280
Minkowski, R
  medium-mass elements, II.1228
  Palomar 200″ Telescope, III.1737
  radio galaxy, III.1776
  radio sources, III.1775
  supernovae, II.1227–8
Minnaert, M, equivalent width, III.1702
Minogin, V G, laser beams, III.1462
Minsky, M
  confocal optical microscopy, III.1422
  confocal scanning laser microscope, III.1450
Misener, superleak flow, II.946
Mistry, N, two-neutrino experiment, II.672
Mitchell, D P, neutron diffraction, I.453
Mitchell, J W
  dislocation lines in silver chloride, III.1526
  solid-state physics, III.1530

Mitra, S K
  rocket studies, III.2001
  The Upper Atmosphere, III.2000
Mittelstaedt, O, speed of light, II.1263
Miyamoto, S, spark chamber, II.753
Moffat, K, protein unfolding, I.438
Mögel, H, sudden ionospheric disturbance (SID), III.1995
Mohorovičić, A, Mohorovičić discontinuity (Moho), III.1954
Mollenauer, L F, soliton-pulse scheme, III.1458, III.1471
Möllenstedt, G, Fresnel electron biprism, III.1585
Møller, C
  meson theory, I.401
  neutron capture experiments, II.1192
  S-matrix, II.680
  symmetrical theory, I.403
  Twelfth Solvay Conference, Brussels, II.646
Monck, W H S, stellar spectra, III.1696
Montroll, E W
  atomic models, II.987
  crystal defects, II.1004
  specific heat of small crystals, II.1005
Moore, C, chemical abundances, III.1702
Moore, W S, whole body imager, III.1924
Morey, G E, amorphous structures, I.477
Morgan, J, plate tectonics, III.1978
Morgan, W W, MKK system, III.1698
Moriya, T, lattice distortions, II.1152
Morley, E W
  Balmer lines, I.90

hydrogen atom, II.641
  optical relativity principle, I.261
Morris, P, nuclear magnetic resonance, II.1175
Mort, J, photocopying, III.1541
Morton, K W, self-avoiding walk (SAW), I.570
Moseley, H G J
  atomic number, I.159–60
  Bohr's theory, I.88
  characteristic x-rays, II.1038
  periodic system, I.362
  periodic table, I.86
  x-ray spectra, I.158
Mosengeil, K, relativistic thermodynamics, I.281
Moss, T S, absorption coefficient of semiconductor, III.1421
Mott, N F, **III.1333**
  Brillouin zone, III.1313
  colour centres, III.1530
  disordered materials, III.1365
  Electron processes in Non-crystalline Materials, III.1369
  paramagnetism, II.1136
  resistivity of dilute alloy, III.1316
  Seebeck coefficient, III.1310
  semiconductors, III.1319, III.1330–4
  solid-state physics, III.1313–14
  The Theory of Atomic Collisions, II.1039
Mottelson, B, II.1216
  collective motion, II.1215
  Coulomb excitation, II.1217
  Coulomb excitation experiments, II.1217
  geometrical collective model, II.1218
  surface oscillations, II.1215
Mould, J, Hubble's constant, III.1793
Moulton, F R
  nebular hypothesis, III.1950
  tidal theory, III.1962
Mouton, H, magneto-optical effect, III.1389
Mow, V C, *Basic Orthopaedic BioMechanics*, III.1896
Mueller, K A, oxygen octahedra, II.1009
Mügge, crystals of copper, gold etc, III.1517
Mukherjee, A K, plastic deformation, III.1534
Mulders, G, equivalent width, III.1702
Müller, A
  Meissner effect, II.957–8
  superconductivity, II.961
Muller, C A, hyperfine line radiation, III.1763
Müller, E, field emission microscope, III.1569
Muller, E W A, solar x-rays, III.1994
Müller, K A, superconductivity, I.490
Mulliken, R
  molecular mechanics, II.1044
  wave mechanics, I.213
Munro, G, ionosphere, III.1984
Murphy, G M
  deuterium, III.1706
  deuteron, II.650, II.1183
Mushotzky, R F, quasars, III.1780
Mussell, L, isocentric couch, III.1873
Mutscheller, A, radiation protection, III.1890
Myers, W D, liquid-drop mass formula, II.1225

Nabarro, F R N
  diffusion creep, III.1534
  dislocations, I.476, III.1526
  physical metallurgy, III.1555
Nagamiya, T, resonance, II.1147

Nakajima, S, pulse oximeter, III.1894
Nambu, Y
  chiral symmetry, II.676
  decay reaction, II.657
  dynamical symmetry breaking, II.957
  pion bound state model, II.756
  quarks, II.687
  SU(3) symmetry, II.709
  Twelfth Solvay Conference, Brussels, II.646
Nasanow, L H, self-consistent renormalized harmonic approximation, II.1004
Nath, N S N, diffraction of light, III.1409
Neddermeyer, S H
  absorption in lead plates, I.385
  cosmic rays, I.386
  elementary particles, I.391
  mesotron, I.231, I.389
  muon, II.650
  shower particles, I.388
Néel, L
  antiferromagnetism, II.1129, II.1132
  ferrimagnetism, II.1157
  neutron diffraction, I.453
Ne'eman, Y, II.678
  isospin, II.661
  strangeness, II.662
  triplets, II.685
Neidigh, R, plasma column, III.1678
Nelmes, R J, neutron diffraction, I.489
Nereson, N, nuclear forces, II.652
Nernst, W H, **I.536**
  calorimetry, II.968–9
  chemical equilibrium, I.535
  First Solvay Conference (1911), I.147
  new physics, I.7
  quantum hypothesis, I.155
  quantum theory, I.82, I.148, I.154
  specific heat, I.152–3, II.970
  thermodynamic theory, III.1699
  two-frequency equation, I.153
Nethercot, A H
  antiferromagnetism, II.1148
  thermal conductivity, II.982
Neugebauer, G
  formation of stars, III.1769
  Orion Nebula, III.1743, III.1767
  sky survey, III.1742–3
Neumann, F
  school of physics, I.15
  semiconductors, III.1419
Neville, F H, **III.1511**
  multiphase alloys, III.1548
  phase diagram, III.1508–9
Newnham, R E, ceramics, III.1538
Newton, I, I.255
  corpuscular hypothesis, I.257
  first law of motion, I.256
  law of gravitation, II.1265
  Opticks, I.76
  particle description of light, I.72
  physical cosmology, III.1721
  Principia, I.255
  rotating bracket, I.256
  second law of motion, II.806
  third law of motion, II.821
Ng, W K, ruby laser, III.1432
Nichols, G A, optical microscopy, III.1406
Nichols, H W, radiowave propagation, III.1990
Nicholson, J W
  angular momentum, I.175
  electron vibration, I.81
  spectra of celestial objects, I.156–7
Nier, A, lead isotopes, III.1960
Nieto, M M, photon mass, III.1452
Niggli, P, x-ray diffraction, I.439
Nishijima, K
  decay reaction, II.657
  isotopic spin multiplets, II.658

Nishikawa, S, electron diffraction, I.463
Nishina, Y
  Dirac equation, I.218
  letter, I.212
  new particles, I.389
Nishizawa, J, semiconductor maser, III.1423
Niu, K, charmed particles, II.700
Noether, F, rigid body, I.279
Nordheim, L
  disordered materials, III.1365
  Schrödinger equations, I.205
Nordsieck, A
  First Shelter Island Conference, II.645
  Master-type equation, I.605
  nuclear exchange potential, I.376
Nordström, G, special-relativistic theories, I.284
Northrop, G A, thermal conductivity, II.982
Northrop, T G
  adiabatic confinement, III.1679–80
  trapped particles in magnetic-mirror fields, III.1683
Nouchi, photon, III.1451–2
Novikov, I D
  black holes, III.1779, III.1780
  structure in the Universe, III.1801
Nowick, A, anelastic relations, III.1523
Nuckolls, J
  direct-drive approach to inertial fusion, III.1676
  fusion power, III.1673
  radiation implosion, III.1674

Obreimow, crystal growth, III.1280
Occhialini, G P S
  cloud chambers, I.371, II.651
  electron–positron pairs, I.231

O'Dell, C R, helium abundance, III.1790
Oehme, R
  parity violation in β-decay, II.665
  singularities in complex angular momentum, II.682
Oersted, H C
  electric current, III.1958
  electromagnetic interactions, III.1953
Ohkawa, T, multipole geometry, III.1645
Ohnuki, Y, quark–lepton analogy, II.700
Okayama, T, slow mesotrons, I.406
Oke, B, supernovae, III.1759
Oke, J, energy in the Sun, III.1748
O'Keefe, J A, optical microscopy, III.1406
Oken, L, *Gesellschaft deutscher Naturforscher und Ärzt*, I.5
Okubo, S, mass relations among baryons and mesons, II.661
Oldham, R D
  earthquake waves, III.1956
  seismic wave paths, III.1954
Olszewski, K, liquefied gases, I.27
O'Neill, E L, communications theory, III.1417
O'Neill, G K, *Storage-ring synchrotron: device for high-energy physics research*, II.739
Onsager, L, **I.551**
  entropy relations, I.600–2
  Fermi surface, III.1351
  ferroelectric transition in KH_2PO_4, I.571
  hypothetical vortices, II.931
  Ising model, I.554–61, II.1018
  Onsager relations, I.616–17, I.626
  quantum theory, III.1350
Oort, J H
  cosmological density, III.1799

hyperfine line radiation, III.1763
neutral hydrogen, III.1763
Palomar 200" Telescope, III.1775
radio astronomy, III.1762
Öpik, E
  giant stars, III.1708–9
  origin of elements, III.1747
  triple-α reaction, III.1746
Oppenheimer, J R
  adiabatic approximation, II.999
  collapsing sphere of dust, I.308
  direct reaction theory, II.1220
  First Shelter Island Conference, II.645
  general-relativistic analysis, III.1712
  mesotron, I.389
  μ-meson, II.652
  muon, II.651
  neutron stars, III.1713
  QED, I.386
  Twelfth Solvay Conference, Brussels, II.646
Orbach, R, fractons, II.1007
Ornstein, L S
  critical correlations, I.563
  pair distribution function, I.547
  spectral line intensity, I.172
Orowan, E
  dislocations, III.1520
  edge dislocation, I.476
Orthmann, W, β-spectrum, I.119
Ortolani, S, ETA-BETA I devices, III.1658
Oseen, C W, wave propagation in cholesteric liquid crystals, III.1407
Osheroff, D, superfluidity, II.950–1
Osmer, P S, quasars, III.1798
Osmond, block copolymers, III.1611
Osterberg, H, thin-film devices, III.1444
Osterbrock, D E, helium problem, III.1790

Ostriker, J P
  binary sources, III.1756
  electron acceleration, III.1752–3
  galactic cannibalism, III.1774
  spiral galaxies, III.1771
Ostwald, W
  dispute with Planck, I.34
  kinetic theory of gases, I.30
  natural phenomena, I.47–8
  quantum theory, I.148
Overbeek, charge effects, III.1611
Overhauser, A W, shell model, I.485, II.997

Pacini, F, neutron stars, III.1752
Pais, A
  First Shelter Island Conference, II.645
  Inward Bound, I.37, II.1183
  Twelfth Solvay Conference, Brussels, II.646
Pancini, E, mesons, II.652
Pankey, supernovae, III.1759
Panofsky, W K H
  electron storage rings, II.739
  Mark III linac, II.737
  proton linear accelerator, II.1205
  proton machines, II.737
  Stanford Linear Accelerator Center (SLAC), II.687–8
Papanastassiou, D A, supernova trigger hypothesis, III.1963
Parisi, G, asymptotic freedom, II.715
Parker, E N
  collisionless shocks, III.1684
  solar wind, III.1685, III.2009
Parker, H M
  dosimetry, III.1881
  radium needles, III.1877
Parker, W H, determination of e/h, II.1268
Parrent, G B, partial coherence, III.1392

Parrish, W, powder photographs, I.436
Paschen, F
  black-body radiation, I.149
  Bohr's theory, I.88
  bolometer, I.23
  far-infrared spectrum, II.982–3
  Zeeman splitting, I.161
Paschos, E A, parton hypothesis, II.690
Pasternack, S, Lamb shift, II.641
Pastukhov, V P, mirror approach, III.1667
Patashinskii, A Z
  correlation components, I.563
  multiple-correlations near critical point, I.562
Patel, C K N, CO_2 laser, III.1442
Paterson, R
  dosimetry, III.1881
  radium needles, III.1877
Patterson, A L, x-ray diffraction pattern, I.441
Patterson, C C, age of the Earth, III.1961
Pauli, W
  atomic dynamics, I.184
  atomic theory, I.179, I.181
  continuous β-spectrum, I.127–8
  Einstein–Podolsky–Rosen (EPR) paradox, I.228
  electromagnetic field, I.391
  electron gas, III.1288
  equivalence relations, I.196
  exclusion principle, I.98–100, I.359, I.528, III.1711
  field quantization, I.283
  Fifth Solvay Conference (1927), I.205
  free particles, I.283
  lattice calculations, I.233
  Lorentz invariance, III.1451
  neutrino, I.371, I.375, III.1706
  non-relativistic theory, I.283
  paramagnetism of normal metals, II.1133
  quantum electrodynamics, I.209, I.362
  quantum mechanics, I.226
  quantum theory, I.180, I.221
  spin-statistics theorem, I.283
  stability of nuclei, I.111–12
  thermal conductivity, II.979–81
  total quantum number, I.162
  two-component wavefunctions, I.277
  wave mechanics, I.206
  Zeeman effect, I.103
Pauling, L
  biomolecular structures, I.497–8
  First Shelter Island Conference, II.645
  inorganic chemistry, I.496
  inorganic compounds, I.492
  pleated sheet structures, I.499
  polypeptide chains, I.497
  quasicrystals, I.481
  residual entropy of ice, I.571
  structure of ice, I.475
  wave mechanics, I.213
Payne, C H
  abundances of elements, III.1701
  Stellar Atmospheres, III.1701
Payne-Gaposchkin, C, spectrum analysis of stars, III.1962
Peacock, J A, radio sources, III.1798
Pearson, impurity effects, III.1281
Pearson, G L, solar cell, III.1421
Pearson, J J, excitations in $s = 1/2$ Heisenberg antiferromagnet, II.1014
Pearson, K, *The Grammar of Science*, I.34
Pease, F G, angular diameter of Betelgeuse, III.1705
Pease, R S
  structure of KH_2PO_4, I.459
  ZETA experiment, III.1642

Pedersen, P O, radio waves,
   III.1992
Peebles, P J E
   clusters of galaxies, III.1771
   cold dark matter, III.1804
   spiral galaxies, III.1771
   synthesis of heavy elements,
      III.1792
   world models, III.1806
Peierls, R E
   exciton absorption spectrum,
      III.1410
   Fermi theory, I.377
   field-strength uncertainty
      product, I.209
   holes, III.2018
   Ising model, I.554
   lattice distortion, II.1000
   meson theory of nuclear forces,
      I.402
   resistivity at low temperatures,
      III.1306
   Schrödinger equation,
      III.1299–300
   theory of nuclear reactions,
      II.1219
   thermal conductivity, II.979–81
   Twelfth Solvay Conference,
      Brussels, II.646
   Umklapp processes, III.1304–6
Peimbert, M, helium abundance,
   III.1790
Pekeris, C L, M_2 tide, II.892
Pelzer, H, lattice deformation,
   III.1321
Penrose, R
   closed trapped surface, III.1778
   global differential geometry,
      I.310
   hot big bang models, III.1778
   quasicrystals, I.482, I.483
   resonance experiments, II.1139
   space-time, I.313, I.318–19
   spinorial techniques, I.277
Penzias, A A
   carbon monoxide molecule,
      III.1763–4
   microwave background
      radiation, III.1790
   microwave radiometer, II.758
   radiation, III.1786
Pepinsky, R
   ferroelectric crystals, III.1538
   XRAC, I.446
Pepper, D M, optical-beam
   transmission, III.1461
Percus, random walk model, I.630
Perkins, D H
   neutrino scattering, II.697
   photographic emulsions, II.653
Perl, M, *Evidence for the existence of
   anomalous lepton production*,
      II.719
Perlman, I
   α-decay spectra, II.1217
   *Nuclear Properties of the Heavier
   Elements*, II.1223
Peron, J, controlled fusion, III.1628
Peronneau, P A, ultrasonic
      imaging, III.1914
Perrin, F
   First Solvay Conference (1911),
      I.147
   Levshin–Perrin law, III.1400
   neutrons, I.122
   relativistic kinematics, I.129
   Twelfth Solvay Conference,
      Brussels, II.646
Perrin, J
   Avogadro's number, III.1610
   Les Atomes, I.43
Pershan, P S, non-linear optics,
   III.1434
Pert, G J, x-ray laser, III.1464
Perutz, M F
   globular proteins, I.501
   haemoglobin, I.500, I.502, I.503
   isomorphous replacement
      method, I.497
   x-ray analysis, III.1514

Peterson, S W, neutron diffraction, I.458–9
Pethick, C J, neutron stars, III.1754
Petit, R, diffraction theory of gratings, III.1466
Petley, B W
  fundamental constants, II.1261
  gravitation, II.1266, II.1267
Petrán, M, confocal scanning laser microscope, III.1450–1
Petschek, H, collisionless shocks, III.1684
Petzoldt, J, positivism, I.319
Pfann, W G, zone-refining, III.1281, III.1543–5
Pfirsch, D, stellarator-type research, III.1655
Pfund, high-reflection coatings, III.1407
Phillips, M, direct reaction theory, II.1220
Phillips, W A, glasses, II.1006
Phragmen, G, single-crystal methods, I.435
Piccard, A, Fifth Solvay Conference (1927), I.205
Piccioni, O, mesons, II.652
Pickering, C, hydrogen lines, I.85
Pickering, E C
  Elements of Physical Manipulation, I.16
  stellar spectra, III.1693, III.1694
Pidd, R W, spark counters, II.753
Pierce, G W
  radio propagation, III.2000
  semiconductors, III.1318
Pierce, J R, photomultipliers, II.750
Piermarini, G, diffraction techniques, I.437
Pierre, F M, charmed meson, II.704
Pike, E R
  imaging theory, III.1467–8
  optical spectrum, III.1454
Pines, D

  electrical discharges in gases, III.1361
  superconductivity, II.937
Pippard, A B
  optics, III.1467
  penetration depth, II.933–4
Pippin, J E, crystal structures, II.1158
Pirani, F A E
  general relativity, I.299
  plane wave solutions, I.312
Piron, C, photon, III.1451
Placzek, G
  light-scattering spectroscopy, III.1408
  phonon dispersion relations, II.993
  transuranic elements, I.131–2
Planck, M, **I.144**
  black-body radiation, I.144, I.524–5, II.973, II.1269, III.1402
  bound system, I.111
  energy density, III.1387
  energy quanta, I.173
  entropy, I.523
  Fifth Solvay Conference (1927), I.205
  finite energy quantum, I.173
  First Solvay Conference (1911), I.147
  oscillators, I.70
  particle dynamics, I.279
  partition function, I.527
  Planck's law, I.67–70
  quantum optics, III.1387
  quantum theory, I.62, I.146–7, I.156, I.522, III.1385
  school of physics, I.15
  Scientific Autobiography, I.34
  second law, I.47
  special-relativistic thermodynamics, I.281
  thermodynamics, I.68
Platzman, R L, continuous spectra

from synchrotron light sources, II.1050
Plücker, J, spectrum of hydrogen, I.76
Podolsky, B, quantum mechanics, I.228
Pohl, R
 Bohr's theory, I.88
 colour centres, III.1529, III.1530
 disordered systems, II.1006
 ionic crystals, III.1291
Pohlhausen, E, free convection, II.843-5
Pohlhausen, K, conduction of heat, II.842
Poincaré, H
 energy conservation, I.61
 First Solvay Conference (1911), I.147
 perturbation theory, III.1830-31
 radio waves, III.1986-7
 recurrence theorem, I.618
 relativity principle, I.265
 relativity theory, I.173
 special theory of relativity, I.262-85
 statistical mechanics, I.524
Pokrovskii, V L
 correlation components, I.563
 multiple-correlations near critical point, I.562
Polanyi, M
 biomolecular structures, I.496
 dislocations, III.1520
 edge dislocation, I.476
 single crystals, III.1526-7
Polder, D, electromagnetic waves, II.1157
Polikanov, S, fission isomers, II.1218
Politzer, H D, running coupling constant, II.713
Pollack, H O
 eigenfunctions and eigenvalues, III.1468

optical imaging, III.1432
Pomeau, Y, delta (δ), III.1849
Pomeranchuk, I Ya
 electron–electron collisons, III.1316
 Pomeranchuk trajectory, II.682
Pontecorvo, B
 neutrinos, II.671
 solar neutrinos, II.757
Popaloizou, J C B, thick discs, III.1781
Porte, G, lamellar phase, III.1608
Porter, A B, imaging theory, III.1391
Porter, C, proton scattering, II.1212
Post, B, Renninger effect, I.450
Post, R F
 Atoms for Peace Conference, III.1639
 Controlled fusion research—an application of the physics of high-temperature plasmas, III.1641
 loss-cone modes of mirror-confined plasmas, III.1662
 mirror machine, III.1679
 Sixteen lectures on controlled thermonuclear reactions, III.1660
Potts, R B, Potts model, I.571
Poulsen, magnetic recording, II.1158
Pound, R V, red shifts, I.291
Powell, C, π-meson, I.231
Powell, C F, photographic emulsions, II.653
Powers, P N, neutron diffraction, I.453
Poynting, J H
 gravitation, II.1266
 theoretical constructs, I.34
Prandtl, L, **II.797**
 boundary-layer equations, II.805, II.806

boundary-layer theory, II.796–8, II.803, II.869
drag due to lift, II.824
drag on a non-lifting body, II.869
elliptic distribution of circulation, II.824
heat transfer, II.842
lift–drag ratios L/D, II.870
perturbation theory, II.829
singular perturbation, II.800
supersonic flow, II.812
triple deck theory, II.829
turbulence, II.815
vortex lines, II.822
vortex wake, II.822
vorticity, II.820
wind-tunnel testing at supersonic speeds, II.879
Prendergast, K H, x-ray sources, III.1755
Prescott, C Y, parity violation, II.698
Press, W H, mass function, III.1804
Preston, T, Zeeman splitting, I.161
Priestley, J, magnesium, III.1359
Prigogine, I, **I.621**
  Brussels school, I.619–22
  Non-equilibrium Statistical Mechanics, I.619
  Twelfth Solvay Conference, Brussels, II.646
Primakoff, H, neutral pion, II.655
Pringle, J E
  accretion discs, III.1964
  thick discs, III.1781
Pringsheim, E, black-body radiation, I.149, III.1387
Prins, J, x-ray diffraction pattern, I.441
Prout, W, atomic species, I.46
Pryce, M H L
  hyperfine structures, II.1139
  quantum theory, III.1413
Purcell, E M
  hyperfine line radiation, III.1763
  nuclear magnetic resonance, II.1094
Pusey, P N, K-distribution, III.1462

Quimby, E
  medical physics, III.1864
  radiotherapy, III.1867
  radium implants, III.1878
Quinn, T J, acoustic interferometer, II.1270

Rabi, I I
  fine/hyperfine structures, II.1047
  First Shelter Island Conference, II.645
  precessional motion, II.1153
Racette, G W, He–Ne laser, III.1435
Radon, J, computed tomography, III.1916
Raff, A D, sea-floor spreading, III.1975
Rainwater, J, II.1216
  quadrupole moment, II.1215
Rajchman, J A, photomultipliers, II.750
Rakshit, H, ionospheric physics, III.2000
Ramachandran, G N, speckle-pattern formation, III.1442
Raman, Sir Chandrasekhara
  Brownian motion, III.1442
  diffraction of light, III.1409
  Lectures in Physical Optics, III.1442
  phase transitions, II.1008
  Raman scattering, II.996, III.1404–6
  secondary scattered radiation, I.172
  speckle, III.1399
Ramberg, E G
  contour phenomena, III.1581–2

magnetic lenses, III.1580
Ramsauer, noble gas atoms, II.1041
Ramsay, Sir William, helium in uranium-bearing mineral, I.58
Rarity, J G, Bell inequalities, III.1471
Rasetti, F
  mesotron lifetime, I.406
  Raman scattering, I.117, II.996
Ratcliffe, J A
  An Introduction to the Ionosphere and Magnetosphere, III.2001
  ionosphere, III.1999
Ray, E, Geiger counter, III.2007
Rayleigh, Lord, **I.21**
  acoustics, III.1910
  aeolian tones, II.850
  black-body radiation, I.30, II.972
  Bohr's theory, I.87
  boundary conditions, II.845
  catastrophe theory, II.798
  Cavendish Laboratory, I.21
  crystal surface, II.1005
  Faraday measurement, II.1269
  grating spectroscopy, III.1391
  law of equal partition of energy, I.143–4
  optics, III.1390
  radio waves, III.1986–7
  random processes, I.586
  senior wrangler, I.11
  shock wave, II.811
  singular perturbation theory, II.811
  sound from flows, II.850
  sound propagation, II.810
  sound waves, II.858
  statistical mechanics, I.173
  The Theory of Sound, I.21, II.808, II.855, III.1910
  wave energy, II.858
  zero-viscosity limit, II.828

Raymond, A L, hexamethylene tetramine, I.495
Raynor, G, physical metallurgy, III.1555
Read, F H, quantum-mechanical reformulation, II.1065
Reber, G
  continuum radio map, III.1762
  radio astronomy, III.1736
Rebka, G A, red shifts, I.291
Rees, M J
  non-thermal phenomena, III.1781
  radio sources, III.1782, III.1783
Regel, A R, energy levels, III.1368
Regener, E, scintillation method, II.1200
Regge, T, Regge poles, II.681–2
Reiche, F, dispersion formula, I.171
Reickhoff, K E, stimulated Brillouin scattering, III.1444
Reid, A
  de Broglie waves, I.171
  electron diffraction, I.463
Reid, J M, pulse-echo ultrasound, III.1911–12
Reimann, B, non-linear analysis, II.857
Reines, F, neutrino, II.663
Reinitzer, F, cholesteric esters, III.1539–40
Reppy, J D, superfluidity, II.932
Retherford, R C, Lamb shift, II.642, II.643
Reusch, rocksalt, III.1517
Revelle, R, Pacific floor measurements, III.1974
Rex, A F, *Maxwell's Demon*, I.576
Reynolds, O
  diffusion of momentum, II.826
  flow in pipes, II.814
  pipe flow, II.825
  Reynolds analogy, II.843
  turbulence, II.798, II.825, III.1834

Rhines, F N
  grain growth, III.1550
  Microstructology—Behavior and Microstructure of Materials, III.1550
Ricci, C G, tensor analysis, I.292
Richards, B, Airy disc, III.1412
Richardson, O W
  electron emission, I.37
  electron optics, III.2000
  Fifth Solvay Conference (1927), I.205
  photoelectric effect, III.1390
  thermionic emission, I.12
Richter, B, electron storage rings, II.739
Richter, R, controlled fusion, III.1628
Riecke
  electrical and thermal conductivities, III.1285
  Hall effect, III.1282
  metallic conduction, III.1284
Riemann, B
  general relativity, III.1722
  propagation speed, II.809
  waveforms, II.810
Risken, H Z
  laser light, III.1443
  laser noise, III.1444
Riste, T, dynamic response of SrTiO$_3$, II.1010
Ritchey, G W, spiral nebulae, III.1717
Ritson, D, b-quark, II.723
Ritter, A, stellar evolution, III.1694–5
Ritz, W
  black-body radiation, I.163
  combination principle, I.25, III.1391
  combination rule, I.186
  spectral frequencies, I.86
Robb, A A, special theory of relativity, I.277–8

Roberts, F, flying-spot microscope, III.1416–17
Roberts, L D, germanium junctions, II.1208
Robertson, H P
  Robertson–Walker metric, III.1729–30
  synchronization of clocks, III.1729
Robertson, J M
  organic chemistry, I.496
  phthalocyanines, I.495
Robinson, D, High Voltage Engineering (HVE), II.1203
Robinson, F N H, anharmonic oscillators, III.1434
Robinson, H
  α-particles, II.1201
  β-ray spectrometer, II.1201
  RaB, RaC and RaE spectra, I.106
Robinson, I K
  plane wave solutions, I.312
  surface crystallography, I.469
Rochester, C D
  elementary particles, II.654
  kaons, II.656
Rochester, G, forked tracks, I.408
Rodionov, S N, tritium-beta experiment, III.1679
Rogers, G L, optical holography, III.1415
Rogerson, J B
  helium problem, III.1790
  interstellar deuterium, III.1791
Rohrer, H, scanning tunnelling microscope, III.1549
Röntgen, W C, x-rays, I.3, I.32, I.53, I.54, I.362, I.424, II.1271, III.1529, III.1857, III.1861
Roosevelt, President, atomic energy research, II.1196
Roozeboom, B, phase diagrams, III.1508
Rosa, E B, speed of light, II.1263

Rosbaud, P, *Progress in Metal Physics*, III.1556
Rose, P H, tandem accelerator, II.1203
Rosen, N, quantum mechanics, I.228
Rosenbluth, M N, **III.1664–5**
  International Summer Course in Plasma Physics 1960, III.1643
  loss-cone modes of mirror-confined plasmas, III.1662
  mirror machine, III.1679
  mirror systems, III.1660
Rosenfeld, L
  gravitational field, I.317
  meson theory, I.401
  symmetrical theory, I.403
  Twelfth Solvay Conference, Brussels, II.646
Rosenhain, W
  amorphous metal hypothesis, III.1518
  An Introduction to the Study of Physical Metallurgy, III.1517
  glass industry, III.1517
  grain boundaries, III.1519, III.1528
  physical metallurgy, III.1516
  slip lines, III.1517
  yield-stresses of single metallic crystals, III.1520
Rosenthal, A H, Sagnac effect, III.1433
Rossby, C G
  potential vorticity concept, II.905
  wave-like current patterns, II.892–5
Rosseland, S, quantum theory, I.179
Rossi, B
  cosmic rays, I.399, II.748
  First Shelter Island Conference, II.645
  nuclear forces, II.652
Rossman, M G
  structure determination, I.445
  virus structures, I.510
Rotblat, angular distributions, II.1220
Roth, W L, superionic conductors, III.1539
Round, H J
  band-structure diagram, III.1417
  yellow light emission, III.1391–2
Roux, D
  lamellar phase, III.1608
  sponge phase, III.1608
Rowan-Robinson, M, radio quasars, III.1797–8
Rowe, H E, non-linear optics, III.1424
Rowland, H, concave spectral gratings, I.54
Royds, T, α-particle, I.59
Rozental, S, meson theory, I.401
Rubbia, C
  field particles W and Z, II.744
  proton–antiproton collider, II.707, II.743
Rubens, H
  black-body radiation, I.149, III.1387
  far-infrared spectrum, II.982–3
  First Solvay Conference (1911), I.147
Rubin, H, magnetic cusp, III.1645
Rubin, V C, rotation curves of galaxies, III.1770
Rubinowicz, A
  polarization of spectral lines, I.178
  selection rules, I.94
Rubinstein, H, Dolen–Horn–Schmid duality, II.683
Rüdenberg, R, electron microscope, III.1573–4
Ruderman, M

indirect-exchange-coupling theory, II.1012
RKKY interaction, II.1164
scalar Higgs fields, III.1806
Rühmkoff, H, higher voltages, I.54
Rumford, Count, *see* Thompson, Sir Benjamin
Runcorn, S K
  continental drift, III.1972
  magnetic North Pole, III.1972
Runge, C, Zeeman splitting, I.161
Rushbrooke, G S, thermodynamic relation, I.560
Ruska, E
  Crantz Colloquium, III.1571–2
  electron microscope, III.1570–1
  electronic lenses, III.1569
  iron-free lenses, III.1568, III.1571
  polepiece lens, III.1567
  transmission electron microscope (TEM), III.1569–70, III.1575–7
Ruska, H, specimen preparation, III.1575–6
Ruskol, E, Moon model, III.1966
Russell, H N
  luminosity–spectral class diagram, III.1696
  photoexcitation, III.1761
  photographic parallax programme, III.1696
  solar atmosphere, III.1701–2
  solar system, III.1962
  tidal theory, III.1951
  white dwarfs, III.1709–10
Russell, J E
  German school system, I.8
  physics in education, I.8
Rutgers, superconductivity, II.924
Rutherford, E (Baron Rutherford of Nelson), **I.55**
  actinium, II.1227
  age of the Earth, III.1694
  α-particles, I.59, I.115, II.638, II.745–6, II.1201

α-rays, I.54, I.58
atomic energy, I.63
atomic nuclei, II.1183
atomic structure, I.74, I.156
β-ray spectrometer, II.1201
β-rays, I.54, I.58
crystal diffraction spectroscopy, II.1207
electrical counting methods, II.1200
electron charge, II.747
electrons, I.64
experimental physics, I.14
First Solvay Conference (1911), I.147
half-life, I.62
hydrogen nucleus, I.109
meeting with Bohr, I.74
Nature of the α-particle, I.58
nuclear structure, I.110, I.113, III.2018
positive ion accelerators, II.1184
quantum conditions, I.359
RaB, RaC and RaE spectra, I.106
Radiations from Radioactive Substances, I.108, I.112, II.1199, II.1200
radioactive decay, III.1949
Radioactive Substances and their Radiations, I.107
Raman effect (combination scattering), III.1404
Silliman lectures, I.52
The distinction between α-rays and β-rays, I.55
thorium emanation, I.59
transformation theory, I.59
wireless transmission, III.1986
Ryle, M
  Cassiopeia A, III.1737
  radio astronomy, III.1738
  radio sources, III.1788
  survey of the sky, III.1788
Ryutov, D, mirror physics, III.1666

Sackur, O, quantum-theoretical equations, I.154
Sadler, C A, x-ray spectroscopy, I.158
Safronov, V, solar system, III.1963
Sagdeev, R Z, fusion-plasma research, III.1642
Sagitov, M V, gravitation, II.1266
Sagnac, G, measurement of rotation, III.1398
Saha, M N
 ionospheric physics, III.2000
 stellar atmospheres, III.1699
 stellar spectra, III.1693
Sakata, S
 fundamental triplet, II.685
 meson theory, I.407
 spin-1 particle, II.654
 two-meson theory, I.408
Sakharov, A D
 baryon–antibaryon asymmetry, III.1806
 baryon asymmetry, II.765
 CP violation, II.675
 Plasma Physics and the Problem of Controlled Thermonuclear Reactions, III.1642
 tokamak, III.1630
Salam, A, II.696
 electroweak unification, II.695
 neutrino, II.667
 Twelfth Solvay Conference, Brussels, II.646
Saleh, B E A, sub-Poissonian Franck–Hertz source, III.1460
Salpeter, E E
 α–α scattering, II.1226
 black holes, III.1779
 nuclear reactions, III.1709
 origin of elements, III.1747
 triple-α reaction, III.1746
Sampson, R A, energy transport, III.1699
Samuelson, E J, K_2CoF_4, II.1016

Sandage, A R
 Evidence for the occurence of violent events in the nuclei of galaxies, III.1777
 Herzsprung–Russell diagrams, III.1793
 Hubble's constant, III.1793
 quasars, III.1777
 redshift–magnitude relation, III.1788, III.1793
 stars in globular clusters, III.1746
 stellar spectra, III.1776
 The ability of the 200" telescope to discriminate between selected world models, III.1793
Sargent, W L W, inverse-Compton catastrophe, III.1781
Satomura, S, Doppler ultrasound, III.1914
Saunders, F A, Russell–Saunders (or L–S) coupling, III.1702
Savedoff, M P, interstellar gas, III.1765
Sayre, D, electron density, I.448
Schafroth, superconductivity, II.937
Schawlow, A L, **III.1428**
 Fabry–Perot resonator, III.1429
 lasers, III.1423, III.1424
 laser spectroscopy, III.1466
 maser, III.1422
 Microwave Spectroscopy, III.1428
Schechter, P, mass function, III.1804
Schechtman, D, quasicrystals, I.481
Scheffer, T J, liquid-crystal optics, III.1452
Schelleng, J C, radiowave propagation, III.1990
Schenectady, R T, electrical discharges in gases, III.1361
Scherk, J, quantum gravity, II.684
Scherrer, P, single crystal, I.434
Scherzer

axially symmetric electron lenses, III.1579
Geometrische Elektronenoptik, III.1578
Scheuer, P A G, radio sources, III.1782, III.1789
Schild, A, general relativity, I.299
Schilling, W, point defects, III.1534
Schlichting, H, Tollmien–Schlichting waves, II.830
Schlier, R J, surface crystallography, I.470
Schlüter, A, stellarator-type research, III.1655
Schmid, C, pion–nucleon charge exchange, II.683
Schmid, E
  plastic deformation, III.1526
  single crystals, III.1527
Schmidt, B V
  Schmidt corrector plate, III.1406
  Schmidt telescopes, III.1745
Schmidt, E, free convection, II.843
Schmidt, G, radioactivity, I.57
Schmidt, G N J, tomato bushy stunt virus (TBSV), I.509
Schmidt, M
  optical spectrum of 3C 273, III.1776
  quasars, III.1798
  radio quasars, III.1797–8
Schmidt, O, monistic theories, III.1963
Schneider, W G
  isotherms of xenon, I.556
  liquid–gas co-existence curve, I.553
Schoenflies, A, symmetry types, I.423
Schönberg, M, Schönberg–Chandrasekhar limit, III.1709
Schottky, W
  colour centres, III.1530
  electric potential, III.1289–90
  vacancies, III.1529
Schrieffer, J R
  BCS theory, II.938
  MOSFET, III.1372
  superconductivity, I.232, II.939, III.1364
Schrödinger, E, **I.193**
  decay law of radioactive substances, I.222
  entangled photon states, III.1394–5
  Fifth Solvay Conference (1927), I.205
  matter waves, I.194
  non-relativistic action function, I.195
  photon mass, III.1452
  quantum mechanics, II.638
  quantum theory, I.174
  relativistic equation, I.283
  wave equations, I.195, I.200
  wave mechanics, I.360
Schubauer, G, boundary layer, II.835
Schubnikow
  crystal growth, III.1280
  zone-melting, III.1281
Schulkes, J A, ferrimagnetism, II.1157
Schulman, J, microemulsions, III.1608
Schulman, L S, 'life' (game), I.629
Schulz, liquid metals, III.1366
Schulz, G J, He atoms, II.1061–2
Schumacher, E E, deformation of lead crystals doped with antimony, III.1544
Schuster, A
  doctorate, I.17
  electrical discharges in gases, III.1985
  energy transport, III.1699
Schwartz, M
  neutrinos, II.671

two-neutrino experiment, II.672
Schwarz, J
  quantum gravity, II.684
  string theories, II.684
Schwarzschild, K
  curvature of space, III.1722
  dynamical laws, I.176
  energy transport, III.1699
  field equations for a point mass, III.1777–8
  Maxwell's equations, I.279
  metric, III.1778
  quantum theory, I.89
  Stark effect, I.162
Schwarzschild, M
  Herzsprung–Russell diagrams, III.1793
  plasma column instabilities, III.1641
  stars in globular clusters, III.1746
Schweber, S
  hierarchical structure, III.2020
  process of emergence, III.2019
Schwers, F, II.972
Schwinger, J
  bosons, II.694–5
  First Shelter Island Conference, II.645
  quantum electrodynamics, II.643, II.740, III.1446
  renormalization, II.646
  temporal behaviour of a general system, I.617
  Twelfth Solvay Conference, Brussels, II.646
Schwitters, R, Mark I Detector, II.703
Scully, M O
  laser light, III.1443
  phaseonium, III.1473
  x-ray lasers, III.1473
Seaborg, G T
  Nuclear Properties of the Heavier Elements, II.1223

  plutonium isotope, II.1223
  transuranic elements, II.1223
Seaton, M J
  chanel coupling formulation, II.1073
  exchanges of angular momentum, II.1060
  R matrices, II.1068
  radial density function, II.1066
Secchi, A, stellar spectra, III.1692–3
Seddon, J A, wave-like patterns, II.897
Seeger, A K, dislocation theory, I.476
Segrè, E
  neutron bombardment, II.1191
  plutonium isotope, II.1223
Seidel, H, optical bistability, III.1449
Seitz, F, **III.1532**
  band structure, III.1314–17
  Brillouin zone, III.1313
  colour centres, III.1531, III.1536
  crystal defects, III.1531
  diffusion, III.1531
  electron band structure of sodium, III.1530
  Imperfections in Nearly Perfect Crystals, III.1535
  materials research, III.1558
  Schrödinger's equation, III.1315
  solid-state physics, III.1313
Selenyi, P, xerography, III.1541
Sen, H K, magneto-ionic theory, III.1992
Sen, N R, ionospheric physics, III.2000
Serber, R
  β-decay, I.401
  direct reaction theory, II.1220
  First Shelter Island Conference, II.645
  mesotron, I.389
  μ-meson, II.652
  muon, II.651

orbit calculations, II.732
Seyfert, C K, spiral galaxies, III.1777
Shaknov, I, scintillation counters, III.1414
Shakura, N, thin accretion discs, III.1779
Sham, L J
　phonon frequency, II.1002
　pseudopotential theory, II.999
Shannon, R R, US optical industry, III.1476
Shapiro, I I, Schwarzschild metric, I.305
Shapiro, J, pion–pion scattering amplitude, II.684
Shapley, H
　Galaxy, III.1714, III.1716, III.1717
　globular clusters, III.1716
　Great Debate, III.1718
　novae, III.1717
　stellar masses, III.1698
Sharp, P F, imaging procedure, III.1920
Sharvin, D Yu, resistance of thin films, III.1370
Sharvin, Yu V, resistance of thin films, III.1370
Shenstone, W A, vitrified silica ware, I.29
Sheppard, R M
　detection of impulses, III.1957
　Earth's core, III.1980
Shimizu, F, self-phase modulation, III.1449
Shklovsky, I S
　Crab Nebula, III.1774
　x-ray sources, III.1755
Shlyaptsev, V, laser amplification, III.1466
Shockley, W B, **III.1324**
　cuprous oxide rectifiers, III.1323
　cyclotron resonance, III.1337–8
　Electrons and Holes in Semiconductors, III.1317
　magnetoresistance, III.1335–6
　photomultipliers, II.750
　transistor, II.744
Shoenberg, D
　de Haas–van Alphen effect, III.1348–52
　superconductivity, II.925
Shore, B W, optical harmonic generation, III.1473
Short, R, sub-Poissonian statistics, III.1460
Shortley, G, *The Theory of Atomic Spectra*, II.1037
Shu, F J, formation of stars, III.1769
Shubnikov, A, superconductivity, II.920
Shuleikin, M V, ionospheric layers, III.1992
Shull, C G
　Fe_3O_4, I.460
　magnetic structures, II.1148
　neutron diffraction, I.453
　powder pattern of MnO, I.460
　spin density, I.462
Shutt, R P
　diffusion chamber, II.752
　heavy particles, II.657
　mesotrons, I.405
Sibbett, W, dye-laser pulses, III.1439
Sidgwick, Faraday measurement, II.1269
Siegbahn, K
　double-focusing beta spectrometer, II.1206
　electron spectrometers, II.1090
Siegbahn, M, x-ray spectra, I.160
Siegert, A J F, Master equation description of gas, I.605
Sievert, R M
　radiation protection, III.1890
　thimble ionization chamber, III.1867

Silin, V P, electrons in metals, III.1363
Silk, J, sound waves, III.1802
Simon, A
  cloud chamber, I.364
  plasma column, III.1678
Simon, Sir Francis
  diamond–graphite equilibrium diagram, I.538
  low-temperature physics, II.920
  third law of thermodynamics, I.537, I.542
  zero-point motion, II.973–4
Simonen, T, TMX experiment, III.1667
Simpson, J A, II.1062
Sinai, Ja, hard spheres system, I.615
Sinitsyn, G V, optical bistability, III.1450
Sitte, K, β-decay theory, I.375
Sitterly, C M, atomic energy levels, II.1093
Skinner, H W B, x-ray emission, III.1361
Sklodowska, M, *see* Curie, M
Slater, J C
  crystal field theory, II.1144
  properties of metals, III.1345
  quantum mechanics, I.211
  quantum theory, I.166, I.180
  Slater determinant, I.215
  Solid State and Molecular Theory, III.1314
  solid-state physics, III.1313
Slee, O B
  radio sources, III.1737
  radio surveys, III.1789
Slepian, D
  eigenfunctions and eigenvalues, III.1468
  optical imaging, III.1432
Slepian, J
  electron emissions, II.749
  secondary emission, III.1399

Slichter, C P, BCS theory, II.940
Slipher, V M, spiral galaxies, III.1725
Sloan, D H, cyclotron, II.736
Slusher, R E, squeezed states of light, III.1469–70
Smale, S, mathematician, III.1840
Smart, J S
  neutron diffraction, I.453
  powder pattern of MnO, I.460
Smekal, A, light-quantum hypothesis, I.172, I.185
Smith, B W, interstellar medium, III.1766
Smith, C S
  A History of Metallography, III.1556
  A Search for Structure, III.1556
  Acta Metallurgica, III.1556
  Institute for the Study of Metals, III.1556
Smith, F G
  Cassiopeia A, III.1737
  radio sources, III.1737
Smith, K, soliton pulse communication, III.1471
Smith, L W, thin-film devices, III.1444
Smith, M G, quasars, III.1798
Smith, N, long-distance communications, III.1998
Smith, P L, *Needs, Analysis and Availability of Data for Space Astronomy*, II.1093
Smith, R A, photoconductivity, III.1327
Smith, T, optical systems, III.1446
Smoluchowski, M S
  dielectric constant, III.1392, III.1396–7
  Maxwell's Demon, I.576
  opalescent bands, I.546
Smoot, G F, cosmic microwave background radiation, III.1804

Smythe, *Atomic Energy for Military Purposes*, II.1223
Snitzer, E, optical-fibre laser, III.1427
Snoek, J L
    ferrimagnetism, II.1155
    strain-aging law, III.1521
Snyder, H
    collapsing sphere of dust, I.308
    general-relativistic analysis, III.1712
Snyder, H S
    accelerators, II.738
    synchrotrons, II.734
Sochor, V, HCN/H$_2$O lasing, III.1443
Soddy, F
    atomic energy, I.63
    radioactive decay, III.1949
    transformation theory, I.59
Soffer, B H, laser development, III.1448
Sohncke, L, symmetry types, I.423
Solomon, J, uncertainty relations, I.228
Solvay, E
    First Solvay Conference (1911), I.147
    new physics, I.7
Somer, J C, imaging, III.1914
Sommerfeld, A, I.180
    astrophysics, I.219
    Atombau and Spektrallinien, I.110
    Atomic Structure and Spectral Lines, I.156
    battle between Boltzmann and Ostwald, I.48
    Bohr's theory, I.88
    Drude–Lorentz model, III.1288
    dynamical quantum theory, I.174
    electric potential, III.1289
    extension of Bohr's model, I.160–1
    fine structure, I.90
    First Solvay Conference (1911), I.147
    free-electron gas hypothesis, I.216
    inner quantum number, I.100–1
    metal electrons, I.215
    Planck's law, I.70
    quantization, I.359
    quantum model, III.1308
    quantum numbers, I.90–2
    quantum rule, I.175
    quantum theory, I.179, III.1285, III.1361
    radio waves, III.1987
    structure factor, I.431
    theory of metals, III.1290
    thermoelectric effects, III.1308
    x-ray diffraction, I.424
    x-ray spectra, I.160
    Zeeman effect, I.102, I.162
Sorby, H
    metallography, III.1507
    microstructure, III.1506
Sorokin, P P
    dye lasers, III.1446
    four-level laser, III.1427
Southworth, G C, waveguide propagation, III.1322
Spain, I L, single-crystal studies, I.437
Spedding, F H, magnetism, II.1143
Spinrad, H, radio galaxies, III.1795
Spitzer, L
    Atoms for Peace Conference, III.1638
    interstellar gas, III.1765
    magnetic fusion, III.1636
    Physics of Fully Ionized Gases, III.1631
    rotational transform, III.1654
    stellarator, III.1630, III.1648, III.1654, III.1655
Sproule, D O, diagnostic ultrasonics, III.1910–11

Sproull, R, materials research, III.1558
Standley, K J
 relaxation time, II.1133
 spinels, II.1156
Stanley, G J, radio sources, III.1737
Stanley, H E, Ising models, I.563
Stark, J
 band spectra, I.77
 chemical and other quanta, I.154
 longitudinal effect, I.162
 quantum physics, I.148, I.155
 Stark effect, I.89
 Zeeman splitting, I.161
Staub, H, element synthesis, II.1226
Staudinger, H, long-chain systems, III.1597
Steenbeck, M, visit to Max Knoll, III.1572–3
Stefan, J, black body, I.66
Stein, P
 computers, III.1843
 mathematician, III.1839
Stein, R F, solar oscillations, III.1750
Steinberger, J
 decay of neutral pion, II.711
 neutral pion, II.655
 two-neutrino experiment, II.672
Steinhardt, P J, world model, III.1807
Steller, J, magnetic mirror, III.1659
Stelson, P H, Coulomb excitation experiments, II.1217
Stephens, W E, element synthesis, II.1226
Stern, J, quantum-theoretical equations, I.154
Stern, O
 electronic orbits, I.164
 ionic crystals, III.1531
 quantum theory, I.221
 specific-heat problem, I.153–4
 surface crystallography, I.468

Sternheimer, R, inner-shell electrons, II.1092
Stevenson, E C
 elementary particles, I.391
 mesotron, I.397
 muon, II.650
 new particles, I.389
Stevenson, M J, four-level laser, III.1427
Stewart, A T, phonons, II.999
Stewart, B
 Earth's magnetic field, III.1621
 geomagnetic variations, III.1984–5, III.1985
 wire-wound resistors, II.1252
Stiles, W G, human eye, III.1407
Stockmayer, W, polymer solutions, III.1597
Stoicheff, B P, stimulated Brillouin scattering, III.1444
Stokes, G G
 molecular vibrations, I.76
 x-rays, I.362
Stoletov, A G, photoelectric effect, III.1390
Stone, J, fibre lasers, III.1427
Stoner, C R, *Short Wave Wireless Communication*, III.1619, III.1620
Stoner, E, white dwarfs, III.1711
Stoner, E C
 electronic levels in atoms, I.168
 spin waves, II.1012
Stoney, G J, estimate for e, I.52
Storey, L R O, whistler studies, III.1995
Stormer, C
 Earth's magnetic field, III.1682
 geomagnetic storms, III.2003
 magnetic-mirror effect, III.1637
 single charged particle, III.2004
Strassmann, F
 elements with $Z = 93$–96, I.130
 neutron bombardment of uranium, II.1195

neutron sources, II.1206
neutron–U collisons, I.130
radiochemical analysis, II.1195
uranium isotope, II.1223
uranium-239, I.132
Stratton, S W, **II.1241**
Straubel, P, imaging theory, III.1418
Strauss, L L
  fusion research, III.1637–40
  Men and Decisions, III.1640
Street, J C
  elementary particles, I.391
  mesotron, I.397
  muon, II.650
  new particles, I.389
Strehl, K, aberration effects, III.1389
Strnat, K J, rare-earth–cobalt alloys, II.1165
Strnat, R M W, rare-earth–cobalt alloys, II.1165
Stroke, G W, holograms, III.1438
Strömgren, B
  interstellar medium, III.1762
  Strömgren spheres, III.1762
Strong, J, anti-reflection coatings, III.1407
Strutinsky, V M, fission isomers, II.1218
Strutt, J W, *see* Rayleigh, Lord
Strutt, R J, thorianite, III.1949
Stueckelberg, E, heavy electrons, I.391
Stuewer, R, nuclear physics, II.1192
Stump, D M, polarized light, III.1412
Sucksmith, W, ferromagnetism, II.1134
Sudarshan, E C G
  neutrinos, II.667
  quasi-classical distribution function, III.1441
  radiation field, III.1388

Suess, E, ocean basins, III.1967
Suess, H E
  chemical elements, III.1747
  nuclear binding energies, II.1210
  shell models, II.1209
Suhl, H, crystal structures, II.1158
Sunyaev, R A
  induced Compton scattering, III.1801
  photon scattering, III.1773
  thin accretion discs, III.1779
Sunyar, A, neutrino helicity, II.668, II.670
Sutherland, P, neutron stars, III.1754
Swedenborg, E, island universes, III.1714
Sweetman, D
  magnetic mirror, III.1659
  neutral beam injection, III.1661
Swiatecki, W T, liquid-drop mass formula, II.1225
Sykes, M E
  critical exponents, I.560
  series expansions, I.557
  site percolation processes, I.572
  two-dimensional lattices, I.573
Synge, E H, optical microscopy, III.1405–6
Szalay, A S, neutrinos, III.1803
Szigeti, B, electronic dipole moment, I.485
Szilard, L
  entropy, I.225
  Maxwell's Demon, I.576
Szöke, A, optical bistability, III.1449
Szymborski, K
  colour centres, III.1529
  solid-state physics, III.1320

Tabor, D, measurement of forces between surfaces, III.1612
Tainter, S, Photophone, III.1422

Talbot, F, daguerreotype process, III.1692
Talleyrand, C M de, metric system, II.1233
Talmi, I, shell model, II.1213
Tamm, I E
  energy loss formula, II.750
  first-order Raman effect, II.996
  nuclear exchange potential, I.376
  Plasma Physics and the Problem of Controlled Thermonuclear Reactions, III.1642
  tokamak, III.1630
  visible light, III.1409
Tammann, G
  binary alloys, I.548
  crystal growth, III.1280
  Hubble's constant, III.1793
  Lehrbuch der Metallographie, III.1554
  liquid crystals, III.1540
  phase diagrams, III.1509, III.1554
Tapster, P R, Bell inequalities, III.1471
Tayler, R J
  helium synthesis, III.1790
  synthesis of heavy elements, III.1792
Taylor, B N
  determination of e/h, II.1268
  magnetic helicity, III.1657
  resonance frequency, II.1095
Taylor, G I, **II.799**
  catastrophe theory, II.798
  diffraction pattern, III.1396
  Diffusion by continuous movements, II.830
  dislocations, III.1520
  edge dislocation, I.476
  equilibrium (wet-bulb) temperature, II.843
  potential-flow forces, II.887
  singular perturbation theory, II.811
  sound waves, II.858
  stabilizing effect of viscosity, II.828
  tidal friction, II.891
  turbulence, II.798, II.832
  vortices, II.833
Taylor, J B
  binary pulsars, III.1757
  plasma pressures, III.1661
Taylor, J H
  binary pulsar, I.298, I.317
  quadrupole radiation formula, I.315
Tebble, R S, domain techniques, II.1155
Teich, M C, sub-Poissonian Franck–Hertz source, III.1460
Teichmann, J
  colour centres, III.1529
  solid-state physics, III.1320
Telegdi, V L, parity violation, II.666
Teller, E
  adiabatic confinement, III.1679–80
  Atoms for Peace Conference, III.1639
  dipole vibration, II.1213–14
  First Shelter Island Conference, II.645
  trapped particles in magnetic-mirror fields, III.1683
Temkin, R J
  amorphous germanium, I.479
  amorphous structure, I.478
Temmer, G M, Coulomb excitation experiments, II.1217
Temperley, H N V, percolation model, I.573
Terhune, R W, Raman Stokes and anti-Stokes rings, III.1436
Ter-Martirosyan, Coulomb excitation, II.1217

Ter-Pogossian, M M, positron emission tomography (PET), III.1920
Tesla, N, radio and high-frequency technology, III.1620
Tharp, M, Mid-Atlantic Ridge, III.1973
Thellung, A, disordered materials, III.1366
Theobald, J P, fission isomers, II.1218
Thiessen, A, direct-drive approach to inertial fusion, III.1676
Thom, R, dynamical systems, III.1850
Thomas, L H, Thomas effect, I.278
Thomas, R G, nuclear reactions, II.1219
Thomas, W, spectral line intensity, I.172
Thompson, Sir Benjamin (Count Rumford), learned societies, I.3
Thompson, G B H, semiconductor lasers, III.1437
Thompson, R, hyperon, II.657
Thompson, S, radioactivity, III.1876
Thompson, W, *see* Lord Kelvin, Lord
Thomson, Sir George
  controlled fusion, III.1636
  de Broglie waves, I.171
  electron diffraction, I.463
  finite electron temperatures, III.1624–5
  toroidal fusion device, III.1634
Thomson, J J, I.50
  atomic model, I.64, I.65
  atomic structure, I.49, I.156
  deflection measurements, II.1271
  determination of e/m, I.53, I.54
  discovery of electron, I.3, III.1565
  electrical conductivity, III.1986
  electrical discharges in gases, III.1985
  electron, III.1857
  electron charge, II.747
  electron diffraction, I.463
  electron emission, I.37
  electronic theory of magnetism, II.1114
  experimental physics, I.14
  finite electron temperatures, III.1624–5
  free electrons, I.54
  gravitation, II.1266
  heat engines, I.534
  line spectra, I.77
  metallic conduction, III.1284
  photoelectron emission, III.1396
  Recollections and Reflections, I.27, I.36
  second wrangler, I.11
  x-ray scattering, I.362
Thomson, R, materials physics, III.1558
Thomson, W, *see* Kelvin, Lord
Thonemann, P C
  fusion reactions, III.1642
  ZETA experiment, III.1641
Thorne, K S, binary sources, III.1756
Thouless, D J, topological long-range order, I.571
Tiktopoulos, singularities in complex angular momentum, II.682
Ting, S C C
  dilepton mass spectrum, II.701
  J resonance, II.702
Tisza, L, two-fluid model, II.927
Tizard, Sir Henry, atomic energy project, II.1198
Todd, A R, nucleic acids, I.504
Todd, Lord, polymerization, III.1550
Todhunter, I, research and genius, I.12

Toennies, J P, helium atom scattering, II.1006
Tolansky, S
  FECO interferometry, III.1412
  microtopography, III.1412
  optical interferometry, III.1411
  rings lens-plate experiment, III.1412
Tollmien, W
  parallel flows, II.828
  stability calculation, II.829
Tolman, R C
  particle dynamics, I.279
  quantum physics, III.1401–2
  spherically symmetric perturbations, III.1801
Tolpȳgo, K B, covalent crystals, II.1002
Tomonaga, S
  β-decay, I.402
  cosmic-ray particles, I.407
  coupling theory, I.407
  quantum electrodynamics, II.643, II.740, III.1446
  renormalization, II.646
  slow mesotrons, I.406
  Twelfth Solvay Conference, Brussels, II.646
Tonks, L, plasma oscillations, III.1624
Totsuji, H, clusters of galaxies, III.1771
Toulouse, percolation processes, I.573
Touschek, B, storage rings, II.740
Townes, C H, **III.1420**
  'dressed' atomic states, III.1421
  electric quadrupole moments, II.1214
  Fabry–Perot resonator, III.1429
  lasers, III.1423
  maser, III.1419, III.1422
  stimulated Brillouin scattering, III.1444
Townsend, A, 'big eddies', II.839

Townsend, J S, ionization detectors, II.746
Toya, T, dispersion curves in sodium, II.999
Trahin, M, perturbation theory, III.1473–4
Treiman, S B
  A Century of Particle Theory, II.770
  β-decay, II.676
  pion decay constant, II.675
Treloar, L R G, rubber elasticity, III.1551
Tremaine, S D
  galactic cannibalism, III.1774
  phase space constraints, III.1804
Trimble, V L, binary sources, III.1756
Trowbridge, J
  electrolytic cells, I.27
  practical classes, I.15
Trump, J G
  electrostatic generator, II.1203
  High Voltage Engineering (HVE), II.1203
Tsoi, V S, bismuth crystal, III.1344–5
Tsu, R
  coupling of quantum wells, III.1459
  quantum-well superlattices, III.1453
Tuck, J L
  magnetic cusp, III.1645
  magnetic pinch, III.1636
Turlay, R, *CP* violation, II.672
Turnbull, D
  age-hardening, III.1533
  nucleation kinetics, III.1547
Turner, R C, flaw detector, III.1911
Tuve, M A
  ionospheric conducting layer, III.1988
  ionospheric plasmas, III.1621
  ionospheric sounding, III.1995

radio receiving equipment, III.1990
van de Graaff generator, II.1203, II.1204
Tweet, D J, surface crystallography, I.469
Twiss, R Q
  optical spectroscopy, III.1421–2
  radio interferometer research, III.1422
Tyson, A, galaxies, III.1797

Uchling, E A, collision term in Boltzmann equation, I.602
Uhlenbeck, G E, **I.594**
  Brownian motion coupled harmonic oscillators, I.608
  collision term in Boltzmann equation, I.602
  Critical Phenomena, I.553
  electron spin, I.101
  First Shelter Island Conference, II.645
  graph theory in statistical mechanics, I.553
  magnetic moment, II.1123
  Master equation, I.593–605
  quantum number, I.168
  random processes, I.628
  spinning electron, I.360
  universality, I.561
Ullmaier, H, point defects, III.1534
Ulrich, R K
  solar model, III.1749
  solar oscillations, III.1750
Underwood, E, *Quantitative Stereology*, III.1549–50
Unsöld, A, hydrogen abundance, III.1701
Upatnieks, J, holography, III.1415, III.1437
Upatnieks, Y J, off-axis optical beam holography, III.1585
Urey, H C
  chemical elements, III.1747
  deuterium, III.1706
  deuteron, II.650, II.1183
  solar system, III.1963

Valasek, J, Rochelle salt, I.489
Vallarta, M S, magnetic-mirror effect, III.1637
Van Allen, J A
  Geiger counter, III.2007
  magnetospheric physics, III.2001
  Radiation Trapped in the Earth's Magnetic Field, III.1683
  rocket sounding experiments, III.1682–3
  Third International Polar Year, III.2006
Van Allen belts, III.1636, III.1683, III.1684
van Cittert, P H, optical coherence, III.1407–8
Van, V, x-ray diffraction, I.498
van de Graaff, R J
  electrostatic generator, II.727–8, II.1186, II.1203
  High Voltage Engineering (HVE), II.1203
van de Hulst, H, radio astronomy, III.1762–3
van der Burg, M G J, field equations, I.313
van der Meer, S
  antiprotons, II.707
  field particles W and Z, II.744
  stochastic cooling, II.743
van der Waals, J D
  doctoral thesis, I.48
  kinetic theory of gases, I.543
  quantum mechanics, II.1042
  spinodal, III.1547–8
Van der Waerden, B, group theory, I.213
van Heel, A C S, fibre optics, III.1419–20
van Hove, L

magnetic correlation theory, II.1016
neutron frequency, III.1418
phonon dispersion relations, II.993
Twelfth Solvay Conference, Brussels, II.646
Van Kampen, N, linear response assumption, I.627
van Laar, J J, statistical mechanics, I.551–4
van Leeuwen, J H
  magnetic properties, II.1115
  theory of metals, III.1286–7
van Maanen, A
  bright spirals M31 and M33, III.1717–18
  nebulae, III.1718
van Rhijn, P J, luminosity function, III.1714
Van Vleck, J H, **II.1131**
  Bohr theory, II.1123–4
  crystal field theory, II.1132
  ferromagnetic resonance, II.1153
  First Shelter Island Conference, II.645
  H–D exchange, II.1140
  magnetic ions, II.1128–9
  molecular orbital theory, II.1143
  Quantum Principles and Line Spectra, II.1131
  rare-earth ions, II.1120
  relaxation theory, II.1146
  spectral line intensity, I.172
  spin–orbit interactions, II.1136
  Theory of Electric and Magnetic Susceptibilities, II.1131
  two-electron system, I.182
van't Hoff, J H, reaction isochore equation, I.535
Varicak, V, kinematic space, I.278
Vashakidze, M A, optical continuum of nebula, III.1774–5

Vassall, A, physics in education, I.9
Vaughan, R A, relaxation time, II.1133
Vavilov, S I
  absorption of light by uranium glass, III.1403
  Milrostructura Sveta, III.1403
  non-linear optics, III.1403
  polarization of fluorescent light, III.1400
  Raman effect, III.1400
  visible light, III.1409
Veksler, V, ion acceleration, II.730
Velikovsky, I, *Worlds in Collision*, III.2013
Veltman, M, electroweak parameters, II.725
Veneziano, G
  Dolen–Horn–Schmid duality, II.683
  scattering amplitude, II.683–4
Venkatesen, T N C, optical bistability, III.1449
Verne, J, science fiction, I.3
Verschaeffelt, J E, classical critical exponents, I.544
Verschaffelt, E, Fifth Solvay Conference (1927), I.205
Verwey, E J W
  charge effects, III.1611
  electrical properties, II.1156
  semiconductors, III.1330
Vigier, J P, matter-wave theory, I.229
Villard, P
  γ-rays, I.54
  ionization, III.1880
  nuclear γ-rays, I.367
Vine, F, magnetic variations of ocean floors, III.1975
Vinen, W F, vibrating-wire experiment, II.931
Vinogradov, A V, laser amplification, III.1466

Virasoro, M, Dolen–Horn–Schmid duality, II.683
Vlasov, A, Landau damping, III.1624, III.1625
Voigt, W
 electron optics, III.2000
 electronic theory of magnetism, II.1114
 magneto-ionic theory, III.1990
 school of physics, I.15
Volkoff, G, neutron stars, III.1713
Volmer, M, nucleation kinetics, III.1547
von Ardenne, scanning transmission microscope (STEM), III.1569, III.1587–9
von Baeyer, O
 β-ray deflection, II.1201
 β-spectrum, I.106
 radioactive decay, I.105
von Borries, B
 polepiece lens, III.1567
 transmission electron microscope (TEM), III.1575–7
von Braun, W, rocket technology, III.1738
von Federov, E, symmetry types, I.423
von Groth, P, crystallography, I.421
von Gutfeld, R J, thermal conductivity, II.982
von Hippel, A, dielectric properties of BaTiO$_3$, I.486
von Humboldt, A, *Gesellschaft deutscher Naturforscher und Ärzt*, I.5
von Kampen, N
 BBKGY hierarchy, I.624
 random walks, I.624
von Kármán, T
 BvK theory, II.985
 crystal lattice vibrations, I.153
 lattice dynamics, I.452
 sound from flows, II.850
 specific heat, II.971–2
 turbulence, II.832
 vibrational specific heat of solids, I.525
von Klitzing, K
 plateaux studies, III.1375
 semiconductors, II.1258
von Laue, M, **I.426**
 crystal structures, III.1514
 electromagnetic wave theory, I.363
 Fresnel's dragging coefficient, I.278
 imaging theory, III.1418
 lattice structure, I.153
 partial coherence, III.1392
 superconductivity, II.918–19
 x-ray diffraction, I.424, I.425
von Neumann, J
 energy surface trajectories, I.599
 First Shelter Island Conference, II.645
 materials research, III.1558
 Mathematical Foundations, I.227
 microphysics, I.225
 non-bounded operators, I.197
 quantum mechanics, I.198, I.206
 Schrödinger equations, I.205
von Rebeur-Paschwitz, E, earthquake vibrations, III.1953–4
von Schweidler, E, decay law of radioactive substances, I.222
von Smoluchowski, M, Brownian motion, I.586
von Weizsäcker, C F
 carbon–nitrogen–oxygen cycle (CNO cycle), III.1707–8
 monistic theories, III.1963
 nuclear energies, II.1191
von Wroblewski, S, liquefied gases, I.27
Voss, H D, *Lightning-induced electron precipitation*, III.1686–7

Wagner, vacancies, III.1529
Wagner, E, x-ray spectra, I.160
Wagoner, R V
  Big Bang, III.1792
  nuclear interactions between light nuclei, III.1791
Wahl, A C, plutonium isotope, II.1223
Wainwright, T E
  autocorrelation function of tagged particle, I.619
  molecular dynamics, I.577
Wakefield, J, holograms, III.1584
Wald, G, rod-vision, III.1416
Wali, K C, *Chandra: a Biography of S Chandrasekhar*, III.1711
Walker, A G
  Robertson–Walker metric, III.1729–30
  synchronization of clocks, III.1729
Walker, Sir Gilbert, systematic shifts of climate, II.907
Walker, J G
  antibunched light, III.1460
  optical data processing, III.1461
Walker, M F, star cluster NGC2264, III.1766–7
Walkinshaw, W, linear accelerators, II.1206
Walls, D F, quantum non-demolition (QND) measurements, III.1469
Walraven, T, Palomar 200" Telescope, III.1775
Walter, W F, germanium junctions, II.1208
Walton, E T S
  disintegration experiments, II.1183, II.1184
  disintegration of lithium, II.729
  electrostatic generators, II.727
  fusion reactions, III.1628
  nuclear reaction, I.285
  nuclear transformation, I.112
  proton acceleration, I.370
Wambacher, H, cosmic-rays tracks, I.404
Wandel, A, quasars, III.1780
Wang, C C, He–Ne laser, III.1435
Wang, Ming Chen
  Brownian motion coupled harmonic oscillators, I.608
  random processes, I.628
Wannier, G
  multi-electron fragmentation, II.1065
  optoelectronic properties, III.1410
  semiconductors, III.1320
Wannier, G H, specific heat, I.555
Warburg, E
  First Solvay Conference (1911), I.147
  quantum number, I.162
Ward, F A B, ionization chambers, II.1200
Ward, J C, quantum theory, III.1413
Ward, W R, Moon model, III.1966
Wardlaw, R S, crystal fields, II.1165
Ware, A A, ZETA experiment, III.1641
Warren, B E, amorphous structure, I.478
Wasserburg, G J, supernova trigger hypothesis, III.1963
Watanabe, Y, semiconductor maser, III.1423
Waterman, P C, electromagnetic scattering, III.1446
Watson, G N, radio waves, III.1987
Watson, J D
  DNA model, I.507
  double-helix structure, I.504
  protein subunits in viruses, I.509
Watson-Watt, R, upper atmosphere, III.1988
Watt, R W, ionosphere, III.1981

Weaire, D, amorphous structures, I.477
Weaver, W, molecular biology, I.496
Weber, A, nucleation kinetics, III.1547
Weber, H F, specific-heat quanta, I.152
Weber, R L, maser principle, III.1417
Weber, W E
  magnetic moments, II.1114
  Zeeman effect, II.1114
Wegener, A L, **III.1968**
  continental drift, III.1967–9
Wegner, E J, universality, I.571
Weidlich, W, laser theory, III.1444
Weigert, F, polarization of fluorescent light, III.1400
Wein, First Solvay Conference (1911), I.147
Weinberg, D L, photon coincidence, III.1435
Weinberg, S, II.696
  Dreams of a Final Theory, II.770
  electroweak unification, II.695
  social construction, I.405
Weinrich, M, parity violation, II.666
Weis, P E, **II.1119**
  Le Magnetisme, II.1119
Weiss, M T, crystal structures, II.1158
Weiss, P
  critical point, I.545
  magnetic fields, II.1117–18
  molecular field hypothesis, I.545
Weissenberg, K, moving-film methods, I.434
Weisskopf, V F
  compound nucleus, II.1219
  electromagnetic field, I.391
  First Shelter Island Conference, II.645

positrons in electron self-energy, II.641
proton scattering, II.1212
quantum theory, I.221
The Joy of Insight, II.770
Welford, W T, compound parabolic concentrator (CPC), III.1446
Wells, H G, science fiction, I.3
Wells, P N T, ultrasonic imaging, III.1914
Wendt, G, longitudinal effect, I.162
Wentzel, G
  asymptotic linkage, I.197
  atomic theory, I.179, I.219
  Fermi field, I.378
  Twelfth Solvay Conference, Brussels, II.646
Wenzel, W A, spark chamber, II.753
West, J, structure of beryl, I.440
Westerhout, G, neutral hydrogen, III.1763
Westgren, J, single-crystal methods, I.435
Weyl, H, I.207
  eigenfunctions with continuous spectra, II.1025
  gauge invariance, I.321
  group theory, I.213
  Weyl's postulate, III.1729
Weymann, R J, electron scattering, III.1801
Whaling, W, resonance state of nucleus, III.1746–7
Wharton, C B, microwave interferometer, III.1659
Wheeler, J A
  First Shelter Island Conference, II.645
  fission process, I.132, II.1195
  hydrogen bomb, III.1636
  S-matrix theory, II.680
  two-photon correlation, III.1413
Whitcomb, R T, area rule, II.881

N81

White, D L, electron–phonon interaction, III.1339
White, J W, neutron beams, I.468
Whitham, G B
  action-conservation principle, II.862
  'equal area' law, II.847
  'exploding wire' phenomenon, II.848
  water motions, II.860
Wick, G C
  nuclear exchange-force model, I.376
  quantum theory, I.221
  Twelfth Solvay Conference, Brussels, II.646
Wideröe, R
  resonant cyclic accelerators, II.729
  The Infancy of Particle Accelerators, II.770
Widom, B
  phase transitions, III.1830
  van der Waals equation, I.561
Wiechert, E
  cathode ray tube, III.1566
  Earth model, III.1953
Wiedlich, W, laser theory, III.1444
Wiegel, red blood cells, I.630
Wieman, C, parity violation, II.699
Wien, W
  atomic vibrations, III.1287
  black-body radiation, I.22, I.53, I.149
  exponential law, I.66–7
Wiener, N
  Brownian process, III.1830
  quantum mechanics, I.192
Wiersma, E C, potassium chromic alum, II.1126
Wiese, W L, *Needs, Analysis and Availability of Data for Space Astronomy*, II.1093
Wightman, A S, Twelfth Solvay Conference, Brussels, II.646

Wigner, E
  band structure, III.1314–17
  electron band structure of sodium, III.1530
  electron gas, III.1361–2
  neutron capture, II.1194
  nuclear resonance, I.126
  reaction cross sections, II.1219
  Schrödinger's equation, III.1315
  Twelfth Solvay Conference, Brussels, II.646
  Wigner function, I.617–18
Wilberforce, S, Darwinian evolution, I.5
Wilczek, F, running coupling constant, II.713
Wild, J J, pulse-echo ultrasound, III.1911–12
Wilkins, M H F, DNA, I.504
Wilkinson, D H
  Ionisation Chambers and Counters, II.1207
  pulse height conversion to pulse time, II.751
  shell-model theory, II.1214
Williams, E J, order–disorder transitions, I.474
Williams, R C, Lamb shift, II.641
Williamson, A, atomic theory, I.46
Wilson, A, science jottings, I.3
Wilson, A J C
  line broadening, I.472
  structure factors, I.439
Wilson, C T R
  alpha/beta particle tracks, II.1200–1
  cloud chamber, I.54
  droplet formation, II.751–2
  Fifth Solvay Conference (1927), I.205
  radiation, I.367
Wilson, H A, electron charge, II.747
Wilson, H F, special orientation relationships, III.1528

Wilson, H R, helical structures, I.498
Wilson, J M
 physics in education, I.9
 senior wrangler, I.11
Wilson, J T, plate tectonics, III.1978
Wilson, K
 computers, III.1834
 field theory, III.1830, III.1833–34
 fixed point, III.1832, III.1833–34
 magnetism, III.1834
 non-perturbative calculations, III.1825
Wilson, K G, **I.565**
 renormalization group, I.564–9
Wilson, R, microwave background radiation, III.1790
Wilson, R R, **II.735**
 Fermilab accelerator, II.734
Wilson, R W
 carbon monoxide molecule, III.1763–4
 microwave radiometer, II.758
 radiation, III.1786
Wilson, W
 electron beams, I.104
 quantum rule, I.175
Winston, R, compound parabolic concentrator (CPC), III.1446
Winther, A, Coulomb excitation experiments, II.1217
Wintner, A, bounded matrices, I.206
Wirtz, C W, spiral galaxies, III.1725
Wisdom, J
 chaotic evolutions, III.1823
 computing, III.1831
Witkowski, J, biomolecular structures, I.496
Wolf, D, melting, III.1533–4
Wolf, E
 Airy disc, III.1412
 coherence properties of partially polarized light, III.1411
 diffraction images, III.1419
 energy transport, III.1459
 optical coherence theory, III.1392, III.1421
 quantum theory of coherence, III.1439
Wolfe, J P, thermal conductivity, II.982
Wolfke, M
 two-stage imaging, III.1583
 x-ray microscope, III.1414
Wollan, E O
 neutron diffraction, I.453
 structure of ice, I.475
Wollaston, W H
 medical physics, III.1857
 solar spectrum, III.1691
 spectral lines, I.156
Wollfson, M M, determinantal inequalities, I.448
Wood, R W
 black-body radiation, I.150, III.1387
 diffraction gratings, III.1397
 direct-drive approach to inertial fusion, III.1676
 fluorescence and photochemistry of materials, III.1401
 Physical Optics, III.1391
 resonance radiation, III.1390
Woodbury, E J, ruby laser, III.1432
Woodruff, D P, surface crystallography, I.466
Woods, A D B, shell model, II.997
Woodward, J J, linear accelerators, II.736
Woolfson, M M
 probability relations, I.449
 structure determination, I.450
Wooster, W, continuous spectrum, I.118, I.126
Wright brothers, powered flight, II.873
Wright, T, island universes, III.1714

N83

Wrinch, D M, Patterson function, I.444
Wu, C S
 parity violation, II.666
 scintillation counters, III.1414
Wu, L-A, squeezed light, III.1471
Wu, S L, gluon jets, II.716
Wu, T T, CP violation, II.673
Wül, B, dielectric properties of BaTiO$_3$, I.486
Wyckoff, R W G
 biomolecular structures, I.496
 crystal structures, III.1513
Wyler, A A, magneto-ionic theory, III.1992
Wynn-Williams, C E
 binary counting circuits, II.748
 ionization chambers, II.1200

Xiao, M, squeezed light, III.1471

Yalow, R, saturation analysis technique, III.1909
Yamaguchi, Y, decay reaction, II.657
Yan, T-M, lepton pair, II.690
Yang, C N
 CP violation, II.673
 isotopic spin symmetry, II.693
 lattice gas model, I.553
 local gauge invariance, II.693
 neutrinos, II.671
 parity conservation, II.665
 parity violation, II.666
 self-interacting fields, II.661
 spin-1 particle, II.654
Yardley, K, x-ray diffraction, I.439
Yariv, A
 guided wave optics, III.1458
 optical waveguides, III.1459
 Quantum Electronics, III.1436
Yeh, Y, Doppler shift, III.1442
Yevick, random walk model, I.630

Yonezawa, F, amorphous structures, I.477, I.479
York, D G, interstellar deuterium, III.1791
York, H
 fusion research, III.1637
 magnetic mirror, III.1659
Yosida, K, RKKY interaction, II.1164
Young, angular distributions, II.1220
Young, F Jr, materials physics, III.1558
Young, J Z, flying-spot microscope, III.1416–17
Young, T
 dimensions of molecules, I.48
 medical physics, III.1857
 wave description of light, I.72
Yuen, H P, minimum-uncertainty packets, III.1469
Yukalov, V I, melting, III.1533–4
Yukawa, H, **I.390**
 exchange forces, I.231
 meson theory, I.378–83, I.389, II.640, II.651–2, II.669
 neutral pions, II.654
 slow mesotrons, I.406
 spin-1 boson, II.692
 Twelfth Solvay Conference, Brussels, II.646
 U-particle hypothesis, I.396
Yuniev, defibrillators, III.1895
Yvon, J, BBKGY hierarchy, I.606, I.623, I.624, I.626

Zachariasen, W H
 amorphous structures, I.477
 extinction, I.451
 structure of metaboric acid, I.449
Zahradnicek, J, gravitation, II.1266
Zanstra, H, planetary nebulae, III.1762

Zavoisky, E, paramagnetic resonance, II.1137–8
Zeeman, P
  magnetism, III.1390
  splitting of spectral lines, I.161
  Zeeman effect, I.54
Zehnder, L, x-ray photograph, I.4
Zeiger, H J, maser, III.1419
Zel'dovich, B Ya, phase conjugation, III.1458
Zeldovich, Ya B
  Big Bang, III.1790
  binary sources, III.1756
  black holes, III.1779, III.1780
  collapsing cloud, III.1803
  cosmological constant, III.1723, III.1806
  induced Compton scattering, III.1801
  neutrinos, III.1803
  photon scattering, III.1773
  photon to baryon–antibaryon ratio, III.1806
  structures in the Universe, III.1801, III.1802
  synthesis of heavy elements, III.1792
  weak vector current, II.668
Zeller, R C, disordered systems, II.1006
Zener, C
  Fermi surface, III.1295
  valence conduction bands, III.1337
Zenneck, J
  radio propagation, III.2000
  radio waves, III.1987
Zermelo, E, process irreversibility, I.524
Zernike, F
  circle polynomials, III.1409
  critical correlations, I.563
  optical coherence, III.1407–9
  pair distribution function, I.547
  phase-contrast microscopy, III.1409–10, III.1419, III.1475
  x-ray diffraction pattern, I.441
Zheludev, N I, polarization symmetry breaking, III.1467
Zherikhin, A, laser amplification, III.1466
Zienau, S, lattice deformation, III.1321
Ziman, J M
  electrons and holes, III.1297
  Electrons and Phonons, III.1305
  lattice conduction of heat in metals, II.981
  pseudopotential concept, III.1367
Zimkina, T M, K-shell absorptions, II.1083
Zimm, B, polymer solutions, III.1597
Zimmerman, G, direct-drive approach to inertial fusion, III.1676
Zinn, W H, neutron diffraction, I.453
Zinn-Justin, J, renormalizability, II.695
Zobernig, G, gluon jets, II.716
Zoll, P
  cardiac pacemakers, III.1895
  defibrillators, III.1895
  electrical stimulation, III.1895
Zupancic, C, Coulomb excitation experiments, II.1217
Zvyagin, B B, oblique texture method, I.464
Zwanziger, singularities in complex angular momentum, II.682
Zweig, G, fractionally charged constituents of matter, II.685
Zwicky, F
  Catalogue of Selected Compact Galaxies and of Post-Eruptive Galaxies, III.1713

clusters of galaxies, III.1731
Coma cluster, III.1732
Schmidt telescope, III.1745
supernovae, III.1712, III.1718, III.1758, III.1759, III.1775

Zworykin, V K
phosphors, III.1536–7
photomultipliers, II.750
synthetic zinc silicate, III.1537
xerography, III.1541

Twentieth Century Physics
Volume III

Twentieth Century Physics
Volume III

Edited by

Laurie M Brown
Northwestern University

Abraham Pais
Rockefeller University
and
Niels Bohr Institute

Sir Brian Pippard
University of Cambridge

Institute of Physics Publishing
Bristol and Philadelphia

and

American Institute of Physics Press
New York

© IOP Publishing Ltd, AIP Press Inc., 1995

All rights reserved. No part of this publication may be reproduced, stored in a retrieval system or transmitted in any form or by any means, electronic, mechanical, photocopying, recording or otherwise, without the prior permission of the publisher. Multiple copying is permitted in accordance with the terms of licences issued by the Copyright Licensing Agency under the terms of its agreement with the Committee of Vice-Chancellors and Principals. Authorization to photocopy items for internal or personal use, or the internal or personal use of specific clients in the USA, is granted by IOP Publishing and AIP Press to libraries and other users registered with the Copyright Clearance Center (CCC) Transaction Reporting Service, providing that the base fee of $19.50 per copy is paid directly to CCC, 27 Congress Street, Salem, MA 01970, USA

British Library Cataloguing-in-Publication Data
A catalogue record for this book is available from the British Library.

In UK and the Rest of the World, excluding North America:
ISBN   0 7503 0353 0   Vol. I
       0 7503 0354 9   Vol. II
       0 7503 0355 7   Vol. III
       0 7503 0310 7   (3 vol. set)

In North America (United States of America, Canada and Mexico):
ISBN   1-56396-047-8   Vol. I
       1-56396-048-6   Vol. II
       1-56396-049-4   Vol. III
       1-56396-314-0   (3 vol. set)

Library of Congress Cataloging-in-Publication Data are available

Published jointly by Institute of Physics Publishing, wholly owned by The Institute of Physics, London, and American Institute of Physics Press, wholly owned by the American Institute of Physics, New York.

Institute of Physics Publishing, Techno House, Redcliffe Way, Bristol BS1 6NX, UK
Institute of Physics Publishing, Suite 1035, The Public Ledger Building, Independence Square, Philadelphia, PA 19106, USA
American Institute of Physics Press, 500 Sunnyside Boulevard, Woodbury, New York 11797-299, USA

Printed and bound in the UK by Bookcraft Ltd, Bath.

CONTENTS

VOLUME III

17	**ELECTRONS IN SOLIDS**	1279
	Sir Brian Pippard	
	17.1 The need for high-quality materials	1280
	17.2 Experimental facts concerning metals	1281
	17.3 Primitive models of metallic conduction	1284
	17.4 Quantum theory of the electron gas	1287
	17.5 Bloch's theorem and its immediate consequences	1291
	17.6 Mechanisms of resistivity	1301
	17.7 Umklapp processes	1304
	17.8 Thermoelectric effects	1306
	17.9 Interlude—Brillouin zones	1311
	17.10 From generalities to particulars; the emergence of solid-state schools	1313
	17.11 Early calculations of band structure	1314
	17.12 Semiconductors—the early years	1317
	17.13 Rectifiers and transistors	1322
	17.14 Photoconductivity	1327
	17.15 Semiconductor physics	1330
	17.16 Hot electrons	1336
	17.17 Metals in the post-war years—the impact of liquid helium	1341
	17.18 Metals as individuals—the Fermi surface programme	1345
	17.19 Beyond the independent-particle model	1359
	17.20 Disordered materials	1365
	17.21 Interference effects	1369
	17.22 The two-dimensional electron gas; the quantum Hall effect	1371
	17.23 Epilogue	1376

18 A HISTORY OF OPTICAL AND OPTOELECTRONIC PHYSICS IN THE TWENTIETH CENTURY   1385
R G W Brown and E R Pike

18.1 Introduction: classical optics up to 1900   1385
18.2 1900–30: early quantum optics   1387
18.3 1930–60: the calm before another storm   1406
18.4 1960–70: the laser and non-linear optics   1424
18.5 1970–94: a wealth of new results and applications   1452
18.6 1994 to 2000: towards the twenty-first century   1472

19 PHYSICS OF MATERIALS   1505
Robert W Cahn

19.1 The establishment of the foundations of a science of materials   1507
19.2 Reference works and data compilations   1509
19.3 Crystal structures   1513
19.4 The birth of physical metallurgy   1516
19.5 The birth of quantitative theory in physical metallurgy   1519
19.6 Point defects and non-metallic materials   1528
19.7 Diffusion   1531
19.8 Electrical and electro-optical ceramics and liquids   1536
19.9 Phase transformations, microstructure and modern instrumentation   1542
19.10 The physics of polymers   1550
19.11 Terms, concepts and institutions   1553

20 ELECTRON-BEAM INSTRUMENTS   1565
T Mulvey

20.1 Early days   1565
20.2 The magnetic electron lens   1566
20.3 Verification of Busch's theory   1569
20.4 The first two-stage electron microscope   1570
20.5 Surpassing the resolving power of the light microscope   1574
20.6 Specimen preparation and radiation damage   1574
20.7 The first serially produced TEM   1575
20.8 Electrostatic electron microscopes   1577
20.9 Abbe's optical theory and the TEM   1578
20.10 Off-axis optical beam holography   1585
20.11 Aberration-free atomic resolution in a TEM by holography   1585
20.12 Scanning electron microscopes   1586
20.13 Conclusion   1590

21	**SOFT MATTER: BIRTH AND GROWTH OF CONCEPTS**	1593
	P G de Gennes	
	21.1 The meaning of 'soft'	1593
	21.2 Polymers	1596
	21.3 Liquid crystals	1603
	21.4 Surfactants	1606
	21.5 Colloids	1610
	21.6 Concluding remarks	1613
22	**PLASMA PHYSICS IN THE TWENTIETH CENTURY**	1617
	Richard F Post	
	22.1 Introduction	1617
	22.2 Plasma physics in the first half of the twentieth century	1619
	22.3 Langmuir and plasma oscillations: Landau and plasma theory	1623
	22.4 The fusion and space-plasma era: circa 1950 to the present	1627
	22.5 Magnetic fusion research post-1960: the long march up the $n\tau T$ slope	1645
	22.6 The growth in importance of theory and computer simulation	1670
	22.7 The other end of the scale: inertial confinement fusion	1673
	22.8 Landmark experiments: general plasma physics	1677
	22.9 Space plasma physics, the ionosphere and beyond	1682
	22.10 Conclusion	1687
23	**ASTROPHYSICS AND COSMOLOGY**	1691
	Malcolm S Longair	
	Part 1 STARS AND STELLAR EVOLUTION UP TO WORLD WAR II	1691
	23.1 The legacy of the nineteenth century	1691
	23.2 The origin of the Hertzsprung–Russell diagram	1692
	23.3 Stellar structure and evolution	1699
	Part 2 THE LARGE-SCALE STRUCTURE OF THE UNIVERSE 1900–39	1714
	23.4 The structure of our Galaxy	1714
	23.5 The great debate	1716
	23.6 The development of relativistic cosmology	1721
	Part 3 THE OPENING UP OF THE ELECTROMAGNETIC SPECTRUM	1734
	23.7 The changing astronomical perspective	1734

Part 4   ASTROPHYSICS AND COSMOLOGY SINCE 1945   1746
23.8   Stars and stellar evolution since 1945   1746
23.9   The physics of the interstellar medium   1761
23.10 The physics of galaxies and clusters of galaxies   1769
23.11 High-energy astrophysics   1774
23.12 Astrophysical cosmology   1784
23.13 The classical cosmological problem   1793
23.14 Galaxy formation   1800
23.15 The very early Universe   1805

24 COMPUTER-GENERATED PHYSICS   1823
Mitchell J Feigenbaum

25 MEDICAL PHYSICS   1855
John R Mallard
25.1   Introduction: chronological evolution of the many branches of the field   1855
25.2   X-rays in diagnosis: diagnostic radiology   1859
25.3   X-rays in treatment: radiotherapy   1863
25.4   Radioactivity in therapy   1876
25.5   Radiation dosimetry   1880
25.6   Radiation biophysics: radiobiology   1884
25.7   Radiation protection   1889
25.8   Medical electronics   1893
25.9   Medical mechanics   1896
25.10 Radioactivity in diagnosis: nuclear medicine   1900
25.11 Medical ultrasonics   1910
25.12 Computed tomography   1916
25.13 Nuclear magnetic resonance imaging   1923
25.14 Conclusion   1933

26 GEOPHYSICS   1943
S G Brush and C S Gillmor
26.1   Introduction   1943
26.2   Origin and age of the Earth (to 1935)   1944
26.3   The Earth's core and magnetism   1952
26.4   Origin and age of the Earth (after 1935)   1960
26.5   The 'revolution in the earth sciences'   1967
26.6   The Earth's upper atmosphere and geospace   1981
26.7   The ionosphere: the early days   1984
26.8   Outwards into the magnetosphere and solar–terrestrial space   2001

27	**REFLECTIONS ON TWENTIETH CENTURY PHYSICS: THREE ESSAYS**	2017
	Historical overview of the twentieth century in physics *Philip Anderson*	2017
	Nature itself *Steven Weinberg*	2033
	Some reflections on physics as a social institution *John Ziman*	2041
	Illustration acknowledgments	A11
	Journal abbreviations	J1
	Subject index	S1
	Name index	N1

VOLUME I

	Preface	ix
	List of contributors	xiii
	Biographical captions	xvii
1	PHYSICS IN 1900 *Sir Brian Pippard*	1
2	INTRODUCING ATOMS AND THEIR NUCLEI *Abraham Pais*	43
3	QUANTA AND QUANTUM MECHANICS *Helmut Rechenberg*	143
4	HISTORY OF RELATIVITY *John Stachel*	249
5	NUCLEAR FORCES, MESONS, AND ISOSPIN SYMMETRY *Laurie M Brown*	357
6	SOLID-STATE STRUCTURE ANALYSIS *William Cochran*	421
7	THERMODYNAMICS AND STATISTICAL MECHANICS (IN EQUILIBRIUM) *Cyril Domb*	521

8	**NON-EQUILIBRIUM STATISTICAL MECHANICS OR *THE VAGARIES OF TIME EVOLUTION*** Max Dresden	585
	Illustration acknowledgments	A1
	Subject index	S1
	Name index	N1

VOLUME II

9	**ELEMENTARY PARTICLE PHYSICS IN THE SECOND HALF OF THE TWENTIETH CENTURY** Val L Fitch and Jonathan L Rosner	635
10	**FLUID MECHANICS** Sir James Lighthill	795
11	**SUPERFLUIDS AND SUPERCONDUCTORS** A J Leggett	913
12	**VIBRATIONS AND SPIN WAVES IN CRYSTALS** R A Cowley and Sir Brian Pippard	967
13	**ATOMIC AND MOLECULAR PHYSICS** Ugo Fano	1023
14	**MAGNETISM** K W H Stevens	1111
15	**NUCLEAR DYNAMICS** David M Brink	1183
16	**UNITS, STANDARDS AND CONSTANTS** Arlie Bailey	1233
	Illustration acknowledgments	A7
	Subject index	S1
	Name index	N1

Chapter 17

ELECTRONS IN SOLIDS

Sir Brian Pippard

Some properties of solids are essentially conferred by the individual atoms or molecules, while others appear only when large numbers are tightly packed together. Thus the electric polarizability of most non-conductive solids is largely due to atomic distortions in the presence of an electric field; solid argon has a higher permittivity (dielectric constant) than gaseous argon solely because there are more atoms per unit volume. The electrical conductivity of a metal or semiconductor, on the other hand, comes about because electrons are readily detached from the closely packed atoms, and roam freely even in the absence of an electric field. This chapter is concerned with the latter class of solids, though the occurrence of electric breakdown in a very strong field brings insulators into the story, if only as minor characters. Two phenomena, ferromagnetism and superconductivity, are sufficiently important to be given separate treatment. What remains is at least as important. Theoretical understanding has often been elusive, and conflicting explanations have often been suggested, but disagreements have not grown into confrontations. The history of solid-state physics may be thought to lack dramatic interest, being a tale of fairly steady progress, with the assimilation of experimental facts into a progressively more comprehensive theoretical framework, and the systematic exploitation of scientific results in the form of commercial products. Its success lies at the heart of modern industrial civilization, which could not have developed, and cannot survive, without solid-state electronics to provide the means of communication and control. The technology, however, is so vast in scope and variety as to deserve a historical account far beyond anything that is possible here, and the emphasis in what follows is on physical ideas.

See also
p 1111
p 913

It is easy to gain the impression from student texts that the subject is almost wholly theoretical; that once one accepts the idea of electrons moving through the crystalline lattice of the positive ions, the details of their behaviour can be deduced by the application of quantum mechanics. To a very limited degree this is true, but one should not overlook the extreme difficulty, especially in the days before computers, of solving any but the simplest quantum-mechanical problems. By good fortune one or two successes were achieved, after the quantum revolution that took place in the years following 1925, with highly simplified models—so highly simplified indeed that without the confirmation of experimental data no one would have dared suggest them as a serious contribution to physics. These models, incorporating the assumption that despite their electrical charges electrons in a metal can move around much like particles in a perfect gas, are to be found far back in history, even before the electron as conceived by Lorentz in 1892, and named by Stoney in 1894, let alone that discovered by J J Thomson in 1897. The early theories, based on classical mechanics and a good measure of wishful thinking, were persuasive enough to encourage continued development, yet in sufficient disagreement with the facts to engender equally dubious alternatives, until in the late 1920s quantum mechanics brought enlightenment.

If we are to understand the events of the first quarter of the twentieth century we must appreciate the experimental information on which the primitive models were based. A large proportion of the available data was of doubtful quality. It was in those times that there arose the legend of how chemists make imprecise measurements on pure samples, while physicists make precise measurements on impure samples. As a preliminary, let us note a few stages by which physicists conquered this weakness.

17.1. The need for high-quality materials [1]

Before the development of x-ray crystallography the word *crystal* probably conjured up in most minds a faceted lump of mineral or salt, rather than an arrangement of atoms. It is not surprising, therefore, that little attention was given to metal crystals, since very few faceted examples were known—notably bismuth which readily crystallizes and can be cleaved to reveal some of its principal planes. Etching of polished metal surfaces showed the microcrystalline structure of cast specimens, but for the most part measurements on metals were made with imperfectly characterized polycrystalline samples. It was only in 1921 that Carpenter and Elam [2] systematically investigated the process of straining, followed by annealing, which caused aluminium plates to grow sizeable crystallites. In 1924 Obreimow and Schubnikow, and in 1925 Bridgman described the growing of single metal crystals by slow cooling of the melt from its lower end. The technique is currently attributed to Bridgman, but would be better ascribed to Tammann whose work Obreimow and Schubnikow acknowledge. Even before this, in

See also p 1526

1918, Czochralski had grown single-crystal wires by slow pulling of a seed out of the melt. His purpose was to study the rate of propagation of the interface between liquid and solid, rather than to produce samples for measurement. The method demanded careful temperature control and very smooth pulling, and seems to have been regarded in those days as too expensive for general use in crystal growing. With the rise of the semiconductor industry it became indispensable.

It is to the semiconductor industry also that we owe our present appreciation of the importance of high purity and precise characterization of samples. The purification and assay of precious metals is an ancient craft which by the end of the nineteenth century had extended its scope to base metals. Thus commercial aluminium, with no more than 0.4% of impurities, was available to Carpenter and Elam, and this was probably quite pure enough for their purpose; but in 1949 Bardeen and Pearson showed that one atom of boron to 10^5 of silicon changes the resistivity of the latter at 80 K by a factor of about one thousand. At lower temperatures the effects of minute amounts of impurity are still more drastic, and chemical purification is inadequate to achieve reproducible results. The process of zone-melting which Pfann first described in 1952 was probably his own invention, but a primitive form is due to Schubnikow, who noticed in 1930 that the resistance of a bismuth rod at a very low temperature is less after several recrystallizations. The *residual resistance* of a sample (strictly its resistivity at zero kelvin) is a sensitive measure of impurity content since the resistivity of a perfectly pure crystal should disappear once scattering of electrons by the thermal vibrations of the lattice has been frozen out. During the slow motion of the liquid–solid interface, as crystallization proceeds, most impurities preferentially remain in the liquid (thus seawater freezes into almost salt-free ice) and are swept out of the growing crystal. Pfann arranged to pass a thin heater along the crystal to move a molten zone through it; after many such passes the impurities were concentrated at the ends, leaving a much purer crystal in the middle.

See also p 1543

17.2. Experimental facts concerning metals [3]

We are not surprised to find, before these developments in technique, considerable disagreement in the early literature on which substances are metals and which semiconductors. The criterion was the variation of resistivity with temperature—in metals the resistivity is low and increases with rise of temperature, in semiconductors it is high and decreases. Most metals present no problem, but even as late as 1931 one can find titanium quoted as a semiconductor and silicon, if pure enough, as a metal. To the experimenters who studied semiconductors before 1939 cuprous oxide was one of the very few in which they had confidence; ten years later germanium and silicon were pre-eminent.

In so far as the resistivity of most metals varies in a very similar way with temperature, being roughly proportional to the absolute temperature except at very low temperatures, it is not a property that tells much about microscopic mechanisms. In the early years of the century much attention was focused on phenomena that seemed at the time by their exceptional character to offer hints, though later they attracted only marginal interest. One of these, however, has occupied a central position ever since its discovery by E H Hall [4], a research fellow at Johns Hopkins University, in 1879 (figure 17.1). An account of the excitement that greeted the discovery is given by Campbell [3], who describes more than 15 theories, including that of Hall himself which is of interest for its assumption, more in keeping with Weber than with Maxwell, that the current is carried by charged particles. If these suffer friction, as does a baseball in flight, and are caused to spin by the magnetic field, then like the baseball they will swerve from their straight trajectory. He is, however, at a loss to understand why the direction of swerve should be different in different metals, even those so nearly alike as iron and nickel. That the Hall effect should be positive in some metals and negative in others presented less of a problem to Riecke and Drude, who developed the first theories of metallic conduction that are recognizable ancestors of those now accepted. For in 1900 it was still possible to suppose that both positively and negatively charged carriers were moving freely in the metal; the sign of the Hall effect was determined by the dominant carrier. This happy illusion was short-lived, however; until the advent of quantum mechanics the Hall effect continued to be one of the vexatious anomalies that kept alive the interest of theorists in metallic conduction.

It did not take long for Hall's announcement to stimulate the discovery of a menagerie of related effects, each named after its discoverer, when heat flow was substituted for electric current and temperature differences for potential differences. It was found that electric currents in the presence of a magnetic field generated a transverse temperature gradient (Ettingshausen) and heat currents a transverse voltage (Nernst); still other combinations were investigated that we need neither specify nor name, and that involved also the thermoelectric effects and their modification by a magnetic field. All these, and their apparently capricious variation from one metal to another, supplemented the challenge of the Hall effect but eventually fell into obscurity, once quantum mechanics had clarified the fundamental structure responsible for all.

Most of these effects were discovered in bismuth, which had been known since 1883 to produce a Hall voltage several thousand times greater than that in gold. In 1884 it was also found to show magnetoresistivity to an exceptional degree, and was brought into use as the sensitive element in small-magnetic-field probes. When Kapitza [5] developed his strong-impulsive-field generator he found the resistance

Electrons in Solids

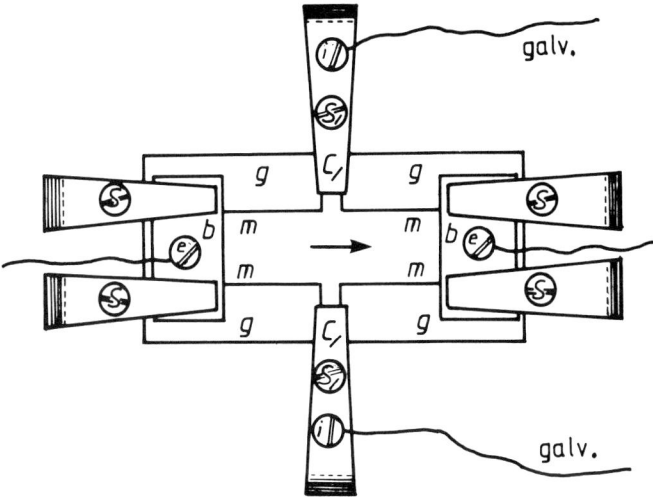

Figure 17.1. *The Hall effect (Hall's [4] picture of his experimental arrangement). A strip of gold foil, with narrow side-arms, is clamped by brass plates to the glass base. Current flows from left to right and the transverse voltage is measured with a moving-magnet galvanometer attached to the leads labelled galv. Hall scraped away the side-arms until, in the absence of a magnetic field, the galvanometer did not respond when the current flowed. On applying a magnetic field normal to the plane of the foil he found a galvanometer deflection.*

of a bismuth wire to be increased 50-fold by a transverse field of 300 kG (30 T) at room temperature, and 1400-fold at the temperature of liquid air. By contrast, a polycrystalline wire of copper exhibited an increase of only 2% at room temperature and 40% in liquid air. In both cases, and with other metals, he concluded that after an initial quadratic rise the resistance settled down to a linear variation with field strength. This was occasionally referred to afterwards as Kapitza's law, but as very few examples have survived later investigation the term has been abandoned. Magnetoresistance has some historical importance, the supposed linear variation very little; it was the fact that magnetoresistance occurred at all that excited the early theorists. After considerable disagreement they came to accept that in their preferred model it should find no place.

We turn now to the Volta effect, or contact potential, and the thermoelectric effects which have at times been assumed to be more closely linked to it than in fact they are. As far back as 1801 Volta discovered that when two different metals are electrically connected they settle down at different potentials, so there is an electric field in the surrounding space. The potential difference may amount to several volts and is easily demonstrated, but cannot be used to generate a current—it is present when the whole system is in equilibrium, unlike the potential difference between the terminals of a Voltaic cell (discovered in 1800) in which the electrodes await the opportunity of reacting chemically with their environment when a current is allowed to flow, and possess a store of chemical energy that can be turned to useful electrical purpose. Of the thermoelectric effects, the Seebeck effect (1822) manifested by the current that flows when a loop, composed of two different metals, has the junctions held at different temperatures, was extensively studied in the nineteenth century and used for temperature measurement and in sensitive radiation detectors. The Peltier heat (1834) liberated or absorbed when a current flows across a junction, and the Thomson heat (1856) liberated or absorbed when a current flows in the presence of a temperature gradient, had been related to the Seebeck effect by W Thomson. His thermodynamical argument is of dubious validity, as he recognized, but is nevertheless now known to give the right answer. The wide variability of these effects from one metal to another, as with the Hall effect, made them attractive to the experimenter and tantalizing to the theorist.

One more experimental observation that excited great interest and much fruitless theorizing was Kamerlingh Onnes' discovery of superconductivity in 1911. This story is told in another chapter and only mentioned here as a further example of information that cried out for, yet resisted, explanation and thus served only to raise doubts about early attempts to understand electrical conduction in metals.

See also p 913

17.3. Primitive models of metallic conduction [6]

Considerable progress was made despite the enigmatic data available. The first substantial theory was set forth by Riecke (1898) with the avowed intention of explaining as many as possible of the results just described. He imagined the interior of a metal to contain many species of particles, with different charges and masses, moving freely as in a gas, apart from frequent collisions with the atoms constituting the rigid metallic structure. His decision to include both positive and negative particles was influenced both by Lorentz's electron theory and by the analogy with electrolytic conductors in which, as Arrhenius had proposed, ions of both sign carried the current. No doubt the occurrence of either sign in the Hall effect also urged him in this direction. He modelled his theory on Clausius's kinetic theory of gases, like him taking all particles of the same species to move at the same speed, proportional to the square root of the absolute temperature, and supposing the path lengths between collisions with fixed atoms to be exponentially distributed—the chance of a path being longer than x is $e^{-x/\ell}$, where ℓ is the mean free path. Two years later Drude and J J Thomson, following Riecke's lead but allowing themselves the further simplification of assuming all paths for a given species to have the same length ℓ, managed thereby to introduce errors of a factor of 2 that

Electrons in Solids

aided the acceptance of their theories by giving better agreement with experiment than there should have been. There is in the work of these two a more intuitive quality that contrasts with Riecke's conscientious analysis.

Riecke calculates the electrical (σ) and thermal (K) conductivities and finds them to be related by a rather complicated formula. Drude proceeds likewise, but notices that if the charges on the particles are not arbitrary but take only the values $\pm e$ a great simplification occurs. With modernized notation, in which k_B stands for Boltzmann's constant, $K/\sigma T = 3k_B^2/e^2$. This agrees so well with experimental data taken around room temperature that it must have been quite a disappointment when Lorentz took proper account of the distribution of free paths and of velocities and replaced the coefficient 3 by 2. Twenty years later Sommerfeld's quantum theory fixed the coefficient at $\pi^2/3$, in excellent agreement with experiment and very close to Drude's lucky shot. The constancy of $K/\sigma T$ had already been noted by Wiedemann and Franz, and its theoretical derivation was the one hopeful sign that the model held promise, at a time when there seemed no way of accounting for the vagaries of thermoelectricity and the Hall effect.

To derive an expression for the Hall effect Riecke starts from Lorentz's expression for the force exerted on a moving charged particle by an electric field $\mathcal{E}$ and a magnetic field B: $F = e(\mathcal{E} + v \times B)$, where v is the velocity of the particle. Because of B, the current does not flow in the same direction as $\mathcal{E}$, so if the geometry of the sample defines the direction of current flow there must be both a component of $\mathcal{E}$ along the current, giving rise to the resistance, and a transverse component, the Hall field $\mathcal{E}_H$. Riecke's expression for $\mathcal{E}_H$ is essentially the same, when translated into modern notation, as the accepted expression for a semiconductor containing both electron and hole carriers. For a single species the Hall field is inversely proportional to the number of carriers, and the sign is determined by the sign of the charge carried. It was Riecke's misfortune that Thomson's discovery of the electron made the assumption of more than one species of carrier hard to maintain. Already in 1901 Richardson, under Thomson's influence, allowed only negative 'corpuscles', i.e. electrons, moving in a background of positive ions, and prevented from escaping by an electrical double layer at the surface which creates a difference $e\Phi$ in potential energy between the inside and the outside. According to the Maxwell–Boltzmann law a tiny fraction of the electrons will be excited thermally to a kinetic energy greater than $e\Phi$, and if they reach the surface will escape. The expression for the number escaping is dominated by the Boltzmann factor $\exp(-e\Phi/k_BT)$, and Richardson's calculation shows that if they are all collected (by a positively charged electrode close at hand) the current will vary as $T^{1/2}\exp(-e\Phi/k_BT)$. In an impressive series of measurements [7] he verified the exponential term—the prefactor $T^{1/2}$ plays so minor a role that it is hard to verify, and in fact is replaced by T^2 in the quantum version. This research provided not only a substantial contribution to the electron theory of metals but a strong theoretical underpinning to the technology of thermionic valves which came into being at the same time. From his results Richardson estimated the work-function $e\Phi$ to be 4.1 electron volts (eV) for platinum and 2.6 eV for sodium, values that he rightly found compatible with such elementary models as he could picture of the atom and the process of ionization. Later the photo-electric effect, which Millikan was the first to study with the requisite attention to surface cleanliness, provided better estimates of Φ. This story, which has more to say about the acceptance of Einstein's photon theory than about metals, is told elsewhere.

See also p 152

The strong arguments in favour of only negative charge carriers did nothing to harm the explanation of the Wiedemann–Franz law, but cast Riecke's explanation of the Hall effect to the winds. From this time until 1929, when Peierls showed how quantum mechanics could resolve the difficulty, there proliferated a succession of theories [3] that had few adherents and do not deserve our attention. It was not simply the occurrence of both signs for the effect that encouraged exotic explanations, but the magnitude also. If the electrons are to be pictured as behaving like atoms of a perfect gas—the model favoured by Lorentz and Richardson among other authorities—they account for the observations only if present in comparable numbers to the atoms (except in bismuth where far fewer are indicated). The high reflectivity of light by metals, as Drude had shown, also favours this view. On the other hand, each electron should contribute $\frac{3}{2}k_B$ to the thermal capacity, and there is no evidence for this at room temperature. So perhaps the number of electrons is much smaller. But then (leaving the Hall effect aside) we find that to account for the magnitude of the electrical conductivity of metals we must suppose the free path between collisions to be something like 50 atomic spacings—an unbelievable result in the light of any contemporary atomic model. These conflicts led Thomson to lose faith in the electron-gas model and to put forward a strange conception of electrons handed on from atom to atom—an attempt to explain away the variable Hall effect that does credit, at least, to his powers of invention.

Bohr's [8] critique (1909 and 1911) of attempts to formulate a consistent theory of metals corrects minor errors made by even such as Lorentz and Poincaré, but though it is a remarkable piece of work for a student it makes no progress in resolving the contradictions. Indeed, it adds one more difficulty to the list by showing that an assembly of charged particles, moving according to classical laws, is not diamagnetic, as had been previously supposed, but is in fact devoid of magnetic properties. Because Bohr's thesis was in Danish, this theorem remained obscure until it was rediscovered in 1919 by van Leeuwen, by whose name it is still known. Bohr's failure to find a publisher for an English version of his thesis indicates that the theory of metals was not at that time a burning issue. Nevertheless, sporadic attempts to resolve the difficulties continued to appear. When the fourth Solvay conference convened [9] in 1924 to discuss electrical conduction in metals, the participants heard masterly summaries by the 71 year old Lorentz of the older theories, and by Bridgman of the *ad hoc* adjustments, mostly flying in the face of reason, by which various optimists, himself included, had attempted to break the impasse. There is no need to reproduce what can be found in the conference proceedings, for most of these attempts follow Thomson in abandoning the concept of a free-electron gas and its one outstanding success—explanation of the Wiedemann–Franz law.

17.4. Quantum theory of the electron gas [10]

So long as the electrons responsible for electrical conduction in a metal were considered to be derived from the atoms by some sort of ionization process, it remained undecided whether their number was constant, or increased with temperature in accordance with Boltzmann's law. Consequently there was no foundation on which to build an explanation of the almost linear variation of resistivity with temperature. But in 1913 Wien [11], and then Keesom [12], considered the consequences of Einstein's quantization of atomic vibrations in a solid. It had been recognized that not all the vibrational energy could be removed by cooling, but that there must remain a zero-point energy $\frac{3}{2}\hbar\omega$ for every atom that vibrated with angular frequency ω. Wien's speculations led him to a set of assumptions that must have seemed wildly improbable to any well-educated critic, and indeed were mentioned without enthusiasm by Lorentz and Bridgman at the 1924 Solvay conference. He imagined that the zero-point vibrations kept the electrons in a constant state of agitation, with so much energy that increasing the temperature had little effect; they would make correspondingly little contribution to the thermal capacity and could be supposed present in numbers comparable to the number of atoms, as required by the Hall effect. Their number being essentially independent of temperature, the increase in resistance with temperature must be attributed to more frequent collisions with the atoms. If it is only the thermally excited vibrations of the atoms that cause them to intercept electron paths (a most unclassical notion, but one already favoured by Kamerlingh Onnes) and if the chance of a collision is proportional to the mean square displacement, the resistivity will be proportional to temperature, only flattening off on cooling, as the last traces of thermal agitation are frozen out.

See also p 915

Though Wien's theory made little impact, it deserves to be remembered because it adumbrates features—zero-point energy of conduction electrons, scattering by thermally excited atomic vibrations only—that proved to be wonderfully justified after 1926, when quantum

> See also
> p 167
> p 527

mechanics could be applied systematically. The first steps in this direction, however, were taken before Heisenberg and Schrödinger had laid the necessary foundations. In 1924 Bose derived Planck's radiation law by considering the radiation in an enclosure as a perfect gas of photons, and inspired Einstein immediately to develop the same idea for a perfect monatomic gas. Einstein's gas differed from the classical gas of Maxwell and Boltzmann only in so far as he assumed, with Bose, that the particles were indistinguishable and that they occupied definite quantum states. Two years later Fermi, inspired by Pauli's exclusion principle, developed an alternative model with the additional assumption that each quantum state could be occupied by one atom only. At low temperatures the difference between the two was highly significant: Einstein's atoms could all pack down into the state of lowest energy (virtually zero energy for an atom in a large container) while Fermi's atoms had to occupy as many different states as there were atoms, and so some inevitably retained considerable zero-point energy. It is interesting that Pauli was at first reluctant to accept Fermi's model, but convinced himself in the end that it must at least apply to an electron gas, since the exclusion principle was an essential feature of the shell model of many-electron atoms. The wave properties of particles were not known to Bose or to Einstein at the time, nor were they used by Fermi; but they are crucial for later developments, and it is in these terms that the fully degenerate state of the Fermi gas, at zero temperature, is explained in Box 17A. At a moderate temperature collisions with the thermally excited atoms can transfer energy amounting to some small multiple of $k_B T$. Those few electrons that lie within this energy range from E_F may be excited, but most are prevented by the exclusion principle from changing their state. Consequently only a fraction of the order T/T_F are excitable and the contribution of the electron gas to the specific heat is correspondingly smaller than what classical theory would predict. In this way one of Wien's bold hypotheses turns out to have hit the mark. At this time there were no measurements available to check the prediction that the electronic contribution to the thermal capacity should be proportional to T. This had to await the development of low-temperature calorimetry.

Once converted to Fermi's view, Pauli lost no time in deducing that, by virtue of the magnetic moment associated with the electron's spin, the electron gas should show a paramagnetic susceptibility independent of temperature, as indeed many metals do. This success of his brilliant former student provided Sommerfeld with the stimulus to redevelop the old Drude–Lorentz model as a Fermi rather than a Boltzmann gas, and it is from here that we may date the inception of the modern theory of metals. Although Sommerfeld's [13] papers appeared 18 months after Schrödinger's first sensational publications, no wave mechanics is involved—the electrons are treated as classical particles subject only to the restrictions of indistinguishability and the exclusion principle.

BOX 17A: FERMI GAS

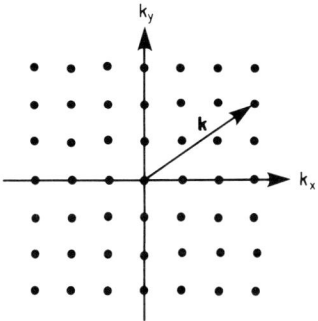

Where classical mechanics used the momentum p to describe the motion of a particle, we use the wave-vector $k = p/\hbar$. A particle contained in a box of volume V can only take certain values of k which are evenly and very closely spaced when V is large enough to contain many atoms. In the simplified two-dimensional diagram each allowed value (*state*) is represented by a point— the line joining the point to the origin represents the vector k. The corresponding three-dimensional diagram is referred to as k-space, and there are $V/8\pi^3$ states per unit volume of k-space. No more than two electrons can occupy each state (the factor 2 arises from electron spin).

At zero temperature a gas of free electrons takes the lowest possible energy, by packing the occupied states in a sphere of radius k_F, centred on the origin. If $k_F^3/3\pi^2 = n$, the number of electrons per unit volume, every state with $k < k_F$ is twice occupied. The electrons then possess kinetic energy ranging from 0 to the *Fermi energy* E_F which is $\hbar^2 k_F^2/2m$. Thus if $n = 8.5 \times 10^{28}$ m^{-3}, as in copper, $E_F = 1.14 \times 10^{-18}$ J or 7 eV; alternatively we may express E_F as $k_B T_F$, where T_F, the *degeneracy temperature*, is 82 000 K in this case.

Sommerfeld takes over from Schottky [14] a ruthlessly simplified version of earlier models, with their assumed variations of electric potential from atom to atom smoothed out into a uniform background. As usual, mutual interactions of the electrons are ignored, so that they may be taken to move freely and independently like molecules in a perfect gas. To be sure, collisions with lattice atoms are necessary to provide a mechanism for resistance, but they are not so frequent as to cast doubt on the general idea of free particles. The positive successes of the Sommerfeld model, as well as the professionalism of his treatment

Twentieth Century Physics

and the high esteem in which he was held, disposed other theorists in its favour, although there was general disapproval of the smoothing away of the individual atoms, in which even Schottky [15], the original proponent, joined. Experimenters were more concerned that it left the enigmas of magnetoresistance and the Hall effect (not to mention superconductivity) still without resolution, though they also were not happy about the simplifying assumptions. Old stagers like Hall [16], then 73, remained wedded to their pre-quantum conceptions, and Barlow [17], a mere novice, was probably not the only one to protest at the cavalier neglect of Coulomb interactions between the electrons themselves, which must surely cast doubt on the analogy with a perfect gas.

This difficulty, though only rarely remarked on in previous treatments, had certainly been appreciated by Lorentz [18]. He noted that collisions between free electrons could not change the total momentum of the electron gas, and therefore did not affect the electrical resistivity; on the other hand, they could destroy the flux of energy carried by electrons in a temperature gradient and thus reduce the thermal conductivity. He concluded that the success of Drude's explanation of the Wiedemann–Franz ratio necessitated belief in non-interacting electrons, however unlikely this seemed. The resolution of the problem came some years after Sommerfeld's paper and will be dealt with in due course. Meanwhile we may note that the objections of experimenters seem to have been ignored by the theorists who pressed ahead to refine the new theory. Sommerfeld himself did not join in; the general questions raised by the new quantum mechanics called for a more fundamental level of enquiry than the theory of metals promised. When in 1931 he and Frank [19] wrote for the newly founded *Reviews of Modern Physics*, and in 1933 when he and Bethe [20] produced the comprehensive *Handbuch* article—the bible to that generation of metal theorists—he was satisfied to present his pioneering contribution and leave later developments to his collaborator. The same must be said of Bethe, once the article was finished, and of Bloch whose contribution to the theory of metals is as significant as any; it was not long before both turned their gaze elsewhere and, apart from occasionally joining in the fashionable sport of demolishing theories of superconductivity, left metals to others.

Nevertheless Sommerfeld's influence as a teacher was pervasive, and his numerous students constituted in a loose way the first research school in the quantum theory of solids. The Americans Houston and Eckart, short-term visitors to Munich, collaborated directly with him in the series of four papers [13] that started the ball rolling. Others whose contributions were more substantial soon moved elsewhere, Bethe to Frankfurt and Peierls to Leipzig where Heisenberg, a former student of Sommerfeld's, encouraged him and Bloch to continue the study of metals. Bloch himself then went to Zürich, where Pauli ran another Sommerfeld outpost, returning to Leipzig in 1929 and being replaced

See also
p 1316
p 1359

as Pauli's assistant by Peierls. Among other major contributors in the first years were Brillouin, an older man who had spent a year with Sommerfeld in 1912 but was then working in Paris, and A H Wilson who left Cambridge to spend a year with Heisenberg in 1931. These were the leaders in the first period, during which emphasis was placed on general principles, with little effort devoted to calculating the properties of particular metals. As the thirties progressed, however, the schools that grew up, in Princeton and Bristol particularly, concerned themselves at least as much with specific problems. At the same time the experimental studies, by Pohl and his Göttingen school, of ionic crystals coloured and made weakly conducting by radiation and heat treatment, began to attract theoretical attention. From this sprang a renewed interest in semiconductors which proved most timely when war and the feverish development of microwave radar created a demand for semiconductor rectifiers. The post-war release into civil life of great numbers of technically skilled young physicists led to an unprecedented surge of research activity, not least in solid-state physics (later, condensed-matter physics). Almost immediately the studies of semiconductors and of metals became separate disciplines, and this was only one early example of the fragmentation into specialisms that has become endemic in science.

17.5. Bloch's theorem and its immediate consequences [21]
Sommerfeld's implausible assumption of a smoothed-out potential was shown to be unnecessary when Bloch [22], in his doctoral thesis, applied wave mechanics to the electron gas. The theorem that goes by his name is not specially original, nor did he attach great importance to it except as a necessary preliminary to the principal topic of his doctoral dissertation. However, by showing that a regular crystalline array of atoms need present no obstacle to the movement of electrons he opened up the systematic investigation of the transport of electricity and heat in solids, and much else besides. The successes of the Sommerfeld model became explicable and, as it turned out later, its failures with regard to the Hall effect and magnetoresistance were remediable; moreover a very simple, and in the main correct, explanation emerged of why some solids are good conductors and others insulators, and what gives semiconductors their peculiar properties. Between December 1928, when Bloch's paper was published, and August 1931, when Wilson [23] submitted his second paper, largely devoted to semiconductors, the intense activity of a few laid the foundations of the modern theory of solids. This brief epoch has been investigated thoroughly by historians and it is enough to give a resumé of the outcome, without which much of what follows would be unintelligible.

Credit, however, must first be given to Bloch's forerunners, even if he was not influenced by them, having already established his own point of view before their work appeared. Earlier in 1928 Houston

[24] had considered how de Broglie waves would interact with the ions of the metal, and had concluded that the wavelets scattered from individual ions would cancel each other out if the crystal were perfectly periodic; but a crystal disturbed by thermal motions would generate a resultant scattered wave. Here was the first indication of how Wien's tentative explanation of the variation of resistivity with temperature might be developed into a rigorous theory. Simultaneously Strutt [25] in Eindhoven, who was interested in the physical applications of Mathieu's differential equation, pointed out that it described the wave mechanics of a particle moving on a line along which the potential energy varied sinusoidally (figure 17.2)—a primitive approximation to the motion of an electron along a regular chain of ions. His diagram shows clearly that there are bands in which the solutions are wave-like and describe unimpeded motion of the particle along the chain, and other bands (unshaded) where no such motion is possible. Both Houston and Strutt discovered important truths, but neither could proceed further, as Bloch did, to obtain a vision of the complete picture. Frenkel also must be considered a significant forerunner. His relative isolation in Leningrad, however, and his idiosyncratic approach greatly lessened his influence, which is hardly to be discerned in the mainstream writings of the German school.

Bloch demonstrated, in much more detail and in three dimensions, what Strutt had demonstrated in one—if the ionic lattice is treated as a periodic variation of potential, $V(r)$, not necessarily sinusoidal, there are solutions of Schrödinger's equation in the form of plane waves that describe an electron travelling in any direction through the crystal without being scattered. The plane-wave solutions are distinctly more complicated than the de Broglie wave of an electron in free space, but a Bloch wave-vector k may still be defined (see Box 17B). It was appreciated from the beginning that exact solutions for any particular choice of $V(r)$ would be rare, and Bloch adopted a well-tested expedient (the *tight-binding* approximation). He assumed that the attractive potential of each atom core was so strong that the wavefunction of a valence electron on one atom hardly overlapped the neighbouring atom, and then chose for the crystal as a whole the superposition of these atomic wavefunctions, but with a phase variation $k \cdot R$ from one atom to the next. The energy associated with this extended wavefunction, calculated by a standard procedure, is much closer to the true energy than the approximate wavefunction is to the unknown true wavefunction—but only for the lowest energy band, e.g. the left-hand branch in figure 17.2. Limited though this result of Bloch's was, it revealed enough to stimulate far-reaching developments. Reminiscing late in life, he confessed that after reaching his conclusions he recast the argument in the group-theoretical form that was then in vogue among real mathematicians, and that Slater castigated as 'gruppenpest'

Electrons in Solids

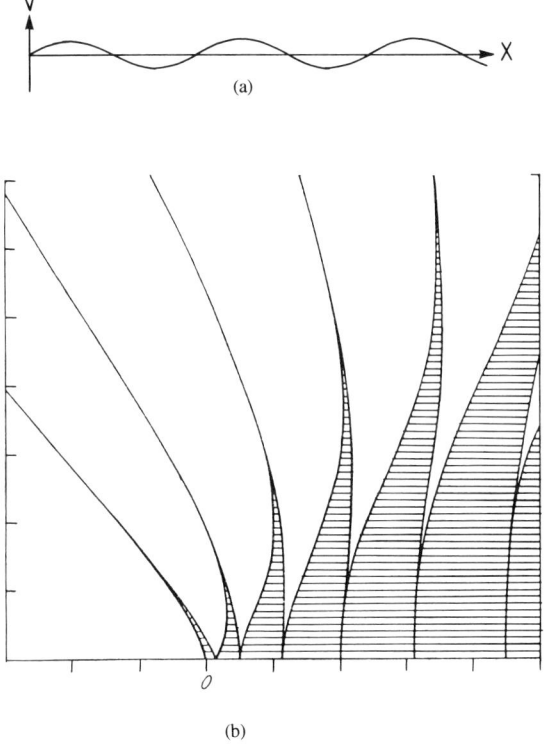

Figure 17.2. Strutt's [25] *model of an electron moving in a periodic one-dimensional potential, $V = V_0 \sin kx$, as in (a). His solution of Schrödinger's equation (b) shows, as shaded regions, the ranges of energy in which the electron can travel along the line. Electron energy is plotted horizontally and the value of V_0 vertically.*

[26]. It did not take long for others to restore the theory to its original comfortable form.

It was probably also the mathematical conventions of the time that caused Bloch to give no diagrams to illustrate his solution, and probably thereby to miss the important insights that Peierls conveyed with simple sketches. In writing his *Handbuch* article Bethe went further and set a skilled mathematical draftsman to work on Bloch's formulae. The result for the face-centred cubic lattice [27] (figure 17.3) gives an impression of how the energy of the electron is related to its wave-vector. For a truly free electron $E = \hbar^2 k^2 / 2m$, and a surface showing all vectors k for which E takes a given value would be a sphere. For motion through the lattice, however, different orientations of k affect the energy differently, and so the surface of constant energy is no longer spherical. For energies near the bottom of the band the surface, surrounding the origin fairly closely, is indeed roughly spherical, but at higher energies it bulges out to contact the hexagonal faces of the Brillouin zone (see Box 17B). When

1293

BOX 17B: BLOCH WAVE

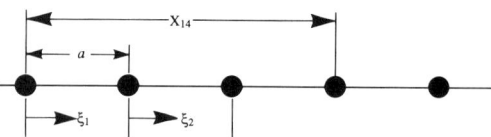

The wave-vector describing an electron moving along a one-dimensional chain of atoms can always be written in the form

$$\psi_m(\xi_m) = \psi_n(\xi_n)e^{ikX_{nm}}$$

which expresses Bloch's result; the form of the wavefunction is the same in every cell (of length a), but there is a progressive phase shift ka from one cell to the next. In the expression above, ξ is measured in each cell from some standard point, and $X_{nm} = (n-m)a$ is the distance between standard points in the nth and mth cells. The form of ψ must be such that it joins continuously from cell to cell.

k is the *Bloch wavenumber*. Whatever phase changes befall ψ in the passage from one cell to a corresponding point in the next cell, we can always add or subtract multiples of 2π so as to cause ka to fall in the range $\pm\pi$; i.e. $-\pi/a < k < \pi/a$,.

In three dimensions each equivalent point can be surrounded by a polyhedron (the *atomic cell*) such that when replicas are stacked to fill all space the atomic lattice is generated. The rhombic dodecahedron on the left, with one atom at its centre, generates in this way a face-centred cubic lattice of points. If r in each cell is measured from the centre, and R_{nm} is the vector joining the centres of the nth and mth cells, Bloch's theorem states that

$$\psi_m(r_m) = \psi_n(r_n)e^{ik \cdot R_{nm}}$$

and k is the *Bloch wave-vector*.

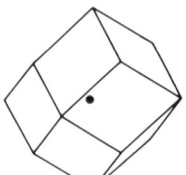
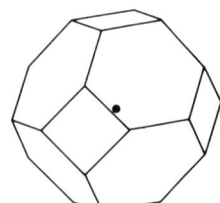

CONTINUED

1294

BOX 17B: CONTINUED

As in one dimension, the values of k can always be constrained to lie within certain limits. For the face-centred cubic lattice, these limits define another polyhedron, the *Brillouin zone*, which is the truncated octahedron on the right. If replicas of this are stacked, their centre points generate a body-centred cubic lattice, which is the *reciprocal lattice* in this case; R_{nm} is the vector joining two reciprocal lattice points.

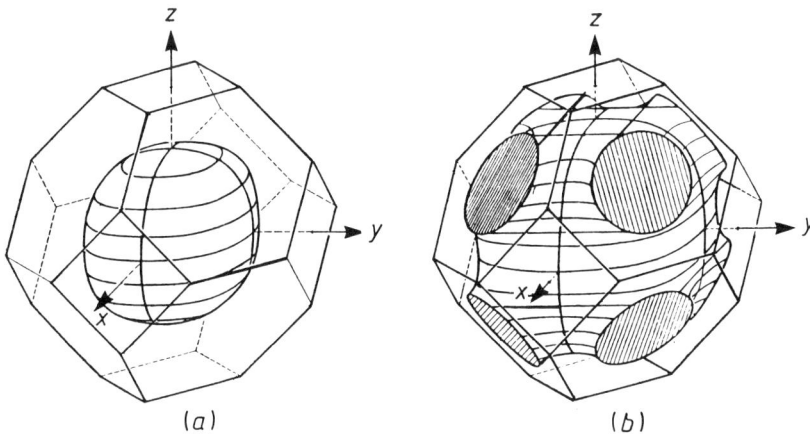

Figure 17.3. *Surfaces of constant energy in k-space for Bloch's tight-binding model of electrons in a face-centred cubic lattice* [20].

this zone is replicated and stacked to fill all k-space, constant-energy surfaces like figure 17.3(b) join to form a three-dimensional network. If the electrons do not interact with one another, they will settle down at the absolute zero to fill all states with energies less than some maximum, the Fermi energy E_F, and thus to occupy evenly all the k-space within the constant-energy surface defined by E_F. This surface gradually came to be called the Fermi surface, the name having been introduced casually by Jones and Zener [28].

What meaning is to be attached to a periodic network representing a surface of constant energy? The answer lies in the point stressed in Box 17B—the Bloch wave-vector only has a meaning when defining the phase difference between points R apart, and this phase difference $k \cdot R$ can be expressed by any number of different k. Equivalent points on the periodically extended energy surface represent equally valid forms of k. In other words, all the replications of the energy surface beyond the Brillouin zone are redundant statements of the information contained in the first zone. They are not, however, trivially redundant since the possibility of replication imposes boundary conditions on its shape—it must meet the faces of the zone in such a way that it joins smoothly to its neighbour. It would become too technical to go more deeply into the restrictions that continuity and crystal symmetry place on the shape; these were elucidated early on by those group-theoretical methods that Slater disliked at the time [29], but to which he was quickly reconciled.

A number of extensions to Bloch's theory were soon developed, and have proved to be fundamental to the understanding of the physics of electrons in solids. They are set out briefly below.

(i) For every periodic crystal structure one can define the unit cell in real space, and a corresponding cell or Brillouin zone in k-space whose replication defines the *reciprocal lattice*; these are the generalizations of the face-centred real-space lattice and body-centred k-space lattice of Box 17B. Just as in real space there is an infinite set of vectors R joining two equivalent points, so in reciprocal space there is a set of reciprocal lattice vectors g. A Bloch wave-vector k can equally well be written as $k+g$.

(ii) In the same way as de Broglie's relation, $\hbar k = p$, relates wavenumber to momentum for a free particle, so for an electron moving through a periodic lattice with Bloch wave-vector k, $\hbar k$ plays a role analogous to momentum. No longer is the energy equal to $\hbar^2 k^2/2m$, and the velocity of the electron is not $\hbar k/m$ but (as Bloch showed by confining the particle in a wave-packet) is given in one dimension by $v = \hbar^{-1} dE/dk$. In three dimensions any component of v is given by $\hbar^{-1}$ times the rate of variation of E with k in that direction. Since E does not change as one moves along a surface of constant energy, v is directed normal to the surface; the closer together the energy surfaces are packed, the greater is the velocity.

(iii) The analogy between $\hbar k$ and momentum was extended by Bloch, who considered acceleration of the electron by an electric field $\mathcal{E}$ and showed that $\hbar \dot{k} = e\mathcal{E}$ and by Peierls and others [30] who showed that to a good approximation the Lorentz force of a magnetic field on a moving charge took the same form as for a free particle, $\hbar \dot{k} = ev \times B$.

With these formalities behind us we turn, in more descriptive mood, to the developments that followed hard on Bloch's theorem. His professor, Heisenberg, had no doubt that this was a significant discovery and advised Peierls, who was also his student and only 21 at the time, to see what it could do for the Hall effect. The advice was sound—Peierls' contributions in the following years entitle him to a high position among the founders of solid-state physics. In a lecture in January 1929, soon afterwards published as a brief note [31], he used

Electrons in Solids

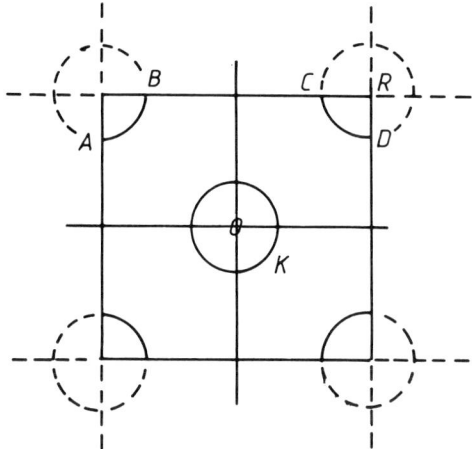

Figure 17.4. *Peierls' [31] diagram of two constant-energy surfaces. The smaller, K, encloses states of lower energy; the other, occupying four corners, is reconstituted as a closed surface when the diagram is replicated, as shown by the broken lines. This surface encloses states of higher energy and gives rise to a Hall effect of reversed sign, as explained in Box 17C.*

a simple two-dimensional version of the tight-binding model to show that a nearly empty zone and one nearly full would give opposite signs of Hall voltage. His diagram is periodically replicated in figure 17.4, as he prescribes, and the argument is carried on in Box 17C, to show how the reversal of sign comes about. For the first time a plausible mechanism had emerged to account for this long-standing mystery. The very helpful representation of the perturbing effect of a field in terms of the hypothetical creation and destruction of a few electrons, so that one can forget about the rest, is implicit in almost every analysis from Peierls onwards, but the explicit diagrammatic representation in Box 17C seems to have been first used by Klemens [32] in 1956, and popularized by Ziman [33].

It was not long before it was realized, though it seems now to be forgotten who made the point explicitly, that the nearly full zone responds to electric and magnetic fields in the same way as would a nearly empty zone of positively charged particles. This inaugurated the concept of *holes* which is central to the understanding of semiconductors. It is worth stressing that the unfortunate choice of the name *hole* led later to not infrequent confusion between a vacant state in an otherwise filled zone (such as is represented in Box 17C by an open circle) and a fictitious positively charged particle in an empty zone. The distinction is clear in the analogous idea introduced by Dirac in his relativistic electron theory, in which all states of negative energy were supposed

1297

BOX 17C: ELECTRONS AND HOLES

An electric field $\mathcal{E}$ shifts the k-value of each negatively charged electron in the opposite direction, so the filled states (shown stippled) inside the Fermi surface move to occupy the broken circle. The change that has occurred is equivalent to creating new electrons on the left and destroying the same number on the right. This is represented in the second diagram by dots outside the Fermi surface (new electrons) and open circles inside (newly vacant states). The arrows show the velocities of electrons in these states. When a magnetic field B is applied, in addition to $\mathcal{E}$, the states are shifted clockwise, to give an upward drift of new electrons and a downward drift of vacancies; both drifts contribute to a downward current, which is the cause of the Hall effect.

The diagrams are now redrawn for a Fermi surface of vacant states, as in the corners of the square in figure 17.4. The velocity is always directed from filled to empty states (lower to higher energy) and the Lorentz force of B now turns the distribution in the opposite sense, to give an upward current. This is Peierls' explanation of sign reversal in the Hall effect.

CONTINUED

BOX 17C: *CONTINUED*

The same effect as in the second row would be obtained if there were positively charged particles whose energy increased in an outward direction as in the row at the top. The filled states are displaced in the same direction as $\mathcal{E}$, and the Lorentz force drives them anticlockwise, to give an upward current as in the second row. The positive particles in this simulation are referred to as *holes*; they are not the same as vacancies. In semiconductors especially, where one finds a small number of vacancies in an otherwise full Brillouin zone, it is convenient to simulate their behaviour with a small number of positive holes in an otherwise empty zone.

filled, but unobservable. If, however, a vacancy was created by, for example, giving an electron enough energy to raise it into a positive energy state, there appeared an electron–positron pair, the positron being, not a vacancy, but a positively charged particle which, alone in the vacuum, simulates the behaviour of an infinite sea of negative-energy electrons, minus one. There is no evidence of any influence of Peierls' (or possibly Heisenberg's) idea of holes on Dirac's thinking.

Shortly after his work on the Hall effect Peierls [34], by now with Pauli in Zürich, analysed the Schrödinger equation for an electron moving in one dimension in a very weak periodic potential. A second simple, but significant, diagram (figure 17.5) illustrated his result. The energy is very nearly a quadratic function of the Bloch wavenumber, but departs from the truly free-electron behaviour, $E = \hbar^2 k^2/2m$, when k approaches π/a. Between 0 and π/a the velocity of the electron, $\hbar^{-1}\,\mathrm{d}E/\mathrm{d}k$, rises at first as for a free electron, but after reaching a maximum drops to zero at the edge of the zone, $k = \pi/a$. There are then a range of energies where the electron cannot move along the lattice, and the next allowed band of energies starts with zero velocity, rising quickly to the free-electron value. If Strutt, who found the succession of allowed and forbidden energy ranges, had only gone into the matter more deeply he would have made this discovery; but he was more interested in mathematical properties than in the physical ideas they might have revealed—in contrast to Peierls at every stage of his long career.

What seems to have delighted Peierls particularly was the realization that the energy discontinuity occurs at the same value of k for both Bloch's tight-binding and his own nearly free model. Moreover the phenomenon is analogous to Bragg reflection of x-rays. An electron, accelerated by an electric field towards the zone boundary, can be

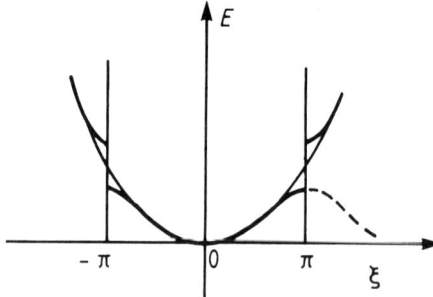

Figure 17.5. Peierls' [34] diagram showing how energy E depends on wavenumber k for an electron moving in a weak periodic lattice potential; $\xi = ka$, where a is the lattice period.

thought either to continue past the boundary in the periodic replication (shown dotted in figure 17.5), or to suffer Bragg reflection from $-\pi/a$ to $+\pi/a$ and continue its acceleration in the central zone. More generally, in three dimensions the faces of the Brillouin zone in figure 17.3 delineate those wave-vectors at which the face-centred cubic lattice causes Bragg reflection; the states within a Brillouin zone are the analogue of one of the bands in the one-dimensional case. This, however, is only the beginning of a development that owes much to Brillouin's studies of wave propagation in crystals, conducted independently in Paris at the same time. His influence was felt most markedly by those who began to enquire into the particular differences between metals, rather than their common characteristics, and we shall leave this till later.

See also p 1311

See also p 1345

Peierls observed that if the electrons entirely fill a zone they cannot conduct electricity; in the language of Box 17C, without a Fermi surface there can be no creation or destruction of electrons. It was left to Wilson, however, to point out the possibility (which Bloch acknowledged he had failed to appreciate) that the bands of forbidden energy at each point on the zone faces might be so wide as to separate the highest level in one zone from the lowest in the next above. Since the number of states in a zone is twice the number of atoms in the sample (in the simplest cases) a solid composed of divalent or quadrivalent atoms might find its valence electrons fitting snugly into filled zones and therefore, like diamond, be an insulator. This could not happen for monovalent (Na, Cu) or trivalent (Al) atoms which should therefore be metallic. If the gaps are not too large, however, the lowest states in the upper zone may lie below the highest in a lower zone—at different values of k, of course. Given this overlap an even-valent element may prove metallic (Mg, Pb). Wilson suggested further that though there may be a gap between the two energy bands, it may be small enough to allow thermal excitation from the lower to the upper at ordinary temperatures, and

Electrons in Solids

that this might account for the existence of elemental semiconductors, if indeed there were any. At that time silicon and germanium were not known to fit the bill, but a few oxides and sulphides were recognized as semiconductors by the accepted criterion of a resistivity that decreased with rising temperature. If thermal excitation is needed to produce free electrons and holes, according to Wilson's suggestion, this behaviour is explicable.

17.6. Mechanisms of resistivity [20, 35]

Once it was realized that quantum mechanics could explain how electrons could move freely through the lattice, albeit with modified dynamical properties, the classical mode of visualizing them as particles tracing out deterministic trajectories was accepted as the sensible way forward. A new set of rules was needed, however, to incorporate the exclusion principle. There is no problem about assuming that the Fermi distribution applies to the electron gas in equilibrium, but the scattering of an electron into another state by an impurity or a lattice vibration must also be considered. In a strictly classical gas the chance that a particle will be scattered from one state of motion into another is determined solely by the nature of the scatterer, which can be represented by the cross-section, or target area, that it presents to the particle. In a Fermi gas, however, scattering cannot occur unless there is a vacant state to receive the scattered particle; the chance of a scattering event taking place is thus proportional both to the cross-section of the scatterer and to the fraction of states that are vacant and available. This new factor was easily included in the theory.

If the scattering agent is an impurity atom, either substituting for one of the lattice atoms or lodged interstitially among them, it may appear so small in comparison with the Bloch wavelength that the scattered wavelet spreads out all round. The electron is almost equally likely to leave in any direction, and forgets its past history. A mean free path between collisions can then be defined or, if preferred, the mean time τ for randomization of motion (relaxation time). Since the Fermi gas, slightly displaced from equilibrium, may be represented as having a few electrons created in some regions of the Fermi surface and an equal number destroyed elsewhere, the assumption of isotropic scattering is equivalent to supposing that the number, n, of these extra particles and vacancies decays exponentially with time: $n \propto e^{-t/\tau}$. This very convenient simplification, introduced by Brillouin [36] in 1930 (if not earlier), was used by Nordheim in his treatment of the resistivity of alloys, and has proved its worth on countless occasions since, even though it may not be fully justified and is sometimes downright misleading.

Such simplifications were not in Bloch's mind when he embarked on the principal theme of his dissertation, which was to calculate the resistivity and its temperature variation. Debye's model (1912) of quantized thermal vibrations, which had been highly successful in accounting for the specific heat of a solid, was the basis of Bloch's conception of the scattering of an electron as a collision with one of Debye's modes; something very similar had served well to describe the Compton effect (1923) as a collision between an electron and an x-ray quantum (photon). But while in the Compton effect it is a photon that is thought of as being scattered by an electron, in metallic conduction the point of view is reversed; what is of interest is the scattering of an electron by a phonon†. As Compton and Debye had assumed in their independent and virtually simultaneous semi-classical analyses, and Bloch confirmed by his quantum-mechanical analysis, both energy and momentum are conserved in the process; more precisely, it is energy and wave-vector that are conserved. The consequence is that a free particle can only create or destroy a quantum of wave motion if it moves faster than the wave. So an electron cannot, by itself, either create or destroy a photon in free space. But an electron in a solid is almost always moving faster than a phonon, and there is no objection to a phonon being destroyed in collision with an electron, nor to an electron spontaneously generating a phonon, always provided that there is an empty state available for the electron to enter. These are the principal processes responsible for the temperature-dependent resistivity.

The most energetic phonons have energy $k_B \Theta_D$, where Θ_D is the characteristic Debye temperature ranging, in metals, from 38 K for caesium to 1460 K for beryllium. Of those whose resistivity had been studied at the time, 19 out of 34 had Θ_D less than 273 K. For these, at room temperature and above, there is enough thermal excitation for any electron near the top of the Fermi distribution to have little trouble finding a vacant state to fall into, if it emits a phonon. Then the rate of scattering will vary in proportion to the thermal energy, which itself is nearly proportional to the temperature; this was Wien's assumption in 1913. At low temperatures only phonons of energy less than a few times $k_B T$ are available to be absorbed, and only low-energy phonons can be excited since the exclusion principle stops an electron from diving deep below the Fermi energy. As a result the wavenumber change suffered by an electron in each process is small compared with its own wavenumber, and it is liable to need many such processes before it forgets its original direction of motion and can be said to have completed a free path. The detailed calculation of this resistive mechanism by the established

† *The photon was introduced by Lewis in 1926 and the phonon by Frenkel in 1932. It is therefore anachronistic to use these terms in connection with the early history of the Compton effect and metallic conduction; but I shall continue to do so. Frenkel's [37] note deserves to be quoted: 'It is not in the least intended to convey hereby the impression that such phonons have a real existence. On the contrary, the possibility of their introduction rather serves to discredit the belief in the real existence of photons.'.*

(Boltzmann equation) method of the time was long and full of hazards. Bloch initially concluded that the resistivity of a metal free from defects would fall to zero at low temperatures as T^3. Others suggested T^4, and eventually Bloch corrected his calculation to T^5. The data available, obtained of course with less than perfect samples which led to the temperature variation at the lower end being masked by a constant component due to defects and impurities, could readily be made to fit any of these proposals. It is now accepted that within the terms of his model Bloch's T^5 law is right, but measurements at ever higher degrees of precision have not surprisingly revealed complications, notably the resistance-minimum phenomenon, the subject of much discussion at a later date.

See also
p 1343

So far, the absolute magnitude of the resistivity has not been mentioned. To calculate this, the strength of the electron–phonon interaction must be known, and this is almost a separate issue; it also was resolved only after several discordant attempts, which were summarized by Bardeen [38] in a paper of 1937, a classic in the field. The essence of the problem is to calculate how the potential due to the ions is modified when an acoustic wave shifts them all in a periodic manner, for the electrons in the perfect lattice have adjusted their wavefunctions to the requirements of the ionic potential, and it is only departures from perfection that give rise to scattering. The first attempts considerably overestimated the scattering. Thus Nordheim assumed that each ion's contribution to the potential moved rigidly with it, so that its perturbing effect was the difference between its potential in the original location and after displacement. Bardeen observed, however, that the electrons could respond collectively to screen the perturbation partially—where ionic displacement results in an excess of positive charge, there will be an increase in the mean electron density to reduce the effect. The partially screened perturbations naturally scatter less than the unscreened, Nordheim-type perturbations, and the magnitude of the resistivity calculated for the alkali metals, especially sodium and potassium, agreed as well with experiment as anyone could have hoped.

With other metals agreement was less satisfactory, as is hardly surprising in view of the simplified model that Bardeen was constrained to adopt if he was to make any progress. To look ahead some years, we may note that his model was in many ways equivalent to that useful, if hypothetical, metal *jellium* (a name apparently introduced in 1953 by Herring [39]) a near relative of Sommerfeld's model. The positive ions in jellium are smeared out into a uniform density which is on the average neutralized by the electrons. When a compressional wave passes through, the positive charge density ceases to be uniform and gives rise, after the collective screening response of the electrons, to the scattering potential for individual electrons. In 1955, when ultrasonic waves had been shown to be strongly attenuated in pure metals at low temperatures (where the electrical conductivity is high) the theory of this process was worked out as a classical calculation of the motion of electrons in the electric field associated with the wave [40]. Rather surprisingly the result, for long wavelengths, agrees completely with Bardeen's quantum calculation—the scattering of electrons by low-frequency phonons and the attenuation of sound waves by electrons are two sides of the same coin. There is some gain in recognizing this, for while it is exceedingly hard to extend the quantum calculation to the case where the electron mean free path is reduced by other scattering processes, no difficulty arises in including these processes in the classical calculation. The results of the latter may therefore be taken over into the quantum context as being, if not strictly correct, at least better than nothing.

To return to Bardeen's paper, his results depart significantly from those for jellium when the phonon wavelength is so short as to be comparable with the lattice spacing. It is then that the Umklapp processes, which had been discovered by Peierls in 1929 and which, since they owe their existence to the periodic crystal structure, are absent in jellium, introduce an additional mode of scattering. Their influence amounts, however, to more than a refinement of the resistivity calculation and deserves separate discussion.

17.7. Umklapp processes

Peierls' [34] discovery arose from his study of heat conduction in insulating crystals, where the heat is carried by thermally excited waves. In terms of phonons the picture is much like that of thermal conduction in a gas though here, especially within Debye's then popular approximation, all phonons described by longitudinal waves have the same velocity, and the velocity of transverse waves is not much different. Anharmonicity of the lattice allows phonons to scatter each other and to generate new frequencies analogous to combination tones in musical acoustics, but so long as the sum of wave-vectors is conserved such scattering does not hinder the flow of energy. In a perfect crystal, therefore, there should be nothing to limit the thermal conductivity—a heat flux once injected would continue indefinitely, though carried by an ever-changing phonon population.

Peierls pointed out, however, that the wave-vector in a discrete lattice is not uniquely defined. The ambiguity is exactly the same as we have already encountered in the definition of the Bloch wave-vector. Just as $k+g$ can represent the same electronic state as k, so a phonon with wave-vector q can also be represented by $q+g$, g being any reciprocal lattice vector. In a collision between phonons (or between a phonon and an electron) it is still essential to conserve wave-vector but the definition of wave-vector is enlarged and may now include a reciprocal lattice vector. If the total wave-vector is changed, in a collision, to the extent of a reciprocal lattice vector, g, the previous condition for energy flux to be

See also p 977

Electrons in Solids

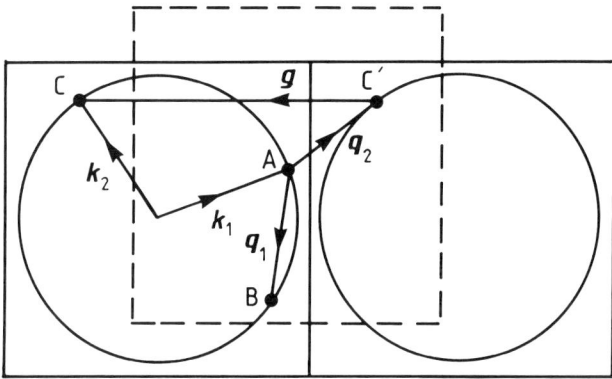

Figure 17.6. *The Umklapp process in electron–phonon scattering. Two replicas are shown of the Fermi surface in its Brillouin zone. An electron at A can be scattered, without invoking a U-process, to any other point on the Fermi surface that lies within the broken square, centred on A, which represents the range of wave-vector q available to the phonons. Thus $A \to B$ is a normal scattering process, needing no help from the lattice. On the other hand $A \to C$ is too large a change to be effected by a normal process, but there is no objection to a transition to the equivalent point C' with the involvement of a shorter wave-vector q_2. Conservation of wave-vector is represented by the equation $k_2 = k_1 + q_2 + g$.*

conserved breaks down, and a thermal resistivity makes its appearance. In such a collision the lattice as a whole recoils with momentum $\hbar g$; this is not directly observed but is important for what follows.

This new process, which Peierls called an Umklappprozess (usually abbreviated to U-process) does not come into play except with phonons of the highest frequency, such as are thermally excited at temperatures not too far below the characteristic Debye temperature. It is the absence of U-processes at lower temperatures that can allow the thermal conductivity of insulating crystals to rise to very high values. A good synthetic white sapphire, for example, at 30 K conducts heat 50 times better than does copper at room temperature.

In jellium, without a lattice structure, U-processes cannot occur, but Bardeen's treatment of the electron–phonon interaction, for all its simplifications, incorporated a lattice at so basic a level that the U-processes emerged automatically. In this case they are intrinsically easier to follow than in a typical multi-phonon collision, since they need involve no more than the change of wave-vector k of an electron during the absorption or emission of a single phonon. Figure 17.6 illustrates how the U-process allows scattering through a wide angle to be attained with a phonon of quite small wave-vector, if the Fermi surface runs close to the zone boundary. Such wide-angle scattering may then persist to lower temperatures than in the absence of U-processes. One has only to examine Ziman's book [33] to see how important U-processes are in the detailed theory of transport in solids.

In 1932 Peierls [41] followed up his first paper by pointing out how these ideas raised serious doubts about Bloch's treatment of resistivity at low temperatures. The variation of ρ as T^5 was derived on the assumption that the phonons scattering the electrons were in thermal equilibrium. But if there are no U-processes, as is the case at low temperatures where the T^5 law holds, the lattice as a whole gains no momentum by recoil. Then the momentum given to the electron gas by the electric field will be transferred to the phonons and will remain with them, being dissipated only by occasional scattering against impurities and imperfections. The very fact that the contribution of the phonons to thermal conductivity is less by several orders of magnitude in a pure metal than in a crystalline insulator shows that the coupling of the phonons to the electrons is enormously stronger than that to defects. It is therefore highly dubious to treat the phonon gas as a system in equilibrium; indeed if there were no defects it is hard to see why there should be any resistivity, electrons and phonons being swept along together by the electric field. Nevertheless the Bloch theory seemed to fit the facts, which led Peierls to suggest that even in the alkalis, the simplest of metals, the Fermi surface might be distorted enough to meet the zone boundaries, when the difficulty would be somewhat alleviated. It was not long before this solution was made to seem unlikely and the problem, when it was not conveniently overlooked, remained an embarrassment for many years. As late as 1960 Ziman [42] was unable to offer a definitive resolution to what even Peierls, the originator, had to confess still puzzled him in 1955. The solution will emerge towards the end of the next section.

17.8. Thermoelectric effects [43]

To proceed further we must examine the thermoelectric effects, and this means stepping back to the beginning of the century and to a scene of confusion, some of it the legacy of Maxwell and his followers. In so far as they sought to avoid introducing the concept of charged particles, and looked to the ether as the medium for all electromagnetic processes, the transport of energy along with charge was foreign to their thought. The Peltier heat arising when a current crossed the junction between two metals was then attributed to an EMF at the junction and this, being measured in microvolts, raised awkward questions about its relation to the contact potential which was of the order of volts. We need not follow the progress and decay of this misapprehension but start the discussion with Lorentz's [44] analysis which was essentially correct, as one would expect from Lorentz, though oversimplified in one significant point. As late as 1919, however, Bridgman [45] felt the need to give a close thermodynamical analysis of the principles, independent of electronic models, in the hope of dispelling prevalent errors. There is little evidence of any further thought being given to

Electrons in Solids

BOX 17D: COLD EMISSION

This simple diagram is Schottky's [14] representation of how the electrical potential varies as one moves into a metal from outside and leaves through the opposite face. The difference between inside and outside is Richardson's Φ and unless an electron acquires kinetic energy in excess of $e\Phi$ it cannot escape. When a positive electrode is brought near the surface the electric field outside changes the potential near the surface. Increasing the field, as in these diagrams, lowers the barrier until a strong enough field enables electrons to leave without the need for kinetic energy; this was Schottky's explanation of cold emission. It depends on the curvature of the step, which he explained as due to the attraction of an electron outside the metal by its electrostatic image. This type of diagram became popular for many purposes where the curvature of the step is irrelevant (e.g. figure 17.7). A more sophisticated use by Schottky will be found in figure 17.11.

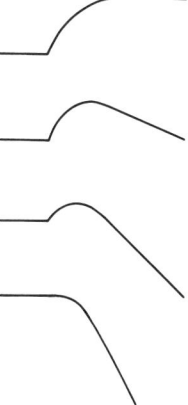

the theoretical issue until it was revived by Sommerfeld [13], whose treatment shows an assurance equalled only by that of Lorentz (and Bohr [8], whose work Sommerfeld may not have known). Some of the misunderstandings might have been avoided by the use of diagrams such as those in Box 17D, which Schottky [14] was perhaps the first to employ, as late as 1923.

Lorentz's treatment of thermoelectric effects can be interpreted in

Twentieth Century Physics

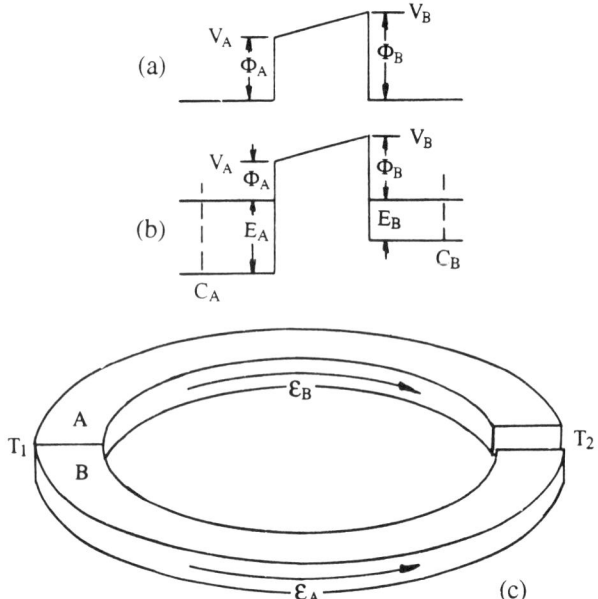

Figure 17.7. *The contact potential and work-function for (a) Lorentz's classical model and (b) Sommerfeld's quantum model. The sharp step at the surface of each metal is produced by an electrical double layer, and there are extra charges on the surfaces to produce the field represented by the sloping potential. If the metals are in intimate contact the charges may merge into an individual double layer. (c) A thermoelectric circuit involving two metals A and B, with internal electric fields $\mathcal{E}_A$ and $\mathcal{E}_B$ accompanying the temperature gradient.*

these terms, as shown in figure 17.7. When two different metals are joined the electrons near the junction almost immediately establish a double layer that ensures no further current flow. In the classical model (*a*), where the electrons have little kinetic energy and lie almost on the bottom of each well, the wells on either side must very nearly match. Then the difference in height, $V_B - V_A$, is the difference in the potential energy of an electron just outside the two metals, and this determines the Volta contact potential, very close to $\Phi_B - \Phi_A$. For a Fermi gas, however, it is the top of the Fermi distributions that must match, as in (*b*), and the energy $e\Phi$ that controls thermionic and cold emission is now the difference in total energy of an electron at the Fermi surface and an electron at rest just outside. Since Lorentz's and Sommerfeld's accounts of the origin of thermoelectric effects run along parallel lines we may proceed immediately to the latter, though still avoiding the detailed mathematical analyses with which both managed to obscure the elementary physics.

When a current is passed across the junction it is as if (see Box 17C) a few electrons near the Fermi level constitute the flow. If they all had

1308

exactly the Fermi energy an electron would leave A and fall naturally into place in B, being accelerated by the double layer from the kinetic energy E_A to the desired energy E_B. There would be no energy imbalance and no Peltier heat generated or absorbed. This is the situation at zero temperature. When the temperature is raised the flow is made up of electrons in a range of a few $k_B T$ around the Fermi energy, and exact energy balance may not be maintained. If, for example, the mobility of electrons in A increases with their energy, while the mobility in B decreases, the electrons crossing the point C_A will carry on the average slightly more energy than those crossing C_B, by an amount measured (probably) in small fractions of $k_B T$, and this energy must be given up at the junction. This is the Peltier heat which, it should now be clear, is quite unrelated to the contact potential. It is, however, more strongly conditioned by the details of electron mobility than Lorentz appreciated. He assumed that all electrons in a given metal had the same mean free path, and was not concerned to refine the argument further.

The need to go deeper was noted by Bohr even in his Master's thesis, and in his doctoral thesis he examined several different scattering models to show how they led to differences in the thermoelectric effects. This was the only result from his theses to be published in English [46] during his lifetime, and it met little response except from Richardson, whose assumptions he was criticizing. Others, not aware of the difficulty, were inclined to assume a straightforward relationship between the thermoelectric effects and the thermodynamical properties of the electron gas in equilibrium. The weakness of the argument is most clearly seen in the Thomson effect. When a current is passed through an unevenly heated metal the electrons travelling from cold to hot have to be given thermal energy from outside; this is the Thomson heat which came to be thought of as the specific heat of electricity. Like the Peltier heat, however, its magnitude depends crucially on the relative importance in the current of the more and the less energetic electrons around the Fermi energy. The specific heat of the electron gas in equilibrium is, at best, only a rough guide to the magnitude.

The Seebeck effect is slightly more complicated, but must be discussed if only because it is the one most frequently measured. Figure 17.7(c) shows a thermoelectric circuit of two metals with their junctions at different temperatures, one of them being opened so that no current flows. At the left-hand junction the Fermi energies of both metals match. If the Fermi energy was constant along each arm, despite the temperature gradient, there would be a match at the right-hand side also; on completing the circuit no current would flow. But normally, with a constant Fermi energy there is a tendency for electrons to diffuse along the temperature gradient—for example, those at the hot end may be more mobile than those at the cold end, and drift towards it faster than those going in the opposite direction. Left to themselves, they will build up space charges until the resulting electric field counteracts the drift. These are the fields shown as $\mathcal{E}_A$ and $\mathcal{E}_B$ in figure 17.7(c), and it is the potential difference due to them at the open circuit that is measured as the thermoelectric (Seebeck) voltage.

Keesom [12] attempted to derive the Seebeck coefficient from properties of the electron gas in equilibrium, starting correctly with the observation that the pressure of the gas is temperature dependent. The electric field in his view supplied the force needed to counter the pressure gradient. But this argument neglects another force—the electrons, in being scattered, exert a force on the scattering centres, and consequently experience an opposite reaction themselves, which must be taken into consideration when calculating $\mathcal{E}$. Bohr in fact had already shown that different scattering models gave, on the one hand, different variations of mean free path with energy and, on the other, different reaction forces. For every model, the Seebeck coefficient was always related to the Thomson and Peltier coefficients by the same relations that Thomson had derived by dubious thermodynamics It was not until Onsager developed his irreversible thermodynamics in 1931 that anything like a rigorous explanation emerged.

See also p 601

This rather detailed analysis of the thermoelectric effects is relevant to Peierls' worry about the equilibrium of the lattice vibrations in the presence of an electric current. Measurements of the Seebeck coefficient of very pure semiconductors and metals, from 1953 onwards, showed clearly that at low temperatures, where U-processes no longer occur, the electron–phonon interaction may become very important. In the presence of a temperature gradient the greater concentration of phonons at the hot end of a sample results in a phonon current that can drag electrons along with it. Consequently an extra electric field is needed to compensate the phonon-induced current, and this will enhance the thermoelectric force. The effect, which was predicted in 1945 by Gurevich, can be very striking (figure 17.8). Nevertheless, even when it produces dramatic thermoelectric effects this phenomenon of *phonon drag* has a remarkably small effect on the electrical resistivity, principally because the observed peaks in figure 17.8, large though they are, are some ten times lower than they would be if the metal were completely free of phonon scattering defects; the effect of phonon drag on the resistance is thereby reduced a 100-fold. We conclude that Peierls was thoroughly justified in his criticism of Bloch's theory, but that the art of material preparation was less far advanced than he had assumed. Indeed, it still is, and Peierls' runaway process has never been demonstrated in practice.

We end the discussion of thermoelectric effects on a more practical note. Mott and Jones [48] express the Seebeck coefficient as proportional to temperature, with a magnitude determined by $(1/\sigma)(d\sigma/dE_F)$—the relative change in conductivity when the number of electrons is altered to change E_F. Leaving aside the correct interpretation of a neat,

Electrons in Solids

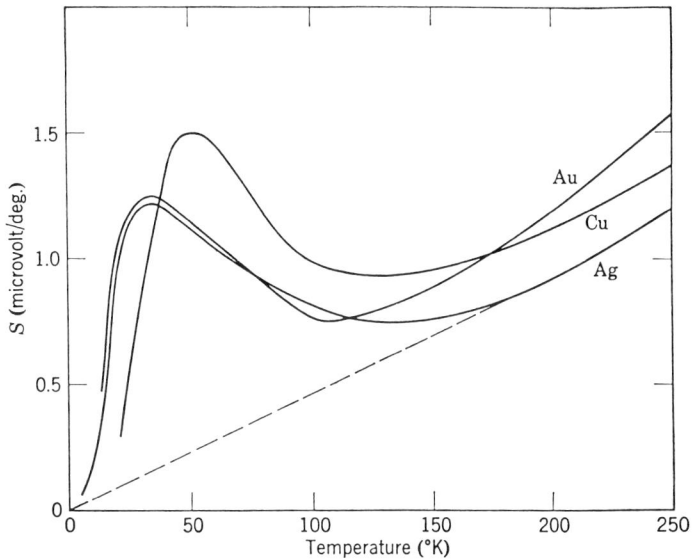

Figure 17.8. *The thermoelectric power of copper, silver and gold, showing the larger humps due to phonon drag* [47].

but not immediately meaningful, formula we may note that when the concentration of electrons and E_F are low, as in bismuth and semiconductors, a given change in E_F makes a proportionately large difference to σ. It was especially Ioffe in Leningrad who saw the potential of semiconductors for refrigerators without moving parts, employing the Peltier effect, and who went to great pains to develop suitable materials. But although he made a pilot model that worked well the system never achieved commercial success. His book [49] is full of information and ideas about how thermoelectricity might be used, some of which have since reached a specialized market.

17.9. Interlude—Brillouin zones

In 1930 Léon Brillouin [50], who was somewhat older than the pioneers of the German school, amused himself (as he admitted) and gave pleasure to others by drawing pictures like that shown in figure 17.9. If one considers the reflection of x-rays by a two-dimensional square lattice, the multitude of lines define the possible values of the wave-vector for which Bragg reflection will occur. These lines are the perpendicular bisectors of lines joining the central point to any other point on the square array of dots (the reciprocal lattice). They also show the values of k at which, according to Peierls' almost free model, the energy of an electron will show a discontinuity. Brillouin pointed out that the whole plane was divided into pieces that could be translated through

1311

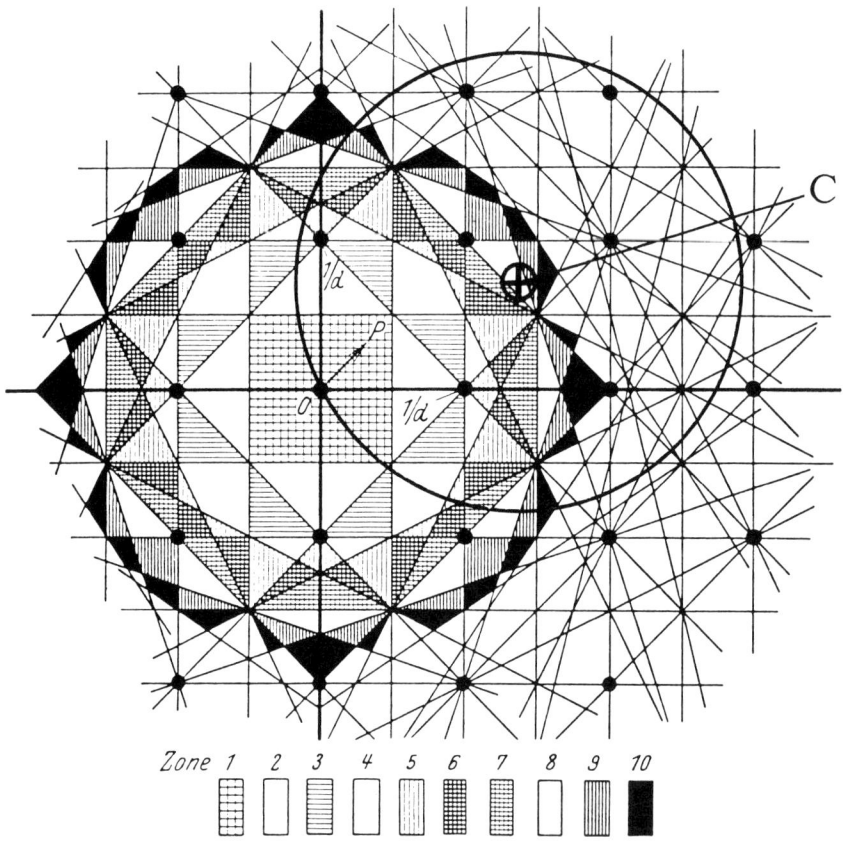

Figure 17.9. Brillouin's [50] *drawing of the dissection of a k-plane by lines of Bragg reflection.*

a reciprocal lattice vector into the central zone (zone 1) so as to fill it perfectly, many times over. Thus the pieces of each shading join up into a square. This is true for any unit cell, in three dimensions as in two, though in three dimensions the shapes can be very complicated. The proof of this assertion seems not to have been given until much later. It is less formidable than it seems at first to decide how to group the bits together, though no rule was given at the time. For example, how can one verify that the point C belongs to the eighth zone as stated by Brillouin? Draw a circle (or sphere in three dimensions) centred on C and passing through O; then the zone number is one more than the number of reciprocal lattice points enclosed.

If it seems odd that the early texts which delighted to reproduce Brillouin's diagrams were silent on such details, the reason probably is that his extended zone picture served rather little useful purpose at the time. The uses to which it was put by Mott and Jones [35], when they sought to make sense of the properties of metals with complex atomic arrangements, were far from transparent. The zones they chose to draw seemed quite different from Brillouin's; and Seitz's [51] formal treatment gave little help. Yet their conception is basically simple, especially in the light of later knowledge. The x-ray reflections of metals with many atoms to the unit cell are mainly rather weak, with a few strong ones. By ignoring all the planes in Brillouin's construction except those that gave strong Bragg reflection Jones (whose ideas inspire this part of the book) could construct a new primitive zone within which there should be no significant discontinuities of energy, and he found in this an explanation for the peculiar crystal structure and properties of bismuth. After years of neglect the 'Jones zone' formed from strong reflecting planes has proved of value in computations of alloy structures. It is also worth remembering that the valuable *Harrison construction*, which we shall meet in due course, is precisely in tune with Brillouin's thought. For the most part, however, Brillouin's scheme was valued as a didactic representation, showing how all the information about Bloch waves and the energies of electrons could be compressed into the first Brillouin zone, with $E_n(k)$ a multivalued function, and the subscript n denoting the zone index. If one wished, this zone and its contours of E_n could be replicated to fill all space. This is what Peierls did; it gave no more information but made transparent the explanation of how electrons behaved when accelerated, and it has been adopted generally since then. Apart from the first zone, most of Brillouin's construction has been dismissed as irrelevant by solid-state physicists.

See also p 1352

17.10. From generalities to particulars; the emergence of solid-state schools [52]

So far we have concentrated on the leaders of the German theoretical school, the paladins of universal law. Their experimental colleagues, concerned less with generalities than with particular examples, and suspicious of theory, had more influence on the study of insulators and semiconductors than on metals. It was largely left to the English-speaking world to pursue the application of theory as well as experiment to specific materials and thus to inaugurate the new science of solid-state physics. In Princeton Eugene Wigner's students, notably Seitz, Bardeen and Herring, led the way (Wigner could never limit himself to solids alone). Slater at MIT had by 1933 already begun a systematic programme to explain chemical bonding in terms of quantum mechanics, when the first paper from Wigner and Seitz inspired him to take in the electronic structure of metals also, and this grew into a major concern of his. Mott's arrival in Bristol, also in 1933, brought into being what soon became, and remained for many years, the premier school of solid-state physics in Britain. In choosing these three schools for special mention I have been influenced by the subject matter of this article, and have ignored the lively research proceeding simultaneously into ferromagnetism and paramagnetic salts, with Stoner in Britain and Van Vleck in the United States as leading exponents.

From the outset the attitudes of Mott's school were in marked contrast to those in Princeton and MIT. Princeton was—perhaps still is—a mandarin among physics departments, the strongest adherent in the New World to the old German ideal of pure scholarship, and the home in exile of Wigner, von Neumann, Einstein, Weyl and others for whom mathematics was the natural language of physics. Mott's approach, however, was always intuitive; he rarely presented his conclusions as the result of mathematical analysis—rather he was inclined to talk around a problem to convince others that the world could not be otherwise than as he saw it. The books of Mott and of Seitz exemplify the difference. Seitz's is a compendium of both basic and esoteric information, presented in formal detail; Mott-and-Jones and Mott-and-Gurney are inspiring works, however irritating for those who seek technical mastery alone.

Slater [53] falls somewhere between these two extremes. Older than Mott, and living during a crucial period in Copenhagen, he was in a position to make fundamental contributions just before and during the rise of quantum mechanics. Indeed he possessed the talent to do so but was no match for Heisenberg and Dirac. It was as a somewhat disappointed man, feeling himself diminished by Bohr's proprietorial attitude, that he took to the application of quantum mechanics instead of its further development. Perhaps in reaction he tended to subordinate his powers of intuition to formal analysis and, later, to elaborate computation; and he always distrusted Mott's informality. Yet when he lectured he concentrated on physical principles which he presented with outstanding lucidity; and he shared with Mott an enormous appetite for detailed knowledge of individual substances and their behaviour, to which the bibliographies of their books bear witness. In the post-war period, as Mott's range broadened, and experience further strengthened his imaginative power and flexibility, Slater gradually became more rigid and impatient of criticism. Their later disagreements should not, however, be taken to diminish the magnitude of their contributions to the developing science of solid-state physics.

17.11. Early calculations of band structure [54]

Wigner and Seitz's [55] landmark paper on sodium has already been mentioned. The primary intention was to explain what holds the metal together, and they showed convincingly that this comes about because the valence electrons have greater freedom in the metal than in the isolated atom. In the lowest state, not only can the wavefunction spread further in the metal than in the isolated atom, meaning that the kinetic energy is less, but the potential energy of the electron in the regions between two atoms is less than at the same distance from the centre of an isolated atom. The consequent lowering of the energy is greatest at the centre of the first Brillouin zone. The extra energy of the other electrons in the band does not outweigh this advantage, so it still requires energy to separate the metal into its constituent atoms. Wigner and Seitz found the binding energy to be greatest when the atomic spacing in the body-centred cubic lattice was 3.6 Å (experimental value 3.8), and was then 1.07×10^5 J mol^{-1} (1.13×10^5); they admitted that the remarkable agreement might be an accident of compensating errors. Indeed, their more searching later study [56] made things rather worse, though by no means bad enough to disparage the attempts; rather, it revealed the great difficulty attending this type of calculation, which to this day remains uncertain in accuracy.

Wigner and Seitz set out to solve Schrödinger's equation for a valence electron in a single atomic cell, in this case a truncated octahedron. The sodium ion at the centre is assumed to have the same electronic configuration as in the free atom, while the single valence electron has the whole cell to itself; its wavefunction must, however, join smoothly to identical wavefunctions in neighbouring cells. They could not avoid approximations in solving the wave equation, but the principal weakness of the calculation is their assumption that the electrons repel one another to the extent that the two valence electrons never occupy the same atomic cell. This turns out to involve little error when the electrons have the same spin and are kept apart by the exclusion principle even without the Coulomb repulsion to help; but the error is greater for electrons of opposite spin, when only the Coulomb repulsion enforces separation and in doing so makes a significant contribution to the energy.

This so-called correlation energy is a skeleton in the solid-state closet. It is the archetypal many-body problem which resists reformulation in terms of a single electron moving in the effective field of all the others. Even if the simple Sommerfeld model is taken as the starting point, it remains almost intractable. It is an odd property of the electron gas, a consequence of the long-range Coulomb force, that it behaves more like a gas of independent particles at high density than at low. Indeed Wigner [57] pointed out that in the low-density gas the electrons incur little penalty in the form of extra kinetic energy if they arrange themselves in a body-centred cubic lattice so as to minimize their electrostatic potential energy†. At high density this is not so; with very little loss of freedom the electrons can arrange to screen each other's Coulomb fields, and so they interact appreciably only during close encounters. For this limiting case Wigner succeeded in developing an approximate theory and suggested a smooth interpolation formula for the correlation energy, covering the

† *Wigner's lattice was anticipated by Lindemann [58] in 1915, but of course he could not develop the idea convincingly by the classical means available to him.*

whole range of electron density. Although the physical principles have been clarified by later work, and although Landau [59] and Anderson [60] have pointed out that there must be a sharp phase change rather than a smooth transition between the low-density and high-density states, Wigner's formula has hardly been improved on and continues to find application in situations very far removed from the idealized Sommerfeld model. On occasion effects attributed to the low-density Wigner lattice are said to have been observed, but they are not altogether convincing. What is undeniable, however, is that Coulomb interactions have far more important consequences in materials where the mobile electrons are only sparsely present, semiconductors and insulators, than in metals.

The many-body problem also arises when a metal contains substitutional impurities of a different atomic number, as if extra charge has been attached to some of the nuclei. If these are negative the electrons in the vicinity are repelled and thus a compensating field is generated that prevents the Coulomb field of the extra charges from extending beyond about 1 Å; similarly a positive charge attracts a screening excess of electrons. The screening idea was not new, but had been introduced by Debye and Hückel [61] in their theory of strong electrolytes. By eliminating the long-range field the ability of the impurity to scatter electrons is greatly reduced, and Mott [62] showed that this concept yielded a reasonable estimate of the resistivity of a dilute alloy. Although an analogous process, by which the Coulomb field of each electron is on the average screened by the others, has often been envisaged, a convincing systematic theory has not emerged. Nevertheless Baber [63] and, independently, Landau and Pomeranchuk [64] made use of this idea to explain away the long-standing riddle of why electron–electron collisions seemed not to occur. Two factors are involved, screening and the exclusion principle. Screening reduces the cross-section for collision to the extent that the mean free path for an electron in sodium might be between 1 nm and 10 nm (10–100 Å) if the electrons were classical particles; this improvement is insufficient to salvage the Wiedemann–Franz law. In a Fermi gas, however, only rather exceptional collisions are allowed, since after collision the electrons must find empty states to occupy. This reduces the number of effective collisions by a factor of about $(T/T_F)^2$, and makes electron–electron collisions in sodium at room temperature 10^3 times less frequent than electron–phonon collisions. It is clear that electron–electron collisions will be hard to detect in sodium and other simple metals, and the Wiedemann–Franz law is safe. Baber, however, was interested in nickel and the other transition metals in which the electrons are very far from free, and here electron–electron collisions may dominate the electrical resistivity at very low temperatures.

Let us briefly return to the developments that followed the papers of Wigner and Seitz. Slater immediately began to extend the calculations

See also p 1290

Electrons in Solids

from the lowest energy state in the valence band of sodium to higher states. His account [65] of the physical ideas is an excellent example of his lucid style of exposition, with the tough mathematics discreetly withheld from the timid reader. He, and others, were struck by how flat the wavefunctions are in the greater part of the regions outside the ion cores—though in the cores they cannot avoid oscillating like the 3s wavefunction of the valence electron in the isolated atom. This led him to try solving Schrödinger's equation in terms of a set of fabricated functions that combined the oscillations inside the cores with plane waves elsewhere—the augmented-plane-wave (APW) method. An alternative procedure was the orthogonalized-plane-wave (OPW) method which built up the wavefunction out of only a few travelling waves, each constructed so as to be intrinsically orthogonal to the core functions, as demanded by Schrödinger's equation. This scheme was devised by Herring, a Princeton student of Wigner's who had migrated to MIT to work with Slater, but also came under Bardeen's influence at Harvard. His profound understanding of physics, as well as his personal achievements, have made him a greatly sought-after counsellor. Neither of these calculational processes can be described in simple language, but like many other difficult topics are treated with clarity, and even occasional humour, in the textbook of Ashcroft and Mermin [66]. Earlier than either, a paper by Hellmann and Kassatotschkin [67] anticipated by nearly 25 years, and ultimately inspired, the idea of the pseudopotential that came to play an important role. It seems, however, to have escaped notice at the time.

At this point we must take note of the growing interest in the physics of semiconductors, which after World War II brought about a revolution in technology and in society as a whole.

17.12. Semiconductors—the early years [68]

In the post-war years semiconductor physics took centre stage for a while, when the invention of the point-contact transistor in 1948 showed where rich rewards were to be harvested. Commercial applications of transistors to hearing aids and portable radios ('trannies'), heralding their invasion of every waking activity, took more than two decades to get under way. By contrast the effect of semiconductors on the research programmes of universities and far-sighted industries was extremely rapid. Shockley's [69] book in 1950 was already a mature account of a substantial body of data and basic theory, and by 1962 some authorities felt that the subject was nearly exhausted [70]. They were mistaken, but they reflected the dismay of many victims of a new phenomenon—a research field being stripped of its best crop before the pioneers had had adequate opportunity to enjoy what they had begun.

Of course the transistor did not spring from nowhere; to find the first hints of this gigantic enterprise we must backtrack at least as far as

1874. In this year two papers appeared independently—Braun attached electrodes to a crystal of galena (lead sulphide), one to the base and the other pressed against the opposite apex, and found conduction to be better in one direction than the other; and Schuster found the same effect when two copper wires, which had long lain around in the laboratory, were lightly pressed together. Schuster detected the effect by incorporating the contact in a circuit driven by a primitive alternator, and finding a steady galvanometer deflection. He may therefore be credited with being the first to rectify an oscillatory signal, though nothing came of this, and only in 1906 did General Dunwoody of the US Army file a patent for the use of carborundum as a radio detector, only a short time after Pickard had filed a patent for the use of silicon [71]. These were the first attempts to profit from an idea that had been tried out a little earlier, independently by Braun and by Pierce in the US, in the hope of replacing the clumsy coherer. By 1915 the crystal and 'cat's whisker' were sufficiently widespread to provide the *Oxford English Dictionary* with its first published usage.

Apart from its use as a radio detector the crystal rectifier interested Pierce as a physicist, but though he made little progress in understanding it, he was able to eliminate one conjectured mechanism. Since all rectifying crystals showed strong thermoelectric effects it was conceivable that the junction with a metal electrode, being heated by the current, generated a voltage that enhanced the ohmic voltage for one direction of current and diminished it for the opposite direction. If this were the process involved, it would take time for the temperature of the junction to settle down, and there would be a lag between the application of the current and the development of the asymmetric voltage. With one of Braun's oscillographs Pierce [72] could detect no such lag. For his part Braun was engaged in showing that there was no transport of sulphide ions in galena and that the current was therefore presumably electronic; moreover he found no evidence for deterioration of the rectifying contact after prolonged current carrying; this was an important practical result.

The physical investigation of semiconductors received great impetus from Presser's selenium rectifier of 1925 and from development of the cuprous oxide rectifier, which was able to handle considerable currents. Grondahl started on this in 1920 but published nothing until 1926 when a workable design could be marketed. His full account [73] contains an excellent bibliography of earlier work on semiconductor rectifiers. In the following years cuprous oxide was adopted as the archetypal semiconductor and was studied systematically, especially in Erlangen and Leningrad. It was Gudden, Pohl's most talented student, who brought to Erlangen the Göttingen tradition of working with meticulously prepared single crystals. As Faraday (1834) and Hittorf (1851) had observed long before, poor conductors tend to conduct better at higher temperatures. With the advantage of a greater range of temperature Gudden showed that the resistivity followed closely the type of behaviour introduced into chemical kinetics by van't Hoff, $\rho \propto e^{\alpha/T}$, with α a constant representing some sort of excitation energy. It seemed to him that between 0.2 and 0.4 eV of energy is needed to detach an electron from its parent site; as the temperature is raised, more manage to free themselves and the resistivity drops.

Gudden also confirmed what Bädecker had noted in 1908, that the conductivity was highly sensitive to the impurity content of the crystal—the more impurity, the better the conduction, with excess of either copper or oxygen counting as impurity. He presented his results in detail to a physics meeting in Bad Elster, in September 1931 [74]. Schottky and five other German physicists also reported their work here, to give an overview of a lively research effort, which provided the basis for A H Wilson's interpretation. A small energy gap allows thermal excitation to produce vacancies in the lower (valence) band and electrons in the upper (conduction) band, both contributing to the conductivity. In addition, impurities of a higher valency than the host may contribute to the conduction band electrons which are only weakly bound to their parents (n-type semiconductor); or, if the valency is lower (p-type), create vacancies in the valence band which are manifested as holes. Somewhat later, Mott and Gurney [75] quantified Wilson's idea with their treatment of a substitutional impurity, such as an antimony atom in germanium, effectively adding one positive charge to a germanium nucleus and one extra electron. In effect the perfect germanium crystal is treated as a modified form of empty space with a rather high dielectric permittivity, through which the electron moves as a free particle, attaching itself to the extra positive charge as if in a Bohr orbit of hydrogen. Because of the dielectric permittivity of the host lattice, and because the curvature d^2E/dk^2† at the bottom of the conduction band is greater than for a free electron, the binding energy is much smaller than in hydrogen. Even at room temperature some of these electrons can escape into the conduction band and make a marked difference to the conductivity. Similarly an impurity of lesser valency (indium in germanium) behaves like a positive hole in Bohr orbit round a negative nucleus.

See also p 1300

If an electron is excited into the conduction band, generating a hole in the valence band, electron and hole attract one another and may form a bound pair, just as (a much later discovery) an electron and a positron combine to form a positronium atom. Energy is needed to separate them into freely moving charge carriers. The idea of a bound pair, now referred to as an exciton, was put forward by Frenkel [76] even before Wilson's theory, but the physical meaning did not become

† In classical and relativistic mechanics dE/dp is the velocity of a particle, and d^2E/dp^2 the reciprocal of its mass. For an electron in a solid the analogy is $\hbar^{-2} d^2E/dk^2 = 1/m^*$, where m^* is the effective mass. Electrons of small effective mass are only weakly bound in a Bohr orbit.

clear until Wannier [77] gave a searching analysis in 1937. In the same way as the centroid of a hydrogen or positronium atom is governed by Schrödinger's equation, so is the centroid of an exciton; its motion through the crystal lattice is described by its own Bloch wave. This is already a more sophisticated description than is afforded by Mott and Gurney's smoothing out of the effects of crystal fields into a permittivity and an effective mass. In its ground state, at zero temperature, a pure insulating crystal has a full valence band and an empty conduction band. The lowest excited state does not correspond to one electron transferred to the conduction band, but to the formation of an exciton at rest; above this are higher excitonic states, analogous to higher Bohr levels in a hydrogen-like atom, which can be recognized in the optical absorption spectrum of the material.

Frenkel's ideas grew from an interest in the absorption of light in transparent ionic crystals, a field that had reached experimental maturity some years before semiconductors. After World War I, Pohl found himself in Göttingen with virtually no equipment, and small chance of acquiring any in the impoverished state of Germany. In these straits he and Gudden decided to look for internal photo-ionization in ionic salts. They were convinced that only well-prepared and well-characterized samples would yield worthwhile results, and it was their work above all that persuaded Mott in Bristol and Seitz in Rochester (later Philadelphia) to shift their theoretical focus from metals to insulators and semiconductors. By 1940 Mott and Gurney's book had firmly established the new field, Mott and Jones having dealt faithfully with metals in 1936. Pohl's own work was summarized very fully by him in 1937 [78], and the findings of a fourth important school, Ioffe's in Leningrad, were presented (in French) in 1935 [79]. A detailed historical survey by Teichmann and Szymborski [80] allows us to pass over this important area of solid-state physics very briefly.

When sodium chloride, to take a specific example, is heated in the presence of sodium vapour it acquires a deep orange–yellow colour, as a result of strong absorption in the green of the spectrum. Between this discovery and its definitive explanation by Pick in 1940 lie nearly 20 years of experiment and hypothesis in Göttingen, with strong emphasis on experiment. What came to be called an F-centre (Farbzentrum) was shown to be a vacant site from which a negative chlorine ion had been extracted, and which served as an attractive centre for an electron. Various types of F-centre, caused by different crystal defects, are well characterized, with definite, if somewhat broad, energy levels that can be studied spectroscopically like the atomic systems they resemble.

At a temperature well below the melting point of the salt the electron may become detached from its F-centre, which leads to the crystal beginning to conduct, in the same way as a semiconductor with impurities. The transparency of the salt, however, also allows

Electrons in Solids

optical ionization of the F-centres, leading to photoconductivity even at low temperatures. This was not observable in the opaque pre-war semiconductors, but a related property—the photo-voltaic effect—had long been known [73]. If a transparent gold film is evaporated onto the surface of cuprous oxide or selenium, to form a rectifying barrier, and light shone through the film, a voltage develops between it and its substrate; electrons excited into the conduction band of the semiconductor have a preferred direction for migration—towards or away from the barrier. Solid-state photosensitive devices are frequently more convenient than vacuum tube photo-cells, and the production of selenium photo-cells became a significant subsidiary of the selenium rectifier industry in Germany. In principle any rectifier can be expected to show the photo-voltaic effect, and quite high efficiencies have been obtained in the conversion of sunlight to electrical power by p–n rectifying junctions of silicon [81]. This is a laboratory development with limited application at present, e.g. to spacecraft and pocket calculators, as the cost is not yet quite competitive enough to make photo-voltaic generation worthwhile on a large scale. Nevertheless it is not a hopeless dream, and if the optimum efficiency ($\sim$25%) can be achieved reliably and cheaply there may be yet another industrial revolution in store, with tropical countries the leading beneficiaries of an environmentally benign source of power.

In 1933 Landau [82] pointed out that an electron in an ionic crystal should polarize the lattice in its neighbourhood, attracting positive ions and repelling negative ions, to create a region of lower potential energy in which it may be trapped in a bound quantum state. This was thought at first a possible mechanism for the F-centres, but by 1940 Mott and Gurney had concluded otherwise, and doubted whether self-trapping had ever been observed; they did not, however, doubt the validity of the basic argument. Landau considered that the relatively slow response of the lattice deformation would tie the electron down, but the idea was laid to rest, for the moment, by Fröhlich, Pelzer and Zienau [83]. The theoretical problem of a moving electron accompanied by its polarization cloud is of the same type as arises in quantum electrodynamics (where it is the vacuum polarization that affects the electron's behaviour) and they adopted the same theoretical techniques. Theirs was, in fact, a very early example of field theory being applied to solid-state physics. The analysis assumed that the polarization extended to a distance considerably exceeding the lattice spacing, and with this reservation the result was unequivocal—electron and polarization can move together with dynamical properties only slightly modified from those of the bare electron; the kinetic energy of ionic motion, as the electron moves, only adds at most 11% to its effective mass. This work on the *large polaron* led Fröhlich to realize that a second electron might fall into the pit dug by the first—that is to say there would be an attractive force between electrons that could counteract their Coulomb repulsion. From this idea came eventually the accepted theory of superconductivity, which is treated elsewhere.

> See also p 936

We shall not carry the story of polarons beyond this point, at which the theory becomes too complicated to allow an account that is both brief and intelligible. It is, however, important to draw attention to Holstein's [84] contribution which clarified the scope of Fröhlich's treatment. If the electrons, before lattice polarization is considered, are rather tightly bound (as in Bloch's first paper) and if the lattice is easily polarized, the resulting polaron is small—hardly larger than the atomic spacing—and in this case virtually immobile, as Landau had conjectured. The electron can now only move by hopping out of its well, leaving the polarized lattice to relax at its own pace, and digging itself another well at its new site.

One other later development of a topic introduced here, but also not to be taken further, concerns the behaviour of excitons when they are present in great numbers, as when a powerful laser is used to generate them. In very pure germanium their lifetime before they recombine—the electron dropping into the vacancy represented by a hole—is long by the standards of solid-state physics, of the order of 1 ms. Furthermore, they can be concentrated by stressing the crystal at one point, when they migrate to the region of maximum stress (minimum energy). Figure 17.10 is a photograph taken with a filter that lets through only the light emitted in recombination. In fact, at this density the excitons dissociate in accordance with a model proposed much earlier by Mott, which will be discussed soon; what is photographed is a neutral plasma of electrons and holes. It has been found that when a certain density is exceeded the plasma condenses into an almost incompressible liquid; at this point there is a change in the quality of the radiation, and it has even been possible to determine the phase boundaries for both dissociation (Mott transition) and condensation [86]. But these beautiful experiments belong to the recent past, and we must return to earlier developments in semiconductors that led to the modern era of solid-state electronics.

> See also p 1334

17.13. Rectifiers and transistors [87]

By the mid-thirties the cat's-whisker rectifier was of interest only to the humblest of radio enthusiasts, but change was imminent. The frequency range to which vacuum tubes could operate as generators was being extended into what is now called the microwave range, above about 10^9 Hz, but the development of amplifiers and rectifiers was lagging. When Southworth [88], at the Bell Telephone Laboratories, began the first serious experiments on waveguide propagation (the theory had been established by Rayleigh in 1894) he was drawn back to the now-obsolete silicon crystal rectifier as the only device of any use at his frequency of 2×10^9 Hz. This was one of the research efforts that started the

Electrons in Solids

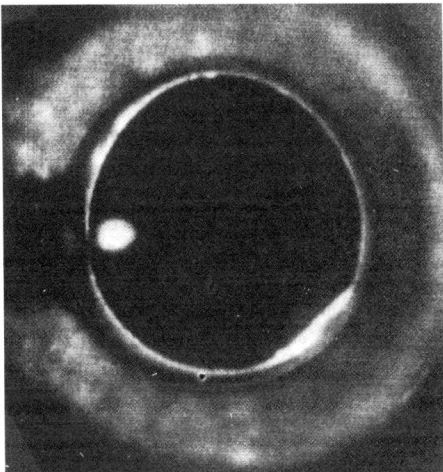

Figure 17.10. *The bright spot on the left is caused by light emitted when electrons and holes recombine in a germanium crystal* [85]. *The diameter of the crystal is 4 mm.*

Bell Laboratories on the course that led to the transistor. It was not, however, Bell that took the lead when radar at centimetric wavelengths was seen as a possibility in the early days of the war in Europe. At that time British radar was further advanced than American, and it was British Thomson–Houston that produced the first cartridges, containing a silicon wafer and cat's whisker, that thereafter played a ubiquitous role as rectifier and mixer.

In the pre-war years Mervin Kelly, a research director at Bell, noting how the bulk and slow action of mechanical relays were holding up the growth of the telephone network, began a programme aimed at replacing them by a solid-state device. In his talented team two members were outstanding—Shockley, a theorist and inventor, and Brattain, an experimenter—and their task was to look into cuprous oxide rectifiers as the basis of a new type of switch. It was not long before they became ambitious to devise semiconductor-based amplifiers, and soon ideas were conceived that only came to fruition after the war; the field-effect transistor, for instance, had occurred to several minds, including Shockley's, from 1930 or even earlier, but was not realized till 1960. The entry of the US into the war interrupted the programme.

During the war the organization of semiconductor research in the US, and control of government funds for the task, were assigned to the MIT Radiation Laboratory [89]; they hived off materials research to industry and universities, including Purdue where a small group under Lark-Horowitz investigated germanium as an alternative to silicon in rectifiers. It was through their work that both elements were finally established as semiconductors; this was not known in Germany until after the war.

William Bradford Shockley

(American, 1910–90)

Shockley's career of innovative research lasted barely 20 years, interrupted by World War II, but for all that his contributions to solid-state physics and to the electronics revolution make him a figure of outstanding importance. Under Slater he gained an early grasp of electron band theory which he took to Bell Telephone Laboratories in 1936. He soon conceived the ambition to find a replacement for the thermionic vacuum tube, and thenceforth until he left in 1955 he was the driving force behind the invention and development of the transistor and other solid-state circuit elements. He was a systematic inventor, with a wide knowledge of physics, whose principles, especially as applied to materials, he understood as well as anybody, and expounded with a clarity that helped unfamiliar notions to be rapidly assimilated and applied. After 1955 he never again touched these heights, and in later life turned his energies towards eugenics and other controversial matters to which he brought high intelligence and somewhat simplistic beliefs concerning racial differences. He could not have chosen a worse time to evangelize for the breeding of superior intellect (defined in his terms) and by the time of his death he had become a target for popular abuse.

Though a rare element, germanium was a useless constituent of flue gases, from which it can be fairly easily separated; it is much easier to purify than silicon, and its capabilities as a rectifier were found to be very attractive. Indeed, in the post-war years it seemed destined to dominate the sophisticated end of the rectifier market (cuprous oxide was good enough for heavy electrical work) until about 1960. By then the purification and production of near-perfect crystals of silicon was driving it out and inaugurating the growth of a vast industry, surely surpassing anything imagined by the inventors of the first transistor barely a dozen years earlier.

In the first wave of enthusiasm for quantum mechanics it was supposed that rectification involved the tunnelling of electrons through an insulating barrier between semiconductor and metal. It was soon realized however, that this would give the wrong direction of easy flow; moreover, the measured capacitance of the junction indicated the presence of a barrier too thick to allow tunnelling. Any barrier must instead be surmounted by electrons excited thermally, and Schottky's first model along these lines was regarded as convincing by Mott and Gurney in 1940. His conjecture of a thick insulating oxide layer was indeed plausible for the cuprous oxide rectifier, but hardly so for germanium and silicon, and in 1942 he made another proposal that has survived. As illustrated in figure 17.11, for an n-type semiconductor a double layer of charge is needed to equalize the Fermi energies of metal and semiconductor at the contact. If, as in the diagram, the metal surface must be negatively charged, it can gain the required number of electrons without appreciably disturbing the copious reservoir of electrons within the metal; the semiconductor, however, has only a sparse distribution of donor impurities which, being raised above the Fermi level, lose their electrons and provide the required positive charge. In the region, the *depletion layer*, the only free charges are those that cross the barrier from the metal and those that enter it from the depths of the semiconductor. In equilibrium there is a small trickle of electrons across the barrier in both directions, and no net current. If the energy on the far right is raised by making the semiconductor negative, more electrons are made available to cross into the metal, and the current rises rapidly; this is the easy direction. There is no corresponding source of electrons when the polarity is reversed. Henisch [90] has given an account of early models and how they were expected to work, and there is no need to take things further here.

It is worth drawing attention to Shockley's exposition which, coming as it did so soon after the point-contact transistor, pays more attention to the details of the metal–semiconductor contact than do later texts, written after the p–n junction rectifier and transistor had ousted their predecessors. He particularly stresses that a typical semiconductor cannot sustain an internal charge imbalance for longer than, say 10^{-11} s; if extra electrons are injected locally into the conduction band of an n-type sample, those already present will rapidly move away and restore the material to its original state. If, on the other hand, electrons are

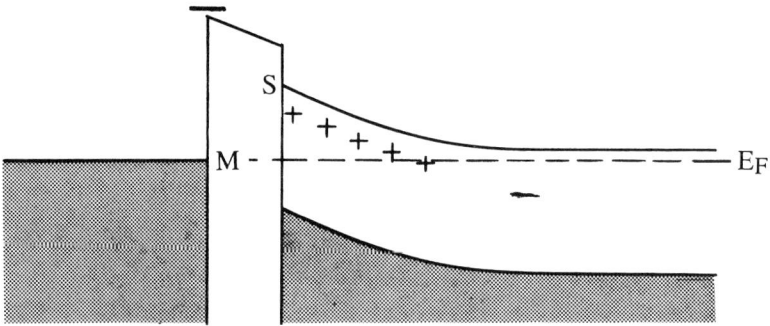

Figure 17.11. *Schottky's model of a rectifying barrier between a metal (M) and a semiconductor (S)* [90]. *The negative charge (−) on the metal surface is matched by a distributed set of ionized donors (+).*

extracted from the valence band (a process usually described as hole injection) electrons in the conduction band will move into the depleted region, to make it electrically neutral, though now with more electrons and some holes as well, so it will conduct better. The importance of this device for modifying carrier concentrations is the reason for the emphasis on minority carriers (e.g. holes in n-type material) in the early papers. One of these [91] describes a beautiful experiment in which short bursts of holes were injected into a rod of n-type germanium to which an electric field was applied. Probes along the rod detected their migration by the change of conductivity as they passed, and so the mobility of the holes could be measured directly. In addition, the attenuation of the signal down the rod revealed their recombination and showed it to be slow. It is the survival and migration of minority carriers injected by one point contact so as to affect the performance of a second contact close at hand that makes possible signal amplification in the point-contact transistor of Bardeen and Brattain [92].

The technological developments that followed the invention of the transistor have been the subject of detailed historical accounts, and need not be elaborated here; but a few comments may be thought relevant. The p–n junction transistor soon superseded the point-contact device, being robust and capable of mass-production in the form of single units that could replace the much larger and shorter-lived vacuum tube amplifier. It was not long before computer builders recognized their merits, and made widely available the means to carry through calculations of a previously disheartening complexity. In 1959 Kilby, of Texas Instruments, applied for a patent covering integrated circuits— at first only two on the same chip, but soon many more. Ten years later the technique of depositing oxide and metal in controlled patterns on silicon surfaces was advanced enough to usher in the era of large-scale integration. Huge memory capacities and logical processing arrays could now be formed on single chips. From 1964 the computing power of a single chip doubled annually, before beginning to slow up around 1976; but it has not levelled off yet. At the same time the cost of a single arithmetical operation fell at a prodigious rate. The effect on daily life, in communications, entertainment and all types of business, needs no comment here, and the character of physical research was affected quite as profoundly. The galvanometer, an almost perfect analogue instrument, was relegated to museums and its place taken by digital instruments directly connected to computers. Henceforth, all too often, measurements were smoothed and fitted to predetermined theoretical formulae without the intervention, fallible or inspired, of critical appraisal. For the first time it was possible to acquire and handle millions of data points, and to contemplate theories that incurred the solution of equations in thousands of independent variables. To set against the loss of chance discoveries, such as in the past initiated important advances, one must grant that little of the physics since about 1970 would have seen the light without the instrumentation made possible by the transistor and its progeny. If it should turn out, as some believe, that this revolution is now reaching its end, and that no comparable enlargement of technical power at an affordable cost comes along, the research physicist may wonder what long-term future can be expected, even while the 'man in the street' may welcome the opportunity to assimilate the changes of a tumultuous epoch.

17.14. Photoconductivity [93]

The invention of the transistor advertised to the physics community that in the Bell Laboratories there was excitement to be had that few universities could match. Several large industrial laboratories in the US took the point, but other countries were still recovering from the war. British industry could indeed have responded positively, had it wished, but preferred to believe it could satisfy its needs by taking licences to use inventions made elsewhere. A considerable, though never fully quantified, emigration of bright young graduates, particularly to Bell, led in the 1960s to serious concern [94] over the 'brain drain', for which no effective countermeasures were found before, in the 1970s, output from the enlarged universities stabilized the position, and indeed gave rise to fears of overproduction of skilled personnel.

Nevertheless, another branch of semiconductor research, originating in military needs, was pursued vigorously in Britain during the post-war period. R A Smith [95], a senior member of the Telecommunications Research Establishment (TRE) which was responsible for airborne radar, was determined not to allow government laboratories to revert to the lamentable state of unpreparedness in which they had languished at the outbreak of war. By encouraging research in new directions he hoped to keep in being a team of first-class scientists, to be on call in time of need. The research topic was ready-made: Germany's scientific war-effort had concentrated less on radar than had the Allies', and had made more progress in using infra-red detectors in combat; the revival of interest in lead sulphide as a photoconductor, under Gudden's influence, may have provided the stimulus [96]. It was the opinion of a British observer [97] in 1947 that with better leadership the Germans might have achieved more; nevertheless, they had led the way, and the challenge was to develop sensitive detectors for wavelengths greater than 5 μm, far in the infra-red, where the radiation from moderately hot bodies is concentrated. Such instruments would also be welcomed in research laboratories, for spectroscopists had begun to appreciate how much information on molecular structure could be gained by exploring this spectral range.

To this end the TRE team began work on photoconductivity in semiconductors, selecting materials with small gaps between the valence and conduction bands, as evident from their photoconductive response to light of low quantum energy. Lead sulphide, and then its selenide and telluride, were fully investigated there, as well as in Cambridge and in the US. At this stage it was noted that two measurements of the energy gap did not always agree, the gap deduced from the temperature variation of the conductivity being smaller than was indicated by the quantum energy needed for photoconduction. The explanation is that the state of maximum energy in the valence band lies at a different point in the Brillouin zone from the state of minimum energy in the conduction band. This does not inhibit thermal excitation by collisions, in which a change in wave-vector is readily accommodated. The velocity of light c is, however, so much greater than electronic velocities that there is very little momentum E/c associated with a photon of energy E and, in effect, wave-vector must be conserved in an optical (*direct* or *vertical*) transition. The minimum energy for photoconduction is therefore determined by the minimum energy difference between states in the two bands with the same value of k.

If there is to be an *indirect* transition at a lower energy, with change of k, there must be a mechanism to supply or take away the difference in wave-vector. Frenkel [76] pointed out in 1931 that a sound wave in the lattice can serve this need, the transition being accompanied by the emission or absorption (if $T \neq 0$) of a relatively slow-moving phonon. Indirect transitions are normally weak but, once looked for, can be detected in the photoconductivity as well as in the impaired transparency of the crystal to radiation of too low a quantum energy to cause a direct transition.

The variation of energy with wavenumber may be quite complicated, as in germanium where the minimum of energy in the conduction band lies neither at the centre nor on a face of the Brillouin zone (figure 17.12). A pioneering attempt by Herman and Callaway [99] to calculate the form

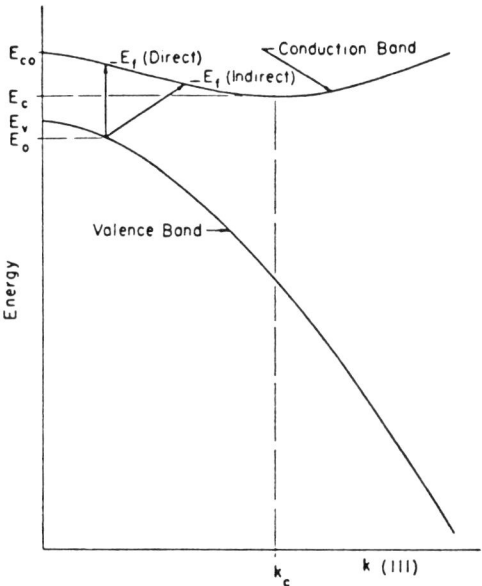

Figure 17.12. *The variation of electron energy with wave-vector along the (111) (cube diagonal) direction in germanium* [98]. *The minimum in the conduction band was inferred from cyclotron resonance measurements (figure 17.15).*

of the valence and conduction bands had hinted at the existence of the minimum in the diagram but it was measurements of cyclotron resonance (to be described shortly) that confirmed it and suggested a refined form of the curve. At about the same time the first experiments with indium antimonide indicated its promising characteristics, and by 1955 Herman [100] had conjectured correctly that in this case the transition of lowest energy would be direct. The energy gap is only $\frac{1}{4}$ eV at zero temperature, so photoconduction should occur as far into the infra-red as 5 μm; at room temperature the gap is even smaller and the range of photoconduction goes out to 7 μm. Since zone refinement of indium antimonide produces crystals of high purity and exact composition, it promised to be an ideal all-purpose semiconductor. In the end, however, advances in the techniques for purifying, crystallizing and doping silicon gradually conquered the market and left indium antimonide as a material of secondary importance except for certain limited applications like photoconductivity, but its thorough study by all available means undoubtedly made a significant contribution to the understanding of semiconductors. When, around 1960, IBM decided to base all its developments on silicon the contest was over, and since then no rival has appeared, though gallium arsenide has established itself, on a smaller scale, for special purposes.

17.15. Semiconductor physics [101]

We shall now forget about applications and concentrate on the physics of semiconductors. After Gudden's [74] 1931 paper on the temperature variation of the resistivity of cuprous oxide, he and his students continued the work, adding other compound semiconductors and studying the Hall effect as well as the resistivity. The diagnostic value of this information is shown in Box 17E. For example, the variation of the Hall constant R_H with impurity content indicates how the concentration of electrons or holes is affected by the impurity. If both electrons and holes are present, R_H expresses the relative importance of the two through a formula that is not so complicated as to preclude interpretation of the measurements. From 1950 the use of liquid helium, to attain temperatures of 4 K and below, became widespread and revealed phenomena that were hidden at room temperature [102]. Earlier results on n-type germanium with small additions of antimony, in which the lowest temperature used was about 90 K, could be satisfactorily interpreted as due to electrons excited from donor atoms or, at higher temperatures, from the valence band. At still lower temperatures, where neither excitation can occur, one would expect the material to be an insulator. But this neglects the inevitable presence in a real sample, however well refined initially, of a few acceptor atoms, e.g. gallium, which can take electrons from donors and leave them ionized at all temperatures. The large Bohr-like orbit of an electron bound to its donor atom may extend as far as an ionized donor in its vicinity, even if there is only one donor to a million germanium atoms; the electron, by hopping to the ionized donor, makes possible a residual conductivity. If no energy is needed for hopping, it can take place even at the lowest temperatures. As far back as 1935 Gudden and Schottky had conjectured this possibility, and it surfaced again sporadically as Fritzsche's [103] account makes clear. In fact the measurements shown in figure 17.13 indicate a small amount of energy needed for hopping, since the resistance continues to rise, albeit not very fast, with falling temperature.

Fritzsche, a student of Pohl's in Göttingen who then joined Lark-Horowitz at Purdue, made a very thorough study of the conductivity of germanium, with carefully controlled amounts of antimony, over a wide temperature range. In addition to clarifying the nature of hopping conduction he revealed another phenomenon that also had been earlier suggested as a possibility, though in a different context. Bloch's theory, as interpreted by Wilson, required an insulator or a semiconductor at zero temperature to have its Brillouin zones either completely full or completely empty; de Boer and Verwey [104] pointed out in 1937 that NiO could not satisfy this condition, but nevertheless was a transparent insulator. Only Mott [105] picked up the problem at the time, and in his first paper on the subject he stressed a distinction in principle between the treatment of dense and rarefied electron assemblies. To give an example

BOX 17E: RESISTIVITY AND HALL EFFECT

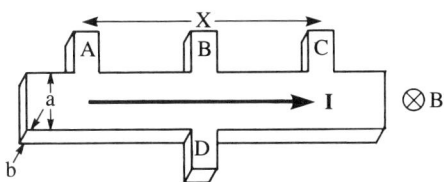

The sample in the diagram has four electrodes arranged to measure the electric field components both parallel and perpendicular to the direction of current flow, in the presence of a transverse magnetic field, B.

The current density $J_x = I/ab$, the field component $E_x = V_{AC}/X$, and the field component $E_y = V_{BD}/a$, V_{AC} and V_{BD} being voltages between the designated electrodes. By suitable measurements, therefore, the quantities on the left can be found.

The longitudinal resistivity, $\rho_{xx} = E_x/J_x$, and the Hall resistivity, $\rho_{yx} = E_y/J_x$.

The mobility, μ, of a group of similar current carriers (electrons or holes) is the mean speed that they acquire in the presence of an electric field of unit strength. It is related to the resistivity ρ_{xx} by the formula $\rho_{xx} = 1/n\mu e$, if only electrons or holes are present.

In a strong magnetic field, $\rho_{yx} = B/ne$, where n is the number of carriers per unit volume.

By measuring ρ_{yx} and ρ_{xx}, therefore, one can determine both n and μ.

If a mixture of electrons and holes are present, the formulae are more involved, and the interpretation of measurements is less assured.

---

not of his choosing, metallic sodium with one valence electron per atom can only have a half-filled Brillouin zone and is therefore, by Bloch's reasoning, a metal. But if we were to separate the atoms to produce, as it were, a cold gas of sodium, each valence electron would remain attached to its atom, since considerable energy is needed to detach it from where it is held by Coulomb attraction, and transfer it to a neutral atom. Hitherto it had been taken for granted that there would be a continuous transition from insulator to good conductor as the atomic separation is decreased, but Mott disputed the assumption, contending that there would be a sharp transition from metallic to insulating behaviour at a certain density. He argued that if one removes a few electrons from their atoms they form a freely moving gas which tends to congregate around

Twentieth Century Physics

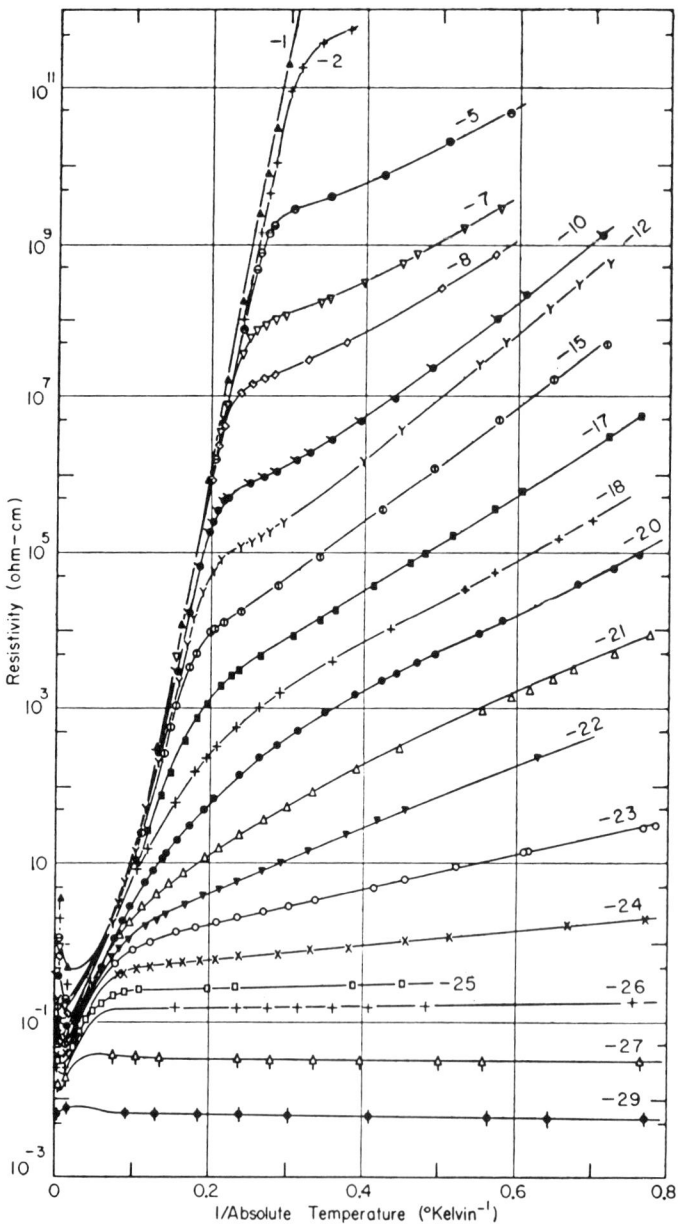

Figure 17.13. The temperature variation of the resistivity of germanium doped with different amounts of antimony [103]. The horizontal scale is $1/T$, so everything occurring above 5 K appears in the left quarter. The vertical scale is logarithmic, with a factor of nearly 10^{15} between top and bottom. The antimony concentrations can be seen in figure 17.14.

Nevill Francis Mott

(British, b 1905)

Mott's research began with the quantum mechanics of scattering and he soon made a reputation by pointing out that the Rutherford law needed modification when identical particles collided. Finding solid-state physics already established in Bristol, when he arrived as first Professor of Theoretical Physics in 1933, he made this his primary interest, as it has remained for the rest of his long, and still active, research career. Every shift of topical emphasis to systems of higher complexity has found him among the leaders,  surrounded always by a cosmopolitan band of disciples. The discovery of high-temperature superconductivity was the latest, and for him the most exciting, of the challenges to his intuitive powers; but it is too soon to know whether his theory will be that one among many speculations to survive. A certain vagueness in daily affairs may be explained by the ability, remarkable in his younger days, to carry many problems in his head at once and to take note privately of what was being said even while thinking of something quite different. A deep concern for social issues and a strong sense of belonging to an ancient culture are factors that led him in middle age to convert from his free-thinking background to the Christian faith, and he has brought an undogmatic lucidity to his discussions of science and religion.

other atoms and reduce the attraction of their ion cores for the valence electrons, thus helping them to become ionized also. But the atoms must be close enough together for the screening effect of a few to be effective in producing so many more ions and free electrons that the process will be cumulative and result in a completely ionized system, that is, a metal. In his view the nickel atoms in NiO were far enough apart for this not to occur.

It is not easy to adjust the separation of sodium atoms, but Mott seized on Fritzsche's measurements as confirmation of his view. As figure 17.14 shows, at 2.5 K the resistivity of germanium is extraordinarily sensitive to impurity concentration over a short range—doubling the antimony content can reduce the resistivity by a factor of 4000, and when the resistivity is low it is almost independent of temperature, like a metal at low temperatures. It must be said that Mott was not the only one to suggest the formation of a metal-like impurity band, but no one else had predicted a sharp transition. Of course, it is not quite sharp in these measurements, but that could hardly be expected with a random distribution of impurities and the inevitable variations of their density in different parts of the sample. The Mott transition has achieved canonical status, and had attracted so much attention that in 1968 a conference lasting $2\frac{1}{2}$ days was entirely devoted to it [106]. The opening address by Mott himself is a good source of historical information on the subject. The same process is responsible for the dissociation of excitons at a high enough density, as noted in connection with figure 17.10.

Enough has been said about the importance of conventional studies of resistivity and Hall constant, and to these we must add magnetoresistance—the change of resistance of a sample in a steady magnetic field. Since it was known from the pre-quantum days that no significant effect could be looked for in an isotropic electron gas, explanations were sought in Bloch theory and in detailed calculations based on rather ill-founded guesses about the shapes of energy surfaces. On the whole, however, magnetoresistance was of minor relevance to semiconductor physics, certainly compared with what it was to reveal about energy surfaces in metals, to which we shall turn later. A much more direct approach to the energy surfaces in semiconductors was provided by a phenomenon new to solid-state physics (though well-known to ionospheric physicists)—cyclotron resonance which, as already mentioned, found immediate application in refining calculations of the band structure of germanium. The idea originated independently with Dorfman and Dingle [107], who pointed out that electrons moving in orbits in a uniform magnetic field can gain energy from an electric field if its direction is caused to rotate in synchronism with the orbital motion. It soon became clear that a very sparse electron gas would be needed to give a readily interpretable result, but suitable conditions could be attained with a semiconductor such as germanium, at a low temperature and in the dark. In their pioneering experiment Dresselhaus, Kip and Kittel [108] not only detected this resonance and thus determined the cyclotron frequency, but showed that the resonant frequencies were different for the two senses of rotation, i.e. for electrons and holes. Since the cyclotron frequency is eB/m^*, the implication is that electrons in the conduction band and holes in the valence band have different effective masses. In fact figure 17.15 shows several resonances, evidence for a more complex band structure than in

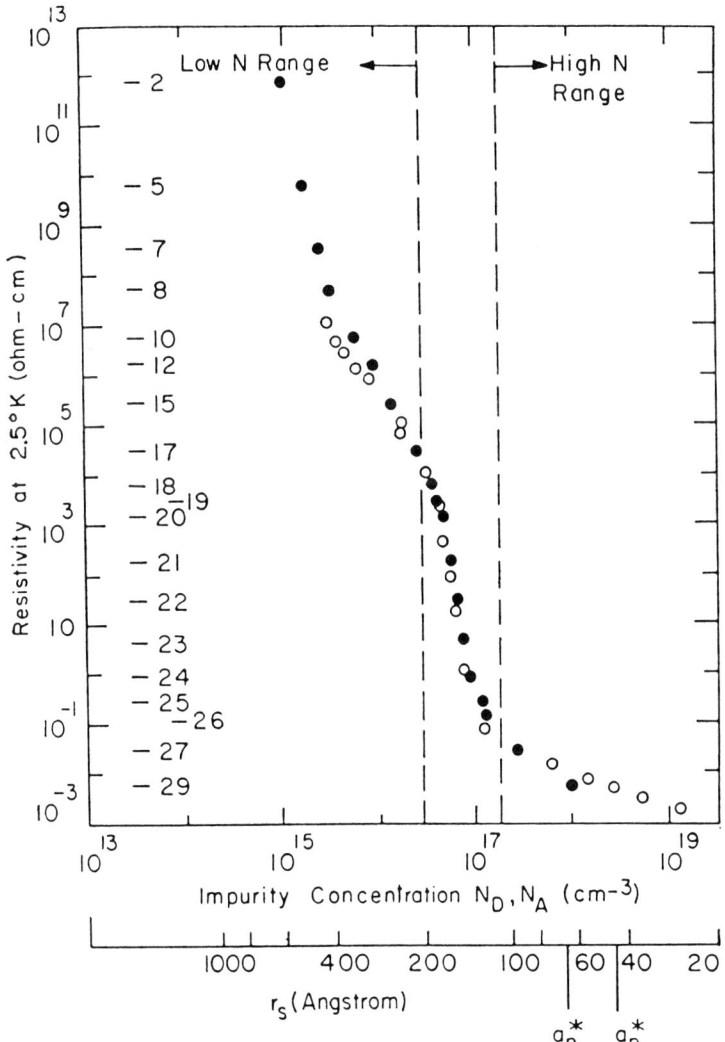

Figure 17.14. *The sharp drop in low-temperature resistivity of germanium around an impurity concentration of 10^{17} atoms cm^{-3}, about two atoms of antimony per million of germanium* [103].

figure 17.12. Later measurements helped still further to substantiate and improve on Herman's calculations.

This work, experimental and theoretical, made it clear that the idea of an electron or hole behaving like a free particle with modified mass m^* is too simple. Not only may m^* change with the orientation of the magnetic field, but the implied quadratic variation of energy with momentum may need to be supplemented with quartic terms. In 1950 Shockley [109]

1335

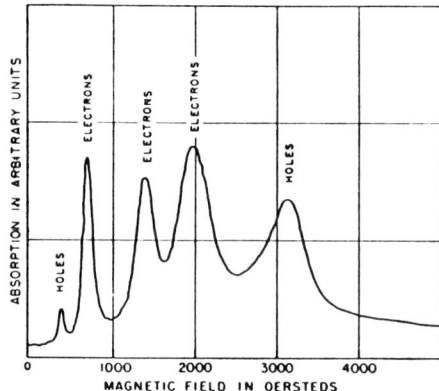

Figure 17.15. *The cyclotron resonance in germanium at 4 K and 24 GHz* [108].

had already been aware of these points as a result of his interest in magnetoresistance. He showed then that the cyclotron frequency may be expressed generally in terms of the cross-sectional area of an energy surface and its variation with energy (figure 17.16). This is probably the first example of truly geometrical, as opposed to algebraic, thinking being applied to the dynamics of electric motion in a lattice. Shockley made clear in conversation that he thought it worth developing further, but nothing happened until a few years later, by which time his idea had been forgotten.

17.16. Hot electrons [110]

Physicists concerned only with metals rarely have to worry about the truth of Ohm's law. Of course, a heavy current can heat a wire enough to change its resistance, but this is a secondary effect. For a significant breakdown of Ohm's law one must impose on the electrons a drift velocity comparable to their random velocities when no current is present. A drift of 1 cm s^{-1} imposed on the electrons in copper, which have random speeds up to 10^8 cm s^{-1}, is no great disturbance, though the resulting current density is 100 A mm^{-2}, more than enough to blow a domestic fuse. The situation is very different in a semiconductor, with far fewer conduction electrons, where the same drift velocity produces only a tiny current and very little heat. By using short pulses of current Ryder in 1951, at Bell, was able to impose drift velocities greater than the initial random velocities, without excessive heating. He found that the current does not continue rising in proportion to the voltage, but eventually reaches saturation (figure 17.17). Shockley lost no time in developing a theory, building on earlier theories of dielectric breakdown, about which something must be said before proceeding with the hot-electron story; it should be remembered that in an insulator, in contrast

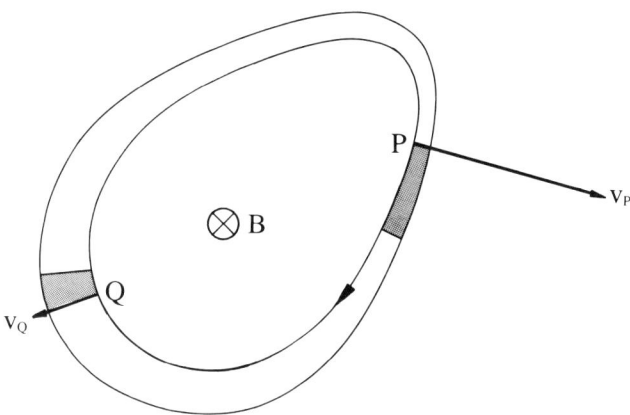

Figure 17.16. Shockley's [109] tube-integral analysis of cyclotron resonance. Two energy contours, sections of energy surfaces in *k*-space by a plane normal to B, are shown. An electron represented by P is caused by the Lorentz force $ev \times B$ to change its state so that the representative point moves round the energy contour at a rate proportional to v. Since the spacing of the contours is inversely proportional to v (the vectors v_P and v_Q are drawn to show this), the point P in its motion sweeps out the intervening area at a constant rate. The equal areas shown shaded at P and Q are swept out in the same time. If the total area between the contours is δA and the energy difference is δE, the time for a complete revolution is $\hbar^2 \, \delta A/eB\, \delta E$, or $(\hbar^2/eB)\, dA/dE$ in the limit of small δE. This result is independent of the shape of the energy contours.

to a semiconductor at ordinary temperatures, there are no electrons (or only very few) naturally present in the conduction band.

Dielectric breakdown has long been known, and was being studied in the early years of the present century, though unsystematically and with little understanding. The first really useful measurements were made in the thirties by von Hippel in Göttingen, where perhaps Pohl's influence led him to work with the alkali halides, and his results were taken in the following decade as touchstones for theory [112]. He himself proposed the avalanche theory—given a strong enough electric field, any electron that happens to appear in the conduction band is accelerated to such an energy, despite collisions, that it can excite another from the valence band; subsequent multiplication by the same process leads eventually to a destructive current density. In 1934 Zener proposed an alternative mechanism by which a strong field causes quantum-mechanical tunnelling of electrons from the valence to the conduction band. This idea has never been strongly favoured, but the avalanche theory was taken up by Fröhlich, and further refined by Seitz, both of whom were satisfied that their estimates of dielectric strength were reasonably in accord with observation; this is not a subject where high precision can be expected in either experiment or theory.

A crucial process in Fröhlich's theory is the interaction of an electron

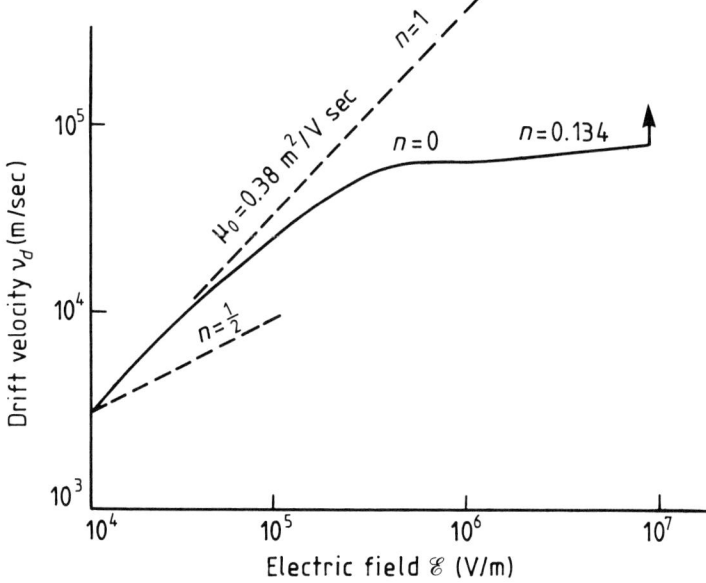

Figure 17.17. *The mean drift velocity of electrons in the conduction band of germanium is proportional to the electric field up to about 3×10^4 V m^{-1}, but then rises more slowly and saturates before a further slow rise and avalanche breakdown at 10^7 V m^{-1} [111]. Note that both scales are logarithmic.*

with the optical phonons—sound waves in which different ions in the unit cell (e.g. Na$^+$ and Cl$^-$) vibrate in opposite senses. These waves frequently show little dispersion, in the sense that it is a good approximation to assume that all vibrate at the same frequency, ω_0, irrespective of wave-vector. An electron accelerated to an energy exceeding $\hbar\omega_0$ is quite likely to excite a phonon and fall to a low energy, only to repeat the operation again and again. If this were inevitable there would be no breakdown. Occasionally, however, an electron manages to achieve a higher energy than $\hbar\omega_0$, and a few runaway electrons continue to be accelerated until they can excite further electrons from the valence band. It is quite tricky to calculate the chance of this outcome, and it is yet another problem to decide how frequently this must occur for the heat generation to make sparkover occur. But Fröhlich and Seitz made satisfactory progress, and it was these ideas that were taken over by Shockley in explaining Ryder's observations on germanium.

The drift velocity of the electrons when the current has become virtually independent of field strength confirms the belief that the excitation of optical phonons is the limiting process. This does not mean, however, that the acoustic phonons, less strongly excited by electrons, are irrelevant. Electrons with insufficient energy to generate an optical phonon can still be scattered by the acoustic phonons, so they pursue random paths on which the electric field superposes a drift. Thus a group of electrons, all starting from rest but taking different paths, soon acquires a distribution of energies, which can be characterized by assigning a temperature to the electron gas. The stronger the field, the wider the spread of energy, and the hotter the gas. By measuring the increased electrical noise Erlbach and Gunn [113] could estimate the temperature; they found that, with a sample immersed in liquid nitrogen and showing no sign of its ionic lattice getting hot, the electron temperature could reach 3500 K or more before optical phonons began to limit the acceleration.

A striking example of phonon interaction with hot electrons was provided by Hutson, McFee and White [114] who studied the attenuation of ultrasonic waves in cadmium sulphide, a piezoelectric crystal in which there is very strong coupling between electrons and acoustic waves. They illuminated the otherwise insulating crystal to introduce conduction electrons in small numbers, yet sufficient to increase the attenuation substantially. On passing a current to cause the electrons to drift in the same direction as the sound waves, they found that when the drift velocity reached, or slightly exceeded, the velocity of sound the attenuation dropped sharply and even became negative, so the crystal burst into spontaneous oscillation. (To see why this happens it may help to remember that a wave always loses energy as it moves through the medium responsible for its dissipation. In this case it is the electron gas rather than the ionic lattice that attenuates the sound wave. If the electrons and the wave both travel from left to right, with the electrons moving faster, then relative to the electrons the wave travels from right to left, and becomes weaker on the left than on the right. The laboratory observer thus sees it becoming stronger as it proceeds to the right.) The amplification sometimes builds up a wave intense enough to shatter the crystal.

This phenomenon and a related one in bismuth [115], neither of them central to the field as a whole, aroused interest at the time, but soon became exhausted as research topics. By that time the basic physics of semiconductors seemed to be well established, and among the large population of solid-state physicists there were many waiting to seize on any novel discovery and work it to death. On the other hand, commercial applications of semiconductors had hardly taken off in 1962, and it was to these that the more far-sighted shifted their attention, and incidentally gave rise to new physics. Some examples may be mentioned briefly before we turn back to the physics of metals, which was taking on new life at this time. At the same time we shall see that the physics of semiconductors is a robust study, for they have shown themselves to be far more versatile than metals in their scope for being manipulated into producing interesting new effects. This is largely a consequence of their sensitivity to doping, by means of which effectively new materials may be produced almost to order. In this respect metals are very inflexible.

There is always a demand for compact, efficient and stable oscillators at as high a frequency as possible, and from the beginning of the semiconductor saga negative-resistance devices were sought that could be used to overcome the inherent losses of resonant systems and permit spontaneous oscillation. One such, probably the earliest, was discovered by Esaki [116] in 1958 when he made a junction between p-type and n-type germanium, both heavily doped. His device differed from the standard rectifying junction in having an exceptionally thin depletion layer through which electrons could tunnel quantum mechanically, between the conduction band on one side to the valence band on the other, so long as no external voltage bias was applied. This is the mechanism, as Esaki remarked, that had been proposed 25 years before to explain the rectifying contact, and abandoned at that time because it predicted the wrong direction of easy conduction. In what he called (conventionally) the reverse direction his diode conducted better than in the 'conventional' forward direction (figure 17.18). The reduction in current comes about when the voltage bias puts the current carriers on each side of the junction at energies in the forbidden gap on the other side. The negative slope between about 0.5 V and 1.5 V can be used to maintain oscillations, and the diode will operate inside a resonant cavity at frequencies as high as 100 GHz, to generate electromagnetic waves of only 3 mm wavelength. Nowadays, however, it is little more than a historical curiosity, having been superseded by more convenient devices with greater power-handling capacity.

A second example of a negative-resistance device, also hailed in its day but now little used, is the Gunn diode [117]. In gallium arsenide there is a shallow energy minimum a little above the bottom of the conduction band. In such a shallow trough the gradient dE/dk is everywhere small, and consequently any electron there moves quite slowly. When the electrons around the bottom of the conduction band are heated by an electric field they may be scattered into this trough, and so, without losing energy, they are slowed down. With increase of field strength this transfer becomes so frequent that the current begins to fall, and the differential conductivity becomes negative. Unlike in a tunnel diode, the seat of the negative resistance is not localized, but extends throughout the semiconductor, and this leads to instability. If we think of ourselves as moving with the mean electron velocity, and imagine a small excess of charge to develop, the electric field it produces causes the neighbouring charges to move towards the excess, rather than away from it as happens in the normal situation of positive resistance. A domain of excess charge quickly builds up, and travels with the mean velocity until it hits an electrode and is dissipated; another domain is then seeded and runs through in the same way. The consequent fluctuations of electric field can be used to drive an oscillatory circuit, and for quite a while the

Electrons in Solids

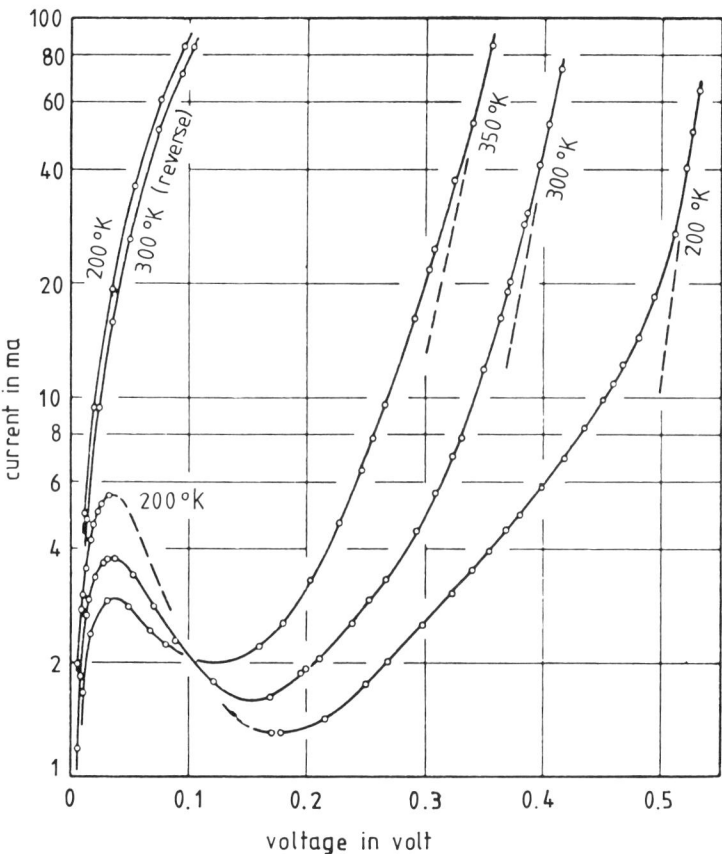

Figure 17.18. *The variation of current (logarithmic vertical scale) with voltage (linear horizontal scale) for a narrow p–n junction in germanium, at three temperatures* [115]. *The negative slope at around 0.1 V can be used to maintain an oscillatory circuit. In the 'reverse' direction the current rises steadily.*

Gunn oscillator was regarded as highly promising. It is still useful as the driver for low-power microwave oscillators in the laboratory but, like the tunnel diode, it has been superseded in many of its original applications.

17.17. Metals in the post-war years—the impact of liquid helium

At the outbreak of World War II there were few laboratories in the world equipped to prepare and use liquid helium, and each had its own home-made liquefier. During the war Collins at MIT developed Kapitza's pioneering expansion liquefier into a marketable product, delivering up to eight litres per hour. By the time he came to write his article [118] for the *Physics Handbook* of 1956 there were 80 models in operation, and since then liquid helium has become a standard laboratory facility.

After about 1960, when commercial production and delivery of the liquid began, it was not even necessary for small users to go to the expense of installing their own machines. The traditional objects of low-temperature research were superconductivity, superfluid helium and paramagnetism, but the importance of liquid helium for the study of semiconductors and of normal (non-superconducting) metals was soon appreciated; programmes were put in hand for measuring electrical and thermal conductivities, and their change in a magnetic field. For this work it is desirable to use the purest available materials, since at 4.2 K (the normal boiling point of helium) the resistance of most samples is determined by the impurity content. Down to the temperature of liquid hydrogen (20 K), scattering by phonons prevents the resistivity of most metals from falling below 1% of its room temperature value, but with some highly refined metals further cooling in liquid helium may produce a many-hundredfold improvement. The mean free path of the conduction electrons is thereby increased from a few tens of nanometres to many micrometres, and this has two significant practical consequences. First, wires or strips may be fabricated that are of large enough cross-section for the material to be of the same quality as in the bulk, yet small enough for collisions of electrons with the boundaries to play a part in determining the resistance. Secondly, when the electron paths in the bulk materials are bent into orbits by a magnetic field, it does not take an enormous field strength to make the orbits small enough to be traversed several times between collisions. Both of these properties allow investigations to be conducted fairly easily that at higher temperatures would be out of the question.

We shall turn to such matters shortly, but must not pass over without comment the considerable volume of work on more down-to-earth properties, such as the temperature variation of electrical and thermal conductivities, aimed at verifying or improving the original Bloch theory, and its refinements by Peierls and the many systematic theorists who came after. The magnitude of this effort is illustrated by the 170-page review article [119] devoted to a single topic—the extent to which the old rule of Matthiessen, concerning the additivity of different mechanisms of resistance, is obeyed, and the often lengthy calculations undertaken to explain rather small discrepancies. The general principles, although elaborated, have survived the tests; it seems that most of the residual discrepancies are irreproducible and only too likely to result from sample imperfections [120].

One important departure from Matthiessen's rule deserves mention, since it stimulated much theoretical work. In 1930 Meissner and Voigt, during a systematic study of the temperature variation of resistivity, had noticed in a few metals (Mg, Mo, Co) that the resistivity fell on cooling, as expected, but reached a minimum in the liquid helium range, below 4 K, and then rose again by about 1%. More systematic studies, especially on the noble metals, confirmed the effect and showed that it occurred when very small amounts of a transition metal (e.g. Fe, Mn) were present in solution [121]. The same samples also exhibited thermopowers many times greater than that of the pure material. The phenomenon excited a number of speculative explanations but remained a mystery until Kondo [122] explained how it arose from the scattering of electrons by the localized spin moments of the solute atoms. This was seized on by the theorists as an opportunity to practise and extend their skill in many-body theory, in consequence of which the explanation of what became known as the Kondo problem remained an almost impenetrable mystery to experimental solid-state physicists. This was one of the earliest examples of a now fairly common schism between theory and experiment, characterized by the reluctance of theorists to explain the underlying physics of their calculations in terms familiar to experimenters. It is not going too far to say that in the early stages of Kondo theory the confused response of seminar speakers to simple questions often led the audience to conclude that the speakers themselves were unable to see beyond the restricted limits of their own formalisms. In such a case I make no apology for passing on with only references to a review article and to texts that do their best to reveal the essential physical ideas [123].

See also p 1167

To turn to more straightforward matters where the electrons may be treated as independent charged particles, the immediate post-war years saw increased interest in measurements on thin films and wires of pure metals at liquid helium temperatures, to study how collisions with the walls affected the resistance. The first observation of what is now called the *size effect*, by Stone in 1898, elicited the first theoretical treatment, by Thomson. Later theories by Fuchs and others, being more firmly anchored in acceptable electron models, can usefully be compared with experiment, but the outcome is not as clear-cut as might have been hoped, as a consequence of such complications as small-angle scattering of the electrons—possibly of only minor importance in the bulk metal, but able to influence the effects of boundary scattering to a significant degree. In the light of Chambers' [124] historical account there is no need to go into further detail.

It is, however, worth commenting on a shift of perspective for which this work can be held responsible. It had been standard practice, following the lead of Lorentz, to begin an analysis of a conduction process by writing down Boltzmann's fundamental differential equation and deriving the appropriate solution. To mathematically trained physicists this was almost second nature, but the appearance on the printed page is discouraging to many experimenters, and the formal process of solution offers little in the way of intuitive insight. When Chambers [125] discussed a particular problem of conduction by a thin wire in a magnetic field, the extra complications of the field and

See also p 587

the boundary conditions made solution of the Boltzmann equation a formidable prospect. He realized, however, that it was only necessary to follow the fortunes of individual electrons at the Fermi energy, to see how far, and in what direction, they travelled between collisions. In effect, he regressed from the sophisticated apparatus of statistical physics to the pictorial methods of primitive kinetic theory, and with his visual imagination thus set free was able to solve the problem. By reworking earlier calculations in his own way he gave impressive testimony to the merits of simplicity. Since then his method, and similar developments, have enabled the mathematically less adept to devise and interpret experimental programmes that might otherwise have seemed too complex to bear thinking about. It must be added that this recipe is not a universal substitute for Boltzmann's equation, though it has a wide range of usefulness.

The particular problem that Chambers tackled was an early example of what grew into a considerable menagerie of experiments involving electrons with long free paths under the influence of a steady magnetic field [126]. Unlike the study of cyclotron resonance in semiconductors, undertaken with the aim of learning about the details of electronic structure in a particular material, the experiments now to be mentioned were for the most part carried out for their own sake, to observe and elucidate attractive and sometimes theoretically challenging phenomena of no very general significance. It was frequently good enough to start with the assumption that the electrons were effectively free, as in Sommerfeld's model, and to disregard the possible complications introduced by the Bloch band structure. Many of the experiments were in fact performed on alkali metals for which it is an excellent approximation. Thus Macdonald, in Oxford, initiated this development in 1949 by measuring the resistance of sodium wires at 4 K in longitudinal and transverse magnetic fields. He found for both arrangements that ultimately the resistance fell with increasing field strength, and his explanation, that the bending of electron paths into helices enabled some to move further without encountering the surface, was well borne out by detailed calculations.

It is easier to develop the theory for a thin plate than for a wire, and Sondheimer predicted that weak oscillations of resistance should be observed as the magnetic field, normal to the plate, is increased. These have been found, sometimes much stronger than originally predicted, to explain which Gurevich had to go beyond the Sommerfeld model and take account of the more sinuous paths followed by electrons in a periodic lattice.

More recently the delicate skill of Sharvin and his colleagues in Moscow has exhibited on a minute scale in metals the phenomena of particle focusing that are exploited in mass spectrometers. For example Tsoi [127] made extremely fine contacts to the surface of a bismuth crystal

Electrons in Solids

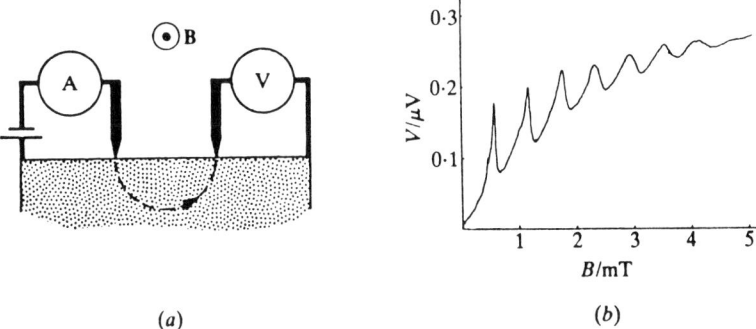

Figure 17.19. (a) A schematic diagram of Tsoi's electron-focusing experiment [127]. Electrons injected from the left-hand point are preferentially focused onto the right-hand point if the magnetic field gives their orbit a diameter equal to the separation of the points or, as (b) shows, a submultiple of this separation.

(figure 17.19) and arranged the magnetic field to focus electrons emitted from one on to the other. It is interesting to see not one spike, but many, in the resistance; apparently the surface was smooth enough to allow a good fraction of the electrons to bounce specularly and, in the stronger field when the orbits are small, to reach the second electrode after several bounces. Obviously the effect depends on electrons not suffering too much random scattering, so the scale of the experiment is determined by the electronic free path, which rarely exceeds 0.1 mm.

17.18. Metals as individuals—the Fermi surface programme [128]

The first theoretical attacks on the problem of what makes one metal differ from another in electronic properties were naturally concentrated on the simplest—the alkali metals [55]—and on the transition metals [129] whose inner, partially occupied, energy bands conferred special properties, notably ferromagnetism in iron. Intermediate between these, and also the subject of attention from 1935 onwards, was copper, whose inner band is full but overlaps the band of conduction electrons. With well-separated bands, as in the alkali metals, it was thought safe to regard the ion cores as rigid sources of electrostatic potential in which the valence electrons, one per atom, moved relatively freely. Not so with copper in which the inner band, with ten electrons per atom, should also be taken into the calculation; but this would have overtaxed computing resources, and various makeshift approximations were adopted. By 1956, after at least six (according to Slater's [130] bibliography) independent attempts by different procedures, it seemed that the results obtained were highly dependent on the assumed form of ionic potential. Prolonged labour produced a very restricted set of numerical values for the energy of a conduction electron at certain selected wavenumbers, and there was only indirect experimental evidence to help decide which, if any, of these results should be believed. Naturally opinions differed; to take a specific example, the sign of the thermoelectric power—opposite to that for the alkali metals—indicated to some that the crystal lattice distorted the constant-energy surfaces so far from the spherical free-electron form that the Fermi surface was pulled into contact with the Brillouin zone boundaries, as in figure 17.3. By contrast Jones [131] was rather committed to belief in the almost free model for the valence electrons, and a nearly spherical Fermi surface, to explain the empirical Hume-Rothery rule†. He was able to make the sign of the thermoelectric power plausible without invoking contact [133], and found support in the calculations of his student Howarth [134] until unfortunately these proved erroneous. All in all, the task of calculating energy surfaces with adequate accuracy, whatever method was used, was disheartening, even to Herman who had enthusiastically set about elucidating the band structure of semiconductors, but eventually retired from the struggle [135].

Soon, however, two independent developments radically changed the situation—experimental methods were discovered that ultimately fixed the shape of the Fermi surface very precisely, and almost simultaneously the power of computers reached a level that allowed electron energies to be calculated at many values of k, with the result that the theoretical shape of the Fermi surface could be predicted [136]. To take the latter first, it became clear, despite previous experience, that the answer was not too sensitive to the choice of ionic potential; moreover the Fermi surface, calculated independently in two ways, agreed encouragingly with the shape newly found by experiment. It was this combination of successes that launched the Fermi surface program and led in due course to an extensive collection of data on almost every metal that could be obtained in pure enough form, and to the critical evaluation of techniques for calculating the energies of electrons in solids.

A convenient starting point for the experimental story is 1940, when H London [137] discovered the *anomalous skin effect* incidentally to measurements of the heat generated by a superconductor in a 1.5 GHz electromagnetic field. Electromagnetic induction prevents a high-frequency field from penetrating more than a little way below the surface of a non-superconducting metal, and the theory of this *skin effect* had long been known. London found that his tin sample at

---

† *Copper alloyed with 2-valent zinc makes brass, and with 4-valent tin makes gun-metal. Hume-Rothery noted that when the concentration that gives an average of 1.33 electrons per atom is exceeded a new crystal structure appears. This concentration of electrons causes the free-electron spherical Fermi surface to touch the zone boundary, and it is tempting to think that this is what makes another structure more favourable. The argument is too simplistic, but Heine [132] has nevertheless shown by a considerably more sophisticated analysis that it enshrines a germ of truth. Whether explained or not, the rule works.*

Electrons in Solids

3.8 K, just above the transition to superconductivity, behaved as though its resistivity was seven times more than at 50 Hz, where the skin depth is much greater, and he drew attention to the analogy with the enhanced DC resistance of thin films due to surface scattering. In view of the equipment available this experiment was a notable achievement, particularly for a man whose lack of manual dexterity was a byword. After the war developments in microwave radar made a repetition quite easy, and I was able to confirm London's result with greater precision [138]. On attempting the theory it became clear that his analogy was not sound. The additional resistance, when the electronic free path is much longer than the skin depth, does not arise from the surface scattering but from the way in which the electric field adjusts itself; thus electrons that enter and leave the skin depth, after colliding only with the surface, take no momentum from the field and hence play no significant part in the conduction process. The only electrons to contribute are those that move so nearly parallel to the surface that they stand a good chance of suffering a collision during their stay in the skin depth. This *ineffectiveness concept* served to describe the results in general terms, though with one undetermined constant whose value was soon fixed after Sondheimer had developed the theory for free electrons in a rigorous fashion and enlisted the aid of a pure mathematician, Reuter, to solve the resulting integral equations to which, in those days, most physicists were strangers [139].

Up to this point the anomalous skin effect was no more than a phenomenon resulting from long free paths, but in 1954 Sondheimer [140] extended the theory to a metal in which the Fermi surface was ellipsoidal rather than spherical. This was a sort of hangover from his student days with Wilson, whose taste was more for soluble models, however artificial, than for the analytical intractability of real energy surfaces. On applying the ineffectiveness concept to this case I was surprised to obtain the same answer and, thus encouraged, wrote down the solution for an arbitrary shape of the Fermi surface [141]. When it turned out that the resistive properties of a pure metal, at a low temperature and a high frequency, were determined (unlike almost every other known property) solely by the geometrical form of the Fermi surface it was obvious that an attempt should be made to discover the form for some chosen metal by appropriate measurements on single crystals, with surfaces cut along different directions so as to make effective the electrons at different parts of the Fermi surface. At the invitation of the Institute for the Study of Metals in Chicago, where the techniques were available to grow and cut slices of copper crystals, and to polish them chemically to the essential smooth and strain-free state, I carried out this task there and was able to devise a Fermi surface that accounted for the measurements [142]. But this surface was pulled out from spherical form to such an extent along the directions of cube diagonals that it would not fit into the

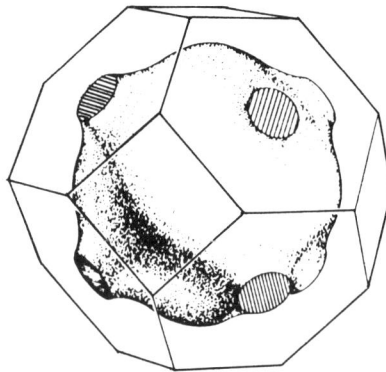

Figure 17.20. *The Fermi surface of copper* [158]. *The shape that was inferred from anomalous-skin-effect measurements had rather smaller contact areas. What is shown here more nearly corresponds to the refined shape given by the de Haas–van Alphen effect.*

Brillouin zone, and it therefore seemed probable that it contacted the zone boundaries (figure 17.20). It was already clear, however, that the method would be hard to make precise, and barely feasible to apply to any more complicated shape. The value of the determination lay less in the result than in the stimulus it gave to making use of a similar geometrical analysis of the far better method employing the *de Haas–van Alphen effect*.

Shoenberg [143] has given an account of the discovery of the effect in 1930, following the observation by Shubnikov and de Haas, in Leiden, that the resistance of a single crystal of bismuth did not vary smoothly with the strength of an applied magnetic field. Instead, particularly at the lowest temperature that they could reach, it showed a slow oscillation with field strength. Soon after, de Haas and van Alphen found a similar effect in the magnetic susceptibility. In both experiments liquid hydrogen was used, and it is surprising that they made only a few preliminary measurements with liquid helium. It was Shoenberg who took up the problem in Cambridge, and more thoroughly in Moscow, where he found a dramatic increase in oscillation amplitude at lower temperatures (figure 17.21). The oscillations became more widely spaced at high fields and, in fact, the theory by then available showed that the period should be constant if the magnetic moment (or torque in this case) is plotted against $1/B$, not B directly. This has since been fully verified, and it is customary to express the periodicity in terms of the change, $\Delta(1/B)$, in $1/B$ in going through one cycle; or, more usually, to define the 'frequency', F, as the reciprocal of $\Delta(1/B)$.

Up to the outbreak of war the de Haas–van Alphen effect (dHvA effect) was an isolated curiosity, peculiar it seemed to the semi-metal bismuth which has only a very small number of electrons in the

Electrons in Solids

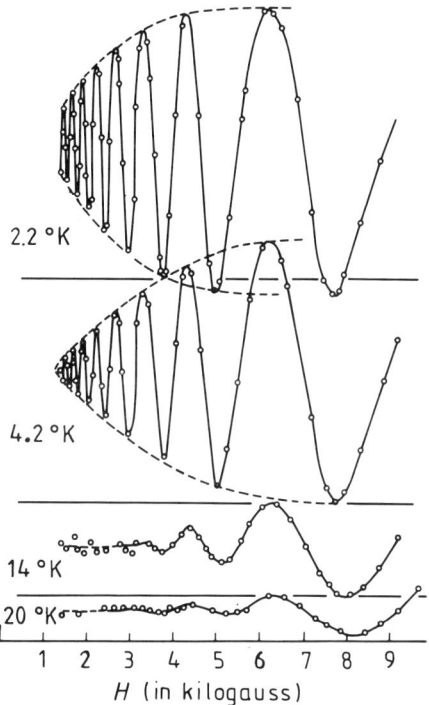

Figure 17.21. *The de Haas–van Alphen effect in a single crystal of bismuth showing how the amplitude of the oscillations increases at lower temperatures. The torque method used by Shoenberg [144] in this early experiment was later replaced by an impulsive magnetic field, and it is not worth describing the precise interpretation of the curves; what is important is the oscillatory character of the magnetic moment.*

conduction band and an equal number of holes in the valence band. It was not expected that the effect would be observable except when, as here, the Fermi surfaces of electrons and holes are very small. However Marcus, his attention drawn to a note in the *Journal of Physics* [145] (which had a brief life as the English translation of a Russian journal) found the effect in zinc. Shoenberg, who had gone on to perform important experiments on superconductors, discovered a renewed interest in the effect, and was soon in thrall to what became a lifelong passion. In Kharkov, where Verkin was the leading experimenter and I M Lifshitz the theoretical inspiration, there began an independent and equally successful search for new examples, but the modest magnetic fields available limited their discoveries to little bits of Fermi surface, as in bismuth and zinc. To detect effects from large surfaces, which most metals must possess, much stronger and highly homogeneous fields are needed. As the last English student of Kapitza, who had returned to Moscow in 1934 and was unable to leave again until 1965, Shoenberg

1349

was familiar with his techniques for generating short and intense pulses of current. The idea appealed, not only as a way of obtaining a stronger field than with an electromagnet, but because, as the field changed, the resulting oscillations of magnetic moment induced a signal in a pick-up coil, from whose frequency F could be determined.

Shoenberg has recounted the pre-war history, from 1930 when Landau predicted the effect, but dismissed its observation as impracticable, through various theoretical treatments to the point when he published Landau's detailed analysis as an appendix to his own paper (Landau himself being imprisoned during Stalin's purge of 1938–9). For the tiny Fermi surfaces in bismuth it was reasonable (as in semiconductors) to assume ellipsoidal energy surfaces, and when the effect was found in other metals the natural tendency was to try to force the results into the same theoretical mould. Even as late as 1954 we find the Procrustean art at work [146], though not without doubts on the part of experimenters and more than doubts from the theorists, Lifshitz in Kharkov and Onsager at Yale. These two had independently devised a more general treatment, though who was first is uncertain since Lifshitz spoke of it to a Ukrainian audience in 1950 but did not publish until 1954, while Onsager, a reluctant author, only published in 1952 ideas that had clearly been in his mind for some time. At the heart of Landau's original analysis was his solution of Schrödinger's equation for a free electron moving in a uniform magnetic field. He had shown that not every energy was allowable, but only those values that led to an integral number of wavelengths in the orbit. It is this orbit quantization that is responsible for the dHvA effect. Onsager and Lifshitz both recognized that this was a problem to which an answer could be given by the old quantum theory. While the newer quantum mechanics ran into difficulties in generalizing the method to electrons in a lattice, the old method rode roughshod over the subtle points. This, said Onsager 'is admittedly a swindle' [147], but it works and has since been thoroughly supported by searching quantum-mechanical treatment. It may be noted that Shockley's tube integrals [109] went half-way to dHvA theory, and his idea was known to Onsager, though very likely not to Lifshitz.

The new geometrical approach, which took as a starting point an arbitrary shape of Fermi surface, without making any assumption about why it took that shape, led to valuable theoretical and experimental developments in the USSR, but not much that involved the dHvA effect, Shoenberg and his students having effectively pre-empted the field for the time being. Even in Cambridge, however, it was only the discovery of the geometrical interpretation of the anomalous skin effect that aroused interest in Onsager's paper, after which geometrical thinking became the order of the day. In Kharkov, Lifshitz and his school applied the same ideas to other phenomena in metals, notably magnetoresistance, and since at this time English translations of Russian papers first began

Electrons in Solids

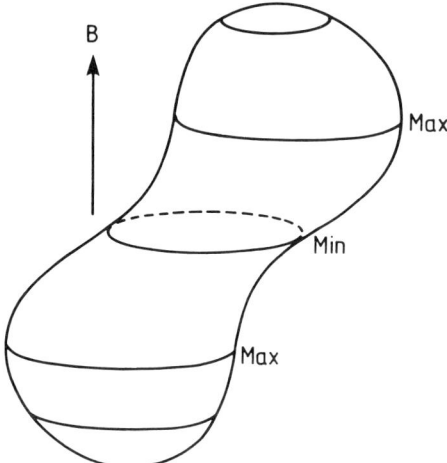

Figure 17.22. *A schematic Fermi surface, showing sections by planes normal to* **B**. *There are two sections of maximal area and one minimal, contributing to the de Haas–van Alphen signal. The other sections make no contribution.*

to appear, an easy and effective interchange of ideas soon developed, though personal contact was still all too rare.

The discovery by Lifshitz and Onsager can be expressed remarkably simply (figure 17.22)—for each orientation of B every extremal cross-section area A_e can give rise to a periodicity in the magnetic moment, $F = \hbar A_e/2\pi e$. Only the area matters, not details of the shape. As this example shows, there may be more than one periodicity, and often many more when the Fermi surface consists of several sheets. By measuring over a range of orientations it is usually possible to decide which periodicities belong to the same sheet. The task is now to construct surfaces that fit the data, normally by devising an expression, containing a small number of adjustable parameters, that may be expected to generate surfaces of the expected type; experimental dHvA data are then used to fix the parameters and check the consistency of the model. This technique of *parametrization* depends for its success on knowing what sort of answer is likely, a matter to which we shall return shortly.

In the anomalous-skin-effect study of copper it was not possible to discern from the measurements alone whether there was contact with the zone boundary. On the other hand, if contact occurs the cross-section has a minimum of area at the boundary, and this should give a dHvA signal; no contact, no signal. Shoenberg therefore set about finding a small single crystal of copper, and eventually acquired a 'whisker' such as is spontaneously extruded from the surface of a stressed metal. At first he was satisfied that there was no neck, but it turned out that he had been misinformed about the crystal orientation of the whisker, and with the

1351

mistake rectified the desired neck signal appeared. In due course silver and gold were found to have Fermi surfaces differing only in detail from that of copper, and Shoenberg's students made careful studies to define the surfaces as precisely as possible.

Computing power was by now sufficient to allow theoretical determinations worth comparing with experiment. Both proceeded in parallel, with the dHvA results serving to validate and refine theoretical techniques. Just at the right time superconducting solenoids appeared. Quite soon they could maintain magnetic fields of 7 T, twice as strong as one can readily get with a good iron-cored electromagnet. More important still, they could be wound to give an extremely uniform field. It was the difficulty of achieving uniformity in 1930 that caused Kapitza to discourage Landau from pursuing his original thoughts on orbit quantization. With superconducting coils there also came big advances in measuring technique, which have been fully described by Shoenberg. All these developments made possible the extension of the programme to more complex metals, whose Fermi surfaces have been determined with high precision.

This experimental precision is not matched by theory, and *a priori* calculation of energy bands remains a relatively imperfect art, albeit that the information obtained is very much better than no information at all. But if the extension of the Fermi surface programme had had to wait on heavy computation in order to explain the results, it would have been far less attractive. Fortunately a unifying concept came to light that satisfied both experimenters and theorists that the energy band structures were not capricious in their variation between metals. On the experimental side it was noticed by Gold [148] that the dHvA results in lead could be explained if one sheet of the Fermi surface had the tubular structure shown schematically in figure 17.23(a). He was led to this possibility by seeing what would happen if the four valence electrons per atom were nearly free. To appreciate the argument let us return to Brillouin's construction (figure 17.9). This is redrawn in figure 17.23(b) with a constant energy contour, i.e. a circle, for a free electron; the various bits of the k-plane are then reconstructed in the central zone, carrying their contour with them, and the pattern is replicated in accordance with Peierls' prescription. The new set of constant-energy contours has sharp cusps, but if there is any lattice potential these will be rounded off. Gold carried out this prescription in the three-dimensional case appropriate to lead, and his tubular structure is the rounded-off version of one of the surfaces that resulted. Systematically developed, it is known as Harrison's construction [149]. Gold's discovery was followed by similar success with other metals. Despite the fact that the electrons are moving in a strong lattice field they behave, paradoxically, as if it is quite weak.

It had been suspected twenty years before, by Hellmann [67] among others, that the electrons might behave as if almost free. In 1958 Phillips

Electrons in Solids

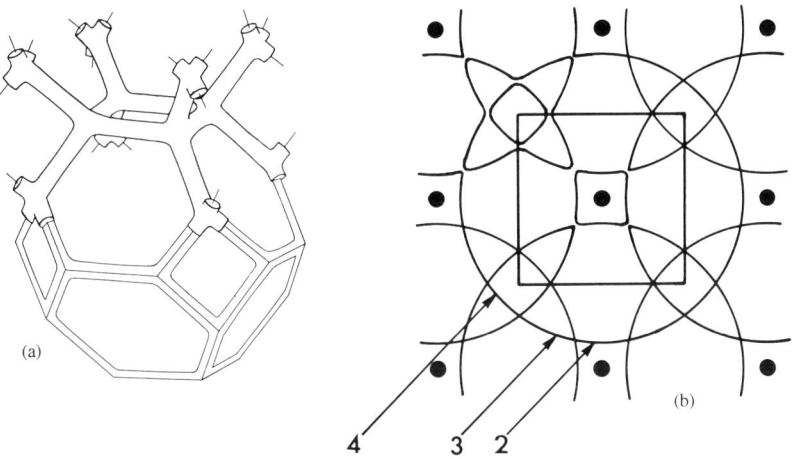

Figure 17.23. *Gold's* [148] *proposal for the Fermi surface of lead, based on the free-electron model. (b) An example of Harrison's construction* [149], *a modification of Brillouin's scheme (figure 17.9). Where Brillouin would have constructed a free-electron Fermi surface centred on the middle of the Brillouin zone (shown as a square), and translated each external piece into the zone, Harrison draws a circle centred on each reciprocal lattice point, and keeps only those parts that lie within the zone. If necessary, periodic extension is used to form closed contours. In this case, rounding of a few sharp cusps at intersections shows how these contours arise when there is a weak lattice potential. The central square contour is derived from the arcs that lie in Brillouin's second zone; the star at the top left-hand corner comes from the third zone, and the square at the corner from the fourth zone. All the other lines are redundant. The central square is a hole contour, and the other two are electron contours.*

[150], alerted by Herring to Hellmann's work, looked more deeply and began to develop the pseudopotential model to which Heine and his research group have made major contributions [151]. The interaction between an atom and its neighbours, in so far as it is mediated by the conduction electrons, can be related to the behaviour of the electron wave in the space between the ion cores. In particular, if a wave is allowed to fall on an ion, all that matters is how much is scattered in various directions, and with what change of phase. The interior process may be very complex, but the final result might frequently have been obtained by replacing the real ionic potential by something far simpler and (as Gold's success shows) far weaker. Thus the computational complexity arising from the real, strong potential can be greatly reduced by replacing it by a *pseudopotential* that has the same effect outside the ion cores. At first this idea was not universally accepted, largely because the critics had in mind phenomena due to processes occurring deep within the ion. Once the limitations were appreciated, however, the pseudopotential took its rightful place as a powerful tool for discussing far more than the shape of the Fermi surface. The original paper by Phillips showed how it was possible to describe the energy bands in diamond, silicon and germanium

by correctly choosing just three parameters to define the pseudopotential for each; indeed, he was able to point out an error in an earlier elaborate computation.

The Fermi surface programme, of which accounts at different levels of complexity have been published, as well as a historical survey, is a fine illustration of the interaction of theory and experiment. Without any guidance from theory, dHvA experiments would not usually have been sufficient for construction of Fermi surfaces. The early successes, however, led to the acceptance of the pseudopotential as a method of parametrizing the metal—the parameters, few in number, are adjusted until the Fermi surface generated by solving Schrödinger's equation fits the dHvA data. This was soon a routine computation, so the construction of Fermi surfaces was left in the hands of the experimental 'fermiologists'. The task of the theorist, starting from more fundamental principles, was then to explain why the fitting parameters take the values they do. In addition, the known parameters can be used to make other calculations, such as of the cohesive energy of different crystalline arrangements. Thus the most stable form can be determined and an explanation given for the changes in crystal structure suffered by many metals with change of temperature and pressure. In other ways, too, knowledge of the differing electronic structure of different metals has thrown light on metallurgical problems, while theorists engaged in devising schemes for calculating electronic structures have been provided with facts against which to test their approximations.

It would be misleading to leave the impression that all metals can be treated by the pseudopotential method. It is very successful with the metals in groups I–III, but, as already noted in passing, the inner 3d shell of electrons in copper lies too near the Fermi level of the conduction band to be treated as a passive source of pseudopotential, and the same is even more emphatically true of the transition elements, rare earths and actinide metals which have unfilled inner shells [152]. For these the Fermi surface, which now has many sheets, is but a part, albeit important, of the experimental data needed to construct a consistent model of the almost free behaviour of the conduction band and the more highly bound electrons in the inner shells. Early in the development of quantum mechanics the relatively large contribution of electrons to the thermal capacity of transition metals, and their strong paramagnetic susceptibility or even ferromagnetism, had encouraged speculation about the narrow and only partially filled inner energy bands responsible for these properties, and later detailed calculations supported the view that their Fermi surface structure was remarkably similar for all, figure 17.24 being typical. The many dHvA frequencies observed in fairly weak fields could be plausibly identified with details of this structure, and there was support from other types of measurements. It is too much to claim that the data alone lead to a unique set of surfaces, but there is general

Electrons in Solids

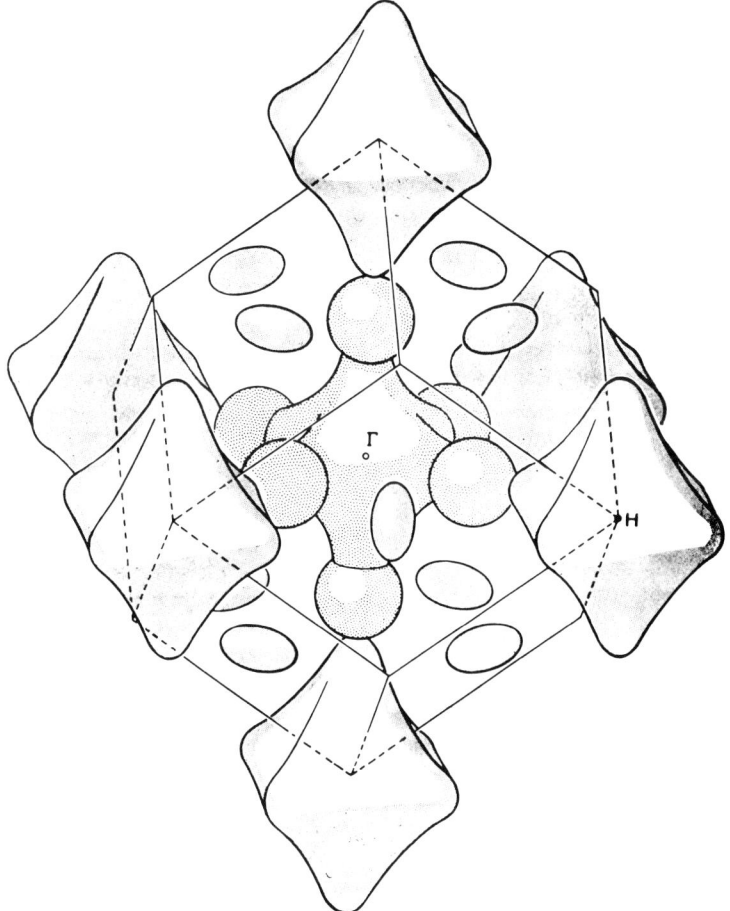

Figure 17.24. *The calculated form of the Fermi surface in tungsten* [153].

consensus that theory and experiment are in sufficient accord to validate the techniques of computation, and hence that they may be applied to even more complex electronic structures for which experimental support is woefully meagre. In recent years the 'heavy-fermion' compounds, of which UPt$_3$ is typical, have been fully investigated. Effective masses of more than a hundred times the electron mass have been observed, though the dHvA signal is so weak and so easily eliminated by thermal effects that it is necessary to use field strengths up to 20 T and cool the sample to a few millidegrees [154].

There are other sources of data that, if less generally applicable than the dHvA effect, have value in particular cases. Although sometimes described merely as instruments for providing detailed information on particular metals, they are phenomena of intrinsic interest and deserve to be appreciated in their own right. It is not possible here to do more than note the principal features of each phenomenon and provide references to accounts that make the historical development fairly clear. At the Cooperstown conference most were discussed fully enough for the published proceedings [128] to serve as a good introduction to this area of physics and its history.

17.18.1. Magnetoresistance [126]

As already noted, Kapitza's [5] supposed law of linear magnetoresistance did nothing to advance understanding. The very existence of a measurable change proved inexplicable according to the primitive models of Lorentz and Sommerfeld, but Peierls and Bethe realized, in about 1930, that non-spherical Fermi surfaces could give an effect [155]. Later, Jones pointed out that although a Sommerfeld-type metal containing only electrons or holes would show no effect, the presence of both made an effect possible—the resistance should rise initially as B^2 before flattening off to a saturation value that might be several times the zero-field value. Indeed, if the numbers of electrons and holes were equal there would be no saturation, and the quadratic rise would continue indefinitely; this was Jones' [156] explanation for the behaviour of bismuth. Much later it was found [157] that with very pure bismuth at liquid helium temperatures, an increase by a factor of more than 10^7 is possible, and his explanation is not doubted.

Little attention was paid to magnetoresistance until it took its place in post-war semiconductor physics as a moderately useful diagnostic tool. For metals it began to assume greater importance with the experimental work of Alekseyevski's Moscow group, and especially with the theoretical work of Lifshitz's Kharkov school. The historical development was presented by Chambers at Cooperstown, where he expounded what has since been regarded as the canonical theory (LAK, short for Lifshitz, Azbel' and Kaganov). Equality of electrons and holes, which is not infrequent in even-valent metals, is only one cause of non-saturating magnetoresistance. Alternatively, the Fermi surface may take a form that allows some electrons, instead of moving in a closed orbit of the sort picked up by the dHvA effect, to follow an *open orbit* that may carry them through the metal some distance, in a sinuous path, before being scattered. That open orbits might occur had been conjectured by Shockley and Chambers, but it was Lifshitz and Peschanskii who presented an elegant topological argument to show precisely when they could be expected and what form they would take. As a result, systematic studies of magnetoresistance were used to reveal the character of particular Fermi surfaces [158]. It is the occurrence of open orbits that is responsible for the extraordinary peaks of resistance in a very pure single-crystal wire of copper as a strong transverse magnetic field is turned about the axis of the wire (figure 17.25). To interpret these results quantitatively the necks, where the Fermi surface contacts

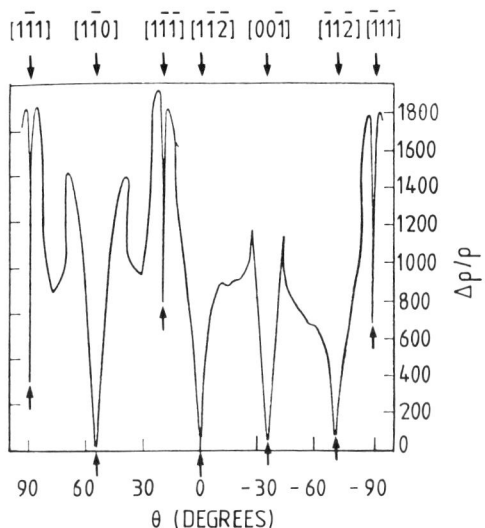

Figure 17.25. *The high-field magnetoresistance of a single-crystal rod of copper* [159]. *As the magnetic field is turned in a plane normal to the rod it passes through various directions of high crystal symmetry (indicated by the arrows) at which the resistance shows sharp minima less than the maxima by a factor of many hundreds.*

the zone boundaries, had to be assumed somewhat larger than was first supposed, and this adjustment has since been verified, and still further refined, by means of the dHvA effect.

17.18.2. Azbel'–Kaner cyclotron resonance (1956)
At almost the same time as Chambers [160] was pointing out that cyclotron resonance, such as had been studied in semiconductors, would not be observed in metals with the magnetic field in the intuitively sensible direction normal to the surface, Azbel' and Kaner [161] suggested directing it parallel to the surface, so that electrons in their orbits could run in and out of the skin depth (figure 17.26). The effect was soon detected, and later, with refined technique, found limited use in determining the velocities of electrons at the Fermi surface. Interesting related effects and extra complications have been discovered and explained, with Moscow the principal centre.

17.18.3. Acoustic resonance (1955) [163]
In 1954 Bömmel observed that ultrasonic waves suffered considerable attenuation in pure metals at low temperatures, which Mason explained as resulting from the very high viscosity of an electron gas when the free path is long. In the megahertz range of frequencies the free path may greatly exceed the sound wavelength, and the theory of electron viscosity takes on the guise of the anomalous skin effect—only electrons

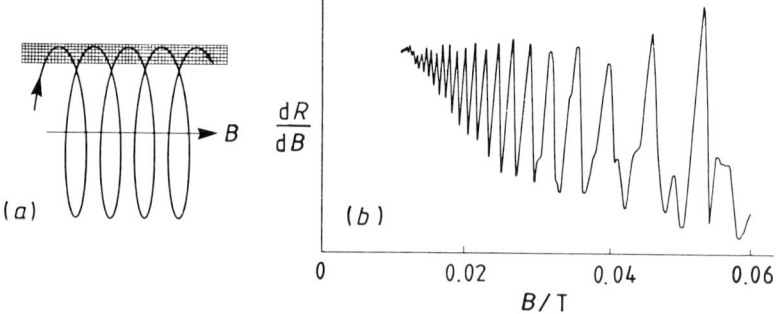

Figure 17.26. *Azbel'–Kaner* [161] *cyclotron resonance in a very pure crystal of tungsten* [162]. *The shaded region in* (a) *indicates the thin surface layer into which the oscillatory field penetrates. With* **B** *parallel to the surface an electron can make regular excursions into this layer. If* **B** *is chosen so that each excursion finds the oscillatory field at the same phase, the electron receives many kicks in the same sense and thus makes a larger than usual contribution to the current. If the cyclotron period is many times the period of oscillation of the field, these resonances occur at fairly close values of* **B**, *as seen in* (b). *The effect is enhanced here by the device of recording, not the surface resistance R directly, but its rate of change with B.*

that move nearly parallel to the wavefronts, so as to be able to 'surf-ride' on the wave, can extract energy from it, but this they do very efficiently. When a magnetic field is applied parallel to the wavefronts the electrons can interact particularly strongly if their orbit dimension matches the wavelength (or a multiple) for then, rather as in Azbel'–Kaner resonance, their interaction is enhanced at every half-orbit. This effect also was first observed by Bömmel, in 1955.

17.18.4. Helicons (1960) [164]

See also p 1988

It has long been known to ionospheric physicists that circularly polarized electromagnetic waves (whistlers) can be propagated in an ionized medium along the direction of a magnetic field. With the weak field of the Earth and a rarefied gas of electrons and ions some of the effects can be striking enough, but in a metal with a high density of electrons, and in a strong magnetic field, they are enormously enhanced. Instead of moving with the velocity of light the waves may travel at no more than a few centimetres per second, so in slabs only millimetres thick electromagnetic resonances may occur at a few Hertz. Aigrain drew attention to such low-frequency resonances in semiconductors, naming them *helicons*, and a little later they were observed in sodium. The theory, for any shape other than an infinite plane slab, is very complicated, and the phenomenon has not found wide use. The wave velocity, however, is determined by the Hall constant, which may therefore be measured without having to make small and fragile samples. It is possible to arrange that the velocity of transverse sound waves coincides with the helicon velocity, to produce a new form of wave motion incorporating both, the energy being carried more or less equally by the lattice vibrations and by the magnetic field. This is but one example, out of many, of strange wave motions in metals.

17.18.5. Magnetic breakdown (1961) [165]
Priestley was surprised to find, in his dHvA study of magnesium, that one of the oscillatory frequencies was too high for the Fermi surface responsible to be contained in the Brillouin zone. Cohen and Falicov explained that if the lattice potential was very weak Bragg reflection might be only partial—an electron reaching a point in its orbit at which reflection should occur had a choice of carrying on or of being reflected. It might thus succeed in completing a free-electron orbit and give rise to the observed high-frequency oscillation. By analogy with Zener's theory of dielectric breakdown, which also is a manifestation of quantum-mechanical tunnelling, they called the effect *magnetic breakdown*. It is not uncommon in metals and can lead to striking oscillatory magnetoresistance, as in figure 17.27 which shows the effect in zinc, and for which an adequate theory has been devised. There are, however, formidable complications to be overcome before a complete theory can be developed, not least that the structure of the quantized energy levels changes infinitely rapidly as the magnetic field is changed, depending on whether B is a rational or irrational multiple of some constant [167]. The problem has some affinity with the fractal structures that have attained prominence in the study of chaos, but virtually nothing has been achieved beyond the most primitive form of the theory.

17.18.6. The Shoenberg magnetic interaction (1962) [168]
The oscillations of magnetization in the dHvA effect can be intense enough to alter the field inside the sample and lead to severe distortions of the oscillations themselves. The most dramatic manifestation occurs in the magnetoresistance of beryllium, where it combines with magnetic breakdown to produce the weird variation shown in figure 17.28.

It will be seen that if we date the modern study of magnetoresistance from the LAK theory of 1955, between this date and 1962 all these new phenomena were uncovered; the inception of the Fermi surface programme also belongs to this brief period. The proceedings of the Cooperstown conference testify to the sense of excitement felt by those who had the good fortune to be involved.

17.19. Beyond the independent-particle model
The experimental Fermi surface programme was mainly conducted in a state of innocence—there seemed to be no cause to doubt that the electrons moved through the lattice independently of one another, even if theorists who had tried to take account of the strong Coulomb interaction

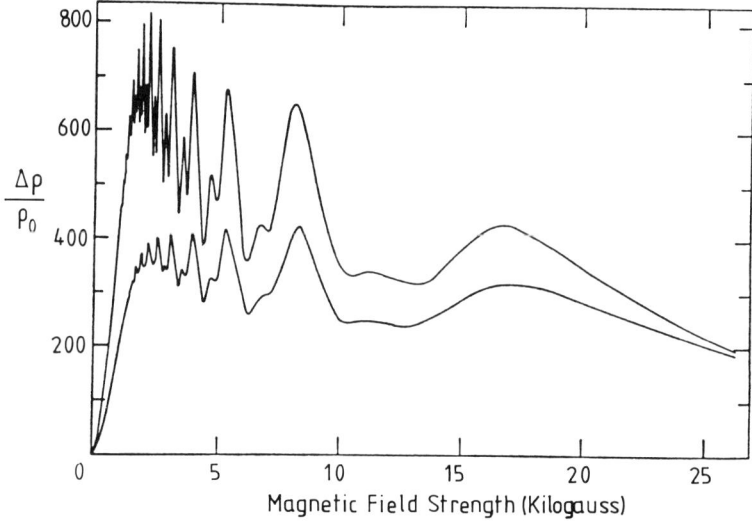

Figure 17.27. Strong oscillations of resistance in a single crystal of zinc, caused by magnetic breakdown [166]. If there were no breakdown the initial quadratic rise of resistance to about 600 times its zero-field value would continue indefinitely, as it does in bismuth. The upper curve was taken with the temperature of the crystal at 1.6 K, the lower at 4.2 K.

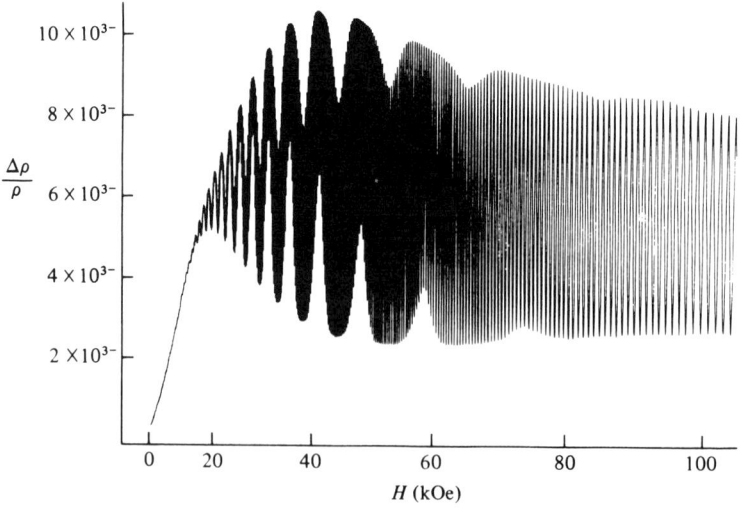

Figure 17.28. The magnetoresistance in a single crystal of beryllium at a temperature of 1.39 K [169], resulting from simultaneous presence of magnetic breakdown and the Shoenberg interaction. At the lower field strengths the rapid oscillations of resistance are not resolved in the recording.

between them had found great difficulty in reaching sensible conclusions. The brilliant success of Sommerfeld's and Bloch's original quantum theories of independent electrons was persuasive enough; and Skinner's [170] pre-war studies of the x-rays emitted, when conduction electrons fell into vacant atomic energy levels, showed a sharp cut-off, consistent with a clear division between filled and empty Bloch states at the Fermi surface. Most published papers of that period ignored the issue, through a few theorists continued to seek the reason for a simplicity that seemed paradoxical.

We have already met the idea of screening, whereby each electron keeps others away from itself and so prevents its Coulomb field from affecting more than its immediate neighbours. This, however correct in essence, could not be regarded as more than a makeshift theory. A respectable theory would treat all electrons as identical and produce without artificiality, it was hoped, a model of independent particles interacting weakly through short-range (i.e. screened) forces. Considerable success along these lines was achieved by Bohm and Pines [171]. For the origins of their model we must go back to studies of electrical discharges in gases, particularly those of Langmuir and his associates at General Electric, Schenectady [172]. It was they who introduced the term *plasma* to denote a neutral, but highly ionized, gas that seemed to them to show some similarity in its properties to a jelly. If neutrality is disturbed by displacing the electrons so as to create excess and deficiency of charge in different regions, the resulting electric field acts as a restoring force; the charge density oscillates at the plasma frequency ω_P until collisions between electrons and ions dissipate the motion. In Langmuir's low-density plasma ω_P was around 10^8 Hz; in most metals, where the electron concentration is far greater, it is over 10^{15} Hz, the frequency of visible or near ultra-violet light, a significant fact in the argument of Bohm and Pines.

See also p 1623

Their discussion leads to the conclusion that the description of the dynamical behaviour of an electron gas can be divided into two parts. As far as the relative motion of electrons when they run close to each other and are repelled by their Coulomb interaction is concerned, they appear to be independent. If, however, they were genuinely independent, accidental fluctuations of charge density would occur all the time. These can be described in terms of quantized plasma oscillations, and now it becomes clear that the quantum energy, $\hbar\omega_P$, is so high that there is no likelihood of their being excited at ordinary temperatures. This means that extended fluctuations in electron density do not occur. Electron motion is coordinated to the extent that they repel one another according to a screened Coulomb law, but they are otherwise independent; there is, however, the additional constraint that they cannot build up extended regions of space-charge. This latter constraint on their freedom is of small importance in a dense electron gas, such as Wigner [57] had already shown to be only slightly affected by Coulomb interactions, but it is significant in more rarefied gases. It also affects the spin paramagnetism with which Pauli initiated the quantum theory of metals. It is difficult to separate the spin contribution from other magnetic effects, but in one or two alkali metals the spin paramagnetic susceptibility has been measured by an ingenious roundabout application of paramagnetic resonance [173]. The results show the expected enhancement in satisfactory agreement with the calculations.

In this theory much depends on deciding how many separate modes of plasma oscillation can exist, for the more there are of these the more constrained is the independent motion of the electrons. Bohm and Pines assumed that only plasma waves with wavelength greater than some cut-off λ_c could exist, and they chose λ_c to minimize the calculated energy of the electron gas. A conceptually more transparent alternative was proposed by Lindhard [174]: short-wavelength plasma waves travel slowly, and only those moving faster than the electron velocity at the Fermi surface can survive, since electrons can surf-ride the slower waves and dissipate their energy. Lindhard's criterion is not in practice greatly different from that of Bohm and Pines, and his analysis was not influential for this particular problem, though the methods he used—extensions of those introduced by his teacher Klein—were later adopted for treating other problems. Lindhard gives credit also to Bloch and Tomonaga for their related ideas, but makes no mention of Landau's [175] earlier and rather obscure discussion of the surf-riding idea, of which he was probably unaware.

Up to this point the theoretical tools employed were of the sort familiar to an earlier generation, but further development involved Feynman diagrams and the apparatus of divergent infinite series. Gell-Mann and Brueckner [176] brought together Bohm–Pines theory and Wigner's analysis of correlation effects; subsequent elaborations, documented by Hubbard [177], provided quantitative stiffening where qualitative ideas had previously lacked rigour. The physical picture, though still incompletely formulated, has survived well.

The study of plasma oscillations was not solely a question of formulating more rigorously the interactions of conduction electrons. Being longitudinal oscillations of charge, with associated electric fields, they are not unlike sound waves and are strongly coupled, through the ionic charges, with sound waves in the metal. This has led to reconsideration [178] of Bardeen's older theory of phonons in sodium, though without any radical change in the results. It was an important stage in his attack on the theory of superconductivity.

A further matter of interest concerning plasma oscillations is the part they play in retarding fast-moving charged particles [179]. Early theories, from Bohr onwards, concentrated on the ejection of bound electrons from atoms as the particle passed by, but when the electrons are free, as in a metal, other mechanisms must be invoked. Lindhard gives a full account of earlier ideas, but rather surprisingly does not pursue the analogy with Čerenkov radiation. A fast-moving charged particle will, in classical terms, generate a shock wave in the plasma; in quantum terms the particle will lose energy in multiples of the quantum $\hbar\omega_P$. The effect is observed in transmission electron microscopes equipped with the means to measure the energy of the emergent electrons. It was first noted in 1941, and left unexplained, by Ruthemann, whose spectrum of energy loss in aluminium shows clearly the excitation of one or more plasmons of 16 eV energy. Since the plasma frequency is different for each material the quantum of energy loss reveals something of the composition of the minute region traversed by the finely focused beam. This is, of course, not the only mechanism of energy loss that has diagnostic value; excitation of core electrons into vacant higher levels is equally specific.

While the implications of Bohm–Pines theory were being worked out, Landau [180] in Moscow was developing a different approach to the problem of interacting particles, extending his earlier ideas inspired by the superfluidity of liquid helium. He was concerned to describe the liquid state of the light isotope of helium, ^{3}He, which like the electron obeys Fermi statistics; consequently his scheme is referred to as Fermi liquid theory. All the essential assumptions, presented dogmatically as if self-evident, have since been justified. The adaptation of the theory to electrons in metals was begun by Silin [181] shortly afterwards. Whereas Bohm and Pines started with a gas of electrons subject to Coulomb interactions, with the intention of deriving a many-body quantum theory of the ground state, Landau's method is to take the ground state as given and proceed immediately to discuss the nature of the lowest excited states, such as would be present at low temperatures and would be responsible for the transport of energy and charge. His fundamental premiss is that these excited states, whose formal description would require immensely complicated wavefunctions involving all the electrons, must have properties closely analogous to individual electrons—they are long-lived, they carry momentum and charge, and they obey Fermi statistics. They are designated *quasi-particles* and interact with one another, but not simply by means of a screened Coulomb interaction. The interactions are collective, in that the velocity of a quasi-particle is determined both by its wave-vector k and by what other quasi-particles are excited. No attempt was made to calculate for any specific metal the magnitude of the predicted effects.

Landau's procedure was characteristic of a man to whom general principles were more important than solving particular problems. He set out to establish a phenomenological theory, that is, a general and necessary framework into which any specific theory must fit, and within which specific measurements and calculations find a consistent

See also p 949

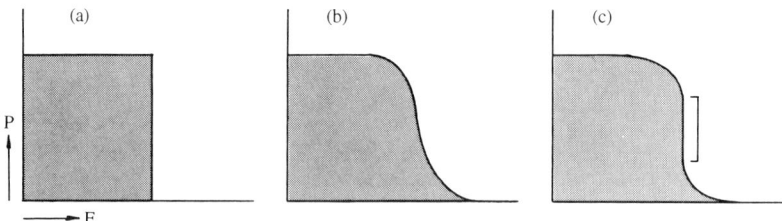

Figure 17.29. *(a) In a Fermi gas of non-interacting particles at zero temperature the chance P of finding a particle in a state of energy E drops sharply to zero at E_F; (b) shows what was believed to be the effect of interactions in smoothing out the transition from full occupation to zero; (c) is Migdal's [182] result, with the residual discontinuity indicated by the bracket.*

description in terms of a limited number of parameters. In contrast to the case for liquid ^{3}He for which new effects like *zero sound* were predicted and later observed, nothing comparable has emerged from the extension to metals, and the direct impact of the theory has been small. On the other hand, the clarification of the concept of quasi-particles has left its mark on subsequent thinking about the quantum mechanics of interacting particles.

Shortly after Landau's first paper on Fermi liquids it received strong support from a general result proved by Migdal [182], also in Moscow. Subject to certain qualifications, the ground state of an assembly of interacting Fermi particles possesses an unexpected property. If we pick particles at random from an assembly of non-interacting particles, we shall find a sharp cut-off in their energy, with none having energy greater than E_F (figure 17.29(a)). With interacting particles it is natural to suppose the cut-off to be smeared out (b), but Migdal showed that there remains a discontinuity (c) which can be taken to define the Fermi energy. It soon became clear, largely through the work of Kohn and Luttinger [183] in the USA, that a Fermi surface in k-space can be defined by this Migdal discontinuity; the quasi-particles of lowest excitation energy (i.e. E_F) have k-values lying on this surface, and it is these quasi-particles that determine such phenomena as the dHvA effect. These results go far towards justifying the innocent assumptions of the fermiologists; one must simply replace the word *electron* by *quasiparticle*.

See also p 937

It is ironic that within a year of Migdal's theorem an exception was published with great acclaim—the theory of superconductivity due to Bardeen, Cooper and Schrieffer. In the ground state of their superconductor there is no Migdal discontinuity, and in addition the first excited states are separated by an energy gap from the ground state. It could be said that Landau anticipated this when he insisted that his theory applied only to Fermi liquids for which no such gap was present. This is an essential part of his basic assumption, shared by Migdal and by Luttinger, that if the Coulomb interaction is imagined to be switched on very gradually, the system will adjust itself smoothly to its new configuration. It is by no means necessary; its denial is at the heart of the Mott transition, and the BCS theory falls for similar reasons outside the scope of Migdal's theorem. There is no general method for determining theoretically whether a given system will obey the theorem, but the empirical evidence of the Fermi surface programme strongly suggests that in non-superconducting metals there is no cause for doubt.

17.20. Disordered materials [184]

The incorporation of imperfections into Bloch's perfect lattice has always been seen as necessary, for without imperfections there is no resistance. The first successes in treating the resistivity of alloys, especially by Nordheim in 1931 and Mott in the years following, are described by Mott and Jones [35]. Their semi-classical methods have, on the whole, proved adequate with metals, but less so with semiconductors and with liquid metals which, of course, are very highly disordered. About 1960 a new phase of theoretical activity began, whose most obvious characteristic is the deployment of fully quantum-mechanical arguments which are either very difficult mathematically or highly intuitive, and sometimes both. Thus Anderson's [185] seminal paper was largely unappreciated for some years by reason of its technical virtuosity, while Mott's [186] apparently simple approach turned out, on closer examination, to depend on an intuition born of long experience, and not generally shared by his colleagues and readers. By surrounding himself, however, with experimenters who put considerable trust in his judgement as to which measurements were most needed, while they patiently digested what they could of his insights, Mott re-established himself (after some years on advisory bodies and in administration) as a potent leader in an exceptionally difficult research field. Anderson also took the lead in suggesting experiments and guiding theorists, and it is certain that without these two disparate pioneers the subject would not have developed so successfully.

The thrust of experiment in these new developments was old-fashioned in principle, however up-to-date the techniques for acquiring large quantities of accurate data. One finds few examples of new phenomena to be exploited, as was so characteristic of the Fermi surface programme. Instead the emphasis was on the classical properties—conductivity, Hall constant, thermoelectricity—and their variation with temperature; and the most hotly debated theoretical issues concerned how these properties were to be related to the configuration of atoms in a disordered system. It is appropriate at this stage to look more closely into the development of ideas on how to treat such problems.

The approach that served Nordheim and Mott well in the thirties had its early critics who pointed out that electrons were frequently assumed to behave as classical particles in circumstances that violated

Heisenberg's uncertainty principle [187]. An electron, if it is to be treated as a particle, must be described by a wave-packet that holds together for longer than the time τ between collisions. This can be achieved only by compounding plane waves taken from an energy range greater than $\hbar/\tau$. So long as this is much less than $k_B T$ no conflict arises in imagining the thermally excited electrons to be well-defined both in energy and position. The criterion, however, is normally not satisfied either at high or at very low temperatures, and seems unnecessarily restrictive. Peierls in 1934 quoted Landau's argument for replacing $k_B T$ by the Fermi energy—if $\hbar/\tau \ll E_F$ the classical theory, in his view, should prove adequate, and this is roughly the same as asking that the mean free path ℓ shall be longer than the de Broglie wavelength. At the time the doubters seem to have been satisfied, though the problem remained in Peierls' mind; colleagues of his in Birmingham took it up again, following a conference in 1956 at which the Japanese theorist Kubo gave a formal expression for the conductivity in terms of exact solutions of Schrödinger's equation for an electron in an imperfect lattice. Since no one has written down any such exact solution, it might seem that the expression is of limited value; it has, however, continued to form the starting point for many theoretical discussions and was indeed used (and probably rediscovered) by Greenwood to provide more substance to Landau's defence of classical methods. Further developments were made by Chester and Thellung [188], who give a full account of the historical background.

By choosing as his starting point exact solutions of Schrödinger's equation, Kubo abandoned the fruitful, if inexact, approach of treating impurities as perturbations of an otherwise perfect lattice. Nordheim had made progress with the theory of resistivity in alloys by assuming the unperturbed state to be a lattice of 'average' atoms, and hence that every atom was in some degree a perturbation. But this is obviously a makeshift. The problem took on new urgency when the electrical properties of liquid metals began to excite interest, for in this case one cannot make the separation into an ideally regular background in which are embedded discrete scattering centres. In a liquid, background and scattering irregularities are inextricable, and demand a new approach.

The study of liquid metals has a long history which can be found in the review articles of Schulz and Cusack, and in Faber's comprehensive treatise [184]. As long ago as 1919 Kent measured the optical reflectivity of a number of liquid elements and alloys, and observed quite good agreement with Drude's classical theory, based on the model of free electrons subject to collisions. At optical frequencies the inertia of the electrons is more effective than collisions in limiting their response to the alternating electrical field, and this was properly taken care of by Drude. Kent noted that the number of electrons needed to explain the results matched the number of valence electrons. It was only later, when the complexities of band theory were appreciated, that this simple result seemed surprising, but it was well confirmed in the fifties by the careful measurements of Schulz and Hodgson. Even more striking, perhaps, were the results on the Hall constant which became fairly reliable after 1960. Until then it had not been well enough realized that unless the geometrical arrangement of electrodes is very carefully thought out, and unusual care taken to achieve a uniform magnetic field, the liquid will be set in motion by the electromagnetic forces and the measurements thereby vitiated. Once these problems had been solved it was found that for many liquid metals the Hall constant took the sign appropriate to electrons, whatever sign it might have in the solid, and that the indicated density of electrons once again matched the density of valence electrons. With the optical properties one might have accepted the explanation that the quantum energy was enough to transcend any band gaps, but this cannot apply to the DC Hall effect. It seemed inescapable that, at least for the valence electrons, no residue of band effects survived melting of the lattice.

The pseudopotential concept arrived most opportunely to resolve the problem. Once it was understood that the complicated Fermi surfaces in solid metals did not demand strong interaction between the electrons and the ions, but could be explained as the consequence of a weak interaction and the Brillouin zone structure, it was clear to Ziman that in the liquid state the electrons were similarly free to wander through a now disordered but still only weakly interacting assembly of ions. The effective number of electrons was then simply the real number, as indicated by experiment. This did not immediately lead to a calculation of the conductivity; for that, one needed to know something about the configuration of the ions. Ziman rediscovered an idea put forward in 1945 by Bhatia and Krishnan, who pointed out that the information required was contained in the x-ray scattering pattern of the liquid. When x-rays fall on a perfect lattice, Bragg reflection sends them into well-defined directions; but a liquid produces no sharp diffraction spots, only diffuse haloes, whose detailed profiles provide just what is needed to describe how a beam of electrons will gradually lose all sense of original direction. The distance in which this happens can be quantitatively related to the mean free path and hence the conductivity. In this way the theory of the conductivity of liquid metals was put on a sound basis, even if it called in the aid of experiment to bypass the formidable problem of the atomic configuration in a liquid.

At the same time rather greater effort was being devoted to the conductivity of disordered, especially amorphous, semiconductors. It had long been known that selenium could be prepared in a glassy form, but with other semiconductors evaporation or sputtering of thin films was the only reliable method; for conductivity measurements this was no handicap. Their behaviour, and even more that of molten semiconductors, presented a problem to those who had accepted Wilson's theory of crystalline semiconductors. If their peculiar properties arise from the proximity of filled valence and empty conduction bands, it is surprising that they remain semiconductors on melting. A plausible explanation was provided by Gubanov, working in Ioffe's Leningrad Institute, a very active centre for this work. In his view long-range order in the atomic arrangement mattered little—it was the arrangement of nearest neighbours that played the key role in determining the general structure of energy levels. A full account of the early phases of this work has been given by Ioffe and Regel [189], who introduce in their article an important distinction between different classes of material. Measurements of electron mobility can be interpreted in terms of electronic free paths, and they point out that if the free path seems to be smaller than the de Broglie wavelength the concept is untenable; for a travelling wave can hardly be defined if it is destroyed before it has a chance to assert its wave-like character. When the apparent free path is as small as this, one must abandon the idea of a mobile electron and attribute conductivity to the hopping of localized electrons from one centre to another. Ioffe and Regel do not remark that their criterion is the same as had been derived on different grounds by Landau [187].

Anderson's 1958 paper demonstrated, for a particular model of a disordered system, that a single electron could be localized without the Coulomb interaction that underlay the Mott transition. There is no implied conflict between alternative explanations—rather, both processes are likely to operate together. Anderson's mechanism is essentially quantum mechanical. A ball bouncing around, without loss of energy, on a random pin-table will wander away from its starting point and in time can travel any distance. On the other hand, a wave scattered by the random array of pins may be confined to a restricted area and never get away, however long one waits. The difference lies in the interference between the multitude of scattered wavelets, which are all phase coherent provided that the pins are fixed. Whatever doubts there may have been about the rigour of Anderson's analysis concerning localization in three dimensions, the situation in one dimension is clear enough—any degree of randomness results in localized wavefunctions. Mott and Twose's [190] original paper made this highly plausible, and subsequent refinements only strengthened the argument [186]. In four dimensions also the result seems clear—there is no localization. In the three-dimensional world, according to Anderson, an electron is neither inevitably non-localized nor inevitably localized; in the impurity band of a semiconductor, probably only the electrons near the lower edge of the band are localized. Appreciation of this point led to the general acceptance of a sharp division—the *mobility edge*—between states in which the electrons are localized and can only move by thermally excited hopping to another localized state, and states that allow freedom of movement, albeit with a very short mean free path. The transition between the two types of behaviour occurs at a point in the band (as we might have come to expect by now) where the mean free path is comparable to the electron wavelength. Precisely what happens at the mobility edge itself has been a matter of controversy, with Mott arguing for a discontinuous change in mobility, and Cohen for something a little smoother. The enormous literature on this and related topics defies fair summary, and the reader who wishes to explore further may start with Mott and Davis's encyclopaedic survey [184], while remembering that on this issue it is inevitably slanted in favour of the authors' interpretations.

17.21. Interference effects [191]

The coherent multiple scattering of electron waves, if insufficient to cause noticeable localization, as is the case in almost every alloy, is not readily discernible in the resistivity. So many scattering processes are involved in a typical sample that the wavelets combine with random phases; the result is as if there were no phase coherence and the electrons behave like balls on a pin-table. When, however, a magnetic field is applied the phase relations are systematically altered, and in such a way as to produce measurable effects under the right conditions. The reason is to be found in a mystifying peculiarity of quantum mechanics. In the absence of a magnetic field the wave-vector of a de Broglie wave, k, is given by $p/\hbar$, p being the momentum mv; in the presence of a field B, however, k is $(mv + eA)\hbar$, A being the vector potential with the property that $B = \text{curl } A$. An electron of given energy, and therefore given velocity v, suffers a change in wavelength when B is applied. Suppose, then, the wave produces a scattered wavelet at a point α, which is subsequently further scattered at β, γ, ... before a remnant returns to α for a second scattering; the phase relationship between the primary wave and the returning secondary wave will be altered by B, since the number of wavelengths contained in the round trip will be altered. Consequently the resultant of all the scattered wavelets will be changed. For any chosen circular tour, $\alpha\beta\gamma\ldots\alpha$, the phase change is $2\pi e\phi/\hbar$, where ϕ is the magnetic flux passing through the polygon $\alpha\beta\gamma\ldots\alpha$. There are, of course, an enormous number of such polygons having different areas and therefore not suffering the same phase changes, so once again in a typical sample the effect will be washed out. But in a very small sample, at a very low temperature, the number of independent paths may be small enough for phase variations to matter. A wire prepared at IBM by the advanced techniques used in fabricating computer circuitry had a cross-section of only 60×38 nm and, as the magnetic field was changed, showed random fluctuations in resistance amounting to a few parts per thousand.

This is perhaps not very spectacular, but the experiment was inspired by earlier theoretical and experimental work in Moscow, and was designed to confirm and extend the more striking phenomenon discovered there—oscillatory variations of resistance with magnetic field. The theoretical prediction, for its part, was inspired by the famous paper of Aharonov and Bohm [192]; this showed how the interference pattern formed by splitting and later recombining an electron beam (as in Young's optical interference experiment) could be displaced if magnetic flux passed between the divided beams, even though it was contained in a shield so that no magnetic field interacted directly with the electrons. The effect, which illustrates how the vector potential A, rather than the field B, controls the de Broglie wavelength, was not strictly a new discovery, since the significance of A had been realized early in the history of quantum mechanics, and was, for example, at the heart of F London's prediction of flux quantization in superconductors. Nevertheless, Aharonov and Bohm deserve the credit for showing how the paradoxical property could be demonstrated explicitly, as indeed it was soon afterwards. Al'tshuler and his colleagues [193] applied the argument to electrons in a small, thin-walled cylinder, in a paper whose physical message is well hidden under mathematical detail. They note that a wavelet scattered at some point α may return to that point after making a circuit of the cylinder. However many scatterings it suffers on the way round, there will be a wavelet making an identical circuit in the reverse sense, and these two will recombine to give a resultant scattered wavelet at α. If there is no magnetic field the phase lengths in opposite senses are identical and they interfere constructively, but flux threading the cylinder makes the phase lengths different and consequently the resultant at α is changed. There are innumerable such pairs of oppositely circulating wavelets, but they all recombine in phase when $B = 0$, and again whenever the flux ϕ is a multiple of $h/2e$; since every circuit has nearly the same area, the oscillatory effects combine in phase, not randomly as in the previous example.

See also p 934

The Sharvins [194], father and son, demonstrated the oscillatory effect on the resistance of a thin film of magnesium evaporated on a 1.5 μm diameter silica thread. The amplitude of the oscillations was only 4 parts per million of the mean resistance. In an experiment at IBM on a minute evaporated ring the oscillations, shown in figure 17.30, had an amplitude of nearly 1/10%. The geometry is different and in this case the period corresponds to a change in ϕ of h/e, since the waves traverse only half the ring before recombining. These experiments have generated successors and, of course, numerous theoretical discussions aimed at quantifying the effects to be observed in *mesoscopic* systems—systems larger than atoms but still very small. They probably have little significance beyond themselves, but deserve notice in their own right as examples of elegant physics.

Electrons in Solids

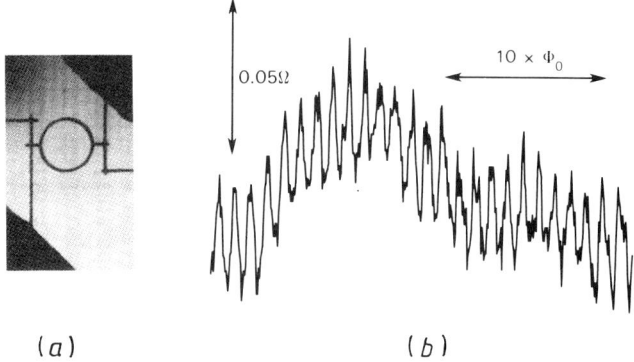

(a)   (b)

Figure 17.30. (a) A microphotograph of an evaporated gold ring of diameter 0.78 μm, with leads to allow its resistance (about 30 Ω) to be measured [195]. As the strength of the magnetic field threading the ring was increased, the resistance oscillated as shown by the trace (b); each cycle corresponds to one more flux quantum h/e through the ring. The temperature was 10 mK.

17.22. The two-dimensional electron gas; the quantum Hall effect [196]

For the topics we have just considered the literature, though extensive, was kept within bounds by virtue of its almost purely academic interest. The same does not apply to this, the last of our topics, which sprang from, and has been nourished by, the electronics industries. Initially the working material was the surface layer of a silicon crystal, coated with a thin layer of silicon dioxide, on top of which a metal electrode was evaporated, as in figure 17.31(a). This is the basic structure of the MOSFET—metal–oxide–semiconductor field-effect transistor—a ubiquitous circuit element in an LSI (large-scale-integration) chip. Later, interest switched to fabricated sandwiches with alternate layers of pure and aluminium-doped gallium arsenide, GaAs. Electrons can be confined to a region very close to one of the interfaces, and the scattering centres in such a device may be so few that the mean free path is 100 μm or more. As a close approximation to an ideal two-dimensional conductor it is the subject of intense study in its own right. But GaAs and its derivatives have such a variety of commercial applications, and such great promise for more, that research into its properties is vigorously pursued in industrial laboratories and, with the support of industry, in the universities. It is the basis of microwave radio transmitters and receivers for low-power communications—even for actuating automatic door-openers, by no means a trivial market. More important is its role as the active component of infra-red lasers in fibre-optic communications and CD players. This presumably explains why about 4000 papers a year are published in physics journals describing work involving GaAs, and why an annual conference has been held for nearly 20 years on this one

1371

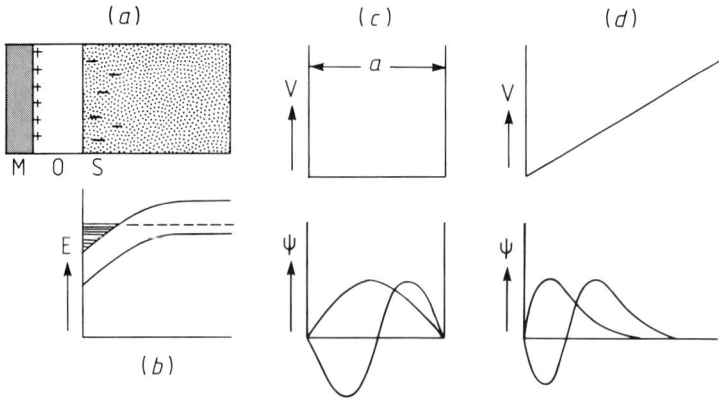

Figure 17.31. (a) The structure of a MOSFET; on the surface of an n-type silicon crystal (stippled) a layer of oxide (plain) is grown, and a metal electrode (shaded) evaporated on top. When the electrode is charged positively the bands are bent, as in (b), to allow negative charge to accumulate below the silicon surface, and form a two-dimensional conductor. (c) A steep-walled potential well (above) allows sinusoidal wavefunctions of which the lowest two are shown below. (d) A triangular well, like that in (b), allows wavefunctions that tail off gradually on one side.

material alone. The volumes [197] containing papers presented there can be consulted to obtain, however superficially, an overview of how the field has developed. No attempt will be made here to epitomize even the principal strands of development. Instead, attention will be confined to one example of pure physics—the quantum Hall effect—which caused surprise and excitement when it was discovered in 1980, and raised a number of theoretical issues that have not yet been resolved.

The two-dimensional electron gas is essentially a quantum concept. If one could make a perfect thin metallic film, with smooth surfaces, a typical wavefunction for an electron in the film would vanish at both surfaces, and the component of wave-vector normal to the film could only take one of the values $n\pi/a$ (figure 17.31(c)). Then, in addition to the energy resulting from motion in the plane of the film, the electron possesses confinement energy $n^2h^2/8ma^2$. When the temperature is so low that $k_B T \ll h^2/8ma^2$, only the lowest state ($n = 1$) is occupied—the electron is still free to run along the film but has no freedom of motion in the normal direction; it behaves strictly as a particle in two-dimensional space. No very stringent condition is implied—at a temperature of 1 K the film can be as much as 100 atoms thick. A more feasible arrangement arose during the early development of the MOSFET, the possibility being foreseen in 1957 by Bardeen's research student, Schrieffer, but discounted at the time as impracticable. Nearly a decade later better understanding of the silicon surface, particularly the benefits of oxidation, made it possible to observe the predicted effects. The situation is shown in

Electrons in Solids

figure 17.31(b), in which the bands have been bent down so far by the field of the positive electrode that there is a thin surface layer—the *inversion layer*—in which conduction electrons are confined. The more positive the electrode, the more electrons in the inversion layer, so with this arrangement the density of the two-dimensional gas can be adjusted from outside, making it a very flexible experimental arrangement. The potential varies almost linearly within the surface (figure 17.31(d)). A very full account [196] has been written of this development up to 1981, and of the properties of the electron gas, ending just as the quantum Hall effect [198] was starting to arouse great interest.

A magnetic field, B, directed normal to the surface, affects conduction far more dramatically than in a three-dimensional metal. In the latter, as shown in figure 17.22, each plane slice of the occupied region of k-space generates an oscillation whose phase is determined by its area; the consequent mutual cancellation of effects is absent in the two-dimensional gas. In particular, if B takes one of the values at which the highest occupied Landau level is exactly full, the gas can still conduct but there are no states to which electrons can be scattered except in higher levels. At a low enough temperature insufficient energy is available and there can be no dissipation. When a current is passed through a strip, no component of electric field appears parallel to the current, but a Hall field normal to the current, being non-dissipative, is not precluded. In the experiments that revealed the periodic vanishing of the resistance, the magnetic field was held constant while the electron density in the film was altered by changing the external electrode potential. What might have been expected, from the argument just offered, is that the resistance would vanish only at certain precisely adjusted electron densities. When, however, the experiment was carried out at a very low temperature and in a strong magnetic field, using the best available material, an extraordinary bonus was apparent (figure 17.32). There are extended ranges of electron density (or electrode potential) in which the resistance vanishes, and within each range the Hall field remains remarkably steady. Moreover the value of the Hall field agrees extremely closely with the simplest theoretical expectation. If one writes the Hall resistance (not resistivity) R_H as the quotient, V/I, of the transverse voltage V between two opposite sides of the strip and the total current I, the shape of the strip and the nature of the material disappear from the expression, leaving R_H as an integral submultiple of h/e^2, or 25812.8 Ω. This discovery immediately raised the hope that the phenomenon could be used to provide an extra item of precision data for determining the values of the fundamental constants. Combined with another relatively recent discovery, the Josephson effect, which yields a very precise measure of h/e, both h and e could be evaluated separately.

But, of course, for this purpose one must be utterly confident that there are no corrections to the simple value h/e^2, and it has indeed been

See also p 1257

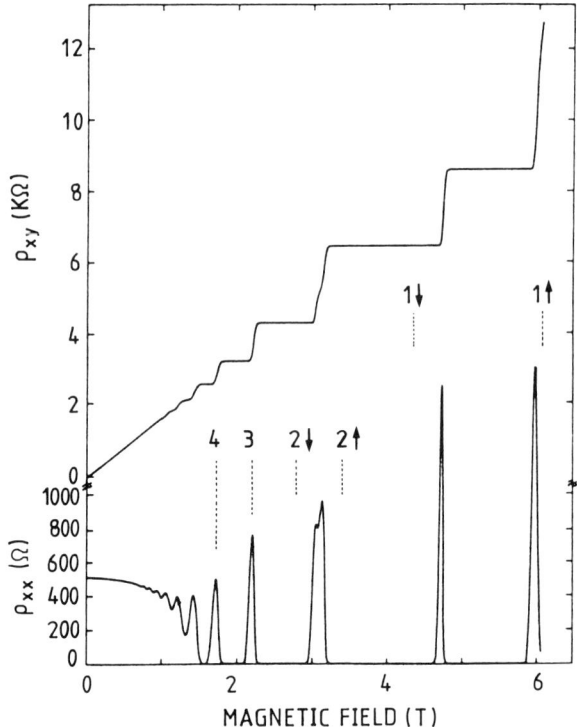

Figure 17.32. *Longitudinal and transverse (Hall) resistance of a two-dimensional electron gas in a heterostructure at a temperature of 8 mK* [199]. *Since it is not so convenient with heterostructures as with* MOSFETS *to change the electron density, the equivalent effect is produced by changing the strength of the magnetic field. The longitudinal resistance (lower curve) vanishes over long stretches of magnetic field strength, during which the Hall resistance is astonishingly constant.*

shown that *if* the resistance vanishes R_H must be given precisely by h/e^2, even though the film is imperfect and contains scattering centres. As to the mechanisms involved, the generally accepted view, after considerable discussion, is that scattering centres give rise to localization of some of the electrons in a quantized Landau level; instead of being a sharp level containing many states of identical energy, it is broadened by scattering, with mobile electrons in the centre and localized electrons at the edges. The plateaux represent the ranges of electrode potential that cause the Fermi level to lie within the bands of localization. It is an extraordinary result that when some of the electrons are rendered useless for conduction the rest acquire sufficient extra mobility to compensate for their loss exactly [200]. No straightforward proof of this has been given, but the technically involved demonstrations have been accepted by those competent to judge them, and certainly experiment shows that the plateaux give a value of h/e^2 that agrees with independent measures to better than one part per million. A readable, and historically informative, account of this work is to be found in von Klitzing's Nobel lecture [199].

The original studies of plateaux with MOSFETs gave nothing as striking as figure 17.32 which was obtained with an epitaxially grown heterostructure. These words need a little explanation. In 1969 Esaki and Tsu, of IBM, proposed to investigate the artificial construction of semiconductors whose properties were graded in layers [201]. For this purpose the technique of molecular beam epitaxy eventually proved the winner, though it involves expensive equipment and (as with all high-level semiconductor research) the most scrupulous standards of cleanliness. The idea is simple enough—in a very high vacuum the constituents of a solid may be evaporated from a heated source and condensed on a cooled substrate so as to settle down, atomic layer by atomic layer, in a very nearly perfect lattice structure. Moreover, after forming a certain number of layers the composition may be altered, say by replacing some of the gallium in GaAs with aluminium. This does not disrupt the crystal structure but it does change the width of the energy gap. In this way it is possible to create within the crystal a smooth non-scattering interface (*heterojunction*) that behaves like the surface of a MOSFET. The oxide layer of the latter is replaced by aluminium-doped GaAs which, having a larger energy gap, does not permit conduction electrons in the pure GaAs to pass the interface. It is also possible to simulate the positively charged external electrode by incorporating donor impurities far enough from the interface that they do not scatter the conduction electrons, yet near enough to produce a triangular potential variation in the GaAs similar to that in figure 17.31(*d*); the number density of conduction electrons in the layer is now fixed, but there is no objection to an external electrode if it is desired to change the density.

The most precise studies of the quantum Hall effect have been made with heterojunctions, but the story does not stop here, since it was observed at Bell in 1982 that there was a subsidiary plateau when the lowest Landau level was only one-third occupied [202]. Since then a fine multiplicity of such rational sub-multiple plateaux have made their appearance, of which figure 17.33 shows a spectacular example. A considerable variety of fanciful explanations have been advanced, but opinion seems to have hardened in favour of a novel type of condensation of the electron gas brought about by Coulomb interaction. This is not the same as the Wigner lattice, which remains elusive. At a conference [204] held in 1991 and devoted to the physics of two-dimensional electron systems the current state of affairs was described by a number of invited speakers, and their papers provide as good a starting point as any for those desirous of knowing more about a highly abstruse topic.

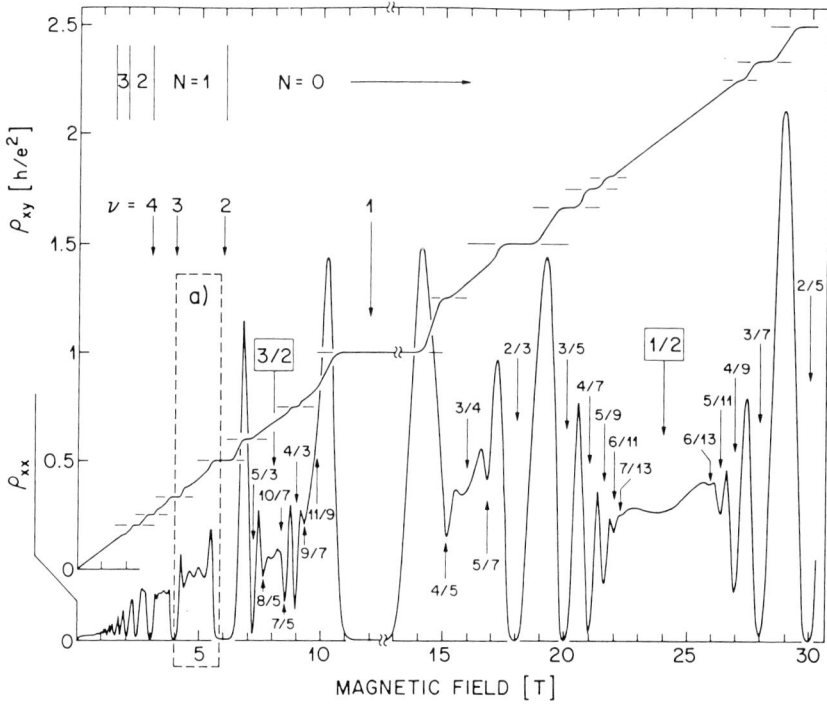

Figure 17.33. *The fractional quantized Hall effect* [203]. *The fractions attached to the minima in the lower curve, which shows the longitudinal resistance of the two-dimensional gas, indicate the fraction of a Landau level that is occupied. Odd denominators are very strongly favoured.*

17.23. Epilogue

As the century draws to its close the electronic properties of solids are involving more people than ever before, while the intellectual skills needed to master the theoretical framework, let alone extend it, are beyond the attainment of a high proportion of those involved. This is not to imply that they are wasting their talents; they are, however, engaged in a very different occupation from what would have attracted a research physicist in the middle years of the century. Such a one, if an experimenter, would have hoped to make a discovery and interpret it, at least qualitatively, in terms of current physical ideas (to overthrow current ideas would already have seemed an overweening ambition). There can be few now who pretend so confident a grasp of their field. To expect anything different is to misunderstand the driving forces behind present research. The armies of workers, grouped around very expensive equipment, are not for the most part seekers after new truths, but aim rather to apply existing knowledge usefully. It is the hope of commercial advantage that loosens the purse-strings. Unexpected

Electrons in Solids

discoveries, such as the fascinating behaviour of two-dimensional, and now one-dimensional and zero-dimensional electron systems, and the equally fascinating properties of heavy-fermion metals and organic conductors, are no longer the primary goal of research, but more like by-products of a vast physics-based industry. By themselves they could not absorb the efforts of a tenth of the work-force of research students, post-doctoral workers and more senior staff who are pouring out the deluge of publications in the field.

Something of this was predictable thirty years ago, though not of course in any detail. Few would have guessed then that so much money would be available or that developments in computers would enlarge the technical range and reduce the cost of experiment so dramatically. But, to speak personally, it now seems to me right to have exhorted younger physicists to look towards applications, rather than pure science, as the likely source of rewarding work [205]. It is hard to believe that as physics develops further it will become simpler again, and we must expect that pure research will continue to be the pursuit of a talented minority—not necessarily the most intelligent or the most imaginative, but those with a peculiar bent. It is all too easy, especially in academic circles, for the academically gifted to parade the notion of their own superiority. The many who can make their mark in other ways should not allow themselves or others to think that they have lost the chance of high distinction—that comes from using one's talent to the full, be it in proving a mathematical theorem, or perfecting an instrument, or teaching, or helping in any way to harness science and technology for the general good.

Acknowledgments

I am indebted to Professor R G Chambers for helpful criticism.

References

[1] 1959 *Methods of Experimental Physics* (New York: Academic), especially p 21 for *Purification* (P H Egli, L R Johnson and W Zimmerman) and p 86 for *Single Crystal Growing* (M Tanenbaum)

[2] Carpenter H C H and Elam C F 1921 *Proc. R. Soc.* A **100** 329

[3] Whittaker E T 1951 *A History of the Theories of Aether and Electricity* 2nd edn (London: Nelson), for Volta and thermoelectric effects
Campbell L L 1923 *Galvanomagnetic and Thermomagnetic Effects* (London: Longmans, Green)

[4] Hall E H 1879 *Am. J. Math.* **2** 287

[5] Kapitza P L 1928 *Proc. R. Soc.* A **119** 358; 1929 *Proc. R. Soc.* A **123** 292

[6] Kaiser W 1987 *Hist. Stud. Phys. Sci.* **17:2** 271

[7] Richardson O W 1903 *Phil. Trans. R. Soc.* A **201** 497

[8] Bohr's Master's and Doctor's dissertations are published with English translations in Nielson J R (ed) 1972 *Niels Bohr Collected Works* vol 1 (Amsterdam: North-Holland).
[9] 1927 *Conductibilité Électrique des Métaux (4th Solvay Conference, 1924)* (Paris: Gauthier-Villars)
[10] Hoddeson L, Braun E, Teichmann J and Weart S (ed) 1992 *Out of the Crystal Maze* (New York: Oxford University Press) ch 2
Eckert M 1987 *Hist. Stud. Phys. Sci.* **17:2** 191
[11] Wien W 1913 *Preuss. Akad. Wiss. Berlin Ber.* **7** 184
[12] Keesom W H 1913 *Phys. Z.* **14** 670
[13] Sommerfeld A 1928 *Z. Phys.* **47** 1; three papers complete the sequence, by W V Houston (p 33), C Eckart (p 38) and finally Sommerfeld again (p 43)
[14] Schottky W 1923 *Z. Phys.* **14** 63
[15] Hoddeson L, Braun E, Teichmann J and Weart S (ed) 1992 *Out of the Crystal Maze* (New York: Oxford University Press) p 104
[16] Hall E H 1928 *Proc. Natl Acad. Sci.* **14** 366, 370
[17] Barlow H M 1929 *Phil. Mag.* **7** 459
[18] Lorentz H A 1905 *Proc. R. Acad. Sci. Amsterdam* **7** 438, 585, 684
[19] Sommerfeld A and Frank N H 1931 *Rev. Mod. Phys.* **3** 1
[20] Sommerfeld A and Bethe H 1933 *Handbuch der Physik* vol 24/2 (Berlin: Springer) p 333
[21] Hoddeson L, Braun E, Teichmann J and Weart S (ed) 1992 *Out of the Crystal Maze* (New York: Oxford University Press)
Mott N F et al 1980 *Proc. R. Soc.* A **371** 3; a symposium on the beginnings of solid-state physics, containing valuable contributions, relevant to this section, by F Bloch, R E Peierls, A H Wilson and H A Bethe
[22] Bloch F 1928 *Z. Phys.* **52** 555
[23] Wilson A H 1932 *Proc. R. Soc.* A **134** 277
[24] Houston W V 1928 *Z. Phys.* **48** 449
[25] Strutt M J O 1928 *Ann. Phys., Lpz.* **86** 319
[26] Slater J C 1975 *Solid State and Molecular Theory: A Scientific Biography* (New York: Wiley) p 60
[27] Sommerfeld A and Bethe H 1933 *Handbuch der Physik* vol 33 (Berlin: Springer) p 40
[28] Jones H and Zener C 1934 *Proc. R. Soc.* A **145** 268
[29] Herring C 1937 *Phys. Rev.* **52** 361
[30] Jones H and Zener C 1934 *Proc. R. Soc.* A **144** 101
[31] Peierls R E 1929 *Phys. Z.* **30** 273
[32] Klemens P G 1956 *Encyclopedia of Physics* vol 14 (Berlin: Springer) p 198
[33] Ziman J M 1960 *Electrons and Phonons* (Oxford: Clarendon) p 386
[34] Peierls R E 1930 *Ann. Phys., Lpz.* **4** 121
[35] Mott N F and Jones H 1936 *The Theory of the Properties of Metals and Alloys* (Oxford: Clarendon) ch 7
[36] Brillouin L 1930 *Les Statistiques Quantiques* vol 2 (Paris: Les Presses Universitaires de France) p 204
[37] Frenkel J 1932 *Wave Mechanics* (Oxford: Clarendon) p 267
[38] Bardeen J 1937 *Phys. Rev.* **52** 688
[39] Herring C 1953 *Structure and Properties of Solid Surfaces* (Chicago, IL: Chicago University Press) p 117
[40] Pippard A B 1955 *Phil. Mag.* **46** 1104
[41] Peierls R E 1932 *Ann. Phys., Lpz.* **12** 154
[42] Ziman J M 1960 *Electrons and Phonons* (Oxford: Clarendon) p 411
[43] Ziman J M 1960 *Electrons and Phonons* (Oxford: Clarendon) p 396

[44] 1927 *Conductibilité Électrique des Métaux (4th Solvay Conference, 1924)* (Paris: Gauthier-Villars) p 9
[45] Bridgman P W 1919 *Phys. Rev.* **14** 306
[46] Bohr N 1912 *Phil. Mag.* **23** 984
[47] MacDonald D K C 1962 *Thermoelectricity* (New York: Wiley) p 71
[48] Mott N F and Jones H 1936 *The Theory of the Properties of Metals and Alloys* (Oxford: Clarendon) p 310
[49] Ioffe A F 1957 *Semiconductor Thermoelements and Thermoelectric Cooling* (London: Infosearch)
[50] Brillouin L 1931 *Die Quantenstatistik* (Berlin: Springer) p 287
[51] Seitz F 1940 *The Modern Theory of Solids* (New York: McGraw-Hill) ch 8
[52] Important texts of the period, in addition to [20], [35] and [51], are:
Fröhlich H 1936 *Elektronentheorie der Metalle* (Berlin: Springer)
Wilson A H 1936 *The Theory of Metals* (Cambridge: Cambridge University Press)
Mott N F and Gurney R W 1940 *Electronic Processes in Ionic Crystals* (Oxford: Clarendon)
[53] Slater J C 1975 *Solid State and Molecular Theory: A Scientific Biography* (New York: Wiley) p 60
Schweber S S 1990 *Hist. Stud. Phys. Sci.* **20** 339
[54] Hoddeson L, Braun E, Teichmann J and Weart S 1992 *Out of the Crystal Maze* (New York: Oxford University Press) ch 3
[55] Wigner E and Seitz F 1933 *Phys. Rev.* **43** 804
[56] Wigner E and Seitz F 1934 *Phys. Rev.* **46** 509
[57] Wigner E 1934 *Phys. Rev.* **46** 1002
[58] Lindemann F A 1915 *Phil. Mag.* **29** 127
[59] Landau L D 1937 *Phys. Z. Sowjet.* **11** 545
[60] Anderson P W 1984 *Basic Notions of Condensed Matter Physics* (Menlo Park, CA: Benjamin/Cummings) p 29
[61] Debye P and Hückel G 1923 *Phys. Z.* **24** 305
[62] See Mott N F and Jones H 1936 *The Theory of the Properties of Metals and Alloys* (Oxford: Clarendon) p 293
[63] Baber W G 1937 *Proc. R. Soc.* A **158** 383
[64] Landau L D and Pomeranchuk I 1936 *Phys. Z. Sowjet.* **10** 649
[65] Slater J C 1934 *Rev. Mod. Phys.* **6** 209
[66] Ashcroft N W and Mermin N D 1976 *Solid State Physics* (New York: Holt, Rinehart and Winston) ch 11
For a full bibliography see Slater J C 1965 *Quantum Theory of Molecules and Solids* vol 2 (New York: McGraw-Hill) p 521.
[67] Hellmann H and Kassatotschkin W 1936 *J. Chem. Phys.* **4** 324
[68] Hoddeson L, Braun E, Teichmann J and Weart S (ed) 1992 *Out of the Crystal Maze* (New York: Oxford University Press) ch 7
[69] Shockley W 1950 *Electrons and Holes in Semiconductors* (New York: Van Nostrand)
[70] 1962 *Proceedings of the International Conference on the Physics of Semiconductors* (London: Institute of Physics and the Physical Society) pp 901, 902
[71] 1906 *US Patent* No 836531 (Pickard) and *US Patent* No 837616 (Dunwoody)
[72] Pierce G W 1909 *Phys. Rev.* **28** 153
[73] Grondahl L O 1933 *Rev. Mod. Phys.* **5** 141
[74] Gudden B 1931 *Phys. Z.* **32** 825
[75] Mott N F and Gurney R W 1940 *Electronic Processes in Ionic Crystals* (Oxford: Clarendon) pp 83, 166
[76] Frenkel J I 1931 *Phys. Rev.* **37** 17

[77] Wannier G 1937 *Phys. Rev.* **52** 191
[78] Pohl R W 1937 *Proc. Phys. Soc.* **49** E3
[79] Ioffe A F 1935 *Actualités Scientifiques et Industrielles* vol 202 (Paris: Hermann)
[80] Hoddeson L, Braun E, Teichmann J and Weart S (ed) 1992 *Out of the Crystal Maze* (New York: Oxford University Press) ch 4
[81] Pankove J I 1971 *Optical Processes in Semiconductors* (Englewood Cliffs, NJ: Prentice-Hall) ch 14
[82] Landau L D 1933 *Phys. Z. Sowjet.* **3** 664
[83] Fröhlich II, Pelzer H and Zienau S 1950 *Phil. Mag.* **41** 221
[84] Holstein T 1959 *Ann. Phys., Lpz.* **8** 343
[85] Wolfe J P, Hansen W L, Haller E E, Markiewicz R S, Kittel C and Jeffries C D 1975 *Phys. Rev. Lett.* **34** 1292
[86] Shah J, Combescot M and Dayem A H 1977 *Phys. Rev. Lett.* **38** 1497
[87] Braun E and Macdonald S 1982 *Revolution in Miniature* 2nd edn (Cambridge: Cambridge University Press)
Hoddeson L, Braun E, Teichmann J and Weart S (ed) 1992 *Out of the Crystal Maze* (New York: Oxford University Press) ch 7
[88] Southworth G C 1950 *Principles and Applications of Waveguide Transmission* (New York: Van Nostrand) pp 8, 648
[89] Torrey H C and Whitmer C A 1948 *Crystal Rectifiers* (New York: McGraw-Hill)
[90] Henisch H K 1949 *Metal Rectifiers* (Oxford: Clarendon)
[91] Haynes J R and Shockley W 1949 *Phys. Rev.* **75** 691
[92] Bardeen J and Brattain W H 1948 *Phys. Rev.* **74** 230
[93] Smith R A, Jones F E and Chasmar R P 1968 *The Detection and Measurement of Infra-red Radiation* 2nd edn (Oxford: Clarendon)
[94] 1962 *Bull. Inst. Phys. Phys. Soc.* **13** 33
[95] Smith S D 1982 *Robert Allan Smith; Biographical Memoirs of Fellows of The Royal Society* (London: The Royal Society) p 479
[96] Fischer F, Gudden B and Treu M 1938 *Phys. Z.* **39** 127
[97] Lovell B (ed) 1947 *Electronics* (London: Pilot) p 124
[98] Hall L H, Bardeen J and Blatt F J 1954 *Phys. Rev.* **95** 559
[99] Herman F and Callaway J 1953 *Phys. Rev.* **89** 518
[100] Herman F 1954 *Phys. Rev.* **95** 847
[101] Burstein E and Egli P H *Advances in Electronics and Electron Physics* vol 7 (New York: Academic) p 1 and Brooks H *Advances in Electronics and Electron Physics* vol 7 (New York: Academic) p 87
[102] Hung C S and Gliessman J R 1950 *Phys. Rev.* **79** 726
[103] Fritzsche H 1958 *J. Phys. Chem. Solids* **6** 69
[104] de Boer J H and Verwey E J W 1937 *Proc. Phys. Soc.* **49** 59
[105] Mott N F 1949 *Proc. Phys. Soc.* A **62** 416
[106] Mott N F 1968 *Rev. Mod. Phys.* **40** 673
[107] Dorfman J 1951 *Dokl. Akad. Nauk* **81** 765
Dingle R B 1952 *Proc. R. Soc.* A **212** 38
[108] Dresselhaus G, Kip A F and Kittel C 1953 *Phys. Rev.* **92** 827
[109] Shockley W 1950 *Phys. Rev.* **79** 191
[110] Shockley W 1951 *Bell System. Tech. J.* **30** 990
Conwell E M 1967 *Solid State Physics Supplement 9* ed F Seitz and H Turnbull (New York: Academic)
[111] Gunn J B 1956 *J. Electron.* **2** 87
[112] O'Dwyer J J 1964 *The Theory of Dielectric Breakdown of Solids* (Oxford: Clarendon)
[113] Erlbach E and Gunn J B 1962 *Proceedings of the International Conference*

Electrons in Solids

       on the *Physics of Semiconductors* (London: Institute of Physics and the Physical Society) p 128; 1962 *Phys. Rev. Lett.* **8** 280
- [114] Hutson A R, McFee J H and White D L 1961 *Phys. Rev. Lett.* **7** 237
- [115] Esaki L 1962 *Phys. Rev. Lett.* **8** 4
- [116] Esaki L 1958 *Phys. Rev.* **109** 603
- [117] Gunn J B 1964 *IBM J. Res. Dev.* **8** 141
  Butcher P N 1967 *Rep. Prog. Phys.* **30** 97
- [118] Collins S C 1956 *Encyclopedia of Physics* vol 14 (Berlin: Springer) p 112
- [119] Bass J 1972 *Adv. Phys.* **21** 421
- [120] van Vucht R J M, van Kempen H and Wyder P 1985 *Rep. Prog. Phys.* **48** 853
- [121] van den Berg G J 1964 *Progress in Low Temperature Physics* vol 4 (Amsterdam: North-Holland) p 194
- [122] Kondo J 1962 *Prog. Theor. Phys.* **28** 846
- [123] Daybell M D and Steyert W A 1968 *Rev. Mod. Phys.* **40** 380
  Dugdale J S 1977 *The Electrical Properties of Metals and Alloys* (London: Arnold) p 189
  Chambers R G 1990 *Electrons in Metals and Semiconductors* (London: Chapman and Hall) p 129
- [124] Chambers R G 1969 *The Physics of Metals* ed J M Ziman (Cambridge: Cambridge University Press) p 175
- [125] Chambers R G 1952 *Proc. R. Soc.* A **202** 378
- [126] Chambers R G 1969 *The Physics of Metals* ed J M Ziman (Cambridge: Cambridge University Press) p 175
  Pippard A B 1989 *Magnetoresistance in Metals* (Cambridge: Cambridge University Press) ch 6
- [127] Tsoi V S 1974 *JETP Lett.* **19** 70
- [128] Harrison W A and Webb M B (ed) 1960 *The Fermi Surface (Cooperstown Conference, 1960)* (New York: Wiley)
  Springford M (ed) 1980 *Electrons at the Fermi Surface* (Cambridge: Cambridge University Press)
  Shoenberg D 1984 *Magnetic Oscillations in Metals* (Cambridge: Cambridge University Press)
  Hoddeson L, Braun E, Teichmann J and Weart S (ed) 1992 *Out of the Crystal Maze* (New York: Oxford University Press) ch 3
- [129] Mott N F and Jones H 1936 *The Theory of the Properties of Metals and Alloys* (Oxford: Clarendon) p 189
- [130] Slater J C 1965 *Quantum Theory of Molecules and Solids* vol 2 (New York: McGraw-Hill) p 303
- [131] See Mott N F and Jones H 1936 *The Theory of the Properties of Metals and Alloys* (Oxford: Clarendon) p 170
- [132] Heine V 1969 *The Physics of Metals* ed J M Ziman (Cambridge: Cambridge University Press) p 51
- [133] Jones H 1955 *Proc. Phys. Soc.* A **68** 1191
- [134] Howarth D J 1959 *Phys. Rev.* **99** 469
- [135] Callaway J 1964 *Energy Band Theory* (New York: Academic) p 168
- [136] Segall B 1961 *Phys. Rev. Lett.* **7** 164; 1962 *Phys. Rev.* **125** 109
  Burdick G A 1961 *Phys. Rev. Lett.* **7** 156
- [137] London H 1940 *Proc. R. Soc.* A **176** 522
- [138] Pippard A B 1947 *Proc. R. Soc.* A **191** 385
- [139] Reuter G E H and Sondheimer E H 1948 *Proc. R. Soc.* A **195** 336
- [140] Sondheimer E H 1954 *Proc. R. Soc.* A **224** 260
- [141] Pippard A B 1954 *Proc. R. Soc.* A **224** 273
- [142] Pippard A B 1957 *Phil. Trans. R. Soc.* A **250** 325

[143] Shoenberg D 1984 *Magnetic Oscillations in Metals* (Cambridge: Cambridge University Press) ch 1
[144] Shoenberg D 1939 *Proc. R. Soc.* A **170** 341
[145] Nakhimovich N M 1941 *J. Phys. USSR* **6** 11
[146] Berlincourt T G 1954 *Phys. Rev.* **94** 1172
[147] 1955 *Les Électrons dans les Métaux (10th Solvay Congress, 1954)* (Brussels: Stoops) p 153
[148] Gold A V 1958 *Phil. Trans. R. Soc.* A **251** 85
[149] Harrison W A and Webb M B (ed) 1960 *The Fermi Surface (Cooperstown Conference, 1960)* (New York: Wiley) p 28
[150] Phillips J C 1958 *Phys. Rev.* **112** 685
[151] Heine V 1969 *The Physics of Metals* ed J M Ziman (Cambridge: Cambridge University Press) ch 1
[152] Mackintosh A R and Andersen O K 1980 *Electrons at the Fermi Surface* (Cambridge: Cambridge University Press) p 149
[153] Cochran J F and Haering R R (ed) 1968 *Solid State Physics* vol 1 (New York: Gordon and Breach) p 146
[154] Julian S R, Teunissen P A A and Wiegers S A J 1992 *Phys. Rev.* B **46** 9821
[155] Peierls R E 1980 *Proc. R. Soc.* A **371** 35
[156] Jones H 1936 *Proc. R. Soc.* A **155** 653
[157] Alers P B and Webber R T 1953 *Phys. Rev.* **91** 1060
[158] Fawcett E 1964 *Adv. Phys.* **13** 139
[159] Klauder J R, Reed W A, Brennert G F and Kunzler J E 1966 *Phys. Rev.* **141** 592
[160] Chambers R G 1956 *Phil. Mag.* **1** 459
[161] Azbel' M Ya and Kaner E A 1956 *Sov. Phys.–JETP* **3** 772
[162] Walsh W M 1968 *Solid State Physics* vol 1, ed J F Cochran and R R Herring (New York: Gordon and Breach) p 158
[163] Pippard A B 1960 *Rep. Prog. Phys.* **23** 176
[164] Petrashov V T 1984 *Rep. Prog. Phys.* **47** 47
Kaner E A and Skobov V G 1968 *Adv. Phys.* **17** 69
[165] Stark R W and Falicov L M 1967 *Progress in Low Temperature Physics* vol 5 (Amsterdam: North-Holland) ch 6
[166] Stark R W 1964 *Phys. Rev.* A **135** 1698
[167] Hofstadter D R 1976 *Phys. Rev.* B **14** 2239
[168] Pippard A B 1980 *Electrons at the Fermi Surface* ed M Springford (Cambridge: Cambridge University Press) ch 4
[169] Reed W A and Condon J H 1970 *Phys. Rev.* B **1** 3504
[170] Skinner H W B 1940 *Phil. Trans. R. Soc.* A **239** 95
[171] Pines D 1955 *Les Électrons dans les Métaux (10th Solvay Congress, 1954)* (Brussels: Stoops) p 9
[172] Tonks L and Langmuir I 1929 *Phys. Rev.* **33** 195
[173] Schumacher R T and Vehse W E 1963 *J. Phys. Chem. Solids* **24** 297
[174] Lindhard J 1954 *Kong. Dansk. Vid. Selsk., Mat.-Fys. Medd.* **28** 8
[175] Landau L D 1946 *J. Phys. USSR* **10** 25
[176] Gell-Mann M and Brueckner K A 1957 *Phys. Rev.* **106** 364
[177] Hubbard J 1958 *Proc. R. Soc.* A **243** 336
[178] Bardeen J and Pines D 1955 *Phys. Rev.* **99** 1140
[179] Pines D 1956 *Rev. Mod. Phys.* **28** 184
[180] Landau L D 1957 *Sov. Phys.–JETP* **3** 920
Wilkins J W 1980 *Electrons at the Fermi Surface* ed M Springford (Cambridge: Cambridge University Press)
[181] Silin V P 1958 *Sov. Phys.–JETP* **6** 387

[182] Migdal A B 1957 *Sov. Phys.–JETP* **5** 333
[183] Luttinger J M 1960 *The Fermi Surface (Cooperstown Conference, 1960)* ed W A Harrison and M B Webb (New York: Wiley) p 2
[184] Mott N F and Davis E A 1979 *Electronic Processes in Non-crystalline Materials* (Oxford: Clarendon)
Faber T E 1972 *Introduction to the Theory of Liquid Metals* (Cambridge: Cambridge University Press)
[185] Anderson P W 1958 *Phys. Rev.* **109** 1492
[186] Mott N F 1967 *Adv. Phys.* **16** 49
[187] See Peierls R E 1955 *Quantum Theory of Solids* (Oxford: Clarendon) p 139
[188] Chester G V and Thellung A 1959 *Proc. Phys. Soc.* **73** 745
[189] Ioffe A F and Regel A R 1960 *Prog. Semicond.* **4** 237
[190] Mott N F and Twose W D 1961 *Adv. Phys.* **10** 107
[191] Aronov A G and Sharvin Yu V 1987 *Rev. Mod. Phys.* **59** 755
Beenakker C W J and van Houten H 1991 *Solid State Physics* vol 44, ed H Ehrenreich and D Turnbull (Boston, MA: Academic) p 1
[192] Aharonov Y and Bohm D 1959 *Phys. Rev.* **115** 485
[193] Al'tshuler B L, Aronov A G and Spivak B Z 1981 *JETP Lett.* **33** 94
[194] Sharvin D Yu and Sharvin Yu V 1981 *JETP Lett.* **34** 272
[195] Webb R A, Washburn S, Umbach C P and Laibowitz R B 1985 *Phys. Rev. Lett.* **54** 2696
[196] Ando T, Fowler A B and Stern F 1982 *Rev. Mod. Phys.* **54** 437
[197] 1972 *Gallium Arsenide and Related Compounds* (Bristol: Institute of Physics)
[198] von Klitzing K, Dorda G and Pepper M 1980 *Phys. Rev. Lett.* **45** 494
[199] von Klitzing K 1986 *Rev. Mod. Phys.* **58** 519
[200] Prange R E 1981 *Phys. Rev.* B **23** 4802; 1990 *Electrons in Metals and Semiconductors* (London: Chapman and Hall) p 199
[201] Esaki L 1985 *Molecular Beam Epitaxy and Heterostructures* ed L L Chang and K Ploog (Dordrecht: Nijhoff) p 1
[202] Tsui D C, Störmer H L and Gossard A C 1982 *Phys. Rev. Lett.* **48** 1559
[203] Willett R, Eisenstein J P, Störmer H L, Tsui D C, Gossard A C and English J H 1987 *Phys. Rev. Lett.* **59** 1776
[204] Physics in two dimensions 1992 *Helv. Phys. Acta* **65** No 2, 3
[205] Pippard A B 1961 *Phys. Today* **14** 38

Chapter 18

A HISTORY OF OPTICAL AND OPTOELECTRONIC PHYSICS IN THE TWENTIETH CENTURY

R G W Brown and E R Pike

18.1. Introduction: classical optics up to 1900

As in other branches of physics, the end of the nineteenth century marks the close of one era (classical optics) and the start of another (quantum optics). The occasion was Max Planck's proposition (1900) that an oscillating electric system does not impart energy to the electromagnetic field in a continuous manner but in finite amounts or 'quanta' of energy $\epsilon = h\nu$. Thus began what Planck would call later [1] '... the long and multiply twisted path to the quantum theory'. It took a long time to mature, but eventually brought him a Nobel Prize (1918).

Between then and 1960, the year of the demonstration of the first laser, many new and apparently disconnected principles of optics were established, but our later discussion of the laser will act as a focus to draw them together. From 1960, there began a still increasing growth of scientific and technological activity in quantum optics, quantum electronics and optoelectronics. The impact on society of this progress is now keenly felt in everyday life, through television, compact-disc recorded sound and video and a wide range of consumer products, office and business systems and military and medical hardware.

The state of classical optics in 1900 is briefly recalled in Chapter 1, and may be assessed by reading the 1902 book of Drude [2]. Classical optics was founded upon the seventeenth century ideas of Descartes concerning transmission of light through the aether, on Snell's law of refraction and Fermat's principle of least time. The interference of light had also been investigated in Hooke's and Boyle's observations of 'Newton's rings', in Newton's studies of colour and in Hooke's and Grimaldi's early studies of diffraction. Hooke had proposed a wave theory of light, later to be improved by Huygens, who also made the fundamental discovery of light polarization. Newton rejected the wave theory of light in favour of his corpuscular theory. The difference of opinion was apparently resolved when, at the beginning of the nineteenth century, the wave theory yielded Young's principle of interference and Fresnel's treatment of diffraction. This century also saw the resolution of many problems in optics by, for instance, Rayleigh, Kirchhoff and of course Maxwell whose celebrated equations for the electromagnetic field encapsulated all of classical optics. However, various questions remained unanswered, especially the reason for the spectral characteristics of black-body radiation. The optical physicist in 1900 knew a great deal, but he still had a few problems to sort out. In 1899, Michelson went so far as to write [3]

> The more important fundamental laws and facts of physical science have all been discovered, and these are so firmly established that the possibility of their ever being supplanted in consequence of new discoveries is exceedingly remote.

Would any one of us say this with respect to the state of physics today, or will it be so totally replaced in the future as has classical physics been in the twentieth century? We think not, but the thought is a sobering one.

In this chapter we can do no more than discuss the historical development of the essential *principles* of optical and optoelectronic physics, with only brief reference to applications, device variations and technology, especially semiconductor optoelectronics. We have chosen only to mention those few items of central importance in optoelectronic devices, leaving device details to be read in the available excellent texts. There are good contemporary texts on all areas of optics and we give a selection in section 18.6.2.

We have chosen to take an essentially chronological approach and produce a history of basic advances which we believe to be the first of its kind for twentieth century optics. A history by separate subject areas such as interferometry, lasers, lenses, etc, can be gleaned from specialist textbooks, and our year by year chronicle gives a fascinating picture of active and continuous development across the broad front line of the subject throughout the twentieth century. We make a feature of referring to the sequence of Nobel Prize awards for optical innovations. This will give an idea of the high profile and esteem in which many of the achievements in optics that we shall be describing have been held.

18.2. 1900–30: early quantum optics

18.2.1. Planck's phenomenology
The solution of the form of the spectral distribution of black-body radiation by Max Planck in 1900 was the first step towards the quantum theory and our understanding of the structure of atoms and molecules [4, 5]. Planck's introduction of energy quanta [6, 7] marked also the start of a new age of optics. It is important for our discussion to understand that Planck created a phenomenological equation that agreed with the whole of the experimentally determined black-body emission spectrum, particularly the new precise results (1900) of Lummer and Pringsheim [8] and of Rubens and Kurlbaum [9]. He had to assume, as an 'act of desperation' as he was later to remark in a letter to R W Wood [10], that the walls of a black body emit or absorb only discrete packets of energy, or quanta. This seemed to imply that these walls were composed of oscillators in thermal equilibrium which possessed discrete energies, restricted to multiples of an energy unit or quantum $h\nu$ rather than the continuous range associated with classical oscillators. By arguing that each field mode could only receive energy from the oscillators in discrete amounts, Planck derived his equation for the energy density $\rho(\nu)$ in a cavity of temperature T:

See also p 146

$$E(\nu)\,d\nu = \frac{8\pi^2\nu^2}{c^3} \frac{h\nu}{\exp(h\nu/kT) - 1}\,d\nu$$

where ν is the oscillator frequency, h is Planck's constant, c is the velocity of light and k is Boltzmann's constant. He did not, at this time or later, wish to suggest that the radiation field was quantized.

There was no immediate appreciation of Planck's result, apart perhaps for a report by Day [11] to the Philosophical Society of Washington. In fact, it was five years before Einstein, to whom Planck's work provided a great stimulus, proposed the quantization of the radiation field itself in one [12] of his three celebrated papers of 1905; and it was ten years before Planck himself published his next paper [13] on the subject in which, still holding Maxwell's equations in great reverence, and not knowing (as we do now) how to have your cake and eat it, he could not bring himself to embrace Einstein's theory of radiation quanta. Indeed later authors have differed in the fine details of interpretation of Planck's work [5]; a completely sound derivation of the black-body law depends on the concept of spontaneous emission which had to wait until Einstein (1916) [14] for its introduction and until Bose (1924) [15] for its consistent derivation, leading to the Bose–Einstein statistics for the photon field.

Even after Bose it was still possible to argue plausibly that the Planck law was entirely due to the quantized matter oscillators and that the field could continue to be unquantized. Another significant outcome of Planck's struggles with the quantum was, fortuitously again, the introduction of zero-point energy into physics. This was done in his 'second theory' of the black-body law of 1911 [16], in which he continued to avoid quantizing the radiation. He allowed his oscillators to build up energy and then release it all in bursts of n quanta with a probability determined by the assumption that the ratio of no emission to emission should be proportional to the intensity of the incident radiation. It is interesting to note that the postulate of a classical zero-point component of the radiation field (stochastic electrodynamics) in the hands of Marshall [17] and others kept alive the hopes of Planck's disciples as late as 1960. It seems that even to the present day Nobel Laureate Willis Lamb still does not believe in photons, but how exactly he understands the photon counting and correlation experiments, which we shall be explaining below, we have not been able to discover, in spite of several interesting discussions with one of the present authors (ERP).

Thus, to be fair to Planck and, indeed, to most of his contemporaries who were slow to take up Einstein's notion of radiation quanta, and to relate him historically to later developments, we should note (dismissing from now on stochastic electrodynamics and other similarly exotic theories) that a theory in which only matter and not radiation is quantized has traditionally been called a 'semi-classical' theory. A significant number of theorists maintained this view into the 1960s, and continued to believe that quantization of radiation is unnecessary to explain all physical effects [18]. Indeed, one of the present authors (ERP) was criticized in print at this time for invoking radiation quanta in an early work on photon-counting statistics, as discussed further in section 18.4.4.

See also p 1439

In modern terminology we can say that semi-classical theory can be used, albeit not entirely rigorously, to 'explain' any optical phenomenon when the quasi-probability 'P representation' of the radiation field, which we shall meet later, introduced independently by Glauber [19] and Sudarshan [20], is everywhere positive and non-singular. The field then has the apparent form of a classical superposition of coherent states. Otherwise we have what is called *non-classical* light. This criterion of a well-behaved P representation is probably not rigorous since it is never entirely clear what is meant by a non-classical field, but it can be taken as a working definition. An alternative definition in terms of inequalities between certain two-point correlation functions of the field can also be used [21]. The exact relation between these two has not been determined. However, given explicit or implicit resolution of such problems, in the circumstances that the P representation is everywhere positive, semi-classical theory can often be invoked. In the end it must be up to the semi-classical practitioner to define criteria as necessary for the particular theory to be used. Just like Gödel's theorem in mathematics, however, semi-classical theory can never be rigorous because implicit assumptions regarding operator ordering have to be made which are not within the scope of the theory itself. It cannot distinguish, for example, between the experimentally distinct expectation values of products of field operators such as E^+E^-, E^-E^+ and $(E^+E^- + E^-E^+)/2$. Planck's black-body law, the Compton effect and even the photoelectric effect, for example, all need quantization of the radiation field if serious contradictions are to be avoided.

A further serious complication for semi-classical theory is the existence of states of the field when photons become, using Schrödinger's term, 'entangled' with each other, another distinct feature of quantum theory, which defies any classical explanation. A third manifestation of the necessity for field quantization is then the violation of the *Bell inequalities* [22], which we will also discuss later, associated with the philosophical topic of the existence of so-called 'local realistic' theories. The first photon-pair correlation experiment, which we will discuss later, where non-classical properties of this type were investigated was not carried out until 1948 by Hanna. Planck, we can therefore say, should be able to rest moderately in peace.

Whilst these early steps towards quantum optics were being taken, the traditional practices of classical optics continued to be pursued. Hartmann [23, 24] explored the use of perforated screens for more accurate testing of wavefront aberrations from large mirror surfaces, a technique to be developed and applied in the 1970s and later [25]. Reflection interference fringes were reported by Lummer in 1901 [26] and also in joint work with Gehrcke in 1903 [27] and the Lummer–Gehrcke plate has featured in many optics textbooks over the years.

Following early observations by Kerr (1875), Cotton and Mouton [28] published their magneto-optical effect in 1907. They found that the birefringence varied as the square of applied magnetic field, in a manner closely similar to Kerr electro-optical double refraction. Faraday [29] (1846) had observed rotation of the plane of polarized light when a magnetic field was applied in the propagation direction. Cotton and Mouton observed birefringence when a magnetic field is applied to a transparent liquid perpendicular to the direction of propagation. Voigt had found the same but weaker effect for vapours in 1902 [30]. We will come back to optics and magnetism later.

The properties of lenses and curved mirrors were also the subject of great interest in the early 1900s. Photography was fashionable and widely practised. The four-element Tessar lens, introduced in 1902 [31] remains a fairly inexpensive lens, giving excellent images over fields up to 60° and for apertures as large as $f/3.5$. It was widely used in mass-market cameras from the 1960s onwards. Strehl's study [32] at this time of the effects of aberration on the Airy pattern formed by a diffraction-limited lens formed the basis of the famous test named after him.

Improvements in lens performance were much sought after, and in 1904 Conrady [33, 34] published his investigations into chromatic aberration, extending the already considerable lens aberration theory dating from Seidel's pioneering work in the previous century. At this time, Wood observed resonance radiation in the excited atoms of a low pressure gas, and the first Köhler-illuminated, ultraviolet-light microscope, with magnification up to 3600, appeared from the Karl Zeiss company [35]. Köhler's illumination system for microscopy (1893) put the specimen in the Fourier plane of the source where translational invariance removes brightness variations of the lamp surface. However, the method was not used for general microscopy until the 1940s when the use of the term *Köhler illumination* became widespread. The method has very recently been extended to electron microscopy [36, 37].

Just prior to the publication by Einstein of his revolutionary ideas, two Nobel Prizes were awarded for achievements related to optics [38], the first in 1902 shared between Zeeman and Lorenz for their investigations of magnetism and the phenomena of radiation, and the second in 1904 to J W Strutt (Lord Rayleigh) who had made various fundamental contributions to optics in the late 1800s.

18.2.2. Einstein's gas of corpuscles

See also p 151

1905 was a momentous year for physics and for optics. Einstein's famous publication [12, 39] concerning the corpuscular nature of light recognized the far-reaching implications of Planck's results from 1900–1. Later, this work would win a Nobel Prize [40]. Einstein showed by thermodynamic arguments that the Planck equation suggests that, at a high enough frequency to satisfy Wien's law, $\rho(\nu) = \alpha \nu^3 \exp(-h\nu/kT)$, light behaves as a gas of independent corpuscles or *quanta*, each of energy $h\nu$. Einstein further noted at low frequencies, where instead the Rayleigh–Jeans law $\rho(\nu) = 8\pi \nu^2 kT/c^2$ holds, that radiation would not behave in this manner.

Using the corpuscular idea, Einstein demonstrated that conservation of energy leads to an equation for the kinetic energy of an emitted photoelectron

$$\frac{mv^2}{2} = h\nu - W$$

where W is the work function, the minimum energy required to liberate the electron from a particular metal. He further showed that the number of emitted electrons is proportional to the incident light intensity, but that their velocity is not related to intensity. These two latter predictions were in agreement with known experimental results on the photoelectric effect, dating back nearly two decades, by Stoletov [41], Ladenburg [42], Lenard [43] and Hallwachs [44–46]. Indeed, the process at that time was widely known as the Hallwachs effect. In the following years Hughes [47], Richardson and Compton [48] and Millikan [49] were to verify Einstein's photoelectric equation to high accuracy; in particular, the linear relation between the potential difference needed to bring a photoelectron to rest and the frequency of the incident light. This work would lead eventually to photo-emission spectroscopy for probing electronic structure [50]. Hallwachs also continued his experiments [51] while Millikan used the photoelectric effect to measure Planck's constant and received a Nobel Prize for this painstaking work in 1923. It is worth noting that the first mention of possible multi-photon effects and the first mention of radiation out of thermodynamic equilibrium (which is necessary for a laser) occur is the 1905 paper of Einstein.

R W Wood published his classic textbook *Physical Optics* in 1905 [52]. Interestingly in this book he describes a method of making radial gradient-index lenses having two plane surfaces, a forerunner of GRIN (gradient-index) lens technology developed in the latter part of the twentieth century [53, 54], especially for use with fibre and waveguide optics. In the field of optical telescope design, Schwarzchild described [55, 56] a class of telescope objectives consisting of two aspheric mirrors and showed that such systems can be made aplanatic. This important result laid the foundations for much telescope design and construction throughout the twentieth century. Also Ritz deduced his combination principle from spectroscopic measurements eight years before Bohr's theory was proposed: an atom in the third excited state (above the ground state) can radiate its energy either in the form of a single light-quantum of frequency v_{30}, or as two quanta, the sum of whose frequencies must be exactly v_{30}.

From the viewpoint of its impact in late twentieth century applied optics, the work of Porter [57] is notable. He explored experimentally Abbe's 30-year-old theory of imaging; the diffraction theory of microscopy where the image was described as a sum of spatial frequencies. The resulting Fourier optics laid the foundations for a great deal of work later in the century concerning image analysis and optical data processing [58].

It was to be another four years before Einstein published further results of fundamental importance to the new optics. In the meantime, some significant events were generated in traditional areas of optical research. Rayleigh [59] calculated (1907) the diffraction of a plane electromagnetic wave by a grating, so providing the basis for modern grating spectroscopy. He developed the dynamical theory of gratings and gave a first mathematical theory of a regular corrugation whose pitch was comparable with the wavelength of light. Rayleigh's theory assumed that the grating depth was small compared with the wavelength. Producing a theory without this restriction occupied theorists throughout the twentieth century, a point we will return to on arriving at 1980. Michelson was awarded the 1907 Nobel Prize for his many contributions to optical instruments for precision spectroscopy and metrology [60], mostly in the late nineteenth century. Perhaps most curiously, Round observed yellow light emission from the semiconductor silicon carbide (SiC) when acted upon by an electric field [61]. It was to be nearly 40 years before the physics of this observation was sorted out, and yet another ten years before the significance of that physics (the p–n junction) was exploited in light-emitting diodes [62].

1907 was also the year in which Max von Laue [63] gave a quantitative measure for partial coherence in terms of a quantity, γ_L, proportional to the square of the time-averaged products of the disturbances at two points in the field (as measured in 1956 by Hanbury Brown and Twiss). Partial coherence, a subject of much important in optics, has been well treated historically by Beran and Parrent [64]; and a comprehensive collection of twentieth century papers on coherence has been compiled by Mandel and Wolf [65]. We shall return many times to developments in optical-coherence theory. However, at this stage we draw the distinction between classical coherence theory (up to 1963) and what followed from Glauber's contribution in that year, which was a quantum theory of optical coherence, essential for the study of many newer optical phenomena.

During 1908 Mie wrote down for the first time a general expression for the scattering of a planar light wave by a homogeneous dielectric sphere of arbitrary size compared with the incident light wavelength [66, 67], a landmark at this time in optical scattering theory which is now known by his name. Although tabulations were made for practical use for various values of the parameters, the modern desk computer has made such calculations easy and the theory is in widespread use. At this time another Nobel Prize was awarded for optics [68], to Lippmann [69, 70] for his method of photographic reproduction of colours based upon the phenomenon of interference within a thick film. The field of light-scattering advanced in another direction when, also in 1908, Smoluchowski [71, 72] published his account of fluctuations in dielectric constant caused by thermal agitations. He addressed in particular the molecular-kinetic theory of the opalescence of gases near critical points but also discussed the effects of roughness and its length-scales on the diffuse reflection of light from surfaces. He was also to make important contributions to the theory of diffusion-controlled aggregation in macromolecular solutions, another area where light-scattering plays an important diagnostic role.

18.2.3. 1909: particles and waves
In 1909, although Einstein had referred to the possibility of non-equilibrium radiation in his earlier paper, there was little motivation for discussing this idea. Indeed, apart from the concept of negative temperature, this was not to come in any detailed fashion until the work of Glauber in 1963. Einstein [73] thus returned to the problem of radiation in thermodynamic equilibrium with its surroundings. Starting

from Planck's equation for radiation density $\rho(\nu)$, and using his previous result for the energy fluctuations of a system in thermal equilibrium

$$\langle E^2 \rangle - \langle E \rangle^2 = \langle [\Delta E]^2 \rangle = kT^2 \frac{\partial \langle E \rangle}{\partial T}$$

Einstein obtained the fluctuation formula

$$\langle [\Delta E(\nu)]^2 \rangle_{\text{total}} = \left(h\nu \rho(\nu) + \frac{c^3}{8\pi \nu^2} \rho^2(\nu) \right) V$$

where

$$\rho(\nu) = \frac{8\pi h\nu^3/c^3}{\exp(h\nu/kT) - 1}$$

is the energy density, V is the volume and $E(\nu)$ is the energy as a function of oscillator frequency ν. The corresponding equation derived from the Rayleigh–Jeans equation when $h\nu \ll kT$ is

$$\langle [\Delta E(\nu)]^2 \rangle_{\text{wave}} = \frac{c^3}{8\pi \nu^2} \rho^2(\nu) V.$$

representing the fluctuations in energies of classical light waves.

From Wien's law, where $h\nu \gg kT$, the particle-like energy fluctuation is calculated to be

$$\langle [\Delta E(\nu)]^2 \rangle_{\text{particle}} = h\nu \rho(\nu) V = h\nu E(\nu)$$

a result characteristic of the classical Poissonian statistics of distinguishable particles. If we re-examine the last three equations, it is easily seen that

$$[\Delta E(\nu)]^2_{\text{total}} = [\Delta E(\nu)]^2_{\text{particle}} + [\Delta E(\nu)]^2_{\text{wave}}$$

and that black-body radiation *simultaneously* exhibits fluctuations characteristic of both particles and waves.

In the high-frequency limit, black-body radiation seems to behave like a gas of independent particles; in the low-frequency limit it seems to behave as a superposition of classical waves; in the intermediate region it has the characteristics of both. Einstein himself looked upon this result as the beginnings of a fusion of the two theories of particles and of waves, which were no longer to be considered to be mutually incompatible. Later developments in the quantum theory were to show that every elementary particle has its corresponding *wave equation*. That the *photon* and its wave should come first was a consequence of the available experimental facts which Planck fitted phenomenologically.

What exactly did Einstein's thermodynamics contribute? Well, he had the answer from Planck so he could work backwards. We know now that in three dimensions there are only two types of statistics, Fermi statistics for particles with half-integral spin and Bose statistics for particles with integral spin (they both go into Boltzmann statistics at sufficiently high temperatures). Maxwell's equations are those of a massless particle of spin one. Thus, on analysis of Planck's energy distribution, Einstein recovered implicitly the thermodynamics of the Bose gas. This was to be made explicit later, of course, by the work of Bose himself.

Optics has played a particularly crucial role in didactic investigations of developments in understanding of twentieth century physics. Wave–particle duality has been the subject of much debate and misunderstanding throughout the twentieth century, with experiments still being performed which probe the same details. We shall therefore spend a little time to discuss the basic ideas. Most physicists have long accepted Bohr's 'Copenhagen interpretation' which was presented at the Fifth Solvay Conference in 1927. This endows quantum theory with the power to calculate all the possible results of any given experiment and their likelihood; however, the actual result of any *particular* experiment, within these possible ones, is not predictable. This says no more, of course, than does classical statistical mechanics. The difference is that in the microscopic world, because of the unavoidable interaction of the experimental apparatus with the measurement and the indistinguishability of particles [74], quantum and classical statistics differ in two ways due, respectively, to *uncertainty* and *entanglement*. For example, if one is predicting the height of a man the interaction with the observer does not change the man's contribution to the statistics of the population. If he were a photon, however, in a photon population, then the act of measurement would kill him off and his contribution to further measurements would cease to exist, thus confusing the true statistics, particularly in a small population. This fundamental difference gives rise to an absolute minimum uncertainty in the values measured related to the mechanics of the interaction between the system and the apparatus. A definitive discussion of the uncertainty principle in the non-relativistic case was given in Heisenberg's little (and apparently nowadays little read!) book *The Physical Principles of the Quantum Theory* in 1930 [75]. Entangled states are multi-particle states that cannot be factored into products of single-particle states in any representation. It was Schrödinger who first used the term entanglement for such a superposition in a multi-particle system. This was in his famous three-part 'cat' paper of 1935 [76] where he poses and resolves the paradox of the cat being half completely dead and half completely alive at the same time by insisting that '... one cannot put the ψ function in place of the physical thing'. He argues at great length that the concept of entanglement lies at the heart of the difference between classical and quantum physics. Entangled photon states came to assume central importance in late twentieth century quantum optics.

To summarize the present-day theory briefly, we note that the description of a system at the quantum level is by means of a 'state function' which lives in a mathematical (Hilbert) space and by means of which the expectation value of any experimental measurements, although not their actual values, may be calculated. In non-relativistic quantum theory this state function can be a square-integrable 'wavefunction' (probability function) in a four-dimensional 'configuration space' or an ensemble of these with given classical probabilities. One can, from the laws of quantum theory or, more crucially for optics since the photon does not have a wavefunction, from relativistic quantum field theory, determine the temporal evolution of this state function after preparation in a given state and so estimate the probability for a later measurement to give any particular value.

To quote Heisenberg [74], who used the configuration-space description, the explanation of an experiment '... requires three distinct steps:

(1) the translation of the initial experimental situation into a probability function;

(2) the following-up of this function in the course of time; and

(3) the statement of a new measurement to be made on the system, the result of which can then be calculated from the probability function'.

So, for example, the famous twentieth century optical problem of explaining the recording of an interference pattern (or perhaps no interference pattern, in the sharper form of two-photon experiments which we discuss later) by light traversing two separate apertures, is answered *not* by saying that a particular photon (as a particle) must have gone through one aperture or the other, or by saying that a wave has gone through both slits and then interferes with itself (either of which leads to contradictions) but by calculation of the possible measurements allowed by the evolution of the state function, which includes the inevitable experimental consequences of destroying individual photons and, for multi-photon states, accounts correctly for any entanglement present.

The most common misinterpretation is to confuse the mathematical space in which the state function is defined (in particular, the configuration space of the wavefunction in non-relativistic quantum mechanics) with real four-dimensional space-time, in other words to do what Schrödinger warns against, namely putting the ψ function in place of the physical thing. The terminology *wave-particle duality* and *complementarity* by encouraging this misinterpretation does not help.

Feynman's [77] opinion, even in the case where an interference pattern is seen, was that this one experiment examines 'a phenomenon which is impossible, *absolutely* impossible, to explain in any classical way, and which has in it the heart of quantum mechanics. In reality, it contains the *only* mystery.'. This is not quite in accordance with the way Schrödinger saw it. Yet a further view was taken by Weinberg [78] who opined that Dirac's treatment of spontaneous emission confirmed the universal character of quantum mechanics. While we are at it we may as well throw in the contribution of Dirac [79] who, late in life, believed that the concept of the probability amplitude is perhaps the most fundamental concept of the quantum theory. The truth is probably that they are all as fundamental as each other and, of course, are all conceptually related.

In the same year (1909) as the paper of Einstein, Taylor [80] (at the suggestion of J J Thomson) performed a simple experiment to investigate whether for extremely weak light sources the individual 'units of energy' which seemed to exist in x-ray and ionization experiments, distributed non-uniformly over the wavefront, would modify ordinary diffraction or interference effects. He photographed the shadow of a needle illuminated by a gas flame through a narrow slit, and inserted smoked-glass filters successively to reduce the light intensity. He estimated the strength of his light source in the weakest case at 5×10^{-6} ergs s^{-1}, which he equated to the strength of a standard candle at one mile. Allowing an exposure time about three months, Taylor created a diffraction pattern of the same form as with stronger illumination. In Thomson's opinion this indicated that the energy units could be no larger than 1.6×10^{-6} ergs if, as he thought, more than one must be present to create an interference pattern. This is, in fact, over a thousand times greater than one of Einstein's quanta.

It is interesting to note that the idea of Einstein's quantum of light seems not to have arrived at Cambridge even by that time. Neither Thomson nor Taylor made any reference to Einstein or Planck's work although they knew from the experiments of Ladenberg that the units of energy varied in direct proportion to the frequency of the light. To quote from Thomson [81] 'The aether has disseminated through it discrete lines of electric force in a state of tension and light consists of transverse vibrations, Röntgen-rays of pulses travelling along these lines.'. With these ideas he explained Lenard's [82] results on photoelectron emission from metals using ultraviolet light, and concluded his paper by noting that '... γ-ray energy units will be very widely separated and will have all the properties of material particles except that they cannot move at any other speed than that of light'.

It was again Einstein [83] who led the way (in 1910) in attacking the physics of a problem that would become important in optics in the future. This time he turned his attention independently to the problem that Smoluchowski [71,72] had also investigated, the effect on light of fluctuations in dielectric constant due to thermal agitations. These works represented the first treatment of light-scattering by a real fluid. The main result is an expression for the extinction coefficient τ, defined as the ratio of the total intensity of light scattered in all directions per unit volume of the diffusing medium to the incident flux density. From the work of Einstein and Smoluchowski τ was found to be given by

$$\tau = \frac{\omega_0^4}{6\pi c^4}\left[k_B T \rho_0^2 \chi_T \left(\frac{\partial \epsilon_0}{\partial \rho}\right)^2_{\rho=\rho_0} + \frac{k_B T^2}{\rho_0 C_v}\left(\frac{\partial \epsilon_0}{\partial T}\right)^2_{T=T_0}\right]$$

where χ_T is the isothermal compressibility, k_B is Boltzmann's constant, T is the equilibrium temperature, ρ_0 is number density, C_v is specific heat at constant volume and ϵ_0 is the dielectric constant. As χ_T tends to infinity near the liquid–gas transition we have the phenomenon of critical opalescence.

The next major step in 'quantum optics' occurred some three years later with the new insights of Niels Bohr, but as ever there continued to be some notable events in established classical optics. Interest in diffraction gratings continued, none stronger than that of Wood, who continued [84] the tradition initiated by Rowland and Michelson, and in 1910 ruled echelette gratings with 2000–3000 grooves per inch. These could send visible light to near 30th order, and later became of great value in infrared spectroscopy.

Other important results were first obtained at this time. Hondros and Debye [85] published a theoretical study of the transmission of light by a dielectric waveguide, and Mauguin [86] considered the propagation of light in a helical structure. Over 50 years later these results found applications in optical fibre communications [87] and in liquid-crystal displays [88]; both are technologies that have dominated late twentieth century optoelectronics and require extensive application of the basic principles of classical optics to complex optical geometries. Such developments are beyond the scope of this chapter and have been summarized elsewhere [89–91].

At this time a landmark step in quantum theory was taken by Bohr [92,93] who combined Planck's quantum ideas with the concept of Rutherford's nuclear atom, as told in Chapter 3. Another major step for twentieth century optics was taken soon after, when Brillouin predicted [94], contrary to prevailing opinion, that scattered light could be shifted in frequency. The story goes that he had his idea on holiday while lying on his back on Monte Rosa looking up at scattering from the sky. He became involved in the war and finished his work in 1922, not without difficulty in understanding his earlier notes. The theory was developed independently by Mandelstam in the Soviet Union, where the phenomenon is called Mandelstam–Brillouin scattering. Landsberg [95] states that Mandelstam had known in 1918 that a spectral shift should occur but his work was not published until 1926 [96]. While trying to verify the effect he discovered 'Raman scattering'.

Mandelstam had understood that the Fourier expansion technique for fluctuations, first used by Einstein, could be related to the acoustic waves

See also p 157

in Debye's theory of specific heat and that their motion would create a frequency shift. Instead, the discovery of the much larger shift due to optic modes deflected him from looking for acoustic-wave scattering and he only got back to it several years later. Landsberg claims [95] that he and Mandelstam finally observed the effect but that, at their suggestion, Rozhdestvenskii at the Institute of Optics in Leningrad, who had a high-resolution echelon grating, put the problem to the young Gross who, by scattering 435.8 nm mercury light at 90° was able to detect Mandelstam–Brillouin spectral shifts of approximately 0.005 nm (6 GHz) from toluene and benzene, and published his first results in 1930 [97].

The Brillouin frequency shift is easily explained as a Doppler shift of scattering from first sound (isentropic pressure waves) excited by thermal fluctuations or, equivalently, as photon–phonon scattering [98]. Much later, with the invention of the laser, these molecular-acoustic effects became much easier to study and Rayleigh, Brillouin and Raman scattering became standard optical techniques in many laboratories. In 1969 [99], with the help of the laser, even scattering from thermally excited second sound in liquid ^{3}He–^{4}He mixtures (propagating isobaric entropy waves) was observed, a feat which, only a few years before, had been predicted to be impossible because of the small frequency shift (77 MHz) and the low scattering intensity (2 photodetection counts/min).

In 1914 Sagnac [100] published the results of his work on the measurement of rotation, started in 1911, by the interference of two counter-propagating light beams in a closed loop. The idea is arguably originally due to Michelson [101]. A slight change in the relative path lengths of each beam, such as may be induced by rotation of the closed loop, results in an interference-fringe shift proportional to the angular speed of rotation, ω. The Sagnac interferometer, particularly in optical-fibre form, later provided the basis for the laser gyroscope which is widely used for aircraft and other navigation.

18.2.4. Spontaneous and induced transitions
In 1916 Einstein made yet another important step forward in the new quantum optics [14, 102]. He had studied the properties of an ensemble of atoms in thermal equilibrium with Wien black-body radiation and, to extend the argument to cover the full range of the Planck equation, he postulated the occurrence of *spontaneous* and *induced* (or *stimulated*) transitions. His idea of spontaneous emission was a process in which an electron loses energy in going from a state of higher energy to one of lower energy analogous to spontaneous radioactive disintegration with a fixed lifetime, A_{mn}^{-1}, characteristic of the two states. The induced transitions were attributed to irradiation with a rate proportional to the density of radiation present. The upward and downward constants of proportionality B_{nm} and B_{mn}, respectively, were related to each other by

the statistical weights of the states $g_n B_{nm} = g_m B_{mn}$ to give the right high-temperature limit, and it then followed that the radiation density obeyed Planck's law

$$\rho = \frac{A_{mn}/B_{mn}}{\exp(h\nu/kT) - 1}$$

where

$$A_{mn} = \frac{8\pi h \nu^3}{c^3} B_{mn}.$$

The A and B terms later became central to the description of laser operation in terms of transition probabilities. The theory of the Compton effect made clear later that the quantum emitted in the stimulated emission process has values of energy and vector momentum given by subtracting the values imparted by recoil from those of the incident quantum. This was unequivocally proved by experiments of Bothe and Geiger [103] and put an end to the alternative radiation theory of Bohr, Kramers and Slater [104].

The diversity and contrast in optical physics activity at this time is perhaps best illustrated by observing that while Einstein was wrestling with A and B coefficients, Twyman and Green were publishing their modification of Michelson's interferometer which was to become so valuable in optical workshops for testing optical component quality [105, 106]. In the Twyman–Green arrangement, the Michelson interferometer is modified to use collimated light so that fringes of equal thickness are obtained from optical elements placed in the non-reference arm of the interferometer. At the same time Millikan published [49] an extensive verification of Einstein's photoelectric equation, having spent ten years of experiment convincing himself of its correctness. Millikan was awarded a Nobel Prize, in part for this work, in 1923, just two years after Einstein's prize.

By 1916–17 most of the basic principles of the new quantum optics had been explored, although there were a number of important steps needed before the laser could become possible. Classical optics was being widely practised throughout this period and continued to grow throughout the twentieth century. A notable example was the completion of the 100″ telescope on Mount Wilson by George E Hale in 1917. It would be 1948 before Mount Palomar received its 200″ Hale telescope, an instrument containing a Pyrex mirror of 18 tons, ground to an accuracy of 50 nm.

18.2.5. From imaging theory to negative dispersion
In 1919 Raman [107] briefly reported observation of an effect we now call 'speckle', the random granular pattern seen in coherent light that has been reflected from a suitably rough surface; also Slepian [108] patented secondary emission as a means for signal amplification, 16 years before

the photomultiplier was demonstrated. In the following year Schriever [109] reported his experimental studies of electromagnetic waves in dielectric wires verifying the earlier Hondros–Debye calculations [85], another step towards fibre optics.

At this time Weigert [110] was studying the polarization of fluorescent light, excited by a linearly polarized beam, in a pioneering experiment which was to give rise to the technique of fluorescence depolarization for the measurement of molecular rotation. Fluorescent light was not then believed to be polarized. Weigert said that '… it did not agree with theory and experience' and he thus took every care to exclude the possibility of his results being due to Tyndall scattering. He used the Stokes technique of complementary filters which, placed together in the incident light path, completely extinguished the beam, but placed respectively in the incident and scattered light paths passed scattered components of shifted frequency. Very interestingly, from the historical point of view concerning the discovery of the Raman effect, Weigert's result was at first disputed by Vavilov and Levshin [111] which elicited a second ironic communication by Weigert [112], pointing out that Vavilov and Levshin were not conversant with this '… well-known and convenient technique … and were probably even using polarization photometers which should only be used for the examination of small polarization effects with great care'. Levshin [113] and Perrin [114] later formulated the *Levshin–Perrin* law for the observed time-dependence of polarization $P(t)$ (a function of viscosity η and molecular volume v), in terms of the polarization P_0 which would be seen at infinite viscosity

$$\frac{1}{P} = \frac{1}{P_0} + \left(\frac{1}{P_0} - \frac{1}{3}\right)\frac{RT}{v\eta}t.$$

See also p 1704

Also in 1920 Michelson [115] reported an interesting development of interferometry. He made his *stellar interferometer* by mounting four mirrors on a girder strapped to the input of the 100" Mount Wilson telescope, creating in effect a very widely spaced two-slit Fizeau interferometer. The fringe visibility of a star so observed varies with mirror separation D, dropping to zero when $D = 1.22\lambda/\alpha$ for a source of angular diameter 2α. The fringes from Betelgeuse disappeared when D was 3 m, corresponding to an angular diameter of 0.047 seconds of arc. By this means, resolving power in optical astronomy had been increased significantly beyond that of a normal telescope.

In 1921, the year of Einstein's Nobel Prize [116], Ladenburg [117] took an important step in the history of the laser, and indeed came close to discovering amplification by stimulated emission. Generalizing the classical Drude formula, he related Bohr's theoretical description of emission and absorption to the number of oscillators present in an ensemble of molecules in thermal equilibrium with radiation. The relationship was formulated in terms of the Einstein A coefficient and was not quite complete, but this was shortly remedied by Kramers [118] and by Kramers and Heisenberg [119], who added negative terms in the dispersion formula to take account of spontaneous emission. Ladenburg and his collaborators worked hard between 1926 and 1930 [120] to verify these negative-dispersion terms. In particular, they found the effect near the red emission line in gaseous neon excited by an electrical discharge [121]. If they had used more current and had had more luck they could have found their dispersion curve becoming negative.

18.2.6. The tale of Miller

The Nobel Prize was awarded to Bohr [38] in 1922 for his study of the structure of atoms and the radiation which emanates from them. Wood [122] described his experiments on the fluorescence and photochemistry of materials including fluorescein (uranine) and rhodamine. Perrin had apparently been the first to associate fluorescence with chemical change in 1918. Wood found fluorescence to vary with temperature and the intensity of exciting light, and in a non-linear manner at higher light intensities.

Miller was in the midst of a repeat performance of the Michelson–Morley aether interferometry experiment, now over a 65 m distance. His data analysis was not showing a null result. Here we have a crucial experiment with real disagreements over the outcome. The Miller experiment is an interesting tale. Back in 1886 Michelson's studies with his interferometer on the effect of motion on the velocity of light had culminated in the famous Michelson–Morley experiment. That experiment had had a profound influence on late nineteenth century and early twentieth century physics as it disproved the existence of the aether. Michelson had constructed his interferometer on a 1.5 m^2 stone block floating on mercury. Seven traverses of each arm resulted in an optical path length of 11 metres, so that the effect of the Earth's motion with rotation of the interferometer should have been about 0.4 fringe displacement, yet no displacement greater than 0.01 fringe was observed, i.e. a null result (which disconcerted Michelson since he had expected to measure the aether drift). This null was disputed by Miller. Whilst Miller failed to find the predicted displacement in his 65 m path system, he did, however, report an apparent small drift and continued to argue for the aether [123]. More sensitive and shorter-path interferometers [124] failed to find any fringe displacement with interferometer rotation, and calculations showed that a temperature change of only 0.001 °C or a pressure change of only 0.05 mm Hg would have caused the Miller observation in his geometry.

Quantum physics was advancing dramatically in this period and two optical applications deserve mention. Duane [125] treated diffraction from a grating using a corpuscular quantum approach and Tolman [126] realized that 'molecules in the upper quantum state may return to the lower quantum state in such a way as to reinforce the primary beam by negative absorption (stimulated emission)'. Yet the laser was not to be invented for another 36 years.

See also p 167

The physics surrounding optics started to change rapidly: Bose [15] succeeded in deriving Planck's black-body radiation spectrum in a logically consistent manner without using classical electrodynamics, on the assumption of complete indistinguishability of the light-quanta, and introduced what came to be known as the *Bose–Einstein* distribution for photons

$$\bar{n}_j = \frac{1}{\exp(h\nu_j/kT) - 1}$$

where $\bar{n}_j$ is the mean number of photons in the jth energy state. His paper was rejected by the *Philosophical Magazine* and he then sent it to Einstein, who translated it and submitted it for him to the *Zeitschrift für Physik* with a note of recommendation.

18.2.7. The new quantum theory

The developments in quantum mechanics in 1925 are fully described in Chapter 3, and we need draw attention to only a few early consequences. Born, Heisenberg and Jordan [127] used harmonic-oscillator decomposition to quantize the radiation field according to matrix mechanics, a consequence of which was the confirmation of Einstein's fluctuation equation, including both wave and particle contributions. Dirac [128], in 1927, described quantization of a free field as an independent dynamical system by representing each field mode as a harmonic oscillator. The rate of spontaneous emission was correctly predicted. Annihilation and creation operators were associated with the mode amplitudes. The single excitation of a mode (i.e. state of the radiation field) represents the presence of a photon in that mode.

We must now discuss the statement of Dirac that a photon can only interfere with itself. His argument was based on the formation of fringes in a Young's double-slit experiment. He reasoned that one rather than two photons must be involved since to form fringes '... sometimes two photons would have to annihilate and sometimes create four which contradicts conservation of energy. Interference between two different photons therefore never occurs.'. Dirac's statement was implicitly only intended to apply to experiments of the double-slit type where the interference is revealed by detecting single photons, not to experiments where correlations between different photons are measured.

His statement has largely gone unchallenged, but a fairly straightforward indication that Dirac's statement is not true is given by the examples of heterodyning of radio waves, or interference between two perfectly coherent laser beams from different lasers. For interference to occur between different photons, the fields in the two beams must each contain amplitudes of photon numbers differing by unity [129]. Thus a state with n photons will not interfere with another n-photon state coming from a different direction, but a state with a quantum superposition of n and $n+1$ photons will interfere with another state with a superposition of m and $m+1$ photons. Coherent states best satisfy this condition and are the only states, apart from a photon interfering with itself, which allow the conventional classical derivation of Young's interference to go through correctly.

In the midst of the huge burst of activity in quantum mechanics, Lewis [130] coined the term 'photon' in 1926 and optical interferometry experiments such as Linnik's studies of surface microtopography were initiated [131]. 1926 also saw perhaps the first non-linear optics experiment. Vavilov and Levshin [132] found a reduction in the absorption of light by uranium glass with increase of intensity, thus violating the law of Bouguer which asserts the independence of absorption coefficient with intensity. This is now known as a photorefractive effect and it was explained as resulting from the depopulation of the ground state by the incident beam. In Vavilov's experiment, light at 454 nm from a high-intensity spark source was passed through either (i) a 33% transmission filter and then uranium glass or (ii) uranium glass and then the same filter. He measured absorptions of 2.576 and 2.544 respectively; a 1.5%±0.3% difference, explained as the effect of a non-linear absorption coefficient. Vavilov himself introduced the term 'non-linear optics'. In his book, *Milrostructura Sveta* (1950), he wrote (translated from the Russian for the authors by Pavel Kornilovitch of King's College London)

> Non-linearity of the medium will be observed not only in absorption. Absorption is related to dispersion, therefore the speed of light in the medium also depends on the light intensity. In general, the dependence on the light intensity, i.e. violation of superposition, will be manifest in other optical properties of the medium: double refraction, dichroism, optical rotation, etc.
>
> An astrophysicist very often is faced with 'non-linear optics' of the medium in his theoretical considerations of conditions inside stars. Because of the enormous density of light at temperatures of millions of degrees, the absorption coefficient and the speed of light must depend on the light intensity very sharply. However, as mentioned above, large violations of superposition and non-linearities could occur even in an optics laboratory, especially in the study of phosphorescent media. It seems that the physical sciences are so accustomed to the linearity of everyday optics that until now no rigorous and formal mathematical apparatus has been developed for solving any 'non-linear' optical problems.

In 1926–27 Kramers and Kronig [133] developed their dispersion relations between absorption χ' and dispersion χ''

$$\chi'(\omega) = \frac{1}{\pi} \text{PV} \int_{-\infty}^{+\infty} \frac{\chi''(\omega')}{\omega' - \omega} \, d\omega'$$

$$\chi''(\omega) = -\frac{1}{\pi} \text{PV} \int_{-\infty}^{+\infty} \frac{\chi'(\omega')}{\omega' - \omega} \, d\omega.$$

The complex susceptibility function $\chi(\omega) = \chi'(\omega) - i\chi''(\omega)$, and PV means the principal value of the following Cauchy integral. It was also at this time that Wiener [134] provided an early relation between coherency and the quantum theory, a subject to be revisited often throughout the century.

18.2.8. Combination scattering—the Raman effect

We now know that what in the West is called the *Raman effect*, and often in the former Soviet Union *combination scattering*, was almost simultaneously discovered by Raman and Krishnan [135, 136] and Mandelstam and Landsberg [137] in 1928. The first definitive measurements of the effect were closer together than the times of publication. New spectral *lines*, shifted from the incident light frequency by an excitation in the optical branch of a crystal of crystalline quartz in the Soviet Union, and by the molecular vibrations of various liquids and vapours in India, were first seen on 21 February by Mandelstam and Landsberg, using a mercury lamp and quartz spectrograph, and on 28 February by Raman and Krishnan, using a direct-vision spectroscope. In his note [138], Raman expresses great surprise that the modified radiation they had reported a few weeks before [135] was now seen to be separated from the incident radiation by a dark band. This had prompted him to make observations with a mercury arc lamp, cut off with a filter above the 436 nm line; he found sharp bright lines at wavelengths beyond the cut-off. In yet a third communication [139] entitled *The optical analogue of the Compton effect* he and Krishnan provide photographic line spectra using a mercury-lamp source. There is some rumour that one of Raman's papers for *Nature* was rejected by a referee who was overruled by the editor. Since records of that time were lost in World War II it is not now possible to know what actually happened [140].

Full credit for the 'simultaneous' discovery was given by Born [141] and by Darwin [142] in the same year and by Rutherford in his November 1929 presidential address to the Royal Society [143]. In their first paper [135] Raman and Krishnan did not report a line spectrum but 'modified radiation of degraded frequency' scattered from a focused beam of sunlight by some 60 different common liquids. They used the technique of complementary filters and stated without further elaboration that spectroscopic confirmation was available. This is difficult to understand.

> See also p 996

We know from Krishnan's diaries that it was only on 7 February of that year that Raman first saw Krishnan's experiments in which polarized fluorescent light was seen scattered from a vessel of liquid pentane. He immediately declared that the result was amazing and on the same day related it to the Kramers–Heisenberg effect [144]. In fact various of his co-workers had been seeing unexplained 'feeble' fluorescence since 1923. Raman had put Krishnan to work in the laboratory because he had been doing theory for a year and prolonged isolation from experiments was not desirable (a sentiment with which we both heartily agree, although prolonged absence from the desk is just as bad, as Raman was to find in a famous controversy with Born—and also Peierls and Landau—who all had the mastery of him when it came to detailed interpretation of the Raman spectra from crystals [145]).

Since only about one part in 10^8 of the incident light was shifted in the Indian experiments, it has been said by Fabelinskii [146] that '... Raman and Krishnan were indeed brave to assert that the light they observed was radiation of altered frequency ... and not luminescence, not the wing of the Rayleigh line and not a consequence of fluorescence of the filters themselves'. This was in 1978 so it is clear that suspicion of Stokes' filter method endured in Moscow long after the original investigations. It was this technique that had also caused Vavilov and Levshin such grief in the Weigert affair. In the event, in 1930, just two years later, fortune favoured the brave and Raman [147] received the Nobel Prize. The Nobel Prize was rarely awarded so quickly; in view of all the circumstances, however, it would be interesting to know why it was not shared with the Russians.

It is curious to note that these important discoveries ('hearing a molecule talk' as one scientist put it at Mandelstam and Landsberg's Moscow seminar) were both accidental in that Mandelstam was looking for Brillouin scattering by acoustic (Debye) waves and Raman was searching for an analogue of the Compton effect. In fact, Smekal [148] and Kramers and Heisenberg [149] had already predicted the effect. Another sad note to this episode is that in France, Rocard [150] had also come to the conclusion that the effect should exist, as well as the Rayleigh-wing scattering due to molecular rotation, but had not published before the experimental discoveries. Cabannes [151], too, had been trying since 1924 to observe combination light-scattering in gases but had too little light intensity to succeed. The advent of the laser gave a tremendous boost to such light-scattering studies, with the help of Loudon's [152] widely cited and timely review of Raman scattering from a number of quantized excitations of crystals.

One other event from 1928 should not go unnoticed. Synge [153] suggested the conventional far-field diffraction barrier in optical microscopy might be overcome by fabricating an optical aperture much smaller than the wavelength of light, and positioning it much closer to the sample than this distance. By using the aperture as a light source, image resolution would be limited to the size of the aperture and not by the wavelength of the light. Synge's idea was actually suggested to him by Einstein. His first thought was to use a total internal reflection surface upon which was placed a sub-microscopic gold particle as a near-field source. He wrote to Einstein who pointed out that the evanescent waves would traverse the gap and suggested using a small hole instead. Synge replied that he had already thought of that but accepted Einstein's criticism, saying that he had misunderstood Rayleigh on total internal reflection [154]. He proposed fabricating such a hole from a metallized glass cone with the coating removed from the tip, a proposal made again in 1986 [155] under the imaginative name of 'optical stethoscopy', by analogy with a medical stethoscope that resolves $\lambda/1000$ by the same principle: unfortunately the name did not stick. Two years earlier 25 nm resolution had been achieved with 488 nm laser light passed through etched 30 nm tips of a quartz crystal [156] and there were several other comparable successes [157–159]. This work of Synge was the first suggestion of near-field, scanning, optical microscopy which is generally credited to O'Keefe in 1956 [160] and to Baez [161] in the same year, who resolved his finger in a near-field experiment at a wavelength of 14 cm. The subject was reborn in 1972 when Ash and Nichols [162] achieved $\lambda/60$ resolution with a near-field scanning hole at 3 cm wavelength, but only became really practical in the 1980s [163, 164] following the development of scanning tunnelling microscopy [165].

As a concluding remark on this period, it can be said that optics provided the first indication of the neutron when Heitler and Herzberg [166] showed that the intensity alternation in the pure rotational Raman spectrum of $^{14}N_2$ [167] could not be explained if nitrogen obeyed Fermi statistics, as was demanded by the electron–proton model of the nucleus.

1900 to 1930 was a truly extraordinary period in the history of optical physics. A revolution in understanding occurred which was not to be fully exploited until the invention of the laser in 1960.

18.3. 1930–60: the calm before another storm

The early 1930s brought a few new developments in classical optical instruments and techniques. Of particular note was the introduction by Schmidt [168] of a telescope comprising a spherical mirror and the aspheric *Schmidt corrector plate* (a shallow toroidal curve) placed at its centre of curvature (see figure 18.1). The performance improvement in wide-field imaging was outstanding and the Schmidt camera has become a most important tool in astronomical research. The first example was built in 1930, and in 1949 the famous 48" Schmidt telescope at the Palomar Observatory in California was completed. Many advances and variations in the design of catadioptric (combination of mirrors and lenses) instruments have occurred throughout the century since this time

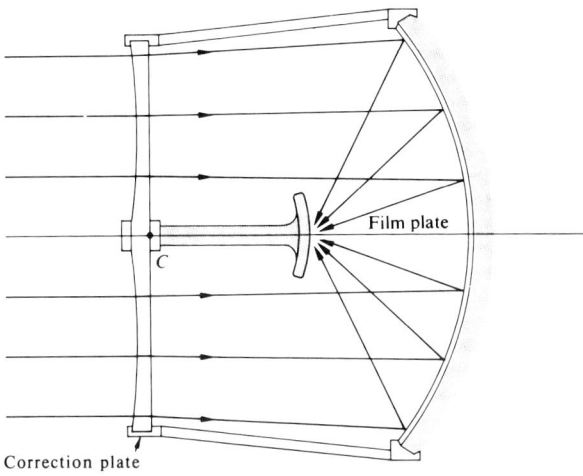

Figure 18.1. *Schmidt corrector telescope.*

[169], and applications of the concept appear in missile trackers and guidance systems, compact commercial telescopes both for the amateur and the professional observer and telephoto camera lenses.

Throughout the early 1930s experiments were conducted by Bauer and Pfund [170] to produce high-reflection coatings for optical surfaces and by Strong [171], a major figure in experimental optics, to produce anti-reflection coatings. Geffken [172] produced metal–dielectric interference filters in 1939. Modern optical instruments using many lens surfaces simply could not operate successfully without the multilayer thin-film optical coatings for which these activities were the foundations. Photographic cameras, lasers and countless other optical systems all depend on high-quality coatings, and the laser gyroscope was to pose an even more stringent requirement in this respect.

In 1933 Oseen [173] improved our understanding of wave propagation in cholesteric liquid crystals; and Stiles and Crawford [174] began to explain the effect named after them, which was still being explored 40 years later [175], that the human eye is directionally sensitive because light entering the centre of the pupil produces a greater response than light entering peripherally.

18.3.1. van Cittert, Zernike and optical coherence

The van Cittert–Zernike [176, 177] theorem was created from research published by van Cittert in 1934 and by Zernike in 1938, and brought a new understanding of coherence in classical optics. The theorem allows the determination of the mutual intensity or field–field correlation function, $J(P_1, P_2)$, and the complex degree of coherence $\mu(P_1, P_2)$

for points P_1 and P_2 on a screen illuminated by an extended quasi-monochromatic primary source of mean wavelength λ. It shows that the mutual intensity is given by

$$J(P_1, P_2) = \int_\sigma I(s) \frac{\exp[ik(R_1 - R_2)]}{R_1 R_2} \, ds$$

where R_1 and R_2 denote the distances between a source point s with intensity $I(s)$ and the points P_1, P_2, respectively, k is $2\pi/\lambda$ and the integral is over the source surface σ. The complex degree of coherence defined by

$$\mu(P_1, P_2) = \frac{J(P_1, P_2)}{\sqrt{I(P_1)}\sqrt{I(P_2)}}$$

takes the value of unity for a fully coherent source and zero for a fully incoherent source. van Cittert was the first to explore these ideas, and Zernike gave a more general treatment applicable to quasi-monochromatic fields produced by any source (coherent, partially coherent or incoherent). Essentially the mutual intensity J is found from a two-dimensional Fourier transform of the intensity distribution across the source. The theorem has found widespread use in statistical optics, image formation and interferometry [178], and has led to significant later developments.

Important new results for light-scattering spectroscopy from Landau and Placzek [179] were published in 1934. The Landau–Placzek 'ratio' applies to a pure simple fluid and relates the relative strengths of scattering of a broadened central line at the incident frequency, caused by entropy fluctuations, and a symmetrically spaced Mandelstam–Brillouin doublet, caused by pressure fluctuations. The ratio of these two intensities was given in terms of the ratio of the specific heats at constant pressure and constant volume, C_p and C_v, respectively, as

$$\frac{I_{\text{doublet}}}{I_{\text{central}}} = \left(\frac{C_p}{C_v} - 1\right).$$

This was also the year when Čerenkov [180] published his observations of radiation from relaxation of non-uniform polarization, now called the Čerenkov effect or, in Russia, the Vavilov–Čerenkov effect. Light is emitted when charged particles, for example electrons, pass through a transparent medium at a velocity greater than the velocity of light in that medium. Vavilov had asked Čerenkov to investigate the visible radiation from radium γ-rays traversing various fluids. Others had observed a weak bluish glow which was initially attributed to fluorescence. Čerenkov observed the radiation, in doubly distilled water in order to eliminate fluorescence, with his own eyes, having had to adapt

A History of Optical and Optoelectronic Physics

for an hour or more in a totally dark room. He found that the visible radiation was polarized along the direction of the incoming radiation; in the following paper, Vavilov [181] postulated that it was fast electrons that were the primary cause of the visible light. A classical theory was given by Tamm and Frank three years later [182].

In fact, directed electromagnetic emission due to the motion of 'electrification' through a dielectric had been predicted by Heaviside in 1899 and also by Sommerfeld in 1904. Single charged particles can be observed by this effect, their speed being related to the angle of the 'bow-wave' emitted; this led to the development of Čerenkov counters widely used in nuclear and space physics. Čerenkov, Frank and Tamm were awarded the Nobel Prize in 1958. A quantum theory of the Čerenkov effect was given by Ginsburg in 1940 [183, 184].

Also at this time, Raman and Nath [185] published their five-paper series describing the diffraction of light by high-frequency sound waves, now known as *Raman–Nath* diffraction. This was an extension to finite systems of the theory of Bragg scattering for infinite, equally spaced scattering planes and is of importance in acousto-optics [186]. Nowadays, acousto-optic diffraction is characterized as either Raman–Nath diffraction or Bragg diffraction, depending on the relation that exists between the acousto-optic cell thickness and the period of the acoustic wave generated by the RF carrier that excites it. For sufficiently thin cells, Raman–Nath diffraction occurs and the cell behaves like a thin phase-grating, generating multiple diffraction orders. For sufficiently thick cells, Bragg diffraction occurs, and, if the acoustic grating is illuminated at the Bragg angle, there is a high diffraction efficiency into a single order.

In 1934–35, when RCA started development of the photomultiplier [187], Zernike [188] introduced a new form of microscopy, *phase-contrast* microscopy. Coherent illumination of transparent objects is employed and an intensity distribution is produced which is directly proportional to the phase changes introduced by the object. This is done by placing a thin phase plate in the back focal plane of the microscope objective in order to advance/retard the phase of the central order with respect to the other diffraction spectra by one-quarter period. The resulting light distribution in the image plane now represents a false amplitude grating and the intensity in the image plane is proportional to the phase change due to the corresponding element of the object. He describes his discovery of phase contrast in an article to *Science* in 1955 [189]. In the 1934 paper he also introduced *Zernike's circle polynomials*, which are orthogonal within the unit circle, and these were later to play an important role in the wave theory of aberrations [190–193]

In a later paper Zernike [177] (1938) not only explored the concept of degree of coherence, as we have discussed in relation to van Cittert, but explored its consequences for microscopy and effects in light propagation. Zernike had tried to interest the famous microscope company of Carl Zeiss (Jena) in the phase-contrast method whilst it was in its infancy in 1932, but met with no enthusiasm from Abbe's former company. He went on to publish the 1934 paper and file a patent. It was only during 1942 that Zeiss supplied their first phase-contrast instrument [189, 194]. Phase-contrast microscopes are nowadays widely used, particularly in biological research.

In the quantum-mechanical description of the photon, the spin angular momentum of a photon is $\pm\hbar$, where the sign determines the handedness. The spin is independent of the energy of the photon. Emission and absorption of a photon entails a change of $\pm\hbar$ in angular momentum. In 1936 Beth [195] made use of this fact in an extremely sensitive experiment involving a torsion pendulum to measure, by direct mechanical means, the angular momentum of photons.

18.3.2. Through World War II
At this stage of the twentieth century, the late 1930s, a second world war was feared and much scientific effort began to be devoted to military problems. Optical research covered plenty of applied interferometry and interferometric spectroscopy. Optics played its role in the military effort through cameras, binoculars, telescopes, rangefinders and other instruments. Throughout the 1930s there had been a steady stream of improvements in photography, new miniature cameras, a move to cellulose acetate film from glass plates and celluloid, developments in fine-grain films and wider use of more sensitive bromide paper for photographic printing. In cinematography, Kodak had introduced 8 mm film for amateurs in 1932, and colour photography was firmly established, firstly with Technicolour and then with three-colour multiple-layer processes such as Kodacolour which used standard cameras [196]. Stroboscopic photography, sequences of pictures taken with exposures of objects by very short flashes of light from electrical discharges, thus 'freezing' object motion, was beginning its development through the work of Edgerton and Killian [197] at the Massachusetts Institute of Technology.

Optoelectronic sources and detectors had not yet started their major development which made components of such vital importance to military hardware from the 1960s. Indeed, at this time Wannier [198] was describing for the first time the basic optoelectronic properties of an exciton, an electon–hole pair bound by Coulomb attraction, that moves through a crystal as if it were a particle in an excited state, to be associated also with the name of Mott. The form of the Wannier exciton absorption spectrum was described by Peierls [199] in the same year.

Optics played a rather different kind of role in the military nuclear effort which was then building up. From the early 1930s to the late 1950s there was considerable activity in the study of nuclear spin, using spectroscopic methods to examine hyperfine structure. Prominent amongst many engaged in this work was Tolansky, who described the optical interferometry techniques required [200]. Photomultiplier tubes were also developed further in this period because of the sharp increase in research in atomic and nuclear physics and their value in scintillation counting [201].

We return now to another step in the history of the laser. In 1940 Fabrikant [202], a physicist in the USSR, wrote in his doctoral dissertation concerning light amplification in a molecular system

> For molecular (atomic) amplification it is necessary that N_2/N_1 be greater than g_2/g_1. Such a situation has not yet been observed in a discharge even though such a ratio of populations is in principle attainable ... Under such conditions we would obtain a radiation output greater than the incident radiation and we could speak of a direct experimental demonstration of the existence of negative absorption.

It is clear that at this time the physics for lasing i.e. amplification of *light*, was understood, though the device-engineering and technology aspects had yet to be worked out. Fabrikant and his colleagues filed a patent in 1951 which was published in 1959 [202], but because of this delay, and no doubt because of the relative isolation of science in the USSR from the West his work had little influence on the development of the laser.

In 1940, Gabor [203] (later to invent holography), patented a new optical concept, the 'super-lens', a pair of (afocal) lens arrays of the sort that later were widely used for compact optics in photocopiers. The lens-array idea had previously been described by Lassus St Genies [204] in 1939.

Starting in 1941 Clark Jones [205, 206] published a series of eight papers describing a new calculus for the treatment of polarization in optical systems. This important work laid the foundations for the matrix treatment of polarization that soon became a standard technique. (Clark Jones also introduced D^*, the detectivity [207] parameter to be widely used in calculations of the performance of infrared imaging systems). The simplicity of the polarization calculus for coherent, polarized beams is most useful for rapid calculations e.g. one plane-polarized light component is just

$$\begin{bmatrix} 1 \\ 0 \end{bmatrix}.$$

For arbitrarily polarized light, the traditional Stokes theory is still required although the introduction of a 4×4 matrix scheme by Mueller in 1943 eased the handling of Stokes vectors. It was some years before Wolf introduced a coherency matrix approach to describe the coherence properties of partially polarized light [208] in which, in particular, he showed how the degree of polarization may be obtained from simple experiments involving a polarization compensator and a polarizer, a technique now commonplace.

In 1943, Hopkins [209] calculated the diffraction pattern of a point object in the focal plane of a lens, which at low aperture is the famous Airy disc pattern, for systems of high relative aperture. Although he did not use a full electromagnetic theory his results are of use for fairly high aperture values. He correctly predicted that for linearly polarized light the Airy pattern would become elliptical as the aperture increased, with the major axis in the direction of the polarization in the object space. As would have been clear from the previous full vector theory of Ignatowsky [210] (unreferenced in Hopkins' paper or the later work of Richards and Wolf [211]), significant departure from the usual Airy form was found for values of numerical aperture larger than about 0.5. Mathematical calculation in optics was prominent at this time; only a year later Luneberg [212] published his landmark text *Mathematical Theory of Optics*.

Multiple-beam interferometry is well known because of the celebrated Fabry–Perot (or Boulouch, see below) interferometer developed at the very end of the nineteenth century. Multiple-beam, thin-film, optical interferometry was developed for surface studies in 1944 by Tolansky who made a simple modification of the standard Newton's rings lens-plate experiment [213]. He coated both contacting surfaces with a high-reflectivity coating. This yielded dramatic improvement in the fringe sharpness and clarity and made possible the new field of optical surface microtopography. In the following year Tolansky extended the technique by inventing FECO interferometry [214] (fringes of equal chromatic order). In this arrangement white light is used in place of monochromatic light; bright sources can be used, and with a spectroscope very sharp fringes are obtained which represent a section through the interfering surfaces. In microtopography, surface-height fluctuations down to 2.5 nm can be measured, a significant improvement on previous optical techniques [215]. A multiple-beam interferogram is shown in figure 18.2.

In 1945 the technical director of the British Broadcasting Corporation approached Hopkins to design a zoom lens for television cameras, which was first used, a few years later, to broadcast a test match from the Lord's cricket ground in London. At about the same time, Hopkins commented briefly on the resolving power of a microscope using polarized light [216], simply drawing attention to the improvement that may thus be gained over unpolarized illumination and to the earlier statement of Stump [217] that this is possible. This was due to the non-circularly symmetric Airy pattern which he had pointed out previously. For unpolarized light Hopkins and Barham [218] were later to introduce a coherence parameter, s, defined as the ratio of the numerical aperture of the condenser to that

Figure 18.2. *Surface microtopography of a diamond face by multiple-beam interferometry.*

of the objective, ranging from zero in the coherent limit when the source is a point, to infinity in the incoherent limit for an extended source. For a single-point image, s is unimportant but for a two-point image they found that an improvement of about 7% in resolution can be achieved when s is approximately 1.45.

The next year saw the publication of a seminal and influential text, Duffieux's [219] *The Fourier Transform and its Applications in Optics*, which became the basis for extensive research in Fourier-transform image processing in the 1960s and 1970s [220]. Duffieux treated the general imaging instrument as a linear filter, established the Fourier-transform relationship between the energy distributions in the image or object plane and in the pupil plane and described the image intensity distribution as a convolution product of the object intensity distribution and an instrumental 'point-spread function' i.e. the impulse response of the system.

Wheeler now proposed [221] a two-photon correlation experiment in which the photons arose from the decay of slow, spin-zero positronium. Conservation of angular momentum predicts that they will be polarized at right angles to each other. This could be measured by using a pair of Geiger counters to detect Compton scattering from solid targets placed in diametrically opposite directions from the source. The detailed quantum theory was provided in the following year by Pryce and Ward [222]. There were two attempts at this experiment a year later: by Hanna at Cambridge (UK) and by Bleuler and Brant at Purdue (Indiana). Hanna [223] found a strong correlation for the perpendicular polarization but it was about 25% lower than the theoretical prediction. Bleuler and Brant's result [224] was 14% too high but within the statistical errors.

A third experiment by Wu and Shaknov in 1950 [225], using scintillation counters, was 2% too high but within a 4% error. All the experiments used a ^{64}Cu source. Other photon-pair correlation experiments were to occur through the rest of the century.

18.3.3. Quantum electrodynamics—the first gauge theory
Stimulated by measurements using new world-war technology, the years 1946–48 saw work on electromagnetic radiation theories which was to provide the foundation of modern theoretical high-energy and particle physics. Quantum electrodynamics (QED) was the first of the so-called gauge field theories which have since evolved into the present-day 'standard model' of theoretical physics, the most recent success of which was the discovery in 1994 of the last remaining quark, the top quark. These developments in QED are fully treated in Chapter 9 and in a recent text [226].

On a less elevated plane, the announcement of the Lau effect [227] was made in 1948; this is self-imaging, Moiré grating interference using an extended white-light source. It has come into prominence (sometimes without attribution to Lau) in recent atom-interferometer experiments. For example, with supersonic atomic beams of sodium of de Broglie wavelength 16 pm in a Lau interferometer, and gratings of 200 nm spacing made at the Cornell University nanofabrication facility, a measurement of the Sagnac effect has been achieved [228] with such sensitivity that the Earth's rotation causes a phase shift of two radians.

1948 was perhaps also the turning point for contact-lens technology. Whilst the development of contact lenses had progressed for over 100 years, the work of Dallos [229] and Bier [230] with scleral lenses and the manufacture of acrylic corneal lenses led to the wearing duration for both types of lens increasing dramatically from a couple of hours to 8–10 hours or more without removal. With other refinements in design and materials, acceptance and widespread use by the public soon followed [231].

18.3.4. The early days of holography
The next major landmark in the history of optics was unusual in that it aroused great interest amongst the general public. When finally developed in its optical form, holography was to become a global fascination. Its foundations can be traced back to x-ray crystallography studies as early as 1920. The work of Wolfke [232] in 1920 and of Bragg [233, 234] in 1939 and 1942 led to the x-ray microscope in which Fourier optics was used to study x-ray films. Starting in 1948, Gabor [235–237] intended to increase the resolution in electron microscopy by similar means, recording scattered or diffracted electrons from an object together with unscattered coherent electrons, and to reconstruct the image from the record by using visible light. Soon after Gabor's

A History of Optical and Optoelectronic Physics

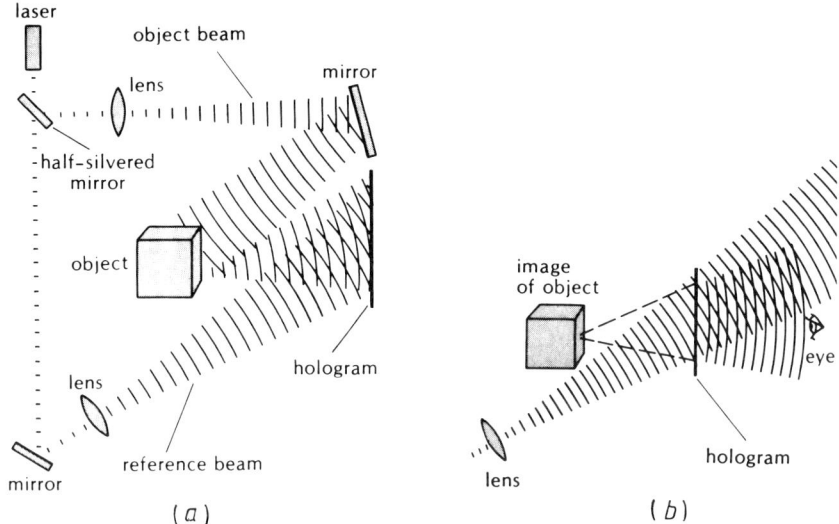

Figure 18.3. *The principle of holography. (a) To make a hologram, the object is illuminated by the object beam, via a half-silvered mirror and an arrangement of mirrors and lenses. All points of the object reflect the illuminating beams and act as sources of spherical waves. The reflected waves and the reference beam interfere at a photographic plate to produce the hologram. (b) To view the hologram, it is illuminated by the laser beam and viewed from the far side.*

work, Rogers [238] demonstrated optical holography by using a high-pressure mercury arc lamp, pre-dating laser holography by a decade. He achieved holograms, hologram copying, three-dimensional image generation, image subtraction, phase holograms and an early attempt at a computed hologram [239]. El-Sum and Kirkpatrick [240] and Lohmann [241] also attempted all-optical holography, but with poor success because of conjugate-image problems, and interest waned. The imagination of the public was captured in the 1960s when Leith and Upatnieks [242] showed how to perform holography in an all-optical configuration using coherent light to allow observation of apparently three-dimensional images for the first time (see figure 18.3). Many developments followed and Gabor received a Nobel Prize in 1971 [243]. We shall discuss further details a little later.

In 1949, Brossel and Kastler [244] published the first of a number of papers that described optical pumping, a technique to redistribute the populations in sub-levels of an excited atomic system. By illuminating the atoms with a specific wavelength, a population inversion could be created. This idea later found application to certain types of laser and won Kastler a Nobel Prize [245] in 1966. A point of historical interest is that Kastler, in a long note just before his death to one of the present authors (ERP), put forward the claim of Boulouch to have invented the Fabry–Perot interferometer (see reference [246]).

1950 saw the publication of Chandrasekhar's book, *Radiative Transfer*

[247], which had a major influence in optical scattering research. Bouwkamp published his solution of the field inside a sub-wavelength hole [248], an important problem in diffraction theory. He showed that Bethe's near-field solution (1944) was seriously in error, but that his far-field results required no corrections.

Occasionally in the scientific literature one encounters an invention that was perhaps not fully understood at the time by its inventor! This seems to have been the case with Fellgett's invention (1951) of multiplex optical spectroscopy [249], a development in high-resolution spectroscopic analysis. Unlike the standard approach using a monochromator whose output spectrum is scanned by a detector and the intensity distribution measured directly, in the multiplex system information from all resolution elements in an interferometer is recorded simultaneously as path length in the interferometer is varied. Then Fourier transformation decodes the interferogram and creates the required spectrum. There is usually a very significant advantage in signal-to-noise ratio which is widely ascribed to the fact that one is observing all frequency components at once. However, we could also say that the ordinary scanning detector sees all time delays at once and, in fact, the two are just a Fourier transform (FT) pair of variables. The FT spectrometer measures the quantity

$$g(\tau) = \frac{1}{\sqrt{2\pi}} \int S(\omega) e^{i\omega\tau} \, d\omega$$

while the monochromator or grating spectrometer measures

$$S(\omega) = \frac{1}{\sqrt{2\pi}} \int g(\tau) e^{-i\omega\tau} \, d\omega.$$

Hence the situation is actually symmetric, and which one has the better signal-to-noise ratio depends on the nature of the noise. If the noise were quasi-sinusoidal in time it would give rise to spurious sharp spectral features in the Fellgett scheme but not in the standard one, while if it were white and impulsive it would average out to a large extent and show the so-called 'Fellgett advantage' in the FT spectrometer, but would give spurious spikes in the spectrum in the classical instrument. The latter type of noise (including typical detector noise) is more common so that the Fellgett advantage normally occurs; FT spectroscopy is widely used, particularly in the infrared where detector noise is unavoidable.

The photochemical basis of rod-vision (the rhodopsin system) was revealed in 1951 when Wald [250] extended his original 1936 discovery. This work provides the basis for our present understanding of the biophysics of visual photoreception [251] and Wald won a Nobel Prize in Medicine in 1967. Young and Roberts constructed a 'flying-spot' microscope, a scanning microscope using the raster of a bright cathode-ray tube as a light source [252]. The rather late publication of work on sheet optical polarizers by Land [253] also occurred at this time. Land deserves far more than just a passing reference to sheet-polarizers as one of his contributions to twentieth century optics. Whilst not, perhaps, a generator of new optical physical principles, he nevertheless made a large impression in the field of applied optics. Besides polarizers, his interests ranged across night goggles, cameras, colour television, holography, vision and many other areas as a most notable inventor. He was particularly famed, of course, for his 1948 invention of the 'instant' camera, in which the photographic print was almost immediately available. An excellent overview of his multifarious activities has appeared recently [254].

It was now that the band-structure diagram of a p–n junction was first used as a model to explain the work of Round [61] and of Lossew in the 1920s and 1930s on silicon carbide; in the same decade Destriau [255] had also observed light emission from ZnS phosphors when excited by an electric field, which he had called *electroluminescence*. In the work of Lehovec, Accardo and Jamgochian [256] electrons, injected across a forward-biased p–n junction, combined with holes in the p-type region of the junction. In the recombination process, the energy lost by the electron is emitted as a photon. Whilst this model has been developed a great deal since then, it still provides our basic understanding of the electroluminescent effect in p–n junctions, one of the most vital components of semiconductor optoelectronics.

18.3.5. Towards masers and lasers

The next few years, leading to the momentous period 1958–62, contained the seeds of new directions in optics and optoelectronics that were to be developed throughout the remainder of the twentieth century.

In 1952 Weber described the maser principle, but without a working device [257]. The basic idea, which is that of the laser also, is to achieve a population inversion between two energy levels of an atom or system by some incoherent-pumping mechanism, and then, either in a long column of the medium or by reflections back and forth in a cavity, to amplify a radiation field at the resonant frequency of the transition by Einstein's stimulated emission mechanism. In this mechanism the presence of the field stimulates the population to drop back into the lower level, thus emitting more radiation coherently into the same field and building it up into a self-sustaining oscillation. We will explain this in more detail when we come to describe the first working laser.

See also p 1424

In the same year Elias and his colleagues published first attempts to connect optics and communications theory [258], later to be more correctly founded by O'Neill [259]. This work was to have important ramifications in optical information processing and imaging. Elias and O'Neill discussed the statistical information theory ideas of Shannon and

Wiener, Fourier theory and the concepts of signal-to-noise ratio in optical systems, autocorrelation functions and optimal filtering.

Following early work by Straubel [260] and Luneberg [261] in the field of imaging, di Francia [262] published his ideas concerning lens-pupil functions (apodization) that yield narrow impulse-response functions, i.e. modifications to the transmission distribution of the lens aperture that yield a point-spread function whose central peak width is narrower than that of the usual Airy pattern, and thus may lead to 'super-resolution'. di Francia showed that the central peak could be as narrow as one wished and could be surrounded by a dark region as wide as one wished. Unfortunately this is at the expense of increased side-lobes outside the 'tailored' region which contain most of the optical power. Imaging theory has pursued super-resolution ideas throughout the remainder of the century and we shall return to this topic later.

di Francia was also soon to be responsible for introducing the Shannon number, $S = X\Omega/\pi$, into the diffraction theory of optical imaging [263] where X is the spatial extent of the object and Ω is π times the pupil diameter divided by (wavelength × focal length). S is a measure of the 'number of degrees of freedom' in the image. It is interesting to note that a similar proposition had been published previously by von Laue [264] in 1914!

Of particular importance to future light-scattering activities, 1952 was the year when Glauber [265, 266] (and also van Hove, in 1954 [267]) described in some detail how the frequency of a neutron was changed when it was scattered by crystals and liquids. The theory was later reformulated by Komarov and Fisher [268].

18.3.6. Exhausting Fourier imaging theory

1953 was the year of coherent imaging. Marechal and Croce [269] demonstrated coherent imaging, especially spatial-filtering improvements to the quality of photographs, which was to become the basis of much coherent image processing in the 1960s and 1970s. They were perhaps the first to show clearly the analogy between spatial frequencies in image formation and temporal frequencies in a communication network. Furthermore, their classic '$4f$' optical filtering geometry has become the standard technique for coherent image manipulation. The basic optics involved is fairly straightforward. Imagine a transparent object located in the front focal plane of a converging lens and illuminated by plane parallel monochromatic light. In the back focal plane of that lens (the so-called transform plane) the Fourier transform of the transparent object is formed. By placing transparent masks of the Fourier transforms of comparison objects or specific mathematical functions in the transform plane and 're-transforming' that plane using a second lens also operated under back/front focal plane conditions, various processes of correlation, convolution, matched filtering, spatial filtering, pattern recognition, etc, may be performed on the original transparent object [58]. When we come to consider the lens more accurately as a band-limited transmission system, rather than an infinite Fourier transformer, it will be clear that we have to consider only those spatial–frequency components that are within the transmission band.

Hopkins [270] now published his comprehensive paper on the Fourier theory of image formation. Using his results of 1951 [271] he calculated imaging properties of an optical system under different conditions of coherence of object illumination for arbitrary aberrations and defocus. Resolution was found to vary with state of partial coherence and with angle of oblique coherent illumination. Known results (of Abbe for example) were recovered in the cases of perfect coherence and complete incoherence (incoherence increases the resolution at the expense of contrast). Linfoot and Wolf [272] were exploring diffraction images in systems with an annular aperture, calculating the three-dimensional light distribution near the focus of an aberration-free optical system. Polarization effects were not addressed in this paper, nor in their later paper [273] in which they calculated the phase distribution near the focus.

The other significant coherent-imaging event of 1953 was the award of a Nobel Prize to Zernike [274] for his demonstration of the phase-contrast method, and especially for his invention of phase-contrast microscopy.

18.3.7. The beginnings of optoelectronics

Semiconductors started to assume serious importance in optics and optoelectronics in 1953, when McKay and McAfee [275, 276] demonstrated electron multiplication in silicon and germanium p–n junctions. The resulting avalanche photodiode has been used for analogue detection in optical communications for some time now, but has only recently begun to take over from the photomultiplier tube for digital photo-detection, thanks mainly to work at RCA Vaudreuil, Canada. Also, in a letter to Teller, and unknown to the maser (and early laser) community, Neumann [277] indicated that one could obtain radiation amplification by stimulated emission in semiconductors, nine years in advance of the first demonstrations of the semiconductor laser.

Gordon, Zeiger and Townes [278] demonstrated maser action for the first time in 1953. This was the big breakthrough that the microwave community had been seeking for many years. It had a dramatic impact and spurred on efforts to find similar action for light. Already Dicke [279] was talking about an 'optical bomb' employing a super-radiant state, i.e. a laser *without mirrors* as a source of radiation, with the emission process occurring coherently in a very short and very intense light burst.

18.3.8. Flexible light pipes—fibre optics

Modern fibre optics could be said to have been started in the next year, when, in consecutive papers van Heel [280] demonstrated 'a new

Charles Hard Townes

(American, b 1915)

Charles Townes was the first scientist to conceive of a workable *microwave amplification system using stimulated emission* (maser). He used ammonia gas as the amplifying medium. He shared the 1964 Nobel Prize with Nikolai Basov and Alexander Prokhorov for work which led to oscillators and amplifiers based on the maser–laser principle.

He received his BS and BA degrees from Furman University, Greenville, South Carolina, his MA from Duke University, North Carolina and his PhD from the California Institute of Technology. During World War II Townes designed radar and navigational devices; later, at Columbia University, he used short-wavelength radar for microwave spectroscopy of atoms and molecules.

From 1961 to 1966 Townes was Provost of the Massachusetts Institute of Technology and since 1967 he has worked at the University of California in microwave and infrared astronomy.

---

method of transporting optical images without aberrations', employing a dielectric waveguide clad with a layer of lower refractive index to reduce evanescent wave losses, and Hopkins and Kapany published a fibre-bundle design of plain fibres small enough in diameter to examine the digestive tract [281]. Both papers mentioned cytoscopic applications but van Heel was more concerned with conquering alignment problems in imaging. Several years earlier Hopkins had been asked if he could improve the design of endoscopes, which at that time were inflexible. Tyndall's nineteenth century demonstration of light passed down a jet of water and along glass tubes inspired him to shine a car lamp down a glass fibre. Cladding greatly reduces the light losses and cladded fibre is now universally used.

In 1954 Bouwkamp's notable review [282] of diffraction theory evaluated earlier twentieth century work and corrected many errors by famous physicists and also Heitler's seminal text *The Quantum Theory of Radiation* was published [283]. In the same year Wolf presented a new optical coherence theory [284], which was a macroscopic theory of interference and diffraction of light from finite sources with a narrow but finite spectral range. He introduced a generalized Huygen's principle for the intensity of such fields and an intensity correlation function (the same as Zernike's degree of coherence) which he related to the spectral intensity of the source. Further, he showed that, in regions where geometrical optics is a valid approximation, the coherence factor itself obeys a simple law of propagation. Previous work by van Cittert, Zernike, Hopkins and Rogers are special cases of Wolf's results. In semiconductor optoelectronics Chapin, Fuller and Pearson [285] demonstrated the first solar cell, and Moss [286] and Burstein [287] studied the absorption coefficient of a semiconductor in relation to its impurity concentration, which later turned out to be relevant to optical bistability. In this process ionized donors or acceptors populate the edges of the conduction or valence bands with electrons or holes so that they cannot take part in an absorption process and this shifts the absorption edge to a higher energy (the Moss–Burstein shift).

18.3.9. *The birth of light-beating spectroscopy*

In 1947 there had appeared two papers [288, 289] which pointed out that optical sources should show similar beating phenomena to those well known in the radio-wave region of the electromagnetic spectrum. The first experimental demonstration occurred in 1955 [290]. When the Zeeman-split Hg 546.1 nm emission line illuminated a square-law photodetector, the two strong optical components beat with each other to produce a component in the photocurrent at about 10^{10} Hz. The widespread application and development of this result with lasers and light-scattering spectroscopy had to wait for a few years. Also in 1955 Lamb collected his Nobel Prize for experimental work on the spectrum of hydrogen. The *Lamb–Retherford* shift of the $2p_{1/2} - 2s_{1/2}$ line, caused by quantum-electrodynamic interactions with vacuum fluctuations, amounts to approximately 1050 MHz. Autler and Townes [291] described 'dressed' atomic states, a similar concept but arising this time from an interaction with an external driving field; this effect would come to prominence in studies of the non-linear optics of atomic systems some 20 years later. The theory of partial coherence was also advanced significantly this year with the independent introduction by Wolf [292], and by Blanc–Lapierre and Dumontet [293], of matrix formulations rigorously applicable to polychromatic fields from any type of source, coherent, partially coherent or incoherent. The degree of coherence was again defined in terms of the cross-correlation of the disturbance at two points in the field.

Optical spectroscopy of a different nature emerged in 1956 with the introduction by Hanbury Brown and Twiss [294] of perhaps the first entirely new physical principle in optical interferometry this century. From their 1954 radio interferometer research [295] they developed a corresponding optical *intensity* interferometry, in which the angular size of stellar objects is obtained. The light from a star is focused by two separated concave mirrors onto two photodetectors, and the correlation of fluctuations in their photocurrents is measured as a function of mirror separation. The two mirrors are in a partially coherent field, so the correlation function for unpolarized light can be written as

$$\langle \Delta I_1(t_1) \Delta I_2(t_1 + \tau) \rangle = \tfrac{1}{2} \bar{I}_1 \bar{I}_2 |\gamma_{1,2}(\tau)|^2$$

where I_1, I_2 are the intensities and $|\gamma_{1,2}(\tau)|^2$ is the mdoulus squared of the degree of coherence; the stellar diameter can be recovered from this function. This technique has been applied in optical astronomy, using collectors of 6.5 m diameter mounted on carriages on a track 188 m in diameter at the Narrabri telescope in Australia, to measure a number of stellar diameters as small as 0.0005 arc sec [296, 297]. This work has led to further developments [298–301], including, in the era of the laser, photon correlation or intensity fluctuation spectroscopy which we shall discuss later.

The earliest confocal optical microscopy patent was filed by Minsky [302, 303] in 1957. The illumination is focused to a small spot on the specimen and scanned in three dimensions. The transmitted (or reflected) light is focused by the objective onto a pinhole in front of a detector. In the patent most of the important features of this type of microscopy were mentioned: improved point resolution and rejection of light-scattered from other than the central focal point. His patent was not published until 1961 but still the work was far in advance of its time. Also in 1957 Hopkins published his paper [304] concerning numerical evaluation of the spatial–frequency response of optical systems, later to be widely used for lens testing by means of optical transfer functions [305–307].

18.3.10. The final pre-laser days

In 1957 Townes, the inventor of the maser (1953), was collaborating with Schawlow at Bell Laboratories on the initial stages of making it operate at optical frequencies. The use of Fabry–Perot [246] resonators was central to their thinking, and the number of excited atoms needed was calculated. The work [308] was circulated to colleagues and to Bell Laboratories Patent Office, which at first refused to patent it because 'optical waves have never been of any importance to communications'. Perhaps their Patent Office had forgotten Alexander Graham Bell and his assistant Sumner Tainter's Photophone experiments in 1880, in which they transmitted a human voice hundreds of metres on a beam of light; the world's first electro-optical system. At Townes' insistence, a patent was filed [309].

Like Townes and Schawlow, Dicke had also been concerned with the Fabry–Perot resonant cavity, having suggested its use for molecular amplification and generation systems in 1956 and having obtained a patent in 1958 [310], the same year as Townes and Schawlow filed their patent and published their results. Interestingly a semiconductor maser patent was filed by Watanabe and Nishizawa [311] in Japan in 1957, the year that Basov [312] started his semiconductor maser activities in the USSR. Townes (1964) and Schawlow (1981) both were awarded Nobel Prizes [313] for different aspects of their laser work. Also in 1958 Prokhorov discussed the possibility of assembling a molecular generator and amplifier for wavelengths shorter than a millimetre, using rotational transitions of ammonia molecules [314]. He and Basov were awarded Nobel Prizes [315] jointly with Townes.

In the same year, there was the prediction of a new kind of quantum semiconductor phenomenon, electro-absorption or photo-assisted tunnelling across the band gap, by Franz [316] and Keldysh [317], which led to a new spectroscopic technique for semiconductors (electro-reflectance) and created new opportunities for optical modulators and detectors. Exploitation of these ideas had to wait nearly another 30 years.

Politics and patents feature prominently in the history of the laser, and the picture is complicated. Bertolotti has assembled many of the details [318], in which Gordon Gould is a central figure. He worked with Townes at Columbia University. His signed laboratory books show that he worked on aspects of the laser invention, including the use of the Fabry–Perot cavity as a laser resonator, and had performed calculations on the feasibility of laser light amplification. In 1959, now at TRG Inc, he filed patents in the USA and UK. Litigation over these patents became a major issue in the laser community and after many years he was granted four US patents for optically pumped laser amplifiers, various laser applications, electric discharge pumping of a laser and use of a Brewster angle window in a laser cavity. As by this time the original laser patents had expired, Gould's royalties arising from the invention of the laser in the end may be far more significant than any others.

Perhaps it is fair to summarize the laser patent problem by giving credit to Townes and Schawlow for describing the physics, and to Gould for various (but not all) device technologies, but many other people throughout the century had also contributed vital ideas towards the invention of the laser, as we have seen.

One or two other events from this epoch are particularly notable. A theory of the formation of excitons with absorption and emission of phonons was published by Elliott [319] and Dresselhaus [320]. Putting exciton theory on a formal theoretical basis extended the previous work of Wannier and helped to open up the field of optical effects in semiconductors. Exciton absorption spectra in important materials such as germanium and silicon were observed at about the same time

[321]. Optical effects in semiconductors have been historically and fully reviewed by Smith [322].

Mandel [323, 324] published his theoretical studies of fluctuations of photon beams, including the semi-classical compound Poisson formula for photon counting statistics now known by his name

$$p_T(n) = \int \frac{\bar{n}^n}{n!} e^{-\bar{n}} P(\bar{n}) \, d\bar{n}$$

where n is photon number and $\bar{n}$ is the mean number of photons counted in time T.

A paper (1959) by Manley and Rowe [325] had a significant impact later in the field of non-linear optics. They derived general conservation relations for parametric interactions involving energy exchange between oscillation fields coupled by a non-linear reactive element. At this time Schawlow [326] concluded that ruby (chromium in corundum) would be a medium for lasing action, but thought mistakenly that the two strongest (691.9 nm and 694.3 nm) lines were not suitable. He also predicted correctly that a laser structure would be simply a rod with one end totally reflecting and the other almost totally reflecting. Everything now pointed to the invention of the laser.

18.4. 1960–70: the laser and non-linear optics

Despite all the research papers and patents, the first working laser was indisputably achieved by Maiman at Hughes Research Laboratory in the middle of 1960, using a crystal of ruby as the lasing medium. Immediately it could be seen that the laser had the expected coherence, narrow linewidth and directionality, although there were practical problems concerning repeatability and stability.

Laser is an acronym for light amplification by stimulated emission of radiation; we start with a simple description of how Maiman's ruby laser worked. He used pink ruby, Al_2O_3 containing about 0.05% by weight of Cr_2O_3. Wrapped around the crystal was a xenon flashlamp, capable of emitting plenty of green light around 550 nm, which chromium ions absorb and by which they are excited to an upper electronic energy level. From this level they can return to their original ground-state by two steps. First they drop to a slightly lower energy level by non-radiative heating of the crystal lattice. The ion can remain at this level, a long-lived metastable state, for a few milliseconds. Unless further excited it will then return to the ground state, emitting a (red) photon of wavelength 694.3 nm. Continuous operation of the flash lamp pumps the chromium ions to the upper energy level; they drop spontaneously to the metastable state, from which random emission of red photons begins. The flashlamp continues to excite more chromium ions and, with a sufficiently strong pump, at some moment the 'bottleneck' at the metastable state causes

Theodore Harold Maiman

(American, b 1927)

Ted Maiman earned his place in the history of modern optics as the inventor of the ruby laser. Maiman's father was an electrical engineer, and Maiman himself graduated in engineering physics from the University of Colorado in 1949. Following his PhD at Stanford, he moved to Miami to join the Hughes Research Laboratories in 1955.

He was particularly interested in the maser, which was developed in the USA and the USSR in 1955. He realized that it should be possible to extend the operating principles of the maser to allow operation at shorter, visible, wavelengths. In 1960, he successfully demonstrated operation of a ruby laser using a 1 cm ruby crystal to produce a pulsed output of red light. Maiman founded his own company, Korad Corporation, in 1962 which became the leading developer and manufacturer of high-power lasers. He has continued to work on lasers and advanced technology, with his own companies in the 1960s and 1970s, and joined TRW Electronics in 1977.

---

there to be more excited than unexcited chromium ions—a population inversion. The randomly (spontaneous) emitted photons, interacting with the other excited chromium atoms, and stimulating emission of identical photons, now start a cascade. A feature of stimulated emission is that the emitted photons are in phase with, have the same polarization and propagate in the same direction as the stimulating photons. The cascade of stimulated photons travels backwards and forwards along the crystal reflected from the crystal end-faces, and builds up to a high intensity by rapid conversion of the energy stored in the metastable levels.

A proportion of this built-up intensity is emitted through the partially reflecting end mirror and forms a coherent laser light pulse. The whole process lasts about 1–2 ms. For a more detailed account, Lengyel's excellent book should be consulted [327].

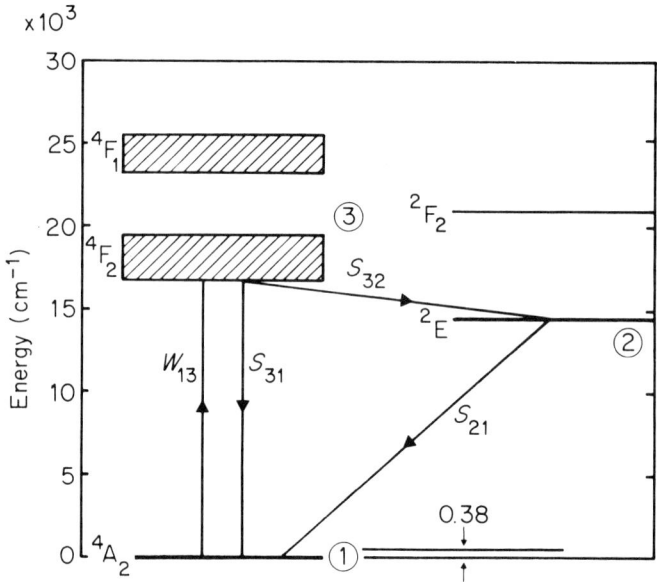

Figure 18.4. *Three-level energy diagram of the ruby laser.*

Not long before his laser paper Maiman described the results of his studies of the fluorescence of ruby [328]. He gave the simplified energy level diagram for triply ionized chromium in corundum shown in figure 18.4. Irradiation of ruby with a green wavelength (550 nm) raises the ions to the $^4F_2$ state, from which they fall to the metastable ^2E state; slow decay by spontaneous emission gives a sharp (R) doublet, 694.3 and 692.9 nm at a temperature of 300 K, causing the observed fluorescence. The fluorescence quantum efficiency was measured to be close to unity.

Maiman calculated and experimentally verified that population changes could be produced in the $^4A_2$ ground state by suitable optical excitation. Experiments with increasing intensity and frequency of a pulsed xenon flash-lamp finally yielded inversion of population and laser emission between the ^2E level and the ground state on 16 May 1960 [329]. It was announced in the *New York Times* on 7 July 1960.

Physical Review Letters rejected Maiman's laser paper, but *Nature* accepted it for publication on 6 August 1960 [330]. More detailed accounts followed in 1961 [331, 332]. The crystal was a 1 cm cube coated on two parallel surfaces with silver and surrounded by a coiled xenon flash lamp. Although passed over by the Nobel Prize committee, years later (April 1987) he won the prestigious third Japan Prize for this achievement.

The invention of the laser revolutionized optics and many other scientific disciplines in the twentieth century. Applications of the laser have proved to be innumerable. Massive effort was put into laser research around the world within a very short space of time from this first demonstration.

Just a few months after the first three-level ruby laser had been developed came the demonstration by Sorokin and Stevenson [333] of the four-level laser using trivalent uranium in calcium fluoride, starting a long list of materials that 'lase'. Schawlow noted later that nearly anything can be made to lase, even jello (jelly in English)! We will not recite every new kind of laser, but note a number of the more significant advances.

The ubiquitous helium–neon laser was also demonstrated just before the end of 1960 by Javan, Bennett Jr and Herriott [334], though not the famous red line. It was an infrared laser at first, with five lines between 1.118 and 1.207 μm. The strongest line produced 15 mW at 1.153 μm. The red 633 nm He–Ne line was first found by White and Rigden [335] in 1962. Here was a new kind of laser, in the gaseous rather than the solid state, and excited by a radio-frequency source.

1960 was without doubt the year of the laser: a great moment in the history of science, the invention of a new type of light source and a new type of light.

The optical-fibre laser was also proposed in 1960, Snitzer [336] noted that fibres of small core diameter, supporting only two modes, could make amplification and laser action possible at low power levels, but that pumping would be difficult. With Koester he succeeded in making an Nd-doped, flash-tube-pumped, glass-fibre amplifier in 1963 [337], achieving a gain of 5×10^4 in 1 m of multimode fibre with a core diameter of 30 μm. The core had a refractive index of 1.54 and the cladding 1.52. He went on to demonstrate with his colleagues [338] in 1969 a single-mode, neodymium-doped, 40 dB fibre power amplifier for an He–Ne laser operating at 1.06 μm. Did they suspect that such fibre amplifiers would be key components in long-distance telephony in the 1990s? In 1972 fibre Raman [339] and Brillouin [340] amplifiers were described and in the following year Stone and Burrus [341] described room-temperature operation of solution-Nd-doped, fused-silica, multimode fibre lasers of diameters from 15 to 800 μm and of 1 cm length. These were end-pumped with pulsed dye or argon-ion lasers.

18.4.1. Non-linear optics takes off

The year following the invention of the ruby laser proved to be the year of non-linear optics that was to provide another landmark in the history of optics. Its birth was heralded by an experiment [342] published in *Physical Review Letters*. Franken and his colleagues demonstrated the generation of radiation at the second-harmonic frequency, i.e. 2ω, in a quartz crystal when it was traversed by a ruby laser pulse of frequency

Arthur L Schawlow

(American, b 1921)

Arthur Schawlow was educated at the University of Toronto, where he received his first degree in 1941 and his PhD in 1949. His research career was firmly established at Bell Telephone Laboratories, and in 1961 he joined Stanford University as Professor of Physics. Schawlow is most noted for his work on the development and use of lasers, although his contributions have been wide-ranging and have also included optical and microwave spectroscopy, and superconductivity. He was co-author of the classic book *Microwave Spectroscopy*, and wrote the first paper describing optical masers, or lasers. He collaborated with Charles Townes in early work on maser principles and is generally credited as co-inventor of the laser. Although he did not share in Townes' Nobel award in 1964, Schawlow did share the 1981 Nobel Prize with Nikolaas Bloembergen for their independent research in laser spectroscopy.

ω. They dispersed the light in a spectrograph and recorded the output on a photographic place. In addition to the over-exposed spot at the ruby wavelength $\lambda = 694.3$ nm, the blue-sensitive plate showed a tiny dark spot corresponding to the ultraviolet wavelength $\lambda = 347$ nm. One of the authors (RGWB) has heard Franken tell with great delight how, during editing, the little spot on the photograph was thought to be a smudge mark, and removed. Thus the published evidence for second-harmonic generation in the paper, wasn't! No erratum was published. Nevertheless, the result was believed and non-linear optics started in earnest. The same team, augmented by Bass, made the first observation of sum-frequency generation. The light from two ruby lasers, with wavelengths differing by one nanometre, was passed through a crystal of triglycine sulphate. Three lines were observed around 347 nm, the middle one being the sum frequency and the two outer ones the second harmonics of the original lines [343].

A matter of a few weeks after publication of Franken's second-harmonic generation (SHG) work, Kaiser and Garrett [344] published observation of two-photon absorption from a ruby-laser pulse in a crystal of CaF_2 doped with Eu^{2+} ions. This observation was some 30 years after the theoretical discussion of two-photon decay in Goeppert-Mayer's PhD thesis at Göttingen, and approximately 11 years after an observation of the effect by Hughes and Grabner, which was explained by Rabi [345]. Again, the high light intensity available from the ruby laser made the experiment much easier. Two-photon absorption terminating in a level of the Eu^{2+} ion was monitored via the fluorescence intensity at $\lambda = 425$ nm of a lower lying level which was populated by relaxation from the excited state. Laser experiments not only opened up two-photon research, but heralded the widespread exploration of multiphoton transitions and their use in high-precision optical spectroscopy [345]. Interestingly, the rate of two-photon absorption from a single laser beam depends on the statistical properties of the light as described by its normalized degree of second-order temporal coherence, $g^{(2)}(\tau)$. For coherent, single-mode light $g^{(2)}(0) = 1$, but for chaotic single-mode light $g^{(2)}(0) = 2$, and thus chaotic light is absorbed at twice the rate of coherent light [346], see also section 18.5.4. The history of these and other early steps in non-linear optics, such as discovery of the basic 'laws', has recently been given by Bloembergen [347].

In addition to these first steps in non-linear optics, there was more fundamental thinking in the physics and operation of lasers. A better understanding of the properties of laser cavities led to improvements in the quality of laser output. An approximate theory of the plane–mirror Fabry–Perot resonator had been achieved by Schawlow and Townes [348] in 1958, but Fox and Li [349, 350] developed the theory to account for the effects of diffraction on the electromagnetic field in Fabry–Perot resonators with square and circular plane mirrors. Boyd, Kogelnik and Gordon [351, 352] studied modes in a resonator comprising two spherical mirrors, demonstrating advantages of alignment stability and lower diffraction losses. The Gaussian field-distribution profile for the lowest order transverse mode TEM_{00q} was derived a little later, as were

Nikolaas Bloembergen

(Dutch, b 1920)

Nikolaas Bloembergen was educated in the Netherlands, studying non-linear saturation phenomena for his PhD in 1948. He moved to Harvard in 1949, where he stayed for his entire working life in the USA.

Although Bloembergen is usually associated with the development of non-linear optics using lasers, his first major contribution came in the early 1950s. Townes' first masers could only work in pulses, as energy levels became depopulated on emission of microwave radiation. Bloembergen devised pumping schemes for three- and four-level masers which allowed continuous operation. Similar pumping schemes were subsequently used to activate all kinds of lasers.

As associate editor of the *Physical Review* and three other journals, and as advisory editor for *Optics Communications* and *Physica*, Bloembergen has had wide-ranging interests and influence. He shared the 1981 Nobel Prize for Physics with Arthur Schawlow for contributions to the development of laser spectroscopy.

the famous Kogelnik–Li equations for transformation and focusing of a Gaussian laser beam by a lens. Some laser mode patterns are shown in figure 18.5.

Another important advance in the subject came from the realization that a time variation of the *quality*, Q, of the laser cavity can result in a massive increase in the power output of pulsed lasers. This is called Q-switching and was proposed by Hellwarth [353] in 1961 and demonstrated by McClung and Hellwarth in 1962 [354]. With the mirrors detached from the ruby rod and a 5 ns nitrobenzene Kerr cell shutter between the ruby and one of the mirrors kept closed, excitation can be built up far in excess of the level reached without the shutter present. The shutter is kept closed until high excitation is reached. When the

A History of Optical and Optoelectronic Physics

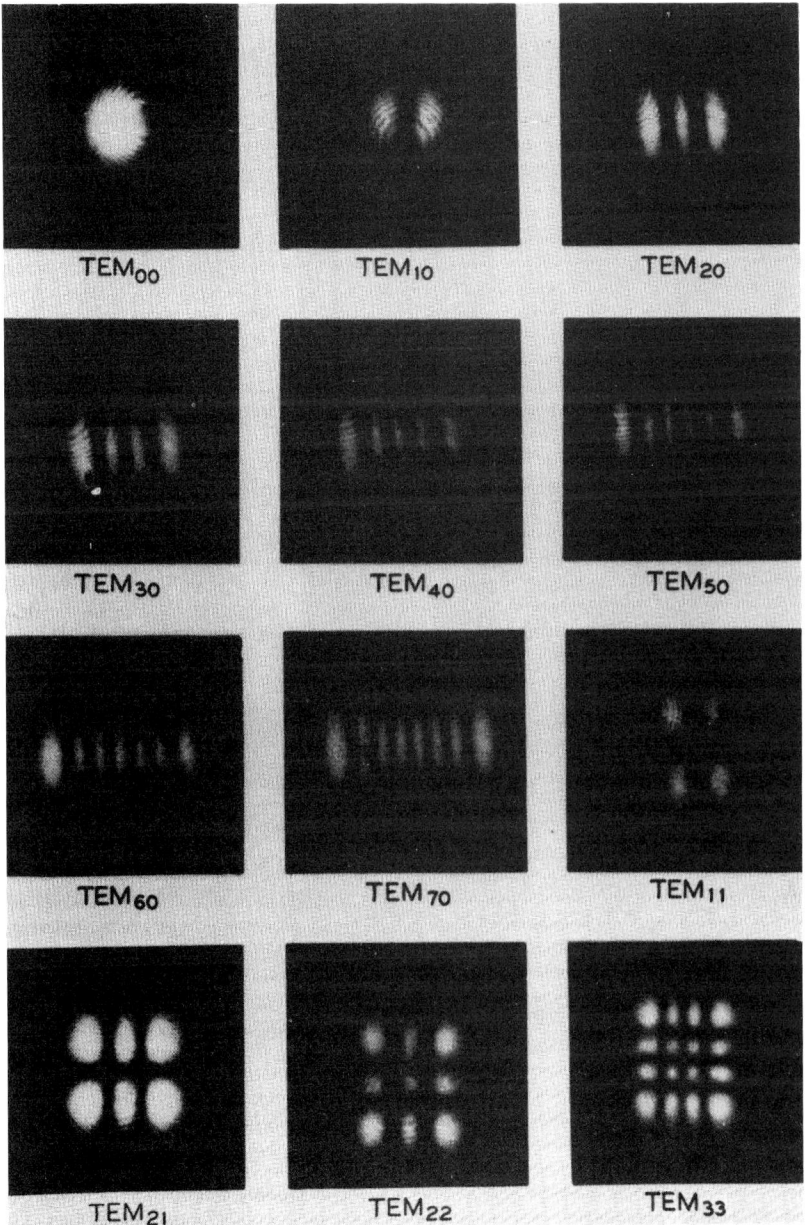

Figure 18.5. *Spatial mode patterns of a laser.*

shutter is opened the radiation builds up rapidly and the excess energy is discharged in an extremely short time, about 0.1 μs, typically with more than 1000 times the power of regular pulses.

Twentieth Century Physics

Whilst using such a laser, Woodbury and Ng [355] at the Hughes Aircraft Company observed both the 694.3 nm ruby-laser line and an additional line at 767 nm with about 20% of the power. They gave no explanation but said that it did not coincide with any of the lines that had been reported in the fluorescence spectrum of ruby. In fact, the difference is an NO vibration frequency (1345 cm^{-1}). Hellwarth and his colleagues quickly found that the line went away when the Q-switch was changed to a potassium dihydrogen phosphate (KDP) crystal. After an intensive period of activity, an important ingredient of which was obtaining a sample of D_6H_6 (which was the only liquid they could find that did not have any absorption at the ruby line and thus ruled out normal three-level laser action), Hellwarth gave the name *stimulated Raman scattering* to the new effect. They placed many liquids in the laser cavity and observed Raman shifts between 785 cm^{-1} in toluene and 3064 cm^{-1} in benzene [356].

18.4.2. A new theory of imaging: eigenfunctions

A significant advance in the theory of optical imaging was made when Slepian and Pollack [357] of the Bell Telephone Laboratories in 1961 finally eschewed the sines and cosines of Fourier, and instead introduced the more relevant prolate spheroidal functions φ_n as basis functions to describe the band-limited transmission of information. It seems odd that the mathematical theory of general linear transformations, which had been around since the end of the nineteenth century, was barely applied to optics until this time. Such is the grip of the Fourier transform, as distinct from the more general Sturm–Liouville theory, particularly in the education of engineers, that it is applied regardless of the nature of the problem at hand and has thus severely delayed further progress. The solution of the difficult coherent-imaging Fredholm integral equation of the first kind

$$I(x) = \int_{-X/2}^{+X/2} \frac{\sin \Omega(x - x')}{(x - x')} O(x') \, dx' \qquad -\frac{X}{2} \leqslant x \leqslant \frac{X}{2}$$

or $I = KO$ in the operator form of a linear transformation, was achieved by finding a differential operator (the first 'miracle' in the field according to Professor Grunbaum) which commuted with the integral operator K and which turned out happily to have known eigenfunctions, φ_n, and eigenvalues, λ_n, so that $\lambda_n \varphi_n = K \varphi_n$. In one dimension this is the same equation as describes the transmission of electrical signals over communication channels. Expansion of both image and object in this eigenfunction basis and use of its orthonormality solved for the first time the inverse problem of finding the object from the measured image. This mathematical solution apparently allows reconstruction of all the experimentally cut-off higher frequencies, which is impossible physically.

The contradiction is resolved by noting that in a perfectly noiseless case and with infinite computing precision one could indeed perform this feat since the Jordon-arc theorem allows the low-frequency image to be continued analytically to all frequencies. In physical terms one sees that the information required is present due to the infinite 'ringing' at the sharp boundaries of the object in this formulation. We will have more to say on this topic shortly.

Two other significant publications from 1961 should be noted. The first is the theory of stimulated emission in semiconductors, elucidated by Bernard and Duraffourg [358] who treated transitions between the conduction and valence bands, and derived the simple conditions under which lasing is possible in a semiconductor p–n junction. The second paper is Mandel's discussion of the concept of cross-spectral purity in coherence theory [359]. In most practical cases, the coherence properties of an extended optical source are position-independent, and the optical fields are then said to be coherence-separable or cross-spectrally pure, allowing the mutual power spectrum of the source to be factored into a product of spatial and spectral components. However, some sources are not cross-spectrally pure, and this leads to interesting interferometric effects.

In 1962 other notable events were reported; Dumke pointed out the importance of using direct energy gap semiconductors such as GaAs to create semiconductor lasers [360] and Rosenthal [361] proposed that a laser should be used to make a gyroscope based on the Sagnac effect. Macek and Davis were the first to publish a successful demonstration of a rotation-sensing, ring-laser gyroscope [362]. The main problem at first was the occurrence of a 'dead zone' at slow rotation rates when scattering from the mirrors and coupling within the gain medium caused the counter-propagating modes to lock together at the same frequency. An expensive programme to develop low-scatter mirrors ensued, but a more ingenious solution has been to use counter-propagating trains of ultra-short pulses which only meet in two places, both in free space. For example, a femtosecond-pulsed, dye-laser ring gyroscope, which can detect non-reciprocal differences in phase of 10^{-6}, has recently been fabricated [363]. This corresponds to rotation rates less than that of the Earth. The instrument has also been used to study the intrinsic response of photodetectors by measuring the change of refractive index due to the generated carriers, to study coherent interactions in a three-level system and to measure small magnetic field splittings in an atomic beam of samarium at fields down to 10^{-9} T.

Just as there was an immediate surge of interest following the announcement of the laser, a surge of a lesser kind followed the first detailed reports of non-linear optics. The phenomena arise from the non-linear response of a medium to electrical and magnetic fields at optical frequencies. The fields must be strong if non-linearity is to become significant, and for optical non-linearity something like 100 kW cm^{-2} is needed, which is unobtainable from standard thermal light sources, but readily obtainable from lasers, even the earliest ones.

In linear optics the electromagnetic field, E, passing through a material exerts a relatively small force on the electrons, resulting in an electrical polarization parallel with and directly proportional to the applied field. In classical terms the polarization $P = \chi^{(1)}E$, where $\chi^{(1)}$ is a dimensionless constant known as the linear susceptibility. In non-linear optics, the non-linear response can usually be described by generalizing the expression for P as a power series in electric field strength E

$$P = \chi^{(1)}E + \chi^{(2)}E^2 + \chi^{(3)}E^3 + \ldots$$

$\chi^{(2)}$ and $\chi^{(3)}$ are known as the second- and third-order non-linear optical susceptibilities. When the vector nature of these fields is taken into account, $\chi^{(1)}$ becomes a second-rank tensor, $\chi^{(2)}$ a third-rank tensor, etc. The low-order tensors and associated symmetry conditions were described at length by Kleinman [364, 365], by Bloembergen [366] and by Garrett and Robinson [367] in terms of anharmonic oscillators. Other papers [368] laid the theoretical foundations of non-linear optical effects such as second-harmonic generation and sum-frequency generation where optical frequencies ω_1 and ω_2 can be added in a non-linear medium. Bloembergen and Pershan [369] set out some of the 'laws' of non-linear optics.

The conversion efficiencies in early non-linear optical experiments were very low but the important idea of 'phase matching' the converted light in a birefringent crystal so that total momentum and energy could be conserved in the interaction [370, 371] gave the subject much greater interest and use. The basic idea is the same as is involved when two acoustic waves (phonons) in a solid combine, through the non-linearity of the medium, to generate a wave at the sum of the frequencies. This is described in Chapter 17, where it is shown how the resultant wave-vector must also be the sum of the constituent wave-vectors if there is to be any significant output. Perfect phase matching can be achieved with a birefringent crystal by polarizing the highest frequency wave in the direction that gives it the lower of the two possible refractive indices. This was achieved by focusing the output of a laser into a single crystal of quartz which was rotated to maximize the intensity of the second-harmonic signal. These experiments used pulsed lasers without precise mode control, but in 1963 Ashkin et al [372] used a continuous-wave He–Ne 1.15 μm laser to give a clear quantitative demonstration of the phase-matching phenomenon in second-harmonic generation.

Unfortunately, with normal dispersion and in isotropic materials, perfect phase matching is impossible, but an alternative technique known as quasi-phase matching has been developed. It corrects the relative phases at regular intervals by means of a structural periodicity built into the non-linear medium [368, 373]. Although enhancement has since been achieved in this way by a number of schemes [374] involving multi-domain and rotationally twinned crystals, stacking of thin plates, and application of periodic electric fields, it had to wait nearly 30 years before it was used to double the frequency of red laser-diode light to blue wavelengths for possible use in optical data-storage systems [375].

Parametric down-conversion, and parametric amplification and oscillation involve the photon-energy conservation process

$$\omega_3 = \omega_2 + \omega_1$$

energy from a 'pump' wave at ω_3 being transferred to lower 'idler' and 'signal' frequencies ω_2 and ω_1 respectively. The basic principles of parametric processes had been described in 1959 [325]. This idea was developed for optics by several groups [368, 376–378] but had to wait until 1964 for the demonstration of an optical amplifier when Wang and Racette [379] amplified light from an He–Ne laser at 633nm, using as a pump a ruby 694.3 nm line doubled in ADP (ammonium dihydrogen phosphate). The parametric medium was a second crystal of ADP. A higher gain was obtained when Akhmanov et al [380] in the following year amplified 1.06 μm Nd:YAG laser radiation, using doubled light of the same frequency as a pump and KDP as the parametric medium. The first optical parametric oscillator (OPO) was also demonstrated in 1965 [381]. This work created the possibility of tuning the frequency of laser emission by using an anisotropic crystal to create $\omega_3 = \omega_1 + \omega_2$, and by varying the crystal orientation. The output frequency may also be tuned by varying the birefringent-crystal temperature. This style of laser was later highly developed by Byer [382] and his colleagues at Stanford University.

Optical parametric oscillation provides an interesting new twist in the tale of lasers and non-linear optics. The laser had made possible observation of non-linear optical effects, which in their turn are used to enhance the performance of a laser itself through sum/difference frequency generation. Spontaneous optical parametric down-conversion later found widespread application as a principal source for quantum-optics experiments with correlated (entangled) photon-pair states [383], leading to further tests of quantum mechanics [384, 385]. This followed a photon coincidence experiment by Burnham and Weinberg [386] who observed virtual simultaneity in the parametric production of optical-photon pairs, i.e. the splitting of a single photon ω_3 into two photons ω_1 and ω_2 of lower energy, by passing a HeCd laser beam at 325 nm through a crystal of ADP.

A number of well-known non-linear optical processes are simply the result of an intensity-dependent refractive index, e.g. self-focusing of

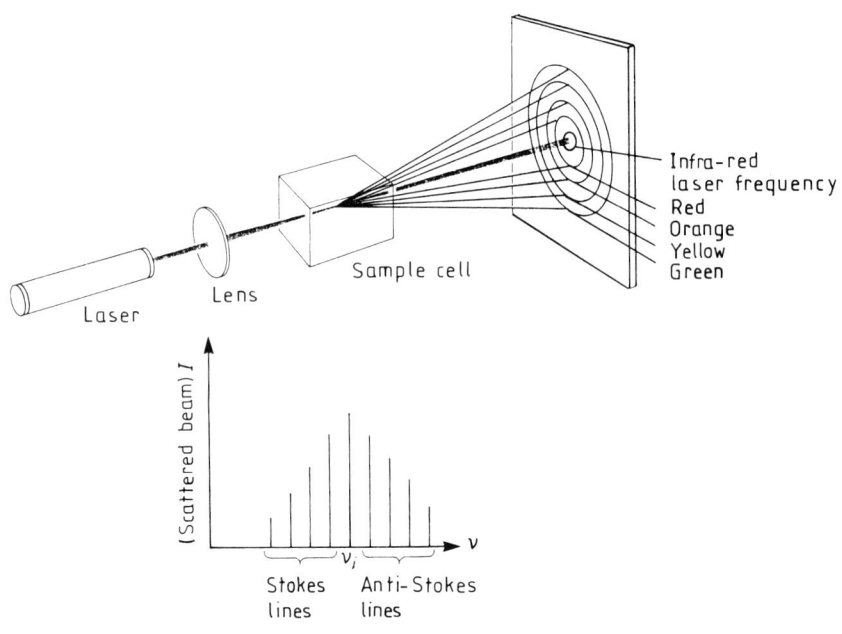

Figure 18.6. *Raman Stokes and anti-Stokes rings.*

light, optical phase conjugation, optical bistability, two-beam coupling, optical-pulse propagation and optical solitons. All these were intensively studied in the 1970s and 1980s, but self-focusing was the first to receive serious attention. It is found, particularly in liquids, that when the power in a laser beam exceeds a critical value its diameter contracts before stabilizing at a reduced value. The self-focusing is due to the dependence of the optical dielectric constant on the square of the electric field and was perhaps first discussed by Askaryan [387] in 1962. The condition for self-focusing is often the same as for stimulated Raman scattering (SRS), which differs from ordinary Raman scattering in that it only occurs above a certain threshold, has directionality similar to that of the exciting beam and is highly monochromatic and efficiently produced. A beautiful colour photograph by R W Terhune of this effect can be found as the frontispiece of the first edition of Yariv's text *Quantum Electronics* [388] which is shown schematically in figure 18.6. Yariv worked with Louisell [389] on the first quantum formulation of non-linear optical interactions. The Hamiltonian introduced in this paper was later used to predict the non-classical effect of light squeezing.

Whilst 1962 brought many new non-linear optical phenomena, interest in the laser was growing rapidly. Besides the demonstration of Q-switching by McClung and Hellwarth [354], the race to demonstrate a

A History of Optical and Optoelectronic Physics

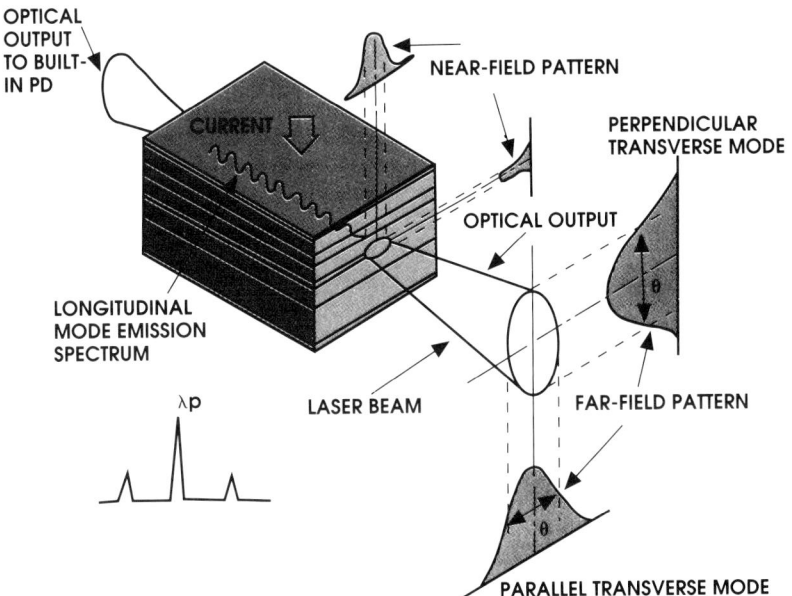

Figure 18.7. *A semiconductor laser diode, showing the structure of the device and the optical output.*

semiconductor laser had a near 'photo-finish' between four teams [390]. General Electric (Schenectady) had a working example in September 1962 [391] and IBM Watson Research Laboratories submitted a paper [392] in early October. Later in October Lincoln Laboratory [393] and General Electric (Syracuse) [394] also reported semiconductor lasers. A schematic heterostructure laser diode is shown in figure 18.7. The basic physics of these devices has been eloquently summarized by Thompson [395].

18.4.3. Practical holography
Another event at this time was to have a major impact in years to come. Leith and Upatnieks [242] demonstrated that Gabor's holography was practical at visible wavelengths in a modified illumination, recording and playback scheme. The breakthrough was their invention of a technique for off-axis reference-beam recording. Reconstruction by illuminating the hologram with the original reference beam now produced image pairs well separated from that beam, so as to avoid overlap which had been the primary reason for failure of earlier experiments. In the Leith–Upatnieks arrangement, figure 18.3, a separate reference beam is derived from the same coherent source that illuminates the object. This reference beam is directed to the photographic plate during the recording process, at an offset-angle θ to the beam that arrives from the object. Both the amplitude and phase of the object wave are encoded into the intensity distribution of the interference fringes recorded on the photograph. It is now assumed that the amplitude transmitted by the processed plate is

linearly related to the intensity in the original interference pattern. To reconstruct the image from the photographic plate (the hologram) it is once more illuminated with the same reference beam that was used to record it. A virtual image of the object is created in its original position and a real image in a conjugate position, both positioned off the axis of the reference beam. This eliminates the problems of overlap encountered by Gabor and earlier researchers.

The power and coherence of lasers made possible high-quality holographic records of deep diffusing objects. At about the same time as Leith and Upatniek's advance, Denisyuk [396] in the USSR formed holograms (volume holograms) by a technique which has some similarities to the Lippmann technique of colour photography [69], the object and reference waves being incident on a photographic emulsion from opposite sides. Recorded interference fringes were therefore layers almost parallel to the plane of the emulsion and spaced by approximately half a wavelength.

In addition to contributions by Stroke [397], Stroke and Labeyrie [398] (who produced white light reflection holograms) and others in developing basic concepts, later technological developments of holography and widespread applications in many branches of optics soon followed; these are described, for example, by Hariharan [399] and Collier et al [239]. Of particular note are holographic interferometry [400] (in which displacements of a rough surface can be measured to a fraction of a visible wavelength), holograms for displays, notably rainbow holograms [401] and three-dimensional, life-size, colour holograms of people and other objects, developed as an art-form [402] that became widely popular and available for purchase. Holograms are now also routinely used, for example, on credit cards as a measure against forgery.

Scientific uses of holography also developed, especially the construction of holographic optical elements [403] which, unlike conventional components, can be made in thin sheets, and computer generated holograms [404] for use in the manipulation of the phase of optical beams and as references for optical testing.

By 1963 the laser, non-linear optics and optical holography were well established as the 'new optics'. The fundamental physics was in reasonable shape to allow many applications in experimental physics, devices and system construction.

At this time we saw the development of rare-earth ion lasers by Johnson [405], for example neodymium (Nd^{3+}) doped into yttrium garnets (the most famous and widely used laser of this type), and also developments of non-linear optical effects such as anti-Stokes Raman emission in the form of a conical shell [406]. Mode-locking [407], an alternative to the Q-switching method for producing high-quality, pulsed laser output, was also developed at this time. In the process of mode-locking the relative phases of the longitudinal modes of a laser are fixed so as to cause emission of a pulse train from the laser cavity. The fixing of the phases can be achieved by devices inside the laser cavity such as an acousto-optic cell, by means of which the effect was first demonstrated in a He–Ne laser [408], or a saturable absorber whose opacity decreases with increasing optical intensity. Passive mode-locking in a ruby laser was devised in the following year [409]. By 1966 laser pulse-widths of picoseconds had been achieved by these methods [410]. Femtosecond laser pulses were generated in 1981 [411] by the technique of *colliding-pulse mode-locking* in which two oppositely directed pulses collide in a thin saturable absorber causing a transient grating of refractive index which synchronizes and shortens both pulses. Another widely used method is the self-mode-locking scheme introduced originally for the titanium-sapphire laser [412], in which the Kerr non-linearity of the gain medium itself was used to produce 60 femtosecond pulses of high power and wide tunability. Adams, Sibbett and Bradley [413] detected a 70 MHz train of picosecond dye-laser pulses with a fast photodiode, and fed its amplified output to the deflection plates of a 'Photochron' streak camera to make an effective 'picosecond oscilloscope' for the study of ultra-fast optical phenomena occurring on the camera photocathode.

At this time Lukosz [414–416] published a series of papers, in which optical images with a resolution exceeding diffraction limits were obtained. These exploited the von Laue (1914) idea of a fixed number of degrees of freedom in a two-dimensional image and used optical tricks to trade between one and two dimensions. Unfortunately, as with apodization, the methods are essentially of curiosity value only.

The classical theory of coherence now played itself out with an elegant and comprehensive survey of fluctuations of light beams, coherence and bunching effects [417], this being the first of two major reviews of the coherence of light with [18]. Both of these works were firmly based on an 'analytic-signal' description of the light field related to its classical wave envelope. Pancharatnam [418], in one of the last papers concerning classical coherence before the quantum theory of coherence was started, addressed effective phase difference and degree of mutual coherence between two polychromatic beams.

18.4.4. *A quantum theory of coherence*
The highlight of 1963 was surely Glauber's introduction of a quantum theory of coherence [19, 419, 420], a very significant step forward, albeit disputed by Mandel and Wolf [421] for a couple of years (on the grounds that field quantization is unnecessary as long as the atoms are quantized) until, in effect, they capitulated.

Glauber's starting point was to separate the electric field operator $E(rt)$ into its positive- and negative-frequency parts

$$E(rt) = E^{(+)}(rt) + E^{(-)}(rt)$$

the positive-frequency part being associated with photon absorption (annihilation) and the negative-frequency part with photon emission (creation) and then to use these to define field correlation functions of arbitrary order. The nth-order correlation function $G^{(n)}$ takes the form of the expectation value

$$G^{(n)}(r_1t_1, \ldots, r_nt_n; r'_nt'_n, \ldots, r'_1t'_1)$$
$$= \text{tr}[\rho E^{(-)}(r_1t_1) \ldots E^{(-)}(r_nt_n) E^{(+)}(r'_nt'_n) \ldots E^{(+)}(r'_1t'_1)]$$

where ρ is the density operator of the radiation field. In the diagonal form where $r_it_i = r'_it'_i$, this has the physical interpretation of the joint probability of photons being detected at the points r_i at times t_i if detectors were present at these places and times. When the field is in a classical statistical superposition of coherent states (to be described below), or, equivalently, if the P representation, (also to be described below), is positive, $G^{(n)}$ is equivalent to the classical expression

$$G^{(n)} = \langle \hat{E}^*(r_1t_1) \ldots \hat{E}^*(r_nt_n) \hat{E}(r'_nt'_n) \ldots \hat{E}(r'_1t'_1) \rangle$$

where $\hat{E}$ is the 'analytic signal' or envelope of the wave-field; a semi-classical description is then possible. We should emphasize again that the lack of necessity to quantize the radiation field even in such cases is only apparent and not rigorously correct. Apart from the introduction of the quantum correlation functions, the second of which, $G^{(2)}$, was to play a central role in the new photon correlation spectroscopy to be discussed below, the main results of Glauber's work include the introduction of coherent states into optics and the quasi-distribution now known as the P representation. Coherent states $|\alpha\rangle$ were defined for the single-mode case on the basis of photon-number states $|n\rangle$, where n is the number of energy quanta in the mode, by the relation

$$|\alpha\rangle = \exp(-|\alpha|^2/2) \sum_{n=0}^{\infty} [\alpha^n/(n!)^{1/2}] |n\rangle$$

where α is an arbitrary complex number. The coherent state is an eigenstate of the photon annihilation operator of the mode. A property of the coherent states is that $G^{(n)}$ separates into a product of $2n$ factors since there are no correlations between photon numbers in such a state. This can be used as an alternative definition of a coherent state.

Description of the general field by a statistical mixture of (over-complete) coherent states led Glauber [19] to the P representation

$$\rho = \int P(\alpha) |\alpha\rangle \langle \alpha| \, d^2\alpha$$

which was also introduced independently by Sudarshan [20]. This is a so-called *quasi-classical* distribution function in the same family as the well-known Wigner function. An excellent summary of the history and development of these ideas has been given by Bertolotti [422]. Sudarshan's [20] paper also introduced the optical equivalence theorem which shows that this 'diagonal' representation of the density operator enables a formulation of a quantum theory of optical coherence in a language formally equivalent in all respects to the language of classical theory as expressed in terms of analytic signals [423] but where probabilities can take on singular or negative values. It is in such cases that we have 'non-classical' light which escapes a semi-classical description.

In 1963–64, Glauber [19,424] also discussed the photodetection process and showed that the probability $p^{(1)}(t)$ of counting a single photon at r between times t_0 and t_1 is given by

$$p^{(1)}(t) = \int_{t_0}^{t_1} S(t-t')G^{(1)}(rt',rt')\,dt'$$

where S is a frequency-dependent sensitivity function. This result was also obtained by Kelley and Kleiner [425] for the broad-band case where $S(t-t') = \delta(t-t')$. This is a fully quantum-mechanical equation since both radiating and detecting systems have been quantized. It neglects depletion of the photon field, a defect which was soon to be remedied by Mollow [426]. There has been some more recent discussion [427–429] of the applicability of this formula to the broad-band case since it relates to $\boldsymbol{E} \cdot \boldsymbol{r}$, the operator for field-energy measurement rather than photon detection and some further frequency factors are thus required which should not fundamentally be ascribed to the detector sensitivity. A basis for photon counting and correlation spectroscopy, however, was now in place, and these subjects were to develop rapidly and widely. The substantial literature has been reviewed by Le Berre [430].

18.4.5. Laser-light-scattering begins
Cummins and his colleagues at Johns Hopkins University published results of two pioneering laser-light-scattering experiments in 1963 and 1964. First [431], they had studied the 'self-beating' fluctuations of light scattered by a suspension of particles in Brownian motion, by recording the frequency spectrum of the detector photocurrent. This spectrum, for a monodisperse suspension of particles, is predicted to be of Lorentzian shape, centred at zero frequency. The smaller the particles, the faster the fluctuations and the broader the spectrum; the linewidth thus permits a measurement of particle size. This appears to be the first experimental use of the fluctuating *speckle pattern*, or *corona* as it used to be called, to determine the size of particles undergoing Brownian

motion, a technique which is now used routinely in many fields of science. The idea, however, must be credited to Sir C V Raman. In a beautiful paper of 1943 Ramachandran [432] first correctly formulated and verified experimentally the basics of speckle-pattern formation and evolution. With a layer of lycopodium powder on a glass slide, in both plane and spherical wavefronts, he demonstrated, by visual classification of speckle-spot intensities, the 1880 exponential law of Rayleigh for the distribution of intensity from a fixed number of randomly phased oscillators. In the conclusion to this paper he says that Raman had suggested to him that Brownian motion could be observed by this intensity-fluctuation technique and that he was proposing to study this possibility experimentally. Apparently he either did not try, or tried and failed, since there seems to be no further record from him. There does, however, appear a description of the phenomenon by Raman himself in his *Lectures in Physical Optics* (1959) [433]. He viewed a mercury lamp behind a small aperture through a thin film of milk, and says

> Bright points of illumination continually appear in the field and others disappear. These changes become less rapid and ultimately stop when the film is dry.

In the second experiment, Yeh and Cummins [434], experimentally investigated the Doppler shift of scattered light by particles suspended in a laminar fluid flow, an experiment which heralded the start of laser velocimetry and anemometry, later to be developed and applied extensively to measure flow in liquids and gases as well as motion and vibration of solids. The experimental arrangement used by Yeh and Cummins is shown in figure 18.8. This type of light-beating spectroscopy was also underway in Benedek's laboratory at MIT [435] where it was used to study critical fluctuations in xenon. The experiment is now being repeated in space by Gammon, a NASA principal investigator, to eliminate the inhomogeneity which gravity induces in the critical state of a fluid.

The laser progressed in 1964. Molecular laser activity started in earnest with the report by Patel, Faust and McFarlane [438] of a CO_2 laser operating at an emission wavelength of 10.6 μm, in the far-infrared, by means of transitions between the vibrational and rotational levels of the molecule. This laser proved to be valuable for industrial and military applications, for reason of its high-power output from relatively small active regions. Even water was found to lase [437] when researchers at SERL (Baldock, UK) tried to make a far-infrared laser with HCl, but forgot to dry it first! Lasing action in the far infrared (sub-mm) region was also discovered by Patel *et al* [438] in neon, when they first found a line at 57.3 μm and later at 68, 85 and 133 μm. The water lines, however, were orders of magnitude stronger. The SERL group also published an HCN laser line at 331 μm [439]. The explanation of the HCN and H_2O

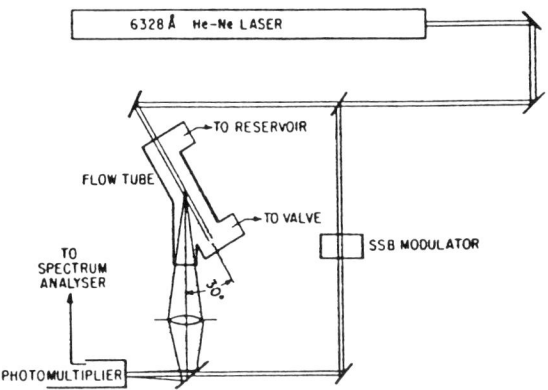

Figure 18.8. *Schematic diagram of the Yeh–Cummins light-beating experiment.*

lasing action took some time to tease out. One of the authors (ERP) has personal experience of the difficulty of this problem and Sochor [440] also failed to identify the correct mechanism. Lide and Maki [441] finally proposed for HCN the enhancement of otherwise small transition probabilities by interaction of nearly degenerate vibration–rotation levels; the resulting Coriolis mixing gives them some of the character of the strongly allowed pure rotational transitions. This explanation was extended to H_2O [442–444] where there exists a perturbative coupling between some near-degenerate high rotational sub-levels of a bending (020) and an asymmetric stretching (001) mode. Such lasers were later applied to radar modelling on scale models of ships and aircraft.

The widely used argon-ion laser was first reported in 1964 [445, 446]. Bridges apparently discovered lasing from argon ions at 488 nm also quite accidentally, but the development of lasers using rare-gas ions, e.g. krypton and xenon [120] was rapid. Blue–green emission argon-ion lasers became another laboratory standard laser to join the less powerful helium–neon lasers already in widespread use.

In 1965 Lamb published his semi-classical theory of the optical maser [447], in which the *Lamb-dip*, at the centre of the line from a gas laser (which varies with the power) was explained as being due to the depletion of the atomic population by waves travelling in both directions along the laser cavity at nearly the same frequency. The first detailed theories of laser light were published by Haken [448] and Risken [449], and later by Scully and Lamb [450] and by Lax and Louisell [451]. Haken considers atoms at random positions interacting with the resonator mode, and adds pumping as an external stochastic process, with losses accounted for phenomenologically. The equations were solved by treating the atom and field quantum operators as *c*-numbers i.e. as expectation values. Linewidths were calculated both below and above threshold and connection was made with Glauber's work on photon statistics. Risken's contribution was a non-linear theory of laser noise. The Haken–Risken results were improved upon by Weidlich and Haake [452] who avoided the c-number treatment of the field, and by Wiedlich *et al* [453] who provided a full quantum theory; the equivalence of this latter theory with that of Scully and Lamb was eventually shown by Casagrande and Lugiato [454]. All these were non-trivial calculations in the quantum theory of open systems. Laser theory determines the number of photons in the lasing mode by considering the transition rates into and out of each level, resulting from the Einstein A and B processes of absorption and emission, and from the loss of photons from the cavity. The lasing atoms are pumped by an external mechanism (electrical discharge or optical pumping, for example) into an excited state of inverted population, and emit radiation into the cavity at a faster rate than they absorb it. The result would diverge but for the cavity loss, and the theory, in the form of a master equation, describes the steady-state balance between these three processes [226].

Four other events of later significance in applied optics occurred in 1964. Osterberg and Smith [455] described experiments with thin-film devices which led to the concept of integrated optics as discussed by Miller [456]. A second type of device combined a holographic spatial filter and an optical correlator [457] in an interferometric technique widely used to record any complex filter for which the point-spread function is known. By recording the interference pattern between the desired filter transfer function and a mutually coherent reference beam, a Fourier hologram of the point-spread function is obtained. The technique was to become of central importance in the development of coherent optical processors, and to lead to extensive applications in experiments concerning optical pattern recognition and image and data/information processing [458, 459].

Thirdly, *stimulated Brillouin scattering* (SBS) was observed by Chiao, Townes and Stoicheff [460]. SBS occurs when the acoustic wave that scatters the optical beam is produced by the optical beam itself. When an intense beam of frequency ω_2 passed through a crystal of quartz or sapphire, a coherent acoustic wave at frequency ω_s was produced within the crystal, while at the same time an optical beam at frequency $\omega_2 - \omega_s$ was generated. The acoustic and scattered optical beams were emitted along specific directions and their generation occurred only above a well-defined input threshold power level. This new non-linear optical effect was later used in phase-conjugation experiments for effective optical-wavefront time reversal and amplification. SBS was also reported in liquids in the same year by Garmire and Townes [461] and by Brewer and Reickhoff [462].

Fourthly, there were two demonstrations of the optical Kerr effect. The original Kerr effect (1875) is the birefringence produced by an electric field, and has been used to Q-switch laser cavities; the optical Kerr effect, on the other hand, uses the electric field of a strong light beam, which can be the beam itself (self-action) or, as in these demonstrations, a pump beam to control the transmission or reflection of a probe beam (the *pump-probe* configuration). Both experiments were of the pump-probe transmission type. The first was by Maker *et al* [463] and the second, which did not make use of lasers, was by Mayer and Gires [464]. The pump-probe optical Kerr effect in specular reflection was not reported until 1992 [465] when a giant non-linear phase shift was seen at an exciton resonance in ZnSe. The same three configurations can be used to produce the inverse Faraday effect. In the Faraday effect (1846), a magnetic field rotates the plane of polarization of a light beam; in the inverse Faraday effect a circularly polarized beam induces a macroscopic magnetic polarization [466]. This was seen in the self-action configuration a year after the first optical Kerr effect measurement. There have been many related developments which may be studied in the original papers [467–478]. Interestingly, *linear* optical activity on reflection was not seen until 1993 [479].

18.4.6. Optics lights up

By 1965 the explosion of activity in optics caused by the invention of the laser is very clear from the research journals, especially those of the leading organization, the Optical Society of America. Many thousands of research publications appeared in the next few decades showing intensive global activity in laser applications and developments. Discerning the advances in fundamental optical physics now becomes difficult as the emphasis is so much on developments and applications.

Important new classes of laser appeared in 1965, including the first chemical laser which used HCl molecules and emitted at 3.7 μm [480]. Chemical lasers were intensively developed by the military for their very high-power output [481]. The metal-vapour laser arrived in the same year, with zinc and cadmium [482], and was to lead to the first demonstration of the helium–cadmium laser in 1967 employing the deep-blue cadmium-ion line at 441.6 nm [483].

At this time, Bloembergen [366] published his text *Nonlinear Optics*, drawing together the physics in this area and presenting non-linear analogues of the linear laws of optics: Brewster's angle, total reflection, surface waves, etc. However, the non-linear analogue of conical refraction had to wait until Schell [484], one of Bloembergen's students, described this in her thesis in the late 1960s.

Another new non-linear effect, stimulated Rayleigh-wing scattering, was reported in 1965 by Mash *et al* [485] and explained by Bloembergen and Lallemand [486] in the following year. The light-scattering here arises from the tendency of anisotropic molecules to become aligned along the electric field vector of an optical wave, so that an induced Rayleigh effect is observed.

In concentrating so much on the new laser and non-linear optical physics that dominated this era, it is important not to forget that more traditional activities in optics were still being vigorously pursued, particularly in applications, optical-system engineering and component developments. Two more basic traditional optical activities, aberrations and scattering theory, were still under investigation. By now, the aberrations of optical systems were clearly the realm of the computer, and much tedious calculation could at last be done by machine. Such calculation was based on the matrix representation of Gaussian optics and techniques of ray-tracing employing matrices developed from the early work of Smith in the 1930s [487, 488]. The aberrations of complex lenses, gratings and holographic optical elements were given much attention. In a review paper of 1965, Welford [489, 490] summarized the basic physics supporting these activities.

Also at this time, a new branch of classical optics was starting to emerge, the optics of non-imaging light concentrators. Primitive forms of concentrator such as the simple cone had been known and used for many years, but the cone is far from ideal, and was now supplanted by the compound parabolic concentrator (CPC), which was capable of coming close to the optimum concentration ratio. CPCs first appeared in the mid-1960s in very different experiments, e.g. for collection of light from Čerenkov counters [491] and in three-dimensional geometries for solar-energy collection [492]. Many developments in CPCs and in efficient solar-energy collection during the next two decades were introduced to a wide optical community by Welford and Winston [493].

Scattering theory was also being performed by computers to eliminate tedious hand-calculation. Mie's equations and their developments, though ideally suited to machine computation, only applied to regular shapes such as spheres, ellipses and cylindrical rods. In 1965 the situation was improved when Waterman [494] explained a matrix formulation of electromagnetic scattering. This led to the T-matrix theory which permits computation of light-scattering patterns by computer for irregular particles of arbitrary shape [495, 496].

Feynman, Schwinger and Tomonaga [497] received the Nobel Prize for their contributions to quantum electrodynamics in 1965. The first dye lasers, discovered by Sorokin and Lankard [498] in 1966, used a solution of chloro-aluminium phthalocyanine, excited by radiation obtained from a giant-pulse ruby laser. This was another important step towards laser light tunable over a broad wavelength range. Since that time many dyes have made this type of laser an important, versatile laboratory tool [499]. A Nobel Prize was awarded to Kastler [500] for his work on optical pumping and Lohmann and Brown [501] made the first computer-generated spatial filter, a cornerstone (with Vander Lugt's [457] work) of future optical image processing and pattern recognition.

18.4.7. Optical-fibre communication

Perhaps the most prescient papers of 1966 were those of Kao and Hockham [502], and Werts [503], who almost simultaneously proposed optical telecommunication using cladded dielectric waveguides or glass optical fibres. Although fibres at this time showed losses exceeding 1000 dB km^{-1}, this was due mainly to impurities from the starting materials and the fabrication process. The crucial breakthrough came in 1970 when Kapron, Keck and Maurer [504] at Corning Glass achieved losses less than 20 dB km^{-1} by a new 'soot' process of chemical vapour deposition. Within two years signals with gigahertz bandwidth had been transmitted over 1 km of fibre in the UK [505], and within another five years telephone circuits were in trial operation in the UK, the USA, Japan and Germany [506]. The British Post Office, for example, started a 140 Mbit s^{-1} link from Stevenage to Hitchin (8 km) while AT&T set up a 44 Mbit s^{-1}, 24-fibre, 2.5 km link in Chicago; a link of 32 Mbits s^{-1} over 53 km without repeater was operational in Japan, and the Deutsches Bundespost set up a 44 Mbit s^{-1} link over 4 km in Berlin. Of the large volume of research we will mention one further breakthrough, heralded by two papers in 1985. A technique of chemical vapour deposition was used to dope silica-glass fibres with rare earths [507], and the first continuous-wave, single-mode, silica-fibre lasers, using up to 100 m of neodymium-doped fibre were described [508]. That same year the Southampton University team filed an erbium-doped optical-fibre amplifier (EDFA) patent [509]. Publications followed in 1987 [510]. The EDFA is now widely used for long-distance telecommunications. Improvements in fibre and components have multiplied to the point where the cost of even trans-oceanic communication is now extremely low—another wonderful story of scientific and technological success for the benefit of the unsuspecting consumer!

Other experiments with important scientific consequences were the demonstrations [511–513] in 1965–66 of single-photon counting which verified Glauber's 1963 theory for various types of light field. Johnson *et al* [513], in particular, had the instrumentation necessary for extremely efficient and accurate experiments. They also made the important step, following Glauber's theory, of autocorrelating the photon pulse train to derive spectral information, as we shall describe later.

In this prolific year Kogelnik and Li [514] explained Gaussian–Hermite (laser) beam propagation through an optical system. The basic equations are

$$w(z) = w_0(z) \left[1 + \left(\frac{\lambda z}{\pi w_0^2} \right)^2 \right]^{1/2}$$

$$d_2 = \left(\frac{(d_1 - f)f^2}{(d_1 - f)^2 + (\pi w_1^2/\lambda)^2} \right) + f$$

$$w_2 = \frac{w_1 f}{[(d_1 - f)^2 + (\pi w_1^2/\lambda)^2]^{1/2}}.$$

The first equation gives the radius w of a laser beam at a distance z from its waist position (the plane where it has its minimum radius, w_0). The second equation gives the position, d_2, of the beam waist, w_2, after a thin lens of focal length f, if the initial distance and waist are d_1 and w_1; the third gives the waist radius, w_2, at the position d_2.

Perhaps the most significant optical event of the next year, 1967, was the ingenious invention of the wavelength-tunable dye laser [515]. Soffer had already made a number of contributions to laser development [516]; now he and McFarland replaced one mirror of a laser-pumped rhodamine 6G laser with a diffraction grating mounted in the Littrow arrangement. The grating was adjusted so that the first-order reflection of the desired wavelength was reflected back upon itself along the axis of the laser, and the tuning was accomplished by rotating the grating. This simple and elegant scheme made the tunable dye laser a standard item of a spectroscopy laboratory [499]. In the late 1980s and the 1990s some of the dye-laser applications were superseded by the titanium–sapphire laser [517], for simple practical reasons.

Tunable coherent radiation became available in a broad spectral range, from the far infrared to vacuum ultraviolet, with increasing use in optical spectroscopy and for measurements which relied upon non-linear optical processes such as *coherent anti-Stokes Raman scattering* (CARS) [518] and multi-photon absorption and ionization. In CARS two laser beams, one of fixed frequency, ω_L, and the other of variable frequency, ω_S, are passed through a medium with an electronic transition of frequency ω. The radiation of frequency $2\omega_L - \omega_S$, produced by four-wave mixing, is monitored. A non-linear interaction gives rise to a polarization wave which imparts a resonant coherent anti-Stokes shift to ω_L when $\omega = \omega_L - \omega_S$. The frequency ω_S is tracked through this region to map out the shape of the Raman line. High-resolution spectroscopy can thus be performed without the use of a spectrometer with the additional advantages that the signal is proportional to the square of the number of scatterers: it is thus much more intense than spontaneous Raman scattering and is highly directional, which allows rejection of diffuse background light.

There were also demonstrations of two new non-linear optical effects, *parametric fluorescence* and *self-phase modulation*. According to the photon energy-level description, an atom first absorbs a photon ω_1 and jumps to a higher virtual level. Decay from this level may be by a two-photon emission process, stimulated either by another optical field ω_2, already present, or arising from spontaneous two-photon emission from the virtual level even if ω_2 is not applied. The fields generated in the latter case are much weaker. This is parametric fluorescence [519].

Self-phase modulation is the change in the phase of an optical pulse due to the non-linearity of the refractive index of the material medium, yet another example of an intensity-dependent refractive index. This was initially studied by Brewer [520] and by Shimizu in 1967 [521] but its importance comes later, in optical soliton propagation for telecommunications.

The years 1968 and 1969 brought to a conclusion perhaps the most momentous decade in the history of optics. The laser and non-linear optics had completely revolutionized the subject.

18.4.8. Integrated optics and optical bistability

1969 saw the beginning of integrated optics [456, 522], and the theory, fabrication and application of guided-wave optical devices. Light is guided along the surface region of a wafer of material by dielectric waveguides situated at or near the wafer surface. The light is confined to a region having a typical dimension, in cross section, of several optical wavelengths. Guided-wave devices can perform passive operations such as beam-splitting at waveguide bifurcations. By fabricating waveguides in active optical materials such as ferroelectrics, electro-optical effects can be exploited to modulate the guided light. By making waveguides in compound semiconductors, passive and active devices can be combined monolithically with lasers and detectors to create integrated optical circuits or 'photonic' integrated circuits. A further extension has been to add optoelectronic devices to such circuits and thus create *optoelectronic integrated circuits*, OEICs.

Certain non-linear optical systems can give more than one output for a given input, because of another non-linear optical effect resulting from an intensity-dependent refractive-index. This is *optical bistability*. 1968 saw an experiment with a neon cell inside a helium–neon laser cavity [523] in which non-linear absorption gave rise to various mode-locking effects. A year later Szöke et al [524] proposed putting the non-linear absorber in a separate Fabry–Perot cavity, and gave a theoretical description which is much simpler than that for the laser cavity. A patent on the same effect was taken out by Seidel [525], and later Szöke filed a patent [526] for the production of short pulses of variable length. Gibbs and Venkatesen demonstrated such optical bistability some years later, and in 1975 [527] they reported bistability in a 2.5 cm cell of sodium vapour at approximately 10^{-6} Torr in a cavity composed of two 90% reflecting mirrors 11 cm apart. They noted that the phenomenon in no way followed the expected behaviour of absorptive bistability and proposed instead a dispersive mechanism. Further results for sodium vapour followed in 1976 [528], and in the following year they took out a patent for optical bistability in indium phosphide [529] for an optical amplifier 'akin to a transistor'. They claimed that the earlier patents had not covered the dispersive effect. In 1978 they patented [530] the use of enhanced non-linearities due to combined absorption and dispersion in ruby, GaAs, GaAsSb and CdS for optical amplification. Soon after they demonstrated the effect in gallium arsenide [531], and initiated extensive worldwide efforts to apply it in optical computing. Semiconductors such as these were known to have high values of $\chi^{(3)}$ [532], but saturation mechanisms just below the band gap enhance these by many orders of magnitude. In 1978 Miller et al [533] at Heriot Watt University (UK) were to observe very large non-linearities in indium antimonide which, in a paper submitted on the same day as that of Gibbs et al [531], to the same journal, gave rise to an *optical transistor* and some logic gates [534]. In the same year Bonifacio and Lugiato gave a comprehensive description of optical bistability in a 'mean-field' approximation [535]. A few years later the Heriot Watt team published an account of their work on all-optical circuits [536] and Gibbs published his book *Optical Bistability* [537].

In 1978, in the USSR, Karpushko and Sinitsyn were investigating optical bistability in thermo-refractive zinc sulphide [538] and in the USA McCall and Gibbs reported thermal bistable effects in a glass filter [539]. Also at this time Wolf described three-dimensional structure determination of semi-transparent objects from holographic data [540], a paper that laid the foundations of diffraction tomography [541].

18.4.9. The confocal microscope
Inspired by the 1961 patent of Minsky [302], and one of Petrán and Hadravský in 1966 [542] (and at the suggestion of Baer to use a laser) in 1969 Davidovits and Egger demonstrated the first confocal scanning laser microscope [543, 544] a prototype using a He–Ne laser in which the laser beam was scanned by moving the objective lens rather than the object as Minsky had done. Petrán, as a research assistant with Egger at Yale, had previously implemented [545] the Petrán and Hadravský scheme using a Nipkow disc of 20 μm thick copper foil of 85 mm diameter rotating at 3 rev s^{-1}. The disc was electrolytically etched with 26 400 holes of 90 μm diameter in an Archimedean spiral. Sunlight was used for illumination; it entered on one side of the disc and the light reflected from the specimen left by the corresponding holes on the opposite diameter. Exposure times ranged from 1 to 4 seconds.

The benefits of scanning microscopy had been pointed out in the prelaser work of Minsky. The application of the laser was very beneficial since the high collimation gives a tight focus of the incident beam. On the other hand, the Petrán and Hadravský scheme had the advantage of many spots of light scanning the specimen in parallel and allowed visual observation at high frame rates. Both methods have undergone major development, in particular, Boyde has brought the use of the latter so-called tandem scanning microscope to a fine art in the laboratory [546].

In 1977 a theory of the new confocal scanning microscopy was published [547] and the imaging properties and applications were discussed [548] a little later. In this arrangement an incident laser beam is highly focused onto the specimen and imaged at high aperture onto a pinhole at the centre of the image plane. A detector behind the pinhole records the signal as the specimen is scanned in three dimensions. Alternatively, in a reflection microscope, the laser beam may be scanned in two transverse dimensions and the specimen in the axial dimension. The theory shows that the transfer function is a product of the usual $J_1(r)$ for coherent imaging with a second $J_1(r)$ for the focused incident beam. If the illuminating lens and the objective lens have the same aperture (as will be the case in reflection) the resulting transfer function of $J_1^2(r)$ shows an improvement in resolution of about 1.4 in the half-width at half-height of the impulse response, as has been demonstrated experimentally [549]. In the incoherent-imaging case a transfer function of $J_1^4(r)$ is achieved with a similar resolution gain.

Both rotating-disc and scanning-beam microscopes are now in wide use, although the laser-beam scanning is usually done before the microscope eyepiece by use of high-speed galvanometric mirrors. The 'sectioning' property has enormous advantages in fluorescence microscopy and in the microscopy of near-transparent media such as the human eye.

18.4.10. The shape of the photon

As a final remark on this decade, we mention the 1969 paper of Amrein [550] in which he dealt with the impossibility (for reasons of gauge and Lorentz invariance first noted by Pauli [551]) of having a point photon, and constructed what he postulated would be the most tightly localized isotropic photon. After Jauch and Piron [552], he called it 'weakly localized' and found that it had power-law tails of energy density falling off asymptotically as the seventh power of distance. With the advent of single-photon experiments it has become experimentally feasible to investigate such localization properties, and new theories with various power-law tails which might be observed in the laboratory have been published recently [553,554]. The experimental measurement of the 'shape of a single photon' is a very difficult task but attempts are being made to see these tails [555]. Recent work by Hellwarth, Nouchi and Adlard (unpublished) disputes the Amrein claim for the most localized photon by putting forward two new explicit solutions with higher power-law tails, of order eight and ten, for the energy density, thus reopening the question of how small a photon can be.

One might wonder further about the fundamentals of the problem of photon localization. Some people have felt that the photon cannot exist in the free Maxwell field. Einstein commented in 1951 that 'All these fifty years of pondering have not brought me any closer to answering the question—what are light-quanta?' [556]: and he is not alone. In spite of many advances, tests of quantum theory and extensive theoretical research, it seems that the photon is still a puzzle.

It should not be forgotten that the photon could have a very small mass; it would then obey the Proca equations. A number of calculations have been performed during this century to put an upper bound on such a mass. In 1943 Schrödinger and McConnell [557] was able to put a bound of the order of 10^{-46} g on the mass of the photon by analysing data on the Earth's magnetic field, while in 1976 Goldhaber and Nieto [557] reduced this limit to 5×10^{-48} g using spacecraft measurements of Jupiter's magnetic field. If the photon had a mass, magnetic (and electric) fields would diminish exponentially with distance; the galactic scale of these experiments makes possible the extreme sensitivity.

18.5. 1970–94: a wealth of new results and applications

From 1970 to the present day (early 1995) optics has been applied to every imaginable problem. Computer-aided lens and system design has been established, catalogue shopping for optical and optoelectronic components has become the norm, with lasers widely available in the public domain. Powerful optical systems are relatively straightforward to design, build and test and we shall discuss this process in the final section of this chapter, but for now we will concentrate on the continuing fundamental developments.

Starting from the early 1970s, the liquid-crystal display rose to dominate optical-display technologies in the 1990s and was studied in universities and industries all over the world. From the optical physicist's viewpoint the need for precise calculation of the optical properties of liquid-crystal optical devices employing different materials and geometries was recognized early on. We have already mentioned two or three developments in the basic understanding of liquid-crystal optics throughout the century. In 1970 Berreman and Scheffer [558] laid the foundations for device calculations through their matrix treatment of liquid-crystal optics, based on Maxwell's equations. They used a 4×4 matrix formulation of the electromagnetic wave equations in stratified media to compute the reflectance and transmittance of single-domain, cholesteric, liquid-crystal films. The importance of their approach was that it can be applied to any system in which the director changes only along one direction, i.e. stratified media, the basis of most electro-optic display cells. Throughout the next 20 and more years there were widespread efforts to elucidate, both theoretically and experimentally, the optical non-linear and electro-optical properties of liquid-crystal materials and devices, driven by the desire to use them in ever more complex and capable liquid-crystal flat-panel displays for televisions and portable computers, etc [91].

Also in 1970, Esaki and Tsu [559] published their seminal theoretical paper on quantum-well superlattices, considering many quantum wells grown on top of one another, with barriers made so thin that tunnelling between wells becomes important, and a superlattice with minibands and minigaps in the band structure is formed. This concept was to prove central to much semiconductor optoelectronic device design in the 1980s and 1990s when fabrication techniques had improved enough to make such structures reliably.

Novel lasers continued to appear throughout the 1970s. The excimer laser was born in 1970, when Basov *et al* [560] demonstrated lasing in Xe_2 molecules at 176 nm. New uses were found for lasers, such as an optical 'bottle', when Ashkin [561, 562] showed that the photon pressure of two opposing argon-ion laser beams (128 mW at 514 nm) could trap 2.68 μm latex spheres in water, a development leading more recently to 'optical tweezers' for manipulating objects under a microscope [563].

Non-linear optics continued to flourish. Lasers were an exciting new way of exploring the rich range of non-linear phenomena possible with two-level atomic systems, for example saturation effects, power broadening, Rabi oscillations, optical Stark shifts and dressed states [564, 565].

18.5.1. Progress in astronomy—laser guide stars

Experiments with laser speckle [566] and other speckle phenomena were also in progress at this time and already an important result for optical astronomy had been obtained. Labeyrie [567] attained diffraction-limited resolution in a large telescope by analysis of speckle patterns in star images, a result which started an intensive period of research. The original images were apparently of poor quality, yet by evaluation of ensemble-averaged autocorrelation functions of many short exposures it proved possible to recover from the speckle interferogram a diffraction-limited autocorrelation function of the stellar field. An outline theory is as follows.

For exposures of less than about 50 ms the blurred image $I_n(x)$ is the convolution of the astronomical object field $O(x)$ with the combined point-spread function $P_n(x)$ of the telescope and a snapshot of the turbulent atmosphere, $I_n(x) = O(x) * P_n(x)$. Fourier transformation of the raw data (the intensities in the image plane of the telescope) $I_n(x)$ gives

$$\tilde{I}_n(u) = \tilde{O}(u)\tilde{P}_n(u).$$

Taking the modulus square of $\tilde{I}_n(u)$ over many frames of data yields

$$\langle|\tilde{I}_n(u)|^2\rangle = |\tilde{O}(u)|^2 \langle|\tilde{P}_n(u)|^2\rangle.$$

The speckle-interferometry transfer function $\langle|\tilde{P}_n(u)|^2\rangle$ is positive and non-zero for all frequencies up to the diffraction limit. Since $\langle|\tilde{P}_n(u)|^2\rangle$ can be found by measurement of an isolated star, the object power spectrum $|\tilde{O}(u)|^2 = \tilde{O}^{(2)}(u)$ can be obtained, providing a diffraction-limited object autocorrelation function $O(x)*O(x)$ after inverse Fourier transformation. Later developments involving measurements of triple-correlations [568] and employing speckle masking and deconvolution [569] led to methods for unravelling true diffraction-limited images from blurred astronomical observations. Speckle-interferometry pictures before and after processing are shown in figure 18.9. Such techniques, however, have largely been superseded by laser beacon and adaptive optic techniques used with multi-element optical telescopes [570–573] developed from the 1970s onwards. The largest of these is the Keck telescope with a 10 m segmented primary mirror. A nearby powerful laser beam illuminates the upper atmosphere causing sodium fluorescence to give the telescope a reference 'guide star' which it can use to adapt its flexible surface to obtain the best image, thus cancelling out the wavefront distortions arising from atmospheric variations lower down. The servo-mechanical system must operate in less than 100 ms in order to 'freeze' these fluctuations. The performance limits of this technique are being fully assessed at the time of writing; they make severe demands on charge-coupled device (CCD) cameras for detection and on the mechanical transducers for driving the mirror segments.

18.5.2. Photon-correlation spectroscopy

The birth of digital photon correlation spectroscopy dates back to a conference paper by Foord et al [574] (Malvern, UK) at about the same time as Benedek [575] described experiments in 'light-beating' spectroscopy by analogue means. Photon counting statistics of the new light sources had been widely explored from 1963, but the Malvern group recognized that much was to be gained from direct electronic computations of Glauber's $G^{(2)}$, the second-order temporal correlations in photon counts, in place of simple photon counting. A theory of optical spectroscopy by 'clipped' digital autocorrelation of photon-counting fluctuations, in which the measured photon correlation function is related to the optical spectrum, was described by Jakeman and Pike in 1969 [576]. The first photon correlator to perform such computations was used in 1970 to measure the size of the molecule haemocyanin [577, 578]. The autocorrelation function for scattering from a monodisperse suspension

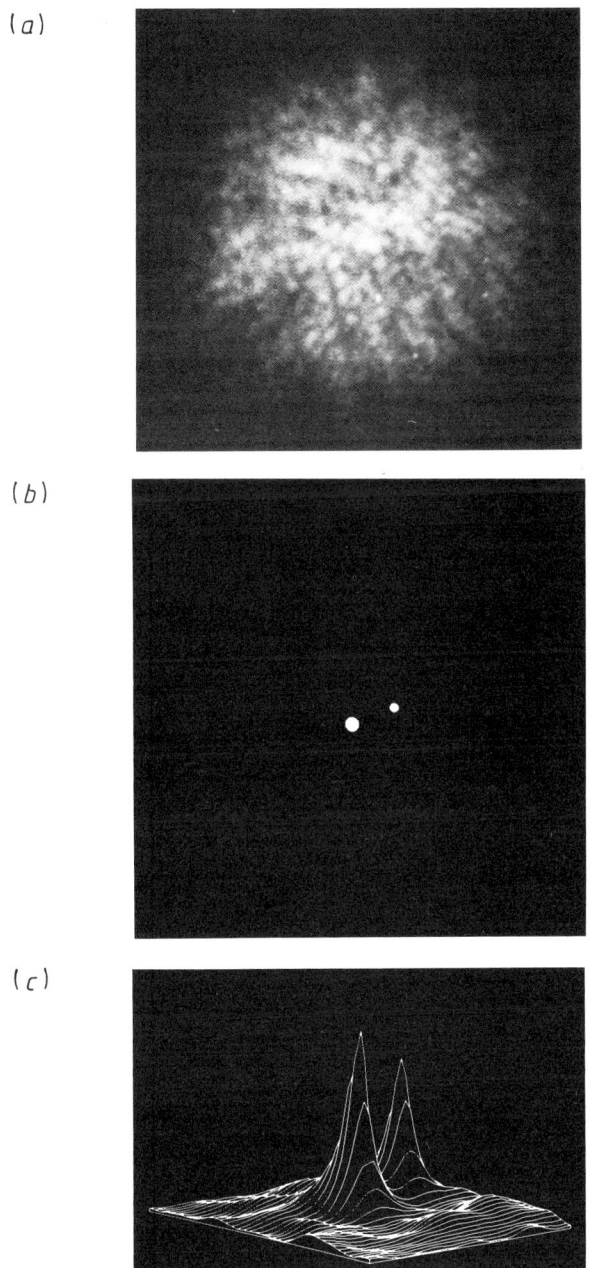

Figure 18.9. *Speckle-masking measurement of the close spectroscopic double star ψ Sagittarii. (a) One of the 150 speckle interferograms evaluated; (b), (c) show the reconstructed true image of ψ Sagittarii.*

is a single exponential function whose decay rate is proportional to the hydrodynamic size of the molecule. The clipping technique was an analogue of the well-known 'hard limiting' of signals borrowed from radar and allowed the high-speed digital multiplications necessary for the formation of the correlation function to be made in real-time.

The photon correlator provided a seven-decade leap in optical resolution, down to a few Hertz, with the implicit cancellation of carrier frequency fluctuations. Commercial construction of digital photon correlators and their worldwide use followed quickly, with application to the analysis of macromolecular suspensions (proteins, enzymes, viruses, polymers, etc), viscosity, thermal diffusion, mutual diffusion, air and fluid velocity flows and turbulence and many other problems with moving atoms, molecules or macroscopic objects [579, 580]. The technique was particularly sensitive for the non-invasive measurement of transonic and supersonic air flows, by making use of the Doppler shifts from natural contaminant particles. By this means accurate flow patterns could be obtained, to compare with the results of the rapidly developing computer codes used in the design of internal-combustion engines, turbocompressors, aerofoils and jet engines (including the Concorde and the Rolls-Royce RB 211) [581–583]. Figure 18.10 shows a fast 'clipped' photon correlator of this era. Some 15 years later the speed of digital electronics had advanced sufficiently for the clipping to be dispensed with, to give some gain in accuracy, and recently the whole operation has been performed with software using a single high-speed CPU chip in the expansion slot of a portable computer [584]. Such a single-board correlator was used, for example, in the recent space-shuttle experiments on critical xenon. Recent thin-film experiments have allowed particle sizing in 'black media', of great interest to the oil and automotive industries for the analysis of soots and asphaltanes [585].

Gabor received his Nobel Prize [586] for the invention and development of holography (perhaps a little late) in 1971.

Electron-beams now began to feature prominently in laser research when Madey [587] proposed a free-electron laser. Coherent radiation was to be generated by a relativistic electron beam moving through a periodic, transverse, DC magnetic field. The electrons 'see' virtual quanta, and Compton-scatter them into real short-wavelength photons. Such lasers have now joined the rest of the laser armoury, and have broad tunability but, as they require large accelerator facilities, they are not convenient for everyday use. A very recent theoretical development by Bonifacio and De Salvo [588] seeks to combine the free-electron laser and conventional gas lasers into a Collective Atomic Recoil Laser. The possibility of building such a laser for the generation of very short wavelengths is being studied at the Stanford Linear Accelerator Laboratory.

In contrast with this type of research, interest in the possibility of optical super-resolution had been growing since the stimulation of

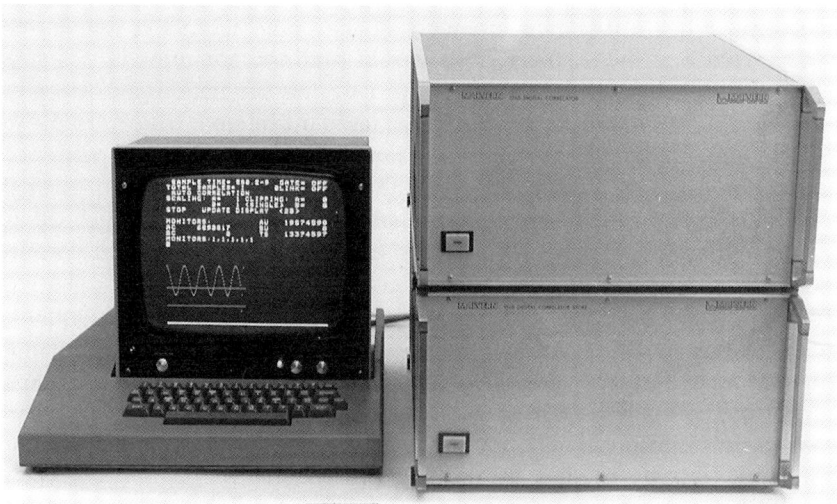

Figure 18.10. *A 100 MHz photon correlator of the early 1970s.*

di Francia's [262, 263] publications, but was still not properly understood. An experimental spur came in 1971 with the demonstration by Smith and Yansen [589] of super-resolving pupil functions by employing three concentric annuli at the lens. In the following year McKechnie [590] described the effect of condenser pupil obstruction on the two-point resolution of a microscope. Improved resolution can be obtained without modifying the point-spread function of the objective and thus degrading image quality, but another decade elapsed before such empirical observations were explained more rigorously.

In 1972 Freedman and Clauser published [591] another in the sequence of experiments, which still continue, to investigate Einstein–Podolsky–Rosen(–Bohm) (EPR(B)) paradoxes in quantum theory. In the EPR(B) *gedanken* experiment a source emits a pair of particles in a singlet state. After the particles have separated and travelled some distance from each other, the correlation properties of their spins (polarization in optics) are measured for different orientations a and b. Each measurement can yield two results, which we designate ± 1. For photons, a measurement along a yields $+1$ if the polarization is parallel to a, and -1 for the opposite condition. For a singlet state, quantum mechanics predicts a correlation between such measurements on the two particles. If $P_{\pm\pm}(a, b)$ are the probabilities of obtaining the result ± 1 along a and along b for the two particles then

$$E(a, b) = P_{++}(a, b) + P_{--}(a, b) - P_{+-}(a, b) - P_{-+}(a, b)$$

is the correlation coefficient of the two measurements. Bell worked out what would happen if such correlations were caused by independent

See also p 229

local properties of the two members of the pair having left the source. He showed that these correlations would then be constrained by certain inequalities that are not always obeyed by the predictions of quantum mechanics. As was shown in 1963 by Glauber, the quantum equivalents of classical probabilities for pure states can be singular and negative, particularly so for number states of the EPR type.

The Freedman–Clauser experiment was a test of local, hidden-variable theories, involving measurement of linear polarization correlations of photons emitted in an atomic cascade of calcium. The observed correlations were markedly stronger than allowed by Bell's inequality with a high statistical accuracy, and provided strong evidence against hidden-variable theories. However, since any tiny loophole to dispute quantum mechanics is always seized upon in this field, it has to be added that this conclusion relies upon an assumption that the polarizers do not somehow amplify the beams! These experiments, confirming the earlier results of Hanna and others, and convincing as they are, still did not end the quest. All other subsequent similar experiments, however, have fortunately also come down firmly on the side of quantum theory!

See also p 1413

Non-linear optics continued to develop, including especially the first studies of harmonic generation of light in (cholesteric) liquid crystals [592] (from which grew an interesting sideline in liquid-crystal research [90, 91]) and the realization that a pulse in an optical fibre can become a solitary wave (soliton) since the wave envelope satisfies the non-linear Schrödinger equation [593]. The confirmatory experiments of Mollenauer *et al* [594] led to the soliton-pulse scheme being demonstrated for telecommunications over thousands of kilometres of optical fibre. It is still unclear whether this mode of operation will supersede the current arrangement which uses incoherent fast pulses.

18.5.3. All done by mirrors
At this time, 1972, another non-linear effect, phase conjugation or optical wavefront reversal in reflection by stimulated Brillouin scattering, was discovered by Zel'dovich *et al* [595] in Moscow, and proved valuable for correction of aberrations in laser-beam wavefronts. The same result can be achieved by a 'dynamic hologram', formed by the incident and pump beams [596–598].

The mathematics of lens aberrations continued to advance when Buchdahl's studies of the Lagrange invariant in relativity theory [599] led him to the first of at least seven papers on the Lagrangean aberration theory of systems lacking all symmetries [600].

One final contribution of note from 1973 was Yariv's paper giving a coupled-mode theory for guided wave optics [601], now the basis for calculation of directional coupling, acousto-optic interactions, distributed feedback lasers and electro-optic modulation in dielectric waveguide geometries.

A History of Optical and Optoelectronic Physics

The early to mid-1970s also saw much study of non-linear optics using a two-level-atom model [602]. The optical Bloch equations are an important approximation for calculations of such atom-field interactions in quantum optics. They are satisfactory when quantum fluctuations are small, and comprise three coupled equations for the field and for the diagonal and off-diagonal atomic operators. In a very recent development [603] it has been shown that extra terms representing local dipole-field effects must be inserted into these equations when the atomic density exceeds about 10^{14} atoms cm^{-3}. In this case some interesting new phenomena arise, including enhancement of inversionless gain and index of refraction, intrinsic optical bistability and what have been christened 'piezophotonic effects', changes in sign of the absorption curve at certain values of the density.

In 1974 Bjorkholm and Ashkin [604] showed very clearly the effects of self-trapping of a laser beam in an atomic sodium vapour. The tendency of diffraction to cause the laser beam to expand is offset by the self-focusing effect and now the laser beam maintains a diameter d over a distance much longer than the usual longitudinal extent (d^2/λ) of the focal region of a beam of that diameter. In 1975 Dingle et al [605] verified, using optical experiments, the formation of minibands in a superlattice, by the coupling of quantum wells as Esaki and Tsu had described in 1970. In 1976 Wolf [606] published his new theory of energy transport in free electromagnetic fields, a rigorous analysis of radiometry for statistically stationary and homogeneous fields. The theory is distinguished from previous work by the use of invariants of electromagnetic correlation tensors for the energy density and Poynting vectors. At this time also, Yariv described how to transmit three-dimensional image information along optical waveguides [607] by non-linear optical mixing.

18.5.4. Photon antibunching

1977 was a busy year, marked in particular by results from the first of a number of 'non-classical-light' experiments, building on the work of Hanbury Brown and Twiss and of Glauber. The value of the normalized equal-time Glauber intensity correlation function at a single point, $g^{(2)}(0)$, is a function of the photon statistics of the field. For a thermal field $g^{(2)}(0) = 2$, but for a coherent field $g^{(2)}(0) = 1$. The enhanced value of $g^{(2)}(0)$ for a thermal field over a coherent field occurs because such photons bunch together with a distribution in each mode

$$p(n) = \bar{n}^n/(1+\bar{n})^{n+1}$$

where n is the photon number and $\bar{n}$ its average. Coherent-state photons are not bunched in this manner, but have a random Poisson distribution. A value of $g^{(2)}(0) < 1$ is not possible for classical populations unless the sampling were to remove individuals from the population once they

See also p 1421

have been counted [608]. Such a sampling scheme 'without replacement' is a feature of photodetection and is built into the definition of $g^{(2)}(\tau)$, since the counted photon is always annihilated. It would be unusual to 'remove' members of a classical population once they had been measured (although this does happen in 'flux' measurements) and this procedure in any case would not give the 'true' statistics. When $g^{(2)}(0) < 1$ the (sub-Poissonian) photon-number distribution is narrower than for a coherent field and only occurs with 'non-classical' light. For certain non-classical light fields it is possible for the photons to be antibunched but not sub-Poissonian. Such a distribution would have $g^{(2)}(0) > 1$ but $g^{(2)}(\tau)$ would exceed the value of $g^{(2)}(0)$ for some positive value of τ.

The first experimental claim of observation of photon antibunching came from Kimble, Dagenais and Mandel [609], a year after its theoretical prediction. They conducted a resonance-fluorescence experiment in which sodium atoms were continuously excited by a dye-laser beam, and reported a value for $g^{(2)}(0) < 1$ with the correlation function increasing in value as the delay time increased from zero.

The critics, without doubting the basic effect, pointed out that the result was masked by the Poisson statistics of atom arrivals in the measurement volume and that the claimed antibunching effect had therefore not strictly been observed [610, 611]. The increasing correlation coefficients with delay, however, are non-classical. In 1983, Short and Mandel [612] observed sub-Poissonian statistics in selected photon *detections* from single atoms, in the same process of resonance fluorescence, but this time by gating the photodetector by emissions from the atom before it entered the sampling region. Although this removes the effect of atom-number fluctuations from the detected light, the light beam itself, as before, is not sub-Poissonian. It was to be eight years before Teich and Saleh [613] created a weakly sub-Poissonian Franck–Hertz source at 253.7 nm from mercury vapour excited by inelastic collisions with a 'quiet' space-charge limited electron beam, and Jakeman and Walker [614], using a Burnham–Weinberg parametric down-conversion arrangement with a feedback loop triggered by a counter in one arm to stop any following events for a short period, produced antibunched light in the other beam. Other demonstrations followed [615, 616].

Another result in non-linear optics was phase conjugation by degenerate four-wave mixing [617–619]. In this scheme four interacting waves exist in a non-linear medium, each of the same frequency (hence degenerate). A lossless medium characterized by a third-order, non-linear susceptibility $\chi^{(3)}$ is illuminated by two strong counter-propagating pump waves E_1 and E_2, and by a signal wave E_3 at the same frequency. The non-linear coupling of these waves creates a new wave E_4 that is the phase conjugate of E_3 [617]. The idea was widely pursued for its ability to remove the effects of aberrations from optical systems [620] and to compensate the atmospheric degradation of optical-beam transmission [621]. An historical survey has been published by Pepper [622].

Another non-linear optical effect, photorefractivity, was the subject of much scrutiny from about 1977 with the publication in western literature of a model by Kukhtarev *et al* [623]. Earlier work had been published in the Soviet Union [624]. The photorefractive effect is the change in refractive index of an optical material that results from the optically induced redistribution of electrons and holes. Unusually it cannot be described by any non-linear susceptibility $\chi^{(n)}$ because, under a wide range of conditions, the change in refractive index in the steady state is independent of the intensity of the light that induces the change. The mechanism involved is that two laser beams create a photorefractive hologram when their interference pattern is incident on a photorefractive crystal. Charge carriers are preferentially excited in the bright fringes and diffuse to regions of low intensity. The electric field of this space charge acts through the linear electro-optic effect to form a volume holographic refractive index grating Δn. Applications such as memories for optical neural network and optical information storage were developed in the 1980s and 1990s.

Around this time, 1978, there was much activity in optical signal and data processing [625, 626] and many scientists were engineering ever more complex optical systems to demonstrate mathematical functionality. Through the 1980s and 1990s this field of activity grew with (smart) spatial-light modulator technology [627], optical computing, interconnects and learning machines, as well as optoelectronic neural networks [628]. There was also lively interest in the reconstruction of objects from the modulus of their Fourier transforms, that is to say, attempts to solve the phase-retrieval problem of optical-coherence theory, and to obtain high-resolution imagery from optical data by iterative processing. Earlier work by Landweber [629] and Fienup [630, 631], was improved by Walker [632] who pointed out that full information on phase was available from the original image plus a second taken through a simple exponential-amplitude filter. 'Blind' deconvolution, in which both image and aberration function are recovered was shown to be possible by Cole as early as 1974 [633] and much more recently by Ayers and Dainty [634]. Motivation for the technique was the recovery of sound from old wax recordings, but it was immediately applied also to blurred images. It may seem like magic to recover both the image and the function which blurred it without knowing either, but the constraints of the imaging process are nearly always sufficient to allow a unique decomposition of the two. It was applied successfully to the unfortunately blurred images of the Hubble space telescope [635] before the remarkable COSTAR mission to correct the optics. In spite of great efforts to define the point-spread function, it could never be specified exactly, and other deconvolution techniques such as the popular Richardson–Lucy iterative deconvolution algorithm [636, 637] always showed some residual rings. The Richardson–Lucy and blind-deconvolution methods have been combined successfully in more recent work [638, 639].

18.5.5. Femtosecond pulses
At the turn of this decade there was also great interest in ultra-short laser pulses and their applications, from non-linear optics to biophysics, especially the development of mode-locking to achieve sub-picosecond pulses [412, 640, 641]. Ultra-short laser pulses were soon a widely available facility [642, 643]. Later developments in mode-locking techniques created pulses of a few femtoseconds, no longer than a few wavelengths of light [644].

At this time also Jakeman and Pusey [645] introduced the K-distribution into the explanation and analysis of certain light-scattering experiments. This two-parameter model allowed description of intensity statistics in the non-Gaussian transition region between log–normal and negative-exponential statistics, and was applied to the scintillation properties of laser beams propagating through atmospheric turbulence (figure 18.11). K-distributions represent a universal class of distributions which generalize the central-limit theorem to the case where the component distributions have infinite variance. They were first noted to occur in six separate and quite different experiments: scattering of laser light from a turbulent layer of nematic liquid crystal, turbulent air and water layers, atmospheric turbulence, scattering (scintillation) of starlight in the upper atmosphere and microwave scattering from a rough sea surface. This gave the clue to the existence of a new central-limit mechanism, and received much attention during the next decade [646].

Trapping of particles by a laser beam, and cooling and storage of atoms, was discussed by Letokhov and Minogin [647] who proposed to use a laser standing wave at a resonance frequency of the atom. Under certain conditions the laser radiation was expected to cool atoms in a low-pressure gas to about 10^{-4} to 10^{-3} K, and because of gradient-force effects the cooled atoms could be held in the light field for a long time (optical molasses) to allow precision spectroscopic studies. Developments and applications of this type of experiment have been widespread [648].

Compact optical discs [649] for recording sound and video started to attract widespread public interest in the late 1970s for the volume of information they could store and the quality of sound that it was possible to reconstruct from them. In 1979 Hopkins [650] modified his earlier diffraction theory [270] of optical images, based on the coherence of object illumination, to provide a theory of laser read-out systems for optical video discs [649].

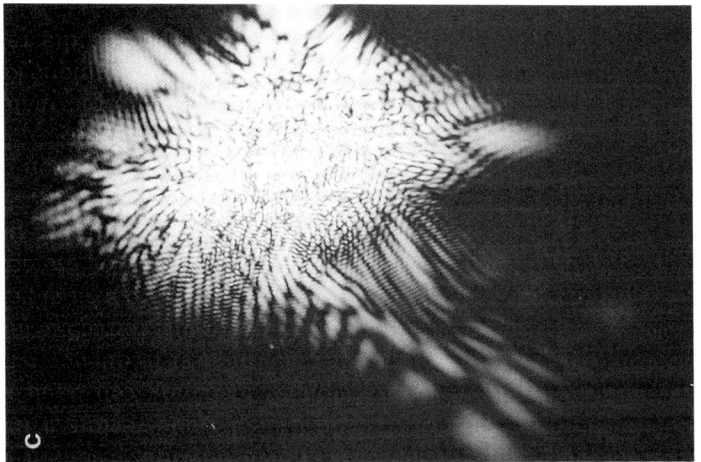
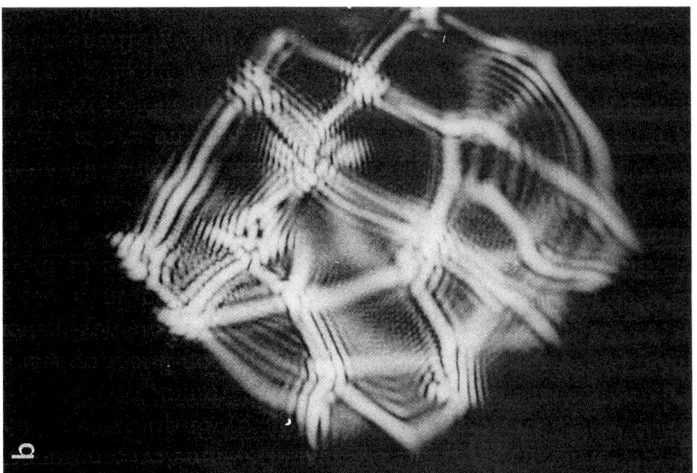
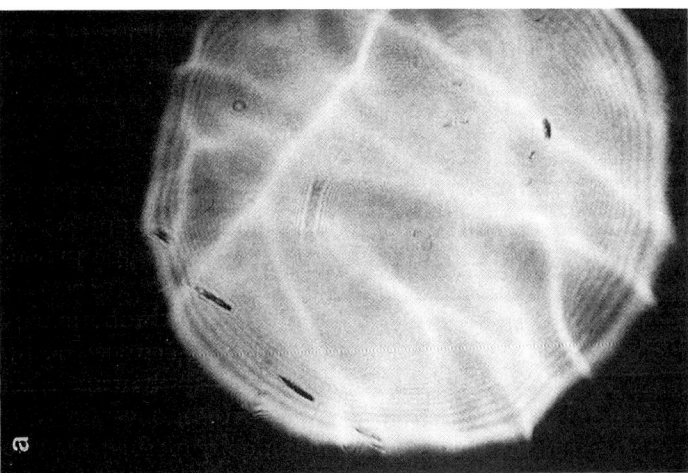

Figure 18.11. *Scintillation in laser-beam propagation. (a) Near field. (b) Caustic region. (c) Far field.*

18.5.6. The impact of the semiconductor laser diode

The double heterostructure laser diode [651] which had been developing since the early 1960s became the workhorse of optical-disc readers after about 1980. Typically, the double heterostructure laser diode comprises a thin (~ 0.1 μm) region of GaAs sandwiched between two regions (n and p type) of AlGaAs, grown epitaxially on a crystalline GaAs substrate. Hundreds of millions of optical-disc systems will be manufactured for music, video and data storage before the end of the twentieth century; lasers and modern optics in compact optical-disc systems (CD) are now household items alongside television. In 1993 over 100 million CD systems were sold globally.

A new concept in laser diodes appeared in 1978–79, the vertical-cavity surface-emitting laser (VCSEL) invented by Iga and colleagues at Tokyo Institute of Technology [652]. Unlike all previous laser diodes, which emitted their light in the plane of the substrate (*edge emitting*), VCSELs emit their light *perpendicular* to the substrate and offer some valuable attributes: low threshold currents, single-mode laser beams of circular cross section, and the potential of fabrication in large arrays, perhaps thousands of lasers per square centimetre. This style of laser is constructed by growing mirror stacks in semiconductor materials, placed either side of an active semiconductor region, to form a distributed Bragg-reflector cavity [653]. The technology of the 1990s allows the practical realization of this new concept.

18.5.7. X-ray lasers

The first reliable reports of an x-ray laser were by Pert and his colleagues at the University of Hull (UK) [654] when they announced laser gain at 18.2 nm in a highly ionized carbon plasma made by vaporizing thin carbon fibres with intense infrared laser pulses. Three-body recombination preferentially fills the higher states and produces a population inversion. While other workers were using foils, Pert realized that an exploding cylindrical expansion cools the plasma much faster than a planar one, enhancing the recombination probability and not populating the ground state [655]. X-ray lasers at shorter wavelengths soon followed [656]. Although unpublished, it is known that a nuclear-bomb-powered x-ray laser was developed around this time by the Lawrence Livermore Laboratories at an underground test site in Nevada. Hora and Miley have reviewed lasers for fusion (including 'Nova' at Lawrence Livermore Laboratory (figure 18.12) and 'Antares' a large CO_2 laser system at Los Alamos Laboratory) and nuclear-pumped lasers [657].

In 1984, at a joint press conference, groups from Lawrence Livermore and Princeton reported new 'laboratory' x-ray lasers. The first group used 260 ps pulses from two beams of the high-power Nova laser directed at a 75 nm-thick film of selenium on a plastic backing; they found strong lines at 20.6 and 20.9 nm [658]. They confirmed

A History of Optical and Optoelectronic Physics

(a)

(b)
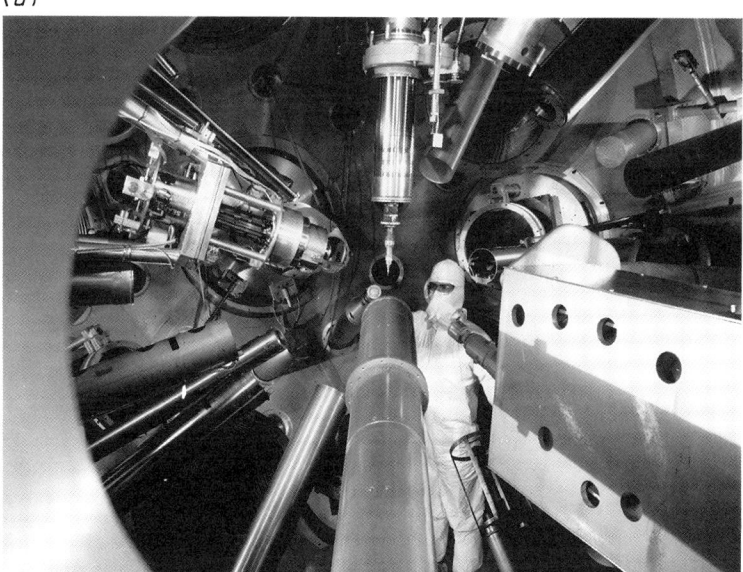

Figure 18.12. *The Nova laser at Lawrence Livermore Laboratory. (a) The main laser hall. (b) The target chamber.*

correct scaling of laser amplification, thus giving perhaps the first incontrovertible evidence for x-ray laser action. The transition which gave rise to the amplified spontaneous emission was $2p^53p$–$2p^53s$. The mechanism had been described by Zherikhin, Koshelev and Letokhov in 1976 [659] and refined by Vinogradov and Shlyaptsev in 1983 [660]. The Princeton group generated 18.2 nm [661] radiation with the same carbon line as Pert, but trapped the plasma in a 90 kG solenoidal magnetic field for long enough to allow recombination. More recent work has moved the limits of laser action down into the 3.5 nm region where there is hope of significant applications to biology by imaging living cellular material.

At this time the planar laser was demonstrated [662] giving a 360° plane coherent emission of 3 mJ pulses from rhodamine 6G dye. Spherical emission lasers were also being considered.

Diffraction was again newsworthy. Petit's 1980 text [663] presented a summary of the rigorous diffraction theory of gratings of all types. Since that time computer software based on the mathematics in this text has been developed and made available by Meystre in a computer program 'Reseau 2000' that is invaluable for diffraction-grating computations.

A rather different type of diffraction study was being conducted by Berry and colleagues [664] at Bristol University who since the mid-1970s had pushed forward the theory of optics on the 'semi-classical' boundary between waves and rays. They investigated so-called 'diffractals' and caustics and introduced catastrophe theory into this area of optics. Their studies centred on the focusing or converging of optical rays passing through random media. The classical perfect point or line foci do not exist in practice because of diffraction, and in any such optical system rays focus to form bright contours or *caustics*. Berry and co-workers explained these caustics and their diffraction patterns in terms of catastrophe theory. Such patterns are commonly produced on the floors of swimming pools when light is focused by the rippling water surface. Scintillation is also used [665] to explain the twinkling of stars and, together with K-distributions, to explain laser-beam break up, figure 18.11, due to propagation through the atmosphere [666].

Bloembergen and Schawlow received their rewards for changing the face of optics in the early 1960s, receiving Nobel Prizes [38] in 1981 for their contributions to laser spectroscopy. Also the irrepressible Hopkins brought together the techniques of ray tracing and diffraction theory to describe the process of image formation in optical systems [667]. His new ray-tracing formulae were applicable to any image-forming system; they avoided previous indeterminacies for infinite pupils and were easily applied in computer programs.

18.5.8. Instabilities and chaos

In the late 1970s and early 1980s interests began to focus on laser instabilities and chaos. Spiking, i.e. unstable emission, had been present in the first experiments with a ruby laser, but now various theorists, notably Ikeda *et al* and Eberly *et al* [668, 669] began to explore conditions for unstable and chaotic emission. The period-doubling route to chaos in lasers was predicted and later observed in an He–Ne laser [670]. Temporal chaos and optical bistability [671] in optical systems was observed in various experiments with single-mode and multimode lasers. Spatial instabilities were studied later, especially transverse modulational instabilities in non-linear media and laser-beam profiles with static, dynamic or chaotic modulation [672,673]. Another spectacular development was the observation of polarization instabilities [674], showing both first- and second-order spontaneous polarization symmetry breaking. Zheludev [675] has reviewed polarization instability and chaos and Knight [345] has given a useful account of optical chaos.

Elsewhere Pippard [676] examined some basic questions of physical and psychological optics at the limits of resolution and contrast, including a demonstration that the eye is at least four times more sensitive to contrast than is implied by the Rayleigh criterion, which he reminds us was no more than an arbitrary convenient rule, unsupported by experiment. We also saw another experimental realization of the EPR *gedanken* experiment and a new theory of resolution in diffraction-limited imaging appeared.

To discuss the famous EPR *gedanken* experiment we return again to higher-order interference experiments, involving correlations of photon detections, which can reveal information about the quantum properties of the electromagnetic field. The old fundamental questions in quantum mechanics of the EPR paradox and the so-called 'Bell inequalities' were addressed in this way by Aspect *et al* [677], confirming the earlier experiments already mentioned. As did Freedman and Clauser, they measured the linear-polarization correlation of pairs of photons emitted in a radiative cascade of a calcium atom, this time using beam-splitting polarizers, which closes the loophole of 'amplifying filters' of the Freedman–Clauser experiments mentioned above. The results again show excellent agreement with the predictions of quantum mechanics and violation of Bell's inequalities. The meaning of the quantum theory thus still attracts interest. Having solved every 'paradox' brought to him some 60 years ago Bohr is probably turning in his grave! Following Bell, Greenburger *et al* [678] have derived an even sharper test in three-photon interferometry where the result is -1 in quantum mechanics and $+1$ in classical theory, and which has already been verified experimentally [385]. Quantum mechanics wins again!

18.5.9. Imaging theory: from eigenfunctions to singular functions
Another advance in 1982 was the appearance of a rigorous mathematical treatment of resolution in diffraction-limited imaging. For many years physicists had proposed methods of attaining *super-resolution* to overcome the classical Airy diffraction limit and Rayleigh criterion [414–416,679–684]. The contribution and advance by Bertero, Pike and co-workers [685,686] was to recognize that object and image domains can

differ significantly from the correspondence determined by geometrical optics, particularly when there are a small number of Nyquist samples (degrees of freedom) in the object, since then diffraction strongly blurs the boundaries of an image. The appropriate mathematics introduces the singular functions and singular values of a first-kind Fredholm equation, and provides a generalization of the eigenfunctions and eigenvalues used by Slepian and Pollack [357]. Using this theory, Bertero and Pike analyse many different cases and find the limits of resolution in each case as a function of noise in the detected image. The Rayleigh criterion could be robustly improved by a factor of two in the coherent-imaging case, and similar significant practical gains could be made in incoherent-imaging systems by the use of new optical arrangements. These effects have since been demonstrated [687] and will be of importance in future high-resolution imaging and microscopy.

Throughout all the new activity in quantum optics, optical components and systems based on lasers, and military applications were being rapidly developed around the world. New optoelectronic devices such as the staircase avalanche photodiode [688] were introduced, employing conduction-band steps in a multilayer geometry to encourage electron ionizations at discrete locations, and to decrease the hole–electron ionization ratio to reduce noise. The bulk of the new optical physics, however, came from the quantum-optics community.

18.5.10. Squeezed light

The experiments on photon bunching, antibunching and sub-Poissonian statistics had investigated intensity (photon-number) fluctuations of the electromagnetic field; in terms of oscillator momentum and position variables p and q in suitable units, these are fluctuations of $p^2 + q^2$. If a harmonic oscillator in its ground state suddenly has its origin displaced and its potential energy lowered it can be forced into a coherent state, in which it has a Gaussian wavefunction which oscillates back and forth between the classical turning points without changing its shape. If, at the same time its frequency is changed, it goes into a *squeezed* state where the width of the Gaussian has an extra 'breathing' motion in phase with the oscillation. For a coherent state $\Delta p = \Delta q$ but for a squeezed state $\Delta q = s^{-1/2}$ and $\Delta p = s^{1/2}$, where s is the *squeezing parameter*, and the uncertainty in one of these variables can be lower than that of a coherent state with a compensating increase in the other to satisfy Heisenberg's uncertainty principle $\Delta p \cdot \Delta q = 1$. During the 1980s considerable theoretical effort was focused on fluctuations in these *quadrature amplitudes* of the electromagnetic field [689]. Since the squeezed states of light have quantum noise fluctuations which can be less than those of the vacuum state of the elecromagnetic field, no semi-classical theory for them exists.

The original paper on squeezing was written by Stoler in 1970 [690] who showed that minimum-uncertainty packets are unitarily equivalent to the coherent states and that coherence is stationary minimality. A discussion of the properties of minimum-uncertainty packets was published soon after by Yuen, who introduced the concept of squeezing as a two-photon coherent state of the radiation field [691]. The experimental demonstration of squeezing had to wait until 1985, but this only came after many theoretical papers [689] and discussions of possible experimental arrangements.

Quantum non-demolition (QND) measurements, which had their origins in the detection of gravitational radiation, were also in progress. The problem of inventing a technically realizable QND scheme had first been posed at least as early as 1974 [692]. Later, in 1980 the problem of gravitational radiation detection was discussed in relation to QND by Braginsky et al [693] and by Caves et al [694]. QND measurements involve careful choice of coupling between the system and the measurement apparatus to avoid the back-action noise of the measurement. One can, for example, construct a scheme where the measurement of the position of a particle is not QND but the measurement of momentum is. Measurement of position then changes momentum which changes subsequent position. However, measurement of momentum changes position which does not affect subsequent momentum. On the other hand, one can postulate a coupling which has the opposite effect [695]. Milburn and Walls proposed a four-wave mixing scheme by which QND may be achieved [696] and other QND experiments soon followed [697, 698]. One of these experiments involved non-linear interactions in an optical fibre that coupled sideband modes of two strong pump waves at different frequencies, while the other was a rather simpler arrangement of optical parametric down-conversion; both achieved substantial noise reductions.

By 1984, work on experimental production of soliton pulses [594] had matured into the first successful soliton laser [699], a totally new device for the production of ultra-short light pulses. Pulse compression of solitons in an optical fibre which is part of the laser's feedback loop, force the laser to produce pulses of definite shape and width, and this was the first source of femtosecond pulses in the near infrared. Interest in solitons has continued to develop both for potential telecommunications use and for the underlying physics [700–705]. At this time another technique with potential for optical logic and computing was being investigated: the quantum confined Stark effect in AlGaAs/GaAs multiquantum wells [706], which was to lead to new bistable optical switching devices (SEEDs) and new architectures for optical computation [707].

The first experimental demonstration of squeezed states of light was published in 1985 by Slusher et al [708], who achieved 7% noise reduction corresponding to nearly 20% squeezing of the vacuum-fluctuation noise power by using non-degenerate, four-wave mixing with sodium atoms in an optical cavity. The following year saw two alternative methods for generation of squeezed light, by four-wave mixing [709] in an optical fibre and by parametric down-conversion [710], a particularly successful technique. Squeezing was later discussed as a possible means of enhanced optical communication [711] and improved interferometry.

Through 1985 and into 1986, Capasso and his colleagues [712] were demonstrating a new type of photoconductivity based on quantum-mechanical tunnelling in superlattices. Since the tunnelling probability of carriers through the barrier layers of a superlattice increases exponentially with decreasing effective mass, electrons are transported through a superlattice much more easily than heavy holes. When light is shone on the superlattice, the photogenerated holes remain localized in the wells while the electrons can tunnel through the barriers. The photoconductivity is controlled by tunnelling and, since the electron mobility perpendicular to the layers depends exponentially on the superlattice thickness, the electron-transit time, the photoconductive gain, and the gain-bandwidth product can be artificially varied over a wide range. This allows greater versatility in device design than is available with standard photoconductors [713].

Sequential resonant tunnelling through a superlattice was also soon observed [714]. With weakly coupled quantum wells because of relatively thick barriers, the electron states are well described by the quasi-eigenstates of the individual wells. If a uniform electric field is now applied and the bias increased, at some point the potential drop across a barrier equals the energy difference between the first two levels of the quantum well. Resonant tunnelling then occurs between the ground state of one well and the first excited state of its neighbour. The excitation energy is carried away by phonons; the process is repeated sequentially through the superlattice. Device applications of this effect in a laser had been proposed 15 years earlier [715], but were not to be realized until 1994 [716].

An especially interesting observation in 1986 was the experimental verification of Berry's topological phase. Two years previously, he had described [717] how the state of a quantum system depends not only on its present location and physical parameters but also on its history: a quantal system in an eigenstate, after being slowly transported around a closed trajectory by varying the parameters in its Hamiltonian, will have acquired a geometrical phase factor in addition to the usual dynamical phase factor. Chiao et al [718] demonstrated that in a helically wound, single-mode, optical fibre a photon of given helicity could be adiabatically transported around a closed path in momentum space. The consequent rotation of the plane of linearly polarized light in the fibre is independent of deformations in the fibre path if the solid angle of the path in momentum space stays constant. This is the topological feature of Berry's phase.

18.5.11. Optical-fibre sensors and telecommunications take hold
Monomode fibre optics enjoyed a vogue from the 1980s onwards, with the ready availability of many types of optical telecommunications fibre, and came into widespread use in the telecommunications industry. Monomode optical-fibre interferometers were applied in optical sensors [719, 720] which showed extraordinarily high sensitivity to various physical parameters such as temperature, pressure, strain, length change, velocity, etc. Many designs have been developed, mostly based on classical interferometers. In 1987, Brown [721] employed the spatial coherence properties of single-mode fibres in photon correlation experiments to create what was later described as 'the ultimate device for dynamic light-scattering' [722]. Combining this with avalanche diodes and solid-state lasers has recently led to a successful commercial photon-correlation spectroscopy (PCS) protein analyser. Throughout the 1980s many new types of monomode-fibre geometries and doping profiles were developed. Optical telecommunications systems using monomode fibres extended the widespread use of multimode fibres [723]. Wavelength demultiplexing techniques were developed to transmit and receive many channels of information simultaneously [724]. Mollenauer and Smith [725] demonstrated soliton pulse communication over 6000 km of monomode fibre in 1988. Hand-in-hand with all this practical optical fibre effort, detailed theoretical calculations were also being undertaken [726].

Activity in purely physics-based optics continued to centre around quantum-optics experiments. Xiao, Wu and Kimble [727] reported precision measurement beyond the shot noise limit in 1987, using squeezed light to achieve a 3 dB improvement of signal-to-noise ratio, an important result for interferometry. Yet further experiments to test the Bell inequalities [728, 729] using parametric down-conversion and two-photon correlation, achieved violations by several standard deviations. Similar but less restrictive experimental results were reported later by Rarity and Tapster [730], who avoided reliance on the polarization of the photons.

New styles of detector and laser were still appearing. In 1988 Levine *et al* [731] demonstrated a superlattice device employing perpendicular transport in a novel infrared detector with spectral response in the 8–12 μm atmospheric window. It employed AlGaAs/GaAs n-type doped quantum wells, with absorption occurring between the ground state of the well and a resonant state or a miniband in the continuum. In 1989 Lallier *et al* [732] made a continuous-wave, single-mode, waveguide laser in lithium niobate and Chandler *et al* [733] made an ion-implanted waveguide laser in Nd:YAG.

Around this time the field of cavity quantum electrodynamics (QED) started to develop as a new line of investigation, the mode structure of the field in a resonator being altered to create novel effects: inhibited or enhanced irreversible spontaneous emission, reversible spontaneous emission, microlaser action and quantum collapse and revival. A laser which works with only one atom in the optical resonator has now been reported [734]. These advances are already showing a practical pay-off; for example micro-disc and single-mode, vertical-cavity semiconductor lasers with ultra-low thresholds and high efficiencies have been demonstrated [735, 736].

With the widespread availability of high-power laser sources, interests in the effects of intense optical fields on matter were pursued. In 1989, optical binding of particles was demonstrated [737], extending the earlier work of Ashkin [561]. Significant forces between dielectric objects can be generated by intense optical fields. Consequently, light waves can serve to bind matter into new organized forms. Perhaps optically induced and organized material structures could be frozen and put to use without light fields?

By now it must be obvious that the growth and diversity of optical sciences in the late 1990s is too great to allow adequate cover in this single chapter. Many exciting avenues of physics such as laser cooling of atoms, enhanced backscattering, femtosecond pulse generation, optical interconnections, non-linear guided wave phenomena, optical mapping of the cortex, new types of tunable laser, etc, are under active investigation around the world. A snapshot of the status of many of these fields was given in a review of the 1980 decade [738], from which the abundance and significance of these activities may be judged. Yearly reviews of some of the optical physics highlights were published throughout the late twentieth century in December issues of *Optics and Photonics News*, and are recommended as overviews of these activities.

18.6. 1994 to 2000: towards the twenty-first century

Theories of cavity QED [739, 740] are so recent that experimental tests are still in progress. For example, it was not until the early 1990s that (vertical) microcavity lasers were constructed to demonstrate control of spontaneous emission, and remarkable capabilities can be expected as we approach the year 2000 [741]. There is eager anticipation of important new optoelectronic devices such as quantum dot and wire lasers [742–745] based on designs incorporating optical and electronic confinement simultaneously [746]. Such structures may be coupled to form quantum cellular automata [747]. Quantum communication and cryptography [748–750] are just starting. The predictions of quantum mechanics are *still* being tested in quantum-optics experiments [751]. Uncertainty also surrounds research into photonic bandgaps [752], the optical equivalent of semiconductor bandgaps produced by three-dimensional Bragg grating structures. There seems to be genuine new optical and device physics to be explored, as witness a recent review of a wide range of forefront optical activities [753].

Another area of optical research offers tantalizing possibilities at present. This is amplification without inversion (AWI) and lasing without inversion (LWI) [754]. Here the whole concept of population inversion that forms the basis of the laser development we have described seems to be turned upside-down! For a so-called three-level 'Λ' system one of the odd-even orthogonal combinations of the two lower energy levels can be populated by an auxiliary strong tuned field leaving the other empty and thus producing an effective inversion. For a three-level 'V' system the mechanism is the old one of Fano resonance [755] which is due to quantum interference of two paths to the same final state. Observation of AWI has been claimed in at least three laboratories [756] and actual LWI is surely not far away. A related effect, which was first seen in the mid-1970s [757], is electromagnetically induced transparency (EIT). A strongly absorbing medium can be made transparent to a suitable probe beam by illuminating it with a pump of the correct frequency. Dalton and Knight [758] pointed out that laser fluctuations can dephase the atomic coherences unless the two beams are critically cross-correlated by, for example, deriving one field from the other by acousto-optic modulation. The quenching of absorption by coherence generated in a ground-state doublet is related to the quenching of spontaneous-emission noise achieved in the so-called correlated-emission laser, in which atomic coherence is generated between a pair of excited states by an RF field [759]. It is also closely related to stimulated electronic resonance Raman transitions. Another interesting coherent-pumping phenomenon pointed out by Scully [760] is that of giant refractive index changes with minimal absorption, creating a medium christened 'phaseonium'. The possibility of achieving LWI in a semiconductor laser has been proposed [761] and will be of considerable importance if achieved effectively. The LWI mechanism may also be of importance to x-ray lasers where conventional inversion has been so difficult to achieve. We recommend the reviews of Kocharovskaya [762] and Scully [763]. A special issue of the journal *Quantum Optics* was also devoted entirely to these questions [756].

Yet another approach to the problem of short-wavelength lasers has been optical harmonic generation by means of 'dressed' photoionization continuum states, sometimes called excess-photon ionization. Work of Shore and Knight [764] shows that the laser responsible for the optical harmonic generation can itself restructure the continuum by a dressing interaction which will induce high harmonics in the polarization (this mechanism is somewhat analogous to the population-trapping effects in bound states which we discussed above). The theory was also considered by Goutier and Trahin [765], using a perturbation theory approach with more limited application.

Various new optoelectronic devices are fascinating at the present time, e.g. semiconductor lasers and LEDs incorporating strain effects [766], master-oscillator power amplifiers (MOPAs) to generate high-power light from laser diodes [767], polymer LEDs [768], multi-colour lasers [769], light-emitting silicon [770, 771], spatial light modulators [772] and microlens arrays [773]. Colour display and other technologies may benefit from these developments [774–776].

The final years of this century could turn out to be as exciting as the opening years with Planck and Einstein.

18.6.1. Applied optics: the impact in science and society

By far the greatest activity in optics and optoelectronics throughout the twentieth century has been in component, device, system and experimental developments and applications. In the latter part of the twentieth century, technological advances have made possible an extraordinarily wide range of uses for optics and the impact has been substantial. So far we have concentrated mainly on the key developments in physics, paying scant attention to the myriad applications. The purpose of this final section is to redress the balance.

The learned societies have led the way in organizing, publishing and disseminating information about optics to scientists around the globe. Most notably the Optical Society of America, formed in 1916, has been the flagship for the scientific advances, both theoretical and experimental, through its broad range of journals and international meetings. A European Optical Society has been founded in recent years. The Society of Photo-Optical Instrumentation Engineers has, especially in the latter part of this century, also been a leader in covering technology and applications of optics and optoelectronics. Optics has played a prominent role in national physics organizations worldwide. Scientists now gather in their thousands to attend international conferences, and the published proceedings of such meetings give a compact, detailed overview of late twentieth century interests in optics and optoelectronics.

We now attempt to give a flavour of the impact of optics in scientific and consumer activities. Inevitably, the laser has revolutionized the situation in the last 30 years of the century, but there are other matters to consider first. From the scientist's viewpoint, the twentieth century has seen major changes in the performance of optical instruments. The telescope has evolved from simple lens or two-mirror arrangements by incorporating Schmidt corrector plates to improve resolution, and active mirror-segment deformation to compensate for tropospheric and atmospheric refractive-index fluctuations. Huge new instruments, nearly 4 m in diameter have been constructed and the Hubble Space Telescope [777] (figure 18.13) has been deployed by NASA, with initial manufacturing errors now happily corrected. Progress has also

See also p 1741

A History of Optical and Optoelectronic Physics

Figure 18.13. *The Hubble Space Telescope.*

been made with ground-based or terrestrial telescopes using multiple apertures. Synthetic aperture optics [778] was studied extensively from the early 1960s, and lately optical aperture synthesis to deliver high-resolution images from coupled telescopes has been pursued [779]. Stellar optical interferometry is now widely practised [780].

The microscope has also been considerably developed this century, with improvements in coated multiple-element lenses, and more recently with the introduction of the confocal laser-beam and Nipkow-disc scanning techniques, and super-resolving schemes. Improvements in computer image-processing techniques have also had considerable impact. Zernike's phase-contrast microscopy was an especially significant advance. Scanning near-field optical microscopy is also under strong development [781, 782].

Optical scientists have benefited enormously from the introduction (in about 1970) of catalogue mail-order shopping for almost any optical or optoelectronic item. The use of computers in the optical sciences has proliferated since the 1960s so that powerful computer software is readily available for lens design, optical-coating design, computer control of component manufacture, and optical experiment design, control and data analysis. Computers and lasers have changed the life of the optical scientist out of all recognition.

For the manufacturer of classical optical and optoelectronic components and systems, there has been huge growth in consumer and business products including lenses and other basic components such as prisms, beam-splitters, filters and the like. Mass-production computer-automated facilities have become well established, especially in Japan, Malaysia, Korea and Taiwan, for photographic cameras and binoculars. Manufacturers have had to adapt to and introduce many new technologies, including semiconductors, glass fibres, graded-index lenses and thin-film coatings. It is now common-place to achieve sub-micron precision in reticles, diffraction gratings, lens mounts and microlens arrays, because of advances in engineering and computer capabilities. A brief survey of the US optical industry activities during much of the twentieth century has been compiled by Shannon [783].

Completely new optoelectronic industries have appeared since the 1960s, as advances in semiconductor physics and engineering engendered a range of technological devices and systems. The liquid-crystal display (LCD) [784], the light-emitting diode (LED) [62], the laser diode (LD) [651], the avalanche photodiode (APD) [785], the charge-coupled-array detector (CCD) [786], the photonic integrated circuit (PIC) and the optoelectronic integrated circuit (OEIC) [787], a large array of photoelectronic imaging devices [788], the solar cell and many other types of light emitter, modulator and detector for all visible and infrared wavelengths have been introduced and mass-produced. Interest in solar cells and non-imaging concentrators expanded greatly after 1970 as these became more efficient [493].

Many of these optoelectronic devices have been crucial to the development of the now vast array of consumer and business system products that are sold worldwide. The television set has come to dominate life in most parts of the developed world. Conventional television based on the 50-year-old technology of the cathode-ray tube will probably be overtaken (in volume manufacture) in the late 1990s by the liquid-crystal television display (figure 18.14). There are competitive types of electronic display, such as electroluminescent panels and plasma devices [789]. Associated with television are the many other useful optoelectronic gadgets now widely available, such as the video recorder and the camcorder [790, 791]. As we approach the year 2000, the consumer is being introduced to new optoelectronic technologies such as facsimile (FAX), laser printers, photocopiers [792, 793], which were previously only affordable in the business world. Electronic imaging is becoming widespread [794]. Virtual reality [795] and 3D television [796, 797] may also be within the grasp of the consumer in the near future.

A History of Optical and Optoelectronic Physics

Figure 18.14. *A 21" (diagonal) liquid-crystal television.*

Central to so many of these important display developments have been the subjects of colour science, photometry, lighting and the use of the CIE (Commission Internationale de l'Éclairage) chart [798]. The physics of digital colour display and printing is now of great importance [799].

Throughout the century the public have had a growing interest in cinema and visual entertainment. Cameras and binoculars have become commonplace [800] with first-class manufacturers in the USA, Europe and the Far East. Recently new optical advances in the recording and projection of very large, high-resolution images have made a lasting impression on all who have seen them [801]. Moving pictures via the telephone are becoming available to the consumer through introduction of the videophone in certain countries. Another kind of display also benefits from twentieth century optics—priceless paintings in art galleries, which can now be cleaned and restored by an excimer laser beam [802].

Military requirements and space exploration have been major driving forces in the development of optics and optoelectronics, especially from around the time of World War II. Very-high-resolution imaging systems have been developed for the visible and infrared areas of the spectrum [803, 804]. Thermal imaging detectors, both cooled and uncooled [805, 806], have become highly developed, of which the 'SPRITE' detector

1477

deserves special mention [807]. The civilian use of hand-held, and now helmet-mounted, pyroelectric vidicon and solid-state, far-infrared detectors by fire-fighters to detect body-heat through smoke is a valuable spin-off from military research and development [808]. Optics has also played a significant role in underwater observations [809]; this application has led to significant effort to develop blue–green lasers which have high transmission in sea water.

The laser has revolutionized the applications of optics since its invention in 1960. Attempts have been made to use lasers for nearly every conceivable optical application and to invent many new applications. The development and application of the laser in the USA has been described by Bromberg [810].

Scientists use lasers to measure a very large range of physical parameters, for example: velocity, size, temperature, pressure, humidity, absorption, level, displacement, vibration frequency, chemical and biological change and composition and electric current. We use lasers for high-resolution microscopy and particle trapping [561]. The basic unit of length, the metre has been defined in terms of a laser transition frequency and measured by optical means with increasing accuracy throughout the century [811]. Currently scientists are engaged in trying to construct laser interferometers to detect gravitational radiation, an experiment of exceptional difficulty to which conventional optics together with quantum optics and quantum electronics offers a potential solution [812, 813].

Industrial consumers have come to use optics and lasers in a very large number of ways, from the very basic measurement of lengths and displacements (metrology) [814] through to measurement of velocities [815, 816] and particle sizes [817] in processes, cutting and welding of materials [818, 819] and isotope separation [820]. Newspapers are printed by lasers in some cities. Semiconductors and other materials are processed by laser [821, 822], e.g. by laser-controlled etching of semiconductors in solutions and by doping of semiconductors using laser-induced chemistry in microelectronics. Aircraft are now navigated by laser gyroscopes [823, 824]. Combustion engines of all kinds are improved by laser velocimetry and CARS experiments [825, 826]. Lidar (light detection and ranging) has become widely used to monitor atmospheric pollution [827]. Recent tests of femtosecond lasers have apparently even suggested their use for control of lightning strikes [828]!

The military and space communities make extensive use of optoelectronics and lasers both in their research and development programmes and in vital applications. Military battlefield uses include rangefinding, tank and rifle sights [829], laser-guided bombs, laser detonation and laser sensor-blinding [830]. Lasers powered by nuclear explosions have been considered [831] and the possibility of a missile-defence laser system was examined in the extraordinarily costly Star

Wars [832] programme in the USA in the 1980s. Secure communications systems employing the highly directional properties of laser beams have been developed for battlefield use and for communication between aircraft and submarines (blue–green wavelengths), as have similar systems for target-vibration signature analysis and identification. In the 1970s the US Navy investigated chemical lasers to defend ships against attack by tactical missiles [833]. It was believed that the Pentagon invested over $1 billion in chemical lasers at that time.

Space-borne uses of lasers include global wind monitoring, long-range communications and scientific experiments. NASA has already flown a space-shuttle 'Lidar In Space Technology Experiment' [834] using an expanded incoherent three-wavelength Nd:YAG laser beam to measure cloud cover and other aerosol and Earth returns. Both NASA [835] and the European Space Agency (ESA) [836] are actively considering space-borne Doppler wind lidar systems for launch in the first decade of the twenty-first century. Early in the history of the laser, the distance of the Moon from the Earth was accurately measured by timing the round-trip time of a laser pulse sent to the Moon and bounced off its surface back to the Earth. This type of laser radar ranging was first demonstrated in May 1962 using a pulsed ruby laser and a 48" telescope at MIT's Lincoln Laboratory [837]. Only a few photons returned into the telescope from each pulse sent to the Moon, but that was sufficient to determine the distance accurately. Reflectors for this purpose were placed by Moon walkers [838] on the Apollo 11 mission. Measurements using these reflectors have shown that tidal dissipation has caused the Moon to move one metre further from the Earth since then. The USSR has also conducted an active Moon-ranging programme [839].

Another use of lasers on a grand scale, which has been under development in the 1980s and 1990s, is laser-fusion-generated electrical power [839–841]. To this end, massive multiple-laser-beam systems have been constructed [839] such as those in the USA (Shiva-Nova), the USSR (Delphin) and Japan (Gekko), but system break-even has yet to be reached. If this can be achieved, lasers may provide a new source of electrical energy in the twenty-first century [842]. As we have seen in our discussion of x-ray lasers, these high-power systems have other scientific uses. To achieve the highest brightness a radar technique known as pulse-compression is used, in which the amplification is carried out on a stretched-out pulse to reduce peak power levels which can damage components. The pulse is sharpened again by a pair of large high-efficiency gratings at the output. In a recent report from the UK Rutherford Appleton Laboratory [843], for example, extremely bright pulses were obtained from their Nd:glass laser Vulcan, by such a compression technique. The laser output was 8 TW in 2.4 ps, which has since been upgraded to 35 TW in 620 fs. 22 J has been deposited on a three-times diffraction-limited spot of 15 μm, giving an irradiance of 10^{19} W cm^{-2}. Such pulses can be used for studies in multi-photon physics, plasma compression, particle acceleration and fusion ignition. Already, using these pulses, electrons have been accelerated to 45 MeV energy, and more than 10^9 neutrons have been obtained from the $D + D$ reaction with a target of deuterated plastic. The radiation pressure (which we discussed previously in connection with the 'optical bottle') in these experiments can be many megabars.

Of more immediate interest to the average person is the widespread use of lasers in medicine [844–846]. Lasers have been used to treat many conditions, for example cancer [847], urological problems such as kidney and gall-stones [848], the welding of burst blood vessels, as scalpels in surgery and fibre endoscopy, port-wine stain removal, de-scaling or unblocking of arteries, repair of detached retinas, dentistry and cosmetic surgery. In the future more complicated laser-optical systems may find regular use such as in retinal blood-flow examination [849, 850] and laser tomography [851–853], techniques developed recently in the research laboratory. Of particular interest in the press in recent years has been the use of lasers to perform corneal re-shaping (by ablation) to correct defective vision without the need for spectacles or contact lenses [846]. Ophthalmic uses of lasers will, no doubt, expand in the future because the eye is the only optical window into the human body and useful information about our health can be gained by retinal observations in this way. The laser has also found many uses in biotechnological applications [854].

Whilst many of the applications of lasers so far described are widely known, they often do not have an immediate impact on everyday life. However, the consumer is using lasers every day in most of the developed nations of the world, many without knowing it. Telephone conversations (and recently videophone and cable television) are now routinely transmitted by laser diode light through optical fibres, inside and between cities and between continents [87, 855–857]. Even a short science-fiction article has been written on how these applications would develop to the year 2003 [858].

Telecommunications by optical-fibre technology were well developed in 1993. We had active fibre devices in the form of lasers, amplifiers, converters and optically activated switches, etc, of remarkable performance. Fibre-amplifier laboratory experiments had demonstrated 10 Gbit s^{-1} transmission over 10 000 km of fibre with effectively zero error and potential for delivery of 12 simultaneous television channels to 39 000 000 customers over a 500 km radius. This implied that nearly the whole UK could be supplied by a single transmitter! Today the Information Super Highway, using a vast network of such telecommunications, promises a revolution in our lives which has been compared with that of the industrial revolution.

More obvious to the consumer, bar-code scanners in supermarkets use scanned red laser light to automate check-out. Compact optical-disc [649] players provide us with high-quality digital audio signals read by laser-diode light from the reflective/diffractive disc. Video programmes for television are recorded and played-back in a similar manner [859], standards for digital video recording are at present being finalized. Mass optical data storage of many gigabytes of information is now becoming available using this and other technologies [860, 861]. The laser is central to the way we record and transmit data, it is central to the way we now live.

Global activity in optics and optoelectronics is a very substantial business. Estimates for Japan alone [862, 863], one of the world's larger manufacturers, indicate that the optoelectronics market in that country was approximately $17 billion in 1979 and $30 billion in 1991. The current prediction is that it will grow to $110 billion in 2000 and to around $260 billion in 2010, but with trading recessions these figures are questionable. Nevertheless, the total world market is of the order of double these figures. Optics is big business. Many today believe that together with life sciences, microelectronics and software, optics and optoelectronics will be a dominant technology in the twenty-first century, with much physics still to be done.

18.6.2. Further reading

More information on related topics can be found in the following general references:

(i) history—[4, 5, 39, 40, 120, 196, 277, 810, 864]
(ii) classical optics—[31, 56, 101, 297, 865]
(iii) optical principles—[56, 865]
(iv) non-linear optics—[866]
(v) quantum optics—[226, 345, 384, 751, 867]
(vi) optoelectronics—[186, 868–870]
(vii) applied optics—[871, 872]
(viii) progress in optics—[873].

Acknowledgments

We have benefited from countless discussions by email, fax, letter, telephone and sometimes even face-to-face with many of the leading characters in our story. We thank them all anonymously. They will know who they are. We apologize in advance for any historical inaccuracies, which we are sure will be brought to our attention! The complex and difficult process of typing and re-typing drafts and amendments to this chapter was performed with great care by Mrs Emma Gilbert, to whom we are also most grateful.

References

[1] Planck M 1958 *Physikalische Abhandlungen und Vorträge* vol III (Braunschweig: Vieweg) p 121
See also Planck M 1960 Nobel Prize address reprinted in *A Survey of Physical Theory* (New York: Dover) p 102.
[2] Drude P K L 1959 *The Theory of Optics* (New York: Dover) (originally published by Longman in 1902)
[3] See for example Corney A 1977 *Atomic and Laser Spectroscopy* (Oxford: Clarendon) p 1
[4] Kangro H 1976 *History of Planck's Radiation Law* (London: Taylor and Francis)
[5] Kuhn T S 1978 *Black-body Theory and the Quantum Discontinuity 1894–1912* (Oxford: Oxford University Press)
[6] Planck M 1900 *Vehr. Deutsch. Phys. Ges.* **2** 202, 237
[7] Planck M 1901 *Ann. Phys., Lpz* **4** 553
[8] Lummer O and Pringsheim E 1900 *Vehr. Deutsch. Phys. Ges.* **2** 163
[9] Rubens H and Kurlbaum F 1900 *Prussian Acad. Wiss.* 929
[10] M Planck, 7 October 1931 letter to R W Wood, in the Archives of the Center for the History of the Philosophy of Physics at the American Institute of Physics, New York City, USA.
[11] Day A L 1902 *Science* **15** 429
[12] Einstein A 1905 *Ann. Phys., Lpz* **17** 132
[13] Planck M 1910 *Ann. Phys., Lpz* **31** 758
[14] Einstein A 1916 *Mitt. Phys. Ges. Zurich* **16** 47 (Engl. Transl. van der Waerden B L (ed) 1967 *Sources of Quantum Mechanics* (Amsterdam: North Holland))
[15] Bose S N 1924 *Z. Phys.* **26** 178 (Engl. Transl. 1976 *Am. J. Phys.* **44** 1056)
[16] Planck M 1911 *Verh. Deutsch. Phys. Ges.* **13** 138
[17] Marshall T W 1963 *Proc. R. Soc.* A **276** 475
[18] Mandel L and Wolf E 1965 *Rev. Mod. Phys.* **37** 231
[19] Glauber R J 1963 *Phys. Rev.* **131** 2766
[20] Sudarshan E C G 1963 *Phys. Rev. Lett.* **10** 277
[21] Loudon R 1985 *Lasers in Applied and Fundamental Research* ed S Stenholm (Bristol: Hilger) p 185
[22] Bell J S 1964 *Physics* **1** 195
[23] Hartmann J 1900 *Z. Instrumentkd.* **20** 47
[24] Hartmann J 1904 *Z. Instrumentkd.* **24** 1
[25] Malacara D (ed) 1992 *Optical Shop Testing* 2nd edn (New York: Wiley)
[26] Lummer O 1901 *Vehr. Deutsch. Phys. Ges.* **3** 85
[27] Lummer O and Gehrcke E 1903 *Ann. Phys., Lpz* **10** 457
[28] Cotton A and Mouton H 1907 *C. R. Acad. Sci., Paris* **145** 229
See also Born M 1933 *Optik* (Berlin: Springer) p 360
[29] Faraday M 1846 *Phil. Trans.* **136** 1
[30] Voigt W 1902 *Ann. Phys., Lpz* **9** 367
See also Voigt W 1908 *Magneto- und Elktro-optik* (Leipzig: Teubner)
[31] Hecht E and Zajac A 1974 *Optics* (Reading, MA: Addison-Wesley)
[32] Strehl K 1903 *Z. Instrumentkd.* **22** 213
[33] Conrady A E 1904 *Mon. Not. R. Astron. Soc.* **64** 182
[34] Welford W T 1986 *Aberrations of Optical Systems* (Bristol: Hilger) p 201
[35] Köhler A 1904 *Z. Wiss. Mikrosk.* **21** 129
[36] Evennett P 1993 *Proc. R. Microsc. Soc.* **28** 180

[37] Probst W, Bauer R, Benner G and Lehman J L 1991 *Proc. 49th Annual Meeting EMSA, San Jose, CA, 1991* (Boston, MA: Electron Microscopy Society of America) p 1010
Benner G and Probst W 1994 *J. Microsc.* **174** 133
[38] Weber R L 1988 *Pioneers of Science* (Bristol: Hilger)
[39] Pais A 1982 *Subtle is the Lord* (Oxford: Oxford University Press) ch 19
[40] Weber R L 1988 *Pioneers of Science* (Bristol: Hilger) p 64
[41] Stoletov A G 1888 *C. R. Acad. Sci., Paris* **106** 1149
[42] Ladenburg E 1903 *Ann. Phys., Lpz* **12** 558
[43] Lenard P 1902 *Ann. Phys., Lpz* **8** 149
[44] Hallwachs W L F 1888 *Ann. Phys. Chem.* **33** 301
[45] Hallwachs W L F 1888 *Ann. Phys. Chem.* **34** 731
[46] Hallwachs W L F 1889 *Ann Phys. Chem.* **37** 666
[47] Hughes A L 1912 *Phil. Trans. R. Soc.* A **212** 205
[48] Richardson O W and Compton K T 1912 *Phil. Mag.* **24** 575
[49] Millikan R A 1916 *Phys. Rev.* **7** 355
[50] Margoritondo G 1988 *Phys. Today* April p 66
[51] Hallwachs W L F 1916 *Handbuch der Radiologie III* ed E Marx (Leipzig)
[52] Wood R W 1988 *Physical Optics* (Washington, DC: Optical Society of America) (first published 1905)
[53] Marchand E W 1978 *Gradient-Index Optics* (New York: Academic)
[54] Moore D T 1980 *Appl. Opt.* **19** 1035
[55] Schwarzchild K 1905 *Astr. Mitt. Kgl. Sternwarte Gött.*
[56] Born M and Wolf E 1975 *Principles of Optics* (Oxford: Pergamon) p 197
[57] Porter A B 1906 *Phil. Mag.* **11** 154
[58] Goodman J W 1968 *Introduction to Fourier Optics* (New York: McGraw-Hill)
[59] Lord Rayleigh (Strutt J W) 1907 *Proc. R. Soc.* A **79** 399
[60] Weber R L 1988 *Pioneers of Science* (Bristol: Hilger) p 31
[61] Round H J 1907 *Electrical World* **19** 309
[62] Bergh A A and Dean P J 1976 *Light Emitting Diodes* (Oxford: Clarendon)
[63] von Laue M 1907 *Ann. Phys., Lpz* **23** 1
[64] Beran M J and Parrent G B 1964 *Theory of Partial Coherence* (Englewood Cliffs, NJ: Prentice-Hall)
[65] Mandel L and Wolf E 1990 *Selected Papers on Coherence and Fluctuations of Light* (Bellingham, WA: SPIE) vols 1 and 2
[66] Mie G 1908 *Ann. Phys., Lpz* **25** 377
[67] Kerker M 1969 *The Scattering of Light and other Electromagnetic Radiation* (New York: Academic) p 54
[68] Weber R L 1988 *Pioneers of Science* (Bristol: Hilger) p 34
[69] Lippmann G 1894 *J. Physique* **3** 97
[70] Collier R J, Burckhardt C B and Lin L H 1971 *Optical Holography* (New York: Academic)
[71] Smoluchowski M 1908 *Ann. Phys., Lpz* **25** 205
[72] Crosignani B, di Porto P and Bertolotti M 1975 *Statistical Properties of Scattered Light* (New York: Academic)
[73] Einstein A 1909 *Phys. Z.* **10** 185
[74] Heisenberg W 1962 *Physics and Philosophy* (New York: Harper and Row)
[75] Heisenberg W 1930 *The Physical Principles of the Quantum Theory* (Chicago: University of Chicago Press)
[76] Schrödinger E 1935 *Naturwissenschaften* **23** 807, 823 and 844 (Engl. Transl. Trimmer J D 1980 *Proc. Am. Phil. Soc.* **124** 323)
[77] Feynman R P, Leighton R B and Sands M 1965 *Lectures on Physics III* (Reading, MA: Addison-Wesley) ch 3
[78] Weinberg S 1977 *Proc. Am. Acad. Arts Sci.* **106** 22

[79]   Dirac P A M 1972 *Fields Quanta* **3** 154
[80]   Taylor G I 1909 *Proc. Camb. Phil. Soc.* **15** 114
[81]   Thompson J J 1907 *Proc. Camb. Phil. Soc.* **14** 41
[82]   Lenard P 1900 *Ann. Phys., Lpz* **1** 486; 1900 *Ann. Phys., Lpz* **3** 298
[83]   Einstein A 1910 *Ann. Phys., Lpz* **33** 1275
[84]   Wood R W 1910 *Phil. Mag.* **20** 770
[85]   Hondros D and Debye P 1910 *Ann. Phys., Lpz* **32** 465
[86]   Mauguin C 1911 *Bull. Soc. Fr. Miner.* **34** 71
[87]   Miller S E and Chynoweth A G (ed) 1979 *Optical Fibre Telecommunications* (New York: Academic) vol 1
[88]   Pankove J I (ed) 1980 *Display Devices* (Berlin: Springer)
       See also Kaneko E 1987 *Liquid crystal TV displays* (Dordrecht: D Reidel)
[89]   Snyder A W and Love J 1983 *Optical Waveguide Theory* (London: Chapman and Hall)
[90]   Janossy I (ed) 1991 *Optical Effects in Liquid Crystals* (Milan: Kluwer Academic)
[91]   Khoo I-C and Wu S-T 1993 *Optics and Non-Linear Optics of Liquid Crystals* (Singapore: World Scientific)
[92]   Bohr N 1913 *Phil. Mag.* **26** 1, 476, 857
[93]   Born M 1969 *Atomic Physics* (London: Blackie)
[94]   Brillouin L 1914 *C. R. Acad. Sci., Paris* **158** 1331; 1922 *Ann. Phys., Paris* **17** 88
[95]   Landsberg G S 1952 *Raman Scattering Spectroscopy* ed K W F Kohlrauch (Russian Transl.) (Moscow: Inostrannoya Literature)
[96]   Mandelstam L I 1926 *Zh. Russ. Fiz.-Khim. Obsch. Fiz.* **58** 381
[97]   Gross E 1930 *Nature* **126** 201
[98]   Chu B 1991 *Laser Light Scattering* 2nd edn (San Diego: Academic) p 54
[99]   Pike E R, Vaughan J M and Vinen W F 1969 *Phys. Lett.* A **30** 373
[100] Sagnac G 1914 *J. Phys. Radium* **4** 177
[101] Hariharan P 1975 *Appl. Opt.* **14** 2319
       See also Hariharan P 1985 *Optical Interferometry* (Sydney: Academic)
[102] Einstein A 1917 *Z. Phys.* **18** 121
[103] Bothe W and Geiger H 1925 *Naturwissenschaften* **13** 440; 1925 *Z. Phys.* **32** 639
[104] Bohr N, Kramers H A and Slater J C 1924 *Phil. Mag.* **47** 785
[105] Twyman F and Green A 1916 *British Patent* No 103,832
[106] Twyman F 1957 *Prism and Lens Making* 2nd edn (London: Hilger and Watts)
[107] Ramaseshan S (ed) 1988 *Scientific Papers of C V Raman III—Optics* (Bangalore: Indian Academy of Sciences) p xi
[108] Slepian J 1919 *US Patent* No 1,450,265
[109] Schriever O 1920 *Ann. Phys., Lpz* **63** 645
[110] Weigert F 1920 *Vehr. Deutsch. Phys. Ges.* **1** 100
[111] Vavilov S I and Levshin W L 1922 *Phys. Z.* **23** 173
[112] Weigert F 1922 *Phys. Z.* **23** 232
[113] Levshin V L 1924 *Z. Phys.* **26** 274
[114] Perrin F 1929 *Ann. Phys., Paris* **12** 169
[115] Michelson A A 1920 *Astrophys. J.* **51** 257
[116] Weber R L 1988 *Pioneers of Science* (Bristol: Hilger) p 64
[117] Ladenburg R W 1921 *Z. Phys.* **4** 451
[118] Kramers H A 1924 *Nature* **133** 673
[119] Kramers H A and Heisenberg W 1925 *Z. Phys.* **31** 681
[120] Bertolotti M 1983 *Masers and Lasers, an Historical Approach* (Bristol: Hilger)
[121] Ladenburg R W and Kopfermann H 1928 *Z. Phys.* **48** 46, 51; 1930 *Z. Phys.* **65** 167; 1928 *Z. Phys. Chem.* **139** 378

[122] Wood R W 1922 *Phil. Mag.* **43** 757
[123] Miller D C 1933 *Rev. Mod. Phys.* **5** 203
[124] Tolansky S 1955 *An Introduction to Interferometry* (London: Longmans) pp 103–4
[125] Duane W 1923 *Proc. Natl Acad. Sci.* **9** 158
[126] Tolman R C 1924 *Phys. Rev.* **23** 693
See also Bertolotti M 1983 *Masers and Lasers, an Historical Approach* (Bristol: Hilger) p 20
[127] Born M, Heisenberg W and Jordan P 1926 *Z. Phys.* **35** 557
[128] Dirac P A M 1927 *Proc. R. Soc.* A **114** 243
[129] Pike E R and Sarkar S 1986 *Frontiers in Quantum Optics* ed E R Pike and S Sarkar (Bristol: Hilger) p 96
[130] Lewis G N 1926 *Nature* **118** 874
[131] Tolansky S 1955 *An Introduction to Interferometry* (London: Longmans) pp 103, 113
See also Linnik V P 1933 *Dokl. Akad. Nauk* **1** 18, 208
[132] Vavilov S I and Levshin V L 1926 *Z. Phys.* **35** 920
[133] Kronig R L 1926 *J. Opt. Soc. Am.* **12** 547
Kramers H A 1927 *Estratto degli Atti del Congresso Internazionale de Fisici Como* vol 2 (Bologna: Zanichelli) p 545
[134] Wiener N 1927–28 *J. Math. Phys. (MIT)* **7** 109
[135] Raman C V and Krishnan K S 1928 *Nature* **121** 501
[136] Raman C V 1928 *Indian J. Phys.* **2** 387
[137] Mandelshtam L I and Landsberg G S 1928 *Zh. Russ. Fiz.-Khim. Obsch. Fiz.* **60** 335
[138] Raman C V 1928 *Nature* **121** 619
[139] Raman C V and Krishnan K S 1928 *Nature* **121** 711
[140] Venkataraman G 1988 *Journey into Light: Life and Science of C V Raman* (Bangalore: Indian Academy of Science in cooperation with the Indian National Science Academy) p 208
[141] Born M 1928 *Naturwissenschaften* **16** 741
[142] Darwin C G 1928 *Nature* **121** 630
[143] Rutherford E 1930 *Proc. R. Soc.* A **126** 184 (Presidential address November 1929)
[144] Venkataraman G 1988 *Journey into Light: Life and Science of C V Raman* (Bangalore: Indian Academy of Science in cooperation with the Indian National Science Academy) p 197
[145] Venkataraman G 1988 *Journey into Light: Life and Science of C V Raman* (Bangalore: Indian Academy of Science in cooperation with the Indian National Science Academy) p 383
[146] Fabelinskii I L 1978 (Engl. Transl.) *Sov. Phys.–Usp.* **21** 780
[147] Weber R L 1988 *Pioneers of Science* (Bristol: Hilger) p 93
[148] Smekal A 1923 *Naturwissenschaften* **11** 873
[149] Kramers H A and Heisenberg W 1925 *Z. Phys.* **31** 681
[150] Rocard Y 1928 *C. R. Acad. Sci., Paris* **186** 1201
[151] Cabannes J 1928 *C. R. Acad. Sci., Paris* **186** 110
[152] Loudon R 1964 *Adv. Phys.* **13** 423
[153] Synge E H 1928 *Phil. Mag.* **6** 356
[154] Einstein's letters at the Hebrew University of Jerusalem, copied with permission to quote to D M Mullen 1990 *Proc. R. Microsc. Soc.* **25** 127
[155] Dürig U, Pohl D W and Rohner F 1986 *J. Appl. Phys.* **59** 3318
[156] Pohl D W, Denk W and Lanz M 1984 *Appl. Phys. Lett.* **44** 651
[157] Lewis A, Isaacson M, Harootunian A and Muray A 1984 *Ultramicroscopy* **13** 227

[158] Massey G A 1984 *Appl. Opt.* **23** 658
[159] Betzig E, Lewis A, Harootunian A, Isaacson M and Kratschmer E 1986 *Biophys. J.* **49** 269
[160] O'Keefe J A 1956 *J. Opt. Soc. Am.* **46** 359
[161] Baez A V 1956 *J. Opt. Soc. Am.* **46** 901
[162] Ash E and Nichols G A 1972 *Nature* **237** 510
[163] Pohl D W, Denk W and Lanz M 1984 *Appl. Phys. Lett.* **44** 651
[164] Betzig E and Trantman J 1992 *Science* **257** 189
[165] Binnig G, Rohrer H, Gerber C H and Weibel E 1982 *Phys. Rev. Lett.* **49** 57
[166] Heitler W and Herzberg G 1929 *Naturwissenschaften* **34** 673
[167] Rasetti F 1929 *Proc. Natl Acad. Sci.* **15** 515
[168] Schmidt B V 1931 *Central-Z. Opt. Mech.* **52** No 2
See also Born M and Wolf E 1975 *Principles of Optics* (Oxford: Pergamon)
[169] Jones L 1995 *Handbook of Optics* vol II, ed M Bass (New York: McGraw-Hill) ch 18
[170] Macleod H A 1969 *Thin Film Optical Filters* (Bristol: Hilger) p 3
[171] Strong J 1936 *J. Opt. Soc. Am.* **26** 73
[172] Geffken W 1939 *Deutsches Reich Patentschrift* No 716,153
[173] Oseen C W 1933 *Trans. Faraday Soc.* **29** 883
[174] Stiles W G and Crawford B H 1933 *Proc. R. Soc.* B **112** 428
[175] Snyder A W and Pask C 1973 *Vision Res.* **13** 1115
[176] van Cittert P H 1934 *Physica* **1** 201
[177] Zernike F 1938 *Physica* **5** 785
[178] Goodman J W 1985 *Statistical Optics* (New York: Wiley)
[179] Landau L D and Placzek G 1934 *Phys. Z. Sov.* **5** 172
[180] Čerenkov P A 1934 *Dokl. Acad. Sci. USSR* **11** 451
[181] Vavilov S I 1934 *Dokl. Acad. Sci. USSR* **11** 457
[182] Tamm I E and Frank I M 1937 *Dokl. Acad. Sci. USSR* **14** 107
[183] Ginsberg V 1940 *Sov. Phys.–JETP* **10** 589
[184] Ginsberg V 1940 *J. Phys. USSR* **2** 441
[185] See Ramaseshan S (ed) 1988 *Scientific Papers of C V Raman III—Optics* (Bangalore: Indian Academy of Sciences); 5 papers by C V Raman and N S N Nath 1935/6
[186] Saleh B E A and Teich M C 1991 *Fundamentals of Photonics* (New York: Wiley) ch 20
[187] Iams H and Salzberg B 1935 *Proc. IRE* **23** 55
[188] Zernike F 1934 *Physica* **1** 689; 1935 *Z. Tech. Phys.* **16** 454
[189] Zernike F 1955 *Science* **121** 345
[190] Mahajan V N 1994 *Opt. Photonics News* S-21
[191] Nijboer B R A 1942 The diffraction theory of aberrations *Thesis* University of Groningen, Holland
[192] Nijboer B R A 1943 *Physica* **10** 679
[193] Nijboer B R A 1947 *Physica* **13** 605
[194] Ferwerda H A 1993 *Opt. Eng.* **32** 3176
[195] Beth R A 1936 *Phys. Rev.* **50** 115
[196] Williams T I 1982 *A Short History of Twentieth Century Technology* (Oxford: Clarendon) ch 26
[197] Edgerton H E and Killian Jr J R (ed) 1979 *Moments of Vision: The Stroboscopic Revolution in Photography* (Cambridge, MA: MIT Press)
[198] Wannier G 1937 *Phys. Rev.* **52** 191
[199] Peierls R 1937 *Ann. Phys.* **13** 905
[200] Tolansky S 1948 *Hyperfine Structure in Line Spectra and Nuclear Spin* (London: Methuen)

[201] Engstrom R W et al 1970 *RCA Photomultiplier Manual* (Harrison, NJ: RCA Corporation)
[202] See Bertolotti M 1983 *Masers and Lasers, an Historical Approach* (Bristol: Hilger) pp 3, 27, 115 and 119
 Fabrikant V A, Vudynskii M M and Butayeva F 1959 *USSR Patent* No 123,209
[203] Gabor D 1940 *UK Patent* No 541,753
[204] Lassus St Genies A H J 1939 *UK Patent* No 506,540
[205] Clark Jones R 1941 *J. Opt. Soc. Am.* **31** 488
[206] Clark Jones R 1956 *J. Opt. Soc. Am.* **46** 126
[207] Clark Jones R 1959 *Proc. IRE* **47** 1495; 1960 *J. Opt. Soc. Am.* **50** 883
[208] Wolf E 1959 *Nuovo Cimento* **13** 1165
[209] Hopkins H H 1943 *Proc. Phys. Soc.* **55** 116
[210] Ignatowsky V S 1919 *Trans. Opt. Inst. Petrograd* **1** paper IV; *Trans. Opt. Inst. Petrograd.* **1** paper V
[211] Richards B and Wolf E 1959 *Proc. R. Soc.* A **253** 358
[212] Luneberg R K 1960 *Mathematical Theory of Optics* (Berkeley, CA: University of California Press) (Mimeographed 1944 Lecture Notes, Brown University, Providence, RI)
[213] Tolansky S 1944 *Phil. Mag.* **35** 120
[214] Tolansky S 1945 *Proc. R. Soc.* A **184** 51
[215] Tolansky S 1948 *Multiple-beam Interferometry of Surfaces and Films* (Oxford: Clarendon)
[216] Hopkins H H 1945 *Nature* **155** 275
[217] Stump D M 1922 *J. R. Microsc. Soc.* 264
[218] Hopkins H H and Barham P M 1950 *Proc. Phys. Soc.* **63** 737
[219] Duffieux P M 1946 (Engl. Transl. 1983 *The Fourier Transform and its Application in Optics* (New York: Wiley))
[220] Reynolds G O, de Velis J B, Parrent Jr G B and Thompson B J 1989 *The New Physical Optics Notebook: Tutorials in Fourier Optics* (Bellingham, WA: SPIE)
[221] Wheeler J A 1946 *Ann. NY Acad. Sci.* **48** 219
[222] Pryce M H L and Ward J C 1946 *Nature* **160** 435
[223] Hanna R C 1948 *Nature* **162** 332
[224] Bleuler E and Brant H L 1948 *Phy. Rev.* **73** 1398
[225] Wu C S and Shaknov I 1950 *Phys. Rev.* **77** 136
[226] Pike E R and Sarkar S 1995 *The Quantum Theory of Radiation* (Oxford: Oxford University Press)
[227] Lau E 1948 *Ann. Phys.* **6** 417
[228] Keith D N, Ekstrom C R, Torchette Q A and Pritchard D E 1991 *Phys. Rev. Lett.* **66** 2693
[229] Dallos J 1946 *Br. J. Opthal.* **30** 607
[230] Bier N 1948 *Optician* **116** 497
[231] Stone J and Phillips A J (ed) 1984 *Contact Lenses* 2nd edn (London: Butterworth)
[232] Wolfke M 1920 *Phys. Z.* **21** 495
[233] Bragg W L 1939 *Nature* **143** 678
[234] Bragg W L 1942 *Nature* **149** 470
[235] Gabor D 1948 *Nature* **161** 777
[236] Gabor D 1949 *Proc. R. Soc.* A **197** 454
[237] Gabor D 1951 *Proc. R. Soc.* B **64** 449
[238] Rogers G L 1952 *Proc. R. Soc. Edin.* A **63** 193
[239] Collier R J, Burkhardt C B and Lin L H 1971 *Optical Holography* (New York: Academic)

[240] El-Sum H M A and Kirkpatrick P 1952 *Phys. Rev.* **85** 763
[241] Lohmann A 1956 *Opt. Acta* **3** 97
[242] Leith E N and Upatnieks J 1962 *J. Opt. Soc. Am.* **52** 1123
See also Born M and Wolf E 1975 *Princples of Optics* (Oxford: Pergamon)
[243] Weber R L 1988 *Pioneers of Science* (Bristol: Hilger) p 221
[244] Brossel J and Kastler A 1949 *C. R. Acad. Sci., Paris* **229** 1213
See also Bertolotti M 1983 *Masers and Lasers, an Historical Approach* (Bristol: Hilger) pp 59 and 72
[245] Weber R L 1988 *Pioneers of Science* (Bristol: Hilger) p 207
[246] Vaughan J M 1989 *The Fabry–Perot Interferometer* (Bristol: Hilger) (in which on p 5 the attribution to M R Boulouch is detailed)
[247] Chandrasekhar S 1950 *Radiative Transfer* (Oxford: Oxford University Press)
[248] Bouwkamp C J 1950 *Philips Res. Rep.* **5** 321, 401
[249] Fellgett P 1951 *Thesis* University of Cambridge, UK; 1958 *J. Phys. Radium* **19** 187, 237
[250] Wald G 1951 *J. Opt. Soc. Am.* **41** 949
[251] Lewis A and Del Priore L V 1988 *Phys. Today* January p 38
[252] Young J Z and Roberts F 1951 *Nature* **167** 231
[253] Land E H 1951 *J. Opt. Soc. Am.* **41** 957
[254] Weeks R F (ed) 1994 *Opt. Photonics News* **5** No 10 (special issue)
[255] Destriau G 1936 *J. Chim. Phys.* **33** 587
[256] Lehovec K, Accardo C A and Jamgochian E 1951 *Phys. Rev.* **83** 603
[257] See Bertolotti M 1983 *Masers and Lasers, an Historical Approach* (Bristol: Hilger) p 74
[258] Elias P, Grey D S and Robinson D Z 1952 *J. Opt. Soc. Am.* **42** 127
See also Elias P 1953 *J. Opt. Soc. Am.* **43** 229
[259] O'Neill E L 1956 *IRE Trans.* **IT-2** 56
[260] Straubel P 1935 *P. Zeeman Verh.* (The Hague: Martinus Nijhoff)
[261] Luneberg R K 1960 *Mathematical Theory of Optics* (Berkeley, CA: University of California Press) (Mimeographed 1944 Lecture Notes Brown University, Providence, RI)
[262] di Francia T 1952 *Nuovo Cimento* **9** 426
[263] di Francia T 1955 *J. Opt. Soc. Am.* **45** 497
[264] von Laue M 1914 *Ann. Phys.* **44** 1197
[265] Glauber R J 1952 *Phys. Rev.* **87** 189
[266] Glauber R J 1954 *Phys. Rev.* **94** 751
[267] van Hove L 1954 *Phys. Rev.* **95** 249
[268] Komarov L I and Fischer I Z 1962 (Engl. Transl. 1963) *Sov. Phys.–JETP* **16** 1358
[269] Marechal A and Croce P 1953 *C. R. Acad. Sci., Paris* **237** 706
[270] Hopkins H H 1953 *Proc. R. Soc.* A **217** 408
[271] Hopkins H H 1951 *Proc. R. Soc.* A **208** 263
[272] Linfoot E H and Wolf E 1953 *Proc. Phys. Soc.* B **66** 145
[273] Linfoot E H and Wolf E 1956 *Proc. Phys. Soc.* B **69** 823
[274] Weber R L 1988 *Pioneers of Science* (Bristol: Hilger) p 147
[275] McKay K G and McAfee K B 1953 *Phys. Rev.* **91** 1079
[276] McKay K G 1954 *Phys. Rev.* **94** 877
[277] Hecht J 1992 *Laser Pioneers* (Boston: Academic)
See also Bertolotti M 1983 *Masers and Lasers, an Historical Approach* (Bristol: Hilger) p 166
[278] Gordon J P, Zeiger H J and Townes C H 1954 *Phys. Rev.* **95** 282
See also Bertolotti M 1983 *Masers and Lasers, an Historical Approach* (Bristol: Hilger) pp 79–81
[279] Dicke R H 1954 *Phys. Rev.* **93** 99

See also Bertolotti M 1983 *Masers and Lasers, an Historical Approach* (Bristol: Hilger) p 114
[280] van Heel A C S 1954 *Nature* **173** 39
[281] Hopkins H H and Kapany N S 1954 *Nature* **173** 39
[282] Bouwkamp C J 1954 *Rep. Prog. Phys.* **XVIII** 35
[283] Heitler W 1954 *The Quantum Theory of Radiation* (Oxford: Clarendon)
[284] Wolf E 1954 *Proc. R. Soc.* A **225** 96
[285] Chapin D M, Fuller C S and Pearson G L 1954 *J. Appl. Phys.* **25** 676
[286] Moss T S 1954 *Proc. Phys. Soc.* B **76** 775
[287] Burstein E 1954 *Phys. Rev.* **93** 632
[288] Gorelik G 1947 *Dokl. Akad. Nauk* **58** 45
[289] Forrester A T, Parkins W E and Gerjuoy E 1947 *Phys. Rev.* **72** 728
[290] Forrester A T, Gudmunsen R A and Johnson P O 1955 *Phys. Rev.* **99** 1691
[291] Autler S H and Townes C H 1955 *Phys. Rev.* **100** 703
[292] Wolf E 1955 *Proc. R. Soc.* A **230** 246
[293] Blanc-Lapierre A and Dumontet P 1955 *Rev. Opt.* **34** 1
[294] Hanbury Brown R and Twiss R Q 1956 *Nature* **178** 1046
[295] Hanbury Brown R and Twiss R Q 1954 *Phil. Mag.* **45** 663
[296] Hanbury Brown R 1974 *The Intensity Inteferometer* (London: Taylor and Francis)
[297] Steel W H 1983 *Interferometry* 2nd edn (Cambridge: Cambridge University Press)
[298] Hanbury Brown R and Twiss R Q 1957 *Proc. R. Soc.* A **242** 300
[299] Hanbury Brown R and Twiss R Q 1957 *Proc. R. Soc.* A **243** 291
[300] Rebka G A and Pound R V 1957 *Nature* **180** 1035
[301] Twiss R Q, Little A G and Hanbury Brown R 1957 *Nature* **180** 324
[302] Minsky M 1961 *US Patent* No 3,013,467, filed 1957
[303] Minsky M 1988 *Scanning* **10** 128
[304] Hopkins H H 1957 *Proc. Phys. Soc.* B **70** 1002
[305] Williams C S and Becklund O A 1989 *Introduction to the Optical Transfer Function* (New York: Wiley)
[306] Baker L R (ed) 1992 *Optical Transfer Function: Measurement* (Bellingham, WA: SPIE)
[307] Baker L R (ed) 1992 *Optical Transfer Function: Foundation and Theory* (Bellingham, WA: SPIE)
[308] See Bertolotti M 1983 *Masers and Lasers, an Historical Approach* (Bristol: Hilger) pp 103–6
[309] Schawlow A L and Townes C H 1960 *US Patent* No 2,929,922, filed 1958
See also Bertolotti M 1983 *Masers and Lasers, an Historical Approach* (Bristol: Hilger) pp 106, 117
[310] Dicke R H 1958 *US Patent* No 2,851,652
See also Bertolotti M 1983 *Masers and Lasers, an Historical Approach* (Bristol: Hilger) p 103
[311] Watanabe Y and Nishizawa J 1960 *Japanese Patent* No 273,217, filed 1957
See also Bertolotti M 1983 *Masers and Lasers, an Historical Approach* (Bristol: Hilger) pp 165, 185
[312] Bertolotti M 1983 *Masers and Lasers, an Historical Approach* (Bristol: Hilger) p 116
[313] Weber R L 1988 *Pioneers of Science* (Bristol: Hilger) p 195
[314] Prokhorov A M 1958 (Engl. Transl.) *Sov. Phys.–JETP* **7** 1140
[315] Weber R L 1988 *Pioneers of Science* (Bristol: Hilger) p 199
[316] Franz W 1958 *Z. Naturforsch.* A **13** 484
[317] Keldysh L V 1958 *Zh. Eksp. Teor. Fiz.* **34** 1138

[318] Bertolotti M 1983 *Masers and Lasers, an Historical Approach* (Bristol: Hilger) p 110
See also Hecht J 1992 *Laser Pioneers* (Boston: Academic) pp 13, 54, 113
[319] Elliott R J 1957 *Phys. Rev.* **108** 1384
[320] Dresselhaus G 1957 *Phys. Rev.* **106** 76
[321] Macfarlane G G, McLean T P, Quarrington J E and Roberts V 1957 *Phys. Rev.* **108** 1377; 1958 *Phys. Rev.* **111** 1245
[322] Smith R A 1978 *Semiconductors* 2nd edn (Cambridge: Cambridge University Press)
[323] Mandel L 1958 *Proc. Phys. Soc.* **72** 1037
[324] Mandel L 1959 *Proc. Phys. Soc.* **74** 233
[325] Manley J M and Rowe H E 1959 *Proc. IRE* **47** 2115
[326] Schawlow A L 1960 *First Int. Quantum Electronics Conf.* (New York: Columbia University Press) p 553
[327] Lengyel B A 1971 *Lasers* 2nd edn (New York: Wiley)
[328] Maiman T H 1960 *Phys. Rev. Lett.* **4** 564
[329] Hecht J 1992 *Laser Pioneers* (Boston: Academic) p 17
[330] Maiman T H 1960 *Nature* **187** 493
See also Hecht J 1992 *Laser Pioneers* (Boston: Academic) p 19
Bertolotti M 1983 *Masers and Lasers, an Historical Approach* (Bristol: Hilger) p 122
[331] Maiman T H 1961 *Phys. Rev.* **123** 1145
[332] Maiman T H, Hoskins R H, D'Haenens I J, Asawa C K and Evtuhov V 1961 *Phys. Rev.* **123** 1151
[333] Sorokin P P and Stevenson M J 1960 *Phys. Rev. Lett.* **5** 557
[334] Javan A, Bennett Jr W R and Herriott D R 1961 *Phys. Rev. Lett.* **6** 106
[335] White A D and Rigden J D 1962 *Proc. IRE* **50** 1697
[336] Snitzer E 1961 *J. Appl. Phys.* **32** 36
[337] Koester C J and Snitzer E 1964 *Appl. Opt.* **3** 1182
[338] Holst G C, Snitzer E and Wallace R 1969 *IEEE J. Quant. Electron.* **5** 342
[339] Stolen R H, Ippen E P and Tynes A R 1972 *Appl. Phys. Lett.* **20** 62
[340] Ippen E P and Stolen R H 1972 *Appl. Phys. Lett.* **21** 539
[341] Stone J and Burrus C A 1973 *Appl. Phys. Lett.* **23** 388
[342] Franken P A, Hill A E, Peters C W and Weinreich G 1961 *Phys. Rev. Lett.* **7** 118
[343] Bass M, Franken P A, Hill A E, Peters C W and Weinreich G 1962 *Phys. Rev. Lett.* **8** 18
[344] Kaiser W and Garrett C G B 1961 *Phys. Rev.* **7** 229
[345] Knight P L 1989 Quantum optics *The New Physics* ed P Davies (Cambridge: Cambridge University Press) ch 10
[346] Teich M C and Wolga G J 1966 *Phys. Rev. Lett.* **16** 625
[347] Bloembergen N 1992 *Huygen's Principle 1690–1990: Theory and Applications* ed H Blok, H A Ferwerda and H K Kuiken (Amsterdam: Elsevier) p 383
[348] Schawlow A L and Townes C H 1958 *Phys. Rev.* **112** 1940
[349] Fox A G and Li T 1960 *Proc. IRE* **48** 1904
[350] Fox A G and Li T 1961 *Bell Syst. Tech. J.* **40** 453
[351] Boyd G D and Gordon J P 1961 *Bell. Syst. Tech. J.* **40** 489
[352] Boyd G D and Kogelnik H 1962 *Bell. Syst. Tech. J.* **41** 1347
[353] Hellwarth R W 1961 *Advances in Quantum Electronics* ed R J Singer (New York: Columbia University Press) p 334
[354] McClung F J and Hellwarth R W 1962 *J. Appl. Phys.* **33** 828
[355] Woodbury E J and Ng W K 1962 *Proc. IRE* **50** 2367
[356] Hellwarth R W 1963 *Phys. Rev.* **130** 1850

Eckhardt G, Hellwarth R W, McClung F J, Schwarz S E, Weiner D and Woodbury E J 1962 *Phys. Rev. Lett.* **9** 455
[357] Slepian D and Pollack H O 1961 *Bell. Syst. Tech. J.* **40** 43
[358] Bernard M G A and Duraffourg G 1961 *Phys. Status Solidi* **1** 669
[359] Mandel L 1961 *J. Opt. Soc. Am.* **51** 1342
[360] Dumke P W 1962 *Phys. Rev.* **127** 1559
[361] Rosenthal A H 1962 *J. Opt. Soc. Am.* **52** 1143
[362] Macek W M and Davis D J M Jr 1963 *Appl. Phys. Lett.* **2** 67
[363] Lai M, Diels J-C and Dennis M 1992 *Opt. Lett.* **17** 1535
[364] Kleinman D A 1962 *Phys. Rev.* **126** 1977
[365] Kleinman D A 1962 *Phys. Rev.* **128** 1761
[366] Bloembergen N 1965 *Nonlinear Optics* (New York: Benjamin)
[367] Garrett C G B and Robinson F N H 1966 *IEEE J. Quant. Electron.* **2** 328
[368] Armstrong J A, Bloembergen N, Ducuing J and Pershan P S 1962 *Phys. Rev.* **127** 1918
Pershan P S 1963 *Phys. Rev.* **130** 919
[369] Bloembergen N and Pershan P S 1962 *Phys. Rev.* **128** 606
[370] Maker P D, Terhune R W, Nisenhoff M and Savage C M 1962 *Phys. Rev. Lett.* **8** 21
[371] Giordmaine J A 1962 *Phys. Rev. Lett.* **8** 19
[372] Ashkin A, Boyd G and Dziedzic J M 1963 *Phys. Rev. Lett.* **11** 14
[373] Franken P A and Ward J F 1963 *Rev. Mod. Phys.* **35** 23
[374] Fejer M M, Magel G A, Jundt D H and Byer R L 1992 *IEEE J. Quant. Electron.* **28** 2631
[375] See, for example, Khurgin J, Colak S, Stolzenberger R and Bhargava R N 1990 *Appl. Phys. Lett.* **57** 2540
[376] Kingston R H 1962 *Proc. IRE* **50** 472
[377] Akhmanov S I and Khokhlov R V 1962 *Sov. Phys.–JETP.* **43** 351
[378] Kroll N M 1962 *Phys. Rev.* **127** 1207
[379] Wang C C and Racette G W 1965 *Appl. Phys. Lett.* **6** 169
[380] Akhmanov S I, Kovrigin A, Piskaras A S, Fadeev V V and Khokhlov R V 1965 *JETP Lett.* **2** 191
[381] Giordmaine J A and Miller R C 1965 *Phys. Rev. Lett.* **14** 973
[382] Yang S T, Eckardt R C and Byer R L 1993 *Opt. Lett.* **18** 971 and references therein
[383] Greenburger D M, Horne M A and Zeilinger A 1993 *Phys. Today* August p 22
[384] Walls D F and Milburn G J 1994 *Quantum Optics* (Berlin: Springer)
[385] Mattle L, Weinfurter H and Zeilinger A 1994 *Proc. EQEC Amsterdam, September, 1994* Paper QTuC4
[386] Burnham D C and Weinberg D L 1970 *Phys. Rev. Lett.* **25** 84
[387] Askaryan G A 1962 (Engl. Transl.) *Sov. Phys.* **15** 1088
[388] Yariv A 1967 *Quantum Electronics* (New York: Wiley) frontispiece and p 398
[389] Yariv A and Louisell W H 1962 *Phys. Rev.* **125** 558
[390] Hecht J 1992 *Laser Pioneers* (Boston: Academic)
See also Bertolotti M 1983 *Masers and Lasers, an Historical Approach* (Bristol: Hilger) p 170
[391] Hall R N, Fenner G E, Kingsley J O, Soltys T J and Carlson R O 1962 *Phys. Rev. Lett.* **9** 366
[392] Nathan M I, Dunke W P, Burns G, Dill Jr F H and Lasher G 1962 *Appl. Phys. Lett.* **1** 62
[393] Keyes R J and Quist T M 1962 *Proc. IRE* **50** 1822
[394] Holonyak Jr N and Bevacqua S F 1962 *Appl. Phys. Lett.* **1** 82

[395] Thompson G B H 1980 *Physics of Semiconductor Laser Devices* (Chichester: Wiley)
[396] Denisyuk Yu N 1962 *Sov. Phys. Dokl.* **7** 543
[397] Stroke G W 1966 *An Introduction to Coherent Optics and Holography* (New York: Academic)
[398] Stroke G W and Labeyrie A E 1966 *Phys. Lett.* **20** 368
[399] Hariharan P 1984 *Optical Holography* (Cambridge: Cambridge University Press)
[400] Powell R L and Stetson K A 1965 *J. Opt. Soc. Am.* **55** 1593
[401] Benton S 1969 *J. Opt. Soc. Am.* **59** 1545
[402] Phillips N J 1983 *Proc. SPIE Int. Soc. Opt. Eng. (USA)* **402** 19
[403] Schwar M R J, Pandya T P and Weinberg F J 1967 *Nature* **215** 239
[404] Lohmann A and Paris D P 1967 *Appl. Opt.* **6** 1739
[405] Johnson L F 1963 *J. Appl. Phys.* **34** 897
[406] Terhune R W 1963 *Bull. Am. Phys. Soc.* **8** 359
[407] Harris S E and Targ R 1964 *Appl. Phys. Lett.* **5** 202
[408] Hargrave L E, Fork R L and Pollack M A 1964 *Appl. Phys. Lett.* **5** 4
[409] Mocker H and Collins R 1965 *Appl. Phys. Lett.* **7** 270
[410] De Maria A J, Stetser D A and Heyman H 1966 *Appl. Phys. Lett.* **8** 22
[411] Fork R L, Greene B I and Shank C V 1981 *Appl. Phys. Lett.* **38** 671
[412] Spence D E, Kean P N and Sibbett W 1991 *Opt. Lett.* **16** 42
[413] Adams M C, Sibbett W and Bradley D 1979 *Adv. Electron. Electron Phys.* **52** 265
[414] Lukosz W and Marchand M 1963 *Opt. Acta* **10** 241
[415] Lukosz W 1966 *J. Opt. Soc. Am.* **56** 1463
[416] Lukosz W 1967 *J. Opt. Soc. Am.* **57** 932
[417] Mandel L 1963 *Progress in Optics II* ed E Wolf (Amsterdam: North-Holland)
[418] Pancharatnam S 1963 *Proc. Indian Acad. Sci.* **57** 231
[419] Glauber R J 1963 *Phys. Rev. Lett.* **10** 84
[420] Glauber R J 1963 *Phys. Rev.* **130** 2529
[421] Mandel L and Wolf E 1968 *J. Phys. A: Gen. Phys.* **1** 625
See also Jakeman E and Pike E R 1968 *J. Phys. A: Gen. Phys.* **1** 627
[422] Bertolotti M 1983 *Masers and Lasers, an Historical Approach* (Bristol: Hilger) pp 217–28
[423] Klauder J R and Sudarshan E C G 1968 *Fundamentals of Quantum Optics* (New York: Benjamin)
[424] Glauber R J 1964 *Quantum Optics and Electronics* ed C DeWitt, A Blandin and C Cohen-Tanoudji Les Houches 1964 (New York: Gordon and Breach) p 65
[425] Kelley P L and Kleiner W H 1964 *Phys. Rev.* A **136** 316
[426] Mollow B R 1968 *Phys. Rev.* **168** 1896
[427] Yurke B 1985 *Phys. Rev.* A **32** 311
[428] Bondurant R S 1985 *Phys. Rev.* A **32** 2797
[429] Drummond P D 1987 *Phys. Rev.* A **35** 4253
[430] Le Berre M 1988 *Photons and Quantum Fluctuations* ed E R Pike and H Walther (Bristol: Hilger) p 31
[431] Cummins H Z, Knable N, Gampel L and Yeh Y 1963 *Appl. Phys. Lett.* **2** 62
[432] Ramachandran G N 1943 *Proc. Indian Acad. Sci.* A **18** 190
[433] Raman C V 1959 *Lectures in Physical Optics Part 1* (Bangalore: Indian Academy of Science) p 160
[434] Yeh Y and Cummins H Z 1964 *Appl. Phys. Lett.* **4** 176
[435] Ford N C Jr and Benedek G B 1965 *Phys. Rev. Lett.* **15** 649
[436] Patel C K N, Faust W L and McFarlane R A 1964 *Bull. Am. Phys. Soc.* **9** 500

[437] Crocker A, Kimmett M F, Gebbie H A and Mathias L E S 1964 *Nature* **201** 250
[438] Patel C K N, Faust W L, McFarlane R A and Garrett C G B 1964 *Appl. Phys. Lett.* **4** 18
[439] Gebbie H A, Stone N W B and Findlay F D 1964 *Nature* **202** 685
[440] Sochor V 1968 *Czech. J. Phys. B* **18** 60
[441] Lide D R Jr and Maki A G 1967 *Appl. Phys. Lett.* **11** 62
[442] Hartmann B and Kleman B 1968 *Appl. Phys. Lett.* **12** 168
[443] Benedict W S 1968 *Appl. Phys. Lett.* **12** 170
[444] Pollack M A and Tomlinson W J 1968 *Appl. Phys. Lett.* **12** 173
[445] Bridges W B 1964 *Appl. Phys. Lett.* **4** 128
[446] Hecht J 1992 *Laser Pioneers* (Boston: Academic)
[447] Lamb W E 1964 *Phys. Rev. A* **134** 1429
[448] Haken H 1964 *Z. Phys.* **181** 96
[449] Risken H Z 1965 *Z. Phys.* **186** 85
[450] Scully M O and Lamb Jr W E 1967 *Phys. Rev.* **159** 208
[451] Lax M and Louisell W 1967 *IEEE J. Quant. Electron.* **3** 47
[452] Weidlich W and Haake F 1965 *Z. Phys.* **185** 30
[453] Weidlich W, Risken H and Haken H 1967 *Z. Phys.* **201** 396
[454] Casagrande F and Lugiato L A 1977 *Phys. Rev. A* **15** 429
[455] Osterberg H and Smith L W 1964 *J. Opt. Soc. Am.* **54** 1073
[456] Miller S E 1969 *Bell Syst. Tech. J.* **48** 2059
[457] Vander Lugt A B 1964 *IEEE Trans. Info. Theor.* **IT-10** 139
[458] Vander Lugt A B 1992 *Optical Signal Processing* (New York: Wiley)
[459] Stark H (ed) 1982 *Applications of Optical Fourier Transforms* (New York: Academic)
[460] Chiao R Y, Townes C H and Stoicheff B P 1964 *Phys. Rev. Lett.* **12** 592
[461] Garmire E and Townes C H 1964 *Appl. Phys. Lett.* **5** 84
[462] Brewer R G and Reickhoff K E 1964 *Phys. Rev. Lett.* **13** 334a
[463] Maker P D, Terhune R W and Savage C M 1964 *Phys. Rev. Lett.* **12** 507
[464] Mayer G and Gires F 1964 *C. R. Acad. Sci., Paris* **258** 2039
[465] Saiki T, Takeuchi K, Kuwata-Gonokami M, Mitsuyu T and Okhawa K 1992 *Appl. Phys. Lett.* **60** 192
[466] van der Ziel J P, Pershan P S and Malmstrom L D 1965 *Phys. Rev. Lett.* **15** 190
[467] Arutyunian V M, Papazyan T A, Adonts C G, Karmenyan A V, Ishkhanyan S P and Khol'ts L 1975 *Sov. Phys.–JETP.* **41** 22
[468] Wieman C and Hänsch T 1976 *Phys. Rev. Lett.* **36** 1170
[469] Popov S V, Zheludev N I and Svirko Yu P 1994 *Opt. Lett.* **19** 13
[470] Kuwata M 1987 *J. Lumin.* **38** 247
[471] Valakh M Ya, Dykman M I, Lisitsa M P, Rudko Yu G and Tarasov G G 1979 *Solid State Commun.* **30** 133
Kovrigin A I, Yakovlev D V, Zhdanov B V and Zheludev N I 1980 *Opt. Commun.* **35** 92
[472] Apanasevitch S, Dovchenko D and Zheludev N I 1987 *Sov. Opt. Spectrosc.* **62** 481
[473] Bungay A R, Popov S V, Zheludev N I and Svirko Yu P 1995 *Opt. Lett.* **20** 356
[474] Akhmanov S A and Zharikov V I 1967 *Sov. Phys.–JETP Lett.* **6** 137
[475] Bairamov B H, Zakharchenya B P, Toporov V V and Kashkhozhev Z M 1973 *Sov. Phys.–Solid State* **15** 1245
[476] Akhmanov S A, Zhdanov B V, Zheludev N I, Kovrigin A I and Kuznetsov V I 1979 *Sov. Phys.–JETP Lett.* **29** 294
[477] Vlasov D V and Zaitseva V P 1971 *Sov. Phys.–JETP Lett.* **14** 171

[478] Zheludev N I and Paraschuk D Yu 1990 *Sov. Phys.–JETP Lett.* **52** 683
[479] Bungay A R, Svirko Yu P and Zheludev N I 1993 *Phys. Rev. Lett.* **70** 3039
[480] Kasper J V V and Pimentel G C 1965 *Phys. Rev. Lett.* **14** 352
[481] Gross R W F and Bott J F 1976 *Handbook of Chemical Lasers* (New York: Wiley)
[482] Silfvast W T, Fowles G R and Hopkins B D 1966 *Appl. Phys. Lett.* **8** 318
[483] Hecht J 1992 *Laser Pioneers* (Boston: Academic)
[484] Schell A J 1977 Thesis Harvard University
See also Schell A J and Bloembergen N 1978 *Phys. Rev.* A **18** 2592
[485] Mash D I, Morozov V V, Starunov V S and Fabelinskii I L 1965 *Sov. Phys.–JETP Lett.* **2** 25
[486] Bloembergen N and Lallemand P 1966 *Phys. Rev. Lett.* **16** 81
[487] Smith T 1930 *Trans. Opt. Soc.* **31** 244
See also Hallbach K 1964 *Am. J. Phys.* **32** 90
[488] Brouwer W 1964 *Matrix Methods in Optical Instrument Design* (New York: Benjamin)
[489] Welford W T 1965 *Progress in Optics IV* ed E Wolf (Amsterdam: North Holland) p 241
[490] Welford W T 1986 *Aberrations of Optical Systems* (Bristol: Hilger)
[491] Hinterberger H and Winston R 1966 *Rev. Sci. Instrum.* **37** 1094
[492] Baranov V K 1966 *Geliotekhnika* **2** 11 (Engl. Transl. *Appl. Solar Energy* **2** 9)
[493] Welford W T and Winston R 1978 *The Optics of Non-Imaging Concentrators* (New York: Academic)
[494] Waterman P C 1965 *Proc. IEEE* **53** 805
[495] Waterman P C 1971 *Phys. Rev.* D **3** 825
[496] Barber P and Yeh C 1975 *Appl. Opt.* **14** 2864
[497] Weber R L 1988 *Pioneers of Science* (Bristol: Hilger) p 201
[498] Sorokin P P and Lankard J R 1966 *IBM J. Res. Dev.* **10** 162
[499] Schäfer F P (ed) 1973 *Dye Lasers* (Berlin: Springer)
[500] Weber R L 1988 *Pioneers of Science* (Bristol: Hilger) p 207
[501] Lohmann A W and Brown B R 1966 *Appl. Opt.* **5** 967
[502] Kao K C and Hockham G A 1966 *Proc. IEEE* **113** 1151
[503] Werts A 1966 *L'Onde Electrique* **46** 967
[504] Kapron F P, Keck D B and Maurer R D 1970 *Appl. Phys. Lett.* **17** 423
[505] Gambling W A 1972 *Opt. Commun.* **6** 317
[506] Sandbank C P 1980 *Optical Fibre Communication Systems* (New York: Wiley)
[507] Poole S B, Payne D N and Fermann M E 1985 *Electron. Lett.* **21** 737
[508] Mears R J, Reekie L, Poole S B and Payne D N 1985 *Electron. Lett.* **21** 738
[509] Mears R J, Payne D N, Poole S B and Reekie L 1985 UK Patent No GB 2,180,392 B, priority date 13 August 1985, filed 13 August 1986
[510] Mears R J, Reekie L, Poole S B and Payne D N 1986 *Tech. Digest OFC'86 Conf., Atlanta* paper TUL15 (Washington DC: Optical Society of America) p 64
Mears R J, Reekie L, Jauncey I M and Payne D N 1987 *Tech. Digest IOOC/OFC Conf., Reno, Nevada* paper W12 (Washington DC: Optical Society of America)
Mears R J, Reekie L, Jauncey I M and Payne D N 1987 *Electron. Lett.* **23** 1026
Desurvire E, Simpson J R and Becker P C 1987 *Opt. Lett.* **12** 888
[511] Arecchi F T 1965 *Phys. Rev. Lett.* **15** 912
[512] Freed C and Haus H A 1966 *Physics of Quantum Electronics (Proc. Conf. San Juan, 1965)* P L Kelley, B Lax and P E Tannenwald (New York: McGraw-Hill)
[513] Johnson F A, McLean T P and Pike E R 1966 *Physics of Quantum Electronics*

(*Proc. Conf. San Juan, 1965*) ed P L Kelley, B Lax and P E Tannenwald (New York: McGraw-Hill)
[514] Kogelnik H and Li T 1966 *Appl. Opt.* **5** 1550
[515] Soffer B H and McFarland B B 1967 *Appl. Phys. Lett.* **10** 266
[516] Soffer B H 1964 *J. Appl. Phys.* **35** 2551
[517] Moulton P F 1986 *J. Opt. Soc. Am.* B **3** 125
[518] Terhune R W 1963 *Bull. Am. Phys. Soc.* **8** 359
[519] Harris S E, Oshman M K and Byer R L 1967 *Phys. Rev. Lett.* **18** 732
[520] Brewer R G 1967 *Phys. Rev. Lett.* **19** 8
[521] Shimizu F 1967 *Phys. Rev. Lett.* **19** 1097
[522] Bass M (ed) 1995 *Handbook of Optics* vol II (New York: McGraw-Hill) ch 6
[523] Lee P H, Schoefer P B and Barker W C 1968 *Appl. Phys. Lett.* **13** 373
[524] Szöke A, Daneu V, Goldhar J and Kurnit N A 1969 *Appl. Phys. Lett.* **15** 376
[525] Seidel H 1969 *US Patent* No 3,610,731, filed 1969
[526] Szöke A 1974 *US Patent* No 3,813,605, filed 1972
[527] Gibbs H M and Vekatesen T N C 1975 *J. Opt. Soc. Am.* **65** 1184
[528] Gibbs H M, McCall S L and Venkatesen T N C 1976 *Phys. Rev. Lett.* **36** 113
[529] Gibbs H M, McCall S L and Venkatesen 1977 *US Patent* No 4,012,699
[530] Gibbs H M, McCall S L and Venkatesen 1978 *US Patent* No 4,121167
[531] Gibbs H M, McCall S L, Venkatesen T N C, Gossard A C, Passner A and Wiegmann W 1979 *Appl. Phys. Lett.* **35** 451
[532] Flytzanis C 1975 *Quantum Electronics* vol 1A, ed H Rabin and C L Tang (New York: Academic) p 9
[533] Miller D A B, Mozolowski M H, Miller A and Smith S D 1978 *Opt. Commun.* **27** 133
[534] Miller D A B, Smith S D and Johnson A M 1979 *Appl. Phys. Lett.* **35** 658
[535] Bonifacio R and Lugiato L A 1978 *Phys. Rev.* A **18** 1129
[536] Smith S D, Janossy I, MacKenzie H A, Matthew J G H, Reid J J E, Taghisadeh M R, Tooley F A P and Walker A C 1985 *Opt. Eng.* **24** 569
[537] Gibbs H M 1985 *Optical Bistability* (Orlando, FL: Academic)
[538] Karpushko F V and Sinitsyn G V 1978 *J. Appl. Spectrosc. USSR* **29** 1323
[539] McCall S L and Gibbs H M 1978 *J. Opt. Soc. Am.* **68** 1378
[540] Wolf E 1969 *Opt. Commun.* **1** 153
[541] Kac A C and Slaney M 1988 *Principles of Computerised Tomographic Imaging* (New York: IEEE)
[542] Petrán M and Hadravský M 1966 *Czech Patent* application No 7,720
[543] Davidovits P and Egger M D 1969 *Nature* **223** 831
[544] Davidovits P and Egger M D 1971 *Appl. Opt.* **10** 1615
[545] Egger M D and Petrán M 1967 *Science* **157** 305
[546] Petrán M, Hadrawský M and Boyde A 1985 *Scanning* **7** 97
[547] Sheppard C J R and Choudhury A 1977 *Opt. Acta* **24** 1051
[548] Wilson T 1980 *Appl. Phys.* **22** 119
[549] Brakenhoff G J, Blom P and Barends P 1979 *J. Microsc.* **117** 219
[550] Amrein W O 1969 *Helv. Phys. Acta* **42** 149
[551] Pauli W 1964 *Collected Scientific Papers* vol II (New York: Interscience) p 608
[552] Jauch J M and Piron C 1967 *Helv. Phys. Acta* **40** 559
[553] Pike E R and Sarkar S 1987 *Phys. Rev.* A **35** 926
[554] Pike E R and Sarkar S 1989 *Quant. Opt.* **1** 61
[555] Pike E R 1991 *ECOOSA 90—Quantum Optics* ed M Bertolotti and E R Pike (Bristol: Institute of Physics) p 188
[556] A Einstein letter to M Besso 12 December 1951, see Pais A 1982 *Subtle is the Lord* (Oxford: Oxford University Press) p 382.

[557] Goldhaber A S and Nieto M M 1971 *Rev. Mod. Phys.* **43** 277
See also Goldhaber A S and Nieto M M 1976 *Sci. Am.* **234** May p 86
[558] Berreman D W and Scheffer T J 1970 *Mol. Cryst. Liq. Cryst.* **11** 395
Berreman D W 1972 *J. Opt. Soc. Am,* **62** 502
[559] Esaki L and Tsu R 1970 *IBM J. Res. Dev.* **14** 61
[560] Basov N G, Danilychev V A, Popov Y M and Khodkevich D D 1970 *Sov. Phys.–JETP. Lett.* **12** 329
[561] Ashkin A 1970 *Phys. Rev. Lett.* **24** 156
[562] Ashkin A and Dziedzic J M 1980 *Appl. Opt.* **19** 660
[563] See, for example, Steubing R W, Cheng S, Wright W H, Numjiri Y and Burns M W 1991 *Cytometry* **12** 505
See also Visscher K and Brakenhoff G J 1991 *Cytometry* **12** 486
[564] Allen L D and Eberly J H 1975 *Optical Resonance and Two Level Atoms* (New York: Wiley)
[565] Boyd R W 1992 *Nonlinear Optics* (San Diego: Academic)
[566] Dainty J C (ed) 1975 *Laser Speckle and Related Phenomena* (Berlin: Springer)
[567] Labeyrie A 1970 *Astron. Astrophys.* **6** 85
[568] Lohmann A W and Wirnitzer B 1984 *Proc. IEEE* **72** 889
[569] Lohmann A W and Weigelt G 1979 *Proc. ESA/ESO Conf. Astronomical Uses of the Space Telescope (Geneva, 1979)* ed F Macchetto *et al* (Geneva: ESA) p 353
[570] Collins G P 1992 *Phys. Today* **45** 17
[571] Davis J 1993 *Laser Focus World* February p 111
[572] Fugate R Q 1993 *Opt. Photonics News* **4** No 6 14
[573] Thompson L A 1994 *Phys. Today* December p 24
[574] Foord R, Jakeman E, Jones R, Oliver C J and Pike E R 1969 *Proc. IRE Conf. Proc.* **14** 271
[575] Benedek G B 1969 *Polarisation, Matière et Rayonnement A Kastler Jubilee Volume* ed Societé Francaise de Physique (Paris: Presses Universitaire de France) pp 49–84
[576] Jakeman E and Pike E R 1969 *J. Phys. A: Gen. Math.* **2** 411
[577] Jakeman E, Jones R, Oliver C J and Pike E R 1970 *RRE Memorandum* No 2621 (Malvern, UK: Ministry of Defence)
[578] Foord R, Jakeman E, Oliver C J, Pike E R, Blagrove R J, Wood E and Peacocke A R 1970 *Nature* **227** 242
[579] Cummins H Z and Pike E R (ed) 1974 *Photon Correlation and Light Beating Spectroscopy* (New York: Plenum)
[580] Cummins H Z and Pike E R (ed) 1977 *Photon Correlation Spectroscopy and Velocimetry* (New York: Plenum)
[581] Abbiss J B, Chubb T W, Mundell A R G, Sharpe P R, Oliver C J and Pike E R 1972 *J. Phys. D: Appl. Phys.* **5** L100
[582] Langdon P 1982 *High Speed Diesel Report* (Brookfield WI: Diesel and Gas Turbine Publications) September/October issue
[583] Abbiss J B, East L F, Nash C R, Parker P, Pike E R and Sawyer W G 1976 *Royal Aircraft Establishment Technical Report* No 75141
[584] Pike E R 1991 *Laboratory and Analysis Technology International* (London: Cornhill) p 57
[585] Institut Francais de Petrole 1993 *Patent Application (France)* No 14,347,000, filed November
[586] Weber R L 1988 *Pioneers of Science* (Bristol: Hilger) p 221
[587] Madey J M J 1971 *J. Appl. Phys.* **42** 1906
[588] Bonifacio R and De Salvo L 1994 *Nucl. Instrum. Methods* **341** 360
[589] Smith A E and Yansen D E 1971 *J. Opt. Soc. Am.* **61** 688
[590] McKechnie T S 1972 *Opt. Acta* **19** 729

[591] Freedman S J and Clauser J F 1972 *Phys. Rev. Lett.* **28** 938
[592] Shelton J W and Shen Y R 1972 *Phys. Rev.* A **5** 1867
[593] Hasegawa A and Tappert F D 1973 *Phys. Lett.* **23** 142
[594] Mollenauer L F, Stolen R H and Gordon J P 1980 *Phys. Rev. Lett.* **45** 1095
[595] Zel'dovich B Ya, Popovichev V I, Ragulskii V V and Faisullov F S 1972 *Sov. Phys.–JETP. Lett.* **15** 109
[596] Gerritsen H J 1967 *Appl. Phys. Lett.* **10** 237
[597] Stepanov B I, Ivakin E V and Rubanov A S 1971 *Sov. Phys. Dokl.* **16** 46
[598] Woerdman J P 1971 *Opt. Commun.* **2** 212
[599] Buchdahl H A 1993 *J. Opt. Soc. Am.* A **10** 524
[600] Buchdahl H A 1972 *J. Opt. Soc. Am.* **62** 1314
[601] Yariv A 1973 *IEEE J. Quant. Electron.* **9** 919
[602] Allen L D and Eberly J H 1975 *Optical Resonance and Two-Level Atoms* (New York: Wiley)
[603] Manka A, Dowling J P, Bowden C M and Fleischauer M 1994 *Quant. Opt.* **6** 371
[604] Bjorkholm J E and Ashkin A 1974 *Phys. Rev. Lett.* **32** 129
[605] Dingle R, Gossard A C and Wiegmann W 1975 *Phys. Rev. Lett.* **34** 1327
[606] Wolf E 1976 *Phys. Rev.* D **13** 869
[607] Yariv A 1976 *J. Opt. Soc. Am.* **66** 301
[608] Jakeman E and Shepherd T J 1984 *J. Phys. A: Math. Gen.* **17** L745
[609] Kimble H J, Dagenais M and Mandel L 1977 *Phys. Rev. Lett.* **39** 691
[610] Jakeman E, Pike E R, Pusey P N and Vaughan J M 1977 *J. Phys. A: Math. Gen.* **10** L257
[611] Agarwal G S, Brown A C, Narducci L M and Vetri G 1977 *Phys. Rev.* A **15** 1613
[612] Short R and Mandel L 1983 *Phys. Rev. Lett.* **51** 384
[613] Teich M C and Saleh B E A 1985 *J. Opt. Soc. Am.* B **2** 275
[614] Jakeman E and Walker J G 1985 *Opt. Commun.* **55** 219
[615] Yamamoto Y, Machida S and Nillson O 1986 *Phys. Rev.* A **34** 4025
[616] Diedrich F and Walther H 1987 *Phys. Rev. Lett.* **58** 203
[617] Bloom D M and Bjorklund G C 1977 *Appl. Phys. Lett.* **31** 592
[618] Hellwarth R W 1977 *J. Opt. Soc. Am.* **67** 1
[619] Yariv A and Pepper D M 1977 *Opt. Lett.* **1** 16
[620] Fisher R A (ed) 1982 *Optical Phase Conjugation* (New York: Academic)
[621] Yariv A 1977 *Opt. Commun.* **21** 49
[622] Pepper D M 1982 *Opt. Eng.* **21** 156
[623] Kukhtarev N V, Markov V B and Odulov S G 1977 *Opt. Commun.* **23** 338
[624] Vinetskii V L and Kukhtarev N V 1975 *Sov. Phys.–Solid State* **16** 2414
[625] Casasent D and Psaltis D 1977 *Proc. IEEE* **65** 770
[626] Casasent D (ed) 1978 *Optical Data Processing* (Berlin: Springer)
[627] Johnson K M, McKnight D J and Underwood I 1993 *IEEE J. Quant. Electron.* **29** 699
[628] Yu F T S and Jutamulia S 1992 *Optical Signal Processing, Computing and Neural Networks* (New York: Wiley)
[629] Landweber L 1951 *Am. J. Math.* **73** 615
[630] Fienup J R 1978 *Opt. Lett.* **3** 27
[631] Fienup J R 1982 *Appl. Opt.* **21** 2758
[632] Walker J G 1981 *Opt. Acta* **28** 735, 1017
[633] Cole E R 1974 *Digital Image Deblurring by Non-linear Homomorphic Filtering* Report No UTEC-CSc-74029 (Salt Lake City, UT: Computer Science Department, University of Utah)
Stockham T G, Cannon T M and Ingebretson R B 1975 *Proc. IEEE* **63** 678
[634] Ayers G R and Dainty J C 1988 *Opt. Lett.* **13** 47

[635] Hanisch R J and White R L (ed) 1994 *The Restoration of HST Images and Spectra-II* (Baltimore, MD: Space Telescope Science Institute)
[636] Richardson W H 1972 *J. Opt. Soc. Am.* **62** 55
[637] Lucy L B 1974 *Astron. J.* **79** 745
[638] Holmes T J 1992 *J. Opt. Soc. Am.* A **9** 1052
[639] Fish D A, Brinicombe A M and Pike E R 1995 *J. Opt. Soc. Am.* A **12** 58
[640] Shank C V, Ippen E P and Shapiro S L (ed) 1978 *Picosecond Phenomena* (New York: Springer)
[641] Bradley D J 1978 *J. Phys. Chem.* **82** 2259
[642] Johnson A M (ed) 1992 *Opt. Photonics News* **3** No 5 (special issue)
[643] Kapteyn H C and Murnane M M 1994 *Opt. Photonics News* **5** 20
[644] Kaiser W 1993 *Ultrashort Laser Pulses* 2nd edn (Berlin: Springer)
[645] Jakeman E and Pusey P N 1978 *Phys. Rev. Lett.* **40** 546
[646] Jakeman E and Tough R J A 1988 *Adv. Phys.* **37** 471
[647] Letokhov V S and Minogin V G 1979 *J. Opt. Soc. Am.* **69** 413
[648] Gilbert S L and Wieman C E 1993 *Opt. Photonics News* **4** 8
[649] Bouwhuis G, Braat J, Huijser A, Pasman J, van Rosmalen G and Schouhamer Immink K 1985 *Principles of Optical Disc Systems* (Bristol: Hilger)
[650] Hopkins H H 1979 *J. Opt. Soc. Am.* **69** 4
[651] Casey H C and Panish M B 1978 *Heterostructure Lasers* (New York: Academic)
[652] Soda H, Iga K, Kitahara C and Suematsu Y 1979 *Japan J. Appl. Phys.* **18** 2329
[653] Jewell J L, Harbison J P, Scherer A, Lee Y H and Florez L T 1991 *IEEE J. Quant. Electron.* **27** 1332
[654] Jacoby D, Pert G J, Ramsden S A, Shorrock L D and Tallents G T 1981 *Opt. Commun.* **37** 193
[655] Matthews D 1985 quoted in *Laser Pioneers* ed J Hecht (revised edition) (New York: Academic) p 275.
[656] Elton R C 1990 *X-Ray Lasers* (New York: Academic)
[657] Hora H and Miley G H 1984 *Laser Interaction and Related Plasma Phenomena* vol 6 (New York: Plenum)
[658] Matthews D L, Hagelstein P L, Rosen M D, Eckhart M J, Ceglio N M, Hazi A U, Medicki H, MacGowan B J, Trebes J E, Whitten B L, Campbell E M, Hatcher C W, Hawryluk A M, Kauffman R L, Pleasance L D, Rambach G, Scofield J, Stone G and Weaver T A 1985 *Phys. Rev. Lett.* **54** 110
[659] Zherikhin A, Koshelev K and Letokhov V 1976 *Sov. J. Quant. Electron.* **6** 82
[660] Vinogradov A V and Shlyaptsev V 1983 *Sov. J. Quant. Electron.* **13** 1511
[661] Suckewer S, Skinner C H, Milchberg H, Keane C and Voorhees D 1985 *Phys. Rev. Lett.* **55** 1753
[662] Horvath Z Gy, Malyutin A A and Kilpio A 1980 *Laser Focus* June p 32
[663] Petit R (ed) 1980 *The Electromagnetic Theory of Gratings* (Berlin: Springer)
[664] Berry M V and Upstill C 1980 *Progress in Optics XVIII* ed E Wolf (Amsterdam: North Holland) p 257
[665] Jakeman E, Pike E R and Pusey P N 1976 *Nature* **263** 215
[666] Jakeman E and Pusey P N 1980 *Inverse Scattering Problems in Optics* ed H P Baltes (Berlin: Springer)
[667] Hopkins H H 1981 *Opt. Acta* **28** 667
[668] Ikeda K, Daida H and Akimoto O 1980 *Phys. Rev. Lett.* **45** 709
[669] Eberly J H, Narozhny N B And Sanchez-Mondragon J J 1980 *Phys. Rev. Lett.* **44** 1323

[670] Weiss C O and King H 1982 *Opt. Commun.* **44** 59
[671] Abraham N B 1983 *Laser Focus* May p 73
[672] Firth W J 1991 Spontaneous spatial patterns in non-linear optics *ECOOSA 90—Quantum Optics* ed M Bertolotti and E R Pike (Bristol: Institute of Physics) p 173
[673] Arecchi F T and Harrison R G (ed) 1987 *Instabilities and Chaos in Quantum Optics* (Berlin: Springer)
[674] Kitano M, Yabudzaki T and Ogawa T 1981 *Phys. Rev. Lett.* **46** 926
[675] Zheludev N I 1989 *Sov. Phys.–Usp.* **32** 357
[676] Pippard A B 1982 *Eur. J. Phys.* **3** 65
[677] Aspect A, Grangier P and Roger G 1982 *Phys. Rev. Lett.* **49** 91
[678] Greenburger D M, Horne M A and Zeilinger A 1993 *Phys. Today* **46** August p 22
[679] di Francia T 1969 *J. Opt. Soc. Am.* **59** 799
[680] Harris J L 1964 *J. Opt. Soc. Am.* **54** 931
[681] McCutchen C W 1967 *J. Opt. Soc. Am.* **57** 1190
[682] Rushforth C K and Harris R W 1968 *J. Opt. Soc. Am.* **58** 539
[683] Gerchberg R W 1974 *Opt. Acta* **21** 709
[684] Pask C 1976 *J. Opt. Soc. Am.* **66** 68
[685] Bertero M and Pike E R 1982 *Opt. Acta* **29** 727
[686] Bertero M, De Mol C, Pike E R and Walker J G 1984 *Opt. Acta* **31** 923
[687] Grochmalicki J, Pike E R and Walker J G 1993 *Pure Appl. Opt.* **2** 1
[688] Capasso F, Tsang W T and Williams G F 1983 *IEEE Trans. Electron. Dev.* **ED-30** 38
[689] Walls D F 1983 *Nature* **306** 141
[690] Stoler D 1970 *Phys. Rev.* D **1** 3217
[691] Yuen H P 1976 *Phys. Rev.* A **13** 2226
[692] Braginsky V B and Vorontsov Y I 1974 *Sov. Phys.–Usp.* **17** 644
[693] Braginsky V B, Vorontsov Y I and Thorne K S 1980 *Science* **209** 547
[694] Caves C M, Thorne K S, Drever R W P, Sandberg V D and Zimmerman M 1980 *Rev. Mod. Phys.* **57** 341
[695] Ozawa M 1988 *Squeezed and Non-classical Light* ed P Tombesi and E R Pike (New York: Plenum) p 263
[696] Milburn G J and Walls D F 1983 *Phys. Rev.* A **28** 2065
[697] Levenson M D, Shelby R M, Reid M and Walls D F 1986 *Phys. Rev. Lett.* **57** 2473
[698] La Porta A, Slusher R E and Yurke B 1989 *Phys. Rev. Lett.* **62** 28
[699] Mollenauer L F and Stolen R H 1984 *Opt. Lett.* **9** 13
[700] Segev M, Crosignani B, Yariv A and Fischer B 1992 *Phys. Rev. Lett.* **68** 923
[701] Crosignani B, Segev M, Engin D Di, Porto P, Yariv A and Salamo G 1993 *J. Opt. Soc. Am.* B **10** 446
[702] Duree G C, Shulz J L, Salamo G J, Segev M, Yariv A, Crosignani B Di, Porto P, Sharp E J and Neurgaonkar R R 1993 *Phys. Rev. Lett.* **71** 533
[703] Swartzlander G A, Andersen D R, Regan J J, Yin H and Kaplan A E 1991 *Phys. Rev. Lett.* **66** 1583
[704] Swartzlander G A and Law C T 1992 *Phys. Rev. Lett.* **69** 2503
[705] Taylor J R (ed) 1992 *Optical Solitons* (Cambridge: Cambridge University Press)
[706] Miller D A B, Chemla D S, Damen T C, Gossard A C, Wiegmann W, Wood T H and Burrus C A 1984 *Phys. Rev. Lett.* **53** 2173

[707] Miller D A B, Chemla D S, Damen T C, Gossard A C, Wiegmann W, Wood T H and Burrus C A 1984 *Appl. Phys. Lett.* **45** 13
[708] Slusher R E, Hollberg L W, Yurke B, Mertz J C and Valley J F 1985 *Phys. Rev. Lett.* **55** 2409
[709] Shelby R M, Levenson M D, Perlmutter S M, DeVoe R G and Walls D F 1986 *Phys. Rev. Lett.* **57** 691
[710] Wu L-A, Kimble H J, Hall J L and Wu H 1986 *Phys. Rev. Lett.* **57** 2520
[711] Slusher R E and Yurke B 1990 *J. Lightwave Technol.* **8** 466
[712] Capasso F, Mohammed K, Cho A Y, Hull R and Hutchinson A L 1986 *Phys. Rev. Lett.* **55** 1152
[713] Capasso F, Mohammed K, Cho A Y, Hull R and Hutchinson A L 1985 *Appl. Phys. Lett.* **47** 420
[714] Capasso F, Mohammed K and Cho A Y 1986 *Appl. Phys. Lett.* **48** 478
[715] Kazarinov R F and Suris R A 1971 *Sov. Phys.–Semicond.* **5** 707
[716] Faist J, Capasso F, Sivco D L, Sitori C, Hutchinson A L and Cho A Y 1994 *Science* **264** 553
[717] Berry M V 1984 *Proc. R. Soc.* A **392** 45
[718] Chiao R Y and Wu Y-S 1986 *Phys. Rev. Lett.* **57** 933
Tomita A And Chao R Y 1986 *Phys. Rev. Lett.* **57** 937
[719] Jackson D A, Dandridge A and Sheem S K 1980 *Opt. Lett.* **5** 139
[720] Jackson D A and Jones J D C 1986 *Opt. Acta* **33** 1469
[721] Brown R G W 1987 *Appl. Opt.* **26** 4846
[722] Ricka J 1993 *Appl. Opt.* **32** 2860
[723] Li T 1993 *Proc. IEEE* **81** 1568
[724] Laude J-P 1993 *Wavelength Division Multiplexing* (New York: Prentice-Hall)
[725] Mollenauer L F and Smith K 1988 *Opt. Lett.* **13** 675
[726] Snyder A W and Love J D 1983 *Optical Waveguide Theory* (London: Chapman and Hall)
[727] Xiao M, Wu L-A and Kimble H J 1987 *Phys. Rev. Lett.* **59** 278
[728] Ou Z Y and Mandel L 1988 *Phys. Rev. Lett.* **61** 50
[729] Shih Y H and Alley C O 1988 *Phys. Rev. Lett.* **61** 2921
[730] Rarity J G and Tapster P R 1990 *Phys. Rev. Lett.* **64** 2495
[731] Levine B F, Bethea C G, Hasnain G, Walker J and Malik R J 1988 *Appl. Phys. Lett.* **53** 296
[732] Lallier E, Pocholle J P, Papuchon M, Grezes-Besset C, Pelletier E, De Micheli M, Li M J, He Q and Ostrowski D B 1989 *Electron. Lett.* **25** 1491
[733] Chandler P J, Field S J, Hanna D C, Shepherd D P, Tropper A C and Zhang L 1989 *Electron. Lett.* **25** 985
[734] Kyungwon An, Childs J J, Dasari R R and Feld M S 1994 *Phys. Rev. Lett.* **73** 3375
[735] Yamamoto Y (ed) 1991 *Coherence, Amplification and Quantum Effects in Semiconductor Lasers* (New York: Wiley)
[736] McCall S L, Levi A F J, Slusher R E, Pearton S J and Logan R A 1992 *Appl. Phys. Lett.* **60** 289
[737] Burns M M, Fournier J M and Golovchenko J A 1989 *Phys. Rev. Lett.* **63** 1233
[738] Guenther B D (ed) 1990 *Opt. Photonics News* **1** No 12 (special issue)
[739] Haroche S and Kleppner D 1989 *Phys. Today* **42** 24
[740] Morin S E, Wu Q and Mossberg T W 1992 *Opt. Photonics News* **3** No 8 8
[741] Slusher R E 1993 *Opt. Photonics News* **4** No 2 8
[742] Kapon E 1992 *Proc. IEEE* **80** 398
[743] Arakawa T, Nishioka M, Nagamune Y and Arakawa Y 1994 *Appl. Phys. Lett.* **64** 2200

[744] Hirayama H, Matsunaga K, Asada M and Suematsu Y 1994 *Electron. Lett.* **30** 142
[745] Chavez-Pirson A, Ando H, Saito H and Kanbe H 1994 *Appl. Phys. Lett.* **64** 1759
[746] Chemla D S (ed) 1993 *Phys. Today* **46** No 6 (special issue)
[747] Lent C S, Tougaw P D and Porod W 1993 *Appl. Phys.* **62** 714
[748] Bennett C H and Brassard G 1984 *Proc. IEEE Int. Conf. on Computers, Systems and Signal Processing (Bangalore)* (New York: IEEE) p 175
[749] Giacobino E, Fabre C and Leuchs G 1989 *Phys. World* February p 31
[750] Ekert A, Rarity J G, Tapster P R and Palma G M 1992 *Phys. Rev. Lett.* **69** 1293
[751] Peřina J, Hradil Z and Jurko B (ed) 1994 *Quantum Optics and the Fundamentals of Physics* (Dordrecht: Kluwer)
[752] Yablonovitch E 1993 *J. Opt. Soc. Am.* B **10** 283
[753] Dainty J C (ed) 1995 *Current Trends in Optics* (New York: Academic)
[754] Hänsch T and Toschek P 1970 *Z. Phys.* **236** 213
Arkhipkin V G and Heller Yu I 1983 *Phys. Lett.* A **98** 12
Kocharovskaya O A and Khanin Ya I 1988 *Sov. Phys.–JETP. Lett.* **48** 630
Harris S 1989 *Phys. Rev. Lett.* **62** 1033
Scully M O, Zhu S-Y and Gavridiles A 1989 *Phys. Rev. Lett.* **62** 2813
[755] Fano U 1961 *Phys. Rev.* **124** 1866
[756] Padmabandu G G and Pilloff H (ed) 1994 *Quant. Opt.* **6** No 4 (special issue)
[757] Arimondo E and Orriols G 1976 *Lett. Nuovo Cimento* **17** 333
Alzetta G, Gozzini A, Moi L and Orriols G 1976 *Nuovo Cimento* B **36** 5
[758] Dalton B J and Knight P L 1982 *Opt. Commun.* **42** 411
[759] Winter M P, Hall J L and Toschek P E 1990 *Phys. Rev. Lett.* **65** 3116
[760] Scully M O 1991 *Phys. Rev. Lett.* **67** 1855
[761] Imamoglu A and Ram R J 1994 *Opt. Lett.* **19** 1744
[762] Kocharovskaya O 1992 *Phys. Rep.* **219** 175
[763] Scully M O 1992 *Phys. Rep.* **219** 191
[764] Shore B W and Knight P L 1987 *J. Phys. B: At. Mol. Phys.* **20** 413
[765] Goutier Y and Trahin M 1982 *IEEE J. Quant. Electron.* **18** 1137
[766] Adams A and O'Reilly E 1992 *Phys. World* October p 43
[767] Welch D 1994 *Phys. World* February p 35
[768] Friend R, Bradley D and Holmes A 1992 *Phys. World* November p 42
[769] Sibbett W and Padgett M 1993 *Phys. World* October p 36
See also Mourou G 1992 *Laser Focus World* June p 51
[770] Canham L T 1992 *Phys. World* March p 41
[771] Iyer S S, Collins R T and Canham L T (ed) 1992 *Light Emission from Silicon (MRS 256)* (Pittsburgh: MRS)
[772] Forrest S R and Hinton H S (ed) 1993 *IEEE J. Quant. Electron.* **29** February (special issue on smart pixels)
[773] Hutley M C (ed) 1991 *Microlens Arrays (IOP Short Meeting Number 30)* (Bristol: Institute of Physics Publishing)
[774] Midwinter J E and Hinton H S (ed) 1992 *Opt. Quant. Electron.* **24** April (special issue on optical interconnects)
[775] Hohn F 1993 *Phys. World* March p 33
[776] Levenson M D 1993 *Phys. Today* July p 28
[777] Hall D (ed) 1982 *The Space Telescope Observatory* (Washington, DC: NASA Scientific and Technical Information Branch I)
[778] Goodman J W 1970 *Progress in Optics VIII* ed E Wolf (Amsterdam: North-Holland)
[779] Connes P, Froehly C and Facq P 1985 *Proc. ESA Colloquium Kilometric*

Optical Arrays in Space (ESA SP-26) (Noordwijk: ESA) p 49
Greenaway A 1991 Meas. Sci. Technol. **2** 1
[780] Armstrong J T, Hutter D J, Johnston K J and Mozurkewich D 1995 Phys. Today **48** No 5 42
[781] Moyer P and van Slambrouck T 1993 Laser Focus World October p 105
[782] Zenhausern F, O'Boyle M P and Wickramsinghe H K 1994 Appl. Phys. Lett. **65** 1623
[783] Shannon R R 1992 Opt. Photonics News **3** No 7 8
[784] Kaneko E 1987 Liquid Crystal TV Displays (Dordrecht: Reidel)
[785] Webb P O, McIntyre R J and Conradi J 1974 RCA Rev. **35** 234
[786] Boyle W S and Smith G E 1976 IEEE Trans. Electron. Dev. **ED-23** 661
Barbe D F (ed) 1980 Charge-coupled Devices (Berlin: Springer)
[787] Koch T L (ed) 1993 Opt. Photonics News **4** No 3 (special issue)
[788] Biberman L M and Nudelman S 1971 Photoelectronic Imaging Devices vol 2 (New York: Plenum)
[789] Whitaker J C 1994 Electronic Displays (New York: McGraw-Hill)
[790] Sandbank C P (ed) 1990 Digital Television (Chichester: Wiley)
[791] Noll A M 1988 Television Technology (Norwood, MA: Artech)
[792] Mort J 1994 Phys. Today **47** No 4 32
[793] Gundlach R W 1990 Technology of our Times ed F Su (Bellingham, WA: SPIE) p 56
[794] Shaw R (ed) 1994 Opt. Photonics News **5** No 1 (special issue)
[795] Helsel S K and Roth J P (ed) 1991 Virtual Reality (Westport: Meckler)
Larijani L C 1994 The Virtual Reality Primer (New York: McGraw-Hill)
[796] Travis A R L 1990 Appl. Opt. **29** 4341
[797] SPIE 1995 Conference 2409A Stereoscopic Displays and Applications VI (Bellingham, WA: SPIE)
[798] Wright W D 1964 The Measurement of Colour (Princeton, NJ: Van Nostrand)
[799] Roetling P (ed) 1992 Phys. Today **45** No 12 (special issue)
[800] Kingslake R 1992 Optics in Photography (Bellingham, WA: SPIE)
[801] See, for example, Baker K and Murray H 1993 Opt. Photonics News **4** No 6 8
[802] Fotakis C 1995 Opt. Photonics News **6** No 5 30
[803] Smith W J 1990 Modern Optical Engineering 2nd edn (New York: McGraw-Hill)
[804] Karim M A (ed) 1992 Electro-Optical Displays (New York: Marcel Dekker)
[805] Hudson R D 1969 Infrared Systems Engineering (New York: Wiley)
[806] Wolfe W L and Zissis G J 1985 The Infrared Handbook (Ann Arbor: ERIM)
[807] Elliott C T 1981 Electron. Lett. **17** 312
[808] Whatmore R W 1991 Rep. Prog. Phys. **49** 1335
[809] Jerlov N G 1976 Marine Optics (Amsterdam: Elsevier)
[810] Bromberg J L 1991 The Laser in America 1950–1970 (Cambridge, MA: MIT Press)
[811] Petley B W 1985 The Fundamental Physical Constants and the Frontier of Measurement (Bristol: Hilger) pp 54–68
[812] Braginsky V B, Vorontsov Y I and Thorne K S 1980 Science **209** 547
See also Walls D F and Milburn G J 1994 Quantum Optics (Berlin: Springer)
Braginsky V B and Vorontsov Y I 1974 Sov. Phys.–Usp. **17** 644
[813] Faber S 1994 New Sci. November p 40
[814] Sona A 1972 Lasers in Metrology Laser Handbook vol 2, ed F T Arecchi and E O Schulz-DuBois (Amsterdam: North-Holland) p 1457
[815] Durst F, Melling A and Whitelaw J H 1981 Principles and Practice of Laser Doppler Anemometry 2nd edn (London: Academic)

[816] Brown R G W and Pike E R 1983 Laser anemometry *Optical Transducers and Techniques in Engineering Measurement* ed A R Luxmore (London: Applied Science)
[817] Barth H G (ed) 1984 *Modern Methods of Particle Size Analysis* (New York: Wiley)
[818] Harry J E and Lunau F W 1972 *IEEE Trans. Ind. Appl.* **IA-8** 418
Lock E V and Hella R A 1974 *IEEE J. Quant. Electron.* **10** 179
[819] Cohen M I and Epperson J P 1968 *Electron Beam and Laser Beam Technology* ed L Martin and A B Elkareh (New York: Academic)
[820] Harrison R G 1979 *Phys. Bull.* **30** 259
[821] Ibbs K G and Osgood R M (ed) 1989 *Laser Chemical Processing for Microelectronics* (Cambridge: Cambridge University Press)
[822] Allen S D (ed) 1992 *Opt. Photonics News* **3** No 6 (special issue)
[823] Burns W K (ed) 1993 *Optical Fibre Rotation Sensing* (New York: Academic)
[824] Lefevre H 1993 *The Fibre Optic Gyroscope* (Boston, MA: Artech)
[825] Levenson M D and Kano S S 1988 *Introduction to Nonlinear Laser Spectroscopy* (San Diego: Academic) p 148
[826] Eckbreth A C 1988 *Laser Diagnostics for Combustion Temperature and Species* (New York: Gordon and Breach)
[827] Measures R M 1984 *Laser Remote Sensing* (New York: Wiley)
[828] Zhao X M and Diels J-C 1993 *Laser Focus World* November p 113
[829] Schagen P and Browning H 1952 *Philips Res. Rep.* No 7 119
[830] Seidel R W 1988 *Phys. Today* October p 36
[831] Robinson Jr C A 1981 *Aviat. Week Space Technol.* **25** February p 23
[832] Hecht J 1984 *Beam Weapons: the Next Arms Race* (New York: Plenum)
[833] Hecht J 1993 *Laser Focus World* December p 91
[834] Winker D M and McCormick M P 1994 *Lidar Techniques for Remote Sensing (SPIE 2310)* (Bellingham, WA: SPIE) p 98
[835] Curran R J et al 1987 LAWS laser atmospheric wind sounder earth observing system *NASA Instrument Panel Report* vol IIg (Washington DC: NASA)
[836] Betout P, Burridge D and Werner Ch 1989 ALADIN atmospheric laser Doppler instrument *ESA Lidar Working Group Report* ESA SP-1112 (Noordwijk: ESA)
[837] Stitch M L 1972 Laser Rangefinding *Laser Handbook* vol 2, ed F T Arecchi and E O Schulz-DuBois (Amsterdam: North-Holland) p 1751
[838] Foller J E and Wampler E J 1970 *Sci. Am.* March p 38
[839] Letokhov V S and Ustinov N D 1983 *Power Lasers and their Applications* (New York: Harwood)
[840] Krupke W F, George E V and Haus R A 1979 Advanced Lasers for Fusion *Laser Handbook* vol 3, ed F T Arecchi and E O Schulz-DuBois (Amsterdam: North-Holland) p 627
Motz H 1979 *The Physics of Laser Fusion* (London: Academic)
[841] Yamanaka C 1991 *Introduction to Laser Fusion* (New York: Gordon and Breach)
[842] Lindl J D, McCrory R L and Campbell E M 1992 *Phys. Today* **45** 32
[843] Danson C N et al 1993 *Opt. Commun.* **103** 392
[844] Rounds D E 1972 Laser Applications to Biology and Medicine *Laser Handbook* vol 2, ed F T Arecchi and E O Schulz-DuBois (Amsterdam: North-Holland) p 1863
[845] Caro R C and Choy D S J (ed) 1992 *Opt. Photonics News* **3** No 10 (special issue)
[846] Deutsch T F 1988 *Phys. Today* October p 56

[847] Morstyn G and Kaye A H (ed) 1990 *Phototherapy of Cancer* (New York: Gordon and Breach)
[848] Arons I J 1993 *Laser Focus World* October p 63
[849] Feke G T and Riva C E 1978 *J. Opt. Soc. Am.* **68** 526
[850] Hill D W, Pike E R and Gardner K 1981 *Trans. Opthal. Soc. UK* **101** 152
[851] Delpy D 1994 *Phys. World* **7** August p 34
[852] Scudder H J 1978 *Proc. IEEE* **66** 628
[853] Weitz D A and Pine D J 1993 Diffusing wave spectroscopy *Dynamic Light Scattering* ed W Brown (Oxford: Clarendon)
[854] Turner A P F, Karube I and Wilson G S (ed) 1987 *Biosensors* (Oxford: Oxford University Press)
[855] O'Mahoney M J 1992 *Opt. Photonics News* **3** No 1 8
[856] Desurvire E 1994 *Erbium-doped Fibre Amplifiers* (New York: Wiley)
[857] Darcie T E 1992 *Opt. Photonics News* **3** No 9 16
[858] Olshansky R 1994 *Opt. Photonics News* **5** No 2 15
[859] Isailovic J 1985 *Videodisc and Optical Memory Systems* (Englewood Cliffs, NJ: Prentice-Hall)
[860] Zech R G 1992 *Opt. Photonics News* **3** No 8 16
[861] Betzig E 1992 *Appl. Phys. Lett.* **61** 142
[862] Kitagawa T (ed) 1992 *OITDA Activity Report 5* (Tokyo: Optoelectronic Industry and Technology Development Association)
[863] Sato T 1992 *Opt. Photonics News* **3** No 11 25
[864] Klein M J 1977 *History of Twentieth Century Physics* ed C Weiner (London: Academic)
[865] Hecht E 1987 *Optics* 2nd edn (Reading, MA: Addison-Wesley)
[866] Boyd R W 1992 *Nonlinear Optics* (San Diego: Academic)
[867] Loudon R 1983 *The Quantum Theory of Light* 2nd edn (Oxford: Oxford University Press)
[868] Sze S M 1981 *Physics of Semiconductor Devices* 2nd edn (New York: Wiley)
[869] Pankove J I 1971 *Optical Processes in Semiconductors* (Englewood Cliffs, NJ: Prentice-Hall)
[870] Bhattacharya P 1994 *Semiconductor Optoelectronic Devices* (Englewood Cliffs, NJ: Prentice-Hall)
[871] Kingslake R (ed) 1965 *Applied Optics and Optical Engineering* vol I (New York: Academic); a series, up to 1992 vol XI
[872] Bass M (ed) 1995 *Handbook of Optics* vols I and II (New York: McGraw-Hill)
[873] Wolf E (ed) 1961 *Progress in Optics I* (Amsterdam: North-Holland); a series up to 1991 vol XXIX

Chapter 19

PHYSICS OF MATERIALS

Robert W Cahn

It has been said that the practice of metallurgy is the second-oldest profession, and indeed, metalworking was a long-established skill at the beginning of the twentieth century. Yet, as the century turned, the scientific understanding of the properties of metals was barely beginning, and other kinds of useful materials were either entirely in the domain of the craftsman or else still in the lap of the future. This is scarcely surprising: useful materials are complex, and understanding them in scientific terms required a willing confrontation of structural complexity. Many novel kinds of material, which are now familiar, had to await the creation of a proper in-depth understanding of those few materials that were in common use in 1900; that was not feasible until the study of simple entities such as isolated atoms had made sufficient progress to smooth the path to the scientific understanding of elaborate assemblies of several different kinds of atom arranged in a multiplicity of crystal structures: indeed at the beginning of the century, some reputable scientists declined to accept the very existence of atoms.

Mastering the physics of materials requires, of course, a general understanding of what distinguishes the properties of different kinds of atom. That, however, is only a beginning. A knowledge of crystal structures is equally necessary and in 1900 that was still 12 years in the future. Paradoxically, imperfections in crystal structures proved just as vital to the creation of a true physics of materials as did the identification of the *ideal* structure which the various kinds of imperfection then impair. To grasp the idea that a missing atom site in a crystal—a *vacancy*—is an entity in its own right, with its own well-defined characteristics, was as difficult an advance as the recognition of the concept of a hole in an electron energy band. Another kind of imperfection which gradually came to be seen as crucial was the *minor impurity*—a chemical defect. The development of the battery of techniques required for a detailed assessment of the nature and concentration of imperfections, structural and chemical alike, gradually came to constitute a subdiscipline in its own right: collectively, the use of such techniques is today called *materials characterization*.

Physicists (and others) seeking to improve useful materials, then, need to be at home with atoms, crystals and imperfections. In addition, they have a special need for two other kinds of insight: they need to understand *phase equilibria*, and they have to be thoroughly at home with *microstructures*. The general theory of phase equilibria is a contribution stemming from physical chemistry and, thanks to the genius of Willard Gibbs in the nineteenth century, this was already in a healthy state at the turn of the century. The expression of that general theory in any particular instance is the *phase* or *equilibrium diagram*, a phase being a region of a body which, when in equilibrium at any particular temperature and pressure, has a constant composition and internal structure. The first really accurate diagrams of this sort were just beginning to be published around 1900. A phase diagram, however, fails entirely to deliver one crucial kind of information: it does tell us what the volume fractions and individual compositions of the different phases are, for a specific overall composition of (say) an alloy at a particular temperature of equilibration, but it cannot tell us anything about the sizes, shapes and mutual disposition and orientations of blocks of those phases, or indeed of the separate crystal grains of any one phase. These issues are the domain of the subsidiary science of *microstructure*, and it is no exaggeration to assert that this is the central 'mystery' of the physics of materials—or perhaps one should say, of physical metallurgy. The study of microstructure by means of the light microscope was already fairly mature by 1900, thanks to the researches of Henry Sorby (1826–1908) in Sheffield thirty years earlier, and indeed several famous metallurgists—Roberts-Austen, Stead and Arnold in England and Osmond in France—were using the microscope to study phase transformations in steels shortly before the end of the nineteenth century. The new century, nevertheless, was to see an extraordinary flowering of this subsidiary science, of microstructure, and it was fated to retain and even enhance its central position in the physics of materials. We shall return to an examination of the role of microstructure in the mature stages of materials physics, and pause here only to point the reader to a superb analysis of Sorby's role in the early scientific study of materials [1]. This book is based on a conference held in 1963, to mark the centenary of Sorby's observation of a Widmanstätten microstructure in a steel; that structure is an array of epitactic layers of one phase precipitated from another, a kind of classical stereotype of metallic two-phase microstructures.

I have called the study of microstructure a 'subsidiary science'; it is only one, albeit the leading one, among the subsidiary sciences which cluster about the physics of materials. I propose to coin a neologism, the *subscidy*, to denote a *sub*sidiary *sci*entific *d*omain, that is, a subsidiary body of scientific understanding concerning some relatively narrow range of phenomena or investigative procedures, one which subsidizes the main scientific structure. Such subscidies deal with matters such as atomic transport through solids (diffusion); the geometry and dynamics of line defects in crystals (dislocations); the structure and local composition of interfaces, such as grain boundaries and free surfaces; the formation and behaviour of materials in forms which are very small in one, two or three dimensions (nanostructured materials, quantum dots); the crystallographic relationships between parent and product phases in a microstructure (this might be termed a subscidy of a subscidy, and there are many of these); and the mechanism and kinetics of the formation of surface layers on materials, for example, oxide films. These are a few of the many subscidies, subsidiary fields of scientific knowledge, in the physics of materials; some of them will feature in the following pages.

19.1. The establishment of the foundations of a science of materials

At the turn of the century, the scientific study of materials (largely metals, in those days) was certainly not recognized by anyone as an independent field of research. In general, it was pursued by physical and inorganic chemists, and in France the apportionment of metallurgy to chemistry has remained in force to this day; this tendency to adhere to long-established ways is perhaps not so surprising in a country in which labels such as *'physiologie animale'* and *'physiologie végétale'* still appear on doors in universities, and where a famous chemist who did not believe in atoms became Foreign Minister (Secretary of State) in 1885 [2]. Physicists had not interested themselves in the study of metals. Henry Sorby, the father of *metallography* (that is, the use of the optical microscope to examine polished and etched cross sections of metals and alloys by reflected light) was an independent, privately educated 'natural philosopher' (he never attended university) and he devoted more time to micrographic study of rocks and minerals than to metals. Indeed, formal crystallography, until the advent of x-ray diffraction, was the province of professional mineralogists, who perforce operated purely on the basis of the external appearance of single-crystal mineralogical specimens.

J Willard Gibbs' theoretical work on phase equilibria was initially appropriated by chemists, though Gibbs himself was a mechanical engineer turned mathematical physicist. In those days, the life of a university professor could be distinctly unrewarding in a pecuniary sense. For his first nine years as professor of mathematical physics at Yale, Gibbs was paid no salary at all, and it was only the attempt of the newly founded John Hopkins University to lure him away that persuaded the miserly authorities at Yale to offer a salary at last—and that did not match John Hopkins' offer! Gibbs, the former mechanical engineer, achieved what one of his biographers called 'a unique mastery of thermodynamics' and applied it to a profound analysis of the equilibria of phases. (Whether this analysis by an engineer/physicist is properly called physics or physical chemistry is an open issue; nineteenth century savants were apt to be cheerfully untroubled by such questions!) In 1876 he published, in an obscure New England journal, what came to be called the *Phase Rule*, as part of a long and difficult—and immortal—theoretical paper 'On the equilibrium of heterogeneous substances' [3]. Even though he circulated reprints liberally, and some distinguished physicists, James Clerk Maxwell in particular, became enthusiastic admirers [4], Gibbs' ideas took a long time to filter through into the consciousness of those who were concerned with the experimental study of phase equilibria. To a large extent this slow recognition was due to the difficult mathematical formulation of Gibbs' ideas. His concept of free energy (or chemical potential) and the related concept of entropy (with its statistical interpretation, soon after, at the hands of Boltzmann), came together to produce Gibbs' simple generalization as to the number of phases that could coexist in equilibrium at any one temperature, in relation to the number of constituents involved. Once this had been expressed in the simple form $F = C + 2 - P$, where C is the number of constituents, P the number of phases and F the number of degrees of freedom, the permissible topography of the phase fields in phase diagrams became clear and a major branch of the modern science of materials gathered momentum. (An example of the operation of this rule refers to a binary alloy system, with $C = 2$. Then, if two phases are to coexist in equilibrium, $F = 2$: this means that the alloy composition and the temperature can be allowed to vary while all the time two phases remain in equilibrium.)

Towards the end of the nineteenth century, the first accurate phase diagrams were published. The Dutch physical chemist Bakhuis Roozeboom of the University of Amsterdam in 1901 published the first phase diagram that was soundly based on the phase rule, for the iron–carbon system (the most important of all metallic phase diagrams). He was quickly followed by a remarkable experimental team of two investigators, Charles Heycock and Francis Neville (see box), both fellows of the Royal Society, who in 1904 jointly published a precocious study of the phase diagram of the copper–tin system—the tin bronzes, encouraged in this enterprise by a long correspondence with Roozeboom. This research culminated in one of the classic papers [5] of early physical metallurgy, in 1904. Figure 19.1 shows their diagram. It was based on the twin techniques of thermal analysis and optical micrography. In thermal analysis, an alloy specimen is slowly cooled from a high temperature, and anomalies in the temperature/time plot pinpoint temperatures at which phase transitions take place. In the other approach, specimens of appropriate compositions are water-quenched from a high temperature in order to preserve the phase structure characteristic of that temperature (this does not always work!), and the specimens are then sectioned, polished and etched to reveal the various phases present. In this, they were following the lead of Sorby's work on steels some decades earlier; Heycock and Neville, however, were the first to marry the micrographic approach to accurate phase diagram construction. In this way, a phase field such as β, which is stable only at high temperature, can be identified. Figure 19.2 shows some of Heycock and Neville's remarkable early micrographs; they could not be improved upon today. A detailed account of Heycock and Neville's lives and collaboration has been published [6], as has an account of the correspondence between them and Roozeboom [7].

Their copper–tin diagram has stood the test of time, even though some fine details have been modified by numerous later investigators. This work, together with the more variegated but much less accurate phase diagram research by Gustav Tammann in Göttingen, stimulated an enormous amount of research on many different binary and ternary systems in the early years of this century. By the 1930s, the findings had become so extensive, with so many contradictions between the work of different investigators who had looked at the same system, that a careful and critical juxtaposition of these researches had become essential. Such a volume, by a German industrial metallurgist, Max Hansen, was published in 1936 [8]. In it, 828 binary systems were critically surveyed; for 456 of these, there was sufficient information to hand to permit a phase diagram to be illustrated. A second edition published in 1958 dealt with 1334 systems and printed 717 diagrams. A number of more elaborate compilations of binary alloy systems have appeared since, especially in the USA. Recently, publication of the first systematic compilation of critically assessed *ternary* metallic phase diagrams has begun in Germany (the venture will require more than a decade and dozens of volumes): the number of ternary systems based on aluminium alone (in four huge volumes) exceeds the total number of all systems examined in the first edition of Hansen's binary book, even though the investigation of a ternary system is a far more time-consuming affair than that of a binary. All this demonstrates the crucial importance, for students of materials, of the researches started by Roozeboom and by Heycock and Neville almost a century ago.

19.2. Reference works and data compilations

This thumbnail sketch of the bibliography of phase diagrams prompts an aside on the broader role of tabulated and graphic compilations in the life of a modern materials scientist. An early 'universal' compilation was published as the *International Critical Tables*. This has long since been relegated to the cellars of most major libraries (a few still resolutely display it!), whereas the evergreen *Handbook of*

Twentieth Century Physics

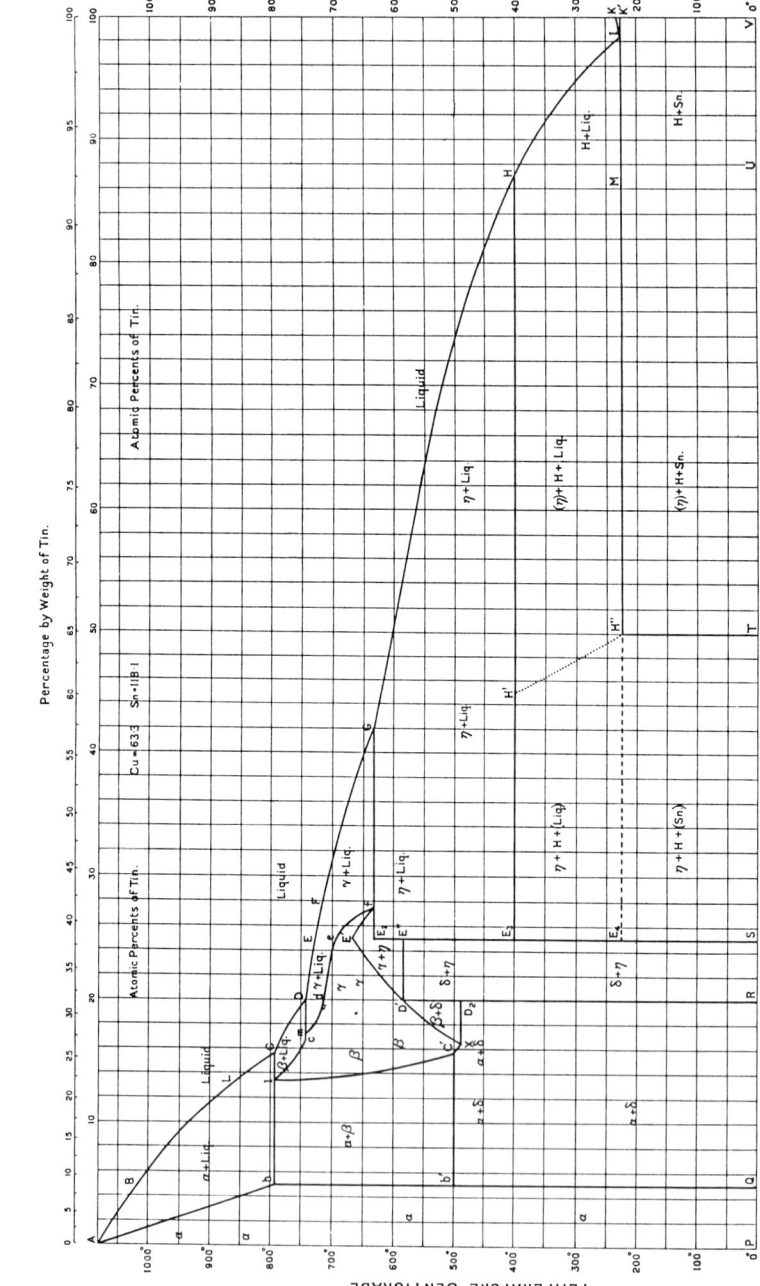

Figure 19.1. Part of the phase diagram of the copper-tin system, as published by Heycock and Neville in 1904.

Pioneers in Metallurgy

Charles Thomas Heycock (British, 1858–1931) and **Francis Henry Neville** (British, 1847–1915) were pioneers in physical metallurgy, before that science had been given a name and respectability. Heycock studied chemistry and Neville mathematics, both at Cambridge where they were to work for the rest of their lives. Their researches on alloy phase equilibria were undertaken over many years in a laboratory in the grounds of Sidney Sussex College, until in 1910 a drunken rowing eight used their paper records to light a celebratory bonfire! Heycock and Neville began by studying the depression of the freezing temperature of various metals when other metals were dissolved in them, testing van't Hoff's theory relating this depression to the latent heat of fusion of the solvent. This work became feasible when they were able to replace mercury thermometers by the platinum resistance thermometer developed independently by H L Callendar and by E H Griffiths, who, indeed, first brought the two investigators together. Over a period of years they focused their attention on the copper–tin system. Both Heycock and Neville were elected to Fellowship of the Royal Society, and Heycock was the first occupant of a new Readership (associate professorship) in metallurgy at Cambridge, endowed by the Worshipful Company of Goldsmiths in 1908; this endowment led directly to the appointment of the first Goldsmiths Professor of Metallurgy in Cambridge in 1932, and the establishment of that discipline as a degree subject.

Twentieth Century Physics

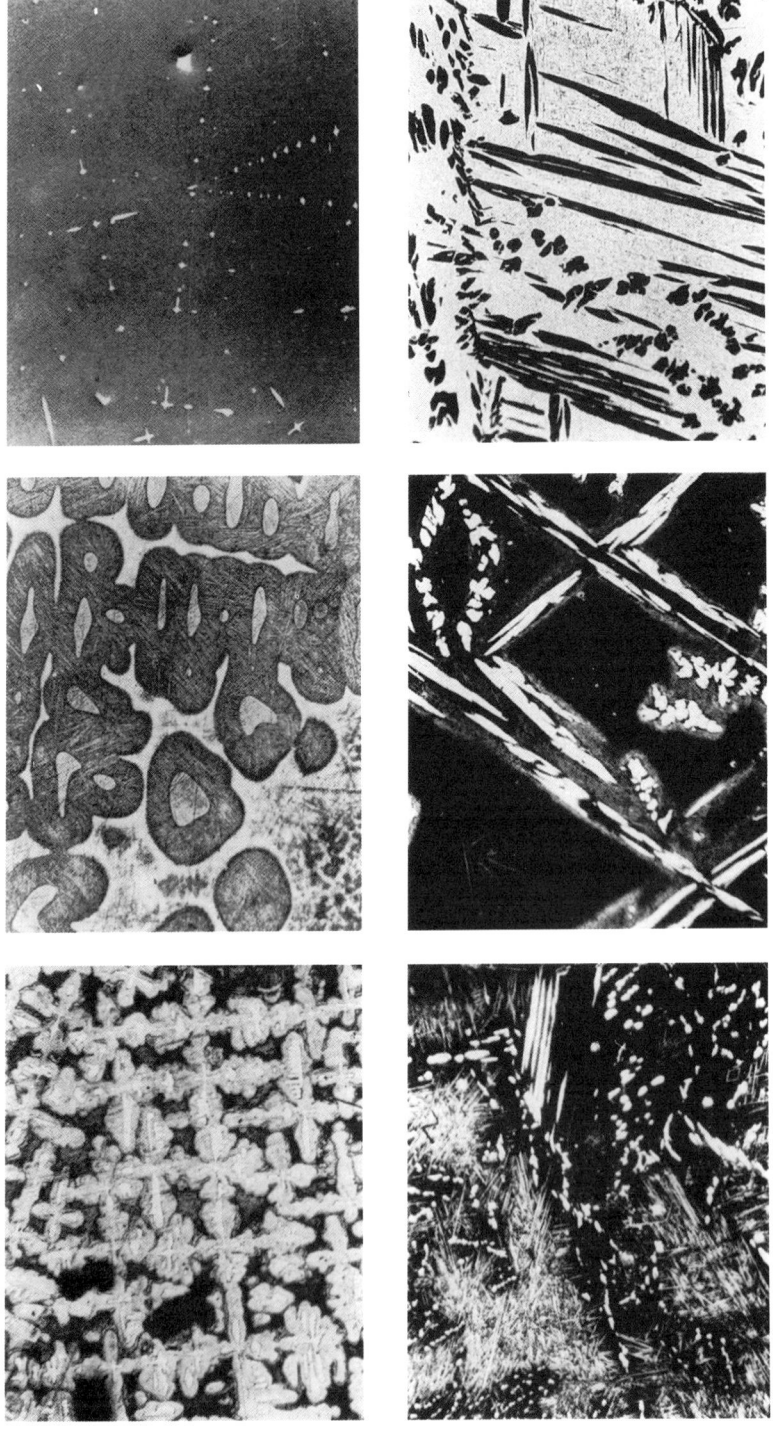

Figure 19.2. A selection of photomicrographs of alloys in the copper–tin system, as published by Heycock and Neville in 1904.

1512

Chemistry and Physics (subtitled *A Ready-Reference Book of Chemical and Physical Data*), first published in 1922 and now in its 73rd edition, remains an indispensable help to researchers on materials as well as many others. It has affectionately been dubbed the 'Rubber Bible' by generations of researchers, because it is published by the Chemical Rubber Company, now retitled the CRC Press, in Florida. Clearly, a systematic procedure for updating such compilations of data is essential and requires formidable organizational skills on the part of the editor. (In more than 70 years, the Rubber Bible has had only three editors). There are other, major data compilations which satisfy the requirement of regular updating, such as the *Landolt–Börnstein Tables* (which, like the Rubber Bible, hovers on the borders of physics and chemistry) and a succession of important handbooks which assembled the crystal structures of tens of thousands of compounds, such as the *Strukturbericht* of the interwar years and a more recent compilation assembled by Ralph Wyckoff [9]. Recently, these printed handbooks have begun to be complemented by computerized databases, on compact disc, tape or accessible 'on-line' via a computer network. For instance, there are several such databases for crystal structures of compounds (a survey has been published recently [10]), and by means of special software, these can be used to calculate bond lengths and interbond angles, statistics of Bravais lattices and space groups, and the like. The most recent database of this kind, the Intermetallic Phases Data Bank, currently has about 55 000 entries and this is beginning to find use in connection with the design of new alloys.

In addition to the compilations that assemble numerical and geometrical data exclusively, there has been a long series of handbooks, which range from collections of definitive review articles (such as the immortal German *Handbuch der Physik* of the interwar years, still cited today), via the frequently revised American *Metals Handbook*, an 11-volume *Encyclopedia of Materials Science and Engineering* (1986–93), and a very recent series of 21 books covering most of the science of materials, *Materials Science and Technology*, to collections of brief essays complemented by extensive numerical tables, such as the successive editions of the British *Metals Reference Book*. The labours of those who edit and those who provide raw material for such books and databases play an essential role in the development of a science of materials in the broadest possible sense.

19.3. Crystal structures

In outlining the scientific prehistory of the modern physics of materials, we focused on phase equilibria and on microstructural examination, with a strong emphasis on metals. These are two legs of the tripod which gave later investigators the elevation to survey extensive terrains and to create a genuine science of solids. The third leg is the study of crystal structures.

There is no need here to go in any detail into the history of x-ray diffraction; that is the province of Chapter 6. Here we will only discuss the matter from the viewpoint of materials physics, for which the events of 1912 were of primordial importance.

Crystallographers working in the context of mineralogy had already, by the end of the nineteenth century, worked out the formal theory of symmetry, lattices, point groups and space groups. Before the crystal structure of sodium chloride had been determined, the theory of the 230 space groups had been worked out. This is one of the most remarkable episodes in science of the general preceding the specific!

This is not the only peculiarity attaching to the beginnings of crystal structure determination. When Max von Laue instructed his assistants, Paul Knipping and Walter Friedrich, to send a beam of x-rays through a crystal of copper sulphate and onto a photographic plate and immediately after did the same with zincblende, they observed the first diffraction spots from a crystal. Laue had been inspired to undertake this experiment by a conversation with Paul Ewald, who had pointed out to him that atoms in a crystal had to be not only periodically arranged, but much closer together than a light wavelength. At the time, the nature of x-rays was still a mystery, and no one knew whether they were to be regarded as made up of particles or of waves. Laue clearly believed, from the way they were produced, that x-rays (*if* they *were* waves) must have a wavelength much shorter than that of visible light. Laue showed to his own and everyone else's satisfaction that indeed x-rays consisted of waves. Nevertheless, the crucial experiments that determined the structures of a number of simple crystals, beginning with sodium chloride, were done by the Braggs, William and Lawrence, father and son, over the following two years. At a meeting in London to celebrate the 40th anniversary of his discovery, Laue remarked, in public, how frustrated he felt afterwards that he had left it to the Braggs to make these epoch-making determinations; he had not made them himself because he was focused, not on the nature of crystals but on the nature of x-rays! By the time he had succeeded in shifting his focus, it was too late. It happens often in the history of physics, and other sciences too, that the fiercely concentrated discoverer of a new phenomenon does not see the consequence that stares him in the face. The minutiae of the events of 1912–14 are beautifully set out in an historical volume published in 1962 [11], while the subtle relations between father and son in the most celebrated of all scientific cooperations between the generations of the same family are memorably described in Gwendolen Caroe's memoir of her father, William Bragg [12]. Another article on 'how Lawrence Bragg invented x-ray analysis', by Max Perutz, has appeared quite recently [13].

It rapidly became clear that most metals had very simple crystal structures, and this helped to account for their ability often to dissolve large percentages of other metals to form solid solutions, something that few minerals or other compounds are able to do. Intermetallic compounds gave the new x-ray crystallographers a great deal to do and coming seriously to grips with this category of material took many long years, and indeed is still going on today. One notable episode concerned the solid solutions between copper and gold, which can be mixed in all proportions to form face-centred cubic solid solutions. A group of Russian chemists, Kurnakov and his colleagues, discovered in 1916 that slowly cooled alloys had anomalously low electrical resistivities at simple compositions (CuAu and Cu_3Au) but the same alloys were no longer anomalous if they were water-quenched from a high temperature; they were quite unable to make any sense of this [14]. The answer came from x-ray diffraction experiments done some years later, in 1925, by a pair of Swedish physicists, Johansson and Linde [15]: they found that the slowly cooled alloys had an *ordered* arrangement of the copper and gold atoms on distinct sublattices, while the alloys quenched from high temperatures had a random distribution of the two species among the lattice sites. This was a very simple idea once it had been pinned down, but was nevertheless so revolutionary that no one had thought of it before (apart from a passing study by a highly original American metallurgist, Edgar Bain, in 1923). It is perhaps poetically just that Bain's name is associated with one of the most controversial of metallurgical phases, bainite in steels, about the precise nature of which controversy is still raging today. The alloy phases of copper and gold were found to order at well-defined temperatures during slow cooling, and such *order–disorder* transformations proved to have an irresistible attraction for students of statistical mechanics. The discovery of atomic order began to inject some rationality into the analysis of intermetallic compounds, which had resisted understanding in terms of the chemical concept of valence. Incidentally, the study of ordered intermetallic compounds has broadened in recent years into an enormous enterprise involving physicists and metallurgists alike, aimed at creating a new category of load-bearing high-temperature materials, for aero-engines in particular. The technology of this field of research has just received a comprehensive and authoritative overview [16], while the equally extensive underlying physics of order–disorder transformations has been more concisely surveyed [17].

See also p 548

Crystallography is a very broad science, stretching from crystal-structure determination to crystal physics (especially the systematic study of anisotropy), the prediction of crystal structures from first principles (very active nowadays, and entirely dependent on the advances in the electron theory of solids treated in Chapter 17 of this book), crystal chemistry and the geometrical study of phase transitions.

In the narrow sense of crystal structure determination it has had a long innings since 1912: no fewer than 26 Nobel Prizes have gone to scientists best described as crystallographers, some of them physicists, some chemists, some biochemists. It is one of those fields where physics and chemistry have become intimately commingled. It has also evinced more than its fair share of quarrelsomeness, since some physicists regard it as a technique rather than a science, while investigators of crystal structures (especially chemists) were for some years inclined to regard anyone who studied other aspects of crystals as second-class citizens. What this shows is that we scientists will never learn: we argue about terminology as though this were an argument about the 'real world', and we cannot be cured of the urge to rank ourselves (or rather, each other) into categories of relative superiority and inferiority.

Returning to metals specifically, the fact that the puzzle concerning copper–gold alloys had been uncovered by inorganic chemists in 1916 and had been resolved by physicists in 1925 is symptomatic of the gradual shift of research on metals from physical and inorganic chemists towards physicists. It is now time to examine this trend.

19.4. The birth of physical metallurgy
In the early years of the twentieth century, the scientific study of materials was virtually restricted to metals and alloys: ceramics were still caught in the craft traditions of earlier centuries, while semiconductors and other 'electronic' materials, polymers and composites were as yet asleep, awaiting the kiss of life from scientists who were then at school.

Those early years, up to 1914, were the time when the study of metals was transformed by a brilliant, energetic and resolute man, Walter Rosenhain. He was an Australian; though he was born in Berlin in 1875, his family emigrated in 1880 to Melbourne and there he was educated as a civil engineer. In 1897 he won a scholarship to take him to Cambridge where he did research with Sir Alfred Ewing (an unusual engineer for his time, with interests in topics such as ferromagnetism). At first he was assigned a problem involving steam jets, but this did not satisfy him. Then Ewing suggested to him that he try to find out how it was possible for a metal to undergo plastic deformation without losing its crystalline structure (which Ewing, but by no means everybody else, recognized metals as having). Thereafter Rosenhain was in his element: he polished the surface of various metals and deformed them slightly, for instance by bending and re-straightening, and examined the deformed metals under a microscope by reflected light—following in Sorby's footsteps. He observed surface steps (figure 19.3) and proved that they were indeed steps by plating the surface with another metal and sectioning the plated assembly. The implication was inescapable: blocks of crystal were sliding over each other along well-defined crystal planes, without losing any cohesion. A series of major papers in the *Philosophical*

Transactions of the Royal Society (see, for example, reference [18]) made this discovery widely known. The phenomenon had in fact been discovered years earlier in Germany by the mineralogist Reusch who examined rocksalt (1867) and another tireless German mineralogist, Mügge, who examined native crystals of copper, gold, etc at about the same time and actually was able to establish the crystallography (slip along ⟨110⟩ on a {111} plane), but it was Rosenhain who was transfixed by his discovery and who in consequence transformed the study of metals. In fact, Rosenhain's observations showed that at intervals along a surface, the directions of the 'slip lines' abruptly changed. This proved that ordinary metals were polycrystalline, made up of small 'grains' typically 0.1 mm across, and this was the deathknell of fanciful ideas still current at that time, such as the notion that a metal began amorphous and gradually crystallized under the influence of fatigue, i.e. under a series of reversing stresses: this idea arose from the fact that a fatigue fracture surface consisted of bright crystalline cleavage facets separated by matte regions. (A paper by Ewing and Humfrey in 1903 got rid of this notion once and for all.) In fact, Rosenhain and Ewing went further than that: they accidentally discovered the phenomenon of recrystallization of plastically deformed metals—the central 'repair mechanism' by which damaged metals recover their pristine physical properties—a topic which has spawned an immense series of researches during the present century.

His work at Cambridge complete, Rosenhain went to work in the British glass industry for some years (where, in his own words, he functioned as 'a tame scientist kept on the premises'); however, his heart was in metallurgy and he set up his own private laboratory at his home, training his wife to prepare specimens for microscopic examination. In 1906 he was appointed Superintendent of the new Department of Metallurgy and Metallurgical Chemistry (the second part of the name was fated to disappear in due course) in the National Physical Laboratory (NPL) at Teddington, near London. For the next 25 years, he used this vantage point to transform research on metals, especially on alloy constitution and its relation to useful properties. His career is described in an excellent biographical article by Kelly [19], prepared for a centenary celebration of Rosenhain's birth.

At the NPL, Rosenhain and his rapidly growing band of collaborators initially concentrated on the development of new aluminium alloys, which involved the perfection of a variety of physical techniques for determining the temperatures of phase transformations, in particular. Gradually, a variety of other metallurgical problems came to be studied by physical methods. An ever-increasing awareness of the importance of the physical approach to the study of metals and alloys led Rosenhain to write his famous text, *An Introduction to the Study of Physical Metallurgy*, published by Constable in London in 1914. This remained popular for

Twentieth Century Physics

Figure 19.3. *Photomicrograph, by reflected light, of glide steps on the surface of plastically deformed lead.*

many years, and a third edition was published in 1934, the year of his death.

In addition to the extensive practical research in his Division, Rosenhain continued to concern himself with the fundamental physics of metals. In his day, that meant issues such as these: what is the structure of the boundaries between the distinct crystal grains in polycrystalline metals (all commercial metals are in fact polycrystalline), and why does metal harden as it is progressively deformed plastically, i.e., why does it work-harden? Rosenhain formulated a generic model, which became known as the amorphous metal hypothesis, according to which grains are held together by 'amorphous cement' at the grain boundaries, and work-hardening is due to the deposition of layers of amorphous material within the slip bands which he had observed. These erroneous ideas he defended with great skill and greater eloquence over many years, against many forceful counterattacks. Metallurgists at last had begun to argue about basics in the way that physicists had long done. Concerning this period and the amorphous grain-boundary cement theory in particular, Rosenhain's biographer has this to say 'The theory was wrong in scientific detail but it was of great utility. It enabled the metallurgist to reason and recognize that at high temperatures grain boundaries are fragile, that heat-treatment involving hot or cold work coupled with annealing can lead to benefits in some instances and to catastrophes such as "hot shortness" in others [this term means brittleness at high temperatures] ... Advances in technology and practice do not always require exact theory. This must always be striven for, it is true, but a hand-waving argument which calls salient facts to attention, if readily grasped in apparently simple terms, can be of great practical utility.' [19].

This controversial claim goes to the heart of the relation between metallurgy as it was, and as it was fated to become under the influence of physical ideas and, more importantly, of the physicist's approach. We turn to this issue next.

19.5. The birth of quantitative theory in physical metallurgy

In astrophysics, reality cannot be changed by anything the observer can do. The classical principle of 'changing one thing at a time' in a scientific experiment, to see what happens to the outcome, has no application to the stars! Therefore, the acceptability of a hypothesis intended to interpret some facet of what is 'out there' depends entirely on rigorous quantitative *self-consistency*. Metallurgists used to think that they were exempt from this rule; while they thought so, physicists did not accept their science as being 'serious'.

The matter was memorably expressed recently in a book, GENIUS— *The Life of Richard Feynman* by James Gleick, 'So many of his witnesses observed the utter freedom of his flights of thought, yet when Feynman talked about his own methods he emphasized not freedom but constraint ... For Feynman the essence of scientific imagination was a powerful and almost painful rule. What scientists create must match reality. It must match what is already known. Scientific imagination, he said, is imagination in a straitjacket ... The rules of harmonic progression made [for Mozart] a cage as unyielding as the sonnet did for Shakespeare. As unyielding and as liberating—for later critics found the creators' genius in the counterpoint of structure and freedom, rigor and inventiveness.'.

This also expresses accurately what was new in the metallurgical breakthroughs of the early 1950s.

As we have seen, Rosenhain fought hard to defend his preferred model of the structure of grain boundaries, based on the notion that layers of amorphous, or glassy, material occupied these discontinuities. The trouble with the battles he fought was twofold: there was no theoretical treatment to predict what properties such a layer would have, for an assumed thickness and composition, and there were insufficient experimental data on the properties of grain boundaries, such as specific energies of grain boundaries. This lack, in turn, was to some degree due to the absence of appropriate experimental techniques of characterization, but not to this alone: no one measured the energy of a grain boundary as a function of the angle of misorientation between the adjacent crystal lattices, not because it was difficult to do, even then, but because metallurgists could not see the point of doing it. Studying a grain boundary *in its own right* was deemed a waste of time—only grain boundaries as they directly affected useful properties such as ductility deserved attention. In other words, the cultivation of subsidies was not yet thought justifiable by most metallurgists.

A look at the history of dislocations is instructive. Dislocations are line defects in crystals, which are now known to be the 'vectors' of plastic deformation. They were first proposed ('invented' would be a better term) in 1934, simultaneously and independently by three people—an applied mathematician, Geoffrey Taylor; an engineer, Egon Orowan; and a physical chemist (later a philosopher!), Michael Polanyi. The invention came, significantly, because of a mismatch between the calculated resistance of a crystal lattice to plastic glide—the process that Rosenhain had first observed in 1898–99—and the actual observed yield-stresses of single metallic crystals. (The fact that anyone at all made single crystals of metals, in order to measure their strength—wholly useless items from an industrialist's viewpoint—showed that the contempt for subsidies was not universal.) The observed strength was hundreds of times smaller than a very simple theoretical approach had predicted, and the solution was to suggest that one block of crystal slid over its neighbour, not all at once but by the progressive displacement of a defect which permitted *localized* glide. The time-honoured analogy is with the problem of moving a large carpet a few inches over a floor: the large friction can only be overcome by pushing a 'ruck' from one side of the carpet to the other.

However, no one could see the postulated dislocations, which were supposed to be long lines, not necessarily straight, but with effective diameters of only a few atoms. Optical microscopes were apparently of no use in searching for these elusive entities. Nevertheless, after World War II dislocations were taken up by some metallurgists, who held them responsible, in a purely handwaving manner and even though there was as yet no evidence at all for their very existence, for a variety of phenomena such as brittle fracture. They were thought by some to explain everything imaginable, and therefore 'respectable' scientists reckoned that they explained nothing.

What was needed was to escape from handwaving. That milestone was passed in 1947 when Alan Cottrell at Birmingham University (which was at that time the principal centre of physical metallurgy research) (see box) formulated a rigorously quantitative theory of the discontinuous yield-stress in mild steel. When a specimen of such a steel is stretched, it behaves elastically until, at a particular stress, it *suddenly* gives way and then continues to deform at a lower stress. If the test is interrupted, then after some minutes holding at ambient temperature the former yield-stress is restored, i.e. the steel strengthens or *strain-ages*. This phenomenon was much debated but not understood at all. Cottrell, influenced by the dislocation theorists Egon Orowan and Frank Nabarro (as set out by Ernest Braun in a recently published history of solid-state physics [20]) came up with a novel model. The essence of Cottrell's idea was given in the abstract of his paper to a conference on dislocations held in Bristol in 1947, as cited by Braun

> It is shown that solute atoms differing in size from those of the solvent [carbon, in fact] can relieve hydrostatic stresses in a crystal and will thus migrate to the regions where they can relieve the most stress. As a result they will cluster round dislocations forming 'atmospheres' similar to the ionic atmospheres of the Debye–Hückel theory of electrolytes. The conditions of formation and properties of these atmospheres are examined and the theory is applied to problems of precipitation, creep and the yield point.

The importance of this advance is hidden in the simple words 'It is shown ...'. Cottrell (later joined by Bruce Bilby in formulating the definitive version of his theory), by precise application of elasticity theory to the problem, was able to work out the concentration gradient across the carbon atmospheres, what determines whether the atmosphere 'condenses' at the dislocation line and thus ensures a well-defined yield-stress, the integrated force holding a dislocation to an atmosphere (which determines the drop in stress after yield has taken place) and, most impressively, he was able to predict the time law governing the reassembly of the atmosphere after the dislocation had been torn away from it by exceeding the yield-stress—that is, the strain-aging kinetics. Thus it was possible to compare accurate measurement with precise theory. The decider was the strain-aging kinetics, because the theory produced the prediction that the fraction of carbon atoms that has rejoined the atmosphere is strictly proportional to $t^{2/3}$, where t is the time of strain-aging after a steel specimen has been taken past its yield-stress.

In 1951, this strain-aging law was checked by Harper [21] by a method which perfectly encapsulates the changes which were transforming physical metallurgy around the middle of the century. It was necessary to measure the change with time of free carbon dissolved in the iron, and to do this in spite of the fact that the solubility of carbon in iron at ambient temperature is only a tiny fraction of one per cent. Harper performed this apparently impossible task by using a torsional pendulum, invented just before World War II by a Dutch physicist, J L Snoek, as shown in figure 19.4(*a*). The specimen is in the form of a wire held under slight tension in the elastic regime, and the inertia arm is sent into

Alan Howard Cottrell
(British, b 1919)

Sir Alan Cottrell is a pioneer who has played a major part in turning physical metallurgy into a respectable scientific discipline, particularly by introducing rigorous theory into the interpretation of the mechanical behaviour of metals and alloys. In 1948 he published a small book, *Theoretical Structural Metallurgy*, which introduced ideas of band theory, statistical thermodynamics (including the thermodynamic interpretation of phase diagrams) and the elastic theory of dislocations to a wide audience, and thereby changed the teaching of physical metallurgy forever. His subsequent, more substantial textbooks built on this crucial beginning.

In 1946/7, when he was a young professor of physical metallurgy at Birmingham University, Cottrell was prevented from doing experiments by the power cuts during that critical winter when coal supplies ran out; universities had to close for weeks because of the lack of heating. Looking for something to do at home, he 'taught himself some elasticity theory and then used this to try to calculate the elastic forces between solute atoms and dislocations'. This period of self-education led straight on to the work described in the text. This small contribution to the history of science comes from Ernest Braun's chapter on mechanical properties of solids in a recent history of solid-state physics [20]. Braun goes on to cite this chill-induced serendipity as confirmation of Egon Orowan's dictum that 'the genesis of dislocation theory provides no exception from the fairly general rule that radical steps in science as in art required in the past some breakdown, malfunctioning, or inefficiency of social life which gave to a few persons opportunities to ramble instead of marching in a prescribed direction'.

free torsional oscillation. The amplitude of oscillation gradually decays because of internal friction, or damping: this damping had been shown to be caused by dissolved carbon (and nitrogen, when that was present also). Roughly speaking, the dissolved carbon atoms, being small, sit in interstitial lattice sites close to an edge of the cubic unit cell of iron, and when that edge is elastically compressed and one perpendicular to it is stretched by an applied stress, then the equilibrium concentrations of carbon in sites along the two cube edges become slightly different: the carbon atoms 'prefer' to sit in sites where the space available is slightly enhanced. After half a cycle of oscillation, the compressed edge becomes stretched and vice versa. When the frequency of oscillation matches the most probable jump frequency of carbon atoms between adjacent sites, then the damping is a maximum. By finding how the temperature of peak damping varies with the (adjustable) pendulum frequency (figure 19.4(b)), the jump frequency and hence the diffusion coefficient can be determined, even below room temperature (figure 19.4(c)). The subtleties of this 'anelastic' technique, and other related ones, were fully set out in 1972 in a classic text by two Americans, Arthur Nowick and Brian Berry [22].

The magnitude of the peak damping is proportional to the amount of carbon in solution. A carbon atom locked in an 'atmosphere' around a dislocation is locked to the stress field of the dislocation and thus cannot oscillate between sites; it does not contribute to the peak damping. By the simple expedient of stretching a steel wire beyond its yield-stress, clamping it into the Snoek pendulum and measuring the decay of the damping coefficient with the passage of time at temperatures near ambient, Harper obtained the experimental plots of figure 19.4(d): here f is the fraction of dissolved carbon which had migrated to the dislocation atmospheres. The $t^{2/3}$ law is perfectly confirmed, and by comparing the slopes of the lines for various temperatures, it was possible to show that the activation energy for strain-aging was identical with that for diffusion of carbon in iron, as determined from figure 19.4(c). After this, Cottrell and Bilby's model for the yield-stress and for strain-aging was generally accepted and so was the existence of dislocations, even though nobody had seen one as yet. Cottrell's 1953 book on dislocation theory [23] marked the coming of age of the subject.

It is worthwhile to present this episode in considerable detail, because it encapsulates very clearly what was new in physical metallurgy in the middle of the century. The elements are: an accurate theory of the effects in question, without disposable parameters; and, to check the theory, the use of a technique of measurement (the Snoek pendulum) which is simple in the extreme in construction but subtle in its quantitative interpretation, so that theory ineluctably comes into the measurement itself. It is quite inconceivable that any handwaver could ever have dreamt up the use of a pendulum to measure dissolved carbon concentrations!

Twentieth Century Physics

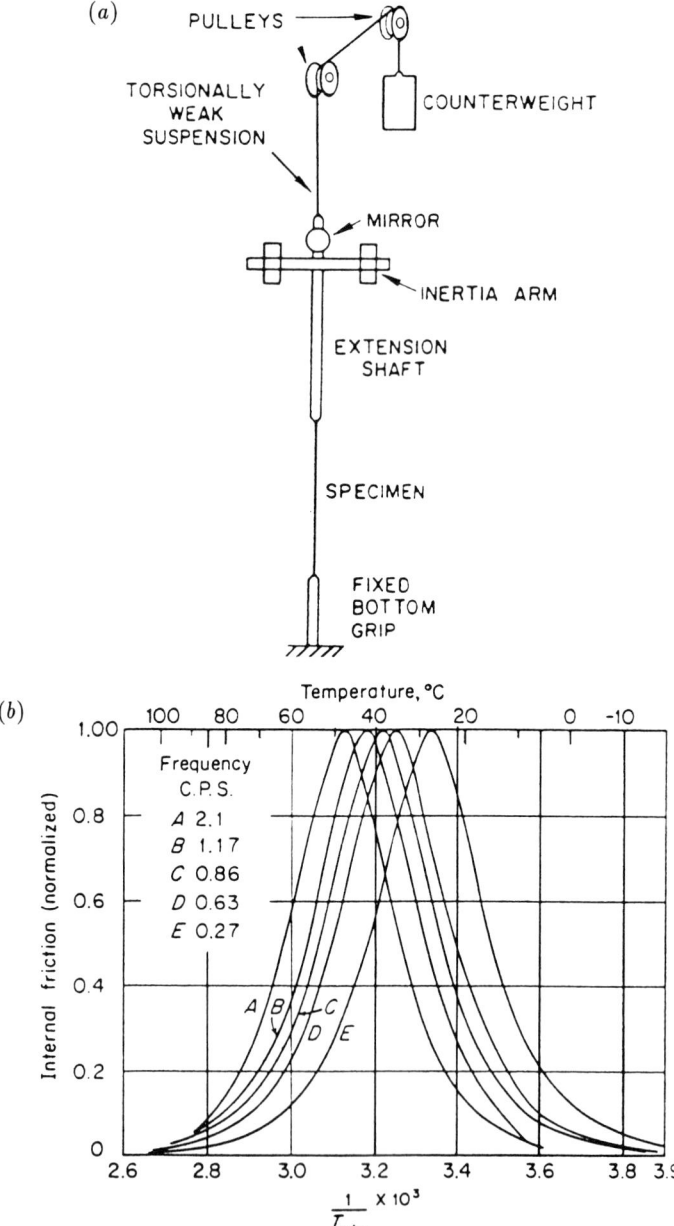

Figure 19.4. Checking the strain-aging law for a dilute iron–carbon alloy. (a) Torsion pendulum ('Snoek pendulum') for measuring free carbon dissolved in iron. (b) Internal friction as a function of temperature for a solid solution of carbon in iron at five different pendulum frequencies (after [24]). (c) Diffusion coefficient of carbon in iron over 14 decades, using the Snoek effect (−30 to 200 °C) and a conventional radioisotope method (400 to 700 °C). (d) The effect of temperature on the strain-aging kinetics, using the Snoek pendulum.

Physics of Materials

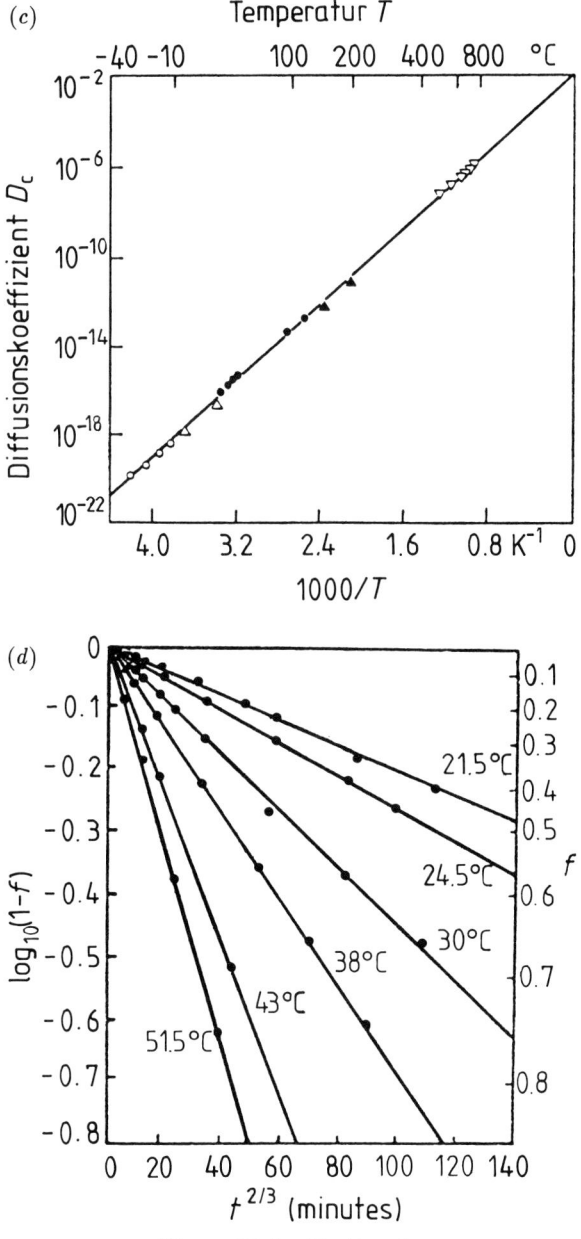

Figure 19.4. *Continued.*

As remarked, in 1952 no one had yet seen a dislocation. Once people had acquired confidence in their presumed reality, from Cottrell, Bilby and Harper's work, it became worthwhile to make a serious effort to see dislocations, in pursuit of that central tenet followed by

1525

physical metallurgists, 'seeing is believing'. Over the next few years, dislocations—or, to be pedantic, their ghosts—were rendered visible by a plethora of means. To cite just two examples: J W Mitchell 'decorated' networks of dislocation lines in silver chloride by using light to print out minute silver particles along the dislocations (1953), and W C Dash exploited the preferential diffusion of copper impurity along dislocation lines in silicon, followed by the use of optical microscopy of thin silicon slices in transmission with infrared light (silicon is transparent to infrared, copper is not). Figure 19.5 shows a celebrated image of a dislocation configuration in silicon made by Dash around 1955. At this time, also, dislocations were first seen in the transmission electron microscope (see Chapter 20 by T Mulvey): here again, it is really a ghost that is seen, because the presence of the defect distorts the lattice and thus locally the intensity of electron diffraction is diminished, and this shows in the image. The direct observation of dislocations, including the electron microscopy, is historically reviewed in Braun's chapter [20] and also in a splendid 1964 book by Amelinckx [25] which is full of beautiful images. An earlier, classic book by Friedel [26] presented both observational evidence and elastic theory of dislocations; it was an instance of the postwar flowering of solid-state physics in France, largely under the influence of Friedel himself; that story has been outlined by him in a recent autobiography [27]. Since these earlier books, Frank Nabarro (a British physicist based in South Africa) has devoted a good deal of his professional life to the study and exposition of the role of dislocations in the physics of materials, and he has brought out a long series of volumes entirely devoted to theory and experiment as they concern dislocations. None of this intense research would ever have materialized without Cottrell's banishment of handwaving.

As we have seen, the invention of dislocations, followed only twenty years later by their visual detection, was the direct outcome of a mismatch between two quantities—the measured yield-stress of a metal crystal and the theoretical 'ideal' value of this quantity. For this to be possible, large metal crystals had to be made, in the form of rods, wires or plates. This began to be done around 1920, either by very slow freezing from one end or else by light deformation followed by exceedingly slow recrystallization. This was an early example of a programme of metallurgical research undertaken purely for the sake of physical understanding. Harold Carpenter, Constance Elam (Mrs Tipper) and Daniel Hanson in England, and Erich Schmid and Walter Boas in Germany were at the forefront of this research; in 1935, Elam [28] and Schmid and Boas [29] simultaneously published influential books on the plastic deformation of these crystals. The subtleties of mechanical behaviour came to be understood primarily through these studies. Incidentally, it is interesting that Michael Polanyi, educated as a physical chemist (and, as already mentioned, later to become

Physics of Materials

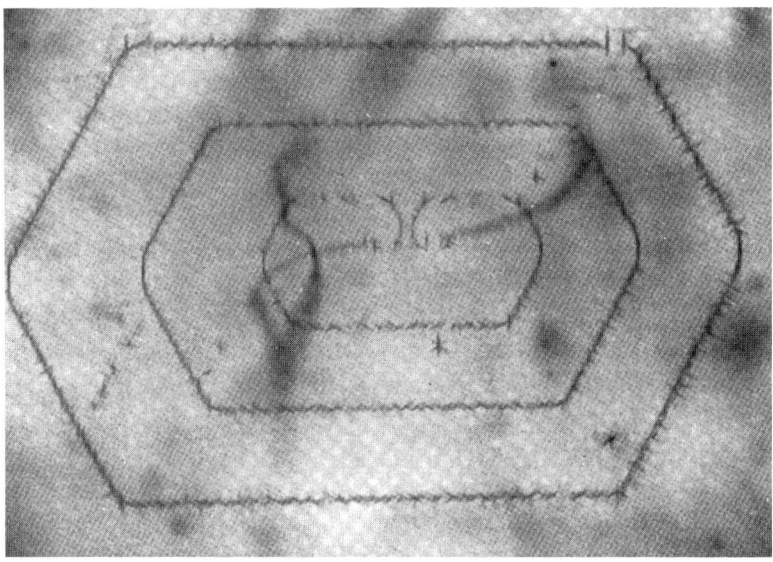

Figure 19.5. *Double-ended dislocation multiplication source in silicon, made visible by copper decoration. (Photograph by W C Dash.)*

first professor of physical chemistry and later a renowned professor of philosophy in Manchester), joined the metal physicist Erich Schmid and the future polymer physicist Hermann Mark in Berlin in the early 1920s to undertake the first study of the mechanical behaviour of metallic (zinc) single crystals [30]. This serves to show that the practice of physics is not restricted to 'paid-up' physicists, and also that good research workers are not content to be typecast.

Other major advances were made by physicists who recognized similar mismatches between quantities predicted by theory and their experimental counterparts. Perhaps the most celebrated was the work in 1947–9 of Charles Frank, then a young physicist newly arrived at Bristol University. He was exposed to colleagues interested in crystal growth and others concerned with dislocations. During a previous stay with Peter Debye in Berlin, Frank had learned that iodine crystals could grow from the vapour at 1% supersaturation whereas the calculations of Frank's Bristol colleague, J Burton, suggested that ~50% supersaturation was needed to nucleate a fresh crystal layer when the last one had covered the entire surface. In 1949, this led Frank to the idea of a mode of growth which would obviate the need to keep nucleating fresh crystal layers: a screw dislocation has the characteristic that the crystal layers surrounding it follow a spiral path, and a crystal growing around the point of emergence of such a dislocation would form a continuous spiral, rather like an old-fashioned castle on a hill with a spiral footpath leading up to it. He worked this out a day or so before one of the Bristol dislocation conferences, in 1949, and hardly had he shown his rough sketch of a predicted 'growth spiral', when a young mineralogist, J Griffin, showed exactly such a configuration on the surface of a beryl crystal. (It turned out later that the spiral steps, not much more than one atomic diameter in height, had become 'decorated' by dirt migrating across the crystal surface till a step was reached, and only thus had they become visible by optical microscopy). Braun [20] gives a step-by-step account of this episode. The Frank/Griffin two-man act was probably the most dramatic instant confirmation of a theory in the history of solid-state physics!

We saw that in the infancy of physical metallurgy, Rosenhain took a dogmatic stance concerning the nature of boundaries between crystal grains in metals, which he thought to have a glassy structure. Following the apotheosis of the quantitative approach in the early 1950s, grain boundaries were re-examined in the light of the hypothesis that they consist of assemblies of dislocations, and this led to an enormous outburst of research, both experimental and theoretical. A further impetus, which has lasted to this day, came from a pioneering geometrical examination in 1955 of 'special orientation relationships' between pairs of adjacent grains, by another team, Kronberg and Wilson [31], from the heroic days of the General Electric (GE) Research Laboratory in New York State. An influential account of the early work was published in 1957 by Donald McLean, a metallurgist who worked in Rosenhain's old laboratory [32], while the extraordinary subtlety of present-day models of such interfaces is revealed in a new multi-author book [33] (the field, like so many others, has expanded beyond the competence of any single scientist!). The understanding of the structure and properties of interfaces—between grains, or between phases liquid, gas and solid—is central to much of modern materials physics.

19.6. Point defects and non-metallic materials

Up to now, the emphasis has been mostly on metallurgy and physical metallurgists. That was where many of the modern concepts in the physics of materials started. However, it would be quite wrong to equate modern materials science with physical metallurgy. For instance, the gradual clarification of the nature of point defects in crystals (an essential counterpart of dislocations, or line defects) came entirely from the concentrated study of ionic crystals, and the study of polymeric materials after World War II began to broaden from being an exclusively chemical pursuit to becoming one of the most fascinating topics of physics research, and that is leaving entirely to one side the huge field of semiconductor physics, dealt with elsewhere in this book. Polymers will be discussed in a later section; here we focus on ionic crystals.

At the beginning of the century, nobody knew that a small proportion of atoms in a crystal are routinely missing, even less that this was not a matter of accident but of thermodynamic equilibrium. The recognition in the 1920s that such 'vacancies' had to exist in equilibrium was due to a school of statistical thermodynamicians including the Russian Frenkel and the Germans, Jost, Wagner and Schottky. A vacancy, moreover, as we know now, is only one kind of 'point defect'; an atom removed for whatever reason from its lattice site can be inserted into a small gap in the crystal structure, and then it becomes an 'interstitial'. Moreover, in insulating crystals a point defect is apt to be associated with a local excess or deficiency of electrons, producing what came to be called 'colour centres', and this can lead to a strong sensitivity to light: an extreme example of this is the photographic reaction in silver halides.

At about the same time as the thermodynamicians came to understand why vacancies had to exist in equilibrium, another group of physicists began a systematic experimental assault on colour centres in insulating crystals: this work was mostly done in Germany, and especially in the famous physics laboratory of Robert Pohl (1884–1976) in Göttingen. A splendid, very detailed account of the slow, faltering approach to a systematic knowledge of the behaviour of these centres has recently been published by Teichmann and Szymborski, as part of the previously mentioned history of solid-state physics [34]. Pohl was a resolute empiricist, and resisted what he regarded as premature attempts by theorists to make sense of his findings. Essentially, his school examined, patiently and systematically, the wavelengths of the optical absorption peaks in synthetic alkali halides to which controlled 'dopants'—small amounts of impurity—had been added. (Another approach was to heat crystals in a vapour of, for instance, an alkali metal.) Work with x-ray irradiation was done also, starting with a precocious series of experiments by Wilhelm Röntgen in the early years of the century; he published an overview in 1921. Other physicists in Germany ignored Pohl's work for many years, or ridiculed it as 'semiphysics' because of the impurities which they thought were bound to vitiate the findings. Several decades were yet to elapse before minor dopants came to the forefront of applied physics in the world of semiconductor devices. Insofar as Pohl permitted any speculation as to the nature of his 'colour centres', he opined that they were of non-localized character, and the adherents of localized and of diffuse colour centres quarrelled fiercely for some years. Even without a theoretical model, Pohl's cultivation of optical spectroscopy, with its extreme sensitivity to minor impurities, led through collaborations to advances in other fields, for instance, the isolation of vitamin D.

See also p 1320

One of the first experimental physicists to work with Pohl on impure ionic crystals was the Hungarian Zoltan Gyulai (1887–1968). He rediscovered colour centres created by x-ray irradiation while working in Göttingen in 1926, and also studied the effect of plastic deformation on the electrical conductivity. Pohl was much impressed by his Hungarian collaborator's qualities, as reported in a small survey of physics in Budapest published recently [35]. This book reveals the astonishing flowering of Hungarian physics during the past century, including the physics of materials, but many of the greatest Hungarian physicists (people like Szilard, Wigner, von Neumann, Teller, von Karman, Gabor, von Hevesy, Kurti) made their names abroad because the unceasing sequence of revolutions and tyrannies made life at home too uncomfortable or even dangerous. However, Gyulai was one of those who returned and he later presided over the influential Roland Eötvös Physical Society in Budapest.

Attempts at a theory of what Pohl's group was discovering started in Russia, whose physicists (notably Yakov Frenkel and Lev Landau) were more interested in Pohl's research than were most of his own compatriots. Frenkel, Landau and Rudolf Peierls, in the early 1930s, favoured the idea of an electron trapped 'by an extremely distorted part of the lattice' which developed into the idea of an 'exciton', an activated atom. Finally, in 1934, Walter Schottky in Germany first proposed that colour centres involved a pairing between an anion vacancy and an extra (trapped) electron—a 'Schottky defect'. (Schottky was a rogue academic who did not like teaching and migrated to industry, where he fastened his teeth on copper oxide rectifiers; thus he approached a fundamental problem in alkali halides via an industrial problem, an unusual sequence at that time!)

At this point, German research with its Russian topdressing was further fertilized by sudden and major input from Britain and especially from the United States. In 1937, at the instigation of Nevill Mott, a physics conference was held in Bristol University, on colour centres (the beginning of a long series of influential physics conferences there, dealing with a variety of topics including dislocations, crystal growth and polymer physics). Pohl delivered a major experimental lecture while R W Gurney and Mott produced a quantum theory of colour centres, leading soon afterwards to their celebrated model of the photographic effect. (This sequence was outlined later by J W Mitchell [36].)

The leading spirit in the United States was Frederick Seitz (see box). He first made his name with his model, jointly with his thesis adviser, Eugene Wigner, for calculating the electron band structure of a simple metal, sodium [37]. Soon afterwards he spent two years working at the General Electric Company's central research centre (the first and at that time the most impressive of the large industrial laboratories in America), and became involved in research on phosphorescent materials ('phosphors') suitable for use as a coating in cathode-ray tubes; to help him in this quest, he began to study Pohl's papers. (These, and other stages in Seitz's life are covered in some autobiographical notes published by the Royal Society [38] and in a very recent full-length autobiography [39].) Conversations with Mott then focused his attention on crystal defects. Many of the people who were to create the theory of colour centres after World War II devoted themselves meanwhile to the improvement of phosphors for radar, before switching to the related topic of radiation damage in relation to the Manhattan Project. After the war, Seitz returned to the problem of colour centres and in 1946 published the first of two celebrated reviews [40], based on his resolute attempts to unravel the nature of colour centres. Theory was now buttressed by purpose-designed experiments: Otto Stern (with two collaborators) was able to show that when ionic crystals had been greatly darkened by irradiation and so were full of colour centres, there was a measurable decrease in density, though by only one part in 10^4! (This remarkably precise measurement of density was achieved by the use of a flotation column, filled with liquid arranged to have a slight gradient of density from top to bottom, and by establishing where the crystal came to rest.)

Vacancies had at last come of age. Following an intense period of research at the heart of which stood Seitz, he published a second review on colour centres [41]. In this review, he distinguished between 12 different types of colour centres, involving single, paired or triple vacancies; many of these later proved to be misidentifications, but nevertheless, in the words of Teichmann and Szymborski, 'it was to Seitz's credit that, starting in the late 1940s, both experimental and theoretical efforts became more convergent and directed to the solution of clearly defined problems' [34]. So, the symbiosis of quantitative theory and experiment got under way at much the same time for metals and for non-metals.

19.7. Diffusion

By the 1940s, as we have seen, the existence of crystal vacancies was well substantiated in insulating crystals, and statistical thermodynamics indicated that they must exist in metals also. In 1942, Huntingdon and Seitz [42] published results of calculations of the formation energies and also of the activation energies for migration of both vacancies and interstitials in metallic copper, and from these calculations it became clear that vacancies must be the 'carriers' of atomic diffusion, rather than interstitials (i.e. atoms squeezed between normal lattice sites). This meant that atoms could move about in a metallic crystal only by exchanging places with adjacent vacancies. Some years later, Huntingdon and Seitz's calculations were supplemented by a series of brilliant experimental studies by Robert Balluffi at the General Electric Laboratory (e.g. Simmons and Balluffi [43]): they combined dilatometry (measurements of changes of length as a function of changing temperature) with precision measurements of lattice parameter (the dimensions of the crystalline unit cell) to measure vacancy concentrations. This was the

Frederick Seitz

(American, b 1911)

Frederick Seitz has played a crucial role in the creation of solid-state physics as well as the wider field of materials science. As a young Stanford graduate, he came to Princeton in 1932 to do research under Edward Condon, a spectroscopist. Condon advised him to work with Eugene Wigner instead, saying 'Solid-state physics is coming, and if you stay with me you'll just do calculations for my book (*Theory of Atomic Spectra*)'. That advice led to the crucial Wigner–Seitz 'cellular model' of 1934–5 which allowed the free-electron theory to be applied to real, not just to idealized metals. Later, he moved on to his famous researches on colour centres in ionic crystals, described in the text. His books (1940 *Modern Theory of Solids*; 1943 *The Physics of Metals*) and the monograph series he founded, *Solid State Physics* 1955 *et seq.*) have been immensely influential in making familiar the concepts of quantum theory and especially band theory as applied to solids. He was professor of physics at Carnegie Institute of Technology from 1942 to 1949 and at the University of Illinois from 1949 to 1965, before moving on to manifold activities on the national scene and finishing as president of a major university.

Before all of these posts, however, as a young physicist, he had spent two years working at the General Electric Central Research Laboratory and that experience clearly influenced his outlook on the physics of materials very deeply: he played a major role in establishing the Materials Research Laboratories which now grace numerous US campuses. In 1993, the Materials Research Society awarded him its highest honour, the von Hippel Award, and in accepting it he spoke with feeling of the unexpected, early role played by the great mathematician, John von Neumann, in pressing for the Materials Research Laboratories to be created. Labels tell one little about scientists' interests!

beginning of a period in which diffusion, and the characteristic formation and migration energies of vacancies, in many metals and alloys were studied intensively. Not in all crystals, however, is diffusion 'carried' by vacancies: it was established some years later that in substances like silicon, diffusion takes place via interstitial atoms.

At the end of the 1940s, it also came to be realized that if excess vacancies were 'frozen' into a metal by quenching it from a high temperature then these extra vacancies would greatly accelerate diffusion at room temperature, until the excess vacancies had diffused away to 'sinks' such as a free surface or the boundaries between crystal grains. In 1950, David Turnbull showed that the phenomenon of age-hardening in certain aluminium alloys could be explained in this way: thus, a binary Al–Cu alloy quenched from ~600 °C hardens slowly at room temperature as excess copper comes out of solid solution in the form of minute copper-rich zones which interfere with the free motion of dislocations. When the measured *equilibrium* diffusion rates in aluminium are extrapolated to room temperature, the predicted age-hardening rate is negligible: it is only the excess vacancies in the quenched alloy that make the process feasible. This kind of non-equilibrium defect population came soon afterwards to be a central concept in the study of radiation damage in nuclear reactor materials: as atoms are knocked off their lattice sites by high-energy photons, neutrons or ions, excess vacancies *and* interstitials are formed in large numbers and lead to numerous anomalous forms of behaviour, including non-equilibrium segregation of solute atoms in alloys which should ideally be of uniform composition, unexpected changes of shape caused by the macroscopic migration of point defects from one side to the other of crystal grains, agglomeration of vacancies to form inconvenient 'voids', and anomalously easy plastic deformation under absurdly small stresses. Grouping of point defects, both vacancies and interstitials, especially in twosomes, has been much examined, and a very recent paper claims to be able to interpret a wide range of properties of both crystals and glasses, including melting, on the basis of the effects of large populations of self-interstitials [44]. (The long sequence of theories of melting, which attempt to come to grips with the awkward fact that while it is easy to supercool melts, it is almost impossible to superheat crystals since melting starts at the surface—as shown by two Dutch physicists in a definitive study [45]—has divided condensed matter theorists into irreconcilable camps; it is a disputatious and wonderfully intriguing field, involving American, Dutch, French, British, Russian, New Zealand and other physicists in great profusion. Two recent papers, by Yukalov [46] and by Wolf *et al* [47], respectively exemplify approaches that take into account the properties of the liquid as well as the solid, and those that focus on imperfections in solids without reference to the free energy of the melt. Much of the theorizing about melting falls into the second category, partly because the rigorous physical modelling of liquids has proved a slow and uphill task [48].)

Returning to radiation damage, this has become a huge field of applied materials research, very much dependent on physical insights about entities such as point defects, single and grouped. A good physically based survey of the field was published by Gittus [49], while a shorter but very up-to-date treatment has just been published by Schilling and Ullmaier [50] (German physicists have been very active in this field). An authoritative recent overview of point defects in metals [51] devotes much of its space to evidence from irradiation experiments, which shows the intimate association between point defects and irradiation. Recently, in an introduction to the 200th volume of the *Journal of Nuclear Materials*, an editor, L K Mansur, opined that 'the cumulative contribution of nuclear materials research, documented in these two hundred volumes [since 1959], lies at the foundation of materials science and engineering and has accelerated the advance of physical science and engineering generally'.

Diffusion in metals and non-metals alike has become a major subscidy in its own right, with numerous subtleties, such as the hindrance of diffusion in atomically ordered intermetallic compounds. (If a nickel vacancy in an ordered alloy such as CuZn changes place with a neighbouring atom it then finds itself in a place reserved for zinc, which incurs a stiff energy penalty.) The result can be seen in figure 19.6, taken from an experimental study of the diffusion of copper and zinc in CuZn both above and below its critical ordering temperature [52]. Diffusion can also lead to a slow change of shape under stress: the process is called *diffusion creep* and its existence and mechanism were first proposed by Frank Nabarro at one of the famous Bristol Physics Conferences in 1947 [53] and further worked out in 1950 by Conyers Herring at Bell Laboratories in America [54]: atoms flow in response to a stress, as seen in figure 19.7, while vacancies flow in the inverse direction. (In an alternative scenario, vacancies diffuse along the grain boundaries rather than through the crystal grains.) Dislocations are not involved. Clearly, the rate of flow is very sensitive to grain diameter, d, varying as $d^{-1/2}$ for the first process and as $d^{-1/3}$ for the alternative process. This understanding has led to the development of *superplastic forming*, a process in which a mastery of microstructural control is exploited to stabilize a very fine-grained structure (typically, 1 μm or even less) in an alloy or ceramic. The material can then be shaped slowly at quite low stresses and intermediate temperatures, and strains of hundreds of per cent can be readily imparted, even to ceramics such as TiO_2 which in the ordinary way are totally brittle. The process, which is in large-scale industrial use, has recently been surveyed by Mukherjee [55]. Superplasticity is a good example of a topic which lies at the borders between physics of materials, metallurgy and engineering, and there are many of these.

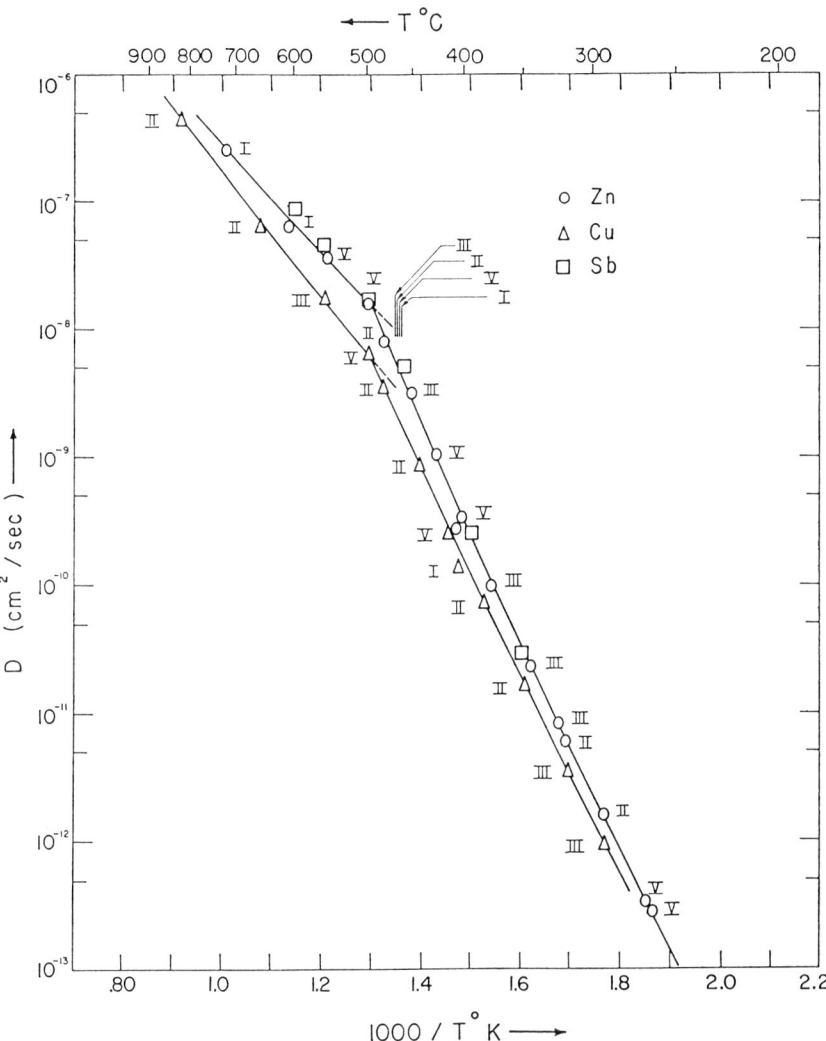

Figure 19.6. *Diffusivities in β-brass, CuZn, as a function of temperature. The discontinuity in slope is close to the critical disordering temperature.*

Diffusion is only one of many processes studied by materials physicists which depend on crystal defects. It is significant that one of the most influential books, published in 1952, at the time when both point and line defects came to be properly understood, was entitled *Imperfections in Nearly Perfect Crystals*. Indeed, in this book Seitz summarized what was known at that time about defects in both metallic and non-metallic crystals [56].

Figure 19.7. *Schematic mechanism of diffusion creep as it operates in superplastic deformation.*

19.8. Electrical and electro-optical ceramics and liquids

The work on colour centres and its consequences for understanding electrically charged defects in insulating and semiconducting crystalline materials helped to stimulate ceramic researches in the electrical/electronic industry. The subject is enormous and here there is space only for a cursory outline of what has happened since 1900.

The main categories of 'electrical/optical ceramics' are as follows: phosphors for TV, radar and oscilloscope screens; voltage-dependent and thermally sensitive resistors; dielectrics, including ferroelectrics; piezoelectric materials, again including ferroelectrics; pyroelectric ceramics; electro-optic ceramics; and magnetic ceramics (which are dealt with elsewhere in this book).

We have already seen that Seitz became motivated to study colour centres during his pre-war sojourn at the General Electric Research Laboratory, where he was exposed to studies of phosphors which could convert the energy in an electron beam into visible radiation, as required for oscilloscopes and television receivers. The term 'phosphor' is used generally for materials which fluoresce and those which phosphoresce (i.e. show persistent light output after the electron stimulus is switched off). Such materials were studied, especially in Germany, early in this century and these early results were assembled by Lenard *et al* [57]. Phosphors were also a matter of acute concern to Vladimir Zworykin (a charismatic Russian immigrant to America); he wanted to inaugurate a television industry in the late 1920s, but failed to persuade his employers, Westinghouse, that this was a realistic objective. According to an intriguing piece of historical research by Notis [58], Zworykin then transferred to another company, RCA, which he was able to persuade to commercialize both television and electron microscopes. For the first of these objectives, he needed a reliable and plentiful material to use as phosphor, with a persistence time of less than 1/30 of a second (at that time, he believed that 30 refreshments of the tube image per second would be essential). Zworykin was fortunate to encounter a ceramic technologist of genius, Hobart Kraner. He had studied crystalline glazes on decorative ceramics (this was an innovation, since most glazes had been glassy), and among these, a zinc silicate glaze [59]. He and others later found that when manganese was added as a nucleation catalyst to encourage crystallization of the glassy precursor, the resulting crystalline glaze was fluorescent. In the meantime, natural zinc silicate, the mineral willemite, was being used as a phosphor, but it was erratic and non-reproducible, and anyway in very short supply. Kraner showed Zworykin that synthetic zinc silicate, Zn_2SiO_4, would serve even better as a phosphor when 'activated' by a 1% manganese addition. This serendipitous development came just when Zworykin needed it, and it enabled him to persuade RCA to proceed with the large-scale manufacture of TV tubes. Kraner, a modest man who published little, did later present a lecture on creativity and the interactions between people needed to stimulate it [60]. The history of materials is full of episodes when the right concatenation of individuals elicited the vitally needed innovation at the right time.

Phosphors to convert x-ray energy into visible light go back to a time soon after x-rays were discovered. Calcium tungstate, $CaWO_4$, was found to be more sensitive to x-rays than the photographic film of that time. Many more efficient phosphors have since been discovered, all doped with rare-earth ions, as recently outlined by an Indian physicist [61]. The early history of all these phosphors, whether for impinging electrons or x-rays, has been surveyed by Harvey [62]. (The generic term for this field of research is 'luminescence', and this is in the title of Harvey's book.)

The relatively simple study of fluorescence and phosphorescence (based on the action of colour centres) has nowadays extended to non-linear optical crystals, in which the refractive index is sensitive to the light intensity (or in the photorefractive variety [63] also to its spatial variation): a range of crystals, the stereotype of which is lithium niobate, is now used.

Ceramic conductors also cover a great range of variety, and a considerable input of fundamental research has been needed to drive them to their present state of subtlety. A good example is the zinc oxide *varistor* (i.e. a voltage-dependent resistor, in a sense an analogue to a photorefractive crystal). A varistor consists of semiconducting ZnO grains separated by a thin intergranular layer rich in bismuth, with a higher resistance than the grains; as voltage increases, increasing areas of intergranular film can participate in the passage of current. These important materials have been described in Japan [64] (a country which has achieved an unchallenged lead in this kind of ceramics, which they call 'functional' or 'fine' ceramics) and in England [65]. This type of microstructure also influences other kinds of conductors, especially those with positive (PTC) or negative (NTC) temperature coefficients of resistivity. For instance, PTC materials [66] have to be impurity-doped polycrystalline ferroelectrics, usually barium titanate (single crystals do not work) and depend on a ferroelectric-to-paraelectric transition in the dopant-rich grain boundaries, which leads to enormous increases in resistivity. Such a ceramic can be used to stabilize temperature.

> See also p 486

Ferroelectrics have just been mentioned. The term is a linguistic curiosity, adapted from 'ferromagnetic'. ('Ferro' here is taken to imply a spontaneous magnetization, or electrification, and those who invented the name chose to forget that 'ferro' actually refers to iron! The corresponding term 'ferroelastic' for non-metallic crystals which display a spontaneous strain is an even weirder linguistic concoction!) Ferroelectric crystals are a large family, the modern archetype of which is barium titanate, $BaTiO_3$, although for years an awkward and unstable organic crystal, Rochelle salt, held sway. These scientifically fascinating crystals are used for their high dielectric constants in capacitors and also for their powerful piezoelectric properties, for instance for sonar. Details would take us too far here, but it can confidently be asserted that ferroelectrics have constituted a major branch of the physics of materials in recent decades. Their complicated scientific history, with many vigorous, even acrimonious controversies, has been excellently mapped out by Cross and Newnham [67]. One of the intriguing pieces of information in this history is that in the 1950s, Bernd Matthias at Bell Laboratories competed with Ray Pepinsky at Pennsylvania State University to see who could discover more novel ferroelectric crystals, just as later he competed again with others to drive up the best superconducting transition temperature in primitive (i.e. metallic) superconductors. Every scientist has his own secret spring of action, if only he has the good fortune to discover it!

Ferroelastic crystals were cited in the last paragraph, in contradistinction to 'paraelastic' (or 'normal') crystals. Ferroelastic crystals, like ferroelectric ones, undergo a phase transition which lowers their symmetry so that they can exist in two mirror orientations with different spontaneous electric or strain features. This kind of crystal has come to be of interest to mineralogists as well as to materials experts: in fact, modern mineralogy is turning more and more towards physics, and recently a theoretical physicist who is also professor of mineralogy in Cambridge has written an enticing book about these recondite materials [68]. (Elastic stresses can also affect phase equilibria in alloys, as is attested by the broad field of research on shape-memory alloys like those in the Ni–Ti system—alloys in which shape change is generated by diffusionless, stress-induced shear transformations; under the right conditions, such alloys can reverse-transform, when heated, back to their pristine shape. The earliest study of this kind of transformation, then called 'thermoelastic transformation', was probably the classical 1949 study [69] by Kurdjumov and Khandros, in Moscow, of certain steels. Kurdjumov, who is still alive in extreme old age, was an extremely influential protagonist of metal physics in the Soviet Academy of Sciences.)

Reverting to functional ceramics, a further large family is that of the *superionic conductors*. This term was introduced by W L Roth (once again, working at the GE Central Laboratory) in 1972 [70]; though his work was published in the *Journal of Solid-State Chemistry*, it could with equal justification have appeared in *Physical Review*, but it is usual with crystallographers that people working in this field are polarized between those who think of themselves as chemists and those who think of themselves as physicists. Superionic conductors are electronically insulating ionic crystals in which either cations or anions move with such ease under the influence of an electric field that the crystals function as efficient conductors in spite of the immobility of electrons. The prototype is a sodium-doped aluminium oxide of formula $NaO.11Al_2O_3$, called beta-alumina. Roth substituted silver for some of the sodium, for the sake of easier x-ray analysis, and found that the silver occupied a minority of certain sites on a particular plane in the crystal structure, leaving many other sites vacant. This configuration is responsible for the extraordinarily high mobility of the silver atoms (or the sodium some of which they replaced); the vacancy-loaded planes have been described as liquid-like. There are now many other superionic conductors and they have important uses as electrolytes in all-solid storage batteries and fuel cells.

One is inclined to think of 'materials' as being solids; when editing an encyclopedia of materials some years ago, I found it required an effort of will to include articles on various aspects of water and on inks. Yet one of the most important families of materials in the general area of consumer electronics are liquid crystals, used in inexpensive displays, for instance in digital watches and calculators. They have a fascinating history as well as deep physics.

Liquid crystals come in several varieties: for the sake of simplified illustration, one can describe them as collections of long molecules tending statistically to lie along a specific direction; there are three types, nematic, cholesteric and smectic, with an increasing measure of order in that sequence, and the variation of degree of alignment as the temperature changes is akin to the behaviour of spins in ferromagnets. The recorded history of these curious materials goes back to 1888, when a 'botanist-cum-chemist', Friedrich Reinitzer, sent some cholesteric esters

See also p 1603

to a 'molecular physicist', Otto Lehmann. They can be considered the joint progenitors of liquid crystals. Reinitzer's compounds showed two distinct melting-temperatures, about 30 K apart. Much puzzlement ensued at a time when the nature of crystalline structure was quite unknown, but Lehmann (who was a single-minded microscopist) and others examined the appearance of the curious phase between the two melting-points, in electric fields and in polarized light. Lehmann concluded that the phase was a form of very soft crystal, or 'flowing crystal'. He was the first to map the curious defect structures (features called 'disclinations' today). Thereupon the famous solid-state chemist, Gustav Tammann, came on the scene. He was an old-style authoritarian and, once established in a prime chair in Göttingen, he refused absolutely to accept the identification of 'flowing crystals' as a novel kind of phase, in spite of the publication by Lehmann in 1904 of a comprehensive book on what was known about them. Ferocious arguments continued for years, as recounted in two instructive historical articles by Kelker [71]. Lehmann, always eccentric and solitary, became more so and devoted his last 20 years to a series of papers on 'Liquid crystals and the theories of life'.

During the first half of this century, progress was mostly made by chemists, who discovered ever new types of liquid crystals. Then the physicists, and particularly theoreticians, became involved and understanding of the structure and properties of liquid crystals advanced rapidly. The principal early input from a physicist came from a French crystallographer, Georges Friedel, grandfather of the Jacques Friedel whom I mentioned earlier in connection with dislocations and French solid-state physics. It was Georges Friedel who invented the nomenclature, nematic, cholesteric and smectic, mentioned above; as Jacques Friedel recounts in his autobiography, family tradition has it that this nomenclature emerged during 'an afternoon of relaxation with his daughters, especially Marie who was a fine Hellenist'! Friedel recognized that the low viscosity of liquid crystals allowed them readily to change their equilibrium state when external conditions were altered, for instance an electric field, and he may thus be regarded as a direct ancestor of the current technological uses of these materials. According to his grandson, Georges Friedel's 1922 survey of liquid crystals [72] is still frequently cited nowadays. The very detailed present understanding of the defect structure and statistical mechanics of liquid crystals is encapsulated in two very recently published second editions of classic books, by de Gennes [73] in Paris and by Chandrasekhar [74] in Bangalore, India. (See also Chapter 21 of this book.)

Liquid-crystal displays depend upon the reorientation of the 'director', the defining alignment vector of a population of liquid-crystalline molecules, by an applied electric field between two glass plates: to apply the field, one uses transparent ceramic conductors,

Physics of Materials

typically tin oxide, of the type mentioned above. Such applications, which are numerous and varied, have been treated in a book series [75].

It is perhaps not too fanciful to compare the stormy history of liquid crystals to that of colour centres in ionic crystals: resolute empiricism followed by fierce strife between rival theoretical schools, until at last a systematic theoretical approach led to understanding and indeed to practical application. In both these domains, it would not be true to say that the empirical approach sufficed to generate practical uses—these in fact had to await the advent of good theory!

In industrial terms, perhaps the most successful of the many innovations that belong in this section is *xerographic* copying or photocopying of documents, together with its offspring, laser-printing the output of computers. This has very recently been reviewed in historical terms by Mort [76]. He explains that 'in the early 1930s, image production using electrostatically charged insulators to attract frictionally charged powders had already been demonstrated'. According to the book on Budapest physics cited earlier [35], this earliest precursor of modern xerography was in fact due to a Hungarian physicist named Pal Selenyi (1884–1954), who between the wars was working in the Tungsram laboratories in Budapest, but apparently the same Zworykin who has already featured in this section, presumably during a visit to Budapest, dissuaded the management from pursuing this invention; apparently he also pooh-poohed a (subsequently successful) electron multiplier invented by another Hungarian physicist, Zoltan Bay (who died very recently). If the book is to be believed, Zworykin must have been an early exponent of the 'not invented here' syndrome of industrial scepticism. Returning to Mort's survey, we learn that the first widely recognized version of xerography was demonstrated by an American, Chester Carlson, in 1938; it was based on amorphous sulphur as the photosensitive receptor and lycopodium powder. It took Carlson six years to raise $3000 of industrial support, and at last, in 1948, a photocopier based on amorphous selenium was announced and took consumers by storm; the market proved to be enormously greater than predicted! Later, selenium was replaced by more reliable synthetic amorphous polymeric films. Mort recounts the substantial part played by John Bardeen, as consultant and as company director, in fostering the early development of practical xerography. Just as the growth of xerographic copying and laser-printing was a physicists' triumph, the development of fax machines was driven by chemistry in the development of modern heat-sensitive papers, most of which have been perfected in Japan.

Xerography depends on amorphous materials, and so do many other modern industrial products. Perhaps the most important of such materials are the optical glasses, which are today available in an astonishing variety of refractive indices, dispersions, absorption spectra, colour response, stress-optical response, etc. Photochromic spectacle lenses constitute one variant which is widely familiar to the public. The development of modern optical glasses [77] has involved both physicists and chemists in such close symbiosis that it is quite impossible to disentangle their contributions, and this is also true of glass fibres for communications [78] which have to be of a purity comparable to that of transistor materials to ensure high transparency. They also have to have a precisely graded range of refractive index across the diameter, which requires sophisticated physical control, and in this respect they are the precursors of a very recent, growing family of materials, mostly ceramic and mostly Japanese, in which thermal shock resistance and other desirable properties are achieved by a progressive change of composition or microstructure across a thick layer; these have been christened 'functionally graded materials'.

19.9. Phase transformations, microstructure and modern instrumentation

At the outset of this chapter, the phase rule, with its concomitant concept, the phase diagram, was presented as one of the pillars of physical metallurgy. Once x-ray diffraction had been born and made crystal structure determination possible, the study of the crystallographic minutiae of solid–solid phase transitions became feasible.

The most important phase transition is that from the liquid to the solid state—solidification. This is a field which has been transformed out of recognition during the past 45 years by physical metallurgists and physicists. The emphasis has been on alloys, with attention to the behaviour of solutes. Since at most compositions, melt and solid in equilibrium have different compositions, there is scope for a great variety of phenomena both in equilibrium and, more particularly, when freezing is too fast for equilibrium to be retained. The non-equilibrium phenomena include such features as dendritic growth (fine solid spikes reaching out into the melt), the trapping of excess amounts of solute in the frozen alloy, and extreme refinement of microstructure. The ramifications of the modern theory of solidification have been set out by an Argentine metallurgist, Heraldo Biloni [79], while the many features of alloys frozen at very high speeds have been set out in a recent monograph [80]. In the first chapter of that volume I have reviewed the history of 'rapid solidification processing', which is a remarkable illustration of new technology deriving from purely scientific curiosity.

Pol Duwez (1907–84) was a Belgian–American physical metallurgist, for many years a professor at the California Institute of Technology. He was fascinated by a very old metallurgist's technique, quenching of hot alloys, typically into water, to preserve metastably at room temperature a non-equilibrium microstructure, notably the hard 'martensite' phase in steels. Duwez examined various exotic ways of enhancing the cooling

Physics of Materials

rate. He also had another obsession of the kind one expects a physicist to evince: he wanted to know why copper and gold, and again gold and silver, form a seamless series of solid solutions in all proportions, while copper and silver do not. Theoretical prediction was that copper and silver should behave like the two complementary pairs, and Duwez supposed that the relative (Gibbs) free energies of a solid solution and of a two-phase structure (as actually found) were delicately balanced and almost equal. If so, it should be possible to 'force' Cu and Ag at the 50/50 composition into a metastable mutual solid solution, and he planned to do this by freezing the melt very rapidly. So he set about inventing ways of quenching a *melt* very fast, whereas hitherto quenching had been done entirely in the solid domain. He succeeded (the technical details do not matter here), was able to make a metastable continuous series of Cu–Ag solid solutions [81], and thereby gave birth to a large new field of useful technology, *rapid solidification processing* of alloys such as tool steels and light aircraft alloys; this technology itself spawned new fields of fundamental materials physics, notably that concerned with metallic glasses made by this approach [82], which then in turn led to the widespread use of glassy Fe–Si–B and similar combinations for transformer laminations. It is worth noting that earlier investigators had developed ways of solidifying alloys rapidly, but only as a means of manufacturing certain shapes cheaply, without any scientific *arrière-pensée*: but it was not until Duwez's *tour de force* had attracted wide attention among good scientists that rapid solidification was taken seriously as an approach that offered scientific as well as technological rewards. Once again, as with colour centres and with liquid crystals, scientific insight, not brute empiricism, led to technological breakthroughs.

An unusual aspect of research into ultra-rapid solidification has been the use of levitated drops of melt, not in contact with any container, as a means to more efficient quenching. The general body of applied physics research into levitation of solids and liquids (which can use electromagnetic fields, concentrated sound waves or laser light, or in appropriate cases can exploit superconductive materials) was exploited here, and specially designed radiofrequency induction coils were used. The whole field has been surveyed by Brandt [83]. Levitation theory is a good example of pure physics being pressed into service to solve specific problems in materials processing.

One crucially important aspect of solidification is forever associated with the name of William Pfann. His great invention, *zone-refining* plus *zone-levelling*, was a crucial component of the manufacture of germanium transistors. Without it, there could have been no transistors! The approach is outlined in Box 19A (see also figure 19.8). Pfann has verbally described what led up to his invention, and his account is preserved in the Bell Laboratory archives. As a youth, he was engaged by Bell

Laboratories as a humble laboratory assistant, beginning with duties such as polishing samples and developing films. He attended evening classes and finally earned a bachelor's degree (in chemical engineering). He records attending a talk by a famous physical metallurgist of the day, Champion Mathewson, who spoke about plastic flow and crystal glide. Like Rosenhain before him, the youthful Pfann was captivated. Then, while still an assistant, he was invited by his boss, E E Schumacher, in the best Bell Laboratory tradition, to 'take half your time and do whatever you want'. Astonished, he remembered Mathewson and chose to study the deformation of lead crystals doped with antimony (as used by the Bell System for cable sheaths). He wanted to make crystals of uniform composition, and promptly invented zone-levelling. (He 'took it for granted that this idea was obvious to everyone, but was wrong'.) Pfann apparently impressed the Bell Director of Research by another piece of technical originality, and was made a fully fledged member of technical staff, though innocent of a doctorate. When William Shockley complained that the available germanium was nothing like pure enough, Pfann, in his own words, 'put my feet up on my desk and tilted my chair back to the window sill for a short nap, a habit then well established. I had scarcely dozed off when I suddenly awoke, brought the chair down with a clack I still remember, and realized that a series of molten zones, passed through the ingot of germanium, would achieve the aim of repeated fractional crystallization'. The new technique was proved and applied in transistor manufacture in a matter of months. Later, he developed the necessary variant that would work for silicon. Pfann has described all this in detail in two editions of his definitive book [84].

The subsequent application of zone-refining to metals, in other laboratories, proved a disappointment, but there was one important exception to this generalization. The General Electric Company in America had wished since the 1920s to manufacture a sealed vacuum circuit breaker for switchgear, with copper electrodes, but it turned out that on each break of current, local heating released too much gas from the copper and the vacuum deteriorated rapidly. Then, in about 1954, F H Horn prepared zone-refined copper and this had a gas content less than one part in 10 million. It proved to be the secret that led to the successful development of a commercial circuit breaker. The detailed history of this development is laid out in a most interesting survey of a range of case-histories in the GE Corporate Research Laboratory in the 1950s and 1960s [85]. It is also worth citing here an excellent biography which sets out how this laboratory, the first American industrial research laboratory and one of the most effective, came to be created [86].

The vast range of solid–solid phase transformations, the study of which is central to physical metallurgy, can only be asserted. There is simply no space here even to outline the range of transformation types (usually subdivided into those that involve atomic diffusion

BOX 19A: ZONE-REFINING

William G Pfann (American, 1917–82) was the begetter of this wonderfully simple and effective method of purifying semiconducting materials.

When the germanium transistor was invented at Bell Laboratories in 1948, it quickly became clear that impurities had to be kept down to the level of one in hundreds of millions of atoms, levels of purity that had never been pursued before; furthermore, the minute residual levels of impurity were required to be uniformly distributed (before dopants were added to shape the transistor). Pfann was ready with the appropriate technique. When a melt is frozen slowly enough to maintain equilibrium, the solid has a different composition from the melt; most commonly, the solid of a binary mixture contains less solute than does the liquid. What Pfann realized was that this feature could be exploited both for purification and for subsequent homogenization. While simple solidification from one end of a rod is an inefficient mode of purification, all that is necessary is to pass a narrow molten zone along a solid rod; the effect is to 'sweep' impurities along the rod (figure 19.8). A simple way to proceed, Pfann found, was to arrange a sequence of ring heaters around a germanium rod and to move the rod slowly through them. At first the rod was held in a crucible, later (especially when silicon began to replace germanium) the surface tension of the thin molten zone was exploited to dispense with any container at all (thereby preventing recontamination). The 'dirty' end of the rod into which most of the impurities had been swept was cut off. Afterwards, molten zones were passed forwards and backwards to homogenize the germanium. Purification was done by zone-refining; homogenization, by zone-levelling. Primitive though it is, in its purification and homogenization performance this technique excelled all other known methods at the time of its introduction. It is much preferable to repeated fractional crystallizations, as used by chemists, because this entails throwing away most of the material between fractionations! Today, the zone method has been replaced by chemical ultrapurification of silicon during its manufacture via a silicon halide gas, but at the time, zone-refining was what made transistors (and, later, integrated circuits) feasible.

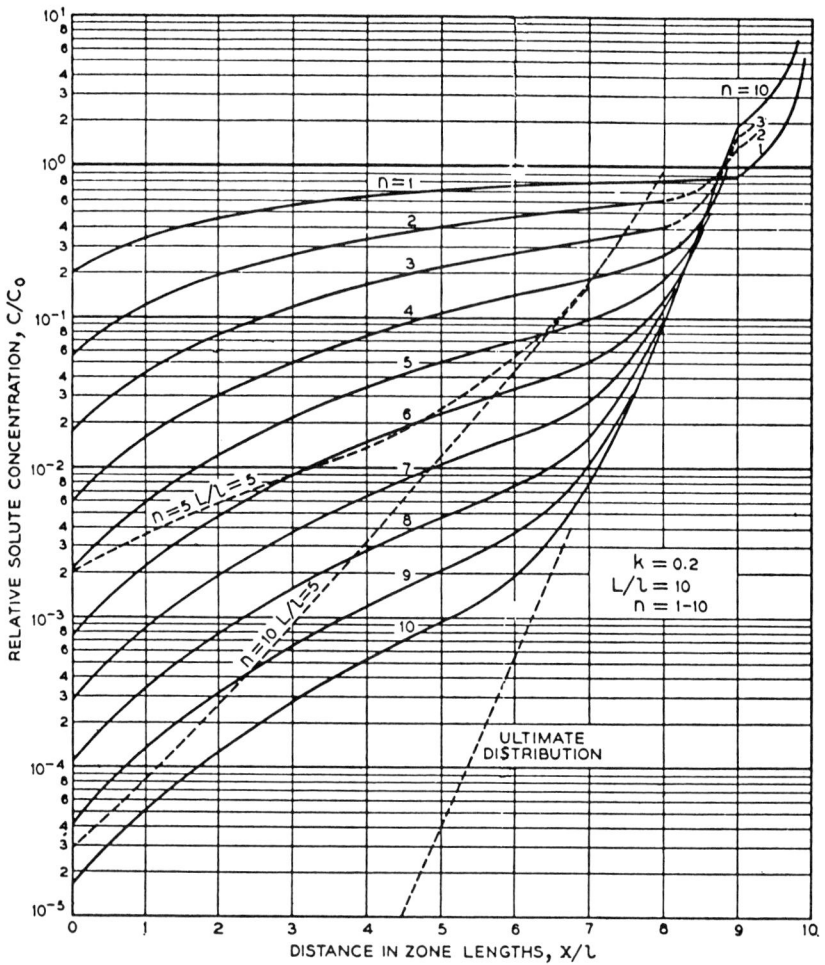

Figure 19.8. *Computed relative solute concentration profile, C/C_0 (where C_0 is the starting concentration of a solute) as a function of the number of passes of a molten zone along a rod of solvent. For the solid curves, the ratio of rod length to zone length is 10; for the dashed curves, the ratio is 5; k is the equilibrium ratio of solubilities of the solute in the solid and the liquid, here taken at the typical value of 0.2. Note the logarithmic concentration scale.*

and others, very rapid, in which there is no time for diffusion and rearrangement of solutes—like the transformations in shape-memory alloys, already mentioned). Historically, in the 1930s, the first sophisticated crystallographic analysis of the orientation relationship between parent and daughter phases was André Guinier's pioneering study of the nature of age-hardening in aluminium–copper alloys [87] (essentially an x-ray study of a precursor or transitional stage on the way to the equilibrium phase) and the long series of studies of diffusive transformations in a wide range of alloys by the school of Charles Barrett and Robert Mehl in Pittsburgh, throughout the 1930s. These last were outlined in 1943 in an early classic book, by the first of these, about the application of crystallography to physical metallurgy [88]. Mehl also began the systematic study of the kinetics of phase transformations, and related processes, including the recrystallization of deformed metals; in this he built on the mathematical analysis of Marvin Avrami, who published a few immensely influential papers (e.g. [89]) and then left the physics of solids entirely. Once kinetics are known in detail, the corresponding mechanism of phase transformation becomes much easier to understand. The extensive studies of age-hardening (also called precipitation-hardening) which followed the invention of electron microscopy to supplement x-ray diffraction were very well summarized in 1968 by John Martin [90]; this unusual book also reprints extracts from a number of the classical papers in this field. The universally acknowledged bible of people working in the broad field of phase transformations in metals and alloys is a substantial theoretical book by an Oxford metallurgist, Jack Christian [91].

One specific aspect of phase transformations has been so influential among physical metallurgists, and also among polymer physicists, that it deserves a specific summary; this is the study of the nucleation and of the spinodal decomposition of phases. The notion of homogeneous nucleation of one phase in another (e.g. of a solid in a supercooled melt) goes back all the way to Gibbs. Minute embryos of different sizes (that is, transient nuclei) constantly form and vanish; when the product phase has a lower free energy than the original phase, as is the case when the latter is supercooled, then some embryos will survive if they reach a size large enough for the gain in volume free energy to outweigh the energy that has to be found to create the sharp interface between the two phases. No less a physicist than Einstein examined the theory of this process with regard to the nucleation of liquid droplets in a vapour phase [92]. Then, after a long period of dormancy, the theory of nucleation kinetics was revived in 1925 by M Volmer and A Weber in Germany [93] and improved further by two German theoretical physicists of note, Richard Becker and Wolfgang Döring [94]. Reliable experimental measurements became possible much later still in 1950, when David Turnbull, at GE, perfected the technique of dividing a melt up into tiny hermetic compartments so that heterogeneous nucleation catalysts were confined to just a few of these; his measurements [95, 96] are still frequently cited.

It took a long time for students of phase transformations to understand clearly that there exists an alternative way for a new phase to emerge by a diffusive process from a parent phase. This process is what the Nobel-prize-winning Dutch physicist Johannes van der Waals (1837–1923), in his doctoral thesis in 1873, first christened the 'spinodal'.

He recognized that a liquid beyond its liquid/gas critical point, having a negative compressibility, was unstable towards *continuous changes*. A negative Gibbs free energy has a similar effect, but this took a very long time to become clear. The matter was at last attacked head-on in 1961 in a famous theoretical paper (based on a 1956 doctoral thesis) by the Swedish metallurgist Mats Hillert [97]: he studied theoretically both atomic segregation and atomic ordering, two alternative diffusional processes, in an unstable metallic solid solution. The issue was taken further by John Cahn and the late John Hilliard in a series of celebrated papers which has caused them to be regarded as the creators of the modern theory of *spinodal decomposition*; first [98] they revived the concept of a *diffuse* interface which gradually thickens as the unstable parent phase decomposes *continuously* into regions of diverging composition (but, typically, of similar crystal structure); later [99], John Cahn generalized the theory to three dimensions. It then emerged that a very clear example of spinodal decomposition in the solid state had been studied in detail as long ago as 1943, at the Cavendish Laboratory by Vera Daniel and Henry Lipson [100], who had examined a copper–nickel–iron ternary alloy. Recently, on an occasion in honour of Mats Hillert, John Cahn [101] mapped out in masterly fashion the history of the spinodal concept and its establishment as a widespread alternative mechanism to classical nucleation in phase transformations, specially of the solid–solid variety. There is an excellent, up-to-date account of the present status of the theory of spinodal decomposition and its relation to experiment and to other branches of physics by Binder [102]. The Hillert/Cahn/Hilliard theory has also proved particularly useful to modern polymer physicists concerned with structure control in polymer blends, since that theory was first applied to these materials in 1979 (see outline by Kyu [103]).

Another feature, which can only be asserted for lack of space to illustrate it, is the central role of microstructural and microcompositional analysis in the study of phase transformations. (This has played an essential part in the experimental study of spinodal decomposition.) Figure 19.2, showing some of Heycock and Neville's early micrographs of multiphase alloys, gives an inkling of the kind of complications that emerge from a microscopic study of such alloys. Both the geometrical disposition and (especially) size scale of dispersed phases, and their composition, crucially influence mechanical properties and also, as we have seen, the electrical behaviour of ceramics. Two instruments have transformed this kind of study. One is the transmission electron microscope, the genesis of which is recounted by T Mulvey elsewhere in this book (Chapter 20); the other, equally important, is the electron microprobe analyser (essentially the brainchild of R Castaing [104] in Paris in 1951–54). This is a device in which a fine focused electron beam strikes a polished specimen surface and the characteristic x-rays emitted are analysed, either by diffraction from an analysing crystal or by direct energy dispersion in a radiation counter. It allows chemical composition in a microstructure to be mapped from point to point by purely physical means, and it represents the high point of a gradual, widespread shift from chemical to physical analysis of composition by materials scientists. (In the 1950s, all metallurgy laboratories employed 'wet' chemical analysts; today they have disappeared.) The electron microprobe analyser has permitted theoretical models of phase transformations, with their quantitative predictions of solute distributions, to be experimentally checked; we have seen how important such checks have been in the history of physical metallurgy. A recent, very up-to-date account of what this crucial instrument can do has been published by Lifshin [105].

The electron microprobe analyser is only one, albeit the most important, of a large number of physical techniques of compositional analysis. Ions, photons, electrons, protons or neutrons can be pressed into service as probes, and any of these can also be measured to give the emitted signal. Some analyse only a nanometre or so at the surface, others average through a large volume; some can resolve and identify at the atomic scale, others average over square centimetres; some have quickly achieved great fame, the most notable being the scanning tunnelling microscope, a device with atomic resolution which won its inventors, G Binnig and H Rohrer, the Nobel Prize within a couple of years, and which has quickly proved to have dozens of unsuspected uses [106]; the Nobel Prize adjudicators certainly showed remarkable foresight. A recent encyclopedia included no fewer than 106 distinct techniques of physical characterization used for materials of various kinds [107], although some of these are used to garner crystallographic as distinct from compositional information, and others, notably thermal analysis—microcalorimetry, thermogravimetry, etc—are used for the precise quantitative study of the temperatures and kinetics of reactions and phase transformations. (To convey a feeling for the scope of this one family of techniques it can be noted, by way of example, that one journal specializing in thermal analysis published 700 papers in one recent year.)

Yet another family of techniques is devoted to the quantitative measurement of microstructure—features such as the grain size distribution, the particle size distribution and the 'mean free path' in two-phase dispersions, the interface area per unit volume, mean curvature of inclusions, porosity in partly sintered materials, fractal dimensions, etc. In broad terms, this involves the determination of three-dimensional features from measurements on two-dimensional sections, and therefore requires sophisticated geometrical and statistical analysis as well as a variety of computer devices devoted to image analysis. The objective is to relate the stereological parameters to properties, both physical and mechanical. An early (1970) classic in this field, called stereology or quantitative metallography, is a theoretical book by an American metallurgist, Ervin Underwood [108]. The most brilliant exponent of this recondite art was another American metallurgist, Frederick Rhines, who showed in 1974 how the topological features of grain growth in a pure metal could be clarified by sophisticated stereological analysis (that topic has exercised a continuing fascination over successive generations of metallurgists and computer modellers). Rhines, who used stereology to good effect in numerous other studies, put a lifetime's experience into a fascinating little monograph somewhat eccentrically titled *Microstructology—Behavior and Microstructure of Materials* [109]. A recent comprehensive overview has been given by the German metallurgist Exner [110].

It is no exaggeration to claim that the huge range of modern methods of characterization represents perhaps the most crucial contribution of physics to the study of useful materials.

19.10. The physics of polymers

See also p 1596

In 1980, Lord Todd, at that time President of the Royal Society of Chemistry in London, was asked what had been chemistry's biggest contribution to society. He felt that despite all the marvellous medical advances, chemistry's biggest contribution was the development of polymerization, according to the preface of a recent book on the history of high-technology polymers [111]. Undoubtedly, modern polymers are a most impressive success story for industrial (and academic) chemists, and polymer chemists (unlike metallurgists) have been awarded Nobel Prizes (but then, sadly, there is no such prize available for prowess in physical metallurgy!). It is all the more ironic that, as recently as the 1920s, the chemical 'establishment' was highly sceptical that high-polymeric molecules existed at all. In the aforementioned book, Melvin Kohan [112], in outlining the history of nylon-66, remarks that when the youthful Wallace Carothers (the inventor of nylon) was put in charge of the Dupont research laboratory in 1926, he and his few colleagues 'viewed a tiny polymer industry heavily dependent on modified natural products, and they saw a technical establishment still not fully convinced by the work of Staudinger, Svedberg, Meyer and others that polymers were valence-bonded, high-molecular-weight entities and not just aggregates of small molecules [although synthetic polymers had been used for almost a century]. Even the definition of the term "polymer" was at issue.'. There were no polymer journals, whereas today there are more than 100, and load-bearing polymers are steadily displacing metals from their place of primacy. The scepticism of conservative chemists at that time is reminiscent of the doubters who saw no need for the 'atomic hypothesis' a century ago, and the equally sceptical metallurgists around 1900 who were not at all convinced that metals consisted of small crystal grains.

However, a close look at history shows that chemists are not the sole begetters of modern polymers. The chemical identity of long-chain molecules, the polymerization process itself, molecular weights and configurations (straight or branched chains, homopolymers, copolymers, block copolymers) are the concern of chemists; but orientation, crystallization, microstructures of polymer blends, viscoelastic properties, strength and stiffness are matters for physicists, and their input has been steadily growing in recent years. Sadly, chemists and physicists are still apt to look on each other with much suspicion in this industry, but perhaps not more so than extractive and physical metallurgists in another field.

The first major contribution of physicists to the understanding of polymer behaviour was the creation of a theory of rubber-like elasticity. Rubbers, natural or synthetic, especially if lightly vulcanized, are capable of enormous elastic strains, quite unlike materials such as metals, and they have other unusual features such as an elastic modulus that rises with increasing temperature. If cooled too much, they lose completely their low modulus and large strain capability. In the early 1970s, the statistical mechanics and thermodynamics of both free and cross-linked assemblies of polymer chains, pioneered by Paul Flory in California, had reached the stage where rubber-like elasticity could be interpreted in quantitative detail. The largest part of the elastic strain is of entropic origin: the more a population of chains is stretched, the smaller its entropy (because there are fewer alternative configurations for a partially uncurled polymer molecule). However, different states of uncurling are also associated with different energies, because of different steric and Coulombic interactions [113]. The main part of the remarkable strain of a stretched rubber band can be regarded as entropy made palpable. The band wishes to return to a macrostate in which it has more alternative configurations at its disposal, i.e. it wants to contract. Flory was in fact a physical chemist, but the statistical mechanics of rubber elasticity was systematically set out by a British physicist, L R G Treloar, in 1975 [114]. The relative contributions of entropic and energetic components to the elastic modulus of a rubber have also been satisfactorily disentangled, in agreement with experimental determinations [115]. Flory's approach also proved to be a necessary precursor to the application of spinodal theory to polymer blends.

The hardest aspect of polymers for a physical metallurgist to envisage is the fact that most commonly they are neither wholly amorphous nor perfectly crystalline, but semicrystalline. For a polymer physicist, it makes sense to say of a polymer that it is 50% crystalline. The development of very strong polymeric fibres, most often made by intense mechanical drawing that aligns the long chains, can be achieved by starting with precursors of different degrees of crystallinity. The physicist Charles Frank, of Bristol (whom we have already encountered in connection with dislocation-aided crystal growth) in 1970 showed how the intrinsic (maximum) stiffness of a stretched polymer chain can be

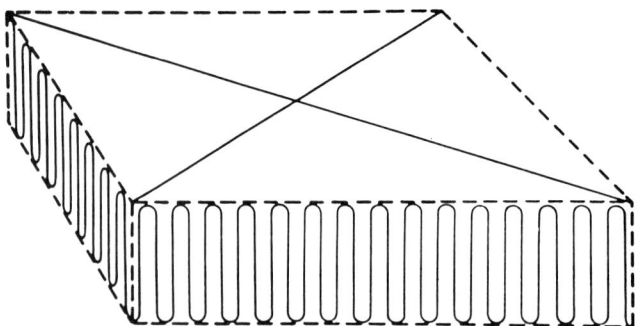

Figure 19.9. *Model of chain-folding in a polymer crystal.*

calculated [116]. Experimental determinations of such stiffness can be made not only by direct mechanical tests but also by x-ray diffraction and from Raman spectroscopic measurements. What is clear is that a mixture of physical theory and experiment has been an essential input in developing such strong and stiff polymeric fibres as Kevlar. The history has been outlined by the English physicist Ian Ward in a *Festschrift* published to mark Frank's 80th birthday [117].

The most striking contribution of physics to the understanding of polymers, however, is undoubtedly the recognition of chain-folding in fully crystallized polymers. Several investigators, notably Erhard Fischer in Germany and Andrew Keller in England, found in 1957, to their intense astonishment, that unbranched polyethylene (the simplest of polymers) can be crystallized from solution to yield thin crystal blades, much thinner than would correspond to the average length of the molecules. The crystal structure was deduced to be as shown in figure 19.9 [118]. Keller made his discovery in Bristol, and in the aforementioned 1991 *Festschrift* [119] he gives a splendid circumstantial account of his discovery and how the special atmosphere of the Bristol physics department made it possible. The discovery was followed by many years of fierce controversy as to how a configuration as shown in the figure could possibly come into existence; this controversy, which has still not quietened 35 years later, is memorably outlined in Keller's 1991 chapter. In a semicrystalline polymer, a single polymer molecule can start in an amorphous region, enter a folded region as in the figure, and emerge from it to re-enter another amorphous zone. It is a long way from the images to which physical metallurgists are accustomed!

Liquid crystals, which we met earlier in this chapter, consist of long molecules. In the extreme case, they are polymeric molecules, and a large family of modern high-performance polymers are in fact liquid-crystal polymers. Studying their morphology has been to a substantial degree the province of physicists, as can be seen by examining bibliographies of papers in the field. A recent survey of these materials has been written by a physicist and a (former) physical metallurgist [120]. More generally, the processing of polymeric liquids requires a sophisticated understanding of the fluid dynamics of anisotropic fluids: substantially aligned chain configurations in amorphous solid polymers can be made by a skilful exploitation of this kind of fluid dynamics [121]. The importance of the spinodal concept in relation to the understanding of polymer blends (i.e. polymer 'alloys') has been emphasized earlier.

These are a few examples, perhaps the most important, of the contributions of physicists during the past three decades to the understanding and the processing of polymers, which many observers regard as destined to become the most important materials of the twenty-first century.

19.11. Terms, concepts and institutions

The field with which this chapter is concerned has been blessed with an extravagant range of names, some applying to the whole, some applying to parts, some including the field as a constituent: metallography, physical metallurgy, metal physics, polymer physics, materials science, applied physics, engineering physics, even (though that name is rarely encountered) physics of materials! Names have caused passions to boil; thus the German metallurgical society (the Deutsche Gesellschaft für Metallkunde) recently went through a prolonged bout of dissension while seeking to broaden its remit: finally, an eminent and combative 92-year-old member, Werner Köster, prevailed against the power of the establishment and the name was changed to Deutsche Gesellschaft für Materialkunde (defeating the more longwinded alternative, Deutsche Gesellschaft für Werkstoffwissenschaften); so, at least the initials remained the same. It is impossible to resist the temptation to quote Shakespeare, who is always apposite. In *Romeo and Juliet*, Juliet (who was much plagued by an excessive preoccupation with family names) bursts out:

> What's in a name? that which we call a rose
> By any other name would smell as sweet.

The naming issue starts with a definition problem: what is a material? One reasonable approach is to say that a material is any useful form of matter, excepting those which are burned, swallowed, injected or spread on farmland. Even that leaves loose ends: is a catalyst a material? Some materials have to be elaborately shaped (e.g. for a mechanical watch); others are used 'as is': a bottle of lubricating oil or a length of fusewire. Some are useful because of their strength, others because of their conductivity, some for their slipperiness, some for their insulating ability. Some need to be ultrapure, some need to be contaminated with dopants, to do what is expected of them. To cover such a multiplicity of nature and end-use, a multiplicity of approaches is unavoidable if understanding and control of their functions is to be achieved. Physics is one such approach, but it can rarely be used in complete isolation.

We have seen in the preceding pages that many who have made substantial contributions to the physics of materials began professional life as physical chemists, as engineers, as metallurgists, as mathematicians, even in one instance as a botanist, and for others, their period as physicists of the solid state was transient indeed. Thus, Ernst Ising, who gave his name to the 'Ising model' in the theory of ferromagnetism with just one paper in 1925 [122], went on to work in the financial sector after he had achieved his doctorate (one hopes that his actuarial problems were more readily soluble than the three-dimensional Ising problem), though 25 years later his name turns up again as the writer of a published letter on 'Goethe as a physicist'! Again, Marvin Avrami, who in 1939 pioneered a general mathematical analysis of nucleation-and-growth phase transformations [89, 123] which is still much used, then promptly moved into astrophysics. Some disciplines worked well together, others maintained a suspicious mutual stance: it would be true to say, I believe, that over the years, solid-state physicists and metallurgists have learnt to work smoothly together. It is only in the academic sphere that members of one of these professions have sometimes been tempted to the conviction that they could do the job of the other better.

We have seen at the outset that metallurgy began with a strong chemical stance and that this insensibly shifted towards a physical outlook as the physics revolution in the early years of this century took hold. This drift can be effectively illustrated by examining the history of one particular institution, at the distinguished German University of Göttingen. In 1903, Gustav Tammann (1861–1939) came to Göttingen from what is now Estonia to take up a new chair of inorganic chemistry; four years later he took over Walter Nernst's chair at the Institute of Physical Chemistry when Nernst moved to Berlin and gradually shifted his interests from inorganic glasses and silicates (and liquid crystals, which he was convinced were imaginary) towards metals, including the determination of numerous phase diagrams. In 1912 he published a *Lehrbuch der Metallographie* (at that time, 'metallography' was a portmanteau word for the study of metals once they had been extracted from their ores). Many excellent German chemical and physical metallurgists were his pupils. In 1938, one of Tammann's pupils, Georg Masing (1885–1957) took over the chair and the Institute which went with it (in Germany, broadly based departments, on the American model, are the exception) and it was renamed the Institute of General Metallurgy (Allgemeine Metallkunde). In 1952, Masing, with his pupil Kurt Lücke, published an influential *Lehrbuch der Metallkunde*. When Masing, in ill health, relinquished his chair in 1955, the Institute fell on hard times and the Faculty resolved to close it down. Thereupon widespread protests came from the German metals industry, until the provincial minister of education intervened and the Institute was reactivated, with a very young (31-year-old) physicist, Peter Haasen (1927–93) in charge. Haasen had taken his doctorate in theoretical physics with Richard Becker and Georg Leibfried. (Becker's and Pohl's physics institutes had made major contributions to materials science.) Haasen very successfully revived the moribund institute, which was at first called Institut für Metallphysik und Allgemeine Metallkunde, but Haasen soon deleted the more technological part of the name! (Not long before his recent death, Haasen told a friend 'I have been body and soul a physicist, physics has been my whole life'.) In 1974, he published a widely praised textbook, *Physikalische Metallkunde*. By the time Haasen retired in 1993, metal physics, notably the study of dislocations in many kinds of material, and of the mechanisms of phase transformations, had made the Institute world-famous. The new professor, appointed in 1994, is Rainer Kirchheim, an experimental metal physicist who has resolved to introduce polymer physics and is thinking of renaming the institute with a name including the term 'Materialphysik'—which would probably be the first academic unit anywhere with 'physics of materials' in its name. This sequence of men, institute names and book titles reveals very clearly the progressive shift of emphasis, over nearly a century, in the study of metals. An equally influential centre of research in physical metallurgy is the Max Planck (formerly Kaiser Wilhelm) Institute of Metals Research in Stuttgart, which under the leadership of a charismatic pupil of Tammann, Werner Köster (1896–1989), was set up in 1933 and has grown to embrace not only metals but ceramics and basic solid-state physics too, in a federal structure. Polymer physics is catered for by a newer Max Planck Institute in Mainz.

In Britain, the metallurgist Daniel Hanson (1892–1953) during the 1930s gradually converted the entirely traditional Department of Metallurgy (part of which turned into the Department of Physical Metallurgy soon after World War II) into an academic pioneer, in which themes such as dislocation theory, statistical thermodynamics, the fundamentals of phase diagrams and the theory of phase transformations for the first time formed part of an undergraduate course in metallurgy. Under the influence in particular of Alan Cottrell and Geoffrey Raynor (1913–83), two young men who were made professors in 1948, and of the young theoreticians Frank Nabarro (b 1916) and John Eshelby (1916–81), the Department became a scientific trendsetter for physical metallurgists the world over, and thereby one of the precursors of the later development of materials science. Cottrell and Raynor were told that one of them was to be professor of physical metallurgy, the other, of metal physics, but it was up to them to decide who was to hold each title: this takes us back to Shakespeare's Juliet!

In America, a place which (about the same time) had a correspondingly great influence of shifting metallurgy towards physics was the Institute for the Study of Metals at the University of Chicago. This was the creation of Cyril Smith (1903–91), a charismatic British metallurgist who became an American citizen, masterminded the metallurgical aspects of the Manhattan Project at Los Alamos and then resolved to help others to undertake the kind of fundamental researches on metals which he had undertaken, under conditions of great difficulty, in an American copper company before World War II. For 15 years from 1946 he and a uniquely able group of physicists, physical chemists and physical metallurgists combined forces to study a great variety of fundamental problems concerning metals and alloys; many visitors (Haasen was one of them) were deeply influenced by what they experienced there. Smith himself studied the development of microstructure and especially the factors governing grain shape in polycrystals, and then moved on under the influence of his historian wife to study the history, first of the science of metals—which he set out in a major book, *A History of Metallography* [124]—and subsequently of materials in a broader sense. The flavour of his interests and style can be appreciated best from a collection of essays, *A Search for Structure* [125]. While Smith was in Chicago, he was involved in a major event, the foundation of a new journal, *Acta Metallurgica*, first published in 1953 and still flourishing today. This came just as the 'quantitative revolution' of the early 1950s was getting strongly under way, and I can still remember clearly the enormous impact it had on young physical metallurgists at the time: its existence was a great encouragement to physically inclined students of metals primarily, and other materials by extension. The new way of studying metals (and, soon enough, other materials too) was consolidated by the new journal, excellently edited by Bruce Chalmers, a British physicist and solidification expert who had settled at Harvard. Other new journals at about the same time, notably *Physics and Chemistry of Solids*, abetted the trend. *Acta Metallurgica* complemented a review journal, *Progress in Metal Physics*, founded four years earlier by a former metal physicist, Paul Rosbaud (1896–1963) who had worked with Erich Schmid in Berlin in the 1920s, had helped establish the metallurgical Institute in Göttingen, had moved to Britain after the war and done much for scientific publishing there [126]. This was probably the first of the numerous present-day review journals in the science of materials. Other very influential journals in the physics of metals include the *Zeitschrift für Metallkunde* (founded 1911, and initially called *Internationale Zeitschrift für Metallographie*) and the *Journal of Materials Research* (founded in 1986 by the Materials Research Society in America).

The Materials Research Society (MRS), founded in the United States in 1973, marks the firm establishment of the modern scientific approach to research on materials, including but by no means restricted to the application of the methods of physics. Recently, an early president of the MRS, Kenneth Jackson (then of Bell Laboratories) outlined the circumstances of the creation of the Society [127]. The Society, which has become in only 20 years the most influential and vivid assembly of scientists concerned with materials (including many physicists), was founded because the existing American national bodies that catered for physicists, chemists, metallurgists, electrochemists and electrical engineers, were not receptive to the research done by their members in industrial laboratories. The way Jackson put it is 'Their work was often regarded as uninteresting to purists in the conventional disciplines and they often failed to achieve recognition for it.'. By this time the broad concept of the *materials scientist* had become common currency (it began in American academia in the 1950s) and, in Jackson's words 'we recognized the need for a forum where materials scientists could come together to interact with a broader range of scientists and technologists than they could at the meetings based on the traditional disciplines, *to address topics which fall into the gaps between the traditional disciplines* [my italics], to provide a coherence to the field of materials science, to provide a home for materials scientists, and to provide proper recognition for the multidisciplinary nature of their contributions'. Indeed, ever since its foundation, the MRS has insisted at its twice-yearly meetings that its members organize symposia mainly on 'current hot topics' that span the traditional disciplines. It is not accidental that their very first symposium, in May 1973, was on *Applications of Phase Transitions in Material Science*. Their monthly *MRS Bulletin* always carries printed symposia on topics that satisfy the above criteria of genuine interdisciplinarity and current importance. Policy in the MRS is very firmly in the hands of its members and their elected representatives, so it seems that the normal hardening of the organizational arteries has been avoided up to now. To what extent materials science by now deserves to be termed a clearly defined discipline in its own right, as well as being the meeting-place of traditional disciplines in the task of solving difficult problems, is a large question which would deserve a separate chapter. The MRS has spawned a number of other national and transnational materials research societies, in Western Europe, India, Japan, China and Mexico, operating on similar principles.

One organizational innovation which did much to foster the development of materials science in general, and the physics of materials specifically, was the creation of the Materials Research Laboratories (MRLs) at American universities. These were—and are—federally funded laboratories necessarily involving multiple academic departments (such as materials science, physics, chemistry, electrical and/or mechanical engineering and chemical engineering), with their own offices, experimental workspace and extensive equipment: existing professors would double up as members of the local MRL, working on diverse problems often in collaboration with colleagues in other fields. The first such laboratories were set up in 1960, after a long gestation beginning with forceful support from the mathematician and polymath John von Neumann (1903–57), who became profoundly convinced of the importance of high-quality research on materials, as recently recounted by Seitz [128]. Von Neumann died before he could pursue this, and it took a great deal more politicking by Seitz and others before the first laboratories were designated. The 25th anniversary of these very effective laboratories was marked by a conference in 1985; in the published form, a historical overview by Robert Sproull does justice to their record [129]. In a broader sense, the developments in American materials science were examined in two major and very interesting publications, in 1974–5 and in 1990, sponsored by the American National Academies. Rather than attempt to survey these here, I refer the reader to my own detailed reviews of these two reports and the 1985 conference report [130].

Within physics itself, at least in America, 'materials physics' has finally become a recognized field of research. This is largely the result of devoted efforts by Robb Thomson (of the National Bureau of Standards, as it was at that time) and Frederick Young Jr (of Oak Ridge National Laboratory). As Thomson has remarked in a private communication 'I have found it desirable and necessary to retain my physics coloration, and have not simply joined up, body and soul, with the metallurgists or ceramists'. At about the time the MRS was formed, Thomson and Young became troubled by the unconcern, as it seemed to them, of the American Physical Society, which about this time stopped having sessions on mechanical properties, and on most crystal defects as well, at its annual March meetings. They campaigned to have a Materials Physics Division set up side by side with the existing Condensed Matter Division. After extensive trials and tribulations, a 'Materials Physics Topical Group' was invented and began operating in 1984; six years later, a fully fledged Materials Physics Division came into existence and now has many hundreds of active members. Its mode of operation is not unlike that of the MRS: focus topics are chosen, organizers picked out and keynote speakers invited, and blank spaces are left to allow last-minute informal discussions to be set up. As Thomson remarks 'one of the other things we were successful about was to pick areas where people outside the narrow physics community would like to participate, and soon we had full sessions on topics which had not been seen in the APS for 20 years'. Competition is a very good thing, not only in the world of business but in the worlds of journal publishing and conference organization also! The (British) Institute of Physics likewise, for a number of years, has had a Materials Group the organizers of which take a relaxed view of what constitutes physics. So, to quote Shakespeare once more, the physics of materials at length has 'a local habitation and a name'.

Many topics in solid-state physics and physical metallurgy that might be thought to belong in this chapter have had to be left without discussion. Some have been omitted because they are too tangential to the practical world of materials; others because they belong more centrally to metallurgy or verge on engineering; yet others, because their practice necessarily entails interdisciplinary input; others again, simply because they are treated elsewhere in this book; finally, others have had to be omitted simply because of the limits on space. In the first category, I can exemplify the (two-dimensional) crystallography of surfaces and the theory of critical phenomena. In the second category, I have omitted the large and important field of mechanical properties and their interpretation in terms of crystal defects. Nevertheless, I cannot in good conscience omit a mention of one of the most steadily influential papers of the century, published in 1920 by Alan Griffith (1893–1963), on the elastic theory of the growth of cracks in brittle solids—a paper still frequently cited today [131]. The theory is based on the balance between changes in the surface energy of the crack faces and in the stored elastic energy, as a crack extends under stress. When Griffith joined Rolls-Royce in 1939, his chief tersely instructed him 'Go on thinking!'.

In the third category, the stereotype could be the 'art and science of growing crystals' (as an early book on the subject was entitled [132]), a skill which requires chemical and physical input in equal measure. Above, I have briefly discussed the role of single crystals of metals in early physical metallurgical research, but have had to leave out discussion of the much larger field of the growth of 'functional' crystals for electronic and electro-optical uses, which now once again involve metallic crystals also. (A very recent piece of research in this field, still in progress in Stuttgart, entails the preparation of beryllium crystals for the monochromatization of synchrotron x-ray beams; the metal for such crystals has to be ultrapurified by chemical means, then made to have precisely the right measure of imperfection to maximize x-ray reflectivity; then designed to be of the right thickness so that unreflected x-rays mostly pass through, thereby preventing heating up from absorbed radiation with the distortion this entails. How should such work be categorized?)

In the fourth category, I have said nothing about the very important topic of modern electron theory of metals or about the whole topic of semiconductors, because other chapters in this work touch upon them. In the final category, perhaps the most important omission forced by lack of space is the broad topic of computer simulation (or modelling), a skill which is a form of physics midway between theory and experiment: it might be termed the scientific equivalent of the spreadsheet in financial analysis, since it is easy to determine quickly the effect on the final outcome of modifying one small feature of the total input. The use of computer modelling in the first-principles prediction of crystal structures has been mentioned in passing, but the use of the approach in predicting binary and ternary phase diagrams and in studying microstructural changes and in interpreting many physical properties, notably by Monte Carlo simulation [133], has had to be left aside. It is one of the many fields of materials physics which is of rapidly growing importance, and now has its own journals—one of the rites of passage for new specialities.

I request the reader's indulgence for the omission of these, and other, topics in materials physics. Fortunately, the field encompasses such an *embarras de richesses* that completeness is unattainable: that is what makes it so endlessly fascinating to its practitioners.

References

[1] Smith C S (ed) 1965 *The Sorby Centennial Symposium on the History of Metallurgy* (New York: Gordon and Breach)
[2] Jacques J 1987 *Berthelot: Autopsie d'un Mythe* (Paris: Belin)
[3] Gibbs J W 1875–8 *Trans. Connecticut Acad. Arts Sci.* **3** 108, 343
[4] Klein M J 1970–80 *Dictionary of Scientific Biography* (entry on J W Gibbs) ed C C Gillispie (New York: Scribner's) p 386
[5] Heycock C T and Neville F H 1904 *Phil. Trans. R. Soc.* A **202** 1
[6] Hunt L B 1980 *The Metallurgist and Materials Technologist* July p 392
[7] Stockdale D 1946 *Metal Progress* p 1183
[8] Hansen M 1936 *Der Aufbau der Zweistofflegierungen (Constitution of Binary Alloys)* (Berlin: Springer)
    Hansen M and Anderko K 1958 *Constitution of Binary Alloys* (New York: McGraw-Hill)
[9] Wyckoff R W G 1963–6 *Crystal Structures* (2nd edn in 5 volumes) (New York: Interscience)
[10] 1987 *Crystallographic Databases—Information Content, Software Systems, Scientific Applications* (Chester: Data Commission of the International Union of Crystallography)
[11] Ewald P P et al 1962 *Fifty Years of X-ray Diffraction* (Utrecht: Oosthoek)
[12] Caroe G M 1978 *William Henry Bragg* (Cambridge: Cambridge University Press)
[13] Perutz M 1990 *The Legacy of Sir Lawrence Bragg* ed J M Thomas and D Phillips (London: The Royal Institution) p 71
[14] Kurnakov N, Zemczuzny A and Zasedelev M 1916 *J. Inst. Metals* **15** 305
[15] Johansson C H and Linde J O 1925 *Ann. Phys.* **78** 439
[16] Westbrook J H and Fleischer R L (ed) 1994 *Intermetallic Compounds: Principles and Practice* (2 volumes) (New York: Wiley) (see especially the historical introduction by Westbrook, pp 3–18)
[17] Cahn R W 1994 *Physics of New Materials* ed F E Fujita (Heidelberg: Springer) p 179
[18] Ewing J A and Rosenhain W 1900 *Phil. Trans. R. Soc.* A **193** 353
[19] Kelly A 1976 *Phil. Trans. R. Soc.* A **282** 5
[20] Braun E 1992 *Out of the Crystal Maze* ed L Hoddesdon et al (Oxford: Oxford University Press) p 340
[21] Harper S 1951 *Phys. Rev.* **83** 709
[22] Nowick A S and Berry B S 1972 *Anelastic Relaxations in Crystalline Solids* (New York: Academic)

[23] Cottrell A H 1953 *Dislocations and Plastic Flow in Crystals* (Oxford: Clarendon)
[24] Wert C and Zener C 1949 *Phys. Rev.* **76** 1169
[25] Amelinckx S 1964 *The Direct Observation of Dislocations* (New York: Academic)
[26] Friedel J 1956 *Les Dislocations* (Paris: Gauthier-Villars)
[27] Friedel J 1994 *Graine de Mandarin* (Paris: Editions Odile Jacob)
[28] Elam C F 1935 *Distortion of Metal Crystals* (Oxford: Clarendon)
[29] Schmid E and Boas W 1935 *Kristallplastizität* (Berlin: Springer)
[30] Mark H, Polanyi M and Schmid E 1927 *Z. Phys.* **12** 58
[31] Kronberg M L and Wilson H F 1955 *Trans. Am. Inst. Mining Metall. Eng.* **185** 501
[32] McLean D 1957 *Grain Boundaries in Metals* (Oxford: Oxford University Press)
[33] Wolf D and Yip S 1992 *Materials Interfaces: Atomic-Level Structure and Properties* (London: Chapman and Hall)
[34] Teichmann J and Szymborski K 1992 *Out of the Crystal Maze* ed L Hoddesdon *et al* (Oxford: Oxford University Press) p 236
[35] Radnai R and Kunfalvi R 1988 *Physics in Budapest* (Amsterdam: North-Holland) pp 64, 74
[36] Mitchell J W 1980 The beginnings of solid state physics *Proc. R. Soc.* A **371** 126
[37] Wigner E and Seitz F 1933 *Phys. Rev.* **43** 804; 1934 *Phys. Rev.* **46** 509
[38] Seitz F 1980 The beginnings of solid state physics *Proc. R. Soc.* A **371** 84
[39] Seitz F 1994 *On the Frontier: My Life in Science* (New York: American Institute of Physics)
[40] Seitz F 1946 *Rev. Mod. Phys.* **18** 384
[41] Seitz F 1954 *Rev. Mod. Phys.* **26** 7
[42] Huntingdon H B and Seitz F 1942 *Phys. Rev.* **61** 315, 325
[43] Simmons R O and Balluffi R W 1960 *Phys. Rev.* **117** 52; 1960 *Phys. Rev.* **119** 600; 1962 *Phys. Rev.* **125** 862; 1963 *Phys. Rev.* **129** 1533
[44] Granato A V 1994 *Phys. Chem. Solids* **55** 931
[45] Frenken J W M and van der Veen J F 1985 *Phys. Rev. Lett.* **54** 134
[46] Yukalov V I 1985 *Phys. Rev.* B **32** 436
[47] Wolf D, Okamoto P R, Yip S, Lutsko J F and Kluge M 1990 *J. Mater. Res.* **5** 286
[48] Barker J A and Henderson D 1976 *Rev. Mod. Phys.* **48** 587
[49] Gittus J 1978 *Irradiation Effects in Crystalline Solids* (London: Applied Science Publishers)
[50] Schilling W and Ullmaier H 1994 *Nuclear Materials, Part II* ed B R T Frost, *Materials Science and Technology* vol 10B, ed R W Cahn, P Haasen and E J Kramer (Weinheim: VCH) p 179
[51] Wollenberger H J 1983 *Physical Metallurgy* 3rd edn, ed R W Cahn (Amsterdam: North-Holland) p 1139
[52] Kuper A B, Lazarus D, Manning J R and Tomizuka C T 1956 *Phys. Rev.* **104** 1536
[53] Nabarro F R N 1948 *Report on a Conference on the Strength of Solids* (London: The Physical Society) p 75
[54] Herring C 1950 *J. Appl. Phys.* **21** 437
[55] Mukherjee A K 1993 *Plastic Deformation and Fracture of Materials* ed H Mughrabi, *Materials Science and Technology* vol 6, ed R W Cahn, P Haasen and E J Kramer (Weinheim: VCH) p 407
[56] Seitz F 1952 *Imperfections in Nearly Perfect Crystals* ed W Shockley (New York: Wiley)

[57] Lenard P, Schmidt F and Tomaschek R 1928 *Handbuch der Experimentalphysik* vol 23 (Leipzig)
[58] Notis M R 1986 *High-Technology Ceramics, Past, Present and Future* vol 3, ed W D Kingery (Westerville, OH: The American Ceramic Society) p 231
[59] Kraner H M 1924 *J. Am. Ceram. Soc.* **7** 868
[60] Kraner H M 1971 *Am. Ceram. Soc. Bull.* **50** 598
[61] Moharil S V 1994 *Bull. Mater. Sci. (Bangalore)* **17** 25
[62] Harvey E N 1957 *History of Luminescence* (Philadelphia: American Philosophical Society)
[63] Agulló-López F 1994 *MRS Bull.* **19** March p 29
[64] Miyayama M and Yanagida H 1988 *Fine Ceramics* ed S Saito (Amsterdam: Elsevier; Tokyo: Ohmsha) p 175
[65] Moulson A J and Herbert J M 1990 *Electroceramics* (London: Chapman and Hall) p 130
[66] Kuwabara M and Yanagida H 1988 *Fine Ceramics* ed S Saito (Amsterdam: Elsevier; Tokyo: Ohmsha) p 286
Moulson A J and Herbert J M 1990 *Electroceramics* (London: Chapman and Hall) p 147
[67] Cross L E and Newnham R E 1986 *High-Technology Ceramics, Past, Present and Future* vol 3, ed W D Kingery (Westerville, OH: The American Ceramic Society) p 289
[68] Salje E K H 1990 *Phase Transitions in Ferroelastic and Co-elastic Crystals. An Introduction for Mineralogists, Material Scientists and Physicists* (Cambridge: Cambridge University Press)
[69] Kurdjumov G, Khandros L G 1949 *Dokl. Akad. Nauk* **66** 211
[70] Roth W L 1972 *J. Solid-State Chem.* **4** 60
[71] Kelker H 1973 *Mol. Cryst. Liq. Cryst.* **21** 1; 1988 *Mol. Cryst. Liq. Cryst.* **165** 1
[72] Friedel G 1922 *Ann. Phys.* **18** 273
[73] de Gennes P G and Prost J 1993 *The Physics of Liquid Crystals* 2nd edn (Oxford: Clarendon)
[74] Chandrasekhar S 1992 *Liquid Crystals* 2nd edn (Cambridge: Cambridge University Press)
[75] Bahadur B (ed) 1991 *Liquid Crystals: Applications and Uses* (3 volumes) (Singapore: World Scientific)
[76] Mort J 1994 *Phys. Today* **47** April p 32
[77] Weber M J 1991 *Glasses and Amorphous Materials* ed J Zarzycki, *Materials Science and Technology* vol 9, ed R W Cahn, P Haasen and E J Kramer (Weinheim: VCH) p 619
[78] MacChesney J B and DiGiovanni D J 1991 *Glasses and Amorphous Materials* ed J Zarzycki, *Materials Science and Technology* vol 9, ed R W Cahn, P Haasen and E J Kramer (Weinheim: VCH) p 753
[79] Biloni H 1983 *Physical Metallurgy* 3rd edn, ed R W Cahn (Amsterdam: North-Holland) p 477
[80] Liebermann H H (ed) 1993 *Rapidly Solidified Alloys* (New York: Marcel Dekker)
[81] Duwez P, Willens R H and Klement W Jr 1960 *J. Appl. Phys.* **31** 36
[82] Cahn R W 1980 *Contemp. Phys.* **21** 43
[83] Brandt E H 1989 *Science* **243** 349
[84] Pfann W G 1958 *Zone Melting* 1st edn (New York: Wiley), 1966 2nd edn
[85] Suits C G and Bueche A M 1967 *Applied Science and Technological Progress* (Washington, DC: National Academy of Sciences) p 308
[86] Wise G 1985 *Whillis R Whitney, General Electric, and the Origins of US Industrial Research* (New York: Columbia University Press)
[87] Guinier A 1938 *Nature* **142** 569

[88] Barrett C S 1943 *The Structure of Metals* 1st edn (New York: McGraw-Hill), 1967 3rd edn (with T B Massalski)
[89] Avrami C 1941 *J. Chem. Phys.* **9** 177
[90] Martin J W 1968 *Precipitation Hardening* (Oxford: Pergamon)
[91] Christian J W 1975 *The Theory of Transformations in Metals and Alloys* 2nd edn (Oxford: Pergamon)
[92] Einstein A 1910 *Ann. Phys.* **33** 1275
[93] Volmer M and Weber A 1925 *Z. Phys. Chem.* **119** 277
[94] Becker R and Döring W 1935 *Ann. Phys. (New Series)* **24** 719
[95] Turnbull D and Cech R E 1950 *J. Appl. Phys.* **21** 804
[96] Turnbull D 1952 *J. Chem. Phys.* **20** 411
[97] Hillert M 1961 *Acta Metall.* **9** 525
[98] Cahn J W and Hilliard J E 1958 *J. Chem. Phys.* **28** 258
[99] Cahn J W 1961 *Acta Metall.* **9** 795
[100] Daniel V and Lipson H 1943 *Proc. R. Soc.* A **181** 368; 1944 *Proc. R. Soc.* A **182** 378
[101] Cahn J W 1991 *Scand. J. Metall.* **20** 9
[102] Binder K 1991 *Phase Transformation in Materials* ed P Haasen, *Materials Science and Technology* vol 5, ed R W Cahn, P Haasen and E J Kramer (Weinheim: VCH) p 405
[103] Kyu T 1993 *Encyclopedia of Materials Science and Engineering* Supplementary vol 3, ed R W Cahn (Oxford: Pergamon) p 1893
[104] Castaing R and Deschamps R 1954 *C. R. Acad. Sci., Paris* **238** 1506
[105] Lifshin E 1993 *Characterization of Materials, Part II* ed E Lifshin, *Materials Science and Technology* vol 2B, ed R W Cahn, P Haasen and E J Kramer (Weinheim: VCH) p 351
[106] DiNardo N J 1993 *Characterization of Materials, Part II* ed E Lifshin, *Materials Science and Technology* vol 2B, ed R W Cahn, P Haasen and E J Kramer (Weinheim: VCH) p 3
[107] Cahn R W and Lifshin E 1993 *Concise Encyclopedia of Materials Characterization* (Oxford: Pergamon)
[108] Underwood E E 1970 *Quantitative Stereology* (Reading, MA: Addison-Wesley)
[109] Rhines F N 1986 *Microstructology—Behavior and Microstructure of Materials* (Stuttgart: Dr Riederer) now (Munich: Hanser)
[110] Exner E E 1993 *Characterization of Materials, Part II* ed E Lifshin, *Materials Science and Technology* vol 2B, ed R W Cahn, P Haasen and E J Kramer (Weinheim: VCH) p 281
[111] Seymour R B and Kirshenbaum G S (ed) 1986 *High-Performance Polymers: Their Origin and Development* (New York: Elsevier)
[112] Kohan M I 1986 *High-Performance Polymers: Their Origin and Development* ed R B Seymour and G S Kirshenbaum (New York: Elsevier) p 19
[113] Flory P J 1969 *Statistical Mechanics of Chain Molecules* (New York: Wiley-Interscience)
[114] Treloar L R G 1975 *The Physics of Rubber Elasticity* (Oxford: Clarendon)
[115] Mark J E 1976 *Macromol. Rev.* **11** 135
[116] Frank F C 1970 *Proc. R. Soc.* A **319** 127
[117] Ward I M 1991 *Sir Charles Frank, OBE, FRS: An Eightieth Birthday Tribute* ed R G Chambers, J E Enderby, A Keller, A R Lang and V W Steeds (Bristol: Hilger) p 322
[118] Keller A 1957 *Phil. Mag.* **11** 1165
[119] Keller A 1991 *Sir Charles Frank, OBE, FRS: An Eightieth Birthday Tribute* ed R G Chambers, J E Enderby, A Keller, A R Lang and V W Steeds (Bristol: Hilger) p 265

[120] Donald A M and Windle A H 1992 *Liquid Crystalline Polymers* (Cambridge: Cambridge University Press)
[121] Mackley M R 1991 *Sir Charles Frank, OBE, FRS: An Eightieth Birthday Tribute* ed R G Chambers, J E Enderby, A Keller, A R Lang and V W Steeds (Bristol: Hilger) p 307
[122] Ising E 1925 *Z. Phys.* **31** 253
[123] Avrami M 1939 *J. Chem. Phys.* **7** 1103
[124] Smith C S 1960 *A History of Metallography* (Chicago, IL: University of Chicago Press)
[125] Smith C S 1981 *A Search for Structure* (Cambridge, MA: MIT Press)
[126] Cahn R W 1994 *Eur. Rev.* **2** 37
[127] Jackson K A 1993 *MRS Bull.* **18** August p 70
[128] Seitz F 1994 *MRS Bull.* **19** March p 60
[129] Sproull R L 1987 *Advancing Materials Research* ed P A Psaras and H D Langford (Washington, DC: National Academy Press) p 25
[130] Cahn R W 1992 *Artifice and Artefacts* (Bristol: Institute of Physics) pp 314, 325 and 348
[131] Griffith A A 1920 *Phil. Trans. R. Soc.* A **221** 163
[132] Gilman J J (ed) 1963 *The Art and Science of Growing Crystals* (New York: Wiley)
[133] Binder K 1992 *Adv. Mater.* **4** 540

Chapter 20

ELECTRON-BEAM INSTRUMENTS

T Mulvey

Electron-beam instruments occupy a wide panorama of scientific and engineering innovation. The present account is mainly concerned with the emergence of basic concepts, invention and innovation rather than the detailed engineering, essential though this is for the user. Many people have made great personal contributions over the years in both innovation and manufacture, so the present account is necessarily selective. The task has been made easier by the publication of two wide-ranging accounts; the first [1] covers the period up to 1939 and the second [2] reviews the scientific and engineering developments up to 1985, including important historical documents in key areas.

20.1. Early days

In 1897, during a regular Friday evening discourse at the Royal Institution of Great Britain, J J Thomson announced the discovery of the electron. The newly discovered 'corpuscle' had an astonishingly high charge to mass ratio, e/m, which Thomson had later confirmed in his cathode ray tube by the ingenious use of superposed magnetic and electric deflecting fields. He surmised correctly that this particle must be of a size and mass extremely small compared with even that of the lightest atom, hydrogen. The existence of particles much smaller than the atom had not previously been suspected. Many German physicists, however, preferred the term Kathodenstrahlung (cathode radiation), emphasizing its possible wave nature, which they could not prove experimentally in a convincing manner, but Thomson's 1897 views prevailed in the end. Belief in the 'wave theory' of the electron was effectively dead in Germany by 1900.

Even after de Broglie's wave theory of 1924 and G P Thomson and A Reid's brilliant demonstration of electron diffraction, in 1927, wave theory played no part in the development of instruments such as the electron microscope. It was the technology of the cathode ray tube developed in this period that cleared the ground for the emergence of electron optical instruments through the efforts of scientifically minded electrical engineers. The physicists had their turn later. Among early discoveries, that of Wiechert in 1899 is especially notable. He found that in a cathode ray tube, completely immersed in the uniform magnetic field of a long solenoid, the diverging electron beam from the cathode could be confined to the paraxial region and even 'concentrated' into a small spot on the distant fluorescent screen by a suitable adjustment of the solenoid current. This converging device does not constitute a lens, however, since a uniform field does not have a unique axis, nor does it have a refractive power that varies with radial height, a distinctive feature of any lens.

20.2. The magnetic electron lens

The first real confrontation between theory and traditional 'know-how' of the behaviour of electron beams in magnetic fields was probably brought on by Dennis Gabor, a gifted Hungarian research student, who had just graduated from the Technische Hochschule (TH) Berlin and was looking for a challenging PhD project. Professor Matthias, Head of the High Voltage Laboratory of the TH Berlin, gave him a place in his research team, specializing in measuring fast electrical transients with high-speed cathode ray oscillographs. He was assigned the task of building a greatly improved instrument that could record the fast electrical surges (*Wanderwelle*) that were then wreaking havoc along the newly installed high-voltage overhead grid lines in Germany. Gabor, without previous experience, introduced many refinements into the design. He replaced the traditional long solenoid, which made access to the electron optical column difficult, by a short solenoid. To reduce stray external magnetic fields that might disturb the beam deflector system, he designed intuitively an innovative iron-shrouded short solenoid that could be conveniently sited between the emitting cathode and the critical operational part of the electron optical column. There was no theory of its action to guide him in his design and, ignoring all received wisdom, he simply placed it where a coil of some 40 mm internal diameter would fit conveniently over the glassware of the tube. The coil was encapsulated in a cylindrical iron shroud, provided with iron end plates. The axial region itself was iron-free. The axial field due to the coil was therefore supplemented by the magnetization of the iron, which at the same time considerably reduced the external magnetic field, killing two birds with one stone. He did not realize at the time, as he later admitted, that he had unknowingly designed and built the first iron-shrouded magnetic

Dennis Gabor

(Hungarian, 1900–79)

Gabor, a gifted son of the Director of Hungarian Coalmines, was educated at home by a succession of French and German governesses and an English tutor. He taught himself physics and mathematics from advanced German textbooks his father let him buy, having trouble later when he went to school and found he knew more than his teacher. When he arrived at the TH Berlin, he was bored; he had already covered the syllabus. He stayed on for a PhD, working on the high-voltage oscillograph, narrowly missing inventing the magnetic electron lens. He regarded himself primarily as an inventor. In 1933, his Jewish background and Nazi hostility caused him to emigrate to Britain. He worked on optics and electron optics at the British Thomson Houston Company Research Laboratory in Rugby. When war was declared on Germany, Gabor was classified as an 'Enemy alien with specialist knowledge'; this saved him from internment in the Isle of Man, but he had to work alone in a hut outside the factory fence. During the Easter holiday of 1947, while idly awaiting a game of tennis, he 'received a sudden vision' of the principle of holography, as he called it, for correcting the spherical aberration of electron microscopes. He received the Nobel Prize in 1971 for this concept but only after his death did holography achieve atomic resolution in the electron microscope. (See reference [3] for more detail.)

electron lens, the forerunner of the modern polepiece lens subsequently patented in 1932 by von Borries and Ruska. Ruska tended to play down the significance of Gabor's iron-shrouded lens on the grounds that it was sheer serendipity! Nevertheless the existence in the same laboratory, however accidentally, of an excellent magnetic lens, must have made it easier for von Borries and Ruska to incorporate what Ruska called the 'finishing touches' of realizing much smaller focal lengths, by inserting iron polepieces of small diameter and spacing into the bores of Gabor's iron-shrouded lens.

I have used modern data to calculate the performance of Gabor's lens, and have found it to measure up well to today's standards, given the constraint of a wide bore. It was a much better one, for example, than either of the two iron-free lenses that Knoll and Ruska used in their first electron microscope. Clearly luck and intuition played a big part here, but had it been used as an objective lens, its resolving power would have exceeded that of the light microscope. Gabor was a profound thinker, but although he struggled to understand how his lens worked, he was only able in the thesis to present a token theory that he found wanting. Nevertheless, he strongly affirmed that his experimental results were essentially correct.

Shortly after presenting his thesis, Gabor saw Hans Busch's seminal papers of 1926 and 1927 on the analogy between the motion of electrons in axially symmetric magnetic or electric fields and the geometrical optics of light rays, an idea which seemed ridiculous at first sight. Busch, like many of the early pioneers, had kept this problem turning over in his mind since 1908 when, as a research student, he had made some measurements on electron-beam concentration in cathode ray tubes using iron-free solenoids. Later, as Professor at Jena, he took it up once more in the early 1920s and examined systematically the rotation of a beam of incoming electrons making a small angle with the direction of the uniform magnetic field. This led him, among other things, to devise in 1922 a considerably improved method for determining the ratio e/m of the electron. Busch realized that the field in a finite solenoid is not uniform, but has a radial component, especially at the ends. He took the decisive step in 1927 of trying to calculate what would happen if the source of electrons lay outside the magnetic field of the solenoid, assuming for simplicity that the solenoid was short compared with its distance from the source. To his astonishment, he found that such a solenoid behaves with electrons in a way very similar to that of a thin glass lens with light. For example, the optical ray equation ($1/u + 1/v = 1/f$) describes how a bundle of paraxial electrons diverging from an 'object' point at a distance u from the magnetic lens forms an electron 'image' at a distance v from the lens. This result was unexpected, since a magnetic field cannot influence the speed of the electron, as glass does with light; only the electron's direction can be changed. In an infinite solenoid, an electron moving parallel to the axis experiences no force, but as it enters a short magnetic lens, the radial component of the field causes it to drift sideways, and this sideways motion interacts with the longitudinal component of field to push the electron towards the axis. The mathematical analysis is subtle and one is not surprised that it took Busch so long to reach a conclusion.

For all his elation, however, he held firmly that experiment is the final arbiter. There was no possibility of setting up a new experiment quickly in Jena, so he dug out the unpublished results that he had obtained as a research student at Göttingen. These results showed, encouragingly at first sight, that as the magnification ratio v/u was increased, the image indeed increased in size. It was, however, not possible to make a direct check on the absolute value of the magnification, because the size of the object (emitting source) was not known accurately and the image size had been judged visually. However, it was relatively easy to compare the ratio of the calculated maximum to minimum image size, as the lens conjugates were varied, to the relevant ratio observed experimentally. According to his theory, the ratio of magnifications should have been just over a hundred to one. The experimentally measured ratio was four to one, a discrepancy of some 25 times. Busch tried hard but was not able to account for it; uncharacteristically, he tended to favour the theory! The theory, at this stage was therefore not definitive, but seemed worth publishing.

It was well received. Gabor himself, who had previously favoured his own experiments rather than his theory in this matter, had no doubts whatever about Busch's theory. He was both elated and shattered by it; he described it as 'not just a surprise but as the throwing of a lighted match into an explosive mixture'. Years later he admitted to the present writer that he kicked himself hard shortly afterwards, for narrowly failing to understand and hence to 'invent' the iron-shrouded magnetic lens and possibly the first electron microscope, when he realized that he had had in his electron column all the components he needed to take a one-stage image of a simple grid. E Ruska [1] has speculated that Busch's dilemma was the main reason why the latter did not put forward in his 1927 paper the possibility of electron optical instruments such as the electron microscope. His theory, of momentous significance for the future of electron optical instruments, unleashed a flurry of activity in Berlin, and elsewhere, in putting together a comprehensive theory of electron optics. Gabor, during the rest of his career, could never get the electron microscope out of his mind and constantly returned to the question of seeing individual atoms in the electron microscope, which he thought, correctly, would be a very demanding project.

20.3. Verification of Busch's theory

At the TH Berlin, Matthias immediately realized the significance of Busch's paper for the development of the high-speed cathode ray oscillograph and the need to resolve the dilemma of Busch's theory and Gabor's experiment. This was the real starting point for the emergence of the new range of electron microscopes that were soon to sprout up in Berlin, including the transmission electron microscope (TEM) of Knoll and Ruska, the emission electron microscope (EEM) of Brüche and Johannson, the scanning transmission microscope (STEM) of von Ardenne, and the field emission microscope of Erwin Müller. Matthias immediately set up a team in his laboratory, led by the young and gifted

Max Knoll, with a remit to investigate and resolve the questions raised by Gabor's new concentration coil and Busch's lens theory, so as to gain a deeper understanding of the high-speed oscillograph. The team under Knoll was wide-ranging and impressive; Freundlich on cathodes, Knoblauch on gas-discharge tubes and Lubszinsky on the screening of beams from stray magnetic fields. Ruska and Knoll concentrated on clarifying the electron optics. Bodo von Borries had his own cathode ray tube project but took a keen interest in the electron optical work and became a close personal friend of Ernst Ruska, playing an indispensable role in the critical days later on when they would work together on the Siemens prototype TEM.

At this point Ruska, although only a final-year-project student, showed his prowess as a careful and gifted experimenter. He had the advantage that he knew how Gabor's lens worked, so was in a strong position to improve it. Nevertheless, for a final-year undergraduate, he showed profound insight in exploring unknown territory in which others had lost their way. He realized that a gas discharge tube as used by Busch and Gabor does not constitute a well-defined source; it is especially difficult to define its axial position, a key requirement for testing Busch's theory. He also needed a reliable method of measuring image size; although conventional fluorescent screens had a high light output they did not have a good resolution. He placed a small aperture of known size in the anode of the discharge tube, so that the 'object' was well defined in position and in lateral extent. In addition, he used visual observation only for setting up the column, not for image measurement. For this he used a separate 'measuring fluorescent screen', in the form of a uranium-glass plate covered with a layer of gold to avoid charging-up effects. This arrangement gave better resolution and the image was photographed by an external camera. The distance between object and image and the position of the coil could be varied and adjusted accurately by sliding vacuum seals. The coil was of narrow axial extent to comply with the assumptions underlying Busch's theory. With this well thought out experiment, Busch's 'thin magnetic lens' theory was completely verified, within the limits of experimental error (some 5%). This constituted a turning point in electron optical instrumentation. The whole of the established theory of geometrical optics now became available for understanding electron devices. Ruska went on to verify Busch's prediction that electrostatic converging devices also behaved like optical lenses.

20.4. The first two-stage electron microscope

Forming an electron image of a metal aperture at a magnification of some 8× with a single lens was a good start, but can an image formed by electrons be further magnified by a second magnetic lens? The first crude two-stage electron microscope shown in figure 20.1 was designed and built in the early part of 1931 by Ruska [1] and his research supervisor

Max Knoll. Both believed exclusively in the corpuscular nature of the electron and in geometrical optics, and had not even heard of Louis de Broglie, so they thought that there would be no theoretical limit to the resolving power of the electron microscope, since the electron was of negligible size compared with the lightest atom. Figure 20.1 shows the brilliance of the design concept, which was mainly carried out by Ruska, building on his previous experience and allowing crucial and interpretable conclusions to be drawn from the experiment. A metal aperture, on the anode of the 50 kV electron source, served as the specimen and defined the specimen plane. A distinctive platinum mesh was placed on this aperture to act as a readily identifiable object. Platinum was chosen because it would stand up well to the intense electron bombardment. The lenses themselves were simple iron-free solenoids, placed outside the vacuum for ease of alignment. Iron shrouds were not employed, although they would have greatly improved the imaging performance. Ruska preferred to use iron-free lenses since he could calculate analytically the axial field distribution of the coil. In this ingeniously conceived column, a second circular aperture was placed in front of the projector lens, onto which the image of the first lens could be focused. This aperture was overlaid by an equally distinctive bronze mesh, which helped to identify the action of the projector lens. Ruska's laboratory notebook records that on 7 April 1931, a two-stage image of the first grid was achieved at a magnification of 17.4×. The image of the second grid also appeared in focus on the fluorescent screen at a magnification of 4.8×, as expected from light optical theory. Because of the low magnification, the effects of lens aberration were negligible, so the results were conclusive and definitive. The principle of the TEM had been unambiguously established.

At the time, neither Knoll nor even Ruska seemed to sense the future commercial prospects of such an instrument. Knoll, in his role as supervisor, was normally assiduous in filing patents for the cathode ray oscillograph, but he appeared to regard the new instrument as mainly illustrating in a striking way the new scientific principles of electron optics. He was later to take the same view when, in 1935 at Telefunken Berlin, he conceived the idea and built the first crude experimental scanning electron microscope (SEM), referred to below. Knoll, perhaps from inexperience (he was 34 at the time, Ruska only 23), did not realize that he was now in a unique position to take out a series of master patents covering a variety of electron optical instruments making use of electron probes, or of transmission instruments such as the compound electron microscope. In fact, he did not concern himself about its commercial future, preferring to offer the newly gained concepts to the wider scientific community. He decided simply to give a survey lecture in the (public) Crantz Colloquium series at the TH Berlin on 4 June 1931, less than eight weeks later.

Twentieth Century Physics

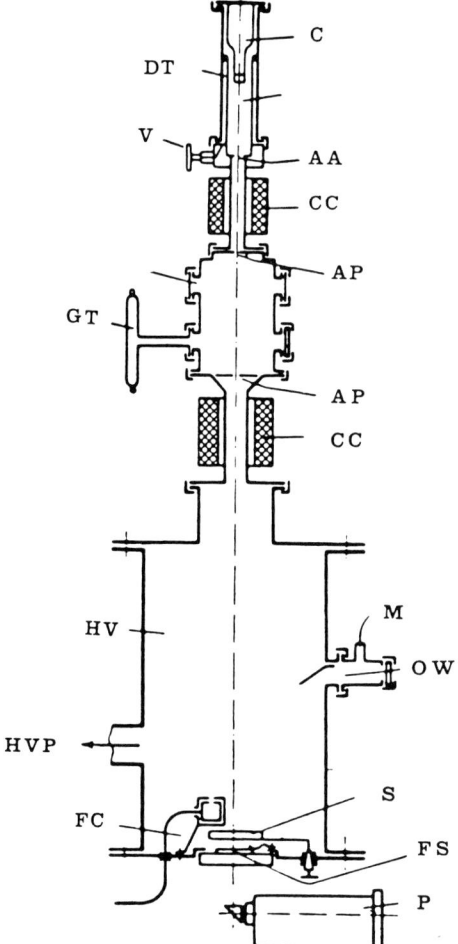

Figure 20.1. *Schematic arrangement of the column of the first two-stage* TEM, *April 1931. Magnification about 13 ×. The electron gun (C) illuminated a metal grid in the object plane (AP) of the first lens coil (CC), which forms an image onto the object plane of the second coil. The image of the second grid is then focused onto the screen S by the second coil. A focused image of the first grid also appeared on the screen, demonstrating two-stage electron imaging. P is the plate camera.*

Events now moved swiftly and in a way that Knoll had not foreseen. In the eight weeks available to him to prepare this new material for the Colloquium, Knoll went to great trouble to bring these advances in the understanding of the cathode ray oscillograph to the attention of the leaders in the field, many of whom were invited to see the equipment and results for themselves. In particular, Knoll gave his friend Dr M Steenbeck at the Siemens-Schuckertwerke, Berlin, a cordial invitation to visit the laboratory. He duly arrived and Knoll and Ruska showed him the equipment and the results obtained. Steenbeck was impressed and disclosed all he had learned to his Chief, Rheinhold Rüdenberg. This event seems to have triggered off something that had been lurking in Rüdenberg's mind. He was then the Chief Electrical Engineer at the Siemens-Schuckertwerke, the heavy electrical plant of Siemens. The related firm Siemens und Halske, where Ruska and von Borries were eventually to build the first serial TEMs, was the light current and scientific instrumentation plant of Siemens. Rüdenberg was also a Visiting Professor at the TH Berlin, well known to Knoll and Ruska and an acknowledged authority on high-voltage oscillographs, holding several patents in this area. He was exceptionally quick on the uptake and had a wide grasp of contemporary science. Significantly he had previously worked in the Patents Department of Siemens-Schuckertwerke.

Up to that point Rüdenberg's ideas, based on the work of Busch and others, appear to have been mainly speculative and unsupported by experiment. That the detailed news of the brilliant Knoll–Ruska experiments came as a powerful trigger to Rüdenberg's mind in the week beginning Monday 25 May 1931 is now clear. Although the manufacture of an electron microscope would not be within the remit of Siemens-Schuckertwerke, Rüdenberg felt that it might well be relevant to the sister firm of Siemens und Halske. On Wednesday 27 May, he called on Dr Fischer at the latter company to discuss the possible commercial significance of an electron microscope for Siemens. After a preliminary discussion, Dr Fisher agreed that the matter was indeed important and called in Dr Abraham, a senior member of the Patent Department of Siemens-Schuckertwerke, who was instructed to draw up forthwith a patent specification for the electron microscope. On purely verbal instructions from Rüdenberg, Abraham drew up the specification and claims himself. So comprehensive was Rüdenberg's briefing, it seems that he only needed to make minor changes to the application. On the following day, Thursday 28 May, Rüdenberg signed the patent agreement. Following standard practice in German firms at the time, the patent was assigned to Siemens-Schuckertwerke. The patent application was received in the German Patent Office on Saturday 30 May 1931, only four days before the Crantz Colloquium, given by Knoll on Thursday 4 June 1931 at the TH Berlin. Rüdenberg attended this colloquium but, understandably, took no part in the discussion. For reasons that have never been adequately explained, the patent was not granted in Germany until 1953, so Knoll and Ruska were blissfully unaware of it until it was first published in France in 1932 and more significantly in the USA in October 1936. By then Rüdenberg, who was of Jewish extraction, had emigrated from Germany and by 1947 had become a US citizen, changing his name to Rudenberg. The Siemens-Schuckertwerke Patent, as a result of World War II, had fallen into the hands of the Custodian for Alien Property. Rudenberg applied for the patent to be re-assigned and granted to himself as inventor instead. The complex legal proceedings and the background to the original submission of the patent in Berlin are fully set out out in reference [4]. The Custodian for Alien Property directed that the Patent be transferred to Rudenberg as the inventor, since in his judgment the matter of the patent lay outside the commercial remit of Siemens-Schuckertwerke and could therefore be regarded as a private invention by Rudenberg, especially as the inventor had not signed the relevant assignment document. The German Patent Office belatedly issued the patent in Germany in 1953, confirming that Rudenberg is the inventor of the electron microscope in patent law, notwithstanding the fact that he played no part in its early development.

20.5. Surpassing the resolving power of the light microscope

In 1932 Knoll's group broke up when Knoll went to Telefunken Berlin and von Borries took a position away from Berlin. Ruska, now on his own in Berlin, was able to get a grant from the Notgemeinschaft der Deutschen Wissenschaft. The goal of the project was to design and construct a greatly improved TEM, operating at 75 kV, that would surpass the resolution limit of the light microscope. Ruska's new microscope employed as many components as possible from his previous instrument. It was a two-stage TEM with a remarkable top magnification of 10 000×, sufficient to record detail down to 10 nm. By the end of 1933, Ruska had obtained a magnification of some 12 000× and a resolution of some 50 nm with a carbon fibre, just slightly better resolution than that of the best light microscopes and much better than that of good routine light microscopes. Experts who came to see this first TEM agreed that it had good resolution, but condemned it out of hand as having, in their opinion, no future as a means of viewing, in 'real' specimens, finer detail than that already visible under the light microscope. It was difficult, at the time, to see that new methods of specimen preparation would soon be developed to exploit the better resolution available and that one would soon learn how to interpret fine detail that had never been seen before in any microscope. No support, however, was forthcoming for TEM and Ruska also had to leave the field and take up a position at Telefunken (Berlin). It seemed like the end of the road for electron microscopy.

20.6. Specimen preparation and radiation damage

The difficulties with specimens turned out eventually to be not so formidable as many had supposed. This was largely due to the work of Marton, a Hungarian physicist working in Brussels. Inspired by Ruska's work, Marton built in 1934 a simple horizontal transmission microscope of modest resolving power. He was optimistic that he could succeed in examining biological specimens in the TEM. His first experiments showed, however, that the 'raw' specimens were quickly burnt up under the beam. Nevertheless, if he used the normal osmium impregnation techniques as used in biological light microscopy, he found he could still see the basic structure of, for example, the leaf of the sundew plant, preserved in skeletal form after the organic material had been destroyed. After some practice in specimen preparation and by applying the laws of electron scattering to thin specimens, he convinced himself that the way forward to true electron microscopy lay in the use of thinner specimens, higher accelerating voltage and internal photography, leading to short exposure times. He constructed an improved TEM, with a vertical column, operating at 80 kV, and provided with specimen and camera air-locks. This allowed many specimens to be investigated in quick succession and photographed internally. This was arguably Marton's greatest contribution, since it came at a critical time. His work prompted light microscopists to adapt their specimen techniques to the TEM, encouraged physicists to contribute to the theory of electron image contrast, and forced manufacturers to provide specimen air-locks and internal cameras with multi-plate facilities. Most importantly, he showed that there were no fundamental reasons why electron microscopy should not be used with run of the mill specimens, given a certain amount of ingenuity in preparing them.

20.7. The first serially produced TEM

Back in Germany, Ernst Ruska and Bodo von Borries continued their interest in further electron microscopy (EM) developments at the TH Berlin and abroad and tried to persuade some firm in Germany to let them build a commercial TEM prototype, suitable for laboratory use, for subsequent serial production. Eventually, in 1937, both Siemens und Halske, Berlin, and Carl Zeiss, Jena, offered to set up a suitable laboratory with the necessary facilities for building a prototype laboratory TEM. They opted for Siemens, since strong support would be needed in the design of the high-stability, high-voltage (100 kV) supplies. The facilities provided by Siemens were not lavish; the site was a disused bakery in Spandau, Berlin, but all the ingredients of success were implanted right from the start. Siemens und Halske at that time already possessed the Rudenberg patents via Siemens-Schuckertwerke; they now acquired the von Borries and Ruska Patent on polepiece lenses. The work at Siemens started in the spring of 1937, with a very young and energetic team, E Ruska, B von Borries, H O Müller and others. They were given a free hand.

Siemens also approved the appointment of Professor Walter Glaser, from the Institute of Theoretical Physics, Prague, to provide the theoretical background in electron optics. Helmut Ruska, the younger brother of Ernst, gave up his career in medicine and took charge of the Applications Laboratory. He was ingenious in devising clever ways round the problems of specimen preparation. By 1939, he had begun to resolve plant viruses, and later human disease viruses.

From the start, two prototype TEMs were built, incorporating the lessons learnt from the experience of Ruska's 1933 TEM. One was set aside for instrumental development, the other for applications work. The instrumental development could run parallel with applications work, each feeding back into the other without interrupting either programme. Progress was extremely rapid and the development work often went on far into the night. By the middle of 1939, the first high-resolution commercial TEM (*Übermikroskop*) was being tested in the factory. This had a thermionic electron gun, replacing the previous low-brightness gas discharge cathode, thus permitting the use of higher magnification. Ruska's intuitive genius now began to blossom in this invigorating atmosphere. He believed from the start that only the objective lens is important in a TEM and that it should have as small a focal length as possible. It was already known that the chromatic aberration coefficient of a magnetic lens is of the same order as the focal length. Fortunately, any reduction in focal length will reduce chromatic and spherical aberration and therefore improve the resolving power. For most practical specimens, which tended to be fairly thick by modern standards, the reduction in chromatic aberration was the more significant. Ultramicrotomes were still several years away. Moreover, for a given maximum magnification, a reduction in objective focal length results in an appreciable reduction in the length of the microscope column, leading to better mechanical stability.

The experience gained from these two prototypes was incorporated in the first Siemens production TEM, with a magnification of $30\,000\times$ and a guaranteed resolution of 7 nm, a remarkable achievement for the first serially produced TEM in the world. This Siemens TEM attracted so much interest that, on receipt of the first order from IG Farben at Höchst, Siemens went ahead to build a batch of ten instruments without waiting for further orders. These instruments set a high standard and made it difficult for other manufacturers to compete. It is remarkable that the development of the TEM in all countries has kept strictly, almost slavishly, to Ruska's initial electron optical design as a PhD student. The tendency was strongly reinforced in the early 1960s when he returned to a suggestion made in 1940 by his former collaborator in Berlin, Walter Glaser, that the best objective lens from the point of view of minimizing spherical aberration would be a strong lens with the specimen in the centre of the lens, the so-called *Einfeldlinse*, or *condenser-objective lens*, since the first part of the lens field focuses the illuminating beam onto the specimen and only the second part forms the image. This would have been a difficult lens to construct and assemble accurately in 1941, so the idea had lain dormant, awaiting improved technology.

The paper describing the high performance of this lens was published by Riecke and Ruska in 1966 and was rapidly adopted in almost every kind of commercial high-resolution electron microscope around the world (cf figure 20.6 later in this chapter), whether TEM, SEM or STEM. Ruska had thus demonstrated that by persistent striving to eliminate instrumental limitations it was possible to see individual atoms. The question nevertheless remained; how is such an image to be interpreted? The combined effects of spherical aberration and diffraction must undoubtedly cause image artefacts to appear at the atomic level of resolution. This became clear in the 1980s, when it was realized that atomic structure could not be deduced intuitively from high-resolution micrographs, but had to be arrived at by trial and error, by image simulation methods or by image processing of a focal series that might allow one to deduce the spherical aberration coefficient and re-synthesize the image.

20.8. Electrostatic electron microscopes

Initially it was believed that electrostatic electron microscopes were the instrument of choice, since no current supplies are needed for the lenses and their focal length does not change when the accelerating voltage is varied. There are, though, operational difficulties and such instruments are no longer manufactured. However, the immense research effort involved was not wasted as it provided the basic understanding of electron wave optics and the instrumentation needed for the correction of spherical aberration. In 1930 the AEG (Allgemeine Elektrizitäts Gesellschaft) Research Institute, Berlin, had set up a wide-ranging study into geometrical electron optics and its industrial applications, such as high-speed cathode ray tubes, and low-energy electron beams. In August 1931 H Johannson joined AEG and began to design an *emission electron microscope*. The results of this investigation were reported by Brüche in 1932 and later by Johannson. This microscope was a single-stage microscope in which the thermionic cathode itself served as the specimen. Electrons from the cathode were accelerated by an apertured plate held at 10–20 V positive with respect to the cathode. A further electrode (anode) accelerated the electrons to some 200 V and focused them onto a fluorescent screen held at anode potential. This innovative lens was known variously as a Johannson-, cathode- or immersion lens.

These contributions established AEG as pioneers of electron microscopy in Berlin. The emission electron microscope created more interest at its birth, since the behaviour of thermionic cathodes could be studied *in situ* with its aid, than it did subsequently, since its resolving power is poor compared with other forms of electron microscope. It also requires a high electric field strength at the surface under examination. Commercial production has now ceased but recently a revival has taken place in the form of the *low energy electron microscope* (LEEM) [5], in which high resolution is not the most important factor, but rather the

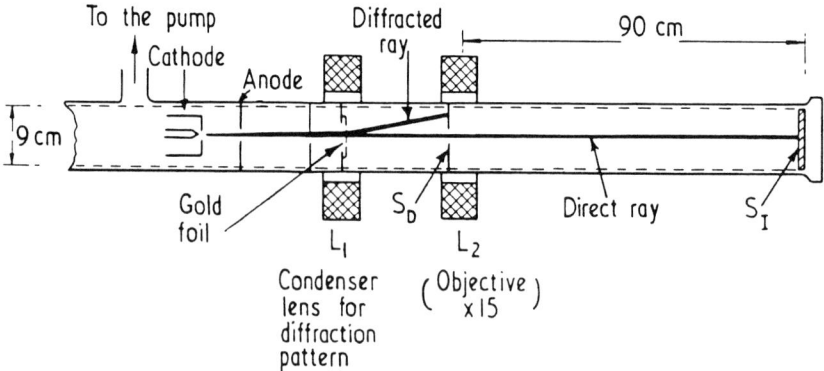

Figure 20.2. *Boersch's experimental* TEM, *magnification* 15 ×, *for investigating Abbe's theory of imaging in the* TEM *with a specimen of gold foil. Lens L_1 can be tilted to place a selected diffracted ray on the axis to form an image on the screen S_I. Selected area diffraction was also demonstrated for the first time.*

ability to observe *in situ* chemical reactions at a well-defined surface. The AEG Institute is also well known for producing the first electron optics monograph *Geometrische Elektronenoptik* by Brüche and Scherzer.

20.9. Abbe's optical theory and the TEM

In this theory of the microscope, a coherent wave passes through the specimen where it is modulated in amplitude and phase. The objective lens forms a *primary image* (diffraction pattern) containing all the information about the specimen, but in a form unfamiliar to the eye. The waves then proceed to form a *secondary image* in the image plane, in a form that can be recognized by the eye. Up to 1936, Abbe's theory had played no part in the development of the electron microscope. In the latter half of 1935, Boersch began an investigation to see whether Abbe's theory holds for the electron microscope. With the simple equipment shown in figure 20.2, consisting of a continuously evacuated cylindrical glass tube with an electron gun at one end and a fluorescent screen S_I at the other, he revolutionized our basic understanding of the electron microscope. Two wire-wound coils, acting as magnetic electron lenses, could be moved in an axial direction over the tube; the image could be photographed externally. The first lens L_1 acted as a 'diffraction' lens, forming a diffraction pattern on an area-selecting aperture S_D (Abbe's primary image) placed in the specimen plane of the objective lens L_2, which had a magnification of only 15×; this was just sufficient for Boersch's purpose. Different modes of imaging were arranged, not by moving the apertures, which were fixed, but by tilting the diffraction lens.

Dark field images and image formation with a given diffracted beam were observed systematically for the first time in the TEM. Selected area diffraction was also demonstrated.

Within the broad limits of this experiment, and to the surprise of many microscopists, the Abbe theory of image formation was found to be valid for the electron microscope. Selected area diffraction was later re-invented, independently, by J B Le Poole around 1944 in the war-time isolation of the Technische Hoogeschool Delft, in his experimental TEM that set the pattern for all future developments by the Philips Company in Eindhoven. Le Poole's PhD design was elegant and definitive, strongly influenced by wave-optical ideas.

While Boersch was explaining the action of the TEM in terms of wave optics, Scherzer published his famous 1936 paper showing that axially symmetric electron lenses differ from optical lenses in that their first-order spherical aberration cannot be negative. All known electron lenses, magnetic and electrostatic, have considerable spherical aberration so it is impossible to correct them by combining them with a lens with negative aberration. This raised grave doubts about the ultimate resolution of the electron microscope; on the wave theory this would now be determined by the combined effects of spherical aberration and electron wavelength, rather than by the size of the electron. This did not matter in 1936, since resolution was clearly limited by instrumental defects. Neither Gabor nor Scherzer shared this view, however. Even ten years later, Gabor, in his well-known monograph [6], estimated on theoretical grounds that the ultimate resolution of the uncorrected TEM would be around 0.8 nm (8 AU) at what he saw as the optimum accelerating voltage of some 60 kV, so that correction of spherical aberration was urgent. His main suggestion was to create space-charge in the lens gap. This would weaken the refractive power of the outer zones of the lens and hence reduce the spherical aberration to zero; up to now, experimental tests have not succeeded.

Scherzer's solution was to abandon axial symmetry. Multipole elements can produce specific amounts of both astigmatism and spherical aberration of either sign. An octupole has no refractive power, but can produce spherical aberration in a straight-line image. A quadrupole can produce an image in the form of two lines at right angles but axially separated from each other. It is thus possible to correct the aberration with octupoles and then reconverge the beams to a point by a further quadrupole. The system was shown to work reasonably well experimentally. The chief disadvantage was that the quadrupoles introduce enormous amounts of spherical aberration. This can be corrected, in principle, by the octupoles, but it is a non-trivial task. After over 25 years of such designs, none has yet reached the stage where it can be fitted advantageously to a commercial electron microscope. However, one of the minor but most fruitful bonuses of Scherzer's work was the electrostatic astigmatism corrector, the *Stigmator*, an eight-pole electric or magnetic device connected as two quadrupoles, each acting on the beam at right angles so that astigmatism can be corrected in azimuth and strength from the control panel. This was to make a dramatic difference in electron microscopy since astigmatism can arise from many causes. It did not save the electrostatic TEM, however, as it proved impossible to increase its accelerating voltage beyond 80 kV, much too low for materials science applications. The electrostatic TEM suddenly became extinct as magnetic TEMs, fitted with Stigmators and working at higher accelerating voltages left them behind.

Hillier and Ramberg's correction system for magnetic lenses was more empirical, but it alerted designers to a serious problem. Eight small iron screws were placed in a ring in the gap spacer of the objective lens and adjusted by trial and error to compensate for image asymmetry. This could not be done *in situ* so the instrument had to be dismantled and reassembled for each successive adjustment! This was clearly only a temporary solution. Nevertheless, Hillier knew how to measure astigmatism quantitatively, with far-reaching consequences. Hillier [7] in 1940, almost simultaneously with Boersch [8] had discovered Fresnel diffraction fringes at the edges of specimens in the electron microscope; these become apparent when the objective lens is slightly out of focus (see figure 20.3). The asymmetry of such fringes at the edges of holes in a thin film is a sensitive indicator of astigmatism, since the appearance of the over-focus and under-focus fringes in a TEM is strikingly different. Hillier was one of the two electron microscopists, at the time, who were familiar with coherent electron beams and Fresnel diffraction effects since he had already built a Boersch projection microscope, but with magnetic lenses, as a high-resolution (50 nm) electron probe microanalyser for light elements, using an electron energy-loss spectrometer (EELS) as the detector.

We may mention in passing that he had also intended to incorporate an x-ray spectrometer for microanalysis, but was not able to obtain industrial approval to develop this. Instead, he took out a patent [9] for electron probe x-ray microanalysis, thereby becoming the inventor, in patent law. It was left to Raymond Castaing [10] in 1949 to establish this technique, from first principles, in the form we now know it. Castaing's main contribution was not in electron optics but in his innovative method of performing accurate chemical analysis at the micrometre level.

Armed with this new Fresnel fringe technique, Hillier and Ramberg [11] were able to demonstrate that, contrary to received wisdom, astigmatism due to manufacturing errors, rather than spherical aberration, had hitherto been the limiting factor for the resolving power of even the best commercial TEMs. This was a breakthrough, since Fresnel tests were straightforward and did not depend on the arbitrary judgment of the operator. Interestingly for the future of holography,

Electron-Beam Instruments

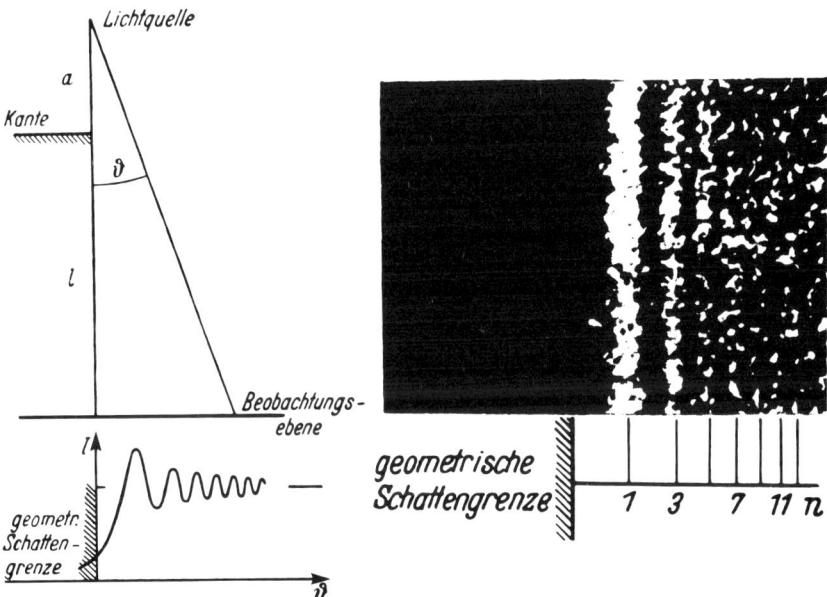

Figure 20.3. *Definitive demonstration [8] of the formation of electron Fresnel diffraction. A point projection electron microscope (870×) projects a shadow image (see figure 20.4) of a sharp edge (Kante) onto the fluorescent screen. Fresnel interference fringes appear outside the geometrical shadow of the edge, as in light optics, thus providing an independent proof of the wave nature of the electron.*

which was to depend strongly on the use of Fresnel fringes, Gabor immediately attacked both Boersch and Hillier for claiming that their published 'edge-fringes' were due to Fresnel diffraction. Instead, he asserted that they must be energy-loss fringes of a type previously published by Hillier himself. Thus in the 1946 edition of his monograph [6], Gabor showed an electron micrograph at 200 000× magnification, sent to him by V K Zworykin, the Head of the RCA Research Division, taken on an RCA TEM. This revealed molecular structure and showed several Fresnel fringes at the edges of a synthetic rubber specimen. After explaining why such fringes must be 'chromatic' fringes, corresponding to the characteristic energy losses in carbon (in multiples of 24 eV), Gabor added a footnote 'These contours were first observed by J Hillier, who instead of connecting them with his observations of 1939, tried to explain them tentatively as Fresnel fringes ... Hillier has not attempted a quantitative explanation, which would have shown that Fresnel fringes cannot account for his observations'.

Gabor, at that time, held the same negative view about Boersch's results. The point is of some historical interest as it confirms that in 1946 Gabor had no inkling about holography. Hillier and Ramberg in their 1947 paper [11] 'Contour phenomena and the attainment of

1581

high resolving power' gave conclusive proof of the nature of these fringes. In the impending second edition of *Electron Microscopy*, Gabor felt compelled to rewrite completely the section on image contrast, omitting the offending passages and accepting completely Boersch and Hillier's explanation of Fresnel diffraction. It seems to the present author that this paper was a trigger for Gabor's unique concept of holography. According to Gabor, the basic concept came to him during the Easter holiday of 1947 (Easter Monday fell on 15 April) as he and his wife were sitting on a bench in the town of Rugby, waiting their turn for a game of tennis. In the vision, which came to him 'effortlessly and without any action on his part', he was told to record, with coherent illumination, the electron wave in amplitude and phase and then reconstruct the wave by illuminating the pattern with coherent light. This would enable the aberration to be corrected. Once converted to the belief in electron Fresnel diffraction and the possibility of phase contrast effects, his powerful insight went to work on Fresnel interference patterns. Boersch's microscope was not a 'shadow microscope', but a Fresnel diffraction microscope, producing a record of the wave leaving the specimen, a point that Boersch himself seems to have overlooked. Gabor now invented *diffraction microscopy*. No one had explained satisfactorily what the 'image' from the projection microscope really was. It was not a shadow but an interferogram! As one moved the specimen away from the focal point more and more fringes appeared around a small specimen detail, revealing more and more of the wave scattered from the specimen. Without ever having seen a projection electron microscope, Gabor calculated the simple case of a perfect lens projecting a Fresnel diffraction image of a single atom on the axis, a short distance away from the focal point. Waves scattered from the atom interfere with the incoming reference beam from the distant point-like source, giving rise to a bright ring on the photographic plate where they are in phase and a dark ring where they are out of phase. The pattern on the photographic plate was identical to that of a zone plate. He then introduced spherical aberration into the projection calculation. This gave a pattern similar to the first but the fringes in the outer zones were slightly crowded together, indicating the greater bending of the wavefront due to spherical aberration. Gabor's genius was to call this record a *hologram*, a 'written record carrying all the information (amplitude and phase) about the wave from the object'. This became his trademark. He realized that the hologram also contained information about the electron optical properties of the device producing the hologram. It was perhaps an even greater stroke of genius when he realized that if one looked through such an electron hologram, illuminated by coherent light, the wave reaching the eye would reproduce a faithful image of the original object on the observer's retina. It followed that Boersch had already produced electron holograms without realizing it. The chief problem was how to

Electron-Beam Instruments

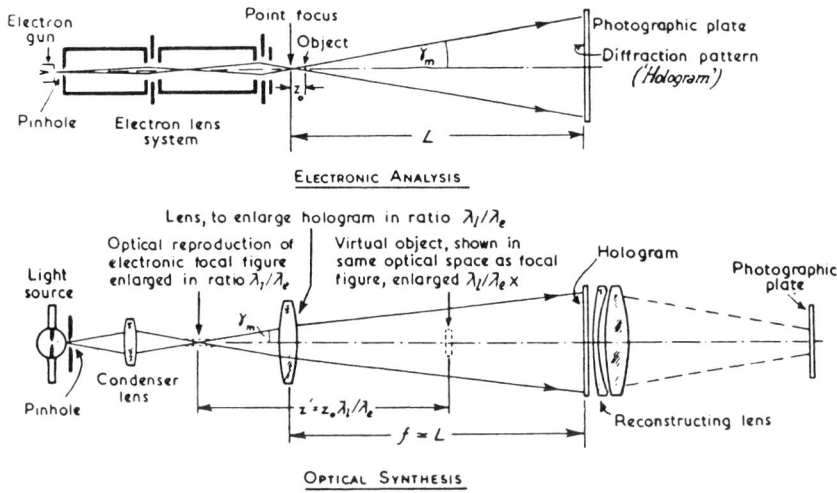

Figure 20.4. *Gabor's 1947 proposal for 'diffraction microscopy' using the ideas of Boersch's electron shadow microscope and Bragg's optical synthesizer* [13].

reconstruct the image from the hologram. In his BTH Company research report on diffraction microscopy, Gabor pointed out that, in principle, the image could be computed from the hologram since if the wave is known in amplitude and phase for a particular plane, it can be calculated for any other.

In 1947 such a calculation was beyond the capacity of any existing calculating machine, but was later to become the method of choice. However, unknown to Gabor, a method of *two-stage imaging* had been around since 1920, when Wolfke [12] put forward the idea of an x-ray microscope, using the two-wavelength principle. Wolfke's idea was to record the x-ray diffraction pattern, corresponding to Abbe's *primary image* on a photographic plate and then obtain the Abbe secondary image of the original object, in visible light, by means of an optical lens system. The paper referred to some experiments carried out by the author but these were not adequately described, and the paper was simply forgotten. Boersch in 1938 and later Bragg in 1942 independently reinvented this method of x-ray microscopy, being unaware of Wolfke's work. Since the photographic plate does not record the phases of the x-ray diffraction pattern, the method will only work for specific crystal structures where the phases are known in advance. This method, therefore, will not work with normal specimens in electron microscopy. Boersch, by 1943 in fact, was well placed to invent electron-beam holography. He had all the tools he needed, including the holograms, but didn't realize that they contained the phase information.

Gabor was strongly impressed by Bragg's reconstructed images, where the phases were known but Gabor had the advantage that the

1583

hologram recorded both amplitude and phase so the correction of spherical aberration would be possible without any prior knowledge. He now wanted to take a hologram [13] in a coherent electron projection microscope and reconstruct it with coherent optical light, correcting residual astigmatism, defocusing errors and spherical aberration optically, as illustrated in figure 20.5. Gabor had moved to Imperial College London as Reader and had persuaded the Director of the AEI Research Laboratory Berkshire, T E Allibone, to set up a small team under the direction of M E Haine, consisting of J Dyson, M E Haine, T Mulvey and J Wakefield with Gabor as consultant. An experimental TEM was converted into a projection electron microscope for taking holograms and an optical bench with coherent monochromatic illumination and variable spherical aberration was constructed by Dyson and Wakefield. Holograms with modest potential resolution were quickly obtained, but it was soon realized that the magnetic projection microscope suffers from a severe restriction of the field of view caused by chromatic aberration. A breakthrough occurred when Haine and Dyson [14] showed that an out-of-focus Fresnel fringe system in a TEM is also a hologram and it does not suffer from a restricted field of view. All that is needed in a holographic TEM is a fine-focus illuminating system and stable electron column and electrical supplies. Thus electron holography could be seen as a normal operational mode of any high-performance TEM. This made both scientific and commercial sense. What was more serious, eventually fatal, was an incorrigible defect in the Gabor 'in-line' hologram, in which the reference beam is in line with the beam from the specimen. A hologram is a generalized zone plate and acts both as a converging and a diverging lens. An observer looking through the hologram therefore sees two objects in line with each other, a real and an imaginary image at different positions on the axis. If he focuses on one of them, he will see the other one out of focus. However, with coherent light, the wanted image will be enveloped in a confusing array of fringes, which will degrade the resolution. The situation can be alleviated somewhat by strongly defocusing the TEM objective, but this leads to excessive exposure times. Holograms requiring several minutes exposure were produced, at a potential resolution of some 0.5 nm (5 AU), but it was not possible to reconstruct them to this accuracy for the above reasons [15]. It was concluded that the Gabor in-line hologram, although a brilliant idea, would not be viable at atomic dimensions. In any case, a field emission gun, not available at the time, was needed to cut down the exposure time by two orders of magnitude. The outstanding problem was to find some way of completely separating the two images. From a manufacturing point of view, the investigation was invaluable since the diagnostic capabilities of holography were to prove indispensible in assessing and improving the performance of high-resolution TEMs in a quantitative way. However, it looked like the end of the road for Gabor holography.

20.10. Off-axis optical beam holography

Help came from an unexpected quarter. Leith and Upatnieks [16] in the USA, realized in 1965 from their experience in radar mapping, which has a strong analogy with holography, that one must always separate the carrier wave and the side bands in a radar transmission in order to obtain good mapping. The waves from the specimen and from the coherent reference source should meet for the first time at the hologram. They therefore produced an off-axis optical hologram by tilting the reference beam with respect to the beam from the specimen, so that two images were completely separated. This solved the problem; the resulting off-axis holograms were striking and had all the properties predicted by Gabor. Optical holography revolutionized optical processing, especially in view of the arrival of the laser, with its high intensity and great coherence. No new principle was involved, but it was a brilliant technical solution.

20.11. Aberration-free atomic resolution in a TEM by holography

The tools for off-axis electron-beam holography had unknowingly been produced in 1953, when Möllenstedt [17] at the AEG Laboratories in Mosbach invented the Fresnel electron biprism, almost by accident. He had placed a thin wire across the objective aperture of an AEG electrostatic TEM in order to obtain dark field images. To his annoyance one day, double images appeared on the fluorescent screen. This was traced to the wire charging up negatively due to hydrocarbon contamination. He immediately asked himself it the images were coherent; if so, he could make a Fresnel biprism. His colleague Düker investigated this and found that such a biprism, with a positively charged wire, could indeed be used as an electron-beam interferometer. It was not recognized immediately that the interferograms it produced were indeed off-axis holograms of the specimen. Later, however, it was possible to think of off-axis holography in the TEM. Möllenstedt, by now Head of Applied Physics at Tübingen University, commissioned his colleague Hannes Lichte to carry out a demanding project in atomic resolution microscopy. A Philips 100 kV EM420 ST equipped with a field emission gun was acquired for this task. The only accessory needed for off-axis holograms was a small Fresnel biprism, placed on the selected area diffraction rod; the normal operation of the instrument was not affected. It was nevertheless an experimental *tour de force* to realize the very high mechanical and electrical stability needed in such a microscope. Off-axis holography exacts a heavy technical price for the correction of spherical aberration, since the basic carrier fringes in the hologram that are modulated by the specimen information have to be about three times finer than the resolution required, namely 0.03 nm for a final resolution of 0.1 nm. Moreover, it was known from the work of Hanssen [18] and his colleagues that the two-wavelength method of reconstruction is inadequate for atomic resolution. It was therefore necessary to digitize the hologram and compute the image, correcting the spherical aberration in the process. The project was successful and the first hologram showing atomic resolution was converted to an aberration-free image by Lichte [19] over the Easter holiday of 1985, the only time that the powerful Martinsried Institute computer needed for this task was available to him. As a result of this and subsequent developments in instrumentation at Tübingen, all the predictions that Gabor made have come true. For example, the really harmful effects of spherical aberration in wrongly interchanging phase and amplitude information are now perfectly clear in practice. After nearly 60 years of intensive theoretical and experimental effort, the central problem in TEM has therefore now been solved. In a more recent development [20] aberration-free imaging has been achieved without the aid of a biprism. This new in-line method has some parallels with the original in-line method of Gabor holography, inasmuch as it employs coherent illumination and it synthesizes, in amplitude and phase, the wave leaving the specimen, which it then corrects for spherical aberration. The method consists in capturing a well-defined focal series of a specimen in a charged coupled device (CCD) camera, which feeds its digital output directly into a computer. Such a series contains all the information about the specimen down to the information limit of the microscope, but in a 'scrambled', i.e. encoded, form. Unwanted non-linear disturbances in the image are first removed by a suitable algorithm and the wave leaving the specimen is recovered by a least squares fit of the data held in the individual micrographs. Correction of defocusing and astigmatism is straightforward, but the correction of spherical aberration needs an extremely accurate estimate of the spherical aberration coefficient of the objective lens. The whole operation is highly computer-intensive but is quite feasible on today's PCs and workstations. The virtue of this method of correcting for spherical aberration is that no modification of the column is needed. These holographic methods undoubtedly simplify the problem of determining structure in the TEM and open an new era in electron microscopy.

20.12. Scanning electron microscopes

In a *scanning electron microscope* (SEM), the column is reversed and so forms a demagnified image (electron probe). This probe is then scanned across the specimen, which can be a solid object or a thin film as in TEM. Electrons scattered by the specimen are collected and form an electrical signal that modulates the beam of a cathode ray tube scanned in synchronism with the specimen probe, thereby forming an image on the screen. The magnification is given by the ratio of the length of the scan on the tube to that on the specimen. The resolution is of the order of the electron probe size. With a solid specimen it is usual to speak

Electron-Beam Instruments

Figure 20.5. *SEM image of a nettle leaf. The remarkable depth of focus gives objects a striking appearance and better resolution compared with light microscopy.*

of SEM; with a thin specimen, the specimen current transmitted can be measured underneath the specimen and this is referred to as scanning transmission EM or STEM. TEM and STEM have comparable resolution in the range 0.1 to 0.2 nm. SEM has a resolution about an order of magnitude worse. The first SEM was conceived and built by Max Knoll at Telefunken Berlin in 1935 (cf references [1,2]). It was a genuine prototype. Although the magnification was around unity, it nevertheless demonstrated the principle behind all modern SEMs. The overriding difficulty at that time was that there was no detector capable of recording single electrons, whereas in TEM the photographic plate fulfilled this role eminently. SEM did not become significant until 1960, when Everhard and Thornley (cf the chapter of Oatley *et al* pp 443–8 in reference [2]) devised their wideband scintillation detector, now in general use.

This paved the way for the development and commercial production of SEM by Oatley and the Cambridge Instrument Company and later by other manufacturers. Figure 20.5, for example, shows a wide field of view image, taken on a modern SEM manufactured by CamScan Ltd, of a nettle leaf, illustrating the remarkable depth of focus and good resolution of the SEM at low magnification, well beyond the capability of a light microscope.

STEM was conceived and realized in prototype form by von Ardenne in Berlin in 1938 (cf von Ardenne's chapter in reference [2]). Although

Twentieth Century Physics

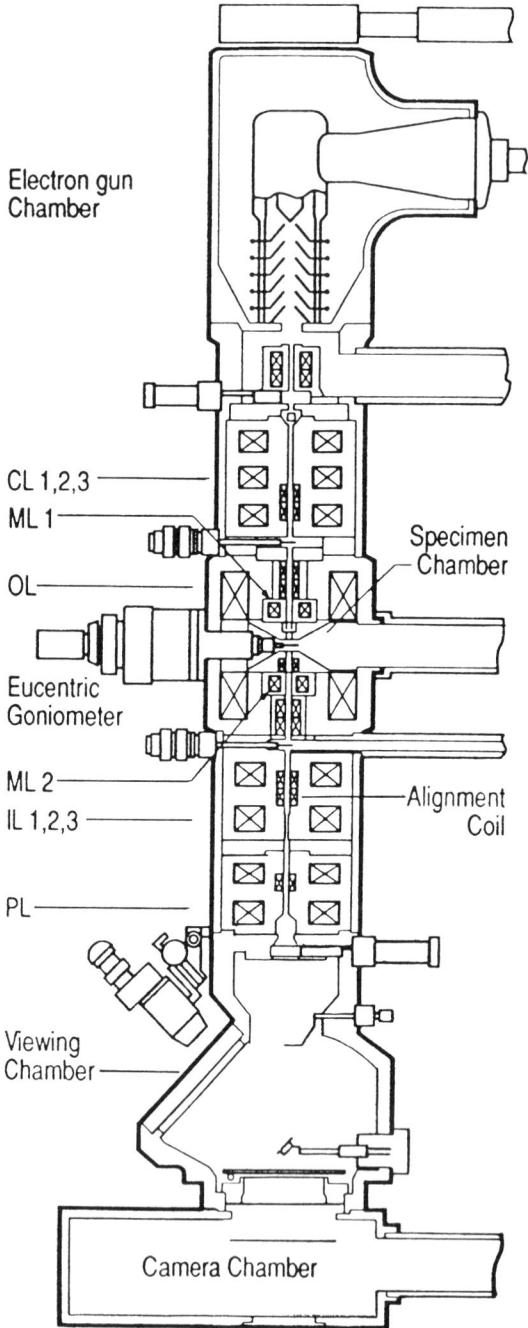

Figure 20.6. *Cross section of the TOPCON EM002B 200 kV column, operating in TEM, STEM and SEM modes. (CL) condenser lens, (ML) mini-lens, (OL) condenser-objective lens, (IL) intermediate lens, (PL) projector lens.*

Electron-Beam Instruments

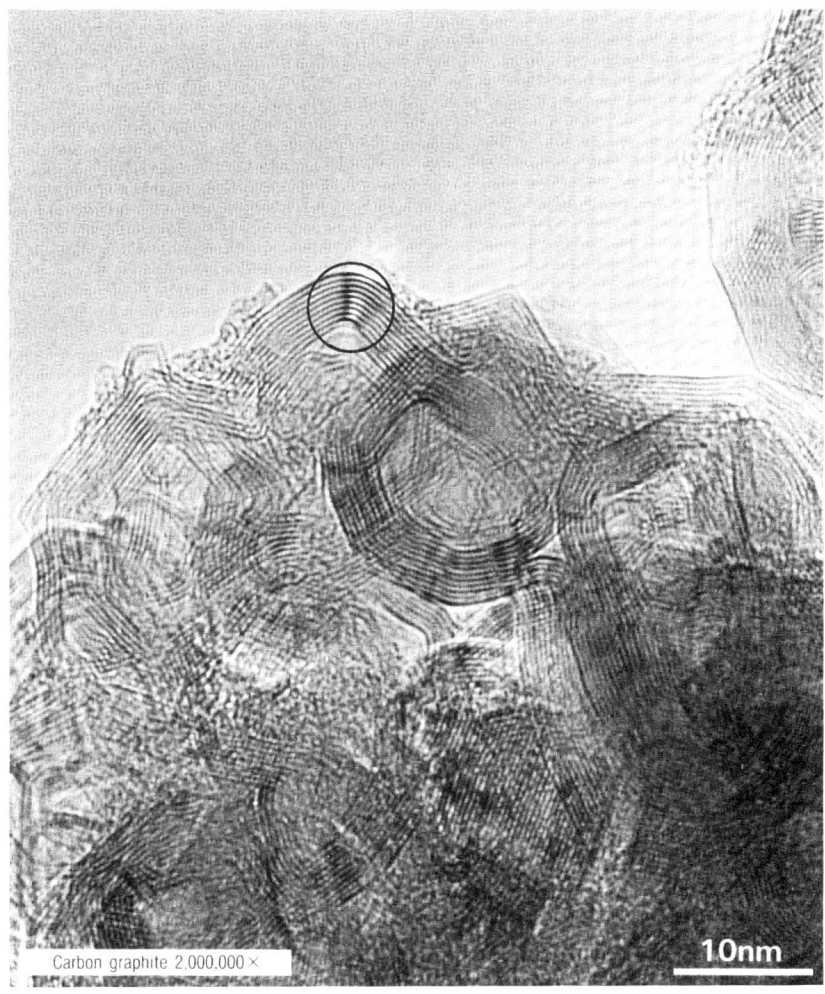

Figure 20.7. *Graphitized carbon, taken in the EM002B, showing atomic planes. Inscribed circle (5 nm) indicates the remarkably small selected area from which electron diffraction patterns and x-ray microanalysis can be obtained.*

he was able to improve the electron optical column by using lenses developed for TEM, there was no visible image, since he was forced to use an inconvenient rotating photographic drum arrangement for recording the transmitted current passing through the specimen. Nevertheless, he established the principle of STEM. For any serial scanning method, the exposure time is orders of magnitude greater than that for parallel imaging in TEM. It was not until the 1960s, when Albert Crewe introduced high brightness field emission sources into STEM, that routine use of STEM became practicable at a resolution equalling that of TEM.

Figure 20.8. *Transverse cross section of a silicon on sapphire (SOS) interface, showing the presence of interesting defects. Micrograph by L A Freeman on the Cambridge University 650 kV TEM.*

Likewise, SEM has also benefited from field emission guns, both in resolution and in its ability to work at low accelerating voltages, even below 1 kV.

20.13. Conclusion

In order to illustrate the enormous scientific and technological progress that has been made in the last 60 years, figure 20.6 shows the cross section of a typical state-of-the-art commercial 200 kV high-resolution analytical column that allows one to work in TEM, STEM and SEM, as well as taking electron diffraction patterns and chemical analyses of selected micro-areas down to some 2 nm in diameter (cf figure 20.7).

This is possible because the complex re-configuration of the column needed for the different tasks can be made rapidly under computer control. The resolution in TEM is 0.2 nm. Images can be acquired on a CCD camera for further processing, including the correction of spherical aberration and astigmatism. Such instruments are now reaching the point where the limits are set by the properties of electrons, rather than by the accuracy of manufacture or the skill of the operator.

There is now a tendency to increase the accelerating voltage of TEMs; this reduces the chromatic aberration and improves the resolving power by a modest amount. 650 kV is about as high as most microscopists would want to go since radiation damage increases rapidly at higher voltages than this. Figure 20.8 shows a transverse section of a silicon-on-sapphire (SOS) interface, showing the presence of many defects thought to be taking up the strain of the 5.5° mismatch between the crystals.

Electron microscopy has now reached an extraordinary state of development in the mere 60 years of its existence.

References

[1] Ruska E 1980 *The Early Development of Electron Lenses and Electron Microscopy* (Stuttgart: Hirzel)
[2] Hawkes P (ed) 1985 *The Beginnings of Electron Microscopy (Advances in Electronics and Electron Physics. Supplement 16)* (New York: Academic) note especially the literature review by P Hawkes p 589; see also Mulvey T (ed) 1996 *The Growth of Electron Microscopy (Advances in Imaging and Electron Physics 95)* (San Diego: Academic)
[3] Mulvey T 1995 *Advances in Imaging and Electron Physics* vol 91, ed P Hawkes (New York: Academic) pp 259–83
[4] Rudenberg vs Clark, Attorney General, Civil Action No 3873 Boston, MA, USA 1947 *Federal Supplement* **72** 381–9
[5] Veneklasen L H 1992 *Rev. Sci. Instrum.* **63** 5513
[6] Gabor D 1946 *The Electron Microscope* 1st edn (London: Hulton) (2nd edn 1948 (London: Electronic Engineering))
[7] Hillier J 1940 *Phys. Rev.* **58** 842
[8] Boersch H 1940 *Naturwissenschaften* **28** 710
[9] Hillier J 1947 US Patent No 2418 029 (applied for in 1943)
[10] Castaing R 1951 *PhD Thesis* University of Paris
[11] Hillier J and Ramberg E G 1947 *J. Appl. Phys.* **18** 48
[12] Wolfke M 1920 *Phys. Z.* **21** 495
[13] Gabor D 1947 *Proc. R. Soc.* A **197** 454
[14] Haine M E and Dyson J 1950 *Nature* **166** 315
[15] Haine M E and Mulvey T 1952 *J. Opt. Soc. Am.* **2** 763
[16] Leith E N and Upatnieks Y J 1962 *J. Opt. Soc. Am.* **52** 112
[17] Möllenstedt G 1991 *Adv. Opt. Electron Microsc.* **13** 1
[18] Hanssen K-J 1986 *Int. Symp. Electr. Optics (Beijing, 1986)* (Beijing: Academica Sinica) p 9
[19] Lichte H 1991 *Adv. Opt. Electron Microsc.* **13** 25
[20] Coene W, Jannssen G, Op de Beek M and van Dyck D 1992 *Phys. Rev. Lett.* **69** 3743–6

Chapter 21

SOFT MATTER: BIRTH AND GROWTH OF CONCEPTS

P G de Gennes

21.1. The meaning of 'soft'

21.1.1. Strong responses

In an electric wrist-watch using a liquid crystal display, the molecules in the display are triggered every second by extremely small electric signals. These liquid crystals are a good illustration of what we call 'soft matter': molecular systems giving large responses for very small perturbations.

The type of perturbation is arbitrary: in the above example it was an electric field, but we can think of magnetic perturbations, of mechanical perturbations (anyone eating a bowl of tapioca can notice that it hardens upon stirring) and of chemical perturbations. A good example of weak but significant chemical action is the *vulcanization of rubbers* invented by Goodyear in 1839 (and described in figure 21.1). Vulcanization is a weak reaction of sulphur on hydrocarbon chains: less than 1% of the carbon atoms are attacked. But the result is dramatic. The system transforms from a liquid into a solid! (a cross-linked system).

Because the reaction level is low, locally this solid is still quite flexible: NMR probes, for instance, would diagnose it as liquid. But at macroscopic scales, the network resists deformations. Goodyear generated a remarkable form of soft solid, commonly known as *natural rubber*.

Another form of perturbation involving soft matter is provided by certain types of *doping*: for instance, take a jar of water and add to it 100 milligrams per litre of long, water-soluble, polymer chains. (The classical, classroom example is polyoxyethylene $(CH_2-CH_2-O)_N$ with $N \gtrsim 10^4$.) The hydrodynamic behaviour of the water is greatly altered [1, 2]! One example, the *tubeless syphon*, is described in figure 21.2, but there are

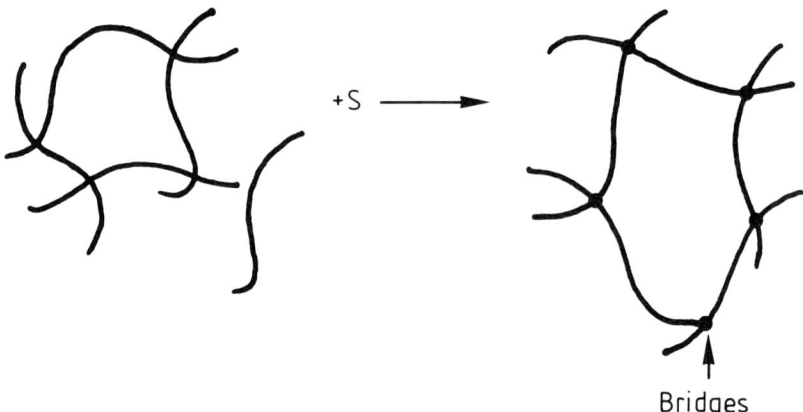

Figure 21.1. *Vulcanization of rubber: a liquid of flexible chains is cross-linked by sulphur atoms. Macroscopically, the system transforms from a liquid to a solid. Microscopically, there is still a lot of motion: the system, as seen via nuclear magnetic resonance, is locally fluid.*

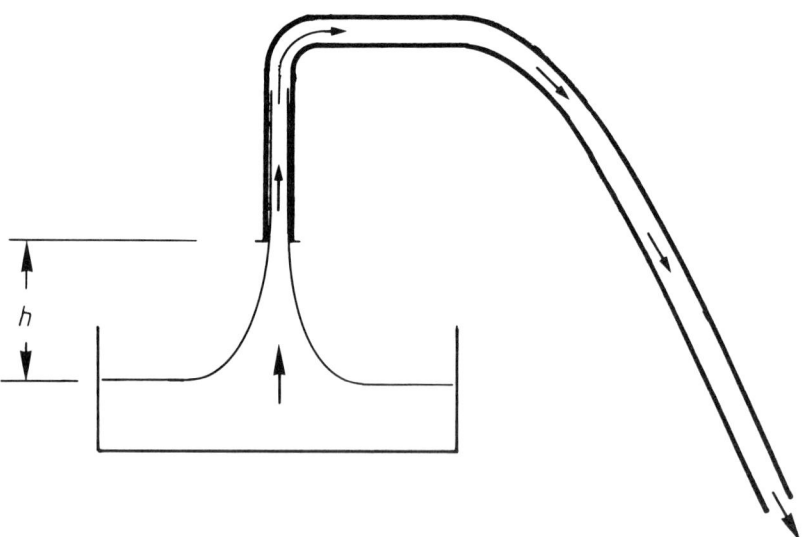

Figure 21.2. *The tubeless syphon. This works with water containing ~ 100 mg l^{-1} of a long-chain, water-soluble, polymer. Typical heights h are in the region of 20 cm.*

many others: for instance, turbulent losses are significantly reduced—an effect which is not yet fully understood, although we have known of it for forty years.

21.1.2. Sources of flexibility

If a condensed matter physicist is asked to imagine a system with strong responses to some perturbation, his first reaction will often involve *critical phenomena*: for instance, near the Curie point T_c of a ferromagnet, the magnetic susceptibility χ is very large.

This is not a very useful answer for most practical purposes because it imposes very stable ambient conditions (temperature, pressure, etc). It is indeed used in solids of high dielectric constant ($BaTiO_3$), together with special metallurgical tricks, aiming for a broader temperature range. But for our soft matter systems, the line of action is different.

One first approach uses systems with a broken (continuous) symmetry. For instance, in an ideal Heisenberg magnet, at $T < T_c$, we have a magnetization vector M_0 of fixed length but arbitrary direction (the broken symmetry is the rotation group). Let us apply a field H_0 along M_0 to stabilize the orientation, and then add a weak normal field H_1. The M aligns along $M_0 + M_1$ and this implies a transverse susceptibility

$$\chi = \frac{M_0}{H_0}$$

which is very large if H_0 is small, at all temperatures below T_c. Certain liquid crystal effects are deeply related to this form of large response.

Another approach is based on *fragile structures*. The weak rubber network of figure 21.1 is one example. Other examples are *gels*, where a flexible network is (usually) swollen by a solvent: gelatin for instance is based on collagen fibres in water, with a certain number of cross-linking points occurring spontaneously between fibres.

Another family of fragile structures is constructed with *molecular sheets*: the basic example is a 'vesicle' as depicted in figure 21.3. In normal conditions, the sheet is fluid and highly deformable, but it is stable: the lipid constituents cannot escape from the sheet. Also, the vesicle is rather impermeable: it can transport solutes like a bag.

Ultimately, we may even say that a long polymer chain is in itself a fragile structure. We know this, for instance, from hydrodynamic experiments in dilute solutions, where the chains break quite easily under shear. The bonds are strong (typically one bond will resist one nanonewton), but in these experiments one bond near the centre of the chemical sequence suffers from friction forces from all the other units in the chain and these add up: a little bit like a rope being pulled at both ends by two groups of children.

Last, but not least, we should notice that the interactions between units (i.e. between lipids in a vesicle, or molecules in a liquid crystal, etc) are rather weak: they are mainly van der Waals interactions. Under room-temperature conditions, these interactions are comparable to the thermal unit kT, thus many soft-matter systems are locally fluid (as, for instance, is a rubber). In the American literature, soft matter is

Twentieth Century Physics

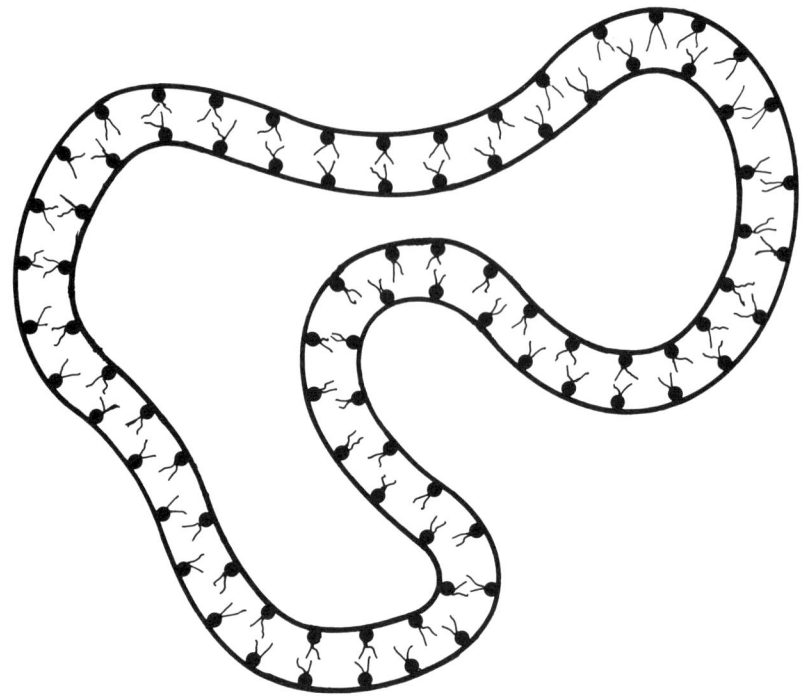

Figure 21.3. *A 'liposome' or 'vesicle' is one example of self assembly, obtained with certain surfactant molecules in water. Here each molecule has a polar head (small circle) and two aliphatic tails. A typical example would be egg lecithin.*

often designated as 'complex fluids', but this denomination is obscure, negative and ultimately incomplete, since rubber is not a fluid at macroscopic scales.

21.2. Polymers

21.2.1. Long-chain systems

A polymer is a long chain, constructed by repetition of a simple, small unit, the 'monomer'. Major examples of synthetic polymers are

$$\text{polyethylene} \quad -\left[CH_2\right]_N-$$

$$\text{polystyrene} \quad -\left[CH_2-\underset{\underset{\bigcirc}{|}}{CH}\right]_N-$$

$$\text{polyvinyl chloride} \quad \left[CH_2-\underset{\underset{Cl}{|}}{CH}\right]_N$$

The 'polymerization index' N is of order 10^3 or more (reaching 10^5 in certain special cases). We should acknowledge here the brilliant work of the chemists, who are able to repeat one elementary operation 10^5 times *without mistakes* (at least in good model systems).

Many objects around us are based on linear chains: wood, foods, textiles, plastics and most of living matter. This preference for linear objects as opposed to branched polymers or other forms such as sheets, etc, is deeply related to the technology of fabrication: in a nylon reaction, as well as in a ribosome, the easiest form of assembly is via a linear sequence of monomers.

21.2.2. Birth of the polymer concept [3]

Natural polymers were used by our ancestors; the first major industrial process using *modified* polymers was the invention of rubber by Goodyear, as already mentioned in the introduction. Later, by suitable modifications, the cellulose from wood became the base for certain artificial fibres, etc. But, during all this nineteenth century period, the concept of long, linear chains was *not recognized*. In fact, many interesting polymer materials were generated, but not seriously studied. This was due to the prevailing dogma: to announce a new chemical product, one had to purify it many times, and ultimately to show its purity by measuring some physical property such as the melting point. Most long-chain systems do not crystallize, and thus do not show a clear melting point; for this reason they were rejected.

It was only around 1920 that H Staudinger was able to prove neatly the existence of long-chain systems—using delicate operations where he synthesized short chains ('oligomers') small enough to be amenable to purification and to the standard dogma. Even Staudinger, in spite of his genius, was still biased by earlier schemes: he was convinced that these chains were *rigid rods*. We recognize now that they are not: in most practical cases, at room temperature, the chains are highly flexible and carry a large entropy. This was first shown by Kuhn, who then was able to understand, from this entropy, the mechanism of rubber elasticity, i.e. when a network of chains is stretched, its entropy decreases and its free energy increases.

21.2.3. Dilute chains

One of the major concepts introduced by Kuhn dealt with the shape of a polymer in very dilute solutions: he described it as a *coil*, or more precisely, as an ideal *random walk*. Assuming that successive units are independent, this leads to a root mean square size

$$R_0 = N^{1/2} a \tag{1}$$

(where a is something like the monomer size). Our understanding of polymer solutions at the Kuhn level then expanded rapidly in the hands of J Kirkwood, W Stockmayer and B Zimm.

Paul Flory

(American, 1910–85)

Paul Flory was born in Sterling, Illinois. He worked first at the Central Research Department of Du Pont, under Wallace Carothers—inventor of nylon and neoprene. In 1948, Peter Debye invited him to give the famous Baker lectures at Cornell on polymer science, causing him to reflect deeply on the whole field of polymer chemistry and physics. One of the outstanding results was the theory of excluded volume effects—the swelling laws for a polymer coil in a good solvent. Another result of Flory's work was his revolutionary notion that chains behave like ideal random walks when they are dense (i.e. in a melt): this was disbelieved at the time and only proven much later by neutron experiments. He also developed the concept of the *Flory temperature*—a temperature for a given solvent/polymer pair at which one polymer coil is nearly ideal.

After moving to Stanford University, Flory pioneered a vast sector of research on the local conformation of polymers (including DNA) and on their ability to make nematic liquids. His two books on polymer science are historical landmarks, and he was awarded the Nobel Prize for Chemistry in 1974.

The next theoretical step is due to Paul Flory [4], who recognized that a chain, in the normal 'good' solvents, tends to avoid itself: the geometrical problem is not a simple random walk, but a *self-avoiding walk*. Flory then constructed an extremely elegant (approximate) description of self-avoiding walks, and came up with a power law for the coil size

$$R_F \simeq aN^\nu \qquad (2)$$

where $\nu = 3/(d+2)$ depends on the dimensionality of space d.

For instance, $\nu = 1$ for $d = 1$ (a self-avoiding chain confined to a line is fully stretched) and $\nu = 1/2$ for $d = 4$ (we recover the Kuhn walk

Soft Matter: Birth and Growth of Concepts

for all $d \geqslant 4$). For the most important case $d = 3$, Flory found $\nu = 0.6$. The difference between this Flory value and the Kuhn prediction is quite important for very long chains: for $N = 10^5$, it implies a factor ~ 3 in size, and is thus very conspicuous in experiments (the size is measured primarily via hydrodynamic properties: a coil in solution is roughly equivalent to a Stokes sphere of radius R_F).

The next major advance concerning the theoretical description of coils is due to S F Edwards: he noticed a deep analogy between one conformation of a long coil and one trajectory for a (non-relativistic) quantum-mechanical particle. The analogue of time t for the particle is the polymerization index N (the number of monomers) for the coil. A certain sum over trajectories is the wavefunction of the particle, the same sum is the statistical weight for the polymer, and, if an external potential is added, both systems are ruled by exactly the same Schrödinger equation [5].

This result was extremely seminal: fifty years of know-how in quantum physics could immediately be transposed to polymer statistics.

21.2.4. Overlapping chains: statics

The Edwards analogy was originally restricted to a single chain: many-chain problems were treated at the Hartree level. Later, a more general relation between the many-chain problems and a certain type of field theory became obvious [6]. This allowed for refined calculations of the exponent ν in equation (3), via renormalization group methods. More importantly, Des Cloizeaux [7] showed that the properties of polymer *solutions*, with overlapping coils, could be analysed neatly in this language, and this was the starting point of many experiments, using neutron scattering (and other techniques), to probe the various spatial scales in a solution.

An important feature of overlapping solutions and particularly of their dense limit (a melt) was already recognized by Flory. Using simple mean-field arguments, he predicted that in this strongly interacting case the chains would return to the ideal form of Kuhn (equation (1)). This was not believed by the polymer community, until, many years later, neutron scattering experiments on isotopic mixtures (of H chains and D chains) in melts proved it unambiguously.

This Flory result may not be completely surprising for a physicist, who knows that an electron gas becomes more ideal when it is dense. There is a rough analogy between melts (where chains cannot intersect) and electron systems (with a Pauli principle), but it is not deep. The Pauli principle forbids two electrons to be at the same place at the same time, while the chain exclusion holds for different 'times' (i.e. the monomer index along both partner chains need not be the same). Thus there is no analogue of a Fermi surface in dense polymer systems. In fact, an adequate interpretation of the Flory result for melts was first provided by

Edwards and is based on screening: the interactions between two units are screened by the ambient chains, and this screening is very strong in dense systems [6–8].

21.2.5. Dynamics of entangled systems

Another important feature of the melts is their mechanical properties: at very low frequencies ω, they behave like a (highly viscous) liquid, while at higher frequencies ($\omega > 1/\tau$), they are elastic like a rubber. The crucial parameter is the relaxation time τ, which describes how one chain 'disentangles' from its neighbours.

A variant of this problem concerns *one* chain moving inside a fixed network: this 'reptation' (snake-like motion) is described in figure 21.4. At any moment of time, the chain is trapped in a certain 'tube' (of length L proportional to N). It diffuses back and forth in this tube, with a diffusion constant D_{tube}. From Einstein's relation, $D_{tube} = kT/\zeta_{tube}$ where ζ is the friction coefficient for the chain (and is proportional to N). Ultimately, the 'reptation time' τ is such that the chain has diffused over a length $\sim L$, and is thus

$$\tau = \frac{L^2}{D_{tube}} \sim N^3. \qquad (3)$$

This N^3 dependence is not very far from the data, which give $\tau \sim N^x$ with x ranging between 3 and 3.5.

After understanding the one-chain problem, the next (difficult) step was to attack the melt, with a large number of entangled chains which are simultaneously moving. Doi and Edwards did this by showing that in a 'monodisperse' melt (i.e. all chains of the same length), the chains C surrounding one particular chain C_0 move so slowly that they still behave like fixed obstacles so the reptation law (equation (3)) still holds.

Starting from this, Doi and Edwards constructed a complete model for the mechanics of entangled systems—indeed it was the first non-trivial microscopic model for a rheological system [9]. A simplified picture of the viscosity is given in [10].

21.2.6. Solid phases: glasses and crystals

When we cool a polymer melt, it usually does not crystallize, but transforms into an amorphous phase—a glass. There are two reasons for this.

(i) Although the monomers are equivalent, they may enter the backbone in different stereochemical patterns. The chain is then stereochemically irregular.

(ii) The kinetics of crystallization are slow for entangled systems.

Soft Matter: Birth and Growth of Concepts

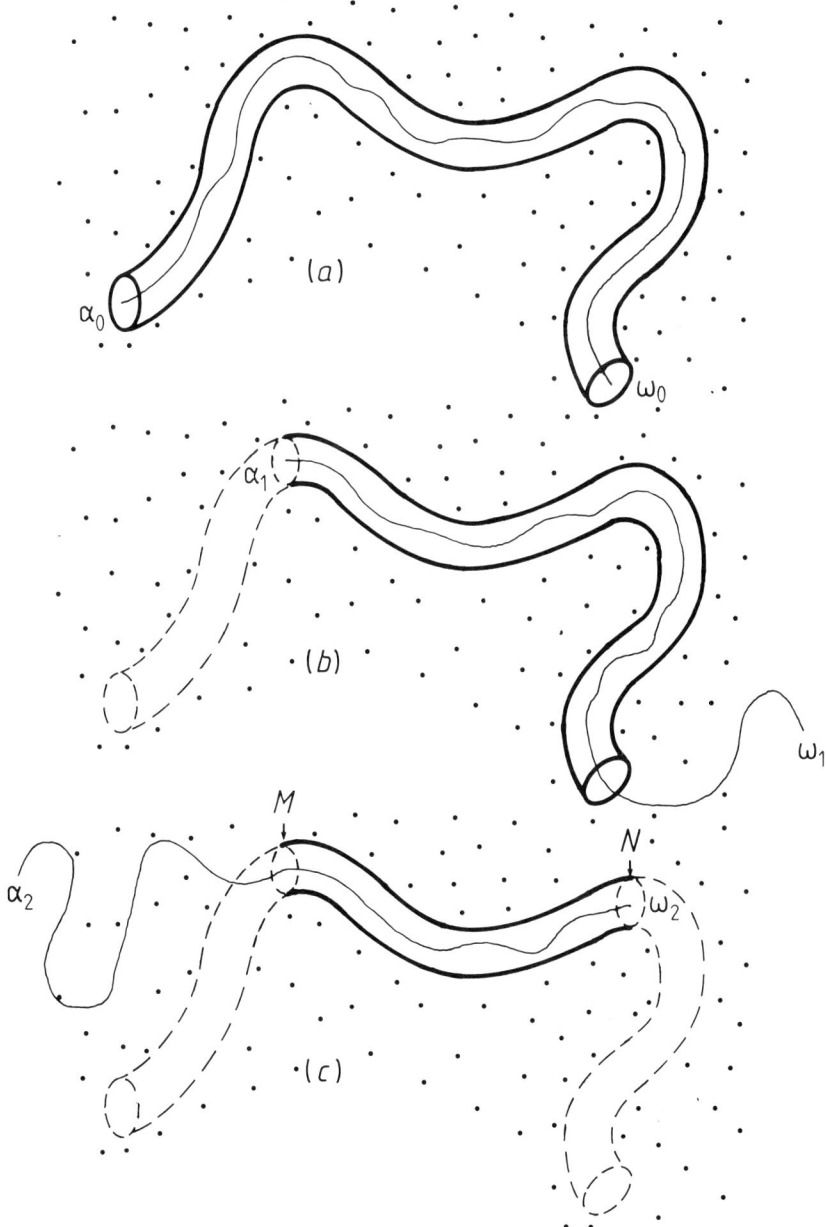

Figure 21.4. *The 'reptation' process for one polymer coil moving in a fixed network (here represented by black dots). At any moment the chain is confined in a certain 'tube' (a concept invented by S F Edwards). The chain escapes by crawling motions: after some time, only the portion MN of the original tube is maintained.*

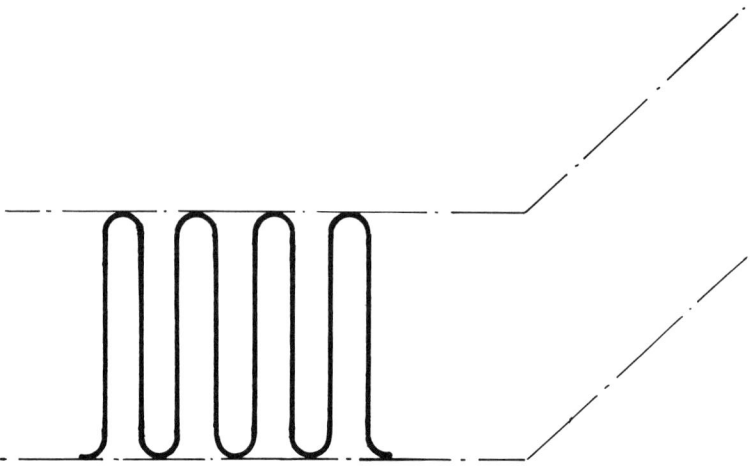

Figure 21.5. *Lateral view of a lamellar crystal of polyethylene with chain folds.*

Typical examples of glassy polymers are polystyrene and polymethyl methacrylate (commercially known as Lucite).

See also p 1552

There are a few favourable cases where the chains are stereochemically regular and may thus crystallize: for instance polyethylene. From dilute solutions of polymers such as polyethylene, one can form platelet crystals: in these platelets the chains are folded (figure 21.5); these folds were a great surprise when they were first found [11]. They are due to kinetic limitations during the coil → crystal transformation [12].

If we cool a *melt*, it will crystallize only partly. This generates an interesting composite of amorphous/crystalline regions: most synthetic textile fibres derive from this principle—the major examples being nylon and polyester (more precisely polyethylene terephthalate), with the formula

$$\left[O-CH_2-CH_2-O-CO-\langle O \rangle-CO \right]_N.$$

21.2.7. Polymer gels

Gels are another interesting form of soft matter. As already mentioned in section 21.1.2, people noticed that gelatin was soluble in water above 37 °C, but that upon cooling it would transform into a transparent gel. Now, we visualize this as a system of peptide chains which, below a certain temperature, can cross-link (by formation of small pieces of multistranded helixes). The net result is an elastic network which is swollen by water: a very soft solid.

Actually, this type of experiment was one of the reasons for the delays in the birth of polymer science [3]. The formation of polymer gels was

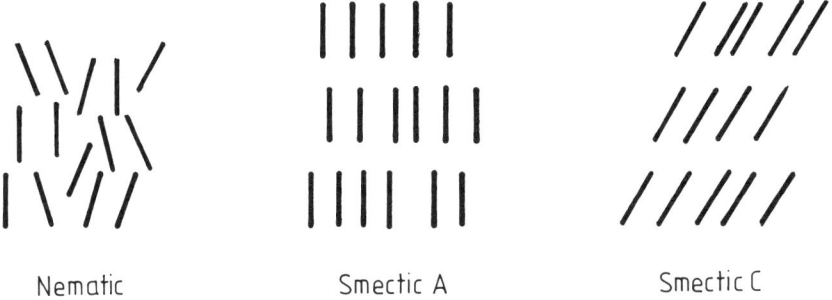

Figure 21.6. *Some examples of liquid crystal phases. The molecules are represented by long rods. Notice that if these rods carry a dipole (along the rod axis), there are as many dipoles 'up' as dipoles 'down'; these phases are not ferroelectric.*

often confused with the aggregation of a suspension of solid particles ('colloid floculation'), and for this reason polymers were considered as colloids, i.e. three-dimensional particles of complex origin, rather than well-defined linear chains. (We come back to the colloid part of the story later in section 21.5.)

In recent years, the phase transformations of gels have provoked great interest: by suitable changes of solvent quality, pH, ionic content, etc, one can provoke an abrupt change in the equilibrium volume of a gel—the kinetics of these changes are still a challenge [13].

21.2.8. Future prospects
In the present period, physical research on polymers is mainly focused on three aspects:

(i) polymers at interfaces [14] (adsorbed, grafted, etc);
(ii) mechanical toughness of bulk phases (glasses, rubbers, etc) [15] and of interfaces (adhesion, tribology) [16];
(iii) possible developments of electrically conducting polymers, especially if these materials become amenable to fibre formation from solutions [17].

More generally, because polymers can be tailored for many purposes, they enter in the 'formulations' of many industrial products: foods, cosmetics, pharmaceuticals, etc. We return to this aspect in section 21.5.

21.3. Liquid crystals
In conventional crystals, the molecules are stacked in a periodic lattice, while in a liquid they are disordered. However, we can build molecules with special shapes (elongated, discoid), plus suitable flexible parts, and generate systems which are in between crystals and liquids. Following their great analyst, Georges Friedel [18], they should be called *mesomorphic phases*—or, more loosely, liquid crystals.

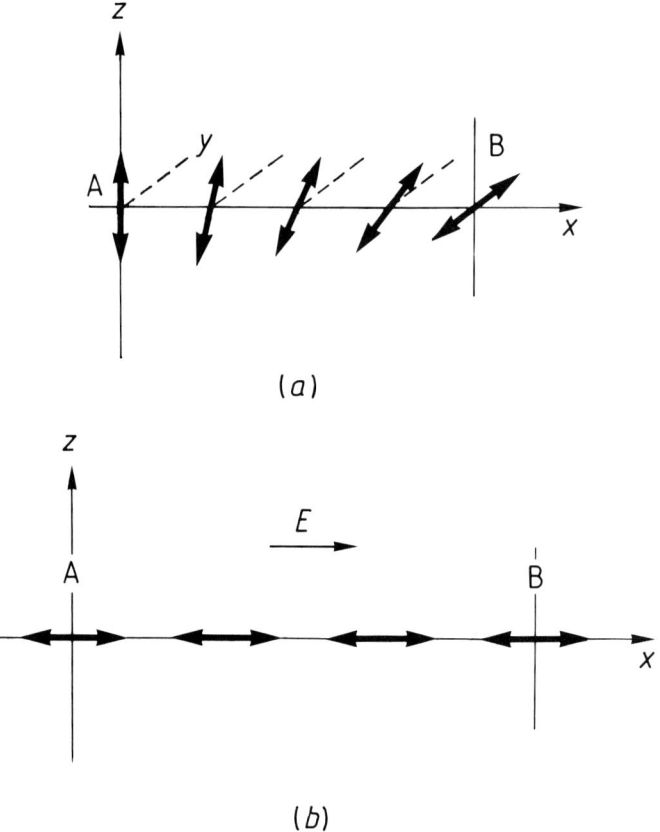

Figure 21.7. *The twisted nematic cell. (a) In zero electric field, the nematic is forced to align parallel to z along the A plate, and parallel to y along the B plate: in between it is helically twisted. A beam of polarized light going from A to B rotates its polarization plane by 90°. (b) Under a field E, the molecules all align along E (i.e. along X). Then the polarization of a light beam going from A to B is unaltered. With parallel polarizers near the cell at A and B, the light beam does not go through in case (a) and does go through in case (b). Most current displays in watches, cars, etc, are based on this idea.*

The main types are shown in figure 21.6. The most fluid systems are the *nematics*, which are optically uniaxial. Their optical axis may be rotated by weak agents: electric fields; magnetic fields; flow; alignment induced by the walls of the container.

One example of the latter case is the twisted nematic cell, which is currently used in display systems (figure 21.7).

The *smectic* phases are also important, in particular for lipid water systems (see section 21.4). One particular smectic phase was invented by R B Meyer in 1975 [19]. This is the so-called chiral smectic C shown in figure 21.8. This phase has a non-zero macroscopic electric moment, and thus couples strongly to electric fields. The next generation of fast

Figure 21.8. *A 'fish analogue' to the C* smectics. The molecules are replaced by fish. There are as many fish with the head up as with the head down, i.e. no dipole along the director axis n. But if the molecules are chiral (e.g. if the fish all carry a dipole along their right eye), these dipoles add up in a direction normal to the plane of this figure: C* has a non-zero electric moment.*

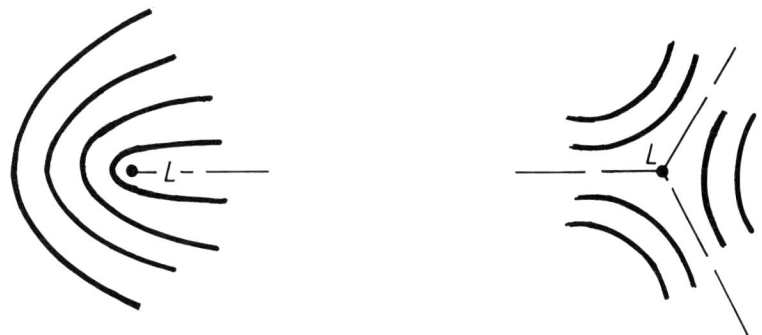

Figure 21.9. *Typical line singularities L in a nematic. Here L is normal to the plane of the sheet. The continuous lines define the local optical axis of the fluid. These singularities were analysed in detail by Charles Frank.*

display systems will probably be based on this family.

With the chiral smectic C system, *a new form of condensed matter* was created, starting from theoretical ideas—the relevant molecules were synthesized a few months later. This is a particular trend of the twentieth century.

An important feature of liquid crystals is their defect structures. Some typical defects are shown in figure 21.9 for nematics and in figure 21.10 for the simplest smectics. In the latter case, it is the observation of the 'focal conic' defects which led G Friedel to predict that smectics were made of liquid, equidistant layers. Working with the microscope at the 100 micron level, he was able to infer the structure at the 10 Å level! More generally, defects provide a specific *signature* for all liquid crystal phases.

Sir Charles Frank

(British, b 1911)

Charles Frank, after graduating from Oxford, showed his exceptional gift for observation while working on the deciphering of strange photographs—which actually represented the first German rocket platforms—during World War II. Since then he has been at Bristol University. Independently of Sakharov, he invented the catalysis of nuclear fusion by mesons, but his principal interest has been in a wide range of problems in condensed matter especially, at first, in the different types of dislocation and their role in crystal growth. He showed how screw dislocations lead to growth spirals on the surface of crystals. He was a pioneer in the field of liquid crystals, elucidating the elasticity of nematics and classifying their topological defects.

Interacting with Andrew Keller he became interested in polymers, and produced a basic model for chain folding in polymer crystals. In his retirement he takes an active interest in geophysics, particularly the behaviour of rocks under pressure and the processes involved in earthquakes.

21.4. Surfactants

21.4.1. Decoration of interfaces

Surfactants are frustrated molecules [20], with one part which likes water (hydrophilic: H), and one part which likes oil (lipophilic: L). Molecules of this type are attracted to an oil–water interface, and also (although less strongly), to an air–water interface: they decrease the corresponding interfacial energy and thus facilitate the making of emulsions and foams.

In many cases, when we increase the concentration of surfactants, we find that some association occurs, in the form of spherical micelles (first described by Hartley in about 1930), on elongated micelles, on lamellar phases, etc (figure 21.11(*a*)). At higher overall surfactant concentrations,

Soft Matter: Birth and Growth of Concepts

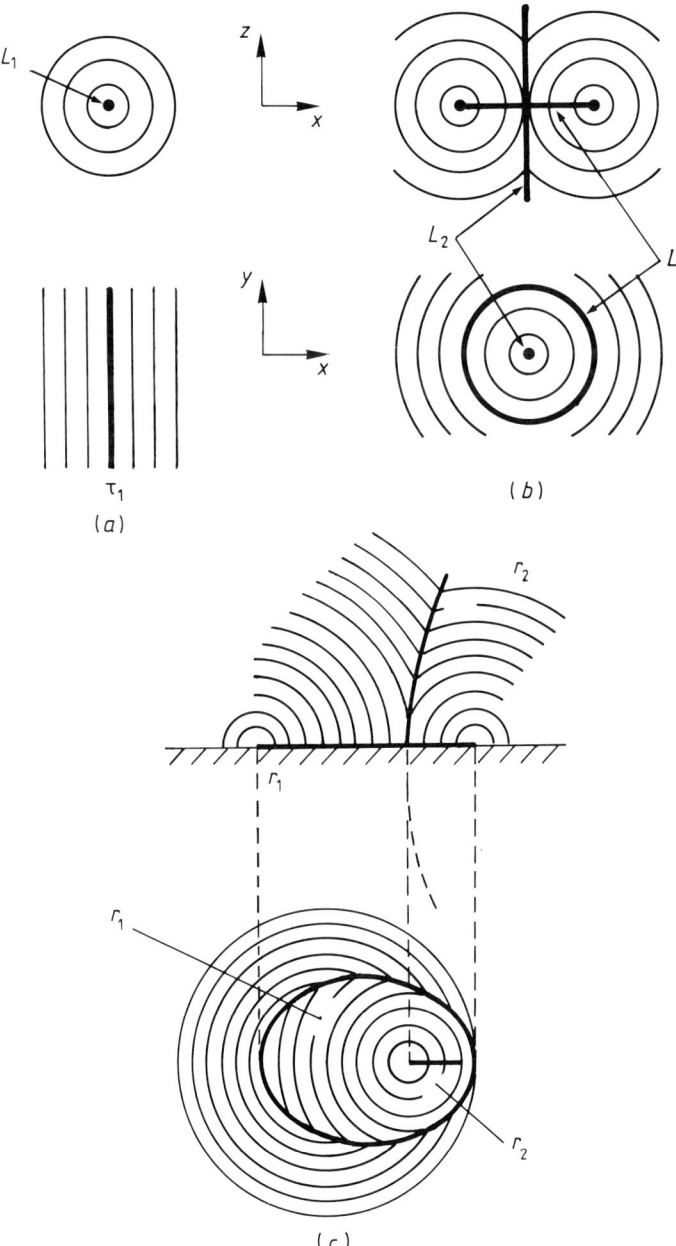

Figure 21.10. *Focal conic textures in a smectic liquid crystal: (a) simple 'jelly-roll' or 'myelinic' arrangement, generating a tube; (b) tube closed into a torus: note the two singular curves (a circle and a straight line); (c) generalization: the circle becomes an ellipse, the straight line becomes a hyperbola.*

their chemical potential is then buffered, and the interfacial tensions become fixed. However, in some favourable cases, one can ensure that no associated phase forms, and reach a vanishing oil–water (O–W) interfacial energy: the system then develops a large O–W area, and builds up a *microemulsion* with characteristic sizes in the region of 500 Å. These microemulsions are thermodynamically stable, fluid and transparent systems. They were first discovered by Hoar and Schulman in 1943 [21] and have a number of technical applications.

21.4.2. An example of associated form: vesicles
Another promising form of surfactant association is the vesicle (or liposome) displayed in figure 21.3, and discovered by Bangham (1964). Liposomes are large (of the order of a few microns), flexible objects, which are almost impermeable for solutes in water: they may possibly become useful as drug carriers (if they are suitably protected) and they have already found some applications in cosmetology.

A liposome is based on essentially insoluble surfactants, and optimizes its area A at fixed surfactant number: the energy is thus stationary with respect to A, or equivalently the surface tension vanishes. This has a number of remarkable consequences.

(i) The shape fluctuations are large [22]. This leads to an interesting repulsion between two neighbouring vesicles, which have restricted fluctuations when their separation becomes small [23].

(ii) If a hole is drilled in a liposome, it heals. This is strictly different from the case of a soap bubble, which has a small but finite surface tension, and which breaks if a local hole is made (figure 21.11(b)).

A number of elegant mechanical experiments on liposomes have been performed by E Evans [24].

21.4.3. Lamellar phases and sponge phases
Clearly, the physics of bilayers, liposomes, etc, is directly related to the statistics of random surfaces [25]. An interesting example of this interrelation was provided by G Porte, D Roux and co-workers [26, 27] and is based on an assembly of bilayers. The most natural association of bilayers is the *lamellar phase* of figure 21.11(b), where the successive sheets make a (relatively trivial) smectic structure: conventional soaps are of this type. However, if the layers are very flexible, we may sometimes obtain a more disordered (isotropic) form, where the sheets fold and achieve many topologies, but always close on themselves (figure 21.12). This is the so-called 'sponge phase' [28, 29]—one of the most provocative forms of soft matter.

Soft Matter: Birth and Growth of Concepts

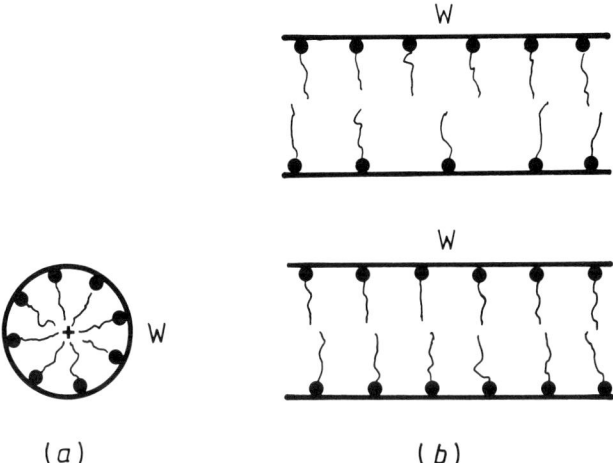

Figure 21.11. *Some aggregation forms of surfactants in water: (a) spherical micelles; (b) lamellar phase (smectic A).*

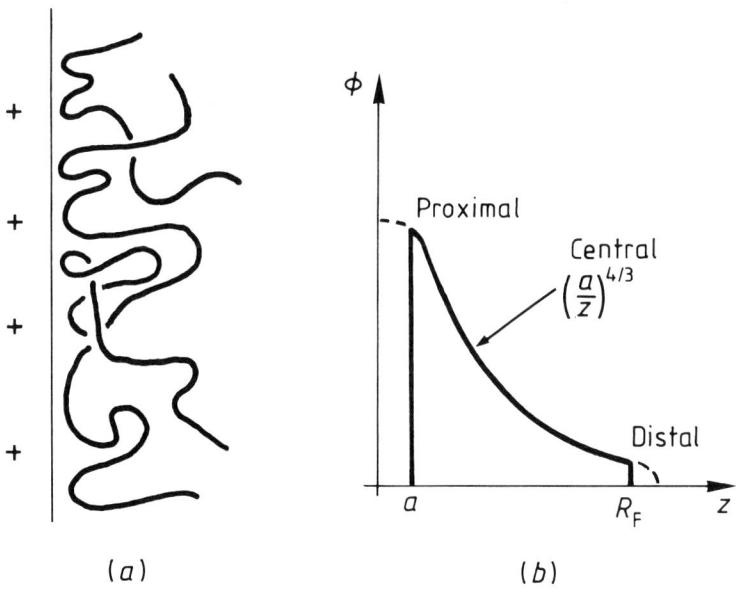

Figure 21.12. *Multichain adsorption from a good solvent: (a) qualitative aspect of the diffuse layer; (b) the concentration profile $\phi(z)$. Note the three regions: (1) proximal (very sensitive to the details of the interactions); (2) central (self-similar); and (3) distal (controlled by a few large loops and tails).*

21.5. Colloids

21.5.1. Definition
Colloid means ultradivided matter: solid grains suspended in a liquid; liquid droplets (e.g. oil) in another liquid (e.g. water); liquid droplets in a gas (aerosols) etc. Many industrial products are based on colloids: for instance, a common form of white paint is based on TiO_2 particles floating in a water matrix (water being preferred to organic solvents for environmental reasons). Typical sizes are in the region of a few microns.

21.5.2. Dilute systems
The volume fraction occupied by the colloidal grains can vary considerably. Very naturally, after a period of purely empirical observations (in the nineteenth century), the attention focused first on *dilute* systems, which are similar to perfect gases—a property used by Jean Perrin to measure Avogadro's number via sedimentation equilibria [30]. The dynamics of individual grains in a solvent was progressively understood through the Stokes friction law, the studies of Einstein and Langevin on diffusion and the Brownian motion data of Perrin. Then attention returned to more concentrated systems, and thus to particle–particle interactions.

21.5.3. Colloid instability and colloid protection
It was recognized early on that colloids tend to be intrinsically *unstable*. This was explained by the Dutch School [31, 32] in terms of long-range van der Waals attractions between particles. Two small spheres of radius R, separated by a small gap $d < R$, have an attraction energy

$$U_{12} = -\frac{A}{12}\frac{R}{d} \qquad (d < R)$$

where A (the Hamaker constant) is a measure of the van der Waals attractive energy between two small molecules upon contact. In room-temperature conditions, the thermal energy kT is comparable to A. If $R \sim 1$ μm and $d = 10$ Å, $|U_{12}|/kT$ is thus of order 100: the grains tend to clump very strongly. One of the basic aims of colloid science is to counteract this attraction—or, in common terminology, to *protect* the colloid.

21.5.4. Charge effects
Protection can be achieved by *electrostatic effects*. In water, for instance, and at neutral pH, many oxides, etc, produce a negative surface charge: this then leads to a long-range repulsion within grains. The range is in fact the screening length, invented by Debye and Hückel. This screening length becomes small if we add salt to the water: thus many colloids

protected via charge effects flocculate (aggregate) in the presence of high salt levels. The theory for all this was constructed in two classical papers by Landau and Derjagin, and Verwey and Overbeek (see [32]).

21.5.5. Protection via surfactants
Cobalt particles ($2R \sim 100$ Å), for instance, can be stabilized in oils using certain surfactants whose polar heads attach to the cobalt surface, while the tails float in oil. These are the basic elements of *ferrofluids*—liquids with strange magneto-hydrodynamic properties.

A vast number of *emulsions* (e.g. droplets of oil in water) are protected by surfactants [33]. These systems are metastable: the interfacial (oil–water) tension does not vanish, but is small. The systems would gain energy by coalescing the droplets, and reducing the oil/water area. They are thus very different from the microemulsions which we mentioned in section 21.4.1.

21.5.6. Protection via polymers
This can occur in these different ways.

(i) Simple adsorption (figure 21.12). For instance, indian ink (in French *encre de Chine* (chinese ink)), is in fact an early example of colloid protection, invented by the ancient Egyptians. To disperse particles of carbon black in water, they added gum arabic to the water: this is a long polysaccharide (polyhyaluronic acid) soluble in water and adsorbed on the carbon surface.

The principles of this stabilization can be understood as follows.

(a) Water must be a *good solvent* of the polymer. This means that any sugar unit along the polymer prefers to be in contact with water, rather than with other sugar units.

(b) The carbon black grain prefers sugar to water. Then, the polymer chains tend to adsorb on the carbon surface, but, because of competition between themselves, they do not stick completely to the grain, instead, they build up a fluffy structure or 'corona' around the grain.

(c) When we force two grains closer together, the coronas overlap: certain sugar units from one corona come into contact with sugar units from the other. The net result from criterion (a) above, is that a repulsive energy builds up: the colloid is protected.

(ii) A second form of polymer protection is based on chemical grafting of chains on the surface (figure 21.13). This generates strong protective layers, but requires delicate use of surface chemistry.

(iii) A third form is based on block copolymers: they are made with two (chemically different) chain portions welded at one end. Osmond (at ICI) patented a number of these systems: for instance, if one part of the chain (the 'anchor') is not water soluble, while the other part (the 'buoy')

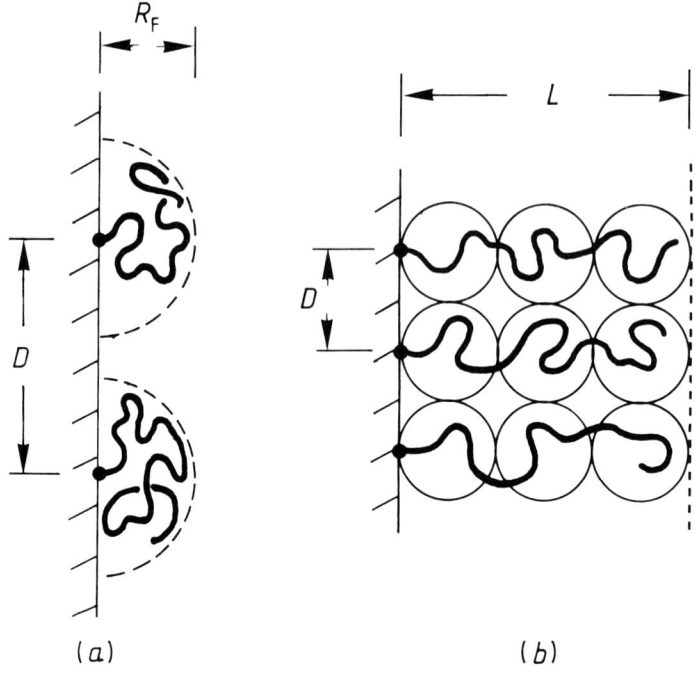

Figure 21.13. *Two types of grafted surfaces: (a) low grating density—the distance between heads D is larger than the coil size R_F (this is the 'mushroom' regime); (b) high grafting density ($D < R_F$) (the 'brush').*

is soluble, the copolymer will often precipitate on the grain surface and generate a protective brush (figure 21.14).

21.5.7. Recent progress and future lines of research
In recent years, these topics have benefited from three main physical advances.

(i) The direct measurement of forces between surfaces, using a machine invented by D Tabor and J Israelashvili [32].

(ii) The use of neutron techniques (small angle scattering or neutron reflectance on flat interfaces) to study the interfacial region [34].

(iii) A number of theoretical reflections on adsorbed and grafted polymer systems [35].

The current problems in colloid science are of two main types.

(i) An extension of the simple ideas sketched above towards multifunctional systems, where, to satisfy different needs, one uses simultaneously a number of additives, e.g. salts, surfactants, polymers.

Soft Matter: Birth and Growth of Concepts

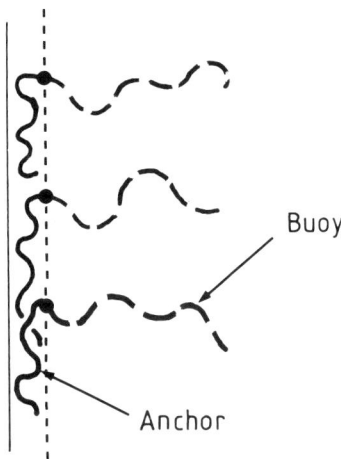

Figure 21.14. *Terminally attached chains via block copolymers AB. The 'anchor' A precipitates against the wall while the 'buoy' B protrudes towards the solution.*

This is the art of 'formulation'—a very important industrial sector for foods, cosmetics, paints, pharmacology, etc.

(ii) An extension to nanoparticles (in the region of 100 Å rather than in the region of a few microns). Here, for instance, an adsorbing polymer coil may be much larger than the supporting grain: systems of this sort are just beginning to be probed and may become important.

21.6. Concluding remarks

Since the days of silex and potteries, hard matter and soft matter have coexisted. The first half of the twentieth century saw a cascade of 'scientific supernovae', such as relativity, quantum mechanics, microscopic physics, which related more directly to 'hard' systems. The second part of the century showed one very brilliant 'supernova' (molecular biology). Another one may be ready to explode (brain function). Some parts of our sky remain dark (e.g. fully developed turbulence). There is also a high noise level in all directions of observation (claimed discoveries which collapse, unrealistic simulations of natural phenomena, etc).

In this stormy world, soft matter, as we defined it, appears as a very small sector, but it represents the *science of everyday life*—and, as such, it should take an increasing share in our educational system. Up to the last century, most children lived in agricultural surroundings, and learned a lot by watching the birds, herding the sheep, repairing tools, etc. Now this experience is lost: our school system ignores it and focuses on abstract principles. We need an education based on simple things.

Insects (temporarily) lost their grasp on the Earth, because of their clumsy, hard, crust. Humans won when their soft hands allowed them to make tools; this ultimately led them to expand their brains, and to think. Soft is beautiful.

References

[1] Ferry J D 1980 *Viscoelastic Properties of Polymers* (New York: Wiley)
[2] Bird R B, Armstrong R and Hassage O 1977 *Dynamics of Polymeric Liquids* (New York: Wiley)
[3] Morawetz H 1985 *Polymers: the Origins and Growth of a Science* (New York: Wiley), an excellent reference for this chapter
[4] Flory P 1971 *Principles of Polymer Chemistry* (Ithaca, NY: Cornell University Press)
[5] Edwards S F 1965 *Proc. Phys. Soc.* **5** 613
[6] de Gennes P G 1985 *Scaling Concepts in Polymer Physics* (Ithaca, NY: Cornell University Press)
[7] Des Cloizeaux J and Jannink G 1987 *Les Polymères en Solution* (Les Ulis: Edition Physique)
[8] Edwards S F 1966 *Proc. Phys. Soc.* **88** 265
[9] Doi M and Edwards S F 1986 *The Theory of Polymer Dynamics* (Oxford: Oxford University Press)
[10] de Gennes P G 1991 *Mater. Res. Soc. Bull.* **I** 20
[11] Keller A 1957 *Phil. Mag.* **2** 1171
[12] Frank F C 1989 *Faraday Discussions* **68** 7
[13] Tanaka T, Sun S T, Hirokawa Y, Kucera J, Hirose Y and Amiya T 1988 *Molecular Conformation and Dynamics of Macromolecules* ed H Nasagawa (Amsterdam: Elsevier) p 203
[14] de Gennes P G 1987 *Adv. Colloid Interface Sci.* **27** 189
[15] Kausch H H (ed) 1983 *Crazing in Polymers* (Berlin: Springer)
[16] Brochard F and de Gennes P G 1992 *Langmuir* **8** 3033
[17] 1989 *Synth. Met. (Symp. Proc.)* C28, this whole issue is a good reference for this topic
[18] Friedel G 1992 *Ann. Phys., Paris* **18** 273
[19] Meyer R B, Liebert L, Keller P and Strzlecki 1975 *J. Physique* **36** L69
[20] Tanford C 1973 *The Hydrophobic Effect* (New York: Wiley)
[21] Hoar T P and Schulman J 1943 *Nature* **152** 102
[22] Brochard F and Lennon J F 1975 This first simple analysis of the fluctuation modes was done for red blood cells *J. Physique* **36** 1035
[23] Helfrich W 1990 *Liquids at Interfaces* ed J Charvolin *et al* (Amsterdam: North-Holland) p 4
[24] Evans E 1991 *Physical Actions in Biological Adhesion (Handbook of Biophysics I)* ed R Lipowsky and E Sackmann
[25] Nelson P *et al* (ed) 1989 *Statistical Mechanics of Membranes and Surfaces* (Singapore: World Scientific)
[26] Porte G, Marignon J, Bassereau P and May R 1988 *J. Physique* **49** 511
[27] Gazeau D, Bellocq A M, Roux D and Zemb T 1989 *Europhys. Lett.* **9** 447
[28] Huse D and Leibler S 1988 *J. Physique* **49** 605
[29] Roux D, Coulon C and Cates M E 1992 *J. Phys. Chem.* **96** 4174
[30] Perrin J 1991 *Les Atomes* (Paris: Flammarion)
[31] 1967 *Molecular Forces* ed Pontifical Academy (Amsterdam: North-Holland)
[32] Israelashvili J N 1985 *Intermolecular and Surface Forces* (New York: Academic)

[33] Becher P 1966 *Emulsions; Theory and Practice* ed R Krieger (New York: Hunlington)
[34] Auroy P, Auvray L and Leger L 1991 *Phys. Rev. Lett.* **66** 719
[35] Milner S, Witten T and Cates M 1988 *Macromolecules* **21** 2610

Chapter 22

PLASMA PHYSICS IN THE TWENTIETH CENTURY

Richard F Post

22.1. Introduction

The designation 'plasma' for a gas composed of charged particles was coined by Irving Langmuir, one of the pioneers in this field of research. The history of twentieth century plasma-physics research as a discipline can be roughly divided into two eras, with a dividing line in time at about mid-century. A further demarcation could be made in terms of the motivations for plasma research. In the first half of this century plasma phenomena generally played a secondary role in other fields of research: i.e. in the study of electrical discharges in gases, and in the investigation of the propagation of radio waves over the surface of the Earth. At or about mid-century some much stronger motivations for theoretical and experimental research into plasma phenomena came to the fore. One of these—controlled fusion research—aimed at an eminently practical and worthwhile goal, the harnessing of nuclear fusion reactions for the generation of electrical power. It was early on perceived that the only foreseeable route to the achievement of the controlled release of fusion power was through the attainment of a thorough understanding of high-temperature plasmas and their interactions with electromagnetic fields.

The issue of plasma temperature, alluded to in the previous paragraph, provides another important dividing line between the early work and the more recent research into the physics of the plasma state. In the early study of gas discharges, mentioned above, the gases involved were only partially ionized, so that the ions and electrons of the plasma were immersed in a background gas of neutral particles against which they continually collided. These neutral particles were in turn in thermal contact with the walls of the discharge chamber. As a result, except on a very transient basis, the kinetic temperature of the interior of the discharge was limited to a few thousand degrees at most. It is therefore no wonder that only hints of the properties of a 'true' plasma (i.e. a gas composed only of charged particles) could be seen in those early experiments.

Contrast this situation with the one encountered in fusion research. Thermonuclear fusion, as its name implies, only occurs when the kinetic temperature of the fusing nuclei is sufficiently high that these nuclei can, when colliding, penetrate their mutual Coulomb barrier deeply enough to permit fusion to occur. To achieve these conditions requires kinetic temperatures in the range of tens to hundreds of millions of degrees kelvin. Long before such temperatures are reached all matter will have become totally ionized, owing to the force of the collisions between its energetic particles. Furthermore, at fusion temperatures the mean free paths for deflecting collisions between the ions and electrons have become orders of magnitude longer than those for the ions and electrons in ordinary gas discharges. In the laboratory one has gone from a situation where the mean free path for collisions can be much smaller than the dimensions of the gas chamber to one where the mean free path is several orders of magnitude larger. This means that in the absence of a non-material means for confining plasma at fusion temperatures such plasmas would disappear in microseconds, their particles having collided with the chamber walls, the electrons and ions then recombining to form neutral atoms. High kinetic temperature, long collisional mean free paths and the concomitant need for confinement of the plasma free from contact with ordinary matter are among the characteristics that distinguish fusion-motivated modern plasma research from pre-midcentury plasma physics.

The other strong motivation for the growth of modern plasma physics as a discipline was to gain an understanding of a wide spectrum of space and astrophysical phenomena. Space physics and astrophysics is a research arena where plasma physics plays a dominant role. As we now know, planet Earth is but a small non-plasma island in a vast sea of plasma. Though tenuous in outer space, plasma is dense and omnipresent in the stars and in their coronas. In fact this 'fourth state of matter'—plasma—is seen as the dominant form of matter in the Universe, and no astrophysical theory that does not include plasma phenomena can hope to represent more than a small fraction of the overall cosmological picture.

The history of plasma physics in the twentieth century starts with a situation where the existence of the plasma state was virtually unknown, and little awareness existed of the potential importance of plasma phenomena in our world and its real importance in the Universe. Beginning with the second half of the century strong motivations for plasma research appeared, marking the real beginning of serious attempts to understand this complex state of matter and its interactions with electromagnetic fields. In effect a new field of physics research was born, with its own list of puzzles and unknowns. In what follows I will be presented with a dilemma in trying to decide what topics and results to discuss, while fitting it all into a relatively few pages of text. Having spent the last 42 years of my professional career in the field of magnetic fusion research, it will be virtually impossible for me to avoid some degree of bias, both in my presentation of fusion research, and in my discussion of other areas of plasma research. To those whose work is thereby inadequately covered, or to those whose work on an important topic preceded that of which I am aware, I apologize. I am alone responsible for the choices made, and for the interpretations of the accompanying research policy decisions that, particularly in recent years, have shaped the directions of plasma research.

22.2. Plasma physics in the first half of the twentieth century

22.2.1. Radio propagation and the ionosphere
It was not until the early decades of this century that evidence, let alone understanding, of plasma-related physical phenomena began to appear. As we have mentioned previously, gas discharges, regimes in which plasma effects might have been seen, though studied by Crookes and others before the turn of the century, were dominated by atomic-physics effects, thereby concealing plasma-related phenomena. As it turned out, therefore, it was in the study of the propagation of radio waves that effects dominated by plasma phenomena were first observed unequivocally and even this observation was somewhat of a fluke.

It is interesting to read accounts of the sequence of events in radio propagation studies from early books on the subject. One such book, *Short Wave Wireless Communication* by A W Ladner and C R Stoner [1], states in its preface 'The discovery, in the third decade of the present century, of the utility of "short" waves for world-wide communication, in which amateur workers can claim a large share of the credit, produced a revolutionary change in the science of wireless and had a profound influence, both technical and economic, on world communication'. Later in the book, a history of the development of wireless communication is given, starting (in 1886) with the experiments of Hertz at very short wavelengths, 35 cm, propagated over short distances, followed by Lodge in 1894 and Marconi in 1896, with distances of 150 yards (Lodge) and 2 miles (Marconi).

In the early years of the twentieth century, as spearheaded by Marconi, there grew up an industry based on wireless transmission, but not at the short wavelengths studied in those early experiments. When long ranges were sought, these transmissions were carried out at the opposite end of the radio spectrum, i.e. at very long wavelengths, up to 10 000 metres (30 kHz). This dramatic shift to long wavelengths was made in response to the requirement for reliable propagation at distances up to 5000 miles. As noted in Ladner and Stoner '... owing to the bad results obtained with waves of 200 metres, shorter waves were not attempted for any except occasional very short-range service'. The 'bad results' referred to that were obtained in these studies early in the century were just the effects we now understand as arising from reflections from the ionospheric plasma, leading to the well-known 'skip' effects encountered in the short-wave bands. Thus it was that early on there was little commercial interest in the short-wave bands, leaving it to the researchers (and to amateur radio enthusiasts) to show their great value for long-range communications. In 1923 Marconi again was involved in pioneering experiments that showed this potential. Using a 1 kW transmitter at Poldhu in Cornwall, England, it was found that reliable transmission could be obtained at a wavelength of 97 metres over a distance of 2300 miles, with much stronger signals, however, during the night-time hours. This range was soon extended, both up in distance and down in wavelength, with strong signals propagating all the way to Australia. By 1926 commercial circuits were opened on wavelengths as short as 26.5 metres, having been preceded by radio amateurs who had made two-way contacts between the United States and Europe in 1923. It is important to note that the name Nikola Tesla is intimately associated with some of the earliest work in radio and high-frequency technology. Some of Tesla's patents in radio technology predated those of Marconi, and Tesla's genius was evident in many aspects of the field. Though flawed, Tesla's ideas for transmitting power by radio waves and his early experiments, showed his awareness of the importance of the ionosphere in radio propagation.

As an additional bit of history, the reasons for the late discovery of the utility of short waves for long-distance communication are described very well by Ladner and Stone

> At this point it seems necessary to give a hint as to why the utility of short waves went so long undiscovered. All the early experiments had shown conclusively that in the neighborhood of a very short-wave transmitting station the attenuation was extremely great. What they had failed to reveal was that in addition to the surface wave which was being investigated and whose characteristics were understood, there was a radiation which entered the ionosphere, and from there was bent towards the Earth. If the bending was sufficiently great, the wave left the layer again and reached the Earth's surface at some point generally very distant from the transmitter. ... Thus in the London–Birmingham tests of 1921 on 15 metres, no doubt strong signals could have been received in many parts of the world, but no one thought of looking for them ....

Looking back now, it seems somewhat surprising that long-distance propagation of short waves was not predicted and discovered earlier than it was, considering that the concept of the Earth having an ionospheric layer had already been inferred, nearly simultaneously, by Heaviside [2] and by Kennelly [3] in 1902. Even earlier, the possible existence of conducting strata in the upper atmosphere had been suggested by Balfour Stewart [4] in formulating a theory to explain the daily variation of the Earth's magnetic field.

The assumption of a conducting ionosphere was forced on the wireless community by the observation of the long-distance propagation of very long-wavelength radio waves, the first band to be exploited commercially. If the Earth were to be modelled solely as a conducting sphere in free space, there would be no way to explain how long-wave radio signals could be found to propagate with modest attenuation over the Earth's surface for thousands of miles. This paradox could only be explained by adopting the hypothesis advanced by Kennelly–Heaviside, that a conducting layer high in the atmosphere had the effect of confining these long radio waves between the Earth's surface and this conducting shell. No one seems to have early on taken the next step: to use the Kennelly–Heaviside layer to predict the 'skip' phenomena associated with waves in the short-wave bands.

One can therefore in hindsight discern that a very practical issue, worldwide communication by wireless, was being dominated by something as seemingly esoteric as plasma-related phenomena in the outer reaches of the Earth's atmosphere. We can also see that if the full import of this fact had been discerned early on, the history of wireless communication might have been very different and with it the history of plasma physics. It is therefore fair to characterize as being among the earliest experiments aimed specifically at plasma phenomena the measurements of the height and electrical properties of the Earth's ionospheric plasmas carried out in the late 1920s and early 1930s by Breit and Tuve [5], Hafstad and Tuve [6], Appleton [7] and others. In some of these experiments, which could be viewed as forerunners to the development of radar during World War II, a short burst of radio-frequency power was emitted vertically from a radio transmitter. After propagating upwards through the atmosphere these radio waves impinged upon, and were reflected downwards by, the ionosphere to a receiver located near to the transmitter. In this way the effective height and the reflection coefficient of the ionosphere was detected. The word 'effective height' is used to describe the distance thus measured, as some of the time delay could be ascribed to the slowing down and reflection of the radio wave as it penetrated the ionosphere. These same types of measurements also turned up the existence, under some conditions, of what appeared to be two 'layers' in the ionosphere, dubbed the 'E' layer and the 'F' layer. The existence of the second layer was deduced from the presence of two reflected signals, separated slightly from each other in time, that were sometimes received. These same types of measurements, interpreted by theory, gave indications of the density of electrons in the ionospheric layers and explained the evident differences in the day–night and wavelength-dependent character of short-wave propagation. Many years later, upon the launching of sounding rockets and instrumented satellites, a more accurate picture of the distribution of plasma in the ionosphere could be deduced, this time from direct measurements.

The theory that allowed a rational explanation for wireless propagation phenomena was worked out in the 1920s and 1930s by several theorists, among them Hartree [8] and Appleton [9]. It is of historical interest that Hartree published, within about a year, three articles, in the same journal (*Proceedings of the Cambridge Philosophical Society*). The first and third one had to do with propagation of electromagnetic waves in a plasma; the second one, in an area where the name of Hartree is well known, was titled *The distribution of charges and current in an atom consisting of many electrons obeying Dirac's equations.*

Considering the fact that the equations needed—Maxwell's equations, the Lorentz force and Newton's equations—were by then old in the field of physics makes it surprising that it was decades after the turn of the century before someone applied them fully to this problem. What is also noteworthy is that it was necessary to include all three of the basic equation sets to explain what was observed in the propagation of radio waves around the world. That is, it was also necessary to take the Earth's magnetic field into account (through the Lorentz force) in order to make sense of what was going on. 'What was going on' included many strange effects, including puzzling changes in polarization and interference, and delay effects that could not be explained by simple models.

Among the early attempts to explain the effect of the ionosphere on radio propagation were the papers by Eccles [10] in 1912, and, filling some missing elements, the paper by Larmor [11] in 1924. Neither of these papers, however, dealt with the important effects on propagation caused by the Earth's magnetic field.

As a prime example of the effect of the Earth's magnetic field on the propagation of radio waves the detection of 'whistler waves' cited in 1919 by Barkhausen [12] can be given. Apparently a complete mystery at the time, they were not analysed theoretically until more than a decade later, by Barkhausen [13] (incorrectly) in 1930 and by Eckersley [14] (correctly) in 1935. These waves are called 'whistlers' because they are picked up (at audio frequencies) sounding like a whistle with a descending tone.

That whistlers were a puzzlement when they were first discovered can be seen from an excerpt from Barkhausen's first (1919) paper, in which he says

> During the war, amplifiers were used extensively on both sides of the front in order to listen in on enemy communications. ...

At certain times a very remarkable whistling note is heard in the telephone. ... These tones were so strong and frequent on many days that at times listening was impossible. ... This phenomenon was certainly related to meteorological influences. ... At first it seems inexplicable how such characteristically weak alternating currents could form periodically in the Earth and in the sea. Possibly further communications from all those who have used such apparatus will contribute to the explanation.

Ten years later Barkhausen, in his 1930 paper, correctly ascribes whistlers as having been launched at the Earth's surface by lightning bolts, and speculates on the role of the ionosphere in their propagation. He, however, then (incorrectly) postulates that their descending tone and their long duration (up to a second) results from multiple reflections from the ionosphere (requiring up to 1000 reflections to explain the data).

The correct explanation, involving the role of the Earth's magnetic field and its effect on waves in the ionospheric plasma, is contained in the paper by Eckersley [14]. As he there details, frequency components of the lightning bolts propagate along the lines of force of the Earth's magnetic field, finally returning to Earth at great distances from their point of launching. Their descending tone character comes about from the fact that the group velocity of a packet of such waves (very much lower than the velocity of electromagnetic waves in free space) decreases with decreasing frequency because of the effect of the Earth's magnetic field on the motion of the electrons in the ionosphere. Thus the waves with the highest frequencies arrive first at the receiving point, followed by the lower-frequency components of the spectrum of waves launched impulsively by the lightning burst.

The experimental and theoretical studies of the ionosphere and its electromagnetic properties that were carried out before mid-century hardly scratched the surface of the subject. With the dawn of near- and far-space exploration in the latter half of the century there came an explosion of understanding of the electrodynamic phenomena that dominate the plasmas surrounding the Earth. We will therefore return to this subject in the latter part of this chapter.

22.3. Langmuir and plasma oscillations: Landau and plasma theory

From the relatively meagre number of publications concerning laboratory plasma experimentation and basic plasma theory that appeared in the first half of the century there are two names that stand out: US physicist Irving Langmuir and Soviet theorist Lev Landau.

Langmuir, in the 1920s, performed critical experiments that elucidated a fundamental property of plasmas—the oscillations of its electrons relative to its ions at what is universally designated as the 'plasma frequency'. Landau, in his pioneering work, in correcting a subtle analytical error made by another Soviet theorist, A Vlasov, arrived at the prediction of a fundamental damping mechanism of those same oscillations, through a process now universally called 'Landau damping'.

Langmuir's plasma oscillations come about from the property of strongly coupled mutual interactions between the electrons and ions that characterizes the plasma state, giving rise to one of its most pronounced characteristics: statistical charge neutrality. Stated simply, except under unusual circumstances, all plasmas strongly tend towards a state of charge neutrality. That is, they will try to preserve, on average, an equal number (chargewise) of electrons or ions in any given volume. The reason for this property is simple: because of the magnitude of the unit electronic charge, the removal of even a small percentage of the electrons or ions from a finite volume of electrically neutral plasma (equal numbers of electronic and ionic charges) at typical fusion or laboratory plasma densities would result in a strong countervailing electric field. For example, at a plasma density of 10^{14} ions and electrons per cm^3 this field would be of order 1×10^6 V cm^{-1} at the surface of a sphere of plasma with a radius of 1 cm if there were to be a 1% departure from charge neutrality within that sphere. Departures from neutrality thus necessarily lead to electric fields that tend to draw back charges (electrons or ions) into those regions of the plasma that depart from neutrality. It follows that if any process displaces a group of electrons from their near-neighbour ions, these electrons will be attracted back towards the ions, creating the classical situation of a mass and a spring, i.e. simple harmonic oscillations. In their simplest form, then, plasma oscillations consist of bulk longitudinal oscillations of the electron plasma cloud with respect to the ions, with a defined frequency (the plasma frequency), but with no defined wavelength.

Langmuir's plasma oscillations, with a frequency varying as the square root of the plasma electron density, occur at millimetre-wave frequencies in laboratory plasmas. Their high frequency reflects directly the strong electrostatic effects arising from departures from charge neutrality. In his investigations of electrical discharges in mercury vapour, Langmuir observed these oscillations and he and Lewi Tonks [15] theorized on their characteristics. According to this first theoretical picture of these oscillations, one that ignored the thermally induced motions of the electrons, the oscillations would occur at only a single frequency, and would therefore not have a wave character, but would instead be a sort of 'body oscillation' of the plasma itself. However, when electron thermal motion is introduced into the theory, it is seen that Langmuir's plasma oscillations take on a dispersive wave-like character, albeit over a limited range of wavelengths and frequencies. The first theoretical treatment that included finite electron temperatures was made by a famous father-and-son team, J J Thomson and G P Thomson, in 1933 [16]. Though their insights were correct, namely that finite temperature introduced a dispersive correction term to Langmuir's plasma oscillations, their numerical coefficient was incorrect, and the correct theory was not presented until years later, arising from the work of Landau [17], and later, that of Bohm and Gross [18], and Bernstein, Greene and Kruskal [19].

The process of Landau damping involves a direct wave–particle interaction without the necessity of randomizing collisions, in contradistinction to the damping of sound waves in ordinary gases where the damping occurs as a result of randomizing collisions between the gas molecules. As such it was something entirely new in the field of gas dynamics—a 'collisionless' damping of waves in a gaseous medium. Furthermore, Landau's treatment of electron wave processes in plasmas is a remarkable example of a case where an analytical subtlety involving the choice of a contour in the complex plane made the difference between an earlier, incorrect, result (Vlasov's) and the correct physical picture. Over-simplified, Vlasov's error was to take the principal value of an integral over the electron distribution function, thereby missing an imaginary term (corresponding to wave damping) that Landau's excursion into the complex plane revealed. Physically, the correct analysis showed the entrapment of electrons in the wave, with the consequent extraction of energy from the wave in the process, whence a damping of wave energy through transfer to a selected group of the electron population. The concept of Landau damping, and of its inverse, Landau 'anti-damping'— unstable growth of plasma waves by transfer of energy from the plasma electrons (or ions) to the wave—is one of the most basic ones needed to understand a wide spectrum of plasma–wave interactions. It therefore stands as a major contribution to plasma theory.

Although in the particular case of plasma oscillations Vlasov had made an error, his equation, the 'Vlasov–Boltzmann equation', is an extremely important one. When properly applied, and when used with the Maxwell and Poisson equations, it completes the fundamental equation set used for virtually all of the sophisticated analyses that have arisen during the incredible flowering of plasma theory that occurred past mid-century as a result of fusion research and space-plasma research.

To conclude this brief summary of plasma-physics research in the first half of the twentieth century, one can say that, while there were preliminary excursions into the unknown continent of plasma phenomena, beyond some basic theory, and a few tantalizing hints from the laboratory, the incredible complexity of the plasma state and its interaction with magnetic and electromagnetic fields was yet to be explored. At the time of writing, as we near the end of that century, we are still a long way from understanding the full nature of the plasma state of matter.

Lev Davidovich Landau

(Soviet, 1908–68)

The name of Lev Landau may be better known for his work in fields of physics other than plasma physics, as, for example, is his work in the field of condensed matter physics, for which he won the Nobel Prize in 1962. He is nevertheless to be honoured in the field of modern plasma physics for a singular and remarkable contribution that he made to the theory of plasmas. This work led to the understanding of a basic process in plasmas, now universally known by the name 'Landau damping'. The term refers to the process by which energy is exchanged between an ensemble of charged particles, i.e. a plasma, and the electric field of a wave propagating through that plasma. En route to the understanding of this process, fundamental to almost every aspect of wave–particle interactions in plasmas, Landau made major contributions to the development of the basic kinetic equations needed to analyse the plasma state, while at the same time discovering a subtle error in earlier attempts at such analyses.

Landau was born in what is now Azerbaijan on January 22, 1908. His father was a petroleum engineer and his mother a physician. He studied at the universities of Baku (1922–24) and Leningrad (1924–27), graduating in 1927. Beginning in 1929 he visited various scientific centres in Europe, during which time he met and began a long-lasting friendship and a working relationship with Nils Bohr. Returning to the Soviet Union in 1932 he headed theoretical groups in Leningrad and Kharkov. In 1937 at the request of Peter Kapitza he moved to Moscow as head of the theoretical division at the Institute of Physical Problems, following which became professor of physics at Moscow State University. During that period he distinguished himself in his collaboration with E M Lifschitz to produce a remarkable set of monographs on theoretical physics, begun in 1938, and published in the 1950s.

CONTINUED

> **CONTINUED**
>
> Lev Landau's career was brought to a tragic and untimely end by a serious automobile accident in 1962. Though he survived the accident, the injuries he suffered were so severe that they precluded any further creative work and he died six years later, mourned by the entire physics community. Landau was indeed a genius in many areas of theoretical physics, but he is especially remembered by the plasma physics community, where his name is forever attached to one of the most basic physical processes in plasmas.

22.4. The fusion and space-plasma era: circa 1950 to the present

Although 'What is past is prologue', no one could have predicted in the 1930s the explosion of interest in plasma-physics-related research that began in mid-century. The first motivator for this mounting interest was the goal of fusion power. Fresh from the sobering scientific success of fission weapons, and motivated by altruism, scientific curiosity and, in some cases perhaps, atonement for feelings of guilt, scientists in the Soviet Union, the United Kingdom, and the United States began to think seriously about ways to achieve the controlled release of fusion energy for the benefit of mankind.

In this context the achievement of fusion power seemed to them an almost ideal goal. The primary fuel for fusion, heavy hydrogen, being a constituent of ordinary water (1 part in 6000 of every atom of hydrogen in water is a deuterium atom), is truly inexhaustible and is universally available. The ashes of the fusion process are ordinary helium, and, with good design of the reactor, induced radioactivity, when it occurs as a by-product of some fusion reactions, can be limited both in amount and half-life. Furthermore, rather than being a runaway process like the fission chain reaction, the fusion flame is one that must be carefully tended, lest it go out. Here indeed was a worthwhile endeavour, one that could earn a lifetime commitment of a professional career.

The example of the controlled release of nuclear energy in fission reactors encouraged these post-war scientists to believe that fusion might as readily be tamed and tapped for useful energy, given a sufficiently astute attack on the problem. They had also, to illuminate their way, the example of the Sun, whose steady energy release could only be explained by the process of thermonuclear fusion. The concept of thermonuclear reactions as an explanation of the sustained energy release was posited by Atkinson and Houtermans in 1929 [20]. This theory, followed as it was in the early 1930s by the discovery in accelerator experiments carried out by Cockroft and Walton [21] and others of fusion reactions between light elements, in particular the heavy isotopes of hydrogen, set the stage for what was to occur 20 years later.

The idea of achieving a controlled release of energy from fusion reactions, achieved 'simply' by creating the right temperatures and densities in a fuel composed of light elements, must have occurred to many physicists soon after the Atkinson and Houtermans' article was published. There is indeed a story that following that event, when the physicist George Gamow, then a citizen of the Soviet Union, gave a talk in Leningrad (now St Petersburg) reviewing the implications of the new theory, the local commissar offered him the use of the Leningrad power grid during the mid-night hours, if he would agree to try to achieve fusion in high-powered electric discharges. Wisely, Gamow declined the offer.

Perhaps the first hint to the public and to those physicists (like myself) outside the *cognoscenti* of both the political implications of controlled fusion, and of its hoped-for achievement by electrical means, came from an unexpected source. On 21 March 1951 the then dictator of Argentina, Juan Peron, and an Austrian physicist in his employ, Ronald Richter, issued an astounding press release. In it they revealed that Peron had initiated a research programme to achieve controlled fusion (through electric discharges as it turned out later) and that on 16 February 'there was held with complete success the first tests which, with the use of this new method, produced controlled liberation of atomic energy'. Upon investigation it was found that Richter's claims were not sustainable, and the matter soon dropped from public view. But, fallacious though its claims were, Peron's announcement had an undoubted impact on the scientific community, causing many of us to begin to think about the fusion challenge, and to speculate on ways to achieve the goal of net fusion power.

Better to understand the fusion challenge as it was perceived in those days, the understanding of this task was the following: to find means to heat a low-density gas of fusion fuel to fusion kinetic temperatures (of order 100×10^6 K), then to confine it without substantial loss of the fuel charge for a long enough time for the fusion energy released to exceed the energy needed to heat the fuel to temperature, taking into account all inefficiencies in the process. The requirement for 'low density' at that time came from the recognition that, except at particle densities which were several orders of magnitude smaller than that of gases at atmospheric densities, the pressure exerted by that gas would exceed the bursting strength of any known 'containment' system.

Early on, in a disarmingly simple calculation, Lawson [22] had stated the net-power requirement in terms of a simple product of the particle density, n, and the energy confinement time, τ, (loss-time constant for the heat content of the fuel—closely related to the loss-time constant for the fuel particles themselves). According to the Lawson criterion, the product $n\tau$ must exceed 10^{14} cm^{-3} s, for even the best of the fusion fuels. It follows then, if particle pressures were to be limited (by strength-of-materials considerations) to at most a few hundred atmospheres, then for a gas at a kinetic temperature of one or two hundred million degrees kelvin, the density would have to be limited to a few times 10^{14} cm^{-3}. According to the Lawson criterion, then, the required confinement time for the plasma fuel ions would have to approach 1 second. But those ions, travelling at mean kinetic speeds of order 10^6 m s^{-1} would, if uncontained, have travelled a distance of the order of the distance between New York and Kansas City! Some non-material means for confining the ions in a very limited space (as compared to their free-flight path length) clearly had to be devised if the fusion furnace were to yield net power. Put another way, in a volume of the order of a cubic metre, the plasma ions would, on average, be executing of order one million bounces against the 'walls' of the containment vessel if confined long enough to satisfy the Lawson criterion. Clearly a successful confining system must also be nearly leak-proof in this sense.

22.4.1. The birth of the magnetic fusion approach

Sometime just before or just after World War II the idea occurred, independently, to physicists in the West and in the Soviet Union, that the only feasible way to achieve the required confinement was to immerse the fusion plasma in a powerful magnetic field. Strong magnetic fields act on the electrons and ions of a plasma to curve their free-flying orbits into tight spirals, thus inhibiting their escape across the field. The first known (to me) clear example of physicists attempting to heat and confine a hydrogen plasma in a magnetic field—as a route to controlled fusion—came, in 1938, in an unlikely setting: the Langley Laboratory of NACA (National Advisory Committee for Aeronautics— the predecessor to NASA), a laboratory dedicated to aerodynamic research. The researchers involved were Arthur Kantrowitz and Eastman Jacobs. Jacobs, Katrowitz's boss, obtained a small grant (5000 dollars) from the director of the laboratory, Dr George Lewis, to build their experiment by arguing that it might someday have applications to aircraft propulsion. That their real goal was fusion was not something that would have been approved at that time and in that setting.

The motivation for their experiment came from an appreciation of the significance of the concept of thermonuclear fusion reactions, as outlined in Hans Bethe's articles on nuclear reactions in the journal *Reviews of Modern Physics* [23]. Kantrowitz's fusion reactor idea was to create a doughnut-shaped magnetic field by winding coils around a toroidal vacuum chamber. They named their device a 'diffusion inhibitor' to disguise its real purpose. Low-density hydrogen gas was then to be introduced into the chamber and ionized and heated by radio-frequency power. The experiment was built, and operated (at night) to look for tell-tale x-ray emission that would signal the achievement of high plasma temperatures. Though a glowing plasma was seen, there were no x-rays. As we now understand, plasma held in by a purely toroidal magnetic field cannot exist in pressure equilibrium but drifts across the confining magnetic field. Later on, the 'tokamak' of Sakharov and Tamm and the 'stellarator' of Lyman Spitzer addressed and solved this problem, but in different ways.

Kantrowitz and Jacobs' experiment was terminated when their boss discovered its true purpose, but it stands as perhaps the first experimental attempt to heat and confine a plasma for the purpose of achieving fusion reactions. Its disappointing results also, perhaps, gave the first experimental evidence of what would become the bane of existence of fusion researchers—plasma instability—and its termination gave the first hint of the way in which fusion research and government research policy would later be intertwined, not always to the benefit of fusion.

The onset of World War II certainly curtailed any serious effort towards fusion, but paradoxically, a war-related effort resulted in another hint of the perverse behaviour of plasmas when immersed in magnetic fields. This effort was the development of the 'calutron' isotope separation at Earnest Lawrence's Berkeley 'Radiation Laboratory'. Here the large magnets of the laboratory's cyclotrons had been pressed into service for research into uranium isotope separation as a part of the Manhattan Project on the atomic bomb. Lawrence's concept entailed the use of ion beams deflected by magnetic fields. In the course of trying to increase the throughput of the devices, clear evidence of unstable behaviour of the ion beams and their accompanying electrons was seen. It was this behaviour that prompted David Bohm, a theorist on the team, to come up with his famous expression for 'Bohm diffusion'. In never-published notes he derived an expression that both estimated the limits on the amplitude of the unstable oscillations of plasmas in magnetic fields and also estimated their consequences in terms of the enhancement of the rate of diffusion of those plasmas across the lines of a confining field (figure 22.1). What was not known at the time is under what circumstances Bohm's result applied and when it did not. Optimists thought it to be merely a curiosity; pessimists thought it dealt a fatal blow to hopes for the magnetic confinement of plasmas for fusion purposes. In fact, in the early years of fusion research Bohm diffusion was often taken as the 'law' to beat, and success was measured in how many 'Bohm times' had been achieved in a given experiment.

To explain the concern raised by Bohm's result one must understand that, early on, so-called 'classical' calculations of the rate of collision-induced diffusion of ions and electrons across a confining magnetic field had been made. These results, exemplified by those given by

Lyman Spitzer in his famous small bible of the early fusion researcher *Physics of Fully Ionized Gases* [24], painted a rosy picture for the future of magnetic fusion. In his book Spitzer summarized much of the then-known theoretical picture of the plasma state, including particularly its predicted rate of diffusion across strong magnetic fields ('strong' in the sense that the orbit diameters of the trapped ions and electrons are small compared with the dimensions of the confining magnetic field). Putting in the numbers appropriate to fusion plasma densities (10^{14} cm^{-3}) and kinetic temperatures (10^8 K) and magnetic field intensities of, say, 5.0 T, Spitzer's equations would predict very small diffusion velocities, of order 1 cm s^{-1}. In fact this classical rate of diffusion is so small that if it were to be achieved in a cylindrical column of fusion plasma, then to meet Lawson's criterion, if the time of loss of the plasma across the field was the controlling factor, the radius of the column would only have to be one or two centimetres! Furthermore, scaling as (radius)^2, one might expect easily to accommodate, to an order of magnitude or two, enhancement (for example, by turbulence) of the rate of diffusion by scaling up to a few centimetres radius. How naive we were!

Contrast the optimistic picture painted by using the classical rate as an index of confinement times with the (too-pessimistic, it turns out) predictions of the Bohm diffusion equation. Here the same plasma numbers result in predicted diffusion times of about a microsecond, six orders of magnitude shorter than the classical result! Furthermore the confinement time became shorter at higher temperatures (where the classical confinement time becomes longer as the temperature increases), and scaled up only as the first power of the magnetic field (where the classical time scales as the square of the confining field).

A perhaps too laconic summary of the course of magnetic fusion research in the last half of this century is that it has been a programme that that has had to live between the rock of Bohm diffusion and the hard place of impatience on the part of the body politic with the fact that no practical fusion reactor has resulted from, now, over 40 years of internationally supported research efforts.

22.4.2. The classified years: Project Sherwood

The years between about 1950 and 1958 saw, in the United States, the United Kingdom and the Soviet Union, a concerted effort, behind each country's shield of classification, to tackle the problem of magnetically confining fusion plasmas. It, perforce, also involved tackling a more fundamental problem in physics—the nature and characteristics of the high-temperature plasma state when strongly coupled to magnetic fields. During this period there were also spawned many ingenious configurations (dubbed 'approaches') of magnetic confining fields, each with its own champions (and detractors). It was during this time that awareness began to grew of how ubiquitous the problem of plasma

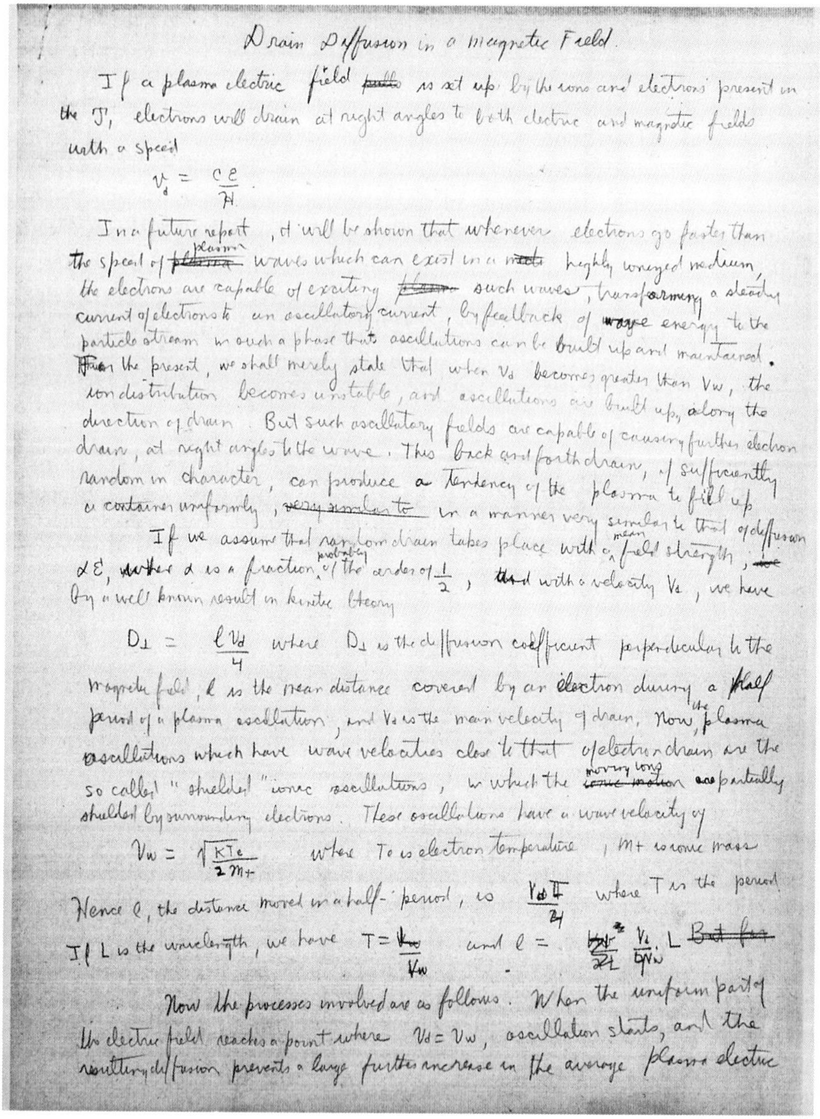

Drain Diffusion in a magnetic Field.

If a plasma electric field is set up by the ions and electrons present in the J', electrons will drain at right angles to both electric and magnetic fields with a speed

$$v_d = \frac{cE}{H}.$$

In a future report, it will be shown that whenever electrons go faster than the speed of plasma waves which can exist in a highly ionized medium, the electrons are capable of exciting such waves, transforming a steady current of electrons to an oscillatory current, by feedback of energy to the particle stream in such a phase that oscillations can be built up and maintained. For the present, we shall merely state that when V_d becomes greater than V_w, the ion distribution becomes unstable, and oscillations are built up, along the direction of drain. But such oscillatory fields are capable of causing further electron drain, at right angles to the wave. This back and forth drain, if sufficiently random in character, can produce a tendency of the plasma to fill up a container uniformly, in a manner very similar to that of diffusion.

If we assume that random drain takes place with a mean field strength, αE, where α is a fraction, probably of the order of $\frac{1}{2}$, and with a velocity V_d, by a well known result in kinetic theory we have

$D_\perp = \frac{l V_d}{4}$ where $D_\perp$ is the diffusion coefficient perpendicular to the magnetic field, l is the mean distance covered by an electron during a half period of a plasma oscillation, and V_d is the mean velocity of drain. Now, the plasma oscillations which have wave velocities close to that of electron drain are the so called "shielded" ionic oscillations, in which the moving ions are partially shielded by surrounding electrons. These oscillations have a wave velocity of

$V_w = \sqrt{\frac{KT_e}{2m_+}}$ where T_e is electron temperature, m_+ is ionic mass

Hence l, the distance moved in a half-period, is $\frac{V_d T}{2}$ where T is the period. If L is the wavelength, we have $T = \frac{L}{V_w}$ and $l = \frac{V_d}{2V_w} L$

Now the processes involved are as follows. When the uniform part of the electric field reaches a point where $V_d = V_w$, oscillation starts, and the resulting diffusion prevents a large further increase in the average plasma electric

Figure 22.1. *Pages from handwritten notes of David Bohm showing his derivation of the famous 'Bohm Diffusion' formula, including the physical arguments he made in carrying out his derivation. This work was never published by Bohm.*

instabilities and plasma turbulence would turn out to be. This awareness came not only from the disappointing results of laboratory experiments, but also from the theoreticians, who became increasingly adept at handling the complexities of plasma theory, particularly when it came to predicting possible new modes of plasma instability.

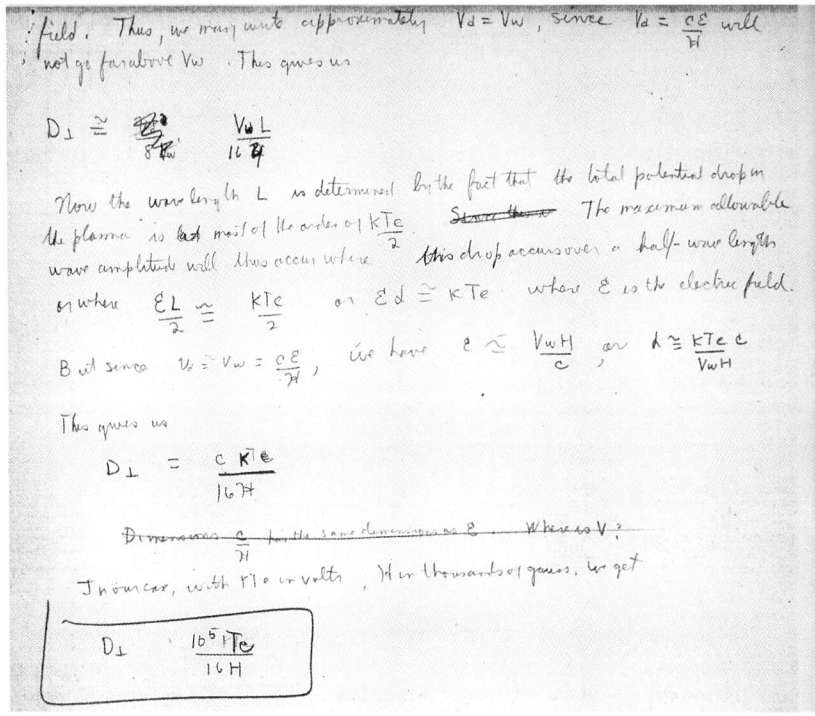

Figure 22.1. *Continued*

In the United States the Atomic Energy Commission placed a 'Secret, Restricted Data' classification on all research on controlled fusion, despite the fact that its goals were entirely peaceful in nature. The rationale for this classification came under discussion at the time of one of the first 'Project Sherwood' meetings that was held, in 1952, in Denver, Colorado, with arguments being made both for and against. In these discussions the, in retrospect naive, assumption was made by those arguing for classification, that small fusion reactors might represent a short-cut to the production of fissionable materials, replacing as neutron sources the huge fission 'piles' then used to produce weapons-grade plutonium. There was even thought, now seen to be even more naive, that miniature hydrogen bombs could be built using knowledge gained from controlled fusion research.

It took only six years of intense effort on fusion research to convince the US fusion community and the Atomic Energy Commission that achieving fusion was a longer term proposition than they had thought, that fusion systems would in all likelihood be large and complex and that the best chances of success lay in declassification, i.e. in involving the wider scientific community. Similar conclusions seem to have been reached by the other parties—the United Kingdom and the Soviet Union.

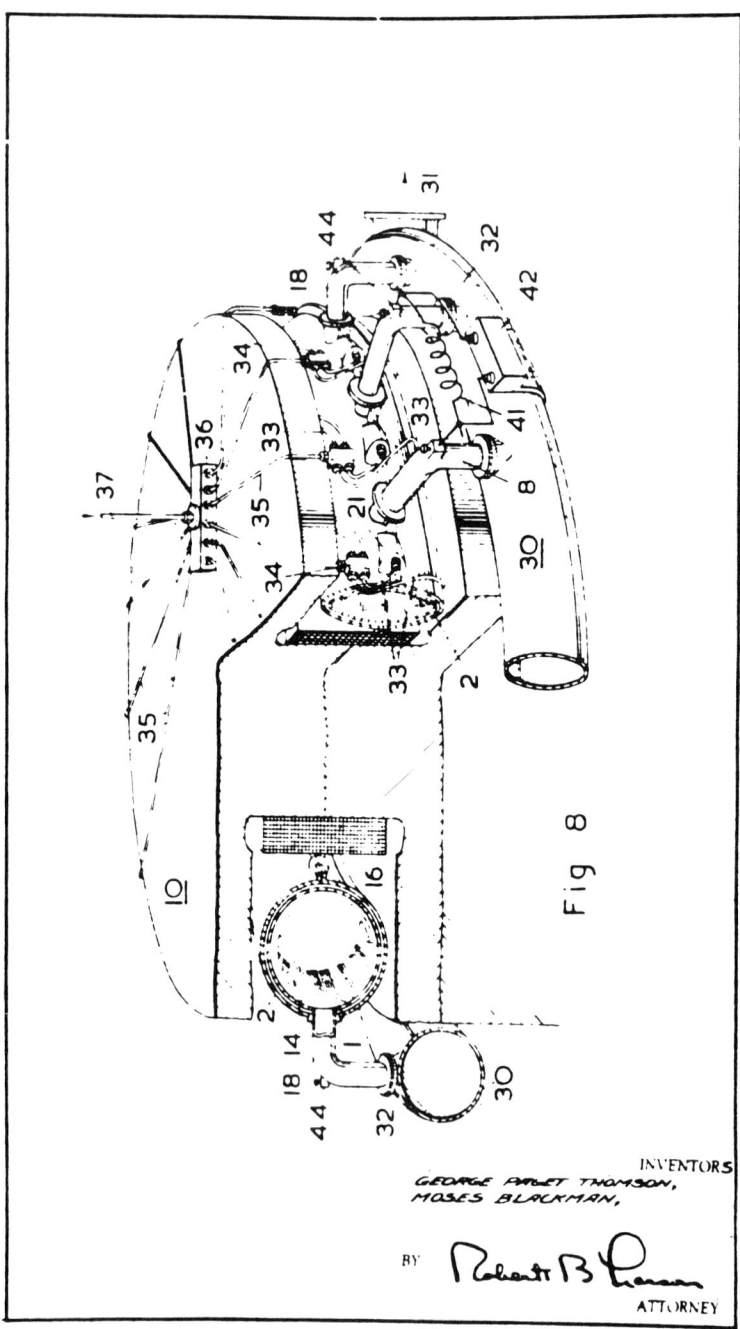

Figure 22.2. *Patent drawing from UK patent application (classified 'Secret') dated 14 January 1952, by Sir George Thomson and Moses Blackman. The application concerns a toroidal fusion device employing travelling radio-frequency waves to induce a pinching current in the plasma.*

Plasma Physics in the Twentieth Century

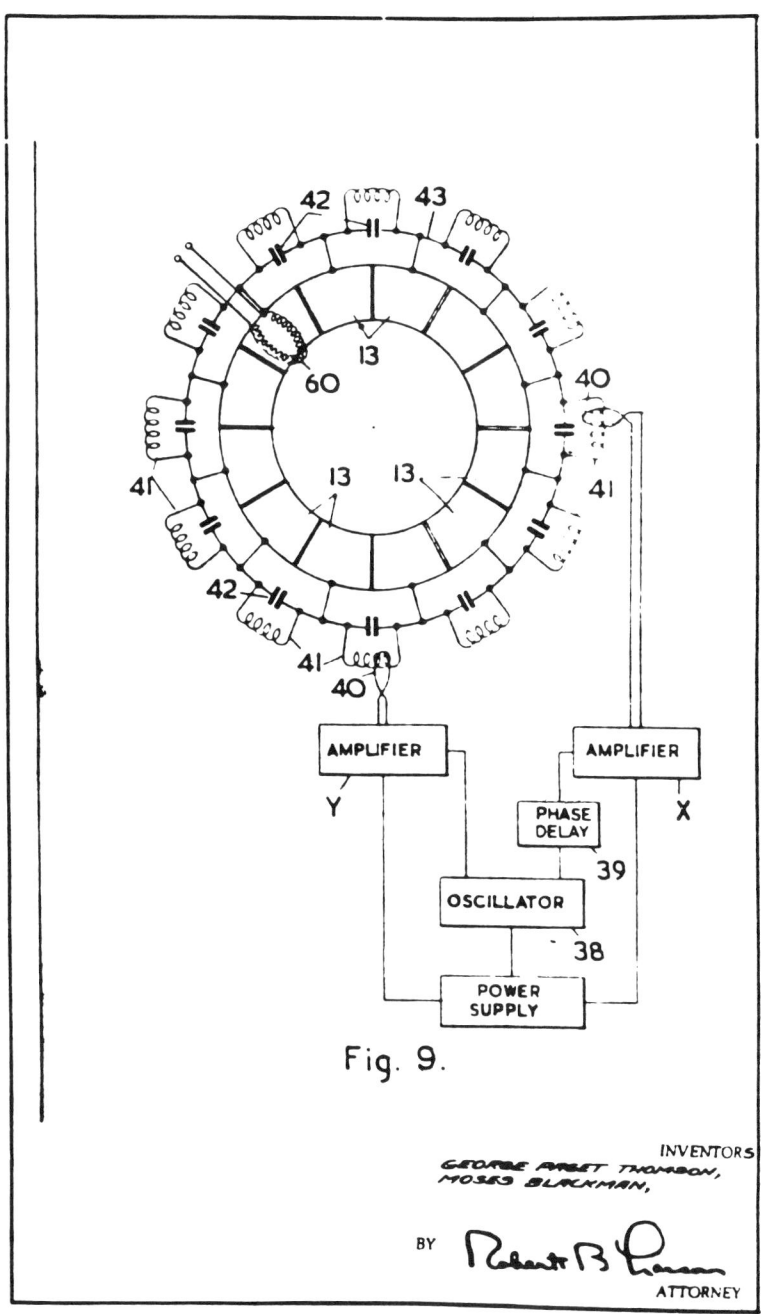

Figure 22.2. *Continued*

In fact the stage had been set for such declassification by an agreement, in 1956, between the United States and the United Kingdom to share their fusion research findings.

Going back to the early 1950s the principal sites for fusion research in the United States were three: Princeton University, Los Alamos and the Livermore Laboratory of the University of California.

In Princeton, Lyman Spitzer proposed to investigate his stellarator approach to magnetic fusion. Aware from theory of the inability of a simple toroidal field (a field produced by winding a coil around a doughnut-shaped chamber) to confine plasma, Spitzer's first proposal was to reshape the doughnut into a 'figure eight'. This ingenious twist was predicted to solve the problem by cancelling the outward drift of the plasma, since 'outward' on one half of the figure-eight was 'inward' on the other. Spitzer's magnetic fusion project was launched within the same buildings as John Wheeler's theoretical group's project which was looking into the physics of the hydrogen bomb. The code name for the group's investigations, Project Matterhorn, was in fact suggested by Spitzer, who had initially planned to participate in its activities. Another of Spitzer's early co-workers on the stellarator project was James Van Allen, who later on became famous for his discovery of the 'Van Allen belts' of plasma particles trapped in the Earth's magnetic field.

In Los Alamos, British physicist James Tuck was looking into a variety of ideas, starting with the 'magnetic pinch', an approach where compression and heating of a plasma to fusion temperatures and densities was to be achieved by passing a huge current through a plasma. Tuck, who had been a part of the English contingent at Los Alamos during the war, had returned to Los Alamos after a stint in England. There he had learned something of the thinking of British physicists, among others, Sir George Thomson, on tackling controlled fusion (figure 22.2). As to the pinch effect, theory showed that the magnetic field caused by this current should cause the plasma column to compress, together with strong heating. This approach, as it turned out, was also one of the ones used in the efforts in the Soviet Union. At Los Alamos, Tuck whimsically named his first pinch machine the Perhapsatron, reflecting no doubt some uncertainty about its likely performance. In its initial form the Perhapsatron was a simple doughnut-shaped vacuum chamber within which a magnetically confined plasma was to be created by inducing a strong current in the plasma by transformer action. No windings around the toroid were believed needed, as the confining magnetic field was to be created entirely by currents within the plasma. One could hardly visualize a simpler or more compact approach to fusion, whence its charm.

At the Livermore Laboratory, preceded by a brief period at Lawrence's Berkeley 'Radiation Laboratory' (Rad Lab), the first small experimental controlled fusion programme was headed by the author.

I had been bitten by the fusion bug by listening to three classified lectures, given at the Rad Lab by Herbert York, who was to become the first director of the new Livermore Laboratory. York had gone to Princeton and Los Alamos to learn of their fusion proposals and had condensed what he had learned, plus some of his own ideas, into these lectures. His intent: to stimulate the interest of members of his audience to join him at the new laboratory, the charter of which was to include research on controlled fusion. Fascinated by the fusion challenge, I wrote a classified memo to York commenting on one of York's ideas, and suggesting some new approaches. Soon after, York invited me to come to Livermore and to set up a small research group to look into what would come to be called the 'mirror machine'. The local code name for the project was 'Arc Research', ostensibly harking back to Lawrence's wartime work on the Calutron at Berkeley. The mirror machine was based on a concept familiar to cosmic-ray physicists: the reflection of spirally charged particles as they move into regions of increasing magnetic field. The magnetic-mirror effect had been discussed early on by the Scandinavian physicist Størmer [25] and later by the Mexican physicist, Vallarta [26], and was described in another early bible of the first fusion researchers, Hannes Alfvén's book *Cosmical Electrodynamics* [27]. My idea was to build a fusion system in the form of a long tube, wrapped with magnet coils, the strength of which would be increased at the ends, forming the magnetic mirrors. Ions and electrons of the plasma, spiralling along the field lines, were then expected to be reflected back at the ends, thus trapping them. The mirror machine was therefore another way of dealing with 'the problem of the ends' (the tendency of plasma to flow freely along field lines, unless this flow is somehow inhibited). This problem was solved in the stellarator and the Perhapsatron by closing the ends back on themselves.

As a hint of things to come, during the early years of the US fusion effort, in the summer of 1955 a remarkable collection of theorists, from the fields of particle physics and from other disciplines, was convened at Los Alamos. The purpose: to advance the field of plasma theory. Out of this summer's activity there came some important theoretical tools for tackling fusion's recondite problems. One of these, the 'CGL' equations (Chew–Goldberger–Low), represented a contribution from these well-known theorists to the analysis of plasma problems.

Looking back on those early years, an 'age of innocence' of the US fusion research effort, it was a period both of intense excitement and of naiveté. It was conducted in an atmosphere of secrecy, excitement, and strong support from the government, here represented by the then chairman of the Atomic Energy Commission, Lewis L Strauss. Strauss had taken on as a personal mission the advancement of fusion research. He hoped to see a practical solution to controlled fusion within his tenure as Chairman. I particularly recall a visit, circa 1953, to the Livermore

(a)

(b)

Figure 22.3. (a) Sir John Cockroft (UK) viewing some of the posters at the exhibition. (b) Group discussion at the exhibition: facing camera—Lyman Spitzer, Princeton University; on his left (with glasses)—R Demirkhanov, USSR; at extreme left of picture—M Gottleib, Princeton University.

Plasma Physics in the Twentieth Century

(c)

(d)

Figure 22.3. *(c) Chairman Lewis L Strauss of US Atomic Energy Commission, viewing the exhibition prepared in response to his request to the US fusion research community. (d) Edward Teller (on right) at the exhibition, facing author R F Post (with briefcase). Other individuals not known.*

Laboratory and to my own experimental area in the corner of a World-War-II-vintage hangar building. Strauss stood in the middle of the area and asked me the direct question: How can I help? How much money can you use? Under Strauss' support, the US fusion effort grew from a sum total of about 1 million dollars for the three fiscal years 1951 to 1953, to 1.7 million dollars in fiscal year 1954, and thence up to an annual budget of about 18 million dollars in 1958, the year of his retirement. Though disappointed that his hoped-for breakthrough had not occurred, Strauss wound up his term by calling for, and seeing to the implementation of, a remarkable showcase of the US fusion effort. This showcase took the form of a very large scientific exhibit, manned by the fusion researchers themselves, at the 1958 'Atoms for Peace' conference in Geneva, Switzerland (figure 22.3).

In his book *Men and Decisions*, Strauss has much to say about his personal interest in fusion. For example, in a section entitled 'Sherwood' he says the following 'One of my first concerns on returning to the Commission was to learn the state of progress in the effort to use the energy of fusion for peaceful purposes ...'. Later on in the section he says 'The Commission unanimously supported my proposal that a more impressive effort should be made ...' and 'The importance of "Sherwood" as the project was called, now conceded to be at least theoretically feasible, can hardly be overstated' and 'I hope to live long enough to see the same natural force which powers the hydrogen bomb tamed for peaceful purposes. A breakthrough could come tomorrow as well as a decade hence. Out of our laboratories may come a discovery as important as the Promethean taming of fire.'.

During Strauss' tenure he had the benefit of some excellent staff work in those that headed the fusion research branch of the Atomic Energy Commission (AEC). One of the first of these was the physicist Amasa S Bishop, whose book *Project Sherwood—The US Program in Controlled Fusion* [28] provides a detailed account of those early years.

22.4.3. Declassification: Geneva 1958 and its consequences
Corresponding to the first full public exposure of previously highly classified fusion research programmes of the Soviet Union, the United Kingdom and the United States, the fusion sections of the 1958 Geneva Conference comprised a remarkable event. In his opening address (read by E I Dobrokhotov) a tone was set by the Soviet physicist Lev Artsimovich that would become a central theme thereafter of fusion research—widespread international cooperation and open exchange of scientific information. He said at the conclusion of his talk 'This problem [fusion] seems to have been created especially for the purpose of developing close cooperation between the scientists and engineers of various countries, working at this problem according to a common

plan, and continuously exchanging their results of their calculations, experiments and engineering developments'.

Mixed together with the enthusiasm and excitement generated with the opening up of fusion research at the Geneva Conference there came also the sobering realization that in none of the fusion plasma experiments that were discussed there did the plasma behave in accordance with the optimistic picture of a quiescent plasma in stable equilibrium with its confining magnetic field. Though they had not been able to talk about it publicly, US and UK scientists had already encountered plasma instabilities in many of their devices, and, in a surprise revelatory talk on 26 April 1956, at the UK Harwell Laboratory, Igor V Kurchatov, after whom the famous Moscow nuclear laboratory was named, specifically alluded to these difficulties. In giving his talk, while flanked by Bulganin and Khrushchev, Kurchatov outlined to his audience the motivations for fusion research and the conditions for fusion, going on to describe Soviet pinch-effect experiments, initiated in 1952, in deuterium plasmas. In these experiments neutrons were observed—supposedly a sign that thermonuclear temperatures had been reached. However, the explanation, he felt, resided in effects caused by dynamic acceleration processes associated with unstable behaviour, rather than from the achievement of high kinetic temperatures. US physicists had also seen pinch-produced neutrons, and were ascribing them to kink-like or sausage-like instabilities of the plasma column, instabilities predicted years earlier by Kruskal and Schwarzschild [29] in the United States.

In the same year, 1956, the author published his first paper on the topic of fusion entitled *Controlled fusion research—an application of the physics of high-temperature plasmas* [30]. In this paper, written under the constraints of classification, I attempted to lay out some of the physics issues that would be encountered in tackling controlled fusion, this for the purpose of interesting the rest of the scientific community in the fusion quest.

Perhaps stimulated to release information prior to Geneva by Kurchatov's Harwell talk, the British made an announcement in 1957 that drew worldwide interest. Based on earlier work (initiated in 1950 and 1951) by A A Ware and P C Thonemann, it described experiments in their large ZETA (Zero-Energy Thermonuclear Apparatus) toroidal pinch experiment, at the time the largest fusion research facility in the western world. In ZETA they too had observed neutrons. Although initially claiming to have reached thermonuclear temperatures by the time of Geneva they realized that spurious acceleration processes, operating on a small fraction of the deuterons in their plasma, had caused the neutrons, and that the actual plasma temperatures and confinement times reached, as limited by radiation losses and turbulence from plasma instabilities, were far below those required for fusion. One year later, in the Geneva

1958 paper on ZETA, UK co-author R S Pease said '... the magnetic field fluctuations, and the current and the voltage transients, suggest that the plasma is by no means stationary'.

At the 1958 Geneva Conference there were some 110 papers presented on theory and experiments on controlled fusion. Scanning them now, one sees: (1) a mixture of optimistic theoretical descriptions of a panoply of confinement geometries, only a few of which have withstood the test of time, (2) experiments, many of which gave evidence of the deviation of plasma behaviour from simple theoretical predictions, and (3) solid theoretical advances in the theory of plasma equilibria in magnetic fields and in basic aspects of the theory of plasma instabilities.

The Geneva Conference also had its share of overly optimistic predictions, a problem that would reappear over the years to follow, perhaps caused by the intense desire to see fusion become a reality, coupled with pressures from the public and from government to 'see some results from all the money spent'. The pronouncements at Geneva 1958 had been preceded by an earlier prediction, by the Indian Physicist H J Bhabha, that startled the attendants of the first, the 1955, Atoms for Peace Conference. In his statement, made at a time when the major powers had disclosed nothing of their efforts towards fusion, Bhabha predicted that power from controlled fusion reactions would be achieved 'in twenty years'. An example of a similar prediction at the Geneva Conference is contained in the conclusion of the keynote talk by UK physicist P C Thonemann. In it he said though '... it is still impossible to answer the question "Can electrical power be generated using the light elements as fuel by themselves?" I believe that this question will be answered in the next decade ...'.

Coincident with the opening of Geneva 1958, the Soviet delegation handed out a remarkable four-volume set of books in Russian. They comprised a collection of technical papers (translated two years later into English and published by Pergamon Press) entitled *Plasma Physics and the Problem of Controlled Thermonuclear Reactions* and edited by senior Soviet physicist, M A Leontovich. The three lead-off papers, written by I E Tamm and A D Sakharov (who was later to become famous in other contexts), can be seen now to be particularly significant. In these papers they define many of the essential features of what would eventually become the 'tokamak' (translated, an abbreviation of 'toroidal chamber, magnetic') approach to fusion, now the dominant one in the international fusion effort. Other papers in the series contained significant articles by other first-rank Soviet fusion-plasma researchers and theorists, for example, Boris B Kadomtsev and Roald Z Sagdeev.

As a result of the scientific disclosures made at Geneva new fusion research efforts appeared in several other countries, notably in Europe and in Japan. These efforts would grow in size over the years until they took a front, even a dominant, role in the pursuit of the fusion goal.

Within two years of Geneva 1958 the Danish Atomic Energy Commission sponsored an 'International Summer Course in Plasma Physics 1960' at Risö, Denmark, where invited lecturers from Europe and the United States discussed advanced theory relevant to fusion plasma research. Edited by Marshall Rosenbluth of the United States, the proceedings of this workshop laid significant parts of the theoretical groundwork for the research to be conducted in the coming years. Ten selected topics were covered, ranging from 'single particle motion' to 'stability' and 'plasma waves and diagnostics'.

The Geneva Conference also contributed to the launching of a continuing series of international conferences on fusion research, sponsored by the International Atomic Energy Agency (IAEA). The first of these, held in September 1961 in Salzburg, Austria, was also full of surprises, achievements and predictions. The growth in size of the fusion effort in a mere three years was reflected in the high quality and the number (115) of papers presented, not counting the 138 submitted, but not presented, papers. Among the surprises and major achievements was the announcement, by Lev Artsimovich, of the results of a watershed experiment by M Ioffe concerning the stabilization of a fundamental instability mode (which I will later discuss) of mirror-type systems, accomplished by reshaping the magnetic field [31]. Among the predictions and pronouncements was a pithy statement by Artsimovich, in his conference summary paper reviewing experimental results, in which he said 'It is now clear to all that our original beliefs that the doors into the desired region of ultra-high temperatures would open smoothly at the first powerful push exerted by the creative energy of physicists, have proved as unfounded as the sinner's hope of entering Paradise without passing through Purgatory. And yet there can be scarcely any doubt that the problem of controlled fusion will eventually be solved. Only we do not know how long we shall have to remain in Purgatory ...'.

An emerging theme, among others, of the Salzburg conference was the increasing sophistication (and proliferation!) of analyses of plasma instabilities. It was clearly recognized by all that the central problem of magnetic confinement fusion was to achieve situations where instabilities and fluctuations were reduced to an acceptably low level, i.e. to a level compatible with achieving confinement satisfying Lawson's criterion.

As a digression concerning the entire field of plasma research, there is a circumstance that delayed the dawn of the modern era of plasma research, particularly high-temperature plasma physics as required in fusion research, beyond the time of emergence of other 'modern' physics disciplines, for example research into the solid state of matter. In the latter discipline Nature provides us with ample material to investigate, and thus experiments can both give important insights and can provide theory with consistent and repeatable checks on its validity. In the case of plasma physics, however, on Earth Nature is not so obliging. Except in the ionosphere and in lightning bolts there is no natural source of plasma available for study. Even in man-made electrical discharges in gases (the earliest plasmas to be studied) as we have earlier mentioned, what went on inside the discharge was much more likely to be dominated by atomic-physics and wall-physics issues than it was by plasma effects. Thus it was that very few of the fascinating complexities associated with the pure plasma state—particularly those involved in its interaction with strong magnetic fields—were discerned in those early laboratory experiments. In the context of fusion plasma research these problems came back to haunt the experimenter, who found that no meaningful results could be obtained unless pristine vacuum conditions and minimal concentrations of higher-Z impurities could be achieved. If this were not the case confinement could be ruined and radiation losses from the plasma (for example, as encountered in the British ZETA experiment) would dominate the energy balance and cool the plasma.

As a matter of fact the serious study in the laboratory of the pure plasma state had to await the development of enabling technologies of the kinds employed (and often, pioneered) in fusion research—strong magnetic fields, high vacuum and intense particle or photon (laser) beam sources. Not only does Nature not provide earthly plasmas for study, but the field of experimental plasma physics was not early on equipped with non-invasive measurement techniques suitable for 'diagnosing' plasma behaviour. These had to be developed in order for the research to proceed. In fact the now-common designation 'plasma diagnostics' applied to the whole spectrum of measuring techniques used in fusion research is a wry admission of the nature of the problems to be addressed. As we have noted, in this field of research exasperating (to the researcher) turbulent and unstable behaviour is more the rule than the exception.

In the development of plasma diagnostic measurements, and in the important atomic-physics issues that accompany all plasma experimentation, even at high temperatures, the contributions of workers in the older field of gas discharges to fusion research should not be ignored. For example, the MIT gaseous electronics group, headed by William P Allis and Sanborn Brown performed many important studies, including the use of microwave techniques to measure electron diffusion rates, years prior to the emergence of magnetic fusion research from classification. Examining the agendas of meetings, such as the 'Gaseous Electronics Conference' in the United States, and the large biennial international meetings 'Ionization Phenomena in Gases' in the period just before fusion was declassified and just after is informative. One finds an appearance of fusion-related topics in the conferences, mixed in with the older topics, such as 'point-to-plane coronal discharges', etc. In the years that followed there has been a considerable cross-fertilization between these two disciplines, with magnetic fusion expertise and technology appearing in such topics as plasma processing of materials, plasma-based switches and many other practical applications of plasma phenomena.

22.5. Magnetic fusion research post-1960: the long march up the $n\tau T$ slope

In the years following 1960 there began a remarkable trek in the quest for success in magnetic plasma confinement. During this trek many of the early approaches fell by the wayside, and one approach, the tokamak, came to dominate the field. Among the early approaches that were abandoned were: the toroidal pinch, of which ZETA was the largest and last example; the 'magnetic cusp' confinement geometry of Grad and Rubin [32], Tuck [33] and others, shot down by its excessive leakage through the cusps; the 'multipole' geometry of Kerst and Ohkawa and their respective co-workers [34], and others, abandoned because of its need for levitated conductors immersed in the plasma; a variety of radio-frequency confinement systems, particularly those investigated in early Soviet work [35]; the 'ASTRON' of Christofilos [36], using a ring of electrons gyrating at relativistic energies to form its confining field; and the linear and toroidal 'theta-pinch' devices, based on the fast magnetic compression of a plasma by a pulsed solenoidal field.

The largest representative of this latter genre was the Los Alamos 'Scyllac' device [37], incorporating one of the largest fast-pulsed condenser banks ever constructed, but it was terminated in 1977. The theta-pinch approach was, in the early days, so popular and so prevalently investigated that it prompted Artsimovich (at the Salzburg Meeting) to say of it, sardonically 'If things go on this way, we shall soon come nearer to the realization of the slogan "Every Hausfrau should have her own theta-pinch".'.

What emerged during this period were four main approaches, and two or three lesser ones, later to be discussed. These four approaches then contended for honours in the stability/confinement competition. They were: the tokamak, the stellarator, the reversed-field pinch and the mirror machine.

In addition to the question of confinement times, another element in the competition was plasma temperature, the *sine qua non* of the fusion objective. Considering the three critical parameters, density, n, confinement time, τ, and temperature, T, what is truly remarkable about the long march that ensued following Geneva 1958 is the many orders-of-magnitude progress in the product of these three that has occurred in the years between then and now. This product, $n\tau T$, so useful as a measure of progress in fusion, has also become something of a two-edged sword. That is, it has been used as a means for the early elimination by policy makers of some approaches, which, given a chance, might have proved in time to be equally as viable as, or possibly even superior to, the then front-runner approaches.

Returning to the early days, remember that fusion researchers were starting with an essentially clean slate in their understanding of the plasma state at that time, and had neither the experimental techniques, the measurement methods, nor the required sophistication of theory at the beginning of the trek. Consider also that the plasma state in its interaction with magnetic fields is far more complex in its behaviour than are fluids and gases in their turbulent behaviour, a subject that is still not adequately understood after more than a century of study. In this light the achievements of fusion plasma physicists in understanding the turbulent behaviour of plasmas and, now, in controlling this behaviour to the point that the fusion goal is in sight, represent a major accomplishment.

In brief, the goals sought in each of the approaches can be simply stated as follows.

(i) To maintain a hot plasma in stable pressure equilibrium with its confining magnetic field, where β, the pressure parameter of the plasma in magnetic fusion parlance, is a substantial fraction of its theoretical limiting value ($\beta = 1.0$).

(ii) To achieve a sufficiently quiescent state of the confined plasma to permit confinement times satisfying the Lawson criterion.

(iii) To achieve sufficiently high efficiencies in the heating of the plasma and in the recovery of fusion energy to yield net fusion power at economically viable levels.

(iv) To accomplish all of the above in a system that is small enough and simple enough to be economically viable.

From the outset, although there were fundamental plasma-physics issues to be addressed, the eventual goal of their research—fusion power—was never forgotten by the fusion researchers. Nor was it by the policy makers, who sometimes used this end to justify less-than-wise means in terminating programmes prematurely.

22.5.1. *The tokamak*

The tokamak is an evolutionary development from the toroidal pinch, a concept which was investigated in early work in the United Kingdom, the Soviet Union and the United States. It (the tokamak) combines the idea from the toroidal pinch of inducing a circulating current in the plasma by transformer action, with the next-simplest—but unstable—concept investigated by Kantrowitz and Jacobs, i.e. wrapping the torus with magnet coils to create a toroidally directed magnetic field. Whereas either magnetic configuration, taken by itself, is unstable, the combined magnetic field, (a strong toroidal field plus a poloidal field, in tokamak jargon) possesses a magneto-hydrodynamic-stable (MHD-stable) equilibrium according to theory. Furthermore, the induced current in the plasma provides a built-in heating mechanism that, in a large enough machine, can carry the plasma temperature to the brink of the thermonuclear regime.

Except for the simpler aspects of its operation, the tokamak has proved, even at the present time, to be intractable to detailed theoretical analysis. Thus it is that the particle and energy confinement properties of the tokamak have had to be determined by analysis of data obtained in, as it turns out, a very large number of tokamaks. These were built on an ever-increasing scale in the hopes of achieving longer and longer confinement times and, with these increased times, increased plasma temperatures. The first tokamaks, those built in the Soviet Union (and one, it turns out, built by B Liley in Australia at about the same time), had major radii of half a metre, plasma radii of 10 cm or so, fields of a few tesla and toroidal currents of order 100 000 A. With these numbers went confinement times of a millisecond or two, plasma densities of order 10^{13} ions cm^{-3}, and plasma ion temperatures of order 100 eV (one million degrees kelvin). Thus $n\tau$, being of order 10^{10}, was four orders of magnitude below the Lawson criterion, with the temperature, T, being another two orders of magnitude below that required for fusion ignition. A long way to go!

Though intractable theoretically, what the tokamak had going for it was the simple fact that its confinement time was found to scale roughly as the square of the plasma radius, as one would expect from a diffusion-limited process. Never mind that the diffusion velocities, as dominated by then unidentified turbulence effects, were orders of magnitude faster than the simple Spitzer classical values. It was clear early on that, based on its empirically determined scaling laws, if the tokamak could be made large enough its confinement times should reach the Lawson criterion values. In a nutshell, therefore, the progress of the tokamak since the earliest days has gone hand-in-hand with the level of its budgetary support, in the sense that progress depended mainly on how large a machine could be financed. To be sure, clever improvements, particularly in the heating and current-drive mechanisms (radio-frequency heating and injection of energetic neutral beams), have occurred along the way, and partial progress has been made in theoretical understanding of the turbulences. Nevertheless, the success of the tokamak, and its present dominance over all other magnetic fusion approaches, has been a direct result of the way in which its confinement time scales with the size of the device.

Between the early 1960s, before the tokamak approach had been accepted internationally, and 1975, when there were more than 20 tokamaks in operation in the world, a paradigm shift in fusion research had occurred. The shift: from a search among many alternatives for a way around the problem of plasma instabilities, to a growing belief, at least on the part of tokamak researchers, that they had the best, perhaps the only, way to the fusion goal. The larger of those tokamaks of 1975 had chamber volumes more than 100 times that of the early tokamaks, coupled with plasma currents in excess of a megampere. Together with these larger dimensions went longer confinement times—many tens of milliseconds, and higher temperatures, approaching 10 million degrees (1 kilovolt kinetic temperature).

As a matter of history, it is worthwhile to point out that much of the political side of the tokamak's success in its early days can be traced directly to the efforts of Lev Artsimovich. After being convinced by their own measurements and by another unusual circumstance (he invited a UK fusion team to bring their advanced laser temperature-measuring apparatus to Moscow to confirm the Soviet measurements), Artsimovich undertook a pilgrimage, in 1969, to the United States to present his case for the tokamak. His discussions convinced US fusion researchers (particularly, those at MIT and Princeton) that they should adopt the Soviet approach. The end result was the initiation of several tokamak projects in the United States, at Oak Ridge, at universities, and, of particular significance, at Princeton. At Princeton, following an initially reluctant reaction to such a major change, they abandoned Spitzer's stellarator approach, in 1970 converting his 'Model C' stellarator into the ST tokamak.

A very informative way to evaluate the emergence of the tokamak as the dominant approach to magnetic fusion research is to examine its history as reported in the series of international fusion meetings sponsored by the IAEA. These meetings, held triennially between 1960 and 1974, and biennially thereafter, are bell-wethers of fusion's progress and trends.

The first meeting at which tokamak experiments were reported was the one in Novosibirsk, Siberia, in 1968. At this meeting there was one paper on tokamak experiments, and three on tokamak theory. All other papers concerned other approaches, such as the pinch, the stellarator, the mirror machine and other, more off-beat, approaches. By 1971 there were eight experimental and 16 theoretical papers on the tokamak, while there were 71 experimental papers on 12 different non-tokamak approaches. In the years to follow the fraction of papers devoted to the tokamak became larger and larger, while the opposite occurred for other approaches, both in the numbers of papers and in the numbers of approaches represented. In 1992, for example, while there were 70 papers on tokamak experiments, there were only 26 on non-tokamak experiments covering only four alternative approaches, representing a complete reversal of the situation that pertained in 1971.

Not only did the ratio of tokamak to non-tokamak papers shift drastically in the two decades following 1971, but also the balance between theory and experiment of tokamak-related papers shifted down to a smaller fraction of theory papers as soon as the 'easy' theory problems (MHD equilibria and stability) were solved and the intractable

Table 22.1. *Experimental and theoretical papers concerning the tokamak given at the IAEA meetings between 1968 and 1992.*

Year	Experimental	Theoretical	Location
1968	1	3	Novosibirsk
1971	8	16	Madison
1974	27	24	Tokyo
1976	30	25	Berchtesgaden
1978	30	30	Innsbruck
1980	28	21	Brussels
1982	37	23	Baltimore
1984	50	20	London
1986	51	22	Kyoto
1988	70	30	Nice
1990	65	27	Washington, DC
1992	70	30	Würzburg

ones were encountered. The historical trends in tokamak-related research are visible in the data of table 22.1 which lists the numbers of experimental papers and theoretical papers concerning the tokamak given at the IAEA meetings between the years 1968 and 1992.

The reports at these same meetings tell a revealing history of the evolution of the tokamak concept itself, from a simple doughnut-shaped plasma of circular cross section, with plasma confinement and heating derived solely from transient currents driven by primitive transformer cores, to tokamaks of non-circular cross section, with megawatt-level neutral beams, microwave and RF power for heating and for driving current in the plasma.

Also gleaned from the IAEA reports are the important contributions to the empirical data base that were made by experiments such as Alcator at MIT. In these experiments it was found (as reported in 1976) that high plasma density enhanced the Lawson $n\tau$ product significantly, presumably because of its damping effects on turbulence. Also a significant gain over the then prevailing trends of energy confinement scaling seen in other tokamak experiments was reported (in 1982) in the ASDEX tokamak at Garching in Germany. By empirically adjusting the heating and the plasma boundary conditions the so-called 'H-mode' ('H' for high-confinement) of operation was found. In H-mode operation the energy confinement time was seen to jump by a factor of two over its normal values, again no doubt as a result of the partial suppression of turbulent modes. Since its first discovery H-mode transitions have been induced in almost all tokamaks where boundary layer control has been effective. Such control is usually established by the use of so-called 'divertors' (localized magnetic field perturbations which divert the escaping plasma into special collectors).

During these same years there also began to appear experiments and theory relating to a very important development: non-inductive current drive. In these approaches tangentially injected neutral beams or specially launched RF waves were employed to augment the normal transformer-action induced currents. These current-drive methods aimed at tackling one of the initial weaknesses of the tokamak for fusion power applications—the fact that in its usual form it is not a steady-state device.

Another development, arising from MHD theory, addressed an additional shortcoming of the tokamak—the rather low value of its beta (relative plasma pressure) parameter. Beta values were limited in early experiments to a per cent or two of the confining magnetic pressure. By evolving away from a simple circular cross section to a generally triangular or a 'D-shaped' plasma cross section, average beta values of 5% or higher were achieved. One can again trace the history of these developments in the tokamak theory papers given in the IAEA meetings, where such papers began to appear in the 1970s.

The evolutionary developments just described, coupled with the larger and larger data base contributed by the dozens of tokamaks that were being built and operated in the 1970s, set the stage for major national and international commitments for very large tokamaks, ones that would take a decade to design and construct. In the United States this commitment was to the TFTR (Tokamak Fusion Test Reactor) at Princeton University (figure 22.4). TFTR was envisaged from the start as a tokamak large enough to produce a significant power yield if operated with 'high octane' fusion fuel, that is, a 50–50 mixture of deuterium and tritium. Similarly, in the European Community, the commitment was made to build an even larger tokamak, the JET (for Joint European Torus) facility, at the Culham Laboratory in the United Kingdom. JET, designed, constructed and operated by a multinational team of scientists and engineers, represents a model of international cooperation in scale and in excellence of its execution (figure 22.5). The Japanese Atomic Energy Agency also undertook the construction of a comparable-size tokamak, JT-60 at Tokai, representing a major commitment by the Japanese to the fusion quest. These commitments, involving, overall, billions of dollars, attest to the importance attached by the major nations to the achievement of fusion power. They also attest to the truly international nature of fusion research, a characteristic which it has consistently exemplified since the 1958 Geneva meeting.

To appreciate the scales to which these major tokamaks have grown, the larger ones of these have chamber dimensions of several metres, corresponding to plasma volumes that are some 3000 times that of the early tokamaks, with correspondingly longer confinement times (approaching a second) and far higher temperatures (hundreds of millions of degrees). These machines, when fine-tuned and heated with megawatts of radio-frequency and neutral beam power, come within

Plasma Physics in the Twentieth Century

Figure 22.4. *Interior view of the TFTR vacuum vessel.*

less than an order of magnitude of fusion 'breakeven', that is, attaining more fusion power release by the plasma than that required to heat it and maintain its temperature. Also, recently, in the TFTR at Princeton, operation with deuterium-tritium (D-T) fuel has yielded up to 10 MW of fusion energy release, a substantial fraction of the input power required to produce and heat the plasma. Similar D-T experiments have also been performed in JET, also with megawatt-level fusion outputs: not yet 'breakeven', but close enough to give increasing confidence that that goal can be reached.

The long march of the tokamak up the fusion trail has been accomplished with no fundamental change in the basic concept since it was first proposed (figure 22.6). To be sure, new techniques have been used and new operating modes have been developed, but the basic magnetic configuration—a strong external toroidal magnetic field upon which is superposed the field from a current in the plasma (many megamperes in the present tokamaks)—has not changed. Nor has there been an essential change in the circumstance that the energy containment time in the tokamak, even in H-mode operation, is degraded, by as yet not understood turbulent processes, by orders of magnitude compared with that expected for a quiescent plasma.

The tokamak, particularly if it is indeed implemented, early in the twenty-first century, in the internationally sponsored ITER (for International Toroidal Experimental Reactor) device, will in all probability demonstrate the scientific feasibility of magnetic fusion. That

Twentieth Century Physics

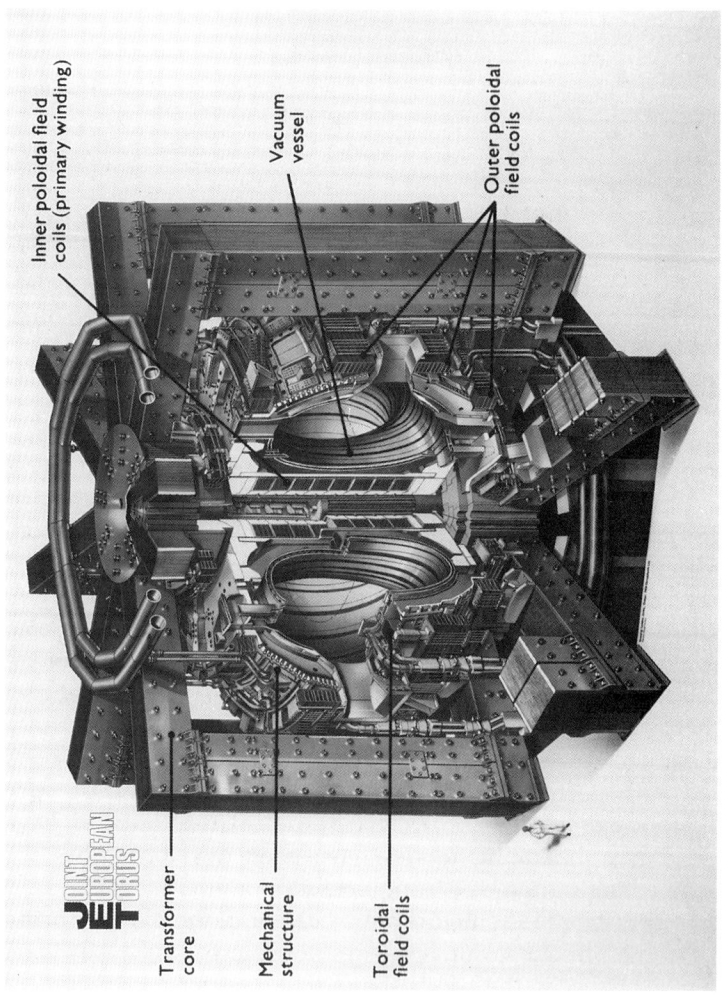

Figure 22.5. Cutaway diagram of the Joint European Torus.

Plasma Physics in the Twentieth Century

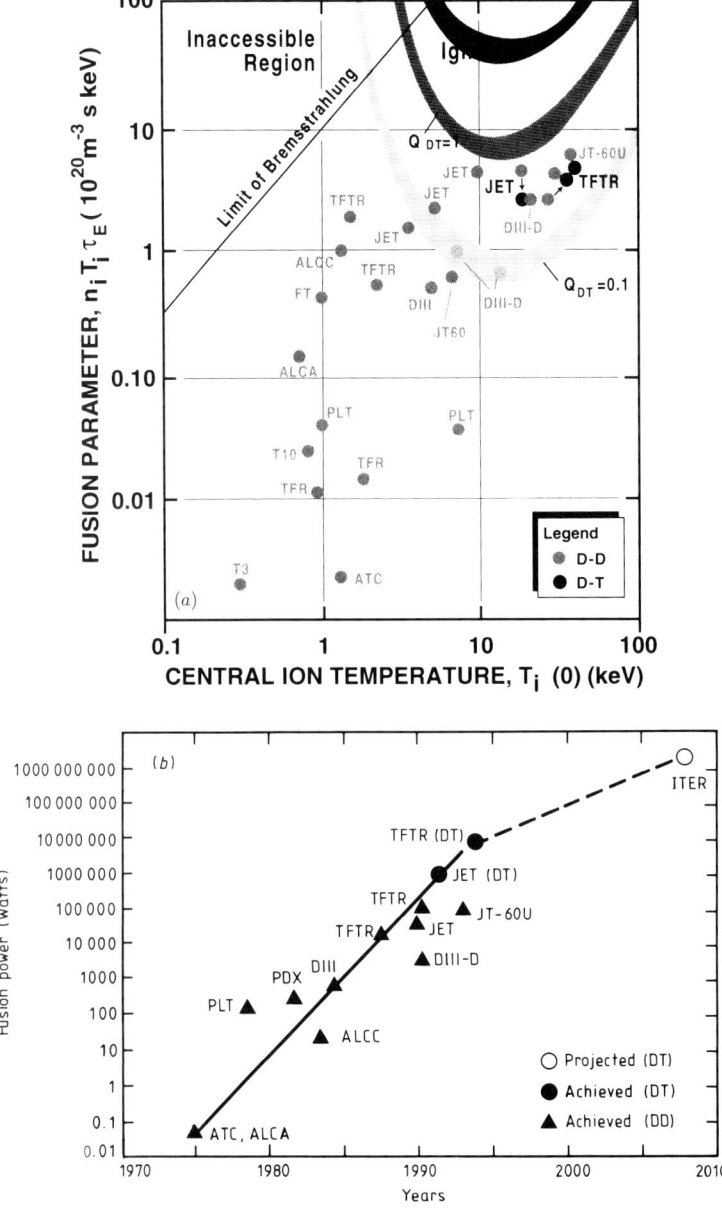

Figure 22.6. *The long march up the fusion trail. (a) Steady-state D-T Lawson diagram, showing central ion temperature against fusion parameter $n\tau T$. The shaded band marked $Q_{DT} = 1$ represents the 'breakeven' zone. (b) Progress in Magnetic Fusion Power. ALCA, Massachusetts Institute of Technology; ATC, Princeton Plasma Physics Laboratory; ALCC, Massachusetts Institute of Technology; PLT, Princeton Large Torus; PDX, Princeton Divertor Experiment; DIII, DIII-D, General Atomics Tokamak Experiments; TFTR, Tokamak Fusion Test Reactor; JET, Joint European Torus; JT-60, Japanese Tokamak; ITER, International Thermonuclear Experimental Reactor.*

is, it will achieve the 'ignited' state of the plasma, i.e. one in which the fusion energy release within the plasma will be sufficient to sustain its temperature. What is far less clear is whether it will point the way to economically feasible fusion power plants. In this sense the tokamak scaling laws, so painstakingly documented over the last three decades, are to some extent also a millstone around its neck. These same laws point to such a large and complicated device in order to produce net power that the economic future of the tokamak is not clear.

22.5.2. The stellarator
The stellarator started as Spitzer's brainchild of a figure-eight chamber around which solenoidal coils were to be wrapped (figure 22.7). However, the stellarator concept has evolved in form far from its original shape, while still retaining an essential feature of Spitzer's insight— the idea of 'rotational transform'. Stated simply, rotational transform corresponds to the idea that in following a magnetic line around the vacuum chamber, upon returning to the cutting plane from which it started the line should not close on itself in a single turn, but should instead go around many times before this occurs. In a simple torus the rotational transform is zero, since each field line is a simple circle, closing on itself in one turn. When this is the case, as it was in the experiment of Kantrowitz and Larson, uncancelled electric fields can build up by particle drifts within the plasma that prevent achieving a stable equilibrium. Rotational transform provides 'communication', i.e. electrical conduction along the field lines, that can neutralize these fields and thus permit an equilibrium state to exist.

In the tokamak, rotational transform is provided by the presence of the internal plasma current. In the stellarator and its progeny it is provided by shaping the external field. At Princeton the evolution of the stellarator concept proceeded from the geometrically intractable figure-eight to a race-track torus around which two types of coils were wound. The first of these was the usual type of winding, producing the main confining field—the toroidal field component. Added to these coils were pairs of helically running coils, like stretched springs, wound around the vacuum chamber. The currents in each pair were oppositely directed, so that the field produced was corkscrew-like. Its effect was to introduce a helical twist to the confining field, i.e. to introduce rotational transform, but now without the necessity (or the desirability) of internal currents in the plasma, the presence of which is essential in the tokamak. The obvious advantage of the stellarator relative to the tokamak is that it can be a steady-state device, since it does not rely on the induction of currents that can decay owing to the finite resistivity of the plasma. The disadvantage is that has a much more complex field, one for which the equilibrium state is much harder to define and to understand than that for the tokamak.

Plasma Physics in the Twentieth Century

Figure 22.7. *Young ladies (attendants at the US exhibit at the 1958 Geneva 'Atoms for Peace' conference, engaged for the exhibit from an interpreter school in Geneva), being briefed on the model 'figure-eight' stellarator (Spitzer's invention).*

Throughout the time that Spitzer's form of the stellarator was under investigation at Princeton its confinement was saddled with a burden, namely, it seemed to follow the Bohm diffusion law, with confinement times not only orders of magnitude less than simple theory, but also varying inversely with the plasma temperature. This behaviour persisted into the large 'Model C' device, at that time the largest fusion research facility in the United States (figure 22.8). At the time of its conversion into a tokamak it was not clear whether Bohm diffusion was endemic to the stellarator concept or whether the particular design was to blame. Work, years later, on modified stellarator geometries has tended to show that well-designed stellarators can be as good plasma confiners as similar-sized tokamaks. However, no stellarator-like devices of a size comparable to the large tokamaks have as yet been built.

Beginning soon after Geneva 1958 another stellarator-type approach was launched in Germany at Garching, near Munich. Here two of the principal theorists involved were Arnold Schlüter and Dieter Pfirsch. In the Garching 'Wendelstein' series of machines an increasingly geometrically sophisticated set of windings was used, ones that combined the production of a strong toroidal field with the simultaneous creation of rotational transform. Other major laboratories in which

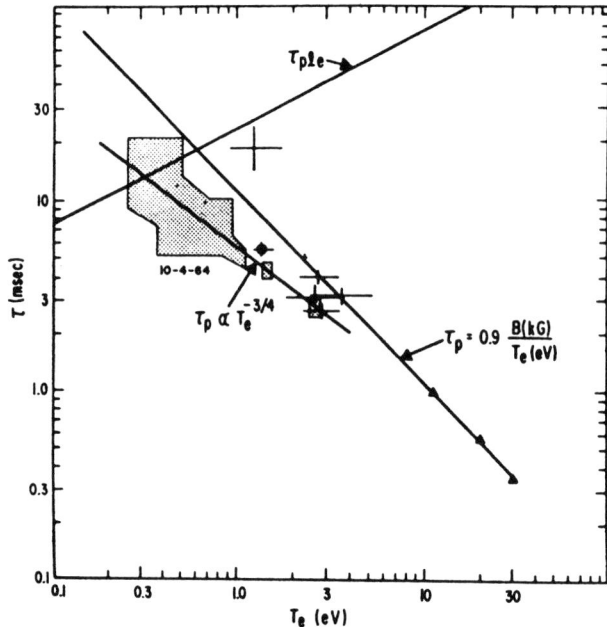

Figure 22.8. *Graph of confinement data from the Princeton Model C stellarator, showing the Bohm-like nature of the confinement, i.e. with a confinement time decreasing with increasing electron temperature (line with data points). The line showing confinement time increasing with temperature represents the prediction of a 'classical' theory, obviously at variance with the data.*

stellarator-type research has been carried out, both before and after the demise of the Princeton Model C, include the Lebedev Physical Institute in Moscow, the Heliotron Center in Kyoto, Japan, and the Oak Ridge National Laboratory.

While the intrinsically steady-state nature of the stellarator concept is a major advantage relative to the tokamak, its toroidal field configuration seems not to exhibit better confinement, and its stable beta (relative plasma pressure) values are apparently even lower than the 5–10% figure that seems to characterize tokamaks.

22.5.3. *The reversed-field pinch*

The reversed-field pinch (RFP) machine is an approach that appears to have been invented as a result of a fortuitous observation. Though it is not easy to exactly pinpoint its origin in time, it seems to have arisen as a result of observations made during the waning days of the British ZETA experiment. In that experiment some field windings had been added, producing a relatively weak toroidal field (relative to the magnetic field from the pinch itself), in hopes of stabilizing the wriggling pinch. What was observed was a surprise: at certain critical values of the plasma current the plasma itself produced an interior toroidal field reversed in direction compared with the external field [38]. At the same time the plasma appeared to become more nearly quiescent, with a consequent improvement in confinement. Thus, instead of overpowering the pinch field to achieve equilibrium and stability, as is done in the tokamak, in the RFP the plasma itself had been coaxed to produce a grossly stable regime, one that might be good enough for fusion purposes. Its major potential advantage was the elimination of the massive toroidal field coils required in the tokamak, with the possibility of building a much smaller and simpler device.

As might be expected, experimentation on the RFP remained for many years a major effort at the UK Culham Laboratory. That Laboratory itself, in its prime, stands as a remarkable example of a government laboratory built initially for the sole purpose of pursuing a broadly based research effort devoted exclusively to magnetic fusion research. The RFP work at Culham followed up on the 1967 hints from ZETA (built at the Harwell Laboratory, but no longer in operation). Under the direction of Hugh Bodin several RFPs were constructed and operated, leading to greatly increased understanding of that approach. It was also at Culham that the most significant theoretical development for the RFP happened—Brian Taylor's theory [39] of the role of magnetic helicity (a measure of the flux linkage between the toroidal and the poloidal field components) in establishing an equilibrium state, i.e. a 'minimum-energy state', in the RFP plasma.

In addition to the RFP work at Culham, significant RFP research was carried out at Los Alamos, where an ambitious programme was laid out, bolstered by economic studies made there by Krakowski [40]. These studies showed that, given a continuation of the favourable scaling law for confinement versus plasma current that was being seen in the experiments, RFP reactors could be built that would have much higher power densities and would be much smaller than comparable tokamak-based systems. Starting with ZT-1 in 1970, the Los Alamos group went on to their ZT-40 and ZT-40M experiments with progressively improved performance, albeit still substantially below that being achieved by tokamaks. However, the empirical scaling law, referred to above, that was found in theirs and others' work showed a steep rise in the $n\tau$ product with current, rising as the $\frac{5}{2}$ power of the current, over a range of an order of magnitude in current, up to 500 kA. Based on these encouraging results Los Alamos proposed and received approval to start construction of a very large RFP, ZT-H. With a toroidal current of 4 MA, ZT-H was predicted to reach confinement times of 75 ms at high plasma density, and at ion temperatures of 400×10^6 K. Construction of this 50 million dollar facility was well along when it was cancelled in 1990, the victim of budget cuts and the major redirection of the US magnetic fusion effort, involving the virtual elimination of all but the tokamak approach.

The other major site for RFP research is the Euratom-CNR facility at Padua in Italy. There Sergio Ortolani led an effort that built and operated the ETA-BETA I devices (1974–77), and the ETA-BETA II devices, followed by the large RFX experiment, now under construction. The Padua experiments shed additional light on the complex processes going on in the RFP. Although Taylor's theory showed the existence of an MHD-stable equilibrium regime, this regime was found, in Padua and in other RFP experiments, not to be a quiescent one. The RFP equilibrium plasma is characterized by a 'burbling' condition, diffusion effects carrying the plasma away from the Taylor state, followed by fluctuation-induced relaxation back to that energetically preferred state. Thus, just as occurs in the tokamak, a macroscopically 'stable' state exists, but it is also microscopically turbulent. Again, as in the tokamak, a purely empirical scaling law points the way to a reactor regime, but in the case of the RFP it is one that projects to a much smaller system than does the tokamak.

In the case of the RFP, as will be seen in the case of mirror systems, next to be discussed, the near total collapse of support for anything but the tokamak that has occurred worldwide, but especially in the United States, means that the potential advantages of magnetic fusion approaches such as the RFP will remain largely unknown until such time that fusion research policies return to a broader-based strategy.

22.5.4. *The magnetic mirror*

The magnetic-mirror approach differs in a fundamental way from the three approaches that have just been described. Each of the former is a variant on a basic toroidal configuration field, differing only in the mix between toroidal and poloidal field components. Each of these approaches is topologically 'endless' in that the only way for a confined gyrating particle to escape confinement is by diffusing across the field lines of the confining field. Thus, provided turbulence levels are not too high, confinement adequate for fusion purposes can always be attained in principle 'simply' by making the device large enough, an approach that might be thought of as following that of the stars themselves. Stars are successful in generating fusion power in their inner regions simply because they are large enough to keep the escape of heat from their interior down to a level that allows their inner temperature, arising from gravitational compression, to rise to fusion levels.

In the magnetic-mirror approach, sketched earlier, the topology is that of an open-ended tube, filled with field lines that emerge only from the ends of the tube. In adopting this topology there are seemingly only two avenues to achieving net fusion power: one either makes the tube so long that the flight time of the plasma along the field lines is longer than that required by the Lawson criterion, or else one finds means to plug the ends so that the ions and electrons bounce back and forth for a long enough time for fusion to occur. The former of these two approaches was in fact considered seriously at Los Alamos, and elsewhere, as a possible fusion power plant version of the theta pinch. The many-kilometre length that was projected to be required was, however, a daunting prospect, so that the idea was dropped.

The mirror idea seems to have arisen at about the same time in the classified programmes in the Soviet Union, the United Kingdom and the United States. In the Soviet Union mirror pioneers included G I Budker and I N Golovin. In the United Kingdom it was D Sweetman who was involved in early experiments at the Aldermaston nuclear-weapons laboratory. The author came to adopt it as his own through having examined another suggestion, made by Herbert York, that the radiation pressure from intense microwave fields should be used to plug the ends of a simple solenoidal field. Although this plugging idea did not appear to be technologically feasible, I suggested to York that magnetic mirrors be used to augment the confinement, a suggestion that led to my joining the new Livermore Laboratory in 1952. In a simple preparatory experiment performed in the Berkeley Laboratory with co-worker J Steller, I had convinced myself that mirrors would indeed improve the confinement of a plasma (produced by a pulsed radio-frequency discharge). This experiment was also, perhaps, one of the first uses of a microwave beam to measure the density of a magnetically confined plasma. The microwave beam idea was then carried forward by Wharton and co-workers [41] to develop the microwave interferometer as a major diagnostic tool in fusion plasma research.

During the classified years at Livermore there followed a series of small experiments, with names such as 'Albedo' (investigating the RF enhancement of mirror reflections) and 'Cucumber' a static mirror field with a mirror ratio of 100 or more, aimed at finding the scaling of mirror losses with the mirror ratio. (The 'mirror ratio' is the ratio of the strength of the magnetic field at the mirror to that at the point mid-way between the mirrors.) The key technical problem in all of our experiments at that time was the creation of a trapped plasma by injection into an evacuated chamber (evacuated to minimize the disastrous hot-ion losses that would arise from charge-exchange processes). What we used was a then-classified technique, the ejection of hot plasma from a deuterated titanium surface by an electrical discharge. Using this technique, in early (unpublished) experiments, Coensgen was able to trap and heat a deuterium plasma (by magnetic compression in a pulsed mirror field). At the peak of the compression he observed the production of neutrons, 'real' ones in the sense that they were sustained in duration and were not accompanied by violent instabilities such as had been encountered in neutron-producing pinch experiments. Years later, theta-pinch experiments were carried out with similar results, providing *ex post facto* validation of Coensgen's earlier results.

By 1954 the mirror approach at Livermore had achieved a sufficient level of scientific credibility to warrant a preliminary assessment of its potential as a fusion power system. During that year the author prepared a series of lectures, issued as a classified report entitled unimaginatively *Sixteen lectures on controlled thermonuclear reactions* [42]. In it I discussed how a long cylindrical mirror system might be filled with hot ions and electrons and how energy might be recovered (by direct conversion through magnetic decompression). A primitive attempt was made to deal with the possibility of MHD instabilities by shaping the magnetic field so as to achieve a near-magnetic-well configuration, but I missed the boat on the actual field configurations that would later be seen to be optimal.

In the report a simple estimate was made of what was from the onset seen as the Sword of Damocles hanging over the mirror approach—mirror losses. Contrary to the toroidal devices, particles are confined in simple mirror systems only if their pitch angles are not too nearly directed along the field lines, since the mirror effect itself depends on the spiralling motion of the particle as it approaches a mirror to produce reflection. Thus if ion–ion scattering, or any other process, such as high-frequency instabilities, deflects the particle so that it aims too nearly along the field lines, then it will be lost upon its next encounter with a mirror. Mirrors of the type then being studied could only rely on near-classical containment and very high temperatures (reduces scattering rates) to achieve a net fusion power. To a crude approximation, one effective collision between a given ion and its companion ions is sufficient to deflect it by 90 degrees, thus into the loss cone. Only at ion kinetic temperatures approaching 100 keV ($\sim 1 \times 10^9$ K) would the energy released by fusion be expected to exceed that required to heat the ions to that temperature. The first rigorous calculation of mirror confinement times was contained in a classified Livermore report, in 1955, by D Judd, W McDonald and M N Rosenbluth [43]. This report and ones that followed in succeeding years [44] demonstrated the marginality of the power balance in simple mirror systems. One of the results obtained showed a perhaps counter-intuitive result: the confinement time in simple mirror systems scales only logarithmically with the mirror ratio, rather than linearly as a simple picture might imply. Furthermore, the confinement time was seen to be essentially independent of the size of the machine, i.e. of the distance between the mirrors, in contrast to the toroidal systems, where confinement time scales up with the square of the radius of the plasma column.

These circumstances, understood early on in the Livermore programme, led to a concentration on achieving quiescent plasmas, and, in time, on achieving high efficiency in the heating and the energy recovery cycles of proposed mirror systems. They also lead to an attitude of 'first things first', i.e. establishing whether stable confinement was possible, following which power-balance issues might be addressed.

In the years following Geneva 1958, the generally encouraging results and the basic simplicity of the mirror concept led to the initiation of mirror research programmes in many countries, France and Japan to name two. Particularly after Ioffe's landmark experiment, researchers were pleased to discover that, contrary to the experience in the stellarator, for example, it was possible to achieve apparently stable plasmas, ones with beta values (plasma pressure relative to the confining magnetic pressure) that were very high compared to those in, say, the tokamak. That magnetic-well-type mirrors could hold in high plasma pressures was in agreement with theory, such as that of R J Hastie and J B Taylor [45] in 1965 who showed that plasma pressures up to the limiting value of $\beta = 1.0$ could still be expected to be stable. This was a startling result indeed, but it was one that would be confirmed experimentally at Livermore, ten years later.

At Livermore the mirror programme evolved into two parallel approaches to the achievement of high ion temperatures at fusion-relevant plasma densities. The first of these, pursued by F Coensgen, represented an extension of the plasma gun idea, coupled with magnetic compression for heating and densification. Here the technological issues were those of pulsed fields and high-speed pumping of neutral gas (to minimize charge-exchange losses). The second approach, led by C Damm, employed the technique of neutral beam injection. The neutral-beam concept, suggested in the early 1950s by W Brobeck and S Colgate, involved the creation of directed beams of energetic hydrogen atoms, created by neutralization of a hydrogen ion beam in a gas cell external to the apparatus. This technique, pioneered at Livermore, would in time be adopted as a preferred heating method in tokamaks around the world. At Livermore it would later become the key to one of the most important results achieved in the mirror programme. Quite independently, neutral beam injection was also being pursued by Sweetman at the Culham Laboratory in the United Kingdom, with results that complemented those being obtained at Livermore.

Early on, that is just before Geneva 1958, there emerged on the scene another group using energetic particle injection to fill a mirror system. Based on an idea attributed to J Luce, a beam of energetic singly ionized hydrogen molecules were injected, at the mid-plane, in orbits that intersected the axis of the machine. At this point they encountered a carbon-arc plasma, which stripped the extra electron and thus produced two hydrogen nuclei, but now with half the radius of curvature in the field so that they could no longer return to the injector snout but were trapped. Through a sequence of increasingly larger machines, starting with the DCX, Oak Ridge was able to build up a high plasma ion temperature, with hundred-keV ions held in for fractions of a second but with densities limited to four orders of magnitude below those believed necessary for fusion power. After minimizing or eliminating obvious sources of losses, such as scattering off the ionizing arc, they still were limited to confinement times of tens of milliseconds, a tiny fraction of the times calculated for mirror end losses by scattering. What they had run into were what were then called 'micro-instabilities', that is, wave–particle instabilities stimulated by the highly ordered (that is, far from Maxwellian) nature of the distribution functions of the trapped ions. This problem would eventually spell the end of the Oak Ridge DCX series of experiments.

In the Soviet Union mirror efforts were also being carried out, with approaches similar to some of those being pursued in the United States. Starting with Budker's original ideas, prior to the 1958 Geneva meeting a very large mirror machine, OGRA, was constructed at the Kurchatov laboratory, under the direction of I N Golovin. OGRA used high-energy ion injection in a manner similar to that employed in the Oak Ridge experiments. Although, like DCX, OGRA was in time decommissioned, Golovin's group has continued to investigate beam-fed mirror experiments for many years, contributing to the knowledge base for mirrors. One of their investigations included one of the earliest known applications of feed-back techniques to the stabilization of plasma instabilities.

At Livermore, in the ALICE neutral-beam injection experiments, the build-up problem was very different from that in the DCXs. Here the key to build-up was deemed to be to employ the self-ionization of excited atom components of the injected beam to trap a seed plasma, upon which build-up could occur if the vacuum were good enough. Thus the thrust was in the direction of achieving better and better vacua, to the point that pressures of order 10^{-10} Torr were achieved, accompanied by build-up of the plasma to densities comparable to those achieved in the DCX. Again, it was found that micro-instabilities stepped in to limit the density.

By 1965 it was recognized that it would be necessary to hit the micro-instability issue head-on if further progress was to be made in the neutral injection approach to steady-state mirror confinement. In response, the author, together with Marshall Rosenbluth, in 1965–66, undertook a systematic study of the so-called 'loss-cone modes' of mirror-confined plasmas [46, 47]. What we found was that there were indeed ways to suppress these modes, based on achieving sufficiently well-randomized trapped-ion distributions and scaling up of the plasma diameter. While the diameters involved were somewhat daunting, corresponding to many ion orbit diameters, they nevertheless provided a path towards better confinement for the mirror programme, one that could be verified by experiment. Based on these results the emphasis shifted in the ALICE (now Baseball) neutral-beam experiments towards achieving better randomized plasmas, through lowering the energy of the injected beams.

A year following the Rosenbluth–Post theoretical analyses, the author conceived of another method to stabilize the most virulent mirror microinstability [48]. This method, involving the introduction of a stream of 'warm plasma' along the field lines, was not tested until nearly ten years later, when the effect was inferred from results in Ioffe's mirror machine and when its efficacy was directly demonstrated by Coensgen in his 2X-IIb mirror experiment, about which more later.

In the intervening years a new situation arose in the US fusion programme. This was the flowering of 'fusion power plant studies', i.e. attempts to predict the size, cost and performance of fusion power plants based on the concepts then under study, in particular the Los Alamos theta-pinch approach, the tokamak and the mirror. The first to fall under the impact of these studies was the Los Alamos Syllac programme, where it appeared that both the stability and confinement problems and the technological difficulty of this approach were such as to preclude its becoming a practical fusion power system. Mirrors did not escape unscathed either, but the electrostatic direct converter, proposed by the author at a Culham fusion conference in 1969 [49] seemed to offer a way out of the mirror's marginal power balance.

Meanwhile, beginning with results presented at the 1968 IAEA fusion conference at Novosibirsk, tokamak research began to gain ground, eventually to the point of almost eliminating every other approach to magnetic fusion. In the United States, by the early 1970s the tokamak had had many successes and was on the way to dominance. But mirrors had not stood still: exploiting the idea of neutral beam injection to the hilt, Coensgen had added 12 husky neutral-beam injectors, each delivering a megawatt of 20 kV neutral beam power, to his new 2XII-B (2X for 'two times' the size of the previous experiment, II for the second experimental apparatus in the sequence, and 'B' for 2XII with beams) (figure 22.9). When even this beam power was not seen to be sufficient to overcome losses from micro-instabilities he soon after added 'warm plasma stabilization', with spectacularly successful results [50]. Not only did the plasma build up to high densities at very high ion temperatures (200×10^6 K), but the confinement times agreed almost exactly with those one would expect theoretically from a completely quiescent mirror-confined plasma. On top of this within a year, that is by 1976, the beta value had been driven off the end of the scale, to a value of 200%, while still maintaining MHD stability! This result, which astounded the fusion community, was the high point in the programme to investigate the simple mirror. It provided the incentive for Washington to take the mirror approach to fusion much more seriously, but at the same time it did not remove the Sword of Damocles—marginal power balance—from dangling, for the moment just out of sight, over the mirror approach.

During the next decade, that is, between 1976 and 1986 the fortunes of

Marshall N Rosenbluth

(American, b 1927)

Marshall N Rosenbluth has a special place in the field of plasma physics, particularly as it relates to the quest for fusion power. Born in Albany, New York, on February 5 1927 he did his undergraduate study at Harvard University, and, in a hint of things to come, in three years obtained his PhD from Columbia University at age 22. Thus it was that in 1950, following a brief period at Stanford University, he joined the Los Alamos Scientific Laboratory as a theorist in their newly founded controlled fusion research program. His entry into the field thus corresponds almost exactly in time to the birth of the modern era of plasma research. Beginning with that, then classified, effort Marshall Rosenbluth began his remarkable career in which his work in plasma theory has impacted almost every area of this field. These contributions were recognized at the time he received the Enrico Fermi Award in 1986. It is appropriate to quote from the citation that accompanied this award, which said, in part

> Marshall Rosenbluth's theoretical contributions have been central to the development of controlled thermonuclear fusion. He is widely considered to be the leading plasma theorist of this generation, including work of central significance for fusion research and development.

CONTINUED

the US mirror programme were to rise and fall. In 1976, a time when the budget level of the magnetic fusion programme was rising, Livermore proposed, and received approval for, the construction of a very large, single-cell mirror machine. It was to be called MFTF (for Mirror Fusion Test Facility), comprised of a 40 ton superconducting magnet coil in the special shape called a 'Yin-Yang coil', after the interlocking nature of its two bent mirror coils [51]. It was to be a machine big enough to hold in multi-hundred kilovolt ions, thus to drive the $n\tau$ product up past 10^{12}.

> **CONTINUED**
>
> The many contributions made by Marshall Rosenbluth were carried out at the various institutions at which he served as a leading theorist. Following Los Alamos in 1956 he moved to the John Jay Hopkins Laboratory of General Atomics, where he led a singularly successful group effort in plasma theory, while at the same time serving as a professor in the department of physics of the University of California at San Diego. His talent was further recognized when he was called to the Institute for Advanced Study in Princeton as a professor in the school of natural sciences, while also having a joint appointment as senior research physicist at the Plasma Physics Laboratory of Princeton University. In 1980 he was appointed as director of the newly formed Institute for Fusion Studies at the University of Texas. In 1987 he returned to the University of California at San Diego, again with a joint association with General Atomics.
>
> Though the name of Marshall Rosenbluth is most closely associated with fusion plasma research, Rosenbluth has made important contributions in other areas, for example in the theory of the scattering of relativistic electrons. He has also served on numerous blue ribbon panels and science advisory groups, while never flagging in his contributions to the scientific literature. To all of these endeavors he brings a remarkable combination of keen physical insight and analytical skills, together with an exceptional ability to work with, and inspire, his myriad co-workers.

This represented a gain of two orders of magnitude over the previous record for mirrors, and thus was a major step towards the satisfaction of the Lawson criterion. This was indeed an endorsement of the mirror concept, but not an unqualified one. Washington was at the same time asking for a clearer route to net fusion power, called 'Q-enhancement' of the mirror (where Q had come to be used to designate the fusion power gain), with $Q > 1$ being the Rubicon for mirrors. Thus it was that, in the same year as the IAEA fusion meeting in Berchtesgaden, Germany, the tandem-mirror concept was invented by T K Fowler and G Logan [52] in the United States and, independently, by Dimov and co-workers [53] in Novosibirsk. Fowler and Logan came to the meeting prepared to reveal their new idea only to discover that Dimov was also going to discuss the identical concept. Dimov was a member of a mirror team in Novosibirsk, founded by one of the pioneers of the mirror concept, Budker. Budker's heritage lived on in the work of top-flight scientists

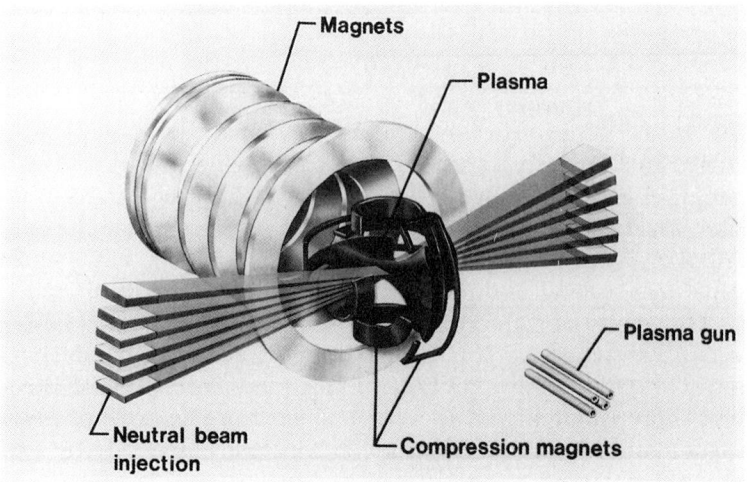

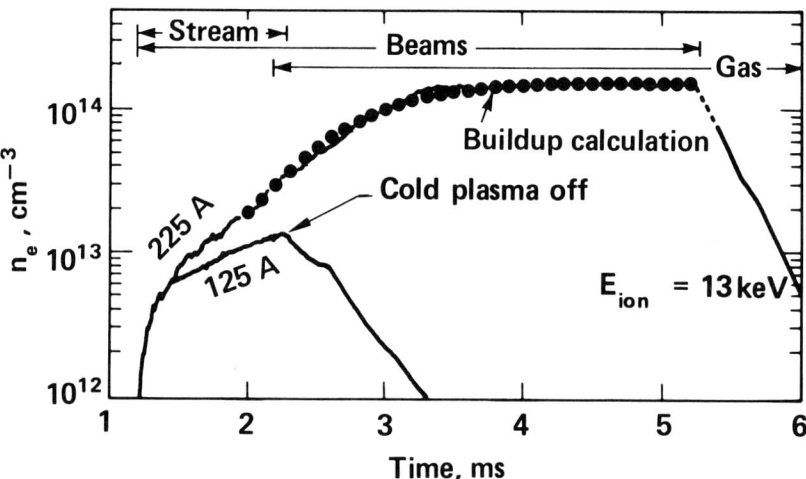

Figure 22.9. *Upper part: schematic drawing of the 2XIIB experiment at Livermore. Lower part: plot of plasma build-up, with and without 'warm plasma' stabilization of high-frequency plasma instability. With stabilization, plasma builds to high density and high beta (140%) in close agreement with the theoretical build-up curve based on solely classical trapping and scattering theory.*

such as Dmitri Ryutov, who has continued to make many theoretical contributions to mirror physics over the years.

The tandem-mirror idea of Dimov, Fowler and Logan built on an earlier idea of Kelley at Oak Ridge. The basic idea was to build a mirror system with a large-volume central cell, flanked on each end by smaller-volume mirror cells. In the tandem-mirror idea the end cells were to be filled with plasma at higher density than that in the central cell. Mirror physics, by then well understood, dictated that the end cells would exist at higher positive potential than the central cell. In this way the end cells comprised electrostatic potential barriers for ions leaking out of the central cells, plugging up the usual mirror leak. The theory of tandem-mirror end losses, as it was first written down by the Russian physicist V P Pastukhov [54], showed an exponential increase in confinement time as compared to that of simple mirrors, thus moving the mirror approach into the 'high-Q' league.

Within an unusually short time, that is about a year, a tandem-mirror experiment, called TMX (for Tandem Mirror Experiment) was proposed by Livermore and received approval by Washington in 1977. At about the same time construction funding was received for MFTF, involving the construction of its huge superconducting coils, the largest yet attempted in the fusion programme. These approvals came against the backdrop of the formation of a blue-ribbon panel, chaired by John S Foster Jr, formed to review the entire fusion programme. Out of concerns that the tokamak might prove too large and complex to become a practical reactor, given the then uncertainty of its scaling laws, they recommended continued support of alternative approaches the major one being the mirror approach. Their recommendations, though they did not result in the hoped-for increase in the fusion budget, helped keep the mirror programme in the running.

By now magnetic fusion research policy was increasingly shaped by the perceived fusion power capabilities and projected economics of the approaches, and less and less by interesting science. Under this pressure ideas for improving the original tandem-mirror concept were being sought, even prior to the advent of the TMX experiment, aimed at improving its perceived economics. Responding to this pressure Logan, in concert with Dave Baldwin, came up with the idea of the 'thermal barrier', a means for enhancing the height of the potential peaks in the end cells while at the same time raising the central plasma density relative to that in the end cells. This idea, seen to improve the economic prospects for tandem mirrors, was eventually incorporated into tandem-mirror experiments, both at Livermore and elsewhere.

Meanwhile the construction of TMX, taking place in the site formerly occupied by Baseball II, was completed in an unusually short time of 18 months. Completed in 1978 it underwent shakedown for the canonical 18 months. Then, in July 1979, the experiments showed unequivocally that the central plasma was being confined for more than 25 ms, ten times as long as it would have been confined without the end plugs. That the plugs really were doing their job was proved by the experimental leader, Tom Simonen, by turning off one plug: the plasma promptly leaked out that end.

At this point in time, the autumn of 1979, the successes in TMX and the increased confidence in the the thermal barrier idea gained from computer simulations led Fowler to propose a bold two-step programme

Twentieth Century Physics

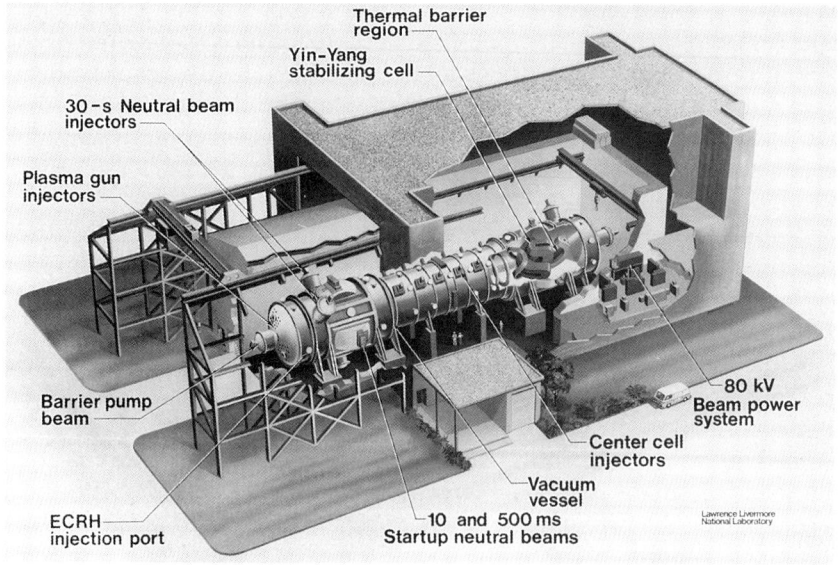

Figure 22.10. *Artist's drawing of the large MFTF-B tandem-mirror experiment constructed at Livermore. Following completion of construction and initial operation the facility was 'mothballed' as a result of budget cutting. Note van (lower right) for scale.*

to propel the mirror idea into the big league, matching in size the largest tokamaks then being considered. He proposed converting the MFTF facility into a 400 million dollar MFTF-B, a large tandem-mirror system with thermal barriers (figure 22.10). This task would be accomplished by building a duplicate of the original MFTF's huge Yin-Yang magnet, then adding a central cell with very large superconducting solenoidal coils. To provide the needed experimental confirmation of the thermal barrier idea, to be accomplished during the construction of MFTF-B, he proposed scrapping TMX and building in the same pit an upgraded tandem mirror, TMX-U.

By 1981 a successful test of the first 150 ton superconducting Yin-Yang magnets of MFTF-B (figure 22.11) was carried out, TMX-U had been constructed and was soon to be operated. By 1982 it was in operation and other tandem-mirror systems, at the University of Wisconsin, MIT and Tsukuba University, Japan were in various stages of construction. By 1983 there were encouraging results from TMX-U. However, soon after troubles began to show up, both on the experimental and the financial front, that would in time prove fatal to the US mirror effort. In 1984 the US magnetic fusion budget reached its peak of about 480 million dollars, after which it declined, by 50 million dollars per year, until it had dropped to 300 million dollars. The commitment to complete

Plasma Physics in the Twentieth Century

Figure 22.11. *The Yin-Yang superconducting mirror-magnet coil of the MFTF-B at the Lawrence Livermore National Laboratory.*

the construction of MFTF-B was honoured, however, so that in 1986 a dedication ceremony of the facility was held, following a successful demonstration of its remarkable engineering aspects—superconducting magnetic fields, cryogenic vacuum pumping system and the multi-megawatt neutral beams and their power supplies. The next day, the project was cancelled by the Department of Energy. The US mirror programme, together with all other so-called 'alternate' approaches were eliminated in response to the draconian reductions in the fusion budget that the Administration and Congress had dictated. Only the tokamak programme remained, apart from a few residual university-level investigations.

Worldwide, a similar attrition had occurred. Apart from the continued effort at Novosibirsk and the very substantial tandem-mirror facility, Gamma-10, at Tsukuba in Japan, mirror research essentially disappeared from the scene. As of this writing, apart from small-scale work, Gamma-10 is the only remaining large-scale mirror experiment in the world. It continues to turn out good results [55], which in another funding climate, would be expected to stimulate others to pursue the mirror concept.

Among the US victims of the budget axe in the post-1986 period was a large RFP experiment at Los Alamos, a just-completed reversed-field theta-pinch experiment near Seattle and numerous smaller experiments on such concepts as the 'spheromak' (a tokamak-like system where the confining field is self-generated by currents in the plasma), investigated at Los Alamos, Princeton and Livermore. Particularly the latter two concepts, coming under the rubric of 'compact toruses', offered the hope of smaller, more compact fusion power systems than the tokamak.

'What goes around, comes around.' As of the writing of this chapter there is beginning to be a reversal of the US budget policies of the 1980s. With a new Administration, and a rethinking of the needs of fusion research there is beginning to be an acceptance, albeit at a small scale, of the need not to put all of ones eggs in the same basket in a research and development effort as difficult and profoundly important for the future as fusion. Thus some of the previously abandoned concepts are being looked at again, at a theoretical or small-experimental level, together with new ones that have appeared as a result of technological advances. It is the author's opinion that this is the right direction for fusion research to go: to have a present front-runner, i.e. the tokamak, but to have smaller, but healthy, components looking at approaches that are distinctly different from the tokamak, so that better routes may be uncovered. Another of humankind's major research efforts—the search for a cure for cancer—is carried out in just this way, and for good reason.

22.6. The growth in importance of theory and computer simulation
Before undertaking a discussion of the history of plasma theory it is important to put such a discussion in the context of our present picture of plasmas and how we think about them. In common with other dualities in physics, for example the dual nature of photons as 'particles' or electromagnetic wave trains, plasmas can be thought of in two ostensibly different ways, each one having its merits, but each not being adequately conceptually described by that representation. One way to think of the plasma state is to think of it as a structureless electrically conducting fluid. The 'magneto-hydrodynamic' (MHD) representation of the plasma state is in this category, and for some plasma phenomena such a representation is a valid one, just as the representation of a gas as a structureless fluid is frequently a useful one. On the other hand, there is no such thing in Nature as a structureless fluid, either non-plasma or plasma. What a plasma really is is a collection of a very large number of individual charged particles, all interacting with each other through mutual Coulomb forces and through the electrical currents associated with their motion, while at the same time interacting with (and thereby modifying) any electromagnetic fields of external origin. It is this property of collective long-range interactions, of the plasma with itself and of the plasma with its electromagnetic environment, that gives rise to the great complexity of its behaviour. This behaviour, far more

complicated than that of ordinary fluids (complex though turbulence-related fluid phenomena are), is what makes the plasma state so difficult to understand in detail. In fact, in the early days of modern plasma research someone, in frustration, put forth an alternate designation for plasma as 'that which cannot be understood'.

Given the growing awareness of the complexity of plasma behaviour that occurred in the early days of fusion research it is not hard to understand the increasing role that analysis and computation has played in the research. Both in fusion and in space-plasma research there has been an exponential growth in theoretical and computational analysis of the plasma state. At first, theoretical models of the plasma were based on extensions of the equations of fluid dynamics by the addition of Maxwell's equations, i.e. early forms of the MHD equations. However, this picture represented a far too simplistic picture of the plasma.

Thus, while both pre-mid-century and post-mid-century plasma physics built on a theoretical base that was largely derived from the classical disciplines of electromagnetic theory and statistical gas-kinetic theory, in the modern era of plasma research the sheer complexity of the phenomena involved required evolutionary changes in the application of these disciplines. Not only were new analytical approaches devised, but also the computer, as a tool in the simulation of complex physical phenomena, was early on brought into the research in major ways. This approach, that is, the use of computation as a research tool, has been particularly evident in fusion research. In its first phases, when this research was classified, many of the researchers involved had participated in the development of nuclear weapons. In weapons research the use of computers to simulate the processes going on in those devices was well established. It was therefore natural that computational methods would from the onset be employed in fusion research. Particularly in the United States and in the European Community and Japan, the use of computers as an aid to understanding plasma behaviour has continued to grow to the point that it has become a dominant factor in the theoretical component of plasma research.

The heavy reliance of plasma researchers on computational methods can be seen as a reflection of the complexity of plasma behaviour. Though the equations to be solved are well known—the laws of motion coupled to Maxwell's equations through the Lorentz force equation—only in a limited number of cases can these equations be solved analytically. As with biological research, the complexity of the phenomena involved is a reflection of the complexity of strongly coupled multi-body systems, not of the basic laws of physics involved.

Returning to the analytical tools that were developed to cope with the complexities of plasma behaviour, it is worthwhile singling out some specific contributions, ones that represented either new insights or insightful improvements in older theories that permitted new solutions to problems. I will list here a few of these, not pretending to be complete.

(i) Landau's theoretical treatment of collisionless damping [17].

(ii) Kruskal and Schwarzchild's prediction of the instability of the magnetic pinch [29].

(iii) Marshall Rosenbluth's work on the application of the Fokker–Planck equation to plasmas, including the 'Rosenbluth potentials' [56].

(iv) The definitive work by Northrop and Teller on the role of the adiabatic invariants of magnetic moment and action in defining particle confinement in magnetic-mirror systems [57].

(v) The 'Grad–Shafranov' equation of plasma pressure equilibrium in a torus [58, 59].

(vi) The paper by Bernstein, Frieman, Kruskal, and Kulsrud on *An energy principle for hydromagnetic stability problems* [60], based in part on an analytical approach suggested by Teller [61].

(vii) The elucidation of plasma wave theory by W P Allis and co-workers [62], followed by the detailed treatment of these waves by Stix [63].

(viii) The treatment of 'finite orbit' corrections to plasma stability theory by Rosenbluth, Krall and Rostoker [64].

(ix) The pioneering paper on velocity–space-plasma instabilities by Harris [65], followed by papers by Mikhailovskii and Timofeev [66] on 'drift-cyclotron' instabilities and by Rosenbluth and Post on 'loss-cone' instabilities in mirror machines [46, 47].

(x) The analysis of the role of 'magnetic wells' in stabilizing MHD instabilities in high-pressure mirror-confined plasmas, exemplified by the calculations of Hastie and Taylor [45].

(xi) The analysis of finite-resistivity instabilities of a sheet pinch by Furth, Killeen and Rosenbluth [67].

(xii) The analysis of 'neo-classical' diffusion in a torus by Galeev and Sagdeev [68].

(xiii) The introduction of the concept of the 'tandem mirror' by Dimov [53] and Fowler and Logan [52].

(xiv) The development of the concept of conservation of magnetic helicity, as presented by J B Taylor in his analysis of the 'reversed-field pinch' approach to magnetic confinement [69].

The above list hardly scratches the surface of what has been a truly prodigious effort in plasma theory in the last 40 years. The strongly international nature of the effort can be seen in the collaborative work that went on, with Soviet, European/UK, US and Japanese workers joining in the effort. This collaboration is clearly discerned, for example, in the results of the several international theory workshops that were held, particularly in the first few years after declassification. An early one of these, the 'Trieste Workshop of 1964' [70], sponsored by the IAEA, contains seminal papers on all aspects of plasma confinement and stability theory, presented by plasma-physics pioneers such as H Furth, M Kruskal, M Rosenbluth, R Sagdeev, A Simon and J B Taylor.

In parallel with analytical advances such as those listed above, there were major advances in the use of computers in simulating plasma behaviour. To mention only a few of the pioneers in this area, there was O Bunemann, who simulated stream-type instabilities and C K Birdsall, who extended this type of code to handle a variety of situations. Another simulation code pioneer, especially in the context of fusion systems, is John Dawson of Princeton (now at the University of California, Los Angeles).

Over the years since computers were first employed in fusion research there have been developed a remarkably broad set of computer codes as aids to the research. Among these codes are ones, such as MAFCO developed for the magnetic fusion programme by Perkins at Livermore, enabling the calculation of magnetic fields from complex coil systems. MAFCO and its successors, EFFI, and others, played an essential role in the design of the coils of the mirror systems at Livermore and in the design of tokamaks, both in the United States and elsewhere. MHD equilibria in tokamaks and related toroidal systems have been treated by many codes, including the PEST code of Princeton. In fact codes have been written to analyse and predict such complex processes as ion–ion and ion–electron scattering processes as described by the Fokker–Planck equation, neutral-beam injection, microwave heating of electrons, RF heating of ions in dense plasmas and the loss of particles caused by cyclotronic instability modes. In laser fusion, multi-element codes have also been written to simulate the implosion, heating and instability processes that occur in that research.

22.7. The other end of the scale: inertial confinement fusion

Not long after Geneva 1958 in the United States, at the Lawrence Livermore National Laboratory some researchers began thinking about a completely different approach to fusion. By then the hydrogen bomb was an awesome reality, one that proved that fusion energy could be released by man-made means. In 1957, Associate Director Harold Brown asked John Nuckolls to explore one conceivable avenue for obtaining power from fusion by detonating small hydrogen bombs in a deep cavity carved out in bedrock and filled with steam. To minimize the scale, cost and radioactive waste Nuckolls examined the following two questions.

(i) What is the smallest possible fusion explosion?

(ii) How can a tiny fusion explosion be ignited without employing a fission explosion to ignite it?

The answer to the first question is exciting and challenging. A milligram of D-T compressed to 1000 times liquid density and ignited by a hot spot contains at a minimum only 20 kJ of energy and produces approximately 100 MJ of fusion yield.

Some months before the invention of the laser, Nuckolls used weapons-design computer codes to make calculations of the compression and ignition of a milligram of deuterium–tritium in a radiation implosion in a 'hohlraum' (today known as 'indirect drive') and began to define the properties of a 'driver' which could heat the tiny hohlraum. Several megajoules of energy were required to drive the hohlraum in these early calculations. During these years, and for many years thereafter the concept of indirect drive as an approach to pellet fusion was classified, only being declassified very recently.

When the laser was invented, it became the leading candidate for inertial confinement fusion (ICF) drivers. Other candidates included charged particle beams. What we are talking about here, of course, is a completely different approach to the fusion confinement problem, one based on inertial effects. That is to say, if a fusion fuel charge can be sufficiently compressed and heated to ignition temperatures in a time short compared with the time for its disassembly then the Lawson criterion time can be reached, not via attaining sufficiently long (seconds) confinement times reached at low plasma densities, but rather through very high densities (greater than solid densities) 'confined' for nanoseconds by inertial effects.

What the fusion researcher is presented with is a circumstance that is not encountered in any other of humankind's means for the generation and use of energy. Between the extremes of fuel densities encountered in fusion approaches, with the tokamak at the lower end and ICF at the higher end, there are more than ten orders of magnitude in fuel density. Since fusion is a binary process—ions colliding with each other—fusion power densities vary as the square of the fuel density. Thus there are more than 20 orders of magnitude between the power density of the tokamak and that of an ICF pellet. For the fusion researcher, this entire range of power densities is available for him to contemplate in some yet-to-be-defined new approach to fusion power.

Returning to the history of inertial fusion at Livermore, in the early 1960s Stirling Colgate made some computer calculations of the implosion and ignition of multi-shell fusion fuel capsules in a hohlraum driven by approximately a megajoule laser pulse. At about the same time Ray Kidder made calculations showing that a capsule containing a milligram of D-T could be directly imploded to ignition by a spherically symmetric laser beam—again using roughly a megajoule of laser light. Meanwhile, Nuckolls extended his indirect-drive calculations showing that a simple droplet of D-T could be ignited by exploiting pulse-shaping of the laser pulses. His motivation: to show that the cost of manufacturing the laser pellet need not stand in the way of commercial power production (other proposed pellets were complicated in structure and difficult to make).

Plasma Physics in the Twentieth Century

Figure 22.12. *Photograph of a typical laser-fusion pellet residing on the head of a common pin. The multiple beams of a laser facility (such as the Livermore Nova) are all focused to converge on this tiny target.*

Nevertheless the economic viability of inertial fusion was not at all clear.

Building on these early studies, the Livermore Laboratory initiated an experimental laser fusion programme in 1962, led by Ray Kidder. The principal goal was to study weapons physics. At the time representing the forefront of laser technology, Kidder set out to build a 12 beam, 1 J, ruby laser system. Beginning in the early 1970s, the laser programme at Livermore grew in laser energy by five orders of magnitude, as the issues turned up in the lower-energy experiments showed the necessity for much higher laser energies in order to approach pellet ignition. In parallel with the evolution of the tokamak, instabilities and lack of perfect spherical symmetry led to the need for ever-larger laser systems in order to reach the desired conditions of temperature and density.

1675

In the late 1960s and early 1970s a laser fusion programme was launched at KMS Fusion under the technical leadership of Keith Breuckner. In the KMS approach, direct-drive implosion symmetry was achieved by use of clamshells, mirrors which reflected light onto the pellet from a pair of solid-state laser beams. The KMS Fusion programme stimulated the progress of AEC laser fusion programmes and a partial declassification of the concept. In this same time period laser fusion programmes were launched at the University of Rochester and at Los Alamos.

In *Nature*, in 1972 [71], there appeared the first unclassified article, by Livermore scientists Nuckolls, Wood, Thiessen and Zimmerman, on the physics of the direct-drive approach to inertial fusion. In it they outlined the concepts of rocket-action (ablation-induced) implosion and compression of a pellet with a temporally shaped laser pulse, including the discussion of symmetry and fluid and plasma instability issues as they were then understood. They also presented computer-based curves of the pellet 'gain' (fusion energy release divided by laser energy input) versus compression, showing that compressions, over liquid density of order 1000, and megajoule-scale lasers would be required to yield a high enough gain (100-fold, in order to allow for laser and energy conversion inefficiencies). Optimistically, compressions of 10 000 scaled to modest gains ($\ll 100$) with lasers smaller than 10 kJ, provided perfect laser plasma coupling and implosion symmetry could be achieved. This paper laid out the task of direct-drive laser fusion, including its major concerns and the potential advantages of short-wavelength lasers. In the intervening 20 years these problems were encountered, and largely solved, by a combination of brute force and extreme subtlety, both in the generation of the laser pulses and in the diagnosing of the implosion process, all carried out on fractional nanosecond time-scales.

According to Nuckolls, Russian scientists told him that in the 1960s Sakharov, who also, as we have noted, did pioneer thinking in magnetic fusion, proposed the use of lasers to ignite small capsules. In other parts of the world other laser-fusion programmes were also initiated, with similar objectives and conclusions. In France, at Limei a programme was initiated in 1962. Starting with fundamental studies of laser–matter interactions, the programme evolved to larger lasers, so that by 1985 a 15 kJ laser had been constructed with the assistance of the Livermore Laboratory. Also, in Japan, the GEKKO series of lasers was built, so that by 1985 a 12-beam system, GEKKO XII, with 30 kJ output had been put in operation.

Over the intervening years larger and larger lasers were built at the Livermore Laboratory under the leadership of John Emmett, culminating (1984) in the largest multi-beam laser system in the world, the Livermore NOVA with its ten beams, and energy of 50 000 Joules. The NOVA laser has in its beam-line a 'frequency tripler', yielding a blue-spectrum light output. The frequency tripler employs non-linear effects in special crystalline material. The motivation: it was found that some of the deleterious effects of plasma instabilities in the pellet are relieved if the laser frequency is increased towards the short-wave end of the optical spectrum. These instabilities spelled the death knell of the Los Alamos approach, which had relied on the use of CO_2 laser output (in the far infrared) to heat and implode the pellet.

Recently the US Department of Energy has supported the construction of the National Ignition Facility, a megajoule-scale glass laser which is projected to achieve ignition and modest gain early in the next decade. Such a facility would be the ICF-equivalent of the large ITER tokamak in the sense of a first demonstration of net fusion power gain, and a preview of an approach to the generation of fusion power.

Although the majority of the work in pellet fusion has concentrated on the use of glass lasers to heat and implode the pellets, there is a consensus that for a power plant a better approach exists, as respects energetic efficiency (glass lasers are quite inefficient in their conversion of electric power to light) and, possibly, capital cost of the driver. This approach is the use of heavy-ion beams (for example mercury ions) accelerated to multi-GeV energies by multiple-beam-line accelerators, which are then focused down on the target. In the United States this work was initiated at the Lawrence Berkeley Laboratory, working in conjunction with the Livermore Laboratory.

Again, we have a situation encountered before. As with the tokamak and with the glass laser, the issue facing the researchers hoping to find a path to commercial fusion power is that of size and cost. In all these cases it is no longer a question as to whether the scientific goal of net fusion power can be achieved; it is now a matter of practicality and cost. It is, however, the author's contention that it is too early to draw final conclusions. In both the case of magnetic fusion and pellet fusion there should be potential variations on their theme, including variants lying intermediate in density between the two, that could result in smaller, simpler and less expensive fusion systems than a simple extrapolation from the present picture of either the tokamak or the pellet fusion approach would represent.

22.8. Landmark experiments: general plasma physics

In the course of the explosion of interest in plasma physics caused by fusion and space research there was also an increase in the level of study of general issues in plasma physics. Unfortunately this research has too often had to ride on the coat-tails of the other two fields, so has frequently suffered from neglect. As a partial summary of the kind of laboratory experiment that has contributed to the general knowledge of plasma (either within or outside the context of fusion research) we will list some that stand out, together with their perceived significance.

Twentieth Century Physics

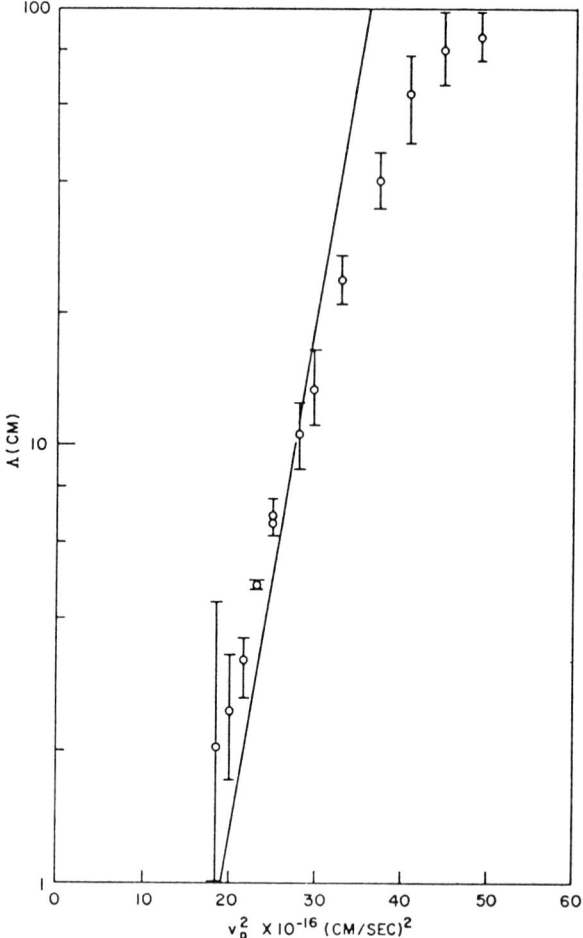

Figure 22.13. *Graph of the logarithm of damping length for plasma waves against the square of the wave phase velocity compared with Landau damping theory. Except at the highest wave velocities the agreement between this experiment and the theory is very close. The deviations at high phase velocity can be explained by effects peculiar to the particular means used to generate the plasma.*

(i) The experimental demonstration of Landau damping of plasma waves by Malmberg, Wharton and Drummond in 1965 [72], showing for the first time in a laboratory experiment the validity of this concept (figure 22.13).

(ii) The experimental and theoretical elucidation, in 1955, by A Simon and R Neidigh of the Oak Ridge Laboratory [73], of a plasma column with a cross-field diffusion rate varying inversely as the square of the magnetic field, rather than inversely linearly as predicted by the turbulence-based Bohm diffusion formula.

EFFECT OF SCATTERING GAS ON CONFINEMENT

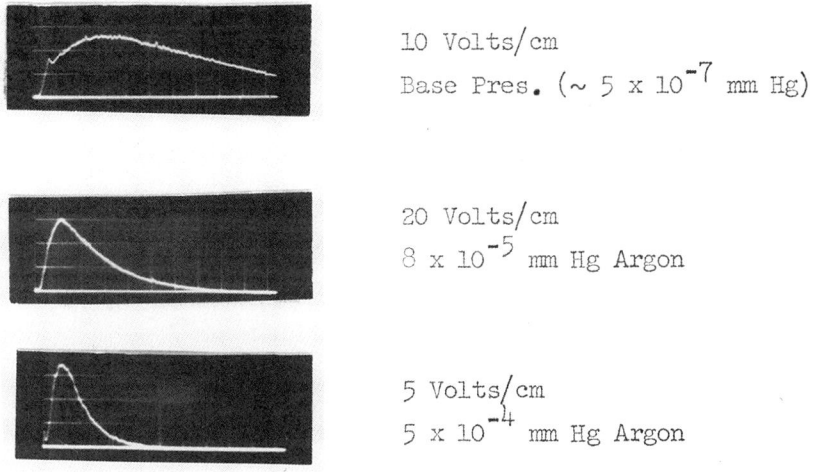

SWEEP TIME 5 MILLISECONDS

Figure 22.14. *Data from the early (1957–59) table-top hot-electron plasma experiments at Livermore, showing the long-time confinement of a plasma column (diameter of order 2 cm) between mirrors, as influenced by scattering losses from a background gas. For the longest decay times the confinement time is more than 10^5 times longer than that predicted by the Bohm diffusion formula.*

(iii) The demonstration of the stable confinement of a 'hot electron' plasma trapped in a mirror machine, by Post, Ellis, Ford and Rosenbluth in 1960 [74], showing confinement times five orders of magnitude longer than the 'Bohm time', apparently rendered stable by finite-orbit effects (figure 22.14).

(iv) The tritium-beta experiment of Rodionov in 1959 [75], and the 'beta-ray' experiment of Gibson, Jordan and Lauer in 1963 [76], demonstrating in the laboratory the validity of the Teller–Northrop theory of adiabatic confinement of particles in mirror systems (figure 22.15).

(v) The mirror experiment by Ioffe's team in 1961 [31], demonstrating the suppression of MHD instabilities in mirror systems by the use of a magnetic-well-type confining field.

(vi) The experiment, in the 'Alice' mirror machine at Livermore, by Damm et al in 1970 [77], showing, in quantitative agreement with theory, the suppression of ion-cyclotron instability modes by electron Landau damping (figure 22.16).

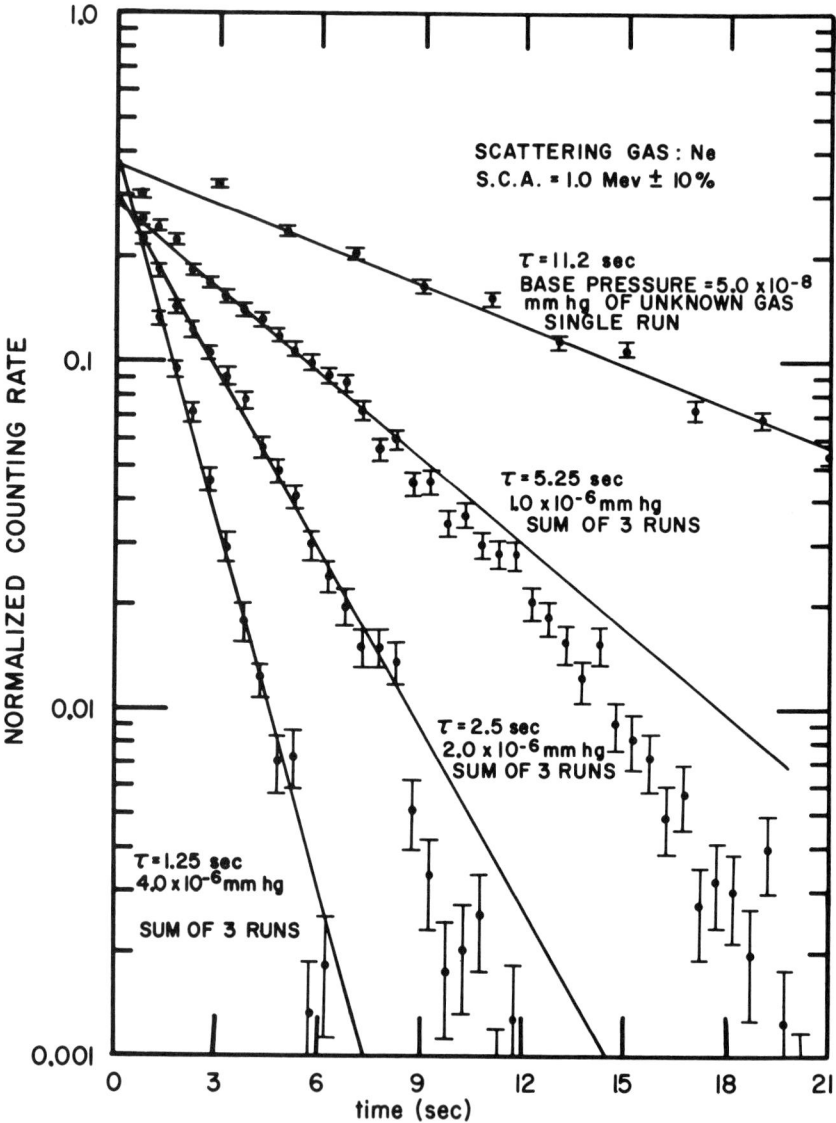

Figure 22.15. *Plots of the rates of escape of mirror-trapped relativistic beta-particles (from the decay of neon-19) as a function of background gas pressure, in the 'beta-ray' experiment of Gibson, Jordon and Lauer [76]. The decay rate shows classical behaviour in agreement with scattering theory. For the longest decays the positrons are undergoing of order 10^8 reflections from the mirrors, demonstrating the validity of the confinement theory of Teller and Northrop based on adiabatic invariants.*

(vii) The demonstration of the critical conditions for the onset of unstable 'drift waves' in a magnetically confined plasma column, carried out in the so-called Q-machines at Princeton and elsewhere.

Plasma Physics in the Twentieth Century

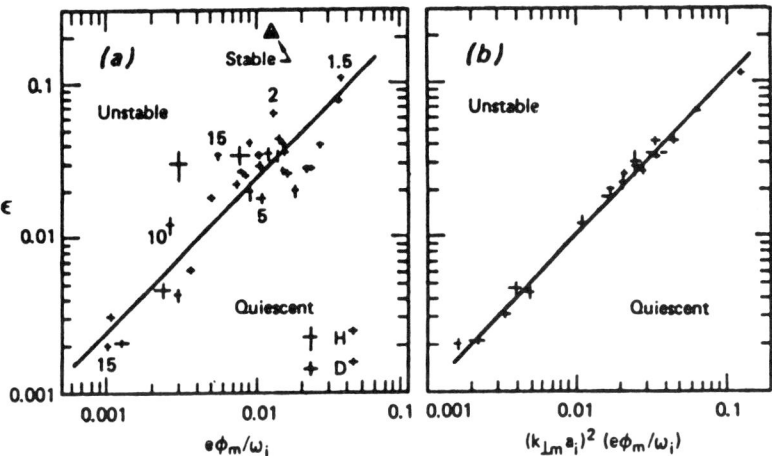

Figure 22.16. *Comparison between experiment and instability theory, showing the role of Landau damping in stabilizing high-frequency wave–particle plasma instabilities in the Livermore Baseball I experiment. Experimental data shows the density threshold of the ion cyclotron instability as a function of the relative confining ambipolar potential for electrons. (a) Data points plotted against $e\phi_m/\omega_i$. (b) Same plot as (a), but including the measured values of $k_\perp a_i$ associated with each data point (determined by mode analysis). The solid line indicates the theoretical prediction based on stabilization by Landau damping; there is close agreement between the experimental values and the theory.*

(viii) The experiments of Malmberg on the magneto-electrostatic confinement of non-neutral electron plasmas [78].

The above list, in those contributions that arose in magnetic fusion research, is conspicuously lean in items from the toroidal fusion approaches. This circumstance can be traced to the fact that in those devices there is always a spectrum of turbulence-related processes going on that makes it difficult to pick out a single effect for investigation in the spirit of landmark results. Nevertheless, the totality of the results from programmes such as the tokamak programme give an overall picture of plasma confinement in toroidal fields which certainly qualifies as a major accomplishment. In the case of the mirror-related experiments, the nature of mirror confinement is such that quiescent or nearly quiescent states can be achieved, ones where clear results can be deduced in the classic manner, for example as in Ioffe's landmark experiment.

Out of the research fields of fusion plasmas and general plasma physics in recent years there has resulted a resurgence of interest in the application of plasma techniques to industrial material processing. Plasmas can be used, for example, to create thin diamond films, to promote certain chemical reactions and to process semiconductor 'chips'. These areas, and their commercial value, have been greatly helped by the technology and the understanding developed in fusion research, added to the knowledge gleaned from the older research field of gas-discharges.

22.9. Space plasma physics, the ionosphere and beyond

The ionosphere, the region surrounding the Earth that introduced researchers to the effects of plasma on the propagation of radio waves, also became the progenitor of space-plasma research studies. Here the possibility of direct measurements of the complex phenomena occurring within the ionosphere had to await the development of rockets with sufficient lift-off capabilities to reach the ionosphere. There were, however, early thoughts and researches relating the ionosphere and the Earth's magnetosphere to phenomena such as the aurorae.

The whole matter of the influence of the Earth's magnetic field on energetic particles coming in from space had been investigated early in the century in studies aimed at understanding the aurorae. These studies may have been stimulated by an even earlier conjecture. As early as 1881 Goldstein [79] had suggested that they might be caused by particles coming from the Sun.

Because of its suspected relevance to the aurorae, the effect of the Earth's magnetic field on incoming particles was examined theoretically by Størmer [80], a Scandinavian who certainly would be aware of their existence. Early in the century (1907 and following) Størmer computed the orbits of incoming charged particles, showing the existence of the magnetic mirror effect in producing 'forbidden zones', regions of latitude where particles below a critical energy could not penetrate, but were instead reflected back into space. Størmer's theoretical concepts were previewed experimentally, in the laboratory, just before the beginning of this century, in the 'Terella' device of Birkeland [81], a small magnetic dipole in the shape of a sphere. Suspended in a vacuum chamber this sphere was bombarded by cathode rays (as electrons were then known). Ionization of the residual gas in the chamber then showed light-filled regions where the electrons could penetrate, and dark regions from which they were excluded, in agreement with Størmer's later findings. Another of Størmer's predictions, the possibility of 'permanently' trapping particles within the Earth's magnetic dipole field would have to wait for nearly 50 years before it could be directly confirmed.

Thus, after a hiatus of many years, the ionosphere again became the focus of space-plasma research in the latter half of the century. Immediately following World War II, when the first high-altitude rockets became available for research use, soundings were taken of the ionosphere, using such rockets as captured German V-2 rockets (reconditioned for their peaceful mission) and the 'Aerobee' sounding rocket. Using two-frequency radio transmissions from these rockets back to Earth, improved measurements were made of the electron densities in the ionosphere (of order 10^4 to 10^5 cm^{-3}) and other important measurements were performed in those early soundings, begun in the United States in 1946. In 1950 the name James Van Allen began to appear in the literature, associated with rocket sounding measurements made on cosmic-ray particles in the ionosphere. His name would, within a few years, become well known, when the mirror-trapped particle 'layers' in the ionosphere were discovered as a result of his research.

In a chapter in the book *Radiation Trapped in the Earth's Magnetic Field* [82] Van Allen describes his findings concerning the belts of trapped energetic charged particles (protons and electrons) surrounding the Earth. Among the remarkable features that his observations revealed was the remarkably long lifetime of the mirror-trapped particles; years for energetic protons and months or more for trapped electrons. These long lifetimes represented an irrefutable confirmation of the theory of Teller and Northrop [57], previously mentioned in connection with the magnetic-mirror fusion programme, dealing with the role of the orbit constants, magnetic moment and longitudinal action, in confining trapped particles in magnetic-mirror fields.

Van Allen discovered his belts in March 1958. A few months prior to that date, October 1957, Nicolas Christofilos of the Lawrence Livermore Laboratory had suggested an audacious experiment to create a man-made radiation belt at high altitudes. His suggestion: detonate a small rocket-launched atomic bomb in the ionosphere and then follow the time history of the energetic electrons released by beta decay at the time of the blast. By April of 1958 the decision was made to proceed with Christofilos' experiment, code-named 'Argus'. It then required the remarkably short time of only four months to plan and execute the experiment, to be coordinated with measurements made by the satellite *Explorer IV*. The satellite recorded and reported to ground stations, around the Earth the electron density in the trapped shell as well as its position and thickness. As reported by Christofilos in his chapter in the above-cited book [82] 'An important observation was that (within the limit of observational error) neither did the electron shell move at all across the magnetic lines nor did the electrons diffuse across these lines.'. Again, a remarkable confirmation of the Teller–Northrop theory. It was also heartening to mirror researchers to find support for their endeavours from such a grand-scale source.

As an addendum to the Argus experiment, it was found that a measurable fraction of the electrons injected into the ionosphere in the Argus experiment could be detected decades after the event! Also, because of the subsequent worldwide ban on the detonation of nuclear explosions in space, Argus was a one-off event, never to be repeated. However, with the increasingly sophisticated instrumentation on satellites other techniques could be used further to probe the mysteries of the ionosphere. In his chapter, Christofilos suggests the injection of radioactive positron or α-particle emitters as an alternative. In later years other techniques, such as barium-vapour clouds, were used.

Having discovered the existence of the Van Allen belts, the issue shifted to their origin. The existence of the Earth's magnetic-mirror fields is a necessary but not a sufficient condition for the existence of trapped particles. However, by 1966 there had already developed a credible theory of the origin of the trapped protons in the inner Van Allen belt as coming from the decay of neutrons of cosmic-ray origin [83]. This theory thus built on the observations of cosmic-ray physicists who had measured the spectrum and the type of incoming cosmic-ray particles. There still remained, however, the issue of the outer belt, and its interaction with the solar wind (the perpetual stream of charged particles coming in from the Sun). Here a whole new set of important issues arose in the contemplation of the impact of the solar wind on the Earth's magnetosphere.

The issue just described has played a fundamental role in plasma physics since its beginning. It is the issue of 'magnetic reconnection', the process by which an abrupt, sometimes chaotic, change in the topology of a magnetic field can take place within a plasma-filled region. In reconnection lines of force 'switch' from one region to another, with consequent changes in the flow of particles in that region. The theory of reconnection built on the foundations made by Alfvén, who had first introduced the fundamental concept in MHD theory of the idea of field lines 'frozen' within a conducting medium—the plasma. In this view, motions, including expansion and compression of the plasma will carry with them the trapped field lines 'frozen' in place by the conducting medium by which they are surrounded. In reconnection this picture breaks down, at least momentarily, when magnetic lines from plasmas impinging on each other must rearrange themselves into a new configuration.

Early work in this field is due to Chapman and Ferraro in 1931 [84]. In this work they discussed the diamagnetic currents that would arise at a plasma boundary. Not until much later, however, would the complex issues involved be understood. A critical feature in this understanding was what are called 'collisionless shocks'. Ordinary shock waves arise from temperature-dependent flow of thermal energy in a gas, coupled through inter-particle collisions. In the ionosphere, however, collisional processes are extremely weak, so that shock waves, if they are to arise, must appeal to electromagnetic coupling effects to appear. Names associated with the theory of collisionless shocks are: space-plasma physicist, E N Parker, fusion researchers, R Kulsrud and I Bernstein, and physicist H Petschek. As can be seen from the names involved, in the arena of space-plasma research there was a strong interplay between magnetic fusion researchers and their counterparts in space-plasma research.

An important element in the ionospheric plasma is the role played by the solar wind. This stream of plasma from the Sun has a major influence on the nature and the shape of the upper portion of the ionosphere. A dynamical theory of the solar wind was proposed in 1958 by E N Parker that explains many of its features [85]. Though incomplete, Parker's model comes closer to fitting with experimental data on the solar wind than an earlier, hydrostatic, model proposed by S Chapman. These and other theoretical models have helped to extend our understanding of the streaming plasmas in interplanetary space, including the dominant role of the Sun in producing them (figure 22.17).

Following the finding of the Van Allen belts, more and more detail as to waves and other transient phenomena in the ionosphere began to be revealed by satellites. In these studies many of the same types of instabilities that were being encountered experimentally, and studied theoretically, in magnetic fusion research, were seen. These included both MHD-types of instabilities, such as were being encountered in mirrors and in the toroidal devices, and velocity–space instabilities involving the interaction of particles and waves. Fast-growing instabilities associated with magnetic reconnections, for example, were seen to play a role in the phenomena of the aurorae, where the sudden dumping of particles into the north and south polar regions causes these awe-inspiring displays. The relationship between the aurorae and so-called 'magnetic storms' associated with changes in the millions-of-amperes ring currents in the ionosphere is being increasingly better understood as a result of theory coupled with satellite observations.

Returning to an earlier reference to the ubiquitous presence of plasma in the Universe, the name of Hannes Alfvén should be specially singled out. Starting in the early 1940s his name is associated with many significant theoretical studies of the subject. He did pioneering work in the theory of MHD waves, and one type of MHD wave carries his name—Alfvén waves. His book *Cosmical Electrodynamics* [27], in addition to being an important source for early fusion researchers, also dealt with a variety of space-plasma topics, such as: solar physics, magnetic storms and aurorae and cosmic rays. In the years following the publication of that book Alfvén continued to contribute to the field of astrophysical plasmas, culminating in his receipt of the Nobel Prize in Physics in 1970, the first one ever awarded in the field of plasma physics.

In our discussion of space-plasma research we have concentrated on the ionosphere and have not discussed the continuum of plasma phenomena in the solar wind between the Sun and the Earth, within the solar corona itself and on into outer space, where plasmas rule supreme and one finds the bizarre plasma effects encountered in pulsars and in novae. Through radio astronomy, and now through space-based telescopes, these fascinating phenomena are yet to be researched.

To conclude this too-brief discussion of space-plasma research, there was made, in the 1980s, a remarkably detailed set of satellite measurements demonstrating directly the effect of lightning bursts on

Twentieth Century Physics

Figure 22.17. *Photograph of the Sun taken in the light of an emission line from hydrogen. The huge arch seen at one point represents light emitted from a plasma being guided up from the Sun's surface along magnetic field lines emerging from large sunspot regions at the base of the arch.*

the trapped electron population in the Van Allen belts. In a way these measurements bring full circle the 'whistler' story that began at the time of World War I. In the 1980s measurements, for example as reported in *Nature* in 1984 by Voss *et al* in their article *Lightning-induced electron precipitation* [86], showed the immediate depletion of the radiation belt, caused by frequency components (at the electron cyclotron frequency) of waves in the ionospheric plasma, waves launched by lightning bolts, when these same waves interact with trapped electrons in the Van Allen belt. The power of resonant wave–particle interactions to perturb the ionospheric plasma was shown by the fact that the energy fluxes of precipitating electrons that were measured were estimated to be up to 100 000 times that of the wave energy flux. The same theory that describes the trapping and loss of particles from fusion mirror machines also describes these losses, yet another example of the cross-fertilization between fusion plasma research and magnetic fusion research.

22.10. Conclusion

As is evident from its history, the discipline of plasma physics has its feet planted both in the applied and in the most basic aspects of physics research. In the applied, fusion research stands as its cornerstone, while in basic science, the dominance of plasma effects in the matter in the rest of our Universe, and even in our understanding of cosmology, is abundantly evident.

In the applied, and in particular in magnetic fusion research, we are far from having an adequate understanding of the complex interplay between plasma and the electromagnetic field. If we had as thorough an understanding of that physics as we do now of the electromagnetic field itself the path towards optimal fusion power systems would be clear, in the light of which our present efforts would seem primitive.

As to the basic aspects of plasma physics, we may never know the full extent of the role of plasma in our Universe. In recent years it has seemed that almost every new discovery in astronomy exposes some new aspect of cosmology where plasma effects are involved in fundamental ways.

Thus it seems to me that over the coming years we can expect that the study of plasmas, both those in space and those in the laboratory, will lead not only to new insights into our Universe, but also, as our understanding grows, to new means of control and practical application of this tantalizingly complex 'fourth state of matter'.

22.10.1. Further reading
More information on the history of fusion research can be found in references [28, 87–89].

Acknowledgments

The author is indebted to the many individuals who provided helpful contributions while the material for this chapter was being assembled. Among them are: T K Fowler, J Nuckolls, T Stix, J B Taylor and M Walt.

References

[1] Ladner A W and Stoner C R 1932 *Short Wave Wireless Communication* (New York: Wiley)
[2] Heaviside O 1902 *Encyclopaedia Britannica* 9th edn, vol 33 (New York: Encyclopaedia Britannica Inc.) p 215
[3] Kennelly A E 1902 *Electrical World Eng.* **15** 473
[4] Stewart B 1902 *Encyclopaedia Britannica* 9th edn, vol 16 (New York: Encyclopaedia Britannica Inc.) p 181
[5] Breit G and Tuve M 1926 *Phys. Rev.* **28** 554
[6] Hafstad L and Tuve M 1926 *Terr. Magn. Atoms. Electr.* **23** March
[7] Appleton E V 1932 *J. Inst. Electr. Eng.* **71** 642
[8] Hartree D R 1931 *Proc. Camb. Phil. Soc.* **27** 143
[9] Appleton E V 1927 *URSI Reports* Washington
[10] Eccles W H 1912 *Proc. R. Soc.* A **87** 79
[11] Larmor J 1924 *Phil. Mag.* **48** 1025
[12] Barkhausen H 1919 *Phys. Z.* **20** 201
[13] Barkhausen H 1930 *Proc. IRE* **18** 1155
[14] Eckersley T L 1935 *Nature* **135** 104
[15] Tonks L and Langmuir I 1929 *Phys. Rev.* **33** 195
[16] Thomson J J and Thomson G P 1933 *Conduction of Electricity Through Gases* 3rd edn, vol 2 (New York: Cambridge University Press)
[17] Landau L D 1946 *J. Phys. USSR* **10** 25
[18] Bohm D and Gross E P 1949 *Phys. Rev.* **75** 1851
[19] Bernstein I B, Greene J M and Kruskal M D 1957 *Phys. Rev.* **108** 546
[20] Atkinson R d'E and Houtermans F G 1929 *Z. Phys.* **54** 656
[21] Cockroft J D and Walton E T S 1932 *R. Soc. Proc.* **137** 229
[22] Lawson J D 1957 *Proc. Phys. Soc.* B **70** 6
[23] Bethe H A 1936 *Rev. Mod. Phys.* **8** 83; 1937 *Rev. Mod. Phys.* **9** 69, 245
[24] Spitzer Jr L 1956 *Physics of Fully Ionized Gaes* (New York: Interscience)
[25] Størmer C 1907 *Arch. Sci. Phys. Genève* **24** 5 113, 221, 317
[26] Vallarta M S 1937 *Nature* **139** 839
[27] Alfvén H 1950 *Cosmical Electrodynamics* (Oxford: Oxford University Press)
[28] Bishop A S 1958 *Project Sherwood—The US Program in Controlled Fusion* (New York: Addison-Wesley)
[29] Kruskal M D and Schwarzchild M 1954 *Proc. R. Soc* A **223** 348
[30] Post R F 1956 *Rev. Mod. Phys.* **28** 338
[31] Gott Y V, Ioffe M S and Telkovsky V G 1963 *Plasma Physics and Controlled Nuclear Fusion Research, Part 3* (Vienna: IAEA) p 1045
[32] Grad H and Rubin H 1958 *Proc. Second United Nations Int. Conf. on the Peaceful Uses of Atomic Energy* vol 31 (Geneva: United Nations) p 190
[33] Tuck J L 1958 *Proc. Second United Nations Int. Conf. on the Peaceful Uses of Atomic Energy* vol 32 (Geneva: United Nations) p 3
[34] Kerst D W *et al* 1971 *Plasma Physics and Controlled Nuclear Fusion Research 1971* vol 1 (Vienna: IAEA) p 3
Ohkawa T, Yoshikawa M, Gilleland J R and Tamano T *Plasma Physics and Controlled Nuclear Fusion Research 1971* vol 1 (Vienna: IAEA) p 15
[35] Vedenov A A *et al* 1958 *Proc. Second United Nations Int. Conf. on the Peaceful Uses of Atomic Energy* vol 32 (Geneva: United Nations) p 239
[36] Christofilos N C 1958 *Proc. Second United Nations Int. Conf. on the Peaceful Uses of Atomic Energy* vol 32 (Geneva: United Nations) p 279
[37] Cantrell E L *et al* 1975 *Plasma Physics and Controlled Fusion Research 1974* vol III (Vienna: IAEA) p 113

[38] Butt E P et al 1966 *Plasma Physics and Controlled Fusion Research* vol II (Vienna: IAEA) p 751
[39] Taylor J B 1974 *Phys. Rev. Lett.* **33** 1139
[40] Krakowski R A et al 1981 *Plasma Physics and Controlled Fusion Research 1980* vol II (Vienna: IAEA) p 607
Hagensen R L, Krakowski R A, Bryne R N and Dobrott D 1983 *Plasma Physics and Controlled Fusion Research* vol I (Vienna: IAEA) p 373
[41] Wharton C B, Howard J C and Heinz O 1958 *Proc. Second United Nations Int. Conf. on the Peaceful Uses of Atomic Energy* vol 32 (Geneva: United Nations) p 388
[42] Post R F 1954 Sixteen lectures on controlled thermonuclear reactions *Lawrence Livermore National Laboratory Report* UCRL-4231
[43] Judd D, McDonald W and Rosenbluth M N 1955 *Conf. on Thermonuclear Reactions, Radiation Laboratory, University of California Berkeley, Report WASH-289* (Washington, DC: Atomic Energy Commission)
[44] See, for example, Fowler T K and Rankin M 1966 *Plasma Phys.* **8** 121
[45] Hastie R J and Taylor J B 1965 *Phys. Fluids* **8** 323
[46] Rosenbluth M N and Post R F 1965 *Phys. Fluids* **8** 547
[47] Post R F and Rosenbluth M N 1966 *Phys. Fluids* **9** 730
[48] Post R F 1967 *Plasma Confined in Open-ended Geometry Proc. Int. Conf. (Gatlinburg, 1967), Report CONF-971127* (Oak Ridge, TN: Oak Ridge National Laboratory) p 309
[49] Post R F 1969 *Nuclear Fusion Reactors Proc. British Nuclear Energy Soc. Conf. (London, 1969)* (Abingdon: Culham Laboratory) p 88
[50] Logan B G et al 1976 *Phys. Rev. Lett.* **37** 1468
[51] Moir R W and Post R F 1969 *Nucl. Fusion* **9** 253
[52] Fowler T K and Logan B G 1977 *Comm. Plasma Phys. Control. Fusion* **2** 167
[53] Dimov G I, Zakaidakov V V and Kishinevskii M E 1976 *Sov. J. Plasma Phys.* **2** 236
[54] Pastukhov V P 1974 *Nucl. Fusion* **14** 3; 1984 *Vopr. Teor. Plazmy* **13** 160
[55] Inutake M et al 1994 *Proc. Int. Conf. on Open Plasma Confinement Systems for Fusion (Novosibirsk, Russia, 1993)* ed A A Kabantsev (Singapore: World Scientific) p 51
Ichimura M 1994 et al *Proc. Int. Conf. on Open Plasma Confinement Systems for Fusion* p 69
Cho T et al 1994 *Proc. Int. Conf. on Open Plasma Confinement Systems for Fusion* p 79
Tamano T et al 1994 *Proc. Int. Conf. on Open Plasma Confinement Systems for Fusion* p 97
[56] Rosenbluth M N, McDonald W M and Judd D C 1957 *Phys. Rev.* **107** 1
[57] Northrop T G and Teller E 1960 *Phys. Rev.* **117** 215
[58] Grad H 1967 *Phys. Fluids* **10** 137
[59] Shafranov V D 1966 *Reviews of Plasma Physics* vol 2 (New York: Consultants Bureau) p 103
[60] Bernstein I B, Frieman E A, Kruskal M D and Kulsrud R M 1958 *Proc. R. Soc.* A **244** 17
[61] Teller E 1954 Paper presented at *Conf. Thermonuclear Reactions (Princeton University, October, 1954)*.
[62] Allis W P, Buchsbaum S J and Bers A 1962 *Waves in Plasmas* (New York: Wiley)
[63] Stix T H 1962 *The Theory of Plasma Waves* (New York: McGraw-Hill)
[64] Rosenbluth M N, Krall N A and Rostoker N 1962 *Nucl. Fusion Suppl.* part 1 143
[65] Harris E G 1959 *Phys. Rev. Lett* **2** 34

[66] Mikhailovskii A B and Timofeev A V 1963 *Sov. Phys.–JETP* **17** 626
[67] Furth H, Killeen J and Rosenbluth M N 1963 *Phys. Fluids* **6** 459
[68] Galeev A A and Sagdeev R Z 1968 *Sov. Phys.–JETP* **26** 233
[69] Taylor J B 1974 *Phys. Rev. Lett.* **33** 1139
[70] 1965 *Plasma Physics* (Vienna: IAEA)
[71] Nuckolls J, Wood L, Thiessen A and Zimmerman G 1972 *Nature* **239** 139
[72] Malmberg J H, Wharton C B and Drummond W E 1957 *Proc. Phys. Soc.* B **70** 6
[73] Simon A 1958 *Proc. Second United Nations Int. Conf. on the Peaceful Uses of Atomic Energy* vol 32 (Geneva: United Nations) p 343
[74] Post R F, Ellis R E, Ford F C and Rosenbluth M N 1960 *Phys. Rev. Lett.* **4** 166
[75] Rodionov S N 1959 *At. Ehnerg.* **6** 623
[76] Gibson G, Jordon W C and Lauer E J 1963 *Phys. Fluids* **6** 116
[77] Damm C C, Foote J H, Hunt A L, Moses K, Post R F and Taylor J B 1970 *Phys. Rev. Lett.* **24** 495
[78] Malmberg J H and Driscoll C F 1980 *Phys. Rev. Lett.* **44** 654
[79] Goldstein S 1881 *Wiedemanns Ann.* **12** 266
[80] Størmer C 1931 *Z. Astrophys.* **3** 31, 227; 1932 *Z. Astrophys.* **4** 290; 1933 *Z. Astrophys.* **6** 333
[81] Birkeland K 1896 *Arch. Sci. Phys.* **1** 497
[82] Van Allen J 1966 *Radiation Trapped in the Earth's Magnetic Field* ed B McCormac (New York: Gordon and Breach)
[83] Lenchek A M and Singer S F 1962 *J. Geophys. Res.* **67** 1263
[84] Chapman S and Ferraro V C A 1931 *Terr. Magn. Atmos. Electr.* **36** 77, 171
[85] Parker E N 1958 *Astrophys. J.* **128** 664
[86] Voss H D *et al* 1984 *Nature* **312** 740
[87] Bromley J L 1982 *Fusion: Science, Politics, and the Invention of a New Energy Source* (Cambridge, MA: MIT Press)
[88] Heppenheimer T A 1984 *The Man-Made Sun: The Quest for Fusion Power* (Boston, MA: Little Brown)
[89] Herman R 1990 *Fusion: The Search for Endless Energy* (Cambridge: Cambridge University Press)

Chapter 23

ASTROPHYSICS AND COSMOLOGY

Malcolm S Longair

PART 1: STARS AND STELLAR EVOLUTION UP TO WORLD WAR II

23.1. The legacy of the nineteenth century

The great revolutions in physics of the early years of the twentieth century have their exact counterparts in the birth of astrophysics and cosmology [1]. Until the late nineteenth century, astrophysics did not exist. Astronomy meant positional astronomy and the techniques of accurate observation had improved steadily since the pioneering efforts of Tycho Brahe in the late seventeenth century. All observations were made by eye using as large telescopes as the astronomers could afford.

The revolution which was to take place at the beginning of the twentieth century can be traced to three key observational and technical advances. The first was the measurement of stellar distances. In 1838, Friedrich Bessel announced the measurement of the trigonometric parallax of the star 61 Cygni which he found to be about one third of an arcsec, corresponding to a distance of 10.3 light-years [2]. His observations showed that the stars are objects similar to our own Sun and set the scale for the Universe of stars. Gradually trigonometric parallaxes became available for a significant number of stars but it was a struggle. By 1900, less than 100 parallaxes had been measured with any accuracy.

The second great development was the birth of astronomical spectroscopy. The early years of the nineteenth century saw a blossoming of experimental spectroscopy. In 1802, spectroscopic measurements of the solar spectrum were made by William Wollaston [3], who observed five strong dark lines but their significance was not appreciated at the time. The breakthrough was due to Joseph Fraunhofer [4] who, in 1814, rediscovered the narrow dark lines which provided precisely defined wavelength standards. He labelled the ten strongest lines in the solar spectrum A, a, B, C, D, E, b, F, G and H and recorded 574 fainter lines between the B and H lines. In the same set of experiments, he found three broad stripes in the spectrum of the bright star Sirius. In 1823, he went on to observe the spectra of the planets and the brightest stars [5], anticipating by about 40 years the next serious attempts to measure the spectra of the stars.

Throughout the 1850s, there was considerable effort in Europe and in the USA to identify the emission lines produced by different substances in flame, spark and arc spectra. The most important work resulted from Robert Bunsen and Gustav Kirchhoff's studies. In Kirchhoff's great papers of 1861–63, entitled *Investigations of the solar spectrum and the spectra of the chemical elements* [6], the solar spectrum was compared with the spark spectra of 30 elements using a four prism arrangement with which it was possible to view simultaneously the spectrum of the element and the solar spectrum. He concluded that the cool, outer regions of the solar atmosphere contained iron, calcium, magnesium, sodium, nickel and chromium and probably cobalt, barium, copper and zinc as well.

The third development was astronomical photography. The discovery of the daguerreotype process was announced almost simultaneously by Daguerre in France and by Fox Talbot in England in 1839. The key development for astronomical photography was the invention in the 1870s of dry collodion plates which allowed long exposures to be made. They were so successful that the search for substitutes for collodion were sought and the result was the invention of the gelatin emulsion in which silver salts are suspended in gelatin. It was found that the speed of the gelatin emulsions could be vastly increased by prolonged exposure to heat or by the addition of ammonia. As a result of these developments, the typical exposure time for terrestrial photography was reduced to about $\frac{1}{15}$ second.

Thus, by the 1880s, the tools needed to begin the detailed study of the spectra of the stars had been developed and the scene was set for the birth of astrophysics as a physical science.

23.2. The origin of the Hertzsprung–Russell diagram

23.2.1. The classification of stellar spectra
In the second half of the nineteenth century, much effort was devoted to the classification of the spectra of large numbers of stars in order to put some order into the diverse features of stellar spectra. The first of these pioneers was the Italian Jesuit priest Father Angelo Secchi who, during the 1860s, classified the spectra of over 4000 stars by visual observation. In the final version of his classification scheme [7], he placed the stars into four classes, corresponding to white or blue stars (class

I), yellow or solar-type stars (class II), red stars with wide absorption bands (class III) and the 'carbon stars', as they are now known, which have 'luminous bands separated by dark intervals', in class IV. The development of photographic spectra greatly enhanced the objectivity of the procedures of classification and, to encompass the wealth of information available, subclasses were added to Secchi's system. The process of stellar classification developed into a major industry and, by 1900, 23 different systems were in use.

While many groups were working on the problem of stellar classification, the whole enterprise was overtaken by the mammoth surveys of stellar spectra which were begun at Harvard under the direction of Edward C Pickering, who was appointed director of the Harvard College Observatory in 1876 [8]. Pickering decided that the most effective means of undertaking spectral studies of very large numbers of stars was to use an objective prism to disperse the images of all the stars in the region of sky under observation. The principal investigators were Williamina P Fleming, Annie Jump Cannon and Antonia C Maury and a large corps of women 'computers'. The team was jokingly referred to as 'Pickering and his harem'.

The first part of the observing programme was completed by January 1889 and consisted of 633 plates containing the spectra of 10 351 stars. The task of examining and classifying the spectra, as well as estimating their magnitudes, was carried out by Fleming. The spectral classification [9] was based upon Secchi's four classes but they were now divided into further subclasses. Class I was divided into four subclasses A, B, C and D, Class II into seven subclasses, E, F, G, H, I, K, L and classes III and IV were renamed M and N. The designation O was used to describe the Wolf-Rayet stars.

The most famous work which led to the standard Harvard classification, published in the Harvard Annals in 1901, was due to Cannon [10]. She based her classifications upon the scheme devised by Fleming with some important amendments. She adopted the suggestion that the O stars are the hottest stars and that the B stars should precede the A stars in the stellar sequence. She dropped a number of the classes so that the basic sequence became O, B, A, F, G, K, M. Stars were placed along this linear sequence on the basis of the presence or absence of different spectral lines, the intention being that the progression of spectral features should be continuous. It was already apparent from the investigations of Norman Lockyer that the sequence was basically a temperature sequence but it would take much further work before the precise relation between spectral type and temperature was established through the work of Saha, Fowler and Milne [11]. The result of the efforts of Pickering, Maury and Cannon was that, by 1912, almost 5000 stars had been classified over the whole sky with classifications far superior to those available previously.

Pickering was, however, already planning an even more ambitious programme, what was to become the Henry Draper (or HD) catalogue. On 11 October 1911, Cannon began the classification of 225 300 stars and completed the task just under four years later [12]. In an extraordinary feat of concentrated effort, she was able to classify spectra at a rate of about 3 per minute and her classifications were repeatable over the years of the survey. Cannon was almost completely deaf throughout her career. In 1938, she was appointed William Cranch Bond Astronomer, one of the first women to receive an appointment from the Harvard Corporation, and she was the first woman to be awarded an honorary degree by Oxford University.

23.2.2. Early theories of stellar structure and evolution

The origin of the theory of stellar structure and evolution can be traced to the understanding of the first law of thermodynamics in the 1850s. The stellar energy source could be attributed to the release of gravitational binding energy. A popular version of the theory invoked meteoritic bombardment of stars as the means of releasing gravitational energy. Hermann von Helmholtz proposed that, rather, the contraction of the Sun itself could provide an enormous reserve of gravitational potential energy. The Sun was conceived of as a convective mass of liquid which gradually contracted and cooled. Energy transport through the Sun was assumed to be by convection. William Thompson (Lord Kelvin) and Helmholtz [13] realized that they could then work out an approximate age for the Sun. This time-scale, which is nowadays referred to as the *Kelvin–Helmholtz time-scale* for cooling of the Sun, or any star, can be roughly estimated by dividing its gravitational potential energy by its present luminosity. For the Sun, this time-scale is about 10^7 years, considerably shorter than the favoured estimates of the age of the Earth from stratigraphic analyses, although these were of dubious reliability. The first reliable estimates of the age of the Earth were made in 1904 by Ernest Rutherford [14], using the relative abundances of radioactive and stable isotopes of the heavy elements.

The first person to investigate the internal structure of the Sun as a gaseous body was the American J Homer Lane [15]. In 1869, he attempted to reproduce its surface properties and to determine the variation of density, temperature and pressure with radius, assuming that the material of the Sun behaved as a perfect gas. He set up the equations of hydrostatic equilibrium and mass conservation but could not reproduce the observed surface properties of the Sun. Although not included in his paper, he was the first to derive the important result that, if a star remains a perfect gas sphere, as it loses energy by radiation and contracts, the temperature *increases* rather than decreases [16].

In the late 1870s, similar calculations were carried out independently by Augustus Ritter at Aachen who identified the initial phase of

Astrophysics and Cosmology

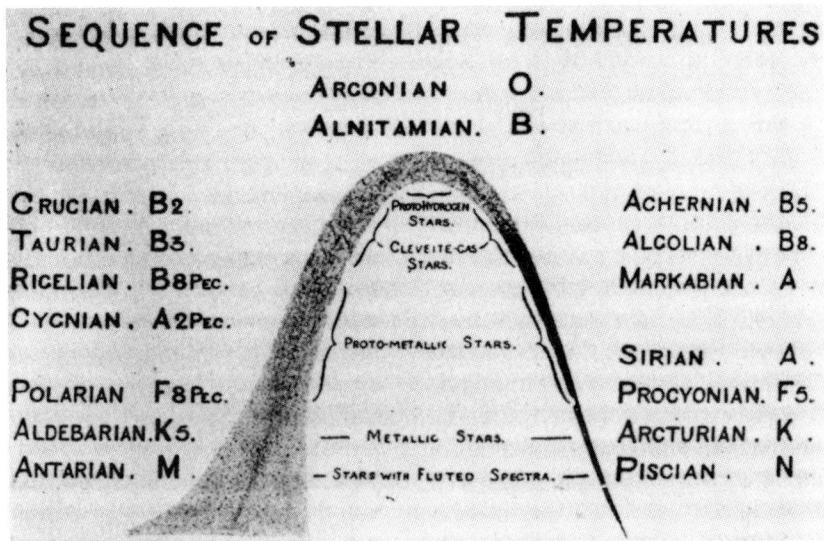

Figure 23.1. *Lockyer's temperature curve for stellar spectral evolution* [19]. *His assignments of the Secchi types to different parts of the arch are indicated.*

evolution of the star as the contraction of a perfect gas sphere, which subsequently cooled according to Kelvin's prescription [17]. The culmination of these physical models for stars was the treatise *Gaskugeln* by Robert Emden [18], which was published in 1907. The intention of the treatise was to attract students at the Technische Hochschule in Munich into theoretical physics by giving as 'practical examples' the internal structure of the stars! Emden showed how the solutions of what became known as the *Lane–Emden equation* describe stars which have a boundary at a finite radius.

Lockyer attempted to identify the various classes of star with an evolutionary picture in the 1880s. Figure 23.1 shows his sketch of his theory of stellar evolution [19]. The evolutionary 'temperature arch' begins at the bottom left which shows the cloud of meteoroids condensing and then contracting so that the star attains maximum temperature at the peak of the arch. The star was then assumed to cool to become a compact red star. Lockyer assigned Secchi classes to the different parts of this evolutionary arch but the reasons for the assignments were obscure. It was implicit in his scheme that there should exist large and small diameter stars at the same temperature. There are similarities to Ritter's more physical picture of stellar evolution.

23.2.3. The Hertzsprung–Russell diagram

With the publication of the first *Henry Draper Catalogue of Stellar Spectra* in 1890 [9], it became possible to make the first tests of the hypothesis that the stars evolve down the spectral sequence from hot stars at A and

B to cool stars at K. By 1893, Monck and Kapteyn [20] had independently concluded that something must be wrong with this picture. They found that the stars with the greatest proper motions, and hence the most nearby and least luminous stars, were not the K and L stars but the F and G stars. This was scarcely consistent with a scheme in which the stars cooled and grew fainter along the spectral sequence.

The solution to this problem was found by the Danish astronomer Ejnar Hertzsprung. He realized that, if the stars radiated like black bodies and their distances were known, it was a straightforward calculation to determine their physical sizes from the Stefan–Boltzmann law. In 1906, he showed that Arcturus had a physical size roughly equal to the diameter of the orbit of Mars, indicating that very large stars exist [21]. With this knowledge, he reinvestigated the data of Monck and Kapteyn but now included the information from Maury's classifications of the high-resolution spectra of bright stars. She had noted that there were three types of spectral lines: in class *a*, the lines were clearly defined and of 'average' width; in class *b*, the lines were much broader and hazy while, in class *c*, the lines were unusually narrrow and sharp. Hertzsprung noted that the *c* stars were distant luminous stars, similar to Arcturus, while the non-*c* stars were nearby objects. The distinction between the classes of what became known as the *dwarf* and the *giant* stars was confirmed by parallax studies [22]. It turned out that among the brightest stars in the sky, there are more giants than dwarfs and this accounted for Monck's and Kapteyn's strange result. The dwarf F and G stars are much less luminous than the giant F and G stars and so the systematic trend of stars becoming intrinsically fainter and redder along the spectral sequence is correct, provided only dwarf stars are selected.

In 1907, Hertzsprung turned his attention to star clusters, for which it can be assumed that all the stars are at the same distance. In 1911, he published the first luminosity–colour diagrams for the Pleiades and Hyades star clusters [23]. In these diagrams there was a prominent continuous sequence of stars which he named the *main sequence* but there was also a very wide range of luminosity among the red stars. These were the first published colour-magnitude diagrams.

Henry Norris Russell arrived at the same diagram by a rather different route. From 1902 to 1905, he worked at the Observatories at Cambridge, England, and, with Arthur Hinks, began one of the first photographic parallax programmes for stars. He returned to Princeton in 1905 and the reduction of the parallax data was completed by 1910. In 1908, Russell had made contact with Pickering who agreed to provide magnitudes and spectra for the 300 stars in the parallax programme. It was immediately apparent that high and low luminosity red stars were present in the sample, a result similar to Hertzsprung's. Russell's famous luminosity–spectral class diagram [24] was published simultaneously in *Nature* and *Popular Astronomy* in 1914

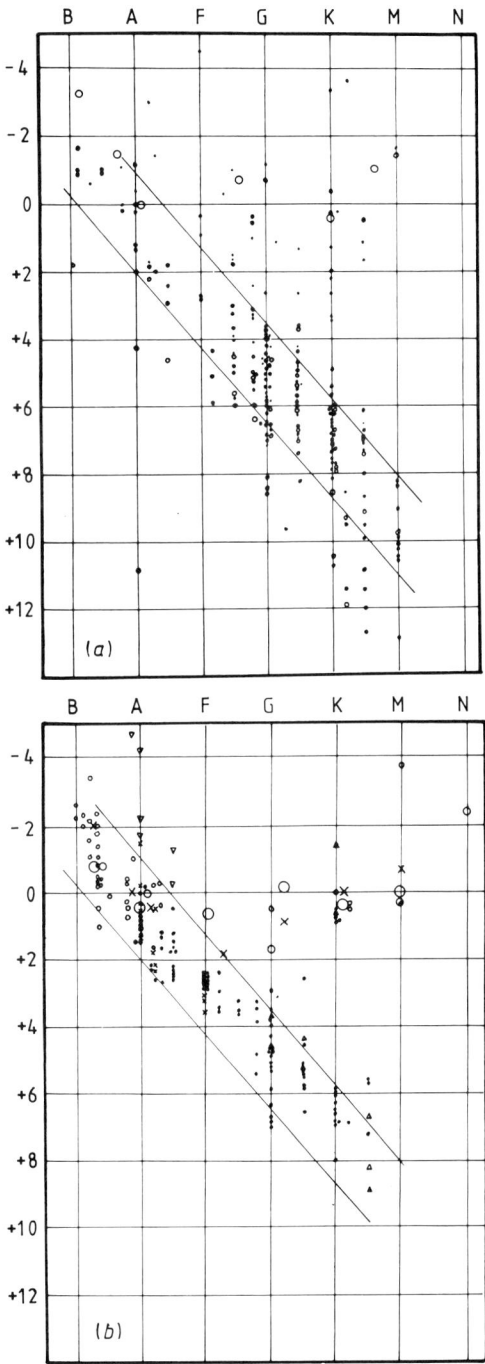

Figure 23.2. The first published 'Russell diagram' showing the relation between absolute magnitude and spectral type [24]. (a) The relation for all nearby stars; (b) the relation derived from studies of four star clusters.

(figure 23.2(a)). The correlation between spectral type and luminosity, indicated by the bounding diagonal lines in figure 23.2(a), corresponds to the *main sequence* described by Hertzsprung. In addition, there are red stars above the main sequence in a region which became known as the *giant branch*. The luminosities of the K and M stars span about 10 astronomical magnitudes, corresponding to a factor of 10 000 in luminosity. Russell's paper also included the luminosity–spectral class diagram for four clusters in which the giant branch is more clearly defined (figure 23.2(b)). The luminosity–spectral class diagram was originally known as the 'Russell diagram' until 1933 when Strömgren introduced the term Hertzsprung–Russell diagram.

Hertzsprung's analysis had shown that features of stellar spectra could be used to determine whether the stars were dwarfs or giants. Independently, Walter Adams and Arnold Kohlschütter [25] discovered other spectral features which could be used as luminosity indicators. By combining data on the parallaxes, proper motions and spectral types, they discovered that within a given spectral class, certain spectral features were sensitive luminosity indicators. In their paper of 1914, they showed that, using these criteria, the absolute magnitudes of stars could be estimated with an accuracy of about 1.5 magnitudes, corresponding to a factor of 4 in intrinsic luminosity. In consequence, the distances of stars could be roughly estimated from the characteristics of their spectra alone. Distances estimated in this way are referred to as *spectroscopic parallaxes*. The luminosity indicators were built into the system of spectral classification, which eventually superseded the Harvard classification system, called the Morgan, Keenan and Kellman (MKK) or Yerkes system [26] published in 1943. In addition to spectral types, the stars were assigned to five luminosity classes from type I, the supergiant stars, to type V, main sequence stars. The MKK system is the basis of the modern system of spectral classification.

One of the other great problems of stellar astronomy is the determination of stellar masses. In 1912, Russell's first graduate student, Harlow Shapley, made a detailed study of the light curves of eclipsing binary stars and from these he was able to demonstrate that the very brightest yellow and red stars were indeed 'giant' stars. Masses were determined for these stars from the parameters of their binary orbits and it was found that the range of luminosities was enormous compared with the range of masses. Russell found at best a weak correlation between luminosity and mass for the stars in his samples [27], consistent with the prevailing view that the red-giant stars represented the earliest phases of stars which contracted to the upper end of the main sequence. The main sequence then represented a cooling sequence as the stars grew older. A relic of these early (and quite incorrect) theories remains in the use of the term *early-type stars* to mean stars in the upper part of the main sequence and *late-type stars* for those on the lower main sequence.

23.3. Stellar structure and evolution

23.3.1. The impact of the new physics
Within ten years, the picture would be changed completely. Contrary to Russell's assertion that the dwarf stars on the main sequence all had more or less the same mass, Jacob Hahn in 1911 showed that there is a correlation between mass and luminosity along the main sequence [28]. By 1919, Hertzsprung [29] had derived an empirical mass–luminosity relation for main-sequence stars, $L \propto M^x$, with $x \approx 7$, somewhat greater than current values, which are about 4 for stars with mass roughly that of the Sun. To rescue the standard theory, the stars would have to lose mass.

A further important development was the idea that energy was transported through the star by radiation rather than by convection. This had first been discussed by R A Sampson in 1894 and studied in more detail by Schuster and Schwarzschild [30] in the early 1900s but had not been generally accepted by astronomers. In 1916, these ideas were revived by Eddington [31] who applied them to radiative transfer in giant stars.

Bohr's theory of atomic structure, published in 1913, had an immediate impact upon astrophysics [32]. The energy levels of atoms and ions in different states of ionization could be determined and this had consequences for the measurement of the temperatures of stellar atmospheres. The earliest attempts to measure the surface temperatures of the stars had assumed that they emit like black bodies, the technique which had been used by Hertzsprung [21] to measure the diameter of Arcturus in 1906. Temperatures had been measured using this technique by a number of workers but it suffered from the problem of the presence of absorption lines in the stellar spectra. To estimate the continuum intensity, observations had to be made between the prominent absorption lines. There remained the problem, however, of the unknown extent to which weak absorption lines depressed the continuum, a phenomenon known as line blanketing.

The first astronomer to apply the idea of using the state of ionization of atoms in stellar atmospheres as a means of measuring temperatures was the Indian astrophysicist Megh Nad Saha [33]. In 1919, Saha visited the German physical chemist Walther Nernst who was working on the thermodynamic theory of the equilibrium state of chemical reactions. Saha acknowledged that this work was the inspiration for his formulation of equilibrium ionization states. He described ionization as 'a sort of chemical reaction, in which we have to substitute ionization for chemical decomposition'. John Eggert, a pupil of Nernst, had already calculated the equilibrium state for eight-times ionized iron in stellar interiors, and Saha applied the same formalism to studies of the solar atmosphere. These considerations led to the famous *Saha equation* which describes the

Arthur Stanley Eddington

(British, 1882–1944)

Arthur Stanley Eddington, born in Kendal in the English Lake District, was the 'most distinguished astrophysicist of his time'. He had an outstanding career at school and university and attained the prestigious rank of first wrangler in only his second year at Cambridge. From 1906 to 1913, he was chief assistant at the Royal Greenwich Observatory, where he learned the techniques of observational astronomy. In 1913, he was appointed Plumian Professor of Astronomy at Cambridge, where he remained for the rest of his life.

His outstanding mathematical ability and his deep insight into physics were employed in tackling the most important problems in astrophysics and cosmology. His greatest contributions were in the astrophysics of the structure and evolution of the stars, which culminated in his great text *The Internal Constitution of the Stars* of 1926. He is remembered as the principal exponent of general relativity in English, his exposition of the theory in *Mathematical Theory of Relativity* of 1923 being regarded by Einstein as the finest presentation of the theory in any language. He was joint leader of the solar eclipse expeditions of 1919 to measure the gravitational deflection of starlight by the Sun. In cosmology, he was a proponent of what are now called the Eddington–Lemaître models of the Universe. In his later years, he devoted an enormous, but unsuccessful, effort to *The Fundamental Theory* in which he attempted to unify quantum theory and relativity and, in the process, to derive the values of the fundamental constants.

Being deeply religious and a Quaker, he declared himself a pacifist during World War I. He never married. He was knighted in 1930 and the Order of Merit was conferred upon him in 1938.

state of ionization of a gas in thermodynamic equilibrium at a given temperature. Boltzmann's equation and the equations of ionization equilibrium are combined in Saha's equation which determines how the state of ionization depends upon both the density and the temperature of the gas. To estimate temperatures, he used the method of 'marginal appearances' of lines based upon the first appearance and disappearance of different spectral lines employed in the Harvard sequence of stellar types. He concluded his paper with the remark that

> the stellar spectra may be regarded as unfolding to us, in an unbroken sequence, the physical processes succeeding each other as the temperature is continually varied from 3000 K to 40 000 K.

In other words, the spectral sequence O, B, A, F, G, K, M is a temperature sequence, the O stars being the hottest and the M stars the coolest.

These concepts were developed by Ralph Fowler and E A Milne [34] who provided a much more complete description of the equilibrium ionization states, including the effects of excited states of atoms and ions. Rather than simply using the first appearance or disappearance of different ions and atoms, they determined the conditions under which the absorption lines would have maximum strength. The way was opened up for precise determination of the abundances of the elements and this task was undertaken by Cecilia Payne, a pupil of Milne's, who carried out these studies at Harvard under Shapley as her PhD project. The most famous aspect of her work was the demonstration that, although the spectra of stars can vary widely, they all have remarkably similar chemical compositions, the principal cause of the differences being the surface temperatures of the stars. In her classic monograph *Stellar Atmospheres* [35], she stated that 'the uniformity of composition of stellar atmospheres appears to be an established fact'. She further showed that these abundances were similar to the terrestrial abundances with the exceptions of the elements hydrogen and helium, which she found to be vastly more abundant in the stars than on Earth. Although she had obtained the correct answer, she did not believe it. She wrote

> Although hydrogen and helium are manifestly very abundant in stellar atmospheres, the actual values derived from the estimates of their marginal appearances are regarded as spurious.

This conclusion reflected the prevailing prejudice. Three years later, in 1928, Albrecht Unsöld [36] showed that hydrogen is indeed very much more abundant than all other elements. This was confirmed by William H McCrea [37] who used the relative intensities of flash spectra to show that the number density of hydrogen atoms at the base of the chromosphere was the same as that found by Unsöld.

Henry Norris Russell was one of the pioneers who tested Saha's theory by comparing the relative intensities of the spectral lines of potassium and rubidium in the solar atmosphere and in sunspots [38]. In 1925, he investigated the problem of the anomalous triplet terms of the alkaline earth metals, calcium, scandium and barium. Russell and Frederick A Saunders [39] developed Alfred Landé's vector model of the atom to take account of the coupling between the spin and orbital angular momentum of the electrons, an interaction which became known as Russell–Saunders or L–S coupling. With this new understanding of atomic spectra, Russell, Walter Adams and Charlotte Moore [40] made a detailed study of the chemical abundances in the solar atmosphere. They used 1288 absorption lines in 228 different multiplets to find the relation between the strengths of the absorption features and the number of absorbing atoms or ions. In Russell's analysis of 1929, the solar abundances for 56 different elements, including hydrogen, and six diatomic molecules were determined [41]. These abundances were all within a factor of two of present estimates.

These procedures were adapted for the determination of the abundances of elements in the stars by Marcel Minnaert and Gerard Mulders [42] in 1930. They introduced the concept of the *equivalent width* of the spectral line, meaning the waveband of continuum radiation of the star which would correspond to the same amount of radiation removed from the continuum, when integrated over the observed line profile. They developed the procedures for relating the equivalent widths of lines in the spectrum to the number of absorbing atoms, taking account of the different processes which broaden the absorption lines, radiation damping, natural damping and thermal broadening. This led to what Minnaert called the *curve of growth* technique for relating the equivalent width of the line to the number of absorbing atoms. The same type of procedure was developed independently by Donald Menzel [43] in 1930 for emission lines. These have become the standard techniques of analysis of the abundances of the elements in stars.

23.3.2. Eddington and the theory of stellar structure
Arthur Stanley Eddington was the central figure in the development of the theory of the internal structure and evolution of the stars. Between 1916 and 1924, he published over a dozen papers which were collected and extended in his great book of 1926 *The Internal Constitution of the Stars* [44]. According to Henry Norris Russell [45], whose theory of stellar evolution was comprehensively demolished by Eddington

> Several investigators—Jeans, Kramers, Eggert—have contributed to this field, but much the largest share is Eddington's.

There is no simpler way of describing Eddington's achievement than to quote from Chandrasekhar's assessment [46]

In the domain of the internal constitution of the stars, Eddington recognized and established the following basic elements of our present understanding:

(1) Radiation pressure must play an increasingly important role in maintaining the equilibrium of stars of increasing mass.

(2) In parts of the star in which radiative equilibrium, as distinct from convective equilibrium, obtains, the temperature gradient is determined jointly by the distribution of the energy sources and of the opacity of the matter to the prevailing radiation field. Precisely

$$\frac{dp_r}{dr} = -\kappa \frac{L(r)}{4\pi cr^2}\rho \qquad p_r = \tfrac{1}{3}aT^4$$

and

$$L(r) = 4\pi \int_0^r \varepsilon\rho r^2\, dr$$

where p_r, κ, ε, and ρ denote, respectively, the radiation pressure, the coefficient of stellar opacity, the rate of energy generation per gram of stellar material, and the density.

(3) The principal physical process contributing to the opacity, κ, is determined by the photo-electric absorption coefficient in the soft x-ray region, that is, by the ionization of the innermost K- and L-shells of the highly ionized atoms.

(4) With electron scattering as the ultimate source of stellar opacity, there is an upper limit to the luminosity, L, that can support a given mass M. The maximum luminosity, set by the inequality $L < 4\pi cGM/\sigma_e$, where σ_e denotes the Thomson scattering coefficient, is now generally referred to as the *Eddington limit*.

(5) In the first approximation, in normal stars (i.e. in stars along the main sequence, the (mass, luminosity, effective temperature)-relation is not very sensitive to the distribution of the energy sources through the star. Therefore, a relation is available for comparison with observations even in the absence of a detailed knowledge of the energy sources of the star.

(6) The burning of hydrogen into helium is the most likely source of stellar energy.

These great insights were not gained without a struggle and many of these issues were the subject of heated debate between Eddington and James Jeans. In his first paper on the internal structure of the stars, Eddington had assumed that the mean atomic mass of the particles was 54, meaning that the star was composed of iron [31]. This was quickly corrected by Jeans who pointed out that, at the high temperature of the

Figure 23.3. *Michelson's stellar interferometer mounted on the 100" telescope on 10 August 1920* [49].

stellar interior, 'a rather extreme state of disintegration is possible' [47]. In Eddington's next paper, a mean atomic weight of 2 was adopted, corresponding to the complete ionization of the atoms, assuming that there is no hydrogen present. Eddington still adhered to the prevailing Russell–Lockyer picture and so applied his theory of radiative transfer to the envelopes of the giant stars which were assumed to be gaseous spheres. In his paper of 1917, he showed that, if the release of gravitational energy was the source of the luminosity of the giant stars, they could not radiate for more that 100 000 years, which is very much less than the age of the Earth [47].

The opportunity of testing Eddington's theory of red giants arose in 1919. Albert A Michelson [48] had been developing the techniques of optical interferometry for almost 30 years and George Ellery Hale, director of the Mount Wilson Observatories, decided that the 100" Hooker telescope should be equipped with a Michelson interferometer to determine the separations of close binary stars, if not the diameters of the stars themselves. Michelson did not know how long the interferometer baseline would have to be but, being aware that the instrument was in the process of construction, Eddington used his theory of the structure of red giants to predict the angular size of Betelgeuse. In the light of this prediction, Michelson built a 6 metre interferometer which was mounted on the top ring of the 100" telescope (figure 23.3). On the

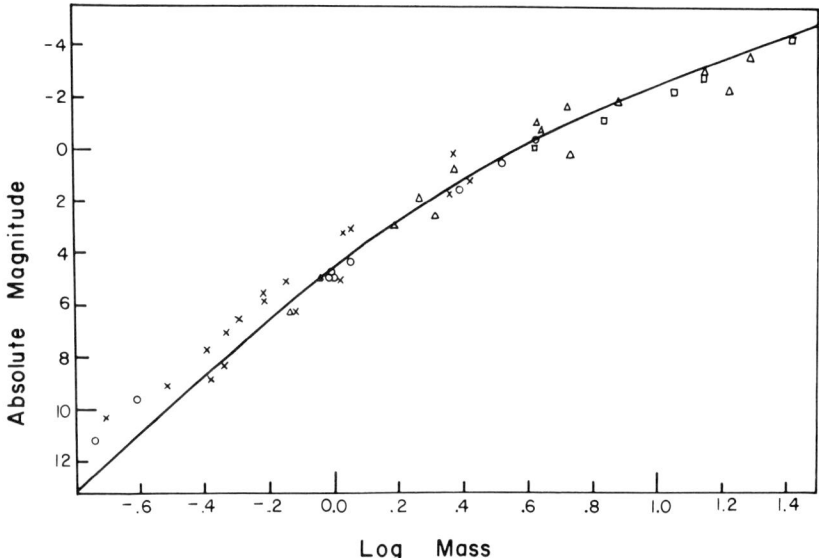

Figure 23.4. *The observed mass–luminosity relation for stars compared with Eddington's theoretical mass–luminosity relation [50]. The solid line shows Eddington's theoretical mass–luminosity relation. The circles, crosses, squares and triangles are data on the masses and luminosities of different types of star.*

night of 13 December 1919, Francis G Pease and J A Anderson measured the angular diameter of Betelgeuse to be 0.047 arcsec, just slightly less than Eddington's prediction [49]. This observation confirmed the large diameters of the red giants. They went on to measure the diameters of five more red giants.

In 1919, Eddington [50] discovered to his surprise that he could account for the observed mass–luminosity relation for main-sequence stars if he applied his theory of stellar structure to the dwarf stars as well (figure 23.4). The implications were profound—the main-sequence stars could not be incompressible liquid spheres but rather gaseous spheres. This cut the foundation from under the standard Russell picture. This conclusion was vigorously opposed by Jeans who argued that the result was spurious since it ignored the process of energy generation inside the stars. As noted by Chandrasekhar, it turns out that the mass–luminosity relation is remarkably independent of the precise process of energy production. Jeans proposed that the source of energy in the Sun was radioactive decay.

In his early papers, Eddington advocated the annihilation of matter as an inexhaustible source of energy for the stars. In 1920, he found a better way in which energy could be generated. In a remarkably prescient paragraph of his Presidential Address to the Mathematical and Physics Section of the British Association for the Advancement of Science at its annual meeting which was held in Cardiff, he stated [51]

> F W Aston's experiments seem to leave no room for doubt that all the elements are constituted out of hydrogen atoms bound together with negative electrons. The nucleus of the helium atom, for example, consists of 4 hydrogen atoms bound with two electrons. But Aston has further shown conclusively that the mass of the helium atom is less than the sum of the masses of the 4 hydrogen atoms which enter into it; ...There is a loss of mass in the synthesis amounting to about 1 part in 120, the atomic weight of hydrogen being 1.008 and that of helium 4. ...Now mass cannot be annihilated, and the deficit can only represent the mass of the electrical energy set free in the transmutation. We can therefore at once calculate the quantity of energy liberated when helium is made out of hydrogen. If 5 per cent of the star's mass consists initially of hydrogen atoms, which are gradually being combined to form more complex elements, the total heat liberated will more than suffice for our demands, and we need look no further for the source of a star's energy.

At that time, this could be no more than a hypothesis but it is indeed the source of energy in the Sun.

23.3.3. *The impact of quantum mechanics and the discovery of new particles*

The resolution of the problem of energy generation in the Sun was one of the immediate fruits of the remarkable theoretical and experimental discoveries in physics of the 1920s and 1930s. These advances were quickly assimilated into astrophysics. Fermi–Dirac statistics [52], discovered in 1926, had application to the equation of state of matter in stars; quantum-mechanical tunnelling was discovered by George Gamow [53] and applied to the inelastic scattering of α-particles by nuclei in 1928, Wolfgang Pauli [54] proposed the existence of the neutrino in 1931 and Enrico Fermi's theory of β-decay was formulated in 1933 [55]. Equally important were the experimental discoveries of this golden age of physics. The discovery of the positron in cosmic-ray cloud chamber experiments was announced by Carl D Anderson [56] in 1931; in the same year, Harold Urey, Ferdinand Brickwedde and George Murphy [57] discovered deuterium through spectroscopic studies of 'distilled' liquid hydrogen; in 1932, the neutron was discovered by James Chadwick [58].

Eddington's proposal that nuclear energy could provide the energy of the Sun was still a matter of speculation. Chandrasekhar visited Copenhagen in 1932 to work with Niels Bohr and his associates and recorded Bohr's attitude [59]

> I cannot be really sympathetic to work in astrophysics because the first question I want to ask when I think of the Sun is where

does the energy come from. You cannot tell me where the energy comes from, so how can I believe all the other things?

Even by then, however, the great discoveries of quantum mechanics were beginning to be applied to the problem of nuclear-energy generation in the Sun.

The problem was that, even at the high temperatures of stellar interiors, the Coulomb repulsion between protons and nuclei is so great that, according to classical physics, this nuclear-energy source cannot be tapped. The solution of the problem had to await Gamow's theory of quantum-mechanical tunnelling. Only one year later, in 1929, Robert Atkinson and Fritz Houtermans [60] applied Gamow's theory to nuclear reactions in the hot central regions of stars. By considering the process of barrier penetration by a Maxwellian distribution of protons, they established two key features of the process of nuclear-energy generation in stars. First, the most effective energy sources involve interactions of nuclei of small electric charge since the Coulomb barriers are smaller than for nuclei with large charges and, second, the particles which penetrate the Coulomb barriers are those few particles in the high-velocity tail of the Maxwellian distribution. As a result, the nuclear reactions can take place at temperatures which are considerably lower than might have been expected. These ideas also suggested why the luminosity of the stars should be a sensitive function of temperature. As the central temperature increases, the rate of barrier penetration increases and so hotter stars should be much more luminous than cooler stars.

By 1931, it had been established that hydrogen was by far the most abundant element in the stars and so Atkinson's objective was to account for the origin of the chemical elements by the successive addition of protons to nuclei. He argued that the formation of helium by the combination of four protons was a very rare process indeed and proposed that, instead, helium could be formed by the successive addition of protons to heavier nuclei which, when they exceeded the mass limit for nuclear stability, would eject α-particles, thus creating helium [61]. This proposal was the precursor of the carbon–nitrogen–oxygen cycle (CNO cycle) which was discovered independently by Carl von Weisäcker and Hans Bethe [62] in 1938. In this famous cycle, carbon acts as a catalyst for the formation of helium through the successive addition of protons accompanied by two β^+ decays as follows

$$^{12}C + p \rightarrow {}^{13}N + \gamma\,; {}^{13}N \rightarrow {}^{13}C + e^+ + \nu_e\,; {}^{13}C + p \rightarrow {}^{14}N + \gamma$$
$$^{14}N + p \rightarrow {}^{15}O + \gamma\,; {}^{15}O \rightarrow {}^{15}N + e^+ + \nu_e\,; {}^{15}N + p \rightarrow {}^{4}He + {}^{12}C.$$

In the meantime, it had become possible to make estimates of the reaction rates for the simplest nuclear reaction, the combination of pairs of protons to form deuterium, which can then combine with other deuterons to form

^{3}He and ^{4}He. The first calculations were carried out by Atkinson [63] in 1936 and were much refined by Bethe and Critchfield in 1938, who combined Fermi's theory of the weak interaction with Gamow's theory of barrier penetration [64]. The principal series of reactions in the proton–proton (or p–p) chain are as follows:

$$p + p \to {}^2H + e^+ + \nu_e \quad : \quad {}^2H + p \to {}^3He + \gamma$$
$$^3He + {}^3He \to {}^4He + 2p.$$

The crucial first reaction in the chain involves a weak interaction in which a positron and neutrino are released in what may be thought of as the transformation of one of the protons into a neutron. This reaction accounts for most of the energy release in the p–p chain but it has never been measured experimentally at the energies of interest for nucleosynthesis in the Sun. Bethe and Critchfield showed that this series of reactions could account for the luminosity of the Sun. In addition, they found that the rate of energy production, ε, of the p–p chain depends upon the central temperature of the star as $\varepsilon \propto T^4$. In 1939, Bethe [62] worked out the corresponding energy production rate for the CNO cycle and found a very much stronger temperature dependence, $\varepsilon \propto T^{17}$. He concluded that the CNO-cycle was dominant in massive stars while the p–p chain was the principal energy source for stars with mass less than about 1.5 times the mass of the Sun. These conclusions have been confirmed by the much more detailed models of stellar structure which became available after World War II and, in particular, with the development of computer codes which have converted the study of stellar structure into one of the most precise of the astrophysical sciences.

23.3.4. The red-giant problem
There remained the problem of accounting for the giant stars which radiate very much greater luminosities than the main-sequence stars at relatively low temperatures. The solution was found in 1938 by the Estonian astrophysicist Ernst Öpik [65]. In the case of stars like the Sun, he realized that nuclear burning takes place within a region which comprises only the central 10% by mass of the star. Energy transport within this core is by radiative transport if the temperature gradient is less than the adiabatic lapse rate and by convection if the temperature gradient exceeds that value. In either case, the net effect is to deplete the available hydrogen fuel within the core of the star. Öpik discussed what happens when the fuel in the core is exhausted. There is no longer pressure support for the central regions of the star and so the core must collapse. The temperatures of the core and the surrounding hydrogen layers increase so that hydrogen burning can take place in a shell about the inert core. Eventually, however, the core collapses until it is hot

Astrophysics and Cosmology

enough for the nuclear reactions to begin the synthesis of carbon from helium nuclei.

Öpik's major discovery was the fact that, when the core of the star collapses, the outer envelope expands to an enormous size. The exact physical cause of the formation of the giant stars is one of the continuing mysteries of stellar structure. Every computation of the evolution of stars onto the giant branch shows that, as hydrogen burning in the core is exhausted, the core collapses and the envelope expands, but the dramatic changes in the star's structure cannot be attributed to any single cause. Many different processes take place simultaneously—the collapse of the central regions, changes in chemical composition of the stellar material with radius and consequently changes in its opacity, the development of extensive convective zones in the stellar envelope and so on. The quantitative theory of red-giant stars needed a more complete theory of the nuclear reactions by which helium nuclei are converted into carbon and this was first discussed by Öpik [66] and then by Salpeter [67] after World War II.

As a result of Öpik's work, there was no need to seek separate physical processes to account for the giant stars—they form quite naturally at the end of the phase of hydrogen burning on the main sequence. The giant phase only lasts a short time compared with the age of stars on the main sequence, because in the giant phase stars burn up the available nuclear fuel at thousands of times the rate at which it is consumed on the main sequence. Thus, the giant phase is a brief final fling before the star settles down to some form of dead star. As Öpik pointed out, this picture was entirely consistent with the observation that the red-giant stars are very much rarer per unit volume of space than the dwarf stars.

The final link in the chain was provided in 1942 by the Brazilian astrophysicist Mario Schönberg and Chandrasekhar [68] who showed that there do not exist stable stellar models in which the helium core contains more than about 10% of the mass of the star. This key result, known as the *Schönberg–Chandrasekhar limit*, explained the formation of red-giant stars during the course of stellar evolution. Stars spend most of their lifetimes on the main sequence, the energy source being the conversion of hydrogen into helium. Hydrogen is depleted in the central regions resulting in the formation of an inert core. When the core grows to 10% of the mass of the star, core collapse and the formation of the red-giant envelope ensues.

23.3.5. White dwarfs

The discovery of white dwarfs is charmingly told in a reminiscence of Henry Norris Russell delivered at a colloquium in Princeton in 1954. In 1910, Russell suggested to Pickering that it would be useful to obtain the spectra of stars for which parallaxes had been measured. Russell's reminiscence continues [69]

> Pickering said 'Well, name one of these stars'. 'Well', said I, 'for example, the faint component of Omicron Eridani'. So Pickering said, 'Well, we make rather a speciality of being able to answer questions like that'. And so we telephoned down to the office of Mrs Fleming and Mrs Fleming said, yes, she'd look it up. In half an hour she came up and said, 'I've got it here, unquestionably spectral type A'. I knew enough, even then, to know what that meant. I was flabbergasted. I was really baffled trying to make out what it meant. ...Well, at that moment, Pickering, Mrs Fleming and I were the only people in the world who knew of the existence of white dwarfs.

The remarkable feature of the faint companion of o-Eridani was that it was a very low-luminosity star and yet it had the type of spectrum which was associated with hot stars on the upper part of the main sequence. Russell included it without comment in his first 'Russell' diagram (figure 23.2(a)), the single A star lying roughly 10 magnitudes below typical main-sequence A stars. Walter Adams [70] drew attention to its remarkable properties in 1914 and discovered another example in the following year, the faint companion of Sirius A known as Sirius B. Eddington realized that these observations implied that white-dwarf stars had to be very dense indeed. Their masses could be determined since both examples are members of binary star systems and their radii could be estimated from Planck's radiation formula and the luminosity of the stars. Their mean density was found to be about 10^8 kg m^{-3}. Eddington [71] argued that there was nothing inherently implausible about such large densities. Matter at high temperatures inside stars would be completely ionized and so there was no reason at that time why the matter could not be compressed to much higher densities than typical terrestrial densities. In fact, he argued that even nuclear densities were quite conceivable. In his paper of 1924, he estimated the gravitational redshift which would be expected from such a compact star according to general relativity and found that it corresponded to a Doppler shift of the spectral lines to longer wavelengths of about 20 km s^{-1}. Adams [72] made very careful spectroscopic observations of Sirius B with the 100" telescope in 1925 and, once account was taken of the orbital motions of the binary stars, a shift of 19 km s^{-1} was measured. Eddington [73] was jubilant

> Prof Adams has thus killed two birds with one stone. He has carried out a new test of Einstein's theory of general relativity, and he has shown that matter at least 2000 times denser than platinum is not only possible, but actually exists in the stellar Universe.

The theory of white dwarfs was one of the first triumphs of the new quantum theory of statistical mechanics as applied to astrophysics. Pauli [74] had enunciated the exclusion principle in 1922 and this led to Fermi–Dirac statistics and the concept of degeneracy pressure. In 1926, Ralph Fowler [75] used these concepts to derive the equation of state of a cold degenerate electron gas and found the important result $p \propto \rho^{5/3}$. This equation of state is independent of the temperature and so the structure of white dwarfs can be derived directly from the Lane–Emden equation. Unlike main-sequence stars, in which pressure support is provided by the thermal pressure of hot gas, the white dwarfs are supported by electron degeneracy pressure. The source of their luminosity is the internal thermal energy with which they were endowed on formation. According to Fowler's picture, the white dwarfs radiate away their internal thermal energy and end up as inert cold stars with all the nuclei and electrons in their lowest ground state.

In 1929, Wilhelm Anderson [76] had shown that the degenerate electrons in the centres of white dwarfs with mass roughly that of the Sun become relativistic. In the extreme relativistic limit, the equation of state of the degenerate electron gas becomes $p \propto \rho^{4/3}$. Again, the result is independent of temperature but the change in the dependence of the pressure upon density from $p \propto \rho^{5/3}$ to $p \propto \rho^{4/3}$ has profound implications. Anderson and Edmund Stoner [77] realized that the consequence was that there do not exist equilibrium configurations for degenerate stars with mass greater than about the mass of the Sun. The most famous analysis of this problem was due to S Chandrasekhar who had begun working on this problem before he arrived to take up his fellowship at Trinity College, Cambridge in 1930. According to Wali's biography [78], he derived the key result while on board the ship *Lloyd Triestino* which was taking him as a 19 year-old from Bombay to London. He found the crucial result that, in the extreme relativistic limit, there is an upper limit to the mass of stable white dwarfs which is $M_{Ch} = 1.46 M_\odot$ for typical stellar material. This mass, M_{Ch}, which depends only upon fundamental constants, is known as the *Chandrasekhar mass* [79]. The cause of the instability is that, in the extreme relativistic limit, both the internal thermal energy U_{th} and the gravitational potential energy U_{grav} of the star depend upon radius in the same way, $U_{th} = \frac{1}{2} U_{grav} \propto R^{-1}$. The gravitational potential energy is proportional to M^2 whereas the internal energy is proportional to M. Therefore, for massive enough stars, the gravitational energy term dominates, causing collapse which cannot be stabilized by the pressure of the degenerate gas, since the two energies always depend upon radius in the same way. The inference is that there is nothing to prevent degenerate stars more massive than M_{Ch} from collapsing to very high densities indeed and possibly to a state of complete gravitational collapse. This conclusion was vigorously challenged by Eddington and led to the famous dispute with Chandrasekhar [80]. Eddington found the idea of complete gravitational collapse unacceptable and believed that there must be some new unspecified physical process which prevented it occurring.

Quite independently, in 1932 Lev Landau [81] had come to the conclusion that gravitational collapse to a singularity should be taken seriously and in 1938 Robert Oppenheimer and Hartland Snyder [82] gave the first general-relativistic analysis of what would be observed in the final stages of gravitational collapse of a pressureless sphere. In their paper, they described many of the key features of what are now termed black holes.

23.3.6. Supernovae and neutron stars

The neutron had been discovered by Chadwick [58] in 1932 and the model of the nucleus consisting of neutrons and protons was quickly adopted, although the problem of how the nucleus could be held together remained to be resolved. The first mention of the possibility of neutron stars appears as a famous 'Additional Remark' to a paper by Walter Baade and Fritz Zwicky [83] in 1934. In that year, they published two papers on the energetics of what they termed 'super-novae'. The extragalactic nature of the 'white nebulae' had been established in the previous decade. Among the objects which played a part in that debate were the novae or 'new stars' which increase rapidly in brightness and then fade away. Some events, apparently of this type, had been observed in nearby galaxies. In their paper, Baade and Zwicky proposed that the population of novae consists of two types, the ordinary novae which are relatively common phenomena, and the 'super-novae', which are very rare but are very energetic indeed. They identified Tycho Brahe's supernova of 1572 and the bright 'nova' observed in the Andromeda Nebula in 1885 as examples of this class of extremely violent explosion. They estimated that the frequency of occurrence of these events was only about once per thousand years per galaxy but, when they occurred, an enormous amount of energy was released, corresponding to a significant fraction of the rest-mass energy of the precursor star. In their second paper, they suggested that such events might be the sources of the cosmic ray particles which had been discovered by Victor Hess [84] in 1912. As an addendum to the second paper they wrote

> With all reserve we advance the view that a super-nova represents the transition of an ordinary star into a *neutron star*, consisting mainly of neutrons. Such a star may possess a very small radius and an extremely high density. As neutrons can be packed much more closely than ordinary nuclei and electrons, the 'gravitational packing' energy in a *cold* neutron star may become very large, and under certain circumstances, may far exceed the ordinary nuclear

packing fractions. A neutron star would therefore represent the most stable configuration of matter as such.

It is best to allow Zwicky to describe how these ideas were received in a quotation from the extraordinary preface to his *Catalogue of Selected Compact Galaxies and of Post-Eruptive Galaxies* [85] of 1968

> In the Los Angeles Times of January 19, 1934, there appeared an insert in one of the comic strips, entitled 'Be Scientific with Ol'Doc Dabble' quoting me as having stated 'Cosmic rays are caused by exploding stars which burn with a fire equal to 100 million suns and then shrivel from $\frac{1}{2}$ million miles diameter to little spheres 14 miles thick', Says Prof. Fritz Zwicky, Swiss Physicist.' This, in all modesty, I claim to be one of the most concise triple predictions ever made in science. More than thirty years were to pass before the statement was proved to be true in every respect.

Indeed, the idea that a neutron star is formed in a supernova explosion was proved to be correct with the discovery of pulsars [86] in 1967.

In the meantime, however, Gamow [87] showed in 1937 that a gas of neutrons could be compressed to a much higher density than a gas of nuclei and electrons, and estimated the probable densities of such stars to be about 10^{17} kg m^{-3}. The issue of the maximum mass of neutron stars was discussed by Landau [88] in 1938 and in much greater detail by Oppenheimer and George Volkoff [89] in the following year. The result they found is not so different from the expression for the upper mass limit for white dwarfs. The physics is the same as in the case of the white dwarfs but now neutron degeneracy pressure holds the star up. Complications arise because it is necessary to take into account the details of the equation of state of neutron matter at nuclear densities and the effects of general relativity can no longer be neglected. They found an upper mass limit of about $0.7M_\odot$. This result is not so different from the best modern estimates which correspond to about $2-3M_\odot$. This is a very much more serious situation than the case of the white dwarfs. The neutron stars are so compact that general relativity is no longer a small correction but is central to the stability of the star. Typically, for neutron stars, the general-relativistic parameter $2GM/Rc^2 \sim 0.3$ and so they have radii which are only about three times the Schwarzschild radius of a black hole of the same mass.

This work created some theoretical interest but little enthusiasm from the observers. The radii of typical neutron stars were expected to be about 10 km and so there was no prospect of detecting significant fluxes of thermal radiation from such tiny stars. Nonetheless, many of the objects which were to play a leading role in the development of high-energy astrophysics in the years following World War II were already in place in the literature, even if there was not a great deal that the astronomers could do about them at that time.

PART 2: THE LARGE-SCALE STRUCTURE OF THE UNIVERSE 1900–39

23.4. The structure of our Galaxy

In the mid-eighteenth century, philosophers such as Emanuel Swedenborg, Thomas Wright, Immanuel Kant and Jean Lambert had speculated that the stars are grouped into *island universes* of which the Milky Way was an example viewed from the inside. These ideas had little physical foundation and were not taken seriously by the astronomers. Towards the end of the eighteenth century, however, William Herschel made the first attempt to define the structure of our Galaxy. He made counts of stars in different directions and, assuming the stars all had the same intrinsic luminosity, he derived his famous picture in which the Milky Way consists of a flattened disc of stars with a diameter about five times its thickness [90]. The fundamental problem with his analysis was that he had no satisfactory way of measuring astronomical distances.

To determine the scale and structure of the system of stars in the Galaxy, Jacobus Kapteyn [91] in 1906 drew up a plan of 206 selected areas in which deep star counts and proper motions were to be measured. By the time the analysis of the star counts was made, it was known that there is a huge range of intrinsic luminosities among random samples of stars and so, to interpret the counts, the distribution of their luminosities in a typical volume of space had to be known—this distribution is known as the stellar *luminosity function*. By 1920, Kapteyn and Pieter J van Rhijn [92] found that the luminosity function could be approximated by a Gaussian distribution with mean absolute magnitude $\bar{M} = 7.7$ and halfwidth of a few magnitudes. Assuming this luminosity function applied to stars throughout the Galaxy, Kapteyn [93] found that it was highly flattened with dimensions 1500 pc perpendicular to the plane and about eight times that size in the Galactic plane (figure 23.5).

Meanwhile, Harlow Shapley was adopting a different approach to the determination of the structure of the Galaxy. At Harvard, Henrietta S Leavitt had been assigned the task of finding the variable stars in the Magellanic Clouds. The advantage of studying systems such as the Magellanic Clouds is that, although their distances may not be known, it is safe to assume that all their members are at the same distances and hence the *relative* luminosities of the stars can be found. Among the 1777 variable stars which Leavitt discovered in the Clouds were a number of Cepheid variables, which have distinctive periodic light curves. In her paper of 1912, the periods and apparent magnitudes of 25 Cepheid variables were reported and the period–luminosity relation displayed for the first time [94] (figure 23.6).

This discovery provided a powerful means of measuring astronomical distances because the Cepheid variables are intrinsically luminous stars

Astrophysics and Cosmology

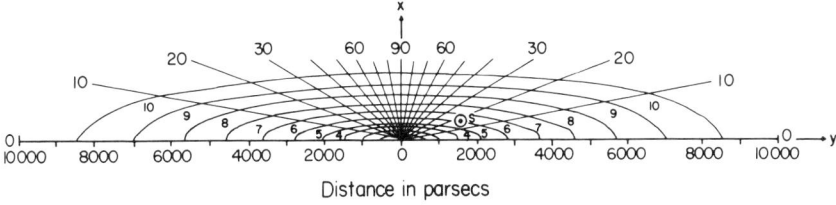

Figure 23.5. *Kapteyn's model for the distribution of stars in the Galaxy [93]. The diagram shows the distribution of stars in a plane perpendicular to the Galactic plane. The curves are lines of constant number density of stars and are in equal logarithmic intervals. The Sun (marked S) is slightly displaced from the centre of the system.*

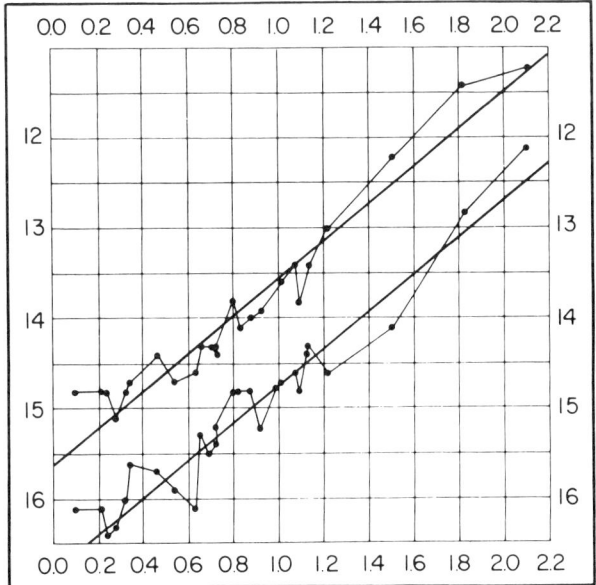

Figure 23.6. *A plot of the period–luminosity relation for the 25 Cepheid variables discovered by Leavitt [94] in the Small Magellanic Cloud. The abscissa is the logarithm of the period of the Cepheid variable in days and the ordinate is the apparent magnitude of the stars. The upper locus is found for the maximum light of the Cepheids and the lower line for their minimum brightnesses.*

and their distinctive light curves can be identified in distant systems. Once the period–luminosity relation had been calibrated locally, the absolute luminosity could be found from the Cepheid's period and hence its distance estimated from its apparent magnitude. This procedure was first carried out by Hertzsprung [95] in 1913 and he derived a distance of 30 000 light years for the Small Magellanic Cloud, the greatest distance of any astronomical object measured at that time, but five times smaller than the best estimate today.

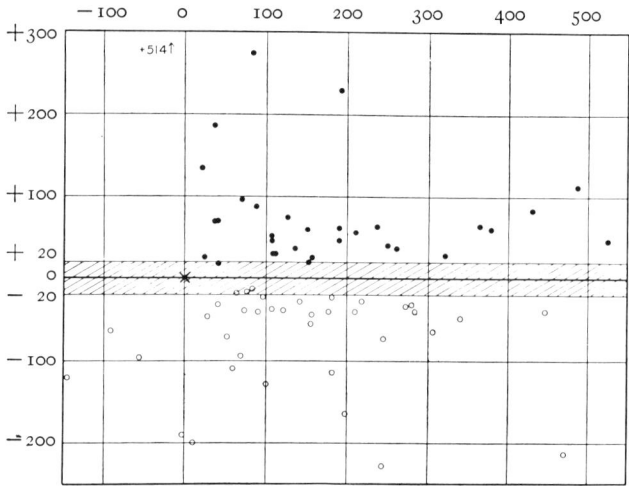

Figure 23.7. *The distribution of globular clusters in the Galaxy according to Shapley's distance measurements* [96]. *The coordinates are distances measured in thousands of light years in the direction of the Galactic centre (abscissa) and perpendicular to the Galactic Plane (ordinate). The Sun is not located at the centre of the system.*

By 1918, Shapley realized that the system of globular clusters provided a means of determining the large-scale structure of our Galaxy. The globular clusters were among the objects generally classed as 'nebulae' but they were clearly resolved into individual stars. In contrast to most of the components of the Milky Way, they extend to high galactic latitudes. Cepheid variables were identified in the globular clusters and the distances found in this way were entirely consistent with observations of giant stars and other characteristic stars found in globular clusters. The scale of the system of globular clusters was enormous and, furthermore, it was not centred upon the Solar System but, rather, most of the clusters are found in a direction centred upon the constellation of Sagittarius. Shapley [96] plotted the distances of the globular clusters and found that the Solar System is located towards the edge of the globular-cluster system (figure 23.7). He estimated the distance to the Galactic centre to be about 50 000 light years, an overestimate because the role of interstellar extinction was not then appreciated. Shapley's picture of the Galaxy differed radically from Kapteyn's Sun-centred Universe, Shapley arguing that Kapteyn's studies referred only to the nearby part of the Galactic system.

23.5. The great debate

These different approaches led to what became known as the *Great Debate* [97]. There were two separate questions to be resolved, the first concerning the scale and structure of our Galaxy and the second concerning the nature of the spiral nebulae. The first question centred on the contrast between Shapley's model of the Galaxy, which had dimension 300 000 light years and in which the Sun is located towards the edge of the system, and the Sun-centred model of Kapteyn in which the Galaxy has dimensions of 30 000 light years.

The second question concerned the issue of whether the spiral nebulae are 'island universes' or constituent members of our own Galaxy. The systematic cataloguing of the nebulae was begun by William Herschel and continued by his son John Herschel who, in 1864, published the *General Catalogue of Nebulae* [98] containing 5079 objects, of which all but 449 were discovered by the Herschels. It provided a large fraction of the entries in the *New General Catalogue of Nebulae and Clusters of Stars* [99] published by J L E Dreyer in 1888. Dreyer produced two supplements to the NGC catalogues known as the *Index Catalogues* [100]. In all, these catalogues contain some 15 000 nebular objects. The process of cataloguing bright nebulae was completed by 1908 but their nature remained a mystery, largely because their distances were unknown. William Huggins and William Miller [101] had shown convincingly from their pioneering spectroscopic observations that some of the nebulae were hot gas clouds, because of the presence of strong emission lines in their spectra, but the nature of the others, 'the white nebulae', was not understood.

In 1917, George W Ritchey [102] discovered by chance a nova in the spiral nebula NGC6946. This led to searches through the plate archives of the major observatories for further examples of novae in spiral nebulae. Heber D Curtis and Shapley [103] announced the discovery of several other novae so that, by the end of 1917, 11 novae were known to have taken place in seven spiral nebulae, four of them having occurred in the Andromeda Nebula. Curtis noted that, at maximum, the novae in our Galaxy typically have apparent magnitudes about 5.5 whereas those in the spiral nebulae were about ten magnitudes fainter. Hence, if they were the same types of object, the spiral nebulae would have to be 100 times more distant than their Galactic counterparts. Shapley drew the same conclusion, estimating the distance of the Andromeda nebula to be 50 times the distance of the nearby novae, that is, about 1 000 000 light years distant.

There were, however, two big problems. The first was that a nova had been observed in the Andromeda Nebula in 1885 which was six magnitudes brighter than the novae which had been used to make the distance estimate. If the Andromeda Nebula really was at a distance of 1 000 000 light years, the nova of 1885 would have been more than 100 times brighter than typical nearby novae. If the nova of 1885 was regarded as a typical nova, the distance of the nebula would have been ten times smaller. A second problem was that Adriaan van Maanen [104] claimed to have measured the proper motions of the arms of the bright spirals M31 and M33 which, at a distance of 1 000 000 light years, would have meant speeds approaching the speed of light.

This was the background to the Great Debate between Shapley and Curtis [105] which was held at the National Academy of Sciences in Washington on 20 April 1920. Shapley took the 'scale of the Universe' to mean the size of the globular cluster system, which he found to have dimensions about 100 000 light years. He accepted van Maanen's observations of the proper motions of the arms of spiral nebulae, which he assumed formed part of an extensive halo about the Galaxy. He also pointed out that the surface brightnesses of the spiral nebulae are very much greater than the surface brightness of the plane of our Galaxy in the vicinity of the Sun and so it was not evident that the spiral nebulae were the same class of object as our Galaxy. There was also the problem that, if Shapley's large dimensions for our Galaxy were adopted, then, even if a distance as large as 1 000 000 light years were adopted for the Andromeda Nebula, our Galaxy would still have been very much larger than the typical spiral nebula and so retain a unique position in the Universe.

Curtis defended the smaller distances inferred from Kapteyn's statistical studies and the 'island universe' picture. He made what turned out to be the correct inference that van Maanen's reported motions of the spiral arms of nebulae were spurious and placed considerable weight upon the use of the novae as distance indicators.

The most serious problem was the neglect of interstellar extinction which affected both Shapley's and Kapteyn's analyses in different ways. Interstellar dust absorption and scattering in the plane of the Galaxy was also responsible for the observation that the spiral nebulae avoid the Milky Way. The central regions of our Galaxy in fact have very similar surface brightnesses to those of the spiral nebulae but interstellar extinction prevents us observing these regions directly in the optical waveband.

Gradually, the origin of the discrepancies between the two pictures was understood. Between 1917 and 1919, the Swedish astronomer Knut Lundmark [106] discovered 22 novae in the Andromeda Nebulae and, if these were assumed to be similar to Galactic novae, a distance of 650 000 light years was found. Lundmark made the distinction between two classes of novae, those used to make the distance measurements belonging to the 'lower class' while novae such as that of 1885 were assigned to the 'upper class' and were to be identified with 'super-novae' by Baade and Zwicky [83] in the 1930s. In 1921, Lundmark made a detailed spectroscopic study of the spiral arms of M33 as well as some of its brightest stars. In 1921, he wrote [107]

> Some objects [in the arms] have a nebular spectrum but most of the objects belonging to the spiral show a strong continuous spectrum without bright lines ... From the spectral evidence, it

seems probable that the spiral nebula consists of ordinary stars, clusters of stars, and some nebular [that is, gaseous] material.

The conclusive proof of the extragalactic nature of the spiral nebulae was provided by Edwin P Hubble [108] in 1925. He discovered 22 Cepheids in M33 and 12 in M31 and they displayed exactly the period–luminosity relation found for Cepheids in the Magellanic Clouds. He was therefore able to make good distance estimates for M31 and M33 of 285 000 light years, much greater than Shapley's largest estimate for the size of our Galaxy.

The extragalactic nature of the spiral galaxies was established and Hubble immediately began to use them as a tool for understanding the large-scale structure of the Universe. He appreciated that the means were now available by which cosmological questions could be addressed by astronomical observation. In the next year, 1926, Hubble [109] published a major study of galaxies which begins with his famous classification scheme, distinguishing between the main classes of ellipticals, normal spirals, barred spirals and irregulars. He presented his classification scheme in the form of a 'tuning-fork' diagram [110] (figure 23.8) which he interpreted as an evolutionary sequence in which the galaxies evolved from spherical elliptical galaxies at the left of the diagram through the sequence of spiral galaxies. This speculation proved to be wholly incorrect but the terms 'early-type' and 'late-type' galaxies are still in use.

Of much greater significance for cosmology was his realization that the number counts of galaxies brighter than a given apparent magnitude provides a test of the homogeneity of the distribution of galaxies in the Universe. If the galaxies are distributed uniformly in space, the number N brighter than the limiting apparent magnitude m is expected to be $\log N = 0.6m +$ (constant). Hubble's galaxy counts extended to an apparent magnitude of 16.7 and increased with increasing apparent magnitude exactly as expected for a uniform distribution.

Next, Hubble worked out the typical masses of the galaxies and hence the mean mass density of the Universe. The value he found was $\rho = 1.5 \times 10^{-28}$ kg m^{-3}. Adopting Einstein's static model for the Universe [109], he found that the radius of curvature of the spherical geometry was 27 000 Mpc and the number of galaxies in the closed Universe 3.5×10^{15}. In the last paragraph of his paper, he noted that the 100" telescope could observe typical galaxies to about 1/600 of the radius of the Universe and bright galaxies such as M31 to several times this distance. He concluded by remarking that

> ... with reasonable increases in the speed of plates and sizes of telescopes, it may become possible to observe an appreciable fraction of the Einstein universe.

Edwin Powell Hubble

(American, 1889–1953)

Hubble's interest in astronomy was stimulated by his studies with George Ellery Hale at the University of Chicago but, having won a Rhodes scholarship, he spent the years 1910–12 studying law at Oxford. There, he won a blue in athletics and fought an exhibition heavyweight boxing match with the French champion, Georges Charpentier. On returning to the USA, he abandoned his career in law and was appointed assistant astronomer  at the Yerkes Observatory. His outstanding skill as an observer led Hale to offer him a position at the Mount Wilson Observatory, which he delayed taking up until after his return with the American Army of Occupation from Germany in 1919.

His skill as an observer enabled him to identify Cepheid variables in the Andromeda Nebula, which, in 1925, led to measurement of its distance. These observations were decisive proof of the extragalactic nature of the spiral nebulae. In 1926, he published the first major survey of the properties of all known types of galaxies, including an estimate of the mean mass density of the Universe and relating this value to the properties of Einstein's static universe. In 1929, he discovered the velocity–distance relation for galaxies within 2 Mpc of our Galaxy. These and his subsequent achievements were brilliantly summarized in his book *The Realm of the Nebulae* of 1936.

During World War II, Hubble was chief of ballistics and director of the Supersonic Wind Tunnel at the Aberdeen Proving Ground in Maryland. After the war, much of his efforts were devoted to the commissioning of the Palomar 200" telescope and he was the first to use it in 1949. As the father of extragalactic astronomy and observational cosmology, his research combined brilliance in observation with physical intuition which led him to be selective in his observations and to grasp the opportunities they presented.

Astrophysics and Cosmology

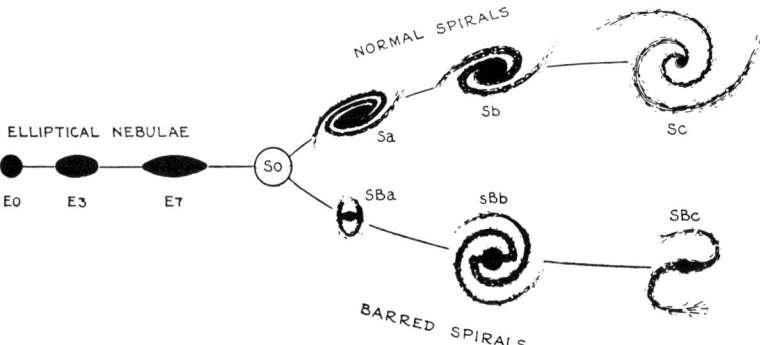

Figure 23.8. *Hubble's 'tuning-fork' diagram illustrating the sequence of nebular types. As Hubble notes in his caption to the diagram which appears in* The Realm of the Nebulae [110] *'The diagram is a schematic representation of the sequences of classification. A few nebulae of mixed types are found between the two sequences of spirals. The transitional stage, S0, is more or less hypothetical. The transition between E7 and SBa is smooth and continuous. Between E7 and Sa, no nebulae are definitely recognized.'. The S0 galaxies were later recognized in photographic surveys of nearby galaxies and may be thought of as disc galaxies with central bulges but without spiral arms.*

Thus, by 1926, the first application of the ideas of relativistic cosmology to the Universe of galaxies had been made. It comes as no surprise that John Ellery Hale began his campaign to raise funds for the construction of the Palomar 200″ telescope in 1928—the study of the Universe of distant galaxies needed the largest telescopes that could be built. In the American tradition of private sponsorship of observational astrophysics, in which the USA had taken a decisive lead, Hale was successful in obtaining a grant of $6 000 000 from the Rockefeller Foundation for the telescope before the year was out.

23.6. The development of relativistic cosmology

23.6.1. Physical cosmology up to the time of Einstein [111]
Isaac Newton appreciated that the unique form of the inverse square law of gravity had important consequences for the large-scale distribution of matter in the Universe. The cosmological problem was addressed in a series of letters between Newton and Richard Bentley [112] and concerned the stability of a universe uniformly filled with stars. The attractive nature of gravity means that matter tends to fall together and Newton was well aware of this problem. His solution was to suppose that the distribution of stars extends infinitely in space so that the net gravitational attraction on any star in a uniform distribution is zero. The problem, which was fully understood by Newton, was that this model is unstable. If any star is slightly perturbed from its equilibrium position,

the attractive force of gravity causes the star to continue to fall in that direction. Newton had to assume that the Universe had been set up in a perfectly balanced state. This remained an unsolved problem.

During the late eighteenth century, non-Euclidean spaces began to be taken seriously by mathematicians who realized that Euclid's fifth postulate, the parallel axiom, might not be essential for the construction of a self-consistent geometry. In his book *On the Principles of Geometry* of 1829, Nicolai Ivanovich Lobachevsky [113] at last solved the problem of the existence of non-Euclidean geometries and showed that Euclid's fifth postulate could not be deduced from the other postulates. Non-Euclidean geometry was placed on a firm theoretical basis by the work of Gauss, Lobachevsky and Bolyai. The generalization of their work to n-dimensional spaces with variable curvature, which proved to be essential for the development of general relativity, was described by Bernhard Riemann [114]. Several authors attempted to measure the curvature of space, for example, Karl Schwarzschild [115] finding a lower limit to the radius of curvature of its geometry of 2500 light years in 1900. Until Einstein's discovery of the general theory of relativity, considerations of the geometry of space and the role of gravity in defining the large-scale structure of the Universe were distinct questions. After 1915, they could no longer be separated.

23.6.2. General relativity and Einstein's universe

The history of the discovery of the general theory of relativity [116] is recounted elsewhere in these volumes (Chapter 4). Perhaps the most remarkable result of the theory was the fact that, because space-time is curved in the vicinity of a massive object, the orbits of planets are expected to precess. In 1859, Le Verrier [117] had discovered that, once account was taken of the influence of the planets, there remained an unexplained component of the precession of the perihelion of Mercury, amounting to about 40 arcsec per century. Einstein showed in 1915 that the precession expected according to the general theory of relativity amounted to 43 arcsec per century, a value in excellent agreement with the present best estimates for the precession.

The theory also predicted the deflection of light by massive bodies because of the curvature of space-time in their vicinity. For the Sun, the predicted deflection of light rays for stars just grazing the limb of the Sun amounted to 1.75 arcsec; according to 'Newtonian' theory, the deflection would have amounted to only half this value. Einstein's prediction provided the stimulus for the famous eclipse expeditions of 1919 led by Eddington and Crommelin [118], one to Sobral in Northern Brazil and the other to the island of Principe, off the coast of West Africa. The results were in agreement with Einstein's prediction, the Sobral result being 1.98 ± 0.12 arcsec and the Principe result 1.61 ± 0.3 arcsec. Because of the technical difficulty of these observations, the precise value of the deflection remained a controversial issue, which was not laid to rest until the development of radio interferometric techniques in the 1970s.

The theory also predicted the gravitational redshift of light originating close to massive compact objects. As already described in section 23.3.5, Adams' careful observations of the spectrum of the white dwarf Sirius B in 1925 showed the expected gravitational redshift [72].

> See also
> p 1710

In 1916, the year after the general theory of relativity was proposed, de Sitter and Ehrenfest [119] had suggested in correspondence that a spherical four-dimensional space-time would eliminate the problem of the boundary conditions at infinity, which pose problems for Newtonian cosmological models. In 1917, Einstein [120] published his famous paper which seemed to resolve the problems inherent in the Newtonian models.

Einstein realized that, in general relativity, he had a theory which could be used to construct models of the Universe as a whole. His reason for taking this issue seriously was his objective of incorporating what he designated *Mach's principle* into the structure of general relativity. By Mach's principle, he meant that the local inertial frame of reference should be defined by the large-scale distribution of matter in the Universe. There were two obstacles to constructing suitable models. The first was that, as appreciated by Newton, the static Newtonian model is unstable, in the sense that local regions collapse under gravity. The second problem concerned the boundary conditions at infinity. Einstein proposed to solve all these problems at one fell swoop by introducing an additional term into the field equations, the infamous *cosmological constant* λ. In Newtonian terms, the cosmological constant corresponds to a repulsive force acting on a test particle at distance r, $f = \frac{1}{3}\lambda r$. Notice that, unlike gravity, the force is independent of the density of matter. As Zeldovich [121] has remarked, the term corresponds to the 'repulsive effect of a vacuum'. The term is negligible on the scale of the solar system and is only appreciable on the largest scales in the Universe [122].

What Einstein effectively did was to add the term $+\lambda$ to Poisson's equation for gravity so that it became, $\nabla f = 4\pi G \rho + \lambda$. There is now a static solution with constant gravitational potential ϕ, $f = -\nabla\phi = 0$ if $\lambda = -4\pi G \rho_0$, where ρ_0 is the density of the static Universe. From Einstein's field equations, it follows that the geometry of the Universe is closed and the radius of curvature of the geometrical sections is $\Re = c/(4\pi G \rho_0)^{1/2}$. This solution eliminated the problem of the boundary conditions at infinity since the geometry is finite and closed. The volume of the spherical geometry is $V = 2\pi^2 \Re^3$ and there is a finite number of galaxies in the Universe. Einstein also believed that he had incorporated Mach's principle into general relativity. The essence of the argument was that static solutions of the field equations did not exist in the absence of matter. The cosmological constant was essential in creating a static, closed model of the Universe.

This was the first fully self-consistent cosmological model but it had been achieved at the cost of introducing the cosmological constant which has remained a thorn in the flesh of cosmologists since its introduction in 1917. Einstein was somewhat uncomfortable about it, acknowledging that the term was 'not justified by our actual knowledge of gravitation' but was merely 'logically consistent'. In 1919, Einstein [123] realized that the term involving the cosmological constant would appear in the field equations of general relativity, quite independent of its cosmological significance. In the solution of the field equations, the λ-term appears as a constant of integration which is normally set equal to zero in standard general relativity.

Cosmologists have taken different views of the λ-term. Willem de Sitter [124] had similar views to Einstein and wrote in 1919 that the term

> ... detracts from the symmetry and elegance of Einstein's original theory, one of whose chief attractions was that it explained so much without introducing any new hypotheses or empirical constant.

Others have regarded it as a constant which appears in the development of general relativity and that its value should be determined by astronomical observation. The inclusion of the term results in a number of quite specific observable astronomical phenomena but there has been no convincing positive evidence that any of these have been observed [122].

23.6.3. De Sitter, Friedman and Lemaître

In the same year that Einstein's first paper on cosmology was published, Willem de Sitter [125] showed that one of Einstein's objectives had not be achieved. He found solutions of Einstein's field equations in the absence of matter, $\rho = p = 0$. Although there is no matter present in the Universe, a test particle still has a perfectly well-defined geodesic along which to travel. As de Sitter asked 'If no matter exists apart from the test body, has this inertia?'. At that time, the principal issues at stake were the origin of inertia and Mach's principle rather than any thought that these considerations might be of relevance to astronomical observation.

In 1922, Lanczos [126] showed that the de Sitter solution could be written alternatively in the form of a metric in which the test particles move apart at an exponentially increasing rate. At almost exactly the same time, Alexander Alexandrovich Freidmann [127] published the first of two classic papers for both static and expanding world models. This work was carried out in Leningrad in the period 1922–24 in the newly-founded Soviet Union. In the first paper of 1922, Friedman found solutions for expanding world models with closed spatial geometries, including those which expand to a maximum radius and eventually collapse to a singularity. In the second paper of 1924, he showed that there exist expanding solutions which are unbounded and which have hyperbolic geometry. These solutions correspond exactly to the standard world models of general relativity and are appropriately known as the *Friedman world models*.

In 1925 Friedman died of typhoid in Leningrad before the fundamental significance of his work was appreciated. The neglect of Friedman's work in those early days is somewhat surprising since Einstein [128] had commented, incorrectly as he admitted, on the first of the two papers in 1923. It was not until Georges Lemaître independently discovered the same solutions in 1927 and then became aware of Friedman's contributions, that the pioneering nature of these papers was appreciated. Lemaître [129] was seeking quite explicitly solutions which had neither of the problems which afflicted either Einstein's universe, which was closed and static, or de Sitter's universe, which was open, empty and expanding.

One of the problems facing the pioneers of relativistic cosmology was the interpretation of the space and time coordinates used in their calculations. De Sitter's solution could be written in apparent stationary form or as an exponentially expanding solution. From the metric, de Sitter had shown that a distance–redshift relation must exist for his model but it was not clear whether or not this was relevant to the observable Universe. The answer came resoundingly in the affirmative with Hubble's famous discovery of the velocity–distance relation for galaxies in 1929 which ushered in a new epoch in astrophysical cosmology.

23.6.4. *The recession of the nebulae*

In 1917, Vesto M Slipher [130] published a paper describing heroic spectroscopic observations of 25 spiral galaxies made with the Lowell Observatory's 24″ telescope. Exposures of 20, 40 and even 80 hours were made to secure these spectra. He found that the velocities of the galaxies inferred from the Doppler shifts of their absorption lines were typically about 570 km s^{-1}, far in excess of the velocity of any known object in our Galaxy. Furthermore, most of the velocities corresponded to the galaxies moving away from the Solar System, that is, the absorption lines were shifted to longer (red) wavelengths and this phenomenon became known as the *redshift* of the galaxies. In his paper, he notes that

> This might suggest that the spiral nebulae are scattering but their distribution on the sky is not in accord with this since they are inclined to cluster.

In 1921, Carl Wilhelm Wirtz [131] searched for correlations between the velocities of the spiral galaxies and other observable properties and concluded that, when the data were averaged in a suitable way

> ... an approximate linear dependence of velocity upon apparent magnitude is visible. This dependence is in the sense that the nearby nebulae tend to approach our Galaxy whereas the distant

ones move away ... The dependence of the magnitudes indicates that the spiral nebulae nearest to us have a lower outward velocity than the distant ones.

By 1929, Hubble [132] had assembled distance estimates of 24 galaxies for which velocities had been measured, all within 2 Mpc of our Galaxy. From these meagre data, Hubble's famous velocity–distance relation was derived (figure 23.9(a)). It is ironic that the main objective of Hubble's paper was not to derive the velocity–distance relation but rather to use the velocities of the galaxies to derive the velocity of the local standard of rest at the Earth relative to the extragalactic nebulae.

With hindsight, it is remarkable that he found the redshift–distance relation from such a nearby sample of galaxies but there was more evidence available even at that time. He noted in his paper that Milton Humason had measured a velocity of 3910 km s^{-1} for the galaxy NGC7619, the brightest galaxy in a cluster. If the velocity–distance relation were correct, the absolute magnitude of this galaxy would be of the same order as those of the brightest galaxies in nearby clusters.

Although he did not write down the famous relation $v = H_0 r$ in this paper, he notes that 'the velocity–distance relation may represent the de Sitter effect'. De Sitter had shown that there would be a redshift of spectral lines which increases with distance but the origin of the effect in terms of the expansion of the system of galaxies was not common knowledge at that time.

The subsequent story is told in Hubble's Silliman Lectures given at Yale University in 1935 and published as the famous and influential monograph *The Realm of the Nebulae* [110] in the following year. The task of extending the measurement of the radial velocities of galaxies to much greater distances was undertaken by Humason using the 100″ Hooker Telescope at Mount Wilson. By 1935, he had measured the velocities of almost 150 galaxies out to distances inferred to be 35 times greater than the distance of the Virgo cluster of galaxies and to radial velocities of 42 000 km s^{-1}, roughly one seventh of the speed of light. Although distances could not be measured directly, Hubble and Humason [133] realized that the luminosity functions of the galaxies in clusters are remarkably similar and so they used the fifth brightest member as measures of the relative distances of the clusters. The resulting redshift–apparent magnitude relation is expected to follow the relation $\log v = 0.2m + \text{constant}$, if the galaxies follow a velocity–distance relation, $v \propto r$. The remarkable results of these arduous observing programmes are shown in figure 23.9(b) and are in excellent agreement with a linear velocity–distance relation.

In Hubble's monograph, he goes on to describe counts of faint galaxies made with the 100″ telescope, which extended to apparent magnitude 21, the faintest counts feasible with the 100″ telescope. Whereas the counts had the expected slope $N(\leqslant m) = 0.6m + \text{constant}$

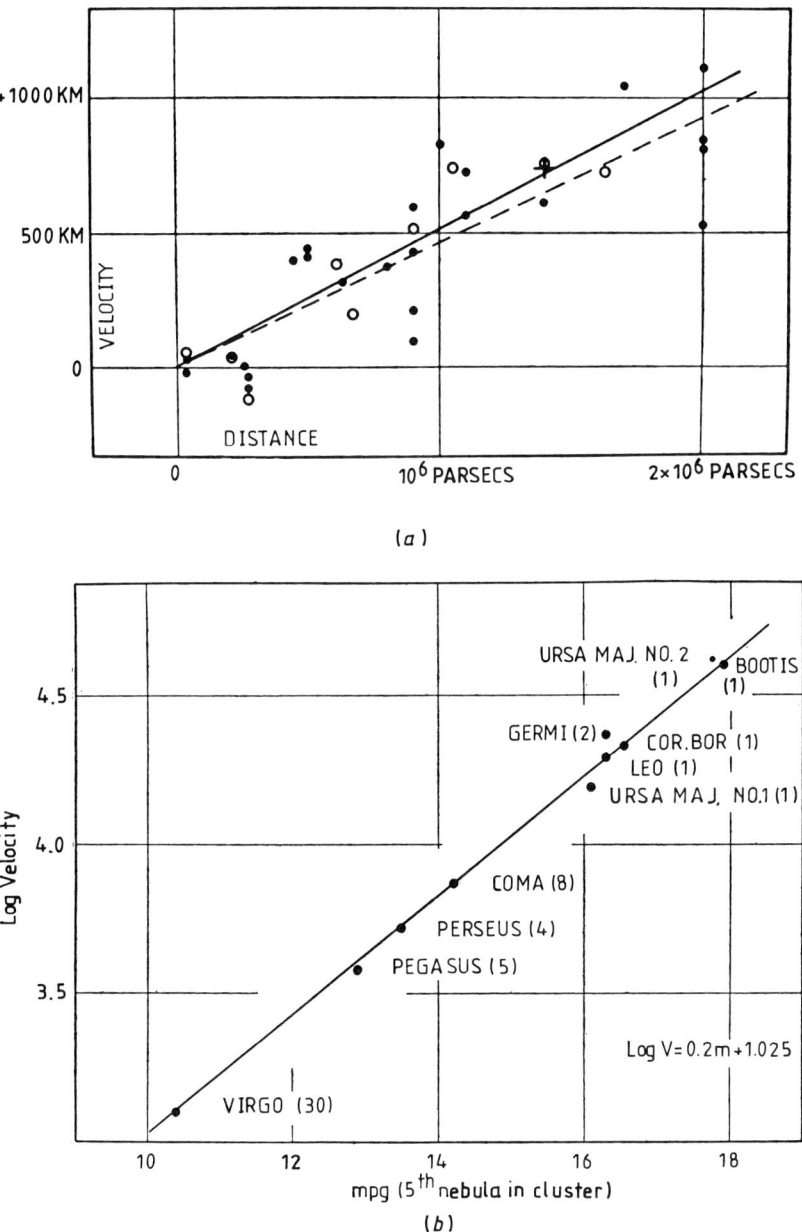

Figure 23.9. (a) Hubble's first version of the velocity–distance relation for nearby galaxies [132]. The filled circles and the full line represent a solution for the solar motion using the nebulae individually; the open circles and the dashed line, a solution combining the nebulae into groups. (b) The velocity–apparent magnitude relation for the fifth brightest member of clusters of galaxies, corrected for galactic obscuration [137]. Each cluster velocity is the mean of the various individual velocities observed in the cluster, the number being indicated by the figure in brackets.

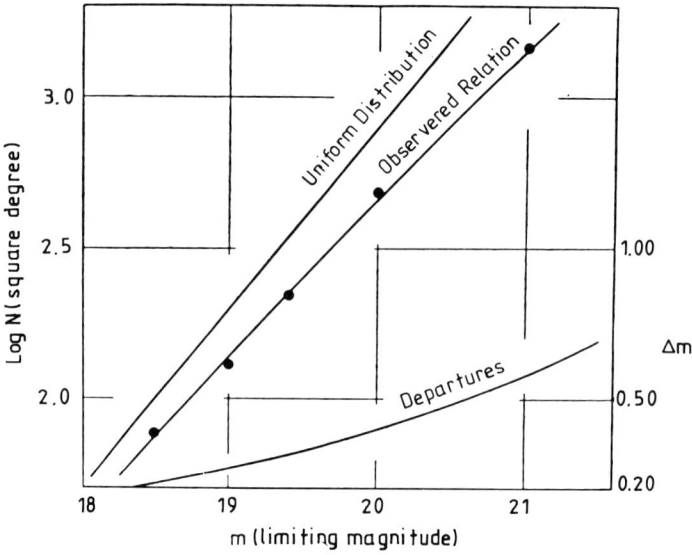

Figure 23.10. Hubble's counts of faint galaxies published in 1936 [110]. The line labelled 'Uniform Distribution' corresponds to the relation $\log N(\geq m) = 0.6m + \text{constant}$. The points represent the observed counts with a best-fitting line shown. At the bottom of the diagram is the difference in magnitude between the uniform counts and the observations which Hubble interpreted as the effect of redshift.

down to 18th magnitude, the counts began to converge at fainter magnitudes (figure 23.10). Hubble correctly concluded that the counts extended to such faint magnitudes, and consequently large distances, that the effects of redshift upon the number counts had to be taken into account. He also correctly concluded that the counts were evidence for the overall homogeneity of the Universe as far the 100″ telescope could observe the galaxies. In the last pages, he speculates that the convergence of the number counts may be associated with the curvature of space. His conclusion that the Universe must have positive curvature was incorrect but this can be attributed to the fact that it took some time before the proper relativistic formulation of the relations between observables and the intrinsic properties of galaxies were worked out.

23.6.5. The Robertson–Walker metric

The discovery of the velocity–distance relation for galaxies acted as a great stimulus to the study of the Friedman models. Of prime importance were the efforts to place the construction of the world models on a firm theoretical foundation. There remained some confusion about the notions of time and distance as used in the cosmological models of general relativity because the field equations could be set up in any frame of reference. The principle of special relativity meant that observers located on galaxies moving relative to one another could not agree on the synchronization of their clocks. By 1935, the problem had been solved independently by Robertson and Walker [134].

In 1932 Hermann Weyl [135] enunciated what is known as *Weyl's postulate*. To eliminate the arbitrariness present in the choice of coordinate frames, Weyl introduced the idea that, to quote Bondi [136]

> The particles of the substratum (representing the nebulae) lie in space-time on a bundle of geodesics diverging from a point in the (finite or infinite) past.

The most important aspect of this statement is the postulate that the geodesics, which represent the world-lines of galaxies, do not intersect, except at a singular point in the past. By the term 'substratum' Bondi meant an imaginary medium which can be thought of as a fluid which defines the overall dynamics of the system of galaxies. The consequence of Weyl's postulate is that there is only one geodesic passing through each point of space-time, except at the origin. Once this postulate is adopted, it becomes possible to assign a notional observer to each world-line and these are known as *fundamental observers* whose clocks measure *cosmic time* from a singular point in the past. One further assumption is essential before the framework of the standard models can be derived. This is known as the *cosmological principle* and it asserts that our Galaxy is not located in a privileged or special position in the Universe. The implication is that our Galaxy is located at a typical point in the Universe and any fundamental observer would observe the same large-scale features of the Universe at the same cosmic time. One requirement is that all observers should observe the same velocity–distance relation at the same cosmic epoch. A second requirement is that, looked at on large enough scales, the Universe should look the same in all directions and that on average the matter should be homogeneously distributed. Hubble's galaxy counts were crucial in showing that the Universe of galaxies indeed seems to be homogeneous since they obeyed accurately the relation $\log N = 0.6m + $ (constant).

Putting these concepts together, Robertson and Walker [134] independently showed that the metric for any isotropically expanding substratum has to have the form

$$ds^2 = dt^2 - \frac{R^2(t)}{c^2}\left(\frac{dr^2}{(1+Kr^2)} + r^2(d\theta^2 + \sin^2\theta\, d\phi^2)\right)$$

where $K = \mathfrak{R}^{-2}$ is the curvature of space at the present epoch, r is a co-moving radial distance coordinate and $R(t)$ is the scale factor which describes how the distance between any two world-lines changes with cosmic time t, and is normalized to unity at the present epoch. This metric is known as the *Robertson–Walker metric*. It contains all the permissible isotropic geometries consistent with the assumptions of isotropy and homogeneity and these are described by the curvature $K = \Re^{-2}$ where $\Re$ is the radius of curvature of the spatial sections of the isotropic curved space. It should be noted that the form of the metric does not depend upon the assumption that the dynamics of the Universe can be described by general relativity. The physics of the expansion has been absorbed into the scale factor $R(t)$. Once this metric of isotropic expanding space-time was derived, it was a relatively straightforward task to use it to derive relations between the intrinsic properties of objects and their observed properties.

23.6.6. Milne–McCrea and Einstein–de Sitter

By far the most important solutions for the scale factor $R(t)$ were those of general relativity, including those with the cosmological constant. There is no simple general closed solution of the field equations and they were the subject of a great deal of study. One of the most important contributions during this period was made by Milne and McCrea [137] in 1934. They showed that, despite the fact that Newtonian mechanics cannot provide a completely self-consistent cosmology, simple ideas from Newtonian physics can provide insight into the physical content of the cosmological models. The important point is that the requirements of isotropy and homogeneity are very powerful constraints upon the properties of the models. In the simplest form of their argument, it can be supposed that our Galaxy is located at the centre of a uniformly expanding sphere. Milne and McCrea showed how it is possible to derive formulae for the scale factor $R(t)$ of exactly the same form as those which come out of the full theory. These results of Milne and McCrea were of considerable importance because they showed that, despite the fact that there are problems with the boundary conditions in the Newtonian model, it can be used successfully on large scales in the Universe and, in particular, on scales less than the radius of curvature of the spatial geometry, it is perfectly adequate to use Newtonian arguments.

By the 1930s, Einstein was regretting the inclusion of the cosmological constant into the field equations. According to George Gamow [138], Einstein stated that the introduction of the cosmological constant was 'the biggest blunder of my life'. In 1932, Einstein and de Sitter [139] showed how one particularly simple solution of the field equations for an expanding universe seemed to be in good accord with observations. They noted that there is one special solution of the equations in which the cosmological constant is zero and the spatial curvature is zero, $\Re \to \infty$, corresponding to Euclidean space sections. This is the famous model often referred to as the *Einstein–de Sitter model* and it has particularly simple dynamics, $R(t) = (t/t_0)^{2/3}$ where $t_0 = (2/3)H_0^{-1}$, H_0 being Hubble's constant, the constant of proportionality in the velocity–distance relation. The model has average density at the present epoch

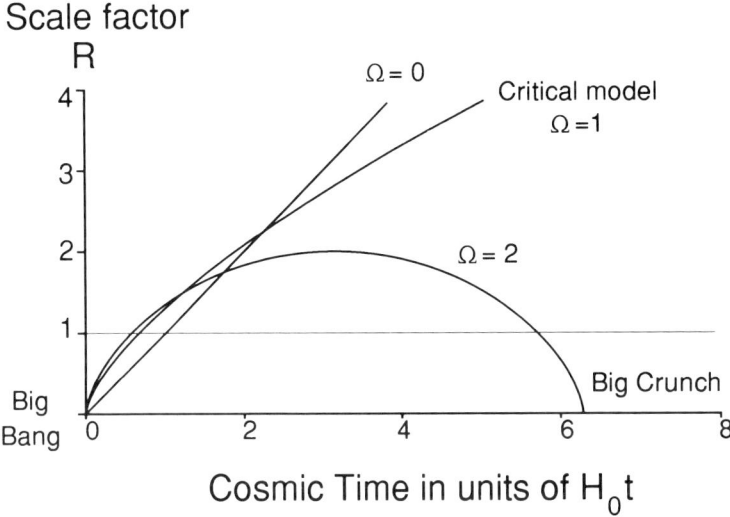

Figure 23.11. *Examples of the dynamics of the standard Friedman world models with $\lambda = 0$. The scale factor $R(t)$ has been normalized to 1 at the present epoch. The critical model, with density parameter $\Omega = 1$ separates the re-collapsing models with $\Omega > 1$ from those which expand forever and which have $\Omega \leqslant 1$.*

$\rho_0 = 3H_0^2/8\pi G$. This density is often referred to as the *critical density* and the Einstein–de Sitter model as the *critical model*, because it separates the ever-expanding models with open, hyperbolic geometries from the models which will eventually collapse to a singularity and which have closed, spherical geometry (figure 23.11). When Einstein and de Sitter inserted the value of Hubble's constant from Hubble's observations, $H_0 = 500$ km s^{-1} Mpc^{-1} into the expression for ρ_0, they found a value of 4×10^{-25} kg m^{-3} for the density of the Universe. Although recognizing that this value was somewhat greater than the value derived by Hubble, they argued that the density was of the correct order of magnitude and, in any case, there might well be a considerable amount of what would now be called 'dark matter' present in the Universe.

Evidence of dark matter in the Universe was not long in coming. In 1937, the remarkable Swiss astronomer Fritz Zwicky [140], working at the Mount Wilson Observatory, made the first studies of rich clusters of galaxies, in particular of the Coma cluster, which is one of the largest regular clusters in the northern sky. The method Zwicky used to estimate the total mass of the cluster had been derived by Eddington in 1916 to estimate the masses of star clusters. Using methods familiar in the theory of gases, Eddington [141] derived the *virial theorem* which relates the total internal kinetic energy T of the stars or galaxies in a cluster to the total gravitational potential energy, $|U|$, if the system is in statistical equilibrium under gravity. The kinetic energy can be written $T = \frac{1}{2}M\langle v^2 \rangle$, where $\langle v^2 \rangle$ is the mean square velocity of the stars or galaxies, and $|U| = GM^2/2R_{cl}$, where R_{cl} is some suitably defined radius which depends upon the distribution of mass in the cluster. For a cluster of stars or galaxies in statistical equilibrium, Eddington showed that $T = \frac{1}{2}|U|$. Therefore, if the cluster is known to be in statistical equilibrium, the total mass of the cluster can be found from the virial theorem, $M \approx 2R_{cl}\langle v^2\rangle/G$.

In rich clusters of galaxies such as the Coma cluster, there is convincing evidence from the radial distribution of galaxies within the cluster that they have reached statistical equilibrium and so good estimates can be made of its mass. Zwicky measured the velocity dispersion of the galaxies in the Coma cluster and, in his paper of 1937, found that there was much more mass in the cluster than could be attributed to the visible parts of galaxies. In solar units, the ratio of mass to optical luminosity of a galaxy such as our own is about 3, whereas for the Coma cluster the ratio was found to be about 500. In other words, there must be about 100 times more dark or hidden matter as compared with visible matter in the cluster. Zwicky's pioneering studies have been confirmed by all subsequent studies of rich clusters of galaxies. The nature of the dark matter remains an open and crucial question for cosmology.

23.6.7. Eddington–Lemaître

Despite Einstein's renunciation of the cosmological constant, this was very far from the end of the story because there remained one very grave problem for the models in which the cosmological constant is set equal to zero. It is a simple calculation to show that, if $\lambda = 0$, the age of the Universe must be less than H_0^{-1}. Using Hubble's estimate of $H_0 = 500$ km s^{-1} Mpc^{-1}, the age of the Universe had to be less than 2×10^9 years old, a figure in conflict with the age of the Earth derived from the ratios of abundances of long-lived radioactive species. The present best estimate for the age of the Earth is about 4.6×10^9 years.

See also p 1960

Eddington and Lemaître [142] immediately recognized that this problem could be eliminated if the cosmological constant is positive. The effect of a positive cosmological constant is to counteract the attractive force of gravity when the Universe grows to a large enough size. Among the solutions, there are special cases corresponding to the Einstein stationary universe but not necessarily at the present epoch. It was possible to conceive of a model which had remained in the static Einstein state for an arbitrarily long period in the past which then began to expand away from that state under the influence of the cosmological term. In this type of *Eddington–Lemaître* model, the age of the Universe could be arbitrarily long. As Eddington [143] expressed it, the Universe would have a 'logarithmic eternity' to fall back on and so resolve the conflict between estimates of Hubble's constant and the age of the

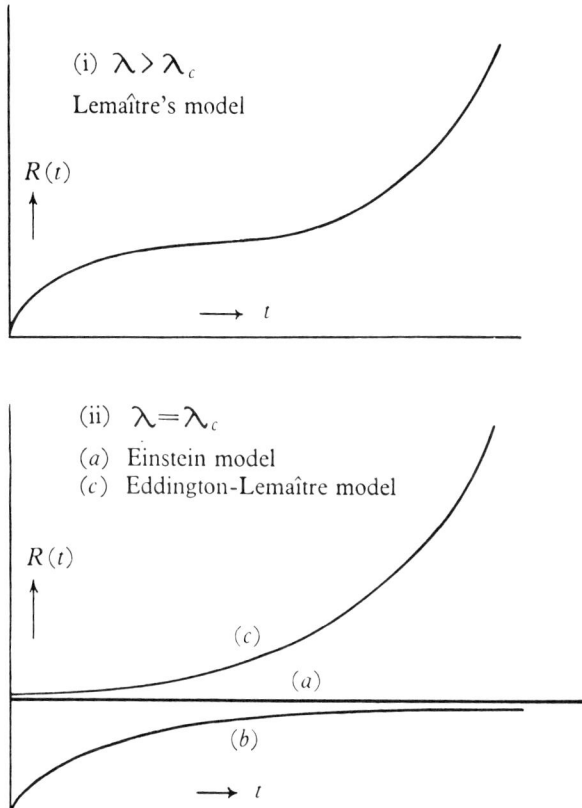

Figure 23.12. Examples of the dynamics of world models in which $\lambda \neq 0$ [144]. The Einstein static model is illustrated by the model for which $R(t)$ is constant for all time. The Eddington–Lemaître model in which the Universe expanded from the Einstein static universe in the infinite past is illustrated as well as other models in which the value of λ is slightly different from the static model.

Earth. Examples of the Eddington–Lemaître models and their close relatives [144] are illustrated in figure 23.12.

23.6.8. The cosmological problem in 1939

Thus, by the end of the 1930s, the basic problems of what I call *classical cosmology* had been clearly identified. The solution of the cosmological problem lay in the determination of the parameters which defined the Friedman world models. This was the goal of the great programmes of observation to be carried out by the 200″ telescope and the subsequent generation of 4 metre class telescopes. The challenge was to measure the parameters which characterize the Universe: Hubble's constant, $H_0 = \dot{R}/R$; the deceleration parameter, $q_0 = -\ddot{R}/R^2$; the curvature of space $\kappa = \Re_0^{-2}$; the mean density of matter in the Universe ρ, in particular, whether or not it attains the critical density ρ_0; the age of the Universe, T_0; and the cosmological constant λ.

These are not all independent. For example, for the models which include the cosmological constant

$$\kappa = \Re^{-2} = \frac{(\Omega - 1) + \frac{1}{3}(\lambda/H_0^2)}{(c/H_0)^2} \qquad q_0 = \frac{\Omega}{2} - \frac{1}{3}\frac{\lambda}{H_0^2}$$

where $\Omega = \rho/\rho_0$ is known as the *density parameter*. If $\lambda = 0$, there is a one-to-one relation between the geometry of the world models, their densities and dynamics, $q_0 = \Omega/2$ and $\kappa = \Re^{-2} = (\Omega - 1)/(c/H_0)^2$.

The determination of these parameters has turned out to be among the most difficult observational challenges in the whole of astronomy and progress has not been as dramatic as the optimists of the 1930s must have hoped. In compensation, completely new vistas were to open up after World War II as the whole of the electromagnetic spectrum became available for astronomical observation.

PART 3: THE OPENING UP OF THE ELECTROMAGNETIC SPECTRUM

23.7. The changing astronomical perspective

Until 1945, astronomy meant optical astronomy [145]. The completion of the 200" telescope in 1949 highlighted the dominance of the USA in observational astrophysics in the period immediately after World War II. Like most of the other major observatories in the USA, the 200" telescope was a private telescope, used more or less exclusively by the astronomers employed at the host institutions, the Mount Wilson Observatory and the Carnegie Institute of Washington. Thus, in the early 1950s, the most important telescope in the world for all types of astrophysical and cosmological research was in the hands of a relatively small group of privileged astronomers.

The astronomical scene was, however, about to change with the development of new ways of tackling astrophysical and cosmological problems. Four reasons can be identified for the major change in outlook of observational and theoretical astrophysicists which has taken place since 1945.

(i) The most important of these has been the expansion of the wavebands available for astronomical observation, which has led to a much more complete description of our physical Universe and to the discovery of new physical phenomena, which are important for fundamental physics as well as astronomy. The capability of making observations from above the Earth's atmosphere opened up the far-infrared, ultraviolet, x-ray and γ-ray wavebands which are inaccessible from the ground.

(ii) There have been remarkable technological developments in the design and construction of telescopes, instruments and detectors for all wavebands. The semiconductor and computer revolutions have been crucial to the advance of both observation and theory.

(iii) There has been a huge increase in activity in the astronomical sciences, at least partly due to the influx of physicists whose research interests and expertise have led them to consider astrophysical problems. By the process of symbiosis between the astrophysical and laboratory sciences, astronomy has assimilated new tools from physics and theoretical physics, most obviously in general relativity and particle physics but also from fields such as chemistry, solid-state physics, plasma physics and superconductivity. Some measure of the increase in activity is provided by the membership of the International Astronomical Union [146] which is open to all professional astronomers and which was founded in 1919. At the first General Assembly held in Rome in 1922, there were just over 200 members from 19 adhering countries. By 1938, the numbers had risen to 550 from 26 countries. The number was roughly the same immediately after World War II but, by the time of the 1991 General Assembly held in Buenos Aires, the membership had risen to about 6700 from 56 adhering countries.

(iv) Astrophysics has become one of the big sciences. The telescopes needed to carry out frontier research have become very complex and costly, international collaboration often being essential to construct and operate them. After World War II, there was a spectacular increase in investment in basic research in the USA, largely stimulated by the huge contributions that research scientists had made during the period of hostilities and the realization of the enormous potential for economic growth as well as strategic defence requirements which the fruits of basic research could bring. The attitude of many of the best research workers had been changed by their war-time experience. To quote Bernard Lovell [147], they adopted an approach to research

> ...utterly different from that deriving from the pre-war environment. The involvement with massive operations had conditioned them to think and behave in ways which would have shocked the pre-war university administrators. All these facts were critical in the large-scale development of astronomy.

The astronomers rode this wave of investment in the fundamental sciences but these initiatives had to be seen in a national or international context rather than as sponsorship by private institutions as had occurred in the USA.

Figure 23.13. *Karl Jansky's radio antenna with which he discovered the radio emission from the Galaxy in 1933.*

23.7.1. Radio astronomy [148]

The expansion of the observable waveband began with Karl Jansky's announcement [149] of the discovery of radio emission from the Galaxy in May 1933. Working at the Bell Telephone Laboratories at Holmdel, New Jersey, Jansky was assigned to the task of identifying naturally occurring sources of radio noise which would interfere with radio transmissions. In what turned out to be a classic series of observations made at the long wavelength of 14.6 metres (20.5 MHz), he discovered the radio emission from the Galaxy (figure 23.13). This discovery was confirmed by Grote Reber, who was a radio engineer and an enthusiastic amateur astronomer. With his home-built radio antenna and receiving system operating at a wavelength of 1.87 metres (160 MHz), he made a radio scan along the plane of the Galaxy which was published in the *Astrophysical Journal* [150] in 1940. Comparison of Jansky's and Reber's observations showed that the emission could not be blackbody radiation, Reber proposing that it was bremsstrahlung. In the immediately following paper in the *Astrophysical Journal*, L G Henyey and P C Keenan [151] showed that, whilst the radiation at 1.87 m might be the bremsstrahlung of gas at 10 000 K, the intensity observed by Jansky at the longer wavelength was far too great for this to be the emission process. Other than this negative conclusion, these observations attracted little attention from professional astronomers. The culmination of Reber's work was the publication of the first map of the radio emission of the Galaxy in the *Astrophysical Journal* [152] in 1944.

The development of radar during World War II had two immediate consequences for radio astronomy. First, sources of radio interference which might confuse radar location had to be identified. In 1942, James S Hey and his colleagues [153] at the Army Operational Research Group in the UK discovered intense radio emission from the Sun which coincided with a period of unusually high sun-spot activity.

Then, in 1946, immediately after the war, his group discovered the first discrete source of radio emission which lay in the constellation of Cygnus, the source which became known as Cygnus A [154]. The second consequence was that the extraordinary research efforts to design powerful radio transmitters and sensitive receivers for radar resulted in new technologies, which were to be exploited by the pioneers of radio astronomy, all of whom came from a background in radar.

Immediately after the war, a number of the radar scientists began the systematic study of the astronomical phenomena discovered, more or less by chance, as a result of the war effort. Further discrete sources of radio emission were discovered and radio interferometry provided the best means of measuring their positions with improved accuracy. In 1948, Martin Ryle and F Graham Smith [155] discovered the most powerful source in the northern hemisphere, Cassiopeia A, and in 1949 the Australian radio astronomers John G Bolton, Gordon J Stanley and O Bruce Slee [156] succeeded in associating three of the discrete radio sources with remarkable nearby astronomical objects, the supernova remnant known as the Crab Nebula and the strange galaxies NGC5128, associated with the source Centaurus A, and M87, associated with Virgo A. In addition to the diffuse radio emission of our own Galaxy, these early surveys established the existence of a population of discrete radio sources, some concentrated towards the plane of the Galaxy but many lying outside it. There was some uncertainty as to whether the isotropic component of the source population was primarily associated with nearby radio stars in our own Galaxy or with distant extragalactic objects [157].

The radio astronomers could not answer this question from the radio observations alone since the radio data did not provide any distance measure for the sources. Distances could only be determined by finding an associated optical object and measuring its distance. In 1951, F Graham Smith [158] measured the positions of the two brightest sources in the northern sky, Cygnus A and Cassiopeia A, with an accuracy of about 1 arcmin and this led to their optical identification by Walter Baade and Rudolph Minkowski [159] using the Palomar 200" Telescope. Cassiopeia A was associated with a young supernova remnant in our own Galaxy, while Cygnus A was associated with a faint and distant galaxy. The latter observation immediately showed that the radio sources could be used for cosmological studies. The cosmological importance of radio astronomical observations of discrete sources was thus apparent by the mid-1950s—fainter radio sources would lie at greater cosmological distances and hence probe the Universe at epochs much earlier than the present [160].

Following a suggestion by Hannes Alfvén and Nicolai Herlofson [161] in 1948, K O Kippenhauer and V L Ginzburg [162] developed the theory that the Galactic radio emission is synchrotron radiation, the emission of ultra-relativistic electrons gyrating in the Galactic magnetic field. By the mid-1950s, the power-law nature of the Galactic radio spectrum and its high degree of polarization were convincing evidence for the correctness of the synchrotron hypothesis. Radio emission is observed throughout the disc of the Galaxy and provides direct evidence for an interstellar flux of very high-energy electrons.

From the early 1950s onwards, radio astronomy developed as a subject in its own right. Large single-dish reflectors were constructed and new interferometric techniques developed. The principle of aperture synthesis, pioneered by Ryle and his colleagues at Cambridge, was of particular importance since it enabled radio astronomers to obtain high angular resolution by adding together coherently the signals from separated telescopes [163]. Such techniques culminated in the construction of large synthesis arrays such as the Very Large Array in the USA and the Australia Telescope. Interferometry was extended to intercontinental baselines, a technique known as very long baseline interferometry (VLBI), resulting in the highest angular resolution, $\approx 10^{-3}$ arcsec, available in any waveband [164]. Radio astronomy resulted in other key discoveries for contemporary astrophysics, neutron stars as the parent bodies of radio pulsars, interstellar molecules through their millimetre line emission and the cosmic microwave background radiation being of special importance.

23.7.2. X- and γ-ray astronomy [165]
Immediately after World War II, those physicists interested in ultraviolet, x-ray and γ-ray astronomy took the first tentative steps into space astronomy. The atmosphere is opaque to all radiation with wavelengths shorter than about 330 nm and so ultraviolet, x-ray and γ-ray astronomy have to be conducted from above the Earth's atmosphere. The German V-2 rocket programme had made enormous strides in rocket technology during the war and the German scientists who had built them, led by Werner von Braun, as well as 300 box cars full of V-2 parts, were taken to the USA where they formed the core of the US Army's rocket programme and were made available for scientific research [166].

One of the prime targets of the early rocket experiments was the ultraviolet and x-ray emission of the Sun which, it was surmised, might be responsible for the ionization of the Earth's ionosphere. The first successful rocket ultraviolet observations of the Sun were made in October 1946 by Herbert Friedman's group at the Naval Research Laboratory [167]. In the following year, they made the first successful x-ray observations of the Sun, confirming the expectation that the Sun's corona is very hot.

The flights of *Sputniks 1* and *2* in late 1957 and Yuri Gagarin's orbital flight in 1961 came as a profound shock to the US administration which realized that the USA had fallen behind the USSR in space technology.

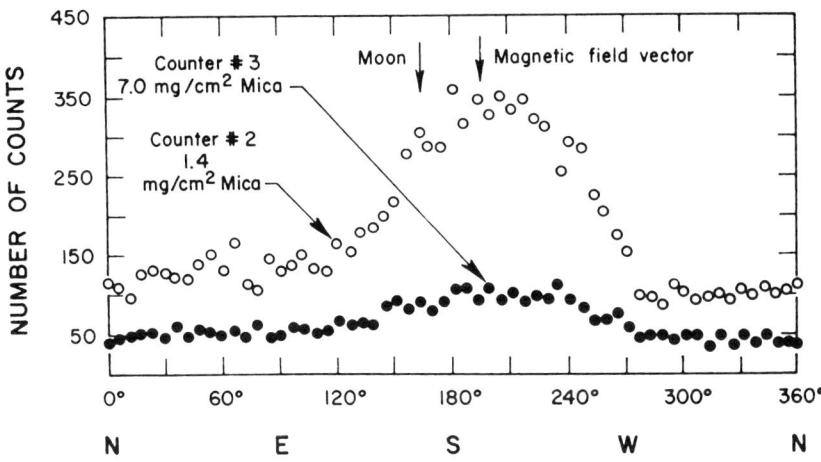

Figure 23.14. *The discovery record of the x-ray source Sco X-1 and of the x-ray background emission by Giacconi and his colleagues [169] in a rocket flight of 1962. The prominent source was observed by both detectors, as was the diffuse background emission.*

The US response was to set up the National Aeronautics and Space Administration (NASA) in July 1958 as a civilian organization to begin the process of catching up with the USSR. As part of that endeavour, the American Science and Engineering group (AS&E) was set up in association with the Massachusetts Institute of Technology to carry out military and civilian contracts. The AS&E group was responsible for the first successful flight, dedicated to searches for extraterrestrial x-rays, which took place in June 1962. In the five minutes of observing time, during which the rocket payload was above the Earth's atmosphere, Giacconi and his colleagues [168] discovered an intense discrete source of emission in the constellation of Scorpius which became known as Sco X-1 (figure 23.14). In addition, an intense background of x-rays was observed which was remarkably uniformly distributed over the sky. These observations were confirmed by other rocket flights by the AS&E group, as well as by Friedman's group at the Naval Research Laboratory (NRL), which discovered x-rays from the supernova remnant, the Crab Nebula [169].

Until 1970, all x-ray astronomy was carried out from rockets, which provided tantalizing glimpses of what was there but the picture was confused. These problems were resolved with the launch in December 1970 of the *UHURU* X-ray Observatory [170], which was the first satellite dedicated to x-ray astronomy and which initiated the successful series of *Explorer* satellites sponsored by NASA. The *UHURU* Observatory conducted the first survey of the x-ray sky and revealed the true nature of the x-ray population. The x-ray sources turned out to include a wide variety of very hot objects—x-ray binaries, supernova remnants, young radio pulsars, active galactic nuclei and the intergalactic gas in clusters of galaxies. The next generation of x-ray space telescopes comprised very large projects and included the High Energy Astrophysical Observatory-A, HEAO-A, launched in August 1977, which carried out a further survey of the x-ray sky. The next x-ray telescope, HEAO-B, also known as the Einstein X-ray Observatory, was launched in November 1978 and not only pioneered the astrophysics of the x-ray sources but also demonstrated that essentially all classes of star are x-ray emitters.

In the early 1960s, there was already interest in γ-ray detectors in space in order to monitor the atmospheric nuclear test-ban treaties concluded between the USA and the USSR. The *Vela* series of satellites was launched for this purpose in the 1960s but there was no intention that they should have any astronomical role. Cosmic γ-rays were first detected in observations made by the *Explorer II* satellite [171] in 1965 but this experiment did little more than show that there existed γ-rays which originated from beyond the Earth's atmosphere. The first important astronomical observations were made by the third Orbiting Solar Observatory (OSO-III) launched in March 1967. The prime discovery of this mission was the detection of γ-rays with energies $E_\gamma > 100$ MeV from the general direction of the Galactic Centre [172]. This γ-ray flux was convincingly interpreted as the γ-ray emission associated with the decay of neutral pions created in collisions between relativistic protons and the cold interstellar gas.

These pioneering observations were followed up by balloon observations but these suffered from severe contamination because of the production of secondary γ-rays by interactions of primary cosmic rays with the nuclei of atoms of the atmosphere. The *Small Astronomical Satellite, SAS-2*, was launched in November 1972 and included an array of spark chambers to detect the electron–positron pairs created when an incoming γ-ray is converted into a pair within the instrument. Although it operated for only eight months and detected about 8000 γ-rays of cosmic origin, it confirmed that there is a general concentration of γ-rays towards the plane of the Galaxy [173]. Discrete sources of γ-rays were present, in particular, two of the sources were associated with the pulsars in the Crab and Vela supernova remnants and evidence was found for diffuse extragalactic γ-ray background radiation. The *SAS-2* mission was followed in 1975 by the equally successful *COS-B* satellite launched by a European consortium. It also consisted of an array of spark chambers sensitive to γ-rays with energies greater than about 70 MeV. It continued to take data continuously for 6.5 years and resulted in a detailed map of the Galactic plane as well as evidence for 24 discrete γ-ray sources [174].

To everyone's surprise, astronomical γ-ray bursts were discovered by the *Vela* satellites, each burst lasting typically less than a minute [175]. During that time, each burst is the brightest source in the sky. The first of these was detected by the *Vela* satellite in 1967 but they were not reported in the scientific literature until 1973. Observations with the Compton Gamma-Ray Observatory, launched in April 1991, have shown that these bursts occur about once per day, apparently randomly over the sky. These sources have proved to be among the most tantalizing discoveries in high-energy astrophysics.

23.7.3. Ultraviolet astronomy and the Hubble Space Telescope

Among the earliest beneficiaries of the opening up of space for astronomy were the ultraviolet astronomers. The central figure in this story is Lyman Spitzer who in 1946 wrote a report on the utilization of space for astronomical purposes for the RAND project of the US Air Force [176]. In 1957, as soon as the USSR space achievements galvanized the US space programme into action, Spitzer and his colleagues planned a series of three space observatories, to be known as the Orbiting Astrophysical Observatories (OAO), which were to be dedicated to spectroscopy in the ultraviolet waveband between 90 to 330 nm.

Unlike the other new astronomical wavebands, the astrophysical objectives of ultraviolet astronomy were very well defined. The resonance transitions of essentially all the common elements were known to lie in the ultraviolet rather than the optical waveband and so the study of the chemical composition of interstellar clouds is very effectively carried out in this waveband. Access to the wavelengths shorter than Lyman-α was of special significance because, among the many resonance lines are those of deuterium which is of cosmological importance. The *Copernicus* satellite (OAO-3), launched in 1972, was the great success of the series [177]. The very high-resolution spectrographs had the capability of exploring the wavebands to the short-wavelength side of the Lyman-α line at 121.6 nm. The abundances of the common elements, as well as deuterium, were measured in the interstellar medium for the first time. The mission also found evidence for the hot component of the interstellar gas through observations of the absorption lines of highly ionized oxygen O^{+5}. The OAO observatories led in turn to the launch of the International Ultraviolet Explorer (IUE) in 1978, which was a joint UK–European Space Agency–NASA project [178].

The IUE was the precursor of the Hubble Space Telescope [179]. Optical astronomers had long been aware of the fact that large aperture telescopes never achieve their theoretical angular resolution because of refractive index fluctuations in the atmosphere which blur the images of stars to typically about 1 arcsec, compared with the diffraction limit of a 4 metre telescope of about 0.03 arcsec. The increase in angular resolution also brings with it an increase in sensitivity for point sources. These problems are largely eliminated by placing the telescope above the Earth's atmosphere. In the 1960s, plans were formulated for the construction of a 3 metre Large Space Telescope, but the diameter was reduced to 2.4 metres when the decision was taken to launch the telescope using the Space Transportation System (STS), more popularly referred to as the *Space Shuttle*. Not only would the telescope be launched by the *Shuttle*, but it would also be regularly serviced by it. The approval process for the Hubble Space Telescope was not straightforward. The biggest problem was the fact that the project was very expensive, the cost estimates being greater than for any pure science programme ever undertaken. International collaboration was secured with the European Space Agency, which negotiated a 15% share in the telescope for European astronomers, and the programme was approved by the Ford administration in 1977.

The programme encountered major technical and financial difficulties within a couple of years of approval and the crisis came in 1981 when the programme almost ran out of money. Managerial changes were made and a new, more realistic budget was set for the programme. The programme was further delayed by the *Challenger* disaster in 1986 but ultimately the telescope was launched in April 1990. Within weeks, it was found that the primary mirror had been figured to the wrong shape, resulting in an unacceptable amount of spherical aberration, which blurred the images so that, without computer enhancement of the data, they were little better than what could be obtained from the ground. The spectacular maintenance and refurbishment mission carried out in December 1993 was a great triumph and restored more or less the full capability of the telescope.

23.7.4. *Infrared astronomy* [180]

The infrared radiation of the Sun was first detected in 1800 by William Herschel [181] in his famous experiments in which he placed thermometers beyond the red end of an optical spectrum of the Sun and found a greater temperature increase than in the red region of the spectrum. The development of infrared astronomy was slow because the photographic process does not work at wavelengths longer than about 1 μm. The first effective infrared detectors for astronomy were only constructed in the 1950s using photoelectric semiconductor materials. The great advantage of working in the infrared waveband is that the interstellar dust, which obscures many of the most interesting regions in gas clouds and galaxies, becomes transparent.

Infrared astronomy was pioneered from the ground using single-element detectors to observe through the atmospheric windows in the near infrared waveband, the available windows occurring at 1.2 (J), 1.65 (H), 2.2 (K), 3.5 (L), 5 (M), 10 (N) and 20 (Q) μm. The earliest observations were made using lead sulphide cells, which were sensitive in the 1 to 4 μm waveband. In the late 1950s and early 1960s, Harold Johnson and his colleagues [182] measured the intensities of several thousands of stars in the J, H and K wavebands. At about the same time, Gerry Neugebauer and Robert Leighton [183] at the California Institute of Technology began a survey of the whole sky at 2.2 μm with a 62" telescope which they built themselves. They surveyed the whole of the sky north of declination $\delta = -33°$ and found 5612 infrared sources. In many cases, the infrared emission was simply the extension of the optical spectra of stars into the infrared waveband but, in addition, many more strong infrared emitters were discovered than had been expected.

Observations at the longer infrared wavelengths were much more difficult than those at 2 μm because of the background of thermal emission from the sky and the telescope. At the longer wavelengths, gallium-doped germanium detectors were pioneered as bolometers by Frank Low [184] in the early 1960s. In 1966 Eric Becklin and Gerry Neugebauer [185] made the first painstaking observations of the Orion Nebula at 1.65, 2.2, 3.5 and 10 μm with the Palomar 200" telescope. To their surprise, a very intense infrared 'star' was discovered, not from the prominent optical nebula but from an obscured area to the north of the four famous trapezium stars, which are responsible for illuminating the nebula. Their preferred interpretation was that the object was a massive protostar which was still enshrouded in a dusty envelope. The dust was responsible for absorbing the energy emitted by the protostar and re-radiating it at far-infrared wavelengths. In another heroic paper of 1968, they made the first infrared observations of the Galactic Centre in the H, K and L wavebands [186]. Because of the decrease in extinction with increasing wavelength, the Galactic Centre region itself was detectable. They found evidence for an increase in the star density towards the Centre as well as a compact region in the Centre itself, coincident with an intense compact radio source.

The great potential of infrared observations led to the construction of telescopes which were optimized for the infrared region of the spectrum. The UK Infrared Telescope (UKIRT) and the NASA Infrared Telescope Facility (IRTF), both located on the summit of Mauna Kea, Hawaii, began operations in the late 1970s and played a major role in making infrared observations an integral part of observational astrophysics. In the mid-1970s, lead sulphide detectors were replaced by the more sensitive indium antimonide detectors. By the mid-1980s, infrared array technology was declassified by the US military agencies and it became possible to build cameras with which to take images in the infrared waveband [187].

The far-infrared regions of the spectrum cannot be observed from the ground because of atmospheric absorption. Pioneering experiments were carried out from high-flying aircraft and from balloon-borne platforms in the 1970s. Frank Low [188] carried out a number of exploratory programmes from a modified executive Lear jet which led in due course to the development by NASA of the Kuiper Airborne Observatory which consists of a Lockheed C-141 transport aircraft with a hole cut in the side to enable observations to be made with a 91 cm diameter telescope at an altitude of about 13 km.

The next natural step was to build a dedicated satellite to undertake a systematic survey of the far-infrared sky. The *Infrared Astronomy Satellite* (*IRAS*), which was an international venture involving the Netherlands, the USA and the UK, was launched in January 1983 and mapped the whole sky in those infrared wavebands which are inaccessible from the ground, namely in bands centred on 12, 25, 60 and 100 μm. These observations have had a major impact upon essentially all branches of astronomy but the outstanding contributions have come in the study of regions of star formation and the realization that the majority of galaxies emit as much radiation in the far-infrared waveband as they do at optical wavelengths [189].

23.7.5. Optical astronomy in the age of the new astronomies
In the 1960s, the construction of a large number of 4 metre class optical telescopes was begun with the intention of providing improved access for many more astronomers to world-ranking observing facilities. Within the USA, the need for national telescopes in addition to the private observatories was recognized with the founding in May 1960 of the Association of Universities for Research in Astronomy (AURA) which had the responsibility of building and operating telescopes for the US astronomical community. The principal facilities were the 4 metre Telescope at the Kitt Peak National Observatory (KPNO) in Arizona, the Sacramento Peak Solar Observatory in New Mexico and the 4 metre telescope at the Cerro Tololo International Observatory (CTIO) at Cerro Tololo in Chile.

Within Europe, it was realized that the many riches of the southern hemisphere, including the Galactic Centre and the Magellanic Clouds, had yet to be explored with large telescopes. The European Southern Observatory (ESO) was set up in 1962 'to establish and operate an astronomical observatory in the southern hemisphere, equipped with powerful instruments with the aim of furthering and organizing collaboration in astronomy'. Operation of the 3.6 metre telescope at La Silla in the Atacama Desert, 600 km north of Santiago, began in 1977.

The UK did not become a member of ESO but joined with Australia to construct the 3.9 metre Anglo-Australian Telescope [190] on Siding Spring Mountain in New South Wales—the telescope was completed in 1975. France, Canada and Hawaii agreed to collaborate in the construction of the 3.6 metre telescope on the summit of Mauna Kea in Hawaii, probably the best and highest all-round site for astronomy in the world, and the telescope began operations in 1979. The UK, Spain and the Netherlands constructed an observatory for the northern skies on the island of La Palma in the Canary Islands, including the 4.2 metre William Herschel Telescope, which was completed in 1987.

Astrophysics and Cosmology

The change was, however, much more than simply an increase in observing time on large telescopes. Perhaps the most dramatic development has been the introduction of highly sensitive electro-optical detectors which, for almost all purposes, have replaced the use of photographic plates. The invention of the charge-coupled device (CCD) in 1969 was of special significance for astronomy [191]. The best photographic plates have quantum efficiencies of only about 1%, whereas a good CCD typically has quantum efficiency of about 70%, the equivalent of increasing the collecting area of the telescope by 70 times. The astronomers realized the potential of these devices for astronomy and they were developed under contract from the Jet Propulsion Laboratory by Texas Instruments who built the first devices specifically for astronomy in 1976. They were adopted as the detectors to be used in the Wide Field Planetary Camera of the Hubble Space Telescope in 1977 and gradually they have been adopted by all major observatories as the prime detector for imaging and for many spectroscopic applications.

Survey work is at the heart of many astronomical projects. The widest field telescopes are the Schmidt telescopes which use an optical design invented by Bernhard Schmidt in 1929 [192]. The use of this type of telescope was pioneered by Zwicky in the 1930s. Immediately after World War II, a large Schmidt telescope of effective aperture 1.2 metres (48") was constructed to support observations made with the 200" telescope at Mount Palomar. The size of each plate was 14 inches, corresponding to about 6° on the sky, and over a period of 8 years, this telescope completed a photographic survey of the complete sky north of declination −20° in blue and red wavebands. Photographic copies of the survey plates were made available to the worldwide community of astronomers and the resulting Palomar Atlas has been one of the most valuable research tools for all astronomers.

> See also p 1406

In the 1960s, the ESO and the UK constructed Schmidt Telescopes in the southern hemisphere to carry out the same type of surveys as had been completed in the northern sky from Mount Palomar. The large sky surveys contained an enormous amount of statistical data of importance for astronomy but the data could only be extracted if high-speed measuring machines were built specially for the purpose. UK astronomers took the lead in these developments with the construction of the COSMOS High-Speed Measuring Machine at the Royal Observatory, Edinburgh and the Automatic Plate Measuring Machine (APM) at Cambridge. These studies have provided many of the most important targets to be observed by the 4 metre class telescopes.

PART 4: ASTROPHYSICS AND COSMOLOGY SINCE 1945

23.8. Stars and stellar evolution since 1945

By 1945, the basic physical processes involved in the evolution of stars was understood and the Hertzsprung–Russell diagrams for stars in clusters provided a test-bed for the theory of stellar evolution. By 1952, Sandage and Schwarzschild [193] had developed evolutionary tracks on the Hertzsprung–Russell diagram for stars in globular clusters and used the main-sequence termination point to estimate the ages of the clusters which turned out to be about 3×10^9 years. There remained, however, an enormous amount of detailed work to be undertaken before a precise comparison between theory and observation could be made. To build detailed models of stars, the equation of state of the material of the star, accurate nuclear reaction rates and the opacity of stellar material for the transfer of radiation have to be known. Hence, the astrophysicists had to have access to a very wide range of data from nuclear, atomic and molecular physics. These data would not have been of much value had it not been for the development of electronic computers which enabled detailed models of the Sun and the stars to be constructed. Thanks to the efforts of pioneers such as L G Henyey, Rudolph Kippenhahn and Icko Iben, the study of the stars has become one of the most precise of the astrophysical sciences [194].

23.8.1. Nucleosynthesis and the origin of the chemical elements

Two of the major problems of stellar astrophysics were closely related. The first concerned the synthesis of the chemical elements and the second the nuclear processes responsible for energy generation, once the stars had moved off the main sequence.

There are no stable isotopes with mass numbers 5 and 8 and, consequently, there is no straightforward way in which protons, neutrons and α-particles can be added successively to helium nuclei as the first of a sequence of reactions which leads to the formation of carbon. The solution was discovered independently by Öpik and Salpeter [195] who pointed out that, when the central temperature of the star reaches about 4×10^8 K, the triple-α reaction, in which three α-particles come together to form carbon, can take place. The cross section for this reaction seemed, however, to be too small to create significant amounts of carbon.

This problem was solved in 1953 by Fred Hoyle [196] who realized that the cross section for the interaction would be increased if there is a resonance associated with formation of ^{12}C in an excited state. Hoyle estimated that the excited state of ^{12}C should occur at about 7.7 MeV, a remarkable prediction in that, at that time, models of nuclei were not sufficiently well developed that any resonance state of any nucleus could be predicted. Ward Whaling and his colleagues [197] were persuaded to search for the resonance and found it at exactly the energy predicted by Hoyle. The inclusion of the carbon resonance increased the cross section for the formation of carbon by the triple-α process by a factor of 10^7. Hoyle went on to show that helium burning takes place at a temperature of 10^8 K, the temperature deduced by Sandage and Schwarzschild for the cores of red-giant stars at the tip of the giant branch.

Öpik and Salpeter realized that once carbon was created, heavier elements such as oxygen and neon could be created by the successive addition of α particles. In his paper of 1954, Hoyle argued that, once the star had exhausted the helium in its core, massive enough stars would continue to contract, increasing the central temperature so that the nuclear burning of ^{12}C into ^{24}Mg would take place and, at a slightly higher temperature, ^{16}O would be converted into ^{32}S. In massive enough stars, the process of nuclear burning would continue until all the nuclear energy resources were used up when the core of the star consists of ^{56}Fe, the element with the greatest nuclear binding energy.

In 1956, Hans Suess and Harold Urey [198] published their detailed analysis of the cosmic abundances of the chemical elements, which fall off rapidly with increasing atomic weight. There are, however, important features of the abundance curves which provide clues to the processes of nucleosynthesis. They drew attention to the overabundance of the 'α-particle' nuclei, such as those with 16, 20, 32 nucleons, as well as the iron group elements and the peaks of stability at $N = 50, 82, 126$, corresponding to the 'magic numbers' of nuclear physics.

The nuclear processes involved in the synthesis of the elements were described in two famous papers published in 1957, by Burbidge, Burbidge, Fowler and Hoyle and by Cameron [199]. In the former paper, eight nuclear processes by which the elements could be synthesized were described. In addition to hydrogen burning, helium burning and the triple α-process, they drew special attention to processes involving the addition of neutrons to pre-existing nuclei, the slow (s) and rapid (r) processes. These reactions provide the means by which nuclei with mass numbers greater than the iron group can be synthesized. In the r-process, several neutrons are added before decay occurs. At high enough temperatures and densities, the inverse β-decay process results in the release of large numbers of neutrons, and supernova explosions were identified as a site in which such reactions could take place. The s-process was believed to occur at an earlier stage in the evolution of stars on the giant branch.

Cameron drew particular attention to the importance of nucleosynthesis in supernova explosions, the process known as *explosive nucleosynthesis*. He realized that different chemical abundances are created if the process of nucleosynthesis takes place in a non-stationary manner, as in the case of supernova explosions. With the development of high-speed computers, it became possible to quantify these predictions and, in 1970,

Arnett and Clayton [200] showed that many of the element abundances could be naturally attributed to explosive nucleosynthesis.

23.8.2. Solar neutrinos

In 1955, Raymond Davis [201] suggested that it might be possible to build detectors to measure the flux of electron neutrinos liberated in the CNO cycle. Because of their very small cross sections for interaction with matter, neutrinos escape essentially unimpeded from their point of origin within the central 10% of the Sun by radius and thus the flux of solar neutrinos provides a direct test of the processes of nucleosynthesis. Davis proposed detecting the solar neutrinos by the nuclear transformations which they would produce in a fluid containing a large number of chlorine atoms. Specifically, the nuclear reaction $^{37}\text{Cl} + \nu_e \rightarrow {}^{37}\text{Ar} + e^-$ has a threshold energy is 0.814 MeV. The argon created in this reaction is radioactive and the amount produced can be measured from the number of radioactive decays of the ^{37}Ar nuclei. Unfortunately, Epstein and Oke [202] had shown that the p–p chain rather than the CNO cycle is the principal source of energy in the Sun.

Neutrinos are, however, emitted in the first reaction of the p–p chain (see section 23.3.3). The first reaction, in which deuterium is formed, is the principal source of neutrinos from the Sun but they are of low energy, the maximum energy being 0.420 MeV, and so could not be detected by a chlorine detector. In 1958, Cameron and Fowler [203] pointed out that more energetic neutrinos are emitted in a rarer side-chain of the main p–p chain.

$$^{7}\text{Be} + p \rightarrow {}^{8}\text{B} + \gamma \; : \; {}^{8}\text{B} \rightarrow {}^{8}\text{Be}^* + e^- + \nu_e \; : \; {}^{8}\text{Be}^* \rightarrow 2\,{}^{4}\text{He}.$$

The electron neutrinos emitted in the decay of ^{8}B nuclei have maximum energy 14.06 MeV and so could be detected in the type of experiment proposed by Davis.

The first detailed predictions of the solar neutrino flux were made by Bahcall [204] in 1964 and, at about the same time, the famous *solar neutrino experiment* was begun by Davis and his colleagues using a 100 000 gallon tank of perchloroethylene, C_2Cl_4, located at the bottom of the Homestake Goldmine in South Dakota. The experiment ran for 20 years and a significant flux of neutrinos was detected, but it corresponded to only about one-quarter of the flux predicted by the standard solar model. This discrepancy is the famous *solar neutrino problem*. The origin of this discrepancy has been one of the most controversial topics in astrophysics since Davis's results were first reported in 1968. Interpreted literally, the result suggests that there must be something wrong either with the nuclear physics or with the astrophysics of the Sun or with both [205].

Confirmation that the flux of high-energy neutrinos indeed originates within the Sun was provided in 1990 by the Kamiokande II neutrino-scattering experiment in which the arrival directions of the incoming

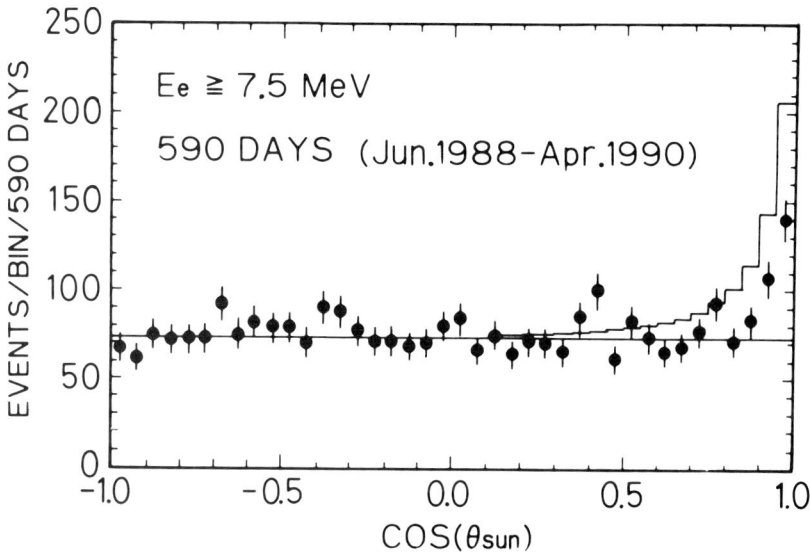

Figure 23.15. *The distribution in $\cos\theta_{Sun}$ for the 590-day sample of neutrinos with $E_e \geq 7.5$ MeV. θ_{Sun} is the angle between the momentum vector of the an electron observed at a given time and the direction of the Sun [206]. The angular resolution of the detector system has been taken into account in calculating the expected distribution of arrival directions of the neutrinos from the Sun which is indicated by the histogram.*

neutrinos were measured [206]. A small but significant excess flux of neutrinos was found in the direction of the Sun (figure 23.15) but it amounted to only about 46% of that expected from the standard solar model of Bahcall and Ulrich.

A key test of the solar models is the detection of the more plentiful low-energy neutrinos from the first reaction of the p–p chain. Gallium has been used as the detector material and, to measure the neutrino flux, the number of radioactive germanium nuclei created by the neutrino interaction, $\nu_e + {}^{71}\text{Ga} \rightarrow e^- + {}^{71}\text{Ge}$, is measured. In June 1992, the first results of the GALLEX experiment were reported [207], the neutrino flux being 83 ± 19 (stat) ± 8 (syst) SNU compared with a value of $132 {}^{+20}_{-17}$ SNU expected from the standard solar models, where 1 SNU = 1 solar neutrino unit. This is a very important result since, although there is still a discrepancy, a neutrino flux of the correct order of magnitude has been detected from the crucial first reaction of the p–p chain.

23.8.3. Helioseismology

The solar neutrinos provided the only method for studying directly the internal structure of the Sun until the discovery of *solar oscillations* with period about five minutes by Leighton and his colleagues [208]. In a prescient paper of 1968, Frazier [209] suggested that the oscillations were

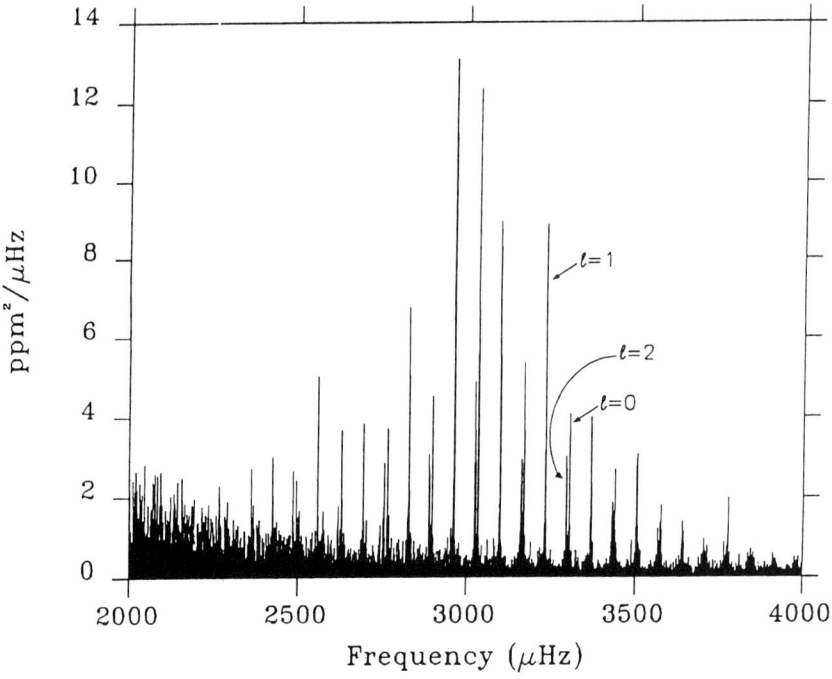

Figure 23.16. *An example of the frequency spectrum of solar oscillations showing some of the normal modes of oscillation of the Sun. These data were derived from 160 days of observation by the IPHIR experiment on board the PHOBOS spacecraft* [212]. *This power spectrum of the low degree p modes shows an alternating pattern of double and single peaks; the double peaks are the $l = 0, 2$ modes and the single peaks the $l = 1$ modes.*

trapped acoustic waves in the outer layers of the Sun but the first proper analyses of the normal modes of oscillation of the Sun were carried out by Ulrich in 1970 and by Leibacher and Stein [210] in 1971. The 'five-minute' oscillations turned out to be standing acoustic waves, confined to the outer layers of the Sun.

The study of the astrophysics of the Sun and the stars was rejuvenated by these discoveries. The low degree modes probe deeply into the interior of the Sun and so could be used to test models of its internal structure. The existence of these low degree modes was deduced by Christensen-Dalsgaard and Gough [211] from the data published by the Birmingham group in 1979. An example of the quality of the data now available is shown in figure 23.16 which shows the power spectrum of the total luminosity of the Sun obtained from the IPHIR experiment on board the *PHOBOS* space probe while in transit between the Earth and Mars in 1988 and 1989 [212].

Because of the similarities of the physical techniques to terrestrial seismology, these studies are referred to as *helioseismology*. The

Astrophysics and Cosmology

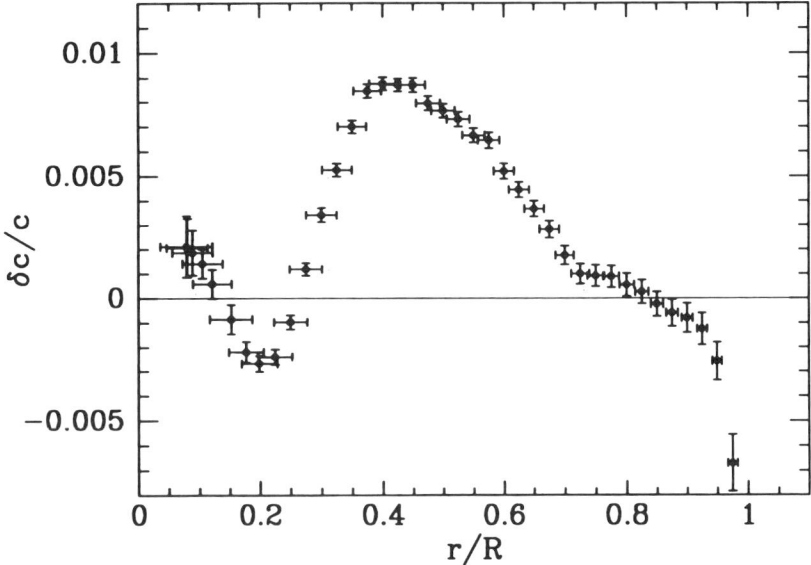

Figure 23.17. *A comparison of the theory and the observations in terms of the fractional deviations of the sound speed inferred from observation (points with error bars) relative to the best-fitting standard model (straight line at $\delta c/c = 0$) of the interior structure of the Sun* [213]. *The agreement is better than 1%.*

helioseismic data can be inverted to derive the variation of the speed of sound within the Sun with radius. As can be seen from figure 23.17, the predictions of the standard solar models agree with the observations to better than 1% throughout the Sun [213]. The observations are not, however, consistent with non-standard models which involve large amounts of core mixing, rapid rotation of the core of the Sun or the presence of weakly interacting massive particles in the core of the Sun, all of which have been proposed to account for the low flux of solar neutrinos. The astrophysics of the standard solar model seems to provide a satisfactory description of the internal structure of the Sun right into its core. If there is a solar neutrino problem, its resolution probably lies with the nuclear physics rather than with the astrophysics.

23.8.4. The discovery of neutron stars

The neutron stars were discovered, more or less by chance, as the parent bodies of radio pulsars by Antony Hewish, Jocelyn Bell and their colleagues in 1967 [214]. Hewish had pioneered the techniques of observing the scintillation of radio sources at long radio wavelengths caused by irregularities in the plasma density along the line of sight to the radio source. In the early 1960s, he used this technique to study the interplanetary plasma and, as a by-product, to discover radio quasars which display strong radio scintillations at low frequencies. He designed and built a large, low-frequency array of collecting area 1.8 hectares operating at a frequency of 81.5 MHz (3.7 m wavelength) in order to record the rapidly fluctuating intensities of bright radio sources on a time-scale of one tenth of a second. This was the key technological development which led to the discovery of radio pulsars [215].

The first sky surveys began in July 1967 and Jocelyn Bell, Hewish's research student, discovered a strange source which seemed to consist entirely of scintillating radio signals (figure 23.18(a)). In November 1967, the source was observed using a recorder with a much shorter time constant and it was found to consist entirely of a series of pulses with pulse period about 1.33 s (figure 23.18(b)). Over the next few months, three further sources were discovered with pulse periods in the range 0.25 to almost 3 s. The name *pulsar* was coined soon after the announcement of the discovery.

Within a year, more than 20 more pulsars were discovered and a flood of theoretical papers appeared. The favoured model for pulsars was described by Thomas Gold in 1968, similar in many respects to Pacini's prescient proposal of 1967, and consisted of an isolated, rotating, magnetized neutron star, in which the magnetic axis of the star and its rotation axis are misaligned [216]. The radio pulses are assumed to originate from beams of radio emission emitted along the magnetic axis. The key observations which supported this picture were the very short stable periods of the pulses and the observation of polarized radio emission within the pulses. Two of the pulsars discovered in 1968 were of special importance. A pulsar of period 0.089 s was discovered in the young supernova remnant in the constellation of Vela and, soon afterwards, a pulsar with a period of only 0.033 s was detected in the centre of the Crab Nebula, the remnant of the supernova which exploded in 1054. Optical pulses from the pulsar with precisely the same pulse period were discovered in 1969 and, within three months, observations from a rocket-borne detector had discovered x-ray pulses as well [217].

The short periods of these pulsars proved beyond any shadow of doubt that the parent bodies of the pulsars were neutron stars, since the break-up rotational speeds of white dwarfs correspond to periods of roughly one second or greater. Furthermore, the formation of neutron stars in supernova explosions was conclusively demonstrated by the coincidence of these short period pulsars with young supernova remnants.

In 1968, Pacini [218] had shown that the magnetic field strengths at the surfaces of the neutron stars had to be enormous, $B \sim 10^6$–10^8 T and, in the next year, the first papers exploring the electrodynamics of pulsars were published. These magnetic fields are so strong that the Lorentz ($v \times B$) force extracts electrons from the surface layers of the neutron star so that electric currents must be present in its magnetosphere [219]. Gunn and Ostriker [220] showed how electrons could be accelerated to

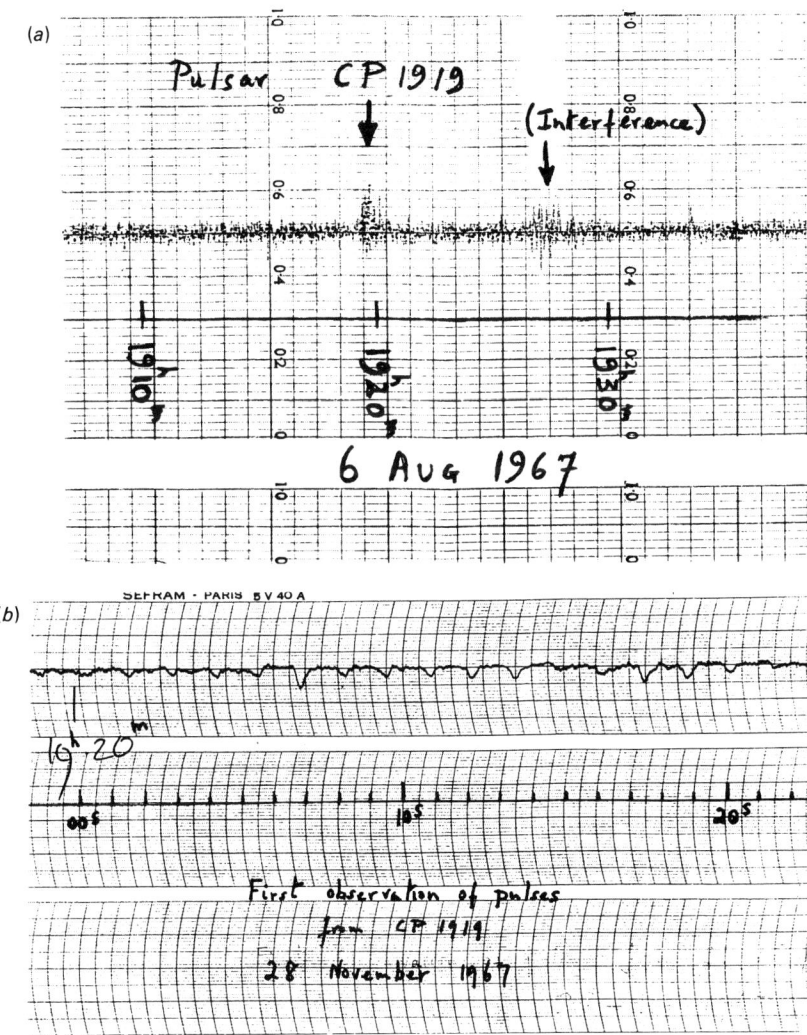

Figure 23.18. *The discovery records of the first pulsar to be discovered, PSR 1919+21* [214]. *(a) The first record of the strange scintillating source labelled CP 1919. Note the subtle differences between the signal from the source and the neighbouring signal due to terrestrial interference. (b) The signals from PSR 1919+21 observed with a recorder of shorter time-constant than the discovery record showing that the signal consists entirely of regularly spaced pulses with period 1.33 s.*

very high energies in the strong electromagnetic waves emitted by the rotating magnetized neutron star.

Studies of the internal structure of the neutron stars were pursued with renewed vigour. The equation of state of neutron matter had been the subject of numerous studies prior to the discovery of pulsars. Much

Twentieth Century Physics

improved calculations were carried out by Baym, Bethe, Pethick and Sutherland [221] in 1971 and these were used to construct the standard models for neutron stars. The equation of state was well understood up to about nuclear densities, $\rho \sim 3 \times 10^{17}$ kg m^{-3}, but there remained uncertainties at higher densities, which may be found in the centres of the most massive neutron stars. The interior is threaded by an intense magnetic field which does not have a strong influence upon the structure of the neutron star but does influence its internal dynamical properties. Long before the pulsars were discovered, theorists had proposed that the interiors of neutron stars should be superfluid. Studies of the properties of the neutron–proton–electron fluid in different regimes inside the neutron stars showed that the inner crust and the neutron liquid phases are superfluid and that the protons are superconducting. In 1969, Baym and co-workers [222] showed that the magnetic field is therefore quantized into vortices which are pinned to the crust of the neutron star.

See also p 930

The link between the interior structure of the star and observable phenomena was provided by the phenomenon *glitches* which were discovered in the Vela pulsar in 1969 [223]. All radio pulsars were known to slow down but, occasionally, discontinuous changes in the slow-down rate were observed in which the period decreased abruptly. One possibility proposed by Baym and co-workers [222] in 1969 was that, as the neutron star slows up, its crust takes up a new shape in a 'star-quake' in which the structure changes discontinuously to take up a new equilibrium configuration. This process cannot, however, be the whole story since, in pulsars such as the Vela pulsar, the glitches occur too frequently. The preferred interpretation is that the glitches may be associated with the unpinning of the vortices with the crust of the neutron star. One important aspect of the glitches is that the recovery of the rotation speed to a steady value is remarkably slow. This is direct observational evidence that a considerable fraction of the moment of inertia of the neutron star must reside in some superfluid form which is only weakly coupled to the crust.

23.8.5. X-ray binaries and the search for black holes

The discovery of binary x-ray sources by the *UHURU* satellite in 1971 was to have a profound influence upon thinking in high-energy astrophysics [224]. In late 1970, the variable source Cygnus X-1 was observed and displayed somewhat random variability on time-scales as short as 100 ms, indicating that the source region must be compact. Equally remarkable were the observations of the source Centaurus X-3 (Cen X-3) made in January 1971. Unlike Cygnus X-1, Cen X-3 emitted regular pulses with a period of about 5 s, longer than that of any known radio pulsar. In May 1971, the pulsation period was found to vary sinusoidally with a period of 2.1 days, indicating that the source was

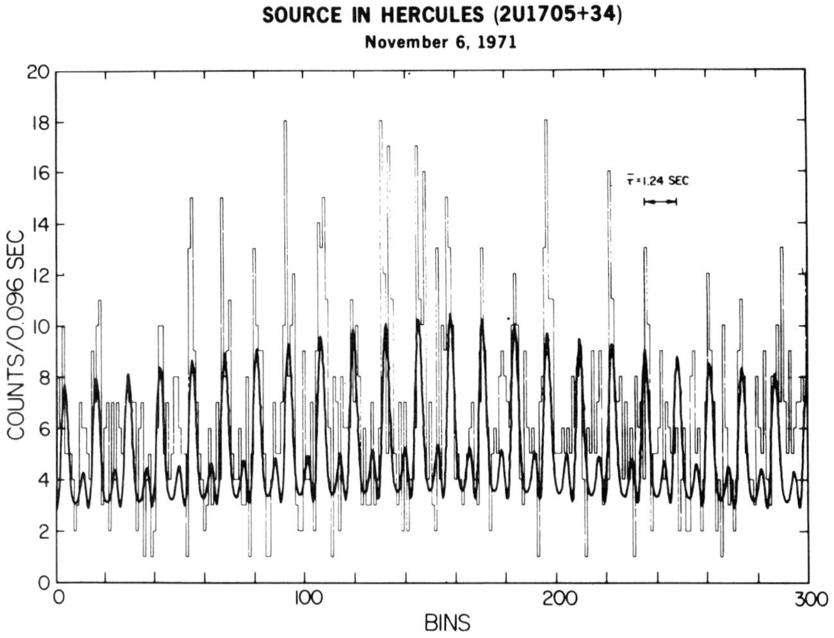

SOURCE IN HERCULES (2U1705+34)
November 6, 1971

Figure 23.19. *The discovery records of the pulsating x-ray source Her X-1* [225]. *The histogram shows the number of counts observed in successive 0.096 s bins. The continuous line shows the best-fitting harmonic curve to the observations, taking account of the varying sensitivity of the telescope as it swept over the source.*

a member of a binary system, the change in period of the pulses being due to the orbital motion of the source. On 6 May 1971, the source disappeared, only to reappear half a day later. This pattern repeated roughly every two days indicating that the x-ray source was being occulted by the primary star of a binary system. With these clues, the primary star was identified with a massive blue star with the same binary period of 2.1 days as the x-ray source. Soon after this discovery, another similar source was discovered [225], Hercules X-1 (Her X-1), which had pulse period 1.24 s and an orbital period of 1.7 days (figure 23.19).

The short pulse period of Her X-1 was strong evidence that the parent body must be a neutron star, similar to those of the radio pulsars. The source of energy for the system was identified as accretion. The idea of accretion as a source of energy for the x-ray sources had already been suggested by Hayakawa and Matsuoka [226] in 1964, who considered normal close binary systems, and by Shklovsky [227] in 1966, who proposed accretion from a binary companion onto a neutron star as the energy source for the brightest x-ray source in the sky Sco X-1. In 1968, Prendergast and Burbidge [228] pointed out that, in the accretion of matter from the primary star onto a compact secondary in a binary system, the accreted matter would necessarily have a considerable amount of angular momentum and so an *accretion disc* would form about the compact star. Accretion of matter from the primary star onto a compact neutron star is a very powerful energy source, up to about 5% of the rest-mass energy of the infalling matter being liberated, roughly an order of magnitude greater than can be liberated by nuclear fusion reactions.

Since the x-ray sources are members of binary systems, the masses of the neutron stars can be estimated using the standard procedures of celestial mechanics. The gratifying result was found that the masses of the seven binary x-ray sources for which this analysis was possible lay in the range $1.2 M_\odot$ to $1.4 M_\odot$, consistent with the upper limit to the masses of neutron stars, which is similar to the Chandrasekhar limit for white dwarfs [229].

In 1973, Margon and Ostriker [230] showed that the luminosities of the binary sources extended up to about $L = 10^{31}$ W which is very close to the Eddington limiting luminosity for accretion onto objects with mass $1 M_\odot$, $L \leqslant 1.3 \times 10^{31} (M/M_\odot)$ W. This result demonstrated that objects existed which could radiate x-rays at luminosities close to the maximum permissible luminosity. Furthermore, it was natural that these sources should emit most of their radiation in the x-ray waveband. Assuming that the x-ray emission originates close to the surface of the neutron star, application of the Stefan–Boltzmann law showed that the temperature of the emitting region had to be greater than about 10^7 K to produce such large luminosities.

From these considerations, a standard model for pulsating binary x-ray sources was developed [231]. In the case of the massive companions, the neutron star is located in a strong stellar wind and the matter within a certain *accretion radius* of the neutron star is accreted onto the star. In the low-mass binary systems, mass transfer takes place through the process of Roche lobe overflow, in which the primary star fills its Roche lobe, the equipotential surface joining the two stars, and so matter attains a lower gravitational potential by collapsing to form an accretion disc about the neutron star. The x-ray pulsations are attributed to accretion onto the poles of the rotating neutron star, the strong non-aligned magnetic field channelling the matter onto the polar regions. Evidence for the presence of strong magnetic fields in the source Her X-1 was found by Trümper and his colleagues [232] in 1978 who identified a cyclotron radiation feature in its x-ray spectrum at about 58 keV, corresponding to a magnetic field strength of about $(4 - 6) \times 10^8$ T.

See also p 306

The next step was to ask whether or not there was any evidence for black holes among the binary x-ray sources. Isolated black holes are very difficult to detect and it is only when they are close to sources of fuel that their presence can be readily detected. In 1965, Zeldovich and Gusyenov [233] had proposed that the observation of x-rays or γ-rays from single-line spectroscopic binaries might be the signature of either a neutron star or black hole. In 1969, Trimble and Thorne [234] investigated whether or not any of the dark companions in known single-line binaries could be massive enough to be black holes but no likely candidates were found and none coincided with known x-ray sources.

The first strong candidate for a black-hole companion was found in 1971 in the bright x-ray source Cyg X-1 which was identified with a ninth magnitude blue supergiant star, which turned out to be the primary star of a binary system with period 5.6 days [235]. Assuming the mass of the supergiant B star was greater than $10M_\odot$, the mass of the invisible companion had to be greater than $3M_\odot$, the most likely values being that the masses were $20M_\odot$ and $10M_\odot$ respectively. The mass of the unseen companion exceeded the upper limit for stability as a neutron star and it was concluded that it must therefore be a black hole.

Since these pioneering studies, three other good examples of x-ray binaries with massive invisible companions have been discovered, the x-ray binary sources LMC X-1, LMC X-3 and A0620-00 [236]. In each of these, the x-ray intensity exhibits short-period variability but no signature of pulsed x-ray emission. The simplest interpretation of the properties of these systems is that they contain black holes.

23.8.6. Radio pulsars and tests of general relativity

The radio pulsars provide some of the very best tests of general relativity. Among the most important systems are the binary pulsars, the first of which was discovered by Russell Hulse and Joseph Taylor [237] in 1974. The system, known as PSR 1913+16, has a binary period of only 7.75 hours and the orbital eccentricity is large, $e = 0.617$. To test general relativity, a precise clock in a rotating frame of reference is required and systems such as PSR 1913+16 are ideal for this purpose. Various parameters of the binary orbit can be measured by precise timing of the arrival times of the radio pulses and these provide estimates of different functions involving the masses of the two neutron stars. Assuming general relativity is the correct theory of gravity, the masses of the two neutron stars, which appear in six independent functions of the orbital parameters, agree with astounding precision. These are the most accurate masses measured for any stars, the values being $1.4411(7)M_\odot$ and $1.3874(7)M_\odot$. There is no evidence for any discrepancy with the expectations of general relativity [238].

A second remarkable measurement has been the rate of loss of rotational energy of the binary system by the emission of gravitational waves. The change of orbital phase of the system PSR 1913+16 has been observed over a period of 17 years and the observed changes agree precisely with the predictions of general relativity [238] (figure 23.20). Although the gravitational waves themselves have not been detected, exactly the correct energy loss rate has been observed. This is a very important result for general relativity since it enables a wide range of alternative theories of gravity to be excluded.

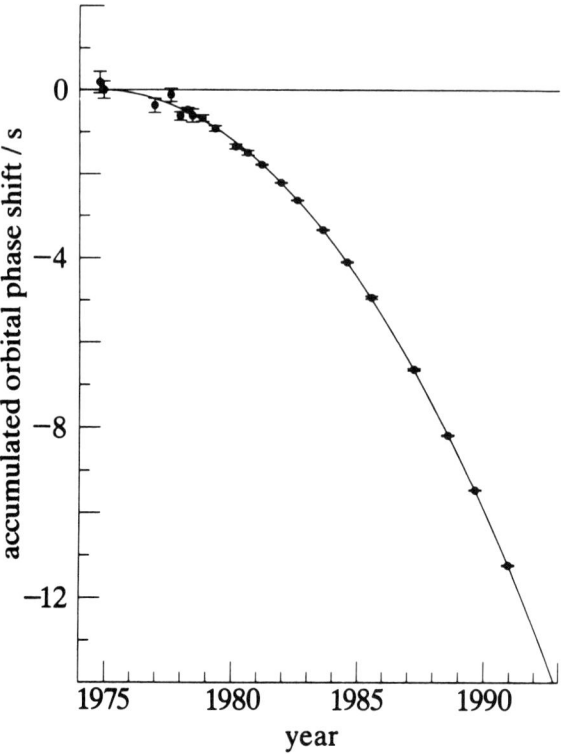

Figure 23.20. *The change of orbital phase as a function of time for the binary neutron star system PSR 1913+16 compared with the expected change due to gravitational radiation energy loss by the binary system* [238].

The technique of accurate pulsar timing can also be used to determine whether or not there is any evidence for the gravitational constant varying with time [238]. These tests are slightly dependent upon the equation of state used to describe the interior of the neutron star but for the range of plausible equations of state, the limits to $\dot{G}/G$ are less than about 10^{-11} per year. Thus, there can have been little change in the value of the gravitational constant over cosmological time-scales.

23.8.7. Supernovae [239]

Fritz Zwicky began the systematic search for supernovae in 1934. In 1936, he supervised the construction of a wide-angle 18″ Schmidt telescope, which was sited at the new observatory at Mount Palomar. The first supernova was discovered in the galaxy NGC4157 in March 1937 and the second in the dwarf spiral galaxy IC4182 on 26 August 1937. Zwicky continued to discover about four supernovae per year in relatively nearby galaxies. With the construction of the 48″ Schmidt telescope at Mount

Palomar in 1949, searches could be made for fainter supernovae and typically about 20 were discovered each year [240].

In 1938, Baade and Zwicky [241] gave the first description of the typical light curve of supernovae which consists of an initial outburst lasting for a few weeks, following which the brightness decreases exponentially with a half-life of about 77 days. In 1941 Minkowski [242] made the important discovery that there are two quite distinct types of supernovae. The spectra of Type I supernovae consist of broad emission bands, the nature of which were not understood until almost 30 years later when Kirshner and Oke [243] showed that they could be interpreted as the superposition of hundreds of lines of Fe^+ and Fe^{++}. No hydrogen lines are observed in their spectra. In contrast, the Type II supernovae show the Balmer series of hydrogen soon after maximum light. A striking feature of the Type I supernovae is that their spectral and luminosity evolution are essentially identical.

In both cases, the energetics of the explosions are so great that they must involve the collapse of the star to some form of compact remnant, either a neutron star or a black hole. The Type II supernovae were identified with the collapse of the central regions of very massive stars, $M \geqslant 8M_\odot$, which have relatively short lifetimes and cannot have travelled far from the regions within which they formed. In contrast, in the case of the Type I supernovae, the most attractive picture is that they are formed by the accretion of mass onto a white dwarf in a binary system. When accretion takes the total mass of the star over the Chandrasekhar limit, collapse to a neutron star must take place. This picture can account for the facts that no hydrogen absorption lines are observed, that their spectra are iron rich and that the supernovae seem to have essentially identical properties. This picture of the formation of neutron stars in binary systems can also account for the high space velocities of radio pulsars if the binary is disrupted in the explosion.

The long exponential decay of supernovae following their initial outbursts was a puzzle, since the characteristic decay time was the same for essentially all supernovae. In 1961, Pankey [244] had suggested in his PhD dissertation that the decay might be associated with the decay of radioactive nuclides created in the explosion and this proposal was put on a proper astrophysical basis by Colgate and McKee [245] in 1969. In the process of collapse to form a neutron star, explosive nucleosynthesis takes place which creates, among other products, the radioactive species ^{56}Ni. This isotope then decays as follows

$$^{56}Ni \xrightarrow{\beta^+} \,^{56}Co \xrightarrow{\beta^+} \,^{56}Fe.$$

The first β-decay has a half-life of only 6.1 days, while the second β-decay, which has a half-life of 77.1 days, is presumed to be the source of energy for the exponential decay of the luminosity of the supernova,

3.5 MeV being liberated in the form of γ-rays in each decay of a ^{56}Co nucleus. The exponential decay of the luminosity of the supernova is attributed to the creation of radioactive nickel in the explosion which is ejected into the expanding envelope of the supernova.

Unquestionably, the most important event in supernova studies this century has been the explosion of a supernova in one of the dwarf companion galaxies of our own Galaxy, the Large Magellanic Cloud. This supernova, known as SN1987A, was first observed optically on 24 February 1987 and reached about third visual magnitude by mid-May 1987 [246]. The supernova coincided precisely with the position of the bright blue supergiant star Sanduleak-69 202 which disappeared following the supernova explosion. This observation indicated that the progenitor of the supernova was a massive early-type B3 star.

One of the pieces of great good fortune was that, at the time of the explosion, neutrino detectors were operational at the Kamiokande experiment in Japan and at the Irvine–Michigan–Brookhaven (IMB) experiment located in an Ohio salt-mine and the signature of the arrival of a burst of neutrinos was convincingly demonstrated in both experiments [247]. Only 20 neutrinos with energies in the range 6 to 39 MeV were detected but they arrived almost simultaneously at the two detectors, the duration of the pulse being about 12 seconds. The supernova was only observed optically some hours after the neutrino pulse, consistent with a picture in which the neutrinos escape more or less directly from the centre of the collapse of the progenitor, whereas the optical light diffuses out through the supernova envelope. This observation, coupled with the measured energies of the neutrinos, enabled a limit of $m_\nu \leqslant 20$ eV to be set to the rest mass of the neutrino. The observation of the neutrino flux from the supernova is uniquely important for the theory of stellar evolution. The neutrino luminosity of the supernova was of the same order as that expected from the formation of a neutron star ($E \approx 10^{46}$ J).

The light curve of the supernova has been followed for almost five years after the initial explosion (figure 23.21). After the initial outburst, the luminosity decayed exponentially with characteristic half-life of 77 days until roughly 800 days after the explosion after which time the rate of decline decreased [246]. To account for the luminosity of the supernova, about 0.07 $M_\odot$ of ^{56}Ni must have been deposited in the supernova envelope and this figure agrees very well with the theoretical expectations of explosive nucleosynthesis. Evidence was found for the γ-ray lines of ^{56}Co within six months of the explosion as well as fine structure lines of cobalt and nickel in the infrared spectrum of the supernova, once the exponential decrease in luminosity began.

The totality of these observations provides direct confirmation for the radioactive theory of the origin of the supernova light curve and the formation of iron peak elements in supernova explosions.

Astrophysics and Cosmology

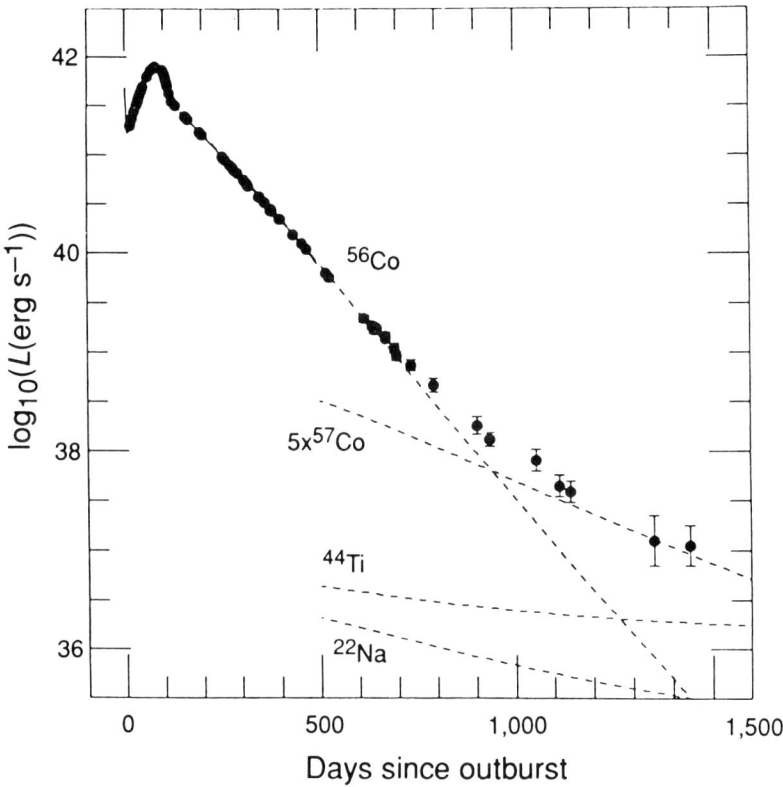

Figure 23.21. *The light curve of SN1987A presented by Chevalier [246]. The ordinate is the bolometric luminosity of the supernova meaning its integrated luminosity in the ultraviolet, optical and infrared wavebands, during the first five years. The energy deposited by radioactive nuclides (broken lines) is based upon the following initial masses: $0.075 M_\odot$ of ^{56}Ni (and subsequently ^{56}Co), $10^{-4} M_\odot$ of ^{44}Ti, $2 \times 10^{-6} M_\odot$ of ^{22}Na and $0.009 M_\odot$ of ^{57}Co, the last being five times the value expected from the solar ratio of $^{56}Fe/^{57}Fe$.*

23.9. The physics of the interstellar medium

By 1939, the existence of various forms of interstellar matter had been established. From the study of interstellar absorption lines and the variation of interstellar extinction with distance, it was known that diffuse gas and dust are present in the interstellar medium [248]. It was recognized in the early 1920s that the central stars of planetary nebulae and the O stars are very hot and so radiate a great deal of energy in the ultraviolet waveband. Russell [249] suggested that the excitation of the emission lines seen in gaseous nebulae and planetary nebulae are due to photoexcitation, and Eddington [250] showed that, as a result, the gas would attain a temperature of about 10 000 K. In 1926, Donald H Menzel [251] applied this model to planetary nebulae, suggesting that the Balmer emission lines were associated with the photoionization of hydrogen by the ultraviolet radiation of the central star followed by recombination of the protons and electrons. This picture was elaborated by Herman Zanstra [252] in 1926 who postulated that the central stars of planetary nebulae had temperatures of about 30 000 K. In a subsequent paper [253], he described the processes of photoionization and recombination of hydrogen, in which each photon with energy greater than 13.6 eV ionizes one hydrogen atom.

The importance of photoionization in relation to regions of ionized gas in the interstellar medium was discussed by the Danish astrophysicist Bengt Strömgren [254] in 1939. He solved the problem of the degree of ionization of the interstellar gas with radial distance from a star and its dependence upon the density of the gas and the temperature of the exciting star. The radius of the ionizing zone is strongly dependent upon the flux of ionizing radiation with wavelength $\lambda \leqslant 91.6$ nm. Strömgren also showed that the regions of ionized gas, often referred to as *Strömgren spheres*, have very sharp edges, in very good agreement with the properties of the regions of ionized hydrogen known as HII regions. The hottest O and B stars are found embedded in these regions of ionized hydrogen.

In 1928, Ira Bowen [255] at last identified the 'nebulium' lines which had baffled spectroscopists since their discovery in 1868. The nebulium lines can be associated with what are now termed *forbidden transitions* between low-lying metastable states of the ions of common elements and the ground state. These lines had not been observed in laboratory experiments because the densities are sufficiently high for the metastable states to be de-excited by collisions.

This picture of the interstellar gas was entirely derived from optical observations. After World War II, observations became possible in the radio, infrared, ultraviolet and x-ray wavebands and revealed the full complexity of the interstellar medium.

23.9.1. Neutral hydrogen and molecular line astronomy

The prediction of the 21 cm line of neutral hydrogen is one of the more remarkable stories attending the birth of radio astronomy [256]. Jan Oort was the director of the Leiden Observatory throughout the period of the German occupation of Holland during World War II and astronomical seminars had to be held in secret in the basement of the Leiden Observatory. Copies of the *Astrophysical Journal* continued to reach Leiden and, in 1944, included Reber's continuum radio map of the galaxy. To quote Hendrik van de Hulst [256]

> In the spring of 1944 Oort said to me: 'We should have a colloquium on the paper by Reber; would you like to study it? And, by the way, radio astronomy can really become very important if there were at least one line in the radio spectrum.

Then we can use the method of differential galactic rotation as we do in optical astronomy.

Van de Hulst [257] took up the challenge and studied the many ways in which atoms, ions and molecules could radiate line emission in the radio waveband. The most significant prediction was that neutral hydrogen should emit hyperfine line radiation at a wavelength of 21.106 cm because of the minute difference in energy when the relative spins of the proton and electron in a hydrogen atom change. Oort and C Alex Muller would probably have been the first to detect the 21 cm line but the first receiver was destroyed in a fire. The first detection of the line was made by Harold I Ewan and Edward M Purcell [258] at Harvard University in 1951 and, six weeks later, Muller and Oort [259] measured the same line. The first maps of the distribution of neutral hydrogen in the Galaxy and the determination of the Galactic rotation curve appeared in 1952. By 1953, neutral hydrogen had been detected in the Magellanic Clouds and, in 1954, high-velocity features in the Galactic Centre and 21 cm absorption spectra were first measured. In 1958, Oort, Kerr and Westerhout [260] published their famous map of the distribution of neutral hydrogen in the plane of the Galaxy, made by combining observations taken in the Netherlands with those taken at Sydney, Australia (figure 23.22). Neutral hydrogen is omnipresent throughout the plane of the Galaxy and extends well beyond the Sun's radius, enabling the Galactic rotation curve to be determined. The rotation curve at and beyond the Sun's orbit about the Galactic Centre is remarkably flat, $v(r) =$ constant, suggesting the presence of dark matter in the halo of the Galaxy [261].

Long before the advent of radio astronomy, it was known that there exist significant abundances of molecules in interstellar space. The molecules CH, CH$^+$ and CN possess electronic transitions in the optical waveband and absorption features associated with these were well known in the spectra of bright stars. The first interstellar molecule to be detected at radio wavelengths was the hydroxyl radical OH, which was observed in absorption against the bright radio source Cassiopeia A at a wavelength of 18 cm in 1963 [262]. Soon afterwards, in 1965, the hydroxyl lines were observed in emission [263]. The surprise was that the sources were very compact and variable in intensity. The brightness temperatures were very great indeed $T_b \geq 10^9$ K, implying that the emission process must involve some form of maser action. This was the beginning of the intensive search for other interstellar molecules. In 1968, ammonia NH$_3$ was detected and in the following year water vapour H$_2$O and formaldehyde H$_2$CO were discovered. All these molecules involve some form of maser action.

A key discovery was the great intensity of the carbon monoxide molecule CO which was first observed by Wilson, Jefferts and Penzias [264] in 1970. CO is one of the simplest molecules which

Twentieth Century Physics

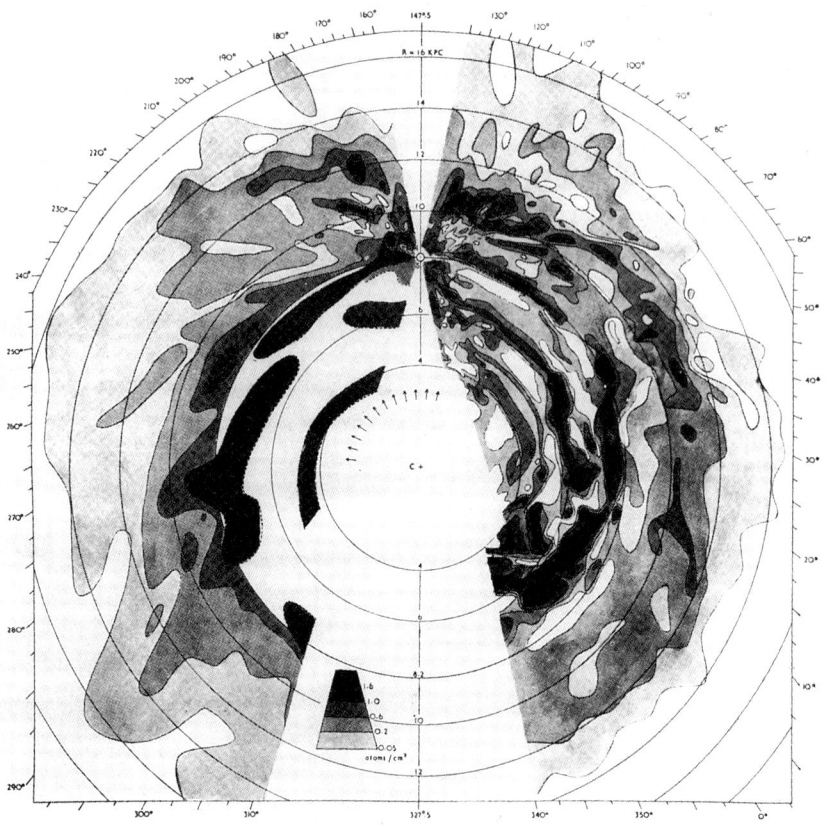

Figure 23.22. *The distribution of atomic hydrogen in the Galactic plane* [260]. *The Sun–centre distance is assumed to be 8 kpc. The numbers around the outside of the figure denote galactic longitudes. In deriving this map, it is assumed that the neutral hydrogen rotates in circular orbits about the Galactic Centre with velocities given by a standard rotation curve.*

emits line radiation by electric dipole transitions between neighbouring rotational states. Since molecular hydrogen has no electric dipole moment, CO is expected to be the most abundant molecule that can be detected by its millimetre and submillimetre line emission and it acts as a tracer for the distribution of molecular hydrogen. Since that time, the number of molecular species known to be present in the interstellar medium has multiplied rapidly, almost 100 different molecules having now been detected, including some remarkable acetylenic chains and some species which are unstable in the laboratory but which can survive in the low density conditions of interstellar space.

The discovery of interstellar molecules came as a surprise because it had been assumed that the ultraviolet radiation in the diffuse interstellar medium would dissociate any but the most tightly bound molecules.

This argument had neglected the shielding role of dust which protects the molecules from ultraviolet radiation. Molecular gas is present throughout the plane of the Galaxy and, as soon as the first maps of its distribution were made, it was found that a large fraction of the molecules belong to *giant molecular clouds*, which typically have masses about $10^6 M_\odot$. By the late 1970s, it was apparent that these clouds contained a great deal of fine-scale structure and these are the sites of star formation. The giant molecular clouds are transparent to millimetre and submillimetre radiation and so the narrow molecular lines are excellent probes of the internal dynamics of the clouds. The presence of so many different molecular species and large densities of interstellar dust gave rise to the new discipline of *interstellar chemistry* [265]. To understand the existence and abundances of the many molecules now observed in molecular clouds requires an understanding of gas-phase reactions and also molecular processes occurring on the surfaces of dust-grains.

23.9.2. *The multiphase interstellar medium*

The theory of the diffuse interstellar medium was the subject of a series of papers by Lyman Spitzer beginning in 1948. In 1950, he and Malcolm Savedoff [266] studied in detail the processes of heating and cooling of the diffuse interstellar gas. In addition to confirming the results of Eddington and Strömgren, they showed that the temperature of the diffuse medium was determined by the balance between heating by the ionization losses of cosmic rays and energy losses associated with the low-lying excited states of CI, CII and SiII. The typical temperature of the diffuse interstellar medium was found to be about 60 K. These low temperatures were confirmed some years later by observations of the 21 cm line of neutral hydrogen.

There remained the problem of confining the cool neutral hydrogen clouds and this was elegantly solved by Field, Goldsmith and Habing [267] in 1969 who considered the thermal stability of a medium heated by interstellar high-energy particles. In 1965, Field [268] had made a detailed study of thermal instabilities in different astrophysical contexts and shown that, in the intermediate range of temperatures between about 100 and 3000 K, there is no thermally stable phase, the stable phases consisting of a low-density, high-temperature gas at $T \simeq 8000$ K in pressure balance with low-temperature gas at $T \simeq 20$ K. This picture became known as the *two-phase model* of the interstellar medium. The inferred temperature of the diffuse neutral hydrogen clouds was too low and one solution was that the interstellar carbon, which is one of the principal coolants of the gas, is depleted relative to its cosmic abundance because it is locked up in interstellar grains.

This picture of the diffuse interstellar medium was soon confronted by observations from the *Copernicus* ultraviolet spectroscopic satellite which was launched in 1972. The depletion of the chemical abundances of the heavy elements in the interstellar medium as compared with their cosmic abundances was confirmed [269]. A very hot component of the medium was detected through observations of the lines of OVI in absorption in the direction of hot stars [270]. Molecular hydrogen was also detected in the direction of reddened stars and it was found that the greater the extinction, the greater the column density of molecular hydrogen, suggesting that its formation is catalysed by dust grains [271].

The discovery of the soft x-ray background from the interstellar medium further complicated the picture [272]. These observations showed a strong anticorrelation between the intensity of the x-ray emission and the column density of neutral hydrogen, the inference being that the soft x-rays suffer photoelectric absorption by interstellar neutral hydrogen. Despite the absorption, this was clear evidence that there exists diffuse soft x-ray background emission from the interstellar medium. In 1974, Cox and Smith [273] provided a convincing explanation for this component. The *UHURU* satellite had shown that supernova remnants are strong x-ray sources in the 1–10 keV waveband and this was naturally interpreted as the bremsstrahlung of gas heated to a very high temperature in the supernova explosion. Cox and Smith found that supernovae occur sufficiently frequently in our Galaxy that old, hot supernova remnants overlap and, by percolation, result in a series of tunnels of hot gas through the interstellar medium. Detailed studies of the local distribution of neutral and hot gas have suggested that the Sun is located in a local hole of hot gas of radius about 50 pc which may have been evacuated by a supernova explosion more than 10^6 years ago [274].

These ideas were synthesized into a picture of the *violent interstellar medium* [275]. The very hot component at 10^6 K, the hot neutral gas at about 10^4 K and the cool diffuse medium at about 100 K are roughly in pressure balance but, in addition, there are the giant molecular clouds. The medium is constantly being buffeted by supernova explosions which may give rise to the formation and cooling of the giant molecular clouds. The gas can also be strongly perturbed by the gravitational influence of spiral arms.

23.9.3. The formation of stars

In 1945, Alfred Joy [276] drew attention to the T-Tauri stars which are variable stars embedded in nebulous clouds of dust and gas. Although there are prominent emission lines in their spectra, these are superimposed upon an absorption line spectrum, consistent with those of stars with mass roughly that of the Sun. In 1947, Viktor Ambartsumian [277] argued that the T-Tauri stars were low-mass main-sequence stars in the process of formation and in 1952 George Herbig [278] suggested that they lay above the main sequence. This idea was proved correct by Merle F Walker [279] who studied the extremely young star cluster NGC2264. In addition to luminous O and B stars, Walker studied the T-Tauri stars and showed that they lay above the main sequence—he interpreted the T-Tauri stars as stars which were contracting towards the main sequence.

The models which Walker had used to study the evolution of pre-main-sequence stars assumed that, as the star contracted quasi-statically towards the main sequence, energy transport through the star was by radiation. This conclusion was shown to be incorrect by Chushiro Hayashi [280] in 1961. Hayashi's analysis concerned the stability of quasi-static models of stars, in which energy transport is by convection rather than by radiation. He had already studied the stability of stars on the giant branch and shown that there is a limiting locus for stellar models when the energy transport by convection extends throughout the whole of the star. This *Hayashi limit* is what eventually stops the expansion of the envelopes of red-giant stars. The Hayashi tracks for stars on the Hertzsprung–Russell diagram are almost vertical and occur well to the right of the diagram, in the region of low surface temperatures, $T \sim 3000$–5000 K. There are no quasi-static solutions for stars to the right of the Hayashi track. Rather, the stars evolve down the Hayashi track until energy transport by radiation becomes the more important process as the central temperature rises. Examples of the evolutionary tracks for stars of different masses worked out by Hayashi in his famous paper are shown in figure 23.23. Hayashi showed that the T-Tauri stars lie precisely in the regions of the Hertzsprung–Russell diagram spanned by his evolutionary tracks.

It was already apparent that star formation was associated with gas clouds in which there are large amounts of dust but the central role of dust in the star-formation process was only appreciated when, in 1965, Eric Becklin and Gerry Neugebauer [281] discovered what they believed to be an infrared protostar in the Orion Nebula. Among the sources they discovered in their pioneering observations at 2.2 μm was one unidentified source which they showed, in the following year, had temperature 700 K, well below the surface temperature of the coolest stars. Although the mapping of regions of star formation with single-element detectors was time consuming, intense, compact, far-infrared sources were found in the vicinity of regions of star formation. The far-infrared luminosities of some of these sources turned out to be enormous, $10^3 - 10^5 \, L_\odot$. It proved to be very difficult to demonstrate that objects like the Becklin–Neugebauer object are indeed protostars. It was quite possible that an O or B star had already formed but that it was still embedded deep within the dense dusty molecular cloud from which it formed, rather than being an object which derived its energy from gravitational collapse of a protostar.

The most important contributions to many aspects of these studies was provided by the *IRAS*, which was launched in 1983. These

Twentieth Century Physics

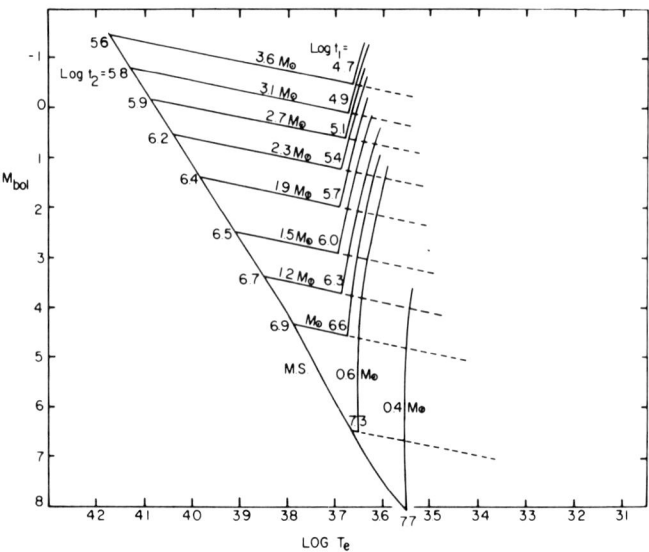

Figure 23.23. *Hayashi's evolutionary tracks and ages for stars with different masses under gravitational contraction* [280]; *the times t_1 and t_2 denote the ages in years at the turning point and at the point where the star joins the main sequence.*

observations confirmed beyond any doubt the significance of dust in the formation of stars [189]. The dust acts as a transformer, absorbing the optical and ultraviolet radiation of protostars and young stellar objects and re-radiating the absorbed energy at the temperature to which the dust is heated. This proves to be an extremely efficient means of getting rid of the gravitational binding energy of matter accreting onto a protostar since the envelope of the star is transparent to radiation in the far-infrared wavebands. Not only were luminous far-infrared sources found in regions of star formation, but lower luminosity sources were discovered as well with luminosities in the range 1–100 $L_\odot$—these are probably the progenitors of solar mass stars. One of the most important early discoveries of the mission was the detection of dust discs around young stars which are interpreted as the material out of which planets are formed.

The theory of star formation has been one of the more contentious areas of modern astrophysics, because it can no longer be assumed that the star is in a quasi-static state whilst evolving from a density enhancement in a giant molecular cloud to a fully fledged main-sequence star. As long ago as 1902, James Jeans [282] had worked out the condition for gravitational collapse, which is that the force of gravity should exceed the force associated with the pressure gradients within the cloud which resist the collapse—this is known as the *Jeans criterion* for collapse. On large enough scales, gravity will always be the dominant force. In the modern version of the theory, the collapsing cloud heats up as it collapses but, so long as the cloud is optically thin to radiation, it can cool by radiation. Once the cloud becomes optically thick to radiation, it begins to heat up and the details of the collapse have to be followed by computer simulations. Among the first of these were calculations by Richard Larson [283] in 1969 who showed that a compact core forms within the collapsing cloud and that the star forms by accretion of matter onto this core. He showed that the dust envelope would absorb the optical and ultraviolet radiation leading to the types of source which had just been discovered by Becklin and Neugebauer. These ideas led in due course to a standard picture of the process of formation of stars due to Frank Shu and his colleagues [284] in 1980. The core of the protostar comes into hydrostatic equilibrium and matter is accreted onto the core through an accretion shock. The binding energy of the accreted matter is released as radiation which is absorbed in the dust envelope, which cannot have a temperature greater than about 1000 K or else the dust grains are evaporated. The dust shell re-radiates away the absorbed energy in the far-infrared waveband.

There must, however, be much more to the story that this. In the late 1970s, maps of young stellar objects were made at millimetre wavelengths and showed that, rather than evidence of infall, the molecular gas in the source L1551 seemed to be be expelled from the source in the form of what became known as a *bipolar outflow* [285]. These molecular outflows have velocities up to about 150 km s^{-1} and extract large momentum and energy fluxes from the star, the total energies corresponding to about 10^{36}–10^{40} J. These outflows are believed to be responsible for a variety of energetic phenomena seen in the vicinity of young stellar objects including the Herbig–Haro objects, high-velocity water-maser sources, shock-excited molecular hydrogen and optically visible jets. The origin of these outflows is not established but they seem to be found wherever there are protostellar or young stellar objects.

23.10. The physics of galaxies and clusters of galaxies

23.10.1. The galaxies

The masses of galaxies range from systems which have masses only about $10^7 M_\odot$ to supergiant elliptical galaxies which, in the most extreme cases, have masses as great as $10^{13} M_\odot$. The dwarf galaxies were recognized by Zwicky [286] in the late 1930s through observations made with the Palomar 18″ Schmidt telescope. Although there are some variations, the distribution of luminosities among galaxies of all types per unit volume can be remarkably well represented by the form of the *luminosity function* introduced by Paul Schechter [287] in 1976

$$\phi(L)\,dL = AL^{-\alpha} \exp(-L/L^*)\,dL.$$

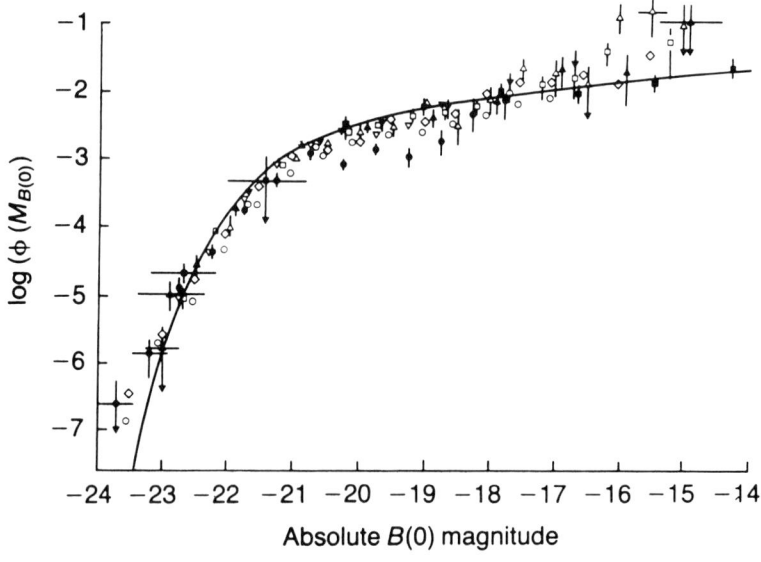

Figure 23.24. *The luminosity function for galaxies derived by Felten [288] in 1977 using data from a large number of different surveys of galaxies. The full curve shows the best-fitting Schechter function to the data.*

In 1977, James Felten [288] surveyed a large number of determinations of the luminosity function for galaxies and showed that they could all be described reasonably well by the function proposed by Schechter (figure 23.24). The value of L^* corresponds to the break in the luminosity function and has value $L^* \approx 10^{10} L_\odot$. The index α was found to have a value of about 0.25, indicating that the luminosity function has a long tail which extends to the dwarf galaxies. Our own Galaxy has a luminosity which is about $0.5 L^*$ and so it is typical of the spiral galaxies found in statistical samples but is by no means among the most luminous of galaxies known.

Of central importance in the study of the physics of galaxies are their mass distributions. These proved to be very demanding programmes observationally but, from the late 1970s onwards, image tubes and CCD detectors became available which greatly increased the capability for taking two-dimensional spectra using long spectrographic slits. Vera Rubin and her colleagues [289] began systematic studies of the rotation curves of galaxies using very long slits which enabled the rotation curves to be determined throughout the body of the galaxy, the narrow emission lines being the preferred tracer of the velocity fields. This work was complemented by observations of the 21 cm line of neutral hydrogen, which enabled the velocity curves of the spirals to be determined to much greater radial distances than the optical observations.

Both types of observation showed that, in the outer regions of galaxies, the velocity curves are generally remarkably flat, $v_{\text{rot}} = $ constant, as far as the rotation curves can be measured. It follows that, if the mass distribution is taken to be spherically symmetric, $M(<r) \propto r$, so that the total mass within radius r increases linearly with distance from the centre. This result contrasts strongly with the variation of the surface brightness of spiral galaxies with distance from the centre which decreases much more rapidly than r^{-2}. The mass to light ratio of the material of the galaxies must increase dramatically in the outer regions of spiral galaxies. The conventional way of expressing this result is to state that there must be a large amount of *dark matter* in the haloes of galaxies. A typical figure for giant spiral galaxies is that they must contain about 10 times as much dark matter as visible matter. Similar results have been found for elliptical galaxies from studies of their globular clusters [290] and from the distribution of their x-ray surface brightnesses [291].

An astrophysical argument for the existence of a dark halo about spiral galaxies was presented in 1973 by Ostriker and Peebles [292], who pointed out that rotating discs are subject to a bar instability unless there is a stabilizing halo present which contains a significant fraction of the mass of the system. Their criterion for stability was that the the ratio of the ordered kinetic energy of the disc T_{orb} to the total potential energy $|U|$ should be less than 0.14 and this result has been confirmed by subsequent analyses. A halo of dark matter about spiral galaxies would provide stabilization of the galactic disc.

23.10.2. Clusters of galaxies

The clustering of galaxies occurs on a very wide range of physical scales, from small groups containing only a few galaxies to the giant clusters and superclusters of galaxies. The clustering properties of galaxies were quantified using correlation functions by Totsuji and Kihara [293] in 1969 and this approach was developed extensively by James Peebles and his colleagues [294] in an important series of papers in the 1970s. The two-point correlation function for galaxies can be written in the form

$$N(r)\, dV = N_0[1 + \xi(r)]\, dV$$

where $N(r)$ is the number density of galaxies at radial distance r from any given galaxy and $\xi(r)$ describes the excess probability, over a uniform distribution N_0, of finding a galaxy at distance r. The function $N(r)$ can be described by a power-law function, $\xi(r) = (r/r_0)^{-\gamma}$ where $\gamma = 1.8$ and $r_0 \approx 8$ Mpc. This function gives a good representation of the clustering of galaxies on scales from about 200 kpc to 10 Mpc. On scales significantly greater than about 10 Mpc, the function decreases more rapidly with increasing physical size.

In fact, the distribution of galaxies is much more complicated than this. In the 1970s, Peebles and his colleagues used the Lick counts of galaxies to demonstrate that, on scales greater than those of clusters of galaxies, the distribution of galaxies has a stringy, cellular appearance [294]. Figure 23.25 shows the local distribution of galaxies derived from a survey of over 14 000 bright galaxies undertaken by Margaret Geller, John Huchra and their colleagues. If the galaxies were uniformly distributed in the local universe, the points would be uniformly distributed over the diagram. It can be seen that there are large 'holes' in which the local number density of galaxies is significantly lower than the mean and there are long 'filaments' of galaxies. The scale of the large holes seen in figure 23.25 is about 30–50 times the scale of a cluster of galaxies. Richard Gott and his colleagues [295] have shown that the topology of the distribution of the galaxies on the large scale is 'sponge-like'. The material of the sponge represents the location of the galaxies and the holes represent the large voids. Both the holes and the distribution of galaxies are continuously connected throughout the local universe. These are the largest known structures in the Universe and one of the great cosmological problems is to reconcile the gross irregularity in the large-scale distribution of galaxies with the remarkable smoothness of the cosmic microwave background radiation.

The richest clusters, such as the Coma cluster of galaxies, have *crossing times*, $t_c = R/v$, which are much less than the age of the Universe. In this expression, v is the mean velocity of a galaxy in the cluster and R is a characteristic size of the cluster. Thus, it is certain that such clusters have come to a state of statistical equilibrium under gravity. It is for this reason that the *virial theorem* [141] can be applied with confidence to these clusters in order to determine their masses. The masses of clusters can range up to $3 \times 10^{15} M_\odot$, very much greater than would be inferred from the light of the galaxies. These observations provide incontrovertible evidence for the presence of dark matter in clusters of galaxies.

One of the most important discoveries of the *UHURU* X-ray Observatory was that some rich clusters of galaxies are intense x-ray sources. In clusters such as the Coma cluster, the x-ray emission is diffuse and fills the core of the cluster [296]. This radiation has been convincingly identified with the bremsstrahlung of hot intra-cluster gas, the clinching piece of evidence being the discovery of emission lines of very highly ionized iron, Fe XXV and Fe XXVI, at 8 keV from the gas by the *Ariel-V* satellite [297]. The intra-cluster gas makes a significant contribution to the total mass of the cluster but it is still an order of magnitude less than that required to bind the cluster gravitationally. The observation of iron emission from the intra-cluster gas indicates that the iron created in the stars in galaxies must be circulated through the intra-cluster medium.

A consequence of the x-ray emitting gas in the cluster is that the hot electrons can scatter photons of the cosmic microwave background radiation which pass through the cluster. This process was first described

Astrophysics and Cosmology

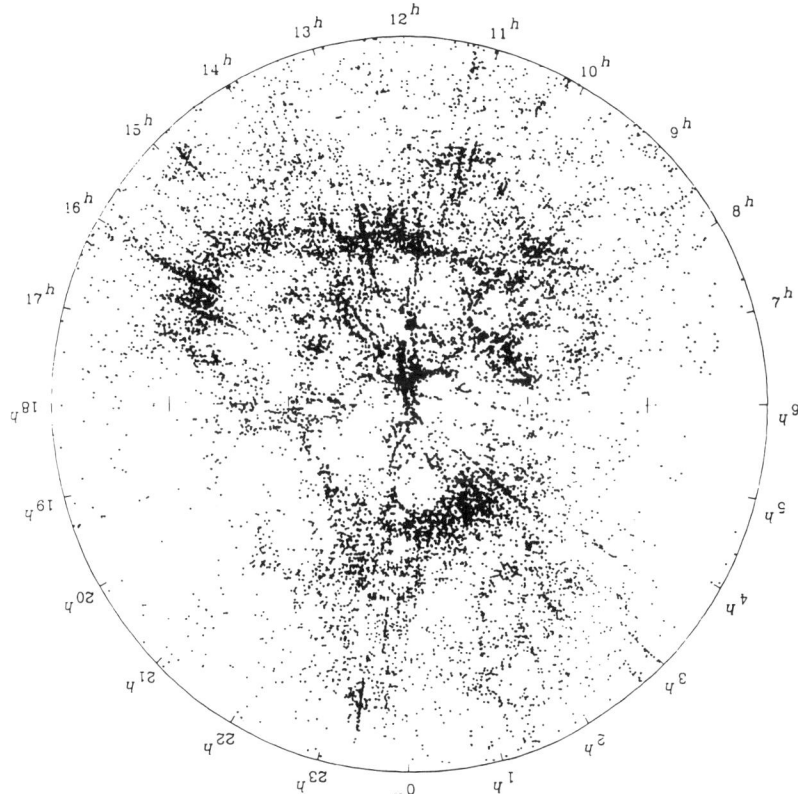

Figure 23.25. *The distribution of galaxies in the nearby Universe as derived from the Harvard–Smithsonian Center for Astrophysics survey of galaxies. The map contains over 14 000 galaxies which form a complete statistical sample around the sky between declinations $\delta = 8.5°$ and $44.5°$. Our Galaxy is located at the centre of the map and the radius of the bounding circle is 150 Mpc. The galaxies within this slice around the sky have been projected onto a plane to show the large-scale features in the distribution of galaxies. Rich clusters of galaxies which are gravitationally bound systems with internal velocity dispersions of about 10^3 km s^{-1} appear as 'fingers' pointing radially towards our Galaxy at the centre of the diagram.*

by Sunyaev and Zeldovich [298] in 1969 and, in it, the low-energy photons of the background radiation are scattered to higher energies by Compton scattering. Consequently, the Planck spectrum of the background radiation is shifted to slightly higher energies, resulting in a decrease in the intensity of the background radiation in the direction of the cluster of galaxies in the Rayleigh–Jeans region of the black-body spectrum. The Sunyaev–Zeldovich effect amounts to only about one part in 3000 of the intensity of the background for a rich cluster of galaxies but it was observed by Birkinshaw and his colleagues [299] in 1990 after many years of difficult observation.

The simplest picture for the dynamical evolution of a cluster of galaxies begins by assuming that the galaxies can be considered to be point masses. When the collapse of protocluster gets underway, large gravitational potential gradients are set up, since the collapse is unlikely to be spherically symmetric, and the system of galaxies relaxes under the influence of these large-scale perturbations. This process was first described by Donald Lynden-Bell [300] in 1967 and is known as *violent relaxation*. He showed that, under the influence of these large potential gradients, the galaxies in the cluster rapidly attain an equilibrium configuration in which galaxies of all masses have the same velocity distribution, consistent with observations of the velocities of galaxies of different masses in clusters. In the process of violent relaxation, the system has to get rid of half of its kinetic energy so that the cluster ends up being bound and satisfying the virial theorem.

Just as in the case of a Maxwellian gas of particles, the galaxies can exchange kinetic energy but now by gravitational encounters. Unlike the case of particles in a gas, the encounters are rather infrequent but the statistical result is the same in that the galaxies tend towards equipartition of energy so that the more massive galaxies slow up and tend to drift towards the centre of the cluster. This process of deceleration is known as *dynamical friction* and was first discussed by Chandrasekhar [301] in the context of the dynamical evolution of star clusters in 1943. This result suggests a reason why, in the regular relaxed clusters, the most massive galaxies are found towards their centres.

The finite sizes of galaxies also have important consequences. Just as in the cases of the peculiar and interacting galaxies, strong tidal forces can cause disruption of galaxies and, in particular, large galaxies tend to disrupt and consume smaller galaxies. This process, described by Ostriker and Tremaine in 1975 [302], is often referred to as *galactic cannibalism* and is likely to be particularly important in clusters of galaxies. The process of galactic cannibalism seems to provide a good explanation of the origin of the huge giant elliptical galaxies observed at the centres of many of the richest clusters of galaxies.

23.11. High-energy astrophysics

23.11.1. Radio astronomy and high-energy astrophysics
The key result of the early history of radioastronomy was the identification of the radio emission of the Galaxy as the synchrotron emission of ultra-high-energy electrons. In 1952, Shklovsky [303] proposed that the same mechanism could account for the radio and optical properties of the continuum emission of the remnant of the supernova of 1954, the Crab Nebula. One consequence of this hypothesis was that the optical continuum of the nebula should be polarized and this was discovered by V A Dombrovski and M A Vashakidze in 1954

Astrophysics and Cosmology

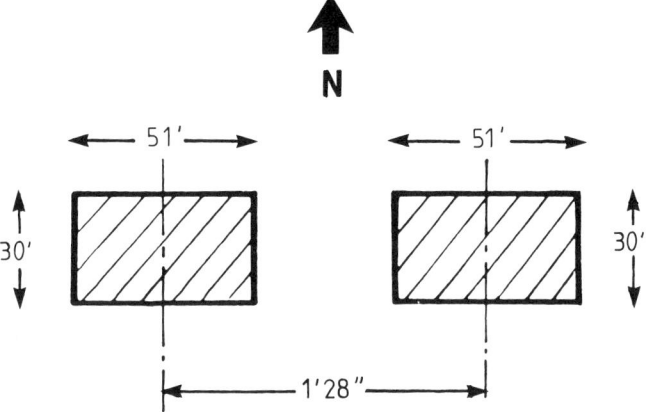

Figure 23.26. *A reconstruction of the radio structure of the radio source Cygnus A from radio interferometric observations by Jennison and Das Gupta [306] at a frequency of 125 MHz.*

and confirmed in 1956 by Oort and Walraven [304], who used superb plates taken with the Palomar 200″ Telescope by Baade in 1955. The famous jet in the nearby giant elliptical galaxy M87 was shown to be polarized by Baade [305] in 1956 and this was interpreted as another example of optical synchrotron radiation.

The identifications of the radio sources Cassiopeia A with a young supernova remnant and Cygnus A with a distant galaxy by Baade and Minkowski [159] in 1954 were crucial. Assuming the radio emission of Cassiopaeia A was synchrotron radiation, huge fluxes of very-high-energy electrons must have been accelerated in the supernova remnant, an idea foreshadowed by Baade and Zwicky's papers [83] of 1934. Equally remarkably, the radio luminosity of the radio galaxy Cygnus A was enormous, more than a million times greater than that of our Galaxy. It must therefore be the source of vast quantities of relativistic material. The radio emission did not, however, originate from the galaxy itself. In 1953, Jennison and Das Gupta [306] at Jodrell Bank used interferometric techniques to show that the radio emission originated from two huge lobes located on either side of the radio galaxy (figure 23.26). Thus, not only must the radio galaxy accelerate an enormous amount of material to relativistic energies, this material has to be ejected into extragalactic space in opposite directions.

To estimate exactly how much energy was necessary to produce the observed flux of synchrotron radiation, Geoffrey Burbidge [307] worked out the minimum energy in high-energy particles and magnetic fields which must be present in the source regions. The energies were enormous, in some sources corresponding to a rest-mass energy of $10^6 M_\odot$. Some astrophysical means had to be found by which a significant fraction of the rest-mass energy of a galaxy could be converted into high-energy particles and magnetic fields.

23.11.2. The discovery of quasars and their close relatives

These radio astronomical discoveries stimulated a great deal of astronomical interest and investment in the construction of radio telescopes. Many of the objects associated with the radio sources were faint galaxies and so accurate radio positions were needed to find the objects associated with the source among the many unrelated stars and galaxies. Many extragalactic radio sources were found to be associated with the most massive and luminous galaxies known which, consequently, could be observed at large redshifts, the largest redshift for any galaxy at that time, $z = 0.46$, being found for the radio galaxy 3C 295 by Rudolf Minkowski [308] in 1960.

By 1962, Matthews and Sandage [309] had identified three of the brightest radio sources, 3C 48, 196 and 286, with 'stars' of an unknown type with strange optical spectra. The breakthrough came in 1962 when Cyril Hazard [310] measured very precisely the radio position of the source 3C 273 by the method of lunar occulation from Australia. The source was identified with what appeared to be a 13th magnitude star. In the same year, Maarten Schmidt [311] obtained the optical spectrum of 3C 273 and identified the Balmer series of hydrogen but with a redshift $z = 0.158$—this was certainly no ordinary star. 3C 273 was the first example of a hyperactive galactic nucleus, its optical luminosity being about 1000 times greater than the luminosity of a galaxy such as our own. Searches through the Harvard plate archives showed that the enormous luminosity of 3C 273 varied on a time-scale of years [312]. Nothing like this had been observed in extragalactic astronomy before. These sources were termed *quasi-stellar radio sources* and within a year this had been abbreviated to the term *quasar*.

One of the first discussions of the implications of these remarkable discoveries was held in Dallas in 1963 at the First Texas Symposium on Relativistic Astrophysics [313]. For the first time, the optical and radio astronomers got together with the theoretical astrophysicists and, in particular, with the general relativists. Perhaps the most important conclusion of the meeting was the realization that the quasars must involve strong gravitational fields and so general relativity must play a central role in understanding their properties. At the closing dinner, Thomas Gold remarked [313]

> Everyone is pleased: the relativists who feel they are being appreciated, who are suddenly experts in a field which they hardly new existed; the astrophysicists for having enlarged their domain, their empire by the annexation of another subject—general relativity.

For some time, there was concern about whether or not the redshifts of the quasars really were of cosmological origin. The short time-scale variability of their enormous luminosities seemed so extreme that some astronomers, the most prominent of whom were Fred Hoyle, Geoffrey Burbidge and Halton Arp [314], suggested that the quasars were actually relatively nearby objects. Although widely discussed, this hypothesis never attracted wide support, largely because no other satisfactory explanation for the origin of the redshifts of the quasars, which had reached redshift $z = 2$ by 1965 [315], was forthcoming.

In many ways, the quasars were discovered too early. It became apparent that they are among the most extreme examples of what are now termed *active galactic nuclei*. The first examples of this class were discovered in the early 1940s by Carl Seyfert [316] who studied the spectra of spiral galaxies which had star-like nuclei. He found that they possessed very intense Balmer and forbidden lines which were very broad, the inferred Doppler broadening corresponding to velocities up to 8500 km s^{-1}. These lines were quite unlike those of HII regions in our Galaxy. In addition, the spectrum of the continuum radiation of the nucleus was quite smooth, unlike the spectrum of starlight.

Seyfert's pioneering work was largely neglected until the 1960s. In 1963, just before the extragalactic nature of the quasars was established, Burbidge, Burbidge and Sandage [317] surveyed a wide range of evidence concerning activity in the nuclei of galaxies in a paper entitled *Evidence for the occurrence of violent events in the nuclei of galaxies*. By this time, it was commonly accepted that the smooth continuum radiation often found in the nuclei of these sources was synchrotron radiation. Radio-quiet counterparts of the quasars were discovered by Sandage [318] in 1965. The similarity between the nuclei of quasars and of the Seyfert galaxies was reinforced by the discovery in 1968 that the nuclear continuum emission of Seyfert galaxies was also variable [319]. It became apparent that there is a continuous sequence of high-energy activity in the nuclei of galaxies, from weak nuclei such as the centre of our own Galaxy to the most extreme examples of quasars, in which the starlight of the galaxy is totally overwhelmed by the intense non-thermal radiation from the nucleus. Among the most extreme examples of active galactic nuclei were the *BL-Lacertae* or *BL-Lac objects* which were discovered as extremely compact and variable radio sources by McLeod and Andrew [320] in 1968.

23.11.3. General relativity and the theory of active galactic nuclei

In parallel with these discoveries of observational astronomy, remarkable progress was made in understanding the role of general relativity, not only in cosmology but also in the ultimate fate of massive stars. The exact solution of the field equations for a point mass in general relativity had been discovered by Karl Schwarzschild [321] in 1916, the year after Einstein's General Theory was published. Schwarzschild did not remark upon the fact that the metric of a point mass, the well-known *Schwarzschild metric*, contains two singularities, one at $r = 0$ and the other at coordinate distance $r = 2GM/c^2$. The second singularity is not a real physical singularity but is associated with the particular choice of coordinate system in which Schwarzschild wrote the metric, as was first demonstrated by Kruskal [322] in the mid-1950s but only published in 1960. The singularity at $r = 0$ is, however, a real physical singularity in space-time.

Symbolic of this turning point in modern astrophysics was the discovery in 1963 by Roy Kerr [323] of the *Kerr metric* which describes the metric of space-time about a rotating black hole and is a generalization of the Schwarzschild solution. In 1965, a further generalization of the Kerr metric was discovered by Newman and his colleagues [324] for the case of a system with finite electric charge by solving the combined Einstein and Maxwell field equations. Only later was it realized that the metric described a rotating black hole with finite electric charge. The Kerr metric depends upon both the mass and angular momentum of the black hole. In 1971, Brandon Carter [325] showed that the only possible solutions for uncharged axisymmetric black holes were the Kerr solutions and in 1972 Stephen Hawking [326] showed that all stationary black holes must be either static or axisymmetric so that the Kerr solutions indeed included all possible forms of black hole. These theorems led to important conclusions about the fate of collapsing bodies in general relativity. No matter how complex the object and its properties before collapse to a black hole, all other properties except its mass, electric charge and angular momentum are radiated away during the collapse, a result often referred to as the *no-hair theorem* for black holes. Thus, as the end points of stellar evolution, black holes are quite remarkably simple objects. Notice that these theorems apply to isolated black holes. Although the black hole cannot possess a magnetic dipole moment, magnetic fields can penetrate into the black hole provided they are firmly attached to the external medium.

One of the key questions addressed by the relativists was whether or not there must be a physical singularity as a result of gravitational collapse. It had been argued by some relativists that the singularity in the Schwarzschild solution was a special case and that, in general, the presence of a singularity might depend upon the initial conditions from which the collapse proceeded. The problem was solved in 1965 by Roger Penrose [327], who showed quite generally that, once a surface from which light cannot escape outwards has formed, what is known as a *closed trapped surface*, there is inevitably a singularity inside that surface. The same techniques were applied to the Universe as a whole by Hawking and Penrose [328] and they were able to show in 1969 that, according to classical general relativity, it is inevitable that there is a singularity at the origin of the Hot Big Bang models of the Universe.

From the point of view of astrophysics, the most important results of the study of black holes, as they were named by Wheeler in 1967, concerned the behaviour of matter in the vicinity of the hole and the maximum amount of gravitational binding energy which could be released when matter falls into the black hole from infinity. The Schwarzschild radius, $r_g = 2GM/c^2$, of a spherically symmetric black hole is the surface of infinite redshift, meaning that radiation emitted from this surface is observed with infinite wavelength at infinity. There is also a last stable circular orbit about this type of black hole at radius $3r_g$. The energy which can be released by an element of mass falling into the black hole corresponds to about 6% of its rest-mass energy. For a maximally rotating Kerr black hole, the surface of infinite redshift shrinks to $r_g = GM/c^2$ and, in the case of co-rotating orbits, up to 42% of the rest-mass energy of the in-falling matter can be released. The rotational energy of the black hole can also be tapped and, in 1969, Penrose [329] showed that up to 29% of the rest-mass energy of a maximally rotating black hole could be made available to the external Universe. These results showed that the accretion of matter onto black holes is potentially an extremely powerful source of energy—nuclear energy, for example, can only release about 0.7% of the rest-mass energy of the matter [330].

Immediately following the discovery of quasars, a plethora of theories appeared in the literature, all of them attempting to account for their huge variations in luminosity on short time-scales. In 1964, Zeldovich and Novikov in Moscow and Salpeter at Cornell [331] pointed out that accretion of matter onto black holes is a very powerful energy source. The matter is most unlikely to fall directly into the black hole since it must acquire some angular momentum and an accretion disc is formed. The energy release is associated with frictional energy losses as the matter transfers its angular momentum outwards. In 1969, the first analysis of thin accretion discs about black holes was carried out by Donald Lynden-Bell [332] who showed how, in principle, they could account for the most extreme active galactic nuclei known at that time. Lynden-Bell assumed that the black holes were of the Schwarzschild variety but Bardeen [333] pointed out in 1970 that the black hole is likely to possess angular momentum as the in-falling matter brings angular momentum with it. The energy release could be correspondingly greater, up to a limit of 42% of the rest-mass energy of the in-falling matter. In 1973, detailed models of accretion discs about black holes were published by Nicolai Shakura and Rashid Sunyaev [334] who showed that, although the nature of the viscosity that is responsible for the outward transport of angular momentum and energy dissipation in the disc is not well understood, many of the properties of thin accretion discs are independent of the viscosity. Accretion models for active galactic nuclei fuelled by accretion discs have become one of the more promising types of model for active galactic nuclei.

One of the most important aspects of these studies was the determination of the masses of the black holes in active galactic nuclei. As early as 1964, Zeldovich and Novikov [335] had pointed out that the masses of quasars had to be very large because of the Eddington limit, which requires that the luminosity of any source cannot be greater than $L = 1.3 \times 10^{31}(M/M_\odot)$ W. Since 3C 273 has a luminosity of at least 10^{40} W, it was evident that the source of energy must have mass greater than $10^9 M_\odot$.

The optical continua of quasars are roughly of power-law form and polarized, indicating that the emission mechanism was probably synchrotron radiation. This radiation is responsible for the excitation of the intense lines emitted from clouds in the vicinity of the nucleus. The power-law form of the continuum spectrum has the advantage that a wide range of different states of ionization would be expected to be present in the clouds, in agreement with the observations. This picture was confirmed by observations of quasars and active galaxies by the International Ultraviolet Observatory (IUE) in the late 1970s and early 1980s. Direct evidence for correlated variability of the nuclear continuum ultraviolet radiation and the strength of the broad line spectrum was found in an intensive observing campaign with the IUE. It was found that there is a delay between an ultraviolet continuum outburst in the nucleus and the excitation of the surrounding clouds. In NGC4151, the delay corresponded to about 10 days and so the distance of the broad emission line regions from the nucleus can only be about a few light days [336]. Combining this dimension with the velocities of the clouds inferred from their Doppler widths, the mass of the central power source was found to be about $10^9 M_\odot$.

In 1986 Wandel and Mushotzky [337] analysed the spectra of quasars and Seyfert galaxies which were known to be intense variable x-ray sources. They made estimates of the masses of the central objects from dynamical and causality arguments, which were in good agreement, and the luminosities of the sources were all significantly less than the corresponding Eddington luminosities (figure 23.27). Thus, for these active galaxies and quasars, there is no problem, in principle, in accounting for their extreme luminosities and the short time-scales of their variability, if it is assumed that the ultimate source of energy is a supermassive black hole in the nucleus.

The case that quasars and active galactic nuclei contain supermassive black holes is remarkably persuasive in general terms but it has proved much more difficult to construct successful models that can account for all their features. As the luminosity of the disc increases, most of the thin-disc approximations break down. An attempt to construct a self-consistent thick-disc model applicable to active galactic nuclei was proposed by Marek Abramowicz and his colleagues [338] in 1978 in which the torus was inflated to such a degree that there were funnels along the axis of the torus which were thought to be relevant to the

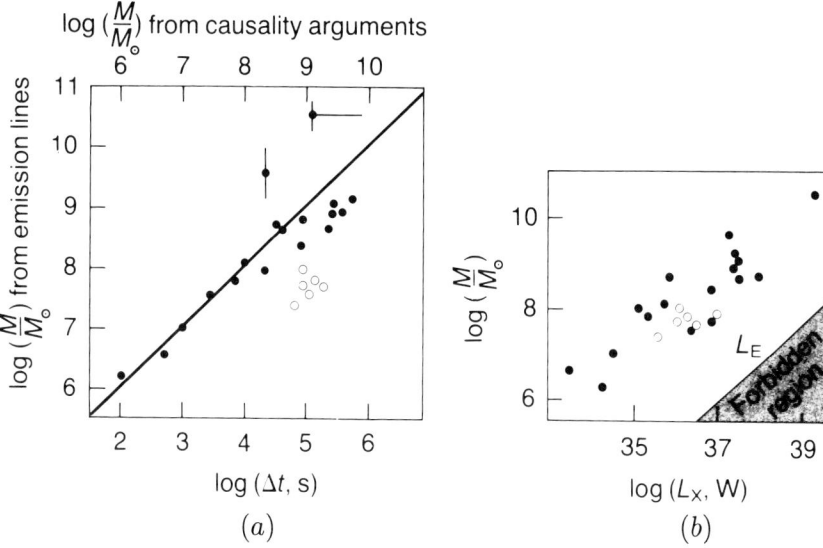

Figure 23.27. (a) Comparison of mass estimates for active galactic nuclei from the variability of their x-ray emission and from dynamical estimates. (b) Comparison of the inferred masses and luminosities with the Eddington limiting luminosity, $L_E = 1.3 \times 10^{31} (M/M_\odot)$ W. All the objects lie well below the Eddington limit [337].

formation of the jets observed to be ejected from many active nuclei. The problem with these thick tori is that they are grossly unstable. In 1984, Papaloizou and Pringle [339] showed that the most unstable modes of thick discs are non-axisymmetric and global in character.

23.11.4. Non-thermal phenomena in active galactic nuclei

The non-thermal properties of active galactic nuclei and the short timescales of their variability created many problems. In 1966, Hoyle, Burbidge and Sargent [340] showed that the quasars are liable to suffer what has been termed the *inverse-Compton catastrophe*. If the variable optical emission is synchrotron radiation, the energy densities of radiation are so great that the relativistic electrons lose more energy by the inverse Compton scattering of the optical photons than by the creation of the optical radiation in the first place, in the process creating intense fluxes of x-rays and γ-rays. There is a critical brightness temperature of about 10^{12} K above which radiation is lost preferentially by inverse Compton scattering rather than by synchrotron radiation. In a remarkably prescient paper of 1966, Martin Rees [341] showed how these problems could be overcome if the source components move out of the nuclear regions at relativistic velocities. Several separate effects contribute to the alleviation of the problems in these models, including the Doppler effect, which increases the observed luminosity of the source as well as aberration effects which make the source appear to be larger than it actually is.

During the late 1960s and early 1970s, the first high-resolution radio maps were made of extragalactic radio sources and these began to reveal the details of their radio structures. The features of particular interest in these maps were the 'hot spots' observed towards the leading edges of the double radio structures. These hot-spots must be temporary phenomena which are continuously supplied with energy from the active galactic nucleus by beams or jets of particles. Variants of this model [342] were proposed by Rees in 1971 and by Peter Scheuer in 1974 and the generic picture of continuous-flow models remains the preferred picture to account for the radio emission of the extended radio sources.

One of the difficulties attending the presence of high-energy electrons in many different astronomical environments was the fact that, as a gas of relativistic particles expands, it loses energy adiabatically. Thus, in supernova remnants such as Cassiopeia A, if the radio-emitting electrons had been injected in the initial explosion, they would have lost all their energy in the adiabatic expansion of the remnant. Similar problems were encountered in accounting for the fluxes of relativistic electrons observed in extended radio sources.

In 1977 and 1978, a number of independent authors [343] showed that *first-order Fermi acceleration* in shock waves is a remarkably effective means of accelerating particles and creating a power-law energy spectrum. If there are high-energy particles present in the vicinity of a shock wave, they are scattered back and forth across the shock wave and, in each crossing in either direction, the particles obtain a fractional increase in energy of order v/c where v is the velocity of the shock wave. There is a probability that the particles are lost from the acceleration region by convection downstream and it is straightforward to show that these two competing effects lead to the formation of a power-law energy spectrum of the form $N(E)\,dE \propto E^{-2}\,dE$. The beauty of this process is that it depends only upon the presence of a strong shock wave. Thus, in supernova shells and at the interface between jets and the interstellar or intergalactic media, where there are strong shock waves, this mechanism provides a means of accelerating the high-energy particles where they are needed. There remain a number of problems with this mechanism—it has proved difficult to obtain spectra which are steeper than $N(E)\,dE \propto E^{-2}\,dE$ and, as applied to supernova remnants, particles cannot be accelerated to energies much greater than 10^{15} eV.

The origin of the strong magnetic field in supernova remnants was solved by Stephen Gull [344] in 1975 when he showed, by numerical simulations, that an expanding supernova shell becomes unstable to the Rayleigh–Taylor instability as the shell decelerates and the resulting turbulence can amplify any seed magnetic field to roughly the values observed. Similar instabilities are expected to take place at the interface between the jets in double radio sources and the intergalactic gas, and can account for the presence of magnetic fields in the source components.

Astrophysics and Cosmology

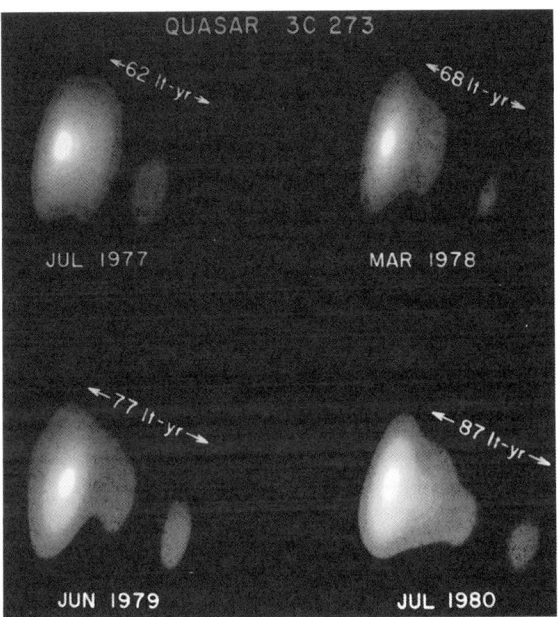

Figure 23.28. *VLBI images of the nucleus of the quasar 3C 273 from 1977 to 1980* [346]. *The radio component is observed to move a distance of 25 light years in 3 years, implying a superluminal velocity of about 8c.*

This picture laid the responsibility for the extended radio sources on jets of material ejected from the active galactic nucleus and direct evidence for these was subsequently found from the observation of radio jets, such as that observed in Cygnus A [345]. Even more remarkable was the evidence derived from VLBI observations of the compact radio cores in active galactic nuclei by Marshall Cohen and his colleagues. By 1980, there was clear evidence that the structures observed in these sources on the scale of milli-arcseconds were not stationary but that the source components appeared to be moving outwards at speeds exceeding the speed of light [346]. In the example of 3C 273 shown in figure 23.28, the fainter component appears to move a distance 25 light years in only 3 years, implying an apparent separation velocity of about eight times the speed of light. Subsequent observations have shown that the phenomenon is remarkably common among the compact radio sources [347].

There immediately followed a flurry of theoretical speculation about the origin of this phenomenon. The simplest explanation, and still the most plausible explanation, was already described by Martin Rees [341] in his papers of 1966—if a source component is ejected at a relativistic velocity from the nucleus at a small angle to the line of sight, the component can have an apparent velocity up to γv where $\gamma = (1 - v^2/c^2)^{-1/2}$ is the Lorentz factor. In addition, the luminosity of the approaching component is enhanced by the Doppler effect and this may account for the one-sidedness of some of the jets observed in active galactic nuclei.

Despite these advances, which enable an empirical picture of the physics of active galaxies and extragalactic radio sources to be derived, the origin of the jets in active nuclei is unclear. One possibility is that they are associated with the funnels which are formed along the rotation axes of thick accretion discs but these are thought to be unstable configurations. Other models ascribe the jets to electromagnetic processes occurring close to the black hole itself and, in the case of the rotating black holes, there is a preferred axis along its axis of rotation.

Much attention has been devoted in the last 10 years to understanding the effects of orientation upon observations of active galaxies. In the case of the superluminal sources, the standard interpretation is that they are only observed when the beam of particles is ejected at an angle close to the line of sight. In 1985 Antonucci and Miller [348] proposed that the distinction between different classes of Seyfert galaxies may be associated with orientation effects. The same idea was proposed by Barthel [349] in 1989 to account for the differences between the radio quasars and the radio galaxies. In his picture, a dust torus obscures the quasar nucleus if its axis is viewed at a large angle to the line of sight, resulting in the observation of a radio galaxy. If the nucleus is observed along the axis of the torus, the nucleus is observed and the intense optical radiation from the nucleus swamps the starlight of the galaxy, resulting in the observation of a radio quasar. The BL-Lac objects, which are the most extreme forms of active galactic nuclei, are thought to be objects which are observed very close to the direction of motion of the relativistic jet and may, in fact, be intrinsically quite weak radio sources.

The most recent results concerning relativistic beaming have come from observations with the Compton Gamma Ray Observatory in 1993 [350]. It has been found that the objects which are the most intense γ-ray sources are those radio quasars which exhibit superluminal motions. The γ-ray luminosities are so great that relativistic bulk motions are necessary to avoid all the γ-rays being degraded by electron–positron pair production, consistent with the observation of superluminal motion of the radio components.

23.12. Astrophysical cosmology

23.12.1. *Gamow and the Big Bang* [351]

In the 1930s, there were two reasons why the synthesis of the chemical elements in the early stages of the Friedman world models was taken seriously. First, the abundances of the elements in stars seemed to be remarkably uniform. The second consideration was that it appeared that the interiors of stars were not hot enough to synthesize the chemical elements. An obvious starting point was to work out the equilibrium abundances of the elements at some high temperature and assume that, if the density and temperature decreased sufficiently rapidly, these abundances would remain 'frozen'.

The first detailed calculations were carried out in 1942 by Chandrasekhar and Henrick [352] who confirmed the expectation of equilibrium theory that, if the elements are in equilibrium at a high temperature, their abundances should be inversely correlated with their binding energies. Chandrasekhar and Henrick found that this was indeed the case for densities of about 10^9 kg m^{-3} and temperatures of about 10^{10} K. There were, however, several gross discrepancies with the observed abundances. The light elements, lithium, beryllium and boron, were vastly overproduced and iron underproduced, as were all the heavier elements with mass numbers greater than about 70. This result was referred to as the 'heavy element catastrophe'. Chandrasekhar and Henrick suggested that some non-equilibrium process was required.

In contrast to this equilibrium picture, Lemaître [353] had proposed that the initial stage of the Friedman models consisted of what he termed a 'primaeval atom', which may be thought of as a sea of neutrons closely packed together, much as in a neutron star. The primaeval neutrons were supposed to disintegrate into the chemical elements as well as the cosmic rays. This was the starting point for George Gamow's attack upon the problem of the origin of the chemical elements. In 1946, he [354] proposed that the synthesis of the chemical elements should take place in the early stages of the Friedman models. He extrapolated the models back to epochs when the densities and temperatures were high enough for nucleosynthesis to take place and found that the time-scale of the Universe at these early stages was too short to establish an equilibrium distribution of the elements.

Ralph Alpher joined Gamow as a research student in 1946 and worked on the problem of primordial nucleosynthesis. Neutron capture cross sections became available in 1946 as a by-product of the nuclear physics programme and these showed the encouraging result that there is an inverse correlation between the neutron capture cross sections and the relative abundances of the elements. In the first calculations, a smooth curve was fitted to these data and it was assumed that the initial conditions consisted of a sea of free neutrons. As protons became available as a result of neutron decay, heavier elements were synthesized by neutron capture. The nuclear reactions were assumed to begin only after the temperature had fallen below that corresponding to the binding energy of deuterium, $kT = 0.1$ MeV, and the Universe was assumed to be static. This theory was published in 1948 by Alpher, Bethe and Gamow [355], Bethe's name being included to complete the $\alpha\beta\gamma$ pun, and they found reasonable agreement with the observed abundances of the elements. The importance of the paper was that it drew attention to the necessity of a hot, dense phase in the early Universe if the elements were to be synthesized cosmologically.

In the same year, Alpher and Robert Herman made improved calculations of primordial nucleosynthesis, including the expansion of the Universe into their calculations. They realized that, at such very high temperatures at early epochs, the Universe was radiation rather than matter dominated and they solved the problem of the subsequent temperature history of the Universe. They came to the far-reaching conclusion that the cooled remnant of the hot early phases should be present in the Universe today and estimated that the temperature of the thermal background should be about 5 K [356]. This radiation was discovered 18 years later by Penzias and Wilson.

There was, however, a major problem with this picture—there are no stable nuclei with mass numbers 5 and 8. Fermi and Turkevich carried out calculations of the evolution of the nuclear abundances of the light elements including 28 nuclear reactions for elements up to mass number 7 in a radiation-dominated, expanding universe and these were published by Alpher and Herman [357] in 1950. These calculations showed that only about one part in 10^7 of the initial mass was converted into elements heavier than helium.

In 1950, another key link in the chain was provided by Hayashi [358] who pointed out that, in the early Universe, at temperatures only ten times greater than that at which the nucleosynthesis took place, the neutrons and protons were brought into thermodynamic equilibrium by the weak interactions

$$e^+ + n \leftrightarrow p + \bar{\nu}_e \qquad \nu_e + n \leftrightarrow p + e^-.$$

Furthermore, at about the same temperature, electron–positron pair production ensures a plentiful supply of positrons and electrons. The result was that, rather than assume arbitrarily that the initial conditions consisted of a sea of neutrons, the equilibrium abundances of protons, neutrons, electrons and all the other constituents of the early Universe could be calculated exactly. In 1953, Alpher, Follin and Herman [359] worked out the evolution of the proton–neutron ratio as the Universe expanded and obtained an answer remarkably similar to modern calculations. They had come very close indeed to the modern picture of the evolution of the early Universe but, before that could happen, steady-state cosmology and the nucleosynthesis of the chemical elements in stars occupied centre stage.

23.12.2. *Steady-state cosmology*

The period immediately after World War II was one of considerable uncertainty concerning the observational, and theoretical foundations of cosmology. There was a time-scale problem because Hubble's estimate of the rate of expansion of the Universe corresponded to a value of H_0^{-1} of only 2×10^9 years. This is the maximum age which any of the Friedman models can have, if the cosmological constant is set equal to zero, and it was known that the age of the Earth was greater than this value.

There were many new ideas in the air. Milne [360] had developed his theory of kinematic relativity, in which he supposed that there are two different times, one associated with dynamical phenomena and another with electromagnetic phenomena. Dirac [361] had been profoundly impressed by coincidences between the very large numbers in physics and the properties of the Universe—for example, the square of the ratio of the strengths of electromagnetic and gravitational forces is roughly equal to the numbers of protons in the Universe. A consequence of his identification of these large numbers was the idea that the gravitational constant should change with time. Eddington [362] had completed his *Fundamental Theory*, in which he attempted to account for the values of the fundamental constants of physics and in which the cosmological constant appeared as a fundamental constant of nature.

It was in this atmosphere that steady-state cosmology was invented by Hermann Bondi, Thomas Gold and Fred Hoyle [363] who published their papers in 1948. Bondi and Gold extended the cosmological principle to what they termed the *perfect cosmological principle* according to which the Universe presents the same large-scale picture to all fundamental observers *at all times*. In this picture, Hubble's constant becomes a fundamental constant of nature, on a par with the charge of the electron and the gravitational constant. They showed that the perfect cosmological principle leads to a unique metric for the dynamics of the Universe with zero spatial curvature. Because of the expansion of the Universe, matter has to be continuously created in order to replace the matter which is dispersing. Hoyle based his theory on a field-theoretical description of the process of the continuous creation of matter, which he described by what he called a *C*-field. The creation rate of matter amounted to only one particle m^{-3} every 300 000 years. A consequence of the theory was that the Universe was infinite in age but the age of typical objects observed in the local universe is only $\frac{1}{3}H_0^{-1}$. Thus, our own Galaxy had to be rather older than the typical object in the Universe but that was not too unreasonable because, if Hubble's constant were 500 km s^{-1} Mpc^{-1}, our own Galaxy would be much larger than other spiral galaxies.

Steady-state cosmology and the theory of continuous creation immediately attracted considerable attention, both within the astronomical community and among the public at large. One important consequence of the development of steady-state cosmology was that Hoyle set about attempting to find an alternative means of understanding the formation of the chemical elements and this was one of the motivations for his remarkable prediction of the carbon resonance [196] and the important paper on the processes of nucleosynthesis in stars by Burbidge, Burbidge, Fowler and Hoyle [199]. Bondi asked what evidence there was for any relics of the hot early phases of the Universe and, with these new results, the abundances of the chemical elements disappeared as evidence.

In the 1950s, two important results were reported of central importance for cosmology. The first concerned the value of Hubble's constant. At the meeting of the International Astronomical Union in Rome in 1952, Walter Baade [364] announced that the distance of the Andromeda Nebula (M31) had been underestimated by a factor of two. The principal indicators used to determine the distance of M31 were the Cepheid variables and Baade discovered that there is a difference in the period–luminosity relations for Cepheids of stellar populations I and II. Using the same type of Cepheid variable in our own Galaxy, in the Magellanic Clouds and in M31, the distance to M31 increased by a factor of two. From the point of view of cosmology, Hubble's constant was reduced to 250 kms^{-1} Mpc^{-1} and hence H_0^{-1} increased to 4×10^9 years. In 1956, Humason, Mayall and Sandage [365] published their redshift–magnitude relation for 474 galaxies and Hubble's constant was revised downwards again to 180 km s^{-1} Mpc^{-1}. These revisions eliminated the discrepancy between the age of the Earth and the age of the Universe according to the standard Friedman models in which the cosmological constant is set equal to zero.

23.12.3. The counts of radio sources

The second piece of evidence came from the surveys of extragalactic radio sources which began in the early 1950s. The central figure in this story was Martin Ryle who was leading the initiatives in radio astronomy at the Cavendish Laboratory in Cambridge. Ryle and Hewish designed and constructed a large four-element interferometer to carry out a new survey of the sky at 81.5 MHz which, being an interferometer, would be sensitive to small angular diameter radio sources. The second Cambridge (2C) survey of radio sources was undertaken in 1954 and the first results published in the following year [366]. Ryle and his colleagues found that the sources were uniformly distributed over the sky and that the numbers of sources increased enormously as the survey extended to fainter and fainter flux densities. In any uniform Euclidean model, the numbers of sources brighter than a given limiting flux density S are expected to follow the relation $N(\geq S) \propto S^{-3/2}$ whereas, at the faintest flux densities, Ryle found an excess of faint sources which could be described by $N(\geq S) \propto S^{-3}$. He concluded that the only reasonable interpretation was that the sources were extragalactic, that they were objects similar in luminosity to Cygnus A, and that there was a much greater number density of sources at large distances than nearby. As Ryle [367] expressed it in his Halley Lecture in Oxford in 1955

> This is a most remarkable and important result, but if we accept the conclusion that most of the radio stars are external to the Galaxy, and this conclusion seems hard to avoid, then there seems no way in which the observations can be explained in terms of a steady-state theory.

These remarkable conclusions came as a surprise to the astronomical community. There was enthusiasm and also some scepticism that such profound conclusions could be drawn from the counts of radio sources, particularly when their physical nature was not understood and only the brightest 20 or so objects had been associated with relatively nearby galaxies.

The Sydney group were making radio surveys of the southern sky at that time with the Mills Cross and found that they could represent the source counts by the relation $N(\geq S) \propto S^{-1.65}$, which they argued was not significantly different from the expectation of uniform world models. In 1957 Bernard Mills and Bruce Slee [368] stated

> We therefore conclude that discrepancies, in the main, reflect errors in the Cambridge catalogue, and accordingly deductions of cosmological interest derived from its analysis are without foundation. An analysis of our results shows that there is no clear evidence for any effect of cosmological importance in the source counts.

The problem with the Cambridge counts was that they extended to surface densities of sources such that the flux densities of the faintest sources were overestimated because of the presence of faint sources in the beam of the telescope, a phenomenon known as confusion. Peter Scheuer [369] devised a statistical procedure for deriving the counts of sources from the survey records themselves without the need to identify individual sources. The technique, which he referred to as the $P(D)$ technique, showed that the slope of the source counts was actually -1.8. Ironically, this result, which is exactly the correct answer, was not trusted, partly because the mathematical techniques used by Scheuer were somewhat forbidding and also because his result differed from the prejudices of both Ryle and Mills. The dispute reached its climax at the Paris Symposium on Radio Astronomy [370] in 1958 and the conflicting positions were not resolved.

The resolution only came with further surveys which were less sensitive to the effects of source confusion and which enabled accurate positions and optical identifications to be made. These showed that Ryle's conclusions of 1955 were basically correct. More radio sources were identified with distant galaxies and the optical identification programmes led to the discovery of quasars in the early 1960s. The source counts showed an excess over the expectations of Euclidean world models [371]. The discrepancies with the Friedman and steady-state models were much greater than this comparison suggested because the predicted source counts converge rapidly as soon as the source populations extend to significant redshifts [372]. By the mid-1960s, the evidence was compelling that there was indeed an excess of sources at large redshifts and this was at variance with the expectations of the steady-state theory.

23.12.4. The helium problem and the discovery of the cosmic microwave background radiation

Helium is one of the more difficult elements to observe astronomically because of its high excitation potential and so can only be observed in hot stars. Osterbrock and Rogerson [373] had shown in 1961 that the abundance of helium seemed to be remarkably uniform wherever it could be observed and corresponded to about 25% by mass. A further important observation was reported by O'Dell, Peimbert and Kinman [374] in 1964 of the helium abundance in a planetary nebula in the old globular cluster M15. Despite the fact that the heavy elements were deficient relative to their cosmic abundances, the helium abundance was still about 25%.

By 1964, it was possible to carry out primordial nucleosynthesis calculations more accurately. Hoyle and Roger Tayler [375] realized that they could undertake much more precise calculations and they obtained the answer that about 25% helium by mass is synthesized in the Big Bang, in remarkable agreement with observation and essentially independent of the matter density of the Universe. One consequence of the Big Bang model, which Hoyle and Tayler did not mention explicitly in their paper, was that the cooled remnant of the thermal radiation present during the very hot early phases should be detectable at centimetre and millimetre wavelengths. Alpher and Herman's prediction had been more or less forgotten when Gamow's theory of primordial nucleosynthesis had failed to account for the creation of the chemical elements. The idea of searching for thermal radiation from the Big Bang was revived in the early 1960s by Yakob Borisevich Zeldovich and his colleagues in Moscow and by Robert Dicke and his colleagues in Princeton [376].

The very next year, in 1965, microwave background radiation was discovered by Arno Penzias and Robert Wilson [377], more or less by accident. They had joined the Bell Telephone Laboratories in the early 1960s with the intention of using the 20 foot horn reflector, which had been built to test telecommunications with the *Echo* satellite, for radio astronomical observations. Having carefully calibrated all parts of the telescope and their 7.35 cm receiver system, they found that there remained about 3.5 ± 1 K excess noise contribution no matter where they pointed the telescope on the sky. Robert Dicke's group in Princeton was attempting exactly the same experiment to detect the cooled remnant of the Big Bang. Discussions with the Princeton group ensued and it was apparent that Penzias and Wilson had discovered what the Princeton physicists were searching for. This was the discovery of the diffuse Cosmic Microwave Background Radiation. Within a few months, the Princeton group [378] had measured a background temperature of 3.0 ± 0.5 K at a wavelength of 3.2 cm, confirming the black-body nature of the background in the Raleigh–Jeans region of the spectrum.

Remarkably, there had been earlier evidence for a diffuse component of millimetre radiation with this radiation temperature from the study of several faint interstellar absorption lines associated with the molecules CH, CH^+ and CN. In the case of CN, for example, absorption was observed from the first rotationally excited state of the molecule as well as the ground state. In 1941, A McKellar [379] had shown that the necessary excitation temperature to populate the first excited state was 2.3 K although the origin of the excitation was then unknown.

Many measurements of the background radiation were made in the following years. At millimetre wavelengths, the observations were very difficult because of atmospheric absorption and a number of balloon observations were made which were broadly consistent with a blackbody radiation spectrum at a temperature of about 2.7 K. The best solution was to carry out the experiments from space and this was brilliantly achieved with the launch of the *Cosmic Background Explorer* (*COBE*) in November 1989. This experiment has shown that the spectrum of the cosmic background radiation is of perfect black-body form, the radiation temperature being 2.725 ± 0.01 K [380].

In 1967, Robert Wagoner, William Fowler and Fred Hoyle [381] repeated the analysis carried out by Hoyle and Tayler but now using cross sections for many more nuclear interactions between light nuclei and knowing that the cosmic microwave background radiation had a radiation temperature of about 2.7 K. These calculations confirmed that about 25% of helium by mass is created by primordial nucleosynthesis and that the figure is remarkably independent of the present density of baryonic matter in the Universe. The abundances of other products of nucleosynthesis, deuterium, ^{3}He and ^{7}Li are, however, sensitive to the mean baryon density in the Universe (figure 23.29). The significance of these elements is that they are not synthesized in stars because they have relatively small nuclear binding energies—deuterium and ^{3}He are destroyed in stars rather than created.

Interstellar absorption lines of deuterium were discovered in the ultraviolet region of the spectrum by John Rogerson and Donald York [382] in 1973 from observations made by the *Copernicus* ultraviolet satellite. An interstellar deuterium abundance of 1.5×10^{-5} by mass relative to hydrogen was found. Subsequent observations showed that the same deuterium abundance is found along the line of sight to other stars [383]. These observations enabled an upper limit to be placed upon

Twentieth Century Physics

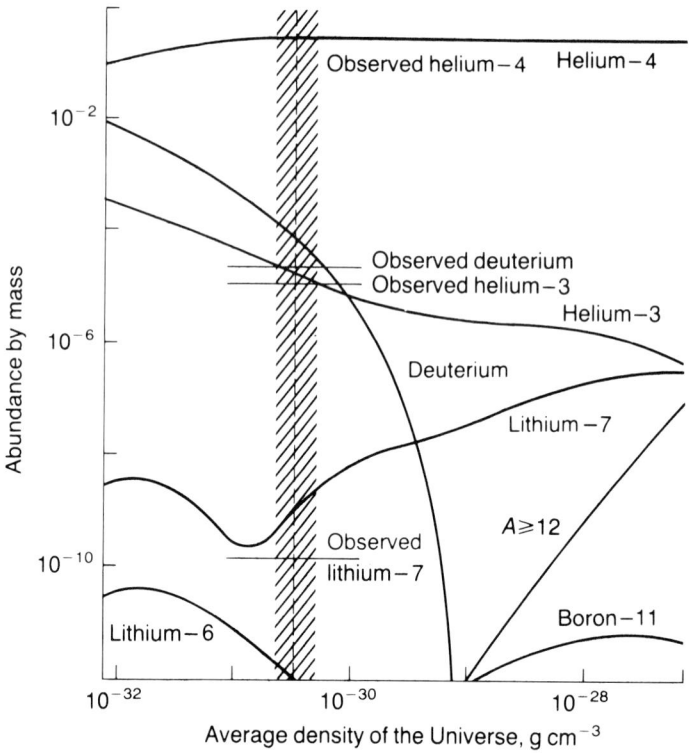

Figure 23.29. *The synthesis of the light elements in the Big Bang according to the calculations of Wagoner, Fowler and Hoyle [381] in 1967.*

the mean baryonic density of the Universe of 1.5×10^{-28} kg m^{-3}—if the mean baryonic density of the Universe were any greater, the primordial deuterium would be underproduced and no other way of creating deuterium astrophysically is known. This upper limit to the baryon density is at least an order of magnitude less than the critical cosmological density.

It was appreciated by Hoyle and Tayler, by Peebles [384] and by the Moscow group working with Zeldovich [385] that the synthesis of the heavy elements provides an important diagnostic tool for the dynamics of the Universe at the epoch of nucleosynthesis. If the Universe expanded too rapidly, the neutron–proton ratio would freeze out at a higher temperature and helium would be overproduced. This result enabled important constraints to be placed upon any variations of the gravitational constant with cosmic epoch, as well as restricting the number of permissible neutrino species to three, a result subsequently confirmed by experiments with the Large Electron–Positron Collider at CERN from studies of the energy widths of the decay products of the Z^0 bosons [386].

Thus, by the late 1960s, there was evidence for two relics of the hot early phases of the Universe, the cosmic microwave background radiation and the cosmic abundances of the light elements. These pieces of evidence supported the standard Big Bang picture, which has been adopted as the standard framework for astrophysical cosmology.

23.13. The classical cosmological problem

Immediately after World War II, the prime instrument for cosmological research was the Palomar 200" telescope which was commissioned in 1948. In 1961, Allan Sandage published a paper entitled *The ability of the 200" telescope to discriminate between selected world models* [387] in which a number of different approaches to determining the basic cosmological parameters listed in section 23.6.8 was discussed. Provided the properties of galaxies do not change with cosmic epoch, several approaches are available for measuring the values of parameters such as Hubble's constant and the deceleration parameter.

23.13.1. Hubble's constant and the age of the universe

Hubble's constant H_0 appears ubiquitously in cosmological formulae. By 1956, Sandage's best estimate was 75 km s^{-1} Mpc^{-1}, the principal reason for this further downward revision being that what had been thought to be the brightest stars in some of the galaxies turned out to be HII regions and star clusters. Also in 1956, Humason, Mayall and Sandage [366] showed that, for galaxies selected at random, there is a large scatter about the mean redshift–magnitude relation because of the breadth of the luminosity function of galaxies, making them unsuitable for determining cosmological parameters. Sandage [388] showed, however, that the absolute magnitudes of the brightest galaxies in clusters have a much narrower dispersion in absolute magnitude, only about 0.3 magnitudes (figure 23.30). Hubble's constant can be found if this relation can be calibrated by measuring the absolute magnitude of the brightest galaxies in nearby rich clusters by techniques which are independent of redshift.

From the 1970s onwards, there has been a continuing controversy over the value of Hubble's constant [389]. In a long series of papers Sandage and Gustav Tammann have found values of Hubble's constant of about 50 km s^{-1} Mpc^{-1} whereas de Vaucouleurs, Aaronson, Mould and their collaborators have found values about 80 km s^{-1} Mpc^{-1}. The nature of the discrepancy can be appreciated from estimates of the distance to the Virgo cluster. If its distance is taken to be 15 Mpc, the higher estimate of H_0 is found, whereas if the distance is taken to be 22 Mpc, values close to 50 km s^{-1} Mpc^{-1} are obtained.

The value of Hubble's constant is closely related to the age of the Universe. The best technique, pioneered by Sandage and Schwarzschild [193] in 1952, involves comparing the Herzsprung–Russell diagrams of the oldest, metal-poor, globular clusters with the

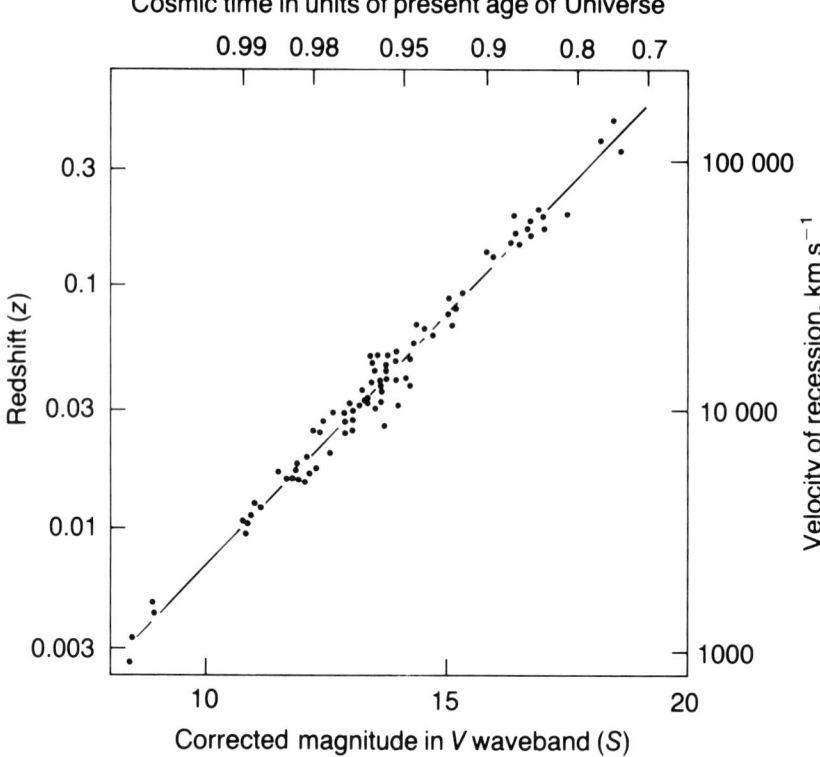

Figure 23.30. *The redshift–V-magnitude relation for the brightest galaxies in clusters presented by Sandage [388] in 1968. The straight line shows that the expected relation if the galaxies all have the same intrinsic luminosity, $m = 5 \log_{10} z + \text{constant}$.*

expectations of the theory of stellar evolution from the main sequence onto the giant branch. An example of what can be achieved is illustrated in figure 23.31 which shows the Herzsprung–Russell diagram for the old globular cluster 47 Tucanae and the predicted isochrones for various assumed ages for the cluster [390]. In this cluster, the abundance of the heavy elements is only 20% of the Solar abundance and the age of the cluster is estimated to be between $(1.2 - 1.4) \times 10^{10}$ years.

For comparison, the time-scale H_0^{-1} for values of Hubble' constant of 50 and 100 km s^{-1} Mpc^{-1} are 2.0×10^{10} and 1.0×10^{10} years respectively. Thus, the lower values of Hubble's constant would be consistent with Friedman world models in which $\lambda = 0$ but the larger values would require λ to be finite and positive or some other modification to the field equations would be needed.

23.13.2. *The deceleration parameter*

The redshift–magnitude relation for the brightest galaxies in clusters displays an impressive linear relation (figure 23.30) but it only extends

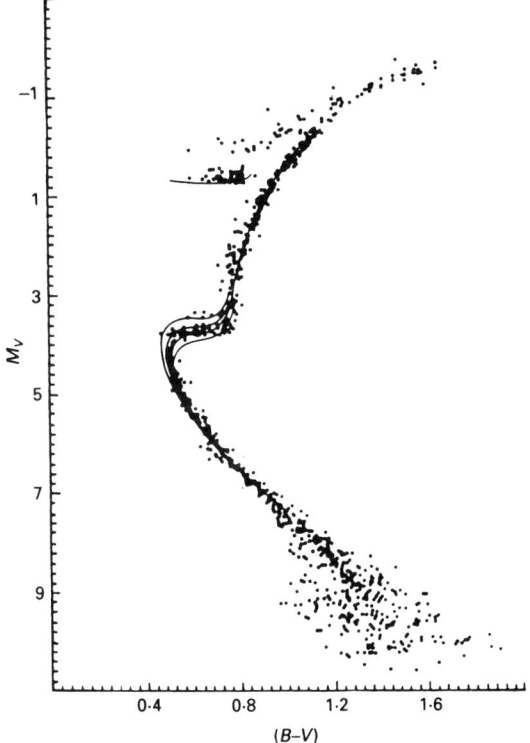

Figure 23.31. *The Herzsprung–Russell diagram for the globular cluster 47 Tucanae derived by John Hesser and his colleagues [390] in 1987. The full curves show isochrones for different assumed ages of the cluster.*

See also p 1697

to redshifts $z \sim 0.5$ at which the differences between the world models are relatively small. Furthermore, corrections have to be made for subtle correlations between the properties of the brightest galaxies in clusters and other properties of the clusters. The uncertainties in the value of q_0 determined by this approach are quite large, Sandage's most recent estimate [391] being $q_0 = 1 \pm 1$.

Another approach to finding galaxies suitable for determining the redshift–magnitude became available in the early 1980s, when the first generation of CCD cameras enabled the identification of essentially all of the bright radio sources in particular areas of sky to be identified with very faint galaxies. These galaxies turned out to have very strong, narrow emission line spectra, and spectroscopy by Hyron Spinrad and his colleagues showed that many of these radio galaxies had very large redshifts. At the same time, infrared photometry of these galaxies became feasible with the development of sensitive indium antimonide detectors. The redshift–magnitude relation for a complete sample of 3CR radio galaxies was determined by Simon Lilly and the author [392]

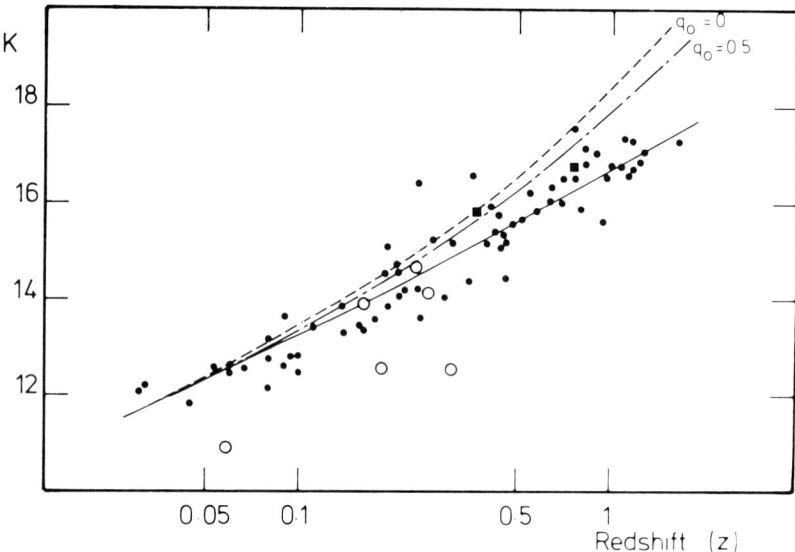

Figure 23.32. *The redshift–K-magnitude relation for a complete sample of radio galaxies from the 3CR catalogue* [392]. *The infrared apparent magnitudes were made at a wavelength of 2.2 μm. The broken curves show the expectations of world models with $q_0 = 0$ and 0.5. The full curve is a best-fitting line for standard world models which include the effects of stellar evolution on the old stellar population of the galaxies. The open circles are radio galaxies in the spectra of which there is contamination by non-thermal optical radiation from the active nuclei.*

at an infrared wavelength of 2.2 μm in 1984 (figure 23.32). It was found that there is a remarkably tight K-magnitude–redshift relation which extended to redshifts of 1.5, but the galaxies at large redshifts were more luminous than expected for world models with $q_0 \sim 0.5$. When the simplest evolutionary correction was made for the increased rate at which stars evolved onto the giant branch at earlier epochs, values of q_0 in the range 0 to 1 were found. There are, however, problems with this simple picture. In the late 1980s, Chambers, Miley and their collaborators used the identification of radio sources with steep radio spectra to find many radio galaxies with redshifts greater than 2. In these radio galaxies, the radio structures are aligned with the optical and infrared images and these have complicated the interpretation of the data [393].

Evidence that the population of galaxies as a whole exhibits evolutionary changes with cosmic epoch has come from the counts of galaxies. Until about 1980, the deepest counts extended to apparent magnitudes of about 22 to 23 and, although there were disagreements between the results of different observers, there was no strong evidence that the counts of galaxies departed from the expectations of uniform world models. In the 1980s, much deeper counts became feasible with the use of CCD cameras on 4 metre class telescopes, the deepest surveys

Astrophysics and Cosmology

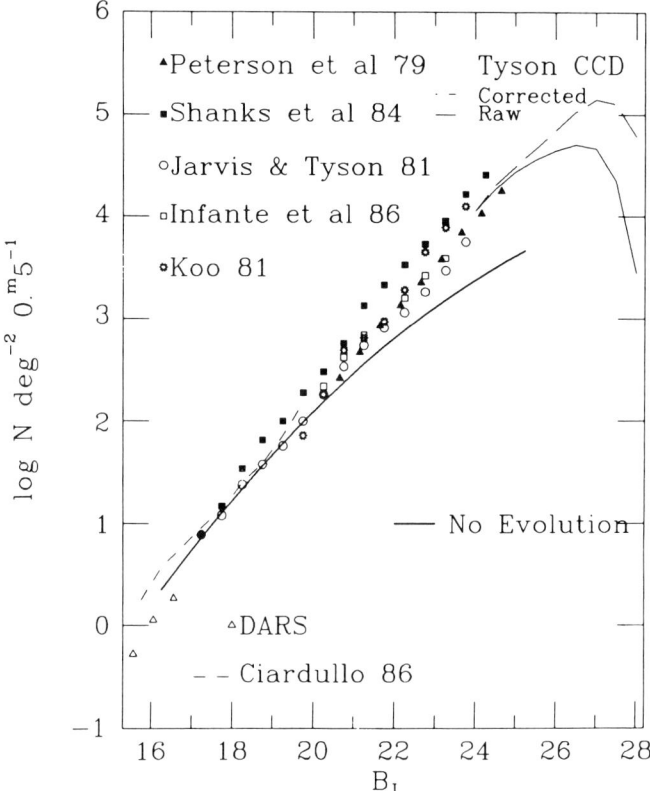

Figure 23.33. *The counts of faint galaxies in the blue (B) waveband compared with the expectations of uniform world models (solid line)* [395].

being carried out by Tyson [394]. It was found that there is an excess of faint galaxies at blue magnitudes greater than about 22 (figure 23.33). In contrast, the counts of galaxies at red and infrared wavelengths showed little evidence of such strong evolution. The nature of these evolutionary changes is not understood.

23.13.3. The cosmological evolution of active galaxies
Evidence that certain classes of extragalactic source display very strong evolutionary changes with cosmic epoch was established in the 1960s. The steep slope of the radio-source counts and the increasing numbers of optical identifications which became available for radio galaxies and quasars could be used to show that the radio-source population displayed very strong effects of cosmological evolution [396]. In 1968, Michael Rowan-Robinson and Maarten Schmidt [397] independently developed a procedure, which became known as the $V/V_{\max}$ test, to show that the radio quasars were located towards the limits of their observable

volumes. Since then, several systematic surveys have been carried out to define the precise form of the cosmological evolution of the radio-source population. The most ambitious study was completed by James Dunlop and John Peacock [398] in 1990 and the nature of the changes of the radio luminosity function of the sources is shown in figure 23.34. It can be seen that the effects of evolution with cosmic epoch are very strong indeed.

By the early 1980s, as a result of systematic surveys of optically selected quasars by Schmidt and Green, Hoag and Smith and Osmer [399], it was clear that the optically selected quasars exhibit strong cosmological evolution, similar in character to the radio quasars. Osmer found that there seemed to be a lack of large redshift objects as compared with the expectations of evolutionary models which could explain the redshift distributions and number counts of objects to a redshift of about 2. In the 1980s, the availability of high-speed measuring machines enabled large complete surveys of quasars to be undertaken. Brian Boyle and his colleagues [400] derived a complete sample of over 400 quasars down to $B = 21$ and redshift $z = 2.2$, all confirmed by spectroscopy. David Koo and his colleagues [401] have searched smaller areas of sky using 4 metre plates and multiwaveband techniques to define complete samples to $B = 22$. Michael Irwin, Richard McMahon and their colleagues [402] perfected the techniques of multicolour searches using high-quality 48" Schmidt plates to find many of the very largest redshift quasars known which now extend out to redshift 4.8.

The first deep surveys of the x-ray sky were carried out by the Einstein X-ray Observatory in 1982 and there was some evidence for an excess of faint x-ray sources associated with active galaxies. Definitive evidence for cosmological evolution of the population of x-ray sources was derived from the deep surveys carried out by the ROSAT satellite, which was launched in 1991. The deep x-ray source counts derived by Gunther Hasinger and his colleagues [403] in 1993 have exactly the same form as the counts of radio sources and quasars, with the slope of the counts exceeding the Euclidean expectation at high flux densities and displaying convergence at the faintest flux densities.

It appears that the populations of active galaxies of all types, which can be observed at large redshifts, exhibit the same form of cosmological evolution. The evolution can be described by assuming that the typical luminosities of the sources change as $L = L_0(1 + z)^3$ in the redshift range $0 < z < 2$. At larger redshifts, the luminosities cannot continue to increase as $(1 + z)^3$, but it is not known whether there is a cut-off or whether the distribution extends to large redshifts with roughly the same co-moving space density at redshift 2. The general conclusion is that the active galaxies must have been considerably more energetic when the Universe was about 20–25% of its present age. The origin of these very strong evolutionary effects is not understood.

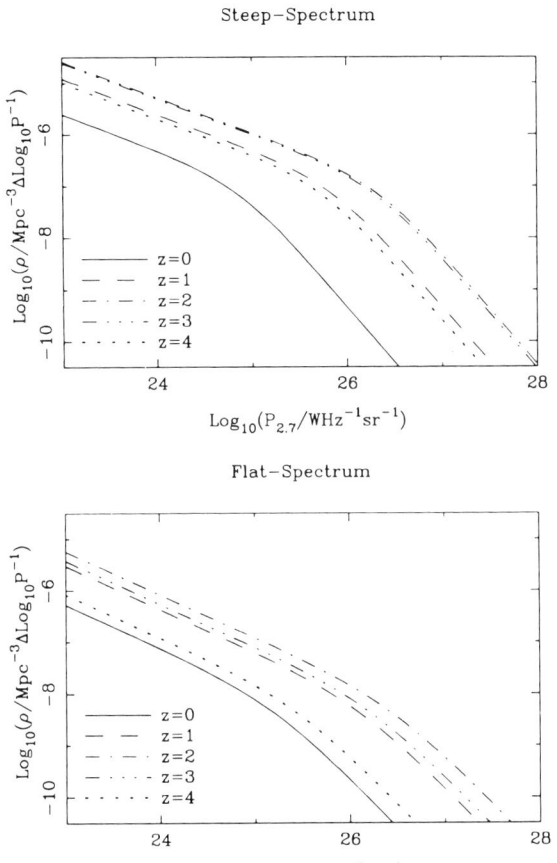

Figure 23.34. *Illustrating the evolution of the luminosity function of extragalactic radio sources with flat and steep radio spectra as a function of redshift z (or cosmic epoch)* [398]. *The luminosity functions shown describe the number of radio sources with different radio luminosities per unit co-moving volume, that is, in a coordinate system which expands with the Universe, so that the figure shows changes over and above the changing density due to the expansion of the Universe.*

23.13.4. The density parameter

The critical cosmological density $\rho_{\rm crit} = 3H_0^2/8\pi G$ depends upon the value of Hubble's constant and so it is convenient to write $H_0 = 100h$ km s^{-1} Mpc^{-1}. Estimates of the average mass density in the form of the visible matter in galaxies were included in Hubble's first major paper [109] on the extragalactic nature of the spiral nebulae in 1926. Hubble's analysis was repeated by Oort [404] in 1958, who found that the average mass density was 3.1×10^{-28} kg m^{-3}, assuming that Hubble's constant was 180 km s^{-1} Mpc^{-1}. This density is not very different from the modern values.

In 1978, Gunn [405] expressed the same result in terms of the necessary mass–luminosity ratio which would be needed if the Universe were to attain the critical density. He found that $(M/L)_{\text{crit}} = 2600h$, very much greater than the values found in our vicinity in the plane of the Galaxy. If account is taken of the dark matter in galaxies, the overall mass–luminosity ratio would be much larger, $M/L \sim 100\text{–}150$. In rich clusters of galaxies such as the Coma cluster, the value of M/L including the dark matter is of the order of 250, but this value is biased towards elliptical and S0 galaxies, which have three times larger values of M/L than the spiral galaxies, which contribute most of the light per unit volume in the Universe at large. These values of M/L are still significantly less than the value needed to close the Universe. Gunn's best estimate of the density parameter for bound systems such as galaxies, groups and clusters of galaxies was about 0.1 and was independent of the value of h.

These results reflect the generally accepted view that, if mass densities are determined for bound systems, the total mass density in the Universe is about a factor of 5 to 10 less than that needed to close the Universe. Therefore, if the Universe were to attain the critical density, the mass would have to be located between the great clusters of galaxies. Evidence that this might be the case has been obtained from studies of the distribution of galaxies on the very largest scales and the observation of their peculiar velocities. In the 1980s, this approach was referred to as the *cosmic virial theorem* [406]. Application to the random velocities of field galaxies suggested that Ω might be larger than 0.2. A similar result was found from studying the in-fall of galaxies into superclusters and other large-scale systems such as the *Great Attractor* [407]. In its most recent incarnation, complete samples of IRAS galaxies have been used to define the local density and velocity fields of galaxies and the overall cosmological density estimated by modelling these distributions [408]. The values of the mean density of the Universe are found to be close to the critical value.

If the Universe is close to the critical density, most of the mass in the Universe must be in some non-baryonic form. The observations of the cosmic deuterium abundance indicate that the present density of baryons cannot be greater than $0.015h^{-2}$ and so, even adopting $h = 0.5$, the mass density of baryons cannot close the Universe.

23.14. Galaxy formation

The standard Friedman world models are isotropic and homogeneous and so the enormous diversity of structure we observe in the Universe is absent. The next step in developing more realistic models of the Universe is to include small perturbations and to study their development under gravity. This problem had been solved by James Jeans [282] in 1902 for the case of a stationary medium as described in section 23.9.3. On large enough scales, the gravitational force of attraction by the matter of the perturbation exceeds the pressure gradients which resist collapse.

The analysis was repeated for the case of spherically symmetric perturbations in an expanding medium in the 1930s by Lemaître and by Tolman [409] and the general solution found by Evgenii Lifshitz [410] in 1946. The condition for gravitational collapse is exactly the same as the Jeans' criterion but, crucially, the growth of the perturbations is no longer exponential but only algebraic. For the critical model, $\Omega_0 = 1$, the density contrast $\Delta = \delta\rho/\rho$ grows with time as $\Delta \propto t^{2/3}$. A similar result is found for radiation dominated universes. The consequence is that the fluctuations from which the large-scale structure of the Universe formed cannot have grown from infinitesimal perturbations. For this reason, Lemaître, Tolman and Lifshitz inferred that galaxies could not be formed by gravitational collapse.

Other authors took the point of view that the solution was to include finite perturbations in the early Universe and then to follow in detail how their mass spectrum would evolve with time. The Moscow school led by Zeldovich, Novikov and their colleagues, and Peebles at Princeton pioneered the study of the development of structure in the Universe in the 1960s. If perturbations on a particular physical scale are tracked backwards into the past, it is found that, at some large redshift, the scale of the perturbation is equal to the horizon scale, that is $r = ct$, where t is the age of the Universe. In 1964, Novikov [411] showed that, to form structures on the scales of galaxies and clusters of galaxies, the density peturbations on the scale of the horizon had to have amplitude $\Delta \sim 10^{-4}$ to guarantee the formation of galaxies by the present epoch.

The discovery of cosmic microwave background radiation in 1965 had an immediate impact upon these studies since the thermal history of the intergalactic gas could be determined in detail. The temperature of the thermal background radiation changes with scale factor as $T = T_0(1+z)$, and so, at a redshift $z \sim 1500$, the temperature of the radiation was about 4000 K, at which there are sufficient photons in the tail of the Planck distribution to ionize all the intergalactic hydrogen. This epoch is referred to as the *epoch of recombination* and, at earlier epochs, the hydrogen is fully ionized. At a slightly earlier epoch, the inertial mass density of the radiation was equal to the mass density of the matter, $\rho c^2 = aT^4$ and, at earlier times, the dynamics of the Universe were radiation-dominated.

The coupling of the matter and radiation by electron scattering was worked out by Weymann [412] in 1966 and in much more detail in 1969 by Zeldovich and Sunyaev [413], who based their analysis upon the theory of induced Compton scattering developed by Kompaneets [414] published in 1956, long after this remarkable classified work had been completed. They showed that, during the radiation-dominated epochs, the matter and radiation are maintained in very close thermal contact by

Compton scattering as long as the intergalactic gas remains ionized. This enabled the speed of sound and hence the Jean's length to be determined at all epochs before the epoch of recombination.

In 1968, Joseph Silk [415] showed that, during the pre-recombination epochs, sound waves in the radiation-dominated plasma are damped by repeated electron scatterings and so perturbations with masses less than about $10^{12} M_\odot$ would be dissipated by the epoch of recombination. All fine-scale structure is wiped out and only large-scale structures on the scale of large galaxies and clusters of galaxies form after recombination. In the early 1970s, Harrison and Zeldovich [416] independently put together information about the spectrum of the initial fluctuations on different physical scales and showed that the observed structures in the Universe could be accounted for if the mass fluctuation spectrum had the form $\Delta(M) \propto M^{-2/3}$ in the very early Universe, corresponding to a power spectrum of initial fluctuations of the form $|\Delta_k|^2 \propto k^n$ with $n = 1$ and amplitude $\sim 10^{-4}$. This spectrum, known as the *Harrison–Zeldovich spectrum*, has the attractive feature that the fluctuations on different mass scales have the same amplitude when they come through the horizon.

In the 1970s, two principal scenarios for the origin of structure in the Universe were developed. The first, known as the *adiabatic* model, was based upon a picture in which the initial perturbations were adiabatic sound waves before recombination and the structure in the Universe was formed by the fragmentation of the large-scale structures which reached amplitude $\delta\rho/\rho \sim 1$ at relatively late epochs. An alternative picture was one in which the perturbations were not sound waves but *isothermal* perturbations in the pre-recombination plasma which were in pressure balance with the background radiation. These would not suffer the damping of small adiabatic mass perturbations and so masses on all scales would survive to the recombination epoch. Galaxies and clusters of galaxies would then form by a process of hierarchical clustering. The adiabatic picture could be thought of as a 'top-down' process in which the largest scale structures form first, whereas the isothermal picture corresponded to a 'bottom-up' process in which small-scale objects come together to form larger structures. In the adiabatic picture, the galaxies, the stars and the chemical elements are all formed at late epochs, whereas in the isothermal picture, the galaxies, stars and heavy elements can begin to form at large redshifts.

A key test of these models is provided by the presence of density fluctuations at the epoch of recombination, which should leave their imprint upon the cosmic microwave background radiation. The principal source of temperature fluctuations on small angular scales was expected to be associated with first-order Doppler scattering due to the collapse of the perturbations [417] and on larger scales due to the gravitational redshift which the photons suffer on escaping from the largest scale perturbations, an effect known as the *Sachs–Wolfe effect* [418]. The predicted amplitudes of the fluctuations in these early theories were $\Delta T/T \sim 10^{-3}-10^{-4}$, well within the capability of isotropy measurements of the cosmic microwave background radiation.

As the limits to the amplitude of fluctuations in the cosmic microwave background radiation continued to improve, the models with low density parameters were in serious conflict with the upper limits to the background fluctuations because, in these, there is very little growth of the perturbations after the epoch of recombination. The limits to the baryonic density from cosmological nucleosynthesis showed that, if the density parameter were 1, most of the matter in the Universe would have to be in some non-baryonic form.

A solution to these problems appeared in 1980 when Lyubimov and his collaborators [419] reported that the electron neutrino had a finite rest mass of about 30 eV. In 1966, Gershtein and Zeldovich [420] had noted that relict neutrinos of finite rest mass could make an appreciable contribution to the mass density of the Universe and, in the 1970s, Marx and Szalay [421] had considered the role of neutrinos of finite rest mass as candidates for the dark matter as well as studying their role in galaxy formation. The intriguing aspect of Lyubimov's result was that, if the relic neutrinos had this rest mass, the Universe would just be closed, $\Omega_0 = 1$.

Zeldovich and his colleagues [422] developed a new version of the adiabatic model in which the Universe was dominated by neutrinos with finite rest mass. The neutrino fluctuations began to grow as soon as they became non-relativistic but, since the neutrinos are weakly interacting, they streamed freely out of the perturbations and so only perturbations on the very largest scale, with masses $M \geqslant 10^{16} M_\odot$ survived to the epoch of recombination. Just as in the adiabatic model, the largest scale perturbations formed first and then the smaller scale structures formed by a process of fragmentation. In 1970, Zeldovich [423] discovered a solution for the non-linear development of a collapsing cloud and he used this to show that the large-scale perturbations would form sheets and pancakes which he believed would resemble the large-scale filamentary structure seen in the distribution of galaxies. The expected amplitude of the fluctuations in the microwave background radiation is much reduced since the fluctuations in the baryonic matter are of low amplitude during the critical phases when the background photons were last scattered. This scenario for galaxy formation became known as the *Hot Dark Matter* picture of galaxy formation, since the neutrinos were still relativistic at relatively late epochs.

There were, however, problems with this picture. It is now believed that Lyubimov's result was erroneous—the present 90% confidence upper limit to the rest mass of the electron neutrino is 7.9 eV [424]. Second, constraints could be set to the mass of the neutrinos, if they were to constitute the dark matter in galaxies, groups and clusters of galaxies. In 1979, Gunn and Tremaine [425] showed how the phase space constraints associated with fermions such as neutrinos could be used to set lower limits to their masses. While 30 eV neutrinos could bind clusters and the haloes of giant galaxies, those needed to bind dwarf galaxies would have to have masses much greater than 30 eV. This was not necessarily a fatal flaw because it could be that some other form of dark matter was present in the haloes of the dwarf galaxies.

There was also the realization about this time that there was a vast range of alternative possibilities for the dark matter which came from theories of particle physics, examples including the axions, supersymmetric particles such as the gravitino or photino and ultra-weakly interacting neutrino-like particles, all of which could indeed be relics of the very early Universe. The period 1980–82 marked the period when the particle physicists began to take the early Universe very seriously as a laboratory for particle physics. In 1982, Peebles [426] introduced the term *cold dark matter* to encompass many of these exotic types of particles suggested by the particle physicists. These weakly interacting massive particles (or WIMPS) may have been created in the very early Universe and would be very cold by now.

The cold-dark-matter scenario [427] is similar in many ways to the isothermal model. Since the matter is very cold, perturbations are not destroyed by free streaming. As in the hot-dark-matter scenario, after the epoch of recombination, the baryonic matter collapses into the growing potential wells in the dark matter. After recombination, galaxies, groups and clusters form by a process of hierarchical clustering. A remarkably useful formalism for the process of hierarchical clustering was described by William Press and Paul Schechter [428] in 1974 which gives a good description of how the mass function of objects of different masses evolves with time for a given input mass function.

These alternative dark matter pictures of galaxy formation have been the subject of a great deal of analysis and computer simulation. The cold-dark-matter scenario can account for many features of the distribution of galaxies in the Universe and it has the great advantage that it can make testable predictions. The only major problem has been to account for the existence of structure on the very largest scales. One of the most important tests has been the prediction of the intensity spectrum of fluctuations in the cosmic microwave background radiation and these were detected in 1992 by the *COBE* satellite. Fluctuations on an angular scale of $10°$ were discovered with amplitude of $\Delta I/I = 10^{-5}$ by George Smoot and his colleagues [429] (figure 23.35). Fluctuations at the same intensity level were observed in 1993 from ground-based observations carried out in Tenerife [430]. The observed intensity fluctuations were about a factor of two greater in amplitude than had been expected according to the standard cold-dark-matter picture. These fluctuations correspond to physical dimensions about ten times the size of the

Astrophysics and Cosmology

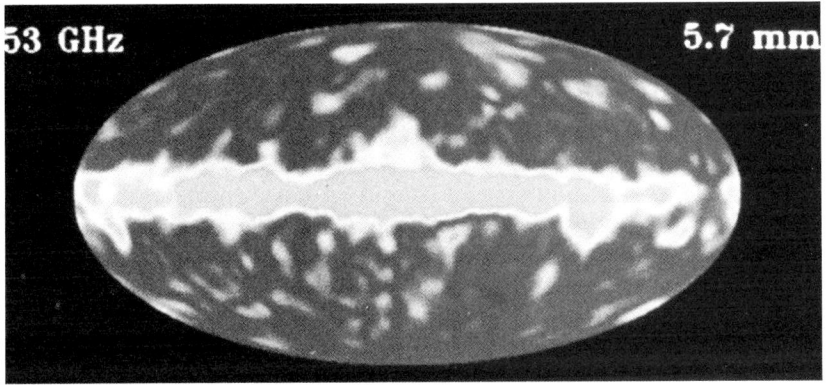

Figure 23.35. *A map of the whole sky in galactic coordinates as observed at a wavelength of 5.7 mm by the COBE satellite [429], once the dipole component associated with the motion of the Earth through the background radiation has been removed. The emission of our Galaxy can be seen as a bright band across the centre of the picture. When averaged statistically over the sky at high latitudes, the intensity fluctuations correspond to a signal of 30 ± 5 μK of cosmological origin.*

largest holes and voids observed in the distribution of galaxies and are associated with the Sachs–Wolfe effect.

The cold-dark-matter picture has become the preferred picture for galaxy formation but it does need some patching up to achieve consistency with all the observations.

23.15. The very early Universe [431]

Despite the success of the standard Big Bang model, it is incomplete in the sense that the initial conditions have to be arranged so that the Universe we observe today is created. The following four pieces of information have to be incorporated as initial conditions.

(i) The Universe must be isotropic.

(ii) There must have been a very small baryon–antibaryon asymmetry in the very early Universe.

(iii) The Universe must have been set up remarkably close to the critical cosmological model, $\Omega = 1$.

(iv) An initial spectrum of fluctuations must have been present from which the present large-scale structure of the Universe formed (see section 23.14).

The first requirement arises from the fact that, further and further into the past, the horizon scale encompassed less and less mass and so why should the Universe be so isotropic on the largest scales today? The second requirement arises from the fact that the photon to baryon ratio today is $N_\gamma/N_B = 4 \times 10^7/\Omega_B h^2$, where Ω_B is the density parameter

1805

in baryons. If photons are neither created or destroyed, this ratio is conserved. At temperatures of about 10^{10} K, electron–positron pair production takes place from the photon field. At an earlier epoch, at a correspondingly higher temperature, baryon–antibaryon pair production takes place with the result that there must be a slight asymmetry in the baryon–antibaryon ratio in the very early Universe. In 1965, Zeldovich [432] showed that, if the Universe were completely symmetric with respect to matter and antimatter, the present day photon to baryon–antibaryon ratio would be about 10^{18}. Various baryon symmetric models of the Universe were proposed by Alfvén and Klein [433] in 1962 and by Omnes [434] in 1969 but none of these convincingly demonstrated how the matter and antimatter could be separated in the early Universe.

The third problem, pointed out by Dicke and Peebles [435] in 1979, arises from the fact that, according to the standard world models, if the Universe were set up with a value of the density parameter differing from the critical value $\Omega = 1$, then it would depart very rapidly from $\Omega = 1$ at later epochs. Since the value of Ω is certainly within a factor of about 10 of the value $\Omega = 1$ now, it must have been extremely close to the critical value in the remote past.

In 1967, Sakharov [436] suggested that the baryon–antibaryon asymmetry might be associated with the type of symmetry-breaking observed in the decays of the K mesons. Similar types of symmetry-breaking occur in grand unified theories of elementary particles at very high temperatures and these have been applied to the physics of the very early Universe. Estimates of the photon to baryon ratio which result from the symmetry breaking are in good general agreement the observed value.

The most important conceptual development came in 1981 with Alan Guth's proposal [437] of the inflationary model for the very early Universe. There had been earlier suggestions foreshadowing his proposal. Zeldovich [121] had noted in 1968 that there is a physical interpretation of the cosmological constant λ associated with zero-point fluctuations of a vacuum. Linde [438] in 1974 and Bludman and Ruderman [439] in 1977 had shown that the scalar Higgs fields, which are introduced to give particles mass, have similar properties to those which would result in a positive cosmological constant.

Guth showed that early exponential expansion of the Universe would solve both the problem of the isotropy of the Universe on a large scale and also drive the Universe towards a flat spatial geometry. The effects of the exponential expansion is to drive neighbouring particles apart at an exponentially increasing rate so that, although the particles were in causal contact in the very early Universe, the exponential inflation quickly drives them far beyond the local horizon and can account for the large-scale isotropy of the Universe by the end of the inflationary epoch. At the end of this phase of exponential inflation, the Universe transforms over into the standard Friedman world model, which, since it has very precisely flat geometry, must have $\Omega = 1$. In Guth's original picture, the change to the Friedman solution took place in a first-order phase transition. The model was revised in 1982 by Linde and by Albrecht and Steinhardt [440] to avoid problems with the original picture, and the transition to the Friedman solutions took place much more slowly and continuously.

Since 1982, the inflationary scenario for the early evolution of the Universe between the epochs when the Universe was 10^{-34} to 10^{-32} s old has been studied very intensively. Among the further successes claimed for the theory has been the realization that quantum fluctuations in the Higgs fields, which drive the inflation, are also amplified in the process of inflation and the Harrison–Zeldovich spectrum comes naturally out the theory. These ideas look promising, although there is not yet a fully satisfactory realization of the physics of the inflationary expansion phase.

The methodological problem with these ideas is that they are based upon extrapolations to energies vastly exceeding those which can possibly be tested in a terrestrial laboratory. Cosmology and particle physics come together in the early Universe and boot-strap their way to a self-consistent solution. This may be the best that we can hope for but it would be preferable to have independent means of testing the theories.

References

[1] There is a wealth of fascinating material on the history of twentieth century astrophysics and cosmology which I have had to condense into a modest space. I have therefore had to be selective in the material included in this survey and have chosen to emphasize the *physics* of astrophysics and cosmology at the expense of purely astronomical themes. With the greatest regret, I have had to omit major areas of astronomy including positional astronomy, planetary astronomy, the history of telescope construction and astronomical instrumentation. Many important astronomical topics are mentioned only briefly. A much expanded version of the text will be published by Cambridge University Press in 1996. In that text, the coverage of the material will be more comprehensive and much more of the astronomical and physical background will be described. In preparing this chapter, I have found the following volumes particularly useful.

Lang K R and Gingerich O (ed) 1979 *A Source Book in Astronomy and Astrophysics, 1900–1975* (Cambridge, MA: Harvard University Press). This volume contains reprints of and brief historical introductions to many of the original articles published between 1900 and 1975 referred to in this survey. All the articles are translated into English.

Gingerich O (ed) 1984 *The General History of Astronomy, vol 4. Astrophysics and Twentieth-Century Astronomy to 1950: Part A* (Cambridge: Cambridge University Press)

Hearnshaw J B 1986 *The Analysis of Starlight: One Hundred and Fifty Years of Astronomical Spectroscopy* (Cambridge: Cambridge University Press)

Bertotti B, Balbinot R, Bergia S and Messina A (ed) 1990 *Modern Cosmology in Retrospect* (Cambridge: Cambridge University Press)

North J D 1965 *The Measure of the Universe* (Oxford: Clarendon)

Gillespie C C (ed) 1981 *Dictionary of Scientific Biography* (New York: Scribner)

I have also assumed some familiarity with astronomical terminology. For more details of the terminology and reviews of many areas of astronomy, the following can be recommended.

Audouze J and Israël (ed) 1988 *The Cambridge Atlas of Astronomy* (Cambridge: Cambridge University Press)

Maran S P (ed) 1992 *The Astronomy and Astrophysics Encyclopedia* (New York: Van Nostrand–Reinhold) and (Cambridge: Cambridge University Press)

[2] Bessel F W 1839 *Astron. Nach.* **16** 64
[3] Wollaston W H 1802 *Phil. Trans. R. Soc.* **92** 365
[4] Fraunhofer's discoveries were first reported in lectures to the Munich Academy of Sciences in 1814 and 1815 and printed in 1817 *Denkschr. Münchener Akad. Wiss.* **5** 193 These papers were published in English in 1823 *Edinburgh Phil. J.* **9** 296 and in 1824 *Edinburgh Phil. J.* **10** 26.
[5] Fraunhofer J 1823 *Gilberts Ann.* **74** 337
[6] Kirchhoff G 1861 *Abh. Berliner Akad.* **62**; Part 1 1862 *Abh. Berliner Akad.* 227; Part 2 1863 *Abh. Berliner Akad.* 225
[7] Classes I to III are described by A Secchi in 1863 *C. R. Acad. Sci., Paris* **93** 364 and Class IV in 1868 *C. R. Acad. Sci., Paris* **66** 124.
[8] The many surveys undertaken by Pickering and his team are described by Hearnshaw J B 1986 *The Analysis of Starlight: One Hundred and Fifty Years of Astronomical Spectroscopy* (Cambridge: Cambridge University Press) ch 5.
[9] Pickering E C 1890 *Harvard Coll. Obs. Ann.* **27** 1. This catalogue was based on the work of Fleming and was entitled *The Draper Catalogue of Stellar Spectra*.
[10] Cannon A J and Pickering E C 1901 *Harvard Obs. Ann.* **28** (Part II) 131
[11] Saha M N 1920 *Phil. Mag.* **40** 479
Fowler R H and Milne E A 1923 *Mon. Not. R. Astron. Soc.* **83** 403; 1924 *Mon. Not. R. Astron. Soc.* **84** 499
[12] The Henry Draper Catalogue was published in *Harvard Ann.* **51** and **55–62** between 1918 and 1924.
[13] von Helmholtz H 1854 Lecture delivered at Königsberg 7 February 1854, published in English in 1856 *Phil. Mag. (series 4)* **11** 489
Thompson W 1854 *Br. Assn. Report Part II*; also 1854 *Phil. Mag.* December
[14] Rutherford E 1907 *J. R. Astron. Soc. Can.* **1** 145
[15] Lane J H 1870 *Am. J. Sci. Arts, 2nd series* **50** 57
[16] These results are contained in Lane's unpublished notes in the US National Archives.
[17] Ritter A 1883 *Wiedemanns Ann.* **20** 137, 897; 1898 *Astrophys. J.* **8** 293
[18] Emden R 1907 *Gaskugeln* (Leipzig: Teubner)
[19] Lockyer's first version of his temperature arch was published in 1887 in *Proc. R. Soc.* **43** 117. He continued to publish versions of the temperature arch, that shown in figure 23.1 dating from 1914.
[20] Monck W H S 1895 *J. Br. Astron. Assoc.* **5** 418; see DeVorkin D 1984 *The General History of Astronomy, vol 4. Astrophysics and Twentieth-Century Astronomy to 1950: Part A* (Cambridge: Cambridge University Press) p 96

[21] See DeVorkin D 1984 *The General History of Astronomy, vol 4. Astrophysics and Twentieth-Century Astronomy to 1950: Part A* (Cambridge: Cambridge University Press) p 97
[22] Hertzsprung E 1905 *Z. Wiss. Photogr.* **3** 429; 1907 *Z. Wiss. Photogr.* **5** 86
[23] Hertzsprung E 1911 *Publ. Astrophys. Obs. Potsdam* **22** No 63
[24] Russell H N 1914 *Popular Astron.* **22** 275, 331; 1914 *Nature* **93** 227, 252 and 281
[25] Adams W S and Kohlschütter A 1914 *Astrophys. J.* **40** 385
[26] Morgan W W, Keenan P C and Kellman E 1943 *Atlas of Stellar Spectra* (Chicago: University of Chicago Press)
[27] See DeVorkin D 1984 *The General History of Astronomy, vol 4. Astrophysics and Twentieth-Century Astronomy to 1950: Part A* (Cambridge: Cambridge University Press) p 102
[28] Halm J 1911 *Mon. Not. R. Astron. Soc.* **71** 610
[29] Hertzsprung E 1919 *Astron. Nach.* **208** 89
[30] Sampson R A 1895 *Mem. R. Astron. Soc.* **51** 123
Schuster A 1902 *Astrophys. J.* **16** 320; 1905 *Astrophys. J.* **21** 1
Schwarzschild K 1906 *Nach. K. Preuss. Akad. Wiss., Göttingen, Math-Phys. Klasse* 1
[31] Eddington A S 1916 *Mon. Not. R. Astron. Soc.* **77** 16
[32] Bohr N 1913 *Phil. Mag.* **23** 1
[33] Saha M N 1920 *Phil. Mag.* **40** 479
[34] Fowler R H and Milne E A 1923 *Mon. Not. R. Astron. Soc.* **83** 403; 1924 *Mon. Not. R. Astron. Soc.* **84** 499
[35] Payne C H 1925 *Harvard College Observatory Monographs, No 1, Stellar Atmospheres* (Cambridge, MA: Harvard University Press)
[36] Unsöld A 1928 *Z. Phys.* **46** 765
[37] McCrea W H 1929 *Mon. Not. R. Astron. Soc.* **89** 483
[38] Russell H N *Astrophys. J.* **55** 119
[39] Russell H N and Saunders F A 1925 *Astrophys. J.* **61** 38
[40] Russell H N, Adams W S and Moore C E 1928 *Astrophys. J.* **68** 1
[41] Russell H N 1929 *Astrophys. J.* **70** 11
[42] Minnaert M and Mulders G 1930 *Z. Astrophys.* **1** 192
[43] Menzel D H 1931 *Publ. Lick Obs.* **17** 230
[44] Eddington A S 1926 *The Internal Constitution of the Stars* (Cambridge: Cambridge University Press)
[45] Russell H N 1925 *Nature* **116** 209
[46] Chandrasekhar S 1983 *Eddington: the Most Distinguished Astrophysicist of his Time* (Cambridge: Cambridge University Press) p 11
[47] Eddington A S 1917 *Mon. Not. R. Astron. Soc.* **77** 596
[48] Michelson A A 1890 *Phil. Mag.* **30** 1
[49] Michelson A A and Pease F G *Astrophys. J.* **53** 249
[50] Eddington A S 1924 *Mon. Not. R. Astron. Soc.* **84** 308
[51] Eddington A S 1920 *Observatory* **43** 353
[52] Fermi E 1926 *Rend. Acc. Lincei* **3** 145
[53] Gamow G 1928 *Z. Phys.* **52** 510
[54] Pauli W 1931 *Phys. Rev.* **38** 579
[55] Fermi E 1934 *Z. Phys.* **88** 161
[56] Anderson C D 1932 *Science* **76** 238
[57] Urey H, Brickwedde F G and Murphy G M 1932 *Phys. Rev.* **39** 164, 864
[58] Chadwick J 1932 *Nature* **129** 312
[59] Chandrasekhar S quoted by Wali K C 1991 *Chandra: a Biography of S Chandrasekhar* (Chicago: University of Chicago Press)
[60] Atkinson R d'E and Houtermans F G 1929 *Z. Phys.* **54** 656

[61] Atkinson R d'E 1931 *Astrophys. J.* **73** 250, 308
[62] von Weizsäcker C F 1937 *Phys. Z.* **38** 176; 1938 *Phys. Z.* **39** 633
Bethe H A 1939 *Phys. Rev.* **55** 434
[63] Atkinson R d'E 1936 *Astrophys. J.* **84** 73
[64] Bethe H A and Critchfield C L 1938 *Phys. Rev.* **54** 248
[65] Öpik E 1938 *Publ. Obs. Astron. Univ. Tartu* **30** 1
[66] Öpik E 1951 *Proc. R. Irish Acad.* **54** 49
[67] Salpeter E E 1952 *Astrophys. J.* **115** 326
[68] Schönberg M and Chandrasekhar S 1942 *Astrophys. J.* **96** 161
[69] See Davis Philip A G and DeVorkin D H (ed) 1977 In Memory of Henry Norris Russell *Dudley Observatory Report* No 13 90, 107
[70] Adams W S 1914 *Publ. Astron. Soc. Pacif.* **26** 198; 1915 *Publ. Astron. Soc. Pacif.* **27** 236
[71] Eddington A S 1924 *Mon. Not. R. Astron. Soc.* **84** 308
[72] Adams W S 1925 *Proc. Natl Acad. Sci.* **11** 382
[73] Eddington A S 1927 *Stars and Atoms* (Oxford: Clarendon) p 52
See also Douglas A V 1956 *The Life of Arthur Stanley Eddington* (London: Nelson) pp 75–8
[74] Pauli W 1925 *Z. Phys.* **31** 765
[75] Fowler R H 1926 *Mon. Not. R. Astron. Soc.* **87** 114
[76] Anderson W 1929 *Z. Phys.* **54** 433
[77] Stoner E C 1929 *Phil. Mag.* **7** 63
[78] Wali K C 1991 *Chandra: a Biography of S Chandrasekhar* (Chicago, IL: University of Chicago Press)
[79] Chandrasekhar S 1931 *Astrophys. J.* **74** 81
[80] See, for example, Chandrasekhar S 1983 *Eddington: the Most Distinguished Astrophysicist of his Time* (Cambridge: Cambridge University Press) p 47
[81] Landau L D 1932 *Phys. Z. Sowjet.* **1** 285
[82] Oppenheimer J R and Snyder H 1939 *Phys. Rev.* **56** 455
[83] Baade W and Zwicky F 1934 *Proc. Natl Acad. Sci.* **20** 254, 259
[84] Hess V F 1912 *Phys. Z.* **13** 1084
[85] Zwicky F 1968 *Catalogue of Selected Compact Galaxies and of Post-Eruptive Galaxies* (Guemlingen, Switzerland: Zwicky)
[86] Hewish A, Bell S J, Pilkington J D, Scott P F and Collins R A 1968 *Nature* **217** 709
[87] Gamow G 1937 *Atomic Nuclei and Nuclear Transformations* (Oxford: Oxford University Press); 1939 *Phys. Rev.* **55** 718
[88] Landau L D 1938 *Nature* **141** 333
[89] Oppenheimer J R and Volkoff G M 1939 *Phys. Rev.* **55** 374
[90] Herschel W 1785 *Phil. Trans. R. Soc.* **75** 213
[91] Kapteyn J C 1906 *Plan of Selected Areas* (Groningen)
[92] Kapteyn J C and van Rhijn P J 1920 *Astrophys. J.* **52** 23
[93] Kapteyn J C 1922 *Astrophys. J.* **55** 302
[94] Leavitt H S 1912 *Harvard Coll. Obs. Circ.* No 173 1
[95] Hertzsprung E 1913 *Astron. Nach.* **196** 201
[96] Shapley H 1918 *Astrophys. J.* **48** 89
[97] This story is told in Berendzen R, Hart R and Seeley D 1976 *Man Discovers the Galaxies* (New York: Science History Publications) and Smith R W 1982 *The Expanding Universe: Astronomy's 'Great Debate' 1900–1913* (Cambridge: Cambridge University Press)
[98] Herschel J 1864 General catalogue of nebulae *Phil. Trans. R. Soc.* **154** 1
[99] Dreyer J L E 1888 New general catalogue of nebulae and clusters of galaxies *Mem. R. Astron. Soc.* **49**

Astrophysics and Cosmology

[100] Dreyer J L E 1895 Index catalogue of galaxies *Mem. R. Astron. Soc.* **51**; 1908 *Mem. R. Astron. Soc.* **59**
[101] Huggins W and Miller W A 1864 *Phil. Trans. R. Soc.* **154** 437
[102] Ritchey G W *Publ. Astron. Soc. Pacif.* **29** 210
[103] Curtis H D 1916 *Publ. Astron. Soc. Pacif.* **29** 180
Shapley H *Publ. Astron. Soc. Pacif.* **29** 213
[104] van Maanen A *Astrophys. J.* **44** 210; this is the first of a series of papers on proper motions in spiral nebulae.
[105] Shapley H 1921 *Bull. Natl Acad. Sci.* **2** 171
Curtis H D 1921 *Bull. Natl Acad. Sci.* **2** 194
[106] Lundmark K 1920 *K. Svenska Vetensk. Akad. Handl.* **60** 63
[107] Lundmark K 1921 *Publ. Astron. Soc. Pacif.* **33** 324
[108] Hubble E P 1925 *Publ. Am. Astron. Soc.* **5** 261
[109] Hubble E P 1926 *Astrophys. J.* **64** 321
[110] Hubble E P 1936 *The Realm of the Nebulae* (New Haven: Yale University Press)
[111] The history of cosmology and the development of relativistic cosmologies is described in the monographs by North J D 1965 *The Measure of the Universe* (Oxford: Clarendon) and Bondi H 1960 *Cosmology* 2nd edn (Cambridge: Cambridge University Press).
[112] A delightful discussion of this correspondence is given by Harrison E R 1987 *Darkness at Night: A Riddle of the Universe* (Cambridge: Cambridge University Press) ch 6.
[113] Lobachevsky N I 1829–30 *On the Principles of Geometry* (Kazan Bulletin)
[114] Riemann B 1854 *Habilitationschrift* (Göttingen: University of Göttingen)
[115] Schwarzschild K 1900 *Viert. Astron. Ges.* **35** 337
[116] Einstein A 1915 *K. Preuss. Akad. Wiss. (Berlin) Sitzungsber.* 844; 1916 *Ann. Phys.* **49** 769
[117] Le Verrier U J J 1859 *C. R. Acad. Sci., Paris* **49** 379
[118] Dyson F W, Eddington A S and Davidson C 1920 *Phil. Trans. R. Soc.* **220** 291
[119] See discussion in North J D 1965 *The Measure of the Universe* (Oxford: Clarendon) p 80.
[120] Einstein A 1917 *K. Preuss. Akad. Wiss. (Berlin) Sitzungsber.* **1** 142
[121] Zeldovich Ya B 1968 *Usp. Fiz. Nauk* **95** 209
[122] Carroll S M, Press W H and Turner E L 1992 *Ann. Rev. Astron. Astrophys.* **30** 499
[123] Einstein A 1919 *K. Preuss. Akad. Wiss. (Berlin) Sitzungsber.* Pt 1 349
[124] de Sitter W 1917 *Proc. Acad. Amst.* **19** 1225
[125] de Sitter W 1917 *Mon. Not. R. Astron. Soc.* **78** 3
[126] Lanczos C 1922 *Phys. Z.* **23** 539
[127] Friedman A A 1922 *Z. Phys.* **10** 377; 1924 *Z. Phys.* **21** 326
[128] Einstein A 1922 *Z. Phys.* **11** 326; 1923 *Z. Phys.* **16** 228
[129] Lemaître G 1927 *Ann. Soc. Scient. Brux.* **A47** 49
[130] Slipher V M 1917 *Proc. Am. Phil. Soc.* **56** 403
[131] Wirtz C W 1921 *Astron. Nach.* **215** 349
[132] Hubble E P 1929 *Proc. Natl Acad. Sci.* **15** 168
[133] Hubble E P and Humason M 1934 *Astrophys. J.* **74** 43
[134] Robertson H P 1935 *Astrophys. J.* **82** 284
Walker A G 1936 *Proc. Lond. Math. Soc. Series 2* **42** 90
[135] Weyl H 1923 *Phys. Z.* **29** 230
[136] Bondi H 1960 *Cosmology* 2nd edn (Cambridge: Cambridge University Press) p 100
[137] Milne E A and McCrea W H 1934 *Q. J. Math.* **5** 64, 73

[138] Gamow G 1970 *My World Line* (New York: Viking) p 44
[139] Einstein A and de Sitter W 1932 *Proc. Natl Acad. Sci.* **18** 213
[140] Zwicky F 1937 *Astrophys. J.* **86** 217
[141] Eddington A S 1915 *Mon. Not. R. Astron. Soc.* **76** 525
[142] Eddington A S 1930 *Mon. Not. R. Astron. Soc.* **90** 669
Lemaître G 1931 *Mon. Not. R. Astron. Soc.* **91** 483
[143] North J D 1965 *The Measure of the Universe* (Oxford: Clarendon) p 125
[144] Bondi H 1960 *Cosmology* 2nd edn (Cambridge: Cambridge University Press) p 84
[145] A survey of astronomical observatories and instrumentation is presented in Part II of 1984 *The General History of Astronomy* (Cambridge: Cambridge University Press).
[146] Information on the development of astronomy internationally can be obtained from the *Proceedings of the General Assemblies of the International Astronomical Union* (Dordrecht: Reidel) which have been held at three-yearly intervals, except during the years of World War II.
[147] Lovell A C B 1987 *Q. J. R. Astron. Soc.* **28** 8
[148] The early history of radio astronomy has been surveyed in Sullivan W T III (ed) 1984 *The Early Years of Radio Astronomy* (Cambridge: Cambridge University Press). Sullivan has also edited a compilation of important early papers on radio astronomy: Sullivan W T III 1982 *Classics in Radio Astronomy* (Dordrecht: Reidel).
[149] Jansky K G 1933 *Proc. Inst. Radio Eng.* **21** 1387
[150] Reber G 1940 *Astrophys. J.* **91** 621
[151] Henyey L G and Keenan P C 1940 *Astrophys. J.* **91** 625
[152] Reber G 1944 *Astrophys. J.* **100** 279
[153] Hey J S 1946 *Nature* **157** 47. This paper reports the observation of intense radio emission associated with a large solar flare which occurred in 1942. The information was declassified after World War II.
[154] Hey J S, Parsons S J and Phillips J W 1946 *Nature* **158** 234
[155] Ryle M and Smith F G 1948 *Nature* **162** 462
[156] Bolton J G, Stanley G J and Slee O B 1949 *Nature* **164** 101
[157] See the discussion on 'The origin of cosmic radio noise' at the *Conference on Dynamics of Ionised Media* held in 1951 at University College, London.
[158] Smith F G 1951 *Nature* **168** 555
[159] Baade W and Minkowski R 1954 *Astrophys. J.* **119** 206
[160] Ryle M 1958 *Proc. R. Soc.* **A248** 289
[161] Alfvén H and Herlofson N 1950 *Phys. Rev.* **78** 616
[162] Kippenhauer K O 1950 *Phys. Rev.* **79** 739
Ginzburg V L 1951 *Dokl. Akad. Nauk* **76** 377
[163] Scheuer P A G 1984 *The Early Years of Radio Astronomy* ed W T Sullivan III (Cambridge: Cambridge University Press) p 249
[164] The first successful VLBI observations are reported in:
Broten N W, Legg T H, Locke J L, McLeish C W, Richards R S, Chisholm R M, Gush H P, Yen J L and Galt J A 1967 *Nature* **215** 38
Bare C, Clark B G, Kellermann K I, Cohen M H and Jauncey D L 1967 *Science* **157** 189
[165] The history of x-ray astronomy is told by Tucker W and Giacconi R 1985 *The X-ray Universe* (Cambridge, MA: Harvard University Press); the development of γ-ray astronomy is described in Ramana Murthy P and Wolfendale A A 1993 *Gamma-ray Astronomy* (Cambridge: Cambridge University Press).
[166] For the history of space exploration, see Rycroft M (ed) 1990 *The Cambridge Encyclopaedia of Space* (Cambridge: Cambridge University Press).

[167] For the early history of ultraviolet observations of the Sun, see Friedman H 1986 *Sun and Earth* (New York: Scientific American Library).
[168] Giacconi R, Gursky H, Paolini F R and Rossi B B 1962 *Phys. Rev. Lett.* **9** 439
[169] Gursky H, Giacconi R, Paolini F R and Rossi B B 1963 *Phys. Rev. Lett.* **11** 530
Bowyer S, Byram E T, Chubb T A and Friedman H 1964 *Science* **146** 912
[170] Giacconi R, Kellogg E, Gorenstein P, Gursky H and Tanenbaum H 1971 *Astrophys. J.* **165** L69
[171] Kraushaar W L, Clark G W, Garmire G P, Borken R, Higbie P and Agogino M 1965 *Astrophys. J.* **141** 845
[172] Clark G W, Garmire G P and Kraushaar W L 1968 *Astrophys. J. Lett.* **153** L203
[173] Fichtel C E, Simpson G A and Thompson D J 1978 *Astrophys. J.* **222** 833
[174] Mayer-Hasselwander et al 1982 *Astron. Astrophys.* **103** 164
[175] Klebesadel R W, Strong I B and Olson R A 1973 *Astrophys. J. Lett.* **182** L85
[176] Smith R W 1989 *The Space Telescope: A Study of NASA, Science, Technology and Politics* (Cambridge: Cambridge University Press) p 30
[177] Rogerson J B, Spitzer L, Drake J F, Dressler K, Jenkins E B Morton D C and York D G 1973 *Astrophys. J.* **181** L97
[178] Kondo Y (ed) 1987 *Exploring the Universe with the IUE Satellite* (Dordrecht: Reidel)
[179] The history of the Hubble Space Telescope project is splendidly told by Smith R W 1989 *The Space Telescope: A Study of NASA, Science, Technology and Politics* (Cambridge: Cambridge University Press). The second edition (1994) includes the problem of the spherical aberration of the primary mirror and its solution using correction optics.
[180] An introduction to the history of infrared astronomy is given by Allen D A 1975 *Infrared: The New Astronomy* (Shaldon, Devon: Keith Reid).
[181] Herschel W 1800 *Phil. Trans. R. Soc.* **90** 255, 284, 293 and 437
[182] Johnson H L, Mitchell R I, Iriate B and Wisniewski W Z 1966 *Commun. Lunar Planetary Lab.* **4** 99
[183] Neugebauer G and Leighton R B 1969 *Two-micron Sky Survey* (NASA SP-3047)
[184] Low F 1961 *J. Opt. Soc. Am.* **51** 1300
[185] Becklin E E and Neugebauer G 1966 *Astrophys. J.* **147** 799
[186] Becklin E E and Neugebauer G 1968 *Astrophys. J.* **151** 145
[187] See the images in Wynn-Williams C G and Becklin E E (ed) 1987 *Infrared Astronomy with Arrays* (Honolulu, HI: Institute for Astronomy, University of Hawaii).
[188] Low F J and Aumann H H 1970 *Astrophys. J. Lett.* **162** L79
Low F J, Aumann H H and Gillespie C M 1970 *Astronaut. Aeronaut.* **7** 26
[189] The first symposium dedicated to the results of the IRAS mission are reported in Israel F P (ed) 1985 *Light on Dark Matter* (Dordrecht: Reidel).
[190] Gascoigne S C B, Proust K M and Robins M O 1990 *The Creation of the Anglo-Australian Telescope* (Cambridge: Cambridge University Press)
[191] Boyle W S and Smith G E 1970 *Bell Syst. Tech. J.* **49** 587
[192] Schmidt B V 1931 *Zentztg Opt. Mechan.* **52** 25
[193] Sandage A R and Schwarzschild M 1952 *Astrophys. J.* **116** 463
[194] An excellent survey of the role of computation in developing models of stellar structure is given by Kippehahn R and Weigert A 1990 *Stellar Structure and Evolution* (Berlin: Springer).
[195] Öpik E 1951 *Proc. R. Irish Acad.* A **54** 49

Salpeter E E *Astrophys. J.* **115** 326
[196] Hoyle F *Astrophys. J. Suppl.* **1** 121
[197] Dunbar D N F, Pixley R E, Wenzel W A and Whaling W 1953 *Phys. Rev.* **92** 649
[198] Suess H E and Urey H C 1956 *Rev. Mod. Phys.* **28** 53
[199] Burbidge E M, Burbidge G R, Fowler W A and Hoyle F 1957 *Rev. Mod. Phys.* **29** 547
Cameron A G W 1957 *Publ. Astron. Soc. Pacif.* **69** 201
[200] Arnett W D and Clayton D D 1970 *Nature* **227** 780
[201] Davis R 1955 *Phys. Rev.* **97** 766
[202] Epstein I 1950 *Astrophys. J.* **112** 207
Oke J B 1950 *J. R. Astron. Soc. Can.* **44** 135
[203] Cameron A G W 1958 *Ann. Rev. Nucl. Sci.* **8** 299
Fowler W A 1958 *Astrophys. J.* **127** 551
[204] Bahcall J N 1964 *Phys. Rev. Lett.* **12** 300
[205] An excellent survey of the solar neutrino problem is given by Bahcall J N 1990 *Neutrino Astrophysics* (Cambridge: Cambridge University Press).
[206] Hirata K S et al 1990 *Phys. Rev. Lett.* **65** 1297
[207] See Hampel W 1994 *Phil. Trans. R. Soc.* A **346** 3
[208] Leighton R B 1960 *Aerodynamic Phenomena in Stellar Atmospheres* ed R N Thomas (Bologna: Zanichelli) p 321
[209] Frazier E N 1968 *Z. Astron.* **68** 345
[210] Ulrich R K 1970 *Astrophys. J.* **162** 993
Leibacher J W and Stein R F 1971 *Astrophys. Lett.* **7** 191
[211] Christensen-Dalsgaard J and Gough D O 1980 *Nature* **288** 544
[212] Toutain T and Frölich C 1992 *Astron. Astrophys.* **257** 287
[213] Courtesy of Professor D O Gough 1993; see also Gough D O 1994 *Phil. Trans. R. Soc.* A **346** 37
[214] Hewish A, Bell S J, Pilkington J D H, Scott P F and Collins R A 1968 *Nature* **217** 709
[215] The history of the discovery of pulsars is described by Hewish A 1986 *Q. J. R. Astron. Soc.* **27** 548
[216] Pacini F 1987 *Nature* **216** 567
Gold T 1968 *Nature* **218** 731
[217] A recent survey of observations and the physics of pulsars, including references to the many original papers is given by Lyne A G and Graham Smith F 1990 *Pulsar Astronomy* (Cambridge: Cambridge University Press).
[218] Pacini F 1968 *Nature* **219** 145
[219] Goldreich P and Julian W H 1969 *Astrophys. J.* **157** 869
[220] Gunn J E and Ostriker J P 1969 *Astrophys. J.* **157** 1395
[221] Baym G Bethe H A and Pethick C J 1971 *Nucl. Phys.* A **175** 225
Baym G, Pethick C J and Sutherland P 1971 *Astrophys. J.* **170** 299
[222] Baym G Pethick C, Pines D and Ruderman M 1969 *Nature* **224** 872
[223] Radhakrishnan V and Manchester R N 1969 *Nature* **222** 228
Reichley P E and Downs G S 1969 *Nature* **222** 229
[224] The history of the UHURU satellite and its discoveries is described by Tucker W and Giacconi R 1985 *The X-ray Universe* (Cambridge, MA: Harvard University Press).
[225] Tananbaum H, Gursky H, Kellogg E M, Levinson R, Schreier E and Giacconi R 1972 *Astrophys. J.* **174** L144
[226] Hayakawa S and Matsuoka M 1964 *Prog. Theor. Phys. Japan Suppl.* **30** 204
[227] Shklovsky I S 1967 *Astron. Zh.* **44** 930
[228] Prendergast K H and Burbidge G R 1968 *Astrophys. J.* **151** L83

[229] Rappaport S A and Joss P C 1983 *Accretion Driven Stellar X-ray Sources* ed W H G Lewin and E P J van den Heuvel p 1
[230] Margon B and Ostriker J P 1973 *Astrophys. J.* **186** 91
[231] For details of the physics of accreting binary systems, see Shapiro S I and Teukolsky S A 1983 *Black Holes, White Dwarfs and Neutron Stars: The Physics of Compact Objects* (New York: Wiley Interscience).
[232] Trümper J, Pietsch W, Reppin C, Voges W, Steinbert R and Kendziorra E 1978 *Astrophys. J. Lett.* **219** L105
[233] Zeldovich Ya B and Gusyenov O H 1965 *Astrophys. J.* **144** 841
[234] Trimble V L and Thorne K S 1969 *Astrophys. J.* **156** 1013
[235] Webster B L and Murdin P 1972 *Nature* **235** 37
[236] McClintock J E 1992 *Proc. Texas ESO/CERN Symp. on Relativistic Astrophysics, Cosmology and Fundamental Particles* ed J D Barrow *et al* (New York: New York Academy of Sciences) p 495
Cowley A P 1992 *Ann. Rev. Astron. Astrophys.* **30** 287
[237] Hulse R and Taylor J 1975 *Astrophys. J. Lett.* **195** L51
[238] Taylor J 1992 *Phil. Trans. R. Soc.* **341** 117
[239] For a comprehensive review of the properties of supernovae, see Trimble V 1982 *Rev. Mod. Phys.* **54** 1183 and 1983 **55** 511.
[240] For an entertaining description of Zwicky's work on supernovae, see Zwicky F 1974 *Supernovae and Supernova Remnants* ed C Battali Cosmovici (Dordrecht: Reidel) p 1.
[241] Baade W and Zwicky F 1938 *Astrophys. J.* **88** 411
[242] Minkowski R 1941 *Publ. Astron. Soc. Pacif.* **53** 224
[243] Kirshner R P and Oke B 1975 *Astrophys. J.* **200** 574
[244] Pankey 1962 *PhD Dissertation* Howard University, Washington, DC
[245] Colgate S and McKee C 1969 *Astrophys. J.* **157** 623
[246] There is a vast literature on SN1987A. Symposia devoted to the supernova include Kafatos M and Michalitsianos A G (ed) 1988 *Supernova 1987A in the Large Magellanic Cloud* (Cambridge: Cambridge University Press) and Danziger I J and Kjär K 1991 *Supernova 1987A and Other Supernovae* (Garching bei München: European Southern Observatory). An excellent summary of the observations up to 1992 is given by Chevalier R 1992 *Nature* **355** 691.
[247] See the excellent discussion of these observations in reference [206].
[248] Plaskett J S and Pearce J A 1933 *Publ. Dom. Astrophys. Obs.* **5** 167
Joy A H 1939 *Astrophys. J.* **89** 356
[249] Russell H N 1921 *Observatory* **44** 72
[250] Eddington A S 1926 *Proc. R. Soc.* A **111** 424
[251] Menzel 1926 *Publ. Astron. Soc. Pacif.* **38** 295
[252] Zanstra H 1926 *Phys. Rev.* **27** 644
[253] Zanstra H 1927 *Astrophys. J.* **65** 50
[254] Strömgren B 1939 *Astrophys. J.* **89** 526
[255] Bowen I S 1928 *Astrophys. J.* **67** 1
[256] Some reminiscences of Oort and Dutch astronomy under the occupation are contained in van Woerden H, Brouw W N and van de Hulst H C (ed) 1980 *Oort and the Universe* (Dordrecht: Reidel).
[257] van de Hulst H C 1945 *Ned. Tijdschr. Natuurkd.* **11** 210
[258] Ewen H I and Purcell E M 1951 *Nature* **168** 356
[259] Muller C A and Oort J H 1951 *Nature* **168** 356
[260] Oort J H, Kerr F J and Westerhout G 1958 *Mon. Not. R. Astron. Soc.* **118** 379
[261] See, for example, Fich M and Tremaine S 1991 *Ann. Rev. Astron. Astrophys.* **29** 409

[262] Weinreb S, Barrett A H, Meeks M L and Henry J C 1963 *Nature* **200** 829
[263] Weaver H, Williams D R W, Dieter N H and Lum W T 1965 *Nature* **208** 29
Weinreb S, Meeks M L, Carter J C, Barrett A H and Rogers A E E 1965 *Nature* **208** 440
[264] Wilson R W, Jefferts K B and Penzias A A 1970 *Astrophys. J. Lett.* **161** L43
[265] An introduction is given by Duley W W and Williams D A 1984 *Interstellar Chemistry* (London: Academic).
[266] Spitzer L and Savedoff M P 1950 *Astrophys. J.* **111** 593
[267] Field G B, Goldsmith D W and Habing H J 1969 *Astrophys. J. Lett.* **55** L149
[268] Field G B 1965 *Astrophys. J.* **142** 531
[269] Field G B 1964 *Astrophys. J.* **187** 453
[270] Rogerson J B, York D G, Drake J F, Jenkins E B, Morton D C and Spitzer L 1973 *Astrophys. J. Lett.* **181** L110
[271] Rogerson J B and York D G 1973 *Astrophys. J. Lett.* **186** L95
[272] Bowyer C S, Field G B and Mack J E 1968 *Nature* **217** 32
[273] Cox D P and Smith B W 1974 *Astrophys. J. Lett.* **189** L105
[274] See, for example, Bruhweiler F C and Vidal-Madjar A 1987 *Exploring the Universe with the IUE Satellite* ed Y Kondo (Dordrecht: Reidel) p 467
[275] McKee C F and Ostriker J P 1977 *Astrophys. J.* **218** 148
[276] Joy A H 1945 *Astrophys. J.* **102** 168
[277] Ambartsumian V A 1947 *Stellar Evolution and Astrophysics* (Yerevan: Armenian Academy of Sciences)
[278] Herbig G H 1952 *J. R. Astron. Soc. Can.* **46** 222
[279] Walker M F 1956 *Astrophys. J. Suppl.* **2** 365
[280] Hayashi C 1961 *Publ. Astron. Soc. Japan* **13** 450
[281] Becklin E E and Neugebauer G 1967 *Astrophys. J.* **147** 799
[282] Jeans J H 1902 *Phil. Trans. R. Soc.* **199** 1
[283] Larson R 1969 *Mon. Not. R. Astron. Soc.* **145** 271 297
[284] Stahler S W, Shu F J and Taam R E 1980 *Astrophys. J.* **241** 641
[285] Snell R L, Loren R B and Plambeck R L 1980 *Astrophys. J. Lett.* **239** L17
[286] Zwicky F 1937 *Astrophys. J.* **217**; 1942 *Phys. Rev.* **61** 489
[287] Schechter P 1976 *Astrophys. J.* **203** 297
[288] Felten J 1977 *Astron. J.* **82** 869
[289] Rubin V C, Ford W K and Thonnard N 1980 *Astrophys. J.* **238** 471
See also Rubin V C 1988 *Large-scale Velocity Fields in the Universe* ed V C Rubin and G V Coyne (Vatican City: Pontificia Academia Scientiarum) p 541
[290] See Trimble V L 1987 *Ann. Rev. Astron. Astrophys.* **25** 425
[291] Fabricant D, Lecar M and Gorenstein P 1980 **241** 552
[292] Ostriker J P and Peebles P J E 1973 *Astrophys. J.* **186** 467
[293] Totsuji H and Kihara T 1969 *Publ. Astron. Soc. Japan* **21** 221
[294] References to the work of Peebles and his colleagues in this area can be found in Peebles' excellent monograph Peebles P J E 1993 *Principles of Physical Cosmology* (Princeton, NJ: Princeton University Press).
[295] Gott J R, Melott A L and Dickinson M 1986 *Astrophys. J.* **306** 341
[296] Gursky H, Kellogg E, Murray S, Leong C, Tananbaum H and Giacconi R *Astrophys. J. Lett.* **167** L81
[297] Mitchell R J, Culhane J L, Davison P J N and Ives J C 1976 *Mon. Not. R. Astron. Soc.* **175** 30P
[298] Sunyaev R A and Zeldovich 1970 *Astrophys. Space Sci.* **7** 20
[299] Birkinshaw M 1990 *The Cosmic Microwave Background: 25 Years Later* ed N Mandolesi and N Vittorio (Dordrecht: Kluwer) p 77
[300] Lynden-Bell D 1967 *Mon. Not. R. Astron. Soc.* **136** 101

[301] Chandrasekhar S 1943 *Astrophys. J.* **97** 251, 259; **98** 270
[302] Ostriker J P and Tremaine S D 1975 *Astrophys. J. Lett.* **202** L113
[303] Shklovsky I S 1953 *Dokl. Akad. Nauk* **90** 983
[304] Dombrovski V A 1954 *Dokl. Akad. Nauk* **94** 1021
Vashakidze M A 1954 *Astron. Tsirk.* No 147
Oort J H and Walraven T 1956 *Bull. Astron. Inst. Neth.* **12** 285
[305] Baade W 1956 *Astrophys. J.* **123** 550
[306] Jennison R C and Das Gupta M K 1953 *Nature* **172** 996
[307] Burbidge G R 1956 *Astrophys. J.* **124** 416; 1959 *Astrophys. J.* **129** 849
[308] Minkowski R 1960 *Astrophys. J.* **132** 908
[309] Matthews T A and Sandage A R 1963 *Astrophys. J.* **138** 30
[310] Hazard C, Mackey M B and Shimmins A J 1963 *Nature* **197** 1037
[311] Schmidt M 1963 *Nature* **197** 1040
[312] Smith H J and Hoffleit D and 1963 *Nature* **198** 650
[313] The proceedings of the First Texas Symposium contain many of the most important papers on radio sources, active galaxies and quasars published up to 1963. Robinson I, Schild A and Schucking E L (ed) 1965 *Quasi-stellar Sources and Gravitational Collapse* (Chicago: University of Chicago Press). A summary of the early observations of quasars is given by G R Burbidge and E M Burbidge 1967 *Quasi-stellar Objects* (San Francisco: Freeman).
[314] The arguments involved in the controversy concerning the origin of quasar redshifts are contained in the volume Field G B, Arp H and Bahcall J N 1973 *The Redshift Controversy* (Reading, MA: Benjamin).
[315] Schmidt M 1965 *Astrophys. J.* **141** 1295
[316] Seyfert C K 1943 *Astrophys. J.* **97** 28
[317] Burbidge E M, Burbidge G R and Sandage A R 1963 *Rev. Mod. Phys.* **35** 947
[318] Sandage A R 1965 *Astrophys. J.* **141** 1560
[319] Fitch W S, Pacholczyk A G and Weymann R J 1968 *Astron. J.* **73** 513
[320] McLeod J M and Andrew B H 1968 *Astrophys. Lett.* **1** 243
[321] Schwarzschild K 1916 *K. Preuss. Akad. Wiss. (Berlin) Sitzungsber.* **1** 189
[322] Kruskal M D 1960 *Phys. Rev.* **119** 1743
[323] Kerr R P 1963 *Phys. Rev. Lett.* **11** 237
[324] Newman E T, Couch K, Chinnapared K, Exton A, Prakash A and Torrence R 1965 *J. Math. Phys.* **6** 918
[325] Carter B 1971 *Phys. Rev. Lett.* **26** 331
[326] Hawking S W 1972 *Commun. Math. Phys.* **25** 152
[327] Penrose R 1965 *Phys. Rev. Lett.* **14** 57
[328] Hawking S W and Penrose R 1969 *Proc. R. Soc.* **A314** 529
[329] Penrose R 1969 *Riv. Nuovo Cimento* **1** 252
[330] The physics of black holes is described in Misner C W, Thorne K S and Wheeler J A 1973 *Gravitation* (San Francisco: Freeman) and Shapiro S I and Teukolsky S A 1983 *Black Holes, White Dwarfs and Neutron Stars: the Physics of Compact Objects* (New York: Wiley Interscience). An excellent review of the history of black holes is presented by Israel W 1987 in *300 Years of Gravitation* ed S W Hawking and W Israel (Cambridge: Cambridge University Press) p 199.
[331] Zeldovich Ya B 1964 *Sov. Phys. Dokl.* **9** 195
Salpeter E E 1964 *Astrophys. J.* **140** 796
[332] Lynden-Bell D 1969 *Nature* **223** 690
[333] Bardeen J M 1970 *Nature* **226** 64
[334] Shakura N and Sunyaev R A 1973 *Astron. Astrophys.* **24** 337
[335] Zeldovich Ya B and Novikov I D 1964 *Sov. Phys. Dokl.* **9** 834

[336] Ulrich M-H et al 1984 *Mon. Not. R. Astron. Soc.* **206** 211
[337] Wandel A and Mushotzky R F 1986 *Astrophys. J.* **306** L63
[338] Abramowicz M A, Jaroszyński M and Sikora M 1978 *Astron. Astrophys.* **63** 221
[339] Papaloizou J C B and Pringle J E 1984 *Mon. Not. R. Astron. Soc.* **208** 721
[340] Hoyle F, Burbidge G R and Sargent W L W 1966 *Nature* **209** 751
[341] Rees M J 1966 *Nature* **211** 468; 1967 *Mon. Not. R. Astron. Soc.* **135** 345
[342] Rees M J 1971 *Nature* **229** 312
Scheuer P A G 1974 *Mon. Not. R. Astron. Soc.* **166** 513
[343] Krymsky G F 1977 *Dokl. Akad. Nauk* **234** 1306
Bell A R 1978 *Mon. Not. R. Astron. Soc.* **182** 147
Axford W I, Leer E and Skandron G 1977 *Proc. 15th Int. Cosmic Ray Conf.* vol 11, p 132
Blandford R D and Ostriker J P 1978 *Astrophys. J.* **221** L29
[344] Gull S F 1975 *Mon. Not. R. Astron. Soc.* **171** 263
[345] Perley R A, Dreher J W and Cowan J J 1984 *Astrophys. J.* **285** L35
[346] Whitney A R, Shapiro I I, Roges A E E, Robertson D S, Knight C A, Clark T A, Goldstein R M, Marandino G E and Vandenberg N R 1971 *Science* **173** 225
Cohen M H, Cannon W, Purcell G H, Shaffer D B, Broderick J J, Kellermann K I and Jauncey D L 1971 *Astrophys. J.* **170** 207
The image shown in figure 23.28 is from Pearson T J, Unwin S C, Cohen M H, Linfield R P, Readhead A C S, Seielstad G A, Simon R S and Walker R C 1982 *Extragalactic Radio Sources* ed D S Heeschen and C M Wade (Dordrecht: Reidel) p 355.
[347] Many important references to superluminal motion can be found in Zensus J A and Pearson T J (ed) 1987 *Superluminal Radio Sources* (Cambridge: Cambridge University Press).
[348] Antonucci R R and Miller J S 1985 *Astrophys. J.* **297** 621
[349] Barthel P D 1989 *Astrophys. J.* **336** 606
[350] Fichtel D 1994 *Frontiers of Space and Ground-Based Astronomy* ed W Wamsteker *et al* (Dordrecht: Kluwer)
[351] The history of Gamow's work on the Big Bang theory is described by Alpher R A and Herman R C in *Modern Cosmology in Retrospect* ed B Bertotti, R Balbinot, S Bergia and A Messina (Cambridge: Cambridge University Press) p 129.
[352] Chandrasekhar S and Henrick L R 1942 *Astrophys. J.* **95** 288
[353] Lemaître G 1931 *Nature* **127** 706
[354] Gamow G 1946 *Phys. Rev.* **70** 572
[355] Alpher R A, Bethe H and Gamow G 1948 *Phys. Rev.* **73** 803
[356] Alpher R A and Herman R C 1948 *Nature* **162** 774
[357] Alpher R A and Herman R C 1950 *Rev. Mod. Phys.* **22** 153
[358] Hayashi C 1950 *Prog. Theor. Phys. Japan* **5** 224
[359] Alpher R A, Follin J W and Herman R C 1953 *Phys. Rev.* **92** 1347
[360] Milne E 1948 *Kinematic Relativity* (Oxford: Clarendon)
[361] Dirac P A M 1937 *Nature* **139** 323
[362] Eddington A S 1946 *Fundamental Theory* ed E Whittaker (Cambridge: Cambridge University Press)
[363] Bondi H and Gold T 1948 *Mon. Not. R. Astron. Soc.* **108** 252
Hoyle F *Mon. Not. R. Astron. Soc.* **108** 372
[364] Baade W 1952 *Trans. IAU* **8** 397
[365] Humason M L, Mayall N U and Sandage A R 1956 *Astron. J.* **61** 97
[366] Shakeshaft J R, Ryle M, Baldwin J E, Elsmore B and Thomson J H 1955 *Mem. R. Astron. Soc.* **67** 106

[367] Ryle M 1955 *Observatory* **75** 137
[368] Mills B and Slee O B 1957 *Aust. J. Phys.* **10** 162
[369] Scheuer P A G 1957 *Proc. Camb. Phil. Soc.* **53** 764
[370] Bracewell R N (ed) 1959 *Paris Symposium on Radio Astronomy* (Stanford: Stanford University Press)
[371] Gower J F R 1966 *Mon. Not. R. Astron. Soc.* **133** 151
[372] Scheuer P A G 1975 *Galaxies and the Universe (Stars and Stellar Systems 9)* ed A Sandage *et al* (Chicago: University of Chicago Press) p 725
Longair M S 1971 *Rep. Prog. Phys.* **34** 1125
[373] Osterbrock D E and Rogerson J B 1961 *Publ. Astron. Soc. Pacif.* **73** 129
[374] O'Dell C R, Peimbert M and Kinman T D 1964 *Astrophys. J.* **140** 119
[375] Hoyle F and Tayler R J 1964 *Nature* **203** 1108
[376] Some appreciation of the scope of Zeldovich's contributions to astrophysics and cosmology can be found in Ostriker J P *et al* (ed) *Selected Works of Yakob Borisevich Zeldovich. Part 2: Astrophysics and Cosmology* (Princeton, NJ: Princeton University Press).
An account of Dicke's role in the revival of the observational study of Gamow's picture of the Big Bang can be found in Peebles P J E 1993 *Principles of Physical Cosmology* (Princeton, NJ: Princeton University Press).
[377] Penzias A A and Wilson R W 1965 *Astrophys. J.* **142** 419
[378] Roll P G and Wilkinson D T 1966 *Phys. Rev. Lett.* **16** 405
[379] McKellar 1941 *Publ. Dom. Astrophys. Obs.* **7** 251
[380] Mather J C *et al* 1990 *Astrophys. J.* **354** L37
[381] Wagoner R V, Fowler W A and Hoyle F 1967 *Astrophys. J.* **148** 3
Wagoner R V *Astrophys. J.* **179** 343
[382] York D G and Rogerson J B 1976 *Astrophys. J.* **203** 378
[383] Vidal-Madjar A, Laurent C, Bonnet R M and York D G 1977 *Astrophys. J.* **211**
[384] Peebles P J E 1966 *Astrophys. J.* **146** 542
[385] Doroshkevich A G, Novikov I D, Sunyaev R A and Zeldovich Ya B 1971 *Highlights of Astronomy* ed C de Jager (Dordrecht: Reidel) p 318
[386] See, for example, Opal Collaboration 1990 *Phys. Lett.* B **240** 497
[387] Sandage A R 1961 *Astrophys. J.* **133** 355
[388] Sandage A R 1968 *Observatory* **88** 91
[389] For a detailed discussion of the different approaches to the determination of Hubble's constant, see Rowan-Robinson M 1985 *The Cosmological Distance Ladder* (New York: W H Freeman and Company); his conclusions are updated in Rowan-Robinson M 1988 *Space Sci. Rev.* **48** 1.
[390] Hesser J E, Harris W E, VandenBerg D A, Allwright J W B, Shott P and Stetson P 1989 *Publ. Astron. Soc. Pacif.* **99** 739
[391] Sandage A R 1994 *The Deep Universe* ed R Kron *et al* (Berlin: Springer)
[392] Lilly S J and Longair M S 1984 *Mon. Not. R. Astron. Soc.* **211** 833
[393] Chambers K C, Miley G K and van Breugel W J M 1987 *Nature* **329** 624
[394] Tyson A 1990 *Galactic and Extragalactic Background Radiation* ed S Bowyer and C Leinert (Dordrecht: Kluwer) p 245
[395] Ellis R 1987 *High Redshift and Primaeval Galaxies* ed J Bergeron *et al* (Gif sur Yvette: Edition Frontières) p 3
[396] Longair M S 1966 *Mon. Not. R. Astron. Soc.* **133** 421
[397] Rowan-Robinson M 1968 *Mon. Not. R. Astron. Soc.* **141** 445
Schmidt M 1968 *Astrophys. J.* **151** 393
[398] Dunlop J S and Peacock J A 1990 *Mon. Not. R. Astron. Soc.* **247** 19
[399] Schmidt M and Green R F 1983 *Astrophys. J.* **269** 352

Hoag A A and Smith M G 1977 *Astrophys. J.* **217** 362
Osmer P S 1982 *Astrophys. J.* **253** 28
[400] Boyle B J, Jones L R, Shanks T, Marano B, Zitelli V and Zamorani G 1991 *Proc. Workshop on The Space Distribution of Quasars: Astronomical Society of the Pacific Conf. Series 21* (San Francisco, CA: Astronomical Society of the Pacific) p 191
[401] Koo D C and Kron R 1988 *Astrophys. J.* **325** 92
[402] Irwin M, McMahon R G and Hazard C 1991 *Proc. Workshop on The Space Distribution of Quasars: Astronomical Society of the Pacific Conf. Series 21* (San Francisco, CA: Astronomical Society of the Pacific) p 117
[403] Hasinger G, Burg R, Giacconi G, Hartner G, Schmidt M, Trümper J and Zamorani G 1993 *Astron. Astrophys.* **275** 1
[404] Oort J 1958 *Solvay Conference on the Structure and Evolution of the Universe* (Brussels: Institut International de Physique Solvay) p 163
[405] Gunn J E 1978 *Observational Cosmology* ed J E Gunn (Geneva: Geneva Observatory Publications) p 1
[406] See, for example, Geller M J and Huchra J 1978 *Astrophys. J.* **221** 1
Davis M and Peebles P J E 1983 *Astrophys. J.* **267** 465
[407] Lynden-Bell D, Faber S M, Burstein D, Davies R L, Dressler A Terlevich R J and Wegner G 1988 *Astrophys. J.* **326** 19
[408] Strauss M A and Davis M 1988 *Large-Scale Motions in the Universe* ed V C Rubin and G Coyne (Vatican City: Pontificia Academia Scientiarum)
[409] Lemaître G 1933 *C. R. Acad. Sci., Paris* **196** 903
Tolman R C 1934 *Proc. Natl Acad. Sci.* **20** 169
[410] Lifshitz E M 1946 *J. USSR Acad. Sci.* **10** 116
[411] Novikov I 1964 *Zh. Eksp. Teor. Fiz.* **46** 686
[412] Weymann R J 1966 *Astrophys. J.* **145** 560
[413] Zeldovich Ya B and Sunyaev R A 1969 *Astrophys. Space Sci.* **4** 301
[414] Kompaneets A 1956 *Zh. Eksp. Teor. Fiz.* **31** 876
[415] Silk J 1968 *Astrophys. J.* **151** 459
[416] Harrison E R 1970 *Phys. Rev.* D **1** 2726
Zeldovich Ya B 1972 *Mon. Not. R. Astron. Soc.* **160** 1P
[417] Sunyaev R A and Zeldovich Ya B 1970 *Astrophys. Space Sci.* **7** 3
[418] Sachs R K and Wolfe A M 1967 *Astrophys. J.* **147** 73
[419] Lyubimov V A, Novikov E G, Nozik V Z, Tretyakov E F and Kozik V S 1980 *Phys. Lett.* B **94** 266
[420] Gershtein S S and Zeldovich Ya B 1966 *Pisma Zh. Eksp. Teor. Fiz.* **4** 174
[421] Marx G and Szalay A S 1972 *Neutrino '72* vol 1 (Budapest: Technoinform) p 191
Szalay A S and Marx G 1976 *Astron. Astrophys.* **49** 437
[422] Zeldovich Ya B and Sunyaev R A 1980 *Pisma Astron. Zh.* **6** 451
Zeldovich Ya B, Doroshkevich A G, Sunyaev R A and Khlopov M Yu 1980 *Pisma Astron. Zh.* **6** 457
Zeldovich Ya B, Doroshkevich A G, Sunyaev R A and Khlopov M Yu 1980 *Pisma Astron. Zh.* **6** 465
[423] Zeldovich Ya B 1970 *Astron. Astrophys.* **5** 84
[424] Particle data group, Hikasa K *et al* 1992 Review of particle properties *Phys. Rev.* D **45** 2.3
[425] Tremaine S and Gunn J E 1979 *Phys. Rev. Lett.* **42** 407
[426] Peebles P J E 1982 *Astrophys. J.* **263** L1
[427] Davis M, Efstathiou G, Frenk C and White S D M 1992 *Nature* **356** 489
[428] Press W H and Schechter P 1974 *Astrophys. J.* **187** 425
[429] Smoot G F *et al* 1992 *Astrophys. J.* **396** L1
[430] Hancock S, Davies R D, Lasenby A N, Guttierrez de la Cruz C M

Watson R A, Rebolo R and Beckman J E 1994 *Nature* **367** 333
[431] The physics of the early universe is splendidly described in Kolb E W and Turner M S 1990 *The Early Universe* (Redwood City, CA: Addison-Wesley).
[432] Zeldovich Ya B 1965 *Adv. Astron. Astrophys.* **3** 241
[433] Alfvén H and Klein O 1962 *Ark. Fyz.* **23** 187
[434] Omnes R 1969 *Phys. Rev. Lett.* **23** 38
[435] Dicke R H and Peebles P J E 1979 *General Relativity: an Einstein Centenary Survey* ed S W Hawking and W Israel (Cambridge: Cambridge University Press) p 504
[436] Sakharov A D 1967 *Pisma Zh. Eksp. Teor. Fiz.* **5** 32
[437] Guth A 1981 *Phys. Rev. D* **23** 347
[438] Linde A D 1974 *Pisma Zh. Eksp. Teor. Fiz.* **19** 183
[439] Bludman S A and Ruderman M 1977 *Phys. Rev. Lett.* **38** 255
[440] Linde A D 1982 *Phys. Lett.* B **108** 389; 1982 *Phys. Lett.* B **129** 177
Albrecht A and Steinhardt P J 1982 *Phys. Rev. Lett.* **48** 1220

Chapter 24

COMPUTER-GENERATED PHYSICS

Mitchell J Feigenbaum

Not so long ago, say 1972, a number of well known physicists took out a full page advert in *The New York Times* strongly criticizing a resurgence of popular interest in astrology and the writings of a certain E Velikovsky that discussed, among other things, an anomalous non-predictable transit of Venus from somewhere-or-other to its present location. The authors trotted out our accumulated wisdom of the orderly certainty of the heavens as well as the surely impossible as dictated by science, which *inter alia*, allowed not at all for the musings of Velikovsky.

As it might have seemed to a sceptic, such confidence was perhaps more an article of faith than a precept of critical thought. Indeed, some twenty years later, much has been observed to vindicate the doubts of the sceptic. In fact, in the last decade, owing to work that I shall ascribe to J Wisdom and J Laskar as exemplars, we now know that Pluto, various planetary moons, and the inclination of the spin axes of the inner planets to the ecliptic (the *obliquities*) are describing disorderly, chaotic evolutions. Indeed, it is probable that the Kirkwood gap in the asteroid belt has been accomplished by that time-honoured opponent of eternal rhythmia, the crass haphazard collisions of these lost asteroids with Earth and Mars in times past.

How has it come to pass in little over twenty years that a world of perfectly tranquil, calculable order has given over to a view so fully injected with doses of chaos? The answers are varied. Mostly, a deep belief in a world view all too easily allows surmise to traduce fact: whatever Velikovsky wrote, our peers, in final analysis, responded all too much in kind. We know much, yet little indeed. Twenty years ago no physicist knew of chaos, and what is more important, of its prevalence.

Next, there were seeds from the past awaiting full germination. Nevertheless, it goes too far to say that our present understanding was adumbrated: the seeds only said to an open mind that things were more complicated, harder, than was generally acknowledged. They should however have fuelled critical scepticism. They did not.

Finally, as many of you already answered to yourselves, there are now computers. But, computers have been with us for close on *fifty* years now. So, we are still left with a genuine conundrum: why is it so obvious now that the classical world is shot with diseases of disorder when no-one believed any such thing only twenty years ago?

I will now offer from my perspective and memory the answers and, more generally, what we have come to know. This subject is basically straightforward because few issues bearing on fundamental new physical theory intrude; rather the question is about how they evolve the equations that physical theory determines. Non-linear dynamics and *chaos* are interesting precisely because the equations develop their solutions in ways qualitatively different from the manner imagined by earlier ignorance. Qualitatively so different that in early numerics and actual physical experiment these behaviours were ignored as various sorts of spurious noise spoiling the results. After all, when one has deep images as to how things should behave, and when difficult-to-control numerical approximations and physical systems display something else, one is prejudiced to believe the former and repeat the latter differently until the mind perceives the desired image. It is very hard to become familiar with seemingly unpatterned new things. This has made me deeply sceptical of research waiting to encounter *emergent behaviour*. After all, if new organizations occur when an object becomes *complex* enough, then one has no need of a computer's service. Rather, one merely needs to look up at the clouds; down at the pebbles and cracks in the pavement and plants and other life; across onto the river, across at the innumerable arbitrarily placed objects in one's environment, not to mention the specific, however peculiarly random, placed hairs on one's head. The world abounds with interacting complex entities, in quantity and quality infinitely surpassing the puny simulations of all the world's computers. It is hard to learn by seeing. So why should the scales miraculously drop off when one stares at a computer screen?

The point is that without a conceptual schema to help one notice what is recurrent, however irregular, no subtle observations are generally possible. Nevertheless, let me admit, fortuities and serendipities do sometimes come about.

The solar system is now obviously infiltrated with chaos because various ideas and insights coalesced to make the phenomenon transparent at last to our minds and visible to our eyes. The path was not straight, and crucial insights came from far afield. Insofar as this is a discussion of how not unimportant ideas came to be known, I shall feel free to comment historically on how my own path intertwined with this effort.

A high-energy physicist wants to understand the fabric of space and time at the smallest scales possible. We have much in the way of theoretical ideas, much that explains experimental data, and much unknown. There are reasons to be dissatisfied with our understanding, even though *approximate* calculations agree with the data. Technically, the extraordinary power of symmetry has made it much easier to relate most experimental facts to just a few irreducible ones, than to account for these latter few. Here actual dynamics is required, and here less is available. Other than through intact symmetry calculations, everything hinges on low-order perturbative calculations. Some of these are controlled, some not. In some cases the best theoretical surmise is that the theory is at most asymptotic. At the height of interest in field theory, one knew that in simple models the renormalized theory was in the orthogonal complement of the perturbation theory. We generally don't know when the theoretical ideas are themselves wrong as opposed to the theory being right and our calculations faulty. While a phenomenological fit to data, good enough to empower predictions, is highly exciting and rewarding, to a critic the theory still feels unfinished. And then too, we have so many mathematical tools at our disposal, its hard not to subsume (by an infinitude of theories) a static set of data. But then too, it is worthy of some deep philosophy to understand why it is humanly almost impossible to find even one!

To my mind, the most significant advance in my career was Wilson's realization (circa 1970) about how one could generally formulate nonperturbative calculations. Spared of details, the idea is how to execute a *fast* algorithm to compute the actual result discretely. Because the idea was deep and naturally expressed in the context of the prevailing mathematics, much conceptual iconography issued out, so that we can have a *sense* for what is apt to be right in a problem.

As for myself in the early 1970s, I didn't care how the particles came out *today*, but sceptical of all our calculations, looked around for ideas to calculate something correctly. Moreover, as all of modern physics is concerned with interacting systems of an infinite number of degrees of freedom, if one could face these questions with adequate tools, one might begin to consider infinitely interconnected sets of neurons, that is, the behaviour of a nervous system.

It is exceedingly difficult to produce an accurate solution to almost any equation. This is true for various reasons, some trivial and some deep. We know we can write down the zeroes of all polynomials of degree four or below. Even if we can't for other polynomials, some version of Newton's method allows us, with pen and paper, to do arbitrarily well although with high mechanical exertion. Similarly, there is a limited list of indefinite integrals we can produce exactly. However, after considerable effort, and paying attention to asymptotic behaviours, we can often meet the challenge. As for definite integrals, there is a bigger list, and even if not analytically soluble, with care towards singularities, we can compute these to any desired accuracy. *Not* that a packaged integrator will do so. These work only for integrands that are relatively singularity-free and strongly bounded. So far there's little *deep* here, although if one wants to establish *truth* it is generally unwise to use an analysis you haven't personally controlled and programmed.

So far this is easy, precisely because only a small number of things need be arithmetically combined. It is easiest if there is just one independent variable, but, when so potentially controllable as the problems above, not much worse in a (small) finite number of dimensions. However, a differential equation for even one variable already pushes us to the *serious*. For non-linear differential equations with nearby singularities, the analysis of the requisite control grows much more difficult, and inherent limitations to precision can become altogether crucial. Here no serious scientist who signs his name on his work should use any packaged program, unless he is *sure* of singularities, convergence and precision. As the number of coupled ordinary differential equations (ODEs) grows, one must be more careful still, and with three non-linear equations, it is axiomatic that if the integration was performed by an inherited package, little credence should be attached to the result. Although anyone who thinks will readily appreciate this, I shall now offer three compelling reasons.

The most elementary is that three coupled non-linear ODEs most generally exhibit chaotic behaviours. Prior to theoretical understanding by the mid-1970s, and despite E Lorenz's seminal discovery a decade before, no such behaviour was generally observed numerically. Now it always is. This is a part of my objection to waiting to see numerics unfold emergent behaviour before our untrained mind and eyes.

Secondly, precision can be crucial. In numerical solution, we are doing a reasonably uncontrolled summation. The terms summed can easily contain a term vastly smaller than the leading term which, when successively summed, can finally grow to overwhelm the presumed leading term. This is endemic in relaxation oscillation with singular perturbations (i.e. almost vanishing highest derivatives) which can appear *dynamically*, rather than explicitly in the equations themselves. The simplest example I know is

$$\ddot{x} + x\dot{x} + x = 0$$

which for $x_0 = 0$, $\dot{x}_0 \gg 1$ is numerically intractable if directly attempted. This should make one shudder when planetary integrations are done for tens of millions of orbits, although reversibility back to initial conditions is highly reassuring.

Thirdly, there is a strong theoretical reason for worry which lies at the heart of Newtonian physics, and which may be expressed briefly in the following way. Galileo discerned the principle of inertia and produced the trajectory for parabolic motion. If one takes the gravitation as constant over strata at successive elevations, one can still write down the trajectory by patching together parabolas, matching y and $\dot{y}$ at each interface. As the strata become infinitesimally thin, this matching poses an infinite number of constraints, and is no longer feasible. However, inertia dictates that $\ddot{y}$ is fixed at each elevation y; Newton is Galileo maintained just locally. But still, it is only the trajectory we care about, and so the ODE *must* be integrated. So long as the pieces smoothly fit together, we can control the integration. But what if we follow Newton not so much as a convenience, but rather because the elementary pieces in each strata are varying almost erratically? Then how do we paste the pieces together with any confidence—how do we control the integration? There are equations which are safe to treat in this way only for a limited stretch. Consider for example a fluid upstream of a cascade, and then downstream below it. Knowing the downstream flow, do we really believe we can determine the upstream smooth behaviour? Certainly if the sheet of water has broken into droplets over the cascade, the fluid equations themselves have failed, and yield a singularity at the moment of break up. Even without this wholesale catastrophe, it is fair to guess that as we analytically propagate the solution backwards up the cascade, we shall encounter a natural boundary beyond which continuation is hopeless. Thus, our very inheritance of the physical theory underlying the ODEs put us on the alert that we may not be able to paste together the pieces in a sufficiently erratic evolution. Should you use someone's integrator here, you are a fool.

Wisdom aside, the first reason is the most galling. When no one knew what almost always happened, no numerical recipe disabused him of his ignorance. Whatever else is to be concluded, it is mandatory to realize that we possess no numerical or other panacea, and that numerics can only *support* our knowledge. On the other hand, it can support it while we are discovering it if we be sufficiently skilful and careful.

This now leaves two last points concerning numerics. It is clear that partial differential equations (PDEs) bring out with a vengeance all the diseases of computing systems of ODEs, unless the PDE is such a tight encoding of (say) a geometrical relationship that good methods can be constructed. In general this is not the case. However, there are systems reminiscent of PDEs, such as cellular automata, for which the numerics are executed exactly. Thus, while the equations of physics are concerned with the continuum for which discrete numerical calculations easily become diseased, these truly discrete systems are natural for a digital computer, and are normally exactly computable. So, the first point is that all the diseases I have mentioned are completely absent in these numerical simulacra of reality. The second point is that usually they are impossible to penetrate analytically. This is in a way the flip side of what I have said above. For the continuum problems, unless we seriously can control the numerics by virtue of analytical forethought, we shall never see the novel phenomena, because the simulation simply fails to produce them, despite their marked presence in a true solution. For these exact discrete systems, we typically see a blithering array of point-to-point erratic variation with no general schema of iconic interpretation: there might be kernels of extraordinary emergent behaviour, but with no reasoned-base perception, who knows how to see it?

So much for the pitfalls of using computers to garner wisdom. From this point onwards, all that is of significance in what I shall describe is embedded in the continuum. As will emerge, it is also in an infinite-dimensional continuum. So, one can neither analytically determine the sought after behaviour, nor can one do other than finite-dimensional numerical calculations. The trick is a synergy: how to use the glimmers of analytical understanding to control and target the numerics in slightly larger arenas, which then improves the analytic conception, and so forth in tentative, gingerly excursions deeper into the unknown. This was the path that led to the theory of universality in period doubling, which I shall now document. It must again be emphasized that when you learn more, just as everything once behaved one way, now they only behave otherwise.

What did I know about computing? Basically, starting in junior high school, I decided that I could calculate the logarithm table myself, and later, the trigonometry tables. I loved Newton's method for solving transcendentals, and in high school I already knew that starting values can make a big difference and lead to erratic non-convergent jumps up to the limit of patience of manual arithmetic. My father showed me his beautiful ivory-on-mahogany slide rule in junior high school, and I quickly realized its idea. I was allowed to use the new Friden calculating machine which, shortly before its transformation into a relic, could also extract square roots. I love numbers and always as an amusement, and more seriously than that, invented new algorithms to calculate them.

The first computer I ever saw was a new IBM at Columbia with magnetic drum memory and patch-board programming capability in 1960. At the same time, in high school, my father's friend and co-worker had given me a Geniac, a set of phenolic-cardboard discs with radial holes, configurable into a sophisticated mechanical Boolean switching array. The games of nim and tick-tack-toe were within its capacity. The various staples and screw heads never made the best of contact, and operation was a touch-and-go enterprise. But not so an included reprint of an article by Shannon on Boolean logic.

I went to CCNY for a degree in electrical engineering, which I completed in 1964. I saw no computer until my last year or so there.

There was a new IBM which took a long spool of punched-hole paper tape which I never used. But in the control systems laboratory there was an analogue computer. That it could draw the phase–space picture of Van der Pol's oscillator in full agreement with the hand-drawn isocline picture was quite a marvel—I wished I had had the astronomical sum of $10 000 to buy one for myself.

I went to MIT as a graduate student, again in electrical engineering, but within a term was in physics pursuing general relativity. Visiting Brooklyn Polytech one day, I saw a new acquisition that looked like a typewriter, but was a programmable digital computer. This was the first computer I ever used, and within an hour had programmed it to take square roots by Newton's method. I almost never saw a computer during my graduate career at MIT. Indeed, as a high-energy theorist in the making, it was made clear to me that only the lower life, such as the star-builders and some nuclear theorists, would ever dirty themselves to use a computer. To the high life this was an instrument to be demeaned and excoriated. As were its users.

As a post-doc at Cornell in 1970 there was a fancy table-top HP computer/calculator. I spent some time mastering the use of this machine. The only other user was Ken Wilson.

After several more years I became a staff member at Los Alamos, and got my first programmable calculator, an original HP65, in December 1974. This was only slightly weaker than the table-top HP—indeed its successor, the HP45, was more powerful than the Cornell table-top machine. The crucial clue to the theory of renormalization in dynamical systems breathed into life in the HP65. At 20 iterations per second and 30 millisecond arithmetic, with a maximum 99 lines of program, this was the state-of-the-art computational power that helped to set the mood that revealed, almost two decades later, the instability of the solar system. How did this happen?

Owning the HP65 had me profligate with my time inventing new numerical algorithms. As a profound technical point, such a device is the device of choice for perfecting numerical methods. The discipline of a very few (five) storage registers and a maximum of 99 instructions spurs the highest level of insightful, tight calculation. The triviality—and paucity—of input/output wastes no time, and diverts from the waste of modern gee-whiz computer junk. The joy is distilled in tight, correct numerical approximation methods. In swift order, I invented new ODE solvers, minimization routines, interpolation methods, etc. For someone who cares for numbers, much of the tedium was eliminated.

When I arrived at Los Alamos, the theory division head, P Carruthers, felt the time was right, and I was the appropriate person, to see if Wilson's *renormalization* group ideas could solve the century and a half old problem of turbulence. In a nutshell, it couldn't—or, so far, hasn't—but led me off in wonderful directions.

> See also p 564

I must now say something about the renormalization group. In the 1950s M Gell-Mann and F Low (my thesis advisor at MIT) noticed that the knowledge from perturbation theory could be extended by paying attention to how renormalizing at different cut-offs, each integrated over more momenta than the former, led to certain functional identities. In about 1969 K Wilson took a different tack towards what this procedure entailed, and realized in collaboration with M Fisher how various anomalous dimensions in field theory, formally related to critical exponents in the statistical mechanics of phase transitions, could be understood to be the scaling ideas in phase transitions as put forward by B Widom, and theoretically highly illuminated by L Kadanoff. This is quite a mouthful. Disregarding the physical meanings, the basic idea is as follows.

Quantum mechanics in the path-integral formulation of Feynman, for imaginary time, is the 'same' as statistical mechanics. Both request the practitioner to perform a certain sum over all the possible configurations that a system might exhibit. In the statistical mechanics of spins (microscopic magnets) on a lattice, one must sum the values of a certain quantity, $e^{-\beta H}$, over all possible ways each of the (innumerable) spins may point. In quantum mechanics, for a propagator (relating the amplitude now to what it was then) the values of $e^{iS/\hbar}$ (S is the action on any particular classical path) when approximated by the values at each discrete time interval τ, leads formally to the same sum as for the magnet. The problem is how to do an infinite-dimensional sum. (It is historically worth noting that these discrete approximations to the paths for imaginary time also precisely agree with N Wiener's 'cylinder' sets for the Brownian process, or in general, for diffusion processes.)

How do we do such a sum? If we can't, then we can't correctly write down the consequences of any of our theories. And in general, we can't. The old approach was by *perturbation theory*. Simple Brownian motion, Gaussian mean field theory for magnets, and free field theory for elementary particles all have exactly the same form with a known completely analytic solution. So, we write the real interacting problem we care about as the easy exact problem plus a correction (the perturbation). We then proceed by writing down a power series for the correction factor from which, if its small and we're lucky—and it turns out neither is generally correct—we can systematically obtain the solution of the correct problem in terms of appropriate averages over the known, unperturbed problem. This is called 'perturbation theory'. That 'theory' appears within the quotes is highly significant, because, in general, the sum makes no sense, and only after exceptional fiddles (renormalization theory) does one argue that the result might actually be correct.

The main clue from the past, owing to Poincaré, is that one must in general expect that the perturbation method couldn't possibly work. In what context did Poincaré reach this result? Why, in planetary dynamics, of course. The perturbation method was invented for this most important of problems, and here Poincaré realized serious new mathematics of a 'qualitative' sort (the introduction of topology) were required. The initial conclusion was that results were diseased beyond cure along the traditional line of thought. When I queried M Hénon on this in 1977 (and he was well in the vanguard of appreciating and spearheading issues of chaotic behaviour in planetary problems) he responded that the best simulations of some 500 years of motion showed nothing to contradict established belief. J Wisdom just short of 10 years ago reached quite different conclusions in computing over several millions of years. (It now appears that order in the solar system is eroded on a time scale of about five million years—not very long for astronomical, or even Earthly, scales.)

So, how does one do these sums correctly?

Wilson's idea was 'divide and conquer'. Consider, for example, an infinite regular one-dimensional line of magnets. Instead of performing the sum of $e^{-\beta H}$ over all possible states of all the magnets, we instead do only a part of the summation, say over the states of the magnets at every *other* site. We still have a lot to sum over. The remaining sum is *not* however over $e^{-\beta H}$ but rather $e^{-\beta H'}$. If we include a rich enough set of different kinds of interactions, H' is the same as H, save for the fact the coupling constants (the coefficients of the different kinds of interactions that comprise H) now have new values that are determined from the original values before the sum, by a certain transformation rule—a 'map'—this is the so-called renormalization group transformation.

Now consider summing over the alternate sites of what is left: now only one quarter of the original sum is left unperformed. However, exactly the same procedure that took us from H to H' now takes us from H' to H''. Thus, this same transformation, say $\tilde{f}$, takes us from the nth partial summation to the $n+1$st always according to the rule $H_{n+1} = \tilde{f}(H_n)$, this is understood as a high-dimensional transformation from the set of coupling constants at level n to those at level $n+1$. Apart from an enlargement of the space of Hs to include many possible coupling constants, this is Kadanoff's idea of 'spin-blocking'. Indeed, for the idea to work there must be an infinite diversity of interactions, and so $\tilde{f}$ is really infinite-dimensional.

There remains a last crucial point: H_{n+1} is *really* different from H_n. This is elementary. By eliminating alternate sites, the distance between two adjacent magnets of H_{n+1} is precisely twice as long as that between the adjacent magnets of H_n. So let us modify $\tilde{f}$ by additionally *reducing* all lengths by a factor of two. We call the result of this total process f and $H_{n+1} = f(H_n)$. Now, H_{n+1} and H_n describe the same kind of magnets spaced the same distance—that is on the same lattices: H_{n+1} and H_n describe possible physics for exactly the same object.

And now the final installment. Recall that Gell-Mann and Low did something that after the fact one could realize was pretty much what Wilson was accomplishing. But they had a punch line. As Gell-Mann told me in 1980 "When me and Francis invented the renormalization group, the only thing we could think of was a fixed point.". This is the cardinal ingredient that Kadanoff missed in finishing his spin-blocking theory. Remember, Gell-Mann and Low by using a succession of cut-offs obtained a mutually consistent relation between the results based upon them. As first appreciated by Wilson, they had effectively argued that the process done again and again, instead of always producing different results, could close the chain by settling down to a constant, reproducing result. This is a *fixed point*. What it meant in the language of Wilson is that each new H_{n+1} isn't all that different from H_n, and in fact, in the limit, H_∞ reproduces itself: $H_\infty = f(H_\infty)$. This mathematically, is the statement of a fixed point: f is a rule of transformation, taking one possible H into another possible H. H_∞ is special: it is a point that f transforms identically into itself.

So what? Well, f is just the rule that accomplishes the elimination (by summing over possibilities) of alternate magnets. We don't need anything further (i.e. no more summing) to get H_∞: $H_\infty = f(H_\infty)$ is an algebraic equation that, given f, we can solve for H_∞. There might be many solutions. However, we know more: f must be *stable* about H_∞. That is, if we add a perturbation to H_∞, the story says that as we keep applying f we must again settle back towards H_∞. That is, the eigenvalues of Df, the infinite-dimensional linearization of f about H_∞, must all produce *contractions*, or lie within the unit circle in the complex plane. There might be several such stable H_∞s. Each is a valid result, but each attracts only *some* of the very original H to itself as we keep applying f to H. (All such original Hs comprise the *basin of attraction* or *universality class* of H_∞.)

So what? Well, remember we had to rescale lengths by a factor of two to construct an f which had a chance of exhibiting a fixed point. Basically, this means that at H_∞ things look the same after a magnification of a factor of two. But then also a factor of 4, 8, 16,.... This means a magnet described by H_∞ exhibits similar fluctuations as to how the magnets are actually deployed, over *all length scales*. This is the hallmark of a (second-order) phase transition. H_∞ is the functional description of a magnet at a phase transition. To know everything about such a magnet we really *never* had to do the whole sum: we simply needed to learn the renormalization rule f.

Now there's a problem. A magnet isn't always at phase transition, but only at the critical temperature. Now, while we wrote $H' = f(H)$, recall that the form of the initial H, H_0, was really βH, with β the inverse of the temperature. So, when we constructed $H_1 = f(\beta H)$, all the coupling constants in H_1 became functions of β—we say they are the 'running' coupling constants to record this fact. It does *not* follow that $H_0 = \beta H$ is attracted to H_∞ if Df happens to have instabilities—eigenvalues outside the unit circle. In this case as we keep applying f to a generic H_0 we will *not* converge towards H_∞. On the contrary, even if we are initially quite close to H_∞, the instabilities cause f to push us *further* away each time. But, should the one parameter β be chosen just right—i.e. at the critical temperature—*and* if there is precisely *one* instability, then that instability can be cancelled, and we will fall into H_∞. This then accounts for all the qualitative facts.

And much more. Imagine that there is just one H_∞. Let us take a whole variety of different minerals, each with a different original H. However, with $H_0 = \beta H$, it is possible that we can find for a particular initial H a particular $\beta(H)$ so that for *this* H_0 we indeed go to H_∞. There is potentially an infinite number of different magnetic materials that experience a phase transition, each at its peculiar critical temperature; but as *all* are governed by exactly the same H_∞, at *criticality* they behave indistinguishably. *This* is the crucial phenomenon that this circle of ideas accounts for: we can start with different sorts of things, yet if they show a certain kind of behaviour (critical phase transition) then they do so quantitatively *identically*. This idea is called *universality* and, when first encountered, is truly remarkable. It says that, for certain regimes, the observed behaviour is altogether independent of microscopic details. You give me almost anything to start off with, I ignore your information, and tell you the right answer—that is, precisely what you will measure!

In particle physics this means you give me a fundamental unrenormalized theory, say quantized Maxwell equations for electricity and magnetism, and I'll give you the renormalized result—i.e. what you actually measure. But then too, you can forget about lots of the exact details of Maxwell—and I'll still give you the same result, and it will be the *correct* result. This is remarkable, and certainly something never to have been guessed from perturbation theory's idea—one exact theory in, one exact result out. Indeed, there is absolutely no way of telling what the initial theory was: anything within the same 'universality class' (by definition) goes to the same fixed point H_∞, and exhibits the same measurable consequences.

So, not only were there clues that the answers were different from what you were guessing but, in a way certainly not anticipated by Poincaré, the entire schema of truth was misdirected.

Before continuing with universality in chaos, there are two last points of Wilson's thoughts I want to discuss. The first is central to the general discussion of this chapter: although the finished results are so clean and elegant and expressible almost exclusively without any equations, that is, emphatically, *not* how Wilson got there. Rather than the iconically simpler magnet, Wilson started with a ϕ^4 field theory and constructed his '*r, u*' model. He deduced the stability of a non-perturbative fixed point only through *numerical* investigations, and only then refined his

1833

thoughts to understand the underlying qualitative schema. Finished thoughts are extraordinary and overwhelmingly impressive. But from whence do they issue? It is amusing to note, and I believe known to all of us who have pitted our wits against the solid inhumanity of a computer, that, as Wilson told me in 1981 "Working too much with a computer makes you less smart than you used to be.". Perhaps this too is the price of wisdom.

The final point pertinent to Wilson's thinking circa 1970 is that H_∞ describes a system with fluctuations at all length scales, that is, a system showing excitations each intimately coupled to the next nearest in scale. Wilson pointed out what a tour de force this was. The classical wisdom is one of quadrature: each part can be solved separately, as though in isolation, and then the different weak interactions added in perturbatively, to produce (rigorously) an arbitrarily accurate result. This is the conventional notion of divide-and-conquer by discerning the significant excitations or mechanisms of a system. What Wilson accomplished was a fast method for that other genre of problem where no separation of scale exists. What problems are like this? Well, evidently the host problem, critical phenomena. But then too, that other phenomenon that defied all classical effort—the turbulence in a fluid that, à la Kolmogorov, exhibits a smooth spectral cascade with every size of eddy intimately interacting with those of adjacent sizes. Wilson, from the beginning, imagined that his ideas could finally deliver the theory of fluid turbulence. It has not. But a curious offshoot did something else: it delivered the first correct theory of the *onset* of turbulence. Wilson's guess had Carruthers set me to turbulence. Now a different story will unfold.

There is a big difference between a statistical magnet and a *deterministic* fluid. The renormalization group idea is a trick to do sums exponentially quickly. From the outset, what is to be summed is clear and prescribed. What does this have to do with the erratic, hard to paste together, solutions of a certain PDE, the Navier–Stokes equations? Whether or not these equations correctly describe an incompressible, dissipative fluid, they display, as is known from analytic guesses and inevitably limited numerical simulations on the world's most powerful computers, behaviours that look pretty much like turbulence. (It is *far* beyond all present computational means to exhibit numerically a simulacrum to three-dimensional fully developed turbulence.)

When things look erratic, the only theoretical tool that exists is statistical. All attempts, starting with O Reynolds in the nineteenth century, have endeavoured to provide a correct statistical description of turbulence. This in no way means the truth is statistical, only that one seeks a limited theoretical result, namely how various measured statistics, elicited from however deterministic an object, should turn out. Since one has no means of knowing, at present, what the actual excitations look like, traditional work has gone vastly further down the statistical path in *presuming* that the causal equations work on a statistical set of initial velocity distributions, this initial set chosen *a priori* with no qualities save for tractable solubility of the required statistics. From the onset, if this sheer guess is inapposite, the results will prove erroneous if not nonsensical, and save for some ingenious attempts at self-consistency, bypass the question of what the equation actually determines. Regrettably, this state of affairs embraces the bulk of effort on the subject of turbulence, with notable exceptions.

In any case, to do Wilson's trick we need to know these complicated jumbled modes that collectively determine the measured outcome of various statistics, typically some correlation functions. But to find the high-frequency modes one has to solve the Navier–Stokes equations, which is precisely what we can't do. So, however much phenomenological thoughts indicated an application of renormalization group ideas, the possibility of doing so is frankly remote.

An essential basis for success is to learn how to find oscillatory modes in a highly non-linear system. So, my first effort at Los Alamos was to understand non-linear oscillations. Thoughts on these matters developed about the time of World War II, but were neither particularly strong nor qualitatively generalizable from one non-linear problem to the next. In a continuum problem an oscillation is a limit cycle—in phase space a loop—or was classically supposed to be only so. This is in fact *very* wrong, and another of Poincaré's *notions*, here quite misapplied. An elementary theorem, the 'Poincaré–Bendixson theorem' indeed establishes this for two *first-order* degrees of freedom. (A system of n degrees of freedom in first-order counting means a coupled system of n first-order ODEs.) The theorem is elementary, especially when a certain device is employed—the idea of a 'Poincaré map'. Let us sketch this out.

We consider two coupled ODEs

$$\dot{x} = u(x, y), \dot{y} = v(x, y).$$

In particular, u and v are independent of t. Starting at an initial point (x_0, y_0) in the (x, y) phase plane, the solution $(x(t), y(t))$ is a curve issuing out of (x_0, y_0). After time T, the point $(x(T), y(T))$ is a smooth function of (x_0, y_0). That is, it is a certain map of the plane to the plane, a two-dimensional f. After time $2T$ the point is just the action of f on the time-T point. So long as we don't care to know just where the point was during an interval of time T, the repeated action of f is all we need care about. That is, once we have integrated the original ODEs over time T and so computed f, we no longer need the original ODEs, and the time-continuous problem has been replaced by the discrete time mapping by f. Such an f is called a 'time-one' map and tells us everything about pictures of the trajectory taken by stroboscopically flashing it every T seconds. Since integration is a smoothing procedure, with u and v continuous, f is differentiable: a small change in (x_0, y_0) determines a small change in $f(x_0, y_0)$ where the point is T seconds later.

However, f isn't *any* smooth function from the plane to the plane. Starting at $t = T$, we can backwards integrate the ODEs to get a unique result at $t = 0$. That is, a time-one map must also be *invertible*. This has a strong consequence. Consider the possibility that f has a fixed point x^*. Then starting at x^*, T seconds later we are back to x^* and indeed we must have been there T seconds earlier, and so for all time, every T seconds we return to x^*. That is, the trajectory describes a closed loop of duration T that it re-describes every interval of time of length T. This is a limit cycle. In particular the trajectory can never produce a piece that is branched like the letter 'γ'.

So far we took a two-dimensional ODE and exactly replaced it with a smooth invertible 2D map that resulted by simply taking snapshots every T seconds. But what is special about T for our ODEs which have no explicit t-dependence? (Such ODEs are 'time homogeneous' or 'autonomous'.) Nothing—we can use any T we want.

Imagine now that the trajectory is oscillating, that is remaining within some finite region of the plane with $x(t)$ and $y(t)$ showing alternating sequences of maxima and minima. Imagine the pattern to be quite irregular. (There will soon be a contradiction.) What is the object we are observing? It is just an irregular clock ticking erratically. Well, rather than an arbitrary T why not use *it* as the clock for the snapshots? Accordingly, each time y achieves a maximum we take a snapshot: we simply use one of the variables as a trigger. Put differently, we take a picture each time $\dot{y} = 0$, or each time the trajectory crosses the curve $v(x, y) = 0$ in that direction that has $\ddot{y} = 0$ (a maximum) or, $\ddot{y} = v_x \dot{x} + v_y \dot{y} = v_x \dot{x} = v_x u < 0$. Since we posit the motion to be oscillatory with y a good variable to observe the oscillations (for example the motion is not $y \equiv$ constant), the orbit must continually cross $v = 0$.

We now immediately generalize the idea of using the internal clock to consider an arbitrary smooth curve which the trajectory regularly crosses transversally (i.e. not tangentially). We call this a 'line of section'. (Had we started with a three or higher d-dimensional system, by generalizing to $\dot{x}_d = 0$, we would have a 'surface of section', a surface one dimension lower than that of the phase space.) Now each time this line is crossed in say the 'down-going' sense we take our picture. But now we have a succession of points on the one-dimensional line, and we have replaced the original 2D continuum in time problem with a 1D discrete map from the line to the line. This construction is called the 'Poincaré map'. The line of section is 1D: we specify any point on it by (say) arc length from some chosen point and call this variable s. From one crossing to the next we have $s' = p(s)$ with p the Poincaré map.

But now p must be well behaved: it must be smooth *and* invertible. This is easy to see. We know the line of section, so that at the point s along it, we know both x_0 and y_0 which serve as initial conditions for the original ODE. It takes some amount of time, $T(x_0, y_0)$, for the orbit to again cross the line of section which is accomplished by the smooth invertible 2D time-one map for $T = T(x_0, y_0)$. This is the new point at $(x(s'), y(s'))$ for $s' = p(s)$. By smoothness of the time-one map, the map p must then be smooth on s. Moreover, given s', we know (x', y') and by the invertibility of the time-one map uniquely determined (x, y) and hence uniquely s, so that p inherits invertibility as well.

What does a smooth invertible map in one dimension look like? The answer: its graph is a smooth *monotone* curve. It may or may not cross the graph of $y = x$, the 45° straight line through the origin. Whenever it does we have $s = p(s)$, a fixed point and hence a limit cycle for the ODEs. If it crosses once, say with slope greater than one, then the next closest crossing must have a slope smaller than one, and so forth.

What does this mean? Consider $s^* = p(s^*)$ and start at $s = s^* + \varepsilon$. Then at the next crossing $s = f(s^* + \varepsilon) \cong f(s^*) + \varepsilon f'(s^*) = s^* + \varepsilon f'(s^*)$. Thus, if $f'(s^*) > 1$, the next point is further from s^* than the first, and we have an *unstable* limit cycle at s^*: starting arbitrarily close to the cycle, we spiral away from it. Reciprocally, for $f'(s^*) < 1$ we have a stable limit cycle. By the monotonicity of p, starting anywhere between two successive fixed points, s systematically moves away from the unstable cycle towards the stable one. So, we now know that whatever the functions u and v of the ODEs, the set of all possible orbits is a set of alternating stable and unstable simple limit cycles. *And nothing else.* No matter what the 2D autonomous system, after a transient, the system beats as a regular clock on a stable limit cycle.

I have done this at great length because it has set up imagery I'll need, and because whichever reader has never encountered these ideas before, that reader should be astounded by the qualitative ease and power to understand what all possible 2D autonomous systems can do despite the fact that for particular (u, v) functions the solution isn't analytically available and uncareful numerical solution can be altogether erroneous.

What does this all mean? For example, if x and y represent the size of populations of a predator and prey species, then in a static world (one for which vegetation grows identically year after year), all that can unfold is a periodic oscillation of the species. In particular, no semblance of anything that can pass for randomness. If the actual behaviour is erratic, no ingenious cooked-up pair of (u, v) can ever give the right result. And if the computer shows buzziness, the answer is simply and manifestly wrong.

In September 1974 when I started thinking about non-linear oscillations, with a small just-acquired familiarity of Poincaré's ideas, it immediately struck me that if simple problems only oscillated, and since the qualitative ideas offered *much* less strong guidance in three and higher dimensions, it was pointless to think about ODEs. And anyway I was not about to use any computer. But as I said much earlier, I already knew that Newton's method could perform in funny, erratic ways even for 1D problems. What was the simplest problem to study that was non-linear? Big surprise, after a linear change of variables

$$x_{n+1} = a + x_n^2 \equiv L(x_n) \tag{1}$$

is the standard form for Newton's method for the zeroes of $u(x)$ with u/u' a quadratic. Notice that there is no 2D autonomous ODE for which this can be its Poincaré map: $a+x^2$ is non-invertible. So whatever fun (1) can indulge in beyond a fixed point is forbidden to any autonomous 2D differential system, and by implication to a 2D invertible mapping. What does (1) correspond to? This I didn't know until several years later.

Certainly, for $a < 1/4$ (1) has a pair of real fixed points. Since $L' = 2x$ and the fixed points are at $2x = 1 \pm \sqrt{1-4a}$, the positive one is unstable and the negative one stable for $a > -3/4$. For $-3/4 < a < 0$, L' is negative, so that the fixed point exhibits alternating, rather than monotone convergence. For $a > 1/4$ no bounded orbits exist, and for $a < -3/4$ both fixed points are unstable. Then what can happen?

Now the point in studying (1) was to understand oscillations more complicated than boring limit cycles (i.e. fixed points). Does (1) do something more interesting? First of all can the map (1) itself oscillate? That is, can I find $y = L(x)$ and $x = L(y)$ for $x \neq y$? Analytically this is easy—and just about as far as analytics allows at this stage. (The finished theory provides new analytics for *infinitely* more complicated behaviours.) We must find a root of $x = L(y) = L(L(x)) \equiv L^2(x)$, the superscript 2 here referring to the second *iterate* of L, not its algebraic square. With L quadratic, this is a quartic which however must factorize: both fixed points also satisfy the quartic. This leaves an elementary quadratic $x^2 + x + (a+1) = 0$ which always has real roots for $a < -3/4$, amusingly just after *both* fixed points became unstable. The stability of this real period-2 oscillation is determined by $|D(L^2)| < 1$ at either point of the cycle. Since $L^{2\prime}(x) = L'(L(x))L'(x) = L'(y)L'(x)$, the stability of each point of a cycle is, as it should be, the same. This readily produces $|a+1| < 1/4$, or $-5/4 < a < -3/4$. Thus as a passes through $-3/4$, the stable one of the two fixed points destabilizes and the newly born orbit of period 2 is born stable.

Are other stable orbits present? What other things can happen for different as? Here analysis is of no direct avail, and one thinks graphically. First of all, for $a > -2$ it isn't hard to see that if x is within $(a, a + a^2)$ it must forever remain within this interval. So, at least for $-2 < a < -5/4$ (1) has bounded oscillatory behaviours more complicated than periodic ones of period 2. To try to understand them, I proceeded graphically.

Computer-Generated Physics

Graphically, period 2 requires the solution of $y = L(x)$ and $x = L(y)$, which amounts to the intersection of the graph of L, a parabola, with that same parabola reflected through $y = x$. This makes it immediately clear that if the parabola lies strictly above $y = x$ (i.e. $a > 1/4$) then no such intersection is possible. (The solutions are two complex conjugate pairs.) It is now elementary to see how the period-2 cycle arises and that for a sufficiently negative it must be unstable.

I proceeded to deduce some graphical methods for yet higher-order orbits, which grew harder and harder to understand, but clearly allowed that all orders of orbits were possible, if out of my control. This was discouraging. I didn't know what connection (1) had to genuine differential systems. On the contrary, I chose to study (1) because it was much simpler than the (impossible) analysis of higher-dimensional ODEs. Should (1) do extraordinary things, would a fluid? Who knew? And if I couldn't penetrate (1)'s complexity, then indeed I had nothing.

At this point fate intruded. Carruthers was promised some ten positions. Arriving at Los Alamos in 1973 he discovered that that promise was to be honoured—provided he fired staff on a 1:1 basis. My position in 1974 was one of these created ones. Well, Carruthers had his eye on several other candidates to make way for his new operation. One was Paul Stein, a mathematician, mostly of combinatorics. So I was sent to talk with him and report back my impressions. Paul and I proved to be lucky men.

After more general discussion, I mentioned to him my surprise at the complications I encountered in looking at (1). It turned out that after working with Ulam, in the previous year he had worked with N Metropolis and another Stein, not about (1) which is concave, but rather, upside-down parabolas. (Notice, by replacing $\xi \equiv -x$, (1) becomes $\xi_{n+1} = (-a) - \xi_n^2$, and $-a$ is positive in the places where exciting things happen. Strictly, he had considered $x' = \lambda x(1 - x)$, which for $x = (\lambda/4)(2\xi - 1)$, $-a = (\lambda/4)(\lambda/4 - 1/2)$ is precisely the same.) And here, in a paper called MSS, had made much conceptual progress—although not exactly proven—in understanding that after period 2 became unstable, then stable period 4 appeared, and again and again an orbit destabilized to be replaced by one of double the period, until finally for an $a > -2$, it had infinitely doubled. What happened just past this point, a_∞, was beyond their comprehension, although they also knew that *inter alia*, for $a < a_\infty$, an orderly pattern of complex periodic orbits appear. Not all of this 'proof' stood up to full scrutiny, and a weaker result along these lines turned out to be known to the Russian Sharkovski more than a decade before, although MSS was unaware of it. Even more striking, MSS asserted that this regular ordering wasn't just the birthright of a parabola, but of virtually anything that looked like a 'hump', and so they named this qualitative ordering the 'U-sequence', with 'U' for 'universal'.

All of this discussion helped save Stein's job. It left me altogether troubled, since (1), whose sole virtue was the possible ease of analytical comprehensibility was growing richer in diseased incomprehensible behaviours, leaving the ground quite tenuous underfoot. And so I decided to look elsewhere. A few months later with the new HP65, there was much to divert in learning how to become a master of the discrete.

The story picks up another half year later, July 1975 in Aspen, Colorado, USA at its physics centre. It turned out that the great mathematician S Smale was also in residence. Smale had rejuvenated the turn of the century French subject of dynamical systems, and to my surprise, was also aware of the infinite doublings of the parabola, and wanted to know about various things past a_∞. One comment, an aside for amusement value, quite struck me. He related that in giving such a talk elsewhere, someone in the audience asked if a_∞ was transcendental. And Smale replied with an amused shrug "Gee, I don't know.". What struck me is that the question should have struck someone as interesting. After all, if a derived from λ, however natural the parameter may appear, by a simple change of coordinates it can be turned into anything else— for example to $b_\infty = 1$ exactly. So, could there be any number in the problem that was actually interesting? We shall soon see.

The arrangement at Aspen is that about three people share a many-bedroom apartment. One of my apartment mates, the evening of Smale's talk, convinced himself that the nth term generated by (1) had to be of the form $(an + b)/(cn + d)$. Despite its patent falsity he was not to be discouraged. So, I decided to see if I could work out how these doublings occur, not qualitatively, but with true quantitatively understanding. The new task I took was to determine from (1), which from this point onwards I used in the upside-down form with a positive ($x' = a - x^2$), an equation for the generating function $\hat{x}(z)$

$$\hat{x}(z) = \sum_{0}^{\infty} x_n z^n$$

which for $\{x_n\}$ bounded is analytic within $|z| = 1$. Clearly, for $x_n \equiv x^*$, $\hat{x} = x^*/(1-z)$, so that $\hat{x}$ has a pole at $z = 1$ and the residue determines the fixed point. Also if $x_n \sim x^* + b\gamma^n$ with $\gamma \equiv L'(x^*)$ and $|\gamma| < 1$, the pattern of convergence to a stable x^*, then there is also a pole at γ^{-1}. Since (1) is just quadratic, a formula for $\hat{x}$ involving just a convolution integral can be written down. From this equation, which resembles a unitarity equation of quantum mechanics, one learns that putting a pole into the integral regenerates a pole, but at the square of the original position. This implies that $\hat{x}$ has poles at γ^{-n} for $n = 0, 1, \ldots$ and

$$\hat{x} = \sum_{0}^{\infty} \frac{r_n}{1 - z\gamma^n}.$$

So long as $\gamma > 0$ ($-1/4 < a < 0$) this amounts to a string of positive geometrically spaced poles starting at $z = 1$ and marching off to infinity. Now at $a = 0$, $x^* = 0$, $\gamma = 0$ and the sum formula above is just the pole at 1 plus a constant. The nearest pole to $z = 1$, at γ^{-1}, is now already at ∞, and $\hat{x}$ at this special *superstable* value ($\gamma = 0$ means convergence faster than any geometric) is the pole at 1 plus an entire function. As a now increases above 0, γ is *negative*, and we have poles along the $+x$-axis at γ^{-2n}, $n = 0, 1, \ldots$ and a string of negative poles at γ^{-2n-1}, $n = 0, 1, \ldots$. As a approaches 3/4, $\gamma \to -1$ and the string of poles becomes a cut, and the string of negative poles also becomes a cut with a new singularity on the circle at $z = -1$. Notice that $(r_0 + r_1 z)/(1 - z^2)$ determines x_n to be period 2, with even iterates at r_0 and odd iterates at r_1. We have thus understood why the bifurcation (a change in the type of orbit) is to a doubled period. Moreover for γ sufficiently close to -1, $\hat{x}$ can be approximately computed directly from the convolution equation, verifying that a singularity of square root type appears at $z = -1$.

As we now increase a further the two cuts wither, leaving behind simple poles at ± 1 and a symmetric pair of geometrically spaced strings of poles on either side of the unit circle. As a increases, the period-2 stability $\gamma_2 \to 0$, and a background entire function replaces all but $1/(1 - z^2)$. Increasing a further now has $\gamma_2 < 0$, a square root appears, and so in addition to the reappeared symmetric strings of poles, there is also a new symmetric set on the *imaginary* axis. Again, as $\gamma_2 \to -1$ the poles coalesce into cuts with two new trailing singularities appearing at $\pm i$. This yields

$$(r_0 + r_1 z + r_2 z^2 + r_3 z^3)/(1 - z^4)$$

the beginning of a period-4 orbit. This story now recurs again and again at each doubling with new diagonal lines of poles appearing each time bisecting the angles between the previous lines of poles. At a_∞, the plane is infinitely bisected. Could we compute the residues of the leading $1/(1 - z^{2^n})$ as $n \to \infty$, we would know quantitatively what this infinitely long orbit at a_∞ looks like.

This was my Aspen result. Several things are missing. Basically, a picture has been provided to determine the entire scenario, *provided* we knew that the doublings would persist, and for an infinite number of times. (The renormalization theory had to be completed before any such thing could be proven.) And to know quantitatively how things worked required a solution to the convolution equation, or as recast given the representation as a sum of poles, solutions to curious equations for the residues. (As I learned a year later this sum is called a 'Julia series' after the great French mathematician early in the century who had apparently tried this same route, and certainly no more successfully.)

So, returning home to Los Alamos in August 1975, I retrieved the HP65 and endeavoured to figure out how to solve the residue equations. After a few days of little progress, I decided instead to find a_n at which a stable 2^n cycle appeared, and then from the cycle itself determine the residues, and see if sense could be made out of how they managed to solve their equation. And so, I first calculated that a_n for which the orbit is superstable, which means $x = 0$ is a point on the orbit. a_n is then a zero of $L^{2^n}(0, a)$. (0 must return to 0 after 2^n steps.) Now this is a polynomial in a of degree essentially 2^{2^n}, which for the 10th doubling is about 10^{300}! Not an analytic task. And then which zero is a_n? The answer: it is the smallest zero larger than all those corresponding to period $1, 2, \ldots, 2^{n-1}$ all of which are also zeroes.

Solving for a zero of a high-order polynomial is a Newton's method procedure, requiring the evaluation of the polynomial at many points before one converges to the solution. For the 5th doubling, $L^{2^5} = L^{32}$ requires 32 calculations of L to get one evaluation of the polynomial. $L = a - x^2$ is, of all equivalent ways of writing a quadratic L, that that is fastest to compute, requiring just 2 arithmetic operations, or 64 operations to get one polynomial value. With looping this was already several seconds of HP65 time, or about half a minute for a Newton's method value of a_5. Since the a_n converge to a_∞, they get very close together for large n. And there are other kinds of orbits of length 2^{2^n} just *above* a_∞. With the time for each erroneous computation becoming minutes, it was critical to guess a good starting approximation. And the slowness of the HP65 gave me ample time to study the previous values to guess the next one.

Because by $n = 4$ it was already clear that a_n was converging to a_∞ geometrically. One sees this by noticing that the difference of successive values decreases by a constant ratio, which quickly appeared to be about 5. By $n = 7$ the next term in the series already solved the equation to machine precision, which was exceeded beyond $n = 8$. But the ratio of differences itself was converging, down to 4.669 before precision deteriorated.

Now this was curious and extraordinary. Much more so than the original idea of better understanding the residue equations. What in this florid calculation was providing geometric convergence—that is *scaling*? As we recall from the renormalization group something should become similar to itself but smaller to lead to a successive geometric convergence. This was not at all obvious and its final understanding was much further away than I guessed.

Something else was curious. As I mentioned in regard to Smale's talk, a_∞ is very uninteresting, mercurially changing upon any reparametrization. But not so for 4.669. In taking a difference, any shift of origin is irrelevant. Then in taking a ratio, any linear change of scale is irrelevant. Since a_n or λ_n or whatever, converges, the relation of a to λ locally becomes linear, and so 4.669 must appear in any reparametrization of a and x of the quadratic provided no malice is exercised to have it non-smooth just at a_∞ or on the orbit of x. So, if

Computer-Generated Physics

any number might actually be interesting in this dynamical story it was 4.669.

I was so struck by this that I spent part of the day trying to see if it was close to various simple combinations of numbers and so forth. Nothing at all clear turned up. Nor did for a month or so.

I spent the first week of October visiting CalTech, as usual without the HP65. As I scratched away over the residue equations, I suddenly was taken aback by a memory. Stein had told me the doubling was the same for anything that looked like a bump. In the MSS paper I had looked over almost a year earlier, I suddenly recalled that $x' = \lambda \sin \pi x$ shows identical qualitative behaviour to (1). Now this was a problem. For (1) the residue picture emerged from the convolution equation for $\hat{x}$. Provided the map was polynomial, something similar, but with multiple convolutions would appear. But for transcendental sin x, no such equation for $\hat{x}$ could be written. (It turns out, as I happily realized only some time after this, that it is easy to produce a general theory for the generating functions for the *residues*.) This meant that even though the picture of criss-crossings of lines of poles was very pretty for the quadratic, the doubling phenomena might find it a superfluous idea, and so the picture need not assist at all.

The day I returned home I decided immediately to check if sin x actually doubled. Indeed it did, but at 1 second per trigonometry the wait was painful. I recalled there was an easy way to guess the next value, and by $n = 4$ again realized there was geometric convergence. By my efforts to fit the ratio, the new result settling down to 4.662 smelled familiar. A quick rummage through my drawer resurrected the sheet with 4.669 for the parabola. Without an instant's hesitation I experienced an overwhelming excitement that I had stumbled upon a piece of the godhead.

I immediately called Stein. No, he didn't know that the doublings converged geometrically and was deeply sceptical that a universal *quantitative* entity could exist. I went over to his office to show him the numbers which had him respond with a repressed anger that I had no right to such suppositions based on just three identical figures. But *twelve* figures would convince him.

Nevertheless, I called my parents that evening and told them that I had discovered something truly remarkable, that, when I had understood it, would make me a famous man.

My colleague, Mark Bolsterli, one of the most knowledgeable computer users, gave me a FORTRAN instruction list book to look at, and told me next morning he'd help me onto a serious Los Alamos computer. His few hours of instruction on the system, editor and the easiest way to get output were extraordinary. I had, under my own steam, 4.6692 by the end of the day. This wasn't my limitation. Rather, for naive iteration, 1/3 of the precision of the machine is all one can do. For single precision CDC at 14 significant figures this was it. So, the next day Lucy Carruthers, another computational expert, gave me the crucial pointers on how to use 29 digit CDC double precision arithmetic. Finally, the next day, some four days after the last encounter, I marched into Stein's office with 4.66920160... agreed to 11 figures for four different problems. This time he concurred, and took out his 'dictionary' of numbers—an ordered list of several hundred pages of decimals and what they represent. By the '9' nothing was close.

So far I have told you about how a number I called δ was born. With this as really the sole clue, I knew it foreshadowed an entire world. After all, one had encountered the phenomenon of independence of the universal critical exponents on a starting Hamiltonian. But here, each starting object was itself like a renormalization group equation with fixed point, etc. So whatever underlay δ was something like a renormalization group on renormalization groups. I never had a doubt of the correctness of this view. But two months were to pass before I knew how to unearth it. There was little to do on a computer at this point. I accomplished just one thing in the week after getting δ universal to 11 digits.

Because δ had to depend upon *something*, whereas hyperbolas etc always gave 4.6692..., within a week I realized it depended upon the order of the maximum of the bump. For $x' = a - |x|^z$ there is a $\delta(z)$ for all $z > 1$ which monotonically increases with z, and $\delta(1) = 2$. Indeed all bumps with local behaviour $|x|^z$ have the same δ for the same z, and nothing else matters. For example $x' = \lambda x(1 - x^2)$ has a quadratic extremum, and so $\delta = 4.669\ldots$. Unless there is a special symmetry, a bump is generically a quadratic bump, and so $\delta(2) = 4.6692\ldots$ is all that one ever sees unless one is maliciously careful. In this sense 4.669... is generically universal to dynamics. I know no one had discovered δ prior to myself, and I believe no one else ever independently did so. (S Grossmann published δ before me, but unmentioned in his paper with Thomae, only after he heard my talk at P Martin's August, 1976 Gordon Conference.)

I say this because it illuminates the issue of how hard it is to see. MSS and several others had a computer estimate of a_∞, but had never looked at the a_ns they had had to calculate to get there. a_∞ seemed important, but certainly wasn't. So far as I can tell, had I not had a training that made me eschew computers, had I not so thoroughly enjoyed the computation and 'meaning' of numbers, had I not been a devotee of the renormalization group, had I not thought about residue equations, and had the HP65 not been so excruciatingly slow, I wouldn't have discovered δ either. It is a *sine qua non* of emergent behaviour. But how do you see it if you don't know what it looks like? After all, fate and luck play all too significant a role.

As I said, the full story didn't unravel immediately. How to find a convergence of a parameter which is just one zero of a polynomial of a stupefyingly high degree, when the only vague theoretical ideas were pertinent to the orbit if you *had* the parameter? By the end of December I grew quite distressed that having this jewel drop into my lap, I could elicit nothing of its genealogy.

Elliott Lieb was taking a tour of the theorists at Los Alamos the last working day of 1975. I told him of the various things I had been working on, and he grew very interested when I told him about δ. I also told him I hadn't a significant clue to understand it.

That weekend everything changed. On a Sunday morning I wrote a letter to an old friend in England, sealed it, and decided, uncharacteristically, to have lunch. I had a green chilliburger. With rippled potato chips.

The rippled chip caught my fancy, and I suddenly knew where the theory came from. The winning icon was the multiple ripples of the chip. You see, if you take a high enough upside-down parabola and iterate it you get *three* bumps, two skewed skinny bumps flanking a fatter symmetric middle bump. $L(0)$, the image of the quadratic maximum, is just the parameter a. For a superstable orbit, a is the largest point on the orbit. For L^{2^n}, a_n is a fixed point situated on a very many bumped curve. This far-to-the-right bump, when composed with itself, since $L^{2^{n+1}} = L^{2^n} \circ L^{2^n}$, becomes three bumps, and the new a_{n+1} is a fixed point on the skinny right-flanking bump. a_n is now no longer on any orbit, but if the bump shrinks faster than the a_ns converge, then a_n is approximately the next nearby fixed point. The difference $a_{n+1} - a_n$ is then the size of the rightmost flanking bump, and δ is just the ratio of the size of a flanking bump compared to the size of the one bump that when composed with itself produced the triple of bumps. So, here was something that reproduced itself on a smaller scale that could account for the geometric ratio δ! And better still, no more arbitrary selection of one zero of an impossible polynomial: the solution lay in the orbit itself, even though δ was about a *parameter*.

I returned home, wrote down some approximate equations, a pair of coupled functional equations, and produced a respectable ball-park value of about 5 for δ. I quickly reopened the letter to my friend Tom, reported that I finally had it and went for a hike. The next day Lieb grew altogether excited and from then on championed my work.

Despite the fact that while almost right, all of this was wrong in that there remained too much undetermined, the crucial idea was that the theory was to be expressed as functional equations, and a fixed point emerged in function space. Moreover it was now clear that not just δ, but all the dynamics as well must be quantitatively universal. This was the great jewel.

Without further input, a quadratic bump iterated doesn't make three bumps of unique fixed scale: rather, details of the bump matter. This was then out of any full control. But then I realized that of all the bumps of

L^{2^n}, one was most special. Should L be symmetric, then the fatter middle bump spawned by the symmetric bump at 0 would also be symmetric, and this significantly cut down on the remaining ambiguities.

Indeed the HP65 revealed that the central bump also rescaled with a geometric factor of α, with $\alpha = -2.502\,907\,875\ldots$. Another universal constant and this one for the actual dynamics! For a superstable orbit of length 2^n the separation between $x = 0$ and its nearest orbital neighbour regularly reduced in scale by a ratio of α, greater than 2 in absolute value, whereas the number of points increased by a factor of exactly 2; this meant that at a_∞ a Cantor set was the 'orbit'. Expressed analytically, the nearest point to $x = 0$ is $L^{2^{n-1}}(0)$, and so, $\alpha^n L^{2^{n-1}}(0) \to$ constant. Thus, at least near $x = 0$, $L^{2^n}(x, a_n)$ converges to something universal. This spelled the end of the use of the HP. It meant L^{2^n} must become everywhere universal (at least near the Cantor set), and justified the big dream that dynamics (forget the misguidance of perturbative wisdom) when appropriately complex in behaviour, knew how to perform independently of details. This is, of course, the most extraordinary discovery I have made in my life.

I now needed wholesale powerful computation. A printout of values made it clear that L^{2^n} magnified by α^n converged to a definite (but *very* complicated) function, and that it was universal over Ls: at least in local clusters of arbitrarily many points the dynamics itself was universal. And so, the mad waste of Silent 700 thermal paper.

Contrary to popular belief, graphics was non-existent at Los Alamos in the 1970s. Specifically, no graphics terminals were allowed outside the 'cleared' area, and even there, only the bomb-builders had these big screens, and these with no interactive digitizing capability. So, to get pictures we had to adopt makeshift methods, with the result that good pictures required feet of printout, and things so intricately bumpy as the universal

$$g_1(x) = \lim \alpha^n L^{2^{n-1}}(x/\alpha^n, a_n)$$

required tens of feet. Because, beyond the visual proof of convergence and universality, I was learning to figure out what equation could possess this horror as a solution.

After an extraordinary amount of analytic/computer effort, some two and a half months, every day, 22 hours a day, until I required medical attention in mid-March, I had produced a theory that read

$$g_1(x) = \alpha g_1(g_1(x/\alpha)) + h(x)$$
$$\frac{\delta}{\alpha} h(x) = h(g_1(x/\alpha)) + g_1'(g_1(x/\alpha))h(x/\alpha)$$

as the functional fixed point of an operator on g_1 and h given by the right-hand sides, with the left-hand sides the results. This operator, T^*, is *stable* over simple g_1 and h as single-bumped input, converging towards the fixed point. To reach this point required a delicate consistency argument for the h equation and the development of algorithms to calculate functional composition operators numerically. And much Silent 700 thermal paper to understand the outcome. While the equations are pretty good—some 10% for $|x|^z$ with $z = 2$—they are approximate, growing accurate for larger z.

These equations were crafted to be totally stable. The fact that $g_1(g_1(0)) = 0$ means that g_1 has resolved the 'Cantor' set at a_∞ into pairs of points, with the pair splitting of $x = 0$ and its neighbour scaled to 1. Thus $g_1(g_1(x))$ has resolved the set better with each point now a fixed point. The magnification of this central bump by α, $g_0(x) \equiv \alpha g_1(g_1(x/\alpha))$, has again the same 'size' as g_1 but is different from it in the same way as $a_0 = 0 \Rightarrow -x^2$ is the same as $a_1 = 1 \Rightarrow 1 - x^2$ for the original parabola. (This tells us $\alpha \approx -2$.) This is so because we haven't yet changed a_n to a_{n+1}. At a_{n+1} this approximation to g_0 has doubled back to g_1. It does so by the addition of h which then is a derivative which knows how much a_n had to change. This is precisely what one does by starting with any L and determining the sequence of superstable parameter values $\{a_n\}$. And universality says that this is completely stable in convergence independent of the initial L and whatever its parametrization, of which the derivative is the initial h. The theory is *totally* stable because it embraces dynamical space and parametric space simultaneously. This is unlike any renormalization group theory, because there we converge towards the fixed point only if the parameter (β) is critical. But the β-axis is altogether boring, whereas here, there is an exceptionally rich parametric structure. These dynamical theories are mathematically much more complex than the field theoretic counterparts.

This stance of full stability in a larger space was technically premature and left the theory of my first paper, over 500 copies of which went out in preprint form after April 1976, only approximate, although conceptually essentially complete. Apart from my conviction that parametric behaviour in this problem was crucial, there was a second reason. At each a_n, 2^n points exist on an orbit, and under continual magnification so strong as to always see the points individually resolved, universality is achieved. But only in local clusters. Now at a_∞ the orbit is not finite. Moreover, the full attractor isn't the orbit, but its closure with cardinality, not of the natural numbers, but of the continuum. I found it hard to believe that magnified infinitely less, so that each bump possessed an infinite number of points of the orbit, would still enjoy universality. Moreover, I knew very well that if the h shift were dropped, the g_1 equation would become *unstable*. The second comment is of course true. (I called this first part of the operation T in my first paper, T^* when the parameter shift was included.) The first is wrong and both misguided.

I gave the first talk on this theory 2 May 1976 at the Institute at Princeton. Yorke and Li's paper "Period Three Implies Chaos" had appeared that very week. (Also unknown to these authors, Sharkovski had fully anticipated their result a decade before.) My colleagues were all practicing high-energy physicists, save for Dyson who found the matter very much to his taste. The most important member of the gathering proved to be my friend Cvitanovic who as a committed high-energy theorist had resisted my attempts that previous December to become involved. He now grew very enthusiastic and started playing with this stuff.

Quite erroneously he convinced himself that if the universal function on composition scaled into itself, then the messy extra h was unnecessary, and just $g(x) = \alpha g(g(x/\alpha))$ without the extra coupled h equation should be all that's necessary. Despite my explanations that that equation would be unstable and g couldn't be g_1, he went off with his HP25 and without seeking g determined that α was then really the right α.

So back at Los Alamos, I now had to find the fixed point of T, with $(Tg)(x) = \alpha g(g(x/\alpha))$ which, unstable, could not be obtained recursively. I quickly learned how to do this, verified to 15 figures that α was indeed α. (Unlike Stein, Cvitanovic's 4 figures I knew to be exact.) But then g wasn't g_1 (of course) and indeed universally resolves clusters at infinitely many points (a Cantor set's worth) on each bump of size 1. And now the progenitor of this all, δ, was nowhere to be found.

But then it became clear that g_0 and g are the opposite ends of a sequence of universal functions g_r, $r = 0, \ldots, \infty$ where g_r resolves the orbit in clusters of 2^r points per unit length bump, and the h equation is precisely the eigenvalue problem (h the eigenfunction) for the infinite-dimensional linearization of T about g, i.e. DT. δ is DT's largest eigenvalue, g is $g(x) = \lim \alpha^n L^{2^n}(x/\alpha^n, a_\infty)$, and g is the unstable 'critical' dynamics, achieved by setting the one parameter a to a_∞. Thus δ is the only eigenvalue of DT outside the unit circle, and numerics now yielded δ to 20 figures.

Very much a renormalization group theory with operator T, the whole parameter axis is almost like the β-axis. Except that it is *very* rich in detail. Unlike spin-blocking, where I can organize the partial summations ('decimations') arbitrarily (even infinitesimally), this structure is so rich that it *must* be organized in only the way of T for the theory to have a fixed point with a solitary 'relevant' (outside the unit circle) eigenvalue. It is also the first exact non-trivial example of a renormalization group, where all of the infinitely many coupling constants (an entire *function*) are rigorously included.

The g_rs are a discrete set of points along the β-axis, and easily generalized to a *continuum* of universal gs. (This, the unstable manifold out of g.) When Cvitanovic became convinced that g wasn't g_1 and that rather than "*The* universal function" there were infinitely many, he regarded the state of affairs as quite distasteful and didn't return to these investigations for several more years.

As I already alluded to, the first 'talk' I gave of this finished theory was at the August 1976 Gordon Conference. I say 'talk' because it was under 10 minutes long. It turned out that several people interested in dynamical systems, including several mathematicians were attending this very *physics* oriented meeting. Martin, who ran it, was sceptical of the relevance of more mathematical offerings, but under protest agreed to a session of ten minute talks. The organizers' response was "So what does it have *really* to do with fluids?".

Most interestingly for myself, B Derrida was there, and mysteriously knew about δ. It turned out that Lieb had called Ruelle in Paris in March telling him of my work, and the fact of δ was passed to Y Pomeau, Derrida's thesis advisor. Derrida and Pomeau stayed only in the parameter space, where they significantly improved upon the understanding of MSS beyond all the period doublings. This resulted in two preprints. However, in an act of highest ethical stance, Pomeau asked me in 1977 when my papers would be published so that they could go ahead and publish theirs. I have never again encountered such exemplary behaviour. Their lectures throughout France in 1976–7 had put universal period doubling 'in the air'.

The second and concluding paper laying out all this structure was shipped as preprints in November 1976 which I brought with me for a week-long set of lectures at Brown University. How well was it received? From the Brown lectures, I came away with a prize in my new acquaintance with L Kadanoff, who was instantly enthused and in the early 1980s pushed these ideas into new arenas. Of course, Lieb was very interested.

But this isn't the full story. The first international talk I gave on the finished theory was at a meeting arranged by N Metropolis in September 1976 in Los Alamos. This produced much warmer interest in Dyson, and to my strong pleasure, from Kac. Although I had no proof that T really did have a fixed point—although my numerics converged with an eigenvalue, highly supportive of rigorous truth—and that DT had one eigenvalue outside the unit circle, to Kac this smelled like convincing if not rigorous proof. (Just after I began my talk, Kac asked from the audience "Sir, is this just numerics or a proof?" "It's not quite a proof." "Is it a reasonable man's proof?" "You'll have to decide for yourself." After the talk, he decided it met the 'reasonable man' criterion and "Let Barry Simon prove it!" he said.) But as I was to learn, I gravely outraged many rank and file mathematicians for exposing and being accredited for the jewel, while the real work of actually proving it would garner no one the deserved credit. This proved to be an exceedingly difficult task which D Sullivan finally completed after a good part of a decade's work in the late 1980s. It turns out that very little extant mathematics exists that can penetrate problems as difficult as period doubling.

By 1978 Kac told me that it made brilliant coffee table conversation, but until some connection with any physical thing could be shown, no one would take it as other than a great curiosity. By and large he was exactly right. Not only must one learn to see, but the sight must grow lucent wherever one looks. While many high-energy theorists found this new behaviour of non-perturbative systems highly interesting, this generally wasn't the case for condensed matter physics.

Many mathematicians didn't believe it. R Bowen found it thoroughly uninteresting at the 1976 Gordon Conference, certainly in contrast to J Guckenheimer's quite elegant proofs of part of MSS. In December 1976 in Rochester, although there are reasons that could have then been clarified, J Moser, after I finally explained that my results were metrically, not qualitatively, universal said "Well then you're wrong." and turned his back on me. In Varenna in 1977 R Thom became totally enthusiastic. So responses were very mixed.

But not for refereed journals. Both papers were rejected, the first after a half-year delay. By then, in 1977, over a thousand copies of the first preprint had been shipped. This has been my full experience. Papers on established subjects are immediately accepted. Every novel paper of mine, without exception, has been rejected by the refereeing process. The reader can easily gather that I regard this entire process as a false guardian and wastefully dishonest.

Anyway, Kac was right. Dynamical systems only became 'science' after Libchaber's measurements in the summer of 1979 showed that a fluid can make a transition to turbulence via period doubling with the generic values of α, δ, and g. Martin, with quite unnecessary magnanimity, publicly retracted his 1976 grumbles, and a new arena opened up for the critical phenomenon theorists who had by then largely consumed what remained of open problems in their subject.

Once 1979 ended, non-perturbative dynamical systems found a place in serious physics departments. Indeed, it had become time for me to leave Los Alamos by 1981, and M Gell-Mann offered me a professorship at CalTech.

Feynman, who had 'eaten alive' two potential hires the previous year, was instructed to be on good behaviour during my colloquium interview. The colloquium proved to be the most enjoyable and electric one in my career: it rapidly devolved into a dialogue between Feynman in the front row and myself. After the talk I went up to his office. "You know, I'm envious of you." he said. "Come on, you of all people can't be envious of *me*." "Well, maybe you're right." he rejoined.

During the ensuing discussion, he related that he had spent the week before my arrival calculating my results on his Commodore Pet computer. (Funny, that hidden away at home, one of the masters of numerical calculation had settled for so simple a machine.) Not that this was the first time that he had considered such problems. Indeed, in 1940 (I questioned if not 1941. No, he said, 1940) he took a part-time employ at The Philadelphia Naval Yard, in anticipation of a war effort. Now, it turned out that naval ordinance used mechanical computers to calculate trajectories à la Galileo. In particular, sin x needed to be calculated. Mechanically, in those days, meant gears. And so, one gear had its teeth spaced according to the derivative of sin x, for which, near 90°, the spacing grew so tight that teeth would be sheared off on the meshing gear. Feynman had immediately hit upon the solution. Namely, n gears in a row, each of them performing the functional nth root of sin x. He had then set about to solve this, and over the remainder of his life returned, from time to time, to playing with functional equations. So much for reminiscences.

Following Libchaber's data, it became necessary to understand how a non-invertible 1D map could precisely show up in an infinite-dimensional fluid. Indeed, 1979 was a banner year. Already in 1978 Pomeau and Derrida had noticed that δ was the usual 4.6692... in the 2D Hénon map, when dissipative. Stein and several other mathematicians told me that $y_{n+1} = x_n$, the way in which Hénon's map becomes 2D, was too trivial a way to be in 2D, and all this was stupid. They indeed proved to be very wrong. In early 1979, under insightful hints from J-P Eckmann, W Franceschini had observed 4.6692... in a five-modal truncation of Navier–Stokes. Somehow, the non-invertible humped 1D map, the Poincaré map of *no* set of ODEs or PDEs was showing up everywhere where it didn't belong.

P Hohenberg faults me for not having guessed this. I wish I could have. I didn't see how it could have been. Anyway, informed numerics were now presaging Libchaber's experimental results.

By Cargèse, 1979 Collet and Eckmann had realized how to explain Franceschini's results. This work is wonderful, and dynamical universality with a vengeance. They realized that in the space of a Poincaré map for *any*-dimensional, but *a fortiori*, dissipative system, a simple but ingenious (motivated by what must happen for Hénon's map to exhibit δ) modification of g had T an operator in the larger space contracting to the 1D T with just one unstable direction, exactly determined by DT to be δ. Thus, it could be argued that while truly no one's Poincaré map, it could effectively be anyone's. And it seemed to exercise this potentiality with exceptional frequency and ease. (Why this prevalence is true is beyond our knowledge.) All their theory says is that g can be naturally embedded in arbitrarily high-dimensional spaces, provided that the dynamics is sufficiently dissipative to be able to relax down to but one (locally) preferred direction. It is 'generically' possible for any dissipative system to do so. But we don't know when. We only know if we observe it in some regime of behaviour to perform several period doublings, then the more we see, the more we are sure it will keep doing so, and then, ineluctably in the usual δ-way.

In fact this is profound. 4.669... is a *very* fast convergence rate. As you vary some parameters, you all too readily can have missed an infinite number of bifurcations. It mightn't matter much to you and your paper, but it matters very much to the physical object that experienced it. What has happened is that its orbit has started knotting itself in ways that now renders it susceptible to all sorts of new instabilities. Without the infinite doublings, with each twisting the orbit in yet a new way, it may have been still quite hardy. Infinitely many important things can happen, and you can miss seeing all of them if you don't know over how small a scale to look. An answer calculated by perturbing what you thought was a simple orbit is very different from that of the linked one. And this raises a deep conceptual problem.

It was always assumed that a fluid starts off flowing in a smooth laminar fashion, but then at high enough Reynolds number begins to writhe in turbulent activity. There was some sort of well defined critical Reynolds number, and one day, we should know how to calculate it.

What I have just explained throws all of this into grave doubt. However well-posed the question 'When will this fluid in this geometrical configuration become turbulent?' may seem to be, it now seems a better guess that it is quite ill-posed, and if insisted upon too strenuously could prove to be undecidable in the sense of Gödel. Because careful, *informed* numerics for high-modal approximations to Navier–Stokes exhibit hosts, if not infinite hosts, of infinite cascades of instabilities. And different diseases live concurrently in different parts of phase space. Is it even mathematically possible to calculate our way analytically beyond this maze to some behaviour we shall finally agree upon to be turbulent? And anyway, at small enough scales, the fluid has become highly spatially disordered long before this gross property becomes apparent, leaving questionable the desideratum that spatial disorder is the signal of 'genuine' turbulence.

It is altogether possible that we might have to settle for a simpler request of our mastery of knowledge over unbridled nature. This alternative is that I *won't* tell just how and when bedlam is to be achieved, but rather, if you tell me what peculiarities you're beginning to notice, I'll tell you quite precisely and just how soon other catastrophes will befall you. Anyway, the best knowledge we now have mitigates for this notion of predictability, and not the classical one. We shall see how well we shall do.

There is much more to this story. For example, that Libchaber's data made it clear how g is to be viewed as embedded in a fluid, and quite precisely just what sort of a Cantor set gs dynamical action produces. Again this is a profound difference from critical phenomena. After H_∞ produces its universal critical exponents, little else is left to mention of its dynamical consequences. Here, g, the analogue of H_∞, or g_r because of the vastly varied dynamics arbitrarily near to g, is just the starting point. Having found T's bizarrely complicated fixed point, we must now actually go ahead and determine dynamically what g makes. It makes a fractal. But no simple geometrical one in the spirit of Mandelbrot. Indeed none of those crafted fractals gives us at all the right idea to view gs with. g, and probably all dynamical equations of nature construct fractals with not just an *infinite* number of local scaling rules, but with *infinite* memory. This means the genealogy of the parents, back to the distant past, matter in the sizes of the children. This is *totally* different from the 'Koch snowflake' and its cousins.

Knowing how things tremble, perhaps in the small, and all hidden away from gross view, is a prerequisite to notice the occurrence. The path to this knowledge is crooked, and every means available required to see just one step ahead. But discovering even so little a piece is exhilarating and shot with the ring of genuine truth. And then too, perhaps the old solar system that sang melodiously (but perhaps boringly) was a more beautiful creation than today's, which to the attuned ear emits a dreadful buzz.

Chapter 25

MEDICAL PHYSICS

John R Mallard

Medical physics is concerned with the application of many branches of physics to medicine. Since most medical physicists are concerned in detail with only one or two branches of the field, no single author can possibly do justice to the whole field as is demanded for this chapter. I have therefore asked experts in various fields each to contribute a section, and I am deeply grateful to:

Professor J R Greening, University of Edinburgh, for section 25.5
Professor Julie Denekamp, University of London and **Professor J F Fowler**, University of Wisconsin, for section 25.6
Dr B Heaton, University of Aberdeen, for section 25.7
Dr A L Evans, West of Scotland Health Boards, for section 25.8
Dr R Wytch, Aberdeen Royal Hospitals, for section 25.9
Professor P N T Wells, University of Bristol, for section 25.11

25.1. Introduction: chronological evolution of the many branches of the field

The twentieth century has been an inspirational period for physics applied to medicine because so many branches of physics have made a real contribution to the diagnosis and treatment of disease and the improvement of health worldwide. It is fascinating to note that much of the early impetus to the natural sciences came from medical practitioners. Etymologically the two subjects share a common root, physics being both the science of medicine and of natural philosophy. One might regard the surge of important contributions by physical scientists to clinical and laboratory medicine, beginning in 1895, as being a *quid pro quo*.

In 1600, William Gilbert, who had read medicine at Cambridge and had become Physician to Queen Elizabeth I, carried out a study of

Twentieth Century Physics

Figure 25.1. *A drawing from the Middle Ages which depicts the conceptual relationship between the patient and modern imagers or 'scanners'.*

the attractive force between lodestones. His widely acclaimed book *De Magnete* helped to achieve some recognition and support for the new sciences which began to grow during the Renaissance. He also reported the extension of the principle of attraction to amber, which was the first essay into electricity. These works led us to electromagnetism, thanks to the research of Faraday and James Clerk Maxwell (Professor of Natural Philosophy at Aberdeen 1856–9), and ultimately to magnetic resonance imaging in 1980 (see section 25.13).

Undoubtedly, the Renaissance and the flowering of art at that time heralded the beginning of medical mechanics (see section 25.9). The exquisite background sketches of Leonardo da Vinci, to cite the most well-known example, show clearly the developing ideas of levers related to the anatomy of the musculature of the human body; and the understanding of simple mechanical bodily functions acted as a source of inspiration for the development of early simple machines and aids for the disabled. A seventeenth century drawing (figure 25.1) even predicts the basic conceptual relationship between the patient and modern body-scanners!

Medical physics itself began to be established in the nineteenth century. One of those of influence was a Scottish physician, Neil Arnott: he was educated in Aberdeen, and had a busy practice in London: in his spare time he gave courses of lectures on natural philosophy and on medical physics. These were enormously popular and formed the basis of his textbook, *Elements of Physics*, published in two volumes in 1825 and 1827. In this book he drew freely on anatomy, physiology and medical practice to illuminate the ideas and methods of physics. There are many other examples of physicians contributing to physics, e.g. J Black (1728–99), T Young (1773–1829), W H Wollaston (1766–1828) and H Helmholtz (1821–94).

However, it was not until the last few years of the nineteenth century that the three discoveries that account for the great majority of the modern applications of medical physics were made. These were the discovery of x-rays by W C Röntgen, Professor of Physics in Wurzburg, Germany, in 1895; of radioactivity by Becquerel in Paris in 1896; and of the electron by J J Thomson in Cambridge in 1897. X-rays have since been used to produce images of the inside of the human body—previously invisible (see section 25.2)—and for the treatment of disease, particularly cancer (see section 25.3). Radionuclides have been used for both of these purposes (see sections 25.4 and 25.10), and in many branches of medical research as 'tracers'. Electrons have provided the basis for the currrent widespread use of electro-medical instrumentation (see section 25.8).

See also p 56

See also p 4

Figure 25.2 shows the year at which each of the fourteen or more branches of today's medical physics and engineering began effectively in the hospitals. It can be seen that medical physics expanded throughout the century as new clinical uses were discovered for the newly emerging branches of physics, and the technology resulting from them.

The use of x-rays to image the structures, organs and tissues inside the body quickly led to the medical profession of diagnostic radiology. x-rays were also used, together with the radiation emitted by radioactive substances, from early in the century to treat malignant tumours, using radiation-induced ionization to kill malignant cells and thereby prevent or reduce the growth of the tumour. This new medical profession—radiotherapy—was fully established by the 1930s. This method of treatment required more precise methods of measuring the amount of radiation being delivered to the tumour (in contrast to the early crude estimates based on the skin redness that developed during treatment), and physicists began to be appointed to leading hospitals from 1912 onwards to develop the techniques of radiation dosimetry (see section 25.5) and to use it directly for the benefit of the patient. By 1939 there were about 30 radiation dosimetrists in the UK. In parallel came the realization that the interaction between ionizing radiation and living tissue was a very complex phenomenon, and radiation biophysics or radiation biology was born (see section 25.6). As the biological effects became more fully understood, the necessity to limit the unplanned and indiscriminate use of ionizing radiation led to the birth of the field of radiation protection (see section 25.7).

Twentieth Century Physics

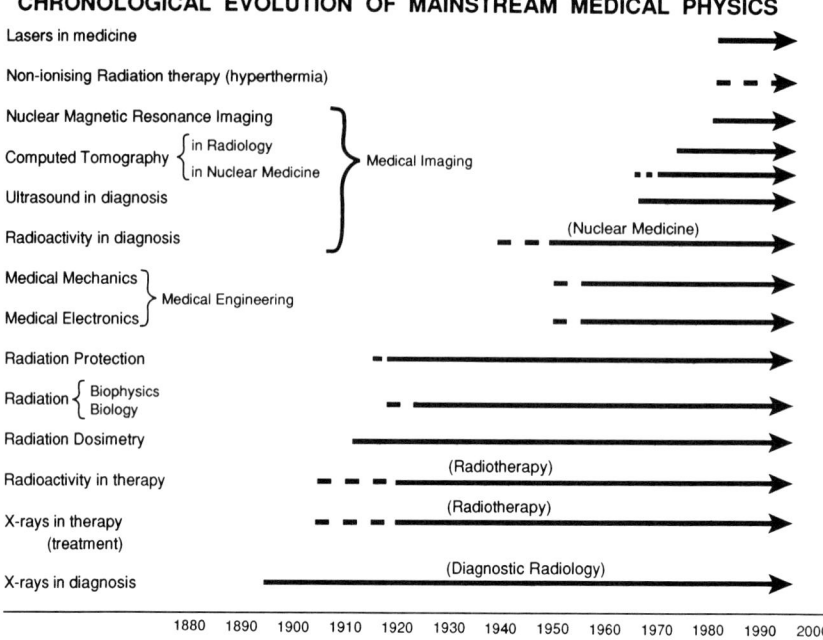

Figure 25.2. *The chronological evolution of mainstream medical physics. The oldest branches are at the bottom, and the most recent additions near the top. Broken lines show periods when the technique is primarily experimental in its application to patients.*

Following World War II there was a tremendous growth in the number, and in the range, of applications of physics to medicine. Artificial radioactivity led not only to large sources of Co-60 for radiotherapy, but also to the administration of safe, trace quantities for diagnostic purposes: the radioactivity becomes localized in an organ (such as the thyroid gland) or a tumour by virtue of the biochemical behaviour of the radioisotope. By early in the 1950s simple methods to image tumours and organs were evolving, and by the mid-1960s 'isotope scanners' were becoming common, to be replaced in the 1970s by gamma cameras. The 'hospital' or medical physicist played a pioneering role in these developments, and the field became known as nuclear medicine (see section 25.10).

The advent of the digital computer led to the digitization of images, which, in turn, made it possible to build up an image of the cross section of the body from imaging information obtained from a series of angles around the outside of the body: this is now known as computed tomography, and was first applied to radionuclide images in the late 1960s and then to x-rays in the early 1970s (see section 25.12). These techniques led to very significant improvements in image contrast and considerable gains in diagnostic accuracy.

Adopting the principles of sonar, ultrasound scanners were developed in the mid-1960s and quickly became a standard method of imaging the foetus in the uterus and a powerful tool for identifying problems during pregnancy—virtually every mother in the developed world has the first glimpse of her child on an ultrasound scanner display. Ultrasound is now a standard technique in radiology for a wide range of diagnostic problems (see section 25.11).

The potential of electron and nuclear magnetic resonance techniques in medicine began to be realized in the early 1970s, and, from 1980, nuclear magnetic resonance imaging rapidly became accepted as a very powerful addition to the diagnostic armoury of the radiologist (see section 25.13).

In contrast to the early days of the twentieth century, when only ionizing radiations were used, virtually every part of the electromagnetic spectrum is now in use in hospitals (see figure 25.3). Medical physics at the end of the twentieth century is a thriving profession which is still expanding.

25.2. X-rays in diagnosis: diagnostic radiology

This account of twentieth century medical physics must begin in 1895, because ionizing radiation provides a large part of the *modus vivendi* of modern medical physics. The present-day collaboration between scientists and medical practitioners began when the famous anatomist Van Koelliker volunteered his hand for the first radiograph taken with Röntgen's new invisible rays, accidentally discovered in the late evening of 8 November 1895. X-rays were very quickly adapted to realize the long-held dream of looking into the human body without opening it. The first clinical radiograph was taken on 13 January 1896, only 66 days after Röntgen's initial discovery. Within a year a thousand articles relating to x-rays were published and organizations of scientists and physicians were being founded to spread the knowledge of and to develop the uses of these new 'miraculous' rays. Notable amongst these was the Röntgen Society, formed in London in 1897.

The Victorians were remarkably quick to exploit the new discovery. X-ray tubes could be bought from an optician in Aberdeen in 1898 [1], and a portable x-ray set was pedalled from hospital to hospital in Berlin in 1906 [2]. Following early use in the Sudan and South America (1898), the widespread and invaluable use of x-rays in World War I earned the lifelong gratitude of innumerable wounded soldiers. As a result, diagnostic radiology quickly became an established profession for medically qualified doctors, who elected in increasing numbers to specialize in the use of these invisible rays. In 1919, one radiologist working for two hours on three afternoons a week coped with all the radiological work at Charing Cross Hospital, London: now there are many specialists working full time. In due course, radiography became

Twentieth Century Physics

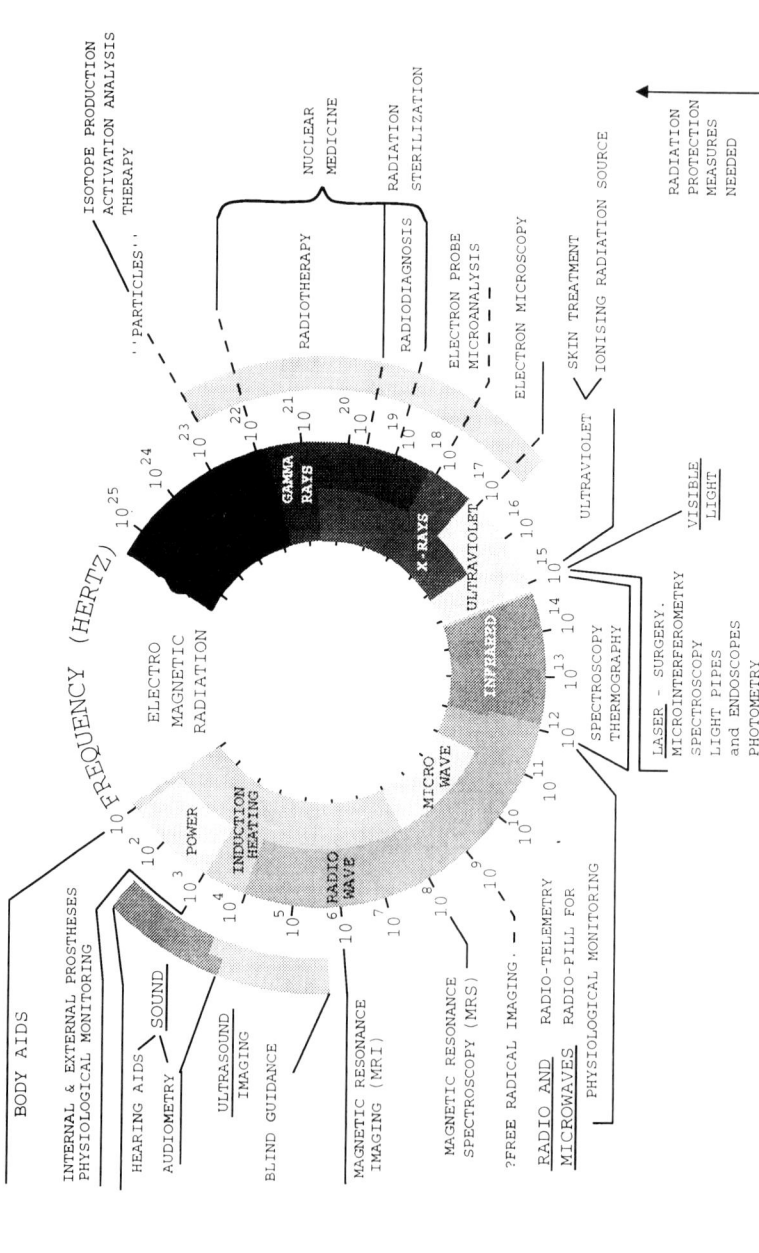

Figure 25.3. *The medical uses of the various parts of the electromagnetic spectrum and the longitudinal wave spectrum. Whereas only the ionizing radiation sections were used in the first half of the century [2], now every part is used.*

a profession for the technical staff who operated the x-ray equipment. Between the two wars, the medical uses of x-rays began to diverge into diagnosis (radiology) and the treatment of tumours (radiotherapy—see section 25.3). The Röntgen Society became the British Institute of Radiology in 1923, bringing together all those associated with ionizing radiations—scientists, clinicians, technicians and industrialists. Röntgen refused to patent his discovery; won the Nobel Prize for it in 1901; and died in 1923 of intestinal cancer, penniless during the economic slump and hyper-inflation in Germany during the early 1920s.

Over the years, the images improved, but not dramatically so. The gradual improvements, which were due almost entirely to improved engineering in the generating, irradiating and recording equipment, were made primarily by physicists and engineers in what became a thriving x-ray industry, responding to the demands of, and collaborating with, the radiologists in the hospitals. Early x-ray tubes—gas tubes excited with a spark induction coil and interrupter—were quickly replaced by the filament-heated cathode high-vacuum tube developed by W D Coolidge of GEC, New York in 1913. The target then became a tungsten button set in a block of copper thermally connected to a radiator. This was generally replaced in the 1930s by rotating anode tubes, first explored by Coolidge in 1915, to dissipate the heat generated by the electron bombardment over a bigger surface area of tungsten target and make possible a much increased intensity of x-ray beam without melting. A modern x-ray set (figure 25.4) uses rotating anode tubes with oil cooling, beams of several different kilovoltages (to allow optimization of the beam energy and penetration for the particular clinical purpose) and a focused electron beam, with several sizes of focal spot on the anode (the smaller the focal spot that can be tolerated, the better the definition of the image for a particular clinical purpose). Such machines, used together with improved photographic films, fluorescent screens and grids of lead strips (set radially to the x-ray source and moved between the patient and the film (Potter–Bucky Diaphragm) to reduce radiation scattered by the patient's tissues) have resulted in the beautifully detailed images that the radiologist takes for granted today.

By the 1930s hospitals had special rooms equipped for diagnostic x-ray work. These contained not only the specialist equipment, but were also provided with radiation-shielded walls and lead-screened cubicles from which the radiographer operated the exposure remotely and was protected from much of the ionizing radiation. These rooms changed little until the advent of more complex equipment such as image intensifiers in the 1960s and 1970s.

Laborious measurements by physicists showed clearly why x-rays were so good at showing up body structures at the x-ray quality of diagnostic radiology (< 150 kVp). Bone has a much higher attenuation coefficient because it contains a high proportion of calcium ($Z = 20$),

Twentieth Century Physics

Figure 25.4. *A modern x-ray tube for diagnostic work up to 140 kV.*

giving it an effective $Z \approx 14$, and hence a high electron density, with predominantly photoelectric effect absorption ($\propto Z^3$). This may be compared with soft tissue muscle ($Z \approx 7.5$) and fat ($Z \approx 6$), with predominantly Compton effect absorption ($\neq Z$) [3]. Selecting the x-ray beam energy (or kilovoltage applied to the x-ray tube) enables one to maximize the difference in intensity recorded on the photographic emulsion (i.e. the contrast): the bony tissue is seen as a shadow. However, it is difficult to image different adjacent soft tissues with only small differences in absorption and scattering coefficients.

Between the wars, physicists began to provide courses to teach the rudiments of radiation physics and x-ray equipment to student radiologists and radiographers. These courses were usually based at the medical schools. In some cases physicists were employed in them separately from the hospitals, and they also taught physics to the undergraduate medical students. Medical physicists employed in hospitals usually had far less to do with diagnostic radiology than with radiotherapy, where their expertise in dosimetry was paramount and of direct clinical importance. This same knowledge, however, has always been relevant to radiation protection, and as these measures have become more and more important in reducing the dose given to the patient during an investigation (see section 25.7), so the physicist has become more and more involved in diagnostic radiology.

Perhaps the biggest advance between the wars and shortly thereafter was the general use of, and gradual improvement in, contrast media, first observed in Philadelphia in 1897. These are compounds containing elements of high atomic number such as barium or iodine: barium is taken by mouth to highlight the stomach and intestines (the barium meal) or as a barium enema; or injected into the blood stream or a strategic blood vessel to display the 'tree' of major and minor vessels branching from it, the artery being chosen to carry the contrast medium to the organ of study. (The first one showed the veins in the hand after an injection of mercury [4]!) This technique is known as angiography, and in the case of a brain tumour, for example, it often shows as an abnormal displacement of blood vessels or an abnormal distribution of them. These techniques considerably increased the range of clinical problems that could be studied and diagnosed by x-rays. They reached their heyday of usage in the 1950s–80s, using image intensifiers that are now up to 40 cm in diameter for coronary angiography, and include 512×512 bit digital acquisition, 50 frames per second recording, with up to 17 000 images possible. Digital subtraction angiography allows images to be subtracted before and after contrast media, to improve perception and detection of abnormal features. Much work is in progress in commercial laboratories to improve x-ray contrast media, one field of particular interest being the development of tissue-specific materials. However, angiographic procedures are usually associated with delivering quite considerable doses of ionizing radiation to the patient and, in addition, there can be severe toxic reactions to the contrast media in some procedures. Magnetic resonance imaging is gradually supplanting them.

Undoubtedly, the most exciting development in x-radiology in the second half of the century has been the introduction of computed tomography; this is discussed in section 25.12.

See also p 1916

25.3. X-rays in treatment: radiotherapy

At the beginning of the century, just five years after the discoveries of Röntgen and Becquerel, the diagnostic use of x-rays, particularly in fracture clinics, was becoming widespread; the damage from overexposure to radiation had been observed; Madame Curie had separated radium from pitchblende; and the first curing of cancer by x-rays had been claimed. The only treatment for cancer at that time was surgery, which was risky, so here was a new 'scientific' method to be explored.

Irradiation with x-rays—radiotherapy—reduces the growth of malignant tumours or kills them, through ionization of the living tissues by the passage of the radiation, a fraction of which is absorbed. This ionization interferes with the reproduction of cells, disturbing their mitosis. It is an extremely complex interaction, which even now is not fully understood: a new specialty, radiation biology, has emerged to study the phenomenon, and recent advances are now improving the

Twentieth Century Physics

effectiveness of modern radiotherapy. Radiation biology is discussed in section 25.6.

> See also p 1886

This ionization interaction occurs in all living tissues, both normal and abnormal; hence the emergence of the field of radiation protection, which has been very important in limiting the ionizing radiation dosage received by staff, and in monitoring the use of radiation for medical purposes. Radiation protection is discussed in section 25.7.

> See also p 1889

It became clear very early in the century that successful treatment depended upon delivering an accurate amount of radiation to the tumour, and it was realized that physicists, with their conceptual knowledge and experimental skills, were needed within the hospital to provide radiation measurements and to help the physician to deliver the correct dose. The first physicist appointed within a hospital was a Dr Russ, who came to the Middlesex Hospital, London, in 1912 from Rutherford's laboratory at Manchester. Ernest Marsden was appointed in June 1914 in Sheffield, also from Manchester—his salary was £275 a year! He was also to care for the 500 mg of radium bromide that was in the possession of the hospital, and the radon that it produced, also used for tumour treatment. Another towering pioneer was Hopwood at St Bartholomew's, London. At the Memorial Hospital in New York, G Failla was appointed in 1915, and Edith Quimby joined him in 1919. One by one, leading hospitals appointed physicists; by World War II there were about 30 of these pioneers in the UK, and perhaps 40 worldwide. The first of the organizations for physicists in medicine—the Hospital Physicist's Association (now the Institution of Physics and Engineering in Medicine and Biology) was formed in 1943 with 53 members. In the USA contributions to the field by pioneers such as Glasser in Cleveland, Failla and Quimby in New York and Landauer in Chicago led to the formation of local hospital physics groups, which culminated in the formation of the American Association of Physicists in Medicine in 1958.

By 1923, an x-ray set operating at 200 kV was installed at Guy's Hospital, London, which, although rudimentary, had all the necessary components of a treatment machine: the high-voltage generator was separate at the back of the room; the evacuated tube was unshielded, but a collimator to define the emergent beam of radiation was used. However, the performance of the sealed-off tubes was poor above 170 kV, and continuously evacuated tubes were used in Sheffield in 1933, as part of the drive for higher voltage equipment of greater reliability [5]. The rooms had to be unusually large for a hospital in those days, and a very popular therapy set became the Newton Victor Maximar which had both the generator and the tube and collimator all in one container. Although it only operated at 140 kV, this was accepted as being sufficiently penetrating to treat fairly superficial tumours, and even though the container was rather unwieldy, it was simple to instal in a medium-sized room. This set was developed in the late 1930s and

Medical Physics

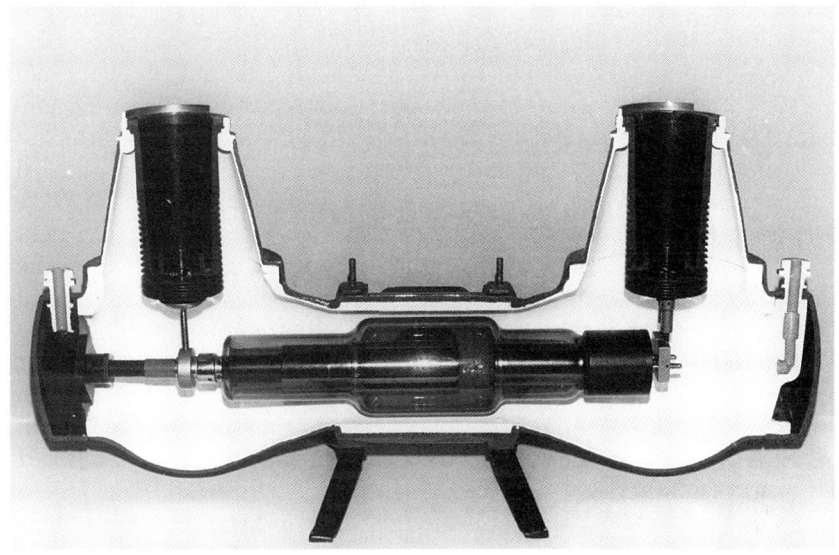

Figure 25.5. *An x-ray tube for radiotherapy at 250 kVp.*

remained in use until the middle 1950s when the main work-horses of most radiotherapy departments became 250 kV machines.

X-ray tubes developed for radiotherapy were very different from those developed for radiology. A therapy tube had a stationary anode which was a massive copper block, in which the tungsten target was embedded, to cope with the kilowatts of heat generated by the bombarding electron beam (figure 25.5). The tube was suspended in oil, which was, in turn, water cooled.

Between the wars, careful measurements by the pioneer hospital physicists established how the intensity of the x-ray beams from the sets varied across the beam and with distance from the anode and depth in the patient. A tank of water was used to copy the absorption and scattering of soft tissue in a patient: the resulting curves are called isodose curves in phantom (figure 25.6). These curves express the result of many different physical phenomena: i.e. the fall-off of intensity of the primary beam of radiation with distance from the anode, which is not a point source; the absorption of the primary beam resulting from interactions with living tissues, which are mainly due to the photoelectric effect and the Compton effect; the scattering of radiation in all directions, which is mainly due to the Compton effect; and the penumbra at the side of the beam arising from the imperfect collimation of the primary beam and radiation scattered from inside the beam.

Also, during this period, the quality of the beam, which depends on the maximum kilovoltage of the electron beam in the tube and its

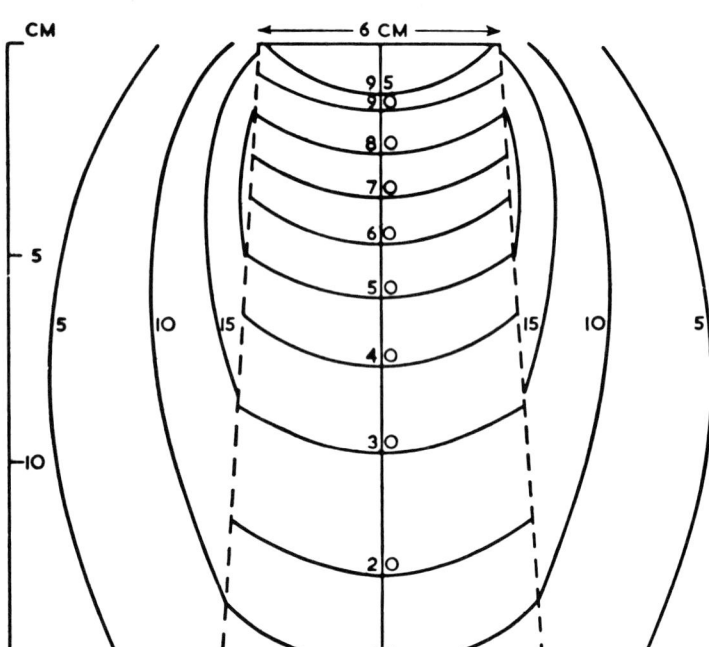

Figure 25.6. *Radiation isodose curves in a water phantom at 220 kVp for a radiation field defined by a collimator with a 6 cm by 10 cm aperture. The x-ray tube focus-to-skin distance (FSD) is 50 cm and the quality of the radiation beam is defined by its half-value layer (HVL) of 1.75 mm of copper.*

energy distribution, began to be described in terms of the thickness of a metal filter which reduced its intensity to one-half [6]. This 'half-value layer' was expressed in millimetres of aluminium for low-energy beams up to about 100 kV or so, and millimetres of copper for higher energies. Materials other than water, such as wood, were used for the phantom, and a 'bolus' of rice, dough or wax was used. Materials which were more 'tissue-equivalent', containing fillers with high atomic number, were developed; examples were 'Lincolnshire bolus' [7] and Mix D wax [8].

The 100% intensity of the beam was measured with a small ionization chamber (thimble chamber) at the surface of the phantom on the central axis of the beam, and included the radiation backscattered from the water tank. Compton's paper on x-ray scattering did not appear until 1923 [9], and the definition of the unit of quantity of radiation—the röntgen—in terms of the ionization of air was not made by the International Commission of Radiation Units until 1928 [10], following a proposal by Duane (of the 1922 Duane–Hunt law of effective wavelength), Professor of Biophysics at Harvard. Standard free-air chambers were produced, and in the early 1930s comparisons were made between standards developed at the National Physical Laboratory (NPL) in the UK and in Germany and the USA. The pioneers, such as Sievert at the Karolinska Institute, Stockholm, Mayneord in London, and Spiers in Leeds, produced a practical chamber—the thimble ionization chamber (early ones were used with a gold-leaf electroscope)—which could be used to measure the radiation in a water phantom, from which the central-axis depth-doses could be determined and accurate isodose curves could be drawn. These defined the treatment beam and allowed a meaningful treatment plan to be evolved. Mayneord and Lamerton's survey of depth-dose data [11], which became a classic of radiotherapy literature, was followed by the work of Quimby in 1944, and the *British Journal of Radiology* (Supplement No 5) [12] was an authoritative compilation of the best available data of the time. National Standards Laboratories offered calibration services, so that the thimble chambers could be accurately related to the free-air values.

The relationship between the ionization in the air in the chamber and the ionization energy deposited in the living tissues has led to the use of several different units over the years; these are discussed in section 25.5. The SI unit of dose in radiotherapy is the gray, named after one of the pioneers, Harold Gray, who entered the field from the Cavendish Laboratory, Cambridge, and whose name also adorns the famous radiobiology laboratory at the Mount Vernon Hospital, London.

The effectiveness of radiotherapy treatment depends upon maximizing the radiation dose to the tumour while minimizing the dose to all the surrounding normal tissues. As radiotherapy developed, multi-beam techniques, in which radiation is incident upon the tumour from several different directions, were developed. Each beam added radiation dose to the tumour, while spreading the dose received by the normal tissues between the tumour and the surface of the patient. Gradually, the practice emerged of developing a radiation treatment plan for each patient; this showed the distribution of radiation in the particular body cross section containing the tumour (figure 25.7). The diagram shows a typical six-field treatment plan at 250 kVp for a tumour in the bladder during the 1950s. At this time, the author did many of these drawings and calculations for each patient, laboriously by hand! In order to minimize the dose of radiation to normal tissue, therapy sets have also been used

William Valentine Mayneord

(British, 1902–88)

William Valentine (Val) Mayneord was born in 1902 and graduated at Birmingham University in 1921, going to St Bartholomew's Hospital, London under Professor Hopwood, where he remained until 1927. He then became physicist at the Royal Cancer Hospital, London (now the Royal Marsden), becoming Professor of Physics applied to Medicine in 1940, and retiring in 1964.

His scientific career included original contributions in a very wide range of subjects, including the first paper reporting chemical carcinogenesis; pioneering work in x-ray and radionuclide dosimetry; pioneering work in radionuclide scanning; extensive studies of environmental radioactivity and the mathematical analysis of radiation carcinogenesis. He participated in many national and international organizations devoted to scientific advance in medicine, and to the safe, effective use of ionizing radiation. He was the first President of the International Organization of Medical Physics, and received many awards and honours, including becoming a CBE in 1957 and Fellow of the Royal Society in 1965. His knowledge of Renaissance art and literature was legendary, and he became a Member of the Scientific Committee of the National Gallery, and a Trustee.

He was a very friendly man with a great sense of humour and great courtesy, but capable of kindly criticism of unworthy behaviour or work. A generation of his students and colleagues have followed in his wake, stimulated by his example.

with a rotational movement, so that the tube moves in an arc centred on the tumour. This is known as rotational therapy: first implemented in 1915 [1], such treatments have usually been regarded as more of esoteric interest than of general use.

Louis Harold Gray

(British, 1905–65)

Louis Harold (Hal) Gray was born in 1905 and graduated from Trinity College, Cambridge in 1927, then working on the absorption of gamma-rays in matter in the magic circle of Rutherford at the Cavendish Laboratory. The Bragg-Gray principle followed, with the concept of absorbed dose, which led eventually to the international unit of radiation dose being called the 'Gray'.

He went to Mount Vernon Hospital, London, in 1933, being amongst the first hospital physicists to be appointed without medical school teaching, and with others, built the first neutron generator for radiobiology in a wooden hut. Using broad bean roots, he developed accurate quantitative methods for testing hypotheses of the action of radiation. After a period at the Medical Research Council Radiotherapeutic Research Unit at Hammersmith Hospital, where he left a legacy of radiobiology and radioisotope research, he returned to Mount Vernon to found the great Radiobiology Research Unit (which is now known as the Gray Laboratory), until his death in 1965. He was a Founder Member of the Hospital Physicists Association, being its Chairman in 1946-7, and President of the British Institute of Radiology in 1950. He was a fellow of the Royal Society, an Honorary Member of the American Radium Society, and Vice-Chairman of the International Commission of Radiation Units, who now award a Gray Medal. The Gray Trust set up by the HPA, BIR and the Association of Radiation Research, sponsors a biennial Gray Conference.

Hal Gray was a man of outstanding intellect and energy. His enthusiasm, kindly interest, resonant voice and laugh are legendary, and he earned the respect and affection of all those who worked with him. He was one of the 'Greats' of this century.

Twentieth Century Physics

(a)

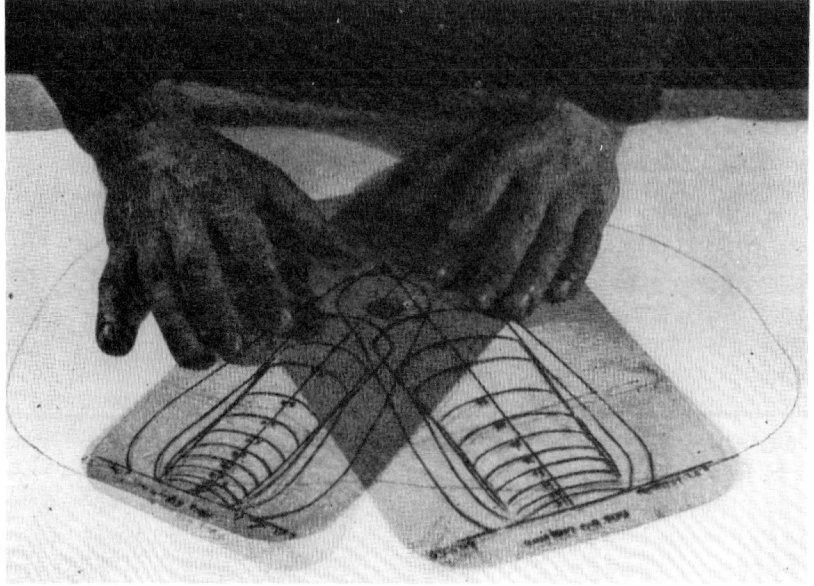

(b)

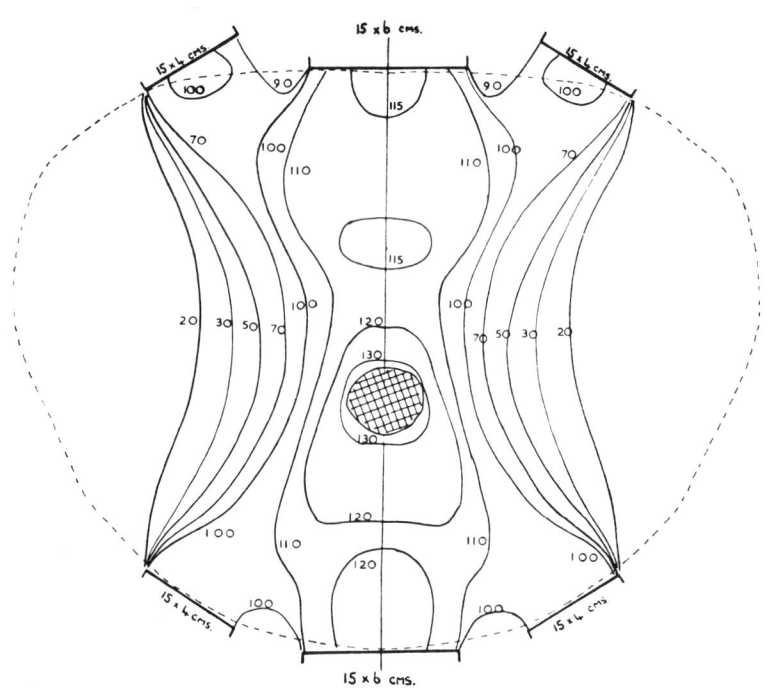

Figure 25.7. (a) The addition of two isodose curves on a body and tumour outline. (b) Final six-field treatment plan at 250 kVp for a bladder tumour, typical of 1950s radiotherapy treatment.

In 1945 most physics services in hospitals consisted of x-ray dose-rate measurements, x-ray treatment planning for radiotherapy patients (usually of the order of 1000 or more new patients a year) and the care of radium and supervision of radium treatments. Hospital physicists were also concerned with radiation protection room design for both radiology and radiotherapy departments. Very few independent hospital physics departments of several staff existed: many of the physics services were provided by physics staff in the medical schools. For the more effective treatment of tumours deep in the body, it was clear that better depth-doses were needed, and a move to higher voltage treatment machines, producing more penetrating radiation, began.

The pioneering supervoltage radiotherapy machines began to be installed just before World War II. Notable amongst these were the Medical Research Council's 500 kV van de Graaff generator at Hammersmith Hospital and the 1 MV continuously evacuated generator at St Bartholomew's Hospital (opened by Rutherford), both built by staff of Metropolitan Vickers Company. In the USA, D W Kerst created the betatron in 1940 at the University of Illinois, his original machine accelerating electrons to 2 MeV, but it was not until 1953 that a 'medical' betatron, manufactured by the Allis-Chalmers Company, with appropriate positioning facilities, was used at the Sloan-Kettering Memorial Hospital, New York. The objective was to generate more penetrating radiation to give a better depth dose to deep-seated tumours (see figure 25.8). The physicists associated with these installations were concerned not only with making these experimental nuclear physics machines work in a clinical environment, which was not an easy task, but also with exploring the changed biological interactions at these considerably higher energies, where the biological effects of a given radiation dose could be expected to be different (see section 25.6).

A quite different attempt to improve the biological effects of radiation on malignant tumours was the use of neutrons. Early attempts in the USA with neutrons were no more successful than the x-ray methods, but further work was performed in the UK in the 1960s and 1970s by the Medical Research Council at Hammersmith Hospital and in Edinburgh, using cyclotrons to generate neutrons by bombarding beryllium targets. Following the experimental irradiation of pigs, in order to verify dose-effect relationships by Fowler (see section 25.6), patients were treated with neutron beams, and early results looked promising, but the type and location of the tumours successfully treated was limited. In general, the use of neutrons is not being continued, but experiments in the neutron irradiation of brain tumours which have concentrated boron-enriched materials are taking place.

The advent of the nuclear reactor led to plentiful supplies of artificial radionuclides in the 1950s. Sources of cobalt-60 (with a half-life of 6 years, emitting 1.2 MeV radiation, eventually reaching several kilocuries),

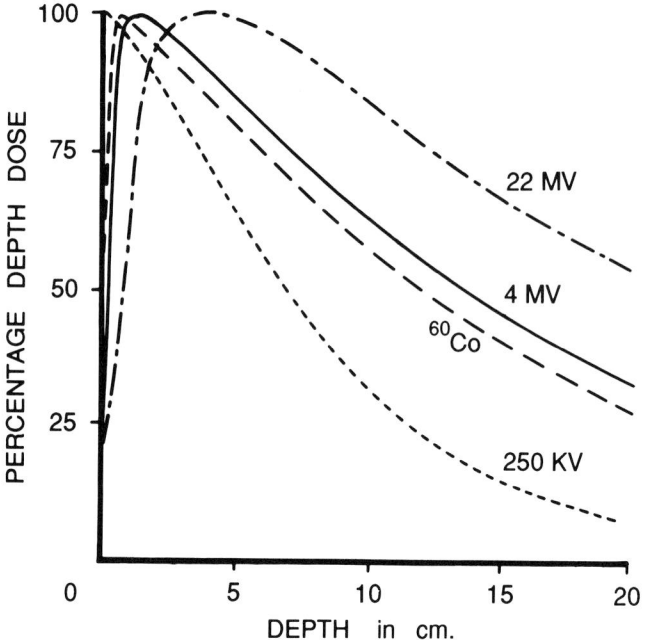

Figure 25.8. *Central-axis depth-dose (CADD) curves for 250 kVp, Co-60 beam unit, 4 MV and 22 MV radiation for typical radiotherapy field sizes and focus–skin distances.*

popularly nicknamed the 'cobalt-bomb', became the work-horse of most radiotherapy departments in the 1960s and 1970s. The first machines contained sources made by Mayneord for neutron irradiation in a reactor, and one of these led to the establishment of telecobalt therapy by Johns in Saskatchewan in 1951 [13]. It is seen (figure 25.8) that higher percentage doses at depth are obtained, because of the greater penetration of the beam and also the greater distance between the source and the patient which became possible. Also wedge fields could be used more than at 250 kV: these are achieved by placing in the beam a wedge of absorbing metal, such as copper, to give a much lower radiation intensity on the side of the beam with the greater thickness of copper than on the other side: they were first developed by Miller at Sheffield in 1944 for 200 kV [14], but at this energy, outputs and depth-dose are too low for them to be truly effective. Wedge fields are of great value in treating tumours in corner sites, e.g. in the head, when two fields at right angles on the corner give a uniform dose distribution to the tumour, while irradiating very little normal tissue (see also figure 25.10 below).

By the 1960s the measurement of radiation-dose distributions in a phantom was being automated, with mechanisms to move the ionization chamber around in the tank of water, which served as the body phantom, used to copy the absorption and scattering of radiation. The movement of the chamber was controlled from outside the treatment room, and its position monitored. Later, automatic mechanisms known as isodose plotters were developed to plot the isodose curves directly.

Before preparing a treatment plan, the radiotherapist must determine precisely the location, shape and size of the tumour; this is not easy. In the case of a brain tumour, for example, x-ray angiographs would previously have been used, perhaps with an isotope brain scan. In the 1990s, x-ray computer tomography (CT) images are used, perhaps in conjunction with magnetic resonance imaging (MRI) and other methods. The directions of beam entry, and the use or otherwise of wedges, are decided in discussion with the hospital physicist. A technician would formerly have drawn up the isodose treatment plan using photographic transparencies of the individual beam isodose curves (see figure 25.7(a): in the 1990s this is all carried out using a computer-based treatment planner.

Figure 25.8 shows typical central-axis depth-dose curves for 250 kVp, 1.2 MeV, 4 MV and 22 MV x-rays. As well as showing the greatly increased penetration and higher percentage dose at depth for the 4 MV and higher-energy radiation, it also shows a very low dose at the surface, so that the maximum dose (100%) is a few millimetres below the surface. This important build-up effect is the result of the much larger number of forwardly directed secondary electrons from Compton interactions which occur immediately beneath the surface of the patient, the intensity of which builds up with depth in the first superficial layers of tissue. This effect becomes much more pronounced at higher beam energies, which give rise to more secondary electrons in the forward direction. This effect results in a low dose delivered to the skin, which leads to a welcome reduction in the reaction of the skin to the treatment: skin reactions are very painful and distressing to the patient, and the traditional severe skin 'burns' of the old 250 kV radiotherapy were a dreadful disadvantage and limited the radiation dose that could be delivered to the tumour.

War-time radar led to the development of microwave electron linear accelerators which have made possible modern radiotherapy treatments of tumours with megavoltage x-rays. The first one was at Hammersmith Hospital, operating at 8 MV, built by Metropolitan Vickers and installed in 1952. The author can remember his boss (L Mussell) taking part in the design of the isocentric couch, which ensured that the radiation was always directed at the tumour whatever the angle of the gantry, and can also remember the first day of treatment with this wonderful new machine! This was an exciting and idealistic time, when scientists and industrialists were anxious to adapt the scientific developments of war to more humane applications. One company developing a therapeutic linear accelerator had their work described as its 'ticket to heaven'. However, with the exception of the leading radiotherapy departments, the Co-60 therapy unit remained the work-horse of most radiotherapy

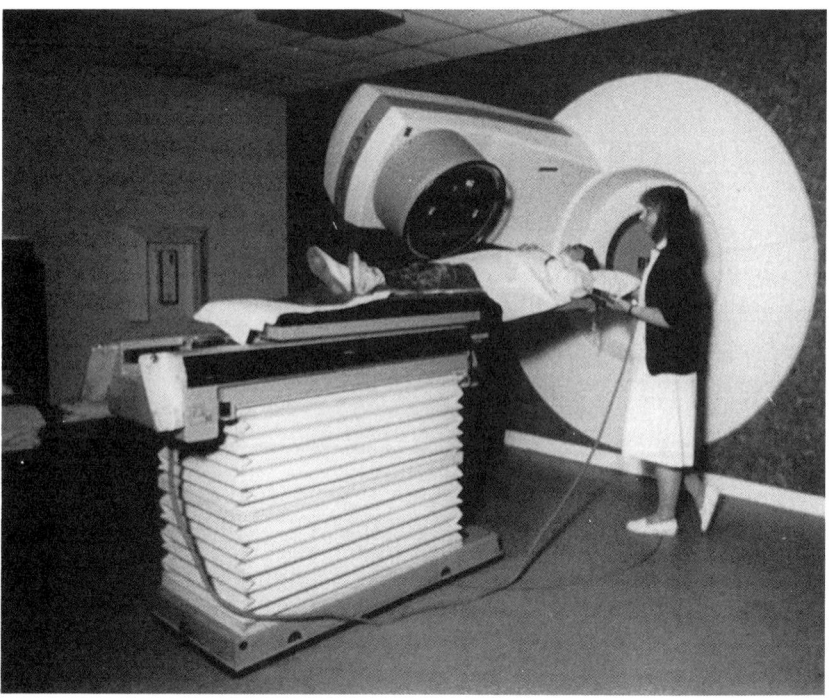

Figure 25.9. *A modern linear accelerator for radiotherapy up to 20 MV. This machine was used for the treatment shown in figure 25.10.*

departments until the 1980s, when linear accelerators became much more reliable and versatile in the hostile clinical environment, partly because of computer-controlled systems. This was also a time when the disadvantages of using cobalt-60 became generally appreciated. These were the problems of day-to-day source manipulation—almost every radiotherapy department has stories of what they had to do when the source 'stuck' between its lead-shielded safe and the treatment head, or when the safety shield would not rotate into the safe position!—and the storage of highly radioactive sources beyond their useful life, which became an embarrassment. Fundamentally, the difficulty is that the radiation from radioactive sources cannot be turned off at will, like an x-ray generator. Most leading hospitals are now equipped with 6–10 MV accelerators, and machines up to 20 MV are now common: these have a selection of energies for the optimum treatment of different depths and sites. Together with the very high-intensity output, the linear electron accelerator has become the machine of choice, such that it is now the work-horse of modern radiotherapy departments, particularly for the treatment of deep-seated tumours. The author entered hospital physics in 1951 when the clinical linear accelerator was just evolving, and he finds it difficult to believe that the end of the century will mark its fiftieth anniversary!

Medical Physics

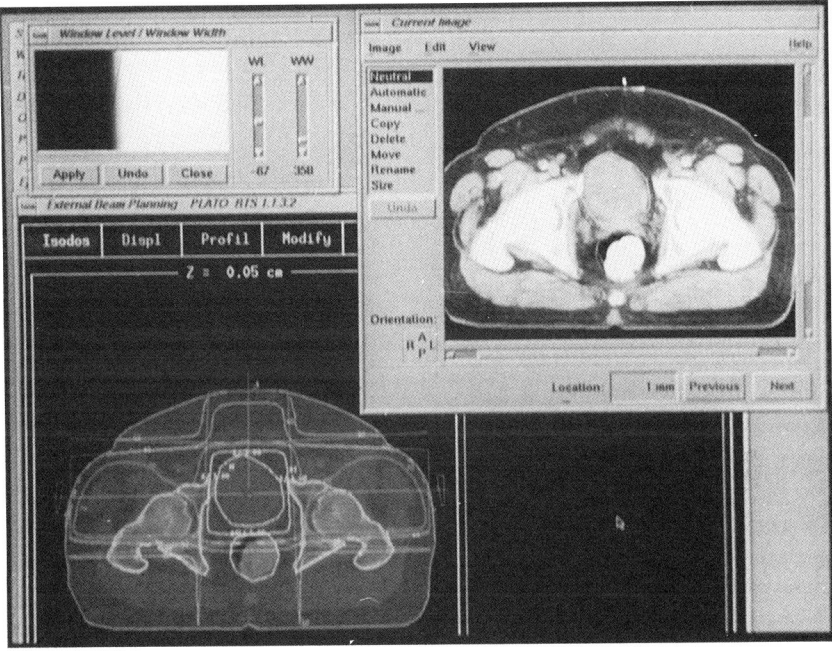

Figure 25.10. *Treatment plan for a bladder tumour using 16 MV x-rays from the linear accelerator shown in figure 25.9. Note that the two fields from the side are wedge fields.*

Many clinical linear accelerators also have the capability of delivering an emergent electron beam instead of x-ray photons. Electron radiotherapy is believed to be advantageous for some treatments, but most centres with this facility only treat a small proportion of patients in this way.

Treatment planning is now all performed by computerized treatment-planning systems. Figure 25.10 shows a plan for treatment of a bladder tumour at 16 MV. The much improved radiation distribution compared with earlier treatments at lower energies can be seen in the very rapid fall-off of intensity outside the tumour volume, and the lower dose delivered to surrounding normal tissues. Of particular note is the fact that the spinal cord is spared unnecessary irradiation by the use of two wedge fields at the side. The tumour receives a very high dose compared with normal tissue, and this, combined with the greater skin-sparing of the higher energy, provides a considerable improvement in safety and comfort for the patient. Currently there is much interest in developing and using three-dimensional plans on a daily basis, with the emphasis on the treatment of the whole tumour volume and little else.

Radiotherapy is gradually becoming much more accepted as a treatment of choice and is being used more frequently, even to the extent that radiotherapy is used instead of radical surgery for early tumours in sites where surgery is notoriously difficult, e.g. the larynx. This much more favourable position is the result of step-by-step contributions from physicists in hospitals and in industry throughout the century. Combined with the fact that tumours are now being found much earlier as a result of improvements in diagnostic methods (which will be discussed in sections 25.10, 25.11, 25.12 and 25.13), the prognosis for a considerable proportion of patients with early malignancy is much brighter than it would have been early in the century.

Following successful treatments of eye tumours with protons of 20 MeV or so at a few centres, it will be interesting to see the clinical results from the Heavy-Ion Medical Accelerator at Chiba, Japan, which will accelerate ions of carbon, neon and silicon to energies between 100 and 800 MeV: these particles have a peak of intense ionization at the end of their path—the Bragg peak, and the energy will be tailored to place this peak at the site of the deep-seated tumour.

25.4. Radioactivity in therapy

Within three months of the discovery of x-rays, radioactivity was discovered virtually simultaneously by Becquerel in Paris and Silvanus Thompson in London, by virtue of the detection of another invisible ray, the gamma-ray. Becquerel is usually credited with this discovery, but his paper appeared rapidly while Thompson's was held up by consultations with the Royal Society! The medical potential of radioactivity was exploited more slowly. Although it was used in the treatment of cancer early in the twentieth century, its diagnostic applications were not developed until the late 1940s (see section 25.10). These applications have developed into the important modern clinical specialty of nuclear medicine.

The Curies discovered radium in pitchblende in 1898, but the production of 100 mg of it, enough for a patient treatment, in 1900 required half a ton of pitchblende to be processed: there is a magnificent photograph of Madame Curie and others standing beside a huge vat of pitchblende in the early 1900s [1]. The diseases first treated with radium were superficial ones: for example, birthmarks at the London Hospital in 1909. Radium tubes with walls of glass, silver or gold were first used in 1910. Radon—or emanation, as it was originally known—is a gas with the advantageous half-life of 3.8 days, compared with the 1620 years of radium. It was first used as an alternative to radium in 1908 in the USA by Duane, and in 1909 in England by Jordan with the radon in small sealed glass or metal containers.

Madame Curie, as Director of the Red Cross Radiology Service during World War I, was responsible for putting on the road 20 'radiological cars' fitted with x-ray sets, nicknamed 'petits Curies', one of which she operated herself; and for the establishment of 200 radiological posts, which had examined more than a million men by the end of the war [1].

The Medical Research Council (MRC) was set up in the UK in 1920, and one of its first tasks was to manage the 2.5 g of radium released from war service, then valued at £72 500. It was to be used for medical research, particularly into the treatment of cancer, and the MRC arranged for it to be distributed amongst nine centres. In 1929 the National Radium Commission established 12 radium centres, to be followed by 10 more. Each centre was to have a full-time physicist with an adequately equipped laboratory and workshop: this helped tremendously to establish physicists in hospitals in the UK, particularly outside London, where the London Medical Schools had previously led the way.

The early development of radium teletherapy machines—machines with the source at a distance—began in 1919 at the Middlesex Hospital, London with 2.5 g of radium provided by the Ministry of Munitions; there were similar developments in Paris, New York and Stockholm. Later designs, often called 'radium bombs', were of higher activity (up to 10 g) and could be used with a variety of radiation beam sizes. Radiation protection measures had to be strengthened: in one unit, made by Bryant Symons Ltd, which was standard equipment for treating head and neck cancers from the late 1930s to the 1960s, the source was housed in a separate lead safe several metres from the teletherapy unit, the source travelling between them by pneumatic transfer.

Although many radium teletherapy units existed during the period 1930–60, the bulk of the work with radium was for surface, interstitial, and intracavitary use in which tumours were treated with radium needles and tubes, and radon 'seeds'. This is now known as brachytherapy (from the Greek *brakhus*, 'short' or 'near to'). Long-handled forceps, lead-shielded benches and lead carrying pots were used, and staff were recommended to have days off and frequent holidays of a month or more! Treatment with radium needles or radon seeds involved their implantation into the tissues to be treated (interstitial therapy, figure 25.11) in accordance with a plan for each individual patient (plans were designed to deliver the required dose to the volume of the tumour as uniformly as possible). Several systems were developed to achieve this, with 'rules' for optimum needle size, activity, separation and geometrical distribution. The most widely adopted system was devised by Parker, a physicist, and Paterson, a radiotherapist, at the famous Christie Institute, Manchester in the mid-1930s [15, 16], based on earlier work by Sievert and Mayneord. The system was based on laborious theoretical calculations of the radiation-exposure distributions around radium sources of simple geometrical shapes, such as a point, line, disc, sphere and cylinder: uniformity of dose was taken as ±10%.

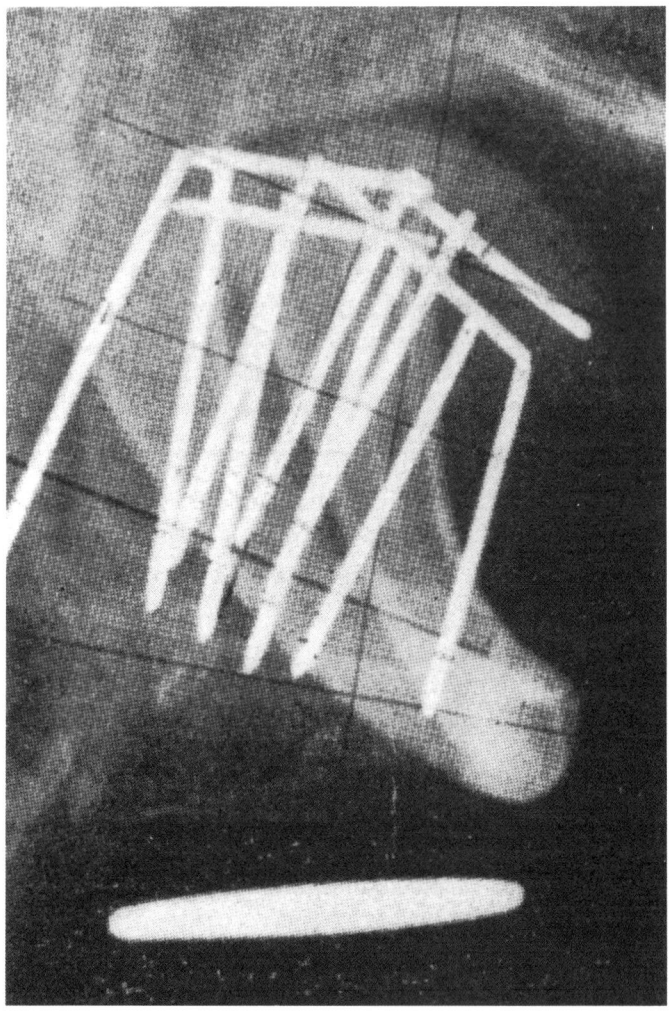

Figure 25.11. *An interstitial radium implant. This is an x-radiograph of a two-plane implant of radium needles which sandwich a tumour of the tongue. To carry out the radiation dose-rate calculations at significant positions in the tissues, it was necessary to reconstruct a 3D model of the implant from stereographic radiographs.*

Another system was that developed by Quimby in New York. On innumerable occasions in the early 1950s, the author calculated into the early hours, aided by pairs of 3D radiographs, in order to report to the therapist the next morning the removal times of the radium 'implants' from that afternoon's theatre session of patients!

Radium needles and tubes were also used in 'moulds' of wax applied to the surface of the skin above the superficial tumour to be treated; and in cork (Paris) or rubber (Manchester) ellipsoids inserted into the vagina, together with a central radium tube inserted into the uterus, for the treatment of cancer of the cervix, which was prevalent.

The outbreak of World War II caused much anxiety within the radium physics community in the UK. Emergency measures were taken to protect radium stocks from bombing raids: for example, in Aberdeen the whole stock of radium was deposited in a cavity at the bottom of a deep granite quarry, and, when it became clear that radiotherapy ought to be restarted, another quarry was found where the radon could be pumped off and supplied not only to the local centre but throughout Scotland and northern England [17]. The more usual method was to put the radium into a bore-hole on the hospital site, which depended on the sub-strata, or to transfer it to more remote sites.

Radium sources have now virtually disappeared from the radiotherapy armamentarium. The nuclear energy programme gave an impetus to the use of artificially produced radioactivity in medicine, and safer and more efficacious substitutes for radium have been introduced, one by one [18]. The very long half-life of radium (1620 years) and the high energy of its gamma-rays (some greater than 1 MeV), presents a greater potential hazard than radioactivity of shorter half-life and lower-energy emissions. In addition, frequent leakage tests had to be carried out on radium sources, because the daughter of radium is the gas radon. The platinum sheath encapsulating the radium suffered much mechanical handling, so that fracture damage and leakage was not uncommon, requiring re-encapsulation. Handling procedures now used for Cs-137 ($t^{1/2} = 30$ years and 0.662 MeV gamma energy) or Ir-192 ($t^{1/2} = 74$ d and 0.613 MeV gamma energy) result in much reduced radiation dose to the handling staff, which includes clinicians, physicists, technicians and nurses. The Manchester system of dosimetry calculations can be applied to these implants, with appropriate modifications. For interstitial implants, radon has been replaced by 2.5 mm gold grains (Au-198: $t^{1/2} = 2.7$ d, 0.412 MeV gamma-ray), and for the special task of destroying the small pituitary gland, pellets of sintered Y-90 oxide emitting 2 MeV beta-rays have been used [19]. Nevertheless, brachytherapy has become less common and is limited to head, neck, breast and gynaecological cancers.

For some treatments with radioactivity, 'afterloading' techniques are now used: empty containers are placed in the optimum positions and the radioactive sources are loaded remotely. Remotely controlled afterloading is popular for treating gynaecological cancer using Co-60 or Cs-137 sources with high dose rates, and equipment designed to facilitate this is available commercially. This technology has led to a renaissance in brachytherapy at other sites, such as the lung (with Ir-192). Advantages include short treatment times and the ready availability of dose distribution software. For teletherapy units radium has been replaced by Co-60 ($t^{1/2} = 5.26$ years, 1.17 and 1.33 MeV gamma-rays) sources which can be produced with high specific activity, so that the source is much smaller physically. These 'cobalt bombs' were discussed in section 25.3.

25.5. Radiation dosimetry

From the very earliest days of the century x-rays and the radiations from radium were being used for the treatment of cancer (and, for better or worse, for the treatment of other diseases as well). Initially a great deal of empiricism was involved, but it soon became apparent that the 'dose' of these radiations had to be controlled within far finer limits than the doses of other medicaments. It was this need for accurate measurement and for the development of the concept of radiation 'dose' that first led to the entry of physicists into a medical environment in appreciable numbers.

Many of the laws of physics, for example all the conservation laws, are special cases of the everyday saying 'You do not get something for nothing'. It was recognized that radiations could not produce biological effects unless some of the radiation energy was absorbed by the biological material. It was a natural thought to define radiation dose in terms of energy absorption per unit volume or unit mass [20, 21]. Unfortunately the amount of radiation used in a daily treatment of a patient would raise the temperature of the tissue by about 0.0005 °C and if this were to be measured with sufficient accuracy the measuring system would have to detect a temperature rise of a few millionths of a °C!

25.5.1. Ionization methods

Physicists turned in other directions. A single alpha particle from radium could produce 100 000 ion pairs when dissipating its energy in a gas. A gamma-ray could release an electron that could produce 10 000 ion pairs, and even a low energy x-ray photon such as is used in radiography could indirectly produce 1000 ion pairs in a gas. The substantial electric charge released by a single ionizing particle, coupled with the sensitivity of even early electroscopes and electrometers, made ionization methods very attractive for the quantative measurement of these radiations.

If ionization methods were to be used, a unit defined in terms of ionization was needed. Villard [22] suggested as a unit for a rather vague 'quantity' of x-rays the amount that would produce 1 esu of electric charge in 1 cm^3 of air under standard conditions of temperature and pressure. This ionization had to be produced in such a way that only air was involved in the absorption of the x-rays and the secondary electrons that produced the ionization. This led to the development at standards laboratories of so-called free-air ionization chambers.

25.5.2. Radium dosimetry

Radium was used for treatment in the form of so-called radium needles. These consisted of hollow tubes of platinum–iridium alloy into which thin-walled tubes containing radium salt were loaded and sealed. Early measurements of the ionization produced in air by known quantities of radium under standard conditions had been made by Eve [23]. However, the initial dosimetry of radium sources for clinical purposes owed much to the Swedish physicist Sievert who, in a series of papers in the 1920s and early 1930s, e.g. [24], calculated the radiation distribution around various distributions of radium sources, also providing tables of integrals (named after him) that allowed for the oblique filtration of the radium in the platinum–iridium sheath. Others, in particular Parker working in Manchester, investigated how best to distribute radium sources in order to produce clinically desirable radiation fields. Parker's data were combined with the clinical judgement of Paterson to produce a simple system of dosimetry that was widely used [25, 26]

25.5.3. Developments in ionization dosimetry
Until the 1930s x-ray and radium dosimetry had followed largely separate paths. New developments were to draw them together.

The First International Congress of Radiology was held in 1925 and set up what is now known as the International Commission on Radiation Units and Measurements (ICRU). The existence of these free-air chambers enabled the ICRU in 1928 to define a quantity called the röntgen to be used with x-radiation that was close to Villard's suggested definition of 1908. In 1937 the ICRU extended the definition to include the gamma-rays from radium. This gave rise to heroic experiments at standards laboratories to measure radium gamma-rays with free-air chambers. At the UK NPL [27] a chamber was constructed having electrodes separated by three metres as the secondary electrons from radium gamma-rays had ranges of metres. At the US National Bureau of Standards (NBS) [28] a smaller chamber containing air at high pressure was constructed (to reduce the range of the electrons).

The ICRU [29] tidied up the definition of the röntgen. Until that time the quantity being measured and the unit of measurement had not been separated. Now a quantity exposure was defined as a measure of x- or gamma-radiation based upon its ability to produce ionization with the röntgen as a unit of exposure (To be precise from 1957 until 1962 the quantity was called exposure dose). The ICRU [30] defined a quantity called absorbed dose which was the (radiation) energy absorbed per unit mass at the point of interest with a unit, the rad, being 100 erg g^{-1}, but in doing so it recognized that ionization methods would need to be used to determine the absorbed dose.

Free-air chambers were of little use for day to day measurements in the field. For this purpose so-called thimble chambers were employed. The name derived from the size and shape of many of them. These chambers had a wall material with an atomic composition as close as possible to that of air and were calibrated against the free-air chambers.

The röntgen required the production of a certain charge per unit mass of air. If the energy, W, needed to release an ion pair in air was known the energy per unit mass of air was known. The energy absorbed per unit mass of some other material placed at the point where the number of röntgens was known could be calculated if the ratio of the mass–energy absorption coefficients of the material and air was known for the x-rays concerned. So the physicists had to measure or calculate the spectral distributions of the x-rays they were using and prepare compilations of the mass–energy absorption coefficients of the materials of interest over a range of energies. Fortunately the ratio of the mass–energy absorption coefficients of air to those of water or soft tissues did not vary rapidly with photon energy so spectral distributions did not have to be accurately known and simpler criteria of x-ray quality could be used. It looked as if the problem of determining energy absorbed per unit mass was solved, at least in principle, but life is not that easy!

The development of the cavity magnetron for radar in World War II made possible the construction of linear accelerators producing electrons with energies (initially) of 4 MeV, and by the mid-1950s these accelerators were in use in hospitals for the treatment of deep-seated tumours. In addition atomic energy programmes made possible the production of large quantities of Co-60 which were put into special containers to produce beams of gamma-rays for medical use. The heroic efforts needed to try to measure radium gamma-rays in röntgens with free-air chambers has already been indicated above. There was no way this would be possible for 4 MeV x-rays. What was the way forward?

Gray [31, 32] showed how the measurement of the energy absorbed by irradiated materials could be achieved if a small gas-filled cavity was introduced into the material of interest. Bragg [33] had earlier presented a qualitative discussion and the theory is called the Bragg–Gray theory. In modern terminology it states that

$$D = sJW/e$$

where D is the radiation energy imparted per unit mass of the cavity wall, s is the ratio of the mass stopping powers of the wall and gas for the charged particles crossing the cavity, J is the charge per unit mass of the cavity gas, W is the energy required to release an ion pair in the gas and e is the electronic charge (the innocent-looking s on the right-hand side of the equation was to engage the attention of physicists for the next 50 years). By equating this expression for D with that derived from an exposure measurement, standards laboratories were able to see that it was possible to produce a cavity-chamber standard of exposure. To do this they had to use a material (graphite was chosen) for which the stopping powers and the mass–energy absorption coefficients relative to air were close to unity and accurately known.

They also had to measure the cavity volume accurately [34]. The NPL used x-rays from a 2 MV van de Graaff generator and other standards laboratories used Co-60 radiation with these cavity exposure standards. At higher radiation energies the concept of exposure broke down owing to difficulty in producing electronic equilibrium and uncertainty in the point of measurement.

Particle accelerators of various kinds were developed to produce beams of x-rays, electrons, protons and neutrons of ever increasing energy to be employed in the treatment of malignant disease. Each extra radiation or increase in energy presented medical physicists with a new challenge. The dosimetry of all these radiations relied on ionization systems that were calibrated against cavity-chamber standards as indicated in the previous paragraph. The calibration was extended to the necessary higher energies using the Bragg–Gray theory, which had been refined over the years by a succession of workers.

25.5.4. Calorimetry

The direct measurement of absorbed dose by calorimetry was a kind of Holy Grail for dosimetrists. As indicated above, the temperature changes to be measured were extremely small and required not only a sensitive measuring system but also extremely careful thermal insulation of the detecting element from its surroundings. Calorimeters were constructed that were rather like a Russian doll in that a thermal screen had a thermal screen inside it and within that thermal screen was the detecting element. All these layers needed to be thermally insulated from each other and yet needed electrical connections for temperature detectors and for calibrating heaters. Most calorimeters were made of graphite and thus determined absorbed dose in that element. The first such calorimeter of reasonable accuracy was that of Genna and Laughlin [35], but the problems were such that it was not until the 1990s that a standards laboratory used a calorimeter as its preferred standard of absorbed dose.

Water, a simple substance with a close approximation to the composition of the more complex material soft tissue, was really the ideal medium in which to determine absorbed dose. Everyone had assumed that the direct calorimetric determination of absorbed dose in water would be invalidated by convection currents. Everyone that is until Domen [36]. He showed that convection was far smaller than had been assumed and also that it could be further reduced by appropriate experimental design. His calorimeter was extremely simple requiring not much more than the placing of a tiny thermistor at the point in a tank of water where the absorbed dose was to be measured (it was not *quite* that simple). Unfortunately the radiation being measured reacts with the water, producing chemical changes which can be exo- or endothermic and the heat of reaction affects the measured result by a few per cent.

The precise nature of this heat defect has yet to be determined, but when it is satisfactorily resolved calorimetry should be opened up to a much greater number of radiation workers.

25.5.5. Chemical dosimetry

Another system used for the dosimetry of high-energy photons and electrons was the oxidation of ferrous sulphate in dilute sulphuric acid solution to ferric sulphate. The solution was about 96% water and was thus well suited to the determination of absorbed dose in water. This system had first been developed by Fricke and his co-workers, e.g. [37]. The ferric ions were measured by ultraviolet spectrophotometry and the energy required for their production by either ionization methods or calorimetry. In the 1970s discrepancies were noticed between the energy required to produce a ferric ion using multimillion volt electrons on the one hand and multimillion volt x-rays on the other. As in the latter case it was again electrons that deposited energy in the dosimeter solution, this was an anomaly in need of explanation, and modifications were found to be needed in the recommended methods of photon dosimetry.

25.5.6. Other methods

Literally hundreds of other chemical reactions have been suggested for dosimetry but very few have found application outside the laboratory of initial use. Other systems use photographic films, thermoluminescence, lyoluminescence, radio-photoluminescence, scintillation detectors, Geiger counters, vacuum chambers, production of electron–hole pairs in silicon and ultraviolet absorption in a wide variety of plastics, but they all need calibration against one of the more fundamental systems discussed above.

The dosimetry of internal radionuclides, either medically administered or adventitiously inhaled or ingested, is discussed in sections 25.7 and 25.10, the problems lying more in the fields of physiology and biochemistry than in physics.

Also for reasons of space there has been no discussion of the problems faced and advances made by medical physicists in transferring the fundamental concepts discussed above to the day-to-day determination of radiation doses to individual patients. These are dealt with in sections 25.3 and 25.4.

The contents of this section are more fully discussed, with scientific detail if required, in a book by Greening [38].

25.6. Radiation biophysics: radiobiology

Radiation biophysics, or radiobiology to give it a broader title, is of immense interest in the fields of cancer induction (carcinogenesis) and cancer treatment (radiotherapy). Because radiation causes damage to the DNA (genetic material) of cells it can cause cell death if the dose is high enough, or a mutation if the damage is slight enough to allow the cell to survive, but with an alteration to its genetic code. Physicists have played a very important role in the evolution of this subject, because of their ingenuity in devising techniques to measure radiation dose and by encouraging their biological colleagues to be very quantitative in their assessment of injury. Some have transmuted and become leading biophysicists and even radiobiologists themselves. This has allowed very precise determination of 'dose response curves', i.e. the relationship between dose of radiation and the amount of damage inflicted. The field has attracted engineers and physicists with a desire to contribute to medicine or basic biological science, and who have the talents and willingness to collaborate or even cross the difficult language and concept barriers of the so-called 'hard' physical and 'soft' biological sciences.

Radiation energy, when deposited in cells, disrupts molecular bonds along the track of each particle. The effect varies with different types of radiation depending upon the density of ionizations, expressed as linear energy transfer (LET). As more and more sophisticated accelerators have been built, biophysicists have gained great insight into the nature of the critical lesions in cells by varying the spatial and temporal distribution of ionizing events, i.e. LET and dose rate. The critical lesion is now known to be a double strand break in the DNA, but not all such breaks are lethal, since many can be repaired. The vast majority of lesions (hundredfold more) are single strand breaks, but these are not serious because the cell can repair them all with great accuracy, using the opposite strand as a template.

Radiation damage is not caused linearly with dose: there is a region at low doses where the cell-killing effect is relatively ineffective (the shoulder) followed by a region where it becomes much more damaging, giving an approximately exponential decrease in cell number with increasing dose. Much of the excitement for biophysicists has been in the interpretation of the mechanisms underlying these non-linear dose-response curves, and in interpreting from them how to make occupational exposure to low doses as safe as possible. After almost half a century of biophysical modelling, there are still heated debates about the interpretation of the relationship between ionizations inflicted and the ultimate injury. There appears to be a linear and a quadratic effect. Is the linear component due to a double strand break from a single track traversal and the quadratic term from two separate tracks happening to occur close enough to interact? Or is all damage due only to track ends, where the ionization density is greatest and the curvilinear response is due to depletion of repair enzymes at higher doses? Why can the cell tolerate some, but not all double strand breaks (DSBs)? These questions and their consequences provide the stimulus for current research activities.

25.6.1. Latency of expression of injury

The thermal energy that is imparted in a lethal dose of radiation (i.e. one that would kill half of a population) is very small, equivalent to a warm cup of tea. This indicates that the target damage (DNA) is very specific. The time at which death occurs is variable, depending upon the natural life span and turnover of cells in different organs/tissues. Huge doses are needed to kill cells outright, i.e. prevent their normal biochemical activities (tens or hundreds of grays). However, a dose of 1–2 Gy can sterilize a cell and make it incapable of further successful divisions. The damage in this case is only expressed when the cells attempt to divide, fail and thereby do not replace the functional differentiated cells. This may occur within a few days in tissues with extensive cell turnover, e.g. those in the gut and hair follicles, or may be delayed weeks (in skin) or months in slowly proliferating tissues, e.g. lung, heart, kidney and most deep organs. Thus, if one wants to describe the LD_{50}, (lethal dose for half the population), one must define whether it is being assessed 1 day, 1 week, 1 month, 1 year, etc, after exposure to the radiation. Often the apparently safe dose decreases as longer observation periods allow more sensitive late responding tissue to express their latent damage.

Time is also important for another reason: the cells and tissues can remove or repair the damaged molecules by chemical, biochemical and physiological processes. This occurs over minutes to hours for biochemistry and days to weeks for physiological compensation. The biophysicist must therefore interface with people from these scientific disciplines, as well as with computer modellers and biostatisticians.

25.6.2. Radiation hazards

The prime risk from low doses of radiation is of mutations which could, in the germ cells, influence future generations, or in somatic cells could give rise to cancers. This is an area of enormous public interest, especially in relation to nuclear power plants, natural radioactivity, e.g. radon, and medical exposure both to patients and to medical staff. Many of the early radiation workers died of radiation-induced cancers, from leukaemias occurring within 5–10 years and solid tumours appearing much later at 20–30 years. This early recognition of its dangers has led to the very effective present control of radiation as an environmental hazard, with physicists providing both the means to monitor dose and the biophysical models to extrapolate and predict the 'maximum permissible doses'. The hazard to future generations is now known to be much less than was at first feared after the nuclear bombs were dropped on Hiroshima and Nagasaki. The lesions produced are probably too severe to be compatible with survival and may lead to abortion of the foetus. Cancer risk estimates are mainly obtained from the survivors of Hiroshima and Nagasaki and of course are limited because of the imprecision in the estimates of the exact dose received by individuals at different places and behind different levels of screening, as well as the speed with which they escaped from the contaminated cities. These data are supplemented by much smaller numbers of individuals exposed in radiation accidents and as time goes on will include data from Chernobyl. Because cancer is such a common disease anyway, affecting about 40% of the population, it is quite difficult to quantify the increase of a few per cent that can be attributed to even a major accident. Opinions are still varied about whether cancer induction is linear with dose, sublinear or even supralinear, because the available data exist only at higher doses than those of interest for setting safe limits for the entire population, or for those occupationally exposed. This is still an active and exciting field of research, with new input now coming from molecular biology techniques. The financial consequences of the linear or non-linear biophysical models are enormous as they dictate the shielding arrangements and clean-up procedures needed in all medical and commercial uses of nuclear establishments, radiation machines or radioactive isotopes.

25.6.3. Radiotherapy

Radiation can kill every single cancer cell in a tumour, but only at the cost of serious damage to the surrounding normal tissues, unless great skill is used. Radiation therapists, radiographers and radiation physicists combine their skills to identify the tumour volume as accurately as possible, then design ways of placing intersecting beams precisely across the tumour, maximizing the dose to tumour and minimizing the dose to surrounding critical structures. Major advances have come from physics and engineering in the design of more sophisticated radiation machines, allowing precise beam localization, excellent dosimetry, multi-leaf flexible collimator design and computer-aided dose planning. In most radiotherapy centres, supervoltage x-rays or electrons are used to give skin sparing and energy deposition at depth. In a few sophisticated centres much more expensive machines are available to deliver protons, π-mesons or heavy charged particles, taking advantage of the Bragg peak to deliver maximum dose as the particles slow down over the last few microns to millimetres. This peak can be scanned across the tumour target volume. In addition to the physical benefits gained from limiting the volume of normal tissue that has to be included in the treatment fields, biophysicists have helped to design and interpret biological experiments which form the scientific basis for the pattern of dose delivery. Experience in the 1930s showed that many small fractions delivered over days or weeks gave much better local tumour control, with less normal tissue injury. It has become common practice to use 2 Gy fractions daily from Monday to Friday until the tolerance dose to the surrounding normal tissue is reached. This is often 60–70 Gy in 6–7 weeks unless a more sensitive organ like lung, kidney, small bowel or spinal cord is included in the treatment field.

Some radiotherapy centres use larger daily doses of 2.5–3 Gy in shorter overall times with similar treatment results. For localized cancers, cure rates of 40–60% are achieved for many common cancers, being greater if early disease is treated and lower for late disease stages. Algorithms have been developed to describe the way the total dose can be altered by varying fraction size, fraction number, the interval between them and the overall time in order to take advantage of biological principles. Again this is an active interface between physicists, biologists and radiotherapists with many exciting new schedules being rigorously tested at present. This must be done in randomized clinical trials, involving several hundred patients receiving the new treatment and several hundred the conventional one. The aim is to prove significant improvements in tumour eradication, without significant increases in late normal tissue damage, which could be life threatening. Each study requires 10–15 years to allow adequate follow up after the necessary patient entry. Such trials require great patience and perseverance as well as many patients, but they are the only way of introducing modified schedules to a treatment which is already capable of curing many patients. It is of course crucial for such long-term studies and for patient safety that the biophysical algorithms are as accurate as possible. The most revolutionary new fractionation schedule being tested is CHART, in which three fractions are given each day, including weekends with at least 6 hours between fractions. If this 12 day intensive treatment regime proves superior to the conventional once-a-day, 6–7 week protracted normal practice, the organization and staffing of radiotherapy departments will need radical alteration. The revolutionary aspect is not just organizational, since the therapists are also reducing the radiation dose to 54 Gy from 66 Gy because the new schedule should give greater radiobiological benefit.

Part of the rationale for CHART is that tumour cells can proliferate extremely rapidly (e.g. every day or two) although in an untreated tumour many of the offspring starve to death as the vascular supply cannot keep pace with their growth. The slow conventional treatment over 6–7 weeks was believed safe because external observation of tumour volume growth indicated doubling times of 2–6 months. It is now known this slow volume doubling is the product of rapid cell production and extensive cell loss—again an area where modelling has benefited from biophysicists' input. As soon as treatment commences, the 'potential growth rate' may become more important as the balance between production and natural loss is disturbed. By a very indirect route, physicists have had a major impact in recent decades, since space technology has been used to create flow cytometers measuring features of thousands of cells per minute. These machines deliver a precise stream of cells through a laser beam and can monitor the proliferation status of each cell by quantifying the intensity of stains for DNA content and for the uptake of DNA precursors, e.g. BrdUdR, provided just before a biopsy is taken. This is a superb example of the interdisciplinary fertilization in cancer research.

Thus the physicists' input into radiobiology and its medical application to radiotherapy is multifaceted. Other new developments include incorporating boronated compounds into tumours and then exposing them to epithermal and thermal neutrons to take advantage of the huge neutron capture cross section. Attempts are being made to develop antibodies which would 'home in' on tumour cells and could be used to carry radioisotopes as nuclear warheads. Since the antibodies usually locate on the cell membrane and the critical target is deep in the nucleus, the precise dose distributions of exotic isotopes with α or particle ranges of a few microns are becoming of increasing interest, requiring more biophysics input.

This exciting field of biophysics and radiobiology has evolved over decades showing the hybrid vigour of cross-discipline fertilization and is still vital and dynamic. Amongst the most famous names at this interface, many of those in radiobiology and radiation biophysics originally came from medical physics or radiation physics. The following list of names (in approximate date order of contributions) is far from complete, but it illustrates the close relationship between medical physics, radiation biophysics and radiobiology: Gray, Lea, Pelc, Alper, Boag, Rotblat, Mayneord, Pollard, Hutchinson, Quimby, Failla, Rossi, Waachsmann, Howard-Flanders, Lamerton, Elkind, Whitmore, Neary, Barendsen, Fowler, Bewley, Steel, Oliver, Liversage, Gilbert, Hall, Field, Hendry, Dewey, Curtis, Raju, Vennart, Martin, Skarsgard, McNally, Goodhead, Turesson, Begg, Wheldon, Dale, Bentzen, Brenner, Joiner and others. Their work, and a much more detailed account of radiobiology than is possible in this short section is presented in [39] and [40].

25.7. Radiation protection

In the hundred years since Röntgen discovered x-rays the thrust of radiation protection has in some ways turned a complete circle. Initially the emphasis was on staff safety in radiology but this changed to accommodate the rapidly expanding nuclear weapon and power programmes. In recent years, however, the emphasis has returned to radiology as the contribution patient radiation doses make to the population radiation dose has been seen to be orders of magnitude greater than any other man-made contribution.

Many people, including myself prior to preparing this section, have the impression from anecdotal stories that an appreciation of the risks involved in working with ionizing radiation and the recommended safety precautions were completely absent in the early years and that only over the last 40 years have we really started to apply ourselves to the problem.

This is not the case and reading through early recommendations on the safety precautions one is often struck by how little has actually changed. What they recommended is, in several areas, still the core of that which is recommended today.

The use of x-rays for both diagnosis and therapy and radium for brachytherapy and teletherapy spread rapidly following their discovery and isolation. By 1915 several countries had formed working groups to look at hazards that were clearly apparent. These groups were drawn together at the Second International Congress of Radiology held in Stockholm in 1928. At this congress, following a proposal by Kaye of the National Physical Laboratory in the UK, the first recommendations for x-ray and radium protection were issued [41, 42]. The unit of the röntgen was also adopted, (subsequently formalized by the International Commission on Radiation Units and Measurements in 1934) without which the advances in protection made over the subsequent years could not have taken place. The röntgen allowed the quantification of exposure dose. At the same congress the International X-ray and Radium Protection Commission was set up. This became better known as the International Commission on Radiological Protection with a role that has changed surprisingly little over the years. It is still an independent body and has freedom over appointments to its various committees. It now provides the core documents and recommendations with regard to operational, managerial and dose limitation concerns in all aspects of worker and population exposure to ionizing radiation. These core documents and recommendations are then introduced into the national legislative framework of each country.

Some things have changed surprisingly little over the years and although the detailed proposals were placed before the commission in 1928 the definitive recommendations of the commission were not finally given until 1934 when a tolerance dose for x-rays of 'about 0.2 international röntgens (r) per day' was recommended for persons in normal health. This exposure, based largely on the work of Mutscheller [43] in the USA and Sievert [44] in Sweden, evolved from the consideration that 1/100th of the skin erythema dose received over a month was tolerable. This tolerable dose, aimed at ensuring that no deterministic effects (considered the most hazardous at the time) occurred, was arrived at in a judgmental and consensual way by physicists, without any biological evidence. (Some people hold the view that the same methodology applies today.) The skill of these physicists is evident when one compares it with current skin dose limits and finds it to be approximately the same. They were astute enough to include the statement that 'no similar tolerance dose is at present available in the case of radium gamma-rays'.

Some of the basic protective structural and operational recommendations can be found in current recommendations, for example the recommendations that 'an x-ray operator should on no account expose himself to a direct beam of x-rays'; 'screening examinations should be conducted as rapidly as possible with minimum intensities and apertures'; 'efficient safeguards should be adopted to avoid the omission of a metal filter in x-ray treatment. To this end, some means of continuously measuring the emergent radiation is recommended'. The shielding recommended for room walls, cubicles and lead/rubber gloves and aprons are the same or close to what we would use today. In fact, because of the much softer x-ray beams used then, this thickness of shielding provided a greater attenuation. The same recommendations cover the use of radium brachytherapy treatments, with again the broad operational appreciation of the problems.

During the period up to 1949 the main role of physicists in medicine was in the field of radiotherapy. Their influence on the protection of staff other than in their own group depended very much on their standing and personality. In the absence of routine personnel dosimetry records it is impossible to even speculate on the doses received by some nurses and radiographers. The implementation of the recommendations in diagnostic x-ray departments depended very much on the professional judgement of the radiographers and radiologists. Anecdotal evidence would indicate that despite the sound recommendations that had been laid down, some people were exposed to high dose rates for long periods of time. It is perhaps not unreasonable to speculate that this state of affairs would have continued longer if it had not been for the impetus that the nuclear programmes gave to biological research and epidemiological studies.

By 1955 the increase in leukaemia among atomic bomb survivors was quite clear and the preliminary findings of an increase in leukaemia in patients following some treatments involving x-rays had also been published. The Medical Research Council (MRC) was requested to look at the hazards to man of nuclear and allied radiation. Although the increase in leukaemia noted above was known about, it was the potential increase in genetic mutations, and hence the gonad dose, which was regarded as the most serious long-term effect. As part of their work the MRC instigated, under the chairmanship of Lord Adrian, what was probably the first national study of the doses given during diagnostic x-ray examinations. The various reports [45–47] of this committee make fascinating reading. The conclusions with regard to the spread of doses given for the same examinations in different hospitals and the recommended solutions for reducing doses were essentially repeated some thirty years later in an essentially similar survey [48, 49]. During the 1950s and 1960s physicists had little influence on the working practices in many x-ray departments. Although the general recommendations were now presented in the form of Codes of Practice, radiographers and radiologists still often tried out equipment on themselves and x-ray engineers could be found checking that the filament was still working by looking up the cone on the x-ray tube which had no added filtration. Radiation doses were, however, routinely monitored in most hospitals from the mid-1950s. Only towards the end of the 1960s did radiation protection officers start to have a real influence in x-ray departments particularly as the updated codes of practice now started to be much more detailed. The stochastic effects of ionizing radiation were now becoming much more apparent. The success in reducing doses to staff in radiology departments since then can be seen by the very small doses they now receive. Patient doses did not receive other than routine consideration until the work on patient doses over the last ten years highlighted once again the high patient doses that were given in some hospitals. This time however the doses have aroused a keen interest and physicists, together with radiologists and radiographers are currently working together to reduce the doses in those hospitals giving high doses. They now work together in the drawing up of the specifications of new equipment part of which involves looking at the radiation dose the equipment gives to patients. However, the scope for major improvements in patient dose reduction in the future is limited. The current modern digital screening equipment and radiography using the fastest film/screen combinations that are available are working very close to the physical limit of quantum mottle (i.e. the statistical variation in x-ray intensity, hence the film blackening, across the x-ray field or radiograph) making further dose reduction unlikely.

The availability of cobalt-60, caesium-137 and iridium-192 from the early 1950s meant that radium could be progressively replaced in radiotherapy departments. New technology meant that afterloading systems could be introduced, eventually reducing the amount of time staff were actively handling the sources to zero as automatically controlled systems were installed. Most radiotherapy departments have now replaced the large cobalt-60 sources with linear accelerators and the doses received by radiotherapy staff are even less than those received by diagnostic radiology staff.

Nuclear medicine protection of both staff and patients benefited from the work of Evans [50] and the nuclear programme. Control has always been applied to the activity of the radiopharmaceutical administered to the patients through, initially informal and latterly formal, certification or authorization procedures. Nuclear medicine departments, like radiotherapy departments always had physicists in direct control of the radioactive materials and it was possible to control the ingestion hazards to a high degree from the very early days. The external hazards were not always controlled to the same degree, particularly the finger doses to radiopharmacists and it is only relatively recently that staff doses have been reduced to the levels found in diagnostic radiology.

In summary, I think that the early pioneers of dose reduction to staff and patients would now be pleased with the levels to which the doses have now been reduced but I suspect that some would be a little sad it took so long in some areas.

25.8. Medical electronics

An abiding image of experimental medical investigation in the early years of the century is that of Eindhoven's patient sitting with his arms and legs in buckets of saline in order to record electrical signals from the heart [51]. Eindhoven was Professor of Physiology at the University of Leiden in Holland when in 1903 he devised the first string galvanometer to detect the small electrical currents originating from the heart. He was subsequently awarded the 1924 Nobel Prize for Physiology or Medicine for his discovery of the electrical properties of the heart through the electrocardiograph. Despite the fact that almost every investigative instrument now has electronic parts, the electrocardiograph perhaps remains the most useful electronic technique in diagnostic medicine: more than 200 million electrocardiograms are recorded each year.

The amplitude and shape of the electrocardiograph signals depend on the sources of the electrical activity in the heart, on the medium conducting the signals to the surface of the body and on the location of the electrodes on the skin. The fundamental source of electrical signals from the heart is cellular. Nerve and muscle cell membranes are excitable and electrical stimulation causes a reversal in ionic flow across the membrane with a consequent reversal in intracellular electrical potential. This electrical activity propagates to neighbouring cells and in muscle cells initiates a contraction. The cells in the heart are organized in such a way that the electrical activity is initiated by pacemaker cells which self-excite at regular intervals. This electrical activity is rapidly conducted throughout the walls of the heart by a low-resistance, high-speed pathway in order that the muscle cells contract in synchrony.

Although simplified modelling allows a gross indication of events in the heart tissue, the interpretation of the electrocardiogram remains empirical despite a century of scientific modelling. Progress in electrocardiographic instrumentation has therefore concentrated on ease of production and on quality of presentation of the electrocardiogram. Modern machines incorporate intelligence to compute a preliminary interpretation as the investigation takes place.

Along with its many uses as a diagnostic instrument, the electrocardiograph has a special utility in the intensive care of coronary patients and as a monitor during surgery. Monitoring the anaesthetized patient concentrates on ensuring adequate ventilation and delivery of oxygen to vital organs such as the brain, the heart and the kidneys. In addition to the ECG, the patient's blood pressure is commonly monitored whilst ventilation is checked by analysis of inspired and expired gases. The oxygen saturation of arterial blood (the percentage of the total haemoglobin that is oxygenated, normally about 95%) can now be measured non-invasively using the pulse oximeter, developed by Nakajima and his colleagues in Japan in 1975 [52]. By measuring the increase in attenuation of light by the pulsatile flow of blood into a finger or an ear at two different light wavelengths, it is possible to calculate the ratio of the absorption coefficients at the two wavelengths. This ratio is empirically related to the oxygen saturation which is calculated by a microprocessor and displayed along with the heart rate.

In all diagnostic instrumentation, new technology is introduced as it becomes available, resulting in smaller, safer and more reliable devices. Instruments for biochemical analysis have been based on novel sensors such as ion-specific sensors, while the recent introduction of catheter-tip charge-coupled cameras is a significant development in endoscopic investigative techniques.

For many years, the most useful therapeutic electrical technique was electrosurgery or surgical diathermy. Although surgical diathermy was for years associated with the name Bovie (an American physicist whose design led to the first commercial unit), his was only one of several teams who experimented with high-frequency currents in the first quarter of the century. In the closing years of the last century, the French physician and physicist d'Arsonval who was director of the Laboratory of Biological Physics of the College de France, Paris, had shown that, provided the frequency was sufficiently high (greater than 10 kHz), alternating electrical currents could flow in the body without causing pain or muscular stimulation [53]. This fact led to the use of high-frequency (100 kHz to 5 MHz) currents to incise tissue and to seal blood vessels during operations. When a fine blade electrode is used, the current density is high and resistance heating results. A large area pad at a remote site allows the same current to complete the circuit at a reduced current density. A continuous waveform is used for cutting whereas for coagulations the duty cycle of the high-frequency current is reduced to 20% or less which produces sealing without excessive damage of tissue.

A more recent thermal knife is the laser scalpel first used clinically in 1965 only a few years after Maiman demonstrated the first laser using a ruby crystal (see Chapter 18). The laser energy is monochromatic and highly collimated so that power densities producing vaporization of tissues can be achieved. The three commonly used lasers, carbon dioxide, argon and Nd:YAG, offer three widely separated wavelengths ranging from the visible to the infrared and their differing absorption characteristics offer differing surgical effects. Their cutting and coagulating properties are used in a wide variety of procedures, particularly in areas where access is difficult such as the throat or the vagina. By using a fibre-optic light guide to convey the laser energy, surgical treatment of internal tissues has become possible via the natural orifices or through 'keyhole' surgery without recourse to large skin incisions and gross displacement of neighbouring tissue and organs.

The Boston physician Paul Zoll was involved in two innovative uses of electrical stimulation which have subsequently developed into life-saving techniques. Defibrillators are instruments that are used to apply an electric shock to the heart in order to convert an ineffective, rapid heart rate back to the normal rhythm. Defibrillation is thought to occur when sufficient heart tissue is depolarized simultaneously so that the heart's own pacemaker (the sinus node) can re-establish the normal sequence or sinus rhythm. The earliest defibrillators used alternating current waveforms, the pioneering clinical applications being performed by Beck and co-workers [54] and by Zoll [55, 56] in the USA. Today's damped sine-wave defibrillator uses a waveform described by the Russian workers Gurvich and Yuniev and is said to produce fewer subsequent arrhythmias. Emergency defibrillation and cardioversion shocks are applied across the chest but in cardioversion the discharge is triggered at a suitable time in the rapid heart cycle by monitoring the electrocardiogram to maximize the chance of a successful conversion to sinus rhythm. Direct defibrillation is sometimes used during cardiac surgery when electrodes are placed on the wall of the heart.

Zoll was also prominent in the introduction of cardiac pacemakers. Although active stimulation of tissue had been known for over half a century, clinical electrical stimulation of the heart to maintain a regular heart rate was not achieved until 1952 when Zoll built the first external pacemaker. This mains-powered apparatus applied painful 100 V impulses to chest surfaces electrodes. Lower voltages and battery-powered stimulators became possible when electrodes were sewn onto the heart wall. The first implantable pacemaker was a rechargeable one used in Sweden by Elmquist and Senning [57], but soon zinc–mercury battery-powered pacemakers became the norm. In the mid-1960s, electrode systems were designed which could be passed via veins and placed against the inside wall of the heart. The short, two-year life of the earliest batteries led to experimentation with nuclear-powered units. These proved heavy and unacceptable because of the potential radiation hazards mainly associated with accidental incineration after death. The advent of the lithium–iodine cell in the 1970s coupled with low-power circuits, now allows lifetimes of up to 10 years to be achieved. Pacemakers now range from the simplest type providing 1 ms, 5 V pulses at heart rates of 70 beats per minute to complex programmable units containing a processor and memory on a chip. Therefore, not only do pacemakers now only stimulate on demand (that is, they sense when the patient's heart rate has decreased) they can also pace the atrium as well as the ventricle to maintain a physiological pacing sequence: some pacemakers can even sense and attempt to defibrillate tachyarrhythmias as they arise.

Therapeutic instruments have also benefited from the availability and reliability of small but powerful electronic devices. Examples are small, portable pumps for drug infusion and many aids for the disabled. A large number of the partially sighted use electronic methods of information retrieval. A range of smaller and more suitable hearing aids are now available whilst a small number of the completely deaf can hear sound as the result of an electronic implant that stimulates the cochlea.

This century, which began with Eindhoven's skilful and painstaking laboratory technique, has therefore seen remarkable developments ranging from the ubiquitous clinical electrocardiograph to the restoration of proper function by the implantation of electronic devices in the human body (for further information see references [58] and [59]).

25.9. Medical mechanics

The contents of this section are more fully discussed in Mow and Hayes *Basic Orthopaedic BioMechanics* [60].

Implant surgery represents one of the major achievements in post-war medicine. The most commonly implanted devices in the human body are either orthopaedic or cardiovascular, others include neurosurgical plastic, maxillofacial and ear, nose and throat implants. Implanted devices have to be designed for extremely high reliability and biocompatibility. A heart valve would have to operate for about 50 million times per year for many years. Typically 10 000 patients per year in the United Kingdom would be fitted with these devices. A total joint replacement of the hip, for example, would be loaded during walking about a million times per year and therefore has to have the appropriate mechanical strength and stiffness to withstand this loading pattern. Approximately 50 000 patients per year in the United Kingdom are fitted with a replacement hip joint. There are many varieties available—approximately 400 different types which can be divided into either cemented or uncemented prostheses. Total joint replacement of the hip is now a very successful operation which gives the patient relief from pain, a stable joint and a mobile joint. The earliest attempts to replace the natural ball and socket joint of the hip produced only limited success. The major step forward came in the early 1960s, when UK surgeon, Professor Sir John Charnley, designed a femoral component which articulated within a plastic cup. Initially, polytetrafluoroethylene (Teflon) was used as the plastic material, but this was soon found to be unsuitable due to excessive wear. However, when ultra-high-molecular-weight polyethylene (UHMWPE) was used together with polymethylmethacrylate (PMMA) bone cement to fix the components in the bone, an extremely effective synthetic joint was created. The Charnley low-friction arthroplasty is still the most common artificial hip joint used in the UK.

Orthopaedic surgeons are now able to replace many joints in the body, in particular the hip, the knee, the shoulder and joints in the finger: other joints such as the elbow and ankle can be replaced but are not particularly successful.

Fractured bones can also be re-united with assorted pieces of metalwork; bone plates, intramedullary nails and external fixators. These devices have certain properties which make them suitable for implantation in the body. They must be biocompatible, which means they are inert, non-toxic, non-carcinogenic, non-allergenic and they must have the appropriate mechanical properties of strength and rigidity, resistance to wear and a good surface finish. There are a whole range of materials that can be implanted into the body: these include stainless steel, titanium alloys and cobalt–chrome alloys. Some of the plastics include ultra-high-molecular-weight polythene, polyurethane and silicone rubbers. The fibrous materials that can be implanted include polyester and carbon fibres, other materials include acrylic bone cement, ceramics and pyrolytic carbon.

Developments in hip design include the use of new materials and improved fixation techniques. The femoral component is often manufactured from titanium or cobalt–chrome–molybdenum alloys. They can be located in the femur by porous coating of the stem surface with beads of titanium or hydroxyapatite. This surface coating provides a roughened surface for osteintegration to occur and this secures the prosthesis in the thigh bone. The acetabular cup can be metal backed with circular grooves so as to provide a screw fit.

25.9.1. Cardiovascular devices
Disorders of the rhythm of the heartbeat can be corrected by the surgical implantation of electrically powered pacemakers so that they can beat 72 times per minute with a regular rhythm. Advances in microelectronics and the development of tissue-compatible materials have improved their reliability and have increased their time of residence inside the body.

Modern anaesthetic equipment makes it possible to perform open surgery safely on the temporarily silent heart, the circulation of blood to the rest of the body being maintained by an extra corporeal blood pump/oxygenator until the heart can resume its function.

The engineering of the pacemaker and the pump/oxygenator has reached a high degree of safety and reliability so that the risk of breakdown is insignificant when judged against the risks associated with the untreated disease.

Major advances in the accuracy of clinical diagnosis have followed recent improvements in the design and construction of diagnostic machines. These include the x-ray display of the chambers of the heart and blood vessels, real-time ultrasonic imaging of the chambers of the beating heart, and pulsed Doppler ultrasound measurements of blood flow within it. Cardiac function and the localization of areas of damage to the heart muscle can be precisely defined using electrocardiography and may be imaged by radionuclide scanning, x-ray computed tomography and nuclear magnetic resonance imaging.

All of these have led to achievement of a high degree of pre-operative diagnosis and more precise pre-operative planning of an operation on the heart.

25.9.2. Artificial heart valves

The ability to safely enter the chambers of the heart has allowed the surgical replacement of defective heart valves. There has been a steady advance during the last 30 years in the design and manufacture of artificial heart valves to meet the engineering specifications of a valve that will withstand the mechanical forces that accompany the opening and the closing of the valve at least 40 million times a year. It must be designed to prevent the accumulation on its working surfaces of accretions of compacted blood cells and clots and also remain securely attached to the heart tissues into which it is sewn. Artificial heart valves in use today are of two types: 'biological' and 'non-biological'. The non-biological or mechanical 'engineering' valves are either 'ball and cage' or 'tilting disc' design and are mostly made of stainless steel. Their design bears no resemblance to the natural valve. 'Biological valves' are constructed of animal tissue (gluteraldehyde-treated porcine cusped heart valves or bovine pericardial tissue) from which a near-replica of the natural valve is tailored. Despite the complexity of the operation for the surgical implantation of an artificial heart valve, it is a remarkably safe procedure with minimal long-term morbidity, although it is known that the valves of biological tissues will degrade and calcify at some time.

25.9.3. Artificial arteries

Degenerative changes to the arteries of the body cause arteriosclerosis, and this results in narrowing of the vessels and a consequent reduction or cessation of flow through them. The coronary arteries supplying the heart are commonly affected. Though woven tubular structures of rigid polymer fibres (polyester or polytetrafluoroethylene) are used frequently to replace or bypass obstructions in the larger peripheral arteries, no satisfactory device is yet available for the smaller vessels (< 6 mm internal diameter) and much work is in progress to meet this need using a variety of novel methods to produce compliant, flexible, microfibrous tubes of elastomeric polymers.

In addition, the material must be blood and tissue compatible, must not degrade in the body and must not yield toxic products. It must also be possible to fabricate it into flexible tubes, the walls of which have an open, porous structure. The engineering contribution to progress in this field lies in the design of an arterial device having mechanical properties similar to those of natural arteries and in the design and construction of machines for their manufacture.

25.9.4. Technology for the disabled

The design of equipment for the disabled, such as artificial limbs (prostheses), splinting and bracing (orthoses) and communication aids has improved significantly during the past half century. There have been many design improvements to both manual and powered wheelchairs and this has allowed many patients to become more mobile. The development of modern lightweight wheelchairs has allowed paraplegic patients to engage in many vigorous sporting activities in which they previously were unable to participate.

The introduction of microtechnology has been a major benefit to the disabled, particularly low-cost microcomputers which have been of great assistance to those with communication difficulties. The use of speech synthesis has allowed many mute people to communicate with the outside world. Perhaps the most notable is Professor Stephen Hawking, Lucasian Professor of Mathematics at Cambridge University, whose contribution to modern science would not have been so evident if modern communication aids had not been available.

Many patients with neurological disorders have such gross movement that they are unable to perform even the simplest manoeuvres. These patients can be trained to operate equipment with specially chosen controls large enough to be handled and tough enough to withstand misuse. There are a whole range of interface devices for controlling various types of microprocessors for environmental, wheelchair and communication purposes.

Advances in materials technology have led to the development of modern ultra-lightweight artificial limbs and the use of computer-aided design and computer-aided manufacture has increased the speed of production and accuracy in the manufacture of these limbs. Orthotics has also benefited from the introduction of new materials technology particularly with the introduction of lightweight thermoplastics and composite materials which are more efficient and much more cosmetic than previously designed splinting devices.

25.9.5. Human movement studies

In order to understand pathological gait it is first necessary to study normal gait, since this provides the standard against which the gait of a patient can be judged. There has been a steady progression from early descriptive studies, through increasingly sophisticated methods of measurement, to mathematical analysis and mathematical modelling. This progression has also been true for recording all aspects of human movement.

It is now commonplace for universities, hospitals, those involved in sports science and ergonomists to have laboratories equipped with force platforms, optoelectronic movement analysis systems, video recording facilities and methods of recording the energy expenditure of movement.

25.9.6. Functional electrical stimulation (FES) of nerve and muscle
In recent years FES has been used for retraining muscles; restoring standing and walking functions to paraplegic patients with spinal cord injuries. Its potential may be much greater and it can be used in both orthopaedic and neurological rehabilitation as an active orthosis. FES uses an electrical pulse or a train of pulses (typically 30 Hz at 120 V with a duration of 10 ms) delivered to the motor points of muscles through nerve fibres to trigger muscle contraction processes in order to induce the movement of a paralysed limb.

The benefits of FES are that it can improve capillary circulation, reduce swelling of a limb, strengthen muscles of a paralysed limb, increase muscle bulk and possibly improve tissue condition, and prevent muscle atrophy through disuse. The reason for the preceding benefits are that movement and tension caused by contraction in muscles stimulate the production of protein and improve the metabolism.

25.10. Radioactivity in diagnosis: nuclear medicine

While section 25.4 was concerned with natural radioactivity used for treatment, this section is concerned with the use of artificially produced radioactivity in diagnostic medicine, which has given birth to the new diagnostic specialty of nuclear medicine. Medical physicists working in hospitals were very much the pioneers of this field. In centres where independent medical physics departments had been set up, these often became the base for the radioisotope service and the focus of the clinical tests that evolved. The reader must forgive the writer if some of what follows becomes a more personal story, because the author has been closely involved in the development of nuclear medicine and with the more recent development of nuclear magnetic resonance imaging. In 1951, reactor-produced iodine-131 was being used for the first time to study, and then to diagnose, abnormalities in the functioning of the thyroid gland in the neck. One of the very early workers in the field, Russell Herbert, showed the author how to measure the radioactivity accumulating in the gland by holding a Geiger–Müller counter over the neck. The I-131, administered as iodide either by mouth or injection, is removed from the blood stream by the thyroid gland where it is metabolized into hormones—mainly thyroxine, which is released into the blood stream to regulate the body's metabolism. The fraction of an administered dose that appeared in the thyroid gland was found to be a measure of the health or otherwise of the gland—an overactive gland collected more I-131 than normal, while an underactive one collected less:

both of these unhappy conditions are very common. The complexities of thyroid function were not fully understood, and the accurate diagnosis and treatment of thyroid-function abnormalities in the individual patient were not possible until artificial radioactivity could be applied to the problem.

The exciting situation of that time, which brought the first glimpses of a new medical field (a field that owed its origins to the high yields of radioisotopes obtained from the new nuclear reactors) is presented in a classic publication by Mayneord [61]. The first Conference on Radioisotope Techniques was sponsored in 1951 by the Atomic Energy Research Establishment at Harwell [62], and was followed by a special issue of the *British Medical Bulletin* devoted to Isotopes in Medicine in 1952 [63]. In addition to the thyroid work, P-32 was used for red blood cell volumes, Na-24 to study muscle clearance, and C-14 labelled compounds for radiocarbon dating. The famous Radiochemical Centre at Amersham was established to supply the burgeoning new endeavour in medicine. The radioisotope tracer technique was beginning to be applied to many different branches of medicine. Physicists found themselves in wards and operating theatres carrying out tests, and patients were being brought to the medical physics departments for measurements. The medical physicists became much better known throughout the hospital, and medical staff came to the departments to discuss results and evolve research projects. They found not only scientists, but also electronic and mechanical workshops and technical staff, all of which had been established to serve radiotherapy and radiology. The support and demand for medical physics services grew both locally and nationally.

The beginning of medical imaging followed quickly at St Bartholomew's Hospital, London. By collimating the counter so that gamma-rays were only counted from about a square centimetre of the neck, moving the counter rectilinearly in centimetre steps over the neck, measuring the counts detected in half minute intervals at each point and writing them down on graph paper, it was possible to draw lines of equal counting rate (isocount lines) which gave a crude image of the thyroid gland showing its shape and size [64] (figure 25.12). If the gland contained a tumour (which would not concentrate the I-131 because it is not normal thyroid tissue), then the tumour would be detected and localized as a 'cold' area, in contrast to the surrounding 'hot' normal tissue (figure 25.12(*b*)). The author carried out innumerable tests of this kind in the early 1950s. Russell Herbert at Liverpool introduced one of the first scintillation detectors to replace the Geiger–Müller counter, using a small calcium tungstate crystal [65].

Ben Cassen in Los Angeles developed the first automatic rectilinear scanner in the mid-1950s [66]. The scintillation counter was moved automatically over the neck in a rectilinear raster and, at the same time, a tapper—rather like a typewriter—moved over paper making a black

Twentieth Century Physics

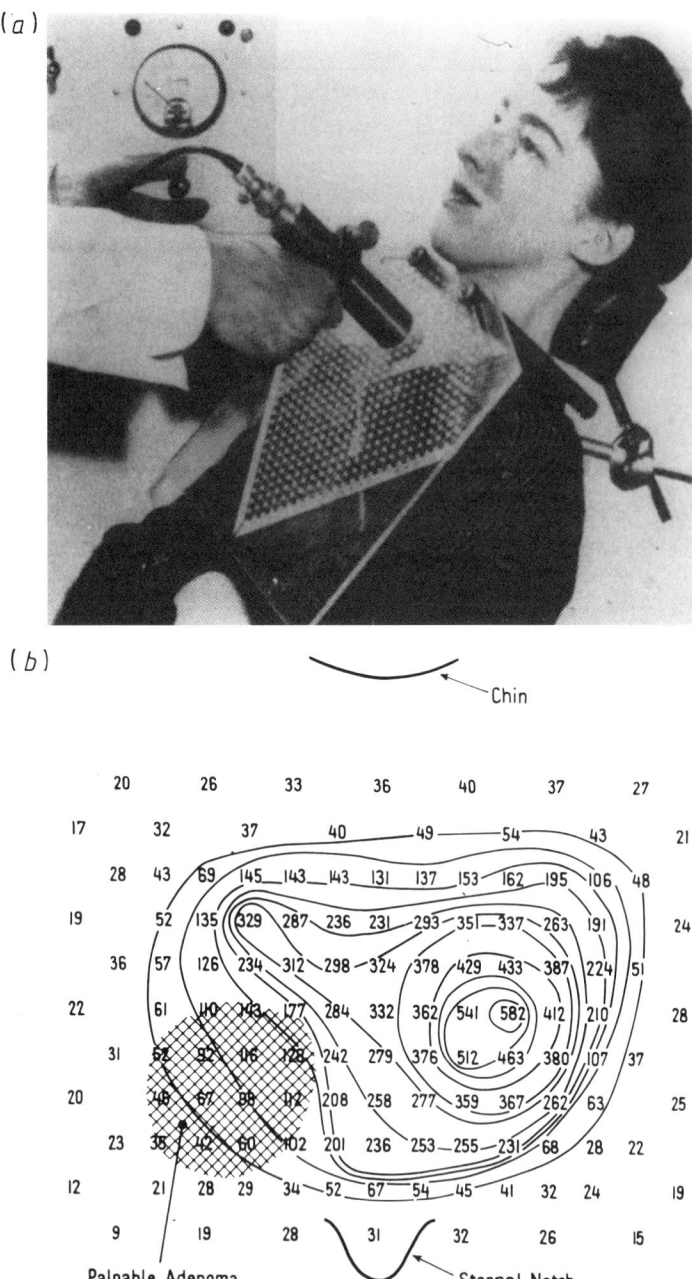

Figure 25.12. (a) Geiger–Müller counter shielded with lead used to detect gamma-rays from I-131 in the thyroid gland. The counter was moved in centimetre steps across the whole gland. (b) Gamma counts per 0.5 min at each point were recorded on graph paper and isocount lines drawn, which indicated the shape and size of the gland and presence of abnormal regions.

Medical Physics

Figure 25.13. *Image of the thyroid gland using I-131, formed with a very early rectilinear scanner and monochrome printer.*

mark every time a chosen number of gamma-rays had been detected. The marks were thus much closer together when the counter was over regions of high radioactivity, and this created an image of the thyroid not unlike that of a newspaper photograph (figure 25.13). However, although the image was better than that obtained from the isocount line method, the quantitative information was lost. In 1957, the author built the first rectilinear scanner in Europe [67], using an Ekco scintillation counter with a half-inch diameter sodium iodide (thallium activated) crystal as detector. In this apparatus, the patient was moved in the raster on a mechanized floating-top couch, mounted on the end of which was the typewriter-like printer mechanism, with a series of coloured tapes placed under the tapper, so that the marks made over higher radioactivity were not only closer together, but were in a series of different colours which were related to the counting rate. It was now once again possible to compare radioactive concentrations in the two lobes of the thyroid. Colour proved to be of even greater value in improving the visual contrast of the 'tumour' relative to the background in situations where the final contrast was minimal on the displays of these primitive machines. The idea of using colour as a quantitative indicator has now been adopted in many different fields. This rectilinear scanner was exhibited at the International Medical Electronics Conference, Olympia, London, 1960.

The detection and localization of brain tumours became possible for the first time using this machine in combination with a gamma-ray contrast provided by detecting the annihilation gamma-rays from the positrons emitted from arsenic-72 and -74 (produced by a cyclotron) [68]. Small, safe, tracer doses of the radioactive arsenic were given to the patient by injection, and a pair of scintillation detectors scanned over the head detecting only the coincident gamma-ray pulses from the positrons (figure 25.14). Although the images were very crude by comparison with those of today, a clinical accuracy of 80% was achieved, far surpassing the 65% of the x-ray angiography of that time (of course, it is much better than that now). The counting sensitivity was very low, and the antero-posterior view was soon found to be useful, in addition to the lateral views, because one could compare the radioactive patterns contra-laterally. This was the beginning of the expansion of the use of radionuclide imaging for many different body systems and organs; and also laid the foundation of the present-day imaging technique known as PET—positron emission tomography.

Cyclotron-produced radionuclides could not, however, be made generally available, and during the early 1960s efforts were made to improve the sensitivity and spatial imaging properties of gamma-ray imaging, particularly for the 360 keV radiation from I-131. Before the detectors for scintillation counters could be made to a diameter of greater than 0.5 inch (1.25 cm), it was sufficient to collimate the radiation with a 0.5 inch diameter cylindrical hole in a block of lead in front of the detector, which was placed as close as possible to the skin. As larger detectors became available, multihole collimators began to appear: the holes were conical, angled to one another so that their central axes pointed to what was known as the focal point, some 10 or 15 cm in front of the collimator. The focal point was the distance of maximum sensitivity to the radioactivity in air, but absorption of the gamma-rays in soft tissue (simulated by a water phantom) led to a falling sensitivity with depth in the patient. During this period, much attention was paid to the design of these multihole collimators to provide maximum sensitivity consistent with obtaining a good spatial resolution, usually described in terms of the FWHM of the point-source spread function at the focal distance. The FWHM was usually of the order of 1–2 cm at 10 cm depth with detectors as large as 5 inches in diameter [69, 70]. Special collimators were then designed to improve the response at greater depth in the body to improve the detection of deeper tumours [69], and some collimators were even made of gold to maximize sensitivity [70]. Brain tumours were

Medical Physics

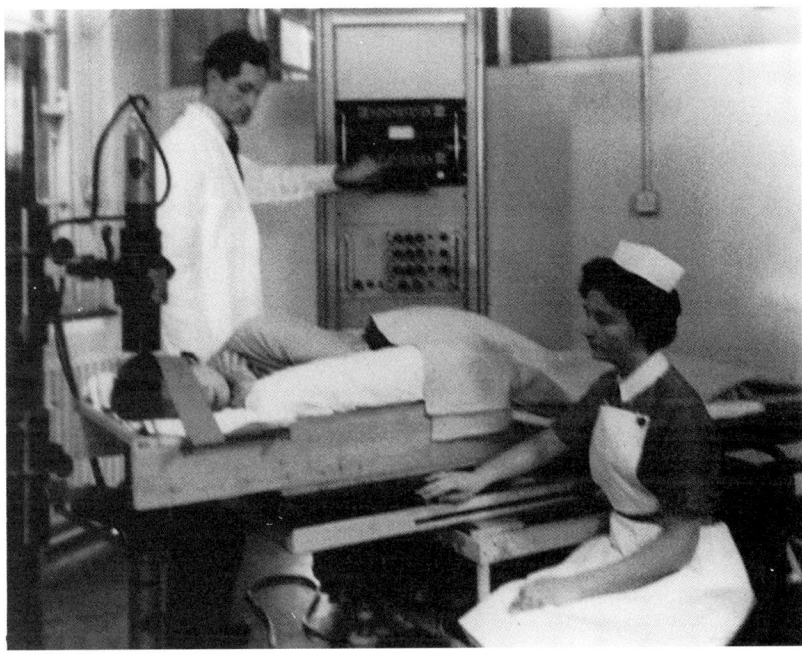

Figure 25.14. *The colour scanner in use to detect and localize a brain tumour. The opposed pair of scintillation counters with 0.5 inch diameter sodium iodide (thallium-activated) detectors, are operated in coincidence to detect the two annihilation gamma-rays from the positrons emitted from As-72 and -74. The rack of electronics shows nucleonics of the 1950s and 1960s, being commercial versions of equipment developed for the nuclear-energy programme at AERE Harwell.*

detected and localized with I-131 labelled human serum albumen (see figure 25.15, which shows that one really is detecting malignant tissue); liver abnormalities with Au-198 colloid; kidney abnormalities with I-131 lipodol; pancreatic abnormalities with Se-75 labelled methionine; thyroid disorders in children with the 2.3 hour half-life, low radiation dose I-132; and others [71, 72]. By this time, the use of radioactive isotopes was impinging significantly on patient management and hospital practice, so much so that, in the USA, these techniques began to be referred to as nuclear medicine. This name gradually came into worldwide parlance, but the author believes this to have been a mistake, because it tainted this most useful and worthy application of physics with the political and ecological overtones of a 'nuclear' activity, and over the years the field has perhaps not received the recognition, resources and funding that it has deserved.

A tremendous improvement came in 1964 with the introduction of technetium-99m [73]. This isomer of 6 hour half-life emits virtually monoenergetic 140 keV photons and delivers only a low radiation dose

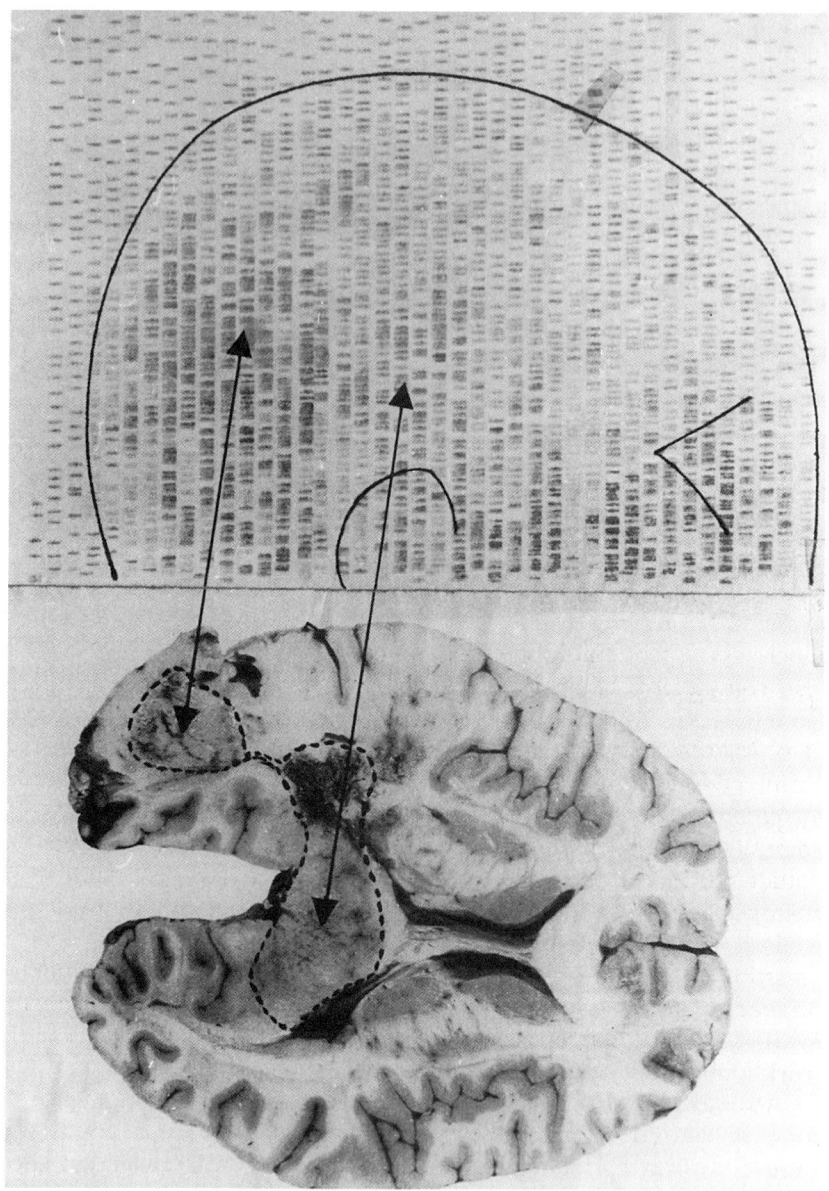

Figure 25.15. *Lateral brain scan (upper image) obtained with 131-I labelled human serum albumen, showing increased counting rate (arrowed) in the occipital region, compared with a histological transverse section of the same brain (lower image) at post mortem showing an occipital glioma infiltrating the posterior half of the corpus callosum. The scan image is spoiled here by being reproduced in monochrome, but the change in shade does indicate the colour change of the original.*

to the patient because it does not emit any β-rays. It also has the tremendous advantage that the septal thickness of collimators for the 140 keV radiation need only be 0.2 mm, whereas for the 360 keV radiation of I-131 the collimators must be 7–8 times thicker [74]. Thus the collimators could be designed with very thin septa, which meant that much more of the detector surface was exposed to the radiation, thereby improving the sensitivity considerably—which, in turn, meant that one could gain some spatial resolution by discarding some sensitivity. Better detection and delineation of brain tumours were thereby achieved, helped by the fact that sodium pertechnetate, the chemical form used for them, had a tumour/normal concentration ratio that was no worse than that of the albumen used formerly, which for favourable tumours is in the range 12–15 [75].

Because technetium is chemically similar to iodine, the whole range of organ-localizing pharmaceuticals became available using Tc-99m. The final advantage was that a molybdenum-99 generator (2.8 day half-life) could be made, from which the Tc-99m could be 'milked' by passing through a weak saline solution: thus a generator could be delivered from the Radiochemical Centre, Amersham, or other supplier, and be available for milking over a week or so, thus making a busy clinical centre much less dependent upon erratic delivery processes. This development also completely changed the routine procedures in busy hospital 'hot laboratories', which became much less dependent upon the presence of physicists, since the main work now became the sterile labelling of various chemicals and pharmaceuticals. Of course this change was aided by commercially available standardized methods of determining the amounts of radioactivity in test tubes and standard hospital dispensing bottles. The equipment usually consisted of a re-entrant ionization chamber and DC amplifier: the radioactive material was placed in the well in the middle of the chamber, which was sealed, usually under pressure, and had two cylindrical electrodes maintained at a voltage difference. Technetium-99m rapidly became the work-horse of nuclear medicine, and today it is still used for over 80% of imaging studies worldwide.

Also at about this time gamma cameras were beginning to be commonly used. The first scintillation crystal, photomultiplier-type gamma camera was developed by Hal Anger in Berkeley, California [76], and was first displayed at the Fifth Annual Meeting of the Society of Nuclear Medicine in June 1958 in Los Angeles. The author enjoyed working with Ekco on the development of the first European one [77]. This had a 5 inch (12.5 cm) diameter, 0.5 inch thick thallium-activated sodium iodide scintillation detector, viewed with seven photomultiplier tubes to provide positional analysis of each scintillation and place a spot in the corresponding position on a storage-tube oscilloscope display. Collimation was provided by a 0.25 inch diameter 'pin-hole' in a shield of lead in front of the detector. The supreme advantage of the gamma camera was that it did not have to move in a rectilinear raster, and so it was able to view the radioactive distribution for the whole period of exposure: it thus had a much greater sensitivity than the scanners, and could form a useful image in a much shorter time. Nevertheless, the first brain image that found a tumour, taken with this first machine, took 20 minutes [77]. The great potential of the gamma camera for dynamic studies was quickly realized and studies of the excretion of the kidneys and blood flow through the heart followed quickly. Nuclear Enterprises of Edinburgh took over the UK gamma-camera development and production, one of the first being in use in Aberdeen in 1967. The company continued progressive development of their camera and by 1979 it had an inherent resolution of 5 mm for 360 keV photons with a uniformity of ±10% [78]. This company made a big impact on the field of radioactive measurement for over 30 years after World War II: a part of the company still survives in England. Picker and IGE in the USA, and Siemens and Philips in Europe, quickly dominated the gamma-camera market; Elscint (Israel), Toshiba (Japan) and others were also very active in the field.

A great fillip to the growth of this field in the 1960s was a series of first-rate international conferences organized by the International Atomic Energy Agency under the rubric of peaceful uses of atomic energy, which brought together the leading protagonists of this new and, to most medical consultants, rather mysterious scientific field in medicine. The first conference was in Vienna in 1959, attended by 36 people; the second one in 1964 in Athens with 162, the third in Salzburg in 1968 with 341, and the fourth in Monte Carlo in 1972 with 484: others followed. By 1972 there were 120 rectilinear scanners and about 30 gamma cameras in routine use in the UK, and, of course, considerably more *per capita* in the USA.

In the early days of nuclear medicine, collaboration with a busy x-radiologist or a physician willing to take an interest in this burgeoning new imaging field was the only way to introduce it in a hospital for use on patients. As the techniques became more clinically useful, however, both radiologists and physicians developed expertise in the field. All major modern hospitals have at least one consultant in nuclear medicine on their staff.

Although the gamma camera was quicker and made dynamic studies possible, it suffered in the early days from serious image distortions, which were not present in the images from rectilinear scanners. Another disadvantage was that the images were not quantitative. It was realized in 1965, however, that the use of two 'crossed' multichannel analysers, one to define $X + \delta X$ and the other $Y + \delta Y$, defines a square element (or pixel) of the image and that pulses passing these analysers would represent the 'counts' in that pixel. Thus digital imaging was born

[79, 80]. One of the very first digital images was shown at the First International Congress of Medical Physics in Harrogate, England in 1965 [79].

It was natural then to connect the gamma camera on-line to a small computer. The DEC PDP8I was originally used, together with a primitive display system of 56×56 elements [80], the computer storing the counts to each element—which then determined the brightness (or colour) of that element on the cathode ray tube display. This made it possible to improve radionuclide imaging of certain difficult organs, such as the pancreas, by using two radionuclides emitting gamma-rays with different energies and subtracting the two images in the computer. Selenium-75 (gamma energies up to 401 keV) methionine localizes in the pancreas, which is a very difficult and elusive organ to image, but is also taken up in the liver, which envelops it: by also making an image of the liver only, with a Tc-99m (gamma 140 keV) labelled colloid, and subtracting the two images, pixel by pixel, one is left with a much improved image of the pancreas [80].

Other types of gamma camera were tried, the author's team making extensive studies of the potential of image intensifiers for radionuclide imaging. The system was very simple in concept, with a large lens to focus the scintillation upon the photocathode, and yielded images of much finer resolution than the Anger-type camera at that time. The clinical realization was defeated, however, by the need for improvement in the size and noise level of the intensifier tubes [81]: when the noise of the tubes was added to the background from the patient, the loss of contrast was too great. The continuing research and development of the multi-national gamma camera manufacturers gradually improved the inherent resolution of the Anger system until it reached, and then surpassed, the performance of the image intensifier systems.

In parallel with the development of medical imaging, there was also progress in the field of biochemical assay. Ekins at the Middlesex Hospital reported in 1959 the assay of trace amounts of the hormone thyroxine, through the use of labelled thyroxine and a specific binding agent, by the saturation analysis technique [82]. At virtually the same time, Yalow and Berson in the USA reported the assay of insulin by a technique that was, in principle, the same as Ekins' but used a binding by antibody to the natural and labelled hormone; they called their method radio-immunoassay. These methods have now developed into major new analytical procedures in diagnosis and physiological research, used throughout the world. Rosalyn Yalow received the Nobel Prize in 1977 in recognition of this major contribution.

The problem of radiation dosimetry of radionuclides distributed internally in the human body is exceedingly complex, and invokes problems lying more in the fields of physiology and biochemistry than in physics. Determinations of radiation dose are required to set the level of radioactivity that can safely be administered for tracer diagnostic tests, and, to ensure that these levels are not exceeded in clinical practice, it is customary for legislation to lay down procedures to be followed. For example, in the UK there is a Radioactive Substances Act, under which consultants in nuclear medicine are licensed and comply with the recommendations of an Administration of Radioactive Substances Advisory Committee. This body has available the most accurate information on the distribution of a particular radiopharmaceutical in the body, together with calculations of the radiation dose deposited in different organs and tissues. The identification of a critical organ to which the largest dose is delivered in general determines the quantity of radiopharmaceutical that can be administered. Different organs are not equally at risk, and appropriate weighting factors have been adopted by the International Commission on Radiation Protection (ICRP). A succinct, comprehensive presentation of this complex field is presented by Dendy [83], with a most useful guide to further reading.

For the macroscopic dosimetry of targeted radionuclide therapy, *in vivo* measurements are now at the millimetre level with techniques such as PET, SPECT and MRI. The range of ionizing particles is important in relation to the target size. The range of Auger electrons in soft tissue is measured in nanometres; for alpha particles it is in micrometres, and for beta particles it is from 0.8 mm for I-131 to 5.0 mm for Y-90. If the range is too great, radiation is deposited outside the target in normal tissues and wasted: if the range is too short, the non-homogeneity of the radionuclide distribution in the target tissue will have a serious effect upon the uniformity of deposition of radiation dose, and some regions may be underdosed while others are overdosed. For I-131 it has been shown that about 1 cubic millimetre is the optimum treatment size (about 1 million cells). The International Commission on Radiation Units and Measurements (ICRU) is presently addressing problems in this field.

This account of the development of nuclear medicine is continued and concluded in section 25.12.

25.11. Medical ultrasonics

The scientific basis of the study of acoustics was established by John Strutt, third Baron Rayleigh [84], who published the two volumes of his book The Theory of Sound in 1877–78. Thus it was, before the end of the first quarter of the twentieth century, that biologists, engineers and physicists had begun to explore the effects of ultrasonic waves on living systems and the feasibility of the detection of underwater obstacles by the reflection of pulsed sound waves had been demonstrated.

25.11.1. The origins of diagnostic ultrasonics
During World War II, experiments carried out independently in the United Kingdom by D O Sproule [85] and in the United States of America

by F A Firestone [86] led to the development of the ultrasonic industrial flaw detector. Using separate quartz transducers as transmitters and receivers, the reflections of brief pulses of megahertz-frequency ultrasound were used to measure the depths of minute fractures within metal castings, by assuming the speed of sound in the structure. It was not until after the end of the war, however, that details of this invention were published.

In the immediate pre- and post-war periods, attempts were made to visualize the internal structures of the body in terms of the two-dimensional distribution of the attenuation of transmitted beams of sound [87]. Encouraging pictures were obtained in which it seemed that the images were related to the internal structure of the head and, particularly, to the fluid-filled ventricles in the brain. Thus it was that research into transmission ultrasonic imaging received active support in the United States, the impetus of which was maintained by the publication of a paper [88] by two very influential physicists, T F Hueter and R H Bolt, which concluded that '... a preliminary evaluation indicates that the echo-reflection method is considerably less promising than the transmission method for general ventriculography, mainly because of the small amount of reflection at the interface between the tissue and the ventricular fluid'. In retrospect, it can be seen that this substantially held up progress in the United States and work there received a further setback when it was demonstrated that the images that had caused so much optimism were in fact due to the effect of the skull itself which coincidentally happened to mimic the shape of the ventricles in its transmission characteristics [89]. There was consternation when 'brain' images were shown to result from scanning an empty skull.

In Scandinavia, a neurosurgeon, Lars Leksell, began experimenting with a British industrial flaw detector in 1950, but this instrument was not sensitive enough to demonstrate echoes from the brain through the intact skull. In 1952, he obtained a more sensitive machine and observed an echo that appeared to originate from the centre of the brain. Whether he or R C Turner in London was actually the first to observe such an echo is not certain; although Turner's work was mentioned in the Annual Report of the British Empire Cancer Campaign in 1952 [90], Leksell did not describe the results of his experiments until 1956 [91].

25.11.2. Two-dimensional pulse-echo imaging
In the early 1950s, the piezoelectric properties of the ferroelectric material lead zirconate titanate were discovered at the National Bureau of Standards in the USA [92]. Probes using lead zirconate titanate instead of quartz had remarkably improved sensitivity and pulse performance; they were soon used in better ultrasonic flaw detectors. Around this time also, J J Wild and J M Reid in Minneapolis and D H Howry in Denver were pursuing the possibility of using pulse-echo ultrasound

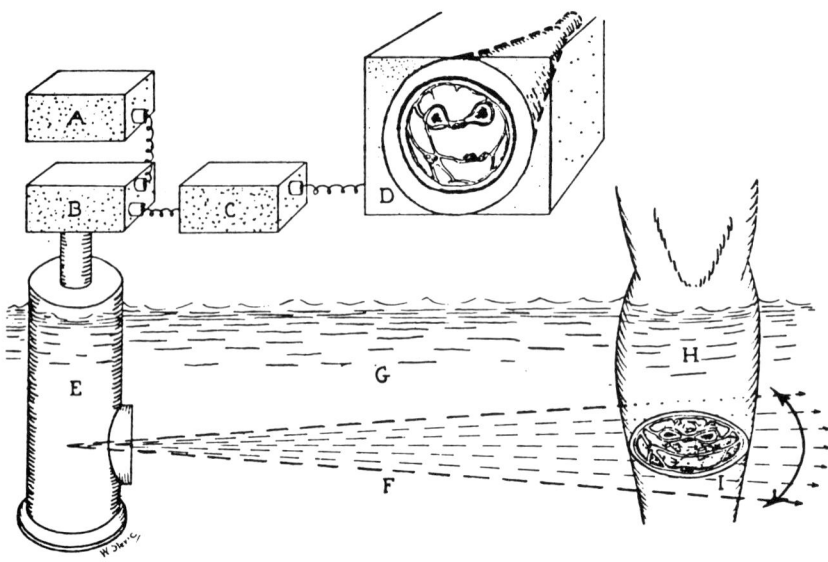

Figure 25.16. *Block diagram of the method used for cross-sectional imaging and described by D H Howry and W R Bliss in the Journal of Laboratory and Clinical Medicine [94]. Assembly E contains an ultrasonic transducer which is mechanically swept through the sector F to produce a cross-sectional image of section I on display D. Assembly A is a timing circuit; the pulse transmitter is in assembly B and assembly C is the receiving amplifier.*

for medical imaging, despite the views of the contemporary scientific establishment in the United States. The Minneapolis group, who for their early experiments used a naval radar trainer which was essentially a water-bath ultrasonic simulator, obtained encouraging results with pathological tissue specimens [93]. They went on to construct two-dimensional imaging machines operating at 15 MHz and produced the first pictures of the internal structures of the body—actually, of the intact female breast. In Denver, Howry [94] constructed the prototype scanner illustrated in figure 25.16 and went on to develop a 2 MHz instrument in which the patient was immersed in a water bath (the central gun turret removed from a scrap B-29 bomber) and, almost single-handedly, he produced images [95] the quality of which was not to be surpassed for some 15–20 years. As an example of what can be achieved nowadays, figure 25.17 is a scan produced by an instrument employing an array of tiny transducer elements operating in real time.

Credit for the realization of the potential clinical value of diagnostic ultrasound belongs to Ian Donald [96], when he was Regius Professor of Midwifery in the University of Glasgow. In 1955, he and his colleagues travelled out of the city in two cars, loaded with fibroids and

Medical Physics

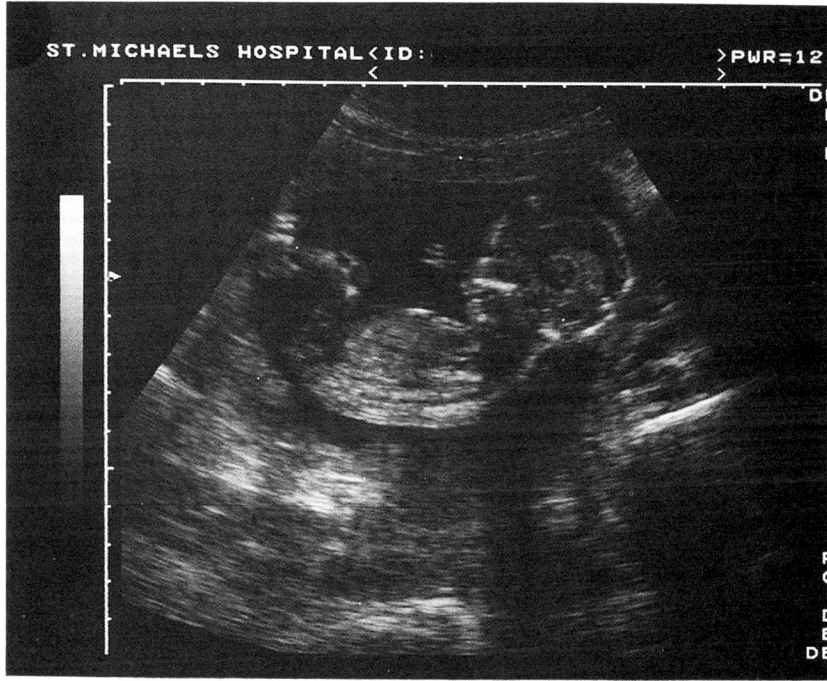

Figure 25.17. *A modern ultrasonic scan showing a foetus in utero at about four months gestation. The scan was made in real time using a curved linear array operating at about 4 MHz. The baby's head can be seen, together with the mouth and nose and there is detail of the brain. Some of the vertebrae can be picked out, together with the leg and even some of the toes.*

a huge ovarian cyst, to a boiler-making factory in Renfrew. There they applied the ultrasonic probe of an industrial flaw detector to the various specimens and observed the echoes that were displayed. Stimulated by the intriguing results that were obtained, Donald began a very fruitful collaboration with T G Brown, an electrical engineer employed by the Kelvin Hughes Company. This led to the development of a two-dimensional pulse-echo ultrasonic scanner in which the transducer was held in contact with the skin, oscillating while it travelled across the scan plane. The machine was demonstrated at the International Conference on Medical Electronics in London in 1960.

25.11.3. Time–position recording

Although the information from two-dimensional ultrasonic images of anatomical structures was soon shown to be of immense clinical value, parallel research in Scandinavia was producing hitherto unobtainable information about the moving structures of the heart. Inge Edler, a cardiologist in Lund, and his colleague C H Hertz, an engineer, showed that it was possible to obtain time–position recordings of the cardiac structures using pulse-echo ultrasound [97]. Much work went into the identification of the anatomical origins of the echoes and the relationship between the characteristics of the waveforms and the functions of the heart valves.

25.11.4. Doppler ultrasound

Meanwhile, in Japan, Shigeo Satomura [98] demonstrated that the frequency shift due to the Doppler effect could be detected in ultrasonic waves reflected by moving structures within the body. It is a fortunate coincidence that the speed of ultrasound in the soft tissues and the ultrasonic frequencies used, which are constrained by considerations of beam formation and attenuation, happen to result in audible Doppler-shift frequencies for targets moving within the body at physiological velocities. This means that the investigator can simply listen to the Doppler signals from the moving structures, using the probe as if it were a directional and very sensitive stethoscope. In the original Japanese work, it was assumed that the Doppler signals originated from the moving walls of the heart; it was soon shown, however, that the weaker signals scattered by blood could also be detected [99].

25.11.5. The modern era of ultrasonic imaging

Although advances in technology and the application of digital techniques have led to improvements in equipment performance which make the instruments of a decade ago almost unrecognizable as the forebears of what is commercially available today, there have really been only three other developments in what might be considered to be fundamental physics. The first of these developments, carried out independently by D W Baker [100] in the USA, P A Peronneau [101] in France and P N T Wells [102] in the UK, was the demonstration that, by combining pulse-echo and Doppler methods, it is possible to separate Doppler signals according to the depths of origin along the ultrasonic beam. Secondly, J C Somer [103] in Utrecht and Nicholaas Bom [104] and his colleagues in Rotterdam were responsible for the introduction of phased and linear array scanners for real-time two-dimensional imaging by the electronic control of the beams produced by stationary transducers. The third development, due to Chihiro Kasai [105] in Japan, was that of real-time two-dimensional flow imaging in which pulse-echo scanning of anatomical structures is combined with estimation of the Doppler frequency shifts across the whole of the two-dimensional scan plane. The pulsed Doppler technique made it possible for the anatomical location of the Doppler sample volume to be identified on the two-dimensional scan, allowing flow characteristics to be studied in detail in peripheral and deep vessels, and within the heart. Colour flow imaging provides the ability rapidly to examine blood flow in complicated anatomical situations; it requires relatively less skill on the part of the user but provides information which is of a more qualitative nature.

25.11.6. Ultrasound in therapy and surgery

In addition to its use in diagnosis, ultrasound has applications in physiotherapy and surgery. In the 1930s, particularly in Germany, there was tremendous enthusiasm for the perceived healing qualities of ultrasonic radiation. Even today, ultrasonic physiotherapy is widely used for the treatment of acute muscular conditions; typically, average intensities of a few watts per square centimetre, often pulsed at millisecond periods, are applied to the affected parts.

At higher intensities, ultrasound can produce damage which may be irreversible. This is exploited in surgical applications. In the 1950s in Indianapolis, an engineer, W J Fry, and his colleagues [106], working with neurosurgeons, developed high-precision equipment for focused ultrasonic surgery. Using brief pulses of 1 MHz ultrasound, trackless lesions could be produced in the brain after the removal of the overlying skull. The research moved along two directions. Firstly, ultrasonic focal lesions were positioned to interrupt nervous pathways in the brains of experimental animals, and the consequent functional changes were related to neuroanatomy. Secondly, attempts were made to ablate parts of the brain responsible for, for example, Parkinsonism and the control of cancer growth. The method has now fallen into disuse, at least partly because of the development of effective drugs.

Other areas in which ultrasound has been used in surgery include the destruction of part of the inner ear in patients with balance disorders and related problems, and for the treatment of laryngeal warts. More recently, the formation of shock waves due to non-linear propagation has been exploited in the destruction of stones without the need for open surgery [107]. Such shock waves can be produced at the focus of a large-aperture source. The shock waves are designed to have sufficient energy to fragment the stone but to be weak enough to avoid damaging the soft tissues in the intervening path.

25.11.7. Medical ultrasound in clinical perspective

In a typical medical teaching hospital today, about 12% of the patients undergoing radiological examinations are imaged by ultrasound. Because ultrasound scanning is rather less labour-intensive than some other studies such as magnetic resonance imaging and computed tomography, it accounts for a little less than 10% of the total workload. The use of ultrasonic physiotherapy is more variable between hospitals, depending on the preferences of the staff, but may account for about 5% of the workload. Surgical applications of ultrasound are still in their infancy but, as minimally invasive procedures become more important, it is certain that ultrasound will emerge as a very significant technique.

25.12. Computed tomography

25.12.1. Computed tomography in nuclear medicine
The advent of the digital computer made possible the development of computed tomography for both radionuclide imaging and x-ray imaging. This was the beginning of high-technology imaging in medicine. In nuclear medicine the technique is known as SPECT (single photon emission computed tomography), and it makes possible the imaging of transverse slices across the body. Until the advent of SPECT in the mid-1960s (it did not become well known and well used until the early 1980s), the only available views of radioactive distributions in the body were those from the front (antero-posterior or AP), the back (postero-anterior or PA), or from the side (left lateral and right lateral). In these views, when trying to image a tumour, for example, one also detects the gamma-rays emitted from the normal tissues between the tumour and the detector; and also, to a lesser extent, from behind the tumour. Figure 25.18 shows that for a tumour which has six times the concentration of radioactivity of the surrounding normal tissues, the contrast in the image is reduced to only 2:1; but the transverse section view through the tumour would give an imaging contrast of 6:1. The mathematics of the method of reconstruction using back projection was originally developed in 1917 by the Polish mathematician Radon in Vienna [108], but it was not until the digital computer, combined with digital imaging, was available, that its use became practicable. Today it is possible, in effect, to divide the body into a series of transverse slices, examine the individual slices for abnormalities and put them back together again.

Tomography was first carried out in nuclear medicine in Philadelphia in 1964 by Kuhl [109], who used an analogue technique. He presented a film of the counter movements and his early clinical images at the IAEA Scanning Conference in Athens to a rapt audience. The first digital system for CT in nuclear medicine (figure 25.19) was built by the team in Aberdeen [110], coordinated by A R Bowley, from 1967 to 1969, and was immediately applied in a clinical environment to the diagnosis of epilepsy [111] by providing a positive diagnosis from imaging to support the symptomatic evidence, which is not always clear-cut. It could never have been expected that our radioisotopes would lead to a step forward in the battle against mental disorders. Four years later, tomography was successfully applied to x-rays by Hounsfield.

For nuclear medicine, in essence, two opposed scintillation counters move laterally across the patient, and their axis is then rotated through 15 degrees (usually smaller rotation angles provide a more realistic computed reconstruction), followed by the next lateral crossing: the sequence is known as translate–rotate. The detectors thus sample the radioactive distribution as a series of linear projections from a series of angles around the patient, all viewing the same transverse section. The

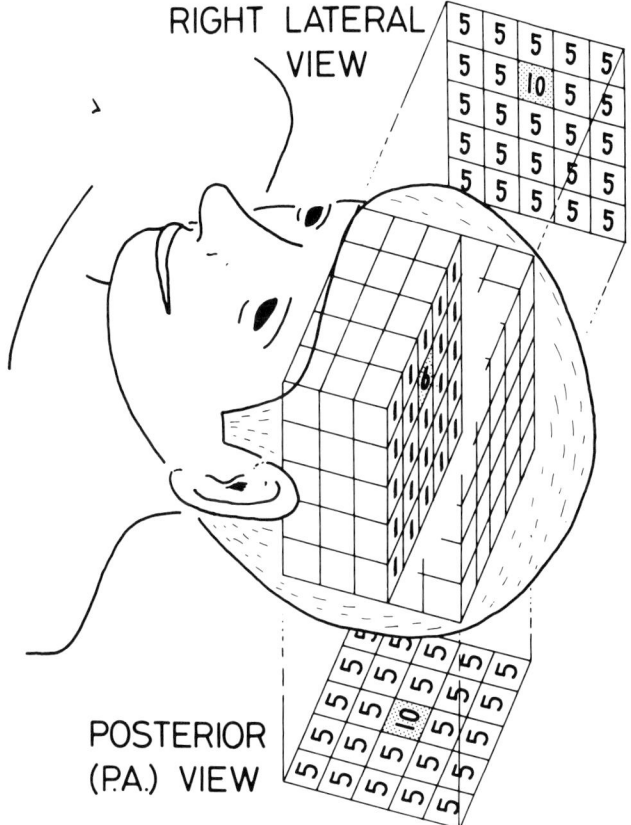

Figure 25.18. *This diagram shows a 'tumour' containing six times the concentration of radioactivity of the surrounding normal tissues. When it is viewed from front or back, and from the sides, as in conventional nuclear medicine imaging with a rectilinear scanner or gamma camera, then the contrast seen in the image is only 2:1. However, if a transverse section view can be made, which contains the tumour in its plane, then the imaging contrast will be 6:1.*

counting rate profiles are back-projected and summed in the computer, each view adding up the tumour hot-spot, while the more random background is averaged, thereby increasing the tumour contrast. The counts to be allocated to each pixel of the transverse section image are obtained from the computer. The problem of high background in the reconstructed image, and its blurring, is tackled by applying a ramp filter in Fourier space. Corrections for attenuation in the body tissues also need to be applied.

Tumours not seen very positively on the conventional AP and lateral views could be clearly detected and localized in the SPECT view (figure 25.20). It gave an improved perception of tumour-to-normal contrast, which led to an increase in brain lesion detection from 85%

Twentieth Century Physics

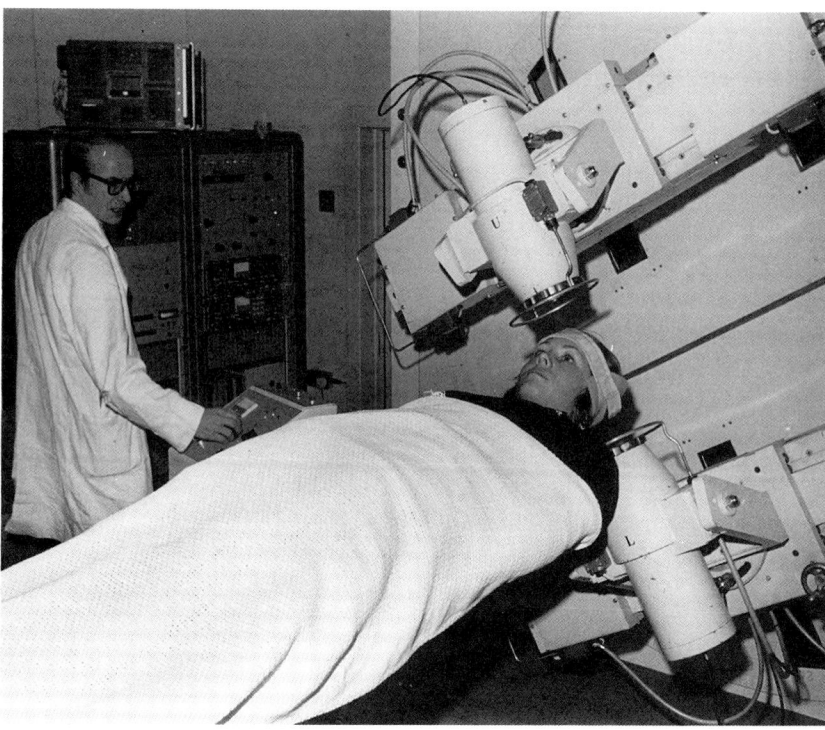

Figure 25.19. *The Aberdeen Section Scanner 1969: the pair of opposed scintillation detectors are seen mounted so that they can be moved to translate across the patient in both directions; both are mounted on a circular gantry which can rotate in increments around the patient.*

to 92%, when the CT view was added to the diagnostic work-up. The improvement was achieved for equivocal cases in which the diagnosis had not hitherto been very certain (the accuracy was four times better for these cases). This scanner, and the gamma-camera version which was built in the mid-1970s [112], was the method of choice for brain lesion detection in Aberdeen for some years, as it was also in other centres where the technique was available. Commercial machines became available in the late 1970s. When x-ray tomography (X-CT) came to Aberdeen in the early 1980s, it became the method of choice for brain lesion detection because of the much improved detail in the images. By 1973, the transverse SPECT view gave great confidence, and much more precise radiotherapy treatments began to be prescribed from the CT isotope scans (this process is now carried out routinely with x-ray CT images, or increasingly with MRI images).

The rotating gamma camera system made possible the imaging of multiple sections simultaneously. It also made possible in principle the imaging of other angular planes, deriving them from a stack of transaxial

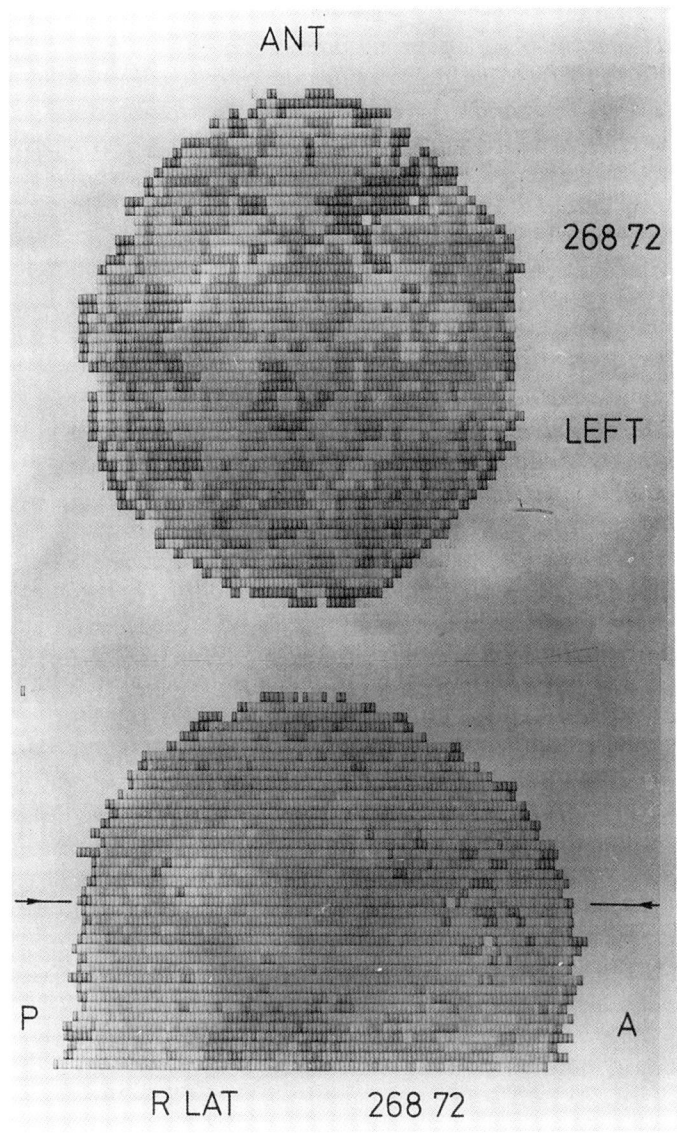

Figure 25.20. *A Tc-99m (sodium pertechnetate) brain image of a patient with a brain tumour which is very clearly perceived on the transverse section* SPECT *image (upper image) but hardly perceived on the conventional lateral view (lower image). On the* SPECT *view, the tumour is seen at depth in the brain (blacker region below centre), clearly separated from a high concentration of radioactivity also seen near the periphery of the head (blacker region on the left side of image): this is the scar tissue from a surgical operation six months before to remove the tumour, and the tumour now seen is a recurrence. On the conventional view, the scar tissue radioactivity overlaps that in the tumour, so that one cannot be sure what has been perceived. The* SPECT *view has made the diagnosis considerably more certain.*

sections: it is now becoming possible to present 3D images from them. Commercial rotating gamma cameras soon became available.

A typical modern gamma camera is entirely digital and has a 15 inch diameter thallium-activated sodium iodide detector which is viewed by 91 photomultiplier tubes—to be compared with the five tubes of the first camera. The images from modern systems are superb (figure 25.21).

During the period of evolution of rectilinear scanners, gamma cameras and SPECT systems, much attention has been paid to the measurement of imaging device performance. An important report sponsored by the ICRU in 1964 [113] led to the universal adoption of line spread function and modulation transfer function. In the early days of radioisotope imaging, it was difficult for the human observer to perceive the 'tumour' in the noisy backgrounds, and considerable interest developed in the evolution of methods to evaluate the images themselves. Psychophysical techniques, based on binary decisions or on ROC (receiver–operator characteristics) have been developed, and these are making possible genuine evaluations of the clinical effectiveness of a particular imaging procedure. This important field is well summarized by Sharp, Chairman of an ICRU Report Committee, in reference [74] and in a recent report [114].

The SPECT technique, together with newer radiopharmaceuticals which have been developed, has made considerable impact on patient management. SPECT is essential for imaging the new cerebral blood flow agents such as 99m-Tc[HM-PAO] and 123-I[IMP] and is the preferred method for imaging the myocardium with 201-Tl. Bone-seeking chemicals (MDP, methylene di-phosphonate) labelled with Tc-99m are used to detect and localize bone malignancies on the rib-cage, pelvis, long bones of the arms and legs, spine and skull, including secondary malignant deposits from primary tumours elsewhere in the body, such as breast and lung (figure 25.21). This has become one of the most important tests that nuclear medicine offers routinely because the results have a significant effect upon patient management (over 2000 a year are performed in Aberdeen). A clinical load of 7000 tests annually from a population of half a million is typical: about 20 different tests are available, and about half of the cases involve SPECT. This is typical of leading teaching hospitals worldwide, and the better general hospitals in the developed world. The current expansion of the field is expected to continue, with the total market for nuclear imaging products projected to grow by 50% to 400 million US dollars by 1997.

In section 25.10 it was seen that, in the very early days, detection of positron annihilation radiation was used for the localization of brain tumours. This technique has been developed by Ter-Pogossian in St Louis [115] to become a powerful clinical research tool, known as positron emission tomography (PET). In the early machines, rows of sodium-iodide crystals and photomultipliers, arranged as a hexagon,

Medical Physics

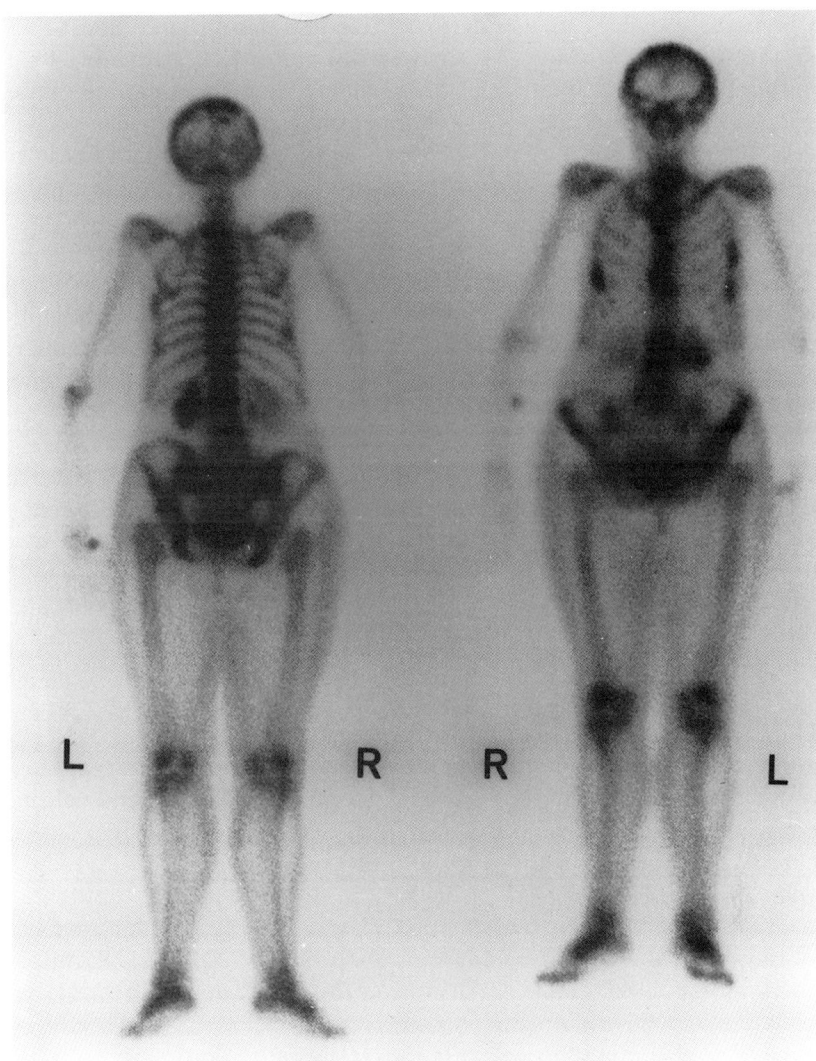

Figure 25.21. *Modern gamma camera image. This has been taken with the bone-seeking Tc-99m labelled di-phosphonate. The bones of the skeleton are clearly seen, with also increased radioactivity (blacker regions) in secondary malignant tumours distributed throughout the ribs, spine, skull and shoulders: these are secondary metastases from a primary tumour in the breast. These images are conventional AP and PA views—they are not tomographs.*

performed a translate–rotate motion, opposing crystals detecting the two annihilation photons. The reconstructed image in the computer showed the distribution in the plane of motion. In a more recent machine (the NeuroECAT III), eight rings of 320 bismuth germanate crystals oscillate

1921

to give fifteen slices of 6 mm FWHM spatial resolution. Short-lived radionuclides are used (C-11, half-life 20 min; N-13, 10 min; O-15, 2.1 min; Fl-18, 1.8 h), all of which are capable of being incorporated into a very wide range of biochemicals—over 400 at a recent count—in specially developed, rapid synthesizing procedures. The isotopes are made by on-site cyclotrons: the first was installed at Hammersmith Hospital in London in 1953, and much of the pioneering clinical work with PET was performed at this site.

Much of the PET clinical work has been concerned with brain imaging, particularly in neurological/psychiatric studies of stroke, epilepsy, schizophrenia and various forms of dementia such as Alzheimer's disease. With research studies using PET and individual diagnosis using SPECT, it is looking more likely that there may be some form of treatment in the future. Also, very excitingly, the PET team in Los Angeles showed the way to study the functioning of the brain itself: using C-11 labelled glucose it is possible to image the areas of the brain that concentrate the material dynamically when the volunteer looks, listens, thinks, remembers and uses various muscles.

There are still only about 100 PET imagers worldwide, largely because of funding difficulties relating to both the imager and the cyclotron, but there has been much growth of interest in this work recently.

25.12.2. Computed tomography in x-radiology
Between the wars many attempts were made to achieve tomographic x-ray images of longitudinal sections by moving the x-ray tube and detector contra-laterally to blur out the image everywhere except in the plane of interest. In radiology, however, the term tomography (from the Greek tomos, section) is now used exclusively in the context of computerized axial tomography (CAT, now usually CT), which was first announced by Godfrey Hounsfield of EMI Ltd, London, at the 1972 British Institute of Radiology Congress [116]. CT has been described as the greatest step forward in radiology since Röntgen's original discovery. Hounsfield shared the 1979 Nobel prize with his collaborator, Cormack, who had carried out theoretical studies for this major revolution in radiology. The first EMI-Scanner was commercially available in 1973 for head scans and by 1975 for body scans. The considerable improvement in contrast made it possible to see for the first time details of some soft-tissue organs using x-rays, and the scanners were in great demand within the radiology community. There have been several types of CT scanners from different manufacturers, including translation–rotation of a single detector and of multi-detector arrays for a single slice; rotation of x-ray tube and detector in an arc; a ring of detectors; and various hybrid systems. Because of the intense commercial development and patent protection, and the direct line of communication already existing between the radiologists and the multinational medical x-ray companies, hospital physicists did

not, in general, play a major part in the introduction of x-ray CT but, nevertheless, the best reviews and books in this field come from medical physicists [117]. Now that physicists play a more central role in radiology departments, partly as a result of the rapid adoption of MRI in radiology, hospital physicists are instrumental in reducing radiation dose to the patient in those CT procedures where this has been found to be excessive.

25.13. Nuclear magnetic resonance imaging

Magnetic resonance imaging for medical applications was first conceived in the early 1960s. At that time, electron spin resonance (ESR) was contributing to frontier research in biochemistry and pharmacology. Measurements were made of ESR signals from normal and malignant tissues [118], which arose from free radicals: they showed that tumours gave rise to a different intensity of signal from normal tissue. So here was an inherent natural contrast between tumours and normal tissue: if a scanner could be built to display the array of ESR signals from the body, one would be able to 'see' tumours and perhaps other things—free radicals are associated with tissue damage—without the need for injections of radioactive material. This potential new imaging technique was reported to the First International Congress of Medical Physics at Harrogate in 1965 [119]. However, ESR imaging presents great difficulties, because the required frequency is too high and is absorbed too readily by aqueous tissues, so that it will not penetrate sufficiently through the body; and even worse, it is heavily scattered, so that any initial free-radical contrast within the body is almost lost when the signals are detected outside.

In the early 1970s, Damadian [120], began measuring the nuclear magnetic resonance parameter T1, the spin-lattice relaxation time, of the water protons in tissue samples. He reported larger values of T1 for malignant tumour tissue than for normal tissue, and he proposed that this difference could be used as the basis for a form of imaging. T1 differences, albeit smaller, were subsequently confirmed by the Aberdeen team [121], who showed that they were related to the water content of the tissues, which varied by about 10%. Because the magnetic resonance of protons occurs at a much lower electromagnetic frequency for a given standing magnetic field strength than ESR, an NMR imager was much more feasible. Further measurements showed that T1 for liver, for example, was one tenth of that of pure water (3.5 s at 24 MHz); values for the spleen were 60% larger; and the brain gave values for white matter that were 40% less than those for grey matter [122]. This suggested that imaging by NMR might show up the different organs, with grey brain differentiated from white—hitherto not possible by imaging—as well as possibly showing tumours and inflammation (for which T1 would be large, because of the high water content). It became possible to predict what a rabbit T1 image might look like [123], and even a human sectional image was drawn predictively [122].

In 1973, Lauterbur [124] suggested a way of forming an NMR image. This was to use a magnetic field gradient across a sample so that one side of it was in higher standing-field strengths, and the protons therefore precessing at higher frequencies, than the other side. Thus a frequency spectrum of the NMR signal coded the strength of the signal with position across the sample: by compounding several spectra taken at a series of angles around the sample, an image could be reconstructed by computed tomography, which was well known by then. The world's first T1-based image of an animal was obtained in this way in March 1974, in Aberdeen [125] (figure 25.22).

For medical diagnostic work, an imager was required that could image the trunk as well as head or limbs, and clearly it needed to measure the relaxation times from voxel to voxel. So, in contrast to most of the other teams, Aberdeen decided to build a whole body imager, which could image T1. Other teams were also searching for magnetic resonance imaging, particularly at the University of Nottingham, where there were three teams—one led by Andrew, who produced the first image of a wrist in 1977 [126]; one by Mansfield, (whose team patented the method of 90° pulses to selectively excite and define a slice in 1974 [127] and who later concentrated on echo-planar fast pulse sequences [128]); and one by Moore, who built a whole body imager in the early 1980s [129]. A team led by Young at GEC, London (who had taken over the medical imaging programme of EMI) first imaged a human head in 1978 [130], and Damadian's team in New York produced the first human thorax section in 1977 [131]. Supported by a grant from the UK Medical Research Council, not granted until 1976, a prototype was built in Aberdeen. It had an air-cooled, four-coil electromagnet (superconducting magnets had not quite arrived), with a standing field of 0.04 T, the highest possible consistent with achieving the necessary field uniformity [122, 123] (figure 25.23). In 1975, Hutchison introduced the inversion-recovery pulse sequence into imaging for obtaining T1-weighted images [132], and by 1979, images were being obtained of volunteers (including the author!) by a line scan technique: these had recognizable shape and showed some internal anatomy, but suffered from serious artefacts due to body movements such as heart beats, which made the images useless for medical purposes [123]. These artefacts were not eliminated until the early months of 1980, when the very first of the 2D Fourier transform methods was introduced, nicknamed 'spin-warp' imaging [133]. A proton density set of pulses gave an image of the distribution of proton concentration throughout a transverse plane across the patient, and a T1 pulse sequence was interleaved, giving another image of the same transverse slice of the patient, but showing the distribution of T1 throughout the plane, pixel by pixel.

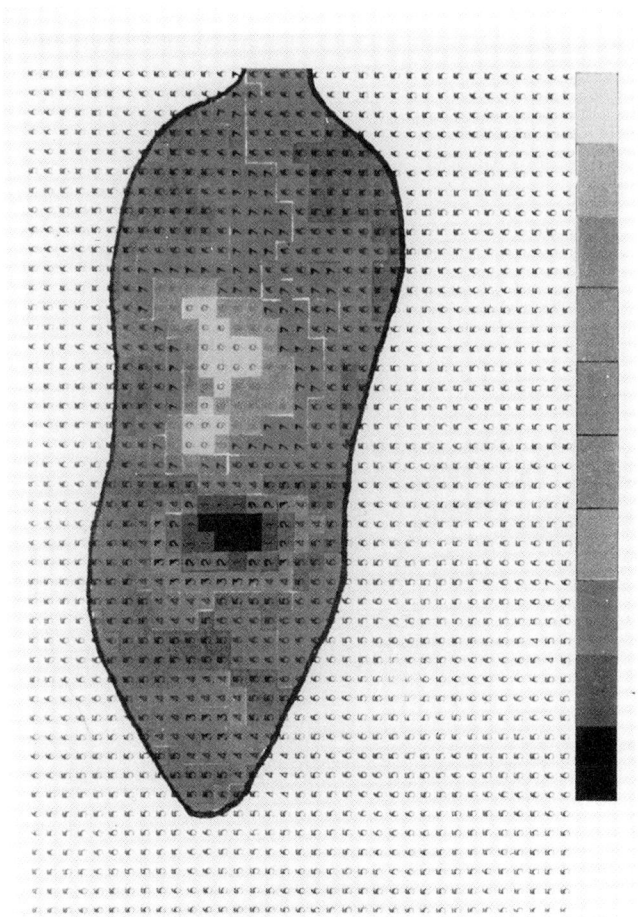

Figure 25.22. The first-ever NMR image of a mouse [125] displaying relaxation time information. The outline of the mouse was shown by NMR intensity signals, related to the densities (or concentrations) of protons from pixel to pixel; and its liver (white) and brain were localized within it by displaying, in colour (originally) for each pixel, the average T1 throughout the thickness of the animal. Excitingly, the long T1 (black) of the oedema around the broken neck (which had to be inflicted to ensure that the animal was completely still during the one hour of the image data acquisition) was imaged so that the very first image had shown a pathological effect. It was also the first-ever quantitative T1 image. The quality of this image is affected by being reproduced only in monochrome.

Twentieth Century Physics

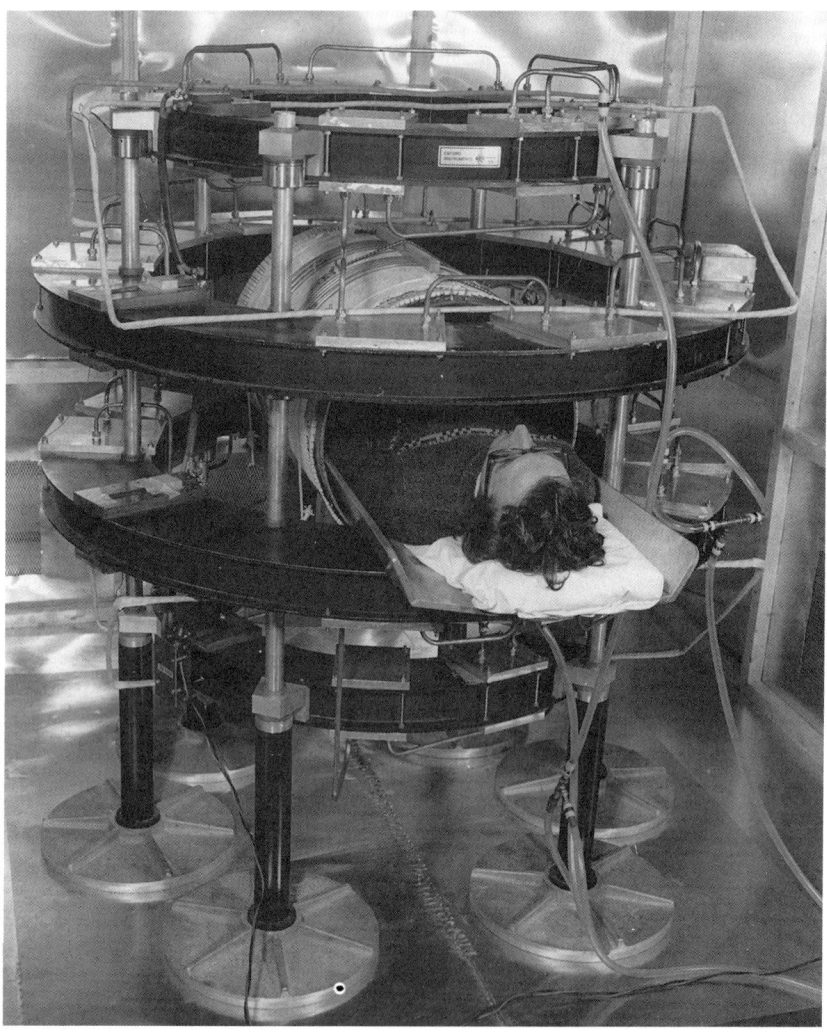

Figure 25.23. *The Aberdeen MK1* NMR *Imager. The four-coil electromagnet providing a vertical standing field of 0.04 T is seen, within which is the cylindrical former surrounding the patient (in this photo Dr J M S Hutchison, the main designer) on which are wound the X and Y field gradient coils, and the* RF *coils.*

The spin-warp technique was the real breakthrough. Following a period of imaging of volunteers, the first patient was imaged on 26 August 1980. Realistic images with startling anatomic detail of immediate clinical usefulness were obtained [134] (figure 25.24): MRI had arrived and was clinically exciting! The machine was quickly in full use to investigate patients, and it was soon found that it was superb for many things other than tumours: e.g. multiple sclerosis (figure 25.25),

Medical Physics

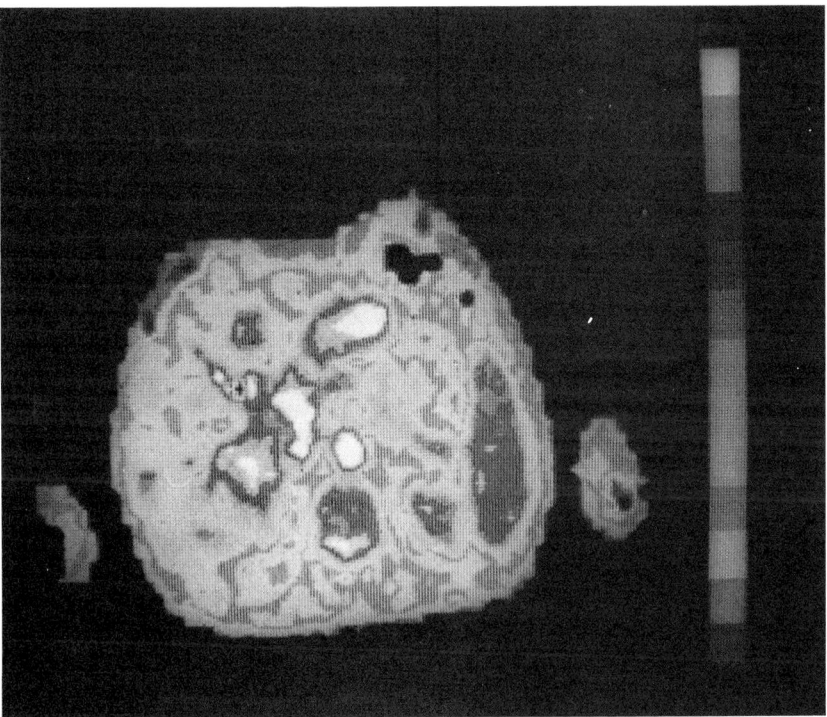

Figure 25.24. *The world's first clinically useful NMR image. It is an abdominal image of a patient with a primary carcinoma of the oesophagus, (imaged on a thoracic section as a result of its longer T1 than the surrounding tissue); massive secondary malignant deposits (T1 > 500 ms) are seen (white areas) in the very swollen liver (T1 = 140 msec); also seen is the spleen (dark ellipse on the right) and the splenic artery; the aorta and spine (in section); cerebrospinal fluid; and also a secondary malignant deposit in the spine not suspected hitherto, and subsequently confirmed using nuclear medicine [134]. The arms at the side of the body are also seen. (This image is spoiled by being reproduced only in monochrome.)*

for which it is now the method of choice. It was clear that NMR imaging gave superb contrast between the soft tissues, a property not possessed by other imaging techniques. Over 900 patients were imaged with this machine over 2.5 years, with many papers published describing world-first clinical series [135, 136]. The Aberdeen team then built a machine of better spatial resolution, which imaged over 9000 patients from 1982 to 1992 [137] (these machines are now in the Science Museum in London and in the Royal Museum in Scotland): it became the basis of a commercial machine produced in Japan, which is now much improved with a permanent magnet.

Medical imaging companies such as Technicare were taking interest from 1978 onwards, and GEC, London, installed their prototype machine 'Neptune' (1500 gauss provided by a superconducting magnet, made by

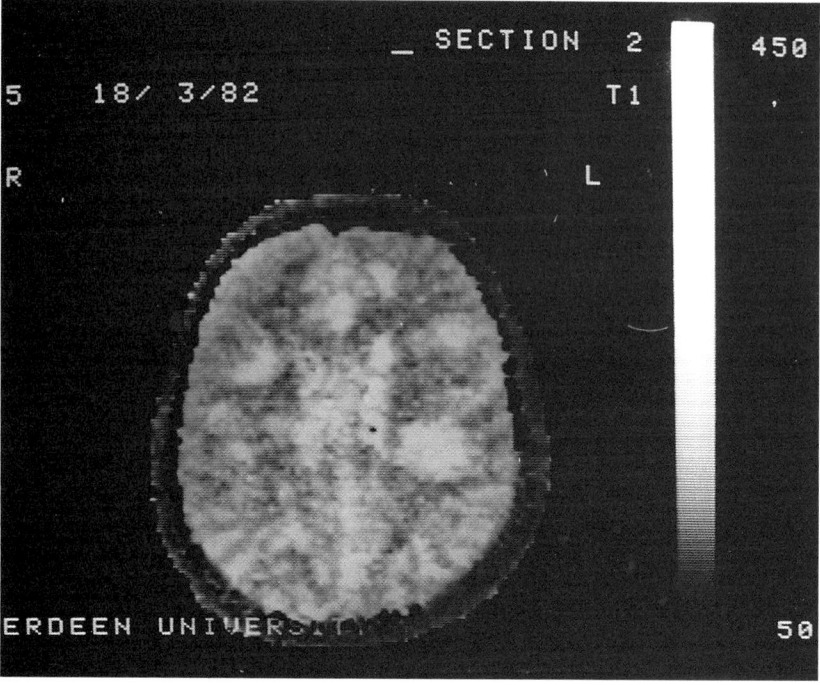

Figure 25.25. *An early image of a section across the head of a patient with multiple sclerosis. The disease removes the fatty-like (low T1) myelin sheath from the nervous tissue, so that where there is disease, the T1 is higher and those areas are seen as whiter blobs in the image.* MRI *is now the method of choice for diagnosing multiple sclerosis and other functional cerebral pathologies.*

Oxford Instrument Co) at Hammersmith Hospital in London in 1981. This trend to higher magnetic fields, and hence greater signal strength, enabled smaller pixels to be used to provide finer spatial resolution: this was just what the radiologists wanted, the images being more akin to the superb detail of their x-radiographs. The 'supercons' also gave a better stability: they were quickly taken up by the competing multinationals. By 1984 fields of 1500 to 5000 gauss became available, and fantastic detail of the soft-tissue brain structure became possible for the first time (figure 25.26): one by one, the major multi-nationals came onstream with superconducting magnet imagers giving up to 1 and 1.5 T, but very expensive indeed at £1m a time and requiring purpose-built suites with high structural installation costs.

The demand was high in the USA and wealthy countries, but much lower in Britain and poorer countries! By 1985 many hundreds of imagers were in use in the USA, Japan and Germany, while the UK had barely ten. The university teams in research laboratories were gradually pushed out of the further development of NMR imaging, as a result

Medical Physics

(a)

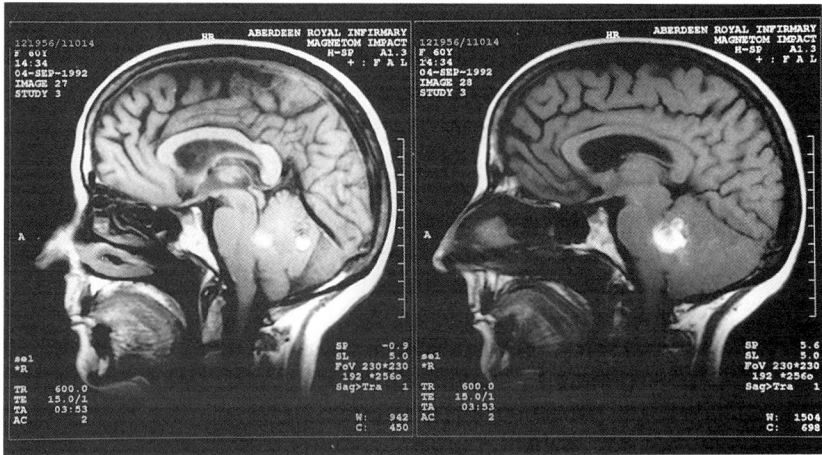

(b)

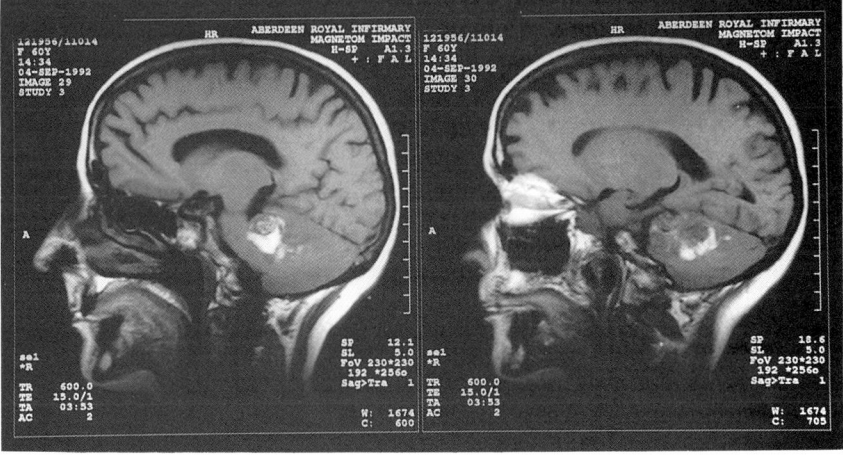

Figure 25.26. *Modern (1992) brain images taken on a commercial machine at 1.0 T. They are four adjacent sagittal slices moving through a cerebellar tumour with oedema (seen as the white area, slightly right of centre) (two in (a) two in (b)). The soft tissue detail is startling, and the capability of seeing the brain section together with the neck and spine is a valuable feature of the technique.*

of the much greater financial and manpower resources that the major multinationals could bring to bear. In the hands of skilled clinical teams worldwide, the new imaging technique was applied to a vast range of clinical problems.

By 1991 there were many thousands of machines worldwide, and by the end of the century more than 200 are expected in the UK. The early problems of using superconducting magnets, such as excessive stray fields, have been overcome, and the frequency of cryogen refills is now yearly. The world is installing about 2000 machines a year, a market of well over 1500m US dollars per annum.

By altering the field gradient directions, one can obtain images of planes at any angle in the body: radiologists usually take the three orthogonal planes, coronal, sagittal and transverse: this is not practicable in other forms of medical imaging. The introduction in 1985 of receiver coils on the surface of the body, first used in magnetic resonance spectroscopy (MRS) in 1980 [138], led to a big improvement in signal strength, and finer detail could be seen. Surface coils are now used routinely for the eye, limb joints and the spine, and mammography with bilateral breast coils is becoming prevalent.

MRI is the most effective method for investigating disorders of the brain and spinal cord [139]. The ability to see the head and neck on one sagittal image is particularly valuable, and is not possible by any other imaging technique. MRI is replacing more invasive investigations in musculo-skeletal diagnosis and has principal uses for oncology, paediatrics and the cardiovascular system. It has a developing role in a large number of other clinical specialties, including ophthalmology, endocrinology and otolaryngology. It is very safe for patients, and the test can be repeated safely, with no demonstrable side-effects; and it does not use ionizing radiation, which can do harm in the course of doing good. It is replacing other expensive, less effective, more invasive tests. It is replacing x-rays for some tests, particularly angiography (the visualization of blood vessels, both arterial and venous) and is also replacing some x-ray CT investigations. It is proving particularly useful for spinal and abdominal imaging, where methods of overcoming tissue movement are particularly helpful. Blood flow is imaged by the inflow/outflow of blood from the imaging plane creating a signal void, and other techniques, such as time-of-flight and phase contrast are used: blood velocity measurements are as accurate as those from ultrasound. MRI is also being used increasingly to analyse sports disorders. Several million patients have now been imaged worldwide, with about 25 000 images a day being performed.

As has been seen, MRI is a very versatile technique with several quite different natural mechanisms for producing imaging contrast: these are the two relaxation times, $T1$ and $T2$, the proton density, motion, and others, which vary from tissue to tissue and in disease, because the water content is related to the biochemical constituents and macromolecules taking part in the living or disease processes [140]. Flow provides another contrast. One can also inject paramagnetic contrast materials which change the magnetic moment of the protons and cellular constituents, and in some pathological conditions increases the imaging contrast: compounds of rare-earth metals such as gadolinium (introduced in 1988) are used, and improved clinical accuracy is reported. Very fast pulse sequences have been developed to give images in milliseconds and real-time movies of the beating heart. MRI techniques sensitive to blood flow, diffusion or relative oxygenation (using the natural difference in paramagnetic susceptibility between de-oxygenated and oxygenated blood) herald the mapping of functional brain activity with a spatial and temporal resolution better than PET.

There is now a slow move down in field strength from the 1 or 1.5 T machines, with field gradients of 15 mT cm^{-1}, which almost became standard in the late 1980s: this is because, although these fields provided fine spatial resolution, the contrast of tumours and other abnormalities was not always as high as expected, and clinical problems of movement artefacts and other losses were more significant. The next generation of machines use rather lower fields, but have increased versatility in pulse sequences and faster imaging times. These machines provide practically as much anatomical detail as those with larger fields (figure 25.26). Multiple slices, acquired in the same times as a single slice in earlier machines, are now standard, so that the body can be seen in section. This is a natural development for 3D imaging.

One of the very recent trends is the evolution of computer procedures to superimpose images from different imaging modes of the same section of a patient, such as PET or SPECT images, which show a specific tissue function, and x-ray CT images which show where the bony structures are, or MRI angiographic images, which show where soft tissues, nerves, blood vessels and tumours are. The overlays must allow for the different geometrical distortions introduced by each different mode. Such overlays are most successful for the head, where bony landmarks can be used as location points, and surgeons are learning to use them as a guide in operations: holograms are being used to present the 3D images to the surgeons.

Although MRI has become the method of choice for the exquisite depiction of cerebral anatomy, MRS, which has always promised biochemical information, has never become a routinely useful clinical examination, despite more than a decade of effort. The 1.5 T machines which became common in the late 1980s were capable of performing MRS as well as very fine spatial resolution MRI. MRS is time consuming and the *in vivo* spectra are difficult to interpret, partly because of decreased sensitivity and low metabolite concentrations. The great complexity of living tissues within a typical imaging voxel presents MRS as a key to its unravelling, but it remains a research tool which has yet to prove its value in daily medicine. 2 T, 4 T and even 10 T systems with 1 metre bore are being developed.

Although MRI can be achieved with nuclei other than protons, but these nuclei remain a research tool only at present, with more interest in F19, because high concentrations can be obtained in blood vessels with liquid perfluorocarbons, as used in artificial blood.

The future will see much improved detail of images acquired at low magnetic field strengths using SQUIDs, which provide a hundred-fold reduction in noise. New designs of magnets will make possible portable bedside imagers: 3D MRI used for stereotactic surgery will become routine: mass screening for vascular plaque and other disorders will come: new contrast agents will be developed which are tissue-specific to home-in on the abnormality sought: a surge of new activity using both MRI and MRS to study and ultimately control degenerative nervous diseases such as MS and the dementias will begin: even the concept of human *in vivo* free radical imaging will come to fruition (using the Oberhauser effect—images of free-radicals passing through the kidneys of a rat have already been obtained [141] and a machine for human whole-body imaging of free-radicals is now being built).

25.13.1. Other medical imaging techniques
Medical imaging also includes new techniques that are either not yet sufficiently developed to be of routine clinical use, or have only a very specialized role.

Thermography presents an image of skin-temperature distribution by imaging the infrared radiation emitted with a cooled semiconductor photon detector. It is being used to assess rheumatoid arthritis, but its early application to the screening of breast cancer is no longer regarded as of clinical value.

Electrical impedance tomography produces 2D or 3D maps of the impedances of the tissues in the body, which, perhaps surprisingly, stretch over at least two orders of magnitude. This requires the solution of the inverse problem from measurements of skin surface potentials resulting from the injection of electric current. Presently the spatial resolution is rather poor: it is best at the surface, but in the centre is only about 10% of the patient diameter. The most promising results occur for dynamic functions such as gastric emptying and lung ventilation.

Biomagnetic imaging detects the weak magnetic fields which result from electrical activity of biological functions, particularly in the brain and the heart. They can be detected by SQUIDs, used as gradiometers to detect the 0.2–50 pT fields outside the body. Typically arrays of 37 SQUIDs are used, and brain-imaging systems are now commercially available. Magnetoencephalography can produce startling dynamic images of brain function in real-time: for example, the activity resulting from a sound signal detected by the ear can be seen arriving at the auditory cortex and then proceeding on to other parts of the brain for interpretation. These dynamic images show how static are PET and MRI images by comparison, even though they may be beautifully detailed spatially.

Diaphanography, or light transmission imaging, can be used to image accessible structures by observing the shadow resulting from infrared illumination of the opposite side. Some biological materials, such as

Figure 25.27. *The memorial in Hamburg to the radiation martyrs.*

haemoglobin, are strong absorbers. The main problem is due to scattered photons, which reduce target detectability. Diaphanography has been evaluated for screening for breast cancer, but has been found to generate an unacceptably high number of false positives.

25.14. Conclusion

This chapter has reviewed the development of the main applications of physics to medicine over the century. The remarkable advancement of the field is perhaps best exemplified by figure 25.3, which shows the electromagnetic spectrum and the various uses of each part of it for medicine. The x-ray and gamma-ray region has been used for radiology and radiotherapy since the beginning of the century. Figure 25.27 shows the memorial in Hamburg to pioneer radiation workers where death could be attributed to radiation overdosage before the effects were more clearly understood. Nuclear medicine, a more recent development, is also based upon gamma-rays. Ultrasound is now a part of radiology, and about a fifth of all imaging tests use ultrasound. The radiofrequency region, not used in medicine in the middle of the century, is now used in magnetic resonance imaging—the present-day leading edge of radiology. Almost every part of the spectrum is now in daily use in medicine—all the result of physicists working in hospitals alongside medical people.

So, modern medical physicists have now to be concerned with all these fields of physics used every day in the hospitals of the world. Medical physics did not exist as a profession a century ago—it is now thriving in every leading hospital, with over 20 000 practitioners worldwide. The birth and growth of the international organizations—the International Organization of Medical Physicists (IOMP), the International Federation of Medical and Biological Engineering (IFMBE), and the European Federation of Medical Physics (EFOMP), which constitute the International Union of Physical and Engineering Sciences in Medicine (IUPESM)—from the 1950s onward, is described in [142]. Medical physicists ensure that all these applications of physics, electronics and computing are properly and accurately used, and they research and innovate to improve the health of the world. One can be sure that there are new fields yet to come. In the same way as a physicist a century ago could not have foreseen what we now have and do, I wish I could be a fly on the wall to see what the next century will bring!

Acknowledgments

I am deeply grateful to my colleagues, listed at the beginning of this chapter for so kindly writing important sections of it. Also I am very grateful to Professor John Laughlin for most helpful notes on the evolution of medical physics in the USA and to Dr John Haggith, Dr Meg Hutchison, Dr Alan Jennings, Professor Harold Miller, Mr Norman Ramsey and Professor Peter Sharp for most helpful notes and comments. Also, I have drawn upon the two lovely books of Dr R F Mould [1], particularly for some of the earlier parts of the century. I am deeply grateful to Mr Raymond Hutcheon of the Department of Bio-Medical Physics and Bio-Engineering, University of Aberdeen for reproducing the illustrations.

References

[1] Mould R F 1980 *A History of X-rays and Radium* (Sutton: IPC)
  See also Mould R F 1993 *A Century of X-rays and Radioactivity in Medicine* (Bristol: Institute of Physics)
[2] Mallard J R 1967 Medical physics—what is it? Hybrid tea—numerically scanning clockwise *Aberdeen Univ. Rev.* **XLII** 12–29
[3] For a more detailed account of the radiation physics of diagnostic radiology see Meredith W J and Massey J B 1972 *Fundamental Physics of Radiology* (Bristol: Wright) or Hine G J and Brownell G L 1956 *Radiation Dosimetry* (New York: Academic).
[4] Mould R F 1980 19th century skiagrams *Hosp. Phys. Assoc. Bull.* (Supplement) September
[5] Miller H 1982 *A Brief History of Medical Physics in Sheffield 1914–1982* (Dept Medical Physics and Clinical Engineering, University of Sheffield)
[6] Christen F T 1913 *Messung and Dosierung der Roentgenstrahlen* (Hamburg: Graefe und Sillem) (abstract in 1913 *Arch. Roentgen Ray* **18** 280)
[7] Lindsay D D and Stern B E 1953 A new tissue-like material for use as bolus *Radiology* **60** 355
[8] Jones D E A and Raine H C 1949 *Br. J. Radiol.* **22** 549
[9] Compton A H and Allison S K 1946 *X-rays in Theory and Experiment* (New York: Van Nostrand)
[10] ICRU 1928 International x-ray unit of intensity *ICRU Report* 2 (Bethesda, MD: ICRU)
[11] Mayneord W V and Lamerton L F 1941 A survey of depth dose data *Br. J. Radiol.* **XIV** 255–64
[12] 1953 Central axis depth dose data *Br. J. Radiol.* Supplement 5
[13] Johns H E and Cunningham J R 1971 *The Physics of Radiology* 3rd edn (Springfield, IL: Thomas)
[14] Ellis F and Miller H 1944 *Br. J. Radiol.* **XVII** 90
[15] Paterson R and Parker H M A dosage system for gamma-ray therapy *Br. J. Radiol.* **7** 592
[16] Meredith W J (ed) 1967 *Radium Dosage: The Manchester System* (Edinburgh: Livingstone)
[17] 1951 *The Radium Commission: A Short History of its Origin and Work 1929–1948* (London: HMSO)
[18] Sinclair W K 1952 Artificial radioactive sources for interstitial therapy *Br. J. Radiol.* **25** 417
[19] Jones E and Mallard J R 1963 The experimental determination of the dose distribution around yttrium-90 sources suitable for pituitary implantation *Phys. Med. Biol.* **8** 59–82
[20] Christen T 1913 *Messung und Dosierung der Röntgenstrahlen* (Hamburg: Graefe und Sillem)
[21] Christen T 1914 *Arch. Roentgen Ray* **19** 210
[22] Villard P 1908 *Arch. d'Elec. Med.* **16** 692
[23] Eve A S 1906 *Phil. Mag.* **12** 189
[24] Sievert R 1932 *Acta. Radiol.* Supplement 14
[25] Paterson R and Parker H M 1934 *Br. J. Radiol.* **7** 592
[26] Paterson R and Parker H M 1938 *Br. J. Radiol.* **11** 252, 313
[27] Kaye G W C and Binks W 1937 *Proc. R. Soc.* A **161** 564
[28] Taylor L S and Singer G 1940 *Am. J. Roentgenol.* **44** 428
[29] ICRU 1957 *ICRU Report* 8 (Washington, DC: ICRU)

[30] ICRU 1954 *Br. J. Radiol.* **27** 243
[31] Gray L H 1929 *Proc. R. Soc.* A **122** 647
[32] Gray L H 1936 *Proc. R. Soc.* A **156** 578
[33] Bragg W H 1912 *Studies in Radioactivity* (London: Macmillan)
[34] Barnard G P, Aston G H, Marsh A R S and Redding K 1964 *Phys. Med. Biol.* **9** 333
[35] Genna S and Laughlin J 1955 *Radiology* **65** 394
[36] Domen S R 1980 *Med. Phys.* **7** 157
[37] Fricke H and Morse S 1927 *Am. J. Roentgenol.* **18** 430
[38] Greening J R 1985 *Fundamentals of Radiation Dosimetry* (Bristol: Hilger)
[39] Hall E J 1994 *Radiobiology for the Radiologist* 4th edn (Philadelphia, PA: Lippincott)
[40] Thames H D and Hendry J H 1987 *Fractionation in Radiotherapy* (London: Taylor and Francis)
[41] 1928 X-ray and radium protection. Recommendations of the 2nd International Congress of Radiology, 1928 *Br. J. Radiol.* **1** 359–63
[42] 1929 X-ray and radium protection. Recommendations of the 2nd International Congress of Radiology, 1928 *Circular No 374* (Bureau of Standards, US Government Printing Office)
[43] Mutscheller A 1925 Physical standards of protection against roentgen ray dangers *Am. J. Roentgenol. Radiat. Ther.* **13** 65–9
[44] Sievert R M 1925 Einage Untersuchungen über Vorrichtungen zum Schutz gegen Rontgenstrahlen *Acta Radiol.* **4** 61–75
[45] 1956 *The Hazards to Man of Nuclear and Allied Radiations Command 9780* (London: HMSO)
[46] 1960 *The Hazards to Man of Nuclear and Allied Radiations Command 1225* (London: HMSO)
[47] 1959 Radiological hazards to patients *Interim Report of Committee* and 1960 *Second Report* (London: HMSO)
[48] Shrimpton P C, Wall B E, Jones D G, Fisher E S, Hillier M C, Kendall E M and Harrison R M 1986 A National survey of doses to patients undergoing a selection of routine x-ray examinations in English hospitals *Chilton NRPB-200* (London: HMSO)
[49] 1990 Patient dose reduction in diagnostic radiology *Doc. NRPB* **1** No 3
[50] Evans R D 1980 Radium poisoning: a review of present knowledge *Health Phys.* **38** 899–905 (read at the Fourth Meeting of the Western Branch, American Public Health Association, 1933)
[51] Eindhoven W 1903 Ein neues Galvanometer *Ann. Phys., Lpz* **12** 1059–71
[52] Nakajima S, Hirai Y, Takase H, Kuse A, Aoyagi S, Kishi M and Yamaguchi K 1975 New pulsed-type earpiece oximeter *Kokyu to Junkan* **23** 709–13
[53] d'Arsonval A 1891 Action physiologique des courants alternatifs *C. R. Séanc. Soc. Biol.* **43** 283–6
[54] Beck C S, Pritchard W H and Feil H S 1947 Ventricular fibrillation of long duration abolished by electric shock *J. Am. Med. Assoc.* **135** 985
[55] Zoll P M 1952 Resuscitation of the heart in ventricular standstill by external electric stimulation *New Engl. J. Med.* **247** 768
[56] Zoll P M, Linenthal A J, Gibson W, Paul M H and Norman L R 1956 Termination of ventricular fibrillation in man by externally applied electric countershock *New Engl. J. Med.* **254** 727
[57] Elmquist R and Senning A 1960 An implantable pacemaker for the heart *Proc. 2nd Int. Conf. Medical Electronics* (London: Iliffe)
[58] Webster J G (ed) 1988 *Encyclopaedia of Medical Devices and Instrumentation* (New York: Wiley)
[59] Geddes L A and Baker L E 1989 *Principles of Applied Biomedical*

[60] Mow V C and Hayes W C 1991 *Basic Orthopaedic BioMechanics* (New York: Raven)
[61] Mayneord W V 1950 Some applications of nuclear physics to medicine *Br. J. Radiol.* Supplement 2
[62] 1953 *Radioisotope Techniques: Proc. Isotope Techniques Conf. (Oxford, July, 1951) vol I, Medical and Physiological Applications* (London: HMSO) pp 1–466
[63] 1952 Isotopes in medicine *Br. Med. Bull.* **8** 115–215
[64] Ansell G and Rotblat J 1948 Radioactive iodine as a diagnostic aid for intrathoracic goitre *Br. J. Radiol.* **21** 552–8
[65] Herbert R J T 1952 Discrimination against noise in scintillation counters *Nucleonics* **10** 37–9
[66] Curtis L and Cassen B 1952 Speeding up and improving contrast of thyroid scintigrams *Nucleonics* **10** 58–9
[67] Mallard J R and Peachey C J 1959 A quantitative automatic body scanner for the localization of radioisotopes *in vivo Br. J. Radiol.* **32** 652
[68] Mallard J R, Fowler J F and Sutton M 1961 Brain tumour detection using radioactive arsenic *Br. J. Radiol.* **34** 562
[69] Mallard J R 1965 Medical radioisotope scanning *Phys. Med. Biol.* **10** 309–34
[70] Gottschalk A and Beck R N (ed) 1968 *Fundamental Problems in Scanning* (Springfield, IL: Thomas)
[71] McCready V R, Taylor D M, Trott N G, Cameron C B, Field E O, French R J and Parker R P (ed) 1967 *Radioactive Isotopes in the Localization of Tumours* (London: Heinemann)
[72] Belcher E H and Vetter H (ed) 1971 *Radioisotopes in Medical Diagnosis* (London: Butterworth)
[73] Harper P V, Beck R, Charleston D and Lathrop K A 1964 Optimisation of a scanning method using technetium-99m *Nucleonics* **22** 50–4
[74] Sharp P F, Dendy P P and Keyes W I 1985 *Radionuclide Imaging Techniques* (New York: Academic)
[75] Matthews C M E and Mallard J R 1965 Distribution of Tc-99m and tumour/brain concentrations in rats *J. Nucl. Med.* **6** 404–8
[76] Anger H O 1958 Scintillation camera *Rev. Sci. Instrum.* **29** 27–33
[77] Mallard J R and Myers M J 1963 The performance and clinical applications of a gamma camera for the visualization of radioactive isotopes *in vivo Br. J. Radiol.* **8** 165–92
[78] Mallard J R and Trott N G 1979 Some aspects of the history of nuclear medicine in the United Kingdom *Semin. Nucl. Med.* **IX** 203–17
[79] Wilks R J and Mallard J R 1966 A small gamma camera—improvements in the resolution, a setting-up procedure and a digital print-out *Int. J. Appl. Radiat. Isot.* **17** 113–9
[80] Mallard J 1987 Hevesy Medal Memorial Lecture—Some call it laziness: I call it deep thought (with apologies to Garfield) *Nucl. Med. Commun.* **8** 691–710
[81] Mitchell J G, Mallard J R, Egerton I B, Caldwell A B, Lakshmanan A V, Nienan C and Turnball W (ed) 1972 Towards a fine-resolution image intensifier gamma camera: the AberGammascope *Medical Radioisotope Scintigraphy* vol 1 (Vienna: IAEA) pp 157–67
[82] Ekins R P 1960 The estimation of thyroxine in human plasma by an electrophoretic technique *Clin. Chem. Acta* **5** 453–9
[83] Dendy P P, Palmer K E and Szaz K F 1989 *Practical Nuclear Medicine* ed P F Sharp, H G Gemmell and F W Smith (Oxford: IRL/Oxford University Press) ch 7

[84] Strutt J W (third Baron Rayleigh) 1877 *The Theory of Sound vol 1* (London: Macmillan); 1878 *The Theory of Sound vol 2* (London: Macmillan)
[85] Desch C H, Sproule D O and Dawson W J 1946 The detection of cracks in steel by means of supersonic waves *J. Iron Steel Inst.* **153** 319
[86] Firestone F A 1946 The supersonic reflectoscope, an instrument for inspecting the interior of solid parts by means of sound waves *J. Acoust. Soc. Am.* **17** 287
[87] Dussik K T, Dussik F and Wyt L 1947 Auf dem Wege zur Hyperphonographie des Gehirnes *Wien Med. Wochschr.* **97** 425
[88] Hueter T F and Bolt R H 1951 An ultrasonic method for outlining the cerebral ventricles *J. Acoust. Soc. Am.* **23** 160
[89] Güttner W, Fielder G and Pätzold J 1952 Über ultraschallabbildungen am Menschlichen Schädel *Acoustica* **2** 148
[90] 1952 *Annual report of the British Empire Cancer Campaign* p 72
[91] Leksell L 1956 Echo-encephalography: detection of intracranial complications following head injury *Acta Chir. Scand.* **110** 301
[92] Jaffe B, Roth R S and Marzullo S 1955 Properties of piezoelectric ceramics in solid-solution series lead titanate-lead zirconate-lead oxide: tin oxide and lead titanate-lead hafnate *J. Res. Natl Bur. Stand.* **55A** 239
[93] Wild J J and Reid J M 1952 Further pilot echographic studies of the histologic structure of tumours of the living intact human breast *Am. J. Pathol.* **28** 839
[94] Howry D H and Bliss W R 1952 Ultrasonic visualization of soft tissue structures of the body *J. Lab. Clin. Med.* **40** 579
[95] Howry D H 1957 Techniques used in the ultrasonic visualization of soft tissues *Ultrasound in Biology and Medicine* ed E Kelly (Washington: American Institute of Biological Sciences) p 49
[96] Donald I 1974 Sonar—the story of an experiment *Ultrasound Med. Biol.* **1** 109
[97] Edler I and Gustafson A 1957 Ultrasonic cardiogram in mitral stenosis *Acta Med. Scand.* **159** 85
[98] Satomura S 1957 Ultrasonic Doppler method for the inspection of cardiac functions *J. Acoust. Soc. Am.* **29** 1181
[99] Kaneko Z, Kotani H, Komuta K and Satomura S 1961 Studies on peripheral circulation by ultrasonic blood-rheograph *Japan. Circ. J.* **25** 203
[100] Baker D W and Watkins D 1970 Pulsed ultrasonic Doppler blood-flow sensing *IEEE Trans. Sonics Ultrason.* **SU-17** 170
[101] Peronneau P A, Hinglais J R, Pellet H M and Leger F 1970 Vélocimètre sanguin par effet Doppler à émission ultra-sonore pulsée *Onde. Élect.* **50** 369
[102] Wells P N T 1969 A range-gated ultrasonic Doppler system *Med. Biol. Eng.* **7** 641
[103] Somer J C 1968 Electronic sector scanning for ultrasonic diagnosis *Ultrasonics* **6** 153
[104] Bom N, Lancée C T, Honkoop J and Hugenholtz P G 1971 Ultrasonic viewer for cross-sectional analysis of moving cardiac structures *Biomed. Eng.* **6** 500
[105] Kasai K, Namekawa K, Koyano A and Omoto R 1985 Real-time two-dimensional blood flow imaging using an autocorrelation technique *IEEE Trans. Sonics Ultrason.* **32** 458
[106] Fry W J, Mosberg W H, Barnard J W and Fry F J 1954 Production of focal destructive lesions in the central nervous system with ultrasound *J. Neurosurg.* **11** 471

[107] Chaussy C 1982 *Extracorporeal Shock Wave Lithotripsy* (Basel: Karger)
[108] Radon J 1917 Über die Bestimmung von Functionen durch ihre Integralwerte langs gewisser Mannigfaltigkeiten *Ber. Verh. Sachs. Akad. Wiss. Lpz Math. Phys. Kl* **69** 262
[109] Kuhl D E 1964 A Cylindrical Radioisotope Scanner for Cylindrical and Section Scanning *Medical Radioisotope Scanning* vol 1 (Vienna: IAEA) pp 273–89
[110] Bowley A R, Taylor C G, Causer D A, Barber D C, Keyes W I, Undrill P E and Mallard J R 1973 A radioisotope scanner for rectilinear, arc, transverse and longitudinal section scanning (ASS—the Aberdeen Section Scanner) *Br. J. Radiol.* **46** 262–71
[111] Choudhury A R, Keyes W I and MacDonald A F 1974 Cerebral scanning including transverse section technique in the investigation of systematic epilepsy *Hans Berger Centenary Symposium on Epilepsy (Edinburgh, 1973)* ed E Harris and D Maudsley (London: Churchill Livingstone) pp 243–9
[112] Chesser R and Gemmell H G 1982 The interfacing of a gamma camera to a DEC gamma-11 data processing system for single photon emission tomography *Phys. Med. Biol.* **27** 437–41
[113] McIntyre W J, Fedoruk S O, Harris C C, Kuhl D E and Mallard J R 1969 Sensitivity and resolution in radioisotope scanning: a report to the International Commission of Radiation Units and Measurements *Medical Radioisotope Scintigraphy* vol 1 (Vienna: IAEA) pp 391–435
Also in 1969 *Nucl. Medizin* **8** 99–146
[114] ICRU Medical imaging: the assessment of image quality *ICRU Report* 54 (Washington, DC: ICRU)
[115] Ter-Pogossian M M 1992 The origins of positron emission tomography *Semin. Nucl. Med.* **22** 140–9
See also Maisey M and Jeffery P 1992 Clinical applications of PET *Br. J. Clin. Prac.* **45** 265–73
[116] Hounsfield G 1973 Computerized transverse axial scanning (tomography). Part 1: Description of system *Br. J. Radiol.* **46** 1016
Ambrose J and Hounsfield G 1973 Part 2: Clinical applications *Br. J. Radiol.* **46** 1023
[117] Pullan B R 1979 The scientific basis of computerized tomography *Recent Advances of Radiology and Medical Imaging* vol 6, ed T Lodge and R Steiner (Edinburgh: Churchill Livingstone) p 1
Webb S (ed) 1988 *The Physics of Medical Imaging* (Bristol: Hilger)
Webb S 1990 *From the Watching of Shadows: the Origins of Radiological Tomography* (Bristol: Hilger)
[118] Cook P D and Mallard J R 1963 An electron spin resonance cavity for the detection of free radicals in the presence of water *Nature* **198** 145–7
Mallard J R and Kent M 1964 Differences observed between electron spin resonance signals from surviving tumour tissues and from their corresponding normal tissues *Nature* **204** 1192; 1966 Electron spin resonance in surviving rat tissues *Nature* **210** 588–91
[119] Mallard J R and Lawn D G 1967 Dielectric absorption of microwaves in human tissues *Nature* **213** 28–30
Mallard J R and Whittingham 1968 *Nature* **218** 366–7
[120] Damadian R 1971 Tumor detection by nuclear magnetic resonance *Science* **171** 1151–3
[121] Gordon R E 1974 Proton NMR relaxation time measurements in some biological tissues *PhD Thesis* University of Aberdeen, UK
[122] Mallard J R, Hutchison J M S, Edelstein W A, Ling R and Foster M A

1979 Imaging by nuclear magnetic resonance and its bio-medical implications *J. Biomed. Eng.* **1** 153–60

[123] Mallard J R, Hutchison J M S, Edelstein W A, Ling C R, Foster M A and Johnson G 1980 *In vivo* NMR imaging in medicine: the Aberdeen approach, both physical and biological *Phil. Trans. R. Soc.* B **289** 519–33

[124] Lauterbur P C 1973 Image formation by induced local interactions: examples employing nuclear magnetic resonance *Nature* **242** 190–1

[125] Hutchison J M S, Mallard J R and Goll G C 1974 *In-vivo* imaging of body structures using proton resonance *Proc. 18th Ampere Conf. (Nottingham, UK, 1974)* ed P S Allen, E R Andrew and C A Bates (Nottingham: University of Nottingham) pp 283–4

[126] Andrew E R 1980 NMR imaging of intact biological systems *Phil. Trans. R. Soc.* B **289** 471–81

[127] Garraway A N, Grannell P K and Mansfield P 1974 Image formation in NMR by a selective irradiative process *J. Phys. C: Solid State Phys.* **7** L457–462

[128] Mansfield P, Morris P G, Ordidge R J, Pykett I L, Bangert V and Coupland R E 1980 Human whole body imaging and detection of breast tumours by NMR *Phil. Trans. R. Soc.* B **289** 503–10

[129] Moore W S and Holland G N 1980 Experimental considerations in implementing a whole body multiple sensitive point nuclear magnetic resonance imaging system *Phil. Trans. R. Soc.* B **289** 511–8

[130] Young I R and Clow H 1978 *New Sci.* November p 588

[131] Damadian R 1980 Field focusing NMR (FONAR) and the formation of chemical images in man *Phil. Trans. R. Soc.* B **289** 489–500

[132] Hutchison J M S 1976 *Imaging by Nuclear Magnetic Resonance Proc 7th L H Gray Conf. (Leeds, 1976)* (Chichester: Wiley) pp 135–41

[133] UK Patent Number 2079946A March 1981. See also Edelstein W A, Hutchison J M S, Johnson G and Redpath T W 1980 Spin-warp NMR imaging and application to human whole-body imaging *Phys. Med. Biol.* **25** 751–6

[134] Mallard J R, Hutchison J M S, Foster M A, Edelstein W A, Ling C R, Smith F W, Selbie R, Johnson G and Redpath T W 1980 Medical imaging by nuclear magnetic resonance—a review of the Aberdeen physical and biological programme *Medical Radionuclide Imaging* (Vienna: IAEA) pp 117–44

See also Smith F W, Mallard J R, Hutchison J M S, Reid A, Johnson J, Redpath T W and Selbie R D 1981 Clinical application of nuclear magnetic resonance *Lancet* January pp 78–9

[135] An example is Pollet J E, Smith F W, Mallard J R, Ah-See A K and Reid A 1981 Whole body nuclear magnetic resonance imaging in medicine: the first report of its use in surgical practice *Br. J. Surg.* **68** 493–4.

[136] Mallard J R 1986 The Wellcome Foundation Lecture 1984. Nuclear magnetic resonance imaging in medicine: medical and biological applications and problems *Proc. R. Soc.* B **226** 391–419

[137] Redpath T W, Hutchison J M S, Eastwood L M, Selbie R D, Johnson G, Jones R A and Mallard J R 1987 A low field imager for clinical use *J. Phys. E: Sci. Instrum.* **20** 1228–34

[138] Ackerman J H, Grove T H, Wong G G, Gadian D G and Radda G 1980 Mapping of metabolites in whole animals by P631 NMR using surface coils *Nature* **283** 167–70

[139] Isherwood I (Chairman) 1992 *Report of the Working Party on the Provision of Magnetic Resonance Imaging Services in the UK* (London: Royal College of Radiologists)

[140] Foster M A 1984 *Magnetic Resonance in Medicine and Biology* (Oxford: Pergamon)
Foster M A and Hutchison J M S 1989 NMR *Imaging* (Oxford: IRL/Oxford University Press)
[141] Lurie D J, Nicholson I, Foster M A and Mallard J R 1990 Free radicals imaged *in-vivo* in the rat by using proton-electron double resonance imaging (PEDRI) *Phil. Trans. R. Soc.* A **333** 453–6
[142] Mallard J R 1994 The birth of the international organizations—with memories *Scope* **3** No 2 25–31

Chapter 26

GEOPHYSICS

S G Brush and C S Gillmor

26.1. Introduction

Geophysics in the twentieth century is such a diverse field and has ties with so many adjacent scientific communities that it is not possible to present a complete history in this chapter. For example, the neutral atmosphere of the Earth, and the oceans, receive little attention here. We have chosen to describe some major contributions to knowledge of the solid Earth (including its origin and age) and of 'geospace', the ionized envelope extending to thousands of miles around the Earth.

See also p 887

'Geophysics' has meant different things in various parts of the world. In many countries the term has included most studies of the Earth that centrally involve physics. In the United States until recent decades, geophysics was a portion of the geological sciences and signified structural studies of the solid Earth. In France, geophysics particularly included geomagnetism. This was true also of Germany and of countries controlled or influenced in the nineteenth century by the German Empire. In the United Kingdom and some Commonwealth countries, upper-atmosphere geophysics was studied in departments of physics, and quite often by mathematicians in departments of 'mathematics and theoretical physics'. Since radio techniques so influenced work in the United States, upper-atmosphere geophysics was studied predominantly in government and industrial laboratories and in electrical engineering departments in universities. With the arrival of the 'space age' in the late 1950s, a considerable portion of upper-atmosphere geophysics research in the United States evolved from physics and applied physics departments performing rocket studies of solar and cosmic-ray physics, but the military interest in defence communications continued to support the large output of research from departments of electrical engineering. In Australia, mixed alliances to American engineering and to British physics saw impressive amounts of upper-atmosphere research from the 1920s onwards both from physics and from electrical engineering departments. Japan conducted upper-atmosphere geophysics research, on the German model, from physics and geomagnetism centres until the conclusion of World War II. Under the American occupation such research shifted to electrical engineering research centres.

Geophysics commingles with physics, the earth and biological sciences and with technology. Ionospheric physics, the physics of the Earth's upper atmosphere, can serve as an example of how geophysics both stimulates and is stimulated by other scientific and technical fields. Beginning in the late 1920s at the Bell Telephone Laboratories, radio astronomy was first developed by ionosphere researchers who were attempting to survey noise sources hampering transatlantic ionospherically propagated radio communications. Other pioneers in radio astronomy were ionospheric workers who studied radio propagation during meteor showers, and radio amateurs who studied non-thermal radio noise bursts from the Sun. Radio astronomy received part of its impetus from numerous British and British Empire ionosphere workers who, after working on radar during World War II, took surplus radar equipment and turned it towards the sky. Except for the Dutch, the early post-war radio astronomers tended to be British and Australians who had some study of the ionosphere in their background. A number of ionospheric workers became connected with operations research and computers during and after the war. This was natural, since geophysics has to solve very difficult differential equations; and as in astronomy, geomagnetism, meteorology, and some other physical sciences, literally years were spent in manual calculations. Ionosphere workers, many of whom were electrical engineers, developed computer techniques and hardware for solving ionospheric problems. For a number of these people the techniques and design of the computer became so interesting they never returned to geophysics.

26.2. Origin and age of the Earth (to 1935) [1][†]

The origin of the solar system, and of the Earth in particular, is the most fundamental problem in geophysics. Apart from the intrinsic fascination of understanding how and when our planet was formed, a plausible solution to that problem could suggest an answer to a more general question: is the formation of a planetary system a normal feature of star formation, or a rare accident? Are planetary systems likely to be so abundant throughout the Universe that the chance of communicating with other civilizations is high, or is life on Earth the lonely exception in a dead galaxy?

At the end of the nineteenth century most astronomers accepted the 'nebular hypothesis' proposed 100 years earlier by the French theorist

† S G Brush is the author of sections 26.2–26.5.

Pierre Simon de Laplace (1749–1827) and the German–British astronomer William Herschel (1738–1822). According to Laplace, the atmosphere of the primaeval Sun extended throughout the entire space now occupied by planetary orbits; it was a hot, luminous rotating cloud of gas, gradually losing its original heat to the cold surrounding space. As the cloud cooled it contracted, rotated faster, and broke up into rings which then condensed to planets. Satellites could be formed by a similar process from clouds surrounding the planets. The remaining central portion would become the Sun.

Herschel suggested that stars are formed by contraction from nebulae like those he observed through his telescope. Thus the Laplace–Herschel nebular hypothesis implied that planetary formation is a universal accompaniment of the birth of stars, and that with sufficiently powerful telescopes we should be able to observe the later stages of the process occurring in our galaxy right now.

The nebular hypothesis was closely connected with nineteenth century geological theories, which generally presumed that the Earth had been formed as a hot fluid ball and then cooled down, solidifying on the outside first. According to this 'contraction theory' the solid crust would not contract as rapidly as the fluid interior, so it would have to wrinkle in order to adjust its diameter to that of the shrinking core; in this way one could explain the formation of mountain ranges and other surface features.

Before about 1840 it was thought that the solid crust is only 50 or 60 miles thick, the rest of the interior being a hot fluid (figure 26.1). That estimate was based on extrapolating the increasing temperature measured as one goes down below the surface, until one gets to a temperature high enough to melt all known rocks. However, the British geophysicist William Hopkins (1793–1866) pointed out that since the melting point probably increases with pressure, the solid crust could be considerably thicker than 50 miles. He thought it even possible that the high pressure at the centre would cause solidification despite the high temperature.

Lord Kelvin [William Thomson] (1824–1907), the most influential British physicist of the nineteenth century, adopted the general scheme of a cooling Earth but attacked two features that geologists relied on for their explanations. First, he estimated the time required to cool down from an initial hot fluid state, using Joseph Fourier's heat conduction theory, and found it to be only 20 to 100 million years compared with the hundreds of millions of years geologists assumed to have been available for slow processes like erosion to produce their observed effects. Second, Kelvin showed that the fluid interior could not be so large nor the crust so thin as geologists believed; if it were, tidal forces would quickly break up the crust and other physical effects that are not observed would be present. Kelvin went to the other extreme and concluded that the entire

Twentieth Century Physics

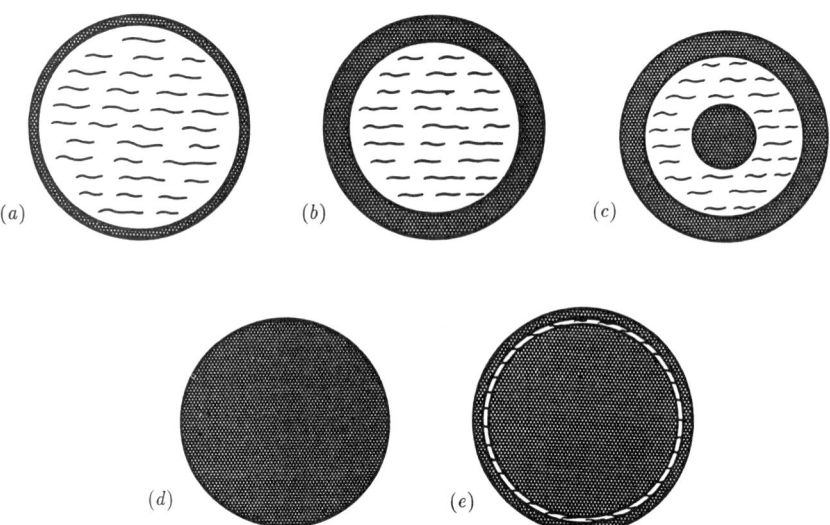

Figure 26.1. (a) Structure of the Earth as imagined by the geologists in the early nineteenth century. They assumed that the observed increase of temperature with depth below the surface could be extrapolated to a depth corresponding to the melting point of rocks, about 50 or 60 miles. Thus a thin solid crust surrounds a hot fluid interior. Lava in volcanic eruptions was thought to come directly from this fluid. (b) Revised model for Earth structure suggested by Hopkins. William Hopkins pointed out in 1839 that since pressure as well as temperature increases with depth below the surface, and the melting temperature probably increases with pressure, rocks would remain solid down to at least several hundred miles. (c) Alternative model suggested by Hopkins. If the melting temperature of the material inside the Earth increases very rapidly with pressure, the central region would be solid, with a liquid region between the solid crust (or 'mantle' as it is now called) and the solid inner core. A century later, a model qualitatively similar to this was revived as a result of the work of Harold Jeffreys and Inge Lehmann (see figure 26.6). (d) Kelvin's model. Lord Kelvin, following an argument of Hopkins, at first considered that the Earth must be entirely solid because one gets the correct values for its rotational properties (precession and nutation) on that assumption. Later he recognized that this argument could be wrong if the fluid interior interacts with the crust so that it rotates in the same way. A better argument is: if the Earth was once a hot fluid sphere, as it cooled solidification would have started at the centre where the pressure was highest. If rocks contract on solidification, then any solid material formed at the outside of the sphere would sink to the centre and could not form a stable solid crust until the entire inside was solid. The best argument (especially against the thin-crust model shown in (a)) is that the same forces that raise tides in the Earth's ocean would raise tides in the interior fluid, either pushing up the crust (so we would observe no net tidal motion in the ocean) or, more likely, would break up the crust. Taken together these arguments persuaded many scientists that the Earth is entirely solid; in the early twentieth century, direct measurements of the rigidity showed that the Earth as a whole is 'as rigid as steel'. (e) Geologists' model of the Earth at the end of the nineteenth century. To explain vulcanism and tectonic processes, geologists (especially British and American) postulated a thin fluid or at least 'plastic' layer beneath the crust. As a concession to Lord Kelvin's arguments (see (d)) they were willing to assume that the rest of the interior is solid.

Geophysics

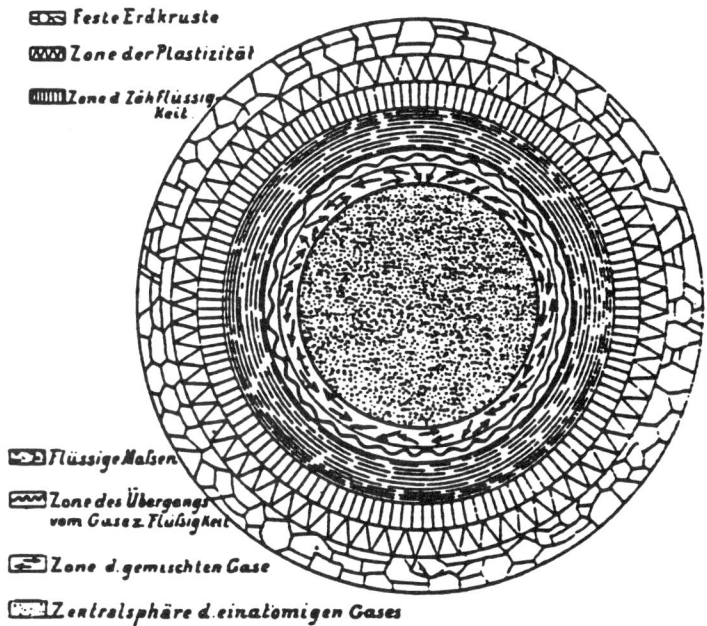

Figure 26.2. *Alternative gaseous-interior model of the Earth. Geophysicists, especially in Germany, proposed that material at the high temperatures and pressure inside the Earth would go through various degrees of plasticity, viscosity (Zähflüssigkeit), fluidity, gaseous fluidity, eventually becoming a supercritical fluid with its molecules broken down into a monatomic gas* [2].

Earth is now solid—'as rigid as steel', but some geologists insisted that a thin fluid or plastic layer beneath the crust was needed to explain mountain-building, volcanoes and crustal movements.

An alternative model, popular in Germany at the end of the nineteenth century, invoked the 'critical point' phenomenon studied by Charles Cagniard de la Tour and Thomas Andrews. At sufficiently high pressures and temperatures, such as those expected to be found inside the Earth, the liquid–gas boundary disappears and one has simply a supercritical fluid (figure 26.2).

Kelvin's limit on the age of the Earth caused difficulties for Charles Darwin's theory of biological evolution. Darwin had assumed that much longer periods of time—of the order of several hundred millions of years—would be available for present-day species (including humans) to emerge by the slow process of natural selection. Thus the late nineteenth century saw a public debate between physicists (led by Kelvin) on one side, and geologists and biologists on the other. Physicists argued that if the findings of other sciences came into conflict with the laws of Nature (as physicists understood those laws), such findings must be invalid. Geologists complained that physicists paid no respect to geological evidence. Yet even though the physicists turned out to be wrong in this particular case, geologists in the twentieth century have accepted physical methods as the most reliable means for establishing the chronology of the Earth's history and the overall structure of the Earth's crust and interior.

Charles Darwin's son, George Howard Darwin (1845–1912), worked with Kelvin and became a leading geophysicist. He was best known for his theory of the evolution of the Moon's orbit. Astronomers had analysed the gradual changes observed in this orbit, known as 'secular acceleration' because the Moon seemed to be moving more rapidly as the centuries went by. According to Newtonian gravitational theory (or Kepler's Third Law), this would mean that the Moon is gradually coming closer to the Earth, but in 1865, the American geophysicist William Ferrel (1817–91) and the French astronomer Charles-Eugène Delaunay (1816–72) proposed that the observed secular acceleration is partly due to a slowing-down of the Earth's rotation, produced by tidal dissipation in the Earth–Moon system. The net result is that over a long period of time the Moon is actually moving more slowly in its orbit, and the Earth–Moon distance is increasing as angular momentum is transferred from the Earth to the Moon. In 1878, G H Darwin showed that this effect could be extrapolated back into the past; about 50 million years ago the Moon's centre would have been no more than 6000 miles from the Earth's surface, and its period of revolution would have coincided with the Earth's period of rotation—about 5.5 hours.

Although the orbit could not be rigorously deduced for earlier times, G H Darwin surmised that the Moon had once been part of the Earth and had been ejected from it in a catastrophic event triggered by the Sun's tidal force. (The force could amplify the free oscillations of the proto-Earth by resonance.) The English geophysicist Osmond Fisher (1817–1914) then suggested that the scar left by the Moon's separation did not completely heal, and that the ocean basins are holes left in the Earth's crust after some flow of the remaining solid towards the original cavity. In this way the birth of the Moon would have produced both the Pacific Ocean basin and the separation of the American continents from Europe and Africa.

The Darwin–Fisher hypothesis attained great popularity in the early twentieth century in the simplified version 'the Moon came out of the Pacific Ocean'. It gradually fell out of favour after 1930 when the British geophysicist Harold Jeffreys (1891–1989) and others presented physical arguments (viscous damping, insufficient angular momentum) to show that the process could not have taken place in the way Darwin proposed.

One alternative was to go back to Laplace's scheme for the formation of planets from a large nebula and postulate an intermediate stage in which each planet condensed from a small nebula while spinning off rings that became satellites. While that might work for the multi-satellite systems of the giant planets it did not offer a plausible explanation of the unique characteristics of the Earth's moon (only one satellite, relatively large but of low density). Another theory, popular in the 1950s and 1960s, was that the Moon was formed elsewhere in the solar system and was subsequently captured gravitationally by the Earth.

At the beginning of the twentieth century the credibility of the Laplace–Herschel–Kelvin scenario for the development and present state of the Earth was assaulted from several directions. First came the discovery of radioactivity, which undermined the assumptions Kelvin had used to estimate the age of the Earth and eventually led to a much better method. By the 1890s Kelvin and other physicists had reduced their estimates to only 24 million years. That figure was consistent with an estimate of the age of the Sun, first suggested by Hermann von Helmholtz, based on the assumption that its heat is derived from gravitational contraction, but after the isolation of radium by Marie and Pierre Curie, it was found that the decay of this radioactive substance produces an enormous amount of heat. Kelvin had assumed no source of heat inside the Earth, but now it appeared that radium generates so much heat that it could replace the heat lost from the Earth's crust by conduction into space; the Earth might actually be warming up rather than cooling down (but this indirectly supported Kelvin's other proposition, that the entire Earth is solid, by suggesting that it is not extremely hot inside). Similarly it was proposed that radium in the Sun could provide enough energy to keep it shining for much longer than 20 million years.

Following the proposal by Ernest Rutherford and Frederick Soddy in 1902 that radioactive decay involves the transmutation of one element into another, Rutherford and others recognized that precise measurements of the relative proportions of these elements could be used to estimate the ages of rocks and hence the minimum age of the Earth's crust. Rutherford's first estimates were in the range of 40 to 500 million years. R J Strutt in 1905 was the first to pass the (American) billion mark when he published an estimate of 2400 million years for a specimen of thorianite, although this was subsequently considered too high. By 1915 the figure of 1600 million years, based on lead/uranium ratios, was widely accepted as the best estimate of the age of the oldest minerals and of the Earth's crust.

Even before radioactivity had cast doubts on the nineteenth century cooling Earth model, the American geologist Thomas Chrowder Chamberlin (1843–1928) had proposed to replace that model by a completely different hypothesis about the origin of the Earth and other planets. Chamberlin first became sceptical about the standard assumption that the Earth had condensed from a hot fluid when he realized that, according to the kinetic theory of gases, not only hydrogen but also oxygen and nitrogen would have escaped from the primaeval Earth (their average molecular speeds would have

1949

exceeded the gravitational escape velocity). With the help of the American astronomer Forest Ray Moulton (1872–1952), he showed that the nebular hypothesis could not satisfactorily explain the physical properties of the solar system. It was already known that the Laplace process was inconsistent with the observed slow rotation of the Sun; Chamberlin and Moulton proved quantitatively that the nebular hypothesis completely fails to account for the distribution of angular momentum in the solar system. Chamberlin proposed that the Earth had been formed by the aggregation of small cold particles, which he called 'planetesimals' (meaning 'infinitesimal planets') and frictional dissipation and differentiation processes would generate enough heat to bring it up to its present temperature. However, Chamberlin agreed with Kelvin that the Earth is now completely solid; this view prevailed until the 1920s (see below) and provided a formidable obstacle to the acceptance of Wegener's continental drift theory.

Having rejected Laplace's theory of the formation of the solar system, Chamberlin still seemed to be attached to Herschel's old idea that the nebulae one sees in the sky are associated with the early stages of stellar evolution. Looking at photographs of spiral nebulae taken by the American astronomer James Keeler, Chamberlin speculated that the two prominent arms belonged to two previously distinct celestial objects. From this thought, and contemplation of solar prominences, he was led to the idea that a planetary system could be generated when another star passed close to the Sun, as explained in figure 26.3.

When it became clear in the 1920s that spiral nebulae are galaxies rather than objects that could be as small as planetary systems, Chamberlin dropped this part of his theory but retained the assumption that two stars interacted in order to release into space the material from which planets formed. As it happened, Harold Jeffreys and the British physicist–astronomer James Hopwood Jeans (1877–1946) independently proposed a similar assumption in 1916 and 1917, respectively. It became known as the 'tidal' theory of the origin of the solar system. But Jeans and Jeffreys rejected Chamberlin's planetesimal hypothesis, retaining instead the older assumption that the planets formed from fluid balls (figure 26.4).

Astronomers realized that any theory which required the close encounter of two stars in order to form planets would entail an extremely small number of planetary systems in the Universe (and an even smaller number of planets that could support the evolution of life). This was consistent with the failure to find convincing evidence for planetary systems surrounding stars other than our own. Jeans seemed to take perverse pleasure in the idea that we are the result of a chance event that has happened only once in the history of the Universe and (because the stars are decaying and thinning out by expansion) will probably never happen again.

Geophysics

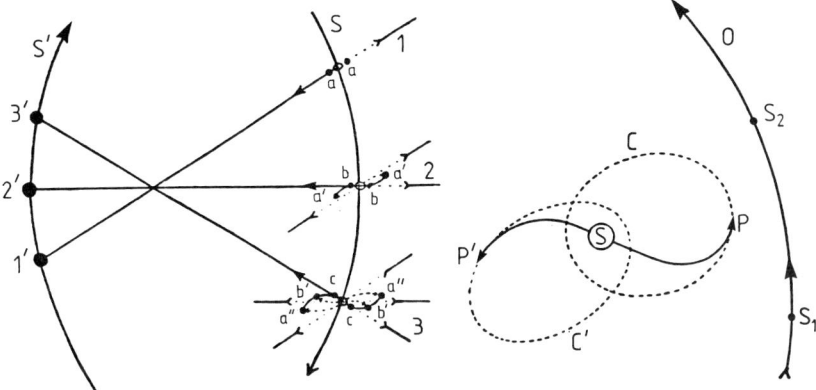

Figure 26.3. *Creation of the solar system by tidal action of a passing star on the Sun (Chamberlin's version). Chamberlin, inspired by Keeler's photographs of spiral nebulae, imagined that such a nebula could be produced when an intruder star S' (in successive positions 1', 2', 3') exerts a tidal force on the Sun S (in successive positions 1, 2, 3). As a result, material is drawn out of the Sun on near and far sides (a). In the second stage, more material (b) has emerged from the Sun, but it will experience a force in a different direction because of the subsequent motion of S' from 1' to 2'; similarly for material (c) emerging in the third stage. The resulting curvature of the filaments produces the spiral-arm appearance. The gaseous filaments condense first to small solid particles (planetesimals) which then aggregate into planets* [3].

The tidal theory, whether the Chamberlin–Moulton or Jeans–Jeffreys version, was generally accepted by astronomers until 1935, even though it was never worked out in sufficient detail to provide a convincing explanation of the quantitative properties of the solar system. Its supporters believed that the tidal theory could overcome the major defect of the nebular hypothesis by showing at least qualitatively how most of the original angular momentum could have been given to the major planets rather than to the Sun.

However, the tidal theory turned out to have serious defects as well. The American astronomer Henry Norris Russell (1877–1957) found two major objections, which he presented in an influential book published in 1935 [5]. First, theories of stellar structure developed by A S Eddington and others in the 1920s indicated that the gases from the interior of the Sun would be at such a high temperature—of the order of a million degrees—that they would dissipate into space before they could condense into planets. Second, a simple dynamical calculation showed that it would be impossible for the tidal encounter to leave enough material in orbits at distances from the Sun corresponding to the giant planets.

There was now *no* satisfactory theory of the origin of the Earth, since the earlier objections to the nebular hypothesis still seemed valid. Moreover, since Jeffreys had shown in 1930 that G H Darwin's hypothesis

1951

Twentieth Century Physics

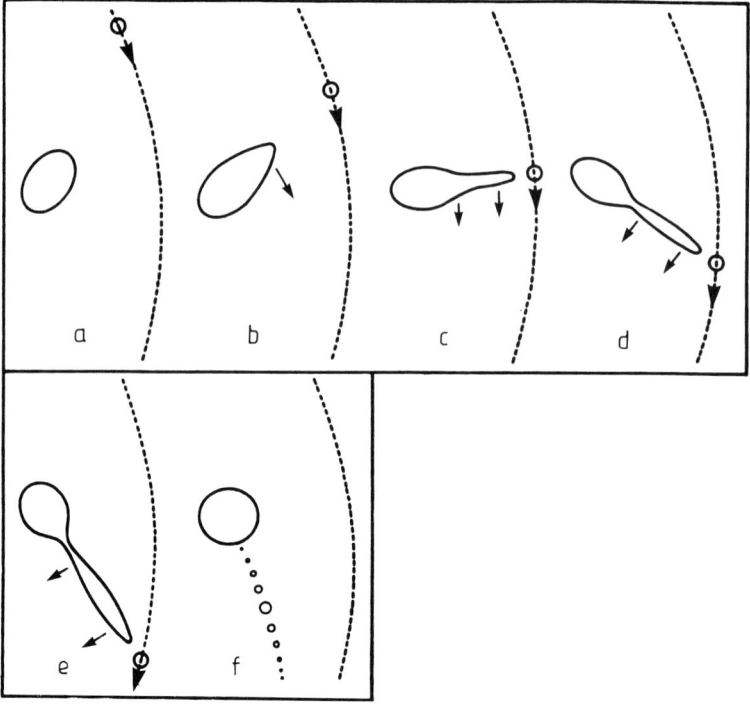

Figure 26.4. *Creation of the solar system by tidal action of a passing star on the Sun (Jeans' version). Jeans imagined that the Sun would first be deformed by the tidal force of the passing star, then a single coherent filament would be detached and subsequently break up into separate fluid blobs which would condense directly to planets. According to Jeans 'This sequence of six diagrams (a, b, c, d, e, f) is intended to suggest the course of events in the Sun, or other star, when its planets were coming into existence. For simplicity the second star is represented throughout by a small circle, although actually this would, of course, also experience deformation and possible break-up'* [4].

about the separation of the Moon from the Earth was inadequate, there was also no satisfactory theory of the origin of the Moon. Before telling how scientists managed to get out of this quandary, let us see how the problem of the Earth's internal structure was revived and resolved.

26.3. The Earth's core and magnetism

Magnetism provided one of the first clues about the nature of the Earth's interior. Before the seventeenth century, the direction of a magnetic compass needle was thought to be determined by influences from the heavens, but in 1600 the British physician William Gilbert (1544–1603) argued that the needle is governed by magnetic forces originating inside the Earth. Using a spherical piece of magnetic iron ore to simulate the Earth, he found that the pattern of magnetic field lines around it matched

1952

the pattern deduced from motions of compass needles at various places on the Earth's surface.

Gilbert's result suggested that the Earth contains a substantial amount of magnetized iron—but is it entirely solid, like a piece of iron ore? New evidence suggested otherwise. As sailors brought back more information about compass readings, it was noticed that the Earth's magnetic field seemed to be changing slowly with time: the field pattern was drifting to the west. This would seem to be impossible if the field was produced by a solid permanent magnet.

To account for the westward drift of the Earth's magnetic field, the British astronomer Edmond Halley (1656–1743) proposed that the field originates in a central core, which is separated from the solid crust by a fluid region. Both crust and core rotate eastward, but the core lags slightly behind the crust so that the magnetic field slips westward relative to the crust.

In the nineteenth century the idea that the Earth's interior is a hot fluid (see above), together with the realization that iron loses its magnetism at high temperatures, seemed to contradict the assumption that terrestrial magnetism originates in a solid iron core. But the discovery of electromagnetic interactions by Oersted, Ampère and Faraday led to the alternative hypothesis that the Earth's magnetic field arises from electrical currents.

After Kelvin had persuaded scientists that the Earth is completely solid, the German geophysicist Emil Wiechert (1861–1928) developed a simple Earth model consisting of a solid iron core and rocky shell. The parameters were chosen to fit measured physical properties of the Earth (density, ellipticity, precession and nutation). The radius of the core was about 4970 km and its density 8.2 (slightly compressed from the normal density of iron, 7.8), with a shell about 1400 km thick having a density of 3.2. Wiechert's model dominated thinking about the Earth's internal structure at the beginning of the twentieth century, but it did not claim to explain the Earth's magnetism.

Further progress came from an unexpected direction. Kelvin had suggested that experiments should be done to measure the tides of the solid Earth—that is, the vertical motion of the Earth's surface associated with lunar and solar positions—in order to determine the rigidity of the Earth. The results seemed to confirm his claim that the Earth is as rigid as steel, but they also led to a discovery that eventually overturned his solid-Earth theory. The first apparatus sensitive enough to measure surface motions accurately, a pendulum constructed by the German astronomer–geophysicist Ernst von Rebeur-Paschwitz (1861–95), was also the first to detect earthquake vibrations that had passed through the inside of the Earth. On 18 April 1889, a major earthquake was registered on the seismograph at the Imperial University of Tokyo; just 64 minutes after the impulse was recorded at Tokyo, Rebeur-Paschwitz observed the disturbance with his instruments at Potsdam and Wilhelmshaven. The seismic waves had gone a distance of more than 5000 miles through the body of the Earth, at an average speed of more than a mile per second.

From that moment on the science of seismology, previously limited to the study of local catastrophes, provided a new method for analysing the structure of the entire Earth. Rebeur-Paschwitz suggested an international network of seismological observatories be set up to record data on standard instruments. After his early death, his dream was realized when it became possible to compare the time it took for seismic waves from an earthquake to arrive at various points on the Earth's surface. The paths taken by the waves, and the variations in speed along the way, could then be reconstructed. Eventually, each depth was found to have a characteristic velocity that could provide information about the nature of the material at that depth.

R D Oldham (1858–1936), an Irish geologist, was one of the first scientists to analyse seismic wave paths inside the Earth. In his report on a 1902 earthquake in Guatemala, Oldham concluded that the Earth has a core in which the seismic wave speed is substantially lower than it is in the surrounding material. He suggested that waves entering the core at an oblique angle change their direction, just as a beam of light is refracted when it enters a medium in which its speed is less than it is in air. A spherical core would bend seismic waves entering at different angles, and they would be bent again on leaving, in such a way that a 'shadow zone' is created (figure 26.5).

Further refinement of seismological analysis led to a very accurate determination of the size of the core by the German–American seismologist Beno Gutenberg (1889–1960) in 1912. Gutenberg established that a sharp change in wave velocity—a decrease of more than 30%—occurs at a depth of 2900 km below the Earth's surface. Thus Gutenberg's core is substantially smaller than Wiechert's.

At about the same time, the Croatian geophysicist Andrija Mohorovičić (1857–1936) found a smaller but significant change in seismic wave speed at a surface that is about 5 km below the ocean floors and 50 km below the continents. The region between this surface and that discovered by Gutenberg is now called the 'mantle' of the Earth. Thus the Mohorovičić discontinuity (the 'Moho') is the boundary between the Earth's crust and mantle, while the Gutenberg discontinuity is the boundary between mantle and core.

Seismologists still accepted Kelvin's view that the entire Earth is solid. They assumed that the changes in seismic wave speed at various depths beneath the Earth's surface are due, not to changes in *physical* state, but to changes in the density of the material resulting from variations in *chemical* composition. Accordingly, the reduced speed in the core was ascribed to its being made of iron, whereas the mantle is made of rocky material.

Geophysics

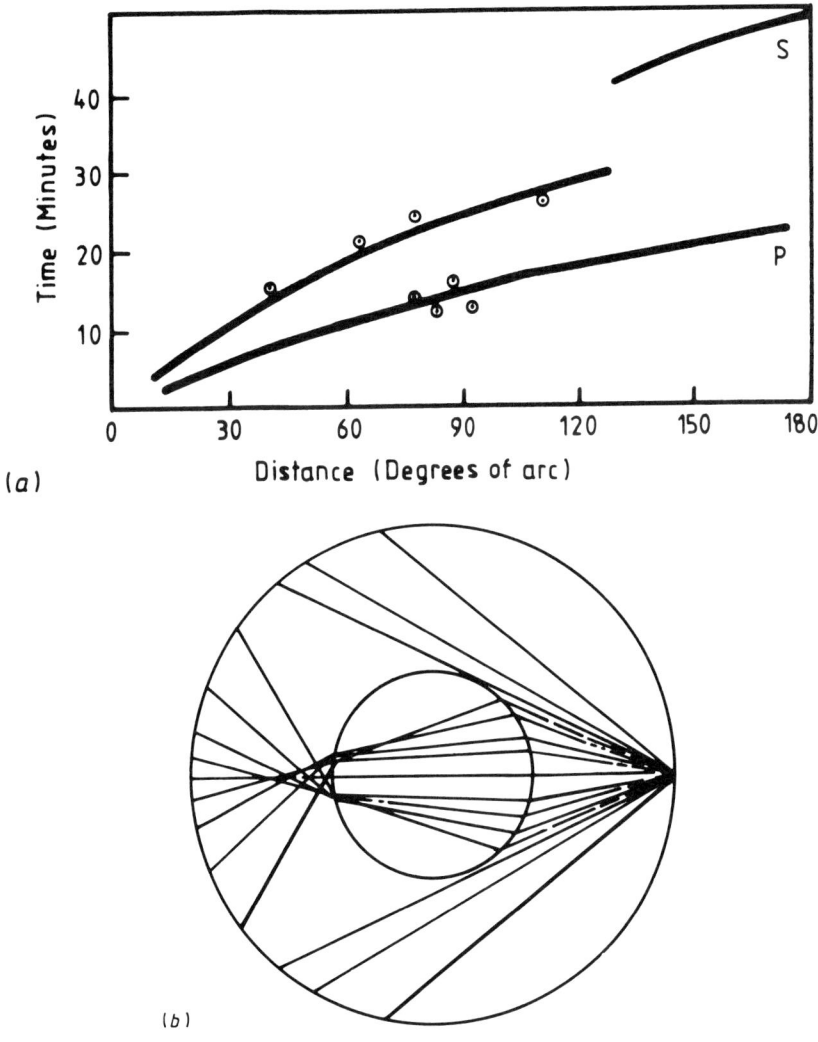

Figure 26.5. *The Earth's core inferred from seismological records. Oldham collected data on the time taken for the first (P) and second (S) phases of preliminary tremors from an earthquake to reach points at different arcual distances on the Earth's surface (a). The resulting curves, especially the break in the upper curve (S waves), could be explained if the seismic waves passing through the Earth's interior were refracted by a core of radius 0.4R in which the speed is 3 km s^{-1} while the speed outside the core (from 0.4 to 1.0R) is 6 km s^{-1} (b) [6]. (Other seismologists later arrived at rather different interpretations of these data, in particular they concluded that S waves do not pass through the core at all; but the qualitative idea of a large central core has survived.)*

Seismological data also contain information about the physical state of the Earth's interior. The theory of wave propagation, developed in the first part of the nineteenth century, provides a way of distinguishing

1955

between fluids and solids. There can be two kinds of wave: *longitudinal* (compression–expansion) and *transverse* (twisting motions, displacement perpendicular to the direction of wave propagation). In seismology these are known as P and S waves, respectively. Longitudinal waves can go through any substance that resists compression and tends to return to its original volume when the excess pressure is removed; this is true of ordinary solids, liquids and gases. Transverse waves can travel only in substances that have elastic resistance to twisting and tend to return to their original shape; this is true only of solids, not fluids [7]. William Hopkins was the first to point out, in 1847, how the theory of wave motion could be applied to earthquake waves.

Oldham had distinguished P and S waves in records of an 1897 earthquake in India, but the evidence in seismograph records was rather difficult to interpret, and several years elapsed before Oldham decided that he had never really seen S waves going through the Earth's core. By 1914 he suspected that the core fails to transmit S waves because it is fluid [8]. Gutenberg, however, was so confident that the core is solid that he published a graph of S wave speeds inside the core, calculated by assuming that they were proportional to the P wave speeds that he had determined there.

By 1925 it was common knowledge among seismologists that no impulses from earthquakes had been observed to pass through the core in the form of S waves. This knowledge might have persuaded them that the core is fluid, and indeed this is the argument usually given in modern textbooks, but a stronger argument was required to overcome Kelvin's doctrine of solidity.

Conclusive proof that the core (or at least a large part of it) must be fluid was provided by Jeffreys in 1926. His best evidence was that the rigidity of the mantle, which could be determined fairly accurately from the speeds of seismic waves, is much greater than the average rigidity of the Earth as a whole, as determined, for example, from tides of the solid Earth. Hence, a compensating region of very low rigidity, i.e. fluid, must exist in the core. Jeffreys showed that all the other evidence, including the fact that S waves never went through the core, agreed with this conclusion.

Jeffreys's new model of the Earth still left the mantle completely solid down to a depth of 2900 km and thus gave no help to advocates of continental drift. Indeed, Jeffreys remained one of the major critics of that hypothesis and of its modern successor, plate tectonics, until his death. Nor did it vindicate imaginative writers, such as Jules Verne and Edgar Rice Burroughs, who postulated open spaces where humans and animals could live inside the Earth.

One further discovery was needed to complete our present picture of the gross physical structure of the Earth. In 1936 the Danish seismologist Inge Lehmann (1888–1993) noted that seismic waves passing close to the

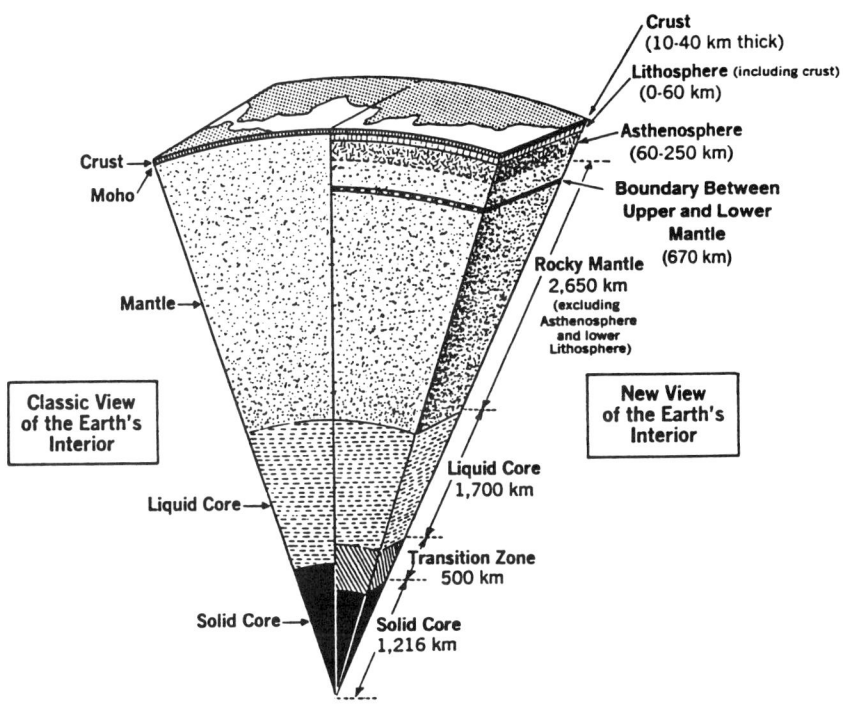

Figure 26.6. *Comparison of a simplified model of the Earth's structure accepted in the 1940s with a more complicated version which was developed in the 1960s and 1970s* [9].

Earth's centre seem to encounter a surface where the velocity jumps suddenly by a small amount. To account for this phenomenon, she proposed that there is an 'inner core' with a radius of about 1400 km (figure 26.6).

The American geophysicist Francis Birch (1903–92) suggested in 1940 that the inner core is solid iron. He conjectured that the melting temperature of iron rises fast enough with pressure to reach about 4000 °C at the centre (exceeding the actual temperature there) and that the Curie temperature increases rapidly enough with pressure to allow the possibility of a ferromagnetic inner core.

Keith Bullen (1906–76), a New Zealand geophysicist, in a number of papers in the 1940s and 1950s, presented evidence that the inner core is solid. One definitive proof would be the detection of impulses that have passed through it as S waves, and Bullen urged a search for such impulses. The first such detection was announced in 1972 by Bruce R Julian, David Davies and Robert M Sheppard, using a seismic array to synthesize data from several stations, but there is still disagreement among seismologists as to whether they interpreted the data correctly, and no other observations have settled the issue. More convincing evidence for the solidity of the inner core came from analysis of the Earth's free oscillations. These oscillations, excited by large earthquakes, were first detected by Hugo Benioff in 1952, and by several seismologists in 1960 and 1968 (these dates correspond to earthquakes in Siberia, Chile and Alaska, respectively). Calculations to infer the size and rigidity of the inner core from data on free oscillations were done by John Derr, Freeman Gilbert, A M Dziewonski, Louis Slichter, Chaim Pekeris and Frank Press.

Until the 1940s, there was very little connection between seismological research on the core (concentrating on its mechanical properties such as rigidity) and research on terrestrial magnetism. It would seem that any theory which ascribed the Earth's magnetic field to electric currents in its interior would involve some assumption about which parts (if any) of the interior are fluid. Yet Harold Jeffreys, in a review of Sydney Chapman's 1936 book on geomagnetism, complained about the 'lack of correspondence between the results and anything we know about the solid earth' [10].

Following George Ellery Hale's discovery of magnetic fields in sunspots at the beginning of the twentieth century, and his investigation of the magnetic structure of the Sun, Joseph Larmor suggested in 1919 that magnetism both in the Sun and the Earth could be due to a 'self-exciting dynamo'. It was known from the work of Oersted and Faraday that an electric current is surrounded by a magnetic field and that electric currents can be generated by the relative motion of a conductor and a magnetic field; perhaps the magnetic field generated by a current of moving electric charges could continually regenerate the current itself without help from outside.

The development of this idea was apparently blocked by T G Cowling's proof in 1933 that an axially symmetric field cannot be self-maintained. The first major theoretical breakthrough was accomplished by the German–American physicist Walter M Elsasser (1904–91) in 1946. His major contribution was the discovery that, starting with the poloidal field characteristic of a magnetic dipole, toroidal ('doughnut-shaped') fields could be generated and amplified by non-uniform fluid motion within the core (figure 26.7). Left to themselves both poloidal and toroidal fields would eventually decay to zero, but a more complicated but plausible pattern of fluid motions could regenerate the poloidal field from the toroidal field. Thus, contrary to inferences from Cowling's theorem, a self-exciting dynamo could be maintained by induction effects in a moving fluid, and could account for the major features of the Earth's magnetic field (including its secular variation) in a physically reasonable way. Elsasser emphasized the connection between the fluidity of the Earth's core as deduced from seismological evidence and the plausibility of a geomagnetic dynamo based on fluid motions.

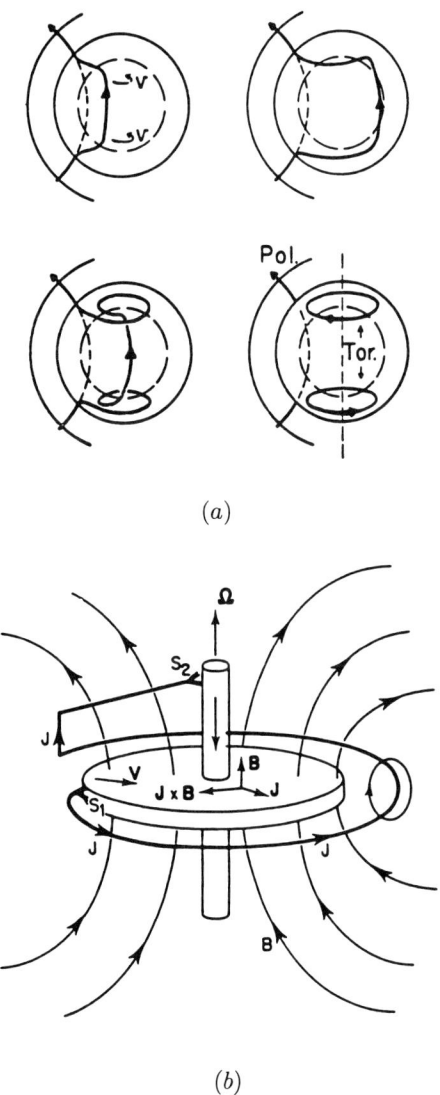

Figure 26.7. Concepts involved in geomagnetic dynamo theory. (a) In a rotating spherical system a magnetic field can be poloidal (lines of force lie in meridional planes) or toroidal (lines of force are circular about the axis) or any combination of these two. Elsasser pointed out that if the lines of force are attached to particles in a fluid which is in non-uniform rotation, an initially poloidal field will be deformed into a toroidal field [11]. (b) The simplest example of a self-exciting dynamo is the homopolar dynamo. Magnetic field B is created by current flowing in the direction shown, in the circular loop running round the periphery of the rotating disc. The current (density J) is generated inductively by the motion of the conducting disc; the velocity v is $\Omega \times r$ where Ω is the angular velocity of the disc. The electrical circuit is completed through sliding contact S_1 and S_2. The Lorentz force $J \times B$ per unit volume on the disc is in the direction shown, opposite to v. The dynamo still requires an external source of energy to keep the disc moving [12].

Twentieth Century Physics

Edward Crisp Bullard (1907–80), a British geophysicist, elaborated Elsasser's theory, supplying an explicit feedback mechanism to regenerate the field. Bullard's model was not entirely satisfactory, and despite considerable efforts by many theorists there is still no precise quantitative theory that explains the behaviour of the Earth's magnetic field. Nevertheless many of its qualitative features are now understood. In particular, the Elsasser–Bullard theory made it plausible that the total dipole field of the Earth could decay to zero (spontaneously or because of external influences) and then re-establish itself in the opposite direction. The theoretical possibility of magnetic reversals and the development of methods for determining when they had occurred in the Earth's past history were crucial to the establishment of plate tectonics in the 1960s (see below).

> See also p 1975

26.4. Origin and age of the Earth (after 1935)

Modern estimates of the age of the Earth, and of the oldest rocks on its surface, depend on nuclear physics. In particular, it was necessary to analyse the composition of rocks and meteorites into the several isotopes of lead, uranium and other elements to measure the decay rates of the radioactive isotopes and to determine the isotopic composition at the time when the rocks (or the Earth) were originally formed.

After the discovery of isotopes in 1913, the radioactive decay series ending in the various (stable) isotopes of lead began to be sorted out. By 1920, the British geophysicist Arthur Holmes (1890–1965) had concluded that the oldest minerals were at least 1600 million years old, and he estimated that the Earth's crust probably was about 2000 million years old.

Further research led to greater and more accurate ages. By 1936 it was understood that two isotopes of lead—206 and 207—were the stable products of two decay series starting with ^{238}U and ^{235}U, respectively. Another isotope, ^{208}Pb, came from a thorium isotope. The fourth stable isotope, ^{204}Pb, was not produced by any such process and was, therefore, called 'non-radiogenic' (like the isotopes of uranium, it was presumably produced before the Earth's formation by nuclear reactions in stars). Using a mass spectroscope, the American physicist Alfred Nier (1911–94) found that these isotopes did not occur in the same proportions in all rocks, even though the average atomic weight of the isotopic mixture was about the same from rock to rock. Assuming that a rock with the highest proportion of non-radiogenic lead would be the closest possible approximation to the primaeval abundances of the other lead isotopes, Nier was able to estimate in 1941 that a particular rock was formed 2570 million years ago [13].

In 1946 Holmes and the German physicist Fritz Georg Houtermans (1903–66) independently pointed out that Nier's method could be extended to give not only the ages of particular rocks but the age of the Earth itself. They obtained values of 2900 million years. Holmes revised his value to 3350 million years in 1947.

Although some scientists pointed out that the available data did not exclude a value for the age of the Earth as high as 5000 million years, the Holmes–Houtermans value of 3000 to 3400 million years was generally accepted until the 1950s. That value was already high enough to produce a conflict with astronomers' estimates, obtained from the Hubble distance–velocity relation for galaxies in an expanding universe, that the Universe itself is only about 2000 million years old. This conflict was one of the reasons for giving serious consideration to the alternative 'steady-state' cosmology, although the age conflict disappeared when astronomers recalibrated their distance scales and boosted the age of the Universe to 10 000 million years or more.

See also p 1725

In 1953, a group of scientists at the University of Chicago and the California Institute of Technology reported that the abundances of the radiogenic lead isotopes in some meteorites were significantly lower than the figures previously considered 'primaeval'. Moreover, the ratio of uranium to lead in these meteorites was extremely low. This meant that little if any of the present amount of ^{206}Pb and ^{207}Pb could be attributed to decay of uranium since the formation of the meteorite. Because it seemed reasonable to suppose that this material was much less affected by chemical differentiation processes than minerals found in the Earth's crust, they concluded that these values were the most appropriate to use for the abundances at the time of the Earth's formation. Geochemist Clair Cameron Patterson (b 1922), one of the members of the group, announced that the Earth was at least 4500 million years old, on the assumption that the primaeval lead isotopic abundances were the same as those found in meteorites. Patterson's special contribution was the development of very accurate methods for measuring small amounts of lead isotopes.

During the next three years, the new value was confirmed by other analyses of lead isotope data and by independent estimates of meteorite ages based on two other radiometric dating techniques (the potassium–argon method and the rubidium–strontium method). In 1956 Patterson proposed the more accurate estimate: 4550 ± 70 million years. By the 1970s most scientists had accepted his estimate, especially after analysis of lunar rocks from the Apollo missions showed that the Moon is also about 4600 million years old.

During the two decades after Patterson first announced his result for the age of the Earth, the earth sciences underwent a major revolution with the advent of plate tectonics, changing most of what was believed about the Earth's past (see below). Yet, remarkably, the most fundamental parameter, its age, has remained fixed—the currently accepted value is within the limits given nearly 40 years ago.

Ideas about the origin of the Earth evolved more erratically during the last few decades: even today we are less certain about *how* the Earth was formed than about *when* it was formed. As described in section 26.2, astronomers and geologists before 1940 considered and rejected two radically different kinds of theory. The Laplace–Herschel nebular hypothesis is an example of a *monistic* theory: the planets formed in the same process as the Sun. The tidal theories of Chamberlin and Moulton, Jeans and Jeffreys are *dualistic* theories: they require the interaction of two stars to produce planets. A monistic origin of our own solar system suggests that many other planetary systems could be formed in a similar way, as part of the general process of star formation; a dualistic origin implies that planetary formation is much rarer because of the low probability that two stars approach closely enough for the required interaction.

New ideas and observations provided a plausible basis for reviving a monistic theory after 1940. For example, the Swedish plasma physicist Hannes Alfvén (b 1908) proposed that the early Sun had a strong magnetic field and was surrounded by an ionized gas. Lines of magnetic force, rotating with the Sun, would be trapped in the ionized gas and transfer angular momentum to it. In this way one might explain why most of the angular momentum of the present solar system is carried by the giant planets, rather than by the Sun as would have been expected on the original nebular hypothesis. Alfvén's 'magnetic braking' mechanism to slow the Sun's rotation was adopted by several theorists in the 1950s and 1960s even though they rejected other aspects of Alfvén's theory of planetary formation.

As Jeans and Henri Poincaré had pointed out early in the twentieth century, if the total mass of the present planets and Sun were uniformly distributed throughout the entire volume of the solar system, the density would be so small that the atoms would simply dissipate into space rather than condensing—but astrochemistry came to the rescue. The British–American astronomer Cecilia Payne-Gaposchkin (1900–79) had demonstrated in 1925 that spectrum analysis of stars, using quantum theory and statistical mechanics, could be used to estimate the abundances of elements in stellar atmospheres, and reached the somewhat hesitant conclusion that the Sun is composed mostly of hydrogen and helium. This contradicted earlier ideas that the Earth and Sun have similar compositions (as one might expect from the tidal theory) but was soon confirmed by H N Russell and other astronomers. If one assumed that the primaeval cloud from which the solar system formed had the same composition as the present-day Sun, it must have been much more massive than the present-day planets and must have subsequently lost most of its hydrogen and helium, at least in the inner solar system. The greater density of such a massive protoplanetary nebula would make it easier for processes such as gravitational instability, viscosity and turbulence to start the condensation of gases and dust.

The revival of monistic theories began with a 1944 paper by C F von Weizsäcker. At the same time Alfvén and Otto Schmidt developed theories that were dualistic insofar as they postulated previously formed clouds of material captured by the Sun, but their theories concentrated on the subsequent development of the solar system and eventually became monistic theories.

In the 1950s, competing theories were proposed by Gerard P Kuiper and Harold C Urey. Kuiper (1905–73), a Dutch–American astronomer, postulated a massive nebula of about one-tenth solar mass (i.e. about 100 times the present mass of the planets) surrounding the Sun, and assumed that it would form large protoplanets by gravitational instability. After the planets formed, the excess material (especially hydrogen and helium) would be blown away by the Sun's radiation pressure. Urey (1893–1981), an American chemist who became a leader in planetary science after World War II, proposed instead that numerous smaller objects (up to lunar size) were first formed and later accumulated into planets. He argued that the high abundance of hydrogen in the primaeval nebula should be taken into account in research on the origin of life; the first organic compounds could have been formed under chemically reducing conditions in the Earth's early atmosphere [14].

Schmidt's theory was developed by Victor Safronov (b 1917) and other Soviet scientists throughout the 1960s and 1970s. It became primarily a model for the accumulation of small solid particles (Chamberlin's planetesimals) from the protoplanetary cloud into planets (figure 26.8). Safronov's model was adopted, with some modifications, by the American geophysicist George W Wetherill (b 1925), who explored its consequences with the help of computer calculations. The Safronov–Wetherill model is now considered the most plausible one for the formation of the terrestrial planets, though it does not yet account quantitatively for their properties.

After the 1950s, a major role in ideas about the origin of the solar system was played by 'isotopic anomalies'. These are discrepancies between the average abundances of isotopes estimated for the solar system and the abundances found in particular meteorites. Although these anomalies have little bearing on most of the traditional problems of planet formation, they were believed to offer important clues to the initial stages of formation and contraction of the solar nebula, relating them to nuclear processes in the Sun and other stars. The best-known example was the 'supernova trigger' hypothesis, based in part on the excess ^{26}Mg found in the Allende meteorite which fell in 1969. The anomaly was established by Typhoon Lee, D A Papanastassiou and G J Wasserburg at Caltech in 1976, and was ascribed to the earlier presence in the solar system of ^{26}Al, which has a half-life of only 700 000 years. It was generally supposed that the ^{26}Al must have been synthesized in a supernova less than a few million years before the formation of the

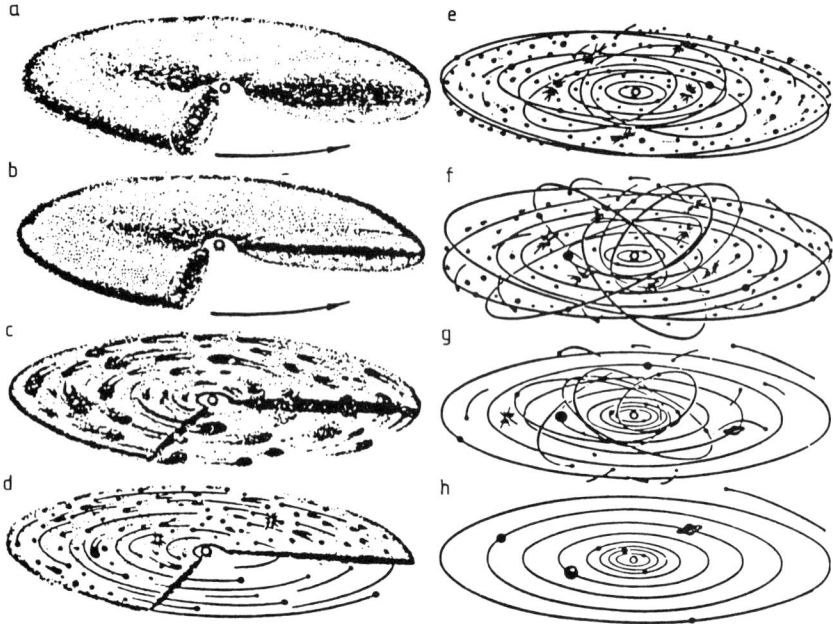

Figure 26.8. *Origin of the solar system according to Schmidt, Safronov and colleagues. The primordial gas–dust cloud begins its evolution somewhat as described by Laplace but does not form rings and large protoplanets at an early stage. Instead, it condenses to small solid particles (planetesimals) which then aggregate to the planets* [15].

solar system. Since a supernova explosion also produces a shock wave that might compress rarefied material to densities high enough for them to become unstable against gravitational collapse, the isotopic anomalies might indicate that the supernova caused the solar system to form.

The most influential advocate of the supernova trigger hypothesis was the Canadian–American astrophysicist Alistair Graham Walter Cameron (b 1925), but Cameron abandoned the hypothesis in 1984 when astrophysical observations indicated that other modes of synthesis were probably responsible for producing ^{26}Al. In the meantime Cameron had developed a quantitative theory of the collapse of the solar nebula and the formation of protoplanets. He argued that turbulent viscosity rather than magnetic braking is primarily responsible for the transfer of angular momentum from Sun to planets. Cameron incorporated into his model the theory of 'accretion discs' developed by D Lynden-Bell and J E Pringle.

Cameron's theoretical work, and research on isotopic anomalies by Donald Clayton and others, conflicted with another hypothesis that was popular in the early 1970s. To explain the specific chemical and physical properties of planets and meteorites, several scientists suggested that all of the material in the solar system (or at least in the region of the terrestrial planets) had been completely vaporized and thoroughly mixed. Thus the solar system was 'born again', preserving no evidence of its earlier history aside from its overall chemical and isotopic composition. As the supposedly homogeneous gas cooled down, its components would condense in a sequence determined by their thermodynamic properties and the pressure–density–temperature profile of the protoplanetary nebula. With some additional assumptions about the relative rates of cooling and aggregation, and about the extent to which thermodynamic equilibrium prevails in the nebula, one could then calculate the chemical compositions of the solid bodies formed at different distances from the Sun.

This 'condensation sequence' model was very attractive to scientists who studied meteorites, but it was the meteoriticists who eventually found evidence to refute it. Isotopic anomalies undermined the assumption that the protoplanetary nebula was well mixed, and it became increasingly difficult to account for the details of meteorite structure by simple condensation from a high-temperature gas. In the late 1970s and 1980s meteoriticists began to favour more complex histories, including the possibility that certain components had been formed elsewhere in the galaxy and survived as interstellar grains through the formation epoch of the solar system.

One might have expected that new results from the space programme would have played an important role in the development of theories of the origin of the solar system in the late twentieth century. After all, that was one of the 'scientific' justifications for funding the programme. In fact the direct contributions were less significant than expected in view of the huge expense of the space programme. Ground-based observations were frequently more decisive. The programme did have an important indirect effect by creating a planetary science community and encouraging the development of sophisticated experimental and computational techniques. The US National Aeronautics and Space Administration (NASA) funded the research of several theorists and enabled them to work out the consequences of their assumptions on computers. Conferences held to discuss plans and results of the space programme provided convenient forums for discussion of the origin of the solar system. New kinds of observations of the Universe beyond our solar system were valuable. Infrared, x-ray and gamma-ray telescopes on artificial satellites placed in orbit outside the Earth's atmosphere, combined with 'high-tech' ground-based observations, provided crucial data on the early stages of star formation and on the abundances of certain isotopes considered significant in planetogonic theories.

In one area the space programme did have a major impact: theories of the origin of the Moon ('selenogony'). Before 1969, there were three competing theories: co-accretion of Earth and Moon (an extension of the nebular hypothesis to satellite formation); fission (G H Darwin's

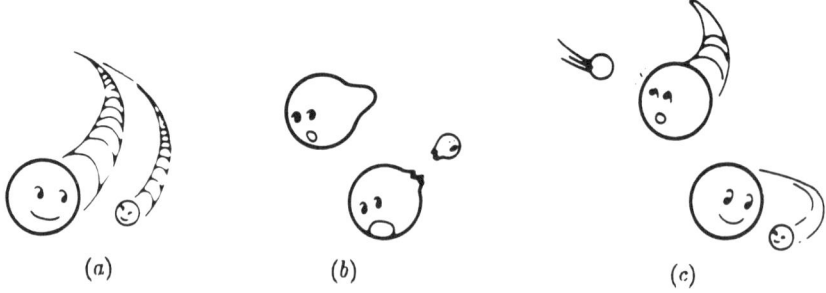

Figure 26.9. Three models for the origin of the Moon, before 1969. (a) Co-accretion: the Moon was formed in orbit around the Earth by the same process as planets formed according to Laplace's nebular hypothesis. This theory was popular in the early and mid-nineteenth century but never worked out in quantitative detail; it was revived in the mid-20th century by Evgenia Ruskol and others. (b) Fission: the Moon was ejected from the Earth through rotational instability. G H Darwin arrived at this hypothesis by calculating the evolution of the lunar orbit as governed by tidal forces, extrapolating back to a time when it was close to Earth and the Moon's period of revolution coincided with the Earth's period of rotation. This theory was popular in the late nineteenth and early twentieth centuries. (c) Capture: the Moon was formed elsewhere in the solar system and then captured by the Earth's gravitational attraction. This theory was worked out quantitatively by Horst Gerstenkorn in the 1950s and strongly advocated by Harold Urey.

hypothesis); and capture of a Moon (formed elsewhere) into orbit around the Earth (figure 26.9). Analysis of the samples brought back by *Apollo 11* and later lunar missions seemed to refute all three theories. This allowed the emergence of a new theory based on an idea that, though suggested earlier by several scientists, was previously considered too implausible to be taken seriously. In 1975 the American astronomers William K Hartmann (b 1939) and Donald R Davis (b 1939) proposed that the Moon was formed from material ejected into a circumterrestrial disc by a large (> 1000 km radius) body that struck the Earth. Cameron, with W R Ward, proposed a similar hypothesis in 1976, postulating an even larger impacting body, comparable in size to Mars (figure 26.10).

The giant-impact theory was made more plausible by the results of the Safronov–Wetherill theory, which suggested that bodies of lunar or martian magnitude were prevalent in the late stages of the formation of the planets. In addition, Mercury's cratered surface, revealed by the unmanned *Mariner 10* spacecraft in 1974, suggested that the terrestrial planets were bombarded by somewhat smaller bodies for hundreds of millions of years after their formation. By 1984 the giant-impact theory seemed to be the best working hypothesis for the origin of the Moon. Planetary scientists since then have been working out its consequences for the early history of the Earth.

Geophysics

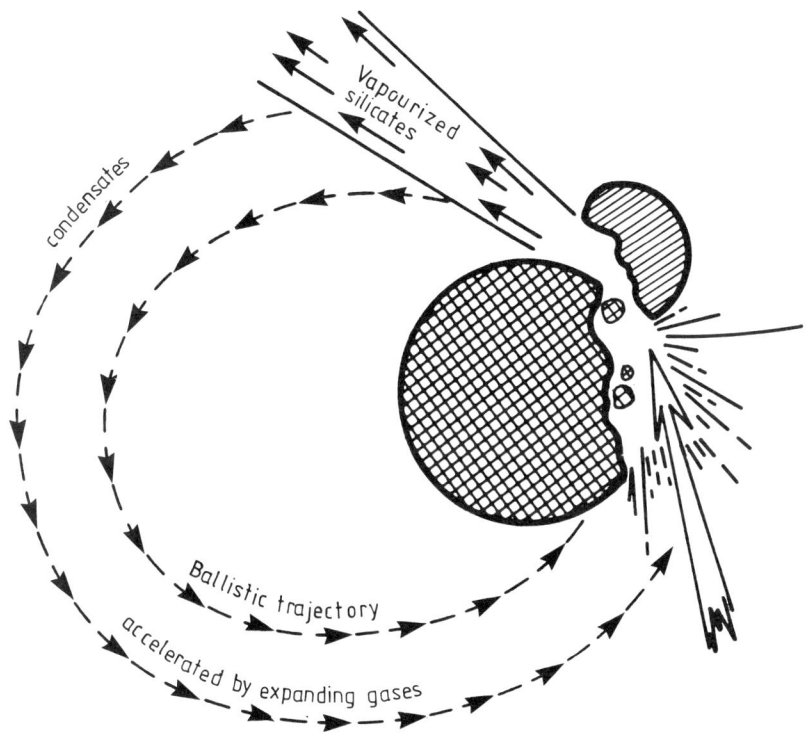

Figure 26.10. *Giant impact model for the formation of the Moon. The impactor may be as large as Mars. Following a glancing blow which ejects silicate material from the Earth's mantle, vapour expands and accelerates condensates into orbits with perigees that clear the Earth's surface (in contrast to the trajectories of solid debris knocked into simple ballistic orbits, which would reaccrete to the Earth). Volatile elements and compounds escape into space; refractory material condenses and eventually forms a single satellite. Tidal interaction with the Earth and its oceans transfers angular momentum to the satellite, pushing its orbit outwards while slowing the Earth's rotation. According to this model half or more of the Moon may come from the impactor, the remainder from the Earth's mantle* [16].

26.5. The 'revolution in the earth sciences' [17]

Alfred Wegener (1880–1930), a German geophysicist, is generally recognized as the founder of the theory of continental drift, though others had suggested on the basis of the shapes of their coastlines that the continents had been split apart from an earlier supercontinent (figure 26.11). Wegener became convinced that Africa and South America had once been joined when he read about the similarity of biological species on these two continents. The similarity was explained by palaeontologists by postulating prehistoric 'land bridges' across the Atlantic. That idea was consistent with Eduard Suess's theory that the ocean basins had been formed by collapse of parts of their floors. But Wegener found this improbable; it seemed to him much simpler to

Alfred Lothar Wegener

(German, 1880–1930)

Wegener's early scientific interests were in meteorology, climatology, geology and glaciology, which he pursued in a very practical and ultimately fatal way throughout his life by undertaking repeated expeditions to the Arctic regions. After receiving his degree in astronomy in 1905, he joined a Danish expedition to Greenland. In 1910, after returning from that expedition, he noticed (as many had done before him) the apparent similarity between the shapes of the coasts of the Atlantic on its west and east sides and wondered whether Europe, Africa and the Americas had once formed a single continent. Using evidence from several sources, he developed the theory of continental drift in a number of articles and a comprehensive book *The Origin of Continents and Oceans*. After serving as a lecturer in meteorology at Marburg, as a junior military officer in World War I and as a meteorologist at the German Marine Observatory near Hamburg he was appointed to a chair of meteorology and geophysics at the University of Graz in 1924. Six years later he disappeared while on an expedition to Greenland. His theory, though hotly debated by geologists and geophysicists during the 1920s, almost disappeared until its revival with the discovery of new evidence and the invention of 'plate tectonics' in the 1960s.

assume that the continents had been part of a single giant land-mass which he called 'Pangaea'. In his book on the origin of continents and oceans (1915), he presented a large quantity of biological and geological evidence for his hypothesis.

The biological evidence was in some cases quite striking. Certain species of earthworm and snail, for example, are found only in western Europe and eastern America; it is hard to believe that they could have swum across the Atlantic. At the other extreme of size, the hippopotamus is found on Madagascar as well as in Africa; could this huge animal have crossed the 250 miles of water in between?

Since geophysicists now believe that Wegener was right, the question arises as to why his theory was generally rejected before the 1960s. One reason was that he did not propose a plausible *mechanism* for the motion of continents; the forces he discussed (connected with the Earth's rotation) seemed much too feeble to push solid continents through the top layer of a solid mantle. However, the theory of continental drift was eventually accepted before any plausible mechanism was established, so this cannot be the whole story.

More significant is the fact that Wegener's *evidence* for the existence of continental drift was not convincing. There were few precise data to establish the geological 'fit' of the western edges of the European and African continents with the eastern edges of the American continent. Few geologists had first-hand knowledge of both sides of the Atlantic Ocean or experience in studying the global questions raised by Wegener's hypothesis. Although continental drift was widely discussed in the 1920s and 1930s, no one made the major research commitment needed to test it.

It has also been suggested that continental drift was not accepted because the evidence for it was biological and geological while the arguments against it were physical. In the early twentieth century physics was the most prestigious science. Despite the fact that Lord Kelvin's physical arguments against the geologists on the age of the Earth had been proved fallacious (section 26.2), both geologists and physicists still believed that physical evidence was more reliable than geological evidence. Geology in the first half of the twentieth century had fallen to a rather low status among the sciences, as compared with physics which was attracting much attention because of its spectacular advances. Biology was scarcely any better off, and arguments based on geographical distribution of species certainly did not carry as much weight as they had in the mid-nineteenth century in connection with the evolution debates.

A few scientists did support Wegener's theory. One was Arthur Holmes, whose work on radiometric dating has already been mentioned. Holmes argued that radioactive minerals inside the Earth would generate enough heat to cause convection currents in the mantle, and these could rise up to the crust and carry the continents along like a conveyor-belt. While Holmes' theory was not accepted either, his widely read books did keep the idea of continental drift alive until sufficient evidence (of the 'right kind') turned up to force other scientists to take it seriously.

The essential physical evidence for continental drift finally came after World War II with the study of magnetism. Physicists such as P M S Blackett (1897–1974), in England, and Walter Elsasser, in the US, took up the old problem of the origin of the Earth's magnetic field.

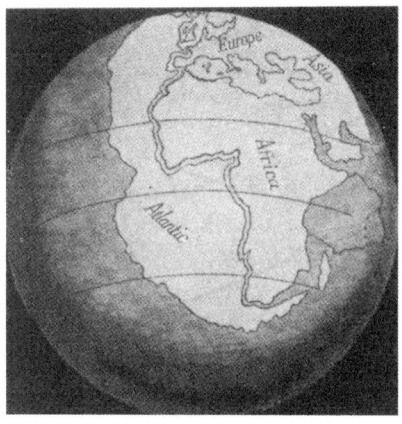 

(a)

Figure 26.11. *Fit of continental edges suggests they may once have been joined. (a) Long before Wegener, scientists imagined that Europe and Africa once formed a single continent with the Americas* [18].

The most obvious explanation, that the Earth contains a permanently magnetized substance like iron in its interior, had long since been rejected because it was known that substances like iron lose their magnetism when heated above the 'Curie temperature', and there was ample evidence that the Earth's interior is hotter than this temperature. Two other possibilities were now investigated: first, that the field is due to a self-sustaining pattern of electric currents; second, that a rotating body acquires a magnetic field as a result of some previously unknown law of physics.

As noted in section 26.3, the first of these two possibilities was successfully worked out theoretically by Elsasser and Bullard. The second was pursued by Blackett using an experimental approach. He borrowed 15.2 kg of gold from the Bank of England and constructed a cylinder which was suspended in his laboratory so that it rotated with the Earth. In order to see whether the rotation of this cylinder produced any magnetic field, he designed a magnetometer sensitive to fields as small as 1×10^{-7} gauss (the Earth itself has a field of about 0.5 gauss at the surface). In 1952 he concluded that the result of the experiment was negative: rotation by itself does not generate a magnetic field.

Blackett's 'negative experiment' had one very important by-product: the world's most sensitive magnetometer, which Blackett and others now used to measure the magnetization of ancient rocks. Rocks containing magnetic iron oxides which have been heated above the Curie temperature and then cooled will acquire a 'remanent magnetism' in the direction of the Earth's magnetic field at the time of cooling. If one notes the precise orientation of the rock in its natural surroundings (and if that

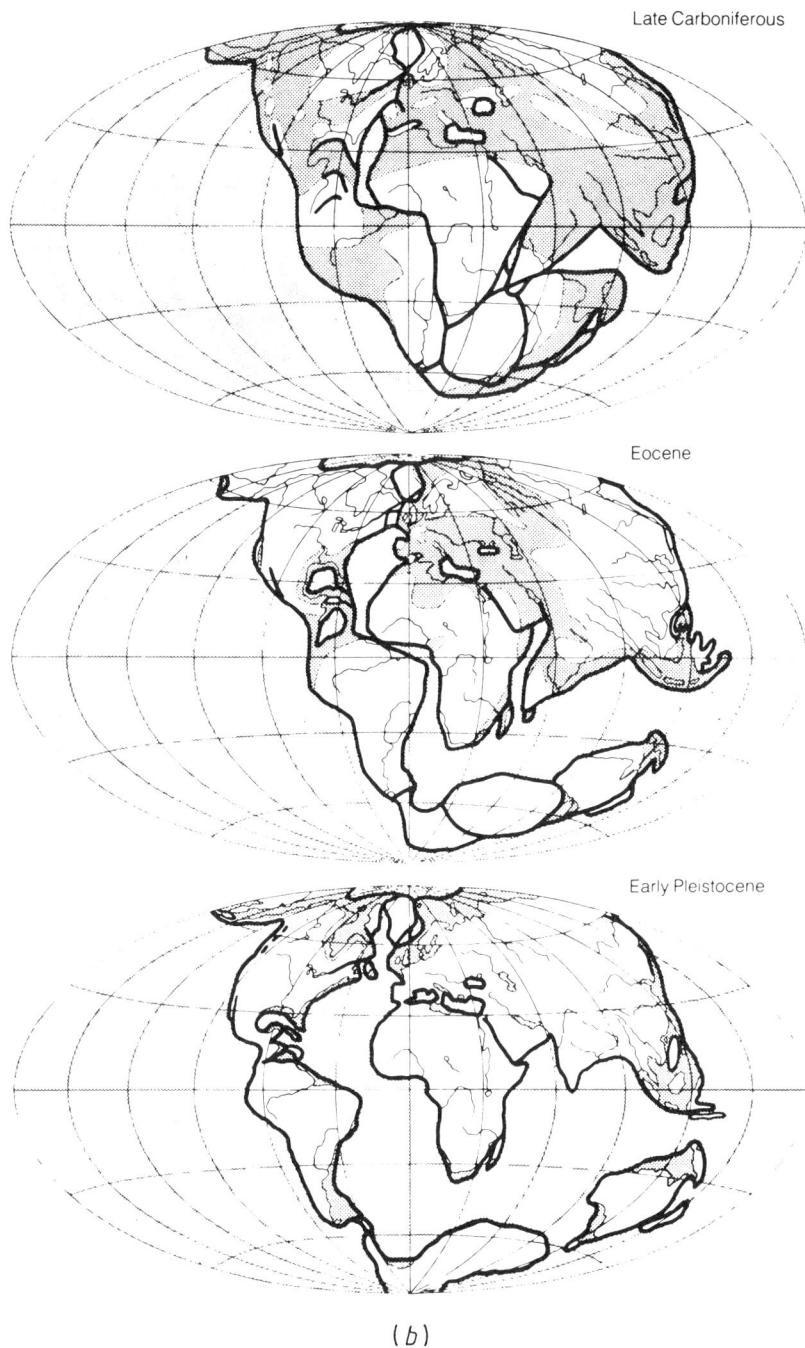

Figure 26.11. *(b) Wegener's reconstruction of the proto-continent Pangaea and the movement of continents after its break-up* [19].

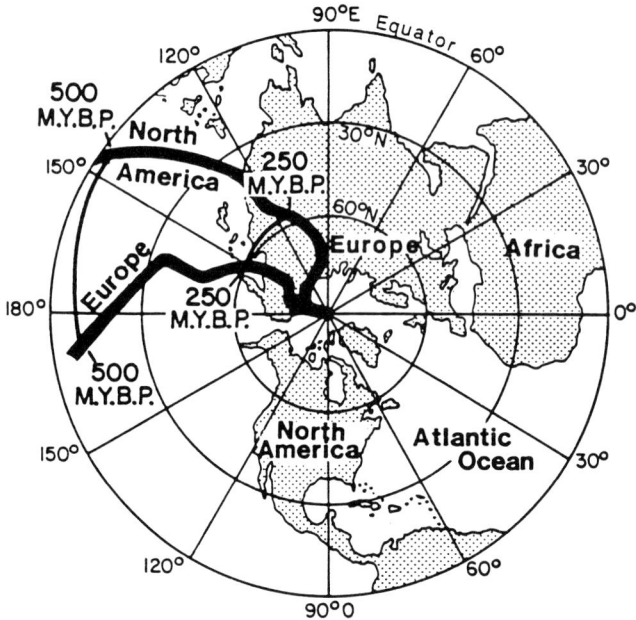

Figure 26.12. *Evidence for continental drift from polar wandering. Past locations of the magnetic North Pole, going back to 500 million years before present, as inferred by Runcorn and his colleagues from magnetic sediments in Europe and North America. If the Earth can have only one magnetic North Pole at a given time, they argued, then the two curves should be made to coincide by moving North America eastward. This would close the Atlantic Ocean and unite North America and Europe to form Pangaea as in figure 26.11* [21].

orientation can be assumed to have remained unchanged since cooling), one can obtain a record of the direction of the Earth's magnetic field in the past [20].

A group of British palaeomagnetists led by Stanley Keith Runcorn (b 1922) was able to show during the 1950s that the magnetic North Pole had wandered from a position near Hawaii about 600 million years ago, through the Pacific to Japan, then across Siberia to its present position, but similar data based on the remanent magnetism of rocks found in America led to a path of polar wandering which was displaced by about 30° in longitude from the one based on British rocks, with the displacement dropping off to zero only for the most recent part of the path. When these paths were plotted on a globe, the most natural explanation seemed to be that America and Europe had been joined together until less than 100 million years ago, and then had drifted apart just as Wegener had asserted (figure 26.12). So, around 1960, Runcorn and other palaeomagnetists became proponents of the theory of continental drift.

About this time geophysicists began to look into the evidence that the Earth's magnetic field had actually reversed itself at various times in the past. Such a reversal had first been reported in 1909 by the French physicist and meteorologist Bernard Brunhes (1867–1910), who measured the magnetization of ancient lava flows in France. Then in 1929 the Japanese geophysicist Motonori Matuyama (1884–1958) found rocks with reverse magnetization in Japan. At that time it seemed absurd to suppose that the Earth's magnetic field could be reversed, so the observations were either ignored or explained as being due to some peculiar property of the magnetic compounds, but now (in 1960) the Elsasser–Bullard dynamo theory made it seem quite plausible that reversals could occur.

New evidence for reversals of the geomagnetic field was obtained by Ian McDougall and his colleagues at the Australian National University, and by Allan Cox, G Brent Dalrymple and Richard Doell at the US Geological Survey in California. By 1963, with the help of the potassium–argon radiometric dating method, Cox, Doell and Dalrymple were able to establish two major periods of reversed magnetism in the past, with briefer reversals within each period. The major periods were named after four major pioneers in the study of the Earth's magnetism, as indicated in figure 26.13.

These epochs, together with the short-period reversals within them, can be represented by a pattern of black and white stripes (figure 26.13(b)) [24].

Palaeomagnetic evidence for wandering of the magnetic poles suggested the possibility of continental drift, but was not by any means sufficient to establish its reality. Nevertheless the time-scale of magnetic reversals did help to provide convincing proof in another way when combined with data from oceanography.

The now-familiar 'Mid-Atlantic Ridge' played a central role in the story. Part of it was first discovered in the 1870s by the British *Challenger* research ship. Further information was obtained in the 1920s and 1930s by 'echo-sounding' (a method, like radar, of determining the distance of solid objects by timing the return trip of waves, in this method, acoustic rather than electromagnetic). Other ridges were found, for example in the Indian Ocean, and it was also discovered that such ridges are split lengthwise by a deep gully (figure 26.14(a)).

In the 1950s the American geologist–oceanographer Marie Tharp made a detailed study of the Mid-Atlantic Ridge using echo-sounding data which indicated a rift valley. She also plotted the locations of a large number of earthquakes and found that almost all of them occurred just below the floor of the rift valley. During the International Geophysical Year (1957–58) extensive studies by research ships from several countries showed that there is a global system of ridges on ocean floors, and that these coincide with major zones of earthquake activity (figure 26.14(b)).

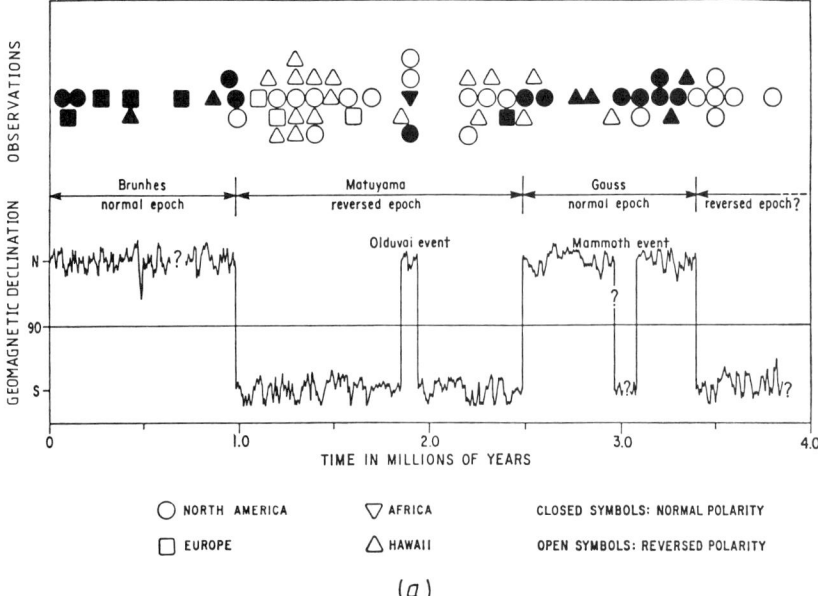

Figure 26.13. *Construction of the time-scale for the orientation of the Earth's magnetic field. (a) 'Magnetic polarities of 64 volcanic rocks and their potassium–argon ages. Geomagnetic declination for moderate latitudes is indicated schematically.'* [22].

Another clue came from the measurement of rates of heat flow just below the ocean floor. It was originally expected that these rates would be much less than those measured on the surface of continents, since the latter were believed to contain large amounts of heat-generating radioactive minerals. But measurements on the Pacific floor, initiated by Bullard and continued by Arthur Maxwell and Roger Revelle, showed that the heat flow there was as great as on continents. The result could have been due either to a greater concentration of radioactive minerals than expected under the ocean floor, or to convective currents bringing heat up from the lower part of the mantle. Bullard, Maxwell and Revelle preferred the latter explanation.

In 1960 the American geologist–geophysicist Harry Hess (1906–69) synthesized the new data on ocean ridges, earthquakes and heat flow with his own observations of 'trenches' found around the rim of the Pacific Ocean [27]. Trenches are regions of the ocean floor which had been found by the Dutch geophysicist Felix Andries Vening Meinesz (1887–1966) in the 1930s to have unusually low gravitational fields. According to Hess, the Earth's mantle (even though 'solid') has large-scale patterns of convection currents: material rises through the ridges and spills out, spreading along the ocean floor. At the same time the floor moves away from the ridges toward the trenches, where it descends into the mantle (figure 26.15).

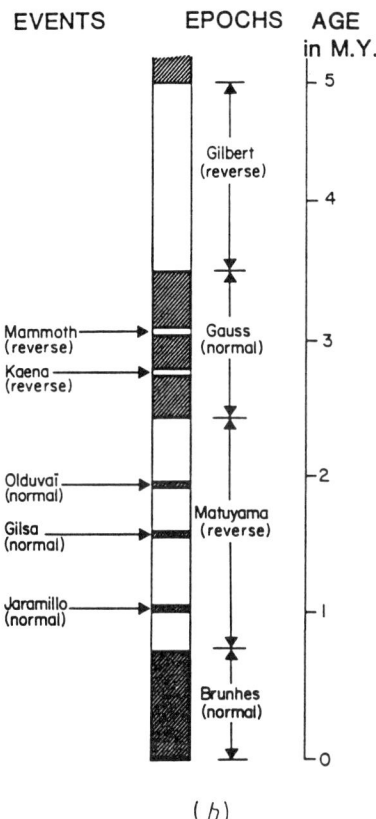

Figure 26.13. *(b) Later simplified version of the geomagnetic reversal time-scale. Dark bands represent periods when the orientation is 'normal' (same as at present); white bands indicate periods when the orientation is reversed. Time is in millions of years before present* [23].

The concept of 'sea-floor spreading' proposed by Hess was soon confirmed in an unexpected way. In 1955, the American geophysicist Arthur D Raff (b 1917) had developed a magnetometer which could be towed behind a research ship to record detailed variations in magnetic intensity, and he soon found a peculiar pattern of alternating 'stripes' of high and low magnetic intensity. The stripes were parallel to the Juan de Fuca Ridge, near Vancouver Island. A similar pattern was found by a British research ship in the Indian Ocean in 1962.

The next breakthrough was made by two British geophysicists, Frederick Vine and Drummond Matthews. Analysing this and other data on magnetic variations of ocean floors, they discovered that stripes of reverse magnetization occur at equal distance on both sides of ridges. Vine and Matthews proposed in 1963 that the pattern of normal and reverse magnetic stripes resulted from the extrusion of mantle material

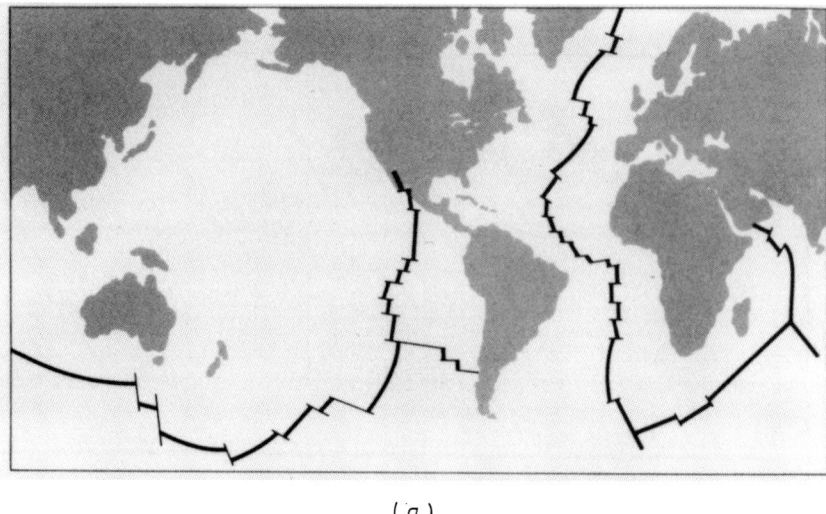

(a)

Figure 26.14. (a) Global pattern of mid-ocean ridges. According to plate tectonics theory, each ridge is a zone of tension along which two plates are being pulled apart [25].

through the ridges, which then acquired a magnetization determined by the Earth's magnetic field at the time it cooled below the Curie temperature [29]. In accordance with Hess's sea-floor spreading theory, the material furthest away from the ridge would be imprinted with the record of the Earth's magnetization at the earliest period, while that nearest the ridge would indicate magnetization in the recent past. By 1966 there was sufficient evidence to show that the pattern of magnetic stripes on both sides going away from an ocean ridge matches the pattern of epochs for the Earth's magnetic field direction as shown in figure 26.13.

According to the Vine–Matthews theory, each part of the sea floor in a strip at a certain distance from an ocean ridge can be associated with a certain period in the Earth's magnetic history—the period when it was extruded through the ridge and then cooled. An independent test of this conclusion was provided by the drilling ship *Glomar Challenger* [30], which took samples of the sediment on the ocean floor; it was found that the age of the bottom layer of sediment, as determined from analysis of its fossils, increased uniformly with distance from the axis of the Mid-Atlantic Ridge, and agreed with that derived from the magnetic variations. The spreading rate was found to be about 2 cm per year in the Atlantic Ocean; it varies from 1 to 5 cm per year in other oceans.

The overall theory developed to account for continental drift, the geographical distribution of earthquake zones and sea-floor spreading was called 'plate tectonics' [31]. It was the product of independent and collaborative work by several scientists in the decade after 1964, led by

Geophysics

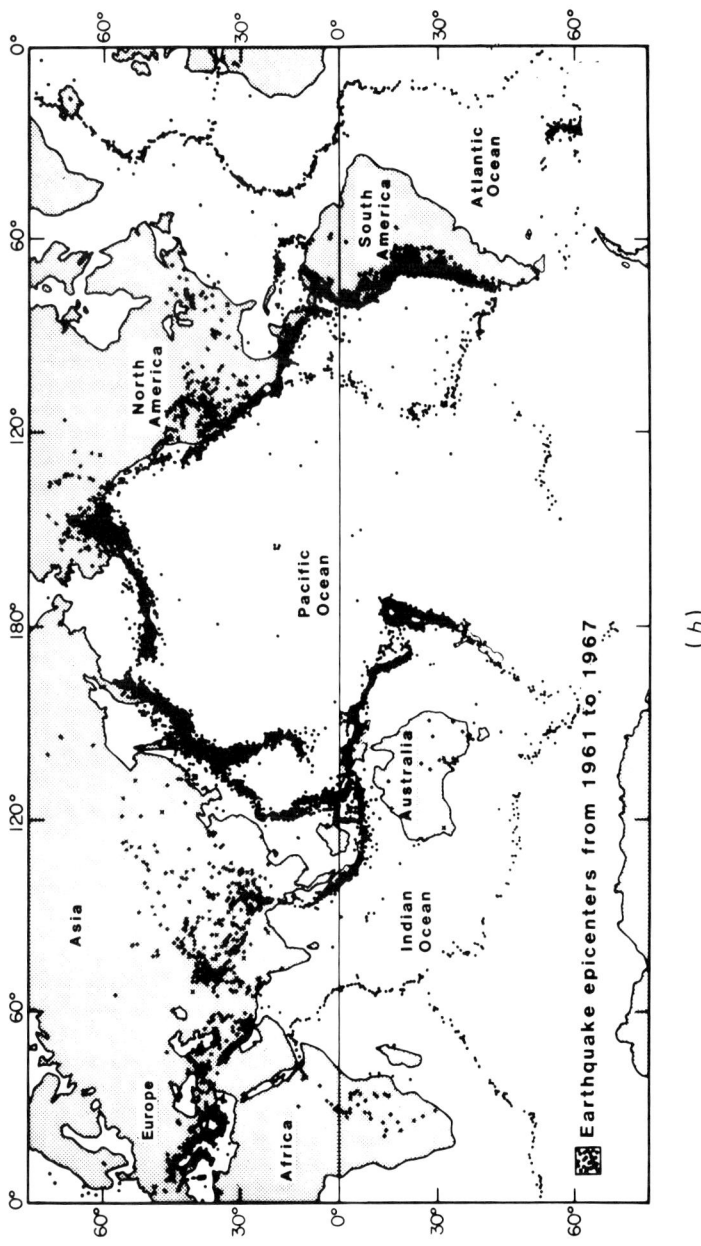

Figure 26.14. (b) Global pattern of earthquake epicentres. Most earthquakes are located along mid-ocean ridges or at trenches that ring the Pacific Ocean; there are two different kinds of plate boundary in plate tectonics theory [26].

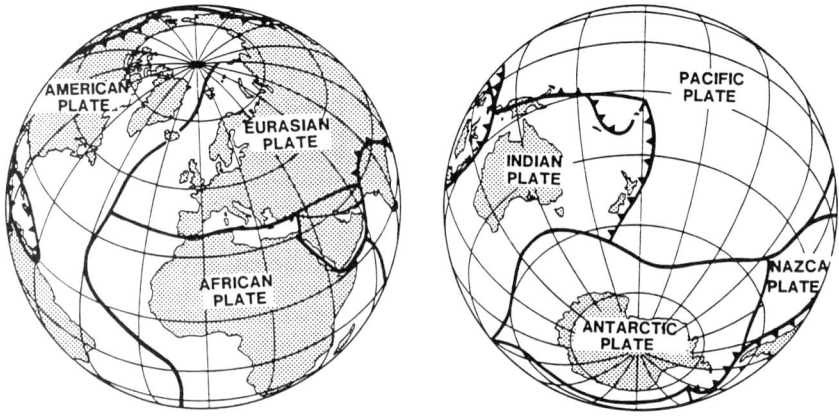

Figure 26.15. *Arrangement of plates on the Earth's surface. Xavier Le Pichon's seven-plate description of the Earth. Later work has refined this model by subdividing some plates* [28].

J Tuzo Wilson, Jason Morgan, Dan McKenzie, and Xavier Le Pichon. According to plate tectonics, the Earth's crust and upper mantle down to a depth of about 60 km are composed of relatively strong rigid rock material, called the 'lithosphere' [32] (figure 26.6), which is split into six large plates and several smaller plates (figure 26.15). Underneath the lithosphere is a weak layer called the 'asthenosphere' [33] composed of rocks which, though solid, are heated close to their melting point and can flow under stress. The asthenosphere goes down to about 250 km below the surface, where one encounters a much harder layer (either a different kind of rock or a pressure-induced denser phase of the same material).

The plates meet along narrow boundary regions which may be of three kinds: (1) ocean ridges (such as the Mid-Atlantic Ridge) where new crust is being created by upwelling of material from the mantle and the sea floor spreads out in both directions from these ridges; (2) trenches, where crust is being forced down under another plate, so that the lithosphere falls into the asthenosphere where it is presumably recycled, eventually to come back through a ridge (figure 26.16); (3) boundaries such as the San Andreas fault in California, where two plates are sliding past each other.

Plate tectonics theory led to the prediction of a new phenomenon, the 'transform fault', which could be recognized by the detailed analysis of earthquake records. In 1965 the Canadian geophysicist J Tuzo Wilson (1908–93) pointed out that a mid-ocean ridge may be offset by a motion of the crust, producing a dislocation that terminates abruptly at the ridge. As material continues to upwell from the ridge the differential motion along the fault will produce a shear which causes further earthquakes;

Geophysics

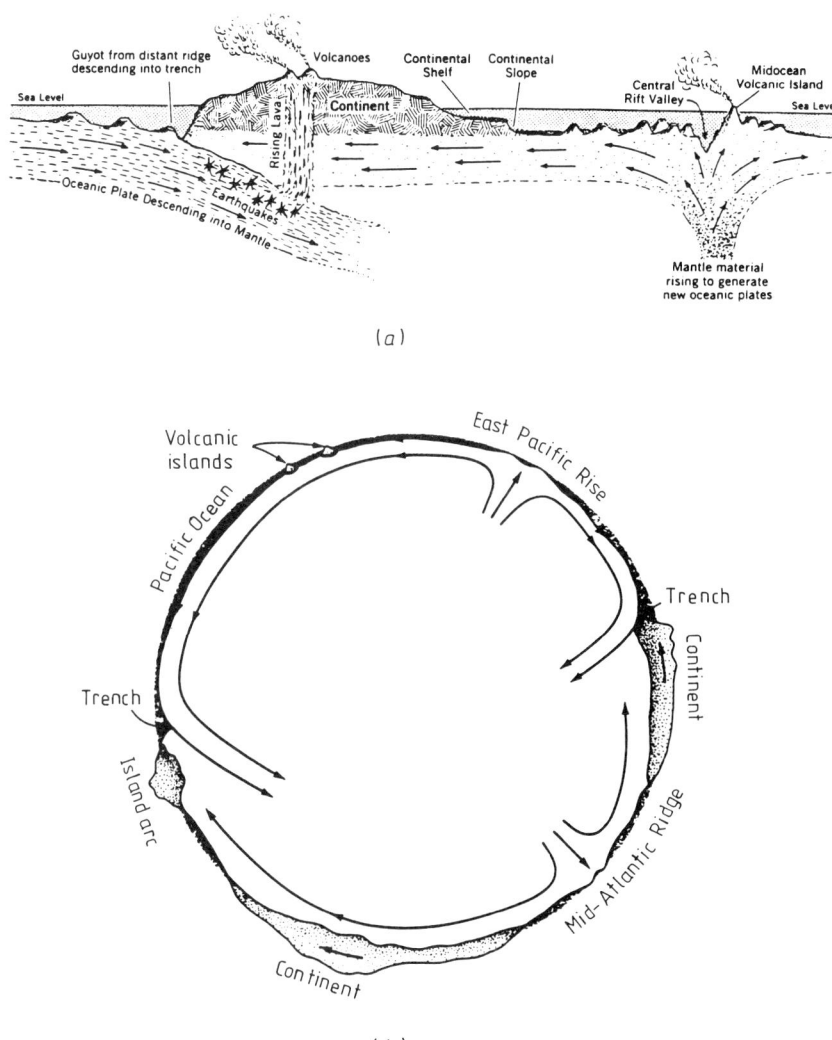

Figure 26.16. *Circulation of material according to plate tectonics. (a) New ocean floor is formed along mid-ocean ridges (on the right), spreads along top of crust carrying the continent with it. A plate covered with old ocean floor may push under another plate at a trench (on the left), diving into the mantle where material will be melted and recycled. Earthquakes and volcanoes are by-products of these processes* [34]. *(b) Equatorial cross section of the Earth showing the distribution of spreading ridges and trenches (called 'subduction zones' in the theory)* [35].

but the shear is in the opposite direction from the motion produced by the dislocation itself (figure 26.17). By looking at the direction of the initial impulse in seismic records, it is possible to distinguish earthquakes produced by such transform faults from those produced by normal or

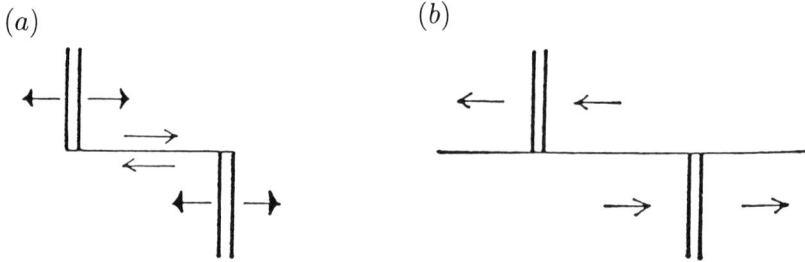

Figure 26.17. *Transform and transcurrent faults. (a) A transform fault (horizontal line) connects offset portions of a mid-ocean ridge (double vertical lines). Solid-headed arrows indicate sea-floor spreading from the ridge, and determine the direction of shear motion along the fault (open arrows). (It is assumed that there is no longer any significant shearing force from the dislocation that originally produced the fault.) (b) A transcurrent fault does not end at the ridge but indicates a continuing shear pushing the two offset parts of the ridge further apart. (It is assumed that there is no significant sea-floor spreading in this case.) Note that the direction of shear (open arrows) is opposite to that in the transform fault. Thus the direction of the initial impulse of an earthquake along this fault will be opposite, so the two kinds of fault can be distinguished in seismic records* [36].

'transcurrent' faults. The confirmation of this prediction was important evidence for the new theory.

The plates move away from ridges and towards trenches, carrying the continents on their backs. While the dense mantle material in the lithosphere can slide under another plate at a trench, the continental rocks are more buoyant, so when two continents collide they will pile up on top and form a mountain range. The most spectacular example of this is the Himalayas, created by the collision of India and Asia.

Expensive technology obviously played a major role in the plate-tectonics revolution, and the financial support of the US Government enabled American scientists to make some of the most important contributions. For the most part this support was given for 'basic research' with no guarantee that there would be any pay-off in practical applications. The National Science Foundation supplied about $68 million from 1967 to 1975 for the ocean-floor sampling project, which included the work done with the *Glomar Challenger*. The Navy has traditionally supported research in oceanography for obvious reasons.

The 'Cold War' between the US and the former USSR, and in particular the nuclear weapons race, advanced geophysics indirectly. During the period of underground testing of atomic bombs, each side wanted to be able to detect the tests of the other side. This required the establishment of large-scale networks of very sensitive seismic instruments; in order to distinguish bomb tests from earthquakes, the characteristics of each had to be precisely investigated. One by-product was the evidence for solidity of the inner core obtained by Julian, Davies and Sheppard (section 26.3); another was a considerable increase in knowledge of seismic activity throughout the world, knowledge which helped to reveal the significance of the ocean ridges.

A possible practical benefit of the plate-tectonics revolution is the ability to predict major earthquakes far enough in advance to allow evacuation and other precautions in vulnerable areas. Greater understanding of the structure and dynamics of the Earth's mantle may assist in the location of accessible mineral deposits and petroleum reservoirs. Further in the future is the possible exploitation of 'geothermal energy'—after we use up all the energy stored in fossil fuels in the crust we should be able to tap the much greater supply of energy involved in the convection currents that move plates around [37].

26.6. The Earth's upper atmosphere and geospace†

The Earth's upper atmosphere and its near-space environment (now commonly termed geospace) comprise regions which begin about 50 km above the surface of the Earth and extend to at least several earth radii. Geophysics deals with all of the states of matter—solid, liquid, gas and plasma—which comprise the Earth, but upper-atmosphere geophysics is characterized by its emphasis on naturally occurring plasma [38]. This upper atmosphere ends, finally, at the outer limits of the region where matter is associated with the Earth rather than with the Sun. The mass of the Earth's upper atmosphere is only a small fraction of a per cent of the mass of the lower neutral atmosphere studied by meteorologists, yet the volume is thousands of times larger. Thus we are considering a huge volume, at laboratory vacuum pressure, primarily acted upon by the terrestrial magnetic field and by energetic radiations from the Sun and from cosmic rays.

Ionization of upper-atmosphere gases, caused mainly by particles and electromagnetic radiation from the Sun, produces electric currents and fields, the reflection of radio waves, the visual and radio aurorae and other phenomena. The plasma recombines or produces other chemical species and releases electromagnetic energy across a wide spectrum from radio waves to light. The geomagnetic field influences most electrical properties of the upper atmosphere and in the outer portions, or magnetosphere, the geomagnetic field causes trapping of energetic particles.

Different terms have been used over the years in the study of the Earth's upper atmosphere and geospace, called here upper-atmosphere geophysics: the *ionosphere*, extending from about 50 to about 1000 km above the Earth's surface, was speculated upon in the nineteenth and early twentieth century and named by Robert Watson Watt in 1926 [39]. The *magnetosphere*, a term coined by Thomas Gold in 1959, extends from the upper part of the ionosphere out to the extent of the Earth's

† C S Gillmor is the author sections 26.6–26.8.

environment that is dominated by the geomagnetic field, the tail of which, however tenuously, extends out beyond the Moon's orbit. The region of the ionosphere out to several earth radii is also sometimes called the *plasmasphere*, thus providing a second name for the inner portion of the magnetosphere—confusing, but typical of scientific fields where more than one specialty can claim expertise. The science historian and philosopher Thomas Kuhn noted this when he related the true story in which a chemist and a physicist were each asked whether a single atom of helium was or was not a molecule. For the chemist the atom of helium was a molecule because it behaved like one with respect to the kinetic theory of gases. For the physicist, on the other hand, the helium atom was not a molecule because it displayed no molecular spectrum [40].

In wishing to unify the physical and chemical study of the upper atmosphere, Sydney Chapman in 1950 [41] termed such study of the ionospheric region, *aeronomy*. For many purposes, the ionosphere and magnetosphere are studied by the same people using similar methods. To present more confusion, perhaps, the Earth's upper-atmospheric regions have been classified differently depending upon whether temperature, composition, state of mixing, gaseous escape or ionization is deemed more central to the problems at hand, or to the scientific community at work. Thus the temperature-classified *troposphere* and *stratosphere* are located lower than, but the *mesosphere* and *thermosphere* overlap, the *ionosphere* (see figure 26.18). In the 1950s and 1960s, most scientists spoke interchangeably of *space physics* and what has been called here *upper-atmosphere geophysics*. As instrumentation advanced and propulsion vehicles became more powerful, space physics came more to identify either those fields of physics and allied areas studied utilizing rockets and satellites, or those physical locations further and further from the Earth's surface. In 1915, 'short' radio waves were waves of less than several hundreds of metres in wavelength; by 1940 'short' radiowaves were less than 40 or perhaps 10 metres in length. Similarly, 'space' as a defining term moved higher and higher above the Earth's surface, from the stratosphere in the early part of this century, to the ionosphere in the inter-war period, to the magnetosphere by about 1960, then to the intra-solar system, then on to 'Deep Space' as the American television science fiction programmes called it beginning in the 1970s.

Similarly, given the numerous close connections between the Sun and the upper atmosphere of the Earth, the community of those scientists studying solar–terrestrial relations, or solar–terrestrial physics (STP) shares many citizens with the upper-atmosphere and geospace community. Over the years as scientists have effectively and *in situ* studied the regions above the Earth, the terms of study have changed. Yet, the strongest tie seems to have been the physics of ionized regions—with the upper-atmosphere and geospace groups closer to

Geophysics

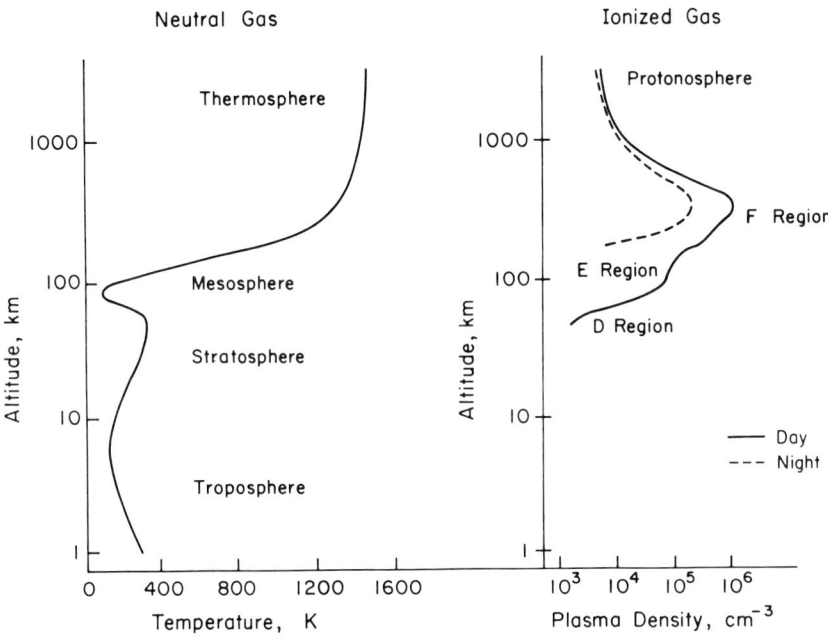

Figure 26.18. *Typical profiles of neutral atmospheric temperature and ionospheric plasma density with the various layers designated* [42].

plasma physics, solar physics and astronomy than to classical neutral air meteorology. This situation has been changing somewhat with the realization that, in terms of theoretical models, upper-atmosphere geophysics has 'seen' phenomena similar to those in meteorology. At an elementary level, for example, there are ionospheric 'storms' and magnetospheric 'sub-storms' with characteristics like storms in neutral air meteorology, and, since the 1960s, some educators and researchers have argued for the recognition of a general *atmospheric physics* or *atmospheric science* as a field that can fruitfully bring together formerly disparate areas. One specific example is the so-called 'winter anomaly' in the ionospheric absorption of radio waves: ionospheric radio absorption is greater in the winter than summer by about a factor of two, and it is much more variable in winter. Following rocket studies and satellite measurements in the late 1960s, it was seen that temperature and wind in the stratosphere affect the lower ionosphere and help produce the 'winter anomalies'. In another, recently discovered connection, lightning discharges ionize the ionosphere, and produce various phenomena. Still another area of study that has developed in the last two decades is that of large-scale drift motions in the upper atmosphere, in some cases linking the neutral atmosphere, the ionosphere and the magnetosphere. Beginning with speculations by

1983

D F Martyn in 1936 and from observations reported by G Munro and by W Beynon in the 1940s and other studies by K Maeda and colleagues, it became recognized that large-scale disturbances in the ionosphere could travel thousands of kilometres and have periods up to one or two days. Such large-scale waves of acoustic and acoustic-gravity form have attracted wide interest both experimentally and theoretically, as, for example, by the work of C O Hines and B H Briggs. In a survey of several hundred ionospheric and magnetospheric scientists conducted in 1972 [43] it was noted by many respondents that the research in the fields of upper-atmosphere geophysics and the lower atmosphere sciences would come closer together due to the importance of coupling mechanisms then presumed to be increasingly important in the dynamics of the Earth's atmosphere. In the same way progress has been made in the promising, but always controversial, area of study of the relations between the Sun and long-term and short-term conditions in the Earth's atmosphere, from the surface troposphere out through the ionosphere and magnetosphere to the magnetopause.

In the concluding sections of this chapter, we shall examine elements of the history of investigations of the ionosphere, the magnetosphere, and solar–terrestrial relations.

26.7. The ionosphere: the early days

An early suggestion of an atmospheric conducting layer was made in connection with geomagnetism by C F Gauss in 1839 when he speculated that daily geomagnetic variations originated from electric currents in the atmosphere. Michael Faraday in 1851 suggested that daily heating of the Earth's atmosphere by the Sun altered the paramagnetism of the atmospheric oxygen and thus could cause the daily geomagnetic variations. Lord Kelvin in 1860 also speculated on the existence of a conducting layer. Electric fields and currents would be generated if winds drove the supposedly conducting layer across the geomagnetic field. This idea was first presented thoroughly by Balfour Stewart of Owens College, Manchester in 1882 in the *Encyclopaedia Britannica* [44]. Stewart dismissed the idea that daily geomagnetic variations could be due to direct solar action. Others had suggested that solar heat produced thermo-electric currents in the Earth's crust and that this caused the daily geomagnetic variations.

Rejecting these theories, Stewart stated that 'the only conceivable magnetic cause capable of operating in such regions must be an electric current ... we know from our study of the aurora that there are such currents in these regions ... convective currents established by the Sun's heating influence in the upper regions of the atmosphere are to be regarded as conductors moving across lines of magnetic force, and are thus the vehicle of electric currents which act upon the magnet'. He imagined the conductivity to be much greater than currently supposed and quoted Hittorf on the subject of conductivity in rarefied gases. In the late nineteenth century, mere low pressure was thought sufficient to ionize a gas and make it conducting. This idea prevailed in the early work of Stewart's student J J Thomson and later in the ideas of Arthur Kennelly. Thus Stewart proposed a dynamo theory of the upper atmosphere. His theory qualitatively explained seasonal and lunar variations and compared wind patterns and geomagnetism.

Another of Stewart's students, Arthur Schuster, developed the dynamo theory in a number of papers. Both Schuster and J J Thomson worked on electrical discharges in gases. Schuster demonstrated that once a gas was ionized, a small potential could maintain a current. These interests in geomagnetism and in electrical conductivity of gases (plasma physics before its time) led Schuster to develop considerably Stewart's dynamo theory, in papers written from 1889 and continuing for a period of 20 years. He quantified Stewart's theory and showed that the majority of the daily geomagnetic variation was due to external causes and that most of the remainder was due to currents within the Earth which were induced by the varying external field. He believed that the conducting layer was in the upper atmosphere and pointed to solar ionizing radiation as cause for the conductivity, and convective motion as cause for the currents. He calculated the ionization and conductivity as a function of the solar zenith angle and, by 1908, had calculated an estimate for the specific conductivity of the layer as 10^{-13} emu and estimated the layer thickness as 300 km. The theories of Stewart and Schuster were further developed by Sydney Chapman and others and considerably influenced upper-atmosphere geophysics and the ionospheric physics which would begin to form in the first decades of the twentieth century.

From the beginnings of radio research in the late nineteenth century, technology would strongly determine the direction and methodologies of the study of geospace. Heinrich Hertz' experiments in Karlsruhe, Germany in 1887–88 demonstrated the propagation of electromagnetic waves in air. Hertz used an unclosed circuit connected to an induction coil to produce the waves and detected them with a small open loop. These experiments with waves of a few centimetres length gained immediate fame for Hertz, and united other workers such as Oliver Lodge into a human wave of experimenters on what would soon be called 'wireless' or 'radio'.

In early attempts to transmit and receive these 'wireless' waves over greater distances, longer antennas and higher transmitting and receiving masts were thought to be necessary. It was widely assumed that radiowaves, like lightwaves, propagated in straight lines. At first, radio was thought useful only as a substitute for lighthouses to warn ships a short distance away. Then it was thought useful for secret naval military uses. Soon a race was on to achieve distance, using very inefficient sparking devices and largely untuned circuits.

Ernest Rutherford as a physics student in 1895 established the New Zealand wireless transmitting record of a couple of hundred metres, even before Guglielmo Marconi began his experiments which would achieve for him the Nobel Prize for physics, which he shared with the radiophysicist K Ferdinand Braun in 1909. In the summer of 1896 Marconi demonstrated wireless communication to British Post Office officials over a distance of a few hundred yards. It seemed after each few succeeding months that Marconi almost doubled his communication distance. By autumn 1899 he had achieved 85 miles, by late 1901, 186 miles. By December 1901 Marconi claimed transatlantic communication (on less than solid evidence), and clearly demonstrated this achievement by February 1902. This feat seemed impossible to explain, unless wireless waves could plough a straight line '100 miles deep' through the Earth or sea; no optical analogy could suggest diffraction around a quarter of the Earth's circumference, for example.

Editors of the scientific and technical press wanted answers, stockholders in telegraph and undersea cable companies wanted answers: the American radio engineer Arthur E Kennelly (born in India of English parents) commented in March 1902 on wireless waves and J J Thomson's researches on electrical conductivity of air at low pressure 'It may be safe to infer ... that at an elevation of about 80 kilometers, or 50 miles, a rarefaction exists which, at ordinary temperatures, accompanies a conductivity to low-frequency alternating currents about 20 times as great as that of ocean water. There is well-known evidence that the waves of wireless telegraphy, propagated through the ether and atmosphere over the surface of the ocean, are reflected by that electrically-conducting surface' [45]. Kennelly hoped soon to see science computing the 'electrical conditions of the upper atmosphere'.

Ever since 1894, the Irish theoretician G F Fitzgerald had been corresponding with the self-taught physicist, engineer and genius Oliver Heaviside about wireless wave propagation around a sphere. They exchanged several letters about this in 1899 in connection with the latest long-distance wireless experiments. Finally, after the Marconi transatlantic feats, Heaviside tucked his speculations into a few comments in an article on 'Telegraphy' he wrote for the *Encyclopaedia Britannica* and which was published in December 1902 'There may possibly be a sufficiently conducting layer in the upper air. If so, the waves will, so to speak, catch on to it more or less. Then the guidance will be by the sea on one side and the upper layer on the other' [46]. Heaviside, a world-renowned expert on telegraph and telephone transmission along cables and transmission lines, seems to be speaking of long-distance radio waves propagating as if along a two-wire transmission line, the sea being one line and the upper conducting layer the other. Several of the world's eminent mathematical physicists [47], including Henri Poincaré (1904), Lord Rayleigh (1904), Jonathan

Zenneck (1907), Arnold Sommerfeld (1909) and G N Watson (1919), attempted to explain this passage of radio waves around a spherical Earth using optical diffraction theory, but physical experience had shown that diffraction could not nearly account for the bending of a radiowave passing from England to Newfoundland.

The first attempt to use quantitative physics to study radio waves propagating in a plasma was made by William H Eccles in 1912. Both Kennelly and Heaviside had assumed an upper *conducting* layer and therefore that the radio waves were *reflected* off the layer rather than bent or refracted while passing through the layer, for it was known that radio waves would not penetrate a conductor. Eccles' theory required him also to consider that the waves did not penetrate far into the layer but reflected at a sharp boundary, otherwise the waves should be attenuated. He assumed the effective charged particles were ions, not electrons. He presented formulas for the velocity and for the absorption of waves in the plasma. Eccles named the posited upper conducting layer the 'Heaviside layer'. It also was known as the 'Kennelly–Heaviside layer', or 'K-H-L', 'upper reflecting layer' or 'upper reflecting surface'. In the next several years Lee de Forest (1913) and Leonard Fuller (1915) in the USA, T L Eckersley (1921) in England and others argued for the existence of an upper ionized layer, following studies of fading and deviating of long-distance radio signals. These studies were not successful, however, in bringing a majority to accept that such an ionized layer was responsible for long-distance radio propagation. For one thing, empirical studies, especially done by Louis W Austin (1911, 1914) for the United States government, had shown that propagation was more successful the longer the wavelength. One problem here was that only extremely long wavelengths (1000 to more than 10 000 metres) were used in the tests.

Following World War I, however, three factors became of importance to the future study of upper-atmosphere geophysics: (1) vacuum tubes, and then quartz crystal oscillating circuits became more readily available; (2) numerous radio amateurs, who had been restricted to the seemingly unimportant 'short wave' lengths of less than 200 metres used low-power homemade equipment to establish contact with each other over thousands of miles; (3) commercial or public broadcasting was established in America and in Europe about 1920–22 and soon a 'boom' in manufacture and sales of radio sets occurred. As the public listened to the radio, solutions to the problems of signal skipping, fading or 'swinging' of signals and 'strays' (atmospheric interference) became of considerable interest to governments and to the public at large.

The astonishing long distances obtained, plus the natural phenomena listed above (skipping, swinging and strays) led some to think again about the Kennelly–Heaviside layer ideas and the possibility of its role in long-distance radio-wave propagation. The British Radio Research Board provided research grants. The Post Office and the National Physical Laboratory offered grants and conducted studies. Similar work was done by the Bureau of Standards in the United States. By the late 1920s, radiophysics research was sponsored by numerous governments around the world. It was in this way that Robert Watson-Watt, Edward V Appleton and others became interested in the upper atmosphere.

Joseph Larmor extended Eccles' theory in the autumn of 1924 by imagining the collision frequencies of the ionized particles in the upper atmosphere to be small enough so that the region acted as a dielectric rather than a conductor, and thus the waves could be bent or refracted back to Earth while passing through the layer. In addition, Larmor assumed the effective particles to be electrons, not ions. A special meeting was held in London in late November 1924 to discuss the 'Ionization in the atmosphere and its influence on the propagation of wireless signals'. In an impressive paper, Edward Appleton added a new parameter, the geomagnetic field, pointing out that a plane polarized wave would be split by the Earth's magnetic field into two oppositely rotating circularly polarized waves. In addition Appleton predicted that there would be less absorption of the radio wave at higher frequencies since there would be many more oscillations of each particle between collisions. He suggested that this could explain propagation success at short wavelengths at night, and a corresponding success with longer wavelengths in daylight. Appleton began to work on what he would call the magneto-ionic theory and by direct means started to test experimentally the existence of the upper-atmosphere conducting layer. His theoretical and experimental work in the physics of the ionosphere over the next 15 years secured for him the Nobel Prize in physics in 1947. Appleton's experimental technique, begun in late 1924, was to compare the strengths of radio waves received at about 100 km from a transmitter, with the transmitter slowly changing frequency. Appleton arranged to use a British Broadcasting Company transmitter during the hours after local midnight. The signal received fluctuated between maxima and minima as the difference in phase between the signal arriving directly along the ground mixed with the signal passing from the transmitter up through the ionosphere and back to Appleton's receiver (figure 26.19). This was analogous to optical wave phenomena and allowed Appleton to calculate the height of reflection or refraction of the signal from the ionosphere. This was similar to the observations made by de Forest and Fuller during World War I. At the Department of Terrestrial Magnetism of the Carnegie Institution of Washington, DC, Gregory Breit and Merle Tuve also sought experimentally to determine the height of the ionospheric conducting layer. They began experiments in late 1924 but gained clear success in mid-1925 only after securing use of a crystal-controlled transmitter at the Naval Research Laboratory, when it was not being used for naval

Geophysics

Figure 26.19. *Edward Appleton with students and staff demonstrating radio receiving apparatus at King's College, Strand, London (undated but probably about 1927).*

communications (figure 26.20). Their method, pulse-echo sounding, was to prove so successful that by 1932 most ionosphere workers around the world were building similar equipment and the technique was to provide the basic model for radar. Their transmitter initially radiated pulses at a frequency of 4 MHz and was received by equipment several miles away which recorded photographically the amplitude and time delay of the echoes. The amplitude could give approximate indications of the signal strength and the pulse delay could be converted into height of the ionospheric layer. The pulse-echo technique evolved to use the cathode ray oscilloscope and to measure the height of ionospheric echoes continuously through time and over a band of frequencies ranging from less than 1 MHz to 20 MHz or more, above which frequencies radio waves usually are no longer refracted from the ionosphere but pass upwards into space. The first automatic multifrequency ionospheric recording (the device became termed the 'ionosonde') was produced by T R Gilliland at the US National Bureau of Standards in April 1933 (figure 26.21). The automatic ionosonde was developed further by the Carnegie Institution of Washington, and then by the Allied Forces in World War II. Several dozen ionosondes were in operation by 1945 and about 200 by the time of the International Geophysical Year 1957–58 [48].

Figure 26.20. *Gregory Breit (left) and Merle A Tuve (right) in the laboratory at the Department of Terrestrial Magnetism, Carnegie Institution of Washington, DC, adjusting radio receiving equipment used in ionospheric pulse sounding experiments (14 February 1927).*

Theory and experiment in early ionospheric physics each led and lagged the other. Almost simultaneously with Appleton, Harold W Nichols and John C Schelleng (1925) of the Bell Telephone Laboratory discussed effects of the terrestrial magnetic field on radiowave propagation. They developed their physics using mathematical matrix methods as utilized earlier, especially by the German optical physicists Paul Drude (1902) and Woldemar Voigt (1908, 1920) and other theoretical physicists. Gregory Breit, Sydney Goldstein and others also published partial solutions to the magneto-ionic theory by 1928. Appleton read two papers in October 1927 at a radiophysics meeting in Washington, DC. Short abstracts of these two papers were published in the conference proceedings in July 1928—Appleton announced that he had found more than one ionized layer, and he discussed the theory of propagation of wireless waves in an ionized medium, extending to the general case of propagation in any direction with respect to the magnetic field.

In preparing this further theoretical work, Appleton was assisted in late 1925 and early 1926 by a young Austrian physicist, Wilhelm Altar. At Appleton's suggestion, Altar took on the project of solving the general case of wave propagation in a magneto-plasma with Appleton daily

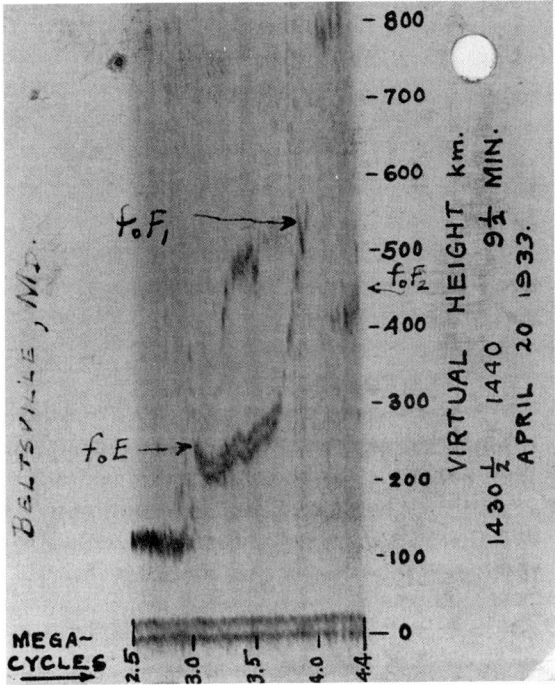

Figure 26.21. *The first automatic multifrequency ionosonde record, made by T R Gilliland of the US National Bureau of Standards at Beltsville, MD, 20 April 1933, over a frequency range of 2.5 to 4.4 MHz. See notation of the 'critical frequencies', $f_0 E$, $f_0 F1$ and $f_0 F2$, indicating reflections from the E and F regions of the ionosphere.*

giving him advice and criticism. Altar prepared a manuscript which presented for the first time the magneto-ionic plasma dispersion relation and which contained, in matrix mathematics termed the dielectric tensor form, the equations for polarization and complex refractive index with and without wave attenuation. Altar's manuscript, prepared in 1925–26 with Appleton's assistance, contains the first derivation for the dispersion relation, or the equation for the wave-normal surface, in the cold-plasma model in plasma physics theory, but the manuscript remained unpublished until 1981. Soon after Altar's work, and quite independently of it, a young German physicist, H Lassen, published (1927) an equivalent theory with the assistance of his mentor at Cologne, Karl Försterling. Altar and Lassen each used matrix inversions which were common in crystal optics theory at the time but less familiar to experimental ionospheric physicists. Altar incorporated complications in his mathematics which would ultimately be needed to cope with magnetoplasmas when both ions and electrons have to be taken into account. This work, however, had little or no influence on the field of plasma physics as it was later to develop and the achievement is sometimes credited to E O Astrom [49].

In spite of Altar's manuscript, Appleton failed to mention it and later sought the help of his old friend Douglas R Hartree (1929, 1931) who convinced Appleton that, following the work of Hendrik Lorentz in his *Theory of Electrons* (1909), a collisional damping term for free electrons, known as the Lorentz polarization term should be included in the magneto-ionic theory. This was later shown to be in error, after much discussion by numerous authors and some experimentation over a dozen years. Interestingly, C G Darwin, a contributor to this debate wrote late in his life (1950) that this problem was not 'one of the exciting growing points in modern physics, but of all the considerable variety of subjects in physics with which I have been concerned in my day, though the present problem may not have been the most important, it has in fact on the whole been the one that caused me most difficulty and puzzlement' [50]. The magneto-ionic theory became the central support for ionospheric research in the theory's modified form as constructed during the 1930s, especially by Henry G Booker, and by Thomas L Eckersley, some of whose work on this was not published for many years. In modern form, H K Sen and A A Wyler (1960) derived the magneto-ionic equations from first principles based on Boltzman's equation. A complete mathematical treatment of the theory was later standardized in a comprehensive work by K G Budden [51].

Early experimental studies essentially exploited radio data to deduce ionosphere and relevant upper-atmosphere physical characteristics using estimates of all factors involved in the production and loss processes in the formation of the ionospheric layers. These included estimates of the solar ultraviolet energy and its effects on the various height regions of the Earth's atmosphere. An early worker in producing such models was E O Hulburt of the Naval Research Laboratory in Washington, DC [52]. Others also worked to produce models of the formation, maintenance and dissipation of the upper atmosphere under solar influence. Models were produced by H Lassen in Germany in 1926 [53] and described in a book by P O Pedersen in Denmark in 1927 [54]. Pedersen, especially, provided expressions for the peak particle concentration and for its dependence on the zenith angle of the solar radiation. These works did not succeed in gaining the attention of sufficient workers, however: the German ionospheric community was still small. Pedersen published in a Danish engineering series and his book was also 'ahead of its time' as were works such as the monograph by the Russian M V Shuleikin [55]. The first model to gain wide attention in accounting for layer formation was one proposed by Sydney Chapman in 1931 [56].

Chapman re-emphasized the points of Pedersen and also calculated the distribution of electrons that would be produced in an equilibrium situation where the rate of production was equal to the rate of loss,

Sydney Chapman

(British, 1888–1970)

Chapman's first research in about 1910, on the theory of gases, began one portion of his widely ranging interests in physics. His other path proceeded to work on solar and lunar influences upon the Earth, especially in geomagnetism, the aurorae, and upper-atmosphere physics. He was active throughout his life as a pacifist, beginning with his exemption from military service in World War I. He was a co-founder of the Chapman–Enskog theory of gases, their viscosity, thermal conductivity and diffusion. This work continued from 1917 until 1940, expanded to include early studies of plasma physics, and resulted in the publication of a standard text. From 1913 until his death, Chapman contributed fundamental studies on geomagnetism, the ionosphere, lunar and solar atmospheric tides, geomagnetic storms, the magnetosphere and the solar wind. Here, too, he produced standard texts. Chapman played one of his most important roles in initiating and guiding the International Geophysical Year (1957–58), the greatest international geophysics co-operation in the twentieth century. After retirement from Oxford in 1953, Chapman as an international science statesman moved to the United States and continued his work from the universities of Alaska and Colorado. Chapman outwardly was reserved, but was a staunch friend and supporter of younger scientists and a man of great integrity. He lived simply and was an absolute enthusiast throughout his life for hiking and swimming.

and showed how the thickness of the electron layer depends on the scale height of the gas that is ionized. He assumed that parallel rays of sunlight of a single colour fell upon an upper atmosphere composed of a single type of molecule in isothermal plane-stratified conditions. The sunlight ionized these molecules, producing electrons and positive ions. Chapman assumed that the atmospheric gas became exponentially rarer with height above the Earth. Once ionized by the sunlight, the electrons remained unattached for some time and then recombined with the positive ions. This compound situation—ionizing radiation penetrating downwards and gradually losing energy, and gas density increasing downwards—would require a definite maximum of electrons produced at some height. Above and below this height the curve of electron density would fall off. Now, many factors actually intervene in the real terrestrial situation. This model was elaborated and termed the 'Chapman layer' model. Today, numerous different ionospheric regions (D, E, F1, F2, etc) at different heights are known to exist at different times and solar conditions. Some of the most difficult problems and some of the most imaginative solutions occupied ionospheric workers during the 1930s and 1940s. For example, Chapman thought the ionizing radiation was corpuscles, particularly at heights of the so-called 'E layer' while Appleton argued for photons as the causative radiation. Solar eclipse measurements in 1933 finally provided clear evidence that both the E and the F1 layers were ionized by photons. In order to explain the production of a layer at any height, one had to know the distribution and ionization cross sections of the different gases in the upper atmosphere. One major aim of ionospheric research was to provide this knowledge. Early ideas drawn from optical meteor trail studies suggested that the temperature above 100 km was about 300 K. Radio data were used to argue that the temperature rose to 1000 K at 300 km and that the important gases below 100 km were molecular oxygen and molecular nitrogen, with atomic oxygen being predominant at higher altitudes. Because ultraviolet radiation is strongly absorbed by air, laboratory measurements on Earth were difficult. By the 1930s, E W A Muller, T L Eckersley and G J Elias had suggested that solar x-rays might produce the E layer of the ionosphere. F Hoyle and D R Bates in 1948 suggested that heavy ions in the hot solar corona would produce sufficient x-rays [57], but confirmation of these ideas was to come only from rocket-borne instruments after World War II [58].

Other natural phenomena occupied the field of ionospheric physics and these studies carried through into the period of rockets, satellites and magnetospheric physics, beginning in the 1950s. In 1933 it was found that one high-powered radio in Luxembourg imposed broadcast modulation upon another broadcast. V A Bailey and D F Martyn in 1934 suggested that the electric field of the disturbing wave locally heated up the electrons in the ionosphere, increasing the collision frequency and the absorption coefficient and affecting the second signal in the region where the two beams intersected. This cross modulation or wave-interaction effect was turned to use as an experimental method. Wave-interaction and ionospheric heating studies remained of interest and have become especially important in geospace plasma physics since the 1970s.

The study of noise or interference has always been of great value to science, and this is especially true of upper-atmospheric physics research. As mentioned above, noise problems led to ionospheric research in the 1920s, and to early radio astronomy in the 1930s. The search to control for noise and scintillation effects first in the ionosphere, then in the solar-system plasma contributed to the discovery of the pulsar in radio astronomy. Other noise problems as experienced on telephone lines revealed naturally dispersive noise tones, whistling atmospherics, or 'whistlers' at very low radio frequencies. T L Eckersley [59] demonstrated that these can result from the frequency dispersion of impulses such as lightning strokes and can reveal the electron content of the ionosphere. Early whistler studies became dormant for two decades until the brilliant investigative work of L R O Storey [60]. This sub-field then became exceedingly active [61] and still is today. The area of very low-, extremely low- and ultra low-frequency radio physics (VLF, ELF and ULF) became of great interest to military communications (submarines and nuclear defence) and revealed itself to be a powerful aid to the study of natural plasmas in the ionosphere and magnetosphere. For example, the first evidence that plasma of ionospheric origin extended to several earth radii was provided using data from ground-based whistler measurements.

It has always been obvious that the Sun is closely involved with upper-atmosphere geophysics and geospace but the exact mechanisms have not been understood. (Very early in this century, Marconi suspected that radio waves propagated further during darkness because then the radio antennas were free of the daytime shield of ionization around the antenna wires caused by the Sun's light!) Several discoveries in the 1930s involved solar action on the upper atmosphere. J H Dellinger (1935) and independently H Mögel (1930) discovered the sudden ionospheric disturbance (SID). Close coincidence in time and intensity were found to exist between bright chromospheric eruptions on the Sun and the attenuation of radio waves propagating over the sunlit portions of the Earth. Radio fade-outs occurred within minutes and could last for hours. Correlation was found also with disturbances to geomagnetic records. Early atomic physics studies suggested the oxidation of atomic oxygen or nitric oxide or ozone in the ionosphere by Lyman-alpha solar radiation as the direct cause, but these ideas were altered following rocket studies of solar x-rays after World War II.

Very exciting opportunities for upper-atmosphere research came with the initiation of rocket and then satellite and space-probe studies. Other new methods evolved from purely ground-based studies or from combinations of ground-based as well as *in situ* instruments. Space will not permit mention of all of the methods, but the powerful technique of incoherent backscatter, or incoherent scatter radar (ISR) will serve here as an excellent illustration. Normal ionospheric sounding as initiated by Breit and Tuve in 1925, depends upon the upward-going radio wave being returned to the ground at a height and radio frequency where the refractive index is zero. The return echo may be coherent, as if reflected from a mirror, or it may be scattered incoherently from a rough surface. However, above a certain frequency, the radio wave is not reflected but passes on upwards into space. Thus, ground-based ionosondes normally yield no data above the height of maximum electron density. Ionosondes flown in satellites and transmitting downwards can provide a 'topside' look down, but such ionosondes also have their limitations.

Each electron in the ionosphere, however, can be considered as a radiator, with a very small scattering cross section, typically about 10^{-19} cm^2. As the electrons are in random thermal motion, each small scattered return echo is added incoherently and the return signal power is proportional to the total number of electrons in the scattering region. The scattering cross section of each electron does not depend upon the frequency of the incident transmitted wave. Therefore higher frequencies can be chosen than would be returned with a normal ionospheric sounder, frequencies which are high enough to suffer very little absorption in the ionosphere. Beginning in the late 1950s, a new type of radio sounding of geospace was developed, that of incoherent scatter radar (ISR). The amount of energy scattered back to the ISR, however, is minuscule. From a range of 300 km in the ionosphere, the energy backscattered to a typical ISR antenna is roughly the same as that from a small copper coin at the same range [62]. Thus the ISR transmitters are very powerful and ISR antennas are among the largest in the world, measuring acres in area. The world's largest single dish antenna, at Arecibo, Puerto Rico, is a 1000 foot spherical section wire dish strung between three mountains. Others are giant steerable parabolas up to 100 metres across, or fields of fixed antennas operated in combination. The transmitters usually have pulse powers in the megawatt range, and typically are used in collaboration with other upper atmosphere observing instruments, such as coherent radars, or ionospheric heating transmitters, and satellite instrumentation.

The principle of ISR was first published by W E Gordon [63] and was almost immediately demonstrated experimentally by K L Bowles. ISR techniques can be used to determine electron density throughout the ionosphere and up to altitudes in the magnetosphere of 7000 km and also to determine electron and ion temperature, ion composition and electric fields. The great promise of this method of studying the upper atmosphere was quickly put into practice. Within about a decade some seven ISR observatories had been built on three continents. There were eight such observatories as of 1989, operating at equatorial, mid-latitude and polar latitudes. One spectacular ISR record of ionospheric plasma irregularities from the observatory at Jicamarca, Peru is shown in figure 26.22. Such studies are frequently coordinated with other ground-based equipment and with satellites or rockets passing overhead. As

Geophysics

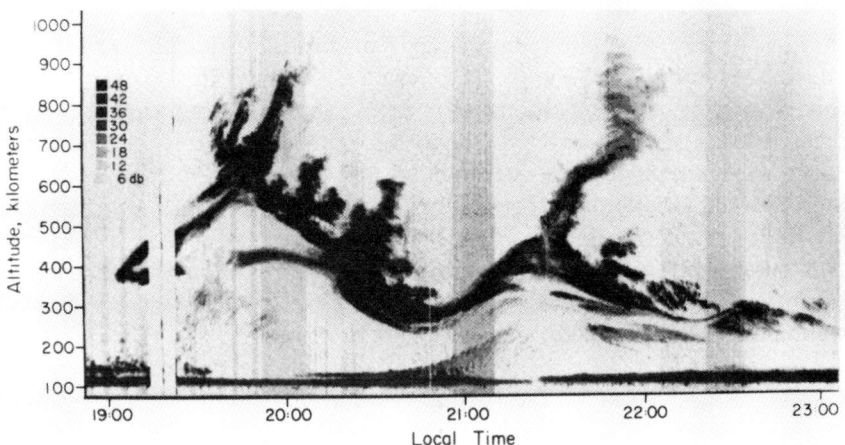

Figure 26.22. *Equatorial spread of F-region ionospheric plasma instability phenomena. Range–time–intensity map displaying the incoherent backscatter power at 3 m wavelengths measured at Jicamarca, Peru. The grey scale is in decibels above the thermal noise level.*

of the mid-1990s, there are literally dozens of techniques for studying geospace, with more than a dozen of these approaches using radio techniques.

In other methods of study, geospace itself has been artificially modified. An early example is the Luxembourg effect, mentioned above, where the ionosphere is in effect heated by the power of the transmitter. Ionospheric heating studies continue today. Of course, many man-made modifications are designed to copy natural phenomena, as are accelerator experiments in high-energy physics. Solar flares and the solar wind in Nature modify geospace. Rockets and satellites have been used to produce artificial aurorae or visible- or radar-trackable trails produced by nitrogen, barium, lithium, caesium or other elements or compounds released into the upper atmosphere. Very low-frequency transmitted signals have been designed to trigger plasma events in the magnetosphere, mimicking natural 'whistler' phenomena. Ion guns have been fired from satellites to create plasma experiments in space, and nuclear devices have been exploded in the ionosphere to create artificial particle belts, aurorae and magnetospheric plasma. The three *Argus* high-altitude nuclear devices detonated in August and September 1958 constituted the first worldwide geophysics experiment, producing an artificial particle radiation belt which was detectable worldwide for several weeks [64].

There are many other interesting topics that could be discussed in this chapter, but one other should be mentioned, as it was to have a great effect on research in the years following World War II. The movement to organize a worldwide network of ionospheric observing stations came from several directions: the tradition in geomagnetism, and in astronomy and meteorology, had established the desirability of worldwide geophysical observing stations. To some extent, ionospheric data-gathering equipment could be situated adjacent to other geophysical equipment, possibly in pre-existing observatories. The by-then powerful radio communications industry, both commercial and military, needed prediction services: which radio frequencies, using what level of power and antenna arrays and operating at what hours of the day would assure reliable regional and worldwide communications? This called for data stations and for prediction models.

The initiation of World War II hastened the assembling of networks and the construction of theoretical radio propagation models. After positing mechanisms for how ionized layers or regions were established and maintained in the upper atmosphere, attempts were made to fit models of electron density profiles as a function of height. An ionospheric sounder on the ground received echoes from layers up to points of maximum electron density. For a unique solution, the electron density must increase monotonically with height. The returned ionospheric echo on any given frequency gives information on the integrated electron content over that path. Before the availability of digital computers, such integral equations could be nearly impossible to solve in practice. It seemed reasonable to try to fit the data to some simple mathematical curve and then adjust the fit. Thus the ionospheric density profiles were approximated over the years by linear, quadratic, exponential, parabolic and 'Chapman' profiles [65], and there was some dispute over which model best fit the greatest number of data.

The National Bureau of Standards (NBS) in Washington had been sending radio prediction forecasts worldwide from radio standard frequency station WWV since 1935. In the mid-1930s there were only three or four stations worldwide making any sort of regular ionospheric sounding data measuring runs and there were only seven such stations by 1939. Most of these had been developed by Lloyd Berkner, formerly of the NBS and then with the Carnegie Institution of Washington. In the years just preceding and during World War II, competitive systems were set up within several countries for purposes of predicting radiowave propagations. In Germany and in Britain, for example, there were both civil and military prediction services. In the United States before the war, Newbern Smith of the National Bureau of Standards had initiated (1937) a prediction system utilizing empirical transmission curves to determine the most efficient frequencies for long-distance communications. During the war the NBS joined the military in what was called the Interservice Radio Propagation Laboratory, headquartered at the NBS in Washington, DC. The British military formed an Inter-Service Ionosphere Bureau

Geophysics

(ISIB) headed by T L Eckersley and G Millington of the Marconi Company in Great Baddow, Essex, England. This group used an approach similar to the Americans. A rival in England to the ISIB was the system developed by Appleton and W J G Beynon at the Radio Research Station at Slough. Though assisted by the American Lloyd Berkner, the Australians first used Smith's curves, then switched to Appleton's approach in predictions at their Ionospheric Prediction Service. Severe war-time communications difficulties were sometimes laid at the doorsteps of the different prediction models. An international Allied radio propagation conference held in Washington in April 1944 improved matters and demonstrated that an effective geophysical prediction system necessitated international operation and much cooperation. By the end of the war there had been more than 40 ionospheric sounding stations operated by the Allies and by the Axis powers. This network of stations served as a main support for ionospheric research planning for the International Geophysical Year 1957–58, during which the number of such stations grew to about 150 [66].

The days of early ionospheric physics research can be spoken of as beginning in about 1902 with the speculations of Kennelly and Heaviside, greatly increasing soon after World War I and lasting until about the end of World War II. Using ground-based equipment, including remote sensing ionospheric radars (the ionosondes), and associated geophysical data from solar and geomagnetic observatories, physicists had constructed reasonable models for much of the ionosphere. J A Ratcliffe (1974) summarized the feelings of many when he wrote 'In 1925 little was known about several phenomena that took part in the formation of the ionosphere. They included: the nature and distribution of the atmospheric gases, the nature of the ionizing radiations, the nature (ionic or electronic) of the ionospheric charged particles, and the mechanism (recombination or attachment) by which electrons were lost. ... Between 1925 and 1955 ground-based radio measurements provided almost the only evidence, and by 1955 they had led to a fairly complete understanding of the ionosphere and of the atmosphere between the heights of about 90 km (the top of the D region) and about 300 km (the peak of the F-layer)' [67]. Beginning in the 1950s, experiments with rockets and satellites, new ground-based techniques of partial reflection, cross-modulation, incoherent scatter, VLF whistler modelling, the impetus of the International Geophysical Year and the infusion of more atomic and plasma physicists enlarged upper-atmosphere geophysics to include the magnetosphere and geospace. Two celebrated physicists commented (1974) upon the comparison of the 'early days' to more recent times 'Except for the discovery of radio waves nothing has had such major direct influence on ionospheric research as the advent of sounding rockets and the advent of satellites' wrote D R Bates [68], and H S W Massey wrote 'Sometimes in the initial

1999

development of a theory it is a disadvantage to know too much and at least the early ionospheric theorists did not suffer in this regard' [69].

As a new scientific specialty is emerging, it can metamorphose from an earlier field or it can combine elements of other scientific areas of study. In upper-atmosphere geophysics, ionospheric studies evolved from electron physics and from radio communications. Thus, one finds citations back to the earlier classic electron and electron optics texts of H A Lorentz, P Drude, W Voigt and O W Richardson. One also finds cited, for instance, the radio propagation and engineering books of G W Pierce and J Zenneck. Early attempts to produce a text or 'Bible' for ionospheric physics such as the book of P O Pedersen [54] were not sufficiently influential and workers continued to utilize the pioneering literature.

Review articles normally appear continuously throughout the evolution of a field, and before the appearance of texts. Science as an institution has long recognized this. For example, the young Michael Faraday in 1821 was put to writing a series of reviews of the new electromagnetism phenomena [70] although he himself had yet to become proficient at the new field. This review-article task was of great benefit to Faraday and indeed changed his career—with results known to the world!

The first standard collections for upper-atmosphere geophysics were written or compiled just before World War II. J A Fleming's edition of *Terrestrial Magnetism and Atmospheric Electricity* [71] formed one volume in a series sponsored by the US National Research Council on the Physics of the Earth. The work was largely written by workers at the Carnegie's Department of Terrestrial Magnetism, with assistance from J Bartels of Berlin, L Vegard of Oslo and others. This was followed by S Chapman and J Bartels' two-volume *Geomagnetism* [72]. Each of these works treated the ionosphere, aurorae, upper-atmosphere and solar–terrestrial relations only in part, but in large and significant part. By 1940 there were numerous other sections of handbooks of physics, geophysics or astrophysics containing material of interest to the upper-atmosphere geophysicist: material linking electromagnetic theory, atomic physics and spectroscopy, aurorae and night sky light, geomagnetism, atmospheric electricity, solar–terrestrial relationships, and radio propagation and communications techniques.

During and immediately following World War II several singly authored texts or monographs appeared. One, S K Mitra's *The Upper Atmosphere* (1947, 1952) [73], became known to almost all workers and was a standard until the appearance of magnetospheric physics. Mitra's volume grew out of a review on ionospheric physics which he published in 1935 and developed into a monograph at the urging of M N Saha. With the aid of his colleagues in Calcutta, especially J N Bhar, N R Sen, H Rakshit and S P Ghosh, Mitra included much material on an ideal model atmosphere, wind systems, geomagnetism, the aurorae, magnetic storms, and especially the ionosphere. The book was a great success and contributed to the numerous students developing into scholars of upper-atmosphere and radio physics both in India and around the world. This volume served both as a text and as research monograph. Mitra was also able to introduce the reader to early rocket studies.

Conference proceedings on the upper atmosphere played a major role beginning in the 1950s. By the late 1950s the ionospheric physics community was producing 500 papers per year and one now saw the appearance of sub-specialty monographs, for example, on whistlers, or sporadic-E or Arctic radio propagation. Actual textbooks began to appear, that is, books designed specifically for the advanced undergraduate or graduate university student. The appearance of textbooks indicates something about the maturity of a scientific specialty. One can then look towards signs of new activity: the magnetosphere began as a small part of ionospheric studies, then it became an equal part, for example, as noted in the title of J A Ratcliffe's *An Introduction to the Ionosphere and Magnetosphere* [74]. Earlier in this chapter the author spoke of hopes of uniting the study of the Sun and its effects upon geospace. Attempts to pull together the entire field of solar effects upon geospace and the Earth's upper atmosphere can be seen in S-I Akasofu and Sydney Chapman's *Solar–Terrestrial Physics* [75], the successor to Chapman and Bartels' *Geomagnetism* (1940) [72]. Akasofu and Chapman saw the field returning to a unity of concerns about solar influences on the Earth's upper atmosphere and electromagnetic environment, and thus centrally involving the study of natural plasmas. For them, the late-nineteenth century studies of geomagnetism and the aurorae, combined with the earlier twentieth century studies of ionospheric physics, aeronomy and cosmic-ray physics, were amplified with the age of satellites and space probes. This should be a message indicating the unity of studies of geospace and the many fascinating problems common also to plasma physics and astrophysics.

26.8. Outwards into the magnetosphere and solar–terrestrial space

As J A Van Allen has written 'The scientific heritage of magnetospheric physics lies principally in studies of geomagnetism, aurorae, and the geophysical aspects of cosmic radiation and solar corpuscular streams' [76]. The rapid growth of the sub-field owes much of its impetus to the new generation of instruments and techniques of space satellites and probes, beginning with rockets in 1946. Early magnetospheric workers felt as if they had broken out of a flat world. W N Hess saw it in 1968 as 'Before exploration of space started, we were standing on the surface of the Earth looking upward. We had balloons to probe up to 30 km [77], and radio waves to study the ionosphere. We could observe meteor trails and aurorae to 100 km and higher, but we were living in essentially a two-dimensional world' [78]. The terrestrial atmosphere was assumed to disappear above a few hundred miles altitude and the external geomagnetic field was assumed to decrease in strength outwards as that of a simple bar magnet. The effects of solar cosmic rays were not recognized or understood, although the Sun's effect on geomagnetic storms had long been a problem of research.

From the late nineteenth century, networks of stations had collaborated on studying the small (about 1% or less) changes in the surface geomagnetic field which occur nearly continuously in time. These changes, most often recorded with ink on paper rolls or on photographic paper, were given names such as 'bays', 'crochets', 'spikes', etc, since the tracings on the records resembled to the human eye the outline, for example, of a marine bay coastline on a map. Making such analogies is common to scientists. When audio observations were made of 'whistlers', researchers named the phenomena to resemble audio sounds, such as 'whistlers', 'tweeks', 'chirps' 'hiss', 'dawn chorus'. When these same phenomena were later recorded on visual trace form, newly discovered features were named again for visual examples, such as 'knee', 'hook', 'nose' or 'cluster'. Study of geomagnetic changes on records revealed that 'storms' occurred, that they could affect large portions of the Earth, and that they seemed to re-occur roughly at periods of 27 days, the synodic period of rotation of the Sun. Certain phenomena, such as the radio fade-outs or SID, occurred in daylight, shortly after solar flares. Most geomagnetic storms, however, seemed to lag solar flares in time by about two days. This work was fully summarized by Chapman and Bartels in 1940.

It was assumed that solar streams of ions or electrons travelled towards the Earth from time to time, at speeds of some hundreds of kilometres per second. T L Eckersley speculated in the late 1920s that such solar streams were involved with the propagation of 'whistlers' and also with the reception of radio echoes which seemed to have inexplicably slow propagation speeds. Chapman thought that solar corpuscular streams were responsible for ionizing the E layer in the ionosphere, but Appleton convinced him that solar radiation was most likely the cause. Thus many specific geophysical events were believed to be linked to sporadically occurring solar streams.

What about a *theory* for the action of the solar stream upon the Earth and its causation of a geomagnetic storm? When the solar stream arrived at the geomagnetic field, it compressed the field and gave rise to an increase in the magnetic field as measured on the Earth's surface. Subsequently, it was argued, the geomagnetic field reversed and decreased and the solar stream then caused a resultant westward-flowing, equatorial ring current around the Earth, and was associated with geomagnetic and auroral activity, especially in the polar regions. As the Sun rotated, the stream of particles would no longer intersect the

Earth's orbit until about 27 days later. More than one storm could be superimposed upon the Earth simultaneously, if there were more than one causative, active solar flare region.

The astrophysicist E A Milne [79] had argued theoretically that, at high levels in the solar atmosphere, upward moving neutral or ionized atoms might be continually accelerated away from the Sun by selective radiation pressure and under appropriate conditions driven at velocities up to 1600 km s^{-1}. (We now know that Milne's argument is incorrect, but it was persuasive in earlier times.) How atoms at such relatively low velocities, of the order of 1000 km s^{-1} could penetrate the Earth's upper atmosphere to ionospheric and auroral heights remained a problem. As Chapman wrote in 1940 [80] 'Unless some undiscovered mechanism exists which imparts much greater velocities to the solar corpuscles, possibly only in the near neighbourhood of the Earth, we must conclude that the Earth's atmosphere is more penetrable, down to 100, 80, or even 70 km, than is indicated by our present information ...'. Since at least 1918, Chapman had written on causes of geomagnetic storms.

In a series of papers beginning in 1930, Chapman and V C A Ferraro [81] argued that the geomagnetic field might influence the solar stream or cloud by carving out a hollow space round the Earth, although the solar stream could reach the lower ionized atmosphere along two 'horns' extending into each polar region. This would account for the initial phase and increase of the geomagnetic field. The main phase of the geomagnetic storm, with resultant radio interference, auroral phenomena, etc, would be produced by charged layers induced on and escaping from the surface of the hollow. A westward flowing equatorial ring current, causing the main phase, was located at a height of several earth radii. Qualitatively, this theory fit the observed facts, though Chapman and Ferraro utilized very simplified idealized cases in their theory. Basically, their approach was correct. Notable alternative models were produced at that time by Alfvén (1939 and subsequently), who also posited the development of a ring current around the Earth and envisioned a weak magnetic field accompanying the solar stream [82]. The problem of explaining geomagnetic storms has continued to this day as an active field of research and will be commented upon below.

Both Chapman and Alfvén were strongly influenced in their theories of the Sun's effects on geomagnetic storms by the earlier auroral studies of Kristian Birkeland and Carl Størmer. Beginning in 1896, Birkeland considered that a stream of electrons from the Sun could reach the geomagnetic field with enough density to produce geomagnetic disturbances. He made experiments wherein cathode rays were projected towards a small magnetized sphere, which he called a 'terrella'. Birkeland (1901, 1908, 1913) [83] demonstrated with beautiful photographs that a toroidal space was obtained around his 'terrella' and that many of the cathode rays (electrons) were directed towards the poles

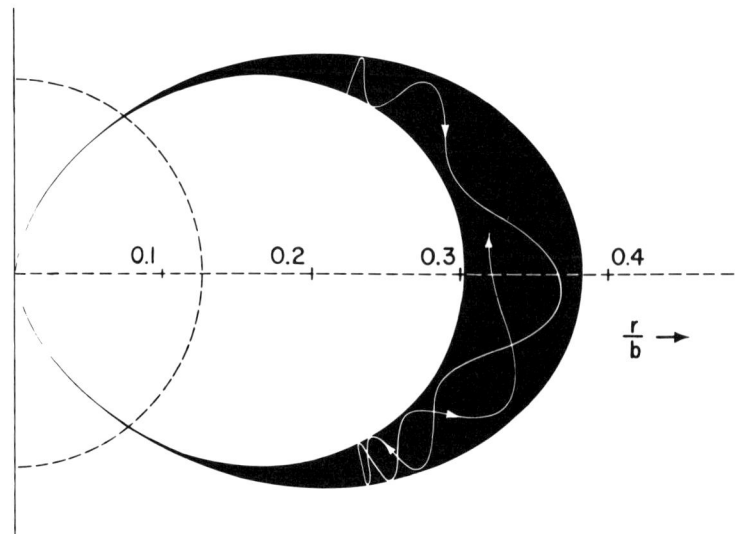

Figure 26.23. *Størmer orbit of a charged particle in the Earth's magnetic field.*

of the sphere, leaving an equatorial belt and the toroidal space relatively free of the particles. Birkeland's theory was criticized by Arthur Schuster and others but it was influential on the Scandanavian school of auroral and magnetic researchers and it was instrumental in inspiring Størmer. In fact, Størmer developed mathematically Birkeland's theory and experiments, reminiscent of the way S D Poisson in the 1820s developed mathematically C A Coulomb's electrical and magnetic experiments.

Størmer dealt with the possible paths of a single charged particle and showed that such a particle from the Sun can reach the Earth only in two narrow zones centred near the polar regions of the Earth. Figure 26.23 illustrates Størmer's 1907 diagram of the meridian projection of the trajectory of an electrically charged particle in a magnetic dipole field [84]. The dark area is the region wherein the particle is trapped. The quantity r/b is the ratio of the radial distance r to b, a parameter in Størmer's model. The dashed semicircle represents the Earth. This single orbit calculation took Størmer's female assistant almost two years to calculate by hand. Størmer devoted many years to studying the motion of electrically charged particles near magnetic dipoles. His work and that of Birkeland would provide inspiration for studies of solar and cosmic-ray particles and their interactions with the Earth's magnetosphere.

Cosmic-ray studies began just before the time of World War I. It was a major growth area in physics, beginning in the late 1920s with questions concerning the nature of the 'penetrating radiation' as it was then called: whether there were particles; if so, whether the particles were charged, whether they were isotropic in their arrival, and what was the energy spectrum of such particles. These questions occupied and frustrated cosmic-ray physicists, since they could not get their instruments in balloons 'above' the neutral atmosphere of the Earth. It would be instrumentation placed in geospace after World War II which would allow further studies of the cosmic rays; new instrumentation which would explore the magnetosphere, assisted by ground-based 'whistler' studies.

From 1946 onwards upper-atmosphere research was conducted using war surplus German V-2 rockets and newer designs such as the US *Aerobee* and *Viking*. Exciting results were obtained, although, as mentioned above, early rocket results were often in error and sometimes confused the situation. Atmospheric pressure, temperature and density were measured. Ozone distribution, electron densities in the D and E regions of the ionosphere were measured and compared with ground-based ionosonde radar results. Cosmic rays were measured. Solar x-ray and ultraviolet measurements made from rockets by Herbert Friedman and his group at the US Naval Laboratory, and by others in the years following 1946, answered questions about ionospheric layer production by solar energy, and opened up the new field of x-ray astronomy. Scientific rocket studies were also underway by the Soviets, Australians, Canadians, English and French. All this stimulated the idea of studying geospace and the Earth from a worldwide platform of ground observatories and from space. The result was a wonderful 18 month collaboration (1957–58) called the International Geophysical Year (IGY), the greatest international scientific expedition in history. The IGY also ushered-in the satellite era.

International expeditions for science, conquest and entertainment were not a new idea. The IGY grew out of earlier Polar Years. The first International Polar Year in 1882–83 was mainly concerned with meteorology and terrestrial magnetism in the Arctic. In the history of science, the first International Polar Year was the first major event in which the fundamental concept was multidisciplinary co-operation on an international scale. Numerous Arctic and Antarctic expeditions were mounted in succeeding years and a Second International Polar Year was initiated in 1932–33, a half century after the first Polar Year. The Second Polar Year plans were hampered by the worldwide economic depression following 1929, so that some expeditions were cancelled and others such as Richard Byrd's Second Antarctic Expedition were delayed for a year. The Second Polar Year also concentrated mostly on meteorology and on the Arctic. In spite of this, considerable upper-atmosphere research had been conducted in Antarctica by 1934, including the first polar ionosonde recordings, 'whistler' studies, and even airborne cosmic-ray studies by A H Compton's group [85].

Lloyd Berkner, ionospheric researcher before World War II, had participated in Byrd's First Antarctic Expedition in 1928–29. Having served in the US Navy during World War II as an engineering and science adviser, Berkner had by the Eisenhower years attained considerable influence in Washington. He was perhaps the most powerful man in science in the United States—advising in the 1950s in high-energy physics, in radio astronomy and geophysics. Berkner was the prime mover for the IGY. It had long been his dream, as well as that of Sydney Chapman and other geophysicists, to follow the Second Polar Year with another programme which would study the South Polar regions in geophysical detail and go beyond the primarily meteorological studies of the other Polar Years with an emphasis on ionospheric, rocket, auroral, airglow and geomagnetic studies.

At a dinner at the home of J A Van Allen in 1950, Berkner proposed a Third International Polar Year for 1957–58, to follow the Second Polar Year by 25 years and also to occur during a sunspot maximum period. (The Second Polar Year occurred nearer a sunspot cycle minimum.) Berkner and Sydney Chapman were primarily responsible for enlarging and developing the idea. The proposal was to exploit the 'rapid advances made since 1933 in scientific, especially ionospheric techniques' as Sydney Chapman said [86]. Here was the hope for international multidisciplinary geophysics to be funded by other than popular fancy. As polar geologist Lawrence Gould had written in 1931, he hoped sincerely 'that the day is past when we must seek support for polar research in terms of spectacular and heroic effort ... the lure of knowledge that is "baited by suffering and death"...' [87]. The idea was sponsored by the International Union for Scientific Radio with the co-operation of the International Astronomical Union and the International Union for Geodesy and Geophysics. The International Council of Scientific Unions formed a special committee in 1952 and several congresses were held to prepare for the IGY.

Particular attention in the IGY was paid to the geographic regions of the Arctic, the Antarctic and the equatorial belt. In terms of funding, special emphasis was given to the Antarctic, ionospheric research, and rocket and satellite studies. As early as 1953, the IGY Secretariat had foreseen the possibility of employing satellites during the IGY, and the IGY symbol designed in 1954 depicted the Antarctic as well as the trajectory of a satellite. The first public announcements of planned satellite launches during IGY were made by the USA and by the USSR in 1955. The IGY discoveries really led to the division of geomagnetism into two different aspects: internal magnetism, relating to the physics of the Earth's interior, and 'external geophysics', illustrating the close connections in geospace between the ionosphere, magnetosphere, and cosmic-ray studies. In the USA programme, satellite work was the largest single budget item, and ionospheric physics was the largest ground-based specialty in terms of funding [88].

The polar regions, in terms of the aurorae, particularly difficult radio propagation conditions, and the possibility of detecting lower-energy charged particles which entered geospace along the magnetic field lines converging at the Poles, had been of interest to upper-atmosphere geophysics for decades. In the early 1950s, rockoon flights in the auroral zone detected low-energy, easily absorbed particles, later found to be electrons. (The rockoon was a balloon which rose to about 20 km altitude and then launched instrument-carrying rockets to an altitude of 90 to 120 km.) Upper-atmosphere geophysics had benefited from rocket studies for a decade and was ready for satellite studies.

The world remembers the launch of the Soviet *Sputnik I* on 4 October 1957, followed by *Sputnik II* with a payload greater than 1000 lb on 3 November. After three failed launch attempts with a Vanguard rocket, the USA launched *Explorer I* on 31 January 1958. This tiny satellite carried a Geiger counter but no tape recorder, so real-time data in one or two minute segments were intercepted by seven stations in the USA and nine others in South America, the Caribbean, Africa, Australia and Asia. The Geiger counter was part of the research of J A Van Allen, George Ludwig and Ernest Ray at the University of Iowa. They received surprising results: near satellite apogee over South America, their Geiger counter data showed zero counts per second, whereas at other times their count rate was an expected 30 counts per second. The interpretation of these puzzling data required some weeks.

Their next successful launch was on 26 March 1958 on the satellite *Explorer III*. They had a Geiger counter aboard as well as a tape recorder, so all data could be recorded and then telemetered to ground stations. On this satellite, their Geiger count-rate went up and down from a normal count of about 20 to a maximum reading of 128 and then to near zero. The astounding finding was that very large particle fluxes had caused the counter electronics to saturate and register zero. In his first public lecture on the satellite results, 1 May 1958, Van Allen emphasized that the large flux was 1000 times larger than the cosmic-ray flux and suggested that the flux was from particles trapped in the Earth's magnetic field. He states that his familiarity with the Størmer orbit paper of 1907 and his own earlier work with magnetic field confinement of trapped particles led him to this idea [89]. He thought that the particles most likely were low-energy electrons, the soft radiation that he had previously observed from rockets launched in the auroral zone. (It was later learned that most of the particles were protons.) *Explorer IV*, launched 26 July 1958 carried four particle detectors with different thresholds. *Explorer IV* was designed not only to measure the natural radiation but also to study the particles artificially placed in the geomagnetic field by the three small nuclear explosions set off from rockets in late August and September 1958 at about 200 to 500 km altitude in the ionosphere as part of the US Department of Defense Project Argus.

Pioneer III was launched by the USA on 6 December 1958 and was

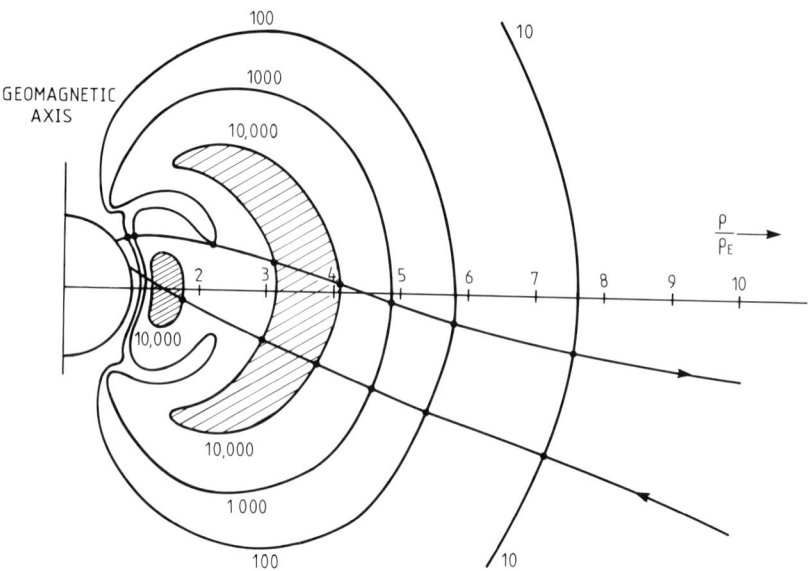

Figure 26.24. Van Allen's original 1959 publication of a sketch of the Earth's inner and outer (Van Allen) radiation belts.

intended to reach the Moon. It did not, but fell back towards the Earth after reaching a distance of over 107 000 km from the Earth. Two Geiger counters were aboard *Pioneer III*. Combining the data from *Explorer IV* and *Pioneer III* allowed Van Allen to produce the first map of the radiation belts, showing the inner and outer zones [90]. Figure 26.24 shows the count rates as contour lines.

The Soviet satellite *Sputnik II* might have detected the radiation belt but the Soviets could not agree on sharing data codes with Australia. Soviet *Sputnik III* and other Soviet deep-space probes, along with further US probes discovered and verified that the inner radiation zone is stable in time and is composed of high-energy protons. The outer zone is quite variable, changing by an order of magnitude in less than a day, and is composed of energetic electrons. These discoveries were made in the Cold War era of science competition. Science, possible science applications and technology were presented as a race between the USA and the USSR or between NATO countries and the Warsaw Pact. In this atmosphere, unfortunately, achievements of some scientists, such as K I Gringauz and V I Krassovskii were downplayed in their own country, since they had 'lost' the race.

What were the weaknesses of the early magnetosphere measurements [91]? There was a failure to identify the species of particles surrounding the Earth, and there was a lack of information on the energy spectrum. Both ground-based and satellite observations contributed to further information concerning the magnetosphere. Don Carpenter, from ground-based whistler measurements, first identified the plasmapause [92]. It was subsequently realized that early Soviet flights as described by K I Gringauz et al [93] had detected the plasmapause and also the solar wind.

What about theory? Following the earlier work of Størmer, Chapman and Ferraro, Alfvén and others, S F Singer in 1957 suggested that solar particles were somehow injected into the geomagnetic field early in the course of a magnetic storm and that such particles were trapped in the Earth's outer atmosphere at altitudes of about six earth radii. These trapped particles moved in helical paths along geomagnetic lines of force back and forth from north pole and south pole and drifted in longitude, causing the ring current. Ludwig Biermann (in 1951 and subsequently) had observed ionized comet tails and concluded that solar gas flowed continuously through the solar system. Eugene Parker developed the theory of gas flow from a solar corona and coined the term 'solar wind' for the continuous flow [94]. Additional evidence for the solar plasma interacting throughout the solar system came from others as well.

What do we now know about geospace, the Earth's ionized and magnetic envelope [95]? The temperature drops with increasing altitude from the Earth's surface to a minimum near 10 km, rises in the stratosphere and drops again due to radiative cooling at the beginning of the mesosphere, at about 80 km altitude. It then rises dramatically to over 1000 K in the thermosphere, the region above about 100 km. Above this altitude the gases are no longer mixed homogeneously but separate out according to their masses. Absorption of solar UV radiation in the thermosphere causes some of the temperature rise and also the ionization of the atmosphere. The ionospheric plasma also is produced in part by energetic particle impact, especially in the polar regions. The various electron density regions (D, E, F1, F2) reach maxima at heights from 80 to 400 km depending upon a complicated combination of ionization rates, local time, solar flux conditions, solar activity, ion and electron dissociation and recombination rates, etc. The peak electron density, in the F region can reach 10^6 cm^{-3} at local noon. Near 100 km altitude N_2 and O_2 dominate. Higher up, NO^+ and O^{2+} dominate until hydrogen dominates in the altitudes above 1000 km. There is a complicated ion chemistry with numerous species, including heavy atomic ions deposited into the ionosphere by meteors. At E region heights of about 100 km, particle energies can vary widely, especially in the auroral regions circling the polar regions. Night-time aurorae are caused by 3–10 keV electrons, though lower-energy electrons and both protons and high-energy electrons cause other phenomena. Storage in the magnetosphere of solar wind energy, and then the release of this energy into the ionosphere adds much complexity to the situation. The coupling mechanisms between the ionosphere and the middle atmosphere below, and particularly the coupling to the magnetosphere above, has received much observational and theoretical attention in the last quarter century. This geospace structure above our heads and its effects on radio signals and polar lights has fascinated us for over a hundred years.

The magnetosphere was revealed in its main outlines in the decade following the late 1950s. An amazing number of additional features and mechanisms were discovered and studied in the succeeding quarter century. The steady solar wind, a fully ionized supersonic collisionless plasma, travels to the Earth. Some of it, as mentioned, enters the ionosphere and upper atmosphere causing the numerous natural phenomena discussed here. Much of the solar plasma is deflected by the geomagnetic field, flowing around the Earth and creating electrical currents. The secondary magnetic field created by these currents counteracts the geomagnetic field on the sunward side of the Earth and increases the magnetic field in certain other locations. A teardrop-shaped closed magnetosphere would result in a simplified model. This might resemble a distorted magnetic dipole. In 1961 Ian Axford and Colin Hines [96] introduced a theory which allowed for solar wind plasma to be coupled to plasma inside the magnetosphere. This would account for some of the observed flow within the magnetosphere. Another theory, involving a partially 'open' magnetosphere was presented by Jim Dungey [97] and involved the idea of geomagnetic field lines which terminate in interplanetary space. Dungey suggested that solar energy could be added to the magnetosphere, in effect, by completing a circuit by reconnection of geomagnetic field lines with field lines in the Earth–Sun space. The interplanetary electric field can then affect the upper atmosphere, entering the ionosphere and the middle atmosphere. In this model, which has been considerably developed, the field lines are swept around the Earth, merging in the anti-sunward 'tail' of the magnetosphere many earth radii behind the Earth. Once reconnection occurs, plasma in the tail flows back towards the Earth.

Pictorial models of geospace, including the ionosphere, the magnetosphere, radiation belts and other aspects, have become so detailed that only a specialist can understand the more elaborate of the pictures. Nevertheless, a somewhat simplified version shown in figure 26.25 serves to give the general outline of what we know of geospace. On the scale of the figure drawing, the various regions D, E, F, etc, of the ionosphere and the auroral zones are immediately adjacent to the outer diameter of the solid Earth. Above this, at distances of from a fraction up to about four Earth radii, lie the trapped radiation belts or zones. Exterior to these are the regions of low-energy plasma; the sunward region is toroidal, with the anti-sunward side having an extended 'tail' in the feature called the plasma sheet. The sheet has a thickness of several Earth radii and a low particle density of about 0.5 cm^{-3}. At the plasmapause, the electron density in the magnetosphere drops abruptly by as much as one or two

Geophysics

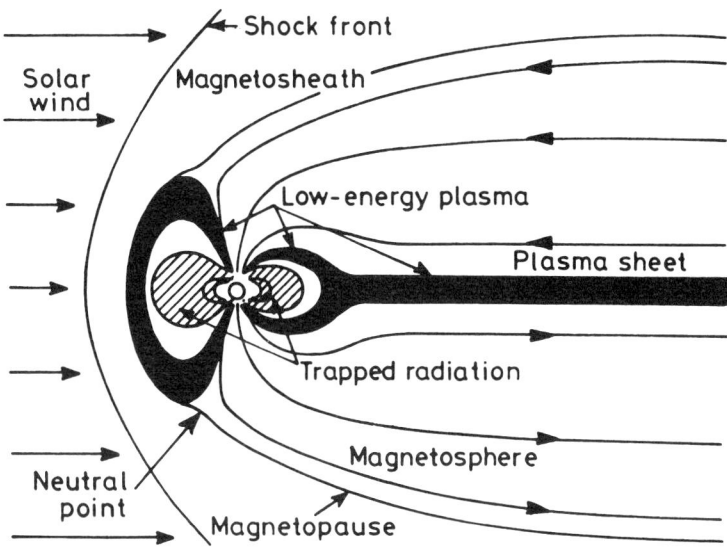

Figure 26.25. *Distribution of the low-energy plasma in relation to other magnetospheric features.*

orders of magnitude but the temperature increases considerably. The interplanetary field is adjacent to the magnetosphere at the turbulent region called the magnetosheath, just outside the magnetopause. Most of the immediate solar-caused phenomena occur along field lines reaching Earth at auroral and subauroral latitudes. Finally, at the Earth's bow wave, there is a bow shock and the supersonic solar plasma creates both a subsonic flow inside the shock layer, and actual entrance of portions of the supersonic wind into the magnetosphere.

Forecasts and hopes today for the study of geospace include: increased interdisciplinary approaches, both in theory, in method, and in techniques and networking of observations; the use of more powerful computers for data analysis and modelling; and the construction of texts and reviews leading toward a synthesis of the plasma physics of our solar system.

Finally, we reach the end of geospace, some Earth radii outside our planet, but we now know that not only geospace, but even the interplanetary medium is greatly affected by solar effects. We find our ionospherists and geomagneticians of several decades ago have evolved to studying our planetary magnetosphere, the interplanetary medium, the atmospheres and magnetospheres of our solar system planets and solar physics. As we began this chapter proclaiming the multidisciplinary morphology of geophysics in discussing its inner-core research, so we find geophysicists merging with solar physicists, planetary geophysicists and astrophysicists at the outer edges of geospace.

References

[1] For further details and references see the forthcoming book on the history of planetary physics by S G Brush, to be published by Cambridge University Press. Some of the material in this article is adapted from that book and used here by permission of the publisher. See also: Brush S G and Landsberg H E 1985 *The History of Geophysics and Meteorology, An Annotated Bibliography* (New York: Garland). S G Brush thanks Rachel Laudan for several useful suggestions. His contribution to this chapter was prepared while he was a Member of the School of Social Science at the Institute for Advanced Study, Princeton, with support from the Andrew W Mellon Foundation and the General Research Board of the University of Maryland, College Park.

[2] Günther S 1897 *Handbuch der Geophysik* (Stuttgart: Enke)

[3] Chamberlin T C 1916 *Origin of the Earth* (Chicago: University of Chicago Press)

[4] Jeans J H 1943 *Endeavour* **2** 3

[5] Russsel H N 1935 *The Solar System and its Origin* (New York: Macmillan)

[6] Oldham R D 1906 *J. Geol. Soc. Lond.* **62** 456

[7] The same reasoning had been used by Augustin Fresnel and Thomas Young, in developing the wave theory of light, to conclude that the lumeniferous aether must be an elastic solid. This was one of the anomalies of nineteenth century aether theory, partly resolved by Maxwell's electromagnetic theory of light and eventually abolished by Einstein's theory of relativity which dispensed with the aether entirely (Chapter 4).

[8] The first published suggestion that observed that seismic velocities imply a fluid or non-rigid core seems to be that of Leonid Leybenson: Leybenson L S 1911 *Trudy Otdel. Fiz. Nauk, Obsh. Lyud. Estestvoz.* **15** No 1

[9] Sullivan W 1991 *Continents in Motion* 2nd edn (New York: American Institute of Physics) p 71

[10] Jeffreys H *Geol. Mag.* **73** 558

[11] Elsasser W M 1955 *Am. J. Phys.* **23** 590

[12] Roberts P H *Geomagnetism* vol 2 (New York: Academic) p 251

[13] For further details see Dalrymple G B 1991 *The Age of the Earth* (Stanford, CA: Stanford University Press).

[14] The famous 1953 experiment by Urey's student, S L Miller, which initiated a new epoch in research on 'chemical evolution', was thus indirectly inspired by the revival of the monistic cosmogony with the help of the Payne–Russell discovery of the high cosmic abundance of hydrogen.

[15] Safronov V S 1987 *Proiskhoshdenie Zemli* (Moscow: Izdatel'stov Nauke) p 15

[16] Wood J A 1986 *Origin of the Moon* ed W K Hartmann *et al* (Houston: Lunar and Planetary Insitute) p 17

[17] For further details and references see:
Frankel H 1979 *Stud. Hist. Phil. Sci.* **10** 21
Glen W 1982 *The Road to Jaramillo* (Stanford, CA: Stanford University Press)
Hallam A 1973 *A Revolution in the Earth Sciences* (Oxford: Clarendon)
Marvin U B 1973 *Continental Drift* (Washington, DC: Smithsonian Institution Press)

[18] Snider-Pelligrini A 1858 *La Création et ses Mystères Devoilés* (Paris: Franck)

[19] Wegener A 1922 *The Origin of Continents and Oceans* (New York: Dutton) (English translation of 4th German edn)

[20] The British have a considerable advantage in this field of 'palaeomagnetism' because strata from every geological period are found relatively undisturbed somewhere in the British Isles.
[21] Allegre C 1988 *The Behavior of the Earth* (Cambridge, MA: Harvard University Press) p 48
[22] Cox A, Doell R and Dalrymple G B 1964 *Science* **144** 1537
[23] Reference [21] p 55
[24] It is instructive to compare the acceptance of magnetic reversals by geophysicists in the 1960s with their refusal at the same time to take seriously the speculations of Immanuel Velikovsky. Velikovsky, a Freudian psychoanalyst who was familiar with a number of ancient Egyptian and Hebrew texts, published a book *Worlds in Collision* (1950) in which he suggested that various events recorded in these texts were associated with astronomical catastrophes. For example, he claimed that the planet Venus had been ejected from Jupiter as a comet that twice came near the Earth, causing such events as the parting of the Red Sea, momentary stopping of the Earth's rotation (when the Sun 'stood still') and reversal of the Earth's magnetic field. Scientists ignored or ridiculed Velikovsky's ideas, and criticized his publisher, Macmillan, for advertising it as a 'science' book, thereby jeopardizing the reputation of legitimate science books. As a result of this disapproval Macmillan transferred the book to another publisher, and the scientists were then accused of censorship and suppression of unorthodox ideas. When a few of Velikovsky's suggestions such as geomagnetic reversals later seemed to have been accepted, his supporters regarded this as a 'confirmation' of his theory and a defeat for the 'scientific establishment', but current understanding of magnetic reversals, as shown in figure 26.13, makes it clear that Velikovsky's claim that such a reversal occurred in biblical times (or any time within the last 10 000 years) is completely wrong. No reversals have been found within the last 700 000 years (the 'Brunhes epoch') in spite of abundant material for examination. Even if there had been such reversals within recorded human history, Velikovsky's evidence would have carried much less weight than that of Brunhes and Matuyama, and even their findings were ignored until they could be fitted into our theoretical understanding of geomagnetism. Science does not progress by accepting at face value every alleged 'observation' or 'experimental fact'—if data do not conform to current theory, in most cases it is better to keep the theory until the evidence against it becomes overwhelming. In the twentieth century, evidence of a physical nature carries the most weight; geological and biological arguments are scarcely heeded, as Wegener found out; and ancient chronicles interpreted by Freudian psychoanalysts have little chance of getting any hearing at all.
[25] Lounsbury J F and Ogden L 1979 *Earth Science* 3rd edn (New York: Harper and Row) p 352
[26] Reference [21] p 83
[27] Hess's paper was circulated to other scientists but was not published until 1962; in the meantime Robert Dietz independently published similar ideas in 1961.
[28] Reference [21] p 100
[29] L W Morley's paper presenting similar ideas was rejected for publication by *Nature* and did not become known until later; see Frankel H 1982 *Hist. Stud. Phys. Sci.* **13** 1.
[30] The name came from 'Global Marine', the company which also operated

the *Glomar Explorer*, used by the US Central Intelligence Agency to recover a sunken Soviet submarine in 1974.
[31] 'Tectonics' is the study of structure; the word come from the same root as 'architecture'—the Greek word 'tekton' meaning carpenter or builder.
[32] 'Lithosphere' (like the more familiar word 'lithograph') comes from the Greek 'lithos' meaning stone.
[33] 'Asthenosphere' is derived from the Greek 'asthenes' meaning weak (same root as 'neurasthenia').
[34] Reference [9] p 75
[35] Reference [21] p 62
[36] Wilson J T 1965 *Science* **150** 482
[37] That project would carry the risks of 'playing with fire' on a global scale. The Earth's magnetic field protects us by deflecting cosmic rays which would otherwise reach the surface and threaten all living things. If disturbances of the solid mantle interact with the geomagnetic dynamo in the core, they could trigger reversals of the Earth's magnetic field, and during the resulting epoch of zero or very small field, these life-saving electric currents would collapse.
[38] One of the most useful texts covering the subject of upper-atmosphere geophysics and one to which I am greatly indebted is Hargreaves J K 1979 *The Upper Atmosphere and Solar–Terrestrial Relations* (Princeton, NJ: Van Nostrand–Reinhold).
[39] Gillmor C S 1976 *Nature* **262** 347
[40] Kuhn T S 1970 *The Structure of Scientific Revolutions* 2nd edn (Chicago) p 50
[41] Chapman S 1950 *J. Atmos. Terr. Phys.* **1** 121
[42] Kelley M C and Heelis R A 1989 *The Earth's Ionosphere: Plasma Physics and Electrodynamics* (San Diego, CA: Academic) p 5
[43] Gillmor C S 1981 *Space Science Comes of Age* ed P A Hanle and C Von Del (Washington, DC: Smithsonian Institution Press) pp 101–14
[44] Balfour S *Encyclopaedia Britannica* vol 16, 9th edn p 181
[45] Kennelly A E 1902 *Electrical World Eng.* **39** 473
[46] Heaviside O 1902 *Encyclopaedia Britannica* vol 33, 10th edn p 215
[47] The paragraphs below are drawn from Gillmor C S 1982 *Proc. Am. Phil. Soc.* **126** 395, and from reference [43].
[48] Gladden S C 1959 *National Bureau of Standards, Technical Note No 28* (US Department of Commerce, Washington, DC)
[49] Gillmor C S 1982 *Proc. Am. Phil. Soc.* **126** 395
[50] Darwin C G 1950 *Proc. Int. Congr. Mathematicians* vol 1 (Providence, RI: American Mathematical Society) p 593
[51] Budden K G 1961 *Radio Waves in the Ionosphere* then enlarged in 1985 as *The Propagation of Radio Waves* (Cambridge: Cambridge University Press).
[52] Waynick A H 1975 *Phil. Trans. R. Soc.* A **280** 11
[53] Lassen H 1926 *Jahrb. Drahtlosen Telegr. Teleph.* **28** 109 and 139
[54] Pedersen P O 1927 *The Propagation of Radio Waves Along the Surface of the Earth and in the Atmosphere* (Danmarks Naturvidenskabelige Samfund, A. Nr. 15a and 15b, Copenhagen), see especially Chapter V
[55] Shuleikin M V 1923 *Propagation of Electromagnetic Energy* (Moscow: First Russian Radio Bureau) (in Russian)
[56] Chapman S 1931 *Proc. Phys. Soc.* **43** 26
[57] Ratcliffe J A 1974 *J. Atmos. Terr. Phys.* **36** 2167
[58] Friedman H 1974 *J. Atmos. Terr. Phys.* **36** 2245
[59] Eckersley T L 1925 *Phil. Mag.* **49** 1250

[60] Storey L R O 1953 *Phil. Trans. R. Soc.* A **246** 113
[61] Helliwell R A 1965 *Whistlers and Related Ionospheric Phenomena* (Stanford, CA)
[62] Kelley M C and Heelis R A 1989 *The Earth's Ionosphere* (San Diego, CA: Academic) p 426
[63] Gordon W E 1958 *Proc. IEEE* **46** 1824
[64] Hess W N 1968 *The Radiation Belt and Magnetosphere* (Waltham, MA: Blaisdell)
[65] Kaur K, Srivastava M P, Nath N and Setty C S G K 1973 *J. Atmos. Terr. Phys.* **35** 1745
[66] Gillmor C S 1982 *Proc. Am. Phil. Soc.* **126** 395 and Gillmor C S 1991 *International Science and National Scientific Identity* ed R W Home and S G Kohlstedt (Deventer: Kluwer) p 181
[67] Ratcliffe J A 1974 *J. Atmos. Terr. Phys.* **36** 2167
[68] Bates D R 1974 *J. Atmos. Terr. Phys.* **36** 2287
[69] Massey H S W 1974 *J. Atmos. Terr. Phys.* **36** 2141
[70] Faraday M 1821–22 *Ann. Phil., New Series* **2** and **3**
[71] Fleming J A 1939 (ed) *Terrestrial Magnetism and Atmospheric Electricity* (New York: McGraw-Hill)
[72] Chapman S and Bartels J 1940 *Geomagnetism* (in 2 volumes) (Oxford: Clarendon)
[73] Mitra S K 1947 *The Upper Atmosphere* 1st edn (Calcutta: The Royal Asiatic Society), 2nd edn 1952
[74] Ratcliffe J A 1972 *An Introduction to the Ionosphere and Magnetosphere* (Cambridge: Cambridge University Press)
[75] Akasofu S-I and Chapman S 1972 *Solar–Terrestrial Physics* (Oxford: Clarendon)
[76] Van Allen J A 1983 *Origins of Magnetospheric Physics* (Washington, DC: Smithsonian Institution Press) p 9
[77] On ballooning, see DeVorkin D H 1989 *Race to the Stratosphere: Manned Scientific Ballooning in America* (New York: Springer)
[78] Hess W N 1968 *The Radiation Belt and Magnetosphere* (Waltham, MA: Blaisdell) p 1
[79] Milne E A 1926 *Mon. Not. R. Astron. Soc.* **86** 459–578
[80] Chapman S and Bartels J 1940 *Geomagnetism* (in 2 volumes) (Oxford: Clarendon) p 810
[81] Chapman S and Ferraro V C A 1930 *Nature* **126** 129
[82] Alfvén H 1939, 1940 *K. Svenska Vetensk. Akad. Handl.* **18** No 3 and No 9
[83] Birkeland K 1901 *Vidensk. Skrifter I. Mat. Naturv. Kl. 1901* 1; 1908 and 1913 *Norwegian Aurora Polaris Expedition 1902–3* vol 1, 1st and 2nd sections (Christiana: Aschehoug)
[84] Størmer C 1907 *Arch. Sci. Phys. Genève* **24** 5, 113, 221 and 317
[85] Gillmor C S 1978 *Upper Atmosphere Research in Antarctica (Antarctic Research Series 29)* ed L J Lanzerotti and C G Park (Washington, DC: American Geophysical Union) p 236
[86] As quoted by Van Allen J A 1983 *Eos* **64** 977.
[87] Gould L 1931 *Cold* (New York: Brewer, Warren and Putnam) p 266
[88] *Int. Geophysical Year General Report No 21* 1965 (Washington, DC: National Academy of Sciences)
[89] Van Allen J A 1983 *Eos* **64** 67
[90] Van Allen J A and L A Frank 1959 *Nature* **184** 219
[91] Here, as in the paragraphs above, I reflect the point of view of Van Allen 1983 *Eos* **64** 977 to whom I am indebted.
[92] Carpenter D L 1963 *J. Geophys. Res.* **68** 1675

[93] Gringauz K I et al 1961 *Artificial Earth Satellites* vol 6, ed L V Kurnosova (New York: Plenum) p 130
[94] Parker E N 1958 *Phys. Fluids* **1** 171
[95] Kelley M C and Heelis R A 1989 *The Earth's Ionosphere* (San Diego, CA: Academic)
[96] Axford W I and Hines C O 1961 *Can. J. Phys.* **39** 1433
[97] Dungey J W 1961 *Phys. Rev. Lett.* **6** 47

Chapter 27

REFLECTIONS ON TWENTIETH CENTURY PHYSICS: THREE ESSAYS

Historical overview of the twentieth century in physics
Philip Anderson

Introduction
To write a philosophical overview of this century of physics is a more than daunting task. It may be that with this century the history of science and technology will be seen to so overshadow and determine the conventional history of the world as to be inextricable from it. The ramifications of physics alone determined the outcome of the century's major war and dominated the politics in the half century since that war, through the physics-based revolution in communications as much as through the revolution in weaponry. With luck the politics of the next century will focus on science-dominated problems: population, energy and global ecology. Technologies based on new science—the Green Revolution, the Pill, increasing control of many diseases, the electronics industry, aerospace, and the many uses of the computer—have dominated world economics and sociology (a wonderful reference on this point is Pico Ayer, *Video Time in Kathmandu*). I also sense seeds of a coming revolution in modes of thinking which certain scientific discoveries—fractals, chaos, complex adaptive systems such as neural nets—are preparing for us. Leaving aside this wider context of physics I turn my gaze inwards, to a great extent, to look at how physics grew and changed, seeing how the world context affects physics and physicists but ignoring the very important feedback loop of how we affect the world.

Even so, I am left with a great variety of choices as to how to structure what I have to say, whether to focus on the great theoretical discoveries such as relativity, the structure of the atom and the nucleus, quantum mechanics, the Standard Model, broken symmetry, chaos, the Big Bang, or on the great technologies such as radar and vacuum tubes, the Bomb(s) and fission, macroscopic coherence and the laser, band theory and semiconductor electronics, diffraction (x-ray and neutrons), radio astronomy, SQUID, interferometry, NMR and MRI, atomic microscopy, computer simulation and computer-aided experimentation. Another focus could be socio-historical: how, when, by whom, and out of what context these developments evolved? I will come back and touch on each of these structurings later, but first I want to talk about two related themes which run through all of physics in this century.

The flight from 'common sense'
The first theme is a complete reversal in our view of nature, much more pervasive than we realize. Maxwell, the ideal nineteenth century physicist, discussed his equations for electromagnetic radiation, not quite with tongue in cheek, in terms of physical entities actually present in the 'ether' to carry his waves about. In his *'anno mirabile'*, 1905, Einstein discovered the physical unreality of the ether, yet expressed his theory of special relativity in terms of very solid and classical 'metre sticks' and 'clocks' and 'observers', and to the end of his life he believed, in some sense, in the reality and primacy of the directly perceived world, and in space and time as he, Albert Einstein, saw them. I would argue that the result of a 'Century of Progress' (the motto of the 1932 Chicago World's Fair) in physics is that no theoretical physicist can take himself seriously if he thinks that way in the 1990s.

Again and again, in many and in very completely structured ways, we are led to conclude that 'what we see' is *not* what is really there. Historically, this revolution began, I think, with Rutherford's discovery of the nuclear atom. Solid bodies were seen not to contain their mass uniformly, but to be mostly empty space. This was soon confirmed by the analysis of x-ray diffraction. But the great wrench in our perceptions of the world was caused by the quantum theory—a dislocation which is yet to be mentally healed even for many physicists. The first problem was the well-known and much-discussed question of measurement theory and the uncertainty principle. However, although less evident, equally important is the realization that the vacuum has properties: it is chock-full of fluctuations which can be felt by particles passing through it, as was proven experimentally in the late 1940s, in a quiet second revolution in our thinking about nature which still engages us. This, perhaps as much as the problem of measurement, left Einstein philosophically at sea: how could a vacuum with properties remain relativistically invariant? Condensed matter physicists, then known as 'solid staters', had taken the idea of a vacuum with properties and turned it around; using Landau's concept of 'elementary excitations', and Heisenberg and Peierls' idea of 'holes', solid matter had taken on the aspect of a vacuum through which excitations travelled like elementary particles, without interacting except with each other. We treated the quantum theory of crystal lattices as though the lattice were itself a vacuum. Soon, Nambu, and then Goldstone, Ward, Salam and Weinberg inverted this idea to invent the field-theoretical version of our concept of 'broken symmetry' that is, there could be a vacuum containing not only fluctuations but actual finite values of the averages of physical quantities, in this case of a quantum field. This left the field theory of the *physical* vacuum having even different symmetries from that of the *real*, underlying theory with its *real* vacuum. Yet another counterintuitive step was necessary before we were able to construct the successful 'standard model' of the elementary particles and forces. This was the introduction of the non-Abelian gauge interactions of quantum chromodynamics, again making the underlying physics of the strong nuclear interactions utterly different from that which is directly measured. Of course, great simplicity and generality derive from these manipulations since there is only *one* type of equation, in which interactions and symmetries are equivalent. The basic postulate of the Standard Model is that all interaction obeys the gauge principle. It is not clear whether we will find still greater dislocations of the conceptual structure of physics as we proceed toward higher energies in the twenty-first century. It is already a serious proposition that the real theory is of strings, not of particles, and has no clear statistics.

Emergence as the God Principle

What this brief jaunt into the philosophical structure tells us is that the structure of physical law can no longer be assumed to correspond in any way with our direct perception of the world. The philosophical phrase is 'emergence at every level'; the properties of space, time and matter which derive from common sense are not the 'true' properties of the underlying theoretical structures as we have come to understand them. This has left physics increasingly estranged from the common wisdom, an estrangement which may have disastrous consequences for both scientists and the lay public. The relatively simple and trivial wrench for the imagination which the beginning student of physics encounters, on being taught to substitute Newtonian intuition for Aristotelian common sense, is as nothing compared to this complete break between direct perception and the underlying theoretical concepts of the physicist. Modern physics has become very remote from the common man or woman.

The second theme is that the process of 'emergence' is in fact the key to the structure of twentieth century science, on all scales. This fact has been emphasized in a very insightful article by Sylvan Schweber in the November 1993 issue of *Physics Today*. Broken symmetry, an emergent property, is also the central concept of solid-state (now 'condensed matter') physics, including the symmetry-changing phase transitions sorted out in the 1970s by Kadanoff, Widom, Fisher and Wilson, as well as the broken symmetries of ferro- and antiferromagnetism and, crucially, the broken gauge symmetries of superfluidity (O Penrose, Onsager and Feynman) and superconductivity (BCS, PWA and Gor'kov), manifesting the quantum dilemma at the macroscopic scale. Emergence is also coming to be understood as the process by which our biological and social world has developed from its physical substrate. As sources of an evolving complexity in living matter, the nineteenth century's 'heat death' and its 'élan vital' are inappropriate descriptions of the apparently inevitable (at least on Earth) emergence of life, first primitive and then, through morphogenesis, with increasing levels of complexity leading on to consciousness, communication, and emerging social complexity. This, then, is the fundamental philosophical insight of twentieth century science: everything we observe emerges from a more primitive substrate, in the precise meaning of the term 'emergent', which is to say obedient to the laws of the more primitive level, but not conceptually consequent from that level. Molecular biology does not violate the laws of chemistry, yet it contains ideas which were not (and probably could not have been) deduced directly from those laws; nuclear physics is assumed to be not inconsistent with QCD, yet it has not even yet been reduced to QCD, etc.

This hierarchical structure is described well in the above-mentioned article by Schweber, which in turn quotes extensively from an earlier (1967–71) piece by the author. His conclusion, as mine, is that philosophically such a structure as the Standard Model, or the laws of chemical bonding, breaks the chain of reductionism and makes further delving into the underlying laws somewhat irrelevant to higher levels of organization. Schweber's article may be seen as a contribution to the great philosophical debate over the SSC and some other large science projects which have dominated the final decade of this century, in rebuttal to such books as *Dreams of a Final Theory* by Weinberg, or *The God Particle* by Lederman, which express a strong belief in the importance and relevance of reductionism. To me it seems to argue effectively that the 'God Principle' of emergence at every level is far more pervasive in our understanding of the universe than any possible 'God Particle' representing a hypothetical milestone in the reduction to ever simpler and more abstract laws of the dynamics of the interiors of subatomic particles.

This mention of the SSC debate is a good point at which to change subjects and have a look at the practical and sociological state of the world of physics, since that debate and its outcome have seemed to symbolize a state of affairs which has been journalistically expressed as 'the end of the age of physics'.

The first half-century and World War II: the triumph of physics
The twentieth century opened with the Western world in the throes of a massive and accelerating technological transformation. Science was

immensely popular, although very few understood that the fundamental discoveries in physics and chemistry of the nineteenth century were behind the rapid development of the internal combustion engine, of electric power and its use, of wireless and telephonic communication, and of flight in the first decades of this century. The glamour figures of science were the practical men, engineers like Edison, Steinmetz, the Wrights, Marconi or chemists like Langmuir. World War I left both of these professions highly regarded as practically useful even in that lethal business, and had little effect on physics except in killing off some of the brighter youth. (It is interesting to note how many of the significant figures of later twentieth century physics, e.g. Wigner, Teller, and Bardeen, started their training in this period as engineers, chemical engineers or chemists.) Chemists and engineers were also the first scientific professionals to be heavily supported by industry. Nonetheless physics shared in the general popular adulation, as Einstein's status as a folk hero attests. The popular appreciation of physics and other sciences was sufficient to attract increasing support for research from private philanthropy: the Nobel Prizes, with their great prestige; the Rockefeller foundation and later others; and private donations which Bohr, Rutherford, and later, and on a larger scale, Millikan and Lawrence were able to attract for research support. Until World War II there was no question of massive governmental support of physics as agricultural science was supported at the Land Grant universities in the USA, or geological science through the Geological Survey (in the USA; similar mechanisms operated elsewhere). Physics was just starting to need the kind of funding only governments could provide. Incidentally, in the decade just before that war another significant innovation took place: some industrial organizations in the USA at least (GE, AT&T, RCA) and some military establishments, seeing the open-ended possibilities of what we would now call 'High Technology', began to fund relatively undirected investigations in physics and other sciences. The seeds of the post-1940 explosion of radar and other technologies were planted primarily in these beginnings.

The events of World War II caused an enormous change throughout the Western world in how physics was perceived and how it was funded. We should keep in mind that a number of physics-related developments were incubated in that war: (i) radar, with a concomitant radical acceleration in all the related electronics and communications fields, such as microwave techniques, solid-state diodes, signal processing, (radar was the practical success of the war, contributing perhaps as much or more than the Allies' phenomenal code-breaking successes to Allied victories); (ii) missile technology, which was developed by the Germans almost exclusively but had little effect on the war's outcome; (iii) jet aircraft and other aerodynamic developments, both sides contributing, with the Germans' superior but desperate effort coming too late; (iv) primitive electronic calculators which had zero operational value; (v) nuclear fission and fusion, which, despite its very high profile, probably had only marginal effect on the war's outcome, as we now know, but has dominated military strategic maneuvering in the postwar half-century. In many countries nuclear fission has contributed significantly to electrical power generation.

Governments emerged from World War II convinced that investments in scientific research were vital to their military strength and were economically valuable, and the major participants on both sides developed, as they recovered economically, systems of national laboratories as well as schemes for supporting science in research universities. Industrial laboratories also enormously expanded and proliferated, although much of that expansion was also funded in the USA by government (mostly military) contracts. Physics was a major beneficiary of all this activity. Where, before the war, the major employment in the profession had been a relatively small number of academic jobs, there now came to be literally tens of thousands of individuals who considered their profession to be research in physics, within academia, at a national laboratory, in a private foundation, or in industry.

For some decades physics more than lived up to these expectations on the part of government. Within less than three years, the first transistor was operating, invented by a group at AT&T Bell Laboratories formed before the war, but expanded specifically to discover semiconductor devices. Equally quickly, the new science of radio astronomy, using radar technology, began to give us the first of several new windows for looking at the universe: a window through which we now see the radiation from the birth of the universe, for one. The fusion bomb made its dreaded appearance, and in combination with emergent ballistic missile and jet technology has held the world in a state of deadly apprehension for half a century. These were only the few of the first fruits of a cornucopia of technological and scientific breakthroughs which transformed our lives and, perhaps as significantly, transformed our understanding of our world, out of all recognition. New semiconductor devices transformed the electronic computer, which, plus an unlikely mix including the sophistication of quantum electronics, leading to the laser, empirical inventions such as xerography, high space technology, and materials technology such as glass fibres, has given us a completely new and rapidly evolving worldwide information network. Physics has transformed other sciences: molecular biology began in departments of physics, physical techniques sparked the plate tectonics revolution. Physics has transformed medical diagnosis with the CAT scan and MRI, and will do more in this area.

The age of Big Science

I do not need to further expand the list of the achievements of the second half of our century. What I do want to do is to discuss some generalities about their history and the sociology which spawned them. My first point is that this period of rapid expansion was also a period of centrifugal fragmentation of physics. A somewhat self-appointed elite had already broken off in the 1930s from the study of atoms and molecules to focus on the nucleus; this was the group who formed the Manhattan Project and its offshoots during the war. After the war they had the choice of two paths. Many remained primarily concerned with weapon design and military technology, and with advising the government, primarily in military matters. Possibly the most visible and controversial example has been Edward Teller. Many others returned to their universities or to national laboratories and followed the natural evolution of their subject from nuclear physics to mesons and other subnuclear particles, or into modern astrophysics, in either case learning to spend money in gigantic amounts and to build giant accelerators. A large coterie followed both paths in different mixes, and formed a most influential group of scientist advisors, including, for instance, a considerable number of members of such groups as JASON and the Institute for Defense Analysis.

It was less than a decade after the war that a definite culture which could be called 'Big Science' grew up, largely out of this nucleus which had been involved in the Manhattan Project. There were other contributors: for instance, the hope of fusion energy by magnetic or inertial confinement came to claim large subventions from the US, British and Soviet Governments. And, at least at first, the large research fission reactors at Brookhaven, Oak Ridge, Harwell, Chalk River, and eventually Grenoble were seen as 'big science' because they cost on a comparable scale, but they were soon taken over as 'user' facilities by large coteries of 'small' scientists. The military fission program, now known to have been callously misrun, fell out of the hands of nuclear physicists, although it still taints their reputation. Finally, so-called 'space science' developed rather independently as a creation of the 'military–industrial complex' on the coat-tails of which many physicists came eventually to ride, with microgravity experiments of uniformly dubious value as well as some quite meaningful astrophysical probes and investigations into solar system physics.

That Big Science culture in the USA, and similar groups elsewhere, tended to have separate, direct access to government and hence to funding sources. It was independent to a great extent of the rest of science, of which it was never a majority component except in funding. In the USA NASA, the Department of Energy (earlier the AEC), and military support operated outside of the standard peer review mechanisms. (The DOE funded other science, but from a separate budget.) The sums involved precluded private support, either by industry, universities, or private foundations; the former came to view Big Science not as an investment but as a 'cash cow', a source of very helpful overhead charges and of enlargement of the bureaucracy, if not of actual profits.

Eventually one could identify an entity (which perhaps was never more than a certain state of mind and set of common interests) which could be thought of as the Big Science subdivision of the military–industrial complex. Whatever one may call it, towards the end of the century a number of controversies or problems arose which indicated that Big Science no longer had unlimited approval from the public nor unlimited access to the public purse. In fact, it is this series of events which have been referred to as 'the end of the age of physics'.

The most conspicuous event has been the collapse of funding for the SSC, but there are a number of related, still unresolved situations:

(i) the US Space Station. This continues, with diminished funding and downsized mission. General scientific opinion is overwhelmingly negative and seems likely to carry the day in the end.

(ii) Fusion. Inertial fusion, the darling of Livermore Laboratories, is being cut back. Magnetic confinement is at last talking about realistic time scales (2040 is the last I heard) and serious engineering questions; clearly it too no longer has a free ride.

(iii) The Laser Interferometric Gravitational 'Observatory' valiantly maintained by the support of one US Senator, seems likely to be another failure hastening the demise of Big Science, though it is not so big. Other Big Science projects (ANS; B-factory) are also in trouble in the USA.

Apparently unrelated, but not necessarily so in the public mind, are two essentially political events: SDI (the Star Wars enterprise) and the end of the Cold War. Physicists of most stripes in the USA opposed—to their great credit—the self-serving, unrealistic, sales pitch for SDI. Nonetheless the public saw it as a promise of safety through high technology, and it was certainly a coterie of the Big Science complex who supported it. The same political mechanism—direct intervention at the White House level—was used by its advocates as was used to support the SSC.

The end of the Cold War is, not incorrectly, seen as an enormous setback for the military–industrial complex. Whatever the personal feelings of the individual scientists involved, Big Science was maintained in its unique position by its ties to that complex, and it will in the end lose power as that complex loses power.

In Europe Big Science still maintains most of its influence. CERN shows no weakening of its stability, and other European large projects continue to thrive. Nonetheless it is possible to detect a public disenchantment with large investments and an increasing concern with economic and political problems which must eventually lead in the same direction as in the USA. The 'age of Big Science' seems likely to end with the twentieth century.

The flowering of 'Small Science'

Big Science is, however, only a fraction of all science, even—or especially—of all fundamental or 'basic' physics. A second type of physicists doing a second kind of physics was given a jump start from wartime advances and postwar funding. The specific nuclei of the resulting growth were distinct from the Manhattan Project core of nuclear, particle, and fusion physics.

(i) The availability of new ranges of coherent sources of electromagnetic radiation, of new types of circuitry, and of ultrasensitive detection, began a continuing trend to cover the entire range of spectroscopy by coherent methods: at first EPR, NMR, and microwave gas spectroscopy. The range took an enormous leap with the invention of the laser in the early 1960s.

(ii) Neutron diffraction and scattering, using the fission reactors left over from the Manhattan Project, and those specially built for the power reactor program, had a large influence on solid-state physics research and formed nuclei of specialists at each of the national centres. These soon became quite intellectually isolated from the Big Science projects, often at the same centres. This isolation was signalled by one of the two or three most conspicuous omissions in the list of Nobel Prizes: neutron diffraction and scattering, allowing the experimental proof of antiferromagnetism and the study of phonon spectra. (In 1994 this omission was finally corrected.)

(iii) Wartime research on semiconductors and other materials, primarily carried out at industrial centres, together with the new sophistication in electronics, led to a great increase in the sophistication and depth of research on semiconductors, which of course spawned the invention of the transistor. This was only the first of a stream of developments of useful solid-state materials and devices, for example, the discovery and development of insulating magnetic materials.

(iv) A very important stimulant to fundamental physics was the development of a liquid helium cryostat affordable by almost any laboratory. This opened up the temperature regime in which quantum effects on almost all properties of condensed matter were of importance. Superconductivity, superfluidity, and fermi surface effects in metals were now open to investigation at many laboratories throughout the world.

(v) As the world's technical sophistication advanced, advances in relatively old kinds of measurement capacities developed: the electron microscope and x-ray diffraction using new, powerful sources being examples which were very important in molecular biophysics and in metallurgy on the atomic scale, as first applications.

The wartime roots of these developments were, on the whole, separate organizationally and geographically from those of Big Science. The twin institutions in the USA whose contributions in wartime had the biggest impact in this area were Bell Laboratories and MIT's Radiation Laboratory with contacts to Harvard's Radio Research Laboratory. It is significant that the Collins liquefier and the Bitter magnet were MIT developments, that the transistor arose at Bell, and that NMR as a useful measurement tool grew out of Harvard. (I emphasize that I am talking not about organizational but about intellectual history. The roles played as government advisers by administrators like Conant of Harvard or Buckley, Baker and Buchsbaum of Bell Laboratories in the arms race and in the government support of both kinds of science are irrelevant to the point.)

From Los Alamos Big Science diffused first to American universities such as Berkeley, Chicago, Princeton, Columbia, Stanford and Cal Tech; similarly small physics tended to spread from Bell Laboratories, Harvard, and MIT to a distinct set of universities, such as Cornell and Illinois, and became strong overseas in the British university system and at Harwell, and eventually Leyden (and Philips Eindhoven and Paris on the continent) as well as Japan, the latter for obvious reasons. Many physics departments accommodated both: Chicago and Harvard, and, at first, Columbia. But it was clear in some physics departments that one group or the other dominated hiring and promotion, often for decades, as it did (to emphasize the Big Science examples) at Cal Tech, UCLA, Columbia (with a separate organizational set-up for some small scientists), Princeton, and Stanford, and on the continent of Europe generally, especially in Italy and Germany, and, when it joined the scientific world, China.

In the USA through the 1950s the American Physical Society tended to have relatively large general meetings with physics of all kinds represented, but by the mid-1960s the March meeting, originally set up to give Small Science a forum, had grown into the equal of any APS meeting and by the 1980s it dwarfed all of the others. Almost in self-defense, the Big Scientists took over the April meeting in Washington, and then began to have separate 'nuclear' and 'particle' divisional meetings. Meanwhile, on the international scene, the International Low Temperature Physics Congress had become a giant convocation of many kinds of Small Science, and other specialities (magnetism, semiconductors) developed their own giant congresses; an entirely separate, very internationalized system of congresses grew up, starting from the Geneva congress where the Cold War opponents first made open contact in what seemed, then, the sensitive fields of nuclear and particle physics.

The growth in these two areas took place in very contrasting ways. The Big Science experimental project has become larger and more collaborative, finally involving, in some cases, literally thousands of scientists on a single experiment. The great majority of these people have no truly independent role to play. On the other hand, theoretical work in this field became more and more speculative, esoteric, and abstract, with only a small fraction of theorists actually involved in detailed interaction with experiment. Some of the best known of present-day theorists, Witten, Penrose, Hawking, Schwarz for a few, have eschewed the prediction of and explanation for experimental fact as a primary goal of theoretical physics.

The evident relevance of much Big Science to the deep, quasireligious human impulse to know the ultimate origin of life and of the universe and the ultimate stuff of which we are made, aside from affording at least some access to public money, also assures it an indefinite supply of eager young recruits, especially to the theoretical effort. These recruits, so numerous as to overwhelm all available research posts even in 'good' times, in turn dominate physics departments in smaller colleges and in the Third World, which helps to perpetuate the supply.

The saturation of Small Science: the end of entitlement
The gigantic growth in Small Science (particularly physics), on the other hand, took place to a large extent because of a perception both in industry and in government that the products created were economically useful. The result was funding on a scale which absorbed essentially all new recruits. Through the first three postwar decades this was unquestionably the case; we hardly need to repeat the litany of practically useful materials and devices which came out during these years. But in this half of physics, too, there were ominous indications of trouble ahead.

It is a great boon that this kind of science can produce both practical, useful devices and methods as well as intellectually exciting scientific knowledge. But this fact can also lead to confusion. The scientists' natural preference is for the more intellectual work which is reinforced by the fact that the pure researcher is more visible, more honoured, more mobile (he publishes in the open literature) and often better paid than the worker on applications, and certainly better paid than the production engineer who supervizes actual manufacture. In academia and national laboratories, and even in some of the great industries, it has seemed as though most departments have been trying to compete with the pure research divisions within the organizations, and with university departments of physics and engineering science, as well, even when the nominal function was to do device development, manufacturing design, or even marketing. The fresh PhD, coming from the relatively 'pure' atmosphere of university research, has found it natural, easy, and often at least financially secure to continue along the lines of his thesis and in the 'research' mode. The sociology, again in the USA, encouraged him to feel this as an entitlement (encouraged, as well, by some unfortunate propaganda from scientific administrators). It is interesting and important to realize that this sociology was also rampant in the Soviet Union and the Eastern Bloc as well as in certain parts of Western Europe, with variations. In the Soviet Union and East Germany, large institutes supported by governments grew up, combining (like our DOE laboratories) military and big space responsibilities with Small Science. Employment in these institutes held great personal advantages relative to the general economy. Reunification, in Germany at least, has found these institutes overpopulated relative to any reasonable mission by factors (conservatively) of 3–10. A similar situation seems to hold in the FSU. In Germany, at least, the cause of the collapse is *not* by any means entirely the abandonment of military and space missions, or of the Cold War mentality, but primarily a serious neglect of the economic, productive aspect of technology; Europe, the USA and the Eastern Bloc differed in degree in this, rather than qualitatively.

Throughout the period the same rapid growth and apparent boundless opportunities meant that Small Science, particularly, seemed an ideal entry point to the advanced societies for immigrants from India, from the Soviet system, and from the Orient. This group of people had an additional motivation to stay in research: visa regulations which allowed for postdoctoral stays and were unfriendly to changes in jobs, and even more so to changes in field or employer. (And often language, or other cultural factors, could make such changes difficult in any case.)

The result was an increasing overcrowding of research as a profession. One may question—and leaders of various kinds did—how much research is too much, or even whether there *could* be too much research, research being viewed as an absolute good. But conditions within the profession, viewed objectively, made it quite clear that aside from the inevitable funding crunch there were dysfunctions in the system.

There was a very sharp change in the nature of a research career. The 'promising' young scientist's publication rate grew by factors of five to ten; the number of applications a young researcher might make for postdoctoral work or an entry-level position from two or three to 50. Senior scientists were overwhelmed with receiving and sending reams of letters of recommendation, which thereupon became meaningless. The numbers of meetings in a given specialized subject, and the number of subjects with a formal meeting list, both grew by factors of ten or more. In many subjects one could 'meet' nearly 52 weeks in the year, somewhere in the world, and leaders in the field were invited to all. Meetings almost inevitably led to publications. Most publications became tactical in this game of jockeying one's way to the top; publications in certain prestige journals were seen as essential entry tickets or score counters rather than as serious means of communication. Great numbers of these publications were about simulations of dubious realism or relevance. Essentially, in the early part of the postwar period the career was science-driven, motivated mostly by absorption with the great enterprise of discovery, and by genuine curiosity as to how nature operates. By the last decade of the century far too many, especially of the young people, were seeing science as a competitive interpersonal game, in which the winner was not the one who was objectively right as to the nature of scientific reality but the one who was successful at getting grants, publishing in PRL, and being noticed in the news pages of *Nature, Science,* or *Physics Today.*

In many subjects the great volume of publications, the fragmentation into self-referential so-called schools who met separately, and a general deterioration in quality which came primarily from excessive specialization and from careerist sociology, meant that quite literally more was worse. This was very obvious in the Soviet Union, in many fields. In the field of high T_c superconductivity with which I am familiar, enhanced national and regional funding has simply proliferated non-communicating schools and subcultures rather than focusing meaningful effort: it seems certain to delay solution of the scientific problem. One wonders whether AIDS could not suffer the same fate. Fortunately in both cases practical, as opposed to scientific, progress is more easily judged and managed.

Observation of the actual working of modern science in this period has led some sociologists to attempt to apply deconstructionist ideas to it: to maintain that the 'truth' is purely sociological and determined by power relationships, not by nature. Even the president of Radcliffe University seemed to be lending her support to this nonsensical point of view. These sociologists are seeing a temporary and, I believe, aberrant *sociological*, rather than scientific, phenomenon; in fact one does not know of any case in which scientific error caused by political, economic or sociological pressure has not eventually been corrected. If this were not so, how could Darwin have prevailed? Or Copernicus? The truth has great power to prevail; it is fixed and solid, while error is variable; reproducibility wins over visions and poltergeists.

To complete the story of the last decades of the twentieth century, eventually the inevitable collapse occurred. It was not through failure of the research engine in the West, but that this engine produced device after device of which Japan—which has not had a great record of original innovation—cornered the great part of the market share and then often lost it to even younger societies. The liquid crystal computer display may have been the straw that broke the camel's back; the entire technology is American, the American market share is approximately zero. Quite sensibly, the American corporations decided, one by one, that it was essential to shift the motivation system away from the entitled 'pure' research career. This coincides with a great wave of immigration from the FSU and Eastern Europe, and with a (however glacial) downsizing of the national laboratories. The US Congress is making threatening noises at the National Science Foundation which supports much university research, but these have not yet resulted in seriously damaging that research. Many students and a few of the sharper postdoctorates are moving out of the research pipeline. Perhaps by the end of the century we will see the 1980s and 1990s as the bad old days and will be back with a reasonably efficient physics establishment. What seems to be the worst danger is allowing the pendulum to swing too far the other way.

Concluding comments
Let me finish this essay with some comments about recent developments and trends in physics, most of which are more optimistic than otherwise.

One healthy trend is growth in the level of complexity of systems which are being studied with the quantitative precision, instrumental ingenuity, and mathematical sophistication which characterize the best of physics. A wonderful indication of this trend was the Nobel Prize award to de Gennes, who has spent decades on 'soft physics', the classical physics of polymers, colloids, liquid crystals, glues, and so on. The field is growing and many promising results are appearing. Biophysics is a mine of unsolved, or only qualitatively, solved problems like biocatalysis, cell microstructure, neural function, and a relatively small coterie are moving in these directions, using very sophisticated technology in many cases.

In the past physics has tended to disengage itself from such fields as molecular biology once they are well started. This has often been a loss both to physics and to the new field, since the attitudes of intellectual rigor and quantitative precision which physics brings can remain relevant even when the matter under study is biological. I hope that physics will more and more retain an interest in these new fields of biophysics.

An explosive advance in non-linear science is leading into new problems and results in driven, complicated non-linear systems. A danger in this area is the temptation to substitute simulation for its own sake for the appropriate subject matter of physics, which is matter and energy in the natural world. Where a simulation or a computation is relevant to the real world, it is physics, but otherwise one tends to lose the guiding criteria of reproducibility and experimental verification which are at the intellectual centre of physics. These fields, both biology and non-linear dynamics, are as yet not promising ones for the 'careerist' sociology, though one finds a certain level of hype surrounding some developments.

The beautiful, flexible kit of mathematical tools which the theoretical physicists have developed, first to build the Standard Model and to solve similar quantum many-body problems in condensed matter, and then in attempts to go farther, should not be allowed to die in the general downsizing of both big and small physics which must occur. (I do not mean here *mathematical* physics, which has tended to become separate from experimental physics and sterile.) The substrate considered may be totally different, but I must believe that it is both beautiful and important that the same mathematical constructs illuminate liquid crystals and quantum chromodynamics; that phase transitions in the early universe are echoed, the one in superconductivity, another in droplet metallization in semiconductors. In theory, more is probably worse and less better, but zero is totally unacceptable. There remain many exciting problems in fundamental physics which will not die with the SSC and its compeers. Big astrophysical science seems set for decades, for instance; and fundamental condensed matter physics is full of problems, if not of solutions.

By the rather mordant picture which I have given of the sociology of physics I do not mean to imply that physics is not intellectually very lively, even in some of the areas one might have expected to have been thoroughly exploited. For instance, it has become increasingly clear during the last decade that the quantum many-body problem encountered in the bizarre new superconductors, in a broad variety of intermetallic compounds with rare-earth atoms, and in layer, surface, and chain structures remains full of surprises, and essentially unsolved. In the opposite case is astrophysics, where each new type of probe leaves us with a new and exciting puzzle to solve, and even the overall structure of the universe remains controversial, much less whether or not there are recognizable traces of phase transitions in the early universe. What is depressing is how few of those who are supposed to be investigating these mysteries are doing so meaningfully, because of the unsatisfactory career-dominated sociology which has arisen.

A third important point is that much of the present malaise in Small Science comes about because in great part it had historically been driven by Cold War and industrial needs for electronics technology. The electronic and communication industry were its chief customers. These industries have correctly and healthily realized that this pipeline of 'hardware' technology is full, and that software and economics will be the limiting factors in future. The 'hardware' problems of the future can be expected to lie in at least two new directions: biomedical technology; and energy and the environment. Great contributions in the past decade have been made by physics to medical diagnostics, and, in a quiet way, to biomaterials. The impact of advanced physics on energy and the environment is only beginning. Satellite studies of all kinds, and fission reactors, are so far all that is of major importance. But solar photovoltaics, batteries, energy storage, transmission, (suitable superconducting cables are not very far away), and of course fusion, are all in the future, as are minerals exploration and trace analysis. These are not fields which have a Bell Laboratories or an IBM to lead the way; new mechanisms to support—and manage—research in these fields will probably evolve.

We conclude, then, with the thought that there is no dearth of new directions for physics and physicists to take. However, there are real problems in the field. One is the increasing estrangement of even the best-educated public from physics which is inherent in the phenomenon of 'emergence'. When we are producing economic goodies for the public, we are tolerated; but we cannot expect them to support our pure curiosity with extraordinary generosity even if we make every effort to explain ourselves, and popularizing and explaining physics will be an increasingly difficult task.

A second problem we have is to survive the present era of overpopulation of research without a complete breakdown in our system of ensuring quality and integrity in the research process, and without having some inappropriate system of standards imposed from outside. Also, scientists must take more responsibility for helping to decide what projects should be funded (even outside their own specialities) and which fields are intellectually sound, as well as relevant to the rest of science. If we do not make more of an effort to regulate ourselves and to curb our tendency to assume an entitlement to go on doing things as we always have, the public tolerance may properly become overstrained.

Nature itself
Steven Weinberg

The state of science at the end of the twentieth century is very different from its condition at the century's beginning. It is not just that we know more now—we have come in this century to understand the very pattern of scientific knowledge. In 1900 many scientists supposed that physics, chemistry and biology each operated under its own autonomous laws. The empire of science was believed to consist of many separate commonwealths, at peace with each other, but separately ruled. A few scientists held fast to Newton's dream of a grand synthesis of all the sciences, but without any clear idea of the terms on which this synthesis would be reached. Today we know that chemical phenomena are what they are because of the physical properties of electrons, electromagnetism, and a hundred or so types of atomic nuclei. Biology of course involves historical accidents in a way that physics and chemistry do not, but the mechanism of heredity which drives biological evolution is now understood in molecular terms, and vitalism, the belief in autonomous biological laws, is safely dead. This has truly been the century of the triumph of reductionism.

The same reductionist tendency is visible *within* physics. This is not a matter of how we carry on the practice of physics, but how we view nature itself. There are many fascinating problems that await solution, some like turbulence left over from the past, and others recently encountered, like high-temperature superconductivity. These problems have to be addressed in their own terms, not by reduction to elementary particle physics. But when these problems are solved, the solution will take the form of a deduction of the phenomenon from known physical principles, such as the equations of hydrodynamics or of electrodynamics, and when we ask why *these* equations are what they are, we trace the answers through many intermediate steps to the same source: the Standard Model of elementary particles. Along with the theory of gravitation and cosmology, the theory of elementary particles thus now constitutes the whole outer frontier of scientific knowledge.

The Standard Model is a quantum field theory. The fundamental ingredients of nature that appear in the underlying equations are fields: the familiar electromagnetic field, and some twenty or so other fields. The so-called elementary particles, like photons and quarks and electrons, are 'quanta' of the fields—bundles of the fields' energy and momentum. The properties of these fields and their interactions are largely dictated by principles of symmetry†, including Einstein's Special Principle of Relativity, together with a principle of 'renormalizability'‡, which dictates that the fields can only interact with each other in certain specially simple ways. The Standard Model has passed every test that can be imposed with existing experimental facilities.

But the Standard Model is clearly not the end of the story. We do not know why it obeys certain symmetries and not others, or why it contains six types of quarks, and not more or less. Beyond this, appearing in the Standard Model there are about 18 numerical parameters (like ratios of quark masses) that must be adjusted by hand to make the predictions of the model agree with experiment. Finally, gravitation cannot be brought into the quantum field theoretic framework of the Standard Model, because gravitational interactions do not satisfy the principle of renormalizability that governs the other interactions.

So now we must make the next step toward a truly unified view of nature. Unfortunately, this next step has turned out to be extraordinarily difficult. We are in a position somewhat like that of Democritus—we can see the outlines of a unifying theory, but it is a theory whose structures become evident only when we examine nature at scales of distance vastly smaller than those accessible in our laboratories. Democritus speculated about atoms, which after over two millennia were discovered to be some ten orders of magnitude smaller (i.e. smaller by a factor of one 10 billionth, or 10^{-10}) than Democritus himself. Today we speculate about a theory that would unify all the forces of nature, and we too can see that the structures of this theory must be very much smaller than the scales of distances that we can study experimentally.

We have two clues to the fundamental scale. One is that the strength of the strong nuclear forces (which hold quarks together inside the particles inside the atomic nucleus) decreases very slowly with decreasing distance, while the strengths of the electromagnetic and weak

---

† *A symmetry principle is a statement that the laws of nature look the same when we change our point of view in some way. The Special Principle of Relativity says that the laws of nature appear the same to observers moving with any constant velocity. There are other spacetime symmetries, that tell us that the laws of nature look the same when we rotate or translate our laboratory, or re-set our clocks. The Standard Model is based on a set of symmetry principles, including these symmetries of spacetime, and also other symmetries that require that the laws of nature take the same form when we make certain changes in the ways that the fields of the theory are labelled.*

‡ *The concept of renormalizability arose from the effort in the late 1940s to make sense of the infinite energies and reaction rates that appeared when calculations in quantum electrodynamics were pushed beyond the first approximation. A theory is said to be renormalizable if these infinities can all be cancelled by suitable redefinitions (or 'renormalizations') of a finite number of parameters of the theory, such as masses and electric charges. Renormalizability is no longer viewed as a fundamental physical requirement needed to eliminate infinities, but as we shall see it survives in the Standard Model for other reasons.*

nuclear forces increase even more slowly, all forces becoming equal at a scale of distances which is currently estimated to be about 10^{-16} of the size of an electron†. The other clue is that gravitation becomes about as strong as the other forces at distances about 10^{-18} times the size of an electron. These two estimates are close enough to encourage us that there really is a final unified theory whose structures become visible at a scale of distances about 10^{-16} to 10^{-18} of the size of an electron. But in a purely numerical sense, we are farther away from being able to observe these structures than Democritus was from being able to observe atoms.

So what are we to do? I am convinced that there is no reason for us to begin to talk about an end to fundamental physics. Nor do I think (though I am not sure) that any great change is needed in the style of physics. There seems to me to be ample hope for progress in the same reductionist mode that has served us so well in this century. To be specific, there are two approaches that are not yet exhausted: a high road, and a low road, 'high' and 'low' here referring to the energy of the processes studied.

One approach, the low road, is to try through the combined efforts of theorists and experimentalists to complete our understanding of physics at energies accessible in our laboratories. The higher the energy, the smaller the scale of the structures that can be studied; with energies no larger than about a trillion electron volts, we cannot directly study structures smaller than about one 10 thousandth the size of the electron. As already mentioned, physics on this scale is known to be well described by the Standard Model. This theory is based in part on a principle of symmetry which prohibits the appearance of any masses for the known elementary particles in the equations of the theory. Something beyond the known particles must break this symmetry. In the original version of the theory, this something was a field pervading the universe, that breaks the symmetry in much the same way that the gravitational field of the Earth breaks the symmetry between up and down. The quanta of this field would show up in our experiments as a particle, known as the Higgs particle. There are other alternatives, but all theories agree that something beyond the known particles of the Standard Model must show up when we study collisions at energies high enough to create masses of the order of a trillion electron volts, a thousand times the mass of the proton.

It is vitally important to learn the details of the symmetry-breaking mechanism, because it is this mechanism that sets the scale of the masses

---

† *The electron does not really have a well-defined size; it appears in our theories as a point particle. I am referring here to what is called the classical radius of the electron, that is, the radius of a sphere with the charge of the electron whose electrostatic energy would equal the energy in the electron's mass. This radius can be taken as representative of the characteristic distances encountered in elementary particle physics; for instance, it is not very different from the size of typical atomic nuclei.*

of the quarks and electron and other known elementary particles; the only mass appearing in the equations of the Standard Model in its simplest version is the Higgs particle mass. From units of mass we can infer units of length, so this is also what sets the scale of lengths characteristic of the elementary particles. Thus we must understand this symmetry-breaking mechanism if we are to solve the 'hierarchy problem', the problem of making sense of the enormous disparity between the lengths encountered in our laboratory, like the size of the electron, and the scale of distances where all the forces become unified.

It was this question that we hoped to settle at the Superconducting Super Collider. Now that the Congress of the United States has decided to cancel the Super Collider, our hopes are riding on plans to build a similar accelerator, the Large Hadron Collider, in Europe. In the mean while, we have a good chance of getting advance information about where the Higgs particle or whatever takes its place might be found. The point is that quantum fluctuations involving the continuous creation and annihilation of this particle (or particles) have a slight effect on quantities that can be measured with present facilities, like the mass of the W particle. We can expect the mass of the W particle to become much more accurately known through experiments at CERN, while experiments at Fermilab are now in the process of providing an accurate value for the mass of the 'top' quark, which also affects the mass of the W particle. When this is done, we will be able to estimate the mass of the Higgs particle (if there is just one), or rule out the possibility of a single Higgs particle at any mass. But we will still need the Large Hadron Collider to pin down the symmetry-breaking mechanism that gives mass to the elementary particles.

There are many other things that might turn up in our experiments while we are waiting for the next generation of large accelerators. These include an electric dipole moment of the neutron or electron, an effect that is expected in some theories of symmetry breaking, or any of the new particles predicted by various theories (supersymmetry, technicolour) which have been suggested in attempts to solve the hierarchy problem. There are also tiny effects like proton decay or neutrino masses that might be directly produced by the physical processes that occur at the very short distance scales of the final unification. These exotic effects would be manifested as additions to the Standard Model that violate the principle of renormalizability mentioned earlier. Though not forbidden, these effects would be proportional to powers of the very small distances from which they arise, and hence very small, which is presumably why they have not yet been detected. But they may yet. There will soon be a new underground facility in Japan known as Super-Kamiokande where proton decay might be discovered, and there are indirect hints from studies of neutrinos from the Sun that neutrinos actually do have small masses. Any or all of these things might turn up in future experiments. But we have been saying this for almost twenty years, since the completion of the Standard Model, and so far none of them have.

Then there is the high road. A fair fraction of particle theorists are now trying to jump over all intermediate steps and proceed directly to the formulation of a final unified theory, without waiting for new data. For a while in the 1970s after the success of the Standard Model it was thought that this final theory might take the form of a quantum field theory, like the Standard Model but simpler and more unified. This hope has been largely abandoned. For one thing, we now understand that *any* physically satisfactory theory will look like a quantum field theory at sufficiently low energy, so the great success of quantum field theory in the Standard Model tells us little about the deeper theory from which it derives. The Standard Model is seen as an 'effective' quantum field theory—the low-energy limit of a quite different fundamental theory. Also, the continued failure to bring gravitation into the ambit of quantum field theory suggests that this fundamental theory is quite different.

The best hope for success along this line is in some sort of string theory. Strings are hypothetical one-dimensional elementary entities, either closed like rubber bands or open like pieces of ordinary string, which like a violin string can vibrate at a great many different frequencies. They are about 10^{-18} times the size of an electron, so when viewed in our laboratories they seem like point particles of various different types, the type depending on the mode in which the string happens to be vibrating. One of the exciting things about string theories is that in some versions they predict a menu of particle types which is impressively similar to what we actually observe in nature. Further, one of the particle types that is predicted by all versions of string theory is the graviton, the quantum of the gravitation field. Thus string theories not only unite gravitation with the rest of elementary particle physics, they explain why gravitation must exist. Finally, the old problem of infinities that stood in the way of a quantum field theory of gravitation is avoided in string theories. The infinities found in our field-theoretic calculations arose from processes in which particles occupy the same point in space. But strings have a finite extension, and therefore two strings can never be brought to zero distance from each other. String theory thus provides our first real candidate for a final unifying theory.

String theory unfortunately has not yet lived up to the great expectations that attended it in the 1980s. There seem to be a great variety of different string theories, and although it is widely suspected that these are really just different solutions of the same universal theory, no one knows what that theory might be. There are formidable mathematical obstacles that stand in the way of solving any of the string theories to find just what it predicts for measurable physical quantities like quark masses, and we cannot tell from first principles which version of string theory is correct; and even if all these obstacles are overcome, we will still be left with the question, why should the real world be described by anything like string theory.

One possible answer to this question is suggested by reference to the historical origins of string theory. In the 1960s, before the advent of the now standard quantum field theory of the strong nuclear forces, many theorists had given up on the idea of describing these forces in terms of any quantum field theory. Instead they sought to calculate the properties of nuclear particles and mesons through a positivistic program, known as 'S-matrix theory' that avoids referring to unobservable quantities like the field of the electron. In this program one proceeds by imposing physically reasonable conditions on observable quantities, specifically on the probabilities of all possible reactions among any numbers of particles. (Among these conditions was the condition that the probabilities of all reactions would always add up to 100%, another requirement that these probabilities depend in a smooth way on the energies and directions of motion of the particles participating in any reaction, yet another requirement regarding the behaviour of these probabilities at very high energy, and finally various symmetry conditions, including the symmetries of spacetime embodied in the Special Principle of Relativity.) It turned out to be extraordinarily difficult to find *any* set of probabilities that satisfied all of these conditions. Finally, by inspired guesswork, a formula for reaction probabilities was found in 1968–9 that seemed to satisfy all these conditions. Shortly after, it was realized that the theory that had been discovered was in fact a theory of strings. It is possible that this history reflects the logical basis of string theory. That is, string theory may ultimately be understood as the only way to satisfy all of the physically reasonable conditions on reaction probabilities, or at least the only way to satisfy these conditions in any theory that includes gravitation.

This point of view has its paradoxical elements. When in *S*-matrix theory we talk of the probabilities of various 'reactions', we have in mind processes in which two or more particles come together after travelling freely over great distances, and then interact, producing new particles which finally travel apart until they are again so far apart that they can no longer interact. This is the paradigmatic experiment of modern elementary particle physics. But this sort of reaction can only occur in a universe that is more or less empty and 'flat'—that is, not filled with a high density of matter, and not pervaded by strong gravitational fields that warp the structure of spacetime. This is indeed the present state of our universe, but it was not so in the early universe, and even today there are objects like black holes where spacetime is grossly curved. It seems odd to take a set of 'reasonable conditions' on reaction probabilities as the fundamental principles of physics, when the reactions to which these conditions refer did not always hold, and do not hold everywhere even today.

Indeed, the fact that our present universe is more or less empty and flat is itself rather paradoxical. In most theories, the quantum fluctuations of various fields would give 'empty' space an energy density so great that the gravitational field it produces would make spacetime grossly curved—so much so, that nothing like an ordinary elementary particle reaction could occur, and no scientists could live to observe it†. This problem is not solved in string theory; most of the large number of proposed string theories predict an enormous vacuum energy density.

To solve the problem of the vacuum energy density, we may need to invoke physical principles that are not only new, but of a new type, different from those that have so far seemed legitimate. This would not be the first time that there have been changes in our ideas of what are permissible fundamental principles. In his 1909 book *The Theory of the Electron*, Lorentz took the opportunity to comment on the difference between Einstein's Special Theory of Relativity, proposed four years earlier, and his own work. Lorentz had tried to use an electromagnetic theory of electron structure to show that matter composed of electrons would when in motion behave in such a way as to make it impossible to detect the effects of its motion on the speed of light, thus explaining the persistent failure to detect any difference in the speed of light along or at right angles to the Earth's motion around the Sun. Einstein in contrast had taken as a fundamental axiom that the speed of light is the same to all observers. Lorentz grumbled that 'Einstein simply postulates what we have deduced, with some difficulty and not always satisfactorily, from the fundamental equations of the electromagnetic field.'. But history was on Einstein's side. From a modern point of view, what Einstein had done was to introduce a principle of symmetry—the invariance of the laws of nature under changes in the velocity of the observer—as one of the fundamental laws of nature. Since Einstein's time, we have become more and more familiar with the idea that symmetry principles of various sorts are legitimate fundamental hypotheses. The Standard Model is largely based on a set of assumed symmetry principles, and string theory can also be viewed in this way. But in the days of Einstein and Lorentz, symmetries were generally regarded as mathematical curiosities, of great value to crystallographers, but hardly worthy to be included among the fundamental laws of nature. It is not surprising that Lorentz was uncomfortable with Einstein's hypothesis of the Principle of Relativity. We too may have to discover new sorts of hypothesis, which may at first seem to us as uncongenial as Einstein's symmetry principle seemed to Lorentz.

We have already encountered such an hypothesis. The so-called Anthropic Principle states that the laws of nature must allow for the appearance of living beings capable of studying the laws of nature. This

---

† *For historical reasons, this is known as the problem of the cosmological constant.*

principle is certainly not widely accepted today, though it provides what is so far the only way we have of solving the problem of a large vacuum energy density. (Too large a vacuum energy density would, depending on its sign, either prevent galaxies from forming, or end the Big Bang too early for life to evolve.) In some cosmological theories, a weak version of the Anthropic Principle would be no more than common sense. If what we now call the laws of nature, including the values of physical constants, vary from place to place in the universe, or from one epoch to another, or from one term in the quantum mechanical wavefunction of the universe to another, then naturally we could only find ourselves in a place or an epoch or a term in the wavefunction hospitable to intelligent life. But in the long run we may have to invoke a stronger form of the Anthropic Principle. When at last we learn the final laws of physics, we will then confront the question, why nature is described by these laws, and none other. There are plenty of imaginable theories that are logically perfectly consistent but nonetheless wrong. For instance, there is nothing *logically* wrong with Newtonian mechanics. Conceivably, the correct final theory is the only logically consistent set of principles consistent with the appearance of intelligent life.

The Anthropic Principle is just one example of an unconventional hypothesis, but one that has already received serious attention. In our search for the final laws of nature, we may in the twenty-first century have to accept the legitimacy of new sorts of fundamental physical principle stranger than any that have so far been imagined.

Some reflections on physics as a social institution
John Ziman

A domain of knowledge
A dictionary definition of 'physics' might start by stating that it is 'a domain of knowledge'. But only philosophers can conceive of knowledge without knowers, and social psychologists inform us that people only come to know the same thing by associating together. The dictionary would further characterize physics as a 'science', or a 'discipline'. It is thus located as a part of one of the major components of modern society—'Science'. This is just a fancy way of insisting that physics is, in many interesting ways, a *social institution*.

Almost all physicists, and the great majority of historians of science, find this concept difficult to grasp. There is not the space here for me to attempt to explain what it implies in principle and to show how it helps our understanding of the past, present and likely future of our great science. A thorough account, moreover, would be lengthened by a detailed repudiation of much zealous silliness that has also arisen out of the 'sociological turn' in contemporary science studies. May I be forgiven for not taking on that thankless task.

This essay has its origins in a chapter dealing with the exact sciences since World War II, for the fourth volume of the *History of the University in Europe* presently being published by the Cambridge University Press. Surprisingly, when I set about writing that chapter, I found that I was saying things that were not entirely trite. Here I simply pick up and develop just a few of the themes that emerge as soon as one begins to look specifically at physics from the same point of view. These observations are quite straightforward: they can be made by anyone familiar with the recent history of physics, without the use of sophisticated metascientific instruments such as the sociology of knowledge. They do not lead to unexpected conclusions; but they do fit together into a framework within which to locate and interpret many apparently unrelated phenomena. They also constitute a stable platform for projecting past and present trends into the future.

A discipline apart
Can the *scope* of physics be defined? I doubt it. The dictionary account would probably create more confusions that it set out to resolve. Knowledge domains can never be precisely delineated. In fact, the

increasing difficulty of differentiating physics from other parts of science is a significant historical phenomenon to which we shall return.

Despite the great depth of its historical roots, physics only began to be treated academically as a distinct science in the nineteenth century. By the beginning of the twentieth century, the division of the natural sciences into separate disciplines was fully established. From then on, their existence as more or less independent social institutions, within and beyond academia, seemed assured.

To start with, it was taken as a fact of life that Chemistry, Physics, Botany, etc, should be taught to undergraduates as if it were a single subject, dealing with a particular aspect of the world according to a unique method. It was the task of the university to turn out 'chemists', 'physicists', 'botanists', *et alia*, who understood things in the appropriate way and knew how to improve that understanding.

At that time, also, the various sciences were carefully differentiated academically from their corresponding technologies. Experimental physicists might eventually become expert at applying their knowledge, but they did not attend the same courses as mechanical or electrical engineers: theoretical physicists might learn how to solve the equations of fluid flow without ever studying the properties of a real propeller or turbine.

Throughout the first half of the twentieth century, the tribal boundaries around physics, and between its 'pure' and 'applied' clans, were sacrosanct within academia, and extended upwards and outwards into society at large. In the UK, for example, it was not until the 1960s that the Physical Society, whose members were mostly university teachers, merged with the Institute of Physics, representing physics graduates working in industry. German industrial firms still do not employ 'physicists' in 'engineering' roles, and government R&D establishments in most countries carefully classify professional physicists separately from chemists, mathematicians, etc, with their own career paths to pursue.

This *disciplinism* was, of course, both a reflection of, and a resource for, professional and organizational differentiation within universities. Like all other academic subjects, physics was segregated into a distinct department, staffed almost entirely by physics graduates all supposedly engaged in original research in physics. Even if they had had the expertise and/or formal qualifications to move to other subjects, university teachers of physics were effectively trapped by their heavy personal investments in highly specialized research.

It was recognized, of course, that certain fields of research lay athwart the regular disciplinary boundaries. But *interdisciplinary* subjects, such as chemical physics, or geophysics, were marginalized both for teaching and research, and often led a precarious institutional existence in small, poorly esteemed sub-departments. It was also understood, though seldom officially acknowledged, that the major scientific disciplines were internally sub-divided into small 'fields' whose inhabitants had little in common. Physics, which had been united in the nineteenth century under the banners of Newton and Maxwell, began to fall apart. By the middle of the twentieth century, it could have been partitioned into those who used Quantum Theory, those who used Relativity, those who used both, and those—still quite a number even in the 1940s and 1950s—who used neither, and did not understand them well enough to teach them properly.

The county and parish boundaries within the land of physics should not, however, be over-emphasized. They were not drawn in the same places, in different countries. There were also outstanding centres, whose research advances set the pace of curriculum development in particular fields, and many, many mediocre departments who lagged far behind them. But translations of such famous works as Dirac's *Quantum Mechanics*, of Courant and Hilbert's *Methoden der Mathematischen Physik* could be used as textbooks throughout the world, showing that there was a common culture of teaching and research.

The biographies of the leading physicists between the World Wars report numerous international visits and meetings. It was evidently quite easy from a scientific point of view, although often still very difficult for other reasons such as language or conditions of employment, for, say, an English physicist to study for a while in a German university, or for physicists from many European countries to take refuge from persecution in academic posts in the United States. Students from all over the world—Russia, India, Argentina, Australia, etc—could come to a centre such as Cambridge or Copenhagen, usually not knowing much about recent research developments but with a sound foundation of established knowledge on which to build. Physics was not yet, perhaps, a *global* social institution, but it was already a multinational association of such institutions with a strong common culture.

Fading frames and dissolving boundaries
But as time went on, the social barriers that previously differentiated physics from the other sciences, and framed them off from the non-academic world, began to fade and dissolve. This shows up in the emergence of new interdisciplinary research areas and trans-sectoral institutions; e.g. centres for research on materials, 'systems', the atmosphere, etc, jointly funded by universities, governments and industrial firms. Beneath the surface there was also much more interpenetration and interdependence of the disciplines, university departments, scientific professions, government agencies and economic sectors that span the many dimensions of R&D.

'Interdisciplinarity' is one of the perennial slogans of advocates of academic reform—and perennially they are disappointed. How was it that it has developed so extensively in the sciences adjacent to physics in the last twenty years? A number of factors have contributed, some internal to the academic enterprise, some stemming from changes in the social, political and economic environment of advanced science.

Pervasive research technologies

The most obvious factor is the diffusion of new research technologies from physics and electronic engineering into all the neighbouring disciplines. This methodological revolution began in the 1920s, with the introduction of electronic measuring instruments into chemical and geological research. This process was hastened by scientists returning to their laboratories with wartime communications and radar experience, until by the 1960s ability to apply these techniques when required was essential for experimental or observational research in all the exact sciences.

The real revolution came in the 1970s, as micro-electronic control, data-processing and digital-computing capabilities were linked to, combined with, and eventually incorporated into every kind of research instrument. This development was irresistible. By definition, all research in the exact sciences involves the acquisition and logical manipulation of quantitative data. This is a task that can be performed by digital micro-electronics with ever-increasing capacity, speed and algorithmic sophistication.

Geographically speaking, this revolution took place somewhat unevenly. European physicists were only marginally involved in the invention and development of the hardware and software that made it possible, and they were relatively slow in adopting the new instruments and techniques that were becoming common in American research laboratories in the 1960s and 1970s. But by the beginning of the 1980s, they had caught up, and one would expect a university physics laboratory in Western Europe to be as well equipped with computers, data terminals and computerized instruments as in the United States. The same cannot be said of physics research in Central and Eastern Europe, where the lag in instrumental capabilities made it almost impossible to keep at the international frontier in experimental research. This material deficiency continues to stunt the growth of the research base in physics throughout the Third World.

The pervasion of all the exact sciences by information technology has brought them much closer together. The linkages with physics are not merely technical, in the sense that the same visual display units, circuit boards and high-level languages are in use in almost every building of the science area of every campus. All the sciences were beginning to learn the same new 'intellectual' procedures, such as computer simulation of spatial phenomena, or topographic reconstruction of observational data. A procedure that had been developed in geophysical prospecting could be just what was needed for non-destructive testing of materials. A massively parallel supercomputer acquired originally for oceanographic studies could later be used for working out the whole history of the universe, of the interaction of quarks inside a nucleus.

It is natural to emphasize the unifying effects of information technology because this technology has become so familiar in everyday life. But a number of other new and powerful research techniques have percolated through the scientific world, far from the fields where they originated. At first, an 'alien' technique may be used unreflectively, simply as a new way of looking at an existing problem. But then an intellectual synergy develops between the fields thus linked, opening up new pathways through departmental and disciplinary barriers.

X-ray diffraction, electron microscopy, laser optics, synchrotron radiation, radioactive isotopes, mass spectrometry—the list is endless—were all conceived and born in the domain of physics. They have grown to maturity and are put to work nowadays in all branches of the natural sciences, from astronomy to zoology. The scanning tunnelling microscope, for example, can provide exquisitely detailed topographic information about a solid surface, whether of a structural alloy, a silicon chip, a meteorite, a mineral or a large organic molecule. A laser beam may be used to measure the creep of continents or catch a chemical reaction in a state of transition. This diffusion of new experimental techniques does not stop at the boundaries of the exact sciences. X-ray optics is exploited to study stars and starfish, macromolecules and membranes. The recent emergence of a whole cluster of generic research technologies is changing the definition of physics as a domain of knowledge and its role as a social institution.

Transdisciplinary disciplines

Whiggish histories of science celebrate the appearance of new scientific tribes that eventually become nations. They describe how the exploration of the natural world extends to new aspects, inviting systematic study. The underlying assumption is that what is then discovered can be characterized according to an innate structure, ranging, as we would say nowadays, from the quark to the quasar in scale, from the particle to the political party in complexity. Each science can accordingly be located uniquely on an abstract map, according to its place in this scheme.

By the middle of the twentieth century, this useful, if philosophically naive scheme was in trouble. All the Dark Continents, the *Terrae Incognitae*, on its map had been traversed, roughly surveyed, and colonized by eager pioneers. At the junctions of the continental plates, buffer zones were established, whose double-barrelled names—Geochemistry, Astrophysics, Geophysics, Chemical Physics, Mathematical Physics, etc—indicated their hybrid status. Since that time, only

Molecular Biology—or should it be labelled Biophysics—has emerged as a brand new science with a previously unexplored subject matter.

This is not to say that physics has not made wonderful progress, with quite amazing discoveries, in the past 50 years. Most observers agree that this has been one of the most productive epochs in its history. It is just to emphasize that these discoveries, unlike those of the previous half century, did not open doors to quite unsuspected realms of being. For example, the search for the fundamental constituents of the natural world has dug down into the atomic nucleus, through the strata of electrons and nucleons, photons and neutrinos, to lay bare the deeper levels of quarks and mesons. But most physicists would regard these discoveries, including their theoretical interpretations and unifications, as successive stages in a grand research programme that was under way when, say, Ernest Rutherford and Niels Bohr were working together in Manchester in 1912. Astronomy, similarly, has obtained striking evidence of a much wider range of objects and phenomena than was imagined in the 1930s, but its general conception of the cosmos is not totally different from what was being taught to students at that time.

With no more worlds to conquer, physics turned at first towards a policy of imperialism. The names given to the hybrid subjects around physics—astrophysics, geophysics, etc—originally indicated the application of physics methods to other aspects of nature. Inspired by naive reductionist fundamentalism, some physicists have tried to enlarge their holdings in these other sciences into take-over bids. This grandiose policy has failed, chiefly because it ignored completely the immense amounts of tacit, non-quantifiable understanding on which the other sciences are based. But the barriers between these traditionally distinct disciplines have been irreparably breached. Thus, the notion that 'all chemistry is really physics' is now largely discounted, even though the chemists themselves celebrate the steady progress that continues to be made on the computation of 'chemical' phenomena—molecular structures, bond strengths, reaction rates, optical properties, etc—using mathematical formulations of 'physical' laws.

A more productive policy, in the long run, has been to move into a higher intellectual dimension. Since the early 1970s, the scientific world has witnessed a new academic phenomenon—the coalescence of elements drawn from a number of existing disciplines into a broad new fields of science. Indeed, these *transdisciplinary* fields are often so wide in their scope that they extend far beyond the natural sciences as traditionally defined. Earth Science includes many geographical aspects of the social sciences as well as Geology and Geophysics. Information Technology is not just advanced engineering, since it depends fundamentally on Mathematics and Physics on the one side, and on Psychology, Sociology and Management Science on the other.

As a case in point, Materials Science extends right across the established exact sciences and their associated technologies. The knowledge and techniques that it derives from Physics and Chemistry are not marginal, but come from the mainstream of research in these traditional disciplines, whilst its applications are radically transforming all branches of Electrical, Mechanical and Structural Engineering. Originally it might have been described as a very loose alliance of recognized sub-disciplines such as Metallurgy, Continuum Mechanics, Crystallography, Polymer Chemistry, and Solid-State Physics. Or one might think of it as a confederation of innumerable research specialities, concerned with a variety of material properties such as Semiconductivity, Magnetism, Crystal Growth, Mechanical Strength, etc. Or it could be defined in practical terms as the search for peculiarly strong, elastic, light, insulating, transparent, superconducting, biologically inert, etc, etc, new materials.

The champions of this new science tend to exaggerate its intellectual coherence. Nevertheless, Materials Science is firmly established as an academic category—and hence as a social institution—that cuts right across the traditional scheme of the exact sciences. In effect, it does not belong to that scheme at all, but presents itself as a member of an alternative classification scheme, where specialized fields of study are defined and aggregated according to a somewhat different principle. A 'matrix structure' is emerging, where the new transdisciplinary 'rows' are more visible and influential in society at large than the old disciplinary 'columns' that sustain them. The authority of Physics is thus being fragmented and shared out across a much wider range of social institutions.

Researching and/or teaching

There is no doubt that the dynamism of an exact science such as physics resides in its research capabilities, rather than in its educational functions. The university tradition depended on a symbiotic relationship between teaching and research, within disciplinary departments or more specialized professorial institutes. The developments described above have put this relationship under very severe stress. The research frontiers have proliferated, diversified, intermingled and moved on, so that they are no longer directly connected with the basic subjects that need to be taught to students.

It is not just that courses and curricula are out of date, and no longer consistent with current knowledge. The point is that knowledge in the physical sciences is cumulative and progressive. New understanding builds on past understanding, but does not supersede it. For example, most theoretical understanding in physics is based on quantum mechanics. This has always been a subtle intellectual discipline, whose principles and procedures take time to grasp. It cannot be simplified and reduced to a school subject, but has to occupy a considerable proportion of every undergraduate course in these disciplines—including the new interdisciplinary subjects, such as materials science, that also rely on it theoretically.

How then can one reconcile the relatively permanent disciplinary structure appropriate for undergraduate teaching with the changing, interdisciplinary, multinational, intersectoral groupings that are now typical of advanced research? One solution is to cut the Gordian knot that ties teaching to research by setting up a separate system of institutes, units, laboratories or centres staffed by full-time research scientists. This policy has been widely followed, especially in countries such as France and Russia, where state-funded national academies have always been active research enterprises.

The sophisticated instrumental requirements and rich technological potentialities of physics would seem to favour this arrangement, even though it severely limits the research opportunities for university teaching staffs; and yet experience in countries such as the USA, the UK, Holland and Sweden shows that these complications can still be handled inside a university framework without separating teaching from research in the working careers of academic scientists.

Finalization

The increasing traffic across the interfaces between the exact sciences is both a cause and a consequence of closer contacts between physics and its associated technologies. But the gap between 'basic' and 'applied' physics has also been closing as a result of *finalization*. This refers to the degree to which research in a particular field can be 'finalized', that is, usefully oriented towards realistic goals.

The original *finalisierung* thesis was put forward by a science studies group in a Max Planck Institute at Starnberg University in West Germany in 1973. In its simplest form, it depends on Thomas Kuhn's account of the life cycle of a field of study, which typically falls into three successive phases. The first phase can be described as *pre-paradigmatic*, in that the field lacks a generally accepted overall explanatory scheme. In these circumstances, there is seldom adequate theoretical support for elaborate programmes of research intended to reach specific practical goals. This was obviously the situation in macroscopic dynamics before, say, the work of Isaac Newton: in most other branches of what we now call classical physics this phase endured until the early nineteenth century.

The *revolutionary* phase is characterized by anomalous observations which do not fit into any existing interpretations, together with a general feeling that conditions exist for a breakthrough into a more coherent theoretical regime. This is evidently a phase where it is still unrealistic to direct research towards designated technological objectives, but where a major effort towards better basic understanding is likely to be very rewarding in the long term. This feeling was, of course, widespread at the beginning of the twentieth century.

What has made physics so exciting in this century is that many major fields have 'had their Kuhnian revolution', and are now effectively finalized. By 1900, classical physics, e.g. Thermodynamics, Electromagnetism, Hydrodynamics, etc, had already established a fundamental paradigm covering most macroscopic systems. Quantum Mechanics and Relativity extended the coverage to most aspects of the microscopic physics of condensed matter, molecules, atoms and nuclei, as well as to most of what was required to explain large-scale macroscopic terrestrial, planetary and stellar phenomena.

Physicists who are nostalgic for their revolutionary past should note that the same transition has taken place in all the exact sciences and in large areas of biology. They may rejoice that High-Energy Physics, Astrophysics and Cosmology are still 'unfinalized', in that there remain fundamental phenomena that are quite unexplored, there is still widespread uncertainty about the theoretical framework within which research ought to be planned, and a broad spectrum of highly speculative projects is being actively pursued, mostly under academic auspices.

This Kuhnian analysis is obviously oversimplified. Scientific knowledge advances on many fronts simultaneously, through innumerable interlocking local cycles of paradigm formation and revision. Even within physics, different fields vary considerably in this dimension, and thus follow different trajectories in academic history. But it is a vulgar error to suppose that the 'normal' science that follows a 'revolution' is intellectually uncreative, or unworthy of the university tradition of the search for truth. On the contrary, finalization opens up new vistas for basic exploration, as well as new opportunities for gainful application.

One effect is that research can be concentrated confidently on innumerable unexplained but important phenomena that have already been observed, in materials and in engineering systems, on Earth and in outer space. A great part of what is reported in these volumes has been revealed by research undertaken in that spirit. Strong theoretical paradigms have also made feasible the most recent development in the *methodology* of physics—that is, realistic computer modelling of natural or artificial systems, simulating observable physical phenomena.

The possibility of carrying out research more systematically, according to a more certain schedule, profoundly affects the social dimensions of the scientific enterprise. Given enough time, and more than enough money, almost any theoretically credible practical objective can be realized. The *management* of the effort becomes more crucial than the conception of its goal. This was demonstrated dramatically during World War II, when fundamental research in nuclear physics turned out to be ripe for exploitation as a military technology. It joined the team already established long ago by other finalized branches of physics, such as classical dynamics (e.g. ballistics), electromagnetism (e.g. telegraphy) and fluid mechanics (e.g. aeronautics).

Similar developments have since taken place in nearly every other field of microscopic physics. Almost all the knowledge of physics that might seem to be needed for any plausible practical application is now thought to be accessible by mission-oriented or strategically programmed research. The scope of these potential applications is rich beyond measure, thus opening a Pandora's box of expectations for power, wealth, welfare—or mere curiosity. In consequence, a social institution that had previously been on the margins of politico-economic life has moved to centre stage. This could not happen without a series of far-reaching structural transformations, which we shall now explore.

Linking the academy with industry
As each field of physics is finalized, it becomes the scene for large scale R&D under industrial and/or governmental auspices. This process was accelerated by the vital part played by physics and physicists in World War II. After the war, the military and peaceful uses of 'atomic energy' and of microelectronics were pursued indefatigably in enormous establishments modelled on Los Alamos, Malvern, Bell Laboratories and other successful laboratory complexes.

'Applied physics' had been a major component of industrial and military R&D since the end of the nineteenth century, but usually as an adjunct to engineering. Now the whole knowledge base of physics began to be thought of as a direct source of technological innovation, and hence worthy of massive support.

The centroid of research effort in many fields has thus moved out of the university into these laboratories, where large resources were available to follow up promising leads. This trend is observable even in research whose proximate motive might best be described as intellectual curiosity. An obvious example of this is in basic semiconductor physics, which began life in university physics laboratories in the 1930s, but where the pace is now set almost entirely by the multinational electronic firms. The same historical development took place 30 years later in laser physics and optics, where academia is no longer the principal site of fundamental research.

At the same time, technological goals gained far greater weight in university physics. The search for knowledge 'for its own sake' was re-oriented (or relabelled!) towards more explicit utilitarian ends by a number of different forces, of which direct industrial funding was not always the most influential. In some countries, university physicists have acquired a regrettable taste for a cut of the defence budget cake, and have learnt to shape their project bids accordingly. In the last two decades, a large proportion of governmental funding of 'academic' physics has come with a tag indicating that it should contribute to the industrial competitiveness of their country on world markets, or some such politico-economic crusade.

The notion of 'strategic' or 'pre-competitive' research provides an umbrella slogan for this development. Apart from Astrophysics, Cosmology, and very High-Energy Particle Physics, it is not implausible to say of almost any physics research project that it might produce knowledge that might improve our basic understanding of a generic technology that could eventually enable a profitable practical invention (etc). This argument is proving a valuable rhetorical resource for university physicists seeking moral justification and financial support for their projects, however distant these may actually be from the marketplace. It could be said that this represents a creative integration of theory with practice, promising substantial benefits to society and demonstrating the vital social role of basic physics, even in its most esoteric and expensive areas. But it also poses a real threat to the future of those branches of science, epitomized by High-Energy Particle Physics, which seem to have no 'strategic' escape from ever more expensive and intellectually baroque ivory towers.

The characteristic research activities and interest of the industrial and academic sectors thus interpenetrate in detail, on university campuses, in science parks, and in corporate R&D laboratories. Take, for example, a typical research question in condensed matter physics: what are the optical properties of an isolated chemical impurity in a glassy material? This question might have been posed in the course of a basic study of the fundamental quantum-mechanical states of electrons in such systems: it might equally well have arisen along the way to developing and manufacturing a more transparent glass fibre for opto-electronic communications. In either case, the actual research would involve exactly the same instruments, specimens, theories, archival sources and technical expertise. Precisely the same investigation might have been mounted in a university department of physics, of materials science or of electronic engineering—or in the R&D laboratory of a multinational electronics firm, or in a research establishment of a national Ministry of Defence. And in every case, serious consideration would be given to whether the research results might give rise to an application that could be patented, or otherwise exploited commercially.

This does not mean that all university work in physics is now directed towards technological invention. 'Curiosity-driven', 'exploratory', 'blue skies' research, is still being done by thousands of academic scientists all over the world. Private sector firms are very far from doing all the basic research they think they might need to support their industrial R&D. The initial training of students and researchers is still primarily a university responsibility, although academic qualifications can often now be gained on the basis of pre-doctoral and post-doctoral research in an industrial or government laboratory. Academia and industry are much more open to each other, and more aware of each others concerns than they used to be, but they are still essentially complementary in their scientific and technological activities.

The fact remains, however, that physics is still not a unitary social institution. Well established academic attitudes and practices—freedom of problem choice and of publication, theoretically defined objectives, reference to disciplinary peer groups, personal autonomy within a collegial structure, and so on—are not easily reconciled with adherence to a programme designed to produce marketable intellectual property relevant to the urgent solution of a practical problem already defined by corporate management. These cultural differences have deeper roots than is usually recognized in public exhortations and subsidies to encourage more effective academy–industry links, and have not been altogether solved by the wide range of new institutions that have been set up in various countries to bridge the divide. The social roles of such bodies as 'Science Parks', 'Fraunhofer Institutes', 'Interdisciplinary Research Centres' and 'Venture Companies' are evolving so rapidly that we still have no feeling for how they should relate to older institutions of the type traditionally gathered under the physics banner.

Collectivization
The *performance* of research in physics has changed even more radically than its subject matter. This change is most economically described as a move from *individual* scientific effort towards much more highly organized *collective* modes of research. The most direct evidence of the change is in the phenomenon of multi-authored scientific papers. Before World War II, the great majority of such papers were published either in the name of single author, or of two scientists in close collaboration—typically a student or junior researcher together with his or her professor.

This did not necessarily mean that people were doing their research entirely alone. In most cases, they would have been students or staff members of a university department or institute, with a dozen or so scientists and technical staff sharing the basic facilities of a research and teaching laboratory. But they would normally have had their own stretch of laboratory bench, and their own personal projects, for which they took individual responsibility from the conception of an investigation, through its performance, to the publication of its results. This applied in principle even in a large 'research school' where almost all the research was actually being carried out under the close supervision of a famous professor, who also claimed much of the credit for its successes.

What we now find is that physics papers that are not exclusively theoretical usually appear above the names of a group of 'authors', ranging in number from two or three up to dozens, or even hundreds. This reflects a reality of actual collaboration, whether in voluntary partnership as professional colleagues or under managerial instruction as members of an organized research team. Although the group may have an easily recognized leader, the research is presented as the collective product of the labour of all its members.

This trend varies in scale and intensity from field to field, but it is not confined to physics. Indeed, it is so pervasive that it cannot be ascribed to any single simple cause. It may even represent a cultural form, diffusing back to Europe from the richer, more business-like universities and research laboratories of the United States. The example of military research during World War II, and of technological R&D in large industrial firms, demonstrated the effectiveness of large research teams focused on the solution of well defined problems. In some fields, this mode of research is almost unavoidable. Even though high-energy physics and cosmology are still 'revolutionary' in spirit, the effective finalization of many technical aspects of their study makes it feasible to mount projects that are so elaborate that they can only be undertaken by large, highly organized groups of very skilled people.

Again, where the science itself has become interdisciplinary or transdisciplinary, research programmes cannot be broken down into independent projects suited to the limited range of expertise of individual researchers. Progress requires the active, day-to-day collaboration of specialists from a number of different scientific traditions, working to a common purpose. Even when the investigation is apparently located within the mainstream of a traditional discipline, it may still require inputs from a number of different technical specialities. Thus, for example, few physicists nowadays are masters of both the experimental and theoretical techniques that contribute to their field, and may thus find it advantageous to publish jointly with colleagues with complementary skills.

A related development has been the rapidly increasing sophistication and cost of research equipment. This is due to a combination of factors that we have already noted—the application of discoveries in one field as research technologies in others (e.g. solid-state detectors in astronomy), systematic redesign and manufacture of novel but very powerful standardized instruments by commercial firms, and, of course, the automatic control of instruments by electronic computers.

It is still just possible in certain fields—for example, in the structural study of unusual forms of condensed matter such as liquid crystals—for a small group of very talented and imaginative scientists to do outstanding research without highly specialized apparatus. But the apparent 'simplicity' of such research is often misleading. As many physicists in Central and Eastern Europe know from experience, routine shortages of very simple, basic equipment—test tubes, fine chemicals, lasers, personal computers, learned journals—can frustrate the most inspired research in the 'sealing wax and string' tradition. In general, even when a particular investigation does not involve the construction and use of one dedicated piece of apparatus, it will usually require the routine use of several elaborate instruments, such as electron microscopes, helium liquefiers, x-ray spectrometers, etc, not to mention access to high-powered computers and databases.

As a result, an effective research institution has to make a heavy investment in such instruments. The actual scale of aggregation necessary to make, say, a university physics department viable is a matter of opinion, but it is now quite clear that this investment is quite uneconomic unless it is shared in use by dozens, if not hundreds, of research scientists. This does not mean that the research itself needs to be organized on the same scale. Quite small research units—e.g. of half a dozen professional researchers, with accompanying assistants, students and technical staff—are still optimal for 'strategic basic research' over wide areas of physics. But access to the necessary instrumental and infrastructural facilities is vital.

Big Physics
The supreme manifestation of collectivization is 'Big Science'. In some research fields, the essential instruments are so large and so indivisibly costly that they are far beyond the personnel and material resources of even the largest university, and can only be provided as national or international 'facilities'. In the public eye, physics is identified with enormous particle accelerators and immensely costly satellite observatories. The fact that very few people can understand the discoveries they make only adds to their symbolic power.

The rise and rise of 'Big Physics' is not only a major factor in public attitudes to science: it is also highly significant within physics itself. It is true that many of the large facilities that seem to dominate physics research are simply sources of special radiation, or very powerful observing instruments, which are mainly used by a number of small research groups working independently. They merely extend to continental distances the established practice of travelling to a special work site to obtain important research material. Thus, a university researcher from, say, Barcelona, may go to Grenoble for a few days, carrying a previously prepared specimen to be placed in a beam of neutrons at one of a number of work stations around the high flux nuclear reactor of the Institut Laue-Langevin. Or two or three astronomers from Boston may spend some time in Hawaii, carrying out a series of observations with one of the telescopes there. The arrangements for running such facilities are often very elaborate and sometimes contentious, but they scarcely impinge directly on the working lives of the scientists who use them.

Even Fermilab and CERN, with their enormous resources and material facilities, do not have large academic staffs of their own. The extraordinarily elaborate and costly experiments that are performed on their immense particle accelerators are planned and carried out by so-called 'collaborations'—that is, by *ad hoc* teams of hundreds of academic scientists drawn from dozens of universities, distributed over a whole continent. A professor from the University of Milano, a lecturer from Dundee, a graduate student from Helsinki, a technician from Vienna might spend a larger proportion of their time working together in Geneva on some feature of an experimental rig. Nevertheless, they would remain employees or students of their home universities, where they would have to return from time to time to carry out their normal academic duties.

The fact is, however, that a large part of the effort that goes into physics is concerned with the design, construction and operation of particle accelerators, research reactors, synchrotron radiation sources, plasma confinement vessels, radio telescopes, space platforms, planetary exploration vehicles, supercomputers, databases, communication networks, etc. Although much of this effort is defined by physicists as 'engineering', or 'management', or 'maintenance services', these features cannot be dismissed from the planning of research, and often determine the nature of the science that is done. The sheer magnitude and complexity of the technology requires a degree of coordination and collaboration that could scarcely have been envisaged by the physicists who pioneered these fields less than 50 years ago.

Internationalization
Originally, all Big Physics facilities were provided and managed nationally, as they still are in many fields in the United States. But elementary economies of scale continually push towards multinational sharing of investments, running costs and usage. Some fields of physics are 'going international' through the development of programmes of research requiring the active collaboration of research groups based in several different countries. Usually this has a geographical rationale—for example, in the study of global environmental phenomena, but sometimes it is done for political reasons, to emphasize transnational solidarity between stronger and weaker scientific communities, or to foster 'precompetitive' research as a basis for marketable technological innovations.

The situation in Europe is particularly instructive. There are only a few institutions like JET—the Joint European Torus—where research scientists from a number of countries are employed full time, under international management. But what one might call a 'European research system', or a 'European science base', is emerging, not as a command structure centred on Brussels but as a loose network of non-governmental organizations such as the European Science Foundation, learned societies such as the Academia Europaea, multilateral intergovernmental agencies such as CERN and the European Space Agency, not to mention the shadowy powers of multinational commercial firms such Siemens,

2055

Airbus Industries, or Shell. When individual research groups collaborate in such projects, they do not merge their legal or administrative identities, and remain firmly rooted in their home institutions. Nevertheless, they have to work together just as closely as if, say, they were American physicists all taking part in a multi-university project funded by a federal agency within the United States.

As we have noted, physics was never strongly culture bound. Each country might have its own style, its own particular interest, its own scientific heroes, but they all belonged in the same intellectual world. In the period since World War II, this international homogeneity in the formal and technical cultures of physics has spread outwards from the written paper, the conference hall and the laboratory bench to other aspects of scientific life. The sheer research technology of physics has forced physicists to become much more *cosmopolitan* in their outlook and working practices.

The high-energy physicists were in the vanguard of this transformation, first by building on their scientific *camaraderie* to create a permanent research facility where they could work together, then by demonstrating the effectiveness of multinational research teams and by establishing data networks linking scientists in their own laboratories, throughout the region. It now seems quite normal for any physicists to be working in a multinational research team, communicating in the *linguafranca* of broken English, and sharing enmity for the obstructive local bureaucracy. A day's scientific business in Paris or Geneva is no more out of the way for a physicist from Cambridge (England) than a meeting in Washington or Chicago would be for a physicist from Cambridge (Massachusetts). An international committee allocating observing time on a European telescope seems just as natural as a national peer review panel doing the same job. If, as is often asserted, it is desirable to create a 'United States of Europe' on a par with the United States of America, then physics has already gone a long way down this path, and has little incentive to turn back.

Boundary Conditions

Science is always revolutionizing itself. For centuries now, each generation of physicists has known so much more than its elders that it has had to completely rewrite its technical manuals, its file of outstanding problems—and sometimes its representations of reality. Cognitive revolution cannot be disassociated from social revolution. Each generation has accordingly had to revise its manuals of scientific etiquette, its file of effective institutional arrangements—and sometimes its representations of communal and societal reality.

Slow but steady change throughout the twentieth century has transformed physics as a social institution. Imagine looking back at quarter-century intervals—for example, over the successive periods punctuated by the beginnings of the two World Wars. The shift from loosely linked individualism to systematic collectivism shows very clearly. The period from 1889 to 1914 sees the consolidation of the German academic machine, which selected scholars competitively on the basis of their research productivity, and provided them individually with facilities, resources and research assistants. From 1914 to 1939 this model spread to other countries, where it hybridized with the Anglo–American departmental system, thus enabling groups of established researchers to pool their efforts. In the next phase, from 1939 to 1964, wartime experience of sophisticated team research encouraged the crystallization of more tightly organized groups within university departments, and also the creation of research establishments and facilities managed along more bureaucratic lines. The latest period, from 1964 to 1989 (another super-political climacteric!) has seen the ascendancy of the 'funding agencies'—governmental, intergovernmental, or quasi-governmental bodies exercising stop–go powers over research projects in the name of peer review panels and other communal groupings.

As we have already seen, similar changes occurred in other dimensions—greater intimacy with other disciplines; spawning, spin-off, capture and colonization of related sub-disciplines; institutional and professional conflation of 'pure' and 'applied' physics in academia and industry; enlargements of organizational scale to continental and global ventures; and so on. Closer study would also reveal a paradoxical change towards less formality, but more specific contractual obligation, in the personal relations between senior physicists, research assistants and students. Both internally and externally, physics as a whole has become a much more tightly connected and socially interactive institution.

What is surprising is not the amount of change, but the continuity and invariance of the cultural norms. After all, physics as a whole has grown by a factor of about a hundred. For every physicist at the beginning of the century, there are at least a hundred now—each spending far more on apparatus and services. No social institution can expand on that scale without radical internal reconstruction. And yet we can easily suppose that Ernest Rutherford could walk into Carlo Rubbia's office, and learn to run CERN in a week—and that Rubbia would acquit himself equally well in Rutherford's Manchester laboratory as of 1910. We can also suppose that, even now, there is a young, unemployed, unqualified theoretical physicist somewhere, who will duly receive recognition, like Albert Einstein, for the sheer genius of his unsolicited papers.

It might be argued that this continuity has only been achieved by staunch defence of the traditional scientific ethos. Certainly, academic physicists still try to follow the norms listed by Robert Merton nearly 50 years ago. But they must also obey the situational logic of promising new research technologies. The circumstances of modern physics research make it almost impossible to carry out many of the specific practices in which those norms were originally embodied—for example, the expression of originality through free choice of research problems, or the evaluation of individuals solely on the basis of their published papers.

It seems to me, rather, that there remains a healthy awareness of the necessary conditions for the advancement of knowledge—social space and opportunity for personal creativity, time for ideas to mature, hospitality to novelty, openness to criticism, respect for demonstrated expertise, and so on. It is heartening to observe how these conditions still prevail within some entirely non-traditional organizations, such as the enormous, highly structured 'collaborations' that carry out what are still quaintly called 'experiments' in high-energy physics.

Left to its own devices, physics would surely go on much as before, transforming itself intellectually and institutionally whilst preserving the inner core of personal motivations and interactions that has made it such an attractive and productive social enterprise. But the internal equations of motion have to satisfy boundary conditions whose effects are penetrating deeper and deeper into the system. Or, to put it another way, the system has expanded until is coming up against its limits to growth, and may be going through a significant, irreversible transition to a new structural phase.

In quantitative terms, the funding of physics research has levelled off, bringing to an end a very long history of continuous growth. Until World War II, basic physics research was largely financed as an adjunct to university teaching, supplemented by private patronage. The contribution of physics to the war effort earned it a privileged place in national budgets. These covered the enormous expenditures required to sustain both a rapidly growing community and rapidly increasing instrumental costs. Until, perhaps, the early 1970s, physicists in economically advanced countries got almost as much from their governments as they asked for—and sometimes rather more than they could use to good effect. In one country after another, however, this generosity came to be questioned by rival supplicants for public funds—other sciences, higher education, health services, and so on. Whenever the local economic climate becomes harsher, the call comes for cuts, especially in those fundamental fields that seemed to cost so much and offer so little practical return. Indeed, in an age that prides itself on its hard-headed, business-like, short-term realism, where every activity is measured by its wealth-producing, value-adding potentialities, it is a tribute to the fascination of basic physics research and discovery that so much is still being spent on high-energy experimental physics, astrophysics and space exploration.

'Steady state' funding is not the only boundary condition on an otherwise highly expansive institution. 'Finalized' physics promises more, and raises higher hopes of practical benefit. Funding bodies want their limited resources to be utilized more efficiently, according to priorities prescribed outside the research world. Researchers are forced to compete more fiercely for these resources, spending much effort in formulating their projects more precisely in advance and accounting for their expenditures in more detail *ex post facto*. A cult of 'evaluation' has developed, applying frequent tests of 'performance' to individuals, research groups, laboratories, universities, nations—even to whole fields of science.

These new conditions are quite reasonable in principle. Physics could not have been allowed to go on expanding at the expense of other important elements of civilized life. Elaborate technical projects and installations do need to be managed professionally. Competitive quality, performance evaluation and accountability are normal requirements for the receipt of public funds. The research priorities of physics professor do not automatically coincide with the needs of society at large.

Unfortunately, these conditions are often imposed and policed by people with limited understanding of the uncertainties, the tacit judgements, the technical virtuosities, the intellectual tensions, the unquantifiable criteria, that actually operate in a high science such as physics. It is a mistake to underestimate the zeal of bureaucrats, the pettiness of accountants, and the populism of politicians. Given free rein, they would take the magic out of physics and reduce it to an orderly system of technical projects and routine investigations. I hate to think what a mess they would make, continually pulling up projects by the roots to see if they were growing, and killing geese that lay golden eggs.

It is not that physicists are much wiser or more virtuous than other people; but at least they know what they are doing in their own domain. Even though this domain can no longer be run as if it were a world apart, its future will depend on the leaders of the physics community becoming aware of the effects of these new forces, and exercising their political and managerial skills to steer the enterprise safely as it forges ahead through very dangerous seas.

ILLUSTRATION ACKNOWLEDGMENTS

Chapter 17

W B Shockley—Fred English Photographers

N F Mott—Courtesy AIP Emilio Segrè Visual Archives

17.2 Reprinted with permission from Strutt, *Ann. der Physik* **86** 319 (1928), figure 1. Copyright 1928 Johann Ambrosius Barth Verlag

17.3 From Mott and Jones, *Properties of Metals and Alloys*, figure 26 a, b, p70, Clarendon Press, Oxford

17.5 Reprinted with permission from Peierls, *Ann. der Physik* **4** 121 (1930), figure 1, Johann Ambrosius Barth Verlag

17.8 From *Thermoelectricity* by D K C MacDonald. Copyright © J Wiley & Sons, Inc. Reprinted by permission of John Wiley & Sons, Inc.

17.9 From *Die Quantenstatistik* by L Brillouin, p278, 1931. Copyright 1931 Springer-Verlag GmbH & Co KG

17.10 Reprinted with permission from Wolfe *et al*, *Phys. Rev. Lett.* **34** 1292 (1975), figure 1. Copyright 1975 The American Physical Society

17.12 Reprinted with permission from Hall *et al*, *Phys. Rev.* **95** 559 (1954), figure 1. Copyright 1954 The American Physical Society

17.13 Reprinted with permission from Fritsche, *J. Phys. Chem. Solids* **6** 69 (1958), figure 2, Pergamon Press, Oxford

17.14 Reprinted with permission from Fritsche, *J. Phys. Chem. Solids* **6** 69 (1958), figure 4, Pergamon Press, Oxford

17.15 Reprinted with permission from Dresselhaus *et al*, *Phys. Rev.* **98** 368 (1958), figure 2. Copyright 1958 The American Physical Society

17.17 From R A Smith, *Semiconductors*, © Cambridge University Press 1978. Reprinted with the permission of Cambridge University Press

17.18 Reprinted with permission from Esaki, *Phys. Rev.* **109** 603 (1958), fig 1. Copyright 1958 The American Physical Society

17.19 A B Pippard, *Magnetoresistance in Metals*, figure 6.14, p222, © Cambridge University Press. Reprinted with the permission of Cambridge University Press

17.20 Reprinted with permission from Fawcett, *Advances in Physics* **13** 139 (1964), Taylor and Francis

17.21 Reprinted with permission from Shoenberg, *Proc. Roy. Soc.* A**170** 341, figure 4 (1939), The Royal Society, London

A11

17.23 Reprinted with permission from A V Gold, *Phil. Trans. Roy. Soc.* A**251** 85, figure 7 (1958) The Royal Society, London

17.24 Reprinted with permission from J F Cochran and R R Haering (ed), *Solid State Physics*, Vol 1, p146 (1968). Copyright 1968 Gordon & Breach Science Publishers

17.25 Reprinted with permission from Klauder *et al*, (1966) *Phys. Rev.* **141**, 592, figure 2a. Copyright 1966 The American Physical Society

17.26 From A B Pippard, *Physics of Vibration* Vol 1, pp 276 and 277, © Cambridge University Press 1989. Reprinted with the permission of Cambridge University Press

17.27 Reprinted with permission from Stark, *Phys. Rev.* A**135** 1698 (1964), figure 13. Copyright 1964 The American Physical Society

17.28 Reprinted with permission from Reed and Condon, *Phys. Rev.* B**1**, 3504 (1970), figure 3. Copyright 1970 The American Physical Society

17.30 Reprinted with permission from R A Webb *et al*, *Phys. Rev. Lett.* **54** 269 (1985), figure 1. Copyright 1985 The American Physical Society

17.32 Reprinted with permission from K von Klitzing, *Rev. Mod. Phys.* **58** 519 (1986), figure 14. Copyright 1986 The American Physical Society

17.33 Reprinted with permission from R Willett *et al*, *Phys. Rev. Lett.* **59** 1776 (1987), figure 1. Copyright 1987 The American Physical Society

Chapter 18

C H Townes—AIP Meggers Gallery of Nobel Laureates. Courtesy MIT

T H Maiman—*Maser to Lasers: An Historical Approach* by M Bertolotti, Adam Hilger, Bristol. Copyright 1983 M Bertolotti

A L Schawlow—Sources for the History of Lasers. Courtesy AIP Emilio Segrè Visual Archives

N Bloembergen—AIP Emilio Segrè Visual Archives

18.1 From Eugene Hecht, *Optics*, Second Edition (p198), © 1987 by Addison-Wesley Publishing Company, Inc. Reprinted by permission of the publisher

18.2 *An Introduction to Interferometry*, Second Edition by S Tolansky, Longman Group Ltd, London

18.3 From *Physics*, Second Edition by H Ohanian, W W Norton & Company, Inc., New York

18.4 *Masers and Lasers: An Historical Approach* by M Bertolotti, Adam Hilger, Bristol. Copyright 1983 M Bertolotti

18.5 Copyright IEEE

18.6 Reproduced with permission from R W Minck, R W Terhune and C C Wang, *Proc. IEEE* **54**, 1357. © 1966 IEEE

18.7 From *The Photonics Design & Applications Handbook 1994*, Laurin Publishing Co, Inc., Pittsfield

18.8 Reprinted with permission from Y Yeh and H Z Cummins, *Appl. Phys. Lett.* **4**, p176 (1964). Copyright 1964 American Institute of Physics

18.9 Reprinted with permission from B Wirmitzer and G Weigelt, *Optics Letters* **8** (1983), p389, Optical Society of American, Washington

18.10 Courtesy of Prof E Jakeman, FRS

18.11 From *Inverse Scattering Problems in Optics* by H P Baltes (ed), figure 3.7 a–c. Copyright Springer-Verlag GmbH & Co KG

18.12 Courtesy of Lawrence Livermore Laboratories, California

18.13 Courtesy of NASA

Illustration Acknowledgments

18.14   Courtesy of Sharp Laboratories of Europe Ltd, Oxford

Chapter 19

19.6   Reprinted with permission from C Herring, 1950, *J. Appl. Phys.* **21** 437. Copyright 1950 American Institute of Physics

19.8   *Whillis R Whitney* by George Wise. Copyright © 1985 by Columbia University Press. Reprinted with permission of the publishers

Chapter 20

D Gabor—AIP Meggers Gallery of Nobel Laureates

Chapter 21

P Flory—Stanford University Press
C Frank—IOP Publishing Ltd, Bristol

Chapter 22

L D Landau—AIP Emilio Segrè Visual Archives
M N Rosenbluth—AIP Emilio Segrè Visual Archives, Physics Today Collection

22.4   Courtesy of the Princeton Plasma Physics Laboratory
22.5   JET Joint Undertaking
22.6   Courtesy of Princeton Plasma Physics Laboratory
22.8   Reprinted with permission from *Plasma Physics and Controlled Fusion Research* **2** 700 (1965) IAEA, Vienna
22.9   Courtesy of Lawrence Livermore National Laboratory
22.10 Courtesy of University of California Lawrence Livermore National Laboratory, Livermore and the US Department of Energy
22.11 Courtesy of Lawrence Livermore National Laboratory
22.12 Courtesy of Lawrence Livermore National Laboratory
22.13 Reprinted with permission from *Phys. Rev. Lett.* **13** 6, p186. Copyright 1964 The American Physical Society
22.14 Courtesy of Lawrence Livermore National Laboratory
22.15 Courtesy of Lawrence Livermore National Laboratory
22.16 Reprinted with permission from *Nuclear Fusion* **27** 10, p1675 (1987) IAEA, Vienna

Chapter 23

A S Eddington—AIP Emilio Segrè Visual Archives
E P Hubble—Hale Observatories, courtesy AIP Emilio Segrè Visual Archives

23.1   Reprinted with permission from J N Lockyer, *Proc. Roy. Soc.* **A43** (1887), 117–56, The Royal Society
23.4   From A S Eddington, 1924, *Mon. Not. R. Astron. Soc.*, **84** 308, Blackwell Science Ltd. Reprinted with permission from Blackwell Science Ltd
23.8   Reprinted with permission from E P Hubble, 1936, *The Realm of the Nebulae*, New Haven: Yale University Press
23.10 Reprinted with permission from E P Hubble, 1936, *The Realm of the Nebulae*, New Haven: Yale University Press
23.12 From H Bondi, 1960 *Cosmology* 2nd edition, p84. Reprinted with permission from Cambridge University Press
23.16 From T Toutain and C Frölich, *Astron. Astrophys.* **257** 287, 1992. Copyright 1992 Springer-Verlag GmbH & Co KG
23.17 Courtesy of Prof D O Gough 1993
23.18 Reprinted with permission from *Nature* **217** 709, A Hewish, S J Bell, J D H Pilkington, P F Scott and R A Collins. Copyright 1968 Macmillan Magazines Ltd
23.20 Reprinted with permission from J Taylor, *Phil. Trans. Roy. Soc. Lond.*, **A341** (1992),

117–234, figure 7, The Royal Society
23.21 From M Kafatos and A G Michalitsianos (ed), 1988 *Supernova 1987A in the Large Magellanic Cloud.* Reprinted with permission from Cambridge University Press
23.22 From J H Oort, F J Kerr and G Westerhout, 1958, *Mon. Not. R. Astron. Soc.,* **118** 379, Blackwell Science Ltd. Reprinted with permission from Blackwell Science Ltd
23.23 Reprinted with permission from C Hayashi, 1961, *Publ. Astron. Soc. Japan* **13** 450, Astronomical Society of Japan
23.25 Courtesy of Drs M Geller and J Huchra, Harvard-Smithsonian Center for Astrophysics
23.26 Reprinted with permission from *Nature* **172** 996, R C Jennison and M K Da Gupta. Copyright 1953 Macmillan Magazines Ltd
23.30 From A R Sandage, 1968, *Observatory* **88** 91. Reprinted with the permission of the Editors of *Observatory*
23.31 From J E Hesser, W E Harris, D A VandenBerg, J W B Allright, P Shott and P Stetson, *Publ. Astron. Soc. Pac.* **99** 739. Reprinted with permission from the Astronomical Society of the Pacific
23.32 From S J Lilly and M S Longair, 1984, *Mon. Not. R. Astron. Soc.,* **211** 833, Blackwell Science Ltd. Reprinted with permission from Blackwell Science Ltd
23.34 From M Rowan-Robinson, 1968, *Mon. Not. R. Astron. Soc.* **141** 445. Reprinted with permission from Blackwell Science Ltd

Chapter 25

W V Mayneord—Courtesy of the Institute of Physical Sciences in Medicine, York

L H Gray—Courtesy of the Institute of Physical Sciences in Medicine, York

Chapter 26

26.4 From *Endeavour* **2** p3 (1943). Elsevier Science Ltd, Oxford
26.6 Reprinted from *Continents in Motion* Second edition, by W Sullivan with the permission of American Institute of Physics. Copyright 1991 American Institute of Physics, New York
26.7 (*a*) Reprinted from *Amer. J. Phys.* **23** 590 with the permission of American Association of Physics Teachers. (*b*) Reprinted from *Geomagnetism* **2** p251 with the permission of Academic Press, New York
26.8 Isdatel'stov Nauke, Moscow
26.9 Lunar and Planetary Institute, Houston
26.10 Lunar and Planetary Institute, Houston
26.12 Reprinted from *The Behavior of the Earth,* by C Allegre, 1988 with the permission of Harvard University Press, Cambridge
26.13 (*a*) Reprinted from A Cox *et al, Science* **144** p1537 (1964). Copyright American Association for the Advancement of Science. (*b*) Reprinted from *The Behavior of the Earth,* by C Allegre, 1988 with the permission of Harvard University Press, Cambridge
26.14 (*a*) Reprinted from *Earth Science* Third edition by J F Lounsbury and L Ogden with the permission of Harper and Row, New York. (*b*) Reprinted from *The Behavior of the Earth,* by C Allegre, 1988 with the permission of Harvard University Press, Cambridge
26.15 Reprinted from *The Behavior of the Earth,* by C Allegre, 1988 with the permission of Harvard University Press, Cambridge

Illustration Acknowledgments

26.16 (a) Reprinted from *Continents in Motion* Second edition, by W Sullivan with the permission of American Institute of Physics. Copyright 1991 American Institute of Physics, New York. (b) Reprinted from *The Behavior of the Earth*, by C Allegre, 1988 with the permission of Harvard University Press, Cambridge

26.17 From *Science* **150** p482 (1965). Copyright American Association for the Advancement of Science

26.19 Camera Portraits, London

26.20 Library of Congress, Washington

26.24 Reprinted with permission from *Nature* **184** p219. Copyright 1959 MacMillan Magazines Ltd, London

JOURNAL ABBREVIATIONS

This listing shows abbreviated journal titles as cited in references, and full journal titles. Many of these journals are no longer active.

Abbreviations	Full Titles
Abh Berliner Akad	Abhandlung Berliner Akademie
Abh Göttingen Ges Wiss	Abhandlung der Göttingen Gesellschaft der Wissenschaften
Abh Preuss Akad Wiss (Berlin)	Königlich Preussische Akademie der Wissenschaften zu Berlin. Abhandlungen
Acad Sci	Academy of Science
Acta Chir Scand	Acta Chirurgica Scandinavica
Acta Cryst	Acta Crystallographica
Acta Math	Acta Mathematica
Acta Metall	Acta Metallurgica
Acta Phys Pol	Acta Physica Polonica
Acta Radiol	Acta Radiologica
Adv Astron Astrophys	Advances in Astronomy and Astrophysics
Adv Catalysis	Advances in Catalysis
Adv Colloid Interface Sci	Advances in Colloid and Interface Science
Adv Electron Electron Phys	Advances in Electronics and Electron Physics
Adv Mater	Advanced Materials
Adv Opt Electron Microsc	Advances in Optical and Electron Microscopy
Adv Phys	Advances in Physics
Adv Struct Res Diffraction Methods	Advances in Structure Research by Diffraction Methods
Adv Virus Res	Advances in Virus Research
Am Ceram Soc Bull	American Ceramic Society Bulletin
Am J Pathol	American Journal of Pathology
Am J Roentgenol	American Journal of Roentgenology

Am J Roentgenol Radiat Ther	American Journal of Roentgenology and Radiation Therapy
Am J Sci	American Journal of Science
Am Math Monthly	American Mathematical Monthly
Angew Chem	Angewandte Chemie
Angew Chem Int Ed Engl	Angewandte Chemie—International edition in English
Ann Chim Phys	Annales de Chimie et Physique
Ann der Phys	Annalen der Physik
Ann NY Acad Sci	Annals of the New York Academy of Sciences
Ann Phil	Annalen der Philosophie
Ann Phys	Annales de Physique
Ann Phys Lpz	Annales de Physique Leipzig
Ann Rev Astron Astrophys	Annual Review of Astronomy and Astrophysics
Ann Rev Biochem	Annual Review of Biochemistry
Ann Rev Nucl Sci	Annual Review of Nuclear Science
Ann Rev Phys Chem	Annual Review of Physical Chemistry
Ann Soc Scient Brux	Annales de la Société Scientifique de Bruxelles
Appl Opt	Applied Optics
Appl Phys Lett	Applied Physics Letters
Arch Elektrotech	Archiv für Elektroteknik
Arch Hist Exact Sci	Archives for the History of Exact Science
Arch Kemi	Arkiv för Kemi
Arch Sci Phys Génève	Archives des Sciences Physique Génève
Arch Sci Phys Nat	Archives des Sciences Physique et Naturelles
Ark Fyz	Arkiv för Matematik, Astronomie och Fysik (Stockholm)
Astron Astrophys	Astronomy and Astrophysics
Astron J	Astronomical Journal
Astron Mitt Königl Sternwarte Göttingen	Astronomische Mitteilungen Königlich Sternwarte Göttingen
Astron Nach	Astronomische Nachrichten
Astron Tsirk	Astronomisk Tidsskrift
Astron Zh	Astronomicheskii Zhurnal
Astronaut Aeronaut	Astronautics and Aeronautics
Astrophys J	Astrophysical Journal
Astrophys J Lett	Astrophysical Journal Letters
Astrophys J Suppl	Astrophysical Journal. Supplement
Astrophys Lett	Astrophysics Letters
Astrophys Space Sci	Astrophysics and Space Science

Atti Congr Int Fis	Atti Congresso Internazionale dei Fisici
Aust J Phys	Australian Journal of Physics
Beitr Phys Atmos	Beiträge zur Physik der Atmosphäre
Bell Syst Tech J	Bell Systems Technical Journal
Ber Deutsch Chem Ges	Berichte der Deutschen Chemischen Geselleschafte
Ber Preuss Akad Wiss	Preussische Akademie der Wissenschaften (Berlin)
Biogr Mem Fell R Soc	Biographical Memoirs of Fellows of the Royal Society
Biophys J	Biophysical Journal
Br J Hist Sci	British Journal of the History of Science
Br J Opthal	British Journal of Opthalmology
Br J Phil Sci	British Journal of the Philosophy of Science
Br J Radiol	British Journal of Radiology
Bull Am Math Soc	Bulletin of the American Mathematical Society
Bull Am Phys Soc	Bulletin of the American Physical Society
Bull Astron Inst Neth	Bulletin of the Astronomical Institute of the Netherlands
Bull Mater Sci (Bangalore)	Bulletin of Material Society (Bangalore)
Bull Natl Acad Sci	Bulletin of the National Academy of Science
Bull Soc Math France	Bulletin de la Société Mathematique de France
Bur Stand Bull	Bureau of Standards Bulletin
Bur Stand J Res	Bureau of Standards Journal of Research
C R Acad Sci Paris	Comptes Rendus de l'Academie des Sciences (Paris)
C R Acad Sci USSR	Comptes Rendus de l'Academie des Sciences USSR
C R Hebd Séanc Acad Sci	Comptes Rendus Hebdomodaires des Séances de l'Academie des Sciences (Paris)
C R Séanc Soc Biol	Comptes Rendus des Séances de la Société de Biologie
Can J Chem	Canadian Journal of Chemistry
Can J Phys	Canadian Journal of Physics
Chem Britain	Chemistry in Britain
Chem Rev	Chemical Review
Clin Chem Acta	Clinica Chemica Acta
Clin Prac	Clinical Practice

Cold Spring Harbor Symp Quant Biol	Cold Spring Harbor Symposium on Quantum Biology
Comments Plasma Phys Controlled Fusion	Comments on Plasma Physics and Controlled Fusion
Commun Lunar Planetary Lab	Communications of the Lunar Planetary Laboratory
Commun Math Phys	Communications in Mathematical Physics
Contemp Phys	Contemporary Physics
Czech J Phys	Czechoslovak Journal of Physics
Dansk Vid Selsk Mat-Fys	Danske Videnskaberne Selskab Mathematisk-Fysiske Meddelanden
Deep Sea Res	Deep Sea Research
Denkschr Münchener Akad Wiss	Bayerische Akademie der Wissenschaften (München). Denkschrift
Doc NRPB	Documents of the National Radiological Protection Board
Dokl Akad Nauk	Doklady Akademii Nauk SSSR
Edinburgh Phil J	Edinburgh Philosophical Journal
Electrical World Eng	Electrical World Engineering
Electron Eng	Electronic Engineering
Electron Lett	Electronics Letters
Elektrochem Z	Elektrochemische Zeitschrift
Encycl Math Wiss	Encyklopädie der Mathematischen Wissenschaften
Ergebn Exakt Naturw	Ergebnisse der Exacten Naturwissenschaften
Ergebn Tech Rontgenk	Ergebnisse der Technischen Röntgentkunde
Eur Rev	European Review
Europhys Lett	Europhysics Letters
Fiz Zh	Fiziologicheskii Zhurnal
Forsch auf d Geb d Ing Wiss	Forschungen auf gem Gebeit der Ingenieurwissenschaften
Fys Tidsskr	Fysisk Tidsskrift
Geol Mag	Geological Magazine
Gilberts Ann	Annalen der Physik (Series 1. Gilberts Annalen)
Handbuch Phys	Handbuch der Physik
Helv Phys Acta	Helvetica Physica Acta
Hist Stud Phys Sci	Historical Studies in the Physical Sciences

IBM J Res Develop	IBM Journal of Research and Development
IEEE J Quant Electron	Institute of Electrical and Electronic Engineers: Journal of Quantum Electronics
IEEE Trans Electron Dev	Institute of Electrical and Electronic Engineers: Transactions on Electron Devices
IEEE Trans Ind Appl	Institute of Electrical and Electronic Engineers: Transactions on Industrial Applications
IEEE Trans Instrum Meas	Institute of Electrical and Electronic Engineers: Transactions on Instrumentation and Measurement
Ind Eng Chem	Industrial and Engineering Chemistry
Indian J Phys	Indian Journal of Physics
IRE Trans	Transactions of the Institute of Radio Engineers
J Acoust Soc Am	Journal of the Acoustic Society of America
J Aero Sci	Journal of Aeronautical Science
J Am Ceram Soc	Journal of the American Ceramic Society
J Am Chem Soc	Journal of the American Chemical Society
J Am Cryst Assoc	Journal of the American Crystallography Association
J Am Math Soc	Journal of the American Mathematical Society
J Appl Cryst	Journal of Applied Crystallography
J Appl Phys	Journal of Applied Physics
J Appl Spectrosc USSR	Journal of Applied Spectroscopy USSR
J Atmos Terr Phys	Journal of Atmospheric and Terrestrial Physics
J Biomed Eng	Journal of Biomedical Engineering
J Biophys Biochem Cytol	Journal de Biophysique et de Biomecanique
J Br Astron Assoc	Journal of the British Astronomical Association
J Chem Inf Computer Sci	Journal of Chemical Information and Computer Sciences
J Chem Phys	Journal of Chemical Physics
J Chem Soc	Journal of the Chemical Society
J de l'Ecole Polytech	Journal de l'Ecole Polytechnique
J Exp Med	Journal of Experimental Medicine
J Fluid Mech	Journal of Fluid Mechanics
J Franklin Inst	Journal of the Franklin Institute

J Geol Soc Lond	Journal of the Geological Society of London
J Geophys Res	Journal of Geophysical Research—Space
J Inst Elec Eng	Journal of Instrumentation and Electrical Engineering
J Inst Metals	Journal of the Institute of Metals
J Iron Steel Inst	Journal of the Iron and Steel Institute
J Lab Clin Med	Journal of Laboratory and Clinical Medicine
J Lightwave Technol	Journal of Lightwave Technology
J Low Temp Phys	Journal of Low Temperature Physics
J Magn Magn Mater	Journal of Magnetism and Magnetic Materials
J Mater Res	Journal of Material Research
J Math Phys	Journal of Mathematical Physics
J Meteorol	Journal of Meteorology
J Meteorol (Japan)	Journal of Meteorology (Japan)
J Microscop	Journal of Microscopy
J Mol Biol	Journal of Molecular Biology
J Neurosurg	Journal of Neurosurgery
J Non-cryst Solids	Journal of Non-crystaline Solids
J Nucl Med	Journal of Nuclear Medicine
J Opt Soc Am	Journal of the Optical Society of America
J Phys A: Math Nucl Gen	Journal of Physics A: Mathematical, Nuclear and General
J Phys B: At Mol Phys	Journal of Physics B: Atomic and Molecular Physics
J Phys C: Solid State Phys	Journal of Physics C: Solid State Physics
J Phys Chem	Journal of Physical Chemistry
J Phys Chem Solids	Journal of Physics and Chemistry of Solids
J Phys E: Sci Instrum	Journal of Physics E: Scientific Instruments
J Phys F: Met Phys	Journal of Physics F: Metal Physics
J Phys Oceanogr	Journal of Physical Oceanography
J Phys (Paris)	Journal de Physique Théorique et Appliquée (Paris)
J Phys Radium	Journal de physique et le radium
J Phys Soc Japan	Journal of the Physical Society of Japan
J Phys Soc Japan (Suppl)	Journal of the Physical Society of Japan (Supplement)
J Phys USSR	Journal of Physics (USSR)
J Physique	Journal de Physique
J Quant Electron	Institute of Electrical and Electronic Engineers: Journal of Quantum Electronics

J R Aeronaut Soc	Journal of the Royal Aeronautical Society
J R Astron Soc Can	Journal of Royal Astronomical Society of Canada
J R Stat Soc	Journal of the Royal Statistical Society
J Res NBS	Journal of Research of the National Bureau of Standards
J Sci Instrum	Journal of Scientific Instruments
J Sound Vib	Journal of Sound and Vibration
J Stat Phys	Journal of Statistical Physics
Jahrb d dtsch Luftfahrtf	Jahrbuch der deutschen Luftfahrtsforschung
Jahrb Drahtlosen Telegr Teleph	Jahrbuch der drahtlosen Telegraphie und Telephonie
Jahrb Rad Elektr	Jahrbuch der Radioaktivität und Elecktronik
JAMA	Journal of the American Medical Association
Japan J Appl Phys	Japanese Journal of Applied Physics
JETP	Journal of Experimental and Theoretical Physics—English translation of Zhurnal Eksperimental'noi i Teoretiskoi Fiziki
JETP Lett	Journal of Experimental and Theoretical Physics: Letters
K Preuss Akad Wiss (Berlin) Sitzungsber	Königlich Preussische Akademie der Wissenschaften zu Berlin Sitzungsberichte
K Preuss Akad Wiss (München) Sitzungsber	Königlich Bayerische Akademie der Wissenschaften (München). Sitzungberichte
K Svenska Vetensk Akad Handl	Künglia Svenska Vetenskakademiens Handlingar
Kgl Danske Vid Selsk Skrifter,	Köngelige Danske Vidsenskabernes Selskab Skrifter
Kgl Ges Wiss Gott	Königlich Gesellschaft der Wissenschaften zu Göttingen
Kon Neder Akad Wet Amsterdam Versl Gewone Vergad Wisen Natuurkd Afd	Verlag van de gewone Vergadering der wis-en natuurkindige Afdeeling, Koniklijke Akademie van Wetenschappen te Amsterdam
Kong Dansk Vid Selsk Matt-fys Medd	Det Köngelige Danske Vidsenskabernes Selskab Matematisk-Fysiske Meddelanden

Koninkl Nederland Akad Wetenschap	Koninklijke Nederlandse Akademie van Wetenschappen te Amsterdam
Liebigs Ann	Justus Liebigs Annalen der Chemie
Lighting Res Technol	Lighting Research and Technology
Macromol Rev	Macromolecular Review
Mat Res Soc Bull	Material Research Society Bulletin
Math Ann	Mathematische Annalen
Math Intell	Mathematical Intelligencer
Meas Sci Technol	Measurement Science and Technology
Med Phys	Medical Physics
Mém Acad R Sci (Paris)	Academie Royale des Sciences (Paris). Mémoires
Mem R Astron Soc	Memoirs of the Royal Astronomical Society
Meteorol Zeitschr	Meteorlogische Zeitschrift
Mitt Phys Ges Zürich	Mitteilungen der Physikalischen Gesellschaft zu Zürich
Mol Cryst Liq Cryst	Molecular Crystals and Liquid Crystals
Mon Not R Astron Soc	Monthly Notices of the Royal Astronomical Society
Mon Weath Rev	Monthly Weather Review
Mon-Ber Akad Wiss Berlin	Königlich Preussische Akademie der Wissenschaften zu Berlin Monatsberichte
MRS Bull	Material Research Society Bulletin
Nach Ges Wiss	Königlich Gesellschaft (Societät) der Wissenschaften zu Göttingen Nachrichten
Nach Ges Wiss Göttingen Math-Phys	Akademie der Wissenschaften zu Göttingen Nachrichten, Mathematisch-physikalische Klasse
Nach Gott Ges	Königlich Gesellschaft (Societät) der Wissenschaften zu Göttingen Nachrichten
Nach Königl Preuss Akad Wiss, Göttingen, Math-Phys Klasse 1	Akademie der Wissenschaften zu Göttingen Nachrichten, Mathematisch-physikalische Klasse 1
Nat Adv Comm Aeronaut Memorandum	National Advisory Commission on Aeronautics. Memorandum
Naturwissenschaften	Die Naturwissenschaften
New England J Med	New England Journal of Medicine
New Sci	New Scientist
Nucl Instrum Methods	Nuclear Instruments and Methods

Nucl Instrum Methods Phys Res	Nuclear Instruments and Methods in Physics Research Section B: Beam interactions with materials and atoms
Nucl Medizin	Nuclear Medizin
Nucl Phys	Nuclear Physics
Nuovo Cim	Il Nuovo Cimento
Nuovo Cim (Suppl)	Nuovo Cimento (Supplement)
Obituary Notices (London)	Obituary Notices of Fellows of the Royal Society (London)
Opt Acta	Optica Acta
Opt Commun	Optics Communications
Opt Eng	Optical Engineering
Opt Lett	Optics Letters
Opt Photonics News	Optics and Photonics News
Opt Spectra	Optics Spectra
Opto-Laser Eur	Opto-Laser Europe
Phil Mag	Philosophical Magazine
Phil Trans R Soc	Philosophical Transactions of the Royal Society of London
Phil Trans R Soc A	Philosophical Transactions of the Royal Society of London Series A—Mathematical and Physical Sciences
Philips Res Rep	Philips Research Reports
Phys Assoc Bull	Physics Association Bulletin
Phys Blätter	Physikalische Blätter
Phys Bull	Physics Bulletin
Phys Fluids	Physics of Fluids
Phys Lett	Physics Letters
Phys Med Biol	Physics in Medicine and Biology
Phys Rep	Physics Reports: Review Section of Physics Letters
Phys Rev	Physical Review
Phys Rev Lett	Physical Review Letters
Phys Scr	Physica Scripta
Phys Today	Physics Today
Phys Z	Physikalische Zeitschrift
Phys Z Sowjet	Physikalische Zeitschrift der Sowjetunion
Pisma Astron Zh	Pis'ma v Astronomicheskii Zhurnal
Pisma Zh Eksp Teor Fiz	Pis'ma v Zhurnal Eksperimental'noi I Teoreticheskoi Fiziki
Pogg Ann	Annalen der Physik (Series 2. Poggendorff's Annalen der Physik und Chemie)

Popular Astron	Popular Astronomy
Proc Akad Sci Amst	Koninklijke Akademie von Wetenschappen te Amsterdam Proceedings
Proc Am Acad Sci	Proceedings of the American Academy of Science
Proc Am Phil Soc	Proceedings of the American Philosophical Society
Proc Astron Soc Pacif	Proceedings of the Astronomical Society of the Pacific
Proc Camb Phil Soc	Proceedings of the Cambridge Philosophical Society
Proc Camb Phil Soc Suppl	Proceedings of the Cambridge Philosophical Society. Supplement
Proc IEEE	Proceedings of the Institute of Electrical and Electronic Engineers
Proc Indian Acad Sci	Proceedings of the Indian Academy of Sciences
Proc Inst Radio Eng	Proceedings of the Institute of Radio Engineers
Proc Ir Acad	Proceedings of the Royal Irish Academy
Proc IRE	Proceedings of the Institute of Radio Engineers
Proc Kon Akad Wetenschap Amsterdam	Koninklijke Akademie von Wetenschappen te Amsterdam Proceedings
Proc Lond Math Soc	Proceedings of the London Mathematics Society
Proc Lond Math Soc Series 2	Proccedings of the London Mathematics Society Series 2
Proc Natl Acad Sci	Proceedings of the National Academy of Sciences
Proc Phys Soc	Proceedings of the Physics Society
Proc Phys-Math Soc Japan	Proceedings of the Physical Mathematics Society of Japan
Proc R Dublin Soc	Proceedings of the Royal Dublin Society
Proc R Inst	Proceedings of the Royal Institute
Proc R Irish Acad	Proceedings of the Royal Irish Academy
Proc R Microsc Soc	Royal Microscopical Society. Proceedings
Proc R Soc	Proceedings of the Royal Society
Proc Virginia Acad Sci	Proceedings of the Virginia Academy of Science
Prog Opt	Progress in Optics
Prog Theor Phys	Progress in Theoretical Physics
Prog Theor Phys Japan	Progress of Theoretical Physics Japan

Prussian Acad Wiss	Königlich Preussische Akademie der Wissenschaften zu Berlin. Abhandlungen
Publ Am Astron Soc	Publications of the American Astronomical Society
Publ Astron Soc Pacif	Publications of the Astronomical Society of the Pacific
Publ Astrophys Obs Potsdam	Publikationen des Astrophysikalischen Observatoriums zu Potsdam
Publ Lick Obs	Publications of the Lick Observatory
Pure Appl Math	Pure and Applied Mathematics
Pure Appl Opt	Pure and Applied Optics
Q J Math	Quarterly Journal of Mathematics
Q J Mech Appl Math	Quarterly Journal of Mechanics and Applied Mathematics
Q J R Astron Soc	Quarterly Journal of the Royal Astronomical Society
Q J R Meteorol Soc	Quarterly Journal of the Royal Meterological Society
Quant Opt	Quantum Optics
Rend Acc Lincei	Accademia Nazionale dei Lincei. Atti Rendiconti Lincei
Rep Prog Phys	Reports on Progress in Physics
Rev Gen Sci Pures Appl	Revue générale de sciences pures et appliquées
Rev Mod Phys	Reviews of Modern Physics
Rev Opt	Reviews of Optometry
Rev Sci Instrum	Review of Scientific Instruments
Rev Scientifique	Revue Scientifique
Ric Scient	Ricerca Scientifica
S B Preuss Akad Wiss	Königlich Preussiche Akademie der Wissenschaften zu Berlin Sitzungsberichte
Scand J Metall	Scandinavian Journal of Metallurgy
Sci Am	Scientific American
Sci Rep Tohoku Univ	Science Reports of the Research Institute Tohoku University
Seminars Nucl Med	Seminars in Nuclear Medicine
SIAM Rev	Society for Industrial and Applied Mathematics Review
Soc Sci Fenn Commentat Phys-Math	Societatis Scientiarum Fennica Commentari (Finnish Academy)

Solid State Chem	Solid State Chemistry
Solid State Commun	Solid State Communications
Solid State Phys	Solid State Physics
Sov J Nucl Phys	Soviet Journal of Nuclear Physics—USSR and Yaderna
Sov J Plasma Phys	Soviet Journal of Plasma Physics
Sov Opt Spectrosc	Optics and Spectroscopy—English translation of Optika i Spektroskopiya
Sov Phys Semiconductor	Fizika i Tekhnika Poluprovodnikov
Sov Phys Solid State	Physics of the Solid State—English translation of Fizika Tverdogo Tela
Sov Phys-JETP	Soviet Physics, JETP
Sov Phys-Usp	Soviet Physics, Uspekhi
Stössen Z Phys	Zeitung für Physik
Stud Hist Phil Sci	Studies in History and Philosophy of Science
Tech Mech u Thermod	Technologie der Mechaniks und Thermodynamics
Terr Mag Atmos Elec	Terrestrial Magnetism and Atmospheric Electronics
Topics Current Chem	Topics in Current Chemistry
Trans Am Electrochem Soc	Transactions of the American Electrochemical Society
Trans Am Inst Mining Metall Eng	Transactions of the American Institute of Mining, Metallurgical and Petroleum Engineers
Trans Am Soc Civ Eng	Transactions of the American Society of Civil Engineers
Trans Connecticut Acad Arts Sci	Transactions of the Connecticut Academy of Arts and Sciences
Trans Farad Soc	Transactions of the Faraday Society
Trans Opt Inst Petrograd	Transactions of the Optical Institute of Petrograd
Trans Opthal Soc UK	Transactions of the Opthalmological Society. UK
Trans R Soc	Transactions of the Royal Society
Trans R Soc Canada	Transactions of the Royal Society of Canada
Trans Soc Nav Arch Mar Eng	Transactions of the Society of Naval Architects and Marine Engineers
Trends Biochem Tech	Trends in Biochemical Technology
Ultrasound Med Biol	Ultrasound in Medicine and Biology

Usp Fiz Nauk	Uspekhi Fizika Nauk (Russian edition of Sov Phys Usp)
Verh Deutsch Phys Ges	Verhandlungen der Deutschen Physikalische Gesellschaft
Verh Ges Deustch Naturf Ärzte	Verhandlungen der Gesellschaft deutsche Naturforscher und Ärte
Verh Naturf Ges Basel	Verhandlungen der Naturforscher Gesellschaft zu Basel
Versl Kon Akad Wetensch Amsterdam	Verlag van de gewone Vergadering der wis-en natuurkindige Afdeeling, Koniklijke Akademie van Wetenschappen te Amsterdam
Vet Akad Arkiv Mat Atr och Fysik	Arkiv für Matematik, Astronomie och Fysik (Stockholm)
Viert Astron Ges	Vierteljahrbschrift der Astronomischen Gesellschaft
Vjschr Naturf Ges Zurich	Vierteljahrschrift der Naturforschenden Geselschaft zu Zürich
Vopr Teor Plazmy	Voprosy Teorii Plazmy
Wiedemann's Ann	Annalen der Physik (Series 3. Wiedemann's Annalen der Physik und Chemie)
Wien Ber	Kaiserliche Akademie der Wissenschaften (Wien) Sitzungsberichte Abteilung
Wien Med Wochschr	Wiener Medizinische Wochenschrift
Wiener Chem Z	Wiener Chemische Zeitschrift
Yad Fiz	Yadernaya Fizika
Z Angew Math Mech	Zeitschrift für Angewandte Mathematik und Mechanik
Z Astrophys	Zeitschrift für Astrophysik
Z Elektrochem	Zeitschrift für Elektrochemie
Z Instrumentkd	Zeitschrift für Instrumentkunde
Z Krist	Zeitschrift für Kristallographie
Z Mech	Zeitschrift für Mechanik
Z Phys	Zeitschrift für Physik
Z Phys Chem	Zeitschrift für Physikalische Chemie
Z Wiss Mikrosk	Zeitschrift für Wissenschaftliche Mikroskopie
Z Wiss Photogr	Zeitschrift für Wissenschaftliche Photographie
Zh Eksp Teor Fiz	Zhurnal Eksperimental'noi i Teoretiskoi Fiziki

SUBJECT INDEX

Note—volume and page numbers in **bold type** refer to items in boxes

A0620-00 x-ray binary source, III.1757
Abbe optical theory, III.1578–84
Abbe primary image, III.1583
Abbe secondary image, III.1583
Abelian gauge theory, II.693, II.714
Aberdeen MK1 NMR Imager, III.1926
Aberdeen Section Scanner, III.1918
Aberration, I.258, I.266, I.269
 see also Lens
Ability of the 200" telescope to discriminate between selected world models, The, III.1793
Abrikosov vortex lattice, II.935
Absorption, III.1404
Absorption coefficient, I.364, I.368, II.1052
Absorption spectra, I.24
Academia Europaea, III.2055
Académie des Sciences, I.6, II.1233
Accelerators, II.657, II.687–8, II.725–44, II.1201–6, II.1225, III.1883, III.1997, III.2036
 colliding beams, II.738–42
 fixed-field alternating-gradient, II.739
 strong focusing, II.733–4
Accretion disc, III.1756, III.1779
Accretion radius, III.1756
Acoustic energy, II.873

Acoustic modes, II.1004
Acoustic paramagnetic resonance (APR), II.1146–7
Acoustic resonance, III.1357–8
Acta Crystallographica, I.511
Acta Metallurgica, III.1556
Actinides, I.97, II.955
 magnetism, II.1166
Actinium, II.1227
Activation energy, III.1531
Active galactic nuclei, III.1777
 and general relativity, III.1777–81
 non-thermal phenomena in, III.1781–4
Adenine, I.504
Adiabatic demagnetization, II.1125–6
Adiabatic model, III.1802
Adiabatic principle, I.92–3
Adler–Weisberger relation, II.679
Administration of Radioactive Substances Advisory Committee, III.1910
AEG (Allgemeine Elektrizitäts Gesellschaft) Research Institute, III.1577–8
Aerobee (rocket), III.1682, III.2005
Aerodynamics, II.821
 heating phenomenon, II.874
 lift theory, II.817

Aeronautics, II.869–87
Aeronomy, III.1982
Aether-drift, I.26
Age-hardening, III.1546, III.1547
AIDS, III.2029
Aircraft, fixed-wing, II.817
Aircraft wings, II.821, II.824
Airy diffraction limit, III.1467
Alcator, III.1649
ALICE neutral-beam injection experiments, III.1662
Alkali metals, III.1345
Allende meteorite, III.1963
Allgemeine Elektrizitäts Gesellschaft Research Institute, see AEG
Allis-Chalmers Company, III.1871
Along-crest energy propagation, II.863–9
α-decay, I.115–16, I.366–7, II.1189
α-particle energies, II.1201
α-particle model, II.1190
α-particles, I.58, I.59, I.104, I.110, I.112, I.113, I.116, I.119, I.122, I.123, I.367, II.638, II.1189, II.1226, III.1706
α-radioactivity, I.52
α-rays, I.37, I.54, I.58, I.104
β-alumina, III.1539
Aluminium alloys, III.1533
Aluminium–copper alloys, III.1533, III.1546
Aluminium–manganese alloy, I.481–3
Alzheimer's disease, III.1922
American Association of Physicists in Medicine, III.1864
American Crystallographic Association, I.512
American Philosophical Society, I.3
American Physical Society, III.1558, III.2026
American Science and Engineering Group (AS&E), III.1739
American Society for X-ray and Electron Diffraction, I.511
Americium, II.1225
Amino acids, I.496, I.497
Amorphous cement, III.1518
Amorphous magnetism, II.1170–1
Amorphous materials, III.1541
Amorphous metal hypothesis, III.1518
Amorphous solids, x-ray diffraction, **I.478**
Amorphous structures, I.477–9
Ampere (unit), II.1235, II.1254, II.1257
  absolute determination, II.1255
Amplification without inversion (AWI), III.1473
Anaesthetic equipment, III.1897
Anderson–Brinkman–Morel (ABM) state, II.953
Andromeda Nebula, III.1717, III.1718, III.1788
Angiography, III.1930
Anglo-Australian Telescope, III.1744
Angstrom (unit), II.1242
Angular momentum, I.81, I.162, I.359, II.818, II.1124, III.1410, III.1756, III.1779, III.1962, III.1964
Angular momentum algebra, **II.1038**
Anharmonic effects, II.1002–4
Anharmonic oscillators, III.1434
Anisotropic superfluid, II.950–1
Annalen der Physik und Chemie, I.7
Annales de chimie et de physique, I.7
Anomalous skin effect, III.1346, III.1347, III.1351, III.1357
Anthracene, crystal structure, I.494
Anthropic Principle, III.2039–40
Antiferroelectric structure, I.487
Antiferromagnetism, I.559, II.1016, II.1113, II.1129–32, II.1147–8
  exchange interactions, II.1143

resonance, II.1151
structures, II.1011
Antimony, III.1332, III.1335, III.1544
Antiparticles, II.638, II.655
Antiprotons, II.655
Antiscreening. *See* Screening and antiscreening
Anyon mechanism, II.962
Applications of Phase Transitions in Material Science, III.1557
Applied optics, III.1474–81
Applied physics, III.2050
Arginine, structure of, I.449–50
Argon, I.75
Argus experiment, III.1683
Army Operational Research Group, III.1736
Arnowitt–Deser–Misner (ADM) mass, I.301
Arsenic-72 and -74, III.1904
Arteries, artificial, III.1898
Arteriosclerosis, III.1898
Artificial limbs, III.1899
ASDEX tokamak, *see* Tokamak
Associated production, II.657
Association of Universities for Research in Astronomy (AURA), III.1744
Asthenosphere, III.1978
Astigmatism, III.1580
ASTRON, III.1645
Astronomical photography, III.1692
Astronomical spectroscopy, III.1691
Astronomical timescales, II.1227
Astronomical Unit (AU), II.1242
Astronomy, I.24, II.757–62, III.1453–4, III.2046
  changing perspective, III.1734–45
  general theory of relativity, I.315–17
  optical, III.1744–5

Astrophysical Journal, III.1762
Astrophysics, I.630, II.757–62, III.1618, III.1691–821, III.2031
  birth of, III.1691
  general theory of relativity, I.315–17, III.1776
  high-energy, III.1774–84
  impact of quantum mechanics, III.1706–8
  since 1945, III.1746–807
  statistical mechanics in, III.1711
Asymmetric unit, I.423
Asymptotic freedom, II.714
Atmosphere, II.899
Atmospheric physics, III.1983
Atmospheric science, III.1983
Atomic bomb, II.1197, III.1630, III.1891, III.1980
Atomic clock, II.1244, II.1247
Atomic dynamics, discretization in, I.184–5
Atomic energy, I.63, I.113, I.132–3, I.285, III.1628
Atomic Energy Research Establishment, III.1901
Atomic model, I.157–8
Atomic number, I.65, I.158–60, I.407, I.429, II.1057, II.1222, II.1223
Atomic physics, I.31, I.358–62, II.755–6, II.1025–49
Atomic pile, *see* Nuclear reactors
Atomic processes, current overviews, II.1101–6
Atomic scale, II.1026–7
Atomic scattering factor, I.431, I.432, I.439, I.453
Atomic spectra, II.1032–9
Atomic standards of measurements, II.1095
Atomic structure, I.63–5, I.156–62, I.211–15, II.1026–7
Atomic systems, optical actions, II.1097–101
Atomic theory, I.43, I.46, I.227

causality in, I.220–3
failure of, I.181–3
principles of, I.176–8
probability in, I.220–3
schools of, I.178–81
Atomic weight, I.65
Atoms, I.43–103
role in metrology and instrumentation, II.1094–7
Atoms for Peace Conference (1955), III.1642
Atoms for Peace Conference (1958), III.1655
AT&T, III.2022
Au$_2$Mn, magnetic moments in, I.462
Aufbauprinzip (building-up principle), I.96–7, I.366
Auger emissions, inner-shell phenomena, II.1090–1
Augmented-plane-wave (APW) method, III.1317
Aurorae, III.2006–7
Auroral phenomena, III.2003
Auswahlprinzip, selection rule, I.178–81
Autocorrelation functions, III.1418
Auto-ionization, II.1061, II.1090, II.1102
Automatic Plate Measuring Machine (APM), III.1745
Avalanche photodiode (APD), III.1419, III.1476
Avogadro's constant, I.150
Avogadro's law, I.46
Avogadro's number, I.43, I.49, I.51, II.1237, II.1269
Azbel'–Kaner cyclotron resonance, III.1357

Balian–Werthamer (BW) state, II.952
Balloon flight, I.368
Balmer formula, I.24, I.78, I.80–5, I.88

Balmer series, I.160
Band gaps, III.1367
Band structure, II.1047, II.1166, III.1314–17, III.1344, III.1530
Band-structure diagram, III.1417
Band theory, II.1165–7, III.1367
Bar-code scanners, III.1480–1
Bardeen–Pines interaction, II.938
Barium, contrast media, III.1863
Barium titanate, III.1538
Baroclinic instability, II.902, II.907
Baroclinic mode, II.896
Baryogenesis, II.761–2
Baryon, II.660, II.661, II.675, II.685, III.1800
Baryon–antibaryon asymmetry, II.760, III.1806
Baryon number, II.658
Basic Orthopaedic BioMechanics, III.1896
BaTiO$_3$, ferroelectric phase transition, I.486
Batteries, I.27
lithium–iodine, III.1895
zinc–mercury, III.1895
BBKGY hierarchy, I.605, I.623
BCS theory, I.489, II.935, II.937, II.937–41, II.961–5
Beevers–Lipson strips, I.446
Beiblatter zu den Annalen der Physik, I.7
Bell inequalities, III.1389
Bell Telephone Laboratories, III.1790, III.1944, III.2022, III.2026
Berkelium, II.1225
Bernoulli's equation, II.822, II.885
Beryl, structure of, I.440
Beryllium, I.119, I.120, III.1360
β-decay, I.106, I.366–7, I.370–2, I.375, I.398–402, II.639, II.664, II.665, II.667, III.1706, III.1707
Fermi's theory, I.374–5
limits on neutrino masses, II.755

Subject Index

β-effect, II.902
β-electrons, I.104
β-particles, I.107, I.109
β-radioactivity, I.52
β-ray spectroscopy, I.103–9
β-rays, I.37, I.54, I.58, I.59, I.104–9
β-spectra, I.106, I.109, I.118–19, I.126–9
β-structures, I.498
Betatron, II.732, II.1204–5, III.1871
B-factories, II.767–8
Big Bang, III.1784–6, III.1790–3, III.2040
'Big Physics', III.2054–5
'Big Science', III.2023–7, III.2054
Billions of electronvolts (GeV), I.358
Binary alloys, critical behaviour of, I.548–9
Binary collisions, I.611, I.612
Binding energy, I.110–12
Biochemical assay, III.1909
Biomedical technology, III.2031
Biomolecular crystals, I.450–3
Biomolecular structures, I.445, I.496–512
Biomolecules, II.1084
Biophysics, III.2030, III.2046
Biot–Savart law, II.818
Bipolar outflow, III.1769
Birge–Bond diagram, II.1268
Bismuth, III.1282–3, III.1339
Bismuth germanate, III.1921
Bistable optical switching devices (SEEDs), III.1469
Bjorken–Glashow hypothesis, II.700
Black body, I.66
Black-body law, I.173, I.177
Black-body radiation, I.22–4, I.30, I.69, I.144, I.148–50, I.174, I.524- 5, II.758, II.972, II.973, II.982, III.1387, III.1393, III.1402

Black holes, I.306–12, II.760–1, III.1754–7, III.1778–80
  no-hair theorem for, III.1778
BL-Lacertae (or BL-Lac) objects, III.1777, III.1784
Bloch equation, II.1174
Bloch functions, II.1166
Bloch theorem, III.1291–301
Bloch theory, III.1342
Bloch wave, **III.1294–5**, III.1320
Bloch wavelength, III.1301
Bloch wavenumber, III.1294, III.1299
Bloch wave-vector, III.1294–6
Block copolymers, III.1611
Blood flow, III.1897–8, III.1914, III.1920
Blood pressure, monitoring, III.1893–4
B-meson, II.721, II.724–5, II.767–8
Bohm diffusion, III.1630, III.1631
Bohm diffusion formula, III.1632
Bohm diffusion law, III.1655
Bohm–Pines theory, III.1362, III.1363
Bohr–Kramers–Slater radiation theory, I.185, I.203
Bohr magneton, I.116
Bohr orbit, III.1319
Bohr radius, I.85, II.1026
Bohr relation, I.90
Bohr–Sommerfeld atomic model, I.175–6
Bohr–Sommerfeld rules, I.92
Bohr–Sommerfeld theory, I.161
Bohr theory, II.1124
Bolometer, I.23
Boltzmann collision number, I.603
Boltzmann combinatorial factor, I.527
Boltzmann constant, I.67, I.150, I.523, I.589, II.1270, III.1285, III.1387, III.1397
Boltzmann distribution, I.542
Boltzmann equation, I.523, I.533,

S5

I.588, I.591, I.593–5, I.597, I.602, I.605, I.608–11, I.617, I.624–6, III.1303, III.1344, III.1701, III.1992
Boltzmann–Grad limit, I.626
Boltzmann law, III.1287
Bond percolation, I.572
Bone fractures, III.1897
Bootstrap programme, II.681
Born–Oppenheimer approximation, II.1043–6
Bose–Einstein condensation, I.530–1, I.537, II.927, II.930
Bose–Einstein distribution, III.1402
Bose–Einstein statistics, I.155, I.533, II.687, III.1387
Bose statistics, I.117, III.1394
Boson, I.382, II.693, II.694, II.699, II.1218–19
Bound scattering length, I.453
Boundary conditions, II.800–1, II.804
Boundary layer, II.804, II.807, II.812–46
Boundary-layer equations, II.804, II.806, II.843
Boundary-layer solution, II.802
Boundary-layer theory, II.797
Boundary-value problem, I.196
Boys' radiometer, I.28
Brachytherapy, III.1877, III.1890, III.1891
Bragg–Gray theory, III.1882
Bragg Law, I.428, I.431, I.469
Bragg peak, III.1887
Bragg reflection, II.990, III.1367
Brain imaging, III.1906, III.1922, III.1929, III.1932
Brain tumour, III.1871, III.1904, III.1905, III.1919
Bravais lattices, I.462, I.466, III.1513
Bravais model, II.984
Breit–Wigner formula, II.1194
Bremsstrahlung, III.1766

Brillouin frequency shift, III.1398
Brillouin scattering, III.1405
Brillouin zone, II.987, II.1011, III.1293, III.1295, III.1296, III.1300, III.1305, III.1311–13, III.1315, III.1328, III.1331, III.1348, III.1359
British Association for the Advancement of Science, I.5, I.52, II.1252, III.1705
British Broadcasting Company (BBC), III.1988
British Institute of Radiology, III.1861
British Journal of Radiology, III.1867
British Medical Bulletin, III.1901
British Radio Research Board, III.1987–8
Brittle fracture, III.1520
Broadcasting, III.1987, III.1994
Brownian motion, I.30, I.43, I.147, I.533, I.592, I.595–7, I.607–8, I.616, I.623, I.628, III.1441–2
Brussels school, I.619–22
BTH Company, III.1583
Bubble chamber, II.659, II.753
Bubble domains, II.1159–60
Building-up principle (*Aufbauprinzip*), I.96–7, I.366
Bulbous bow, II.885
Bulk-carriers, II.885
Bunsen burner, I.19, I.75
Buoyancy force, II.844
Bureau International de l'Heure (BIH), II.1247, II.1248
Bureau International des Poids et Mesures, I.20
Burgers vector, I.475–6
Busch's lens theory, III.1570
  verification of, III.1569–70

Cabibbo–Kobayashi–Maskawa (CKM) matrix, II.674, II.723–5
Caesium, I.75

Caesium-137, III.1892
Caesium chloride, I.429
Caesium clock, II.1245–8
Calcite, I.430, I.485
Calcium tungstate, III.1537, III.1901
Californium, II.1225, II.1227
Calorimetry, II.968, III.1883–4
Calutron isotope separation, III.1630
Cambridge Database System, I.512
Cambridge Electron Accelerator (CEA), II.741
Cambridge Structural Database, I.512
Cambridge University, I.9–15, I.17, I.20, I.21, I.54
Cancer, III.1857, III.1886, III.1917
Candela, II.1251
Canonical ensemble, I.526
Cantor set, III.1846, III.1847, III.1852
Capacitance, II.1255
Capacitors, II.1256
Carbon, formation of, III.1746
Carbon bond, II.1043
Carbon dioxide, atmospheric, II.908
Carbon monoxide in interstellar medium, III.1763–4
Carbon–nitrogen–oxygen cycle (CNO cycle), III.1707–8
Cardiac function, III.1898
Cardiac pacemaker, III.1895
Cardiovascular devices, III.1897–8
Carl Zeiss, III.1410
Carnot cycle, I.536–7, II.1249
Carnot principle, I.60
Cascade showers, I.387
Cassiopeia A, III.1737, III.1775, III.1782
Catalogue of Selected Compact Galaxies and of Post-Eruptive Galaxies, III.1713
Cathode ray, I.27, I.36, I.52, I.53

Cathode-ray oscilloscope, III.1989
Cathode-ray tube, III.1530, III.1568, III.1909
Cauchy condition, I.484
Cauchy integral, III.1404
Cauchy model, II.967
Caustics, III.1466
Cavendish Laboratory, Cambridge, I.17, I.20, I.21, I.54
Cavity radiation, I.144, I.145
CCDM, II.1240
Cellular automata, I.629–31
Cellulose fibres, III.1597
Centaurus X-3 (Cen X-3), III.1754
Centimetre–gram–second (CGS) system, II.1252
Central-axis depth-dose (CADD) curves, III.1872, III.1873
Centrosymmetric structure, I.448, I.449
Cepheids, III.1714, III.1716, III.1719
Ceramics, III.1534
  conductors, III.1537, III.1540
  crystalline glazes on, III.1537
  electrical, III.1536–42
  electro-optical, III.1536–42
  functional (or 'fine'), III.1538
Čerenkov counter, II.749–51, III.1446
Čerenkov detector, II.758, II.765
Čerenkov effect, III.1408
Čerenkov radiation, III.1363
CERN, III.2024, III.2036, III.2054, III.2055, III.2057
Cerro Tololo International Observatory (CTIO), III.1744
Cesium-137, III.1879
C-field, III.1787
CGS units, II.1235
Challenger research ship, III.1973
Challenger space disaster, III.1742
Chandrasekhar limit, III.1759
Chandrasekhar mass, III.1711

Channel classification, prototype of, II.1060–3
Channel coupling, empirical parametrization, II.1073
Chaos, I.593, III.1466–7, III.1823, III.1826, III.1831, III.1848
Chaotic motions, II.815
Chapman–Enskog method, I.597–8
Chapman–Kolmógoroff equation, I.596
Chapman–Kolmógoroff–Smoluchowski–Bachelier equation, I.604
Chapman layer model, III.1994
Characteristic impedance of free space, II.1261
Charge-coupled devices (CCD), II.1248, III.1454, III.1476, III.1586, III.1590, III.1745, III.1770, III.1796
Charged particles
  detection of, II.1200
  measuring devices and detectors, II.1206–8
Charm hypothesis, II.699–700, II.702–4
Charmed particles, II.700–1, II.704–6, **II.705**
Charmonium, II.701–2, II.704
Charnley low-friction arthroplasty, III.1896
CHART fractionation schedule, III.1888
Chemical constant, I.154
Chemical defect, III.1505
Chemical elements, I.45
  formation of, III.1787
  origin of, III.1746–7, III.1785
  synthesis of, III.1747, III.1784–5
Chemical potential, I.532
Chemical quanta, I.154–5
Chemical Rubber Company (now CRC Press), III.1513
Chemical spectroscopy, I.19

Chew–Goldberger–Low (CGL) equations, III.1637
Chew–Low theory, II.680
Chiral invariance, II.676
Chiral smectic C, III.1604, III.1605
Chiral symmetry, II.676
Cholesteryl iodide, I.444
CIE (Commission Internationale de l'Éclairage) chart, II.1251, III.1477
Cinematography, III.1410
Clarendon Laboratory, I.20
Classical statistics, I.525–8
Clausius–Mossotti law, I.485, I.546
Climate, II.899–912
Clocks, II.1242–6
Close collisions, II.1076, II.1077
  by ions and atoms, II.1041–2
  of slow electrons, II.1040–1
Closed trapped surface, III.1778
Cloud chamber, I.54, I.367, I.384, I.388, I.389, I.396, I.408, II.120, II.751–2, II.1200
Cloudy crystal ball model, II.1212
Clusters, II.1084
  masses, III.1772
Cobalt-60, III.1858, III.1873, III.1874, III.1879, III.1892
Cobalt-bomb, III.1872
Cobalt particles, III.1611
Cobalt structure, I.472
COBE satellite, III.1804
Cockcroft–Walton generator, II.639, II.727
Co–Cr–Mo alloys, III.1897
CODATA, II.1261, II.1262, II.1266, II.1268–72
Coherence theory, III.1408, III.1439–41
  cross-spectral purity in, III.1433
Coherent anti-Stokes Raman scattering (CARS), III.1448
Coherent imaging, III.1418
Coherent potential approximation, II.1006

Coherent regeneration, II.674
Cohesive energy, I.484–5
Cold dark matter, III.1804, III.1805
Cold emission, **III.1307**
Collagen, structure of, I.499
Collective Atomic Recoil Laser, III.1456
Collectivization, III.2052–4
Colliding-pulse mode-locking, III.1439
Collision sequences, I.612
Collisions between atoms or ions, II.1039–42, II.1074–81
　collision speed comparable to electron velcities, II.1079–80
　highly stripped ions, II.1080–1
　projectiles faster than atomic electrons, II.1076–7
　projectiles slower than atomic electrons, II.1077–9
Colloids
　definition, III.1610–13
　dilute systems, III.1610
　future lines of research, III.1612–13
　instability, III.1610
　protection, III.1610, III.1611
　recent progress, III.1612–13
Colour centres, III.1529, III.1531, III.1541
Colour flow imaging, III.1914
Coma cluster, III.1732, III.1772, III.1800
Combination scattering, III.1404–6
Combinatorial analysis, I.526
Combinatorial formula, I.527–8
Commission Internationale de l'Éclairage chart, *see* CIE
Communications
　radio, III.1985–6, III.2000
　theory, III.1417
Compact optical-discs, III.1462, III.1464
Complementarity, I.201–4, I.223–5, III.1395

Complex fluids, III.1596
Complex susceptibility function, III.1404
Complexions, I.523, I.526
Compositeness, II.762–3
Compound nucleus, II.1219
Compound parabolic concentrator (CPC), III.1446
Compton effect, I.120, I.163–72, I.362–6, II.638, III.1302, III.1399, III.1404, III.1862, III.1865
Compton Gamma-Ray Observatory, III.1741, III.1784
Compton scattering, I.218, I.365, I.368, II.991, II.1057, III.1413, III.1773, III.1801
Compton wavelength, II.756, II.1026
Computational fluid dynamics (CFD), II.881–2, II.906–7
Computed tomography (CT), III.1858, III.1873, III.1898, III.1916–23
Computer-aided design (CAD), III.1899
Computer-aided manufacture (CAM), III.1899
Computer analysis, II.906
Computer-generated physics, III.1823–53
Computer graphics, I.628
Computer simulation, I.619, I.628, III.1670–3, III.2044
Computerized treatment-planning systems, III.1875
Computers, I.593, I.622
　early, III.1828–9
　in medical physics, III.1858
　in optical sciences, III.1475–6
　supercomputer, III.2045
Concave spectral gratings, I.54
Concorde, II.877, II.878

S9

Condensation sequence model, III.1965
Condensed matter, II.756, III.2051
Condenser-objective lens, Einfeldlinse, III.1576
Condensing gas, statistical mechanics, I.551–4
Conduction band, III.1319, III.1329, III.1334, III.1340, III.1354
Conduction electrons, II.1168, III.1362
Conductors, magnetism, II.1133–6
Conference on Radioisotope Techniques, III.1901
Configuration space, III.1395
Confocal microscope, III.1450–1
Connected clusters, I.572
Conservation laws, I.374
Conserved vector current (CVC), II.667–9
Constant-energy surfaces, III.1297
Constant-Q technique, II.994
Consultative Committee for Electricity (CCE), II.1254, II.1256–60, II.1262, II.1271, II.1273
Consultative Committee for Photometry (CCP), II.1251
Consultative Committee on the Definition of the Second (CCDS), II.1247
Consultative Committee on Thermometry (CCT), II.1250
Contact-lens technology, III.1414
Contact potential, III.1283, III.1308
Continental drift, III.1967, III.1969, III.1972, III.1973
Continuous phase transitions, II.1009, II.1015
Continuous random network model, I.477
Continuum problems, III.1828
Contraction theory, III.1945

Controlled fusion research, III.1617, III.1633
Controlled fusion research—an application of the physics of high-temperature plasmas, III.1641
Convergent–divergent nozzle, II.879, II.880
Cooling of atoms, II.1099
Cooper pairs, II.938–42, II.946–8, II.950–2, II.954, II.956, II.962–4
Coordinated Universal Time (UTC), II.1247, II.1248
Copernicus satellite, III.1741, III.1765, III.1791
Copper, I.430, III.1515, III.1531, III.1543
Copper–gold alloys, III.1516
Copper oxide superconductors, II.959
Copper–silver solid solution, III.1543
Copper–tin diagram, III.1509
Corona, III.1441
Coronary arteries, III.1898
Coronium, I.76, I.81
Correlation diagram, II.1044
Correlation energy, III.1315
Correspondence principle, I.86
COS-B satellite, III.1740
Cosmic Background Explorer (COBE), II.758, III.1791
Cosmic microwave background, II.758
Cosmic rays, I.364, I.367–70, II.655–7, II.692
  intensity measurement, I.384
  penetrating radiation, I.388, I.404–5
  physics, II.760, III.2004–5
  soft and hard components, I.383–94
  see also Mesotron

Cosmical Electrodynamics, III.1636, III.1685
Cosmological constant, III.1723, III.1724, III.1730, III.1734, III.2039
Cosmological principle, III.1729
Cosmological problem (1939), III.1733–4
Cosmology, II.757–62, III.1691–821
  astrophysical, III.1784–93
  birth of, III.1691
  physical, III.1721–2
  relativistic, III.1721–34
  research since 1945, III.1746–807
  steady-state, III.1786–8
COSMOS High-Speed Measuring Machine, III.1745
COSTAR mission, III.1461
Coster–Kronig transitions, II.1089
Coulomb attraction, III.1331
Coulomb excitation, II.1217
Coulomb field, II.1217, III.1316
Coulomb force, I.123, I.376, III.1315
Coulomb interactions, II.1040, II.1125, III.1290, III.1316, III.1359, III.1362–4, III.1375, III.1551
Coulomb law, I.116, II.641, III.1361
Coulomb repulsion, III.1322
Covering lattice, I.573
Cowling's theorem, III.1958
CP violation, II.671–5
Crab Nebula, III.1737, III.1739, III.1752, III.1774
Critical behaviour, I.568
  binary alloys, I.548–9
Critical current, II.916
Critical density, III.1731
Critical fields, II.920
Critical fluctuations, II.1010, II.1016
Critical magnetic field, II.916
Critical model, III.1731
Critical opalescence, I.546–7

Critical phenomena, I.570, III.1595
Critical point, I.543, III.1947
Critical temperature, I.545, I.547, II.975, II.976
Critical velocity, II.928
Cross-spectral purity in coherence theory, III.1433
Cryostats, helium, I.436–7, III.2025
Crystal boundaries, III.1528
Crystal defects, III.1320, III.1531, III.1535
Crystal diffraction spectroscopy, II.1207
Crystal field, rare-earth elements, II.1164–5
Crystal field splittings, II.1164
Crystal field theory, II.1126–9, II.1136, II.1139, II.1141, II.1144–5
Crystal growth, I.466
Crystal lattice, III.1520
Crystal morphology, I.421
Crystal rectifier, III.1318
Crystal spectrometer, I.430, I.433
Crystal structure, I.423, III.1513–16, III.1529
  analysis, I.484–96
  and magnetic properties, II.1156–8
  determination methods, I.439–50
  faults in layer, I.472–3
  imperfections in, III.1505–6
Crystal surfaces, *see* Surface crystallography
Crystalline cleavage facets, III.1517
Crystalline virus, I.437
Crystallites, III.1280
Crystallographic analysis, III.1546
Crystallographic Data Centres, I.512
Crystallography, I.421–4, III.1515–17, III.1547
  history of, I.512
Crystallography in North America, I.512

Crystals
  dielectric properties, I.485–6
  dislocations, I.475–7
  doubly refracting, I.485
  imperfect, I.471–84
  optical properties, I.485–6
  see also under specific crystal
    types
CuAu, III.1515
Cu$_3$Au, III.1515
Cu–Ni–Fe alloy, III.1548
Curie law, I.540–1, II.1116, II.1133
Curie point, I.544–6
Curie temperature, I.545–6, I.549,
  I.550, I.559, III.1970, III.1976
Curie–Weiss Δs, II.1128–9
Curie–Weiss law, II.1116, II.1120,
  II.1128
Curium, II.1225
Curve of growth technique,
  III.1702
CuZn, III.1534
Cyclotrons, II.729–30, II.1199,
  II.1201, II.1206, II.1221,
  III.1871
  frequency, III.1334, III.1336
  resonance, III.1336, III.1337,
    III.1357
Cygnus A, III.1737, III.1775,
  III.1783, III.1788
Cygnus X-1, III.1754, III.1757
Cytosine, I.504

Daguerreotype process, III.1692
d'Alembert's Paradox, II.796,
  II.803–8
d'Alembert's Theorem, II.803–8,
  II.824, II.870, II.886
Danish Atomic Energy
  Commission, III.1643
Dark matter, II.761, II.768
Darwin–Fisher hypothesis, III.1948
Data collection, II.906
Data processing, thermodynamics
  of, I.576

de Broglie relation, I.457, III.1296
de Broglie waves, I.199, I.387,
  II.946, II.975, II.976, II.991,
  III.1292
Debye characteristic temperature,
  I.452
Debye equivalent, II.977
Debye formula, II.973
Debye function, II.972
Debye–Hückel theory, III.1521
Debye temperature, II.972, II.975,
  II.976, III.1305
Deceleration parameter, III.1794–7
Deep inelastic scattering, II.714–15
Defibrillation, III.1895
de Haas–van Alphen effect,
  II.1166, III.1348–56
δ-function, I.607
δ-rays, I.37
De Magnete, I.544, III.1856
Dementias, III.1932
Dendritic growth, III.1542
Density matrix, I.602, I.617
Density of states, II.986, II.1006
Density parameter, III.1734,
  III.1799–800
Deoxyribonucleic acid (DNA),
  I.504, I.507
Depletion layer, III.1325
Detectors, II.744–54
Deterministic hypothesis, I.222
Deuterium, I.112, I.370, III.1627,
  III.1706, III.1800
Deuterium-tritium, III.1651,
  III.1674
Deuteron, I.370, I.403
Deutsche Gesellschaft für
  Metallkunde, III.1553
Deutsche Gesellschaft für
  Naturforscher und Ärzte,
  I.47
Deutsche Gesellschaft für
  Technische Röntgenkunde,
  I.511

Subject Index

Diabatic potential energy functions, II.1079
Diagnostic machines, III.1897
Diagnostic medicine, III.1900–10
Diagnostic radiology, III.1859–63
Diagnostic ultrasonics, III.1910–11
Diamagnetism, II.1113–15
Diamond, I.429
Diamond–graphite equilibrium diagram, I.539
Diaphanography (light transmission imaging), III.1932–3
Dictionary of Applied Physics, II.977
Dictionary of Scientific Biography, I.16
Dielectric breakdown, III.1336–7
Dielectric constant, III.1595
Dielectric properties of crystals, I.485–6
Dielectric response, II.1055
Dielectric waveguide, III.1397, III.1420
Differential equations, III.1826
Diffractals, III.1466
Diffraction gratings, III.1401–2, III.1466
Diffraction microscopy, III.1582
Diffraction pattern, III.1412
Diffraction theory, III.1420
Diffractometer, I.435
Diffusion, II.818–19, III.1531–5, III.1548
  by continuous movements, II.830
  Fick's law of, I.597, I.601
Diffusion coefficient, I.597
Diffusion–convection balance for scalar quantities, II.841–6
Diffusion creep, III.1534
Diffusion inhibitor, III.1629
Digital subtraction angiography, III.1863
Diketopiperazine, I.464
Dimensional analysis, II.838

Diopside, structure of, I.440
Dipole–dipole interaction, II.987
Dipole vibration, II.1213
Dirac delta-function, I.206
Dirac equation, **I.361**, I.362
Dirac–Jordan transformation theory, I.198–201
Dirac theory, II.651
Directional distribution patterns, **II.1035**
Disabled persons, technology for, III.1899
Disciplinism, III.2042
Discretization process, I.629
Dislocations, I.475–7, III.1507, III.1520, III.1521, III.1525, III.1526, III.1528
Disordered materials, III.1365–9
Dispersion, III.1404
  curves, II.988–90, II.998
  formulae, I.171–2
  relations, II.680
  theory, I.184–5
Dispersive waves, II.856–60
Displacement law, I.66
Distorted wave Born approximation (DWBA), II.1220, II.1221
Distribution of charges and current in an atom consisting of many electrons obeying Dirac's equations, The, III.1622
D-lines, I.66
DNA, *see* Deoxyribonucleic acid (DNA)
Dolen–Horn–Schmid duality, II.683
Doping (and dopants), III.1529, III.1544, III.1593
Doppler effect, I.269, I.270, II.1094, III.1781, III.1784
Doppler shift, III.1398, III.1710
Doppler ultrasound, III.1914
Double-helix structure, I.506

Double strand breaks (DSBs), III.1885
Dreams of a Final Theory, III.2020
Drell–Yan process, II.701–2
Drift chambers, II.749, II.754
Drude–Lorentz model, III.1288
Duality, II.682–4
Dulong and Petit's law, II.970, II.972, II.980
Dwarf stars, III.1696, III.1698
Dye molecules, II.1087
Dynamic hologram, III.1458
Dynamical electroweak symmetry breaking, II.763
Dynamical friction, III.1774
Dynamical symmetry breaking, II.957
Dynamical systems, III.1850
   classification of, I.614–16
Dynamo theory, III.1985

Earth
   core, III.1952–60
   core size, III.1954
   fluid envelope, II.887–909
   gaseous-interior model, III.1947
   Laplace–Herschel–Kelvin scenario, III.1949
   liquid–gas boundary, III.1947
   magnetic field, III.1622–3, III.1952–60, III.1973
   origin and age (post-1935), III.1960–6
   origin and age (pre-1935), III.1944–52
   primaeval, III.1949
   structure, III.1946, III.1954, III.1956, III.1957
Earth–Moon system, III.1948
Earth sciences, III.1967, III.2046
Earthquakes, III.1956, III.1978–80
   zones, III.1976
Echo-sounding data, III.1973
École Polytechnique, I.15

Eddington–Lemaître model, III.1732–3
Eddington luminosities, III.1780
Edge dislocation, I.475–6
Education, I.8–16
   curricula, I.8
   higher, women in, I.8, I.10
EFFI, III.1673
Ehrenfest dog-flea model, I.618
Eigenfunctions, III.1432–7
Eigenvalue problem, III.1848
EIH method, I.303
Einfeldlinse (condenser-objective lens), III.1576
Einstein–Bohr frequency condition, I.359
Einstein–Bose statistics, I.602–3
Einstein–de Sitter model, III.1730–1
Einstein–Podolsky–Rosen(–Bohm) (EPR(B)) paradoxes, I.228–9, III.1457
Einstein X-ray Observatory, III.1798
Einsteinium, II.1224
E layer, III.1994, III.2002
Elasser–Bullard theory, III.1960
Elastic constants, II.983
Elastic scattering, II.687
Elasticity, I.484–5
Electric charges, I.51
Electric fields, I.27, I.261, III.1283, III.1362, III.1434, III.1435
Electric polarization, II.1031
Electrical conduction in metals, III.1284–7
Electrical conductivity, III.1279, III.1530
Electrical counting methods, II.1200
Electrical impedance tomography, III.1932
Electrical stimulation, III.1895
Electrical surges (*Wanderwelle*), III.1566

Electrical units and standards, II.1252–60
Electricity, I.51, I.52
Electricity supply industry, II.1255
Electro-absorption, III.1423
Electrocardiogram/graph, III.1893, III.1896
Electrodynamics, I.88, I.266, I.267, I.270
  in special theory of relativity, I.257, I.280–1
Electroluminescence, III.1417
Electrolysis, I.51
Electromagnetic field, I.358, III.1434, III.2039
Electromagnetic gauge invariance, II.693
Electromagnetic induction, I.266, II.1112
Electromagnetic interactions, I.391
Electromagnetic radiation, I.197, II.1264, III.2018, III.2025
Electromagnetic spectrum, III.1734–45
Electromagnetic theory, I.362
Electromagnetic waves, I.260, III.1391, III.1622, III.1985
Electromagnetism, II.640, II.693, III.1856, III.2000
Electron beams, III.1456, III.1565–91
Electron charge/mass ratio (e/m_e), II.1271
Electron–deuteron scattering, II.698
Electron density, I.440, I.446, I.448, I.452, I.454, I.502, III.1316
  regions, III.2009, III.2010
Electron diffraction, I.463–5, I.481
Electron distribution, I.452
Electron–electron collisions, II.739–40, III.1316
Electron energy
  LEED experiments, I.466
  RHEED experiments, I.468

Electron energy-loss spectrometer (EELS), III.1580
Electron $g - 2$ factor, II.647
Electron gas, III.1361, III.1362, III.1371–5
  quantum theory of, III.1287–91
Electron–hole pairs, II.999
Electron layer, III.1993
Electron magnetic moment, II.644–5
Electron magnetic resonance, III.1859
Electron mass, II.1271
Electron microprobe analyser, III.1548–9
Electron microscopes, I.481, II.931, III.1526, III.1569, III.1570–4
  see also under specific types of electron microscope
Electron mobility, III.1368
Electron multiplication, III.1419
Electron–neutrino pair, II.639
Electron pair, II.1028
Electron paramagnetic resonance (EPR), II.1137–52
  3d ions, II.1139–41
  4f ions, II.1141–3
  gedanken experiment, III.1457, III.1467
  iron group ions, II.1137–9
  rare earths, II.1141–3
Electron–phonon collisions, III.1316
Electron–phonon coupling, II.1000
Electron–phonon interaction, II.941, II.999, III.1303
Electron–phonon scattering, III.1305
Electron physics, III.2000
Electron–positron annihilation, II.690–1, II.715, II.717, II.740
Electron–positron collisions, II.691, II.711, II.740, II.741, II.768
Electron–positron pair production, III.1784, III.1786, III.1806

S15

Electron probe x-ray microanalysis, III.1580
Electron–proton collider (HERA), II.766–7
Electron scattering, II.1033, II.1221–2
  cross section, III.1996
Electron scattering factor, I.463–4
Electron spin, I.215
Electron spin resonance (ESR), II.1137–8, III.1923–33
Electron synchrotrons, II.731–2
Electron theory, I.30
  of metals, I.215–17
Electronic devices, therapeutic, III.1896
Electronics, medical. *See* Medical electronics
Electronics technology, III.2031
Electrons, I.36, I.51, I.53, I.59, I.64, I.77, I.81–90, I.95–9, I.115, I.158, I.167–8, I.369, I.377, I.457, II.637–8, II.641, **III.1298–9**
  free, I.372
  hot, III.1336–41
  in medical physics, III.1857
  in solids, III.1279–383
  see also Valence electrons
Electronvolt (eV), I.358
Electrostatic effects, charge effects, III.1610
Electrostatic electron microscopes, III.1577–8
Electrostatic generators, II.727–9
Electrosurgery, III.1894
Electroweak theory, II.640, II.762
Electroweak unification, II.692–709
Elementary charge (e), II.1271
Elementary excitations, II.928, II.929
Elementary particles, I.115, I.231–2, I.370–1, I.391, I.408–9
  in special theory of relativity, I.282–3
  physics of, II.635–794
  Standard Model of, III.2030, III.2033–9
  see also under specific particles
Elementary phenomena, II.1029–32
Elements, *see* Chemical elements
Elements of Physics, III.1857
El Niño/Southern oscillation phenomenon (ENSO), II.908
Elsasser–Bullard dynamo theory, III.1973
EMI-Scanner, III.1922
Emission electron microscope (EEM), III.1569, III.1577
Emulsions, III.1611
Encyclopaedia Britannica, I.36
Encyclopedia of Materials Science and Engineering, III.1513
Endocrinology, III.1930
Energy bands, III.1300
Energy density, III.1393
Energy fluctuation, I.163
Energy gap, II.932, II.952, III.1329, III.1433
Energy loss, 'hidden', II.846–9
Energy principle for hydromagnetic stability problems, An, III.1672
Enskog time, I.619
Entanglement, III.1394
Entropy, I.47, I.523, I.601
  calorimetric and statistical estimates, I.539–40
Eötvös experiments, I.286–7
Ephemeris Time, II.1247
Epilepsy, III.1922
  diagnosis, III.1916
Epoch of recombination, III.1801, III.1803
Equations of motion, I.48, I.186, I.303
Equilibrium angular momentum, II.932
Equilibrium state, II.920

Equilibrium theory, III.1785
Equipartition theorem, I.5, I.523, I.528
Equivalence principle, I.286–91
Equivalent width concept, III.1702
Erbium-doped optical-fibre amplifier (EDFA), III.1447
Ergodic surface, I.525
Ergodic theorems, I.599, I.626
Essen Ring, II.1244
ETA-BETA I, III.1658
ETA-BETA II, III.1658
Euler–Lagrange equation, I.195
Euratom-CNR facility, Padua, III.1658
European Federation of Medical Physics (EFOMP), III.1934
European Optical Society, III.1474
European research system (or science base), III.2055
European Science Foundation, III.2055
European Southern Observatory (ESO), III.1744
European Space Agency, III.1742, III.2055
Europium sulphide, I.559
Ewald transformation, II.988
Exchange interactions, II.1129, II.1143-4
Excitation channels, II.1059-73
Exciton absorption spectra, III.1423
Exciton theory, III.1423
Exclusion, **II.1028–9**
Exclusion principle, I.97–100, I.168, I.603, II.686, II.1116
Exluded volume, I.569
Experimental facilities, I.24–9
Experimental physics, I.13, I.14
Explorer I, III.2007
Explorer II, III.1740
Explorer III, III.2007
Explorer IV, III.1683, III.2007, III.2008

Explorer satellites, III.1739
Explosive nucleosynthesis, III.1747–8
Explosive showers, I.387
Exponential law, I.67
Extinction coefficient, III.1396
Extremely low-frequency radio physics, III.1995
Eye tumours, III.1876
Eyewall, II.904

F1 layer, III.1994
Fabry–Perot étalons, II.1240
Fabry–Perot interferometer, I.19, III.1412
Fabry–Perot resonators, III.1422, III.1423, III.1429
Fabry–Perot-type beam receiver, I.315
Face-centred-cubic lattice, I.557, III.1295
Facsimile (FAX), III.1476
Far infrared radiation, II.1054
Faraday (unit), I.51, II.1269
Faraday's constant, I.150, II.1237
Faraday's law, I.52
Fast charged particles, II.1039–40
Fast collisions, II.1076
Fast processes, II.1085
F-centre (Farbzentrum), III.1320–1
Fe–Ag–B alloys, III.1543
FECO interferometry, III.1412
Fellgett advantage, III.1416
Fellows of the Royal Society, I.10, I.12
Femtosecond pulses, III.1462
Fermi–Dirac statistics, I.117, I.123, I.155, I.531, I.533, I.603, II.687, III.1706, III.1711
Fermi distribution, II.1168, III.1301, III.1302, III.1308
Fermi energy, III.1295, III.1302, III.1309, III.1364
Fermi-field theory, I.375–8, I.382, I.392, I.393, I.404

Fermi gas, II.1209, **III.1289**, III.1301, III.1308, III.1316
Fermi particles, I.603
Fermi statistics, II.1190, III.1394
Fermi surface, II.999, II.1002, III.1305, III.1308, III.1345–59, III.2025
Fermi theory, II.651
Fermilab, III.2054
Fermions, II.676, II.694, II.955, II.1168–70, III.1355, III.1804
Fermium, II.1224
Ferrimagnetic insulators, II.1156
Ferrimagnetism, II.1113, II.1152–65
  magnetic properties, II.1157
Ferroelastic crystals, III.1538
Ferroelectricity, I.486–9, III.1538
Ferrofluids, III.1611
Ferromagnetic model, I.570
Ferromagnetic resonance, II.1152–3
Ferromagnetism, I.559, I.562, II.1015, II.1113, II.1120–3, II.1134–6, II.1152–65, III.1279
Feynman diagrams, **I.361**
Fibre-optic light, III.1894
Fibre optics, III.1419–21
  see also Optical fibre
Fick's law of diffusion, I.597, I.601
Field cycling behaviour, II.1121
Field emission microscope, III.1569
Field ion microscope, I.50
Fierz interference, II.664
Finalization, III.2048–50
  pre-paradigmatic phase, III.2048
  revolutionary phase, III.2048–9
Fine structure, **II.1036–7**
  constant, II.640, II.1026, II.1262, II.1271
Finite energy quantum, I.173
Finnegans Wake, II.685
First Antarctic Expedition, III.2005
First International Congress of Medical Physics, III.1909, III.1923
First International Congress of Radiology, III.1881
First law of thermodynamics, I.374, I.521
First-order Fermi acceleration, III.1782
First-order melting transition, I.577
First Texas Symposium on Relativistic Astrophysics, III.1776
Fischer–Riesz theorem, I.206
Fitzgerald–Lorentz contraction, I.32, I.264
Fixed point, I.567, III.1832
Fixed-wing aircraft, II.817
Flash-photolysis, II.1085
Flash-spectroscopy, II.1085
Flight efficiency, II.869–73
Flood waters, II.897
Flory temperature, III.1598
Flow
  in pipes, II.814–15, II.825
  resistance, II.815
Fluctuation equation, III.1393, III.1402
Fluid envelope, II.860, II.887–909
Fluid mechanics, II.795–912
Fluid particles, I.572
Fluids, microscopic critical behaviour, I.546–7
Fluorescence, III.1537
Fluorite, I.430
Flying-spot microscope, III.1416
Focal conic defects, III.1605
Fokker–Planck equation, I.595–7, I.624, I.627
Forbidden transitions, III.1762
Forced convection, II.843, II.844
Form factors, II.660, II.687
FORTRAN, III.1843
Fortschritte der Physik, I.7
Foucault's pendulum, I.2–3
Fourier imaging theory, III.1418–19

Fourier series, I.441, I.623
Fourier synthesis, I.501
Fourier techniques, reconstructed (111) surface, I.469
Fourier-transform, III.1418, III.1924
  image processing, III.1413
  spectrometer, III.1416
  spectroscopy, II.1055
Fourier's law of thermal conduction, I.601
Fractals, I.574, I.630
Fractured bones, III.1897
Fragile structures, III.1595
Fragmentation, II.1102, II.1103
  dynamics of, II.1105–6
Franck–Condon rule, II.1046
Franck–Hertz experiment, I.89
Free convection, II.843, II.844
Free-electron gas hypothesis, I.216
Free electrons, I.372
Free-energy phase diagram, I.538
Freedman–Clauser experiment, III.1458, III.1467
Frequency
  dissemination of, II.1248–9
  standards, II.1243–6
Fresnel biprism, III.1585
Fresnel diffraction fringes, III.1580–2
Friedel's law, I.440, I.464, I.495
Friedman world model, III.1725, III.1731, III.1733, III.1784, III.1807
Froome interferometer, II.1264
Frustration, II.1170
FSU, III.2028
Fullerene crystals, II.956
Functional electrical stimulation (FES), III.1900
Functionally graded materials, III.1542
Fundamental observers, III.1729
Fundamental physics, III.2034–5
Fundamental Theory, The, III.1700, III.1787

Fusion bomb, III.2022
Fusion research, III.1617, III.1618, III.1627–9, III.2024
  declassification, III.1640–5
  principal sites for, III.1636
  see also Magnetic fusion research
FWHM, III.1904, III.1922

Gadolinium, magnetic properties, II.1162
Galactic cannibalism, III.1774
Galactic Centre, III.1740, III.1744, III.1763
Galaxy(ies), III.1769–71
  active, III.1797–8
  clusters of, III.1771–4
  crossing times, III.1772
  dark halo, III.1771
  dark matter, III.1771
  distribution of, III.1771–2
  elliptical, III.1771
  formation, III.1800–5
  luminosity function for, III.1769–70
  masses, III.1769
  structure of, III.1714–21
  velocity curves, III.1771
  violent events in nuclei of, III.1777
Galena crystal, III.1318
Galilean transformation law, I.257
Galilei–Newtonian kinematics, I.257
GALLEX experiment, III.1749
Gallium, I.75, III.1749
Gallium arsenide (GaAs), III.1371, III.1375, III.1464
Gamma-10, III.1669
Gamma camera, III.1907–9, III.1918, III.1920, III.1921
Gamma-decay, I.106
Gamma-radioactivity, I.53
Gamma-ray astronomy, III.1738–41
Gamma-ray detectors, III.1740

Gamma-ray imaging, III.1904
Gamma-ray wavelengths, II.1207
Gamma-rays, I.54, I.367, I.457,
    II.1207, III.1784, III.1879,
    III.1902
Gamow–Teller type interactions,
    II.664
Gargamelle Collaboration, II.697
Gas constant (R), II.1269
Gas discharges, I.27
Gas of corpuscles, III.1390–2
Gaseous Electronics Conference,
    III.1644
Gases, I.47
    ionization phenomena in,
        III.1644
    kinetic theory of, I.30, I.35, I.46,
        I.522, I.523, I.587, III.1284
    liquid, I.28, I.543–4
    specific heat, I.528–30
Gaskugeln, III.1695
Gauge theory, III.1414
Gaussian field-distribution profile,
    III.1429
Gaussian–Hermite (laser) beam
    propagation, III.1447
Geiger(–Müller) counter, I.108,
    II.747, II.1200, II.1207,
    III.1413, III.1900–2, III.2008
Geiger–Nuttall relation, I.116
GEKKO series of lasers, III.1676
Gell-Mann–Okubo mass relation,
    II.661
General Catalogue of Nebulae,
    III.1717
General Conference on Weights
    and Measures (CGPM),
    II.1234–6, II.1247, II.1248,
    II.1251, II.1257, II.1262,
    II.1264
General Electric Company (GEC)
    Research Laboratories,
    II.1204
General Electric (GE) Research
    Laboratory, New York State,
    III.1528
General theory of relativity, I.249,
    I.286–320, III.1722–4
    alternative formulations and
        foundations, I.298–9
    alternative scenarios, I.295–6
    and active galactic nuceli,
        III.1777–81
    approximation schemes, I.302
    astronomical and astrophysical
        applications and tests,
        I.315–17, III.1776
    classical tests, I.303–6
    equivalence principle, I.286–91
    exact solutions, I.302
    field equations, I.293–5, I.313
    gravitational energy, I.300–1
    later work on, I.296–8
    metric-tensor field, I.291–3
    physical interpretation, I.301
    Schwarzschild solution, I.303–6
    software programs, I.302
    tests, III.1757–8
Generalized form factor, II.1040
Genetic code, I.507
Geneva Conference (1958),
    III.1640–5, III.1642, III.1643
Geniac, III.1828
*GENIUS—The Life of Richard
    Feynman*, III.1519
Geology, III.2046
Geomagnetic dynamo theory,
    III.1958–9
Geomagnetic field, reversals of,
    III.1973
Geomagnetic field lines, III.2010
Geomagnetic storm, III.2002–3
Geomagnetism, III.2000, III.2001
Geometrical collective model,
    II.1218
Geophysical observing stations,
    III.1998
Geophysics, III.1943–2016, III.2046
    history, III.1943–4
    use of term, III.1943

S20

Georan, II.1242
Geospace, III.1981, III.1997–2001
 future study propsects, III.2011
 pictorial models, III.2010
Germanium, I.75, I.470, II.982,
 II.1208, III.1281, III.1301,
 III.1319, III.1322, III.1324,
 III.1332–6, III.1340, III.1544,
 III.1749
Gesellschaft deutscher Naturforscher und Ärzte, I.5
Giant branch, III.1698
Giant dipole resonance (GDR), II.1213–14
Giant molecular clouds, III.1765
Giant stars, III.1696, III.1698, III.1708–9
Gibbs ensembles, I.531–3
Gibbs free energy, III.1543, III.1548
Gibbs function, I.549
Gibbs Paradox, I.528
Ginzburg–Landau theory, I.568, II.934, II.935, II.940
Glashow, Iliopoulos and Maiani (GIM) mechanism, II.700
Glashow model, II.694
Glashow–Weinberg–Salam theory, II.695, II.697, II.698, II.707
Glass industry, III.1517
Glasses, III.1533, III.1600–2
 phonons in, II.1004–7
Glitches, III.1754
Global Positioning Satellite System, II.1249
Glomar Challenger, III.1976, III.1980
Gluons, II.636, II.637, II.684, II.714, II.716
Glutamine, structure of, I.449
God Particle, The, III.2020
God principle, III.2019–20
Gold, III.1515, III.1543
Gold-198, III.1905
Goodyear, III.1593, III.1597
Göttingen quantum mechanics, I.183–96

Gradient-index lenses, III.1391
Grain boundaries, III.1518–20, III.1528
Grain size distribution, III.1549
Grand canonical ensemble, I.532
Grand partition function (GPF), I.533
Grand unified theory, I.323, II.763–4
Grating spectroscopy, III.1391
Gravitation, II.757–62, III.2037
 and quantum theory, I.318
 in special theory of relativity, I.283–4
 Newtonian constant of, I.28
 theory of, I.283–4, II.762
Gravitational collapse, I.306–12
Gravitational constant (G), II.1265–7
Gravitational energy, I.300–1
Gravitational field equations, I.253
Gravitational radiation, I.312–15
Gravitational waves, I.312, I.315
Gravity fields
 Moon, II.888
 Sun, II.888
Gray (unit), III.1867
Grazing collisions by ions and atoms, II.1041–2
Great Debate, III.1716–21
Green statistics, II.687
Greenhouse effect, II.899
Green's function, I.617, I.623–4
Ground-state energy, II.1045
Group-theoretical model, I.211–15
Group velocity, II.856
Guanine, I.504
Guided wave optics, III.1458
Guy's Hospital, III.1864
Gyromagnetic effect, II.1153
Gyromagnetic ratio, II.1237, II.1257
Gyroscope
 femtosecond-pulsed dye-laser ring, III.1433

S21

rotation-sensing ring-laser, III.1433

Habilitation thesis, I.184
Hadron colliders, II.707–8, II.742–4, II.766–7
Hadrons, II.686, II.690–1, II.699–700, II.704, II.713, II.715–17
Haemoglobin, I.500, I.502–3
Hafnium, I.100
Half-life concept, I.59
Hall constant, III.1334, III.1367
Hall effect, I.9, II.1096, III.1282, III.1284–6, III.1296, III.1298, III.1299, **III.1331**
Hall field, III.1373
Hall potential, II.1096
Hall resistance, II.1258
Hambergite, structure of, I.451
Hamilton–Jacobi equation, I.176, I.194
Hamilton–Jacobi theory, I.169, I.191
Hamiltonian, I.566–9, I.612
Hamiltonian matrix, I.190
Hamiltonian scheme, I.189
Hammersmith Hospital, III.1871, III.1873, III.1922
Handbook of Chemistry and Physics, III.1509–13
Handbuch der Physik, III.1513
Harker–Kasper inequality, I.447, I.448
Harmonic oscillator, II.973
Harrison construction, III.1313
Harrison–Zeldovich spectrum, III.1802, III.1807
Harvard Radio Research Laboratory, III.2026
Hayashi limit, III.1767
Heart valves, artificial, III.1898
Heat capacity, I.78–80
Heat conduction theory, III.1945
Heat Theorem, II.968, II.975

Heat-treatment, III.1519
Heaviside layer, III.1987
Heavy-atom derivatives, I.501
Heavy-atom method, I.444
Heavy element catastrophe, III.1785
Heavy fermions, II.955, II.1168–70
Heavy hydrogen, III.1627, III.1628
Heavy-Ion Medical Accelerator, Chiba, Japan, III.1876
Heisenberg–Dirac (H–D) theory, II.1128–9, II.1148–9
Heisenberg exchange integral, I.213
Heisenberg magnet, III.1595
Heisenberg n–p model, I.382
Heisenberg systems, II.1014
Helicity, II.667
Helicons, III.1358–9
Helimagnetic structure, I.462
Helioseismology, III.1749–51
Helium, I.58, I.75, I.81, I.101–2, I.110, I.182, I.211, II.974, II.1004, III.1707
  and discovery of cosmic microwave background radiation, III.1790–3
  cryostats, I.436–7, III.2025
  liquid, II.913–15, II.921–32, III.1341–5
  superfluid phases, II.949–55
  synthesis, II.1226
α-helix, I.498, I.508
Helmholtz free energy, I.527, I.532
Henry Draper Catalogue of Stellar Spectra, III.1694, III.1695
HERA (electron–proton collider), II.766–7
Hercules X-1 (Her X-1), III.1755
Hermann–Mauguin symbol, I.439
Hermite functions, I.625
Hermite polynomials, I.625
Hermitian affine connection, I.322
Hermitian matrices, I.190
Hermitian operators, I.205

Hertzsprung–Russell diagram, III.1692-8, III.1746, III.1793-4
Her X-1, III.1756
Hess–Fairbank effect, II.944-6, II.955, II.956
Hg–Ba–Ca–Cu–O compound, II.959
Hierarchical clustering, III.1804
Higgs boson, II.695, II.762
Higgs fields, III.1807
Higgs mechanism, II.694
Higgs particle, III.2036
High Energy Astrophysical Observatory A and B (HEAO-A and B), III.1740
High Energy Particle Physics, III.2051
High-energy physics, III.2056, III.2058
High-resolution images, III.1477
High-temperature superconductor, I.489-90
High Voltage Engineering (HVE), II.1203
Highest occupied molecular orbital (HOMO), II.1083
Hilbert conditions, I.293
Hilbert space, II.1025, III.1395
Hip replacement, III.1896
History of Metallography, A, III.1556
History of the University in Europe, III.2041
H-mode operation, III.1649
Hohlraumstrahlung, I.66
Holes, **III.1289-9**
  concept of, III.1297
Holograms, III.1415, III.1438, III.1582-4
Holography, III.1437-9, III.1580-1
  history of, III.1414-17
  off-axis optical beam, III.1585
  principle of, III.1415
Hooker 100" telescope, III.1704, III.1726

Hospital Physicist's Association, III.1864
Hot Big Bang models of Universe, III.1778
Hot Dark Matter, III.1803, III.1804
Hot electrons, III.1336-41
  phonon interaction with, III.1339
Hot shortness, III.1519
Hot-wire techniques, II.831
Hovercraft, II.886
HP65 programmable calculator, III.1829, III.1842, III.1846
H-theorem, I.35
Hubbard U, II.1166
Hubble space telescope, III.1461, III.1474, III.1741-2, III.1745
Hubble's constant, III.1730, III.1731, III.1787, III.1788, III.1793-4, III.1799
Human movement studies, III.1899-900
Hume-Rothery rule, III.1346
Hydraulic jump, II.858-9
Hydrodynamic forces on ships, II.882
Hydrofoil boat, II.886
Hydrogen, I.64, I.77, I.78, I.80-6, I.160, I.359
  molecular, I.181, II.1045, **II.1048**, II.1069-73, III.1766
  ortho, **II.1048**
  ortho–para, I.540
  para, **II.1048**
Hydrogen atoms, I.90-2, I.211, I.360, I.570, II.641
Hydrogen bombs, III.1633, III.1673
Hydrogen bond, II.1043
Hydrogen isotope, I.370
Hydrogen spectrum, I.90
Hydroxyapatite, III.1897
Hyperfine interactions, II.649-50
Hyperfine structure, **II.1036-7**, II.1214
Hyperon, II.657

Hypersonic aerodynamics, II.879–80
Hyperspherical coordinates, II.1103–5
Hypothesis of energy elements, I.145
Hysteresis, II.918, II.1121

Ice, structure of, I.474–5
Illinois University, III.1871
Illustrated London News, I.2
Image processing, Fourier-transform, III.1413
Imaging theory, III.1399–401, III.1467–8
IMB-3 Detector, II.759
Imperfections in Nearly Perfect Crystals, III.1535
Implant surgery, III.1896
Incoherent backscatter, III.1995
Incoherent scatter radar (ISR), III.1995, III.1996
Independent-particle model, III.1359–65
Index Catalogues, III.1717
Indian ink, III.1611
Indirect-exchange-coupling theory, II.1012
Indium, I.75
Indium antimonide (InSb), reconstructed (111) surface, I.469
Induced transitions, III.1398–9
Induction coil, I.54
Ineffectiveness concept, III.1347
Inelastic neutron scattering, II.1162
Inelastic scattering, II.687–90, II.697, II.1221
Inequality relations, I.447
Inertial confinement fusion (ICF), III.1673–7
Inflation, II.761–2
Information retrieval, electronic methods, III.1896
Information technology, III.2044–6

Infrared absorption, II.988
Infrared astronomy, III.1742–4
Infrared Astronomy Satellite (IRAS), III.1744
Infrared radiation, laser sources of, II.1054
Infrared reflectivity, II.1003
Inner quantum number, I.100
Inner-shell phenomena, II.1089–92
  Auger emissions, II.1090–1
  spectra, II.1038–9
  x-ray studies, II.1089–90
Inorganic compounds, structure analysis, I.492–3
Inorganic Crystal Structure Database, I.512
In-plane x-ray diffraction, I.466
Instant camera, III.1417
Institut für Metallphysik und Allgemeine Metallkunde, III.1555
Institute for Defense Analysis, III.2023
Institute for the Study of Metals, University of Chicago, III.1556
Institute Laue-Langevin, III.2054
Institute of General Metallurgy (Allgemeine Metallkunde), III.1554–5
Institute of Metals Research, III.1555
Institute of Physics, III.1558, III.2042
Institution of Physics and Engineering in Medicine and Biology, III.1864
Instrumentation, role of atoms, II.1094–7
Integrated intensity, **I.434**
Integrated optics, III.1449–50
Interaction II, I.391, I.393, I.402
Interaction III, I.392, I.397
Interdisciplinarity, III.2042–4
Interference effects, III.1369–70

Subject Index

Interference pattern, III.1395
Interferometric spectroscopy, III.1410
Interferometry, I.25, I.26, III.1410, III.1738
  see also Optical interferometry techniques
Intermediate state, II.924
Intermediate valence, II.1168–70
Intermetallic compounds, III.1515
Intermetallic Phases Data Bank, III.1513
Internal Constitution of the Stars, The, III.1700, III.1702
Internal conversion, I.107
Internal field, I.545
Internal gravity waves, II.866
Internal pressure, I.543
International Ampere, II.1254, II.1257
International Astronomical Union (IAU), II.1242, II.1247
International Atomic Energy Agency (IAEA), III.1643, III.1648, III.1649, III.1908
  Scanning Conference, III.1916
International Atomic Time (TAI), II.1247, II.1248
International Bureau of Weights and Measures (BIPM), II.1234, II.1236, II.1240, II.1248, II.1251, II.1257, II.1259, II.1272
International Centre for Diffraction Data, I.512
International Commission on Radiation Protection (ICRP), III.1910
International Commission on Radiation Units and Measurements (ICRU), III.1867, III.1881, III.1910, III.1920
International Commission on Radiological Protection, III.1890
International Committee of Weights and Measures (CIPM), II.1234, II.1240, II.1242, II.1247, II.1249–51, II.1257, II.1264, II.1271
International Conference on Medical Electronics, III.1913
International Council for Scientific Unions, III.2006
International Critical Tables, II.1267, III.1509
International Earth Rotation Service (IERS), II.1248
International Federation of Medical and Biological Engineering (IFMBE), III.1934
International Geophysical Year (IGY), III.1973, III.1989, III.2005–6
International Medical Electronics Conference, III.1904
International Ohm, II.1254, II.1257
International Organization of Medical Physicists (IOMP), III.1934
International Polar Year (1882–83), III.2005
International Practical Temperature Scale (IPTS-48), II.1250
International System of Units (SI system), II.1235
International Tables for the Determination of Crystal Structures, I.511
International Tables for X-ray Crystallography, I.439
International Temperature Scale, II.1250
International Toroidal Experimental Reactor) device, see ITER

S25

International Ultraviolet Explorer, III.1741
International Ultraviolet Observatory, III.1780
International Union for Geodesy and Geophysics, III.2006
International Union for Scientific Radio, III.2006
International Union of Crystallography, I.511–12
International Union of Physical and Engineering Sciences in Medicine (IUPESM), III.1934
International Union of Pure and Applied Physics (IUPAP), II.1268
International Union of Radio Science (URSI), II.1273
International Units, II.1255, II.1260
International Volt, II.1254, II.1257
International Weights and Measures Organization, II.1234
International X-ray and Radium Protection Commission, III.1890
Internationale Zeitschrift für Metallographie, III.1556
Internationalization, III.2055–6
Inter-Service Ionosphere Bureau (ISIB), III.1998–9
Interservice Radio Propagation Laboratory, III.1998
Interstellar chemistry, III.1765
Interstellar medium, III.1761–9
  soft x-ray background, III.1766
  two-phase model, III.1765
  violent, III.1766
Interstitials, III.1529, III.1531, III.1533
Introduction to the Ionosphere and Magnetosphere, An, III.2001
Introduction to the Study of Physical Metallurgy, An, III.1517

Inverse-Compton catastrophe, III.1781
Inverse-Compton scattering, III.1781
Inverse-square law, II.1266
Inversion layer, III.1373
Investigations of the solar spectrum and the spectra of the chemical elements, III.1692
Iodine, contrast media, III.1863
Iodine-123, III.1920
Iodine-131, III.1900, III.1903–6
Iodine-132, III.1905
Ion–atom collisions, see Collisions between atoms or ions
Ion guns, III.1997
Ionic crystals, II.997–8, III.1528–9, III.1531, III.1541
Ionization chamber (thimble chamber), III.1866–7, III.1881
Ionization detectors, II.746–9
Ionization energy, III.1867
Ionization equilibrium, III.1701
Ionization Phenomena in Gases, III.1644
Ionization threshold, II.1069–73
Ionization thresholds, spectral connections at, II.1058–9
Ionizing radiation, III.1857, III.1889
Ionosonde, III.1989, III.1991
Ionosphere, III.1619–23, III.1682–7, III.1981–2001
  'E' layer, III.1621
  electron content, III.1995
  'F' layer, III.1621
Ionospheric physics, III.1944
IRAS satellite, III.1767
Iridium-192, III.1879, III.1892
Iron–carbon alloy, III.1524
Iron-shrouded lens, III.1567
Irreducible cluster integrals, I.552
Irvine–Michigan–Brookhaven (IMB) experiment, III.1760
Ising model, I.554–61, III.1554

Subject Index

Isobars, II.680
Isomeric levels, II.1210
Isomorphous replacement method, I.445
Isospin, **I.395**
Isothermal compressibility, I.546, III.1397
Isotope effect, II.933
Isotopes, I.107
Isotopic charge, II.676
ITER (International Toroidal Experimental Reactor) device, III.1651
ITS-27 temperature scale, II.1250
ITS-90 temperature scale, II.1250, II.1260

J/ψ, II.701–2
Jacobians, I.12
Japanese Atomic Energy Agency, III.1650
JASON, III.2023
Jeans criterion for collapse, III.1768
Jellium, III.1303–5
JET (Joint European Torus), III.1650, III.1651, III.2055
Jets, II.715–16
jj-coupling shell model, II.1209, II.1210
Johnson noise fluctuations, I.601
Joint European Torus, *see* JET
Joint replacement, III.1896
Jones zone, III.1313
Josephson constant, II.1258, II.1271
Josephson effect, II.941–2, II.1096, II.1257, II.1258
Josephson junction, II.1095
Joule–Thomson effect, I.28
Journal of Applied Crystallography, I.511
Journal of Materials Research, III.1556
Journal of Nuclear Materials, III.1534
Julia series, III.1841

Kamioka detector, II.765
Kamiokande neutrino-scattering experiment, II.759, III.1748
Kaons, II.650–4, II.656–7, II.660, II.678
Kapitza's law, III.1283
Kármán vortex street, II.851
K-distributions, III.1466
Kelvin (degree), II.1250
Kelvin Circulation Theorem, II.817, II.822
Kelvin–Helmholtz time-scale, III.1694
Kennedy–Thorndike experiment, I.285
Kennelly–Heaviside layer, III.1621, III.1987
Kerr metric, III.1778
Kerst–Serber theory, II.1206
Kevlar polymeric fibres, III.1552
Keyhole surgery, III.1895
KHF_2, structure of, I.458
KH_2PO_4
  ferroelectricity, I.488
  nuclear density, I.461
  structure of, I.459
Kiloelectronvolt (keV), I.358
Kilogram (unit), II.1233–8
Kinematic shock wave, II.899
Kinematic wave, II.898
Kinematics, relativistic, I.364
Kinetic constant, I.154
Kinetic energy, I.90, II.837
Kinetic equations, I.608–11
Kinetic theory of gases, I.30, I.35, I.46, I.522, I.523, I.587, III.1284
Kirchhoff's function, I.148
Kirchhoff's law, I.66
Kitt Peak National Observatory (KPNO), III.1744
Klein–Gordon equation, I.209, I.283
Klein–Gordon theory, I.209
Klein–Nishina formula, I.218, I.365

S27

Klystrons, II.1206
K-meson, see Kaon
KMS Fusion, III.1676
$KNbO_3$ ferroelectric phase transition, I.486
Knowledge domains, III.2041
Kogelnik–Li equations, III.1430
Köhler illumination, III.1390
Kolmógoroff–Sinai entropy, I.615
Kolmógorov dissipation length, II.838
Kondo effect, II.1167–9
Kondo problem, III.1343
Kondo theory, III.1343
Korteweg–de Vries equation, II.859
Kramers equation, I.607–8
Kramers–Heisenberg effect, III.1405
Kronecker delta-function, I.208
Krypton, I.75
K systems, I.615
Kuhnian revolution, III.2049
Kuiper Airborne Observatory, III.1743
k-values, III.1364

Laboratory of Biological Physics of the College de France, III.1894
LAK theory, III.1359
Lamb-dip, III.1443
Lamb–Retherford shift, III.1421
Lamb shift, II.641–3, II.649
Lamellar phases, III.1608
Laminar flow, II.827
Landau anti-damping, III.1625
Landau critical velocity, II.928
Landau damping, III.1624, III.1625
Landau expansion, II.1009
Landau–Ginzburg–Wilson Hamiltonian, I.568
Landau theory of second-order transitions, I.549–51, II.1009
Landolt–Börnstein series, II.1165
Landolt–Börnstein Tables, III.1513

Lane–Emden equation, III.1695, III.1711
Langevin equations, I.595–7, I.608, I.614, I.624, I.626, I.628
Langevin function, I.548, II.1116
Langley's bolometer, I.23
Laplace–Herschel nebular hypothesis, III.1945, III.1962
Large Electron–Positron Collider (LEP), II.741, III.1792
Large Hadron Collider, III.2036
Large Magellanic Cloud, II.758, III.1760
Large Space Telescope, III.1741
Large-scale-integration, see LSI
Laser anemometry, III.1442
Laser beam
  particle trapping, III.1462
  self-trapping, III.1459
  wavefronts, III.1458
Laser diode, III.1476, III.1480
Laser-fusion, III.1676
Laser-fusion-generated electrical power, III.1479
Laser-fusion pellet, III.1675
Laser interferometer gravitational-wave observatory (LIGO) project, I.298, I.317
Laser Interferometric Gravitational Observatory, III.2024
Laser-light-scattering, III.1441–5
Laser-printing, III.1541
Laser read-out systems, III.1462
Laser sources of infrared radiation, II.1054
Laser speckle, III.1453
Laser standards, II.1242
Laser velocimetry, III.1442
Lasers, III.1417–18, III.1424–52
  applications, III.1478–9
  argon, III.1894
  argon-ion, III.1453
  carbon dioxide, III.1442, III.1677, III.1894

chemical, III.1445
continuous-wave single-mode silica-fibre, III.1447
excimer, III.1453
free-electron, III.1456
GEKKO series, III.1676
glass, III.1677
HCN and H_2O action, III.1442–3
helium–cadmium, III.1435, III.1445
helium–neon, III.1427, III.1435, III.1439
history of, III.1411, III.1423
in medicine, III.1480
instabilities, III.1466–7
molecular, III.1442
Nd:glass, III.1479
Nd:YAG, III.1435, III.1471, III.1479, III.1894
noise, III.1444
Nova, III.1464, III.1676
rare-earth ion, III.1438
rare-gas ions, III.1443
ruby, III.1424, III.1426, III.1429, III.1430, III.1432, III.1479, III.1894
soliton, III.1469
spatial mode patterns, III.1431
titanium–sapphire, III.1448
tunable, III.1472
wavelength-tunable dye, III.1448
wire, III.1472
x-ray, III.1464–6, III.1479
see also under specific types and applications
Lasing without inversion (LWI), III.1473
Lattice conduction, II.981
Lattice dynamics, II.967–91, II.997–1007
Lattice gas model, I.553
Lattice gauge theory, II.756
Lattice vibrations, I.452
Lau effect, III.1414
Lau interferometer, III.1414

Laue equations for X-ray diffraction, **I.427**
Laue photographs, I.458
Lead, II.999
 crystals, III.1544
 isotopes, III.1960, III.1961
Lead sulphide detectors, III.1743
Learned societies, I.3
Lehrbuch der Metallographie, III.1554
Length, II.1239–42
Lens aberrations, III.1458
Lens-pupil functions, III.1418
Lenses
 contact, III.1414
 gradient-index, III.1391
 iron-free, III.1568
 magnification ratio, III.1569
 properties of, III.1389
 see also under specific types of lens
Lenz–Ising model. See Ising model
Lepton number, II.664, II.755
Leptons, II.635–6, II.692–5, II.700, II.704, II.719–25
Leukaemia, III.1891
Levitation theory, III.1543
Levshin–Perrin law, III.1400
Liapunov functions, I.621–2
Lidar systems, III.1479
Lift–drag ratio, II.870, II.871, II.881
Light, speed of, I.260, I.261, I.266, II.1240, II.1262–4
Light-beating spectroscopy, III.1421
Light-emitting diode (LED), III.1474, III.1476
Light microscope, resolving power, III.1574
Light-quantum hypothesis, I.71–2, I.151–2, I.167–8, I.185, I.363
Light rays, I.260
Light-scattering spectroscopy, III.1408
Light transformations by atoms, II.1101

Light transmission imaging (diaphanography), III.1932–3
Light waves, I.258
Lightning-induced electron precipitation, III.1686
Line broadening, I.471–2
Line of section, III.1836
Linear accelerators, II.734–8, II.1205, III.1874, III.1875
Linear energy transfer (LET), III.1885
Linear response theory, I.626–7
Linnean system, I.47
Liouville equation, I.593, I.602, I.605–7, I.619, I.620
Liposomes, III.1608
Liquid crystal display (LCD), III.1476, III.1540–1, III.1593, III.2029
Liquid crystal television display, III.1476
Liquid crystals, II.1084, III.1539–42, III.1552–3, III.1603–5, III.2031
Liquid-drop model, II.1186–91, II.1213
Liquid gases, I.28
 critical point, I.543–4
Liquid helium cryostat, III.2025
Liquid metals, III.1366–7
Lithosphere, III.1978
Liverpool Astronomical Society, I.24
Lloyd Triestino, III.1711
LMC X-1, III.1757
LMC X-3, III.1757
Local gauge invariance, II.693
Local gauge transformation, II.693
Lockheed C-141 transport aircraft, III.1743
Lodestone, II.1112, II.1156, III.1856
London Chemical Society, I.46
London equation, II.925, **II.926–7**
Long-chain systems, III.1596–7
Longitudinal waves, III.1956
Long-range collisions, II.1076

Long-range effects, **II.1066–7**
Long-range interactions, II.1063–5
Long-range parameters, **II.1074–5**
Lorentz contraction, I.264
Lorentz–Einstein formula, I.285
Lorentz electron model, I.32
Lorentz factor, III.1783
Lorentz force, III.1622
Lorentz force equation, III.1671
Lorentz–Rayleigh model, I.629
Lorentz transformation, I.264, I.265, I.272, I.274, I.275, I.281
Loschmidt's constant, I.150
Low-energy electron diffraction (LEED), I.466
Low-energy electron microscope (LEEM), III.1577
Lowest unoccupied molecular orbital (LUMO), II.1083
LSI (large-scale-integration), III.1371
Luminescence, III.1537
Luminosity function for galaxies, III.1769–70
Luminosity–spectral class diagram (Russell diagram), III.1696, III.1698, III.1710
Luminous stars, III.1696
Lummer–Gehrcke plate, III.1389
Lummer–Gehrcke spectrograph, I.19

Mach number, II.813, II.814, II.821–4, II.880
Mach's principle, III.1723
Macroscopic wavefunctions, II.943
Macrostate, I.523, I.526
MAFCO, III.1673
Magellanic Clouds, II.758, III.1714, III.1715, III.1719, III.1744, III.1760, III.1788
Magic numbers, II.1225
Magnesium, formation of, III.1747
Magnesium oxide (MgO) crystals, I.453

Subject Index

Magnetic anisotropy, II.1134, II.1154
Magnetic behaviour, types of, II.1112
Magnetic breakdown, III.1359
Magnetic bubbles, II.1161
Magnetic compass, III.1952
Magnetic dipole, II.1111, III.1958, III.2004
  interactions, II.1011
Magnetic domain, II.1154–5
  patterns, II.1135
Magnetic electron lens, III.1566–9
Magnetic equation of state, I.545
Magnetic excitons, II.1162
Magnetic field, I.27, I.102, I.108, I.261, I.541, II.1111, II.1125, II.1126, II.1132, II.1172, III.1282, III.1345, III.1352, III.1358, III.1373
  Earth's, III.1622–3, III.1952–60, III.1973
  in sunspots, III.1958
Magnetic flux, II.919, II.1121
Magnetic fusion research, III.1619, III.1629–31
  classification era, III.1631–40
  post-1960, III.1645–70
Magnetic induction, II.1111
Magnetic insulators, co-operative phenomena in, II.1147–8
Magnetic intensity, II.1111
Magnetic interactions, II.1010
Magnetic lens, III.1568, III.1580
Magnetic materials, I.559
Magnetic mirror, III.1658–70
Magnetic models, I.570
Magnetic moments, I.376, II.1115, II.1117, II.1123, II.1166–7
  in Au_2Mn, I.462
Magnetic monopoles, II.766
Magnetic phase transitions, II.1014–18
Magnetic pinch, III.1636
Magnetic properties, II.1112
  and crystal structures, II.1156–8
Magnetic recording, II.1158–9
Magnetic recording, II.1158
Magnetic resonance, I.541–2
Magnetic resonance imaging (MRI), II.1176, III.1873, III.1910, III.1923-33
Magnetic resonance spectroscopy (MRS), III.1930, III.1931
Magnetic saturation, II.1145
Magnetic spectrometers, II.1206–8
Magnetic structures, I.462, I.463, II.991
Magnetic susceptibility, I.545, I.557, I.558, III.1595
Magnetism, I.65, II.1111–81, III.1832
  conductors, II.1133–6
  developments—prior to twentieth century, II.1111–13
  developments—1900–25, II.1113–23
  developments—1925–50, II.1123–5
  developments—1950 onwards, II.1136–7
  molecular field theory, II.1014
  quantum concepts, II.1115–16
Magnetization, I.555, I.559
  intensity of, II.1111
  spontaneous, I.573, II.1118
Magnetoencephalography, III.1932
Magneto-hydrodynamic (MHD) theory, III.1648, III.1650, III.1658, III.1663, III.1670, III.1672, III.1673, III.1684, III.1685
Magneto-ionic theory, III.1990, III.1992
Magnetometer, III.1970
Magneto-optical effect, III.1389
Magnetopause, III.2011
Magnetoresistance, III.1356–7
Magnetosphere, III.1752, III.1981, III.1983, III.1984, III.2001–11

S31

Magnons, II.1162
Manchester Literary and Philosophical Society, I.3
Mandelstam–Brillouin doublet, III.1408
Mandelstam–Brillouin scattering, III.1397
Mandelstam–Brillouin spectral shifts, III.1398
Manganese fluoride, I.559
Manganese oxide (MnO), powder pattern, I.460
Manhattan Project, III.1531, III.1556, III.1630, III.2023, III.2025
Markovian kinetic equation, I.620
Masers, III.1417–18, III.1422
Mass, II.1236–8
Mass absorption anomaly, I.398
Mass–energy absorption coefficients, III.1882
Mass–energy equivalence relation, I.285
Mass spectrographs, I.111
Massachusetts Institute of Technology, I.16
Massey criterion, II.1077
Master equation, I.593, I.605–7, I.618
Master function, I.606
Master-oscillator power amplifiers (MOPAs), III.1474
Materials
  characterization, III.1506
  high-quality, III.1280–1
  physics of, III.1505–64
  science, III.2046–7
  technology, III.2022
  terminology, III.1553
  see also under specific materials
Materials Research Laboratories (MRLs), III.1557
Materials Research Society (MRS), III.1556–7

Materials Science and Technology, III.1513
Mathematical Theory of Relativity, III.1700
Mathematical theory of speculation, I.586
Mathematical tools, III.2030
Mathematics Tripos, I.10, I.13
Matrix mechanics, I.360
Matter, I.44, I.47
  condensed, II.756
  constituents of, I.357–8
  new forms of, II.650–62
  thermodynamic properties, I.542
Max Planck Institute, III.1555
Maxwell–Boltzmann distribution, I.589–91, I.598
Maxwell–Boltzmann law, III.1285
Maxwell–Boltzmann statistics, I.527
Maxwell equations, I.33, I.260–1, I.264, I.265, I.270, I.279, I.283, III.1394, III.1622, III.1625, III.1671, III.2018
Maxwellian distribution of protons, III.1707
Maxwellian electromagnetism, I.29
Maxwell's Demon, I.576
Maxwell's electromagnetic theory of light, I.26
Maxwell's theory, I.29, I.321
Mean field approximations, I.554
Mean free path, I.522, III.1549
Mean square radius of gyration, I.569
Mean thermal energy, I.585
Measurement
  applications, II.1240–2
  capacities, III.2025
Mechanical oscillograph, I.26
Mechanical principle of equivalence, I.287–8
Mechanocaloric effect, II.922
Medical electronics, III.1893–6
Medical imaging, III.1901, III.1912

Subject Index

new techniques, III.1932–3
Medical mechanics, III.1896–900
Medical physics, III.1855–941, III.2022
  branches of, III.1855–9
  chronological evolution, III.1855–9
  computers in, III.1858
  modern applications, III.1857
Medical Research Council (MRC), III.1871, III.1877, III.1891, III.1924
Medical ultrasound, III.1910–15
  clinical perspective, III.1915
Medicine
  lasers in, III.1480
  nuclear, III.1858, III.1892, III.1900–10
Meetings, III.2028
Meissner (or Meissner–Ochsenfeld) effect, II.919–20, II.924, II.941, II.957
Mekometer, II.1240
Men and Decisions, III.1640
Mesomorphic phases, III.1603
Meson Factory, II.1205
Mesons, I.378–83, I.389–94, I.402–4, II.640, II.643, II.661, II.686, II.703, II.721
  B-meson, II.721, II.724–5, II.767–8
Mesophere, III.1982
Mesoscopic systems, III.1370
Mesotrons, I.389, I.404–5
  cosmic-ray, I.394–9
  decay, I.399–402
  decay versus capture, I.405–7
  lifetime, I.397–8, I.406
  nuclear interaction, I.407
Metaboric acid, structure of, I.449
Metal–dielectric interference filters, III.1407
Metallography, III.1468, III.1507, III.1554
Metallurgy, III.1505, III.1554
  pioneers in, **III.1511**
Metals
  electrical conduction in, III.1284–7
  electron theory, I.215–17
  groups I–III, III.1354
  individual properties, III.1345–59
  liquid, III.1366–7
  phonon dispersion curves, II.999
  phonons in, II.998–9
  physics of, III.1518
  post-war research, III.1341–5
  resistivity variation with temperature, III.1281–2
Metals Crystallographic Data File, I.512
Metals Handbook, III.1513
Metals Reference Book, III.1513
Metalworking, III.1505
Metastable effect, II.946
Metastable equilibrium, I.530
Metastable states, II.918
Meteorites, III.1965
Methoden der Mathematischen Physik, III.2043
Methylene di-phosphonate (MDP), III.1920
Metre (unit), II.1235, II.1239
Metre Convention, II.1236
Metric system, II.1233, II.1235
Metrically transitive systems, I.599
Metrology, II.1233, II.1273
  role of atoms, II.1094–7
Metropolitan Vickers Company, III.1871, III.1873
MFTF (Mirror Fusion Test Facility), III.1664, III.1668, III.1669
Michelson interferometer, III.1704
Michelson–Morley experiment, I.32, I.262, I.264, I.266, III.1401
Michelson's interferometer, III.1399

Micro-canonical ensemble, I.532
Microphysics, I.210–33
Microscopes, I.26, III.1475
  see also under specific types of microscope
Microscopic physics, I.358
Microstructology—Behavior and Microstructure of Materials, III.1550
Microstructure, III.1542–3
  quantitative measurement of, III.1549
  study of, III.1506–7
  see also under specific materials
Microtechnology, III.1899
Microwave electron linear accelerators, III.1873
Microwave radiation, II.1053–4
Microwaves
  applications, II.1273
  frequency standards, II.1055
  parametric amplifier for, II.1158
Mid-Atlantic Ridge, III.1973, III.1976
Middlesex Hospital, III.1877
Milky Way, III.1714, III.1716, III.1718
Miller indices, I.421
Millions of electronvolts (MeV), I.358
Mineralogy, III.1514
Minkowski space, I.292
Minor impurity, III.1505
Mirror Fusion Test Facility, *see* MFTF
Mirror machine, III.1648, III.1662
Mirrors, properties of, III.1389
MIT Radiation Laboratory, III.2026
Mixed state, II.920
MKSA system, II.1235, II.1257
Mobility edge, III.1368
Models, I.603–5, I.618–19, I.628–9
Mohorovičić discontinuity ('Moho'), III.1954
Moiré grating interference, III.1414

Molecular beam analysis, II.1033
Molecular biology, I.496, I.497, III.2020, III.2030, III.2046
Molecular bonds, II.1042–9
Molecular collisions, I.588, II.827
Molecular dissociation, II.1067
Molecular dynamics, I.577
Molecular field hypothesis, I.545
Molecular fluids, I.557
Molecular gases, I.35
Molecular hydrogen, I.181, II.1045, **II.1048**
  in interstellar medium, III.1766
  photoabsorption by, II.1069–73
Molecular line astronomy, III.1762–5
Molecular physics, I.358–62, II.1025–49, II.1081–9
Molecular resonances, II.1084
Molecular sheets, III.1595
Molecular spectra, II.1046–7
Molecular spectroscopy, I.529, II.1085–7
Molecular structure, I.211–15
Molecular systems, perturbations in, III.1593
Molecular theory, I.43
Molecular viscosity, II.826
Molecules, I.45, I.47–9, I.81–2, I.85, I.587, I.588
Molybdenum-99, III.1907
Momentum space, I.567
Momentum transport, II.853
Monomers, III.1596
Monotone curve, III.1837
Monsoon phenomena, II.907–8
Monte Carlo methods, I.572, I.628, III.1560
Moon
  formation of, III.1967
  gravity fields, II.888
  orbit, III.1948
  origin of, III.1952, III.1965, III.1966

Morgan, Keenan and Kellman (MKK) system, III.1698
MOSFET, III.1371, III.1372, III.1375
Moss–Burstein shift, III.1421
Mott transition, III.1365, III.1368
Mount Palomar, III.1745, III.1758
Mount Vernon Hospital, III.1867
Moving-coil experiment, II.1238
MSS (paper), III.1839, III.1843, III.1849, III.1850
MULTAN (multiple tangent formula method), I.450
Multi-channel formulation, II.1067–73
Multi-electron fragmentation, II.1065–7
Multi-electron spectra, II.1035–7
Multi-particle system, III.1394
Multiphase interstellar medium, III.1765–6
Multi-photon processes, II.1099–100
Multiple-beam interferometry, III.1412
Multiple collisions, I.611
Multiple-laser-beam systems, III.1479
Multiple sclerosis (MS), III.1928, III.1932
Multiplex optical spectroscopy, III.1416
Multiply connected graphs, I.552
Multi-TeV hadron colliders, II.767
Muon $g - 2$ factor, II.647–9
Muons, II.640, II.650–4
Myoglobin, I.445
  structure, I.500–1

Nambu–Goldstone boson, II.694, II.695
Nambu–Goldstone mode, II.694
Nambu–Jona-Lasinio model, II.756
$NaNbO_3$, dielectric and crystallographic properties, I.487

Naphthalene, crystal structure, I.494
National Institute of Standards and Technology (NIST), II.1235
National laboratories, role of, II.1254–5
National Measurement Accreditation Service (NAMAS), II.1272
National Physical Laboratory (NPL), I.20, II.1234, II.1243, III.1517, III.1867, III.1881, III.1890, III.1988
National Radium Commission, III.1877
National Science Foundation, III.1980
National Standards Laboratories, III.1867
Natural philosophy, I.13
Natural rubber, III.1593
Natural Sciences Tripos, I.10–14
Nature, I.5, I.6, I.96, I.116, I.129
Naval Research Laboratory (NRL), III.1739, III.1988, III.1992
Navier–Stokes equations, III.1834, III.1835
Near-infrared radiation, II.1053–4
Near-space environment, III.1981
Nebular hypothesis, III.1944–5, III.1950
Nebulium, I.75, I.81
Néel temperature, II.1129, II.1151, II.1152, II.1159
Negative dispersion, III.1399–401
Negative-resistance device, III.1340
Negative temperatures, I.541–2
Neodymium, III.1438
  paramagnetic susceptibility, II.1162
Neodymium–iron borates, II.1165
Neon, I.75, III.1747
Neptunium, II.1223

Nernst–Lindemann formula, II.972
NeuroECAT III, III.1921
Neurological disorders, III.1899
Neutral currents, II.695–7, II.699
Neutral hydrogen, III.1762–5
Neutral kaons, II.658–9
Neutral weak currents, II.695
Neutrino masses, β-decay limits on, II.755
Neutrinoless double-β decay, II.755
Neutrinos, I.129, I.370, I.371, II.639, II.640, II.663–4, II.669–71, II.758–60, III.1748–9
  helicity, **II.670**
  oscillations, II.763
  supernova, II.758
  see also Solar neutrinos
Neutron beams, I.468
Neutron bombardment, I.122
Neutron capture, II.1193, II.1223, II.1227, III.1785
Neutron decay, III.1785
Neutron diffraction, I.452–63, I.487, III.2025
Neutron flux, II.994–6
Neutron guide, I.458
Neutron irradiation, I.130, III.1871, III.1872
Neutron–neutron forces, II.1190
Neutron–proton forces, II.1190
Neutron–proton model, I.371–3, II.1190
Neutron–proton ratio, III.1792
Neutron scattering, I.453, II.991–6, II.1011, II.1166, III.2025
Neutron sources, II.1199, II.1206
Neutron stars, III.1712–13, III.1751–4, III.1756
Neutrons, I.112, I.118–26, I.369–71, I.379, II.639, II.991
  lifetime and decay properties, II.754–5
New General Catalogue of Nebulae and Clusters of Stars, III.1717

Newton Victor Maximar, III.1864
Newtonian constant of gravitation, I.28
Newtonian mechanics, I.29, I.36, I.257, I.261
Newton's equations, III.1622
Newton's rings, III.1386
NGC2264 star cluster, III.1767
$NH_4H_2PO_4$, antiferroelectricity, I.488
Nicholson–Bohr quantum rule, I.175
Nickel–titanium alloys, III.1539
Nobel Prizes, III.2021
Noble metals, III.1343
No-hair theorem for black holes, III.1778
Noise problems in upper-atmospheric physics research, III.1995
Noise radiation, II.872–3
Non-abelian gauge theory, II.693
Non-classical light, III.1388
Non-crystalline solids, I.471–84
Non-dimensional parameters, II.812–16
Non-equilibrium theory, I.630
Non-imaging light concentrators, III.1446
Non-linear differential equations, III.1826
Non-linear effects, II.808, II.846–69
Non-linear equations, III.1826
Non-linear field equations, II.800
Non-linear field theories, II.798
Non-linear optics, II.1100, III.1424–53, III.1458–61
Non-linear oscillations, III.1837–8
Non-linear relations, I.233
Non-linear science, III.2030
Non-linear systems, III.1835, III.2030
Non-linear theory, II.810
Non-linearity, II.856–60, III.1403
Non-metallic materials, III.1528–31

Non-perturbative calculations, III.1825
Non-stationary state, II.1029
Nordström's theory, I.295
Nova laser, III.1464, III.1676
Nuclear atom, discovery of, III.2018
Nuclear binding, I.111, II.1208, II.1210
Nuclear bombs, III.1886
Nuclear demagnetization, II.1175
Nuclear dynamics
  as many-body problem, II.1191–5
  background, II.1183–91
  developments in, II.1183–232
  effects of World War II, II.1196–9
  technical advances, II.1199–208
Nuclear energy, I.107, II.1191, III.1879
Nuclear fission, I.126, I.129–33, II.1195, II.1197
Nuclear forces, I.123, I.378–83, I.402–4, II.1208
  fundamental theories, I.375–83
Nuclear fusion, *see* Fusion research
Nuclear magnetic moments, I.116–17, II.1171–2
Nuclear magnetic resonance (NMR), II.1094, II.1171–5, III.1859, III.2026
Nuclear magnetic resonance (NMR) imaging, III.1898, III.1923–33
Nuclear magnetism, II.1171–5
Nuclear masses, II.1186–91
Nuclear matter, density of, II.1189, II.1190
Nuclear medicine, III.1858, III.1892, III.1900–10
Nuclear physics, I.115, I.119, II.754–62, III.1871
  defence laboratories, II.1196
  research laboratories, II.1198

Nuclear reactions, I.123–6, II.1219–22, III.1629
  direct reaction mechanism, II.1220–1
  nucleus theory of, II.1192
Nuclear reactors, II.1197, III.1871
Nuclear scattering, II.1012, II.1219–22
Nuclear size, I.116
Nuclear spin, I.117, I.541, II.1171, III.1410–11
Nuclear statistics, I.117–18
Nuclear systematics, I.366–7
Nuclear volume, II.1189, II.1208
Nuclear weapons, III.1980
Nucleation process, III.1547
Nucleic acids, I.503–4, I.507
Nucleon isobars, II.659
Nucleon–nucleon force, II.1212
Nucleons, II.639, II.658, II.660, II.675, II.1218
Nucleosynthesis, II.758–60, III.1746–7, III.1792
Nucleotide chains, I.506
Nucleus, I.103–33, I.370
  collective motion in, II.1213–19
  compound, II.1219
  Heisenberg's n–p model, I.371–3
  low-energy collective modes, II.1214–18
  models, I.109–15
  shell structure in, II.1208–13
Numerical calculations, III.1828
Numerical solution, III.1826

Oak Ridge DCX experiments, III.1661–2
Ocean current, II.862
  patterns, II.892, II.896
Ocean engineering, II.869–87
Ocean ridges, III.1978, III.1981
Ocean tides, II.888
Ocean waves, II.857, II.860
Offshore structures, II.886–7
OGRA mirror machine, III.1662

Ohm
  absolute determination, II.1255
  international, II.1254, II.1257
  monitoring, II.1258
  reference standards, II.1256
Ohm's law, III.1336
Oligonucleotide structures, I.507
On the Principles of Geometry, III.1722
Onsager–Feynman quantization condition, II.944–5
Onsager regression hypothesis, I.616
Onsager relations, I.600–2, I.615–16, I.626–7
Ophthalmology, III.1930
Optical actions, atomic systems, II.1097–101
Optical astronomy, III.1744–5
Optical bistability, III.1449–50
Optical coatings, III.1407
Optical coherence theory, III.1392, III.1407–10, III.1461
Optical comparators, II.1240
Optical computing, III.1461
Optical fibre
  communication, III.1447–9
  sensors, III.1471–2
  telecommunications, III.1471–2, III.1480
  see also Fibre optics
Optical filtering, III.1418
Optical glasses, III.1541
Optical goniometer, I.421
Optical imaging, III.1432–7
Optical information processing, III.1417
Optical instruments and techniques, I.26, III.1406–24
Optical interferometry techniques, III.1411, III.1422
Optical Kerr effect, III.1444–5
Optical maser, III.1443
Optical micrography, III.1508

Optical microscopes, III.1405, III.1422, III.1520, III.1526
Optical parametric oscillator (OPO), III.1435
Optical phonons, III.1338
Optical physics, III.1385–504
Optical properties, crystals, I.485–6
Optical pumping, II.1097–9, III.1415
Optical research, World War II, III.1410–14
Optical sciences, computers in, III.1475–6
Optical signal and data processing, III.1461
Optical Society of America, III.1445, III.1474
Optical spectroscopy, III.1421–2
Optical stethoscopy, III.1406
Optical systems
  polarization in, III.1411–12
  signal-to-noise ratio in, III.1418
Optical theory, Abbe's, III.1578–84
Optical transistor, III.1450
Optical video discs, III.1462
Optics
  classical, III.1385–6
  in special theory of relativity, I.257
  linear, III.1434
  new results and applications (1970–94), III.1452–72
  non-linear, III.1427–32
  towards the twenty-first century, III.1472–81
Optimal filtering, III.1418
Optoelectronic communications, III.2051
Optoelectronic devices, III.1472, III.1476
Optoelectronic industries, III.1476
Optoelectronic integrated circuits (OEICs), III.1449, III.1476
Optoelectronic neural networks, III.1461

Subject Index

Optoelectronic physics, III.1385–504
Optoelectronics
 early research, III.1419
 semiconductors, III.1417
Orbital magnetic moments, II.1012
Orbiting Astrophysical Observatories (OAO), III.1741
Orbiting Solar Observatory (OSO-III), III.1740
Order–disorder transition, I.459, I.473–4, I.578, II.1009, II.1016, III.1515
 lambda transition, II.915, II.922–4, II.927, II.944
Ordinary differential equations (ODEs), III.1826–7, III.1829, III.1835- 9, III.1851
Organic compounds, structure analysis, I.493–6
Orion Nebula, III.1767
Orthogonalized-plane-wave (OPW) method, III.1317
Orthopaedic surgery, III.1896–7
Orthotics, III.1899
Oscillation, II.896, II.907–8, II.973
Oscillator strengths, II.1058
Otolaryngology, III.1930
Out of the Crystal Maze, I.512
Oxalic acid, I.452, I.454
Oxford University, I.20
Oxygen, formation of, III.1747

Pacemakers, III.1895, III.1897
Pacific floor, measurements, III.1974
Padé Approximant, I.558
Pair connectedness, I.573
Pair distribution function, I.547
Palatini variational method, I.299
Palomar Atlas, III.1745
Palomar telescope, III.1721, III.1775, III.1793
Pangaea (giant land-mass), III.1968

Paraelastic crystals, III.1538
Parallel plate ionization chamber, I.54
Parallel shear flow, II.828
Paramagnetic insulators, II.1116–20
Paramagnetic materials, I.459
Paramagnetic salts, I.540–1
Paramagnetism, II.1015, II.1113, II.1119, II.1125–33, III.1354, III.1362
Parametric fluorescence, III.1448–9
Parametrization technique, III.1351
Parastatistics, II.687
Paris Symposium on Radio Astronomy, III.1789
Parity, II.654, **II.1028–9**
 conservation, II.665
 violation, II.639, II.698, II.699
Partial coherence, III.1392
Partial conservation of axial current (PCAC), II.676
Partial differential equations (PDEs), III.1827, III.1834, III.1851
Particle accelerators, *see* Accelerators
Particle content, II.763
Particle energies, II.1201
Particle physics, III.1833, III.2038
 see also Elementary particles; and under specific particle types
Particle size distribution, III.1549
Partition function, I.527, I.532
Parton hypothesis, II.689–90, II.715
Paschen–Wien law, I.149
Patterson functions, I.441–5, I.495, I.501
Pauli exclusion principle, I.97–100, I.168, I.603, II.686, II.1116
Pauli spin-statistics theorem, I.283
Peltier heat, III.1284
Penicillin, structure of, I.496
Penning traps, II.1099

S39

Penrose tiling, I.483
Percolation processes, I.572–4
Perfect cosmological principle, III.1787
Period–luminosity relation, III.1714–15
Periodic table, I.95–7, I.359, II.635
Permanent magnets, II.1165, II.1238
Permittivity of free space, II.1261
Perovskites, I.486–7, I.489, II.1156
Perturbation
  in molecular systems, III.1593
  singular theory, II.800–3, II.829
  strong responses to, III.1593–694
  theory, I.360, III.1830, III.1833
PET (positron emission tomography), III.1904, III.1910, III.1920, III.1922, III.1931
Phase angles, I.439, I.442, I.450, I.502
Phase conjugation, III.1460
Phase-contrast microscopy, III.1409–10
Phase diagrams, III.1506, III.1508, III.1509, III.1542
  free-energy, I.538
Phase equilibria, III.1506
Phase matching, III.1434
Phase Rule, III.1508, III.1542
Phase transformations, I.542–74, I.577, II.1007–10, III.1542–50
  microstructural and microcompositional analysis in, III.1548–50
Phaseonium, III.1473
Philosophical Magazine, I.7
Philosophical Transactions of the Royal Society, I.7, III.1516–17
PHOBOS space probe, III.1750
Phonon dispersion curves, II.1000, II.1002
Phonon drag, III.1310, III.1311
Phonon exchange, II.953

Phonon interaction with hot electrons, III.1339
Phonons, II.981, II.982, II.990, II.998–9, II.1004, III.1302, III.1338
  in crystals containing defects and in glasses, II.1004–7
Phosphorescent materials (phosphors), III.1403, III.1530, III.1536, III.1537
Phosphors, *see* Phosphorescent materials (phosphors)
Photoabsorption
  by molecular hydrogen, II.1069–73
  spectra, II.1052, II.1053, II.1054, II.1056
Photo-assisted tunnelling, III.1423
Photocathode, III.1439
Photo-cells, III.1321
Photochromic spectacle lenses, III.1542
Photochron streak camera, III.1439
Photoconductivity, III.1321, III.1327–9
Photocopying, III.1541
Photodetection process, III.1441
Photodetectors, III.1433
Photodiode, III.1439
Photoelectric effect, I.37, III.1286
Photoelectronic imaging devices, III.1476
Photo-emission spectroscopy, III.1391
Photographic plates, II.1200
Photography, III.1389, III.1392, III.1410
Photoionization, II.1052, III.1762
Photometry, II.1251–2
Photomultiplier tubes, II.1207, III.1409, III.1411, III.1920
Photon antibunching, III.1459–62
Photon bunching, III.1468
Photon-correlation spectroscopy (PCS), III.1454–8

protein analyser, III.1471
Photon counting statistics, III.1424
Photon energy, II.1052
  conservation process, III.1435
Photonic bandgaps, III.1472
Photonic integrated circuit (PIC), III.1476
Photons, I.117, I.386, II.637, II.638, II.641, II.693, II.1028, III.1302, III.1393, III.1402, III.1403, III.1410, III.1413, III.1451–2, III.1994
Photophone experiments, III.1422
Photoprocesses, II.1102
Photorefractivity, III.1461
Photosensitive devices, III.1321
Photo-voltaic effect, III.1321
Physical constants, II.1233, II.1260–72
  linked sets, II.1267–72
Physical metallurgy
  birth of, III.1516–19
  quantitative theory, III.1519–28
Physical Review, The, I.6
Physical Society, III.2042
Physics
  as domain of knowledge, III.2041
  as social institution, III.2041, III.2056
  definition, III.2041
  methodology of, III.2049
  scope of, III.2041
  sociology of, III.2031
Physics and Chemistry of Solids, III.1556
Physics of Fully Ionized Gases, III.1631
Physikalische Gesellschaft, I.88
Physikalische Metallkunde, III.1555
Physikalische-Technische Reichsanstalt (PTR), I.7, I.20, I.69, I.107, I.916, II.1234, II.1254
Physikalische Zeitschrift, I.6, I.7

Piezoelectric properties, III.1538
Piezophotonic effects, III.1459
π-meson, *see* Pions
Pioneer III, III.2007–8
Pion–nucleon scattering, II.659
Pion–pion scattering, II.681
Pions, I.408, II.650–5, II.658, II.665, II.675–6, II.711
Pipes, flow in, II.825
Pippard coherence length, II.947
Pitchblende, III.1876
Planck constant, I.67, I.152, I.359, I.360, II.638, II.1026, II.1271, III.1387
Planck energy, II.972
Planck formula, II.970, II.971
Planck law, I.70, I.151, I.363, I.527, III.1399
Planetary dynamics, III.1830–1
Planetary systems, III.1950
Planets
  formation, III.1948–9, III.1963
  infinitesimal, III.1950
Plasma, III.1361
  diagnostic measurements, III.1644
  frequency, III.1623
  instability, III.1630
  oscillations, III.1362, III.1623–5
  radio wave propagation in, III.1987
  temperature, III.1617
  theory, III.1623–5
Plasma physics, III.1617–90, III.1991–2, III.1994
  developments—circa 1950 to the present, III.1627–45
  developments—first half of twentieth century, III.1619–25
  history of, III.1617, III.1618
  landmark experiments, III.1677–81
  research, III.1617, III.1618
  space, III.1682–7

Plasma Physics and the Problem of Controlled Thermonuclear Reactions, III.1642
Plasmapause, III.2009, III.2010
Plasmasphere, III.1982
Plastic deformation, III.1520, III.1526, III.1530
Plate tectonics, III.1956, III.1961, III.1976, III.1978–80, III.2022
Platinum–iridium alloy, II.1237, II.1239
Pleated sheet structures, I.500
Plunging breakers, II.862
Plutonium, II.1223–5
  discovery, II.1197
  weapons-grade, III.1633
p–n junction, III.1325, III.1417
Poincaré–Bendixson theorem, III.1835
Poincaré map, III.1836, III.1851
Poincaré recurrence theorem, I.618
Poincaré–Zermelo recurrence conflict, I.592
Point defects, III.1528–31, III.1533, III.1534
Poisson equation, III.1625
Poisson formula, III.1424
Poissonian statistics, III.1393
Polar Years, III.2005
Polaritons, II.988, II.990
Polarization in optical systems, III.1411–12
Polarization rules, I.94–5
Polarized light microscopy, III.1412
Polarizing microscope, I.421
Polarons, III.1321
Polepiece lens, III.1567
Polonium, I.119, II.1200, II.1206
Polycrystal Book Service, I.512
Polycrystalline metals, III.1518
Polycrystalline samples, III.1280
Polyethylene, III.1602
Polymerization, III.1550, III.1551
Polymerization index, III.1597

Polymers, II.1084, III.1596–603
  blends, III.1553
  chain-folding, III.1552
  colloid protection, III.1611
  configurations, I.569–70
  crystallization, III.1602
  dynamics of entangled systems, III.1600
  fibres, III.1898
  future prospects, III.1603
  gels, III.1602–3
  high-technology, III.1550
  modified, III.1597
  overlapping, III.1599
  physics of, III.1550–3
  reptation process, III.1601
  semicrystalline nature of, III.1551
  solid phases, III.1600
  solutions, III.1599
  statics, III.1599
  structure, I.569–70
Poly-τ-methyl-l-glutamate, I.498
Polymethylmethacrylate (PMMA), III.1602, III.1896
Polyoxyethylene, III.1593
Polypeptide chains, I.496, I.497, I.499–501, I.510
Polysaccharides, III.1611
Polystyrene, III.1602
Polytetrafluoroethylene (Teflon), III.1896
Pomeranchuk trajectory (Pomeron), II.682
Pomeron (Pomeranchuk trajectory), II.682
Positive electron, I.109
Positively charged light particle, I.231
Positron emission tomography, *see* PET
Positronium, II.645–6, II.649, II.650, II.702
Positrons, I.370, I.371, I.377, II.638
Post Office, III.1988

Potassium dihydrogen phosphate (KDP), I.570–1, III.1432
Potential-energy curves, II.1045
Potential flow, II.803, II.804, II.887
Potential vorticity, II.895
Potts model, I.571
Powder diffraction, I.453
Powder photographs, I.435
Poynting's vector, I.33
Precious metals, III.1281
Precipitation-hardening, see Age-hardening
P representation, III.1440
Primaeval atom, III.1785
Primary image, III.1578
Primordial nucleosynthesis, III.1785, III.1786
Princeton Model C stellarator, III.1656
Principal quantum number, I.84, I.91, I.215
Principle of complementarity, I.197
Principle of Relativity, III.2039
Probabilities in quantum physics, I.93–4
Probability amplitude manipulation, II.1064
Probability relations, I.449
Proceedings of the Royal Society, I.7
Programmable calculators, III.1829
Project Sherwood, III.1631–40
Project Sherwood—The US Program in Controlled Fusion, III.1640
Projection electron microscope, III.1584
Proportional counters, II.1207
Protein crystallography, I.445, I.502
Protein Data Bank, I.512
Proteins, I.496
  fibrous, I.497, I.499
  globular, I.497
Proton–antiproton colliders, II.743–4

Proton decay, II.764–5
Proton–electron (PE) model, I.109–10
Proton linear accelerators, II.1205
Proton–neutron (PN) system, I.123
Proton–proton chain, III.1708
Proton–proton colliders, II.742–3
Proton–proton collisions, I.377
Proton synchrotrons, II.732–3
Protons, I.113, I.115, I.118, I.370, II.639, II.701
  in medical physics, III.1876
  Maxwellian distribution of, III.1707
Prototype phenomena, II.1060–3
Pseudopotential, III.1353, III.1367
PSR 1913+16 binary pulsar, III.1757
Publications, III.2028
Pulsars, III.1752
Pulse-echo imaging, III.1911–13
Pulse-echo technique, III.1989
Pump/oxygenator, III.1897
Pump-probe configuration, III.1445
P–V–T relations, I.28
Pyrites, I.430

Q-switching, III.1430, III.1436, III.1438
Quadrature amplitudes, III.1468
Quadrupole moments, I.403, II.1214, II.1215
Qualitative quantum mechanics, I.116
Quanta, I.378, II.638, III.1390
  and radiation, I.146–55
  empirical foundations, I.145–72
  specific heat, I.152–4
Quantitative metallography, III.1549
Quantitative theory, I.44
Quantum chemistry, theory and computations, II.1087–8

Quantum chromodynamics (QCD),
  I.322–3, II.636, II.692,
  II.709–19, II.756, II.769,
  III.2020, III.2031
Quantum communication, III.1472
Quantum conditions, I.359
Quantum cryptography, III.1472
Quantum dot, III.1472
Quantum dynamics, I.72–89
Quantum electrodynamics (QED),
  I.358, **I.361**, I.362, I.374, I.378,
  I.385, I.386, I.391, II.640–50,
  II.646, II.740, II.769, III.1414,
  III.1446, III.1472
Quantum energy, III.1361
Quantum field theories (QFTs),
  I.206–9, I.318, **I.380–1**, I.617,
  II.640, III.2033, III.2037,
  III.2038
Quantum formula, I.156
Quantum gravity, I.317–19
Quantum Hall effect, II.1096–7,
  II.1257, II.1259, III.1372,
  III.1375
Quantum mechanical effects,
  I.163–72
Quantum mechanical many-body
  problem, II.1212
Quantum mechanical parameters,
  II.1063
Quantum mechanical
  reformulation, II.1065
Quantum mechanical scattering
  theory, I.217–19
Quantum mechanical
  transformation theory, I.199
Quantum mechanical tunnelling,
  III.1706, III.1707
Quantum mechanical
  wavefunction, II.934
Quantum mechanics, I.30, I.61,
  I.65, I.72, I.115, I.125,
  I.143–248, I.359, I.360,
  I.602–3, I.627–8, II.638, II.917,
  II.1026, II.1034, II.1058,
  III.1280, III.1325, III.1403,
  III.1467, III.1830, III.2049
  applications, I.211–20, I.229
  as reformulated classical
    mechanics, I.189–91
  aspects of theory beyond,
    I.232–3
  causality in, I.220–9
  complementarity in, I.220–9
  completeness and reality of,
    I.225–9
  in astrophysics, III.1706–8
  mathematical foundation,
    I.196–208
  new structures within 'standard
    theory', I.230–2
  non-relativistic, I.220
  origin and completion of, I.173
  physical interpretation, I.196–208
  present situation, I.229–30
  reality in, I.220–9
Quantum Mechanics, III.2043
Quantum non-demolition (QND)
  measurements, III.1469
Quantum numbers, I.84, I.90–100,
  I.157, I.162, I.182, I.215,
  II.657, II.658, II.1034, II.1036,
  II.1124
Quantum of circulation, II.931
Quantum optics, III.1387–406,
  III.1468, III.1471, III.1472
Quantum phenomena, I.253
Quantum physics, I.65–72, I.89
  probabilities in, I.93–4
Quantum postulate, I.70, I.81
Quantum statistics, I.525–8, I.537
Quantum-theoretical
  reformulation and
  non-commuting variables,
  I.186–9
Quantum theory, I.37, I.54, I.62,
  I.65, I.68, I.71, I.80, I.88,
  I.89, I.101–3, I.524–31, I.537,
  III.1285, III.1387, III.1393,

Subject Index

III.1395, III.1402–4, III.1962, III.2018
and gravitation, I.318
evolution, I.144–5
in special theory of relativity, I.282–3
of electron gas, III.1287–91
of ions, II.1123–5
of measurement, I.223–5, I.233
with classical statistics, I.527
Quarkonium decays, II.716–17
Quarks, II.635–6, II.677, II.684–7, II.699, II.719–25
  b-quarks, II.722–3
  charmed, II.716
  colours, II.687, II.691, II.709, II.711–13
  flavours, II.687
  free, II.691–2
  statistics, II.709–10
Quartz clock, II.1244, II.1247, II.1248
Quasars, III.1776–7, III.1798
  discovery of, III.1789
Quasi-classical distribution function, III.1441
Quasicrystals, I.481–4
Quasiparticles, III.1363, III.1364
Quasi-phase matching, III.1434–5
Quasi-stellar radio sources, III.1776
Quenching, III.1533, III.1542

Radar, III.1944
Radial distribution function, I.480
Radiation, II.1028–32
  and quanta, I.146–55
  see also under specific types of radiation
Radiation accidents, III.1887
Radiation actions, II.1049–59
  spectrum of, II.1055–8
Radiation biology, III.1857
Radiation biophysics, III.1857, III.1884–9

Radiation damage, III.1531, III.1534, III.1884–5
Radiation-dose distributions, III.1872
Radiation dosimetry, III.1880–4, III.1887–9, III.1890
  calorimetry in, III.1883–4
  chemical, III.1884
  ionization methods, III.1880–3
  radionuclides, III.1909
Radiation energy, III.1885
Radiation hazards, III.1886–7, III.1895
Radiation injury, latency of expression, III.1886
Radiation isodose curves, III.1865
Radiation protection, III.1857, III.1871, III.1889–93
Radiation shielding, III.1891
Radiation Trapped in the Earth's Magnetic Field, III.1683
Radiationless transitions, II.1086–7
Radiative instability, I.83
Radio, III.1987
Radio astronomy, III.1736–8, III.1774–6, III.1944
Radio communications, III.1985–6, III.2000
Radio galaxies, III.1795, III.1797
Radio interference, III.2003
Radio propagation, III.1619–23, III.2000, III.2007
Radio pulsars, III.1752, III.1754, III.1757–8
Radio quasars, III.1751, III.1797
Radio research, III.1985
Radio scintillations, III.1751
Radio signals, III.1987, III.1988
Radio sources, III.1776, III.1782, III.1788–90, III.1797–9
Radio telescopes, III.1776
Radio waves, III.1985–8
  attenuation of, III.1995
  propagation, III.1990

Radioactive decay, I.93, I.106, II.639, III.1960
Radioactive isotopes, II.1222
Radioactive Substances Act, III.1910
Radioactive tracers, III.1858
Radioactivity, I.36, I.37, I.49, I.52–3, I.56–63
  artificially produced, III.1900–10
  in medical physics, III.1857
Radiobiology, III.1884–9
Radiochemical Centre, III.1901, III.1907
Radiofrequency energy, II.1057
Radiofrequency spectroscopy, II.1047
Radiography, III.1878
Radiometric dating, III.1969, III.1973
Radionuclides, III.1871, III.1904
  dosimetry, III.1884
  imaging, III.1916
  in medical physics, III.1857
  radiation dosimetry, III.1909
  scanning, III.1898
Radiotherapy, III.1857, III.1858, III.1862–80
  computer-aided dose planning, III.1887
Radium, I.58, I.61, I.104–8, III.1876–9
  bomb, III.1877
  dosimetry, III.1880–1
  implants, III.1878
  needles, III.1877, III.1878
Radon, II.1206, III.1876, III.1879, III.1916
  seeds, III.1877
Raman effect, I.163, I.171–2, II.996, III.1404–6
Raman–Nath diffraction, III.1409
Raman scattering, I.117, II.988, II.990, II.996–7, II.1011, II.1013, II.1397, III.1436
RAND project, III.1741

Random close packing (RCP), I.577
Random loose packing (RLP), I.577
Random processes, I.595–7, I.600
Random walks, I.569–70, I.586, I.624, I.630, III.1597
Rapid solidification processing, III.1543
Rare-earth elements, I.97, II.955
  crystal fields, II.1164–5
  electron paramagnetic resonance (EPR), II.1141–3
Rare-earth ion lasers, III.1438
Rare-earth metals and alloys, I.463, II.1012, II.1168
  magnetic properties, II.1160–5
Rayleigh criterion, III.1467
Rayleigh–Einstein–Jeans (REJ) law, I.69
Rayleigh–Jeans equation, III.1393
Rayleigh–Jeans law, III.1390
Rayleigh–Lorentz models, I.628
Rayleigh number, II.844–5
Rayleigh scattering, II.1057
Rayleigh–Taylor instability, III.1782
Reactive collisions, II.1048–9, II.1088–9
Real space, I.567
  renormalization, I.568, I.569
Realm of the Nebulae, The, III.1720, III.1726
Reciprocal lattice, III.1295, III.1296
Recrystallization, III.1526
Rectifiers, III.1322–7
Red blood cells, I.630
Red giants, III.1704, III.1708–9
Red stars, III.1696
Redshift–distance relation, III.1726
Redshift–magnitude relation, III.1794–7
Redshifts, III.1725, III.1777, III.1798
Reduced radial distribution function, I.481
Reflection high-energy electron diffraction (RHEED), I.466

Subject Index

Refractive index, I.546, III.1449
Regge poles, II.681–2, II.718
Regge trajectories, II.682
Relativistic cosmology, III.1721–34
Relativistic kinematics, I.364
Relativistic statistical mechanics, I.282
Relativistic thermodynamics, I.281–2
Relativistic velocity (kinematic) space, I.278
Relativity, III.2049
  kinematics, I.161
  philosophical treatments and popular reactions, I.319–20
  theory of, I.62, I.249–356, II.1266
  see also General theory of relativity; Special theory of relativity
Relaxation processes, II.1132–3, II.1145–6
Renninger effect, I.465
Renormalizability, II.694, II.695, III.2034, III.2036
Renormalization, I.564–9, II.646–7, II.713, II.756, II.1009, II.1015, III.1829–30, III.1832, III.1834, III.1844, III.1848
Renormalization rule, III.1832
Research
  career prospects, III.2028
  funding, III.2058
  laboratories, III.2050
  linking academy with industry, III.2050–1
  management, III.2049
  needs, I.19–22
  performance, III.2052
  physics, I.16–18
  strategic (or pre-competitive), III.2051
  technologies, III.2044–5
Researching and/or teaching, III.2047–8
Residual resistance, III.1281

Resistivity, **III.1331**, III.1334
  absolute magnitude of, III.1303
  mechanisms of, III.1301–4
  temperature variation of, III.1302, III.1342
Resonance
  effects, II.1059–73
  excitation, I.125
  phenomena, II.659–60, II.1029, II.1060–3, II.1067–73
Resonant absorption of radiation energy, II.1057
Resonant electron excitations, II.1082–4
Reststrahlen bands, II.983
Reversed-field pinch (RFP), III.1656–8
Reviews of Modern Physics, III.1290
Revue Scientifique, I.6
Reynolds number, II.807, II.815, II.816, II.825, II.828, II.834, II.837, II.841, II.844, II.850, III.1852
Reynolds stress, II.826, II.827, II.830, II.837, II.853
R-factor, I.450–1
RF SQUID ring, II.942
Ribonucleic acid (RNA), I.504, I.509
Richardson–Lucy iterative deconvolution algorithm, III.1462
Riemann tensor, I.292, I.312
Riemannian metric, I.286
Riemannian space-time, I.322
River water, II.897–8
RKKT theory, II.1170
RKKY interaction, II.1164
R matrices, II.1068–9, **II.1070**
RNA, see Ribonucleic acid (RNA)
Robertson–Walker metric, III.1728–30
Rochelle salt, III.1538
  ferroelectricity, I.489
Rockets, III.1682, III.1738, III.1997, III.1999, III.2005

Rocks
  ages of, III.1960–1
  magnetization, III.1970
  remanent magnetism of, III.1970–2
Rod-vision, III.1416
Röntgen (unit), III.1881–2, III.1890
Röntgen Society, III.1861
Rossby waves, II.893, II.894, II.902
Rotating-disc microscopes, III.1451
Rotational motion, II.1216
Rotational spectra, II.1217
Rotational therapy, III.1868
Rotons, II.928
Royal Greenwich Observatory, II.1248
Royal Institution of Great Britain, I.3, I.27
Royal Society, II.1233
  Fellows of, I.10, I.12
  Philosophical Transactions, I.7, III.1516–17
  Proceedings of, I.7
'Rubber Bible', III.1513
Rubbers, III.1551
  vulcanization of, III.1551, III.1593
Rubidium, I.75
Running coupling constant, II.713
Russell diagram, III.1696, III.1698, III.1710
Russell–Saunders (or L–S) coupling, III.1702
Rydberg constant, I.157–8, II.1271
Rydberg energy, II.1026
Rydberg formula, II.1034

Sachs–Wolfe effect, III.1802, III.1805
Sacramento Peak Solar Observatory, III.1744
Safronov–Wetherill model, III.1963
Safronov–Wetherill theory, III.1966
Sagnac effect, III.1433
Sagnac interferometer, III.1398

Saha equation, III.1699, III.1701
St Bartholomew's Hospital, III.1871, III.1901
Salicylic acid, I.443
Salt, *see* Sodium chloride
San Andreas fault, III.1978
Sanduleak-69 202, III.1760
Satellites, III.1683, III.1741, III.1765, III.1791, III.1997, III.1999, III.2031
  see also under named satellites
Saturation analysis technique, III.1909
Scalar quantities,
  diffusion-convection balance for, II.841–6
Scale factor, III.1730, III.1731
Scaling, II.688–9
  empirical derivation, I.561–4
  equation of state, I.568
  violation, II.714–15
Scandium, I.75
Scanning electron microscope (SEM), III.1451, III.1571, III.1586–90
Scanning transmission electron microscope (STEM), III.1569, III.1587–90
Scattering phenomena, I.217–19
Schizophrenia, III.1922
Schlieren photography, II.863, II.880
Schmidt camera, III.1406
Schmidt corrector plate, III.1406
Schmidt telescope, III.1406, III.1745, III.1758
Schönberg–Chandrasekhar limit, III.1709
Schottky defect, III.1530
Schrödinger equation, I.196, I.205, I.209, I.225, I.602, II.798, II.862, III.1292, III.1293, III.1299, III.1315, III.1317, III.1320, III.1366, III.1458
Schrödinger wavefunction, I.174

Schrödinger wave mechanics, I.183–96
Schwarzschild metric, III.1778
Schwarzschild singularity, I.306
Science, I.6
Science Abstracts, I.1, I.7
Scientific American, I.5, I.6
Scintillation counters, II.749–51, III.1901, III.1904, III.1916
Scintillation crystal, III.1907
Scintillation detectors, II.1207–8, III.1901, III.1904
Sco X-1 x-ray source, III.1739, III.1755
Screening and antiscreening, II.1091–2, III.1361, III.1363
Screw dislocation, I.476
Scyllac device, III.1645
SDI (Star Wars enterprise), III.2024
Sea-floor spreading, III.1975, III.1976
Sea transport, II.882–7
Search for Structure, A, III.1556
Second (unit), II.1235, II.1247–8
Second Antarctic Expedition, III.2005
Second-harmonic generation (SHG), III.1429
Second International Congress of Radiology, III.1890
Second International Polar Year (1932–33), III.2005
Second law of thermodynamics, I.46, I.68–9, I.374, I.521, I.524, I.534- 5, I.575
Second-order transitions, I.549–51
Secondary image, III.1578
Seebeck coefficient, III.1310
Seebeck effect, III.1284, III.1309
Seebeck voltage, III.1310
Seismology, III.1954, III.1955, III.1956, III.1980
Selection rules (*Auswahlprinzip*), I.94, I.178–81
Selenium, III.1541

Selenium-75, III.1905, III.1909
Selenium rectifier, III.1318
Self-avoiding walk (SAW), I.569–70, III.1598
Self-consistency, III.1519
Self-interstitials, III.1533
Self-phase modulation, III.1448–9
Self-similarity, I.574
Semi-classical theory, III.1388–9
Semiconductors, II.1002, III.1297, III.1301, III.1323, III.1367–8, III.1419, III.2025
   crystalline, III.1368
   detectors, II.1207
   devices, III.2022
   industry, III.1281
   laser diode, III.1464
   lasers, III.1474
   molten, III.1368
   optoelectronics, III.1417
   particle detectors, II.1208
   physical investigation, III.1318
   physics, III.1330–6
   post-war years, III.1317–22
   stimulated emission in, III.1433
   yellow light emission from, III.1392
Semi-leptonic decays, II.675
Senior Wrangler, I.10–11
Sequential resonant tunnelling, III.1470
Seyfert galaxies, III.1777, III.1780, III.1784
Shallow-water theory, II.890, II.892, II.896
Shannon number, III.1418
Shape-memory alloys, III.1539
Shell model, II.997, II.1214
Shell structure, II.1027, II.1208–13
Ship design, II.884
Ship hulls, II.884, II.885
Ship model testing, II.886
Ship–wave pattern, II.884
Ships, hydrodynamic forces on, II.882

Shock waves, II.798, II.808–12, II.847–9, II.872–82
Shoenberg magnetic interaction, III.1359
Short-range interactions, II.1063–5
Short-range parameters, II.1073, **II.1074–5**
Short Wave Wireless Communication, III.1619
Shortt clock, II.1243
Shubnikov groups, I.462
Sidereal time, II.1247
Siegbahn double-focusing beta spectrometer, II.1206
Siemens-Schuckertwerke, III.1573, III.1574
Signal-to-noise ratio in optical systems, III.1418
Silicon, I.470, III.1281, III.1301, III.1526
Silicon on sapphire (SOS) interface, III.1590
Silver, solid solutions, III.1543
Silver chloride, III.1526
Silver halides, III.1529
Single crystals, I.435–7, I.506, III.1280–1, III.1318, III.1360, III.1520
Single dish antenna, Arecibo, Puerto Rico, III.1996
Single photon emission computed tomography, *see* SPECT
Singular perturbation theory, II.800–3, II.829
Singularities, I.306–12
Site percolation, I.572
Sixteen lectures on controlled thermonuclear reactions, III.1660
Size effect, III.1343
Skin effect, II.1157
Skin-temperature distribution, III.1932
SLAC-LBL Collaboration, II.701
Slender-body aerodynamics, II.877

Sliding charge density-wave conduction, II.936
Sloan-Kettering Memorial Hospital, III.1871
Small Astronomical Satellite (SAS-2), III.1740
Small Magellanic Cloud, III.1715
Small Science, III.2025–31
S-matrix theory, II.680, III.2038
Smectic phases, III.1604
Smith chart, II.680
Snoek pendulum, III.1523
Society of Nuclear Medicine, III.1907
Society of Photo-Optical Instrumentation Engineers, III.1474
Sociological phenomena, III.2029
Sociology of physics, III.2031
Sodium, I.65, II.999, III.1530
Sodium chloride, I.429, I.430, III.1514
Sodium iodide
 crystals, II.1207, III.1903
 scintillation counters, II.1210
Sodium pertechnetate, III.1919
Sodium potassium tartrate, *see* Rochelle salt
Soft matter, III.1593–615
 use of term, III.1593–696
Soft-mode theory, II.1007–8
Solar cell, III.1476
Solar energy, III.1446, III.2005
Solar flares, III.1997
Solar gas, III.2009
Solar neutrinos, II.757, II.1227, III.1748–9
Solar oscillations, III.1749
Solar particles, III.2009
Solar radiation, II.899, II.900
Solar spectrum, spectroscopic measurements of, III.1691–2
Solar stream, III.2002, III.2003
Solar system, III.1824, III.1831, III.1950

Subject Index

dualistic theory, III.1962, III.1963
isotopic anomalies, III.1963–4
monistic theory, III.1962, III.1963
origin of, III.1965
tidal theory, III.1950, III.1951, III.1952
Solar–terrestrial physics (STP), III.1982
Solar–Terrestrial Physics, III.2001
Solar–terrestrial space, III.2001–11
Solar wind, III.1997, III.2009
Solar zenith angle, III.1985
Solenoids, iron-free, III.1568
Solid–solid phase transformations, III.1542, III.1544
Solid solutions, III.1543
Solid-state physics, II.1165–6, II.1169, III.1279, III.1526, III.1529, III.2019, III.2025
  emergence of schools, III.1313–14
Solid-state structure analysis, I.421–519
Solidification, III.1542
Solids
  electrons in, III.1279–383
  properties of, III.1279
Soliton pulses, III.1469
Sorbonne, I.15
Sound from flows, II.850–6
Sound intensity, II.851
Sound waves, II.798, II.846, II.857, II.873, II.950
  interaction of, **II.978–9**
Southern oscillation, II.907–8
Space exploration, III.1479
Space groups, III.1513
Space physics, III.1618, III.1982
Space plasma physics, III.1682–7
Space-probe studies, III.1995
Space research, III.1943, III.1965, III.2007–9, III.2024
Space Shuttle, III.1742
Space technology, III.2022
Space-time, I.251, III.1730

Space Transportation System (STS), III.1742
Spark chambers, II.753–4
Spatial-light modulator technology, III.1461
Special Principle of Relativity, III.2034
Special theory of relativity, I.54, I.249, III.2039
  alternate formulations and formalisms, I.273–8
  causal formulations, I.277–8
  constraints, I.249–51
  continuum mechanics, I.279–80
  core principles, I.249
  electrodynamics in, I.257, I.280–1
  elementary particles in, I.282–3
  experimental tests and applications, I.284–5
  formulation, I.262–73
  formulation without light postulate, I.277
  four-dimensional formulation, I.274–6
  gravitation theories in, I.283–4
  later development, I.273
  origins of, I.254–62
  particle dynamics, I.279
  quantum theory in, I.282–3
  relativistic statistical mechanics, I.282
  relativistic thermodynamics, I.281–2
  relativistic velocity (kinematic) space, I.278
  rigid motions, I.279–80
  space-time structure, I.251
  spinor formalism, I.277
  twistor formalism, I.277
Specific gravity, I.46, II.967
Specific heat, I.550, I.555, I.558–60, II.967–73
  lattice theories, II.982–91
  of gases, I.528–30
  quanta, I.152–4

S51

Speckle interferometry, III.1454
Speckle pattern, III.1441
Speckle phenomena, III.1453
SPECT (single photon emission computed tomography), III.1910, III.1916, III.1917, III.1918, III.1920, III.1931
Spectral analysis, I.156
Spectral connection at ionization thresholds, II.1058–9
Spectral density, I.66, I.600
Spectral fingerprints, II.1092–4
Spectral frequencies, I.77
Spectral function, I.69
Spectral lines, I.156–62
Spectral numerology, I.77
Spectrometer, II.992, II.994
Spectroscopy, I.24, I.74–8, II.1039
  equipment, I.24
  interferometric, III.1410
  parallaxes, III.1698
Speculation, mathematical theory of, I.586
Speed of light, I.260, I.261, I.266, II.1240, II.1262–4
Sphalerite, I.429
Spheromak, III.1670
Spilling breakers, II.862
Spin, I.100–1, **I.395**, I.402, II.654
Spin-blocking, III.1831
Spin-fluctuation-exchange, II.953
Spin-Hamiltonian, II.1139–41, II.1151
Spin-lattice relaxation time, III.1923
Spin model, I.564
Spin–orbit coupling, II.1012, II.1126, II.1127, II.1136, II.1210
Spin–orbit interaction, II.1127, II.1212
Spin-warp technique, III.1924, III.1926
Spin waves, II.1010–14, II.1150–2

Spinels, crystallographic data, II.1156
Spinodal, III.1547–8
Spiral nebulae, III.1716–21
Sponge model, II.921
Sponge phases, III.1608
Spontaneous emission, I.93–4
Spontaneous transitions, III.1398–9
SPRITE detector, III.1477
Sputnik I, III.1738, III.2007
Sputnik II, III.1738, III.2007, III.2008
Sputnik III, III.2008
Squeezed light, III.1468–71
Squeezing parameter, III.1468
SQUIDs, III.1932
SSC debate, III.2020, III.2024, III.2031
Stability/instability, II.828–9
Stable fixed points, I.567
Stacking faults, I.473
Standard Model of elementary particles, III.2030, III.2033–9
Standards, II.1233, II.1234, II.1236–60
Stanford Linear Accelerator Center (SLAC), II.687–8, II.741
Star clusters, III.1696
Stark effect, I.89, I.161, III.1469
Stars, III.1691–713
  early-type, III.1698
  formation, III.1766–9, III.1945
  interstellar medium, III.1761–9
  island universes, III.1714, III.1717, III.1718
  late-type, III.1698
  luminosities, III.1698
  main-sequence, III.1698, III.1705
  mass–luminosity relation, III.1705
  photographic parallax programme, III.1696
  research since 1945, III.1746–60
  see also Stellar
Stationary states, I.83, I.86
Stationary waves, II.875

Statistical mechanics, I.523, I.524, I.542, I.586, III.1830, III.1962
  condensing gas, I.551–4
  development of, I.591
  in astrophysics, III.1711
  method of mean values, I.535
  models in, I.603–5
  non-equilibrium, I.585–633
Steering of atoms, II.1099
Stefan–Boltzmann law, III.1696, III.1756
Stellar aberration, I.258, I.259
Stellar Atmospheres, III.1701
Stellar distances, measurement of, III.1691
Stellar evolution, III.1691–713
  research since 1945, III.1746–60
Stellar interferometer, III.1400
Stellar luminosity function, III.1714
Stellar masses, determination of, III.1698
Stellar optical interferometry, III.1475
Stellar spectra, classification of, III.1692–4
Stellar structure, III.1694–5, III.1699
  theory of, III.1702–5
Stellarator, III.1648, III.1654–6
Stereology, III.1549
Stern–Gerlach effect, I.163–72, I.202
Stern–Gerlach magnets, II.1032, II.1047
Stigmator electrostatic astigmatism corrector, III.1580
Stimulated Brillouin scattering (SBS), III.1444
Stimulated Raman scattering (SRS), III.1432, III.1436
Stimulated Rayleigh-wing scattering, III.1445
Stokes theory, III.1411
Storage-ring synchrotron, II.739

Strain-aging, III.1521, III.1523
'Strange particles', II.655–9, II.677–9
Stratosphere, II.900, III.1982
String theory, II.682–4, II.756, II.766, III.2037–9
Stripping reactions, II.1220
Stroke, III.1922
Strömgren spheres, III.1762
Strong coupling theory, I.407
Strong interactions, II.680–4
Strong responses to perturbations, III.1593–694
Strongly correlated electron systems, II.1168
Strouhal number, II.850
Structure analysis, I.450–3
Structure-factor formula, I.433
Structure factors, I.439, I.441, I.447, I.449
Structure functions, II.690
Structure Reports, I.511, I.512
Strukturbericht, I.511, III.1513
SU(2), II.660, II.694, II.725
SU(3), II.660–2, II.677, II.679, II.684
Subscidy (*subsidiary scientific domain*), III.1507, III.1520
Substance X, I.64
Sudden ionospheric disturbance (SID), III.1995
Sulphur, III.1541
  formation of, III.1747
Sum rules, II.1058–9
Sun
  gravity fields, II.888
  internal structure, III.1694
  luminosity of, III.1708
  magnetic structure, III.1958
  see also Solar
Sunspots, magnetic fields in, III.1958
Sunyaev–Zeldovich effect, III.1773
Supercomputer, III.2045
Superconducting alloys, II.920

Superconducting magnets,
 III.1927, III.1929
Superconducting Super Collider,
 III.2036
Superconductivity, I.232, I.489–92,
 III.1279, III.1347, III.2025
  BCS theory and its aftermath,
   II.937–41
  developments—pre-1933,
   II.915–17
  developments—1945, II.917–21
  developments—1945–56, II.932
  discovery of, II.915
  high-temperature, II.957–65
  modern unified picture, II.943–9
  new developments, II.949
  pre-BCS attempts at microscopic
   theory, II.936–7
  related phenomena, II.955–7
  theoretical developments
   1933–45, II.923–9
  type-I, II.920, II.948
  type-II, II.920–1, II.941, II.948
Supercritical fluid, III.1947
Superfluid density, II.944
Superfluidity, II.913–15, II.922,
 III.2025
  irrotational component, II.929
  modern unified picture, II.943–9
  new developments, II.949
  related phenomena, II.955–7
Superionic conductors, III.1539
Super-Kamiokande detector,
 II.765, II.768, III.2036
Superlattices, III.1470
Supernova neutrinos, II.758
Supernova trigger hypothesis,
 III.1963–4
Supernovae, II.1227–8, III.1712–13,
 III.1718, III.1758–60, III.1766,
 III.1963–4
  light curve, III.1760
  SN1987A, III.1760
Superplasticity, III.1534
Super-resolution, III.1467

Supersonic aerofoil, II.880
Supersonic boom, II.876
Supersonic speeds, II.872, II.874,
 II.877, II.879
Superstring theories, II.684
Supersymmetry, II.762–3
Supra-heat-conductivity, II.922
Surface crystallography, I.465–70
Surface of section, III.1836
Surface waves, II.883
Surfactants, III.1606–8
 colloid protection, III.1611
Surgery
 diathermy, III.1894
 ultrasound in, III.1915
Symmetry, I.430, I.462, **II.1028–9**
Symmetry breaking, II.694, II.762
Symmetry elements, I.423
Symmetry principle, III.2034,
 III.2035, III.2039
Synchrocyclotrons, II.730–1, II.1201
Synchrotron light, II.1050–3
 spectroscopy, II.1052
Synchrotrons, II.639, II.732–3
 DESY, I.436
Synthetic sapphire crystal, II.982
Systems development, III.2030

Tandem accelerator, II.1203
Tandem-mirror facility, III.1669
Tandem Mirror Experiment, see
 TMX
Tandem scanning microscope,
 III.1451
Tankers, II.885
Taylor thickness, II.811
Teaching and/or researching,
 III.2047–8
Technetium-99m, III.1905–7,
 III.1909, III.1919
Technicare Neptune, III.1927
Technische Hoogeschool Delft,
 III.1579
Teflon (polytetrafluoroethylene),
 III.1896

Telecobalt therapy, III.1872
Telecommunication networks, II.906
Telecommunications by optical-fibre technology, III.1480
Telecommunications Research Establishment (TRE), III.1327–8
Telescopes, III.1734, III.1735, III.1738
  see also under specific types and applications
Teletherapy, III.1877, III.1879, III.1890
Television cameras, III.1412
Television industry, III.1476, III.1536
Teller–Northrop theory, III.1683
Temperature coefficients of resistivity, III.1538
Temperature gradients, II.901
Temperature measurement, II.1249–51
Temperature scales, II.1249–51
  ITS-90, II.1260
Temperature variation of resistivity, III.1302, III.1342
Terrella (magnetized sphere), III.2003
Terrestrial Magnetism and Atmospheric Electricity, III.2000
Tetramethyl manganese ammonium chloride (TMMC), II.1014
TFTR, see Tokamak Fusion Test Reactor (TFTR)
Thallium, I.75
Theorem of equipartition of energy, see Equipartition theorem
Theoretical Structural Metallurgy, III.1522

Theory of the Electron, The, III.1992, III.2039
Thermal analysis, III.1508
Thermal capacity, II.967, III.1354
Thermal conductivity, II.977–82, III.1285, III.1290
Thermal energy, I.66
Thermal equilibrium, I.530
Thermal expansion, II.975–7
Thermal vibrations, II.984
Thermal wind equation, II.901
Thermals, II.845
Thermionic emission, I.37
Thermodynamic equilibrium, III.1391
Thermodynamic internal energy, I.532
Thermodynamics, I.5, I.35, I.149
  as science, I.521
  first law of, I.374, I.521
  nineteenth century background, I.521–4
  of data processing, I.576
  of heterogeneous substances, I.532
  relativistic, I.281–2
  second law of, I.46, I.68–9, I.374, I.521, I.524, I.534–5, I.575
  special-relativistic, I.281
  third law of, I.535–7, I.537–42, II.968
Thermoelastic transformation, III.1539
Thermoelectric effects, III.1283–4, III.1306–11
Thermoelectric (Seebeck) voltage, III.1310
Thermography, III.1932
Thermometers, II.1249
Thermonuclear explosion, II.1224
Thermonuclear fusion, III.1618, III.1629
Thermonuclear reactions, II.1226
Thermos flask, I.27
Thermosphere, III.1982, III.2009

Thimble chamber (ionization chamber), III.1866–7, III.1881
Thin-film devices, III.1444
Thin films, II.1170–1
Third International Polar Year (1957–58), III.2006
Third law of thermodynamics, II.968
　applications, I.537–42
　historical survey, I.535–7
Thomas–Fermi method, I.464, II.1190, II.1191
Thomson moving-magnet galvanometer, I.23
Thomson scattering, I.65
Thomson–Lampard theorem, II.1255
Thorium, I.59, I.60
Threshold effects, II.1063–5, **II.1066–7**
Thymine, I.504
Thyroid gland, III.1900–3
Tidal currents, II.888
Tidal kinetic energy, II.891
Tide gauges, II.891–2
Tide-raising forces, II.892, II.896
Tide tables, II.890, II.892
Tight-binding approximation, III.1292
Tight-binding model, III.1295, III.1297
Time
　dissemination of, II.1248–9
　measurement, II.1242–6
Time–position recording, III.1913–14
Time-resolved crystallography, I.438
Time-scales, II.1247–8
Tin oxide, III.1541
Titanium and titanium alloys, III.1281, III.1897
Tl–Ba–Ca–Cu–O compound, II.959
Tl$_2$Sr$_2$Cu$_3$O$_{10}$, I.492

TMX (Tandem Mirror Experiment), III.1667–8
Tobacco mosaic virus (TMV), I.508
Tokamak Fusion Test Reactor (TFTR), III.1646–54, III.1663, III.1677
Tollmien–Schlichting waves, II.832
Tomato bushy stunt virus (TBSV), I.509, I.510
Tomography. *See* Computed tomography
TOPCON EM002B 200 kV column, III.1588
Torsional oscillation, III.1523
Transcurrent faults, III.1980
Transdisciplinary disciplines, III.2045–7
Transform faults, III.1979
Transformation laws, I.269
Transformation theory, I.59, I.60, I.360
Transistors, III.1322–7
　commercial applications, III.1317
　invention of, III.1317
Transition
　abrupt, II.833, II.834
　amplitudes, I.185, **II.1064**
　restrained, II.833, II.834
　state, II.1049
　taxonomy, II.832–41
　types of, II.825–32
Transition metals, III.1343, III.1354
Transmission electron microscope (TEM), III.1526, III.1548, III.1569, III.1570, III.1573
　aberration-free atomic resolution, III.1585–6
　accelerating voltage, III.1590
　and Abbe's optical theory, III.1578–84
　astigmatism corrector, III.1580
　commercial prototype, III.1575–6
　high-resolution commercial, III.1576
　holographic, III.1584

over-focus and under-focus fringes, III.1580
RCA, III.1581
resolution, III.1590
selected area diffraction, III.1579
serially produced, III.1575–7
specimen preparation and radiation damage, III.1574–5
two-stage, III.1574
vertical column, III.1575
Transonic flight, II.882
Transport coefficients, I.597–8, I.612, I.614
Transport parameters, I.593, I.614
Transport phenomena, I.592
Transuranium elements, II.1222–5
Transverse temperature gradient, III.1282
Transverse voltage, III.1282
Transverse waves, III.1956
Trapping of atoms, II.1099
Triplet coupling, II.1086
Triplets (antitriplets), II.684–5
Triton, II.1208
Tropical cyclones, II.904–6
Troposphere, II.900, III.1982, III.1984
T-Tauri stars, III.1766–7
Tubeless syphon, III.1593–4
Tumours. *See* Cancer; Radiotherapy
Tungsten, III.1355
Tunnel effect, I.217
Turbo-fan engine, II.873
Turbulence, II.843, III.1834
  isotropic, II.832
  taxonomy, II.832–41
  types of, II.825–32
Turbulent flow, II.825
Turnip yellow mosaic virus (TYMV), I.510
Twinned crystals, III.1435
Twisted nematic cell, III.1604
Two-fluid hydrodynamics, II.929
Two-level-atom model, III.1459

UHURU satellite, III.1739, III.1754, III.1766
UK Infrared Telescope (UKIRT), III.1743
Ultra high-molecular-weight polyethylene (UHMWPE), III.1896
Ultra low-frequency radio physics, III.1995
Ultrasonic imaging, III.1914–15
Ultrasonic industrial flaw detector, III.1911
Ultrasonic probe, III.1913
Ultrasonic radiation, therapeutic, III.1915
Ultrasound
  diagnostic, III.1910–11
  in medicine, III.1910–15
  in surgery, III.1915
  in therapy, III.1915
  scanning, III.1859, III.1913, III.1915
Ultraviolet astronomy, III.1741–2
Ultraviolet energy, III.1992
Ultraviolet light, II.1050–3, III.1396
Ultraviolet radiation, III.1994, III.2009
Umklapp processes, II.980, III.1304–6
Uncertainty relations, I.174, I.201–4, I.209, I.228, III.1394
Uniaxial ferroelectrics, II.1009
Unified field theories, I.320–3
Unit cell, I.423
Unitary symmetry, II.660–2
Units, II.1236–60
Universal gas constant, I.150
Universal Time (UT), II.1247
Universality, I.549–51, I.563, II.664, II.677, III.1833
  empirical derivation, I.561–4
Universe
  age of, III.1732, III.1793–4
  dark matter, III.1731
  dynamics of, III.1730, III.1787

S57

Einstein's, III.1722–4
expansion of, III.1786, III.1787, III.1806
geometry of, III.1723
Hot Big Bang models of, III.1778
hot early phases, III.1788, III.1793
large-scale structure, III.1714–34, III.1805
mean baryonic density, III.1791–2
mean density, III.1800
mean mass density, III.1719
temperature history, III.1786
total mass density, III.1800
very early, III.1805–7
U-particles, I.393, I.394, I.396, I.397
Upper atmosphere, III.1981–4
geophysics, III.1943–4, III.1982
ionization of gases, III.1981
Upper Atmosphere, The, III.2000
Upsilon, II.721
Upstream wakes, II.868–9
U-quanta, I.378, I.382, I.389, I.391
Uracil, I.504
Uranium, I.56, I.60, I.61, I.106, II.1199, II.1223, II.1227, III.1630, III.1960
Uranium-235, I.131–2
US Atomic Energy Commission, III.1633
US Bureau of Standards, III.1988
US National Aeronautics and Space Administration (NASA), III.1739, III.1965
US National Bureau of Standards (NBS), I.20, II.1235, II.1236, III.1881, III.1998
U-sequence, III.1839
UTC, *see* Coordinated Universal Time (UTC)
Utrecht School, I.629

Vacancies, III.1505, III.1529, III.1531, III.1533
Vacuum polarization, II.641, II.1031
Vacuum technology, I.27
Vacuum tubes, I.53, III.1987
Valence band, III.1317, III.1319, III.1326, III.1329, III.1334
Valence electrons, III.1314, III.1317, III.1366
Van Allen belts, III.1683–4
van de Graaff generator, II.727–8, II.1203, III.1871, III.1883
van der Waals equation, I.28, I.547, I.552, I.561
van der Waals law, II.973
van der Waals theory, I.555, I.557
Varistors, III.1537
Väsälä–Brunt frequency, II.864
V–A theory, II.667–9
Vavilov–Čerenkov effect, III.1408
Vela satellite, III.1740, III.1741
Velocity–distance relation, III.1726, III.1727, III.1728
Velocity–position space (μ-space), I.587, I.590
Veneziano amplitude, II.684
Vertical-cavity surface-emitting laser (VCSEL), III.1464
Very high-resolution imaging systems, III.1477
Very long baseline interferometry (VLBI), III.1738
Very low-frequency radio physics, III.1995
Very low temperatures, I.540–1
Vesicles, III.1608
Vibration, II.967–1010
Vibrational excitation, II.1073
Vibrational motion, II.1044
Vibrational specific heat of solids, I.525
Vibrational transitions, II.1073
Videophone, III.1477
Viking (rocket), III.2005
Vine–Matthews theory, III.1976

Subject Index

Virial theorem, II.1026, III.1731, III.1772, III.1800
Virtual reality, III.1476
Viruses, I.507–11
  rod-shaped, I.508
  spherical, I.509
Viscosity, II.807, II.813–14
  stabilizing effect of, II.828
Viscous diffusion, II.827
Vitamin B_{12}, structure of, I.495
Vitamin D, isolation of, III.1529
Vlasov–Boltzmann equation, III.1625
Volt
  absolute determination, II.1256
  reference standards, II.1256
Volta effect, III.1283
Voltaic cell, III.1284
von Klitzing constant, II.1259, II.1271
von Neumann equation, I.602, I.612
Vortex lines, II.820, II.822, II.823
Vortex-wake, II.876
Vorticity, II.816–24, II.889, II.890, II.892, II.894, II.901
V-particles, I.408–9
Vulcanization, III.1551, III.1593

Wake fluid, II.821
Wakes, II.812–16
Wanderwelle, electrical surges, III.1566
Water-bath ultrasonic simulator, III.1912
Water waves, II.798
Watson–Crick model, I.504
Wave, *see* under specific wave types
Wave action, II.861
Wave energy, II.862
Wave equation, I.195, I.283, III.1393
Wave frequency, II.863
Wave functions, I.204, **I.361**, I.617, II.941–3, II.949, II.1124
Wave generation, II.846–69, II.882–5, II.900
Wave-interaction effect, III.1994
Wave mechanics, I.174, I.191–6, I.217, I.360
Wave momentum, II.866
Wave numbers, II.838, II.839, II.861, II.864, II.866–8, II.885, II.893, II.984
Wave packets, II.836, III.1366
Wave–particle duality, I.168–71, I.364, III.1394, III.1395
Wave propagation, II.846–69, II.877, III.1407, III.1955
Wave theory, I.258, I.363
Waveforms, II.848, II.858, II.861
Wavelength, II.883, II.902
  standards, II.1239–40, II.1242
Wave-making resistance, II.867
Wavevectors, II.988–90, II.1001
W-bosons, II.637, II.706–9
Weak interactions, II.663–71, II.692–3
Weak universality, II.664
Weather, II.899–912
  forecasting, II.888, II.899, II.906–9
Weber bar, I.314
Weinberg–Salam model, II.695, II.762
Wet-bulb temperature, II.843
Weyl's postulate, III.1729
Wheatstone bridge, I.22
Whiskers, III.1351
Whistler waves, III.1622, III.1623, III.1997, III.2005
  measurements, III.2009
  modelling, III.1999
Whistling atmospherics, III.1995
White dwarfs, III.1709–12
Wide Field Planetary Camera, III.1745

Wiedemann–Franz law, III.1286, III.1287, III.1316
Wiedemann–Franz ratio, III.1290
Wien–Planck spectral equation, I.69
Wiener–Khintchine theorem, I.600–2
Wien's law, I.69, I.151, I.524, III.1393
Wigner function, I.617–18, III.1441
William Herschel Telescope, III.1744
Wilson cloud chamber, *see* Cloud chamber
Wilson's scaling method, I.440, I.445, I.447
Wind effects, II.901, II.903–4
Wind instruments, II.851–2
Wind speed, II.843
Wind speed/sound speed ratio, II.813
Wind tree model, I.603
Wind-tunnels, II.830, II.879
Winding number, II.955
Wireless. *See* Radio
Wireless telegraphy, III.1986
Wollaston wire, II.831
Women in higher education, I.8, I.10
Work-function, III.1286, III.1308
Work-hardening, III.1518
World Directory of Crystallographers, I.511
W-particle, III.2036
Wranglers, I.10–12
Wurtzite, I.429

Xenon, I.75, I.556
Xerographic copying, III.1541
X-ray absorption, II.1051
X-ray analysis, III.1514
X-ray Analysis Group, I.511
X-ray angiography, III.1904
X-ray apparatus, II.1051
X-ray astronomy, III.1738–41, III.2005
X-ray binaries, III.1754–7
X-ray contrast media, III.1862–3
X-ray crystallography, I.446, I.469, I.484, III.1515
X-ray diffraction, I.30, I.424–32, I.456, I.457, I.484, I.498, II.983, III.1583
  amorphous solids, **I.478**
  in-plane, I.466
  Laue's equations for, **I.427**
X-ray dose-rate measurements, III.1871
X-ray emission, III.1630
X-ray equipment, III.1861
X-ray imaging, III.1916, III.1922–3
X-ray intensities, I.440
X-ray irradiation, III.1529
X-ray lasers, III.1464–6
X-ray measurements, II.1010
X-ray microscope, III.1583
X-ray optics, I.484
X-ray photons, I.363
X-ray pulsations, III.1756
X-ray pulses, III.1752
X-ray scattering, I.165, I.362, I.365, I.453, II.990, III.1867
X-ray sets, III.1864, III.1877
X-ray sky, III.1798
X-ray sources, III.1755–7
X-ray spectra, I.86, I.158–60
X-ray spectrometer, III.1580
X-ray spectroscopy, I.158–60, I.429, I.484
X-ray studies, inner-shell phenomena, II.1089–90
X-ray telescopes, III.1740
X-ray tomography (X-CT), III.1918
X-ray tubes, III.1859, III.1861, III.1862, III.1865
X-rays, I.3, I.32, I.36, I.52, I.53, I.65, I.362–6, I.421–4, II.1050–3, III.1514
  discovery, I.424

in diagnosis, III.1859–63
in medical physics, III.1857
see also Radiotherapy

Yang–Mills theory, II.692–4, II.713, II.714, II.763
YBa$_2$Cu$_3$O$_7$, II.959
YBa$_2$Cu$_3$O$_{7-\delta}$, I.490–1
Yerkes system, III.1698
Yield-stress, III.1520–1, III.1523
Yin-Yang coil, III.1664
Young's double-slit experiment, III.1402
Yttrium–iron–garnet, II.1159
Yukawa meson theory, I.378–83, II.651–2

Z-bosons, II.637, II.706–9
Zeeman effect, I.7, I.36, I.52, I.54, I.99–103, I.160–2, I.182, I.183, I.184, I.211, II.1114, II.1115, II.1149
Zeeman splittings, II.1120

Zeeman sub-state, II.953
Zeitschrift für Kristallographie, I.511
Zeitschrift für Metallkunde, III.1556
Zenith angle, III.1992
Zernike's circle polynomials, III.1409
Zero-point motion, II.973–5, II.1004
Zero-resistance state, II.961
Zero sound, II.950, III.1364
ZETA (Zero-Energy Thermonuclear Apparatus) toroidal pinch experiment, III.1641–2, III.1645, III.1657
Zhukovski's theorem, II.822
Zinc, III.1360
Zinc oxide, III.1537
Zinc sulphide scintillators, II.1200
Zincblende, III.1514
Zone-levelling, III.1543, III.1544
Zone-refining, III.1543, III.1544, **III.1545**
Zweig's rule, II.685

NAME INDEX

Note—volume and page numbers in **bold type** refer to items in boxes

Aaronson, M, Hubble's constant, III.1793
Abbott, E A, *Flatland*, I.5
Abelson, P H, transuranium element, II.1223
Åberg, T, secondary excitations, II.1090
Abragam, A
  hyperfine structures, II.1139
  nuclear magnetic resonance, II.1175
  resonance research, II.1143
Abraham, electron microscope, III.1573
Abraham, M
  special-relativistic theories, I.284
  stress–energy tensor, I.280, I.281
Abramowicz, M A, self-consistent thick-disc model, III.1780
Abrikosov, A A, superconductivity, II.935
Accardo, C A, p–n junction, III.1417
Adams, M C, dye-laser pulses, III.1439
Adams, W C
  A stars, III.1710
  chemical abundances, III.1702
  luminosity indicators, III.1698
Adams, W S, Sirius B, III.1710, III.1723

Ademollo, M, Dolen–Horn–Schmid duality, II.683
Adlard, photon, III.1451–2
Adler, S, soft photon emission, II.679
Aharonov, Y
  field strength, II.693
  interference patterns, III.1370
Akasofu, S-I, *Solar–Terrestrial Physics*, III.2001
Akhmanov, S I, Nd:YAG laser, III.1435
Albrecht, A, world model, III.1807
Alder, B J
  autocorrelation function of tagged particle, I.619
  molecular dynamics, I.577
Alder, K, Coulomb excitation experiments, II.1217
Alexander, S, fractons, II.1007
Alfvén, H
  baryon symmetric models, III.1806
  Cosmical Electrodynamics, III.1637, III.1685
  Galactic radio emission, III.1737–8
  MHD theory, III.1684, III.1685
  monistic theory, III.1962, III.1963
  ring current around the Earth,

III.2003
Allen, J F, superleak flow, II.946
Allis, W P, electron diffusion rates, III.1644
Alpher, R A
　abundances of elements, III.1785–6
　black-body radiation, II.758
　primordial nucleosynthesis, III.1785, III.1786, III.1790
　proton–neutron ratio, III.1786
Als-Nielsen, J, order–disorder transition in β-brass, II.1009, II.1016
Altar, W, wave propagation, III.1990–1
Altarelli, G, asymptotic freedom, II.715
Al'tshuler, B L, scattered wavelets, III.1370
Alvarez, L W
　neutron diffraction, I.459
　particle interactions, II.659
　proton linear accelerator, II.1205
　proton machines, II.737
　60" cyclotron, II.1202
　tandem accelerator, II.1203
Ambartsumian, V A, T-Tauri stars, III.1766
Amelinckx, S, dislocations, III.1526
Ampère, A M
　electromagnetic interactions, III.1953
　magnetism, I.65
Amrein, W O, point photon, III.1451
Anaxagoras, matter, I.44
Anaximenes, air as basic substance, I.358
Anderson, C D, **I.386**
　absorption in lead plates, I.385
　electron–positron pairs, I.231
　elementary particles, I.391
　mesotron, I.231, I.389
　muon, II.650
　positron, III.1706
　shower particles, I.388
　standard model, I.231
　Wilson cloud chamber, I.371
Anderson, J A, angular diameter of Betelgeuse, III.1705
Anderson, J L, relativistic thermodynamics, I.281
Anderson, P W
　bosons, II.695
　disordered materials, III.1365
　electron density, III.1316
　exchange interactions, II.1144
　glasses, II.1006
　phase transitions, II.1008
　phonon exchange, II.953
Anderson, W C
　speed of light, II.1263
　white dwarfs, III.1711
Andrade, E N, crystal diffraction spectroscopy, II.1207
Andrew, E R
　magnetic resonance imaging, III.1924
　nuclear magnetic resonance, II.1175
Andrews, T
　critical point in carbon dioxide, I.28
　critical point phenomenon, III.1947
Anger, H, gamma camera, III.1907
Ångström, A J
　normal solar spectrum, II.1242
　spectrum of hydrogen, I.76
Antonucci, R R, Seyfert galaxies, III.1784
Appleton, E V
　radio receiving apparatus, III.1989
　solar radiation, III.2002
　upper atmosphere, III.1988
　wireless propagation, III.1622
Araki, G
　cosmic-ray particles, I.407

slow mesotrons, I.406
Argos, P, structure determination, I.445
Arima, A, collective motion in nuclei, II.1218
Aristotle, matter, I.44
Armenteros, R, hyperon, II.657
Armstrong, H E, organic compounds, I.493
Arndt, U W, single-crystal oscillation, I.437
Arnett, W D, element abundances, III.1748
Arnott, N
  Elements of Physics, III.1857
  medical physics, III.1856–7
Arp, H, quasars, III.1777
Artsimovich, L
  Atoms for Peace Conference, III.1640
  mirror-type systems, III.1643
  tokamak, III.1648
Asaro, F, alpha-decay spectra, II.1217
Ash, E, optical microscopy, III.1406
Ashkin, A
  argon-ion laser beams, III.1453
  He–Ne laser, III.1434
  optical binding of particles, III.1472
  self-trapping of laser beam, III.1459
Askaryan, G A, self-focusing, III.1436
Aspect, A, Bell inequalities, III.1467
Astbury, W T
  DNA, I.504
  fibrous proteins, I.497
  x-ray diffraction, I.439
Aston, F W
  isotope masses, I.111–12
  stellar energy source, III.1706

Astrom, E O, plasma physics, III.1992
Atkinson, R d'E
  nuclear reactions, II.1225, III.1707
  thermonuclear reactions, III.1627
Auger, P, neutrons, I.122
Austin, L W, radio propagation, III.1987
Autler, S H, 'dressed' atomic states, III.1421
Avenarius, critical opalescence, I.546
Avery, Q T, DNA, I.504
Avogadro, A, hypothesis, I.46
Avrami, M, phase transformations, III.1554
Axford, I, solar wind plasma, III.2010
Ayers, G R, blind deconvolution, III.1461
Azbel, M Ya
  canonical theory, III.1356
  cyclotron resonance, III.1357

Baade, W
  Andromeda Nebula, III.1788
  medium-mass elements, II.1228
  Palomar 200" Telescope, III.1737
  radio sources, III.1775
  supernovae, II.1227–8, III.1712, III.1718, III.1759, III.1775
Bacher, R F
  atomic energy levels, II.1093
  nuclear structure theory, II.1208
  shell model, II.1209
  Thomas–Fermi model, II.1191
Back, E
  total quantum number, I.162
  Zeeman splitting, I.161
Bacon, G E
  neutron diffraction, I.456
  structure of KH_2PO_4, I.459
Badash, L, I.29

Baez, A V, optical microscopy, III.1406
Bahcall, J N
  solar model, III.1749
  solar neutrino flux, III.1748
Bailey, V A, broadcast modulation, III.1994
Bain, E, bainite, III.1515
Bainbridge, K
  H-particle–electron model, I.112
  mass spectrometry, II.1187
Baker, D W, ultrasonic imaging, III.1914
Baker, G A, Padé Approximant, I.558
Baldwin, D, thermal barrier, III.1667
Baldwin, G C, dipole vibration, II.1213
Balescu, R, *Equilibrium and Non-equilibrium Statistical Mechanics*, I.620
Balluffi, R, diffusion, III.1531
Balmer, J, R value, I.77–8
Banerjee, M K, proton inelastic scattering, II.1221
Bangham, vesicles (or liposomes), III.1608
Barber, C, electron storage rings, II.739
Bardakci, K, singularities in complex angular momentum, II.682
Bardeen, J, **II.939**
  BCS theory, II.938
  impurity effects, III.1281
  interatomic forces, II.999
  ion potential, III.1303
  point-contact transistor, III.1326
  solid-state physics, III.1313
  superconductivity, I.232, II.937, III.1364
  theory of phonons, III.1362
  transistor, II.744
  xerography, III.1541
Bardeen, J M, black holes, III.1779
Barham, P M, coherence parameter, III.1412
Barkhausen, H
  magnetic flux, II.1121
  radio waves, III.1622
  whistler waves, III.1622–3
Barkla, C G
  x-ray scattering, I.362, I.424
  x-ray spectroscopy, I.158
Barlow, W, crystal structure, I.423–4
Barnes, W H, structure of ice, I.474
Barrett, C, diffusive transformations, III.1547
Bartels, J, *Geomagnetism*, III.2000
Barthel, P D, radio quasars and radio galaxies, III.1784
Bartlett, J A, shell models, II.1209
Barut, A O, singularities in complex angular momentum, II.682
Basov, N G
  excimer laser, III.1453
  semiconductor maser, III.1423
Bass, M, sum-frequency generation, III.1429
Bate, G, magnetic materials, II.1159
Bates, D R
  ionospheric research, III.1999
  probability of electronic excitations, II.1080
  solar x-rays, III.1994
Bauer, high-reflection coatings, III.1407
Baxter, R J, ferroelectric model, I.571
Bay, Z, electron multiplier, III.1541
Baym, G, neutron stars, III.1754
Bear, R S, three-stranded coiled-coil model, I.499
Beck, C S, defibrillators, III.1895
Beck, G, β-decay theory, I.372, I.375

Beck, K, single crystals of ferromagnets, II.1123
Becker, H, neutron, I.119
Becker, R, nucleation kinetics, III.1547
Becklin, E E
  formation of stars, III.1769
  Orion Nebula, III.1743, III.1767
Becquerel, A H
  radioactivity, I.37, III.1857, III.1876
  uranic rays, I.53, I.56–7
Bednorz, J G
  Meissner effect, II.957–8
  superconductivity, I.490, II.961
Beevers, C A
  Rochelle salt, I.489
  structure determination, I.441
Behn, U, specific heat, II.968
Bell, A G, Photophone, III.1422
Bell, F O, DNA, I.504
Bell, J S
  EPR paradox, I.229
  inequalities, III.1389
Bell, P R, scintillators, II.750
Bell, R J, amorphous SiO_2, I.478–9
Bell, S J
  neutron stars, III.1751
  sky surveys, III.1752
Bémont, G, radium, I.58
Bénard, H
  sound from flows, II.850
  surface tension, II.845
Benedek, G
  'light-beating' spectroscopy, III.1454
  spontaneous magnetization, I.559
Benioff, H, earthquake oscillations, III.1958
Bennet, W R Jr, helium–neon laser, III.1427
Bennett, F, tandem accelerator, II.1203

Bennewitz, K, zero-point motion, II.973–4
Bentley, R, physical cosmology, III.1721
Beran, M J, partial coherence, III.1392
Berkner, L
  IGY, III.2006
  ionospheric research, III.2005–6
  ionospheric sounding, III.1998
  Third International Polar Year, III.2006
Berlinger, W, oxygen octahedra, II.1009
Berman, R
  diamond–graphite equilibrium diagram, I.538
  thermal conductivity, II.981
Bermen, S M, hadrons, II.690
Bernal, J D
  amorphous structure, I.479
  crystal oscillation, I.434
  globular proteins, I.497
  haemoglobin, I.497
  random loose packing (RLP) and random close packing (RCP), I.577
  residual entropy of ice, I.571
  stereochemical formulae, I.495
  structure of ice, I.474
  tobacco mosaic virus (TMV), I.508
Bernard, M G A, stimulated emission in semiconductors, III.1433
Bernoulli, D
  Bernoulli equation, II.806
  gas pressure, I.47
Bernstein, I B
  collisionless shocks, III.1684
  plasma oscillations, III.1625
Berreman, D W, liquid-crystal optics, III.1452
Berry, B, anelastic relations, III.1523

Berry, M V
  diffractals and caustics, III.1466
  topological phase, III.1470
Berson, S, saturation analysis technique, III.1909
Bertero, M, imaging theory, III.1467–8
Bertolotti, M, lasers, III.1423
Berzelius, J J, nomenclature for elements, I.45
Besicovich, A S, quasicrystals, I.482
Bessel, F, stellar distances, III.1691
Beth, R A, angular momentum of photons, III.1410
Bethe, H A
  abundances of elements, III.1785–6
  bremsstrahlung, I.385
  carbon–nitrogen–oxygen cycle (CNO cycle), III.1707–8
  face-centred cubic lattice, III.1293
  fast charged particles, II.1040
  Fermi theory, I.377
  First Shelter Island Conference, II.645
  mesons, I.401, I.403, II.654
  near-field/far field solution, III.1416
  neutron capture, II.1193
  neutron stars, III.1754
  nuclear physics, II.1184
  nuclear reactions, I.132, II.1192, III.1629
  nuclear structure theory, II.1208
  origin of the elements, II.1225
  pp process, II.1226
  quantum fluctuations, II.1013–14
  shell model, II.1209
  spacing of states in highly excited nucleus, II.1213
  theory of metals, III.1290
  Thomas–Fermi model, II.1191
Twelfth Solvay Conference, Brussels, II.646
  weak-interaction-treatment, II.1040
Beynon, W, ionosphere, III.1984
Bhabha, H J
  cosmic rays, I.393
  fusion reactions, III.1642
  showers, I.388
  U-particles, I.393
Bhar, J N, ionospheric physics, III.2000
Bhatia, pseudopotential concept, III.1367
Bieler, É S
  α-particles, I.114
  anomalous scattering, I.366
Bier, N, scleral lenses, III.1414
Biermann, L, ionized comet tails, III.2009
Bijvoet, J M
  dextro (d) and laevo (l) compounds, I.495
  strychnine, I.495
Bilby, B A, yield-stress and strain-aging, III.1523
Bilderback, D, area detector, I.438
Billy, H, Renninger effect, I.450
Biloni, H, solidification, III.1542
Binnig, G
  scanning tunnelling microscope, III.1549
  tunnelling electron microscope, I.466
Birch, F, Earth's core, III.1957
Birdsall, C K, stream-type instabilities, III.1673
Birge, R T
  Birge–Bond diagram, II.1268
  constants, II.1271
  least-squares adjustment, II.1270
  speed of light, II.1263
Birkeland, K
  geomagnetic storms, III.2003
  Terella device, III.1682

Birkhoff, G D, energy surface trajectories, I.599
Birkinshaw, M, Sunyaev–Zeldovich effect, III.1773
Bishop, A S, *Project Sherwood—The US Program in Controlled Fusion*, III.1640
Bitter, F, antiferromagnetism, II.1129
Bjerrum, N
  Applications of the quantum hypothesis to molecular spectra, I.82
  molecules, I.81–2
  quantum theory, I.155
Bjorken, B J
  charmed quark, II.700
  quark–lepton analogy, II.700
Bjorken, J D
  hadrons, II.690
  parton hypothesis, II.690
  point-like constituents, II.689
Bjorkholm, J E, self-trapping of laser beam, III.1459
Bjornholm, S, fission isomers, II.1218
Black, J, medical physics, III.1857
Black, J C, ultrasonic properties, II.1006
Blackett, P M S, **I.400**
  cloud chambers, II.651
  Earth's magnetic field, III.1969–70
  electron–positron pairs, I.231
  nuclear reactions, II.1201
  penetrating radiation, I.388
  The craft of experimental physics, II.745
  Wilson cloud chamber, I.371
Blackman, M
  BvK theory, II.984
  perturbation theory, II.1002
  specific heat, II.985–7
  toroidal fusion device, III.1634

Blair, J S
  alpha particles, II.1221
  diffraction theory, II.1221
Blanc-Lapierre, A, polychromatic fields, III.1421
Blau, M, cosmic-rays tracks, I.404
Bleaney, B, resonance research, II.1143
Bleuler, E, two-photon correlation, III.1413
Blewett, J P
  accelerators, II.725–7
  Particle Accelerators, II.1201
Bliss, W R, cross-sectional imaging, III.1912
Bloch, F
  electron gas, III.1291–301
  gyroscopic motion, II.1172
  magnetism, II.1137
  neutron diffraction, I.453, I.459
  nuclear magnetic resonance, II.1094
  quantum theory, I.221, III.1361
  spin waves, II.1010–11
Block, M, parity states, II.665
Block, S, diffraction techniques, I.437
Bloembergen, N, **III.1430**
  anharmonic oscillators, III.1434
  laser spectroscopy, III.1466
  non-linear optics, II.1100, III.1434
  Nonlinear Optics, III.1445
Blow, D M, structure determination, I.445
Bludman, S A, scalar Higgs fields, III.1806
Boas, W, plastic deformation, III.1526
Bobeck, A H, bubble domains, II.1160
Bodin, H, reversed-field pinch (RFP) research, III.1657
Boersch, electron holograms, III.1582–3

Bogoliubov, N N, **I.610**
  BBKGY hierarchy, I.606, I.623, I.624, I.626
  kinetic equation, I.609
  properties of dilute Bose gas, II.930
Bohm, D
  Bohm diffusion, III.1630, III.1631, III.1632
  electrical discharges in gases, III.1361
  EPR paradox, I.229
  field strength, II.693
  First Shelter Island Conference, II.645
  interference patterns, III.1370
  plasma oscillations, III.1625
Bohr, A, II.1216
  compound nucleus theory, II.1219
  Coulomb excitation experiments, II.1217
  geometrical collective model, II.1218
  quadrupole moment, II.1215
Bohr, H, quasicrystals, I.482
Bohr, J, Fourier techniques, I.469
Bohr, N, **I.79**, I.180
  atomic theory, I.157, II.1117, III.1401
  β-decay theory, I.372
  β-spectra, I.126–7
  Balmer formula, I.80, I.82
  Bohr radius, I.85
  building-up principle (*Aufbauprinzip*), I.96–7
  Cockcroft–Walton accelerator, II.1188
  complementarity theory, I.224
  compound nucleus, II.1192, II.1194
  correspondence principle, I.86
  dumb-bell model, I.181–2
  dynamical quantum theory, I.174
  Einstein–Podolsky–Rosen (EPR) paradox, I.228
  electron collisions, II.1042
  fast charged particles, II.1039
  Fifth Solvay Conference (1927), I.205
  fission process, II.1195
  high-speed β-particles, I.107
  hydrogen atom, I.82–6, I.91
  impact of theory, I.86–9
  Light and life, I.224
  liquid drop, I.131, II.1213
  m-values, I.95
  magnetic properties, II.1115
  mechanical forces, I.78
  'new hypothesis', I.84
  nuclear collective motions, II.1213
  nuclear fission, I.126
  nuclear reactions, I.123–6
  On the constitution of atoms and molecules, I.157
  On the general theory, I.176
  On the quantum theory of line spectra, I.94
  periodic system, I.97
  periodic table, I.95
  principal quantum number, I.91
  principle of complementarity, I.197
  quantum conditions, I.359
  quantum dynamics, I.72–82
  quantum hypothesis, I.93
  quantum mechanics, I.225, I.226, I.228
  quantum optics, III.1397
  quantum theory, I.80, I.166, I.175, I.221, II.1115, III.1397
  radiative instability, I.83
  spacing of states in highly excited nucleus, II.1213
  spectra, I.77
  superconductivity, II.917
  theory of atomic structure, III.1699

theory of metals, III.1286
theory of nuclear reactions, II.1192
Twelfth Solvay Conference, Brussels, II.646
uranium-235, I.131–2
x-ray spectra, I.86
Bolsterli, M, computers, III.1843
Bolt, R H, ventriculography, III.1911
Bolton, J G, radio sources, III.1737
Boltzmann, L E, **I.31**
  battle with Ostwald, I.48
  black body, I.66
  black-body radiation, I.149
  energy surface trajectories, I.590
  entropy, I.47
  equipartition theorem, I.67
  H-theorem, I.35
  kinetic theory of gases, I.587
  Liouville theorem, I.525
  process irreversibility, I.524
  second law of thermodynamics, I.524
  specific heat, II.970
  statistical mechanics, I.173, I.523
Bom, N, imaging, III.1914
Bömmel, acoustic resonance, III.1357–8
Bond, W N, constants, II.1267–8
Bondi, H
  chemical elements, III.1788
  field equations, I.313
  plane wave solutions, I.312
  steady-state cosmology, III.1787
Bonhoeffer, K F, molecular hydrogen, II.1048
Bonifacio, R, Collective Atomic Recoil Laser, III.1456
Booker, H G, magneto-ionic theory, III.1992
Bormann, E, decay law of radioactive substances, I.222
Born, M, I.180, **II.985**
  adiabatic approximation, II.999

Atomic Physics, I.210
BBKGY hierarchy, I.606, I.623, I.624, I.626
crystal lattice vibrations, I.153
discrete quantum mechanics, I.184
Dynamical Theory of Crystal Lattices, II.988
Fifth Solvay Conference (1927), I.205
harmonic-oscillator decomposition, III.1402
lattice dynamics, I.452
Lattice theory of rigid bodies, I.189
matrix mechanics, I.360
method of long waves, II.1003–4
On quantum mechanics, I.184
On the quantum mechanics of collision processes, I.198
perturbation theory, II.1002
properties of fluids, I.553
quantum mechanics, I.185, II.638
quantum theory, I.181, I.188
Raman effect, II.996, III.1404
rigid body, I.279
Schrödinger's wavefunction, I.174
self-consistent renormalized harmonic approximation, II.1004
specific heat, II.971–2
spectral line intensity, I.172
The structure of the atom, I.189
theory of ionic crystals, II.997
vibrational specific heat of solids, I.525
Bose, S N
  black-body radiation, III.1402
  light-quanta, I.167
  Planck's law, I.527
  radiation law, III.1288
Bothe, W
  coincident counting, I.364
  cosmic rays, II.748
  neutron, I.119

x-ray scattering, I.166
Bouguer, gravitation, II.1265
Bouwkamp, C J, diffraction theory, III.1420
Bowen, I S, nebulium lines, III.1762
Bowen, R, Gordon Conference (1976), III.1850
Bowles, K L, incoherent scatter radar (ISR), III.1996
Bowley, A R, nuclear medicine, III.1916
Boyd, G D, resonator modes, III.1429
Boyde, A, tandem scanning microscope, III.1451
Boyle, B J
  Newton's rings, III.1386
  quasars, III.1798
Boys, C V
  fine silica fibres, I.28
  gravitation, II.1266
  radiometer, I.6, I.23, I.28
  Soap Bubbles and the Forces that Mould Them, I.5
Bozorth, R M, x-ray study of KHF$_2$, I.458
Bradley, A W, single-crystal methods, I.435
Bradley, D, dye-laser pulses, III.1439
Bradley, J, stellar aberration, I.259
Bragg, Sir William (Henry), I.429
  α-particles, I.104
  Bragg–Gray theory, III.1882
  crystal spectrometer, I.430
  naphthalene, I.494
  reflection method, I.158
  structure factors, I.439
  x-ray diffraction, I.426, I.429
  x-rays, I.363
Bragg, Sir (William) Lawrence, **I.429**
  binary alloys, I.548
  doubly refracting crystal, I.485

Fifth Solvay Conference (1927), I.205
Fourier series method, I.440
inorganic compounds, I.492
order–disorder transitions, I.473
reflection method, I.158
sodium chloride, I.493
structure of beryl, I.440
Twelfth Solvay Conference, Brussels, II.646
x-ray analysis, I.496
x-ray diffraction, I.426, I.427–8, I.431
x-ray microscope, III.1414
Braginsky, V B, quantum non-demolition (QND) measurements, III.1469
Brahe, T
  supernova, III.1712
  supernovae, II.1228
Brandt, E H, levitation theory, III.1543
Brant, H L, two-photon correlation, III.1413
Brattain, W H
  point-contact transistor, III.1326
  transistor, II.744
Braun, E
  growth spiral, III.1528
  solid-state physics, III.1521
Braun, F, cathode-ray oscillograph, I.26, I.27
Braun, K F, wireless communication, III.1986
Bravais, A, symmetry types, I.423
Breit, G
  First Shelter Island Conference, II.645
  ionospheric conducting layer, III.1988
  ionospheric plasmas, III.1621
  ionospheric sounding, III.1995
  magneto-ionic theory, III.1990
  neutron capture, II.1194
  nuclear resonance, I.126

radio receiving equipment,
  III.1990
 reaction cross sections, II.1219
 Tesla coil, II.1185
Bretherton, F P, momentum
  transfer, II.867
Breuckner, K, laser fusion
  programme, III.1676
Brewer, R G
 self-phase modulation, III.1449
 stimulated Brillouin scattering,
  III.1444
Brickwedde, F G
 deuterium, III.1706
 deuteron, II.650, II.1183
Bridges, W B, argon-ion laser,
  III.1443
Bridgman, P W
 crystal growth, III.1280
 thermoelectric effects, III.1306
Briggs, B H, ionosphere, III.1984
Brill, R, biomolecular structures,
  I.497
Brillouin, L
 asymptotic linkage, I.197
 Brillouin lines and Brillouin
  zones, I.216, III.1311–13
 Fifth Solvay Conference (1927),
  I.205
 First Solvay Conference (1911),
  I.147
 Maxwell's Demon cannot operate,
  I.576
 quantum theory, III.1397
Brindley, G W, zero-point motion,
  II.974
Brinkman, W F, phonon exchange,
  II.953
Broadbent, S R, percolation
  processes, I.572
Brobeck, W, neutral beam
  injection, III.1661
Brockhouse, B N, **II.993**
 dispersion curves of germanium,
  II.1002

interatomic forces, II.999
neutron spectroscopy, I.486
phonon dispersion relations,
  II.993–4
phonons, II.999
shell model, II.997
spin-wave energies, II.1012
triple-axis crystal spectrometer,
  II.992
Broer, L J F, nuclear resonance,
  II.1172
Bromberg, J L, lasers, III.1478
Bronsteyn, M P, linearized
  quantum gravity, I.317
Brossel, J, optical pumping,
  III.1415
Brown, B R, computer-generated
  spatial filter, III.1447
Brown, H, fusion power, III.1673
Brown, R G W, spatial coherence,
  III.1471
Brown, S, electron diffusion rates,
  III.1644
Brown, T G, pulse-echo ultrasonic
  scanner, III.1913
Brüche
 emission electron microscope
  (EEM), III.1569
 Geometrische Elektronenoptik,
  III.1578
Brueckner, K A, correlation effects,
  III.1362
Brunhes, B, Earth's magnetic field,
  III.1973
Brush, S G
 Lenz–Ising model, I.554
 planetary physics, III.2012
Buchdahl, H A, Langrangean
  aberration theory, III.1458
Buckingham, M J, λ-point of
  liquid He^4, I.559
Budden, K G, radio waves, III.1992
Budker, G I
 cooling method, II.743
 magnetic mirror, III.1659

mirror concept, III.1665
mirror machine, III.1662
storage rings, II.741
Buechner, W W, electrostatic generator, II.1203
Buerger, M J, precession camera, I.434
Bullard, E C
  Earth's magnetic field, III.1960
  Pacific floor measurements, III.1974
Bullen, K, Earth's core, III.1957
Bunemann, O, stream-type instabilities, III.1673
Bunn, C W, difference synthesis, I.452
Bunsen, R
  Bunsen burner, I.75–6
  spectral emission lines, III.1692
  spectral lines, I.156
Burbidge, E M
  Evidence for the occurrence of violent events in the nuclei of galaxies, III.1777
  nucleosynthesis in stars, III.1788
  supernovae, II.1227, II.1228
  synthesis of elements, III.1747
Burbidge, G R
  Evidence for the occurrence of violent events in the nuclei of galaxies, III.1777
  inverse-Compton catastrophe, III.1781
  nucleosynthesis in stars, III.1788
  quasars, III.1777
  radio sources, III.1775
  supernovae, II.1227, II.1228
  synthesis of elements, III.1747
  x-ray sources, III.1755
Burger, R, Thermos flask, I.27
Burgers, W G, dislocation theory, I.476
Burke, P G, *R*-matrix calculations, II.1069

Burnham, D C, photon coincidence, III.1435
Burns, G, superconductors, I.490
Burrow, P D, molecular resonances, II.1084
Burrows, angular distributions, II.1220
Burrus, C A, fibre lasers, III.1427
Burstein, E, absorption coefficient of semiconductor, III.1421
Busch, H
  electron motion, III.1568–9
  thin magnetic lens theory, III.1570
Buschow, K H J, permanent magnet materials, II.1165
Butler, C C
  elementary particles, II.654
  forked tracks, I.408
  kaons, II.656
Butler, S T, theory for (d,p) reactions, II.1220
Byer, R L, laser frequency, III.1435
Byrd, R
  First Antarctic Expedition, III.2005
  Second Antarctic Expedition, III.2005

Cabannes, J, combination light-scattering in gases, III.1405
Cabibbo, N, charge-changing weak current, II.677
Cady, properties of quartz, II.1244
Cagniard de la Tour, C, critical point phenomenon, III.1947
Cahan, D, measurement, I.15
Cahn, J, spinodal concept, III.1548
Callan, C, point-like constituents, II.690
Callaway, J, valence/conduction bands, III.1328–9
Cameron, A G W

explosive nucleosynthesis, III.1747
pp chain, II.1227
solar neutrinos, III.1748
supernova trigger hypothesis, III.1964
synthesis of elements, III.1747
Campbell, L L, resistivity variation with temperature, III.1282
Cannizzaro, S
  Karlsruhe congress, I.46
  molecules, I.45
Cannon, A J, stellar spectra, III.1693, III.1694
Capasso, F, photoconductivity, III.1470
Carathéodory, C
  second law of thermodynamics, I.534
  temperature concept, I.534
Carlson, C, xerography, III.1541
Carnot, S, second law of thermodynamics, I.521
Caroe, O, *William Henry Bragg*, III.1514
Carothers, W, polymer industry, III.1550
Carpenter, D, whistler measurements, III.2009–10
Carpenter, H C H
  crystal growth, III.1280
  plastic deformation, III.1526
Carrington, A, vibrational spectrum, II.1067
Carruthers, L, computational expert, III.1844
Cartan, E, non-symmetric connection, I.321
Carter, B, black holes, III.1778
Casagrande, F, laser theory, III.1444
Casimir, H B G, two-fluid model, II.924, II.925
Caspar, D L D
  tobacco mosaic virus (TMV), I.508
  tomato bushy stunt virus (TBSV), I.510
Cassels, J M, critical fluctuations, II.1016
Cassen, B
  automatic rectilinear scanner, III.1901
  isotope invariance, I.378
Castaing, R
  electron microprobe analyser, III.1548
  electron probe x-ray microanalysis, III.1580
Cauchy, A, elasticity of crystals, I.424
Caullery, M
  École Polytechnique, I.15
  laboratory equipment, I.19
Cave, H M, ionization chambers, II.1200
Cavendish, gravitation, II.1266
Caves, C M, quantum non-demolition (QND) measurements, III.1469
Cercignani, C
  BBKGY hierarchy, I.626
  The Boltzmann Equation and its Applications, I.626
Čerenkov, P A
  charged particles, II.750
  radiation from relaxation of non-uniform polarization, III.1408
Chadwick, Sir James, I.120
  α-particles, I.114, II.1190
  anomalous scattering, I.366
  continuous spectrum, I.107–8, I.118
  neutron, I.119–20, I.122, I.369, I.370, II.639, II.1183, III.1706, III.1712
  positron, II.650

Possible existence of a neutron,
  I.120
*Radiations from Radioactive
  Substances*, I.108, I.112,
  II.1199, II.1200
Chalmers, B, new journal, III.1556
Chamberlin, T C
  nebular hypothesis, III.1950
  tidal theory, III.1962
Chambers, K C, radio galaxies,
  III.1796
Chambers, R G
  boundary scattering, III.1343–4
  cyclotron resonance, III.1357
  electron free paths, III.1344
Chandler, P J, waveguide laser,
  III.1471
Chandrasekhar, S
  astrophysics, III.1706
  degenerate stars, III.1711
  dynamical friction, III.1774
  *Eddington: the Most Distinguished
    Astrophysicist of his Time*,
    III.1702–3
  equilibrium theory, III.1785
  liquid crystals, III.1540
  Radiative Transfer, III.1415–16
  Schönberg–Chandrasekhar limit,
    III.1709
Chandresekhar, S, random walks,
  I.608
Chapin, D M, solar cell, III.1421
Chapman, S, **III.1993**
  diamagnetic currents, III.1684
  geomagnetic field, III.2003
  geomagnetic storms, III.2003
  Geomagnetism, III.2000
  ionospheric layers, III.1992–4
  Solar–Terrestrial Physics, III.2001
  statistical mechanics, I.598
  upper atmosphere, III.1982,
    III.1985
Charnley, Sir John, hip
  replacement, III.1896
Charpak, G
  drift chambers, II.754
  multi-wire proportional
    chambers, II.749
Cherwell, Lord, *see* Lindemann, F
  A (later Lord Cherwell)
Chester, G V, disordered
  materials, III.1366
Chevalier, R, light curve of
  SN1987A, III.1761
Chew, G F
  pion–nucleon resonance, II.680
  pion–pion scattering, II.681
  Regge trajectories, II.682
  Twelfth Solvay Conference,
    Brussels, II.646
Chiao, R Y
  optical fibres, III.1470
  stimulated Brillouin scattering,
    III.1444
Chodorow, M, high-power pulsed
  klystron, II.736
Christensen-Dalsgaard, J, low
  degree modes, III.1750
Christenson, J, CP violation, II.672
Christian, J, phase transformations,
  III.1547
Christoffel, tensor analysis, I.292
Christofilos, N C
  alternating-gradient focusing,
    II.734
  Argus experiment, III.1683
  ASTRON, III.1645
Christy, R F
  meson theory, I.407
  supernovae, II.1228
Cini, M, Twelfth Solvay
  Conference, Brussels, II.646
Clark Jones, R, polarization in
  optical systems, III.1411
Clauser, J F, Einstein–Podolsky
  –Rosen(–Bohm) (EPR(B))
  paradoxes, III.1457
Clausius, R
  heat engines, I.534

kinetic theory of gases, I.522–3, I.587, III.1284
On the kind of motion we call heat, I.47
second law of thermodynamics, I.521
virial theorem, I.543
Clayton, D D
　element abundances, III.1748
　isotopic anomalies, III.1964
Clifford, W K, atoms, I.31
Cline, D, proton–antiproton colliders, II.743
Clothier, W K, voltage measurement, II.1256
Cochran, W
　determinational inequalities, I.448
　helical structure, I.498
　phase transitions, II.1008
　probability relations, I.449
　pseudopotential theory, II.999
　shell model, II.997, II.1002
　structure determination, I.452
Cockcroft, Sir John
　atomic energy project, II.1198
　Atoms for Peace Conference, III.1638
　disintegration experiments, II.1184
　disintegration of lithium, II.729
　electrostatic generators, II.727
　fusion reactions, III.1628
　nuclear disintegration, II.1183
　nuclear reaction, I.285
　nuclear transformation, I.112
　proton acceleration, I.370
Coensgen, F
　deuterium plasma, III.1659
　neutral-beam injectors, III.1663
　plasma gun, III.1661
Cohen, E G D, collision configurations, I.611
Cohen, M
　Landau phonon–roton dispersion relation, II.994
　Landau spectrum, II.930
Cohen, M H, galactic nuclei, III.1783
Colclough, A R, gas constant (R), II.1269
Cole, E R, blind deconvolution, III.1461
Coleman, S, running coupling constant, II.713
Coles, D
　law of the wake, II.840
　law of the wall, II.840
Colgate, S
　neutral beam injection, III.1661
　neutron star, III.1759
Colle, R, molecular dissociation, II.1067
Collier, R J, holography, III.1438
Collins, M F, spin-wave modes, II.1151
Collins, S C, liquid helium, III.1341
Compton, A H, I.166
　cloud chamber, I.364
　cosmic-ray studies, III.2005
　Fifth Solvay Conference (1927), I.205
　Secondary radiation produced by x-rays, I.165
　x-ray scattering, I.362, I.364, III.1867
　x-rays, I.363
Compton, K T, photoelectric effect, III.1390
Condon, E U
　α-decay, I.115, I.217
　atomic energy levels, II.1093
　electrostatic generators, II.727
　isotope invariance, I.378
　Nature, I.116
　quantum-mechanical theory of alpha decay, II.1184
　The Theory of Atomic Spectra, II.1037

Conrady, A E, chromatic aberration, III.1390
Conversi, M, mesons, II.652
Conway, 'life' (game), I.629
Cook, A H, measurements in physic, II.1273
Coolidge, W D, cathode high-vacuum tube, III.1861
Cooper, L
  BCS theory, II.938
  superconductivity, I.232, II.939, III.1364
Cooper, M J, anharmonicity in BaF_2, I.452
Corey, R B
  biomolecular structures, I.497–8
  pleated sheet structures, I.499
  polypeptide chains, I.497
Cork, C, spark chamber, II.753
Cornford, F M, Cambridge University, I.19
Coster, D, hafnium, I.100
Cotton, Sir Alan, magneto-optical effect, III.1389
Cottrell, Sir Alan, **III.1522**
  discontinuous yield-stress in mild steel, III.1520–1
  dislocation theory, III.1523
  physical metallurgy, III.1555
  yield-stress and strain-aging, III.1523
Coulomb, C A, electrical and magnetic experiments, III.2004
Coulson, C H, computations, II.1087
Courant, *Methoden der Mathematischen Physik*, III.2043
Courant, E D
  accelerators, II.738
  synchrotrons, II.734
Cowan, C, neutrino, II.663
Cowling, T G, axially symmetric field, III.1958

Cox, A, geomagnetic field, III.1973
Cox, D P, interstellar medium, III.1766
Craik, D J, domain techniques, II.1155
Crasemann, B, secondary excitations, II.1090
Crawford, B H, human eye, III.1407
Crewe, A, high brightness field emission sources, III.1589
Crick, F H C
  DNA model, I.507
  double-helix structure, I.504
  helical structure, I.498
  polypeptides, I.498
  protein subunits in viruses, I.509
Critchfield, pp process, II.1226
Croat, J J, neodymium–iron borates, II.1165
Croce, P, coherent imaging, III.1418
Crommelin, eclipse expeditions, III.1722
Cronin, J W
  CP violation, II.672
  spark chamber, II.753
Crookes, W
  cathode rays, II.1271
  phosphorescence of zinc sulphide, II.745
Cross, L E, ceramics, III.1538
Crowfoot, D M
  globular proteins, I.497
  tomato bushy stunt virus (TBSV), I.509
Cummerow, R L, electron spin resonance, II.1138
Cummins, H Z, Doppler shift, III.1442
Curie, I, radioactivity, I.121
Curie, M, I.57
  atoms, I.49
  Fifth Solvay Conference (1927), I.205

First Solvay Conference (1911), I.147
  radioactivity, I.56–8, I.60
  radium, I.6, II.1272, III.1863, III.1876
  uranium, I.61
Curie, P
  diamagnetism, II.1116
  ferromagnets, II.1118
  magnetism, I.545, II.1113
  Nobel Prize lecture, I.63
  radioactivity, I.56–8
  radium, III.1876
  uranium, I.61
Curien, H, x-ray scattering, II.990
Curle, N
  jet-edge system, II.852
  sound waves, II.851
Curtis, H D
  Great Debate, III.1718
  novae, III.1717
Cusack, N E, liquid metals, III.1366
Cvejanovic, quantum-mechanical reformulation, II.1065
Cvitanovic, P, universal function, III.1848
Czochralski, single-crystal wires, III.1281

Dabbs, J W T, germanium junctions, II.1208
Dagenais, M, photon antibunching, III.1460
Daguerre, L-J-M, daguerreotype process, III.1692
Dainty, J C, blind deconvolution, III.1461
Dällenbach, W, invariant averaging procedure, I.281
Dallos, J, scleral lenses, III.1414
Dalrymple, G B, geomagnetic field, III.1973
Dalton, B J, laser fluctuations, III.1473
Dalton, J
  learned societies, I.3
  New System of Chemical Philosophy, I.44
Damadian, R, nuclear magnetic resonance, III.1923
Damm, C C
  Alice mirror machine, III.1679
  neutral beam injection, III.1661
Danby, G, two-neutrino experiment, II.672
Daniel, V, copper–nickel–iron alloy, III.1548
Darmois, gravitational field, I.314
Darrow, K K, First Shelter Island Conference, II.645
d'Arsonval, surgical diathermy, III.1894
Darwin, C G
  biological evolution, III.1947
  dispersion formula, I.171
  extinction, I.451
  integrated intensity formula, I.434
  magneto-ionic theory, III.1992
  Raman effect (combination scattering), III.1404
  relativistic hydrogen spectrum, I.218
  statistical mechanics, I.535
  structure factor, I.431
Darwin, G H
  Moon–Earth hypothesis, III.1951–2
  Moon model, III.1966
  Moon's orbit, III.1948
Das Gupta, M K, Cygnus A, III.1775
Dash, W C, diffusion of copper impurity, III.1526
Dashen, R, axial–vector coupling constant and pion–nucleon scattering, II.679
Daunt, energy gap, II.932

Davidovits, confocal scanning laser microscope, III.1450
Davies, D
  detection of impulses, III.1957
  Earth's core, III.1980
da Vinci, L, medical mechanics, III.1856
Davis, D J M Jr, ring-laser gyroscope, III.1433
Davis, D R, Moon model, III.1966
Davis, E A, *Electron processes in Non-crystalline Materials*, III.1369
Davis, R, solar neutrinos, II.757, III.1748
Davisson, C
  electron diffraction, I.463
  electron reflection, I.170
Davy, Sir Humphry, learned societies, I.3
Dawson, B, temperature factor, I.452
Dawson, J, fusion systems, III.1673
Day, A L, report to Philosophical Society of Washington, III.1387
Dean, P, amorphous SiO_2, I.479
de Boer, J H
  critical temperature, II.975
  electrical properties, II.1156
  law of corresponding states, II.974
  semiconductors, III.1330
Debrenne, P, cryostat, I.437
de Broglie, L, I.170
  elementary particles, I.282–3
  Fifth Solvay Conference (1927), I.205
  First Solvay Conference (1911), I.147
  matter-wave theory, I.174, I.229
  On the general definition of the correspondence between wave and (particle) motion, I.169
  wave mechanics, I.227
  wave–particle duality, I.168–9
de Broglie, M
  crystal oscillation, I.434
  x-ray spectra, I.160
Debye, P, **II.971**
  adiabatic demagnetization, II.1125
  dielectric waveguide, III.1397
  dynamical laws, I.176
  electron screening, III.1316
  energy spectrum, I.153
  Fifth Solvay Conference (1927), I.205
  Planck's law, I.70
  screening length, III.1610
  single crystal, I.434
  specific heat, II.971–2, II.977, II.980, II.982, II.1011, III.1398
  thermal conductivity, II.977–8
  thermal vibration, I.431
  vibrational specific heat of solids, I.525
  x-ray scattering, I.165
  x-rays, I.363
  Zeeman effect, I.102
de Forest, L
  triode, I.37
  upper ionized layer, III.1987
de Gennes, P G
  liquid crystals, II.1084, III.1540
  SAW model, I.570
  soft physics, III.2030
de Haas, W J
  Kondo effect, II.1167
  potassium chromic alum, II.1126
de Hevesy, G, quantum theory, I.180
Dehmelt, H G, charged particles, II.1099
Delaunay, C-E, Earth–Moon system, III.1948
Delchar, T A, surface crystallography, I.466
Dellinger, J H, sudden ionospheric disturbance (SID), III.1995

Demirkhanov, R, Atoms for Peace Conference, III.1638
Democritus, atoms, I.43, III.2034–5
Denisyuk, Yu N, holograms, III.1438
Dennison, D M, wavefunctions, I.529
Derjagin, B V, charge effects, III.1611
Derrida, B, delta (δ), III.1849
De Rújula, A, charmed particles, II.704
des Cloizeaux, J
　excitations in $s = 1/2$ Heisenberg antiferromagnet, II.1014
　polymer solutions, III.1599
De Salvo, L, Collective Atomic Recoil Laser, III.1456
Desclaux, J P, shell electrons, II.1091
de Sitter, W
　cosmological constant, III.1724
　double-star systems, I.285
　Einstein–de Sitter model, III.1730–1
　spherical four-dimensional space-time, III.1723
　Universe, III.1724
Destriau, G, electroluminescence, III.1417
Deutsch, M, positronium, II.646
de Vaucouleurs, G, Hubble's constant, III.1793
de Vries, G, weak non-linear and weak dispersive effects, II.859
Devonshire, A F, ferroelectric transition in $BaTiO_3$, II.1007
Dewar, J, liquefied gases, I.28
Dick, B J, shell model, I.485, II.997
Dicke, R H
　Big Bang, III.1790
　Fabry–Perot resonant cavity, III.1423
　optical bomb, III.1419
　solar oblateness, I.303
　world models, III.1806
Dickinson, R G, hexamethylene tetramine, I.495
Dietrich, O W, order–disorder transition in β-brass, II.1009, II.1016
di Francia, T, lens-pupil functions, III.1418, III.1457
Dillon, J F
　bubble domains, II.1159
　domain techniques, II.1155
Dimov, G I, tandem-mirror concept, III.1665–6
Dingle, R, superlattice minibands, III.1459
Dirac, P A M, I.189
　atomic mechanics, I.183
　electromagnetic field, I.358
　electron spin, II.1125
　electron statistics, I.528
　Fifth Solvay Conference (1927), I.205
　four-component wavefunctions, I.277
　free particles, I.283
　particle dynamics, I.279
　properties of the Universe, III.1787
　quantization of free field, III.1402
　quantum electrodynamics (QED), I.360, I.362
　quantum mechanics, I.188, I.198, I.211
　Quantum Mechanics, III.2043
　relativistically invariant theory, I.283
　transformation theory, I.201, I.360
　Twelfth Solvay Conference, Brussels, II.646
　wavefunctions, I.200
Dixon, J M, crystal fields, II.1165

Doell, R, geomagnetic field, III.1973
Doi, M
  entangled systems, III.1600
  monodisperse melt, III.1600
Dolen, R, pion–nucleon charge exchange, II.683
Domb, C
  critical behaviour, I.556, I.560
  critical exponents, I.560
  magnetic field at critical point, I.562
  order–disorder transformation, I.578
  site percolation processes, I.572
Dombrovski, V A, optical continuum of nebula, III.1774–5
Domen, S R, calorimetry, III.1883
Donald, I, diagnostic ultrasound, III.1912–13
Donder, Th De, Fifth Solvay Conference (1927), I.205
Döring, W, nucleation kinetics, III.1547
Dorsey, N E, speed of light, II.1263
Drell, S D, lepton pair, II.690
Dresden, M, game rules, I.629
Dresselhaus, G
  phonons, III.1423
  resonant frequencies, III.1334
Dreyer, J L E
  Index Catalogues, III.1717
  New General Catalogue of Nebulae and Clusters of Stars, III.1717
Droste, J, field equations, I.303
Drude, P K L
  electron optics, III.2000
  electron vibration, II.983
  Hall effect, III.1282
  magneto-ionic theory, III.1990
  metallic conduction, III.1284
  school of physics, I.15
  The Theory of Optics, III.1385
Dryden, H L
  closed-circuit wind-tunnels, II.830–1
  turbulence, II.798
Duane, W
  quantum physics, III.1401–2
  radiation quanta, I.165
  radon, III.1876
  röntgen, III.1867
Duffieux, P M, *The Fourier Transform and its Applications in Optics*, III.1413
Dumke, P W, semiconductor lasers, III.1433
DuMond, J W M, curved crystal focusing spectrometer, II.1207
Dumontet, P, polychromatic fields, III.1421
Dungey, J, magnetosphere, III.2010
Dunlop, J S, radio sources, III.1798
Dunning, J R, fission experiments, II.1196
Dunoyer, L, surface crystallography, I.468
Duraffourg, G, stimulated emission in semiconductors, III.1433
Duwez, P, quenching of hot alloys, III.1542–3
Dyson, F J
  spin-wave theory, II.1011
  Twelfth Solvay Conference, Brussels, II.646
Dyson, J
  Fresnel fringe system, III.1584
  holograms, III.1584

Eberly, J H, unstable and chaotic emission, III.1466
Eccles, W H, radio waves, III.1622, III.1987
Eckart, C
  continuous media, I.280
  equivalence relations, I.196

Eckart, M, theory of metals, III.1290
Eckersley, T L
  long-distance communications, III.1999
  magneto-ionic theory, III.1992
  solar x-rays, III.1994
  upper ionized layer, III.1987
  whistler waves, III.1622–3, III.1995, III.2002
Eddington, A S, **III.1700**
  eclipse expeditions, III.1722
  Eddington–Lemaître model, III.1732–3
  energy transport, III.1699
  fine structure constant, II.1262
  Fundamental Theory, III.1787
  gravitational waves, I.312
  internal structure of the stars, III.1703
  interstellar medium, III.1761
  mass–luminosity relation, III.1705
  Mathematical Theory of Relativity, III.1700
  red giants, III.1704
  The Fundamental Theory, III.1700
  The Internal Constitution of the Stars, III.1700, III.1702
  theory of stellar structure, III.1702–6
  virial theorem, III.1731–2
  white dwarfs, III.1710
Edgerton, H E, stroboscopic photography, III.1410
Edler, I, time–position recordings, III.1913–14
Edwards, O S, stacking faults, I.473
Edwards, S F
  entangled systems, III.1600
  monodisperse melt, III.1600
  quantum-mechanical particles, III.1599

Egger, confocal scanning laser microscope, III.1450
Ehlers, J, general relativity, I.299
Ehrenberg, W, LEED experiments, I.466
Ehrenfest, P
  adiabatic principle, I.92–3
  dog-flea model, I.604, I.618
  Fifth Solvay Conference (1927), I.205
  higher-order transitions, I.549
  periodic systems, I.177
  phase transitions, II.923–4
  spherical four-dimensional space-time, III.1723
  statistical mechanics, I.586
  thermodynamics and irreversibility, I.599
  wind tree model, I.603
Ehrenfest, T
  statistical mechanics, I.586
  thermodynamics and irreversibility, I.599
Eindhoven, W, medical electronics, III.1893
Einstein, A, **I.250**
  anno mirabile, III.2018
  binding energy, I.110–11
  black-body radiation, I.163
  Bohr's theory, I.88
  Brownian motion, I.586
  conservation of energy, III.1390
  cosmological constant, III.1723, III.1730
  dielectric constant, III.1396–7
  Einstein–de Sitter model, III.1730–1
  energy surface trajectories, I.590
  Fifth Solvay Conference (1927), I.205
  First Solvay Conference (1911), I.147
  fluctuations, I.533–4
  gas of corpuscles, III.1390–2
  general theory of relativity,

I.251, I.253, I.286–320,
III.1722, III.1723
gravitation, I.292
heat capacity, I.78
light-quanta, I.54, I.71, I.151–2,
I.167, I.173, I.363, III.1396,
III.1452
nucleation of liquid droplets,
III.1547
opalescent bands, I.546
optical microscopy, III.1406
quanta, II.969
quantum concept, I.71, I.147,
I.228, I.585, III.1398–9
quantum physics, I.93–4
radiation in thermodynamic
equilibrium, III.1392–3
radiation law, I.152
relativity theory, I.34, I.61, I.173
space-time structure, I.251
special theory of relativity, I.54,
I.249, I.251, I.262–85, III.2039
specific heat, I.153–4, II.970
statistical mechanics, I.173
stress–energy tensor, I.281
unified-field theory, I.232
Ekins, R P, saturation analysis
technique, III.1909
Elam, C I
crystal growth, III.1280
plastic deformation, III.1526
Elias, G J, solar x-rays, III.1994
Elias, P, communications theory,
III.1417
Eliashberg, G M, electron–phonon
interaction, II.941
Elliott, R J, phonons, III.1423
Ellis, C D
continuous spectrum, I.118, I.126
internal conversion, I.107
*Radiations from Radioactive
Substances*, I.108, I.112,
II.1199, II.1200
Ellis, R E, mirror machine, III.1679
Elsasser, W M

Earth's magnetic field,
III.1969–70
electron reflection, I.170
geomagnetic dynamo, III.1958
neutron diffraction, I.453
shell models, II.1209
toroidal (doughnut-shaped)
fields, III.1958
Elster, phosphorescence of zinc
sulphide, II.745
El-Sum, H M A, holography,
III.1415
Emanuel, K A, tropical cyclones,
II.904
Emden, R, *Gaskugeln*, III.1695
Emmons, H W
laminar flow, II.825
randomly originating turbulent
spots, II.835
Empedocles, elements, I.44
Enskog, D, statistical mechanics,
I.598
Eötvös, R V
equivalence principle, I.286
gravitation, II.1267
Epicurus, atoms, I.44
Epstein, J B, energy in the Sun,
III.1748
Epstein, P S
dynamical laws, I.176
quantum theory, I.89
Stark effect, I.162
Ernst, M, non-linearity, I.625
Ertel, H, tropical cyclones, II.905
Esaki, L
coupling of quantum wells,
III.1459
germanium, III.1340
quantum-well superlattices,
III.1453
Eshelby, J, physical metallurgy,
III.1555
Essam, J W
critical exponents, I.560
pair connectedness, I.573

two-dimensional lattices, I.573
Essen, L, speed of light, II.1263
Estermann, I, surface crystallography, I.468
Eucken, A, II.972
  calorimeter, II.968
  liquid-air temperature, II.977
  quantum mechanics, I.146
  specific-heat problem, I.153–4
Evans, E, liposomes, III.1608
Eve, A S
  ionizing agents, I.367
  radium, III.1881
  secondary x-rays, I.362
Ewald, P P
  crystal structures, III.1514
  Fifty Years of X-ray Diffraction, I.511
  reciprocal lattice, I.428
  x-ray diffraction, I.424
Ewan, H I, hyperfine line radiation, III.1763
Ewing, Sir Alfred, plastic deformation, III.1516
Ewing, D H, compound nucleus, II.1219

Faber, T E, liquid metals, III.1366
Fabrikant, V A, light amplification, III.1411
Failla, G, medical physics, III.1864
Fairbank, W M
  λ-point of liquid He^4, I.559
  superfluidity, II.932
Fankuchen, I, tobacco mosaic virus (TMV), I.508
Fano, U, resonance, III.1473
Faraday, M
  Chemical History of a Candle, I.5
  conductivity, III.1318
  disliquifying point, I.543
  electric current, III.1958
  electrical properties, II.1113
  electrolysis, I.51
  electromagnetic interactions, III.1953
  electromagnetism, III.1856, III.2000
  experimental physics, I.14
  geomagnetic variations, III.1984
  learned societies, I.3
  magnetic properties, II.1112, II.1113
Farnsworth, H G, surface crystallography, I.470
Faust, W L, CO_2 laser, III.1442
Feher, E R, crystal field theory, II.1144
Fellgett, P, multiplex optical spectroscopy, III.1416
Felten, J, luminosity function, III.1770
Fermi, E, **II.1187**
  β-decay theory, I.374–5, I.375
  electron statistics, I.528
  neutron bombardment, I.122
  neutron capture, II.1186, II.1193
  neutron diffraction, I.456
  neutron irradiation, I.129–30
  neutron sources, II.1199, II.1206
  quantum mechanics, II.1031
  quantum states, III.1288
  Raman scattering, II.996
  synchrocyclotron, II.733
  Tentativo di una teoria della emissione di raggi β, I.129
  theory of β-decay, III.1706
Fermi, L, *Atoms in the Family*, II.1191
Ferraro, V C A
  diamagnetic currents, III.1684
  geomagnetic field, III.2003
Ferrel, W, Earth–Moon system, III.1948
Feshbach, H
  compound nucleus, II.1219
  First Shelter Island Conference, II.645

projection-operator technique, II.1221
proton scattering, II.1212
Feynman, R P, **II.648**
  First Shelter Island Conference, II.645
  Landau phonon–roton dispersion relation, II.994
  Landau spectrum, II.930
  neutrinos, II.667
  parity states, II.665
  point-like constituents, II.689
  quantum electrodynamics, II.643, II.740, III.1446
  quantum mechanics, III.1395, III.1830
  renormalization, II.646
  scientific imagination, III.1519
  Twelfth Solvay Conference, Brussels, II.646
  weak vector current, II.668
Field, G B, interstellar medium, III.1765
Fienup, J R, optical data processing, III.1461
Fierz, M, free particles, I.283
Finch, J F, polio virus, I.510
Finkelstein, D, Schwarzschild singularity, I.308
Finney, J L, amorphous structure, I.479
Firestone, F A, ultrasonic flaw detector, III.1911
Fischer, E, polyethylene, III.1552
Fischer, K, AgCl, II.998
Fisher, I Z, neutron frequency, III.1418
Fisher, M E
  covering lattice, I.573
  critical exponents, I.560
  field theory, III.1830
  pair distribution function, I.547
Fisher, O, Earth–Moon system, III.1948
Fitch, V, *CP* violation, II.672

Fitzgerald, G F
  Fitzgerald–Lorentz contraction, I.264
  wireless wave propagation, III.1986
Fizeau, A H L, velocity of light, I.266
Fleming, J A, *Terrestrial Magnetism and Atmospheric Electricity*, III.2000
Flory, P, **III.1598**
  excluded volume, I.569
  polymer chains, III.1551
  polymer solutions, III.1599
  self-avoiding walk, III.1598
Fock, V
  configuration (Fock)-space method, I.208
  equations of motion, I.303
  self-gravitating systems, I.313
Foëx, G, *Le Magnetisme*, II.1119
Foley, electric quadrupole moments, II.1214
Follin, J W, proton–neutron ratio, III.1786
Fomichev, V I, K-shell absorptions, II.1083
Foord, R, photon correlation spectroscopy, III.1454
Ford, F C, mirror machine, III.1679
Försterling, K, wave propagation, III.1991
Fortuin, C, bond percolation model, I.573
Foucault, J B L, pendulum, I.2, I.3
Fowler, R H
  electron gas, III.1711
  Fifth Solvay Conference (1927), I.205
  high-density matter in stars, I.215
  ionization, III.1701
  quantum theory, I.180
  residual entropy of ice, I.571
  statistical mechanics, I.535

Statistical Mechanics, I.533
Statistical Thermodynamics, I.537
stellar spectra, III.1693
structure of ice, I.474
white dwarf stars, I.531
Fowler, T K, tandem-mirror concept, III.1665–6
Fowler, W A
 Big Bang, III.1792
 nuclear interactions between light nuclei, III.1791
 nucleosynthesis in stars, III.1788
 origin of the elements, II.1225
 pp chain, II.1227
 solar neutrinos, III.1748
 supernovae, II.1227, II.1228
 synthesis of elements, III.1747
Fox, A G, Fabry–Perot resonator, III.1429
Franceschini, W, delta (δ), III.1851
Franck, J
 Franck–Hertz experiment, I.89–90
 quantum theory, I.179
 scattering experiments, I.217
 stationary electron orbits, I.158
Frank, A, rare-earth ions, II.1120
Frank, Sir Charles, **III.1606**
 crystal growth, III.1527
 crystallinity, III.1551–2
 line singularities in nematics, III.1605
Frank, I, energy loss formula, II.750
Frank, I M, visible light, III.1409
Frank, N H, theory of metals, III.1290
Franken, P A, ruby laser, III.1427–9
Franklin, B, learned societies, I.3
Franklin, R E
 DNA, I.504
 tobacco mosaic virus (TMV), I.508

Franz, W, photo-assisted tunnelling, III.1423
Fraunhofer, J
 solar spectrum, III.1691–2
 spectral lines, I.156
Frautschi, S C, Regge trajectories, II.682
Frazier, E N, solar oscillations, III.1749–50
Freedman, S J, Einstein–Podolsky–Rosen(–Bohm) (EPR(B)) paradoxes, III.1457
Freeman, A J, rare-earth elements, II.1164
Freidmann, A A, world models, III.1724–5
Frenkel, J
 phonon, III.1302
 semiconductors, III.1319–20
 sound waves, III.1328
 superconductivity, II.917, II.936
 vacancies, III.1529
Fresnel, A
 diffraction, III.1386
 wave description of light, I.72
Freud, P von, energy complexes, I.301
Freundlich, E F
 general relativity, I.305
 red shift, I.289
Fricke, H, chemical dosimetry, III.1884
Friedel, G
 focal conic defects, III.1605
 liquid crystals, III.1540
 mesomorphic phases, III.1603
Friedman, H
 Crab Nebula, III.1739
 rocket experiments, III.1738
 rocket studies, III.2005
 world models, III.1784, III.1807
Friedman, J I, parity violation, II.666
Friedrich, W
 crystal structures, III.1514

x-ray diffraction, I.425
Frisch, O R
  fission hypothesis, II.1195
  neutron–U collisions, I.130
Fritzsch, H, quarks, II.711
Fritzsche, H, semiconductors, III.1330
Fröhlich, H
  electron–phonon interactions, III.1337–8
  heavy electrons, I.392
  lattice deformation, III.1321
  meson theory, I.394
  metallic conductivity, I.216
  superconductivity, II.936
Froome, K D, speed of light, II.1263
Fry, D W, linear accelerators, II.1206
Fry, W J, ultrasonic surgery, III.1915
Fuchs, K, properties of fluids, I.553
Fukui, S, spark chamber, II.753
Fuller, C S, solar cell, III.1421
Fuller, L, upper ionized layer, III.1987
Furnas, T C, powder photographs, I.436
Furry, W, positrons in electron self-energy, II.641

Gabor, D, **III.1567**
  diffraction microscopy, III.1582
  electron microscope, III.1414–15, III.1581
  holography, III.1456, III.1582–4
  magnetic electron lens, III.1566–9
  super-lens, III.1411
Gagarin, Y, orbital flight, III.1738
Gaillard, J-M, two-neutrino experiment, II.672
Galileo Gallilea, I.254
  mechanical principal of relativity, I.254
  principle of inertia, III.1827
  speed of light, II.1262
Galison, P, magnetic moment and angular momentum, II.1123
Gamow, G
  abundances of elements, III.1785–6
  α-decay, I.115, I.217, II.1184
  α-particles, II.1189
  α-scattering, II.1189
  Atomic Nuclei and Nuclear Transformations, II.1225
  Big Bang, III.1784–6
  black-body radiation, II.758
  Constitution of Atomic Nuclei and Radioactivity, I.115
  cosmological constant, III.1730
  electrostatic generators, II.727
  fusion reactions, III.1628
  heavier elements, II.1226, II.1227
  liquid drop, I.126, II.1186
  neutrons, III.1713
  quantum-mechanical tunnelling, III.1706
  The Constitution of Atomic Nuclei and Radioactivity, II.1186
Garmire, E, stimulated Brillouin scattering, III.1444
Garrett, C G B
  anharmonic oscillators, III.1434
  two-photon absorption, III.1429
Garwin, R, parity violation, II.666
Gauss, C F, geomagnetism, III.1984
Geffken, W, metal–dielectric interference filters, III.1407
Gehrcke, E, Lummer–Gehrcke plate, III.1389
Geiger, H
  α-particles, I.113, II.745–6
  α-scattering, II.1200
  β-spectrum, I.107
  coincident counting, I.364
  electrical counting methods, II.1200

electron charge, II.747
Nature of the α-particle, I.58
x-ray scattering, I.166
Geissler, J, improved vacua, I.54
Geitel, phosphorescence of zinc sulphide, II.745
Geller, M, bright galaxies, III.1772
Gell-Mann, M, II.678
  axial coupling constant, II.676
  axial–vector coupling constant and pion–nucleon scattering, II.679
  baryons, II.661
  β-decay, II.677
  correlation effects, III.1362
  isotopic spin multiplets, II.658
  mass relations among baryons and mesons, II.661
  neutrinos, II.667
  quark–lepton analogy, II.700
  quarks, II.711
  renormalization group, III.1830, III.1832
  strangeness, II.662
  SU(3) triplets, II.685
  The Quark and the Jaguar, II.770
  Twelfth Solvay Conference, Brussels, II.646
  unitary symmetry, II.677
  weak hadronic current, II.678
  weak vector current, II.668
Genna, S, calorimetry, III.1883
Georgi, H, charmed particles, II.704
Gerlach, W, beam splitting, I.165
Germain, G, structure determination, I.450
Germer, L H, de Broglie waves, I.171
Gershtein, S S
  neutrinos, III.1803
  weak vector current, II.668
Gerstenkorn, H, Moon model, III.1966

Ghosh, S P, ionospheric physics, III.2000
Giacconi, R, Sco X-1, III.1739
Giaever, I, energy gap, II.933
Giauque, W F
  adiabatic demagnetization, II.1125
  gadolinium sulphate, II.1126
Gibbs, H M
  optical bistability, III.1449
  Optical Bistability, III.1450
  thermal bistable effects, III.1450
Gibbs, J W, **I.522**
  canonical ensemble, I.526
  Elementary Principles of Statistical Mechanics, I.586, I.593
  ensembles, I.531–3
  heterogeneous systems, I.34–5
  identical particles, I.528
  partition function, I.527
  phase equilibria, III.1507–8
  PhD, I.17
  statistical mechanics, I.173, I.523, I.606
  thermodynamics, I.521
Gibson, angular distributions, II.1220
Gibson, G, beta-ray experiment, III.1679–80
Gilbert, W
  De Magnete, I.544, III.1855–6
  Earth's magnetism, III.1952–3
  electromagnetism, II.1252
  magnetic forces, III.1952
  magnetism, II.1111
Gilliland, T R, ionosonde, III.1989, III.1991
Ginsburg, V, Čerenkov effect, III.1409
Gintzon, E L
  high-power pulsed klystron, II.736
  linear accelerators, II.736
Ginzburg, D M
  energy gap, II.932

superconductivity, II.934–5
Ginzburg, V L, Galactic radio emission, III.1737–8
Giorgi, G, MKSA system, II.1235
Gires, F, pump-probe configuration, III.1445
Gittelman, B, electron storage rings, II.739
Gittus, J, radiation damage, III.1534
Glaser, D A, bubble chamber, II.659, II.753
Glaser, W, electron optics, III.1575
Glashow, S L, II.696
 charmed particles, II.704
 charmed quark, II.700, II.705
 neutral currents, II.700
 quark–lepton analogy, II.700
 SU(2) x U(1) group, II.694
Glauber, R J
 neutron frequency, III.1418
 photodetection, III.1441
 quantum theory of coherence, III.1439–41
 radiation field, III.1388
Glazebrook, R T (later Sir Richard), **II.1253**
Gleick, J, *GENIUS—The Life of Richard Feynman*, III.1519
Gold, A V, de Haas–van Alphen effect, III.1352
Gold, T
 magnetosphere, III.1981
 pulsars, III.1752
 Relativistic Astrophysics Symposium, III.1776
 steady-state cosmology, III.1787
Goldberg, H, triplets, II.685
Goldberg, J, conservation laws, I.301
Goldberger, M L
 beta-decay, II.676
 pion decay constant, II.675
 Twelfth Solvay Conference, Brussels, II.646

Goldhaber, A S, photon mass, III.1452
Goldhaber, G, charmed meson, II.704
Goldhaber, M
 dipole vibration, II.1213–14
 neutrino helicity, II.668, II.670
 shell model, II.1215
Goldman, I M, dielectric properties of BaTiO$_3$, I.486
Goldschmidt, V M
 First Solvay Conference (1911), I.147
 inorganic compounds, I.492
Goldsmith, D W, interstellar medium, III.1765
Goldsmith, H H, neutron sources, II.1199
Goldstein, S
 energetic particles, III.1682
 magneto-ionic theory, III.1990
Goldstone, J, general theorem, II.676
Golovin, I N
 beam-fed mirror experiments, III.1662
 magnetic mirror, III.1659
Goodman, C, Coulomb excitation experiments, II.1217
Goodman, P, electron diffraction, I.464
Gordon, J P
 maser, III.1419
 relativistic hydrogen spectrum, I.218
 resonator modes, III.1429
Gordon, W E
 incoherent scatter radar (ISR), III.1996
 Schrödinger method, I.365
Gor'kov, L, BCS theory, II.940
Gorter, C J
 magnetic field, II.1133
 nuclear resonance, II.1172
 relaxation time, II.1133

superconductivity, II.921
Twelfth Solvay Conference,
  Brussels, II.646
two-fluid model, II.924, II.925
Gorter, E W, ferrimagnetism,
  II.1157
Görtler, H, boundary layers,
  II.832–3
Gott, J R, distribution of galaxies,
  III.1772
Gottleib, M, Atoms for Peace
  Conference, III.1638
Goudsmit, S
  atomic energy levels, II.1093
  electron spin, I.101
  magnetic moment, II.1123
  quantum number, I.168
  spinning electron, I.360
Gough, D O, low degree modes,
  III.1750
Gould, G, lasers, III.1423
Gould, L, polar research, III.2006
Goulianos, K, two-neutrino
  experiment, II.672
Goutier, Y, perturbation theory,
  III.1473–4
Gouy, A, gravity effects, I.553
Gowing, M, *Britain and Atomic
  Energy 1939–45*, II.1198
Grad, H
  magnetic cusp, III.1645
  non-linearity, I.625
Gratias, D, quasicrystals, I.481
Gray, L H, **III.1869**
  Bragg–Gray theory, III.1882
  radiotherapy, III.1867
Green, H S, parastatistics, II.687
Green, M B, string theories, II.684
Green, M S, collision
  configurations, I.611
Green, R F, quasars, III.1798
Greenberg, O W
  quarks, II.687
  SU(3) symmetry, II.709

Greenburger, D M, three-photon
  interferometry, III.1467
Greene, C H
  critical behaviours, II.1067
  R matrices, II.1069
Greene, J M, plasma oscillations,
  III.1625
Greening, J R, dosimetry, III.1884
Gregory, Sir Richard, *Nature*, I.6
Greinacher, H, α-particle, II.748
Gribov, V N, singularities
  in complex angular
  momentum, II.682
Griffin, J, growth spiral, III.1528
Griffith, A A, crack growth,
  III.1559
Griffiths, J H E, ferromagnetic
  resonance, II.1152–3
Griffiths, R B, smoothness
  postulate, I.563
Gringauz, K I
  plasmapause, III.2009
  solar wind, III.2009
Grodzins, L, neutrino helicity,
  II.668, II.670
Grondahl, L O
  photo-voltaic effect, III.1321
  semiconductors, III.1318
Gross, D
  point-like constituents, II.690
  running coupling constant,
    II.713
Gross, E, spectral shift, III.1398
Gross, E P, plasma oscillations,
  III.1625
Grossmann, M, gravitation, I.292
Groth, crystallography, I.424
Grüneisen, E, thermal expansion,
  II.975–7
Gubanov, long-range order,
  III.1368
Guckenheimer, J, MSS, III.1850
Gudden, B
  photo-ionization, III.1320
  single crystals, III.1318–19

temperature variation of resistivity, III.1330
Guen, H S, BBKGY hierarchy, I.606, I.623, I.624, I.626
Guggenheim, E A
 coexistence curves, I.555
 Statistical Thermodynamics, I.537
Guggenheimer, J, shell models, II.1209
Guinier, A, age-hardening in aluminium–copper alloys, III.1546
Gull, S F, supernova remnants, III.1782
Gunn, J E
 electron acceleration, III.1752–3
 mass–luminosity ratio, III.1800
 phase space constraints, III.1804
Gurney, R
 α-decay, I.115, I.217, II.1184
 colour centres, III.1530
 electrostatic generators, II.727
 Nature, I.116
 semiconductors, III.1319
 solid-state physics, III.1314
Gurvich, defibrillators, III.1895
Gusyenov, O H, binary sources, III.1756
Gutenberg, B
 earthquake waves, III.1956
 seismological analysis, III.1954
Guth, A, Universe model, III.1806
Guye, Ch E, Fifth Solvay Conference (1927), I.205
Gyulai, Z, colour centres, III.1529–30

Haake, F, laser theory, III.1444
Haas, A E
 Bohr radius, I.81
 quantum formula, I.156
Haasen, P, *Physikalische Metallkunde*, III.1555
Habgood, H W, isotherms of xenon, I.556

Habing, H J, interstellar medium, III.1765
Hadravský, M, confocal scanning laser microscope, III.1450–1
Hafstad, L, ionospheric plasmas, III.1621
Hahn, E L, resonant absorption, II.1174
Hahn, J, stellar mass, III.1699
Hahn, O, I.105
 β-ray deflection, II.1201
 β-spectrum, I.104, I.105, I.106
 elements with $Z = 93$–96, I.130
 neutron bombardment, II.1194, II.1195
 neutron sources, II.1206
 neutron–U collisions, I.130
 uranium isotope, II.1223
 uranium-239, I.132
Haine, M E, Fresnel fringe system, III.1584
Haldane, F D M, $s = 1$ Heisenberg chains, II.1014
Hale, G E, magnetic fields in sunspots, III.1958
Hale, J E, Palomar 200" telescope, III.1721
Hall, E H
 physics in education, I.9
 practical classes, I.15
 resistivity variation with temperature, III.1282
Halley, E, Earth's magnetic field, III.1953
Halliday, D, electron spin resonance, II.1138
Hallwachs, W L F, photoelectric effect, III.1390–1
Halperin, B I, ultrasonic properties, II.1006
Halpern, O
 neutron diffraction, I.459
 quantum theory, I.183
Hamilton, W R, dynamical laws, I.176

Hammersley, J M
  bond percolation processes, I.572
  percolation processes, I.572
  self-avoiding walk (SAW), I.570
Han, M Y
  quarks, II.687
  SU(3) symmetry, II.709
Hanbury Brown, R
  optical spectroscopy, III.1421
  radio interferometer research, III.1422
Haneman, D, surface crystallography, I.470
Hanna, R C, two-photon correlation, III.1413
Hansen, M, binary systems, III.1509
Hansen, W W
  electron linear accelerators, II.1205
  linac, II.736
  linear accelerators, II.736
Hanson, D
  physics of materials, III.1555
  plastic deformation, III.1526
Hanssen, K-J, atomic resolution, III.1585–6
Hara, Y, quark–lepton analogy, II.700
Hariharan, P, holography, III.1438
Harker, D
  inequality relations, I.447
  Patterson function, I.443
Harper, S, strain-aging law, III.1521
Harrison, E R, structures in the Universe, III.1802
Harrison, S C, tomato bushy stunt virus (TBSV), I.510
Harte, B, *Plain Language from Truthful James*, I.6
Hartley, spherical micelles, III.1606
Hartmann, J, wavefront aberrations, III.1389

Hartmann, W K, Moon model, III.1966
Hartree, D R
  magneto-ionic theory, III.1992
  multi-electron spectra, II.1036
  self-consistent field method, I.432, I.433
  wireless propagation, III.1622
Harvey, E N, luminescence, III.1537
Hasenöhrl, First Solvay Conference (1911), I.147
Hasinger, G, x-ray source counts, III.1798
Hastie, R J, plasma pressures, III.1661
Hauptman, H
  determinantal inequalities, I.448
  tangent formula, I.449
  The Solution of the Phase Problem. I. The Centrosymmetric Crystal, I.449
Hauser, W, compound nucleus, II.1219
Haüy, R J, lattice theory, I.422–3
Hawking, S
  black holes, I.317, II.761, III.1778
  communication aids, III.1899
  hot big bang models, III.1778
Haxel, O
  nuclear binding energies, II.1210
  shell models, II.1209
Hayakawa, S, x-ray sources, III.1755
Hayashi, C, quasi-static models of stars, III.1767
Hayes, W, two-neutrino experiment, II.672
Hayes, W C, *Basic Orthopaedic BioMechanics*, III.1896
Hazard, C, radio sources, III.1776
Heaviside, O
  ionospheric layer, III.1621
  radio waves, III.1987

wireless wave propagation, III.1986
Hebel, P Zum, BCS theory, II.940
Heisenberg, W K, I.189, **I.373**
  atomic theory, I.179
  cosmic-ray review, I.230
  discrete quantum mechanics, I.184
  dispersion formula, I.172
  doublets, I.102
  Einstein–Podolsky–Rosen (EPR) paradox, I.228
  electron spin, II.1125
  equation of motion, I.186
  exchange forces, I.231
  exchange integral, I.213
  Fermi-field theory, I.377
  ferromagnetism, I.554
  Fifth Solvay Conference (1927), I.205
  harmonic-oscillator decomposition, III.1402
  holes, III.2018
  lattice calculations, I.233
  meson theory, I.404
  microphysical phenomena, I.224
  neutron–neutron, neutron–proton forces, II.1190
  neutron–proton model, II.1184
  nuclear charge-exchange force, I.376
  nuclear forces, I.123
  nuclear volume, II.1190
  nucleus, I.371–3, II.1191
  periodic system, I.178
  QED, I.362
  quantum mechanics, I.226, I.358, II.638, II.1058–9
  quantum theory, I.144, I.180, I.187, I.221
  relativity, I.359
  spontaneous emission, III.1401
  superconductivity, II.932, II.934, II.936
  The Physical Principles of the Quantum Theory, III.1394
  theory of explosive showers, I.387–8
  total quantum number, I.162
  Twelfth Solvay Conference, Brussels, II.646
  two-electron system, I.213
  uncertainty relations, I.174
  unified-field theory, I.232
Heitler, W
  bremsstrahlung, I.385
  cosmic rays, I.393
  covalent molecular bond, II.1042
  heavy electrons, I.392
  meson theory, I.394
  nuclear statistics, I.117
  showers, I.388
  The Quantum Theory of Radiation, II.1031, III.1420–1
  Twelfth Solvay Conference, Brussels, II.646
Heller, P, spontaneous magnetization, I.559
Hellmann, H, pseudopotential model, III.1352–3
Hellwarth, R W
  photon, III.1451–2
  Q-switching, III.1430, III.1432, III.1436
Helm, G, *Energetik*, I.48
Helmholtz, H von
  age of the Sun, III.1949
  electricity, I.52
  laboratory equipment, I.20
  medical physics, III.1857
  Physikalisch-Technische Reichsanstalt, I.20, II.1254
  Sun's time-scale, III.1694
Henisch, H K, semiconductors, III.1325
Henkel, superfluidity, II.932
Hénon, M, chaotic behaviour, III.1831

Henrick, L R, equilibrium theory, III.1785
Henriot, E, Fifth Solvay Conference (1927), I.205
Henyey, L G
  radio astronomy, III.1736
  study of the stars, III.1746
Herbert, R, thyroid gland, III.1900
Herbert, R J T, scintillation detector, III.1901
Herbig, G H, T-Tauri stars, III.1766
Herbst, J F, neodymium–iron borates, II.1165
Herglotz, G
  elastic media, I.280
  rigid body, I.279
Herlofson, N, Galactic radio emission, III.1737–8
Herman, F
  photoconductivity, III.1329
  valence/conduction bands, III.1328–9
Herman, R C
  black-body radiation, II.758
  primordial nucleosynthesis, III.1786, III.1790
  proton–neutron ratio, III.1786
Herring, C
  diffusion creep, III.1534
  jellium, III.1303–4
  pseudopotential model, III.1352–3
  solid-state physics, III.1313
  spin waves, II.1012
Herriott, D R, helium–neon laser, III.1427
Herschel, J, *General Catalogue of Nebulae*, III.1717
Herschel, W
  Galaxy, III.1714
  infrared radiation, III.1742
  nebulae, III.1717, III.1945
Hertz, C H, time–position recordings, III.1913–14
Hertz, G
  Franck–Hertz experiment, I.89–90
  scattering experiments, I.217
  stationary electron orbits, I.158
Hertz, H, electromagnetic waves, III.1985
Hertzsprung, E
  diameter of Arcturus, III.1699
  main sequence, III.1698
  mass–luminosity relation, III.1699
  Small Magellanic Cloud, III.1715
  star clusters, III.1696
  stellar sizes, III.1696
Herzberg, G, **II.1072**
  ionization threshold of H_2, II.1069–72
  molecular mechanics, II.1044
  nuclear statistics, I.117
  spectra of diatomic molecules, II.1046
Herzen, E
  Fifth Solvay Conference (1927), I.205
  First Solvay Conference (1911), I.147
Herzfeld, K F, vibration of alkali halide, I.485
Herzog, R O, fibrous proteins, I.497
Hess, G B, superfluidity, II.932
Hess, H
  ocean ridges, III.1974
  sea-floor spreading, III.1976
Hess, V F, I.369
  balloon flight, I.368
  cosmic ray particles, III.1712
  radiation, II.639
Hess, W N, magnetospheric physics, III.2001
Hevesy, G
  Bohr's theory, I.88
  hafnium, I.100
Hewish, A
  neutron stars, III.1751

survey of the sky, III.1788
Hey, J S, radio astronomy, III.1736
Heycock, C T, **III.1511**
  multiphase alloys, III.1548
  phase diagram, III.1508–9
Heydenburg, M P, Coulomb excitation experiments, II.1217
Heyl, P R, gravitation, II.1266
Higgs, P W
  bosons, II.695
  dynamical symmetry breaking, II.957
Higinbotham, W A, diode coupling, II.748
Hilbert, D, I.207
  gravitational field, I.314
  Hilbert's space, II.1025
  infinite quadratic forms and matrices, I.191–2
  Methoden der Mathematischen Physik, III.2043
  Schrödinger equations, I.205
  theory of linear integral equations, I.174
  variational principle, I.294
Hildebrand, R, charge-independence, II.654
Hill, R D, shell model, II.1215
Hillert, M
  atomic segregation/atomic ordering, III.1548
  spinodal concept, III.1548
Hilliard, J, spinodal decomposition, III.1548
Hillier, J
  contour phenomena, III.1581–2
  Fresnel diffraction, III.1582
  magnetic lenses, III.1580
Hines, C O
  ionosphere, III.1984
  solar wind plasma, III.2010
Hinks, A, photographic parallax programme, III.1696
Hirakawa, K, K_2CoF_4, II.1016

Hisano, K, fine structure, II.1003
Hittorf, J W
  conductivity, III.1318
  rarefied gases, III.1985
Hoag, A A, quasars, III.1798
Hoar, T P, microemulsions, III.1608
Hockham, G A, optical telecommunication, III.1447
Hodgkin, D C, penicillin, I.496
Hofmann, S, superheavy elements, II.1225
Hofstadter, R
  electron scattering, II.1221
  nuclear structure studies, II.1206
  sodium iodide crystals, II.1207
  sodium iodide scintillator, II.750–1
Hohenberg, P
  delta (δ), III.1851
  phonon frequency, II.1002
Holborn, L, black-body radiation, I.22
Holmes, A
  Earth's crust, III.1960
  radioactive minerals, III.1969
Holmes, K C, tobacco mosaic virus (TMV), I.508
Holstein, T, lattice polarization, III.1322
Holt, angular distributions, II.1220
Honda, K, ferromagnetism, II.1134
Hondros, D, dielectric waveguide, III.1397
Hooke, R, Newton's rings, III.1386
Hopkins, H H
  coherence parameter, III.1412
  diffraction pattern of a point object, III.1412
  fibre optics, III.1420
  image formation, III.1419
  polarized light, III.1412
  spatial-frequency response of optical systems, III.1422
Hopkins, W

Earth structure, III.1946
earthquake waves, III.1956
Hopkinson, J, *critical temperature*, I.545
Hopwood, medical physics, III.1864
Hora, H, laser systems, III.1464
Horn, D, pion–nucleon charge exchange, II.683
Horn, F H, zone-refined copper, III.1544
Horne, R W, adenovirus, I.510
Horton, self-consistent renormalized harmonic approximation, II.1004
Horwitz, L, BBKGY hierarchy, I.623
Hoselet, First Solvay Conference (1911), I.147
Houston, W V
 de Broglie waves, III.1291–2
 Lamb shift, II.641
 theory of metals, III.1290
Houtermans, F
 nuclear reactions, II.1225
 nuclear relations, III.1707
Houtermans, F G
 age of the Earth, III.1960–1
 thermonuclear reactions, III.1627
Howry, D H
 cross-sectional imaging, III.1912
 prototype scanner, III.1912
 pulse-echo ultrasound, III.1911–12
Hoxton, L G, diffusion cloud chamber, II.752
Hoyle, F
 Big Bang, III.1792
 excited state of ^{12}C, III.1746
 helium synthesis, III.1790
 inverse-Compton catastrophe, III.1781
 nuclear interactions between light nuclei, III.1791
 nucleosynthesis in stars, III.1788
 quasars, III.1777
 red giant stars, II.1226–7
 solar x-rays, III.1994
 steady-state cosmology, III.1787
 supernovae, II.1227, II.1228
 synthesis of elements, III.1747, III.1792
Huang, K
 dipole–dipole forces, II.988
 Dynamical Theory of Crystal Lattices, II.988
 Raman effect, II.996
Hubbard, J
 correlation effects, III.1362
 Hubbard U, II.1166
Hubble, E P, **III.1720**
 counts of faint galaxies, III.1728
 spiral nebulae, III.1719
 The Realm of the Nebulae, III.1720, III.1726
 tuning-fork diagram, III.1721
 velocity–distance relation, III.1726, III.1727
Huchra, J, bright galaxies, III.1772
Hückel, double/triple bonds, II.1043
Hückel, E, wave mechanics, I.213
Hückel, G
 electron screening, III.1316
 screening length, III.1610
Hudson, R P, magnetic cooling, II.1126
Hueter, T F, ventriculography, III.1911
Huggins, M L, chain conformations, I.497
Huggins, W, nebulae, III.1717
Hughes, A L, photoelectric effect, III.1390
Hughes, E W, least squares method, I.452
Hughes, V W, parity violation, II.698
Hughes, W, Rochelle salt, I.489
Hugoniot, A, conservation equations, II.812

Hulburt, E O, solar ultraviolet energy, III.1992
Hull, A W, single crystal, I.434
Hulse, J M
  binary pulsar, I.317
  quadrupole radiation formula, I.315
Hulse, R, binary pulsars, I.298, III.1757
Hulthen, magnetization, II.1131–2
Humason, M L
  galaxy NGC7619, III.1726
  redshift–magnitude relation, III.1788, III.1793
Hümmer, K, Renninger effect, I.450
Hund, F
  analysis of spectra, II.1124
  electron reflection, I.170
  molecular mechanics, II.1044
  quantum theory, I.145
  rare-earth ions, II.1120
  tunnel effect, I.217
  wave mechanics, I.213
Hunter, D L, magnetic field at critical point, I.562
Huntingdon, H B, diffusion, III.1531
Hutson, A R, electron–phonon interaction, III.1339
Hüttel, A, speed of light, II.1263
Huus, T, Coulomb excitation experiments, II.1217
Huxley, T H, Darwinian evolution, I.5
Huygens, C
  light polarization, III.1386
  wave description of light, I.72
  wave theory, I.258
Hyde, E K, *Nuclear Properties of the Heavier Elements*, II.1223

Iachello, F, collective motion in nuclei, II.1218
Iben, I, study of the stars, III.1746

Ibser, H W, neutron sources, II.1199
Ignatowsky, V S, vector theory, III.1412
Ikeda, H, K_2CoF_4, II.1016
Ikeda, K, unstable and chaotic emission, III.1466
Iliopoulos, J
  charmed quark, II.700, II.705
  neutral currents, II.700
Inglis, D R, Cranking Formula, II.1218
Ingram, V, sickle cell anaemia, I.502
Inoue, T, meson theory, I.407, I.408
Inui, T, bulbous bow, II.885
Ioffe, A F
  energy levels, III.1368
  Semiconductor Thermoelements and Thermoelectric Cooling, III.1311
Ioffe, M
  mirror experiment, III.1679
  mirror-type systems, III.1643
Iona, M, KCl, II.988
Irwin, M, redshift quasars, III.1798
Ishiwara, J, quantum rule, I.175
Ising, E
  Ising model, III.1554
  magnetic moment, I.554
Ising, G, linear accelerator, II.736
Israelashvili, J N, measurement of forces between surfaces, III.1612
Iwanenko, D
  neutrons, I.122
  nuclear exchange potential, I.376
Iyengar, P K, dispersion curves of germanium, II.1002

Jackson, K, Materials Research Society (MRS), III.1557
Jacobi, C G J, dynamical laws, I.176

Jacobs, E
  magnetic fusion, III.1629–30
  tokamak, III.1646
Jacobsen, E H, acoustic paramagnetic resonance (APR), II.1146
Jakeman, E
  antibunched light, III.1460
  K-distribution, III.1462
  optical spectrum, III.1454
James, R W
  sodium chloride crystals, I.433
  structure factors, I.439
  zero-point motion, II.974
Jamgochian, E, p–n junction, III.1417
Jamieson, J C, x-ray powder diffraction, I.437
Jancke, W, fibrous proteins, I.497
Jánossy, L, forked tracks, I.408
Jansky, K, radio astronomy, III.1736
Jauch, J M, photon, III.1451
Javan, A, helium–neon laser, III.1427
Jeans, First Solvay Conference (1911), I.147
Jeans, J H
  black-body radiation, I.163
  galaxy formation, III.1800
  gravitational collapse, III.1768
  internal structure of the stars, III.1703
  spiral nebulae, III.1950
  statistical mechanics, I.173
  tidal theory, III.1962
Jefferts, K B, carbon monoxide molecule, III.1763–4
Jeffreys, H
  Earth model, III.1956
  Earth–Moon system, III.1948
  Earth's core, III.1956
  geomagnetism, III.1958
  spiral nebulae, III.1950
  tidal theory, III.1962

Jenkin, F, wire-wound resistors, II.1252
Jennison, R C, Cygnus A, III.1775
Jensen, J H D, II.1211
  Breit–Wigner formula, II.1194
  nuclear binding energies, II.1210
  shell model, II.1209, II.1210, II.1214, II.1220
Johannson, H, emission electron microscope (EEM), III.1569, III.1577
Johansson, C H, crystal structure, III.1515
Johns, H E, telecobalt therapy, III.1872
Johnson, F A, optical-fibre communication, III.1447
Johnson, F M, antiferromagnetism, II.1148
Johnson, J B, Johnson noise fluctuations, I.601
Johnson, L F, rare-earth ion lasers, III.1438
Johnson, M H, neutron diffraction, I.459
Johnson, W H, metre standards, II.1239
Joliot-Curie, F, I.121
  radioactivity, I.121, I.375, II.1186, II.1222
Joliot-Curie, I, I.121
  neutron bombardment, II.1194
  radioactivity, I.375, II.1186, II.1222
Jona-Lasinio, G
  dynamical symmetry breaking, II.957
  pion bound state model, II.756
Jones, H
  Brillouin zone, III.1313
  disordered materials, III.1365
  Fermi surface, III.1295
  Seebeck coefficient, III.1310
  solid-state physics, III.1314
Jordan, P

field quantization, I.283
harmonic-oscillator
  decomposition, III.1402
matrix mechanics, I.198, I.360
Physics of the Twentieth Century,
  I.210
quantization of waves, I.208
quantum electrodynamics, I.209
transformation scheme, I.201
Jordan, W C, beta-ray experiment,
  III.1679–80
Josephson, B, Josephson effect,
  II.941–2
Jost, vacancies, III.1529
Joule, J P
  first law of thermodynamics,
    I.521
  learned societies, I.3
Joy, A H, T-Tauri stars, III.1766
Joyce, G S, long-range and
  short-range forces, I.563
Julian, B R
  detection of impulses, III.1957
  Earth's core, III.1980
Jungnickel, C, organized research,
  I.16
Justi, superconductivity, II.921
Jüttner, Maxwell–Boltzmann
  distribution law, I.282

Kadanoff, L P
  block spin variables, I.567
  delta (δ), III.1849
  hypothesis of universality, I.563
  phase transitions, III.1830
  scaling properties, I.563–4
  spin-blocking, III.1831, III.1832
  universality, I.571
Kadomtsev, B B, fusion-plasma
  research, III.1642
Kae, M, Ehrenfest dog-flea model,
  I.618
Kaganov, canonical theory,
  III.1356

Kaiser, W, two-photon absorption,
  III.1429
Kalckar, F
  liquid drop, I.126, II.1213
  spacing of states in highly
    excited nucleus, II.1213
Källén, G, Twelfth Solvay
  Conference, Brussels, II.646
Kallman, H
  fluorescent light pulses, II.1207
  scintillation, II.750
Kaluza, T, general relativity and
  electromagnetism, I.321–2
Kamerlingh Onnes, H, **II.914**
  critical point, I.545
  First Solvay Conference (1911),
    I.147
  liquid helium, II.913
  superconductivity, I.489, II.915,
    III.1284
Kaner, E A, cyclotron resonance,
  III.1357
Kant, I, island universes, III.1714
Kantrowitz, A
  magnetic fusion, III.1629–30
  tokamak, III.1646
Kao, K C, optical
  telecommunication, III.1447
Kapany, N S, fibre optics, III.1420
Kapitza, P L
  helium liquefier, II.920
  law of linear magnetoresistance,
    III.1356
  strong-impulsive-field generator,
    III.1282–3
  superfluidity, II.922
  superleak flow, II.946
Kapron, F P, chemical vapour
  deposition, III.1447
Kapteyn, J C
  Galaxy, III.1714, III.1715
  luminosity function, III.1714
  stellar spectra, III.1696
Kapur, P L, theory of nuclear
  reactions, II.1219

Name Index

Karagioz, O V, gravitation, II.1266
Karle, I L, structure of arginine dihydrate, I.449–50
Karle, J
  determinantal inequalities, I.448
  structure of arginine dihydrate, I.449–50
  tangent formula, I.449
  The Solution of the Phase Problem. I. The Centrosymmetric Crystal, I.449
Karpushko, F V, optical bistability, III.1450
Kasai, C, flow imaging, III.1914
Kasper, J S
  crystal structure of dekaborane, I.447
  inequality relations, I.447
Kasteleyn, P W, bond percolation model, I.573
Kastler, A, optical pumping, II.1097, III.1415, III.1447
Kasuya, T, RKKY interaction, II.1164
Kay, W, scintillation counting, I.113
Kaya, S, ferromagnetism, II.1134
Kaye, radiation protection, III.1890
Kayser, H
  handbook of spectroscopy, I.76
  history of spectroscopy, I.24
Keck, D B, chemical vapour deposition, III.1447
Keeler, J, spiral nebulae, III.1950
Keenan, P C
  MKK system, III.1698
  radio astronomy, III.1736
Keesom, A P, liquid helium, II.914
Keesom, W H
  atomic vibrations, III.1287
  liquid helium, II.914, II.922
  Seebeck coefficient, III.1310
Keffer, F, spin-wave modes, II.1151

Kekulé, A, Karlsruhe conference, I.45
Keldysh, L V, photo-assisted tunnelling, III.1423
Kelker, H, liquid crystals, III.1540
Kellar, J N, line broadening, I.472
Keller, A, polyethylene, III.1552
Kellermann, E W, NaCl, II.987–9
Kellers, C F, λ-point of liquid He^4, I.559
Kellcy, P L, photodetection, III.1441
Kellman, E, MKK system, III.1698
Kelly, A, Walter Rosenhain, III.1517–19
Kelly, M, cuprous oxide rectifiers, III.1323
Kelvin, Lord
  conducting layer, III.1984
  cooling Earth, III.1945
  Earth structure, III.1946
  Earth's rigidity, III.1953
  gaseous molecule, I.48
  law of equal partition of energy, I.143–4
  Maxwell's Demon, I.576
  papers and patents, I.2
  radium atoms, I.62
  second law of thermodynamics, I.521
  second wrangler, I.11
  Seebeck effect, III.1284
  ship–wave pattern, II.883
  Sun's time-scale, III.1694
  temperature scales, II.1249
Kemmer, N
  meson theory, I.394
  new particles, I.393
  quantum field theory interaction, I.378
  symmetric theory, I.403
  U-particles, I.393
Kendrew, J C
  globular proteins, I.501

isomorphous replacement method, I.500
phase angles, I.502
Kennedy, J W, plutonium isotope, II.1223
Kennelly, A E
  ionosphere, III.1621, III.1985
  radio waves, III.1986, III.1987
Kepler, J, supernovae, II.1228
Kerr, A, electro-optical double refraction, III.1389
Kerr, F J, neutral hydrogen, III.1763
Kerr, R, Kerr metric, III.1778
Kerst, D W
  betatron, II.732, II.1204, III.1871
  fixed-field alternating-gradient accelerators, II.739
  high-intensity beams, II.738
  multipole geometry, III.1645
  orbit calculations, II.732
Keuffel, J W, spark counters, II.753
Kevles, D, physics research, I.16
Khalatnikov, I M
  curvature singularities, I.310
  Stochasticity in relativistic cosmology, I.630
Khandros, L G, metal physics, III.1539
Khintchine, A I, *Mathematical Formulation of Statistical Mechanics*, I.614
Khoklov, stochastic particle acceleration, I.630
Khriplovich, I B, Yang–Mills theory, II.713
Kibble, B P
  fundamental constants, II.1260
  NPL moving-coil experiment, II.1238
Kidder, R
  laser fusion programme, III.1675
  radiation implosion, III.1674
Kihara, T, clusters of galaxies, III.1771

Kikuchi, S, electron diffraction, I.463
Kilby, J St C, integrated circuits, III.1326
Killian, J R Jr, stroboscopic photography, III.1410
Kim, Y K, shell electrons, II.1091
Kimble, H J
  photon antibunching, III.1460
  squeezed light, III.1471
Kinman, T D, helium abundance, III.1790
Kinoshita, S, charged particle tracks, II.1200
Kip, A F, resonant frequencies, III.1334
Kippenhahn, R, study of the stars, III.1746
Kippenhauer, K O, Galactic radio emission, III.1737–8
Kirchheim, R, physics of materials, III.1555
Kirchhoff, G R
  black-body radiation, I.22, I.66, I.148
  collaboration with Bunsen, I.75–6
  induction coil, I.54
  Investigations of the solar spectrum and the spectra of the chemical elements, III.1692
  quantum theory, I.65
  radiation at constant temperature, I.521
  school of physics, I.15
  spectral lines, I.156
Kirkpatrick, H A, curved crystal focusing spectrometer, II.1207
Kirkpatrick, P, holography, III.1415
Kirkwood, J
  BBKGY hierarchy, I.606, I.623, I.624, I.626
  polymer solutions, III.1597

Kirshner, R P, supernovae, III.1759
Kittel, C
 antiferromagnetic resonance, II.1151
 ferromagnetic resonance, II.1153
 ferromagnets, II.1147
 resonance, II.1147
 resonant frequencies, III.1334
 RKKY interaction, II.1164
 spin waves, II.1012
Klaiber, G S, dipole vibration, II.1213
Kleeman, R, α-particles, I.104
Klein, M J, black-body radiation, I.22
Klein, O
 baryon symmetric models, III.1806
 Dirac electrons, I.218
 Dirac equation, I.218
 quantum theory, I.179
 Twelfth Solvay Conference, Brussels, II.646
Kleiner, W H
 crystal field theory, II.1144
 photodetection, III.1441
Kleinman, D A, anharmonic oscillators, III.1434
Klemens, P G, electrons and holes, III.1297
Klug, A
 polio virus, I.510
 tobacco mosaic virus (TMV), I.508
 turnip yellow mosaic virus, I.510
Knight, P L
 laser fluctuations, III.1473
 optical harmonic generation, III.1473
Knipping, P
 crystal structures, III.1514
 x-ray diffraction, I.425
Knoll, M
 Crantz Colloquium, III.1571–2
 electron microscope, III.1571
 high-speed oscillograph, III.1570
 iron-free lenses, III.1568
 scanning electron microscope, III.1587
 transmission electron microscope (TEM), III.1569–70
Knudsen, M
 Fifth Solvay Conference (1927), I.205
 First Solvay Conference (1911), I.147
Kobayashi, M
 CP violation, II.701
 quarks, II.720, II.723
Kocharovskaya, O, x-ray lasers, III.1473
Koehler, T R, self-consistent renormalized harmonic approximation, II.1004
Koehler, W C
 neutron diffraction, I.459
 rare-earth metals, I.463
Koester, C J, optical-fibre laser, III.1427
Kogelnik, H
 Gaussian–Hermite (laser) beam propagation, III.1447
 resonator modes, III.1429
Kogut, J, hadrons, II.690
Kohan, M, *High Performance Polymers*, III.1550
Köhler, A, illumination system for microscopy, III.1390
Kohlhörster, W, balloon flights, I.368
Kohlrausch, F, *Leitfaden der Praktischen Physik*, I.15
Kohlschütter, A, luminosity indicators, III.1698
Kohn, W
 Fermi surface, III.1364
 phonon frequency, II.1002
 wavevectors, II.999

Kolmogoroff, A N, *K* systems, I.615
Kolmogorov, A N, turbulence, II.798, II.837
Komarov, L I, neutron frequency, III.1418
Kompaneets, A, induced Compton scattering, III.1801
Kondo, J, electron scattering, III.1343
Koo, D C, sky surveys, III.1798
Koppe, superconductivity, II.932, II.934, II.936
Korteweg, D J, weak non-linear and weak dispersive effects, II.859
Koschmieder, E L, free convection, II.846
Koshelev, K, laser amplification, III.1466
Kossel, W
  periodic table, I.95
  x-ray spectra, I.160
Kosterlitz, J M, topological long-range order, I.571
Krakowski, R A, reversed-field pinch (RFP) research, III.1657
Kramers, H A
  asymptotic linkage, I.197
  atomic theory, I.197
  correspondence principle, I.86
  dispersion formula, I.171, I.172, I.185
  dispersion relations, III.1404
  Fifth Solvay Conference (1927), I.205
  Kramers' theorem, II.1125
  particle escape probability, I.607–8
  periodic table, I.96
  potassium chromic alum, II.1126
  quantum theory, I.166, I.179
  specific heat, I.555
  spontaneous emission, III.1401
  two-electron system, I.182

Kraner, H
  crystallization, III.1537
  synthetic zinc silicate, III.1537
Kratzer, A, atomic theory, I.179
Krishnan, K S
  pseudopotential concept, III.1367
  Raman effect (combination scattering), III.1404–6
  Raman scattering, II.996
Kronberg, M L, special orientation relationships, III.1528
Kronig, R L
  dispersion relations, III.1404
  magnetic field, II.1133
  nuclear magnetic moments, I.116
  nuclear spin, I.117
  superconductivity, II.917
Kruger, R, semiconducting crystals, II.1006
Kruskal, M D
  coordinate system, III.1778
  plasma column instabilities, III.1641
  plasma oscillations, III.1625
Kubo, R, **I.613**
  lattice impurities, III.1366
  spin waves, II.1011
  transport coefficients, I.612, I.626–7
Kuhl, D E, tomography, III.1916
Kuhn
  random walk, III.1597
  rubber elasticity, III.1597
Kuhn, T S
  black-body radiation, I.22, II.973
  helium atom, III.1982
  life cycle of field of study, III.2048
  quanta, II.969
Kuhn, W, spectral line intensity, I.172
Kuiper, G P, solar system, III.1963
Kukhtarev, N V, photorefractivity, III.1461

Kukich, superfluidity, II.932
Kulsrud, R, collisionless shocks, III.1684
Kunsman, C H, electron reflection, I.170
Kunze, P, charged particles, I.388
Kurchatov, I V, fusion research, III.1641
Kurdjumov, G, metal physics, III.1539
Kurlbaum, F, black-body radiation, I.149, III.1387
Kurnakov, crystal structure, III.1515
Kurti, N, low-temperature physics, II.920
Kusaka, S, meson theory, I.407
Kyu, T, polymer blends, III.1548

Labeyrie, A E
  holograms, III.1438
  optical astronomy, III.1453
Laborde, radium, I.61
Ladenburg, E, photoelectric effect, III.1390
Ladenburg, R W
  dispersion formula, I.171
  emission and absorption, III.1400
Ladner, A W, *Short Wave Wireless Communication*, III.1619, III.1620
Lallemand, P, stimulated Rayleigh-wing scattering, III.1445
Lallier, E, waveguide laser, III.1471
Lamb, W E
  First Shelter Island Conference, II.645
  hydrogen spectrum, III.1421
  Lamb shift, II.641–3
  laser light, III.1443
  Master-type equation, I.605
  optical maser, III.1443
  photon counting, III.1388
Lambert, J, island universes, III.1714
Lamerton, L F, depth-dose data, III.1867
Lanczos, C
  field equations, III.1724
  matrix mechanics, I.192
  quantum theory, I.196
Land, E H, sheet optical polarizers, III.1417
Landau, L D, **III.1626–7**
  abrupt and restrained transition, II.833–4
  charge effects, III.1611
  damping theory, III.1678
  diamagnetism, II.1133–4
  electron density, III.1316
  electron–electron collisons, III.1316
  energy complexes, I.301
  field-strength uncertainty product, I.209
  gravitational collapse, III.1712
  interacting particles, III.1363
  ionic crystals, III.1321
  light-scattering spectroscopy, III.1408
  magnetization, II.1134
  neutrino, II.667
  neutron stars, III.1713
  non-linear stability theory, II.826
  phenomenological theory, II.1007
  plasma oscillations, III.1623–5
  quantum liquid, II.928
  second-order transitions, I.549–51
  self-gravitating systems, I.313
  spectrum, II.930
  spin-1 particle, II.654
  superconductivity, II.924, II.934–6
  superfluidity, II.929
  surf-riding, III.1362

two-fluid hydrodynamics, II.929, II.944
zero sound, II.950
Landauer, R, thermodynamics of data processing, I.576
Landé, A
  quantum numbers, I.101
  two-electron system, I.182
  Zeeman effect, I.162
Landsberg, G S
  Raman effect (combination scattering), III.1404–6
  Raman scattering, II.996
  spectral shift, III.1397, III.1398
Landweber, L, optical data processing, III.1461
Lane, A M, nuclear reactions, II.1219
Lane, J H, sun's structure, III.1694
Lange, L, inertial frame of reference, I.255
Langenberg, D N, Determination of e/h, II.1268
Langevin, P
  Boltzmann relation, I.545
  diamagnetism, II.1117
  Fifth Solvay Conference (1927), I.205
  First Solvay Conference (1911), I.147
  internal energy, I.111
  stochastic processes, I.596
  Zeeman effect/diamagnetism, II.1114
Langley, S P, black-body radiation, I.22
Langmuir, I
  Fifth Solvay Conference (1927), I.205
  plasma oscillations, III.1623–5
Langsdorf, A Jr, diffusion cloud chamber, II.752
Lankard, J R, dye lasers, III.1446
Laplace, P S
  black holes, II.760

central force programme, I.321
fluid-mechanical analysis, II.889
nebular hypothesis, III.1945
sound waves, II.808
Laporte, O, atomic theory, I.179
Larmor, Sir Joseph
  aether theory, I.5
  radio propagation, III.1622
  self-exciting dynamo, III.1958
  upper atmosphere, III.1988
Larson, R, formation of stars, III.1769
Laskar, J, chaotic evolutions, III.1823
Lassen, H
  upper atmosphere, III.1992
  wave propagation, III.1991
Lassettre, E, molecular resonances, II.1084
Lassus St Genies, A H J, lens arrays, III.1411
Latham, R, critical fluctuations, II.1016
Lau, E, Lau effect, III.1414
Laub, J, stress–energy tensor, I.281
Lauer, E J, beta-ray experiment, III.1679–80
Laughlin, J, calorimetry, III.1883
Launay, J M, vibrational excitation, II.1089
Lauterbur, P C, NMR image, III.1924
Laval, J, x-ray scattering, II.990
Lawrence, E O, I.371, **II.1185**
  calutron isotope separation, III.1630
  cyclotron, II.639, II.729, II.736, II.1201
  direct reaction theory, II.1220
Lawson, J D
  fusion reactions, III.1628–9
  $n\tau$ product, III.1649
Lax, M, laser light, III.1443
Leavitt, H S, Magellanic Clouds, III.1714

Le Berre, M, *Photons and Quantum Fluctuations*, III.1441
Lebowitz, stochastic dynamics, I.629
Lederman, L M
　Drell–Yan process, II.701
　high-mass lepton pairs, II.721
　parity violation, II.666
　two-neutrino experiment, II.672
Lee, B W, renormalizability, II.695
Lee, E W, transverse Kerr effect, II.1155
Lee, T, supernova trigger hypothesis, III.1963
Lee, T D
　lattice gas model, I.553
　neutrinos, II.671
　parity conservation, II.665
　parity violation, II.666
Lee, Y T, cross-beam collision products, II.1089
Leff, H S, *Maxwell's Demon*, I.576
Legvold, S, magnetic configurations, II.1162
Lehmann, H, haemoglobin, I.502
Lehmann, I, Earth structure, III.1956-7
Lehmann, O
　cholesteric esters, III.1540
　liquid crystals, III.1540
Lehovec, K, p–n junction, III.1417
Leibacher, J W, solar oscillations, III.1750
Leighton, R B
　sky survey, III.1743
　solar oscillations, III.1749–50
Leith, E N
　holography, III.1415, III.1437
　off-axis optical beam holography, III.1585
Leksell, L, flaw detector, III.1911
Lemaître, G
　Eddington–Lemaître model, III.1732–3
　primaeval atom, III.1785

Schwarzschild horizon, I.307
　spherically symmetric perturbations, III.1801
　Universe, III.1725
Lenard, P
　photoelectric effect, I.152, III.1390
　photoelectron emission, III.1396
Lengyels, B A, *Lasers*, III.1425
Lenz, W
　atomic theory, I.179
　ferromagnetism, I.554
Leontovich, M A, *Plasma Physics and the Problem of Controlled Thermonuclear Reactions*, III.1642
Le Pichon, X
　Earth's surface, III.1978
　plate tectonics, III.1978
Le Poole, J B, selected area diffraction, III.1579
Leprince-Ringuet, L, K-meson (or kaon), II.656
Letokhov, V S
　laser amplification, III.1466
　laser beams, III.1462
Le Verrier, U J J, precession of perihelion of Mercury, III.1722
Levi-Civita, T, tensor analysis, I.292
Levine, B F, superlattice device, III.1471
Levinson, C A, proton inelastic scattering, II.1221
Levshin, V L
　absorption of light by uranium glass, III.1403
　Levshin–Perrin law, III.1400
　polarization of fluorescent light, III.1400
　Raman effect, III.1400
Levy, H A, neutron diffraction, I.458–9
Lévy, M

axial coupling constant, II.676
β-decay, II.677
weak hadronic current, II.678
Lewis, G N
  direct reaction theory, II.1220
  particle dynamics, I.279
  photons, I.358, III.1302, III.1403
Lhéritier, M, K-meson (or kaon), II.656
Li, T
  Fabry–Perot resonator, III.1429
  Gaussian–Hermite (laser) beam propagation, III.1447
Libchaber, A, dynamical systems3.1852
Lichte, H, atomic resolution microscopy, III.1585
Lichten, W, potential curves for Ar_2^+ complex, II.1079
Lide, D R Jr, HCN laser line, III.1443
Lieb, E H
  delta (δ), III.1845
  percolation model, I.573
  two-dimensional ferroelectric models, I.570–1
Lifshin, E, electron microprobe analyser, III.1549
Lifshitz, E M
  curvature singularities, I.310
  energy complexes, I.301
  galaxy formation, III.1801
  self-gravitating systems, I.313
  Statistical Physics, I.550
Lifshitz, I M
  canonical theory, III.1356
  crystal defects, II.1004
  crystal surface, II.1005
  Fermi surface, III.1351
  quantum theory, III.1350
Liftshitz, E, magnetization, II.1134
Liley, B, tokamak, III.1647
Lilly, S J, radio galaxies, III.1795–6
Lima de Faria, J, *Historical Atlas of Crystallography*, I.511–12

Linde, A D
  scalar Higgs fields, III.1806
  world model, III.1807
Linde, J O, crystal structure, III.1515
Lindemann, F A (later Lord Cherwell)
  Clarendon Laboratory, II.920
  crystal lattice, III.1315
  First Solvay Conference (1911), I.147
  two-frequency equation, I.153
Lindhard, J
  electron gas, III.1362
  Kubo procedure, I.626
Linfoot, E H, diffraction images, III.1419
Lippmann, G, interference within a thick film, III.1392
Lipscomb, W N, boron hydrides, I.493
Lipson, H, copper–nickel–iron alloy, III.1548
Lipson, H S
  stacking faults, I.473
  structure determination, I.441
Livingston, M S
  accelerators, II.725–7, II.738, II.1201
  cyclotron, II.736
  direct reaction theory, II.1220
  nuclear physics, II.1184
  Particle Accelerators, II.1201
  synchrotrons, II.734
Lobachevsky, N I, *On the Principles of Geometry*, III.1722
Lockyer, Sir Norman
  Nature, I.6
  spectrograph, I.25
  stellar evolution, III.1695
  stellar spectra, III.1693
Lodge, O J, obituary of George Fitzgerald, I.29
Logan, B G, thermal barrier, III.1667

Logan, G, tandem-mirror concept, III.1665–6
Lohmann, A W
  computer-generated spatial filter, III.1447
  holography, III.1415
London, F
  covalent molecular bond, II.1042
  Josephson junction, II.1095
  penetration depth, II.935
  superconductivity, II.924–5
  Superfluids, II.934
London, H
  anomalous skin effect, III.1346–7
  superconductivity, II.921, II.924–5
  superfluidity, II.932
  thermoelectric effects, II.922
Long, R R, upstream wakes, II.868
Longair, M S, radio galaxies, III.1795–6
Longuet-Higgins, M S, wave energy, II.860–1
Lonsdale, K
  hexamethyl benzene, I.495
  x-ray scattering, II.990
Lonzarich, G G, saturation magnetization, II.1167
Lorentz, H A, **I.18**
  anomalous Zeeman effect, I.102
  electric dipoles, II.1117
  electrodynamics of moving bodies, I.261
  electromagnetic theory, I.63
  electron collisions, III.1290
  electron optics, III.2000
  electron theory, II.1114, III.1284
  Fifth Solvay Conference (1927), I.205
  First Solvay Conference (1911), I.147
  general theories of relativity, I.34
  induced precession, II.1115
  magnetism, III.1390

Maxwell equations, I.280
microscopic electron theory, I.281
quantum theory, I.145, I.585
random processes, I.586
special theory of relativity, I.34, I.262–85
splitting of spectral lines, I.161
statistical mechanics, I.173
The Theory of the Electron, III.1992, III.2039
thermoelectric effects, III.1306, III.1307–8, III.1308
Loschmidt, J
  Avogadro's number, II.1269
  dimensions of molecules, I.48
Lossew, band-structure diagram, III.1417
Loudon, R, virtual electronic transitions, II.996–7
Louisell, W, laser light, III.1443
Love, A E H
  crystal surface, II.1005
  The Mathematical Theory of Elasticity, II.967
Lovelace, C, pion–pion scattering amplitude, II.684
Loveland, W D, transuranic elements, II.1223
Lovell, B, development of astronomy, III.1735
Low, F E
  bolometers, III.1743
  electric quadrupole moments, II.1214
  pion–nucleon resonance, II.680
  pion–pion scattering, II.681
  renormalization group, III.1830
Lowde, R, neutron diffraction, I.456
Lowman, C E
  magnetic materials, II.1159
  Magnetic Recording, II.1158
Lowry, R A, gravitation, II.1266

Lu, K T, photoabsorption spectrum of xenon, II.1073
Luce, J, energetic particle injection, III.1661
Lücke, K, *Lehrbuch der Metallkunde*, III.1554
Lucretius, atoms, I.44
Ludwig, G, Geiger counter, III.2007
Lugiato, L A, laser theory, III.1444
Lukirskii, A P
  bremsstrahlung, II.1050
  x-ray apparatus, II.1051
Lukosz, W, optical images, III.1439
Lummer, O R, black-body radiation, I.22, I.149, III.1387
Lummer, U, reflection interference fringes, III.1389
Lunbeck, R J, critical temperature, II.975
Lundmark, K, Andromeda Nebulae, III.1718
Luneberg, R K
  imaging theory, III.1418
  Mathematical Theory of Optics, III.1412
Luttinger, J M, Fermi surface, III.1364
Lyddane, R H, vibration of alkali halide, I.485
Lynden-Bell, D
  accretion discs, III.1964
  black holes, III.1779
  violent relaxation, III.1774
Lynn, J E, fission isomers, II.1218
Lyons, H, properties of quartz, II.1244
Lyubimov, V A, electron neutrino, III.1803

McAfee, K B, semiconductors, III.1419
McCall, S L, thermal bistable effects, III.1450
McClelland, C L, Coulomb excitation experiments, II.1217
McClung, F J, Q-switching, III.1430, III.1436
McConnell, A J, photon mass, III.1452
McCormmach, R, organized research, I.16
McCrea, W H, cosmological models, III.1730
Macdonald, resistance of sodium wires, III.1344
McDonald, W, mirror systems, III.1660
MacDougall, D P, gadolinium sulphate, II.1126
McDougall, I, geomagnetic field, III.1973
Macek, W M, ring-laser gyroscope, III.1433
McFarland, B B, laser development, III.1448
McFarlane, C G B, CO_2 laser, III.1442
McFee, J H, electron–phonon interaction, III.1339
McGowan, F K, Coulomb excitation experiments, II.1217
Mach, E
  atoms, I.48
  kinetic theory of gases, I.30
  Mach number, II.813
MacInnes, D, First Shelter Island Conference, II.645
McIntyre, P, proton–antiproton colliders, II.743
McKay, K G, semiconductors, II.1208, III.1419
McKee, C, neutron star, III.1759
McKellar, A, excited states, III.1791
McKenzie, D, plate tectonics, III.1978

Mackintosh, A R, exchange interactions, II.1164
McLennan, liquid helium, II.922
McMahon, R G, redshift quasars, III.1798
McMillan, E M
  electron synchrotron, II.731
  frequency modulation, II.1202
  ion acceleration, II.730
  neptunium, II.1224
  60″ cyclotron, II.1202
  transuranium element, II.1223
Madansky, L, spark counters, II.753
Madelung, E
  electron vibration, II.983–4
  quantum theory, I.227
Madey, J M J, free-electron laser, III.1456
Maeda, K, ionosphere, III.1984
Maiani, L
  charmed quark, II.700, II.705
  neutral currents, II.700
Maiman, T H, **III.1425**
  ruby laser, III.1424–6, III.1894
Majorana, E
  deuteron, I.376
  nuclear stability, II.1190
Maker, P D, pump-probe configuration, III.1445
Maki, A G, HCN laser line, III.1443
Maki, Z, quark–lepton analogy, II.700
Mallard, J R, gamma camera, III.1907
Malmberg, J H, magneto-electrostatic confinement, III.1681
Mandel, L
  coherence theory, III.1433
  optical coherence, III.1392
  photon antibunching, III.1460
  photon beams, III.1424

  quantum theory of coherence, III.1439
  sub-Poissonian statistics, III.1460
Mandelbrot, B
  fractals, I.574
  self-similarity, I.574
Mandelstam, L I
  first-order Raman effect, II.996
  Mandelstam–Brillouin scattering, III.1397
  Raman effect (combination scattering), III.1404–6
  Raman scattering, II.996
  secondary scattered radiation, I.172
Mandelstam, S, Twelfth Solvay Conference, Brussels, II.646
Manley, J M, non-linear optics, III.1424
Mansfield, P
  magnetic resonance imaging, III.1924
  nuclear magnetic resonance, II.1175
Mansur, L K, nuclear materials research, III.1534
Marconi, G
  electromagnetic waves, I.29
  radio waves, III.1995
  telegraph signals, I.54
  wireless transmission, III.1619–20, III.1986
Marechal, A, coherent imaging, III.1418
Margon, B, binary sources, III.1756
Mark, H
  biomolecular structures, I.497
  single crystals, III.1527
Marsden, E
  α-particles, I.113
  α-scattering, II.1200
  medical physics, III.1864
Marshak, R E
  Conceptual Foundations of Modern Particle Physics, II.770

First Shelter Island Conference, II.645
mesons, II.654
neutrinos, II.667
Marshall, L, neutron diffraction, I.456
Marshall, T W, stochastic electrodynamics, III.1388
Martin, J, x-ray diffraction, III.1547
Martin, P, temporal behaviour of a general system, I.617
Marton, transmission microscope, III.1574–5
Martyn, D F
 broadcast modulation, III.1994
 ionosphere, III.1984
Marx, G, neutrinos, III.1803
Mascart, E, motion of the Earth, I.258
Mash, D I, stimulated Rayleigh-wing scattering, III.1445
Mashkevich, V S, covalent crystals, II.1002
Masing, G, *Lehrbuch der Metallkunde*, III.1554
Maskawa, T
 CP violation, II.701
 quarks, II.720, II.723
Massey, Sir Harrie, **II.1024**
 ionospheric research, III.1999
 The Theory of Atomic Collisions, II.1024, II.1039
Mathewson, C, plastic flow and crystal glide, III.1544
Matsuoka, M, x-ray sources, III.1755
Matthews, D, magnetic variations of ocean floors, III.1975
Matthews, T A, stellar spectra, III.1776
Matthias, B T
 ferroelectric crystals, III.1538
 Matthias' rules, I.489

Matuyama, M, Earth's magnetic field, III.1973
Mauguin, C, helical structure, III.1397
Maurer, R D, chemical vapour deposition, III.1447
Maxwell, A, Pacific floor measurements, III.1974
Maxwell, J C
 Atom, I.76
 atoms, I.51
 electrolysis, I.51
 electromagnetic waves, I.260
 electromagnetism, III.1856
 equal area construction, I.544
 equipartition of energy, I.523
 field theory, I.321
 kinetic theory of gases, I.587
 molecules, I.47
 new fields of research, I.29
 second law of thermodynamics, I.523
 second wrangler, I.11
 sodium yellow line, II.1239
 statistical mechanics, I.173
 Theory of Heat, I.575
 wire-wound resistors, II.1252
Mayall, N U, redshift–magnitude relation, III.1788, III.1793
Mayer, J E, statistical mechanics, I.551–4
Mayer, M G, II.1211
 shell model, II.1209, II.1210, II.1214, II.1220
Mayneord, W V, **III.1868**
 depth-dose data, III.1867
 nuclear medicine, III.1901
 thimble ionization chamber, III.1867
Meggers, W F, **II.1030–1**
 atomic energy levels, II.1093
Mehl, R, diffusive transformations, III.1547
Meinesz, F A V, ocean floor, III.1974

Meissner, W, superconductivity, II.916–19, II.946
Meitner, L, I.105
  β-decays, I.118
  β-ray deflection, II.1201
  β-spectrum, I.104, I.105, I.119
  elements with $Z = 93$–96, I.130
  neutron bombardment, II.1194
  neutron sources, II.1206
  quantum theory, I.221
  uranium isotope, II.1223
  uranium-239, I.132
Melvill, T, discrete spectra, I.74–5
Mendeleev, D I
  absolute boiling temperature, I.543
  Avogadro's law, I.46
  periodic table, I.96
Mendelssohn, K
  energy gap, II.932
  low-temperature physics, II.920
Mendenhall, T C, squared paper, I.16
Menter, J M, electron microscope, I.477
Menzel, D H
  abundances of elements, III.1702
  planetary nebulae, III.1761
Merton, R, norms of scientific ethos, III.2057–8
Merz, W, dielectric constants, I.487
Metzner, A W K, field equations, I.313
Meyer, K H, biomolecular structures, I.497
Meyer, R B, smectic phase, III.1604
Michell, J, gravitation, II.1266
Michelson, A A, **I.33**
  Balmer lines, I.90
  cadmium red line, II.1239–40
  future discoveries, I.29
  hydrogen atom, II.641
  interferometer, I.25
  Lightwaves and their Uses, I.26
  optical instruments, III.1391
  optical interferometry, III.1704
  optical relativity principle, I.261
  spectroscopy, I.25
  speed of light, II.1263
  stellar interferometer, III.1400
  Studies in Optics, I.26
  Zeeman splitting, I.161
Mie, G
  non-linear field theories, I.232
  special-relativistic gravitation theory, I.284, I.321
Migdal, A B
  electron–phonon interaction, II.941
  interacting Fermi particles, III.1364
  phonons, II.999
Migneco, E, fission isomers, II.1218
Milburn, G J, quantum non-demolition (QND) measurements, III.1469
Miley, G H, laser systems, III.1464
Miley, G K, radio galaxies, III.1796
Miller, D A B, indium antimonide, III.1450
Miller, D C, Michelson–Morley experiment, III.1401
Miller, H, wedge fields, III.1872
Miller, J S, Seyfert galaxies, III.1784
Miller, S E, integrated optics, III.1444
Miller, W, nebulae, III.1717
Miller, W H, Miller indices, I.421
Millikan, R A
  absorption data, I.368
  constituents of matter, I.357
  electron charge, II.747
  mesotron, I.389
  new cosmic-ray particle, I.396
  oil drop method, II.1271
  photoelectric effect, I.363, III.1390–1
Millington, G, long-distance communications, III.1999

Mills, B, radio surveys, III.1789
Mills, R L
  isotopic spin symmetry, II.693
  local gauge invariance, II.693
  self-interacting fields, II.661
Milne, E A
  cosmological models, III.1730
  electrons in metals, I.531
  ionization, III.1701
  solar atmosphere, III.2003
  stellar spectra, III.1693
  theory of kinematic relativity, III.1787
Minkowski, H, **I.252**
  four-dimensional formulation, I.274
  particle dynamics, I.279
  phenomenological electrodynamics, I.280
  relativistic vacuum electrodynamics, I.280
Minkowski, R
  medium-mass elements, II.1228
  Palomar 200" Telescope, III.1737
  radio galaxy, III.1776
  radio sources, III.1775
  supernovae, II.1227–8
Minnaert, M, equivalent width, III.1702
Minogin, V G, laser beams, III.1462
Minsky, M
  confocal optical microscopy, III.1422
  confocal scanning laser microscope, III.1450
Misener, superleak flow, II.946
Mistry, N, two-neutrino experiment, II.672
Mitchell, D P, neutron diffraction, I.453
Mitchell, J W
  dislocation lines in silver chloride, III.1526
  solid-state physics, III.1530

Mitra, S K
  rocket studies, III.2001
  The Upper Atmosphere, III.2000
Mittelstaedt, O, speed of light, II.1263
Miyamoto, S, spark chamber, II.753
Moffat, K, protein unfolding, I.438
Mögel, H, sudden ionospheric disturbance (SID), III.1995
Mohorovičić, A, Mohorovičić discontinuity (Moho), III.1954
Mollenauer, L F, soliton-pulse scheme, III.1458, III.1471
Möllenstedt, G, Fresnel electron biprism, III.1585
Møller, C
  meson theory, I.401
  neutron capture experiments, II.1192
  S-matrix, II.680
  symmetrical theory, I.403
  Twelfth Solvay Conference, Brussels, II.646
Monck, W H S, stellar spectra, III.1696
Montroll, E W
  atomic models, II.987
  crystal defects, II.1004
  specific heat of small crystals, II.1005
Moore, C, chemical abundances, III.1702
Moore, W S, whole body imager, III.1924
Morey, G E, amorphous structures, I.477
Morgan, J, plate tectonics, III.1978
Morgan, W W, MKK system, III.1698
Moriya, T, lattice distortions, II.1152
Morley, E W
  Balmer lines, I.90

hydrogen atom, II.641
  optical relativity principle, I.261
Morris, P, nuclear magnetic resonance, II.1175
Mort, J, photocopying, III.1541
Morton, K W, self-avoiding walk (SAW), I.570
Moseley, H G J
  atomic number, I.159–60
  Bohr's theory, I.88
  characteristic x-rays, II.1038
  periodic system, I.362
  periodic table, I.86
  x-ray spectra, I.158
Mosengeil, K, relativistic thermodynamics, I.281
Moss, T S, absorption coefficient of semiconductor, III.1421
Mott, N F, **III.1333**
  Brillouin zone, III.1313
  colour centres, III.1530
  disordered materials, III.1365
  Electron processes in Non-crystalline Materials, III.1369
  paramagnetism, II.1136
  resistivity of dilute alloy, III.1316
  Seebeck coefficient, III.1310
  semiconductors, III.1319, III.1330–4
  solid-state physics, III.1313–14
  The Theory of Atomic Collisions, II.1039
Mottelson, B, II.1216
  collective motion, II.1215
  Coulomb excitation, II.1217
  Coulomb excitation experiments, II.1217
  geometrical collective model, II.1218
  surface oscillations, II.1215
Mould, J, Hubble's constant, III.1793
Moulton, F R
  nebular hypothesis, III.1950
  tidal theory, III.1962
Mouton, H, magneto-optical effect, III.1389
Mow, V C, *Basic Orthopaedic BioMechanics*, III.1896
Mueller, K A, oxygen octahedra, II.1009
Mügge, crystals of copper, gold etc, III.1517
Mukherjee, A K, plastic deformation, III.1534
Mulders, G, equivalent width, III.1702
Müller, A
  Meissner effect, II.957–8
  superconductivity, II.961
Muller, C A, hyperfine line radiation, III.1763
Müller, E, field emission microscope, III.1569
Muller, E W A, solar x-rays, III.1994
Müller, K A, superconductivity, I.490
Mulliken, R
  molecular mechanics, II.1044
  wave mechanics, I.213
Munro, G, ionosphere, III.1984
Murphy, G M
  deuterium, III.1706
  deuteron, II.650, II.1183
Mushotzky, R F, quasars, III.1780
Mussell, L, isocentric couch, III.1873
Mutscheller, A, radiation protection, III.1890
Myers, W D, liquid-drop mass formula, II.1225

Nabarro, F R N
  diffusion creep, III.1534
  dislocations, I.476, III.1526
  physical metallurgy, III.1555
Nagamiya, T, resonance, II.1147

Nakajima, S, pulse oximeter, III.1894
Nambu, Y
  chiral symmetry, II.676
  decay reaction, II.657
  dynamical symmetry breaking, II.957
  pion bound state model, II.756
  quarks, II.687
  SU(3) symmetry, II.709
  Twelfth Solvay Conference, Brussels, II.646
Nasanow, L H, self-consistent renormalized harmonic approximation, II.1004
Nath, N S N, diffraction of light, III.1409
Neddermeyer, S H
  absorption in lead plates, I.385
  cosmic rays, I.386
  elementary particles, I.391
  mesotron, I.231, I.389
  muon, II.650
  shower particles, I.388
Néel, L
  antiferromagnetism, II.1129, II.1132
  ferrimagnetism, II.1157
  neutron diffraction, I.453
Ne'eman, Y, II.678
  isospin, II.661
  strangeness, II.662
  triplets, II.685
Neidigh, R, plasma column, III.1678
Nelmes, R J, neutron diffraction, I.489
Nereson, N, nuclear forces, II.652
Nernst, W H, **I.536**
  calorimetry, II.968–9
  chemical equilibrium, I.535
  First Solvay Conference (1911), I.147
  new physics, I.7
  quantum hypothesis, I.155
  quantum theory, I.82, I.148, I.154
  specific heat, I.152–3, II.970
  thermodynamic theory, III.1699
  two-frequency equation, I.153
Nethercot, A H
  antiferromagnetism, II.1148
  thermal conductivity, II.982
Neugebauer, G
  formation of stars, III.1769
  Orion Nebula, III.1743, III.1767
  sky survey, III.1742–3
Neumann, F
  school of physics, I.15
  semiconductors, III.1419
Neville, F H, **III.1511**
  multiphase alloys, III.1548
  phase diagram, III.1508–9
Newnham, R E, ceramics, III.1538
Newton, I, I.255
  corpuscular hypothesis, I.257
  first law of motion, I.256
  law of gravitation, II.1265
  Opticks, I.76
  particle description of light, I.72
  physical cosmology, III.1721
  Principia, I.255
  rotating bracket, I.256
  second law of motion, II.806
  third law of motion, II.821
Ng, W K, ruby laser, III.1432
Nichols, G A, optical microscopy, III.1406
Nichols, H W, radiowave propagation, III.1990
Nicholson, J W
  angular momentum, I.175
  electron vibration, I.81
  spectra of celestial objects, I.156–7
Nier, A, lead isotopes, III.1960
Nieto, M M, photon mass, III.1452
Niggli, P, x-ray diffraction, I.439
Nishijima, K
  decay reaction, II.657
  isotopic spin multiplets, II.658

Nishikawa, S, electron diffraction, I.463
Nishina, Y
　Dirac equation, I.218
　letter, I.212
　new particles, I.389
Nishizawa, J, semiconductor maser, III.1423
Niu, K, charmed particles, II.700
Noether, F, rigid body, I.279
Nordheim, L
　disordered materials, III.1365
　Schrödinger equations, I.205
Nordsieck, A
　First Shelter Island Conference, II.645
　Master-type equation, I.605
　nuclear exchange potential, I.376
Nordström, G, special-relativistic theories, I.284
Northrop, G A, thermal conductivity, II.982
Northrop, T G
　adiabatic confinement, III.1679–80
　trapped particles in magnetic-mirror fields, III.1683
Nouchi, photon, III.1451–2
Novikov, I D
　black holes, III.1779, III.1780
　structure in the Universe, III.1801
Nowick, A, anelastic relations, III.1523
Nuckolls, J
　direct-drive approach to inertial fusion, III.1676
　fusion power, III.1673
　radiation implosion, III.1674

Obreimow, crystal growth, III.1280
Occhialini, G P S
　cloud chambers, I.371, II.651
　electron–positron pairs, I.231

O'Dell, C R, helium abundance, III.1790
Oehme, R
　parity violation in β-decay, II.665
　singularities in complex angular momentum, II.682
Oersted, H C
　electric current, III.1958
　electromagnetic interactions, III.1953
Ohkawa, T, multipole geometry, III.1645
Ohnuki, Y, quark–lepton analogy, II.700
Okayama, T, slow mesotrons, I.406
Oke, B, supernovae, III.1759
Oke, I, energy in the Sun, III.1748
O'Keefe, J A, optical microscopy, III.1406
Oken, L, *Gesellschaft deutscher Naturforscher und Ärzt*, I.5
Okubo, S, mass relations among baryons and mesons, II.661
Oldham, R D
　earthquake waves, III.1956
　seismic wave paths, III.1954
Olszewski, K, liquefied gases, I.27
O'Neill, E L, communications theory, III.1417
O'Neill, G K, *Storage-ring synchrotron: device for high-energy physics research*, II.739
Onsager, L, **I.551**
　entropy relations, I.600–2
　Fermi surface, III.1351
　ferroelectric transition in KH_2PO_4, I.571
　hypothetical vortices, II.931
　Ising model, I.554–61, II.1018
　Onsager relations, I.616–17, I.626
　quantum theory, III.1350
Oort, J H
　cosmological density, III.1799

hyperfine line radiation, III.1763
neutral hydrogen, III.1763
Palomar 200" Telescope, III.1775
radio astronomy, III.1762
Öpik, E
  giant stars, III.1708–9
  origin of elements, III.1747
  triple-α reaction, III.1746
Oppenheimer, J R
  adiabatic approximation, II.999
  collapsing sphere of dust, I.308
  direct reaction theory, II.1220
  First Shelter Island Conference, II.645
  general-relativistic analysis, III.1712
  mesotron, I.389
  μ-meson, II.652
  muon, II.651
  neutron stars, III.1713
  QED, I.386
  Twelfth Solvay Conference, Brussels, II.646
Orbach, R, fractons, II.1007
Ornstein, L S
  critical correlations, I.563
  pair distribution function, I.547
  spectral line intensity, I.172
Orowan, E
  dislocations, III.1520
  edge dislocation, I.476
Orthmann, W, β-spectrum, I.119
Ortolani, S, ETA-BETA I devices, III.1658
Oseen, C W, wave propagation in cholesteric liquid crystals, III.1407
Osheroff, D, superfluidity, II.950–1
Osmer, P S, quasars, III.1798
Osmond, block copolymers, III.1611
Osterberg, H, thin-film devices, III.1444
Osterbrock, D E, helium problem, III.1790

Ostriker, J P
  binary sources, III.1756
  electron acceleration, III.1752–3
  galactic cannibalism, III.1774
  spiral galaxies, III.1771
Ostwald, W
  dispute with Planck, I.34
  kinetic theory of gases, I.30
  natural phenomena, I.47–8
  quantum theory, I.148
Overbeek, charge effects, III.1611
Overhauser, A W, shell model, I.485, II.997

Pacini, F, neutron stars, III.1752
Pais, A
  First Shelter Island Conference, II.645
  Inward Bound, I.37, II.1183
  Twelfth Solvay Conference, Brussels, II.646
Pancini, E, mesons, II.652
Pankey, supernovae, III.1759
Panofsky, W K H
  electron storage rings, II.739
  Mark III linac, II.737
  proton linear accelerator, II.1205
  proton machines, II.737
  Stanford Linear Accelerator Center (SLAC), II.687–8
Papanastassiou, D A, supernova trigger hypothesis, III.1963
Parisi, G, asymptotic freedom, II.715
Parker, E N
  collisionless shocks, III.1684
  solar wind, III.1685, III.2009
Parker, H M
  dosimetry, III.1881
  radium needles, III.1877
Parker, W H, determination of e/h, II.1268
Parrent, G B, partial coherence, III.1392

Name Index

Parrish, W, powder photographs, I.436
Paschen, F
  black-body radiation, I.149
  Bohr's theory, I.88
  bolometer, I.23
  far-infrared spectrum, II.982–3
  Zeeman splitting, I.161
Paschos, E A, parton hypothesis, II.690
Pasternack, S, Lamb shift, II.641
Pastukhov, V P, mirror approach, III.1667
Patashinskii, A Z
  correlation components, I.563
  multiple-correlations near critical point, I.562
Patel, C K N, CO_2 laser, III.1442
Paterson, R
  dosimetry, III.1881
  radium needles, III.1877
Patterson, A L, x-ray diffraction pattern, I.441
Patterson, C C, age of the Earth, III.1961
Pauli, W
  atomic dynamics, I.184
  atomic theory, I.179, I.181
  continuous β-spectrum, I.127–8
  Einstein–Podolsky–Rosen (EPR) paradox, I.228
  electromagnetic field, I.391
  electron gas, III.1288
  equivalence relations, I.196
  exclusion principle, I.98–100, I.359, I.528, III.1711
  field quantization, I.283
  Fifth Solvay Conference (1927), I.205
  free particles, I.283
  lattice calculations, I.233
  Lorentz invariance, III.1451
  neutrino, I.371, I.375, III.1706
  non-relativistic theory, I.283
  paramagnetism of normal metals, II.1133
  quantum electrodynamics, I.209, I.362
  quantum mechanics, I.226
  quantum theory, I.180, I.221
  spin-statistics theorem, I.283
  stability of nuclei, I.111–12
  thermal conductivity, II.979–81
  total quantum number, I.162
  two-component wavefunctions, I.277
  wave mechanics, I.206
  Zeeman effect, I.103
Pauling, L
  biomolecular structures, I.497–8
  First Shelter Island Conference, II.645
  inorganic chemistry, I.496
  inorganic compounds, I.492
  pleated sheet structures, I.499
  polypeptide chains, I.497
  quasicrystals, I.481
  residual entropy of ice, I.571
  structure of ice, I.475
  wave mechanics, I.213
Payne, C H
  abundances of elements, III.1701
  Stellar Atmospheres, III.1701
Payne-Gaposchkin, C, spectrum analysis of stars, III.1962
Peacock, J A, radio sources, III.1798
Pearson, impurity effects, III.1281
Pearson, G L, solar cell, III.1421
Pearson, J J, excitations in $s = 1/2$ Heisenberg antiferromagnet, II.1014
Pearson, K, *The Grammar of Science*, I.34
Pease, F G, angular diameter of Betelgeuse, III.1705
Pease, R S
  structure of KH_2PO_4, I.459
  ZETA experiment, III.1642

N57

Pedersen, P O, radio waves, III.1992
Peebles, P J E
 clusters of galaxies, III.1771
 cold dark matter, III.1804
 spiral galaxies, III.1771
 synthesis of heavy elements, III.1792
 world models, III.1806
Peierls, R E
 exciton absorption spectrum, III.1410
 Fermi theory, I.377
 field-strength uncertainty product, I.209
 holes, III.2018
 Ising model, I.554
 lattice distortion, II.1000
 meson theory of nuclear forces, I.402
 resistivity at low temperatures, III.1306
 Schrödinger equation, III.1299–300
 theory of nuclear reactions, II.1219
 thermal conductivity, II.979–81
 Twelfth Solvay Conference, Brussels, II.646
 Umklapp processes, III.1304–6
Peimbert, M, helium abundance, III.1790
Pekeris, C L, M_2 tide, II.892
Pelzer, H, lattice deformation, III.1321
Penrose, R
 closed trapped surface, III.1778
 global differential geometry, I.310
 hot big bang models, III.1778
 quasicrystals, I.482, I.483
 resonance experiments, II.1139
 space-time, I.313, I.318–19
 spinorial techniques, I.277
Penzias, A A
 carbon monoxide molecule, III.1763–4
 microwave background radiation, III.1790
 microwave radiometer, II.758
 radiation, III.1786
Pepinsky, R
 ferroelectric crystals, III.1538
 XRAC, I.446
Pepper, D M, optical-beam transmission, III.1461
Percus, random walk model, I.630
Perkins, D H
 neutrino scattering, II.697
 photographic emulsions, II.653
Perl, M, *Evidence for the existence of anomalous lepton production*, II.719
Perlman, I
 α-decay spectra, II.1217
 Nuclear Properties of the Heavier Elements, II.1223
Peron, J, controlled fusion, III.1628
Peronneau, P A, ultrasonic imaging, III.1914
Perrin, F
 First Solvay Conference (1911), I.147
 Levshin–Perrin law, III.1400
 neutrons, I.122
 relativistic kinematics, I.129
 Twelfth Solvay Conference, Brussels, II.646
Perrin, J
 Avogadro's number, III.1610
 Les Atomes, I.43
Pershan, P S, non-linear optics, III.1434
Pert, G J, x-ray laser, III.1464
Perutz, M F
 globular proteins, I.501
 haemoglobin, I.500, I.502, I.503
 isomorphous replacement method, I.497
 x-ray analysis, III.1514

Peterson, S W, neutron diffraction, I.458–9
Pethick, C J, neutron stars, III.1754
Petit, R, diffraction theory of gratings, III.1466
Petley, B W
 fundamental constants, II.1261
 gravitation, II.1266, II.1267
Petrán, M, confocal scanning laser microscope, III.1450–1
Petschek, H, collisionless shocks, III.1684
Petzoldt, J, positivism, I.319
Pfann, W G, zone-refining, III.1281, III.1543–5
Pfirsch, D, stellarator-type research, III.1655
Pfund, high-reflection coatings, III.1407
Phillips, M, direct reaction theory, II.1220
Phillips, W A, glasses, II.1006
Phragmen, G, single-crystal methods, I.435
Piccard, A, Fifth Solvay Conference (1927), I.205
Piccioni, O, mesons, II.652
Pickering, C, hydrogen lines, I.85
Pickering, E C
 Elements of Physical Manipulation, I.16
 stellar spectra, III.1693, III.1694
Pidd, R W, spark counters, II.753
Pierce, G W
 radio propagation, III.2000
 semiconductors, III.1318
Pierce, J R, photomultipliers, II.750
Piermarini, G, diffraction techniques, I.437
Pierre, F M, charmed meson, II.704
Pike, E R
 imaging theory, III.1467–8
 optical spectrum, III.1454
Pines, D
 electrical discharges in gases, III.1361
 superconductivity, II.937
Pippard, A B
 optics, III.1467
 penetration depth, II.933–4
Pippin, J E, crystal structures, II.1158
Pirani, F A E
 general relativity, I.299
 plane wave solutions, I.312
Piron, C, photon, III.1451
Placzek, G
 light-scattering spectroscopy, III.1408
 phonon dispersion relations, II.993
 transuranic elements, I.131–2
Planck, M, **I.144**
 black-body radiation, I.144, I.524–5, II.973, II.1269, III.1402
 bound system, I.111
 energy density, III.1387
 energy quanta, I.173
 entropy, I.523
 Fifth Solvay Conference (1927), I.205
 finite energy quantum, I.173
 First Solvay Conference (1911), I.147
 oscillators, I.70
 particle dynamics, I.279
 partition function, I.527
 Planck's law, I.67–70
 quantum optics, III.1387
 quantum theory, I.62, I.146–7, I.156, I.522, III.1385
 school of physics, I.15
 Scientific Autobiography, I.34
 second law, I.47
 special-relativistic thermodynamics, I.281
 thermodynamics, I.68
Platzman, R L, continuous spectra

from synchrotron light sources, II.1050
Plücker, J, spectrum of hydrogen, I.76
Podolsky, B, quantum mechanics, I.228
Pohl, R
  Bohr's theory, I.88
  colour centres, III.1529, III.1530
  disordered systems, II.1006
  ionic crystals, III.1291
Pohlhausen, E, free convection, II.843–5
Pohlhausen, K, conduction of heat, II.842
Poincaré, H
  energy conservation, I.61
  First Solvay Conference (1911), I.147
  perturbation theory, III.1830–31
  radio waves, III.1986–7
  recurrence theorem, I.618
  relativity principle, I.265
  relativity theory, I.173
  special theory of relativity, I.262–85
  statistical mechanics, I.524
Pokrovskii, V L
  correlation components, I.563
  multiple-correlations near critical point, I.562
Polanyi, M
  biomolecular structures, I.496
  dislocations, III.1520
  edge dislocation, I.476
  single crystals, III.1526–7
Polder, D, electromagnetic waves, II.1157
Polikanov, S, fission isomers, II.1218
Politzer, H D, running coupling constant, II.713
Pollack, H O
  eigenfunctions and eigenvalues, III.1468

optical imaging, III.1432
Pomeau, Y, delta (δ), III.1849
Pomeranchuk, I Ya
  electron–electron collisons, III.1316
Pomeranchuk trajectory, II.682
Pontecorvo, B
  neutrinos, II.671
  solar neutrinos, II.757
Popaloizou, J C B, thick discs, III.1781
Porte, G, lamellar phase, III.1608
Porter, A B, imaging theory, III.1391
Porter, C, proton scattering, II.1212
Post, B, Renninger effect, I.450
Post, R F
  Atoms for Peace Conference, III.1639
  Controlled fusion research—an application of the physics of high-temperature plasmas, III.1641
  loss-cone modes of mirror-confined plasmas, III.1662
  mirror machine, III.1679
  Sixteen lectures on controlled thermonuclear reactions, III.1660
Potts, R B, Potts model, I.571
Poulsen, magnetic recording, II.1158
Pound, R V, red shifts, I.291
Powell, C, π-meson, I.231
Powell, C F, photographic emulsions, II.653
Powers, P N, neutron diffraction, I.453
Poynting, J H
  gravitation, II.1266
  theoretical constructs, I.34
Prandtl, L, **II.797**
  boundary-layer equations, II.805, II.806

boundary-layer theory, II.796–8, II.803, II.869
drag due to lift, II.824
drag on a non-lifting body, II.869
elliptic distribution of circulation, II.824
heat transfer, II.842
lift–drag ratios L/D, II.870
perturbation theory, II.829
singular perturbation, II.800
supersonic flow, II.812
triple deck theory, II.829
turbulence, II.815
vortex lines, II.822
vortex wake, II.822
vorticity, II.820
wind-tunnel testing at supersonic speeds, II.879
Prendergast, K H, x-ray sources, III.1755
Prescott, C Y, parity violation, II.698
Press, W H, mass function, III.1804
Preston, T, Zeeman splitting, I.161
Priestley, J, magnesium, III.1359
Prigogine, I, **I.621**
 Brussels school, I.619–22
 Non-equilibrium Statistical Mechanics, I.619
 Twelfth Solvay Conference, Brussels, II.646
Primakoff, H, neutral pion, II.655
Pringle, J E
 accretion discs, III.1964
 thick discs, III.1781
Pringsheim, E, black-body radiation, I.149, III.1387
Prins, J, x-ray diffraction pattern, I.441
Prout, W, atomic species, I.46
Pryce, M H L
 hyperfine structures, II.1139
 quantum theory, III.1413
Purcell, E M

hyperfine line radiation, III.1763
nuclear magnetic resonance, II.1094
Pusey, P N, K-distribution, III.1462

Quimby, E
 medical physics, III.1864
 radiotherapy, III.1867
 radium implants, III.1878
Quinn, T J, acoustic interferometer, II.1270

Rabi, I I
 fine/hyperfine structures, II.1047
 First Shelter Island Conference, II.645
 precessional motion, II.1153
Racette, G W, He–Ne laser, III.1435
Radon, J, computed tomography, III.1916
Raff, A D, sea-floor spreading, III.1975
Rainwater, J, II.1216
 quadrupole moment, II.1215
Rajchman, J A, photomultipliers, II.750
Rakshit, H, ionospheric physics, III.2000
Ramachandran, G N, speckle-pattern formation, III.1442
Raman, Sir Chandrasekhara
 Brownian motion, III.1442
 diffraction of light, III.1409
 Lectures in Physical Optics, III.1442
 phase transitions, II.1008
 Raman scattering, II.996, III.1404–6
 secondary scattered radiation, I.172
 speckle, III.1399
Ramberg, E G
 contour phenomena, III.1581–2

magnetic lenses, III.1580
Ramsauer, noble gas atoms, II.1041
Ramsay, Sir William, helium in uranium-bearing mineral, I.58
Rarity, J G, Bell inequalities, III.1471
Rasetti, F
  mesotron lifetime, I.406
  Raman scattering, I.117, II.996
Ratcliffe, J A
  An Introduction to the Ionosphere and Magnetosphere, III.2001
  ionosphere, III.1999
Ray, E, Geiger counter, III.2007
Rayleigh, Lord, **I.21**
  acoustics, III.1910
  aeolian tones, II.850
  black-body radiation, I.30, II.972
  Bohr's theory, I.87
  boundary conditions, II.845
  catastrophe theory, II.798
  Cavendish Laboratory, I.21
  crystal surface, II.1005
  Faraday measurement, II.1269
  grating spectroscopy, III.1391
  law of equal partition of energy, I.143–4
  optics, III.1390
  radio waves, III.1986–7
  random processes, I.586
  senior wrangler, I.11
  shock wave, II.811
  singular perturbation theory, II.811
  sound from flows, II.850
  sound propagation, II.810
  sound waves, II.858
  statistical mechanics, I.173
  The Theory of Sound, I.21, II.808, II.855, III.1910
  wave energy, II.858
  zero-viscosity limit, II.828

Raymond, A L, hexamethylene tetramine, I.495
Raynor, G, physical metallurgy, III.1555
Read, F H, quantum-mechanical reformulation, II.1065
Reber, G
  continuum radio map, III.1762
  radio astronomy, III.1736
Rebka, G A, red shifts, I.291
Rees, M J
  non-thermal phenomena, III.1781
  radio sources, III.1782, III.1783
Regel, A R, energy levels, III.1368
Regener, E, scintillation method, II.1200
Regge, T, Regge poles, II.681–2
Reiche, F, dispersion formula, I.171
Reickhoff, K E, stimulated Brillouin scattering, III.1444
Reid, A
  de Broglie waves, I.171
  electron diffraction, I.463
Reid, J M, pulse-echo ultrasound, III.1911–12
Reimann, B, non-linear analysis, II.857
Reines, F, neutrino, II.663
Reinitzer, F, cholesteric esters, III.1539–40
Reppy, J D, superfluidity, II.932
Retherford, R C, Lamb shift, II.642, II.643
Reusch, rocksalt, III.1517
Revelle, R, Pacific floor measurements, III.1974
Rex, A F, *Maxwell's Demon*, I.576
Reynolds, O
  diffusion of momentum, II.826
  flow in pipes, II.814
  pipe flow, II.825
  Reynolds analogy, II.843
  turbulence, II.798, II.825, III.1834

Rhines, F N
  grain growth, III.1550
  Microstructology—Behavior and Microstructure of Materials, III.1550
Ricci, C G, tensor analysis, I.292
Richards, B, Airy disc, III.1412
Richardson, O W
  electron emission, I.37
  electron optics, III.2000
  Fifth Solvay Conference (1927), I.205
  photoelectric effect, III.1390
  thermionic emission, I.12
Richter, B, electron storage rings, II.739
Richter, R, controlled fusion, III.1628
Riecke
  electrical and thermal conductivities, III.1285
  Hall effect, III.1282
  metallic conduction, III.1284
Riemann, B
  general relativity, III.1722
  propagation speed, II.809
  waveforms, II.810
Risken, H Z
  laser light, III.1443
  laser noise, III.1444
Riste, T, dynamic response of SrTiO$_3$, II.1010
Ritchey, G W, spiral nebulae, III.1717
Ritson, D, b-quark, II.723
Ritter, A, stellar evolution, III.1694–5
Ritz, W
  black-body radiation, I.163
  combination principle, I.25, III.1391
  combination rule, I.186
  spectral frequencies, I.86
Robb, A A, special theory of relativity, I.277–8

Roberts, F, flying-spot microscope, III.1416–17
Roberts, L D, germanium junctions, II.1208
Robertson, H P
  Robertson–Walker metric, III.1729–30
  synchronization of clocks, III.1729
Robertson, J M
  organic chemistry, I.496
  phthalocyanines, I.495
Robinson, D, High Voltage Engineering (HVE), II.1203
Robinson, F N H, anharmonic oscillators, III.1434
Robinson, H
  α-particles, II.1201
  β-ray spectrometer, II.1201
  RaB, RaC and RaE spectra, I.106
Robinson, I K
  plane wave solutions, I.312
  surface crystallography, I.469
Rochester, C D
  elementary particles, II.654
  kaons, II.656
Rochester, G, forked tracks, I.408
Rodionov, S N, tritium-beta experiment, III.1679
Rogers, G L, optical holography, III.1415
Rogerson, J B
  helium problem, III.1790
  interstellar deuterium, III.1791
Rohrer, H, scanning tunnelling microscope, III.1549
Röntgen, W C, x-rays, I.3, I.32, I.53, I.54, I.362, I.424, II.1271, III.1529, III.1857, III.1861
Roosevelt, President, atomic energy research, II.1196
Roozeboom, B, phase diagrams, III.1508
Rosa, E B, speed of light, II.1263

Rosbaud, P, *Progress in Metal Physics*, III.1556
Rose, P H, tandem accelerator, II.1203
Rosen, N, quantum mechanics, I.228
Rosenbluth, M N, **III.1664–5**
　International Summer Course in Plasma Physics 1960, III.1643
　loss-cone modes of mirror-confined plasmas, III.1662
　mirror machine, III.1679
　mirror systems, III.1660
Rosenfeld, L
　gravitational field, I.317
　meson theory, I.401
　symmetrical theory, I.403
　Twelfth Solvay Conference, Brussels, II.646
Rosenhain, W
　amorphous metal hypothesis, III.1518
　An Introduction to the Study of Physical Metallurgy, III.1517
　glass industry, III.1517
　grain boundaries, III.1519, III.1528
　physical metallurgy, III.1516
　slip lines, III.1517
　yield-stresses of single metallic crystals, III.1520
Rosenthal, A H, Sagnac effect, III.1433
Rossby, C G
　potential vorticity concept, II.905
　wave-like current patterns, II.892–5
Rosseland, S, quantum theory, I.179
Rossi, B
　cosmic rays, I.399, II.748
　First Shelter Island Conference, II.645
　nuclear forces, II.652
Rossman, M G
　structure determination, I.445
　virus structures, I.510
Rotblat, angular distributions, II.1220
Roth, W L, superionic conductors, III.1539
Round, H J
　band-structure diagram, III.1417
　yellow light emission, III.1391–2
Roux, D
　lamellar phase, III.1608
　sponge phase, III.1608
Rowan-Robinson, M, radio quasars, III.1797–8
Rowe, H E, non-linear optics, III.1424
Rowland, H, concave spectral gratings, I.54
Royds, T, α-particle, I.59
Rozental, S, meson theory, I.401
Rubbia, C
　field particles W and Z, II.744
　proton–antiproton collider, II.707, II.743
Rubens, H
　black-body radiation, I.149, III.1387
　far-infrared spectrum, II.982–3
　First Solvay Conference (1911), I.147
Rubin, H, magnetic cusp, III.1645
Rubin, V C, rotation curves of galaxies, III.1770
Rubinowicz, A
　polarization of spectral lines, I.178
　selection rules, I.94
Rubinstein, H, Dolen–Horn–Schmid duality, II.683
Rüdenberg, R, electron microscope, III.1573–4
Ruderman, M

indirect-exchange-coupling theory, II.1012
RKKY interaction, II.1164
scalar Higgs fields, III.1806
Rühmkoff, H, higher voltages, I.54
Rumford, Count, *see* Thompson, Sir Benjamin
Runcorn, S K
  continental drift, III.1972
  magnetic North Pole, III.1972
Runge, C, Zeeman splitting, I.161
Rushbrooke, G S, thermodynamic relation, I.560
Ruska, E
  Crantz Colloquium, III.1571–2
  electron microscope, III.1570–1
  electronic lenses, III.1569
  iron-free lenses, III.1568, III.1571
  polepiece lens, III.1567
  transmission electron microscope (TEM), III.1569–70, III.1575–7
Ruska, H, specimen preparation, III.1575–6
Ruskol, E, Moon model, III.1966
Russell, H N
  luminosity–spectral class diagram, III.1696
  photoexcitation, III.1761
  photographic parallax programme, III.1696
  solar atmosphere, III.1701–2
  solar system, III.1962
  tidal theory, III.1951
  white dwarfs, III.1709–10
Russell, J E
  German school system, I.8
  physics in education, I.8
Rutgers, superconductivity, II.924
Rutherford, E (Baron Rutherford of Nelson), **I.55**
  actinium, II.1227
  age of the Earth, III.1694
  α-particles, I.59, I.115, II.638, II.745–6, II.1201
  α-rays, I.54, I.58
  atomic energy, I.63
  atomic nuclei, II.1183
  atomic structure, I.74, I.156
  β-ray spectrometer, II.1201
  β-rays, I.54, I.58
  crystal diffraction spectroscopy, II.1207
  electrical counting methods, II.1200
  electron charge, II.747
  electrons, I.64
  experimental physics, I.14
  First Solvay Conference (1911), I.147
  half-life, I.62
  hydrogen nucleus, I.109
  meeting with Bohr, I.74
  Nature of the α-particle, I.58
  nuclear structure, I.110, I.113, III.2018
  positive ion accelerators, II.1184
  quantum conditions, I.359
  RaB, RaC and RaE spectra, I.106
  Radiations from Radioactive Substances, I.108, I.112, II.1199, II.1200
  radioactive decay, III.1949
  Radioactive Substances and their Radiations, I.107
  Raman effect (combination scattering), III.1404
  Silliman lectures, I.52
  The distinction between α-rays and β-rays, I.55
  thorium emanation, I.59
  transformation theory, I.59
  wireless transmission, III.1986
Ryle, M
  Cassiopeia A, III.1737
  radio astronomy, III.1738
  radio sources, III.1788
  survey of the sky, III.1788
Ryutov, D, mirror physics, III.1666

Sackur, O, quantum-theoretical equations, I.154
Sadler, C A, x-ray spectroscopy, I.158
Safronov, V, solar system, III.1963
Sagdeev, R Z, fusion-plasma research, III.1642
Sagitov, M V, gravitation, II.1266
Sagnac, G, measurement of rotation, III.1398
Saha, M N
 ionospheric physics, III.2000
 stellar atmospheres, III.1699
 stellar spectra, III.1693
Sakata, S
 fundamental triplet, II.685
 meson theory, I.407
 spin-1 particle, II.654
 two-meson theory, I.408
Sakharov, A D
 baryon–antibaryon asymmetry, III.1806
 baryon asymmetry, II.765
 CP violation, II.675
 Plasma Physics and the Problem of Controlled Thermonuclear Reactions, III.1642
 tokamak, III.1630
Salam, A, II.696
 electroweak unification, II.695
 neutrino, II.667
 Twelfth Solvay Conference, Brussels, II.646
Saleh, B E A, sub-Poissonian Franck–Hertz source, III.1460
Salpeter, E E
 α–α scattering, II.1226
 black holes, III.1779
 nuclear reactions, III.1709
 origin of elements, III.1747
 triple-α reaction, III.1746
Sampson, R A, energy transport, III.1699
Samuelson, E J, K_2CoF_4, II.1016

Sandage, A R
 Evidence for the occurence of violent events in the nuclei of galaxies, III.1777
 Herzsprung–Russell diagrams, III.1793
 Hubble's constant, III.1793
 quasars, III.1777
 redshift–magnitude relation, III.1788, III.1793
 stars in globular clusters, III.1746
 stellar spectra, III.1776
 The ability of the 200" telescope to discriminate between selected world models, III.1793
Sargent, W L W, inverse-Compton catastrophe, III.1781
Satomura, S, Doppler ultrasound, III.1914
Saunders, F A, Russell–Saunders (or L–S) coupling, III.1702
Savedoff, M P, interstellar gas, III.1765
Sayre, D, electron density, I.448
Schafroth, superconductivity, II.937
Schawlow, A L, **III.1428**
 Fabry–Perot resonator, III.1429
 lasers, III.1423, III.1424
 laser spectroscopy, III.1466
 maser, III.1422
 Microwave Spectroscopy, III.1428
Schechter, P, mass function, III.1804
Schechtman, D, quasicrystals, I.481
Scheffer, T J, liquid-crystal optics, III.1452
Schelleng, J C, radiowave propagation, III.1990
Schenectady, R T, electrical discharges in gases, III.1361
Scherk, J, quantum gravity, II.684
Scherrer, P, single crystal, I.434
Scherzer

axially symmetric electron lenses, III.1579
Geometrische Elektronenoptik, III.1578
Scheuer, P A G, radio sources, III.1782, III.1789
Schild, A, general relativity, I.299
Schilling, W, point defects, III.1534
Schlichting, H, Tollmien–Schlichting waves, II.830
Schlier, R J, surface crystallography, I.470
Schlüter, A, stellarator-type research, III.1655
Schmid, C, pion–nucleon charge exchange, II.683
Schmid, E
  plastic deformation, III.1526
  single crystals, III.1527
Schmidt, B V
  Schmidt corrector plate, III.1406
  Schmidt telescopes, III.1745
Schmidt, E, free convection, II.843
Schmidt, G, radioactivity, I.57
Schmidt, G N J, tomato bushy stunt virus (TBSV), I.509
Schmidt, M
  optical spectrum of 3C 273, III.1776
  quasars, III.1798
  radio quasars, III.1797–8
Schmidt, O, monistic theories, III.1963
Schneider, W G
  isotherms of xenon, I.556
  liquid–gas co-existence curve, I.553
Schoenflies, A, symmetry types, I.423
Schönberg, M, Schönberg–Chandrasekhar limit, III.1709
Schottky, W
  colour centres, III.1530
  electric potential, III.1289–90
  vacancies, III.1529
Schrieffer, J R
  BCS theory, II.938
  MOSFET, III.1372
  superconductivity, I.232, II.939, III.1364
Schrödinger, E, **I.193**
  decay law of radioactive substances, I.222
  entangled photon states, III.1394–5
  Fifth Solvay Conference (1927), I.205
  matter waves, I.194
  non-relativistic action function, I.195
  photon mass, III.1452
  quantum mechanics, II.638
  quantum theory, I.174
  relativistic equation, I.283
  wave equations, I.195, I.200
  wave mechanics, I.360
Schubauer, G, boundary layer, II.835
Schubnikow
  crystal growth, III.1280
  zone-melting, III.1281
Schulkes, J A, ferrimagnetism, II.1157
Schulman, J, microemulsions, III.1608
Schulman, L S, 'life' (game), I.629
Schulz, liquid metals, III.1366
Schulz, G J, He atoms, II.1061–2
Schumacher, E E, deformation of lead crystals doped with antimony, III.1544
Schuster, A
  doctorate, I.17
  electrical discharges in gases, III.1985
  energy transport, III.1699
Schwartz, M
  neutrinos, II.671

two-neutrino experiment, II.672
Schwarz, J
  quantum gravity, II.684
  string theories, II.684
Schwarzschild, K
  curvature of space, III.1722
  dynamical laws, I.176
  energy transport, III.1699
  field equations for a point mass, III.1777–8
  Maxwell's equations, I.279
  metric, III.1778
  quantum theory, I.89
  Stark effect, I.162
Schwarzschild, M
  Herzsprung–Russell diagrams, III.1793
  plasma column instabilities, III.1641
  stars in globular clusters, III.1746
Schweber, S
  hierarchical structure, III.2020
  process of emergence, III.2019
Schwers, F, II.972
Schwinger, J
  bosons, II.694–5
  First Shelter Island Conference, II.645
  quantum electrodynamics, II.643, II.740, III.1446
  renormalization, II.646
  temporal behaviour of a general system, I.617
  Twelfth Solvay Conference, Brussels, II.646
Schwitters, R, Mark I Detector, II.703
Scully, M O
  laser light, III.1443
  phaseonium, III.1473
  x-ray lasers, III.1473
Seaborg, G T
  Nuclear Properties of the Heavier Elements, II.1223

plutonium isotope, II.1223
  transuranic elements, II.1223
Seaton, M J
  chanel coupling formulation, II.1073
  exchanges of angular momentum, II.1060
  R matrices, II.1068
  radial density function, II.1066
Secchi, A, stellar spectra, III.1692–3
Seddon, J A, wave-like patterns, II.897
Seeger, A K, dislocation theory, I.476
Segrè, E
  neutron bombardment, II.1191
  plutonium isotope, II.1223
Seidel, H, optical bistability, III.1449
Seitz, F, **III.1532**
  band structure, III.1314–17
  Brillouin zone, III.1313
  colour centres, III.1531, III.1536
  crystal defects, III.1531
  diffusion, III.1531
  electron band structure of sodium, III.1530
  Imperfections in Nearly Perfect Crystals, III.1535
  materials research, III.1558
  Schrödinger's equation, III.1315
  solid-state physics, III.1313
Selenyi, P, xerography, III.1541
Sen, H K, magneto-ionic theory, III.1992
Sen, N R, ionospheric physics, III.2000
Serber, R
  β-decay, I.401
  direct reaction theory, II.1220
  First Shelter Island Conference, II.645
  mesotron, I.389
  μ-meson, II.652
  muon, II.651

orbit calculations, II.732
Seyfert, C K, spiral galaxies, III.1777
Shaknov, I, scintillation counters, III.1414
Shakura, N, thin accretion discs, III.1779
Sham, L J
　phonon frequency, II.1002
　pseudopotential theory, II.999
Shannon, R R, US optical industry, III.1476
Shapiro, I I, Schwarzschild metric, I.305
Shapiro, J, pion–pion scattering amplitude, II.684
Shapley, H
　Galaxy, III.1714, III.1716, III.1717
　globular clusters, III.1716
　Great Debate, III.1718
　novae, III.1717
　stellar masses, III.1698
Sharp, P F, imaging procedure, III.1920
Sharvin, D Yu, resistance of thin films, III.1370
Sharvin, Yu V, resistance of thin films, III.1370
Shenstone, W A, vitrified silica ware, I.29
Sheppard, R M
　detection of impulses, III.1957
　Earth's core, III.1980
Shimizu, F, self-phase modulation, III.1449
Shklovsky, I S
　Crab Nebula, III.1774
　x-ray sources, III.1755
Shlyaptsev, V, laser amplification, III.1466
Shockley, W B, **III.1324**
　cuprous oxide rectifiers, III.1323
　cyclotron resonance, III.1337–8
　Electrons and Holes in Semiconductors, III.1317
　magnetoresistance, III.1335–6
　photomultipliers, II.750
　transistor, II.744
Shoenberg, D
　de Haas–van Alphen effect, III.1348–52
　superconductivity, II.925
Shore, B W, optical harmonic generation, III.1473
Short, R, sub-Poissonian statistics, III.1460
Shortley, G, *The Theory of Atomic Spectra*, II.1037
Shu, F J, formation of stars, III.1769
Shubnikov, A, superconductivity, II.920
Shuleikin, M V, ionospheric layers, III.1992
Shull, C G
　Fe$_3$O$_4$, I.460
　magnetic structures, II.1148
　neutron diffraction, I.453
　powder pattern of MnO, I.460
　spin density, I.462
Shutt, R P
　diffusion chamber, II.752
　heavy particles, II.657
　mesotrons, I.405
Sibbett, W, dye-laser pulses, III.1439
Sidgwick, Faraday measurement, II.1269
Siegbahn, K
　double-focusing beta spectrometer, II.1206
　electron spectrometers, II.1090
Siegbahn, M, x-ray spectra, I.160
Siegert, A J F, Master equation description of gas, I.605
Sievert, R M
　radiation protection, III.1890
　thimble ionization chamber, III.1867

Silin, V P, electrons in metals, III.1363
Silk, J, sound waves, III.1802
Simon, A
  cloud chamber, I.364
  plasma column, III.1678
Simon, Sir Francis
  diamond–graphite equilibrium diagram, I.538
  low-temperature physics, II.920
  third law of thermodynamics, I.537, I.542
  zero-point motion, II.973–4
Simonen, T, TMX experiment, III.1667
Simpson, J A, II.1062
Sinai, Ja, hard spheres system, I.615
Sinitsyn, G V, optical bistability, III.1450
Sitte, K, β-decay theory, I.375
Sitterly, C M, atomic energy levels, II.1093
Skinner, H W B, x-ray emission, III.1361
Sklodowska, M, see Curie, M
Slater, J C
  crystal field theory, II.1144
  properties of metals, III.1345
  quantum mechanics, I.211
  quantum theory, I.166, I.180
  Slater determinant, I.215
  Solid State and Molecular Theory, III.1314
  solid-state physics, III.1313
Slee, O B
  radio sources, III.1737
  radio surveys, III.1789
Slepian, D
  eigenfunctions and eigenvalues, III.1468
  optical imaging, III.1432
Slepian, J
  electron emissions, II.749
  secondary emission, III.1399

Slichter, C P, BCS theory, II.940
Slipher, V M, spiral galaxies, III.1725
Sloan, D H, cyclotron, II.736
Slusher, R E, squeezed states of light, III.1469–70
Smale, S, mathematician, III.1840
Smart, J S
  neutron diffraction, I.453
  powder pattern of MnO, I.460
Smekal, A, light-quantum hypothesis, I.172, I.185
Smith, B W, interstellar medium, III.1766
Smith, C S
  A History of Metallography, III.1556
  A Search for Structure, III.1556
  Acta Metallurgica, III.1556
  Institute for the Study of Metals, III.1556
Smith, F G
  Cassiopeia A, III.1737
  radio sources, III.1737
Smith, K, soliton pulse communication, III.1471
Smith, L W, thin-film devices, III.1444
Smith, M G, quasars, III.1798
Smith, N, long-distance communications, III.1998
Smith, P L, *Needs, Analysis and Availability of Data for Space Astronomy*, II.1093
Smith, R A, photoconductivity, III.1327
Smith, T, optical systems, III.1446
Smoluchowski, M S
  dielectric constant, III.1392, III.1396–7
  Maxwell's Demon, I.576
  opalescent bands, I.546
Smoot, G F, cosmic microwave background radiation, III.1804

Smythe, *Atomic Energy for Military Purposes*, II.1223
Snitzer, E, optical-fibre laser, III.1427
Snoek, J L
  ferrimagnetism, II.1155
  strain-aging law, III.1521
Snyder, H
  collapsing sphere of dust, I.308
  general-relativistic analysis, III.1712
Snyder, H S
  accelerators, II.738
  synchrotrons, II.734
Sochor, V, HCN/H$_2$O lasing, III.1443
Soddy, F
  atomic energy, I.63
  radioactive decay, III.1949
  transformation theory, I.59
Soffer, B H, laser development, III.1448
Sohncke, L, symmetry types, I.423
Solomon, J, uncertainty relations, I.228
Solvay, E
  First Solvay Conference (1911), I.147
  new physics, I.7
Somer, J C, imaging, III.1914
Sommerfeld, A, I.180
  astrophysics, I.219
  Atombau and Spektrallinien, I.110
  Atomic Structure and Spectral Lines, I.156
  battle between Boltzmann and Ostwald, I.48
  Bohr's theory, I.88
  Drude–Lorentz model, III.1288
  dynamical quantum theory, I.174
  electric potential, III.1289
  extension of Bohr's model, I.160–1
  fine structure, I.90
  First Solvay Conference (1911), I.147
  free-electron gas hypothesis, I.216
  inner quantum number, I.100–1
  metal electrons, I.215
  Planck's law, I.70
  quantization, I.359
  quantum model, III.1308
  quantum numbers, I.90–2
  quantum rule, I.175
  quantum theory, I.179, III.1285, III.1361
  radio waves, III.1987
  structure factor, I.431
  theory of metals, III.1290
  thermoelectric effects, III.1308
  x-ray diffraction, I.424
  x-ray spectra, I.160
  Zeeman effect, I.102, I.162
Sorby, H
  metallography, III.1507
  microstructure, III.1506
Sorokin, P P
  dye lasers, III.1446
  four-level laser, III.1427
Southworth, G C, waveguide propagation, III.1322
Spain, I L, single-crystal studies, I.437
Spedding, F H, magnetism, II.1143
Spinrad, H, radio galaxies, III.1795
Spitzer, L
  Atoms for Peace Conference, III.1638
  interstellar gas, III.1765
  magnetic fusion, III.1636
  Physics of Fully Ionized Gases, III.1631
  rotational transform, III.1654
  stellarator, III.1630, III.1648, III.1654, III.1655
Sproule, D O, diagnostic ultrasonics, III.1910–11

Sproull, R, materials research, III.1558
Standley, K J
  relaxation time, II.1133
  spinels, II.1156
Stanley, G J, radio sources, III.1737
Stanley, H E, Ising models, I.563
Stark, J
  band spectra, I.77
  chemical and other quanta, I.154
  longitudinal effect, I.162
  quantum physics, I.148, I.155
  Stark effect, I.89
  Zeeman splitting, I.161
Staub, H, element synthesis, II.1226
Staudinger, H, long-chain systems, III.1597
Steenbeck, M, visit to Max Knoll, III.1572–3
Stefan, J, black body, I.66
Stein, P
  computers, III.1843
  mathematician, III.1839
Stein, R F, solar oscillations, III.1750
Steinberger, J
  decay of neutral pion, II.711
  neutral pion, II.655
  two-neutrino experiment, II.672
Steinhardt, P J, world model, III.1807
Steller, J, magnetic mirror, III.1659
Stelson, P H, Coulomb excitation experiments, II.1217
Stephens, W E, element synthesis, II.1226
Stern, J, quantum-theoretical equations, I.154
Stern, O
  electronic orbits, I.164
  ionic crystals, III.1531
  quantum theory, I.221
  specific-heat problem, I.153–4
  surface crystallography, I.468

Sternheimer, R, inner-shell electrons, II.1092
Stevenson, E C
  elementary particles, I.391
  mesotron, I.397
  muon, II.650
  new particles, I.389
Stevenson, M J, four-level laser, III.1427
Stewart, A T, phonons, II.999
Stewart, B
  Earth's magnetic field, III.1621
  geomagnetic variations, III.1984–5, III.1985
  wire-wound resistors, II.1252
Stiles, W G, human eye, III.1407
Stockmayer, W, polymer solutions, III.1597
Stoicheff, B P, stimulated Brillouin scattering, III.1444
Stokes, G G
  molecular vibrations, I.76
  x-rays, I.362
Stoletov, A G, photoelectric effect, III.1390
Stone, J, fibre lasers, III.1427
Stoner, C R, *Short Wave Wireless Communication*, III.1619, III.1620
Stoner, E, white dwarfs, III.1711
Stoner, E C
  electronic levels in atoms, I.168
  spin waves, II.1012
Stoney, G J, estimate for e, I.52
Storey, L R O, whistler studies, III.1995
Stormer, C
  Earth's magnetic field, III.1682
  geomagnetic storms, III.2003
  magnetic-mirror effect, III.1637
  single charged particle, III.2004
Strassmann, F
  elements with $Z = 93$–96, I.130
  neutron bombardment of uranium, II.1195

neutron sources, II.1206
neutron–U collisons, I.130
radiochemical analysis, II.1195
uranium isotope, II.1223
uranium-239, I.132
Stratton, S W, **II.1241**
Straubel, P, imaging theory, III.1418
Strauss, L L
　fusion research, III.1637–40
　Men and Decisions, III.1640
Street, J C
　elementary particles, I.391
　mesotron, I.397
　muon, II.650
　new particles, I.389
Strehl, K, aberration effects, III.1389
Strnat, K J, rare-earth–cobalt alloys, II.1165
Strnat, R M W, rare-earth–cobalt alloys, II.1165
Stroke, G W, holograms, III.1438
Strömgren, B
　interstellar medium, III.1762
　Strömgren spheres, III.1762
Strong, J, anti-reflection coatings, III.1407
Strutinsky, V M, fission isomers, II.1218
Strutt, J W, *see* Rayleigh, Lord
Strutt, R J, thorianite, III.1949
Stueckelberg, E, heavy electrons, I.391
Stuewer, R, nuclear physics, II.1192
Stump, D M, polarized light, III.1412
Sucksmith, W, ferromagnetism, II.1134
Sudarshan, E C G
　neutrinos, II.667
　quasi-classical distribution function, III.1441
　radiation field, III.1388

Suess, E, ocean basins, III.1967
Suess, H E
　chemical elements, III.1747
　nuclear binding energies, II.1210
　shell models, II.1209
Suhl, H, crystal structures, II.1158
Sunyaev, R A
　induced Compton scattering, III.1801
　photon scattering, III.1773
　thin accretion discs, III.1779
Sunyar, A, neutrino helicity, II.668, II.670
Sutherland, P, neutron stars, III.1754
Swedenborg, E, island universes, III.1714
Sweetman, D
　magnetic mirror, III.1659
　neutral beam injection, III.1661
Swiatecki, W T, liquid-drop mass formula, II.1225
Sykes, M E
　critical exponents, I.560
　series expansions, I.557
　site percolation processes, I.572
　two-dimensional lattices, I.573
Synge, E H, optical microscopy, III.1405–6
Szalay, A S, neutrinos, III.1803
Szigeti, B, electronic dipole moment, I.485
Szilard, L
　entropy, I.225
　Maxwell's Demon, I.576
Szöke, A, optical bistability, III.1449
Szymborski, K
　colour centres, III.1529
　solid-state physics, III.1320

Tabor, D, measurement of forces between surfaces, III.1612
Tainter, S, Photophone, III.1422

Talbot, F, daguerreotype process, III.1692
Talleyrand, C M de, metric system, II.1233
Talmi, I, shell model, II.1213
Tamm, I E
  energy loss formula, II.750
  first-order Raman effect, II.996
  nuclear exchange potential, I.376
  Plasma Physics and the Problem of Controlled Thermonuclear Reactions, III.1642
  tokamak, III.1630
  visible light, III.1409
Tammann, G
  binary alloys, I.548
  crystal growth, III.1280
  Hubble's constant, III.1793
  Lehrbuch der Metallographie, III.1554
  liquid crystals, III.1540
  phase diagrams, III.1509, III.1554
Tapster, P R, Bell inequalities, III.1471
Tayler, R J
  helium synthesis, III.1790
  synthesis of heavy elements, III.1792
Taylor, B N
  determination of e/h, II.1268
  magnetic helicity, III.1657
  resonance frequency, II.1095
Taylor, G I, **II.799**
  catastrophe theory, II.798
  diffraction pattern, III.1396
  Diffusion by continuous movements, II.830
  dislocations, III.1520
  edge dislocation, I.476
  equilibrium (wet-bulb) temperature, II.843
  potential-flow forces, II.887
  singular perturbation theory, II.811
  sound waves, II.858
  stabilizing effect of viscosity, II.828
  tidal friction, II.891
  turbulence, II.798, II.832
  vortices, II.833
Taylor, J B
  binary pulsars, III.1757
  plasma pressures, III.1661
Taylor, J H
  binary pulsar, I.298, I.317
  quadrupole radiation formula, I.315
Tebble, R S, domain techniques, II.1155
Teich, M C, sub-Poissonian Franck–Hertz source, III.1460
Teichmann, J
  colour centres, III.1529
  solid-state physics, III.1320
Telegdi, V L, parity violation, II.666
Teller, E
  adiabatic confinement, III.1679–80
  Atoms for Peace Conference, III.1639
  dipole vibration, II.1213–14
  First Shelter Island Conference, II.645
  trapped particles in magnetic-mirror fields, III.1683
Temkin, R J
  amorphous germanium, I.479
  amorphous structure, I.478
Temmer, G M, Coulomb excitation experiments, II.1217
Temperley, H N V, percolation model, I.573
Terhune, R W, Raman Stokes and anti-Stokes rings, III.1436
Ter-Martirosyan, Coulomb excitation, II.1217

Ter-Pogossian, M M, positron emission tomography (PET), III.1920
Tesla, N, radio and high-frequency technology, III.1620
Tharp, M, Mid-Atlantic Ridge, III.1973
Thellung, A, disordered materials, III.1366
Theobald, J P, fission isomers, II.1218
Thiessen, A, direct-drive approach to inertial fusion, III.1676
Thom, R, dynamical systems, III.1850
Thomas, L H, Thomas effect, I.278
Thomas, R G, nuclear reactions, II.1219
Thomas, W, spectral line intensity, I.172
Thompson, Sir Benjamin (Count Rumford), learned societies, I.3
Thompson, G B H, semiconductor lasers, III.1437
Thompson, R, hyperon, II.657
Thompson, S, radioactivity, III.1876
Thompson, W, see Lord Kelvin, Lord
Thomson, Sir George
  controlled fusion, III.1636
  de Broglie waves, I.171
  electron diffraction, I.463
  finite electron temperatures, III.1624–5
  toroidal fusion device, III.1634
Thomson, J J, I.50
  atomic model, I.64, I.65
  atomic structure, I.49, I.156
  deflection measurements, II.1271
  determination of e/m, I.53, I.54
  discovery of electron, I.3, III.1565
  electrical conductivity, III.1986
  electrical discharges in gases, III.1985
  electron, III.1857
  electron charge, II.747
  electron diffraction, I.463
  electron emission, I.37
  electronic theory of magnetism, II.1114
  experimental physics, I.14
  finite electron temperatures, III.1624–5
  free electrons, I.54
  gravitation, II.1266
  heat engines, I.534
  line spectra, I.77
  metallic conduction, III.1284
  photoelectron emission, III.1396
  Recollections and Reflections, I.27, I.36
  second wrangler, I.11
  x-ray scattering, I.362
Thomson, R, materials physics, III.1558
Thomson, W, see Kelvin, Lord
Thonemann, P C
  fusion reactions, III.1642
  ZETA experiment, III.1641
Thorne, K S, binary sources, III.1756
Thouless, D J, topological long-range order, I.571
Tiktopoulos, singularities in complex angular momentum, II.682
Ting, S C C
  dilepton mass spectrum, II.701
  J resonance, II.702
Tisza, L, two-fluid model, II.927
Tizard, Sir Henry, atomic energy project, II.1198
Todd, A R, nucleic acids, I.504
Todd, Lord, polymerization, III.1550
Todhunter, I, research and genius, I.12

N75

Toennies, J P, helium atom scattering, II.1006
Tolansky, S
 FECO interferometry, III.1412
 microtopography, III.1412
 optical interferometry, III.1411
 rings lens-plate experiment, III.1412
Tollmien, W
 parallel flows, II.828
 stability calculation, II.829
Tolman, R C
 particle dynamics, I.279
 quantum physics, III.1401–2
 spherically symmetric perturbations, III.1801
Tolpȳgo, K B, covalent crystals, II.1002
Tomonaga, S
 β-decay, I.402
 cosmic-ray particles, I.407
 coupling theory, I.407
 quantum electrodynamics, II.643, II.740, III.1446
 renormalization, II.646
 slow mesotrons, I.406
 Twelfth Solvay Conference, Brussels, II.646
Tonks, L, plasma oscillations, III.1624
Totsuji, H, clusters of galaxies, III.1771
Toulouse, percolation processes, I.573
Touschek, B, storage rings, II.740
Townes, C H, **III.1420**
 'dressed' atomic states, III.1421
 electric quadrupole moments, II.1214
 Fabry–Perot resonator, III.1429
 lasers, III.1423
 maser, III.1419, III.1422
 stimulated Brillouin scattering, III.1444
Townsend, A, 'big eddies', II.839

Townsend, J S, ionization detectors, II.746
Toya, T, dispersion curves in sodium, II.999
Trahin, M, perturbation theory, III.1473–4
Treiman, S B
 A Century of Particle Theory, II.770
 β-decay, II.676
 pion decay constant, II.675
Treloar, L R G, rubber elasticity, III.1551
Tremaine, S D
 galactic cannibalism, III.1774
 phase space constraints, III.1804
Trimble, V L, binary sources, III.1756
Trowbridge, J
 electrolytic cells, I.27
 practical classes, I.15
Trump, J G
 electrostatic generator, II.1203
 High Voltage Engineering (HVE), II.1203
Tsoi, V S, bismuth crystal, III.1344–5
Tsu, R
 coupling of quantum wells, III.1459
 quantum-well superlattices, III.1453
Tuck, J L
 magnetic cusp, III.1645
 magnetic pinch, III.1636
Turlay, R, *CP* violation, II.672
Turnbull, D
 age-hardening, III.1533
 nucleation kinetics, III.1547
Turner, R C, flaw detector, III.1911
Tuve, M A
 ionospheric conducting layer, III.1988
 ionospheric plasmas, III.1621
 ionospheric sounding, III.1995

radio receiving equipment, III.1990
van de Graaff generator, II.1203, II.1204
Tweet, D J, surface crystallography, I.469
Twiss, R Q
  optical spectroscopy, III.1421–2
  radio interferometer research, III.1422
Tyson, A, galaxies, III.1797

Uchling, E A, collision term in Boltzmann equation, I.602
Uhlenbeck, G E, **I.594**
  Brownian motion coupled harmonic oscillators, I.608
  collision term in Boltzmann equation, I.602
  Critical Phenomena, I.553
  electron spin, I.101
  First Shelter Island Conference, II.645
  graph theory in statistical mechanics, I.553
  magnetic moment, II.1123
  Master equation, I.593–605
  quantum number, I.168
  random processes, I.628
  spinning electron, I.360
  universality, I.561
Ullmaier, H, point defects, III.1534
Ulrich, R K
  solar model, III.1749
  solar oscillations, III.1750
Underwood, E, *Quantitative Stereology*, III.1549–50
Unsöld, A, hydrogen abundance, III.1701
Upatnieks, J, holography, III.1415, III.1437
Upatnieks, Y J, off-axis optical beam holography, III.1585
Urey, H C
  chemical elements, III.1747
  deuterium, III.1706
  deuteron, II.650, II.1183
  solar system, III.1963

Valasek, J, Rochelle salt, I.489
Vallarta, M S, magnetic-mirror effect, III.1637
Van Allen, J A
  Geiger counter, III.2007
  magnetospheric physics, III.2001
  Radiation Trapped in the Earth's Magnetic Field, III.1683
  rocket sounding experiments, III.1682–3
  Third International Polar Year, III.2006
  Van Allen belts, III.1636, III.1683, III.1684
van Cittert, P H, optical coherence, III.1407–8
Van, V, x-ray diffraction, I.498
van de Graaff, R J
  electrostatic generator, II.727–8, II.1186, II.1203
  High Voltage Engineering (HVE), II.1203
van de Hulst, H, radio astronomy, III.1762–3
van der Burg, M G J, field equations, I.313
van der Meer, S
  antiprotons, II.707
  field particles W and Z, II.744
  stochastic cooling, II.743
van der Waals, J D
  doctoral thesis, I.48
  kinetic theory of gases, I.543
  quantum mechanics, II.1042
  spinodal, III.1547–8
Van der Waerden, B, group theory, I.213
van Heel, A C S, fibre optics, III.1419–20
van Hove, L

magnetic correlation theory, II.1016
neutron frequency, III.1418
phonon dispersion relations, II.993
Twelfth Solvay Conference, Brussels, II.646
Van Kampen, N, linear response assumption, I.627
van Laar, J J, statistical mechanics, I.551–4
van Leeuwen, J H
　magnetic properties, II.1115
　theory of metals, III.1286–7
van Maanen, A
　bright spirals M31 and M33, III.1717–18
　nebulae, III.1718
van Rhijn, P J, luminosity function, III.1714
Van Vleck, J H, **II.1131**
　Bohr theory, II.1123–4
　crystal field theory, II.1132
　ferromagnetic resonance, II.1153
　First Shelter Island Conference, II.645
　H–D exchange, II.1140
　magnetic ions, II.1128–9
　molecular orbital theory, II.1143
　Quantum Principles and Line Spectra, II.1131
　rare-earth ions, II.1120
　relaxation theory, II.1146
　spectral line intensity, I.172
　spin–orbit interactions, II.1136
　Theory of Electric and Magnetic Susceptibilities, II.1131
　two-electron system, I.182
van't Hoff, J H, reaction isochore equation, I.535
Varicak, V, kinematic space, I.278
Vashakidze, M A, optical continuum of nebula, III.1774–5

Vassall, A, physics in education, I.9
Vaughan, R A, relaxation time, II.1133
Vavilov, S I
　absorption of light by uranium glass, III.1403
　Milrostructura Sveta, III.1403
　non-linear optics, III.1403
　polarization of fluorescent light, III.1400
　Raman effect, III.1400
　visible light, III.1409
Veksler, V, ion acceleration, II.730
Velikovsky, I, *Worlds in Collision*, III.2013
Veltman, M, electroweak parameters, II.725
Veneziano, G
　Dolen–Horn–Schmid duality, II.683
　scattering amplitude, II.683–4
Venkatesen, T N C, optical bistability, III.1449
Verne, J, science fiction, I.3
Verschaeffelt, J E, classical critical exponents, I.544
Verschaffelt, E, Fifth Solvay Conference (1927), I.205
Verwey, E J W
　charge effects, III.1611
　electrical properties, II.1156
　semiconductors, III.1330
Vigier, J P, matter-wave theory, I.229
Villard, P
　γ-rays, I.54
　ionization, III.1880
　nuclear γ-rays, I.367
Vine, F, magnetic variations of ocean floors, III.1975
Vinen, W F, vibrating-wire experiment, II.931
Vinogradov, A V, laser amplification, III.1466

Virasoro, M, Dolen–Horn–Schmid duality, II.683
Vlasov, A, Landau damping, III.1624, III.1625
Voigt, W
  electron optics, III.2000
  electronic theory of magnetism, II.1114
  magneto-ionic theory, III.1990
  school of physics, I.15
Volkoff, G, neutron stars, III.1713
Volmer, M, nucleation kinetics, III.1547
von Ardenne, scanning transmission microscope (STEM), III.1569, III.1587–9
von Baeyer, O
  β-ray deflection, II.1201
  β-spectrum, I.106
  radioactive decay, I.105
von Borries, B
  polepiece lens, III.1567
  transmission electron microscope (TEM), III.1575–7
von Braun, W, rocket technology, III.1738
von Federov, E, symmetry types, I.423
von Groth, P, crystallography, I.421
von Gutfeld, R J, thermal conductivity, II.982
von Hippel, A, dielectric properties of BaTiO$_3$, I.486
von Humboldt, A, *Gesellschaft deutscher Naturforscher und Ärzt*, I.5
von Kampen, N
  BBKGY hierarchy, I.624
  random walks, I.624
von Kármán, T
  BvK theory, II.985
  crystal lattice vibrations, I.153
  lattice dynamics, I.452
  sound from flows, II.850

specific heat, II.971–2
turbulence, II.832
vibrational specific heat of solids, I.525
von Klitzing, K
  plateaux studies, III.1375
  semiconductors, II.1258
von Laue, M, **I.426**
  crystal structures, III.1514
  electromagnetic wave theory, I.363
  Fresnel's dragging coefficient, I.278
  imaging theory, III.1418
  lattice structure, I.153
  partial coherence, III.1392
  superconductivity, II.918–19
  x-ray diffraction, I.424, I.425
von Neumann, J
  energy surface trajectories, I.599
  First Shelter Island Conference, II.645
  materials research, III.1558
  Mathematical Foundations, I.227
  microphysics, I.225
  non-bounded operators, I.197
  quantum mechanics, I.198, I.206
  Schrödinger equations, I.205
von Rebeur-Paschwitz, E, earthquake vibrations, III.1953–4
von Schweidler, E, decay law of radioactive substances, I.222
von Smoluchowski, M, Brownian motion, I.586
von Weizsäcker, C F
  carbon–nitrogen–oxygen cycle (CNO cycle), III.1707–8
  monistic theories, III.1963
  nuclear energies, II.1191
von Wroblewski, S, liquefied gases, I.27
Voss, H D, *Lightning-induced electron precipitation*, III.1686–7

Wagner, vacancies, III.1529
Wagner, E, x-ray spectra, I.160
Wagoner, R V
  Big Bang, III.1792
  nuclear interactions between light nuclei, III.1791
Wahl, A C, plutonium isotope, II.1223
Wainwright, T E
  autocorrelation function of tagged particle, I.619
  molecular dynamics, I.577
Wakefield, J, holograms, III.1584
Wald, G, rod-vision, III.1416
Wali, K C, *Chandra: a Biography of S Chandrasekhar*, III.1711
Walker, A G
  Robertson–Walker metric, III.1729–30
  synchronization of clocks, III.1729
Walker, Sir Gilbert, systematic shifts of climate, II.907
Walker, J G
  antibunched light, III.1460
  optical data processing, III.1461
Walker, M F, star cluster NGC2264, III.1766–7
Walkinshaw, W, linear accelerators, II.1206
Walls, D F, quantum non-demolition (QND) measurements, III.1469
Walraven, T, Palomar 200" Telescope, III.1775
Walter, W F, germanium junctions, II.1208
Walton, E T S
  disintegration experiments, II.1183, II.1184
  disintegration of lithium, II.729
  electrostatic generators, II.727
  fusion reactions, III.1628
  nuclear reaction, I.285
  nuclear transformation, I.112

proton acceleration, I.370
Wambacher, H, cosmic-rays tracks, I.404
Wandel, A, quasars, III.1780
Wang, C C, He–Ne laser, III.1435
Wang, Ming Chen
  Brownian motion coupled harmonic oscillators, I.608
  random processes, I.628
Wannier, G
  multi-electron fragmentation, II.1065
  optoelectronic properties, III.1410
  semiconductors, III.1320
Wannier, G H, specific heat, I.555
Warburg, E
  First Solvay Conference (1911), I.147
  quantum number, I.162
Ward, F A B, ionization chambers, II.1200
Ward, J C, quantum theory, III.1413
Ward, W R, Moon model, III.1966
Wardlaw, R S, crystal fields, II.1165
Ware, A A, ZETA experiment, III.1641
Warren, B E, amorphous structure, I.478
Wasserburg, G J, supernova trigger hypothesis, III.1963
Watanabe, Y, semiconductor maser, III.1423
Waterman, P C, electromagnetic scattering, III.1446
Watson, G N, radio waves, III.1987
Watson, J D
  DNA model, I.507
  double-helix structure, I.504
  protein subunits in viruses, I.509
Watson-Watt, R, upper atmosphere, III.1988
Watt, R W, ionosphere, III.1981

Weaire, D, amorphous structures, I.477
Weaver, W, molecular biology, I.496
Weber, A, nucleation kinetics, III.1547
Weber, H F, specific-heat quanta, I.152
Weber, R L, maser principle, III.1417
Weber, W E
  magnetic moments, II.1114
  Zeeman effect, II.1114
Wegener, A L, **III.1968**
  continental drift, III.1967–9
Wegner, E J, universality, I.571
Weidlich, W, laser theory, III.1444
Weigert, F, polarization of fluorescent light, III.1400
Wein, First Solvay Conference (1911), I.147
Weinberg, D L, photon coincidence, III.1435
Weinberg, S, II.696
  Dreams of a Final Theory, II.770
  electroweak unification, II.695
  social construction, I.405
Weinrich, M, parity violation, II.666
Weis, P E, **II.1119**
  Le Magnetisme, II.1119
Weiss, M T, crystal structures, II.1158
Weiss, P
  critical point, I.545
  magnetic fields, II.1117–18
  molecular field hypothesis, I.545
Weissenberg, K, moving-film methods, I.434
Weisskopf, V F
  compound nucleus, II.1219
  electromagnetic field, I.391
  First Shelter Island Conference, II.645

positrons in electron self-energy, II.641
proton scattering, II.1212
quantum theory, I.221
The Joy of Insight, II.770
Welford, W T, compound parabolic concentrator (CPC), III.1446
Wells, H G, science fiction, I.3
Wells, P N T, ultrasonic imaging, III.1914
Wendt, G, longitudinal effect, I.162
Wentzel, G
  asymptotic linkage, I.197
  atomic theory, I.179, I.219
  Fermi field, I.378
  Twelfth Solvay Conference, Brussels, II.646
Wenzel, W A, spark chamber, II.753
West, J, structure of beryl, I.440
Westerhout, G, neutral hydrogen, III.1763
Westgren, J, single-crystal methods, I.435
Weyl, H, I.207
  eigenfunctions with continuous spectra, II.1025
  gauge invariance, I.321
  group theory, I.213
  Weyl's postulate, III.1729
Weymann, R J, electron scattering, III.1801
Whaling, W, resonance state of nucleus, III.1746–7
Wharton, C B, microwave interferometer, III.1659
Wheeler, J A
  First Shelter Island Conference, II.645
  fission process, I.132, II.1195
  hydrogen bomb, III.1636
  S-matrix theory, II.680
  two-photon correlation, III.1413
Whitcomb, R T, area rule, II.881

N81

White, D L, electron–phonon interaction, III.1339
White, J W, neutron beams, I.468
Whitham, G B
  action-conservation principle, II.862
  'equal area' law, II.847
  'exploding wire' phenomenon, II.848
  water motions, II.860
Wick, G C
  nuclear exchange-force model, I.376
  quantum theory, I.221
  Twelfth Solvay Conference, Brussels, II.646
Wideröe, R
  resonant cyclic accelerators, II.729
  The Infancy of Particle Accelerators, II.770
Widom, B
  phase transitions, III.1830
  van der Waals equation, I.561
Wiechert, E
  cathode ray tube, III.1566
  Earth model, III.1953
Wiedlich, W, laser theory, III.1444
Wiegel, red blood cells, I.630
Wieman, C, parity violation, II.699
Wien, W
  atomic vibrations, III.1287
  black-body radiation, I.22, I.53, I.149
  exponential law, I.66–7
Wiener, N
  Brownian process, III.1830
  quantum mechanics, I.192
Wiersma, E C, potassium chromic alum, II.1126
Wiese, W L, *Needs, Analysis and Availability of Data for Space Astronomy*, II.1093
Wightman, A S, Twelfth Solvay Conference, Brussels, II.646

Wigner, E
  band structure, III.1314–17
  electron band structure of sodium, III.1530
  electron gas, III.1361–2
  neutron capture, II.1194
  nuclear resonance, I.126
  reaction cross sections, II.1219
  Schrödinger's equation, III.1315
  Twelfth Solvay Conference, Brussels, II.646
Wigner function, I.617–18
Wilberforce, S, Darwinian evolution, I.5
Wilczek, F, running coupling constant, II.713
Wild, J J, pulse-echo ultrasound, III.1911–12
Wilkins, M H F, DNA, I.504
Wilkinson, D H
  Ionisation Chambers and Counters, II.1207
  pulse height conversion to pulse time, II.751
  shell-model theory, II.1214
Williams, E J, order–disorder transitions, I.474
Williams, R C, Lamb shift, II.641
Williamson, A, atomic theory, I.46
Wilson, A, science jottings, I.3
Wilson, A J C
  line broadening, I.472
  structure factors, I.439
Wilson, C T R
  alpha/beta particle tracks, II.1200–1
  cloud chamber, I.54
  droplet formation, II.751–2
  Fifth Solvay Conference (1927), I.205
  radiation, I.367
Wilson, H A, electron charge, II.747
Wilson, H F, special orientation relationships, III.1528

Wilson, H R, helical structures, I.498
Wilson, J M
    physics in education, I.9
    senior wrangler, I.11
Wilson, J T, plate tectonics, III.1978
Wilson, K
    computers, III.1834
    field theory, III.1830, III.1833–34
    fixed point, III.1832, III.1833–34
    magnetism, III.1834
    non-perturbative calculations, III.1825
Wilson, K G, **I.565**
    renormalization group, I.564–9
Wilson, R, microwave background radiation, III.1790
Wilson, R R, **II.735**
    Fermilab accelerator, II.734
Wilson, R W
    carbon monoxide molecule, III.1763–4
    microwave radiometer, II.758
    radiation, III.1786
Wilson, W
    electron beams, I.104
    quantum rule, I.175
Winston, R, compound parabolic concentrator (CPC), III.1446
Winther, A, Coulomb excitation experiments, II.1217
Wintner, A, bounded matrices, I.206
Wirtz, C W, spiral galaxies, III.1725
Wisdom, J
    chaotic evolutions, III.1823
    computing, III.1831
Witkowski, J, biomolecular structures, I.496
Wolf, D, melting, III.1533–4
Wolf, E
    Airy disc, III.1412
    coherence properties of partially polarized light, III.1411
    diffraction images, III.1419
    energy transport, III.1459
    optical coherence theory, III.1392, III.1421
    quantum theory of coherence, III.1439
Wolfe, J P, thermal conductivity, II.982
Wolfke, M
    two-stage imaging, III.1583
    x-ray microscope, III.1414
Wollan, E O
    neutron diffraction, I.453
    structure of ice, I.475
Wollaston, W H
    medical physics, III.1857
    solar spectrum, III.1691
    spectral lines, I.156
Wollfson, M M, determinantal inequalities, I.448
Wood, R W
    black-body radiation, I.150, III.1387
    diffraction gratings, III.1397
    direct-drive approach to inertial fusion, III.1676
    fluorescence and photochemistry of materials, III.1401
    Physical Optics, III.1391
    resonance radiation, III.1390
Woodbury, E J, ruby laser, III.1432
Woodruff, D P, surface crystallography, I.466
Woods, A D B, shell model, II.997
Woodward, J J, linear accelerators, II.736
Woolfson, M M
    probability relations, I.449
    structure determination, I.450
Wooster, W, continuous spectrum, I.118, I.126
Wright brothers, powered flight, II.873
Wright, T, island universes, III.1714

Wrinch, D M, Patterson function, I.444
Wu, C S
  parity violation, II.666
  scintillation counters, III.1414
Wu, L-A, squeezed light, III.1471
Wu, S L, gluon jets, II.716
Wu, T T, *CP* violation, II.673
Wül, B, dielectric properties of BaTiO$_3$, I.486
Wyckoff, R W G
  biomolecular structures, I.496
  crystal structures, III.1513
Wyler, A A, magneto-ionic theory, III.1992
Wynn-Williams, C E
  binary counting circuits, II.748
  ionization chambers, II.1200

Xiao, M, squeezed light, III.1471

Yalow, R, saturation analysis technique, III.1909
Yamaguchi, Y, decay reaction, II.657
Yan, T-M, lepton pair, II.690
Yang, C N
  CP violation, II.673
  isotopic spin symmetry, II.693
  lattice gas model, I.553
  local gauge invariance, II.693
  neutrinos, II.671
  parity conservation, II.665
  parity violation, II.666
  self-interacting fields, II.661
  spin-1 particle, II.654
Yardley, K, x-ray diffraction, I.439
Yariv, A
  guided wave optics, III.1458
  optical waveguides, III.1459
  Quantum Electronics, III.1436
Yeh, Y, Doppler shift, III.1442
Yevick, random walk model, I.630

Yonezawa, F, amorphous structures, I.477, I.479
York, D G, interstellar deuterium, III.1791
York, H
  fusion research, III.1637
  magnetic mirror, III.1659
Yosida, K, RKKY interaction, II.1164
Young, angular distributions, II.1220
Young, F Jr, materials physics, III.1558
Young, J Z, flying-spot microscope, III.1416–17
Young, T
  dimensions of molecules, I.48
  medical physics, III.1857
  wave description of light, I.72
Yuen, H P, minimum-uncertainty packets, III.1469
Yukalov, V I, melting, III.1533–4
Yukawa, H, **I.390**
  exchange forces, I.231
  meson theory, I.378–83, I.389, II.640, II.651–2, II.669
  neutral pions, II.654
  slow mesotrons, I.406
  spin-1 boson, II.692
  Twelfth Solvay Conference, Brussels, II.646
  U-particle hypothesis, I.396
Yuniev, defibrillators, III.1895
Yvon, J, BBKGY hierarchy, I.606, I.623, I.624, I.626

Zachariasen, W H
  amorphous structures, I.477
  extinction, I.451
  structure of metaboric acid, I.449
Zahradnicek, J, gravitation, II.1266
Zanstra, H, planetary nebulae, III.1762

Zavoisky, E, paramagnetic resonance, II.1137–8
Zeeman, P
  magnetism, III.1390
  splitting of spectral lines, I.161
  Zeeman effect, I.54
Zehnder, L, x-ray photograph, I.4
Zeiger, H J, maser, III.1419
Zel'dovich, B Ya, phase conjugation, III.1458
Zeldovich, Ya B
  Big Bang, III.1790
  binary sources, III.1756
  black holes, III.1779, III.1780
  collapsing cloud, III.1803
  cosmological constant, III.1723, III.1806
  induced Compton scattering, III.1801
  neutrinos, III.1803
  photon scattering, III.1773
  photon to baryon–antibaryon ratio, III.1806
  structures in the Universe, III.1801, III.1802
  synthesis of heavy elements, III.1792
  weak vector current, II.668
Zeller, R C, disordered systems, II.1006
Zener, C
  Fermi surface, III.1295
  valence conduction bands, III.1337
Zenneck, J
  radio propagation, III.2000
  radio waves, III.1987
Zermelo, E, process irreversibility, I.524
Zernike, F
  circle polynomials, III.1409
  critical correlations, I.563
  optical coherence, III.1407–9
  pair distribution function, I.547
  phase-contrast microscopy, III.1409–10, III.1419, III.1475
  x-ray diffraction pattern, I.441
Zheludev, N I, polarization symmetry breaking, III.1467
Zherikhin, A, laser amplification, III.1466
Zienau, S, lattice deformation, III.1321
Ziman, J M
  electrons and holes, III.1297
  Electrons and Phonons, III.1305
  lattice conduction of heat in metals, II.981
  pseudopotential concept, III.1367
Zimkina, T M, K-shell absorptions, II.1083
Zimm, B, polymer solutions, III.1597
Zimmerman, G, direct-drive approach to inertial fusion, III.1676
Zinn, W H, neutron diffraction, I.453
Zinn-Justin, J, renormalizability, II.695
Zobernig, G, gluon jets, II.716
Zoll, P
  cardiac pacemakers, III.1895
  defibrillators, III.1895
  electrical stimulation, III.1895
Zupancic, C, Coulomb excitation experiments, II.1217
Zvyagin, B B, oblique texture method, I.464
Zwanziger, singularities in complex angular momentum, II.682
Zweig, G, fractionally charged constituents of matter, II.685
Zwicky, F
  Catalogue of Selected Compact Galaxies and of Post-Eruptive Galaxies, III.1713

clusters of galaxies, III.1731
Coma cluster, III.1732
Schmidt telescope, III.1745
supernovae, III.1712, III.1718, III.1758, III.1759, III.1775

Zworykin, V K
phosphors, III.1536–7
photomultipliers, II.750
synthetic zinc silicate, III.1537
xerography, III.1541

To I.G.R.
From J.A.Z.R.
29.1.2001